# MyMathLab® = Your Resource for Success

## In the lab, at home,...

- Access videos, PowerPoint® slides, and animations.
- Complete assigned homework and quizzes.
- Learn from your own personalized Study Plan.
- Print out the *Student Workbook for Elementary Algebra for College Students* for additional practice.
- Explore even more tools for success.

## ...and on the go.

 Download the free Pearson eText App to access the full eText on your Apple® or Android™ device. MyMathLab log-in required.

 Use your Chapter Test as a study tool! Chapter Test Prep Videos show step-by-step solutions to all Chapter Test exercises. Access these videos in MyMathLab or by scanning the code.

Scan the code or go to: www.youtube.com/user/AngelElemInterAlg

## Don't Miss Out! Log In Today.

MyMathLab delivers proven results in helping individual students succeed. It provides engaging experiences that personalize, stimulate, and measure learning for each student. And, it comes from a trusted partner with educational expertise and an eye on the future.

To learn more about how MyMathLab combines proven learning applications with powerful assessment, visit **www.mymathlab.com**

VIDEOS • POWERPOINT SLIDES • ANIMATIONS • HOMEWORK • QUIZZES • PERSONALIZED STUDY PLAN • TOOLS FOR SUCCESS

To my brother, Jerry
Allen R. Angel


To my brothers and sisters,
Nancy, Steve, Cathy, Rod, and Mary Jo
And to the memory of Dave
Dennis C. Runde

# Contents

## 1 Real Numbers   1

## 2 Solving Linear Equations   91

## 3 Applications of Algebra   165

# 7  Rational Expressions and Equations     379

# 8  Functions and Their Graphs     449

# 9  Systems of Linear Equations     517

# 10  Inequalities in One and Two Variables     585

# 11 Roots, Radicals, and Complex Numbers 623

# 12 Quadratic Functions 691

# 13 Exponential and Logarithmic Functions 771

# Preface

This book was written for college students and other adults who have never been exposed to algebra or who have been exposed but need a refresher course. Our primary goal was to write a book that students can read, understand, and enjoy. To achieve this goal we have used short sentences, clear explanations, and many detailed, worked-out examples. We have tried to make the book relevant to college students by using practical applications of algebra throughout the text.

The many factors that contributed to the success of the previous editions have been retained. In preparing this revision, we considered the suggestions of instructors and students throughout the country. The *Principles and Standards for School Mathematics*, prepared by the National Council of Teachers of Mathematics (NCTM), and *Beyond Crossroads: Implementing Mathematics Standards in the First Two Years of College*, by the American Mathematical Association of Two-Year Colleges (AMATYC) together with advances in technology, influenced the writing of this text.

## New to This Edition

One of the most important features of the text is its readability. The book is very readable for students of all reading skill levels. The Fourth Edition continues this emphasis and has been revised with a focus on improving accessibility and addressing the learning needs and styles of today's students. To this end, the following changes have been made:

### Content Changes

- Discussions throughout the text have been thoroughly revised for brevity and accessibility. Whenever possible, a visual example or diagram is used to explain concepts and procedures.
- **Understanding Algebra** is a new feature appearing in the margin throughout the text. Placed at key points, **Understanding Algebra** draws students' attention quickly to the important concepts and facts that they need to master.
- The pedagogical use of color has been enhanced and now includes a color-coded system for variables and notation to support a more visual approach.
- Exercise sets now begin with new **Warm-up Exercises**—and include an emphasis on vocabulary. These exercises are great as a warm-up to the homework or as a 5-minute quiz. The Concept/Writing exercises (formerly found at the start of the exercise sets) are now located after the Problem Solving section in the exercise sets.
- Exercises and applications have been updated throughout.

## Enhancements to Resources

- The Chapter Test Prep Video and Lecture Series Videos are now captioned in both English and Spanish. The videos are available in MyMathLab. The Chapter Test Prep videos are also available on YouTube.
- MyMathLab and MathXL have been significantly updated including:

  1. A substantial increase in exercises coverage
  2. Suggested Assignments in homework builder
  3. New study skills and math-reading connections coverage

## Features of the Text

### Full-Color Format

Color is used pedagogically in the following ways:

- Important definitions and procedures are color screened.
- Color screening or color type is used to make other important items stand out.
- Artwork is enhanced and clarified with use of multiple colors.
- The full-color format allows for easy identification of important features by students.
- The full-color format makes the text more appealing and interesting to students.

### Accuracy

Accuracy in a mathematics text is essential. To ensure accuracy in this book, math teachers from around the country have read the pages carefully for typographical errors and have checked all the answers.

### Connections

Many of our students do not thoroughly grasp new concepts the first time they are presented. In this text we encourage students to make connections. That is, we introduce a concept, then later in the text briefly reintroduce it and build upon it. Often an important concept is used in many sections of the text. Important concepts are also reinforced throughout the text in the Cumulative Review Exercises and Cumulative Review Tests.

### Chapter Opening Application

Each chapter begins with a real-life application related to the material covered in the chapter. By the time students complete the chapter, they should have the knowledge to work the problem.

## Goals of This Chapter

This feature on the chapter opener page gives students a preview of the chapter and also indicates where this material will be used again in other chapters of the book. This material helps students see the connections among various topics in the book and the connection to real-world situations.

## The Use of Icons

At the beginning of each exercise set the icons for MathXL®, MathXL®, and for MyMathLab, MyMathLab®, are illustrated to remind students of these homework resources.

## Keyed Section Objectives

Each section opens with a list of skills that the student should learn in that section. The objectives are then keyed to the appropriate portions of the sections with blue numbers such as **1**.

## Problem Solving

Pólya's five-step problem-solving procedure is discussed in Section 1.2. Throughout the book, problem solving and Pólya's problem-solving procedure are emphasized.

## Practical Applications

Practical applications of algebra are stressed throughout the text. Students need to learn how to translate application problems into algebraic symbols. The problem-solving approach used throughout this text gives students ample practice in setting up and solving application problems. The use of practical applications motivates students.

## Detailed, Worked-Out Examples

A wealth of examples have been worked out in a step-by-step, detailed manner. Important steps are highlighted in color, and no steps are omitted until after the student has seen a sufficient number of similar examples.

## Now Try Exercises

In each section, after each example, students are asked to work an exercise that parallels the example given in the text. These Now Try Exercises make the students *active*, rather than passive, learners and they reinforce the concepts as students work the exercises. Through these exercises, students have the opportunity to immediately apply what they have learned. After each example, Now Try Exercises are indicated in green type such as Now Try Excercise 27. They are also indicated in green type in the exercise sets, such as 27.

## Study Skills Section

Students taking this course may benefit from a review of important study skills. Such study skills are essential for success in mathematics. Section 1.1, the first section of the text, discusses such study skills. This section should be very beneficial for your students and should help them to achieve success in mathematics.

## Understanding Algebra

The new **Understanding Algebra** boxes appear in the margin throughout the text. Placed at key points, **Understanding Algebra** helps students focus on the important concepts and facts that they need to master.

## Helpful Hints

The Helpful Hint boxes offer useful suggestions for problem solving and other varied topics. They are set off in a special manner so that students will be sure to read them.

## Avoiding Common Errors

Common student errors are illustrated. Explanations of why the shown procedures are incorrect are given. Explanations of how students may avoid such errors are also presented.

## Exercise Sets

The exercise sets are broken into three main categories: Warm-Up Exercises, Practice the Skills, and Problem Solving. Many exercise sets also contain Concept/Writing, Challenge Problems, and/or Group Activities. Each exercise set is graded in difficulty. The early problems help develop the student's confidence, and then students are eased gradually into the more difficult problems. A sufficient number and variety of examples are given in each section for the student to successfully complete even the more difficult exercises. The number of exercises in each section is more than ample for student assignments and practice.

## Warm-Up Exercises

Exercise sets now begin with new Warm-Up Exercises. These fill-in-the-blank exercises include an emphasis on vocabulary. They serve as a great warm-up to the homework exercises or as 5-minute quizzes.

## Problem-Solving Exercises

These exercises help students become better thinkers and problem solvers. Many of these exercises involve real-life applications of algebra. It is important for students to be able to apply what they learn to real-life situations. Many problem-solving exercises help with this.

## Concept/Writing Exercises

Most exercise sets include exercises that require students to write out the answers in words. These exercises improve students' understanding and comprehension of the material. Many of these exercises involve problem solving and conceptualization and help develop better reasoning and critical thinking skills. These exercises are located following the Problem-Solving exercises within the end-of-section exercise sets.

## Challenge Problems

These exercises, which are part of many exercise sets, provide a variety of problems. Many were written to stimulate student thinking. Others provide additional applications of algebra or present material from future sections of the book so that students can see and learn the material on their own before it is covered in class. Others are more challenging than those in the regular exercise set.

## Cumulative Review Exercises

All exercise sets (after the first two) contain questions from previous sections in the chapter and from previous chapters. These Cumulative Review Exercises will reinforce topics that were previously covered and help students retain the earlier material, while they are learning the new material. For the students' benefit, Cumulative Review Exercises are keyed to the section where the material is covered, using brackets, such as [3.4].

## Group Activities

Many exercise sets have group activity exercises that lead to interesting group discussions. Many students learn well in a cooperative learning atmosphere, and these exercises will get students talking mathematics to one another.

## Mid-Chapter Tests

In the middle of each chapter is a Mid-Chapter Test. Students should take each Mid-Chapter Test to make sure they understand the material presented in the chapter up to that point. In the student answers, brackets such as [2.3] are used to indicate the section where the material was first presented.

## Chapter Summary

At the end of each chapter is a comprehensive chapter summary that includes important chapter facts and examples illustrating these important facts.

## Chapter Review Exercises

At the end of each chapter are review exercises that cover all types of exercises presented in the chapter. The review exercises are keyed using color numbers and brackets, such as [1.5], to the sections where the material was first introduced.

## Chapter Practice Tests

The comprehensive end-of-chapter practice test will enable the students to see how well they are prepared for the actual class test. The section where the material was first introduced is indicated in brackets in the student answers.

## Cumulative Review Tests

These tests, which appear at the end of each chapter after the first, test the students' knowledge of material from the beginning of the book to the end of that chapter. Students can use these tests for review, as well as for preparation for the final exam. These exams, like the Cumulative Review Exercises, will serve to reinforce topics taught earlier. In the answer section, after each answer, the section where that material was covered is given using brackets.

## Answers

The *odd answers* are provided for the exercise sets. *All answers* are provided for the Cumulative Review Exercises, Mid-Chapter Test, Chapter Review Exercises, Chapter Practice Tests, and Cumulative Review Tests. Answers are not provided for the Group Activity exercises since we want students to reach agreement by themselves on the answers to these exercises.

# Prerequisite

This text assumes no prior knowledge of algebra. However, a working knowledge of arithmetic skills is important. Fractions are reviewed early in the text, and decimals and percent are reviewed in Appendix A.

# Modes of Instruction

The format and readability of this book lend it to many different modes of instruction. The constant reinforcement of concepts will result in greater understanding and retention of the material by your students.

The features of the text and the large variety of supplements available make this text suitable for many types of instructional modes, including

- lecture
- hybrid or blended courses
- distance learning
- self-paced instruction
- modified lecture
- cooperative or group study
- learning laboratory

# Student and Instructor Resources

## STUDENT RESOURCES

| | |
|---|---|
| **Student Solutions Manual**<br>Provides complete worked-out solutions to<br>• the odd-numbered section exercises<br>• all exercises in the Mid-Chapter Tests, Chapter Reviews, Chapter Practice Tests, and Cumulative Review Tests | **Worksheets for Classroom or Lab Practice**<br>• Extra practice exercises for every section of the text with ample space for students to show their work |
| **Lecture Videos**<br>• For each section of the text, there are about 20 minutes of lecture.<br>• Captioned in English and Spanish<br>• Available in MyMathLab® | **Chapter Test Prep Videos**<br>• Step-by-step solutions to every exercise in each Chapter Practice Test<br>• Available in MyMathLab®<br>• Available on YouTube at http://www.youtube.com/user/AngelElemInterAlg |

## INSTRUCTOR RESOURCES

| | |
|---|---|
| **Annotated Instructor's Edition**<br>Contains all the content found in the student edition, plus the following:<br>• Answers to exercises on the same text page with graphing answers in the Graphing Answer section at the back of the text<br>• Instructor Example provided in the margin paired with each student example | **Instructor's Resource Manual with Tests and Mini-Lectures**<br>• Mini-lectures for each text section<br>• Several forms of test per chapter (free response and multiple choice)<br>• Answers to all items<br>• Available for download from the IRC and in MyMathLab® |
| | **TestGen®**<br>• Available for download from the IRC |
| **Instructor's Solutions Manual**<br>• Available for download from the IRC and in MyMathLab® | **Online Resources**<br>• MyMathLab® (access code required)<br>• MathXL® (access code required) |

# Acknowledgments

We thank our spouses, Kathy Angel and Kris Runde, for their support and encouragement throughout the project. We are grateful for their wonderful support and understanding while we worked on the book.

We also thank our children: Robert and Steven Angel and Alex, Nick, and Max Runde. They also gave us support and encouragement and were very understanding when we could not spend as much time with them as we wished because of book deadlines. Special thanks to daughter-in-law, Kathy; mother-in-law, Patricia, and father-in-law, Scott. Without the support and understanding of our families, this book would not be a reality.

We want to thank Rafiq Ladhani and his team at Edumedia for accuracy reviewing the pages and checking all answers.

We thank Larry Gilligan, University of Cincinnati, and Donna Petrie, Monroe Community College, for their contributions to the series.

Many people at Pearson deserve thanks, including all those listed on the copyright page. In particular, we thank Michael Hirsch, Editor in Chief; Matt Summers, Editorial Project Manager; Michelle Renda, Executive Marketing Manager; Patty Bergin, Senior Production Manager; and Barbara Atkinson, Senior Designer.

We would like to thank the following reviewers and focus group participants for their thoughtful comments and suggestions:

Laura Adkins, *Missouri Southern State College, MO*
Darla Aquilar, *Pima Community College, AZ*
Arthur Altshiller, *Los Angeles Valley College, CA*
Frances Alvarado, *University of Texas–Pan American, TX*
Jose Alvarado, *University of Texas–Pan American, TX*
Jacob Amidon, *Cayuga Community College, NY*
Bhagirathi Anand, *Long Beach City College, CA*
Ben Anderson, *Darton College, GA*
Sheila Anderson, *Housatonic Community College, CT*
Peter Arvanites, *State University of New York–Rockland Community College, NY*
Jannette Avery, *Monroe Community College, NY*
Mary Lou Baker, *Columbia State Community College, TN*
Linda Barton, *Ball State University, IN*
Jon Becker, *Indiana University, IN*
Sharon Berrian, *Northwest Shoals Community College, AL*
Paul Boisvert, *Oakton Community College, IL*
Dianne Bolen, *Northeast Mississippi Community College, MS*
Julie Bonds, *Sonoma State University, CA*
Beverly Broomell, *Suffolk County Community College, NY*
Clark Brown, *Mojave Community College, AZ*
Connie Buller, *Metropolitan Community College, NE*
Lavon Burton, *Abilene Christian University, TX*
Marc Campbell, *Daytona Beach Community College, FL*
Mitzi Chaffer, *Central Michigan University, MI*
Terry Cheng, *Irvine Valley College, CA*
Julie Chesser, *Owens Community College, OH*
Kim Christensen, *Maple Woods Community College, MO*

Barry Cogan, *Macomb Community College, MI*
Pat C. Cook, *Weatherford College, TX*
Ted Corley, *Arizona State University and Glendale Community College, AZ*
Charles Curtis, *Missouri Southern State College, MO*
Joseph de Guzman, *Riverside City College (Norco), CA*
Lisa Delong Cuneo, *Penn. State University–Dubois, PA*
Stephan Delong, *Tidewater Community College, VA*
Deborah Doucette, *Erie Community College (North), NY*
Marla Dresch Butler, *Gavilan Community College, CA*
William Echols, *Houston Community College, TX*
Gary Egan, *Monroe Community College, NY*
Mark W. Ernsthausen, *Monroe Community College, NY*
Elizabeth Farber, *Bucks County Community College, PA*
Dale Felkins, *Arkansas Technical University, AR*
Warrene Ferry, *Jones County Junior College, MS*
Christine Fogal, *Monroe Community College, NY*
Reginald Fulwood, *Palm Beach Community College, FL*
Gary Glaze, *Spokane Falls Community College, WA*
James Griffiths, *San Jacinto College, TX*
Susan Grody, *Broward Community College, FL*
Kathy Gross, *Cayuga Community College, NY*
Abdollah Hajikandi, *State University of New York–Buffalo, NY*
Cynthia Harrison, *Baton Rouge Community College, LA*
Mary Beth Headlee, *State College of Florida, FL*
Richard Hobbs, *Mission College, CA*
Joe Howe, *St. Charles Community College, MO*
Laura L. Hoye, *Trident Technical College, SC*
Barbara Hughes, *San Jacinto Community College (Central), TX*
Kelly Jahns, *Spokane Community College, WA*
Mary Johnson, *Inver Hills Community College, MN*
Cheryl Kane, *University of Nebraska–Lincoln, NE*
Judy Kasabian, *El Camino College, CA*
Jane Keller, *Metropolitan Community College, NE*
Mike Kirby, *Tidewater Community College, VA*
Maryanne Kirkpatrick, *Laramie County Community College, WY*
Marcia Kleinz, *Atlantic Cape Community College, NJ*
William Krant, *Palo Alto College, TX*
Gayle L. Krzemien, *Pikes Peak Community College, CO*
Shannon Lavey, *Cayuga Community College, NY*
Mitchel Levy, *Broward Community College, FL*
Mitzi Logan, *Pitt Community College, NC*
Jason Mahar, *Monroe Community College, NY*
Kimberley A. Martello, *Monroe Community College, NY*
Constance Meade, *College of Southern Idaho, ID*
Lynette Meslinsky, *Erie Community College, NY*
Shywanda Moore, *Meridian Community College, MS*
Elizabeth Morrison, *Valencia Community College, FL*
Catherine Moushon, *Elgin Community College, IL*
Elsie Newman, *Owens Community College, OH*
Charlotte Newsom, *Tidewater Community College, VA*
Kathy Nickell, *College of DuPage, IL*
Charles Odion, *Houston Community College, TX*
Jean Olsen, *Pikes Peak Community College, CO*
Shelle Patterson, *Moberly Area Community College, MO*
Patricia Pifko, *Housatonic Community College, CT*

Jeanne Pirie, *Erie Community College (North), NY*
David Price, *Tarrant County College, TX*
Elise Price, *Tarrant County College, TX*
Dennis Reissig, *Suffolk County Community College, NY*
Linda Retterath, *Mission College, CA*
Dale Rohm, *University of Wisconsin–Stevens Point, WI*
Behnaz Rouhani, *Athens Technical College, GA*
Troy Rux, *Spokane Falls Community College, WA*
Hassan Saffari, *Prestonburg Community College, KY*
Brian Sanders, *Modesto Junior College, CA*
Glenn R. Sandifer, *San Jacinto Community College, (Central), TX*
Rebecca Schantz, *Prairie State College, IL*
Cristela Sifuentez, *University of Texas–Pan American, TX*
Rick Silvey, *St. Mary College, KS*
Julia Simms, *Southern Illinois University–Edwardsville, IL*
Linda Smoke, *Central Michigan University, MI*
Jed Soifer, *Atlantic Cape Community College, NJ*
Richard C. Stewart, *Monroe Community College, NY*
Elizabeth Suco, *Miami–Dade College, FL*
Fereja Tahir, *Illinois Central College, IL*
Harold Tanner, *Orangeburg–Calhoun Technological College, SC*
Dale Thielker, *Ranken Technological College, MO*
Burnette Thompson, Jr., *Houston Community College, TX*
Mary Vachon, *San Joaquin Delta College, CA*

Andrea Vorwark, *Maple Woods Community College, MO*
Ken Wagman, *Gavilan Community College, CA*
Patrick Ward, *Illinois Central College, IL*
Robert E. White, *Allan Hancock College, CA*
Cindy Wilson, *Henderson State University, AZ*
Ronald Yates, *Community College of Southern Nevada, NV*

## Focus Group Participants

Linda Barton, *Ball State University, IN*
Karen Egedy, *Baton Rouge Community College, LA*
Daniel Fahringer, *Harrisburg Area Community College, PA*
Sharon Hansa, *Longview Community College, MO*
Cynthia Harrison, *Baton Rouge Community College, LA*
Judy Kasabian, *El Camino College, CA*
Mark Molino, *Erie Community College, NY*
Kris Mudunuri, *Long Beach City College, CA*
Fred Peskoff, *Borough of Manhattan Community College, NY*
David Price, *Tarrant County College, TX*
Elise Price, *Tarrant County College, TX*
Adrian Ranic, *Erie Community College, NY*
Dale Siegel, *Kingsborough Community College, NY*
Christopher Yarish, *Harrisburg Area Community College, PA*

# To the Student

Algebra is a course that requires active participation. You must read the text and pay attention in class, and, most important, you must work the exercises. The more exercises you work, the better.

The text was written with you in mind. Short, clear sentences are used, and many examples are given to illustrate specific points. The text stresses useful applications of algebra. Hopefully, as you progress through the course, you will come to realize that algebra is not just another math course that you are required to take, but a course that offers a wealth of useful information and applications.

This text makes full use of color. The different colors are used to highlight important information. Important procedures, definitions, and formulas are placed within colored boxes.

The boxes marked **Understanding Algebra** should be studied carefully. They emphasize concepts and facts that you need to master to succeed. **Helpful Hints** should be studied carefully, for they stress important information. Be sure to study **Avoiding Common Errors** boxes. These boxes point out common errors and provide the correct procedures for doing these problems.

After each example you will see a Now Try Exercise reference, such as Now Try Exercise 27. The exercise indicated is very similar to the example given in the book. You may wish to try the indicated exercise after you read the example to make sure you truly understand the example. In the exercise set, the Now Try Exercises are written in green, such as 27.

Some questions you should ask your professor early in the course include: What supplements are available for use? Where can help be obtained when the professor is not available? Supplements that may be available include the Student Solutions Manual; the Lecture Series Videos; the Chapter Test Prep Videos; MathXL®; and MyMathLab®. All these items are discussed under the heading of Supplements in Section 1.1 and listed in the Preface.

You may wish to form a study group with other students in your class. Many students find that working in small groups provides an excellent way to learn the material. By discussing and explaining the concepts and exercises to one another, you reinforce your own understanding. Once guidelines and procedures are determined by your group, make sure to follow them.

One of the first things you should do is to read Section 1.1, Study Skills for Success in Mathematics. Read this section slowly and carefully, and pay particular attention to the advice and information given. Occasionally, refer back to this section. This could be the most important section of the book. Pay special attention to the material on doing your homework and on attending class.

At the end of all exercise sets (after the first two) are

**Cumulative Review Exercises**. You should work these problems on a regular basis, even if they are not assigned. These problems are from earlier sections and chapters of the text, and they will refresh your memory and reinforce those topics. If you have a problem when working these exercises, read the appropriate section of the text or study your notes that correspond to that material. The section of the text where the Cumulative Review Exercise was introduced is indicated by brackets, [ ], to the left of the exercise. After reviewing the material, if you still have a problem, make an appointment to see your professor. Working the Cumulative Review Exercises throughout the semester will also help prepare you to take your final exam.

Near the middle of each chapter is a **Mid-Chapter Test**. You should take each Mid-Chapter Test to make sure you understand the material up to that point. The section where the material was first introduced is given in brackets after the answer in the answer section of the book.

At the end of each chapter are a **Chapter Summary, Chapter Review Exercises**, a **Chapter Practice Test**, and a **Cumulative Review Test**. Before each examination you should review this material carefully and take the Chapter Practice Test (you may want to review the *Chapter Test Prep Video* also). If you do well on the Chapter Practice Test, you should do well on the class test. The questions in the Review Exercises are marked to indicate the section in which that material was first introduced. If you have a problem with a Review Exercise question, reread the section indicated. You may also wish to take the Cumulative Review Test that appears at the end of every chapter (starting with Chapter 2).

In the back of the text there is an **answer section** that contains the answers to the *odd-numbered* exercises, including the Challenge Problems. Answers to *all* Cumulative Review Exercises, Mid-Chapter Tests, Chapter Review Exercises, Chapter Practice Tests, and Cumulative Review Tests are provided. Answers to the Group Activity exercises are not provided, for we wish students to reach agreement by themselves on answers to these exercises. The answers should be used only to check your work. For the Mid-Chapter Tests, Chapter Practice Tests, and Cumulative Review Tests, after each answer the section number where that type of exercise was covered is provided.

We have tried to make this text as clear and error free as possible. No text is perfect, however. If you find an error in the text, or an example or section that you believe can be improved, we would greatly appreciate hearing from you. If you enjoy the text, we would also appreciate hearing from you. You can submit comments to math@pearson.com, subject for Allen Angel and Dennis Runde.

**Allen R. Angel**
**Dennis C. Runde**

# 1 Real Numbers

## Goals of This Chapter

This chapter will provide you with the foundation that you need in order to succeed in this course and all other mathematics courses you will take. Learning proper study skills is the first step in building this foundation. *Please read Section 1.1 carefully and follow the advice given.* The emphasis of this chapter is to provide you with an understanding of the real number system.

In this chapter, you will learn a five-step problem-solving procedure that will be used throughout the book. Once you have learned the material in this chapter, you will be able to tackle the subsequent chapters in the book with confidence.

**A college education is worth money!** The amount of average annual income increases dramatically as one's education increases. For example, in 2007 someone with a bachelor's degree earned more than twice as much as a person without a high school diploma. In Exercise 43 on page 19, we will see how to analyze pictorial data to calculate the financial advantages of a college education.

© Andrew Rich/iStockphoto

# 1.1 Study Skills for Success in Mathematics

**1** Recognize the goals of this text.

**2** Learn proper study skills.

**3** Prepare for and take exams.

**4** Learn to manage time.

This section is extremely important. Take the time to read it carefully and follow the advice given.

Most of you taking this course fall into one of three categories: (1) those who did not take algebra in high school, (2) those who took algebra in high school but did not understand the material, or (3) those who successfully completed algebra in high school but have been out of school for some time and need to take the course again. Whichever the case, you will need to acquire study skills for mathematics courses.

Before we discuss study skills, we will present the goals of this text. These goals may help you realize why certain topics are covered in the text and why they are covered as they are.

## **1** Recognize the Goals of This Text

The goals of this text include:

1. Presenting traditional algebra topics
2. Preparing you for more advanced mathematics courses
3. Building your confidence in, and your enjoyment of, mathematics
4. Improving your reasoning and critical thinking skills
5. Increasing your understanding of how important mathematics is in solving real-life problems
6. Encouraging you to think analytically, so that you will feel comfortable translating real-life problems into mathematical equations, and then solving the problems.

In addition to teaching you the mathematical content, our goals are to teach you to be more *mathematically literate*, which is also called *quantitatively literate*. We wish to teach you to *communicate mathematically*, to teach you to *understand and interpret data* in a variety of formats, to teach you measurement and geometric concepts, to teach you to *reason more logically*, and to teach you to be able to represent real world applications mathematically, which is called *modeling*. Throughout the book we will strive to increase your mathematical understanding to help you become more successful in mathematics, in your future job, and throughout life.

We also realize that some of you may have some mathematics anxiety. We have written the book to try to help you overcome that anxiety by building your confidence in mathematics.

It is important to realize that this course is the foundation for more advanced mathematics courses. A thorough understanding of algebra will make it easier for you to succeed in later mathematics courses and in life.

## **2** Learn Proper Study Skills

***Have a Positive Attitude*** You may be thinking to yourself, "I hate math," or "I wish I did not have to take this class." You may have heard of "math anxiety" and feel you fit this category. The first thing to do to be successful in this course is to change your attitude to a more positive one. You must be willing to give this course, and yourself, a fair chance.

Based on past experiences in mathematics, you may feel that this is difficult. However, mathematics is something you need to work at. Many of you are more mature now than when you took previous mathematics courses. Your maturity and desire to learn are extremely important and can make a tremendous difference in your ability to succeed in mathematics. I believe you can be successful in this course, but you also need to believe it.

***Prepare for and Attend Class*** To be prepared for class, you need to do your homework assignments completely. If you have difficulty with the homework, or some of the concepts, write down questions to ask your instructor. If you were given a reading assignment, read the appropriate material carefully before class.

After the material is explained in class, read the corresponding sections of the text slowly and carefully, word by word.

You should plan to attend every class. Generally, the more absences you have, the lower your grade will be. Every time you miss a class, you miss important information. If you must miss a class, contact your instructor ahead of time, and get the reading assignment and homework. If possible, before the next class, try to copy a friend's notes to help you understand the material you missed.

*In algebra and other mathematics courses, the material you learn is cumulative. The new material is built on material that was presented previously. You must understand each section before moving on to the next section, and each chapter before moving on to the next chapter. Therefore, do not let yourself fall behind.* Seek help as soon as you need it—do not wait! You will greatly increase your chance of success in this course by following the study skills presented in this section.

While in class, pay attention to what your instructor is saying. If you don't understand something, ask your instructor to repeat the material. If you don't ask questions, your instructor will not know that you have a problem understanding the material.

In class, take careful notes. Write numbers and letters clearly, so that you can read them later. Make sure your *x*'s do not look like *y*'s and vice versa. It is not necessary to write down every word your instructor says. Copy the major points and the examples that do not appear in the text. You should not be taking notes so frantically that you lose track of what your instructor is saying.

**Read the Text** Mathematics textbooks should be read slowly and carefully, word by word. If you do not understand something, reread that material. It is a good idea to read with a pencil in your hand, making notes as you proceed.

Don't panic! As you read the examples, notice that the "flow" is basically downward. It is a challenge but try to understand the reasons for each step. This downward movement is a sequence of steps that takes a problem from statement toward its solution. Each step is important to understand. If you have trouble with the rationale for a step, you should ask your instructor for clarification.

When you come across a new concept or definition, you may wish to underline or highlight it so that it stands out. Then it will be easier to find later. Also, work the **Now Try Exercises** that appear in the text following each example. The Now Try Exercises are designed so that you have the opportunity to immediately apply new ideas. Make notes of things you do not understand to ask your instructor.

There are numerous boxes in the left margin marked **Understanding Algebra**. These boxes give alternative wording and additional illustration of important concepts. You may want to give these special attention as you read and see how they help with topics in the text and examples.

This textbook has other special features to help you. I suggest that you pay particular attention to these highlighted features, including the **Avoiding Common Errors** boxes, the **Helpful Hint** boxes, and important procedures and definitions identified by color. The Avoiding Common Errors boxes point out the most common errors made by students. Read and study this material very carefully and make sure that you understand what is explained. If you avoid making these common errors, your chances of success in this and other mathematics classes will be increased greatly. The Helpful Hints offer many valuable techniques for working certain problems. They may also present some very useful information or show an alternative way to work a problem.

**Do the Homework** *Two very important commitments that you must make to be successful in this course are attending class and doing your homework regularly.* Your assignments must be worked conscientiously and completely. Do your homework as soon as possible, so the material presented in class will be fresh in your mind. It is through doing homework that you truly learn the material. While working homework you will become aware of the types of problems that you need further help with. If you do not work the assigned exercises, you will not know what questions to ask in class.

When you do your homework, make sure that you write it neatly and carefully. Pay particular attention to copying signs and exponents correctly.

Don't forget to check the answers to your homework assignments. This book contains the answers to the odd-numbered exercises in the back of the book. In addition, the answers to all the Cumulative Review Exercises, Mid-Chapter Tests, Chapter Review Exercises, Chapter Practice Tests, and Cumulative Review Tests are in the back of the book. The section number where the material is first introduced is provided next to the exercises for the Cumulative Review Exercises and Chapter Review Exercises. The section number where the material is first introduced is provided with the answers in the back of the book for the Mid-Chapter Tests, Chapter Practice Tests, and Cumulative Review Tests. Answers to the Group Activity Exercises are not provided because we want you to arrive at the answers as a group.

Ask questions in class about homework problems you don't understand. You should not feel comfortable until you understand all the concepts needed to work every assigned problem successfully.

**Study for Class**   Study in the proper atmosphere, in an area where you will not be constantly disturbed, so that your attention can be devoted to what you are reading. The area where you study should be well ventilated and well lit. You should have sufficient desk space to spread out all your materials. Your chair should be comfortable. You should try to minimize distractions while you are studying. You should not study for hours on end. Short study breaks are a good idea.

Before you begin studying, make sure that you have all the materials you need (pencils, markers, calculator, etc.). You may wish to highlight the important points covered in class or in the book.

It is recommended that students study and do homework for at least two hours for each hour of class time. Some students require more time than others. It is important to spread your studying time out over the entire week rather than studying during one large block of time.

When studying, you should not only understand how to work a problem but also know *why* you follow the specific steps you do to work the problem. If you do not have an understanding of why you follow the specific process, you will not be able to transfer the process to solve similar problems.

This book has Mid-Chapter Tests in the middle of each chapter. These exercises reinforce material presented in the first half of the chapter. They will also help you determine if you need to go back and review the topics covered in the first half of the chapter. For any of the Mid-Chapter Test questions that you get incorrect, turn to the section provided with the answers in the back of the book and review that section. This book also has Cumulative Review Exercises at the end of every section after Section 1.2. These exercises reinforce material presented earlier in the course, and you will be less likely to forget the material if you review it repeatedly throughout the course. The exercises will also help prepare you for the final exam. Even if these exercises are not assigned for homework, I urge you to work them as part of your studying process.

## 3 Prepare for and Take Exams

If you study a little bit each day you should not need to cram the night before an exam. Begin your studying early. If you wait until the last minute, you may not have time to seek the help you may need if you find you cannot work a problem.

To prepare for an exam:

1. Read your class notes.
2. Review your homework assignments.
3. Study formulas, definitions, and procedures you will need for the exam.
4. Read the Avoiding Common Errors boxes and Helpful Hint boxes carefully.
5. Read the summary at the end of each chapter.
6. Work the Chapter Review Exercises at the end of each chapter. If you have difficulties, restudy those sections. If you still have trouble, seek help.

7. Work the Mid-Chapter Test and the Chapter Practice Test.

8. Rework quizzes previously given if the material covered in the quizzes will be included on the test.

9. If your exam is a cumulative exam, work the Cumulative Review Test.

10. Now, if you can arrange it, you may want to consider a session of study with a partner or group from your class. With a partner, you can construct a sample test to take to simulate your actual test and help alleviate test anxiety. Try these steps:

   a) Using three-by-five-inch index cards, go through the text and select representative problems—writing the question on one side of the card and the answer or page reference on the other. Choose questions you think will most likely be asked; don't choose easy problems. Have your study partner do the same thing. Probably 20 to 25 good, representative questions should do it.

   b) Here is the key: *shuffle the cards*. One thing that makes tests more difficult than homework is that homework problems are often of the same type and knowing how to start the problem is not too difficult. But test questions are all mixed up and to simulate that, shuffle the cards.

   c) You take your partner's test—be sure to give yourself the same amount of time your instructor will give you—and your partner takes your test. Try to avoid distractions (music, food, etc.). Grade your partner's test and have your partner grade your test. Then study weak areas and repeat the process if necessary.

**Prepare for Midterm and Final Exam**  When studying for a comprehensive midterm or final exam follow the procedures discussed for preparing for an exam. However, also:

1. Study all your previous tests and quizzes carefully. Make sure that you have learned to work the problems that you may have previously missed.

2. Work the Cumulative Review Test at the end of each chapter. These tests cover the material from the beginning of the book to the end of that chapter.

3. If your instructor has given you a worksheet or practice exam, make sure that you complete it. Ask questions about any problems you do not understand.

4. Begin your studying process early so that you can seek all the help you need in a timely manner.

© Lisa F. Young/Shutterstock

**Take an Exam**  Make sure you get sufficient sleep the night before the test. Arrive at the exam site early so that you have a few minutes to relax before the exam. If you rush into the exam, you will start out nervous and anxious. After you are given the exam, you should do the following:

1. Carefully write down any formulas or ideas that you want to remember.

2. Look over the entire exam quickly to get an idea of its length. Also make sure that no pages are missing.

3. Read the test directions carefully.

4. Read each question carefully. Show all of your work. Answer each question completely, and make sure that you have answered the specific question asked.

5. Work the questions you understand best first; then go back and work those you are not sure of. Do not spend too much time on any one problem or you may not be able to complete the exam. Be prepared to spend more time on problems worth more points.

6. Attempt each problem. You may get at least partial credit even if you do not obtain the correct answer. If you make no attempt at answering the question, you will lose full credit.

7. Work carefully step by step. Copy all signs and exponents correctly when working from step to step, and make sure to copy the original question from the test correctly.

8. Write clearly so that your instructor can read your work. If your instructor cannot read your work, you may lose credit. When appropriate, make sure that your final answer stands out by placing a box around it.

9. If you have time, check your work and your answers.

10. Do not be concerned if others finish the test before you or if you are the last to finish. Use any extra time to check your work.

Stay calm when taking your test. Do not get upset if you come across a problem you can't figure out right away. Go on to something else and come back to that problem later.

## 4 Learn to Manage Time

As mentioned earlier, it is recommended that students study and do homework for at least two hours for each hour of class time. Finding the necessary time to study is not always easy. The following are some suggestions that you may find helpful.

1. Plan ahead. Determine when you will study and do your homework. Do not schedule other activities for these periods. Try to space these periods evenly over the week.

2. Be organized, so that you will not have to waste time looking for your books, your pencil, your calculator, or your notes.

3. If you are allowed to use a calculator, use it for tedious calculations.

4. When you stop studying, clearly mark where you stopped in the text.

5. Try not to take on added responsibilities. You must set your priorities. If your education is a top priority, as it should be, you may have to reduce time spent on other activities.

6. If time is a problem, do not overburden yourself with too many courses.

*Use Supplements* This text comes with a large variety of supplements. Find out from your instructor early in the semester which supplements are available and might be beneficial for you to use. Supplements should not replace reading the text, but should be used to enhance your understanding of the material. If you miss a class, you may want to review the video on the topic you missed before attending the next class.

The supplements available are: the Student Solutions Manual which works out the odd section exercises as well as all the end-of-chapter exercises; the Lecture Series Videos, which show about 20 minutes of lecture per section; the Chapter Test Prep Video, which works out every problem in every Chapter Practice Test; MathXL® MathXL®, a powerful online tutorial and homework system; MyMathLab® MyMathLab, the online course which houses MathXL. The Lecture Series Videos and Chapter Test Prep Videos are available through MyMathLab. The Chapter Test Prep Videos are also available on YouTube (search Angel "Elementary and Intermediate Algebra" and click on "Channels").

*Seek Help* Be sure to get help as soon as you need it! Do not wait! In mathematics, one day's material is usually based on the previous day's material. So, if you don't understand the material today, you may not be able to understand the material tomorrow.

Where should you seek help? There are often a number of resources on campus. Try to make a friend in the class with whom you can study. Often, you can help one another. You may wish to form a study group with other students in your class. Discussing the concepts and homework with your peers will reinforce your own understanding of the material.

You should know your instructor's office hours, and you should not hesitate to seek help from your instructor when you need it. Make sure you read the assigned material and attempt the homework before meeting with your instructor. Come prepared with specific questions to ask.

There are often other sources of help available. Many colleges have a mathematics lab or a mathematics learning center where tutors are available. Ask your instructor early in the semester where and when tutoring is available. Arrange for a tutor as soon as you need one.

*A Final Word* You can be successful at mathematics if you attend class regularly, pay attention in class, study your text carefully, do your homework daily, review regularly, and seek help as soon as you need it. Good luck in your course and remember: *Mathematics is not a spectator sport!*

## EXERCISE SET 1.1   MathXL®   MyMathLab®

*Do you know:*

1. your professor's name and office hours?
2. your professor's office location and telephone number?
3. where and when you can obtain help if your professor is not available?
4. the name and phone number of a friend in your class?
5. what supplements are available to assist you in learning?
6. if your instructor is recommending the use of a particular calculator?
7. when you can use your calculator in this course?
8. if your instructor is requiring the use of MyMathLab?

*If you do not know the answers to questions 1–8, you should find out as soon as possible.*

9. What are your goals for this course?
10. What are your reasons for taking this course?
11. List the things you need to do to prepare properly for class.

12. Are you beginning this course with a positive attitude? It is important that you do!
13. For each hour of class time, how many hours outside of class are recommended for studying and doing homework?
14. Explain how a mathematics text should be read.
15. Two very important commitments that you must make to be successful in this course are **a)** doing homework regularly and completely and **b)** attending class regularly. Explain why these commitments are necessary.
16. When studying, you should not only understand how to work a problem, but also why you follow the specific steps you do. Why is this important?
17. Have you given any thought to studying with a friend or a group of friends? Can you see any advantages in doing so? Can you see any disadvantages in doing so?
18. Write a summary of the steps you should follow when taking an exam.

# 1.2 Problem Solving

1 Learn the five-step problem-solving procedure.
2 Solve problems involving bar, line, and circle graphs.
3 Solve problems involving statistics.

## 1 Learn the Five-Step Problem-Solving Procedure

One of the main reasons we study mathematics is to use it to solve real-life problems. To solve most real-life problems mathematically, we need to be able to express the problem in mathematical symbols. We will spend a great deal of time explaining how to express real-life applications mathematically.

You can approach any problem using the general five-step **problem-solving procedure** developed by George Pólya (1887–1985) in his book *How to Solve It*.

**Guidelines for Problem Solving**

1. **Understand the problem.**
   - Read the problem *carefully* at least twice. In the first reading, get a general overview of the problem. In the second reading, determine (*a*) exactly what you are being asked to find and (*b*) what information the problem provides.
   - Make a list of the given facts. Determine which are pertinent to solving the problem.
   - Determine whether you can substitute smaller or simpler numbers to make the problem more understandable.
   - If it will help you organize the information, list the information in a table.
   - If possible, make a sketch to illustrate the problem. Label the information given.

2. **Translate the problem to mathematical language.**
   - This will generally involve expressing the problem in terms of an algebraic expression or equation. (We will explain how to express application problems as equations in Chapter 3.)
   - Determine whether there is a formula that can be used to solve the problem.

3. **Carry out all necessary calculations.**

4. **Check the answer obtained in step 3.**
   - Ask yourself, "Does the answer make sense?" "Is the answer reasonable?" If the answer is not reasonable, recheck your method for solving the problem and your calculations.
   - Check the solution in the original wording of the problem if possible.

5. **Make sure you have answered the question.**
   - State the answer clearly.

**Understanding Algebra**

An *expression* is a collection of numbers, letters, grouping symbols, and operations.

In step 2 we use the words *algebraic expression*. An **algebraic expression**, sometimes simply referred to as an **expression**, is a general term for any collection of numbers, letters (called variables), grouping symbols such as parentheses ( ) or brackets [ ], and **operations** (such as addition, subtraction, multiplication, and division). In this section we will not be using variables, so we will discuss their use later.

Examples of Expressions

$$3 + 4, \qquad 6(12 \div 3), \qquad (2)(7)$$

The following examples show how to apply the guidelines for problem solving. In some problems it may not be possible or necessary to list every step in the procedure. If you need to review procedures for adding, subtracting, multiplying, or dividing decimal numbers, or if you need a review of percents, read Appendix A before proceeding.

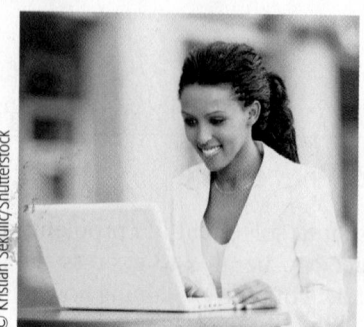

EXAMPLE 1 **Buying Games** Georgia May is deciding which would be less expensive, buying her son's birthday presents on eBay or buying them at a local toy store. Founded in 1995, eBay is The World's Online Marketplace® for the sale of goods and services by a diverse community of individuals and small businesses. The eBay community includes more than 100 million registered members from around the world. The local toy store is only minutes from Georgia's house. Therefore, the cost of gasoline for her car will not factor into her decision. On eBay, the three games Georgia would like to purchase cost $5.99, $9.95, and $19.95. Shipping costs for the games would total $11.10. There would be no sales tax on this purchase. At the local toy store, the total cost for the same three games would be $57.89 plus 8.25% sales tax.

a) Which would be less expensive for Georgia, purchasing the games on eBay or at the local toy store?

b) How much would Georgia save by making the less expensive purchase?

*Solution*

**a) Understand the problem**  A careful reading of the problem shows that the task is to determine if it would be less expensive for Georgia to purchase the games on eBay or at the local toy store. Make a list of all the information given and determine which information is needed to solve the problem.

| Information Given | Pertinent to Solving the Problem? |
|---|---|
| eBay includes more than 100 million registered members | no |
| $5.99, $9.95, and $19.95 cost of the games on eBay | yes |
| $11.10 shipping costs on eBay | yes |
| no sales tax on the eBay purchase | yes |
| cost of gasoline not a factor in Georgia's decision | yes |
| $57.89 cost of the games at the local toy store | yes |
| 8.25% sales tax at the local toy store | yes |

To determine whether eBay or a local store would be the better choice for purchasing the games, it is *not* necessary to know that eBay includes more than 100 million registered members. Solving this problem involves:

- calculating the total cost of games on eBay (including shipping)
- calculating the total cost of games at local store (including sales tax)

To calculate sales tax, you need to determine 8.25% of the cost of the games at the local toy store. When performing calculations, numbers given as percents are changed to decimal numbers. So we will use 0.0825 for 8.25%.

**Translate the problem into mathematical language**

total cost of games on eBay = cost of each individual game + shipping costs

total cost of games at local toy store = total cost of games + 8.25% sales tax

**Carry out the calculations**

total cost of games on eBay = $5.99 + $9.95 + $19.95 + $11.10 = $46.99

total cost of games at local toy store = $57.89 + 0.0825($57.89)

= $57.89 + $4.78 = $62.67

**Check the answer**   The total costs of $46.99 and $62.67 are reasonable based on the information given.

**Answer the question asked**   It would be less expensive for Georgia to purchase the games on eBay.

**b) Understand**  To determine how much Georgia would save by making the less expensive purchase, you need to subtract the total cost of the games on eBay from the total cost of the games at the local toy store.

**Translate**

total cost at toy store − total cost on eBay = amount Georgia would save

**Carry Out**   $62.67 − $46.99 = $15.68

**Check**   The answer $15.68 seems reasonable.

**Answer**   Georgia would save $15.68 by purchasing the games on eBay.

Now Try Exercise 27

## EXAMPLE 2  Processor Speed

In April 2008, IBM began shipping its fastest processor, the Power6, that could perform 5.0 billion operations per second (5.0 gigahertz, symbolized 5.0 GHz). How many operations could it perform in 0.4 second?

*Solution*   Understand   We are given the name of the processor, a speed of about 5.0 billion (5,000,000,000) operations per second, and 0.4 second. To determine the answer to this problem, the name of the processor, Power6, is not needed.

To obtain the answer, will we need to multiply or divide? Often a fairly simple problem seems more difficult because of the numbers involved. When very large or very small numbers make the problem confusing, try substituting commonly used numbers in the problem to determine your problem-solving strategy. Suppose the problem said the processor can perform 6 operations per second. How many operations can it perform in 2 seconds? Simply multiply $6 \times 2$ to get 12. Since we multiplied to obtain this answer, we also will need to multiply to obtain the answer to the given problem.

Translate
number of operations in 0.4 second = 0.4(number of operations per second)

Carry Out

$$= 0.4(5{,}000{,}000{,}000)$$
$$= 2{,}000{,}000{,}000$$

Check   The answer, 2,000,000,000 operations, is less than the 5,000,000,000 operations per second, which makes sense because the processor is operating for less than a second.

Answer   In 0.4 second, the processor can perform about 2,000,000,000 operations.

Now Try Exercise 21

## EXAMPLE 3  Medical Insurance

Brook Matthews' medical insurance policy is similar to that of many workers. Her policy requires that she pay the first $100 of medical expenses each calendar year (called a deductible). After the deductible is paid, she pays 20% of the medical expenses (called a co-payment) and the insurance company pays 80%. (There is a maximum co-payment of $600 that she must pay each year. After that, the insurance company pays 100% of the fee schedule.) On January 1, Brook sprained her ankle playing tennis. She went to the doctor's office for an examination and X rays. The total bill of $325 was sent to the insurance company.

**a)** How much of the bill will Brook be responsible for?

**b)** How much will the insurance company be responsible for?

*Solution*   **a)** Understand   First we list all the *relevant* given information.

Given Information

$100 deductible

20% co-payment after deductible

80% paid by insurance company after deductible

$325 doctor bill

All the other information is not needed to solve the problem. Brook will be responsible for the first $100 and 20% of the remaining balance. The insurance company will be responsible for 80% of the balance after the deductible. Before we can find what Brook owes, we need to first find the balance of the bill after the deductible. The balance of the bill after the deductible is $325 − $100 = $225.

Translate
Brook's responsibility = deductible + 20% of balance of bill after the deductible

**Understanding Algebra**

When solving word problems, it is often helpful to solve a simpler problem first.

**Understanding Algebra**

Notice that the answer to the problem in Example 2 is a complete English sentence, not simply a number, and the number has a unit, operations.

Carry Out      Brook's responsibility = 100 + 20%(225)

$$= 100 + 0.20(225)$$

$$= 100 + 45$$

$$= 145$$

Check and Answer   The answer appears reasonable. Brook will be responsible for $145.

**b)** The insurance company will be responsible for 80% of the balance after the deductible.

insurance company's responsibility = 80% of balance after deductible

$$= 0.80(225)$$

$$= 180$$

Thus, the insurance company is responsible for $180. This checks because the sum of Brook's responsibility and the insurance company's responsibility is equal to the doctor's bill.

$$\$145 + \$180 = \$325$$

We could have also found the answer to part **b)** by subtracting Brook's responsibility from the total amount of the bill, but to give you more practice with percents we decided to show the solution as we did.

Now Try Exercise 33

### Understanding Algebra

Bar graphs, line graphs, circle graphs (or pie charts) are ways of pictorially representing data.

## 2 Solve Problems Involving Bar, Line, and Circle Graphs

Problem solving often involves understanding and reading graphs and sets of data (or numbers). To work Example 4, you must interpret a bar graph and work with data.

EXAMPLE 4 **Walking It Off** Experts suggest that people walk 10,000 steps daily. Depending on stride length, each mile ranges between 2000 and 2500 steps. **Figure 1.1** is a **bar graph** that shows the number of steps it takes to burn off calories from a garden salad with fat-free dressing, a 12-ounce can of soda, a doughnut, and a cheeseburger.

**a)** Using the bar graph in **Figure 1.1**, estimate the number of steps it takes to burn off calories from a cheeseburger.

**b)** If Cliff Jackson can walk a mile in 2000 steps, how many miles will he have to walk in order to burn off the calories from the cheeseburger he ate for lunch?

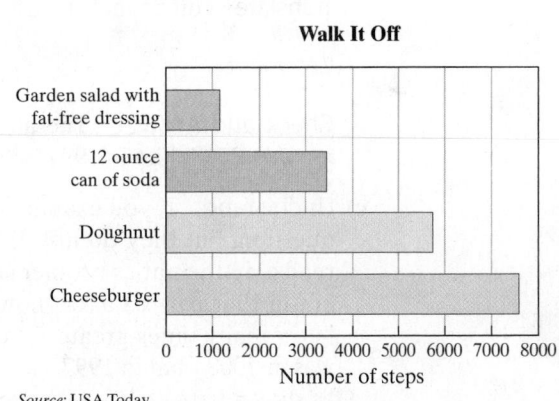

**FIGURE 1.1**      *Source*: USA Today

*Solution*

a) Using the bar to the right of Cheeseburger in **Figure 1.1**, we estimate that the number of steps it takes to burn off calories from a cheeseburger is about 7600.

b) Understand   Since it takes Cliff 2000 steps to walk a mile, it follows that he would need to take 4000 steps to walk 2 miles, 6000 steps to walk 3 miles, and so on. To determine how many miles Cliff will have to walk in order to burn off the calories from the cheeseburger, we need to divide as follows.

Translate     $\text{miles to walk} = \dfrac{\text{number of steps to burn off calories}}{2000}$

Carry Out     $\text{miles to walk} = \dfrac{7600}{2000} = 3.8$

Check and Answer   The answer appears reasonable. Cliff will need to walk 3.8 miles in order to burn off the calories from the cheeseburger that he ate for lunch.

Now Try Exercise 35

In Example 5, we will use the symbol $\approx$, which is read "**is approximately equal to.**" If, for example, the answer to a problem is 34.12432, we may write the answer as $\approx 34.1$.

**Understanding Algebra**

The symbol $\approx$ means "approximately equal to." So we can write $\dfrac{1}{3} \approx 0.33$, for example.

**Super Bowl 30-second Commercial Cost**

*Source*: www.swivel.com/data_sets/show/1002826

**FIGURE 1.2**

**EXAMPLE 5   The Super Bowl**   The **line graph** in **Figure 1.2** shows the cost of a 30-second commercial during Super Bowls from 1997 to 2008. The advertising prices are set by the TV network.

a) Estimate the cost of 30-second commercials in 1997 and 2008.

b) How much more was the cost for a 30-second commercial in 2008 than in 1997?

c) How many times greater was the cost of a 30-second commercial in 2008 than in 1997?

*Solution*

a) When reading a line graph where the line has some thickness, as in **Figure 1.2**, we will use the center of the line to make our estimate. By observing the dashed lines on the graph, we can estimate that the cost of a 30-second commercial was about $1.2 million (or $1,200,000) in 1997 and about $2.7 million (or $2,700,000) in 2008.

b) We use the problem-solving procedure to answer the question.

Understand   To determine how much more the cost of a 30-second commercial was in 2008 than in 1997, we need to subtract.

Translate     difference in cost = cost in 2008 − cost in 1997

Carry Out                    = $2,700,000 − $1,200,000 = $1,500,000

Check and Answer   The answer appears reasonable. The cost was $1,500,000 more in 2008 than in 1997.

c) Understand   If you examine parts **b)** and **c)**, they may appear to ask the same question, but they do not. In Section 1.1, we indicated that it is important to read a mathematics problem carefully, word by word. The two parts are different in that part **b)** asks "how much more was the cost" whereas part **c)** asks "how many <u>times</u> greater." To determine the number of times greater the cost was in 2008 than in 1997, we need to divide the cost in 2008 by the cost in 1997, as shown below.

Translate     $\text{number of times greater} = \dfrac{\text{cost in 2008}}{\text{cost in 1997}}$

Carry Out   $\text{number of times greater} = \dfrac{2{,}700{,}000}{1{,}200{,}000} \approx 2.25$

Check and Answer   By observing the graph, we see that the answer is reasonable. The cost of a 30-second commercial during the Super Bowl in 2008 was about 2.25 times the cost in 1997.

Now Try Exercise 37

EXAMPLE 6  **Stay-at-Home Parents Figure 1.3** is a **circle graph** that shows the reasons why married mothers with children under the age of 15 have stayed out of the labor force for the past year. Use **Figure 1.3** to determine the number of married mothers with children under the age of 15 who have stayed out of the labor force for the following reasons: to care for home and family, ill/disabled, retired, going to school, could not find work, and other.

**Reasons for Married Mothers with Children Under**
**Age 15 Staying Out of the Labor Force for the Past Year**

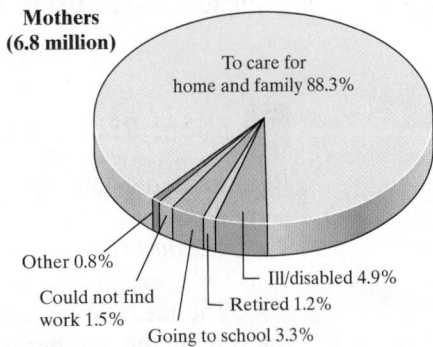

FIGURE 1.3    *Source*: U.S. Census Bureau

*Solution*   Understand There were 6.8 million married mothers with children under the age of 15 who were out of the labor force. Of these mothers, 88.3% were out of the labor force to care for home and family. To determine the number of married mothers with children under the age of 15 out of the labor force to care for home and family, we need to find 88.3% of the total number of mothers. To do this, we multiply as follows.

Translate

$$\begin{pmatrix} \text{to care for home} \\ \text{and family} \end{pmatrix} = \begin{pmatrix} \text{percent out of the labor force} \\ \text{to care for home and family} \end{pmatrix}\begin{pmatrix} \text{total number out} \\ \text{of the labor force} \end{pmatrix}$$

Carry Out

$\text{to care for home and family} = 0.883(6.8\text{ million})$

$\qquad\qquad\qquad\qquad\qquad\quad = 6.0044\text{ million}$

Thus, 6.0044 million married mothers with children under the age of 15 were out of the labor force to care for home and family.

   To find the number who were out of the labor force for being ill/disabled, we do a similar calculation.

$\text{ill/disabled} = 0.049(6.8\text{ million})$

$\qquad\qquad\quad = 0.3332\text{ million}$

We do similar calculations to find the number who were retired, going to school, could not find work, or for other reasons were out of the labor force.

$$\text{retired} = 0.012(6.8 \text{ million})$$
$$= 0.0816 \text{ million}$$
$$\text{going to school} = 0.033(6.8 \text{ million})$$
$$= 0.2244 \text{ million}$$
$$\text{could not find work} = 0.015(6.8 \text{ million})$$
$$= 0.102 \text{ million}$$
$$\text{other reasons} = 0.008(6.8 \text{ million})$$
$$= 0.0544 \text{ million}$$

**Check**  If we add the six amounts, we obtain the 6.8 million total. Therefore, our answer is correct.

6.0044 million + 0.3332 million + 0.0816 million + 0.2244 million
+ 0.102 million + 0.0544 million = 6.8 million

**Answer**  The number of married mothers with children under the age of 15 who were out of the labor force were as follows: 6.0044 million to care for home and family; 0.3332 million were ill/disabled; 0.0816 million were retired; 0.2244 million were going to school; 0.102 million could not find work; for other reasons, 0.0544 million.

Now Try Exercise 39

## 3  Solve Problems Involving Statistics

Because understanding statistics is so important in our society, we will now discuss certain statistical topics and use them in solving problems.

The *mean* and *median* are two **measures of central tendency**, which are also referred to as *averages*. An average is a value that is representative of a set of data (or numbers).

The **mean** of a set of data is determined by adding all the values and dividing the sum by the number of values. For example, to find the mean of 6, 9, 3, 12, 12, we do the following.

$$\text{mean} = \frac{6 + 9 + 3 + 12 + 12}{5} = \frac{42}{5} = 8.4$$

We divided the sum by 5 since there are five values. The mean is the most commonly used average and it is generally what is thought of when we use the word *average*.

Another average is the median. The **median** is the value in the middle of a set of **ranked data**. The data may be ranked from smallest to largest or largest to smallest. To find the median of 6, 9, 3, 12, 12, we can rank the data from smallest to largest as follows.

$$3, 6, 9, 12, 12$$
$$\uparrow$$
$$\text{Middle value}$$

The value in the middle of the ranked set of data is 9. Therefore, the median is 9. Note that half the values will be above the median and half will be below the median.

If there is an even number of pieces of data, the median is halfway between the two middle pieces. For example, to find the median of 3, 12, 5, 12, 17, 9, we can rank the data as follows.

$$3, 5, 9, 12, 12, 17$$
$$\uparrow$$
$$\text{Middle values}$$

Since there are six pieces of data (an even number), we find the value halfway between the two middle pieces, the 9 and the 12. To find the median, we add these values and divide the sum by 2.

$$\text{median} = \frac{9 + 12}{2} = \frac{21}{2} = 10.5$$

Thus, the median is 10.5. Note that half the values are above and half are below 10.5.

---

**Understanding Algebra**

The *mean* is found by dividing the sum of the data by the number of data points. The *median* is the middle score of the ranked data.

**EXAMPLE 7  The Mean Grade**  Alfonso Ramirez's first six exam grades are 90, 87, 76, 84, 78, and 62.

**a)** Find the mean for Alfonso's six grades.

**b)** If one more exam is to be given, what is the minimum grade that Alfonso can receive to obtain at least a B average (a mean average of 80 or better)?

**c)** If there is only one more exam, is it possible for Alfonso to obtain an A average (90 or better)? Explain.

*Solution*

**a)** To obtain the mean, we add the six grades and divide by 6.

$$\text{mean} = \frac{90 + 87 + 76 + 84 + 78 + 62}{6} = \frac{477}{6} = 79.5$$

**b)** We will show the problem-solving steps for this part of the example.

Understand   For the mean average of seven exams to be 80, the total points for the seven exams must be 7(80) or 560. The minimum grade needed can be found by subtracting the sum of the first six grades from 560.

Translate

$$\text{minimum grade needed} = 560 - \text{sum of first six exam grades}$$

Carry Out
$$= 560 - (90 + 87 + 76 + 84 + 78 + 62)$$
$$= 560 - 477$$
$$= 83$$

Check   We can check to see that a seventh grade of 83 gives a mean of 80 as follows.

$$\text{mean} = \frac{90 + 87 + 76 + 84 + 78 + 62 + 83}{7} = \frac{560}{7} = 80$$

Answer   A seventh grade of 83 or higher will result in at least a B average.

**c)** We can use the same reasoning as in part **b)**. For a 90 average, the total points that Alfonso will need to attain is $90(7) = 630$. Since his total points are 477, he will need $630 - 477$ or 153 points to obtain an A average. Since the maximum number of points available on most exams is 100, Alfonso would not be able to obtain an A in the course.

*Now Try Exercise 41*

---

# EXERCISE SET 1.2

MathXL®   MyMathLab®
MathXL®     MyMathLab

## Warm-Up Exercises

*Fill in the blanks with the appropriate word, phrase, or symbol(s) from the following list.*

| | | | |
|---|---|---|---|
| expression | measures of central tendency | mean | median |
| equation | approximately equal to | grouping symbols | problem solving |
| checking | circle graphs | understanding the problem | |

**1.** The _____ of the data 2, 4, 7, 8, 9 is 7.

**2.** A general collection of numbers, symbols, and operations is called a(n) _____ .

**3.** The symbol ≈ means _____ .

**4.** The _____ of the data 2, 4, 7, 8, 9 is 6.

**5.** One of the five important steps in problem solving, seeing if your answer makes sense, is referred to as _____ a problem.

**6.** The mean and median are types of averages, also called _____.

**7.** Graphical representation of data includes bar graphs, line graphs, and _____.

**8.** Parentheses and brackets are examples of _____.

**9.** In this book we use Pólya's five-step approach for _____.

**10.** Reading a problem at least twice, making a list of facts, and making a sketch are the problem-solving step called _____.

## Practice the Skills

**11. Test Grades** Jenna Webber's test grades are 78, 97, 59, 74, and 74. For Jenna's grades, determine the **a)** mean and **b)** median.

**12. Bowling Scores** William Krant's bowling scores for five games were 161, 131, 187, 163, and 145. For William's games, determine the **a)** mean and **b)** median.

**13. Electric Bills** The Malones' electric bills for January through June, 2006, were $96.56, $108.78, $87.23, $85.90, $79.55, and $65.88. For these bills, determine the **a)** mean and **b)** median.

**14. Grocery Bills** Antoinette Payne's monthly grocery bills for the first five months of 2006 were $204.83, $153.85, $210.03, $119.76, and $128.38. For Antoinette's grocery bills, determine the **a)** mean and **b)** median.

**15. Dry Summers** The following figure shows the 10 driest summers in the Southeast from 1895 through 2004. Determine the **a)** mean and **b)** median inches of rainfall for the 10 years shown.

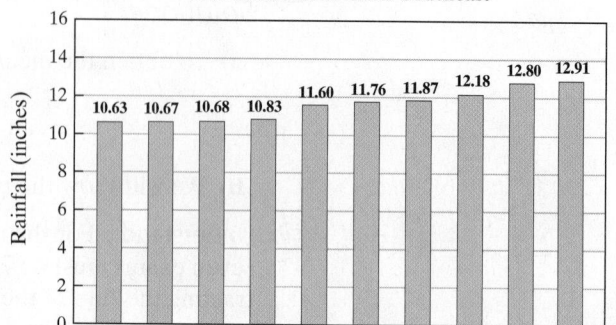

**Driest Summers in the Southeast**

*Source:* Gloria Forthun, Southeast Regional Climate Center
Records are from 1895 through 2004. The Southeast is Va., N.C., S.C., Ga., Fla., and Ala.

**16. Homes for Sale** Eight homes are for sale in a community. The sale prices are $124,100, $175,900, $142,300, $164,800, $146,000, $210,000, $112,200, and $153,600. Determine the **a)** mean and **b)** median sale price of the eight homes.

## Problem Solving

**17. Commissions** Barbara Riedell earns a 5% commission on appliances she sells. Her sales last week totaled $9400. Find her week's earnings.

**18. Empire State Building** May 1, 1931, was the opening day of the Empire State Building. It stands 1454 feet or 443 meters high. Use this information to determine the approximate number of feet in a meter.

**19. Sales Tax    a)** The sales tax in Jefferson County is 8%. What was the sales tax that Scott Reed paid on a used car that cost $16,700 before tax?

**b)** What is the total cost of the car including tax?

**20. Checking Account** The balance in Debbie Ogilvie's checking account is $312.60. She purchased five DVDs at $17.11 each including tax. If she pays by check, what is the new balance in her checking account?

**21. Computer Processor** Suppose a computer processor can perform about 2.3 billion operations per second. How many operations can it perform in 0.7 second?

**22. Buying a Computer** Pat Sullivan wants to purchase a computer that sells for $950. He can either pay the total amount at the time of purchase or agree to pay the store $200 down and $33 per month for 24 months.

**a)** If he pays the down payment and monthly charge, how much will he pay for the computer?

**b)** How much money can he save by paying the total amount at the time of purchase?

**23. Energy Values** The following table gives the approximate energy values of some foods and the approximate energy consumption of some activities, in kilojoules (kJ). Determine how long it would take for you to use up the energy from the following.

**a)** a hamburger by running

**b)** a chocolate milkshake by walking

**c)** a glass of skim milk by cycling

| Energy Value, Food | (kJ) | Energy Consumption, Activity | (kJ/min) |
|---|---|---|---|
| Chocolate milkshake | 2200 | Walking | 25 |
| Fried egg | 460 | Cycling | 35 |
| Hamburger | 1550 | Swimming | 50 |
| Strawberry shortcake | 1440 | Running | 80 |
| Glass of skim milk | 350 | | |

A green numbered exercise, such as 21, indicates a Now Try Exercise.

24. **Jet Ski** The rental cost of a jet ski from Don's Ski Rental is $20 per half-hour, and the rental cost from A. J.'s Ski Rental is $50 per hour. Suppose you plan to rent a jet ski for 3 hours.
   a) Which is the better deal?
   b) How much will you save?

25. **Gas Mileage** When the odometer in Tribet LaPierre's car reads 16,741.3, he fills his gas tank. The next time he fills his tank it takes 10.5 gallons, and his odometer reads 16,935.4. Determine the number of miles per gallon that his car gets.

26. **Income Taxes** The federal income tax rate schedule for a *joint return* in 2007 is illustrated in the following table.

| Adjusted Gross Income | Taxes |
|---|---|
| $0–$15,650 | 10% of income |
| $15,650–$63,700 | $1565 + 15% in excess of $15,650 |
| $63,700–$128,500 | $8772.50 + 25% in excess of $63,700 |
| $128,500–$195,850 | $24,972.50 + 28% in excess of $128,500 |
| $195,850–$349,700 | $43,830 + 33% in excess of $195,850 |
| $349,700 and up | $94,601 + 35% in excess of $349,700 |

*Source:* www.irs.gov

   a) If the Donovins' adjusted gross income in 2007 was $34,612, determine their taxes.
   b) If the Ortegas' 2007 adjusted gross income was $75,610, determine their taxes.

27. **Buying Tires** Eric Weiss purchased four tires through the Internet. He paid $62.30 plus $6.20 shipping and handling per tire. There was no sales tax on this purchase. When he received the tires, Eric had to pay $8.00 per tire for mounting and balancing. At a local tire store, his total cost for the four tires with mounting and balancing would have been $425 plus 8% sales tax. How much did Eric save by purchasing the tires through the Internet?

28. **Baseball Salaries** Jason Giambi, of the New York Yankees, was the highest paid professional baseball player in 2007, according to *USA Today*. He earned $23.4 million. Bartolo Colon, then of the Los Angeles Angels, was the highest paid pitcher, earning $16 million. In 2007, Giambi batted 254 times and Colon pitched 99 innings. Determine approximately how much more Colon received per inning pitched than Giambi did per at bat.

29. **Balance** Consider the figure shown. Assuming the green and red blocks have the same weight, where should a single green block, ■, be placed to make the scale balanced? Explain how you determined your answer.

30. **Taxi Ride** A taxicab charges $2 upon a customer's entering the taxi, then 30 cents for each $\frac{1}{4}$ mile traveled and 20 cents

for each 30 seconds stopped in traffic. David Lopez takes a taxi ride for a distance of 3 miles where the taxi spends 90 seconds stopped in traffic. Determine David's cost of the taxi ride.

 31. **Leaky Faucet** A faucet that leaks 1 ounce of water per minute wastes 11.25 gallons in a day.
   a) How many gallons of water are wasted in a (non-leap) year?
   b) If water costs $5.20 per 1000 gallons, how much additional money per year is being spent on the water bill?

32. **Conversions** a) What is 1 mile per hour equal to in feet per hour? One mile contains 5280 feet.
   b) What is 1 mile per hour equal to in feet per second?
   c) What is 60 miles per hour equal to in feet per second?

33. **Medical Insurance** Mel LeBar's medical insurance policy requires that he pay a $150 deductible each calendar year. After the deductible is paid, he pays 20% of the medical expenses and the insurance company pays 80%. On January 1, Mel's daughter accidentally closed the car door on his finger. He went to the doctor's office for an examination and X rays. The total bill of $365 was sent to the insurance company. If Mel had not as yet paid any of the deductible,
   a) how much of the bill will Mel be responsible for?
   b) how much will the insurance company be responsible for?

34. **Insurance** Drivers under the age of 25 who pass a driver education course generally have their auto insurance premium decreased by 10%. Most insurers will offer this 10% deduction until the driver reaches 25. A particular driver education course costs $70. Don Beville, who just turned 18, has auto insurance that costs $630 per year.
   a) Excluding the cost of the driver education course, how much would Don save in auto insurance premiums, from the age of 18 until the age of 25, by taking the driver education course?
   b) What would be his net savings after the cost of the course?

35. **Math Scores** The bar graph below shows some 2006 scores from the Program for International Student Assessment, a test for 15-year-olds. The scale was constructed so that an average score is 500.

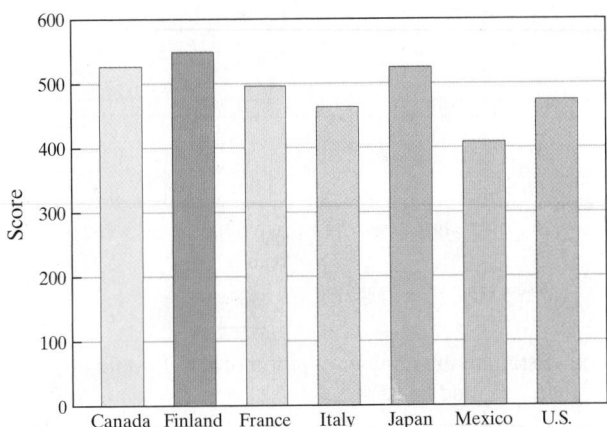

*Source: New York Times*

**a)** Which country had the highest math score shown? Estimate the score.

**b)** Which country had the lowest math score shown? Estimate the score.

**c)** Estimate the difference between the scores for Finland and Mexico.

**36. Commercial Airlines** The bar graph shows the total scheduled passengers on U.S. commercial airlines from 1999 to 2007 with the forecasted totals through 2016. From the bar graph, estimate

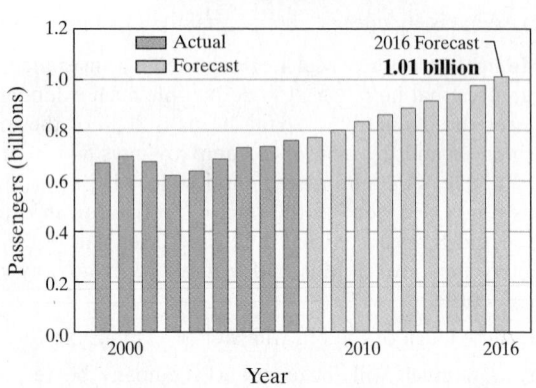

**Total Scheduled Passengers on U.S. Commercial Airlines**

*Source*: Federal Aviation Administration

**a)** the total number of scheduled passengers on U.S. commercial airlines in 2000.

**b)** the forecasted total number of scheduled passengers on U.S. commercial airlines in 2015.

**c)** how many times greater is the forecasted total number of scheduled passengers on U.S. commercial airlines in 2015 than in 2000?

**37. Motorcycle Sales** The line graph below shows the U.S. sales of new motorcycles from 1992 to 2007.

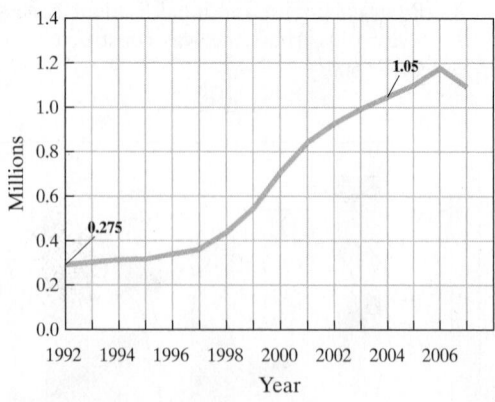

**U.S. Sales of New Motorcycles**

*Source*: Motorcycle Industry Council

**a)** Estimate the number of new motorcycles sold in the U.S. in 1992 and in 2004.

**b)** How many more new motorcycles were sold in the U.S. in 2004 than in 1992?

**c)** How many times greater was the number of new motorcycles sold in 2007 than in 1992?

**38. SAT Math Scores** The line graph below shows the number of freshmen students at the College of Applied Science with high school math SAT scores over 700.

**a)** During which consecutive years did the number of students remain constant?

During which consecutive years did the number of freshmen with high school math SAT scores over 700

**b)** increase the most?

**c)** decrease the most?

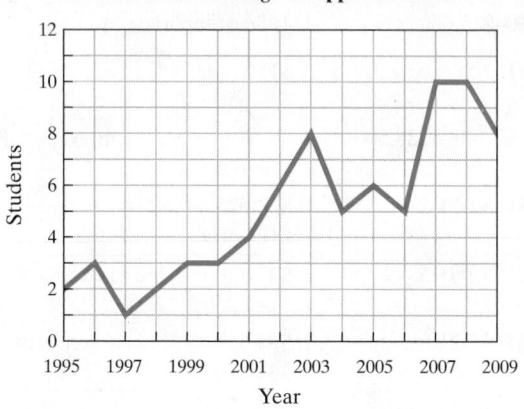

**Number of College Freshman with SAT Math Scores over 700 — College of Applied Science**

**39. Adoption** Approximately 1.7 million U.S. households included adopted children. The circle graph below shows the percent of these households that had one adopted child, two adopted children, and three or more adopted children. Estimate the number of U.S. households that had

**a)** one adopted child.

**b)** two adopted children.

**c)** three or more adopted children.

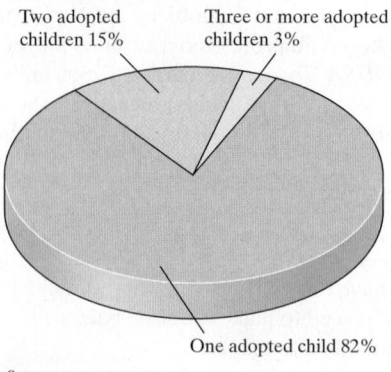

**U.S. Households with Adopted Children (1.7 million households)**

Two adopted children 15%

Three or more adopted children 3%

One adopted child 82%

*Source*: www.census.gov

**40. Jeopardy!** As of this writing, Ken Jennings, a software engineer from Salt Lake City, Utah, is the second highest record holder for the most money ever won on a television game show. The circle graph on the next page shows the outcome of the 160 Daily Doubles that Ken attempted on his *Jeopardy!* run. Use the circle graph to determine

**a)** the number of Daily Doubles Ken answered correctly.

**b)** the number of Daily Doubles Ken answered incorrectly.

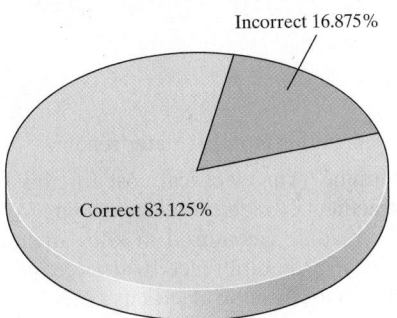

**Ken Jennings's Jeopardy Daily Doubles
(160 Total)**

Incorrect 16.875%

Correct 83.125%

*Source:* www.tvgameshows.net/jenningswinnings.htm

**41. Test Grades** A mean of 60 on all exams is needed to pass a course. On his first five exams Lamond Paine's grades are 50, 59, 67, 80, and 56.

**a)** What is the minimum grade that Lamond can receive on the sixth exam to pass the course?

**b)** An average of 70 is needed to get a C in the course. Is it possible for Lamond to get a C? If so, what is the minimum grade that Lamond can receive on the sixth exam?

© Shutterstock

**42. Test Grades** A mean of 80 on all exams is needed to earn a B in a course. On her first four exams, Heather Feldman's grades are 95, 88, 82, and 85.

**a)** What is the minimum grade that Heather can receive on the fifth exam to earn a B in the course?

**b)** A mean of 90 is needed to earn an A in the course. Is it possible for Heather to get an A? If so, what is the minimum grade that Heather can receive on the fifth exam?

**43. Level of Education** The following bar graph shows the median annual earnings by level of education in 2007 in the United States. Which degree has median annual earnings that are approximately 2.2 times the median annual earnings of someone with less than a high school diploma?

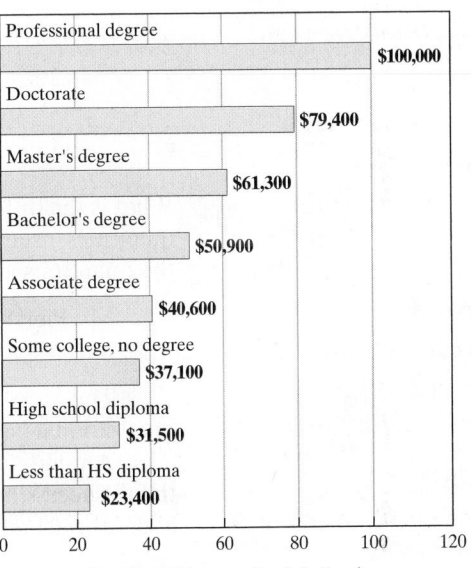

**Median Annual Earnings
by Level of Education**

Professional degree — $100,000

Doctorate — $79,400

Master's degree — $61,300

Bachelor's degree — $50,900

Associate degree — $40,600

Some college, no degree — $37,100

High school diploma — $31,500

Less than HS diploma — $23,400

Earnings (thousands of dollars)

*Source:* College Board

**44. Exams** Mike Ambrosino's mean average on six exams is 78. Find the sum of his scores.

**45. Construct Data** Construct a set of five pieces of data with a mean of 70 and no two values the same.

## Concept/Writing Exercises

**46.** Suppose a set of data of 10 numbers has a mean of 6. An 11th number is added to the data and it has a value of 5. Will the mean of the new data set increase or decrease? Explain.

**47.** Consider the set of data 2, 3, 5, 6, 70. Without doing any calculations, can you determine whether the mean or the median is greater? Explain your answer.

## Challenge Problem

**48. Reading Meters** The figure shows how to read an electric meter.

**Step 1.** Start with the dial on the right. If the arrow falls between two numbers, use the smaller number on the

dial (except when the dial is between 9 and 0, then use the 9). Notice the arrows above the meters indicate the direction the dial is moving (clockwise, then counterclockwise).

**Step 2.** If the pointer is directly on a number, check the dial to the *right* to make sure it has passed 0 and is headed toward 1. If the dial to the right has not passed 0, use the next lower number. The number on the meters on the right is 16064.

*Source:* Southern California Edison, Understanding Your Electricity Bill

Suppose your previous month's reading was as shown on the left, and this month's meter reading is as shown below.

a) Determine this month's meter reading.

b) Determine your electrical cost for this month by first subtracting last month's meter reading from this month's meter reading (measured in kilowatt hours), and then multiplying the difference by the cost per kilowatt hour of electricity. Assume electricity costs 24.3 cents per kilowatt hour.

# 1.3 Fractions

1  Learn multiplication symbols and recognize factors.

2  Simplify fractions.

3  Multiply fractions.

4  Divide fractions.

5  Add and subtract fractions.

6  Change mixed numbers to fractions and vice versa.

What is the difference between arithmetic and algebra? When doing arithmetic, all the quantities used in the calculations are known. In algebra, however, one or more of the quantities are unknown and must be found. Consider the following:

Mr. Piersma has 1 gallon of paint. In order to paint his bedroom, he needs 3 gallons of paint. How many additional gallons does he need?

This is an example of an algebraic problem. The *unknown* quantity is the number of additional gallons of paint needed. Mr. Piersma needs 2 more gallons of paint.

An understanding of decimal numbers (see Appendix A) and fractions is essential to success in algebra. You will need to know how to simplify a fraction and how to add, subtract, multiply, and divide fractions.

## 1  Learn Multiplication Symbols and Recognize Factors

In algebra we often use letters called **variables** to represent numbers. Letters commonly used as variables are *x, y,* and *z,* but other letters can be used as variables. Variables are usually shown in italics. So that we do not confuse the variable $x$ with the multiplication sign $\times$ we often use different notation to indicate multiplication.

### Understanding Algebra

$5x$ means 5 times $x$
and
$rs$ means $r$ times $s$

**Multiplication Symbols**

If $a$ and $b$ represent any two mathematical quantities, then each of the following may be used to indicate the product of $a$ and $b$ ("$a$ times $b$").

$$ab \qquad a \cdot b \qquad a(b) \qquad (a)b \qquad (a)(b)$$

Examples

| 3 times 4 may be written: | 3 times $x$ may be written: | $x$ times $y$ may be written: |
|---|---|---|
| | $3x$ | $xy$ |
| 3(4) | 3($x$) | $x$($y$) |
| (3)4 | (3)$x$ | ($x$)$y$ |
| (3)(4) | (3)($x$) | ($x$)($y$) |

Now we will introduce the term *factors,* which we shall be using throughout the text.

## Understanding Algebra

A *factor* is a number or expression that is multiplied by another number or expression.

### Factors

The numbers or variables that are multiplied in a multiplication problem are called **factors**.

If $a \cdot b = c$, then $a$ and $b$ are *factors* of $c$.

For example, in $3 \cdot 5 = 15$, the numbers 3 and 5 are factors of the product 15. In $2 \cdot 15 = 30$, the numbers 2 and 15 are factors of the product 30. Note that 30 has many other factors. Since $5 \cdot 6 = 30$, the numbers 5 and 6 are also factors of 30. Since $3x$ means 3 times $x$, both the 3 and the $x$ are factors of $3x$.

## 2 Simplify Fractions

Now we have the necessary information to discuss **fractions**. The top number of a fraction is called the **numerator**, and the bottom number is called the **denominator**. In the fraction $\frac{3}{5}$, the 3 is the numerator and the 5 is the denominator.

## Understanding Algebra

The fraction $\frac{3}{5}$ means the same as 3/5 or 3 ÷ 5

### Helpful Hint

Consider the fraction $\frac{3}{5}$. There are equivalent methods of expressing this fraction, as illustrated below.

$$\frac{3}{5} = 3/5 = 3 \div 5 = 5\overline{)3}$$

In general, $\frac{a}{b} = a/b = a \div b = b\overline{)a}$

## Understanding Algebra

*GCF* stands for "greatest common factor." It is the largest number that divides evenly into the two given numbers. The GCF of 12 and 18 is 6.

A fraction is **simplified**, or **reduced to its lowest terms**, when the numerator and denominator have no common factors other than 1. To simplify a fraction, follow these steps.

### To Simplify a Fraction

1. Find the largest number that will divide (without remainder) both the numerator and the denominator. This number is called the **greatest common factor** (GCF).
2. Then divide both the numerator and the denominator by the greatest common factor.

*If you do not remember how to find the greatest common factor of two or more numbers, read Appendix B.*

**EXAMPLE 1** Simplify  **a)** $\frac{10}{25}$  **b)** $\frac{6}{18}$.

*Solution*

**a)** The largest number that divides both 10 and 25 is 5. Therefore, 5 is the greatest common factor. Divide both the numerator and the denominator by 5 to simplify the fraction.

$$\frac{10}{25} = \frac{10 \div 5}{25 \div 5} = \frac{2}{5}$$

**b)** The largest number that divides both 6 and 18 is 6. Divide both the numerator and the denominator by 6.

$$\frac{6}{18} = \frac{6 \div 6}{18 \div 6} = \frac{1}{3}$$

Note in part **b)** that both the numerator and denominator could have been written with a factor of 6. Then the common factor 6 could be divided out.

$$\frac{6}{18} = \frac{1 \cdot \cancel{6}}{3 \cdot \cancel{6}} = \frac{1}{3}$$

Now Try Exercise 11

When you work with fractions you should always simplify your answers.

### 3 Multiply Fractions

To multiply two or more fractions, multiply their numerators together and multiply their denominators together.

**To Multiply Fractions**

$$\frac{a}{b} \cdot \frac{c}{d} = \frac{ac}{bd}$$

**EXAMPLE 2** Multiply $\frac{3}{13}$ by $\frac{5}{11}$.

*Solution* $\quad \frac{3}{13} \cdot \frac{5}{11} = \frac{3 \cdot 5}{13 \cdot 11} = \frac{15}{143}$

Now Try Exercise 39

Before multiplying fractions, to help avoid having to simplify an answer, we often divide both a numerator and a denominator by a common factor.

**EXAMPLE 3** Multiply **a)** $\frac{8}{17} \cdot \frac{5}{16}$ **b)** $\frac{27}{40} \cdot \frac{16}{9}$.

*Solution*

**a)** Since the numerator 8 and the denominator 16 can both be divided by the common factor 8, we divide out the 8 first. Then we multiply.

$$\frac{8}{17} \cdot \frac{5}{16} = \frac{\overset{1}{\cancel{8}}}{17} \cdot \frac{5}{\underset{2}{\cancel{16}}} = \frac{1 \cdot 5}{17 \cdot 2} = \frac{5}{34}$$

**b)**

$$\frac{27}{40} \cdot \frac{16}{9} = \frac{\overset{3}{\cancel{27}}}{40} \cdot \frac{16}{\underset{1}{\cancel{9}}} \qquad \text{Divide both 27 and 9 by 9.}$$

$$= \frac{\overset{3}{\cancel{27}}}{\underset{5}{\cancel{40}}} \cdot \frac{\overset{2}{\cancel{16}}}{\underset{1}{\cancel{9}}} \qquad \text{Divide both 40 and 16 by 8.}$$

$$= \frac{3 \cdot 2}{5 \cdot 1} = \frac{6}{5}$$

Now Try Exercise 41

### Understanding Algebra

When we write 1, 2, 3, 4, ..., the three dots, called an *ellipsis,* indicate the pattern continues indefinitely.

The numbers 0, 1, 2, 3, 4, ... are called **whole numbers**. The whole numbers continue indefinitely. Thus, the numbers 468 and 5043 are also whole numbers. To multiply a whole number by a fraction, write the whole number with a denominator of 1 and then multiply.

EXAMPLE 4 **Lawn Mower Engine** Some engines run on a mixture of gas and oil. A particular lawn mower engine requires a mixture of $\frac{5}{64}$ gallon of oil for each gallon of gasoline used. A lawn care company wishes to make a mixture for this engine using 12 gallons of gasoline. How much oil must be used?

*Solution*   We must multiply 12 by $\frac{5}{64}$ to determine the amount of oil that must be used. First we write 12 as $\frac{12}{1}$, then we divide both 12 and 64 by their greatest common factor, 4, as follows.

$$12 \cdot \frac{5}{64} = \frac{12}{1} \cdot \frac{5}{64} = \frac{\overset{3}{\cancel{12}}}{1} \cdot \frac{5}{\underset{16}{\cancel{64}}} = \frac{3 \cdot 5}{1 \cdot 16} = \frac{15}{16}$$

Thus, $\frac{15}{16}$ gallon of oil must be added to the 12 gallons of gasoline to make the proper mixture.

Now Try Exercise 89

### 4   Divide Fractions

To divide one fraction by another, invert the divisor (the second fraction if written with $\div$) and proceed as in multiplication.

> **To Divide Fractions**
>
> $$\frac{a}{b} \div \frac{c}{d} = \frac{a}{b} \cdot \frac{d}{c} = \frac{ad}{bc}$$

Sometimes, rather than being asked to add, subtract, multiply, or divide, you may be asked to *evaluate* an expression. To **evaluate** an expression means to obtain the answer to the problem using the operations given.

EXAMPLE 5   Evaluate   **a)** $\frac{3}{5} \div \frac{5}{6}$    **b)** $\frac{3}{8} \div 12$.

*Solution*

**a)** $\frac{3}{5} \div \frac{5}{6} = \frac{3}{5} \cdot \frac{6}{5} = \frac{3 \cdot 6}{5 \cdot 5} = \frac{18}{25}$

**b)** Write 12 as $\frac{12}{1}$. Then invert the divisor and multiply.

$$\frac{3}{8} \div 12 = \frac{3}{8} \div \frac{12}{1} = \frac{\overset{1}{\cancel{3}}}{8} \cdot \frac{1}{\underset{4}{\cancel{12}}} = \frac{1}{32}$$

Now Try Exercise 47

### 5   Add and Subtract Fractions

*Fractions that have the same* (or a *common*) *denominator can be added or subtracted.* To add or subtract fractions with the same denominator, add or subtract the numerators and keep the common denominator.

> **To Add and Subtract Fractions**
>
> $$\frac{a}{c} + \frac{b}{c} = \frac{a+b}{c} \quad \text{or} \quad \frac{a}{c} - \frac{b}{c} = \frac{a-b}{c}$$

EXAMPLE 6 **a)** Add $\dfrac{6}{15} + \dfrac{2}{15}$.    **b)** Subtract $\dfrac{8}{13} - \dfrac{5}{13}$.

*Solution*

**a)** $\dfrac{6}{15} + \dfrac{2}{15} = \dfrac{6+2}{15} = \dfrac{8}{15}$    **b)** $\dfrac{8}{13} - \dfrac{5}{13} = \dfrac{8-5}{13} = \dfrac{3}{13}$

Now Try Exercise 55

*To add (or subtract) fractions with unlike denominators, we must first rewrite each fraction with a common denominator.* The smallest number that is divisible by two or more denominators is called the **least common denominator** or **LCD**. *If you have forgotten how to find the least common denominator, review Appendix B now.*

EXAMPLE 7 Add $\dfrac{1}{2} + \dfrac{1}{5}$.

*Solution*    We cannot add these fractions until we rewrite them with a common denominator. Since the lowest number that both 2 and 5 evenly divide into is 10, we will first rewrite both fractions with the least common denominator of 10.

$$\dfrac{1}{2} = \dfrac{1}{2} \cdot \dfrac{5}{5} = \dfrac{5}{10} \quad \text{and} \quad \dfrac{1}{5} = \dfrac{1}{5} \cdot \dfrac{2}{2} = \dfrac{2}{10}$$

Now we add.

$$\dfrac{1}{2} + \dfrac{1}{5} = \dfrac{5}{10} + \dfrac{2}{10} = \dfrac{7}{10}$$

Now Try Exercise 63

Multiplying both the numerator and denominator by the same number is the same as multiplying by 1. Thus, the value of the fraction does not change.

EXAMPLE 8 How much larger is $\dfrac{3}{4}$ inch than $\dfrac{2}{3}$ inch?

*Solution*    To find the difference, we need to subtract $\dfrac{2}{3}$ inch from $\dfrac{3}{4}$ inch.

$$\dfrac{3}{4} - \dfrac{2}{3}$$

The least common denominator is 12. Therefore, we rewrite both fractions with a denominator of 12.

$$\dfrac{3}{4} = \dfrac{3}{4} \cdot \dfrac{3}{3} = \dfrac{9}{12} \quad \text{and} \quad \dfrac{2}{3} = \dfrac{2}{3} \cdot \dfrac{4}{4} = \dfrac{8}{12}$$

Now we subtract.

$$\dfrac{3}{4} - \dfrac{2}{3} = \dfrac{9}{12} - \dfrac{8}{12} = \dfrac{1}{12}$$

Thus, $\dfrac{3}{4}$ inch is $\dfrac{1}{12}$ inch greater than $\dfrac{2}{3}$ inch.

Now Try Exercise 75

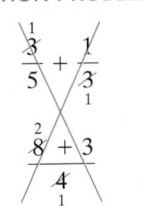

**Avoiding Common Errors**

It is important to remember that dividing out a common factor in the numerator of one fraction and the denominator of a different fraction can be performed only when multiplying fractions. *This process cannot be performed when adding or subtracting fractions.*

CORRECT
MULTIPLICATION PROBLEMS

$$\dfrac{\overset{1}{\cancel{3}}}{5} \cdot \dfrac{1}{\underset{1}{\cancel{3}}}$$

$$\dfrac{\overset{2}{\cancel{8}} \cdot 3}{\underset{1}{\cancel{4}}}$$

INCORRECT
ADDITION PROBLEMS

$$\dfrac{\overset{1}{\cancel{3}}}{5} + \dfrac{1}{\underset{1}{\cancel{3}}}$$

$$\dfrac{\overset{2}{\cancel{8}} + 3}{\underset{1}{\cancel{4}}}$$

---

**Understanding Algebra**

The mixed number $2\frac{3}{4}$ means $2 + \frac{3}{4}$.

---

## 6  Change Mixed Numbers to Fractions and Vice Versa

Consider the number $5\frac{2}{3}$. This is an example of a **mixed number**. A mixed number consists of a whole number followed by a fraction. The mixed number $5\frac{2}{3}$ means $5 + \frac{2}{3}$. We can change $5\frac{2}{3}$ to a fraction as follows.

$$5\frac{2}{3} = \boxed{5} + \frac{2}{3} = \boxed{\frac{15}{3}} + \frac{2}{3} = \frac{15 + 2}{3} = \frac{17}{3}$$

Notice that we expressed the whole number, 5, as a fraction with a denominator of 3, then added the fractions.

**EXAMPLE 9**   Change $7\frac{3}{8}$ to a fraction.

*Solution*

$$7\frac{3}{8} = \boxed{7} + \frac{3}{8} = \boxed{\frac{56}{8}} + \frac{3}{8} = \frac{56 + 3}{8} = \frac{59}{8}$$

Now Try Exercise 25

Now consider the fraction $\frac{17}{3}$. We convert it to a mixed number as follows.

$$\frac{17}{3} = \boxed{\frac{15}{3}} + \frac{2}{3} = \boxed{5} + \frac{2}{3} = 5\frac{2}{3}$$

Notice we wrote $\frac{17}{3}$ as a sum of two fractions, each with the denominator of 3. The first fraction being added, $\frac{15}{3}$, is the equivalent of the largest integer that is less than $\frac{17}{3}$.

**EXAMPLE 10**   Change $\frac{43}{6}$ to a mixed number.

*Solution*   $\dfrac{43}{6} = \boxed{\dfrac{42}{6}} + \dfrac{1}{6} = \boxed{7} + \dfrac{1}{6} = 7\dfrac{1}{6}$

Now Try Exercise 31

**Helpful Hint**

Notice in Example 10 the fraction $\frac{43}{6}$ is simplified because the greatest common divisor of the numerator and denominator is 1. Do not confuse simplifying a fraction with changing a fraction with a value greater than 1 to a mixed number. The fraction $\frac{43}{6}$ can be converted to the mixed number $7\frac{1}{6}$. However, $\frac{43}{6}$ is a simplified fraction.

Now we will work examples that contain mixed numbers.

**FIGURE 1.4**

**EXAMPLE 11** **Plumbing** To repair a plumbing leak, a coupling $\frac{1}{2}$ inch long is glued to a piece of plastic pipe. After the gluing, the piece of plastic pipe showing is $2\frac{9}{16}$ inches, as shown in **Figure 1.4**. How long is the combined length?

*Solution* Understand and Translate   We need to add $2\frac{9}{16}$ inches and $\frac{1}{2}$ inch to obtain the combined length. After we write both fractions with a common denominator, we add the numbers.

Carry Out

$$
\text{original numbers}\left\{
\begin{array}{r}
2\frac{9}{16} \\
+\ \frac{1}{2} \\
\hline
\end{array}
\right. \rightarrow
\left.
\begin{array}{r}
2\frac{9}{16} \\
+\ \frac{8}{16} \\
\hline
2\frac{17}{16}
\end{array}
\right\}
\text{fractions rewritten with common denominator of 16}
$$

Since $2\frac{17}{16} = 2 + \frac{17}{16} = 2 + 1\frac{1}{16} = 3\frac{1}{16}$, the sum is $3\frac{1}{16}$.

Check and Answer   The answer appears reasonable. Thus, the total length is $3\frac{1}{16}$ inches.

*Now Try Exercise 93*

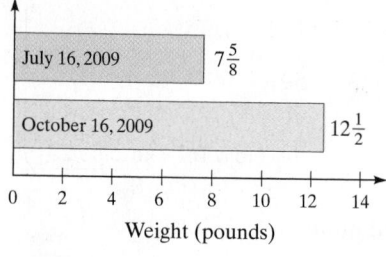

**FIGURE 1.5**

**EXAMPLE 12** **Gaining Weight** The graph in **Figure 1.5** shows the weight of the Smallwoods' baby boy, Jonathan, on July 16, 2009, and on October 16, 2009. How much weight did Jonathan gain during this time period?

*Solution* Understand and Translate   To find the increase, we need to subtract Jonathan's weight on July 16, 2009, from his weight on October 16, 2009. We will subtract vertically.

Carry Out

$$
\begin{array}{r}
12\frac{1}{2} \\
-\ 7\frac{5}{8} \\
\hline
\end{array}
\quad \rightarrow \quad
\begin{array}{r}
12\frac{4}{8} \\
-\ 7\frac{5}{8} \\
\hline
\end{array}
$$

Since we wish to subtract $\frac{5}{8}$ from $\frac{4}{8}$, and $\frac{5}{8}$ is greater than $\frac{4}{8}$, we write $12\frac{4}{8}$ as $11\frac{12}{8}$. To get $11\frac{12}{8}$, we take 1 unit from the number 12 and write it as $\frac{8}{8}$. This gives $11 + 1 + \frac{4}{8} = 11 + \frac{8}{8} + \frac{4}{8} = 11 + \frac{12}{8} = 11\frac{12}{8}$. Now we subtract as follows.

$$
\begin{array}{r}
12\frac{1}{2} \\
-\ 7\frac{5}{8} \\
\hline
\end{array}
\ \rightarrow \
\begin{array}{r}
12\frac{4}{8} \\
-\ 7\frac{5}{8} \\
\hline
\end{array}
\ \rightarrow \
\begin{array}{r}
11\frac{12}{8} \\
-\ 7\frac{5}{8} \\
\hline
4\frac{7}{8}
\end{array}
$$

Check and Answer   By examining the graph, we see that the answer is reasonable. Thus, Jonathan gained $4\frac{7}{8}$ pounds during this time period.

*Now Try Exercise 85*

Although it is not necessary to change mixed numbers to fractions when adding or subtracting mixed numbers, it is necessary to change mixed numbers to fractions when multiplying or dividing mixed numbers. We illustrate this procedure in Example 13.

4 ft

$10\frac{2}{3}$ ft

**FIGURE 1.6**

**EXAMPLE 13  Cutting Wood**  A carpenter is cutting a rectangular piece of wood 4 feet wide by $10\frac{2}{3}$ feet long into four equal strips, as shown in **Figure 1.6**. Find the dimensions of each strip.

*Solution*   Understand and Translate   We know from **Figure 1.6** that one side will have a width of 4 feet. To find the length of the strips, we need to divide $10\frac{2}{3}$ by 4.

Carry Out

$$10\frac{2}{3} \div 4 = \frac{32}{3} \div \frac{4}{1} = \frac{\overset{8}{\cancel{32}}}{3} \cdot \frac{1}{\underset{1}{\cancel{4}}} = \frac{8}{3} = 2\frac{2}{3}$$

Check and Answer   If you multiply $2\frac{2}{3}$ by 4 you obtain the original length, $10\frac{2}{3}$.

Thus, the calculation is correct. The dimensions of each strip will be 4 feet by $2\frac{2}{3}$ feet.

Now Try Exercise 87

---

# EXERCISE SET 1.3      MathXL®    MyMathLab®
MathXL®      MyMathLab

## Warm-Up Exercises

*Fill in the blanks with the appropriate word, phrase, or symbol(s) from the following list.*

| variables | factors | LCD | added or subtracted |
| GCF | mixed number | ellipsis | $\frac{1}{6}$ |
| $\frac{2}{3}$ | denominator | $\frac{3}{2}$ | |

1. When two fractions are being _____ we rewrite them so that they both have the same (common) denominator.

2. $5 + \frac{1}{3}$ is usually written as $5\frac{1}{3}$, which is called a _____.

3. Letters that represent numbers are called _____.

4. In the expression 2, 4, 6, 8, ..., the three dots, called an _____, signify the sequence continues indefinitely.

5. $\frac{1}{3} \div \frac{1}{2} =$ _____.

6. Numbers or variables that are multiplied together are called _____.

7. In the fraction $\frac{3}{4}$, 4 is called the _____.

8. 15 is the _____ of 30 and 75.

9. To perform the division $\frac{4}{7} \div \frac{2}{3}$ we rewrite it as $\frac{4}{7}$ _____.

10. 40 is the _____ of the fractions $\frac{3}{8}$ and $\frac{7}{10}$.

## Practice the Skills

*Simplify each fraction. If a fraction is already simplified, so state.*

11. $\frac{10}{15}$       12. $\frac{40}{10}$       13. $\frac{6}{24}$       14. $\frac{19}{25}$       15. $\frac{36}{76}$       16. $\frac{16}{72}$

17. $\frac{18}{42}$       18. $\frac{60}{105}$       19. $\frac{18}{49}$       20. $\frac{100}{144}$       21. $\frac{12}{25}$       22. $\frac{42}{138}$

*Convert each mixed number to a fraction.*

**23.** $2\frac{13}{15}$     **24.** $15\frac{1}{3}$     **25.** $7\frac{2}{3}$     **26.** $14\frac{3}{4}$     **27.** $3\frac{5}{18}$     **28.** $2\frac{2}{9}$     **29.** $9\frac{6}{17}$     **30.** $3\frac{3}{32}$

*Write each fraction as a mixed number.*

**31.** $\frac{7}{4}$     **32.** $\frac{18}{7}$     **33.** $\frac{13}{4}$     **34.** $\frac{9}{2}$     **35.** $\frac{32}{7}$     **36.** $\frac{110}{20}$     **37.** $\frac{86}{14}$     **38.** $\frac{72}{14}$

*Find each product or quotient. Simplify the answer.*

**39.** $\frac{1}{3}\cdot\frac{4}{5}$     **40.** $\frac{6}{13}\cdot\frac{7}{17}$     **41.** $\frac{5}{12}\cdot\frac{4}{15}$     **42.** $\frac{36}{48}\cdot\frac{16}{45}$

**43.** $\frac{3}{4}\div\frac{1}{2}$     **44.** $\frac{15}{16}\cdot\frac{4}{3}$     **45.** $\frac{3}{8}\div\frac{3}{4}$     **46.** $\frac{3}{8}\cdot\frac{10}{11}$

**47.** $\frac{10}{3}\div\frac{5}{9}$     **48.** $\frac{5}{9}\div30$     **49.** $\frac{15}{4}\cdot\frac{2}{3}$     **50.** $\frac{5}{12}\div\frac{4}{3}$

**51.** $5\frac{3}{8}\div1\frac{1}{4}$     **52.** $\left(2\frac{1}{5}\right)\left(\frac{7}{8}\right)$     **53.** $\frac{28}{13}\cdot\frac{2}{7}$     **54.** $4\frac{4}{5}\div\frac{8}{15}$

*Add or subtract. Simplify each answer.*

**55.** $\frac{3}{8}+\frac{2}{8}$     **56.** $\frac{18}{36}-\frac{5}{36}$     **57.** $\frac{3}{14}-\frac{1}{14}$     **58.** $\frac{1}{4}+\frac{3}{4}$

**59.** $\frac{4}{5}+\frac{6}{15}$     **60.** $\frac{7}{8}-\frac{5}{6}$     **61.** $\frac{9}{17}+\frac{2}{34}$     **62.** $\frac{3}{7}+\frac{17}{35}$

**63.** $\frac{1}{3}+\frac{1}{4}$     **64.** $\frac{1}{6}-\frac{1}{18}$     **65.** $\frac{7}{12}-\frac{2}{9}$     **66.** $\frac{3}{7}+\frac{5}{12}$

**67.** $3\frac{1}{8}-\frac{5}{12}$     **68.** $5\frac{6}{7}+4\frac{5}{8}$     **69.** $6\frac{1}{3}-3\frac{1}{2}$     **70.** $2\frac{3}{8}+3\frac{3}{4}$

**71.** $9\frac{2}{5}-6\frac{1}{2}$     **72.** $4\frac{5}{9}-\frac{7}{8}$     **73.** $5\frac{9}{10}+3\frac{1}{3}$     **74.** $8\frac{2}{7}-3\frac{1}{3}$

**75.** How much larger is $\frac{5}{6}$ mile than $\frac{3}{8}$ mile?     **76.** How much larger is $\frac{1}{5}$ meter than $\frac{1}{7}$ meter?

## Problem Solving

**77. Height Gain** The following graph shows Rebecca Bersagel's height, in inches, on her 8th and 12th birthdays. How much had Rebecca grown in the 4 years?

**Rebecca Bersagel's Height**

**78. Road Paving** The following graph shows the progress of the Davenport Paving Company in paving the Memorial Highway. How much of the highway was paved from June through August?

**Highway Paved in Selected Months**

*In many problems you will need to subtract a fraction from 1, where 1 represents "the whole" or the "total amount." Exercises 79–82 are answered by subtracting the given fraction from 1.*

**79. Putting Success** Lamont is a local golf pro and last year he made $\frac{46}{55}$ of all his putts from within six feet. What fraction of his putts from within six feet did he miss?

**80. Global Warming** The probability that an event does not occur may be found by subtracting the probability that the event does occur from 1. If the probability that global warming is occurring is $\frac{7}{9}$, find the probability that global warming is not occurring.

**81. Time to BS Degree** A recent study by the non-profit Education Trust found that $\frac{37}{100}$ of first-time freshmen at four-year schools earned their bachelor's degrees in four years. What fraction of freshmen did not finish their bachelor's degree in 4 years?

**Freshman Finishing Bachelor's Degree**

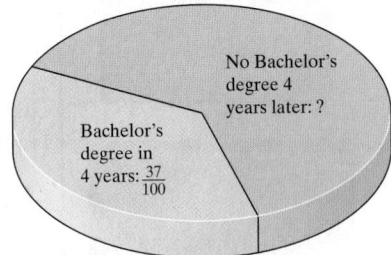

*Source*: The Education Trust

**82. Home Heating** The following circle graph shows the fraction of U.S. homes that used electricity to heat their homes. Determine the fraction of U.S. homes that did not use electricity.

**How Americans Heat Their Homes**

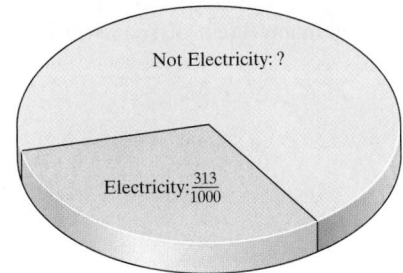

*Source*: www.census.gov

*In Exercises 83–98, answer the questions asked.*

**83. Python's Growth** An albino python at Cypress Gardens, Florida, was 3 feet, $3\frac{1}{4}$ inches when born. Its present length is 15 feet, $2\frac{1}{2}$ inches. How much has it grown since birth?

Albino python

**84. Cream Pie** A Boston cream pie weighs $1\frac{5}{16}$ pounds. If the pie is to be divided equally among 6 people, how much will each person get?

**85. Running** Denise started a running program in January when she could run a mile in $10\frac{1}{2}$ minutes. After 6 months, Denise could run a mile in $8\frac{1}{5}$ minutes. By how many minutes did Denise improve in 6 months?

**86. Baking Turkey** The instructions on a turkey indicate that a 12- to 16-pound turkey should bake at 325°F for about 22 minutes per pound. Josephine Nickola is planning to bake a $13\frac{1}{2}$-pound turkey. Approximately how long should the turkey be baked?

**87. Wood Cut** Debbie Anderson cuts a piece of wood measuring $3\frac{1}{8}$ inches into two equal pieces. How long is each piece?

**88. Pants Inseam** The inseam on a new pair of pants is 32 inches. If Don O'Neal's inseam is $29\frac{3}{8}$ inches, by how much will the pants need to be shortened?

**89. Drug Amount** A nurse must give $\frac{1}{16}$ milligram of a drug for each kilogram of patient mass. If Mr. Krisanda has a mass of 80 kilograms, find the amount of the drug Mr. Krisanda should be given.

**90. Chopped Onions** A recipe for pot roast calls for $\frac{1}{4}$ cup chopped onions for each pound of beef. For $5\frac{1}{2}$ pounds of beef, how many cups of chopped onions are needed?

**91. Shampoo** A bottle of shampoo contains 15 fluid ounces. If Tierra Bentley uses $\frac{3}{8}$ of an ounce each time she washes her hair, how many times can Tierra wash her hair using this bottle?

**92. Fencing** Matt Mesaros wants to fence in his backyard as shown. The three sides to be fenced measure $16\frac{2}{3}$ yards, $22\frac{2}{3}$ yards, and $14\frac{1}{8}$ yards.

**a)** How much fence will Matt need?

**b)** If Matt buys 60 yards of fence, how much will be left over?

**93. Windows** An insulated window for a house is made up of two pieces of glass, each $\frac{1}{4}$-inch thick, with a 1-inch space between them. What is the total thickness of this window?

**94. Truck Weight** A flatbed tow truck weighing $4\frac{1}{2}$ tons is carrying two cars. One car weighs $1\frac{1}{6}$ tons, the other weighs $1\frac{3}{4}$ tons. What is the total weight of the tow truck and the two cars?

**95. Cutting Wood** A 28-inch length of wood is to be cut into $4\frac{2}{3}$ inch strips. How many whole strips can be made? Disregard loss of wood due to cuts made.

**96. Fasten Bolts** A mechanic wishes to use a bolt to fasten a piece of wood $4\frac{1}{2}$ inches thick to a metal tube $2\frac{1}{3}$ inches thick. If the thickness of the nut is $\frac{1}{8}$ inch, find the length of the shaft of the bolt so that the nut fits flush with the end of the bolt (see the figure below).

**97. Entertainment Center** Scott Morningstar just bought a high definition television that will sit atop his entertainment credenza. The TV measures $36\frac{1}{2}$ inches high. Its stand is $14\frac{1}{8}$ inches high and the credenza is $31\frac{3}{4}$ inches high.

**a)** If Scott has 8-foot ceilings in his house, will there be sufficient room for this purchase?

**b)** Find the total height of the TV, the stand, and the credenza.

**98. Soda** If five 2-liter bottles of soda are split evenly among 30 people, how many liters of soda will each person get?

## Concept/Writing Exercises

**99.** Another way of saying "find the LCD of the 3 fractions $\frac{1}{6}, \frac{2}{3}, \frac{7}{10}$" is to say "find the least common multiple (LCM) of 6, 3, and 10." Write a few sentences to describe how to go about finding the LCM of 6, 3, and 10.

**100.** Explain how to simplify a fraction.

## Challenge Problems

**101.** Add or subtract the following fractions using the rule discussed in this section. Your answer should be a single fraction, and it should contain the symbols given in the exercise.

**a)** $\dfrac{*}{a} + \dfrac{?}{a}$

**b)** $\dfrac{\odot}{?} - \dfrac{\square}{?}$

**c)** $\dfrac{\triangle}{\square} + \dfrac{4}{\square}$

**d)** $\dfrac{x}{3} - \dfrac{2}{3}$

**e)** $\dfrac{12}{x} - \dfrac{4}{x}$

**102.** Multiply the following fractions using the rule discussed in this section. Your answer should be a single fraction and it should contain the symbols given in the exercise.

**a)** $\dfrac{\triangle}{a} \cdot \dfrac{\square}{b}$

**b)** $\dfrac{6}{3} \cdot \dfrac{\triangle}{\square}$

**c)** $\dfrac{x}{a} \cdot \dfrac{y}{b}$

**d)** $\dfrac{3}{8} \cdot \dfrac{4}{y}$

**e)** $\dfrac{3}{x} \cdot \dfrac{x}{y}$

**103. Drug Dosage** An allopurinol pill comes in 300-milligram doses. Dr. Muechler wants a patient to get 450 milligrams each day by cutting the pills in half and taking one-half pill three times a day. If he wants to prescribe enough pills for a 6-month period (assume 30 days per month), how many pills should he prescribe?

## Group Activity

*Discuss and answer Exercise 104 as a group.*

**104. Potatoes** The following table gives the amount of each ingredient recommended to make 2, 4, and 8 servings of instant mashed potatoes.

| Servings | 2 | 4 | 8 |
|---|---|---|---|
| Water | $\frac{2}{3}$ cup | $1\frac{1}{3}$ cups | $2\frac{2}{3}$ cups |
| Milk | 2 tbsp | $\frac{1}{3}$ cup | $\frac{2}{3}$ cup |
| Butter* | 1 tbsp | 2 tbsp | 4 tbsp |
| Salt† | $\frac{1}{4}$ tsp | $\frac{1}{2}$ tsp | 1 tsp |
| Potato flakes | $\frac{2}{3}$ cup | $1\frac{1}{3}$ cup | $2\frac{2}{3}$ cups |

*or margarine

†Less salt can be used if desired.

Determine the amount of potato flakes and milk needed to make 6 servings by the different methods described. When working with milk, 16 tbsp = 1 cup.

a) Group member 1: Determine the amounts of potato flakes and milk needed to make 6 servings by multiplying the amounts for 2 servings by 3.

b) Group member 2: Determine the amounts by adding the amounts for 2 servings to the amounts for 4 servings.

c) Group member 3: Determine the amounts by finding the average (mean) of 4 and 8 servings.

d) As a group, determine the amounts by subtracting the amounts for 2 servings from the amounts for 8 servings.

e) As a group, compare your answers from parts **a)** through **d)**. Are they all the same? If not, can you explain why?

## Cumulative Review Exercises

[1.1] **105.** What is your instructor's name and office hours?

[1.2] **106.** What is the mean of 9, 8, 15, 32, 16?

[1.3] **107.** What is the median of 9, 8, 15, 32, 16?

[1.3] **108.** What are variables?

# 1.4 The Real Number System

**1** Identify sets of numbers.

**2** Know the structure of the real numbers.

This section introduces you to different sets of numbers and to the structure of the real number system.

## Understanding Algebra

A *set* is a collection of elements—in this book, usually numbers. The set of natural numbers less than 6, for example, is {1, 2, 3, 4, 5}.

## 1 Identify Sets of Numbers

A **set** is a collection of **elements** listed within braces. The set $\{a, b, c, d, e\}$ consists of five elements, namely $a, b, c, d,$ and $e$. A set that contains no elements is called an **empty set** (or **null set**). The symbols { } or ∅ are used to represent the empty set.

   Two important sets are the natural numbers and the whole numbers. The whole numbers were introduced earlier.

$$\text{Natural numbers:} \quad \{1, 2, 3, 4, 5, \dots\}$$

$$\text{Whole numbers:} \quad \{0, 1, 2, 3, 4, 5, \dots\}$$

An aid in understanding sets of numbers is a number line (**Fig. 1.7**).

FIGURE 1.7

The number line continues indefinitely in both directions. The numbers to the right of 0 are positive and those to the left of 0 are negative. Zero is neither positive nor negative (**Fig. 1.8**).

FIGURE 1.8

**Figure 1.9** illustrates the natural numbers marked on a number line. The natural numbers are also called the **counting numbers** or the **positive integers**.

FIGURE 1.9

Another important set of numbers is the integers.

Integers:   $\{\ldots, -5, -4, -3, -2, -1, 0, 1, 2, 3, 4, 5, \ldots\}$

Negative integers        Positive integers

The integers consist of the negative integers, 0, and the positive integers. The integers are marked on the number line in **Figure 1.10**.

FIGURE 1.10

Fractions and certain decimal numbers do not belong to the set of integers but they do belong to the set of rational numbers. The set of **rational numbers** consists of all the numbers that can be expressed as a quotient (or a ratio) of two integers, with the denominator not 0.

Rational numbers:   {quotient of two integers, denominator not 0}

All integers are rational numbers since they can be written with a denominator of 1. For example, $3 = \frac{3}{1}$, $-12 = \frac{-12}{1}$, and $0 = \frac{0}{1}$. All fractions containing integers in the numerator and denominator (with the denominator not 0) are rational numbers. For example, the fraction $\frac{3}{5}$ is a quotient of two integers and is a rational number.

When a fraction that is a ratio of two integers is converted to a decimal number by dividing the numerator by the denominator, the quotient will always be either a *terminating decimal number,* such as 0.3 and 3.25, or a *repeating decimal number* such as 0.3333... and 5.2727.... The three dots at the end of a number indicate that the numbers continue to repeat in the same manner indefinitely. All terminating decimal numbers and all repeating decimal numbers are rational numbers and can be expressed as a quotient of two integers. For example, $0.3 = \frac{3}{10}$, $3.25 = \frac{325}{100}$, $0.3333\ldots = \frac{1}{3}$, and $5.2727\ldots = \frac{522}{99}$. Some rational numbers are illustrated on the number line in **Figure 1.11**.

FIGURE 1.11

> **Understanding Algebra**
>
> *Rational numbers* can be expressed as fractions where the numerator and denominator are integers.
> or
> *Rational numbers* can be expressed as decimals that either terminate or repeat.

> **Understanding Algebra**
>
> *Irrational numbers* are numbers that have infinite, non-repeating decimal representations.

Some numbers are not rational. Numbers such as the square root of 2, written $\sqrt{2}$, are not rational numbers. Any number that can be represented on the number line that is not a rational number is called an **irrational number**. Irrational numbers are non-terminating, non-repeating decimal numbers. For example, $\sqrt{2}$ cannot be expressed exactly as a decimal number. Irrational numbers can only be *approximated* by decimal numbers. $\sqrt{2}$ is *approximately* 1.41. Thus, we may write $\sqrt{2} \approx 1.41$. Some irrational numbers are illustrated on the number line in **Figure 1.12**.

FIGURE 1.12

## 2 Know the Structure of the Real Numbers

Any number that is either rational or irrational is called a **real number** and the number line is often referred to as a **real number line.**

The symbol $\mathbb{R}$ is used to represent the set of real numbers. The natural numbers, the whole numbers, the integers, the rational numbers, and the irrational numbers are all real numbers. **Figure 1.13** illustrates the relationships between the various sets of numbers within the set of real numbers.

Consider the natural number 5. If we follow the branches in **Figure 1.13a** upward, we see that the number 5 is also a whole number, an integer, a rational number, and a real number. Now consider the number $\frac{1}{2}$. It belongs to the noninteger rational numbers. If we follow the branches upward, we can see that $\frac{1}{2}$ is also a rational number and a real number.

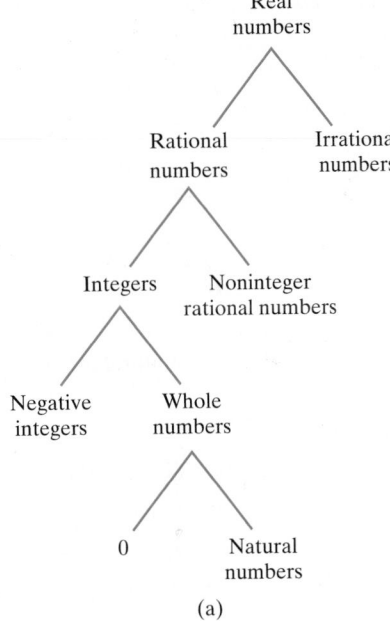

| Real Numbers | |
|---|---|
| Rational numbers | Irrational numbers |
| (Integers and noninteger rational numbers) | (Certain* square roots and other special numbers) |
| $-12$   $4$   $0$ | $\sqrt{2}$      $\sqrt{5}$ |
| $\frac{3}{8}$   $\frac{1}{3}$   $-1.24$ | $\pi$      $\sqrt{12}$ |
| $-1\frac{3}{5}$   $-2.463$ | |

*Other higher roots like $\sqrt[3]{2}$ and $\sqrt[4]{5}$ are also irrational numbers.

(a)                               (b)

**FIGURE 1.13**

---

**EXAMPLE 1**   Consider the following set of numbers.

$$\left\{ \sqrt{11}, -\frac{3}{8}, -0.6, -4, 7\frac{1}{2}, 12, 0, \sqrt{5}, -48, 3.9, -\sqrt{7} \right\}$$

List the elements of the set that are

**a)** natural numbers.      **b)** whole numbers.      **c)** integers.

**d)** rational numbers.      **e)** irrational numbers.      **f)** real numbers.

*Solution*   We will list the elements from left to right as they appear in the set. However, the elements may be listed in any order.

**a)** 12

**b)** 12, 0

**c)** $-4, 12, 0, -48$

**d)** $-\frac{3}{8}, -0.6, -4, 7\frac{1}{2}, 12, 0, -48, 3.9$

**e)** $\sqrt{11}, \sqrt{5}, -\sqrt{7}$

**f)** $\sqrt{11}, -\frac{3}{8}, -0.6, -4, 7\frac{1}{2}, 12, 0, \sqrt{5}, -48, 3.9, -\sqrt{7}$

Now Try Exercise 51

# EXERCISE SET 1.4     MathXL®   MyMathLab®
MathXL®      MyMathLab

## Warm-Up Exercises

*Fill in the blanks with the appropriate word, phrase, or symbol(s) from the following list.*

| | | | |
|---|---|---|---|
| set | irrational numbers | rational numbers | the empty set |
| $\{1, 2, 3\}$ | counting numbers | whole numbers | real number line |
| integers | $\{\ldots, -3, -2, -1\}$ | $\pi$ | $\dfrac{1}{2}$ |

1. $\sqrt{5}$ and $\sqrt{7}$ are examples of _____.

2. The set of negative integers is _____.

3. Another name for the positive integers is the set of _____.

4. The set $\{\ldots, -2, -1, 0, 1, 2, 3, \ldots\}$ is more commonly referred to as the set of _____.

5. The set of real numbers can be displayed pictorially as a _____.

6. The symbol $\varnothing$ is used to denote _____.

7. $\{0, 1, 2, 3, \ldots\}$ is called the set of _____.

8. Numbers that can be expressed as a fraction having integer numerator and non-zero integer denominator are called _____.

9. An example of a real number that is not a rational number is _____.

10. In general, a collection of elements is called a _____.

## Practice the Skills

*In Exercises 11–16, list each set of numbers.*

11. Integers

12. Counting numbers

13. Whole numbers

14. Positive integers

15. Negative integers

16. Natural numbers

*In Exercises 17–48, indicate whether each statement is true or false.*

17. 0 is a whole number.

18. $-1$ is a negative integer.

19. $-7.3$ is a real number.

20. $\dfrac{3}{5}$ is an integer.

21. 0.6 is an integer.

22. 0 is a natural number.

23. $\sqrt{2}$ is a rational number.

24. $\sqrt{3}$ is a real number.

25. $-\dfrac{1}{5}$ is a rational number.

26. $-2\dfrac{1}{3}$ is a rational number.

27. 0 is a rational number.

28. 9.2 is a rational number.

29. $4\dfrac{5}{8}$ is an irrational number.

30. 0 is not a positive number.

31. $-\sqrt{5}$ is an irrational number.

32. Every counting number is a rational number.

33. The symbol $\varnothing$ is used to represent the empty set.

34. Every integer is negative.

35. Every real number is a rational number.

36. Every negative integer is a real number.

37. Every rational number is a real number.

38. When zero is added to the set of counting numbers, the set of whole numbers is formed.

39. Some real numbers are not rational numbers.

40. Some irrational numbers are not real numbers.

41. Every negative number is a negative integer.

42. All real numbers can be represented on a number line.

43. The symbol $\mathbb{R}$ is used to represent the set of real numbers.

44. Any number to the left of zero on a number line is a negative number.

45. Every number greater than zero is a positive integer.

46. Irrational numbers cannot be represented on a number line.

47. When the negative integers, the positive integers, and 0 are combined, the integers are formed.

48. The natural numbers, counting numbers, and positive integers are different names for the same set of numbers.

49. **Hotels** Some buildings in Europe have elevators that list negative numbers for floors below the ground level. For example, a floor might be designated as $-2$. In many countries, floor number 13 is omitted because of superstition. Considering the numbers $-2$ and 13, list those that are

    a) positive integers.

    b) rational numbers.

    c) real numbers.

    d) whole numbers.

50. **Address** We generally think of house addresses as being integers greater than 0. Have you ever seen a house number that was not an integer greater than 0? There are quite a few such addresses in cities and towns across America. The house

numbers 0 and $2\frac{1}{2}$ appear on Legare Street in Charleston, SC. Considering the numbers 0 and $2\frac{1}{2}$, list those that are

© Allen R. Angel

**a)** integers

**b)** rational numbers

**c)** real numbers

**51.** Consider the following set of numbers.

$$\left\{-\frac{5}{7}, 0, -2, 3, 6\frac{1}{4}, \sqrt{7}, -\sqrt{3}, 1.63, 77\right\}$$

List the numbers that are

**a)** positive integers.

**b)** whole numbers.

**c)** integers.

**d)** rational numbers.

**e)** irrational numbers.

**f)** real numbers.

**52.** Consider the following set of numbers.

$$\left\{-6, 7, 12.4, -\frac{9}{5}, -2\frac{1}{4}, \sqrt{3}, 0, 9, \sqrt{7}, 0.35, \frac{22}{7}\right\}$$

List the numbers that are

**a)** positive integers.

**b)** whole numbers.

**c)** integers.

**d)** rational numbers.

**e)** irrational numbers.

**f)** real numbers.

## Problem Solving

*Give three examples of numbers that satisfy the given conditions.*

**53.** An integer but not a negative integer.

**54.** A real number but not an integer.

**55.** An irrational number and a negative number.

**56.** A real number and a rational number.

**57.** A rational number but not a natural number.

**58.** An integer and a rational number.

**59.** A negative integer and a rational number.

**60.** A negative integer and a real number.

**61.** A real number but not a positive rational number.

**62.** A rational number but not a negative number.

**63.** A real number but not an irrational number.

**64.** A negative number but not a negative integer.

*Three dots inside a set indicate that the set continues in the same manner. For example, $\{1, 2, 3, \ldots, 84\}$ is the set of natural numbers from 1 up to and including 84. In Exercises 65 and 66, determine the number of elements in each set.*

**65.** $\{8, 9, 10, 11, \ldots, 94\}$

**66.** $\{-4, -3, -2, -1, 0, 1, \ldots, 64\}$

## Challenge Problems

*The diagrams in Exercises 67 and 68 are called **Venn diagrams**. Venn diagrams are used to illustrate sets. For example, in the diagrams, circle A contains all the elements in set A, and circle B contains all the elements in set B. For each diagram, determine **a)** set A, **b)** set B, **c)** the set of elements that belong to both set A and set B, and **d)** the set of elements that belong to either set A or set B.*

**67.**

**68.**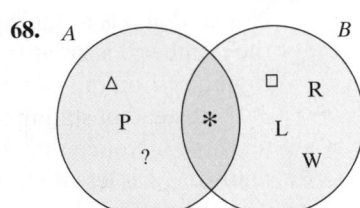

**69.** Consider the sets $A = \{1, 2, 3, 4\}$ and $B = \{1, 2, 3, 4, \ldots\}$.

   **a)** Explain the difference between set $A$ and set $B$.

   **b)** How many elements are in set $A$?

   **c)** How many elements are in set $B$?

   **d)** Set $A$ is an example of a *finite set*. Can you guess the name given to a set like set $B$?

**70.** How many decimal numbers are there

   **a)** between 1.0 and 2.0?

   **b)** between 1.4 and 1.5? Explain your answer.

**71.** How many fractions are there

   **a)** between 1 and 2?

   **b)** between $\frac{1}{3}$ and $\frac{1}{5}$? Explain your answer.

## Group Activity

*Discuss and answer Exercise 72 as a group.*

**72.** Set $A$ **union** set $B$, symbolized $A \cup B$, consists of the set of elements that belong to set $A$ or set $B$ (or both sets). Set $A$ **intersection** set $B$, symbolized $A \cap B$, consists of the set of elements that both set $A$ and set $B$ have in common. Note that the elements that belong to both sets are listed only once in the union of the sets.

   Consider the pairs of sets below.

   Group member 1:    $A = \{2, 3, 4, 6, 8, 9\}$    $B = \{1, 2, 3, 5, 7, 8\}$

   Group member 2:    $A = \{a, b, c, d, g, i, j\}$    $B = \{b, c, d, h, m, p\}$

   Group member 3:    $A = \{\text{red, blue, green, yellow}\}$    $B = \{\text{pink, orange, purple}\}$

   **a)** Group member 1: Find the union and intersection of the sets marked Group member 1.

   **b)** Group member 2: Find the union and intersection of the sets marked Group member 2.

   **c)** Group member 3: Find the union and intersection of the sets marked Group member 3.

   **d)** Now as a group, check each other's work. Correct any mistakes.

   **e)** As a group, using group member 1's sets, construct a Venn diagram like those shown in Exercises 67 and 68.

## Cumulative Review Exercises

[1.3]  **73.** Convert $5\frac{4}{5}$ to a fraction.

**74.** Write $\frac{16}{3}$ as a mixed number.

**75.** Subtract $\frac{7}{8} - \frac{1}{3}$.

**76.** Divide $\frac{3}{5} \div 6\frac{3}{4}$.

# 1.5 Inequalities

**1** Determine which is the greater of two numbers.

**2** Find the absolute value of a number.

## 1 Determine Which Is the Greater of Two Numbers

The number line, which shows numbers increasing from left to right, can be used to explain inequalities (see **Fig. 1.14**). When comparing two numbers, *the number to the right on the number line is the greater number, and the number to the left is the lesser number.* The symbol $>$ is used to represent the words "is greater than." The symbol $<$ is used to represent the words "is less than."

FIGURE 1.14

The statement that the number 3 is greater than the number 2 is written $3 > 2$. Notice that 3 is to the right of 2 on the number line in **Figure 1.14**. The statement that the number 0 is greater than the number $-1$ is written $0 > -1$. Notice that 0 is to the right of $-1$ on the number line.

   Instead of stating that 3 is greater than 2, we could state that 2 is less than 3, written $2 < 3$. Notice that 2 is to the left of 3 on the number line. The statement that the number $-1$ is less than the number 0 is written $-1 < 0$. Notice that $-1$ is to the left of 0 on the number line.

**EXAMPLE 1** Insert either $>$ or $<$ in the shaded area between each pair of numbers to make a true statement.

**a)** $-4$  $-2$    **b)** $-\dfrac{3}{2}$ ▨ $2.5$    **c)** $\dfrac{1}{2}$ ▨ $\dfrac{1}{4}$    **d)** $-2$ ▨ $4$

*Solution*    The points given are shown on the number line (**Fig. 1.15**).

**FIGURE 1.15**

**a)** $-4 < -2$; notice that $-4$ is to the left of $-2$.

**b)** $-\dfrac{3}{2} < 2.5$; notice that $-\dfrac{3}{2}$ is to the left of $2.5$.

**c)** $\dfrac{1}{2} > \dfrac{1}{4}$; notice that $\dfrac{1}{2}$ is to the right of $\dfrac{1}{4}$.

**d)** $-2 < 4$; notice that $-2$ is to the left of $4$.

Now Try Exercise 35

**EXAMPLE 2** Insert either $>$ or $<$ in the shaded area between each pair of numbers to make a true statement.

**a)** $-3$ ▨ $3$    **b)** $-3$ ▨ $-4$    **c)** $-4$ ▨ $0$    **d)** $-1.08$ ▨ $-1.8$

*Solution*    The numbers given are shown on the number line (**Fig. 1.16**).

**FIGURE 1.16**

**a)** $-3 < 3$; notice that $-3$ is to the left of $3$.

**b)** $-3 > -4$; notice that $-3$ is to the right of $-4$.

**c)** $-4 < 0$; notice that $-4$ is to the left of $0$.

**d)** $-1.08 > -1.8$; notice that $-1.08$ is to the right of $-1.8$.

Now Try Exercise 43

## 2  Find the Absolute Value of a Number

The **absolute value** of a number can be considered the distance between the number and 0 on a number line. Thus, the absolute value of 3, written $|3|$, is 3 since it is 3 units from 0 on a number line. Similarly, the absolute value of $-3$, written $|-3|$, is also 3 since $-3$ is 3 units from 0. See **Figure 1.17**.

$$|3| = 3 \quad \text{and} \quad |-3| = 3$$

**FIGURE 1.17**

Since the absolute value of a number measures the distance (without regard to direction) of a number from 0 on the number line, *the absolute value of every number will be either positive or zero.*

| Number | Absolute Value of Number |
|--------|--------------------------|
| 6 | $|6| = 6$ |
| $-6$ | $|-6| = 6$ |
| 0 | $|0| = 0$ |
| $-\dfrac{1}{2}$ | $\left|-\dfrac{1}{2}\right| = \dfrac{1}{2}$ |

### Understanding Algebra

The absolute value of a real number will always be either zero or positive.

The negative of the absolute value of a nonzero number will always be a negative number. For example,

$$-|2| = -(2) = -2 \quad \text{and} \quad -|-3| = -(3) = -3$$

**EXAMPLE 3** Insert either $>$, $<$, or $=$ in each shaded area to make a true statement.

**a)** $|3|$ ▨ $3$     **b)** $|-2|$ ▨ $|2|$     **c)** $-2$ ▨ $|-4|$

**d)** $|-2|$ ▨ $|-4|$     **e)** $\left|-\dfrac{2}{5}\right|$ ▨ $|-0.42|$

*Solution*

**a)** $|3| = 3$.     **b)** $|-2| = |2|$, since both $|-2|$ and $|2|$ equal 2.

**c)** $-2 < |-4|$, since $|-4| = 4$.

**d)** $|-2| < |-4|$, since $|-2| = 2$ and $|-4| = 4$.

**e)** When an absolute value contains a fraction, we can compare it to an absolute value containing a decimal number by rewriting the fraction as a decimal number and comparing the absolute values of the decimal numbers.

$$\left|-\dfrac{2}{5}\right| = |-0.40| = 0.40$$

$$|-0.42| = 0.42$$

Therefore, $\left|-\dfrac{2}{5}\right| < |-0.42|$ because $0.40 < 0.42$.

Now Try Exercise 57

The concept of absolute value is very important in higher-level mathematics courses.

---

## EXERCISE SET 1.5    MathXL®    MyMathLab®

### Warm-Up Exercises

*Fill in the blanks with the appropriate word, phrase, or symbol(s) from the following list.*

| | | | |
|---|---|---|---|
| $|-4|$ | 0 | $|6 - (-4)|$ | distance |
| is greater than | $-4$ | is less than | $|a|$ |
| positive number | negative number | True | False |

**1.** Regardless of the value of $a$, the value of $|a| - |a|$ is _____.

**2.** The symbol $<$ means _____.

**3.** The absolute value of the number $a$ is expressed as _____.

**4.** If we write $x > 0$, alternatively we could say that $x$ is a _____.

**5.** (True or False) If $a$ and $b$ are real numbers and $a < b$, then $b > a$. _____

**6.** The symbol $>$ means _____.

**7.** The distance between 6 and –4 on the number line can be expressed as _____.

**8.** The distance the number –4 is from zero can be expressed as _____.

**9.** The negative of the absolute value of a nonzero number will always be a _____.

**10.** The absolute value of a number represents its _____ from 0 on a real number line.

## Practice the Skills

*Evaluate.*

**11.** $|7|$

**12.** $|-6|$

**13.** $|-15|$

**14.** $|0|$

**15.** $-|0|$

**16.** $|54|$

**17.** $-|-5|$

**18.** $-|92|$

**19.** $-|26|$

**20.** $-|-34|$

*Insert either $<$ or $>$ in each shaded area to make a true statement.*

**21.** 36 ▨ 24

**22.** 4 ▨ –2

**23.** –4 ▨ 0

**24.** –6 ▨ –4

**25.** $\dfrac{1}{2}$ ▨ $-\dfrac{2}{3}$

**26.** $\dfrac{3}{5}$ ▨ $\dfrac{4}{5}$

**27.** 0.7 ▨ 0.8

**28.** –0.2 ▨ –0.4

**29.** $-\dfrac{1}{2}$ ▨ –1

**30.** –0.14 ▨ –0.92

**31.** –5 ▨ 5

**32.** $-\dfrac{3}{4}$ ▨ –1

**33.** –2.1 ▨ –2

**34.** –1.83 ▨ –1.82

**35.** $\dfrac{4}{5}$ ▨ $-\dfrac{4}{5}$

**36.** –9 ▨ –12

**37.** $-\dfrac{3}{8}$ ▨ $\dfrac{3}{8}$

**38.** –4.09 ▨ –5.3

**39.** 0.49 ▨ 0.43

**40.** –1.0 ▨ –0.7

**41.** –0.086 ▨ –0.095

**42.** $\dfrac{1}{2}$ ▨ $-\dfrac{1}{2}$

**43.** 0.001 ▨ 0.002

**44.** –0.006 ▨ –0.007

**45.** $\dfrac{5}{8}$ ▨ 0.6

**46.** 2.7 ▨ $\dfrac{10}{3}$

**47.** $-\dfrac{4}{3}$ ▨ $-\dfrac{2}{3}$

**48.** $\dfrac{19}{2}$ ▨ $\dfrac{17}{2}$

**49.** –0.8 ▨ $-\dfrac{3}{5}$

**50.** –0.7 ▨ –0.2

**51.** 0.3 ▨ $\dfrac{1}{3}$

**52.** $\dfrac{9}{20}$ ▨ 0.42

**53.** $-\dfrac{17}{30}$ ▨ $-\dfrac{16}{20}$

**54.** $\dfrac{13}{15}$ ▨ $\dfrac{8}{9}$

**55.** $-(-6)$ ▨ $-(-5)$

**56.** $-\left(-\dfrac{12}{13}\right)$ ▨ $\dfrac{7}{8}$

*Insert either $<$, $>$, or $=$ in each shaded area to make a true statement.*

**57.** 5 ▨ $|-2|$

**58.** $|-12|$ ▨ $|-13|$

**59.** $\dfrac{3}{4}$ ▨ $|-4|$

**60.** $|-4|$ ▨ –3

**61.** $|0|$ ▨ $|-4|$

**62.** $|-2.1|$ ▨ $|-1.8|$

**63.** 4 ▨ $\left|-\dfrac{9}{2}\right|$

**64.** $|-5|$ ▨ $-|-6|$

**65.** $\left|-\dfrac{4}{5}\right|$ ▨ $\left|-\dfrac{5}{4}\right|$

**66.** $\left|\dfrac{2}{5}\right|$ ▨ $|-0.40|$

**67.** $|-4.6|$ ▨ $\left|-\dfrac{23}{5}\right|$

**68.** $\left|-\dfrac{8}{3}\right|$ ▨ $|-3.5|$

*Insert either $>$, $<$, or $=$ in each shaded area to make a true statement.*

**69.** $\dfrac{2}{3}+\dfrac{2}{3}+\dfrac{2}{3}+\dfrac{2}{3}$ ▨ $4\cdot\dfrac{2}{3}$

**70.** $\dfrac{3}{4}+\dfrac{3}{4}$ ▨ $\dfrac{3}{4}\cdot\dfrac{3}{4}$

**71.** $\dfrac{1}{2}\cdot\dfrac{1}{2}$ ▨ $\dfrac{1}{2}\div\dfrac{1}{2}$

**72.** $5\div\dfrac{2}{3}$ ▨ $\dfrac{2}{3}\div5$

**73.** $\dfrac{7}{8}-\dfrac{1}{2}$ ▨ $\dfrac{7}{8}\div\dfrac{1}{2}$

**74.** $3\dfrac{1}{5}+\dfrac{1}{3}$ ▨ $3\dfrac{1}{5}\cdot\dfrac{1}{3}$

*Arrange the numbers from smallest to largest.*

**75.** $0.46, \dfrac{4}{9}, |-5|, -|-1|, \dfrac{3}{7}$

**76.** $-\dfrac{3}{4}, -|0.6|, -\dfrac{5}{9}, -1.74, |-1.9|$

**77.** $\dfrac{2}{3}, 0.6, |-2.6|, \dfrac{19}{25}, \dfrac{5}{12}$

**78.** $-|-5|, |-9|, \left|-\dfrac{12}{5}\right|, 2.7, \dfrac{7}{12}$

## Problem Solving

**79.** What numbers are 4 units from 0 on a number line?

**80.** What numbers are 100 units from 0 on the number line?

*In Exercises 81–88, give three real numbers that satisfy all the stated criteria. If no real numbers satisfy the criteria, so state and explain why.*

81. less than 4 and greater than 8

82. greater than 4 and less than 6

83. less than −2 and greater than −6

84. greater than −5 and greater than −9

85. greater than −3 and greater than 3

86. less than −3 and less than 3

87. greater than $|-2|$ and less than $|-6|$

88. greater than $|-3|$ and less than $|3|$

89. a) Consider the word *between*. What does this word mean?

   b) List three real numbers between 4 and 6.

   c) Is the number 4 between the numbers 4 and 6? Explain.

   d) Is the number 5 between the numbers 4 and 6? Explain.

   e) Is it true or false that the real numbers between 4 and 6 are the real numbers that are both greater than 4 and less than 6? Explain.

90. **Property Crime Rates** The following line graph shows the U.S. property crime rates per 1000 households from 1973 to 2005.

**Property Crime Rates**

*Note*: Property crimes include burglary, theft, and motor vehicle theft.

*Source*: www.ojp.usdoj.gov

Estimate the year(s) when the property crime rate was

a) first less than 400 per 1000 households.

b) first less than 200 per 1000 households.

c) greater than 400 per 1000 households and less than 600 per 1000 households.

91. **Peanuts** The following bar graph shows the percent of the recommended daily allowance of certain nutrients in 1 ounce of dry-roasted, salted peanuts. Use the bar graph to determine which nutrients provide

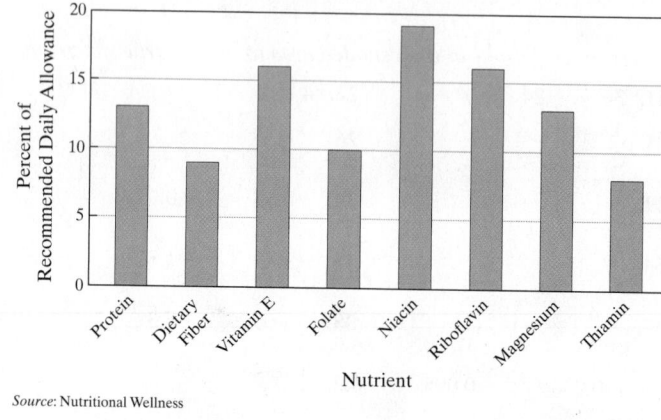

*Source*: Nutritional Wellness

a) less than 10% of the recommended daily allowance.

b) greater than 15% of the recommended daily allowance.

## Concept/Writing Exercises

92. Are there any real numbers whose absolute value is not a positive number? Explain your answer.

93. Will $|a| - |a| = 0$ always be true for any real number $a$? Explain.

94. Suppose $a$ and $b$ represent any two real numbers. Suppose $a > b$ is true. Will $|a| > |b|$ also be true? Explain and give an example to support your answer.

95. Suppose $a$ and $b$ represent any two real numbers. Suppose $|a| > |b|$ is true. Will $a > b$ also be true? Explain and give an example to support your answer.

## Challenge Problems

96. A number greater than 0 and less than 1 (or between 0 and 1) is multiplied by itself. Will the product be less than, equal to, or greater than the original number selected? Explain why this is always true.

97. A number between 0 and 1 is divided by itself. Will the quotient be less than, equal to, or greater than the original number selected? Explain why this is always true.

98. What two numbers can be substituted for $x$ to make $|x| = 3$ a true statement?

99. Are there any values for $x$ that would make $|x| = -5$ a true statement? Explain.

100. a) To what is $|x|$ equal if $x$ represents a real number greater than or equal to 0?

   b) To what is $|x|$ equal if $x$ represents a real number less than 0?

   c) Fill in the following shaded areas to make a true statement.

   $$|x| = \begin{cases} \phantom{xx} , x \geq 0 \\ \phantom{xx} , x < 0 \end{cases}$$

## Group Activity

*Discuss and answer Exercise 101 as a group.*

**101. a)** Group member 1: Draw a number line and mark points on the line to represent the following numbers.

$$|-2|, \quad -|3|, \quad -\left|\frac{1}{3}\right|$$

**b)** Group member 2: Do the same as in part **a)**, but on your number line mark points for the following numbers.

$$|-4|, \quad -|2|, \quad \left|-\frac{3}{5}\right|$$

**c)** Group member 3: Do the same as in parts **a)** and **b)**, but mark points for the following numbers.

$$|0|, \quad \left|\frac{16}{5}\right|, \quad -|-3|$$

**d)** As a group, construct one number line that contains all the points listed in parts **a)**, **b)**, and **c)**.

## Cumulative Review Exercises

[1.3] **102.** Add $2\frac{3}{5} + 3\frac{1}{3}$.

[1.4] **103.** List the set of integers.

**104.** List the set of whole numbers.

**105.** Consider the following set of numbers.

$$\left\{5, -2, 0, \frac{1}{3}, \sqrt{3}, -\frac{5}{9}, 2.3, \pi\right\}$$

List the numbers that are

**a)** natural numbers.

**b)** whole numbers.

**c)** integers.

**d)** rational numbers.

**e)** irrational numbers.

**f)** real numbers.

## Mid-Chapter Test: 1.1–1.5

*To find out how well you understand the chapter material to this point, take this brief test. The answers, and the section where the material was initially discussed, are given in the back of the book. Review any questions that you answered incorrectly.*

**1.** For each hour of class time, how many hours outside of class are recommended for studying and doing homework?

**2. Phone Bills** Jason Wisely's monthly phone bills for the first six months of 2009 were $78.83, $96.57, $62.23, $88.79, $101.75, and $55.62. For Jason's phone bills, determine the **a)** mean and **b)** median.

**3. Checking Account** The balance in Elizabeth Mater's checking account is $652.70. She deposited $230.75 and then purchased three books at $19.62 each, including tax. If she pays by check, what is the new balance in her checking account?

**4. Boat Rental** The rental cost of a boat from Natwora's Boat Rental is $7.50 per 15 minutes, and the rental cost from Gurney's Boat Rental is $18 per half hour. Suppose you plan to rent a boat for 4 hours.

**a)** Which is the better deal?

**b)** How much will you save?

**5. Water Rate** The water rate in Livingston County is $1.85 per 1000 gallons of water used. What is the water bill for a resident of Livingston County who uses 33,700 gallons of water?

*Perform the indicated operation.*

**6.** $\frac{3}{7} \cdot \frac{7}{18}$

**7.** $\frac{9}{16} \div \frac{15}{13}$

**8.** $\frac{5}{8} + \frac{3}{5}$

**9.** $6\frac{1}{4} - 3\frac{1}{5}$

**10. Garden** Justin Calhoun wants to fence in his rectangular garden to keep out deer. His garden measures $14\frac{2}{3}$ feet by $12\frac{1}{2}$ feet. How much fence will Justin need?

© Sarah Joos/Shutterstock

*In Exercises 11–15, indicate whether each statement is true or false.*

**11.** 0 is a natural number.

**12.** $-8.6$ is a real number.

**13.** $3\frac{2}{3}$ is an irrational number.

**14.** Every integer is a real number.

**15.** Every whole number is a counting number.

**16.** Evaluate $-\left|-\frac{7}{10}\right|$.

*Insert either* <, >, *or* = *in each shaded area to make a true statement.*

**17.** −0.005 ▢ −0.006     **18.** $\dfrac{7}{8}$ ▢ $\dfrac{5}{6}$     **19.** $|-9|$ ▢ $|-19|$     **20.** $\left|-\dfrac{3}{8}\right|$ ▢ $|-0.375|$

# 1.6 Addition of Real Numbers

1 Add real numbers using a number line.

2 Add fractions.

3 Identify opposites.

4 Add using absolute values.

## Understanding Algebra

With the exception of 0, *any number without a sign in front of it is positive.* For example, 3 means +3 and 5 means +5.

There are many practical uses for negative numbers. A submarine diving below sea level, a bank account that has been overdrawn, a business spending more than it earns, and a temperature below zero are some examples.

The four basic **operations** of arithmetic are addition, subtraction, multiplication, and division. Both positive and negative numbers can be added, subtracted, multiplied, and divided. In this section, we discuss the operation of addition.

### 1 Add Real Numbers Using a Number Line

To add numbers, we make use of a number line. Represent the first number to be added (first *addend*) by an arrow starting at 0. The arrow is drawn to the right if the number is positive. If the number is negative, the arrow is drawn to the left. From the tip of the first arrow, draw a second arrow to represent the second addend. The second arrow is drawn to the right or left, as just explained. The sum of the two numbers is found at the tip of the second arrow.

**EXAMPLE 1**  Evaluate $3 + (-4)$ using a number line.

*Solution*    *Always begin at 0.* Since the first addend, the 3, is positive, the first arrow starts at 0 and is drawn 3 units to the right (**Fig. 1.18**).

## Understanding Algebra

When adding a positive number and a negative number, the result may be either positive or negative.

$3 + (-4) = -1$

$6 + (-1) = 5$

**FIGURE 1.18**

**FIGURE 1.19**

The second arrow starts at 3 and is drawn 4 units to the left, since the second addend is negative (**Fig. 1.19**). The tip of the second arrow is at −1. Thus,

$$3 + (-4) = -1$$

Now Try Exercise 27

**EXAMPLE 2**  Evaluate $-4 + 2$ using a number line.

*Solution*    Begin at 0. Since the first addend is negative, −4, the first arrow is drawn 4 units to the left. From there, since 2 is positive, the second arrow is drawn 2 units to the right. The second arrow ends at −2 (**Fig. 1.20**).

**FIGURE 1.20**

$$-4 + 2 = -2$$

Now Try Exercise 37

EXAMPLE 3  Evaluate $-3 + (-2)$ using a number line.

*Solution*    Start at 0. Since both numbers being added are negative, both arrows will be drawn to the left (**Fig. 1.21**).

FIGURE 1.21

$$-3 + (-2) = -5$$

Now Try Exercise 39

In Example 3, we can think of the expression $-3 + (-2)$ as combining a *loss* of 3 and a *loss* of 2 for a total *loss* of 5, or $-5$.

EXAMPLE 4  Add $5 + (-5)$ using a number line.

*Solution*    The first arrow starts at 0 and is drawn 5 units to the right. The second arrow starts at 5 and is drawn 5 units to the left. The tip of the second arrow is at 0. Thus, $5 + (-5) = 0$ (**Fig. 1.22**).

FIGURE 1.22

$$5 + (-5) = 0$$

Now Try Exercise 31

EXAMPLE 5  **Below-Zero Temperatures** At the beginning of the five o'clock news broadcast on a winter day in Rochester, New York, the chief meteorologist reported that the temperature was three degrees below zero Fahrenheit. During the weather segment twenty minutes later, the chief meteorologist stated that the temperature had dropped four degrees since the beginning of the news broadcast. Find the temperature at 5:20 P.M.

*Solution*    A vertical number line (**Fig. 1.23**) may help you visualize this problem.

$$-3 + (-4) = -7°F$$

Now Try Exercise 117

FIGURE 1.23

## 2  Add Fractions

To add fractions, where one or more of the fractions is negative, we use the same general procedure discussed in Section 1.3. Whenever the denominators are not the same, we will find the least common denominator (LCD) and then we obtain the answer by adding the numerators while keeping the LCD.

For example, suppose after obtaining a common denominator, we have $-\dfrac{19}{29} + \dfrac{13}{29}$.

To obtain the numerator of the answer, we may add $-19 + 13$ on the number line to obtain $-6$. The denominator of the answer is the common denominator, 29. Thus, the answer is $-\dfrac{6}{29}$.

$$-\frac{19}{29} + \frac{13}{29} = \frac{-19 + 13}{29} = \frac{-6}{29} = -\frac{6}{29}$$

Let's look at one more example. Suppose after obtaining the LCD, we have $-\dfrac{7}{40} + \left(-\dfrac{5}{40}\right)$. We add $-7 + (-5)$ on the number line to obtain $-12$. The denominator of the answer is 40. Thus, the answer before being simplified is $-\dfrac{12}{40}$. The final answer simplifies to $-\dfrac{3}{10}$. We show these calculations as follows:

$$-\frac{7}{40} + \left(-\frac{5}{40}\right) = \frac{-7 + (-5)}{40} = \frac{-12}{40} = -\frac{3}{10}$$

**EXAMPLE 6** Add $\dfrac{7}{16} + \left(-\dfrac{2}{3}\right)$.

*Solution*    The LCD is 48. Changing each fraction to a fraction with a denominator of 48 yields

$$\frac{7}{16} \cdot \frac{3}{3} + \left(-\frac{2}{3}\right) \cdot \frac{16}{16}$$

$$\text{or} \quad \frac{21}{48} + \left(-\frac{32}{48}\right)$$

To add these fractions, we keep the LCD and add the numerators to get

$$\frac{7}{16} + \left(-\frac{2}{3}\right) = \frac{21}{48} + \left(-\frac{32}{48}\right) = \frac{21 + (-32)}{48}$$

Now we add $21 + (-32)$ on a number line to get the numerator of the fraction, $-11$; see **Figure 1.24**.

**FIGURE 1.24**

Thus, $\dfrac{7}{16} + \left(-\dfrac{2}{3}\right) = \dfrac{21}{48} + \left(-\dfrac{32}{48}\right) = \dfrac{21 + (-32)}{48} = -\dfrac{11}{48}$.

*Now Try Exercise 77*

**EXAMPLE 7** Add $-\dfrac{7}{8} + \left(-\dfrac{3}{40}\right)$.

*Solution*    The LCD is 40. Rewriting the first fraction with the LCD gives the following.

$$-\frac{7}{8} + \left(-\frac{3}{40}\right) = -\frac{7}{8} \cdot \frac{5}{5} + \left(-\frac{3}{40}\right)$$

$$= -\frac{35}{40} + \left(-\frac{3}{40}\right) = \frac{-35 + (-3)}{40}$$

Now we add $-35 + (-3)$ to get the numerator of the fraction, $-38$; see **Figure 1.25**

**FIGURE 1.25**

Thus, $-\dfrac{7}{8} + \left(-\dfrac{3}{40}\right) = \dfrac{-35}{40} + \left(\dfrac{-3}{40}\right) = \dfrac{-35 + (-3)}{40} = -\dfrac{38}{40} = -\dfrac{19}{20}$.

*Now Try Exercise 85*

---

**Helpful Hint**

When the numbers being added are large, you are not expected to actually mark and count the units. For example, when determining $21 + (-32)$ you will not need to count 32 units to the left from 21 to obtain the answer $-11$.

In objective 3, we will show you how to obtain $21 + (-32) = -11$ without having to draw number lines. We present adding on a number line here to help you understand the concept of addition of signed numbers, and to help you in determining, without doing any calculations, whether the sum of two signed numbers will be a positive number, a negative number, or zero.

---

## 3  Identify Opposites

Now let's consider **opposites**, or **additive inverses**.

**Opposites (or Additive Inverses)**

Any two numbers whose sum is zero are said to be **opposites**, or **additive inverses**, of each other. In general, if we let $a$ represent any real number, then its opposite is $-a$ and $a + (-a) = 0$.

In Example 4, the sum of 5 and $-5$ is 0. Thus, $-5$ is the opposite of 5 and 5 is the opposite of $-5$.

**EXAMPLE 8**  Find the opposite of each number.

**a)** 3     **b)** $-4$     **c)** $-\dfrac{7}{8}$

*Solution*

**a)** The opposite of 3 is $-3$, since $3 + (-3) = 0$.

**b)** The opposite of $-4$ is 4, since $-4 + 4 = 0$.

**c)** The opposite of $-\dfrac{7}{8}$ is $\dfrac{7}{8}$, since $-\dfrac{7}{8} + \dfrac{7}{8} = 0$.

*Now Try Exercise 15*

---

**Understanding Algebra**

The *additive inverse* of a number has the property that the sum of the number and its additive inverse is 0. So, $(-5)$ is the additive inverse, or *opposite,* of 5 because $(-5) + 5 = 0$.

## 4  Add Using Absolute Values

Recall that the absolute value of a nonzero number will always be positive.

**Adding Real Numbers with the Same Sign**

To add real numbers with the same sign (either both positive or both negative), add their absolute values. The sum has the same sign as the numbers being added.

**EXAMPLE 9**  Add $4 + 8$.

*Solution*   Since both numbers have the same sign, both positive, we add their absolute values: $|4| + |8| = 4 + 8 = 12$. Since both numbers being added are positive, the sum is positive. Thus, $4 + 8 = 12$.

*Now Try Exercise 49*

**EXAMPLE 10**  Add $-6 + (-9)$.

*Solution*   Since both numbers have the same sign, both negative, we add their absolute values: $|-6| + |-9| = 6 + 9 = 15$. Since both numbers being added are negative, their sum is negative. Thus, $-6 + (-9) = -15$.

*Now Try Exercise 51*

## Understanding Algebra

The sum of two positive numbers will always be positive and the sum of two negative numbers will always be negative.

### Adding Two Signed Numbers with Different Signs

To add two signed numbers with different signs, one positive and the other negative, subtract the smaller absolute value from the larger absolute value. The answer has the sign of the number with the larger absolute value.

**EXAMPLE 11** Add $10 + (-6)$.

*Solution*    The two numbers being added have different signs, so we subtract the smaller absolute value from the larger: $|10| - |-6| = 10 - 6 = 4$. Since $|10|$ is greater than $|-6|$ and the sign of 10 is positive, the sum is positive. Thus, $10 + (-6) = 4$.

*Now Try Exercise 53*

**EXAMPLE 12** Add $12 + (-18)$.

*Solution*    The numbers being added have different signs, so we subtract the smaller absolute value from the larger: $|-18| - |12| = 18 - 12 = 6$. Since $|-18|$ is greater than $|12|$ and the sign of $-18$ is negative, the sum is negative. Thus, $12 + (-18) = -6$.

*Now Try Exercise 55*

**EXAMPLE 13** Add $-21 + 20$.

*Solution*    The two numbers being added have different signs, so we subtract the smaller absolute value from the larger: $|-21| - |20| = 21 - 20 = 1$. Since $|-21|$ is greater than $|20|$, the sum is negative. Therefore, $-21 + 20 = -1$.

*Now Try Exercise 61*

Now let's look at some additional examples that contain fractions and decimal numbers.

**EXAMPLE 14** Add $-\dfrac{3}{5} + \dfrac{4}{7}$.

*Solution*    We can write each fraction with the least common denominator, 35.

$$-\frac{3}{5} + \frac{4}{7} = -\frac{3}{5} \cdot \frac{7}{7} + \frac{4}{7} \cdot \frac{5}{5}$$

$$= \frac{-21}{35} + \frac{20}{35} = \frac{-21 + 20}{35}$$

Since $|-21|$ is greater than $|20|$, the final answer will be negative. In Example 13, we found that $-21 + 20 = -1$. Thus, we can write

$$\frac{-21}{35} + \frac{20}{35} = \frac{-21 + 20}{35} = \frac{-1}{35} = -\frac{1}{35}$$

Thus, $-\dfrac{3}{5} + \dfrac{4}{7} = -\dfrac{1}{35}$.

*Now Try Exercise 87*

Examples 15 and 16 contain decimal numbers. If you have forgotten how to perform the basic operations of addition, subtraction, multiplication, and division of decimal numbers, read Appendix A now.

EXAMPLE 15   Add $-37.45 + (-26.98)$.

*Solution*   Since both numbers have the same sign, both negative, we add their absolute values: $|-37.45| + |-26.98| = 37.45 + 26.98$.

$$
\begin{array}{r}
37.45 \\
+ \ 26.98 \\
\hline
64.43
\end{array}
$$

Since two negative numbers are being added, the sum is negative. Therefore, $-37.45 + (-26.98) = -64.43$.

Now Try Exercise 67

EXAMPLE 16   **Net Profit or Loss**   The B.J. Donaldson Printing Company had a loss of $4005.69 for the first 6 months of the year and a profit of $29,645.78 for the second 6 months of the year. Find the net profit or loss for the year.

*Solution*   Understand and Translate   This problem can be represented as $-4005.69 + 29,645.78$. Since the numbers have different signs, we subtract the smaller absolute value from the larger.

Carry Out   $|29,645.78| - |-4005.69| = 29,645.78 - 4005.69$

$$
\begin{array}{r}
29,645.78 \\
- \quad 4005.69 \\
\hline
25,640.09
\end{array}
$$

Since $|29,645.78|$ is greater than $|-4005.69|$ and the sign of 29,645.78 is positive, the sum is positive. Thus, $-4005.69 + 29,645.78 = 25,640.09$.

Check and Answer   The answer is reasonable. Thus, the net profit for the year was $25,640.09.

Now Try Exercise 121

## Understanding Algebra

The sum of two signed numbers with different signs may be either positive or negative. The sign of the sum will be the same as the sign of the number with the larger absolute value.

$$13 + (-45) = -32$$

$$(-17) + 20 = 3$$

> **Helpful Hint**
>
> Architects often make a scale model of a building before starting construction of the building. This model helps them visualize the project and often helps them avoid problems.
>
> Mathematicians also construct models. A *mathematical model* may be a physical representation of a mathematical concept. It may be as simple as using tiles or chips to represent specific numbers. For example, below we use a model to help explain addition of real numbers. This may help some of you understand the concepts better.
>
> We let a red chip represent $+1$ and a green chip represent $-1$.
>
> ● $= +1$       ● $= -1$
>
> If we add $+1$ and $-1$, or a red and a green chip, we get 0.
> Now consider the addition problem $3 + (-5)$. We can represent this as
>
> $$\underbrace{● \ ● \ ●}_{3} \ + \ \underbrace{● \ ● \ ● \ ● \ ●}_{-5}$$
>
> If we remove 3 red chips and 3 green chips, or three zeros, we are left with 2 green chips, which represents a sum of $-2$. Thus, $3 + (-5) = -2$,
>
> $$\cancel{●} \ \cancel{●} \ \cancel{●} \ + \ \cancel{●} \ \cancel{●} \ \cancel{●} \ ● \ ●$$
>
> Now consider the problem $-4 + (-2)$. We can represent this as
>
> $$\underbrace{● \ ● \ ● \ ●}_{-4} \ + \ \underbrace{● \ ●}_{-2}$$
>
> Since we end up with 6 green chips, and each green chip represents $-1$, the sum is $-6$. Therefore, $-4 + (-2) = -6$.

# EXERCISE SET 1.6    MathXL®    MyMathLab®
MathXL®    MyMathLab

## Warm-Up Exercises

*Fill in the blanks with the appropriate word, phrase, or symbol(s) from the following list.*

| | | | | |
|---|---|---|---|---|
| opposites | additive inverse | common denominator | addends | sum |
| absolute value | −8 | numerator | negative | 8 |
| positive | | | | |

1. The sum of two negative numbers is always _____ .

2. Another expression for "opposite" of a real number is _____ .

3. The expression $|x|$ is read "the _____ of $x$."

4. The sum of two positive numbers is always _____ .

5. In the statement $(-8) + 5 = -3$, the number $-3$ is called the _____ of $-8$ and 5.

6. In the statement $(-8) + 5 = -3$, the numbers $-8$ and 5 are called _____ .

7. $-|-8| =$ _____ .

8. $|-8| =$ _____ .

9. When adding two fractions with different signs, we first find the _____ .

10. Two numbers that add up to zero are _____ of each other.

*In Exercises 11 and 12, are the calculations shown correct? If not, explain why not.*

11. $\dfrac{-5}{12} + \dfrac{9}{12} = \dfrac{-5 + 9}{12} = \dfrac{4}{12} = \dfrac{1}{3}$

12. $\dfrac{-6}{70} + \left(\dfrac{-9}{70}\right) = \dfrac{-6 + (-9)}{70} = \dfrac{-15}{70} = -\dfrac{3}{14}$

## Practice the Skills

*Write the opposite of each number.*

13. 19

14. −7

15. −28

16. 3

17. 0

18. $-3\dfrac{1}{2}$

19. $\dfrac{5}{3}$

20. $-\dfrac{1}{4}$

21. $2\dfrac{3}{5}$

22. −1

23. 3.72

24. −0.721

*Add.*

25. $5 + 16$

26. $-8 + 2$

27. $4 + (-3)$

28. $9 + (-12)$

29. $-4 + (-2)$

30. $-3 + (-5)$

31. $6 + (-6)$

32. $-8 + 8$

33. $-4 + 4$

34. $-6 + 6$

35. $-8 + (-2)$

36. $6 + (-5)$

37. $-7 + 3$

38. $-6 + 9$

39. $-8 + (-5)$

40. $0 + (-3)$

41. $0 + 0$

42. $0 + (-0)$

43. $-8 + 0$

44. $-9 + 13$

45. $-18 + (-9)$

46. $-7 + 7$

47. $-33 + (-31)$

48. $-27 + (-9)$

49. $7 + 9$

50. $12 + 3$

51. $-8 + (-4)$

52. $-25 + (-36)$

53. $6 + (-3)$

54. $52 + (-25)$

55. $13 + (-19)$

56. $34 + (-40)$

57. $180 + (-220)$

58. $-452 + 312$

59. $-11 + (-20)$

60. $-33 + (-92)$

61. $-67 + 28$

62. $183 + (-183)$

63. $184 + (-93)$

64. $-19 + 176$

65. $80.5 + (-90.4)$

66. $-24.6 + (-13.9)$

67. $-124.7 + (-19.3)$

68. $106.3 + (-110.9)$

69. $-123.56 + (-18.35)$

70. $-72.79 + 33.47$

71. $-99.36 + 45.71$

72. $-84.15 + (-29.98)$

*Add.*

73. $\dfrac{3}{5} + \dfrac{1}{7}$

74. $\dfrac{5}{8} + \dfrac{3}{5}$

75. $\dfrac{5}{12} + \dfrac{6}{7}$

76. $\dfrac{2}{9} + \dfrac{3}{10}$

77. $-\dfrac{8}{11} + \dfrac{4}{5}$

78. $-\dfrac{4}{9} + \dfrac{5}{27}$

79. $-\dfrac{7}{10} + \dfrac{11}{90}$

80. $\dfrac{8}{9} + \left(-\dfrac{1}{3}\right)$

81. $-\dfrac{7}{30} + \left(-\dfrac{4}{5}\right)$

82. $-\dfrac{7}{9} + \left(-\dfrac{1}{5}\right)$

83. $\dfrac{9}{25} + \left(-\dfrac{3}{50}\right)$

84. $-\dfrac{1}{15} + \left(-\dfrac{5}{6}\right)$

85. $-\dfrac{4}{5} + \left(-\dfrac{5}{75}\right)$

86. $\dfrac{5}{36} + \left(-\dfrac{5}{24}\right)$

87. $-\dfrac{9}{24} + \dfrac{5}{7}$

88. $-\dfrac{9}{40} + \dfrac{4}{15}$

89. $-\dfrac{5}{12} + \left(-\dfrac{3}{10}\right)$

90. $\dfrac{7}{16} + \left(-\dfrac{5}{24}\right)$

91. $-\dfrac{13}{14} + \left(-\dfrac{7}{42}\right)$

92. $-\dfrac{11}{27} + \left(-\dfrac{7}{18}\right)$

*In Exercises 93–108,* **a)** *determine by observation whether the sum will be a positive number, zero, or a negative number;* **b)** *find the sum.*

**93.** 587 + (−197)

**94.** −140 + (−629)

**95.** −84 + (−289)

**96.** −647 + 352

**97.** −947 + 495

**98.** 762 + (−762)

**99.** −496 + (−804)

**100.** −354 + 1090

**101.** −375 + 263

**102.** 1127 + (−84)

**103.** −1833 + (−2047)

**104.** −426 + 572

**105.** 3124 + (−2013)

**106.** −9095 + (−647)

**107.** −1025 + (−1025)

**108.** 7513 + (−4361)

*Indicate whether each statement is true or false.*

**109.** The sum of two negative numbers is always a negative number.

**110.** The sum of a negative number and a positive number is sometimes a negative number.

**111.** The sum of two positive numbers is never a negative number.

**112.** The sum of a positive number and a negative number is always a negative number.

**113.** The sum of a positive number and a negative number is always a positive number.

**114.** The sum of a number and its opposite is always equal to zero.

## Problem Solving

*Write an expression that can be used to solve each problem and then solve.*

**115. Credit Card** David Nurkiewicz owed $94 on his credit card. He charged another item costing $183. Find the amount that David owed.

**116. Charge Card** Mrs. Chu charged $142 worth of goods on her charge card. Find her balance after she made a payment of $87.

**117. Football** A football team lost 18 yards on one play and then lost 3 yards on the following play. What was the total loss in yardage?

**118. Overdrawn Checking Account** Mrs. Jahn is unaware that her checking account has been overdrawn by $56. While shopping, she writes a check for $162. Find the total amount by which Mrs. Jahn has overdrawn her account.

**119. Drilling for Water** A company is drilling a well. During the first week they drilled 27 feet, and during the second week they drilled another 34 feet before they struck water. How deep is the well?

**120. Coffee Bar** The Frenches opened a coffee bar. Their income and expenses for their first three months of operation are shown in the following graph.

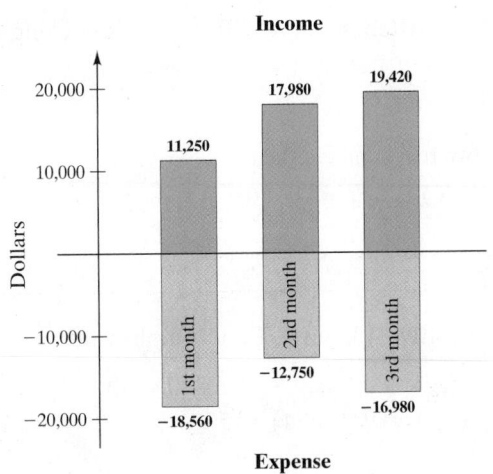

**a)** Find the net profit or loss (the sum of income and expenses) for the first month.

**b)** Find the net profit or loss for the second month.

**c)** Find the net profit or loss for the third month.

**121. High Mountain** The Web site *www.guinnessworldrecords.com* lists Mauna Kea in Hawaii as the tallest mountain in the world when measured from its base to its peak. The base of Mauna Kea is 19,684 feet below sea level. The total height of the mountain from its base to its peak is 33,480 feet. How high is the peak of Mauna Kea above sea level?

Hawaii's Mauna Kea is the world's highest mountain measured from the seafloor to its peak.

**122. Net Profit or Loss** The Crafty Scrapbook Company had a loss of $3000 for the first 4 months of the year and a profit of $37,400 for the last 8 months of the year. Find the net profit or loss for the year.

**123. Surplus and Deficit** The following graph shows the surplus or deficit for Gilmar Publishing Company in Cincinnati, Ohio, for the years 1996 through 2008.

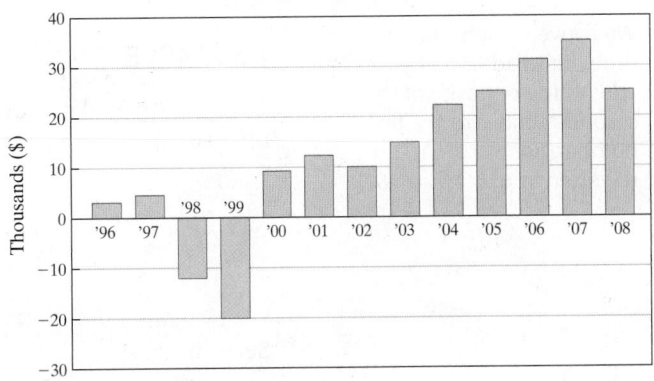

*Source*: www.gilmarpublishing.com

**a)** Estimate the surplus or deficit for Gilmar Publishing in 1998.

**b)** Estimate the surplus or deficit for Gilmar Publishing for each of the years 2006, 2007, and 2008. Then estimate the surplus or deficit from 2006 through 2008 by adding these three estimates.

**124. Stocks** The following chart shows percent changes from the first quarter of 2007 through the first quarter of 2008 for the mutual fund Fidelity Puritan Fund.

Determine the percent change for the fund from the first quarter of 2007 through the first quarter of 2008 by adding the individual percents.

| Fidelity Puritan Fund | Percent change from previous quarter |
|---|---|
| 1st quarter 2007 | 4.2% |
| 2nd quarter 2007 | 5.2% |
| 3rd quarter 2007 | 0.2% |
| 4th quarter 2007 | −13.5% |
| 1st quarter 2008 | −3.0% |

## Challenge Problems

*Evaluate each exercise by adding the numbers from left to right. We will discuss problems like this shortly.*

**125.** $(-8) + (-6) + (-12)$
**126.** $5 + (-7) + (-8)$
**127.** $29 + (-46) + 37$
**128.** $4 + (-5) + 6 + (-8)$
**129.** $(-12) + (-10) + 25 + (-3)$
**130.** $(-4) + (-2) + (-15) + (-27)$
**131.** $\frac{1}{2} + \left(-\frac{1}{3}\right) + \frac{1}{5}$
**132.** $-\frac{3}{8} + \left(-\frac{2}{9}\right) + \left(-\frac{1}{2}\right)$

*Find the following sums. Explain how you determined your answer. (Hint: Pair small numbers with large numbers from the ends inward.)*

**133.** $1 + 2 + 3 + \cdots + 10$
**134.** $1 + 2 + 3 + \cdots + 20$

## Cumulative Review Exercises

[1.3] **135.** Multiply $\left(\frac{4}{7}\right)\left(2\frac{3}{8}\right)$.
**136.** Subtract $3 - \frac{5}{16}$.

[1.4] **137.** True or False: Every number less than zero is a negative integer.

[1.5] *Insert either $<$, $>$, or $=$ in each shaded area to make a true statement.*

**138.** $|-3|$ ▨ $2$
**139.** $8$ ▨ $|-12|$

# 1.7 Subtraction of Real Numbers

**1** Subtract numbers.
**2** Subtract numbers mentally.
**3** Evaluate expressions containing more than two numbers.

### Understanding Algebra

We convert a subtraction problem to addition by adding the opposite of the second number to the first number. So, $6 - (-10)$ is rewritten as $6 + 10 = 16$.

## 1 Subtract Numbers

Any subtraction problem can be rewritten as an addition problem using the additive inverse.

**To Subtract Real Numbers**

In general, if $a$ and $b$ represent any two real numbers, then
$$a - b = a + (-b)$$

**EXAMPLE 1** Evaluate $9 - (+4)$.

*Solution* We are subtracting a positive 4 from 9. To accomplish this, we add the opposite of +4, which is −4, to 9.

$$9 - (+4) = 9 + (-4) = 5$$

Subtract  Positive 4  Add  Negative 4

We evaluated $9 + (-4)$ using the procedures for *adding* real numbers presented in Section 1.6.

Now Try Exercise 13

## Understanding Algebra

The parts of a subtraction problem:

$$16 \leftarrow \text{minuend}$$
$$- \underline{12} \leftarrow \text{subtrahend}$$
$$4 \leftarrow \text{difference}$$

Often in a subtraction problem, when the number being subtracted is a positive number, the + sign preceding the number being subtracted is not shown. For example, in the subtraction $9 - 4$,

$$9 - \boxed{4} \text{ means } 9 - \boxed{(+4)}$$

Thus, to evaluate $9 - 4$, we must add the opposite of 4, which is $-4$, to 9.

$$9 - 4 = 9 + (-4) = 5$$

Subtract  Positive 4  Add  Negative 4

This procedure is illustrated in Example 2.

**EXAMPLE 2** Evaluate $5 - 3$. ────────

*Solution*   We must subtract a positive 3 from 5. To change this problem to an addition problem, we add the opposite of 3, which is $-3$, to 5.

Subtraction    Addition
problem       problem
$$5 - 3 = 5 + (-3) = 2$$

Subtract  Positive 3  Add  Negative 3

Now Try Exercise 15

**EXAMPLE 3** Evaluate. ────────

**a)** $4 - 9$     **b)** $-4 - 2$

*Solution*

**a)** Add the opposite of 9, which is $-9$, to 4.

$$4 - 9 = 4 + (-9) = -5$$

**b)** Add the opposite of 2, which is $-2$, to $-4$.

$$-4 - 2 = -4 + (-2) = -6$$

Now Try Exercise 17

In Example 4, we subtract numbers that contain decimal points.

**EXAMPLE 4** Evaluate $16.32 - 18.75$. ────────

*Solution*   Add the opposite of 18.75, which is $-18.75$, to 16.32.

$$16.32 - 18.75 = 16.32 + (-18.75) = -2.43$$

Now Try Exercise 53

In Examples 5 and 6, we will show how to subtract a negative number.

**EXAMPLE 5** Evaluate $4 - (-2)$. ────────

*Solution*   We are asked to subtract a negative 2 from 4. To do this, we add the opposite of $-2$, which is 2, to 4.

$$4 - (-2) = 4 + 2 = 6$$

Subtract  Negative 2  Add  Positive 2

Now Try Exercise 19

## Understanding Algebra

Whenever we subtract a negative number, we *always* replace the two negative signs with a plus sign.

$$6 - (-9) = 6 + 9$$

---

**Helpful Hint**

By examining Example 5, we see that

$$4 - (-2) = 4 + 2$$

Two negative     Plus
signs together

---

EXAMPLE 6 Evaluate. ───

  **a)** $7 - (-5)$      **b)** $-15 - (-12)$

*Solution*

  **a)** Since we are subtracting a negative number, adding the opposite of $-5$, which is 5, to 7 will result in the two negative signs being replaced by a plus sign.

$$7 - (-5) = 7 + 5 = 12$$

  **b)** $-15 - (-12) = -15 + 12 = -3$

Now Try Exercise 29

---

**Helpful Hint**

We will now indicate how we may illustrate subtraction using colored chips. Remember from the preceding section that a red chip represents $+1$ and a green chip $-1$.

● = +1     ● = −1

Consider the subtraction problem $2 - 5$. If we change this to an addition problem, we get $2 + (-5)$. We can then add, as was done in the preceding section. The figure below shows that $2 + (-5) = -3$.

● ● + ● ● ● ● ●

Now consider $-2 - 5$. This means $-2 + (-5)$, which can be represented as follows:

● ● + ● ● ● ● ●

Thus, $-2 - 5 = -7$.
Now consider the problem $-3 - (-5)$. This can be rewritten as $-3 + 5$, which can be represented as follows:

● ● ● + ● ● ● ● ●

Thus, $-3 - (-5) = 2$.
Some students still have difficulty understanding why when you subtract a negative number you obtain a positive number. Let's look at the problem $3 - (-2)$. This time we will look at it from a slightly different point of view. Let's start with 3:

● ● ●

From this we wish to subtract a negative 2. To the $+3$ shown above we will add two zeros by adding two $+1 - 1$ combinations. Remember, $+1$ and $-1$ sum to 0.

$\underbrace{● ● ●}_{+3}$ + $\underbrace{● ●}_{0}$ + $\underbrace{● ●}_{0}$

Now we can subtract or "take away" the two $-1$'s as shown:

● ● ● + ● ● + ● ●

From this we see that we are left with $3 + 2$ or 5. Thus, $3 - (-2) = 5$.

EXAMPLE 7 Subtract 12 from 3.

**Solution**

$$3 - 12 = 3 + (-12) = -9$$

Now Try Exercise 57

**Helpful Hint**

Example 7 asked us to "subtract 12 from 3." The correct way of writing this is $3 - 12$. Notice that the number following the word "from" is our starting point. That is where the calculation begins. For example:

Subtract 2 from 7 means $7 - 2$.      From 7, subtract 2 means $7 - 2$.

Subtract 5 from $-1$ means $-1 - 5$.      From $-1$, subtract 5 means $-1 - 5$.

Subtract $-4$ from $-2$ means $-2 - (-4)$.      From $-2$, subtract $-4$ means $-2 - (-4)$.

Subtract $-3$ from 6 means $6 - (-3)$.      From 6, subtract $-3$ means $6 - (-3)$.

Subtract $a$ from $b$ means $b - a$.      From $a$, subtract $b$ means $a - b$.

EXAMPLE 8 Subtract 5 from 5.

**Solution**

$$5 - 5 = 5 + (-5) = 0$$

Now Try Exercise 59

EXAMPLE 9 Subtract $-6.48$ from $4.25$.

**Solution**

$$4.25 - (-6.48) = 4.25 + 6.48 = 10.73$$

Now Try Exercise 61

Now we will perform subtraction problems that contain fractions.

EXAMPLE 10 Subtract $\dfrac{5}{9} - \dfrac{13}{15}$.

**Solution**    Begin by changing the subtraction problem to an addition problem.

$$\frac{5}{9} - \frac{13}{15} = \frac{5}{9} + \left(-\frac{13}{15}\right)$$

Now rewrite the fractions with the LCD, 45, and add the fractions as was done in the last section.

$$\frac{5}{9} + \left(-\frac{13}{15}\right) = \frac{5}{9} \cdot \frac{5}{5} + \left(-\frac{13}{15}\right) \cdot \frac{3}{3}$$

$$= \frac{25}{45} + \left(-\frac{39}{45}\right) = \frac{25 + (-39)}{45} = \frac{-14}{45} = -\frac{14}{45}$$

Thus, $\dfrac{5}{9} - \dfrac{13}{15} = -\dfrac{14}{45}$.

Now Try Exercise 85

**EXAMPLE 11** Subtract $-\dfrac{7}{18}$ from $-\dfrac{9}{15}$.

*Solution*    This problem is written $-\dfrac{9}{15} - \left(-\dfrac{7}{18}\right)$.
We can simplify this as follows.

$$-\frac{9}{15} - \left(-\frac{7}{18}\right) = -\frac{9}{15} + \frac{7}{18}.$$

The LCD of 15 and 18 is 90. Rewriting the fractions with a common denominator gives

$$-\frac{9}{15} \cdot \frac{6}{6} + \frac{7}{18} \cdot \frac{5}{5} = -\frac{54}{90} + \frac{35}{90} = \frac{-54 + 35}{90}$$

$$= \frac{-19}{90} = -\frac{19}{90}.$$

Now Try Exercise 87

Let us now look at some applications that involve subtraction.

**EXAMPLE 12** **A Gift Card Balance** Brigitte Martineau's gift card indicates a balance of $86.23 before she makes purchases totaling $127.49. Find the amount that Brigitte owes the cashier after she uses her gift card.

*Solution*    Understand and Translate    We can obtain the amount that Brigitte owes by subtracting 127.49 from 86.23.

Carry Out    $86.23 - 127.49 = 86.23 + (-127.49) = -41.26$

Check and Answer    The negative indicates a deficit, which is what we expect. Therefore, Brigitte owes the cashier $41.26.

Now Try Exercise 131

**EXAMPLE 13** **Temperature Difference** On January 30, 2008, the high temperature for the day in Laredo, Texas, was 86°F. On the same day, the low temperature in Wahpeton, North Dakota, was −26°F. Find the difference in their temperatures.

*Solution*    Understand and Translate    The word *difference* in the example title indicates subtraction. We can obtain the difference in their temperatures by subtracting as follows.

Carry Out    $86 - (-26) = 86 + 26 = 112$

Check and Answer    Therefore, the high temperature in Laredo is 112°F greater than the low temperature in Wahpeton.

Now Try Exercise 135

Laredo, Texas

**EXAMPLE 14** **Measuring Snow** A kindergarten class in Richfield, Minnesota, has a snow gauge placed outside its window that is left untouched for two days. Suppose that on the first day $6\dfrac{3}{8}$ inches of snow falls. On the second day, no snow falls, but $1\dfrac{1}{2}$ inches of the first day's snowfall melts. How much snow remains after the second day?

*Solution*    Understand and Translate    From the first amount, $6\dfrac{3}{8}$ inches, we must subtract $1\dfrac{1}{2}$ inches.

Carry Out    We begin by changing the subtraction problem to an addition problem. We then change the mixed numbers to fractions, and then rewrite each fraction with the LCD, 8.

$$6\frac{3}{8} - 1\frac{1}{2} = 6\frac{3}{8} + \left(-1\frac{1}{2}\right)$$

$$= \frac{51}{8} + \left(-\frac{3}{2}\right)$$

$$= \frac{51}{8} + \left(-\frac{3}{2}\right) \cdot \frac{4}{4}$$

$$= \frac{51}{8} + \left(-\frac{12}{8}\right)$$

$$= \frac{51 + (-12)}{8} = \frac{39}{8} \quad \text{or} \quad 4\frac{7}{8}$$

**Check and Answer** Thus, after the second day there were $4\frac{7}{8}$ inches of snow remaining. Based upon the numbers given in the problem, the answer seems reasonable.

*Now Try Exercise 133*

**EXAMPLE 15** Evaluate. ——————————————

**a)** $15 + (-4)$ **b)** $-16 - 3$ **c)** $19 + (-14)$

**d)** $7 - (-9)$ **e)** $-9 - (-3)$ **f)** $8 - 13$

*Solution* Parts **a)** and **c)** are addition problems, whereas the other parts are subtraction problems. We can rewrite each subtraction problem as an addition problem to evaluate.

**a)** $15 + (-4) = 11$ **b)** $-16 - 3 = -16 + (-3) = -19$

**c)** $19 + (-14) = 5$ **d)** $7 - (-9) = 7 + 9 = 16$

**e)** $-9 - (-3) = -9 + 3 = -6$ **f)** $8 - 13 = 8 + (-13) = -5$

*Now Try Exercise 31*

## 2 Subtract Numbers Mentally

In the previous examples, we rewrote subtraction problems as addition problems. We did this because we know how to add real numbers. After this chapter, when we work out a subtraction problem, we will not show this step. *You need to practice and thoroughly understand how to add and subtract real numbers. When asked to evaluate an expression like* $-4 - 6$*, you need to be able to compute the answer mentally.*

Let's evaluate a few subtraction problems without showing the process of changing the subtraction to addition.

**EXAMPLE 16** Evaluate. ——————————————

**a)** $-7 - 5$ **b)** $4 - 12$ **c)** $18 - 25$ **d)** $-20 - 12$

*Solution*

**a)** $-7 - 5 = -12$ **b)** $4 - 12 = -8$ **c)** $18 - 25 = -7$ **d)** $-20 - 12 = -32$

*Now Try Exercise 33*

In Example 16 **a)**, we may have reasoned that $-7 - 5$ meant $-7 + (-5)$, which is $-12$, but we did not need to show it.

**EXAMPLE 17** Evaluate $-\frac{3}{5} - \frac{7}{8}$. ——————————————

*Solution* Write each fraction with the LCD, 40.

$$-\frac{3}{5} \cdot \frac{8}{8} - \frac{7}{8} \cdot \frac{5}{5} = -\frac{24}{40} - \frac{35}{40} = \frac{-24 - 35}{40} = -\frac{59}{40} = -1\frac{19}{40}$$

*Now Try Exercise 77*

Notice in Example 17, when we had $-\dfrac{24}{40} - \dfrac{35}{40}$, we could have written $\dfrac{-24 + (-35)}{40}$, but at this time we elected to write it as $\dfrac{-24 - 35}{40}$. Since $-24 - 35$ is $-59$, the answer is $-\dfrac{59}{40}$ or $-1\dfrac{19}{40}$.

## 3 Evaluate Expressions Containing More Than Two Numbers

In evaluating expressions involving more than one addition and subtraction, work from left to right unless parentheses or other grouping symbols appear.

> **Understanding Algebra**
>
> When performing a string of additions and subtractions, always work from left to right.

**EXAMPLE 18** Evaluate.

**a)** $9 - 12 + 3$     **b)** $-7 - 15 - 6$     **c)** $-5 + 1 - 8$

*Solution*    We work from left to right.

**a)** $\underbrace{9 - 12}\ + 3$
$= -3\ \ + 3$
$= 0$

**b)** $\underbrace{-7 - 15}\ - 6$
$= -22\ \ - 6$
$= -28$

**c)** $\underbrace{-5 + 1}\ - 8$
$= -4\ \ - 8$
$= -12$

Now Try Exercise 119

> **Understanding Algebra**
>
> Whenever we see an expression of the form $a + (-b)$, we can write the expression as $a - b$.
>
> $2 + (-5) = 2 - 5$
>
> Whenever we see an expression of the form $a - (-b)$, we can rewrite it as $a + b$.
>
> $6 - (-8) = 6 + 8$

**Rewriting Expressions**

In general, for any real numbers $a$ and $b$,

$$a + (-b) = a - b, \text{ and}$$
$$a - (-b) = a + b$$

Using the given information, the expression $9 + (-12) - (-8)$ may be simplified to $9 - 12 + 8$.

**EXAMPLE 19**

**a)** Evaluate $-5 - (-9) + (-12) + (-3)$.

**b)** Simplify the expression in part **a)**.

**c)** Evaluate the simplified expression in part **b)**.

*Solution*

**a)** We work from left to right. The shading indicates the additions being performed to get to the next step.

$$-5 - (-9) + (-12) + (-3) = -5 + 9 + (-12) + (-3)$$
$$= 4 + (-12) + (-3)$$
$$= -8 + (-3)$$
$$= -11$$

**b)** The expression simplifies as follows:

$$-5 - (-9) + (-12) + (-3) = -5 + 9 - 12 - 3$$

**c)** Evaluate the simplified expression from left to right. Begin by adding $-5 + 9$ to obtain 4.

$$-5 + 9 - 12 - 3 = 4 - 12 - 3$$
$$= -8 - 3$$
$$= -11$$

When you come across an expression like the one in Example 19 **a)**, you should simplify it as we did in part **b)** and then evaluate the simplified expression.

Now Try Exercise 129

# EXERCISE SET 1.7  MathXL®  MyMathLab®
MathXL®  MyMathLab

## Warm-Up Exercises

*Fill in the blanks with the appropriate word, phrase, or symbol(s) from the following list.*

| | | | | |
|---|---|---|---|---|
| opposite | subtrahend | $-a - b$ | $a + b$ | difference |
| zero | left to right | minuend | $a + (-b)$ | $a - b$ |
| $-a + b$ | | | | |

1. In the equation $4 - 7 = -3$, 4 is called the _____.

2. In the equation $4 - 7 = -3$, 7 is called the _____.

3. In the equation $4 - 7 = -3$, $-3$ is called the _____.

4. $a - b$ could be rewritten as _____.

5. When subtracting a number, we add its _____.

6. When a number is subtracted from itself, the result is _____.

7. When many numbers are being added and subtracted, we always work from _____.

8. The opposite of the number $a + b$ is _____.

9. $-a - (-b)$ could be rewritten as _____.

10. $a - (-b)$ could be rewritten as _____.

*In Exercises 11 and 12, are the following calculations correct? If not, explain why.*

11. $\dfrac{4}{9} - \dfrac{3}{7} = \dfrac{28}{63} - \dfrac{27}{63} = \dfrac{28 - 27}{63} = \dfrac{1}{63}$

12. $-\dfrac{5}{12} - \dfrac{7}{9} = -\dfrac{15}{36} - \dfrac{28}{36} = \dfrac{-15 - 28}{36} = -\dfrac{43}{36}$

## Practice the Skills

*Evaluate.*

13. $8 - (+2)$
14. $17 - (+8)$
15. $12 - 5$
16. $9 - 4$

17. $8 - 9$
18. $-6 - 3$
19. $9 - (-3)$
20. $17 - (-5)$

21. $-8 - 8$
22. $-4 - (-3)$
23. $0 - 9$
24. $19 - (-9)$

25. $8 - 8$
26. $10 - 10$
27. $-3 - 1$
28. $-5.7 - (-3.1)$

29. $-8 - (-5)$
30. $4 - 9$
31. $6 - (-3)$
32. $6 - 10$

33. $-9 - 11$
34. $37 - 40$
35. $0 - (-9.8)$
36. $-6.3 - 4.7$

37. $-4.8 - (-5.1)$
38. $-4 - (-4)$
39. $44 - 7$
40. $9 - 9$

41. $-8 - (-12)$
42. $-6 - (-2)$
43. $18 - (-4)$
44. $-25 - 16$

45. $-9 - 2$
46. $-85 - (-8)$
47. $-90.7 - 40.3$
48. $-52.6 - 37.9$

49. $-45 - 39$
50. $-500 - (-400)$
51. $70 - (-70)$
52. $130 - (-90)$

53. $42.3 - 49.7$
54. $81.3 - 92.5$
55. $-7.85 - (-3.92)$
56. $-12.43 - (-9.57)$

57. Subtract 15 from 4.
58. Subtract 7 from 1.
59. Subtract 21 from 21.
60. Subtract 13 from 13.

61. Subtract $-12.4$ from $-6.3$.
62. Subtract 17.3 from $-9.8$.
63. Subtract $-7.9$ from 10.3.

64. Subtract $-23$ from $-23$.
65. Subtract 24 from 13.
66. Subtract $-11.7$ from $-5.2$.

67. Subtract 7.8 from $-10.3$.
68. Subtract $-9.6$ from 3.4.

*Evaluate.*

69. $\dfrac{5}{9} - \dfrac{1}{8}$
70. $\dfrac{4}{5} - \dfrac{5}{6}$
71. $\dfrac{8}{15} - \dfrac{7}{45}$
72. $\dfrac{5}{12} - \dfrac{7}{8}$

73. $-\dfrac{7}{10} - \dfrac{5}{12}$
74. $-\dfrac{1}{4} - \dfrac{2}{3}$
75. $-\dfrac{4}{15} - \dfrac{3}{20}$
76. $-\dfrac{5}{4} - \dfrac{7}{11}$

77. $-\dfrac{7}{12} - \dfrac{5}{40}$
78. $-\dfrac{5}{6} - \dfrac{3}{32}$
79. $\dfrac{5}{8} - \dfrac{6}{48}$
80. $\dfrac{17}{18} - \dfrac{13}{20}$

81. $-\dfrac{4}{9} - \left(-\dfrac{3}{5}\right)$
82. $\dfrac{5}{20} - \left(-\dfrac{1}{8}\right)$
83. $\dfrac{3}{16} - \left(-\dfrac{5}{8}\right)$
84. $-\dfrac{5}{12} - \left(-\dfrac{3}{8}\right)$

85. Subtract $\dfrac{7}{9}$ from $\dfrac{4}{7}$.
86. Subtract $\dfrac{7}{15}$ from $\dfrac{5}{8}$.
87. Subtract $-\dfrac{3}{10}$ from $-\dfrac{5}{12}$.
88. Subtract $-\dfrac{5}{16}$ from $-\dfrac{9}{10}$.

*In Exercises 89–106,* **a)** *determine by observation whether the difference will be a positive number, zero, or a negative number;* **b)** *find the difference; and* **c)** *examine your answer to part* **b)** *to see whether it is reasonable and makes sense.*

**89.** 378 − 279

**90.** 483 − 569

**91.** −482 − 137

**92.** 178 − (−377)

**93.** 843 − (−745)

**94.** 864 − (−762)

**95.** −408 − (−604)

**96.** −623 − 111

**97.** −1024 − (−576)

**98.** −104.7 − 27.6

**99.** 165.7 − 49.6

**100.** −40.2 − (−12.6)

**101.** Subtract 364 from 295.

**102.** Subtract −433 from −932.

**103.** Subtract 647 from −1023.

**104.** Subtract 2432 from −4120.

**105.** Subtract −7.62 from −7.62.

**106.** Subtract 36.7 from −103.2.

*Evaluate.*

**107.** 7 + 5 − (+8)

**108.** 15 − (+9) − (+5)

**109.** −6 + (−6) + 16

**110.** 9 − 4 + (−2)

**111.** −13 − (+5) + 3

**112.** 7 − (+4) − (−3)

**113.** −9 − (−3) + 4

**114.** 15 + (−7) − (−3)

**115.** 5 − (−9) + (−1)

**116.** 12 + (−5) − (−4)

**117.** 17 + (−8) − (+14)

**118.** −7 + 6 − 3

**119.** −36 − 5 + 9

**120.** 45 − 3 − 7

**121.** −12 + 7 − 9

**122.** −2 − 7 − 13

**123.** 25 − 19 + 3

**124.** −4 − 1 + 5

**125.** −4 − 6 + 5 − 7

**126.** −9 − 3 − (−4) + 5

**127.** 17 + (−3) − 9 − (−7)

**128.** 32 + 5 − 7 − 12

**129.** −9 + (−7) + (−5) − (−3)

**130.** 6 − 9 − (−3) + 12

## Problem Solving

**131.** **Lands' End** The Lands' End catalog department had 300 ladies' blue cardigan sweaters in stock on December 1. By December 9, the department had taken orders for 343 of the sweaters.

**a)** How many sweaters were on back order?

**b)** If the catalog department wanted 100 sweaters in addition to those already ordered, how many sweaters would it need to back order?

© Iya Cattel/iStockphoto.com

**132.** **Leadville, Co** According to the *Guinness Book of World Records,* the city with the greatest elevation in the United States is Leadville, Colorado, at 10,152 feet. The city with the lowest elevation in the United States, at 184 feet below sea level, is Calipatria, California. What is the difference in the elevation of these cities?

**133.** **Measuring Rainfall** At Kim Christensen's house, a rain gauge is placed in the yard and is left untouched for 2 days. Suppose that on the first day, $2\frac{1}{4}$ inches of rain falls. On the second day, no rain falls, but $\frac{3}{8}$ inch of the first day's rainfall evaporates. How much water remains in the gauge after the second day?

**134.** **Death Valley** A medical supply package is dropped into Death Valley, California, from a helicopter 1605.7 feet above sea level. The package lands at a location in Death Valley 267.4 feet below sea level. What vertical distance did the package travel?

**135.** **Temperature Change** The greatest change in temperature ever recorded within a 24-hour period occurred at Browning, Montana, on January 23, 1916. The temperature fell from 44°F to −56°F. How much did the temperature drop?

**136.** **Going Home** Two college students are driving on an expressway, going home for spring break. Shawntoya travels 58.5 miles in 1 hour. Marcelino travels 67.3 miles in 1 hour.

**a)** If Shawntoya and Marcelino start at the same parking lot and travel in opposite directions, how far apart will they be in 1 hour?

**b)** If Shawntoya and Marcelino start at the same parking lot and travel in the same direction, how far apart will they be in 1 hour?

**137.** **Golf** The chart below shows some final scores at the Masters golf tournament, held in Augusta, Georgia, in 2008.

| Golfer | Score (above or below par) |
|---|---|
| T. Immelman | −8 |
| T. Woods | −5 |
| S. Cink | −4 |
| P. Harrington | −2 |
| M. Jimenez | −1 |
| V. Singh | +1 |
| H. Stenson | +2 |
| A. Cabrera | +4 |
| K. Choi | +10 |

*Source:* www.masters.org

a) If par for the Masters is 288 strokes, determine T. Immelman's score in 2008.

b) What was the difference in the strokes between K. Choi and S. Cink in 2008?

**138. Inseam** Christine Henry purchases a new pair of pants whose inseam is $32\frac{1}{2}$ inches. If $2\frac{3}{4}$ inches are cut from the inseam of the pants, what will be the new inseam of the pants?

## Concept/Writing Exercises

**139.** Your friend is having trouble in algebra class distinguishing among these three expressions: $x - y$, $y - x$, and $x - (-y)$. Using 3 for $x$ and 8 for $y$, explain in writing why all three expressions are different.

**140.** Simplify $3 - (-9) + (-4)$ by eliminating two signs next to one another and replacing them with a single sign. Explain how you determined your answer and then evaluate your answer.

## Challenge Problems

*Find each sum.*

**141.** $1 - 2 + 3 - 4 + 5 - 6 + 7 - 8 + 9 - 10$

**142.** $1 - 2 + 3 - 4 + 5 - 6 + \cdots + 99 - 100$

**143.** Consider a number line.

a) What is the distance, in units, between $-11$ and $-3$?

b) Write a subtraction problem to represent this distance (the distance is to be positive).

**144. Stock** Amy Tait buys a stock for $50. Will the stock be worth more if it decreases by 10% and then increases by 10%, or if it increases by 10% and then decreases by 10%, or will the value be the same either way?

**145. Rolling Ball** A ball rolls off a table and follows the path indicated in the figure. Suppose the maximum height reached by the ball on each bounce is 1 foot less than on the previous bounce.

a) Determine the total vertical distance traveled by the ball.

b) If we consider the ball moving in a downward direction as negative, and the ball moving in an upward direction as positive, what was the net vertical distance traveled (from its starting point) by the ball?

## Cumulative Review Exercises

[1.4] **146.** List the set of counting numbers.

**147.** Explain the relationship between the set of rational numbers, the set of irrational numbers, and the set of real numbers.

[1.5] *Insert either $>$, $<$, or $=$ in each shaded area to make the statement true.*

**148.** $|-3|$ ▨ $-5$

**149.** $-|-9|$ ▨ $-|-5|$

[1.7] **150.** Subtract $\frac{7}{8}$ from $\frac{5}{6}$.

# 1.8 Multiplication and Division of Real Numbers

**1** Multiply numbers.

**2** Divide numbers.

**3** Remove negative signs from denominators.

**4** Evaluate divisions involving 0.

## **1** Multiply Numbers

The following rules are used in determining the sign of the product when two numbers are multiplied.

### The Sign of the Product of Two Real Numbers

1. The product of two numbers with **like** signs is a **positive** number.

2. The product of two numbers with **unlike** signs is a **negative** number.

The product of two positive numbers or two negative numbers will be a positive number. The product of a positive number and a negative number will be a negative number.

**EXAMPLE 1** Evaluate.

a) $4(-5)$    b) $(-6)(7)$    c) $(-9)(-3)$

*Solution*

a) Since the numbers have unlike signs, the product is negative.
$$4(-5) = -20$$

b) Since the numbers have unlike signs, the product is negative.
$$(-6)(7) = -42$$

c) Since the numbers have like signs, both negative, the product is positive.
$$(-9)(-3) = 27$$

Now Try Exercise 17

**EXAMPLE 2** Evaluate.

a) $(-8)(5)$    b) $(-4)(-8)$    c) $0(6)$
d) $0(-2)$    e) $4.2(-9.7)$    f) $(-3.62)(-6.18)$

*Solution*

a) $(-8)(5) = -40$    b) $(-4)(-8) = 32$    c) $0(6) = 0$
d) $0(-2) = 0$    e) $4.2(-9.7) = -40.74$    f) $(-3.62)(-6.18) = 22.3716$

*Note that zero multiplied by any real number equals zero.*

Now Try Exercise 29

**Helpful Hint**

At this point some students begin confusing problems like $-2 - 3$ with $(-2)(-3)$ and problems like $2 - 3$ with $2(-3)$. If you do not understand the difference between problems like $-2 - 3$ and $(-2)(-3)$, make an appointment to see your instructor as soon as possible.

| Subtraction Problems | Multiplication Problems |
|---|---|
| $-2 - 3 = -5$ | $(-2)(-3) = 6$ |
| $2 - 3 = -1$ | $(2)(-3) = -6$ |

**EXAMPLE 3** Evaluate.

a) $\left(\dfrac{-1}{8}\right)\left(\dfrac{-3}{5}\right)$    b) $\left(\dfrac{3}{20}\right)\left(\dfrac{-3}{10}\right)$

*Solution*

a) $\left(\dfrac{-1}{8}\right)\left(\dfrac{-3}{5}\right) = \dfrac{(-1)(-3)}{8(5)} = \dfrac{3}{40}$    b) $\left(\dfrac{3}{20}\right)\left(\dfrac{-3}{10}\right) = \dfrac{3(-3)}{20(10)} = -\dfrac{9}{200}$

Now Try Exercise 41

Sometimes you may be asked to perform more than one multiplication in a given problem. When this happens, the sign of the final product can be determined by counting the number of *negative* numbers being multiplied. *The product of an even number of negative numbers will always be positive. The product of an odd number of negative numbers will always be negative.*

EXAMPLE 4  Evaluate.

**a)** $(-5)(-3)(1)(-4)$

**b)** $(-2)(-4)(-1)(3)(-4)$

*Solution*

**a)** Since there are an odd number of negative numbers, the product will be negative, as illustrated.

$$(-5)(-3)(1)(-4) = (15)(1)(-4)$$
$$= (15)(-4)$$
$$= -60$$

**b)** Since there are an even number of negative numbers, the product will be positive, as illustrated.

$$(-2)(-4)(-1)(3)(-4) = (8)(-1)(3)(-4)$$
$$= (-8)(3)(-4)$$
$$= (-24)(-4)$$
$$= 96$$

Now Try Exercise 35

## 2  Divide Numbers

### Understanding Algebra

In the division problem

$$(24) \div (4) = 6$$

24 is called the *dividend*, 4 is called the *divisor*, and 6 is called the *quotient*.

The rules for dividing numbers are very similar to those used in multiplying numbers.

---

**The Sign of the Quotient of Two Real Numbers**

1. The quotient of two numbers with **like** signs is a **positive** number.
2. The quotient of two numbers with **unlike** signs is a **negative** number.

---

Therefore, the quotient of two positive numbers or two negative numbers will be a positive number. The quotient of a positive number and a negative number will be a negative number.

EXAMPLE 5  Evaluate.

**a)** $\dfrac{10}{-5}$

**b)** $\dfrac{-45}{5}$

**c)** $\dfrac{-36}{-6}$

*Solution*

**a)** Since the numbers have unlike signs, the quotient is negative.

$$\frac{10}{-5} = -2$$

### Understanding Algebra

A fraction is really a division problem:

$$(24) \div (4) = 6$$

is the same as

$$\frac{24}{4} = 6$$

**b)** Since the numbers have unlike signs, the quotient is negative.

$$\frac{-45}{5} = -9$$

**c)** Since the numbers have like signs, both negative, the quotient is positive.

$$\frac{-36}{-6} = 6$$

Now Try Exercise 55

**EXAMPLE 6** Evaluate.

**a)** $-16 \div (-2)$    **b)** $\dfrac{-2}{3} \div \dfrac{-5}{7}$

*Solution*

**a)** Since the numbers have like signs, both negative, the quotient is positive.

$$-16 \div (-2) = \frac{-16}{-2} = 8$$

**b)** Invert the *divisor*, $\dfrac{-5}{7}$, and then multiply.

$$\frac{-2}{3} \div \frac{-5}{7} = \left(\frac{-2}{3}\right)\left(\frac{7}{-5}\right) = \frac{-14}{-15} = \frac{14}{15}$$

Now Try Exercise 77

**EXAMPLE 7** Evaluate, rounding your answer to the nearest hundredth when appropriate.

**a)** $-18.86 \div 4.1$    **b)** $\dfrac{-27.2}{-2.6}$

*Solution*

**a)** Since the numbers have unlike signs, the quotient is negative.

$$\frac{-18.86}{4.1} = -4.6$$

**b)** Since the numbers have like signs, both negative, the quotient is positive.

$$\frac{-27.2}{-2.6} \approx 10.46$$

The answer in part **b)** was rounded to two decimal places, or hundredths. If you have forgotten how to round decimal numbers, review Appendix A now.

Now Try Exercise 69

> **Helpful Hint**
>
> For multiplication and division of two real numbers:
>
> $$(+)(+) = + \quad \left.\begin{array}{l} \end{array}\right\} \quad \frac{(+)}{(+)} = + \quad \left.\begin{array}{l} \end{array}\right\}$$
> $$(-)(-) = + \qquad\qquad \frac{(-)}{(-)} = +$$
>
> Like signs give positive products and quotients.
>
> $$(+)(-) = - \quad \left.\begin{array}{l} \end{array}\right\} \quad \frac{(+)}{(-)} = - \quad \left.\begin{array}{l} \end{array}\right\}$$
> $$(-)(+) = - \qquad\qquad \frac{(-)}{(+)} = -$$
>
> Unlike signs give negative products and quotients.

## 3 Remove Negative Signs from Denominators

We now know that the quotient of a positive number and a negative number is a negative number. The fractions $-\dfrac{3}{4}, \dfrac{-3}{4}$, and $\dfrac{3}{-4}$ all represent the same negative number, negative three-fourths.

**The Quotient of a Positive Number and a Negative Number**

If $a$ and $b$ represent any real numbers, $b \neq 0$, then

$$\frac{a}{-b} = \frac{-a}{b} = -\frac{a}{b}$$

In mathematics we generally do not write a fraction with a negative sign in the denominator. When a negative sign appears in a denominator, we can move it to the numerator or place it in front of the fraction. For example, the fraction $\frac{5}{-7}$ should be written as either $\frac{-5}{7}$ or $-\frac{5}{7}$.

EXAMPLE 8   Evaluate $\frac{3}{7} \div \left(\frac{-12}{35}\right)$.

*Solution*

$$\frac{3}{7} \div \left(\frac{-12}{35}\right) = \frac{\overset{1}{3}}{\underset{1}{7}} \cdot \left(\frac{\overset{5}{35}}{\underset{4}{-12}}\right) = \frac{1(5)}{1(-4)} = \frac{5}{-4} = -\frac{5}{4}$$

Now Try Exercise 73

The operations on real numbers are summarized in **Table 1.1**.

**TABLE 1.1   Summary of Operations on Real Numbers**

| Signs of Numbers | Addition | Subtraction | Multiplication | Division |
|---|---|---|---|---|
| Both Numbers Are Positive | Sum Is Always Positive | Difference May Be Either Positive or Negative | Product Is Always Positive | Quotient Is Always Positive |
| Examples | | | | |
| 6 and 2 | $6 + 2 = 8$ | $6 - 2 = 4$ | $6 \cdot 2 = 12$ | $6 \div 2 = 3$ |
| 2 and 6 | $2 + 6 = 8$ | $2 - 6 = -4$ | $2 \cdot 6 = 12$ | $2 \div 6 = \frac{1}{3}$ |
| One Number Is Positive and the Other Number Is Negative | Sum May Be Either Positive or Negative | Difference May Be Either Positive or Negative | Product Is Always Negative | Quotient Is Always Negative |
| Examples | | | | |
| 6 and $-2$ | $6 + (-2) = 4$ | $6 - (-2) = 8$ | $6(-2) = -12$ | $6 \div (-2) = -3$ |
| $-6$ and 2 | $-6 + 2 = -4$ | $-6 - 2 = -8$ | $-6(2) = -12$ | $-6 \div 2 = -3$ |
| Both Numbers Are Negative | Sum Is Always Negative | Difference May Be Either Positive or Negative | Product Is Always Positive | Quotient Is Always Positive |
| Examples | | | | |
| $-6$ and $-2$ | $-6 + (-2) = -8$ | $-6 - (-2) = -4$ | $-6(-2) = 12$ | $-6 \div (-2) = 3$ |
| $-2$ and $-6$ | $-2 + (-6) = -8$ | $-2 - (-6) = 4$ | $-2(-6) = 12$ | $-2 \div (-6) = \frac{1}{3}$ |

## 4  Evaluate Divisions Involving 0

What is $\frac{0}{1}$ equal to? Note that $\frac{6}{3} = 2$ because $3 \cdot 2 = 6$. We can follow the same procedure to determine the value of $\frac{0}{1}$. Suppose that $\frac{0}{1}$ is equal to some number, which we will designate by ? .

$$\text{If } \frac{0}{1} = \boxed{?} \quad \text{then} \quad 1 \cdot \boxed{?} = 0$$

Since only $1 \cdot 0 = 0$, the ? must be 0. Thus, $\frac{0}{1} = 0$. Using the same technique, we can show that zero divided by any nonzero number is zero.

### Zero Divided by a Nonzero Number

If $a$ represents any real number except 0, then

$$0 \div a = \frac{0}{a} = 0$$

Now what is $\frac{1}{0}$ equal to?

$$\text{If } \frac{1}{0} = \boxed{?} \quad \text{then} \quad 0 \cdot \boxed{?} = 1$$

But since 0 multiplied by any number will be 0, there is no value that can replace ? . We say that $\frac{1}{0}$ is **undefined**. Using the same technique, we can show that any real number, except 0, divided by 0 is undefined.

### Division by Zero

If $a$ represents any real number except 0, then

$$a \div 0 \quad \text{or} \quad \frac{a}{0} \quad \text{is } \textbf{undefined}$$

What is $\frac{0}{0}$ equal to?

$$\text{If } \frac{0}{0} = \boxed{?} \quad \text{then} \quad 0 \cdot \boxed{?} = 0$$

Since the product of any number and 0 is 0, the ? can be replaced by any real number. Therefore, the quotient $\frac{0}{0}$ cannot be determined, and so there is no answer. Thus, we will not use it in this course.[*]

### Summary of Division Involving 0

If $a$ represents any real number except 0, then

$$\frac{0}{a} = 0 \quad \text{and} \quad \frac{a}{0} \text{ is undefined}$$

EXAMPLE 9  Indicate whether each quotient is 0 or undefined.

**a)** $\frac{0}{2}$    **b)** $\frac{5}{0}$    **c)** $\frac{0}{-4}$    **d)** $\frac{-2}{0}$

*Solution*    The answer to parts **a)** and **c)** is 0. The answer to parts **b)** and **d)** is undefined.

Now Try Exercise 95

---

[*]At this level, some professors prefer to call $\frac{0}{0}$ *indeterminate* while others prefer to call $\frac{0}{0}$ *undefined*. In higher-level mathematics courses, $\frac{0}{0}$ is sometimes referred to as an *indeterminate form*.

# EXERCISE SET 1.8    MathXL®  MyMathLab®

## Warm-Up Exercises

*Fill in the blanks with the appropriate word, phrase, or symbol(s) from the following list.*

| | | | | |
|---|---|---|---|---|
| positive | negative | infinity | product | indeterminate form |
| zero | 63 | undefined | $-\dfrac{a}{b}$ | $\dfrac{a}{b}$ |

$-63$

1. The product of a positive real number with a negative real number is _____.

2. 16 divided by 0 is _____.

3. 0 divided by 8 is _____.

4. The fraction $\dfrac{a}{-b}$ may be rewritten as _____.

5. The fraction $-\dfrac{a}{-b}$ may be rewritten as _____.

6. If $x$ is 9 and $y$ is $-7$ then the value of $xy$ is _____.

7. The product of two negative numbers is a _____ number.

8. 0 divided by 0 is an _____.

9. If $x$ is 9 and $y$ is $-7$ then the value of $x(-y)$ is _____.

10. When two real numbers are multiplied, the result is called the _____ of the two numbers.

*Determine the sign of each product.*

11. $(8)(4)(-5)$

12. $(-9)(-12)(20)$

13. $(-102)(-16)(24)(19)$

14. $(1054)(-92)(-16)(-37)$

15. $(-40)(-16)(30)(50)(-13)$

16. $(-1)(3)(-462)(-196)(-312)$

## Practice the Skills

*Find each product.*

17. $(-5)(-4)$

18. $-4(2)$

19. $5(-3)$

20. $6(-2)$

21. $(-8)(-20)$

22. $(-3)(2)$

23. $-2.1(6)$

24. $-1(8.7)$

25. $6(7)$

26. $-9(-4)$

27. $9(-9)$

28. $(7)(-8)$

29. $(-5)(-6)$

30. $0(-5)$

31. $(-9)(0)(-6)$

32. $5(-4)(2)$

33. $(21)(-1)(4)$

34. $2(8)(-1)(-3)$

35. $-1(-3)(3)(-8)$

36. $(2)(-4)(-5)(-1)$

37. $(-4)(5)(-7)(10)$

38. $(-3)(2)(5)(3)$

39. $(-1)(3)(0)(-7)$

40. $(-6)(6)(4)(-4)$

*Find each product.*

41. $\left(\dfrac{-1}{2}\right)\left(\dfrac{3}{5}\right)$

42. $\left(\dfrac{1}{3}\right)\left(\dfrac{-3}{5}\right)$

43. $\left(\dfrac{-5}{9}\right)\left(\dfrac{-7}{15}\right)$

44. $\left(\dfrac{4}{5}\right)\left(\dfrac{-3}{10}\right)$

45. $\left(\dfrac{6}{-3}\right)\left(\dfrac{4}{-2}\right)$

46. $\left(\dfrac{9}{-10}\right)\left(\dfrac{6}{-7}\right)$

47. $\left(\dfrac{3}{4}\right)\left(\dfrac{-2}{15}\right)$

48. $\left(\dfrac{-9}{10}\right)\left(\dfrac{7}{-8}\right)$

*Find each quotient.*

49. $\dfrac{-42}{6}$

50. $25 \div 5$

51. $-16 \div (-4)$

52. $\dfrac{-18}{9}$

53. $\dfrac{-36}{-9}$

54. $\dfrac{30}{-6}$

55. $\dfrac{36}{-2}$

56. $\dfrac{-15}{-1}$

57. $\dfrac{-19.8}{-2}$

58. $-15.6/(-3)$

59. $40/(-4)$

60. $\dfrac{63}{-7}$

61. $\dfrac{-66}{2}$

62. $\dfrac{-25}{-5}$

63. $\dfrac{48}{-12}$

64. $\dfrac{-10}{10}$

65. $-64.8 \div (-4)$

66. $-86.4/(-2)$

67. Divide 0 by 4.

68. Divide 26 by $-13$.

69. Divide 30.8 by $-5.6$.

70. Divide $-67.64$ by 7.6.

71. Divide $-30$ by $-5$.

72. Divide $-36$ by $-6$.

*Find each quotient.*

73. $\dfrac{3}{12} \div \left(\dfrac{-5}{8}\right)$

74. $4 \div \left(\dfrac{-6}{13}\right)$

75. $\dfrac{-5}{12} \div (-3)$

76. $\dfrac{6}{15} \div \left(\dfrac{7}{30}\right)$

77. $\dfrac{-15}{21} \div \left(\dfrac{-15}{21}\right)$

78. $\dfrac{-4}{9} \div \left(\dfrac{-6}{7}\right)$

79. $(-12) \div \dfrac{5}{12}$

80. $\dfrac{-16}{3} \div \left(\dfrac{5}{-9}\right)$

*Evaluate.*

**81.** $-4(8)$

**82.** $\dfrac{-18}{-2}$

**83.** $\dfrac{-100}{-5}$

**84.** $-50 \div (-10)$

**85.** $-7(2)$

**86.** $6.4(-8)$

**87.** $27.9 \div (-3)$

**88.** Divide 130 by $-10$.

**89.** $-100 \div 5$

**90.** $4(-2)(-1)(-5)$

**91.** Divide $-90$ by $-90$.

**92.** $(6)(1)(-3)(4)$

*Indicate whether each quotient is 0 or undefined.*

**93.** $0 \div 8.6$

**94.** $\dfrac{-2.7}{0}$

**95.** $\dfrac{5}{0}$

**96.** $\dfrac{0}{1}$

**97.** $0 \div (-7)$

**98.** $\dfrac{6}{0}$

**99.** 8 divided by 0

**100.** 0 divided by 12

*In Exercises 101–116,* **a)** *determine by observation whether the product or quotient will be a positive number, zero, a negative number, or undefined;* **b)** *find the product or quotient if it exists;* **c)** *examine your answer in part* **b)** *to see whether it is reasonable and makes sense.*

**101.** $92(-38)$

**102.** $-168 \div 42$

**103.** $-240/15$

**104.** $0/12$

**105.** $243 \div (-27)$

**106.** $(323)(-115)$

**107.** $(-49)(-126)$

**108.** $(1530)(0)$

**109.** $0 \div 5335$

**110.** $-86.4 \div (-36)$

**111.** $8.2 \div 0$

**112.** $-37.74 \div 0$

**113.** $8 \div (2.5)$

**114.** $(1.1)(9.72)(6.3)$

**115.** $(-3.0)(4.2)(-18)$

**116.** $-288.86/1.43$

*Indicate whether each statement is true or false.*

**117.** The product of two negative numbers is a negative number.

**118.** The product of a positive number and a negative number is a negative number.

**119.** The quotient of two numbers with unlike signs is a positive number.

**120.** The quotient of two negative numbers is a positive number.

**121.** The product of an even number of negative numbers is a positive number.

**122.** Zero divided by 1 is 1.

**123.** The product of an odd number of negative numbers is a negative number.

**124.** Six divided by 0 is 0.

**125.** Zero divided by 1 is undefined.

**126.** The product of 0 and any real number is 0.

**127.** Five divided by 0 is undefined.

**128.** Division by 0 does not result in a real number.

## Problem Solving

**129. Football** A high school football team is penalized three times, each time with a loss of 15 yards, or $-15$ yards. Find the total loss due to penalties.

**130. Submarine Dive** A submarine is at a depth of $-160$ feet (160 feet below sea level). It dives to 3 times that depth. Find its new depth.

**131. Credit Card** Leona De Vito's balance on her credit card is $-\$520$ (she owes \$520). She pays back $\dfrac{1}{5}$ of this balance.

**a)** How much did she pay back?

**b)** What is her new balance?

**132. Money Owed** Brian Philip owes his Dad \$500. After he makes four payments of \$40 each, how much will he still owe?

**133. Garage Sale** Four sisters made a total of \$775.40 at a garage sale. After they each give their husbands \$50 and split the remaining amount equally, how much will each woman receive?

**134. Wind Chill** On Monday in Minneapolis the wind chill temperature was $-30°$F. On Tuesday the wind chill temperature was only $\dfrac{1}{3}$ of what it was on Monday. What was the wind chill temperature on Tuesday?

**135. Test Score** Because of incorrect work, Josue Nunez lost 4 points on each of the five questions on his math test.

**a)** How many points did Josue lose altogether?

**b)** If the maximum score possible was 100%, what is Josue's test score?

**136. Lab Work** Jack's job is to monitor the temperature of a superheated piece of metal for ten hours in the lab. He observed that the metal cooled 15° each hour for ten hours.

**a)** What number represents the total drop in temperature?

**b)** If the temperature of the metal was originally 678°, what was it at the end of the ten hours?

**137. Heart Rate** The Johns Hopkins Medical Letter states that to find a person's *target heart rate* in beats per minute, follow this procedure. Subtract the person's age from 220, then multiply this difference by 60% and 75%. The difference multiplied by 60% gives the lower limit and the difference multiplied by 75% gives the upper limit.

**a)** Find the target heart rate range of a 50-year-old.

**b)** Find your own target heart rate.

## Challenge Problems

*We will learn in the next section that* $2^3 = 2 \cdot 2 \cdot 2$ *and* $x^n = \underbrace{x \cdot x \cdot x \cdots \cdot x}_{n \text{ factors of } x}$.

*Use this information to evaluate each expression. (n factors of x)*

**138.** $3^4$  

**139.** $(-5)^3$  

**140.** $\left(\dfrac{2}{3}\right)^3$  

**141.** $1^{100}$  

**142.** $(-1)^{81}$

**143.** Will the product of $(-1)(-2)(-3)(-4) \cdots (-10)$ be a positive number or a negative number? Explain how you determined your answer.

**144.** Will the product of $(1)(-2)(3)(-4)(5)(-6) \cdots (33)(-34)$ be a positive number or a negative number? Explain how you determined your answer.

## Group Activity

*Discuss and answer Exercise 145 as a group, according to the instructions.*

**145. a)** Each member of the group is to do this procedure separately. At this time do not share your number with the other members of your group.
  1. Choose a number between 2 and 10.
  2. Multiply your number by 9.
  3. Add the two digits in the product together.
  4. Subtract 5 from the sum.
  5. Now choose the corresponding letter of the alphabet that corresponds with the difference found. For example, 1 is a, 2 is b, 3 is c, and so on.
  6. Choose a *one-word* country that starts with that letter.
  7. Now choose a *one-word* animal that starts with the last letter of the country selected.
  8. Finally, choose a color that starts with the last letter of the animal chosen.

**b)** Now share your final answer with the other members of your group. Did you all get the same answer?

**c)** Most people will obtain the answer *orange*. As a group, write a paragraph or two explaining why.

## Cumulative Review Exercises

[1.5] **146.** Insert either $<$, $>$, or $=$ in the shaded area to make a true statement.
$$|-3.6| \quad\rule{1.5em}{0.8em}\quad |-2.7|$$

[1.6] **147.** Add $-\dfrac{7}{12} + \left(-\dfrac{1}{10}\right)$.

[1.7] **148.** Subtract $-18$ from $-20$.

**149.** Evaluate $6 - 3 - 4 - 2$.

**150.** Evaluate $5 - (-2) + 3 - 7$.

# 1.9 Exponents, Parentheses, and the Order of Operations

1 Learn the meaning of exponents.

2 Evaluate expressions containing exponents.

3 Learn the difference between $-x^2$ and $(-x)^2$.

4 Learn the order of operations.

5 Learn the use of parentheses.

6 Evaluate expressions containing variables.

### Understanding Algebra

Exponents are simply a shorthand notation for repeated multiplication. For example, $3^4$ means $3 \cdot 3 \cdot 3 \cdot 3 = 81$.

## 1 Learn the Meaning of Exponents

In the expression $4^2$, the 4 is called the **base**, and the 2 is called the **exponent**. The number $4^2$ is read "4 squared" or "4 to the second power" and means

$$\underbrace{4 \cdot 4}_{2 \text{ factors of } 4} = 4^{\overset{\displaystyle\text{base}}{2}} \leftarrow \text{exponent}$$

The number $4^3$ is read "4 cubed" or "4 to the third power" and means

$$\underbrace{4 \cdot 4 \cdot 4}_{3 \text{ factors of } 4} = 4^3$$

In general, the number $b$ to the $n$th power, written $b^n$, means

$$\underbrace{b \cdot b \cdot b \cdots \cdot b}_{n \text{ factors of } b} = b^n$$

Thus, $b^4 = b \cdot b \cdot b \cdot b$ or $bbbb$ and $x^3 = x \cdot x \cdot x$ or $xxx$.

## 2 Evaluate Expressions Containing Exponents

Let's evaluate some expressions that contain exponents.

**EXAMPLE 1** Evaluate.  **a)** $3^2$  **b)** $2^5$  **c)** $1^5$  **d)** $(-6)^2$  **e)** $(-2)^3$  **f)** $\left(\dfrac{2}{3}\right)^2$

*Solution*

**a)** $3^2 = 3 \cdot 3 = 9$

**b)** $2^5 = 2 \cdot 2 \cdot 2 \cdot 2 \cdot 2 = 32$

**c)** $1^5 = 1 \cdot 1 \cdot 1 \cdot 1 \cdot 1 = 1$ (1 raised to any power equals 1; why?)

**d)** $(-6)^2 = (-6)(-6) = 36$

**e)** $(-2)^3 = (-2)(-2)(-2) = -8$

**f)** $\left(\dfrac{2}{3}\right)^2 = \left(\dfrac{2}{3}\right)\left(\dfrac{2}{3}\right) = \dfrac{4}{9}$

Now Try Exercise 19

Now Try Exercise 19

> **Understanding Algebra**
>
> If no exponent is written, it is assumed to be 1.
>
> $6^1 = 6$   and   $x^1 = x$

It is not necessary to write exponents of 1. For example, when writing $xxy$, we write $x^2y$ and not $x^2y^1$. *Whenever we see a variable or number without an exponent, we always assume that the variable or number has an exponent of 1.*

### Examples of Exponential Notation

**a)** $xyxx = x^3y$

**b)** $xyzzy = xy^2z^2$

**c)** $3aabbb = 3a^2b^3$

**d)** $5xyyyy = 5xy^4$

**e)** $4 \cdot 4rrs = 4^2r^2s$

**f)** $5 \cdot 5 \cdot 5mmn = 5^3m^2n$

Notice in parts **a)** and **b)** that the order of the factors does not matter.

> **Helpful Hint**
>
> Note that $x + x + x + x + x + x = 6x$ and $x \cdot x \cdot x \cdot x \cdot x \cdot x = x^6$. Be careful that you do not get addition and multiplication confused.

> **Understanding Algebra**
>
> An exponent refers only to the number or variable that directly precedes it unless parentheses are used to indicate otherwise.
>
> No parentheses:   Parentheses:
>
> $-4^2$         $(-5)^2$
>
> $-(4)(4) = -16$   $(-5)(-5) = 25$

## 3 Learn the Difference Between $-x^2$ and $(-x)^2$

*An exponent refers only to the number or variable that directly precedes it unless parentheses are used to indicate otherwise.* For example, in the expression $3x^2$, only the $x$ is squared. In the expression $-x^2$, only the $x$ is squared. We can write $-x^2$ as $-1x^2$ because any real number may be multiplied by 1 without affecting its value.

$$-x^2 = -1x^2$$

By looking at $-1x^2$ we can see that only the $x$ is squared, not the $-1$. If the entire expression $-x$ were to be squared, we would need to use parentheses and write $(-x)^2$. Note the difference in the following two examples:

$$-x^2 = -(x)(x)$$
$$(-x)^2 = (-x)(-x)$$

Consider the expressions $-3^2$ and $(-3)^2$. How do they differ?

$$-3^2 = -(3)(3) = -9$$
$$(-3)^2 = (-3)(-3) = 9$$

> **Helpful Hint**
>
> The expression $-x^2$ is read "negative $x$ squared," or "the opposite of $x$ squared." The expression $(-x)^2$ is read "negative $x$, quantity squared."

**EXAMPLE 2** Evaluate.     **a)** $-5^2$     **b)** $(-5)^2$     **c)** $-2^3$     **d)** $(-2)^3$

*Solution*

**a)** $-5^2 = -(5)(5) = -25$          **b)** $(-5)^2 = (-5)(-5) = 25$

**c)** $-2^3 = -(2)(2)(2) = -8$          **d)** $(-2)^3 = (-2)(-2)(-2) = -8$

Now Try Exercise 11

**EXAMPLE 3** Evaluate.     **a)** $-2^4$     **b)** $(-2)^4$

*Solution*

**a)** $-2^4 = -(2)(2)(2)(2) = -16$          **b)** $(-2)^4 = (-2)(-2)(-2)(-2) = 16$

Now Try Exercise 23

## 4  Learn the Order of Operations

Now that we have introduced exponents we can present the **order of operations**. Can you evaluate $2 + 3 \cdot 4$? Is it 20? Or is it 14? To answer this, you must know the order of operations to follow when evaluating a mathematical expression.

> **Understanding Algebra**
>
> *Grouping symbols* can include combinations of parentheses, brackets, braces, and even fraction bars. For example,
>
> $$\frac{-4(5 + 2) - 8}{19 + 3(-5)}$$
>
> evaluates to $\dfrac{-4(7) - 8}{19 - 15}$ or $\dfrac{-36}{4}$
>
> or simply $-9$.

**Order of Operations: To Evaluate Mathematical Expressions, Use the Following Order**

1. First, evaluate the information within **parentheses** ( ), brackets [ ], or braces { }. These are **grouping symbols**, for they group information together. A fraction bar, —, also serves as a grouping symbol. If the expression contains nested grouping symbols (one pair of grouping symbols within another pair), evaluate the information in the innermost grouping symbols first.

2. Next, evaluate all **exponents**.

3. Next, evaluate all **multiplications** or **divisions** in the order in which they occur, working from left to right.

4. Finally, evaluate all **additions** or **subtractions** in the order in which they occur, working from left to right.

Some students remember the word PEMDAS or the phrase "Please Excuse My Dear Aunt Sally" to help them remember the order of operations. PEMDAS helps them remember the order: **P**arentheses, **E**xponents, **M**ultiplication, **D**ivision, **A**ddition, **S**ubtraction. Remember, this does not imply multiplication before division or addition before subtraction.

We can now evaluate $2 + 3 \cdot 4$. Since multiplications are performed before additions,

$$2 + 3 \cdot 4 \quad \text{means} \quad 2 + (3 \cdot 4) = 2 + 12 = 14$$

## 5  Learn the Use of Parentheses

Grouping symbols may be used (1) to change the order of operations to be followed in evaluating an algebraic expression or (2) to help clarify the understanding of an expression.

To evaluate the expression $2 + 3 \cdot 4$, we would normally perform the multiplication, $3 \cdot 4$, first. If we wished to have the addition performed before the multiplication, we could indicate this by placing parentheses around $2 + 3$:

$$(2 + 3) \cdot 4 = 5 \cdot 4 = 20$$

Sometimes it may be necessary to use more than one set of grouping symbols to indicate the order to be followed when evaluating an expression. When one set of grouping symbols is within another set of grouping symbols, we call these **nested grouping symbols**. Whenever we are given an expression with nested grouping symbols, we always evaluate the numbers in the *innermost grouping symbols first*. Color shading is used in the following examples to indicate the order in which the expression is evaluated.

$$6[2 + 3(4 + 1)] = 6[2 + 3(5)] = 6[2 + 15] = 6[17] = 102$$

$$4[3(6 - 4) \div 6] = 4[3(2) \div 6] = 4[6 \div 6] = 4[1] = 4$$

$$\{2 + [(8 \div 4)^2 - 1]\}^2 = \{2 + [2^2 - 1]\}^2 = \{2 + [4 - 1]\}^2 = \{2 + 3\}^2 = 5^2 = 25$$

> **Helpful Hint**
>
> If parentheses are not used to change the order of operations, multiplications and divisions are always performed before additions and subtractions. When a problem has only multiplications and divisions, work from left to right. Similarly, when a problem has only additions and subtractions, work from left to right.

## EXAMPLE 4 Evaluate $6 + 3 \cdot 5^2 - 4$.

*Solution* Colored shading is used to indicate the order in which the expression is to be evaluated.

$$
\begin{aligned}
6 + 3 \cdot 5^2 - 4 \quad & \text{Exponent} \\
= 6 + 3 \cdot 25 - 4 \quad & \text{Multiply.} \\
= 6 + 75 - 4 \quad & \text{Add.} \\
= 81 - 4 \\
= 77
\end{aligned}
$$

Now Try Exercise 51

## EXAMPLE 5 Evaluate $-7 + 2[-6 + (36 \div 3^2)]$.

*Solution*

$$
\begin{aligned}
-7 + 2[-6 + (36 \div 3^2)] \quad & \text{Exponent} \\
= -7 + 2[-6 + (36 \div 9)] \quad & \text{Divide.} \\
= -7 + 2[-6 + 4] \quad & \text{Add.} \\
= -7 + 2[-2] \quad & \text{Multiply.} \\
= -7 - 4 \\
= -11
\end{aligned}
$$

Now Try Exercise 65

## EXAMPLE 6 Evaluate $(8 \div 2) + 7(5 - 2)^2$.

*Solution*

$$
\begin{aligned}
(8 \div 2) + 7(5 - 2)^2 \quad & \text{Parentheses} \\
= 4 + 7(3)^2 \quad & \text{Exponent} \\
= 4 + 7 \cdot 9 \quad & \text{Multiply.} \\
= 4 + 63 \\
= 67
\end{aligned}
$$

Now Try Exercise 71

**EXAMPLE 7** Evaluate $-8 - 81 \div 9 \cdot 2^2 + 7$.

*Solution*

$$
\begin{aligned}
&\phantom{=}\; -8 - 81 \div 9 \cdot \boxed{2^2} + 7 && \text{Exponent} \\
&= -8 - \boxed{81 \div 9} \cdot 4 + 7 && \text{Divide.} \\
&= -8 - \boxed{9 \cdot 4} + 7 && \text{Multiply.} \\
&= \boxed{-8 - 36} + 7 && \text{Subtract.} \\
&= -44 + 7 \\
&= -37
\end{aligned}
$$

Now Try Exercise 69

**EXAMPLE 8** Evaluate.  **a)** $-4^2 + 6 \div 3$      **b)** $(-4)^2 + 6 \div 3$

*Solution*

**a)**
$$
\begin{aligned}
&\phantom{=}\; - \boxed{4^2} + 6 \div 3 && \text{Exponent} \\
&= -16 + \boxed{6 \div 3} && \text{Divide.} \\
&= -16 + 2 \\
&= -14
\end{aligned}
$$

**b)**
$$
\begin{aligned}
&\phantom{=}\; \boxed{(-4)^2} + 6 \div 3 && \text{Exponent} \\
&= 16 + \boxed{6 \div 3} && \text{Divide.} \\
&= 16 + 2 \\
&= 18
\end{aligned}
$$

Now Try Exercise 75

**EXAMPLE 9** Evaluate $\dfrac{3}{8} - \dfrac{2}{5} \cdot \dfrac{1}{12}$.

*Solution*   First perform the multiplication.

$$
\begin{aligned}
&\phantom{=}\; \frac{3}{8} - \left( \frac{\overset{1}{2}}{5} \cdot \frac{1}{\underset{6}{12}} \right) && \text{Multiply.} \\
&= \frac{3}{8} - \frac{1}{30} && \text{Subtract.} \\
&= \frac{45}{120} - \frac{4}{120} \\
&= \frac{41}{120}
\end{aligned}
$$

Now Try Exercise 81

**EXAMPLE 10** Write the following statements as mathematical expressions using parentheses and brackets and then evaluate: Multiply 12 by 3. Add 8 to this product. Subtract 7 from this sum. Divide this difference by 6.

*Solution*

$$
\begin{aligned}
& 12 \cdot 3 && \text{Multiply 12 by 3.} \\
& (12 \cdot 3) + 8 && \text{Add 8.} \\
& [(12 \cdot 3) + 8] - 7 && \text{Subtract 7.} \\
& \{[(12 \cdot 3) + 8] - 7\} \div 6 && \text{Divide the difference by 6.}
\end{aligned}
$$

Now evaluate.

$$\{[(\ 12 \cdot 3\ ) + 8] - 7\} \div 6$$
$$= \{[36 + 8] - 7\} \div 6$$
$$= \{44 - 7\} \div 6$$
$$= 37 \div 6$$
$$= \frac{37}{6}$$

Now Try Exercise 125

As shown in Example 10, sometimes brackets, [ ], and braces, { }, are used in place of parentheses to help avoid confusion. If only parentheses had been used, the preceding expression would appear as $(((12 \cdot 3) + 8) - 7) \div 6$.

## 6 Evaluate Expressions Containing Variables

Now we will evaluate some expressions for given values of the variables.

**EXAMPLE 11** Evaluate $5x - 4$ when $x = 3$.

*Solution*    Substitute 3 for $x$ in the expression.

$$5x - 4 = 5(3) - 4 = 15 - 4 = 11$$

Now Try Exercise 105

**EXAMPLE 12** Evaluate **a)** $x^2$ **b)** $-x^2$ and **c)** $(-x)^2$ when $x = 3$.

*Solution*    Substitute 3 for $x$.

**a)** $x^2 = 3^2 = 3(3) = 9$            **b)** $-x^2 = -3^2 = -(3)(3) = -9$

**c)** $(-x)^2 = (-3)^2 = (-3)(-3) = 9$

Now Try Exercise 95

**EXAMPLE 13** Evaluate **a)** $y^2$ **b)** $-y^2$ and **c)** $(-y)^2$ when $y = -4$.

*Solution*    Substitute $-4$ for $y$.

**a)** $y^2 = (-4)^2 = (-4)(-4) = 16$     **b)** $-y^2 = -(-4)^2 = -(-4)(-4) = -16$

**c)** $(-y)^2 = [-(-4)]^2 = (4)^2 = 16$

Now Try Exercise 97

Note that $-x^2$ will always be a negative number for any nonzero value of $x$, and $(-x)^2$ will always be a positive number for any nonzero value of $x$.

## Understanding Algebra

It is easy to confuse the expressions $(-6)^2$ and $-6^2$. $(-6)^2$ means $(-6) \cdot (-6) = 36$. The expression $-6^2$ represents the opposite of $6^2$. So, $-6^2 = -36$.

**Avoiding Common Errors**

The expression $-x^2$ means $-(x^2)$. When asked to evaluate $-x^2$ for any real number $x$, many students will incorrectly treat $-x^2$ as $(-x)^2$. For example, to evaluate $-x^2$ when $x = 5$,

| CORRECT | INCORRECT |
|---|---|
| $-5^2 = -(5^2) = -(5)(5)$ | $-5^2 = (-5)(-5)$ |
| $= -25$ | $= 25$ |

EXAMPLE 14   Evaluate $2x^2 + 4x + 1$ when $x = \dfrac{1}{4}$.

*Solution*   Substitute $\dfrac{1}{4}$ for each $x$ in the expression, then evaluate using the order of operations.

$$2x^2 + 4x + 1 = 2\left(\dfrac{1}{4}\right)^2 + 4\left(\dfrac{1}{4}\right) + 1 \quad \text{Substitute.}$$

$$= 2\left(\dfrac{1}{16}\right) + 4\left(\dfrac{1}{4}\right) + 1 \quad \text{Exponent}$$

$$= \dfrac{1}{8} + 1 + 1 \quad \text{Multiply.}$$

$$= \dfrac{1}{8} + 2 \quad \text{Add.}$$

$$= 2\dfrac{1}{8}$$

Now Try Exercise 113

EXAMPLE 15   Evaluate $-y^2 + 3(x + 2) - 5$ when $x = -3$ and $y = -2$.

*Solution*   Substitute $-3$ for each $x$ and $-2$ for each $y$, then evaluate using the order of operations.

$$
\begin{aligned}
-y^2 + 3(x + 2) - 5 &= -(-2)^2 + 3(-3 + 2) - 5 \quad &&\text{Substitute.}\\
&= -(-2)^2 + 3(-1) - 5 \quad &&\text{Parentheses}\\
&= -(4) + 3(-1) - 5 \quad &&\text{Exponent}\\
&= -4 - 3 - 5 \quad &&\text{Multiply.}\\
&= -7 - 5 \quad &&\text{Subtract, left to right.}\\
&= -12
\end{aligned}
$$

Now Try Exercise 121

# EXERCISE SET 1.9

MathXL®   MyMathLab®
MathXL®   MyMathLab

## Warm-Up Exercises

*Fill in the blanks with the appropriate word, phrase, or symbol(s) from the following list.*

base

left to right

exponent

right to left

grouping symbols

exponent expressions

innermost

1. When an expression has only additions and subtractions, it is evaluated from _____.

2. After evaluating grouping symbols, the next order of operation is to evaluate _____.

3. Parentheses, brackets, and braces are examples of _____.

4. In the expression $7^5$, 7 is called the _____.

5. In the expression $7^5$, 5 is called the _____.

6. When grouping symbols are nested, begin evaluating at the _____ group.

## Practice the Skills

*Evaluate.*

7. $5^2$

8. $2^3$

9. $1^7$

10. $4^1$

11. $-8^2$

12. $7^3$

13. $(-3)^2$

14. $-6^3$

15. $(-1)^3$

16. $2^5$

17. $-10^2$

18. $5^3$

19. $(-9)^2$

20. $(-3)^3$

21. $3^3$

22. $-7^2$

23. $(-4)^4$

24. $-4^4$

25. $-2^4$

26. $3^2(4)^2$

**27.** $\left(\dfrac{3}{4}\right)^2$      **28.** $\left(\dfrac{5}{8}\right)^3$      **29.** $\left(-\dfrac{1}{2}\right)^5$      **30.** $\left(-\dfrac{2}{3}\right)^4$

**31.** $5^2 \cdot 3^2$      **32.** $(-1)^4(2)^4$      **33.** $4^3 \cdot 3^2$      **34.** $(-2)^3(-1)^9$

*In Exercises 35–46,* **a)** *determine by observation whether the answer should be positive or negative and explain your answer;* **b)** *evaluate the expression; and* **c)** *determine whether your answer in part* **b)** *is reasonable and makes sense.*

**35.** $7^3$      **36.** $4^6$      **37.** $6^4$      **38.** $-2^5$

**39.** $(-3)^5$      **40.** $10^3$      **41.** $(-5)^4$      **42.** $(1.3)^3$

**43.** $-(-9)^2$      **44.** $(-3.3)^3$      **45.** $-\left(\dfrac{3}{8}\right)^2$      **46.** $\left(-\dfrac{3}{4}\right)^3$

*Evaluate.*

**47.** $3 + 3 \cdot 6$      **48.** $7 - 5^2 + 8$      **49.** $6 - 6 + 8$

**50.** $(8^2 \div 4) - (20 - 4)$      **51.** $-7 + 2 \cdot 6^2 - 8$      **52.** $6 + 2 \cdot 3^2 - 10$

**53.** $-3^3 + 27$      **54.** $(-2)^3 + 8 \div 4$      **55.** $(4 - 5) \cdot (5 - 1)^2$

**56.** $-10 - 6 - 3 - 2$      **57.** $3 \cdot 7 + 4 \cdot 2$      **58.** $4^2 - 3 \cdot 4 - 6$

**59.** $5 - 2(7 + 5)$      **60.** $8 + 3(6 + 4)$      **61.** $-32 - 5(7 - 10)^2$

**62.** $-40 - 3(4 - 8)^2$      **63.** $\dfrac{3}{4} + 2\left(\dfrac{1}{5}\right)^2$      **64.** $-\dfrac{2}{3} - 3\left(\dfrac{3}{4}\right)^2$

**65.** $-4 + 3[-1 + (12 \div 2^2)]$      **66.** $-2 + 4[-3 + (48 \div 4^2)]$      **67.** $(6 \div 3)^3 + 4^2 \div 8$

**68.** $4 + (4^2 - 13)^4 - 3$      **69.** $-7 - 48 \div 6 \cdot 2^2 + 5$      **70.** $-7 - 56 \div 7 \cdot 2^2 + 4$

**71.** $(9 \div 3) + 4(7 - 2)^2$      **72.** $(12 \div 4) + 5(6 - 4)^2$      **73.** $[4 + ((5 - 2)^2 \div 3)^2]^2$

**74.** $(20 \div 5 \cdot 5 \div 5 - 5)^2$      **75.** $(-3)^3 + 8 \div 2$      **76.** $-3^3 + 8 \div 2$

**77.** $2[1.55 + 5(3.7)] - 3.35$      **78.** $(8.4 + 3.1)^2 - (3.64 - 1.2)$      **79.** $\left(\dfrac{2}{5} + \dfrac{3}{8}\right) - \dfrac{3}{20}$

**80.** $\left(\dfrac{5}{6} \cdot \dfrac{4}{5}\right) + \left(\dfrac{2}{3} \cdot \dfrac{5}{8}\right)$      **81.** $\dfrac{3}{4} - 4 \cdot \dfrac{5}{40}$      **82.** $\dfrac{1}{8} - \dfrac{1}{4} \cdot \dfrac{3}{2} + \dfrac{3}{5}$

**83.** $\dfrac{4}{5} + \dfrac{3}{4} \div \dfrac{1}{2} - \dfrac{2}{3}$      **84.** $\dfrac{12 - (4 - 6)^2}{6 + 4^2 \div 2^2}$      **85.** $\dfrac{-4 - [2(9 \div 3) - 5]}{6^2 - 3^2 \cdot 7}$

**86.** $\dfrac{[(7 - 3)^2 - 4]^2}{9 - 16 \div 8 - 4}$      **87.** $\dfrac{-[4 - (6 - 12)^2]}{[(9 \div 3) + 4]^2 + 2^2}$      **88.** $\dfrac{[(5 - (3 - 7)) - 2]^2}{2[(16 \div 2^2) - (8 \cdot 4)]}$

**89.** $\{5 - 2[4 - (6 \div 2)]^2\}^2$      **90.** $\{-6 - [3(16 \div 4^2)^2]\}^2$      **91.** $-\{4 - [-3 - (2 - 5)]^2\}$

**92.** $3\{4[(3 - 4)^2 - 3]^3 - 1\}$      **93.** $\{4 - 3[2 - (9 \div 3)]^2\}^2$      **94.** $2\{5[(4 - 6)^3 - 1]^2 - 3\}$

*Evaluate* **a)** $x^2$, **b)** $-x^2$, *and* **c)** $(-x)^2$ *for the following values of x.*

**95.** $5$      **96.** $8$      **97.** $-2$      **98.** $-5$

**99.** $6$      **100.** $7$      **101.** $-\dfrac{1}{3}$      **102.** $\dfrac{3}{4}$

*Evaluate each expression for the given value of the variable or variables.*

**103.** $x + 6; x = -2$      **104.** $2x - 4x + 5; x = 3$      **105.** $-7z - 3; z = 6$

**106.** $3(x - 2); x = 5$      **107.** $a^2 - 6; a = -3$      **108.** $b^2 - 8; b = 5$

**109.** $3p^2 - 6p - 4; p = 2$      **110.** $2r^2 - 5r + 3; r = 1$      **111.** $-4x^2 - 2x + 1; x = -1$

**112.** $-t^2 - 4t + 5; t = -4$      **113.** $-x^2 - 2x + 5; x = \dfrac{1}{2}$      **114.** $2x^2 - 4x - 10; x = \dfrac{3}{4}$

**115.** $4(3x + 1)^2 - 6x; x = 5$      **116.** $3n^2(2n - 1) + 5; n = -4$

**117.** $r^2 - s^2; r = -2, s = -3$      **118.** $p^2 - q^2; p = 5, q = -3$

**119.** $5(x - 6y) + 3x - 7y; x = 1, y = -5$      **120.** $4(x + y)^2 + 2(x + y) + 3; x = 2, y = 4$

**121.** $3(x - 4)^2 - (3y - 4)^2; x = -1, y = -2$      **122.** $6x^2 + 3xy - y^2; x = 2, y = -3$

## Problem Solving

*Write the following statements as mathematical expressions using parentheses and brackets, and then evaluate.*

**123.** Multiply 6 by 3. From this product, subtract 4. From this difference, subtract 2.

**124.** Add 4 to 9. Divide this sum by 2. Add 10 to this quotient.

**125.** Multiply 10 by 4. Add 9 to this product. Subtract 6 from this sum. Divide this difference by 7.

**126.** Multiply 6 by 3. To this product, add 27. Divide this sum by 8. Multiply this quotient by 10.

**127.** Add $\frac{4}{5}$ to $\frac{3}{7}$. Multiply this sum by $\frac{2}{3}$.

**128.** Multiply $\frac{3}{8}$ by $\frac{4}{5}$. To this product, add $\frac{7}{120}$. From this sum, subtract $\frac{1}{60}$.

**129.** For what value or values of $x$ does $-(x^2) = -x^2$?

**130.** For what value or values of $x$ does $x = x^2$?

**131. Road Trip** If a car travels at 65 miles per hour, the distance it travels in $t$ hours is $65t$. Determine how far a car traveling at 65 miles per hour travels in 2.5 hours.

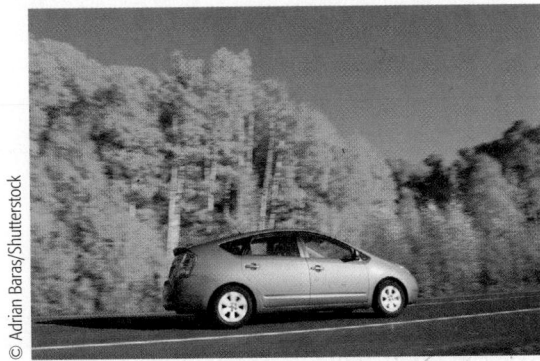

© Adrian Baras/Shutterstock

**132. Sales Tax** If the sales tax on an item is 8%, the sales tax on an item costing $d$ dollars can be found by the expression $0.08d$. Determine the sales tax on a scrapbook that costs $19.99.

**133. Projectile Height** An object is projected upward with an initial velocity of 48 feet per second from the top of a 70-foot building. The height of the object above the ground at any time $t$, in seconds, can be found by the expression $-16t^2 + 48t + 70$. Determine the height of the object after 2 seconds.

**134. Car Cost** If the sales tax on an item is 8%, then the total cost of an item $c$, including sales tax, can be found by the expression $c + 0.08c$. Find the total cost of a car that costs $17,000.

**135. Volume** In a gravel pit, gravel often forms the shape of a cone as it comes off of a conveyor belt. The volume of a cone is represented by the expression $\frac{\pi r^2 h}{3}$. If the radius of the cone is 4 feet and its height is 3 feet, what is its volume to the nearest cubic foot?

**136.** Rework problem 135 for a radius of 10 feet and a height of 9 feet.

## Challenge Problems

**137. Grass Growth** The rate of growth of grass in inches per week depends on a number of factors, including rainfall and temperature. For a certain region of the country, the growth per week can be approximated by the expression $0.2R^2 + 0.003RT + 0.0001T^2$, where $R$ is the weekly rainfall, in inches, and $T$ is the average weekly temperature, in degrees Fahrenheit. Find the amount of growth of grass for a week in which the rainfall is 2 inches and the average temperature is 70°F.

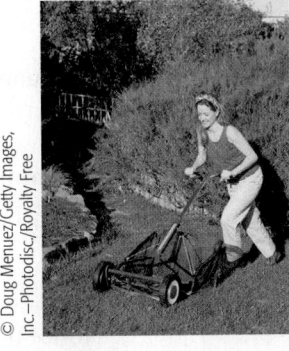

© Doug Menuez/Getty Images, Inc.–Photodisc/Royalty Free

*Insert one pair of parentheses to make each statement true.*

**138.** $14 + 6 \div 2 \times 4 = 40$

**139.** $12 - 4 - 6 + 10 = 24$

**140.** $24 \div 6 \div 2 + 2 = 1$

## Group Activity

*Discuss and answer Exercises 141–144 as a group, according to the instructions. Each question has four parts. For parts* **a)**, **b)**, *and* **c)**, *simplify the expression and write the answer in exponential form. Use the knowledge gained in parts* **a)**–**c)** *to answer part* **d)**. *(General rules that may be used to solve exercises like these will be discussed in Chapter 6.)*

**a)** *Group member 1: Do part* **a)** *of each exercise.*

**b)** *Group member 2: Do part* **b)** *of each exercise.*

**c)** *Group member 3: Do part* **c)** *of each exercise.*

**d)** *As a group, answer part* **d)** *of each exercise. You may need to make up other examples like parts* **a)**–**c)** *to help you answer part* **d)**.

**141. a)** $2^2 \cdot 2^3$  **b)** $3^2 \cdot 3^3$  **c)** $2^3 \cdot 2^4$  **d)** $x^m \cdot x^n$

**142. a)** $\frac{2^3}{2^2}$  **b)** $\frac{3^4}{3^2}$  **c)** $\frac{4^5}{4^3}$  **d)** $\frac{x^m}{x^n}$

**143. a)** $(2^3)^2$  **b)** $(3^3)^2$  **c)** $(4^2)^2$  **d)** $(x^m)^n$

**144. a)** $(2x)^2$  **b)** $(3x)^2$  **c)** $(4x)^3$  **d)** $(ax)^m$

## Cumulative Review Exercises

[1.2]  **145. Dogs** The graph shows the number of dogs in various houses selected at random in a neighborhood.

**Dogs in Selected Houses**

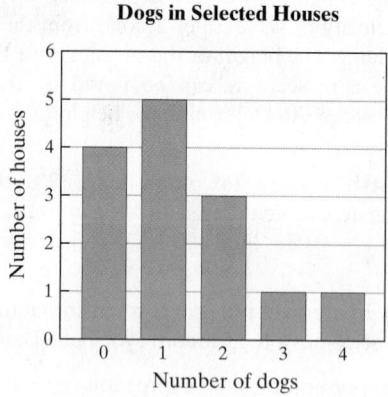

**a)** How many houses have two dogs?

**b)** Make a chart showing the number of houses that have no dogs, one dog, two dogs, and so on.

**c)** How many dogs in total are there in all the houses?

**d)** Determine the mean number of dogs in all the houses surveyed.

**146. Taxi Cost** Yellow Cab charges $2.40 for the first $\frac{1}{2}$ mile plus 20 cents for each additional $\frac{1}{8}$ mile or part thereof. Find the cost of a 3-mile trip.

[1.6]  **147.** Add $-\frac{7}{12} + \frac{4}{9}$.

[1.8]  **148.** Divide $\left(\frac{-5}{7}\right) \div \left(\frac{-3}{14}\right)$.

# 1.10 Properties of the Real Number System

1  Learn the commutative property.

2  Learn the associative property.

3  Learn the distributive property.

4  Learn the identity properties.

5  Learn the inverse properties.

Here, we introduce various properties of the real number system.

## 1  Learn the Commutative Property

The **commutative property of addition** states that the order in which any two real numbers are added does not matter.

**Commutative Property of Addition**

If $a$ and $b$ represent any two real numbers, then

$$a + b = b + a$$

Notice that the commutative property involves a change in *order*. For example,

$$4 + 3 = 3 + 4$$
$$7 = 7$$

The **commutative property of multiplication** states that the order in which any two real numbers are multiplied does not matter.

**Commutative Property of Multiplication**

If $a$ and $b$ represent any two real numbers, then

$$a \cdot b = b \cdot a$$

For example,

$$6 \cdot 3 = 3 \cdot 6$$
$$18 = 18$$

*The commutative property does not hold for subtraction or division.* For example,
$4 - 6 \neq 6 - 4$ and $6 \div 3 \neq 3 \div 6$.

**Understanding Algebra**

Two numbers can be added in either order or multiplied in either order. Thus, those operations are *commutative*. Neither subtraction nor division are commutative operations.

## 2 Learn the Associative Property

The **associative property of addition** states that, in the addition of three or more numbers, parentheses may be placed around any two adjacent numbers without changing the results.

### Associative Property of Addition

If $a$, $b$, and $c$ represent any three real numbers, then

$$(a + b) + c = a + (b + c)$$

Notice that the associative property involves a change of *grouping*. For example,

$$(3 + 4) + 5 = 3 + (4 + 5)$$
$$7 + 5 = 3 + 9$$
$$12 = 12$$

In this example, the 3 and 4 are grouped together on the left, and the 4 and 5 are grouped together on the right.

The **associative property of multiplication** states that, in the multiplication of three or more numbers, parentheses may be placed around any two adjacent numbers without changing the results.

### Associative Property of Multiplication

If $a$, $b$, and $c$ represent any three real numbers, then

$$(a \cdot b) \cdot c = a \cdot (b \cdot c)$$

For example,

$$(6 \cdot 2) \cdot 4 = 6 \cdot (2 \cdot 4)$$
$$12 \cdot 4 = 6 \cdot 8$$
$$48 = 48$$

Since the associative property involves a change of grouping, when the associative property is used, the content within the parentheses changes.

*The associative property does not hold for subtraction or division.* For example, $(4 - 1) - 3 \neq 4 - (1 - 3)$ and $(8 \div 4) \div 2 \neq 8 \div (4 \div 2)$.

Often when we add numbers we group the numbers so that we can add them easily. For example, when we add $70 + 50 + 30$ we may first add the $70 + 30$ to get 100. We are able to do this because of the commutative and associative properties.

$$(70 + 50) + 30 = 70 + (50 + 30) \quad \text{Associative property of addition}$$
$$= 70 + (30 + 50) \quad \text{Commutative property of addition}$$
$$= (70 + 30) + 50 \quad \text{Associative property of addition}$$
$$= 100 + 50 \quad \text{Addition facts}$$
$$= 150$$

Notice in the second step that the same numbers remained in parentheses but the order of the numbers changed, $50 + 30$ to $30 + 50$. Since this step involved a change in order (and not grouping), this is the commutative property of addition.

## 3 Learn the Distributive Property

A very important property of the real numbers is the **distributive property of multiplication over addition**. We often shorten the name to the **distributive property**.

**Distributive Property**

If $a$, $b$, and $c$ represent any three real numbers, then

$$a(b + c) = ab + ac$$

For example, if we let $a = 2$, $b = 3$, and $c = 4$, then

$$2(3 + 4) = (2 \cdot 3) + (2 \cdot 4)$$
$$2 \cdot 7 = 6 + 8$$
$$14 = 14$$

Therefore, we may either add first and then multiply, or multiply first and then add. Another example of the distributive property is

$$2(x + 3) = 2 \cdot x + 2 \cdot 3 = 2x + 6$$

The distributive property can be expanded in the following manner:

$$a(b + c + d + \cdots + n) = ab + ac + ad + \cdots + an$$

For example, $3(x + y + 5) = 3x + 3y + 15$.

**Helpful Hint**

The *commutative property* changes *order*.

The *associative property* changes *grouping*.

The *distributive property* involves *two operations,* usually multiplication and addition.

**EXAMPLE 1**  Name each property illustrated.

a) $4 + (-2) = -2 + 4$     b) $5(r + s) = 5 \cdot r + 5 \cdot s = 5r + 5s$

c) $x \cdot y = y \cdot x$     d) $(-12 + 3) + 4 = -12 + (3 + 4)$

*Solution*

a) Commutative property of addition

b) Distributive property

c) Commutative property of multiplication

d) Associative property of addition

Now Try Exercise 29

**Helpful Hint**

Do not confuse the distributive property with the associative property of multiplication. Make sure you understand the difference.

| Distributive Property | Associative Property of Multiplication |
|---|---|
| $3(4 + x) = 3 \cdot 4 + 3 \cdot x$ | $3(4 \cdot x) = (3 \cdot 4)x$ |
| $= 12 + 3x$ | $= 12x$ |

For the distributive property to be used, there must be two *terms* within parentheses, separated by a plus or minus sign as in $3(4 + x)$.

## 4 Learn the Identity Properties

Now we will discuss the **identity properties**. When the number 0 is added to any real number, the real number is unchanged. For example, $5 + 0 = 5$ and $0 + 5 = 5$. For this reason we call 0 the **identity element of addition** or **additive identity**. When any real number is multiplied by 1, the real number is unchanged. For example, $7 \cdot 1 = 7$ and $1 \cdot 7 = 7$. For this reason we call 1 the **identity element of multiplication** or **multiplicative identity**.

---

**Identity Properties**

If $a$ represents any real number, then

$$a + 0 = a \quad \text{and} \quad 0 + a = a \qquad \text{Identity property of addition}$$

and

$$a \cdot 1 = a \quad \text{and} \quad 1 \cdot a = a \qquad \text{Identity property of multiplication}$$

---

We often use the identity properties without realizing we are using them. For example, when we reduce $\dfrac{15}{50}$, we may do the following:

$$\frac{15}{50} = \frac{3 \cdot 5}{10 \cdot 5} = \frac{3}{10} \cdot \frac{5}{5} = \frac{3}{10} \cdot 1 = \frac{3}{10}$$

When we showed that $\dfrac{3}{10} \cdot 1 = \dfrac{3}{10}$, we used the identity property of multiplication.

## 5 Learn the Inverse Properties

The last properties we will discuss in this chapter are the **inverse properties**. Numbers like 3 and $-3$ are *opposites* or *additive inverses* because $3 + (-3) = 0$ and $-3 + 3 = 0$. Any two numbers whose sum is 0 are called *additive inverses* of each other. In general, for any real number $a$ its additive inverse is $-a$.

Numbers like 4 and $\dfrac{1}{4}$ are *reciprocals* or *multiplicative inverses* because $4 \cdot \dfrac{1}{4} = 1$ and $\dfrac{1}{4} \cdot 4 = 1$. Any two numbers whose product is 1 are called *multiplicative inverses* of each other. In general, for any real number $a$, its multiplicative inverse is $\dfrac{1}{a}$. The inverse properties are summarized below.

---

**Inverse Properties**

If $a$ represents any real number, then

$$a + (-a) = 0 \quad \text{and} \quad -a + a = 0 \qquad \text{Inverse property of addition}$$

and

$$a \cdot \frac{1}{a} = 1 \quad \text{and} \quad \frac{1}{a} \cdot a = 1 \, (a \neq 0) \qquad \text{Inverse property of multiplication}$$

---

We often use the inverse properties without realizing we are using them. For example, to evaluate the expression $6x + 2$ when $x = \dfrac{1}{6}$, we may do the following:

$$6x + 2 = 6\left(\frac{1}{6}\right) + 2 = 1 + 2 = 3$$

When we multiplied $6\left(\dfrac{1}{6}\right)$ and replaced it with 1, we used the inverse property of multiplication. We will be using both the identity and inverse properties throughout the book, although we may not specifically refer to them by name.

EXAMPLE 2  Name each property illustrated.

**a)** $2(x + 6) = (2 \cdot x) + (2 \cdot 6) = 2x + 12$    **b)** $3x \cdot 1 = 3x$

**c)** $(3 \cdot 6) \cdot 5 = 3 \cdot (6 \cdot 5)$    **d)** $y \cdot \dfrac{1}{y} = 1$

**e)** $2a + (-2a) = 0$    **f)** $3y + 0 = 3y$

### Solution

**a)** Distributive property
**b)** Identity property of multiplication
**c)** Associative property of multiplication
**d)** Inverse property of multiplication
**e)** Inverse property of addition
**f)** Identity property of addition

Now Try Exercise 35

EXAMPLE 3  In parts **a)**–**f)**, the name of a property is given followed by part of an equation. Complete the equation, to the right of the equals sign, to illustrate the given property.

**a)** Associative property of multiplication
$(5 \cdot 4) \cdot 7 =$

**b)** Inverse property of addition
$3c + (-3c) =$

**c)** Identity property of multiplication
$6y \cdot 1 =$

**d)** Distributive property
$3(x + 5) =$

**e)** Identity property of addition
$2a + 0 =$

**f)** Inverse property of multiplication
$b \cdot \dfrac{1}{b} =$

### Solution

**a)** $5 \cdot (4 \cdot 7)$   **b)** $0$   **c)** $6y$   **d)** $3x + 15$   **e)** $2a$   **f)** $1$

Now Try Exercise 55

# EXERCISE SET 1.10   MathXL®   MyMathLab®

## Warm-Up Exercises

*Fill in the blanks with the appropriate word, phrase, or symbol(s) from the following list.*

associative property of addition
commutative property of addition
additive identity
distributive property of multiplication over addition
multiplicative inverse

associative property of multiplication
commutative property of multiplication
multiplicative identity
additive inverse
zero

undefined

**1.** $(3 + 4) + 10$ is the same as $3 + (4 + 10)$ illustrates the _____.

**2.** $(-7)(-19)$ equals $(-19)(-7)$ illustrates the _____.

**3.** A number without a multiplicative inverse is _____.

**4.** $(a \cdot b) \cdot c$ is the same as $a \cdot (b \cdot c)$ illustrates the _____.

**5.** $-2a$ is the _____ of $2a$.

**6.** $\dfrac{1}{2}$ is the _____ of 2.

**7.** When $6(2 + 4)$ is rewritten as $12 + 24$, the _____ is illustrated.

**8.** $x + y = y + x$ illustrates the _____.

**9.** When the number 1 multiplies a number, no change occurs. Thus, 1 is called the _____.

**10.** When 0 is added to a number, there is no change. Thus, 0 is called the _____.

## Practice the Skills

*In Exercises 11–22, for the given expression, determine* **a)** *the additive inverse, and* **b)** *the multiplicative inverse.*

**11.** 6

**12.** 5

**13.** −3

**14.** −7

**15.** $x$

**16.** $z$

**17.** 1.6

**18.** −0.125

**19.** $\dfrac{1}{5}$

**20.** $\dfrac{1}{8}$

**21.** $-\dfrac{5}{6}$

**22.** $-\dfrac{2}{9}$

## Practice the Skills

*Name each property illustrated.*

**23.** $6(x + 7) = 6x + 42$

**24.** $3 + y = y + 3$

**25.** $(x + 3) + 5 = x + (3 + 5)$

**26.** $1(x + 3) = (1)(x) + (1)(3) = x + 3$

**27.** $5 \cdot y = y \cdot 5$

**28.** $-4x + 4x = 0$

**29.** $p \cdot (q \cdot r) = (p \cdot q) \cdot r$

**30.** $2(x + 4) = 2x + 8$

**31.** $4(d + 3) = 4d + 12$

**32.** $3 + (4 + t) = (3 + 4) + t$

**33.** $3z \cdot 1 = 3z$

**34.** $0 + 3y = 3y$

**35.** $2y \cdot \dfrac{1}{2y} = 1$

**36.** $x \cdot y = y \cdot x$

*In Exercises 37–58, the name of a property is given followed by part of an equation. Complete the equation, to the right of the equals sign, to illustrate the given property.*

**37.** commutative property of addition

$-4 + 1 =$

**38.** inverse property of addition

$(-7a) + 7a =$

**39.** associative property of multiplication

$-6 \cdot (4 \cdot 2) =$

**40.** associative property of addition

$-5 + (6 + 8) =$

**41.** distributive property

$-2(x + y) =$

**42.** distributive property

$4(x + 3) =$

**43.** commutative property of multiplication

$x \cdot y =$

**44.** identity property of multiplication

$(1)\left(-\dfrac{1}{3}b\right) =$

**45.** commutative property of addition

$4x + 3y =$

**46.** associative property of multiplication

$-9 \cdot (3 \cdot 8) =$

**47.** associative property of addition

$(a + b) + 3 =$

**48.** commutative property of multiplication

$(x + 2)3 =$

**49.** associative property of addition

$(3x + 4) + 6 =$

**50.** commutative property of addition

$3(x + y) =$

**51.** commutative property of multiplication

$3(m + n) =$

**52.** associative property of multiplication

$(3x)y =$

**53.** distributive property

$4(x + y + 3)$

**54.** distributive property

$3(x + y + 2) =$

**55.** inverse property of addition

$3n + (-3n) =$

**56.** identity property of addition

$0 + 2x =$

**57.** identity property of multiplication

$\left(\dfrac{5}{2}n\right)(1) =$

**58.** inverse property of multiplication

$\left(\dfrac{x}{2}\right)\left(\dfrac{2}{x}\right) =$

## Problem Solving

*Indicate whether the given processes are commutative. That is, does changing the order in which the actions are done result in the same final outcome? Explain each answer.*

**59.** Putting sugar and then cream in coffee; putting cream and then sugar in coffee.

**60.** Applying suntan lotion and then sunning yourself; sunning yourself and then applying suntan lotion.

**61.** Putting your contacts in and then washing your hair; washing your hair and then putting your contacts in.

**62.** Putting on your sweater and then your coat; putting on your coat and then your sweater.

**63.** Writing on the chalkboard and then erasing the chalkboard; erasing the chalkboard and then writing on the chalkboard.

**64.** Getting your hands dirty and then washing your hands; washing your hands and then getting your hands dirty.

*In Exercises 65–70, indicate whether the given processes are associative. For a process to be associative, the final outcome must be the same when the first two actions are performed first or when the last two actions are performed first. Explain each answer.*

**65.** Cleaning the kitchen sink, dusting the bedroom furniture, and doing the laundry.

**66.** In a store, buying cereal, soap, and dog food.

**67.** Turning on a DVD player, inserting a DVD, and watching the DVD.

**68.** Putting on a shirt, a tie, and a sweater.

**69.** Starting a car, moving the shift lever to drive, and then stepping on the gas.

**70.** Putting cereal, milk, and sugar in a bowl.

© Rohit Seth/Shutterstock

**71.** The commutative property of addition is $a + b = b + a$. Explain why $(3 + 4) + x = x + (3 + 4)$ also illustrates the commutative property of addition.

**72.** The commutative property of multiplication is $a \cdot b = b \cdot a$. Explain why $(3 + 4) \cdot x = x \cdot (3 + 4)$ also illustrates the commutative property of multiplication.

## Challenge Problems

**73.** Consider $x + (3 + 5) = x + (5 + 3)$. Does this illustrate the commutative property of addition or the associative property of addition? Explain.

**74.** Consider $x + (3 + 5) = (3 + 5) + x$. Does this illustrate the commutative property of addition or the associative property of addition? Explain.

**75.** Consider $x + (3 + 5) = (x + 3) + 5$. Does this illustrate the commutative property of addition? Explain.

**76.** The commutative property of multiplication is $a \cdot b = b \cdot a$. Explain why $(3 + 4) \cdot (5 + 6) = (5 + 6) \cdot (3 + 4)$ also illustrates the commutative property of multiplication.

## Cumulative Review Exercises

[1.3] **77.** Add $2\frac{3}{5} + \frac{2}{3}$.

**78.** Subtract $3\frac{5}{8} - 2\frac{3}{16}$.

[1.6] **79.** Add $102.7 + (-113.9)$.

**80.** Write the opposite of $\frac{7}{8}$.

# Chapter 1 Summary

| IMPORTANT FACTS AND CONCEPTS | EXAMPLES |
|---|---|

**Section 1.2**

**Guidelines for Problem Solving**

1. **Understand the problem.**

2. **Translate the problem to mathematical language.**

3. **Carry out the mathematical calculations necessary to solve the problem.**

4. **Check the answer obtained in step 3.**

5. **Make sure you have answered the question.**

See page 8 for more details on problem solving.

Jacob Thomas can pay either $725 cash for a desk or pay $300 down and $20 a month for 24 months. How much money can he save by paying cash?

*Solution:* Understand We need to determine how much he would pay if he paid $300 down and $20 a month for 24 months. This amount would then be compared to $725.

Translate

$$\begin{pmatrix} \text{amount if} \\ \text{paid over 24} \\ \text{months} \end{pmatrix} = \$300 + \begin{pmatrix} \text{additional} \\ \text{amount paid} \\ \text{over 24 months} \end{pmatrix}$$

Carry Out

$$= \$300 + \$20(24)$$
$$= \$300 + \$480$$
$$= \$780$$

The difference in the amounts is $780 - 725 = 55$.

Check and Answer

The answer appears reasonable. Jacob can save $55 by paying the total amount at the time of purchase.

An **algebraic expression** is a general term for any collection of numbers, variables, grouping symbols, and operations.

$$8(24 \div 4), 3x - 5, a^2 - 6$$

**Bar Graph**

**Top 10 Community-based Organization Websites by U.S. Marketshare Visits (%) July 2008**

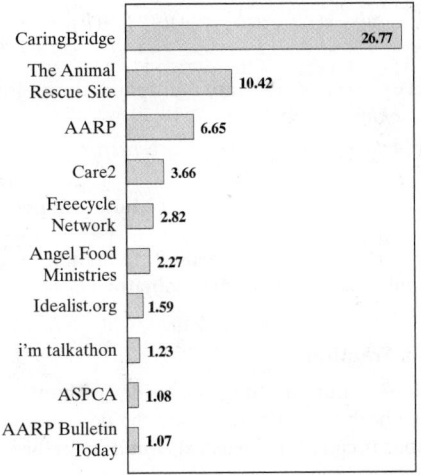

*Source*: www.marketingcharts.com

**Line Graph**

**Indiana SAT Math Scores**

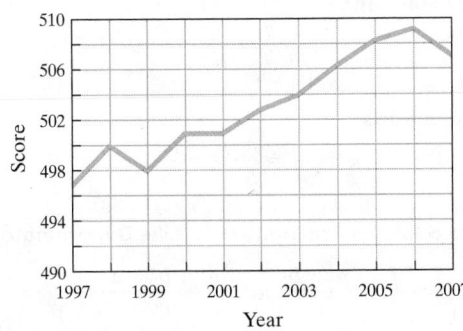

*Source*: www.doe.state.in.us

| IMPORTANT FACTS AND CONCEPTS | EXAMPLES |
|---|---|

### Section 1.2 (continued)

**Circle Graph**

**New Housing Units in the United States
2008**

*Source*: www.census.gov

---

The symbol $\approx$ is read **"is approximately equal to."**

$$57.91536 \approx 57.92$$

---

**Measures of central tendency** or **averages** are values that are representative of a set of data.

The **mean** of a set of data is determined by adding all the values and dividing the sum by the number of values.

The **median** is the value in the middle of a set of **ranked data**.

The mean and median are averages.

The mean of Adam Michael's test grades of 82, 95, 76, 92, and 88 is
$$\frac{82 + 95 + 76 + 92 + 88}{5} = \frac{433}{5} = 86.6.$$

The median of Adam's five test grades given above is 88 (circled).
76, 82, ⑧⑧, 92, 95

### Section 1.3

**Multiplication Symbols**

If $a$ and $b$ represent any two mathematical quantities, then each of the following may be used to indicate the product of $a$ and $b$ ("$a$ times $b$").

$$ab \quad a \cdot b \quad a(b) \quad (a)b \quad (a)(b)$$

5 times $z$ may be written:

$$5z, \quad 5 \cdot z, \quad 5(z), \quad (5)z \text{ or } (5)(z)$$

---

The numbers or variables that are multiplied in a multiplication problem are called **factors**.

    If $a \cdot b = c$, then $a$ and $b$ are *factors* of $c$.

In $7 \cdot 9 = 63$, the numbers 7 and 9 are factors of 63.

---

The numbers $0, 1, 2, 3, 4, \ldots$ are called **whole numbers**.

12, 105, and 0 are whole numbers.

---

The top number of a fraction is called the **numerator,** and the bottom number is called the **denominator**.

In the fraction $\frac{6}{11}$, the 6 is the numerator and the 11 is the denominator.

---

**To Simplify a Fraction**

1. Find the largest number that will divide (without remainder) both the numerator and the denominator. This number is called the **greatest common factor** (GCF).

2. Then divide both the numerator and the denominator by the greatest common factor.

The GCF of 36 and 48 is 12. Therefore,

$$\frac{36}{48} = \frac{36 \div 12}{48 \div 12} = \frac{3}{4}$$

---

**To Multiply Fractions**

$$\frac{a}{b} \cdot \frac{c}{d} = \frac{ac}{bd}$$

$$\frac{3}{4} \cdot \frac{3}{5} = \frac{9}{20}$$

---

**To Divide Fractions**

$$\frac{a}{b} \div \frac{c}{d} = \frac{a}{b} \cdot \frac{d}{c} = \frac{ad}{bc}$$

$$\frac{9}{7} \div \frac{4}{5} = \frac{9}{7} \cdot \frac{5}{4} = \frac{45}{28}$$

---

**To Add and Subtract Fractions with Like Denominators**

$$\frac{a}{c} + \frac{b}{c} = \frac{a + b}{c} \quad \text{or} \quad \frac{a}{c} - \frac{b}{c} = \frac{a - b}{c}$$

1. $\dfrac{3}{13} + \dfrac{7}{13} = \dfrac{3 + 7}{13} = \dfrac{10}{13}$

2. $\dfrac{4}{5} - \dfrac{1}{5} = \dfrac{4 - 1}{5} = \dfrac{3}{5}$

| IMPORTANT FACTS AND CONCEPTS | EXAMPLES |
|---|---|

**Section 1.3 (continued)**

| | |
|---|---|
| The smallest number that is divisible by two or more denominators is called the **least common denominator** or **LCD**. | In the fractions $\frac{7}{9}$ and $\frac{2}{5}$, 45 is the LCD. |
| **To add (or subtract) fractions with unlike denominators,** first rewrite each fraction with a common denominator. | Subtract $\frac{7}{9} - \frac{2}{5}$. $$\frac{7}{9} - \frac{2}{5} = \frac{7}{9} \cdot \frac{5}{5} - \frac{2}{5} \cdot \frac{9}{9}$$ $$= \frac{35}{45} - \frac{18}{45} = \frac{17}{45}$$ |
| A **mixed number** consists of a whole number followed by a fraction. | $7\frac{4}{5}$ is a mixed number. |
| Change a mixed number to a fraction. Change a fraction to a mixed number. | $$6\frac{2}{3} = 6 + \frac{2}{3} = \frac{18}{3} + \frac{2}{3} = \frac{20}{3}$$ $$\frac{52}{7} = \frac{49}{7} + \frac{3}{7} = 7 + \frac{3}{7} = 7\frac{3}{7}$$ |

**Section 1.4**

| | |
|---|---|
| A **set** is a collection of **elements** listed within braces. | The set $\{2, 4, 6, 8\}$ consists of four elements, namely 2, 4, 6, and 8. |
| A set that contains no elements is called an **empty set** (or **null set**). | The set of dogs that can fly is an empty set. |
| **Natural numbers:** $\{1, 2, 3, 4, 5, \ldots\}$ The natural numbers are also called the **positive integers** or the **counting numbers**. | 23, 16, and 1231 are natural numbers. |
| **Whole numbers:** $\{0, 1, 2, 3, 4, 5, \ldots\}$ | 35, 0, and 257 are whole numbers. |
| **Integers:** $\{\underbrace{\ldots, -3, -2, -1}_{\text{Negative integers}}, 0, \underbrace{1, 2, 3 \ldots}_{\text{Positive integers}}\}$ | $-101$, 0, and 236 are integers. |
| **Rational numbers:** $\{$quotient of two integers, denominator not 0$\}$ | $-3, 2.8, 9\frac{1}{2}, 0, 15, -\frac{9}{13}$, and $-3.6$ are rational numbers. |
| **Irrational numbers:** $\{$numbers that can be represented on the number line that are not rational numbers$\}$ | $\sqrt{13}$ and $-\sqrt{7}$ are irrational numbers. |
| **Real numbers:** $\{$all numbers that can be represented on a number line$\}$ | $5.7, -\frac{3}{8}, -16, 0, \sqrt{5}, 3\frac{1}{7}, -2.1, -\sqrt{2}$, and 31 are real numbers. |

**Section 1.5**

| | |
|---|---|
| The symbol $>$ is used to represent the words "is greater than." | **1.** $3.8 > 3.08$ |
| The symbol $<$ is used to represent the words "is less than." | **2.** $-7 < -1$ |
| The **absolute value** of a number can be considered the distance between the number and 0 on a number line. | **1.** $\lvert 12 \rvert = 12$ |
| The absolute value of every number will be either *positive* or *zero*. | **2.** $\lvert 0 \rvert = 0$ **3.** $\lvert -100 \rvert = 100$ |

**Section 1.6**

| | |
|---|---|
| **Add Using a Number Line:** | Add $5 + (-7)$. |
| Represent the first number to be added (first *addend*) by an arrow starting at 0 on the number line. The arrow is drawn to the right if the number is positive, and to the left if the number is negative. From the tip of the first arrow, draw a second arrow to represent the second addend. The second arrow is drawn to the right or left, as just explained. The sum of the two numbers is found at the tip of the second arrow. |  $$5 + (-7) = -2$$ |

| IMPORTANT FACTS AND CONCEPTS | EXAMPLES |
|---|---|

### Section 1.6 (continued)

| | |
|---|---|
| Any two numbers whose sum is zero are said to be **opposites** (or **additive inverses**) of each other. | **1.** The opposite of 8 is $-8$.<br><br>**2.** The opposite of $-\dfrac{5}{6}$ is $\dfrac{5}{6}$. |
| **Add Using Absolute Value:**<br><br>**To add real numbers with the same sign,** add their absolute values. The sum has the same sign as the numbers being added. | Add $-4 + (-9)$.<br><br>$$\lvert -4 \rvert + \lvert -9 \rvert = 4 + 9 = 13$$<br><br>Since both numbers being added are negative, the sum is negative. Thus, $-4 + (-9) = -13$. |
| **To add two signed numbers with different signs,** subtract the smaller absolute value from the larger absolute value. The answer has the sign of the number with the larger absolute value. | Add $-25 + 10$.<br><br>$$\lvert -25 \rvert - \lvert 10 \rvert = 25 - 10 = 15$$<br><br>Since $\lvert -25 \rvert$ is greater than $\lvert 10 \rvert$, the sum is negative. Thus, $-25 + 10 = -15$. |

### Section 1.7

| | |
|---|---|
| **To Subtract Real Numbers**<br><br>In general, if $a$ and $b$ represent any two real numbers, then<br><br>$$a - b = a + (-b)$$<br><br>In evaluating expressions involving more than one addition and subtraction, work from left to right unless parentheses or other grouping symbols appear. | **1.** $8 - (+4) = 8 + (-4) = 4$<br>**2.** $-7 - 5 = -7 + (-5) = -12$<br>**3.** $-12 - (-6) = -12 + 6 = -6$<br>**4.** $-8 + 5 - 11 = -3 - 11 = -14$ |

### Section 1.8

| | |
|---|---|
| **The Sign of the Product of Two Real Numbers**<br><br>**1.** The product of two numbers with **like** signs is a **positive** number.<br><br>**2.** The product of two numbers with **unlike** signs is a **negative** number. | **1.** $-8(-7) = 56$<br><br>**2.** $5(-6) = -30$ |
| **The Sign of the Quotient of Two Real Numbers**<br><br>**1.** The quotient of two numbers with **like** signs is a **positive** number.<br><br>**2.** The quotient of two numbers with **unlike** signs is a **negative** number. | **1.** $\dfrac{-81}{-9} = 9$<br><br>**2.** $\dfrac{-35}{7} = -5$ |
| **The Quotient of a Positive Number and a Negative Number**<br><br>If $a$ and $b$ represent any real numbers, $b \neq 0$, then<br><br>$$\frac{a}{-b} = \frac{-a}{b} = -\frac{a}{b}$$ | $$\frac{3}{-8} = \frac{-3}{8} = -\frac{3}{8}$$ |
| **Summary of Division Involving 0**<br><br>If $a$ represents any real number except 0, then<br><br>$$\frac{0}{a} = 0 \qquad \frac{a}{0} \text{ is undefined}$$ | **1.** $\dfrac{0}{12} = 0$<br><br>**2.** $\dfrac{5}{0}$ is undefined |

### Section 1.9

| | |
|---|---|
| In general, the number $b$ to the $n$th power, written $b^n$, means<br><br>$$\underbrace{b \cdot b \cdot b \cdot \cdots \cdot b}_{n \text{ factors of } b} = b^n$$ | $$x^5 = x \cdot x \cdot x \cdot x \cdot x$$ |

| IMPORTANT FACTS AND CONCEPTS | EXAMPLES |
|---|---|

### Section 1.9 (continued)

**Order of Operations**

1. First, evaluate the information within **parentheses** ( ), brackets [ ], or braces { }. These are **grouping symbols**. A fraction bar, $-$, also serves as a grouping symbol. If the expression contains nested grouping symbols, evaluate the information in the innermost grouping symbols first.

2. Next, evaluate all **exponents**.

3. Next, evaluate all **multiplications** or **divisions** in order from left to right.

4. Finally, evaluate all **additions** or **subtractions** in order from left to right.

$$-16 - 3(6 - 8)^2$$
$$= -16 - 3(-2)^2$$
$$= -16 - 3(4)$$
$$= -16 - 12$$
$$= -28$$

### Section 1.10

**Commutative Property of Addition**

If $a$ and $b$ represent any two real numbers, then
$$a + b = b + a$$

$$9 + 6 = 6 + 9$$
$$15 = 15$$

**Commutative Property of Multiplication**

If $a$ and $b$ represent any two real numbers, then
$$a \cdot b = b \cdot a$$

$$6 \cdot 7 = 7 \cdot 6$$
$$42 = 42$$

**Associative Property of Addition**

If $a$, $b$, and $c$ represent any three real numbers, then
$$(a + b) + c = a + (b + c)$$

$$(1 + 2) + 3 = 1 + (2 + 3)$$
$$3 + 3 = 1 + 5$$
$$6 = 6$$

**Associative Property of Multiplication**

If $a$, $b$, and $c$ represent any three real numbers, then
$$(a \cdot b) \cdot c = a \cdot (b \cdot c)$$

$$(5 \cdot 4) \cdot 3 = 5 \cdot (4 \cdot 3)$$
$$20 \cdot 3 = 5 \cdot 12$$
$$60 = 60$$

**Distributive Property**

If $a$, $b$, and $c$ represent any three real numbers, then
$$a(b + c) = ab + ac$$

$$4(y + 9) = 4 \cdot y + 4 \cdot 9 = 4y + 36$$

**Identity Properties**

If $a$ represents any real number, then

1. $a + 0 = a$  and  $0 + a = a$  Identity property of addition

and

2. $a \cdot 1 = a$  and  $1 \cdot a = a$  Identity property of multiplication

1. $15 + 0 = 0 + 15 = 15$

2. $8 \cdot 1 = 1 \cdot 8 = 8$

**Inverse Properties**

If $a$ represents any real number, then

1. $a + (-a) = 0$  and  $-a + a = 0$  Inverse property of addition

and

2. $a \cdot \dfrac{1}{a} = 1$  and  $\dfrac{1}{a} \cdot a = 1 (a \neq 0)$  Inverse property of multiplication

1. $9 + (-9) = -9 + 9 = 0$

2. $6 \cdot \dfrac{1}{6} = \dfrac{1}{6} \cdot 6 = 1$

## Chapter 1 Review Exercises

[1.2]   *Solve.*

**1. Final Exam Review** Eraj Basnayake bought doughnuts for his math classes for their final exam review sessions. His first class ate 32 doughnuts. His second class ate 29 doughnuts. His third class ate 36 doughnuts, and his fourth class ate 31 doughnuts. If he purchased 13 dozen doughnuts, how many did he have left over?

© debr22pics/Shutterstock

**2. Inflation** Assume that the rate of inflation for tuition for a course is 7% per year for the next two years. What will tuition for the course cost in two years if the tuition cost is $500 today?

**3. Sales Tax** The sales tax in Cattaraugus County is 8.25%.
   **a)** What was the sales tax that Andrew Jacob paid on a laptop computer that cost $899.99 before tax?
   **b)** What is the total cost of the laptop computer including tax?

**4. Plasma TV** Dan Marketos wants to purchase a high definition plasma TV for $3000. He can either pay the total amount at the time of purchase, or he can agree to pay the store $400 down and $225 a month for 12 months. How much money can he save by paying the total amount at the time of purchase?

**5. Test Grades** On Angie Smajstrla's first five exams her grades were 75, 79, 86, 88, and 64. Find the **a)** mean and **b)** median of her grades.

**6. Little League** The number of points scored by a Little League baseball team in its last six games were 21, 3, 17, 10, 9, and 6. Find the **a)** mean and **b)** median of the points.

**7. Commute Times** Recently, the U.S. Census Bureau released a report that included the following bar graph. The bar graph shows some of the longest average commute-to-work times

in the United States. Estimate the average one-way commute time in
   **a)** New York.
   **b)** Illinois.

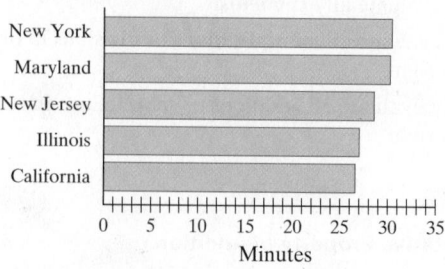

*Source*: US Census Bureau

**8. Freshman Majors** The graphs below show the majors of the 2006 and the 2009 freshmen classes at the University of Cincinnati's College of Applied Science.
   **a)** If there were 900 freshmen in 2006, determine the number of Information Technology majors.
   **b)** If there were 1100 freshmen in 2009, determine the number of Sports Administration majors.

**College of Applied Sciences**
**Major of Entering Freshman**

**Entering Class: 2006**
10%, 35%, 30%, 25%

**Entering Class: 2009**
10%, 45%, 15%, 30%

□ Engineering Technology     □ Information Technology
□ Sports Administration     □ All Others

*Source*: www.uc.edu

[1.3]   *Perform each indicated operation. Simplify your answers.*

**9.** $\dfrac{3}{5} \cdot \dfrac{5}{6}$     **10.** $3\dfrac{5}{7} + 2\dfrac{1}{3}$     **11.** $\dfrac{5}{12} \div \dfrac{3}{5}$     **12.** $\dfrac{5}{6} + \dfrac{1}{3}$     **13.** $3\dfrac{1}{6} - 1\dfrac{1}{4}$     **14.** $7\dfrac{3}{8} \div \dfrac{5}{12}$

[1.4]   **15.** List the set of natural numbers.
**16.** List the set of whole numbers.
**17.** List the set of integers.
**18.** Describe the set of rational numbers.
**19.** Consider the following set of numbers.

$$\left\{ 3, -5, -12, 0, \dfrac{1}{2}, -0.62, \sqrt{7}, 426, -3\dfrac{1}{4} \right\}$$

List the numbers that are
   **a)** positive integers.
   **b)** whole numbers.
   **c)** integers.
   **d)** rational numbers.
   **e)** irrational numbers.
   **f)** real numbers.

**20.** Consider the following set of numbers.

$$\left\{ -2.3, -8, -9, 1\dfrac{1}{2}, \sqrt{2}, -\sqrt{2}, 1, -\dfrac{3}{17} \right\}$$

List the numbers that are
   **a)** natural numbers.
   **b)** whole numbers.
   **c)** negative integers.
   **d)** integers.
   **e)** rational numbers.
   **f)** irrational numbers.
   **g)** real numbers.

**[1.5]**   *Insert either* $<, >,$ *or* $=$ *in each shaded area to make a true statement.*

**21.** $-7$ ▓ $-5$          **22.** $-2.6$ ▓ $-3.6$          **23.** $0.50$ ▓ $0.509$          **24.** $4.6$ ▓ $4.06$

**25.** $-6.3$ ▓ $-6.03$          **26.** $5$ ▓ $|-3|$          **27.** $\left|-\dfrac{9}{2}\right|$ ▓ $|-4.5|$          **28.** $|-10|$ ▓ $|-7|$

**[1.6–1.7]**   *Evaluate.*

**29.** $-9 + (-5)$          **30.** $-6 + 6$          **31.** $0 + (-3)$          **32.** $-10 + 4$

**33.** $-8 - (-2)$          **34.** $-2 - (-4)$          **35.** $4 - (-4)$          **36.** $12 - 12$

**37.** $2 - 7$          **38.** $7 - (-7)$          **39.** $0 - (-4)$          **40.** $-7 - 5$

**41.** $\dfrac{4}{3} - \dfrac{3}{4}$          **42.** $\dfrac{1}{2} + \dfrac{3}{5}$          **43.** $\dfrac{5}{9} - \dfrac{3}{4}$          **44.** $-\dfrac{5}{7} + \dfrac{3}{8}$

**45.** $-\dfrac{5}{12} - \dfrac{5}{6}$          **46.** $-\dfrac{6}{7} + \dfrac{5}{12}$          **47.** $\dfrac{2}{9} - \dfrac{3}{10}$          **48.** $\dfrac{5}{12} - \left(-\dfrac{3}{5}\right)$

*Evaluate.*

**49.** $9 - 4 + 9$          **50.** $-8 - 9 + 14$          **51.** $-5 - 4 - 3$          **52.** $-2 + (-3) - 2$

**53.** $17 - (+4) - (-3)$          **54.** $6 - (-2) + 3$

**[1.8]**   *Evaluate.*

**55.** $7(-9)$          **56.** $(-8.2)(-3.1)$          **57.** $(-4)(-5)(-6)$          **58.** $\left(\dfrac{3}{5}\right)\left(\dfrac{-2}{7}\right)$

**59.** $\left(\dfrac{10}{11}\right)\left(\dfrac{3}{-5}\right)$          **60.** $\left(\dfrac{-5}{8}\right)\left(\dfrac{-3}{7}\right)$          **61.** $0\left(\dfrac{4}{9}\right)$          **62.** $(-4)(-6)(-2)(-3)$

*Evaluate.*

**63.** $45 \div (-3)$          **64.** $12 \div (-2)$          **65.** $-14.72 \div 4.6$          **66.** $-37.41 \div (-8.7)$

**67.** $-88 \div (-11)$          **68.** $-4 \div \left(\dfrac{-4}{9}\right)$          **69.** $\dfrac{28}{-3} \div \left(\dfrac{9}{-2}\right)$          **70.** $\dfrac{14}{3} \div \left(\dfrac{-6}{5}\right)$

*Indicate whether each quotient is* $0$ *or undefined.*

**71.** $0 \div 5$          **72.** $0 \div (-6)$          **73.** $-12 \div 0$          **74.** $-4 \div 0$          **75.** $\dfrac{8.3}{0}$          **76.** $\dfrac{0}{-9.8}$

**[1.6–1.8, 1.9]**   *Evaluate.*

**77.** $-5(3 - 8)$          **78.** $2(4 - 8)$          **79.** $(3 - 6) + 4$

**80.** $(-4 + 3) - (2 - 6)$          **81.** $[6 + 3(-2)] - 6$          **82.** $(-5 - 3)(4)$

**83.** $[12 + (-4)] + (6 - 8)$          **84.** $9[3 + (-4)] + 5$          **85.** $-4(-3) + [4 \div (-2)]$

**86.** $(-3 \cdot 4) \div (-2 \cdot 6)$          **87.** $(-3)(-4) + 6 - 3$          **88.** $[-2(3) + 6] - 4$

**[1.9]**   *Evaluate.*

**89.** $-6^2$          **90.** $(-6)^2$          **91.** $2^4$          **92.** $(-3)^3$

**93.** $(-1)^9$          **94.** $(-2)^5$          **95.** $\left(\dfrac{-4}{5}\right)^2$          **96.** $\left(\dfrac{2}{5}\right)^3$

**97.** $5^3 \cdot (-2)^2$          **98.** $(-2)^4\left(\dfrac{1}{2}\right)^2$          **99.** $\left(-\dfrac{2}{3}\right)^2 \cdot 3^3$          **100.** $(-4)^3(-2)^2$

*Evaluate.*

**101.** $-5 + 3 \cdot 4$          **102.** $4 \cdot 6 + 4 \cdot 2$          **103.** $(3.7 - 4.1)^2 + 6.2$

**104.** $10 - 36 \div 4 \cdot 3$          **105.** $6 - 3^2 \cdot 5$          **106.** $[6.9 - (3 \cdot 5)] + 5.8$

**107.** $\dfrac{6^2 - 4 \cdot 3^2}{-[6 - (3 - 4)]}$          **108.** $\dfrac{4 + 5^2 \div 5}{6 - (-3 + 2)}$          **109.** $3[9 - (4^2 + 3)] \cdot 2$

**110.** $(-3^2 + 4^2) + (3^2 \div 3)$          **111.** $2^3 \div 4 + 6 \cdot 3$          **112.** $(4 \div 2)^4 + 4^2 \div 2^2$

**113.** $(8 - 2^2)^2 - 4 \cdot 3 + 10$          **114.** $4^3 \div 4^2 - 5(2 - 7) \div 5$          **115.** $-\{-4[27 \div 3^2 - 2(4 - 2)]\}$

**116.** $2\{4^3 - 6[4 - (2 - 4)] - 3\}$

*Evaluate each expression for the given values.*

**117.** $3x - 7; x = 4$          **118.** $6 - 4x; x = -5$          **119.** $2x^2 - 5x + 3; x = 6$

**120.** $5y^2 + 3y - 2; y = -1$          **121.** $-x^2 + 2x - 3; x = -2$          **122.** $-x^2 + 2x - 3; x = 2$

**123.** $-3x^2 - 5x + 5; x = 1$          **124.** $-x^2 - 8x - 12y; x = -3, y = -2$

**[1.6–1.9]**   **a)** *Evaluate each expression, and* **b)** *check to see whether your answer is reasonable.*

**125.** $278 + (-493)$          **126.** $324 - (-29.6)$          **127.** $\dfrac{-17.28}{6}$

**128.** $(-62)(-1.9)$          **129.** $(-4)^8$          **130.** $-(4.2)^3$

[1.10] *Name each indicated property.*

**131.** $(7 + 4) + 9 = 7 + (4 + 9)$

**132.** $-5(a + 2) = -5a - 10$

**133.** $6x + 3x = 3x + 6x$

**134.** $(x + 4)3 = 3(x + 4)$

**135.** $4(x + 3) = 4x + 12$

**136.** $(x + 7) + 4 = x + (7 + 4)$

**137.** $8b \cdot 1 = 8b$

**138.** $-8y + 8y = 0$

## Chapter 1 Practice Test

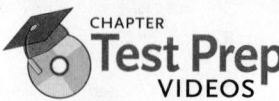

*Chapter Test Prep Videos provide fully worked-out solutions to any of the exercises you want to review. Chapter Test Prep Videos are available via MyMathLab®, or on YouTube (search "Angel Elementary and Intermediate Algebra" and click on "Channels").*

1. **Shopping** While shopping, Mia Nguyen purchases two half-gallons of milk for $1.30 each, one Boston cream pie for $4.75, and three 2-liter bottles of soda for $1.10 each.
   a) What is her total bill before tax?
   b) If there is a 7% sales tax on the bottles of soda, how much is the sales tax?
   c) How much is her total bill including tax?
   d) How much change will she receive from a $50 bill?

2. **TV Commercials** The following line graph shows the cost for a 30-second commercial during the same time slot on the same channel for 12 consecutive years. How many times greater was the cost for a 30-second commercial in the twelfth year than in the first year?

**Cost of a 30-Second Commercial**

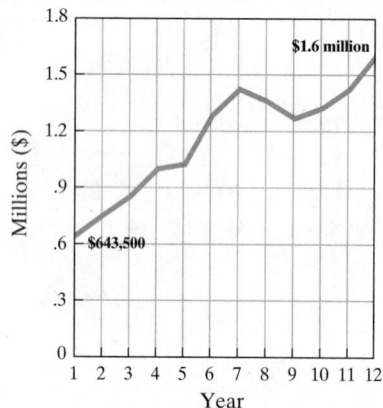

3. **Radio** The following graph shows the median number of listeners for various radio stations during a specific time.

**Median Number of Listeners**

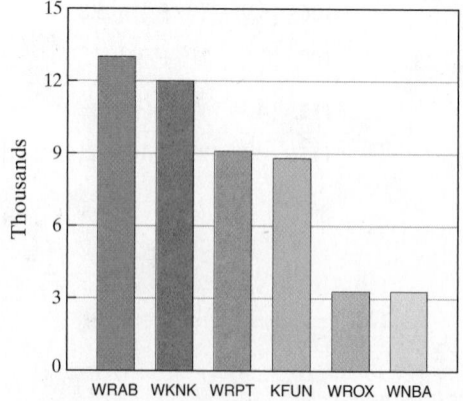

a) Determine the median number of listeners who listened to WRAB during this time.

b) The median number of listeners who listened to KFUN during this time was approximately 8.8 thousand. Explain what this means.

4. Consider the following set of numbers.
$$\left\{ -6, 42, -3\frac{1}{2}, 0, 6.52, \sqrt{5}, \frac{5}{9}, -7, -1 \right\}$$

List the numbers that are
a) natural numbers.
b) whole numbers.
c) integers.
d) rational numbers.
e) irrational numbers.
f) real numbers.

*Insert either $<$, $>$, or $=$ in each shaded area to make a true statement.*

**5.** $-9.9$ ▨ $-9.09$

**6.** $|-3|$ ▨ $|-2|$

*Evaluate.*

**7.** $-7 + (-8)$

**8.** $-6 - 5$

**9.** $15 - 12 - 17$

**10.** $(-4 + 6) - 3(-2)$

**11.** $(-4)(-3)(2)(-1)$

**12.** $\left( \dfrac{-2}{9} \right) \div \left( \dfrac{-7}{8} \right)$

**13.** $\left( -18 \cdot \dfrac{1}{2} \right) \div 3$

**14.** $-\dfrac{3}{8} - \dfrac{4}{7}$

**15.** $-6(-2 - 3) \div 5 \cdot 2$

**16.** $\left( -\dfrac{2}{3} \right)^5$

**17.** $[6 + ((9 - 3)^2 \div 18)^2]^2$

**18.** Explain why $-x^2$ will always be a negative value for any nonzero real number selected for $x$.

*Evaluate the expression for the given values.*

**19.** $5x^2 - 8$; $x = -3$

**20.** $-x^2 - 6x + 3$; $x = -2$

**21.** $6x - 3y^2 + 4$; $x = 3, y = -2$

**22.** $-x^2 + xy + y^2$; $x = 1, y = -2$

*Name each indicated property.*

**23.** $x + 3 = 3 + x$

**24.** $4(x + 9) = 4x + 36$

**25.** $(2 + x) + 4 = 2 + (x + 4)$

# 2 Solving Linear Equations

## Goals of This Chapter

The major emphasis of this chapter is to teach you how to solve linear equations. *To be successful in solving linear equations, you need to have a thorough understanding of adding, subtracting, multiplying, and dividing real numbers.* This material was discussed in Chapter 1.

The first four sections of this chapter will give you the building blocks you will need for solving linear equations. Section 2.5 combines the material previously presented to teach you how to solve a variety of linear equations. In the last few sections of this chapter, you will learn about formulas, ratios, and proportions.

You will be using principles learned in this chapter throughout the book and in real life.

**There are many algebraic formulas** that are very useful in everyday life. For example, we can use algebra to determine the amount we owe on a loan, or to convert dollars to other currencies. Algebraic formulas are also useful in sports and recreation. In Example 7 on page 136, we use algebraic formulas to find the area and circumference of a pizza.

© dovgan/Shutterstock

# 2.1 Combining Like Terms

1 Identify terms.

2 Identify like terms.

3 Combine like terms.

4 Use the distributive property.

5 Remove parentheses when they are preceded by a plus or minus sign.

6 Simplify an expression.

## Understanding Algebra

*Terms* are parts of an expression that get added or subtracted. The expression $2x - 3y - 5$ has 3 terms:

$$2x - 3y - 5$$

terms

$2x$ is a term, $-3y$ is a term, and $-5$ is a term.

## Understanding Algebra

*Terms* are added (or subtracted); *factors* are multiplied.

So, in $4x + 7y$, there are two terms (they are $4x$ and $7y$). The 4 and $x$ are factors of the first term and 7 and $y$ are factors of the second.

## Understanding Algebra

A *coefficient* is the numerical factor of a term. If none is shown, it is understood that the coefficient is a "1".

## 1 Identify Terms

In Section 1.3, we indicated that letters called **variables** are used to represent numbers. A variable can represent a variety of different numbers.

As was indicated in Chapter 1, an **expression**, or **algebraic expression,** is a collection of numbers, variables, grouping symbols, and operation symbols.

### Examples of Expressions

$$7, \quad x^2 - 6, \quad 4x - 3, \quad 2(x + 5) + 6, \quad \frac{x + 3}{4}$$

When an algebraic expression consists of several parts, the parts that are *added* are called the **terms** of the expression. Consider the expression $2x - 3y - 5$. The expression can be written as $2x + (-3y) + (-5)$, and so the expression $2x - 3y - 5$ has three terms: $2x$, $-3y$, and $-5$. The expression $3x^2 + 2xy + 5(x + y)$ also has three terms: $3x^2, 2xy,$ and $5(x + y)$.

When listing the terms of an expression, it is not necessary to list the + sign at the beginning of a term.

| Expression | Terms |
|---|---|
| $-2x + 3y - 8$ | $-2x, 3y, -8$ |
| $3y^2 - 2x + \frac{1}{2}$ | $3y^2, -2x, \frac{1}{2}$ |
| $7 + x + 4 - 5x$ | $7, x, 4, -5x$ |
| $3(x - 1) - 4x + 2$ | $3(x - 1), -4x, 2$ |
| $\frac{x + 4}{3} - 5x + 3$ | $\frac{x + 4}{3}, -5x, 3$ |

The numerical part of a term is called its **numerical coefficient** or simply its **coefficient.** In the term $6x$, the 6 is the numerical coefficient. Note that $6x$ means the variable $x$ is multiplied by 6.

| Term | Numerical Coefficient |
|---|---|
| $5x$ | $5$ |
| $7$ | $7$ |
| $-\frac{1}{2}x$ | $-\frac{1}{2}$ |
| $4(x - 3)$ | $4$ |
| $\frac{2x}{3}$ | $\frac{2}{3}$, since $\frac{2x}{3}$ means $\frac{2}{3}x$ |
| $\frac{x + 4}{3}$ | $\frac{1}{3}$, since $\frac{x + 4}{3}$ means $\frac{1}{3}(x + 4)$ |

*Whenever a term appears without a numerical coefficient, we assume that the numerical coefficient is 1.*

### Examples

| | |
|---|---|
| $x$ means $1x$ | $-x$ means $-1x$ |
| $x^2$ means $1x^2$ | $-x^2$ means $-1x^2$ |
| $xy$ means $1xy$ | $-xy$ means $-1xy$ |
| $(x + 2)$ means $1(x + 2)$ | $-(x + 2)$ means $-1(x + 2)$ |

If an expression has a term that is a number (without a variable), we refer to that number as a **constant term**, or simply a **constant**. In the expression $x^2 + 3x - 4$, the $-4$ is a constant term, or a constant.

## 2  Identify Like Terms

**Like terms** are terms that have the same variables with the same exponents, respectively. Constants, such as 4 and $-6$, are like terms. Some examples of like terms and unlike terms follow. Note that if two terms are like terms, only their numerical coefficients may differ.

> ### Understanding Algebra
>
> In the expression $2x - 3y - 8x$, we refer to the $2x$ term and the $-8x$ term as *like terms* because their variable parts are identical.

| Like Terms | Unlike Terms | |
|---|---|---|
| $3x, \quad -4x$ | $3x, \quad 2$ | ⟵ (One term has a variable, the other is a constant.) |
| $4y, \quad 6y$ | $3x, \quad 4y$ | ⟵ (Variables differ) |
| $5, \quad -6$ | $x, \quad 3$ | ⟵ (One term has a variable, the other is a constant.) |
| $3(x + 1), \quad -2(x + 1)$ | $2x, \quad 3xy$ | ⟵ (Variables differ) |
| $3x^2, \quad 4x^2$ | $3x, \quad 4x^2$ | ⟵ (Exponents differ) |
| $5ab, \quad 2ab$ | $4a, \quad 2ab$ | ⟵ (Variables differ) |

**EXAMPLE 1**   Identify any like terms.

**a)** $2x + 3x + 4$    **b)** $2x + 3y + 2$    **c)** $x + 3 + y - \dfrac{1}{2}$   **d)** $x + 3x^2 - 4x^2$

**e)** $5x - x + 6$    **f)** $3 - 2x + 4x - 6$   **g)** $12 + x^2 - x + 7$

*Solution*

**a)** $2x$ and $3x$ are like terms.

**b)** There are no like terms.

**c)** $3$ and $-\dfrac{1}{2}$ are like terms.

**d)** $3x^2$ and $-4x^2$ are like terms.

**e)** $5x$ and $-x$ (or $-1x$) are like terms.

**f)** $3$ and $-6$ are like terms; $-2x$ and $4x$ are like terms.

**g)** $12$ and $7$ are like terms.

NowTry Exercise 3

## 3  Combine Like Terms

We often need to simplify expressions by combining like terms. To **combine like terms** means to add or subtract the like terms in an expression. To combine like terms, we can use the procedure that follows.

> **To Combine Like Terms**
>
> 1. Determine which terms are like terms.
> 2. Add or subtract the coefficients of the like terms.
> 3. Multiply the number found in step 2 by the common variable(s).

Examples 2 through 7 illustrate this procedure.

**EXAMPLE 2**   Combine like terms: $5x + 4x$.

*Solution*   $5x$ and $4x$ are like terms with the common variable $x$. Since $5 + 4 = 9$, then $5x + 4x = 9x$.

NowTry Exercise 9

**EXAMPLE 3**  Combine like terms: $\dfrac{3}{5}x - \dfrac{2}{3}x$. 

*Solution*  Since $\dfrac{3}{5} - \dfrac{2}{3} = \dfrac{9}{15} - \dfrac{10}{15} = -\dfrac{1}{15}$, then $\dfrac{3}{5}x - \dfrac{2}{3}x = -\dfrac{1}{15}x$.

*Now Try Exercise 15*

**EXAMPLE 4**  Combine like terms: $6.47b - 8.39b$. 

*Solution*  Since $6.47 - 8.39 = -1.92$, then $6.47b - 8.39b = -1.92b$.

*Now Try Exercise 35*

**EXAMPLE 5**  Combine like terms: $3x + x + 5$. 

*Solution*  The $3x$ and $x$ are like terms.

$$3x + x + 5 = 3x + 1x + 5 = 4x + 5$$

*Now Try Exercise 19*

Because of the commutative property of addition, the order of the terms in the answer is not critical. Thus, $5 + 4x$ is also an acceptable answer to Example 5. When writing answers, we generally list the terms containing variables in alphabetical order from left to right, and list the constant term last.

The commutative and associative properties of addition will be used to rearrange the terms in Examples 6 and 7.

**EXAMPLE 6**  Combine like terms: $3b + 6a - 5 - 2a$. 

*Solution*  The only like terms are $6a$ and $-2a$.

$$3b + 6a - 5 - 2a = 6a - 2a + 3b - 5 \quad \text{Rearrange terms.}$$
$$= 4a + 3b - 5 \quad \text{Combined like terms.}$$

*Now Try Exercise 21*

**EXAMPLE 7**  Combine like terms: $-2x^2 + 3y - 4x^2 + 3 - y + 5$. 

*Solution*

$$-2x^2 \quad \text{and} \quad -4x^2 \quad \text{are like terms.}$$
$$3y \quad \text{and} \quad -y \quad \text{are like terms}$$
$$3 \quad \text{and} \quad 5 \quad \text{are like terms.}$$

Grouping the like terms together gives

$$-2x^2 + 3y - 4x^2 + 3 - y + 5 = -2x^2 - 4x^2 + 3y - y + 3 + 5$$
$$= -6x^2 + 2y + 8$$

*Now Try Exercise 29*

## 4  Use the Distributive Property

We introduced the distributive property in Section 1.10. Because this property is so important, we will study it again. But before we do, let's briefly review the subtraction of real numbers. Recall from Section 1.7 that

$$6 - 3 = 6 + (-3)$$

In general,

> ### Subtraction of Real Numbers
>
> For any real numbers $a$ and $b$,
> $$a - b = a + (-b)$$

We will use the fact that $a + (-b)$ means $a - b$ in discussing the distributive property.

> ### Distributive Property
>
> For any real numbers $a$, $b$, and $c$,
> $$a(b + c) = ab + ac$$

> **Understanding Algebra**
>
> We use the distributive property when combining like terms:
>
> $$5x + 4x = (5 + 4)x = 9x$$
> or
> $$7y - 3y = (7 - 3)y = 4y$$

**EXAMPLE 8**  Use the distributive property to remove parentheses.

**a)** $2(x + 4)$        **b)** $-5(p + 3)$

*Solution*

**a)** $2(x + 4) = 2x + 2(4) = 2x + 8$

**b)** $-5(p + 3) = -5p + (-5)(3) = -5p + (-15) = -5p - 15$

Note in part **b)** that, instead of leaving the answer $-5p + (-15)$, we wrote it as $-5p - 15$, which is the preferred form of the answer.

Now Try Exercise 61

**EXAMPLE 9**  Use the distributive property to remove parentheses.

**a)** $3(x - 2)$        **b)** $-2(4x - 3)$

*Solution*

**a)** By the definition of subtraction, we write $x - 2$ as $x + (-2)$.

$$3(x - 2) = 3[x + (-2)] = 3x + 3(-2)$$
$$= 3x + (-6)$$
$$= 3x - 6$$

**b)** $-2(4x - 3) = -2[4x + (-3)] = -2(4x) + (-2)(-3) = -8x + 6$

Now Try Exercise 63

The distributive property is very important to the study of algebra. Study the Helpful Hint that follows.

With a little practice, you will be able to eliminate some of the intermediate steps when you use the distributive property. When using the distributive property, there are eight possibilities with regard to signs. Study and understand the eight possibilities that follow.

Positive Coefficient

**a)** $2(x) = 2x$

$2(x + 3) = 2x + 6$

$2(+3) = +6$

**b)** $2(x) = 2x$

$2(x - 3) = 2x - 6$

$2(-3) = -6$

**c)** $2(-x) = -2x$

$2(-x + 3) = -2x + 6$

$2(+3) = +6$

**d)** $2(-x) = -2x$

$2(-x - 3) = -2x - 6$

$2(-3) = -6$

Negative Coefficient

**e)** $(-2)(x) = -2x$

$-2(x + 3) = -2x - 6$

$(-2)(+3) = -6$

**f)** $(-2)(x) = -2x$

$-2(x - 3) = -2x + 6$

$(-2)(-3) = +6$

**g)** $(-2)(-x) = 2x$

$-2(-x + 3) = 2x - 6$

$(-2)(+3) = -6$

**h)** $(-2)(-x) = 2x$

$-2(-x - 3) = 2x + 6$

$(-2)(-3) = +6$

## Understanding Algebra

Remember, the distributive property is used for multiplying over a sum of 2 or more terms. This is sometimes referred to as *expanding* an expression. So, $2(x - 3y + 9)$ can be rewritten as $2x - 6y + 18$

The distributive property can be expanded as follows:

$$a(b + c + d + \cdots + n) = ab + ac + ad + \cdots + an$$

### Examples of the Expanded Distributive Property

$$3(x + y + z) = 3x + 3y + 3z$$
$$2(x + y - 3) = 2x + 2y - 6$$

**EXAMPLE 10**  Use the distributive property to remove parentheses.

**a)** $4(x - 3)$     **b)** $-6(5y - 1)$     **c)** $-\dfrac{1}{2}(4r + 5)$     **d)** $-7(2x + 4y - 9z)$

*Solution*

**a)** $4(x - 3) = 4x - 12$     **b)** $-6(5y - 1) = -30y + 6$

**c)** $-\dfrac{1}{2}(4r + 5) = -2r - \dfrac{5}{2}$     **d)** $-7(2x + 4y - 9z) = -14x - 28y + 63z$

Now Try Exercise 79

The distributive property can also be used from the right, as in Example 11.

**EXAMPLE 11**  Use the distributive property to remove parentheses from the expression $(2x - 8y)4$.

*Solution*   We distribute the 4 on the right side of the parentheses over the terms within the parentheses.

$$(2x - 8y)4 = 2x(4) - 8y(4)$$
$$= 8x - 32y$$

Now Try Exercise 83

Example 11 could have been rewritten as $4(2x - 8y)$ by the commutative property of multiplication, and then the 4 could have been distributed from the left to obtain the same answer, $8x - 32y$.

Students sometimes try to use the distributive property when it cannot be used. For the distributive property to be used, there must be a + or − between the terms *within parentheses* and the terms within parentheses must be *multiplied* by some number or expression. Study the following correct simplifications carefully.

$$4(2xy) = 8xy \quad \text{(Distributive property is not used)}$$
$$4(2x + y) = 8x + 4y \quad \text{(Distributive property is used)}$$
$$(2x + y) - 4 = 2x + y - 4 \quad \text{(Distributive property is not used)}$$
$$(2x + y)(-4) = -8x - 4y \quad \text{(Distributive property is used)}$$

## 5 Remove Parentheses When They Are Preceded by a Plus or Minus Sign

In the expression $(4x + 3)$, how do we remove parentheses? Recall that the coefficient of a term is assumed to be 1 if none is shown. Therefore, we may write

$$(4x + 3) = 1(4x + 3)$$
$$= 1(4x) + (1)(3)$$
$$= 4x + 3$$

**Understanding Algebra**

When no sign or a plus sign precedes parentheses, the parentheses may be removed without having to change the expression inside parentheses.

**Other Examples**
$$(x + 3) = x + 3$$
$$(2x - 3) = 2x - 3$$
$$+(2x - 5) = 2x - 5$$
$$+(x + 2y - 6) = x + 2y - 6$$

Now consider the expression $-(4x + 3)$. How do we remove parentheses in this expression? Here, the coefficient in front of the parentheses is $-1$, so each term within the parentheses is multiplied by $-1$.

$$-(4x + 3) = -1(4x + 3)$$
$$= -1(4x) + (-1)(3)$$
$$= -4x + (-3)$$
$$= -4x - 3$$

Thus, $-(4x + 3) = -4x - 3$. *When a minus sign precedes parentheses, the signs of all the terms within the parentheses are changed when the parentheses are removed.*

**Understanding Algebra**

Think of the sign preceding a term as being part of that term. So when we rewrite $-(2x - 3)$, we multiply $-1$ by $2x$ and $-1$ by $-3$ to obtain $-2x + 3$.

**Examples**
$$-(x + 4) = -x - 4$$
$$-(-2x + 3) = 2x - 3$$
$$-(5x - y + 3) = -5x + y - 3$$
$$-(-4c - 3d - 5) = 4c + 3d + 5$$

## 6 Simplify an Expression

Combining what we learned in the preceding discussions, we have the following procedure for **simplifying an expression.**

**To Simplify an Expression**

1. Use the distributive property to remove any parentheses.
2. Combine like terms.

**EXAMPLE 12** Simplify $6 - (2x + 3)$.

**Solution**

$$6 - (2x + 3) = 6 - 2x - 3 \qquad \text{Use the distributive property.}$$
$$= -2x + 3 \qquad \text{Combined like terms.}$$

*Note:* $3 - 2x$ is the same as $-2x + 3$; however, we generally write the term containing the variable first.

Now Try Exercise 89

**EXAMPLE 13** Simplify $-\left(\dfrac{2}{3}x - \dfrac{1}{4}\right) + 3x$.

**Solution** 
$$-\left(\frac{2}{3}x - \frac{1}{4}\right) + 3x = -\frac{2}{3}x + \frac{1}{4} + 3x \qquad \text{Distributive property}$$
$$= -\frac{2}{3}x + 3x + \frac{1}{4} \qquad \text{Rearranged terms.}$$
$$= -\frac{2}{3}x + \frac{9}{3}x + \frac{1}{4} \qquad \text{Wrote } x \text{ terms with the LCD, 3.}$$
$$= \frac{7}{3}x + \frac{1}{4} \qquad \text{Combined like terms.}$$

Now Try Exercise 101

Notice in Example 13 that $\dfrac{7}{3}x$ and $\dfrac{1}{4}$ could not be combined because they are not like terms.

**EXAMPLE 14** Simplify $\dfrac{3}{4}x + \dfrac{1}{3}(5x - 2)$.

**Solution**

$$\frac{3}{4}x + \frac{1}{3}(5x - 2) = \frac{3}{4}x + \frac{1}{3}(5x) + \frac{1}{3}(-2) \qquad \text{Distributive property}$$
$$= \frac{3}{4}x + \frac{5}{3}x - \frac{2}{3}$$
$$= \frac{9}{12}x + \frac{20}{12}x - \frac{2}{3} \qquad \text{Wrote } x \text{ terms with the LCD, 12.}$$
$$= \frac{29}{12}x - \frac{2}{3} \qquad \text{Combined like terms.}$$

Now Try Exercise 103

**EXAMPLE 15** Simplify $3(2a - 5) - 3(b - 6) - 4a$.
**Solution**

$$3(2a - 5) - 3(b - 6) - 4a = 6a - 15 - 3b + 18 - 4a \qquad \text{Distributive property}$$
$$= 6a - 4a - 3b - 15 + 18 \qquad \text{Rearranged terms.}$$
$$= 2a - 3b + 3 \qquad \text{Combined like terms.}$$

Now Try Exercise 107

> **Helpful Hint**
>
> Keep in mind the difference between the concepts of *term* and *factor*. When two or more expressions are **multiplied**, each expression is a **factor** of the product. For example, since $4 \cdot 3 = 12$, the 4 and the 3 are factors of 12. Since $3 \cdot x = 3x$, the 3 and the $x$ are factors of $3x$. Similarly, in the expression $5xyz$, the $5$, $x$, $y$, and $z$ are all factors.
>
> In an expression, the parts that are **added** are the **terms** of the expression. For example, the expression $2x^2 + 3x - 4$ has three terms, $2x^2$, $3x$, and $-4$. Note that the terms of an expression may have factors. For example, in the term $2x^2$, the 2 and the $x^2$ are factors because they are multiplied.

# EXERCISE SET 2.1   MathXL®   MyMathLab®

## Warm-Up Exercises

*Fill in the blanks with the appropriate word, phrase, or symbol(s) from the following list.*

terms          $4x - 6y - 18$          like terms          unlike terms          $-4x + 6y + 18$

constant          factors          coefficient          variable

**1.** In the expression $5x - 3y + 17 - 2x$, the 17 is called a _____ term.

**2.** When we apply the distributive property to $-2(2x - 3y - 9)$, we obtain _____.

**3.** In the expression $5x - 3y + 17 - 2x$, the $5x$ and $-2x$ are called _____.

**4.** In the expression $5x - 3y + 17 - 2x$, the $-3$ is called the _____ of the second term.

**5.** In the expression $5x - 3y + 17 - 2x$, the $5x$ and $-3y$ are called _____.

**6.** In the expression $12x + 17$, the 12 and $x$ are _____ of the first term.

**7.** In the expression $-4x^2 + 17x - 90$, the $-4x^2, 17x$, and $-90$ are called _____.

**8.** In the expression $17x$, the $x$ is called a _____.

## Practice the Skills

*Combine like terms when possible. If not possible, rewrite the expression as is.*

**9.** $6x + 8x$
**10.** $4x - 5x$
**11.** $3x + 6$
**12.** $4x + 3y$
**13.** $y + 3 + 4y$
**14.** $4x - 7x + 4$
**15.** $\frac{3}{4}a - \frac{6}{11}a$
**16.** $\frac{3}{4}p - \frac{2}{7}p$
**17.** $2t - 6x + 5t$
**18.** $-7 - 4m - 66$
**19.** $-2w - 3w + 5$
**20.** $-8y - 4y - 7$
**21.** $-x + 2 - x - 2$
**22.** $8x - 2y - 1 - 3x$
**23.** $3 + 6x - 3 - 6x$
**24.** $y - 2y + 5$
**25.** $5 + 2t - 4t + 16$
**26.** $5s - 3s - 2s$
**27.** $4p - 6 - 16p - 2$
**28.** $-6t + 5 + 2t - 9$
**29.** $3x^2 - 9y^2 + 7x^2 - 5 - y^2 - 2$
**30.** $-4x^2 - 6y - 3x^2 + 6 - y - 1$
**31.** $-2x + 4x - 8$
**32.** $4 - x + 4x - 8$
**33.** $b + 4 + \frac{3}{5}$
**34.** $\frac{3}{4}y + 2 + y$
**35.** $5.1n + 6.42 - 4.3n$
**36.** $2x^2 + 3y^2 + 14x + 5y^2$
**37.** $\frac{1}{2}a + 3b + 1$
**38.** $x + \frac{1}{2}y - \frac{3}{8}y$
**39.** $13.4x + 1.2x + 8.3$
**40.** $-4x^2 - 3.1 - 5.2$
**41.** $-x^2 + 2x^2 + y$
**42.** $1 + x^2 + 6 - 3x^2$
**43.** $2x - 7y - 5x + 2y$
**44.** $3x - 7 - 9 + 4x$
**45.** $4 - 3n^2 + 9 - 2n$
**46.** $9x + y - 2 - 4x$
**47.** $-19.36 + 40.02x + 12.25 - 18.3x$
**48.** $52x - 52x - 63.5 - 63.5$
**49.** $\frac{3}{5}x - 3 - \frac{7}{4}x - 2$
**50.** $\frac{1}{2}y - 4 + \frac{3}{4}x - \frac{1}{5}y$
**51.** $5w^3 + 2w^2 + w + 3$
**52.** $4p^2 - 3p^2 + 2p - 5p$
**53.** $2z - 5z^3 - 2z^3 - z^2$
**54.** $5ab - 3ab$
**55.** $6x^2 - 6xy + 3y^2$
**56.** $x^2 - 3xy - 2xy + 6$
**57.** $4a^2 - 3ab + 6ab + b^2$
**58.** $4b^2 - 8bc + 5bc + c^2$

*Use the distributive property to remove parentheses.*

**59.** $5(x + 2)$
**60.** $2(-y + 5)$
**61.** $5(x + 4)$
**62.** $-2(y + 8)$
**63.** $3(x - 6)$
**64.** $-2(x - 4)$
**65.** $-\frac{1}{2}(2x - 4)$
**66.** $-4(x + 6)$
**67.** $1(-4 + x)$
**68.** $\frac{2}{3}(m - 18)$
**69.** $\frac{4}{5}(s - 5)$
**70.** $5(x - y + 5)$
**71.** $-0.3(3x^2 + 5)$
**72.** $-(x - 3)$
**73.** $-\frac{1}{3}(3r - 12)$
**74.** $-2(x + y - z)$
**75.** $0.7(2x + 0.5)$
**76.** $-(x + 4y)$

**77.** $-(-x + y)$        **78.** $(3x + 4y - 6)$        **79.** $-(2x + 4y - 8)$
**80.** $-3(2a + 3b - 7)$        **81.** $1.1(3.1x - 5.2y + 2.8)$        **82.** $-4(-2m - 3n + 8)$
**83.** $(2x - 9y)5$        **84.** $(8b - 1)7$        **85.** $(r + 3s - 19)$
**86.** $(-p + 2q - 3)$        **87.** $-3(-x + 2y + 4)$        **88.** $2.3(1.6x + 5.1y - 4.1)$

*Simplify.*

**89.** $5 - (3x + 4)$        **90.** $7 - (2y - 9)$        **91.** $-2(3 - x) + 7$
**92.** $-(3t - 3) + 5$        **93.** $6x + 2(4x + 9)$        **94.** $3(x + y) + 2y$
**95.** $2(x - y) + 2x + 3$        **96.** $6 + (x - 5) + 3x$        **97.** $4(2c - 3) - 3(c - 4)$
**98.** $4 + (2y + 2) + y$        **99.** $8x - (x - 3)$        **100.** $-(x - 5) - 3x + 4$
**101.** $-\left(\dfrac{3}{4}x - \dfrac{1}{3}\right) + 2x$        **102.** $-\left(\dfrac{7}{8}x - \dfrac{1}{2}\right) - 3x$        **103.** $\dfrac{2}{3}x + \dfrac{1}{2}(5x - 4)$
**104.** $\dfrac{4}{5}x + \dfrac{1}{7}(3x - 1)$        **105.** $-(3s + 4) - (s + 2)$        **106.** $6 - 2(2w + 3) + 5w$
**107.** $4(x - 1) + 2(3 - x) - 4$        **108.** $4(3b - 2) - 5(c - 4) - 6b$        **109.** $4(m + 3) - 4m - 12$
**110.** $-3(a + 2b) + 3(a + 2b)$        **111.** $0.4 - (y + 5) + 0.6 - 2$        **112.** $4 - (2 - x) + 3x$
**113.** $4 + (3x - 4) - 5$        **114.** $2y - 6(y - 2) + 3$        **115.** $4(x + 2) - 3(x - 4) - 5$
**116.** $6 - (a - 5) - (2b + 1)$        **117.** $-0.2(6 - x) - 4(y + 0.4)$        **118.** $-5(2y - 8) - 3(1 + x) - 7$
**119.** $-6x + 7y - (3 + x) + (x + 3)$        **120.** $3(t - 2) - 2(t + 4) - 6$        **121.** $\dfrac{1}{2}(x + 3) + \dfrac{1}{3}(3x + 6)$
**122.** $\dfrac{2}{3}(r - 2) - \dfrac{1}{2}(r + 4)$

## Problem Solving

*If* $\square + \square + \square + \odot + \odot$ *can be represented as* $3\square + 2\odot$, *write an expression to represent each of the following.*

**123.** $\square + \ominus + \ominus + \square + \ominus$

**124.** $\otimes + \dot{\smile} + \otimes + \dot{\smile} + \dot{\smile} + \dot{\smile}$

**125.** $x + y + \triangle + \triangle + x + y + y$

**126.** $2 + x + 2 + \ominus + \ominus + 2 + y$

*In Exercises 127 and 128, consider the following. The positive factors of 6 are 1, 2, 3, and 6 since*

$$1 \cdot 6 = 6$$
$$2 \cdot 3 = 6$$
$$\uparrow \ \uparrow$$
factors

**127.** List all the positive factors of 18.
**128.** List all the positive factors of 24.

## Concept/Writing Exercises

**129. a)** When a minus sign precedes an expression within parentheses, explain how to remove parentheses.
  **b)** Write $-(x - 8)$ without parentheses,

**130. a)** What are like terms? Determine whether the following are like terms. If not, explain why.
  **b)** $3x, 4y$        **c)** $7, -2$
  **d)** $5x^2, 5x$        **e)** $4x, -5xy$

## Challenge Problems

*Simplify.*

**131.** $4x^2 + 5y^2 + 6(3x^2 - 5y^2) - 4x + 3$        **132.** $2x^2 - 4x + 8x^2 - 3(x + 2) - x^2 - 2$

**133.** $2[3 + 4(x - 5)] - [2 - (x - 3)]$        **134.** $\dfrac{1}{4}\left[3 - 2(y + 1)\right] - \dfrac{1}{3}\left[2 - (y - 6)\right]$

## Cumulative Review Exercises

[1.5]    *Evaluate.*
  **135.** $|-7|$
  **136.** $-|-16|$
[1.7]    **137.** Evaluate $-4 - 13 - (-6)$.

[1.9]    **138.** Write a paragraph explaining the order of operations.
  **139.** Evaluate $-x^2 + 5x - 6$ when $x = -1$.

# 2.2 The Addition Property of Equality

1. Identify linear equations.
2. Check solutions to equations.
3. Identify equivalent equations.
4. Use the addition property to solve equations.
5. Solve equations by doing some steps mentally.

## 1 Identify Linear Equations

A statement that shows two algebraic expressions are equal is called an **equation**. For example, $4x + 3 = 2x - 4$ is an equation. In this chapter, we learn to solve **linear equations** in one variable.

### Linear Equation

A **linear equation** in one variable is an equation that can be written in the form

$$ax + b = c$$

where $a$, $b$, and $c$ are real numbers and $a \neq 0$.

Examples of Linear Equations

$$x + 4 = 7$$
$$2x - 4 = 6$$

## 2 Check Solutions to Equations

### Solution to an Equation

The **solution to an equation** is the number or numbers that when substituted for the variable or variables make the equation a true statement.

The solution to $x + 4 = 7$ is 3 because when $x$ is 3, the statement is true: 7 equals 7. A solution can be **checked** by substituting your answer for the variable in the original equation. If the result is true, your solution is correct. If the result is false, then you need to go back and find your error. Try to check all your solutions as this will improve your algebra skills.

To check whether 3 is the solution to $x + 4 = 7$, we substitute 3 for each $x$ in the equation.

Check
$$x = 3$$
$$x + 4 = 7$$
$$3 + 4 \overset{?}{=} 7$$
$$7 = 7 \quad \text{True}$$

Since the check results in a true statement, 3 is a solution.

EXAMPLE 1  Consider the equation $2x - 4 = 6$. Determine whether 3 is a solution.

*Solution*   To determine whether 3 is a solution to the equation, we substitute 3 for $x$.

Check
$$x = 3$$
$$2x - 4 = 6$$
$$2(3) - 4 \overset{?}{=} 6$$
$$6 - 4 \overset{?}{=} 6$$
$$2 = 6 \quad \text{False}$$

Since we obtained a false statement, 3 is not a solution.

Now Try Exercise 11

Now check to see if 5 is a solution to the equation in Example 1. Your check should show that 5 is a solution.

---

### Understanding Algebra

An *equation* is a statement that two expressions are equal.

$$2x - 3 = 7$$
$$y - 4 = 12 \text{ and}$$
$$6t + 14 = 2.3$$

are all *linear equations in one variable* in which the variable is $x$, $y$, and $t$, respectively.

### Understanding Algebra

A *solution* to an equation is a number that makes the equation a true statement.

### Understanding Algebra

We use the symbol $\overset{?}{=}$ in the process of checking a solution — we are questioning whether a statement is true.

**EXAMPLE 2** Determine whether 18 is a solution to the following equation.

$$3x - 2(x + 3) = 12$$

*Solution*    To determine whether 18 is a solution, we substitute 18 for each $x$ in the equation.

Check                                                 $x = 18$

$$3x - 2(x + 3) = 12$$
$$3(18) - 2(18 + 3) \stackrel{?}{=} 12$$
$$3(18) - 2(21) \stackrel{?}{=} 12$$
$$54 - 42 \stackrel{?}{=} 12$$
$$12 = 12 \quad \text{True}$$

Since we obtained a true statement, 18 is a solution.

Now Try Exercise 15

**EXAMPLE 3** Determine whether $-\dfrac{3}{2}$ is a solution to the following equation.

$$3(n + 3) = 6 + n$$

*Solution*    Substitute $-\dfrac{3}{2}$ for each $n$ in the equation.

Check                                                 $n = -\dfrac{3}{2}$

$$3(n + 3) = 6 + n$$
$$3\left(-\frac{3}{2} + 3\right) \stackrel{?}{=} 6 + \left(-\frac{3}{2}\right)$$
$$3\left(-\frac{3}{2} + \frac{6}{2}\right) \stackrel{?}{=} \frac{12}{2} - \frac{3}{2}$$
$$3\left(\frac{3}{2}\right) \stackrel{?}{=} \frac{9}{2}$$
$$\frac{9}{2} = \frac{9}{2} \qquad \text{True}$$

Thus, $-\dfrac{3}{2}$ is a solution.

Now Try Exercise 21

## 3 Identify Equivalent Equations

Left side of equation = Right side of equation

**FIGURE 2.1**

### Understanding Algebra

The strategy for solving a linear equation is to *isolate* the variable on one side of the = sign.

To **solve an equation** means to find its solution. *To solve an equation, it is necessary to get the variable alone on one side of the equals sign. When we get an equation in this form, we say that we isolated the variable.* To isolate the variable we make use of two properties: the addition and multiplication properties of equality. Look first at **Figure 2.1.**

Think of an equation as a balanced statement whose left side is balanced by its right side. When solving an equation, we must make sure that the equation remains balanced at all times. *We ensure that an equation always remains balanced by doing the same thing to both sides of the equation.* For example, if we add a number to the left side of the equation, we must add exactly the same number to the right side. If we multiply the right side of the equation by some number, we must multiply the left side by the same number.

When we add the same number to both sides of an equation or multiply both sides of an equation by the same nonzero number, we do not change the solution to the

## Understanding Algebra

Think of an equation as a balance scale. When a solution checks, the scale balances:

But when a solution doesn't check, the sides are not in balance:

equation. Two or more equations with the same solution are called **equivalent equations.** The equations $2x - 4 = 2$, $2x = 6$, and $x = 3$ are equivalent, since the solution to each is 3.

Check    $x = 3$

$$2x - 4 = 2 \qquad\qquad 2x = 6 \qquad\qquad x = 3$$
$$2(3) - 4 \overset{?}{=} 2 \qquad\qquad 2(3) \overset{?}{=} 6 \qquad\qquad 3 = 3 \quad \text{True}$$
$$6 - 4 \overset{?}{=} 2 \qquad\qquad 6 = 6 \quad \text{True}$$
$$2 = 2 \quad \text{True}$$

*When solving an equation, we use the addition and multiplication properties to express a given equation as equivalent equations until we obtain the solution.*

### 4  Use the Addition Property to Solve Equations

Now we are ready to define the **addition property of equality.**

---

**Addition Property of Equality**

If $a = b$ then $a + c = b + c$ for any real numbers $a$, $b$, and $c$.

---

This property means that the same number can be added to both sides of an equation.

To isolate the variable when solving equations of the form $x + a = b$, *we add* $(-a)$ *to both sides of the equals sign.* This isolates the variable $x$ on the left side of the equation.

## Understanding Algebra

Since we can think of an equation as a scale we are trying to keep in balance, adding the same quantity to both sides preserves the balance.

Now, add 4 to both sides:
$x - 4 + 4 = -3 + 4$ or
simply $x = 1$.

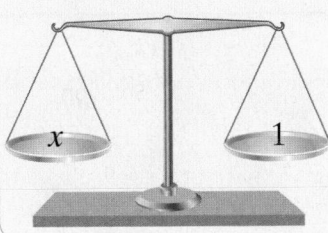

| Equation | To Solve, Use the Addition Property to Eliminate the Number . . . | . . . By Adding the Opposite to Both Sides |
|---|---|---|
| $x - 4 = -3$ | $-4$ | $4$ |
| $x + 5 = 9$ | $5$ | $-5$ |
| $-3 = k + 7$ | $7$ | $-7$ |
| $-5 = x - 4$ | $-4$ | $4$ |
| $-6.25 = y + 12.78$ | $12.78$ | $-12.78$ |

Now let's work some examples.

---

EXAMPLE 4  Solve the equation $x - 4 = -3$.

*Solution*    To isolate the variable, $x$, we must eliminate the $-4$ from the left side of the equation. To do this we add 4, the opposite of $-4$, to *both sides* of the equation.

$$x - 4 = -3$$
$$x - 4 + 4 = -3 + 4 \qquad \text{Add 4 to both sides.}$$
$$x + 0 = 1$$
$$x = 1$$

Note how the process helps to isolate $x$.

Check
$$x - 4 = -3$$
$$1 - 4 \overset{?}{=} -3$$
$$-3 = -3 \quad \text{True}$$

Now Try Exercise 27

---

Space limitations prevent us from showing all checks. However, *you should check all of your answers.*

**EXAMPLE 5** Solve the equation $x + 5 = 9$.

*Solution*    To solve this equation, we must eliminate the 5 from the left side of the equation. To do this, we add $-5$, the opposite of 5, to *both sides* of the equation.

$$x + 5 = 9$$
$$x + 5 + (\boxed{-5}) = 9 + (\boxed{-5}) \quad \text{Add } -5 \text{ to both sides.}$$
$$x + 0 = 4$$
$$x = 4$$

Now Try Exercise 29

### Understanding Algebra

"Subtract 5 from both sides" is equivalent to saying "Add $(-5)$ to both sides."

In Example 5, we added $-5$ to both sides of the equation. We know that $5 + (-5) = 5 - 5$. Thus, adding a negative 5 to both sides of the equation is equivalent to subtracting a 5 from both sides of the equation. *Since subtraction is defined in terms of addition, the addition property also allows us to subtract the same number from both sides of the equation.* Thus, Example 5 could have also been worked as follows:

$$x + 5 = 9$$
$$x + 5 \boxed{- 5} = 9 \boxed{- 5} \quad \text{Subtract 5 from both sides.}$$
$$x + 0 = 4$$
$$x = 4$$

Rather than adding a negative number to both sides of the equation, we will subtract a number from both sides of the equation.

**EXAMPLE 6** Solve the equation $-3 = k + 7$.

*Solution*    We must isolate the variable, $k$, which is on the right side of the equals sign.

$$-3 = k + 7$$
$$-3 \boxed{- 7} = k + 7 \boxed{- 7} \quad \text{Subtract 7 from both sides.}$$
$$-10 = k + 0$$
$$-10 = k$$

Check
$$-3 = k + 7$$
$$-3 \overset{?}{=} -10 + 7$$
$$-3 = -3 \qquad \text{True}$$

Now Try Exercise 25

### Helpful Hint

Remember that our goal in solving an equation is to get the variable alone on one side of the equation. To do this, we add or subtract *the number on the same side of the equation as the variable* to or from both sides of the equation.

| Equation | Must Eliminate | Number to Add (or Subtract) to (or from) Both Sides of the Equation | Correct Results | Solution |
|---|---|---|---|---|
| $x - 5 = 8$ | $-5$ | add 5 | $x - 5 + 5 = 8 + 5$ | $x = 13$ |
| $x - 3 = -12$ | $-3$ | add 3 | $x - 3 + 3 = -12 + 3$ | $x = -9$ |
| $2 = x - 7$ | $-7$ | add 7 | $2 + 7 = x - 7 + 7$ | $9 = x$ or $x = 9$ |
| $x + 12 = -5$ | $+12$ | subtract 12 | $x + 12 - 12 = -5 - 12$ | $x = -17$ |
| $6 = x + 4$ | $+4$ | subtract 4 | $6 - 4 = x + 4 - 4$ | $2 = x$ or $x = 2$ |
| $13 = x + 9$ | $+9$ | subtract 9 | $13 - 9 = x + 9 - 9$ | $4 = x$ or $x = 4$ |

Notice that under the *Correct Results* column, when the equation is simplified by combining terms, the $x$ will become isolated because the sum of a number and its opposite is 0, and $x + 0$ equals $x$.

**EXAMPLE 7** Solve the equation $-5 = x - 4$.

*Solution*    The variable, $x$, is on the right side of the equation. By adding 4 to both sides of the equation, we eliminate $-4$ from the right and $x$ is isolated on the right side:

$$-5 = x - 4$$
$$-5 + 4 = x - 4 + 4 \quad \text{Add 4 to both sides.}$$
$$-1 = x + 0$$
$$-1 = x$$

Thus, the solution is $-1$.

Now Try Exercise 35

**EXAMPLE 8** Solve the equation $-6.25 = y + 12.78$.

*Solution*    The variable, $y$, is on the right side of the equation. To isolate the variable, subtract 12.78 from both sides of the equation.

$$-6.25 = y + 12.78$$
$$-6.25 - 12.78 = y + 12.78 - 12.78 \quad \text{Subtract 12.78 from both sides.}$$
$$-19.03 = y + 0$$
$$-19.03 = y$$

Thus, the solution is $-19.03$.

Now Try Exercise 65

**Avoiding Common Errors**

When solving an equation, our goal is to isolate the variable on one side of the equals sign. Consider the equation $x + 3 = -4$. How do we solve it?

CORRECT

Remove the 3 from the left side of the equation.
$$x + 3 = -4$$
$$x + 3 - 3 = -4 - 3$$
$$x = -7$$
Variable is now isolated.

INCORRECT

Remove the $-4$ from the right side of the equation.
$$x + 3 = -4$$
$$x + 3 + 4 = -4 + 4$$
$$x + 7 = 0$$
Variable is not isolated.

Remember, use the addition property to *remove the number that is on the same side of the equals sign as the variable.*

## 5  Solve Equations by Doing Some Steps Mentally

Consider the following two problems.

**a)**
$$x - 5 = 12$$
$$x - 5 + 5 = 12 + 5$$
$$x + 0 = 12 + 5$$
$$x = 17$$

**b)**
$$15 = x + 3$$
$$15 - 3 = x + 3 - 3$$
$$15 - 3 = x + 0$$
$$12 = x$$

When you feel comfortable using the addition property of equality, you may wish to do some of the steps mentally to reduce some of the written work. For example, the preceding two problems may be shortened as follows:

**Shortened Form**

a)
$$x - 5 = 12$$
$$x - 5 + 5 = 12 + 5 \quad \longleftarrow \boxed{\text{Do this step mentally.}}$$
$$x = 12 + 5$$
$$x = 17$$

Shortened Form
$$x - 5 = 12$$
$$x = 12 + 5$$
$$x = 17$$

b)
$$15 = x + 3$$
$$15 - 3 = x + 3 - 3 \quad \longleftarrow \boxed{\text{Do this step mentally.}}$$
$$15 - 3 = x$$
$$12 = x$$

Shortened Form
$$15 = x + 3$$
$$15 - 3 = x$$
$$12 = x$$

# EXERCISE SET 2.2    MathXL®   MyMathLab®

## Warm-Up Exercises

*Fill in the blanks with the appropriate word, phrase, or symbol(s) from the following list.*

| equation | solutions | True | linear | 2 | opposite |
| 6 | isolate | equivalent | False | check | |

1. To solve the equation $6 = x - 2$, first add _____ to both sides of the equation.

2. Numbers which, when substituted for the variable, make the equation a true statement are called _____.

3. The strategy for solving a linear equation in one variable includes a process to _____ the variable.

4. The symbol $\stackrel{?}{=}$ is used to _____ the solution in the given equation.

5. The equation $x - 3 = -19$ is an example of a _____ equation in the variable $x$.

6. Two equations with the same solutions are called _____ equations.

7. A statement of equality between two expressions, like $x + 7 = 19$, is called an _____.

8. To solve the equation $x - 6 = -2$, first add _____ to both sides of the equation.

9. When solving the equation $x + a = b$, the _____ of $a$ is added to both sides.

10. (True or False) The equations $x + 19 = -14$ and $x + 60 = 27$ are equivalent equations. _____

## Practice the Skills

11. Is $x = 2$ a solution of $4x - 3 = 5$?

12. Is $x = -6$ a solution of $2x + 1 = x - 5$?

13. Is $x = -3$ a solution of $2x - 5 = 5(x + 2)$?

14. Is $x = 1$ a solution of $2(x - 3) = -3(x + 1)$?

15. Is $p = -15$ a solution of $2p - 5(p + 7) = 10$?

16. Is $k = -2$ a solution of $5k - 6(k - 1) = 8$?

17. Is $x = 3.4$ a solution of $3(x + 2) - 3(x - 1) = 9$?

18. Is $x = \frac{3}{4}$ a solution of $x + 5 = 5x + 2$?

19. Is $x = \frac{1}{2}$ a solution of $4x - 4 = 2x - 2$?

20. Is $x = \frac{1}{3}$ a solution of $7x + 3 = 2x + 5$?

21. Is $x = \frac{11}{2}$ a solution of $3(x + 2) = 5(x - 1)$?

22. Is $h = 3$ a solution of $-(h - 5) - (h - 6) = 3h - 4$?

*Solve each equation and check your solution.*

23. $x + 2 = 7$
24. $x - 4 = 13$
25. $-6 = x + 1$
26. $-5 = x + 4$
27. $x - 4 = -8$
28. $x - 16 = -12$
29. $t + 9 = 52$
30. $x + 8 = 17$
31. $-6 + w = 9$
32. $3 = 7 + t$
33. $27 = x + 16$
34. $50 = x - 35$
35. $-18 = x - 14$
36. $-4 = w + 5$
37. $9 + x = 4$
38. $x + 29 = -29$
39. $4 + x = -8$
40. $9 = x - 3$
41. $7 + r = -23$
42. $a - 5 = -9$
43. $8 = 8 + v$
44. $9 + x = 12$
45. $7 + x = -50$
46. $-19 = 8 + t$
47. $12 = 16 + x$
48. $62 = z - 15$
49. $15 + y = -50$
50. $-20 = 4 + x$

**51.** $-15 + x = -15$    **52.** $8 = 8 + x$    **53.** $5 = x - 12$    **54.** $-12 = 20 + c$

**55.** $-50 = x - 24$    **56.** $-29 + x = -15$    **57.** $43 = 15 + p$    **58.** $-25 = 74 + x$

**59.** $40.2 + d = -5.9$    **60.** $-27.23 + x = 9.77$    **61.** $-37 + x = 9.5$    **62.** $7.2 + x = 7.2$

**63.** $x - 8.77 = -17$    **64.** $6.1 + x = 10.2$    **65.** $9.32 = x + 3.75$    **66.** $-5.62 = y + 11.39$

## Problem Solving

**67.** Do you think the equation $x + 1 = x + 2$ has a real number as a solution? Explain. (We will discuss equations like this in Section 2.5.)

**68.** Do you think the equation $x + 4 = x + 4$ has more than one real number as a solution? If so, how many solutions does it have? Explain. (We will discuss equations like this in Section 2.5.)

## Concept/Writing Exercises

**69.** To solve a linear equation, we "isolate the variable." Explain what this means.

**70.** Explain why these three equations are equivalent:
$$2x + 3 = 5, \qquad 2x = 2, \qquad x = 1$$

## Challenge Problems

*We can solve equations that contain unknown symbols. Solve each equation for the symbol indicated by adding or subtracting a symbol to or from both sides of the equation. Explain each answer. (Remember that to solve the equation you want to isolate the symbol you are solving for on one side of the equation.)*

**71.** $x - \triangle = \square$, for $x$

**72.** $\square + \odot = \triangle$, for $\odot$

**73.** $\odot = \square + \triangle$, for $\square$

**74.** $\square = \triangle + \odot$, for $\odot$

## Group Activity

*Discuss and answer Exercise 75 as a group.*

**75.** Consider the equation $2(x + 3) = 2x + 6$.

**a)** Group member 1: Determine whether 4 is a solution to the equation.

**b)** Group member 2: Determine whether $-2$ is a solution to the equation.

**c)** Group member 3: Determine whether 0.3 is a solution to the equation.

**d)** Each group member: Select a number not used in parts **a)–c)** and determine whether that number is a solution to the equation.

**e)** As a group, write what you think is the solution to the equation $2(x + 3) = 2x + 6$ and write a paragraph explaining your answer.

## Cumulative Review Exercises

[1.6]    *Add.*

**76.** $-\dfrac{7}{15} + \dfrac{5}{6}$

**77.** $-\dfrac{11}{12} + \left(-\dfrac{3}{8}\right)$

[2.1]    *Simplify.*

**78.** $4x + 3(x - 2) - 5x - 7$

**79.** $-(2t + 4) + 3(4t - 7) - 3t$

# 2.3 The Multiplication Property of Equality

1 Identify reciprocals.

2 Use the multiplication property to solve equations.

3 Solve equations of the form $-x = a$.

4 Do some steps mentally when solving equations.

## 1 Identify Reciprocals

Recall that two numbers are **reciprocals** of each other when their product is 1. Some examples of numbers and their reciprocals follow.

| Number | Reciprocal | Product |
|--------|-----------|---------|
| 2 | $\dfrac{1}{2}$ | $(2)\left(\dfrac{1}{2}\right) = 1$ |
| $-\dfrac{3}{5}$ | $-\dfrac{5}{3}$ | $\left(-\dfrac{3}{5}\right)\left(-\dfrac{5}{3}\right) = 1$ |
| $-1$ | $-1$ | $(-1)(-1) = 1$ |

The reciprocal of a positive number is a positive number and the reciprocal of a negative number is a negative number. Note that 0 has no reciprocal.

## 2 Use the Multiplication Property to Solve Equations

In Section 2.2, we used the addition property of equality to solve equations of the form $x + a = b$, where $a$ and $b$ represent real numbers. In this section, we use the multiplication property of equality to solve equations of the form $ax = b$, where $a$ and $b$ represent real numbers.

> **Helpful Hint**
>
> It is important that you recognize the difference between equations like $x + 2 = 8$ and $2x = 8$. In $x + 2 = 8$, the 2 is a *term* that is being added to $x$, so we use the addition property to solve the equation. In $2x = 8$, the 2 is a *factor* of $2x$. The 2 is the coefficient multiplying the $x$, so we use the multiplication property to solve the equation.

> **Multiplication Property of Equality**
>
> If $a = b$, then $a \cdot c = b \cdot c$ for any real numbers $a$, $b$, and $c$.

This property means that both sides of an equation can be multiplied by the same nonzero number. *The multiplication property can be used to solve equations of the form $ax = b$.* We can isolate the variable in equations of this form by multiplying both sides of the equation by the reciprocal of $a$, which is $\dfrac{1}{a}$. Doing so makes the numerical coefficient of the variable, $x$, become 1.

| Equation | To Solve, Use the Multiplication Property to Change the Term . . . | . . . By Multiplying Both Sides by the Reciprocal |
|---|---|---|
| $4x = 9$ | $4x$ to $1x$ | $\dfrac{1}{4}$ |
| $-5x = 20$ | $-5x$ to $1x$ | $-\dfrac{1}{5}$ |
| $15 = \dfrac{1}{2}x$ | $\dfrac{1}{2}x$ to $1x$ | $2$ |
| $7 = -\dfrac{3}{5}x$ | $-\dfrac{3}{5}x$ to $1x$ | $-\dfrac{5}{3}$ |

Now let's work some examples.

**EXAMPLE 1** Solve the equation $9x = 63$.

**Solution** To isolate the variable, $x$, we must change the $9x$ on the left side of the equals sign to $1x$. To do this, we multiply both sides of the equation by the reciprocal of 9, which is $\dfrac{1}{9}$.

$$9x = 63$$

$$\dfrac{1}{9} \cdot 9x = \dfrac{1}{9} \cdot 63 \qquad \text{Multiply both sides by } \dfrac{1}{9}.$$

$$\dfrac{1}{\underset{1}{\cancel{9}}} \cdot \overset{1}{\cancel{9}}x = \dfrac{1}{\underset{1}{\cancel{9}}} \cdot \overset{7}{\cancel{63}} \qquad \text{Divide out the common factors.}$$

$$1x = 7$$

$$x = 7$$

Now Try Exercise 9

Notice in Example 1 that $1x$ is replaced by $x$ in the last step. Usually we do this step mentally.

**EXAMPLE 2** Solve the equation $\dfrac{x}{2} = 4$.

*Solution* Since dividing by 2 is the same as multiplying by $\dfrac{1}{2}$, the equation $\dfrac{x}{2} = 4$ is the same as $\dfrac{1}{2}x = 4$. We will therefore multiply both sides of the equation by the reciprocal of $\dfrac{1}{2}$, which is 2.

$$\frac{x}{2} = 4$$

$$\overset{1}{\cancel{2}}\left(\frac{x}{\underset{1}{\cancel{2}}}\right) = 2 \cdot 4 \qquad \text{Multiply both sides by 2.}$$

$$x = 2 \cdot 4$$

$$x = 8$$

Now Try Exercise 11

Now Try Exercise 11

> **Understanding Algebra**
>
> $\dfrac{x}{2}$ is the same as $\dfrac{1}{2}x$.

**EXAMPLE 3** Solve the equation $\dfrac{2}{3}x = 6$.

*Solution* The reciprocal of $\dfrac{2}{3}$ is $\dfrac{3}{2}$. We multiply both sides of the equation by $\dfrac{3}{2}$.

$$\frac{2}{3}x = 6$$

$$\frac{3}{2} \cdot \frac{2}{3}x = \frac{3}{2} \cdot 6 \qquad \text{Multiply both sides by } \frac{3}{2}.$$

$$1x = 9$$

$$x = 9$$

We will show a check of this solution.

Check

$$\frac{2}{3}x = 6$$

$$\frac{2}{3}(9) \overset{?}{=} 6$$

$$6 = 6 \qquad \text{True}$$

Now Try Exercise 49

> **Understanding Algebra**
>
> *Strategy for Equation-Solving:*
>
> 1. Isolate the variable on one side of the equation.
> 2. Be sure the coefficient of the variable is a "1" by multiplying both sides of the equation by the reciprocal of the coefficient of the variable.

In Example 1, we multiplied both sides of the equation $9x = 63$ by $\dfrac{1}{9}$ to isolate the variable. We could have also isolated the variable by dividing both sides of the equation by 9, as follows:

$$9x = 63$$

$$\frac{\overset{1}{\cancel{9}}x}{\underset{1}{\cancel{9}}} = \frac{\overset{7}{\cancel{63}}}{\underset{1}{\cancel{9}}} \qquad \text{Divide both sides by 9.}$$

$$x = 7$$

We can do this because dividing by 9 is equivalent to multiplying by $\dfrac{1}{9}$. *Since division can be defined in terms of multiplication $\left(\dfrac{a}{b} \text{ means } a \cdot \dfrac{1}{b}\right)$, the multiplication property also allows us to divide both sides of an equation by the same nonzero number.*

**EXAMPLE 4** Solve the equation $8w = 3$.

*Solution*   To solve the equation we divide both sides of the equation by 8.

$$8w = 3$$

$$\frac{8w}{8} = \frac{3}{8} \qquad \text{Divide both sides by 8.}$$

$$w = \frac{3}{8}$$

Now Try Exercise 33

**EXAMPLE 5** Solve the equation $-15 = -3z$.

*Solution*   To isolate $z$, we divide both sides of the equation by $-3$.

$$-15 = -3z$$

$$\frac{-15}{-3} = \frac{-3z}{-3} \qquad \text{Divide both sides by } -3.$$

$$5 = z$$

Now Try Exercise 21

**EXAMPLE 6** Solve the equation $0.24x = 1.20$.

*Solution*   Divide both sides of the equation by 0.24 to isolate the variable $x$.

$$0.24x = 1.20$$

$$\frac{0.24x}{0.24} = \frac{1.20}{0.24} \qquad \text{Divide both sides by 0.24.}$$

$$1x = 5$$

$$x = 5$$

Now Try Exercise 35

**Helpful Hint**

When solving an equation of the form $ax = b$, we can isolate the variable by

1. multiplying both sides of the equation by the reciprocal of $a$, $\frac{1}{a}$, as was done in Examples 1, 2, and 3, or

2. dividing both sides of the equation by $a$, as was done in Examples 4, 5, and 6.

Either method may be used to isolate the variable. However, if the equation contains a fraction, or fractions, you will arrive at a solution more quickly by multiplying by the reciprocal of $a$. This is illustrated in Examples 7 and 8.

**EXAMPLE 7** Solve the equation $-2x = \frac{3}{5}$.

*Solution*   Since this equation contains a fraction, we will isolate the variable by multiplying both sides of the equation by $-\frac{1}{2}$, which is the reciprocal of $-2$.

$$-2x = \frac{3}{5}$$

$$\left(-\frac{1}{2}\right)(-2x) = \left(-\frac{1}{2}\right)\left(\frac{3}{5}\right) \qquad \text{Multiply both sides by } -\frac{1}{2}.$$

$$1x = \left(-\frac{1}{2}\right)\left(\frac{3}{5}\right)$$

$$x = -\frac{3}{10}$$

Now Try Exercise 39

**EXAMPLE 8** Solve the equation $-6 = -\dfrac{3}{5}x$.

*Solution*   Since this equation contains a fraction, we will isolate the variable by multiplying both sides of the equation by the reciprocal of $-\dfrac{3}{5}$, which is $-\dfrac{5}{3}$.

$$-6 = -\frac{3}{5}x$$

$$\left(-\frac{5}{3}\right)(-6) = \left(-\frac{5}{3}\right)\left(-\frac{3}{5}x\right) \quad \text{Multiply both sides by } -\frac{5}{3}.$$

$$10 = 1x$$

$$10 = x$$

Now Try Exercise 57

In Example 8, the equation was written as $-6 = -\dfrac{3}{5}x$. This equation is equivalent to the equations $-6 = \dfrac{-3}{5}x$ and $-6 = \dfrac{3}{-5}x$. Can you explain why? All three equations have the same solution, 10.

## 3   Solve Equations of the Form $-x = a$

When solving an equation, we may obtain an equation like $-x = 7$. This is not a solution since $-x = 7$ means $-1x = 7$. The solution to an equation is of the form $x =$ some number. When an equation is of the form $-x = 7$, we can solve for $x$ by multiplying both sides of the equation by $-1$, as illustrated in the following example.

**EXAMPLE 9** Solve the equation $-x = 7$.

*Solution*   $-x = 7$ means that $-1x = 7$. We are solving for $x$, not $-x$. We can multiply both sides of the equation by $-1$ to isolate $x$ on the left side of the equation.

$$-x = 7$$

$$-1x = 7$$

$$(-1)(-1x) = (-1)(7) \quad \text{Multiply both sides by } -1.$$

$$1x = -7$$

$$x = -7$$

Check

$$-x = 7$$

$$-(-7) \overset{?}{=} 7$$

$$7 = 7 \quad \text{True}$$

Thus, the solution is $-7$.

Now Try Exercise 23

### Understanding Algebra

For any real number $a$, if $-x = a$, then $x = -a$. For example,

if $-x = 7$, then $x = -7$.

If $-x = -2$, then $x = 2$.

Example 9 may also be solved by dividing both sides of the equation by $-1$. Whenever we have the negative of a variable equal to a quantity, as in Example 9, we can solve for the variable by multiplying or dividing both sides of the equation by $-1$.

**EXAMPLE 10** Solve the equation $-x = -5$.

*Solution*

$$-x = -5$$

$$-1x = -5$$

$$(-1)(-1x) = (-1)(-5) \quad \text{Multiply both sides by } -1.$$

$$1x = 5$$

$$x = 5$$

Now Try Exercise 25

## 4 Do Some Steps Mentally When Solving Equations

When you feel comfortable using the multiplication property, you may wish to do some of the steps mentally to reduce some of the written work. Now we present two examples worked out in detail, along with their shortened form.

**EXAMPLE 11** Solve the equation $-3t = -21$.

*Solution*

$$-3t = -21$$
$$\frac{-3t}{-3} = \frac{-21}{-3} \quad \leftarrow \text{Do this step mentally.}$$
$$t = \frac{-21}{-3}$$
$$t = 7$$

SHORTENED FORM

$$-3t = -21$$
$$t = \frac{-21}{-3}$$
$$t = 7$$

Now Try Exercise 61

**EXAMPLE 12** Solve the equation $\frac{1}{5}x = 20$.

*Solution*

$$\frac{1}{5}x = 20$$
$$5\left(\frac{1}{5}x\right) = 5(20) \quad \leftarrow \text{Do this step mentally.}$$
$$x = 5(20)$$
$$x = 100$$

SHORTENED FORM

$$\frac{1}{5}x = 20$$
$$x = 5(20)$$
$$x = 100$$

Now Try Exercise 63

### Helpful Hint

The **addition property** is used to solve equations of the form $x + a = b$. The *addition property* is used when a number is *added to or subtracted from* a variable.

$$x + 3 = -6$$
$$x + 3 - 3 = -6 - 3$$
$$x = -9$$

$$x - 5 = -2$$
$$x - 5 + 5 = -2 + 5$$
$$x = 3$$

The **multiplication property** is used to solve equations of the form $ax = b$. It is used when a variable is *multiplied* or *divided* by a number.

$$3x = 6$$
$$\frac{3x}{3} = \frac{6}{3}$$
$$x = 2$$

$$\frac{x}{2} = 4$$
$$2\left(\frac{x}{2}\right) = 2(4)$$
$$x = 8$$

$$\frac{2}{5}x = 12$$
$$\left(\frac{5}{2}\right)\left(\frac{2}{5}x\right) = \left(\frac{5}{2}\right)(12)$$
$$x = 30$$

# EXERCISE SET 2.3

MathXL®    MyMathLab®

## Warm-Up Exercises

*Fill in the blanks with the appropriate word, phrase, or symbol(s) from the following list.*

| $\frac{x}{3}$ | $\frac{3}{x}$ | $\frac{1}{3}$ | $-1$ | check |
| 1 | multiplicative inverse | 3 | isolate | $-\frac{7}{x}$ |

**1.** The expression $\frac{1}{3}x$ can be rewritten as _____.

**2.** To solve the equation $3x = 5$, multiply both sides of the equation by _____.

**3.** To solve the equation $\frac{x}{3} = 7$, we multiply both sides of the equation by _____.

**4.** The strategy for solving a linear equation in one variable includes a process to _____ the variable.

**5.** The symbol $\overset{?}{=}$ is used to _____ the solution in the given equation.

**6.** Another way of saying "reciprocal" is _____.

**7.** The reciprocal of $\frac{x}{-7}$ is _____.

**8.** Two numbers equal to their own reciprocal are _____ and _____.

## Practice the Skills

*Solve each equation and check your solution.*

**9.** $5x = 20$

**10.** $5x = 50$

**11.** $\frac{x}{3} = 7$

**12.** $\frac{y}{5} = 3$

**13.** $-4x = 12$

**14.** $8 = 16y$

**15.** $\frac{x}{4} = -20$

**16.** $\frac{x}{3} = -3$

**17.** $\frac{x}{5} = 4$

**18.** $-7t = 49$

**19.** $-27n = 81$

**20.** $\frac{x}{8} = -3$

**21.** $-7 = 3r$

**22.** $16 = -4y$

**23.** $-x = 13$

**24.** $-x = 9$

**25.** $-x = -8$

**26.** $-x = -15$

**27.** $-\frac{w}{3} = -10$

**28.** $-4 = \frac{c}{7}$

**29.** $4 = -12x$

**30.** $12y = -15$

**31.** $-\frac{x}{3} = -2$

**32.** $-\frac{a}{8} = -7$

**33.** $43t = 26$

**34.** $-24x = -18$

**35.** $-4.2x = -8.4$

**36.** $-3.88 = 1.94y$

**37.** $3x = \frac{3}{5}$

**38.** $7x = -7$

**39.** $5x = -\frac{3}{8}$

**40.** $-2b = -\frac{4}{5}$

**41.** $16 = -\frac{x}{4}$

**42.** $\frac{c}{9} = 0$

**43.** $-\frac{b}{4} = -60$

**44.** $-x = -\frac{5}{9}$

**45.** $\frac{x}{5} = -9$

**46.** $-3r = 0$

**47.** $5 = \frac{x}{4}$

**48.** $-8 = \frac{x}{-5}$

**49.** $\frac{3}{5}d = -30$

**50.** $\frac{2}{7}x = 7$

**51.** $\frac{y}{-2} = 0$

**52.** $-6x = \frac{5}{2}$

**53.** $\frac{-7}{8}w = 0$

**54.** $-x = \frac{15}{8}$

**55.** $\frac{1}{5}x = 4.5$

**56.** $-\frac{1}{4}x = \frac{3}{4}$

**57.** $-4 = -\frac{2}{3}z$

**58.** $-9 = \frac{-5}{3}n$

**59.** $-1.4x = 28.28$

**60.** $-0.42x = -2.142$

*Solve each equation by doing some steps mentally. Check your solution.*

**61.** $-8x = -56$

**62.** $-9x = -45$

**63.** $\frac{2}{3}x = 6$

**64.** $\frac{1}{3}x = 15$

## Concept/Writing Exercises

**65. a)** Explain the difference between $5 + x = 10$ and $5x = 10$.
  **b)** Solve $5 + x = 10$.
  **c)** Solve $5x = 10$.

**66. a)** Explain the difference between $3 + x = 6$ and $3x = 6$.
  **b)** Solve $3 + x = 6$.
  **c)** Solve $3x = 6$.

**67.** Consider the equation $\frac{2}{3}x = 4$. This equation could be solved by multiplying both sides of the equation by $\frac{3}{2}$, the reciprocal of $\frac{2}{3}$, or by dividing both sides of the equation by $\frac{2}{3}$. Which method do you feel would be easier? Explain your answer. Find the solution to the equation.

**68.** Consider the equation $4x = \frac{3}{5}$. Would it be easier to solve this equation by dividing both sides of the equation by 4 or by multiplying both sides of the equation by $\frac{1}{4}$, the reciprocal of 4? Explain your answer. Find the solution to the problem.

**69.** Consider the equation $\frac{3}{7}x = \frac{4}{5}$. Would it be easier to solve this equation by dividing both sides of the equation by $\frac{3}{7}$ or by multiplying both sides of the equation by $\frac{7}{3}$, the reciprocal of $\frac{3}{7}$? Explain your answer. Find the solution to the equation.

## Challenge Problems

**70.** Consider the equation $\square \odot = \triangle$.
  a) To solve for $\odot$, what symbol do we need to isolate?
  b) How would you isolate the symbol you specified in part **a)**?
  c) Solve the equation for $\odot$.

**71.** Consider the equation $\ddot\smile = \triangle \square$.
  a) To solve for $\square$, what symbol do we need to isolate?
  b) How would you isolate the symbol you specified in part **a)**?
  c) Solve the equation for $\square$.

**72.** Consider the equation $\# = \dfrac{\ddot\smile}{\triangle}$.
  a) To solve for $\ddot\smile$, what symbol do we need to isolate?
  b) How would you isolate the symbol you specified in part **a)**?
  c) Solve the equation for $\ddot\smile$.

## Cumulative Review Exercises

[1.7]  **73.** Subtract $-4$ from $-8$.
[1.8]  **74.** Evaluate $(-3)(-2)(5)(-1)$.
[1.9]  **75.** Evaluate $4^2 - 2^3 \cdot 6 \div 3 + 6$.
[1.10] **76.** Name the property illustrated.
   $2 + (4 + y) = (2 + 4) + y$
[2.2]  **77.** Solve the equation $-48 = x + 9$.

# 2.4 Solving Linear Equations with a Variable on Only One Side of the Equation

1 Solve linear equations with a variable on only one side of the equals sign.

2 Solve equations containing decimal numbers or fractions.

## 1 Solve Linear Equations with a Variable on Only One Side of the Equals Sign

In this section, we will use *both* the addition and multiplication properties of equality to solve linear equations in which the variable appears on only one side of the equals sign.

No one method is the "best" to solve all linear equations. But the following general procedure can be used to solve linear equations when the variable appears on only one side of the equation.

### To Solve Linear Equations with a Variable on Only One Side of the Equals Sign

1. If the equation contains fractions, multiply *both* sides of the equation by the least common denominator (LCD). This will eliminate the fractions from the equation.
2. Use the distributive property to remove parentheses.
3. Combine like terms on the same side of the equals sign.
4. Use the addition property to obtain an equation with the term containing the variable on one side of the equals sign and a constant on the other side. This will result in an equation of the form $ax = b$.
5. Use the multiplication property to isolate the variable.
6. Check the solution in the original equation.

### Understanding Algebra

As we solve more complicated equations, these six steps will provide the strategy for *isolating the variable.*

When solving an equation you should always check your solution. To conserve space, we will not show all checks.

Consider the equation $2x + 4 = 10$, which contains no fractions, no parentheses, and no like terms on the same side of the equals sign. Therefore, we start with step 4.

$$2x + 4 = 10$$
$$2x + 4 - 4 = 10 - 4 \quad \text{Addition property (add } (-4) \text{ to both sides).}$$
$$2x = 6 \quad \text{x-term is now isolated on the left.}$$

Now we use the multiplication property (step 5) to solve for $x$.

$$2x = 6$$

$$\frac{2x}{2} = \frac{6}{2} \quad \text{Multiplication property (divide both sides by 2).}$$

$$1x = 3 \quad \text{We need not write the coefficient ``1.''}$$

$$x = 3 \quad \text{Final answer}$$

The solution to the equation $2x + 4 = 10$ is 3. Check this solution:

$$2x + 4 = 10$$

$$2(3) + 4 = 10$$

$$6 + 4 = 10$$

$$10 = 10 \quad \text{True}$$

**EXAMPLE 1** Solve the equation $5x - 7 = 13$. —————————

*Solution*   Since the equation contains no fractions nor parentheses, and since there are no like terms to be combined, we start with step 4.

Step 4
$$5x - 7 = 13$$
$$5x - 7 + 7 = 13 + 7 \quad \text{Add 7 to both sides.}$$
$$5x = 20$$

Step 5
$$\frac{5x}{5} = \frac{20}{5} \quad \text{Divide both sides by 5.}$$
$$x = 4$$

Step 6   Check
$$5x - 7 = 13$$
$$5(4) - 7 \stackrel{?}{=} 13$$
$$20 - 7 \stackrel{?}{=} 13$$
$$13 = 13 \quad \text{True}$$

Since the check is true, the solution is 4. Note that after completing step 4, we obtain $5x = 20$, which is an equation of the form $ax = b$. After completing step 5, we obtain the answer in the form $x = $ some real number.

Now Try Exercise 7

**EXAMPLE 2** Solve the equation $-2r - 6 = -3$. —————————

*Solution*

Step 4
$$-2r - 6 = -3$$
$$-2r - 6 + 6 = -3 + 6 \quad \text{Add 6 to both sides.}$$
$$-2r = 3$$

Step 5
$$\frac{-2r}{-2} = \frac{3}{-2} \quad \text{Divide both sides by } -2.$$
$$r = -\frac{3}{2}$$

Step 6   Check
$$-2r - 6 = -3$$
$$-2\left(-\frac{3}{2}\right) - 6 \stackrel{?}{=} -3$$
$$3 - 6 \stackrel{?}{=} -3$$
$$-3 = -3 \quad \text{True}$$

The solution is $-\dfrac{3}{2}$.

Now Try Exercise 15

Note that checks are always made with the *original* equation. In some of the following examples, the check will be omitted to save space. You should check all of your answers.

**EXAMPLE 3** Solve the equation $16 = 4x + 6 - 2x$.

*Solution*   Again we must isolate the variable, $x$. Since the right side of the equation has two like terms containing the variable, $x$, we will first combine these like terms.

| Step 3 | $16 = 4x + 6 - 2x$ | |
| | $16 = 2x + 6$ | Like terms were combined. |
| Step 4 | $16 - 6 = 2x + 6 - 6$ | Subtract 6 from both sides. |
| | $10 = 2x$ | |
| Step 5 | $\dfrac{10}{2} = \dfrac{2x}{2}$ | Divide both sides by 2. |
| | $5 = x$ | |

*Now Try Exercise 29*

The preceding solution can be condensed as follows.

$$16 = 4x + 6 - 2x$$
$$16 = 2x + 6 \quad \text{Like terms were combined.}$$
$$10 = 2x \quad \text{6 was subtracted from both sides.}$$
$$5 = x \quad \text{Both sides were divided by 2.}$$

**EXAMPLE 4** Solve the equation $5x - 2(x + 4) = 3$.

*Solution*

| Step 2 | $5x - 2(x + 4) = 3$ | |
| | $5x - 2x - 8 = 3$ | Distributive property was used. |
| Step 3 | $3x - 8 = 3$ | Like terms were combined. |
| Step 4 | $3x - 8 + 8 = 3 + 8$ | Add 8 to both sides. |
| | $3x = 11$ | |
| Step 5 | $\dfrac{3x}{3} = \dfrac{11}{3}$ | Divide both sides by 3. |
| | $x = \dfrac{11}{3}$ | |

*Now Try Exercise 57*

The solution to Example 4 can be condensed as follows:

$$5x - 2(x + 4) = 3$$
$$5x - 2x - 8 = 3 \quad \text{Distributive property was used.}$$
$$3x - 8 = 3 \quad \text{Like terms were combined.}$$
$$3x = 11 \quad \text{8 was added to both sides.}$$
$$x = \dfrac{11}{3} \quad \text{Both sides were divided by 3.}$$

**EXAMPLE 5** Solve the equation $3p - (2p + 5) = 7$.

*Solution*
$$3p - (2p + 5) = 7$$
$$3p - 2p - 5 = 7 \quad \text{Distributive property was used.}$$
$$p - 5 = 7 \quad \text{Like terms were combined.}$$
$$p = 12 \quad \text{5 was added to both sides.}$$

*Now Try Exercise 61*

## 2 Solve Equations Containing Decimal Numbers or Fractions

Example 6 illustrates two methods to solve an equation that contains decimal numbers.

**EXAMPLE 6** Solve the equation $x + 1.24 - 0.07x = 4.96$.

*Solution* We will work this example using two methods. In method 1, we work with decimal numbers throughout the solving process. In method 2, we multiply both sides of the equation by a power of 10 to change the decimal numbers to integers.

Method 1

$$x + 1.24 - 0.07x = 4.96$$

$$0.93x + 1.24 = 4.96 \qquad \text{Like terms were combined;} \\ 1x - 0.07x = 0.93x.$$

$$0.93x + 1.24 - 1.24 = 4.96 - 1.24 \qquad \text{Subtract 1.24 from both sides.}$$

$$0.93x = 3.72$$

$$\frac{0.93x}{0.93} = \frac{3.72}{0.93} \qquad \text{Divide both sides by 0.93.}$$

$$x = 4$$

Method 2 We can eliminate the decimal numbers from the equation by multiplying both sides of the equation by 10 if the decimal numbers are given in tenths, by 100 if the decimal numbers are given in hundredths, and so on. In Example 6, since the decimal numbers are in hundredths, you can eliminate the decimals from the equation by multiplying both sides of the equation by 100. This alternate method would give the following.

$$x + 1.24 - 0.07x = 4.96$$

$$100(x + 1.24 - 0.07x) = 100(4.96) \qquad \text{Multiply both sides of equation by 100.}$$

$$100(x) + 100(1.24) - 100(0.07x) = 496 \qquad \text{Distributive property}$$

$$100x + 124 - 7x = 496$$

$$93x + 124 = 496 \qquad \text{Like terms were combined.}$$

$$93x = 372 \qquad \text{124 was subtracted from both sides.}$$

$$x = 4 \qquad \text{Both sides were divided by 93.}$$

Study both methods provided to see which method you prefer.

Now Try Exercise 33

Often, the first step in solving equations containing fractions is to multiply both sides of the equation by the LCD to eliminate the fractions from the equations. Examples 7–9 illustrate this procedure.

**EXAMPLE 7** Solve $\frac{1}{5}(x + 1) = 1$.

*Solution* The LCD of the fraction is 5. We will begin by multiplying both sides of the equation by the LCD. This step will eliminate fractions from the equation.

$$\frac{1}{5}(x + 1) = 1$$

Step 1 
$$5\left[\frac{1}{5}(x + 1)\right] = 5 \cdot 1 \qquad \text{Multiply both sides by the LCD, 5.}$$

$$5\left(\frac{1}{5}\right)(x + 1) = 5$$

$$x + 1 = 5$$

Step 4 
$$x = 4 \qquad \text{1 was subtracted from both sides.}$$

Step 6   Check

$$\frac{1}{5}(x + 1) = 1$$

$$\frac{1}{5}(4 + 1) \stackrel{?}{=} 1$$

$$\frac{1}{5}(5) \stackrel{?}{=} 1$$

$$1 = 1 \quad \text{True}$$

The solution is 4.

Now Try Exercise 37

Example 7 could also be written as $\dfrac{x + 1}{5} = 1$. To solve this equation, we would begin by multiplying both sides of the equation by the LCD, 5, as follows.

$$\frac{x + 1}{5} = 1$$

$$5\left(\frac{x + 1}{5}\right) = 5 \cdot 1$$

$$x + 1 = 5$$

$$x = 4$$

**EXAMPLE 8**  Solve the equation $\dfrac{d}{2} + 3d = 14$.

*Solution*   Multiply both sides of the equation by the LCD, 2. This step will eliminate fractions from the equation.

Step 1   $$2\left(\frac{d}{2} + 3d\right) = 2 \cdot 14 \quad \text{Multiply both sides by the LCD, 2.}$$

Step 2   $$2\left(\frac{d}{2}\right) + 2 \cdot 3d = 2 \cdot 14 \quad \text{Distributive property}$$

$$d + 6d = 28$$

Step 3   $$7d = 28 \quad \text{Like terms were combined.}$$

Step 5   $$d = 4 \quad \text{Both sides were divided by 7.}$$

Step 6   Check   $$\frac{d}{2} + 3d = 14$$

$$\frac{4}{2} + 3(4) \stackrel{?}{=} 14$$

$$2 + 12 \stackrel{?}{=} 14$$

$$14 = 14 \quad \text{True}$$

Now Try Exercise 81

**EXAMPLE 9**  Solve the equation $\dfrac{1}{5}x - \dfrac{3}{8}x = \dfrac{1}{10}$.

*Solution*   The LCD of 5, 8, and 10 is 40. Multiply both sides of the equation by 40 to eliminate fractions from the equation.

$$\frac{1}{5}x - \frac{3}{8}x = \frac{1}{10}$$

Step 1   $$40\left(\frac{1}{5}x - \frac{3}{8}x\right) = 40\left(\frac{1}{10}\right) \quad \text{Multiply both sides by the LCD, 40.}$$

Step 2   $$40\left(\frac{1}{5}x\right) - 40\left(\frac{3}{8}x\right) = 40\left(\frac{1}{10}\right) \quad \text{Distributive property}$$

$$8x - 15x = 4$$

Step 3                            $-7x = 4$          Like terms were combined.

Step 5                            $x = -\dfrac{4}{7}$          Both sides were divided by $-7$.

Step 6   Check                   $\dfrac{1}{5}x - \dfrac{3}{8}x = \dfrac{1}{10}$

$\dfrac{1}{5}\left(-\dfrac{4}{7}\right) - \dfrac{3}{8}\left(-\dfrac{4}{7}\right) \overset{?}{=} \dfrac{1}{10}$          Substitute $-\dfrac{4}{7}$ for each $x$.

$-\dfrac{4}{35} + \dfrac{3}{14} \overset{?}{=} \dfrac{1}{10}$          Divided out common factors, then multiplied fractions.

$-\dfrac{8}{70} + \dfrac{15}{70} \overset{?}{=} \dfrac{7}{70}$          Wrote each fraction with the LCD, 70.

$\dfrac{7}{70} = \dfrac{7}{70}$          True

Now Try Exercise 91

---

### Helpful Hint

Some of the most commonly used terms in algebra are "evaluate," "simplify," "solve," and "check." Make sure you understand what each term means and when each term is used.

Evaluate:   To *evaluate an expression* means to find its numerical value.

Evaluate          $16 \div 2^2 + 36 \div 4$

$= 16 \div 4 + 36 \div 4$

$= 4 + 36 \div 4$

$= 4 + 9$

$= 13$

Evaluate          $-x^2 + 3x - 2$ when $x = 4$

$= -4^2 + 3(4) - 2$

$= -16 + 3(4) - 2$

$= -16 + 12 - 2$

$= -4 - 2$

$= -6$

Simplify:   To *simplify an expression* means to perform the operations and combine like terms.

Simplify          $3(x - 2) - 4(2x + 3)$

$3(x - 2) - 4(2x + 3) = 3x - 6 - 8x - 12$

$= -5x - 18$

Note that when you simplify an expression containing variables you do not generally end up with just a numerical value unless all the variable terms happen to add to zero.

Solve:   To *solve an equation* means to find the value or the values of the variable that make the equation a true statement.

Solve          $2x + 3(x + 1) = 18$

$2x + 3x + 3 = 18$

$5x + 3 = 18$

$5x = 15$

$x = 3$

Check:    To *check the proposed solution to an equation,* substitute the value in the original equation. If this result is true, then the answer checks. For example, to check the solution to the equation just solved, we substitute 3 for $x$ in the original equation.

Check

$$2x + 3(x + 1) = 18$$
$$2(3) + 3(3 + 1) \overset{?}{=} 18$$
$$2(3) + 3(4) \overset{?}{=} 18$$
$$6 + 12 \overset{?}{=} 18$$
$$18 = 18 \quad \text{True}$$

Since we obtained a true statement, the 3 checks.

*It is important to realize that expressions may be evaluated or simplified (depending on the type of problem) and equations are solved and then checked.*

# EXERCISE SET 2.4    MathXL®    MyMathLab®

## Warm-Up Exercises

*Fill in the blanks with the appropriate word, phrase, or symbol(s) from the following list.*

expression              equation              variable was $x$
LCD                     a power of ten        multiplication property of equality
to isolate the variable

1. The _____ allows us to multiply or divide both sides of an equation by the same nonzero number.

2. The fundamental strategy in solving a linear equation is _____.

3. To solve an equation involving decimal numbers, you may want to first multiply both sides of the equation by _____ to change the decimal numbers into integers.

4. To solve an equation involving fractions, you may want to first multiply both sides of the equation by the fractions' _____ to change the fractions into integers.

5. Something that might get evaluated or simplified is called an _____.

6. Something that might get solved is called an _____.

## Practice the Skills

*Solve each equation.*

7. $5x - 6 = 19$

8. $2t - 4 = 8$

9. $-4w - 9 = 11$

10. $-4x + 6 = 20$

11. $3x + 6 = 12$

12. $6 - 3x = 18$

13. $5x - 2 = 10$

14. $-2t + 9 = 21$

15. $-5k - 9 = -19$

16. $-4x - 7 = -6$

17. $12 - x = 5$

18. $-3x - 3 = -12$

19. $8 + 3x = 19$

20. $-2x + 7 = -10$

21. $16x + 5 = -14$

22. $19 = 25 + 4x$

23. $-4.2 = 3x + 31.8$

24. $-24 + 16x = -24$

25. $7r - 16 = -2$

26. $-2w + 4 = -8$

27. $60 = -5s + 9$

28. $15 = 7x + 1$

29. $14 = 5x + 8 - 3x$

30. $15 = 6x - 3 + 3x$

31. $2.3x - 9.34 = 6.3$

32. $x + 0.05x = 21$

33. $0.91y + 2.25 - 0.01y = 5.85$

34. $0.15 = 0.05x - 1.35 - 0.20x$

35. $28.8 = x + 1.40x$

36. $8.40 = 2.45x - 1.05x$

37. $\frac{1}{7}(x + 6) = 4$

38. $\frac{m - 6}{5} = 2$

39. $\frac{d + 3}{7} = 9$

**40.** $\frac{1}{5}(x + 2) = -3$

**41.** $\frac{1}{3}(t - 7) = -7$

**42.** $\frac{2}{3}(n - 3) = 8$

**43.** $\frac{3}{4}(x - 5) = -12$

**44.** $\frac{1}{4} = \frac{z + 1}{4}$

**45.** $\frac{x + 4}{7} = \frac{2}{7}$

**46.** $\frac{4x + 5}{6} = \frac{7}{2}$

**47.** $\frac{3}{4} = \frac{4m - 5}{6}$

**48.** $\frac{5}{6} = \frac{5t - 4}{2}$

**49.** $4(n + 3) = 12$

**50.** $3(x - 2) = 12$

**51.** $-2(x - 3) = 26$

**52.** $5(3 - x) = 15$

**53.** $-4 = -(x + 7)$

**54.** $-3(2 - 3x) = 9$

**55.** $12 = 4(x - 3)$

**56.** $-2(x + 8) - 5 = 1$

**57.** $2x - 3(x + 5) = 6$

**58.** $5(3x + 1) - 12x = -2$

**59.** $-3r - 4(r + 2) = 11$

**60.** $9 = -2(a - 3)$

**61.** $x - 3(2x + 3) = 36$

**62.** $3y - (y + 5) = 9$

**63.** $5x + 3x - 4x - 7 = 9$

**64.** $4(x + 2) = 13$

**65.** $0.7(x - 3) = 1.4$

**66.** $21 + (c - 9) = 24$

**67.** $2.5(4q - 3) = 0.5$

**68.** $0.1(2.4x + 5) = 1.7$

**69.** $3 - 2(x + 3) + 2 = 1$

**70.** $2(3x - 4) - 4x = 12$

**71.** $1 + (x + 3) + 6x = 6$

**72.** $5x - 2x + 7x = -81$

**73.** $4.85 - 6.4x + 1.11 = 25.8$

**74.** $5.76 - 4.24x - 1.9x = 27.864$

**75.** $7 = 8 - 5(m + 3)$

**76.** $4 = \frac{3t + 1}{7}$

**77.** $9 = \frac{2s + 9}{5}$

**78.** $12 = \frac{4d - 1}{3}$

**79.** $x + \frac{2}{3} = \frac{3}{5}$

**80.** $n - \frac{1}{4} = \frac{1}{2}$

**81.** $\frac{r}{3} + 2r = 6$

**82.** $\frac{x}{4} - 6x = 23$

**83.** $\frac{3}{7} = \frac{3t}{4} + 1$

**84.** $\frac{5}{8} = \frac{5t}{6} + 2$

**85.** $\frac{1}{2}r + \frac{1}{5}r = 7$

**86.** $\frac{x}{3} - \frac{3x}{4} = \frac{1}{12}$

**87.** $\frac{2}{8} + \frac{3}{4} = \frac{w}{5}$

**88.** $\frac{x}{4} - \frac{x}{6} = \frac{1}{4}$

**89.** $\frac{1}{2}x + 5 = \frac{1}{8}$

**90.** $\frac{4}{5} + n = \frac{1}{3}$

**91.** $\frac{4}{5}s - \frac{3}{4}s = \frac{1}{10}$

**92.** $\frac{1}{3}x - \frac{3}{4}x = \frac{1}{5}$

**93.** $\frac{4}{9} = \frac{1}{3}(n - 7)$

**94.** $-\frac{3}{8} = \frac{1}{8} - \frac{2x}{7}$

**95.** $-\frac{3}{5} = -\frac{1}{6} - \frac{3}{4}q$

**96.** $-\frac{3}{5} = -\frac{1}{6} - \frac{5}{4}m$

## Concept/Writing Exercises

**97. a)** Explain why it is easier to solve the equation $3x + 2 = 11$ by first subtracting 2 from both sides of the equation rather than by first dividing both sides of the equation by 3.

   **b)** Solve the equation.

**98. a)** Explain why it is easier to solve the equation $5x - 3 = 12$ by first adding 3 to both sides of the equation rather than by first dividing both sides of the equation by 5.

   **b)** Solve the equation.

## Challenge Problems

*For Exercises 99–101, solve the equation.*

**99.** $3(x - 2) - (x + 5) - 2(3 - 2x) = 18$

**100.** $-6 = -(x - 5) - 3(5 + 2x) - 4(2x - 4)$

**101.** $4[3 - 2(x + 4)] - (x + 3) = 13$

**102.** Solve the equation $\square \ddot{\smile} - \triangledown = @$ for $\ddot{\smile}$.

## Group Activity

*In Chapter 3, we will discuss procedures for writing application problems as equations. Let's look at an application now.*

**Birthday Party** John Logan purchased 2 large chocolate bars and a birthday card. The birthday card cost $3. The total cost was $9. What was the price of a single chocolate bar?

*This problem can be represented by the equation $2x + 3 = 9$, which can be used to solve the problem. Solving the equation we find that $x$, the price of a single chocolate bar, is $3.*

*For Exercises 103 and 104, each group member should do parts* **a)** *and* **b)**. *Then do part* **c)** *as a group.*

a) *Obtain an equation that can be used to solve the problem.*

b) *Solve the equation and answer the question.*

c) *Compare and check each other's work.*

**103. Stationery** Eduardo Verner purchased three boxes of stationery. He also purchased wrapping paper and thank-you cards. If the wrapping paper and thank-you cards together cost $6, and the total he paid was $42, find the cost of a box of stationery.

**104. Candies** Mahandi Ison purchased three rolls of peppermint candies and the local newspaper. The newspaper cost 50 cents. He paid $2.75 in all. What did a roll of candies cost?

## Cumulative Review Exercises

[1.4] **105.** True or false: Every real number is a rational number.

[1.9] **106.** Evaluate $[5(2 - 6) + 3(8 \div 4)^2]^2$.

[2.2] **107.** To solve an equation, what do you need to do to the variable?

[2.3] **108.** To solve the equation $7 = -4x$, would you add 4 to both sides of the equation or divide both sides of the equation by $-4$? Explain your answer.

## Mid-Chapter Test: 2.1–2.4

*To find out how well you understand the chapter material to this point, take this brief test. The answers, and the section where the material was initially discussed, are given in the back of the book. Review any questions that you answered incorrectly.*

*In Exercises 1 and 2, combine like terms.*

1. $5x - 9y - 12 + 4y - 7x + 6$

2. $\dfrac{2}{5}x - 8 - \dfrac{3}{4}x + \dfrac{1}{2}$

*In Exercises 3 and 4, use the distributive property to remove parentheses.*

3. $-4(2a - 3b + 16)$

4. $1.6(2.1x - 3.4y - 5.2)$

5. Simplify $5(t - 3) - 3(t + 7) - 2$.

6. Is $x = 2$ a solution of $3(x - 4) = -2(x + 1)$?

7. Is $p = \dfrac{2}{5}$ a solution of $7p - 3 = 2p - 5$?

*In Exercises 8–10, solve each equation and check your solution.*

8. $x - 5 = -9$

9. $120 + x = -40$

10. $-16 = 7 + y$

11. When solving the equation $\dfrac{x}{4} = 5$, what would you do to isolate the variable? Explain.

*In Exercises 12–15, solve each equation and check your solution.*

12. $4 = 12y$

13. $\dfrac{x}{8} = 3$

14. $-\dfrac{x}{5} = -2$

15. $-x = \dfrac{3}{7}$

*In Exercises 16–20, solve each equation.*

16. $6x - 3 = 12$

17. $-4 = -2w - 7$

18. $\dfrac{3}{8} = \dfrac{4n - 1}{6}$

19. $-5(x + 4) - 7 = 3$

20. $8 - 9(y + 4) + 6 = -2$

# 2.5 Solving Linear Equations with the Variable on Both Sides of the Equation

1 Solve equations with the variable on both sides of the equals sign.

2 Solve equations containing decimal numbers or fractions.

3 Identify identities and contradictions.

## 1 Solve Equations with the Variable on Both Sides of the Equals Sign

The equation $4x + 6 = 2x + 4$ contains the variable, $x$, on both sides of the equals sign. To solve equations of this type, we rewrite the equation so that all terms containing the variable are on only one side of the equals sign, and all terms not containing the variable are on the other side of the equals sign. The following steps in the procedure are only guidelines to use to solve these types of equations. For example, there may be times when you may choose to use the distributive property, step 2, before multiplying both sides of the equation by the LCD (step 1). We will illustrate this variation in Examples 9 and 10.

### Understanding Algebra

As the equations get more complicated, our ultimate goal remains the same: *to isolate the variable on one side of the equation.*

---

**To Solve Linear Equations with the Variable on Both Sides of the Equals Sign**

1. **Fractions:** If the equation contains fractions, multiply **both** sides of the equation by the least common denominator (LCD). This will eliminate fractions from the equation.
2. **Parentheses:** Use the distributive property to remove parentheses.
3. **Like terms:** Combine like terms on each side of the equals sign.
4. **Addition property:** Use the addition property to rewrite the equation with all terms containing the variable on one side of the equals sign and all terms not containing the variable on the other side of the equals sign. It may be necessary to use the addition property twice to accomplish this goal.
5. **Multiplication property:** Use the multiplication property to isolate the variable.
6. **Check:** Check the solution in the original equation.

---

Remember: *our goal in solving a linear equation is to isolate the variable, that is, to obtain an equation in which the variable is alone on one side of the equation.* For example, consider the equation $3x + 4 = x + 12$. It contains no fractions or parentheses nor are there any like terms on the same side of the equals sign. Therefore, we start with step 4 above and use the addition property twice. Notice in the steps outlined below that we subtract $x$ from both sides and then subtract 4 from both sides:

$$3x + 4 = x + 12$$

$$3x \; - x \; + 4 = x \; - x \; + 12 \qquad \text{Addition property (subtract } x \text{ from both sides)}$$

$$2x + 4 = 12 \qquad \text{Variable appears only on left side of equation.}$$

Notice that the variable, $x$, now appears on only one side of the equation. However, $+4$ still appears on the same side.

$$2x + 4 = 12$$

$$2x + 4 \; - 4 = 12 \; - 4 \qquad \text{Addition property (subtract 4 from both sides)}$$

$$2x = 8 \qquad \text{Term with } x \text{ is now isolated (in the form } ax = b).$$

Finally, to find $x$, we use the multiplication property:

$$2x = 8$$

$$\frac{\overset{1}{\cancel{2}}x}{\cancel{2}} = \frac{\overset{4}{\cancel{8}}}{\cancel{2}} \qquad \text{Multiplication property (divide both sides by 2)}$$

$$x = 4 \qquad \text{Goal accomplished: } x \text{ is isolated.}$$

The solution to the equation is 4.

**EXAMPLE 1** Solve the equation $4x + 6 = 2x + 4$.

*Solution*   We start by getting all the terms with the variable on one side of the equals sign and all terms without the variable on the other side. The terms with the variable may be collected on either side of the equals sign. We will illustrate two methods of solving this equation.

**Method 1**   Isolate the variable term on the left.

$$4x + 6 = 2x + 4$$

Step 4 $\qquad 4x - 2x + 6 = 2x - 2x + 4 \qquad$ Subtract $2x$ from both sides.

$$2x + 6 = 4$$

Step 4 $\qquad 2x + 6 - 6 = 4 - 6 \qquad$ Subtract 6 from both sides.

$$2x = -2$$

Step 5 $\qquad \dfrac{2x}{2} = \dfrac{-2}{2} \qquad$ Divide both sides by 2.

$$x = -1$$

**Method 2**   Isolate the variable term on the right.

$$4x + 6 = 2x + 4$$

Step 4 $\qquad 4x - 4x + 6 = 2x - 4x + 4 \qquad$ Subtract $4x$ from both sides.

$$6 = -2x + 4$$

Step 4 $\qquad 6 - 4 = -2x + 4 - 4 \qquad$ Subtract 4 from both sides.

$$2 = -2x$$

Step 5 $\qquad \dfrac{2}{-2} = \dfrac{-2x}{-2} \qquad$ Divide both sides by $-2$.

$$-1 = x$$

The same answer is obtained whether we collect the terms with the variable on the left or right side.

Step 6   Check

$$4x + 6 = 2x + 4$$

$$4(-1) + 6 \overset{?}{=} 2(-1) + 4$$

$$-4 + 6 \overset{?}{=} -2 + 4$$

$$2 = 2 \qquad \text{True}$$

Since the check is true, the solution is $-1$.

Now Try Exercise 19

**EXAMPLE 2** Solve the equation $2x - 3 - 5x = 13 + 4x - 2$.

*Solution*   We will choose to collect the terms containing the variable on the right side of the equation in order to create a positive coefficient of $x$. Since there are like terms *on the same side of the equals sign*, we will begin by combining these like terms.

Step 3 $\qquad 2x - 3 - 5x = 13 + 4x - 2$

$\qquad\qquad -3x - 3 = 4x + 11 \qquad$ Combined like terms.

Step 4 $\qquad -3x + 3x - 3 = 4x + 3x + 11 \qquad$ Add $3x$ to both sides.

$\qquad\qquad -3 = 7x + 11$

Step 4 $\qquad -3 - 11 = 7x + 11 - 11 \qquad$ Subtract 11 from both sides.

$\qquad\qquad -14 = 7x$

Step 5 $\qquad \dfrac{-14}{7} = \dfrac{7x}{7} \qquad$ Divide both sides by 7.

$\qquad\qquad -2 = x$

Step 6    Check

$$2x - 3 - 5x = 13 + 4x - 2$$
$$2(-2) - 3 - 5(-2) \stackrel{?}{=} 13 + 4(-2) - 2$$
$$-4 - 3 + 10 \stackrel{?}{=} 13 - 8 - 2$$
$$-7 + 10 \stackrel{?}{=} 5 - 2$$
$$3 = 3 \qquad \text{True}$$

Since the check is true, the solution is $-2$.

NowTry Exercise 29

The solution to Example 2 could be condensed as follows:

$$2x - 3 - 5x = 13 + 4x - 2$$

| | |
|---|---|
| $-3x - 3 = 4x + 11$ | Combined like terms. |
| $-3 = 7x + 11$ | Added $3x$ to both sides. |
| $-14 = 7x$ | Subtracted 11 from both sides. |
| $-2 = x$ | Divided both sides by 7. |

We solved Example 2 by moving the terms containing the variable to the right side of the equation. Now rework the problem by moving the terms containing the variable to the left side of the equation. You should obtain the same answer.

**EXAMPLE 3**  Solve the equation $2(p + 3) = -3p + 10$.

*Solution*

| | | |
|---|---|---|
| | $2(p + 3) = -3p + 10$ | |
| Step 2 | $2p + 6 = -3p + 10$ | Distributive property was used. |
| Step 4 | $2p + 3p + 6 = -3p + 3p + 10$ | Add $3p$ to both sides. |
| | $5p + 6 = 10$ | |
| Step 4 | $5p + 6 - 6 = 10 - 6$ | Subtract 6 from both sides. |
| | $5p = 4$ | |
| Step 5 | $\dfrac{5p}{5} = \dfrac{4}{5}$ | Divide both sides by 5. |
| | $p = \dfrac{4}{5}$ | |

The solution is $\dfrac{4}{5}$.

NowTry Exercise 27

The solution to Example 3 could be condensed as follows:

| | | |
|---|---|---|
| | $2(p + 3) = -3p + 10$ | |
| | $2p + 6 = -3p + 10$ | Distributive property was used. |
| | $5p + 6 = 10$ | Added $3p$ to both sides. |
| | $5p = 4$ | Subtracted 6 from both sides. |
| | $p = \dfrac{4}{5}$ | Divided both sides by 5. |

**Understanding Algebra**

In Example 3, the variable could also have been isolated on the right side by first subtracting $2p$:

$$2p + 6 = -3p + 10$$
$$6 = -5p + 10$$
$$-4 = -5p$$
$$\dfrac{-4}{-5} = p$$
$$\dfrac{4}{5} = p$$

**EXAMPLE 4**  Solve the equation $2(x - 5) + 3 = 3x + 9$.

*Solution*

| | | |
|---|---|---|
| | $2(x - 5) + 3 = 3x + 9$ | |
| Step 2 | $2x - 10 + 3 = 3x + 9$ | Distributive property was used. |
| Step 3 | $2x - 7 = 3x + 9$ | Combined like terms. |
| Step 4 | $-7 = x + 9$ | Subtracted $2x$ from both sides. |
| Step 4 | $-16 = x$ | Subtracted 9 from both sides. |

The solution is $-16$.

NowTry Exercise 35

**EXAMPLE 5** Solve the equation $7 - 2x + 5x = -2(-3x + 4)$.

*Solution*

$$7 - 2x + 5x = -2(-3x + 4)$$

| Step 2 | $7 - 2x + 5x = 6x - 8$ | Used the distributive property. |
| Step 3 | $7 + 3x = 6x - 8$ | Combined like terms. |
| Step 4 | $7 = 3x - 8$ | Subtracted $3x$ from both sides. |
| Step 4 | $15 = 3x$ | Added 8 to both sides. |
| Step 5 | $5 = x$ | Divided both sides by 3. |

The solution is 5.

Now Try Exercise 63

## 2 Solve Equations Containing Decimal Numbers or Fractions

Now we will solve an equation that contains decimal numbers. We will illustrate two procedures for solving Example 6.

**EXAMPLE 6** Solve the equation $5.74x + 5.42 = 2.24x - 9.28$.

*Solution*

**Method 1** Notice that there are no like terms on the same side of the equals sign that can be combined.

$$5.74x + 5.42 = 2.24x - 9.28$$

| Step 4 | $5.74x - 2.24x + 5.42 = 2.24x - 2.24x - 9.28$ | Subtract $2.24x$ from both sides. |
| | $3.50x + 5.42 = -9.28$ | |
| Step 4 | $3.50x + 5.42 - 5.42 = -9.28 - 5.42$ | Subtract 5.42 from both sides. |
| | $3.50x = -14.70$ | |
| Step 5 | $\dfrac{3.50x}{3.50} = \dfrac{-14.70}{3.50}$ | Divide both sides by 3.50. |
| | $x = -4.20$ | |

The solution is $-4.20$.

**Method 2** Since the given equation has numbers given in hundredths, we will multiply both sides of the equation by 100.

$$5.74x + 5.42 = 2.24x - 9.28$$

| | $100(5.74x + 5.42) = 100(2.24x - 9.28)$ | Multiply both sides by 100. |
| | $100(5.74x) + 100(5.42) = 100(2.24x) - 100(9.28)$ | Distributive property |
| | $574x + 542 = 224x - 928$ | |
| Step 4 | $574x + 542 - 542 = 224x - 928 - 542$ | Subtract 542 from both sides. |
| | $574x = 224x - 1470$ | |
| Step 4 | $574x - 224x = 224x - 224x - 1470$ | Subtract $224x$ from both sides. |
| | $350x = -1470$ | |
| Step 5 | $\dfrac{350x}{350} = \dfrac{-1470}{350}$ | Divide both sides by 350. |
| | $x = -4.20$ | |

Notice we obtain the same answer using either method. You may use either method to solve equations of this type.

Now Try Exercise 25

### Understanding Algebra

To eliminate decimal numbers from an equation, multiply by a power of 10:

| If equation contains numbers to . . . | . . . then multiply both sides by |
| --- | --- |
| tenths | 10 |
| hundredths | 100 |
| thousandths | 1000 |

and so on.

Now let's solve some equations that contain fractions.

EXAMPLE 7  Solve the equation $\frac{1}{2}a = \frac{3}{4}a + \frac{1}{5}$.

*Solution*   The least common denominator is 20. Begin by multiplying both sides of the equation by the LCD.

$$\frac{1}{2}a = \frac{3}{4}a + \frac{1}{5}$$

Step 1        $20\left(\frac{1}{2}a\right) = 20\left(\frac{3}{4}a + \frac{1}{5}\right)$        Multiply both sides by the LCD, 20.

Step 2        $10a = \overset{5}{20}\left(\frac{3}{4}a\right) + \overset{4}{20}\left(\frac{1}{\underset{1}{5}}\right)$        Distributive property

              $10a = 15a + 4$

Step 4        $-5a = 4$        Subtracted $15a$ from both sides.

Step 5        $a = -\frac{4}{5}$        Divided both sides by $-5$.

Step 6   Check

$$\frac{1}{2}a = \frac{3}{4}a + \frac{1}{5}$$

$$\frac{1}{2}\left(-\frac{4}{5}\right) \overset{?}{=} \frac{3}{4}\left(-\frac{4}{5}\right) + \frac{1}{5}$$

$$-\frac{2}{5} \overset{?}{=} -\frac{3}{5} + \frac{1}{5}$$

$$-\frac{2}{5} = -\frac{2}{5}$$        True

The solution is $-\frac{4}{5}$.

Now Try Exercise 43

## Understanding Algebra

To eliminate fractions from equations, multiply both sides of the equation by the LCD of the fractions in the equation.

## Understanding Algebra

It is very helpful to place each side of the equation in parentheses before multiplying both sides by the LCD.

> **Helpful Hint**
>
> The equation in Example 7, $\frac{1}{2}a = \frac{3}{4}a + \frac{1}{5}$, could have been written as $\frac{a}{2} = \frac{3a}{4} + \frac{1}{5}$ because $\frac{1}{2}a$ is the same as $\frac{a}{2}$, and $\frac{3}{4}a$ is the same as $\frac{3a}{4}$. You would solve the equation $\frac{a}{2} = \frac{3a}{4} + \frac{1}{5}$ the same way you solved the equation in Example 7.

EXAMPLE 8  Solve the equation $\frac{x}{4} + 3 = 2(x - 2)$.

*Solution*   We will begin by multiplying both sides of the equation by the LCD, 4.

$$\frac{x}{4} + 3 = 2(x - 2)$$

$$4\left(\frac{x}{4} + 3\right) = 4[2(x - 2)]$$        Multiply both sides by the LCD, 4.

$$4\left(\frac{x}{4}\right) + 4(3) = 4[2(x - 2)]$$        Used the distributive property (on left).

$$x + 12 = 8(x - 2)$$

$$x + 12 = 8x - 16$$        Used the distributive property (on right).

$$12 = 7x - 16$$        Subtracted $x$ from both sides.

$$28 = 7x$$        Added 16 to both sides.

$$4 = x$$        Divided both sides by 7.

A check will show that 4 is the solution.

Now Try Exercise 61

In Example 8, we began the solution by multiplying both sides of the equation by the LCD. In Example 9, we will solve the same equation, but this time we will begin by using the distributive property.

**EXAMPLE 9** Solve the equation in Example 8, $\frac{x}{4} + 3 = 2(x - 2)$, by first using the distributive property.

*Solution*   Begin by using the distributive property.

$$\frac{x}{4} + 3 = 2(x - 2)$$

$$\frac{x}{4} + 3 = 2x - 4 \qquad \text{Used the distributive property (on right).}$$

$$4\left(\frac{x}{4} + 3\right) = 4(2x - 4) \qquad \text{Multiply both sides by the LCD, 4.}$$

$$4\left(\frac{x}{4}\right) + 4(3) = 4(2x) - 4(4) \qquad \text{Distributive property (on left and right)}$$

$$x + 12 = 8x - 16$$

$$12 = 7x - 16 \qquad x \text{ was subtracted from both sides.}$$

$$28 = 7x \qquad 16 \text{ was added to both sides.}$$

$$4 = x \qquad \text{Both sides were divided by 7.}$$

The solution is 4.

Now Try Exercise 65

Notice that we obtained the same answer in Examples 8 and 9.

**EXAMPLE 10** Solve the equation $\frac{1}{2}(2x + 3) = \frac{2}{3}(x - 6) + 4$.

Notice that this equation contains one term on the left side of the equals sign and two terms on the right side of the equals sign.

*Solution*   We will work this problem by first using the distributive property.

$$\frac{1}{2}(2x + 3) = \frac{2}{3}(x - 6) + 4$$

$$\frac{1}{2}(2x) + \frac{1}{2}(3) = \frac{2}{3}(x) - \frac{2}{3}(\overset{2}{6}) + 4 \qquad \text{Distributive property (on left and right)}$$

$$x + \frac{3}{2} = \frac{2}{3}x - 4 + 4$$

$$x + \frac{3}{2} = \frac{2}{3}x \qquad \text{Combined like terms.}$$

$$6\left(x + \frac{3}{2}\right) = 6\left(\frac{2}{3}x\right) \qquad \text{Multiply both sides by the LCD, 6.}$$

$$6x + 6\left(\frac{3}{2}\right) = 6\left(\frac{2}{3}x\right) \qquad \text{Used the distributive property.}$$

$$6x + \overset{3}{6}\left(\frac{3}{2}\right) = \overset{2}{6}\left(\frac{2}{3}x\right)$$

$$6x + 9 = 4x$$

$$2x + 9 = 0 \qquad \text{Subtracted } 4x \text{ from both sides.}$$

$$2x = -9 \qquad \text{Subtracted 9 from both sides.}$$

$$x = -\frac{9}{2} \qquad \text{Divided both sides by 2.}$$

Check   Substitute $-\dfrac{9}{2}$ for each $x$ in the equation.

$$\frac{1}{2}(2x + 3) = \frac{2}{3}(x - 6) + 4$$

$$\frac{1}{2}\left[2\left(-\frac{9}{2}\right) + 3\right] \stackrel{?}{=} \frac{2}{3}\left(-\frac{9}{2} - 6\right) + 4$$

$$\frac{1}{2}[-9 + 3] \stackrel{?}{=} \frac{2}{3}\left(-\frac{9}{2} - \frac{12}{2}\right) + 4$$

$$\frac{1}{2}[-6] \stackrel{?}{=} \frac{2}{3}\left(-\frac{21}{2}\right) + 4$$

$$-3 \stackrel{?}{=} -7 + 4$$

$$-3 = -3 \qquad\qquad \text{True}$$

The solution is $-\dfrac{9}{2}$.

Now Try Exercise 75

In Example 10, we began by using the distributive property. We could have also begun by multiplying both sides of the equation by the LCD, 6, before using the distributive property. Work Example 10 again now by first multiplying both sides of the equation by the LCD, 6.

Example 10 could have also been written as $\dfrac{2x + 3}{2} = \dfrac{2(x - 6)}{3} + 4$. If you were given the equation in this form, you could begin by using the distributive property on $2(x - 6)$ or you could begin by multiplying both sides of the equation by the LCD, 6. Because this is just another way of writing the equation in Example 10, the answer would be $-\dfrac{9}{2}$.

We will discuss solving equations containing fractions in more detail later in the book.

## 3  Identify Identities and Contradictions

Thus far all the equations we have solved have had a single value for a solution. Additionally, there are two other types of equations based on their set of solutions. All three are listed below:

> **Types of Equations**
>
> A **conditional equation** is true for a specific value(s) of the variable.
> An **identity** is true for all values of the variable.
> A **contradiction** is not true for any value of the variable.

**Understanding Algebra**

The solution in Example 11 could have been stopped at "$3x - 5 = 3x - 5$." Since one side is identical to the other side, the equation is true for infinitely many values of $x$. We say the solution is "all real numbers."

**EXAMPLE  11**  Solve the equation $5x - 5 - 2x = 3(x - 2) + 1$.

*Solution*

$$5x - 5 - 2x = 3(x - 2) + 1$$

$$5x - 5 - 2x = 3x - 6 + 1 \qquad \text{Used the distributive property.}$$

$$3x - 5 = 3x - 5 \qquad\qquad \text{Combined like terms.}$$

Since the same expression appears on both sides of the equals sign, the statement is true for infinitely many values of $x$. If we continue to solve this equation further, we might obtain

$$3x - 5 = 3x - 5$$
$$3x = 3x \qquad \text{Added 5 to both sides.}$$
$$0 = 0 \qquad \text{Subtracted } 3x \text{ from both sides.}$$

The answer is: all real numbers.

NOTE: *When solving an equation that is always true, like the equation in Example 11, write your answer as "all real numbers."*

Now Try Exercise 47

**EXAMPLE 12** Solve the equation $-2x + 5 + 3x = 5x - 4x + 7$.

*Solution*

$$-2x + 5 + 3x = 5x - 4x + 7$$
$$x + 5 = x + 7 \qquad \text{Combined like terms.}$$
$$x - x + 5 = x - x + 7 \qquad \text{Subtract } x \text{ from both sides.}$$
$$5 = 7 \qquad \text{False}$$

The answer is: no solution.

NOTE: *When solving an equation that is never true, like the equation in Example 12, write your answer as "no solution."*

Now Try Exercise 31

## Understanding Algebra

When solving an equation, if you obtain an obviously false statement, as in Example 12, the equation has *no solution*. We state the answer formally as *"no solution."*

### Helpful Hint

When solving equations, remember two important things: (1) your goal is to isolate the variable, and (2) whatever you do to one side of the equation you must also do to the other side. That is, you must treat both sides of the equation equally.

# EXERCISE SET 2.5    MathXL®   MyMathLab®

## Warm-Up Exercises

*Fill in the blanks with the appropriate word, phrase, or symbol(s) from the following list.*

| | | | |
|---|---|---|---|
| identity | conditional equation | contradiction | True |
| specific value(s) of the variable | least common denominator | power of 10 | False |
| all values of the variable | no real numbers | isolate the variable | |

**1.** If a linear equation has infinitely many solutions, it is called a(n) _____ .

**2.** If a linear equation has a specific solution, it is called a(n) _____ .

**3.** To eliminate fractions from an equation, multiply both sides of the equation by the _____ .

**4.** If there is no solution to a linear equation, it is called a(n) _____ .

**5.** A conditional equation is true for _____ .

**6.** An identity is true for _____ .

**7.** A contradiction is true for _____ .

**8.** The ultimate goal in the process of solving linear equations is to _____ .

**9.** (True or False) The equation $2x + 0 = 0$ is an example of a contradiction. _____

**10.** To eliminate decimal numbers from an equation, multiply both sides of the equation by a _____ .

## Practice the Skills

*Solve each equation.*

**11.** $3x = -2x + 10$

**12.** $y + 4 = 2y - 9$

**13.** $-4x + 10 = 6x$

**14.** $3a = 4a + 8$

**15.** $5x + 3 = 6$

**16.** $-6x = 2x + 16$

**17.** $21 - 6p = 3p - 2p$

**18.** $8 - 3x = 4x + 50$

**19.** $2x - 8 = 3x - 6$

**20.** $5x + 7 = 3x + 5$

**21.** $6 - 2y = 9 - 8y + 6y$

**22.** $-4 + 2y = 2y - 6 + y$

**23.** $124.8 - 9.4x = 4.8x + 32.5$

**24.** $9 - 0.5x = 4.5x + 8.5$

**25.** $0.62x - 0.65 = 9.75 - 2.63x$

**26.** $8.71 - 2.44x = 11.02 - 5.74x$

**27.** $5x + 3 = 2(x + 6)$

**28.** $x - 14 = 3(x + 2)$

**29.** $4y - 2 - 8y = 19 + 5y - 3$

**30.** $3x - 5 + 9x = 2 + 4x + 9$

**31.** $2(x - 2) = 4x - 6 - 2x$

**32.** $4r = 10 - 2(r - 4)$

**33.** $-(w + 2) = -6w + 11$

**34.** $7(-3m + 5) = 3(10 - 6m)$

**35.** $-3(2t - 5) + 5 = 3t + 13$

**36.** $4(x - 3) + 2 = 2x + 12$

**37.** $\dfrac{a}{5} = \dfrac{a - 3}{2}$

**38.** $\dfrac{b}{16} = \dfrac{b - 6}{4}$

**39.** $\dfrac{n}{10} = 9 - \dfrac{n}{5}$

**40.** $6 - \dfrac{x}{4} = \dfrac{x}{8}$

**41.** $\dfrac{7}{2} - \dfrac{x}{3} = 2x$

**42.** $\dfrac{x}{4} - 3 = -2x$

**43.** $\dfrac{5}{8} + \dfrac{1}{4}a = \dfrac{1}{2}a$

**44.** $\dfrac{3}{4}x + \dfrac{1}{2} = \dfrac{1}{2}x$

**45.** $0.1(x + 10) = 0.3x - 4$

**46.** $5(3.2x - 3) = 2(x - 4)$

**47.** $2(x + 4) = 4x + 3 - 2x + 5$

**48.** $3(y - 1) + 9 = 8y + 6 - 5y$

**49.** $5(3n + 3) = 2(5n - 4) + 6n$

**50.** $-4(-3z - 5) = -(10z + 8) - 2z$

**51.** $-(3 - p) = -(2p + 3)$

**52.** $12 - 2x - 3(x + 2) = 4x + 6 - x$

**53.** $-(x + 4) + 5 = 4x + 1 - 5x$

**54.** $18x + 3(4x - 9) = -6x + 45$

**55.** $35(2x - 1) = 7(x + 4) + 3x$

**56.** $10(x - 10) + 5 = 5(2x - 20)$

**57.** $0.4(x + 0.7) = 0.6(x - 4.2)$

**58.** $0.5(6x - 8) = 1.4(x - 5) - 0.2$

**59.** $\dfrac{3}{5}x - 2 = x - \dfrac{1}{2}$

**60.** $\dfrac{3}{5}x + 4 = \dfrac{1}{5}x + 5$

**61.** $\dfrac{y}{5} + 2 = 3(y - 4)$

**62.** $2(x - 4) = \dfrac{x}{5} + 10$

**63.** $12 - 3x + 7x = -2(-5x + 6)$

**64.** $-2x - 3 - x = -3(-2x + 7)$

**65.** $3(x - 6) - 4(3x + 1) = x - 22$

**66.** $-2(-3x + 5) + 6 = 4(x - 2)$

**67.** $5 + 2x = 6(x + 1) - 5(x - 3)$

**68.** $4 - (6x + 6) = -(-2x + 10)$

**69.** $7 - (-y - 5) = 2(y + 3) - 6(y + 3)$

**70.** $12 - 6x + 3(2x + 3) = 2x + 5$

**71.** $\dfrac{3}{5}(x - 6) = \dfrac{2}{3}(3x - 5)$

**72.** $\dfrac{1}{2}(2d + 4) = \dfrac{1}{3}(4d - 4)$

**73.** $\dfrac{3(2r - 5)}{5} = \dfrac{3r - 6}{3}$

**74.** $\dfrac{3(x - 4)}{4} = \dfrac{5(2x - 3)}{3}$

**75.** $\dfrac{2}{7}(5x + 4) = \dfrac{1}{2}(3x - 4) + 1$

**76.** $\dfrac{5}{12}(x + 2) = \dfrac{2}{3}(2x + 1) + \dfrac{1}{6}$

**77.** $\dfrac{a - 5}{2} = \dfrac{3a}{4} + \dfrac{a - 25}{6}$

**78.** $\dfrac{a - 7}{3} = \dfrac{a + 5}{2} - \dfrac{7a - 1}{6}$

## Concept/Writing Exercises

**79. a)** Construct a *conditional equation* containing three terms on the left side of the equals sign and two terms on the right side of the equals sign.

  **b)** Explain how you know your answer to part **a)** is a conditional equation.

  **c)** Solve the equation.

**80. a)** Construct a *conditional equation* containing two terms on the left side of the equals sign and three terms on the right side of the equals sign.

  **b)** Explain how you know your answer to part **a)** is a conditional equation.

  **c)** Solve the equation.

**81. a)** Construct an *identity* containing three terms on the left side of the equals sign and two terms on the right side of the equals sign.

  **b)** Explain how you know your answer to part **a)** is an identity.

  **c)** What is the solution to the equation?

**82. a)** Construct an *identity* containing two terms on the left side of the equals sign and three terms on the right side of the equals sign.

  **b)** Explain how you know your answer to part **a)** is an identity.

  **c)** What is the solution to the equation?

**83. a)** Construct a *contradiction* containing three terms on the left side of the equals sign and two terms on the right side of the equals sign.

  **b)** Explain how you know your answer to part **a)** is a contradiction.

  **c)** What is the solution to the equation?

**84. a)** Construct a *contradiction* containing three terms on the left side of the equals sign and four terms on the right side of the equals sign.

  **b)** Explain how you know your answer to part **a)** is a contradiction.

  **c)** What is the solution to the equation?

## Challenge Problems

**85.** Solve the equation $5\text{✳} - 1 = 4\text{✳} + 5\text{✳}$ for ✳.

**86.** Solve the equation $2\triangle - 4 = 3\triangle + 5 - \triangle$ for △.

**87.** Solve the equation $3\text{☺} - 5 = 2\text{☺} - 5 + \text{☺}$ for ☺.

**88.** Solve $-2(x + 3) + 5x = 3(4 - 2x) - (x + 2)$.

**89.** Solve $4 - [5 - 3(x + 2)] = x - 3$.

## Group Activity

*Discuss and answer Exercise 90 as a group. In the next chapter, we will be discussing procedures for writing application problems as equations. Let's get some practice now.*

**90. Chocolate Bars** Consider the following word problem. Mary Kay purchased two large chocolate bars. The total cost of the two chocolate bars was equal to the cost of one chocolate bar plus $6. Find the cost of one chocolate bar.

**a)** Each group member: Represent this problem as an equation with the variable $x$.

**b)** Each group member: Solve the equation you determined in part **a)**.

**c)** As a group, check your equation and your answer to make sure that it makes sense.

## Cumulative Review Exercises

[1.5] **91.** Evaluate.

     **a)** $|4|$      **b)** $|-7|$      **c)** $|0|$

[1.9] **92.** $\dfrac{x^a}{x^b}$

[2.1] **93.** Explain the difference between factors and terms.

**94.** Simplify $2(x - 3) + 4x - (4 - x)$.

[2.4] **95.** Solve $2(x - 3) + 4x - (4 - x) = 0$.

**96.** Solve $(x + 4) - (4x - 3) = 16$.

# 2.6 Formulas

**1** Use the simple interest formula and the distance formula.

**2** Use geometric formulas.

**3** Solve for a variable in a formula.

> **Formula**
>
> A **formula** is an equation commonly used to express a specific relationship mathematically.

For example, the formula for the area of a rectangle is

$$\text{area} = \text{length} \cdot \text{width} \quad \text{or} \quad A = lw$$

To **evaluate a formula,** substitute the appropriate numerical values for the variables and perform the indicated operations.

## 1 Use the Simple Interest Formula and the Distance Formula

A formula commonly used in banking is the **simple interest formula.**

> **Simple Interest Formula**
>
> $$\text{interest} = \text{principal} \cdot \text{rate} \cdot \text{time} \quad \text{or} \quad i = prt$$

### Understanding Algebra

Formulas are used in business, science, and many other disciplines. A formula is an equation that states a relationship involving several variables.

This formula is used to determine the simple interest, $i$, earned on some savings accounts, or the simple interest an individual must pay on certain loans. In the simple interest formula $i = prt$, $p$ is the principal (the amount invested or borrowed), $r$ is the interest rate in decimal form, and $t$ is the amount of time of the investment or loan.

**EXAMPLE 1** **Auto Loan** To buy a car, Mary Beth Orrange borrowed $10,000 from a bank for 3 years. The bank charged 5% simple annual interest for the loan. How much interest will Mary Beth owe the bank?

*Solution* Understand and Translate Since the bank charged simple interest, we use the simple interest formula to solve the problem. We are given that the rate, $r$, is 5%, or 0.05 in decimal form. The principal, $p$, is $10,000 and the time, $t$, is 3 years. We substitute these values in the simple interest formula and solve for the interest, $i$.

Carry Out

$$i = prt$$
$$i = 10{,}000(0.05)(3)$$
$$i = 1500$$

Check There are various ways to check this problem. First ask yourself "Is the answer realistic?" $1500 is a realistic answer. The interest on $10,000 for 1 year at 5% is $500. Therefore for 3 years, an interest of $1500 is correct.

Answer Mary Beth will pay $1500 interest. After 3 years, when she repays the loan, she will pay the principal, $10,000, plus the interest, $1500 for a total of $11,500.

Now Try Exercise 83

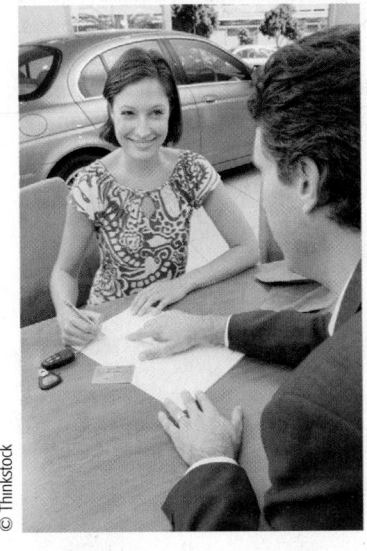

© Thinkstock

**EXAMPLE 2** **Savings Account** John Starmack invests $4000 in a savings account that earns simple interest for 2 years. If the interest earned from the account is $500, find the rate.

*Solution* Understand and Translate We use the simple interest formula, $i = prt$. We are given the principal, $p$, the time, $t$, and the interest, $i$. We are asked to find the rate, $r$. We substitute the given values in the simple interest formula and solve the resulting equation for $r$.

$$i = prt$$
$$500 = 4000(r)(2)$$

Carry Out
$$500 = 8000r$$
$$\frac{500}{8000} = \frac{8000r}{8000}$$
$$0.0625 = r$$

Check and Answer The simple interest rate of 0.0625 or 6.25% per year is realistic. If we substitute $p = \$4000$, $r = 0.0625$ and $t = 2$, we obtain the interest, $i = \$500$. Thus, the answer checks. The simple interest rate is 6.25%.

Now Try Exercise 85

Another important formula is the distance formula.

**Distance Formula**

$$\text{distance} = \text{rate} \cdot \text{time} \quad \text{or} \quad d = r \cdot t$$

**EXAMPLE 3** **Auto Race** At a NASCAR auto race, Dale Earnhardt, Jr., completed the race in 3.2 hours at an average speed of 156.25 miles per hour. Determine the distance of the race.

*Solution* Understand and Translate We are given the rate, 156.25 miles per hour, and the time is 3.2 hours. We are asked to find the distance.

$$\text{distance} = \text{rate} \cdot \text{time}$$

Carry Out
$$= (156.25)(3.2) = 500$$

Answer Thus, the distance of the race was 500 miles.

Now Try Exercise 89

Let's look at the units in Example 3. The rate is given in miles per hour and the time is given in hours. If we analyze the units (a process called *dimensional analysis*), we see that the answer is given in miles.

$$\text{distance} = \text{rate} \cdot \text{time}$$
$$= \frac{\text{miles}}{\text{hour}} \cdot \text{hour}$$
$$= \text{miles}$$

Now we will discuss geometric formulas that will be used throughout the book.

## 2  Use Geometric Formulas

The **perimeter**, $P$, is the sum of the lengths of the sides of a figure. Perimeters are measured in the same common unit as the sides. For example, perimeter may be measured in centimeters, inches, or feet. The **area, $A$,** is the measure of the amount of surface within the figure's boundaries. Areas are measured in square units. For example, area may be measured in square centimeters, square inches, or square feet. **Table 2.1** gives the formulas for finding the areas and perimeters of triangles and quadrilaterals. **Quadrilateral** is a general name for a four-sided figure.

In **Table 2.1**, the letter $h$ is used to represent the *height* of the figure. In the figure of the trapezoid, the sides $b$ and $d$ are called the *bases* of the trapezoid. In the triangle, the side labeled $b$ is called the *base* of the triangle.

**TABLE 2.1     Formulas for Areas and Perimeters of Quadrilaterals and Triangles*

| Figure | Sketch | Area | Perimeter |
|---|---|---|---|
| **Square** | $s$ | $A = s^2$ | $P = 4s$ |
| **Rectangle** | $w$, $l$ | $A = lw$ | $P = 2l + 2w$ |
| **Parallelogram** | $h$, $w$, $l$ | $A = lh$ | $P = 2l + 2w$ |
| **Trapezoid** | $b$, $a$, $h$, $c$, $d$ | $A = \frac{1}{2}h(b + d)$ | $P = a + b + c + d$ |
| **Triangle** | $a$, $h$, $c$, $b$ | $A = \frac{1}{2}bh$ | $P = a + b + c$ |

EXAMPLE 4  **Building an Exercise Area**  Dr. Alex Taurke, a veterinarian, decides to fence in a large rectangular area in the yard behind his office for exercising dogs that are boarded overnight. The part of the yard to be fenced in will be 40 feet long and 23 feet wide (see **Fig. 2.2**).

**a)** How much fencing is needed?

**b)** How large, in square feet, will the fenced-in area be?

*Solution*

**a)** Understand  To find the amount of fencing required, we need to find the perimeter of the rectangular area to be fenced in. To find the perimeter, $P$,

*See Appendix C for additional information on geometry and geometric figures.

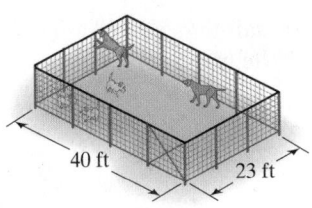

**FIGURE 2.2**

substitute 40 for the length, *l*, and 23 for the width, *w*, in the perimeter formula, $P = 2l + 2w$.

$$P = 2l + 2w$$

Carry Out   $$P = 2(40) + 2(23) = 80 + 46 = 126$$

Check and Answer   By looking at **Figure 2.2**, we can see that a perimeter of 126 feet is a reasonable answer. Thus, 126 feet of fencing will be needed to fence in the area for the dogs to exercise.

**b)** To find the fenced-in area, substitute 40 for the length and 23 for the width in the formula for the area of a rectangle. Since we are multiplying an amount measured in feet by a second amount measured in feet, the answer will be in square feet (or ft²).

$$A = lw$$
$$= 40(23) = 920 \text{ square feet (or 920 ft}^2)$$

Based upon the data given, an area of 920 ft² is reasonable. The area to be fenced in will be 920 square feet.

Now Try Exercise 91

**EXAMPLE 5   Panoramic Photo**   Heather Hunter enlarges rectangular panoramic photos, like the one shown below. One of her enlarged panoramic photos has a perimeter of 116 inches and a length of 40 inches. Find the width of the photo.

© Allen R. Angel

*Solution*   Understand and Translate   The perimeter, *P*, is 116 inches and the length, *l*, is 40 inches. Substitute these values into the formula for the perimeter of a rectangle and solve for the width, *w*.

$$P = 2l + 2w$$
$$116 = 2(40) + 2w$$

Carry Out   $$116 = 80 + 2w$$

$$116 - 80 = 80 - 80 + 2w \quad \text{Subtract 80 from both sides.}$$

$$36 = 2w$$

$$\frac{36}{2} = \frac{2w}{2} \quad\quad\quad \text{Divide both sides by 2.}$$

$$18 = w$$

Check and Answer   By comparing the length and width, you should realize that the dimensions of a length of 40 inches and a width of 18 inches is reasonable. The answer is, the width of the photo is 18 inches.

Now Try Exercise 25

**EXAMPLE 6 Sailboat** A small sailboat has a triangular sail that has an area of 30 square feet and a base of 5 feet (see **Fig. 2.3**). Determine the height of the sail.

*Solution* Understand and Translate We use the formula for the area of a triangle given in **Table 2.1** on page 134.

**FIGURE 2.3**

$$A = \frac{1}{2}bh$$

$$30 = \frac{1}{2}(5)h$$

Carry Out

$$2 \cdot 30 = 2 \cdot \frac{1}{2}(5)h \qquad \text{Multiply both sides by 2.}$$

$$60 = 5h$$

$$\frac{60}{5} = \frac{5h}{5} \qquad \text{Divide both sides by 5.}$$

$$12 = h$$

Check and Answer The height of the triangle is 12 feet. By looking at **Figure 2.3**, you may realize that a sail 12 feet tall and 5 feet wide at the base is reasonable. Thus, the height of the sail is 12 feet.

NowTry Exercise 93

Another figure that we see and use daily is the circle. The **circumference**, *C*, is the length (or perimeter) of the curve that forms a circle. The **radius**, *r*, is the line segment from the center of the circle to any point on the circle (**Fig. 2.4a**). The **diameter** of a circle is a line segment through the center whose endpoints both lie on the circle (**Fig. 2.4b**). *Note that the length of the diameter is twice the length of the radius.*

The formulas for both the area and the circumference of a circle are given in **Table 2.2**.

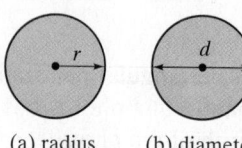

(a) radius   (b) diameter

**FIGURE 2.4**

## Understanding Algebra

$\pi$ has an infinite, non-repeating decimal expansion. It is sometimes approximated by $\frac{22}{7}$ or by 3.14.

Note: If you use 3.14 for $\pi$, your answers might vary slightly from the answers given in this book.

**TABLE 2.2** Formulas for Circles

| Circle | Area | Circumference |
|---|---|---|
| | $A = \pi r^2$ | $C = 2\pi r$ |

The value of **pi**, symbolized by the Greek lowercase letter $\pi$, is an irrational number that cannot be exactly expressed as a decimal number or a numerical fraction. Pi is *approximately* 3.14. Here it is approximated to 100 digits:

$\pi \approx$ 3.14159 26535 89793 23846 26433 83279 50288 41971 69399 37510 58209 74944 59230 78164 06286 20899 86280 34825 34211 7068...

**EXAMPLE 7 Pizza** A Pizza Hut large pizza has a diameter of 14 inches. Determine the area and circumference of the pizza.

*Solution* The radius is half the diameter, so $r = \frac{14}{2} = 7$ inches.

$A = \pi r^2$ $\qquad\qquad\qquad\qquad$ $C = 2\pi r$

$A = \pi (7)^2$ $\qquad\qquad\qquad\quad$ $C = 2\pi (7)$

$A = \pi (49)$ $\qquad\qquad\qquad\quad$ $C \approx 43.98$ inches

$A \approx 153.94$ square inches

To obtain our answers 153.94 and 43.98, we used the $\boxed{\pi}$ key on a calculator and rounded our final answer to the nearest hundredth. If you do not have a calculator with a $\boxed{\pi}$ key and use 3.14 for $\pi$, your answer for the area would be 153.86.

NowTry Exercise 95

**Table 2.3** below gives formulas for finding the volume of certain three-dimensional figures. **Volume** is the measure of the space occupied by a figure, and it is measured in cubic units, such as cubic centimeters or cubic feet.

EXAMPLE 8  **Spaceship Earth**  The inside of Spaceship Earth at Epcot Center in Disney World, Florida, is a sphere with a diameter of 165 feet (see photo). Determine the volume of Spaceship Earth.

*Solution*    Understand and Translate    **Table 2.3** gives the formula for the volume of a sphere. The formula involves the radius. Since the diameter is 165 feet, its radius is $\dfrac{165}{2} = 82.5$ feet.

$$V = \frac{4}{3}\pi r^3$$

Carry Out        $V = \dfrac{4}{3}\pi(82.5)^3 = \dfrac{4}{3}\pi(561{,}515.625) \approx 2{,}352{,}071.15$

Check and Answer   The volume inside the sphere is very large, so the answer of about 2,352,071.15 cubic feet is reasonable.

Now Try Exercise 103

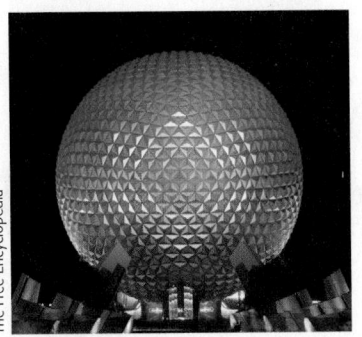

Spaceship Earth

**Understanding Algebra**

Legend for **Table 2.3**

$w$ = width

$l$ = length

$h$ = height

$r$ = radius

**TABLE 2.3    Formulas for Volumes of Three-Dimensional Figures**

| Figure | Sketch | Volume |
|---|---|---|
| Rectangular solid | | $V = lwh$ |
| Right circular cylinder | | $V = \pi r^2 h$ |
| Right circular cone | | $V = \dfrac{1}{3}\pi r^2 h$ |
| Sphere | | $V = \dfrac{4}{3}\pi r^3$ |

## 3  Solve for a Variable in a Formula

Often in this course and in other mathematics and science courses, you will be given an equation or formula solved for one variable and have to solve it for a different variable.

To solve for a variable in a formula, treat each of the quantities, except the one for which you are solving, as if they were constants. Then solve for the desired variable by isolating it on one side of the equation.

EXAMPLE 9  **Distance Formula**  Solve the distance formula $d = rt$ for $t$.

*Solution*    We must get $t$ all by itself on one side of the equals sign. Since $t$ is multiplied by $r$, we divide both sides of the equation by $r$ to isolate the $t$.

$$d = rt$$

$$\frac{d}{r} = \frac{\cancel{r}t}{\cancel{r}} \qquad \text{Divide both sides by } r.$$

$$\frac{d}{r} = t$$

Therefore, $t = \dfrac{d}{r}$.

Now Try Exercise 45

EXAMPLE 10  **Perimeter of Rectangle**  The formula for the perimeter of a rectangle is $P = 2l + 2w$. Solve this formula for the length, $l$.

*Solution*    We must get $l$ all by itself on one side of the equation.

$$P = 2l + 2w$$

$$P - 2w = 2l + 2w - 2w \qquad \text{Subtract } 2w \text{ from both sides.}$$

$$P - 2w = 2l$$

$$\frac{P - 2w}{2} = \frac{2l}{2} \qquad \text{Divide both sides by 2.}$$

$$\frac{P - 2w}{2} = l \quad \left( \text{or} \quad l = \frac{P}{2} - w \right)$$

Now Try Exercise 53

Some formulas contain fractions. When a formula contains a fraction, we can eliminate the fraction by multiplying both sides of the equation by the least common denominator.

EXAMPLE 11  The formula for the area of a triangle is $A = \dfrac{1}{2}bh$. Solve this formula for $h$.

*Solution*    We begin by multiplying both sides of the equation by the LCD, 2, to eliminate the fraction.

$$A = \frac{1}{2}bh$$

$$2 \cdot A = 2 \cdot \frac{1}{\cancel{2}}bh \qquad \text{Multiply both sides by 2.}$$

$$2A = bh$$

$$\frac{2A}{b} = \frac{\cancel{b}h}{\cancel{b}} \qquad \text{Divide both sides by } b \text{ to isolate } h.$$

$$\frac{2A}{b} = h$$

Thus, $h = \dfrac{2A}{b}$.

Now Try Exercise 51

### Write Equations in $y = mx + b$ Form

When discussing graphing later in this book, we will need to solve many equations for the variable $y$, and write the equation in the form $y = mx + b$, where $m$ and $b$ represent real numbers. Examples of equations in this form are $y = 2x + 4$, $y = -\dfrac{1}{2}x - 3$, and $y = \dfrac{4}{5}x + \dfrac{1}{3}$.

**EXAMPLE 12** Solve the equation $6x + 3y = 12$ for $y$. Write the answer in $y = mx + b$ form.

*Solution*   Begin by isolating the term containing the variable $y$.

$$6x + 3y = 12$$

$$6x - 6x + 3y = 12 - 6x \qquad \text{Subtract } 6x \text{ from both sides.}$$

$$3y = 12 - 6x$$

$$\frac{3y}{3} = \frac{12 - 6x}{3} \qquad \text{Divide both sides by 3.}$$

$$y = \frac{12 - 6x}{3}$$

$$y = \frac{12}{3} - \frac{6x}{3} \qquad \text{Wrote as two fractions.}$$

$$y = 4 - 2x$$

$$y = -2x + 4$$

Now Try Exercise 67

---

**Helpful Hint**

Notice that in Example 12, when we obtained $y = \dfrac{12 - 6x}{3}$, we had solved the equation for $y$ since the $y$ was isolated on one side of the equation. When we wrote the answer as $y = -2x + 4$, we wrote the equation in $y = mx + b$ form.

---

**EXAMPLE 13** Solve the equation $y - \dfrac{1}{3} = \dfrac{1}{4}(x - 6)$ for $y$. Write the answer in $y = mx + b$ form.

*Solution*   Multiply both sides of the equation by the LCD, 12.

$$y - \frac{1}{3} = \frac{1}{4}(x - 6)$$

$$12\left(y - \frac{1}{3}\right) = 12 \cdot \frac{1}{4}(x - 6) \qquad \text{Multiply both sides by 12.}$$

$$12y - 4 = 3(x - 6) \qquad \text{Distributive property used on left.}$$

$$12y - 4 = 3x - 18 \qquad \text{Distributive property used on right.}$$

$$12y = 3x - 14 \qquad \text{Added 4 to both sides.}$$

$$y = \frac{3x - 14}{12} \qquad \text{Divided both sides by 12.}$$

$$y = \frac{3x}{12} - \frac{14}{12} \qquad \text{Write as two fractions.}$$

$$y = \frac{1}{4}x - \frac{7}{6}$$

Now Try Exercise 79

In Example 13, you may wish to use the distributive property on the right side of the equation before multiplying both sides of the equation by the LCD, 12. Try working the example using this method now to see which procedure you prefer.

# EXERCISE SET 2.6    MathXL®   MyMathLab®

## Warm-Up Exercises

*Fill in the blanks with the appropriate word, phrase, or symbol(s) from the following list.*

| | | |
|---|---|---|
| formula | quadrilateral | rate · time |
| time | evaluating | cubic feet |

| | |
|---|---|
| radius | π | diameter |
| square inches | decimal | |

1. A four-sided figure is called a _____ .
2. Half the diameter is the _____ .
3. An equation used to express a specific relationship mathematically is called a _____ .
4. Interest is found by multiplying the principal times the rate times the _____ .
5. A possible unit of measure for volume is _____ .

6. A possible unit of measure for area is _____ .
7. When the circumference of a circle is divided by its diameter, the result is _____ .
8. Distance = _____ .
9. The process of substituting values and performing indicated operations on a formula is called _____ it.
10. When calculating simple interest, the rate should be expressed as a _____ .

## Practice the Skills

*Use the formula to find the value of the variable indicated. Round decimal answers to the nearest hundredth.*

11. $d = rt$ (distance formula); find $d$ when $r = 60$ and $t = 8$.
12. $P = 4s$ (perimeter of a square); find $P$ when $s = 6$.
13. $A = lw$ (area of a rectangle); find $A$ when $l = 12$ and $w = 8$.
14. $A = s^2$ (area of a square); find $A$ when $s = 9$.
15. $i = prt$ (simple interest formula); find $i$ when $p = 2000$, $r = 0.06$, and $t = 3$.
16. $c = 2.54i$ (to change inches to centimeters); find $c$ when $i = 12$.
17. $P = 2l + 2w$ (perimeter of a rectangle); find $P$ when $l = 8$ and $w = 10$.
18. $f = 1.47m$ (to change speed from mph to ft/sec); find $f$ when $m = 60$.
19. $A = \pi r^2$ (area of a circle); find $A$ when $r = 10$.
20. $A = \dfrac{m + n}{2}$ (mean of two values); find $A$ when $m = 16$ and $n = 56$.
21. $A = \dfrac{a + b + c}{3}$ (mean of three values); find $A$ when $a = 72$, $b = 81$, and $c = 93$.

22. $p = i^2 r$ (formula for finding electrical power); find $r$ when $p = 2000$ and $i = 4$.
23. $z = \dfrac{x - m}{s}$ (statistics formula for finding the $z$-score); find $z$ when $x = 100$, $m = 80$, and $s = 10$.
24. $A = \dfrac{1}{2}bh$ (area of a triangle); find $b$ when $A = 30$ and $h = 10$.
25. $P = 2l + 2w$ (perimeter of a rectangle); find $l$ when $P = 28$ and $w = 6$.
26. $A = P(1 + rt)$ (banking formula to find the amount in an account); find $r$ when $A = 1050$, $t = 1$, and $P = 1000$.
27. $V = \pi r^2 h$ (volume of a cylinder); find $h$ when $V = 678.24$ and $r = 6$.
28. $V = \dfrac{4}{3}\pi r^3$ (volume of a sphere); find $V$ when $r = 8$.
29. $B = \dfrac{703w}{h^2}$ (for finding body mass index); find $w$ when $B = 24$ and $h = 61$.
30. $S = C + rC$ (for determining selling price when an item is marked up); find $S$ when $C = 160$ and $r = 0.12$ (or 12%).

*In Exercises 31–36, use* **Tables 2.1, 2.2,** *and* **2.3** *to find the formula for the area or volume of the figure. Then determine either the area or volume.*

31.

32.

33.

34.

35.

36.

*In Exercises 37 and 38, use the formula $C = \frac{5}{9}(F - 32)$ to find the Celsius temperature (C) equivalent to the given Fahrenheit temperature (F).*

**37.** $F = 50°$ 

**38.** $F = 86°$

*In Exercises 39 and 40, use the formula $F = \frac{9}{5}C + 32$, to find the Fahrenheit temperature (F) equivalent to the given Celsius temperature (C).*

**39.** $C = 25°$ 

**40.** $C = -40°$

*In Exercises 41–44, find the missing quantity. Use the ideal gas law, $P = KT/V$, where P is pressure, T is temperature, V is volume, and K is a constant.*

**41.** $T = 20, K = 2, V = 1$ 

**42.** $P = 80, T = 100, V = 5$

**43.** $T = 30, P = 3, K = 0.5$ 

**44.** $P = 100, K = 2, V = 6$

*In Exercises 45–66, solve for the indicated variable.*

**45.** $A = lw$, for $l$

**46.** $P = 4s$, for $s$

**47.** $d = rt$, for $t$

**48.** $C = \pi d$, for $d$

**49.** $i = prt$, for $t$

**50.** $V = lwh$, for $l$

**51.** $A = \frac{1}{2}bh$, for $b$

**52.** $E = IR$, for $I$

**53.** $P = 2l + 2w$, for $w$

**54.** $PV = KT$, for $T$

**55.** $3 - 2r = n$, for $r$

**56.** $4m + 5n = 25$, for $n$

**57.** $y = mx + b$, for $b$

**58.** $y = mx + b$, for $x$

**59.** $d = a + b + c$, for $b$

**60.** $ax + by = c$, for $y$

**61.** $ax + by + c = 0$, for $y$

**62.** $V = \pi r^2 h$, for $h$

**63.** $V = \frac{1}{3}\pi r^2 h$, for $h$

**64.** $A = \frac{m + 2d}{3}$, for $d$

**65.** $A = \frac{m + d}{2}$, for $m$

**66.** $L = \frac{c + 2d}{4}$, for $d$

*In Exercises 67–82, solve each equation for y. Write the answer in $y = mx + b$ form. See Examples 12 and 13.*

**67.** $2x + y = 8$

**68.** $6x + 2y = -12$

**69.** $-3x + 3y = -18$

**70.** $-2y + 4x = -8$

**71.** $4x = 6y - 8$

**72.** $15 = 3y - x$

**73.** $5y = -10 + 3x$

**74.** $-2y = -3x - 18$

**75.** $-6y = 15 - 3x$

**76.** $-12 = -2x - 3y$

**77.** $-10 = -x - 2y$

**78.** $4x + 3y = 20$

**79.** $y + 3 = -\frac{1}{3}(x - 4)$

**80.** $y - 3 = \frac{2}{3}(x + 4)$

**81.** $y - \frac{1}{5} = 2\left(x + \frac{1}{3}\right)$

**82.** $y + 5 = \frac{3}{4}\left(x + \frac{1}{2}\right)$

## Problem Solving

*In Exercises 83–86, use the simple interest formula.*

**83. Auto Loan** Thang Tran decided to borrow $6000 from Citibank to help pay for a car. His loan was for 3 years at a simple interest rate of 8%. How much interest will Thang pay?

**84. Simple Interest Loan** Holly Broesamle lent her brother $4000 for a period of 2 years. At the end of the 2 years, her brother repaid the $4000 plus $640 interest. What simple interest rate did her brother pay?

*In Exercises 87–90, use the distance formula.*

**87. Average Speed** On her way from Omaha, Nebraska, to Kansas City, Kansas, Peg Hovde traveled 180 miles in 3 hours. What was her average speed?

**85. Savings Account** Mary Seitz invested a certain amount of money in a savings account paying 3% simple interest per year. When she withdrew her money at the end of 3 years, she received $450 in interest. How much money did Mary place in the savings account?

**86. Savings Account** Peter Ostroushko put $6000 in a savings account earning $3\frac{1}{2}$% simple interest per year. When he withdrew his money, he received $840 in interest. How long had he left his money in the account?

**88. Walk** Lisa Feintech went for a walk where she walked at an average speed of 3.4 miles per hour for 2 hours. How far did she walk?

**89. Fastest Car** The fastest speed recorded on land was about 763.2 miles per hour by a jet-powered car called ThrustSSC. If, during the speed trial, the car traveled for 0.01 hour, how far had the car traveled?

The ThrustSSC

**90. Fastest Plane** The fastest aircraft is the Lockheed SR-71 Blackbird. If, during the trial run, the plane covered a distance of 660 miles in 0.3 hours, determine the plane's average speed.

*Use the formulas given in* **Tables 2.1, 2.2,** *and* **2.3** *to work Exercises 91–104.*

**91. DVD Player** A portable DVD player has a screen with a length of 8 inches and a width (or height) of 6 inches. Determine the area of the screen.

**92. Television** A plasma television has a screen with a length of 34.9 inches and a width (or height) of 19.6 inches. Determine the perimeter of the screen.

**93. Yield Sign** A yield traffic sign is triangular with a base of 36 inches and a height of 31 inches. Find the area of the sign.

**94. Fencing** Milt McGowen has a rectangular lot that measures 100 feet by 60 feet. If Milt wants to fence in his lot, how much fencing will he need?

**95. Swimming Pool** A circular above-ground swimming pool has a diameter of 24 feet. Determine the circumference of the pool.

**96. Living Room Table** A round living room table top has a diameter of 3 feet. Find the area of the table top.

**97. Kite** Below we show a kite. Determine the area of the kite.

**98. Laptop** The carrying case for a laptop computer measures 17 inches long by 14 inches wide by 6 inches deep. Determine the volume of the case.

**99. Trapezoidal Sign** Canter Martin made a sign to display at a baseball game. The sign was in the shape of a trapezoid. Its bases are 4 feet and 3 feet, and its height is 2 feet. Find the area of the sign.

**100. Jacuzzi** The inside of a circular jacuzzi is 8 feet in diameter. If the water inside the jacuzzi is 3 feet deep, determine, in cubic feet, the volume of water in the jacuzzi.

**101. Banyan Tree** The largest banyan tree in the continental United States is at the Edison House in Fort Myers, Florida. The circumference of the aerial roots of the tree is 390 feet. Find the *diameter* of the aerial roots to the nearest tenth of a foot.

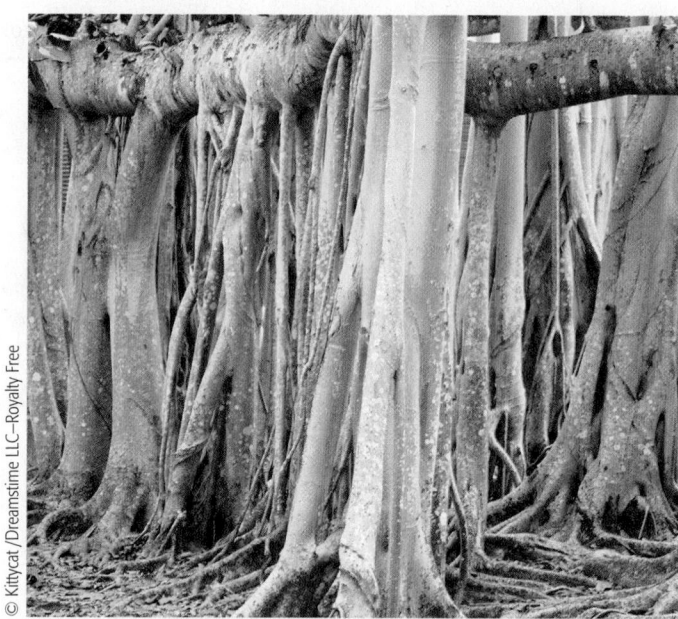

**102. Amphitheater** The seats in an amphitheater are inside a trapezoidal area as shown in the figure. The bases of the trapezoidal area are 80 feet and 200 feet, and the height is 100 feet. Find the area of the floor occupied by seats.

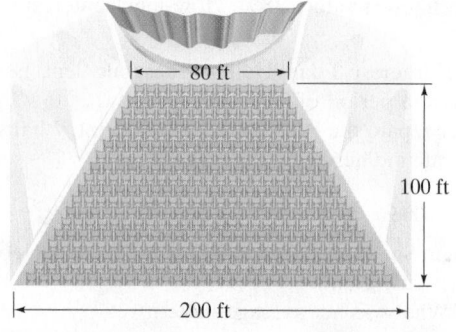

**103. Basketball** Find the volume of a basketball if its diameter is 9 inches.

**104. Oil Drum** Roberto Sanchez has an empty oil drum that he uses for storage. The oil drum is 4 feet high and has a diameter of 24 inches. Find the volume of the drum in cubic feet.

**105. Body Mass Index** A person's body mass index (BMI) is found by multiplying a person's weight, $w$, in pounds by 703, then dividing this product by the square of the person's height, $h$, in inches.

a) Write a formula to find the BMI.

b) Brandy Belmont is 5 feet 3 inches tall and weighs 135 pounds. Find her BMI.

**106. Body Mass Index** Refer to Exercise 105. Mario Guzza's weight is 162 pounds, and he is 5 feet 7 inches tall. Find his BMI.

## Challenge Problems

**107. Cereal Box** A cereal box is to be made by folding the cardboard along the dashed lines as shown in the figure on the right.

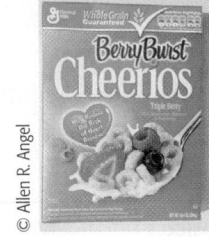

© Allen R. Angel

a) Using the formula

$$\text{volume} = \text{length} \cdot \text{width} \cdot \text{height}$$

write an equation for the volume of the box.

b) Find the volume of the box when $x = 7$ cm.

c) Write an equation for the surface area of the box.

d) Find the surface area when $x = 7$ cm.

## Concept/Writing Exercises

**108.** When using the distance formula, what happens to the distance if the rate is doubled and the time is halved? Explain.

**109.** When using the simple interest formula, what happens to the simple interest if both the principal and rate are doubled but the time is halved? Explain.

**110.** Consider the formula for the area of a square, $A = s^2$. If the length of the side of a square, $s$, is doubled, what is the change in its area? Explain.

**111.** Consider the formula for the volume of a cube, $V = s^3$. If the length of the side of a cube, $s$, is doubled, what is the change in its volume? Explain.

**112.** Which would have the greater area, a square whose side has a length of $s$ inches, or a circle whose diameter has a length of $s$ inches? Explain, using a sketch.

**113.** Which would have the greater area, a square whose diagonal has a length of $s$ inches, or a circle whose diameter has a length of $s$ inches? Explain, using a sketch.

## Group Activity

**114. Square Face on Cube** Consider the following photo. The front of the figure is a square with a smaller black square painted on the center of the larger square. Suppose the length of one side of the larger square is $A$, and the length of one side of the smaller (black) square is $B$. Also the thickness of the block is $C$.

© Allen R. Angel

a) Group member one: Determine an expression for the surface area of the black square.

b) Group member two: Determine an expression for the surface area of the larger square (which includes the smaller square).

c) Group member three: Determine the surface area of the larger square minus the black square (the purple area shown).

d) As a group, write an expression for the volume of the entire solid block.

e) As a group, determine the volume of the entire solid block if its length is 1.5 feet and its width is 0.8 feet.

## Cumulative Review Exercises

[1.7]  **115.** Evaluate $-\dfrac{4}{15} + \dfrac{3}{5}$.

**116.** Evaluate $-6 + 7 - 4 - 3$.

[1.9]  **117.** Evaluate. $[4(12 \div 2^2 - 3)^2]^2$.

[2.4]  **118.** Solve the equation $\dfrac{r}{2} + 2r = 20$.

# 2.7 Ratios and Proportions

1  Understand ratios.

2  Solve proportions using cross-multiplication.

3  Solve applications.

4  Use proportions to change units.

5  Use proportions to solve problems involving similar figures.

## 1  Understand Ratios

### Ratio

A **ratio** is a quotient of two quantities.

Ratios provide a way to compare two numbers or quantities. The ratio of the number $a$ to the number $b$ may be written

$$a \text{ to } b, \quad a:b, \quad \text{or} \quad \frac{a}{b}$$

where $a$ and $b$ are called the **terms of the ratio.** Notice that the symbol : can be used to indicate a ratio.

**EXAMPLE 1**  **Favorite Movies**  Several children in sixth grade were asked to name their favorite movie of 2007. The results are indicated on the graph in **Figure 2.5.**

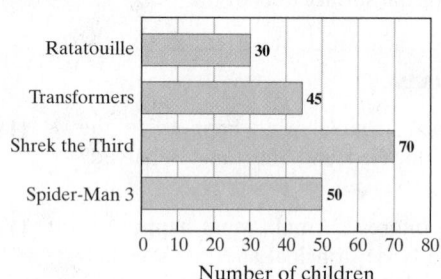

**Favorite Movies**

FIGURE 2.5

**a)** Find the ratio of the number of children who selected *Shrek the Third* to those who selected *Transformers*. Then write the ratio in lowest terms.

**b)** Find the ratio of the number of children who selected *Spider-Man 3* to the total number in the survey.

*Solution*    We will use our five-step problem-solving procedure.

**a)** Understand and Translate    The ratio we are seeking is

Number who selected *Shrek the Third* : Number who selected *Transformers*

Carry Out    We substitute the appropriate values into the ratio.

This gives

$$70:45$$

To write the ratio in *lowest terms*, we simplify by dividing each number in the ratio by 5, the greatest number that divides both terms in the ratio. This gives

$$14:9$$

Check and Answer    Our division is correct. The ratio is $14:9$.

**b)** We use the same procedure as in part **a)**. Fifty children selected *Spider-Man 3*. There were $50 + 70 + 45 + 30$ or 195 children surveyed. Thus, the ratio is $50:195$, which simplifies to

$$10:39$$

### Understanding Algebra

A *ratio* is another word for a fraction. It can be denoted 3 different ways: $\dfrac{a}{b}$, $a:b$, or with the word "to" as "$a$ to $b$."

Now Try Exercise 33

The answer in Example 1, part **a)** could have also been written $\frac{14}{9}$ or 14 to 9. The answer in part **b)** could have also been written $\frac{10}{39}$ or 10 to 39.

EXAMPLE 2  **Cholesterol Level**  There are two types of cholesterol: low-density lipoprotein, (LDL—considered the harmful type of cholesterol) and high-density lipoprotein (HDL—considered the healthful type of cholesterol). Some doctors recommend that the ratio of LDL to HDL be less than or equal to 4 : 1. Mr. Suarez's cholesterol test showed that his LDL was 167 milligrams per deciliter, and his HDL was 40 milligrams per deciliter. Is Mr. Suarez's ratio of LDL to HDL less than or equal to the recommended 4 : 1 ratio?

*Solution*    Understand    We need to determine if Mr. Suarez's LDL to HDL ratio is less than or equal to 4 : 1.

Translate    Mr. Suarez's LDL to HDL ratio is 167 : 40. To make the second term equal to 1, we divide both terms in the ratio by the second term, 40.

Carry Out
$$\frac{167}{40} : \frac{40}{40}$$
$$\text{or } 4.175 : 1$$

Check and Answer    Our division is correct. Therefore, Mr. Suarez's ratio is not less than or equal to the desired 4 : 1 ratio.

Now Try Exercise 87

EXAMPLE 3  **Gas–Oil Mixture**  Some power equipment, such as chainsaws and blowers, use a gas–oil mixture to run the engine. The instructions on a particular chainsaw indicate that 5 gallons of gasoline should be mixed with 40 ounces of special oil to obtain the proper gas–oil mixture. Find the ratio of gasoline to oil in the proper mixture.

*Solution*    Understand    To express these quantities in a ratio, both quantities must be in the same units. We can either convert 5 gallons to ounces or 40 ounces to gallons.

Translate    Let's change 5 gallons to ounces. Since there are 128 ounces in 1 gallon, 5 gallons of gas equals 5(128) or 640 ounces. The ratio we are seeking is

$$\text{ounces of gasoline : ounces of oil}$$

Carry Out
$$640 : 40$$
$$\text{or } 16 : 1 \qquad \text{Divide both terms by 40 to simplify.}$$

Check and Answer    Our simplification is correct. The correct ratio of gas to oil for this chainsaw is 16 : 1.

Now Try Exercise 23

EXAMPLE 4  **Gear Ratio**  The *gear ratio* of two gears is defined as

$$\text{gear ratio} = \frac{\text{number of teeth on the driving gear}}{\text{number of teeth on the driven gear}}$$

Find the gear ratio of the gears shown in **Figure 2.6.**

*Solution*    Understand and Translate    To find the gear ratio we need to substitute the appropriate values.

Carry Out    $\text{gear ratio} = \dfrac{\text{number of teeth on driving gear}}{\text{number of teeth on driven gear}} = \dfrac{60}{8} = \dfrac{15}{2}$

Thus, the gear ratio is 15 : 2. Gear ratios are generally given as some quantity to 1. If we divide both terms of the ratio by the second term, we will obtain a ratio of some number to 1. Dividing both 15 and 2 by 2 gives a gear ratio of 7.5 : 1.

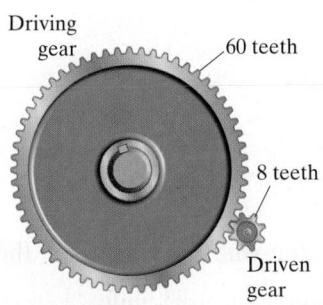

Driving gear — 60 teeth

8 teeth

Driven gear

**FIGURE 2.6**

**Check and Answer**    The gear ratio is 7.5 : 1. This means that as the driving gear goes around once the driven gear goes around 7.5 times. (A typical first gear ratio on a passenger car may be 3.545 : 1.)

Now Try Exercise 27

## 2 Solve Proportions Using Cross-Multiplication

### Proportion

A **proportion** is a special type of equation. It is a statement of equality between two ratios.

One way of denoting a proportion is $a : b = c : d$, which is read "$a$ is to $b$ as $c$ is to $d$." In this text we write proportions as

$$\frac{a}{b} = \frac{c}{d}$$

### Understanding Algebra

When two ratios are set equal, we call it a *proportion*. For example, consider $\frac{3}{12} = \frac{1}{4}$. The numbers 3 and 4 are called the *extremes* and 12 and 1 are called the *means* of the proportion. *The product of the means* $(12 \cdot 1 = 12)$ *always equals the product of the extremes* $(3 \cdot 4 = 12)$.

The $a$ and $d$ are referred to as the **extremes**, and the $b$ and $c$ are referred to as the **means** of the proportion. In Sections 2.4 and 2.5, we solved equations containing fractions by multiplying both sides of the equation by the LCD to eliminate fractions. For example, for the proportion

$$\frac{x}{3} = \frac{35}{15}$$

$$15\left(\frac{x}{3}\right) = 15\left(\frac{35}{15}\right) \quad \text{Mulyiply both sides by the LCD, 15.}$$

$$5x = 35$$

$$x = 7$$

Another method that can be used to solve proportions is **cross-multiplication**.

### Cross-Multiplication

If $\dfrac{a}{b} = \dfrac{c}{d}$, then $ad = bc$.

*The product of the extremes, ad, is equal to the product of the means, bc.*

If any three of the four quantities of a proportion are known, the fourth quantity can easily be found.

**EXAMPLE 5**    Solve $\dfrac{x}{3} = \dfrac{35}{15}$ for $x$ by cross-multiplying.

*Solution*

$$\frac{x}{3} = \frac{35}{15}$$

$$x \cdot 15 = 3 \cdot 35$$

$$15x = 105$$

$$x = \frac{105}{15} = 7$$

Check

$$\frac{x}{3} = \frac{35}{15}$$

$$\frac{7}{3} \overset{?}{=} \frac{35}{15}$$

$$\frac{7}{3} = \frac{7}{3} \quad \text{True}$$

Now Try Exercise 37

Earlier, we solved the proportion $\dfrac{x}{3} = \dfrac{35}{15}$ by multiplying both sides of the equation by 15. In Example 5, we solved the same proportion using cross-multiplication. Notice we obtained the same solution, 7, in each case.

**EXAMPLE 6** Solve $\dfrac{-8}{3} = \dfrac{64}{x}$ for $x$ by cross-multiplying.

*Solution*

$$\dfrac{-8}{3} = \dfrac{64}{x}$$

$$-8 \cdot x = 3 \cdot 64$$

$$-8x = 192$$

$$\dfrac{-8x}{-8} = \dfrac{192}{-8}$$

$$x = -24$$

Check

$$\dfrac{-8}{3} = \dfrac{64}{x}$$

$$\dfrac{-8}{3} \overset{?}{=} \dfrac{64}{-24}$$

$$\dfrac{-8}{3} \overset{?}{=} \dfrac{8}{-3}$$

$$\dfrac{-8}{3} = \dfrac{-8}{3} \quad \text{True}$$

Now Try Exercise 41

## **3** Solve Applications

Often, practical problems can be solved using proportions. Below are specific directions for translating application problems into proportions.

> **To Solve Problems Using Proportions**
>
> 1. Understand the problem.
> 2. Translate the problem into mathematical language.
>    a) First, represent the unknown quantity by a variable.
>    b) Second, set up the proportion by listing the given ratio on one side of the equals sign, and the variable and the other given quantity on the other side of the equals sign. When setting up the proportion, the same respective quantities should occupy the same respective positions on the left and the right. For example, an acceptable proportion might be
>
> $$\text{Given ratio} \left\{ \dfrac{\text{miles}}{\text{hour}} = \dfrac{\text{miles}}{\text{hour}} \right.$$
>
> 3. Carry out the mathematical calculations necessary to solve the problem.
>    a) Once the proportion is correctly written, drop the units and cross-multiply.
>    b) Solve the resulting equation and label with the appropriate units.
> 4. Check the answer obtained in step 3.
> 5. Make sure you have answered the question.

### Understanding Algebra

When solving applied problems involving proportions, it is crucial that each ratio has the same units! For example, if one ratio is given in miles/hour and the second ratio is given in feet/hour, one of the ratio's units must be changed before setting up the proportion.

**EXAMPLE 7** **Painting** A gallon of paint will cover a surface area of 575 square feet.

**a)** How many gallons of paint are needed to cover a house with a surface area of 6525 square feet?

**b)** If a gallon of paint costs $24.99, what will it cost to paint the house?

*Solution*

**a)** Understand   The given ratio is 1 gallon per 575 square feet. The unknown quantity is the number of gallons necessary to cover 6525 square feet.

Translate   Let $x$ = number of gallons of paint needed.

$$\text{Given ratio} \left\{ \dfrac{1 \text{ gallon}}{575 \text{ square feet}} = \dfrac{x \text{ gallons}}{6525 \text{ square feet}} \right. \begin{array}{l} \leftarrow \text{Unknown} \\ \leftarrow \text{Given quantity} \end{array}$$

Note how the amount, in gallons, and the area, in square feet, are given in the same relative positions.

© iofoto/Shutterstock

Carry Out

$$\frac{1}{575} = \frac{x}{6525}$$

$1(6525) = 575x$    Cross-multiply.

$6525 = 575x.$    Solve.

$$\frac{6525}{575} = x$$

$$11.3 \approx x$$

Check   Both ratios in the proportion, $\frac{1}{575}$ and $\frac{11.3}{6525}$, have approximately the same value of 0.00173. Thus, the answer of about 11.3 gallons checks.

Answer   The amount of paint needed to cover an area of 6525 square feet is about 11.3 gallons.

**b)** Assuming the painter only buys full gallons of paint, he will need to buy 12 gallons in order to paint the house. Since each gallon costs $24.99, the cost to paint the house is found by multiplication.

$$12 \times 24.99 = \$299.88$$

The cost to paint the house is $299.88.

Now Try Exercise 61

© Allen R. Angel

EXAMPLE 8 At a charity luncheon, the guests meet and get autographs from members of the New York Yankees. If a particular player signs, on the average, 33 autographs in 4 minutes, how much time must be allowed for him to sign 350 autographs?

*Solution*   The unknown quantity is the time needed for the player to sign 350 autographs. We are given that, on the average, he signs 33 autographs in 4 minutes. We will use this given ratio in setting up our proportion.

Translate   We will let $x$ represent the time to sign 350 autographs.

$$\text{Given ratio} \left\{ \frac{33 \text{ autographs}}{4 \text{ minutes}} = \frac{350 \text{ autographs}}{x \text{ minutes}} \right.$$

Carry Out

$$\frac{33}{4} = \frac{350}{x}$$

$$33x = 4(350)$$

$$33x = 1400$$

$$x = \frac{1400}{33} \approx 42.4$$

Check and Answer   Both ratios in the proportion, $\frac{33}{4}$ and $\frac{350}{42.4}$, have approximately the same value of 8.25. Thus, about 42.4 minutes would be needed for the player to sign the 350 autographs.

Now Try Exercise 55

EXAMPLE 9 **Drug Dosage** A nurse is to give a patient 250 milligrams of the drug simethicone. The drug is available only in a solution whose concentration is 40 milligrams of simethicone per 0.6 milliliter of solution. How many milliliters of solution should the nurse give the patient?

*Solution*   Understand and Translate   We can set up the proportion using the medication on hand as the given ratio and the number of milliliters needed to be given as the unknown.

$$\text{Given ratio} \left\{ \frac{40 \text{ milligrams}}{0.6 \text{ milliliter}} = \frac{250 \text{ milligrams}}{x \text{ milliliters}} \right. \quad \begin{matrix} \longleftarrow \text{ Desired medication} \\ \longleftarrow \text{ Unknown} \end{matrix}$$

Carry Out

$$\frac{40}{0.6} = \frac{250}{x}$$

$$40x = 0.6(250) \qquad \text{Cross-multiply.}$$

$$40x = 150 \qquad \text{Solve.}$$

$$x = \frac{150}{40} = 3.75$$

Check and Answer   The nurse should administer 3.75 milliliters of the simethicone solution.

Now Try Exercise 69

**Avoiding Common Errors**

When you are setting up a proportion, it does not matter which unit in the given ratio is in the numerator and which is in the denominator as long as the units in the other ratio are *in the same relative position*. For example,

$$\frac{60 \text{ miles}}{1.5 \text{ hours}} = \frac{x \text{ miles}}{4.2 \text{ hours}} \quad \text{and} \quad \frac{1.5 \text{ hours}}{60 \text{ miles}} = \frac{4.2 \text{ hours}}{x \text{ miles}}$$

will both give the same answer of 168 (try it and see). When setting up the proportion, set it up so that it makes the most sense to you. Notice that when setting up a proportion containing different units, the same units should not be multiplied by themselves during cross-multiplication.

CORRECT

$$\frac{\text{miles}}{\text{hour}} = \frac{\text{miles}}{\text{hour}}$$

INCORRECT

$$\frac{\text{miles}}{\text{hour}} \diagdown = \diagup \frac{\text{hour}}{\text{miles}}$$

## 4 Use Proportions to Change Units

Proportions can also be used to convert from one quantity to another. For example, you can use a proportion to convert a measurement in feet to a measurement in meters, or to convert from pounds to kilograms.

EXAMPLE 10   **Converting Kilometers to Miles**   There are approximately 1.6 kilometers in 1 mile. What is the distance, in miles, of 78 kilometers?

*Solution*   Understand and Translate   We know that 1 mile ≈ 1.6 kilometers. We use this known fact in one ratio of our proportion. In the second ratio, we set the quantities with the same units in the same respective positions. The unknown quantity is the number of miles, which we will call $x$.

$$\text{Known ratio} \left\{ \frac{1 \text{ mile}}{1.6 \text{ kilometers}} = \frac{x \text{ miles}}{78 \text{ kilometers}} \right.$$

Note that both numerators contain the same units, and both denominators contain the same units.

Carry Out    Now solve for $x$ by cross-multiplying.

$$\frac{1}{1.6} = \frac{x}{78}$$

$$1(78) = 1.6x \quad \text{Cross-multiply.}$$

$$78 = 1.6x \quad \text{Solve.}$$

$$\frac{78}{1.6} = \frac{1.6x}{1.6}$$

$$48.75 = x$$

Check and Answer    Thus, 78 kilometers equals about 48.75 miles.

Now Try Exercise 75

EXAMPLE 11    **Exchanging Currency**    When people travel to a foreign country they often need to exchange currency. Donna Boccio visited Cancun, Mexico. She stopped by a local bank and was told that $1 U.S. could be exchanged for 10.96 pesos.

a) How many pesos would she get if she exchanged $150 U.S.?

b) Later that same day, Donna went to the city market where she purchased a ceramic figurine. The price she negotiated for the figurine was 245 pesos. Using the exchange rate given, determine the cost of the figurine in U.S. dollars.

*Solution*

a) Understand    We use the fact that $1 U.S. can be exchanged for 10.96 Mexican pesos. We use this known fact for one ratio in our proportion. In the second ratio, we set the quantities with same units in the same respective positions.

Translate    The unknown quantity is the number of pesos, which we shall call $x$.

$$\text{Given ratio} \left\{ \frac{\$1\text{ U.S.}}{10.96\text{ pesos}} = \frac{\$150\text{ U.S.}}{x\text{ pesos}} \right.$$

Note that both numerators contain U.S. dollars and both denominators contain pesos.

Carry Out    $$\frac{1}{10.96} = \frac{150}{x}$$

$$1x = 10.96(150)$$

$$x = 1644$$

Check and Answer    Thus, $150 U.S. could be exchanged for 1644 Mexican pesos.

b) Understand and Translate    We use the same given ratio that we used in part a). Now we must find the equivalent in U.S. dollars of 245 Mexican pesos. Let's call the equivalent U.S. dollars $x$.

$$\text{Given ratio} \left\{ \frac{\$1\text{ U.S.}}{10.96\text{ pesos}} = \frac{\$x\text{ U.S.}}{245\text{ pesos}} \right.$$

Carry Out    $$\frac{1}{10.96} = \frac{x}{245}$$

$$1(245) = 10.96x$$

$$245 = 10.96x$$

$$22.35 \approx x$$

Check and Answer    The cost of the figurine in U.S. dollars is $22.35.

Now Try Exercise 85

> **Helpful Hint**
>
> Some of the problems we have just worked using proportions could have been done without using proportions. However, when working problems of this type, students often have difficulty in deciding whether to multiply or divide to obtain the correct answer. By setting up a proportion, you may be better able to understand the problem and have more success in obtaining the correct answer.

## 5  Use Proportions to Solve Problems Involving Similar Figures

### Understanding Algebra

Two figures are called *similar* if they have the same shape but (possibly) different size.

> **Similar Figures**
>
> Two figures are said to be **similar** when their corresponding angles are equal and their corresponding sides are in proportion.

Proportions can be used to solve problems involving similar figures.

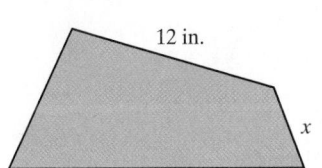

**EXAMPLE 12** The figures to the left are similar. Find the length of the side indicated by the *x*.

*Solution*   We set up a proportion of corresponding sides to find the length of side *x*.

5 inches and 12 inches are corresponding sides of similar figures. $\longrightarrow \dfrac{5}{2} = \dfrac{12}{x}$

2 inches and *x* are corresponding sides of similar figures.

$$\dfrac{5}{2} = \dfrac{12}{x}$$

(Lengths from smaller figure / Lengths from larger figure)

$$5x = 24$$

$$x = \dfrac{24}{5} = 4.8$$

Thus, the side indicated by *x* is 4.8 inches in length.

Now Try Exercise 51

Note in Example 12 that the proportion could have also been set up as

$$\dfrac{5}{12} = \dfrac{2}{x}$$

because one pair of corresponding sides is in the numerators and another pair is in the denominators.

**EXAMPLE 13** Triangles *ABC* and *AB'C'* are similar triangles. Find the length of side *AB'*.

*Solution*   We set up a proportion of corresponding sides to find the length of side *AB'*. We will let *x* represent the length of side *AB'*.

$$\dfrac{\text{length of } AB}{\text{length of } BC} = \dfrac{\text{length of } AB'}{\text{length of } B'C'}$$

Now we insert the proper values and solve for the variable, *x*.

$$\dfrac{15}{9} = \dfrac{x}{7.2}$$

$$(15)(7.2) = 9x$$

$$108 = 9x$$

$$12 = x$$

Thus, the length of side *AB'* is 12 inches.

Now Try Exercise 53

# EXERCISE SET 2.7    MathXL®   MyMathLab®
MathXL®    MyMathLab

## Warm-Up Exercises

*Fill in the blanks with the appropriate word, phrase, or symbol(s) from the following list.*

ratio                cross-multiplying            similar              $c:d$                    $d:c$

units               proportion                  means               extremes

1. A _____ is a statement of equality between two ratios.

2. In the proportion $\frac{a}{b} = \frac{p}{q}$, $a$ and $q$ are called _____.

3. In the proportion $\frac{a}{b} = \frac{p}{q}$, $b$ and $p$ are called _____.

4. A method for solving proportions is _____.

5. Two figures that have equal corresponding angles and have their corresponding sides in proportion are called _____.

6. Another way of expressing the ratio $\frac{c}{d}$ is _____.

7. When solving an applied problem involving proportions, it is crucial that the two ratios have the same _____.

8. The quotient of two quantities is called a _____.

*In Exercises 9–12, is the proportion set up correctly?*

9. $\dfrac{\text{gal}}{\text{min}} = \dfrac{\text{gal}}{\text{min}}$

10. $\dfrac{\text{mi}}{\text{hr}} = \dfrac{\text{mi}}{\text{hr}}$

11. $\dfrac{\text{ft}}{\text{sec}} = \dfrac{\text{sec}}{\text{ft}}$

12. $\dfrac{\text{tax}}{\text{cost}} = \dfrac{\text{cost}}{\text{tax}}$

## Practice the Skills

*The results of a mathematics examination are 6 A's, 4 B's, 9 C's, 3 D's, and 2 F's. Write the following ratios in lowest terms.*

13. A's to C's

14. F's to total grades

15. D's to A's

16. Grades better than C to total grades

17. Total grades to D's

18. Grades better than C to grades less than C

*Determine the following ratios. Write each ratio in lowest terms.*

19. 7 gallons to 4 gallons

20. 50 cents to 60 dollars

21. 5 ounces to 15 ounces

22. 18 liters to 24 liters

23. 3 hours to 30 minutes

24. 6 feet to 4 yards

25. 7 dimes to 12 quarters

26. 26 ounces to 4 pounds

*In Exercises 27 and 28, find the gear ratio. Write the ratio as some quantity to 1. (See Example 4.)*

27. Driving gear, 40 teeth; driven gear, 5 teeth

28. Driving gear, 30 teeth; driven gear, 8 teeth

*In Exercises 29–32, **a)** Determine the indicated ratio, and **b)** write the ratio as some quantity to 1.*

29. **American Consumers**  Each year the average American consumer drinks approximately 50 gallons of soft drinks compared to 26 gallons of coffee, 23 gallons of milk, and less than 10 gallons of fruit juices. What is the ratio of the number of gallons of soft drinks consumed to the number of gallons of milk consumed?

30. **Mail Letter**  In January 2009, the cost to mail a one-ounce letter was 42 cents and the cost to mail a two-ounce letter was 59 cents. What is the ratio of the cost to mail a one-ounce letter to the cost to mail a two-ounce letter?

31. **Minimum Wage**  The minimum wage in the United States in 1996 was \$4.75 per hour and the minimum wage in 2009 was \$7.25 per hour. What is the ratio of the U.S. minimum wage in 2009 to the U.S. minimum wage in 1996?

32. **Population**  The United States population in 1990 was about 249 million, and the population in 2008 was about 305 million. What is the ratio of the U.S. population in 2008 to the U.S. population in 1990?

*Exercises 33–36 show graphs. For each exercise, find the indicated ratio.*

33. **Traveling the Toll Roads**  See figure on p. 153.

   a) Determine the ratio of the toll rate on the Delaware Turnpike to the toll rate on the Garden State Parkway.

   b) Determine the ratio of the toll rate on the NJ Turnpike to the toll rate on PA Route 66.

**How Toll Rates Compare**

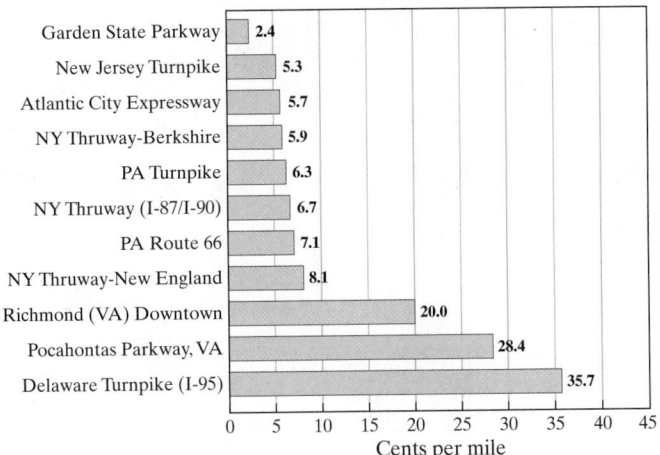

*Source*: Pennsylvania Turnpike Commission, 2008

**34. Number of Passports Processed**

**a)** Estimate the ratio of passports processed in the United States in 1999 to those processed in the United States in 2007.

**b)** Estimate the ratio of passports processed in the United States in 2002 to those processed in the United States in 2004.

**Passports Processed in the U.S.**
Fiscal year (Oct.1–Sept. 30)

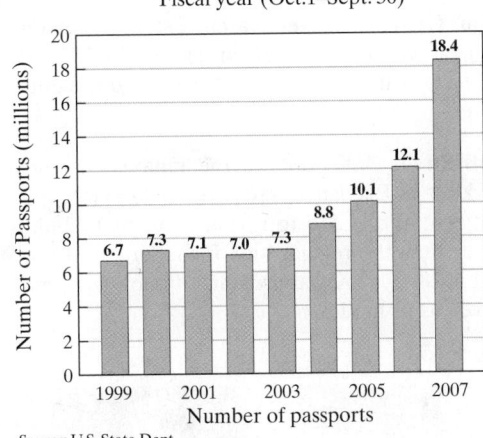

*Source*: U.S. State Dept.

**35. Favorite Doughnut**

**a)** Determine the ratio of people whose favorite doughnut is glazed to people whose favorite doughnut is filled.

**b)** Determine the ratio of people whose favorite doughnut is frosted to people whose favorite doughnut is plain.

**Favorite Doughnut Flavors**

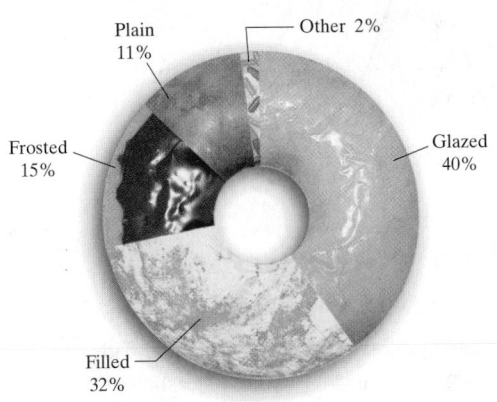

*Source:* The Heller Research Group

**36. Commuting to Work**

**a)** Determine the ratio of workers who drove alone to work to those who carpooled to work.

**b)** Determine the ratio of workers who walked to work to those who worked at home.

**Commuting to Work, 2006**

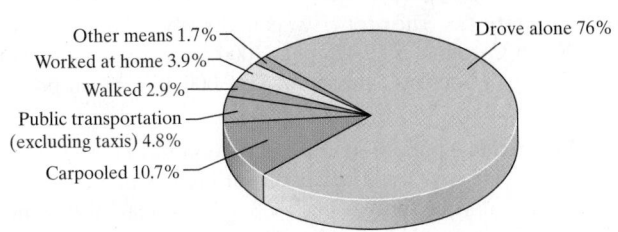

*Source*: www.factfinder.census.gov

*Solve each proportion for the variable by cross-multiplying.*

**37.** $\dfrac{x}{3} = \dfrac{20}{5}$

**38.** $\dfrac{x}{8} = \dfrac{24}{48}$

**39.** $\dfrac{5}{3} = \dfrac{75}{a}$

**40.** $\dfrac{x}{3} = \dfrac{90}{30}$

**41.** $\dfrac{-7}{3} = \dfrac{21}{p}$

**42.** $\dfrac{-12}{13} = \dfrac{36}{x}$

**43.** $\dfrac{15}{45} = \dfrac{x}{-6}$

**44.** $\dfrac{y}{6} = \dfrac{7}{42}$

**45.** $\dfrac{3}{z} = \dfrac{-1.5}{27}$

**46.** $\dfrac{3}{12} = \dfrac{-1.4}{z}$

**47.** $\dfrac{9}{12} = \dfrac{x}{8}$

**48.** $\dfrac{2}{20} = \dfrac{x}{200}$

*The following figures are similar. For each pair, find the length of the side indicated by x.*

**49.**

3 in.

8 in.

12 in.      x

**50.**

2 ft     1.8 ft

0.8 ft     x

**51.**

**52.**

**53.**

**54.**

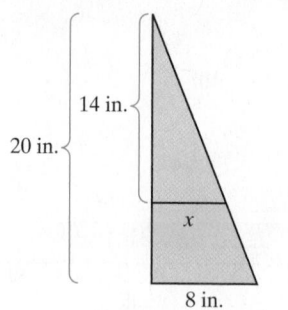

## Problem Solving

*In Exercises 55–74, write a proportion that can be used to solve the problem. Then solve the equation to obtain the answer.*

**55. Washing Clothes** A bottle of liquid Tide contains 100 fluid ounces. If one wash load requires 4 ounces of the detergent, how many washes can be done with one bottle of Tide?

**56. Laying Cable** A telephone cable crew is laying cable at a rate of 42 feet an hour. How long will it take them to lay 252 feet of cable?

**57. Truck Mileage** A 2009 Toyota Tacoma is rated to get 26 miles per gallon. How far can it travel on 15.5 gallons of gas?

**58. Purchasing Stock** If 2 shares of stock can be purchased for $38.25, how many shares can be purchased for $344.25?

**59. Model Train** A model train set is in a ratio of 1 : 20. That is, one foot of the model represents 20 feet of the original train. If a caboose is 30 feet long, how long should the model be?

**60. Property Tax** The property tax in the city of Hendersonville, North Carolina, is $9.475 per $1000 of assessed value. If the Estever's house is assessed at $145,000, how much property tax will they owe?

**61. Insecticide Application** The instructions on a bottle of liquid insecticide say "use 3 teaspoons of insecticide per gallon of water." If your sprayer has an 8-gallon capacity, how much insecticide should be used to fill the sprayer?

**62. Spreading Fertilizer** If a 40-pound bag of fertilizer covers 5000 square feet, how many pounds of fertilizer are needed to cover an area of 26,000 square feet?

**63. Blue Heron** The photograph shows a blue heron. If the blue heron that measures 3.5 inches in the photo is actually 3.75 feet tall, approximately how long is its beak if it measures 0.4 inch in the photo?

**64. Maps** On a map, 0.5 inch represents 22 miles. What will be the length on a map that corresponds to a distance of 55 miles?

**65. Onion Soup** A recipe for 6 servings of French onion soup requires $1\frac{1}{2}$ cups of thinly sliced onions. If the recipe were to be made for 15 servings, how many cups of onions would be needed?

**66. John Grisham Novel** Karen Estes is currently reading a John Grisham novel. If she reads 72 pages in 1.3 hours, how long will it take her to read the entire 656-page novel?

**67. Wall Street Bull** Suppose the famous bull near the New York Stock Exchange (see photo below) is a replica of a real bull in a ratio of 2.95 to 1. That is, the metal bull is 2.95 times larger than the regular bull. If the length of the Wall Street bull is 28 feet long, approximately how long is the bull that served as its model?

**68. Flood** When they returned home from vacation, the Duncans had a foot of water in their basement. They contacted their fire department, which sent equipment to pump out the water. After the pump had been on for 30 minutes, 3 inches of water had been removed. How long, from the time they started pumping, did it take to remove all the water from the basement?

**69. Drug Dosage** A nurse must administer 220 micrograms of atropine sulfate. The drug is available in solution form. The concentration of the atropine sulfate solution is 400 micrograms per milliliter. How many milliliters should be given?

**70. Dosage by Body Surface** A doctor asks a nurse to administer 0.7 gram of meprobamate per square meter of body surface. The patient's body surface is 0.6 square meter. How much meprobamate should be given?

**71. Reading a Novel** Mary read 40 pages of a novel in 30 minutes. If she continues reading at the same rate, how long will it take her to read the entire 760-page book?

**72. Swimming Laps** Jason Abbott swims 3 laps in 2.3 minutes. Approximately how long will it take him to swim 30 laps if he continues to swim at the same rate?

**73. Prader-Willi Syndrome** It is estimated that each year in the United States about 1 in every 12,000 (1 : 12,000) people is born with a genetic disorder called Prader-Willi syndrome. If there were approximately 4,315,000 births in the United States in 2007, approximately how may children were born with Prader-Willi syndrome?

**74. Scrapbooking** Penelope Penna completed 4 scrapbook pages in 20.5 minutes. Approximately how long will it take her to complete 36 scrapbook pages if she continues to complete the scrapbook at the same rate?

*In Exercises 75–86, use a proportion to make the conversion. Round your answers to two decimal places.*

**75.** Convert 78 inches to feet.

**76.** Convert 22,704 feet to miles (5280 feet = 1 mile).

**77.** Convert 26.1 square feet to square yards (9 square feet = 1 square yard).

**78.** Convert 146.4 ounces to pounds.

**79. Newborn** One inch equals 2.54 centimeters. Find the length of a newborn, in inches, if it measures 50.8 centimeters.

**80. Distance** One mile equals approximately 1.6 kilometers. Find the distance, in kilometers, from San Diego, California, to San Francisco, California—a distance of 520 miles.

San Francisco, California

**81. Baseball** Todd Helton holds the major league record for most doubles, 59, in a 162-game season. In the first 60 games of a season, how many doubles would a player need to hit to be on schedule to break Helton's record?

**82. Topsoil** A 40-pound bag of topsoil covers 12 square feet (one inch deep). How many pounds of the topsoil are needed to cover 350 square feet (one inch deep)?

**83. Interest on Savings** Jim Chao invests a certain amount of money in a savings account. If he earned $110.52 in 180 days, how much interest would he earn in 500 days assuming the interest rate stays the same?

**84. Gold** If gold is selling for $834 per 480 grains (a troy ounce), what is the cost per grain?

**85. Mexican Pesos** Suppose that the exchange rate from U.S. dollars to Mexican pesos is $1 per 10.1 pesos. How many pesos would Elizabeth Averbeck receive if she exchanged $200 U.S.?

**86. Currency Exchange** When John Gauzer visited the United States from Canada, he exchanged $10.56 Canadian for $10 U.S. If he exchanges his remaining $600 Canadian for U.S. dollars, how much more in dollars will he receive?

**87. Cholesterol** See Example 2. Ken Rauch's LDL is 127 milligrams per deciliter (mg/dL). His HDL is 60 mg/dL. Is Mr. Rauch's ratio of LDL to HDL less than or equal to the 4 : 1 recommended level?

**88. Cholesterol**

**a)** Another ratio used by some doctors when measuring cholesterol level is the ratio of total cholesterol to HDL.* Is this ratio increased or decreased if the total cholesterol remains the same but the HDL is increased? Explain.

**b)** Doctors recommend that the ratio of total cholesterol to HDL be less than or equal to 4.5 : 1. If Mike's total cholesterol is 220 mg/dL and his HDL is 50 mg/dL, is his ratio less than or equal to 4.5 : 1? Explain.

*Total cholesterol includes both LDL and HDL, plus other types of cholesterol.

## Concept/Writing Exercises

**89.** For the proportion $\dfrac{a}{b} = \dfrac{c}{d}$, if $a$ increases while $b$ and $d$ stay the same, what must happen to $c$? Explain.

**90.** For the proportion $\dfrac{a}{b} = \dfrac{c}{d}$, if $a$ and $c$ remain the same while $d$ decreases, what must happen to $b$? Explain.

## Challenge Problems

**91. Wear on Tires**  A new Goodyear tire has a tread of about 0.34 inch. After 5000 miles the tread is about 0.31 inch. If the legal minimum amount of tread for a tire is 0.06 inch, how many more miles will the tires last?

**92. Apple Pie**  The recipe for the filling for an apple pie calls for

12 cups sliced apples          $\dfrac{1}{4}$ teaspoon salt

$\dfrac{1}{2}$ cup flour          $1\dfrac{1}{2}$ cups sugar

1 teaspoon nutmeg          2 tablespoons butter or margarine

1 teaspoon cinnamon

Determine the amount of each of the other ingredients that should be used if only 8 cups of apples are available.

**93. Insulin**  Insulin comes in 10-cubic-centimeter (cc) vials labeled in the number of units of insulin per cubic centimeter. Thus, a vial labeled U40 means there are 40 units of insulin per cubic centimeter of fluid. If a patient needs 25 units of insulin, how many cubic centimeters of fluid should be drawn up into a syringe from the U40 vial?

## Group Activity

*Discuss and answer Exercises 94 and 95 as a group.*

**94. a)**  Each group member: Find the ratio of your height to your arm span (finger tips to finger tips) when your arms are extended horizontally outward. You will need help from your group in getting these measurements.

**b)**  If a box were to be drawn about your body with your arms extended, would the box be a square or a rectangle? If a rectangle, would the longer length be your arm span or your height measurement? Explain.

**c)**  Compare these results with other members of your group.

**d)**  What one ratio would you use to report the height to arm span for your group as a whole? Explain.

**95.**  A special ratio in mathematics is called the *golden ratio*. Do research in a history of mathematics book or on the Internet, and as a group write a paper that explains what the golden ratio is and why it is important.

## Cumulative Review Exercises

[1.10]  *Name each illustrated property.*

    **96.** $x + 3 = 3 + x$

    **97.** $3(xy) = (3x)y$

    **98.** $2(x - 3) = 2x - 6$

[2.5]  **99.** Solve $3(4x - 3) = 6(2x + 1) - 15$

[2.6]  **100.** Solve $y = mx + b$ for $m$.

# Chapter 2 Summary

| IMPORTANT FACTS AND CONCEPTS | EXAMPLES |
|---|---|
| **Section 2.1** | |
| The **terms** of an expression are the parts that are added. | $2x^2 - 3xy + 5$ has 3 terms: $2x^2$, $-3xy$, and 5. |
| The numerical part of a term is called its **numerical coefficient.** | The numerical coefficient of $\dfrac{3x}{4}$ is $\dfrac{3}{4}$. |
| A **constant** is a term that is a number without a variable. | In $3x^2 + 2x - 7$, the $-7$ is a constant. |
| **Like terms** have the same variables with the same exponents.<br><br>**To Combine Like Terms**<br><br>1. Determine which terms are like terms.<br>2. Add or subtract the coefficients of the like terms.<br>3. Multiply the number found in step 2 by the common variable(s). | $7x$ and $x$;  $6y^2$ and $2y^2$;  $3(x + 4)$ and $-8(x + 4)$<br><br>$$-3x^2 + 4y - 7x^2 + 6 - y - 9$$ $$= -3x^2 - 7x^2 + 4y - y + 6 - 9$$ $$= -10x^2 + 3y - 3$$ |

| IMPORTANT FACTS AND CONCEPTS | EXAMPLES |
|---|---|

### Section 2.1 (cont.)

**Distributive Property**

For any real numbers $a$, $b$, and $c$,

$$a(b + c) = ab + ac$$

$$-5(3r - 6) = -15r + 30$$

---

**To Simplify an Expression**

**1.** Use the distributive property to remove any parentheses.

**2.** Combine like terms.

$$2(3c - 1) - 5(c + 4) - 6$$
$$= 6c - 2 - 5c - 20 - 6$$
$$= c - 28$$

---

When two or more expressions are multiplied, each expression is a **factor** of the product.

Since $7 \cdot 8 = 56$, the 7 and the 8 are factors of 56.

### Section 2.2

A **linear equation** in one variable is an equation that can be written in the form

$$ax + b = c$$

where $a$, $b$, and $c$ are real numbers and $a \neq 0$.

$$9x - 2 = 16$$

---

The **solution to an equation** is the number or numbers that when substituted for the variable or variables make the equation a true statement.

The solution to an equation may be **checked** by substituting the value that is believed to be the solution for the variable in the original equation.

The solution to $2x + 3 = 9$ is 3.

To check whether $-2$ is the solution to $-7x + 1 = 15$:

$$-7x + 1 = 15$$
$$-7(-2) + 1 \overset{?}{=} 15$$
$$14 + 1 \overset{?}{=} 15$$
$$15 = 15 \quad \text{True}$$

Thus, $-2$ is the solution.

---

Two or more equations with the same solution are called **equivalent equations.**

$-4x = 12$, $2x - 3 = -9$, and $x = -3$ are equivalent equations

---

**Addition Property of Equality**

If $a = b$, then $a + c = b + c$ for any real numbers $a$, $b$, and $c$.

Solve the equation $x - 9 = -2$.

$$x - 9 = -2$$
$$x - 9 \;+\; 9 = -2 \;+\; 9$$
$$x = 7$$

### Section 2.3

Two numbers are **reciprocals** of each other when their product is 1.

3 and $\frac{1}{3}$ are reciprocals since $3 \cdot \frac{1}{3} = 1$.

---

**Multiplication Property of Equality**

If $a = b$, then $a \cdot c = b \cdot c$ for any real numbers $a$, $b$, and $c$.

Solve the equation $\frac{3}{7}x = 6$.

$$\frac{3}{7}x = 6$$
$$\frac{7}{3} \cdot \frac{3}{7}x = \frac{7}{3} \cdot 6$$
$$x = 14$$

| IMPORTANT FACTS AND CONCEPTS | EXAMPLES |
|---|---|

### Section 2.4

**To Solve Linear Equations with a Variable on Only One Side of the Equals Sign**

1. If the equation contains fractions, multiply *both* sides of the equation by the least common denominator (LCD).

2. Use the distributive property to remove parentheses.

3. Combine like terms on the same side of the equals sign.

4. Use the addition property to obtain an equation with the term containing the variable on one side of the equals sign and a constant on the other side.

5. Use the multiplication property to isolate the variable.

6. Check the solution in the original equation.

Solve the equation $3(x - 5) - 6x = -2$.

$$3(x - 5) - 6x = -2$$
$$3x - 15 - 6x = -2$$
$$-3x - 15 = -2$$
$$-3x - 15\ \boxed{+\ 15} = -2\ \boxed{+\ 15}$$
$$-3x = 13$$
$$\frac{-3x}{-3} = \frac{13}{-3}$$
$$x = -\frac{13}{3}$$

A check will show that $-\dfrac{13}{3}$ is the solution.

### Section 2.5

**To Solve Linear Equations with the Variable on Both Sides of the Equals Sign**

1. If the equation contains fractions, multiply *both* sides of the equation by the LCD.

2. Use the distributive property to remove parentheses.

3. Combine like terms on the same side of the equals sign.

4. Use the addition property to rewrite the equation with all terms containing the variable on one side of the equals sign and all terms not containing the variable on the other side of the equals sign.

5. Use the multiplication property to isolate the variable.

6. Check the solution in the original equation.

Solve the equation $9 - 3x - 2(x + 5) = 4x + 7 - x$.

$$9 - 3x - 2(x + 5) = 4x + 7 - x$$
$$9 - 3x - 2x - 10 = 4x + 7 - x$$
$$-5x - 1 = 3x + 7$$
$$-5x\ \boxed{+\ 5x}\ - 1 = 3x\ \boxed{+\ 5x}\ + 7$$
$$-1 = 8x + 7$$
$$-1\ \boxed{-\ 7} = 8x + 7\ \boxed{-\ 7}$$
$$-8 = 8x$$
$$\frac{-8}{8} = \frac{8x}{8}$$
$$-1 = x$$

A check will show that $-1$ is the solution.

---

A **conditional equation** is an equation that has a single value for a solution.

An **identity** is an equation that is true for infinitely many values of the variable.

A **contradiction** is an equation that has no solution.

$3x - 2 = 8$ is a conditional equation since its solution is $\dfrac{10}{3}$.

$-4(x + 3) = -5x - 12 + x$ is an identity because the equation is true for all real numbers.

$-9x + 7 + 6x = -5x + 1 + 2x$ is a contradiction because the equation is never true and has no solution.

### Section 2.6

**Simple Interest Formula**

$$\text{interest} = \text{principal} \cdot \text{rate} \cdot \text{time} \quad \text{or} \quad i = prt$$

Determine the interest earned on a \$5000 investment at 3% simple interest for 2 years.

$$i = prt$$
$$i = 5000(0.03)(2)$$
$$i = \$300$$

**Distance Formula**

$$\text{distance} = \text{rate} \cdot \text{time} \quad \text{or} \quad d = r \cdot t$$

Timothy John completed a snowmobile race in 2.4 hours at an average speed of 75 miles per hour. Determine the distance of the race.

$$d = rt$$
$$d = (75)(2.4)$$
$$d = 180 \text{ miles}$$

| IMPORTANT FACTS AND CONCEPTS | EXAMPLES |
|---|---|

**Section 2.6 (cont.)**

**Area** is the measure of the amount of surface within the figure's boundaries. Areas are measured in square units.

**Perimeter** is the sum of the lengths of the sides of a figure. Perimeters are measured in the same common unit as the sides.

Formulas for areas and perimeters of quadrilaterals and triangles can be found in **Table 2.1** on page 134.

Determine the perimeter and area of the following trapezoid.

$$P = a + b + c + d$$

$$P = 5 + 6 + 10.4 + 18.6$$

$$P = 40 \text{ ft}$$

$$A = \frac{1}{2}h(b + d)$$

$$A = \frac{1}{2}(4)(6 + 18.6)$$

$$A = 49.2 \text{ ft}^2$$

---

The **circumference** of a circle is the length (or perimeter) of the curve that forms a circle.

Formulas for the area and circumference of a circle can be found in **Table 2.2** on page 136.

Determine the area and circumference of the following circle.

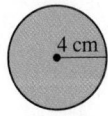

$$A = \pi r^2$$

$$A = \pi(4)^2$$

$$A = \pi(16)$$

$$A \approx 50.27 \text{ cm}^2$$

$$C = 2\pi r$$

$$C = 2\pi(4)$$

$$C = 8\pi$$

$$C \approx 25.13 \text{ cm}$$

---

**Volume** may be considered the space occupied by a figure. Volume is measured is cubic units.

Volume formulas can be found in **Table 2.3** on page 137.

Determine the volume of the following figure.

$$V = \frac{1}{3}\pi r^2 h$$

$$V = \frac{1}{3}\pi(3)^2(7)$$

$$V = \frac{1}{3}\pi(9)(7)$$

$$V = 21\pi$$

$$V \approx 65.97 \text{ m}^3$$

| IMPORTANT FACTS AND CONCEPTS | EXAMPLES |
|---|---|

**Section 2.6 (cont.)**

| | |
|---|---|
| To **solve for a variable in a formula,** treat each of the quantities, except the one for which you are solving, as if they were constants. Then solve for the desired variable by isolating it on one side of the equation. | Solve $V = lwh$, for $h$. $$V = lwh$$ $$\frac{V}{lw} = \frac{lwh}{lw}$$ $$\frac{V}{lw} = h$$ |

**Section 2.7**

| | |
|---|---|
| A **ratio** is a quotient of two quantities. | $3$ to $5$,    $3:5$,    $\dfrac{3}{5}$ |
| A **proportion** is a statement of equality between two ratios. In the proportion $\dfrac{a}{b} = \dfrac{c}{d}$, the $a$ and $d$ are called the **extremes**, and the $b$ and $c$ are called the **means** of the proportion. | $\dfrac{7}{10} = \dfrac{21}{30}$ <br> 7 and 30 are the extremes. 10 and 21 are the means. |
| **Cross-Multiplication** <br> If $\dfrac{a}{b} = \dfrac{c}{d}$, then $ad = bc$. | Solve $\dfrac{-9}{2} = \dfrac{126}{x}$ for $x$ by cross-multiplying. $$\frac{-9}{2} = \frac{126}{x}$$ $$-9 \cdot x = 2 \cdot 126$$ $$-9x = 252$$ $$\frac{-9x}{-9} = \frac{252}{-9}$$ $$x = -28$$ |
| **To Solve Problems Using Proportions** <br><br> 1. Understand the problem. <br><br> 2. Translate the problem into mathematical language. <br><br> 3. Carry out the mathematical calculations necessary to solve the problem. <br><br> 4. Check the answer obtained in step 3. <br><br> 5. Make sure you have answered the question. <br><br> See page 147 for more details on proportions. | Melanie Jo can type 40 words per minute. If she types for 20.5 minutes, how many words will she type? $$\frac{40 \text{ words}}{1 \text{ minute}} = \frac{x \text{ words}}{20.5 \text{ minutes}}$$ $$40(20.5) = 1(x)$$ $$820 = x$$ Melanie Jo will type 820 words. |
| **Similar figures** are figures whose corresponding angles are equal and whose corresponding sides are in proportion. | These two figures are similar. <br> 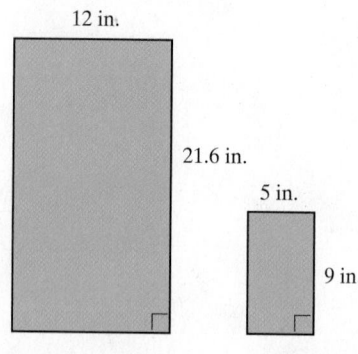 |

## Chapter 2 Review Exercises

[2.1] *Use the distributive property to simplify.*

**1.** $3(x + 8)$

**2.** $5(x - 2)$

**3.** $-2(x + 4)$

**4.** $-(x + 2)$

**5.** $-(m + 8)$

**6.** $-4(4 - x)$

**7.** $5(5 - p)$

**8.** $6(4x - 5)$

**9.** $-5(5t - 5)$

**10.** $4(-x + 3)$

**11.** $\frac{1}{2}(2x + 4)$

**12.** $-\frac{1}{3}(3 + 6y)$

**13.** $-(x + 2y - z)$

**14.** $-3(2a - 5b + 7)$

[2.1] *Simplify.*

**15.** $10q - 6q$

**16.** $5 - 3y + 3$

**17.** $1 + 3x + 2x$

**18.** $-2x - x + 3y$

**19.** $4m + 2n + 4m + 6n$

**20.** $9x + 3y + 2$

**21.** $6x - 2x + 3y + 6$

**22.** $x + 8x - 9x + 3$

**23.** $-4x^2 - 8x^2 + 3$

**24.** $-2(3a^2 - 4) + 6a^2 - 8$

**25.** $2x + 3(x + 4) - 5$

**26.** $-4 + 2(3 - 2b) + b$

**27.** $6 - (-7x + 6) - 7x$

**28.** $2(2x + 5) - 10 - 4$

**29.** $-6(4 - 3x) - 18 + 4x$

**30.** $4y - 3(x + y) + 6x^2$

**31.** $\frac{1}{4}d + 2 - \frac{3}{5}d + 5$

**32.** $3 - (a - b) + (a - b)$

**33.** $\frac{5}{6}x - \frac{1}{3}(2x - 6)$

**34.** $\frac{2}{3} - \frac{1}{4}n - \frac{1}{3}(n + 2)$

[2.2–2.5] *Solve.*

**35.** $-3x = -3$

**36.** $t + 6 = -7$

**37.** $x - 4 = 7$

**38.** $\frac{x}{3} = -9$

**39.** $5x + 1 = 12$

**40.** $14 = 3 + 2x$

**41.** $4c + 11 = -21$

**42.** $9 - 2a = 15$

**43.** $-x = -12$

**44.** $3(x - 2) = 6$

**45.** $-12 = 3(2x - 8)$

**46.** $4(6 + 2x) = 0$

**47.** $-6n + 2n + 6 = 0$

**48.** $-3 = 3w - (4w + 6)$

**49.** $6 - (2n + 3) - 4n = 6$

**50.** $4x + 6 - 7x + 9 = 18$

**51.** $5 + 3(x - 1) = 3(x + 1) - 1$

**52.** $8.4r - 6.3 = 6.3 + 2.1r$

**53.** $19.6 - 21.3t = 80.1 - 9.2t$

**54.** $0.35(c - 5) = 0.45(c + 4)$

**55.** $0.2(x + 6) = -0.3(2x - 1)$

**56.** $-2.3(x - 8) = 3.7(x + 4)$

**57.** $\frac{p}{3} + 2 = \frac{1}{4}$

**58.** $\frac{d}{6} + \frac{1}{7} = 2$

**59.** $\frac{3}{5}(r - 6) = 3r$

**60.** $\frac{2}{3}w = \frac{1}{6}(w - 2)$

**61.** $8x - 5 = -4x + 19$

**62.** $-(w + 2) = 2(3w - 6)$

**63.** $2x + 6 = 3x + 9 - 3$

**64.** $-5a + 3 = 2a + 10$

**65.** $5p - 2 = -2(-3p + 6)$

**66.** $3x - 12x = 24 - 9x$

**67.** $4(2x - 3) + 4 = 8x - 8$

**68.** $4 - c - 2(4 - 3c) = 3(c - 4)$

**69.** $2(x + 7) = 6x + 9 - 4x$

**70.** $-5(3 - 4x) = -6 + 20x - 9$

**71.** $4(x - 3) - (x + 5) = 0$

**72.** $-2(4 - x) = 6(x + 2) + 3x$

**73.** $\frac{x + 3}{2} = \frac{x}{2}$

**74.** $\frac{y}{7} = \frac{y - 5}{2}$

**75.** $\frac{1}{5}(3s + 4) = \frac{1}{3}(2s - 8)$

**76.** $\frac{2(2t - 4)}{5} = \frac{3t + 6}{4} - \frac{3}{2}$

**77.** $\frac{2}{5}(2 - x) = \frac{1}{6}(-2x + 2)$

**78.** $\frac{x}{4} + \frac{x}{6} = \frac{1}{2}(x + 3)$

[2.6]   *Use the formula to find the value of the variable indicated.*

**79.** $y = mx + b$ (slope-intercept form of a line); find $m$ when $y = 7$, $x = 2$, and $b = 1$.

**80.** $A = \frac{1}{2}h(b + d)$ (area of a trapezoid); find $A$ when $h = 12$, $b = 3$, and $d = 5$.

*Determine the area or volume of the figure.*

**81.**
3 cm
8 cm

**82.**
2 in.

*Solve for the indicated variable.*

**83.** $P = 2l + 2w$, for $l$

**84.** $y - y_1 = m(x - x_1)$, for $m$

**85.** $-x + 3y = 2$, for $y$

**86. Spring Break** Yong Wolfer traveled to Florida for spring break at an average speed of 61.7 miles per hour for 5 hours. How for did he travel?

**87. Flower Garden** Chrishawn Miller has a rectangular flower garden that measures 20 feet by 12 feet. What is the area of Chrishawn's flower garden?

**88. Tuna Fish** Find the volume of a tuna fish can if its diameter is 4 inches and its height is 2 inches.

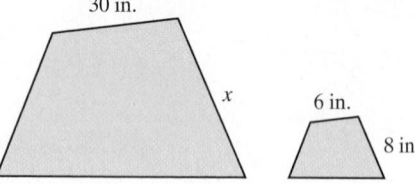

© Steve Mason/Getty Images, Inc.—Photodisc/Royalty Free

[2.7]   *Determine the following ratios. Write each ratio in lowest terms.*

**89.** 15 inches to 20 inches

**90.** 80 ounces to 12 pounds

**91.** 4 minutes : 40 seconds

*Solve each proportion.*

**92.** $\dfrac{x}{4} = \dfrac{8}{16}$

**93.** $\dfrac{5}{20} = \dfrac{x}{80}$

**94.** $\dfrac{3}{t} = \dfrac{15}{45}$

**95.** $\dfrac{20}{45} = \dfrac{15}{q}$

**96.** $\dfrac{6}{5} = \dfrac{-12}{x}$

**97.** $\dfrac{b}{6} = \dfrac{8}{-3}$

**98.** $\dfrac{-7}{9} = \dfrac{-12}{y}$

**99.** $\dfrac{x}{-15} = \dfrac{30}{-5}$

*The following pairs of figures are similar. For each pair, find the length of the side indicated by x.*

**100.**
30 in.
$x$
6 in.
8 in.

**101.**
3.5 ft
$x$
2 ft
7 ft

*Set up a proportion and solve each problem.*

**102. Boat Trip** A boat travels 40 miles in 1.8 hours. If it travels at the same rate, how long will it take for it to travel 140 miles?

**103. Washing Dishes** If Adam Kloza can wash 12 dishes in 3.5 minutes, how many dishes can he wash in 21 minutes?

**104. Copy Machine** If a copy machine can copy 20 pages per minute, how many pages can be copied in 22 minutes?

**105. Map Scale** If the scale of a map is 1 inch to 60 miles, what distance on the map represents 380 miles?

**106. Model Car** Bryce Winston builds a model car to a scale of 1 inch to 1.5 feet. If the completed model is 10.5 inches, what is the size of the actual car?

© Thinkstock

**107. Money Exchange** Suppose that one U.S. dollar can be exchanged for 9.165 Mexican pesos. Find the value of 1 peso in terms of U.S. dollars.

**108. Ketchup** If a machine can fill and cap 80 bottles of ketchup in 50 seconds, how many bottles of ketchup can it fill and cap in 2 minutes?

## Chapter 2 Practice Test

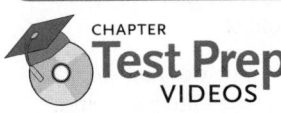

*Chapter Test Prep Videos provide fully worked-out solutions to any of the exercises you want to review. Chapter Test Prep Videos are available via MyMathLab®, or on You Tube (search "Angel Elementary and Intermediate Algebra" and click on "Channels").*

*Use the distributive property to simplify.*

**1.** $-3(4 - 2x)$

**2.** $-(x + 3y - 4)$

*Simplify.*

**3.** $5x - 8x + 4$

**4.** $4 + 2x - 3x + 6$

**5.** $-y - x - 4x - 6$

**6.** $a - 2b + 6a - 6b - 3$

**7.** $2x^2 + 3 + 2(3x - 2)$

*Solve Exercises 8–16.*

**8.** $2.4x - 6.3 = 3.3$

**9.** $\frac{5}{6}(x - 2) = x - 3$

**10.** $2x - 3(-2x + 4) = -13 + x$

**11.** $3x - 4 - x = 2(x + 5)$

**12.** $-3(2x + 3) = -2(3x + 1) - 7$

**13.** $ax + by + c = 0$, for $x$

**14.** $-6x + 5y = -2$, for $y$

**15.** $\frac{1}{7}(2x - 5) = \frac{3}{8}x - \frac{5}{7}$

**16.** $\frac{9}{x} = \frac{3}{-15}$

**17.** What do we call an equation that has
  **a)** exactly one solution,
  **b)** no solution,
  **c)** all real numbers as its solution?

**18.** The following figures are similar. Find the length of side $x$.

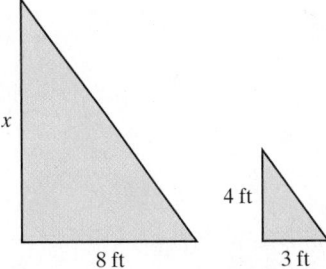

**19. Simple Interest Loan** Laura Hoye lent her sister $2000 for a period of 1 year. At the end of 1 year, her sister repaid the $2000 plus $80 interest. What simple interest rate did her sister pay?

**20. Peanut Butter Pie** A peanut butter pie has a diameter of 9 inches. Determine the circumference of the pie.

**21. Travel Time** While traveling, you notice that you traveled 25 miles in 35 minutes. If your speed does not change, how long will it take you to travel 125 miles?

## Cumulative Review Test

*Take the following test and check your answers with those given in the back of the book. Review any questions that you answered incorrectly. The section where the material was covered is indicated after the answer.*

1. Multiply $\dfrac{52}{15} \cdot \dfrac{10}{13}$

2. Divide $\dfrac{5}{24} \div \dfrac{2}{9}$

3. Insert $<$, $>$, or $=$ in the shaded area to make a true statement: $|-2|$ ▨ $1$.

4. Evaluate $-5 - (-4) + 12 - 8$.

5. Subtract $-6$ from $-7$.

6. Evaluate $20 - 6 \div 3 \cdot 2$.

7. Evaluate $3[6 - (4 - 8^2)] - 30$.

8. Evaluate $-2x^2 - 6x + 8$ when $x = -2$.

9. Name the illustrated property.

$$-5(x - 3y - 4z) = -5x + 15y + 20z$$

*Simplify.*

10. $8x + 2y + 4x - y$

11. $9 - \dfrac{2}{3}x + 16 + \dfrac{3}{4}x$

12. $3x^2 + 5 + 4(2x - 7)$

*Solve.*

13. $7t + 3 = -4$

14. $\dfrac{1}{4}x = -11$

15. $4(x - 2) = 5(x - 1) + 3x + 2$

16. $\dfrac{3}{4}n - \dfrac{1}{5} = \dfrac{2}{3}n$

17. $A = \dfrac{a + b + c}{3}$, for $b$

18. $\dfrac{40}{30} = \dfrac{3}{x}$

19. **Trampoline** A circular trampoline has a diameter of 22 feet. Determine the area of the trampoline.

20. **Earnings** If Samuel earns \$10.50 after working for 2 hours mowing a lawn, how much does he earn after 8 hours?

# 3

# Applications of Algebra

## Goals of This Chapter

The emphasis of this chapter is to get you to express real-world problems mathematically and solve them. This process of representing real-life situations mathematically is called *modeling*. For many students, this chapter is the most important chapter in the entire text. The material presented in this chapter will not only help you succeed in this course, but will help you succeed throughout life!

**The mixture of chemicals** in the laboratory is an important process. In some cases, algebra can be used to determine the amounts of the chemicals needed for a specific process. In Example 6 on page 202 you will see how to determine the mixture proportions for a particular solution.

© Olivier Le Queinec/Shutterstock

# 3.1 Changing Application Problems into Equations

1 Translate phrases into mathematical expressions.

2 Express the relationship between two related quantities.

3 Write expressions involving multiplication.

4 Translate applications into equations.

## 1 Translate Phrases into Mathematical Expressions

**Helpful Hint**

**Study Tip**

It is important that you prepare for this chapter carefully. Make sure you read the book and work the examples carefully. *Attend class every day, and most of all, work all the exercises assigned to you.*

As you read through the examples in the rest of the chapter, think about how they can be expanded to other, similar problems. For example, in Example 1 **a)** we will state that the distance, $d$, increased by 10 miles, can be represented by $d + 10$. You can generalize this to other, similar problems. For example, a weight, $w$, increased by 15 pounds, can be represented as $w + 15$.

One practical advantage of knowing algebra is that you can use it to solve everyday problems by first *translating application problems into mathematical language*. One purpose of this section is to help you take an application problem, also referred to as a *word* or *verbal problem*, and write it as a mathematical equation.

Often the most difficult part of solving an application problem is translating it into an equation. Before you can translate a problem into an equation, you must understand the meaning of certain words and phrases and how they are expressed mathematically. **Table 3.1** is a list of selected words and phrases and the operations they imply. We used the variable $x$. However, any variable could have been used.

**TABLE 3.1**

| Word or Phrase | Operation | Statement | Algebraic Form |
|---|---|---|---|
| Added to<br>More than<br>Increased by<br>The sum of | Addition | 8 *added to* a number<br>6 *more than* a number<br>A number *increased by* 3<br>*The sum of* a number and 4 | $x + 8$<br>$x + 6$<br>$x + 3$<br>$x + 4$ |
| Subtracted from<br>Less than<br>Decreased by<br>The difference between | Subtraction | 6 *subtracted from* a number<br>2 *less than* a number<br>A number *decreased by* 5<br>*The difference between* a number and 9 | $x - 6$<br>$x - 2$<br>$x - 5$<br>$x - 9$ |
| Multiplied by<br>The product of<br>Twice a number, 3 times a number, etc.<br>Of, when used with a percent or fraction | Multiplication | A number *multiplied by* 6<br>*The product of* 4 and a number<br>*Twice a number*<br>20% *of* a number | $6x$<br>$4x$<br>$2x$<br>$0.20x$ |
| Divided by<br><br>The quotient of | Division | A number *divided by* 8<br><br>*The quotient of* a number and 6 | $\dfrac{x}{8}$<br><br>$\dfrac{x}{6}$ |

Often a statement contains more than one operation. The following chart provides some examples of this.

| Statement | Algebraic Form |
|---|---|
| Four more than twice a number | $\underset{\text{Twice a number}}{\underline{2x}} + 4$ |
| Five less than 3 times a number | $\underset{\text{Three times a number}}{\underline{3x}} - 5$ |
| Three times the sum of a number and 8 | $3(\underset{\text{The sum of a number and 8}}{\underline{x + 8}})$ |
| Twice the difference between a number and 4 | $2(\underset{\text{The difference between a number and 4}}{\underline{x - 4}})$ |

Subtraction is not commutative. That is, $a - b \neq b - a$. Therefore, you must be very careful when writing expressions involving subtraction. Study the following examples.

5 less than 3 times a number

| CORRECT | INCORRECT |
|---------|-----------|
| $3x - 5$ | ~~$5 - 3x$~~ |

5 subtracted from 3 times a number

| CORRECT | INCORRECT |
|---------|-----------|
| $3x - 5$ | ~~$5 - 3x$~~ |

the difference between $3x$ and 5

| CORRECT | INCORRECT |
|---------|-----------|
| $3x - 5$ | ~~$5 - 3x$~~ |

> ## Understanding Algebra
>
> Subtraction is *not* commutative. For example, $6 - 9$ is not equal to $9 - 6$. Thus, $x - 7$ is different from $7 - x$, so beware!

Often an algebraic expression can be written in several different ways.

| Algebraic Expression | Statements |
|---|---|
| $2x + 3$ | Three more than twice a number<br>The sum of twice a number and 3<br>Twice a number, increased by 3<br>Three added to twice a number |
| $3x - 4$ | Four less than 3 times a number<br>Three times a number, decreased by 4<br>The difference between 3 times a number and 4<br>Four subtracted from 3 times a number |

**EXAMPLE 1**  Express each statement as an algebraic expression.

a) The distance, $d$, increased by 10 miles

b) Six times the height, $h$

c) Eight less than twice the area, $a$

d) Four pounds more than 5 times the weight, $w$

*Solution*

a) $d + 10$    b) $6h$    c) $2a - 8$    d) $5w + 4$

*Now Try Exercise 13*

## 2  Express the Relationship between Two Related Quantities

When two numbers are related to each other, we will often represent one number as $x$ and the other number as an expression containing $x$.

| Statement | One Number | Other Number |
|---|---|---|
| John is 5 years older than Mary | Let $x$ = Mary's age | Let $x + 5$ = John's age |
| Mike's age now and Mike's age in 8 years | Let $x$ = Mike's age now | Let $x + 8$ = Mike's age in 8 years |
| The first number is 6 times the second number | Let $x$ = second number | Let $6x$ = the first number |
| The first number is 12% less than the second number | Let $x$ = second number | Let $x - 0.12x$ = the first number |

> ## Understanding Algebra
>
> It is important to first determine what quantity we are representing with a letter. Then we should express it literally as "Let $x$ = "

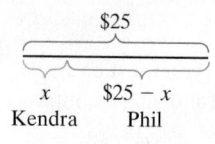

**FIGURE 3.1**

Now let's consider a problem in which one quantity is divided into two parts. For example, suppose $25 is divided between Kendra and Phil.

| If Kendra gets . . . | Then Phil gets . . . |
| --- | --- |
| $20 | $25 − $20 or $5 |
| $15 | $25 − $15 or $10 |
| $8 | $25 − $8 or $17 |
| $2 | $25 − $2 or $23 |

In general, if we let $x =$ the amount Kendra gets, then $25 - x$ is the amount Phil gets. Note that the sum of $x$ and $25 - x$ is 25. (See **Figure 3.1**.)

## Understanding Algebra

In general, if $T$ represents the total to be divided into two parts, then if one part is called $x$, the other part will be $T - x$. See **Figure 3.2**.

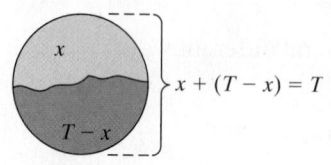

**FIGURE 3.2**

**EXAMPLE 2** For each part, determine what to let $x =$.

**a)** Mary weighs 15 pounds more than Sue.

**b)** The length of a rectangle is 4 inches more than twice its width.

**c)** Joe earns $56.20 less than Larry.

**d)** The theater sold 525 tickets. Some were adults' tickets and some were children's tickets.

*Solution*    In general, when writing expressions to represent word problems, if a quantity $A$ is expressed in terms of a quantity $B$, then we let $x =$ quantity $B$.

**a)** Since Mary's weight is expressed in terms of Sue's weight, we let $x =$ Sue's weight.

**b)** Since the length of a rectangle is expressed in terms of its width, we let $x =$ width of the rectangle.

**c)** Since the amount Joe earns is expressed in terms of what Larry earns, we let $x =$ amount Larry makes.

**d)** Here, since neither quantity of adults' tickets nor children's tickets is expressed in terms of the other's quantity, we can let $x =$ number of adults' tickets sold or let $x =$ number of children's tickets sold. If we let $x =$ number of adults' tickets sold, then $525 - x$ will equal the number of children's tickets sold. If we let $x =$ number of children's tickets sold, then $525 - x$ will equal the number of adults' tickets sold.

Now Try Exercise 37

**EXAMPLE 3** For each relationship, select a variable to represent one quantity and state what that variable represents. Then express the second quantity in terms of the variable selected.

**a)** The Dukes scored 12 points more than the Tigers.

**b)** An adult robin is 4.3 times the weight of a baby robin.

**c)** Bill and Mary share $75.

**d)** Kim has 7 more than 5 times the amount Sylvia has.

**e)** The length of a rectangle is 3 feet less than 4 times its width.

*Solution*    To express the relationships, we must first decide which quantity we will let the variable represent. To give you practice with variables other than $x$, we will select different letters to represent the variable.

**a)** Since the number of points scored by the Dukes is expressed in terms of the number of points scored by the Tigers, we will select the variable $t$.

Let $t =$ number of points scored by the Tigers.

Then $t + 12 =$ number of points scored by the Dukes.

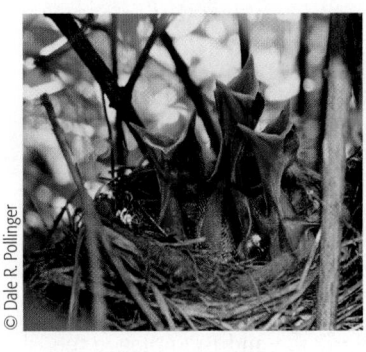

**b)** The weight of an adult robin is given in terms of the weight of a baby robin.

Let $w$ = weight of a baby robin.

Then $4.3w$ = weight of an adult robin.

**c)** We are not told how much of the $75 each person receives. In this case we can let the variable represent the amount either person receives. We will let $a$ represent the amount Bill receives.

Let $a$ = amount Bill receives.

Then $75 - a$ = amount Mary receives.

**d)** The amount Kim has is given in terms of the amount Sylvia has.

Let $s$ = amount Sylvia has.

Then $5s + 7$ = amount Kim has.

**e)** The length of the rectangle is given in terms of the width of the rectangle.

Let $w$ = width of the rectangle.

Then $4w - 3$ = length of the rectangle.

Now Try Exercise 47

## 3 Write Expressions Involving Multiplication

Consider the statement "the cost of 3 items at $5 each." The cost would be 3 times $5 and we could express the cost as $3 \cdot 5$ or $3(5)$. Now consider the statement "the cost of $x$ items at $5 each." The cost would be $x$ times $5 and we could express the cost as $x \cdot 5$ or $x(5)$. It is customary to write this product as $5x$. Thus, the cost of $x$ items at $5 each is represented as $5x$.

Finally, consider the statement "the cost of $x$ items at $y$ dollars each." We write the cost of $x$ items at $y$ dollars as $\boldsymbol{xy}$.

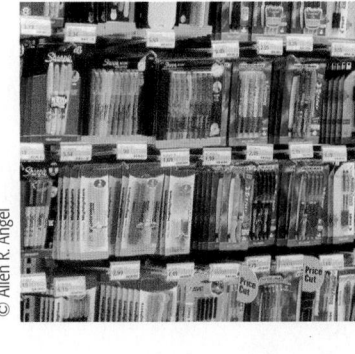

**EXAMPLE 4** Write each statement as an algebraic expression.

**a)** The cost of purchasing $x$ pens at $2 each
**b)** A 5% commission on $x$ dollars in sales
**c)** The dollar amount earned in $h$ hours if a person earns $6.50 per hour
**d)** The number of cents in $q$ quarters
**e)** The number of ounces in $x$ pounds

*Solution*

**a)** We can reason like this:

| 1 pen would cost | $1(2)$ dollars | $= \$2$ |
| 2 pens would cost | $2(2)$ dollars | $= \$4$ |
| 3 pens would cost | $3(2)$ dollars | $= \$6$ |
| $\vdots$ | $\vdots$ | $\vdots$ |
| $x$ pens would cost | $x(2)$ dollars | or $\boldsymbol{2x}$ dollars |

Thus the cost would be $2x$ dollars.

**b)** A 5% commission on $1 sales would be $0.05(1)$, on $2 sales $0.05(2)$, on $3 sales $0.05(3)$, on $4 sales $0.05(4)$, and so on. Therefore, the commission on sales of $x$ dollars would be $0.05(x)$ or $0.05x$.

*Note*: If you need a review on changing a percent to a decimal number, review Appendix A.

**c)** In one hour the person would earn $1(\$6.50)$. In two hours the person would earn $2(\$6.50)$, and in $h$ hours the person would earn $h(\$6.50)$ or $\$6.50h$.

**d)** We know that each quarter is worth 25 cents. Thus, one quarter is 1(25) cents. Two quarters is 2(25) cents, and so on. Therefore, $q$ quarters is $q(25)$ cents or $25q$ cents.

**e)** Each pound is equal to 16 ounces. Using the same reasoning as in part **d)**, we see that $x$ pounds is $16x$ ounces.

Now Try Exercise 71

**EXAMPLE 5  Truck Rental** Maria Mears rented a truck for 1 day. She paid a daily fee of $88 and a mileage fee of 75 cents per mile. Write an expression that represents her total cost when she drives $x$ miles.

*Solution*    Maria's total cost consists of two parts, the daily fee and the mileage fee. Notice the daily fee is given in terms of dollars, and the mileage fee is given in cents. When writing an expression to represent the total cost, we want the units to be the same. Therefore, we will use a mileage fee of $0.75 per mile, which is equal to 75 cents per mile.

Let $x$ = number of miles driven.

Then $0.75x$ = cost of driving $x$ miles.

$\overbrace{\phantom{...........}}^{\text{total cost}}$

daily fee + mileage fee

88    +    0.75$x$

> ## Understanding Algebra
>
> Caution! Be sure that the units are consistent throughout the problem!

Thus, the expression that represents Maria's total cost is $88 + 0.75x$.

Now Try Exercise 75

**EXAMPLE 6  Write a Sum or Difference** In a bus the number of males was 3 more than twice the number of females. Write an expression for

**a)** the sum of the number of males and females

**b)** the difference between the number of males and females

**c)** the difference between the number of females and males

*Solution*    Since the number of males is expressed in terms of the number of females, we let the variable represent the number of females. We will choose $x$ to represent the variable.

Let $x$ = number of females.

Then $2x + 3$ = number of males.

**a)** The expression for the sum of the number of males and females is

$\overbrace{(2x + 3)}^{\substack{\text{number} \\ \text{of males}}} + \overbrace{x}^{\substack{\text{number} \\ \text{of females}}}$

**b)** The expression for the difference between the number of males and females is

$\overbrace{(2x + 3)}^{\substack{\text{number} \\ \text{of males}}} - \overbrace{x}^{\substack{\text{number} \\ \text{of females}}}$

**c)** The expression for the difference between the number of females and males is

$\overbrace{x}^{\substack{\text{number} \\ \text{of females}}} - \overbrace{(2x + 3)}^{\substack{\text{number} \\ \text{of males}}}$

Notice that parentheses are needed around the $2x + 3$ since both terms $2x$ and 3 are being subtracted.

Now Try Exercise 87

> **Helpful Hint**
>
> ## Using Parentheses When Writing Expressions
>
> **Sum:** When writing the sum of two quantities, parentheses may be used to help in the under-standing of the problem, but they are not necessary.
>
> | Examples | Answers |
> |---|---|
> | Find the sum of $x$ and $2x - 3$. | $x + (2x - 3)$  or $x + 2x - 3$ |
> | Find the sum of $3c - 4$ and $c + 5$. | $(3c - 4) + (c + 5)$ or $3c - 4 + c + 5$ |
>
> **Difference:** When writing the difference of two quantities, when only *a single term* is being subtracted, parentheses may be used in the understanding of the problem, but they are not necessary.
>
> | Examples | Answers |
> |---|---|
> | Subtract $r$ from $3r - 2$. | $(3r - 2) - r$ or $3r - 2 - r$ |
> | Find the difference between $2s + 6$ and $s$. | $(2s + 6) - s$ or $2s + 6 - s$ |
>
> When writing the difference of two quantities, *when two or more terms are being subtracted, parentheses* **must** *be placed around all the terms being subtracted,* since all the terms are being subtracted and not just the first term.
>
> | Examples | Answers |
> |---|---|
> | Subtract $x + 2$ from $3x$. | $3x - (x + 2)$ |
> | Subtract $3t - 4$ from $5t$. | $5t - (3t - 4)$ |
> | Subtract $r - 5$ from $2r + 3$. | $(2r + 3) - (r - 5)$ or $2r + 3 - (r - 5)$ |
> | Find the difference between $6$ and $m + 3$. | $6 - (m + 3)$ |
> | Find the difference between $4n - 9$ and $2n - 3$. | $(4n - 9) - (2n - 3)$ or $4n - 9 - (2n - 3)$ |

## Expressions Involving Percent

Example 7 involves percent. Whenever we perform a calculation involving percent, we change the percent to a decimal number or a fraction first.

When shopping we may see a "25% off" sign. We assume that this means 25% off of the *original cost*, even though this is not stated. If we let $c$ represent the original cost, then 25% of the original cost would be represented as $0.25c$. Twenty-five percent off the original cost means the original cost, $c$, decreased by 25% of the original cost. Twenty five percent off the original cost would be represented as $c - 0.25c$.

$$25\% \text{ off the original cost}$$

$$c - 0.25c$$

Original cost ——— decreased by ——— 25% of the original cost

Now let's work an example involving percent.

**EXAMPLE 7**  Write each statement as an algebraic expression.

**a)** The cost of a pair of boots, $c$, increased by 6%

**b)** The population in the town of Brooksville, $p$, decreased by 12%

## Solution

**a)** The question asks for the cost increased by 6%. We assume that this means the cost increased by 6% of the original cost. Therefore, the answer is $c + 0.06c$.

**b)** Using the same reasoning as in part **a)**, the answer is $p - 0.12p$.

Now Try Exercise 79

## Understanding Algebra

If the sales tax rate is 6%, the *tax* on an item costing *x* dollars is $0.06x$.

The *total amount*, which is the cost of the item plus the tax on the item, is represented as $x + 0.06x$, which can also be written as $1.06x$.

### Avoiding Common Errors

In Example 7 **a)** we asked you to represent a cost, $c$, increased by 6%. Note, the answer is $c + 0.06c$. Often, students write the answer to this question as $c + 0.06$. It is important to realize that a percent of a quantity must always be a percent multiplied by some number or letter. Some phrases involving the word percent and the correct and incorrect interpretations follow.

| PHRASE | CORRECT | INCORRECT |
|---|---|---|
| A $7\frac{1}{2}$ % sales tax on $c$ dollars | $0.075c$ | ~~0.075~~ |
| The cost, $c$, increased by a $7\frac{1}{2}$ % sales tax | $c + 0.075c$ | ~~$c + 0.075$~~ |
| The cost, $c$, reduced by 25% | $c - 0.25c$ | ~~$c - 0.25$~~ |

## 4  Translate Applications into Equations

When writing application problems as equations, the word *is* often means *is equal to* and is represented by an equals sign. Some examples of statements written as equations follow.

| Statement | Equation |
|---|---|
| Six times a number *is* 42. | $6x = 42$ |
| Five more than twice a number *is* 4. | $2x + 5 = 4$ |
| A number decreased by 4 *is* 3 more than twice the number. | $x - 4 = 2x + 3$ |
| The sum of a number and the number increased by 4 *is* 60. | $x + (x + 4) = 60$ |
| Twice the difference of a number and 3 *is* the sum of the number and 20. | $2(x - 3) = x + 20$ |
| A number increased by 15% *is* 120. | $x + 0.15x = 120$ |
| Six less than three times a number *is* one-fourth the number. | $3x - 6 = \frac{1}{4}x$ |

Now let's work some examples where we write equations.

### EXAMPLE 8  Translate Words into Equations  Write each problem as an equation.

**a)** A New York City subway car has 36 seats. The number of seats in $s$ subway cars *is* 180.

**b)** The number of cents in $d$ dimes *is* 120.

**c)** The cost of $x$ gallons of gasoline at $4.20 per gallon *is* $35.20.

## Solution

**a)** 1 subway car has 36 seats, 2 subway cars have 72 seats, and $s$ subway cars have $36s$ seats. Since there are 180 seats, the equation is $36s = 180$.

**b)** The number of cents in $d$ dimes is $d(10)$ or $10d$. Since the number of cents in $d$ dimes *is* 120, the equation is $10d = 120$.

**c)** Using similar reasoning as in parts **a)** and **b)**, the equation is $4.20x = 35.20$.

Now Try Exercise 109

> **Helpful Hint**
>
> In a written expression certain other words may be used in place of *is* to represent the equals sign. Some of these are *will be, was, yields,* and *gives.* For example,
>
> "When 4 is added to a number, the sum *will be* 20" can be expressed as $x + 4 = 20$.
>
> "Six subtracted from a number *was* $\frac{1}{2}$ the number" can be expressed as $x - 6 = \frac{1}{2}x$.
>
> "A rental car cost \$75 per day. The cost for renting the car for $x$ days *was* \$150" can be expressed as $75x = 150$.

**EXAMPLE 9  Translate Words into an Equation**  Write the problem as an equation.

One number is 4 less than twice the other. Their sum is 14.

*Solution*

$$\text{Let } x = \text{one number.}$$
$$\text{Then } 2x - 4 = \text{second number.}$$

Now we write the equation using the information given.

$$\text{first number } + \text{ second number} = 14$$
$$x + (2x - 4) = 14$$

Now Try Exercise 99

> ## Understanding Algebra
>
> If $x$ is an integer, then $x + 1$ represents the *next* **consecutive integer.**
>
> If $y$ is an even integer, then $y + 2$ represents the *next* **consecutive even integer**.
>
> If $z$ is an odd integer, then $z + 2$ represents the *next* **consecutive odd integer**.

In Example 10 we will use the term *consecutive even integers*.

**EXAMPLE 10  Consecutive Even Integers**  Write the problem as an equation.

For two consecutive even integers, the sum of the smaller and 3 times the larger is 22.

*Solution*   First, we express the two consecutive even integers in terms of the variable.

$$\text{Let } x = \text{smaller consecutive even integer.}$$
$$\text{Then } x + 2 = \text{larger consecutive even integer.}$$

Now we write the equation using the information given.

$$\text{smaller } + 3 \text{ times the larger} = 22$$
$$x + 3(x + 2) = 22$$

Now Try Exercise 107

**EXAMPLE 11  Translate Words into an Equation**  Write the problem as an equation.

One train travels 3 miles more than twice the distance another train travels. The total distance traveled by both trains is 800 miles.

*Solution*   First express the distance traveled by each train in terms of the variable.

$$\text{Let } x = \text{distance traveled by one train.}$$
$$\text{Then } 2x + 3 = \text{distance traveled by second train.}$$

Now write the equation using the information given.

$$\text{distance of train 1 } + \text{ distance of train 2} = \text{total distance}$$
$$x + (2x + 3) = 800$$

Now Try Exercise 121

EXAMPLE 12 **Translate Words into an Equation** Write the problem as an equation.

Lori Soushon is 4 years older than 3 times the age of her son Ron. The difference in Lori's age and Ron's age is 26 years.

*Solution* Since Lori's age is given in terms of Ron's age, we will let the variable represent Ron's age.

$$\text{Let } x = \text{Ron's age.}$$
$$\text{Then } 3x + 4 = \text{Lori's age.}$$

We are told that the difference in Lori's age and Ron's age is 26 years. The word *difference* indicates subtraction. Since Lori is older than Ron, we must subtract Ron's age from Lori's age to get a positive number.

$$\text{Lori's age } - \text{ Ron's age } = 26$$
$$(3x + 4) - x = 26$$

Now Try Exercise 117

Example 13 will involve percent.

EXAMPLE 13 **Translate Words into an Equation.** Write the problem as an equation.

The 2009 property tax for Danielle's house was 3.9% greater than her property tax in 2008. Her property tax in 2009 was $4008.

*Solution* In this example, we will choose to use the variable $t$, for tax. Since the 2009 property tax is based upon the 2008 property tax, we will let the variable represent the 2008 property tax.

$$\text{Let } t = 2008 \text{ property tax.}$$
$$\text{Then } \underbrace{t + 0.039t}_{\text{this represents the 3.9% increase}} = 2009 \text{ property tax.}$$

Since Danielle's 2009 property tax *was* $4008, the equation we write is $t + 0.039t = 4008$.

Now Try Exercise 123

**Helpful Hint**

It is important that you understand this section and work all your assigned homework problems. You will use the material learned in this section in the next three sections and throughout the book.

# EXERCISE SET 3.1   MathXL®  MyMathLab®

## Warm-Up Exercises

*Fill in the blanks with the appropriate word, phrase, or symbol(s) from the following list.*

| | | | | | |
|---|---|---|---|---|---|
| $2c + 7$ | $2c - 7$ | $c + 0.07$ | $c + 7$ | $c + 2$ | $c + 0.07c$ |
| $7 - c$ | $c - 0.07c$ | equals | $0.07c$ | $c - 7$ | $c + 3$ |

**1.** If $c$ represents the cost of an item, then the 7% tax on that item is represented by _____.

**2.** Consider the statement "Barry is seven years older than Chuck." If $c$ represents Chuck's age, then _____ is the expression that represents Barry's age.

**3.** A seven-foot board is cut into two pieces. If the length of one piece is represented by $c$, then _____ represents the length of the other piece.

**4.** If $c$ represents an odd number, the next consecutive odd number is represented by _____.

**5.** If *c* represents the dollar cost of an item, and then the cost is increased by 7¢, the new cost is represented by _____ .

**6.** If *c* represents the cost of an item, and then the cost is decreased by 7%, the new cost is represented by _____ .

**7.** Simone is 7 years older than twice Carla's age. If *c* represents Carla's age, _____ represents Simone's age.

**8.** A first number is 7 less than two times a second number. If *c* represents the second number, then the first number is represented by _____ .

**9.** When translating a problem, the word "is" gets translated to an _____ symbol.

**10.** If *c* represents the cost of an item and if the sales tax is 7%, then the amount you need to pay for the item is represented by _____ .

## Practice the Skills

*In Exercises 11–30, express the statement as an algebraic expression. See Example 1.*

**11.** The height, *h*, increased by 4 inches

**12.** The weight, *w*, increased by 20 pounds

**13.** The age, *a*, decreased by 5 years

**14.** The time, *t*, decreased by 3 hours

**15.** Five times the height, *h*

**16.** Seven times the length, *l*

**17.** Twice the distance, *d*

**18.** Three times the rate, *r*

**19.** One-half the age, *a*

**20.** One-third the weight, *w*

**21.** Five subtracted from *r*

**22.** Nine subtracted from *p*

**23.** *m* subtracted from 12

**24.** *n* subtracted from 4

**25.** Eight pounds more than twice the weight, *w*

**26.** Six inches more than 3 times the height, *h*

**27.** Four years less than 5 times the age, *a*

**28.** One mile more than $\frac{1}{2}$ the distance, *d*

**29.** One-third the weight, *w*, decreased by 9 pounds

**30.** One-fifth the height, *h*, increased by 2 feet

*In Exercises 31–44, determine what x =. See Example 2.*

**31.** Paul is 4 inches taller than Sonya.

**32.** The length of a rectangle is 5 inches greater than its width.

**33.** The length of Tortuga Beach is 60 feet shorter than the length of Jones Beach.

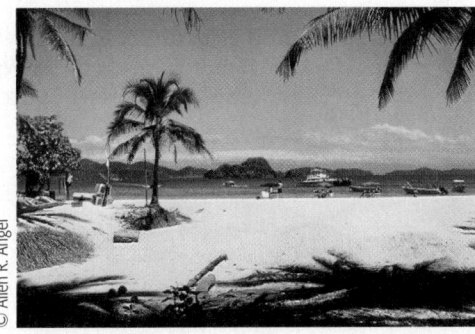

© Allen R. Angel

**34.** Wilma ran 4 miles per hour faster than Natasha.

**35.** The United States won 3 times the number of medals that Finland won.

**36.** The distance to Georgia is $\frac{1}{2}$ the distance to Tennessee.

**37.** The Cadillac costs $200 more than twice the cost of the Chevy.

**38.** Noah received 25 more votes than 3 times the number of votes that Tawnya received.

**39.** June's grade was 2 points less than twice Teri's grade.

**40.** Alberto's salary was $2000 greater than 4 times Nick's salary.

**41.** $60 is divided between Kristen and Yvonne.

**42.** Drawka has 25 marbles. They are either red or blue marbles.

**43.** Together Don and Angela weigh 270 pounds.

**44.** Together Oliver and Dalane have 1053 clients.

*In Exercises 45–56, select a variable to represent one quantity and state what that variable represents. Express the second quantity in terms of the variable selected. See Example 3. Note that the variable you select may be different than the variable used in the answers in the back of the book.*

**45.** The table costs 5 times as much as the chair.

**46.** Joan's house is 810 square feet larger than Alfredo's house.

**47.** The area of the living room is 20 square feet greater than twice the area of the kitchen.

**48.** The amount in Darla's savings account is $250 less than 4 times the amount in Carmen's savings account.

**49.** The length of the rectangle is 2 inches less than 6 times the width.

**50.** A total of $600 is to be divided between Evita and Brian.

**51.** A total of 20 medals were won by Sweden and Brazil.

**52.** A movie theatre sold a total of 220 more adults' tickets than children's tickets.

**53.** Mike's age is 2 years more than $\frac{1}{2}$ George's age.

**54.** The book *Golden Angels* sold 4 copies less than 5 times the amount the book *Flycatcher* sold.

**55.** Jan and Edward used two different treadmills for exercise. The total distance walked between them was 6.4 miles.

**56.** On a 540-mile trip Cheng and Elsie shared the driving.

*In Exercises 57–84, write the indicated expression. See Examples 3–7.*

**57. Age** Dan Graber is *n* years old now. Write an expression that represents his age in 8 years.

**58. Speed-Reading** John Debruzzi used to read *p* words per minute. After taking a speed-reading course, his speed increased by 60 words per minute. Write an expression that represents his new reading speed.

**59. Motorcycle** Melissa Blum is selling her motorcycle. She was asking *x* dollars for the motorcycle but has cut the price in half. Write an expression that represents the new price.

**60. Age** Cathy Bennett's son is one-third as old as Cathy, *c*. Write an expression for her son's age.

**61. Age** Gayle Krzemien's age is one less than twice Mary Lou Baker's age, *a*. Write an expression for Gayle's age.

**62. Calories** The calories in a serving of mixed nuts is 280 calories less than twice the number of calories in a serving of cashew nuts, *c*. Write an expression for the number of calories in a serving of mixed nuts.

**63. Temperature** The average daily temperature in Jacksonville, Florida, in July is 30° less than twice its average daily temperature in January, *t*. Write an expression for the average daily temperature in July.

**64. Population** In 2008, the population of China was 100 million more than 1.1 times the population of India, *p*. Write an expression for the population of China in millions.

**65. Weight** Anika Angel weighed *p* pounds at birth. At age 6 months her weight was 2.3 pounds less than twice her birth weight. Write an expression for Anika's weight at age 6 months.

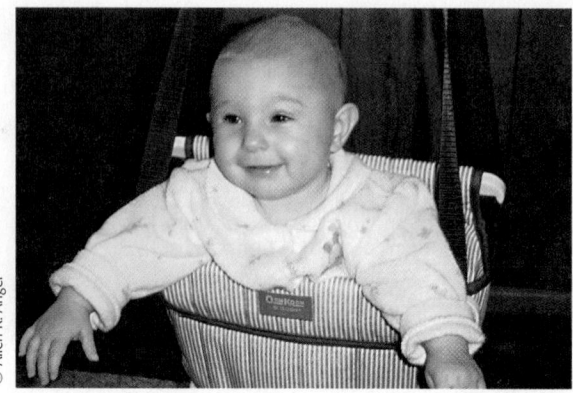

See Exericse 65.

**66. Population Increase** The city of Clarkville has a population of 4000. If the population increases by 300 people per year, write an expression that represents the population after *n* years.

**67. Profits** Monica and Julia share in the profits of a toy store. If the total profit is $80,000 and *m* is the amount Monica receives, write an expression for the amount Julia receives.

**68. Charity Event** A total of 83 men and women attended a charity event. If the number of men who attended is *m*, write an expression for the number of women who attended.

**69. Home Runs** Hank Aaron had 673 less than twice the number of home runs Babe Ruth had, *r*. Write an expression for the number of home runs Hank Aaron had.

**70.** In 2007, the country with the highest average life expectancy was Andorra and the country with the lowest average life expectancy was Swaziland. The average life expectancy in Andorra was 21.1 years greater than twice that in Swaziland, *s*. Write an expression for the average life expectancy in Andorra. (*Source:* CIA World Factbook).

**71. Money** Carolyn Curley found that she had *x* dimes in her handbag. Write an expression that represents this quantity of money in cents.

**72. Weight** Jason Mahar's weight is *w* pounds. Write an expression that represents his weight in ounces.

**73. Money** Susan Grady has *d* dollars in her purse. Write an expression that represents this quantity of money in cents.

**74. Soil** A total of six hundred pounds of soil is put onto two trucks. If *a* pounds of soil is placed onto one truck, write an expression for the amount of soil placed onto the other truck.

**75. Truck Rental** Bob Melina rented a truck for a trip. He paid a daily fee of $45 and a mileage fee of 40 cents a mile. Write an expression that represents his total cost when he travels *x* miles in one day.

**76. Soil Delivery** Mary Vachon had topsoil delivered to her house. The total cost included a delivery charge of $48 plus $60 per cubic yard of soil. Write an expression for the total cost if Mary has *x* cubic yards of soil delivered.

**77. Sales Increase** Barry Cogan is a sales representative for a medical supply company. His 2010 sales increased by 20% over his 2009 sales, *s*. Write an expression for his 2010 sales.

**78. Salary Increase** Charles Idion, an engineer, had a salary increase of 15% over last year's salary, *s*. Write an expression for this year's salary.

**79. Electricity Use** Jean Olson's electricity use in 2009 decreased by 12% from her 2008 electricity use, *e*. Write an expression for her 2009 electricity use.

**80. Shirt Sale** At a 25% off everything sale, Bill Winchief purchased a new shirt. If *c* represents the original cost, write an expression for the sale price of the shirt.

**81. Car Cost** The cost of a new car purchased in Collier County included a 7% sales tax. If *c* represents the cost of the car before tax, write an expression for the total cost, including the sales tax.

**82.** The number of rolls of film sold in 2009 was 74.5% fewer than the number of rolls sold in 1999. Write an expression for the number of rolls sold in 2009 if *r* represents the number of rolls sold in 1999. (*Source:* Photo Marketing Association)

**83.** In 2007, the state with the greatest median age was Maine and the state with the lowest was Utah. The median age in Utah was 30.7% less than that of Maine. If *m* represents the median age in Maine, write an expression for the median age in Utah. (*Source:* U.S. Census Bureau)

**84. Growth in North Las Vegas** North Las Vegas, Nevada, according to the U.S. Census Bureau, had the greatest population growth from 2005 to 2006 of any major city in the United States. Its population in 2006 was 11.9% greater than its population in 2005. If *p* represents North Las Vegas's population in 2005, write an expression for its 2006 population.

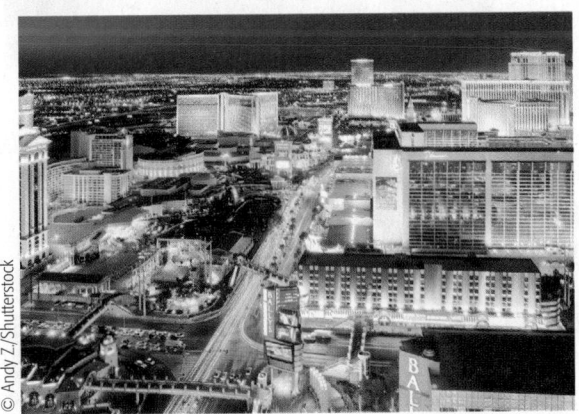

North Las Vegas

*In Exercises 85–98, write the indicated expression. You will need to select the variable to use. Note that the variable you use may be different than the variable used in the answer section. See Example 6.*

**85. Weight** Jennifer's weight is 15 pounds more than Frieda's weight. Write an expression for the sum of their weights.

**86. Length** The Chestnut-mandibled toucan is 3 inches longer than the keel-billed toucan (shown). Write an expression for the sum of their lengths.

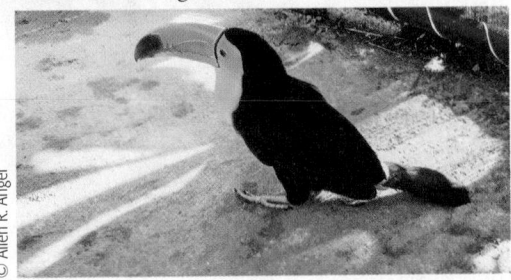

**87. Height** Armando is taller than his son Luis. Armando is 1 inch less than twice Luis's height. Write an expression for the *difference* in Armando's and Luis's heights.

**88. Profits** A company's profits in 2009 were $100 less than the company's profits in 2010. Write an expression for the *difference* in the 2010 and 2009 profits.

**89. Numbers** A first number is 40 less than 3 times a second number. If the second number is represented by *x*, write an expression for the first number subtracted from the second number.

**90. Numbers** A first number is 16 less than twice a second number. If the second number is represented by *n*, write an expression for the first number subtracted from the second number.

**91. Weight** Jill's four-year-old child weighs 3 pounds less than twice the weight of her two-year-old child. Write an expression for the sum of their weights.

**92. Assessed Value** The assessed value of Glen Sandifer's house is $200 more than twice that of Olga Thompson's house. Write an expression for the sum of the assessed values.

**93. Land Area** The largest state in area is Alaska and the smallest is Rhode Island. The area of Alaska is 462 square miles more than 479 times the area of Rhode Island. Write an expression for the sum of the areas of the two states.

**94. Black Bear** The number of black bear sightings in 2009 in Ocala National Forest was 6 less than twice the number of sightings in 2005. Write an expression for the number of 2005 sightings subtracted from the number of 2009 sightings.

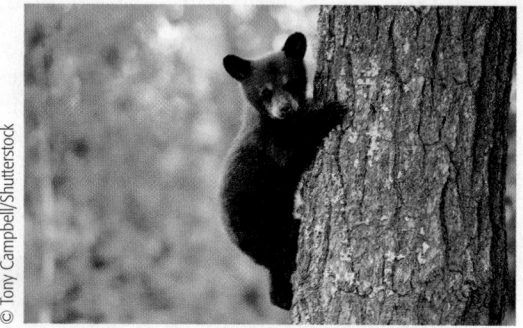

**95. Stocks** The price of Apple Computer stock is 39.4% greater than the cost of Research in Motion stock. Write an expression for the sum of the prices of Research in Motion and Apple Computer stocks.

**96. Car Cost** The cost of a 2009 Chevrolet Corvette LT-1 coupe increased by 2.1% over the cost of a 2008 model. Write an expression for the sum of the costs of a 2008 and a 2009 Corvette.

**97. Bank Assets** The 2008 assets of Fifth Third Bancorp were 11.2% higher than their 2007 assets. Write an expression for the 2008 assets subtracted from the 2007 assets.

**98. Spam** The number of pieces of spam (or junk mail) that Laura Hoye received in 2009 was 12% less than the number she received in 2008. Write an expression for the difference between the number of pieces of spam she received in 2008 and 2009.

*In Exercises 99–130, write an equation to represent the problem. See Examples 8–13.*

**99. Two Numbers** One number is 4 times another. The sum of the two numbers is 20.

**100. Age** Marie is 6 years older than Denise. The sum of their ages is 48.

**101. Consecutive Integers** The sum of two consecutive integers is 41.

**102. Even Integers** The product of two consecutive even integers is 74.

**103. Numbers** Twice a number decreased by 8 is 12.

**104. Consecutive Integers** For two consecutive integers, the sum of the smaller and twice the larger is 29.

**105. Numbers** One-fifth of the sum of a number and 10 is 150.

**106. Numbers** One-third of the sum of a number and 12 is 5.

**107. Even Integers** For two consecutive even integers, the sum of the smaller and twice the larger is 22.

**108. Odd Integers** For two consecutive odd integers, the sum of 3 times the smaller and the larger is 14.

**109. Earnings** John Jones earns $12.50 per hour. If he works $h$ hours his earnings will be $150.

**110. Employees** The T. W. Wilson company plans to increase its number of employees by 20 per year. The increase in the number of employees in $t$ years will be 120.

**111. Top Soil** Abe Mantell purchased $x$ bags of topsoil at a cost of $2.99 a bag. The total he paid was $17.94.

**112. Plants** Mark Ernsthausen purchased $p$ plants at a nursery for $5.99 each. The total he paid was $65.89.

**113. Quarters** The number of cents in $q$ quarters is 175.

**114. Seconds** The number of seconds in $m$ minutes is 480.

**115. Age** Darta Aguilar is 1 year older than twice Julie Chesser's age. The sum of their ages is 52.

**116. Horses** Marc Campbell owns more horses than Selina Jones owns. The number of horses that Marc owns is 3 less than five times the number that Selina owns. The difference in the number of horses owned by Marc and Selina is 5.

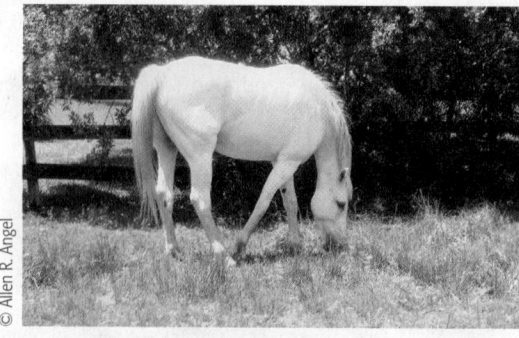

**117. Baseball Cards** Jakob Meyer owns 300 more than twice the number of baseball cards that Saul Gonzales owns. The difference in the number of cards that Jakob and Saul own is 420.

**118. Jogging** David Ostrow jogs 5 times as far as Jennifer Freer. The total distance traveled by both people is 8 miles.

**119. Amtrak** An Amtrak train travels 4 miles less than twice the distance traveled by a Southern Pacific train. The total distance traveled by both trains is 890 miles.

**120. Wagon Ride** On a wagon ride the number of girls was 6 less than twice the number of boys. The total number of boys and girls on the wagon was 18.

**121. Distance Walked** Donna Douglas walked 2 miles less than 3 times as far as Malik Oamar walked. Together they walked a total of 12.6 miles.

**122. Distance** Lilia Orlova rollerbladed 4 miles less than twice the distance she ran. The total distance she traveled was 15 miles.

**123. Viper** The cost of a new Dodge Viper increased by 0.2% over last year's model. The new price is $86,460.

**124. Income** Dan Tadeo's 2009 income was 4.6% greater than his 2008 income. His income in 2009 was $56,900.

**125. Population** The population of the town of Tom's Valley decreased by 1.9%. The population after the decrease was 12,087.

**126. DVD Player** At the Better Buy Warehouse, Anne Long purchased a DVD player that was reduced by 10% for $208.

**127. New Car** Carlotta Diaz bought a new car. The cost of the car plus a 7% sales tax was $32,600.

**128. Sport Coat** David Gillespie purchased a sport coat at a 25% off sale. He paid $195 for the sport coat.

**129. Cost of Meal** Beth Rechsteiner ate at a steakhouse. The cost of the meal plus a 15% tip was $42.50.

**130. Railroad** In a narrow gauge railway, the distance between the tracks is about 64% of the distance between the tracks in a standard gauge railroad. The difference in the distances between the tracks in the two types of railroads is about 1.67 feet.

## Challenge Problem

**131. Time**

a) Write an algebraic expression for the number of seconds in $d$ days, $h$ hours, $m$ minutes, and $s$ seconds.

b) Use the expression found in part **a)** to determine the number of seconds in 4 days, 6 hours, 15 minutes, and 25 seconds.

## Group Activity

*Exercises 132 and 133 will help prepare you for the next section, where we set up and solve application problems. Discuss and work each exercise as a group. For each exercise, write down the quantity you are being asked to find and represent this quantity with a variable. Then write an equation containing your variable that can be used to solve the problem. Do not solve the equation.*

**132. Water Usage** An average bath uses 30 gallons of water and an average shower uses 6 gallons of water per minute. How long a shower would result in the same water usage as a bath?

**133. Salary Plans** An employee has a choice of two salary plans. Plan A provides a weekly salary of $200 plus a 5% commission on the employee's sales. Plan B provides a weekly salary of $100 plus an 8% commission on the employee's sales. What must be the weekly sales for the two plans to give the same weekly salary?

## Cumulative Review Exercises

[1.9]  **134.** Evaluate $3[(4 - 16) \div 2] + 5^2 - 3$.

[2.6]  **135.** $P = 2l + 2w$; find $l$ when $P = 40$ and $w = 5$.

**136.** Solve $3x - 2y = 6$ for $y$.

[2.7]  **137.** Solve the proportion $\dfrac{3.6}{x} = \dfrac{10}{7}$.

**138.** Write the ratio 26 ounces to 4 pounds in lowest terms.

# 3.2 Solving Application Problems

1 Use the problem-solving procedure.

2 Set up and solve number application problems.

3 Set up and solve application problems involving money.

4 Set up and solve applications concerning percent.

## 1 Use the Problem-Solving Procedure

The general problem-solving procedure given in Section 1.2 can be used to solve all types of verbal problems. Below, we present the **five-step problem-solving procedure** again so you can easily refer to it. We have included some additional information under steps 1 and 2, since in this section we are going to emphasize translating application problems into equations.

### Problem-Solving Procedure for Solving Applications

1. **Understand the problem.** Identify the quantity or quantities you are being asked to find.
2. **Translate the problem into mathematical language (express the problem as an equation).**
   a) Choose a variable to represent one quantity, *and write down exactly what it represents.* Represent any other quantity to be found in terms of this variable.
   b) Using the information from step a), write an equation that represents the application.
3. **Carry out the mathematical calculations (solve the equation).**
4. **Check the answer (using the *original* application).**
5. **Answer the question asked.**

Sometimes we will combine two steps in the problem-solving procedure when it helps to clarify the explanation. We may not show the check of a problem to save space. Even if we do not show a check, you should check the problem yourself and make sure your answer is reasonable and makes sense.

## 2 Set Up and Solve Number Application Problems

The examples presented here involve information and data but do not contain percents.

EXAMPLE 1 **An Unknown Number** Two subtracted from 4 times a number is 10. Find the number.

*Solution* Understand   To solve this problem, we need to express the statement given as an equation. We are asked to find the unknown number.

Translate   Let $x$ = the unknown number. Now write the equation.

$$\underbrace{\text{2 subtracted from 4 times a number}}_{4x - 2} \quad \overset{\text{is}}{\underset{=}{\downarrow}} \quad \overset{10}{\underset{10}{\downarrow}}$$

Carry Out
$$4x = 12$$
$$x = 3$$

Check   Substitute 3 for the number in the original problem, two subtracted from 4 times a number is 10.

$$4(3) - 2 \overset{?}{=} 10$$
$$10 = 10 \quad \text{True}$$

Answer   Since the solution checks, the unknown number is 3.

Now Try Exercise 7

EXAMPLE 2  **Number Problem**  The sum of two numbers is 26. Find the two numbers if the larger number is 2 less than three times the smaller number.

*Solution*  Understand  We are given that "the larger number is 2 less than three times the smaller number." Notice that the larger number is expressed in terms of the smaller number. Therefore, we will let the variable represent the smaller number.

Translate                        Let $x$ = smaller number.

                            Then $3x - 2$ = larger number.

The sum of the two numbers is 26. Therefore, we write the equation

$$\text{smaller number} + \text{larger number} = 26$$
$$x + (3x - 2) = 26$$

Carry Out   Now we solve the equation.

$$4x - 2 = 26$$
$$4x = 28$$
$$x = 7$$

The smaller number is 7. Now we find the larger number.

$$\text{larger number} = 3x - 2$$
$$= 3(7) - 2 \quad \text{Substitute 7 for } x.$$
$$= 19$$

The larger number is 19.

Check   The sum of the two numbers is 26.

$$7 + 19 \stackrel{?}{=} 26$$
$$26 = 26 \quad \text{True}$$

Answer   The two numbers are 7 and 19.

Now Try Exercise 13

> **Helpful Hint**
>
> When reading a word problem, ask yourself, "How many answers are required?" In Example 2, the question asked for the two numbers. The answer is 7 and 19. It is important that you read the question and identify what you are being asked to find. If the question had asked "Find the *smaller* of the two numbers if the larger number is 2 less than three times the smaller number," then the answer would have been only the 7. If the question had asked to find the *larger* of the two numbers, then the answer would have been only 19. *Make sure you answer the question asked in the problem.*

EXAMPLE 3  **2008 Summer Olympics**  In the 2008 Olympics in Beijing, China, the United States won the most medals and China won the second greatest number of medals. The United States won 90 less than twice the number of medals won by China. If the difference between the number of medals won by the United States and China was 10, determine the number of medals won by the United States.

*Solution*  Understand  The word *difference* in the problem indicates that this problem will involve subtraction. We are asked to find the number of medals won by the United States. Since the number of medals won by the United States is given in terms of the number of medals won by China, we will let the variable represent medals won by China. We will use the variable $c$.

Translate                        Let $c$ = medals won by China.

                            Then $2c - 90$ = medals won by the United States.

© Pete Niesen/Shutterstock

Since we are dealing with positive amounts, we must subtract the smaller quantity from the larger. Since the difference in medals between the United States and China is 10, we write the following equation.

$$\underbrace{\text{number of medals won by U.S.}}_{2c-90} - \underbrace{\text{number of medals won by China}}_{c} = 10$$

Carry Out
$$2c - 90 - c = 10$$
$$c - 90 = 10$$
$$c = 100$$

Check and Answer   Remember $c$ represents the number of medals won by China. We are asked to find the number of medals won by the United States, which we have represented as $2c - 90$. Now, we substitute the known value of 100 for $c$ in the expression $2c - 90$: $2(100) - 90 = 110$. *The answer is that the United States won 110 medals*. Notice that the difference in the number of medals won by the United States and China is $110 - 100 = 10$, so the answer checks.

Now Try Exercise 27

**EXAMPLE 4** **Bicycles** The Chain Wheel Drive Bicycle Company presently manufactures 800 bicycles a month. Each month after this month the company plans to increase production by 150 bicycles a month until its monthly production reaches 1700 bicycles. How long will it take the company to reach its production goal?

*Solution*   Understand   We are asked to find the *number of months* that it will take for the company's production to reach 1700 bicycles a month. Next month its production will increase by 150 bicycles. In two months, its production will increase by $2(150)$ over the present month's production. In $n$ months, its production will increase by $n(150)$ or $150n$. We will use this information when we write the equation to solve the problem.

Translate
Let $n$ = number of months.

Then $150n$ = increase in production over $n$ months.

$$(\text{present production}) + \left(\begin{array}{c}\text{increased production}\\\text{over } n \text{ months}\end{array}\right) = \text{future production}$$

Carry Out
$$800 + 150n = 1700$$
$$150n = 900$$
$$n = \frac{900}{150}$$
$$n = 6 \text{ months}$$

Check and Answer   As a check, let's list the number of bicycles produced this month and for the next 6 months.

| Presently | Next month | Month 2 | Month 3 | Month 4 | Month 5 | Month 6 |
|---|---|---|---|---|---|---|
| ↓ | ↓ | ↓ | ↓ | ↓ | ↓ | ↓ |
| 800 | 950 | 1100 | 1250 | 1400 | 1550 | 1700 |

Thus, in 6 months the company will produce 1700 bicycles per month.

Now Try Exercise 23

## 3 Set Up and Solve Application Problems Involving Money

When setting up an equation that involves money, you must make sure that all the monetary units entered into the equation are the same, either all dollars or all cents. When pieces of information are given in both dollars and cents, we generally convert the amount given in cents to an equivalent amount of dollars. For example, when renting a truck the cost may be $50 a day plus 90 cents a mile. When writing the equation, we would write the 90 cents a mile as $0.90 a mile. The cost of traveling $x$ miles at 90 cents a mile would be written $0.90x$.

**EXAMPLE 5  Grub Problem**  Part of Kim Martello's lawn was destroyed by grubs. She decided to purchase new sod to lay down. The cost of the sod is 45 cents per square foot plus a delivery charge of $59. If the total cost of delivery plus the sod was $284, how many square feet of sod was delivered?

*Solution*    Understand    The total cost consists of two parts, a cost of 45 cents per square foot of sod, plus a delivery charge of $59. We need to determine the number of square feet of sod that will result in a total cost of $284.

Translate

$$\text{Let } x = \text{number of square feet of sod.}$$
$$\text{Then } 0.45x = \text{cost of } x \text{ square feet of sod.}$$

$$\text{sod cost} + \text{delivery cost} = \text{total cost}$$

$$0.45x + 59 = 284 \qquad \text{Subtract 59 from both sides.}$$

Carry Out

$$0.45x = 225$$

$$\frac{0.45x}{0.45} = \frac{225}{0.45}$$

$$x = 500$$

Check    The cost of 500 square feet of sod at 45 cents a square foot is $500(0.45) = \$225$. Adding the $225 to the delivery cost of $59 gives $284, so the answer checks.

Answer    Five hundred square feet of sod was delivered.

Now Try Exercise 35

**EXAMPLE 6  Photo Printer**  Elsie Newman is going to purchase a photo printer to print pictures from her digital camera. She is considering a Hewlett-Packard (HP) printer and a Lexmark printer. The HP printer costs $419 and the cost for the ink and paper is 14 cents per photo printed. The Lexmark printer costs $299 and the cost for the ink and paper is 18 cents per photo. How many photos would need to be printed for the total cost of the printers, ink, and paper to be the same?

*Solution*    Understand    The HP printer has a greater initial cost ($419 versus $299); however, its cost per photo printed is less (14 cents versus 18 cents). We are asked to find the number of photos printed so that the total cost of the two printers will be the same.

Translate

$$\text{Let } n = \text{number of photos.}$$
$$\text{Then } 0.14n = \text{cost for printing } n \text{ photos with the HP printer}$$
$$\text{and } 0.18n = \text{cost for printing } n \text{ photos with the Lexmark printer.}$$

$$\text{total cost of HP printer} = \text{total cost of Lexmark printer}$$

$$\binom{\text{initial}}{\text{cost}} + \binom{\text{cost}}{\text{for } n \text{ photos}} = \binom{\text{initial}}{\text{cost}} + \binom{\text{cost}}{\text{for } n \text{ photos}}$$

$$419 + 0.14n = 299 + 0.18n$$

Carry Out

$$120 + 0.14n = 0.18n \qquad \text{299 was subtracted from both sides.}$$

$$120 = 0.04n \qquad \text{0.14n was subtracted from both sides.}$$

$$\frac{120}{0.04} = \frac{0.04n}{0.04}$$

$$3000 = n$$

Check and Answer    The total cost would be the same when 3000 photos were printed. We will leave the check of this answer for you.

Now Try Exercise 39

## Understanding Algebra

Recall, the word *percent* means *per hundred*. So, 30% of 80 is 0.30 × 80 = 24.

In application problems involving percent, we are always taking percents *of* quantities. Remember, "of" means multiply.

## 4 Set Up and Solve Applications Concerning Percent

Now we'll look at some application problems that involve percent. Remember that a percent is always a percent of something. Thus if the cost of an item, *c*, is increased by 8%, we would represent the new cost as $c + 0.08c$, and not $c + 0.08$. See the Avoiding Common Errors box on page 172.

EXAMPLE 7 **Water Bike Rental** At a beachfront hotel, the cost for a water bike rental is $30 per half hour, which includes a $7\frac{1}{2}$% sales tax. Find the cost of the rental before tax.

*Solution*    Understand    We are asked to find the cost of the water bike rental before tax. The cost of the rental before tax plus the tax on the water bike must equal $30.

Translate              Let $x$ = cost of the rental before tax.
                   Then $0.075x$ = tax on the rental.

(cost of the water bike rental before tax) + (tax on the rental) = 30

$$x + 0.075x = 30$$

Carry Out                                      $$1.075x = 30$$

$$x = \frac{30}{1.075}$$

$$x \approx 27.91$$

Check and Answer    A check will show that if the cost of the rental is $27.91, the cost of the rental including a $7\frac{1}{2}$% tax is about $30.

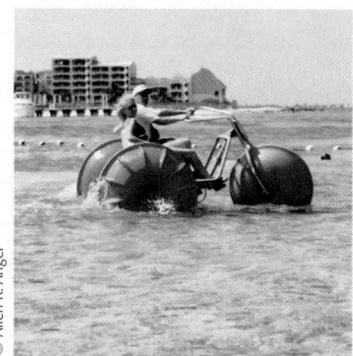

Now Try Exercise 47

EXAMPLE 8 **Caloric Intake** If the caloric intake for men in the United States increased 10% from 1971 to 2009 and reached 2695 calories in 2009, determine the average caloric intake for men in 1971.

*Solution*    Understand    We represent the 2009 caloric intake in terms of the unknown 1971 caloric intake. We use $c$ to represent the 1971 caloric intake and the 2009 caloric intake can be expressed in terms of $c$.

Translate                   Let $c$ = the 1971 caloric intake.
                 Then $c + 0.10c$ = the 2009 caloric intake.

Since the 2009 caloric intake is 2695 calories, we set up the following equation.

2009 caloric intake    is    2695
        ↓              ↓     ↓

$$\overbrace{c + 0.10c} = 2695$$

Carry Out                    $$1.10c = 2695$$

$$c = \frac{2695}{1.10}$$

$$c = 2450$$

Check and Answer    Since $c$ represents the 1971 caloric intake, and it is less than the 2009 caloric intake, our answer is reasonable. The 1971 caloric intake for men was 2450 calories.

Now Try Exercise 49

EXAMPLE 9 **Salary Plans** Jeanne Pirie has accepted a position selling medical supplies and equipment. During her first year, she is given a choice of salary plans. Plan 1 is a $450 weekly base salary plus a 3% commission of weekly sales. Plan 2 is a straight 10% commission of weekly sales. What weekly amount of sales, in dollars, would result in Jeanne receiving the same salary from both plans?

*Solution*  Understand  We are asked to find the *amount of sales*, in dollars, that will result in the same total salary from both plans. To solve this problem, we write expressions to represent the salary from each of the plans and set the salaries equal to one another.

Translate

Let $x$ = amount of sales in dollars.

Then $0.03x$ = commission from plan 1 sales

and $0.10x$ = commission from plan 2 sales.

$$\text{salary from plan 1} = \text{salary from plan 2}$$

$$\text{base salary} + 3\% \text{ commission} = 10\% \text{ commission}$$

$$450 + 0.03x = 0.10x$$

Carry Out

$$450 = 0.07x$$

$$\text{or} \quad 0.07x = 450$$

$$\frac{0.07x}{0.07} = \frac{450}{0.07}$$

$$x \approx 6428.57$$

Check   We will leave it up to you to show that sales of $6428.57 result in Jeanne receiving the same weekly salary from both plans.

Answer   Jeanne's weekly salary will be the same from both plans if she sells $6428.57 worth of medical supplies and equipment.

Now Try Exercise 59

---

### Helpful Hint

Here are some suggestions if you find you are having some difficulty with application problems.

1. Instructor—Make an appointment to see your instructor. Make sure you have read the material in the book and attempted all the homework problems. Go with specific questions for your instructor.

2. Tutoring—If your college learning center offers free tutoring, you may wish to take advantage of tutoring.

3. Study Group—Form a study group with classmates. Exchange phone numbers and e-mail addresses. You may be able to help one another.

4. Student's Solutions Manual—If you get stuck on an exercise you may want to use the Student's Solutions Manual to help you understand a problem. Do not use the Solutions Manual in place of working the exercises. In general, the Solutions Manual should be used only to check your work.

5. MyMathLab—MyMathLab provides exercises correlated to the text. In addition, online tools such as video lectures, animations, and a multimedia textbook are available to help you understand the material.

6. MathXL®—MathXL is a powerful online homework, tutorial, and assessment system correlated specifically to this text. You can take chapter tests in MathXL and receive a personalized study plan based on your test results. The study plan links directly to tutorial exercises for the objectives you need to study or retest.

7. Pearson Tutor Center—Once the program has been initiated by your instructor, you can get individual tutoring by phone, fax, or e-mail.

It is important that you keep trying! Remember, the more you practice, the better you will become at solving application problems.

# EXERCISE SET 3.2    MathXL®    MyMathLab®
MathXL®    MyMathLab

## Warm-Up Exercises

*Fill in the blanks with the appropriate word, phrase, or symbol(s) from the following list.*

$2x + (2x + 2) = 20$          $x + (x + 2) = 20$          $x + (2x) = 20$

$x + (2x + 2) = 20$          $(2x + 10) - x = 20$          $x + (10x - 2) = 20$

$(2x + 2) - x = 20$

1. The statement "a number ($x$) plus two more than twice that number is 20" can be represented by the equation _____.

2. The statement "a number ($x$) plus twice that number is 20" can be represented by the equation _____.

3. The statement "the sum of two consecutive odd numbers is 20" can be represented by the equation _____.

4. The statement "the sum of two numbers is 20 and the larger number is 2 less than ten times the smaller number ($x$)" can be represented by the equation _____.

5. The statement "the difference of two numbers is 20 and the larger number is 2 more than twice the smaller number ($x$)" can be represented by the equation _____.

6. The statement "the difference of two numbers is 20 and the larger number is 10 more than twice the smaller number ($x$)" can be represented by the equation _____.

## Practice the Skills/Problem Solving

*Exercises 7–32 involve finding a number or numbers. Review Examples 1–4, then set up an equation that can be used to solve the problem. Solve the equation and **answer the question asked**.*

7. **Unknown Number** Three subtracted from 4 times a number is 17. Find the number.

8. **Unknown Number** Five subtracted from 6 times a number is 13. Find the number.

9. **Consecutive Integers** The sum of two consecutive integers is 87. Find the numbers.

10. **Consecutive Integers** The sum of two consecutive integers is 113. Find the numbers.

11. **Odd Integers** The sum of two consecutive odd integers is 96. Find the numbers.

12. **Even Integers** The sum of two consecutive even integers is 146. Find the numbers.

13. **Sum of Numbers** One number is 3 more than twice a second number. Their sum is 27. Find the numbers.

14. **Sum of Numbers** One number is 5 less than 3 times a second number. Their sum is 43. Find the numbers.

15. **Difference of Numbers** The larger of two numbers is 4 less than five times the smaller. When the smaller number is subtracted from the larger, the difference is 4. Find the two numbers.

16. **Sum of Numbers** One number is 2 less than 3 times a second number. Their sum is 26. Find the numbers.

17. **Difference of Numbers** The larger of two integers is 8 less than twice the smaller. When the smaller number is subtracted from the larger, the difference is 17. Find the two numbers.

18. **Facing Pages** The sum of the two facing page numbers in an open book is 145. What are the page numbers?

19. **Grandma's Gifts** Grandma gave some baseball cards to Richey and some to Erin. She gave 3 times the amount to Erin as she did to Richey. If the total amount she gave to both of them was 260 cards, how many cards did she give to Richey?

20. **Ski Shop** The Alpine Valley Ski Shop sold 6 times as many downhill skis as cross-country skis. Determine the number of pairs of cross-country skis sold if the difference in the number of pairs of downhill and cross-country skis sold is 1800.

21. **Animal Art** Joseph Murray built a horse for display at a baseball stadium. It took him 1.4 hours more than twice the number of hours to attach the baseball gloves to the horse than to build the horse. If the total time it took him to build the horse and attach the gloves was 32.6 hours, how long did it take to attach the gloves?

© Allen R. Angel

The author, Allen R. Angel, is shown in this photo.

22. **Candle Shop** A candle shop makes 60 candles per week. It plans to increase the number of candles it makes by 8 per week until it reaches a production of 132 candles per week. How many weeks will it take for the shop to reach its production schedule?

23. **Collecting Frogs** Mary Shapiro collects ceramic and stuffed frogs. She presently has 422 frogs. She wishes to add 6 a week to her collection until her collection reaches a total of 500 frogs. How long will it take Mary's frog collection to reach 500 frogs?

24. **Population** The town of Dover currently has a population of 6500. If its population is increasing at a rate of 1200 people per year, how long will it take for the population to reach 20,600?

25. **Circuit Boards** The FGN Company produces circuit boards. It now has 4600 employees nationwide. It wishes to reduce

the number of employees by 250 per year through retirements, until its total employment is 2200. How long will this take?

26. **Computers** The CTN Corporation has a supply of 3600 computers. It wishes to ship 120 computers each week until its supply drops to 2000. How long will this take?

27. **Tornados** According to The Weather Channel, the greatest number of tornados in the United States occurs in June and the fewest number occurs in December. The average number of tornados in June is 16 less than 11 times the average number of tornados in December. If the difference between the average number of tornados in June and December is 204, determine the average number of tornados in December and June.

© Thinkstock

28. **Watching TV** According to the Kaiser Family Foundation, in the United States, the amount of time per day spent by 8–18-year-olds watching television is 16 minutes more than 5 times the number of minutes they spend reading. If the total amount of time per day reading and watching television is 274 minutes, determine the number of minutes spent watching television.

29. **Housekeepers** The average hourly wage paid to hotel housekeepers in New York City is $1.46 more than twice the average wage paid to hotel housekeepers in New Orleans. Determine the average hourly wage paid to housekeepers in New York City if the difference in their average hourly wages is $8.10.

30. **Albums** According to *Billboard Magazine*, the top-selling album of all time is Michael Jackson's *Thriller* and the second biggest selling album was AC/DC's *Back in Black*. The sum of the sales of those two titles is 150 million albums. *Thriller* sold 24 million albums more than twice that of *Back in Black*. Determine the sales of both albums.

© AP Wide World Photos

31. **Pork Production Costs** According to *Pig International*, a 2008 study showed that pork production was most expensive in Japan. In Japan, the cost of producing a kilogram of pork was 14 cents less than 3 times what it cost to produce a kilogram of pork in the United States. If the sum of the Japan cost and the U.S. cost was $2.94, determine the cost for each country.

32. **Oil Consumption** According to the U.S. Energy Information Administration, oil and liquid fuel consumption was projected to be 4.8% higher in 2015 than in 2008. If the sum of consumption for these two years is 209.6 quadrillion Btu, determine the projected consumption for 2015.

*Exercises 33–46 involve money. Read Examples 5–6, then set up an equation that can be used to solve the problem. Solve the equation and* **answer the question asked.**

33. **Gasoline** Luvia Rivera has only $48 to purchase gasoline. If gasoline costs $3.84 per gallon, determine how many gallons of gasoline Luvia can purchase.

34. **Truck Rental** Carol Battle rents a truck for one day and pays $50 per day plus 30 cents a mile. How far can Carol drive in one day if she has only $92?

35. **Copy Machine** Yamil Bernz purchased a copy machine for $2100 and a one-year maintenance protection plan that costs 2 cents per copy made. If he spends a total of $2462 in a year, which includes the cost of the machine and the copies made, determine the number of copies he made.

36. **Gym Membership** At Goldies Gym there is a one-time membership fee of $300 plus dues of $40 per month. If Carlos Manieri has spent a total of $700 for Goldies Gym, how long has he been a member?

37. **Television** Miles Potier's Time Warner cable bill costs $72.68 per month plus $3.95 for each On Demand movie he watches that month. If his cable bill for December was $96.38, determine the number of On Demand movies he watched in December.

38. **Hardwood Floors** Ruth Zasada is having hardwood floors installed in her living room. The cost for the material is $2840 plus an installation charge of $1.90 per square foot. If the total cost for the material plus installation is $5120, determine the area of her living room.

© minik/Shutterstock

39. **Truck Rental** Howard Sporn is considering two companies from which to rent a truck. American Truck Rental charges $20 per day and 25 cents a mile. SavMor Truck Rental charges $35 a day and 15 cents a mile. How far would Howard need to drive in one day for the both companies to have the same total cost?

**40. Washing Machines** Scott Montgomery is considering two washing machines, a Kenmore® and a Neptune®. The Neptune costs $454 while the Kenmore costs $362. The energy guides indicate that the Kenmore will cost an estimated $84 per year to operate and the Neptune will cost an estimated $38 per year to operate. How long will it be before the total cost is the same for both washing machines?

**41. Salaries** Brooke Mills is being recruited by a number of high-tech companies. Data Technology Corporation has offered her an annual salary of $40,000 per year plus a $2400 increase per year. Nuteck has offered her an annual salary of $49,600 per year plus an $800 increase per year. In how many years will the salaries from the companies be the same?

**42. Racquet Club** The Coastline Racquet Club has two payment plans for its members. Plan 1 has a monthly fee of $20 plus $8 per hour for court time. Plan 2 has no monthly fee, but court time is $16.25 per hour. If court time is rented in 1-hour intervals, how many hours would you have to play per month so that plan 1 becomes a better buy?

**43. Printers** Hector Hanna will purchase one of two laser printers, a Hewlett-Packard (HP) or a Lexmark®. The HP costs $499 and the Lexmark costs $419. Suppose, because of the price of the ink cartridges, the cost of printing a page on the HP is $0.06 per page and the cost of printing a page on the Lexmark is $0.08 per page. How many pages would need to be printed for the two printers to have the same total cost?

**44. Satellite or Cable** Sean Stewart is deciding whether to select a satellite receiver or cable for his television programming. The satellite receiver costs $298.90 and the monthly charge is $68.70. With cable there is no initial cost to purchase equipment, but the monthly charge for comparable channels is $74.80. After how many months will the total cost of the two systems be equal?

© Neale Cousland/Shutterstock

**45. Newsletter** Neil Simpson had a professional organization newsletter printed and sent out to all the members. The total cost included a $600 printing cost plus a 39 cents mailing cost for each envelope. If the total cost was $1380, determine how many newsletters were mailed.

**46. Patio Resurfacing** Elizabeth Chu is having her patio resurfaced using concrete pavers. She is considering two companies for the job. A & E Pavers charges $1500 for the pavers plus $40 per hour for labor to install the pavers. The Jerilyn Fairman Company charges $1800 for the pavers plus $25 per hour to install the pavers. How many hours of labor would result in the same total cost with both companies?

*Exercises 47–68 involve percents. Read Examples 7–9, then set up an equation that can be used to solve the problem. Solve the equation and* **answer the question asked**.

**47. Airfare** The airfare for a flight from Amarillo, Texas, to New Orleans cost $295.34, which includes a 7% sales tax. What is the cost of the flight before tax? *Source:* expedia.com

© Nick Rowe/Getty Images, Inc.–Photodisc/Royalty Free

**48. New Car** Yoliette Fournier purchased a new car. The cost of the car, including a 7.5% sales tax was $24,600. What was the cost of the car before tax?

**49. Salary Increase** Zhen Tong just received a job offer that will pay him 30% more than his present job does. If the salary at his new job will be $30,200, determine his present salary.

**50. New Headquarters** Tarrach and Associates plans on increasing the size of its headquarters by 20%. If its new headquarters is to be 14,200 square feet, determine the size of its present headquarters.

**51. Auto Exports** According to the Federal Reserve Bank of Chicago, in 2007 exports of new light vehicles increased 41%

over 1996 in the United States. If in 2007 about 1.7 million light vehicles were exported, determine the number of vehicles exported in 1996.

© minik/Shutterstock

**52. Retirement Income** Ray and Mary Burnham have decided to retire. They estimate their annual income after retirement will be reduced by 15% from their pre-retirement income. If they estimate their retirement income to be $42,000, determine their pre-retirement income.

**53. Autographs** A tennis star was hired to sign autographs at a convention. She was paid $3000 plus 3% of all admission fees collected at the door. The total amount she received for the day was $3750. Find the total amount collected at the door.

**54. Sale** At a 1-day 20% off sale, Jane Demsky purchased a hat for $25.99. What is the regular price of the hat?

**55. Wage Cut** A manufacturing plant is running at a deficit. To avoid layoffs, the workers agree on a temporary wage cut of 2%. If the average salary in the plant after the wage cut is $38,600, what was the average salary before the wage cut?

**56. Teachers** During the 2006 contract negotiations, the city school board approved a 5% pay increase for its teachers effective in 2007. If Dana Frick, a first-grade teacher, projects his 2007 annual salary to be $46,400, what was his 2006 salary?

**57. Earnings and Education** The U.S. Census Bureau reported that in 2005, graduates with an associate's degree earned an average of 24.6% less than graduates with a bachelor's degree. If, in 2005, the average graduate with an associate's degree earned $37,600, determine the average salary of a graduate with a bachelor's degree.

**58. Sales Volume** Mona Fabricant receives a weekly salary of $350. She also receives a 6% commission on the total sales she makes. What must her sales be in a week, if she is to make a total of $710?

**59. Salary Plans** Vince McAdams, a salesman, is given a choice of two salary plans. Plan 1 is a weekly salary of $600 plus 2% commission of sales. Plan 2 is a straight commission of 10% of sales. How much in sales must Vince make in a week for both plans to result in the same salary?

**60. Area** The Johnson Performing Arts Center has increased in size. The area of the new building is 42% larger than the area of the original building. If the area of the new building is 56,000 square feet, determine the area of the original building.

**61. Book** The number of pages in the third edition of a book was 4% less than the number of pages in the second edition. If the number of pages in the third edition is 480, determine the number of pages in the second edition.

**62. Financial Planning** Belen Poltorade, a financial planner, is offering her customers two financial plans for managing their assets. With plan 1 she charges a planning fee of $1000 plus 1% of the assets she will manage for the customers. With plan 2 she charges a planning fee of $500 plus 2% of the assets she will manage. How much in customer assets would result in both plans having the same total fees?

**63. Salary Plans** Becky Schwartz, a saleswoman, is offered two salary plans. Plan 1 is $400 per week salary plus a 2% commission of sales. Plan 2 is a $250 per week salary plus a 16% commission of sales. How much would Becky need to make in sales for the salary to be the same from both plans?

**64. Art Show** Bill Rush wants to rent a building for a week to show his artwork and has been offered two rental plans. Plan 1 is a rental fee of $500 plus 3% of the dollar sales he makes.

Plan 2 is $100 plus 15% of the dollar sales he makes. What dollar sales would result in both plans having the same total cost?

© Keith Levit/Shutterstock

**65. Eating Out** After Linda Kodama is seated in a restaurant, she realizes that she has only $30. From this $30 she must pay a 7% tax and she wishes to leave a 15% tip on the price of the meal before tax. What is the maximum price for a meal that she can afford to pay?

**66. Membership Fees** The Holiday Health Club has reduced its annual membership fee by 10%. In addition, if you sign up on a Monday, the Club will take an additional $20 off the already reduced price. If Jorge Sanchez purchases a year's membership on a Monday and pays $250, what is the regular membership fee?

© Andresr/Shutterstock

**67. Estate** Phil Dodge left an estate valued at $140,000. In his will, he specified that his wife will get 25% more of his estate than his daughter. How much will his wife receive?

**68. Charitable Giving** Charles Ford made a $200,000 cash contribution to two charities, the American Red Cross and the United Way. The amount received by the American Red Cross was 30% greater than the amount received by the United Way. How much did the United Way receive?

**69. Oil Use** According to the International Energy Agency, the demand for oil in North America is far greater than in any other region. In 2007, North America used an average of 5.1 million gallons of oil per day less than twice that used by Europe, the second greatest user. The total used by both regions in one day was 40.8 million gallons. Determine the number of gallons of oil used in one day in Europe in 2007.

**70.** Refer to Exercise 69. Determine the number of gallons of oil used in one day in North America in 2007.

## Concept/Writing Exercises

**71.** Outline the five-step problem-solving procedure we use.

**72.** Explain the concept of *percent increase* to a friend. If you made $8 per hour last month and got a 15% increase, what is your new hourly wage?

## Challenge Problems

**73. Average Value** To find the *average* of a set of values, you find the sum of the values and divide the sum by the number of values.

**a)** If Paul Lavenski's first three test grades are 74, 88, and 76, write an equation that can be used to find the grade that Paul must get on his fourth exam to have an 80 average.

**b)** Solve the equation from part **a)** and determine the grade Paul must receive.

**74. Driver Education** A driver education course costs $45 but saves those under age twenty-five 10% of their annual insurance premiums until they reach age twenty-five. Scott Day has just turned 18, and his insurance costs $600 per year.

**a)** How long will it take for the amount saved from insurance to equal the price of the course?

**b)** Including the cost of the course, when Scott turns 25, how much will he have saved?

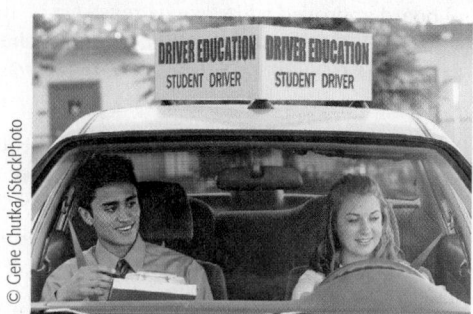
© Gene Chutka/iStockPhoto

## Cumulative Review Exercises

[1.9]   **75.** Evaluate $4[(4 - 6) \div 2] + 3^2 - 1$.

[1.10]  **76.** Name the following property: $3x + 4 = 4 + 3x$.

[2.6]   **77.** Solve the formula $A = \frac{1}{2}bh$ for $h$.

[2.7]   **78.** Solve the proportion $\frac{4.5}{6} = \frac{9}{x}$.

## Mid-Chapter Test: 3.1–3.2

*To find out how well you understand the chapter material to this point, take this brief test. The answers and the section where the material was initially discussed are given in the back of the book. Review any questions you answered incorrectly.*

*In Exercises 1–6, express each statement as an algebraic expression.*

**1.** Six times the weight, $w$

**2.** Five inches more than 3 times the height, $h$

**3.** Represent the cost, $c$, increased by 20%, as a mathematical expression.

**4.** Dennis Donahue rents a truck for $60 per day plus 95 cents per mile, $m$. Write an expression for the total cost of the rental for one day.

**5.** Write an expression for the number of cents in $n$ half-dollars.

**6.** Twenty-five dollars is divided between Amy Keyser and Sherry Norris. If Amy gets $x$ dollars, how much will Sherry get?

**7.** Explain why the cost, $c$, of an item at a 25% off sale is not $c - 25$. Write the correct algebraic expression for the cost of an item at a 25% off sale.

**8.** In the statement, determine what $x =$.

A Gaudy Leaf Frog is 2 centimeters longer than 3 times the length of a Poison Dart Frog (see photos).

© Allen R. Angel

Gaudy Leaf Frog

© Allen R. Angel

Poison Dart Frog

**9.** Select a variable to represent one quantity and state what that variable represents. Express the second quantity in terms of the variable selected.

The distance Mary traveled is 6 miles more than 4 times the distance Pedro traveled.

**10.** The value of a car in 2005 was 18% less than its value in 2006. Write an expression for the difference in the value of the car from 2006 to 2005.

*In Exercises 11 and 12, write the problem as an equation. Do not solve.*

**11.** The population of Cedar Oaks increased by 12%. The population after the increase was 38,619.

**12.** For two consecutive odd integers, the sum of the smaller and 3 times the larger is 26.

In Exercises 13–20, write an equation that can be used to solve the problem. Solve the equation and answer the question asked.

13. **Consecutive Integers** The sum of two consecutive integers is 93. Find the numbers.

14. **Numbers** The larger of two integers is one less than 3 times the smaller. When the smaller number is subtracted from the larger, the difference is 7. Find the numbers.

15. **Candy** A candy manufacturer presently produces 240 boxes of candy a day and wants to increase production by 20 boxes per day until it produces 600 boxes of candy per day. How many days will it take for production to reach 600 boxes per day?

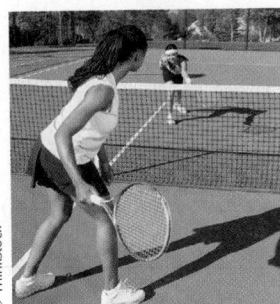

© Thinkstock

See Exercise 16.

16. **Tennis** Kristina Schmid is considering joining one of two tennis clubs. At Dale's Tennis Club the monthly fee is $90, and court time is $4 per hour. At Abel's Tennis Club the monthly fee is $30, but court time is $8 per hour. How many hours in a month would Kristina need to play for the total cost to be the same with both clubs?

17. **Television** The cost of a television plus a 7% sales tax is $749. Find the cost of the television before tax.

18. **Clients** Anita and Betty together have a total of 600 clients. If Anita has 12 more than twice the number of clients Betty has, determine the number of clients each person has.

19. **Truck Rental** A truck cost $36 a day plus 18 cents a mile to rent. If the total cost for a one-day rental is $45.36, how many miles were driven?

20. **Salary Plans** A salesman is offered two salary plans. Plan 1 is $200 per week plus 8% commission of the dollar sales he makes. Plan 2 is $300 per week plus 6% of the dollar sales he makes. How much in sales must the salesman make in a week for the two plans to have the same total salary?

# 3.3 Geometric Problems

1 Solve geometric problems.

## 1 Solve Geometric Problems

This section serves two purposes. One is to reinforce the geometric formulas introduced in Section 2.6. The second is to reinforce procedures for setting up and solving verbal problems discussed in Sections 3.1 and 3.2.

**EXAMPLE 1  Sandbox** Christine O'Connor is planning to build a sandbox for her daughter. She has 30 feet of lumber with which to build the perimeter. What should be the dimensions of the rectangular sandbox if the length is to be 3 feet longer than the width (**Fig. 3.3**)?

*Solution*   Understand   We are asked to find the dimensions of the sandbox that Christine plans to build. The perimeter of the sandbox will be 30 feet. Since the length is given in terms of the width, we will let $w$ represent the width. Then we can express the length in terms of $w$. To solve this problem, we use the formula for the perimeter of a rectangle, $P = 2l + 2w$, where $P = 30$ feet.

Translate                          Let $w$ = width of the sandbox.

Then $w + 3$ = length of the sandbox.

$$P = 2l + 2w$$

Carry Out

$$30 = 2(w + 3) + 2w$$
$$30 = 2w + 6 + 2w$$
$$30 = 4w + 6$$
$$24 = 4w$$
$$6 = w$$

$w$

$l = w + 3$

FIGURE 3.3

### Understanding Algebra

*Recall some useful geometry formulas:*

| | |
|---|---|
| Rectangle | Area: $A = l \cdot w$ |
| | Perimeter: $P = 2l + 2w$ |
| Triangle | Area: $A = \frac{1}{2}b \cdot h$ |
| | Sum of the measures of the interior angles = 180° |
| Quadrilateral | Sum of the measures of the interior angles = 360° |
| Circle | Area $= \pi r^2$ |
| | Circumference $= 2\pi r$ |

## Understanding Algebra

*Additional geometry facts:*

Equilateral triangle: has 3 equal sides and 3 equal angles (all 60°)

Isosceles triangles: has 2 equal sides

Congruent triangles: corresponding sides and corresponding angles are equal

Similar triangles: corresponding sides are in proportion and corresponding angles are equal

Polygon with *n* sides: sum of the measures of the interior angles is $(n-2)180°$

Complementary angles: angles whose measures add to 90°

Supplementary angles: angles whose measures add to 180°

FIGURE 3.4

The width is **6 feet.** Since the length is 3 feet longer than the width, the length is $6 + 3 = 9$ feet.

**Check**   We will check the solution by substituting the appropriate values in the perimeter formula.

$$P = 2l + 2w$$
$$30 \overset{?}{=} 2(9) + 2(6)$$
$$30 = 30 \qquad \text{True}$$

**Answer**   The width of the sandbox will be 6 feet and the length will be 9 feet.

*Now Try Exercise 23*

A triangle that contains two sides of equal length is called an **isosceles triangle.** In isosceles triangles, the angles opposite the two sides of equal length have equal measures.

**EXAMPLE 2   Corner Lot**  Mr. and Mrs. Harmon Katz have a corner lot that is in the shape of an isosceles triangle. Two angles of their triangular lot are the same and the third angle is 30° greater than the other two. Find the measure of all three angles (see **Fig. 3.4**).

*Solution*   **Understand**   To solve this problem, you must know that the sum of the angles of any triangle measures 180°. We are asked to find the measure of each of the three angles, where the two smaller angles have the same measure. We will let *x* represent the measure of the smaller angles, and then we will express the larger angle in terms of *x*.

**Translate**          Let $x$ = the measure of each smaller angle.

Then $x + 30$ = the measure of the larger angle.

sum of the 3 angles = 180

**Carry Out**          $x + x + (x + 30) = 180$
$$3x + 30 = 180$$
$$3x = 150$$
$$x = \frac{150}{3} = 50$$

The two smaller angles are each 50°. The larger angle is $x + 30°$ or $50° + 30° = 80°$.

**Check and Answer**   Since $50° + 50° + 80° = 180°$, the answer checks. The two smaller angles are each 50° and the larger angle is 80°.

*Now Try Exercise 11*

Recall from Section 2.6 that a quadrilateral is a four-sided figure. Quadrilaterals include squares, rectangles, parallelograms, and trapezoids. The sum of the measures of the angles of any quadrilateral is 360°.

FIGURE 3.5

**EXAMPLE 3   Water Trough**  Sarah Fuqua owns horses and uses a water trough whose ends are trapezoids. The measure of the two bottom angles of the trapezoid are the same, and the measure of the two top angles are the same. The bottom angles measure 15° less than twice the measure of the top angles. Find the measure of each angle.

*Solution*   **Understand**   To help visualize the problem, we draw a picture of the trapezoid, as in **Figure 3.5**. We use the fact that the sum of the measures of the four angles of a quadrilateral is 360°.

Translate          Let $x$ = the measure of each of the two smaller angles.

Then $2x - 15$ = the measure of each of the two larger angles.

$$\left(\begin{array}{c}\text{measure of the}\\\text{two smaller angles}\end{array}\right) + \left(\begin{array}{c}\text{measure of the}\\\text{two larger angles}\end{array}\right) = 360$$

Carry Out

$$x + x + (2x - 15) + (2x - 15) = 360$$
$$x + x + 2x - 15 + 2x - 15 = 360$$
$$6x - 30 = 360$$
$$6x = 390$$
$$x = 65$$

Each smaller angle is 65°. Each larger angle is $2x - 15 = 2(65) - 15 = 115°$.

**Check and Answer**   Since 65° + 65° + 115° + 115° = 360°, the answer checks. Each smaller angle is 65° and each larger angle is 115°.

Now Try Exercise 27

**EXAMPLE 4  Fenced-In Area**   Ronald Yates recently started an ostrich farm. He is separating the ostriches by fencing in three equal rectangular areas, as shown in **Figure 3.6**. The length of the fenced-in area, $l$, is to be 30 feet greater than the width and the total amount of fencing available is 660 feet. Find the length and width of the fenced-in area.

*Solution*   Understand   The fencing consists of four pieces of fence of length $w$, and two pieces of fence of length $l$.

Translate          Let $w$ = width of fenced-in area.

Then $w + 30$ = length of fenced-in area.

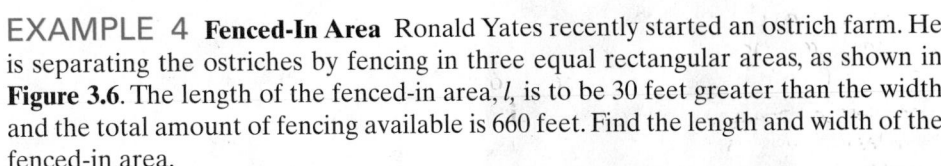

$$\left(\begin{array}{c}\text{4 pieces of fence}\\\text{of length } w\end{array}\right) + \left(\begin{array}{c}\text{2 pieces of fence}\\\text{of length } w + 30\end{array}\right) = 660$$

**FIGURE 3.6**

Carry Out

$$4w + 2(w + 30) = 660$$
$$4w + 2w + 60 = 660$$
$$6w + 60 = 660$$
$$6w = 600$$
$$w = 100$$

Since the width is 100 feet, the length is $w + 30$ or $100 + 30$ or 130 feet.

**Check and Answer**   Since $4(100) + 2(130) = 660$, the answer checks. The width of the fenced-in area is 100 feet and the length is 130 feet.

Now Try Exercise 37

# EXERCISE SET 3.3          MathXL®   MyMathLab®

## Warm-Up Exercises

*Fill in the blanks with the appropriate word, phrase, or symbol(s) from the following list.*

| | | | | | |
|---|---|---|---|---|---|
| 360° | 180° | 90° | $2l + 12$ | 12 | equilateral |
| 10 | quadrilateral | $l \cdot w$ | 45° | isosceles | |

1. In a rectangle, the length is 1 inch more than twice the width and the perimeter is 62 inches. Then the width is _____ inches.

2. The area of a rectangle is 120 square inches. If the length is 10 inches, then the width is _____ inches.

3. The sum of the measures of the interior angles of a quadrilateral is _____.

4. The sum of the measures of the angles of a triangle is _____.

5. A triangle with two equal sides is called _____.

6. A triangle with three equal sides is called _____.

7. The area of a rectangle of length $l$ and width $w$ is _____.

8. A four-sided figure is called a _____.

**9.** If one angle of an isosceles triangle is 90°, then each other angle must measure _____ .

**10.** The perimeter of a rectangle of length *l* and width 6 is _____ .

## Practice the Skills/Problem Solving

*Solve the following geometric problems.*\*

**11. Isosceles Triangle** In an isosceles triangle, one angle is 42° greater than the other two equal angles. Find the measure of all three angles. See Example 2.

**12. Triangular Building** This building in New York City, referred to as the Flatiron Building, has a perimeter in the shape of an isosceles triangle. If the shortest side of the triangle is 50 feet shorter than the two longer sides, and the perimeter is 196 feet, determine the length of the three sides of the triangle.

© Allen R. Angel

**13. A Special Triangle** An **equilateral triangle** is a triangle that has three sides of the same length. The perimeter of an equilateral triangle is 34.5 inches. Find the length of each side.

**14. Equilateral Triangle** The perimeter of an equilateral triangle is 48.6 centimeters. Find the length of each side. See Exercise 13.

**15. Complementary Angles** Two angles are **complementary angles** if the sum of their measures is 90°. Angle *A* and angle *B* are complementary angles, and angle *A* is 21° more than twice angle *B*. Find the measures of angle *A* and angle *B*.

Complementary Angles

**16. Complementary Angles** Angles *A* and *B* are complementary angles, and angle *B* is 14° less than angle *A*. Find the measures of angle *A* and angle *B*. See Exercise 15.

**17. Supplementary Angles** Two angles are **supplementary angles** if the sum of their measures is 180°. Angle *A* and angle *B* are supplementary angles, and angle *B* is 8° less than three times angle *A*. Find the measures of angle *A* and angle *B*.

Supplementary Angles

**18. Supplementary Angles** Angles *A* and *B* are supplementary angles and angle *A* is 2° more than 4 times angle *B*. Find the measures of angle *A* and angle *B*. See Exercise 17.

**19. Vertical Angles** When two lines intersect, the opposite angles are called **vertical angles**. Vertical angles have equal measures. Determine the measures of the vertical angles indicated in the following figure.

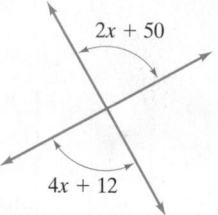
$2x + 50$
$4x + 12$

**20. Vertical Angles** A pair of vertical angles is indicated in the following figure. Determine the measure of the vertical angles indicated. See Exercise 19.

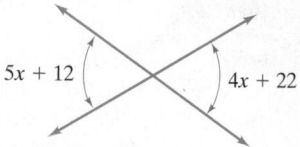
$5x + 12$    $4x + 22$

**21. Unknown Angles** One angle of a triangle is 10° greater than the smallest angle, and the third angle is 30° less than twice the smallest angle. Find the measures of the three angles.

**22. Unknown Angles** One angle of a triangle is 20° larger than the smallest angle, and the third angle is 6 times as large as the smallest angle. Find the measures of the three angles.

**23. Dimensions of Rectangle** The length of a rectangle is 6 feet more than its width. What are the dimensions of the rectangle if the perimeter is 44 feet?

**24. Dimensions of Rectangle** The perimeter of a rectangle is 120 feet. Find the length and width of the rectangle if the length is twice the width.

**25. Tennis Court** The length of a regulation tennis court is 6 feet greater than twice its width. The perimeter of the court is 228 feet. Find the length and width of the court.

Length
Width

\*See Appendix C for more material on geometry.

26. **Patio** Rikki Blair is building a rectangular patio. The perimeter of the patio is to be 96 feet. Determine the dimensions of the patio if the length is to be 6 feet less than twice the width.

27. **Parallelogram** In a parallelogram the opposite angles have the same measures. Each of the two larger angles in a parallelogram is 20° less than 3 times the smaller angles. Find the measure of each angle.

28. **Parallelogram** The two smaller angles of a parallelogram have equal measures, and the two larger angles each measure 27° less than twice each smaller angle. Find the measure of each angle.

29. **Rhombus** A rhombus is a parallelogram with four equal sides. Each of the two larger angles of a rhombus is 5 times as large as the two smaller angles. Find the measure of each of the four angles.

30. **Rhombus** Each of the two larger angles of a rhombus are 20° less than four times the two smaller angles. Find the measure of each of the four angles.

31. **Quadrilateral** The measure of one angle of a quadrilateral is 10° greater than the smallest angle; the third angle is 14° greater than twice the smallest angle; and the fourth angle is 21° greater than the smallest angle. Find the measures of the four angles of the quadrilateral.

32. **Quadrilateral** The measure of one angle of a quadrilateral is twice the smallest angle; the third angle is 20° greater than the smallest angle; and the fourth angle is 20° less than twice the smallest angle. Find the measures of the four angles of the quadrilateral.

33. **Building a Bookcase** A bookcase is to have four shelves, including the top, as shown. The height of the bookcase is to be 3 feet more than the width. Find the width and height of the bookcase if only 30 feet of lumber is available.

34. **Bookcase** A bookcase is to have four shelves as shown. The height of the bookcase is to be 2 feet more than the width, and only 20 feet of lumber is available. What should be the width and height of the bookcase?

35. **American Flag** The dimensions of the American flag appear in the figure below. The perimeter of this particular flag is 580 inches.

a) Determine its length and width.

b) What is the width of each of the stripes?

c) The (left-to-right) width of the blue rectangle is always 76% of the vertical height of the flag. How many inches is that blue width?

d) The vertical height of the blue rectangle is always 53.85% of the vertical height of the flag. How many inches is the blue height?

e) What is the area of this flag?

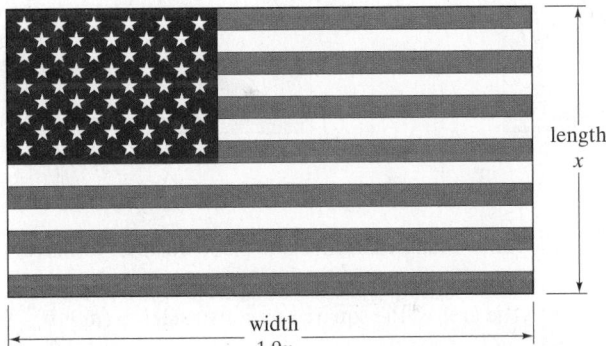

*Source:* www.usflag.org

36. **Storage Shelves** Carlotta Perez plans to build storage shelves as shown. She has only 45 feet of lumber for the entire unit and wishes the width to be 3 times the height. Find the width and height of the unit.

37. **Fenced-In Area** A rectangular area is to be fenced in along a straight river bank as illustrated. The length of the fenced-in area is to be 5 feet greater than the width, and the total amount of fencing to be used is 71 feet. Find the width and length of the fenced-in area.

38. **Gardening** Trina Zimmerman is placing a border around and within a garden where she intends to plant flowers (see the figure). She has 60 feet of bordering, and the length of the garden is to be 2 feet greater than the width. Find the length and width of the garden. The red shows the location of all the bordering in the figure.

## Challenge Problems

39. One way to express the area of the figure on the right is $(a + b)(c + d)$. Can you determine another expression, using the area of the four rectangles, to represent the area of the figure?

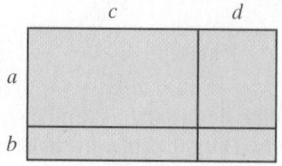

## Group Activity

*Discuss and answer Exercise 40 as a group.*

40. Consider the four pieces shown. Two are squares and two are rectangles.

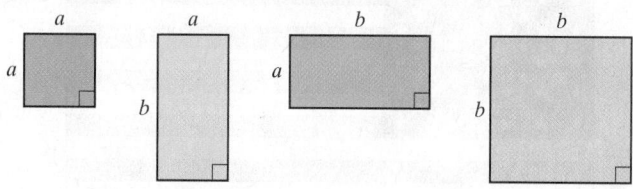

   a) Individually, rearrange and place the four pieces together to form one square.

   b) The area of the square you constructed is $(a + b)^2$. Write another expression for the area of the square by adding the four individual areas.

   c) Compare your answers. If each member of the group did not get the same answers to parts **a)** and **b)**, work together to determine the correct answer.

   d) Answer the following question as a group. If $b$ is twice the length of $a$, and the perimeter of the square you created is 54 inches, find the length of $a$ and $b$.

   e) Use the values of $a$ and $b$ found in part **d)** to find the area of the square you created.

   f) Use the values of $a$ and $b$ found in part **d)** to find the areas of the four individual pieces that make up the large square.

   g) Does the sum of the areas of the four pieces found in part **f)** equal the area of the large square found in part **e)**? Is this what you expected? Explain.

## Cumulative Review Exercises

*Insert either $>$, $<$, or $=$ in each shaded area to make the statement true.*

[1.5]   41. $-|-6|$ ▨ $|-4|$

   42. $|-3|$ ▨ $-|3|$

[1.7]   43. Evaluate $-8 - (-2) + (-4)$.

[2.1]   44. Simplify $-7y + x - 3(x - 2) + 2y$.

[2.6]   45. Solve $6x + 3y = 9$ for $y$.

# 3.4 Motion, Money, and Mixture Problems

1. Solve motion problems involving two rates.

2. Solve money problems.

3. Solve mixture problems.

## Understanding Algebra

Motion problems involve *rates*. Units of rates include miles per hour (mph), feet per second (ft/s) and meters per second (m/s).

The formula we use in motion problems is:

distance = rate × time

or $d = r \cdot t$.

We now discuss three additional types of applications: motion, money, and mixture problems. These problems are grouped in the same section because, as you will learn shortly, you use the same general multiplication procedure to solve them. We begin by discussing motion problems.

### 1 Solve Motion Problems Involving Two Rates

A **motion problem** is one in which an object is moving at a specific rate for a specific period of time. Examples of motion problems include a car traveling at a constant speed or a person walking at a constant speed or a boat being rowed at a constant speed. In this section, we will discuss motion problems that involve *two rates*, such as two trains traveling at different speeds. We will use the distance formula, distance = rate × time, and construct tables like the following one to organize the information. The formula at the top of the table shows how the distance in the last column is calculated.

| | Rate | × | Time | = | Distance |
|---|---|---|---|---|---|
| Item | Rate | | Time | | Distance |
| Item 1 | | | | | distance 1 |
| Item 2 | | | | | distance 2 |

Examples 1 and 2 illustrate the procedure used.

**EXAMPLE 1  Camping Trip**  Maryanne and Paul Justinger and their son Danny are on a canoe trip on the Erie Canal. Danny is in one canoe and Paul and Maryanne are in a second canoe. Both canoes start at the same time from the same point and travel in the same direction. The parents paddle their canoe at 2 miles per hour and their son paddles his canoe at 4 miles per hour. In how many hours will the two canoes be 5 miles apart?

*Solution*   Understand and Translate   We are asked to find the time it takes for the canoes to become separated by 5 miles. We construct a table to aid us in setting up the problem.

Let $t$ = time when canoes are 5 miles apart.

Draw a sketch to help visualize the problem (**Fig. 3.7**). When the two canoes are 5 miles apart, each has traveled for the same number of hours, $t$.

| Canoe | Rate | Time | Distance |
|---|---|---|---|
| Parents | 2 | $t$ | $2t$ |
| Son | 4 | $t$ | $4t$ |

Since the canoes are traveling in the same direction, the distance between them is found by subtracting the distance traveled by the slower canoe from the distance traveled by the faster canoe.

$$\begin{pmatrix} \text{distance traveled} \\ \text{by faster canoe} \end{pmatrix} - \begin{pmatrix} \text{distance traveled} \\ \text{by slower canoe} \end{pmatrix} = 5 \text{ miles}$$

$$4t \quad - \quad 2t \quad = 5$$

Carry Out

$$2t = 5$$

$$t = 2.5$$

**Answer**   After 2.5 hours the two canoes will be 5 miles apart.

(a) Beginning of trip

(b) After $t$ hours

**FIGURE 3.7**

Now Try Exercise 7

**EXAMPLE 2  Paving Roads** Two highway paving crews are 20 miles apart working toward each other. One crew paves 0.4 mile of road per day more than the other crew, and the two crews meet after 10 days. Find the rate at which each crew paves the road.

*Solution*    Understand and Translate    We are asked to find the two rates. We are told that both crews work for 10 days.

$$\text{Let } r = \text{rate of slower crew.}$$
$$\text{Then } r + 0.4 = \text{rate of faster crew.}$$

We make a sketch (**Fig. 3.8**) and set up a table of values.

Rate: $r$        Rate: $r + 0.4$

|← 20 miles →|
covered in 10 days

**FIGURE 3.8**

| Crew | Rate | Time | Distance |
|------|------|------|----------|
| Slower | $r$ | 10 | $10r$ |
| Faster | $r + 0.4$ | 10 | $10(r + 0.4)$ |

The total distance covered by both crews is 20 miles. Since the crews are moving in opposite directions, the distance between them is found by adding the two distances.

$$\left(\begin{array}{c}\text{distance covered}\\\text{by slower crew}\end{array}\right) + \left(\begin{array}{c}\text{distance covered}\\\text{by faster crew}\end{array}\right) = 20 \text{ miles}$$

Carry Out

$$10r \quad + \quad 10(r + 0.4) \quad = 20$$
$$10r + 10r + 4 = 20$$
$$20r + 4 = 20$$
$$20r = 16$$
$$\frac{20r}{20} = \frac{16}{20}$$
$$r = 0.8$$

Answer   The slower crew paves 0.8 mile of road per day and the faster crew paves $r + 0.4$ or $0.8 + 0.4 = 1.2$ miles of road per day.

Now Try Exercise 23

### Understanding Algebra

Notice that in Example 2, the rate, $r$, is measured in *miles* (of road) *per day*.

---

**Helpful Hint**

When working with two different moving items, if the items are moving in the same direction, the solution will involve subtracting the smaller distance from the larger distance as in Example 1. If the items are moving in opposite directions, the solution will involve adding the distances together as in Example 2.

### Understanding Algebra

*Reminder:*

interest = principal × rate × time

### Understanding Algebra

In money problems, *"rate"* applies to rate of interest and is expressed as a decimal.

In context, it is easily distinguished from *rate* used in motion problems.

## 2  Solve Money Problems

One type of money problem involves simple interest. When working with simple interest problems involving two amounts, we can use a table like the one below.

|  | Principal × | Rate × | Time = | Interest |
|---------|-----------|------|------|----------|
| Account | Principal | Rate | Time | Interest |
| Account 1 |  |  |  | interest 1 |
| Account 2 |  |  |  | interest 2 |

**EXAMPLE 3  Investments** Carmine DeSanto has $15,000 to invest. He is considering two investments. One is a loan he can make to another party that pays him 8% simple interest for a year. A second investment is a 1-year certificate of deposit that pays 5%. Carmine decides that he wants to place some money in each investment, and he wants to earn a total of $1125 interest in 1 year from the two investments. How much money should Carmine put in each investment?

*Solution*   Understand and Translate

Let $x$ = amount to be invested at 5%.

Then $15{,}000 - x$ = amount to be invested at 8%.

We use the simple interest formula, interest = principal · rate · time, to solve this problem.

| Account | Principal | Rate | Time | Interest |
|---------|-----------|------|------|----------|
| CD | $x$ | 0.05 | 1 | $0.05x$ |
| Loan | $15{,}000 - x$ | 0.08 | 1 | $0.08(15{,}000 - x)$ |

Since the sum of the interest from the two investments is $1125, we write the equation

$$\left(\begin{array}{c}\text{interest from}\\ \text{5\% CD}\end{array}\right) + \left(\begin{array}{c}\text{interest from}\\ \text{8\% investment}\end{array}\right) = \text{total interest}$$

$$0.05x + 0.08(15{,}000 - x) = 1125$$

Carry Out
$$0.05x + 0.08(15{,}000) - 0.08(x) = 1125$$
$$0.05x + 1200 - 0.08x = 1125$$
$$-0.03x + 1200 = 1125$$
$$-0.03x = -75$$
$$x = \frac{-75}{-0.03} = 2500$$

Check and Answer   Thus, $2500 should be invested at 5% interest. The amount to be invested at 8% is

$$15{,}000 - x = 15{,}000 - 2500 = 12{,}500$$

The total amount invested is $2500 + $12,500 = $15,000, which checks with the information given.

Now Try Exercise 33

In Example 3, we let $x$ represent the amount invested at 5%. If we had let $x$ represent the amount invested at 8%, the answer would not have changed. Rework Example 3 now, letting $x$ represent the amount invested at 8%.

In other types of problems involving two amounts of money, we generally set up similar tables, as illustrated in the next example.

**EXAMPLE 4   Rocking Chairs** Johnson's Patio Furniture Store sells two types of rocking chairs. The single-person rocking chair sells for $130 each and the two-person rocking chair sells for $240 each. On a given day 10 rocking chairs were sold for a total of $1740. Determine the number of single-person and the number of two-person rocking chairs that were sold.

*Solution*   Understand and Translate   We are asked to find the number of each type of rocking chair sold.

Let $x$ = number of single–person rocking chairs sold.

Then $10 - x$ = number of two–person rocking chairs sold.

The income received from the sale of the single-person rocking chairs is found by multiplying the number of single-person rocking chairs sold by the cost of a single-person rocking chair. The income received from the sale of the two-person rocking chairs is found by multiplying the number of two-person rocking chairs sold by the cost of a two-person rocking chair.

$$\left(\begin{array}{c}\text{Number of} \\ \text{Rocking chairs}\end{array}\right) \times \left(\begin{array}{c}\text{Cost of} \\ \text{Rocking chairs}\end{array}\right) = \left(\begin{array}{c}\text{Income from} \\ \text{Rocking chairs}\end{array}\right)$$

| Rocking Chair | Number of Rocking Chairs | Cost | Income from Rocking Chairs |
|---|---|---|---|
| Single | $x$ | 130 | $130x$ |
| Double | $10 - x$ | 240 | $240(10 - x)$ |

$$\left(\begin{array}{c}\text{income from} \\ \text{single-person} \\ \text{rocking chairs}\end{array}\right) + \left(\begin{array}{c}\text{income from} \\ \text{two-person} \\ \text{rocking chairs}\end{array}\right) = \text{total income}$$

$$130x + 240(10 - x) = 1740$$

Carry Out
$$130x + 2400 - 240x = 1740$$
$$-110x + 2400 = 1740$$
$$-110x = -660$$
$$x = \frac{-660}{-110} = 6$$

**Check and Answer**   Six single-person rocking chairs and $10 - 6$ or 4 two-person rocking chairs were sold.

Check
$$\text{income from 6 single-person rocking chairs} = \underline{\phantom{00}780}$$
$$\text{income from 4 two-person rocking chairs} = \underline{\phantom{00}960}$$
$$\text{total} = 1740 \quad \text{True}$$

Now Try Exercise 43

## 3  Solve Mixture Problems

Any problem in which two or more quantities are combined to produce a single quantity or a single quantity is separated into two or more quantities may be considered a **mixture problem**.

Mixture problems in this section will generally be one of two types. In one type, we will mix two solids, as illustrated in **Figure 3.9a**, and be concerned about the value or cost of the mixture. In the second type, we will mix two liquids or solutions, as illustrated in **Figure 3.9b**, and be concerned about the content or strength of the mixture.

Type 1

Mixing solids together

Type 2

Mixing liquids, or solutions, together

(a) Concerned about the value or the cost of the mixture

(b) Concerned about the content or the strength of the mixture

**FIGURE 3.9**

As we did with motion problems involving two rates and with money problems, we will use a table to help analyze mixture problems.

When we construct a table for mixture problems, our table will generally have three rows. One row will be for each of the two individual items being mixed, and the third row will be for the mixture of the two items.

If we know the total weight of two items is 10 pounds and one item weighs $x$ pounds, then the second item weighs $(10 - x)$ pounds.

Reason:

item 1 + item 2 = total
$x$     + $(10 - x)$ =  10

## Type 1—Mixing Solids

When working with mixture problems involving solids, we generally use the fact that the value (or cost) of one part of the mixture plus the value (or cost) of the second part of the mixture is equal to the total value (or total cost) of the mixture.

When we are combining two solid items and are interested in the *value* of the mixture, the following table, or a variation of it, is often used.

Quantity × Price (per unit) = Value of Item

| Item | Quantity | Price | Value of Item |
|------|----------|-------|---------------|
| Item 1 | | | value of item 1 |
| Item 2 | | | value of item 2 |
| Mixture | | | value of mixture |

When we use this table, we generally use the following formula to solve the problem.

value of item 1 + value of item 2 = value of mixture

Now let's look at a mixture problem where we discuss the value or cost of the mixture.

EXAMPLE 5 **Grass Seed** Scott's Family grass seed sells for $2.65 per pound, and Scott's Spot Filler grass seed sells for $2.30 per pound. How many pounds of each should be mixed to get a 10-pound mixture that sells for $2.40 per pound?

*Solution* Understand and Translate We are asked to find the number of pounds of each type of grass seed.

Let $x$ = number of pounds of Family grass seed.

Then $10 - x$ = number of pounds of Spot Filler grass seed.

We make a sketch of the situation (**Fig. 3.10**), then construct a table.

FIGURE 3.10

The cost or value of the seeds is found by multiplying the number of pounds by the price per pound.

| Type of Seed | Number of Pounds | Cost per Pound | Cost of Seed |
|--------------|------------------|----------------|--------------|
| Family | $x$ | 2.65 | $2.65x$ |
| Spot Filler | $10 - x$ | 2.30 | $2.30(10 - x)$ |
| Mixture | 10 | 2.40 | $2.40(10)$ |

$$\left(\begin{array}{c}\text{cost of}\\\text{Family Seed}\end{array}\right) + \left(\begin{array}{c}\text{cost of Spot}\\\text{Filler Seed}\end{array}\right) = \text{cost of mixture}$$

$$2.65x + 2.30(10 - x) = 2.40(10)$$

Carry Out $\quad 2.65x + 23.0 - 2.30x = 24.0$

$$0.35x + 23.0 = 24.0$$

$$0.35x = 1.00$$

$$x \approx 2.86$$

Answer Thus, about 2.86 pounds of the Family grass seed must be mixed with $10 - x$ or $10 - 2.86 = 7.14$ pounds of the Spot Filler grass seeds to make a mixture that sells for $2.40 a pound.

Now Try Exercise 47

### Type 2—Mixing Solutions

We generally solve mixture problems involving solutions by using the fact that the amount of one part of the mixture plus the amount of the second part of the mixture is equal to the total amount of the mixture.

When working with solutions, we use the following formula: *amount of substance in the solution = quantity of solution × strength of solution (in percent written as a decimal)*. When we are mixing two quantities and are interested in the *composition* of the mixture, we generally use the following table or a variation of the table.

|  | Quantity × | Strength = | Amount of Substance |
|---|---|---|---|
| Solution | Quantity | Strength | Amount of Substance |
| Solution 1 |  |  | amount of substance in solution 1 |
| Solution 2 |  |  | amount of substance in solution 2 |
| Mixture |  |  | amount of substance in mixture |

When using this table, we generally use the following formula to solve the problem.

$$\left(\begin{array}{c} \text{amount of substance} \\ \text{in solution 1} \end{array}\right) + \left(\begin{array}{c} \text{amount of substance} \\ \text{in solution 2} \end{array}\right) = \left(\begin{array}{c} \text{amount of substance} \\ \text{in mixture} \end{array}\right)$$

Let us now look at an example of a mixture problem where two solutions are combined.

EXAMPLE 6 **Mixing Acid Solutions** Mr. Dave Lumsford needs a 10% acetic acid solution for a chemistry experiment. After checking the store room, he finds that there are only 5% and 20% acetic acid solutions available. Mr. Lumsford decides to make the 10% solution by combining the 5% and 20% solutions. How many liters of the 5% solution must he add to 8 liters of the 20% solution to get a solution that is 10% acetic acid?

*Solution*    Understand and Translate    We are asked to find the number of liters of the 5% acetic acid solution to mix with 8 liters of the 20% acetic acid solution.

Let $x$ = number of liters of 5% acetic acid solution.

Let's draw a sketch of the problem (**Fig. 3.11**).

FIGURE 3.11

The amount of acid in a given solution is found by multiplying the number of liters by the percent strength.

| Solution | Liters | Strength | Amount of Acetic Acid |
|---|---|---|---|
| 5% | $x$ | 0.05 | $0.05x$ |
| 20% | 8 | 0.20 | $0.20(8)$ |
| Mixture | $x + 8$ | 0.10 | $0.10(x + 8)$ |

$$\begin{pmatrix} \text{amount of acid} \\ \text{in 5\% solution} \end{pmatrix} + \begin{pmatrix} \text{amount of acid} \\ \text{in 20\% solution} \end{pmatrix} = \begin{pmatrix} \text{amount of acid} \\ \text{in 10\% mixture} \end{pmatrix}$$

$$0.05x \quad + \quad 0.20(8) \quad = 0.10(x + 8)$$

**Carry Out**
$$0.05x + 1.6 = 0.10x + 0.8$$
$$0.05x + 0.8 = 0.10x$$
$$0.8 = 0.05x$$
$$\frac{0.8}{0.05} = x$$
$$16 = x$$

**Answer**   Sixteen liters of 5% acetic acid solution must be added to the 8 liters of 20% acetic acid solution to get a 10% acetic acid solution. The total number of liters that will be obtained is $16 + 8$ or 24.

Now Try Exercise 55

**EXAMPLE 7**   Nicole Pappas, a medical researcher, has 40% and 5% solutions of phenobarbital. How much of each solution must she mix to get 0.6 liter of a 20% phenobarbital solution?

*Solution*   Understand and Translate   We are asked to find how much of the 40% and 5% phenobarbital solutions must be mixed to get 0.6 liter of a 20% solution. We can choose to let $x$ be the amount of either the 40% or the 5% solution. We will choose as follows:

Let $x$ = number of liters of the 40% solution.

Then $0.6 - x$ = number of liters of the 5% solution.

Remember from Section 3.1 that if a total of 0.6 liter is divided in two, if one part is $x$, the other part is $0.6 - x$.

Let's draw a sketch of the problem (**Fig. 3.12**).

FIGURE 3.12

The amount of phenobarbital in a given solution is found by multiplying the number of liters by the percent strength.

| Solution | Liters | Strength | Amount of Phenobarbital |
|----------|--------|----------|--------------------------|
| 40%      | $x$    | 0.40     | $0.40x$                  |
| 5%       | $0.6 - x$ | 0.05  | $0.05(0.6 - x)$          |
| Mixture  | 0.6    | 0.20     | $0.6(0.20)$              |

$$\begin{pmatrix} \text{amount of phenobarbital} \\ \text{in 40\% solution} \end{pmatrix} + \begin{pmatrix} \text{amount of phenobarbital} \\ \text{in 5\% solution} \end{pmatrix} = \begin{pmatrix} \text{amount of phenobarbital} \\ \text{in mixture} \end{pmatrix}$$

$$0.40x \quad + \quad 0.05(0.6 - x) \quad = (0.6)(0.20)$$

Carry Out

$$0.40x + 0.03 - 0.05x = 0.12$$
$$0.35x + 0.03 = 0.12$$
$$0.35x = 0.09$$
$$x \approx 0.26$$

**Answer** Since the answer was less than 0.6 liter, the answer is reasonable. About 0.26 liter of the 40% solution must be mixed with about $0.6 - x = 0.60 - 0.26 = 0.34$ liter of the 5% solution to get 0.6 liter of the 20% mixture.

Now Try Exercise 63

In Example 7, we chose to let $x$ = number of liters of the 40% solution. We could have selected to let $x$ = number of liters of the 5% solution. Then $0.6 - x$ would be the number of liters of the 40% solution. Had you worked the problem out like this, you would have found that $x$ was approximately 0.34 liter. Try reworking Example 7 now letting $x$ = number of liters of the 5% solution.

**EXAMPLE 8** An orange punch contains 4% orange juice. If 5 ounces of water is added to 8 ounces of the punch, determine the percent of orange juice in the mixture.

*Solution*    Understand and Translate    We are asked to find the percent of orange juice in the mixture.

Let $x$ = percent of orange juice in the mixture.

We will again set up a table.

| Solution | Ounces | Percent of Juice | Amount of Juice |
|----------|--------|------------------|-----------------|
| Punch | 8 | 0.04 | 8(0.04) |
| Water | 5 | 0.00 | 5(0.00) |
| Mixture | 13 | $x$ | 13$x$ |

$$\left(\begin{array}{c}\text{amount of juice}\\\text{in punch}\end{array}\right) + \left(\begin{array}{c}\text{amount of juice}\\\text{in water}\end{array}\right) = \left(\begin{array}{c}\text{amount of juice}\\\text{in mixture}\end{array}\right)$$

$$8(0.04) \quad + \quad 5(0.00) \quad = 13x$$

Carry Out

$$0.32 + 0.00 = 13x$$
$$0.32 = 13x$$
$$0.025 \approx x$$

**Answer** Therefore, the percent of juice in the mixture is about 2.5%.

Now Try Exercise 59

# EXERCISE SET 3.4    MathXL®    MyMathLab®
MathXL®        MyMathLab

## Warm-Up Exercises

*Fill in the blanks with the appropriate word, phrase, or symbol(s) from the following list.*

| | | | |
|---|---|---|---|
| $d = r \cdot t$ | adding | subtracting | multiplying |
| $i = p \cdot r \cdot t$ | $8 - x$ | the percent | |

**1.** Solving a motion problem when the two items are traveling in the same direction usually involves _____ the distances.

**2.** In a mixture problem, if there is a total of 8 liters and the amount of one unknown is $x$, then the amount of the other unknown is _____.

**3.** A formula important in the solution of motion problems is _____.

**4.** Solving a money problem may include using the formula _____.

**5.** Solving a motion problem when the two items are traveling in different directions usually involves _____ the distances.

**6.** To find the amount of alcohol in an 8-liter solution, we multiply the quantity of solution times _____ of alcohol in the solution.

## Practice the Skills/Problem Solving

*In Exercises 7–66, set up an equation that can be used to solve each problem. Solve the equation, and answer the question.*

*In Exercises 7–32, solve the motion problem. See Examples 1 and 2.*

7. **Ferries** Two high-speed ferries leave at the same time from Ft. Myers, Florida, going to Key West, Florida. The first ferry, the *BigCat*, travels at 34 miles per hour. The second ferry, the *Atlantic Cat*, travels at 28 miles per hour. In how many hours will the two ferries be 6 miles apart?

8. **Trains** Two trains in New York City start at the same station going in the same direction on sets of parallel tracks. The local train stops often and averages 18.4 miles per hour. The express train stops less frequently and averages 30.2 miles per hour. In how many hours will the two trains be 5.9 miles apart?

9. **Horseback Riding** Two friends, Jodi Cotton and Abe Mantell, go horseback riding on the same trail in the same direction. Jodi's horse travels at 8 miles per hour while Abe's horse travels at a slower pace. After 2 hours they are 4 miles apart. Find the speed at which Abe's horse is traveling.

10. **Camel Riding** In the Outback in Australia, Betty Sue Adams and Carl Minieri go camel riding in the same direction along the same path. Betty Sue's camel travels at 6 miles per hour while Carl's camel travels at a slower pace. After 3 hours they are 2.4 miles apart. Find the speed of Carl's camel.

11. **Airplanes** A Jet Blue airplane leaves Chicago for New York at the same time a Southwest airplane leaves New York for Chicago. The distance from Chicago to New York is 821 miles. If the Jet Blue plane travels at 560 miles per hour and the Southwest plane travels at 580 miles per hour, how long into their flights will the two planes pass each other?

12. **Walking** Barb Dansky and Sandy Spears are at opposite ends of a shopping mall 4780.4 feet apart walking toward each other. If Barb walks 1.5 feet per second and Sandy walks 2.2 feet per second, how long will it be before they meet?

13. **Walkie-Talkies** Willie and Shanna Johnston have walkie-talkies that have a range of 16.8 miles. Willie and Shanna start at the same point and walk in opposite directions. If Willie walks 3 miles per hour and Shanna walks 4 miles per hour, how long will it take before they are out of range?

14. **Blue Angels** At a Navy Blue Angel air show two F/A-18 Hornet jets travel toward each other, both at a speed of 1000 miles per hour. After they pass each other, if they were to keep flying at the same speed in opposite directions, how long would it take for them to be 500 miles apart?

15. **Product Testing** The Goodyear Tire Company is testing new tires by placing them on a machine that can simulate the tires riding on a road. First, the machine runs the tires for 7.2 hours at 60 miles per hour. Then the tires are run for 6.8 hours at a different speed. After this 14-hour period, the machine indicates that the tires have traveled the equivalent of 908 miles. Find the second speed to which the machine was set.

16. **Ski Lifts** To get to the top of Whistler Mountain, people must use two different ski lifts. The first lift travels 4 miles per hour for 0.2 hours. The second lift travels for 0.3 hours to reach the top of the mountain. If the total distance traveled up the mountain is 1.2 miles, find the average speed of the second ski lift.

17. **Earthquakes** Earthquakes generate circular *p*-waves and *s*-waves, which travel outward (see the figure). Suppose the *p*-waves have a velocity of 3.6 miles per second and the *s*-waves have a velocity of 1.8 miles per second. How long after the earthquake will *p*-waves and *s*-waves be 80 miles apart?

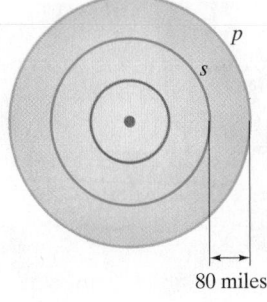

80 miles

18. **Navigating O'Hare Airport** Sadie Bragg and Jose Cruz are in Chicago's O'Hare airport walking between terminals. Sadie walks on the moving walkway (like a flat escalator moving along the floor). Her speed (relative to the ground) is 220 feet per minute. Jose starts walking at the same time and walks alongside the walkway at a speed of 100 feet per minute. How long have they been walking when Sadie is 600 feet ahead of Jose?

19. **Disabled Boat** Two Coast Guard cutters are 225 miles apart traveling toward each other, one from the east and the other from the west, searching for a disabled boat. The eastbound cutter travels 5 miles per hour faster than the westbound cutter. If the two cutters pass each other after 3 hours, find the average speed of each cutter.

20. **Snowplowing** Two snowplowing crews are clearing snow from a 2.56-mile runway at an airport. Crew A starts at one end of the runway and crew B starts at the other end. They start cleaning at the same time and they work toward each other. Crew A clears the runway at a speed of 0.4 miles per hour faster than crew B. If they meet 0.5 hours after they start, find the average speed of each snowplow.

21. **Round Trip** Samia Metwali walks for a time at 4 miles per hour, then slows down and walks at 3.2 miles per hour. The total distance she walked was 6 miles. If she walked for 0.5 hour more at 3.2 miles per hour than she did at 4 miles per hour, determine the time Samia walked at 4 miles per hour.

22. **Visit to Grandchild** Chuck Neumann drove from his house in Auburn Hills, Michigan, to visit his grandchild in Pasadena, Texas, a distance of 1343 miles. Part of the way he drove at 60 miles per hour and part of the way he drove at 70 miles per hour. If he drove for 0.5 hour more at 60 miles per hour than he did at 70 miles per hour, determine the time Chuck drove at 60 miles per hour.

23. **Paving Road** Two crews are laying blacktop on a road. They start at the same time at opposite ends of a 12-mile road and work toward one another. One crew lays blacktop at an average rate of 0.75 mile a day faster than the other crew. If the two crews meet after 3.2 days, find the rate of each crew.

24. **Beach Clean-Up** On Earth Day, two groups of people clean a 7-mile stretch of Myrtle Beach. Auturo's group and Jane's group start at the same time at opposite ends of the beach and walk toward each other. Auturo's group is traveling at a rate of 0.5 mile per hour faster than Jane's group, and they meet in 2 hours. Find the speed of each group.

25. **Sailing** Two sailboats are 9.8 miles apart and sailing toward each other. The larger boat, the *Pythagoras*, sails 4 miles per hour faster than the smaller boat, the *Apollo*. The two boats pass each other after 0.7 hour. Find the speed of each boat.

© EyeWire Collection/Getty Images–Photodisc-Royalty Free

26. **Exercising** Dien and Phuong Vu belong to a health club and exercise together regularly. They start running on two treadmills at the same time. Dien's machine is set for 6 miles per hour and Phuong's machine is set for 4 miles per hour. When they finish, they compare the distances and find that together they have run a total of 11 miles. How long had they run?

27. **Traffic Jam** Betty Truitt drives for a number of hours at 70 miles per hour. When traffic slows, she drives at 50 miles per hour. She travels at 50 miles per hour for 0.5 hour longer than she traveled at 70 miles per hour. The difference in the distance traveled at 50 mph and 70 mph is 5 miles. Determine how long Betty traveled at 50 miles per hour.

© Ke Wang/Shutterstock

28. **Salt Mine** The ore at a mine must travel on two different conveyer belts to be loaded onto a train. The second conveyer belt travels at a rate of 0.6 foot per second faster than the first conveyer belt. The ore travels 180 seconds on the first belt and 160 seconds on the second belt. If the total distance traveled by the ore is 1116 feet, determine the speed of the second belt.

29. **Ironman Triathlon** A triathlon consists of three parts: swimming, cycling, and running. Participants in the Ironman Triathlon in Canada must swim, cycle, and run certain distances. The 2008 women's winner was Australia's Belinda Granger. She swam at an average of 2.64 miles per hour for 0.91 hours then cycled at an average of 22.96 miles per hour for 4.88 hours. Finally, she ran at an average of 7.62 miles per hour for 3.44 hours.

   a) Estimate the distance that Belinda swam.

   b) Estimate the distance that Belinda cycled.

   c) Estimate the distance that Belinda ran.

   d) Estimate the total distance covered during the triathlon.

   e) Find the winning time of the triathlon.

30. **Ironman Triathlon** Refer to Exercise 29, about the Ironman Triathlon. The 2008 men's winner of the Ironman was Bryan Rhodes of New Zealand. He swam at an average speed of 2.93 miles per hour for 0.82 hour, then cycled at an average speed of 23.85 miles per hour for 4.70 hours, then ran at an average speed of 8.94 miles per hour for 2.93 hours. Answer the questions in Exercise 29 **a)–e)** using Bryan's data.

31. **Bicycling** Dom Palmo leaves his house at noon and rides his bicycle south at a uniform rate. His wife, Sue, leaves at the same time heading due north. Dom's rate is 10 kilometers per hour faster than Sue's rate. At 5 PM they are 160 kilometers apart. Find the rate of travel for each cyclist.

32. **Auto Travel** Two cars are 45 miles apart traveling toward one another. One is traveling at 70 mph and the other at 65 mph. How long will it take them to meet?

*In Exercises 33–46, solve the money problem. See Examples 3 and 4.*

33. **Simple Interest** Paul and Donna Petrie invested $12,000, part at 5% simple interest and the rest at 7% simple interest for a period of 1 year. How much did they invest at each rate if their total annual interest from both investments was $800? (Use interest = principal·rate·time.)

34. **Simple Interest** Jerry Correa invested $7000, part at 8% simple interest and the rest at 5% simple interest for a period of 1 year. If he received a total annual interest of $476 from both investments, how much did he invest at each rate?

35. **Simple Interest** Aleksandra Tomich invested $6000, part at 6% simple interest and part at 4% simple interest for a period of 1 year. How much did she invest at each rate if each account earned the same interest?

36. **Simple Interest** Susan Foreman invested $12,500, part at 7% simple interest and part at 6% simple interest for a period of 1 year. How much was invested at each rate if each account earned the same interest?

37. **Simple Interest** Míng Wang invested $10,000, part at 4% and part at 5% simple interest for a period of 1 year. How much was invested in each account if the interest earned in the 5% account was $320 greater than the interest earned in the 4% account?

38. **Simple Interest** Sharon Sledge invested $20,000, part at 5% and part at 7% simple interest for a period of 1 year. How much was invested in each account if the interest earned in the 7% account was $440 greater than the interest earned in the 5% account?

39. **Rate Increase** Patricia Burgess knows that at some point during the calendar year her basic monthly telephone rate increased from $17.10 to $18.40. If she paid a total of $207.80 for basic telephone service for the calendar year, in what month did the rate increase take effect?

40. **Cable TV** Violet Kokola knows that her subscription rate for the basic tier of cable television increased from $18.20 to $19.50 at some point during the calendar year. She paid a total of $230.10 for the year to the cable company. Determine the month of the rate increase.

41. **Wages** Mihály Sarett holds two part-time jobs. One job, at Home Depot, pays $6.50 an hour and the second job, at a veterinary clinic, pays $7.00 per hour. Last week Mihály worked a total of 18 hours and earned $122.00. How many hours did Mihály work at each job?

42. **Rock and Roll Hall of Fame** At the Rock and Roll Hall of Fame in Cleveland, Ohio, adult admission is $22 and child admission is $13. During one day, a total of 1551 adult and child admissions were collected and $27,543 in admission fees was collected. How many child admissions were collected?

Rock and Roll Hall of Fame

43. **Baseball Hall of Fame** At the Baseball Hall of Fame in Cooperstown, New York, adult admission is $16.50 and child admission is $6.00. During one day a total of 2100 adult and child admissions were collected and $25,200 in admission fees was collected. How many adult admissions were collected?

Baseball Hall of Fame

44. **Ticket Sales** At a movie theater an evening show cost $7.50 and a matinee cost $4.75. On one day there was one matinee and one evening showing. On that day a total of 310 tickets were sold, which resulted in ticket sales of $2022.50. How many people went to the matinee and how many went to the evening show?

45. **Stock Purchase** Suppose Nike stock is selling at $78 a share and Kellogg stock is selling at $33 a share. Mike Moussa has a maximum of $10,000 to invest. He wishes to purchase five times as many shares of Kellogg as of Nike. Only whole shares of stock can be purchased.
   a) How many shares of each will he purchase?
   b) How much money will be left over?

46. **Stock Purchase** Suppose Wal-Mart stock is selling at $59 a share and Mattel stock is selling at $28 a share. Amy Waller has a maximum of $6000 to invest. She wishes to purchase four times as many shares of Wal-Mart as of Mattel. Only whole shares of stock can be purchased.
   a) How many shares of each will she purchase?
   b) How much money will be left over?

*In Exercises 47–66, solve the mixture problem. See Examples 5–8.*

47. **Nut Shop** Jean Valjean owns a nut shop where walnuts cost $6.80 per pound and almonds cost $6.40 per pound. Jean gets an order that specifically requests a 30-pound mixture of walnuts and almonds that will cost $6.65 per pound. How many pounds of each type of nut should Jean mix to get the desired mixture?

48. **Grass Seed** Scott's® Family grass seed sells for $2.45 per pound and Scott's Spot Filler grass seed sells for $2.10 per

pound. How many pounds of each should be mixed to get a 10-pound mixture that sells for $2.20 per pound?

49. **Topsoil** Strained topsoil sells for $160 per cubic yard and unstrained topsoil sells for $120 per cubic yard. How many cubic yards of each should be mixed to make 8 cubic yards of a mixture that sells for $150 per cubic yard?

50. **Bird Food** At Agway Gardens, bird food is sold in bulk. In one barrel are sunflower seeds that sell for $1.80 per pound.

In a second barrel is cracked corn that sells for $1.40 per pound. If a mixture is made by taking 2.5 pounds of the sunflower seeds and 1 pound of the cracked corn, what should the mixture cost per pound?

© Margaret M. Stewart/Shutterstock

**51. Bulk Candies** A grocery store sells certain candies in bulk. The Good and Plenty cost $2.49 per pound and Sweet Treats cost $2.89 per pound. If Jane Strange takes 3 scoops of Good and Plenty and mixes it with 5 scoops of Sweet Treats, how much per pound should the mixture sell for? Assume each scoop contained the same weight of candy.

© Tiggy Gallery/Shutterstock

**52. Coffee House** Ruth Cordeff runs a coffee house where chocolate almond coffee beans sell for $7.00 per pound and hazelnut coffee beans sell for $6.10 per pound. A customer asks Ruth to make a 6-pound mixture using the two kinds of beans. How many pounds of each should be used if the mixture is to cost $6.40 per pound?

**53. Beef Wellington** To make beef Wellington, Chef Ramon uses a marinade that is a blend of two red wines. He mixes 5 liters of a wine that is 12% alcohol by volume with 2 liters of a wine that is 9% alcohol by volume. Determine the alcohol content of the mixture.

**54. Pharmacy** Susan Staples, a pharmacist, has a 60% solution of sodium iodite. She also has a 25% solution of the same drug. She gets a prescription calling for a 40% solution of the drug. How much of each solution should she mix to make 0.5 liter of the 40% solution?

**55. Sulfuric Acid** In chemistry class, Todd Corbin has 1 liter of a 20% sulfuric acid solution. How much of a 12% sulfuric acid solution must he mix with the 1 liter of 20% solution to make a 15% sulfuric acid solution?

**56. Paint** Nick Cooper has two cans of white paint, one with a 2% yellow pigment and the other with a 5% yellow pigment. Nick wants to mix the two paints to get paint with a 4% yellow pigment. How much of the 5% yellow pigment paint should be mixed with 0.4 gallon of the 2% yellow pigment paint to get the desired paint?

**57. Mouthwash** The label on the Listerine® Cool Mint Antiseptic mouthwash says that it is 21.6% alcohol by volume. The label on the Scope Original Mint mouthwash says that it is 15.0% alcohol by volume. If Hans mixes 6 ounces of the Listerine with 4 ounces of the Scope, what is the percent alcohol content of the mixture?

**58. Clorox** Clorox® bleach is 5.25% sodium hypochlorite. The instructions on the Clorox bottle say to add 1 cup of Clorox to 4 cups (a quart) of water. Find the percent of sodium hypochlorite in the mixture.

**59. Orange Juice** Mary Ann Terwilliger has made 6 quarts of an orange juice punch for a party. The punch contains 12% orange juice. She feels that she may need more punch, but she has no more orange juice so she adds $\frac{1}{2}$ quart of water to the punch. Find the percent of orange juice in the new mixture.

**60. Insecticide** The active ingredient in one type of Ortho® insect spray is 50% malathion. The instructions on the bottle say to add 1 fluid ounce to a gallon (128 fluid ounces) of water. What percent of malathion will be in the mixture?

© Alistair Scott/Shutterstock

**61. Hawaiian Punch** The label on a 12-ounce can of frozen concentrate Hawaiian Punch® indicates that when the can of concentrate is mixed with 3 cans (36 ounces) of cold water, the resulting mixture is 10% juice. Find the percent of pure juice in the concentrate.

**62. Salt Concentration** Suppose the dolphins at Sea World must be kept in salt water with an 0.8% salt content. After a week of warm weather, the salt content has increased to 0.9% due to water evaporation. How much water with 0% salt content must be added to 50,000 gallons of the 0.9% salt water to lower the salt concentration to 0.8%?

**63. Bleach** Clorox® bleach is 5.25% sodium hypochlorite and swimming pool shock treatment is 10.5% sodium hypochlorite. How much of each item must be mixed to get 6 cups of a mixture that is 7.2% sodium hypochlorite?

**64. Antifreeze** Prestone® antifreeze contains 12% ethylene glycose, and Zerex® antifreeze contains 9% ethylene glycose. How much of each type antifreeze should be mixed to get 10 quarts of a mixture that is 10% ethylene glycose?

**65. Alcohol Solution** How many pints of a 12% isotrophic alcohol solution should Jenny Crawford mix with 15 pints of a 5% isotrophic alcohol solution to get a mixture that is 8% isotrophic alcohol?

**66. Plant Food** Miracle-Gro® All Purpose liquid plant food has 12% nitrogen. Miracle-Gro Quick Start liquid plant food has 4% nitrogen. If 2 cups of the All Purpose plant food are mixed with 3 cups of the Quick Start plant food, determine the percent of nitrogen in the mixture.

## Challenge Problems

**67. Fat Albert** The home base of the Navy's Blue Angels is in Pensacola, Florida. The Angels spend winters in El Centro, California. Assume they fly at about 900 miles per hour when they fly from Pensacola to El Centro in their F/A-18 Hornets. On every trip, their C-130 transport (affectionately called Fat Albert) leaves before them, carrying supplies and support personnel. The C-130 generally travels at about 370 miles per hour. On a trip from Pensacola to El Centro, how long before the Hornets leave should Fat Albert leave if it is to arrive 3 hours before the Hornets? The flying distance between Pensacola and El Centro is 1720 miles.

## Group Activity

*Discuss and answer Exercises 68 and 69 as a group.*

**68. Race Horse** According to the *Guinness Book of World Records*, the fastest race horse speed recorded was by a horse called Big Racket on February 5, 1945, in Mexico City, Mexico. Big Racket ran a $\frac{3}{4}$-mile race in 62.41 seconds. Find Big Racket's speed in miles per hour. Round your answer to the nearest hundredth.

**69. Garage Door Opener** An automatic garage door opener is designed to begin to open when a car is 100 feet from the garage. At what rate will the garage door have to open if it is to raise 6 feet by the time a car traveling at 4 miles per hour reaches it? (1 mile per hour $\approx 1.47$ feet per second.)

## Cumulative Review Exercises

[1.3] **70. a)** Divide $2\frac{3}{4} \div 1\frac{5}{8}$.

**b)** Add $2\frac{3}{4} + 1\frac{5}{8}$.

[2.5] **71.** Solve the equation $6(x - 3) = 4x - 18 + 2x$.

[2.7] **72.** Solve the proportion $\frac{6}{x} = \frac{72}{9}$.

[3.2] **73. Consecutive Integers** The sum of two consecutive integers is 77. Find the numbers.

# Chapter 3 Summary

| IMPORTANT FACTS AND CONCEPTS | EXAMPLES |
|---|---|
| **Section 3.1** | |
| $a$ subtracted from $b$ means $b - a$.<br>The difference between $a$ and $b$ means $a - b$. | 3 subtracted from $5x$ is $5x - 3$.<br>The difference between 3 and $5x$ is $3 - 5x$. |
| If $T$ represents the total amount to be divided in two parts, and if $x$ represents one of the parts, then $T - x$ represents the other part. | If \$800 is divided between Jay and Rose and if Rose gets $x$ dollars, then Jay gets $800 - x$ dollars. |
| A percent is always a percent of some number. | An 8% sales tax on $r$ dollars is $0.08r$.<br>The cost of an item $c$ increased by 5% is $c + 0.05c$. |
| When writing equations, the words *is, was, will be, yields, gives,* often mean =. | "Two less than 5 times a number *is* the number increased by 3" can be expressed as $5x - 2 = x + 3$. |
| **Consecutive integers**, such as 23 and 24, differ by 1 unit.<br>**Consecutive even integers**, such as 24 and 26, differ by 2 units.<br>**Consecutive odd integers**, such as 23 and 25, differ by 2 units. | $x$ and $x + 1$ represent consecutive integers.<br>$x$ and $x + 2$ represent consecutive even integers.<br>$x$ and $x + 2$ represent consecutive odd integers. |
| **Section 3.2** | |
| **Problem-Solving Procedure**<br><br>1. Understand the problem.<br>2. Translate the problem into mathematical language.<br>3. Carry out the mathematical calculation.<br>4. Check the answer.<br>5. Answer the question asked.<br><br>(See page 180 for more detailed information.) | Three subtracted from 4 times a number is 17. Find the number.<br>**Solution: Understand** We need to express the information given as an equation. We are asked to find the unknown number.<br>**Translate** Let $x =$ the unknown number, then we can write the equation<br>$$4x - 3 = 17$$<br>**Carry out**<br>$$4x = 20$$<br>$$x = 5$$<br>**Check** Substitute 5 for the unknown number<br>$$4(5) - 3 = 17$$<br>$$17 = 17 \quad \text{True}$$<br>**Answer** The unknown number is 5. |
| If you are having difficulty with this section, seek help. | See Helpful Hint on page 185 for possible sources for help. |

| IMPORTANT FACTS AND CONCEPTS | EXAMPLES |
|---|---|

### Section 3.3

An **isosceles triangle** has two sides of the same length. The angles opposite the sides of equal length have equal measures.

Two angles of an isosceles triangle are each 30° greater than the smallest angle. Find the measures of the three angles.

**Solution:**      Let $x$ = the smallest angle.

Then $x + 30$ = the measure of each larger angle.

$$x + (x + 30) + (x + 30) = 180$$
$$3x + 60 = 180$$
$$3x = 120$$
$$x = 40$$

The smallest angle is 40°, so the two larger angles are 40° + 30° or 70°. Note that 40° + 70° + 70° = 180°, so the answer checks.

### Section 3.4

A **motion problem** can involve two rates.

Peter and Paul go walking. They start at the same point at the same time and walk in the same direction. Peter walks at 4 mph and Paul walks at 3.5 mph. In how many hours will they be 1 mile apart?

**Solution:**   Let $t$ = time when they are 1 mile apart.

$$4t - 3.5t = 1$$
$$0.5t = 1$$
$$t = 2$$

In 2 hours they will be 1 mile apart.

A **money problem** can involve two rates of interest or two different costs.

John sold 12 paintings in one day. Some sold at $80 and the others sold at $125. If he collected a total of $1275, how many of each type did he sell?

**Solution:**     Let $x$ = number of $80 paintings.

Then $12 - x$ = number of $125 paintings.

$$80x + 125(12 - x) = 1275$$
$$80x + 1500 - 125x = 1275$$
$$-45x + 1500 = 1275$$
$$-45x = -225$$
$$x = 5$$

Five $80 paintings and 12 – 5 or 7 $125 paintings were sold.

A **mixture problem** may involve mixing different strengths or different types of solutions, or mixing solid items, such as nuts.

How many liters of an 8% acetic acid solution must be mixed with 6 liters of a 15% acetic acid solution to get a 10% acetic acid solution?

**Solution:**   Let $x$ = number of liters of the 8% solution.

$$0.08x + 0.15(6) = 0.10(x + 6)$$
$$0.08x + 0.9 = 0.10x + 0.6$$
$$0.9 = 0.02x + 0.6$$
$$0.3 = 0.02x$$
$$15 = x$$

Fifteen liters of the 8% solution must be mixed with the 6 liters of the 15% solution.

## Chapter 3 Review Exercises

[3.1]

1. **Age** Charles's age is 7 more than 3 times Norman's age, $n$. Write an expression for Charles's age.

2. **Gasoline** The cost of a gallon of gasoline in California is 1.2 times the cost of a gallon of gasoline in Georgia, $g$. Write an expression for the cost of a gallon of gasoline in California.

3. **Dress** Write an expression for the cost of a dress, $d$, reduced by 25%.

4. **Pounds** Write an expression for the number of ounces in $y$ pounds.

5. **Money** Two hundred dollars is divided between Jishing Wang and Norma Agras. If Jishing gets $x$ dollars, write an expression for the amount Norma gets.

6. **Age** Mario is 6 years older than seven times Dino's age. Select a variable to represent one quantity and state what the variable represents. Express the second quantity in terms of the variable selected.

7. **Robberies** The number of robberies in 2009 was 12% less than the number of robberies in 2008. Write an expression for the difference in the number of robberies between 2008 and 2009.

8. **Numbers** The smaller of two numbers is 24 less than 3 times the larger. When the smaller is subtracted from the larger, the difference is 8. Write an equation to represent this information. Do not solve the equation.

[3.2]  *In Exercises 9–18, set up an equation that can be used to solve the problem. Solve the equation and answer the question.*

9. **Numbers** One number is 8 more than the other. Find the two numbers if their sum is 74.

10. **Consecutive Integers** The sum of two consecutive integers is 237. Find the two integers.

11. **Numbers** The larger of two integers is 3 more than 5 times the smaller integer. Find the two numbers if the smaller subtracted from the larger is 31.

12. **New Car** Shaana recently purchased a new car. What was the cost of the car before tax if the total cost including a 7% tax was $23,260?

13. **Bagels** A bakery currently ships 520 bagels per month to various outlets. They wish to increase the shipment of bagels by 20 per month until reaching a shipment level of 900 bagels. How long will this take?

14. **Salary Comparison** Irene Doo, a salesperson, receives a salary of $600 per week plus a 3% commission on all sales she makes. Her company is planning on changing her salary to $500 per week, plus an 8% commission on all sales she makes. What weekly dollar sales would she have to make for the total salaries from each plan to be the same?

15. **Sale Price** During a going-out-of-business sale, all prices were reduced by 20%. If, during the sale Kathy Golladay purchased an HDTV stand for $495, what was the original price of the HDTV stand?

See Exercise 15.

16. **Landscaping** Two Brothers Nursery charges $400 for a tree and $45 per hour to plant the tree. ABC Nursery charges $200 for the same size tree and $65 per hour to plant the tree. How many hours for planting would result in the total price for both landscapers being the same?

17. **House Price** According to the National Association of Realtors, the median sale price for single-family homes increased 11.7% from June 2003 to January 2005. If the median price was $191,000 in 2005, find the median price in 2003.

18. **Tax Refund** According to the Internal Revenue Service, the average refund in 2007 was $3 less than 4 times the 1980 average refund. If the average refund in 2007 was $2733, find the average refund in 1980.

[3.3]  *Solve each problem.*

19. **Unknown Angles** One angle of a triangle measures 10° greater than the smallest angle, and the third angle measures 10° less than twice the smallest angle. Find the measures of the three angles.

20. **Unknown Angles** One angle of a trapezoid measures 10° greater than the smallest angle; a third angle measures five times the smallest angle; and the fourth angle measures 20° greater than four times the smallest angle. Find the measure of the four angles.

21. **Garden** Steve Rodi has a rectangular garden whose length is 4 feet longer than its width. The perimeter of the garden is 70 feet. Find the width and length of the garden.

22. **Designing a House** The figure on the right shows plans for a rectangular basement of a house. The dots represents poles in the ground and the lines represent string that has been attached to the poles. The length of the basement is to be 30 feet greater than the width. If a total of 310 feet of string was used to mark off the rooms, find the width and length of the basement.

See Exercise 22.

23. **Rhombus** The two larger angles of a rhombus are each 3 times the measure of the two smaller angles. Find the measure of each angle.

24. **Bookcase** Wade Ellis is building a bookcase with 4 shelves. The length of the bookcase is to be twice the height, as illustrated. If only 20 feet of lumber is available, what should be the length and height of the bookcase?

[3.4]  *Solve each problem.*

25. **Jogging** Two joggers follow the same route. Harold Lowe jogs at 8 kilometers per hour and Susan Karney Fackert at 6 kilometers per hour. If they leave at the same time, how long will it take for them to be 4 kilometers apart?

26. **Trains Leaving** Two trains going in opposite directions leave from the same station on parallel tracks. One train travels at 50 miles per hour and the other at 60 miles per hour. How long will it take for the trains to be 440 miles apart?

27. **Pittsburgh Incline** The Duquesne Incline in Pittsburgh, Pennsylvania, is shown on the next page. The two cars start at the same time at opposite ends of the incline and travel toward each other at the same speed. The length of the incline is 400 feet and the time it takes for the cars to be at the

halfway point is about 22.73 seconds. Determine, in feet per second, the speed the cars travel.

**28. Savings Accounts** Martha Goshaw wishes to place part of $12,000 into a savings account earning 8% simple interest and part into a savings account earning $7\frac{1}{4}$% simple interest. How much should she invest in each if she wishes to earn $900 in interest for the year?

**[3.2–3.4].**   *Solve each problem.*

**33. Numbers** The sum of two consecutive odd integers is 208. Find the two integers.

**34. Television** What is the cost of a television before tax if the total cost, including a 6% tax, is $477?

**35. Medical Supplies** Mr. Chang sells medical supplies. He receives a weekly salary of $300 plus a 5% commission on the sales he makes. If Mr. Chang earned $900 last week, what were his sales in dollars?

**36. Triangle** One angle of a triangle is 8° greater than the smallest angle. The third angle is 4° greater than twice the smallest angle. Find the measure of the three angles of the triangle.

**37. Increase Staff** The Darchelle Leggett Company plans to increase its number of employees by 25 per year. If the company now has 427 employees, how long will it take before they have 627 employees?

**38. Parallelogram** The two larger angles of a parallelogram each measure 40° greater than the two smaller angles. Find the measure of the four angles.

**39. Copy Centers** Copy King charges a monthly fee of $20 plus 4 cents per copy. King Kopie charges a monthly fee of $25 plus 3 cents a copy. How many copies made in a month would result in both companies charging the same amount?

**40. Swimming** Rita Gonzales and Jim Ham are going swimming in Putnam Lake. They start swimming in the same direction at the same time. Rita swims at 1 mph and Jim swims at a slower pace. After 0.5 hour they are 0.2 mile apart. Find the speed that Jim is swimming.

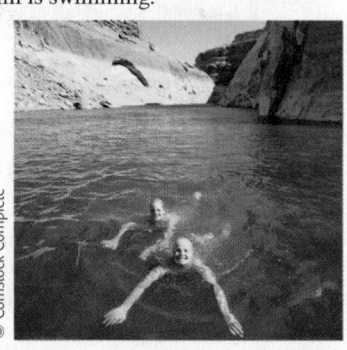

**29. Savings Accounts** Aimee Tait invests $4000 into two savings accounts. One account pays 3% simple interest and the other account pays 3.5% simple interest. If the interest earned in the account paying 3.5% simple interest is $94.50 more than the interest earned in the 3% account, how much was invested in each account?

**30. Holiday Punch** Marcie Waderman is having a holiday party at her house. She made 2 gallons of a punch solution that contains 2% alcohol. How much pure punch must Marcie add to the punch to reduce the alcohol level to 1.5%?

**31. Wind Chimes** Alan Carmell makes and then sells wind chimes. He makes two types, a smaller one that sells for $8 and a larger one that sells for $20. At an arts and crafts show he sells a total of 30 units, and his total receipts were $492. How many of each type of chime did he sell?

**32. Acid Solution** Bruce Kennan, a chemist, wishes to make 2 liters of an 8% acid solution by mixing a 10% acid solution and a 5% acid solution. How many liters of each should he use?

**41. Butcher** A butcher combined ground beef that cost $3.50 per pound with ground beef that cost $4.10 per pound. How many pounds of each were used to make 80 pounds of a mixture that sells for $3.65 per pound?

**42. Speed Traveled** Two brothers who are 230 miles apart start driving toward each other at the same time. The younger brother travels 5 miles per hour faster than the older brother, and the brothers meet after 2 hours. Find the speed traveled by each brother.

**43. Acid Solution** How many liters of a 30% acid solution must be mixed with 2 liters of a 12% acid solution to obtain a 15% acid solution?

**44. Fencing** Kathy Tomaino is partitioning a rectangular yard using fencing, as shown below. If the length is 1.5 times the width and if only 96 feet of fencing is available, find the length and width of the partitioned area.

**45. Acid Solution** Six liters of a 3% sulfuric acid solution is mixed with an 8% sulfuric acid solution. If the mixture is a 4% sulfuric acid solution, determine how many liters of the 8% sulfuric acid solution were used in the mixture.

## Chapter 3 Practice Test

*Chapter Test Prep Videos provide fully worked-out solutions to any of the exercises you want to review. Chapter Test Prep Videos are available via MyMathLab®, or on YouTube (search "Angel Elementary and Intermediate Algebra" and click on "Channels").*

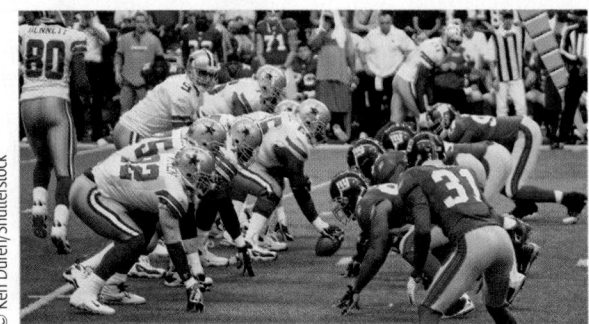

1. **Money** Five hundred dollars was divided between Boris and Monique. If Monique received *n* dollars, write an expression for the amount Boris received.
2. **Restaurants** The money Sally Sestini earned was $6000 more than twice what William Rowley earned. Write an expression for what Sally earned.
3. Write an expression for the number of seconds in *t* minutes.
4. **Season Tickets** The cost of a season ticket for the Dallas Cowboys football games increased by 6% from the previous year's cost, *c*. Write an expression for the cost of a season ticket for this year.

*In Exercises 5–9, select a variable to represent one quantity and state what it represents. Express the second quantity in terms of the variable selected.*

5. **Tic Tacs** The number of packages of peppermint flavored Tic Tacs® sold was 105 packages less than 7 times the number of packages of orange flavored Tic Tacs sold.

6. **Men and Women** At a play there were 600 men and women.
7. **Numbers** The larger of two numbers is 1 less than twice the smaller number. Write an expression for the smaller number subtracted from the larger number
8. **Liquid Tide** The number of fluid ounces in a large bottle of Tide® is 18 fluid ounces more than the amount in a smaller bottle. Write an expression for the sum of the amounts in a small and large bottle.
9. **Nuts** The cost of a can of Planters® Deluxe Nuts is 84% greater than the cost of a can of Planters Peanuts. Write an expression for the difference in cost between the Deluxe Nuts and the Peanuts.

*In Exercises 10–25, set up an equation that can be used to solve the problem. Solve the problem and answer the question asked.*

10. **Integers** The sum of two integers is 158. Find the two integers if the larger is 10 less than twice the smaller.
11. **Consecutive Odd Integers** For two consecutive odd integers, the sum of the smaller and 4 times the larger is 33. Find the integers.
12. **Numbers** One number is 12 less than 5 times the other. Find the two numbers if their sum is 42.
13. **Lawn Furniture** Dona Bishop purchased a set of lawn furniture. The cost of the furniture, including a 6% tax, was $2650. Find the cost of the furniture before tax.
14. **Eating Out** Mark Sullivan has only $40. He wishes to leave a 15% tip. Find the price of the most expensive meal that he can order.
15. **Business Venture** Julie Burgmeier receives twice the profit in a business venture than Peter Ancona does. If the profit for the year was $120,000, how much will each receive?
16. **Snowplowing** William Echols is going to hire a snow plowing service. Elizabeth Suco charges an annual fee of $80, plus $5 each time she plows. Jon Wilkins charges an annual fee of $50, plus $10 each time he plows. How many times would the snow need to be plowed for the cost of both plans to be the same?
17. **Laser Printers** A Delta laser printer cost $499. The cost of printing each page of text is 1 cent. A TexMar laser printer cost $350. The cost of printing each page is 3 cents. How many pages would need to be printed for the total cost of both printers to be the same?

18. **Triangle** A triangle has a perimeter of 75 inches. Find the three sides if one side is 15 inches larger than the smallest side, and the third side is twice the smallest side.
19. **Flag** Paul Murphy's flag has a perimeter of 28 feet. Find the dimensions of the flag if the length is 4 feet less than twice its width.
20. **Trapezoid** The two larger angles of a trapezoid measure 3° more than twice the two smaller angles. Determine the four angles of the trapezoid.
21. **Laying Cable** Ellis and Harlene Matza are digging a shallow 67.2-foot-long trench to lay electrical cable to a new outdoor light fixture they just installed. They start digging at the same time at opposite ends of where the trench is to go, and dig toward each other. Ellis digs at a rate of 0.2 foot per minute faster than Harlene, and they meet after 84 minutes. Find the speed at which each digs.
22. **Running** Alice and Bonnie start running at the same time from the same point and run in the same direction. Alice runs at 8 mph while Bonnie runs at a slower pace. After 2 hours they are 4 miles apart. Determine the speed at which Bonnie is running.
23. **Bulk Candy** A candy shop sells candy in bulk. In one bin is Jelly Belly candy, which sells for $2.20 per pound, and in a second bin is Kits, which sells for $2.75 per pound. How much of each type should be mixed to obtain a 3-pound mixture that sells for $2.40 per pound?

**24. Salt Solution** How many liters of 20% salt solution must be added to 60 liters of 40% salt solution to get a solution that is 35% salt?

**25. Acid Solution** A chemist wishes to make 3 liters of a 6% acid solution by mixing an 8% acid solution and a 5% acid solution. How many liters of each should she mix?

## Cumulative Review Test

*Take the following test and check your answers with those given in the back of the book. Review any questions that you answered incorrectly. The section where the material was covered is indicated after the answer.*

**1. Social Security** The following circle graph shows what the typical retiree receives in social security, as a percent of their total income.

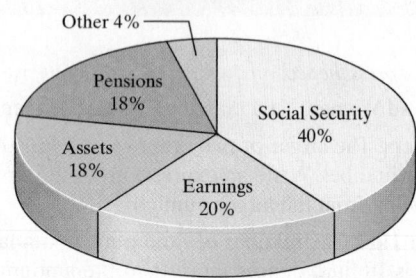

**Where Social Security Recipients Get Their Income**

Other 4%
Pensions 18%
Assets 18%
Earnings 20%
Social Security 40%

*Source:* Newsweek

If Emily receives $40,000 per year, and her income is typical of all social security recipients, how much is she receiving in social security?

**2. Olympic Gold Medals** The following graph illustrates the number of gold medals awarded in the 2008 Beijing Olympics for the top 5 countries.

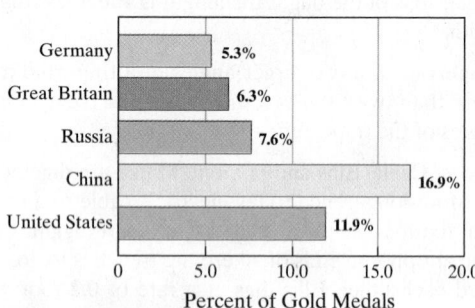

**Percent of Gold Medals Awarded in 2008 Olympics**

Germany        5.3%
Great Britain    6.3%
Russia          7.6%
China          16.9%
United States   11.9%

Percent of Gold Medals

*Source:* results.beijing2008.cn

**a)** What is the percentage difference between the number of gold medals awarded to China and Germany?

**b)** If 302 gold medals were awarded in the 2008 Olympics, how many were awarded to the United States?

**3. Carbon Dioxide Levels** David Warner, an environmentalist, was checking the level of carbon dioxide in the air. On five readings, he got the results indicated in the table above on the right.

| Test | Carbon Dioxide (parts per million) |
|------|-----------------------------------|
| 1    | 5                                 |
| 2    | 6                                 |
| 3    | 8                                 |
| 4    | 12                                |
| 5    | 5                                 |

**a)** Find the mean level of carbon dioxide detected.

**b)** Find the median level of carbon dioxide detected.

**4.** Evaluate $\dfrac{5}{12} \div \dfrac{3}{4}$.

**5.** How much larger is $\dfrac{2}{3}$ inch than $\dfrac{1}{8}$ inch?

**6. a)** List the set of natural numbers.

   **b)** List the set of whole numbers.

   **c)** What is a rational number?

**7. a)** Evaluate $|-4|$.

   **b)** Which is greater, $|-5|$ or $|-3|$? Explain.

**8.** Evaluate $2 - 6^2 \div 2 \cdot 2$.

**9.** Simplify $4(2x - 3) - 2(3x + 5) - 6$.

*In Exercises 10–13, solve the equation.*

**10.** $5x - 6 = x + 14$

**11.** $6r = 2(r + 3) - (r + 5)$

**12.** $2(x + 5) = 3(2x - 4) - 4x$

**13.** $\dfrac{4.8}{x} = -\dfrac{3}{5}$

**14.** If $A = \pi r^2$, find $A$ when $r = 6$.

**15.** Consider the equation $4x + 8y = 16$.

   **a)** Solve the equation for $y$ and write the equation in $y = mx + b$ form.

   **b)** Find $y$ when $x = -4$.

**16.** Solve the formula $P = 2l + 2w$ for $w$.

**17. Gas Needed** If Lisa Shough's car can travel 50 miles on 2 gallons of gasoline, how many gallons of gas will it need to travel 225 miles?

**18. Calling Plan** Lori Sypher is considering two cellular telephone plans. Plan A has a monthly charge of $19.95 plus 35 cents per minute. Plan B has a monthly charge of $29.95 plus 10 cents per minute. How long would Lori need to talk in a month for the two plans to have the same total cost?

**19. Sum of Numbers** The sum of two numbers is 29. Find the two numbers if the larger is 11 greater than twice the smaller.

**20. Quadrilateral** One angle of a quadrilateral measures 5° larger than the smallest angle; the third angle measures 50° larger than the smallest angle; and the fourth angle measures 25° greater than 4 times the smallest angle. Find the measure of each angle of the quadrilateral.

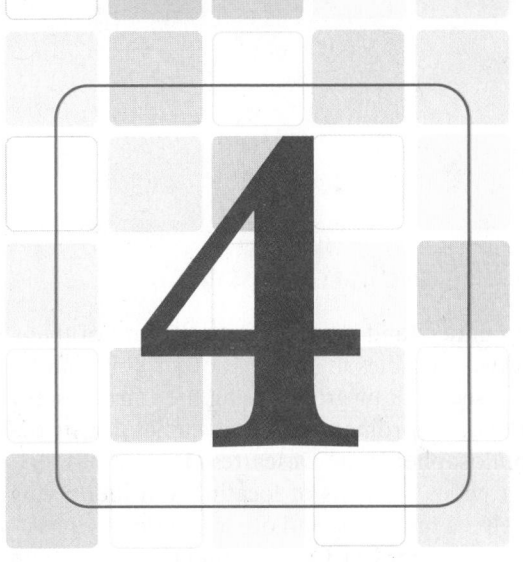

# 4
# Graphing Linear Equations

## Goals of This Chapter

In this chapter you will learn how to graph linear equations. The graphs of linear equations are straight lines. Graphing is one of the most important topics in mathematics.

The two most important concepts we will discuss in this chapter are slopes and functions. Functions are a unifying concept in mathematics.

This chapter contains concepts that are central to mathematics. If you plan to take more mathematics courses, graphs and functions will probably be a significant part of those courses.

© Monkey Business Images/Dreamstime LLC–Royalty Free

**We see graphs** daily. They are very important in both mathematics and in everyday living. Graphs are used to display information. For example, in Exercise 75 on page 232, we use a graph to illustrate the total monthly cost for telephone calls.

# 4.1 The Cartesian Coordinate System and Linear Equations in Two Variables

René Descartes

1 Plot points in the Cartesian coordinate system.

2 Determine whether an ordered pair is a solution to a linear equation.

## 1  Plot Points in the Cartesian Coordinate System

Many algebraic relationships are easier to understand if we can see a picture of them. A **graph** shows the relationship between two variables in an equation. In this chapter we discuss several procedures that can be used to draw graphs using the **Cartesian (or rectangular) coordinate system**. The Cartesian coordinate system is named for its developer, the French mathematician and philosopher René Descartes (1596–1650).

The Cartesian coordinate system provides a means of locating and identifying points just as the coordinates on a map help us find cities and other locations. Consider the map of the Great Smoky Mountains (see **Fig. 4.1**). Can you find Cades Cove on the map? If we tell you that it is in grid A3, you can probably find it much more quickly and easily.

FIGURE 4.1

The Cartesian coordinate system is a grid system, like that of a map, except that it is formed by two axes (or number lines) drawn perpendicular to each other. The two intersecting axes form four **quadrants**, numbered I through IV in **Figure 4.2**.

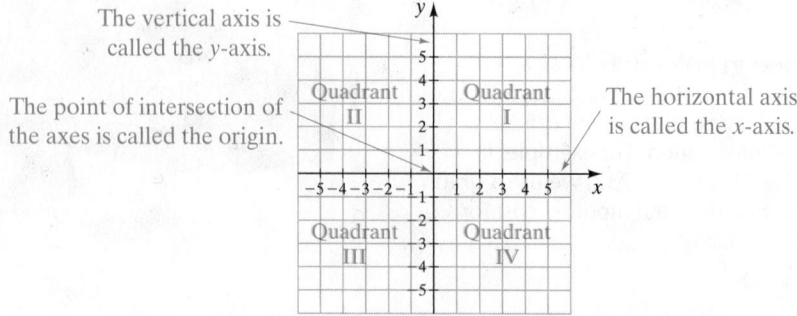

The vertical axis is called the *y*-axis.

The point of intersection of the axes is called the origin.

The horizontal axis is called the *x*-axis.

FIGURE 4.2

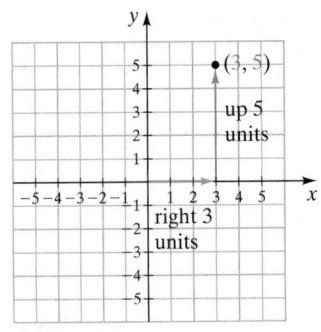

**FIGURE 4.3**

To locate a point in the Cartesian coordinate system, we will use an **ordered pair** of the form $(x, y)$. The first number, $x$, is called the **x-coordinate** and the second number, $y$, is called the **y-coordinate**.

To plot the point $(3, 5)$, start at the origin,

$$(3, 5)$$

$$\begin{cases} x\text{-coordinate is } 3 \longrightarrow \uparrow \uparrow \longleftarrow y\text{-coordinate is } 5 \\ \text{means} \qquad\qquad\qquad\qquad \text{means} \\ \text{go } right \text{ 3 units} \qquad\qquad \text{then go } up \text{ 5 units} \end{cases}$$

To plot the point, $(-4, -2)$, start at the origin,

$$(-4, -2)$$

$$\text{go } left \text{ 4 units} \longrightarrow \uparrow \quad \uparrow \longleftarrow \text{then go } down \text{ 2 units}$$

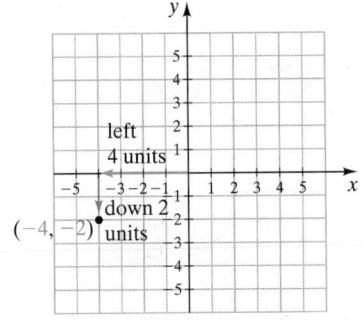

**FIGURE 4.4**

**EXAMPLE 1** Plot each point on the same axes.

a) $A(5, 3)$     b) $B(2, 4)$     c) $C(-3, 1)$

d) $D(4, 0)$     e) $E(-2, -5)$     f) $F(0, -3)$

g) $G(0, 2)$     h) $H\left(6, -\dfrac{9}{2}\right)$     i) $I\left(-\dfrac{3}{2}, -\dfrac{5}{2}\right)$

*Solution*   The first number in each ordered pair is the $x$-coordinate and the second number is the $y$-coordinate. The points are plotted in **Figure 4.5**.

> ## Understanding Algebra
>
> Remember that the order of the numbers in an *ordered pair* is very important. For example, (3, 5) is a different point than (5, 3). The *x*-coordinate is always listed first in an ordered pair.

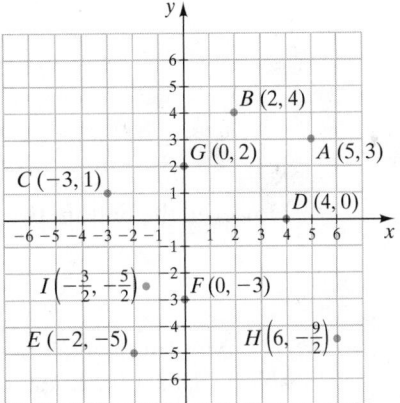

**FIGURE 4.5**

Note that when the $x$-coordinate is 0, as in Example 1 **f)** and 1 **g)**, the point is on the $y$-axis. When the $y$-coordinate is 0, as in Example 1 **d)**, the point is on the $x$-axis.

Now Try Exercise 23

**EXAMPLE 2** List the ordered pairs for each point shown in **Figure 4.6**.

**FIGURE 4.6**

*Solution*    Remember to give the *x*-value first in the ordered pair.

| Point | Ordered Pair |
|-------|--------------|
| A | $(-6, -1)$ |
| B | $(-2, 3)$ |
| C | $(0, 2)$ |
| D | $(5, -1)$ |
| E | $(-4, 0)$ |
| F | $(6, 1)$ |
| G | $(0, 0)$ |

Now Try Exercise 21

## 2  Determine Whether an Ordered Pair Is a Solution to a Linear Equation

### Linear Equations in Two Variables

A **linear equation in two variables** is an equation that can be put in the form

$$ax + by = c$$

where $a, b$, and $c$ are real numbers.

The graphs of equations of the form $ax + by = c$ are straight lines. For this reason such equations are called linear. A linear equation in the form $ax + by = c$ is said to be in **standard form.**

### Examples of Linear Equations

$$4x - 3y = 12$$
$$y = 5x + 3$$
$$x - 3y + 4 = 0$$

Note in the examples that only the equation $4x - 3y = 12$ is in standard form. However, the bottom two equations can be written in standard form, as follows:

$$y = 5x + 3 \qquad\qquad x - 3y + 4 = 0$$
$$-5x + y = 3 \qquad\qquad x - 3y = -4$$

Most of the equations we have discussed thus far have contained only one variable. Consider the linear equation in *one* variable, $2x + 3 = 5$. What is its solution?

$$2x + 3 = 5$$
$$2x = 2$$
$$x = 1$$

This equation has only one solution, 1.

Check
$$2x + 3 = 5$$
$$2(1) + 3 \overset{?}{=} 5$$
$$5 = 5 \quad \text{True}$$

Now consider the linear equation in *two* variables, $y = x + 1$. Since the equation contains two variables, its solutions must contain two numbers, one for each variable. One pair of numbers that satisfies this equation is $x = 1$ and $y = 2$. To see that this is true, we substitute both values into the equation.

Check
$$y = x + 1$$
$$2 \overset{?}{=} 1 + 1$$
$$2 = 2 \qquad \text{True}$$

We write this answer as an ordered pair by writing the $x$- and $y$-values within parentheses separated by a comma. Therefore, one solution to this equation is the ordered pair $(1, 2)$. The equation $y = x + 1$ has other solutions.

| Solution | Solution | Solution |
|---|---|---|
| $x = 2, y = 3$ | $x = -3, y = -2$ | $x = -\dfrac{1}{3}, y = \dfrac{2}{3}$ |

Check  
$y = x + 1$  
$3 \overset{?}{=} 2 + 1$  
$3 = 3$   True

$y = x + 1$  
$-2 \overset{?}{=} -3 + 1$  
$-2 = -2$   True

$y = x + 1$  
$\dfrac{2}{3} \overset{?}{=} -\dfrac{1}{3} + 1$  
$\dfrac{2}{3} = \dfrac{2}{3}$   True

Solution Written as an Ordered Pair

$(2, 3)$        $(-3, -2)$        $\left(-\dfrac{1}{3}, \dfrac{2}{3}\right)$

How many possible solutions does the equation $y = x + 1$ have? The equation $y = x + 1$ has an unlimited or *infinite number* of possible solutions. Since it is not possible to list all the specific solutions, the solutions are illustrated with a graph.

> **Graph of an Equation**
>
> A **graph** of an equation in two variables is an illustration of the set of points whose coordinates satisfy the equation.

**Figure 4.7a** shows the points $(2, 3)$, $(-3, -2)$, and $\left(-\dfrac{1}{3}, \dfrac{2}{3}\right)$ plotted in the Cartesian coordinate system. **Figure 4.7b** shows a straight line drawn through the three points. Arrowheads are placed at the ends of the line to show that the line continues in both directions. Every point on this line will satisfy the equation $y = x + 1$, so this graph illustrates all the solutions of $y = x + 1$. The ordered pair $(1, 2)$, which is on the line, also satisfies the equation.

In **Figure 4.7b**, what do you notice about the points $(2, 3)$, $(1, 2)$, $\left(-\dfrac{1}{3}, \dfrac{2}{3}\right)$, and $(-3, -2)$? You probably noticed that they are in a straight line. A set of points that are on a line are said to be **collinear**.

## Understanding Algebra

The equation $y = x + 1$ has an infinite number of solutions. Each solution is represented by a point on the line in **Figure 4.7b**. Also, each point on the line represents a solution to the equation $y = x + 1$.

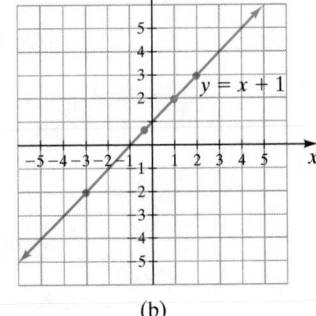

FIGURE 4.7        (a)        (b)

**EXAMPLE 3**  Determine whether the three points appear to be collinear.

**a)** $(2, 7), (0, 3)$, and $(-2, -1)$

**b)** $(0, 5), \left(\dfrac{5}{2}, 0\right)$, and $(5, -5)$

**c)** $(-2, -5), (0, 1)$, and $(6, 8)$

*Solution*   We plot the points to determine whether they appear to be collinear. The solution is shown in **Figure 4.8**.

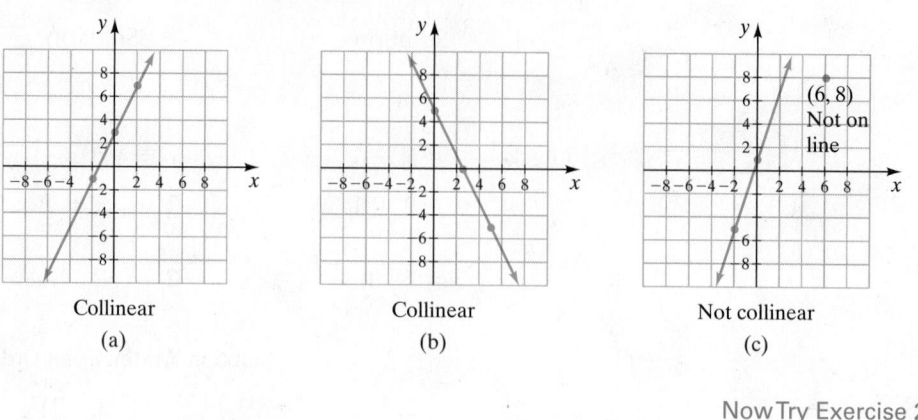

FIGURE 4.8

|  Collinear  |  Collinear  |  Not collinear  |
| (a) | (b) | (c) |

Now Try Exercise 27

To graph an equation, you will need to determine ordered pairs that satisfy the equation and then plot the points.

> **Helpful Hint**
>
> Only two points are needed to graph a linear equation because the graph of every linear equation is a straight line. However, if you graph a linear equation using only two points and you have made an error in determining or plotting one of those points, your graph will be wrong and you will not know it. In **Figures 4.9a and b** we plot only two points to show that if only one of the two points plotted is incorrect, the graph will be wrong. In both **Figures 4.9a and b** we use the ordered pair $(-2, -2)$. However, in **Figure 4.9a** the second point is $(1, 2)$, while in **Figure 4.9b** the second point is $(2, 1)$. Notice how the two graphs differ.
>
> If you use at least three points to plot your graph, as in **Figure 4.7b** on page 219, and they appear to be collinear, you probably have not made a mistake.
>
>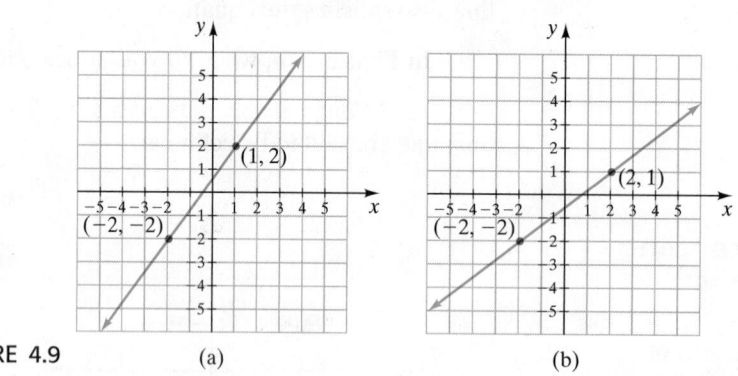
>
> FIGURE 4.9          (a)                    (b)

## EXAMPLE 4

**a)** Determine which of the following ordered pairs satisfy the equation $2x + y = 4$.

$$(2, 0), (0, 4), (3, 1), (-1, 6)$$

**b)** Plot all the points that satisfy the equation on the same axes and draw a straight line through the points.

**c)** What does this line represent?

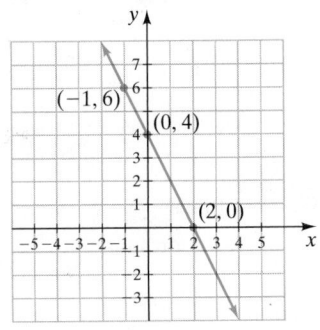

FIGURE 4.10

**Solution**

**a)** We substitute values for $x$ and $y$ into the equation $2x + y = 4$ and determine whether they check.

Check

$(2, 0)$

$2x + y = 4$

$2(2) + 0 \overset{?}{=} 4$

$4 = 4$   True

$(0, 4)$

$2x + y = 4$

$2(0) + 4 \overset{?}{=} 4$

$4 = 4$   True

$(3, 1)$

$2x + y = 4$

$2(3) + 1 \overset{?}{=} 4$

$7 = 4$   False

$(-1, 6)$

$2x + y = 4$

$2(-1) + 6 \overset{?}{=} 4$

$4 = 4$   True

The ordered pairs $(2, 0)$, $(0, 4)$, and $(-1, 6)$ satisfy the equation. The ordered pair $(3, 1)$ does not satisfy the equation.

**b)** **Figure 4.10** shows the three points that satisfy the equation. A line drawn through the three points shows that they appear to be collinear. Note that the ordered pair $(3, 1)$, which is not a solution to the equation, is not a point on this line.

**c)** The line represents all solutions of $2x + y = 4$. The coordinates of every point on this line satisfy the equation $2x + y = 4$.

Now Try Exercise 31

## Using Your Graphing Calculator ▪▪▪▪▪

Throughout this book, we will introduce some of the basic uses of the graphing calculator. More detailed information is available in your calculator manual. A **window** is the screen in which a graph is displayed. In this book all windows and keystrokes will be from a TI-84 Plus graphing calculator. **Figure 4.11** shows the standard window and **Figure 4.12** shows the standard window settings of the **TI-84** Plus.

FIGURE 4.11

```
WINDOW
 Xmin=-10
 Xmax=10
 Xscl=1
 Ymin=-10
 Ymax=10
 Yscl=1
 Xres=1
```

FIGURE 4.12

To graph the equation $y = -2x + 3$, press

Y= ‖ (-) ‖ 2 ‖ X,T,θ,n ‖ + ‖ 3 ‖ GRAPH

The graph of the equation then appears in the window as shown in **Figure 4.13**.

FIGURE 4.13

# EXERCISE SET 4.1

MathXL®     MyMathLab®

MathXL®     MyMathLab

## Warm-Up Exercises

*Fill in the blanks with the appropriate word, phrase, or symbol(s) from the following list.*

| | | | |
|---|---|---|---|
| line | solution | Roman numerals | $y$-axis | linear |
| collinear | $x$-axis | Cartesian | $x$-coordinate | $y$-coordinate |

1. In an ordered pair the _____ is listed first and the _____ is listed second.
2. The _____ is the horizontal axis.
3. The _____ is the vertical axis.
4. A _____ equation can be put into the form $ax + by = c$.

5. The graph of a linear equation is a _____.
6. A _____ to an equation with two variables is an ordered pair that makes the equation true.
7. The quadrants of the Cartesian coordinate system are labeled using _____.
8. Three or more points that are on the same line are called _____.

## Practice the Skills

*Indicate the quadrant in which each of the points belongs.*

9. $(-2, 5)$     10. $(-4, 4)$     11. $(5, -6)$     12. $(2, -3)$
13. $(3, 6)$     14. $(4, 30)$     15. $(-17, -87)$     16. $(63, 47)$
17. $(-124, -132)$     18. $(75, 200)$     19. $(-8, 42)$     20. $(76, -92)$

21. List the ordered pairs corresponding to each point.

22. List the ordered pairs corresponding to each point.

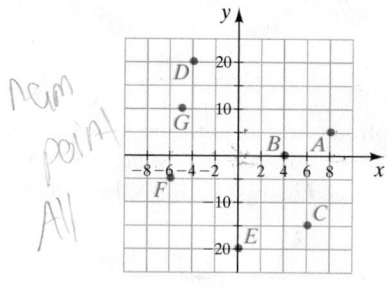

*Plot each point on the same axes.*

23. $A(3, 2), B(-4, 1), C(0, -3), D(-2, 0), E(-3, -4), F\left(-4, -\dfrac{5}{2}\right)$

24. $A(-3, -1), B(2, 0), C(3, 2), D\left(\dfrac{1}{2}, -4\right), E(-4, 2), F(0, 5)$

25. $A(4, 0), B(-1, 3), C(2, 4), D(0, -2), E(-3, -3), F(2, -3)$

26. $A(-3, 4), B(2, 3), C(0, 3), D(-1, 0), E(-2, -2), F(2, -4)$

*Plot the following points. Then determine whether they appear to be collinear.*

27. $A(1, -1), B(5, 3), C(-3, -5), D(0, -2), E(2, 0)$     28. $A(1, -1), B(3, 5), C(0, -3), D(-2, -7), E(2, 1)$

29. $A(1, 5), B\left(-\dfrac{1}{2}, \dfrac{1}{2}\right), C(0, 2), D(-5, -3), E(-2, -4)$     30. $A(1, -2), B(0, -5), C(4, 1), D(-1, -8), E\left(\dfrac{1}{2}, -\dfrac{7}{2}\right)$

*In Exercises 31–36,* **a)** *determine which of the four ordered pairs does not satisfy the given equation.* **b)** *Plot all the points that satisfy the equation on the same axes and draw a straight line through the points.*

31. $y = x + 2$,     **a)** $(2, 4)$     **b)** $(-2, 0)$     **c)** $(-1, 5)$     **d)** $(0, 2)$
32. $2x + y = -4$,     **a)** $(-2, 0)$     **b)** $(2, 3)$     **c)** $(0, -4)$     **d)** $(-1, -2)$
33. $3x - 2y = 6$,     **a)** $(4, 0)$     **b)** $(2, 0)$     **c)** $\left(\dfrac{2}{3}, -2\right)$     **d)** $\left(\dfrac{4}{3}, -1\right)$
34. $4x - 3y = 0$,     **a)** $(3, 4)$     **b)** $(-3, -4)$     **c)** $(0, 0)$     **d)** $(2, 5)$

**35.** $\frac{1}{2}x + 4y = 4$,          **a)** $(-2, 3)$          **b)** $\left(2, \frac{3}{4}\right)$          **c)** $(0, 1)$          **d)** $\left(-4, \frac{3}{2}\right)$

**36.** $y = \frac{1}{2}x + 2$,          **a)** $(0, 2)$          **b)** $(-4, 3)$          **c)** $(-2, 1)$          **d)** $(4, 4)$

## Problem Solving

*Consider the linear equation $y = 3x - 4$. In Exercises 37–40, find the value of y that makes the given ordered pair a solution to the equation.*

**37.** $(2, y)$          **38.** $(-1, y)$          **39.** $(0, y)$          **40.** $(3, y)$

*Consider the linear equation $2x + 3y = 12$. In Exercises 41–44, find the value of x that makes the given ordered pair a solution to the equation.*

**41.** $(x, 0)$          **42.** $(x, -2)$          **43.** $\left(x, \frac{11}{3}\right)$          **44.** $\left(x, \frac{22}{3}\right)$

**45. Longitude and Latitude** Another type of coordinate system that is used to identify a location or position on Earth's surface involves *latitude* and *longitude*. On a globe, the longitudinal lines are lines that go from top to bottom; on a world map they go up and down. The latitudinal lines go around the globe, or left to right on a world map. The locations of Hurricane Georges and Tropical Storm Hermine are indicated on the map on the right.

*Source*: National Weather Service

**a)** Estimate the latitude and longitude of Hurricane Georges.

**b)** Estimate the latitude and longitude of Tropical Storm Hermine.

**c)** Estimate the latitude and longitude of the city of Miami.

**d)** Use either a map or a globe to estimate the latitude and longitude of your college.

## Concept/Writing Exercises

**46.** What is the value of $y$ at the point where a straight line crosses the $x$-axis? Explain.

**47.** What is the value of $x$ at the point where a straight line crosses the $y$-axis? Explain.

**48.** What does the graph of a linear equation illustrate?

**49.** Why are arrowheads added to the ends of graphs of linear equations?

**50.** How many solutions does a linear equation in two variables have?

## Group Activity

*In Section 4.2 we discuss how to find ordered pairs to plot when graphing linear equations. Let's see if you can draw some graphs now. Individually work parts **a)** through **c)** in Exercises 51–54.*

**a)** *Select any three values for x and find the corresponding values of y.*

**b)** *Plot the points (they should appear to be collinear).*

**c)** *Draw the graph.*

**d)** *As a group, compare your answers. You should all have the same lines.*

**51.** $y = 2x$          **52.** $y = x$

**53.** $y = 2x + 1$          **54.** $y = -2x$

## Cumulative Review Exercises

[1.3]　**55.** Subtract $\dfrac{7}{12} - \dfrac{2}{9}$.

[1.9]　**56.** Evaluate **a)** $x^2$, **b)** $-x^2$, and **c)** $(-x)^2$ when $x = -3$.

[2.5]　**57.** Solve the equation $\dfrac{1}{2}(x - 3) = \dfrac{1}{3}x + 2$.

[2.6]　**58.** Solve the equation $3x - 2y = 4$ for $y$.

[3.2]　**59.** One number is 14 more than another number. The sum of the two numbers is 60. Find the numbers.

[3.4]　**60.** Two boats leave the dock at the same time headed in the same direction. Sandy's boat travels at 34 miles per hour and Tanner's boat travels at 28 miles per hour. In how many hours will the two boats be 12 miles apart?

# 4.2 Graphing Linear Equations

1 Graph linear equations by plotting points.

2 Graph linear equations of the form $ax + by = 0$.

3 Graph linear equations using the $x$- and $y$-intercepts.

4 Graph horizontal and vertical lines.

5 Study applications of graphs.

Now we are ready to graph linear equations.

## 1 Graph Linear Equations by Plotting Points

Graphing by plotting points is the most versatile method of graphing because we can also use it to graph other types of equations.

### To Graph Linear Equations by Plotting Points

1. **Solve for $y$:** Solve the equation for the variable $y$. That is, get $y$ by itself on the left side of the equation.

2. **Substitute a number in for $x$:** Select a number and substitute it into the equation for $x$ and find the corresponding value for $y$. Record the ordered pair $(x, y)$.

3. **Repeat step 2:** Select two different values for $x$. This will give you two additional ordered pairs.

4. **Plot the points:** Plot the three ordered pairs. The three points should be collinear. If they are not, recheck your work.

5. **Draw the line:** With a straightedge, draw a straight line through the three points. Draw an arrowhead on each end of the line.

### Helpful Hints

- If you have forgotten how to solve an equation for $y$, review Section 2.6.
- When selecting numbers for $x$, you should select numbers that result in integer values for $y$.
- Select numbers for $x$ that are small enough so that the ordered pairs obtained can be plotted easily on the axes.
- A good number to select for $x$ is always 0.

EXAMPLE 1　Graph $y = 2x + 4$.

*Solution*　The equation is already solved for $y$. We will arbitrarily select the numbers $-1$, 0, and 1 and substitute them into the equation for $x$ and find the corresponding values for $y$. The calculations that follow show that when $x = -1$, $y = 2$, when $x = 0$, $y = 4$, and when $x = 1$, $y = 6$.

| $x$ | $y = 2x + 4$ | Ordered Pair |
|-----|--------------|--------------|
| $-1$ | $y = 2(-1) + 4 = 2$ | $(-1, 2)$ |
| $0$ | $y = 2(0) + 4 = 4$ | $(0, 4)$ |
| $1$ | $y = 2(1) + 4 = 6$ | $(1, 6)$ |

| $x$ | $y$ |
|-----|-----|
| $-1$ | 2 |
| 0 | 4 |
| 1 | 6 |

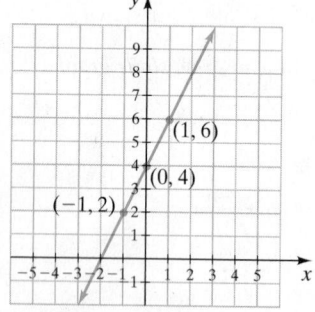

FIGURE 4.14

We then plot the three ordered pairs on the same axes **(Fig. 4.14)**. Connect the points with a straight line and place the arrowheads at the ends of the line.

Now Try Exercise 25

If we select any point on this line, the ordered pair represented by that point will be a solution to the equation $y = 2x + 4$. Similarly, any solution to the equation will be represented by a point on the line. Notice that $(-2, 0)$ and $(2, 8)$ are points on the line (see **Figure 4.15**). Let's verify that they are solutions to the equation

Check $(-2, 0)$          Check $(2, 8)$

$y = 2x + 4$              $y = 2x + 4$

$0 = 2(-2) + 4$           $8 = 2(2) + 4$

$0 = -4 + 4$              $8 = 4 + 4$

$0 = 0$  True            $8 = 8$  True

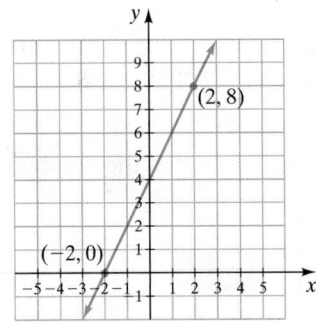

**FIGURE 4.15**

**EXAMPLE 2**  Graph $3y = 5x - 6$.

*Solution*   We begin by solving the equation for $y$.

$$3y = 5x - 6$$

$$y = \frac{5x - 6}{3} \qquad \text{Divided both sides by 3.}$$

$$y = \frac{5x}{3} - \frac{6}{3} \qquad \text{Wrote as two fractions.}$$

$$y = \frac{5}{3}x - 2$$

Now we can see that if we select values for $x$ that are multiples of the denominator, 3, the values we obtain for $y$ will be integers. Let's select the values $-3, 0$, and $3$ for $x$.

$$y = \frac{5}{3}x - 2$$

Let $x = -3$.   $y = \frac{5}{3}(-3) - 2 = -5 - 2 = -7$

Let $x = 0$.    $y = \frac{5}{3}(0) - 2 = 0 - 2 = -2$

Let $x = 3$.    $y = \frac{5}{3}(3) - 2 = 5 - 2 = 3$

| $x$ | $y$ |
|-----|-----|
| $-3$ | $-7$ |
| $0$ | $-2$ |
| $3$ | $3$ |

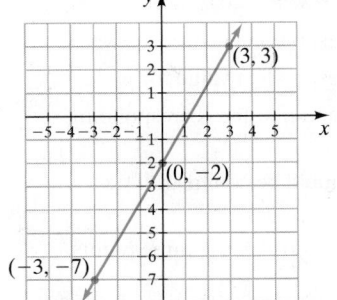

**FIGURE 4.16**

Finally, we plot the points and draw the straight line (**Figure 4.16**).

Now Try Exercise 31

## 2 Graph Linear Equations of the Form $ax + by = 0$

**EXAMPLE 3**  Graph $2x + 5y = 0$.

*Solution*   We begin by solving the equation for $y$.

$$2x + 5y = 0$$

$$5y = -2x$$

$$y = -\frac{2x}{5} \quad \text{or} \quad y = -\frac{2}{5}x$$

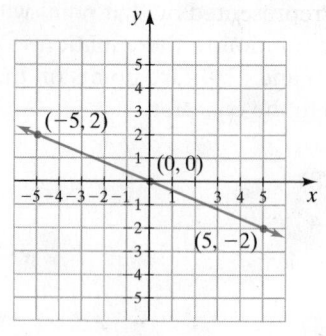

**FIGURE 4.17**

If we select values for $x$ that are multiples of 5, we will get integer values for $y$. We will arbitrarily select the values $x = -5$, $x = 0$, and $x = 5$.

$$y = -\frac{2}{5}x$$

Let $x = -5$. $\qquad y = \left(-\frac{2}{5}\right)(-5) = 2$

Let $x = 0$. $\qquad y = \left(-\frac{2}{5}\right)(0) = 0$

Let $x = 5$. $\qquad y = -\frac{2}{5}(5) = -2$

| $x$ | $y$ |
|-----|-----|
| $-5$ | $2$ |
| $0$ | $0$ |
| $5$ | $-2$ |

Now we plot the points and draw the graph **(Fig. 4.17)**.

Now Try Exercise 37

Note that the graph in Example 3 passes through the origin.

> ### Understanding Algebra
>
> The graph of every linear equation with a constant of 0 (equations of the form $ax + by = 0$) will pass through the origin.

### 3  Graph Linear Equations Using the $x$- and $y$-Intercepts

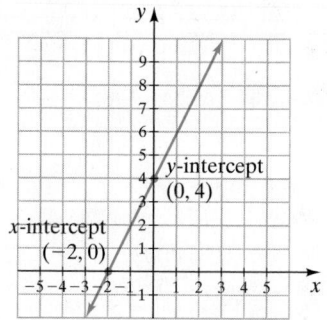

**FIGURE 4.18**

> ### $x$- and $y$-intercepts
>
> - The **$x$-intercept** is the point at which a graph crosses the $x$-axis.
> - The **$y$-intercept** is the point at which a graph crosses the $y$-axis.

Consider the graph in **Figure 4.18**, which is the graph we drew in Example 1. Note that the line crosses the $x$-axis at $-2$. Therefore, the $x$-intercept is $(-2, 0)$. Also note that the line crosses the $y$-axis at 4. Therefore, the $y$-intercept is $(0, 4)$.

It is often convenient to graph linear equations by finding the $x$- and $y$-intercepts.

> ### Understanding Algebra
>
> In general terms, the $x$-intercept is $(x, 0)$ and the $y$-intercept is $(0, y)$.

> ### To Graph Linear Equations Using the $x$- and $y$-intercepts
>
> 1. **Find the $y$-intercept.** Set $x$ equal to 0 and find the corresponding value for $y$.
> 2. **Find the $x$-intercept.** Set $y$ equal to 0 and find the corresponding value for $x$.
> 3. **Determine a checkpoint.** Select a nonzero value for $x$ and find the corresponding value for $y$.
> 4. **Plot the intercepts and checkpoint.** The three points should be collinear. If not, check your work.
> 5. **Draw the line.** Using a straightedge, draw a straight line through the three points. Draw an arrowhead on each end of the line.

**EXAMPLE 4** Graph $3y = 6x + 12$ using the x- and y-intercepts.

*Solution*    To find the y-intercept, set $x = 0$ and find the corresponding value of $y$.

$$3y = 6x + 12$$
$$3y = 6(0) + 12$$
$$3y = 0 + 12$$
$$3y = 12$$
$$y = \frac{12}{3} = 4$$

Therefore, the $y$-intercept is $(0, 4)$. To find the $x$-intercept, set $y = 0$ and find the corresponding value of $x$.

$$3y = 6x + 12$$
$$3(0) = 6x + 12$$
$$0 = 6x + 12$$
$$-12 = 6x$$
$$\frac{-12}{6} = x$$
$$-2 = x$$

Therefore, the $x$-intercept is $(-2, 0)$.

Next, to find a checkpoint, we will select a nonzero value for $x$ and find the corresponding value for $y$. We will select $x = 2$.

$$x = 2$$
$$3y = 6x + 12$$
$$3y = 6(2) + 12$$
$$3y = 12 + 12$$
$$3y = 24$$
$$y = \frac{24}{3} = 8$$

Therefore, the checkpoint is $(2, 8)$. Now plot the $y$-intercept, $(0, 4)$, the $x$-intercept, $(-2, 0)$, and the checkpoint $(2, 8)$ (Fig. 4.19). Since the three points appear to be collinear, draw the straight line through all three points.

Now Try Exercise 55

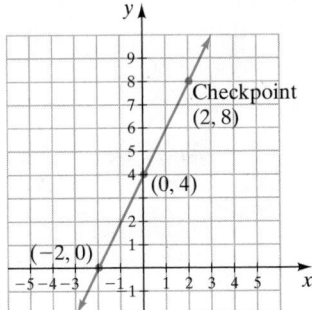

FIGURE 4.19

**EXAMPLE 5** Graph $2x + 5y = 12$ using the $x$- and $y$-intercepts.

*Solution*

| Find $y$-intercept | Find $x$-intercept | Checkpoint |
|---|---|---|
| Let $x = 0$. | Let $y = 0$. | Let $x = 1$. |
| $2x + 5y = 12$ | $2x + 5y = 12$ | $2x + 5y = 12$ |
| $2(0) + 5y = 12$ | $2x + 5(0) = 12$ | $2(1) + 5y = 12$ |
| $0 + 5y = 12$ | $2x + 0 = 12$ | $2 + 5y = 12$ |
| $5y = 12$ | $2x = 12$ | $5y = 10$ |
| $y = \dfrac{12}{5}$ | $x = 6$ | $y = 2$ |

The three ordered pairs are $\left(0, \dfrac{12}{5}\right)$, $(6, 0)$, and $(1, 2)$.

The three points appear to be collinear. Draw a straight line through all three points (**Fig. 4.20**).

Now Try Exercise 45

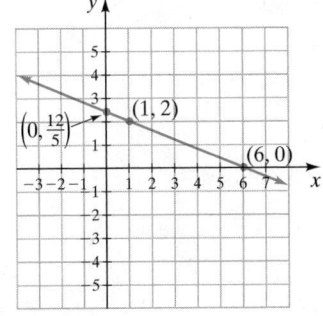

FIGURE 4.20

**EXAMPLE 6** Graph $y = 20x + 60$ using the $x$- and $y$-intercepts.

*Solution*

| Find $y$-intercept | Find $x$-intercept | Checkpoint |
|---|---|---|
| Let $x = 0$. | Let $y = 0$. | Let $x = 3$. |
| $y = 20x + 60$ | $y = 20x + 60$ | $y = 20x + 60$ |
| $y = 20(0) + 60$ | $0 = 20x + 60$ | $y = 20(3) + 60$ |
| $y = 60$ | $-60 = 20x$ | $y = 60 + 60$ |
| | $-3 = x$ | $y = 120$ |

**FIGURE 4.21**

The three ordered pairs are $(0, 60)$, $(\phantom{xx}, 0)$, and $(3, 120)$. Since the values of $y$ are large, we let each interval on the $y$-axis be 15 units rather than 1 (**Fig. 4.21**). Now we plot the points and draw the graph.

Now Try Exercise 59

When selecting the scales for your axes, you should realize that different scales will result in the same equation having a different appearance. Consider the graphs shown in **Figure 4.22**. Both graphs represent the same equation, $y = x$. In **Figure 4.22a** both the $x$- and $y$-axes have the same scale. In **Figure 4.22b**, the $x$- and $y$-axes do not have the same scale. Both graphs are correct in that each represents the graph of $y = x$. The difference in appearance is due to the difference in scales on the $x$-axis.

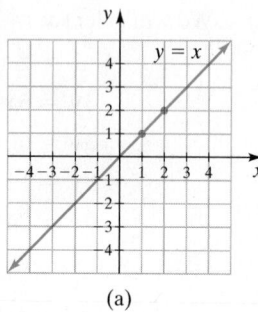

**FIGURE 4.22**

(a)                    (b)

## Using Your Graphing Calculator ▮▮▮▯▯

We will find the intercepts of the graph of $y = 4x - 5$. First, graph the function by pressing the keys shown below. The graph is shown in **Figure 4.23a**.

$$\boxed{Y=}\;\boxed{4}\;\boxed{X,T,Q,n}\;\boxed{-}\;\boxed{5}\;\boxed{GRAPH}$$

(a)                    (b)

**FIGURE 4.23**

To get the $y$-intercept, press $\boxed{TRACE}$. The $y$-intercept is $(0, -5)$ as shown in **Figure 4.23b**. To get the $x$-intercept, we use the "zero" feature by first pressing the keys

$$\boxed{2nd}\;\boxed{TRACE}\;\boxed{2}$$

You are asked for a left-bound. Move the flashing cursor left or right by using the keys $\boxed{<}$ or $\boxed{>}$. Move the cursor to the left of the $x$-intercept and press $\boxed{ENTER}$.

Next, you are asked for a right-bound. Move the cursor to the right of the $x$-intercept and press $\boxed{ENTER}$. Finally, you are asked for a guess. Move the cursor close to the $x$-intercept and press $\boxed{ENTER}$. The screen is shown in **Figure 4.24**. The $x$-intercept is $(1.25, 0)$.

**FIGURE 4.24**

## 4 Graph Horizontal and Vertical Lines

When a linear equation contains only one variable, its graph will be either a horizontal or a vertical line, as is explained in Examples 7 and 8.

**EXAMPLE 7** Graph $y = 3$.

*Solution*    This equation can be written as $y = 3 + 0x$. Thus, for any value of $x$ selected, $y$ will be 3. The graph of $y = 3$ is illustrated in **Figure 4.25**.

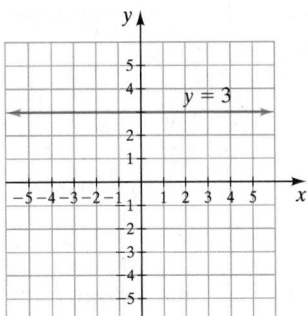

**FIGURE 4.25**

Now Try Exercise 23

**Horizontal Line**

The graph of an equation of the form $y = b$ is a **horizontal line** whose $y$-intercept is $(0, b)$.

**EXAMPLE 8** Graph $x = -2$.

*Solution*    This equation can be written as $x = -2 + 0y$. Thus, for any value of $y$ selected, $x$ will have a value of $-2$. The graph of $x = -2$ is illustrated in **Figure 4.26**.

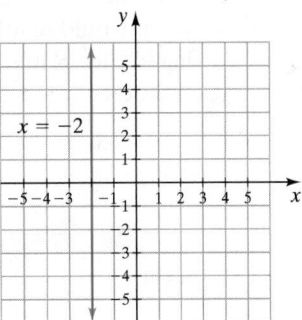

**FIGURE 4.26**

Now Try Exercise 21

**Vertical Line**

The graph of an equation of the form $x = a$ is a **vertical line** whose $x$-intercept is $(a, 0)$.

## 5 Study Applications of Graphs

Let's look at an application of graphing a linear equation.

**EXAMPLE 9** **Weekly Salary** Carol Waters accepted a position as a sales representative at a furniture store where she is paid a weekly salary plus a commission on her sales. She will receive a salary of $300 per week plus a 7% commission on all her sales, $s$.

  **a)** Write an equation for the salary Carol will receive, $R$, in terms of the sales, $s$.

  **b)** Graph the salary for sales of $0 up to and including $20,000.

**c)** From the graph, estimate Carol's salary if her weekly sales are $15,000.

**d)** From the graph, estimate the sales needed for Carol to earn a weekly salary of $900.

*Solution*

**a)** Since $s$ is the amount of sales, a 7% commission on $s$ dollars in sales is $0.07s$.

$$\text{salary received} = \$300 + \text{commission}$$
$$R = 300 + 0.07s$$

**b)** We select three values for $s$ and find the corresponding values of $R$.

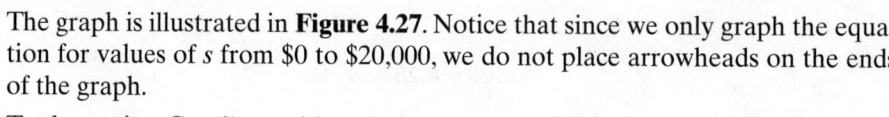

| $s$ | $R$ |
|---|---|
| 0 | 300 |
| 10,000 | 1000 |
| 20,000 | 1700 |

$R = 300 + 0.07s$

Let $s = 0$.    $R = 300 + 0.07(0) = 300$

Let $s = 10,000$.    $R = 300 + 0.07(10,000) = 1000$

Let $s = 20,000$.    $R = 300 + 0.07(20,000) = 1700$

**FIGURE 4.27**

The graph is illustrated in **Figure 4.27**. Notice that since we only graph the equation for values of $s$ from $0 to $20,000, we do not place arrowheads on the ends of the graph.

**c)** To determine Carol's weekly salary on sales of $15,000, locate $15,000 on the sales axis. Then draw a vertical line up to where it intersects the graph, the *purple* line in **Figure 4.27**. Now draw a horizontal line across to the salary axis. Since the horizontal line crosses the salary axis at about $1350, weekly sales of $15,000 would result in a weekly salary of about $1350. We can find the exact salary by substituting 15,000 for $s$ in the equation $R = 300 + 0.07s$ and finding the value of $R$. Do this now.

**d)** To find the sales needed for Carol to earn a weekly salary of $900, we find $900 on the salary axis. We then draw a horizontal line from that point to the graph, as shown with the *green* line in **Figure 4.27**. We then draw a vertical line from the point of intersection of the graph to the sales axis. Thus, sales of about $8600 per week would result in a salary of $900. We can find an exact answer by substituting 900 for $R$ in the equation $R = 300 + 0.07s$ and solving the equation for $s$. Do this now.

Now Try Exercise 75

# EXERCISE SET 4.2

MathXL®    MyMathLab®

## Warm-Up Exercises

*Fill in the blanks with the appropriate word, phrase, or symbol(s) from the following list.*

| | | | | |
|---|---|---|---|---|
| $(x, 0)$ | horizontal | $y$-intercept | $x$-axis | diagonal |
| $(0, y)$ | $x$-intercept | vertical | $y$-axis | |

**1.** The point at which a graph crosses the _____ is called the $x$-intercept.

**2.** The point at which a graph crosses the _____ is called the $y$-intercept.

**3.** To find the _____ of a graph of a linear equation, set $x$ equal to 0 and find the corresponding value for $y$.

**4.** To find the _____ of a graph of a linear equation, set $y$ equal to 0 and find the corresponding value for $x$.

**5.** The $x$-intercept of a graph will have coordinates _____.

**6.** The $y$-intercept of a graph will have coordinates _____.

**7.** The graph of an equation of the form $y = b$ is a _____ line.

**8.** The graph of an equation of the form $x = a$ is a _____ line.

## Practice the Skills

*Find the missing coordinate if the ordered pair is to be a solution to the equation $3x + y = 9$.*

**9.** $(2, ?)$

**10.** $(-2, ?)$

**11.** $(?, -6)$

**12.** $(?, -9)$

**13.** $(?, 0)$

**14.** $\left(\dfrac{3}{2}, ?\right)$

*Find the missing coordinate in the given solutions for $3x - 2y = 8$.*

**15.** $(4, ?)$

**16.** $(0, ?)$

**17.** $(?, 0)$

**18.** $\left(?, -\dfrac{5}{2}\right)$

**19.** $(-4, ?)$

**20.** $(?, -3)$

*Graph each equation.*

**21.** $x = -3$

**22.** $x = \dfrac{3}{2}$

**23.** $y = 4$

**24.** $y = -\dfrac{5}{3}$

*Graph by plotting points. Plot at least three points for each graph.*

**25.** $y = 3x - 1$

**26.** $y = -x + 3$

**27.** $y = 4x - 2$

**28.** $y = x - 4$

**29.** $x + 2y = 6$

**30.** $-3x + 3y = 6$

**31.** $3x - 2y = 4$

**32.** $3x - 2y = 6$

**33.** $4x + 3y = -9$

**34.** $6y - 12x = 18$

**35.** $6x + 5y = 30$

**36.** $2x + 3y = -6$

**37.** $-4x + 5y = 0$

**38.** $3x + 2y = 0$

**39.** $y = -20x + 60$

**40.** $2y - 100x = 50$

**41.** $y = \dfrac{4}{3}x$

**42.** $y = -\dfrac{3}{5}x$

**43.** $y = \dfrac{1}{2}x + 4$

**44.** $y = -\dfrac{2}{5}x + 2$

*Graph using the x- and y-intercepts.*

**45.** $y = 3x + 3$

**46.** $y = -3x + 6$

**47.** $y = -4x + 2$

**48.** $y = -2x + 5$

**49.** $y = 4x + 16$

**50.** $y = -5x + 4$

**51.** $4y + 6x = 24$

**52.** $4x = 3y - 9$

**53.** $\dfrac{1}{2}x + 2y = 4$

**54.** $x + \dfrac{1}{2}y = 2$

**55.** $12x - 24y = 48$

**56.** $25x + 50y = 100$

**57.** $8y = 6x - 12$

**58.** $6y = -4x + 12$

**59.** $y = 15x + 45$

**60.** $y = -10x + 30$

**61.** $\dfrac{1}{3}x + \dfrac{1}{4}y = 12$

**62.** $\dfrac{1}{4}x - \dfrac{2}{3}y = 60$

**63.** $\dfrac{1}{2}x = \dfrac{2}{5}y - 80$

**64.** $\dfrac{2}{3}y = \dfrac{5}{4}x + 120$

*Write the equation represented by the given graph.*

**65.**

**66.**

**67.**

**68.**

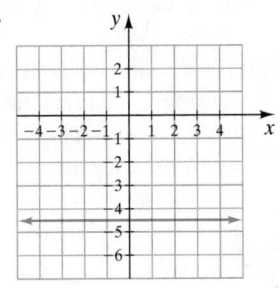

## Problem Solving

**69.** What is the value of $a$ if the graph of $ax + 2y = 15$ is to have an $x$-intercept of $(3, 0)$?

**70.** What is the value of $a$ if the graph of $ax - 4y = 8$ is to have an $x$-intercept of $(4, 0)$?

**71.** What is the value of $b$ if the graph of $3x + by = 14$ is to have a $y$-intercept of $(0, 7)$?

**72.** What is the value of $b$ if the graph of $4x + by = 15$ is to have a $y$-intercept of $(0, -3)$?

*The bar graphs in Exercises 73 and 74 display information. State whether the graph displays a linear relationship. Explain your answer.*

**73.**

**74.**

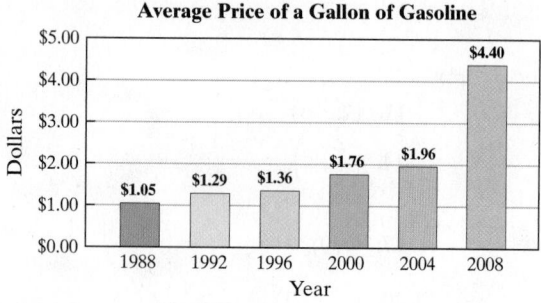

**Average Price of a Gallon of Gasoline**

Source: American Petroleum Institute

*Review Example 9 before working Exercises 75–80.*

**75. Telephone Calls** Alexus Judd's telephone plan consists of a monthly fee of $15 plus 10 cents per minute for long-distance calls made.

**a)** Write an equation for the total monthly cost, $C$, when $n$ minutes are used for long-distance calls.

**b)** Graph the equation for up to and including 100 minutes of long-distance calls made.

**c)** Estimate the total monthly cost if 40 minutes of long-distance calls are made.

**d)** If the total monthly bill is $25, estimate the number of minutes used for long-distance calls.

**76. Distance Traveled** Distance traveled is calculated using the formula,

$$\text{distance} = \text{rate} \cdot \text{time or } d = rt.$$

Assume the rate of a car is a constant 30 miles per hour.

**a)** Write an equation for the distance, $d$, in terms of time, $t$.

**b)** Graph the equation for times of 0 to 20 hours inclusive.

**c)** Estimate the distance traveled in 12 hours.

**d)** If the distance traveled is 150 miles, estimate the time traveled.

**77. Truck Rental** Lynn Brown needs a large truck to move some furniture. She found that the cost, $C$, of renting a truck is $40 per day plus $1 per mile, $m$.

**a)** Write an equation for the cost in terms of the miles driven.

**b)** Graph the equation for values up to and including 100 miles.

**c)** Estimate the cost of driving 60 miles in one day.

**d)** Estimate the miles driven if the cost for one day is $70.

**78. Simple Interest** Simple interest is calculated by the simple interest formula,

$$\text{interest} = \text{principal} \cdot \text{rate} \cdot \text{time or } I = prt.$$

Suppose the principal is $10,000 and the rate is 5%.

**a)** Write an equation for simple interest in terms of time.

**b)** Graph the equation for times of 0 to 20 years inclusive.

**c)** What is the simple interest for 10 years?

**d)** If the simple interest is $500, find the length of time.

**79. Video Store Profit** The weekly profit, $P$, of a video rental store can be approximated by the formula $P = 1.5n - 200$, where $n$ is the number of videos rented weekly.

**a)** Draw a graph of profit in terms of video rentals for up to and including 1000 videos.

**b)** Estimate the weekly profit if 500 videos are rented.

**c)** Estimate the number of videos rented if the week's profit is $1000.

**80. Playing Tennis** The cost, $C$, of playing tennis in the Downtown Tennis Club includes an annual $200 membership fee plus $10 per hour, $h$, of court time.

**a)** Write an equation for the annual cost of playing tennis at the Downtown Tennis Club in terms of hours played.

**b)** Graph the equation for up to and including 300 hours.

**c)** Estimate the cost for playing 50 hours in a year.

**d)** If the annual cost for playing tennis was $1700, estimate how many hours of tennis were played.

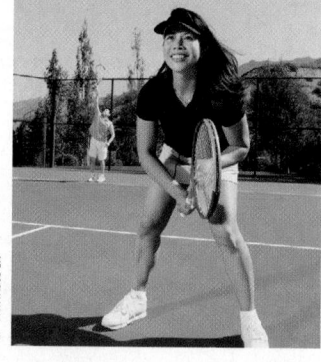

© Thinkstock

*Determine the coefficients to be placed in the shaded areas so that the graph of the equation will be a line with the x- and y-intercepts specified. Explain how you determined your answer.*

**81.** ▢$x$ + ▢$y$ = 6; $x$-intercept at 2; $y$-intercept at 3

**82.** ▢$x$ + ▢$y$ = 18; $x$-intercept at −3, $y$-intercept at 6

**83.** ▢$x$ − ▢$y$ = −12; $x$-intercept at −2, $y$-intercept at 3

**84.** ▢$x$ − ▢$y$ = 30; $x$-intercept at −10, $y$-intercept at −15

## Challenge Problems

**85.** Consider the following equations: $y = 2x - 1$, $y = -x + 5$.

**a)** Carefully graph both equations on the same axes.

**b)** Determine the point of intersection of the two graphs.

**c)** Substitute the values for $x$ and $y$ at the point of intersection into each of the two equations and determine whether the point of intersection satisfies each equation.

**d)** Do you believe there are any other ordered pairs that satisfy both equations? Explain your answer. (We will

study equations like these, called systems of equations, in Chapter 8.)

**86.** In Chapter 10 we will be graphing quadratic equations. The graphs of quadratic equations are *not* straight lines. Graph the quadratic equation $y = x^2 - 4$ by selecting values for $x$. Find the corresponding values of $y$, then plot the points. Make sure you plot a sufficient number of points to get an accurate graph.

## Group Activity

*Discuss and answer Exercise 87 as a group.*

**87.** Let's study the graphs of the equations $y = 2x + 4$, $y = 2x + 2$, and $y = 2x - 2$ to see how they are similar and how they differ. Each group member should start with the same axes.

  **a)** Group member 1: Graph $y = 2x + 4$.

  **b)** Group member 2: Graph $y = 2x + 2$.

  **c)** Group member 3: Graph $y = 2x - 2$.

  **d)** Now transfer all three graphs onto the same axes. (You can use one of the group members' graphs or you can construct new axes.)

  **e)** Explain what you notice about the three graphs.

  **f)** Explain what you notice about the $y$-intercepts.

## Cumulative Review Exercises

[1.9] **88.** Evaluate $2[6 - (4 - 5)] \div 2 - 8^2$.

[2.6] **89.** Find the circumference and area of a circle with a radius of 3 inches.

[2.7] **90. House Cleaning** According to the instructions on a bottle of concentrated household cleaner, 8 ounces of the cleaner should be mixed with 3 gallons of water. If your bucket holds only 2.5 gallons of water, how much cleaner should you use?

[3.2] **91. Refrigerator Purchase** Kristin Runde purchased a new refrigerator. The cost of the refrigerator, includ- ing a 6.5% sales tax, was $1491. What was the price of the refrigerator before tax?

See Exercise 91.

# 4.3 Slope of a Line

1 Find the slope of a line.

2 Recognize positive and negative slopes.

3 Examine the slopes of horizontal and vertical lines.

4 Examine the slopes of parallel and perpendicular lines.

### Understanding Algebra

Slope is often described with the phrase *"rise over run."*

## 1 Find the Slope of a Line

The *slope of a line* is a measure of the *steepness* of the line. The slope of a line is an important concept in many areas of mathematics. A knowledge of slope is helpful in understanding linear equations.

**Slope of a Line**

- The **slope of a line,** $m$, is the ratio of the vertical change, or *rise*, to the horizontal change, or *run*, between any two selected points on the line.

- $m = \text{slope} = \dfrac{\text{vertical change}}{\text{horizontal change}} = \dfrac{\text{rise}}{\text{run}}$

**Helpful Hint**

*Slope*

We often come across slope in everyday life. A road may have a slope (called the *grade*) of 8%. A roof may have a slope (called the *pitch*) of $\frac{6}{15}$. Suppose a road has an 8% grade. Since $8\% = \frac{8}{100}$, this means the road rises 8 feet for each 100 feet of horizontal length. A roof pitch of $\frac{6}{15}$ means the roof rises 6 feet for each 15 feet of horizontal length.

As an example of slope, consider the line that goes through the two points $(1, 2)$ and $(3, 6)$ (see **Fig. 4.28a**). From **Figure 4.28b**, we can see that the vertical change is $6 - 2$, or 4 units. The horizontal change is $3 - 1$, or 2 units.

FIGURE 4.28        (a)                                    (b)

$$m = \text{slope} = \frac{\text{vertical change}}{\text{horizontal change}} = \frac{\text{rise}}{\text{run}} = \frac{4}{2} = 2$$

Thus, the slope of the line through these two points is 2. By examining the line connecting these two points, we can see that as the graph moves up 2 units it moves to the right 1 unit (**Fig. 4.29**).

Now we present the formula to find the slope of a line given any two points $(x_1, y_1)$ and $(x_2, y_2)$ on the line. Look at **Figure 4.30**. The **rise** is the difference between $y_2$ and $y_1$ and the **run** is the difference between $x_2$ and $x_1$.

FIGURE 4.29

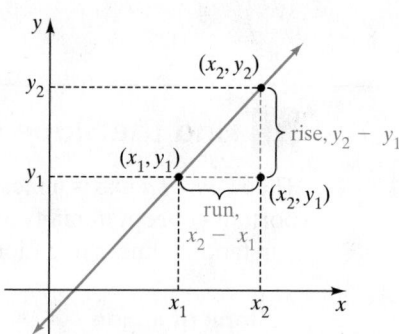

FIGURE 4.30

## Understanding Algebra

The Greek letter delta, $\Delta$, is often used to represent the phrase "the change in." So, $\Delta y$ represents "the change in $y$" and $\Delta x$ represents "the change in $x$" and the formula for slope can be given as

$$m = \frac{\Delta y}{\Delta x} = \frac{y_2 - y_1}{x_2 - x_1}$$

### Slope of a Line Through the Points $(x_1, y_1)$ and $(x_2, y_2)$

$$m = \text{slope} = \frac{\text{vertical change}}{\text{horizontal change}} = \frac{\text{rise}}{\text{run}} = \frac{y_2 - y_1}{x_2 - x_1}$$

### Helpful Hint

It makes no difference which two points on a line are selected when finding the slope of the line. It also makes no difference which point you choose as $(x_1, y_1)$ or $(x_2, y_2)$.

EXAMPLE 1   Find the slope of the line through the points $(-6, 1)$ and $(3, 5)$.

*Solution*    We will designate $(-6, 1)$ as $(x_1, y_1)$ and $(3, 5)$ as $(x_2, y_2)$.

$$m = \frac{y_2 - y_1}{x_2 - x_1}$$

$$= \frac{5 - 1}{3 - (-6)}$$

$$= \frac{5 - 1}{3 + 6} = \frac{4}{9}$$

Thus, the slope is $\frac{4}{9}$.

If we had designated $(3, 5)$ as $(x_1, y_1)$ and $(-6, 1)$ as $(x_2, y_2)$, we would have obtained the same results.

$$m = \frac{y_2 - y_1}{x_2 - x_1}$$

$$= \frac{1 - 5}{-6 - 3} = \frac{-4}{-9} = \frac{4}{9}$$

Now Try Exercise 13

**Avoiding Common Errors**

Students sometimes subtract the $x$'s and $y$'s in the slope formula in the wrong order. For instance, using the problem in Example 1:

$$m = \frac{y_2 - y_1}{x_1 - x_2} = \frac{5 - 1}{-6 - 3} = \frac{4}{-9} = -\frac{4}{9}$$

Notice that subtracting in this incorrect order results in a negative slope, when the actual slope of the line is positive. The same sign error will occur each time subtraction is done incorrectly in this manner.

## 2  Recognize Positive and Negative Slopes

The slope of a line that is neither horizontal nor vertical is either positive or negative. See the lines in **Figure 4.31**.

Positive slope

Negative slope

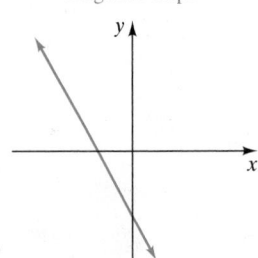

**FIGURE 4.31**

- Line rises from left to right
- As $x$ increases, $y$ increases

(a)

- Line falls from left to right
- As $x$ increases, $y$ decreases

(b)

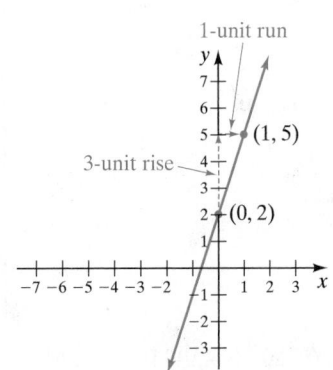

**FIGURE 4.32**

**EXAMPLE 2**  Consider the line in Figure 4.32.

**a)** Determine the slope of the line by observing the rise and the run between the points $(1, 5)$ and $(0, 2)$ on the graph.

**b)** Calculate the slope of the line using the two given points.

*Solution*

**a)** Notice that the slope is positive since the line rises from left to right. From **Figure 4.32** we see that the rise is $+3$ units and the run is $+1$ unit. Thus, the slope of the line is $\frac{3}{1}$ or 3.

**b)** We will use the ordered pairs $(1, 5)$ and $(0, 2)$.

Let $(x_2, y_2)$ be $(1, 5)$. Let $(x_1, y_1)$ be $(0, 2)$.

$$m = \frac{y_2 - y_1}{x_2 - x_1} = \frac{5 - 2}{1 - 0} = \frac{3}{1} = 3$$

Note that the slope obtained in part **b)** agrees with the slope obtained in part **a)**. If we had designated $(1, 5)$ as $(x_1, y_1)$ and $(0, 2)$ as $(x_2, y_2)$, the slope would not have changed. Try it and see that you will still obtain a slope of 3.

Now Try Exercise 25

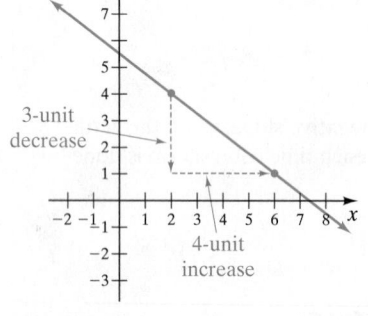

**FIGURE 4.33**

**EXAMPLE 3** Find the slope of the line in **Figure 4.33** using the vertical change and horizontal change between the two points shown.

*Solution*    Since the graph falls from left to right, the line has a negative slope. From **Figure 4.33** we see that the vertical change between the two given points is $-3$ units and the horizontal change between the two given points is 4 units. Thus, the slope is $\dfrac{-3}{4}$ or $-\dfrac{3}{4}$.

Now Try Exercise 29

Using the two points shown in **Figure 4.33** and the definition of slope, calculate the slope of the line in Example 3. You should obtain the same answer.

**EXAMPLE 4** **Driving to a Concert** Boyd Williamson is driving to see an AC/DC concert. The graph shown in **Figure 4.34** shows the time in hours along the horizontal axis and the distance in miles along the vertical axis.

**FIGURE 4.34**

**a)** Determine the slope of the red line segment.

**b)** Determine the slope of the blue line segment.

*Solution*

**a)** We can choose any two points on the red line segment. We will use the points $(0, 0)$ as $(x_1, y_1)$ and $(1, 55)$ as $(x_2, y_2)$.

$$m = \frac{y_2 - y_1}{x_2 - x_1} = \frac{55 - 0}{1 - 0} = \frac{55}{1} = 55$$

Thus, the slope of the red line segment is 55.

**b)** We will use the points $(2, 110)$ as $(x_1, y_1)$ and $(3, 180)$ as $(x_2, y_2)$.

$$m = \frac{y_2 - y_1}{x_2 - x_1} = \frac{180 - 110}{3 - 2} = \frac{70}{1} = 70$$

Thus, the slope of the blue line segment is 70.

Now Try Exercise 71

## 3 Examine the Slopes of Horizontal and Vertical Lines

Consider the graph of $y = 5$ (**Fig. 4.35**). What is its slope?

The graph is parallel to the $x$-axis and goes through the points $(2, 5)$ and $(6, 5)$. Select $(6, 5)$ as $(x_2, y_2)$ and $(2, 5)$ as $(x_1, y_1)$. Then the slope of the line is

$$m = \frac{y_2 - y_1}{x_2 - x_1} = \frac{5 - 5}{6 - 2} = \frac{0}{4} = 0$$

Since there is no change in $y$, this line has a slope of 0. Note that *any* two points on the line would yield the same slope, 0.

FIGURE 4.35

> ### Slope of a Horizontal Line
>
> Every horizontal line has a slope of 0.

Consider the graph of $x = 2$ (**Fig. 4.36**). What is its slope?

The graph is parallel to the $y$-axis and goes through the points $(2, 1)$ and $(2, 4)$. Select $(2, 4)$ as, $(x_2, y_2)$ and $(2, 1)$ as $(x_1, y_1)$. Then the slope of the line is

$$m = \frac{y_2 - y_1}{x_2 - x_1} = \frac{4 - 1}{2 - 2} = \frac{3}{0}$$

We learned in Section 1.8 that $\frac{3}{0}$ is undefined. Thus, we say that the slope of this line is undefined.

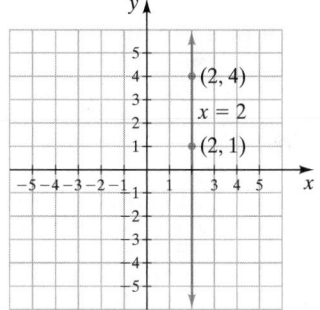

FIGURE 4.36

> ### Slope of a Vertical Line
>
> The slope of any vertical line is undefined.

## 4 Examine the Slopes of Parallel and Perpendicular Lines

> ### Parallel Lines
>
> Two lines are **parallel** when they lie in the same plane but do not intersect.

**Figure 4.37** illustrates two parallel lines.

If we compute the slope of line 1 using the given points, we obtain a slope of 3. If we compute the slope of line 2, we obtain a slope of 3. Notice both lines have the same slope. Any two nonvertical lines that have the same slope are parallel lines.

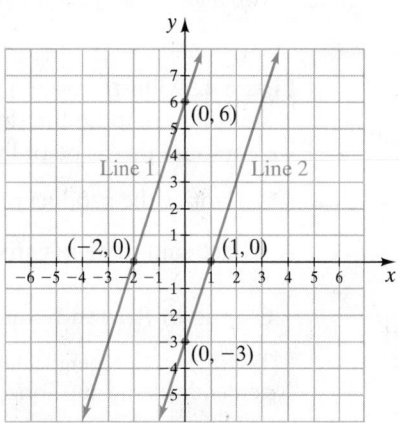

FIGURE 4.37

### Slopes of Parallel Lines

Two lines are parallel when they have the same slope and different *y*-intercepts. Any two vertical lines are parallel to each other.

## EXAMPLE 5

**a)** Draw a line with a slope of $\frac{1}{2}$ through the point $(2, 3)$.

**b)** On the same set of axes, draw a line with a slope of $\frac{1}{2}$ through the point $(-1, -3)$.

**c)** Are the two lines in parts **a)** and **b)** parallel? Explain.

*Solution*

**a)** Place a dot at $(2, 3)$. Because the slope is a positive $\frac{1}{2}$, from the point $(2, 3)$ move *up* 1 unit and to the *right* 2 units to get a second point. Draw a line through the two points; see the blue line in **Figure 4.38**.

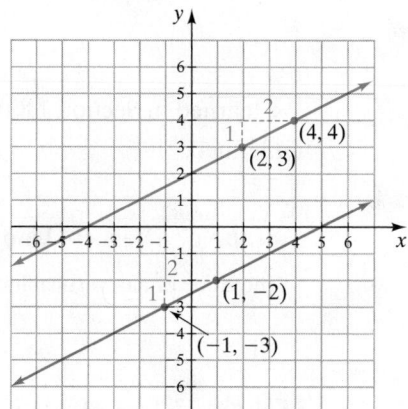

**FIGURE 4.38**

**b)** Place a dot at $(-1, -3)$. From the point $(-1, -3)$ move up 1 unit and to the right 2 units to get a second point. Draw a line through the two points; see the red line in **Figure 4.38**.

**c)** The lines appear to be parallel on the graph. Since both lines have the same slope, $\frac{1}{2}$, they are parallel lines.

Now Try Exercise 73

### Perpendicular Lines

Two lines are **perpendicular** when they intersect and form right $(90°)$ angles.

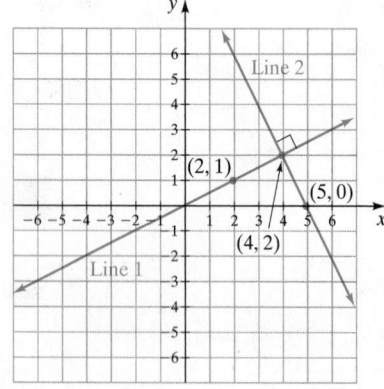

**FIGURE 4.39**

**Figure 4.39** illustrates two perpendicular lines.

If we compute the slope of line 1 using the given points, we obtain a slope of $\frac{1}{2}$. If we compute the slope of line 2 using the given points, we obtain a slope of $-2$.

Notice the product of their slopes, $\frac{1}{2}(-2)$, is $-1$. Any two numbers whose product is $-1$ are said to be **negative reciprocals** of each other.

### Slopes of Perpendicular Lines

Two lines are perpendicular when their slopes are negative reciprocals. Any vertical line is perpendicular to any horizontal line.

EXAMPLE 6

a) Draw a line with a slope of $-3$ through the point $(2, 3)$.

b) On the same set of axes, draw a line with a slope of $\dfrac{1}{3}$ through the point $(-1, -3)$.

c) Are the two lines in parts **a)** and **b)** perpendicular? Explain.

*Solution*

a) Place a dot at $(2, 3)$. A slope of $-3$ means $\dfrac{-3}{1}$. So, from the point $(2, 3)$ move *down* 3 units and to the *right* 1 unit to get a second point. Draw a line through the two points; see the blue line in **Figure 4.40**.

b) Place a dot at $(-1, -3)$. The slope is $\dfrac{1}{3}$, so from this point move up 1 unit and to the *right* 3 units. Draw a line through the two points; see the red line in **Figure 4.40**.

**Understanding Algebra**

In general, if $m$ represents a number, its negative reciprocal will be $-\dfrac{1}{m}$ because
$$m\left(-\dfrac{1}{m}\right) = -1.$$

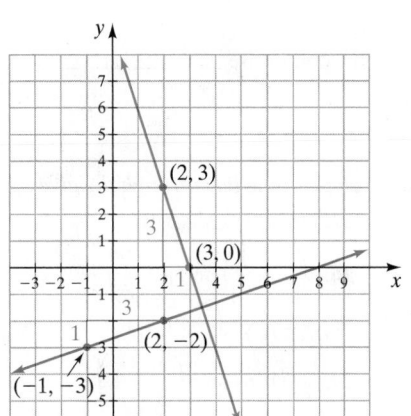

**FIGURE 4.40**

c) The lines appear to be perpendicular on the graph. To determine if they are perpendicular, multiply the slopes of the two lines together.

$$(-3)\left(\dfrac{1}{3}\right) = -1$$

Since the slopes are negative reciprocals, the two lines are perpendicular.

Now Try Exercise 75

EXAMPLE 7   If $m_1$ represents the slope of line 1 and $m_2$ represents the slope of line 2, determine if line 1 and line 2 are parallel, perpendicular or neither.

a) $m_1 = \dfrac{5}{6}, m_2 = \dfrac{5}{6}$      b) $m_1 = \dfrac{2}{5}, m_2 = 4$      c) $m_1 = \dfrac{3}{5}, m_2 = -\dfrac{5}{3}$

*Solution*

a) Since the slopes are the same, both $\dfrac{5}{6}$, the lines are parallel.

b) Since the slopes are not the same, the lines are not parallel. Since $m_1 \cdot m_2 = \left(\dfrac{2}{5}\right)(4) \neq -1$, the slopes are not negative reciprocals and the lines are not perpendicular. Thus the answer is neither.

c) Since the slopes are not the same, the lines are not parallel. Since $m_1 \cdot m_2 = \dfrac{3}{5}\left(-\dfrac{5}{3}\right) = -1$, the slopes are negative reciprocals and the lines are perpendicular.

Now Try Exercise 53

# EXERCISE SET 4.3    MathXL°    MyMathLab°
MathXL°    MyMathLab

## Warm-Up Exercises

*Fill in the blanks with the appropriate word, phrase, or symbol(s) from the following list.*

| | | | |
|---|---|---|---|
| horizontal | over | vertical | equal |
| perpendicular | run | opposites | ratio |
| rise | parallel | product | negative reciprocals |

1. The vertical change between two points on a line is known as the _____ .

2. The horizontal change between two points on a line is known as the _____ .

3. Slope is the _____ of the vertical change to the horizontal change.

4. Slope is often described with the phrase rise _____ run.

5. Every _____ line has a slope of 0.

6. Every _____ line has an undefined slope.

7. Two lines are _____ when they lie in same plane but do not intersect.

8. Two lines are _____ when they intersect and form right angles.

9. The slopes of parallel lines are _____ .

10. The slopes of perpendicular lines are _____ .

## Practice the Skills

*Using the slope formula, find the slope of the line through the given points.*

11. $(2, 1)$ and $(5, 7)$

12. $(-2, 3)$ and $(3, 8)$

13. $(8, 0)$ and $(4, -2)$

14. $(-3, 2)$ and $(-6, 1)$

15. $(3, 5)$ and $(-1, 5)$

16. $\left(-4, \frac{1}{2}\right)$ and $\left(2, \frac{1}{2}\right)$

17. $(5, -6)$ and $(8, -3)$

18. $(9, 3)$ and $(5, -6)$

19. $(6, 4)$ and $(6, 2)$

20. $(-7, 8)$ and $(3, -1)$

21. $(6, 0)$ and $(-2, 3)$

22. $(-2, 3)$ and $(-2, -5)$

23. $\left(0, \frac{5}{2}\right)$ and $\left(-\frac{3}{4}, 2\right)$

24. $(-1, 8)$ and $\left(\frac{1}{3}, -1\right)$

*By observing the vertical and horizontal change of the line between the two points indicated, determine the slope of each line.*

25.

26.

27.

28.

29.

30.

31.

32.

33.

**34.**

**35.**

**36.**

**37.**

**38.**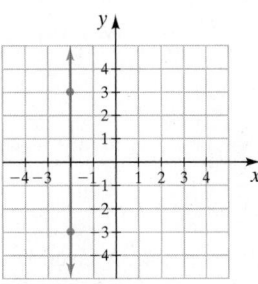

*In Exercises 39–48, graph the line with the given slope that goes through the given point.*

**39.** Through $(3, -1)$ with $m = 2$.

**40.** Through $(-1, -2)$ with $m = -2$

**41.** Through $(0, -2)$ with $m = \dfrac{1}{2}$

**42.** Through $(0, 2)$ with $m = -\dfrac{1}{2}$

**43.** Through $(0, 0)$ with $m = -\dfrac{1}{3}$

**44.** Through $(-3, 4)$ with $m = -\dfrac{2}{3}$

**45.** Through $(-3, 2)$ with $m = 0$

**46.** Through $(-1, 3)$ with $m = 0$

**47.** Through $(2, -2)$ with slope undefined

**48.** Through $(-1, 5)$ with slope undefined

*In Exercises 49–64, $m_1$ represents the slope of line 1, and $m_2$ represents the slope of the distinct line, line 2. Indicate whether line 1 and line 2 are parallel, perpendicular, or neither.*

**49.** $m_1 = 5, m_2 = 5$

**50.** $m_1 = -1, m_2 = -1$

**51.** $m_1 = 7, m_2 = -7$

**52.** $m_1 = \dfrac{3}{2}, m_2 = -4$

**53.** $m_1 = \dfrac{2}{3}, m_2 = -\dfrac{3}{2}$

**54.** $m_1 = \dfrac{1}{4}, m_2 = -4$

**55.** $m_1 = 6, m_2 = \dfrac{2}{3}$

**56.** $m_1 = -\dfrac{1}{3}, m_2 = -3$

**57.** $m_1 = \dfrac{1}{4}, m_2 = 4$

**58.** $m_1 = 6, m_2 = -\dfrac{1}{6}$

**59.** $m_1 = 0, m_2 = 0$

**60.** $m_1 = 0, m_2 = -\dfrac{2}{5}$

**61.** $m_1$ is undefined, $m_2$ is undefined

**62.** $m_1 = 0, m_2$ is undefined

**63.** $m_1$ is undefined, $m_2 = 0$

**64.** $m_1 = 5, m_2 = -5$

**65.** The slope of a given line is 3. If a line is to be drawn parallel to the given line, what will be its slope?

**66.** The slope of a given line is $-2$. If a line is to be drawn parallel to the given line, what will be its slope?

**67.** The slope of a given line is $-4$. If a line is to be drawn perpendicular to the given line, what will be its slope?

**68.** The slope of a given line is 5. If a line is to be drawn perpendicular to the given line, what will be its slope?

## Problem Solving

*In Exercises 69 and 70, determine which line (the first or second) has the greater slope. Explain your answer. Notice that the scales on the x- and y-axes are different.*

**69.**

**70.**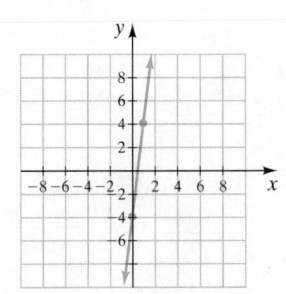

*In Exercises 71 and 72, find the slope of the line segments indicated in* **a)** *red and* **b)** *blue.*

**71.**

**Driving to a Baseball Game**

**72.**

**Stopping Distance on Wet Pavement for Midsize Car**

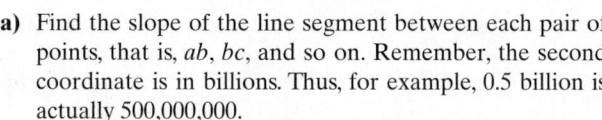
*Source:* Automobile Association of America

**73.** A given line goes through the points $(2, 6)$ and $(4, -2)$. If a line is to be drawn parallel to the given line, what will be its slope?

**74.** A given line goes through the points $(-2, 5)$ and $(4, 7)$. If a line is to be drawn parallel to the given line, what will be its slope?

**75.** A given line goes through the points $(1, -7)$ and $(2, 1)$. If a line is to be drawn perpendicular to the given line, what will be its slope?

**76.** A given line goes through the points $(-3, 0)$ and  If a line is to be drawn perpendicular to the given line, what will be its slope?

## Challenge Problems

**77.** A line contains the points $(1, b)$ and $(4, 9)$. If the line has slope $m = 2$, what is the value of $b$?

**78.** A line contains the points $(1, 2)$ and $(a, -4)$. If the line has slope $m = 3$, what is the value of $a$?

**79.** A quadrilateral (a four-sided figure) has four vertices (the points where the sides meet). Vertex $A$ is at $(0, 1)$, vertex $B$ is at $(6, 2)$, vertex $C$ is at $(5, 4)$, and vertex $D$ is at $(1, -1)$.

**a)** Graph the quadrilateral in the Cartesian coordinate system.

**b)** Find the slopes of sides $AC$, $CB$, $DB$, and $AD$.

**c)** Do you think this figure is a parallelogram? Explain.

**80.** **Population** The following graph shows the world's population estimated to the year 2016.

**World Population Growth**

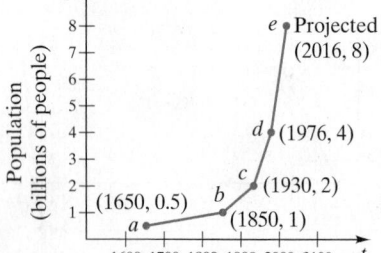

Year

**a)** Find the slope of the line segment between each pair of points, that is, $ab$, $bc$, and so on. Remember, the second coordinate is in billions. Thus, for example, 0.5 billion is actually 500,000,000.

**b)** Would you say that this graph represents a linear equation? Explain.

**81.** Consider the graph below.

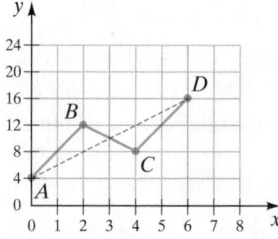

**a)** Determine the slope of each of the three solid blue lines.

**b)** Determine the average of the three slopes found in part **a)**.

**c)** Determine the slope of the red dashed line from $A$ to $D$.

**d)** Determine whether the slope of the red dashed line from $A$ to $D$ is the same as the (mean) average of the slopes of the three solid blue lines.

**e)** Explain what this example illustrates.

## Group Activity

*Discuss and answer Exercise 82 as a group, according to the instructions.*

**82.** The slope of a hill and the slope of a line both measure steepness. However, there are several important differences.

    **a)** As a group, explain how you think the slope of a hill is determined.

    **b)** Is the slope of a line, graphed in the Cartesian coordinate system, measured in any specific unit?

    **c)** Is the slope of a hill measured in any specific unit?

© Ija Masik/Shutterstock

## Cumulative Review Exercises

[1.9] **83.** Evaluate $4x^2 + 9x + \dfrac{x}{3}$ when $x = 0$.

[2.3] **84. a)** If $-x = -\dfrac{5}{2}$, what is the value of $x$?

      **b)** If $8x = 0$, what is the value of $x$?

[2.4] **85.** Solve $\dfrac{2}{3}x + \dfrac{1}{7}x = 4$.

[2.6] **86.** Solve the equation $d = a + b + c$ for $c$.

[4.2] **87.** Find the $x$- and $y$-intercepts for the line whose equation is $5x - 3y = 30$.

## Mid-Chapter Test: 4.1–4.3

*To find out how well you understand the chapter material to this point, take this brief test. The answers, and the section where the material was initially discussed, are given in the back of the book. Review any questions that you answered incorrectly.*

**1.** In which quadrant does the point $(3, -4)$ belong?

**2.** Plot the points $A(2, 6), B(-3, 1), C(-5, -2), D(0, -4), E(4, -7)$ on the same axes.

**3.** Determine which of the three ordered pairs does not satisfy the equation $\dfrac{1}{3}x + y = -2$.

    **a)** $(3, -3)$     **b)** $(0, 2)$     **c)** $(-6, 0)$

**4.** Find the value of $y$ that makes the ordered pair $(-1, y)$ a solution to the linear equation $y = 5x + 1$.

**5.** Find the value of $x$ that makes the ordered pair $(x, 2)$ a solution to the equation $3x - 4y = 1$.

**6.** What does the graph of an equation illustrate?

*Graph each equation.*

**7.** $x = \dfrac{5}{2}$

**8.** $y = -2$

*Graph by plotting points.*

**9.** $y = 3x + 1$

**10.** $y = -\dfrac{1}{2}x + 4$

*Graph using the x- and y-intercepts.*

**11.** $3x - 4y = 12$

**12.** $\dfrac{1}{2}x + \dfrac{1}{5}y = 10$

*Find the slope of the line through each pair of points.*

**13.** $(-1, 5)$ and $(6, 3)$     **14.** $(4, 2)$ and $(7, 2)$

**15.** $(-3, 0)$ and $(-3, 5)$

*Graph the line with the given slope that goes through the given point.*

**16.** Through $(-2, 3)$ with $m = -\dfrac{1}{2}$

**17.** Through $(4, 1)$ with $m = \dfrac{3}{5}$.

*Indicate whether the lines with the following slopes are parallel, perpendicular, or neither.*

**18.** $m_1 = 5$ and $m_2 = \dfrac{1}{5}$

**19.** $m_1 = \dfrac{6}{7}$ and $m_2 = -\dfrac{7}{6}$

**20. Interest** The following graph illustrates the interest obtained when $1000 is invested for 1 year at various interest rates from 0% to 10%. Determine the slope of the line in the graph.

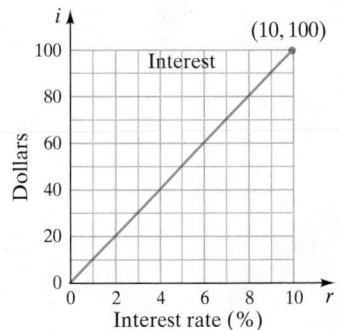

# 4.4 Slope-Intercept and Point-Slope Forms of a Linear Equation

1 Write a linear equation in slope-intercept form.

2 Graph a linear equation using the slope and $y$-intercept.

3 Use the slope-intercept form to determine the equation of a line.

4 Use the point-slope form to determine the equation of a line.

5 Compare the three methods of graphing linear equations.

In Section 4.1 we introduced the *standard form* of a linear equation, $ax + by = c$. In this section we introduce two more forms, the slope-intercept form and the point-slope form.

## 1 Write a Linear Equation in Slope-Intercept Form

A very important form of a linear equation is the **slope-intercept form, $y = mx + b$**. The graph of an equation of the form $y = mx + b$ will always be a straight line with a **slope of $m$** and a **$y$-intercept $(0, b)$**.

### Slope-Intercept Form of a Linear Equation

$$y = mx + b$$

where $m$ is the slope, and $(0, b)$ is the $y$-intercept of the line.

Slope        $y$-intercept is $(0, b)$

$$y = mx + b$$

| Equations in Slope-Intercept Form | Slope, $m$ | $y$-Intercept $(0, b)$ |
|---|---|---|
| $y = 4x - 6$ | $4$ | $(0, -6)$ |
| $y = \dfrac{1}{2}x + \dfrac{3}{2}$ | $\dfrac{1}{2}$ | $\left(0, \dfrac{3}{2}\right)$ |
| $y = -5x + 3$ | $-5$ | $(0, 3)$ |

### Writing an Equation in Slope-Intercept Form

To write a linear equation in slope-intercept form, solve the equation for $y$.

Once the equation is solved for $y$, the numerical coefficient of the $x$-term will be the slope, and the constant term will give the $y$-intercept.

**EXAMPLE 1** Write the equation $-3x + 4y = 8$ in slope-intercept form. State the slope and $y$-intercept.

**Solution**    To write this equation in slope-intercept form, we solve the equation for $y$.

$$-3x + 4y = 8$$
$$4y = 3x + 8 \qquad \text{Added } 3x \text{ to both sides.}$$
$$y = \frac{3x + 8}{4} \qquad \text{Divided both sides by 4.}$$
$$y = \frac{3}{4}x + \frac{8}{4} \qquad \text{Wrote as two fractions.}$$
$$y = \frac{3}{4}x + 2 \qquad \text{Equation in slope-intercept form.}$$

$$y = \frac{3}{4}x + 2$$
$$m = \frac{3}{4} \qquad b = 2$$

Thus, the slope is $\dfrac{3}{4}$, and the $y$-intercept is $(0, 2)$.

Now Try Exercise 11

**EXAMPLE 2** Determine whether the two equations represent lines that are parallel, perpendicular, or neither.

**a)** $2x + y = 9$

$2y = -4x + 5$

**b)** $3x - 2y = 7$

$6y + 4x = -6$

*Solution*    Two lines that have the same slope are parallel lines, and two lines whose slopes are negative reciprocals are perpendicular lines. We can determine the slope of each line by solving each equation for $y$. The coefficient of the $x$ term will be the slope.

**a)**  $2x + y = 9$

$$y = -2x + 9$$

$$m = -2$$

$2y = -4x + 5$

$$y = \frac{-4x + 5}{2}$$

$$y = -2x + \frac{5}{2}$$

$$m = -2$$

Since both equations have the same slope, $-2$, the equations represent lines that are parallel. Notice the equations represent two different lines because their $y$-intercepts are different.

**b)**  $3x - 2y = 7$

$$-2y = -3x + 7$$

$$y = \frac{-3x + 7}{-2}$$

$$y = \frac{3}{2}x - \frac{7}{2}$$

$$m = \frac{3}{2}$$

$6y + 4x = -6$

$$6y = -4x - 6$$

$$y = \frac{-4x - 6}{6}$$

$$y = -\frac{2}{3}x - 1$$

$$m = -\frac{2}{3}$$

The slope of one line is $\frac{3}{2}$ and the slope of the other line is $-\frac{2}{3}$. Since the slopes are negative reciprocals, the lines are perpendicular.

Now Try Exercise 39

## 2  Graph a Linear Equation Using the Slope and $y$-Intercept

In Section 4.2 we discussed graphing a linear equation by (1) plotting points and by (2) using the $x$- and $y$-intercepts. Now we present a third method. This method makes use of the slope and the $y$-intercept. We graph equations using the slope and $y$-intercept in a manner very similar to the way we worked Examples 5 and 6 in Section 4.3.

---

**To Graph Linear Equations Using the Slope and $y$-intercept**

1. If necessary, solve the equation for $y$.
2. Plot the $y$-intercept.
3. Use the slope to find two more points on the line.
4. Using a straightedge, draw a straight line through the three points. Draw an arrowhead on each end of the line.

---

**EXAMPLE 3** Graph $-3x + 4y = 8$ by using the slope and $y$-intercept.

*Solution*    In Example 1 we solved $-3x + 4y = 8$ for $y$. We found that

$$y = \frac{3}{4}x + 2$$

The slope of the line is $\frac{3}{4}$ and the $y$-intercept is $(0, 2)$. We plot the $y$-intercept at 2 on

the $y$-axis (**Fig. 4.41**). Now we use the slope, $\frac{3}{4}$, to find a second point. Since the slope

is positive, we move 3 units up and 4 units to the right to find a second point at $(4, 5)$.
We continue this process to obtain a third point at $(8, 8)$. Now we draw a straight line
through the three points.

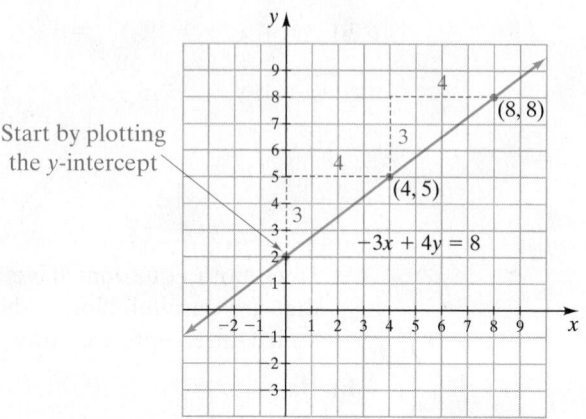

FIGURE 4.41

Now Try Exercise 19

**EXAMPLE 4**  Graph $5x + 3y = 12$ by using the slope and $y$-intercept.

*Solution*    First, solve the equation for $y$.

$$5x + 3y = 12$$
$$3y = -5x + 12$$
$$y = \frac{-5x + 12}{3}$$
$$= -\frac{5}{3}x + 4$$

Thus, the slope is $-\frac{5}{3}$ and the $y$-intercept is $(0, 4)$. Plot the $y$-intercept at 4 on the

$y$-axis (**Fig. 4.42**). Then, since $-\frac{5}{3} = \frac{-5}{3}$, move 5 units down and 3 units to the right to

determine the next point. You can follow this procedure again to obtain a third point.
Finally, draw the straight line between the plotted points.

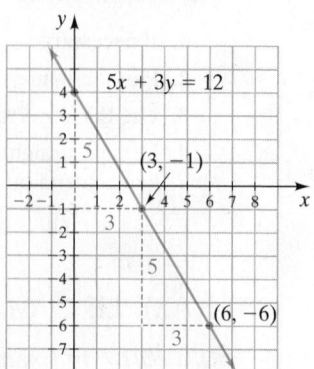

FIGURE 4.42

Now Try Exercise 21

### 3 Use the Slope-Intercept Form to Determine the Equation of a Line

We can also use the slope-intercept form to write the equation of a given line. To do so, we need to determine the slope, $m$, and the $y$-intercept, $(0, b)$. We then can substitute $m$ and $b$ into $y = mx + b$ to get the equation of the given line.

**EXAMPLE 5**  Determine the equation of the line shown in **Figure 4.43.**

*Solution*    The graph shows that the $y$-intercept is at $-5$ so $b = -5$. Since the line falls from left to right it has a negative slope. Looking at the point $(0, -5)$ and $(-2, 1)$, we can see that the rise is 6 and the run is $-2$. So,

$$m = \frac{\text{rise}}{\text{run}} = \frac{6}{-2} = -3$$

Substituting $-3$ for $m$ and $-5$ for $b$ into the slope-intercept form gives us the equation $y = -3x - 5$, which is the equation of the line shown.

Now Try Exercise 29

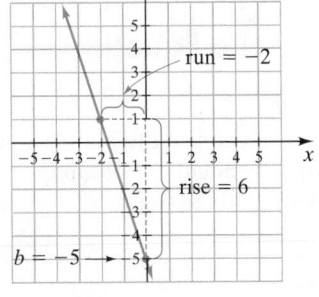

**FIGURE 4.43**

**EXAMPLE 6  Artistic Vases** Kris, a pottery artist, makes ceramic vases that he sells at art shows. His business has a fixed monthly cost (booth rental, advertising, cell phone, etc.) and a variable cost per vase made (cost of materials, cost of labor, cost for kiln use, etc.). The total monthly cost for making $x$ vases is illustrated in the graph in **Figure 4.44**.

**FIGURE 4.44**

**a)** Find the equation of the total monthly cost when $x$ vases are made.

**b)** Use the equation found in part **a)** to find the total monthly cost if 30 vases are made.

**c)** Use the graph in **Figure 4.44** to see whether your answer in part **b)** appears correct.

*Solution*

**a)** Understand and Translate    Notice that the vertical axis is cost, $C$, and not $y$. Since $y$ is replaced by $C$, we will use $C = mx + b$, in which $b$ is where the graph crosses the vertical or $C$-axis. Note that the graph crosses the vertical axis at 200. Thus, $b$ is 200. Now use the points $(0, 200)$ and $(50, 1450)$ to find the slope.

Carry Out
$$m = \frac{y_2 - y_1}{x_2 - x_1}$$
$$= \frac{1450 - 200}{50 - 0} = \frac{1250}{50} = 25$$

Answer    The slope is 25. So $m = 25$ and $b = 200$ and the equation in slope-intercept form is

$$C = mx + b$$
$$= 25x + 200$$

**b)** To find the monthly cost when 30 vases are made, we substitute 30 for $x$.

$$C = 25x + 200$$
$$= 25(30) + 200$$
$$= 750 + 200 = 950$$

The monthly cost when 30 vases are made is $950.

**c)** If we draw a vertical line up from 30 on the $x$-axis (the red line), we see that the corresponding cost is about $950. Thus, our answer in part **b)** appears correct.

Now Try Exercise 65

## Understanding Algebra

Forms of Linear Equations

1. Standard Form

   $ax + by = c$

2. Slope-Intercept Form

   $y = mx + b$

3. Point-Slope Form

   $y - y_1 = m(x - x_1)$

## 4  Use the Point-Slope Form to Determine the Equation of a Line

When the slope of a line and a point on the line are known, we can use the point-slope form to determine the equation of the line. The **point-slope form** can be obtained by beginning with the slope between any selected point $(x, y)$ and a fixed point $(x_1, y_1)$ on a line.

$$m = \frac{y - y_1}{x - x_1} \quad \text{or} \quad \frac{m}{1} = \frac{y - y_1}{x - x_1}$$

Now cross-multiply to obtain

$$m(x - x_1) = y - y_1 \quad \text{or} \quad y - y_1 = m(x - x_1)$$

---

**Point-Slope Form of a Linear Equation**

$$y - y_1 = m(x - x_1)$$

where $m$ is the slope of the line and $(x_1, y_1)$ is a point on the line.

---

EXAMPLE 7  Write an equation, in slope-intercept form, of the line that goes through the point $(1, 3)$ and has a slope of 2.

*Solution*  Since we are given a point on the line and the slope of the line, we begin by writing the equation in point-slope form. Since the slope is 2 and the point on the line is $(1, 3)$, we have $m = 2$, $x_1 = 1$, and $y_1 = 3$. We substitute into the point-slope form and solve for y as follows.

$$y - y_1 = m(x - x_1)$$
$$y - 3 = 2(x - 1) \qquad \text{Equation in point-slope form}$$
$$y - 3 = 2x - 2 \qquad \text{Distributive property}$$
$$y = 2x + 1 \qquad \text{Equation in slope-intercept form}$$

Now Try Exercise 51

EXAMPLE 8  Write an equation, in slope-intercept form, of the line that goes through the point $(1, -2)$ and has a slope of $\frac{3}{4}$.

*Solution*  We will begin with the point-slope form of a line, where $m = \frac{3}{4}$, $x_1 = 1$, and $y_1 = -2$.

**Understanding Algebra**

To simplify an equation that contains a fraction, multiply both sides of the equation by the denominator of the fraction.

$$y - y_1 = m(x - x_1)$$

$$y - (-2) = \frac{3}{4}(x - 1) \qquad \text{Equation in point-slope form}$$

$$4(y + 2) = 4 \cdot \frac{3}{4}(x - 1) \quad \text{Multiply both sides by 4.}$$

$$4y + 8 = 3(x - 1) \qquad \text{Distributive property (left side of equation)}$$

$$4y + 8 = 3x - 3 \qquad \text{Distributive property (right side of equation)}$$

$$4y = 3x - 11 \qquad \text{Subtracted 8 from both sides.}$$

$$y = \frac{3x - 11}{4} \qquad \text{Divided both sides by 4.}$$

$$y = \frac{3}{4}x - \frac{11}{4} \qquad \text{Equation in slope-intercept form}$$

Now Try Exercise 53

**Helpful Hints**

We have discussed three forms of a linear equation. We summarize the three forms below. It is important that you memorize these forms.

| Standard Form | Examples |
|---|---|
| $ax + by = c$ | $2x - 3y = 8$ |
|  | $-5x + y = -2$ |

| Slope-Intercept Form | Examples |
|---|---|
| $y = mx + b$ | $y = 2x - 5$ |
| $m$ is the slope, $(0, b)$ is the $y$-intercept | $y = -\frac{3}{2}x + 2$ |

| Point-Slope Form | Examples |
|---|---|
| $y - y_1 = m(x - x_1)$ | $y - 3 = 2(x + 4)$ |
| $m$ is the slope, $(x_1, y_1)$ is a point on the line | $y + 5 = -4(x - 1)$ |

We now discuss how to use the point-slope form to determine the equation of a line when two points on the line are known.

**EXAMPLE 9** Find an equation of the line through the points $(-1, 3)$ and $(-5, 1)$. Write the equation in slope-intercept form.

*Solution*   To use the point-slope form, we must first find the slope of the line through the two points.

$$m = \frac{y_2 - y_1}{x_2 - x_1} = \frac{1 - 3}{-5 - (-1)} = \frac{1 - 3}{-5 + 1} = \frac{-2}{-4} = \frac{1}{2}$$

Thus $m = \frac{1}{2}$. We can use either point in determining the equation of the line.

This example will be worked out using each of the points to show that the solutions obtained are identical.

Using the point $(-1, 3)$ as $(x_1, y_1)$,

$$y - y_1 = m(x - x_1)$$

$$y - 3 = \frac{1}{2}[x - (-1)]$$

$$y - 3 = \frac{1}{2}(x + 1)$$

$$2 \cdot (y - 3) = 2 \cdot \frac{1}{2}(x + 1) \quad \text{Multiply both sides by the LCD, 2.}$$

$$2y - 6 = x + 1$$

$$2y = x + 7$$

$$y = \frac{x + 7}{2} \quad \text{or} \quad y = \frac{1}{2}x + \frac{7}{2}$$

Using the point $(-5, 1)$ as $(x_1, y_1)$,

$$y - y_1 = m(x - x_1)$$

$$y - 1 = \frac{1}{2}[x - (-5)]$$

$$y - 1 = \frac{1}{2}(x + 5)$$

$$2 \cdot (y - 1) = 2 \cdot \frac{1}{2}(x + 5) \quad \text{Multiply both sides by the LCD, 2.}$$

$$2y - 2 = x + 5$$

$$2y = x + 7$$

$$y = \frac{x + 7}{2} \quad \text{or} \quad y = \frac{1}{2}x + \frac{7}{2}$$

Note that the equations are identical.

Now Try Exercise 57

---

**Helpful Hints**

In the Exercises, you will be asked to write a linear equation in slope-intercept form. Even though you will eventually write the equation in slope-intercept form, you may need to start your work with the point-slope form. Below we indicate the initial form to use to solve the problem.

1. Begin with the **slope-intercept form** if you know
   • The slope of the line and the $y$-intercept
2. Begin with the **point-slope form** if you know
   a) The slope of the line and a point on the line other than the $y$-intercept or
   b) Two points on the line (first find the slope, then use the point-slope form)

---

**5** **Compare the Three Methods of Graphing Linear Equations**

We have discussed three methods to graph a linear equation: (1) plotting points, (2) using the $x$- and $y$-intercepts, and (3) using the slope and $y$-intercept. No single method is always the easiest to use. If the equation is given in slope-intercept form, $y = mx + b$, then graphing by plotting points or by using the slope and $y$-intercept might be easier. If the equation is given in standard form, $ax + by = c$, then graphing using the intercepts might be easier.

**EXAMPLE 10** Graph $3x - 2y = 8$

a) by plotting points;

b) using the $x$- and $y$-intercepts;

c) using the slope and $y$-intercept.

*Solution*    For parts a) and c) we will write the equation in slope-intercept form by solving for $y$.

$$3x - 2y = 8$$

$$-2y = -3x + 8$$

$$y = \frac{-3x + 8}{-2} = \frac{3}{2}x - 4$$

a) **Plotting Points** We substitute values for $x$ and find the corresponding values of $y$. Three ordered pairs are indicated in the following table. Next we plot the ordered pairs and draw the graph (**Fig. 4.45**).

$$y = \frac{3}{2}x - 4$$

| $x$ | $y$ |
|---|---|
| 0 | $-4$ |
| 2 | $-1$ |
| 4 | 2 |

FIGURE 4.45

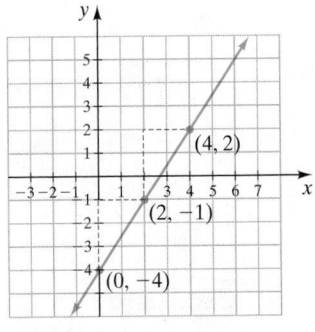

FIGURE 4.46

b) **Intercepts** We find the $x$- and $y$-intercepts and a checkpoint. Then we plot the points and draw the graph (**Fig. 4.46**).

$$3x - 2y = 8$$

| $x$-Intercept | $y$-Intercept | Checkpoint |
|---|---|---|
| Let $y = 0$. | Let $x = 0$. | Let $x = 2$. |
| $3x - 2y = 8$ | $3x - 2y = 8$ | $3x - 2y = 8$ |
| $3x - 2(0) = 8$ | $3(0) - 2y = 8$ | $3(2) - 2y = 8$ |
| $3x = 8$ | $-2y = 8$ | $6 - 2y = 8$ |
| $x = \dfrac{8}{3}$ | $y = -4$ | $-2y = 2$ |
|  |  | $y = -1$ |

The three ordered pairs are $\left(\dfrac{8}{3}, 0\right)$, $(0, -4)$, and $(2, -1)$.

c) **Slope and $y$-intercept** The $y$-intercept is $(0, -4)$; therefore, we place a point at $-4$ on the $y$-axis. Since the slope is $\dfrac{3}{2}$, we obtain a second point by moving 3 units up and 2 units to the right. The graph is illustrated in **Figure 4.47**. Notice that we get the same line by all three methods.

*Now Try Exercise 27*

FIGURE 4.47

---

# EXERCISE SET 4.4

MathXL®    MyMathLab®

MathXL®    MyMathLab

## Warm-Up Exercises

*Fill in the blanks with the appropriate word, phrase, or symbol(s) from the following list.*

| numerator | $y$-intercept | standard | point-slope | slope |
|---|---|---|---|---|
| slope-intercept | denominator | point | $y$ | |

1. The _____ form of a linear equation is $y - y_1 = m(x - x_1)$.

2. The _____ form of a linear equation is $ax + by = c$.

3. The _____ form of a linear equation is $y = mx + b$.

4. To write a linear equation in slope-intercept form, solve the equation for _____.

5. When graphing a linear equation using the slope and $y$-intercept, you first plot the _____.

6. To simplify an equation that contains a fraction, multiply both sides of the equation by the _____ of the fraction.

**7.** If you are asked to find the equation of a line and if you know the _____ and the $y$-intercept, you should use the slope-intercept form.

**8.** If you are asked to find the equation of a line and if you know the _____ and the slope, you should use the point-slope form.

## Practice the Skills

*Determine the slope and y-intercept of the line represented by the given equation.*

**9.** $y = 3x + 1$

**10.** $y = -3x + 17$

**11.** $4x - 3y = 21$

**12.** $7x = 5y + 25$

*Determine the slope and y-intercept of the line represented by each equation. Graph the line using the slope and y-intercept.*

**13.** $y = x - 3$

**14.** $y = -x + 5$

**15.** $y = 3x + 2$

**16.** $3x + y = 4$

**17.** $y = 2x$

**18.** $y = -4x$

**19.** $-2x + y = -3$

**20.** $3x + 3y = 9$

**21.** $5x - 2y = 10$

**22.** $-x + 2y = 8$

**23.** $6x + 12y = 18$

**24.** $16y = 8x + 32$

**25.** $-6x + 2y - 8 = 0$

**26.** $4x = 6y + 9$

**27.** $3x = 2y - 4$

**28.** $20x = 80y + 40$

*Determine the equation of each line.*

**29.**

**30.**

**31.**

**32.**

**33.**

**34.**

**35.**

**36.**
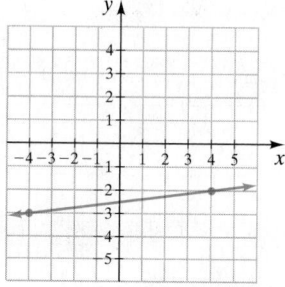

*Determine whether each pair of lines are parallel, perpendicular, or neither.*

**37.** $y = 3x - 2$
$\quad y = 3x + 2$

**38.** $2x + 3y = 1$
$\quad y = -\dfrac{2}{3}x + 3$

**39.** $4x + 2y = 7$
$\quad 4x = 8y + 12$

**40.** $3x - 5y = 7$
$\quad 5y + 3x = 4$

**41.** $3x + 5y = 9$
$\quad 6x = -10y + 9$

**42.** $8x + 2y = 10$
$\quad x - 7 = 4y$

**43.** $y = \dfrac{1}{2}x - 2$
$\quad 2y = 6x + 9$

**44.** $3y - 4 = -5x$
$\quad y = -\dfrac{5}{3}x - 3$

**45.** $5y = 2x + 9$
$\quad -10x = 4y + 11$

**46.** $3x - 9y = 21$
$\quad -3x + 9y = 27$

**47.** $3x + 7y = 21$
$\quad 7x + 3y = 21$

**48.** $5x - 6y = 18$
$\quad -6x + 5y = 10$

## Problem Solving

*Write the equation of each line, with the given properties, in slope-intercept form.*

**49.** Slope = 3, through $(0, 2)$

**50.** Slope = 2, through $(4, 3)$

**51.** Slope = $-3$, through $(-4, 5)$

**52.** Slope = $-3$, through $(2, 0)$

**53.** Slope = $\frac{1}{2}$, through $(-1, -3)$

**54.** Slope = $-\frac{2}{3}$, through $(4, -5)$

**55.** Slope = $\frac{2}{3}$, $y$-intercept is $(0, 6)$

**56.** Slope = $\frac{1}{9}$, $y$-intercept is $\left(0, -\frac{2}{5}\right)$

**57.** Through $(-4, -2)$ and $(-2, 4)$

**58.** Through $(7, 4)$ and $(6, 3)$

**59.** Through $(-6, 9)$ and $(8, -12)$

**60.** Through $(3, 0)$ and $(-3, 5)$

**61.** Through $(10, 3)$ and $(0, -2)$

**62.** Through $(-6, -2)$ and $(5, -3)$

**63.** Slope = $7.4$, $y$-intercept is $(0, -4.5)$

**64.** Slope = $-\frac{7}{8}$, $y$-intercept is $\left(0, -\frac{3}{10}\right)$

**65. Weight Loss Clinic** Stacy Best owns a weight loss clinic. She charges her clients a one-time membership fee. She also charges per pound of weight lost. Therefore, the more successful she is at helping clients lose weight, the more income she will receive. The following graph shows a client's cost of losing weight.

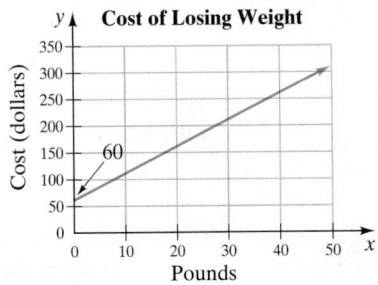

**a)** Find the equation that represents the cost for a client who loses $x$ pounds.

**b)** Use the equation found in part **a)** to determine the cost for a client who loses 30 pounds.

**66. Submarine Submerges** A submarine is submerged below sea level. Tom Johnson, the captain, orders the ship to dive slowly. The following graph illustrates the submarine's depth at a time $t$ minutes after the submarine begins to dive.

See Exercise 66.

**a)** Find the equation that represents the depth at time $t$.

**b)** Use the equation found in part **a)** to find the submarine's depth after 30 minutes.

## Concept/Writing Exercises

**67.** Suppose that you were asked to write the equation of a line with the properties given below. Which form of a linear equation—standard form, slope-intercept form, or point-slope form—would you start with? Explain your answer.

**a)** The slope of the line and the $y$-intercept of the line

**b)** The slope and a point on the line

**c)** Two points on the line

**68.** Consider the two equations $20x - 30y = 50$ and $-20x + 30y = 40$.

**a)** When these equations are graphed, will the two lines have the same slope? Explain how you determined your answer.

**b)** When these two equations are graphed, will they be parallel lines?

**69.** Assume the slope of a line is 2 and two points on the line are $(-5, -4)$ and $(3, 12)$.

**a)** If you use $(-5, -4)$ as $(x_1, y_1)$ and then $(3, 12)$ as $(x_1, y_1)$ will the appearance of the two equations be the same in point-slope form? Explain.

**b)** Find the equation, in point-slope form, using $(-5, -4)$ as $(x_1, y_1)$.

**c)** Find the equation, in point-slope form, using $(3, 12)$ as $(x_1, y_1)$.

**d)** Write the equation obtained in part **b)** in slope-intercept form.

e) Write the equation obtained in part **c)** in slope-intercept form.

f) Are the equations obtained in parts **d)** and **e)** the same? If not, explain why.

**70.** Assume the slope of a line is $-3$ and two points on the line are $(-1,8)$ and $(2,-1)$.

  a) If you use $(-1,8)$ as $(x_1, y_1)$ and then $(2,-1)$ as $(x_1, y_1)$ will the appearance of the two equations be the same in point-slope form? Explain.

  b) Find the equation, in point-slope form, using $(-1,8)$ as $(x_1, y_1)$.

c) Find the equation, in point-slope form, using $(2,-1)$ as $(x_1, y_1)$.

d) Write the equation obtained in part **b)** in slope-intercept form.

e) Write the equation obtained in part **c)** in slope-intercept form.

f) Are the equations obtained in parts **d)** and **e)** the same? If not, explain why.

## Challenge Problems

**71. Unit Conversions** The following graph shows the approximate relationship between speed in miles per hour and feet per second.

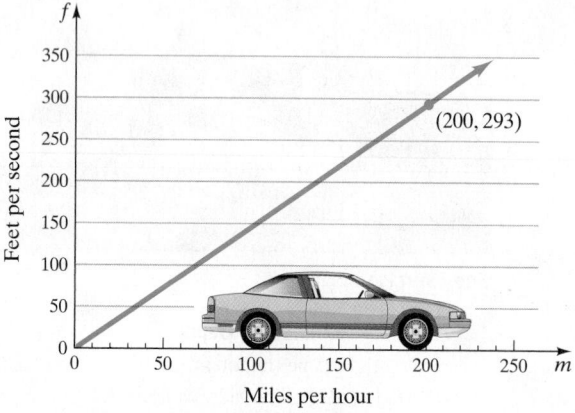

Miles per hour

a) Determine the slope of the line.

b) Determine the equation of the line.

c) At the 2008 Daytona 500, Ryan Newman, the winner, had an average speed of 152.7 miles per hour. Use the equation you obtained in part **b)** to determine the speed in feet per second.

d) Use the graph to estimate a speed of 100 miles per hour in feet per second.

e) Use the graph to estimate a speed of 80 feet per second in miles per hour.

**72. Temperature** The following graph shows the relationship between Fahrenheit temperature and Celsius temperature.

a) Determine the slope of the line.

b) Determine the equation of the line in slope-intercept form.

c) Use the equation (or formula) you obtained in part **b)** to find the Fahrenheit temperature when the Celsius temperature is $20°$.

d) Use the graph to estimate the Celsius temperature when the Fahrenheit temperature is $100°$.

e) Estimate the Celsius temperature that corresponds to a Fahrenheit temperature of $0°$.

**73.** Determine the equation of the line with $y$-intercept at 5 that is parallel to the line whose equation is $2x + y = 6$. Explain how you determined your answer.

**74.** Will a line through the points $(60, 30)$ and $(20, 90)$ be parallel to the line with $x$-intercept at 2 and $y$-intercept at 3? Explain how you determined your answer.

**75.** Write an equation of the line parallel to the graph of $3x - 4y = 6$ that passes through the point $(-8, -1)$.

**76.** Determine the equation of the straight line that intersects the greatest number of shaded points on the following graph.

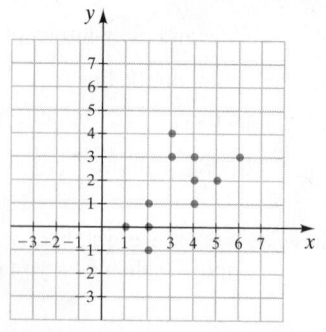

## Group Activity

*Discuss and answer Exercise 77 as a group, according to the instructions.*

**77.** Consider the equation $-3x + 2y = 4$.

**a)** Group member 1: Explain how to graph this equation by plotting points. Then graph the equation by plotting points.

**b)** Group member 2: Explain how to graph this equation using the intercepts. Then graph the equation using the intercepts.

**c)** Group member 3: Explain how to graph this equation using the slope and $y$-intercept. Then graph the equation using the slope and $y$-intercept.

**d)** As a group, compare your graphs. Did you all obtain the same graph? If not, determine why.

## Cumulative Review Exercises

[1.5]   **78.** Insert either $>$, $<$, or $=$ in the shaded area to make the statement true: $|-4|$ ▨ $|-9|$.

[1.8] *Indicate whether each statement is true or false.*

**79.** The product of two negative numbers is always a positive number.

**80.** The sum of two negative numbers is always a negative number.

**81.** The difference of two negative numbers is always a negative number.

**82.** The quotient of two negative numbers is always a negative number.

[1.9]   **83.** Evaluate $4^3$.

[2.6]   **84.** Solve $i = prt$ for $r$.

# Chapter 4 Summary

| IMPORTANT FACTS AND CONCEPTS | EXAMPLES |
|---|---|

### Section 4.1

| The **Cartesian coordinate system** is formed by two axes drawn perpendicular to each other. The point of intersection is called the **origin**. The horizontal axis is called the **x-axis**. The vertical axis is called the **y-axis.** **Ordered pairs** are of the form $(x, y)$. | **Cartesian coordinate system**  |
| A **linear equation in two variables** is an equation that can be put in the form $$ax + by = c$$ where $a, b$, and $c$ are real numbers. This form is also called the **standard form** for a linear equation. | $3x + 7y = 2, \quad -2x - y = 9$ |
| A **graph** of an equation in two variables is an illustration of a set of points whose coordinates satisfy the equation. Points that lie in a straight line are **collinear**. | Every point on the graph satisfies the equation $y = 2x - 1$. The points on the graph are collinear and the graph is a straight line.  |

| IMPORTANT FACTS AND CONCEPTS | EXAMPLES |
|---|---|

### Section 4.2

**To Graph Linear Equations by Plotting Points**

1. Solve the linear equation for the variable $y$.
2. Select a value for the variable $x$. Substitute this value in the equation for $x$ and find the corresponding value of $y$. Record the ordered pair $(x, y)$.
3. Repeat step 2 with two different values of $x$.
4. Plot the three ordered pairs.
5. Draw a straight line through the three points. Draw an arrowhead on each end of the line.

Graph $y = -x + 2$.

**Table**

| $x$ | $y$ |
|---|---|
| $-1$ | 3 |
| 0 | 2 |
| 2 | 0 |

---

The **$x$-intercept** is the point where the graph crosses the $x$-axis.
The **$y$-intercept** is the point where the graph crosses the $y$-axis.

On the graph above, the $x$-intercept is $(2, 0)$ and the $y$-intercept is $(0, 2)$.

---

**To Graph Linear Equations Using the $x$- and $y$-Intercepts**

1. Find the $y$-intercept by setting $x$ in the given equation equal to 0 and finding the corresponding value of $y$.
2. Find the $x$-intercept by setting $y$ in the given equation equal to 0 and finding the corresponding value of $x$.
3. Determine a checkpoint by selecting a nonzero value for $x$ and finding the corresponding value of $y$.
4. Plot the $y$-intercept, the $x$-intercept, and the checkpoint.
5. Draw a straight line through the three points. Draw an arrowhead on each end of the line.

Graph $4x + 2y = 8$ using the $x$- and $y$-intercepts.
Let $x = 0$:

$$4(0) + 2y = 8$$
$$2y = 8$$
$$y = 4.$$

$y$-intercept: $(0, 4)$
Let $y = 0$:

$$4x + 2(0) = 8$$
$$4x = 8$$
$$x = 2$$

$x$-intercept: $(2, 0)$

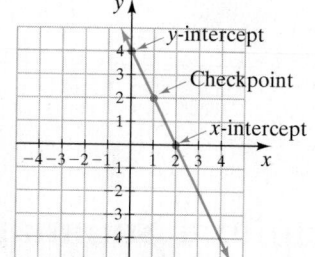

Checkpoint: $(1, 2)$

---

**Horizontal Line**

The graph of an equation of the form $y = b$ is a horizontal line whose $y$-intercept is $(0, b)$.

**Vertical Line**

The graph of an equation of the form $x = a$ is a vertical line whose $x$-intercept is $(a, 0)$.

Graph $y = 2$.

Graph $x = -4$.

---

### Section 4.3

The **slope of a line** is a ratio of the vertical change to the horizontal change between any two selected points on the line.
The slope of a line through the points $(x_1, y_1)$ and $(x_2, y_2)$ is

$$m = \text{slope} = \frac{\text{change in } y \text{ (vertical change)}}{\text{change in } x \text{ (horizontal change)}} = \frac{\text{rise}}{\text{run}} = \frac{y_2 - y_1}{x_2 - x_1}$$

The slope of the line through $(-1, 3)$ and $(5, 7)$ is

$$m = \frac{7 - 3}{5 - (-1)} = \frac{4}{6} = \frac{2}{3}$$

| IMPORTANT FACTS AND CONCEPTS | EXAMPLES |
|---|---|

**Section 4.3 (Cont.)**

| | |
|---|---|
| A line where the value of $y$ increases as $x$ increases has a **positive slope**.<br><br>A line where the value of $y$ decreases as $x$ increases has a **negative slope**.<br><br>A horizontal line has a **slope of 0**.<br><br>The slope of a vertical line is **undefined**. |  <br><br>Positive slope (rises to right)   Negative slope (falls to right)<br><br> <br><br>Slope is 0. (horizontal line)   Slope is undefined. (vertical line) |
| Two nonvertical lines with the same slope and different $y$-intercepts are **parallel lines**. Any two vertical lines are parallel to each other. | The graphs of the equations $y = 2x + 3$ and $y = 2x + 4$ are parallel lines since the graphs have the same slope, 2, and different $y$-intercepts. |
| Two lines whose slopes are negative reciprocals of each other are **perpendicular lines**. Any vertical line is perpendicular to any horizontal line. | The graphs of the equations $y = 2x + 4$ and $y = -\dfrac{1}{2}x + 3$ are perpendicular lines since the slopes of the graphs are negative reciprocals of each other. |

**Section 4.4**

| | |
|---|---|
| **Slope-Intercept Form of a Linear Equation**<br><br>$$y = mx + b$$<br><br>where $m$ is the slope, and $(0, b)$ is the $y$-intercept of the line. | The graph of $y = 3x - 4$ has a slope of 3 and a $y$-intercept of $(0, -4)$.<br><br>The equation of a line with a slope of $-\dfrac{1}{2}$ and a $y$-intercept of $(0, 6)$ is $y = -\dfrac{1}{2}x + 6$. |
| **To Graph Linear Equations Using the Slope and y-Intercept**<br><br>1. If necessary, solve the equation for $y$.<br>2. Plot the $y$-intercept.<br>3. Use the slope to find two more points on the line.<br>4. Using a straightedge, draw a straight line through the three points. Draw an arrowhead on each end of the line. | $2x + 4y = 8$ written in slope-intercept form is $y = -\dfrac{1}{2}x + 2$.<br><br>The slope is $-\dfrac{1}{2}$ and the $y$-intercept is $(0, 2)$.<br><br>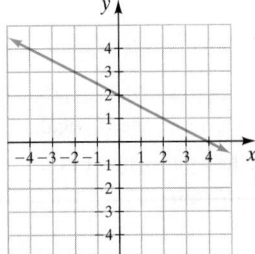 |
| **Point-Slope Form of a Linear Equation**<br><br>$$y - y_1 = m(x - x_1)$$<br><br>where $m$ is the slope of the line and $(x_1, y_1)$ is a point on the line. | The graph of $y - 7 = \dfrac{5}{6}(x - 3)$ has a slope of $\dfrac{5}{6}$ and contains the point $(3, 7)$. |

## Chapter 4 Review Exercises

[4.1]

1. Plot each ordered pair on the same axes.
   a) $A(5, 3)$    b) $B(0, 6)$    c) $C\left(5, \frac{1}{2}\right)$
   d) $D(-4, 3)$    e) $E(-6, -1)$    f) $F(-2, 0)$

2. Determine whether the following points are collinear.

   $$(7, 1), (6, 8), (-2, 0), (4, 5)$$

3. Which of the following ordered pairs satisfy the equation $2x + 3y = 9$?
   a) $(3, 1)$    b) $\left(5, -\frac{1}{3}\right)$
   c) $(-2, 4)$    d) $\left(2, \frac{5}{3}\right)$

[4.2]

4. Find the missing coordinate in the following solutions to $3x - 2y = 8$.
   a) $(-2, ?)$    b) $(0, ?)$
   c) $(?, 5)$    d) $(?, 0)$

*Graph each equation using the method of your choice.*

5. $y = 4$          6. $x = 2$          7. $y = 3x$          8. $y = 2x - 1$
9. $y = -2x + 5$    10. $2y + x = 8$    11. $-2x + 3y = 6$    12. $5x + 2y + 10 = 0$
13. $5x + 10y = 20$    14. $\frac{2}{3}x = \frac{1}{4}y + 20$

[4.3]   Find the slope of the line through the given points.

15. $(6, -4)$ and $(1, 5)$          16. $(-4, -6)$ and $(8, -7)$          17. $(-2, -3)$ and $(-4, 1)$
18. What is the slope of a horizontal line?          19. What is the slope of a vertical line?
20. Define the slope of a straight line.

*Find the slope of each line.*

21.

22.

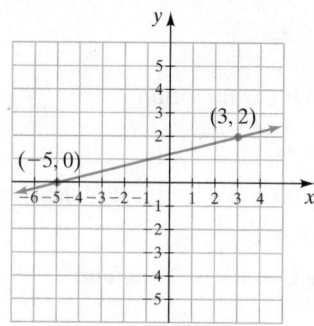

*Assume that line 1 and line 2 are distinct lines. If $m_1$ represents the slope of line 1 and $m_2$ represents the slope of line 2, determine if line 1 and line 2 are parallel, perpendicular, or neither.*

23. $m_1 = \frac{7}{8}, m_2 = -\frac{7}{8}$          24. $m_1 = -3, m_2 = \frac{1}{3}$

25. David Nezelek is riding his bicycle in a charity fund raiser. The graph below shows the time in hours along the horizontal axis and the distance in miles along the vertical axis.

   Find the slope of the line segment in
   a) red          b) blue

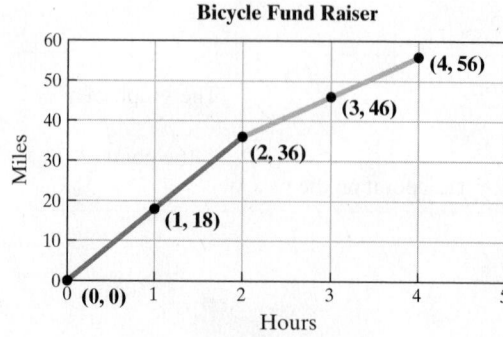

[4.4]   *Determine the slope and y-intercept of the graph of each equation.*

26. $6x + 7y = 21$          27. $2x + 7 = 0$          28. $4y + 12 = 0$

*Write the equation of each line.*

**29.**

**30.**

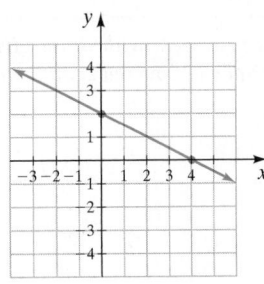

*Determine whether each pair of lines is parallel, perpendicular, or neither.*

**31.** $y = 2x - 7$
      $3y = 6x + 18$

**32.** $2x - 3y = 15$
      $3x + 2y = 12$

*Find the equation of each line with the given properties.*

**33.** Slope $= 3$ through $(2, 7)$

**34.** Slope $= -\dfrac{2}{3}$, through $(3, 2)$

**35.** Slope $= 0$, through $(6, 2)$

**36.** Slope is undefined, through $(4, 1)$

**37.** Through $(-2, 4)$ and $(0, -3)$

**38.** Through $(-5, -2)$ and $(-5, 3)$

## Chapter 4 Practice Test

*Chapter Test Prep Videos provide fully worked-out solutions to any of the exercises you want to review. Chapter Test Prep Videos are available via MyMathLab®, or on You Tube* (search "Angel Elementary and Intermediate Algebra" and click on "Channels").*

**1.** What is a graph?

**2.** In which quadrants do the following points lie?

  **a)** $(3, -5)$

  **b)** $\left(-2, \dfrac{1}{2}\right)$

**3. a)** What is the standard form of a linear equation?

  **b)** What is the slope-intercept form of a linear equation?

  **c)** What is the point-slope form of a linear equation?

**4.** Which of the following ordered pairs satisfy the equation $3y = 5x - 9$?

  **a)** $(4, 2)$        **b)** $\left(\dfrac{9}{5}, 0\right)$

  **c)** $(-1, -10)$     **d)** $(0, -3)$

**5.** Find the slope of the line through the points $(-2, 5)$ and $(4, -3)$.

**6.** Find the slope and $y$-intercept of $4x - 9y = 15$

**7.** Write an equation of the graph in the accompanying figure.

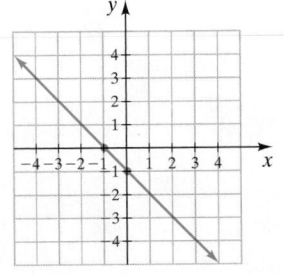

**8.** Graph $x = -4$.

**9.** Graph $y = 2$.

**10.** Graph $y = 3x - 2$ by plotting points.

**11. a)** Solve the equation $3x - 6y = 12$ for $y$.

  **b)** Graph the equation by plotting points.

**12.** Graph $3x + 5y = 15$ using the intercepts.

**13.** Write, in slope-intercept form, an equation of the line with a slope of 4 passing through the point $(2, -5)$

**14.** Write, in slope-intercept form, an equation of the line passing through the points $(3, -1)$ and $(-4, 2)$.

**15.** Determine whether the following equations represent parallel lines. Explain how you determined your answer.

$$2y = 3x - 6 \quad \text{and} \quad y - \frac{3}{2}x = -5$$

**16.** Graph $y = 3x - 4$ using the slope and $y$-intercept.

**17.** Graph $4x - 2y = 6$ using the slope and $y$-intercept.

## Cumulative Review Test

*Take the following test and check your answers with those given in the back of the book. Review any questions that you answered incorrectly. The section where the material was covered is indicated after the answer.*

1. Write the set of
   a) natural numbers.
   b) whole numbers.

2. Evaluate $4 - 9 - (-10) + 13$.

3. Evaluate $(8 \div 4)^3 + 9^2 \div 3$.

4. Evaluate $(10 \div 5 \cdot 5 + 5 - 5)^2$.

5. Name each indicated property.
   a) $3(x + 2) = 3x + 3 \cdot 2$
   b) $a + b = b + a$

6. Solve $2x + 5 = 3(x - 5)$.

7. Solve $3(x - 1) - (x + 4) = 2x - 7$

8. Solve the proportion $\dfrac{2}{20} = \dfrac{x}{200}$.

9. **Chicken Soup** At Tsong Hsu's Grocery Store, 3 cans of chicken soup sell for $1.50. Find the cost of 8 cans.

10. Solve $v = lwh$ for $w$

11. **Numbers** Eleven increased by twice a number is 19. Find the number.

12. **Rectangle** The length of a rectangle is 3 more than twice the width. Find the length and width of the rectangle if its perimeter is 36 feet.

13. **Running** Two runners start at the same point and run in opposite directions. One runs at 6 mph and the other runs at 8 mph. In how many hours will they be 28 miles apart?

14. Give three ordered pairs that satisfy the equation $2x + 4y = 8$.

15. Graph $y = 3x - 5$ by plotting points.

16. Graph $6x - 3y = -12$ using the intercepts.

17. Find the slope of the line through the points $(5, -6)$ and $(6, -5)$.

18. The slope of a given line is $-7$. If a line is to be drawn parallel to the given line, what will be its slope?

19. Graph $y = \dfrac{2}{3}x - 3$ using the slope and $y$-intercept.

20. Write the equation, in point-slope form, of the line with a slope of 3 passing through the point $(5, 2)$.

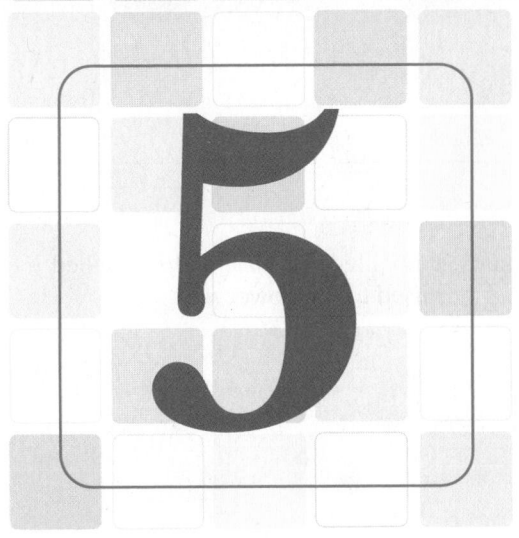

# 5 Exponents and Polynomials

## Goals of This Chapter

The major emphasis of this chapter is to teach you how to work with exponents, polynomials, and scientific notation. You must understand the rules of exponents presented in the first two sections in order to be successful with the remaining material in the chapter.

   You will use the rules learned in this chapter throughout the book, especially in Chapter 6, when factoring is introduced.

**Many areas of** science and technology deal with very small and very large numbers. A *googol* is a very large number, 1 followed by 100 zeros. The word googol was first coined by mathematician Edward Kasner (suggested by his 9-year old nephew). Kasner also proclaimed that the number of grains of sand on the beach at Coney Island was less than a googol. In Exercise 94 on page 290, we will examine the scientific notation that allows us to deal more readily with large numbers like a googol in this chapter.

© Bryan Busovicki/Shutterstock

# 5.1 Exponents

1  Review exponents.

2  Learn the rules of exponents.

3  Simplify an expression before using the expanded power rule.

## Understanding Algebra

For now, think of exponents as repeated multiplication. For example, $x^3$ means $x \cdot x \cdot x$ so $x$ is used as a factor 3 times. In $x^3$, the $x$ is called the *base* and the 3 is called the *exponent*.

### 1  Review Exponents

Recall from Section 1.9, in the expression $x^n$, $x$ is called the **base** and $n$ is called the **exponent**. $x^n$ is read "$x$ to the $n$th power" or "$x$ raised to the power $n$."

$$x^2 = \underbrace{x \cdot x}_{2 \text{ factors of } x}$$

$$x^4 = \underbrace{x \cdot x \cdot x \cdot x}_{4 \text{ factors of } x}$$

$$x^m = \underbrace{x \cdot x \cdot x \cdot \cdots \cdot x}_{m \text{ factors of } x}$$

**EXAMPLE 1**  Write $xxxxyyy$ using exponents.

*Solution*

$$\underbrace{xxxx}_{\substack{4 \text{ factors} \\ \text{of } x}} \quad \underbrace{yyy}_{\substack{3 \text{ factors} \\ \text{of } y}} = x^4 y^3$$

Now Try Exercise 9

Remember:

$$\left.\begin{array}{l} x = 1x \\ x^2 y = 1x^2 y \end{array}\right\}$$ When a term containing a variable is given without a numerical coefficient, the numerical coefficient is assumed to be 1.

$$\left.\begin{array}{l} x = x^1, \\ xy = x^1 y^1, \\ x^2 y = x^2 y^1, \\ 2xy^2 = 2x^1 y^2 \end{array}\right\}$$ When a variable or numerical value is given without an exponent, the exponent of that variable or numerical value is assumed to be 1.

### 2  Learn the Rules of Exponents

**EXAMPLE 2**  Multiply $x^4 \cdot x^3$.

*Solution*

$$\overbrace{x \cdot x \cdot x \cdot x}^{x^4} \cdot \overbrace{x \cdot x \cdot x}^{x^3} = x^7$$

Now Try Exercise 11

Example 2 illustrates the **product rule for exponents**.

## Understanding Algebra

When multiplying expressions with the same base, we add the exponents:

$$y^2 \cdot y^4 = y^6$$

> **Product Rule for Exponents**
>
> $$x^m \cdot x^n = x^{m+n}$$
>
> When multiplying expressions with the same base, we keep the base and **add** the exponents.

In Example 2, we showed that $x^4 \cdot x^3 = x^7$. This problem could also be done using the product rule: $x^4 \cdot x^3 = x^{4+3} = x^7$.

EXAMPLE 3  Multiply each expression using the product rule.

**a)** $3^2 \cdot 3$     **b)** $2^4 \cdot 2^2$     **c)** $x \cdot x^4$     **d)** $x^3 \cdot x^6$     **e)** $y^4 \cdot y^7$

*Solution*

**a)** $3^2 \cdot 3 = 3^2 \cdot 3^1 = 3^{2+1} = 3^3$ or 27     **b)** $2^4 \cdot 2^2 = 2^{4+2} = 2^6$ or 64

**c)** $x \cdot x^4 = x^1 \cdot x^4 = x^{1+4} = x^5$     **d)** $x^3 \cdot x^6 = x^{3+6} = x^9$

**e)** $y^4 \cdot y^7 = y^{4+7} = y^{11}$

Now Try Exercise 19

## Understanding Algebra

Throughout this section, whenever we are working with an expression with a variable in the denominator, we will assume that the denominator does not equal zero. When we are introducing rules of exponents, we will remind you of this. For example, in the Quotient Rule, we will write $x \neq 0$.

**Avoiding Common Errors**

Note in Example 3 **a)** that $3^2 \cdot 3^1$ is $3^3$ and not $9^3$. *When multiplying powers of the same base, do not multiply the bases.*

| CORRECT | INCORRECT |
|---|---|
| $3^2 \cdot 3^1 = 3^3$ | $3^2 \cdot 3^1 = 9^3$ |

Note also in Example 3 **b)** that $2^4 \cdot 2^2$ is $2^6$ and not $2^8$. *When multiplying powers of the same base, do not multiply the exponents.*

| CORRECT | INCORRECT |
|---|---|
| $2^4 \cdot 2^2 = 2^6$ | $2^4 \cdot 2^2 = 2^8$ |

Example 4 will help you understand the **quotient rule for exponents**.

EXAMPLE 4  Divide $x^5 \div x^3$.

*Solution*

$$\frac{x^5}{x^3} = \frac{\overset{1}{\cancel{x}} \cdot \overset{1}{\cancel{x}} \cdot \overset{1}{\cancel{x}} \cdot x \cdot x}{\underset{1}{\cancel{x}} \cdot \underset{1}{\cancel{x}} \cdot \underset{1}{\cancel{x}}} = \frac{1x^2}{1} = x^2$$

Now Try Exercise 23

When dividing expressions with the same base, keep the base and *subtract* the exponent in the denominator from the exponent in the numerator.

## Understanding Algebra

When dividing expressions with the same base, we subtract the exponents:

$$t^7 \div t^4 = \frac{t^7}{t^4} = t^3, t \neq 0$$

**Quotient Rule for Exponents**

$$\frac{x^m}{x^n} = x^{m-n}, \qquad x \neq 0$$

When dividing expressions with the same base, we keep the base and *subtract* the exponent in the denominator from the exponent in the numerator.

In Example 4, we showed that $\dfrac{x^5}{x^3} = x^2$. This problem could also be done using the quotient rule: $\dfrac{x^5}{x^3} = x^{5-3} = x^2$.

EXAMPLE 5  Divide each expression using the quotient rule.

**a)** $\dfrac{3^5}{3^2}$     **b)** $\dfrac{z^8}{z^2}$     **c)** $\dfrac{x^{12}}{x^5}$     **d)** $\dfrac{y^{10}}{y^8}$     **e)** $\dfrac{6^4}{6}$

*Solution*

**a)** $\dfrac{3^5}{3^2} = 3^{5-2} = 3^3$ or 27     **b)** $\dfrac{z^8}{z^2} = \dfrac{z^8}{z^2} = z^{8-2} = z^6$     **c)** $\dfrac{x^{12}}{x^5} = x^{12-5} = x^7$

**d)** $\dfrac{y^{10}}{y^8} = y^{10-8} = y^2$     **e)** $\dfrac{6^4}{6} = \dfrac{6^4}{6^1} = 6^{4-1} = 6^3$ or 216

Now Try Exercise 25

**Avoiding Common Errors**

Note in Example 5 **a)** that $\frac{3^5}{3^2}$ is $3^3$ and not $1^3$. *When dividing powers of the same base, do not divide out the bases.*

CORRECT

$$\frac{3^3}{3^1} = 3^2 \text{ or } 9$$

INCORRECT

$$\frac{\cancel{3}^3}{\cancel{3}^1} \cancel{=} 1^2$$

Also note in Example 5 **b)** that $\frac{z^8}{z^2} = z^6$ and not $z^4$. *When dividing powers of the same base, do not divide the exponents.*

The answer to Example 5 **c)**, $\frac{x^{12}}{x^5}$, is $x^7$. We obtained this answer using the quotient rule. This answer could also be obtained by dividing out the common factors in both the numerator and denominator as follows.

$$\frac{x^{12}}{x^5} = \frac{(\cancel{x} \cdot \cancel{x} \cdot \cancel{x} \cdot \cancel{x} \cdot \cancel{x}) \cdot x \cdot x \cdot x \cdot x \cdot x \cdot x \cdot x}{(\cancel{x} \cdot \cancel{x} \cdot \cancel{x} \cdot \cancel{x} \cdot \cancel{x})} = x^7$$

We divided out the product of five $x$'s, which is $x^5$. We can indicate this process in shortened form as follows.

$$\frac{x^{12}}{x^5} = \frac{\overset{1}{\cancel{x^5}} \cdot x^7}{\underset{1}{\cancel{x^5}}} = x^7$$

## Understanding Algebra

To divide expressions with the same base when the denominator's exponent is greater than the numerator's, divide out the common factors:

$$t^4 \div t^7 = \frac{t^4}{t^7} = \frac{\overset{1}{\cancel{t^4}}}{\cancel{t^4} \, t^3} = \frac{1}{t^3}, \ t \neq 0$$

To simplify an expression like $\frac{x^5}{x^{12}}$, where the exponent in the denominator is greater than the exponent in the numerator, we divide out common factors (in this case, the common factor is $x^5$).

Note: Numerator and denominator must have the same base

$$\frac{x^5}{x^{12}} = \frac{\overset{1}{\cancel{x^5}}}{\underset{1}{\cancel{x^5}} \cdot x^7} = \frac{1}{x^7}$$

We will now simplify some expressions by dividing out common factors.

## EXAMPLE 6

Simplify each expression by dividing out a common factor in both the numerator and denominator.

**a)** $\dfrac{x^9}{x^{12}}$     **b)** $\dfrac{y^4}{y^9}$

*Solution*

**a)** Since the numerator is $x^9$, we write the denominator with a factor of $x^9$. Since $x^9 \cdot x^3 = x^{12}$, we rewrite $x^{12}$ as $x^9 \cdot x^3$.

$$\frac{x^9}{x^{12}} = \frac{\overset{1}{\cancel{x^9}}}{\underset{1}{\cancel{x^9}} \cdot x^3} = \frac{1}{x^3}$$

**b)** $\dfrac{y^4}{y^9} = \dfrac{\overset{1}{\cancel{y^4}}}{\underset{1}{\cancel{y^4}} \cdot y^5} = \dfrac{1}{y^5}$

Now Try Exercise 27

In the next section, we will show another way to simplify expressions like $\dfrac{x^9}{x^{12}}$ by using the negative exponent rule.

Example 7 leads us to our next rule, the **zero exponent rule**.

EXAMPLE 7  Divide $\dfrac{x^3}{x^3}$.

*Solution*    By the quotient rule,

$$\frac{x^3}{x^3} = x^{3-3} = x^0$$

However,

$$\frac{x^3}{x^3} = \frac{1x^3}{1x^3} = \frac{1 \cdot \cancel{x} \cdot \cancel{x} \cdot \cancel{x}}{1 \cdot \cancel{x} \cdot \cancel{x} \cdot \cancel{x}} = \frac{1}{1} = 1$$

Since $\dfrac{x^3}{x^3} = x^0$ and $\dfrac{x^3}{x^3} = 1$, then $x^0$ must equal 1.

Now Try Exercise 29

**Zero Exponent Rule**

$$x^0 = 1, \qquad x \neq 0$$

Any real number, except 0, raised to the zero power equals 1. Note that $0^0$ is undefined.

EXAMPLE 8  Simplify each expression. Assume $x \neq 0$ and $z \neq 0$.

**a)** $3^0$        **b)** $x^0$        **c)** $3x^0$        **d)** $(3x)^0$        **e)** $4x^2y^3z^0$

*Solution*

**a)** $3^0 = 1$

**b)** $x^0 = 1$

**c)** $3x^0 = 3(x^0)$        Remember, the exponent refers only to the immediately
$\qquad = 3 \cdot 1 = 3$        preceding symbol unless parentheses are used.

**d)** $(3x)^0 = 1$

**e)** $4x^2y^3z^0 = 4x^2y^3 \cdot 1 = 4x^2y^3$

Now Try Exercise 37

**Avoiding Common Errors**

An expression raised to the zero power is not equal to 0; it is equal to 1.

| CORRECT | INCORRECT |
|---------|-----------|
| $x^0 = 1$ | $x^0 = 0$ |
| $5^0 = 1$ | $5^0 = 0$ |

The **power rule** will be explained with the aid of Example 9.

EXAMPLE 9  Simplify $(x^3)^2$.

*Solution*

$$(x^3)^2 = \underbrace{x^3 \cdot x^3}_{\text{2 factors of } x^3} = x^{3+3} = x^6$$

Now Try Exercise 45

## Power Rule for Exponents

$$(x^m)^n = x^{m \cdot n}$$

When raising an exponential expression to a power, keep the base and *multiply* the exponents.

**EXAMPLE 10** Simplify each expression.

**a)** $(x^3)^5$     **b)** $(3^4)^2$     **c)** $(y^5)^7$

*Solution*

**a)** $(x^3)^5 = x^{3 \cdot 5} = x^{15}$     **b)** $(3^4)^2 = 3^{4 \cdot 2} = 3^8$     **c)** $(y^5)^7 = y^{5 \cdot 7} = y^{35}$

Now Try Exercise 51

**Helpful Hint**

Students often confuse the product and power rules. Note the difference carefully.

| Product Rule | Power Rule |
|---|---|
| $x^m \cdot x^n = x^{m+n}$ | $(x^m)^n = x^{m \cdot n}$ |
| $2^3 \cdot 2^5 = 2^{3+5} = 2^8$ | $(2^3)^5 = 2^{3 \cdot 5} = 2^{15}$ |

The power rule can be extended to include powers of products and powers of quotients. Consider $(4x^2)^3$, the product $4x^2$ raised to the power 3:

$$
\begin{aligned}
(4x^2)^3 &= 4x^2 \cdot 4x^2 \cdot 4x^2 \\
&= (4 \cdot 4 \cdot 4)(x^2 \cdot x^2 \cdot x^2) \\
&= 4^3(x^2)^3 \qquad \text{Observe both 4 and } x^2 \text{ are raised to the power 3.} \\
&= 64x^6
\end{aligned}
$$

This leads us to the **power of a product rule:***

## Power of a Product Rule

$$(xy)^n = x^n y^n$$

When a product is raised to a power, each factor in the product is raised to that power.

**EXAMPLE 11** Simplify each expression.

**a)** $(a^2b^3)^4$     **b)** $\left(-\dfrac{2}{3}x^6y\right)^3$

*Solution*

**a)** $(a^2b^3)^4 = (a^2)^4 \cdot (b^3)^4$     Raise each factor to the power 4.

$\qquad\qquad = a^8 b^{12}$

**b)** $\left(-\dfrac{2}{3}x^6y\right)^3 = \left(-\dfrac{2}{3}\right)^3 \cdot (x^6)^3 \cdot (y)^3$     Raise each factor to the power 3.

$\qquad\qquad = -\dfrac{8}{27}x^{18}y^3$     Because $\left(-\dfrac{2}{3}\right)^3 = \left(-\dfrac{2}{3}\right) \cdot \left(-\dfrac{2}{3}\right) \cdot \left(-\dfrac{2}{3}\right)$

Now Try Exercise 59

---

*When we write exponential expressions, it is always understood that the values of variables we use in definitions are always restricted to avoid $0^0$. Thus, when we write $(xy)^n$, for example, $n$ cannot be zero when either $x$ or $y$ is zero.

Just like we extended the power rule to a product, we can apply similar reasoning to extend the power rule to a quotient. Consider $\left(\dfrac{5}{x^3}\right)^2$, the quotient $\dfrac{5}{x^3}$ raised to the power 2:

$$\left(\frac{5}{x^3}\right)^2 = \frac{5}{x^3} \cdot \frac{5}{x^3}$$

$$= \frac{5 \cdot 5}{x^3 \cdot x^3}$$

$$= \frac{5^2}{(x^3)^2} \quad \text{Observe both the numerator and denominator are raised to the power 2.}$$

$$= \frac{25}{x^6}$$

This leads us to the **power of a quotient rule:**

---

### Power of a Quotient Rule

$$\left(\frac{x}{y}\right)^n = \frac{x^n}{y^n}, \qquad y \neq 0$$

When a quotient is raised to a power, both the numerator and the denominator are raised to that power.

---

**EXAMPLE 12** Simplify each expression.

**a)** $\left(\dfrac{p^2}{q^3}\right)^4$   **b)** $\left(-\dfrac{x^2}{z^3}\right)^5$

*Solution*

**a)** $\left(\dfrac{p^2}{q^3}\right)^4 = \dfrac{(p^2)^4}{(q^3)^4}$   Raise numerator and denominator to the power 4.

$$= \frac{p^8}{q^{12}}$$

**b)** $\left(-\dfrac{x^2}{z^3}\right)^5 = \left(\dfrac{-x^2}{z^3}\right)^5$

$$= \frac{(-x^2)^5}{(z^3)^5} \qquad \text{Raise numerator and denominator to the power 5.}$$

$$= \frac{-x^{10}}{z^{15}} \ \text{ or } \ -\frac{x^{10}}{z^{15}} \qquad \text{Either answer is acceptable.}$$

Now Try Exercise 85

We now combine the previous two rules, the power rule for products and the power rule for quotients, into the **expanded power rule**.

---

### Expanded Power Rule for Exponents

$$\left(\frac{ax}{by}\right)^m = \frac{a^m x^m}{b^m y^m}, \quad b \neq 0, y \neq 0$$

Every factor within parentheses is raised to the power outside the parentheses when the expression is simplified.

---

**EXAMPLE 13** Simplify $\left(\dfrac{2r^2 s}{t^4}\right)^3$.

*Solution*

$$\left(\frac{2r^2 s}{t^4}\right)^3 = \frac{2^3 (r^2)^3 s^3}{(t^4)^3} \qquad \begin{array}{l}\text{Expanded power rule; each factor}\\ \text{is raised to the power 3.}\end{array}$$

$$= \frac{8r^6 s^3}{t^{12}}$$

Now Try Exercise 87

## 3 Simplify an Expression Before Using the Expanded Power Rule

Whenever we have an expression raised to a power, it helps to simplify the expression in parentheses before using the expanded power rule.

**EXAMPLE 14** Simplify $\left(\dfrac{9x^3 y^2}{3xy^2}\right)^3$.

*Solution*    We first simplify the expression within parentheses by dividing out common factors.

$$\left(\frac{9x^3 y^2}{3xy^2}\right)^3 = \left(\frac{9}{3} \cdot \frac{x^3}{x} \cdot \frac{y^2}{y^2}\right)^3 = (3x^2)^3$$

Now we use the expanded power rule to simplify further.

$$(3x^2)^3 = 3^3 (x^2)^3 = 27x^6$$

Thus, $\left(\dfrac{9x^3 y^2}{3xy^2}\right)^3 = 27x^6$.

Now Try Exercise 93

### Helpful Hint

**Study Tip**

Be very careful when writing exponents. Since exponents are generally smaller than regular text, take your time and write them clearly, and position them properly. If exponents are not written clearly it is very easy to confuse exponents such as 2 and 3, or 1 and 4, or 0 and 6. If you write down or carry an exponent from step to step incorrectly, you will obtain an incorrect answer.

**EXAMPLE 15** Simplify $\left(\dfrac{25x^4 y^3}{5x^2 y^7}\right)^4$.

*Solution*    Begin by simplifying the expression within parentheses.

$$\left(\frac{25x^4 y^3}{5x^2 y^7}\right)^4 = \left(\frac{25}{5} \cdot \frac{x^4}{x^2} \cdot \frac{y^3}{y^7}\right)^4 = \left(\frac{5x^2}{y^4}\right)^4$$

Now use the expanded power rule to simplify further.

$$\left(\frac{5x^2}{y^4}\right)^4 = \frac{5^4 (x^2)^4}{(y^4)^4} = \frac{625x^8}{y^{16}}$$

Thus, $\left(\dfrac{25x^4 y^3}{5x^2 y^7}\right)^4 = \dfrac{625x^8}{y^{16}}$.

Now Try Exercise 95

**Avoiding Common Errors**

Students sometimes make errors in simplifying expressions containing exponents. One of the most common errors follows.

CORRECT

$$\dfrac{4}{2x} = \dfrac{\overset{2}{\cancel{4}}}{\underset{1}{\cancel{2}x}} = \dfrac{2}{x}$$

$$\dfrac{x}{xy} = \dfrac{\overset{1}{\cancel{x}}}{\underset{1}{\cancel{x}y}} = \dfrac{1}{y}$$

$$\dfrac{5x^3y^2}{y^2} = \dfrac{5x^3\,\overset{1}{\cancel{y^2}}}{\underset{1}{\cancel{y^2}}} = 5x^3$$

INCORRECT

$$\dfrac{4}{x+2} = \dfrac{\overset{2}{\cancel{4}}}{x+\cancel{2}} = \dfrac{2}{x+1}$$

$$\dfrac{x}{x+y} = \dfrac{\overset{1}{\cancel{x}}}{\cancel{x}+y} = \dfrac{1}{1+y}$$

$$\dfrac{5x^3+y^2}{y^2} = \dfrac{5x^3+\overset{1}{\cancel{y^2}}}{\underset{1}{\cancel{y^2}}} = 5x^3+1$$

The simplifications on the right side are not correct because only common *factors* can be divided out (remember, factors are multiplied together). In the first denominator on the right, $x + 2$, the $x$ and 2 are terms, not factors, since they are being added. Similarly, in the second denominator, $x + y$, the $x$ and the $y$ are terms, not factors, since they are being added. Also, in the numerator $5x^3 + y^2$, the $5x^3$ and $y^2$ are terms, not factors. No common factors can be divided out in the fractions on the right.

**EXAMPLE 16**  Simplify $(2a^5b^3)^5 (3a^2b)$.

*Solution*  First simplify $(2a^5b^3)^5$ by using the expanded power rule.

$$(2a^5b^3)^5 = 2^5 a^{5\cdot5} b^{3\cdot5} = 32a^{25}b^{15}$$

Now use the product rule to simplify further.

$$(2a^5b^3)^5 (3a^2b) = (32a^{25}b^{15})(3a^2b^1)$$
$$= 32 \cdot 3 \cdot a^{25} \cdot a^2 \cdot b^{15} \cdot b^1$$
$$= 96a^{25+2}b^{15+1}$$
$$= 96a^{27}b^{16}$$

Thus, $(2a^5b^3)^5 (3a^2b) = 96a^{27}b^{16}$.

Now Try Exercise 125

**Summary of the Rules of Exponents Presented in This Section**

1. **Product Rule**

$$x^m \cdot x^n = x^{m+n}$$

2. **Quotient Rule**

$$\dfrac{x^m}{x^n} = x^{m-n}, \quad x \neq 0$$

3. **Zero Exponent Rule**

$$x^0 = 1, \quad x \neq 0$$

4. **Power Rule**

$$(x^m)^n = x^{m\cdot n}$$

5. **Power of a Product Rule**

$$(xy)^n = x^n y^n$$

6. **Power of a Quotient Rule**

$$\left(\dfrac{x}{y}\right)^n = \dfrac{x^n}{y^n}, \quad y \neq 0$$

7. **Expanded Power Rule**

$$\left(\dfrac{ax}{by}\right)^m = \dfrac{a^m x^m}{b^m y^m}, \quad b \neq 0, \quad y \neq 0$$

# EXERCISE SET 5.1

MathXL®   MyMathLab®
MathXL®    MyMathLab

## Warm-Up Exercises

*Fill in the blanks with the appropriate word, phrase, or symbol(s) from the following list.*

| | | | |
|---|---|---|---|
| 1 | product rule for exponents | power rule for exponents | addition rule for exponents |
| exponent | base | quotient rule for exponents | expanded power rule for exponents |
| 0 | $p^3q^4r^5$ | $\dfrac{p^3r^5}{q^4}$ | |

**1.** The only value of $x$ for which $x^0$ is not equal to 1 is _____.

**2.** $(x^2)^8 = x^{16}$ illustrates the _____.

**3.** In the expression $t^p$, $t$ is called the _____.

**4.** In the expression $t^p$, $p$ is called the _____.

**5.** $\left(\dfrac{3x^2}{2y^3}\right)^2 = \dfrac{9x^4}{4y^6}$ illustrates the _____.

**6.** If there is no exponent written for a variable, it is assumed to be the number _____.

**7.** $x^2 \cdot x^8 = x^{10}$ illustrates the _____.

**8.** $\dfrac{x^8}{x^3} = x^5$, $x \neq 0$ illustrates the _____.

**9.** When $pppqqqqrrrrr$ is rewritten using exponents, it is _____.

**10.** When $\dfrac{pppprrrrr}{qqqq}$ is rewritten using exponents, it is _____.

## Practice the Skills

*Multiply.*

**11.** $x^6 \cdot x^4$

**12.** $x^6 \cdot x$

**13.** $-z^4 \cdot z$

**14.** $t^7 \cdot t^2$

**15.** $y^3 \cdot y^2$

**16.** $4^2 \cdot 4^3$

**17.** $3^2 \cdot 3^3$

**18.** $-x^3 \cdot x^4$

**19.** $z^3 \cdot z^5$

**20.** $2^4 \cdot 2^2$

*Divide.*

**21.** $\dfrac{6^2}{6}$

**22.** $\dfrac{x^4}{x^3}$

**23.** $\dfrac{x^{10}}{x^3}$

**24.** $\dfrac{y^9}{y}$

**25.** $\dfrac{3^6}{3^2}$

**26.** $\dfrac{4^5}{4^3}$

**27.** $\dfrac{y^4}{y^6}$

**28.** $\dfrac{a^7}{a^9}$

**29.** $\dfrac{c^4}{c^4}$

**30.** $\dfrac{5^4}{5^4}$

**31.** $\dfrac{q^3}{q^9}$

**32.** $\dfrac{x^9}{x^{13}}$

*Simplify.*

**33.** $x^0$

**34.** $5^0$

**35.** $3x^0$

**36.** $-7x^0$

**37.** $4(5d)^0$

**38.** $-2(8t)^0$

**39.** $-9(-4y)^0$

**40.** $-(-x)^0$

**41.** $6x^3y^2z^0$

**42.** $-5xy^2z^0$

**43.** $-8r(st)^0$

**44.** $-3(a^2b^5c^3)^0$

*Simplify.*

**45.** $(x^4)^2$

**46.** $(a^5)^3$

**47.** $(x^5)^5$

**48.** $(y^5)^2$

**49.** $(x^3)^1$

**50.** $(x^6)^{20}$

**51.** $(x^4)^3$

**52.** $(x^5)^4$

**53.** $(n^6)^3$

**54.** $(1.3x)^2$

**55.** $(-2w^2)^3$

**56.** $(-3x)^2$

**57.** $(-3t^3)^3$

**58.** $(-xy)^4$

**59.** $(4x^3y^2)^3$

**60.** $(3a^2b^4)^3$

*Simplify.*

**61.** $\left(\dfrac{x}{3}\right)^2$

**62.** $\left(\dfrac{-2}{x}\right)^3$

**63.** $\left(\dfrac{y}{x}\right)^4$

**64.** $\left(\dfrac{2}{y}\right)^4$

**65.** $\left(\dfrac{-6}{x}\right)^3$

**66.** $\left(\dfrac{4m}{n}\right)^3$

**67.** $\left(\dfrac{2x}{y}\right)^3$

**68.** $\left(\dfrac{3s}{t^2}\right)^2$

**69.** $\left(\dfrac{4p}{5}\right)^2$

**70.** $\left(\dfrac{2y^3}{x}\right)^4$

**71.** $\left(\dfrac{3x^4}{y}\right)^3$

**72.** $\left(\dfrac{-4x^2}{5}\right)^2$

*Simplify.*

**73.** $\dfrac{a^8 b}{ab^4}$

**74.** $\dfrac{x^3 y^5}{x^7 y}$

**75.** $\dfrac{5x^{12} y^2}{10xy^9}$

**76.** $\dfrac{10x^3 y^8}{2xy^{10}}$

**77.** $\dfrac{30y^5 z^3}{5yz^6}$

**78.** $\dfrac{3ab}{27a^3 b^4}$

**79.** $\dfrac{35x^4 y^9}{15x^9 y^{12}}$

**80.** $\dfrac{6m^3 n^9}{9m^7 n^{12}}$

**81.** $-\dfrac{36xy^7 z}{12x^4 y^5 z}$

**82.** $\dfrac{4x^4 y^7 z^3}{32x^5 y^4 z^9}$

**83.** $-\dfrac{6x^2 y^7 z}{3x^5 y^9 z^6}$

**84.** $-\dfrac{25x^4 y^{10}}{30x^3 y^7 z}$

*Simplify.*

**85.** $\left(\dfrac{10x^4}{5x^6}\right)^3$

**86.** $\left(\dfrac{4x^4}{8x^8}\right)^3$

**87.** $\left(\dfrac{6y^6}{2y^3}\right)^3$

**88.** $\left(\dfrac{4xy^5}{y}\right)^3$

**89.** $\left(\dfrac{6a^2 b^4}{3a^7 b^9}\right)^0$

**90.** $\left(\dfrac{16y^6}{24y^{10}}\right)^3$

**91.** $\left(\dfrac{x^4 y^3}{x^2 y^5}\right)^2$

**92.** $\left(\dfrac{2x^7 y^2}{4xy}\right)^3$

**93.** $\left(\dfrac{9y^2 z^7}{18y^9 z}\right)^4$

**94.** $\left(\dfrac{y^7 z^5}{y^8 z^4}\right)^{10}$

**95.** $\left(\dfrac{25s^4 t}{5s^6 t^4}\right)^3$

**96.** $\left(\dfrac{-64xy^6}{32xy^9}\right)^4$

*Simplify.*

**97.** $(3xy^4)^2$

**98.** $(4ab^3)^3$

**99.** $(5ab^3)(b)$

**100.** $(6xy^5)(3x^2 y^4)$

**101.** $(-2xy)(3xy)$

**102.** $(-3x^4 y^2)(5x^2 y)$

**103.** $(5x^2 y)(3xy^5)$

**104.** $(-5xy)(-2xy^6)$

**105.** $(-3p^2 q)^2 (-p^2 q)$

**106.** $(2c^3 d^2)^2 (3cd)^0$

**107.** $(7r^3 s^2)^2 (9r^3 s^4)^0$

**108.** $(3x^2)^4 (2xy^5)$

*Simplify.*

**109.** $(-x)^2$

**110.** $(2xy^4)^3$

**111.** $\left(\dfrac{x^5 y^5}{xy^5}\right)^3$

**112.** $(2x^2 y^5)(3x^5 y^4)^3$

**113.** $(2.5x^3)^2$

**114.** $(-3a^2 b^3 c^4)^3$

**115.** $\dfrac{x^9 y^3}{x^2 y^7}$

**116.** $(xy^4)(xy^4)^3$

**117.** $\left(-\dfrac{m^4}{n^3}\right)^3$

**118.** $\left(-\dfrac{12x}{16x^7 y^2}\right)^2$

**119.** $(-6x^3 y^2)^3$

**120.** $(3x^6 y)^2 (4xy^8)^0$

**121.** $(-2x^4 y^2 z)^3$

**122.** $\left(\dfrac{z}{4}\right)^3$

**123.** $(9r^4 s^5)^3$

**124.** $(5x^4 z^{10})^2 (2x^2 z^8)$

**125.** $(4x^2 y)(3xy^2)^3$

**126.** $\dfrac{x^2 y^6}{x^4 y}$

**127.** $(1.3x^2 y^4)^2$

**128.** $\left(\dfrac{-3x^3}{4}\right)^3$

**129.** $(x^7 y^5)(xy^2)^4$

**130.** $(4c^3 d^2)(2c^5 d^3)^2$

**131.** $\left(\dfrac{-x^4 z^7}{x^2 z^5}\right)^4$

**132.** $(x^4 y^6)^3 (3x^2 y^5)$

*Study the Avoiding Common Errors box on page 269. Simplify the following expressions by dividing out common factors. If the expression cannot be simplified by dividing out common factors, so state.*

**133.** $\dfrac{a + b}{b}$

**134.** $\dfrac{xy}{x}$

**135.** $\dfrac{y^2 + 3}{y}$

**136.** $\dfrac{a + 9}{3}$

**137.** $\dfrac{6yz^4}{yz^2}$

**138.** $\dfrac{x}{x + 1}$

**139.** $\dfrac{a^2 + b^2}{a^2}$

**140.** $\dfrac{x^4}{x^2 y}$

## Problem Solving

**141.** What is the value of $a^3 b$ if $a = 2$ and $b = 5$?

**142.** What is the value of $xy^2$ if $x = -3$ and $y = -4$?

**143.** What is the value of $(xy)^0$ if $x = -5$ and $y = 3$?

**144.** What is the value of $(xy)^0$ if $x = 2$ and $y = 4$?

**145.** Consider the expression $(-9x^4 y^6)^8$. When the expanded power rule is used to simplify the expression, will the *sign* of the simplified expression be positive or negative? Explain how you determined your answer.

**146.** Consider the expression $(-x^5 y^7)^9$. When the expanded power rule is used to simplify the expression, will the *sign* of the simplified expression be positive or negative? Explain how you determined your answer.

*Write an expression for the total area of the figure or figures shown.*

**147.**

**148.**

**149.**

**150.**

## Concept/Writing Exercises

**151.** Explain the difference between the product rule and the power rule for exponents.

**152.** Explain the zero exponent rule. What restrictions are there on the base?

## Challenge Problems

*Simplify.*

**153.** $(3yz^2)^2\left(\dfrac{2y^3z^5}{10y^6z^4}\right)^0(4y^2z^3)^3$

**154.** $\left(\dfrac{3x^4y^5}{6x^6y^8}\right)^3\left(\dfrac{9x^7y^8}{3x^3y^5}\right)^2$

## Group Activity

*Discuss and answer Exercise 155 as a group, according to the instructions.*

**155.** In the next section we will be working with negative exponents. To prepare for that work, use the expression $\dfrac{3^2}{3^3}$ to work parts **a)** through **c)**. Work parts **a)** through **d)** individually, then part **e)** as a group.

   **a)** Divide out common factors in the numerator and denominator and determine the value of the expression.

   **b)** Use the quotient rule on the given expression and write down your results.

   **c)** Write a statement of equality using the results of part **a)** and **b)** above.

   **d)** Repeat parts **a)** through **c)** for the expression $\dfrac{2^3}{2^4}$.

   **e)** As a group, compare your answers to parts **a)** through **d)**, then write an exponential expression for $\dfrac{1}{x^m}$.

## Cumulative Review Exercises

[1.9] **156.** Evaluate $3^4 \div 3^3 - (5-8) + 7$.

[2.1] **157.** Simplify $-4(x-3) + 5x - 2$.

[2.5] **158.** Solve the equation
$2(x+4) - 3 = 5x + 4 - 3x + 1$.

[2.6] **159. Rectangle a)** Use the formula $P = 2l + 2w$ to find the length of the sides of the rectangle shown if the perimeter of the rectangle is 26 inches.

   **b)** Solve the formula $P = 2l + 2w$ for $w$.

$x + 5$

$x$

# 5.2 Negative Exponents

1 Understand the negative exponent rule.

2 Simplify expressions containing negative exponents.

## 1 Understand the Negative Exponent Rule

One additional rule that involves exponents is the negative exponent rule.

**EXAMPLE 1** Simplify $\dfrac{x^3}{x^5}$ by **a)** using the quotient rule and **b)** dividing out common factors.

*Solution*

**a)** By the quotient rule,

$$\frac{x^3}{x^5} = x^{3-5} = x^{-2}$$

**b)** By dividing out common factors,

$$\frac{x^3}{x^5} = \frac{\cancel{x} \cdot \cancel{x} \cdot \cancel{x}}{\cancel{x} \cdot \cancel{x} \cdot \cancel{x} \cdot x \cdot x} = \frac{1}{x^2}$$

Now Try Exercise 35

In Example 1, we see that $\dfrac{x^3}{x^5}$ is equal to both $x^{-2}$ and $\dfrac{1}{x^2}$. Therefore, $x^{-2}$ must equal $\dfrac{1}{x^2}$. That is, $x^{-2} = \dfrac{1}{x^2}$. This is an example of the **negative exponent rule**.

---

**Negative Exponent Rule**

$$x^{-m} = \frac{1}{x^m}, \qquad x \neq 0$$

When a variable or number is raised to a negative exponent, the expression may be rewritten as 1 divided by the variable or number raised to that positive exponent.

---

Examples

$$x^{-6} = \frac{1}{x^6} \qquad 4^{-2} = \frac{1}{4^2} = \frac{1}{16}$$

$$y^{-7} = \frac{1}{y^7} \qquad 5^{-3} = \frac{1}{5^3} = \frac{1}{125}$$

## Understanding Algebra

Negative exponents have nothing to do with the sign of a quantity—they simply indicate "reciprocal":

$$(6)^{-2} = (6^2)^{-1} = \frac{1}{6^2} = \frac{1}{36}$$

**Avoiding Common Errors**

Students sometimes believe that a negative exponent automatically makes the value of the expression negative. This is not true.

| EXPRESSION | CORRECT | INCORRECT | ALSO INCORRECT |
|---|---|---|---|
| $3^{-2}$ | $\dfrac{1}{3^2} = \dfrac{1}{9}$ | $\cancel{-3^2}$ | $\cancel{-\dfrac{1}{3^2}}$ |
| $x^{-3}$ | $\dfrac{1}{x^3}$ | $\cancel{-x^3}$ | $\cancel{-\dfrac{1}{x^3}}$ |

To help you see that the negative exponent rule makes sense, consider the following sequence of exponential expressions and their corresponding values.

$$2^3 = 8, \quad 2^2 = 4, \quad 2^1 = 2, \quad 2^0 = 1, \quad 2^{-1} = \frac{1}{2^1} \text{ or } \frac{1}{2}, \quad 2^{-2} = \frac{1}{2^2} \text{ or } \frac{1}{4}, \quad 2^{-3} = \frac{1}{2^3} \text{ or } \frac{1}{8}$$

Note that each time the exponent decreases by 1, the value of the expression is halved. For example, when we go from $2^3$ to $2^2$, the value of the expression goes from $8$ to $4$. If we continue decreasing the exponents beyond $2^0 = 1$, the next exponent in the pattern is $-1$. If we take half of 1 we get $\frac{1}{2}$. This pattern illustrates that $x^{-m} = \frac{1}{x^m}$.

## 2 Simplify Expressions Containing Negative Exponents

Generally, *when you are asked to simplify an exponential expression your final answer should contain no negative exponents.* You may simplify exponential expressions using the negative exponent rule and the rules of exponents as shown in the next examples.

**EXAMPLE 2** Use the negative exponent rule to write each expression with a positive exponent. Simplify the expressions further when possible.

    **a)** $y^{-5}$     **b)** $x^{-4}$     **c)** $2^{-3}$     **d)** $6^{-1}$     **e)** $-5^{-3}$     **f)** $(-5)^{-3}$

*Solution*

**a)** $y^{-5} = \dfrac{1}{y^5}$                 **b)** $x^{-4} = \dfrac{1}{x^4}$

**c)** $2^{-3} = \dfrac{1}{2^3} = \dfrac{1}{8}$         **d)** $6^{-1} = \dfrac{1}{6}$

**e)** $-5^{-3} = -\dfrac{1}{5^3} = -\dfrac{1}{125}$      **f)** $(-5)^{-3} = \dfrac{1}{(-5)^3} = \dfrac{1}{-125} = -\dfrac{1}{125}$

Now Try Exercise 11

**EXAMPLE 3** Use the negative exponent rule to write each expression with a positive exponent.

    **a)** $\dfrac{1}{x^{-2}}$        **b)** $\dfrac{1}{4^{-1}}$

*Solution*     First use the negative exponent rule on the denominator. Then simplify further.

   **a)** $\dfrac{1}{x^{-2}} = \dfrac{1}{\frac{1}{x^2}} = 1 \div \dfrac{1}{x^2} = \dfrac{1}{1} \cdot \dfrac{x^2}{1} = x^2$     **b)** $\dfrac{1}{4^{-1}} = \dfrac{1}{\frac{1}{4}} = 1 \div \dfrac{1}{4} = \dfrac{1}{1} \cdot \dfrac{4}{1} = 4$

Now Try Exercise 15

## Understanding Algebra

Negative exponent in denominator:

$$\frac{3}{x^{-4}} = 3 \div x^{-4} = 3 \div \frac{1}{x^4}$$

$$= 3 \cdot \frac{x^4}{1} = 3x^4$$

(Recall: dividing by a quantity is the same as multiplying by its reciprocal.)

### Helpful Hint

From Examples 2 and 3, we can see that when a factor is moved from the denominator to the numerator or from the numerator to the denominator, the sign of the *exponent* changes.

$$x^{-4} = \frac{1}{x^4} \qquad\qquad \frac{1}{x^{-4}} = x^4$$

$$3^{-5} = \frac{1}{3^5} \qquad\qquad \frac{1}{3^{-5}} = 3^5$$

Now let's look at additional examples that combine two or more of the rules presented so far.

**EXAMPLE 4** Simplify.  **a)** $\left(z^{-5}\right)^4$  **b)** $\left(4^2\right)^{-3}$

*Solution*

**a)** $\left(z^{-5}\right)^4 = z^{(-5)(4)}$   Power rule

$\quad\quad\quad\;\; = z^{-20}$

$\quad\quad\quad\;\; = \dfrac{1}{z^{20}}$   Negative exponent rule

**b)** $\left(4^2\right)^{-3} = 4^{(2)(-3)}$   Power rule

$\quad\quad\quad\;\; = 4^{-6}$

$\quad\quad\quad\;\; = \dfrac{1}{4^6}$   Negative exponent rule

Now Try Exercise 25

**EXAMPLE 5** Simplify.  **a)** $x^3 \cdot x^{-5}$  **b)** $3^{-4} \cdot 3^{-7}$

*Solution*

**a)** $x^3 \cdot x^{-5} = x^{3+(-5)}$   Product rule

$\quad\quad\quad\;\; = x^{-2}$

$\quad\quad\quad\;\; = \dfrac{1}{x^2}$   Negative exponent rule

**b)** $3^{-4} \cdot 3^{-7} = 3^{-4+(-7)}$   Product rule

$\quad\quad\quad\;\; = 3^{-11}$

$\quad\quad\quad\;\; = \dfrac{1}{3^{11}}$   Negative exponent rule

Now Try Exercise 51

**Avoiding Common Errors**

What is the sum of $3^2 + 3^{-2}$ ? Look carefully at the correct solution.

CORRECT

$3^2 + 3^{-2} = 9 + \dfrac{1}{9}$

$\quad\quad\quad\; = 9\dfrac{1}{9}$

INCORRECT

$3^2 + 3^{-2} = 0$

$3^2 + 3^{-2} = 3^0 = 1$

Note that $3^2 \cdot 3^{-2} = 3^{2+(-2)} = 3^0 = 1$.

**EXAMPLE 6** Simplify.  **a)** $\dfrac{b^{-5}}{b^{13}}$  **b)** $\dfrac{6^{-8}}{6^{-5}}$

*Solution*

**a)** $\dfrac{b^{-5}}{b^{13}} = b^{-5-13}$   Quotient rule

$\quad\quad\; = b^{-18}$

$\quad\quad\; = \dfrac{1}{b^{18}}$   Negative exponent rule

**b)** $\dfrac{6^{-8}}{6^{-5}} = 6^{-8-(-5)}$   Quotient rule

$\quad\quad\; = 6^{-8+5}$

$\quad\quad\; = 6^{-3}$

$\quad\quad\; = \dfrac{1}{6^3} \text{ or } \dfrac{1}{216}$   Negative exponent rule

Now Try Exercise 81

Read the following Helpful Hint carefully.

> **Helpful Hint**
>
> An alternative way to simplify a fraction involving a variable with a negative exponent in the numerator is to move the variable with the negative exponent from the numerator to the denominator and change the sign of the exponent:
>
> $$\frac{x^{-4}}{x^5} = \frac{1}{x^5 \cdot x^4} = \frac{1}{x^{5+4}} = \frac{1}{x^9}$$
>
> Similarly, if we have a fraction involving a variable with a negative exponent in the denominator we can move the variable with the negative exponent from the denominator to the numerator and change the sign of the exponent:
>
> $$\frac{y^3}{y^{-7}} = y^3 \cdot y^7 = y^{3+7} = y^{10}$$
>
> Now consider a division problem where a number or variable has a negative exponent in both its numerator and its denominator, such as in Example 6 **b)**. Another way to simplify such an expression is to move the variable with the more negative exponent from the numerator to the denominator, or from the denominator to the numerator, and change the sign of the exponent from negative to positive. For example,
>
> $$\frac{x^{-8}}{x^{-3}} = \frac{1}{x^8 \cdot x^{-3}} = \frac{1}{x^{8-3}} = \frac{1}{x^5} \qquad \text{Note that } -8 < -3.$$
>
> $$\frac{y^{-4}}{y^{-7}} = y^7 \cdot y^{-4} = y^{7-4} = y^3 \qquad \text{Note that } -7 < -4.$$

## EXAMPLE 7  Simplify.

**a)** $7x^4 \left(6x^{-9}\right)$      **b)** $\dfrac{16r^3s^{-3}}{8rs^2}$      **c)** $\dfrac{2x^2y^5}{8x^7y^{-3}}$

*Solution*

**a)** $7x^4 \left(6x^{-9}\right) = 7 \cdot 6 \cdot x^4 \cdot x^{-9} = 42x^{-5} = \dfrac{42}{x^5}$

**b)** $\dfrac{16r^3s^{-3}}{8rs^2} = \dfrac{16}{8} \cdot \dfrac{r^3}{r} \cdot \dfrac{s^{-3}}{s^2}$

$\qquad\qquad = 2 \cdot r^2 \cdot \dfrac{1}{s^5} = \dfrac{2r^2}{s^5}$    $s^{-3}$ was rewritten in the denominator as $s^3$

**c)** $\dfrac{2x^2y^5}{8x^7y^{-3}} = \dfrac{2}{8} \cdot \dfrac{x^2}{x^7} \cdot \dfrac{y^5}{y^{-3}}$

$\qquad\qquad = \dfrac{1}{4} \cdot \dfrac{1}{x^5} \cdot y^8 = \dfrac{y^8}{4x^5}$    $y^{-3}$ was rewritten in the numerator as $y^3$

Now Try Exercise 121

## EXAMPLE 8  Simplify.

**a)** $\left(5x^{-3}\right)^{-2}$    **b)** $\left(-5x^{-3}\right)^{-2}$      **c)** $\left(-5x^{-3}\right)^{-3}$

*Solution*    Begin by using the expanded power rule.

**a)** $\left(5x^{-3}\right)^{-2} = 5^{-2}x^{(-3)(-2)}$

$\qquad\qquad = 5^{-2}x^6$

$\qquad\qquad = \dfrac{1}{5^2}x^6$

$\qquad\qquad = \dfrac{x^6}{25}$

**b)** $(-5x^{-3})^{-2} = (-5)^{-2}x^{(-3)(-2)}$

$\qquad\qquad = \dfrac{1}{(-5)^2}x^6$

$\qquad\qquad = \dfrac{x^6}{25}$

**c)** $(-5x^{-3})^{-3} = (-5)^{-3}x^{(-3)(-3)}$

$\qquad\qquad = \dfrac{1}{(-5)^3}x^9$

$\qquad\qquad = \dfrac{1}{-125}x^9$

$\qquad\qquad = -\dfrac{x^9}{125}$

Now Try Exercise 105

**Avoiding Common Errors**

Can you explain why the simplification on the right is incorrect?

| CORRECT | INCORRECT |
|---|---|
| $\dfrac{x^3 y^{-2}}{w} = \dfrac{x^3}{wy^2}$ | $\dfrac{x^3 + y^{-2}}{w} \ne \dfrac{x^3}{w + y^2}$ |

The simplification on the right is incorrect because in the numerator $x^3 + y^{-2}$, the $y^{-2}$ is *not a factor*, it is a *term*.

**EXAMPLE 9**   Simplify $\left(\dfrac{2}{3}\right)^{-2}$.

**Solution**   By the expanded power rule, we may write

$$\left(\dfrac{2}{3}\right)^{-2} = \dfrac{2^{-2}}{3^{-2}} = \dfrac{\dfrac{1}{2^2}}{\dfrac{1}{3^2}} = \dfrac{1}{2^2}\cdot\dfrac{3^2}{1} = \dfrac{3^2}{2^2} = \dfrac{9}{4}$$

Now Try Exercise 95

If we examine the results of Example 9, we see that

$$\left(\dfrac{2}{3}\right)^{-2} = \dfrac{3^2}{2^2} = \left(\dfrac{3}{2}\right)^2.$$

This example illustrates that $\left(\dfrac{a}{b}\right)^{-m} = \left(\dfrac{b}{a}\right)^m$ when $a \ne 0$ and $b \ne 0$. Thus, for example,

$\left(\dfrac{3}{4}\right)^{-5} = \left(\dfrac{4}{3}\right)^5$ and $\left(\dfrac{5}{9}\right)^{-3} = \left(\dfrac{9}{5}\right)^3$. We can summarize this information as follows.

**A Fraction Raised to a Negative Exponent Rule**

For a fraction of the form $\dfrac{a}{b}$, $a \ne 0$ and $b \ne 0$, $\left(\dfrac{a}{b}\right)^{-m} = \left(\dfrac{b}{a}\right)^m$.

**EXAMPLE 10**   Simplify.    **a)** $\left(\dfrac{4}{5}\right)^{-3}$    **b)** $\left(\dfrac{x^5}{y^7}\right)^{-4}$

**Solution**   We use the above rule to simplify.

**a)** $\left(\dfrac{4}{5}\right)^{-3} = \left(\dfrac{5}{4}\right)^3 = \dfrac{5^3}{4^3} = \dfrac{125}{64}$    **b)** $\left(\dfrac{x^5}{y^7}\right)^{-4} = \left(\dfrac{y^7}{x^5}\right)^4 = \dfrac{y^{7\cdot4}}{x^{5\cdot4}} = \dfrac{y^{28}}{x^{20}}$

Now Try Exercise 97

EXAMPLE 11 Simplify.     **a)** $\left(\dfrac{x^2y^{-3}}{z^4}\right)^{-5}$     **b)** $\left(\dfrac{2x^{-3}y^2z}{x^2}\right)^2$

*Solution*

**a)** We will work part **a)** using two different methods. In method 1, we begin by using the expanded power rule. In method 2, we use a fraction raised to a negative exponent rule before we use the expanded power rule. You may use either method.

Method 1

$$\left(\frac{x^2y^{-3}}{z^4}\right)^{-5} = \frac{x^{2(-5)}y^{(-3)(-5)}}{z^{4(-5)}} \qquad \text{Expanded power rule}$$

$$= \frac{x^{-10}y^{15}}{z^{-20}} \qquad \text{Multiplied exponents}$$

$$= \frac{y^{15}z^{20}}{x^{10}} \qquad \text{Negative exponent rule}$$

Method 2

$$\left(\frac{x^2y^{-3}}{z^4}\right)^{-5} = \left(\frac{z^4}{x^2y^{-3}}\right)^5 \qquad \left(\frac{a}{b}\right)^{-m} = \left(\frac{b}{a}\right)^m$$

$$= \left(\frac{y^3z^4}{x^2}\right)^5 \qquad \begin{array}{l}\text{Simplified expression within}\\\text{parentheses}\end{array}$$

$$= \frac{y^{3\cdot5}z^{4\cdot5}}{x^{2\cdot5}} \qquad \text{Expanded power rule}$$

$$= \frac{y^{15}z^{20}}{x^{10}} \qquad \text{Multiplied exponents}$$

**b)** First simplify the expression within parentheses, then square the results. To simplify, we note that $\dfrac{x^{-3}}{x^2}$ becomes $\dfrac{1}{x^5}$.

$$\left(\frac{2x^{-3}y^2z}{x^2}\right)^2 = \left(\frac{2y^2z}{x^5}\right)^2 = \frac{2^2y^{2\cdot2}z^{1\cdot2}}{x^{5\cdot2}} = \frac{4y^4z^2}{x^{10}}$$

Now Try Exercise 125

**Understanding Algebra**

Remember: a negative exponent indicates "reciprocal."

$$\left(\frac{2x^2}{3y^4}\right)^{-3} = \left(\frac{3y^4}{2x^2}\right)^3 = \frac{27y^{12}}{8x^6}$$

---

**Summary of Rules of Exponents**

1. **Product Rule**      $x^m \cdot x^n = x^{m+n}$

2. **Quotient Rule**     $\dfrac{x^m}{x^n} = x^{m-n}, \qquad x \neq 0$

3. **Zero Exponent Rule**     $x^0 = 1, \qquad x \neq 0$

4. **Power Rule**     $(x^m)^n = x^{m \cdot n}$

5. **Power of a Product Rule**     $(xy)^n = x^n y^n$

6. **Power of a Quotient Rule**     $\left(\dfrac{x}{y}\right)^n = \dfrac{x^n}{y^n}, \qquad y \neq 0$

7. **Expanded Power Rule**     $\left(\dfrac{ax}{by}\right)^m = \dfrac{a^m x^m}{b^m y^m}, \qquad b \neq 0, y \neq 0$

8. **Negative Exponent Rule**     $x^{-m} = \dfrac{1}{x^m}, \qquad x \neq 0$

9. **Fraction Raised to a Negative Exponent Rule**     $\left(\dfrac{a}{b}\right)^{-m} = \left(\dfrac{b}{a}\right)^m, \qquad a \neq 0, b \neq 0$

# EXERCISE SET $5.2$    MathXL®   MyMathLab®

## Warm-Up Exercises

*Fill in the blanks with the appropriate word, phrase, or symbol(s) from the following list.*

$4x^4$                    $x^2$                    $-2^4$                    $\dfrac{1}{x^2}$

$(-2)^4$                  $2^{-4}$                  $1$                      $x^4$

$0$                       $\dfrac{1}{x^4}$          $\dfrac{1}{4x^4}$        $-2x$

1. The simplified from of $x^{-2}$ is _____ .
2. The simplified form of $(x^{-2})^{-1}$ is _____ .
3. Which is the largest: $-2^4$, $(-2)^4$, or $2^{-4}$? _____
4. Which number is negative: $-2^4$, $(-2)^4$, or $2^{-4}$? _____

5. The simplified form of $\dfrac{x^2}{x^{-2}}$ is _____ .
6. The expression $x^{-2}$ is undefined if $x$ has the value _____ .

7. The simplified form of $\dfrac{x^{-2}}{x^2}$ is _____ .
8. The simplified form of $\left(\dfrac{2x^2y^{-4}}{y^{-4}}\right)^{-2}$ is _____ .
9. The simplified form of $\left(\dfrac{x^{-2}y^{-4}}{2y^{-4}}\right)^{-2}$ is _____ .
10. Which of the following is in simplified form: $-2x$, $x^{-2}$, or $\left(\dfrac{1}{x}\right)^{-2}$ ? _____

## Practice the Skills

*Simplify.*

11. $x^{-6}$
12. $y^{-5}$
13. $5^{-1}$
14. $7^{-2}$

15. $\dfrac{1}{t^{-3}}$
16. $\dfrac{1}{b^{-4}}$
17. $\dfrac{1}{a^{-1}}$
18. $\dfrac{1}{y^{-4}}$

19. $\dfrac{1}{6^{-2}}$
20. $\dfrac{1}{4^{-3}}$
21. $(x^{-2})^{10}$
22. $(m^{-5})^{-2}$

23. $(y^{-5})^4$
24. $(a^5)^{-4}$
25. $(x^4)^{-2}$
26. $(x^{-9})^{-2}$

27. $(3^{-2})^{-1}$
28. $(2^{-3})^2$
29. $y^4 \cdot y^{-2}$
30. $x^{-3} \cdot x^1$

31. $x^7 \cdot x^{-5}$
32. $d^{-3} \cdot d^{-4}$
33. $3^{-2} \cdot 3^4$
34. $6^{-3} \cdot 6^6$

35. $\dfrac{r^5}{r^6}$
36. $\dfrac{x^2}{x^{-1}}$
37. $\dfrac{p^0}{p^{-3}}$
38. $\dfrac{x^{-2}}{x^5}$

39. $\dfrac{x^{-7}}{x^{-3}}$
40. $\dfrac{z^{-11}}{z^{-12}}$
41. $\dfrac{3^2}{3^{-1}}$
42. $\dfrac{4^2}{4^{-1}}$

43. $5^{-3}$
44. $x^{-7}$
45. $\dfrac{1}{z^{-9}}$
46. $\dfrac{1}{3^{-3}}$

47. $(p^{-4})^{-6}$
48. $(x^{-3})^{-4}$
49. $(y^{-2})^{-3}$
50. $z^9 \cdot z^{-12}$

51. $x^3 \cdot x^{-7}$
52. $x^{-3} \cdot x^{-5}$
53. $x^{-8} \cdot x^{-7}$
54. $8^{-3} \cdot 8^3$

55. $-4^{-2}$
56. $(-4)^{-2}$
57. $-(-4)^{-2}$
58. $-2^{-3}$

59. $(-2)^{-3}$
60. $-(-2)^{-3}$
61. $(-6)^{-2}$
62. $-6^{-2}$

63. $\dfrac{x^{-5}}{x^5}$
64. $\dfrac{y^6}{y^{-8}}$
65. $\dfrac{n^{-5}}{n^{-7}}$
66. $\dfrac{3^{-4}}{3}$

67. $\dfrac{9^{-3}}{9^{-3}}$
68. $(7q^5r^2)^0$
69. $(2^{-1} + 3^{-1})^0$
70. $(3^{-1} + 4^2)^0$

71. $\dfrac{2}{2^{-5}}$
72. $(z^{-5})^{-9}$
73. $(x^{-4})^{-2}$
74. $(x^{-7})^0$

**75.** $(x^0)^{-2}$

**76.** $(12^{-1})^{-2}$

**77.** $2^{-3} \cdot 2$

**78.** $7^5 \cdot 7^{-3}$

**79.** $7^{-5} \cdot 7^3$

**80.** $\dfrac{z^{-3}}{z^{-7}}$

**81.** $\dfrac{x^{-1}}{x^{-4}}$

**82.** $\dfrac{r^6}{r}$

**83.** $(4^2)^{-1}$

**84.** $(2^{-2})^{-2}$

**85.** $\dfrac{5}{5^{-2}}$

**86.** $\dfrac{x^6}{x^7}$

**87.** $\dfrac{3^{-4}}{3^{-2}}$

**88.** $x^{-10} \cdot x^8$

**89.** $\dfrac{8^{-1}}{8^{-1}}$

**90.** $2x^{-1}y$

**91.** $(-6x^2)^{-2}$

**92.** $(-3z^3)^{-2}$

**93.** $3x^{-2}y^2$

**94.** $-5x^4y^{-2}$

**95.** $\left(\dfrac{1}{2}\right)^{-2}$

**96.** $\left(\dfrac{3}{5}\right)^{-2}$

**97.** $\left(\dfrac{5}{4}\right)^{-3}$

**98.** $\left(\dfrac{3}{5}\right)^{-3}$

**99.** $\left(\dfrac{c^4}{d^2}\right)^{-2}$

**100.** $\left(\dfrac{x^2}{y}\right)^{-2}$

**101.** $-\left(\dfrac{r^4}{s}\right)^{-4}$

**102.** $-\left(\dfrac{m^3}{n^4}\right)^{-5}$

**103.** $-7a^{-3}b^{-4}$

**104.** $(3x^2y^3)^{-2}$

**105.** $(4x^5y^{-3})^{-3}$

**106.** $2w(3w^{-5})$

**107.** $(3z^{-4})(6z^{-5})$

**108.** $2x^5(3x^{-6})$

**109.** $4x^4(-2x^{-4})$

**110.** $(9x^5)(-3x^{-7})$

**111.** $(4x^2y)(3x^3y^{-1})$

**112.** $(7a^{-6}b^{-1})(a^9b^0)$

**113.** $(-5y^2)(4y^{-3}z^5)$

**114.** $(-3y^{-2})(5x^{-1}y^3)$

**115.** $\dfrac{24d^{12}}{3d^8}$

**116.** $\dfrac{8z^{-4}}{32z^{-2}}$

**117.** $\dfrac{36x^{-4}}{9x^{-2}}$

**118.** $\dfrac{18m^{-3}n^0}{6m^5n^9}$

**119.** $\dfrac{3x^4y^{-2}}{6y^3}$

**120.** $\dfrac{16x^{-7}y^{-2}}{4x^5y^2}$

**121.** $\dfrac{32x^4y^{-2}}{4x^{-2}y^0}$

**122.** $\dfrac{21x^{-3}z^2}{7xz^{-3}}$

**123.** $\left(\dfrac{5x^4y^{-7}}{z^3}\right)^{-2}$

**124.** $\left(\dfrac{b^4c^{-2}}{2d^{-3}}\right)^{-1}$

**125.** $\left(\dfrac{2r^{-5}s^9}{t^{12}}\right)^{-4}$

**126.** $\left(\dfrac{5m^{-1}n^{-3}}{p^2}\right)^{-3}$

**127.** $\left(\dfrac{x^3y^{-4}z}{y^{-2}}\right)^{-6}$

**128.** $\left(\dfrac{3p^{-1}q^{-2}r^3}{p^2}\right)^3$

**129.** $\left(\dfrac{p^6q^{-3}}{4p^8}\right)^2$

**130.** $\left(\dfrac{x^{12}y^5}{y^{-3}z}\right)^{-4}$

## Problem Solving

**131. a)** Does $p^{-1}q^{-1} = \dfrac{1}{pq}$ ?

**b)** Does $p^{-1} + q^{-1} = \dfrac{1}{p+q}$ ?

**132. a)** Does $\dfrac{x^{-1}y^2}{z} = \dfrac{y^2}{xz}$ ?

**b)** Does $\dfrac{x^{-1} + y^2}{z} = \dfrac{y^2}{x+z}$ ?

*Evaluate.*

**133.** $4^2 + 4^{-2}$

**134.** $8^2 + 8^{-2}$

**135.** $5^3 + 5^{-3}$

**136.** $6^{-3} + 6^3$

*Evaluate.*

**137.** $5^0 - 3^{-1}$

**138.** $4^{-1} - 3^{-1}$

**139.** $2^{-3} - 2^3 \cdot 2^{-3}$

**140.** $2 \cdot 4^{-1} + 4 \cdot 3^{-1}$

**141.** $2 \cdot 4^{-1} - 4 \cdot 3^{-1}$

**142.** $2 \cdot 4^{-1} - 3^{-1}$

**143.** $3 \cdot 5^0 - 5 \cdot 3^{-2}$

**144.** $10^2 + 10^{-2}$

*Determine the number that when placed in the shaded area makes the statement true.*

**145.** $3^{\blacksquare} = \dfrac{1}{9}$

**146.** $\dfrac{1}{2^{\blacksquare}} = 64$

**147.** $\dfrac{1}{6^{\blacksquare}} = 216$

**148.** $4^{\blacksquare} = \dfrac{1}{256}$

## Concept/Writing Exercises

**149.** Explain the difference between the product rule and the power rule. Give an example of each.

**150. a)** For what value of $x$ is $x^0 \neq 1$?

**b)** Write $pppqqqqrrrrr$ using exponents.

## Challenge Problems

*In Exercises 151–153, determine the number (or numbers) that when placed in the shaded area (or areas) make the statement true.*

**151.** $(x^{\blacksquare}y^3)^{-2} = \dfrac{x^4}{y^6}$

**152.** $(x^4y^{-3})^{\blacksquare} = \dfrac{y^9}{x^{12}}$

**153.** $(\blacksquare\, x^{\blacksquare}y^{-2})^3 = \dfrac{8}{x^9y^6}$

**154.** For any nonzero real number $a$, if $a^{-1} = x$, describe the following in terms of $x$.

    **a)** $-a^{-1}$         **b)** $\dfrac{1}{a^{-1}}$

**155.** Consider $(3^{-1} + 2^{-1})^0$. We know this is equal to 1 by the zero exponent rule. Determine the error in the following calculation. Explain your answer.

$$(3^{-1} + 2^{-1})^0 = (3^{-1})^0 + (2^{-1})^0$$
$$= 3^{-1(0)} + 2^{-1(0)}$$
$$= 3^0 + 2^0$$
$$= 1 + 1 = 2$$

## Group Activity

*Discuss and answer Exercise 156 as a group.*

**156.** Often problems involving exponents can be done in more than one way. Consider

$$\left(\frac{3x^2y^3}{x}\right)^{-2}$$

    **a)** Group member 1: Simplify this expression by first simplifying the expression within parentheses.

    **b)** Group member 2: Simplify this expression by first using the expanded power rule.

    **c)** Group member 3: Simplify this expression by first using the negative exponent rule.

    **d)** Compare your answers. If you did not all get the same answers, determine why.

    **e)** As a group, decide which method—**a)**, **b)**, or **c)**—was the easiest way to simplify this expression.

## Cumulative Review Exercises

[2.7] **157. Racing** If a race car travels 104 miles in 52 minutes, how far will it travel in 93 minutes (assuming all conditions stay the same)?

© Kevin L. Chesson/Shutterstock

[3.2] **158. Even Integers** The sum of two consecutive even integers is 190. Find the numbers.

[3.3] **159. Shed** Michael Beattie is building a rectangular shed. The perimeter of the shed is to be 56 feet. Determine the dimensions of the shed if the length is to be 8 feet less than twice the width.

[3.4] **160. Simple Interest** Mia Kattee invested $9000, part at 3% and part at 4% simple interest for a period of one year. How much was invested in each account if the interest earned in the 3% account was $32 greater than the interest earned in the 4% account?

[5.1] **161.** Simplify $(6xy^5)(3x^2y^4)$.

# 5.3 Scientific Notation

1 Convert numbers to and from scientific notation.

2 Recognize numbers in scientific notation with a coefficient of 1.

3 Do calculations using scientific notation.

## 1 Convert Numbers to and from Scientific Notation

Because it is difficult to work with and compute with really large or really small numbers, we often express such numbers using exponents. For example, the projected world population in 2050 of 8,909,000,000 people could be written as $8.909 \times 10^9$. An influenza virus with diameter 0.0000001 meter could be written as $1.0 \times 10^{-7}$ meter. Numbers such as $8.909 \times 10^9$ and $1.0 \times 10^{-7}$ are in a form called *scientific notation*.

### Scientific Notation

A number written in **scientific notation** is written as a number greater than or equal to 1 and less than 10 ($1 \le a < 10$) multiplied by some power of 10. The exponent on the 10 must be an integer.

Examples of Numbers in Scientific Notation

$$1.2 \times 10^6$$
$$3.762 \times 10^3$$
$$8.07 \times 10^{-2}$$
$$1.0 \times 10^{-5}$$

This is how we change the number 68,400 to scientific notation.

$$68,400 = 6.84 \times 10,000 \qquad \text{to go from 68,400 to 6.84 the decimal point was moved}$$
$$\text{4 places to the left}$$
$$= 6.84 \times 10^4 \qquad \text{Note that } 10,000 = 10 \cdot 10 \cdot 10 \cdot 10 = 10^4.$$

Therefore, $68,400 = 6.84 \times 10^4$. Note that the exponent on the 10, the 4, is the same as the number of places the decimal point was moved to the left.

Following are general steps to write a number in scientific notation.

### To Write a Number in Scientific Notation

1. Move the decimal point in the original number to the right of the first nonzero digit. This will give a number greater than or equal to 1 and less than 10.

2. Count the number of places you moved the decimal point to obtain the number in step 1. If the original number was 10 or greater, the count is considered positive. If the original number was less than 1, the count is considered negative.

3. Multiply the number obtained in step 1 by 10 raised to the count (power) found in step 2.

EXAMPLE 1   Write the following numbers in scientific notation.

**a)** 18,500      **b)** 0.0000416      **c)** 3,721,000      **d)** 0.0093

*Solution*

**a)** The original number is greater than 10; therefore, the exponent is positive. The decimal point in 18,500 belongs after the last zero.

$$18,500. = 1.85 \times 10^4$$
$$\text{4 places}$$

**b)** The original number is less than 1; therefore, the exponent is negative.

$$0.0000416 = 4.16 \times 10^{-5}$$
$$\text{5 places}$$

**c)** $3,721,000 = 3.721 \times 10^6$
$$\text{6 places}$$

**d)** $0.0093 = 9.3 \times 10^{-3}$

      3 places

Now Try Exercise 13

When we write a number in scientific notation, we are allowed to leave our answer with a negative exponent, as in parts **b)** and **d)** of Example 1.

Now we explain how to write a number in scientific notation as a number without exponents, or in decimal form.

---

### To Convert a Number from Scientific Notation to Decimal Form

1. Observe the exponent of the power of 10.
2. **a)** If the exponent is positive, move the decimal point in the number (greater than or equal to 1 and less than 10) to the right the same number of places as the exponent. It may be necessary to add zeros to the number. This will result in a number greater than or equal to 10.
   **b)** If the exponent is 0, do not move the decimal point. Drop the factor $10^0$ since it equals 1. This will result in a number greater than or equal to 1 but less than 10.
   **c)** If the exponent is negative, move the decimal point in the number to the left the same number of places as the exponent. It may be necessary to add zeros to the number. This will result in a number less than 1.

---

**EXAMPLE 2** Write each number without exponents.

**a)** $2.9 \times 10^4$        **b)** $6.28 \times 10^{-3}$        **c)** $7.95 \times 10^8$

*Solution*

**a)** Move the decimal point four places to the right (because the exponent, 4, is positive).

$$2.9 \times 10^4 = 2.9 \times 10{,}000 = 29{,}000$$

**b)** Move the decimal point three places to the left (because the exponent, –3, is negative)

$$6.28 \times 10^{-3} = 0.00628$$

**c)** Move the decimal point eight places to the right.

$$7.95 \times 10^8 = 795{,}000{,}000$$

Now Try Exercise 29

## 2 Recognize Numbers in Scientific Notation with a Coefficient of 1

The metric system is used throughout most of the world as the main system of measurement. Prefixes indicate a power of 10 and are always used with some type of base unit.

For example, a *millimeter* is $\dfrac{1}{1000}$ meter. A *megagram* is 1,000,000 grams, and so on. The following table illustrates the meaning of some prefixes.*

---

*There are other prefixes not listed. For example, centi is $10^{-2}$ or 0.01 times the base unit.

## Understanding Algebra

Some of the base units of the metric system include:

| Unit | Abbreviation | Measure |
|------|--------------|---------|
| meter | m | length |
| gram | g | mass |
| liter | l | volume |
| hertz | Hz | frequency |
| byte | B | computer memory |

| Prefix | Meaning | Symbol | Meaning as a Decimal Number |
|--------|---------|--------|------------------------------|
| nano | $10^{-9}$ | n | $\dfrac{1}{1,000,000,000}$ or 0.000000001 |
| micro | $10^{-6}$ | $\mu$ | $\dfrac{1}{1,000,000}$ or 0.000001 |
| milli | $10^{-3}$ | m | $\dfrac{1}{1000}$ or 0.001 |
| base unit** | $10^{0}$ | | 1 |
| kilo | $10^{3}$ | k | 1000 |
| mega | $10^{6}$ | M | 1,000,000 |
| giga | $10^{9}$ | G | 1,000,000,000 |

** The base unit is not a prefix. We included this row to include $10^{0}$ in the chart.

Note that if no numerical coefficient is indicated, it is assumed to be 1. Thus, $10^{-3}$ means $1.0 \times 10^{-3}$ and $10^{9}$ means $1.0 \times 10^{9}$. The table below shows the translation of "an 80 giabyte hard drive", "a 50 micrometer measurement", and "a 325 milligram prescription".

| Written Value | Meaning | Scientific Notation | In Decimal Form |
|---------------|---------|---------------------|------------------|
| 80 gigabytes 80 GB | $80 \times 10^{9}$ bytes | $8.0 \times 10^{10}$ | 80,000,000,000 bytes |
| 50 micrometers 50 $\mu$m | $50 \times 10^{-6}$ meters | $5.0 \times 10^{-5}$ | 0.00005 meter |
| 325 milligrams 325 mg | $325 \times 10^{-3}$ grams | $3.25 \times 10^{-1}$ | 0.325 gram |

In **Figure 5.1**, we see that the frequency of FM radio and VHF TV is about $10^{8}$ hertz ($10^{8}$ Hz). Thus, the frequency of FM radio is $10^{8} = 1.0 \times 10^{8} = 100,000,000$ hertz. This number, one hundred million hertz, can also be expressed as $100 \times 10^{6}$ or 100 megahertz, 100 MHz.

FIGURE 5.1

## EXAMPLE 3

a) Write 52 kilograms without the metric prefix "kilo" and then express the answer in scientific notation.

b) Write 183 nanoseconds without the metric prefix "nano" and then express the answer in scientific notation.

### Solution

a) 52 kilograms (52 kg) $= 52 \times 10^{3}$ grams $= 52,000$ grams $= 5.2 \times 10^{4}$ grams

b) 183 nanoseconds (183 ns) $= 183 \times 10^{-9}$ seconds $= 0.000000183$ second $= 1.83 \times 10^{-7}$ second

Now Try Exercise 45

## 3  Do Calculations Using Scientific Notation

We can use the rules of exponents presented in Sections 5.1 and 5.2 when working with numbers written in scientific notation.

**EXAMPLE 4**  Multiply $(4.2 \times 10^6)(2.0 \times 10^{-4})$. Write the answer in decimal form.

*Solution*  By the commutative and associative properties of multiplication we can rearrange the expression as follows.

$$(4.2 \times 10^6)(2.0 \times 10^{-4}) = (4.2 \times 2.0)(10^6 \times 10^{-4})$$

$$= 8.4 \times 10^{6+(-4)} \qquad \text{Product rule was used.}$$

$$= 8.4 \times 10^2 \qquad \text{Scientific notation}$$

$$= 840 \qquad \text{Decimal form}$$

Now Try Exercise 55

**EXAMPLE 5**  Divide $\dfrac{3.2 \times 10^{-6}}{5.0 \times 10^{-3}}$. Write the answer in scientific notation.

*Solution*
$$\frac{3.2 \times 10^{-6}}{5.0 \times 10^{-3}} = \left(\frac{3.2}{5.0}\right)\left(\frac{10^{-6}}{10^{-3}}\right)$$

$$= 0.64 \times 10^{-6-(-3)} \qquad \text{Quotient rule was used.}$$

$$= 0.64 \times 10^{-6+3}$$

$$= 0.64 \times 10^{-3}$$

$$= 6.4 \times 10^{-4} \qquad \text{Scientific notation}$$

The answer to Example 5 in decimal form would be 0.00064.

Now Try Exercise 59

## Using Your Calculator

What will your calculator show when you multiply very large or very small numbers? The answer depends on whether your calculator has the ability to display an answer in scientific notation. On calculators without the ability to express numbers in scientific notation, you will probably get an error message because the answer will be too large or too small for the display. For example, on a calculator without scientific notation:

$$8000000 \;\boxed{\times}\; 600000 \;\boxed{=}\; \boxed{\text{Error}}$$

On scientific calculators and graphing calculators, the answer to this example might be displayed in the following ways.

Possible displays

$$8000000 \;\boxed{\times}\; 600000 \;\boxed{=}\; \boxed{4.8 \qquad 12}$$

$$8000000 \;\boxed{\times}\; 600000 \;\boxed{=}\; \boxed{4.8 \qquad ^{12}}$$

$$8000000 \;\boxed{\times}\; 600000 \;\boxed{=}\; \boxed{4.8\text{E}12}$$

Each answer means $4.8 \times 10^{12}$. Let's look at one more example.

Possible displays

$$0.0000003 \;\boxed{\times}\; 0.004 \;\boxed{=}\; \boxed{1.2 \qquad -9}$$

$$0.0000003 \;\boxed{\times}\; 0.004 \;\boxed{=}\; \boxed{1.2 \qquad ^{-9}}$$

$$0.0000003 \;\boxed{\times}\; 0.004 \;\boxed{=}\; \boxed{1.2\text{E}-9}$$

Each display means $1.2 \times 10^{-9}$. On some calculators you will press the $\boxed{\text{ENTER}}$ key instead of the $\boxed{=}$ key. The TI-84 Plus graphing calculator displays answers using E, such as 4.8E12.

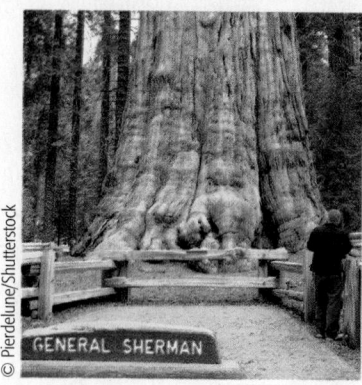
© Pierdelune/Shutterstock

GENERAL SHERMAN

**EXAMPLE 6** **Biggest Trees** The largest known tree (by volume) is a sequoia, called *General Sherman*, and has an estimated weight of $1.20 \times 10^7$ pounds. The largest coast redwood tree, called *Lost Monarch*, has an estimated weight of $1.11 \times 10^6$ pounds.

**a)** How much greater is the weight of the sequoia than the weight of the redwood?

**b)** How many times greater is the weight of the sequoia than the weight of the redwood?

*Solution*

**a)** Understand   We need to subtract $1.11 \times 10^6$ from $1.20 \times 10^7$. To add or subtract numbers in scientific notation, we generally make the exponents on the 10's the same. This will allow us to add or subtract the numerical values preceding the base while maintaining the common base and exponent on the base.

Translate   We can write $1.20 \times 10^7$ as $12.00 \times 10^6$. Now we subtract as follows.

Carry Out
$$
\begin{array}{r}
12.00 \times 10^6 \\
-1.11 \times 10^6 \\
\hline
10.89 \times 10^6 \quad \text{or } 1.089 \times 10^7
\end{array}
$$

Notice that we did not subtract the $10^6$'s. Think, of them as the unit "millions". In other words, 12.00 million minus 1.11 million is 10.89 million.

Check   We can check by writing the numbers out in decimal form:
$$
\begin{array}{r}
12,000,000 \\
-1,110,000 \\
\hline
10,890,000 \quad \text{or } 1.089 \times 10^7
\end{array}
$$

Answer   The largest sequoia is $1.089 \times 10^7$ pounds heavier than the largest coast redwood.

**b)** Understand   In part **b)** we are asked to find the *number of times* greater the weight of the sequoia is than the weight of the redwood. To find the number of times greater, we perform division.

Translate   Divide the estimated weight of the sequoia by the estimated weight of the redwood.

Carry Out
$$
\frac{1.20 \times 10^7}{1.11 \times 10^6} = \frac{1.20}{1.11} \times \frac{10^7}{10^6}
$$
$$
\approx 1.08 \times 10^{7-6} \quad \text{Quotient rule was used.}
$$
$$
\approx 1.08 \times 10
$$
$$
\approx 10.8
$$

Check   We can check by writing the numbers out in decimal form:
$$
\frac{12,000,000}{1,110,000} \approx 10.8
$$

Answer   The largest sequoia is 10.8 times heavier than the largest coast redwood.

Now Try Exercise 75

**EXAMPLE 7** **Fastest Computer** As of June 2008, the fastest computer in the world was the IBM "Roadrunner" located at the Los Alamos National Laboratory in New Mexico. It can perform $1.026 \times 10^{15}$ operations per second, which equates to it taking 0.000000000000000975 second to do one calculation. How long would it take this computer to perform 700 billion (700,000,000,000) calculations? *Source:* www.top500.org

*Solution*

**Understand**  The computer could perform 1 calculation in $1(0.000000000000000975)$ second, 2 calculations in $2(0.000000000000000975)$ second, 3 calculations in $3(0.000000000000000975)$ second, and 700 billion calculations in $700{,}000{,}000{,}000$ $(0.000000000000000975)$ second.

**Translate**  The number $0.000000000000000975$ written in scientific notation is $9.75 \times 10^{-16}$. To compute $700{,}000{,}000{,}000(0.000000000000000975)$, we convert each calculation to scientific notation:

**Carry Out**
$$700{,}000{,}000{,}000(0.000000000000000975) = (7.0 \times 10^{11})(9.75 \times 10^{-16})$$
$$= (7.0 \times 9.75)(10^{11} \times 10^{-16})$$
$$= 68.25 \times 10^{-5}$$
$$= 6.825 \times 10^{-4}$$
$$= 0.0006825$$

**Answer**  The IBM Roadrunner would take about $0.0006825$ of a second to perform 700 billion calculations.

The IBM Roadrunner computer

© Courtesy of IBM

Now Try Exercise 79

# EXERCISE SET 5.3

MathXL®  MyMathLab®

## Warm-Up Exercises

*Fill in the blanks with the appropriate word, phrase, or symbol(s) from the following list.*

| | | |
|---|---|---|
| $5.12 \times 10^{-4}$ | $5.12 \times 10^{-5}$ | integer |
| $5.12 \times 10^{5}$ | $5.12 \times 10^{-6}$ | 10 |
| $1 \le a < 10$ | 11 | $-11$ |

1. The number $0.0000512$ expressed in scientific notation is _____ .

2. If the positive number $a \times 10^{b}$ is written in scientific notation, then $a$ must be _____ .

3. If the positive number $a \times 10^{b}$ is greater than 10, then $b$ must be an _____ .

4. When $51.2 \times 10^{-5}$ is written in scientific notation, it is _____ .

5. When $0.512 \times 10^{-5}$ is written in scientific notation, it is _____ .

6. The number $512{,}000$ expressed in scientific notation is _____ .

7. When $6.00 \times 10^{4}$ is multiplied by $5.00 \times 10^{6}$ and put into scientific notation, the power of 10 will be _____ .

8. When $2.00 \times 10^{-4}$ is multiplied by $5.20 \times 10^{-8}$ and put into scientific notation, the power of 10 will be _____ .

## Practice the Skills

*Express each number in scientific notation.*

9. $20{,}000$
10. $1270$
11. $450$
12. $0.0125$

13. $350{,}000$
14. $3{,}610{,}000$
15. $7950$
16. $0.000089$

17. $0.053$
18. $19{,}000$
19. $0.000726$
20. $0.00000186$

21. $5{,}260{,}000{,}000$
22. $0.0075$
23. $0.00000914$
24. $74{,}100$

25. $220{,}300$
26. $0.08$
27. $0.005104$
28. $416{,}000$

*Express each number in decimal form (without exponents).*

29. $4.3 \times 10^{4}$
30. $1.63 \times 10^{-4}$
31. $9.32 \times 10^{-6}$
32. $6.15 \times 10^{5}$

33. $2.13 \times 10^{-5}$
34. $7.26 \times 10^{-6}$
35. $6.25 \times 10^{5}$
36. $4.6 \times 10^{1}$

**37.** $9.0 \times 10^6$      **38.** $6.475 \times 10^1$      **39.** $5.35 \times 10^2$      **40.** $3.14 \times 10^{-2}$

**41.** $7.73 \times 10^{-7}$      **42.** $6.201 \times 10^{-4}$      **43.** $1.0 \times 10^4$      **44.** $7.13 \times 10^{-4}$

*In Exercises 45–52, write the quantity without metric prefixes and then write the quantity in scientific notation. See Example 3.*

**45.** 8 micrometers      **46.** 29 micrograms      **47.** 125 gigawatts      **48.** 8.7 nanoseconds

**49.** 15.3 kilometers      **50.** 80.2 megahertz      **51.** 48.2 millimeters      **52.** 3.12 milligrams

*Perform each indicated operation and express each number in decimal form (without exponents).*

**53.** $(3.0 \times 10^2)(3.0 \times 10^5)$      **54.** $(2.0 \times 10^{-3})(3.0 \times 10^2)$      **55.** $(2.7 \times 10^{-6})(9.0 \times 10^4)$

**56.** $(1.3 \times 10^{-8})(1.74 \times 10^6)$      **57.** $(1.6 \times 10^{-2})(4.0 \times 10^{-3})$      **58.** $(4.0 \times 10^5)(1.2 \times 10^{-4})$

**59.** $\dfrac{3.9 \times 10^{-5}}{3.0 \times 10^{-2}}$      **60.** $\dfrac{6.0 \times 10^{-3}}{3.0 \times 10^1}$      **61.** $\dfrac{7.5 \times 10^6}{3.0 \times 10^3}$

**62.** $\dfrac{1.4 \times 10^7}{4.0 \times 10^8}$      **63.** $\dfrac{2.0 \times 10^4}{8.0 \times 10^{-2}}$      **64.** $\dfrac{1.6 \times 10^4}{8.0 \times 10^{-3}}$

*Perform each indicated operation by first converting each number to scientific notation. Write the answer in scientific notation.*

**65.** $(700,000)(8,000,000)$      **66.** $(0.003)(0.00015)$      **67.** $(0.0004)(320)$

**68.** $(67,000)(200,000)$      **69.** $\dfrac{5,600,000}{8000}$      **70.** $\dfrac{0.00004}{200}$

**71.** $\dfrac{0.00035}{0.000002}$      **72.** $\dfrac{150,000}{0.0005}$

**73.** List the following numbers from smallest to largest: $7.3 \times 10^2$, $3.3 \times 10^{-4}$, $1.75 \times 10^6$, $5.3$.

**74.** List the following numbers from smallest to largest: $4.8 \times 10^5$, $3.2 \times 10^{-1}$, $4.6$, $8.3 \times 10^{-4}$.

## Problem Solving

*In Exercises 75–92, write the answer in decimal form (without exponents) unless asked to do otherwise.*

**75. Population** In 2008, the U.S. population was about $3.05 \times 10^8$ people and the world population was about $6.72 \times 10^9$ people.

**a)** How many people lived outside the United States in 2008?

**b)** How many times greater is the world population than the U.S. population?

**76. Diapers** Laid end to end the 18 billion disposable diapers thrown away in the United States each year would reach the moon and back seven times (seven round trips).

**a)** Write 18 billion in scientific notation.

**b)** If the distance from Earth to the moon is $2.38 \times 10^5$ miles, what is the length of all these diapers placed end to end? Write your answer in scientific notation and as a number without exponents.

**77. Niagara Falls** A treaty between the United States and Canada requires that during the tourist season a minimum of 100,000 cubic feet of water per second flows over Niagara Falls (another 130,000 to 160,000 cubic feet/sec is diverted for power generation). Find the minimum volume of water that will flow over the falls in a 24-hour period during the tourist season.

© Canada Bureau of Travel

**78. PGA Tour** The top five money leaders from the 2007 PGA Tour are listed above and to the right.

| Golfer | Winnings |
| --- | --- |
| Tiger Woods | $10,867,052 |
| Phil Mickelson | $5,819,988 |
| Vijay Singh | $4,728,376 |
| Steve Stricker | $4,663,077 |
| K. J. Choi | $4,587,859 |

*Source:* www.pgatour.com

**a)** Write, in scientific notation, Tiger Woods's and Vijay Singh's approximate money total.

**b)** Determine, in scientific notation, the average daily amount that Tiger Woods earned on the PGA tour in 2007.

**79. Computer Speed** If a computer can do a calculation in 0.000002 second, how long, in seconds, would it take the computer to do 8 trillion (8,000,000,000,000) calculations? Write your answer in scientific notation.

**80. Fortune Cookies** Steven Yang, who runs M & Y Trading Company, a San Francisco Company, prints about 90% of all the $1.02 \times 10^9$ fortunes in fortune cookies in America in any given year.

**a)** How many fortunes does Yang print in a year?

**b)** How many fortunes does Yang print in a day?

**81. Movies** The gross ticket sales of the top five movies in the United States as of August 28, 2008, are listed below.

| Movie | Year Released | Approximate U.S. Gross Ticket Sales |
|---|---|---|
| **1.** *Titanic* | 1997 | $601,000,000 |
| **2.** *The Dark Knight* | 2008 | $518,000,000 |
| **3.** *Star Wars Ep. IV: A New Hope* | 1977 | $461,000,000 |
| **4.** *Shrek 2* | 2004 | $437,000,000 |
| **5.** *E.T.* | 1982 | $435,000,000 |

*Source:* http://www.imdb.com/boxoffice/

**a)** How much greater was the gross ticket sales of *Titanic* than *E.T.*? Write your answer in scientific notation.

**b)** How many times greater was the gross ticket sales of *Titanic* than *E.T.*?

**82. Melanoma** The number of new cases of melanoma has increased from $5.5 \times 10^4$ in 2003 to about $6.25 \times 10^4$ in 2008.

**a)** How much greater was the number of cases of melanoma in 2008 than in 2003?

**b)** How many times greater was the number of cases of melanoma in 2008 than in 2003?

*Source:* National Cancer Institute

**83. World's Richest** The worth of Warren Buffett (the world's richest person in 2007) was $6.2 \times 10^{10}$. The worth of Charles Ergen (the world's eighty-seventh richest person in 2007) was $9.5 \times 10^9$.

*Source:* www.forbes.com

**a)** How much greater was the worth of Warren Buffett than the worth of Charles Ergen in 2007?

**b)** How many times greater was the worth of Warren Buffett than the worth of Charles Ergen in 2007?

**84. Light from the Sun** The sun is $9.3 \times 10^7$ miles from Earth. Light travels at a speed of $1.86 \times 10^5$ miles per second. How long, in both seconds and in minutes, does it take light from the sun to reach Earth?

**85. Astronomy** The mass of Earth, Earth's moon, and the planet Jupiter are listed below.

Earth:   5,794,000,000,000,000,000,000,000 metric tons

Moon:   73,400,000,000,000,000,000 metric tons

Jupiter:   1,899,000,000,000,000,000,000,000,000 metric tons

**a)** Write the mass of Earth, the moon, and Jupiter in scientific notation.

**b)** How many times greater is the mass of Earth than the mass of the moon?

**c)** How many times greater is the mass of Jupiter than the mass of Earth?

**86. Missing Persons** Of the nearly 47,600 active adult missing-person FBI cases, about 53% were men. Determine, in scientific notation, the approximate number of active adult missing-person FBI cases that were men. *Source: Rochester Democrat & Chronicle*

**87. Camera Products and Services** The circle graph below shows the percentage of revenue from digital products and services and from traditional products and services during the first quarter of 2008 for Kodak. The company's total revenue for the first quarter of 2008 was $2,093,000,000. How much revenue was from digital products and services? Write your answer in scientific notation.

**Kodak First Quarter 2008 Revenues**

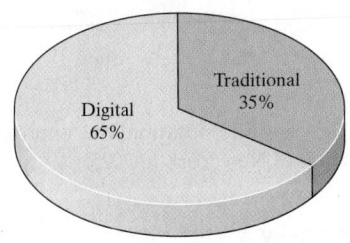

*Source:* www.bnet.com

**88. Chesapeake Bay Oysters** The following bar graph shows the approximate oyster population in the Chesapeake Bay.

**Chesapeake Bay Oyster Population**

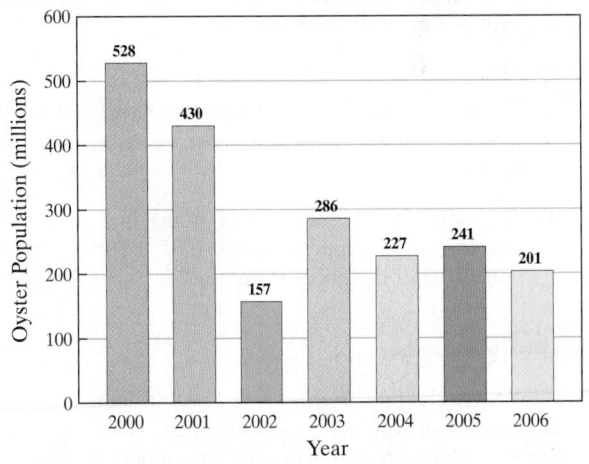

*Source:* Maryland Department of Natural Resources

**a)** Give, in scientific notation, the approximate population of oysters in 2000.

**b)** Give, in scientific notation, the approximate population of oysters in 2006.

**c)** Determine, in scientific notation, the approximate difference between the population of oysters in 2006 and in 2000.

**89. Population Projection** The following bar graph shows the population projections for certain states for the year 2025.

**Projection of Total Population: 2025**

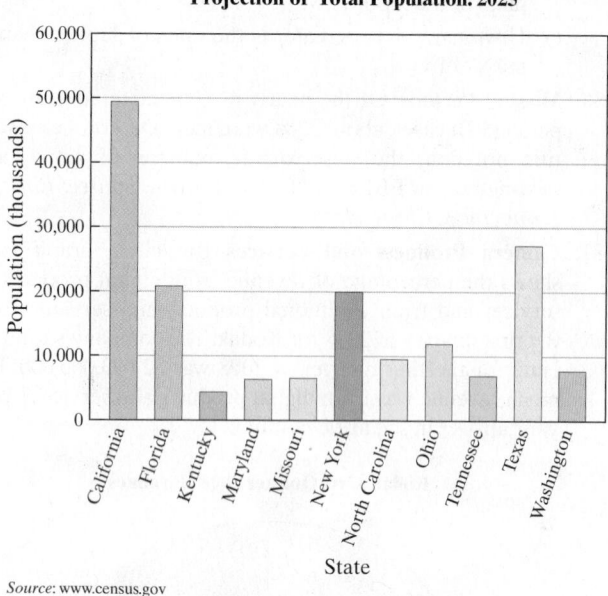

*Source*: www.census.gov

**a)** Write, in scientific notation, the approximate projected population of New York in 2025.

**b)** Determine, in scientific notation, the approximate sum of the projected populations of the two most populated states in 2025.

**c)** Determine, in scientific notation, the approximate difference between the projected population of the two most populated states in 2025.

**90. The Spice Girls** The Spice Girls hold the Guinness World Record as having the highest ever annual earnings by a girl band. The four group members had a total income of $49,000,000 in 1998. If each member of the band received an equal amount, determine, in scientific notation, the amount earned by each group member in 1998.

**91. Structure of Matter** An article in *Scientific American* states that physicists have created a *standard model* that describes the structure of matter down to $10^{-18}$ meters. If this number is written without exponents, how many zeroes would there be to the right of the decimal point?

**92. A Large Number** Avogadro's number, named after the nineteenth-century Italian chemist Amedeo Avogadro, is roughly $6.02 \times 10^{23}$. It represents the number of atoms in 12 grams of pure carbon.

**a)** If this number were written out in decimal form, how many digits would it contain?

**b)** What is the number of atoms in 1 gram of pure carbon? Write your answer in scientific notation.

## Concept/Writing Exercises

**93.** Describe the form of a number given in scientific notation.

**94.** A **googol** is the number 1 followed by 100 zeros and is the largest number many calculators can handle. From what we learned in this section, a googol can be written as $10^{100}$ or, in scientific notation, as $1.0 \times 10^{100}$. What is a googol raised to the fifth power in scientific notation?

**95.** See Exercise 94.

**a)** Estimate the number of seconds that have occurred since January 1, 1900. Is that number more or less than a googol?

**b)** The total number of ways of filling in a nine-by-nine *Sudoku* grid is approximately $6.67 \times 10^{21}$. How many times greater is a googol than this?

**96.** The five planets closest to the sun are, in order, Mercury, Venus, Earth, Mars, and Jupiter. Their average distances from the sun are given below:

| Planet | Average Distance to the Sun in Miles |
| --- | --- |
| Mercury | 35,983,610 |
| Venus | 67,232,360 |
| Earth | 92,957,100 |
| Mars | 141,635,300 |
| Jupiter | 484,632,000 |

In scientific notation, give the ratio of Jupiter's average distance to the sun to Mercury's average distance to the sun.

## Challenge Problems

**97. The Movie Contact** In the movie *Contact*, Jodie Foster plays an astronomer who makes the statement "There are 400 billion stars out there just in our universe alone. If only one out

© NOAO/NSF

of a million of those had planets, and if just one out of a million of those had life, and if just one out of a million of those had intelligent life, there would be literally millions of civilizations out there." Do you believe this statement is correct? Explain your answer.

**98.** How many times smaller is 1 nanosecond than one millisecond?

**99.** How many times, either greater or smaller, is $10^{-12}$ meters than $10^{-18}$ meters?

**100. Light Year** Light travels at a speed of $1.86 \times 10^5$ miles per second. A **light year** is the distance light travels in one year. Determine the number of miles in a light year.

## Group Activity

*Discuss and answer Exercise 101 as a group.*

**101. A Million versus a Billion**  Do you have any idea of the difference in size between a million (1,000,000), a billion (1,000,000,000), and a trillion (1,000,000,000,000)?

**a)** Write a million, a billion, and a trillion in scientific notation.

**b)** Group member 1: Determine how long it would take to spend a million dollars if you spent $1000 a day.

**c)** Group member 2: Repeat part **b)** for a billion dollars.

**d)** Group member 3: Repeat part **b)** for a trillion dollars.

**e)** As a group, determine how many times greater a billion dollars is than a million dollars.

## Cumulative Review Exercises

[1.9] **102.** Evaluate $4x^2 + 3x + \dfrac{x}{2}$ when $x = 0$.

[2.3] **103. a)** If $-x = -\dfrac{3}{2}$, what is the value of $x$?

  **b)** If $5x = 0$, what is the value of $x$?

[2.5] **104.** Solve the equation $2x - 3(x - 2) = x + 2$.

[5.1] **105.** Simplify $\left( \dfrac{-2x^5 y^7}{8x^8 y^3} \right)^3$.

---

### Mid-Chapter Test: 5.1–5.3

*To find out how well you understand the chapter material to this point, take this brief test. The answers, and the section where the material was initially discussed, are given in the back of the book. Review any questions that you answered incorrectly.*

*Simplify.*

**1.** $y^{11} \cdot y^{20}$

**2.** $\dfrac{x^{13}}{x^{10}}$

**3.** $(-3x^5 y^7)(-2xy^6)$

**4.** $\dfrac{6a^{12} b^8}{9a^7 b^2}$

**5.** $(-4x^2 y^4)^3$

**6.** $\left( \dfrac{-5s^4 t^6}{10s^6 t^3} \right)^2$

**7.** $(7x^9 y^5)(-3xy^4)^2$

**8.** $\dfrac{p^{-3}}{p^5}$

**9.** $x^{-4} \cdot x^{-6}$

**10.** $(3^{-1} + 5^2)^0$

**11.** $\left( \dfrac{3}{7} \right)^{-2}$

**12.** $(8x^{-2} y^5)(4x^3 y^{-6})$

**13.** $\dfrac{6m^{-4} n^{-7}}{2m^0 n^{-1}}$

**14.** $\left( \dfrac{2x^{-3} y^{-4}}{x^3 y z^{-2}} \right)^{-2}$

**15. a)** Describe how to write a number 10 or greater in scientific notation.

  **b)** Describe how to write a number less than 1 in scientific notation.

**16.** Express 6,540,000,000 in scientific notation.

**17.** Express $3.27 \times 10^{-5}$ in decimal form (without exponents).

**18.** Write 18.9 kilometers without metric prefixes.

**19.** Multiply $(3.4 \times 10^{-6})(7.0 \times 10^3)$ and express the answer in decimal form (without exponents).

**20.** Divide $\dfrac{0.00006}{200}$ by first converting to scientific notation. Write the answer in scientific notation.

---

# 5.4 Addition and Subtraction of Polynomials

**1** Identify polynomials.

**2** Add polynomials.

**3** Subtract polynomials.

**4** Subtract polynomials in columns.

## **1** Identify Polynomials

**A polynomial in $x$** is an expression that is the sum of a finite number of terms of the form $ax^n$, for any real number $a$ and any *whole number n.*

| Examples of Polynomials | Not Polynomials | |
|---|---|---|
| $8x$ | $4x^{1/2}$ | (Fractional exponent) |
| $\dfrac{1}{3}x - 4$ | $3x^2 + 4x^{-1} + 5$ | (Negative exponent) |
| $x^2 - 2x + 1$ | $4 + \dfrac{1}{x}$ | $\left( \dfrac{1}{x} = x^{-1}, \text{negative exponent} \right)$ |

A polynomial is written in **descending order** when the exponents on the variable decrease from left to right.

### Example of Polynomial in Descending Order

$$2x^4 + 4x^2 - 6x + 3$$

The constant term 3 is last because it can be written as $3x^0$. Remember that $x^0 = 1$.

A polynomial can be in more than one variable. For example, $3xy + 2$ is a polynomial in two variables, $x$ and $y$.

A polynomial with one term is called a **monomial**. A **binomial** is a two-termed polynomial. A **trinomial** is a three-termed polynomial. Polynomials containing more than three terms usually are not given special names.

| Type of Polynomial | Number of Terms | Examples |
|---|---|---|
| Monomial | One | $8, \ 4x, \ -6x^2$ |
| Binomial | Two | $x + 5, \ x^2 - 6, \ 4y^2 - 5y$ |
| Trinomial | Three | $x^2 - 2x + 3, \ 3z^2 - 6z + 7$ |

### Degree

The **degree of a term in one variable** is the exponent on the variable in that term. The degree of $2x^3$ is 3.

The **degree of a term in two or more variables** is the sum of the exponents on the variables. The degree of $4x^2y^3$ is 5.

The **degree of a polynomial** is the same as that of its highest-degree term.

| Term | Degree of Term | |
|---|---|---|
| $4x^2$ | 2 | |
| $2x^2y^5$ | 7 | $(2 + 5 = 7)$ |
| $-23a^3b^4c$ | 8 | $(3 + 4 + 1 = 8)$ |
| $-5x$ | 1 | $(-5x$ is $-5x^1)$ |
| 3 | 0 | $(3$ is $3x^0)$ |

| Polynomial | Degree of Polynomial | |
|---|---|---|
| $8x^3 + 2x^2 - 3x + 4$ | 3 | $(8x^3$ is highest-degree term.$)$ |
| $x^2 - 4$ | 2 | $(x^2$ is highest-degree term.$)$ |
| $6x - 5$ | 1 | $(6x$ or $6x^1$ is highest-degree term.$)$ |
| 4 | 0 | $(4$ or $4x^0$ is highest-degree term.$)$ |
| $x^2y^4 + 2x + 3$ | 6 | $(x^2y^4$ is highest-degree term.$)$ |

### 2 Add Polynomials

In Section 2.1, we stated that like terms are terms having the same variables and the same exponents. That is, like terms may differ only in their numerical coefficients.

### Examples of Like Terms

$$3, \ -5$$
$$2x, \ x$$
$$-2x^2, \ 4x^2$$
$$3y^2, \ 5y^2$$
$$3xy^2, \ 5xy^2$$

**To Add Polynomials**

To add polynomials, combine the like terms of the polynomials.

**EXAMPLE 1** Add $(4x^2 + 6x + 3) + (2x^2 + 5x - 1)$.

**Solution** Remember that $(4x^2 + 6x + 3) = 1(4x^2 + 6x + 3)$ and $(2x^2 + 5x - 1) = 1(2x^2 + 5x - 1)$.

$$(4x^2 + 6x + 3) + (2x^2 + 5x - 1)$$
$$= 1(4x^2 + 6x + 3) + 1(2x^2 + 5x - 1)$$
$$= 4x^2 + 6x + 3 + 2x^2 + 5x - 1 \qquad \text{Used the distributive property to remove parentheses.}$$
$$= \underline{4x^2 + 2x^2} + \underline{6x + 5x} + \underline{3 - 1} \qquad \text{Rearranged terms.}$$
$$= 6x^2 + 11x + 2 \qquad \text{Combined like terms.}$$

Now Try Exercise 67

**EXAMPLE 2** Add $(5a^2 + 3a + b) + (a^2 - 7a + 3)$.

**Solution**
$$(5a^2 + 3a + b) + (a^2 - 7a + 3)$$
$$= 5a^2 + 3a + b + a^2 - 7a + 3 \qquad \text{Removed parentheses.}$$
$$= \underline{5a^2 + a^2} + \underline{3a - 7a} + b + 3 \qquad \text{Rearranged terms.}$$
$$= 6a^2 - 4a + b + 3 \qquad \text{Combined like terms.}$$

Now Try Exercise 75

**EXAMPLE 3** Add $(3x^2y - 4xy + y) + (x^2y + 2xy + 3y)$.

**Solution**
$$(3x^2y - 4xy + y) + (x^2y + 2xy + 3y)$$
$$= 3x^2y - 4xy + y + x^2y + 2xy + 3y \qquad \text{Removed parentheses.}$$
$$= \underline{3x^2y + x^2y} \ \underline{- 4xy + 2xy} + \underline{y + 3y} \qquad \text{Rearranged terms.}$$
$$= 4x^2y - 2xy + 4y \qquad \text{Combined like terms.}$$

Now Try Exercise 77

Usually, when we add polynomials, we will do so horizontally as in Examples 1 through 3. Sometimes, it is easier to add polynomials in columns.

**To Add Polynomials in Columns**

1. Arrange polynomials in descending order, one under the other with like terms in the same columns.
2. Add the terms in each column.

**EXAMPLE 4** Add $5x^2 - 9x - 3$ and $-3x^2 - 4x + 6$ using columns.

**Solution**
$$\begin{array}{r} 5x^2 - 9x - 3 \\ -3x^2 - 4x + 6 \\ \hline 2x^2 - 13x + 3 \end{array}$$

Now Try Exercise 83

**EXAMPLE 5** Add $5w^3 + 2w - 4$ and $2w^2 - 6w - 3$ using columns.

**Solution** Since the polynomial $5w^3 + 2w - 4$ does not have a $w^2$ term, we will add the term $0w^2$ to the polynomial. This procedure helps in aligning like terms.

$$\begin{array}{r} 5w^3 + 0w^2 + 2w - 4 \\ 2w^2 - 6w - 3 \\ \hline 5w^3 + 2w^2 - 4w - 7 \end{array}$$

Now Try Exercise 85

## 3  Subtract Polynomials

### To Subtract Polynomials

1. Use the distributive property to remove parentheses. This will have the effect of changing the sign of *every* term within the parentheses of the polynomial being subtracted.
2. Combine like terms.

**EXAMPLE 6**  Subtract $(3x^2 - 2x + 5) - (x^2 - 3x + 4)$.

*Solution*    $(3x^2 - 2x + 5)$ means $1(3x^2 - 2x + 5)$ and $(x^2 - 3x + 4)$ means $1(x^2 - 3x + 4)$. We use this information in the solution, as shown below.

$$(3x^2 - 2x + 5) - (x^2 - 3x + 4) = 1(3x^2 - 2x + 5) - 1(x^2 - 3x + 4)$$

$$= 3x^2 - 2x + 5 - x^2 + 3x - 4 \qquad \text{Removed parentheses.}$$

$$= \underline{3x^2 - x^2} \; \underline{- 2x + 3x} \; \underline{+ 5 - 4} \qquad \text{Rearranged terms.}$$

$$= \qquad 2x^2 \qquad + \quad x \qquad + \quad 1 \qquad \text{Combined like terms.}$$

Now Try Exercise 97

**EXAMPLE 7**  Subtract $(-3x^2 - 5x + 3)$ from $(x^3 + 2x + 6)$.

*Solution*    $$(x^3 + 2x + 6) - (-3x^2 - 5x + 3)$$

$$= x^3 + 2x + 6 + 3x^2 + 5x - 3 \qquad \text{Removed parentheses.}$$

$$= x^3 + 3x^2 \; \underline{+ 2x + 5x} \; \underline{+ 6 - 3} \qquad \text{Rearranged terms.}$$

$$= x^3 + 3x^2 + \quad 7x \qquad + \quad 3 \qquad \text{Combined like terms.}$$

Now Try Exercise 107

---

### Understanding Algebra

Remember, when a negative sign precedes parentheses, the sign of each term within parentheses is changed:

$$-(-3x^2 - 5x + 3)$$

becomes

$$+3x^2 + 5x - 3$$

---

### Avoiding Common Errors

One of the most common mistakes occurs when subtracting polynomials. When subtracting one polynomial from another, *the sign of each term in the polynomial being subtracted must be changed, not just the sign of the first term.*

| CORRECT | INCORRECT |
|---|---|
| $6x^2 - 4x + 3 - (2x^2 - 3x + 4)$ | $6x^2 - 4x + 3 - (2x^2 - 3x + 4)$ |
| $= 6x^2 - 4x + 3 \; - \; 2x^2 \; + \; 3x \; - \; 4$ | $= 6x^2 - 4x + 3 - 2x^2 - 3x + 4$ |
| $= 4x^2 - x - 1$ | $= 4x^2 - 7x + 7$ |

Do not make this mistake!

---

## 4  Subtract Polynomials in Columns

### To Subtract Polynomials in Columns

1. Write *the polynomial being subtracted* below the polynomial from which it is being subtracted. List like terms in the same column.
2. *Change the sign of each term* in the polynomial being subtracted. (This step can be done mentally, if you like.)
3. Add the terms in each column.

**EXAMPLE 8** Subtract $(2x^2 - 4x + 6)$ from $(4x^2 + 5x + 8)$ using columns.

*Solution*  Align like terms in columns (step 1).

$$\begin{array}{r} 4x^2 + 5x + 8 \\ -(2x^2 - 4x + 6) \end{array}\quad\text{Align like terms.}$$

Change *all* signs in the second row (step 2); then add (step 3).

$$\begin{array}{r} 4x^2 + 5x + 8 \\ -\;2x^2 + 4x - 6 \\ \hline 2x^2 + 9x + 2 \end{array}\quad\begin{array}{l}\text{Changed all signs.}\\ \text{Added terms.}\end{array}$$

Now Try Exercise 113

**EXAMPLE 9** Subtract $(2x^2 - 6)$ from $(-3x^3 + 4x - 3)$ using columns.

*Solution*  To help align like terms, write each expression in descending order. If any power of $x$ is missing, write that term with a numerical coefficient of 0.

$$-3x^3 + 4x - 3 = -3x^3 + 0x^2 + 4x - 3$$
$$2x^2 - 6 = 2x^2 + 0x - 6$$

Align like terms.

$$\begin{array}{r} -3x^3 + 0x^2 + 4x - 3 \\ -(2x^2 + 0x - 6) \end{array}$$

Change all signs in the second row; then add the terms in each column.

$$\begin{array}{r} -3x^3 + 0x^2 + 4x - 3 \\ -\;2x^2 - 0x + 6 \\ \hline -3x^3 - 2x^2 + 4x + 3 \end{array}$$

Now Try Exercise 115

**NOTE:** You may find that you can change the signs mentally and can therefore align and change the signs in one step.

---

# EXERCISE SET 5.4

MathXL®  MyMathLab®

MathXL®      MyMathLab

## Warm-Up Exercises

*Fill in the blanks with the appropriate word, phrase, or symbol(s) from the following list.*

| | | | |
|---|---|---|---|
| $3x - 4$ | 3 | 2 | 1 |
| 5 | 6 | $-5x^3 + 3x + 4$ | $3x + 4$ |
| 7 | True | False | |

**1.** When $3x - 5x^3 + 4$ is written in descending order, it is _____ .

**2.** $(4x^3 + 10x - 1) - (4x^3 + 7x + 3) = $ _____ .

**3.** $(2x^3 - 8x + 1) + (-2x^3 + 11x + 3) = $ _____ .

**4.** A trinomial is a polynomial with _____ term(s).

**5.** A monomial is a polynomial with _____ term(s).

**6.** A binomial is a polynomial with _____ term(s).

**7.** The degree of the term $6x^2y^3$ is _____ .

**8.** The degree of the polynomial $x^6 - 8x^7 + 4$ is _____ .

**9.** (True or False) The expression $\frac{1}{2}y^3 - 7y + 8$ is a polynomial in $y$. _____

**10.** (True or False) The expression $\frac{1}{2y^3} - 7y + 8$ is a polynomial in $y$. _____

---

## Practice the Skills

*Indicate the degree of each term.*

**11.** $x^3$

**12.** $6x^2$

**13.** $-2x$

**14.** $-56t^9$

**15.** $7s^4$

**16.** $4s^7$

**17.** $x^5$

**18.** $z^{11}$

**19.** $-3b^8$

**20.** $5a^4$

**21.** $x^2y$

**22.** $a^4b^3$

**23.** $3r^2s^8$

**24.** $6m^5n^8$

**25.** $-8x^3y^5z$

**26.** $-12p^4q^7r$

*Indicate which expressions are polynomials. If the polynomial has a specific name—monomial, binomial, or trinomial—give that name.*

**27.** $2x^2 - 6x + 7$

**28.** $x^2 + 3$

**29.** $-6$

**30.** $4x^{-2}$

**31.** $9a^4 - 5$

**32.** $7x + 8$

**33.** $8x^9$

**34.** $3x^{1/2} + 2x$

**35.** $a^{-1} + 4$

**36.** $x^3 - 8x^2 + 8$

**37.** $6n^3 - 5n^2 + 4n - 3$

**38.** $10x^2$

**39.** $4 - 2b^2 - 5b$

**40.** $5p^{-3}$

**41.** $\dfrac{2}{3}x^2 - \dfrac{1}{x}$

**42.** $0.6r^4 - \dfrac{1}{2}r^3 - 0.4r^2 - \dfrac{1}{3}$

*Express each polynomial in descending order. If the polynomial is already in descending order, so state. Give the degree of each polynomial.*

**43.** $18$

**44.** $4 + 5x$

**45.** $-4 + x^2 - 2x$

**46.** $6x - 5$

**47.** $x + 3x^2 - 8$

**48.** $4 - 3p^3$

**49.** $-a - 3$

**50.** $2x^2 + 5x - 8$

**51.** $6w^2 - 5w + 9$

**52.** $15$

**53.** $-4 + x - 3x^2 + 4x^3$

**54.** $1 - x^3 + 3x$

**55.** $5x + 3x^2 - 6 - 2x^4$

**56.** $-3r - 5r^2 + 2r^4 - 6$

*Add.*

**57.** $(9x - 2) + (4x - 7)$

**58.** $(5x - 6) + (2x - 3)$

**59.** $(-3x + 8) + (2x + 3)$

**60.** $(-7x - 9) + (-2x + 9)$

**61.** $(t + 7) + (-3t - 8)$

**62.** $(4x - 3) + (3x - 3)$

**63.** $(x^2 + 2.6x - 3) + (4x + 3.8)$

**64.** $(-4p^2 - 3p - 2) + (-p^2 - 4)$

**65.** $(4m - 3) + (5m^2 - 4m + 7)$

**66.** $(-x^2 - 2x - 4) + (4x^2 + 3)$

**67.** $(2x^2 - 3x + 5) + (-x^2 + 6x - 8)$

**68.** $(x^2 - 6x + 7) + (-x^2 + 3x + 5)$

**69.** $(-x^2 - 4x + 8) + \left(5x - 2x^2 + \dfrac{1}{2}\right)$

**70.** $(8x^2 + 3x - 5) + \left(x^2 + \dfrac{1}{2}x + 2\right)$

**71.** $(5.2n^2 - 6n + 1.7) + (3n^2 + 1.2n - 2.3)$

**72.** $(8x^2 + 4) + (-2.6x^2 - 5x - 2.3)$

**73.** $(-7x^3 - 3x^2 + 4) + (4x + 5x^3 - 7)$

**74.** $(6x^3 - 4x^2 - 7) + (3x^2 + 3x - 3)$

**75.** $(8x^2 + 2x - y) + (3x^2 - 9x + 5)$

**76.** $(-7a^2 + 3a - b) + (4a^2 - 2a - 8)$

**77.** $(2x^2y + 2x - 3) + (3x^2y - 5x + 5)$

**78.** $(x^2y + x - y) + (2x^2y + 2x - 6y + 3)$

*Add using columns.*

**79.** Add $8x - 7$ and $3x + 4$.

**80.** Add $-x + 5$ and $-4x - 5$.

**81.** Add $4y^2 - 2y + 4$ and $3y^2 + 1$.

**82.** Add $6m^2 - 2m + 1$ and $-10m^2 + 8$.

**83.** Add $-x^2 - 3x + 3$ and $5x^2 + 5x - 7$.

**84.** Add $-2s^2 - s + 5$ and $3s^2 - 6s$.

**85.** Add $2x^3 + 3x^2 + 6x - 9$ and $7 - 4x^2$.

**86.** Add $-3x^3 + 3x + 9$ and $2x^2 - 4$.

**87.** Add $4n^3 - 5n^2 + n - 6$ and $-n^3 - 6n^2 - 2n + 8$.

**88.** Add $7x^3 + 5x - 6$ and $3x^3 - 4x^2 - x + 8$.

*Subtract.*

**89.** $(4x - 4) - (2x + 2)$

**90.** $(6x - 5) - (2x - 3)$

**91.** $(-2x - 3) - (-5x - 7)$

**92.** $(10x - 3) - (-2x + 7)$

**93.** $(-r + 5) - (2r + 5)$

**94.** $(4x + 8) - (3x + 9)$

**95.** $(-y^2 + 4y - 5.2) - (5y^2 + 2.1y + 7.5)$

**96.** $(9x^2 + 7x - 5) - (3x^2 + 3.5)$

**97.** $(5x^2 - x - 1) - (-3x^2 - 2x - 5)$

**98.** $(-a^2 + 3a + 12) - (-4a^2 - 3)$

**99.** $(-4.1n^2 - 3n) - (2.3n^2 - 9n + 7.6)$

**100.** $(7x - 0.6) - (-2x^2 + 4x - 8)$

**101.** $(8x^3 - 2x^2 - 4x + 5) - (5x^2 + 8)$

**102.** $\left(9x^3 - \dfrac{1}{5}\right) - (x^2 + 5x)$

**103.** $(2x^3 - 4x^2 + 5x - 7) - \left(3x + \dfrac{3}{5}x^2 - 5\right)$

**104.** $(-3x^2 + 4x - 7) - \left(x^3 + 4x^2 - \dfrac{3}{4}x\right)$

**105.** Subtract $(7x + 4)$ from $(8x + 2)$.

**106.** Subtract $(-4x + 7)$ from $(-3x - 9)$.

**107.** Subtract $(5x - 6)$ from $(2x^2 - 4x + 8)$.

**108.** Subtract $(3x^2 - 5x - 3)$ from $(-x^2 + 3x + 10)$.

**109.** Subtract $(-2c^2 + 7c - 7)$ from $(-5c^3 - 6c^2 + 7)$.

**110.** Subtract $(4x^3 - 6x^2)$ from $(3x^3 + 5x^2 + 9x - 7)$.

*Subtract using columns.*

**111.** Subtract $(3x - 3)$ from $(6x + 5)$.

**112.** Subtract $(6x + 8)$ from $(2x - 5)$.

**113.** Subtract $(2a^2 + 3a - 9)$ from $(5a^2 - 13a + 19)$.

**114.** Subtract $(4x^2 - 7x + 3)$ from $(8x^2 + 5x - 9)$.

**115.** Subtract $(6x^2 - 1)$ from $(7x^2 - 3x - 4)$.

**116.** Subtract $(5n^3 + 7n - 9)$ from $(2n^3 - 6n + 3)$.

**117.** Subtract $(5x^2 + 4)$ from $(x^2 + 4)$.

**118.** Subtract $(-5m^2 + 6m)$ from $(m - 6)$.

**119.** Subtract $(x^2 + 6x - 7)$ from $(4x^3 - 6x^2 + 7x - 9)$.

**120.** Subtract $(2x^3 + 4x^2 - 9x)$ from $(-5x^3 + 4x - 12)$.

## Problem Solving

**121.** Make up your own addition problem where the sum of two binomials is $-5x - 1$.

**122.** Make up your own addition problem where the sum of two trinomials is $3x^3 - 2x - 4$.

**123.** Make up your own subtraction problem where the difference of two trinomials is $7x - 3$.

**124.** Make up your own subtraction problem where the difference of two trinomials is $-x^2 + 4x - 5$.

**125.** When two binomials are added, will the sum always, sometimes, or never be a binomial? Explain your answer and give examples to support your answer.

**126.** When one binomial is subtracted from another, will the difference always, sometimes, or never be a binomial? Explain your answer and give examples to support your answer.

**127.** When two trinomials are added, will the sum always, sometimes, or never be a trinomial? Explain your answer and give examples to support your answer.

**128.** When one trinomial is subtracted from another, will the difference always, sometimes, or never be a trinomial? Explain your answer and give examples to support your answer.

**129.** Write a fourth-degree trinomial in the variable $x$ that has neither a second-degree term nor a constant term.

**130.** Write a sixth-degree trinomial in the variable $x$ that has no fifth-, fourth-, first-, or zero-degree terms.

*Write a polynomial that represents the area of each figure shown.*

**131.**

**132.**

**133.**

**134.**

## Concept/Writing Exercises

**135.** Is it possible to have a fourth-degree trinomial in $x$ that has no third-, second-, or zero-degree terms and contains no like terms? Explain.

**136.** Is it possible to have a fifth-degree trinomial in $x$ that has no fourth-, third-, second-, or first-degree terms and contains no like terms? Explain.

## Challenge Problems

*Simplify.*

**137.** $(3x^2 - 6x + 3) - (2x^2 - x - 6) - (x^2 + 7x - 9)$

**138.** $3x^2y - 6xy - 2xy + 9xy^2 - 5xy + 3x$

**139.** $4(x^2 + 2x - 3) - 6(2 - 4x - x^2) - 2x(x + 2)$

## Group Activity

*Discuss and answer Exercise 140 as a group.*

**140.** Make up a trinomial, a binomial, and a different trinomial such that (first trinomial) + (binomial) − (second trinomial) = 0.

## Cumulative Review Exercises

[2.7] **141.** Solve $\dfrac{5}{9} = \dfrac{2.5}{x}$ by cross-multiplying.

[3.1] **142.** Linda Rudolph is $n$ years old now. Write an expression that represents her age 5 years ago.

[4.2] **143.** Graph the equation $3y = 9$.

[4.2] **144.** Graph $y = -2x - 4$.

[5.1] **145.** Simplify $\left( \dfrac{4x^3 y^5}{12x^7 y^4} \right)^3$.

[5.3] **146.** Write $3.5 \times 10^{-2}$ without using exponents.

# 5.5 Multiplication of Polynomials

1 Multiply a monomial by a monomial.

2 Multiply a polynomial by a monomial.

3 Multiply binomials using the distributive property.

4 Multiply binomials using the FOIL method.

5 Multiply binomials using formulas for special products.

6 Multiply any two polynomials.

## 1 Multiply a Monomial by a Monomial

To multiply two monomials, multiply their coefficients and use the product rule of exponents to determine the exponents on the variables.

**EXAMPLE 1** Multiply.

**a)** $(7x^3)(6x^5)$     **b)** $(4b^2)(-9b^7)$

*Solution*

**a)** $(7x^3)(6x^5) = 7 \cdot 6 \cdot x^3 \cdot x^5 = 42x^{3+5} = 42x^8$

**b)** $(4b^2)(-9b^7) = (4)(-9) \cdot b^2 \cdot b^7 = -36b^{2+7} = -36b^9$

Now Try Exercise 15

**EXAMPLE 2** Multiply $(5x^2 y)(8x^5 y^4)$.

*Solution*   Remember that when a variable is given without an exponent we assume that the exponent on the variable is 1.

$$(5x^2 y)(8x^5 y^4) = 40x^{2+5} y^{1+4} = 40x^7 y^5$$

Now Try Exercise 19

**EXAMPLE 3** Multiply.

**a)** $6xy^2 z^5 (-3x^4 y^7 z)$                    **b)** $(-4x^4 z^9)(-3xy^7 z^3)$

*Solution*

**a)** $6xy^2 z^5 (-3x^4 y^7 z) = -18x^5 y^9 z^6$

**b)** $(-4x^4 z^9)(-3xy^7 z^3) = 12x^5 y^7 z^{12}$

Now Try Exercise 21

## 2 Multiply a Polynomial by a Monomial

To multiply a polynomial by a monomial, we use the distributive property presented earlier.

$$a(b + c) = ab + ac$$

The distributive property can be expanded to

$$a(b + c + d + \cdots + n) = ab + ac + ad + \cdots + an$$

**Understanding Algebra**

When multiplying a monomial by a polynomial, the monomial is multiplied by each term of the polynomial:

$3x^2(2x^3 - 5x + 7)$

$3x^2(2x^3) + 3x^2(-5x) + 3x^2(7)$

$6x^5 - 15x^3 + 21x^2$

EXAMPLE 4  Multiply $3x(2x^2 + 4)$.

*Solution*
$$3x(2x^2 + 4) = (3x)(2x^2) + (3x)(4)$$
$$= 6x^3 + 12x$$

Now Try Exercise 29

EXAMPLE 5  Multiply $-3n(4n^2 - 2n - 1)$.

*Solution*  $-3n(4n^2 - 2n - 1) = (-3n)(4n^2) + (-3n)(-2n) + (-3n)(-1)$
$$= -12n^3 + 6n^2 + 3n$$

Now Try Exercise 33

EXAMPLE 6  Multiply $5x^2(4x^3 - 2x + 7)$.

*Solution*  $5x^2(4x^3 - 2x + 7) = (5x^2)(4x^3) + (5x^2)(-2x) + (5x^2)(7)$
$$= 20x^5 - 10x^3 + 35x^2$$

Now Try Exercise 37

EXAMPLE 7  Multiply $2x(3x^2y - 6xy + 5)$.

*Solution*  $2x(3x^2y - 6xy + 5) = (2x)(3x^2y) + (2x)(-6xy) + (2x)(5)$
$$= 6x^3y - 12x^2y + 10x$$

Now Try Exercise 39

In Example 8, we perform a multiplication where the monomial is placed to the right of the polynomial. Each term of the polynomial is multiplied by the monomial, as illustrated in the example.

EXAMPLE 8  Multiply $(3x^3 - 2xy + 3)4x$.

*Solution*  $(3x^3 - 2xy + 3)4x = (3x^3)(4x) + (-2xy)(4x) + (3)(4x)$
$$= 12x^4 - 8x^2y + 12x$$

Now Try Exercise 41

The problem in Example 8 could be written as $4x(3x^3 - 2xy + 3)$ by the commutative property of multiplication, and then simplified as in Examples 4 through 7.

## 3  Multiply Binomials Using the Distributive Property

Now we will discuss multiplying a binomial by a binomial. Before we explain how to do this, consider the multiplication problem $43 \cdot 12$.

$$
\begin{array}{rl}
& 43 \quad \longleftarrow \text{ Multiplicand} \\
& \underline{12} \quad \longleftarrow \text{ Multiplier} \\
2(4) \longrightarrow & 86 \quad \longleftarrow 2(3) \\
1(4) \longrightarrow & \underline{43} \quad \longleftarrow 1(3) \\
& 516 \quad \longleftarrow \text{ Product}
\end{array}
$$

Note how the 2 multiplies both the 3 and the 4, and the 1 also multiplies both the 3 and the 4. That is, every digit in the multiplier multiplies every digit in the multiplicand. We can also illustrate the multiplication process as follows.

$$(43)(12) = (40 + 3)\ (10 + 2)$$
$$= (40 + 3)\ (10)\ +\ (40 + 3)\ (2)$$
$$= (40)\ (10)\ +\ (3)\ (10)\ +\ (40)\ (2)\ +\ (3)\ (2)$$
$$=\quad 400\quad +\quad 30\quad +\quad 80\quad +\quad 6$$
$$= 516$$

## Understanding Algebra

When multiplying two polynomials, every term in the first polynomial multiplies every term in the second polynomial.

$$(2x + 5)(3x + 10)$$

$$(2x + 5)3x + (2x + 5)10$$

$$6x^2 + 15x + 20x + 50$$

$$6x^2 + 35x + 50$$

Whenever any two polynomials are multiplied, the same process must be followed. That is, *every term in one polynomial must multiply every term in the other polynomial.*

Consider multiplying $(a + b)(c + d)$. Treating $(a + b)$ as a single term and using the distributive property, we get

$$(a + b)\ (c + d) = (a + b)\ c + (a + b)\ d$$

Using the distributive property a second time gives

$$= ac + bc + ad + bd$$

Notice how each term of the first polynomial was multiplied by each term of the second polynomial, and all the products were added to obtain the answer.

EXAMPLE 9   Multiply $(3x + 2)(x - 5)$.

*Solution*   $(3x + 2)\ (x - 5) = (3x + 2)\ x + (3x + 2)\ (-5)$
$$= 3x(x) + 2(x) + 3x(-5) + 2(-5)$$
$$= 3x^2\ + 2x - 15x\ - 10$$
$$= 3x^2\ - 13x - 10$$

Now Try Exercise 43

Note that after performing the multiplication, like terms must be combined.

EXAMPLE 10   Multiply $(x - 4)(y + 3)$.

*Solution*   $(x - 4)\ (y + 3) = (x - 4)\ y + (x - 4)\ 3$
$$= xy - 4y + 3x - 12$$

Now Try Exercise 65

## Understanding Algebra

| F | O | I | L |
|---|---|---|---|
| First | Outer | Inner | Last |

When multiplying two binomials, multiply the **f**irst terms of each, the **o**uter terms of each, the **i**nner terms of each, and the **l**ast terms of each.

## 4  Multiply Binomials Using the FOIL Method

A commonly used method to multiply two binomials is the **FOIL method**. This procedure also results in each term of one binomial being multiplied by each term in the other binomial. The FOIL method is not actually a different method used to multiply binomials but rather an acronym to help students remember to correctly apply the distributive property.

**The FOIL Method**

Consider $(a + b)(c + d)$.

**F**    stands for **first**—multiply the first terms of each binomial together:

$$\overset{\overset{\text{F}}{\overbrace{\quad\quad}}}{(a + b)}(c + d) \qquad \text{product } ac$$

**O**    stands for **outer**—multiply the two outer terms together:

$$\overset{\overset{\text{O}}{\overbrace{\quad\quad\quad\quad}}}{(a + b)(c + d)} \qquad \text{product } ad$$

**I**    stands for **inner**—multiply the two inner terms together:

$$(a + \overset{\overset{\text{I}}{\overbrace{\quad}}}{b)(c} + d) \qquad \text{product } bc$$

**L**    stands for **last**—multiply the last terms together:

$$(a + b)(c + \overset{\overset{\text{L}}{\overbrace{\quad}}}{d)} \qquad \text{product } bd$$

The product of the two binomials is the sum of these four products.

$$(a + b)(c + d) = ac + ad + bc + bd$$

**EXAMPLE  11**  Using the FOIL method, multiply $(2x - 3)(x + 4)$.

**Solution**

$$(2x - 3)(x + 4)$$

$$\overset{\text{F}}{\phantom{=}} \quad\ \overset{\text{O}}{\phantom{=}} \quad\ \overset{\text{I}}{\phantom{=}} \quad\quad \overset{\text{L}}{\phantom{=}}$$
$$= (2x)(x) + (2x)(4) + (-3)(x) + (-3)(4)$$
$$= \quad 2x^2 \quad + \quad 8x \quad - \quad 3x \quad - \quad 12$$
$$= 2x^2 + 5x - 12$$

Thus, $(2x - 3)(x + 4) = 2x^2 + 5x - 12$.

Now Try Exercise 45

**EXAMPLE  12**  Multiply $(4 - 2x)(6 - 5x)$.

**Solution**

$$(4 - 2x)(6 - 5x)$$

$$\overset{\text{F}}{\phantom{=}} \quad\ \overset{\text{O}}{\phantom{=}} \quad\quad \overset{\text{I}}{\phantom{=}} \quad\quad\ \overset{\text{L}}{\phantom{=}}$$
$$= 4(6) + 4(-5x) + (-2x)(6) + (-2x)(-5x)$$
$$= \quad 24 \quad - \quad 20x \quad - \quad 12x \quad + \quad 10x^2$$
$$= 10x^2 - 32x + 24$$

Thus, $(4 - 2x)(6 - 5x) = 10x^2 - 32x + 24$.

Now Try Exercise 63

EXAMPLE 13 Multiply $(4p + 5)(4p - 5)$.

*Solution*

$$\begin{array}{cccccccc} & & & \text{F} & \text{O} & \text{I} & \text{L} \\ (4p + 5)(4p - 5) & = & (4p)(4p) & + & (4p)(-5) & + & (5)(4p) & + & (5)(-5) \\ & = & 16p^2 & & - & 20p & & + & 20p & & - & 25 \\ & = & 16p^2 - 25 \end{array}$$

Thus, $(4p + 5)(4p - 5) = 16p^2 - 25$.

Now Try Exercise 57

## 5 Multiply Binomials Using Formulas for Special Products

Example 13 illustrated a special product, the product of the sum and difference of the same two terms.

> **Special Product: Product of the Sum and Difference of the Same Two Terms**
>
> $$\underbrace{(a + b)}_{\substack{\text{sum of} \\ \text{2 terms}}}\underbrace{(a - b)}_{\substack{\text{difference} \\ \text{2 terms}}} = a^2 - b^2 \text{ where } a \text{ and } b \text{ represent terms.}$$

The expression on the right side of the equals sign is called **the difference of two squares**. Since multiplication is commutative, $(a + b)(a - b) = a^2 - b^2$ can also be written $(a - b)(a + b) = a^2 - b^2$.

EXAMPLE 14 Use the rule for finding the product of the sum and difference of two quantities to multiply each expression.

**a)** $(x + 5)(x - 5)$    **b)** $(2x + 3)(2x - 3)$    **c)** $(3x - 2y)(3x + 2y)$

*Solution*

**a)** If we let $x = a$ and $5 = b$, then

$$\begin{array}{ccc} (a + b)(a - b) & = & a^2 - b^2 \\ \downarrow\;\;\downarrow\;\downarrow\;\;\downarrow & & \downarrow\;\;\;\;\downarrow \\ (x + 5)(x - 5) & = & (x)^2 - (5)^2 \\ & = & x^2 - 25 \end{array}$$

**b)**

$$\begin{array}{ccc} (a + b)(a - b) & = & a^2 - b^2 \\ \downarrow\;\;\downarrow\;\downarrow\;\;\downarrow & & \downarrow\;\;\;\;\downarrow \\ (2x + 3)(2x - 3) & = & (2x)^2 - (3)^2 \\ & = & 4x^2 - 9 \end{array}$$

**c)**

$$\begin{array}{ccc} (a - b)(a + b) & = & a^2 - b^2 \\ \downarrow\;\;\downarrow\;\downarrow\;\;\downarrow & & \downarrow\;\;\;\;\downarrow \\ (3x - 2y)(3x + 2y) & = & (3x)^2 - (2y)^2 \\ & = & 9x^2 - 4y^2 \end{array}$$

Now Try Exercise 77

Example 14 could also have been done using the FOIL method.

EXAMPLE 15  Using the FOIL method, multiply $(x + 3)^2$.

*Solution*   $(x + 3)^2 = (x + 3)(x + 3)$

$$\begin{array}{cccc} F & O & I & L \end{array}$$
$$= x(x) + x(3) + 3(x) + (3)(3)$$
$$= x^2 + 3x + 3x + 9$$
$$= x^2 + 6x + 9$$

Now Try Exercise 59

Example 15 illustrates the **square of a binomial**, another special product.

---

**Special Product: Square of Binomial Formulas**

$$(a + b)^2 = (a + b)(a + b) = a^2 + 2ab + b^2$$
$$(a - b)^2 = (a - b)(a - b) = a^2 - 2ab + b^2$$

To square a binomial, add the square of the first term, twice the product of the terms, and the square of the second term.

---

EXAMPLE 16  Use the square of a binomial formula to multiply each expression.

**a)** $(x + 5)^2$     **b)** $(2x - 3)^2$     **c)** $(3r + 2s)^2$     **d)** $(x - 3)(x - 3)$

*Solution*

**a)** If we let $x = a$ and $5 = b$, then

$$(a + b)(a + b) = a^2 + 2\,a\,b + b^2$$

$$(x + 5)^2 = (x + 5)(x + 5) = (x)^2 + 2(x)(5) + (5)^2$$
$$= x^2 + 10x + 25$$

**b)**

$$(a - b)(a - b) = a^2 - 2\,a\,b + b^2$$

$$(2x - 3)^2 = (2x - 3)(2x - 3) = (2x)^2 - 2(2x)(3) + (3)^2$$
$$= 4x^2 - 12x + 9$$

**c)**

$$(a + b)(a + b) = a^2 + 2\,a\,b + b^2$$

$$(3r + 2s)^2 = (3r + 2s)(3r + 2s) = (3r)^2 + 2(3r)(2s) + (2s)^2$$
$$= 9r^2 + 12rs + 4s^2$$

**d)**

$$(a - b)(a - b) = a^2 - 2\,a\,b + b^2$$

$$(x - 3)(x - 3) = (x - 3)(x - 3) = (x)^2 - 2(x)(3) + (3)^2$$
$$= x^2 - 6x + 9$$

Now Try Exercise 83

Example 16 could also have been done using the FOIL method.

**Avoiding Common Errors**

| CORRECT | INCORRECT |
|---|---|
| $(a + b)^2 = a^2 + 2ab + b^2$ | $(a + b)^2 = a^2 + b^2$ |
| $(a - b)^2 = a^2 - 2ab + b^2$ | $(a - b)^2 = a^2 - b^2$ |

*Do not forget the middle term when you square a binomial.*

$$(x + 2)^2 \neq x^2 + 4$$
$$(x + 2)^2 = (x + 2)(x + 2)$$
$$= x^2 + 4x + 4$$

## 6  Multiply Any Two Polynomials

When multiplying any two polynomials, each term of one polynomial must be multiplied by each term of the other polynomial. For example.

$$(3x + 2)(4x^2 - 5x - 3)$$

$$= 3x\,(4x^2 - 5x - 3)\ + 2\,(4x^2 - 5x - 3) \qquad \text{Use the distributive property.}$$

$$= 12x^3 - 15x^2 - 9x + 8x^2 - 10x - 6$$

$$= 12x^3 - 7x^2 - 19x - 6$$

Thus, $(3x + 2)(4x^2 - 5x - 3) = 12x^3 - 7x^2 - 19x - 6$.

Alternatively, you may prefer to multiply a polynomial by a polynomial using a vertical procedure. On page 299, we showed that when multiplying the number 43 by the number 12, we multiply each digit in the number 43 by each digit in the number 12. Review that example now. We can follow a similar procedure, being careful, however, to align like terms in the same column when performing the individual multiplications.

**EXAMPLE 17**  Multiply $(3x + 4)(2x + 5)$.

*Solution*    First write the polynomials one beneath the other.

$$\begin{array}{r} 3x + 4 \\ \underline{2x + 5} \end{array}$$

Next, multiply each term in $(3x + 4)$ by 5.

$$\begin{array}{r} 3x + 4 \\ \underline{2x + 5} \\ 5(3x + 4) \longrightarrow\ \ 15x + 20 \end{array}$$

Next, multiply each term in $(3x + 4)$ by $2x$ and align like terms.

$$\begin{array}{r} 3x + 4 \\ \underline{2x + 5} \\ 15x + 20 \\ 2x(3x + 4) \longrightarrow\ \ \underline{6x^2 + \ 8x \qquad} \\ 6x^2 + 23x + 20 \end{array} \quad \text{Added like terms in columns.}$$

Now Try Exercise 55

The same answer for Example 17 would be obtained using the FOIL method.

**EXAMPLE 18** Multiply $(4y + 3)(2y^2 - 7y - 5)$.

*Solution*   For convenience, we place the shorter expression on the bottom, as illustrated.

$$
\begin{array}{r}
2y^2 - 7y - 5 \\
4y + 3 \\
\hline
6y^2 - 21y - 15 \\
8y^3 - 28y^2 - 20y \\
\hline
8y^3 - 22y^2 - 41y - 15
\end{array}
$$

Multiplied the top polynomial by 3.
Multiplied the top polynomial by $4y$; aligned like terms.
Added like terms in columns.

*Now Try Exercise 95*

**EXAMPLE 19** Multiply $(x^2 - 3x + 2)(2x^2 - 3)$.

*Solution*
$$
\begin{array}{r}
x^2 - 3x + 2 \\
2x^2 - 3 \\
\hline
-3x^2 + 9x - 6 \\
2x^4 - 6x^3 + 4x^2 \\
\hline
2x^4 - 6x^3 + x^2 + 9x - 6
\end{array}
$$

Multiplied the top polynomial by $-3$.
Multiplied the top polynomial by $2x^2$; aligned like terms.
Added like terms in columns.

*Now Try Exercise 103*

**EXAMPLE 20** Multiply $(3x^3 - 2x^2 + 4x + 6)(x^2 - 5x)$.

*Solution*
$$
\begin{array}{r}
3x^3 - 2x^2 + 4x + 6 \\
x^2 - 5x \\
\hline
-15x^4 + 10x^3 - 20x^2 - 30x \\
3x^5 - 2x^4 + 4x^3 + 6x^2 \\
\hline
3x^5 - 17x^4 + 14x^3 - 14x^2 - 30x
\end{array}
$$

Multiplied the top polynomial by $-5x$.
Multiplied the top polynomial by $x^2$; aligned like terms.
Added like terms in columns.

*Now Try Exercise 105*

# EXERCISE SET 5.5        MathXL®   MyMathLab®

## Warm-Up Exercises

*Fill in the blanks with the appropriate word, phrase, or symbol(s) from the following list.*

difference of two squares     trinomial     FOIL     False     True
$4t^2 + 4t + 1$     $4t^2 + 1$     $4t^2 - 1$     $4t^2 - 4t + 1$     align

1. When $(2x - 5)(3x + 4)$ is multiplied and simplified, the resulting expression is a(n) _____.

2. When $(2t - 1)^2$ is multiplied and simplified, the result is _____.

3. (True or False) $(x + 5)^2 = x^2 + 25$. _____

4. When $(2t + 1)(2t - 1)$ is multiplied and simplified, the result is _____.

5. When multiplying polynomials vertically, it is important to _____ like terms.

6. When $(2t + 1)^2$ is multiplied and simplified, the result is _____.

7. The result of multiplying $(x + 4)(x - 4)$ is an expression called the _____.

8. When multiplying two binomials, a commonly used method is _____.

## Practice the Skills

*Multiply.*

9. $(3x^2)(5x^3)$

10. $(-3x^2)(-10x^3)$

11. $(6t^7)(-5t^3)$

12. $(-4z^{12})(5z^3)$

13. $4x^3(-5x^3)$

14. $(-13y^4)(-10y^3)$

15. $(3x^4)(-8x^2)$

16. $(-7p^5)(-2p^3)$

17. $5x^3y^5(4x^2y)$

**18.** $-5x^2y^4\,(2x^3y^2)$

**19.** $(4xy^6)(-7x^2y^9)$

**20.** $(4a^3b^7)(6a^2b)$

**21.** $9xy^6\,(6x^5y^8)$

**22.** $(6m^3n^4)(3n^5)$

**23.** $(6x^2y)\left(\dfrac{1}{2}x^4\right)$

**24.** $\dfrac{3}{4}x(8x^2y^3)$

**25.** $(3.3x^4)(1.8x^4y^3)$

**26.** $(2.3x^5)(4.1x^2y^4)$

*Multiply.*

**27.** $9(x-6)$

**28.** $4(x+3)$

**29.** $-3x(2x-2)$

**30.** $-4p(-3p+6)$

**31.** $-2(8y+5)$

**32.** $2x(x^2+3x-1)$

**33.** $-2x(x^2-2x+5)$

**34.** $-6c(-3c^2+5c-6)$

**35.** $5x(-4x^2+6x-4)$

**36.** $(3x^2+x-6)x$

**37.** $0.5x^2\,(x^3-6x^2-1)$

**38.** $2.3b^2\,(2b^2-b+3)$

**39.** $0.3x(2xy+5x-6y)$

**40.** $-\dfrac{1}{2}x^3\,(2x^2+4x-6y^2)$

**41.** $(x^2-4y^3-3)y^4$

**42.** $\dfrac{1}{4}y^4\,(y^2-12y+4x)$

*Multiply.*

**43.** $(5x-2)(x+4)$

**44.** $(2x-3)(x+5)$

**45.** $(2t+5)(3t-6)$

**46.** $(4a-1)(a+4)$

**47.** $(2x-4)(2x+4)$

**48.** $(4+5w)(3+w)$

**49.** $(8-5x)(6+x)$

**50.** $(-x+3)(2x+5)$

**51.** $(6x-1)(-2x+5)$

**52.** $(7n-3)(2n+1)$

**53.** $(x-2)(4x-2)$

**54.** $(2x+3)(x+5)$

**55.** $(3k-6)(4k-2)$

**56.** $(3d-5)(4d-1)$

**57.** $(x-2)(x+2)$

**58.** $(3x-8)(2x+3)$

**59.** $(2x-3)(2x-3)$

**60.** $(7x+3)(2x+4)$

**61.** $(6z-4)(7-z)$

**62.** $(6-2m)(5m-3)$

**63.** $(9-2x)(7-4x)$

**64.** $(2-5x)(7-2x)$

**65.** $(x+7)(y-3)$

**66.** $(z+2y)(4z-3)$

**67.** $(2x-3y)(3x+2y)$

**68.** $(2x+3)(2y-5)$

**69.** $(9x+y)(4-3x)$

**70.** $(2x-0.1)(x+2.4)$

**71.** $(x+0.6)(x+0.3)$

**72.** $(3x-6)\left(x+\dfrac{1}{3}\right)$

**73.** $(x+4)\left(x-\dfrac{1}{2}\right)$

**74.** $(2y-4)\left(\dfrac{1}{2}x-1\right)$

*Multiply using a special product formula.*

**75.** $(x+8)(x-8)$

**76.** $(x+3)^2$

**77.** $(3x-8)(3x+8)$

**78.** $(r-4)(r-4)$

**79.** $(x+y)^2$

**80.** $(2x-7)(2x+7)$

**81.** $(x-0.2)^2$

**82.** $(a+3b)(a-3b)$

**83.** $(4x+5)(4x+5)$

**84.** $(5x+4)(5x-4)$

**85.** $(0.4x+y)^2$

**86.** $\left(x-\dfrac{1}{2}y\right)^2$

**87.** $(4c-5d)(4c+5d)$

**88.** $(4+3w)(4-3w)$

**89.** $(-2x+6)(-2x-6)$

**90.** $(-3m+2n)(-3m-2n)$

**91.** $(7s-3t)^2$

**92.** $(7a+2)^2$

*Multiply.*

**93.** $(4m+3)(4m^2-5m+6)$

**94.** $(x+4)(3x^2+4x-1)$

**95.** $(3x+2)(4x^2-x+5)$

**96.** $(x-1)(3x^2+3x+2)$

**97.** $(-2x^2-4x+1)(7x-3)$

**98.** $(4x^2+9x-2)(x-2)$

**99.** $(a+b)(a^2-ab+b^2)$

**100.** $(a-b)(a^2+ab+b^2)$

**101.** $(3t^2 - 2t + 4)(2t^2 + 3t + 1)$

**102.** $(x^2 - 2x + 3)(x^2 - 4)$

**103.** $(x^2 - x + 3)(x^2 - 2x)$

**104.** $(6x + 4)(2x^2 + 2x - 4)$

**105.** $(2x^3 - 6x^2 + x - 3)(x^2 + 4x)$

**106.** $(3y^3 + 4y^2 - y + 7)(y^2 - 5y)$

*Determine the cube of each expression by writing the expression as the square of an expression multiplied by another expression. For example, $(x + 3y)^3 = (x + 3y)^2 (x + 3y)$.*

**107.** $(b - 1)^3$

**108.** $(x + 2)^3$

**109.** $(3a - 5)^3$

**110.** $(2z + 3)^3$

## Concept/Writing Exercises

**111.** Will the product of a monomial and a monomial always be a monomial? Explain your answer.

**112.** Will the product of a monomial and a binomial ever be a trinomial? Explain your answer.

**113.** Will the product of two binomials after like terms are combined always be a trinomial? Explain your answer.

**114.** Will the product of any polynomial and a binomial always be a polynomial? Explain.

## Problem Solving

*Consider the multiplications in Exercises 115 and 116. Determine the exponents to be placed in the shaded areas.*

**115.** $3x^2 (2x^{\boxed{\phantom{2}}} - 5x^{\boxed{\phantom{2}}} + 3x^{\boxed{\phantom{2}}}) = 6x^8 - 15x^5 + 9x^3$.

**116.** $4x^3 (x^{\boxed{\phantom{2}}} + 2x^{\boxed{\phantom{2}}} - 5x^{\boxed{\phantom{2}}}) = 4x^7 + 8x^5 - 20x^4$.

**117.** Suppose that one side of a rectangle is represented as $x + 2$ and a second side is represented as $2x + 1$.

   **a)** Express the area of the rectangle in terms of $x$.

   **b)** Find the area if $x = 4$ feet.

   **c)** What value of $x$, in feet, would result in the rectangle being a square? Explain how you determined your answer.

**118.** Consider the figure below.

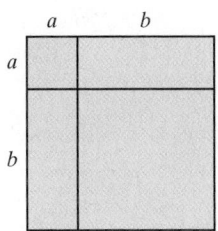

   **a)** Write an expression for the length of the top.

   **b)** Write an expression for the length of the left side.

   **c)** Is this figure a square? Explain.

   **d)** Express the area of this square as the square of a binomial.

   **e)** Determine the area of the square by summing the areas of the four individual pieces.

   **f)** Using the figure and your answer to part **e)**, complete the following.
$$(a + b)^2 = ?$$

**119.** Suppose that a rectangular solid has length $x + 5$, width $3x + 4$, and height $2x - 2$ (see the figure).

   **a)** Write a polynomial that represents the area of the base by multiplying the length by the width.

   **b)** The volume of the figure can be found by multiplying the area of the base by the height. Write a polynomial that represents the volume of the figure.

   **c)** Using the polynomial in part **b)**, find the volume of the figure if $x$ is 4 feet.

   **d)** Using the binomials given for the length, width, and height, find the volume if $x$ is 4 feet.

   **e)** Are your answers to parts **c)** and **d)** the same? If not, explain why.

## Challenge Problems

*Multiply.*

**120.** $(2x^3 - 6x^2 + 5x - 3)(3x^3 - 6x + 4)$

**121.** $\left(\dfrac{1}{2}x + \dfrac{2}{3}\right)\left(\dfrac{2}{3}x - \dfrac{2}{5}\right)$

## Group Activity

**122.** Consider the trinomial $2x^2 + 7x + 3$.

   **a)** As a group, determine whether there is a maximum number of pairs of binomials whose product is $2x^2 + 7x + 3$. That is, how many different pairs of binomials can go in the shaded areas?
$$2x^2 + 7x + 3 = (\boxed{\phantom{xx}})(\boxed{\phantom{xx}})$$

   **b)** Individually, find a pair of binomials whose product is $2x^2 + 7x + 3$.

   **c)** Compare your answer to part **b)** with the other members of your group. If you did not all arrive at the same answer, explain why.

## Cumulative Review Exercises

[2.5] **123.** Solve the equation $3(x + 7) - 5 = 3x - 17$.

[3.3] **124. Complementary Angles** Angles $C$ and $D$ are complementary angles, and angle $D$ is 16° less than angle $C$. Find the measures of angle $C$ and angle $D$.

[5.1] **125.** Simplify $\left(\dfrac{3xy^4}{6y^6}\right)^4$.

[5.1–5.2] **126.** Evaluate the following.

a) $-6^3$

b) $6^{-3}$

[5.4] **127.** Subtract $4x^2 - 4x - 9$ from $-x^2 - 6x + 5$.

# 5.6 Division of Polynomials

**1** Divide a polynomial by a monomial.

**2** Divide a polynomial by a binomial.

**3** Check division of polynomial problems.

**4** Write polynomials in descending order when dividing.

Now let's see how to divide polynomials.

## **1** Divide a Polynomial by a Monomial

> **To Divide a Polynomial by a Monomial**
>
> To divide a polynomial by a monomial, divide each term of the polynomial by the monomial.

**EXAMPLE 1** Divide.

a) $\dfrac{4x + 20}{4}$    b) $\dfrac{9x^2 - 6x}{3x}$

*Solution*

a) $\dfrac{4x + 20}{4} = \dfrac{4x}{4} + \dfrac{20}{4} = x + 5$

b) $\dfrac{9x^2 - 6x}{3x} = \dfrac{9x^2}{3x} - \dfrac{6x}{3x} = 3x - 2$

*Now Try Exercise 17*

> **Avoiding Common Errors**
>
> | CORRECT | INCORRECT |
> |---|---|
> | $\dfrac{x + 2}{2} = \dfrac{x}{2} + \dfrac{2}{2} = \dfrac{x}{2} + 1$ | $\dfrac{x + \overset{1}{\cancel{2}}}{\underset{1}{\cancel{2}}} = \dfrac{x + 1}{1} = x + 1$ |
> | $\dfrac{x + 2}{x} = \dfrac{x}{x} + \dfrac{2}{x} = 1 + \dfrac{2}{x}$ | $\dfrac{\overset{1}{\cancel{x}} + 2}{\underset{1}{\cancel{x}}} = \dfrac{1 + 2}{1} = 3$ |
>
> Can you explain why the procedures on the right are not correct?

**EXAMPLE 2** Divide $\dfrac{4t^5 - 6t^4 + 8t - 3}{2t^2}$.

*Solution*

$$\dfrac{4t^5 - 6t^4 + 8t - 3}{2t^2} = \dfrac{4t^5}{2t^2} - \dfrac{6t^4}{2t^2} + \dfrac{8t}{2t^2} - \dfrac{3}{2t^2}$$

$$= 2t^3 - 3t^2 + \dfrac{4}{t} - \dfrac{3}{2t^2}$$

*Now Try Exercise 37*

**EXAMPLE 3** Divide $\dfrac{3x^3 - 6x^2 + 4x - 1}{-3x}$.

*Solution*    A negative sign appears in the denominator. Usually, it is easier to divide if the divisor is positive. We multiply both numerator and denominator by $-1$ to get a positive denominator.

$$\frac{(-1)(3x^3 - 6x^2 + 4x - 1)}{(-1)(-3x)} = \frac{-3x^3 + 6x^2 - 4x + 1}{3x}$$

$$= \frac{-3x^3}{3x} + \frac{6x^2}{3x} - \frac{4x}{3x} + \frac{1}{3x}$$

$$= -x^2 + 2x - \frac{4}{3} + \frac{1}{3x}$$

Now Try Exercise 41

## 2  Divide a Polynomial by a Binomial

We divide a polynomial by a binomial in much the same way as we perform long division.

**EXAMPLE 4** Divide $\dfrac{x^2 + 6x + 8}{x + 2}$.    ← Dividend
    ← Divisor

*Solution*    Rewrite the division problem in the following form:

$$x + 2 \overline{)x^2 + 6x + 8}$$

Divide $x^2$ (the first term in the dividend) by $x$ (the first term in the divisor).

$$\frac{x^2}{x} = x$$

Place the quotient, $x$, above the like term containing $x$ in the dividend.

$$\begin{array}{r} x \phantom{+6x+8} \\ x + 2 \overline{)x^2 + 6x + 8} \end{array}$$

Next, multiply the $x$ by $x + 2$ as you would do in long division and place the terms of the product under their like terms.

$$\begin{array}{r} \text{Times} \qquad x \\ x + 2 \overline{)x^2 + 6x + 8} \\ \text{Equals} \quad x^2 + 2x \quad \longleftarrow \ x(x+2) \end{array}$$

Now subtract $x^2 + 2x$ from $x^2 + 6x$. When subtracting, remember to change the sign of the terms being subtracted and then add the like terms.

$$\begin{array}{r} x \phantom{+6x+8} \\ x + 2 \overline{)\ x^2 + 6x + 8} \\ \underline{x^2 + 2x} \\ 4x \end{array}$$

Next, bring down the 8, the next term in the dividend.

$$\begin{array}{r} x \phantom{+6x+8} \\ x + 2 \overline{)x^2 + 6x + 8} \\ \underline{x^2 + 2x} \\ 4x + 8 \end{array}$$

Now divide $4x$, the first term at the bottom, by $x$, the first term in the divisor.

$$\frac{4x}{x} = +4$$

Write the $+4$ in the quotient above the constant in the dividend.

$$\begin{array}{r} x + 4 \\ x + 2\overline{)x^2 + 6x + 8} \\ \underline{x^2 + 2x} \\ 4x + 8 \end{array}$$

Multiply the $x + 2$ by 4 and place the terms of the product under their like terms.

$$\begin{array}{r} \text{Times} \\ x + 4 \\ x + 2\overline{)x^2 + 6x + 8} \\ \underline{x^2 + 2x} \\ \text{Equals} \quad 4x + 8 \\ 4x + 8 \leftarrow 4(x + 2) \end{array}$$

Now subtract.

$$\begin{array}{r} x + 4 \quad \leftarrow \text{Quotient} \\ x + 2\overline{)x^2 + 6x + 8} \\ \underline{x^2 + 2x} \\ 4x + 8 \\ \underline{4x + 8} \\ 0 \quad \leftarrow \text{Remainder} \end{array}$$

Thus,

$$\frac{x^2 + 6x + 8}{x + 2} = x + 4$$

There is no remainder.

Now Try Exercise 43

**EXAMPLE 5** Divide $\dfrac{3x^2 + x - 12}{x + 2}$.

*Solution*   First write the problem in the following form:

$$x + 2\overline{)3x^2 + x - 12}$$

Since $3x^2$ divided by $x$ is $3x$, place $3x$ above the $x$-term in the dividend.

$$\begin{array}{r} 3x \\ x + 2\overline{)3x^2 + x - 12} \end{array}$$

Then multiply $3x(x + 2)$ and write the product $3x^2 + 6x$ as shown below. Then subtract to get a difference of $-5x$.

$$\begin{array}{r} 3x \\ x + 2\overline{)\ 3x^2 + \ x - 12} \\ \underline{3x^2 + 6x} \\ -5x \end{array}$$

Next bring down the $-12$. Then divide $-5x$ by $x$, which gives $-5$. Place $-5$ over $-12$ in the dividend, as shown below. Then multiply $-5(x + 2)$. Write the product, $-5x - 10$, below the $-5x - 12$. Then subtract to get a remainder of $-2$.

$$\begin{array}{r} 3x - 5 \\ x + 2\overline{)3x^2 + \ x - 12} \\ \underline{3x^2 + 6x} \\ -5x - 12 \\ \underline{-5x - 10} \\ -2 \end{array}$$

When there is a remainder, as in this example, list the quotient plus the remainder above the divisor. Thus,

$$\frac{3x^2 + x - 12}{x + 2} = 3x - 5 + \frac{-2}{x + 2} = 3x - 5 - \frac{2}{x + 2}$$

Now Try Exercise 45

**EXAMPLE 6** Divide $\dfrac{6x^2 - 5x + 5}{2x + 3}$.

**Solution**

$$\frac{6x^2}{2x} \qquad \frac{-14x}{2x}$$

$$\begin{array}{r} 3x - 7 \\ 2x + 3 \overline{\smash{\big)}\ 6x^2 - 5x + 5} \\ \underline{6x^2 + 9x} \qquad \longleftarrow 3x(2x+3) \\ -14x + 5 \\ \underline{+14x + 21} \qquad \longleftarrow -7(2x+3) \\ 26 \qquad \longleftarrow \text{Remainder} \end{array}$$

Thus, $\dfrac{6x^2 - 5x + 5}{2x + 3} = 3x - 7 + \dfrac{26}{2x + 3}$.

Now Try Exercise 57

## 3 Check Division of Polynomial Problems

The answer to a division problem can be checked. Consider the division problem $13 \div 5$.

$$\begin{array}{r} 2 \\ 5 \overline{\smash{\big)}\ 13} \\ \underline{10} \\ 3 \end{array}$$

Note that the divisor times the quotient, plus the remainder, equals the dividend:

$$(\text{divisor} \times \text{quotient}) + \text{remainder} = \text{dividend}$$
$$(5 \cdot 2) + 3 \overset{?}{=} 13$$
$$10 + 3 \overset{?}{=} 13$$
$$13 = 13 \qquad \text{True}$$

This same procedure can be used to check all division problems.

---

**To Check Division of Polynomials**

$$(\text{divisor} \times \text{quotient}) + \text{remainder} = \text{dividend}$$

---

Let's check the answer to Example 6. The divisor is $2x + 3$, the quotient is $3x - 7$, the remainder is 26, and the dividend is $6x^2 - 5x + 5$.

**Check**  $(\text{divisor} \times \text{quotient}) + \text{remainder} = \text{dividend}$

$$(2x + 3)(3x - 7) + 26 \overset{?}{=} 6x^2 - 5x + 5$$

$$(6x^2 - 5x - 21) + 26 \overset{?}{=} 6x^2 - 5x + 5$$

$$6x^2 - 5x + 5 = 6x^2 - 5x + 5 \qquad \text{True}$$

## 4   Write Polynomials in Descending Order When Dividing

When dividing a polynomial by a binomial, both the polynomial and binomial should be listed in descending order. If there is no term for a certain power, we write that term with a numerical coefficient of 0 as a placeholder. This will help keep like terms aligned. For example, to divide $(6x^2 + x^3 - 4)/(x - 2)$, we begin by writing $(x^3 + 6x^2 + 0x - 4)/(x - 2)$.

EXAMPLE 7   Divide $(-x + 9x^3 - 28)$ by $(3x - 4)$.

*Solution*   First we rewrite the dividend in descending order to get $(9x^3 - x - 28) \div (3x - 4)$. Since there is no $x^2$ term in the dividend, we will add $0x^2$ to help align like terms.

### Understanding Algebra

When a term in "missing", we can insert it with a zero coefficient as a placeholder.

For example, $9x^3 - x - 28$ has no $x^2$ term. But we can write the expression as $9x^3 + 0x^2 - x - 28$ so terms align in the division process.

$$\frac{9x^3}{3x} \quad \frac{12x^2}{3x} \quad \frac{15x}{3x}$$

$$
\begin{array}{r}
3x^2 + 4x + 5 \\
3x - 4{\overline{\smash{\big)}\,9x^3 + 0x^2 - x - 28}} \\
\underline{9x^3 - 12x^2} \quad\quad\quad\quad\quad \leftarrow 3x^2(3x-4) \\
12x^2 - x \\
\underline{12x^2 - 16x} \quad\quad\quad \leftarrow 4x(3x-4) \\
15x - 28 \\
\underline{15x - 20} \quad \leftarrow 5(3x-4) \\
-8 \quad \leftarrow \text{Remainder}
\end{array}
$$

Thus, $\dfrac{-x + 9x^3 - 28}{3x - 4} = 3x^2 + 4x + 5 - \dfrac{8}{3x - 4}$. Check this division yourself using the procedure discussed on page 311.

Now Try Exercise 55

---

# EXERCISE SET 5.6    MathXL®   MyMathLab®

## Warm-Up Exercises

*Fill in the blanks with the appropriate word, phrase, or symbol(s) from the following list.*

| | | | |
|---|---|---|---|
| divisor | $\dfrac{6x^2 - 7x - 20}{(2x - 5)} = (3x + 4)$ | remainder | dividend |
| quotient | | descending order | ascending order |
| $6x^3 + 8x^2 + 16 = (3x^2 - 2x + 4)(2x + 4)$ | $6x^3 + 8x^2 + 0x + 16$ | True | False |

1. The multiplication statement $(2x - 5)(3x + 4) = 6x^2 - 7x - 20$, when written as a division problem, is _____.

2. In the division statement $(2x^2 + 11x + 12) \div (x + 5) = (2x + 1) + \dfrac{7}{x + 5}$, the quantity $2x^2 + 11x + 12$ is called the _____.

3. In the division statement $(2x^2 + 11x + 12) \div (x + 5) = (2x + 1) + \dfrac{7}{x + 5}$, the quantity $x + 5$ is called the _____.

4. In the division statement $(2x^2 + 11x + 12) \div (x + 5) = (2x + 1) + \dfrac{7}{x + 5}$, the quantity $2x + 1$ is called the _____.

5. In the division statement $(2x^2 + 11x + 12) \div (x + 5) = (2x + 1) + \dfrac{7}{x + 5}$, the quantity $7$ is called the _____.

6. The division statement $(6x^3 + 8x^2 + 16) \div (3x^2 - 2x + 4) = 2x + 4$, when written as a multiplication problem is _____.

**7.** To prepare to do the division problem $\dfrac{6x^3 + 8x^2 + 16}{3x^2 - 2x + 4}$, the numerator $6x^3 + 8x^2 + 16$ is better written as _____.

**8.** When dividing a polynomial by a binomial, the terms of each should be written in _____.

**9.** (True or False) $\dfrac{2x + 8}{2} = x + 8$. _____

**10.** (True or False) $\dfrac{y + 5}{y} = 1 + \dfrac{5}{y}$. _____

*Rewrite each multiplication problem as a division problem. There is more than one correct answer.*

**11.** $(x - 7)(x + 6) = x^2 - x - 42$

**12.** $(x + 3)(3x - 1) = 3x^2 + 8x - 3$

**13.** $(2x + 3)(x + 1) = 2x^2 + 5x + 3$

**14.** $(2x - 5)(x + 1) = 2x^2 - 3x - 5$

**15.** $(2x + 3)(2x - 3) = 4x^2 - 9$

**16.** $(3n + 4)(n - 5) = 3n^2 - 11n - 20$

## Practice the Skills

*Divide.*

**17.** $\dfrac{3t + 6}{3}$

**18.** $\dfrac{4x - 6}{2}$

**19.** $\dfrac{4n + 10}{4}$

**20.** $(-3x - 8) \div 4$

**21.** $\dfrac{7x + 6}{3}$

**22.** $\dfrac{5x - 10}{5}$

**23.** $\dfrac{-6x + 4}{2}$

**24.** $\dfrac{-5a + 4}{-3}$

**25.** $\dfrac{-9x - 3}{-3}$

**26.** $\dfrac{8x - 3}{-8}$

**27.** $\dfrac{2x + 16}{4}$

**28.** $\dfrac{2p - 3}{2p}$

**29.** $\dfrac{4 - 10w}{-4}$

**30.** $\dfrac{6 - 5x}{-3x}$

**31.** $(4x^2 + 8x - 12) \div 4x^2$

**32.** $\dfrac{12x^2 - 6x + 3}{3}$

**33.** $\dfrac{-4x^5 + 6x + 8}{2x^2}$

**34.** $\dfrac{6t^2 + 3t + 8}{2}$

**35.** $(x^5 + 3x^4 - 3) \div x^3$

**36.** $(6x^2 - 7x + 9) \div 3x$

**37.** $\dfrac{6x^5 - 4x^4 + 12x^3 - 5x^2}{2x^3}$

**38.** $\dfrac{9x^2 + 18x - 7}{-9}$

**39.** $\dfrac{8k^3 + 6k^2 - 8}{-4k}$

**40.** $\dfrac{-12x^4 + 6x^2 - 15x + 4}{-3x}$

**41.** $\dfrac{12x^5 + 3x^4 - 10x^2 - 9}{-3x^2}$

**42.** $\dfrac{-15m^3 - 6m^2 + 15}{-5m^3}$

*Divide.*

**43.** $\dfrac{x^2 + 4x + 3}{x + 1}$

**44.** $(2x^2 + 3x - 35) \div (x + 5)$

**45.** $\dfrac{5y^2 - 34y - 7}{y - 7}$

**46.** $\dfrac{2p^2 - 7p - 15}{p - 5}$

**47.** $\dfrac{6x^2 + 16x + 8}{3x + 2}$

**48.** $\dfrac{3r^2 + 5r - 8}{r - 1}$

**49.** $\dfrac{x^2 - 16}{-4 + x}$

**50.** $\dfrac{6t^2 - 7t - 20}{3t + 4}$

**51.** $(2x^2 + 7x - 18) \div (2x - 3)$

**52.** $(4a^2 - 25) \div (2a - 5)$

**53.** $\dfrac{x^2 - 36}{x - 6}$

**54.** $\dfrac{9x^2 - 16}{3x - 4}$

**55.** $\dfrac{-x + 9x^3 - 16}{3x - 4}$

**56.** $\dfrac{10x + 3x^2 + 6}{x + 2}$

**57.** $\dfrac{6x + 8x^2 - 12}{2x + 3}$

**58.** $\dfrac{x^3 + 5x^2 + 2x - 8}{x + 2}$

**59.** $\dfrac{7x^3 + 28x^2 - 5x - 20}{x + 4}$

**60.** $\dfrac{2x^3 - 3x^2 - 3x + 6}{x - 1}$

**61.** $\dfrac{2t^3 - 4t^2 + 12}{t - 2}$

**62.** $\dfrac{2x^3 + 6x - 4}{x + 4}$

**63.** $(w^3 - 8) \div (w - 3)$

**64.** $\dfrac{x^3 + 8}{x + 2}$

**65.** $\dfrac{x^3 - 27}{x - 3}$

**66.** $\dfrac{x^3 + 64}{x + 4}$

**67.** $\dfrac{4x^3 - 5x}{2x - 1}$

**68.** $\dfrac{9x^3 - x + 3}{3x - 2}$

**69.** $\dfrac{-m^3 - 6m^2 + 2m - 3}{m - 1}$

**70.** $\dfrac{-x^3 + 3x^2 + 14x + 16}{x + 3}$     **71.** $\dfrac{4t^3 - t + 4}{t + 2}$     **72.** $\dfrac{9n^3 - 6n + 4}{3n - 3}$

## Concept/Writing Exercises

**73.** When dividing a binomial by a monomial, must the quotient be a binomial? Explain and give an example to support your answer.

**74.** When dividing a trinomial by a monomial, must the quotient be a trinomial? Explain and given an example to support your answer.

## Problem Solving

**75.** If the divisor is $x + 4$, the quotient is $2x + 3$, and the remainder is 4, find the dividend (or the polynomial being divided).

**76.** If the divisor is $2x - 3$, the quotient is $3x - 1$, and the remainder is $-2$, find the dividend.

**77.** If a polynomial of degree 4 in $x$ is divided by a polynomial of degree 1 in $x$, what will be the degree of the quotient? Explain.

**78.** If a polynomial of degree 2 in $x$ is divided by a polynomial of degree 1 in $x$, what will be the degree of the quotient? Explain.

*Determine the monomial to be placed in the shaded area to make a true statement. Explain how you determined your answer.*

**79.** $\dfrac{16x^4 + 20x^3 - 4x^2 + 12x}{\rule{1.5cm}{0.35cm}} = 4x^3 + 5x^2 - x + 3$

**80.** $\dfrac{9x^5 - 6x^4 + 3x^2 + 12}{\rule{1.5cm}{0.35cm}} = 3x^3 - 2x^2 + 1 + \dfrac{4}{x^2}$

*Determine the exponents to be placed in the shaded areas to make a true statement. Explain how you determined your answer.*

**81.** $\dfrac{8x^{\blacksquare} + 4x^{\blacksquare} - 20x^{\blacksquare} - 5x^{\blacksquare}}{2x^2} = 4x^3 + 2x - 10 - \dfrac{5}{2x}$

**82.** $\dfrac{15x^{\blacksquare} + 25x^{\blacksquare} + 5x^{\blacksquare} + 10x^{\blacksquare}}{5x^2} = 3x^5 + 5x^4 + x^2 + 2$

## Challenge Problems

*Divide. The quotients in Exercises 83 and 84 will contain fractions.*

**83.** $\dfrac{3x^3 - 5}{3x - 2}$     **84.** $\dfrac{4x^3 - 4x + 6}{2x + 3}$     **85.** $\dfrac{3x^2 + 6x - 10}{-x - 3}$

## Group Activity

*Discuss and answer Exercises 86 and 87 as a group. Determine the polynomial that when substituted in the shaded area results in a true statement. Explain how you determined your answer.*

**86.** $\dfrac{\rule{1cm}{0.35cm}}{x + 4} = x + 2 + \dfrac{2}{x + 4}$

**87.** $\dfrac{\rule{1cm}{0.35cm}}{x + 3} = x + 1 - \dfrac{1}{x + 3}$

## Cumulative Review Exercises

[1.4] **88.** Consider the set of numbers

$$\left\{ 2, -5, 0, \sqrt{7}, \dfrac{2}{5}, -6.3, \sqrt{3}, -\dfrac{23}{34} \right\}.$$

List those that are

**a)** natural numbers;

**b)** whole numbers;

**c)** rational numbers;

**d)** irrational numbers;

**e)** real numbers.

[1.8] **89. a)** To what is $\dfrac{0}{1}$ equal?

**b)** How do we refer to an expression like $\dfrac{1}{0}$?

[1.9] **90.** Give the order of operations to be followed when evaluating a mathematical expression.

[2.5] **91.** Solve the equation $2(x + 3) + 2x = x + 4$.

[3.2] **92. Sale** At a 30% off sale Jennifer Lucking purchased a sweater for $27.65. What was the original price of the sweater?

[5.2] **93.** Simplify $\dfrac{x^9}{x^{-4}}$.

# Chapter 5 Summary

| IMPORTANT FACTS AND CONCEPTS | EXAMPLES |
|---|---|

### Section 5.1

| | |
|---|---|
| In the expression $x^n$, $x$ is called the **base** and $n$ is called the **exponent**. | base $\rightarrow 3^4 \nwarrow$ exponent |

| **Rules of Exponents** | Simplify. |
|---|---|
| **1. Product Rule** $\quad x^m \cdot x^n = x^{m+n}$ | **1.** $x^5 \cdot x^4 = x^{5+4} = x^9$ |
| **2. Quotient Rule** $\quad \dfrac{x^m}{x^n} = x^{m-n}, \qquad x \neq 0$ | **2.** $\dfrac{x^{12}}{x^7} = x^{12-7} = x^5$ |
| **3. Zero Exponent Rule** $\quad x^0 = 1, \qquad x \neq 0$ | **3.** $(-3ab^4)^0 = 1$ |
| **4. Power Rule** $\quad (x^m)^n = x^{m \cdot n}$ | **4.** $(x^6)^3 = x^{6 \cdot 3} = x^{18}$ |
| **5. Power of a Product Rule** $\quad (xy)^n = x^n y^n$ | **5.** $(5t)^2 = 5^2 t^2 = 25t^2$ |
| **6. Power of a Quotient Rule** $\left(\dfrac{x}{y}\right)^n = \dfrac{x^n}{y^n}, \quad y \neq 0$ | **6.** $\left(\dfrac{x}{y}\right)^6 = \dfrac{x^6}{y^6}$ |
| **7. Expanded Power Rule** $\left(\dfrac{ax}{by}\right)^m = \dfrac{a^m x^m}{b^m y^m}, \quad b \neq 0, \quad y \neq 0$ | **7.** $\left(\dfrac{4x}{5y}\right)^2 = \dfrac{4^2 x^2}{5^2 y^2} = \dfrac{16x^2}{25y^2}$ |

### Section 5.2

| | |
|---|---|
| **Negative Exponent Rule** $$x^{-m} = \frac{1}{x^m}, \qquad x \neq 0$$ | $$x^{-2} = \frac{1}{x^2}$$ $$\frac{1}{y^{-6}} = y^6$$ |
| **A Fraction Raised to a Negative Exponent Rule** For a fraction of the form $\dfrac{a}{b}$, $a \neq 0$ and $b \neq 0$, $\left(\dfrac{a}{b}\right)^{-m} = \left(\dfrac{b}{a}\right)^m$ | $\left(\dfrac{7}{8}\right)^{-2} = \left(\dfrac{8}{7}\right)^2 = \dfrac{8^2}{7^2} = \dfrac{64}{49}$ |

### Section 5.3

| | |
|---|---|
| Each number written in **scientific notation** is written as a number greater than or equal to 1 and less than 10 multiplied by some power of 10. | $1.3 \times 10^7$ $4.76 \times 10^{-2}$ |
| **To Write a Number in Scientific Notation** 1. Move the decimal point in the original number to the right of the first nonzero digit. 2. Count the number of places you moved the decimal point in step 1. If the original number was 10 or greater, the count is positive. If the original number was less than 1, the count is negative. 3. Multiply the number obtained in step 1 by 10 raised to the count (power) found in step 2. | $25{,}700 = 2.57 \times 10^4$ $0.0000346 = 3.46 \times 10^{-5}$ |
| **To Convert a Number from Scientific Notation to Decimal Form** 1. Observe the exponent of the power of 10. 2. **a)** If the exponent is positive, move the decimal point in the number to the right the same number of places as the exponent. **b)** If the exponent is 0, do not move the decimal point. **c)** If the exponent is negative, move the decimal point in the number to the left the same number of places as the exponent. | $9.8 \times 10^6 = 9{,}800{,}000$ $5.17 \times 10^{-3} = 0.00517$ |

| IMPORTANT FACTS AND CONCEPTS | EXAMPLES |
|---|---|

**Section 5.4**

| | |
|---|---|
| A **polynomial in** $x$ is an expression containing the sum of a finite number of terms of the form $ax^n$, for any real number $a$ and any whole number $n$. | $\frac{1}{5}x - 2$ and $x^2 - 4x + 7$ are both polynomials in $x$. |
| A polynomial is written in **descending order** when the exponents on the variable decrease from left to right. | $5x^4 - 3x^3 + 7x^2 - 6x + 9$ is written in descending order. |
| A **monomial** is a polynomial with one term. | $-7y^2$ is a monomial. |
| A **binomial** is a two-termed polynomial. | $x^2 - 8$ is a binomial. |
| A **trinomial** is a three-termed polynomial. | $4z^2 - 9z + 1$ is a trinomial. |
| The **degree of a term** of a polynomial in **one variable** is the exponent on the variable in that term. | $2y^6$ has degree six. |
| The **degree of a term** of a polynomial in **two or more variables** is the sum of the exponents on those variables. | $3x^2y^5$ has degree seven. |
| The **degree of a polynomial** is the same as that of its highest-degree term. | $9x^3 + 2x^2 - 5x + 4$ has degree three. |

**To Add Polynomials**

To add polynomials, combine the like terms of the polynomials.

$$(3x^2 - 9x + 4) + (2x^2 - 3x - 5) = \underbrace{3x^2 + 2x^2}\ \underbrace{-9x - 3x}\ \underbrace{+ 4 - 5}$$
$$= \quad 5x^2 \qquad - 12x \qquad - 1$$

**To Subtract Polynomials**

1. Use the distributive property to remove parentheses.
2. Combine like terms.

$$(9a^2 - 6a + 1) - (a^2 - 5a - 3)$$
$$= 9a^2 - 6a + 1 - a^2 + 5a + 3$$
$$= \underbrace{9a^2 - a^2}\ \underbrace{-6a + 5a}\ \underbrace{+ 1 + 3}$$
$$= \quad 8a^2 \qquad - a \qquad + 4$$

**Section 5.5**

**FOIL Method to Multiply Two Binomials (First, Outer, Inner, Last)**

$$(a + b)(c + d)$$

$$\overset{\text{F} \qquad \text{O} \qquad \text{I} \qquad \text{L}}{(3x - 5)(x + 4) = (3x)(x) + (3x)(4) + (-5)(x) + (-5)(4)}$$
$$= \quad 3x^2 \quad + \quad 12x \quad - \quad 5x \quad - \quad 20$$
$$= \quad 3x^2 + 7x - 20$$

**Product of Sum and Difference of the Same Two Terms (also called the difference of two squares):**

$$(a + b)(a - b) = a^2 - b^2$$

$$(y + 6)(y - 6) = (y)^2 - (6)^2$$
$$= y^2 - 36$$

**Square of a Binomial**

$$(a + b)^2 = a^2 + 2ab + b^2$$

$$(a - b)^2 = a^2 - 2ab + b^2$$

$$(x + 7)^2 = (x)^2 + 2(x)(7) + (7)^2$$
$$= x^2 + 14x + 49$$
$$(z - 3)^2 = (z)^2 - 2(z)(3) + (3)^2$$
$$= z^2 - 6z + 9$$

**To Multiply Any Two Polynomials**

To multiply any two polynomials, each term of one polynomial must multiply each term of the second polynomial.

$$(x^2 + 3x + 5)(x - 2) \text{ or}$$

$$\begin{array}{r} x^2 + 3x + 5 \\ x - 2 \\ \hline -2x^2 - 6x - 10 \\ x^3 + 3x^2 + 5x \\ \hline x^3 + \ x^2 - \ x - 10 \end{array}$$

| IMPORTANT FACTS AND CONCEPTS | EXAMPLES |
|---|---|

**Section 5.6**

| | |
|---|---|
| **To Divide a Polynomial by a Monomial**<br>To divide a polynomial by a monomial, divide each term of the polynomial by the monomial. | $\dfrac{6x + 24}{6} = \dfrac{6x}{6} + \dfrac{24}{6} = x + 4$ |
| **To Divide a Polynomial by a Binomial**<br>To divide a polynomial by a binomial we perform division in much the same way as we perform long division. | $\dfrac{x^2 - 4x + 3}{x + 2}$<br><br>$\begin{array}{r} x - 6 \\ x + 2\overline{)x^2 - 4x + 3} \\ \underline{x^2 + 2x} \\ -6x + 3 \\ \underline{-6x - 12} \\ 15 \end{array}$<br><br>$\dfrac{x^2 - 4x + 3}{x + 2} = x - 6 + \dfrac{15}{x + 2}$ |

## Chapter 5 Review Exercises

[5.1]   *Simplify.*

**1.** $x^5 \cdot x^3$

**2.** $x^2 \cdot x^4$

**3.** $3^2 \cdot 3^3$

**4.** $2^4 \cdot 2$

**5.** $\dfrac{x^4}{x}$

**6.** $\dfrac{a^5}{a^5}$

**7.** $\dfrac{5^5}{5^3}$

**8.** $\dfrac{4^4}{4}$

**9.** $\dfrac{x^6}{x^8}$

**10.** $\dfrac{y^4}{y}$

**11.** $x^0$

**12.** $7y^0$

**13.** $(-6z)^0$

**14.** $6^0$

**15.** $(5x)^2$

**16.** $(3a)^3$

**17.** $(-3x)^3$

**18.** $(6s)^3$

**19.** $(2x^2)^4$

**20.** $(-t^4)^6$

**21.** $(-p^8)^4$

**22.** $\left(-\dfrac{2x^3}{y}\right)^2$

**23.** $\left(-\dfrac{5y^2}{2b}\right)^2$

**24.** $6x^2 \cdot 4x^3$

**25.** $\dfrac{16x^2y}{4xy^2}$

**26.** $2x(3xy^3)^3$

**27.** $\left(\dfrac{9x^2y}{3xy}\right)^2$

**28.** $(2x^2y)^3\,(3xy^4)$

**29.** $4x^2y^3\,(2x^3y^4)^2$

**30.** $3c^2\,(2c^4d^3)$

**31.** $\left(\dfrac{9a^3b^2}{3ab^7}\right)^3$

**32.** $\left(\dfrac{21x^4y^3}{7y^2}\right)^3$

[5.2]   *Simplify.*

**33.** $b^{-9}$

**34.** $3^{-3}$

**35.** $5^{-2}$

**36.** $\dfrac{1}{z^{-2}}$

**37.** $\dfrac{1}{x^{-7}}$

**38.** $\dfrac{1}{4^{-2}}$

**39.** $y^5 \cdot y^{-8}$

**40.** $x^{-2} \cdot x^{-3}$

**41.** $p^{-6} \cdot p^4$

**42.** $a^{-2} \cdot a^{-3}$

**43.** $\dfrac{m^5}{m^{-5}}$

**44.** $\dfrac{x^5}{x^{-2}}$

**45.** $\dfrac{x^{-3}}{x^3}$

**46.** $(3x^4)^{-2}$

**47.** $(4x^{-3}y)^{-3}$

**48.** $(-2m^{-3}n)^2$

**49.** $6y^{-2} \cdot 2y^4$

**50.** $(-5y^{-3}z)^3$

**51.** $(-4x^{-2}y^3)^{-2}$

**52.** $2x(3x^{-2})$

**53.** $(5x^{-2}y)(2x^4y)$

**54.** $4y^{-2}\,(3x^2y)$

**55.** $4x^5\,(6x^{-7}y^2)$

**56.** $\dfrac{6xy^4}{2xy^{-1}}$

**57.** $\dfrac{12x^{-2}y^3}{3xy^2}$

**58.** $\dfrac{49x^2y^{-3}}{7x^{-3}y}$

**59.** $\dfrac{4x^8y^{-2}}{8x^7y^3}$

**60.** $\dfrac{36x^4y^7}{9x^5y^{-3}}$

[5.3]   *Express each number in scientific notation.*

**61.** 1,720,000

**62.** 0.153

**63.** 0.00763

**64.** 47,000

**65.** 5760

**66.** 0.000314

*Express each number without exponents.*

**67.** $7.5 \times 10^{-3}$

**68.** $6.52 \times 10^{-4}$

**69.** $8.9 \times 10^{6}$

**70.** $5.12 \times 10^{4}$

**71.** $3.14 \times 10^{-5}$

**72.** $1.103 \times 10^{7}$

*Write each quantity as a base unit without metric prefixes and then write the quantity in scientific notation.*

**73.** 92 milliliters

**74.** 6 gigameters

**75.** 12.8 micrograms

**76.** 19.2 kilograms

*Perform each indicated operation and write your answer without exponents.*

**77.** $(2.5 \times 10^{2})(3.4 \times 10^{-4})$

**78.** $(4.2 \times 10^{-3})(3.0 \times 10^{5})$

**79.** $(3.5 \times 10^{-2})(7.0 \times 10^{3})$

**80.** $\dfrac{7.94 \times 10^{6}}{2.0 \times 10^{-2}}$

**81.** $\dfrac{1.5 \times 10^{-2}}{5.0 \times 10^{2}}$

**82.** $\dfrac{6.5 \times 10^{4}}{2.0 \times 10^{6}}$

*Convert each number to scientific notation. Then calculate. Express your answer in scientific notation.*

**83.** $(14{,}000)(260{,}000)$

**84.** $(0.00053)(40{,}000)$

**85.** $(12{,}500)(400{,}000)$

**86.** $\dfrac{250}{500{,}000}$

**87.** $\dfrac{0.000068}{0.02}$

**88.** $\dfrac{850{,}000}{0.025}$

**89. Milk Tank** A milk tank holds $6.4 \times 10^{6}$ fluid ounces of milk. If one gallon is $1.28 \times 10^{2}$ fluid ounces, determine the number of gallons of milk the tank holds.

**90. Social Security** In 2008 there was about $4.3 trillion in the social security trust fund.

  **a)** Write this amount without scientific notation.

  **b)** Using scientific notation, determine the annual amount of interest obtained in a year if the interest rate is 2.5% per year. Write the answer in scientific notation.

[5.4]   *Indicate whether each expression is a polynomial. If the polynomial is not written in descending order, rewrite it in descending order. If the polynomial has a specific name, give that name. State the degree of each polynomial.*

**91.** $x^{-4} - 8$

**92.** 7

**93.** $x^{2} - 4 + 3x$

**94.** $-3 - x + 4x^{2}$

**95.** $4x^{1/2} - 6$

**96.** $13x^{3} - 4$

**97.** $x - 4x^{2}$

**98.** $y^{5} + y^{-3} - 9$

**99.** $2x^{3} - 7 + 4x^{2} - 3x$

[5.4–5.6]   *Perform each indicated operation.*

**100.** $(x + 8) + (4x - 11)$

**101.** $(2d - 3) + (5d + 7)$

**102.** $(-x - 10) + (-2x + 5)$

**103.** $(-3x^{2} + 9x + 5) + (-x^{2} + 2x - 12)$

**104.** $(-m^{2} + 5m - 8) + (6m^{2} - 5m - 2)$

**105.** $(6.2p - 4.3) + (1.9p + 7.1)$

**106.** $(-6y - 7) - (-3y + 8)$

**107.** $(4x^{2} - 9x) - (3x + 15)$

**108.** $(5a^{2} - 6a - 9) - (2a^{2} - a + 12)$

**109.** $(x^{2} + 7x - 3) - (x^{2} + 3x - 5)$

**110.** $(-2x^{2} + 8x - 7) - (3x^{2} + 12)$

**111.** $\dfrac{1}{7}x(21x + 21)$

**112.** $-3x(5x + 4)$

**113.** $3x(2x^2 - 4x + 7)$

**114.** $-c(2c^2 - 3c + 5)$

**115.** $-7b(-4b^2 - 3b - 5)$

**116.** $(x + 4)(x + 5)$

**117.** $(3x + 6)(-4x + 1)$

**118.** $(-5x + 3)^2$

**119.** $(6 - 2x)(2 + 3x)$

**120.** $(r + 5)(r - 5)$

**121.** $(x - 1)(3x^2 + 4x - 6)$

**122.** $(3x + 1)(x^2 + 2x + 4)$

**123.** $(-4x + 2)(3x^2 - x + 7)$

**124.** $\dfrac{2x + 4}{2}$

**125.** $\dfrac{12y + 18}{3}$

**126.** $\dfrac{8x^2 + 4x}{x}$

**127.** $\dfrac{6x^2 + 9x - 4}{3}$

**128.** $\dfrac{6w^2 - 5w + 3}{3w}$

**129.** $\dfrac{16x^6 - 8x^5 - 3x^3 + 1}{4x}$

**130.** $\dfrac{8m - 4}{-2}$

**131.** $\dfrac{5x^3 + 10x + 2}{2x^2}$

**132.** $\dfrac{5x^2 - 6x + 15}{3x}$

**133.** $\dfrac{x^2 + x - 12}{x - 3}$

**134.** $\dfrac{5x^2 + 28x - 10}{x + 6}$

**135.** $\dfrac{6n^2 + 19n + 3}{6n + 1}$

**136.** $\dfrac{4x^3 + 12x^2 + x - 12}{2x + 3}$

**137.** $\dfrac{4x^2 - 12x + 9}{2x - 3}$

## Chapter 5 Practice Test

*Chapter Test Prep Videos provide fully worked-out solutions to any of the exercises you want to review. Chapter Test Prep Videos are available via MyMathLab®, or on YouTube™ (search "Angel Elementary and Intermediate Algebra" and click on "Channels").*

*Simplify each expression.*

**1.** $5x^4 \cdot 3x^2$

**2.** $(3xy^2)^3$

**3.** $\dfrac{24p^7}{3p^2}$

**4.** $\left(\dfrac{3x^2 y}{6xy^3}\right)^3$

**5.** $(2x^3 y^{-2})^{-2}$

**6.** $(4x^0)(3x^2)^0$

**7.** $\dfrac{30x^6 y^2}{45x^{-1} y}$

*Convert each number to scientific notation and then determine the answer. Express your answer in scientific notation.*

**8.** $(285{,}000)(50{,}000)$

**9.** $\dfrac{0.0008}{4000}$

*Determine whether each expression is a polynomial. If the polynomial has a specific name, give that name.*

**10.** $4x$

**11.** $-8c + 5$

**12.** $x^{-2} + 4$

**13.** Write the polynomial $-5 + 6x^3 - 2x^2 + 5x$ in descending order, and give its degree.

*In Exercises 14–24, perform each indicated operation.*

**14.** $(6x - 4) + (2x^2 - 5x - 3)$

**15.** $(y^2 - 7y + 3) - (4y^2 - 5y - 2)$

**16.** $(4x^2 - 5) - (x^2 + x - 8)$

**17.** $-5d(-3d + 8)$

**18.** $(5x + 8)(3x - 4)$

**19.** $(9 - 4c)(5 + 3c)$

**20.** $(3x - 5)(2x^2 + 4x - 5)$

**21.** $\dfrac{16x^2 + 8x - 4}{4}$

**22.** $\dfrac{-12x^2 - 6x + 5}{-3x}$

**23.** $\dfrac{8x^2 - 2x - 15}{2x - 3}$

**24.** $\dfrac{12x^2 + 7x - 12}{4x + 5}$

**25. Half-Life** The half-life of an element is the time it takes one half the amount of a radioactive element to decay. The half-life of carbon 14 ($C^{14}$) is 5730 years. The half-life of uranium 238 ($U^{238}$) is $4.46 \times 10^9$ years.

© Don Farrall/Getty Images, Inc.–Photodisc/Royalty Free

**a)** Write the half-life of $C^{14}$ in scientific notation.

**b)** How many times longer is the half-life of $U^{238}$ than $C^{14}$?

# Cumulative Review Test

*Take the following test and check your answers with those given in the back of the book. Review any questions that you answered incorrectly. The section where the material was covered is indicated after the answer.*

**1.** Evaluate $12 + 8 \div 2^2 + 3$.

**2.** Simplify $7 - (2x - 3) + 2x - 8(1 - x)$.

**3.** Evaluate $-4x^2 + x - 7$ when $x = -2$.

**4.** Solve $\dfrac{5}{8} = \dfrac{5t}{6} + 2$

**5.** Solve $3x + 5 = 4(x - 2)$.

**6.** Solve the equation $3(x + 2) + 3x - 5 = 4x + 1$.

**7.** Solve the equation $3x - 2 = y - 7$ for $y$.

**8.** Find the slope of the line through the points $(1, 3)$ and $(5, 1)$.

**9.** Determine whether the following pair of lines is parallel.
$$3x - 5y = 7$$
$$5y + 3x = 2$$

**10.** Simplify $\left(\dfrac{5xy^{-3}}{x^{-2}y^5}\right)^2$.

**11.** Write the polynomial $-5x + 2 - 7x^2$ in descending order and give the degree.

*Perform each indicated operation.*

**12.** $(x^2 + 4x - 3) + (2x^2 + 5x + 1)$

**13.** $(6a^2 + 3a + 2) - (a^2 - 3a - 3)$

**14.** $(5t - 3)(2t - 1)$

**15.** $(2x - 1)(3x^2 - 5x + 2)$

**16.** $\dfrac{10d^2 + 12d - 8}{4d}$

**17.** $\dfrac{6x^2 + 11x - 10}{3x - 2}$

**18.** **Chicken Soup**  At Art's Grocery Store, three cans of chicken soup sell for \$1.25. Find the cost of eight cans.

**19.** **Average Speed**  Bob Dolan drives from Jackson, Mississippi, to Tallulah, Louisiana, a distance of 60 miles. At the same time, Nick Reide starts driving from Tallulah to Jackson along the same route. If Bob and Nick meet after 0.5 hour and Nick's average speed was 7 miles per hour greater than Bob's, find the average speed of each car.

**20.** **Rectangle**  The length of a rectangle is 2 less than 3 times the width. Find the dimensions of the rectangle if its perimeter is 28 feet.

# 6 Factoring

## Goals of This Chapter

The major emphasis of this chapter is to teach you how to factor polynomials. Factoring polynomials is the reverse process of multiplying polynomials.

In the first five sections of this chapter, you will learn how to factor a monomial from a polynomial, factor by grouping, factor trinomials of the form $ax^2 + bx + c$ when $a = 1$ and $a \neq 1$, and factor by using special factoring formulas. In the last two sections of this chapter, you will learn how to solve quadratic equations using factoring and how to solve applications of quadratic equations.

It is essential that you have a thorough understanding of factoring, especially Sections 6.3 through 6.5, to complete Chapter 7 successfully.

© Allen R. Angel

**What do Donald Duck, the** Scarecrow from *The Wizard of Oz*, and a baseball diamond all have in common? The answer is the Pythagorean Theorem! Donald Duck is on a postage stamp talking about the Pythagorean Theorem, the Scarecrow talks about the Pythagorean Theorem in the movie *The Wizard of Oz* (see Exercise 54 on page 371), and the Pythagorean Theorem can be used to find the distance between home plate and second base in a baseball diamond (see Exercise 45 on page 370).

# 6.1 Factoring a Monomial from a Polynomial

1. Identify factors.

2. Determine the greatest common factor of two or more numbers.

3. Determine the greatest common factor of two or more terms.

4. Factor a monomial from a polynomial.

## Understanding Algebra

Factoring is the reverse process of multiplication.

*Multiplying*:

$3x(2x^2 + 5) \overset{\text{becomes}}{\rightarrow} 6x^3 + 15x$

*Factoring*:

$6x^3 + 15x \overset{\text{becomes}}{\rightarrow} 3x(2x^2 + 5)$

## 1 Identify Factors

In Chapter 5, you learned how to multiply polynomials. In this chapter, we focus on factoring, the reverse process of multiplication.

> **Factor**
>
> To **factor an expression** means to write the expression as a product of its factors.
>     In general, if $a \cdot b = c$, then $a$ and $b$ are called **factors** of $c$.

For example, in Section 5.5 we showed that $-12n^3 + 6n^2 + 3n$ can also be written as $-3n(4n^2 - 2n - 1)$. We say that $(-3n)$ and $(4n^2 - 2n - 1)$ are *factors* of $-12n^3 + 6n^2 + 3n$.

$3 \cdot 5 = 15$; so 3 and 5 are factors of 15.

$x^3 \cdot x^4 = x^7$; so $x^3$ and $x^4$ are factors of $x^7$.

$x(x + 2) = x^2 + 2x$; so $x$ and $x + 2$ are factors of $x^2 + 2x$.

$(x - 1)(x + 3) = x^2 + 2x - 3$; so $x - 1$ and $x + 3$ are factors of $x^2 + 2x - 3$.

**EXAMPLE 1** List the factors of $6x^3$.

*Solution*

| Factors | Factors |
|---|---|
| $1 \cdot 6x^3 = 6x^3$ | $x \cdot 6x^2 = 6x^3$ |
| $2 \cdot 3x^3 = 6x^3$ | $2x \cdot 3x^2 = 6x^3$ |
| $3 \cdot 2x^3 = 6x^3$ | $3x \cdot 2x^2 = 6x^3$ |
| $6 \cdot x^3\ = 6x^3$ | $6x \cdot x^2\ = 6x^3$ |

The factors of $6x^3$ are $1, 2, 3, 6, x, 2x, 3x, 6x, x^2, 2x^2, 3x^2, 6x^2, x^3, 2x^3, 3x^3,$ and $6x^3$. The opposite (or negative) of each of these factors is also a factor, but these opposites are generally not listed unless specifically requested.

Now Try Exercise 7

Here are examples of multiplying and factoring. Notice again that factoring is the reverse process of multiplying.

| Multiplying | Factoring |
|---|---|
| $3(2x + 5) = 6x + 15$ | $6x + 15 = 3(2x + 5)$ |
| $4y(y - 7) = 4y^2 - 28y$ | $4y^2 - 28y = 4y(y - 7)$ |
| $(x + 1)(x + 3) = x^2 + 4x + 3$ | $x^2 + 4x + 3 = (x + 1)(x + 3)$ |

## 2 Determine the Greatest Common Factor of Two or More Numbers

To factor a monomial from a polynomial, we make use of the *greatest common factor* (*GCF*). If you wish to see additional material on obtaining the GCF, you may read Appendix B, where we also discuss finding the GCF.

Before we examine how to find the GCF, a discussion of **prime numbers** is appropriate.

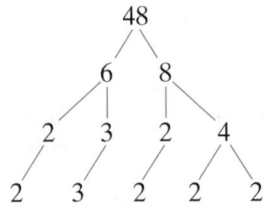

FIGURE 6.1

### Prime Numbers and Composite Numbers

A **prime number** is an integer greater than 1 that has exactly two factors, itself and 1.
The first 15 prime numbers are:

$$2, 3, 5, 7, 11, 13, 17, 19, 23, 29, 31, 37, 41, 43, 47$$

A **composite** number is an integer greater than 1 that is not prime.
The first 15 composite numbers are:

$$4, 6, 8, 9, 10, 12, 14, 15, 16, 18, 20, 21, 22, 24, 25$$

The number 1 is neither prime nor composite.

Examples 2 and 3 present the procedure for writing composite numbers as the product of prime numbers.

EXAMPLE 2  Write 48 as a product of prime numbers. ⎯⎯

*Solution*    Select any two numbers whose product is 48. Two possibilities are $6 \cdot 8$ and $4 \cdot 12$, but there are other choices. Continue breaking down the factors until all the factors are prime, as illustrated in **Figure 6.1**.
Note that no matter how you select your initial factors,

$$48 = 2 \cdot 2 \cdot 2 \cdot 2 \cdot 3 = 2^4 \cdot 3$$

Now Try Exercise 9

In Example 2, we found that $48 = 2 \cdot 2 \cdot 2 \cdot 2 \cdot 3 = 2^4 \cdot 3$. The $2 \cdot 2 \cdot 2 \cdot 2 \cdot 3$ or $2^4 \cdot 3$ may also be referred to as the **prime factorization** of 48.

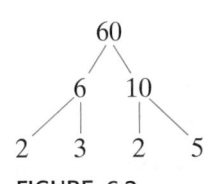

FIGURE 6.2

EXAMPLE 3  Write 60 as a product of its prime factors. ⎯⎯

*Solution*    One way to find the prime factors is shown in **Figure 6.2**. Therefore, $60 = 2 \cdot 2 \cdot 3 \cdot 5 = 2^2 \cdot 3 \cdot 5$.

Now Try Exercise 11

The **greatest common factor (GCF)** of two numbers is the greatest number that divides evenly into each of the numbers. The greatest common factor of 6 and 8 is 2 because 2 is the largest number that divides into both 6 and 8. To find the GCF of two or more numbers, a method involving the numbers' prime factorizations is used.

### To Determine the GCF of Two or More Numbers

1. Write each number as a product of prime factors.
2. Determine the prime factors common to all the numbers.
3. Multiply the common factors found in step 2. The product of these factors is the GCF.

EXAMPLE 4  Determine the greatest common factor of 48 and 60. ⎯⎯

*Solution*    From Examples 2 and 3, we know that

Step 1

$$48 = 2 \cdot 2 \cdot 2 \cdot 2 \cdot 3 = 2^4 \cdot 3$$
$$60 = 2 \cdot 2 \cdot 3 \cdot 5 = 2^2 \cdot 3 \cdot 5$$

Step 2    The common factors are circled. Two factors of 2 and one factor of 3 are common to both numbers. The product of these factors is the GCF of 48 and 60:

Step 3                    $\text{GCF} = 2 \cdot 2 \cdot 3 = 12$

The GCF of 48 and 60 is 12. Twelve is the greatest number that divides evenly into both 48 and 60.

Now Try Exercise 15

**EXAMPLE 5** Determine the GCF of 18 and 24.

*Solution*

$$18 = 2 \cdot 3 \cdot 3 = 2 \cdot 3^2$$
$$24 = 2 \cdot 2 \cdot 2 \cdot 3 = 2^3 \cdot 3$$

One factor of 2 and one factor of 3 are common to both 18 and 24.

$$\text{GCF} = 2 \cdot 3 = 6$$

Now Try Exercise 19

## 3  Determine the Greatest Common Factor of Two or More Terms

Consider the terms $x^3, x^4, x^5,$ and $x^6$. The GCF of these terms is $x^3$, since $x^3$ is the largest number of $x$'s common to all four terms. We can illustrate this by writing the terms in factored form, with $x^3$ as one factor.

$$x^3 = x^3 \cdot 1$$
$$x^4 = x^3 \cdot x$$
$$x^5 = x^3 \cdot x^2$$
$$x^6 = x^3 \cdot x^3$$

GCF of all four terms is $x^3$.

Notice that $x^3$ evenly divides all four terms,

$$\frac{x^3}{x^3} = 1, \quad \text{and} \quad \frac{x^4}{x^3} = x, \quad \text{and} \quad \frac{x^5}{x^3} = x^2, \quad \text{and} \quad \frac{x^6}{x^3} = x^3.$$

**EXAMPLE 6** Determine the GCF of the terms $m^9, m^5, m^7,$ and $m^4$.

*Solution*   The GCF is $m^4$ because $m^4$ is the largest factor common to all the terms.

Now Try Exercise 21

**EXAMPLE 7** Determine the GCF of the terms $x^2y^3, x^3y^2$ and $xy^4$.

*Solution*

$$x^2y^3 = x \cdot x \cdot y \cdot y \cdot y$$
$$x^3y^2 = x \cdot x \cdot x \cdot y \cdot y$$
$$xy^4 = x \cdot y \cdot y \cdot y \cdot y$$

One $x$ is common to each term ↑   ↑ Two $y$'s are common to each term

$$x^1 \quad y^2$$

Thus, the GCF of the three terms is $xy^2$

Now Try Exercise 29

### Greatest Common Factor of Two or More Terms

To determine the GCF of two or more terms

1. Find the GCF of the numerical coefficients of the terms.
2. Find the largest power of each variable that is common to all of the terms.
3. The GCF is the product of the number from step 1 and the variable expressions from step 2.

**EXAMPLE 8** Determine the GCF of each group of terms.

**a)** $18y^2, 15y^3, 27y^5$    **b)** $-20x^2, 12x, 40x^3$    **c)** $5s^4, s^7, s^3$

*Solution*

**a)** The GCF of 18, 15, and 27 is 3. The GCF of $y^2, y^3,$ and $y^5$ is $y^2$. Therefore, the GCF of the three terms is $3y^2$.

**b)** The GCF of $-20$, $12$, and $40$ is $4$. The GCF of $x^2$, $x$, and $x^3$ is $x$. Therefore, the GCF of the three terms is $4x$.

**c)** The GCF of $5$, $1$, and $1$ is $1$. The GCF of $s^4$, $s^7$, and $s^3$ is $s^3$. Therefore, the GCF of the three terms is $1s^3$, which we write as $s^3$.

Now Try Exercise 33

**EXAMPLE 9**   Determine the GCF of $48x^2yz$ and $60x^3y^3$.

*Solution*

1. The numerical coefficients are $48$ and $60$. From Example 4, the GCF of $48$ and $60$ is $12$.

2. The variables $x$ and $y$ are common to both terms. The variable $z$ does not appear in the second term. The largest power of $x$ common to both terms is $x^2$. The largest power of $y$ common to both terms is $y$.

3. Thus, the GCF is the product of $12$, $x^2$, and $y$, or $12x^2y$.

Now Try Exercise 25

**EXAMPLE 10**   Determine the GCF of each pair of terms.

**a)** $a(a-6)$ and $3(a-6)$

**b)** $t(t+4)$ and $t+4$

**c)** $3(p+q)$ and $4p(p+q)$

*Solution*

**a)** The GCF is $(a-6)$.

**b)** $t+4$ can be written as $1(t+4)$. Therefore, the GCF of $t(t+4)$ and $1(t+4)$ is $t+4$.

**c)** The GCF is $(p+q)$.

Now Try Exercise 39

**Understanding Algebra**

Remember, factoring is the reverse process of multiplication.
    Also, it is important to factor out the *greatest* common factor—not just a common factor!

## 4   Factor a Monomial from a Polynomial

Factoring is the reverse process of multiplying factors. As mentioned earlier, to *factor an expression* means to write the expression as a product of its factors.

> **To Factor a Monomial from a Polynomial**
>
> 1. Determine the greatest common factor of all terms in the polynomial.
> 2. Write each term as the product of the GCF and its other factor.
> 3. Use the distributive property to factor out the GCF.

**Understanding Algebra**

With multiplication, we usually see the distributive property used as follows:

$4(x+2) = 4 \cdot x + 4 \cdot 2 = 4x+8$

But, with factoring, we reverse the process as follows:

$4x+8 = 4 \cdot x + 4 \cdot 2 = 4(x+2)$

In step 3 of the process, where we indicate that we use the distributive property, we factor the GCF out of the terms in the polynomial. For example, to factor $4 \cdot x + 4 \cdot 2$, we factor out the GCF, $4$, to write $4(x+2)$.

**EXAMPLE 11**   Factor $6x + 18$.

*Solution*   The GCF is $6$.

$$6x + 18 = \boxed{6} \cdot x + \boxed{6} \cdot 3 \qquad \text{Write each term as a product of the GCF and its other factor.}$$

$$= \boxed{6}(x+3) \qquad \text{Distributive property}$$

Now Try Exercise 49

> **Helpful Hint**
>
> Checking a factoring problem involves two steps. First, multiply the factored result—you should obtain the original expression. Second, be sure each factor in your factored result cannot be factored further. For example, if we factor $3x^2 + 6x$ as $x(3x + 6)$, the first part of the check—multiplying $x$ by $(3x + 6)$—does in fact yield $3x^2 + 6x$. However, there is still a common factor, 3, in $(3x + 6)$, which needs to be factored out. The correct answer is that $3x^2 + 6x$ factors as $3x(x + 2)$. So, be sure the factors in your result do not have any remaining common factors.

**EXAMPLE 12**  Factor $15x - 20$.

*Solution*   The GCF is 5.

$$15x - 20 = \boxed{5} \cdot 3x - \boxed{5} \cdot 4$$
$$= \boxed{5}(3x - 4)$$

The factored binomial no longer has any common factors and we can check that the factoring is correct by multiplying.

Now Try Exercise 51

**EXAMPLE 13**  Factor $6y^2 + 9y^5$.

*Solution*   The GCF is $3y^2$.

$$6y^2 + 9y^5 = \boxed{3y^2} \cdot 2 + \boxed{3y^2} \cdot 3y^3$$
$$= \boxed{3y^2}(2 + 3y^3)$$

The factored binomial no longer has any common factors and we can check that the factoring is correct by multiplying.

Now Try Exercise 59

**EXAMPLE 14**  Factor $8q^3 - 20q^2 - 12q$.

*Solution*   The GCF is $4q$.

$$8q^3 - 20q^2 - 12q = \boxed{4q} \cdot 2q^2 - \boxed{4q} \cdot 5q - \boxed{4q} \cdot 3$$
$$= \boxed{4q}(2q^2 - 5q - 3)$$

*Check*   $4q(2q^2 - 5q - 3) = 8q^3 - 20q^2 - 12q$

Now Try Exercise 81

**EXAMPLE 15**  Factor $35x^2 - 25x + 5$.

*Solution*   The GCF is 5.

$$35x^2 - 25x + 5 = \boxed{5} \cdot 7x^2 - \boxed{5} \cdot 5x + \boxed{5} \cdot 1$$
$$= \boxed{5}(7x^2 - 5x + 1)$$

**Understanding Algebra**

In Example 15, notice we rewrite the constant term 5 as $5 \cdot 1$ so that when 5 is factored out, a 1 remains.

The factored trinomial no longer has any common factors and we can check that the factoring is correct by multiplying.

Now Try Exercise 85

**EXAMPLE 16**  Factor $4x^3 + x^2 + 8x^2y$.

*Solution*   The GCF is $x^2$.

$$4x^3 + x^2 + 8x^2y = \boxed{x^2} \cdot 4x + \boxed{x^2} \cdot 1 + \boxed{x^2} \cdot 8y$$
$$= \boxed{x^2}(4x + 1 + 8y)$$

The factored trinomial no longer has any common factors and we can check that the factoring is correct by multiplying.

Now Try Exercise 89

*Notice in Examples 15 and 16 that when one of the terms is itself the GCF, we express it in factored form as the product of the term itself and 1.*

**EXAMPLE 17** Factor $x(5x - 2) + 7(5x - 2)$.

*Solution*   The GCF of $x(5x - 2)$ and $7(5x - 2)$ is $(5x - 2)$. Factoring out the GCF gives

$$x(5x - 2) + 7(5x - 2) = (5x - 2)(x + 7)$$

This factored result no longer has any common factors and we can check that the factoring is correct by multiplying.

Now Try Exercise 95

**EXAMPLE 18** Factor $4x(3x - 5) - 7(3x - 5)$.

*Solution*   The GCF of $4x(3x - 5)$ and $-7(3x - 5)$ is $(3x - 5)$. Factoring out the GCF gives

$$4x(3x - 5) - 7(3x - 5) = (3x - 5)(4x - 7)$$

This factored result no longer has any common factors and we can check that the factoring is correct by multiplying.

Recall from Section 1.10 that the commutative property of multiplication states that the order in which any two real numbers are multiplied does not matter. Therefore, $(3x - 5)(4x - 7)$ can also be written $(4x - 7)(3x - 5)$. In the book, we will place the common factor on the left.

Now Try Exercise 97

**EXAMPLE 19** Factor $2x(x + 3) - 5(x + 3)$.

*Solution*   The GCF of $2x(x + 3)$ and $-5(x + 3)$ is $(x + 3)$. Factoring out the GCF gives

$$2x(x + 3) - 5(x + 3) = (x + 3)(2x - 5)$$

The factored result no longer has any common factors and we can check that the factoring is correct by multiplying.

Now Try Exercise 93

*Whenever you are factoring a polynomial by any of the methods presented in this chapter, the first step will always be to see if there is a common factor (other than 1) to all the terms in the polynomial. If so, factor the greatest common factor from each term using the distributive property.*

---

# EXERCISE SET 6.1

MathXL®   MyMathLab®

MathXL®       MyMathLab

## Warm-Up Exercises

*Fill in the blanks with the appropriate word, phrase, or symbol(s) from the following list.*

| | | | | |
|---|---|---|---|---|
| composite | $4x^4$ | $2x^2y^3$ | 1 | 2 |
| 12 | prime | 6 | $40x^3y^6$ | multiplying |

**1.** The GCF of 24 and 42 is _____ .

**2.** The positive integer that is neither prime nor composite is the number _____ .

**3.** The GCF of $4x^2y^6$, $10x^3y^4$, and $20x^3y^3$ is _____ .

**4.** Factoring is the reverse process of _____ .

**5.** The only even prime number is _____ .

**6.** An even integer greater than 2 is an example of a _____ number.

**7.** The positive factors of _____ are 1, 2, 3, 4, 6, and 12.

**8.** An integer greater than 1 that is divisible only by itself and 1 is called a _____ number.

## Practice the Skills

*Write each number as a product of prime numbers.*

**9.** 20          **10.** 120          **11.** 90          **12.** 540          **13.** 248          **14.** 144

*Determine the greatest common factor for each pair of numbers.*

**15.** $40, 56$          **16.** $45, 27$          **17.** $70, 98$          **18.** $120, 96$          **19.** $80, 126$          **20.** $88, 160$

*Determine the greatest common factor for each group of terms.*

**21.** $x^5, x^3, x^2$

**22.** $y^3, y^5, y^2$

**23.** $3x, 6x^2, 9x^3$

**24.** $6p, 4p^2, 8p^3$

**25.** $a, ab, ab^2$

**26.** $x, y, z$

**27.** $q^3r, q^2r^2, qr^4$

**28.** $4x^2y^2, 3xy^4, 2xy^2$

**29.** $x^3y^7, x^7y^{12}, x^5y^5$

**30.** $6x, 12y, 18x^2$

**31.** $-3, 20x, 30x^2$

**32.** $24s^5, 6r^3s^3, 15rs^2$

**33.** $9x^3y^4, 8x^2y^4, 12x^4y^2$

**34.** $16x^9y^{12}, 8x^5y^3, 20x^4y^2$

**35.** $40x^3, 27x, 30x^4y^2$

**36.** $6p^4q^3, 9p^2q^5, 9p^4q^2$

**37.** $8(x - 4), 7(x - 4)$

**38.** $4(x - 5), 3x(x - 5)$

**39.** $x^2(2x - 3), 5(2x - 3)$

**40.** $x(9x - 3), 9x - 3$

**41.** $3w + 5, 6(3w + 5)$

**42.** $b(b + 3), b + 3$

**43.** $x - 4, y(x - 4)$

**44.** $3y(x + 2), 3(x + 2)$

**45.** $3(x - 1), 5(x - 1)^2$

**46.** $5(n + 2), 7(n + 2)^2$

**47.** $(x - 9)(x + 6), (x - 9)(x + 3)$

**48.** $(a + 4)(a - 3), 5(a - 3)$

*Factor the GCF from each term in the expression.*

**49.** $4x - 80$

**50.** $4x + 2$

**51.** $15x - 5$

**52.** $12x + 15$

**53.** $7q + 28$

**54.** $3t^2 - 10t$

**55.** $9x^2 - 12x$

**56.** $24y - 6y^2$

**57.** $7x^5 - 9x^4$

**58.** $9x + 27x^3$

**59.** $3x^5 - 12x^2$

**60.** $26p^2 - 8p$

**61.** $36x^{12} + 24x^8$

**62.** $45y^{12} + 30y^{10}$

**63.** $27y^{15} - 9y^3$

**64.** $30w^5 + 25w^3$

**65.** $y + 6x^3y$

**66.** $4x^2y - 6x$

**67.** $7a^4 + 3a^2$

**68.** $3x^2y + 6x^2y^2$

**69.** $16xy^2z + 4x^3y$

**70.** $48m^4n^2 - 16mn^2$

**71.** $80x^5y^3z^4 - 36x^2yz^3$

**72.** $56xy^5z^{13} - 24y^4z^2$

**73.** $25x^2yz^3 + 25x^3yz$

**74.** $13y^5z^3 - 11xy^2z^5$

**75.** $19x^4y^{12}z^{13} - 8x^5y^3z^9$

**76.** $16r^4s^5t^3 - 20r^5s^4t$

**77.** $8c^2 - 4c - 32$

**78.** $x^3 - 4x^2 - 3x$

**79.** $9x^2 + 18x + 3$

**80.** $4x^2 + 8x + 24$

**81.** $4x^3 - 8x^2 + 12x$

**82.** $12a^3 - 16a^2 - 4a$

**83.** $40b^2 - 48c + 24$

**84.** $5x^3 - xy^2 + x$

**85.** $15p^2 - 6p + 9$

**86.** $45y^3 - 63y^2 + 27y$

**87.** $9a^4 - 6a^3 + 3ab$

**88.** $45v^4w^2 + 10v^2x - 20vw^5$

**89.** $8x^2y + 12xy^2 + 5xy$

**90.** $52x^2y^2 + 16xy^3 + 26z$

**91.** $x(x - 7) + 6(x - 7)$

**92.** $9x(3x - 4) - 4(3x - 4)$

**93.** $3b(a - 2) - 4(a - 2)$

**94.** $3x(7x + 1) - 2(7x + 1)$

**95.** $4x(2x + 1) + 1(2x + 1)$

**96.** $4m(5m - 1) - 3(5m - 1)$

**97.** $5x(2x + 1) + 2x + 1$

**98.** $3x(4x - 5) + 4x - 5$

**99.** $3c(6c + 7) - 2(6c + 7)$

**100.** $5t(t - 2) - 3(t - 2)$

## Problem Solving

*Factor each expression, if possible. Treat the unknown symbol as if it were a variable.*

**101.** $12\nabla - 6\nabla^2$

**102.** $3\star + 6$

**103.** $12\square^3 - 4\square^2 + 4\square$

**104.** $\copyright + 11\triangle$

## Concept/Writing Exercises

**105.** Explain how to check a factoring problem.

**106.** What is the greatest common factor of two or more numbers?

## Challenge Problems

**107.** Factor $6x^5(2x + 7) + 4x^3(2x + 7) - 2x^2(2x + 7)$.

**108.** Factor $4x^2(x - 3)^3 - 6x(x - 3)^2 + 4(x - 3)$.

**109.** Factor $x^2 + 2x + 3x + 6$. (*Hint:* Factor the first two terms, then factor the last two terms, then factor the resulting two terms. We will discuss factoring problems of this type in Section 6.2.)

## Cumulative Review Exercises

[2.1] **110.** Simplify $2x - (x - 5) + 4(3 - x)$.

[2.5] **111.** Solve the equation $4 + 3(x - 8) = x - 4(x + 2)$.

[2.6] **112.** Solve the equation $4x - 5y = 20$ for $y$.

**113. Volume** Find the volume of the cone shown below.

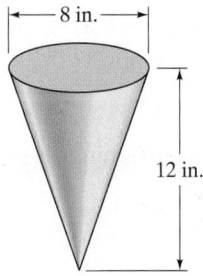

[3.2] **114. Numbers** The sum of two numbers is 41. Find the two numbers if the larger number is one less than twice the smaller number.

[5.1] **115.** Simplify $\left(\dfrac{3x^2y^3}{2x^5y^2}\right)^2$.

# 6.2 Factoring by Grouping

**1** Factor a polynomial containing four terms by grouping.

## **1** Factor a Polynomial Containing Four Terms by Grouping

It may be possible to factor a polynomial containing four or more terms by removing common factors from groups of terms. This process is called **factoring by grouping**, which is illustrated in Example 1.

**EXAMPLE 1** Factor $ax + ay + bx + by$ by grouping.

*Solution*    There is no factor (other than 1) common to all four terms. However, $a$ is common to the first two terms and $b$ is common to the last two terms. Factor $a$ from the first two terms and $b$ from the last two terms.

$$ax + ay + bx + by = a(x + y) + b(x + y)$$

This factoring gives two terms, and $(x + y)$ is common to both terms. Proceed to factor $(x + y)$ from each term, as shown below.

$$a(x + y) + b(x + y) = (x + y)(a + b)$$

Notice that when $(x + y)$ is factored out we are left with $a + b$, which becomes the other factor. Thus, $ax + ay + bx + by = (x + y)(a + b)$.

Now Try Exercise 7

### Understanding Algebra

$(ax + ay) \xrightarrow{\text{factor out } a} a(x + y)$

$(bx + by) \xrightarrow{\text{factor out } b} b(x + y)$

$a(x + y) + b(x + y) \xrightarrow{\text{factor out } (x+y)}$

$\qquad\qquad (x + y)(a + b)$

**To Factor a Four-Term Polynomial Using Grouping**

1. Determine whether there are any factors common to all four terms. If so, factor the greatest common factor from each of the four terms.
2. If necessary, arrange the four terms so that the first two terms have a common factor and the last two have a common factor.
3. Use the distributive property to factor each group of two terms.
4. Factor the greatest common factor from the results of step 3.

**EXAMPLE 2**   Factor $x^2 + 3x + 4x + 12$ by grouping.

**Solution**    No factor is common to all four terms. However, you can factor $x$ from the first two terms and 4 from the last two terms.

$$x^2 + 3x + 4x + 12 = x(x + 3) + 4(x + 3)$$

$x + 3$ is common

$$x(x + 3) + 4(x + 3) = (x + 3)(x + 4)$$

Thus, $x^2 + 3x + 4x + 12 = (x + 3)(x + 4)$

Now Try Exercise 11

In Example 2, the $3x$ and $4x$ are like terms. However, if we were to combine them we would not be able to factor the four terms by grouping.

**EXAMPLE 3**   Factor $15x^2 + 10x + 12x + 8$ by grouping.

**Solution**

$$15x^2 + 10x + 12x + 8 = 5x(3x + 2) + 4(3x + 2)$$

Factor $5x$ from the first two terms and 4 from the last two terms.

$$= (3x + 2)(5x + 4)$$

Now Try Exercise 17

*A factoring-by-grouping problem can be checked by multiplying the factors using the FOIL method. If you have not made a mistake, your result will be the polynomial you began with. Here is a check of Example 3.*

Check                               F          O          I          L

$$(3x + 2)(5x + 4) = (3x)(5x) + (3x)(4) + (2)(5x) + (2)(4)$$

$$= 15x^2 + 12x + 10x + 8$$

$$= 15x^2 + 10x + 12x + 8$$

We can write $12x + 10x$ as $10x + 12x$ because of the commutative property of addition. Since this is the polynomial we started with, the factoring is correct.

**EXAMPLE 4**   Factor $15x^2 + 12x + 10x + 8$ by grouping.

**Solution**    $15x^2 + 12x + 10x + 8 = 3x(5x + 4) + 2(5x + 4)$

$$= (5x + 4)(3x + 2)$$

Now Try Exercise 19

Notice that Example 4 is the same as Example 3 with the two middle terms interchanged. The answers to Examples 3 and 4 are equivalent since only the order of the factors are changed.

**EXAMPLE 5** Factor $x^2 - 3x + x - 3$ by grouping.

*Solution*   In the first two terms, $x$ is the common factor. Is there a common factor in the last two terms? Yes; remember that 1 is a factor of every term. Factor 1 from the last two terms.

$$x^2 - 3x + x - 3 = x^2 - 3x + 1 \cdot x - 1 \cdot 3$$
$$= x(x - 3) + 1(x - 3)$$
$$= (x - 3)(x + 1)$$

Note that $x - 3$ was expressed as $1 \cdot x - 1 \cdot 3 = 1(x - 3)$.

Now Try Exercise 21

> **Understanding Algebra**
>
> In Example 5, notice that $x - 3$ can be rewritten as $1 \cdot x - 1 \cdot 3$ and then as $1(x - 3)$.

**EXAMPLE 6** Factor $6x^2 - 3x - 2x + 1$ by grouping.

*Solution*   When $3x$ is factored from the first two terms, we get

$$6x^2 - 3x - 2x + 1 = 3x(2x - 1) - 2x + 1$$

What should we factor from the last two terms? We rewrite $-2x + 1$ as $-1(2x - 1)$.
   *Whenever we wish to change the sign of each term of an expression, we can factor out a negative number from each term. In this case, we factor out $-1$.*

$$-2x + 1 = -1(2x - 1)$$

Next, we have

$$3x(2x - 1) - 2x + 1 = 3x(2x - 1) - 1(2x - 1)$$

Now we factor out the common factor $(2x - 1)$.

$$3x(2x - 1) - 1(2x - 1) = (2x - 1)(3x - 1)$$

Now Try Exercise 23

> **Understanding Algebra**
>
> It is important to recognize that $2x - 1$ is the opposite of $-2x + 1$. So, we can replace $-2x + 1$ with the equivalent $-1(2x - 1)$.

**EXAMPLE 7** Factor $q^2 + 3q - q - 3$ by grouping.

*Solution*
$$q^2 + 3q - q - 3 = q(q + 3) - q - 3 \qquad \text{Factored out } q.$$
$$= q(q + 3) - 1(q + 3) \qquad \text{Factored out } -1.$$
$$= (q + 3)(q - 1) \qquad \text{Factored out } (q + 3).$$

Now Try Exercise 25

**EXAMPLE 8** Factor $3x^2 - 6x - 4x + 8$ by grouping.

*Solution*
$$3x^2 - 6x - 4x + 8 = 3x(x - 2) - 4(x - 2)$$
$$= (x - 2)(3x - 4)$$

Note: $-4x + 8 = -4(x - 2)$.

Now Try Exercise 27

> **Helpful Hint**
>
> When factoring four terms by grouping, if the coefficient of the third term is positive, as in Examples 2 through 5, you will generally factor out a positive coefficient from the last two terms. *If the coefficient of the third term is negative*, as in Examples 6 through 8, *you will generally factor out a negative coefficient from the last two terms.* The sign of the coefficient of the third term in the expression *must be included* so that the factoring results in two terms. For example,
>
> $$2x^2 + 8x + 3x + 12 = 2x(x + 4) + 3(x + 4) = (x + 4)(2x + 3)$$
> $$3x^2 - 15x - 2x + 10 = 3x(x - 5) - 2(x - 5) = (x - 5)(3x - 2)$$

When factoring four terms by grouping, the two middle terms need not be like terms. This is illustrated in Examples 9 and 10.

EXAMPLE 9   Factor $xy + 3x - 2y - 6$ by grouping. ───

*Solution*   This problem contains two variables, $x$ and $y$. Factor $x$ from the first two terms and $-2$ from the last two terms.

$$xy + 3x - 2y - 6 = x(y + 3) - 2(y + 3)$$
$$= (y + 3)(x - 2) \quad \text{Factored out } (y + 3).$$

Now Try Exercise 41

EXAMPLE 10   Factor $2xy + 4y + 3x + 6$. ───

*Solution*   We will factor out $2y$ from the first two terms and 3 from the last two terms.

$$2xy + 4y + 3x + 6 = 2y(x + 2) + 3(x + 2)$$

Factor out the common factor $(x + 2)$ from each term on the right.

$$2y(x + 2) + 3(x + 2) = (x + 2)(2y + 3)$$

Check                          F         O         I         L

$$(x + 2)(2y + 3) = (x)(2y) + (x)(3) + (2)(2y) + (2)(3)$$
$$= 2xy + 3x + 4y + 6$$
$$= 2xy + 4y + 3x + 6$$

Now Try Exercise 29

If Example 10 were given as $2xy + 3x + 4y + 6$, would the results be the same? Try it and see.

EXAMPLE 11   Factor $15a^2 - 10ab + 12ab - 8b^2$. ───

*Solution*   Factor $5a$ from the first two terms and $4b$ from the last two terms.

$$15a^2 - 10ab + 12ab - 8b^2 = 5a(3a - 2b) + 4b(3a - 2b)$$
$$= (3a - 2b)(5a + 4b)$$

Now Try Exercise 31

EXAMPLE 12   Factor $3x^2 - 15x + 6x - 30$. ───

*Solution*   *The first step in any factoring problem is to determine whether all the terms have a common factor. If so, we factor out the greatest common factor (GCF).* In this polynomial, 3 is the GCF. Therefore, we begin by factoring out the 3.

$$3x^2 - 15x + 6x - 30 = 3(x^2 - 5x + 2x - 10)$$

Now we factor the expression in parentheses by grouping. We factor out $x$ from the first two terms and 2 from the last two terms.

$$3(x^2 - 5x + 2x - 10) = 3[x(x - 5) + 2(x - 5)]$$
$$= 3[(x - 5)(x + 2)]$$
$$= 3(x - 5)(x + 2)$$

Thus, $3x^2 - 15x + 6x - 30 = 3(x - 5)(x + 2)$.

Now Try Exercise 49

# EXERCISE SET 6.2   MathXL®   MyMathLab®

## Warm-Up Exercises

*Fill in the blanks with the appropriate word, phrase, or symbol(s) from the following list.*

| | | | |
|---|---|---|---|
| FOIL | $-1(x + 2)$ | $-1(2x - 1)$ | factoring by grouping |
| $(x + 2)$ | $2(x + 2)$ | $(2x - 1)(x + 3)$ | $(2x - 1)(x - 3)$ |

**1.** The technique used to factor $cx + dx + cy + dy$ is called _____.

**2.** When factoring $6x^2 - 3x - 2x + 1$, the last two terms may be rewritten as _____.

**3.** When $2x^2 + 6x - x - 3$ is factored, the result is _____.

**4.** When factoring $x^2 + 2x - x - 2$, the last two terms may be rewritten as _____.

**5.** A factoring-by-grouping problem can be checked by multiplying the result using the _____ method.

**6.** In $x^2 + 2x + 4x + 8$, there is a common factor in the first two terms and in the last two terms. That common factor is _____.

## Practice the Skills

*Factor by grouping.*

**7.** $x^2 + 3x + 2x + 6$

**8.** $x^2 + 7x + 3x + 21$

**9.** $t^2 + 5t + 4t + 20$

**10.** $x^2 - x + 3x - 3$

**11.** $x^2 + 2x + 5x + 10$

**12.** $y^2 - 7y + 5y - 35$

**13.** $c^2 - 4c + 7c - 28$

**14.** $r^2 - 4r + 6r - 24$

**15.** $4x^2 - 6x + 6x - 9$

**16.** $4b^2 - 10b + 10b - 25$

**17.** $3x^2 + 9x + x + 3$

**18.** $a^2 + a + 3a + 3$

**19.** $6x^2 + 3x - 2x - 1$

**20.** $5x^2 + 30x - 3x - 18$

**21.** $8x^2 + 32x + x + 4$

**22.** $9w^2 - 6w - 6w + 4$

**23.** $12t^2 - 8t - 3t + 2$

**24.** $12x^2 + 42x - 10x - 35$

**25.** $x^2 + 9x - x - 9$

**26.** $35x^2 - 40x + 21x - 24$

**27.** $6p^2 + 15p - 4p - 10$

**28.** $10c^2 + 25c - 6c - 15$

**29.** $x^2 + 2xy - 3xy - 6y^2$

**30.** $x^2 - 3xy + 4xy - 12y^2$

**31.** $3x^2 + 2xy - 9xy - 6y^2$

**32.** $3x^2 - 18xy + 4xy - 24y^2$

**33.** $10x^2 - 12xy - 25xy + 30y^2$

**34.** $6a^2 - 3ab + 4ab - 2b^2$

**35.** $x^2 - bx - ax + ab$

**36.** $x^2 + bx + ax + ab$

**37.** $xy + 9x - 5y - 45$

**38.** $x^2 - 2x + ax - 2a$

**39.** $a^2 + 3a + ab + 3b$

**40.** $3x^2 - 15x - 2xy + 10y$

**41.** $xy - x + 5y - 5$

**42.** $y^2 - yb + ya - ab$

**43.** $12 + 8y - 3x - 2xy$

**44.** $7y - 49 - xy + 7x$

**45.** $z^3 + 5z^2 + z + 5$

**46.** $x^3 - 3x^2 + 2x - 6$

**47.** $x^3 - 5x^2 + 8x - 40$

**48.** $y^3 - 3y + 2y^2 - 6$

**49.** $2x^2 - 12x + 8x - 48$

**50.** $3x^2 - 3x - 3x + 3$

**51.** $4x^2 + 8x + 8x + 16$

**52.** $3z^4 - 3z^3 - 7z^3 + 7z^2$

**53.** $6x^3 + 9x^2 - 2x^2 - 3x$

**54.** $9x^3 + 6x^2 - 45x^2 - 30x$

**55.** $p^3 - 6p^2q + 2p^2q - 12pq^2$

**56.** $18x^2 + 27xy + 12xy + 18y^2$

*Rearrange the terms so that the first two terms have a common factor and the last two terms have a common factor (other than 1). Then factor by grouping. There may be more than one way to arrange the factors. However, the answer should be equivalent regardless of the arrangement selected.*

**57.** $5x + 5y + xy + 25$

**58.** $5m + 2w + mw + 10$

**59.** $6x + 5y + xy + 30$

**60.** $ax - 10 - 5x + 2a$

**61.** $ax + by + ay + bx$

**62.** $ax - 21 - 3a + 7x$

**63.** $rs - 42 + 6s - 7r$

**64.** $ca - 2b + 2a - cb$

**65.** $dc + 3c - ad - 3a$

**66.** $ac - bd - ad + bc$

## Problem Solving

**67.** If you know that a polynomial with four terms is factorable by a specific arrangement of the terms, then will *any* arrangement of the terms be factorable by grouping? Explain, and support your answer with an example.

*Factor each expression, if possible. Treat the unknown symbol as if it were a variable.*

**68.** $\varnothing^2 + 3\varnothing + 4\varnothing + 12$             **69.** $\odot^2 + 3\odot - 5\odot - 15$           **70.** $\Delta^2 + 2\Delta - \Delta + 6$

## Concept/Writing Exercises

**71.** A polynomial of four terms is factored by grouping and the result is $(x - 2y)(x - 3)$. Find the polynomial that was factored and explain how you determined the answer.

**72.** A polynomial of four terms is factored by grouping and the result is $(x - 2)(x + 4)$. Find the polynomial that was factored, and explain how you determined the answer.

## Challenge Problems

*In Section 6.4, we will factor trinomials of the form $ax^2 + bx + c$, $a \neq 1$, using grouping. To do this we rewrite the middle term of the trinomial, bx, as a sum or difference of two terms. Then we factor the resulting polynomial of four terms by grouping. For Exercises 73–78,* **a)** *rewrite the trinomial as a polynomial of four terms by replacing the bx-term with the sum or difference given.* **b)** *Factor the polynomial of four terms. Note that the factors obtained are the factors of the trinomial.*

**73.** $2x^2 - 11x + 15, -11x = -5x - 6x$     **74.** $3x^2 + 10x + 8, 10x = 4x + 6x$     **75.** $2x^2 - 11x + 15, -11x = -6x - 5x$

**76.** $3x^2 + 10x + 8, 10x = 6x + 4x$      **77.** $4x^2 - 17x - 15, -17x = 3x - 20x$     **78.** $4x^2 - 17x - 15, -17x = -20x + 3x$

*Factor each expression, if possible. Treat the unknown symbols as if they were variables.*

**79.** $\star\odot + 3\star + 2\odot + 6$                 **80.** $2\Delta^2 - 4\Delta\star - 8\Delta\star + 16\star^2$

## Cumulative Review Exercises

[2.5]   **81.** Solve $5 - 3(2x - 7) = 4(x + 5) - 6$.

[3.4]   **82. Mixture.** To celebrate the tenth anniversary of their candy store's opening, Harlan and Shelley Bricker decide to create a special candy mixture containing jelly beans and gumdrops. The jelly beans sell for $6.25 per pound and the gumdrops sell for $2.50 per pound. How many pounds of each type of candy will be needed to make a 50-pound mixture that will sell for $4.75 per pound?

[5.6]   **83.** Divide $\dfrac{15x^3 - 6x^2 - 9x + 5}{3x}$.

       **84.** Divide $\dfrac{a^2 - 16}{a + 4}$.

See Exercise 82.

# 6.3 Factoring Trinomials of the Form $ax^2 + bx + c$, $a = 1$

1   Factor trinomials of the form $ax^2 + bx + c$, where $a = 1$.

2   Remove the greatest common factor from a trinomial.

### Understanding Algebra

We factor trinomials of the form $ax^2 + bx + c$ in this section and the next section.

In this section, $a = 1$ so the trinomials appear as $x^2 + bx + c$. In the next section, $a \neq 1$ and a different strategy will be employed.

## 1   Factor Trinomials of the Form $ax^2 + bx + c$, where $a = 1$

Now we discuss how to factor trinomials of the form $ax^2 + bx + c$, where $a$, the numerical coefficient of the squared term, is 1. Examples of such trinomials are

$$x^2 + 7x + 12 \qquad\qquad x^2 - 2x - 24$$
$$a = 1, b = 7, c = 12 \qquad a = 1, b = -2, c = -24$$

Recall that factoring is the reverse process of multiplication. We can show with the FOIL method of multiplying binomials that

$$(x + 3)(x + 4) = x^2 + 7x + 12 \quad \text{and} \quad (x - 6)(x + 4) = x^2 - 2x - 24$$

Therefore, $x^2 + 7x + 12$ and $x^2 - 2x - 24$ factor as follows:

$$x^2 + 7x + 12 = (x + 3)(x + 4) \quad \text{and} \quad x^2 - 2x - 24 = (x - 6)(x + 4)$$

In general, when we factor a trinomial of the form $x^2 + bx + c$ we will get a pair of binomial factors as follows:

$$x^2 + bx + c = (x + \blacksquare)(x + \blacksquare)$$

Numbers go here that
add up to $b$ and
multiply to $c$.

To determine the numbers to place in the shaded areas, try different sets of factors of the constant, $c$. We multiply each pair of factors using the FOIL method, and continue until we find the pair whose sum of the products of the outer and inner terms is the same as the $x$-term in the trinomial. This method for factoring is called **trial and error**.

**EXAMPLE 1**   Factor $x^2 + 7x + 12$ by trial and error.

*Solution*    Begin by listing the factors of 12. Then list the possible factors of the trinomial, and the products of these factors. Finally, determine which, if any, of these products gives the correct middle term, $7x$.

| Factors of 12 | Possible Factors of Trinomial | Product of Factors |
|---|---|---|
| $(1)(12)$ | $(x + 1)(x + 12)$ | $x^2 + 13x + 12$ |
| $(2)(6)$ | $(x + 2)(x + 6)$ | $x^2 + 8x + 12$ |
| $(3)(4)$ | $(x + 3)(x + 4)$ | $x^2 + 7x + 12$ |
| $(-1)(-12)$ | $(x - 1)(x - 12)$ | $x^2 - 13x + 12$ |
| $(-2)(-6)$ | $(x - 2)(x - 6)$ | $x^2 - 8x + 12$ |
| $(-3)(-4)$ | $(x - 3)(x - 4)$ | $x^2 - 7x + 12$ |

In the last column, we find the trinomial we are seeking in the third line. Thus,

$$x^2 + 7x + 12 = (x + 3)(x + 4)$$

Now Try Exercise 17

Now let's consider an easier way to factor $x^2 + 7x + 12$. In Section 5.5, we illustrated how the FOIL method is used to multiply two binomials. Let's multiply $(x + 3)(x + 4)$ using the FOIL method.

$$
\begin{aligned}
(x + 3)(x + 4) &= x^2 + 4x + 3x + 12 \\
&= x^2 + 7x + 12
\end{aligned}
$$

We see that $(x + 3)(x + 4) = x^2 + 7x + 12$.

Note that the *sum of the outer and inner terms is 7x and the product of the last terms is 12*. To factor $x^2 + 7x + 12$ , we look for two numbers whose product is 12 and whose sum is 7. We list the factors of 12 first and then list the sum of the factors.

| Factors of 12 | Sum of Factors |
|---|---|
| $(1)(12) = 12$ | $1 + 12 = 13$ |
| $(2)(6) = 12$ | $2 + 6 = 8$ |
| $(3)(4) = 12$ | $3 + 4 = 7$ |
| $(-1)(-12) = 12$ | $-1 + (-12) = -13$ |
| $(-2)(-6) = 12$ | $-2 + (-6) = -8$ |
| $(-3)(-4) = 12$ | $-3 + (-4) = -7$ |

The only factors of 12 whose sum is a positive 7 are 3 and 4. The factors of $x^2 + 7x + 12$ will therefore be $(x + 3)$ and $(x + 4)$.

$$x^2 + 7x + 12 = (x + 3)(x + 4)$$

In the previous illustration, all the possible factors of 12 were listed so that you could see them. However, once you find the specific factors you are seeking when working a problem, you need go no further.

---

### To Factor Trinomials of the Form $ax^2 + bx + c$, where $a = 1$

1. Find two numbers whose product equals the constant, $c$, and whose sum equals the coefficient of the $x$-term, $b$.

2. Use the two numbers found in step 1, including their signs, to write the trinomial in factored form. The trinomial in factored form will be

$$(x + \text{first number})(x + \text{second number})$$

---

The sign of the constant, $c$, is the key in finding the sign of the two numbers to be placed in parentheses. See the Helpful Hint below.

---

**Helpful Hint**

To factor a trinomial of the form $x^2 + bx + c$, first observe the sign of the constant.

**a)** If the constant, $c$, is positive, both numbers in the factors will have the same sign, either both positive or both negative. If $b$ is positive, both factors will contain positive numbers, and if $b$ is negative, both factors will contain negative numbers.

Example:

$$x^2 + 7x + 12 = (x + 3)(x + 4)$$

The coefficient, $b$, is positive     The constant, $c$, is positive     Positive  Positive

Both factors have positive numbers.

Example:

$$x^2 - 5x + 6 = (x - 2)(x - 3)$$

The coefficient, $b$, is negative     The constant, $c$, is positive     Negative  Negative

Both factors have negative numbers.

**b)** If the constant is negative, the two numbers in the factors will have opposite signs. That is, one number will be positive and the other number will be negative.

Example:

$$x^2 + x - 6 = (x + 3)(x - 2)$$

The coefficient, $b$, is positive     The constant, $c$, is negative     Positive  Negative

One factor has a positive number and the other factor has a negative number.

Example:

$$x^2 - 3x - 10 = (x + 2)(x - 5)$$

The coefficient, $b$, is negative     The constant, $c$, is negative     Positive  Negative

One factor has a positive number and the other factor has a negative number.

We will use this information as a starting point when factoring trinomials.

**EXAMPLE 2** Consider a trinomial of the form $x^2 + bx + c$. Use the signs of $b$ and $c$ given below to determine the signs of the numbers in the factors.

a) $b$ is negative and $c$ is positive

b) $b$ is negative and $c$ is negative

c) $b$ is positive and $c$ is negative

d) $b$ is positive and $c$ is positive

*Solution*   In each case we look at the sign of the constant, $c$, first.

a) Since $c$ is positive, both numbers must have the same sign. Since $b$ is negative, both factors will contain negative numbers.

b) Since $c$ is negative, one factor will contain a positive number and the other will contain a negative number.

c) Since $c$ is negative, one factor will contain a positive number and the other will contain a negative number.

d) Since $c$ is positive, both numbers must have the same sign. Since $b$ is positive, both factors will contain positive numbers.

*Now Try Exercise 1*

> **Understanding Algebra**
>
> To factor $x^2 + bx + c$, first observe the sign of $c$, and then observe the sign of $b$:
> $x^2 + 11x + 18 \rightarrow (x + 9)(x + 2)$
> $x^2 - 11x + 18 \rightarrow (x - 9)(x - 2)$
>
> if $c$ is positive, numbers have same sign
>
> $x^2 - 7x - 18 \rightarrow (x - 9)(x + 2)$
> $x^2 + 7x - 18 \rightarrow (x + 9)(x - 2)$
>
> if $c$ is negative, numbers have different signs

**EXAMPLE 3** Factor $x^2 + x - 6$.

*Solution*   We must find two numbers whose product is the constant, $-6$, and whose sum is the coefficient of the $x$-term, 1. Remember that $x$ means $1x$. Since the constant is negative, one number must be positive and the other negative. We now list the factors of $-6$ and look for the two factors whose sum is 1.

| Factors of $-6$ | Sum of Factors |
|---|---|
| $1(-6) = -6$ | $1 + (-6) = -5$ |
| $2(-3) = -6$ | $2 + (-3) = -1$ |
| $3(-2) = -6$ | $3 + (-2) = 1$ |
| $6(-1) = -6$ | $6 + (-1) = 5$ |

The numbers 3 and $-2$ have a product of $-6$ and a sum of 1. Thus, the factors are $(x + 3)$ and $(x - 2)$.

$$x^2 + x - 6 = (x + 3)(x - 2)$$

The order of the factors is not crucial. Therefore, $x^2 + x - 6 = (x - 2)(x + 3)$ is also an acceptable answer.

*Now Try Exercise 19*

> **Understanding Algebra**
>
> Use FOIL to check the answer to Example 3.
>
> $(x + 3)(x - 2) = x^2 - 2x + 3x - 6$
> $\qquad\qquad\quad = x^2 + x - 6$
>
> Since the product of factors is identical to the original trinomial, the factoring is correct.

**EXAMPLE 4** Factor $x^2 - x - 6$.

*Solution*   The factors of $-6$ are illustrated in Example 3. The factors whose product is $-6$ and whose sum is $-1$ are 2 and $-3$.

| Factors of $-6$ | Sum of Factors |
|---|---|
| $2(-3) = -6$ | $2 + (-3) = -1$ |

Therefore,

$$x^2 - x - 6 = (x + 2)(x - 3)$$

*Now Try Exercise 25*

**EXAMPLE 5** Factor $x^2 - 5x + 6$.

*Solution*   We must find two numbers whose product is 6 and whose sum is $-5$. Since the constant, 6, is positive, both numbers must have the same sign. Since the

**Understanding Algebra**

To factor a trinomial, write the trinomial in descending order. To factor $-5x + x^2 + 6$, we must first rewrite it as $x^2 - 5x + 6$.

coefficient of the $x$-term, $-5$, is negative, both numbers must be negative. We now list the negative factors of 6 and look for the pair whose sum is $-5$.

| Factors of 6 | Sum of Factors |
|---|---|
| $(-1)(-6)$ | $-1 + (-6) = -7$ |
| $(-2)(-3)$ | $-2 + (-3) = -5$ |

The factors of 6 whose sum is $-5$ are $-2$ and $-3$.

$$x^2 - 5x + 6 = (x - 2)(x - 3)$$

Now Try Exercise 29

**EXAMPLE 6** Factor $r^2 + 2r - 24$.

*Solution*    We must find the two factors of $-24$ whose sum is 2. Since the constant is negative, one factor will be positive and the other factor will be negative.

| Factors of $-24$ | Sum of Factors |
|---|---|
| $(1)(-24)$ | $1 + (-24) = -23$ |
| $(2)(-12)$ | $2 + (-12) = -10$ |
| $(3)(-8)$ | $3 + (-8) = -5$ |
| $(4)(-6)$ | $4 + (-6) = -2$ |
| $(6)(-4)$ | $6 + (-4) = 2$ |

Since we have found the two numbers, 6 and $-4$, whose product is $-24$ and whose sum is 2, we need go no further.

$$r^2 + 2r - 24 = (r + 6)(r - 4)$$

Now Try Exercise 33

**EXAMPLE 7** Factor $x^2 - 8x + 16$.

*Solution*    We must find the factors of 16 whose sum is $-8$. Both factors must be negative. The two factors whose product is 16 and whose sum is $-8$ are $-4$ and $-4$.

$$x^2 - 8x + 16 = (x - 4)(x - 4)$$
$$= (x - 4)^2$$

Now Try Exercise 41

**EXAMPLE 8** Factor $x^2 - 11x - 60$.

*Solution*    We must find two numbers whose product is $-60$ and whose sum is $-11$. Since the constant is negative, one number must be positive and the other negative. The desired numbers are $-15$ and 4 because $(-15)(4) = -60$ and $-15 + 4 = -11$.

$$x^2 - 11x - 60 = (x - 15)(x + 4)$$

Now Try Exercise 47

Not all trinomials are factorable using integer coefficients as Example 9 will show.

**Prime Polynomials**

A **prime polynomial** is a polynomial that cannot be factored using only integer coefficients.

For example, $x^2 + 3x + 8$ cannot be factored using integer coefficients because there are no two integers whose product is 8 and whose sum is 3. If asked to factor it, the correct answer is to say it is **prime**.

**EXAMPLE 9** Factor $x^2 + 5x + 12$.

*Solution*   Let's first find the two numbers whose product is 12 and whose sum is 5. Since both the constant and the coefficient of the $x$-term are positive, the two numbers must also be positive.

| Factors of 12 | Sum of Factors |
|---|---|
| (1)(12) | $1 + 12 = 13$ |
| (2)(6) | $2 + 6 = 8$ |
| (3)(4) | $3 + 4 = 7$ |

Note that there are no two integers whose product is 12 and whose sum is 5. Thus, $x^2 + 5x + 12$ is prime.

*Now Try Exercise 31*

In Examples 10 and 11, we will factor trinomials of the form $x^2 + bxy + cy^2$, where $b$ and $c$ are real numbers. An example of a trinomial in this form is $x^2 - 2xy - 15y^2$. When factoring trinomials of this form, the factors must be of the form as follows.

$$x^2 + bxy + cy^2 = (x + \blacksquare y)(x + \blacksquare y)$$

Numbers go here that
• add up to $b$
• multiply to $c$

> **Understanding Algebra**
>
> When factoring a trinomial of the form $x^2 + bx + c$, there is at most one pair of factors whose product is $x^2 + bx + c$. For example, when factoring $x^2 - 12x + 32$, the only factors are $(x - 4)$ and $(x - 8)$.

**EXAMPLE 10** Factor $x^2 + 3xy + 2y^2$.

*Solution*   In this problem, the second term contains two variables, $x$ and $y$. The product of the first terms of the factors we are looking for must be $x^2$, and the product of the last terms of the factors must be $2y^2$.

We must find two numbers whose product is 2 (from $2y^2$) and whose sum is 3 (from $3xy$). The two numbers are 1 and 2. Thus,

$$x^2 + 3xy + 2y^2 = (x + 1y)(x + 2y) = (x + y)(x + 2y)$$

*Now Try Exercise 65*

**EXAMPLE 11** Factor $x^2 - 2xy - 15y^2$.

*Solution*   Find two numbers whose product is $-15$ and whose sum is $-2$. The numbers are $-5$ and 3. The last terms must be $-5y$ and $3y$ to obtain $-15y^2$.

$$x^2 - 2xy - 15y^2 = (x - 5y)(x + 3y)$$

*Now Try Exercise 69*

> **Understanding Algebra**
>
> To factor $x^2 + 2xy + y^2$, we need to find two numbers whose product is 1 and whose sum is 2. The numbers are 1 and 1. So,
>
> $x^2 + 2xy + y^2 = (x + 1y)(x + 1y)$
>
> which we write as $(x + y)^2$.

## 2 Remove the Greatest Common Factor from a Trinomial

Sometimes each term of a trinomial has a common factor. When this occurs, factor out the greatest common factor (GCF) first, as explained in Section 6.1. *The first step in any factoring problem is to factor out the GCF.* After factoring out the GCF, you should factor the remaining trinomial further, if possible, until it is completely factored.

> **Understanding Algebra**
>
> *ALWAYS* factor out the GCF first.

**EXAMPLE 12** Factor $2x^2 + 2x - 12$.

*Solution*   Since 2 is the GCF, we factor it out.

$$2x^2 + 2x - 12 = 2(x^2 + x - 6)$$

Now we factor the remaining trinomial $x^2 + x - 6$ into $(x + 3)(x - 2)$. Thus,

$$2x^2 + 2x - 12 = 2(x + 3)(x - 2).$$

Note that the trinomial $2x^2 + 2x - 12$ is now completely factored into *three* factors: two binomial factors, $x + 3$ and $x - 2$, and a monomial factor, 2. After 2 has been factored out, it plays no part in the factoring of the remaining trinomial.

*Now Try Exercise 71*

**EXAMPLE 13**   Factor $3n^3 + 24n^2 - 60n$.

**Solution**   The GCF is $3n$. After factoring out the $3n$, we factor the remaining trinomial.

$$3n^3 + 24n^2 - 60n = 3n(n^2 + 8n - 20) \quad \text{Factored out the GCF.}$$
$$= 3n(n + 10)(n - 2) \quad \text{Factored the remaining trinomial.}$$

*Now Try Exercise 79*

# EXERCISE SET 6.3   MathXL®   MyMathLab®

## Warm-Up Exercises

*Fill in the blanks with the appropriate word, phrase, or symbol(s) from the following list.*

| | |
|---|---|
| $2x^2 + 18$ | prime |
| $2(x - 5)(x - 1)$ | $2(x - 2)(x - 3)$ |
| one positive and one negative | $(x - y)(x - 5y)$ |

| | |
|---|---|
| $2x^2 - 12x + 18$ | factor out the GCF |
| $(2x - 4)(x - 3)$ | $y(x - 1)(x - 5)$ |
| both positive | both negative |

1. When the trinomial $x^2 - 20x + 36$ is factored, the signs that will appear in the binomial factors are _____.

2. A polynomial that cannot be factored using integer coefficients is said to be _____.

3. When $2x^2 - 10x + 12$ is factored, the result is _____.

4. The first step in factoring is to _____.

5. When $2x^2 - 12x + 10$ is factored, the result is _____.

6. When $x^2 - 6xy + 5y^2$ is factored, the result is _____.

7. When the trinomial $x^2 + 20x + 36$ is factored, the signs that will appear in the binomial factors are _____.

8. When the trinomial $x^2 + 16x - 36$ is factored, the signs that will appear in the binomial factors are _____.

9. When $x^2y - 6xy + 5y$ is factored, the result is _____.

10. When the factored expression $2(x - 3)^2$ is multiplied out, the result is _____.

## Practice the Skills

*Factor each polynomial. If the polynomial is prime, so state.*

11. $x^2 + 11x + 24$
12. $t^2 + 10t + 21$
13. $y^2 - 10y + 16$
14. $x^2 - 11x + 24$
15. $x^2 - 7x + 10$
16. $x^2 + 8x + 15$
17. $x^2 + 6x + 8$
18. $x^2 - 3x + 2$
19. $x^2 + 5x - 24$
20. $x^2 - x - 12$
21. $x^2 + 4x - 6$
22. $y^2 - 6y + 8$
23. $y^2 - 13y + 12$
24. $x^2 + 3x - 54$
25. $a^2 - 2a - 8$
26. $p^2 + 3p - 10$
27. $r^2 - 2r - 15$
28. $x^2 - 6x + 8$
29. $b^2 - 11b + 18$
30. $x^2 + 11x - 30$
31. $x^2 - 8x - 15$
32. $x^2 - 8x + 7$
33. $q^2 + 4q - 45$
34. $x^2 + 10x + 25$
35. $x^2 - 7x - 30$
36. $b^2 - 9b - 36$
37. $x^2 + 4x + 4$
38. $x^2 - 4x + 4$
39. $s^2 - 8s + 16$
40. $u^2 + 2u + 1$
41. $p^2 - 12p + 36$
42. $x^2 - 10x - 25$
43. $-18w + w^2 + 45$
44. $-11x + x^2 + 10$
45. $10x - 39 + x^2$
46. $-3x + 8 + x^2$
47. $x^2 - x - 20$
48. $t^2 - 28t - 60$
49. $y^2 + 13y + 40$
50. $r^2 + 14r + 48$
51. $x^2 + 12x - 64$
52. $x^2 - 18x + 80$
53. $s^2 + 14s - 24$
54. $x^2 - 13x + 36$
55. $x^2 - 20x + 64$
56. $x^2 + 19x + 48$
57. $a^2 - 20a + 99$
58. $x^2 + 5x - 24$
59. $x^2 + 2 + 3x$
60. $m^2 - 11 - 10m$
61. $7w - 18 + w^2$
62. $30 + y^2 - 13y$
63. $x^2 - 8xy + 15y^2$
64. $x^2 - 2xy + y^2$
65. $m^2 - 6mn + 9n^2$
66. $b^2 - 2bc - 3c^2$
67. $x^2 + 8xy + 12y^2$
68. $x^2 + 16xy - 17y^2$
69. $m^2 - 5mn - 24n^2$
70. $c^2 + 2cd - 24d^2$

*Factor completely.*

**71.** $6x^2 - 30x + 24$

**72.** $2a^2 - 12a - 32$

**73.** $5x^2 + 20x + 15$

**74.** $4x^2 + 12x - 16$

**75.** $2x^2 - 18x + 40$

**76.** $3y^2 - 33y + 54$

**77.** $b^3 - 7b^2 + 10b$

**78.** $c^3 + 8c^2 - 48c$

**79.** $3z^3 - 21z^2 - 54z$

**80.** $3x^3 - 36x^2 + 33x$

**81.** $x^3 + 8x^2 + 16x$

**82.** $2x^3y - 12x^2y + 10xy$

**83.** $7a^2 - 35ab + 42b^2$

**84.** $3x^3 + 3x^2y - 18xy^2$

**85.** $3r^3 + 6r^2t - 24rt^2$

**86.** $r^2s + 7rs^2 + 12s^3$

**87.** $x^4 - 4x^3 - 21x^2$

**88.** $2z^5 + 14z^4 + 12z^3$

## Problem Solving

**89.** The first two columns in the following table describe the signs of the coefficient of the $x$-term and constant term of a trinomial of the form $x^2 + bx + c$. Determine whether the third column should contain "both positive," "both negative," or "one positive and one negative."

| Sign of Coefficient of $x$-term | Sign of Constant Term | Sign of Constant Terms in the Binomial Factors |
|---|---|---|
| − | + | |
| − | − | |
| + | − | |
| + | + | |

**90.** Assume that a trinomial of the form $x^2 + bx + c$ is factorable. Determine whether the constant terms in the factors are "both positive," "both negative," or "one positive and one negative" for the given signs of $b$ and $c$.

**a)** $b < 0, c > 0$

**b)** $b > 0, c > 0$

**c)** $b > 0, c < 0$

**d)** $b < 0, c < 0$

**91.** Write a trinomial whose binomial factors contain constant terms whose sum is $-12$ and have a product of 32. Show the factoring of the trinomial.

**92.** Write a trinomial whose binomial factors contain constant terms whose sum is 5 and have a product of 4. Show the factoring of the trinomial.

**93.** Write a trinomial whose binomial factors contain constant terms whose sum is $-2$ and have a product of $-35$. Show the factoring of the trinomial.

**94.** Write a trinomial whose binomial factors contain constant terms whose sum is 5 and have a product of $-14$. Show the factoring of the trinomial.

## Concept/Writing Exercises

**95.** On an exam, a student factored $2x^2 - 6x + 4$ as $(2x - 4)(x - 1)$. Even though $(2x - 4)(x - 1)$ does multiply out to $2x^2 - 6x + 4$, why did his or her professor deduct points?

**96.** Explain how to determine the factors when factoring a trinomial of the form $x^2 + bx + c$.

## Challenge Problems

*Factor.*

**97.** $x^2 + 0.6x + 0.08$

**98.** $x^2 - 0.5x - 0.06$

**99.** $x^2 + \dfrac{2}{5}x + \dfrac{1}{25}$

**100.** $x^2 - \dfrac{2}{7}x + \dfrac{1}{49}$

**101.** $x^2 - 24x - 256$

**102.** $x^2 + 5x - 300$

## Cumulative Review Exercises

[2.5] **103.** Solve the equation $4(2x - 4) = 5x + 11$.

[3.4] **104.** **Mixing Solutions** Karen Moreau, a chemist, mixes 4 liters of an 18% acid solution with 1 liter of a 26% acid solution. Find the strength of the mixture.

[4.2] **105.** Graph $y = 3x - 2$.

[5.5] **106.** Multiply $(2x^2 + 5x - 6)(x - 2)$.

[5.6] **107.** Divide $3x^2 - 10x - 10$ by $x - 4$.

[6.2] **108.** Factor $20x^2 + 8x - 15x - 6$ by grouping.

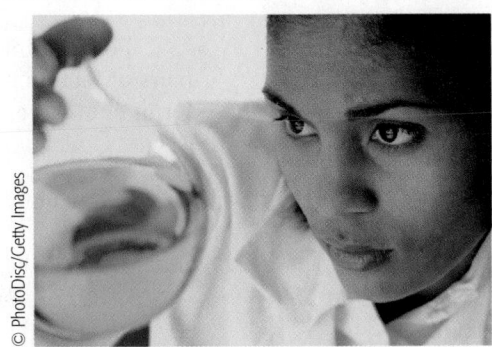

© PhotoDisc/Getty Images

See Exercise 104.

# 6.4 Factoring Trinomials of the Form $ax^2 + bx + c, a \neq 1$

1 Factor trinomials of the form $ax^2 + bx + c, a \neq 1$, by trial and error.

2 Factor trinomials of the form $ax^2 + bx + c, a \neq 1$, by grouping.

### An Important Note

In this section, we discuss two methods of factoring trinomials of the form $ax^2 + bx + c$, $a \neq 1$. That is, we will be factoring trinomials whose squared term has a numerical coefficient not equal to 1, after removing any common factors. Examples of trinomials with $a \neq 1$ are

$$\overbrace{\qquad\qquad}^{a \neq 1}$$
$$2x^2 + 11x + 12 \,(\text{Notice } a = 2) \qquad 4x^2 - 3x + 1(\text{Notice } a = 4)$$

The methods we discuss are (1) **factoring by trial and error** and (2) **factoring by grouping**. We present two different methods for factoring these trinomials because some students, and some instructors, prefer the first method, while others prefer the second method. You may use either method unless your instructor asks you to use a specific method. We will use the same examples to illustrate both methods so that you can make a comparison. Each method is treated independently of the other. Factoring by trial and error was introduced in Section 6.3 and factoring by grouping was introduced in Section 6.2.

## 1 Factor Trinomials of the Form $ax^2 + bx + c, a \neq 1$, by Trial and Error

Recall that factoring is the reverse of multiplying. Consider the product of the following two binomials:

$$\begin{array}{cccc} \text{F} & \text{O} & \text{I} & \text{L} \end{array}$$
$$(2x + 3)(x + 5) = 2x(x) + (2x)(5) + 3(x) + 3(5)$$
$$= 2x^2 + 10x + 3x + 15$$
$$= 2x^2 + 13x + 15$$

Note that $2x^2 + 13x + 15$ in factored form is $(2x + 3)(x + 5)$.

$$2x^2 + 13x + 15 = (2x + 3)(x + 5)$$

When factoring a trinomial of the form $ax^2 + bx + c$ by trial and error, the product of the $x$-terms in the binomial factors must equal the first term of the trinomial, $ax^2$. Also, the product of the constants in the binomial factors, including their signs, must equal the constant, $c$, of the trinomial.

Product of constants in factors must equal $c$.

$$ax^2 + bx + c = (x\text{-term} + 1\text{st constant})(x\text{-term} + 2\text{nd constant})$$

Product of $x$-terms in factors must equal $ax^2$.

For example, when factoring the trinomial $2x^2 + 7x + 6$, each of the following pairs of factors has a product of the first terms equal to $2x^2$ and a product of the last terms equal to 6.

### Understanding Algebra

$$\begin{array}{cccc} \text{F} & \text{O} & \text{I} & \text{L} \end{array}$$

Use FOIL to multiply $(2x + 3)(x + 5)$.

Product of FIRST terms is $2x^2$.

$$2x \cdot x = 2x^2$$

Sum of OUTER and INNER products is $13x$.

$$2x \cdot 5 + 3 \cdot x = 13x$$

Product of LAST terms is 15.

$$3 \cdot 5 = 15$$

$$(2x + 3)(x + 5) = 2x^2 + 13x + 15$$

| Trinomial | Possible Factors | Product of First Terms | Product of Last Terms |
|---|---|---|---|
| $2x^2 + 7x + 6$ | $(2x + 1)(x + 6)$ | $2x(x) = 2x^2$ | $1(6) = 6$ |
| | $(2x + 2)(x + 3)$ | $2x(x) = 2x^2$ | $2(3) = 6$ |
| | $(2x + 3)(x + 2)$ | $2x(x) = 2x^2$ | $3(2) = 6$ |
| | $(2x + 6)(x + 1)$ | $2x(x) = 2x^2$ | $6(1) = 6$ |

How do we determine which is the correct factoring of the trinomial $2x^2 + 7x + 6$? The key lies in the $x$-term. We need to find the pair of factors whose sum of the products of the outer and inner terms is equal to the $x$-term of the trinomial.

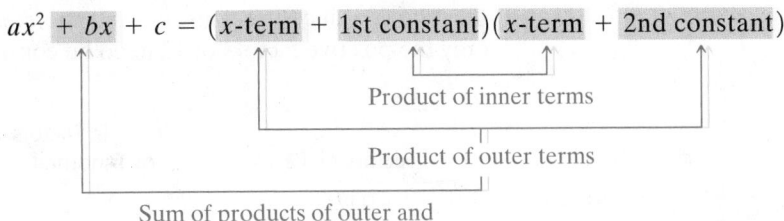

$$ax^2 + bx + c = (x\text{-term} + 1\text{st constant})(x\text{-term} + 2\text{nd constant})$$

Product of inner terms

Product of outer terms

Sum of products of outer and inner terms must equal $bx$.

Now look at the possible pairs of factors we obtained for $2x^2 + 7x + 6$ to see if any yield the correct $x$-term, $7x$.

| Trinomial | Possible Factors | Product of the First Terms | Product of the Last Terms | Sum of the Products of Outer and Inner Terms |
|---|---|---|---|---|
| $2x^2 + 7x + 6$ | $(2x + 1)(x + 6)$ | $2x^2$ | 6 | $2x(6) + 1(x) = 13x$ |
| | $(2x + 2)(x + 3)$ | $2x^2$ | 6 | $2x(3) + 2(x) = 8x$ |
| | $(2x + 3)(x + 2)$ | $2x^2$ | 6 | $2x(2) + 3(x) = 7x$ |
| | $(2x + 6)(x + 1)$ | $2x^2$ | 6 | $2x(1) + 6(x) = 8x$ |

Since $(2x + 3)(x + 2)$ yields the correct $x$-term, $7x$, the factors of the trinomial $2x^2 + 7x + 6$ are $(2x + 3)$ and $(x + 2)$.

$$2x^2 + 7x + 6 = (2x + 3)(x + 2)$$

We can check this factoring using the FOIL method.

Check

$$\begin{aligned} (2x + 3)(x + 2) &= \overset{F}{2x(x)} + \overset{O}{2x(2)} + \overset{I}{3(x)} + \overset{L}{3(2)} \\ &= 2x^2 + 4x + 3x + 6 \\ &= 2x^2 + 7x + 6 \end{aligned}$$

Since we obtained the original trinomial, our factoring is correct.

## Understanding Algebra

Notice that $2x^2 + 13x + 15$ factors as $(2x + 3)(x + 5)$. The product $(2x + 5)(x + 3)$ will yield a $2x^2$ term and a 15 term, but the middle term is $11x$ and not $13x$!

> **Helpful Hint**
>
> When factoring a trinomial of the form $ax^2 + bx + c$, remember that the sign of the constant, $c$, and the sign of the $x$-term, $bx$, offer valuable information. When factoring a trinomial by trial and error, first check the sign of the constant. If it is positive, the signs in both factors will be the same as the sign of the $x$-term. If the constant is negative, one factor will contain a plus sign and the other a minus sign.

Now we outline the procedure to factor trinomials of the form $ax^2 + bx + c$, $a \neq 1$, by trial and error. Keep in mind that the more you practice, the better you will become at factoring.

> **To Factor Trinomials of the Form $ax^2 + bx + c, a \neq 1$, by Trial and Error**
>
> 1. Factor out the greatest common factor (GCF), if any.
> 2. Write all pairs of factors of the coefficient of the squared term, $a$.
> 3. Write all pairs of factors of the constant term, $c$.
> 4. Try combinations of these factors until the correct middle term, $bx$, is found.

**EXAMPLE 1** Factor $3x^2 + 20x + 12$.

*Solution*  We first determine that there is no GCF to factor out. Since the first term is $3x^2$, the factors will be of the form $(3x + \quad)(x + \quad)$. Now we must find the numbers to place in the shaded areas. The product of the last terms in the factors must be 12. Since the constant and the coefficient of the $x$-term are both positive, only the positive factors of 12 need be considered.

| Factors of 12 | Possible Factors of Trinomial | Sum of the Products of the Outer and Inner Terms |
|---|---|---|
| 1(12) | $(3x + 1)(x + 12)$ | $37x$ |
| 2(6) | $(3x + 2)(x + 6)$ | $20x$ |
| 3(4) | $(3x + 3)(x + 4)$ | $15x$ |
| 4(3) | $(3x + 4)(x + 3)$ | $13x$ |
| 6(2) | $(3x + 6)(x + 2)$ | $12x$ |
| 12(1) | $(3x + 12)(x + 1)$ | $15x$ |

Since the product of $(3x + 2)$ and $(x + 6)$ yields the correct $x$-term, $20x$, they are the correct factors.

$$3x^2 + 20x + 12 = (3x + 2)(x + 6)$$

*Now Try Exercise 5*

**EXAMPLE 2** Factor $5x^2 - 7x - 6$.

*Solution*  There is no GCF to factor out. Since the first term is $5x^2$, one factor must contain a $5x$ and the other an $x$. List the factors of $-6$ and look for the pair of factors that yields $-7x$.

| Factors of $-6$ | Possible Factors | Sum of the Products of the Outer and Inner Terms |
|---|---|---|
| $-1(6)$ | $(5x - 1)(x + 6)$ | $29x$ |
| $-2(3)$ | $(5x - 2)(x + 3)$ | $13x$ |
| $-3(2)$ | $(5x - 3)(x + 2)$ | $7x$ |
| $-6(1)$ | $(5x - 6)(x + 1)$ | $-x$ |

Since we did not obtain the desired quantity, $-7x$, by writing the negative factor with the $5x$, we will now try listing the negative factor with the $x$.

| Factors of $-6$ | Possible Factors | Sum of the Products of the Outer and Inner Terms |
|---|---|---|
| $1(-6)$ | $(5x + 1)(x - 6)$ | $-29x$ |
| $2(-3)$ | $(5x + 2)(x - 3)$ | $-13x$ |
| $3(-2)$ | $(5x + 3)(x - 2)$ | $-7x$ |
| $6(-1)$ | $(5x + 6)(x - 1)$ | $x$ |

We see that $(5x + 3)(x - 2)$ gives the $-7x$ we are looking for. Thus,

$$5x^2 - 7x - 6 = (5x + 3)(x - 2)$$

Again we listed all the possible combinations for you to study.

*Now Try Exercise 9*

**Helpful Hint**

In Example 2, we were asked to factor $5x^2 - 7x - 6$. When we considered $-3(2)$ in the first set of possible factors, we obtained

| Factors of $-6$ | Possible Factors | Sum of the Products of the Outer and Inner Terms |
|---|---|---|
| $-3(2)$ | $(5x - 3)(x + 2)$ | $7x$ |

Later in the solution we tried the factors $3(-2)$ and obtained the correct answer.

| | Possible Factors | Sum of the Products of the Outer and Inner Terms |
|---|---|---|
| $3(-2)$ | $(5x + 3)(x - 2)$ | $-7x$ |

When factoring a trinomial with a *negative constant*, if you obtain the $x$-term whose sign is the opposite of the one you are seeking, *reverse the signs on the constants* in the factors. This should give you the set of factors you are seeking.

### EXAMPLE 3 Factor $8x^2 + 33x + 4$.

*Solution*    There is no GCF. Since the first term is $8x^2$, the possible factors may be of the form $(8x\ \ )(x\ \ )$ or $(4x\ \ )(2x\ \ )$. When this situation occurs, we will generally start with the middle-size pair of factors. Thus, we begin with $(4x\ \ )(2x\ \ )$. If this pair does not lead to the solution, we will then try $(8x\ \ )(x\ \ )$. We now list the factors of the constant, 4. Since all signs are positive, we list only the positive factors of 4.

**Understanding Algebra**

In Example 3, another reason we can eliminate the factors

$(4x + 1)(2x + 4)$
$(4x + 2)(2x + 2)$
$(4x + 4)(2x + 1)$
$(8x + 2)(x + 2)$
$(8x + 4)(x + 1)$

is that each product has a binomial that contains a common factor.

Since the original trinomial did not have any common factors, there can be no common factors in the binomial factors.

| Factors of 4 | Possible Factors | Sum of the Products of the Outer and Inner Terms |
|---|---|---|
| $1(4)$ | $(4x + 1)(2x + 4)$ | $18x$ |
| $2(2)$ | $(4x + 2)(2x + 2)$ | $12x$ |
| $4(1)$ | $(4x + 4)(2x + 1)$ | $12x$ |

Since we did not obtain the correct factors with $(4x\ \ )(2x\ \ )$, we now try $(8x\ \ )(x\ \ )$.

| Factors of 4 | Possible Factors | Sum of the Products of the Outer and Inner Terms |
|---|---|---|
| $1(4)$ | $(8x + 1)(x + 4)$ | $33x$ |
| $2(2)$ | $(8x + 2)(x + 2)$ | $18x$ |
| $4(1)$ | $(8x + 4)(x + 1)$ | $12x$ |

Since the product of $(8x + 1)$ and $(x + 4)$ yields the correct $x$-term, $33x$, they are the correct factors.

$$8x^2 + 33x + 4 = (8x + 1)(x + 4)$$

*Now Try Exercise 19*

### EXAMPLE 4 Factor $25t^2 - 10t + 1$.

*Solution*    The factors must be of the form $(25t\ \ )(t\ \ )$ or $(5t\ \ )(5t\ \ )$. We will start with the middle-size factors $(5t\ \ )(5t\ \ )$. Since the constant is positive and the coefficient of the $x$-term is negative, both factors must be negative.

| Factors of 1 | Possible Factors | Sum of the Products of the Outer and Inner Terms |
|---|---|---|
| $(-1)(-1)$ | $(5t - 1)(5t - 1)$ | $-10t$ |

Since we found the correct factors, we can stop.

$$25t^2 - 10t + 1 = (5t - 1)(5t - 1) = (5t - 1)^2$$

*Now Try Exercise 13*

**EXAMPLE 5** Factor $2x^2 + 3x + 7$.

**Solution** The factors will be of the form $(2x \quad)(x \quad)$. We need only consider the positive factors of 7.

| Factors of 7 | Possible Factors | Sum of the Products of the Outer and Inner Terms |
|---|---|---|
| 1(7) | $(2x + 1)(x + 7)$ | $15x$ |
| 7(1) | $(2x + 7)(x + 1)$ | $9x$ |

Since we have tried all possible combinations and we have not obtained the $x$-term, $3x$, this trinomial *cannot be factored using integer coefficients*. Thus, the trinomial $2x^2 + 3x + 7$ is a *prime polynomial*.

Now Try Exercise 17

**EXAMPLE 6** Factor $6a^2 + 19ab + 3b^2$.

**Solution** This trinomial is different from the other trinomials in that the last term is not a constant but contains $b^2$. The factoring process is the same, except that the second term of both factors will contain $b$. Consider factors of the form $(3a \quad)(2a \quad)$. If we cannot find the factors, then we try factors of the form $(6a \quad)(a \quad)$.

| Factors of 3 | Possible Factors | Sum of the Products of the Outer and Inner Terms |
|---|---|---|
| 1(3) | $(3a + b)(2a + 3b)$ | $11ab$ |
| 3(1) | $(3a + 3b)(2a + b)$ | $9ab$ |
| 1(3) | $(6a + b)(a + 3b)$ | $19ab$ |
| 3(1) | $(6a + 3b)(a + b)$ | $9ab$ |

$$6a^2 + 19ab + 3b^2 = (6a + b)(a + 3b)$$

**Check** $(6a + b)(a + 3b) = 6a^2 + 18ab + ab + 3b^2 = 6a^2 + 19ab + 3b^2$

Now Try Exercise 55

**EXAMPLE 7** Factor $6x^2 - 13xy - 8y^2$.

**Solution** We begin with factors of the form $(3x \quad)(2x \quad)$. If we cannot find the solution from these, we will try $(6x \quad)(x \quad)$. Since the last term, $-8y^2$, is negative, one factor will contain a plus sign and the other will contain a minus sign.

| Factors of $-8$ | Possible Factors | Sum of the Products of the Outer and Inner Terms |
|---|---|---|
| 1(−8) | $(3x + y)(2x - 8y)$ | $-22xy$ |
| 2(−4) | $(3x + 2y)(2x - 4y)$ | $-8xy$ |
| 4(−2) | $(3x + 4y)(2x - 2y)$ | $2xy$ |
| 8(−1) | $(3x + 8y)(2x - y)$ | $13xy$ |

We are looking for $-13xy$. When we considered $8(-1)$, we obtained $13xy$. As explained in the Helpful Hint on page 345, if we reverse the signs of the numbers in the factors, we will obtain the factors we are seeking.

$(3x + 8y)(2x - y)$   Product is $6x^2 + 13xy - 8y^2$

$(3x - 8y)(2x + y)$   Product is $6x^2 - 13xy - 8y^2$

Therefore, $6x^2 - 13xy - 8y^2 = (3x - 8y)(2x + y)$.

Now Try Exercise 57

**EXAMPLE 8** Factor $6x^3 + 15x^2 - 36x$.

*Solution*    *The first step in any factoring problem is to factor out the GCF. In this example, $3x$ is common to all three terms. We begin by factoring out the $3x$. Then we continue factoring by trial and error.*

$$6x^3 + 15x^2 - 36x = 3x(2x^2 + 5x - 12)$$
$$= 3x(2x - 3)(x + 4)$$

Now Try Exercise 47

**2** **Factor Trinomials of the Form $ax^2 + bx + c, a \neq 1$, by Grouping**

The steps in the following box give the procedure for factoring trinomials by grouping.

> **To Factor Trinomials of the Form $ax^2 + bx + c, a \neq 1$, by Grouping**
>
> 1. Factor out the greatest common factor, if any.
> 2. Find two numbers whose product is equal to the product of $a$ times $c$, and whose sum is equal to $b$.
> 3. Rewrite the middle term, $bx$, as the sum or difference of two terms using the numbers found in step 2.
> 4. Factor by grouping as explained in Section 6.2.

This process will be made clearer in Example 9. We will rework Examples 1 through 8 here using factoring by grouping. Example 9, which follows, is the same trinomial given in Example 1. After you study this method and try some exercises, you will gain a feel for which method you prefer using.

**EXAMPLE 9** Factor $3x^2 + 20x + 12$.

*Solution*

1. There is no GCF to factor out.

$$a = 3 \quad b = 20 \quad c = 12$$

2. We must find two numbers whose product is $a \cdot c$ and whose sum is $b$. We must therefore find two numbers whose product equals $3 \cdot 12 = 36$ and whose sum equals 20. Only the positive factors of 36 need be considered since all signs of the trinomial are positive.

| Factors of 36 | Sum of Factors |
|:---:|:---:|
| (1)(36) | $1 + 36 = 37$ |
| (2)(18) | $2 + 18 = 20$ |
| (3)(12) | $3 + 12 = 15$ |
| (4)(9) | $4 + 9 = 13$ |
| (6)(6) | $6 + 6 = 12$ |

The desired factors are 2 and 18.

3. Rewrite $20x$ as the sum or difference of two terms using the values found in step 2. Therefore, we rewrite $20x$ as $2x + 18x$.

$$3x^2 + 20x + 12$$

$$= 3x^2 + 2x + 18x + 12$$

## Understanding Algebra

In Example 9, when listing the four terms, we could have written the middle terms as $18x + 2x$.

$$3x^2 + 18x + 2x + 12$$
$$3x(x + 6) + 2(x + 6)$$
$$(x + 6)(3x + 2)$$

This is an equivalent answer.

**4.** Now factor by grouping. Start by factoring out a common factor from the first two terms and a common factor from the last two terms. This procedure was discussed in Section 6.2.

$$\overbrace{3x^2 + 2x}^{x \text{ is common factor}} + \overbrace{18x + 12}^{6 \text{ is common factor}}$$
$$= x(3x + 2) + 6(3x + 2)$$
$$= (3x + 2)(x + 6)$$

*Now Try Exercise 7*

## Understanding Algebra

When trying pairs of factors to obtain the middle term, if you obtain the opposite of the coefficient you are seeking, reverse the signs in the factors. This should give you the coefficient you are seeking. If you try to factor $5x^2 - 7x - 6$ as $(x + 2)(5x - 3)$, for example, but see that that product is $5x^2 + 7x - 6$, reverse the signs to obtain $(x - 2)(5x + 3) = 5x^2 - 7x - 6$.

**EXAMPLE 10** Factor $5x^2 - 7x - 6$.

*Solution*   There are no common factors other than 1.

$$a = 5, \quad b = -7, \quad c = -6$$

The product of $a$ times $c$ is $5(-6) = -30$. We must find two numbers whose product is $-30$ and whose sum is $-7$.

| Factors of $-30$ | Sum of Factors |
|---|---|
| $(-1)(30)$ | $-1 + 30 = 29$ |
| $(-2)(15)$ | $-2 + 15 = 13$ |
| $(-3)(10)$ | $-3 + 10 = 7$ |
| $(-5)(6)$ | $-5 + 6 = 1$ |
| $(-6)(5)$ | $-6 + 5 = -1$ |
| $(-10)(3)$ | $-10 + 3 = -7$ |
| $(-15)(2)$ | $-15 + 2 = -13$ |
| $(-30)(1)$ | $-30 + 1 = -29$ |

Rewrite the middle term of the trinomial, $-7x$, as $-10x + 3x$.

$$5x^2 - 7x - 6$$
$$= 5x^2 - 10x + 3x - 6 \qquad \text{Now factor by grouping.}$$
$$= 5x(x - 2) + 3(x - 2)$$
$$= (x - 2)(5x + 3)$$

*Now Try Exercise 31*

In Example 10, we could have expressed the $-7x$ as $3x - 10x$ and obtained the same answer. Try working Example 10 by rewriting $-7x$ as $3x - 10x$.

**Helpful Hint**

In Example 10 we were looking for two factors of $-30$ whose sum was $-7$. When we considered the factors $-3$ and $10$, we obtained a sum of 7, which is the opposite of $-7$. When trying pairs of factors to obtain the middle term, if you obtain the opposite of the coefficient you are seeking, reverse the signs in the factors. This should give you the coefficient you are seeking.

## Using Your Graphing Calculator

In Example 10, we began factoring $5x^2 - 7x - 6$ by finding two numbers whose product equals $-30$ and whose sum equals $-7$. The graphing calculator can be used to help find those numbers.

We can represent two numbers whose product is $-30$ by $x$ and $-\dfrac{30}{x}$. Notice $x\left(-\dfrac{30}{x}\right) = -30$. Let $Y_1 = -\dfrac{30}{x}$, the second

factor in $x\left(-\dfrac{30}{x}\right)$. Let $Y_2 = -\dfrac{30}{x} + x$, the sum of the factors. Use the TABLE feature to create

pairs of numbers whose product is $-30$ (the product of columns $X$ and $Y_1$ is $-30$) and whose sum is in column $Y_2$ (see **Figure 6.3**). Look down the $Y_2$ column until you find the sum you are looking for, $-7$.

Use the value for $X$, 3, and the value for $Y_1$, $-10$, to rewrite the middle term of the trinomial, $-7x$, as $3x - 10x$. Continue factoring by grouping.

| X | Y1 | Y2 |
|---|-----|------|
| 1 | -30 | -29 |
| 2 | -15 | -13 |
| 3 | -10 | -7 |
| 4 | -7.5 | -3.5 |
| 5 | -6 | -1 |
| 6 | -5 | 1 |
| 7 | -4.286 | 2.7143 |

X=3

FIGURE 6.3

---

**EXAMPLE 11** Factor $8x^2 + 33x + 4$.

*Solution*  There is no GCF to factor out. We must find two numbers whose product is $8 \cdot 4$ or 32 and whose sum is 33. The numbers are 1 and 32.

| Factors of 32 | Sum of Factors |
|---------------|----------------|
| $(1)(32)$     | $1 + 32 = 33$  |

Rewrite $33x$ as $32x + x$. Then factor by grouping.

$$8x^2 + 33x + 4$$
$$= 8x^2 + 32x + x + 4$$
$$= 8x(x + 4) + 1(x + 4)$$
$$= (x + 4)(8x + 1)$$

Now Try Exercise 25

### Understanding Algebra

In Example 11 we could also have written:

$$8x^2 + 33x + 4$$
$$= 8x^2 + x + 32x + 4$$
$$= x(8x + 1) + 4(8x + 1)$$
$$= (8x + 1)(x + 4)$$

---

**EXAMPLE 12** Factor $25t^2 - 10t + 1$.

*Solution*  There is no GCF to factor out. We must find two numbers whose product is $25 \cdot 1$ or 25 and whose sum is $-10$. Since the product of $a$ times $c$ is positive and the coefficient of the $t$-term is negative, both numerical factors must be negative.

| Factors of 25 | Sum of Factors |
|---------------|----------------|
| $(-1)(-25)$   | $-1 + (-25) = -26$ |
| $(-5)(-5)$    | $-5 + (-5) = -10$  |

The desired factors are $-5$ and $-5$.

$$25t^2 - 10t + 1$$
$$= 25t^2 - 5t - 5t + 1 \qquad \text{Rewrote } -10t \text{ as } -5t - 5t.$$
$$= 5t(5t - 1) - 5t + 1$$
$$= 5t(5t - 1) - 1(5t - 1) \qquad \text{Rewrote } -5t + 1 \text{ as } -1(5t - 1).$$
$$= (5t - 1)(5t - 1) \text{ or } (5t - 1)^2$$

Now Try Exercise 29

---

**Helpful Hint**

When attempting to factor a trinomial, if there are no two integers whose product equals $a \cdot c$ and whose sum equals $b$, the trinomial is prime.

**EXAMPLE 13** Factor $2x^2 + 3x + 7$.

*Solution*    There are no common factors other than 1. We must find two numbers whose product is 14 and whose sum is 3. We need consider only positive factors of 14. Why?

| Factors of 14 | Sum of Factors |
|---------------|----------------|
| (1)(14)       | $1 + 14 = 15$  |
| (2)(7)        | $2 + 7 = 9$    |

Since there are no factors of 14 whose sum is 3, we conclude that this trinomial cannot be factored with integers. This is an example of a *prime polynomial*.

Now Try Exercise 21

**EXAMPLE 14** Factor $6a^2 + 19ab + 3b^2$.

*Solution*    There is no GCF to factor out. This trinomial contains two variables. It is factored in basically the same manner as the previous examples. Find two numbers whose product is $6 \cdot 3$ or 18 and whose sum is 19. The two numbers are 18 and 1.

$$6a^2 + 19ab + 3b^2$$
$$= 6a^2 + 18ab + ab + 3b^2$$
$$= 6a(a + 3b) + b(a + 3b)$$
$$= (a + 3b)(6a + b)$$

Now Try Exercise 61

**EXAMPLE 15** Factor $6x^2 - 13xy - 8y^2$.

*Solution*    There is no GCF to factor out. Find two numbers whose product is $6(-8)$ or $-48$ and whose sum is $-13$. Since the product is negative, one factor must be positive and the other negative. Some factors are given below.

| Product of Factors | Sum of Factors        |
|--------------------|-----------------------|
| (1)(−48)           | $1 + (-48) = -47$     |
| (2)(−24)           | $2 + (-24) = -22$     |
| (3)(−16)           | $3 + (-16) = -13$     |

There are many other factors, but we have found the pair we were looking for. The two numbers whose product is $-48$ and whose sum is $-13$ are 3 and $-16$.

$$6x^2 - 13xy - 8y^2$$
$$= 6x^2 + 3xy - 16xy - 8y^2$$
$$= 3x(2x + y) - 8y(2x + y)$$
$$= (2x + y)(3x - 8y)$$

Check    $(2x + y)(3x - 8y)$

$$
\begin{array}{cccc}
\text{F} & \text{O} & \text{I} & \text{L} \\
\end{array}
$$
$$= (2x)(3x) + (2x)(-8y) + (y)(3x) + (y)(-8y)$$
$$= 6x^2 - 16xy + 3xy - 8y^2$$
$$= 6x^2 - 13xy - 8y^2$$

Now Try Exercise 63

Remember that in any factoring problem our first step is to factor the GCF from each term. We then continue to factor the trinomial, if possible.

**EXAMPLE 16**  Factor $6x^3 + 15x^2 - 36x$.

*Solution*   The factor $3x$ is common to all three terms. Factor the $3x$ from each term of the polynomial.

$$6x^3 + 15x^2 - 36x = 3x(2x^2 + 5x - 12)$$

Now continue by factoring $2x^2 + 5x - 12$. The two numbers whose product is $2(-12)$ or $-24$ and whose sum is 5 are 8 and $-3$.

$$3x(2x^2 + 5x - 12)$$
$$= 3x(2x^2 + 8x - 3x - 12)$$
$$= 3x[2x(x + 4) - 3(x + 4)]$$
$$= 3x(x + 4)(2x - 3)$$

NowTry Exercise 49

**Helpful Hint**

**Which Method Should You Use to Factor a Trinomial?**

If your instructor asks you to use a specific method, you should use that method. If your instructor does not require a specific method, you should use the method you feel most comfortable with. You may wish to start with the trial-and-error method if there are only a few possible factors to try. If you cannot find the factors by trial and error or if there are many possible factors to consider, you may wish to use the grouping procedure. With time and practice you will learn which method you feel most comfortable with and which method gives you greater success.

# EXERCISE SET 6.4    MathXL®  MyMathLab®

## Warm-Up Exercises

*Fill in the blanks with the appropriate word, phrase, or symbol(s) from the following list.*

last

factor out the GCF

first

outer

factoring by grouping

inner

1. The two methods of factoring the trinomial $ax^2 + bx + c$, $a \neq 1$ are trial and error and _____.

2. When factoring the trinomial $ax^2 + bx + c$ into two binomials, the product of the _____ terms of the binomials must equal $ax^2$.

3. The first step in factoring is to _____.

4. When factoring the trinomial $ax^2 + bx + c$ into two binomials, the product of the _____ terms of the binomials must equal $c$.

## Practice the Skills

*Factor completely. If the polynomial is prime, so state.*

5. $2x^2 + 11x + 5$

6. $2x^2 + 9x + 4$

7. $3x^2 + 14x + 8$

8. $7x^2 + 37x + 10$

9. $5x^2 - 9x - 2$

10. $3y^2 + 17y + 10$

11. $3r^2 + 13r - 10$

12. $3x^2 - 2x - 8$

13. $4z^2 - 12z + 9$

14. $4n^2 - 9n + 5$

15. $6z^2 + z - 12$

16. $5m^2 - 17m + 6$

17. $5a^2 - 12a + 6$

18. $2x^2 - x - 6$

19. $8x^2 + 19x + 6$

20. $6y^2 - 11y + 4$

21. $3x^2 + 11x + 4$

22. $3a^2 + 7a - 20$

23. $5y^2 - 16y + 3$

24. $5x^2 + 2x + 9$

25. $7x^2 + 43x + 6$

26. $7x^2 - 8x + 1$

27. $4x^2 + 4x - 15$

28. $15x^2 - 19x + 6$

29. $49t^2 - 14t + 1$

30. $16z^2 - 8z + 1$

31. $5z^2 - 6z - 8$

32. $3z^2 - 11z - 6$

33. $4y^2 + 5y - 6$

34. $5y^2 - 3y - 1$

35. $10x^2 - 27x + 5$

36. $6a^2 + 7a - 10$

37. $10d^2 - 7d - 12$

38. $6x^2 + 13x + 3$

39. $8x^2 - 46x - 12$

40. $12x^2 - 13x - 35$

**41.** $10t + 3 + 7t^2$

**42.** $n - 30 + n^2$

**43.** $6x^2 + 16x + 10$

**44.** $12z^2 + 32z + 20$

**45.** $6x^3 - 5x^2 - 4x$

**46.** $8x^3 + 8x^2 - 6x$

**47.** $12x^3 + 28x^2 + 8x$

**48.** $18x^3 - 21x^2 - 9x$

**49.** $4x^3 - 2x^2 - 12x$

**50.** $300x^2 - 400x - 400$

**51.** $48c^2 + 8c - 16$

**52.** $28x^2 - 28x + 7$

**53.** $4p - 12 + 8p^2$

**54.** $72 + 3r^2 - 30r$

**55.** $8c^2 + 41cd + 5d^2$

**56.** $8x^2 - 8xy - 6y^2$

**57.** $15x^2 - xy - 6y^2$

**58.** $2x^2 - 7xy + 3y^2$

**59.** $12x^2 + 10xy - 8y^2$

**60.** $12a^2 - 34ab + 24b^2$

**61.** $7p^2 + 13pq + 6q^2$

**62.** $24x^2 - 92x + 80$

**63.** $6m^2 - mn - 2n^2$

**64.** $8m^2 + 4mn - 4n^2$

**65.** $8x^3 + 10x^2y + 3xy^2$

**66.** $8a^2b + 10ab^2 + 3b^3$

**67.** $4x^4 + 8x^3y + 3x^2y^2$

**68.** $26u^2v + 6uv^2 + 24u^3$

## Problem Solving

*For Exercises 69–74, write a polynomial whose factors are listed.*

**69.** $3x + 1, x - 7$

**70.** $6y - 5, 4y - 3$

**71.** $5, x + 3, 2x + 1$

**72.** $3, 2x + 3, x - 4$

**73.** $t^2, t + 4, 3t - 1$

**74.** $5x^2, 3x - 7, 2x + 3$

**75. a)** If you know one binomial factor of a trinomial, explain how you can use division to find the second binomial factor of the trinomial (see Section 5.6).

**b)** One factor of $18x^2 + 93x + 110$ is $3x + 10$. Use division to find the second factor.

**76.** One factor of $30x^2 - 17x - 247$ is $6x - 19$. Find the other factor.

## Concept/Writing Exercises

**77.** Explain the relationship between the process of factoring trinomials and the process of multiplying binomials.

**78.** When factoring a trinomial of the form $ax^2 + bx + c$,

**a)** What must the product of the first terms of the binomial factors equal?

**b)** What must the product of the constants in the binomial factors equal?

## Challenge Problems

*Factor each trinomial.*

**79.** $18x^2 + 9x - 20$

**80.** $9p^2 - 104p + 55$

**81.** $15x^2 - 124x + 160$

**82.** $16x^2 - 62x - 45$

**83.** $105a^2 - 220a - 160$

**84.** $72x^2 + 417x - 420$

**85.** Two factors of $6x^3 + 235x^2 + 2250x$ are $x$ and $3x + 50$. Determine the other factor. Explain how you determined your answer.

**86.** Two factors of the polynomial $2x^3 + 11x^2 + 3x - 36$ are $x + 3$ and $2x - 3$. Determine the third factor. Explain how you determined your answer.

## Cumulative Review Exercises

[1.9] **87.** Evaluate $-x^2 - 4(y + 3) + 2y^2$ when $x = -3$ and $y = -5$.

[2.6] **88. Daytona 500** Ryan Newman won the 2008 Dayton 500 in a time of about 3.27 hours. If the race covered 507.5 miles, find the average speed of Newman's car.

[6.1] **89.** Factor $36x^4y^3 - 12xy^2 + 24x^5y^6$.

[6.3] **90.** Factor $b^2 + 4b - 96$.

See Exercise 88. Ryan Newman after winning the Daytona 500.

Mid-Chapter Test: 6.1–6.4

*To find out how well you understand the chapter material to this point, take this brief test. The answers, and the section where the material was initially discussed, are given in the back of the book. Review any questions that you answered incorrectly.*

1. How may any factoring problem be checked?

2. Determine the greatest common factor of $18xy^2, 27x^3y^4$, and $12x^2y^3$.

*In Exercises 3–5, factor the GCF from each term in the expression.*

3. $4a^2b^3 - 24a^3b$

4. $5c(d - 6) - 3(d - 6)$

5. $7x(2x + 9) + 2x + 9$

*In Exercises 6–10, factor by grouping.*

6. $x^2 + 4x + 7x + 28$

7. $x^2 + 5x - 3x - 15$

8. $6a^2 + 15ab - 2ab - 5b^2$

9. $5x^2 - 2xy - 45x + 18y$

10. $8x^3 + 4x^2 - 48x^2 - 24x$

*In Exercises 11–20, factor each polynomial completely. If the polynomial is prime, so state.*

11. $x^2 - 10x + 21$

12. $t^2 + 9t + 20$

13. $p^2 - 3p - 8$

14. $x^2 + 16x + 64$

15. $m^2 - 4mn - 45n^2$

16. $3x^2 + 17x + 10$

17. $4z^2 - 11z + 6$

18. $3y^2 + 13y + 6$

19. $9x^2 - 6x + 1$

20. $6a^2 + 3ab - 3b^2$

# 6.5 Special Factoring Formulas and a General Review of Factoring

1 Factor the difference of two squares.

2 Factor the sum and difference of two cubes.

3 Learn the general procedure for factoring a polynomial.

The special formulas, for certain types of factoring problems, we focus on in this section are the *difference of two squares, the sum of two cubes, and the difference of two cubes. You will need to memorize the three highlighted formulas in this section* so that you can use them whenever you need them.

## 1 Factor the Difference of Two Squares

Consider the binomial $x^2 - 9$. Note that each term of the binomial can be expressed as the square of some expression.

$$x^2 - 9 = x^2 - 3^2$$

This is an example of a **difference of two squares**.

**Understanding Algebra**

$x^2 - y^2$   Difference of two squares; is factorable

$x^2 + y^2$   Sum of two squares; is *not* factorable

> **Difference of Two Squares**
> $$a^2 - b^2 = (a + b)(a - b)$$

**EXAMPLE 1** Factor $x^2 - 9$.

*Solution*    If we write $x^2 - 9$ as a difference of two squares, we have $(x)^2 - (3)^2$. Using the difference of two squares formula, where $a$ is replaced by $x$ and $b$ is replaced by 3, we obtain the following:

$$a^2 - b^2 = (a + b)(a - b)$$
$$x^2 - 9 = (x)^2 - (3)^2 = (x + 3)(x - 3)$$

Thus, $x^2 - 9 = (x + 3)(x - 3)$.

Now Try Exercise 13

**EXAMPLE 2** Factor using the difference of two squares formula.

**a)** $x^2 - 16$          **b)** $25x^2 - 4$          **c)** $36x^2 - 49y^2$

*Solution*

**a)** $x^2 - 16 = (x)^2 - (4)^2$
$$= (x + 4)(x - 4)$$
**b)** $25x^2 - 4 = (5x)^2 - (2)^2$
$$= (5x + 2)(5x - 2)$$
**c)** $36x^2 - 49y^2 = (6x)^2 - (7y)^2$
$$= (6x + 7y)(6x - 7y)$$

Now Try Exercise 21

**EXAMPLE 3** Factor each difference of two squares.

**a)** $16x^4 - 9y^4$                          **b)** $x^6 - y^4$

*Solution*

**a)**          $16x^4 - 9y^4 = (4x^2)^2 - (3y^2)^2$     Rewrote as a difference of two squares.

$$= (4x^2 + 3y^2)(4x^2 - 3y^2)$$     Difference of two squares formula was used.

**b)** Rewrite $x^6$ as $(x^3)^2$ and $y^4$ as $(y^2)^2$, then use the difference of two squares formula.
$$x^6 - y^4 = (x^3)^2 - (y^2)^2$$
$$= (x^3 + y^2)(x^3 - y^2)$$

Now Try Exercise 29

**EXAMPLE 4** Factor $9x^2 - 36y^2$ using the difference of two squares formula.

*Solution*

$$9x^2 - 36y^2 = 9(x^2 - 4y^2)$$     Factor out the GCF, 9

$$9(x^2 - 4y^2) = 9[(x)^2 - (2y)^2]$$     Rewrote as difference of two squares.

$$= 9(x + 2y)(x - 2y)$$     Difference of two squares formula was used.

Now Try Exercise 23

Notice in Example 4 that $9x^2 - 36y^2$ is the difference of two squares, $(3x)^2 - (6y)^2$. If you factor this difference of squares without first factoring out the GCF, 9, the factoring may be more difficult. After you factor this difference of squares you will need to factor out the GCF, 3, from each binomial factor, as illustrated below.
$$9x^2 - 36y^2 = (3x)^2 - (6y)^2$$
$$= (3x + 6y)(3x - 6y)$$
$$= 3(x + 2y)3(x - 2y)$$
$$= 9(x + 2y)(x - 2y)$$

We obtain the same answer as we did in Example 4. However, since we did not factor out the common factor 9 first, we had to work a little harder to obtain the answer.

**EXAMPLE 5** Factor $z^4 - 16$ using the difference of two squares formula.

*Solution*     We rewrite $z^4$ as $(z^2)^2$ and 16 as $(4)^2$, then use the difference of two squares formula.
$$z^4 - 16 = (z^2)^2 - (4)^2$$
$$= (z^2 + 4)(z^2 - 4)$$

Notice that the second factor, $z^2 - 4$, is also the difference of two squares. To complete the factoring, we use the difference of two squares formula again to factor $z^2 - 4$.
$$= (z^2 + 4)(z^2 - 4)$$
$$= (z^2 + 4)(z + 2)(z - 2)$$

Now Try Exercise 35

**Avoiding Common Errors**

The difference of two squares can be factored. However, a sum of two squares, where there is no common factor to the two terms, cannot be factored using real numbers.

| CORRECT | INCORRECT |
|---|---|
| $a^2 - b^2 = (a + b)(a - b)$ | ~~$a^2 + b^2 = (a + b)(a + b)$~~ |

## 2  Factor the Sum and Difference of Two Cubes

Consider the product of $(a + b)(a^2 - ab + b^2)$.

$$
\begin{array}{r}
a^2 \ - ab \ + b^2 \\
a \ + b \\
\hline
a^2b - ab^2 + b^3 \qquad \leftarrow b(a^2 - ab + b^2) \\
a^3 - a^2b + ab^2 \qquad\qquad \leftarrow a(a^2 - ab + b^2) \\
\hline
a^3 \qquad\qquad\qquad + b^3 \quad \leftarrow \text{Sum of terms}
\end{array}
$$

Thus, $(a + b)(a^2 - ab + b^2) = a^3 + b^3$. Since factoring is the opposite of multiplying, we may factor $a^3 + b^3$ as follows:

$$a^3 + b^3 = (a + b)(a^2 - ab + b^2)$$

We see, using the same procedure, that $a^3 - b^3 = (a - b)(a^2 + ab + b^2)$. The expression $a^3 + b^3$ is a sum of two cubes and the expression $a^3 - b^3$ is a difference of two cubes.

**Sum of Two Cubes**

$$a^3 + b^3 = (a + b)(a^2 - ab + b^2)$$

**Difference of Two Cubes**

$$a^3 - b^3 = (a - b)(a^2 + ab + b^2)$$

Note that the trinomials $a^2 - ab + b^2$ and $a^2 + ab + b^2$ cannot be factored further. Now let's solve some factoring problems using the sum and the difference of two cubes.

**EXAMPLE 6**  Factor $x^3 + 8$.

*Solution*   We rewrite $x^3 + 8$ as a sum of two cubes: $x^3 + 8 = (x)^3 + (2)^3$. Using the sum of two cubes formula, where $a$ is replaced by $x$ and $b$ is replaced by 2, we get

$$
\begin{array}{c}
a^3 \ + \ b^3 = (a + b)(a^2 - a \cdot b + b^2) \\
\downarrow \quad\quad \downarrow \quad\quad \downarrow \quad\quad \downarrow \quad \downarrow \quad\quad \downarrow \quad \downarrow \quad\quad \downarrow \\
x^3 + 8 = (x)^3 + (2)^3 = (x + 2)(x^2 - x \cdot 2 + 2^2) \\
= (x + 2)(x^2 - 2x + 4)
\end{array}
$$

You can check the factoring by multiplying $(x + 2)(x^2 - 2x + 4)$. If factored correctly, the product of the factors will equal the original expression, $x^3 + 8$. Try it and see.

Now Try Exercise 41

**Helpful Hint**

To help remember the signs in the sum or difference of two cubes formulas, consider

$$a^3 + b^3 = (a + b)(a^2 - ab + b^2)$$

↑ Same sign       Always
 Opposite sign    positive

$$a^3 - b^3 = (a - b)(a^2 + ab + b^2)$$

↑ Same sign       Always
 Opposite sign    positive

**EXAMPLE 7** Factor $y^3 - 125$.

**Solution**   We rewrite $y^3 - 125$ as a difference of two cubes: $(y)^3 - (5)^3$. Using the difference of two cubes formula, where $a$ is replaced by $y$ and $b$ is replaced by 5, we get

$$a^3 - b^3 = (a - b)(a^2 + a \cdot b + b^2)$$

$$y^3 - 125 = (y)^3 - (5)^3 = (y - 5)(y^2 + y \cdot 5 + 5^2)$$

$$= (y - 5)(y^2 + 5y + 25)$$

Now Try Exercise 43

**EXAMPLE 8** Factor $64a^3 - b^3$.

**Solution**   We rewrite $64a^3 - b^3$ as a difference of two cubes. Since $(4a)^3 = 64a^3$, we write

$$64a^3 - b^3 = (4a)^3 - (b)^3$$

$$= (4a - b)[(4a)^2 + (4a)(b) + b^2]$$

$$= (4a - b)(16a^2 + 4ab + b^2)$$

Now Try Exercise 49

**EXAMPLE 9** Factor $8r^3 + 27s^3$.

**Solution**   We rewrite $8r^3 + 27s^3$ as a sum of two cubes. Since $8r^3 = (2r)^3$ and $27s^3 = (3s)^3$, we write

$$8r^3 + 27s^3 = (2r)^3 + (3s)^3$$

$$= (2r + 3s)[(2r)^2 - (2r)(3s) + (3s)^2]$$

$$= (2r + 3s)(4r^2 - 6rs + 9s^2)$$

Now Try Exercise 53

---

**Avoiding Common Errors**

Recall that $a^2 + b^2 \neq (a + b)^2$ and $a^2 - b^2 \neq (a - b)^2$. The same principle applies to the sum and difference of two cubes.

| CORRECT | INCORRECT |
|---|---|
| $a^3 + b^3 = (a + b)(a^2 - ab + b^2)$ | $a^3 + b^3 = (a + b)^3$ |
| $a^3 - b^3 = (a - b)(a^2 + ab + b^2)$ | $a^3 - b^3 = (a - b)^3$ |

Since $(a + b)^3 = (a + b)(a + b)(a + b)$, it cannot possibly equal $a^3 + b^3$. Also, since $(a - b)^3 = (a - b)(a - b)(a - b)$, it cannot possibly equal $a^3 - b^3$.

It may be easier to see that, for example, $a^3 + b^3 = (a + b)(a^2 - ab + b^2)$ and not $(a + b)^3$ by substituting numbers for $a$ and $b$. Suppose $a = 3$ and $b = 4$, then

$$a^3 + b^3 = (a + b)(a^2 - ab + b^2)$$

$$3^3 + 4^3 = (3 + 4)[3^2 - 3(4) + 4^2]$$

$$27 + 64 = 7(13)$$

$$91 = 91$$

$$\text{but}\quad a^3 + b^3 \neq (a + b)^3$$

$$3^3 + 4^3 \neq (3 + 4)^3$$

$$91 \neq 343$$

---

## 3  Learn the General Procedure for Factoring a Polynomial

In this chapter, we have presented several methods of factoring. We now combine techniques from this and previous sections to give you an overview of a general factoring procedure.

## Understanding Algebra

Be careful! $16x^2 - 64$ is the difference of two squares. There is, however, a GCF, 16. Factor the 16 out first then use the formula:

$$16x^2 - 64 = 16(x^2 - 4)$$
$$= 16(x + 2)(x - 2)$$

Here is a general procedure for factoring any polynomial:

### General Procedure for Factoring a Polynomial

1. If all the terms of the polynomial have a greatest common factor other than 1, factor it out.
2. If the polynomial has two terms (or is a binomial), determine whether it is a difference of two squares or a sum or a difference of two cubes. If so, factor using the appropriate formula.
3. If the polynomial has three terms, factor the trinomial using the methods discussed in Sections 6.3 and 6.4.
4. If the polynomial has more than three terms, try factoring by grouping.
5. As a final step, examine your factored polynomial to determine whether the terms in any factors have a common factor. If you find a common factor, factor it out at this point.

### EXAMPLE 10   Factor $3x^4 - 27x^2$.

*Solution*   First determine whether the terms have a greatest common factor other than 1. Since $3x^2$ is common to both terms, factor it out.

$$3x^4 - 27x^2 = 3x^2(x^2 - 9)$$
$$= 3x^2(x + 3)(x - 3) \quad \text{Difference of two squares formula was used.}$$

Note that $x^2 - 9$ is a difference of two squares.

Now Try Exercise 69

### EXAMPLE 11   Factor $2m^2n^2 + 6m^2n - 36m^2$.

*Solution*   Begin by factoring the GCF, $2m^2$, from each term. Then factor the remaining trinomial.

$$2m^2n^2 + 6m^2n - 36m^2 = 2m^2(n^2 + 3n - 18)$$
$$= 2m^2(n + 6)(n - 3)$$

Now Try Exercise 81

### EXAMPLE 12   Factor $15c^2d - 10cd + 20d$.

*Solution*

$$15c^2d - 10cd + 20d = 5d(3c^2 - 2c + 4)$$

Since $3c^2 - 2c + 4$ cannot be factored, we stop here.

Now Try Exercise 77

### EXAMPLE 13   Factor $3xy + 6x + 3y + 6$.

*Solution*   Always begin by determining whether all the terms in the polynomial have a common factor. In this example, 3 is the GCF. Factor 3 from each term.

$$3xy + 6x + 3y + 6 = 3(xy + 2x + y + 2)$$

Now factor by grouping.

$$= 3[x(y + 2) + 1(y + 2)]$$
$$= 3(y + 2)(x + 1)$$

Now Try Exercise 79

In Example 13, what would happen if we forgot to factor out the common factor 3? Let's rework the problem without first factoring out the 3, and see what happens. Factor $3x$ from the first two terms, and 3 from the last two terms.

$$3xy + 6x + 3y + 6 = 3x(y + 2) + 3(y + 2)$$
$$= (y + 2)(3x + 3)$$

In step 5 of the general factoring procedure above, we are reminded to examine the factored polynomial to see whether the terms in any factor have a common factor. If we study the factors, we see that the factor has the common factor 3. If we factor out the 3 from $3x + 3$, we will obtain the same answer obtained in Example 13.

$$(y + 2)(3x + 3) = 3(y + 2)(x + 1)$$

**EXAMPLE 14** Factor $12x^2 + 12x - 9$.

*Solution*    First factor out the GCF, 3. Then factor the remaining trinomial by one of the methods discussed in Section 6.4 (either by grouping or trial and error).

$$12x^2 + 12x - 9 = 3(4x^2 + 4x - 3)$$
$$= 3(2x + 3)(2x - 1)$$

Now Try Exercise 59

**EXAMPLE 15** Factor $2x^4y + 54xy$.

*Solution*    First factor out the GCF, $2xy$.

$$2x^4y + 54xy = 2xy(x^3 + 27)$$
$$= 2xy(x + 3)(x^2 - 3x + 9)$$

Note that $x^3 + 27$ is a sum of two cubes.

Now Try Exercise 95

# EXERCISE SET 6.5

MathXL®    MyMathLab®

MathXL®    MyMathLab

## Warm-Up Exercises

*Fill in the blanks with the appropriate word, phrase, or symbol(s) from the following list.*

the sum of two squares

the difference of two cubes

the sum of two cubes

factor out the greatest common factor

prime polynomial

use a formula

the difference of two squares

call your instructor

**1.** $x^2 - 1$ is an example of a binomial that is _____.

**4.** $x^3 + 1$ is an example of a binomial that is _____.

**2.** $x^3 - 1$ is an example of a binomial that is _____.

**5.** The first step in any factoring problem is to _____.

**3.** $x^2 + 1$ is an example of a binomial that is _____.

**6.** $x^2 + x - 7$ is an example of a _____.

*In Exercises 7–12, the binomial is a sum of squares. There is no formula for factoring the sum of squares. However, sometimes a common factor can be factored out from a sum of squares. Factor those polynomials that are factorable. If the polynomial is not factorable, write the word prime.*

**7.** $x^2 + 9$

**8.** $4y^2 + 1$

**9.** $9b^2 + 81$

**10.** $16s^2 + 64t^2$

**11.** $16m^2 + 36n^2$

**12.** $9y^2 + 16z^2$

## Practice the Skills

*Factor each difference of two squares.*

**13.** $y^2 - 25$

**14.** $t^2 - 4$

**15.** $81 - z^2$

**16.** $64 - z^2$

**17.** $x^2 - 49$

**18.** $c^2 - d^2$

**19.** $x^2 - y^2$

**20.** $16x^2 - 9$

**21.** $9y^2 - 25z^2$

**22.** $64z^2 - 9$

**23.** $64a^2 - 36b^2$

**24.** $100x^2 - 81y^2$

**25.** $36 - 49x^2$

**26.** $100 - y^4$

**27.** $z^4 - 81x^2$

**28.** $9x^4 - 81y^2$

**29.** $25x^4 - 49y^4$

**30.** $4x^4 - 25y^4$

**31.** $36m^4 - 49n^2$

**32.** $10x^2 - 160$

**33.** $2x^4 - 50y^2$

**34.** $4x^3 - xy^2$

**35.** $5x^4 - 405$

**36.** $36x^4 - 4y^2$

*Factor each sum or difference of two cubes.*

**37.** $x^3 + y^3$

**38.** $a^3 - b^3$

**39.** $x^3 - y^3$

**40.** $a^3 + b^3$

**41.** $x^3 + 64$

**42.** $x^3 - 8$

**43.** $x^3 - 27$

**44.** $a^3 + 27$

**45.** $a^3 + 1$

**46.** $t^3 - 1$

**47.** $27x^3 - 1$

**48.** $64y^3 + 125$

**49.** $27a^3 - 125$

**50.** $125 + q^3$

**51.** $27 - 8y^3$

**52.** $8 + 27y^3$

**53.** $64m^3 + 27n^3$

**54.** $64x^3 - 125y^3$

**55.** $8a^3 - 27b^3$

**56.** $27c^3 + 125d^3$

*Factor completely.*

**57.** $4t^2 - 24t + 36$

**58.** $3x^2 - 9x - 12$

**59.** $50x^2 - 10x - 12$

**60.** $3x^2 - 48$

**61.** $2d^2 + 16d + 32$

**62.** $3x^2 + 9x + 12x + 36$

**63.** $5x^2 - 10x - 15$

**64.** $3xy - 6x + 9y - 18$

**65.** $5x^2 - 20$

**66.** $x^2y + 2xy - 6xy - 12y$

**67.** $2x^2 - 50$

**68.** $4a^2y - 64y^3$

**69.** $2x^2y - 18y$

**70.** $3x^3 - 147x$

**71.** $3x^3y^2 + 3y^2$

**72.** $x^4 - 125x$

**73.** $2x^3 - 16$

**74.** $x^3 - 27y^3$

**75.** $18x^2 - 50$

**76.** $54a^3 - 16$

**77.** $6t^2r - 15tr + 21r$

**78.** $12n^2 + 4n - 16$

**79.** $6x^2 - 4x + 24x - 16$

**80.** $4ab^2 + 4ab - 24a$

**81.** $2rs^2 - 10rs - 48r$

**82.** $4x^4 - 26x^3 + 30x^2$

**83.** $4x^2 + 5x - 6$

**84.** $12a^2 + 36a + 27$

**85.** $25b^2 - 100$

**86.** $3b^2 - 75c^2$

**87.** $a^5b^2 - 4a^3b^4$

**88.** $12x^2 + 36x - 3x - 9$

**89.** $5x^4 + 10x^3 + 5x^2$

**90.** $3c^6 + 12c^4d^2$

**91.** $x^3 + 25x$

**92.** $8y^2 - 23y - 3$

**93.** $y^4 - 16$

**94.** $36a^2 - 15ab - 6b^2$

**95.** $16m^3 + 250$

**96.** $2ab - 3b + 4a - 6$

**97.** $ac + 2a + bc + 2b$

**98.** $x^3 - 100x$

**99.** $9 - 9y^4$

## Problem Solving

*Factor each expression. Treat the unknown symbols as if they were variables.*

**100.** $\blacklozenge\text{✳} + 2\blacklozenge + \text{☺}\text{✳} + 2\text{☺}$

**101.** $2\blacklozenge^6 + 4\blacklozenge^4\text{✳}^2$

**102.** $4\blacklozenge^2\text{✳} - 6\blacklozenge\text{✳} - 20\text{✳}\blacklozenge + 30\text{✳}$

## Concept/Writing Exercises

**103.** Explain why the sum of two squares, $a^2 + b^2$, cannot be factored using real numbers.

**104.** Have you ever seen the proof that 1 is equal to 2? Here it is.

Let $a = b$, then square both sides of the equation:

$$a^2 = b^2$$

$$a^2 = b \cdot b$$

$$a^2 = ab \qquad \text{Substituted because } a = b.$$

$$a^2 - b^2 = ab - b^2 \qquad \text{Subtract } b^2 \text{ from both sides of the equation.}$$

$$(a + b)(a - b) = b(a - b) \qquad \text{Factored both sides of the equation.}$$

$$\frac{(a + b)\cancel{(a - b)}}{\cancel{(a - b)}} = \frac{b\cancel{(a - b)}}{\cancel{(a - b)}} \qquad \text{Divide both sides of the equation by } (a - b) \text{ and divide out common factors.}$$

$$a + b = b$$

$$b + b = b \qquad \text{Substituted because } a = b.$$

$$2b = b$$

$$\frac{\overset{1}{\cancel{2b}}}{\underset{1}{\cancel{b}}} = \frac{\overset{1}{\cancel{b}}}{\underset{1}{\cancel{b}}} \qquad \text{Divide both sides of the equation by } b.$$

$$2 = 1$$

Obviously, $2 \neq 1$. Therefore, we must have made an error somewhere. Can you find it?

## Challenge Problems

**105.** Factor $x^6 - 27y^9$.

**107.** Factor $x^2 - 6x + 9 - 4y^2$. (*Hint:* Write the first three terms as the square of a binomial.)

**109.** Factor $x^2 + 10x + 25 - y^2 + 4y - 4$. (*Hint:* Group the first three terms and the last three terms.)

**106.** Factor $x^6 + 1$.

**108.** Factor $x^6 - y^6$. (*Hint:* Factor initially as the difference of two squares.)

## Cumulative Review Exercises

[2.5] **110.** Solve $7x - 2(x + 6) = 2x - 5$.

[2.6] **111. Area** Use the formula $A = \dfrac{1}{2}h(b + d)$ to find $h$ in the following trapezoid if the area of the trapezoid is 36 square inches.

6 in.

$h$

12 in.

[5.1] **112.** Simplify $-9\,(a^3b^2c^6)^0$.

[5.1] **113.** Simplify $\left(\dfrac{4x^4y}{6xy^5}\right)^3$.

[5.2] **114.** Simplify $a^{-4}a^{-7}$.

# 6.6 Solving Quadratic Equations Using Factoring

1 Recognize quadratic equations.

2 Solve quadratic equations using factoring.

## 1 Recognize Quadratic Equations

In this section, we introduce **quadratic equations**, which are equations that contain a second-degree term and no term of a higher degree.

> **Quadratic Equation**
>
> Quadratic equations have the form
> $$ax^2 + bx + c = 0$$
> where $a$, $b$, and $c$ are real numbers, $a \neq 0$.

### Examples of Quadratic Equations
$$x^2 + 4x - 12 = 0$$
$$2x^2 - 5x = 0$$
$$3x^2 - 2 = 0$$

Quadratic equations like these, in which one side of the equation is written in descending order and the other side of the equation is 0, are said to be in **standard form**.

Some quadratic equations can be solved by factoring. To solve a quadratic equation by factoring, we use the **zero-factor property**.

> **Zero-Factor Property**
>
> If $ab = 0$, then $a = 0$ or $b = 0$.

**Understanding Algebra**

If you multiply any number by 0, the product is 0. That is, if $a = 0$ or $b = 0$, then $ab = 0$. The reverse is also true. If a product equals 0, at least one of its factors must be 0.

We now illustrate how the zero-factor property is used in solving equations.

**EXAMPLE 1** Solve the equation $(x + 3)(x + 4) = 0$.

*Solution*    Since the product of the factors equals 0, according to the zero-factor property, one or both factors must equal 0. Set each factor equal to 0, and solve each resulting equation.

$$
\begin{array}{ccc}
x + 3 = 0 & \text{or} & x + 4 = 0 \\
x + 3 - 3 = 0 - 3 & & x + 4 - 4 = 0 - 4 \\
x = -3 & & x = -4
\end{array}
$$

Thus, if $x$ is either $-3$ or $-4$, the product of the factors is 0. The solutions to the equation are $-3$ and $-4$.

Check $\qquad\qquad x = -3 \qquad\qquad\qquad\qquad\qquad x = -4$

$$(x + 3)(x + 4) = 0 \qquad\qquad\qquad (x + 3)(x + 4) = 0$$
$$(-3 + 3)(-3 + 4) \stackrel{?}{=} 0 \qquad\qquad (-4 + 3)(-4 + 4) \stackrel{?}{=} 0$$
$$0(1) \stackrel{?}{=} 0 \qquad\qquad\qquad\qquad -1(0) \stackrel{?}{=} 0$$
$$0 = 0 \quad \text{True} \qquad\qquad\qquad\qquad 0 = 0 \quad \text{True}$$

Now Try Exercise 7

**EXAMPLE 2** Solve the equation $(3x - 2)(4x + 1) = 0$.

*Solution*  Set each factor equal to 0 and solve for $x$.

$$3x - 2 = 0 \qquad \text{to} \qquad 4x + 1 = 0$$
$$3x = 2 \qquad\qquad\qquad 4x = -1$$
$$x = \frac{2}{3} \qquad\qquad\qquad x = -\frac{1}{4}$$

The solutions to the equation are $\frac{2}{3}$ and $-\frac{1}{4}$.

Now Try Exercise 11

## 2 Solve Quadratic Equations Using Factoring

**General Rules to Solve a Quadratic Equation Using Factoring**

1. Write the equation in standard form with the squared term having a positive coefficient. This will result in one side of the equation being 0.
2. Factor the side of the equation that is not 0.
3. Set each factor *containing a variable* equal to 0 and solve each equation.
4. Check each solution found in step 3 in the *original* equation.

Now we will use factoring to solve quadratic equations.

**EXAMPLE 3** Solve the equation $3x^2 = 12x$.

*Solution*

$$3x^2 = 12x$$
$$3x^2 - 12x = 12x - 12x \qquad \text{Subtract } 12x \text{ from both sides to make one side 0.}$$
$$3x^2 - 12x = 0$$
$$3x(x - 4) = 0 \qquad\qquad\qquad \text{Factored out the GCF, } 3x.$$

Now set each factor equal to 0.

$$3x = 0 \qquad \text{or} \qquad x - 4 = 0$$
$$x = \frac{0}{3} \qquad\qquad\qquad x = 4$$
$$x = 0$$

The solutions to the quadratic equation are 0 and 4. Check by substituting $x = 0$, then $x = 4$, in $3x^2 = 12x$.

Now Try Exercise 47

**EXAMPLE 4** Solve the equation $x^2 + 10x + 28 = 4$.

*Solution*

$$x^2 + 10x + 28 = 4$$
$$x^2 + 10x + 24 = 0 \qquad\qquad \text{Subtracted 4 from both sides.}$$
$$(x + 4)(x + 6) = 0 \qquad\qquad \text{Factored}$$
$$x + 4 = 0 \qquad \text{or} \qquad x + 6 = 0 \qquad \text{Set each factor equal to 0.}$$
$$x = -4 \qquad\qquad\qquad x = -6$$

The solutions are $-4$ and $-6$. We will check these values in the original equation.

Check

$$x = -4$$
$$x^2 + 10x + 28 = 4$$
$$(-4)^2 + 10(-4) + 28 \overset{?}{=} 4$$
$$16 - 40 + 28 \overset{?}{=} 4$$
$$-24 + 28 \overset{?}{=} 4$$
$$4 = 4 \quad \text{True}$$

$$x = -6$$
$$x^2 + 10x + 28 = 4$$
$$(-6)^2 + 10(-6) + 28 \overset{?}{=} 4$$
$$36 - 60 + 28 \overset{?}{=} 4$$
$$-24 + 28 \overset{?}{=} 4$$
$$4 = 4 \quad \text{True}$$

Now Try Exercise 23

**EXAMPLE 5** Solve the equation $4y^2 + 5y - 20 = -11y$.

*Solution*    Since all terms are not on the same side of the equation, add $11y$ to both sides of the equation.

$$4y^2 + 5y - 20 = -11y$$
$$4y^2 + 16y - 20 = 0 \qquad \text{Added } 11y \text{ to both sides.}$$
$$4(y^2 + 4y - 5) = 0 \qquad \text{Factored out GCF, 4.}$$
$$4(y + 5)(y - 1) = 0 \qquad \text{Factored}$$
$$y + 5 = 0 \quad \text{or} \quad y - 1 = 0 \quad \text{Set each factor equal to 0.}$$
$$y = -5 \qquad\qquad y = 1$$

*Since 4 is a factor that does not contain a variable, we do not set it equal to 0.* The solutions to the quadratic equation are $-5$ and 1.

Now Try Exercise 25

**EXAMPLE 6** Solve the equation $-x^2 + 5x + 6 = 0$.

*Solution*    When the squared term is negative, we generally make it positive by multiplying both sides of the equation by $-1$.

$$-1(-x^2 + 5x + 6) = -1 \cdot 0$$
$$x^2 - 5x - 6 = 0$$
$$(x - 6)(x + 1) = 0$$
$$x - 6 = 0 \quad \text{or} \quad x + 1 = 0 \qquad \text{Set each factor equal to 0.}$$
$$x = 6 \qquad\qquad x = -1$$

A check using the original equation will show that the solutions are 6 and $-1$.

Now Try Exercise 33

## Understanding Algebra

To solve $x^2 = 8x$ it is *not* correct to divide both sides of the equation by $x$. Since it is a quadratic equation, we solve it this way:

$$x^2 = 8x$$
$$x^2 - 8x = 0$$
$$x(x - 8) = 0$$
$$x = 0 \quad \text{or} \quad x = 8$$

**Avoiding Common Errors**

Be careful not to confuse factoring a polynomial with using factoring as a method to solve an equation.

| CORRECT | INCORRECT |
|---|---|
| Factor:  $x^2 + 3x + 2$ | Factor:  $x^2 + 3x + 2$ |
| $(x + 2)(x + 1)$ | $(x + 2)(x + 1)$ |
|  | ~~$x + 2 = 0$  or  $x + 1 = 0$~~ |
|  | ~~$x = -2$        $x = -1$~~ |

The expression $x^2 + 3x + 2$ is a polynomial, not an equation. Since it is not an equation, it cannot be solved.

CORRECT

Solve:    $x^2 + 3x + 2 = 0$
$$(x + 2)(x + 1) = 0$$
$$x + 2 = 0 \quad \text{or} \quad x + 1 = 0$$
$$x = -2 \qquad\qquad x = -1$$

**EXAMPLE 7**  Solve the equation $x^2 = 49$.

*Solution*

$$x^2 = 49$$
$$x^2 - 49 = 0 \qquad \text{49 was subtracted from both sides.}$$
$$(x + 7)(x - 7) = 0 \qquad \text{Factored using the difference of two squares.}$$
$$x + 7 = 0 \quad \text{or} \quad x - 7 = 0 \quad \text{Set each factor equal to 0.}$$
$$x = -7 \qquad\qquad x = 7$$

The solutions are $-7$ and $7$.

Now Try Exercise 45

**EXAMPLE 8**  Solve the equation $(x - 3)(x + 1) = 5$.

*Solution*

$$(x - 3)(x + 1) = 5$$
$$x^2 - 2x - 3 = 5 \qquad \text{Factors were multiplied.}$$
$$x^2 - 2x - 8 = 0 \qquad \text{Wrote the equation in standard form.}$$
$$(x - 4)(x + 2) = 0 \qquad \text{Trinomial was factored.}$$
$$x - 4 = 0 \quad \text{or} \quad x + 2 = 0 \quad \text{Zero-factor property}$$
$$x = 4 \qquad\qquad x = -2$$

The solutions are $4$ and $-2$. We will check these values in the original equation.

Check

$$x = 4 \qquad\qquad\qquad\qquad x = -2$$
$$(x - 3)(x + 1) = 5 \qquad\qquad (x - 3)(x + 1) = 5$$
$$(4 - 3)(4 + 1) \stackrel{?}{=} 5 \qquad\qquad (-2 - 3)(-2 + 1) \stackrel{?}{=} 5$$
$$1(5) \stackrel{?}{=} 5 \qquad\qquad\qquad (-5)(-1) \stackrel{?}{=} 5$$
$$5 = 5 \quad \text{True} \qquad\qquad\qquad 5 = 5 \quad \text{True}$$

Now Try Exercise 51

> **Helpful Hint**
>
> In Example 8, you might have been tempted to start the problem by writing
>
> $$x - 3 = 5 \quad \text{or} \quad x + 1 = 5.$$
>
> This would lead to an incorrect solution. Remember, the zero-factor property only holds when one side of the equation is equal to 0. In Example 8, once we obtained $(x - 4)(x + 2) = 0$, we were able to use the zero-factor property.

# EXERCISE SET 6.6

MathXL®   MyMathLab®
MathXL®    MyMathLab

## Warm-Up Exercises

*Fill in the blanks with the appropriate word, phrase, or symbol(s) from the following list.*

| | | | |
|---|---|---|---|
| $x = 0$ or $x = 2$ | $x = -2$ or $x = 2$ | $x = 3$ or $x = -2$ or $x = 2$ | $x = -2$ or $x = 3$ |
| standard form | zero-factor property | $x = -2$ or $x = 0$ | |

**1.** "If $ab = 0$ then either $a = 0$ or $b = 0$" is called the

_____.

**2.** When $x^2 + 12 = 7x$ is rewritten as $x^2 - 7x + 12 = 0$, we say that the quadratic equation is now in _____.

**3.** The solution to $x^2 = 2x$ is _____.

**4.** The solution to $x^2 + 2x = 0$ is _____.

**5.** The solution to $3(x + 2)(x - 2) = 0$ is _____.

**6.** The solution to $(x + 1)(x - 2) = 4$ is _____.

## Practice the Skills

*Solve.*

**7.** $(x + 8)(x - 7) = 0$

**8.** $-2x(x + 9) = 0$

**9.** $7x(x - 8) = 0$

**10.** $(t + 3)(t + 5) = 0$

**11.** $(3x + 7)(2x - 11) = 0$

**12.** $(3x - 2)(x - 5) = 0$

**13.** $x^2 - 16 = 0$

**14.** $y^2 - 9 = 0$

**15.** $x^2 - 12x = 0$

**16.** $9x^2 + 27x = 0$

**17.** $x^2 + 7x = 0$

**18.** $a^2 - 4a - 12 = 0$

**19.** $x^2 - 8x + 16 = 0$

**20.** $x^2 + 12x + 36 = 0$

**21.** $x^2 + 12x = -20$

**22.** $3y^2 - 4 = -4y$

**23.** $x^2 + 12x + 22 = 2$

**24.** $3x^2 = -21x - 18$

**25.** $2x^2 + 3x - 24 = 5x$

**26.** $x^2 = 4x + 21$

**27.** $23p - 24 = -p^2$

**28.** $3x^2 - 9x - 30 = 0$

**29.** $33w + 90 = -3w^2$

**30.** $t^2 + 44 + 15t = 0$

**31.** $-2x - 15 = -x^2$

**32.** $-9x + 20 = -x^2$

**33.** $-x^2 + 29x + 30 = 0$

**34.** $12y - 11 = y^2$

**35.** $-15 = 4m^2 + 17m$

**36.** $z^2 + 8z = -16$

**37.** $9p^2 = -21p - 6$

**38.** $2x^2 - 5 = 3x$

**39.** $3r^2 + 13r = 10$

**40.** $3x^2 = 7x + 20$

**41.** $4x^2 + 4x - 48 = 0$

**42.** $6x^2 - 7x - 5 = 0$

**43.** $8x^2 + 2x = 3$

**44.** $2x^2 + 4x - 6 = 0$

**45.** $c^2 = 64$

**46.** $2n^2 + 36 = -18n$

**47.** $2x^2 = 50x$

**48.** $4x^2 - 25 = 0$

**49.** $x^2 = 100$

**50.** $3x^2 - 48 = 0$

**51.** $(x - 2)(x - 1) = 12$

**52.** $(x + 2)(x + 5) = -2$

**53.** $(3x + 2)(x + 1) = 4$

**54.** $(x - 1)(2x - 5) = 9$

**55.** $2(a^2 + 9) = 15a$

**56.** $x(x + 5) = 6$

## Problem Solving

*In Exercises 57–60, create a quadratic equation with the given solutions. Explain how you determined your answers.*

**57.** $6, -4$

**58.** $-3, -5$

**59.** $6, 0$

**60.** $0, -9$

**61.** The solutions to a quadratic equation are $\frac{1}{2}$ and $-\frac{1}{3}$.
   **a)** Give two possible factors with integer coefficients that were set equal to 0 to obtain these solutions.
   **b)** Write a quadratic equation whose solutions are $\frac{1}{2}$ and $-\frac{1}{3}$.

**62.** The solutions to a quadratic equation are $\frac{2}{3}$ and $-\frac{3}{4}$.
   **a)** Give two possible factors with integer coefficients that were set equal to 0 to obtain these solutions.
   **b)** Write a quadratic equation whose solutions are $\frac{2}{3}$ and $-\frac{3}{4}$.

## Concept/Writing Exercises

**63.** Carefully observe the quadratic equations $x^2 - 7x + 12 = 0$, $2x^2 - 14x + 24 = 0$, and $3x^2 - 21x + 36 = 0$. Explain how you know that all three equations have the same solution.

**64.** Write the steps to the method you would use to solve $(x + 1)(x - 2) = 4$ as if you were writing an email to a friend taking an algebra course.

## Challenge Problems

**65.** Solve the equation $(2x - 3)(x - 4) = (x - 5)(x + 3) + 7$.

**66.** Solve the equation $(x - 3)(x - 2) = (x + 5)(2x - 3) + 21$.

**67.** Solve the equation $x(x - 3)(x + 2) = 0$.

**68.** Solve the equation $x^3 - 10x^2 + 24x = 0$.

## Cumulative Review Exercises

[1.7]    **69.** Subtract $\frac{3}{5} - \frac{2}{9}$.

[2.5]    **70. a)** What is the name given to an equation that has an infinite number of solutions?

   **b)** What is the name given to an equation that has no solution?

[2.7]  **71. Cypress Gardens** At Cypress Gardens, there is a long line of people waiting to go through the entrance. If 160 people are admitted in 13 minutes, how many people will be admitted in 60 minutes? Assume the rate stays the same.

[5.1]  **72.** Simplify $\left(\dfrac{3p^5q^7}{p^9q^8}\right)^2$.

[5.4]  *Identify the following as a monomial, binomial, trinomial or not a polynomial. If an expression is not a polynomial, explain why.*

**73.** $2x$          **74.** $x - 3$

**75.** $\dfrac{1}{x}$          **76.** $x^2 - 6x + 9$

Cypress Gardens; see Exercise 71.

# 6.7 Applications of Quadratic Equations

1  Solve applications by factoring quadratic equations.

2  Learn the Pythagorean Theorem.

## 1  Solve Applications by Factoring Quadratic Equations

In Section 6.6, we learned how to solve quadratic equations by factoring. In this section, we will discuss and solve application problems that require solving quadratic equations to obtain the answer. In Example 1, we will solve a problem involving a relationship between two numbers.

**EXAMPLE 1  Number Problem** The product of two numbers is 78. Find the two numbers if one number is 7 more than the other.

*Solution*  Understand and Translate   Our goal is to find the two numbers.

Let $x$ = smaller number.

$x + 7$ = larger number.

$x(x + 7) = 78$  "Product" refers to multiplication.

Carry Out

$$x^2 + 7x = 78$$
$$x^2 + 7x - 78 = 0$$
$$(x - 6)(x + 13) = 0$$
$$x - 6 = 0 \quad \text{or} \quad x + 13 = 0$$
$$x = 6 \qquad\qquad x = -13$$

Remember that $x$ represents the smaller of the two numbers. This problem has two possible solutions.

|  | Solution 1 | Solution 2 |
|---|---|---|
| Smaller number | 6 | −13 |
| Larger number | $x + 7 = 6 + 7 = 13$ | $x + 7 = -13 + 7 = -6$ |

Thus, the two possible solutions are 6 and 13, and −13 and −6.

| Check | 6 and 13 | −13 and −6 |
|---|---|---|
| Product of the two numbers is 78. | $6 \cdot 13 = 78$ | $(-13)(-6) = 78$ |
| One number is 7 more than the other number. | 13 is 7 more than 6. | −6 is 7 more than −13. |

Answer   One solution is: smaller number 6, larger number 13. A second solution is: smaller number −13, larger number −6. You must give both solutions. If the question had stated "the product of two *positive* numbers is 78," the only solution would be 6 and 13.

Now Try Exercise 13

Now let us work an application problem involving geometry.

**EXAMPLE 2  Advertising**  The marketing department of a large publishing company is planning to make a large rectangular sign to advertise a new book at a convention. They want the length of the sign to be 3 feet longer than the width (**Fig. 6.4**). Signs at the convention may have a maximum area of 54 square feet. Find the length and width of the sign if the area is to be 54 square feet.

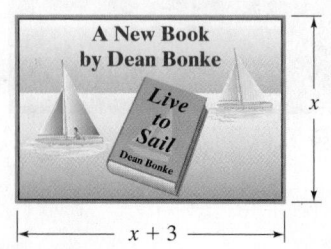

*Solution*  Understand and Translate  We need to find the length and width of the sign. We will use the formula for the area of a rectangle.

**FIGURE 6.4**

$$\text{Let } x = \text{width.}$$
$$x + 3 = \text{length.}$$
$$\text{area} = \text{length} \cdot \text{width}$$
$$54 = (x + 3)x$$

Carry Out
$$54 = x^2 + 3x$$
$$0 = x^2 + 3x - 54 \quad \text{Standard quadratic form}$$
$$\text{or} \quad x^2 + 3x - 54 = 0 \quad \text{Also standard quadratic form}$$
$$(x - 6)(x + 9) = 0$$
$$x - 6 = 0 \quad \text{or} \quad x + 9 = 0$$
$$x = 6 \qquad\qquad x = -9 \quad \text{Reject } -9 \text{ because width cannot be negative.}$$

Check and Answer  Since the width of the sign cannot be a negative number, the only solution is

$$\text{width} = x = 6 \text{ feet}, \quad \text{length} = x + 3 = 6 + 3 = 9 \text{ feet}$$

The area, length · width, is 54 square feet, and the length is 3 feet more than the width, so the answer checks.

Now Try Exercise 21

**EXAMPLE 3  Earth's Gravitational Field**  In Earth's gravitational field, the distance, $d$, in feet, that an object falls $t$ seconds after it has been released is given by the formula $d = 16t^2$. While at the top of a roller coaster, a rider's eyeglasses slide off his head and fall out of the cart. How long does it take the eyeglasses to reach the ground 64 feet below?

*Solution*  Understand and Translate  Substitute 64 for $d$ in the formula and then solve for $t$.

$$d = 16t^2$$
$$64 = 16t^2$$

Carry Out
$$\frac{64}{16} = t^2$$
$$4 = t^2$$

Now subtract 4 from both sides of the equation and write the equation with 0 on the right side to put the quadratic equation in standard form.

$$4 - 4 = t^2 - 4$$
$$0 = t^2 - 4$$
$$\text{or} \quad t^2 - 4 = 0$$
$$(t + 2)(t - 2) = 0$$
$$t + 2 = 0 \quad \text{or} \quad t - 2 = 0 \quad \text{Reject } -2 \text{ because time cannot be negative.}$$
$$t = -2 \qquad\qquad t = 2$$

Check and Answer  Since $t$ represents the number of seconds, it must be a positive number. Thus, the only possible answer is 2 seconds. It takes 2 seconds for the eyeglasses (or any other object falling under the influence of gravity) to fall 64 feet.

Now Try Exercise 25

Pythagoras of Samos

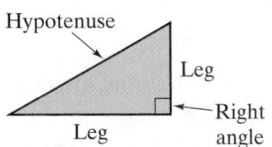

FIGURE 6.5

## 2 Learn the Pythagorean Theorem

We now introduce the Pythagorean Theorem, which describes an important relationship between the length of the sides of a right triangle. The Pythagorean Theorem is named after Pythagoras of Samos ($\approx$569 B.C.–475 B.C.) who was born in Samos, Ionia. Pythagoras is often described as the first pure mathematician. Now let's discuss the Pythagorean Theorem.

A **right triangle** is a triangle that contains a right, or 90°, angle (**Fig. 6.5**). The two shorter sides of a right triangle are called the **legs** and the largest side, which is always opposite the right angle, is called the **hypotenuse**. The **Pythagorean Theorem** expresses the relationship between the lengths of the legs of a right triangle and its hypotenuse.

> **Pythagorean Theorem**
>
> The square of the hypotenuse of a right triangle is equal to the sum of the squares of the two legs.
>
> $$(\text{leg})^2 + (\text{leg})^2 = (\text{hypotenuse})^2$$
>
> If $a$ and $b$ represent the legs, and $c$ represents the hypotenuse, then
>
> $$a^2 + b^2 = c^2$$

When you use the Pythagorean Theorem, it makes no difference which leg you designate as $a$ and which leg you designate as $b$, but the hypotenuse, the longest side, is always designated as $c$.

**EXAMPLE 4** **Verifying Right Triangles** Determine if a right triangle can have the following sides.

**a)** 3 inches, 4 inches, 5 inches        **b)** 2 inches, 5 inches, 7 inches

*Solution*

**a)** Understand    To determine if a right triangle can have the sides given, we will use the Pythagorean Theorem.

Translate    We must always select the largest side to represent the hypotenuse, $c$. We will designate the length of leg $a$ to be 3 inches and the length of leg $b$ to be 4 inches. The length of the hypotenuse, $c$, will be 5 inches. See **Figure 6.6**.

$$a^2 + b^2 = c^2$$
$$3^2 + 4^2 \stackrel{?}{=} 5^2$$

Carry Out
$$9 + 16 \stackrel{?}{=} 25$$
$$25 = 25 \quad \text{True}$$

Check and Answer    Since using the Pythagorean Theorem results in a true statement, a right triangle can have the given sides.

**b)** We will let leg $a$ have a length of 2 inches, leg $b$ have a length of 5 inches, and the hypotenuse, $c$, have a length of 7 inches.

$$a^2 + b^2 = c^2$$
$$2^2 + 5^2 \stackrel{?}{=} 7^2$$
$$4 + 25 \stackrel{?}{=} 49$$
$$29 = 49 \quad \text{False}$$

Since 29 is not equal to 49, the Pythagorean Theorem does not hold for these lengths. Therefore, no right triangle can have sides with lengths of 2 inches, 5 inches, and 7 inches.

Now Try Exercise 27

## Understanding Algebra

The proof of the Pythagorean Theorem is often expressed visually:

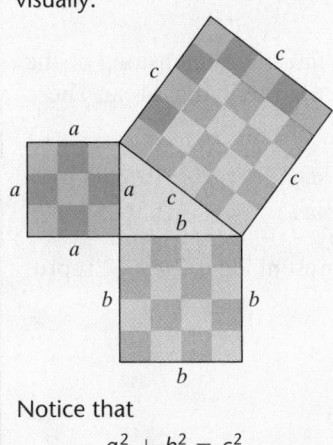

Notice that
$$a^2 + b^2 = c^2$$
$$3^2 + 4^2 = 5^2$$

FIGURE 6.6

**FIGURE 6.7**

**EXAMPLE 5** **Using the Pythagorean Theorem** One leg of a right triangle is 7 feet longer than the other leg. The hypotenuse is 13 feet. Find the dimensions of the right triangle.

*Solution*   Understand and Translate   We will first draw a diagram of the situation. See **Figure 6.7**.

Now we will use the Pythagorean Theorem to determine the dimensions of the right triangle.

$$a^2 + b^2 = c^2$$
$$x^2 + (x + 7)^2 = 13^2$$

Carry Out

$$x^2 + (x^2 + 14x + 49) = 169$$
$$2x^2 + 14x - 120 = 0$$
$$2(x^2 + 7x - 60) = 0$$
$$2(x + 12)(x - 5) = 0$$
$$x + 12 = 0 \quad \text{or} \quad x - 5 = 0$$
$$x = -12 \qquad\qquad x = 5$$

Check and Answer   Since a length cannot be a negative number, the only answer is 5. The dimensions of the right triangle are: one leg is 5 feet, the other leg is $x + 7$ or 12 feet, and the hypotenuse is 13 feet. Since $5^2 + 12^2 = 25 + 144 = 169$, which is $13^2$, the answer checks.

Now Try Exercise 35

**EXAMPLE 6** **Sand and Water Table** Clayton Jackson is building a rectangular table for his son to hold sand in one area and water in a different area (**Fig. 6.8**). The length of the table will be 2 feet less than twice the width. He is placing a divider that is 5 feet in length along the diagonal of the table to separate the sand from the water. Find the dimensions of the table.

*Solution*   Understand and Translate   Our goal is to find the dimensions of the table. **Figure 6.8** shows a right triangle. Therefore, we will use the Pythagorean Theorem to answer the question.

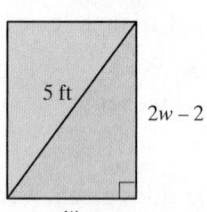

**FIGURE 6.8**

$$\text{Let } w = \text{width of the table.}$$
$$\text{Then } 2w - 2 = \text{length of the table.}$$

Now we use the Pythagorean Theorem. We will let $w$ represent leg $a$, $2w - 2$ represent leg $b$, and 5 represent the hypotenuse, $c$.

$$a^2 + b^2 = c^2$$
$$w^2 + (2w - 2)^2 = 5^2$$

Carry Out

$$w^2 + 4w^2 - 8w + 4 = 25$$
$$5w^2 - 8w + 4 = 25$$
$$5w^2 - 8w - 21 = 0$$
$$(5w + 7)(w - 3) = 0$$
$$5w + 7 = 0 \quad \text{or} \quad w - 3 = 0$$
$$5w = -7 \qquad\qquad w = 3$$
$$w = -\frac{7}{5}$$

Check and Answer   Since the width of the table cannot be a negative number, the only solution is 3. Therefore, the width of the table is 3 feet. The length of the table is $2w - 2 = 2(3) - 2 = 6 - 2 = 4$ feet.

Now Try Exercise 37

In the exercise set, we use the terms consecutive integers, consecutive even integers, and consecutive odd integers. Recall from Section 3.1 that **consecutive integers** may be represented as $x$ and $x + 1$. **Consecutive even** or **consecutive odd integers** may be represented as $x$ and $x + 2$.

# EXERCISE SET 6.7

MathXL® MyMathLab®
MathXL® MyMathLab

## Warm-Up Exercises

*Fill in the blanks with the appropriate word, phrase, or symbol(s) from the following list.*

| | | |
|---|---|---|
| 90° | hypotenuse | the Pythagorean Theorem |
| 45° | legs | 10 | 5 |

**1.** The longest side of a right triangle is called its _____ .

**2.** "The hypotenuse squared is equal to the sum of the squares of the other two sides" is called _____ .

**3.** Every right triangle contains exactly one angle of measure _____ .

**4.** The two shorter sides of a right triangle are called its _____ .

## Practice the Skills

*In Exercises 5–8, determine the value of the question mark.*

**5.**

**6.**

**7.**

**8.**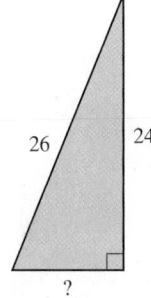

*In Exercises 9–12, a and b represent two legs of a right triangle and c represents the hypotenuse. Determine the value of the question mark (?).*

**9.** $a = 24, c = 26, b = ?$   **10.** $a = 12, b = 16, c = ?$   **11.** $a = 15, b = 36, c = ?$   **12.** $b = 20, c = 25, a = ?$

## Problem Solving

*Express each problem as an equation, then solve.*

**13. Product of Numbers** The product of two positive numbers is 221. Determine the two numbers if one is 4 more than the other.

**14. Positive Numbers** The product of two positive numbers is 245. Determine the two numbers if one number is 5 times the other.

**15. Positive Numbers** The product of two positive numbers is 84. Find the two numbers if one number is 2 more than twice the other.

**16. Consecutive Integers** The product of two consecutive positive integers is 56. Find the two integers.

**17. Consecutive Even Integers** The product of two consecutive positive even integers is 288. Find the two integers.

**18. Consecutive Odd Integers** The product of two consecutive positive odd integers is 143. Determine the two integers.

**19. Area of Rectangle** The area of a rectangle is 36 square feet. Determine the length and width if the length is 4 times the width.

**20. Rectangular Scrapbook** A scrapbook page has an area of 180 square inches. Find the length and width if the width is 3 inches less than the length.

**21. Rectangular Garden** Maureen Woolhouse has a rectangular garden whose width is two-thirds its length. If its area is 150 square feet, determine the length and width of the garden.

**22. Buying Wallpaper** Alejandro Ibanez wishes to buy a wallpaper border to go along the top of one wall in his living room. The length of the wall is 7 feet greater than its height.

**a)** Find the length and height of the wall if the area of the wall is 120 square feet.

**b)** What is the length of the border he will need?

**c)** If the border costs $4 per linear foot, how much will the border cost?

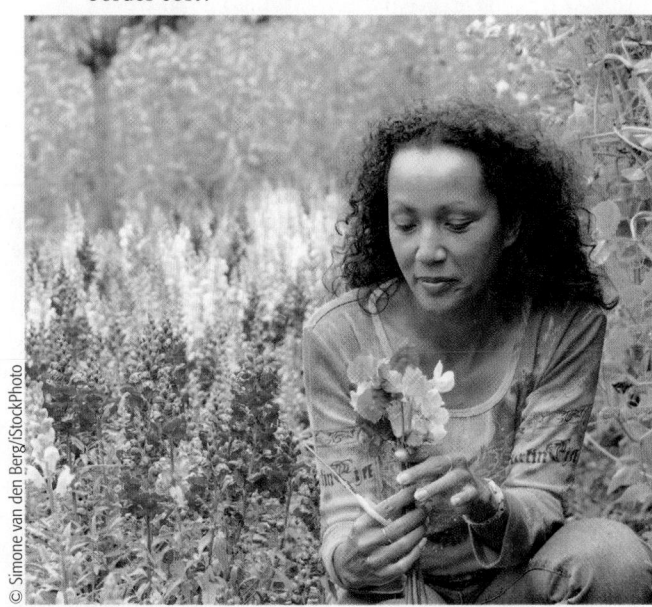
© Simone van den Berg/iStockPhoto

See Exercise 21.

**23. Square** If each side of a square is increased by 4 meters, the area becomes 81 square meters. Determine the length of a side of the original square.

**24. Sign** If the length of the sign in Example 2 is to be 2 feet longer than the width and the area is to be 35 square feet, determine the dimensions of the sign.

**25. Dropped Egg** How long would it take for an egg dropped from a helicopter to fall 256 feet to the ground? See Example 3.

**26. Falling Rock** How long would it take a rock that falls from a cliff 400 feet above the sea to hit the sea? See Example 3.

*In Exercises 27–30, determine if a right triangle can have the following sides where a and b represent the legs and c represents the hypotenuse. Explain your answer.*

**27.** $a = 7, b = 24, c = 25$   **28.** $a = 16, c = 20, b = 22$   **29.** $a = 9, b = 40, c = 41$   **30.** $a = 13, b = 18, c = 28$

*In Exercises 31–34, find the value of the question mark.*

**31.**

**32.**

**33.**

**34.**
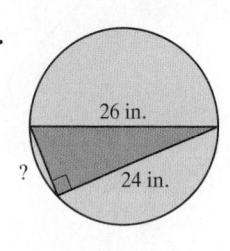

**35. Triangle** One leg of a right triangle is 2 feet longer than the other leg. The hypotenuse is 10 feet. Find the lengths of the three sides of the triangle.

**36. Triangle** One leg of a right triangle is two inches more than twice the other leg. The hypotenuse is 13 inches. Find the lengths of the three sides of the triangle.

**37. Artwork** Rachel bought a framed piece of artwork as a souvenir from her trip to Disney World. The diagonal of the frame is 15 inches. If the length of the frame is 3 inches greater than its width, find the dimensions of the frame.

**38. Laptop** The top of a new experimental rectangular laptop computer has a diagonal of 17 inches. If the length of the computer is 1 inch less than twice its width, find the dimensions of the computer.

**39. Rectangular Garden** Mary Ann Tuerk has constructed a rectangular garden. The length of the garden is 3 feet more than three times its width. The diagonal of the garden is 4 feet more than three times the width. Find the length and width of the garden.

**40. Height of Tree** A tree is supported by ropes. One rope goes from the top of the tree to a point on the ground. The height of the tree is 4 feet more than twice the distance between the base of the tree and the rope anchored in the ground. The length of the rope is 6 feet more than twice the distance be-

tween the base of the tree and the rope anchored in the ground. Find the height of the tree.

**41. Book Store** A book store owner finds that her daily profit, $P$, is approximated by the formula $P = x^2 - 15x - 50$, where $x$ is the number of books she sells. How many books must she sell in a day for her profit to be $400?

**42. Water Sprinklers** The cost, $C$, for manufacturing $x$ water sprinklers is given by the formula $C = x^2 - 27x - 20$. Determine the number of water sprinklers manufactured at a cost of $70.

**43. Sum of Numbers** The sum, $s$, of the first $n$ even numbers is given by the formula $s = n^2 + n$. Determine $n$ for the given sums:
**a)** $s = 20$        **b)** $s = 90$

**44. Telephone Lines** For a switchboard that handles $n$ telephone lines, the maximum number of telephone connections, $C$, that it can make simultaneously is given by the formula

$$C = \frac{n(n - 1)}{2}.$$

**a)** How many telephone connections can a switchboard make simultaneously if it handles 15 lines?

**b)** How many lines does a switchboard have if it can make 55 telephone connections simultaneously?

**45. Backyard Baseball** In baseball, the infield is called the "diamond"; it is actually a square. But the DeVito brothers had to improvise in their backyard for infield practice. They made the distance from home to first base 60 feet, the distance from first base to second base 80 feet, the distance from second base to third base 60 feet, and the distance from third base to home 80 feet as depicted in the rectangle below. How far is the distance from second base to home plate?

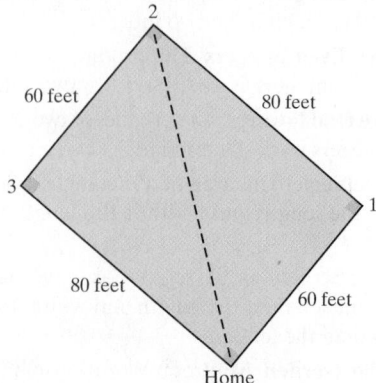

**46.** Refer to Exercise 45. If they had moved all the bases counterclockwise (so that home is where first is in the diagram and first is where second is in the diagram, etc.), would the distance from home to second base change?

## Challenge Problems

**47. Area** Determine the area of the rectangle.

30 ft

18 ft

**48. Area** Determine the area of the circle.

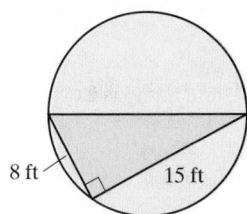

8 ft    15 ft

**49.** Solve the equation $x^3 - 4x^2 - 32x = 0$.

**50.** Solve the equation $x^3 + 2x^2 - 15x = 0$.

**51.** Create an equation whose solutions are $-2, 0,$ and $3$. Explain how you determined your answer.

**52. Numbers** The product of two numbers is $-63$. Determine the numbers if their sum is $-2$.

**53. Numbers** The sum of two numbers is 9. The sum of the squares of the two numbers is 45. Determine the two numbers.

**54. Wizard of Oz** Near the end of the movie *The Wizard of Oz*, when the scarecrow receives his diploma, he starts talking rapidly and seems to impress us with his newfound intelligence. One of the things that the scarecrow *attempts* to say is the Pythagorean Theorem. The only problem is that he says it incorrectly! Rent or borrow a videotape or DVD of the movie and write down exactly what the scarecrow says about the Pythagorean Theorem. Then see if you can rewrite the script to make it mathematically correct.

© Allen R. Angel

## Group Activity

*Discuss and solve Exercises 55 and 56 in groups.*

**55. Cost and Revenue** The break-even point for a manufacturer occurs when its cost of production, $C$, is equal to its revenue, $R$. The cost equation for a company is $C = 2x^2 - 20x + 600$ and its revenue equation is $R = x^2 + 50x - 400$, where $x$ is the number of units produced and sold. How many units must be produced and sold for the manufacturer to break even? There are two values.

**56. Cannonball** When a certain cannon is fired, the height, in feet, of the cannonball at time $t$ can be found by using the formula $h = -16t^2 + 128t$.

**a)** Determine the height of the cannonball 3 seconds after being fired.

**b)** Determine the time it takes for the cannonball to hit the ground. (*Hint:* What is the value of $h$ at impact?)

## Cumulative Review Exercises

[3.1] **57.** Express the statement "seven less than three times a number" as a mathematical expression.

[5.4] **58.** Subtract $x^2 - 4x + 6$ from $3x + 2$.

[5.5] **59.** Multiply $(3x^2 + 2x - 4)(2x - 1)$.

[5.6] **60.** Divide $\dfrac{6x^2 - 19x + 15}{3x - 5}$ by dividing the numerator by the denominator.

[6.4] **61.** Divide $\dfrac{6x^2 - 19x + 15}{3x - 5}$ by factoring the numerator and dividing out common factors.

# Chapter 6 Summary

| IMPORTANT FACTS AND CONCEPTS | EXAMPLES |
|---|---|
| **Section 6.1** | |
| To **factor an expression** means to write the expression as a product of its factors. | $x^2 + 2x - 35 = (x + 7)(x - 5)$ |
| If $a \cdot b = c$, then $a$ and $b$ are **factors** of $c$. | 6 and 4 are factors of 24. |
| The **greatest common factor (GCF)** of two or more numbers is the greatest number that divides into all the numbers. | The GCF of 36 and 48 is 12. |
| A **prime number** is an integer greater than 1 that has exactly two factors, itself and 1. A positive integer (other than 1) that is not prime is called **composite**. The number 1 is neither prime nor composite; it is called a **unit**. | 23 is a prime number. 72 is a composite number. |

| IMPORTANT FACTS AND CONCEPTS | EXAMPLES |
|---|---|

### Section 6.1 (cont.)

**To Determine the GCF of Two or More Numbers**

1. Write each number as a product of prime factors.
2. Determine the prime factors common to all the numbers.
3. Multiply the common factors found in step 2. The product is the GCF.

$$40 = 2^3 \cdot 5 \quad \text{and} \quad 140 = 2^2 \cdot 5 \cdot 7$$

The GCF of 40 and 140 is $2^2 \cdot 5 = 4 \cdot 5 = 20$.

**To Determine the Greatest Common Factor of Two or More Terms**

1. Find the GCF of the numerical coefficients of the terms.
2. Find the largest power of each variable that is common to all of the terms.
3. The GCF is product of the number from step 1 and the variable expressions from step 2.

The GCF of $6xy^2$, $4x^3y^4$, and $8x^2y^3$ is $2xy^2$.

**To Factor a Monomial from a Polynomial**

1. Determine the greatest common factor of all terms in the polynomial.
2. Write each term as the product of the GCF and its other factor.
3. Use the distributive property to factor out the GCF.

$$6a^4 + 27a^3 - 18a^2 = 3a^2(2a^2 + 9a - 6)$$

### Section 6.2

**To Factor a Four-Term Polynomial Using Grouping**

1. If all the terms have a GCF other than 1, factor it out.
2. If necessary, arrange the four terms so that the first two terms have a common factor and the last two have a common factor.
3. Use the distributive property to factor each group of two terms.
4. Factor the GCF from the results of step 3.

$$xy + 5x - 3y - 15 = x(y + 5) - 3(y + 5)$$
$$= (y + 5)(x - 3)$$

### Section 6.3

**To Factor Trinomials of the Form $ax^2 + bx + c$, where $a = 1$**

1. Find two numbers whose product equals the constant, $c$, and whose sum equals the coefficient of the $x$-term, $b$.
2. Use the two numbers found in step 1, including their signs, to write the trinomial in factored form. The trinomial in factored form will be
$$(x + \text{first number})(x + \text{second number}).$$

Factor $x^2 + 5x - 36$.

The two numbers whose product is $-36$ and whose sum is 5 are 9 and $-4$.

Therefore,
$$x^2 + 5x - 36 = (x + 9)(x - 4).$$

A **prime polynomial** is a polynomial that cannot be factored using only integer coefficients

$x^2 - 7x + 11$ is a prime polynomial.

### Section 6.4

**To Factor Trinomials of the Form $ax^2 + bx + c$, $a \neq 1$, by Trial and Error**

1. If all the terms have a GCF other than 1, factor it out.
2. Write all pairs of factors of the coefficient of the squared term, $a$.
3. Write all pairs of factors of the constant term, $c$.
4. Try various combinations of these factors until the correct middle term, $bx$, is found.

Factor $3x^2 - 2x - 8$.

| Factors of $-8$ | Possible Factors | Sum of the Products of the Outer and Inner Terms |
|---|---|---|
| $-1(8)$ | $(3x - 1)(x + 8)$ | $23x$ |
| $-2(4)$ | $(3x - 2)(x + 4)$ | $10x$ |
| $-4(2)$ | $(3x - 4)(x + 2)$ | $2x$ |
| $-8(1)$ | $(3x - 8)(x + 1)$ | $-5x$ |
| $4(-2)$ | $(3x + 4)(x - 2)$ | $-2x$ |

Therefore, $3x^2 - 2x - 8 = (3x + 4)(x - 2)$.

| IMPORTANT FACTS AND CONCEPTS | EXAMPLES |
|---|---|

**Section 6.4 (cont.)**

**To Factor Trinomials of the Form $ax^2 + bx + c$, $a \neq 1$, by Grouping**

1. If all the terms have a GCF other than 1, factor it out.

2. Find two numbers whose product is equal to the product of $a$ times $c$, and whose sum is equal to $b$.

3. Rewrite the middle term, $bx$, as the sum or difference of two terms using the numbers found in step 2.

4. Factor by grouping as explained in Section 6.2.

Factor $4x^2 + 19x - 30$.

$$ac = 4(-30) = -120$$

Two numbers whose product is $ac$, or $-120$, and whose sum is $b$, or 19, are 24 and $-5$.

$$\overset{19x}{\overbrace{\phantom{xxxxxx}}}$$
$$4x^2 + 19x - 30 = 4x^2 + 24x - 5x - 30$$
$$= 4x(x + 6) - 5(x + 6)$$
$$= (x + 6)(4x - 5)$$

**Section 6.5**

**Difference of Two Squares**

$$a^2 - b^2 = (a + b)(a - b)$$

$$y^2 - 49 = (y + 7)(y - 7)$$

**Sum of Two Cubes**

$$a^3 + b^3 = (a + b)(a^2 - ab + b^2)$$

$$8p^3 + q^3 = (2p)^3 + (q)^3$$
$$= (2p + q)[(2p)^2 - (2p)(q) + (q)^2]$$
$$= (2p + q)(4p^2 - 2pq + q^2)$$

**Difference of Two Cubes**

$$a^3 - b^3 = (a - b)(a^2 + ab + b^2)$$

$$8p^3 - q^3 = (2p)^3 - (q)^3$$
$$= (2p - q)[(2p)^2 + (2p)(q) + (q)^2]$$
$$= (2p - q)(4p^2 + 2pq + q^2)$$

**General Procedure for Factoring a Polynomial**

1. If all the terms have a GCF other than 1, factor it out.

2. If the polynomial has two terms, determine whether it is a difference of two squares or a sum or a difference of two cubes. If so, factor using the appropriate formula from Section 6.5.

3. If the polynomial has three terms, factor the trinomial using the methods discussed in Sections 6.3 and 6.4.

4. If the polynomial has more than three terms, try factoring by grouping as discussed in Section 6.2.

5. Examine your factored polynomial to determine whether the terms in any factors have a common factor. If you find a common factor, factor it out.

Factor $3x^2 - 48$.
$$3x^2 - 48 = 3(x^2 - 16)$$
$$= 3(x + 4)(x - 4)$$

**Section 6.6**

**Quadratic Equation**

Quadratic equations have the form

$$ax^2 + bx + c = 0$$

where $a$, $b$, and $c$ are real numbers, $a \neq 0$.

$6x^2 - 7x + 3 = 0$ is a quadratic equation in **standard form**.

**Zero-Factor Property**

If $ab = 0$, then $a = 0$ or $b = 0$.

If $(x + 3)(x - 1) = 0$, then $x + 3 = 0$ or $x - 1 = 0$.

| IMPORTANT FACTS AND CONCEPTS | EXAMPLES |
|---|---|

**Section 6.6 (cont.)**

**To Solve a Quadratic Equation Using Factoring**

1. Write the equation in standard form with the squared term having a positive coefficient.
2. Factor the side of the equation that is not 0.
3. Set each factor *containing a variable* equal to 0 and solve each equation.
4. Check each solution found in step 3 in the *original* equation.

Solve the equation $x^2 - 3x - 52 = -12$.

$$x^2 - 3x - 52 = -12$$
$$x^2 - 3x - 40 = 0$$
$$(x - 8)(x + 5) = 0$$
$$x - 8 = 0 \quad \text{or} \quad x + 5 = 0$$
$$x = 8 \qquad\qquad x = -5$$

**Section 6.7**

A **right triangle** is a triangle that contains a right, or 90°, angle.

Right angle

**Pythagorean Theorem**

If $a$ and $b$ represent the legs of a right triangle, and $c$ represents the hypotenuse, then

$$a^2 + b^2 = c^2$$

Hypotenuse
Leg
Right angle
Leg

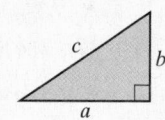
$c$
$b$
$a$

Determine the value of $x$.

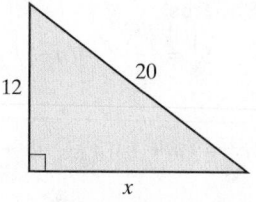
12
20
$x$

$$12^2 + x^2 = 20^2$$
$$144 + x^2 = 400$$
$$x^2 = 256$$
$$x = 16$$

## Chapter 6 Review Exercises

[6.1]   *Find the greatest common factor for each set of terms.*

**1.** $3y^5, y^4, 2y^3$

**2.** $3p, 6p^2, 9p^3$

**3.** $18c^4, 12c^2, 30c^5$

**4.** $20x^2y^3, 25x^3y^4, 10x^5y^2z$

**5.** $9xyz, 12xz, 36, x^2y$

**6.** $9st, 16s^2t, 24s, s^3t^3$

**7.** $8(x - 3), x - 3$

**8.** $x(x + 5), x + 5$

*Factor each expression. If an expression is prime, so state.*

**9.** $7x - 35$

**10.** $35x - 5$

**11.** $24y^2 - 4y$

**12.** $55p^3 - 20p^2$

**13.** $60a^2b - 36ab^2$

**14.** $9xy - 36x^3y^2$

**15.** $20x^3y^2 + 8x^9y^3 - 16x^5y^2$

**16.** $24x^2 - 13y^2 + 6xy$

**17.** $14a^2b - 7b - a^3$

**18.** $x(5x + 3) - 2(5x + 3)$

**19.** $3t(t - 1) + 4(t - 1)$

**20.** $2x(4x - 3) + 4x - 3$

[6.2]   *Factor by grouping.*

**21.** $x^2 + 6x + 2x + 12$

**22.** $x^2 - 5x + 4x - 20$

**23.** $y^2 - 6y - 6y + 36$

**24.** $3xy + 3x + 2y + 2$

**25.** $4a^2 - 4ab - a + b$

**26.** $2x^2 + 12x - x - 6$

**27.** $x^2 + 3x - 2xy - 6y$

**28.** $5x^2 - xy + 20xy - 4y^2$

**29.** $4x^2 + 12xy - 5xy - 15y^2$

**30.** $6a^2 - 10ab - 3ab + 5b^2$

**31.** $pq - 3q + 4p - 12$

**32.** $3x^2 - 9xy + 2xy - 6y^2$

**33.** $7a^2 + 14ab - ab - 2b^2$

**34.** $8x^2 - 4x + 6x - 3$

[6.3]   *Factor completely. If an expression is prime, so state.*

**35.** $x^2 + 5x + 6$

**36.** $x^2 + 4x - 15$

**37.** $x^2 + 11x + 18$

**38.** $n^2 + 3n - 40$

**39.** $b^2 + b - 20$

**40.** $x^2 - 15x + 56$

**41.** $c^2 - 10c - 20$

**42.** $y^2 - 10y - 22$

**43.** $x^3 - 17x^2 + 72x$

**44.** $t^3 - 5t^2 - 36t$

**45.** $x^2 - 2xy - 15y^2$

**46.** $4x^3 + 32x^2y + 60xy^2$

[6.4]   *Factor completely. If an expression is prime, so state.*

**47.** $2x^2 - x - 15$

**48.** $6x^2 - 29x - 5$

**49.** $4x^2 - 9x + 5$

**50.** $5m^2 - 14m + 8$

**51.** $16y^2 + 8y - 3$

**52.** $5x^2 - 32x + 12$

**53.** $2t^2 + 14t + 9$

**54.** $5x^2 + 37x - 24$

**55.** $6s^2 + 13s + 5$

**56.** $6x^2 + 11x - 10$

**57.** $12x^2 + 2x - 4$

**58.** $25x^2 - 30x + 9$

**59.** $9x^3 - 12x^2 + 4x$

**60.** $18x^3 + 12x^2 - 16x$

**61.** $4a^2 - 16ab + 15b^2$

**62.** $16a^2 - 22ab - 3b^2$

[6.5]   *Factor completely.*

**63.** $x^2 - 100$

**64.** $x^2 - 36$

**65.** $3x^2 - 48$

**66.** $81x^2 - 9y^2$

**67.** $81 - a^2$

**68.** $64 - x^2$

**69.** $16x^4 - 49y^2$

**70.** $64x^6 - 49y^6$

**71.** $a^3 + b^3$

**72.** $x^3 - y^3$

**73.** $x^3 - 1$

**74.** $x^3 + 8$

**75.** $a^3 + 27$

**76.** $b^3 - 64$

**77.** $125a^3 + b^3$

**78.** $27 - 8y^3$

**79.** $3x^3 - 192y^3$

**80.** $27x^4 - 75y^2$

[6.1–6.5]   *Factor completely.*

**81.** $x^2 - 14x + 48$

**82.** $3x^2 - 18x + 27$

**83.** $5q^2 - 5$

**84.** $8x^2 + 16x - 24$

**85.** $4y^2 - 36$

**86.** $x^2 - 6x - 27$

**87.** $9x^2 - 6x + 1$

**88.** $7x^2 + 25x - 12$

**89.** $6b^3 - 6$

**90.** $x^3y - 27y$

**91.** $a^2b - 2ab - 15b$

**92.** $6x^3 + 30x^2 + 9x^2 + 45x$

**93.** $x^2 - 4xy + 3y^2$

**94.** $3m^2 + 2mn - 8n^2$

**95.** $4x^2 + 12xy + 9y^2$

**96.** $25a^2 - 49b^2$

**97.** $xy - 7x + 2y - 14$

**98.** $16y^5 - 25y^7$

**99.** $6x^2 + 5xy - 21y^2$

**100.** $4x^3 + 18x^2y + 20xy^2$

**101.** $16x^4 - 8x^3 - 3x^2$

**102.** $d^4 - 16$

[6.6]   *Solve.*

**103.** $x(x + 9) = 0$

**104.** $(a - 2)(a + 6) = 0$

**105.** $(x + 5)(4x - 3) = 0$

**106.** $x^2 + 7x = 0$

**107.** $6x^2 + 30x = 0$

**108.** $6x^2 + 18x = 0$

**109.** $r^2 + 9r + 18 = 0$

**110.** $x^2 - 3x = -2$

**111.** $x^2 - 12 = -x$

**112.** $15x + 12 = -3x^2$

**113.** $x^2 - 6x + 8 = 0$

**114.** $3p^2 + 6p = 45$

**115.** $8x^2 - 3 = -10x$

**116.** $3p^2 - 11p = 4$

**117.** $4x^2 - 16 = 0$

**118.** $49x^2 - 100 = 0$

**119.** $8x^2 - 14x + 3 = 0$

**120.** $-48x = -12x^2 - 45$

[6.7]

**121.** State the Pythagorean Theorem.

**122.** What is the longest side of a right triangle called?

*In Exercises 123 and 124, determine the value of the question mark (?).*

**123.**

**124.**

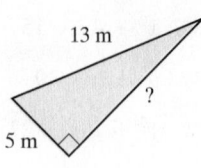

*Express each problem as an equation, then solve.*

**125. Product of Integers** The product of two consecutive positive odd integers is 99. Determine the two integers.

**126. Product of Integers** The product of two positive integers is 56. Determine the integers if the larger is 6 more than twice the smaller.

**127. Area of Rectangle** The area of a rectangle is 180 square feet. Determine the length and width of the rectangle if the length is 3 feet greater than the width.

**128. Right Triangle** One leg of a right triangle is 7 feet longer than the other leg. The hypotenuse is 9 feet longer than the shortest leg. Find the lengths of the three sides of the triangle.

**129. Square** The length of each side of a square is made smaller by 4 inches. If the area of the resulting square is 25 square inches, determine the length of a side of the original square.

**130. Table** Brian has a rectangular table. The length of the table is 2 feet greater than the width of the table. A diagonal across the table is 4 feet greater than the width of the table. Find the length of the diagonal across the table.

**131. Falling Pear** How long would it take a pear that falls off a 16-foot tree to hit the ground?

**132. Baking Cookies** The Pine Hills Neighborhood Association has determined that the cost, $C$, to make $x$ dozen cookies can be estimated by the formula $C = x^2 - 79x + 20$. If they have \$100 to be used to make the cookies, how many dozen cookies can the association make to sell at a fund-raiser?

© Jaimie Duplass/Shutterstock

## Chapter 6 Practice Test

*Chapter Test Prep Videos provide fully worked-out solutions to any of the exercises you want to review. Chapter Test Prep Videos are available via* MyMathLab®, *or on* You Tube *(search "Angel Elementary and Intermediate Algebra" and click on "Channels").*

**1.** Determine the greatest common factor of $9y^5, 15y^3$, and $27y^4$.

**2.** Determine the greatest common factor of $8p^3q^2, 32p^2q^5$, and $24p^4q^3$.

*Factor completely.*

**3.** $5x^2y^3 - 15x^5y^2$

**4.** $8a^3b - 12a^2b^2 + 28a^2b$

**5.** $4x^2 - 20x + x - 5$

**6.** $a^2 - 4ab - 5ab + 20b^2$

**7.** $r^2 + 5r - 24$

**8.** $25a^2 - 5ab - 6b^2$

**9.** $4x^2 - 16x - 48$

**10.** $2y^3 - y^2 - 3y$

**11.** $12x^2 - xy - 6y^2$

**12.** $x^2 - 9y^2$

**13.** $x^3 - 64$

*Solve.*

**14.** $(6x - 5)(x + 3) = 0$

**15.** $x^2 - 6x = 0$

**16.** $x^2 = 64$

**17.** $x^2 + 18x + 81 = 0$

**18.** $x^2 - 7x + 12 = 0$

**19.** $x^2 + 6 = -5x$

**20. Right Triangle** Find the length of the side indicated with a question mark.

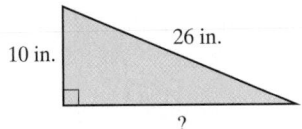

10 in.    26 in.

?

**21. Right Triangle** In a right triangle, one leg is 2 feet less than twice the length of the smaller leg. The hypotenuse is 2 feet more than twice the length of the smaller leg. Determine the hypotenuse of the triangle.

**22. Product of Integers** The product of two positive integers is 36. Determine the two integers if the larger is 1 more than twice the smaller.

**23. Consecutive Even Integers** The product of two positive consecutive even integers is 168. Determine the integers.

**24. Rectangle** The area of a rectangle is 24 square meters. Determine the length and width of the rectangle if its length is 2 meters greater than its width.

**25. Fallen Object** How long would it take for an object dropped from a hot air balloon to fall 1600 feet to the ground?

© Thomas Barrat/Shutterstock

## Cumulative Review Test

*Take the following test and check your answers with those given in the back of the book. Review any questions that you answered incorrectly. The section where the material was covered is indicated after the answer.*

**1.** Evaluate $4 - 5(2x + 4x^2 - 21)$ when $x = -4$.

**2.** Evaluate $5x^2 - 3y + 7(2 + y^2 - 4x)$ when $x = 3$ and $y = -2$.

**3. Motel Room** The cost of a motel room including a 12% state tax and 3% county tax is $103.50. Determine the cost of the room before tax.

**4.** Consider the set of numbers

$$\left\{ -6, -0.2, \frac{3}{5}, \sqrt{7}, -\sqrt{2}, 7, 0, -\frac{5}{9}, 1.34 \right\}$$

List the elements that are

**a)** natural numbers.

**b)** rational numbers.

**c)** irrational numbers.

**d)** real numbers.

**5.** Which is greater, $|-8|$ or $-|8|$? Explain your answer.

**6.** Solve the equation $4x - 2 = 4(x - 7) + 2x$ for $x$.

**7.** Solve the proportion $\dfrac{5}{12} = \dfrac{8}{x}$ for $x$ by cross-multiplying.

**8. Painting** A gallon of paint covers 825 square feet. How much paint is needed to cover a house with a surface area of 5775 square feet?

**9.** Solve the equation $4x + 3y = 7$ for $y$. Write your answer in $y = mx + b$ form.

**10. Acid Solution** How many liters of a 10% acid solution must be mixed with three liters of a 4% acid solution to get an 8% acid solution?

**11. Cross-Country Skiing** Two cross-country skiers follow the same trail in a local park. Brooke Stoner skis at a rate of 8 kilometers per hour and Bob Thoresen skis at a rate of 4 kilometers per hour. How long will it take Brooke to catch Bob if she leaves 15 minutes after he does?

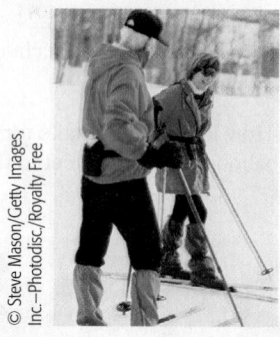

© Steve Mason/Getty Images, Inc.–Photodisc/Royalty Free

**12.** Graph $y = -\dfrac{3}{5}x + 1$ using the slope and *y*-intercept.

**13.** Find an equation of a line through the points $(3, 7)$ and $(-2, 4)$. Write the equation in slope-intercept form.

**14.** Simplify $(2x^{-3})^{-2}(4x^{-3}y^2)^3$.

**15.** Subtract $(4x^3 - 3x^2 + 7)$ from $(x^3 - x^2 + 6x - 5)$.

*Perform the operations indicated.*

**16.** $(3x - 2)(x^2 + 5x - 6)$

**17.** $\dfrac{x^2 - 2x + 6}{x + 3}$

**18.** Factor $qr + 2q - 8r - 16$ by grouping.

**19.** Factor $5x^2 - 7x - 6$.

**20.** Factor $7y^3 - 63y$.

# 7 Rational Expressions and Equations

## Goals of This Chapter

You worked with rational numbers when you worked with fractions in arithmetic. Fractions that contain variables are often referred to as rational expressions. The same basic procedures that you used with arithmetic fractions will be used with rational expressions. You might wish to review Section 1.3 since the material presented in this chapter builds upon the procedures presented there.

Equations that contain rational expressions are called rational equations. We will solve rational equations in Section 7.6.

*To be successful in this chapter, you need to have a complete understanding of factoring, which was presented in Chapter 6. Sections 6.3 and 6.4 are especially important.*

© Konstantin Shevtsov/Shutterstock

**When two people paint** a house, the job gets done faster than if each person painted the house separately. In Exercise 31 on page 430, we will see how algebra can be used to determine how much faster the job gets done.

379

# 7.1 Simplifying Rational Expressions

1. Determine the values for which a rational expression is defined.
2. Understand the three signs of a fraction.
3. Simplify rational expressions.
4. Factor a negative 1 from a polynomial.

## 1 Determine the Values for Which a Rational Expression Is Defined

We begin this chapter by defining a *rational expression*.

**Rational Expression**

A **rational expression** is an expression of the form $\frac{p}{q}$, where $p$ and $q$ are polynomials and $q \neq 0$.

### Examples of Rational Expressions

$$\frac{4}{5}, \quad \frac{x+3}{x}, \quad \frac{x^2+8x}{x-3}, \quad \frac{x}{x^2-4}$$

The denominator of a rational expression cannot equal 0 since division by 0 is not defined.

| Expression | Defined | Undefined |
|---|---|---|
| $\frac{x+3}{x}$ | for all real numbers except $0$ | when $x=0$ |
| $\frac{x^2+8x}{x-3}$ | for all real numbers except $3$ | when $x=3$ |
| $\frac{x}{x^2-4}$ | for all real numbers except $2$ and $-2$ | when $x=2$ or $x=-2$ |

*Whenever a rational expression has a variable in the denominator, we always assume that the value or values of the variable that make the denominator 0 are excluded.*

One method that can be used to determine the value or values of the variable that are excluded is to set the denominator equal to 0 and then solve the resulting equation for the variable.

EXAMPLE 1  Determine the value or values of the variable for which the rational expression is defined.

**a)** $\dfrac{7}{x-4}$    **b)** $\dfrac{3x+4}{2x-7}$    **c)** $\dfrac{x+5}{x^2+6x-7}$

*Solution*

**a)** To determine the conditions for which the denominator is zero, we set $x-4$ equal to 0.

$$x - 4 = 0$$
$$x = 4$$

We say $\dfrac{7}{x-4}$ is defined for all real numbers except 4, or simply $x \neq 4$.

**b)** In $\dfrac{3x+4}{2x-7}$, we set the denominator $2x-7$ equal to 0 to find values we need to exclude.

$$2x - 7 = 0$$
$$2x = 7$$
$$x = \frac{7}{2}$$

Thus when we consider the rational expression $\dfrac{3x+4}{2x-7}$, we say $x$ cannot equal $\dfrac{7}{2}$, or $x \neq \dfrac{7}{2}$.

**c)** Again, we set the denominator of $\dfrac{x+5}{x^2+6x-7}$ equal to zero.

$$x^2 + 6x - 7 = 0$$
$$(x+7)(x-1) = 0$$
$$x + 7 = 0 \quad \text{or} \quad x - 1 = 0$$
$$x = -7 \qquad\qquad x = 1$$

Thus, for $\dfrac{x+5}{x^2+6x-7}$, we say $x \neq -7$ and $x \neq 1$.

Now Try Exercise 15

## Understanding Algebra

Since changing any two of a fraction's three signs does not change the fraction, we can write the following:

$$\frac{3}{-4} = \frac{-3}{4} = -\frac{3}{4}$$

$$\frac{-2x}{-3y^2} = -\frac{-2x}{3y^2} = \frac{2x}{3y^2}$$

$$\frac{-4a}{-5b} = \frac{-4a}{5b} = -\frac{4a}{5b}$$

## 2 Understand the Three Signs of a Fraction

Three **signs** are associated with any fraction: the sign of the numerator, the sign of the denominator, and the sign of the fraction.

Sign of fraction $\longrightarrow +\dfrac{-a}{+b}$ — Sign of numerator — Sign of denominator

Whenever any of the three signs is omitted, we assume it to be positive. For example,

$$\frac{a}{b} \quad \text{means} \quad +\frac{+a}{+b}$$

$$\frac{-a}{b} \quad \text{means} \quad +\frac{-a}{+b}$$

$$-\frac{a}{b} \quad \text{means} \quad -\frac{+a}{+b}$$

### Negative Fractions

Changing any two of the three signs of a fraction does not change the value of a fraction.

$$\frac{-a}{b} = -\frac{a}{b} = \frac{a}{-b}$$

Generally, we do not write a fraction with a negative denominator. For example, the expression $\dfrac{2}{-5}$ would be written as either $\dfrac{-2}{5}$ or $-\dfrac{2}{5}$. The expression $\dfrac{x}{-(4-x)}$ can be written $\dfrac{x}{x-4}$ since $-(4-x) = -4 + x$ or $x - 4$.

## 3 Simplify Rational Expressions

A rational expression is **simplified** or **reduced to its lowest terms** when the numerator and denominator have no common factors other than 1. The fraction $\dfrac{9}{12}$ is not simplified because 9 and 12 both contain the common factor 3. When the 3 is factored out, the simplified fraction is $\dfrac{3}{4}$.

$$\frac{9}{12} = \frac{\overset{1}{\cancel{3}} \cdot 3}{\underset{1}{\cancel{3}} \cdot 4} = \frac{3}{4}$$

Not simplified $\longrightarrow$ $\longleftarrow$ Simplified

The rational expression $\dfrac{ab - b^2}{2b}$ is not simplified because both the numerator and denominator have a common factor, $b$. To simplify this expression, factor $b$ from each term in the numerator, then divide it out.

$$\frac{ab - b^2}{2b} = \frac{\overset{1}{b}(a - b)}{2\underset{1}{b}} = \frac{a - b}{2}$$

Thus, $\dfrac{ab - b^2}{2b}$ becomes $\dfrac{a - b}{2}$ when simplified.

## Understanding Algebra

We say that the simplified form of
$$\frac{ab - b^2}{2b} \text{ is } \frac{a - b}{2}.$$
We should keep in mind that in both original and simplified expressions, $b$ cannot equal 0 because that would make the original denominator 0.

### To Simplify Rational Expressions

1. Factor both the numerator and denominator as completely as possible.
2. Divide out any factors common to both the numerator and denominator.

**EXAMPLE 2**  Simplify $\dfrac{5x^3 + 10x^2 - 35x}{10x^2}$.

*Solution*
$$\frac{5x^3 + 10x^2 - 35x}{10x^2} \qquad \leftarrow \text{Observe that the GCF is } 5x.$$

$$= \frac{\overset{1}{5x}(x^2 + 2x - 7)}{5\underset{1}{x} \cdot 2x} \qquad \leftarrow \begin{array}{l}\text{Factor } 5x \text{ from the numerator and the} \\ \text{denominator and divide it out.}\end{array}$$

$$= \frac{x^2 + 2x - 7}{2x}$$

Now Try Exercise 33

### Helpful Hint

In Example 2 we *simplified* a polynomial divided by a monomial using factoring. In Section 5.6 we *divided* polynomials by monomials by writing each term in the numerator over the expression in the denominator. For example,

$$\frac{5x^3 + 10x^2 - 35x}{10x^2} = \frac{5x^3}{10x^2} + \frac{10x^2}{10x^2} - \frac{35x}{10x^2}$$

$$= \frac{x}{2} + 1 - \frac{7}{2x}$$

The answer above, $\dfrac{x}{2} + 1 - \dfrac{7}{2x}$, is equivalent to the answer $\dfrac{x^2 + 2x - 7}{2x}$, which was obtained by factoring in Example 2. We show this below.

$$\frac{x}{2} + 1 - \frac{7}{2x}$$

$$= \frac{x}{x} \cdot \frac{x}{2} + \frac{2x}{2x} \cdot 1 - \frac{7}{2x} \qquad \text{Write each term with LCD } 2x.$$

$$= \frac{x^2}{2x} + \frac{2x}{2x} - \frac{7}{2x}$$

$$= \frac{x^2 + 2x - 7}{2x}$$

When asked to *simplify* an expression we will factor numerators and denominators, when possible, then divide out common factors. This process was illustrated in Example 2 and will be further illustrated in Examples 3 through 5.

**EXAMPLE 3** Simplify $\dfrac{x^2 - x - 12}{x + 3}$.

**Solution**   Factor the numerator; then divide out the common factor.

$$\frac{x^2 - x - 12}{x + 3} = \frac{\overset{1}{\cancel{(x + 3)}}(x - 4)}{\underset{1}{\cancel{x + 3}}} = x - 4$$

Now Try Exercise 35

**EXAMPLE 4** Simplify $\dfrac{r^2 - 25}{r - 5}$.

**Solution**   Factor the numerator; then divide out common factors.

$$\frac{r^2 - 25}{r - 5} = \frac{(r + 5)\overset{1}{\cancel{(r - 5)}}}{\underset{1}{\cancel{r - 5}}} = r + 5$$

Now Try Exercise 61

**EXAMPLE 5** Simplify $\dfrac{3x^2 - 10x - 8}{x^2 + 3x - 28}$.

**Solution**   Factor both the numerator and denominator, then divide out common factors.

$$\frac{3x^2 - 10x - 8}{x^2 + 3x - 28} = \frac{(3x + 2)\overset{1}{\cancel{(x - 4)}}}{(x + 7)\underset{1}{\cancel{(x - 4)}}} = \frac{3x + 2}{x + 7}.$$

Note that $\dfrac{3x + 2}{x + 7}$ cannot be simplified any further.

Now Try Exercise 41

---

**Avoiding Common Errors**

*Remember: Only common factors can be divided out from expressions.*

CORRECT

$$\frac{\overset{5}{\cancel{20}}\,\overset{x}{\cancel{x^2}}}{\underset{1}{\cancel{4}}\,\underset{1}{\cancel{x}}} = 5x$$

INCORRECT

In the denominator of the example on the left, $4x$, the 4 and $x$ are factors since they are *multiplied* together. The 4 and the $x$ are also both factors of the numerator $20x^2$, since $20x^2$ can be written $4 \cdot x \cdot 5x$.

Some students incorrectly divide out *terms*. In the expression $\dfrac{x^2 - 20}{x - 4}$, the $x$ and $-4$ are *terms* of the denominator, not factors, and therefore cannot be divided out.

---

## 4 Factor a Negative 1 from a Polynomial

*Recall from Section 6.2 that when −1 is factored from a polynomial, the sign of each term in the polynomial changes.*

Examples

$$-3x + 10 = -1(3x - 10) = -(3x - 10)$$
$$5 - 2x = -1(-5 + 2x) = -(2x - 5)$$
$$-2x^2 + 3x - 4 = -1(2x^2 - 3x + 4) = -(2x^2 - 3x + 4)$$

Whenever the terms in a numerator and denominator differ only in their signs, we can factor out −1 from either the numerator or denominator and then divide out the common factor.

## Understanding Algebra

An expression divided by its *additive inverse* (or *opposite*) is −1:

$$\frac{a - b}{b - a} = -1 \quad \text{(provided } a \neq b\text{)}$$

**EXAMPLE 6**   Simplify $\dfrac{3x - 8}{8 - 3x}$.

*Solution*   Since each term in the numerator differs only in sign from its like term in the denominator, we will factor −1 from each term in the denominator.

$$\frac{3x - 8}{8 - 3x} = \frac{3x - 8}{-1(-8 + 3x)}$$

$$= \frac{\overset{1}{\cancel{3x - 8}}}{-\underset{1}{\cancel{(3x - 8)}}}$$

$$= -1$$

Now Try Exercise 43

**EXAMPLE 7**   Simplify $\dfrac{4n^2 - 23n - 6}{6 - n}$.

## Understanding Algebra

In Example 7 we simplified $\dfrac{4n^2 - 23n - 6}{6 - n}$ to $-(4n + 1)$.

Notice $n \neq 6$ in the original expression. Therefore, it is assumed that $n \neq 6$ in the simplified expression, $-(4n + 1)$, as well.

*Solution*   $\dfrac{4n^2 - 23n - 6}{6 - n} = \dfrac{(4n + 1)(n - 6)}{6 - n}$   The terms in $n - 6$ differ only in sign from the terms in $6 - n$.

$$= (4n + 1)(-1) \quad \text{Replaced } \frac{n - 6}{6 - n} \text{ with } -1.$$

$$= -(4n + 1)$$

Note that $-4n - 1$ is also an acceptable answer.

Now Try Exercise 49

# EXERCISE SET 7.1

MathXL®   MyMathLab®

MathXL®       MyMathLab

## Warm-Up Exercises

*Fill in the blanks with the appropriate word, phrase, or symbol(s) from the following list.*

| −4 | reduced to lowest terms | $\dfrac{4}{x}$ | $4x - 1$ | $x - 4$ |
| 1 | rational expression | 4 | $\dfrac{5 - 2x^2}{2x^2}$ | $\dfrac{-4}{x}$ |
| −1 | | | | |

**1.** An expression of the form $\dfrac{p}{q}$ where $p$ and $q$ are polynomials ($q \neq 0$) is called a(n) _____.

**2.** The expression $\dfrac{x - 1}{4 - x}$ is defined for all values of $x$ except _____.

**3.** When a fraction's numerator and denominator have no common factor (other than 1), we say the fraction is _____.

**4.** When $\dfrac{25 - 10x^2}{10x^2}$ is simplified, the result is _____.

**5.** When $\dfrac{x^2 - x - 12}{x + 3}$ is simplified, the result is _____.

**6.** If $x \neq 8$, then $-\dfrac{8 - x}{x - 8}$ simplifies to _____.

**7.** If $x \neq 2$, then $\dfrac{x - 2}{2 - x}$ simplifies to _____.

**8.** The expression $\dfrac{4}{-x}$ can be written as _____.

**9.** When $\dfrac{4x^2 - 13x - 3}{x - 3}$ is simplified, the result is _____.

**10.** The expression $\dfrac{2x}{x + 4}$ is defined for all values of $x$ except _____.

## Practice the Skills

*Determine the value or values of the variable for which each expression is defined.*

**11.** $\dfrac{5}{x-1}$

**12.** $\dfrac{x+1}{4-x}$

**13.** $\dfrac{x-2}{x}$

**14.** $\dfrac{8}{r+3}$

**15.** $\dfrac{7}{4n-16}$

**16.** $\dfrac{7}{2x-3}$

**17.** $\dfrac{x+4}{x^2-4}$

**18.** $\dfrac{9}{x^2+4x-5}$

**19.** $\dfrac{2x-3}{2x^2-9x+9}$

**20.** $\dfrac{x^2+2}{2x^2-13x+15}$

**21.** $\dfrac{x}{x^2+36}$

**22.** $\dfrac{4}{9+x^2}$

**23.** $\dfrac{p+8}{4p^2-25}$

**24.** $\dfrac{10}{9r^2-16}$

*Simplify the following rational expressions that contain a monomial divided by a monomial. This material was covered in Sections 5.1 and 5.2, and it will help prepare you for the next section.*

**25.** $\dfrac{8x^3y}{24x^2y^5}$

**26.** $\dfrac{18x^3y^2}{30x^4y^5}$

**27.** $\dfrac{(2a^4b^5)^3}{2a^{12}b^{20}}$

**28.** $\dfrac{(5r^2s^3)^2}{(3r^4s)^3}$

*Simplify.*

**29.** $\dfrac{5x}{x+xy}$

**30.** $\dfrac{12t}{3t+9}$

**31.** $\dfrac{5x+15}{x+3}$

**32.** $\dfrac{3x^2+6x}{3x^2+9x}$

**33.** $\dfrac{x^3+6x^2+7x}{2x}$

**34.** $\dfrac{x^2y^2-2xy+5y}{y}$

**35.** $\dfrac{r^2-r-2}{r-2}$

**36.** $\dfrac{b-6}{b^2-8b+12}$

**37.** $\dfrac{x^2+2x}{x^2+4x+4}$

**38.** $\dfrac{x^2+3x-18}{4x-12}$

**39.** $\dfrac{z^2-10z+25}{z^2-25}$

**40.** $\dfrac{k^2-6k+9}{k^2-9}$

**41.** $\dfrac{x^2-2x-3}{x^2-x-6}$

**42.** $\dfrac{4x^2-12x-40}{2x^2-16x+30}$

**43.** $\dfrac{4x-8}{8-4x}$

**44.** $\dfrac{10a-6}{3-5a}$

**45.** $\dfrac{x^2-2x-8}{4-x}$

**46.** $\dfrac{7-s}{s^2-12s+35}$

**47.** $\dfrac{x^2+3x-18}{-2x^2+6x}$

**48.** $\dfrac{3p^2-13p-10}{3p+2}$

**49.** $\dfrac{2x^2+5x-3}{1-2x}$

**50.** $\dfrac{x^2-25}{x^2-2x-15}$

**51.** $\dfrac{m-2}{4m^2-13m+10}$

**52.** $\dfrac{2x^2-13x+21}{(x-3)^2}$

**53.** $\dfrac{x^2-25}{(x+5)^2}$

**54.** $\dfrac{16x^2+24x+9}{4x+3}$

**55.** $\dfrac{6x^2-13x+6}{3x-2}$

**56.** $\dfrac{6t^2-7t-5}{3t-5}$

**57.** $\dfrac{x^2-3x+4x-12}{x+4}$

**58.** $\dfrac{x^2-2x+4x-8}{2x^2+3x+8x+12}$

**59.** $\dfrac{2x^2-8x+3x-12}{2x^2+8x+3x+12}$

**60.** $\dfrac{x^3-125}{x^2-25}$

**61.** $\dfrac{a^3-8}{a-2}$

**62.** $\dfrac{y^3+1}{y+1}$

**63.** $\dfrac{9s^2-16t^2}{3s-4t}$

**64.** $\dfrac{a+6b}{a^2-36b^2}$

**65.** $\dfrac{6x+9y}{2x^2+xy-3y^2}$

**66.** $\dfrac{3k^2+6kr-9r^2}{k^2+5kr+6r^2}$

## Problem Solving

*Simplify the following expressions, if possible. Treat the unknown symbol as if it were a variable.*

**67.** $\dfrac{3☺}{15}$

**68.** $\dfrac{☺}{☺+8☺^2}$

**69.** $\dfrac{7\Delta}{14\Delta+63}$

**70.** $\dfrac{\Delta^2 + 2\Delta}{\Delta^2 + 4\Delta + 4}$

**71.** $\dfrac{3\Delta - 4}{4 - 3\Delta}$

**72.** $\dfrac{(\Delta - 3)^2}{\Delta^2 - 6\Delta + 9}$

*Determine the denominator that will make each statement true. Explain how you obtained your answer.*

**73.** $\dfrac{x^2 - x - 6}{\blacksquare} = x - 3$

**74.** $\dfrac{2x^2 + 11x + 12}{\blacksquare} = 2x + 3$

*Determine the numerator that will make the statement true. Explain how you obtained your answer.*

**75.** $\dfrac{\blacksquare}{x + 4} = x + 5$

## Concept/Writing Exercises

**76.** Explain how to simplify a rational expression.

**77.** Explain how to determine the value or values of the variable that makes a rational expression undefined.

## Challenge Problems

*In Exercises 78 and 79,* **a)** *determine the value or values that x cannot represent.* **b)** *Simplify the expression.*

**78.** $\dfrac{x - 4}{2x^2 - 5x - 8x + 20}$

**79.** $\dfrac{x + 5}{2x^3 + 7x^2 - 15x}$

*Simplify. Explain how you determined your answer.*

**80.** $\dfrac{\frac{1}{6}x^5 - \frac{2}{3}x^4}{x^4}$

**81.** $\dfrac{\frac{1}{5}x^5 - \frac{2}{3}x^4}{\frac{1}{5}x^5 - \frac{2}{3}x^4}$

**82.** $\dfrac{\frac{1}{5}x^5 - \frac{2}{3}x^4}{\frac{2}{3}x^4 - \frac{1}{5}x^5}$

## Group Activity

*Discuss and answer Exercise 83 as a group.*

**83. a)** As a group, determine the values of the variable where the expression $\dfrac{x^2 - 25}{x^3 + 2x^2 - 15x}$ is undefined.

**b)** As a group, simplify the rational expression.

**c)** Group member 1: Substitute 6 in the *original expression* and evaluate.

**d)** Group member 2: Substitute 6 in the *simplified expression* from part **b)** and compare your result to that of Group member 1.

**e)** Group member 3: Substitute −2 in the original expression and in the simplified expression in part **b)**. Compare your answers.

**f)** As a group, discuss the results of your work in parts **c)–e)**.

**g)** Now, as a group, substitute −5 in the original expression and in the simplified expression. Discuss your results.

**h)** Is $\dfrac{x^2 - 25}{x^3 + 2x^2 - 15x}$ always equal to its simplified form for *any* value of $x$? Explain your answer.

## Cumulative Review Exercises

[2.6]  **84.** Solve the formula $z = \dfrac{x - y}{4}$ for $y$.

[3.3]  **85. Triangle** Find the measures of the three angles of a triangle if one angle is 30° greater than the smallest angle, and the third angle is 10° greater than 3 times the smallest angle.

[5.1]  **86.** Simplify $\left(\dfrac{5x^2y^2}{9x^4y^3}\right)^2$.

[5.4]  **87.** Subtract $3x^2 - 4x - 8 - (-5x^2 + 6x + 11)$.

[6.3]  **88.** Factor $3a^2 - 6a - 72$ completely.

[6.7]  **89.** Find the length of the hypotenuse of the right triangle.

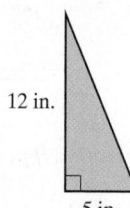

12 in.

5 in.

# 7.2 Multiplication and Division of Rational Expressions

1 Multiply rational expressions.

2 Divide rational expressions.

## Understanding Algebra

Multiplying fractions in algebra is just like multiplying fractions in arithmetic.

Simplify by dividing out common factors first, then multiply numerators and multiply denominators.

**Arithmetic:**

$$\frac{2}{3} \times \frac{6}{7} = \frac{2 \times \overset{2}{\cancel{6}}}{\underset{1}{\cancel{3}} \times 7} = \frac{4}{7}$$

**Algebra:**

$$\frac{x}{3} \cdot \frac{6}{7} = \frac{x \cdot \overset{2}{\cancel{6}}}{\underset{1}{\cancel{3}} \cdot 7} = \frac{2x}{7}$$

## 1 Multiply Rational Expressions

In Section 1.3 we reviewed multiplication of numerical fractions. Recall that to multiply two fractions we multiply their numerators together and multiply their denominators together.

### To Multiply Two Fractions

$$\frac{a}{b} \cdot \frac{c}{d} = \frac{a \cdot c}{b \cdot d}, \quad b \neq 0 \quad \text{and} \quad d \neq 0$$

**EXAMPLE 1** Multiply $\left(\dfrac{3}{5}\right)\left(\dfrac{-2}{9}\right)$.

*Solution*   First divide out common factors; then multiply.

$$\frac{\overset{1}{\cancel{3}}}{5} \cdot \frac{-2}{\underset{3}{\cancel{9}}} = \frac{1 \cdot (-2)}{5 \cdot 3} = -\frac{2}{15}$$

Now Try Exercise 7

The same principles apply when multiplying rational expressions containing variables. Before multiplying, you should first divide out any factors common to both a numerator and a denominator.

### To Multiply Rational Expressions

1. Factor all numerators and denominators completely.
2. Divide out common factors.
3. Multiply numerators together and multiply denominators together.

**EXAMPLE 2** Multiply $\dfrac{3x^2}{2y} \cdot \dfrac{4y^3}{3x}$.

*Solution*   This problem can be represented as

$$\frac{3x^2}{2y} \cdot \frac{4y^3}{3x} = \frac{3xx}{2y} \cdot \frac{4yyy}{3x}$$

$$= \frac{\overset{1}{\cancel{3}}\overset{1}{\cancel{x}}x}{2y} \cdot \frac{4yyy}{\underset{1}{\cancel{3}}\underset{1}{\cancel{x}}} \qquad \text{Divide out the 3's and } x\text{'s.}$$

$$= \frac{\overset{1}{\cancel{3}}\overset{1}{\cancel{x}}x}{\underset{1}{\cancel{2}}\underset{1}{\cancel{y}}} \cdot \frac{\overset{2}{\cancel{4}}\overset{1}{\cancel{y}}yy}{\underset{1}{\cancel{3}}\underset{1}{\cancel{x}}} \qquad \begin{array}{l}\text{Divide both the 4 and the 2}\\\text{by 2, and divide out the } y\text{'s.}\end{array}$$

Now we multiply the remaining numerators together and the remaining denominators together.

$$= \frac{2xy^2}{1} \quad \text{or} \quad 2xy^2$$

Now Try Exercise 19

Rather than illustrating this entire process when multiplying rational expressions, we will often proceed as follows:

$$\frac{3x^2}{2y} \cdot \frac{4y^3}{3x}$$

$$= \frac{\overset{1}{\cancel{3}}\overset{x}{\cancel{x^2}}}{\underset{1}{\cancel{2}}\underset{1}{\cancel{y}}} \cdot \frac{\overset{2}{\cancel{4}}\overset{y^2}{\cancel{y^3}}}{\underset{1}{\cancel{3}}\underset{1}{\cancel{x}}} = 2xy^2$$

**EXAMPLE 3** Multiply $-\dfrac{5y^2}{2x^3}\cdot\dfrac{3x^2}{7y^2}$.

*Solution*

$$-\dfrac{5\overset{1}{y^2}}{2x^3}\cdot\dfrac{3\overset{1}{x^2}}{7y^2}=-\dfrac{15}{14x}$$

Now Try Exercise 53

## Understanding Algebra

When we divide out common factors, it is not necessary to write the 1's as was done in Examples 2 and 3.

From this point on, instead of showing

$$-\dfrac{5y^{2\,1}}{2x_x^3}\cdot\dfrac{3x^{2\,1}}{7y^2_1}=-\dfrac{15}{14x}$$

we will generally show

$$-\dfrac{5y^2}{2x_x^3}\cdot\dfrac{3x^2}{7y^2}=-\dfrac{15}{14x}.$$

**EXAMPLE 4** Multiply $(x-6)\cdot\dfrac{4}{x^3-6x^2}$.

*Solution*

$$(x-6)\cdot\dfrac{4}{x^3-6x^2}=\dfrac{x-6}{1}\cdot\dfrac{4}{x^2(x-6)}=\dfrac{4}{x^2}$$

Now Try Exercise 27

**EXAMPLE 5** Multiply $\dfrac{(x+2)^2}{6x^2}\cdot\dfrac{3x}{x^2-4}$.

*Solution*

$$\dfrac{(x+2)^2}{6x^2}\cdot\dfrac{3x}{x^2-4}=\dfrac{(x+2)(x+2)}{6x^2}\cdot\dfrac{3x}{(x+2)(x-2)}$$

$$=\dfrac{(x+2)(x+2)}{\underset{2}{6}\underset{x}{x^2}}\cdot\dfrac{3x}{(x+2)(x-2)}=\dfrac{x+2}{2x(x-2)}$$

Now Try Exercise 61

In the answer to Example 5 we could have multiplied the factors in the denominator to get $\dfrac{x+2}{2x^2-4x}$. This is also a correct answer. In this section we will leave rational answers with the numerator as a polynomial (in unfactored form) and the denominators in factored form, as was given in Example 5.

**EXAMPLE 6** Multiply $\dfrac{b-4}{5b}\cdot\dfrac{10b}{4-b}$.

*Solution*

$$\dfrac{b-4}{5b}\cdot\dfrac{\overset{2}{10b}}{4-b}=\dfrac{2(b-4)}{4-b}.$$

Next, we rewrite $4-b$ as $-1(b-4)$. Thus,

$$\dfrac{2(b-4)}{4-b}=\dfrac{2(b-4)}{-1(b-4)}=-2$$

Now Try Exercise 23

### Helpful Hint

When only the signs differ in a numerator and denominator in a multiplication problem, factor out $-1$ *from either the numerator or denominator*; then divide out the common factor.

$$\dfrac{a-b}{x}\cdot\dfrac{y}{b-a}=\dfrac{a-b}{x}\cdot\dfrac{y}{-1(a-b)}=-\dfrac{y}{x}$$

**EXAMPLE 7** Multiply $\dfrac{3x + 2}{2x - 1} \cdot \dfrac{4 - 8x}{3x + 2}$.

**Solution**

$$\dfrac{3x + 2}{2x - 1} \cdot \dfrac{4 - 8x}{3x + 2} = \dfrac{3x + 2}{2x - 1} \cdot \dfrac{4(1 - 2x)}{3x + 2} \qquad \text{Factor.}$$

$$= \dfrac{\cancel{3x + 2}}{2x - 1} \cdot \dfrac{4(1 - 2x)}{\cancel{3x + 2}}. \qquad \text{Divide out common factors.}$$

Note that the factor $(1 - 2x)$ in the numerator of the second fraction differs only in sign from $2x - 1$, the denominator of the first fraction. We will therefore factor $-1$ from each term of the $(1 - 2x)$ in the numerator of the second fraction.

$$= \dfrac{\cancel{3x + 2}}{2x - 1} \cdot \dfrac{4(-1)(2x - 1)}{\cancel{3x + 2}} \qquad \text{Factored } -1 \text{ from the second numerator.}$$

$$= \dfrac{\cancel{3x + 2}}{\cancel{2x - 1}} \cdot \dfrac{-4\cancel{(2x - 1)}}{\cancel{3x + 2}} \qquad \text{Divide out common factors.}$$

$$= \dfrac{-4}{1} = -4$$

Now Try Exercise 25

**EXAMPLE 8** Multiply $\dfrac{2x^2 + 5x - 12}{6x^2 - 11x + 3} \cdot \dfrac{3x^2 + 2x - 1}{x^2 + 5x + 4}$.

**Solution** Factor all numerators and denominators, and then divide out common factors.

$$\dfrac{2x^2 + 5x - 12}{6x^2 - 11x + 3} \cdot \dfrac{3x^2 + 2x - 1}{x^2 + 5x + 4} = \dfrac{(2x - 3)(x + 4)}{(2x - 3)(3x - 1)} \cdot \dfrac{(3x - 1)(x + 1)}{(x + 1)(x + 4)}$$

$$= \dfrac{\cancel{(2x - 3)}\,\cancel{(x + 4)}}{\cancel{(2x - 3)}\,\cancel{(3x - 1)}} \cdot \dfrac{\cancel{(3x - 1)}\,\cancel{(x + 1)}}{\cancel{(x + 1)}\,\cancel{(x + 4)}} = 1$$

Now Try Exercise 67

**EXAMPLE 9** Multiply $\dfrac{2x^3 - 18x^2 + 16x}{6y^2} \cdot \dfrac{-2y}{3x^2 - 3x}$.

**Solution**

$$\dfrac{2x^3 - 18x^2 + 16x}{6y^2} \cdot \dfrac{-2y}{3x^2 - 3x} = \dfrac{2x(x^2 - 9x + 8)}{6y^2} \cdot \dfrac{-2y}{3x(x - 1)}$$

$$= \dfrac{2x(x - 8)(x - 1)}{6y^2} \cdot \dfrac{-2y}{3x(x - 1)}$$

$$= \dfrac{\cancel{2}x(x - 8)\cancel{(x - 1)}}{\underset{3\ y}{\cancel{6}\,\cancel{y^2}}} \cdot \dfrac{-2\cancel{y}}{3\cancel{x}\cancel{(x - 1)}}$$

$$= \dfrac{-2(x - 8)}{9y} = \dfrac{-2x + 16}{9y}$$

Now Try Exercise 29

**EXAMPLE 10** Multiply $\dfrac{x^2 - y^2}{x + y} \cdot \dfrac{x + 2y}{2x^2 - 3xy + y^2}$.

**Solution**

$$\dfrac{x^2 - y^2}{x + y} \cdot \dfrac{x + 2y}{2x^2 - 3xy + y^2} = \dfrac{(x + y)(x - y)}{x + y} \cdot \dfrac{x + 2y}{(2x - y)(x - y)}$$

$$= \dfrac{\cancel{(x + y)}\,\cancel{(x - y)}}{\cancel{x + y}} \cdot \dfrac{x + 2y}{(2x - y)\cancel{(x - y)}}$$

$$= \dfrac{x + 2y}{2x - y}$$

Now Try Exercise 33

## 2 Divide Rational Expressions

Recall that to divide one fraction by a second fraction, we multiply the first fraction by the reciprocal of the second fraction (or by the reciprocal of the divisor).

**To Divide Two Fractions**

$$\frac{a}{b} \div \frac{c}{d} = \frac{a}{b} \cdot \frac{d}{c} = \frac{ad}{bc}, \quad b \neq 0, \quad d \neq 0, \quad \text{and} \quad c \neq 0$$

**EXAMPLE 11** Divide.

**a)** $\dfrac{2}{7} \div \dfrac{9}{7}$    **b)** $\dfrac{3}{4} \div \dfrac{10}{6}$

*Solution*

**a)** $\dfrac{2}{7} \div \dfrac{9}{7} = \dfrac{2}{\overset{}{\underset{1}{7}}} \cdot \dfrac{\overset{1}{7}}{9} = \dfrac{2 \cdot 1}{1 \cdot 9} = \dfrac{2}{9}$

**b)** $\dfrac{3}{4} \div \dfrac{10}{6} = \dfrac{3}{\underset{2}{4}} \cdot \dfrac{\overset{3}{6}}{10} = \dfrac{3 \cdot 3}{2 \cdot 10} = \dfrac{9}{20}$

*Now Try Exercise 13*

> ### Understanding Algebra
>
> Dividing fractions in algebra is just like dividing fractions in arithmetic. Multiply the first fraction by the reciprocal of the second fraction, dividing out any common factors.
>
> *Arithmetic:*
>
> $$\frac{2}{3} \div \frac{6}{7} = \frac{2}{3} \cdot \frac{7}{6} = \frac{\overset{1}{2} \cdot 7}{3 \cdot \underset{3}{6}} = \frac{7}{9}$$
>
> *Algebra:*
>
> $$\frac{x}{3} \div \frac{5}{6} = \frac{x}{3} \cdot \frac{\overset{2}{6}}{5} = \frac{2x}{5}$$

The same principles are used to **divide rational expressions**.

**To Divide Rational Expressions**

Multiply the first fraction by the reciprocal of the second fraction.

**EXAMPLE 12** Divide $\dfrac{8x^3}{z} \div \dfrac{5z^3}{3}$.

*Solution*    $\dfrac{8x^3}{z} \div \dfrac{5z^3}{3} = \dfrac{8x^3}{z} \cdot \dfrac{3}{5z^3}$    Multiply $\dfrac{8x^3}{z}$ by $\dfrac{3}{5z^3}$, the reciprocal of $\dfrac{5z^3}{3}$.

$$= \dfrac{24x^3}{5z^4}$$    There are no common factors to divide out.

*Now Try Exercise 35*

**EXAMPLE 13** Divide $\dfrac{x^2 - 9}{x + 4} \div \dfrac{x - 3}{x + 4}$.

*Solution*

$$\frac{x^2 - 9}{x + 4} \div \frac{x - 3}{x + 4} = \frac{x^2 - 9}{x + 4} \cdot \frac{x + 4}{x - 3}$$    Multiply the first fraction by the reciprocal of the second fraction.

$$= \frac{(x + 3)(x - 3)}{x + 4} \cdot \frac{x + 4}{x - 3}$$    Factor, and divide out common factors.

$$= x + 3$$

*Now Try Exercise 49*

EXAMPLE 14 Divide $\dfrac{-1}{2x - 3} \div \dfrac{8}{3 - 2x}$.

*Solution*

$$\dfrac{-1}{2x - 3} \div \dfrac{8}{3 - 2x} = \dfrac{-1}{2x - 3} \cdot \dfrac{3 - 2x}{8}$$

Multiply the first fraction by the reciprocal of the second fraction.

$$= \dfrac{-1}{2x - 3} \cdot \dfrac{-1(2x - 3)}{8}$$

Factor out $-1$, then divide out common factors.

$$= \dfrac{(-1)(-1)}{(1)(8)} = \dfrac{1}{8}$$

Now Try Exercise 41

EXAMPLE 15 Divide $\dfrac{w^2 - 11w + 30}{w^2} \div (w - 5)^2$.

*Solution*   $(w - 5)^2$ means $\dfrac{(w - 5)^2}{1}$.

$$\dfrac{w^2 - 11w + 30}{w^2} \div (w - 5)^2 = \dfrac{w^2 - 11w + 30}{w^2} \cdot \dfrac{1}{(w - 5)^2}$$

$$= \dfrac{(w - 6)(w - 5)}{w^2} \cdot \dfrac{1}{(w - 5)(w - 5)}$$

$$= \dfrac{w - 6}{w^2(w - 5)}$$

Now Try Exercise 43

EXAMPLE 16 Divide $\dfrac{12x^2 - 22x + 8}{7x} \div \dfrac{3x^2 + 2x - 8}{2x^2 + 4x}$.

*Solution*

$$\dfrac{12x^2 - 22x + 8}{7x} \div \dfrac{3x^2 + 2x - 8}{2x^2 + 4x} = \dfrac{12x^2 - 22x + 8}{7x} \cdot \dfrac{2x^2 + 4x}{3x^2 + 2x - 8}$$

$$= \dfrac{2(6x^2 - 11x + 4)}{7x} \cdot \dfrac{2x(x + 2)}{(3x - 4)(x + 2)}$$

$$= \dfrac{2(3x - 4)(2x - 1)}{7x} \cdot \dfrac{2x(x + 2)}{(3x - 4)(x + 2)}$$

$$= \dfrac{4(2x - 1)}{7} = \dfrac{8x - 4}{7}$$

Now Try Exercise 45

# EXERCISE SET 7.2

MathXL®   MyMathLab®

MathXL®   MyMathLab

## Warm-Up Exercises

*Fill in the blanks with the appropriate word, phrase, or symbol(s) from the following list.*

| | | | |
|---|---|---|---|
| $\dfrac{10}{x - 4}$ | factor | reciprocal | additive inverse |
| $\dfrac{10}{x^3 - 4}$ | divide out | $x + 4$ | $x - 3$ |

**1.** When multiplying two rational expressions, the first step should be to _____ numerators and denominators.

**2.** When dividing two fractions, we multiply the first fraction by the _____ of the second fraction.

3. When multiplying two rational expressions, prior to multiplying the numerators and the denominators, you should _____ common factors.

4. When $\dfrac{10}{x^3 - 4x^2}$ is multiplied by $x^2$, the result is _____.

5. When $\dfrac{x^2 - 16}{x - 2}$ is divided by $\dfrac{x - 4}{x - 2}$, the result is _____.

6. When $\dfrac{x^2 + x - 12}{x + 5}$ is divided by $\dfrac{x^2 - x - 20}{x^2 - 25}$, the result is _____.

## Practice the Skills

*Multiply or divide as indicated.*

7. $\left(\dfrac{1}{5}\right)\left(\dfrac{15}{19}\right)$

8. $\left(\dfrac{3}{8}\right)\left(-\dfrac{9}{33}\right)$

9. $\left(\dfrac{6}{8}\right)\left(-\dfrac{10}{14}\right)$

10. $\left(\dfrac{7}{9}\right)\left(\dfrac{81}{98}\right)$

11. $\left(-\dfrac{4}{11}\right)\left(-\dfrac{55}{64}\right)$

12. $\left(-\dfrac{12}{13}\right)\left(-\dfrac{65}{42}\right)$

13. $\dfrac{3}{7} \div \dfrac{5}{7}$

14. $\dfrac{3}{8} \div \dfrac{15}{44}$

15. $-\dfrac{2}{9} \div \dfrac{32}{39}$

16. $\left(-\dfrac{3}{4}\right) \div \left(-\dfrac{15}{16}\right)$

*Multiply.*

17. $\dfrac{6x}{4y} \cdot \dfrac{y^2}{12}$

18. $\dfrac{15x^3 y^2}{2z} \cdot \dfrac{2z}{5xy^3}$

19. $\dfrac{14x^2}{y^4} \cdot \dfrac{5x^2}{y^2}$

20. $\dfrac{7n^3}{16m} \cdot \dfrac{-4}{21m^2 n^3}$

21. $\dfrac{6x^5 y^3}{5z^3} \cdot \dfrac{6x^4}{5yz^4}$

22. $\dfrac{x^2 - 4}{x^2 - 16} \cdot \dfrac{x - 4}{x - 2}$

23. $\dfrac{3x - 2}{3x + 2} \cdot \dfrac{x - 1}{1 - x}$

24. $\dfrac{m - 5}{2m + 5} \cdot \dfrac{8m}{-m + 5}$

25. $\dfrac{x^2 + 7x + 6}{x + 6} \cdot \dfrac{1}{x + 1}$

26. $\dfrac{b^2 + 7b + 12}{6b} \cdot \dfrac{b^2 - 4b}{b^2 - b - 12}$

27. $\dfrac{a}{a^2 - b^2} \cdot \dfrac{a + b}{a^2 + ab}$

28. $\dfrac{t^2 - 36}{t^2 + t - 30} \cdot \dfrac{t - 5}{4t}$

29. $\dfrac{6x^2 - 14x - 12}{6x + 4} \cdot \dfrac{2x + 4}{2x^2 - 2x - 12}$

30. $\dfrac{2x^2 - 9x + 9}{8x - 12} \cdot \dfrac{4x}{x^2 - 3x}$

31. $\dfrac{3x^2 - 13x - 10}{x^2 - 2x - 15} \cdot \dfrac{x^2 + x - 2}{3x^2 - x - 2}$

32. $\dfrac{2t^2 - t - 6}{2t^2 - 3t - 2} \cdot \dfrac{2t^2 - 5t - 3}{2t^2 + 11t + 12}$

33. $\dfrac{x + 9}{x - 3} \cdot \dfrac{x^3 - 27}{x^2 + 3x + 9}$

34. $\dfrac{x^3 + 8}{x^2 - x - 6} \cdot \dfrac{x + 5}{x^2 - 2x + 4}$

*Divide.*

35. $\dfrac{12x^3}{y^2} \div \dfrac{3x}{y^3}$

36. $\dfrac{9x^3}{5} \div \dfrac{1}{20x^2}$

37. $\dfrac{15xy^2}{4z} \div \dfrac{5x^2 y^2}{12z^2}$

38. $\dfrac{36y}{5z^2} \div \dfrac{3xy}{2z}$

39. $\dfrac{11xy}{7ab^2} \div \dfrac{6xy}{7}$

40. $3xz \div \dfrac{6xy}{z}$

41. $\dfrac{12r + 6}{r} \div \dfrac{2r + 1}{r^3}$

42. $\dfrac{x - 3}{10y^2} \div \dfrac{x^2 - 9}{5xy}$

43. $\dfrac{x^2 + 11x + 18}{x} \div \dfrac{x + 2}{x}$

44. $\dfrac{1}{x^2 + 7x - 18} \div \dfrac{1}{x^2 - 17x + 30}$

45. $\dfrac{x^2 - 12x + 32}{x^2 - 6x - 16} \div \dfrac{x^2 - x - 12}{x^2 - 5x - 24}$

46. $\dfrac{a - b}{9a + 9b} \div \dfrac{a^2 - b^2}{a^2 + 2a + 7}$

47. $\dfrac{2x^2 + 9x + 4}{x^2 + 7x + 12} \div \dfrac{2x^2 - x - 1}{(x + 3)^2}$

48. $\dfrac{a^2 - b^2}{9} \div \dfrac{3a - 3b}{27x^2}$

49. $\dfrac{x^2 - y^2}{x^2 - 2xy + y^2} \div \dfrac{x + y}{y - x}$

50. $\dfrac{9x^2 - 9y^2}{18x^2 y^2} \div \dfrac{3x + 3y}{12x^2 y^5}$

51. $\dfrac{5x^2 - 4x - 1}{5x^2 + 6x + 1} \div \dfrac{x^2 - 5x + 4}{x^2 + 2x + 1}$

52. $\dfrac{7n^2 - 15n + 2}{n^2 + n - 6} \div \dfrac{n^2 - 3n - 10}{n^2 - 2n - 15}$

*Perform each indicated operation.*

53. $\dfrac{11z}{6y^2} \cdot \dfrac{24x^2 y^4}{11z}$

54. $\dfrac{5z^3}{7} \cdot \dfrac{9x^2}{15z}$

55. $\dfrac{63a^2 b^3}{20c^3} \cdot \dfrac{4c^4}{9a^3 b^5}$

56. $\dfrac{-2xw}{y^5} \div \dfrac{8x^2}{y^6}$

57. $\dfrac{-xy}{a} \div \dfrac{-2ax}{6y}$

58. $\dfrac{27x}{8y^2} \div 3x^2 y^2$

59. $\dfrac{64m^6}{21x^5 y^7} \cdot \dfrac{14x^{12} y^5}{16m^5}$

60. $\dfrac{-18x^2 y}{11z^2} \cdot \dfrac{22z^3}{x^2 y^5}$

61. $\dfrac{(x + 3)^2}{5x^2} \cdot \dfrac{10x}{x^2 - 9}$

62. $\dfrac{1}{4x - 3} \cdot (24x - 18)$

63. $\dfrac{1}{5x^2 y^2} \div \dfrac{1}{35x^3 y}$

64. $\dfrac{x^2 y^5}{3z} \div \dfrac{7z}{2x}$

**65.** $\dfrac{(4m)^2}{8n^3} \div \dfrac{m^6 n^8}{2}$

**66.** $\dfrac{11r^5 s^2}{(r^2 s^3)^3} \cdot \dfrac{6r^4}{4s}$

**67.** $\dfrac{r^2 + 5r + 6}{r^2 + 9r + 18} \cdot \dfrac{r^2 + 4r - 12}{r^2 - 5r + 6}$

**68.** $\dfrac{z^2 - z - 20}{z^2 - 3z - 10} \cdot \dfrac{(z + 2)^2}{(z + 4)^2}$

**69.** $\dfrac{x^2 - 12x + 36}{x^2 - 8x + 12} \div \dfrac{x^2 - 7x + 12}{x^2 - 6x + 8}$

**70.** $\dfrac{p^2 - 5p + 6}{p^2 - 10p + 16} \div \dfrac{p^2 + 2p}{p^2 - 6p - 16}$

**71.** $\dfrac{2w^2 + 3w - 35}{w^2 - 7w - 8} \cdot \dfrac{w^2 - 5w - 24}{w^2 + 8w + 15}$

**72.** $\dfrac{3z^2 - 4z - 4}{z^2 - 4} \cdot \dfrac{2z^2 + 5z + 2}{2z^2 - 3z - 2}$

**73.** $\dfrac{q^2 - 11q + 30}{2q^2 - 7q - 15} \div \dfrac{q^2 - 2q - 24}{q^2 - q - 20}$

**74.** $\dfrac{2x^2 - 19x + 24}{x^2 - 12x + 32} \div \dfrac{2x^2 + x - 6}{x^2 + 7x + 10}$

**75.** $\dfrac{4n^2 - 9}{9n^2 - 1} \cdot \dfrac{3n^2 - 2n - 1}{2n^2 - 5n + 3}$

**76.** $\dfrac{2z^2 + 9z + 9}{4z^2 - 9} \div \dfrac{(z + 3)^2}{(2z - 3)^2}$

## Problem Solving

*Perform each indicated operation. Treat $\Delta$ and ☺ as if they were variables.*

**77.** $\dfrac{6\Delta^2}{13} \cdot \dfrac{13}{36\Delta^5}$

**78.** $\dfrac{\Delta - 7}{2\Delta + 5} \cdot \dfrac{3\Delta}{-\Delta + 7}$

**79.** $\dfrac{\Delta - ☺}{9\Delta - 9☺} \div \dfrac{\Delta^2 - ☺^2}{\Delta^2 + 2\Delta☺ + ☺^2}$

**80.** $\dfrac{\Delta^2 - ☺^2}{\Delta^2 - 2\Delta☺ + ☺^2} \div \dfrac{\Delta + ☺}{☺ - \Delta}$

*For each equation, fill in the shaded area with a binomial or trinomial to make the statement true. Explain how you determined your answer.*

**81.** $\dfrac{\boxed{\phantom{x}}}{x + 2} = x + 3$

**82.** $\dfrac{x + 5}{\boxed{\phantom{x}}} = \dfrac{1}{x - 5}$

**83.** $\dfrac{\boxed{\phantom{x}}}{x - 6} = x + 2$

**84.** $\dfrac{\boxed{\phantom{x}}}{x^2 - 7x + 10} = \dfrac{1}{x - 2}$

**85.** $\dfrac{\boxed{\phantom{x}}}{x^2 - 4} \cdot \dfrac{x + 2}{x - 1} = 1$

**86.** $\dfrac{x + 4}{x^2 + 9x + 20} \cdot \dfrac{\boxed{\phantom{x}}}{x - 2} = 1$

## Concept/Writing Exercises

**87.** Explain how to multiply two rational expressions.

**88.** Explain how to divide two rational expressions.

## Challenge Problems

*Simplify.*

**89.** $\left(\dfrac{x^2 - x - 6}{2x^2 - 9x + 9} \div \dfrac{x^2 + x - 12}{x^2 + 3x - 4}\right) \cdot \dfrac{2x^2 - 5x + 3}{x^2 + x - 2}$

**90.** $\left(\dfrac{x^2 + 4x + 3}{x^2 - 6x - 16} \div \dfrac{x^2 + 5x + 6}{x^2 - 9x + 8}\right) \cdot \left(\dfrac{x^2 - 1}{x^2 + 4x + 4}\right)$

*For Exercises 91 and 92, determine the polynomials that when placed in the shaded areas make the statement true. Explain how you determined your answer.*

**91.** $\dfrac{\boxed{\phantom{x}}}{\boxed{\phantom{x}}} \cdot \dfrac{x^2 + 3x - 4}{x^2 - 4x + 3} = \dfrac{x - 2}{x - 5}$

**92.** $\dfrac{\boxed{\phantom{x}}}{x^2 + x - 2} \cdot \dfrac{x^2 + 6x + 8}{\boxed{\phantom{x}}} = \dfrac{x + 3}{x + 5}$

## Group Activity

**93.** Consider the three problems that follow:

**1.** $\left(\dfrac{x + 2}{x - 3}\right) \div \left(\dfrac{x^2 - 5x + 6}{x - 2} \cdot \dfrac{x + 2}{x - 3}\right)$

**2.** $\left(\dfrac{x + 2}{x - 3} \div \dfrac{x^2 - 5x + 6}{x - 2}\right) \cdot \left(\dfrac{x + 2}{x - 3}\right)$

**3.** $\left(\dfrac{x + 2}{x - 3}\right) \div \left(\dfrac{x^2 - 5x + 6}{x - 2}\right) \cdot \left(\dfrac{x + 2}{x - 3}\right)$

**a)** Without working the problem, decide as a group which of the problems will have the same answer. Explain.

**b)** Individually, simplify each of the three problems.

**c)** Compare your answers to part **b)** with those of the other members of your group. If you did not get the same answers, determine why.

## Cumulative Review Exercises

[3.4]    **94. White Water Rafting** Jon and Louis leave Idaho Falls paddling downstream at an average speed of 15 miles per hour toward Pocatello. On the return trip, paddling against the current, they average 5 miles per hour. If the trip back to Idaho Falls took 2 hours longer than the trip out, find the time it took them to reach Pocatello.

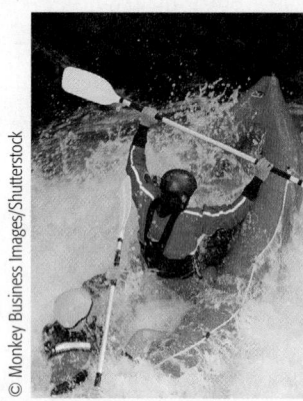

© Monkey Business Images/Shutterstock

[5.5]   **95.** Multiply $(4x^3y^2z^4)(3xy^3z^7)$.

[5.6]   **96.** Divide $\dfrac{4x^3 - 5x}{2x - 1}$.

[6.4]   **97.** Factor $6x^2 - 18x - 60$.

[6.6]   **98.** Solve $3x^2 - 9x - 30 = 0$.

# 7.3 Addition and Subtraction of Rational Expressions with a Common Denominator and Finding the Least Common Denominator

1  Add and subtract rational expressions with a common denominator.

2  Find the least common denominator.

## 1  Add and Subtract Rational Expressions with a Common Denominator

Recall that when adding or subtracting two fractions with a common denominator we add or subtract the numerators while keeping the common denominator.

### To Add or Subtract Two Fractions with a Common Denominator

Adding: $\dfrac{a}{c} + \dfrac{b}{c} = \dfrac{a + b}{c}, c \neq 0$    Subtracting: $\dfrac{a}{c} - \dfrac{b}{c} = \dfrac{a - b}{c}, c \neq 0$

EXAMPLE 1   **a)** Add $\dfrac{7}{16} + \dfrac{8}{16}$.    **b)** Subtract $\dfrac{4}{9} - \dfrac{1}{9}$.

*Solution*

**a)** $\dfrac{7}{16} + \dfrac{8}{16} = \dfrac{7 + 8}{16} = \dfrac{15}{16}$    **b)** $\dfrac{4}{9} - \dfrac{1}{9} = \dfrac{4 - 1}{9} = \dfrac{3}{9} = \dfrac{1}{3}$

Now Try Exercise 13

When adding or subtracting rational expressions, we use the same principles that we use when adding or subtracting fractions.

### To Add or Subtract Rational Expressions with a Common Denominator

1. Add or subtract the numerators.
2. Place the sum or difference of the numerators found in step 1 over the common denominator.
3. Simplify the fraction if possible.

**EXAMPLE 2** Add $\dfrac{3}{x-4} + \dfrac{x+8}{x-4}$.

*Solution*
$$\frac{3}{x-4} + \frac{x+8}{x-4} = \frac{3+(x+8)}{x-4} = \frac{x+11}{x-4}$$

Now Try Exercise 19

**EXAMPLE 3** Add $\dfrac{2x^2+7}{x+3} + \dfrac{6x-7}{x+3}$.

*Solution*
$$\frac{2x^2+7}{x+3} + \frac{6x-7}{x+3} = \frac{(2x^2+7)+(6x-7)}{x+3}$$
$$= \frac{2x^2+7+6x-7}{x+3}$$
$$= \frac{2x^2+6x}{x+3}$$

Now factor $2x$ from each term in the numerator and simplify.

$$= \frac{2x\cancel{(x+3)}}{\cancel{x+3}} = 2x$$

Now Try Exercise 21

**EXAMPLE 4** Add $\dfrac{x^2+2x-2}{(x+5)(x-2)} + \dfrac{5x+12}{(x+5)(x-2)}$.

*Solution*
$$\frac{x^2+2x-2}{(x+5)(x-2)} + \frac{5x+12}{(x+5)(x-2)} = \frac{(x^2+2x-2)+(5x+12)}{(x+5)(x-2)} \qquad \text{Write as a single fraction.}$$
$$= \frac{x^2+2x-2+5x+12}{(x+5)(x-2)} \qquad \begin{array}{l}\text{Parentheses}\\\text{were removed in}\\\text{the numerator.}\end{array}$$
$$= \frac{x^2+7x+10}{(x+5)(x-2)} \qquad \begin{array}{l}\text{Like terms}\\\text{were combined.}\end{array}$$
$$= \frac{\cancel{(x+5)}(x+2)}{\cancel{(x+5)}(x-2)} \qquad \begin{array}{l}\text{Factor, divide}\\\text{out common}\\\text{factor.}\end{array}$$
$$= \frac{x+2}{x-2}$$

Now Try Exercise 27

    *When subtracting rational expressions, be sure to subtract the entire numerator of the fraction being subtracted.*

> **Avoiding Common Errors**
>
> Consider the subtraction
> $$\frac{4x}{x-2} - \frac{2x+1}{x-2}$$
> Many people begin problems of this type incorrectly. Here are the correct and incorrect ways of working this problem.
>
> | CORRECT | INCORRECT |
> |---|---|
> | $\dfrac{4x}{x-2} - \dfrac{2x+1}{x-2} = \dfrac{4x-(2x+1)}{x-2}$ | $\dfrac{4x}{\cancel{x-2}} - \dfrac{\cancel{2x+1}}{x-2} \neq \dfrac{\cancel{4x-2x+1}}{\cancel{x-2}}$ |
> | $\qquad = \dfrac{4x-2x-1}{x-2}$ | |
> | $\qquad = \dfrac{2x-1}{x-2}$ | |
>
> *Note that the entire numerator of the second fraction, not just the first term, must be subtracted.* Also note that the sign of *each* term of the numerator being subtracted will change when the parentheses are removed.

**EXAMPLE 5** Subtract $\dfrac{x^2 - 6x + 3}{x^2 + 7x + 12} - \dfrac{x^2 - 8x - 5}{x^2 + 7x + 12}.$

*Solution*

$$\dfrac{x^2 - 6x + 3}{x^2 + 7x + 12} - \dfrac{x^2 - 8x - 5}{x^2 + 7x + 12} = \dfrac{(x^2 - 6x + 3) - (x^2 - 8x - 5)}{x^2 + 7x + 12} \quad \text{Write as a single fraction.}$$

$$= \dfrac{x^2 - 6x + 3 - x^2 + 8x + 5}{x^2 + 7x + 12} \quad \text{Removed parentheses.}$$

$$= \dfrac{2x + 8}{x^2 + 7x + 12} \quad \text{Combined like terms.}$$

$$= \dfrac{2\cancel{(x + 4)}}{(x + 3)\cancel{(x + 4)}} \quad \text{Factor, divide out common factor.}$$

$$= \dfrac{2}{x + 3}$$

Now Try Exercise 43

**EXAMPLE 6** Subtract $\dfrac{6r}{r - 5} - \dfrac{4r^2 - 17r + 15}{r - 5}.$

*Solution*

$$\dfrac{6r}{r - 5} - \dfrac{4r^2 - 17r + 15}{r - 5} = \dfrac{6r - (4r^2 - 17r + 15)}{r - 5} \quad \text{Write as a single fraction.}$$

$$= \dfrac{6r - 4r^2 + 17r - 15}{r - 5} \quad \text{Parentheses were removed.}$$

$$= \dfrac{-4r^2 + 23r - 15}{r - 5} \quad \text{Like terms were combined.}$$

$$= \dfrac{-(4r^2 - 23r + 15)}{r - 5} \quad \text{Factored out } -1$$

$$= \dfrac{-(4r - 3)\cancel{(r - 5)}}{\cancel{r - 5}} \quad \text{Factor, divide out common factor.}$$

$$= -(4r - 3) \quad \text{or} \quad -4r + 3$$

Now Try Exercise 31

## 2  Find the Least Common Denominator

To add two fractions with unlike denominators, we first obtain the least common denominator. Now we explain how to find the *least common denominator* for rational expressions.

**EXAMPLE 7** Add $\dfrac{4}{7} + \dfrac{2}{3}.$

*Solution*   The least common denominator (LCD) of the fractions $\dfrac{4}{7}$ and $\dfrac{2}{3}$ is 21 because it is the smallest number that is divisible by both denominators, 7 and 3. Rewrite each fraction so that its denominator is 21.

$$\dfrac{4}{7} + \dfrac{2}{3} = \dfrac{3}{3} \cdot \dfrac{4}{7} + \dfrac{2}{3} \cdot \dfrac{7}{7}$$

$$= \dfrac{12}{21} + \dfrac{14}{21}$$

$$= \dfrac{26}{21}$$

Now Try Exercise 95

**To Find the Least Common Denominator of Rational Expressions**

1. Factor each denominator completely. Any factors that occur more than once should be expressed as powers. For example, $(x - 3)(x - 3)$ should be expressed as $(x - 3)^2$.
2. List all different factors (other than 1) that appear in any of the denominators. When the same factor appears in more than one denominator, write that factor with the highest power that appears on it.
3. The least common denominator is the product of all the factors listed in step 2.

**EXAMPLE 8** Find the least common denominator.

$$\frac{1}{7} + \frac{1}{y}$$

*Solution*   The only factor (other than 1) of the first denominator is 7. The only factor (other than 1) of the second denominator is $y$. The LCD is therefore $7 \cdot y = 7y$.

Now Try Exercise 53

**EXAMPLE 9** Find the LCD.

$$\frac{8}{x^2} - \frac{3}{5x}$$

*Solution*   The factors that appear in the denominators are 5 and $x$. List each factor with its highest power. The LCD is the product of these factors.

Highest power of 5 ⟶    ⟵ Highest power of $x$

$$\text{LCD} = 5^1 \cdot x^2 = 5x^2$$

Now Try Exercise 59

**EXAMPLE 10** Find the LCD.

$$\frac{11}{18x^3y} + \frac{5}{27x^2y^3}$$

*Solution*   We first write the denominators in factored form. To determine the LCD, we multiply the highest power of each factor.

$18x^3y = \boxed{2} \cdot 3^2 \cdot \boxed{x^3} \cdot y$    The prime factorization of 18 is $2 \cdot 3 \cdot 3$ or $2 \cdot 3^2$

$27x^2y^3 = \qquad 3^3 \cdot x^2 \cdot \boxed{y^3}$    The prime factorization of 27 is $3 \cdot 3 \cdot 3$ or $3^3$

Highest power of 3 ⟶    ⟵ Highest power of $x$

$$\text{LCD} = 2 \cdot 3^3 \cdot x^3 \cdot y^3$$

Highest power of 2 ⟶    ⟵ Highest power of $y$

**Understanding Algebra**

Numerical coefficients like those in Example 10 need to be factored also. To help you write numbers in their prime factored form, you may want to review Section 6.1 or consult Appendix B.

Thus, the LCD is $2 \cdot 3^3 \cdot x^3 \cdot y^3 = 54x^3y^3$

Now Try Exercise 61

**EXAMPLE 11** Find the LCD.

$$\frac{9}{x} - \frac{2z}{x + 3}$$

*Solution*   The factors in the denominators are $x$ and $x + 3$. *Note that the $x$ in the second denominator, $x + 3$, is a term, not a factor.*

$$\text{LCD} = x(x + 3)$$

Now Try Exercise 65

**EXAMPLE 12** Find the LCD. ────────────────

$$\frac{7}{3x^2 - 6x} + \frac{8x^2}{x^2 - 4x + 4}$$

*Solution*   Factor both denominators.

$$\frac{7}{3x^2 - 6x} + \frac{8x^2}{x^2 - 4x + 4} = \frac{7}{3x(x - 2)} + \frac{8x^2}{(x - 2)(x - 2)}$$

$$= \frac{7}{3x(x - 2)} + \frac{8x^2}{(x - 2)^2}$$

The factors in the denominators are $3, x$, and $x - 2$. List the highest power of each of these factors.

$$\text{LCD} = 3 \cdot x \cdot (x - 2)^2 = 3x(x - 2)^2.$$

Now Try Exercise 85

**EXAMPLE 13** Find the LCD. ────────────────

$$\frac{11x}{x^2 - x - 12} - \frac{6x^2}{x^2 - 7x + 12}$$

*Solution*   Factor both denominators.

$$\frac{11x}{x^2 - x - 12} - \frac{6x^2}{x^2 - 7x + 12} = \frac{11x}{(x + 3)(x - 4)} - \frac{6x^2}{(x - 3)(x - 4)}$$

The factors in the denominators are $x + 3$, $x - 4$, and $x - 3$.

$$\text{LCD} = (x + 3)(x - 4)(x - 3)$$

*Although $x - 4$ is a common factor of each denominator, the highest power of that factor that appears in each denominator is 1.*

Now Try Exercise 81

**EXAMPLE 14** Find the LCD. ────────────────

$$\frac{6w}{w^2 - 14w + 45} + w + 8$$

*Solution*   Factor the denominator of the first term.

$$\frac{6w}{w^2 - 14w + 45} + w + 8 = \frac{6w}{(w - 5)(w - 9)} + w + 8$$

Since the denominator of $w + 8$ is 1, the expression can be rewritten as

$$\frac{6w}{(w - 5)(w - 9)} + \frac{w + 8}{1}$$

The LCD is therefore $1(w - 5)(w - 9)$ or simply $(w - 5)(w - 9)$.

Now Try Exercise 89

# EXERCISE SET 7.3    MathXL®   MyMathLab®

MathXL®        MyMathLab

## Warm-Up Exercises

*Fill in the blanks with the appropriate word, phrase, or symbol(s) from the following list.*

| $\dfrac{3x - 2}{y}$ | True | False | $\dfrac{3x + 2}{y}$ | $12x^3$ | $x + 4$ | $x^3$ |
|---|---|---|---|---|---|---|

**1.** The result of the subtraction problem $\dfrac{4x}{y} - \dfrac{(x - 2)}{y}$ is

_____ .

**2.** (True or False) When adding two fractions with the same denominator, we add the numerators and we add the denominators. _____

**3.** (True or False) When adding two rational expressions, after the addition takes place, there may remain the possibility of further simplification. _____

**4.** The LCD of $4x^2, 6x^3$, and $12x$ is _____ .

*Determine the LCD to be used to perform each indicated operation. Explain how you determined the LCD. Do not perform the operations.*

**5.** $\dfrac{9}{x+6} - \dfrac{2}{x}$

**6.** $\dfrac{10}{x-2} + \dfrac{3}{7}$

**7.** $\dfrac{2}{x+3} + \dfrac{1}{x} + \dfrac{1}{4}$

**8.** $\dfrac{6}{x-3} + \dfrac{1}{x} - \dfrac{1}{8}$

*In Exercises 9–12,* **a)** *explain why the expression on the left side of the not-equal-to sign is not equal to the expression on the right side.* **b)** *Show what the expression on the right side should be for it to be equal to the one on the left.*

**9.** $\dfrac{4x-3}{5x+4} - \dfrac{2x-9}{5x+4} \neq \dfrac{4x-3-2x-9}{5x+4}$

**10.** $\dfrac{5x}{2x-3} - \dfrac{-3x-7}{2x-3} \neq \dfrac{5x+3x-7}{2x-3}$

**11.** $\dfrac{8x-2}{x^2-4x+3} - \dfrac{3x^2-4x+5}{x^2-4x+3} \neq \dfrac{8x-2-3x^2-4x+5}{x^2-4x+3}$

**12.** $\dfrac{2x+5}{x^2-6x} - \dfrac{-x^2+3x+6}{x^2-6x} \neq \dfrac{2x+5+x^2+3x+6}{x^2-6x}$

## Practice the Skills

*Add or subtract.*

**13.** $\dfrac{1}{7} + \dfrac{4}{7}$

**14.** $\dfrac{8}{5} - \dfrac{6}{5}$

**15.** $\dfrac{5r+2}{4} - \dfrac{3}{4}$

**16.** $\dfrac{3x+6}{2} - \dfrac{x}{2}$

**17.** $\dfrac{2}{x} + \dfrac{x+4}{x}$

**18.** $\dfrac{3x+1}{x+1} + \dfrac{6x+8}{x+1}$

**19.** $\dfrac{6}{n+1} + \dfrac{n+2}{n+1}$

**20.** $\dfrac{7}{t-2} - \dfrac{t+4}{t-2}$

**21.** $\dfrac{x}{x-3} + \dfrac{4x+9}{x-3}$

**22.** $\dfrac{4x-3}{x-7} - \dfrac{2x+8}{x-7}$

**23.** $\dfrac{4t+7}{5t^2} - \dfrac{3t+4}{5t^2}$

**24.** $\dfrac{3w+6}{w^2+2w+1} + \dfrac{-2w-5}{w^2+2w+1}$

**25.** $\dfrac{5x+4}{x^2-x-12} + \dfrac{-4x-1}{x^2-x-12}$

**26.** $\dfrac{-x-6}{x^2-16} + \dfrac{2(x+5)}{x^2-16}$

**27.** $\dfrac{2m+5}{(m+4)(m-3)} - \dfrac{m+1}{(m+4)(m-3)}$

**28.** $\dfrac{x^2+3x}{(x+6)(x-3)} - \dfrac{x+15}{(x+6)(x-3)}$

**29.** $\dfrac{2p-6}{p-5} - \dfrac{p+6}{p-5}$

**30.** $\dfrac{x^2-6}{3x} - \dfrac{x^2+4x-11}{3x}$

**31.** $\dfrac{x^2+4x+1}{x+2} - \dfrac{5x+7}{x+2}$

**32.** $\dfrac{-4x+2}{3x+6} + \dfrac{4(x-1)}{3x+6}$

**33.** $\dfrac{3x+13}{2x+10} - \dfrac{2(x+4)}{2x+10}$

**34.** $\dfrac{x^2}{x+4} - \dfrac{16}{x+4}$

**35.** $\dfrac{b^2-2b-2}{b^2-b-6} + \dfrac{b-4}{b^2-b-6}$

**36.** $\dfrac{4x+17}{3-x} - \dfrac{3x+20}{3-x}$

**37.** $\dfrac{t-3}{t+3} - \dfrac{-3t-15}{t+3}$

**38.** $\dfrac{x+8}{3x+2} - \dfrac{x+8}{3x+2}$

**39.** $\dfrac{3x^2+15x}{x^3+2x^2-8x} + \dfrac{2x^2+5x}{x^3+2x^2-8x}$

**40.** $\dfrac{x^2-12}{x+5} - \dfrac{13}{x+5}$

**41.** $\dfrac{3x^2-9x}{4x^2-8x} + \dfrac{3x}{4x^2-8x}$

**42.** $\dfrac{x^3-10x^2+35x}{x(x-6)} - \dfrac{x^2+5x}{x(x-6)}$

**43.** $\dfrac{3x^2-4x+6}{3x^2+7x+2} - \dfrac{10x+11}{3x^2+7x+2}$

**44.** $\dfrac{x^2-2}{x^2+6x-7} - \dfrac{-4x+19}{x^2+6x-7}$

**45.** $\dfrac{x^2+3x-6}{x^2-5x+4} - \dfrac{-2x^2+4x-4}{x^2-5x+4}$

**46.** $\dfrac{4x^2+15}{9x^2-64} - \dfrac{x^2-x+39}{9x^2-64}$

**47.** $\dfrac{5x^2+30x+8}{x^2-64} + \dfrac{x^2+19x}{x^2-64}$

**48.** $\dfrac{20x^2+8x+1}{6x^2+x-2} - \dfrac{8x^2-9x-5}{6x^2+x-2}$

*Find the least common denominator for each expression.*

**49.** $\dfrac{x}{5} + \dfrac{x+4}{5}$

**50.** $\dfrac{2+r}{17} - \dfrac{12}{17}$

**51.** $\dfrac{3}{n} + \dfrac{1}{9n}$

**52.** $\dfrac{6}{x+1} - \dfrac{4}{7}$

**53.** $\dfrac{3}{5x} + \dfrac{7}{3}$

**54.** $\dfrac{1}{8} + \dfrac{1}{z}$

**55.** $\dfrac{6}{p} + \dfrac{9}{p^3}$

**56.** $\dfrac{2x}{x+3} + \dfrac{6}{x-9}$

**57.** $\dfrac{m+3}{3m-4} + m$

**58.** $\dfrac{x+4}{2x} + \dfrac{8}{7x}$

**59.** $\dfrac{t}{6t} + \dfrac{4}{t^2}$

**60.** $\dfrac{x}{5x^2} + \dfrac{9}{7x^3}$

**61.** $\dfrac{x+1}{12x^2y} - \dfrac{7}{9x^3}$

**62.** $\dfrac{-3}{8x^2y^2} + \dfrac{5}{12x^4y^5}$

**63.** $\dfrac{4}{2r^4s^5} - \dfrac{5}{9r^3s^7}$

**64.** $\dfrac{5}{4w^5z^4} + \dfrac{4}{9wz^2}$

**65.** $\dfrac{3}{m} - \dfrac{17m}{m+2}$

**66.** $\dfrac{x-3}{17} - \dfrac{6}{x-5}$

**67.** $\dfrac{5x-2}{x^2+x} - \dfrac{13}{x}$

**68.** $\dfrac{3t}{t-5} + \dfrac{2}{5-t}$

**69.** $\dfrac{n}{4n-1}+\dfrac{n-8}{1-4n}$    **70.** $\dfrac{3}{-2a+3b}-\dfrac{10}{2a-3b}$    **71.** $\dfrac{3}{4k-5r}-\dfrac{10}{-4k+5r}$    **72.** $\dfrac{p}{4p^2+2p}-\dfrac{7}{2p+1}$

**73.** $\dfrac{4}{2q^2+2q}-\dfrac{5}{9q}$    **74.** $\dfrac{10}{(x+4)(x+2)}-\dfrac{6+x}{x+2}$    **75.** $\dfrac{21}{24x^2y}+\dfrac{x+4}{15xy^3}$    **76.** $\dfrac{p^2+4}{p^2-25}+\dfrac{9}{p-5}$

**77.** $\dfrac{11}{3x+12}+\dfrac{3x+1}{2x+4}$    **78.** $6x^2+\dfrac{8x}{x-7}$    **79.** $\dfrac{9x+4}{x+1}-\dfrac{2x-6}{x+8}$

**80.** $\dfrac{x+3}{x^2+11x+18}-\dfrac{x^2-11}{x^2-3x-10}$    **81.** $\dfrac{x-2}{x^2-5x-24}+\dfrac{3}{x^2+11x+24}$    **82.** $\dfrac{6n}{n^2-4}-\dfrac{n-3}{n^2-5n-14}$

**83.** $\dfrac{5}{(a-4)^2}-\dfrac{a+2}{a^2-7a+12}$    **84.** $\dfrac{3x+5}{x^2-1}+\dfrac{x^2-18}{(x+1)^2}$    **85.** $\dfrac{9x}{x^2+6x+5}-\dfrac{5x^2}{x^2+4x+3}$

**86.** $\dfrac{6x+5}{x+2}+\dfrac{3x}{(x+2)^2}$    **87.** $\dfrac{3x-5}{x^2-6x+9}+\dfrac{3}{x-3}$    **88.** $\dfrac{2n+11}{(n+5)(n+2)}-\dfrac{3n-5}{(n-3)(n+5)}$

**89.** $\dfrac{8x^2}{x^2-7x+6}+x-9$    **90.** $\dfrac{2x-1}{x^2-25}+x-10$    **91.** $\dfrac{t-1}{3t^2+10t-8}-\dfrac{11}{3t^2+11t-4}$

**92.** $\dfrac{-4x+9}{2x^2+5x+2}+\dfrac{x^2}{3x^2+4x-4}$    **93.** $\dfrac{3x-1}{4x^2+4x+1}+\dfrac{x^2+x-9}{8x^2+10x+3}$    **94.** $\dfrac{3x+7}{6x^2+11x-10}+\dfrac{x^2-8}{9x^2-12x+4}$

*Add or subtract using the technique from Example 7.*

**95.** $\dfrac{1}{7}+\dfrac{2}{5}$    **96.** $\dfrac{3}{8}+\dfrac{1}{4}$    **97.** $\dfrac{2}{9}+\dfrac{3}{4}$    **98.** $\dfrac{5}{6}-\dfrac{1}{3}$    **99.** $\dfrac{5}{9}-\dfrac{1}{2}$    **100.** $\dfrac{6}{5}-\dfrac{3}{10}$

## Problem Solving

*List the polynomial to be placed in each shaded area to make a true statement. Explain how you determined your answer.*

**101.** $\dfrac{x^2-6x+3}{x+3}+\dfrac{\blacksquare}{x+3}=\dfrac{2x^2-5x-6}{x+3}$    **102.** $\dfrac{4x^2-6x-7}{x^2-4}-\dfrac{\blacksquare}{x^2-4}=\dfrac{2x^2+x-3}{x^2-4}$

**103.** $\dfrac{-x^2-4x+3}{2x+5}+\dfrac{\blacksquare}{2x+5}=\dfrac{5x-7}{2x+5}$    **104.** $\dfrac{-3x^2-9}{(x+4)(x-2)}-\dfrac{\blacksquare}{(x+4)(x-2)}=\dfrac{x^2+3x}{(x+4)(x-2)}$

*Find the least common denominator of each expression.*

**105.** $\dfrac{3}{☺}+\dfrac{4}{5☺}$    **106.** $\dfrac{5}{8\Delta^2☺^2}+\dfrac{6}{5\Delta^4☺^5}$

**107.** $\dfrac{8}{\Delta^2-9}-\dfrac{2}{\Delta+3}$    **108.** $\dfrac{6}{\Delta+3}-\dfrac{\Delta+5}{\Delta^2-4\Delta+3}$

## Concept/Writing Exercises

**109.** Explain how to add or subtract rational expressions with a common denominator.

**110.** Explain how to find the least common denominator of two rational expressions.

## Challenge Problems

*Perform each indicated operation.*

**111.** $\dfrac{4x-1}{x^2-25}-\dfrac{3x^2-8}{x^2-25}+\dfrac{8x-7}{x^2-25}$    **112.** $\dfrac{x^2-8x+2}{x+7}+\dfrac{2x^2-5x}{x+7}-\dfrac{3x^2+7x+10}{x+7}$

*Find the least common denominator for each expression.*

**113.** $\dfrac{17}{6x^5y^9}-\dfrac{9}{2x^3y}+\dfrac{6}{5x^{12}y^2}$    **114.** $\dfrac{2x}{x-3}-\dfrac{3}{x^2-9}+\dfrac{5}{x+3}$

**115.** $\dfrac{3x}{x^2-x-12}+\dfrac{2}{x^2-6x+8}+\dfrac{3}{x^2+x-6}$    **116.** $\dfrac{9}{x^2-4}-\dfrac{8}{3x^2+5x-2}+\dfrac{7}{3x^2-7x+2}$

## Cumulative Review Exercises

[1.3] **117.** Subtract $4\frac{3}{5} - 2\frac{5}{9}$.

[2.5] **118.** Solve $6x + 4 = -(x + 2) - 3x + 4$.

[2.7] **119. Hummingbird Food** The instructions on a bottle of concentrated hummingbird food indicate that 6 ounces of the concentrate should be mixed with 1 gallon (128 ounces) of water. If you wish to mix the concentrate with only 48 ounces of water, how much concentrate should you use?

[3.2] **120. Tennis Club** A Tennis Club has two payment plans. Plan 1 is a yearly membership fee of $250 plus $5.00 per hour for use of the tennis court. Plan 2 is an annual membership fee of $600 with no charge for court time. How many hours would Malcolm Wu have to play in a year to make the cost of plan 1 equal to the cost of plan 2?

© Michael Woodruff/Shutterstock

[4.3] **121.** What is the slope of the line in the graph below?

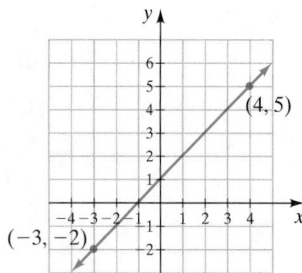

[5.3] **122.** Use scientific notation to evaluate $\dfrac{840{,}000{,}000}{0.0021}$. Leave your answer in scientific notation.

[6.6] **123.** Solve $2x^2 - 3 = x$.

# 7.4 Addition and Subtraction of Rational Expressions

**1** Add and subtract rational expressions.

In Section 7.3 we discussed how to add and subtract rational expressions with a common denominator. Now we discuss adding and subtracting rational expressions that are not given with a common denominator.

## **1** Add and Subtract Rational Expressions

The method used to add and subtract rational expressions with unlike denominators is outlined in Example 1.

**EXAMPLE 1** Add $\dfrac{7}{x} + \dfrac{6}{y}$.

*Solution*   First we determine the LCD as outlined in Section 7.3.

$$LCD = xy$$

We write each fraction with the LCD. We do this by multiplying *both* the numerator and denominator of each fraction by any factors needed to obtain the LCD.

$$\frac{7}{x} + \frac{6}{y} = \frac{7}{x} \cdot \frac{y}{y} + \frac{6}{y} \cdot \frac{x}{x}$$

$\left\{\begin{array}{l} \text{Multiply } \dfrac{7}{x} \text{ by } \dfrac{y}{y} \text{ and} \\[2mm] \text{multiply } \dfrac{6}{y} \text{ by } \dfrac{x}{x}. \\[2mm] \text{Values do not change since } \dfrac{x}{x} = 1 \text{ and } \dfrac{y}{y} = 1. \end{array}\right.$

$$= \frac{7y}{xy} + \frac{6x}{xy}$$

## Understanding Algebra

The concept of adding rational expressions is the same as adding fractions in arithmetic. Rewrite the fractions as equivalent fractions with the least common denominator.

**Arithmetic:**

$$\frac{1}{8} + \frac{1}{2} = \frac{1}{8} + \frac{4}{8} = \frac{1+4}{8} = \frac{5}{8}$$

**Algebra:**

$$\frac{3}{x^2} + \frac{2}{x^3} = \frac{3x}{x^3} + \frac{2}{x^3} = \frac{3x+2}{x^3}$$

Now we add the numerators, and keep the LCD, $xy$.

$$\frac{7y}{xy} + \frac{6x}{xy} = \frac{7y+6x}{xy} \quad \text{or} \quad \frac{6x+7y}{xy}$$

Now Try Exercise 7

### To Add or Subtract Two Rational Expressions with Unlike Denominators

1. Determine the LCD.
2. Rewrite each fraction as an equivalent fraction with the LCD. This is done by multiplying both the numerator and denominator of each fraction by any factors needed to obtain the LCD.
3. Add or subtract the numerators while maintaining the LCD.
4. When possible, factor the remaining numerator and simplify the fraction.

**EXAMPLE 2** Add $\dfrac{1}{4x^2y} + \dfrac{3}{14xy^3}$.

**Solution**    The LCD is $28x^2y^3$. We must write each fraction with the denominator $28x^2y^3$.

$$\frac{1}{4x^2y} + \frac{3}{14xy^3} = \frac{1}{4x^2y} \cdot \boxed{\frac{7y^2}{7y^2}} + \frac{3}{14xy^3} \cdot \boxed{\frac{2x}{2x}}$$

$$\left\{ \begin{array}{l} \text{Multiply } \dfrac{1}{4x^2y} \text{ by } \dfrac{7y^2}{7y^2} \text{ and} \\[2mm] \text{multiply } \dfrac{3}{14xy^3} \text{ by } \dfrac{2x}{2x} \\[2mm] \text{to obtain the LCD of } 28x^2y^3. \end{array} \right.$$

$$= \frac{7y^2}{28x^2y^3} + \frac{6x}{28x^2y^3}$$

$$= \frac{7y^2 + 6x}{28x^2y^3} \quad \text{or} \quad \frac{6x + 7y^2}{28x^2y^3}$$

Now Try Exercise 15

**EXAMPLE 3** Add $\dfrac{3}{x+2} + \dfrac{5}{x}$.

**Solution**    The LCD is $x(x+2)$.

$$\frac{3}{x+2} + \frac{5}{x} = \frac{3}{x+2} \cdot \boxed{\frac{x}{x}} + \frac{5}{x} \cdot \boxed{\frac{x+2}{x+2}}$$

$$\left\{ \begin{array}{l} \text{Multiply } \dfrac{3}{x+2} \text{ by } \dfrac{x}{x} \text{ and} \\[2mm] \text{multiply } \dfrac{5}{x} \text{ by } \dfrac{x+2}{x+2} \\[2mm] \text{to obtain the LCD of } x(x+2). \end{array} \right.$$

$$= \frac{3x}{x(x+2)} + \frac{5(x+2)}{x(x+2)} \qquad \begin{array}{l}\text{Each fraction has been written as an} \\ \text{equivalent fraction with the LCD of} \\ x(x+2)\end{array}$$

$$= \frac{3x}{x(x+2)} + \frac{5x+10}{x(x+2)} \qquad \text{Distributive property was used.}$$

$$= \frac{3x + (5x+10)}{x(x+2)} \qquad \text{Rewrote as a single fraction.}$$

$$= \frac{3x + 5x + 10}{x(x+2)} \qquad \begin{array}{l}\text{Parentheses were removed in the} \\ \text{numerator.}\end{array}$$

$$= \frac{8x + 10}{x(x+2)} \qquad \begin{array}{l}\text{Like terms were combined in the} \\ \text{numerator.}\end{array}$$

Now Try Exercise 25

> **Helpful Hint**
>
> Look at the answer to Example 3, $\dfrac{8x + 10}{x(x + 2)}$. Notice that the numerator could have been factored to obtain $\dfrac{2(4x + 5)}{x(x + 2)}$. Also notice that the denominator could have been multiplied to get $\dfrac{8x + 10}{x^2 + 2x}$. All three of these answers are equivalent and each is correct. In this section, *when writing answers, unless there is a common factor in the numerator and denominator we will leave the numerator in unfactored form and the denominator in factored form.* If both the numerator and denominator have a common factor, we will factor the numerator and simplify the fraction.

**EXAMPLE 4**   Subtract $\dfrac{w}{w - 7} - \dfrac{6}{w - 4}$.

*Solution*   The LCD is $(w - 7)(w - 4)$.

$$\frac{w}{w - 7} - \frac{6}{w - 4} = \frac{w - 4}{w - 4} \cdot \frac{w}{w - 7} - \frac{6}{w - 4} \cdot \frac{w - 7}{w - 7}$$

Multiply $\dfrac{w}{w - 7}$ by $\dfrac{w - 4}{w - 4}$ and multiply $\dfrac{6}{w - 4}$ by $\dfrac{w - 7}{w - 7}$.

$$= \frac{w(w - 4)}{(w - 4)(w - 7)} - \frac{6(w - 7)}{(w - 4)(w - 7)}$$

Rewrote each fraction as an equivalent fraction with the LCD.

$$= \frac{w^2 - 4w}{(w - 4)(w - 7)} - \frac{6w - 42}{(w - 4)(w - 7)}$$

Distributive property was used in the numerators.

$$= \frac{(w^2 - 4w) - (6w - 42)}{(w - 4)(w - 7)}$$

Wrote as a single fraction.

$$= \frac{w^2 - 4w - 6w + 42}{(w - 4)(w - 7)}$$

Parentheses were removed in the numerator.

$$= \frac{w^2 - 10w + 42}{(w - 4)(w - 7)}$$

Like terms were combined in the numerator.

Now Try Exercise 29

> **Avoiding Common Errors**
>
> Remember: When subtracting fractions, the subtraction symbol changes the sign of each term in the numerator following it.
>
> CORRECT
>
> $$\frac{5x}{x + 2} - \frac{x - 3}{x + 2} = \frac{5x - x + 3}{x + 2} = \frac{4x + 3}{x + 2}$$
>
> INCORRECT
>
> $$\frac{5x}{x + 2} - \frac{x - 3}{x + 2} = \frac{5x - x - 3}{x + 2} = \frac{4x - 3}{x + 2}$$

**EXAMPLE 5**   Subtract $\dfrac{x + 2}{x - 4} - \dfrac{x + 3}{x + 4}$.

*Solution*   The LCD is $(x - 4)(x + 4)$.

$$\frac{x + 2}{x - 4} - \frac{x + 3}{x + 4} = \frac{x + 4}{x + 4} \cdot \frac{x + 2}{x - 4} - \frac{x + 3}{x + 4} \cdot \frac{x - 4}{x - 4}$$

$$= \frac{(x + 4)(x + 2)}{(x + 4)(x - 4)} - \frac{(x + 3)(x - 4)}{(x + 4)(x - 4)}$$

Rewrote each fraction as an equivalent fraction with the LCD.

Use the FOIL method to multiply each numerator.

$$= \frac{x^2 + 6x + 8}{(x+4)(x-4)} - \frac{x^2 - x - 12}{(x+4)(x-4)}$$

$$= \frac{(x^2 + 6x + 8) - (x^2 - x - 12)}{(x+4)(x-4)} \qquad \text{Wrote as a single fraction.}$$

$$= \frac{x^2 + 6x + 8 - x^2 + x + 12}{(x+4)(x-4)} \qquad \begin{array}{l}\text{Parentheses were removed in the} \\ \text{numerator.}\end{array}$$

$$= \frac{7x + 20}{(x+4)(x-4)} \qquad \begin{array}{l}\text{Like terms were combined in the} \\ \text{numerator.}\end{array}$$

*Now Try Exercise 37*

Consider the problem

$$\frac{4}{x-2} + \frac{x+3}{2-x}$$

Because the denominators are opposites, we can proceed by converting one of the denominators to the other by multiplying by $-1$. See the Helpful Hint below.

### Understanding Algebra

The expressions $a - b$ and $b - a$ are opposites and, therefore, differ only by a factor of $-1$. Thus,

$$(b-a)(-1) = -b + a = a - b.$$

> **Helpful Hint**
>
> When adding or subtracting fractions whose denominators are opposites, multiply both the numerator *and* the denominator of *either* fraction by $-1$. Then both fractions will have the same denominator.
>
> $$\frac{x}{a-b} + \frac{y}{b-a} = \frac{x}{a-b} + \frac{y}{b-a} \cdot \frac{-1}{-1}$$
>
> $$= \frac{x}{a-b} + \frac{-y}{a-b}$$
>
> $$= \frac{x-y}{a-b}$$

**EXAMPLE 6**  Add $\dfrac{4}{x-2} + \dfrac{x+3}{2-x}$.

**Solution**   Since the denominators differ only in sign, we may multiply both the numerator and the denominator of either fraction by $-1$.

$$\frac{4}{x-2} + \frac{x+3}{2-x} = \frac{4}{x-2} + \frac{x+3}{2-x} \cdot \frac{-1}{-1} \qquad \begin{array}{l}\text{Multiply numerator and denominator} \\ \text{of the second fraction by } -1.\end{array}$$

$$= \frac{4}{x-2} + \frac{(-x-3)}{x-2} \qquad \text{Notice } (2-x)(-1) = x - 2.$$

$$= \frac{4 + (-x-3)}{x-2} \qquad \text{Wrote as a single fraction.}$$

$$= \frac{4 - x - 3}{x-2} \qquad \begin{array}{l}\text{Parentheses were removed in the} \\ \text{numerator.}\end{array}$$

$$= \frac{-x + 1}{x-2} \qquad \begin{array}{l}\text{Like terms were combined in the} \\ \text{numerator.}\end{array}$$

*Now Try Exercise 31*

Let's work another example where the denominators differ only in sign.

**EXAMPLE 7**  Subtract $\dfrac{a-9}{3a-4} - \dfrac{2a-5}{4-3a}$.

**Solution**   The denominators of the two fractions differ only in sign.

$$\frac{a-9}{3a-4} - \frac{2a-5}{4-3a} = \frac{a-9}{3a-4} - \frac{2a-5}{4-3a} \cdot \frac{-1}{-1} \qquad \begin{array}{l}\text{Multiply numerator and denominator of} \\ \text{the second fraction by } -1.\end{array}$$

$$= \frac{a-9}{3a-4} - \frac{(-2a+5)}{3a-4} \qquad \text{The common denominator is } 3a-4.$$

$$= \frac{(a-9)-(-2a+5)}{3a-4} \qquad \text{Wrote as a single fraction.}$$

$$= \frac{a-9+2a-5}{3a-4} \qquad \text{Parentheses were removed in the numerator.}$$

$$= \frac{3a-14}{3a-4} \qquad \text{Like terms were combined in the numerator.}$$

**Now Try Exercise 33**

**EXAMPLE 8**  Add $\dfrac{3}{x^2+5x+6} + \dfrac{1}{3x^2+8x-3}$.

*Solution*

$$\frac{3}{x^2+5x+6} + \frac{1}{3x^2+8x-3} = \frac{3}{(x+2)(x+3)} + \frac{1}{(3x-1)(x+3)}$$

The LCD is $(x+2)(x+3)(3x-1)$.

$$= \frac{3x-1}{3x-1} \cdot \frac{3}{(x+2)(x+3)} + \frac{1}{(3x-1)(x+3)} \cdot \frac{x+2}{x+2}$$

$$= \frac{9x-3}{(3x-1)(x+2)(x+3)} + \frac{x+2}{(3x-1)(x+2)(x+3)}$$

$$= \frac{(9x-3)+(x+2)}{(3x-1)(x+2)(x+3)}$$

$$= \frac{9x-3+x+2}{(3x-1)(x+2)(x+3)}$$

$$= \frac{10x-1}{(3x-1)(x+2)(x+3)}$$

**Now Try Exercise 55**

**EXAMPLE 9**  Subtract $\dfrac{5}{x^2-5x} - \dfrac{x}{5x-25}$.

*Solution*

$$\frac{5}{x^2-5x} - \frac{x}{5x-25} = \frac{5}{x(x-5)} - \frac{x}{5(x-5)}$$

The LCD is $5x(x-5)$.

$$= \frac{5}{5} \cdot \frac{5}{x(x-5)} - \frac{x}{5(x-5)} \cdot \frac{x}{x}$$

$$= \frac{25}{5x(x-5)} - \frac{x^2}{5x(x-5)}$$

$$= \frac{25-x^2}{5x(x-5)} \qquad \text{Factor the numerator.}$$

$$= \frac{(5-x)(5+x)}{5x(x-5)}$$

$$= \frac{-1(x-5)(x+5)}{5x(x-5)} \qquad 5-x = -1(x-5)$$

$$= \frac{-1\cancel{(x-5)}(x+5)}{5x\cancel{(x-5)}} \qquad \text{Simplify.}$$

$$= \frac{-1(x+5)}{5x} \quad \text{or} \quad -\frac{x+5}{5x}$$

**Now Try Exercise 65**

**Avoiding Common Errors**

1. A common error in an addition or subtraction problem is to add or subtract the numerators and the denominators. Here is one such example.

CORRECT

$$\frac{1}{x} + \frac{x}{1} = \frac{1}{x} + \frac{x}{1} \cdot \boxed{\frac{x}{x}}$$

$$= \frac{1}{x} + \frac{x^2}{x}$$

$$= \frac{1 + x^2}{x} \quad \text{or} \quad \frac{x^2 + 1}{x}$$

INCORRECT

Remember that to add or subtract fractions you must first have a common denominator. Then you add or subtract the numerators while maintaining the common denominator.

2. Another common mistake is to treat an addition or subtraction problem as a multiplication problem. You can divide out common factors only when *multiplying* expressions, not when adding or subtracting them.

CORRECT

$$\frac{1}{x} \cdot \frac{x}{1} = \frac{1}{\cancel{x}} \cdot \frac{\overset{1}{\cancel{x}}}{1}$$

$$= 1 \cdot 1 = 1$$

INCORRECT

$$= 1 + 1 = 2$$

---

# EXERCISE SET 7.4    MathXL®    MyMathLab®

## Warm-Up Exercises

*Fill in the blanks with the appropriate word, phrase, or symbol(s) from the following list.*

$24z^2$    you are multiplying by 1    $x$    $\dfrac{y + 5}{12z^2}$

$\dfrac{3yz + 10}{12z^2}$    $12z^2$    opposites    the LCD

1. When adding two fractions whose denominators are _____, multiply the numerator and denominator of either fraction by −1.

2. The first step in adding two rational expressions with unlike denominators is to determine _____.

3. The value of a fraction does not change when you multiply the numerator and denominator by the same quantity because _____.

4. When adding $\dfrac{2}{x^2 - 4x} + \dfrac{5}{x - 4}$ the numerator and denominator of the second fraction should be multiplied by _____.

5. When adding $\dfrac{y}{4z} + \dfrac{5}{6z^2}$, the LCD is _____.

6. $\dfrac{y}{4z} + \dfrac{5}{6z^2} =$ _____.

---

## Practice the Skills

*Add or subtract.*

7. $\dfrac{2}{x} + \dfrac{5}{y}$

8. $\dfrac{5}{x} - \dfrac{1}{y}$

9. $\dfrac{5}{t^2} + \dfrac{1}{2t}$

10. $7 - \dfrac{1}{x^2}$

11. $3 + \dfrac{8}{x}$

12. $\dfrac{5}{6y} + \dfrac{3}{5y^2}$

13. $\dfrac{2}{x^2} + \dfrac{3}{5x}$

14. $\dfrac{6}{x} - \dfrac{5}{x^2}$

15. $\dfrac{9}{4x^2y} + \dfrac{3}{5xy^2}$

16. $\dfrac{7}{12x^4y} - \dfrac{1}{5x^2y^3}$

17. $4y + \dfrac{x}{y}$

18. $x + \dfrac{2x}{y}$

19. $\dfrac{3a - 1}{2a} + \dfrac{5}{3a}$

20. $\dfrac{11}{n} + 5$

21. $\dfrac{6x}{y} + \dfrac{2y}{xy}$

22. $\dfrac{4}{5p} - \dfrac{5}{2p^2}$

**23.** $\dfrac{9}{b} - \dfrac{4}{5a^2}$

**24.** $\dfrac{x-3}{x} - \dfrac{7}{6x}$

**25.** $\dfrac{4}{x} + \dfrac{9}{x-3}$

**26.** $10 - \dfrac{3}{x-3}$

**27.** $\dfrac{9}{p+3} + \dfrac{2}{p}$

**28.** $\dfrac{2a}{a+b} + \dfrac{a-b}{a}$

**29.** $\dfrac{5}{d+1} - \dfrac{d}{3d+5}$

**30.** $\dfrac{2}{x-3} - \dfrac{5}{x-1}$

**31.** $\dfrac{8}{p-3} + \dfrac{2}{3-p}$

**32.** $\dfrac{3}{n-5} - \dfrac{7}{5-n}$

**33.** $\dfrac{9}{x+7} - \dfrac{5}{-x-7}$

**34.** $\dfrac{6}{7x-1} - \dfrac{3}{1-7x}$

**35.** $\dfrac{8}{a-2} + \dfrac{a}{2a-4}$

**36.** $\dfrac{4}{y-1} + \dfrac{3}{y+1}$

**37.** $\dfrac{x+5}{x-5} - \dfrac{x-5}{x+5}$

**38.** $\dfrac{x+7}{x+3} - \dfrac{x+3}{x+7}$

**39.** $\dfrac{5}{6n+3} - \dfrac{2}{n}$

**40.** $\dfrac{x}{4x-4} - \dfrac{1}{3x}$

**41.** $\dfrac{3}{2w+10} + \dfrac{6}{w+2}$

**42.** $\dfrac{5k}{4k-8} - \dfrac{k}{k+2}$

**43.** $\dfrac{z}{z^2-16} + \dfrac{4}{z+4}$

**44.** $\dfrac{5}{(x+4)^2} + \dfrac{3}{x+4}$

**45.** $\dfrac{x+2}{x^2-4} - \dfrac{2}{x+2}$

**46.** $\dfrac{3}{(x-2)(x+3)} + \dfrac{5}{(x+2)(x+3)}$

**47.** $\dfrac{3r+4}{r^2-10r+24} - \dfrac{2}{r-6}$

**48.** $\dfrac{x+9}{x^2-3x-10} - \dfrac{2}{x-5}$

**49.** $\dfrac{x^2-3}{x^2+2x-8} - \dfrac{x-4}{x+4}$

**50.** $\dfrac{x+8}{x^2-4x+4} - \dfrac{x+1}{x-2}$

**51.** $\dfrac{x-6}{x^2+10x+25} + \dfrac{x-3}{x+5}$

**52.** $\dfrac{x}{x^2-xy} - \dfrac{y}{xy-x^2}$

**53.** $\dfrac{5}{a^2-9a+8} - \dfrac{6}{a^2-6a-16}$

**54.** $\dfrac{4}{a^2+2a-15} - \dfrac{1}{a^2-9}$

**55.** $\dfrac{2}{x^2+6x+9} + \dfrac{7}{x^2+x-6}$

**56.** $\dfrac{x}{2x^2+7x-4} + \dfrac{4}{x^2-x-20}$

**57.** $\dfrac{x}{2x^2+7x+3} - \dfrac{5}{3x^2+7x-6}$

**58.** $\dfrac{x}{6x^2+7x+2} + \dfrac{5}{2x^2-3x-2}$

**59.** $\dfrac{x}{4x^2+11x+6} - \dfrac{2}{8x^2+2x-3}$

**60.** $\dfrac{x}{5x^2-9x-2} - \dfrac{1}{3x^2-7x+2}$

**61.** $\dfrac{3w+12}{w^2+w-12} - \dfrac{2}{w-3}$

**62.** $\dfrac{5x+10}{x^2-5x-14} - \dfrac{4}{x-7}$

**63.** $\dfrac{4r}{2r^2-10r+12} + \dfrac{4}{r-2}$

**64.** $\dfrac{6m}{3m^2-24m+48} - \dfrac{2}{m-4}$

**65.** $\dfrac{4}{x^2-4x} - \dfrac{x}{4x-16}$

**66.** $\dfrac{6}{x^2-6x} - \dfrac{x}{6x-36}$

## Problem Solving

*For what value(s) of x is each expression defined?*

**67.** $\dfrac{8}{x} + 6$

**68.** $\dfrac{6}{x-1} - \dfrac{5}{x}$

**69.** $\dfrac{3}{x-4} + \dfrac{7}{x+6}$

**70.** $\dfrac{4}{x^2-9} - \dfrac{9}{x+3}$

*Add or subtract. Treat the unknown symbols as if they were variables.*

**71.** $\dfrac{3}{\Delta-2} - \dfrac{4}{2-\Delta}$

**72.** $\dfrac{\Delta}{2\Delta^2+7\Delta-4} + \dfrac{2}{\Delta^2-\Delta-20}$

## Challenge Problems

*Under what conditions is each expression defined? Explain your answers.*

**73.** $\dfrac{5}{a+b} + \dfrac{4}{a}$

**74.** $\dfrac{x+2}{x+5y} - \dfrac{y-2}{3x}$

*Perform each indicated operation.*

**75.** $\dfrac{x}{x^2 - 9} + \dfrac{2x}{x + 3} + \dfrac{2x^2 - 5x}{9 - x^2}$

**76.** $\dfrac{8x + 9}{x^2 + x - 6} + \dfrac{x}{x + 3} - \dfrac{5}{x - 2}$

**77.** $\dfrac{x + 6}{4 - x^2} - \dfrac{x + 3}{x + 2} + \dfrac{x - 3}{2 - x}$

**78.** $\dfrac{3x - 1}{x + 2} + \dfrac{x}{x - 3} - \dfrac{4}{2x + 3}$

**79.** $\dfrac{2}{x^2 - x - 6} + \dfrac{3}{x^2 - 2x - 3} + \dfrac{1}{x^2 + 3x + 2}$

**80.** $\dfrac{3x}{x^2 - 4} + \dfrac{4}{x^3 + 8}$

## Group Activity

*Discuss and answer Exercise 81 as a group.*

**81. a)** As a group, find the LCD of

$$\frac{x + 3y}{x^2 + 3xy + 2y^2} + \frac{y - x}{2x^2 + 3xy + y^2}$$

   **b)** As a group, perform the indicated operation, but do not simplify your answer.

   **c)** As a group, simplify your answer.

   **d)** Group member 1: Substitute 2 for $x$ and 1 for $y$ in the fraction on the left in part **a)** and evaluate.

   **e)** Group member 2: Substitute 2 for $x$ and 1 for $y$ in the fraction on the right in part **a)** and evaluate.

   **f)** Group member 3: Add the numerical fractions found in parts **d)** and **e)**.

   **g)** Individually, substitute 2 for $x$ and 1 for $y$ in the expression obtained in part **b)** and evaluate.

   **h)** Individually, substitute 2 for $x$ and 1 for $y$ in the expression obtained in part **c)**, evaluate, and compare your answers.

   **i)** As a group, discuss what you discovered from this activity.

   **j)** Do you think your results would have been similar for any numbers substituted for $x$ and $y$ (for which the denominator is not 0)? Why?

## Cumulative Review Exercises

[2.7] **82. White Pass Railroad** The White Pass Railroad is a narrow gauge railroad that travels slowly through the mountains of Alaska. If the train travels 22 miles in 0.8 hours, how long will it take to travel 42 miles? Assume the train travels at the same rate throughout the trip.

[4.3] **83.** A line passes through the points $(-4, 6)$ and $(3, -2)$. Find the slope of the line.

[5.6] **84.** Divide $(8x^2 + 6x - 15) \div (2x + 3)$.

[7.2] **85.** Multiply $\dfrac{x^2 + xy - 6y^2}{x^2 - xy - 2y^2} \cdot \dfrac{y^2 - x^2}{x^2 + 2xy - 3y^2}$.

© Allen R. Angel

## Mid-Chapter Test: 7.1–7.4

*To find out how well you understand the chapter material to this point, take this brief test. The answers, and the section where the material was initially discussed, are given in the back of the book. Review any questions that you answered incorrectly.*

*Determine the value or values of the variable where each expression is defined.*

**1.** $\dfrac{9}{3x - 2}$

**2.** $\dfrac{2x + 1}{x^2 - 5x - 14}$

*Simplify each rational expression.*

**3.** $\dfrac{9x + 18}{x + 2}$

**4.** $\dfrac{2x^2 + 13x + 15}{3x^2 + 14x - 5}$

**5.** $\dfrac{25r^2 - 36t^2}{5r - 6t}$

*Multiply or divide as indicated.*

**6.** $\dfrac{15x^2}{2y} \cdot \dfrac{4y^4}{5x^5}$

**7.** $\dfrac{m - 3}{m + 4} \cdot \dfrac{m^2 + 8m + 16}{3 - m}$

**8.** $\dfrac{x^3 + 27}{x^2 - 2x - 15} \cdot \dfrac{x^2 - 7x + 10}{x^2 - 3x + 9}$

**9.** $\dfrac{5x - 1}{x^2 + 11x + 10} \div \dfrac{10x - 2}{x^2 + 17x + 70}$

**10.** $\dfrac{5x^2 + 7x + 2}{x^2 + 6x + 5} \div \dfrac{7x^2 - 39x - 18}{x^2 - x - 30}$

*Add or subtract as indicated.*

**11.** $\dfrac{x^2}{x + 6} - \dfrac{36}{x + 6}$

**12.** $\dfrac{2x^2 - 2x}{2x + 5} + \dfrac{x - 15}{2x + 5}$

**13.** $\dfrac{3t^2 - t}{4t^2 - 9t + 2} - \dfrac{3t + 4}{4t^2 - 9t + 2}$

*Find the least common denominator.*

**14.** $\dfrac{2m}{6m^2 + 3m} + \dfrac{m + 7}{2m + 1}$

**15.** $\dfrac{9x + 8}{2x^2 - 5x - 12} + \dfrac{2x + 3}{x^2 - 9x + 20}$

*For Exercises 16–19, add or subtract as indicated.*

**16.** $\dfrac{x + 1}{2x} + \dfrac{4x - 3}{5x}$

**17.** $\dfrac{2a + 5}{a + 3} - \dfrac{3a + 1}{a - 4}$

**18.** $\dfrac{x^2 + 5}{2x^2 + 13x + 6} + \dfrac{3x - 1}{2x + 1}$

**19.** $\dfrac{x}{x^2 + 3x + 2} - \dfrac{4}{x^2 - x - 6}$

**20.** To add the rational expressions $\dfrac{7}{x + 1} + \dfrac{8}{x}$, Samuel Ditsi decided to add both numerators and then add both denominators to get $\dfrac{7 + 8}{(x + 1) + x}$, which simplified to $\dfrac{15}{2x + 1}$. This procedure is wrong. Why is it wrong? Explain your answer. Then add the rational expressions $\dfrac{7}{x + 1} + \dfrac{8}{x}$ correctly.

# 7.5 Complex Fractions

**1** Simplify complex fractions by combining terms.

**2** Simplify complex fractions using multiplication first to clear fractions.

## **1** Simplify Complex Fractions by Combining Terms

### Complex Fraction

A **complex fraction** is one that has a fraction in its numerator or its denominator or in both its numerator and denominator.

### Examples of Complex Fractions

$$\dfrac{\frac{3}{5}}{7} \qquad \dfrac{\frac{x + 9}{x}}{4x} \qquad \dfrac{\frac{x}{y}}{x + 1} \qquad \dfrac{\frac{a + b}{a}}{\frac{a - b}{b}}$$

The expression above the **main fraction bar** is the numerator of the complex fraction, and the expression below the main fraction bar is the denominator of the complex fraction.

Numerator of complex fraction $\left\{ \begin{array}{l} a + b \\ \overline{a} \end{array} \right.$

$\longleftarrow$ Main fraction bar

Denominator of complex fraction $\left\{ \begin{array}{l} a - b \\ \overline{b} \end{array} \right.$

There are two methods to simplify complex fractions. The first reinforces many of the concepts used in this chapter because we may need to add, subtract, multiply, and divide simpler fractions as we simplify the complex fraction. Many students prefer to use the second method because the answer may be obtained more quickly.

### Method 1 — To Simplify a Complex Fraction by Combining Terms

1. Add or subtract the fractions in both the numerator and denominator of the complex fraction to obtain single fractions in both the numerator and the denominator.

2. Multiply the fraction in the numerator by the reciprocal of the fraction in the denominator.

3. Simplify further if possible.

## Understanding Algebra

The complex fraction

$$\dfrac{\frac{a + b}{c}}{\frac{d + e}{f}}$$

is another way to represent the division

$$\dfrac{a + b}{c} \div \dfrac{d + e}{f}.$$

We perform the division by multiplying the first fraction by the reciprocal of the second fraction:

$$\dfrac{a + b}{c} \cdot \dfrac{f}{d + e}.$$

**EXAMPLE 1**  Simplify $\dfrac{\frac{ab^2}{c^3}}{\frac{a}{bc^2}}$.

**Solution**   Since both numerator and denominator are already single fractions, we begin with step 2.

$$\frac{\frac{ab^2}{c^3}}{\frac{a}{bc^2}} = \frac{ab^2}{c^3} \div \frac{a}{bc^2}$$   A fraction bar means "divided by."

$$= \frac{ab^2}{c_c^3} \cdot \frac{bc^2}{a}$$   Multiply by $\frac{bc^2}{a}$, the reciprocal of $\frac{a}{bc^2}$.

$$= \frac{b^3}{c}$$   Both $a$ and $c^2$ were factored out of the numerator and denominator.

Thus the expression simplifies to $\dfrac{b^3}{c}$.

Now Try Exercise 11

**EXAMPLE 2**  Simplify $\dfrac{a + \frac{1}{x}}{x + \frac{1}{a}}$.

**Solution**   Step 1 is to express the numerator and denominator as single fractions. To obtain one fraction in the numerator, we notice that the LCD is $x$. Multiply $a$ by $\frac{x}{x}$.

$$a + \frac{1}{x} = a \cdot \frac{x}{x} + \frac{1}{x} = \frac{ax + 1}{x}$$   We use this as the numerator.

To obtain one fraction in the denominator, we notice that the LCD is $a$. Multiply $x$ by $\frac{a}{a}$.

$$x + \frac{1}{a} = x \cdot \frac{a}{a} + \frac{1}{a} = \frac{ax + 1}{a}$$   We use this as the denominator.

So

$$\frac{a + \frac{1}{x}}{x + \frac{1}{a}} = \frac{\frac{ax + 1}{x}}{\frac{ax + 1}{a}} = \frac{ax + 1}{x} \div \frac{ax + 1}{a}$$

$$= \frac{ax + 1}{x} \cdot \frac{a}{ax + 1} = \frac{a}{x}$$

Now Try Exercise 25

In Example 4 we will rework Example 2 using the second method.

## 2  Simplify Complex Fractions Using Multiplication First to Clear Fractions

**Method 2 — To Simplify a Complex Fraction Using Multiplication First**

1. Find the least common denominator of *all* the denominators appearing in the complex fraction.

2. Multiply both the numerator and denominator of the complex fraction by the LCD found in step 1.

3. Simplify when possible.

EXAMPLE 3  Simplify $\dfrac{\frac{2}{3} + \frac{1}{5}}{\frac{4}{5} - \frac{1}{3}}$.

*Solution*   The denominators in the complex fraction are 3 and 5. Multiply both the numerator and denominator of the complex fraction by 15, the LCD of the complex fraction.

$$\frac{\frac{2}{3} + \frac{1}{5}}{\frac{4}{5} - \frac{1}{3}} = \frac{15}{15} \cdot \frac{\left(\frac{2}{3} + \frac{1}{5}\right)}{\left(\frac{4}{5} - \frac{1}{3}\right)} = \frac{15\left(\frac{2}{3}\right) + 15\left(\frac{1}{5}\right)}{15\left(\frac{4}{5}\right) - 15\left(\frac{1}{3}\right)}$$

Now simplify.

$$= \frac{10 + 3}{12 - 5} = \frac{13}{7}$$

Now Try Exercise 9

## Understanding Algebra

The LCD of the complex fraction

$$\frac{\frac{a + b}{c}}{\frac{d + e}{f}}$$

is *cf* (*c* from the numerator and *f* from the denominator). To eliminate fractions within the complex fraction, we multiply both the numerator and the denominator by *cf*:

$$\frac{cf}{cf} \cdot \frac{\left(\frac{a + b}{c}\right)}{\left(\frac{d + e}{f}\right)} = \frac{(a + b)f}{(d + e)c}$$

Now we will rework Example 2 using method 2.

EXAMPLE 4  Simplify $\dfrac{a + \frac{1}{x}}{x + \frac{1}{a}}$.

*Solution*   The denominators in the complex fraction are *x* and *a*. Multiply both the numerator and denominator of the complex fraction by *ax*, the LCD of the complex fraction.

$$\frac{a + \frac{1}{x}}{x + \frac{1}{a}} = \frac{ax}{ax} \cdot \frac{\left(a + \frac{1}{x}\right)}{\left(x + \frac{1}{a}\right)} = \frac{a^2x + a}{ax^2 + x}$$

$$= \frac{a\,\cancel{(ax + 1)}}{x\,\cancel{(ax + 1)}} = \frac{a}{x}$$

Now Try Exercise 25

Note that the answers to Examples 2 and 4 are the same.

EXAMPLE 5  Simplify $\dfrac{y^2}{\frac{1}{x} + \frac{1}{y}}$.

*Solution*   The denominators in the complex fraction are *x* and *y*. Multiply both the numerator and denominator of the complex fraction by *xy*, the LCD of the complex fraction.

$$\frac{y^2}{\frac{1}{x} + \frac{1}{y}} = \frac{xy}{xy} \cdot \frac{y^2}{\left(\frac{1}{x} + \frac{1}{y}\right)}$$

$$= \frac{xy^3}{xy\left(\frac{1}{x}\right) + xy\left(\frac{1}{y}\right)}$$

$$= \frac{xy^3}{y + x}$$

Now Try Exercise 33

> **Helpful Hint**
>
> We have presented two methods for simplifying complex fractions. Which method should you use? Although either method can be used to simplify complex fractions, most students prefer to use method 1 when both the numerator and denominator consist of a single term, as in Example 1. When the complex fraction has a sum or difference of expressions in either the numerator or denominator, as in Examples 2, 3, 4, or 5, most students prefer to use method 2.

## EXERCISE SET 7.5    MathXL®   MyMathLab®

### Warm-Up Exercises

*Fill in the blanks with the appropriate word, phrase, or symbol(s) from the following list.*

complex fraction          denominator          numerator          5          $\dfrac{5}{3}$          $x^2 + 5x + 6$

**1.** The numerator of the complex fraction $\dfrac{\frac{5}{3}}{x^2 + 5x + 6}$ is _____.

**2.** A fraction that contains a fraction in its numerator or its denominator or in both is called a _____.

**3.** The numerator of the complex fraction $\dfrac{5}{\frac{3}{x^2 + 5x + 6}}$ is _____.

**4.** To simplify a complex fraction using multiplication, multiply both the numerator and _____ of the complex fraction by the LCD of the complex fraction.

### Practice the Skills

*Simplify.*

**5.** $\dfrac{1 + \frac{2}{3}}{5 + \frac{1}{3}}$

**6.** $\dfrac{2 + \frac{4}{5}}{1 - \frac{9}{16}}$

**7.** $\dfrac{2 + \frac{3}{8}}{1 + \frac{1}{3}}$

**8.** $\dfrac{\frac{1}{4} + \frac{5}{6}}{\frac{2}{3} + \frac{3}{5}}$

**9.** $\dfrac{\frac{2}{3} + \frac{1}{4}}{\frac{5}{6} - \frac{1}{3}}$

**10.** $\dfrac{\frac{5}{8} + \frac{1}{3}}{\frac{11}{12} - \frac{1}{6}}$

**11.** $\dfrac{\frac{xy^2}{7}}{\frac{3}{x^2}}$

**12.** $\dfrac{\frac{11a}{b^3}}{\frac{b^2}{4}}$

**13.** $\dfrac{\frac{6a^2b}{7}}{\frac{9ac^2}{b^2}}$

**14.** $\dfrac{\frac{18x^4}{5y^4z^5}}{\frac{9xy^2}{15z^5}}$

**15.** $\dfrac{a - \frac{a}{b}}{3 + a}$ all over $b$ — $\dfrac{a - \frac{a}{b}}{\frac{3 + a}{b}}$

**16.** $\dfrac{a + \frac{2}{b}}{\frac{a}{b}}$

**17.** $\dfrac{\frac{9}{t} + \frac{3}{t^2}}{3 + \frac{1}{t}}$

**18.** $\dfrac{\frac{4}{a} + \frac{1}{2a}}{a + \frac{a}{2}}$

**19.** $\dfrac{5 - \frac{1}{x}}{4 - \frac{1}{x}}$

**20.** $\dfrac{\frac{2x}{x - y}}{\frac{x^2}{y}}$

**21.** $\dfrac{\frac{m}{n} - \frac{n}{m}}{\frac{m + n}{n}}$

**22.** $\dfrac{5}{\frac{1}{x} + y}$

**23.** $\dfrac{\frac{a^2}{b} - b}{\frac{b^2}{a} - a}$

**24.** $\dfrac{\frac{1}{x^2} - \frac{8}{x}}{3 + \frac{1}{x^2}}$

**25.** $\dfrac{2 - \frac{a}{b}}{\frac{a}{b} - 2}$

**26.** $\dfrac{\frac{x}{y} - 9}{\frac{-x}{y} + 9}$

**27.** $\dfrac{\frac{4}{t^2} + \frac{4}{t}}{\frac{4}{t} + \frac{4}{t^2}}$

**28.** $\dfrac{\frac{a^2 - b^2}{a}}{\frac{a + b}{a^4}}$

**29.** $\dfrac{\frac{1}{a} - \frac{1}{b}}{\frac{1}{ab}}$

**30.** $\dfrac{\frac{1}{a} - \frac{1}{b}}{\frac{1}{a} + \frac{1}{b}}$

**31.** $\dfrac{\frac{a}{b} + \frac{1}{a}}{\frac{b}{a} + \frac{1}{a}}$

**32.** $\dfrac{\frac{2}{a} + \frac{3}{b}}{\frac{1}{a}}$

**33.** $\dfrac{x}{\dfrac{1}{x} - \dfrac{1}{y}}$

**34.** $\dfrac{\dfrac{1}{a} + \dfrac{1}{b}}{ab}$

**35.** $\dfrac{\dfrac{5}{a} + \dfrac{5}{a^2}}{\dfrac{5}{b} + \dfrac{5}{b^2}}$

**36.** $\dfrac{\dfrac{x}{y} - \dfrac{2}{x}}{\dfrac{y}{x} + \dfrac{1}{y}}$

## Problem Solving

*For the complex fractions in Exercises 37–40,*

**a)** *Determine which of the two methods discussed in this section you would use to simplify the fraction. Explain why.*

**b)** *Simplify by the method you selected in part **a)**.*

**c)** *Simplify by the method you did not select in part **a)**. If your answers to parts **b)** and **c)** are not the same, explain why.*

**37.** $\dfrac{5 + \dfrac{3}{5}}{\dfrac{1}{8} - 4}$

**38.** $\dfrac{\dfrac{x + y}{x^3} - \dfrac{1}{x}}{\dfrac{x - y}{x^5} + 5}$

**39.** $\dfrac{\dfrac{x - y}{x + y} + \dfrac{6}{x + y}}{2 - \dfrac{7}{x + y}}$

**40.** $\dfrac{\dfrac{25}{x - y} + \dfrac{2}{x + y}}{\dfrac{5}{x - y} - \dfrac{3}{x + y}}$

*In Exercises 41 and 42,* **a)** *write the complex fraction, and* **b)** *simplify the complex fraction.*

**41.** The numerator of the complex fraction consists of one term: 5 is divided by $12x$. The denominator of the complex fraction consists of two terms: 4 divided by $3x$ is subtracted from 8 divided by $x^2$.

**42.** The numerator of the complex fraction consists of two terms: 3 divided by $2x$ is subtracted from 6 divided by $x$. The denominator of the complex fraction consists of two terms: the sum of $x$ and the quantity 1 divided by $x$.

## Concept/Writing Exercises

**43.** Explain what a complex fraction is.

**44. a)** Select the method you prefer to simplify complex fractions and then write down a step-by-step procedure for simplifying complex fractions using that method.

**b)** Using your answer to part **a)**, simplify $\dfrac{\dfrac{4}{x} - \dfrac{3}{y}}{x + \dfrac{1}{y}}$.

## Challenge Problems

*Simplify. (Hint: Refer to Section 5.2, which discusses negative exponents.)*

**45.** $\dfrac{x^{-1} + y^{-1}}{3}$

**46.** $\dfrac{x^{-1} + y^{-1}}{y^{-1}}$

**47.** $\dfrac{x^{-1} + y^{-1}}{x^{-1}y^{-1}}$

**48.** $\dfrac{x^{-2} - y^{-2}}{y^{-1} - x^{-1}}$

**49. Jack** The efficiency of a jack, $E$, is expressed by the formula

$$E = \dfrac{\dfrac{1}{2}h}{h + \dfrac{1}{2}},$$ where $h$ is determined by the pitch of the jack's

thread. Determine the efficiency of a jack if $h$ is

**a)** $\dfrac{2}{3}$  **b)** $\dfrac{4}{5}$

Pitch

*Simplify.*

**50.** $\dfrac{\dfrac{x}{y} + \dfrac{y}{x} + \dfrac{2}{x}}{\dfrac{x}{y} + y}$

**51.** $\dfrac{\dfrac{a}{b} + b - \dfrac{1}{a}}{\dfrac{a}{b^2} - \dfrac{b}{a} + \dfrac{3}{a^2}}$

**52.** $\dfrac{x}{4 + \dfrac{x}{1 + x}}$

## Cumulative Review Exercises

[2.5] **53.** Solve the equation

$$2x - 8(5 - x) = 9x - 3(x + 2).$$

[5.4] **54.** What is a polynomial?

[6.3] **55.** Factor $x^2 - 13x + 40$.

[7.4] **56.** Subtract $\dfrac{x}{3x^2 + 17x - 6} - \dfrac{2}{x^2 + 3x - 18}$.

# 7.6 Solving Rational Equations

1 Solve rational equations with integer denominators.

2 Solve rational equations where a variable appears in a denominator.

## 1 Solve Rational Equations with Integer Denominators

A **rational equation** is one that contains one or more rational expressions. A rational equation may contain rational coefficients, such as $\frac{1}{2}x + \frac{3}{5}x = 8$ or $\frac{x}{2} + \frac{3x}{5} = 8$. A rational equation may also have a variable in a denominator, such as $\frac{4}{x-2} = 5$.

The emphasis of this section will be on solving rational equations where a variable appears in a denominator.

### To Solve Rational Equations

1. Determine the least common denominator (LCD) of all fractions in the equation.
2. Multiply *both* sides of the equation by the LCD. *This will result in every term in the equation being multiplied by the LCD.*
3. Remove any parentheses and combine like terms on each side of the equation.
4. Solve the equation using the properties discussed in earlier chapters.
5. Check your solution in the *original* equation.

The purpose of multiplying both sides of the equation by the LCD (step 2) is to eliminate all fractions from the equation. After both sides of the equation are multiplied by the LCD, the resulting equation should contain no fractions.

EXAMPLE 1   Solve $\frac{t}{4} - \frac{t}{5} = 1$ for $t$. ─────────

*Solution*   The LCD of 4 and 5 is 20. Multiply both sides of the equation by 20.

$$\frac{t}{4} - \frac{t}{5} = 1$$

$$20\left(\frac{t}{4} - \frac{t}{5}\right) = 20 \cdot 1 \qquad \text{Multiply both sides by the LCD, 20.}$$

$$20\left(\frac{t}{4}\right) - 20\left(\frac{t}{5}\right) = 20 \qquad \text{Distributive property}$$

$$5t - 4t = 20$$

$$t = 20$$

Check      $$\frac{t}{4} - \frac{t}{5} = 1$$

$$\frac{20}{4} - \frac{20}{5} \overset{?}{=} 1$$

$$5 - 4 \overset{?}{=} 1$$

$$1 = 1 \quad \text{True}$$

The solution is 20.

Now Try Exercise 13

EXAMPLE 2   Solve $\frac{x-5}{30} = \frac{4}{5} - \frac{x-1}{10}$. ─────────

*Solution*   Multiply both sides of the equation by the LCD, 30.

$$\frac{x-5}{30} = \frac{4}{5} - \frac{x-1}{10}$$

$$30\left(\frac{x-5}{30}\right) = 30\left(\frac{4}{5} - \frac{x-1}{10}\right) \qquad \text{Multiply both sides by the LCD, 30.}$$

$$x - 5 = 30\left(\frac{4}{5}\right) - 30\left(\frac{x-1}{10}\right) \qquad \text{Distributive property}$$

$$x - 5 = 24 - 3(x-1)$$

$$x - 5 = 24 - 3x + 3 \qquad \text{Distributive property}$$

$$x - 5 = -3x + 27 \qquad \text{Combined like terms.}$$

$$4x - 5 = 27 \qquad \qquad 3x \text{ was added to both sides.}$$

$$4x = 32 \qquad \qquad 5 \text{ was added to both sides.}$$

$$x = 8 \qquad \qquad \text{Both sides were divided by 4.}$$

A check will show that the answer is 8.

Now Try Exercise 27

## 2 Solve Rational Equations Where a Variable Appears in a Denominator

When solving a rational equation where a variable appears in any denominator, you *must* check your answer. *Whenever a variable appears in any denominator of a rational equation, it is necessary to check your answer in the original equation. If the answer obtained makes any denominator equal to zero, that value is not a solution to the equation.* Such values are called **extraneous roots** or **extraneous solutions**.

**EXAMPLE 3** Solve $4 - \dfrac{5}{x} = \dfrac{3}{2}$.

*Solution*    Multiply both sides of the equation by the LCD, $2x$.

$$2x\left(4 - \frac{5}{x}\right) = \left(\frac{3}{2}\right) \cdot 2x \qquad \text{Multiply both sides by the LCD, } 2x.$$

$$2x(4) - 2x\left(\frac{5}{x}\right) = \left(\frac{3}{2}\right) \cdot 2x \qquad \text{Distributive property}$$

$$8x - 10 = 3x$$

$$5x - 10 = 0 \qquad \qquad 3x \text{ was subtracted from both sides.}$$

$$5x = 10 \qquad \qquad 10 \text{ was added to both sides.}$$

$$x = 2$$

Check        $4 - \dfrac{5}{x} = \dfrac{3}{2}$

$$4 - \frac{5}{2} \stackrel{?}{=} \frac{3}{2}$$

$$\frac{8}{2} - \frac{5}{2} \stackrel{?}{=} \frac{3}{2}$$

$$\frac{3}{2} = \frac{3}{2} \qquad \text{True}$$

Since 2 does check, it is the solution to the equation.

Now Try Exercise 17

**EXAMPLE 4**   Solve $\dfrac{p-5}{p+3} = \dfrac{1}{5}$.

**Solution**   The LCD is $5(p+3)$. Multiply both sides of the equation by the LCD.

$$5(p+3) \cdot \frac{(p-5)}{p+3} = \frac{1}{5} \cdot 5(p+3)$$

$$5(p-5) = 1(p+3)$$

$$5p - 25 = p + 3$$

$$4p - 25 = 3$$

$$4p = 28$$

$$p = 7$$

A check will show that 7 is the solution.

Now Try Exercise 43

In Section 2.7 we illustrated that proportions of the form

$$\frac{a}{b} = \frac{c}{d}$$

can be cross-multiplied to obtain $a \cdot d = b \cdot c$. Example 4 is a proportion and can also be solved by cross-multiplying, as we will do in Example 5.

**EXAMPLE 5**   Use cross-multiplication to solve $\dfrac{9}{x+1} = \dfrac{5}{x-3}$.

**Solution**                $$\frac{9}{x+1} = \frac{5}{x-3}$$

$$9(x-3) = 5(x+1) \qquad \text{Cross-multiplied.}$$

$$9x - 27 = 5x + 5 \qquad \text{Distributive property was used.}$$

$$4x - 27 = 5$$

$$4x = 32$$

$$x = 8$$

A check will show that 8 is the solution to the equation.

Now Try Exercise 41

The following examples involve quadratic equations. Recall from Section 6.6 that quadratic equations have the form $ax^2 + bx + c = 0$, where $a \neq 0$.

**EXAMPLE 6**   Solve $x + \dfrac{12}{x} = -7$.

**Solution**                $$x + \frac{12}{x} = -7$$

$$x \cdot \left( x + \frac{12}{x} \right) = -7 \cdot x \qquad \text{Multiply both sides by } x.$$

$$x(x) + x\left( \frac{12}{x} \right) = -7x \qquad \text{Distributive property was used.}$$

$$x^2 + 12 = -7x$$

$$x^2 + 7x + 12 = 0 \qquad \text{7x was added to both sides.}$$

$$(x+3)(x+4) = 0 \qquad \text{Factored.}$$

$$x + 3 = 0 \quad \text{or} \quad x + 4 = 0 \qquad \text{Zero-factor property}$$

$$x = -3 \qquad\qquad x = -4$$

Check

$$x = -3$$

$$x + \frac{12}{x} = -7$$

$$-3 + \frac{12}{-3} \stackrel{?}{=} -7$$

$$-3 + (-4) \stackrel{?}{=} -7$$

$$-7 = -7 \quad \text{True}$$

$$x = -4$$

$$x + \frac{12}{x} = -7$$

$$-4 + \frac{12}{-4} \stackrel{?}{=} -7$$

$$-4 + (-3) \stackrel{?}{=} -7$$

$$-7 = -7 \quad \text{True}$$

The solutions are $-3$ and $-4$.

Now Try Exercise 57

**EXAMPLE 7**  Solve $\dfrac{x^2 - 2x}{x - 6} = \dfrac{24}{x - 6}$.

*Solution*   If we try to solve this equation using cross-multiplication we will get a cubic equation. We will solve this equation by multiplying both sides of the equation by the LCD, $x - 6$.

$$\frac{x^2 - 2x}{x - 6} = \frac{24}{x - 6}$$

$$\cancel{x - 6} \cdot \frac{x^2 - 2x}{\cancel{x - 6}} = \frac{24}{\cancel{x - 6}} \cdot \cancel{x - 6} \qquad \text{Multiply both sides by the LCD, } x - 6.$$

$$x^2 - 2x = 24$$

$$x^2 - 2x - 24 = 0 \qquad \qquad 24 \text{ was subtracted from both sides.}$$

$$(x + 4)(x - 6) = 0 \qquad \qquad \text{Factored}$$

$$x + 4 = 0 \quad \text{or} \quad x - 6 = 0 \qquad \text{Zero-factor property}$$

$$x = -4 \qquad \qquad x = 6$$

Check

$$x = -4$$

$$\frac{x^2 - 2x}{x - 6} = \frac{24}{x - 6}$$

$$\frac{(-4)^2 - 2(-4)}{-4 - 6} \stackrel{?}{=} \frac{24}{-4 - 6}$$

$$\frac{16 + 8}{-10} \stackrel{?}{=} \frac{24}{-10}$$

$$\frac{24}{-10} = \frac{24}{-10} \quad \text{True}$$

$$x = 6$$

$$\frac{x^2 - 2x}{x - 6} = \frac{24}{x - 6}$$

$$\frac{6^2 - 2(6)}{6 - 6} \stackrel{?}{=} \frac{24}{6 - 6}$$

$$\frac{24}{0} = \frac{24}{0}$$

↑         ↑

Since the denominator is 0, and we cannot divide by 0, 6 is not a solution.

Since $\dfrac{24}{0}$ is not a real number, 6 is an extraneous solution. Thus, this equation has only one solution, $-4$.

Now Try Exercise 47

**Understanding Algebra**

When multiplying both sides of an equation by a variable expression, checking answers is crucial to determine if any of your answers are extraneous solutions that must be eliminated from the final answer.

**Helpful Hint**

Remember, when solving a rational equation in which a variable appears in a denominator, you must check *all* your answers to make sure that none is an extraneous root. If any of your answers make any denominator 0, that answer is an extraneous root and not a true solution.

**EXAMPLE 8**  Solve $\dfrac{5w}{w^2 - 4} + \dfrac{1}{w - 2} = \dfrac{4}{w + 2}$.

*Solution*   First factor $w^2 - 4$.

$$\frac{5w}{(w + 2)(w - 2)} + \frac{1}{w - 2} = \frac{4}{w + 2}$$

Multiply both sides of the equation by the LCD, $(w + 2)(w - 2)$.

$$(w + 2)(w - 2)\left[\frac{5w}{(w + 2)(w - 2)} + \frac{1}{w - 2}\right] = \frac{4}{w + 2} \cdot (w + 2)(w - 2)$$

$$(w + 2)(w - 2) \cdot \frac{5w}{(w + 2)(w - 2)} + (w + 2)(w - 2) \cdot \frac{1}{w - 2} = \frac{4}{w + 2} \cdot (w + 2)(w - 2)$$

$$\cancel{(w + 2)}\cancel{(w - 2)} \cdot \frac{5w}{\cancel{(w + 2)}\cancel{(w - 2)}} + (w + 2)\cancel{(w - 2)} \cdot \frac{1}{\cancel{w - 2}} = \frac{4}{\cancel{w + 2}} \cdot \cancel{(w + 2)}(w - 2)$$

$$5w + (w + 2) = 4(w - 2)$$

$$6w + 2 = 4w - 8$$

$$2w + 2 = -8$$

$$2w = -10$$

$$w = -5$$

A check will show that $-5$ is the solution to the equation.

Now Try Exercise 65

### Helpful Hint

Some students confuse adding and subtracting rational expressions with solving rational equations. When adding or subtracting rational expressions, we must rewrite each expression with a common denominator. When solving a rational equation, we multiply both sides of the equation by the LCD to eliminate fractions from the equation. Consider the following two problems. Note that the one on the right is an equation because it contains an equals sign. We will work both problems. The LCD for both problems is $x(x + 4)$.

**Understanding Algebra**

When adding and subtracting rational expressions, we usually end up with an algebraic expression.

When solving rational equations, the solution, if one exists, will be a numerical value or values.

| Adding Rational Expressions | Solving Rational Equations |
|---|---|
| $\dfrac{x + 2}{x + 4} + \dfrac{3}{x}$ | $\dfrac{x + 2}{x + 4} = \dfrac{3}{x}$ |

We rewrite each fraction with the LCD, $x(x + 4)$.

$$= \frac{x}{x} \cdot \frac{x + 2}{x + 4} + \frac{3}{x} \cdot \frac{x + 4}{x + 4}$$

$$= \frac{x(x + 2)}{x(x + 4)} + \frac{3(x + 4)}{x(x + 4)}$$

$$= \frac{x^2 + 2x}{x(x + 4)} + \frac{3x + 12}{x(x + 4)}$$

$$= \frac{x^2 + 2x + 3x + 12}{x(x + 4)}$$

$$= \frac{x^2 + 5x + 12}{x(x + 4)}$$

We eliminate fractions by multiplying both sides of the equation by the LCD, $x(x + 4)$.

$$(x)(x + 4)\left(\frac{x + 2}{x + 4}\right) = \frac{3}{x} (x)(x + 4)$$

$$x(x + 2) = 3(x + 4)$$

$$x^2 + 2x = 3x + 12$$

$$x^2 - x - 12 = 0$$

$$(x - 4)(x + 3) = 0$$

$$x - 4 = 0 \quad \text{or} \quad x + 3 = 0$$

$$x = 4 \qquad\qquad x = -3$$

The numbers 4 and $-3$ on the right will both check and are thus solutions to the equation.

# EXERCISE SET 7.6

MathXL°   MyMathLab°
MathXL®         MyMathLab

## Warm-Up Exercises

*Fill in the blanks with the appropriate word, phrase, or symbol(s) from the following list.*

LCD                        check                    rational expression              No                    $(x + 1)(x - 3)$

rational equation          multiplied               $x + 1$                          Yes

1. When solving rational equations, both sides of the equation are _____ by the LCD.

2. When solving rational equations with variable denominators, it is very important to _____ your answers.

3. $\dfrac{x}{2} - \dfrac{x}{3} + \dfrac{5}{2x + 7}$ is an example of a _____ .

4. The first step in solving a rational equation is determining the _____ of all fractions.

5. $\dfrac{x}{2} - \dfrac{x}{3} = \dfrac{5}{2x + 7}$ is an example of a _____ .

6. (Yes or No)  Is  $x = 2$  a solution to the equation $\dfrac{3}{x - 2} + \dfrac{2}{x + 2} = \dfrac{4x}{x^2 - 4}$? _____

7. (Yes or No)  Is  $x = 1$  a solution to the equation $\dfrac{3}{x - 2} + \dfrac{2}{x + 2} = \dfrac{7x}{x^2 - 4}$? _____

8. The first step in solving the equation $\dfrac{9}{x + 1} = \dfrac{5}{x - 3}$ is to multiply both sides of the equation by the algebraic expression _____ .

## Practice the Skills

*Solve each equation and check your solution. See Examples 1 and 2.*

9. $\dfrac{x}{3} + \dfrac{x}{2} = 10$

10. $\dfrac{x}{3} - \dfrac{x}{2} = 10$

11. $\dfrac{y}{6} - \dfrac{y}{4} = \dfrac{1}{2}$

12. $\dfrac{x}{4} - \dfrac{x}{6} = \dfrac{1}{2}$

13. $\dfrac{x}{3} - \dfrac{x}{4} = 1$

14. $\dfrac{t}{5} - \dfrac{t}{6} = 2$

15. $\dfrac{r}{6} = \dfrac{r}{4} + \dfrac{1}{3}$

16. $\dfrac{n}{5} = \dfrac{n}{6} + \dfrac{2}{3}$

17. $\dfrac{z}{2} + 6 = \dfrac{z}{5}$

18. $\dfrac{3w}{5} - 6 = w$

19. $\dfrac{z}{6} + \dfrac{2}{3} = \dfrac{z}{5} - \dfrac{1}{3}$

20. $\dfrac{m - 2}{6} = \dfrac{2}{3} + \dfrac{m}{12}$

21. $d + 7 = \dfrac{3}{2}d + 5$

22. $\dfrac{q}{5} + \dfrac{q}{2} = \dfrac{21}{10}$

23. $3k + \dfrac{1}{6} = 4k - 4$

24. $\dfrac{p}{4} + \dfrac{1}{4} = \dfrac{p}{3} - \dfrac{1}{2}$

25. $\dfrac{n + 6}{3} = \dfrac{5(n - 8)}{10}$

26. $\dfrac{3(x - 6)}{5} = \dfrac{4(x + 2)}{8}$

27. $\dfrac{x - 5}{15} = \dfrac{3}{5} - \dfrac{x - 4}{10}$

28. $\dfrac{z + 4}{6} = \dfrac{3}{2} - \dfrac{2z + 2}{12}$

29. $\dfrac{-p + 1}{4} + \dfrac{13}{20} = \dfrac{p}{5} - \dfrac{p - 1}{2}$

30. $\dfrac{1}{10} - \dfrac{n + 1}{6} = \dfrac{1}{5} - \dfrac{n + 10}{15}$

31. $\dfrac{d - 3}{4} + \dfrac{1}{15} = \dfrac{2d + 1}{3} - \dfrac{34}{15}$

32. $\dfrac{t + 4}{5} = \dfrac{5}{8} + \dfrac{t + 7}{40}$

*Solve each equation and check your solution. See Examples 3–8.*

33. $2 + \dfrac{3}{x} = \dfrac{11}{4}$

34. $3 - \dfrac{1}{x} = \dfrac{14}{5}$

35. $7 - \dfrac{5}{x} = \dfrac{9}{2}$

36. $4 + \dfrac{3}{z} = \dfrac{9}{2}$

37. $\dfrac{4}{n} - \dfrac{3}{2n} = \dfrac{1}{2}$

38. $\dfrac{5}{3x} + \dfrac{2}{x} = 1$

39. $\dfrac{x - 1}{x - 5} = \dfrac{4}{x - 5}$

40. $\dfrac{2x + 3}{x + 2} = \dfrac{3}{2}$

41. $\dfrac{5}{a + 3} = \dfrac{4}{a + 1}$

42. $\dfrac{5}{x + 2} = \dfrac{1}{x - 4}$

43. $\dfrac{y + 3}{y - 3} = \dfrac{6}{4}$

44. $\dfrac{x}{x + 6} = \dfrac{2}{5}$

45. $\dfrac{2x - 3}{x - 4} = \dfrac{5}{x - 4}$

46. $\dfrac{3}{x} + 9 = \dfrac{3}{x}$

47. $\dfrac{x^2}{x - 3} = \dfrac{9}{x - 3}$

48. $\dfrac{x^2}{x + 5} = \dfrac{25}{x + 5}$

49. $\dfrac{n - 3}{n + 2} = \dfrac{n + 4}{n + 10}$

50. $\dfrac{x + 5}{x + 1} = \dfrac{x - 6}{x - 3}$

51. $\dfrac{1}{r} = \dfrac{3r}{8r + 3}$

52. $\dfrac{1}{r} = \dfrac{2r}{r + 15}$

53. $\dfrac{k}{k + 2} = \dfrac{3}{k - 2}$

54. $\dfrac{3a - 2}{2a + 2} = \dfrac{3}{a - 1}$

55. $\dfrac{4}{r} + r = \dfrac{20}{r}$

56. $a + \dfrac{5}{a} = \dfrac{14}{a}$

57. $x + \dfrac{20}{x} = -9$

58. $x - \dfrac{32}{x} = 4$

59. $\dfrac{3y - 2}{y + 1} = 4 - \dfrac{y + 2}{y - 1}$

60. $\dfrac{2b}{b + 1} = 2 - \dfrac{5}{2b}$

61. $\dfrac{1}{x + 3} + \dfrac{1}{x - 3} = \dfrac{-5}{x^2 - 9}$

62. $\dfrac{t + 2}{t - 5} - \dfrac{3}{4} = \dfrac{6}{t - 5}$

63. $\dfrac{x}{x - 3} + \dfrac{3}{2} = \dfrac{3}{x - 3}$

64. $\dfrac{y}{2y + 2} + \dfrac{2y - 16}{4y + 4} = \dfrac{y - 3}{y + 1}$

65. $\dfrac{3}{x - 5} - \dfrac{4}{x + 5} = \dfrac{11}{x^2 - 25}$

66. $\dfrac{2n^2 - 15}{n^2 + n - 6} = \dfrac{n + 1}{n + 3} + \dfrac{n - 3}{n - 2}$

**67.** $\dfrac{3x}{x^2 - 9} + \dfrac{1}{x - 3} = \dfrac{3}{x + 3}$

**68.** $\dfrac{3}{x + 3} + \dfrac{5}{x + 4} = \dfrac{12x + 7}{x^2 + 7x + 12}$

**69.** $\dfrac{1}{y - 1} + \dfrac{1}{2} = \dfrac{2}{y^2 - 1}$

**70.** $\dfrac{2y}{y + 2} = \dfrac{y}{y + 3} - \dfrac{3}{y^2 + 5y + 6}$

**71.** $\dfrac{3t}{6t + 6} + \dfrac{t}{2t + 2} = \dfrac{2t - 3}{t + 1}$

**72.** $\dfrac{2}{x - 2} - \dfrac{1}{x + 1} = \dfrac{2}{x^2 - x - 2}$

## Problem Solving

*In Exercises 73–78, determine the solution by observation. Explain how you determined your answer.*

**73.** $\dfrac{3}{x - 2} = \dfrac{x - 2}{x - 2}$

**74.** $\dfrac{1}{2} + \dfrac{x}{2} = \dfrac{5}{2}$

**75.** $\dfrac{x}{x - 6} + \dfrac{x}{x - 6} = 0$

**76.** $\dfrac{x}{4} + \dfrac{3x}{4} = x$

**77.** $\dfrac{x - 2}{3} + \dfrac{x - 2}{3} = \dfrac{2x - 4}{3}$

**78.** $\dfrac{3}{x} - \dfrac{1}{x} = \dfrac{2}{x}$

**79. Optics**  A formula frequently used in optics is

$$\frac{1}{p} + \frac{1}{q} = \frac{1}{f}$$

where $p$ represents the distance of the object from a mirror (or lens), $q$ represents the distance of the image from the mirror (or lens), and $f$ represents the focal length of the mirror (or lens). If a mirror has a focal length of 10 centimeters, how far from the mirror will the image appear when the object is 30 centimeters from the mirror?

© Kellie. L. Folkerts/Shutterstock

## Challenge Problems

**80. a)** Explain why the equation $\dfrac{x^2}{x - 3} = \dfrac{9}{x - 3}$ cannot be solved by cross-multiplying using the material presented in the book.

   **b)** Solve the equation given in part **a)**.

**81.** Solve the equation $\dfrac{x - 4}{x^2 - 2x} = \dfrac{-4}{x^2 - 4}$

**82. Electrical Resistance**  In electronics the total resistance $R_T$, of resistors wired in a parallel circuit is determined by the formula

$$\frac{1}{R_T} = \frac{1}{R_1} + \frac{1}{R_2} + \frac{1}{R_3} + \cdots + \frac{1}{R_n}$$

where $R_1, R_2, R_3, \ldots, R_n$ are the resistances of the individual resistors (measured in ohms) in the circuit.

   **a)** Find the total resistance if two resistors, one of 200 ohms and the other of 300 ohms, are wired in a parallel circuit.

   **b)** If three identical resistors are to be wired in parallel, what should be the resistance of each resistor if the total resistance of the circuit is to be 300 ohms?

**83.** Can an equation of the form $\dfrac{a}{x} + 1 = \dfrac{a}{x}$ have a real number solution for any real number $a$? Explain your answer.

## Group Activity

*Discuss and answer Exercise 84 as a group.*

**84. a)**  As a group, discuss two different methods you can use to solve the equation $\dfrac{x + 3}{5} = \dfrac{x}{4}$.

   **b)**  Group member 1: Solve the equation by obtaining a common denominator.
   Group member 2: Solve the equation by cross-multiplying.
   Group member 3: Check the results of group member 1 and group member 2.

   **c)**  Individually, create another equation by taking the reciprocal of each term in the equation in part **a)**. Compare your results. Do you think that the reciprocal of the answer you found in part **b)** will be the solution to this equation? Explain.

   **d)**  Individually, solve the equation you found in part **c)** and check your answer. Compare your work with the other group members. Was the conclusion you came to in part **c)** correct? Explain.

   **e)**  As a group, solve the equation $\dfrac{1}{x} + \dfrac{1}{3} = \dfrac{2}{x}$. Check your result.

   **f)**  As a group, create another equation by taking the reciprocal of each term of the equation in part **e)**. Do you think that the reciprocal of the answer you found in part **e)** will be the solution to this equation? Explain.

   **g)**  Individually, solve the equation you found in part **f)** and check your answer. Compare your work with the other group members. Did your group make the correct conclusion in part **f)**? Explain.

   **h)**  As a group, discuss the relationship between the solution to the equation $\dfrac{7}{x - 9} = \dfrac{3}{x}$ and the solution to the equation $\dfrac{x - 9}{7} = \dfrac{x}{3}$. Explain your answer.

## Cumulative Review Exercises

[3.2]  **85. Internet Plans**  An Internet service offers two plans for its customers. One plan includes 5 hours of use and costs $7.95 per month. Each additional minute after the 5 hours costs $0.15. The second plan costs $19.95 per month and provides unlimited Internet access. How many hours would Jake LaRue have to use the Internet monthly to make the second plan the less expensive?

   **86. Filling a Hot Tub**  How long will it take to fill a 600-gallon hot tub if water is flowing into the hot tub at a rate of 4 gallons a minute?

[3.3]  **87. Supplementary Angles**  Two angles are supplementary angles if the sum of their measures is 180°. Find the two supplementary angles if the smaller angle is 30° less than half the larger angle.

© Mcinich/Dreamstime LLC–Royalty Free

See Exercise 86.

[5.6]  **88.** Multiply $(3.4 \times 10^{-5})(2 \times 10^{13})$.

[6.6]  **89.** Explain the difference between a linear equation and a quadratic equation, and give an example of each.

# 7.7 Rational Equations: Applications and Problem Solving

1  Set up and solve applications containing rational expressions.

2  Set up and solve motion problems.

3  Set up and solve work problems.

$\frac{3}{5}x - 5$

$x$

**FIGURE 7.1**

## 1  Set Up and Solve Applications Containing Rational Expressions

Many applications of algebra involve rational equations. After we represent the application as an equation, we solve the rational equation as we did in Section 7.6.

The first type of application we will consider is a *geometry problem*.

**EXAMPLE 1**  **A New Rug**  Mary and Larry Armstrong are interested in purchasing a carpet whose area is 60 square feet. Determine the length and width if the width is 5 feet less than $\frac{3}{5}$ of the length. See **Figure 7.1**.

*Solution*

Understand and Translate

Let $x =$ length.

Then $\dfrac{3}{5}x - 5 =$ width.

$$\text{area} = \text{length} \cdot \text{width}$$

$$60 = x\left(\frac{3}{5}x - 5\right)$$

Carry Out

$$60 = \frac{3}{5}x^2 - 5x$$

$$5\,(60) = 5\left(\frac{3}{5}x^2 - 5x\right) \qquad \text{Multiply both sides by 5.}$$

$$300 = 3x^2 - 25x \qquad \text{Distributive property was used.}$$

$$0 = 3x^2 - 25x - 300 \qquad \text{Subtracted 300 from both sides.}$$

$$\text{or} \quad 3x^2 - 25x - 300 = 0$$

$$(3x + 20)(x - 15) = 0 \qquad \text{Factored.}$$

$$3x + 20 = 0 \qquad \text{or} \quad x - 15 = 0 \qquad \text{Zero-factor property}$$

$$3x = -20 \qquad\qquad\qquad x = 15$$

$$x = -\frac{20}{3}$$

**Check and Answer**   Since the length of a rectangle cannot be negative, we can eliminate $-\dfrac{20}{3}$ as an answer to our problem.

$$\text{length} = x = 15 \text{ feet}$$
$$\text{width} = \frac{3}{5}(15) - 5 = 4 \text{ feet}$$

**Check**
$$a = lw$$
$$60 \stackrel{?}{=} 15(4)$$
$$60 = 60 \qquad \text{True}$$

Therefore, the length is 15 feet and the width is 4 feet.

Now Try Exercise 5

Now we will work with a problem that expresses the relationship between two numbers. Problems like this are sometimes referred to as *number problems*.

**EXAMPLE 2  Reciprocals**  One number is 4 times another number. The sum of their reciprocals is $\dfrac{5}{2}$. Determine the numbers.

*Solution*    Understand and Translate

$$\text{Let } x = \text{first number.}$$
$$\text{Then } 4x = \text{second number.}$$

The reciprocal of the first number is $\dfrac{1}{x}$ and the reciprocal of the second number is $\dfrac{1}{4x}$. The sum of their reciprocals is $\dfrac{5}{2}$, thus, $\dfrac{1}{x} + \dfrac{1}{4x} = \dfrac{5}{2}$.

Carry Out
$$\frac{1}{x} + \frac{1}{4x} = \frac{5}{2}$$
$$4x\left(\frac{1}{x} + \frac{1}{4x}\right) = 4x\left(\frac{5}{2}\right) \qquad \text{Multiply both sides by the LCD, } 4x.$$
$$4x\left(\frac{1}{x}\right) + 4x\left(\frac{1}{4x}\right) = 10x \qquad \text{Distributive property}$$
$$4 + 1 = 10x$$
$$5 = 10x$$
$$\frac{5}{10} = x$$
$$\frac{1}{2} = x$$

**Check**   The first number is $\dfrac{1}{2}$. The second number is therefore $4x = 4\left(\dfrac{1}{2}\right) = 2$.

Let's now check if the sum of the reciprocals is $\dfrac{5}{2}$. The reciprocal of $\dfrac{1}{2}$ is 2. The reciprocal of 2 is $\dfrac{1}{2}$. The sum of the reciprocals is

$$2 + \frac{1}{2} = \frac{4}{2} + \frac{1}{2} = \frac{5}{2}$$

**Answer**   Since the sum of the reciprocals is $\dfrac{5}{2}$, the two numbers are 2 and $\dfrac{1}{2}$.

Now Try Exercise 11

## Understanding Algebra

The distance formula is usually written as:

$$\text{distance} = \text{rate} \cdot \text{time}$$

However, it is sometimes convenient to solve the formula for *time*:

$$\frac{\text{distance}}{\text{rate}} = \frac{\text{rate} \cdot \text{time}}{\text{rate}}$$

$$\frac{\text{distance}}{\text{rate}} = \text{time}$$

or   $\text{time} = \dfrac{\text{distance}}{\text{rate}}$

## 2 Set Up and Solve Motion Problems

In Chapter 3 we discussed *motion problems*. Recall that

$$\text{distance} = \text{rate} \cdot \text{time}$$

If we solve this equation for time, we obtain

$$\text{time} = \frac{\text{distance}}{\text{rate}} \quad \text{or} \quad t = \frac{d}{r}$$

This equation is useful in solving motion problems when the total time of travel for two objects or the time of travel between two points is known.

EXAMPLE 3 **Canoeing** Cindy Kilborn went canoeing in the Colorado River. The current in the river was 2 miles per hour. If it took Cindy the same amount of time to travel 10 miles downstream as 2 miles upstream, determine the speed at which Cindy's canoe would travel in still water.

*Solution*   Understand and Translate

Let $r =$ the canoe's speed in still water.

Then $r + 2 =$ the canoe's speed traveling downstream (with current)

and $r - 2 =$ the canoe's speed traveling upstream (against current.)

| Direction | Distance | Rate | Time |
|---|---|---|---|
| Downstream | 10 | $r + 2$ | $\dfrac{10}{r + 2}$ |
| Upstream | 2 | $r - 2$ | $\dfrac{2}{r - 2}$ |

Since the time it takes to travel 10 miles downstream is the same as the time to travel 2 miles upstream, we set the times equal to each other and then solve the resulting equation.

$$\text{time downstream} = \text{time upstream}$$

$$\frac{10}{r + 2} = \frac{2}{r - 2}$$

Carry Out
$$10(r - 2) = 2(r + 2) \qquad \text{Cross-multiplied.}$$
$$10r - 20 = 2r + 4$$
$$8r = 24$$
$$r = 3$$

Check and Answer   A check will show that 3 satisfies the equation. Thus, the canoe would travel at 3 miles per hour in still water.

Now Try Exercise 15

EXAMPLE 4 **Scenic Route** Shelby Kaylor drives along Route 72 in Oahu, Hawaii. Because of the beautiful scenery she drives an average of 20 miles per hour. Then she drives inland and averages 65 miles per hour. If the total distance she drove was 100 miles and the total time she drove was 3.5 hours, how long did she drive at each speed?

*Solution*   Understand and Translate

Let $d =$ distance traveled at 20 miles per hour.

Then $100 - d =$ distance traveled at 65 miles per hour.

| Direction | Distance | Rate | Time |
|---|---|---|---|
| Shoreline | $d$ | 20 | $\dfrac{d}{20}$ |
| Inland | $100 - d$ | 65 | $\dfrac{100 - d}{65}$ |

Since the total time spent driving is 3.5 hours, we write

time along shoreline + time inland = 3.5 hours

$$\frac{d}{20} + \frac{100 - d}{65} = 3.5$$

Carry Out

$$260\left(\frac{d}{20} + \frac{100 - d}{65}\right) = 260(3.5)$$

Multiply both sides by the LCD, 260.

$$\overset{13}{\cancel{260}}\left(\frac{d}{20}\right) + \overset{4}{\cancel{260}}\left(\frac{100 - d}{65}\right) = 910$$

Distributive property

$$13d + 4(100 - d) = 910$$
$$13d + 400 - 4d = 910$$
$$9d + 400 = 910$$
$$9d = 510$$
$$d = \frac{510}{9}$$
$$d \approx 57$$

**Answer**   Remember that the question asked us to find the *time* spent traveling at each speed. The variable $d$ represents the distance traveled at 20 miles per hour. To find the time traveled and to answer the question asked, we need to evaluate $\frac{d}{20}$ and $\frac{100 - d}{65}$ for $d = 57$.

Time at 20 mph

$$\frac{d}{20} = \frac{57}{20} \approx 2.9$$

Time at 65 mph

$$\frac{100 - d}{65} = \frac{100 - 57}{65} = \frac{43}{65} \approx 0.6$$

Thus, Shelby drove about 2.9 hours along the shoreline and about 0.6 hours inland. The total time was 2.9 + 0.6 or 3.5 hours.

Now Try Exercise 17

EXAMPLE 5   **Distance of a Race**   At a fund-raising race participants can either bike, walk, or run. Kim Clark, who rode a bike, completed the entire distance of the race with an average speed of 16 kilometers per hour (kph). Steve Schwartz, who jogged, completed the entire distance with an average speed of 5 kph. If Kim completed the race in 2.75 hours less time than Steve did, determine the distance the race covered.

*Solution*   Understand and Translate   Let $d =$ the distance from the start to the finish of the race. Then we can construct the following table. To determine the time, we divide the distance by the rate.

| Person | Distance | Rate | Time |
|--------|----------|------|------|
| Kim | $d$ | 16 | $\frac{d}{16}$ |
| Steve | $d$ | 5 | $\frac{d}{5}$ |

We are given that Kim completed the race in 2.75 hours less time than Steve did. Therefore, to make Kim's and Steve's times equal, we need to subtract 2.75 hours from Steve's time (or add 2.75 hours to Kim's time).

Time for Kim = Time for Steve − 2.75 hours

$$\frac{d}{16} = \frac{d}{5} - 2.75$$

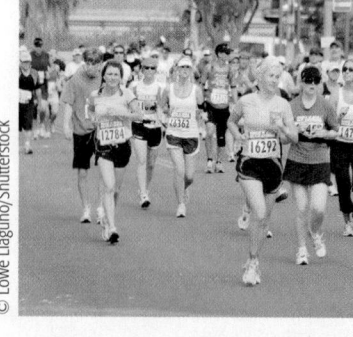

Carry Out
$$80\left(\frac{d}{16}\right) = 80\left(\frac{d}{5} - 2.75\right)$$   Multiply both sides by the LCD, 80.

$$5d = 80\left(\frac{d}{5}\right) - 80(2.75)$$   Distributive property was used.

$$5d = 16d - 220$$

$$-11d = -220$$

$$d = 20$$

Check and Answer   To check this answer we will determine the times it took Kim and Steve to complete the race and see if the difference between the times is 2.75 hours. To determine the times, divide the distance, 20 kilometers, by the rate.

$$\text{Kim's time} = \frac{d}{16} = \frac{20}{16} = 1.25 \text{ hours}$$

$$\text{Steve's time} = \frac{d}{5} = \frac{20}{5} = 4 \text{ hours}$$

Since $4 - 1.25 = 2.75$ hours, the answer checks. Therefore the distance the race covered is 20 kilometers.

Now Try Exercise 25

## 3 Set Up and Solve Work Problems

When two machines or two people work together to get a job done, the situation leads to solving a *work problem*. To solve work problems, we use the fact summarized in the following diagram:

$$\left(\begin{array}{c}\text{part of task done}\\\text{by one person}\\\text{or machine}\end{array}\right) + \left(\begin{array}{c}\text{part of task done}\\\text{by second person}\\\text{or machine}\end{array}\right) = \left(\begin{array}{c}1\\(\text{one completed})\\\text{task}\end{array}\right)$$

To determine the part of the task done by each person or machine, we use the following formula:

**rate of work · time worked = part of task completed**

An important step in solving these problems is determining the *rate* of work.
   Consider the following examples:

- If Joe can do a task by himself in 5 hours, his *rate* is $\frac{1}{5}$ of the task per hour.

- If Yoko can do a task by herself in 4 hours, her *rate* is $\frac{1}{4}$ of the task per hour.

- If a pump can empty a large tank in 10 hours, its *rate* is $\frac{1}{10}$ of a tank per hour.

In general, if a task can get done in $x$ hours, then the *rate* for completing that task is $\frac{1}{x}$ of the task per hour.

> ### Understanding Algebra
>
> Problems where two or more people or machines work together to complete a task are referred to as *work problems*

> ### Understanding Algebra
>
> If JoAnn can perform a task in 6 hours, then her *rate* of work is $\frac{1}{6}$ task per hour.
>
>    If she then works $t$ hours, the *amount* of work she has completed is represented as $\frac{1}{6} \cdot t$ or $\frac{t}{6}$ of the task.

EXAMPLE 6   **Plowing a Field**   Bob can plow a field by himself in 20 hours. His wife, Mary, can plow the same field by herself in 30 hours. How long will it take them to plow the field if they work together?

*Solution*   Understand and Translate   Let $t$ = the time, in hours, for Bob and Mary working together to plow the field. We will construct a table to help us in finding the part of the task completed by Bob and Mary in $t$ hours.

© Doug Menuez/Getty Images, Inc.–Photodisc./Royalty Free

| Person | Rate of Work (part of the task completed per hour) | Time Worked | Part of Task |
|---|---|---|---|
| Bob | $\dfrac{1}{20}$ | $t$ | $\dfrac{t}{20}$ |
| Mary | $\dfrac{1}{30}$ | $t$ | $\dfrac{t}{30}$ |

$$\left(\begin{array}{c}\text{part of the field plowed}\\\text{by Bob in } t \text{ hours}\end{array}\right) + \left(\begin{array}{c}\text{part of the field plowed}\\\text{by Mary in } t \text{ hours}\end{array}\right) = 1(\text{entire field plowed})$$

$$\frac{t}{20} + \frac{t}{30} = 1$$

**Carry Out**   Now multiply both sides of the equation by the LCD, 60.

$$60\left(\frac{t}{20} + \frac{t}{30}\right) = 60 \cdot 1$$

$$\overset{3}{60}\left(\frac{t}{20}\right) + \overset{2}{60}\left(\frac{t}{30}\right) = 60 \qquad \text{Distributive property}$$

$$3t + 2t = 60$$

$$5t = 60$$

$$t = 12$$

**Answer**   Thus, Bob and Mary working together can plow the field in 12 hours. We leave the check for you.

Now Try Exercise 27

**Helpful Hint**

In Example 6, Bob could plow the field by himself in 20 hours, and Mary could plow the field by herself in 30 hours. We determined that together they could plow the field in 12 hours. Does this answer make sense? Since you would expect the time to plow the field together to be less than the time either of them could plow it alone, the answer makes sense.

EXAMPLE 7  **Storing Wine**  At a winery in Napa Valley, California, one pipe can fill a tank with wine in 3 hours and another pipe can empty the tank in 5 hours. If the valves to both pipes are open, how long will it take to fill the empty tank?

*Solution*   Understand and Translate   Let $t =$ amount of time to fill the tank with the valves to both pipes open.

| Pipe | Rate of Work | Time | Part of Task |
|---|---|---|---|
| Pipe filling tank | $\dfrac{1}{3}$ | $t$ | $\dfrac{t}{3}$ |
| Pipe emptying tank | $\dfrac{1}{5}$ | $t$ | $\dfrac{t}{5}$ |

As one pipe is filling, the other is emptying the tank. Therefore, instead of adding the parts of the task, we will subtract the parts of the task.

$$\left(\begin{array}{c}\text{part of tank}\\\text{filled in } t \text{ hours}\end{array}\right) - \left(\begin{array}{c}\text{part of tank}\\\text{emptied in } t \text{ hours}\end{array}\right) = 1(\text{total tank filled})$$

$$\frac{t}{3} - \frac{t}{5} = 1$$

Carry Out

$$15\left(\frac{t}{3} - \frac{t}{5}\right) = 15 \cdot 1 \quad \text{Multiply both sides by the LCD, 15.}$$

$$\overset{5}{\cancel{15}}\left(\frac{t}{\cancel{3}}\right) - \overset{3}{\cancel{15}}\left(\frac{t}{\cancel{5}}\right) = 15 \quad \text{Distributive property was used.}$$

$$5t - 3t = 15$$
$$2t = 15$$
$$t = 7\frac{1}{2}$$

**Check and Answer**  The tank will be filled in $7\frac{1}{2}$ hours. This answer is reasonable because we expect it to take longer than 3 hours when the tank is being drained at the same time.

Now Try Exercise 35

**EXAMPLE 8 Cleaning Service** Linda and John Franco own a house cleaning service. When Linda cleans Damon's house by herself, it takes 7 hours. When Linda and John work together, they can clean the house in 4 hours. How long will it take John to clean the house by himself?

*Solution*  Let $t$ = time for John to clean the house by himself. Then John's rate is $\frac{1}{t}$. Since Linda can clean the house by herself in 7 hours, her rate is $\frac{1}{7}$ of the job per hour. In the table below, we use the fact that together they can clean the house in 4 hours.

| Worker | Rate of Work | Time | Part of Task |
|--------|--------------|------|--------------|
| Linda | $\frac{1}{7}$ | 4 | $\frac{4}{7}$ |
| John | $\frac{1}{t}$ | 4 | $\frac{4}{t}$ |

$$\left(\begin{array}{c}\text{part of house}\\\text{cleaned by Linda}\end{array}\right) + \left(\begin{array}{c}\text{part of house}\\\text{cleaned by John}\end{array}\right) = 1$$

$$\frac{4}{7} + \frac{4}{t} = 1$$

Carry Out

$$7t\left(\frac{4}{7} + \frac{4}{t}\right) = 7t \cdot 1 \quad \text{Multiply both sides by the LCD, } 7t.$$

$$\cancel{7}t\left(\frac{4}{\cancel{7}}\right) + 7\cancel{t}\left(\frac{4}{\cancel{t}}\right) = 7t \quad \text{Distributive property was used.}$$

$$4t + 28 = 7t$$
$$28 = 3t$$
$$\frac{28}{3} = t$$
$$9\frac{1}{3} = t$$

**Understanding Algebra**

In Example 8, notice that if two people were to work at Linda's rate, they would clean the house in $\frac{1}{2} \times 7 = \frac{7}{2}$, or $3\frac{1}{2}$, hours.

Since Linda and John together clean the house in 4 hours, we know that John cleans at a slower rate than Linda!

This is one way to determine that our answer makes sense.

**Check and Answer**  Thus, it takes John $9\frac{1}{3}$ hours, or 9 hours 20 minutes, to clean the house by himself. This answer is reasonable because we expect it to take longer for John to clean the house by himself than it would for Linda and John working together.

Now Try Exercise 39

EXAMPLE 9 **Thank-You Notes** Peter and Kaitlyn Kewin are handwriting thank-you notes to guests who attended their 20th wedding anniversary party. Kaitlyn by herself could write all the notes in 6 hours and Peter could write all the notes by himself in 10 hours. After Kaitlyn has been writing thank-you notes for 4 hours by herself, she must leave town on business. Peter then continues the task of writing the thank-you notes. How long will it take Peter to finish writing the remaining notes?

*Solution*   Understand and Translate   Let $t$ = time it will take Peter to finish writing the notes.

| Person | Rate of Work | Time | Part of Task |
|--------|--------------|------|--------------|
| Kaitlyn | $\dfrac{1}{6}$ | 4 | $\dfrac{4}{6} = \dfrac{2}{3}$ |
| Peter | $\dfrac{1}{10}$ | $t$ | $\dfrac{t}{10}$ |

$$\left( \begin{array}{c} \text{part of notes written} \\ \text{by Kaitlyn} \end{array} \right) + \left( \begin{array}{c} \text{part of notes written} \\ \text{by Peter} \end{array} \right) = 1$$

$$\frac{2}{3} \qquad + \qquad \frac{t}{10} \qquad = 1$$

Carry Out   $30\left(\dfrac{2}{3} + \dfrac{t}{10}\right) = 30 \cdot 1$   Multiply both sides by the LCD, 30.

$$\overset{10}{\cancel{30}}\left(\frac{2}{\cancel{3}_{1}}\right) + \overset{3}{\cancel{30}}\left(\frac{t}{\cancel{10}_{1}}\right) = 30 \qquad \text{Distributive property}$$

$$20 + 3t = 30$$

$$3t = 10$$

$$t = \frac{10}{3} \quad \text{or} \quad 3\frac{1}{3}$$

Answer   Thus, it will take Peter $3\dfrac{1}{3}$ hours to complete the notes.

Now Try Exercise 37

# EXERCISE SET 7.7    MathXL®    MyMathLab®
MathXL®        MyMathLab

## Warm-Up Exercises

*Fill in the blanks with the appropriate word, phrase, or symbol(s) from the following list.*

1 hour         $\dfrac{1}{5}$         $\dfrac{1}{3}$         $\dfrac{\text{distance}}{\text{rate}}$         $\dfrac{\text{rate}}{\text{distance}}$         3         1 complete task

**1.** In most work problems, one side of the equation is 1. The 1 represents _____ .

**2.** If Harlan Bricker can paint a room in 3 hours by himself, then his rate of work is _____ of the task per hour.

**3.** If Michelle Slocum can complete a task in 5 hours, then _____ represents the part of the task completed by her in one hour.

**4.** If we solve the equation distance = rate · time for time, we get, time = _____ .

## Practice the Skills/Problem Solving

*In Exercises 5–42, solve the problem and answer the question.*

Geometry Problems; see Example 1.

5. **Packaging Computers** The Phillips Paper Company makes rectangular pieces of cardboard for packing computers. The sheets of cardboard are to have an area of 99 square inches, and the length of a sheet is to be 5 inches more than $\frac{2}{3}$ its width. Determine the length and width of the cardboard to be manufactured. Use $A = l \cdot w$.

6. **Carry-on Luggage** On most airlines carry-on luggage can have a maximum width of 10 inches with a maximum volume of 3840 cubic inches (see the figure). If the height of the luggage is $\frac{2}{3}$ the length, determine the dimensions of the largest piece of carry-on luggage. Use $V = lwh$.

7. **Triangles of Dough** Pillsbury Crescent Rolls are packaged in tubes that contain perforated triangles of dough. The base of the triangular piece of dough is about 5 centimeters more than its height. Determine the base and height of a piece of dough if the area is about 42 square centimeters. Use $A = \frac{1}{2}bh$.

© Nikola Spasenoski/Shutterstock

8. **Yield Sign** Yield right of way signs used in the United States are triangles. The area of the sign is about 558 square inches. The height of the sign is about 5 inches less than its base. Determine the length of the base of a yield right of way sign.

9. **Triangular Garden** A triangular area is 20 square feet. Find the base of the triangular area if the height is 1 foot more than $\frac{1}{2}$ the base. Use $A = \frac{1}{2}bh$.

10. **Roofing** One side of the roof of Shelby Kaylor's house is in the shape of a trapezoid and has an area of 200 square feet. If the height of the trapezoid is $\frac{1}{4}$ the sum of the 2 bases, determine the height of the trapezoid. Use $A = \frac{1}{2}h(b + d)$.

Number Problems; see Example 2.

11. **Difference of Numbers** One number is 9 times larger than another. The difference of their reciprocals is 1. Determine the two numbers.

12. **Sum of Numbers** One number is 3 times larger than another. The sum of their reciprocals is $\frac{4}{3}$. Determine the two numbers.

13. **Increased Numerator** The numerator of the fraction $\frac{3}{4}$ is increased by an amount so that the value of the resulting fraction is $\frac{5}{2}$. Determine the amount by which the numerator was increased.

14. **Decreased Denominator** The denominator of the fraction $\frac{8}{21}$ is decreased by an amount so that the value of the resulting fraction is $\frac{1}{2}$. Determine the amount by which the denominator was decreased.

Motion Problems; see Examples 3–5.

15. **Paddleboat Ride** In the Mississippi River near New Orleans, the Creole Queen paddleboat travels 6 miles upstream (against the current) in the same amount of time it travels 12 miles downstream (with the current). If the current of the river is 3 miles per hour, determine the speed of the Creole Queen in still water.

16. **Kayak Ride** Kathy Boothby-Sestak can paddle her kayak 6 miles per hour in still water. It takes her as long to paddle 5 miles upstream as 10 miles downstream in the Wabash River near Lafayette, Indiana. Determine the river's current.

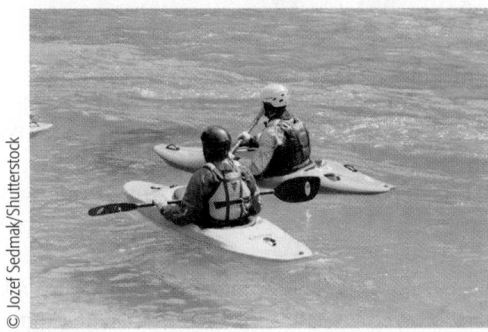

© Jozef Sedmak/Shutterstock

17. **Trolley Ride** A trolley travels in one direction at an average of 12 miles per hour, then turns around and travels on the same track in the opposite direction at 12 miles per hour. If the total time traveling on the trolley is $2\frac{1}{2}$ hours, how far did the trolley travel in one direction?

18. **Motorcycle Trip** Brandy Dawson and Jason Dodge start a motorcycle trip at the same point north of Fort Worth, Texas. Both are traveling to San Antonio, Texas, a distance of about 400 kilometers. Brandy rides 30 kilometers per hour faster than Jason does. When Brandy reaches her destination, Jason has only traveled to Austin, Texas, a distance of about 250 kilometers. Determine the approximate speed of each motorcycle.

19. **Jet Flight** Elenore Morales traveled 1600 miles by commercial jet from Kansas City, Missouri, to Spokane, Washington. She then traveled an additional 500 miles on a private

propeller plane from Spokane to Billings, Montana. If the speed of the jet was 4 times the speed of the propeller plane and the total time in the air was 6 hours, determine the speed of each plane.

20. **Exercise Regimen** Chris Barker walks a distance of 2 miles on an indoor track and then jogs at twice his walking speed for another 2 miles. If the total time spent on the track was one hour, determine the speeds at which he walks and jogs.

21. **No Wake Zone** Alisha is traveling by motorboat from her dock to Paradise Island. While she is in a no wake zone, her average speed is 4 miles per hour. Once she leaves the no wake zone, her average speed is 28 miles per hour. If the total distance traveled from her dock to the island is 36.6 miles and the total time of the trip is 1.7 hours, determine the distance from her dock to the end of the no wake zone and the distance from the end of the no wake zone to Paradise Island.

22. **The Tail of the Dragon** US 129 in Tennessee is a very popular road for motorcyclists because one stretch of that road has 318 curves and is called The Tail of the Dragon. Larry Gilligan rode his motorcycle in one direction and averaged 22 miles per hour. On the return trip it was raining and he only averaged 11 miles per hour. If the round trip took 1.5 hours, what is the length of The Tail of the Dragon?

© Karkas/Shutterstock

23. **Headwind and Tailwind** A Boeing 747 flew from San Francisco to Honolulu, a distance of 2900 miles. Flying with the wind, it averaged 600 miles per hour. When the wind changed from a tailwind to a headwind, the plane's speed dropped to 550 miles per hour. If the total time of the trip was 5 hours, determine the length of time it flew at each speed.

24. **Thalys Train** The Thalys train in Europe has been known to travel an average 240 kilometers per hour (kph). Prior to using the Thalys trains in Europe, trains traveled an average speed of 120 kph. If a Thalys train traveling from Brussels to Amsterdam can complete its trip in 0.88 hour less time than an older train, determine the distance from Brussels to Amsterdam.

© Allen R. Angel

Thalys train

25. **Water Skiers** At a water show a boat pulls a water skier at a speed of 30 feet per second. When it reaches the end of the lake, more skiers are added to be pulled by the boat, so the boat's speed drops to 25 feet per second. If the boat traveled the same distance on both trips and the trip back with the additional skiers took 8 seconds longer than the trip with the single skier, how far, in feet, in one direction, had the boat traveled?

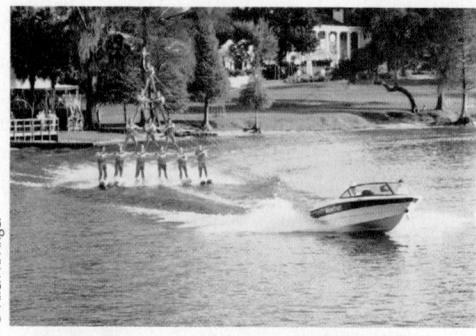
© Allen R. Angel

26. **Cross-Country Skiing** Alana Bradley and her father Tim begin skiing the same cross-country ski trail in Elmwood Park in Sioux Falls, South Dakota, at the same time. If Alana, who averages 9 miles per hour, finishes the trail 0.25 hours sooner than her father, who averages 6 miles per hour, determine the length of the trail.

Work Problems; see Examples 6–9.

27. **Wallpaper** Reynaldo and Felicia Fernandez decide to wallpaper their family room. Felicia, who has wallpapering experience, can wallpaper the room in 6 hours. Reynaldo can wallpaper the same room in 8 hours. How long will it take them to wallpaper the family room if they work together?

28. **Conveyor Belt** At a salt mine, one conveyor belt requires 20 minutes to fill a large truck with ore. A second conveyor belt requires 30 minutes to fill the same truck with ore. How long would it take if both conveyor belts were working together to fill the truck with ore?

29. **Picking Peaches** In a peach orchard in Williamson, New York, Gary Rominger can load his truck with peaches in 6 hours. His friend, Alex Taurke, takes twice as long to load Gary's truck with peaches. How long will it take them working together to load the truck with peaches?

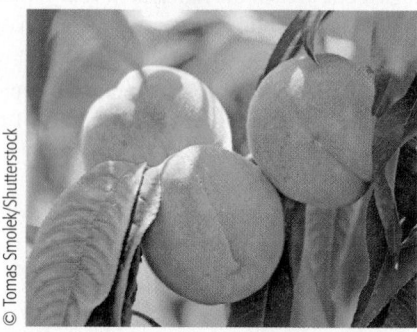
© Tomas Smolek/Shutterstock

30. **Watering Plants** In a small nursery, Becky Hailey can water all the plants in 30 minutes. Her co-worker, Karen Grizzaffi, can water all the plants in 20 minutes. How long will it take them working together to water the plants?

31. **Painting a Room** Eric Kweeder can paint a room in 60 minutes. His brother, Jessup, can paint the same room in 40 minutes. How long will it take them working together to paint the room?

**32. Tree Chipping** The ClearCut Tree Service has two models of chippers, the "Pirate" and the "Ninja." The Pirate can chip a 20-ton load of trees in 2 hours while the Ninja can chip a 20-ton load of trees in 4 hours. How long will it take the two chippers working simultaneously to chip the 20-ton load of trees?

© Susan Law Cain/Shutterstock

**33. Hot Tub** Pam and Loren Fornieri know that their hot tub can be filled in 40 minutes and drained completely in 60 minutes. If the water is turned on and the drain is left open, how long would it take the tub to fill completely?

**34. Filling a Tank** During a rainstorm, the rain is flowing into a large holding tank. At the rate the rain is falling, the empty tank would fill in 8 hours. At the bottom of the tank is a spigot to dispense water. Typically, it takes about 12 hours with the spigot wide open to empty the water in a full tank. If the tank is empty and the spigot has been accidentally left open, and the rain falls at the constant rate, how long would it take for the tank to fill completely?

**35. Payroll Checks** At the Community Savings Bank, it takes a computer 40 minutes to process and print payroll checks. When a second computer is used and the two computers work together, the checks can be processed and printed in 24 minutes. How long would it take the second computer by itself to process and print the payroll checks?

**36. Flowing Water** When the water is turned on and passes through a small hose, a pool can be filled in 6 hours. When the water is turned on at two spigots and passes through both the small hose and a large hose, the pool can be filled in 2 hours. How long would it take to fill the pool using only the large hose?

**37. Digging a Trench** A construction company with two backhoes has contracted to dig a long trench for drainage pipes. The larger backhoe can dig the entire trench by itself in 12 days. The smaller backhoe can dig the entire trench by itself in 15 days. The large backhoe begins working on the trench by itself, but after 5 days it is transferred to a different job and the smaller backhoe begins working on the trench. How long will it take for the smaller backhoe to complete the job?

**38. Delivery of Food** Ian and Nicole Murphy deliver food to various restaurants. If Ian drove the entire trip, the trip would take about 10 hours. If Nicole drove the entire trip, the trip would take about 8 hours. After Nicole had been driving for 4 hours, Ian takes over the driving. About how much longer will Ian drive before they reach their final destination?

**39. Snowstorm** Following a snowstorm, Ken and Bettina Reeves must clear their driveway and sidewalk. Ken can clear the snow by himself in 4 hours, and Bettina can clear the snow by herself in 6 hours. After Bettina has been working for 3 hours, Ken is able to join her. How much longer will it take them working together to remove the rest of the snow?

© Julien/Shutterstock

**40. Photocopying** The College of Applied Science just added two new photocopier machines to its business office and retained one older machine. The older machine, working alone, can copy a full set of diplomas in 3 hours. Each of the two new machines, working alone, can copy a full set of diplomas in 2 hours.

a) How long would it take the three copiers working together to copy a full set of diplomas?

b) How long would it take the three copiers working together to copy 100 full sets of diplomas?

**41. Skimming Oil** A boat designed to skim oil off the surface of the water has two skimmers. One skimmer can fill the boat's holding tank in 60 hours while the second skimmer can fill the boat's holding tank in 50 hours. There is also a valve in the holding tank that is used to transfer the oil to a larger vessel. If no new oil is coming into the holding tank, a full holding tank of skimmed oil can be transferred to a larger tank in 30 hours. If both skimmers begin skimming and the valve on the holding tank is opened, how long will it take for the empty holding tank on the boat to fill?

**42. Flower Garden** Bob can plant a flower garden by himself in 8 hours. Mary can plant the same garden by herself in 10 hours, and Gloria can plant the same garden by herself in 12 hours. How long would it take them working together to plant the garden?

© Doug Menuez/Getty Images, Inc—Photodisc/Royalty Free

## Challenge Problems

**43. Reciprocal of a Number** If 2 times a number is added to 3 times the reciprocal of the number, the answer is 7. Determine the number(s).

**44. Determine a Number** The reciprocal of the difference of a certain number and 5 is twice the reciprocal of the difference of twice the number and 10. Determine the number(s).

**45. Picking Blueberries** Ed and Samantha Weisman, whose parents own a fruit farm, must each pick the same number of pints of blueberries each day during the season. Ed picks an average of 8 pints per hour, while Samantha picks an average of 4 pints per hour. If Ed and Samantha begin picking blueberries at the same time, and Samantha finishes 1 hour after Ed, how many pints of blueberries must each pick?

**46. Sorting Mail** A mail processing machine can sort a large bin of mail in 1 hour. A newer model can sort the same quantity of mail in 30 minutes. If they operate together, how long will it take them to sort the bin of mail?

## Cumulative Review Exercises

[2.1]   **47.** Simplify $\frac{1}{2}(x + 3) - (2x + 15)$.

[4.2]   **48.** Graph $x = -2$.

[7.2]   **49.** Divide $\dfrac{x^2 - 14x + 48}{x^2 - 5x - 24} \div \dfrac{2x^2 - 13x + 6}{2x^2 + 5x - 3}$.

[7.4]   **50.** Subtract $\dfrac{x}{6x^2 - x - 15} - \dfrac{5}{9x^2 - 12x - 5}$.

# 7.8 Variation

**1** Solve direct variation problems.

**2** Solve inverse variation problems.

**3** Solve joint variation problems.

**4** Solve combined variation problems.

**Variation equations** show how one quantity changes in relation to another quantity or quantities. In this section we will discuss three types of variation: *direct*, *inverse*, and *joint* variation.

## **1** Solve Direct Variation Problems

**Direct variation** involves two variables that increase together or decrease together. For example, consider a car traveling 80 miles per hour on an interstate highway. The car travels

- 80 miles in 1 hour,
- 160 miles in 2 hours,
- 240 miles in 3 hours, and so on.

As the *time* increases, the *distance* also increases.
     The formula used to calculate distance traveled is

$$\text{distance} = \text{rate} \cdot \text{time}$$

Since the rate in the example above is constant, the formula can be written

$$d = 80t$$

We say distance *varies directly* as time or that distance is *directly proportional* to time.

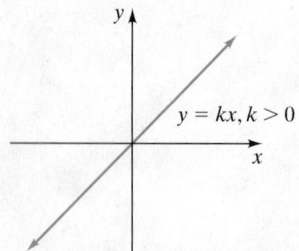

FIGURE 7.2

> **Direct Variation**
>
> If a variable $y$ varies directly as a variable $x$, then
>
> $$y = kx$$
>
> where $k$ is the **constant of proportionality** or the **variation constant**.

The graph of $y = kx$, $k > 0$, is always a straight line that goes through the origin (see **Fig. 7.2**). The slope of the line depends on the value of $k$.

EXAMPLE 1 **Circle** The circumference of a circle, $C$, is directly proportional to (or varies directly as) its radius, $r$. Write the equation for the circumference of a circle if the constant of proportionality, $k$, is $2\pi$.

*Solution*

$$C = kr \qquad \text{\small $C$ varies directly as $r$}$$
$$C = 2\pi r \qquad \text{\small Constant of proportionality is $2\pi$}$$

Now Try Exercise 11

EXAMPLE 2 **Administering a Drug** The amount, $a$, of the drug theophylline given to patients is directly proportional to the patient's mass, $m$, in kilograms.

a) Write the variation equation.

b) If 150 mg is given to a boy whose mass is 30 kg, find the constant of proportionality.

c) How much of the drug should be given to a patient whose mass is 62 kg?

*Solution*

a) We are told the two quantities are directly proportional. We therefore set up a direct variation equation.

$$a = km$$

b) **Understand and Translate** To determine the value of the constant of proportionality, we substitute the given values for the amount and mass. We then solve for $k$.

$$a = km$$
$$150 = k(30) \qquad \text{\small Substitute the given values.}$$

Carry Out $\qquad\qquad\qquad 5 = k$

Answer Thus, $k = 5$ mg. Five milligrams of the drug should be given for each kilogram of a person's mass.

c) **Understand and Translate** Now that we know the constant of proportionality we can use it to determine the amount of the drug to use for a person's mass. We set up the variation equation and substitute the values of $k$ and $m$.

$$a = km$$
$$a = 5(62) \qquad \text{\small Substitute the given values.}$$

Carry Out $\qquad\qquad\qquad a = 310$

Answer Thus, 310 mg of theophylline should be given to a person whose mass is 62 kg.

Now Try Exercise 57

## Understanding Algebra

Direct variation involves two variables that increase together or decrease together.

The phrases

- $y$ varies directly as $x$, and
- $y$ is directly proportional to $x$,

are represented by the direct variation eqation

$$y = kx.$$

EXAMPLE 3 $y$ varies directly as the square of $z$. If $y$ is 80 when $z$ is 20, find $y$ when $z$ is 45.

*Solution* Since $y$ varies directly as the *square of z*, we begin with the formula $y = kz^2$. Since the constant of proportionality is not given, we must first find $k$ using the given information.

$$y = kz^2$$
$$80 = k(20)^2 \qquad \text{\small Substitute the given values.}$$
$$80 = 400k \qquad \text{\small Solve for $k$.}$$
$$\frac{80}{400} = \frac{400k}{400}$$
$$0.2 = k$$

We now use $k = 0.2$ to find $y$ when $z$ is 45.

$$y = kz^2$$
$$y = 0.2(45)^2 \qquad \text{Substitute the given values.}$$
$$y = 405$$

Thus, when $z$ equals 45, $y$ equals 405.

Now Try Exercise 35

## 2 Solve Inverse Variation Problems

**Inverse variation** involves two variables in which one variable increases as the other decreases and vice versa. For example, consider traveling 120 miles in a car. If the car is traveling

- 30 miles per hour, the trip takes 4 hours,
- 40 miles per hour, the trip takes 3 hours,
- 60 miles per hour, the trip takes 2 hours, and so on.

As the *rate* increases, the *time* to travel 120 miles decreases.

The formula used to calculate time given the distance and the rate is

$$\text{time} = \frac{\text{distance}}{\text{rate}}$$

Since the distance in the examples above is constant, the formula can be rewritten

$$\text{time} = \frac{120}{\text{rate}}$$

We say time *varies inversely* as rate or that time is *inversely proportional* to rate.

---

**Inverse Variation**

If a variable $y$ varies inversely as a variable $x$, then

$$y = \frac{k}{x} \qquad (\text{or } xy = k)$$

where $k$ is the constant of proportionality.

---

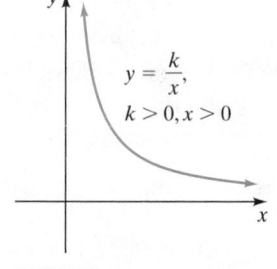

$$y = \frac{k}{x},$$
$$k > 0, x > 0$$

**FIGURE 7.3**

The graph of $y = \dfrac{k}{x}$, for $k > 0$ and $x > 0$, will have the shape illustrated in **Figure 7.3**. The graph of an inverse variation equation is not defined at $x = 0$ because 0 is not in the domain of the function $y = \dfrac{k}{x}$.

EXAMPLE 4  **Melting Ice**  The amount of time, $t$, it takes a block of ice to melt in water is inversely proportional to the water's temperature, $T$.

**a)** Write the variation equation.

**b)** If a block of ice takes 15 minutes to melt in 60°F water, determine the constant of proportionality.

**c)** Determine how long it will take a block of ice of the same size to melt in 50°F water.

*Solution*

**a)** The hotter the water temperature, the shorter the time for the block of ice to melt. The inverse variation is

$$t = \frac{k}{T}$$

## Understanding Algebra

Inverse variation involves two variables in which one variable increases as the other decreases and vice versa.

The phrases

- *y* varies inversely as *x*, and
- *y* is inversely proportional to *x*,

are represented by the inverse variation equation

$$y = \frac{k}{x}.$$

**b) Understand and Translate**   To determine the constant of proportionality, we substitute the values for the temperature and time and solve for *k*.

$$t = \frac{k}{T}$$

$$15 = \frac{k}{60} \qquad \text{Substitute the given values.}$$

Carry Out $$\qquad 900 = k$$

**Answer**   The constant of proportionality is 900.

**c) Understand and Translate**   Now that we know the constant of proportionality, we can use it to determine how long it will take for the same size block of ice to melt in 50°F water. We set up the proportion, substitute the values for *k* and *T*, and solve for *t*.

$$t = \frac{k}{T}$$

$$t = \frac{900}{50} \qquad \text{Substitute the given values.}$$

Carry Out $$\qquad t = 18$$

**Answer**   It will take 18 minutes for the block of ice to melt in the 50°F water.

Now Try Exercise 61

**EXAMPLE 5   Lighting**   The illuminance, $I$, of a light source varies inversely as the square of the distance, $d$, from the source. Assuming that the illuminance is 75 units at a distance of 4 meters, find the formula that expresses the relationship between the illuminance and the distance.

*Solution*   Understand and Translate   Since the illuminance varies inversely as the *square* of the distance, the general form of the equation is

$$I = \frac{k}{d^2}$$

To find $k$, we substitute the given values for $I$ and $d$.

$$75 = \frac{k}{4^2} \qquad \text{Substitute the given values.}$$

Carry Out $$\qquad 75 = \frac{k}{16} \qquad \text{Solve for } k.$$

$$(75)(16) = k$$

$$1200 = k$$

**Answer**   The formula is $I = \dfrac{1200}{d^2}$.

Now Try Exercise 65

## 3 Solve Joint Variation Problems

**Joint variation** involves a variable that varies directly as the product of two or more other variables.

**Joint Variation**

If $y$ varies jointly as $x$ and $z$, then

$$y = kxz$$

where $k$ is the constant of proportionality.

**EXAMPLE 6  Area of a Triangle**  The area, $A$, of a triangle varies jointly as its base, $b$, and height, $h$. If the area of a triangle is 48 square inches when its base is 12 inches and its height is 8 inches, find the area of a triangle whose base is 15 inches and height is 40 inches.

*Solution*  Understand and Translate  First write the joint variation equation and then substitute the known values and solve for $k$.

$$A = kbh$$

$$48 = k(12)(8) \qquad \text{Substitute the given values.}$$

Carry Out

$$48 = k(96) \qquad \text{Solve for } k.$$

$$\frac{48}{96} = k$$

$$k = \frac{1}{2}$$

Now solve for the unknown area using the given values.

$$A = kbh$$

$$= \frac{1}{2}(15)(40) \qquad \text{Substitute the given values.}$$

$$= 300$$

Answer  The area of the triangle is 300 square inches.

Now Try Exercise 69

**Summary of Variation Equations**

| DIRECT | INVERSE | JOINT |
|--------|---------|-------|
| $y = kx$ | $y = \dfrac{k}{x}$ | $y = kxz$ |

## 4  Solve Combined Variation Problems

In **combined variation**, direct and inverse variation occurs among three or more variables at the same time.

**EXAMPLE 7  Pretzel Shop**  The owners of an Auntie Anne's Pretzel Shop find that their weekly sales of pretzels, $S$, vary directly as their advertising budget, $A$, and inversely as their pretzel price, $P$. When their advertising budget is $400 and the price is $1, they sell 6200 pretzels.

**a)** Write a variation equation expressing $S$ in terms of $A$ and $P$. Include the value of the constant.

481

**b)** Find the expected sales if the advertising budget is $600 and the price is $1.20.

*Solution*

**a)** Understand and Translate    We begin with the equation

$$S = \frac{kA}{P}$$

$$6200 = \frac{k(400)}{1} \qquad \text{Substitute the given values.}$$

Carry Out $\qquad\qquad 6200 = 400k \qquad\qquad \text{Solve for } k.$

$$15.5 = k$$

Answer    Therefore, the equation for the sales of pretzels is $S = \frac{15.5A}{P}$.

**b)** Understand and Translate    Now that we know the combined variation equation, we can use it to determine the expected sales for the given values.

$$S = \frac{15.5A}{P}$$

$$= \frac{15.5(600)}{1.20} \qquad \text{Substitute the given values.}$$

Carry Out $\qquad\qquad\qquad = 7750$

Answer    They can expect to sell 7750 pretzels.

Now Try Exercise 71

**EXAMPLE 8  Electrostatic Force**  The electrostatic force, $F$, of repulsion between two positive electrical charges is jointly proportional to the two charges, $q_1$ and $q_2$, and inversely proportional to the square of the distance, $d$, between the two charges. Express $F$ in terms of $q_1$, $q_2$, and $d$.

*Solution*

$$F = \frac{kq_1q_2}{d^2}$$

Now Try Exercise 75

# EXERCISE SET 7.8

MathXL®   MyMathLab®

MathXL®      MyMathLab

## Warm-Up Exercises

*Fill in the blanks with the appropriate word, phrase, or symbol(s) from the following list.*

| joint | direct | combined | inverse | constant | $y = kx$ | $y = \dfrac{k}{x}$ |

**1.** In _____ variation, direct and inverse variation occurs among three or more variables at the same time.

**2.** _____ variation involves two variables in which one variable increases as the other decreases and vice versa.

**3.** _____ variation involves two variables that increase together or decrease together.

**4.** _____ variation involves a variable that varies directly as the product of two or more other variables.

**5.** When $y$ varies directly as $x$, we write $y = kx$, where $k$ is the _____ of proportionality.

**6.** If $y$ varies inversely as $x$, then the variation equation is _____ .

## Practice the Skills

*In Exercises 7–24, determine whether the variation between the indicated quantities is direct or inverse.*

7. The speed of a car and the distance traveled in one hour.

8. The number of pages Tom can read in a 2-hour period and his reading speed

9. The speed of an athlete and the time it takes him to run a 10-kilometer race

© Lowe Llaguno/Shutterstock

10. Barbara's weekly salary and the amount of money withheld for state income taxes

11. The radius of a circle and its area

12. The side of a cube and its volume

13. The radius of a balloon and its volume

14. The diameter of a circle and its circumference

15. The diameter of a hose and the volume of water coming out of the hose

16. The air temperature and the thickness of the clothing needed for a person to stay warm

17. The time it takes an ice cube to melt in water and the temperature of the water

18. The distance between two cities on a map and the actual distance between the two cities

19. The area of a lawn and the time it takes to mow the lawn

20. The displacement, measured in liters, of an engine and the horsepower of the engine

© Selena/Shutterstock

21. The speed of the typist and the time taken to type a 500-word essay

22. The number of calories eaten and the amount of exercise required to burn off those calories

23. The light illuminating an object and the distance the light is from the object

24. The number of calories in a cheeseburger and the size of the cheeseburger

*For Exercises 25–32, **a)** write the variation equation and **b)** find the quantity indicated.*

25. $x$ varies directly as $y$. Find $x$ when $y = 12$ and $k = 6$.

26. $t$ varies inversely as $r$. Find $t$ when $r = 20$ and $k = 60$.

27. $y$ varies directly as $R$. Find $y$ when $R = 180$ and $k = 1.7$.

28. $x$ varies inversely as $y$. Find $x$ when $y = 25$ and $k = 5$.

29. $R$ varies inversely as $W$. Find $R$ when $W = 160$ and $k = 8$.

30. $L$ varies inversely as the square of $P$. Find $L$ when $P = 4$ and $k = 100$.

31. $A$ varies directly as $B$ and inversely as $C$. Find $A$ when $B = 12$, $C = 4$, and $k = 3$.

32. $A$ varies jointly as $R_1$ and $R_2$ and inversely as the square of $L$. Find $A$ when $R_1 = 120$, $R_2 = 8$, $L = 5$, and $k = \dfrac{3}{2}$.

*For Exercises 33–42, **a)** write the variation equation and **b)** find the quantity indicated.*

33. $x$ varies directly as $y$. If $x$ is 12 when $y$ is 3, find $x$ when $y$ is 5.

34. $Z$ varies directly as $W$. If $Z$ is 7 when $W$ is 28, find $Z$ when $W$ is 140.

35. $y$ varies directly as the square of $R$. If $y$ is 5 when $R$ is 5, find $y$ when $R$ is 10.

36. $P$ varies directly as the square of $Q$. If $P$ is 32 when $Q$ is 4, find $P$ when $Q$ is 7.

37. $S$ varies inversely as $G$. If $S$ is 12 when $G$ is 0.4, find $S$ when $G$ is 5.

38. $C$ varies inversely as $J$. If $C$ is 7 when $J$ is 0.7, find $C$ when $J$ is 12.

39. $x$ varies inversely as the square of $P$. If $x$ is 4 when $P$ is 5, find $x$ when $P$ is 2.

40. $R$ varies inversely as the square of $T$. If $R$ is 3 when $T$ is 6, find $R$ when $T$ is 2.

41. $F$ varies jointly as $M_1$ and $M_2$ and inversely as $d$. If $F$ is 20 when $M_1 = 5$, $M_2 = 10$, and $d = 0.2$, find $F$ when $M_1 = 10$, $M_2 = 20$, and $d = 0.4$.

42. $F$ varies jointly as $q_1$ and $q_2$ and inversely as the square of $d$. If $F$ is 8 when $q_1 = 2$, $q_2 = 8$, and $d = 4$, find $F$ when $q_1 = 28$, $q_2 = 12$, and $d = 2$.

## Problem Solving

43. Assume $a$ varies directly as $b$. If $b$ is doubled, how will it affect $a$? Explain.

44. Assume $a$ varies directly as $b^2$. If $b$ is doubled, how will it affect $a$? Explain.

**45.** Assume $y$ varies inversely as $x$. If $x$ is doubled, how will it affect $y$? Explain.

**46.** Assume $y$ varies inversely as $a^2$. If $a$ is doubled, how will it affect $y$? Explain.

*In Exercises 47–52, use the formula $F = \dfrac{km_1m_2}{d^2}$.*

**47.** If $m_1$ is doubled, how will it affect $F$?

**48.** If $m_1$ is quadrupled and $d$ is doubled, how will it affect $F$?

**49.** If $m_1$ is doubled and $m_2$ is halved, how will it affect $F$?

**50.** If $d$ is halved, how will it affect $F$?

**51.** If $m_1$ is halved and $m_2$ is quadrupled, how will it affect $F$?

**52.** If $m_1$ is doubled, $m_2$ is quadrupled, and $d$ is quadrupled, how will it affect $F$?

*In Exercises 53 and 54, determine if the variation is of the form $y = kx$ or $y = \dfrac{k}{x}$, and find $k$.*

**53.**

| $x$ | $y$ |
|-----|-----|
| 2 | $\dfrac{5}{2}$ |
| 5 | 1 |
| 10 | $\dfrac{1}{2}$ |
| 20 | $\dfrac{1}{4}$ |

**54.**

| $x$ | $y$ |
|-----|-----|
| 6 | 2 |
| 9 | 3 |
| 15 | 5 |
| 27 | 9 |

**55. Profit** The profit from selling lamps is directly proportional to the number of lamps sold. When 150 lamps are sold, the profit is $2542.50. Find the profit when 520 lamps are sold.

**56. Profit** The profit from selling stereos is directly proportional to the number of stereos sold. When 65 stereos are sold, the profit is $4056. Find the profit when 80 stereos are sold.

**57. Antibiotic** The recommended dosage, $d$, of the antibiotic drug vancomycin is directly proportional to a person's weight. If Phuong Kim, who is 132 pounds, is given 2376 milligrams, find the recommended dosage for Nathan Brown, who weighs 172 pounds.

**58. Dollars and Pesos** American dollars vary directly to Mexican pesos. The more dollars you convert the more pesos you receive. Last week, Carlos Manuel converted $275 into 3507 pesos. Today, he received $400 from his aunt. If the conversion rate is unchanged when he converts the $400 into pesos, how many pesos (round your answer to the nearest peso) will he receive?

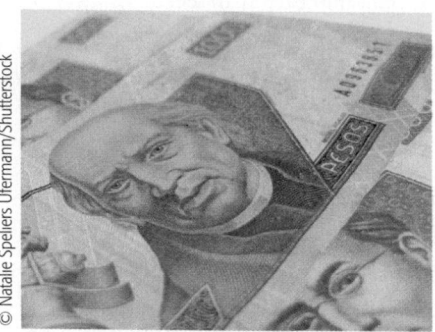

**59. Hooke's Law** Hooke's law states that the length a spring will stretch, $S$, varies directly with the force (or weight), $F$, attached to the spring. If a spring stretches 1.4 inches when 20 pounds is attached, how far will it stretch when 15 pounds is attached?

**60. Distance** When a car travels at a constant speed, the distance traveled, $d$, is directly proportional to the time, $t$. If a car travels 150 miles in 2.5 hours, how far will the same car travel in 4 hours?

**61. Pressure and Volume** The volume of a gas, $V$, varies inversely as its pressure, $P$. If the volume, $V$, is 800 cubic centimeters when the pressure is 200 millimeters (mm) of mercury, find the volume when the pressure is 25 mm of mercury.

**62. Building a Brick Wall** The time, $t$, required to build a brick wall varies inversely as the number of people, $n$, working on it. If it takes 8 hours for five bricklayers to build a wall, how long will it take four bricklayers to build a wall?

**63. Running a Race** The time, $t$, it takes a runner to cover a specified distance is inversely proportional to the runner's speed. If Jann Avery runs at an average of 6 miles per hour, she will finish a race in 2.6 hours. How long will it take Jackie Donofrio, who runs at 5 miles per hour, to finish the same race?

**64. Pitching a Ball** When a ball is pitched in a professional baseball game, the time, $t$, it takes for the ball to reach home plate varies inversely with the speed, $s$, of the pitch.* A ball pitched at 90 miles per hour takes 0.459 second to reach the plate. How long will it take a ball pitched at 75 miles per hour to reach the plate?

---

*A ball slows down on its way to the plate due to air resistance. For a 95-mph pitch, the ball is about 8 mph faster when it leaves the pitcher's hand than when it crosses the plate.

**65. Intensity of Light** The intensity, $I$, of light received at a source varies inversely as the square of the distance, $d$, from the source. If the light intensity is 20 foot-candles at 15 feet, find the light intensity at 10 feet.

**66. Tennis Ball** When a tennis player serves the ball, the time it takes for the ball to hit the ground in the service box is inversely proportional to the speed the ball is traveling. If Andy Roddick serves at 122 miles per hour, it takes 0.21 second for the ball to hit the ground after striking his racquet. How long will it take the ball to hit the ground if he serves at 80 miles per hour?

**67. Stopping Distance** Assume that the stopping distance of a van varies directly with the square of the speed. A van traveling 40 miles per hour can stop in 60 feet. If the van is traveling 56 miles per hour, what is its stopping distance?

**68. Falling Rock** A rock is dropped from the top of a cliff. The distance it falls in feet is directly proportional to the square of the time in seconds. If the rock falls 4 feet in $\frac{1}{2}$ second, how far will it fall in 3 seconds?

**69. Volume of a Pyramid** The volume, $V$, of a pyramid varies jointly as the area of its base, $B$, and its height, $h$ (see the figure). If the volume of the pyramid is 160 cubic meters when the area of its base is 48 square meters and its height is 10 meters, find the volume of a pyramid when the area of its base is 42 square meters and its height is 9 meters.

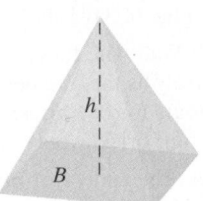

**70. Mortgage Payment** The monthly mortgage payment, $P$, you pay on a mortgage varies jointly as the interest rate, $r$, and the amount of the mortgage, $m$. If the monthly mortgage payment on a $50,000 mortgage at a 7% interest rate is $332.50, find the monthly payment on a $66,000 mortgage at 7%.

**71. DVD Rental** The weekly DVD rentals, $R$, at Hollywood Video vary directly with their advertising budget, $A$, and inversely with the daily rental price, $P$. When their advertising budget is $400 and the rental price is $2 per day, they rent 4600 DVDs per week. How many DVDs would they rent per week if they increased their advertising budget to $500 and raised their rental price to $2.50?

**72. Electrical Resistance** The electrical resistance of a wire, $R$, varies directly as its length, $L$, and inversely as its cross-sectional area, $A$. If the resistance of a wire is 0.2 ohm when the length is 200 feet and its cross-sectional area is 0.05 square inch, find the resistance of a wire whose length is 5000 feet with a cross-sectional area of 0.01 square inch.

**73. Weight of an Object** The weight, $w$, of an object in Earth's atmosphere varies inversely with the square of the distance, $d$, between the object and the center of Earth. A 140-pound person standing on Earth is approximately 4000 miles from Earth's center. Find the weight (or gravitational force of attraction) of this person at a distance 100 miles from Earth's surface.

**74. Wattage Rating** The wattage rating of an appliance, $W$, varies jointly as the square of the current, $I$, and the resistance, $R$. If the wattage is 3 watts when the current is 0.1 ampere and the resistance is 100 ohms, find the wattage when the current is 0.4 ampere and the resistance is 250 ohms.

**75. Phone Calls** The number of phone calls between two cities during a given time period, $N$, varies directly as the populations $p_1$ and $p_2$ of the two cities and inversely as the distance, $d$, between them. If 100,000 calls are made between two cities 300 miles apart and the populations of the cities are 60,000 and 200,000, how many calls are made between two cities with populations of 125,000 and 175,000 that are 450 miles apart?

© Thinkstock

**76. Water Bill** In a specific region of the country, the amount of a customer's water bill, $W$, is directly proportional to the average daily temperature for the month, $T$, the lawn area, $A$, and the square root of $F$, where $F$ is the family size, and inversely proportional to the number of inches of rain, $R$.

In one month, the average daily temperature is 78°F and the number of inches of rain is 5.6. If the average family of four, who has 1000 square feet of lawn, pays $68 for water, estimate the water bill in the same month for the average family of six, who has 1500 square feet of lawn.

**77. Intensity of Illumination** An article in the magazine *Outdoor and Travel Photography* states, "If a surface is illuminated by a point-source of light (a flash), the intensity of illumination produced is inversely proportional to the square of the distance separating them."

If the subject you are photographing is 4 feet from the flash, and the illumination on this subject is $\frac{1}{16}$ of the light of the flash, what is the intensity of illumination on an object that is 7 feet from the flash?

© Photos.com

**78. Force of Attraction** One of Newton's laws states that the force of attraction, $F$, between two masses is directly proportional to the masses of the two objects, $m_1$ and $m_2$, and inversely proportional to the square of the distance, $d$, between the two masses.

**a)** Write the formula that represents Newton's law.

**b)** What happens to the force of attraction if one mass is doubled, the other mass is tripled, and the distance between the objects is halved?

**79. Pressure on an Object** The pressure, $P$, in pounds per square inch (psi) on an object $x$ feet below the sea is 14.70 psi plus the product of a constant of proportionality, $k$, and the number of feet, $x$, the object is below sea level (see the figure). The 14.70 represents the weight, in pounds, of the column of air (from sea level to the top of the atmosphere) standing over a 1-inch-by-1-inch square of ocean. The $kx$ represents the weight, in pounds, of a column of water 1 inch by 1 inch by $x$ feet.

**a)** Write a formula for the pressure on an object $x$ feet below sea level.

**b)** If the pressure gauge in a submarine 60 feet deep registers 40.5 psi, find the constant $k$.

**c)** A submarine is built to withstand a pressure of 160 psi. How deep can the submarine go?

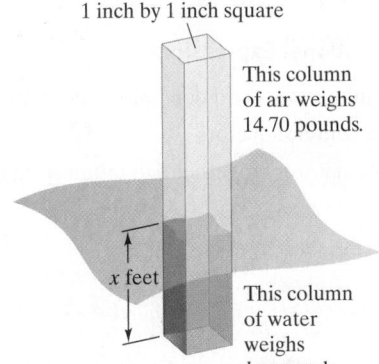

1 inch by 1 inch square

This column of air weighs 14.70 pounds.

$x$ feet

This column of water weighs $kx$ pounds.

## Cumulative Review Exercises

[5.6]  **80.** Divide $\dfrac{8x^2 + 6x - 21}{4x + 9}$.

[6.1]  **81.** Factor $y(z - 2) + 8(z - 2)$.

[6.6]  **82.** Solve $3x^2 - 24 = -6x$.

[7.2]  **83.** Multiply $\dfrac{x + 8}{x - 3} \cdot \dfrac{x^3 - 27}{x^2 + 3x + 9}$.

# Chapter 7 Summary

| IMPORTANT FACTS AND CONCEPTS | EXAMPLES |
|---|---|
| **Section 7.1** | |
| A **rational expression** is an expression of the form $\dfrac{p}{q}$, where $p$ and $q$ are polynomials and $q \neq 0$. | $\dfrac{x + 2}{x}$ and $\dfrac{x^2 + x}{x - 1}$ are rational expressions. |
| Whenever we have a rational expression containing a variable in the denominator, we always assume that the value or values of the variable that make the denominator 0 are excluded. | The rational expression $\dfrac{x - 5}{x - 3}$ is defined for all real numbers except 3. |
| A rational expression is simplified or reduced to its lowest terms when the numerator and denominator have no common factors other than 1. | $\dfrac{1}{8}$ and $\dfrac{x^2 + x + 1}{2x^2 + 3x + 7}$ are expressions reduced to lowest terms. |
| **To Simplify Rational Expressions** <br> 1. Factor both the numerator and denominator completely. <br> 2. Divide out common factors. | $\dfrac{12x^2 - 11x - 5}{9x^2 + 6x + 1} = \dfrac{\cancel{(3x + 1)}(4x - 5)}{\cancel{(3x + 1)}(3x + 1)} = \dfrac{4x - 5}{3x + 1}$ |

| IMPORTANT FACTS AND CONCEPTS | EXAMPLES |
|---|---|

## Section 7.2

**To Multiply Two Fractions**

$$\frac{a}{b} \cdot \frac{c}{d} = \frac{a \cdot c}{b \cdot d}, \quad b \neq 0 \text{ and } d \neq 0$$

$$\frac{1}{3} \cdot \frac{4}{5} = \frac{1 \cdot 4}{3 \cdot 5} = \frac{4}{15}$$

**To Multiply Rational Expressions**

1. Factor all numerators and denominators completely.
2. Divide out common factors.
3. Multiply numerators together and multiply denominators together.

$$\frac{4x^3}{7y^2} \cdot \frac{14y^3}{6x} = \frac{\overset{2}{\cancel{4}}\,\overset{x^2}{\cancel{x^3}} \cdot \overset{2}{\cancel{14}}\,\overset{y}{\cancel{y^3}}}{\cancel{7}y^2 \cdot \underset{3}{\cancel{6}}\,\cancel{x}} = \frac{4x^2 y}{3}$$

**To Divide Two Fractions**

$$\frac{a}{b} \div \frac{c}{d} = \frac{a}{b} \cdot \frac{d}{c} = \frac{ad}{bc}, \quad b \neq 0, \ d \neq 0, \text{ and } c \neq 0$$

$$\frac{3}{5} \div \frac{6}{7} = \frac{\overset{1}{\cancel{3}}}{5} \cdot \frac{7}{\underset{2}{\cancel{6}}} = \frac{7}{10}$$

**To Divide Rational Expressions**

Multiply the first fraction by the reciprocal of the second fraction.

$$\frac{x+4}{x+3} \div \frac{3x+12}{x+3} = \frac{\cancel{x+4}}{\cancel{x+3}} \cdot \frac{\cancel{x+3}}{3\cancel{(x+4)}} = \frac{1}{3}$$

## Section 7.3

**To Add or Subtract Two Fractions**

$$\frac{a}{c} + \frac{b}{c} = \frac{a+b}{c}, c \neq 0 \qquad \frac{a}{c} - \frac{b}{c} = \frac{a-b}{c}, c \neq 0$$

$$\frac{3}{11} + \frac{4}{11} = \frac{7}{11}, \qquad \frac{18}{19} - \frac{5}{19} = \frac{13}{19}$$

**To Add or Subtract Rational Expressions with a Common Denominator**

1. Add or subtract the numerators.
2. Place the sum or difference of the numerators over the common denominator.
3. Simplify the fraction if possible.

$$\frac{2}{x-3} + \frac{x+5}{x-3} = \frac{2+x+5}{x-3} = \frac{x+7}{x-3}$$

**To Find the Least Common Denominator of Rational Expressions**

1. Factor each denominator completely.
2. List all different factors of each denominator. When the same factor appears in more than one denominator, write that factor with the highest power that appears on it.
3. The least common denominator is the product of all the factors listed in step 2.

Find the least common denominator.

$$\frac{1}{9x^3y^4} + \frac{1}{6x^5y^3}$$

$$9x^3y^4 = 3 \cdot 3x^3y^4$$

$$6x^5y^3 = 3 \cdot 2x^5y^3$$

LCD is $2 \cdot 3^2 x^5 y^4 = 18x^5 y^4$.

## Section 7.4

**To Add or Subtract Two Rational Expressions with Unlike Denominators**

1. Determine the LCD.
2. Rewrite each fraction as an equivalent fraction with the LCD.
3. Add or subtract the numerators while maintaining the LCD.
4. When possible, factor the remaining numerator and simplify the fraction.

$$\frac{9}{m} + \frac{5}{m-1} = \frac{m-1}{m-1} \cdot \frac{9}{m} + \frac{5}{m-1} \cdot \frac{m}{m}$$

$$= \frac{9(m-1)}{m(m-1)} + \frac{5m}{m(m-1)}$$

$$= \frac{9m-9+5m}{m(m-1)}$$

$$= \frac{14m-9}{m(m-1)}$$

## Section 7.5

A **complex fraction** is one that has a fraction in its numerator or its denominator or in both its numerator and denominator.

$$\frac{\frac{2}{3}}{\frac{4}{7}}, \quad \frac{\frac{1}{x} + \frac{1}{y}}{\frac{1}{a} + \frac{1}{b}}$$

| IMPORTANT FACTS AND CONCEPTS | EXAMPLES |
|---|---|

### Section 7.5 (cont.)

**Method 1—To Simplify a Complex Fraction by Combining Terms**

1. Add or subtract the fractions in both the numerator and denominator of the complex fraction to obtain single fractions in both.
2. Multiply the fraction in the numerator by the reciprocal of the fraction in the denominator.
3. Simplify further if possible.

$$\frac{1 + \frac{1}{x}}{x} = \frac{\frac{x}{x} + \frac{1}{x}}{x} = \frac{\frac{x+1}{x}}{x}$$

$$= \frac{x+1}{x} \cdot \frac{1}{x} = \frac{x+1}{x^2}$$

**Method 2—To Simplify a Complex Fraction Using Multiplication First**

1. Find the LCD of *all* the denominators appearing in the complex fraction.
2. Multiply both the numerator and denominator of the complex fraction by the LCD found in step 1.
3. Simplify when possible.

$$\frac{1 + \frac{1}{x}}{x} = \frac{x}{x} \cdot \frac{1 + \frac{1}{x}}{x} = \frac{x(1) + x\left(\frac{1}{x}\right)}{x(x)} = \frac{x+1}{x^2}$$

Note: LCD = $x$.

### Section 7.6

A **rational equation** is an equation that contains one or more rational expressions.

**To Solve Rational Equations**

1. Determine the LCD of all fractions in the equation.
2. Multiply *both* sides of the equation by the LCD.
3. Remove any parentheses and combine like terms on each side of the equation.
4. Solve the equation using the properties discussed in earlier chapters.
5. Check your solution in the *original* equation.

$$\frac{1}{3}x - \frac{1}{7}x = 10, \qquad x + \frac{9}{x} = \frac{1}{3}$$

$$\frac{x}{5} - \frac{x}{8} = 1$$

$$40\left(\frac{x}{5} - \frac{x}{8}\right) = 40\,(1)$$

$$8x - 5x = 40$$

$$3x = 40$$

$$x = \frac{40}{3}$$

A check shows that $\frac{40}{3}$ is the solution.

### Section 7.7

**Applications**

A **geometry problem** involves geometric figures and formulas.

A rectangle has an area of 70 square meters. Find the dimensions if the width is 3 meters shorter than the length.

The answer is 7 meters by 10 meters.

A **motion problem** involves distance, rate, and time and uses the formula

$$\text{distance} = \text{rate} \cdot \text{time}$$

or

$$\text{time} = \frac{\text{distance}}{\text{rate}}$$

or

$$\text{rate} = \frac{\text{distance}}{\text{time}}$$

A cyclist can travel 20 miles with the wind to his back in the same time he can travel 12 miles going into the wind. If the wind is blowing at 2 miles per hour, find the speed of the cyclist without any wind.

The answer is 8 miles per hour.

A **work problem** involves two or more machines or people working together to complete a specific task.

Tom can paint a room in 6 hours and Bill can paint the same room in 4 hours. How long will it take them working together to paint this room?

The answer is 2.4 hours.

### Section 7.8

**Variation equations** show how one quantity changes in relation to another quantity or quantities.

| IMPORTANT FACTS AND CONCEPTS | EXAMPLES |
|---|---|
| **Section 7.8 (cont.)** | |
| **Direct Variation** <br><br> If a variable $y$ varies directly as a variable $x$, then $y = kx$, where $k$ is the constant of proportionality. | $y = 3x$ |
| **Inverse Variation** <br><br> If a variable $y$ varies inversely as a variable $x$, then <br><br> $$y = \dfrac{k}{x} \, (\text{or } xy = k)$$ <br><br> where $k$ is the constant of proportionality. | $y = \dfrac{3}{x}$ |
| **Joint Variation** <br><br> If $y$ varies jointly as $x$ and $z$, then <br><br> $$y = kxz$$ <br><br> where $k$ is the constant of proportionality. | $y = 3xz$ |

## Chapter 7 Review Exercises

[7.1]    *Determine the values of the variable for which the following expressions are defined.*

**1.** $\dfrac{5}{2x - 38}$

**2.** $\dfrac{2x + 1}{x^2 - 8x + 15}$

**3.** $\dfrac{7x - 1}{5x^2 + 4x - 1}$

*Simplify.*

**4.** $\dfrac{y}{xy - 8y}$

**5.** $\dfrac{x^3 + 5x^2 + 12x}{x}$

**6.** $\dfrac{9x^2 + 3xy}{3x}$

**7.** $\dfrac{x^2 + 2x - 8}{x - 2}$

**8.** $\dfrac{a^2 - 81}{a - 9}$

**9.** $\dfrac{-2x^2 + 7x + 4}{x - 4}$

**10.** $\dfrac{b^2 - 7b + 10}{b^2 - 3b - 10}$

**11.** $\dfrac{4x^2 - 11x - 3}{4x^2 - 7x - 2}$

**12.** $\dfrac{2x^2 - 21x + 40}{4x^2 - 4x - 15}$

[7.2]    *Multiply.*

**13.** $\dfrac{5a^2}{6b} \cdot \dfrac{2}{4a^2 b}$

**14.** $\dfrac{30x^2 y^3}{3z} \cdot \dfrac{6z^3}{5xy^3}$

**15.** $\dfrac{20a^3 b^4}{7c^3} \cdot \dfrac{14c^7}{5a^5 b}$

**16.** $\dfrac{2}{x - 4} \cdot \dfrac{4 - x}{9}$

**17.** $\dfrac{-m + 4}{15m} \cdot \dfrac{10m}{m - 4}$

**18.** $\dfrac{a - 2}{a + 3} \cdot \dfrac{a^2 + 4a + 3}{a^2 - a - 2}$

*Divide.*

**19.** $\dfrac{9x^6}{y^2} \div \dfrac{x^4}{4y}$

**20.** $\dfrac{5xy^2}{z} \div \dfrac{x^4 y^2}{4z^2}$

**21.** $\dfrac{6a + 6b}{a^2} \div \dfrac{a^2 - b^2}{a^2}$

**22.** $\dfrac{1}{a^2 + 8a + 15} \div \dfrac{8}{a + 5}$

**23.** $(t + 8) \div \dfrac{t^2 + 5t - 24}{t - 3}$

**24.** $\dfrac{x^2 + xy - 2y^2}{2y} \div \dfrac{x + 2y}{12y^2}$

[7.3]    *Add or subtract.*

**25.** $\dfrac{n}{n + 5} - \dfrac{2}{n + 5}$

**26.** $\dfrac{4x}{x + 7} + \dfrac{28}{x + 7}$

**27.** $\dfrac{5x - 4}{x + 8} + \dfrac{44}{x + 8}$

**28.** $\dfrac{7x - 3}{x^2 + 7x - 30} - \dfrac{3x + 9}{x^2 + 7x - 30}$

**29.** $\dfrac{5h^2 + 12h - 1}{h + 5} - \dfrac{h^2 - 5h + 14}{h + 5}$

**30.** $\dfrac{6x^2 - 4x}{2x - 3} - \dfrac{-3x + 12}{2x - 3}$

*Find the least common denominator for each expression.*

**31.** $\dfrac{a}{8} + \dfrac{5a}{3}$

**32.** $\dfrac{10}{x+3} + \dfrac{2x}{x+3}$

**33.** $\dfrac{10}{4xy^3} - \dfrac{11}{10x^2y}$

**34.** $\dfrac{6}{x-3} - \dfrac{2}{x}$

**35.** $\dfrac{8}{n+5} + \dfrac{2n-3}{n-4}$

**36.** $\dfrac{5x-12}{x^2+2x} - \dfrac{4}{x+2}$

**37.** $\dfrac{2r+1}{r-s} - \dfrac{6}{r^2-s^2}$

**38.** $\dfrac{3x^2}{x-9} + 10x^3$

**39.** $\dfrac{19x-5}{x^2+2x-35} + \dfrac{-10x+1}{x^2+9x+14}$

**[7.4]** *Add or subtract.*

**40.** $\dfrac{5}{3y^2} + \dfrac{y}{2y}$

**41.** $\dfrac{3x}{xy} + \dfrac{1}{4x}$

**42.** $\dfrac{5x}{3xy} - \dfrac{6}{x^2}$

**43.** $7 - \dfrac{2}{x+2}$

**44.** $\dfrac{x-y}{y} - \dfrac{x+y}{x}$

**45.** $\dfrac{7}{x+4} + \dfrac{2}{x}$

**46.** $\dfrac{2}{3x} - \dfrac{3}{3x-6}$

**47.** $\dfrac{1}{(z+5)} + \dfrac{9}{(z+5)^2}$

**48.** $\dfrac{x+2}{x^2-x-6} + \dfrac{x-3}{x^2-8x+15}$

**[7.2–7.4]** *Perform each indicated operation.*

**49.** $\dfrac{x+4}{x+6} - \dfrac{x-5}{x+2}$

**50.** $2 + \dfrac{x}{x-4}$

**51.** $\dfrac{a+2}{b} \div \dfrac{a-2}{5b^2}$

**52.** $\dfrac{x+5}{x^2-9} + \dfrac{2}{x+3}$

**53.** $\dfrac{6p+12q}{p^2q} \cdot \dfrac{p^5}{p+2q}$

**54.** $\dfrac{8}{(x+2)(x-3)} - \dfrac{6}{(x-2)(x+2)}$

**55.** $\dfrac{x+7}{x^2+9x+14} - \dfrac{x-10}{x^2-49}$

**56.** $\dfrac{x-y}{x+y} \cdot \dfrac{xy+x^2}{x^2-y^2}$

**57.** $\dfrac{3x^2-27y^2}{30} \div \dfrac{(x-3y)^2}{6}$

**58.** $\dfrac{a^2-11a+30}{a-6} \cdot \dfrac{a^2-8a+15}{a^2-10a+25}$

**59.** $\dfrac{a}{a^2-1} - \dfrac{3}{3a^2-2a-5}$

**60.** $\dfrac{2x^2+6x-20}{x^2-2x} \div \dfrac{x^2+7x+10}{2x^2-8}$

**[7.5]** *Simplify each complex fraction.*

**61.** $\dfrac{5+\frac{1}{3}}{\frac{3}{4}}$

**62.** $\dfrac{1+\frac{5}{8}}{3-\frac{9}{16}}$

**63.** $\dfrac{\frac{12ab}{9c}}{\frac{4a}{c^2}}$

**64.** $\dfrac{\frac{18x^4y^2}{9xy^5}}{4z^2}$

**65.** $\dfrac{a-\frac{a}{b}}{1+\frac{a}{b}}$

**66.** $\dfrac{r^2+\frac{7}{s}}{s^2}$

**67.** $\dfrac{\frac{3}{x}+\frac{2}{x^2}}{5-\frac{1}{x}}$

**68.** $\dfrac{\frac{x}{x+y}}{\frac{x^2}{4x+4y}}$

**69.** $\dfrac{\frac{9}{x}}{\frac{9}{x^2}}$

**70.** $\dfrac{\frac{1}{a}+3}{\frac{1}{a}+\frac{3}{a}}$

**71.** $\dfrac{\frac{1}{x^2}-\frac{1}{x}}{\frac{1}{x^2}+\frac{1}{x}}$

**72.** $\dfrac{\frac{8x}{y}-x}{\frac{y}{x}-1}$

**[7.6]** *Solve.*

**73.** $\dfrac{5}{8} = \dfrac{10}{x+3}$

**74.** $\dfrac{x}{4} = \dfrac{x-3}{2}$

**75.** $\dfrac{12}{n} + 2 = \dfrac{n}{4}$

**76.** $\dfrac{10}{m} + \dfrac{3}{2} = \dfrac{m}{10}$

**77.** $\dfrac{-4}{d} = \dfrac{3}{2} + \dfrac{4-d}{d}$

**78.** $\dfrac{1}{x-7} + \dfrac{1}{x+7} = \dfrac{1}{x^2-49}$

**79.** $\dfrac{x-3}{x-2} + \dfrac{x+1}{x+3} = \dfrac{2x^2 + x + 1}{x^2 + x - 6}$

**80.** $\dfrac{a}{a^2 - 64} + \dfrac{4}{a+8} = \dfrac{3}{a-8}$

**81.** $\dfrac{d}{d-4} - 4 = \dfrac{4}{d-4}$

[7.7]   *Solve.*

**82. Sandcastles** It takes John and Amy Brogan 6 hours to build a sandcastle. It takes Paul and Cindy Carter 4 hours to make the same sandcastle. How long will it take all four people together to build the sandcastle?

**83. Filling a Pool** One hose can fill a swimming pool in 7 hours. A second hose can siphon all the water out of a full pool in 12 hours. How long will it take to fill the pool if, while one hose is filling the pool, the other hose is siphoning water from the pool?

**84. Sum of Numbers** One number is six times as large as another. The sum of their reciprocals is 7. Determine the numbers.

**85. Rollerblading and Bicycling** Robert Johnston can travel 3 miles on his rollerblades in the same time Tran Lee can travel 8 miles on his mountain bike. If Tran's speed on his bike is 3.5 miles per hour faster than that of Robert on his rollerblades, determine Robert's and Tran's speeds.

[7.8]

**86.** $W$ is directly proportional to the square of $L$ and inversely proportional to $A$. If $W = 4$ when $L = 2$ and $A = 10$, find $W$ when $L = 5$ and $A = 20$.

**87.** $z$ is jointly proportional to $x$ and $y$ and inversely proportional to the square of $r$. If $z = 12$ when $x = 20$, $y = 8$, and $r = 8$, find $z$ when $x = 10$, $y = 80$, and $r = 3$.

**88. Drug Dosage** The recommended dosage, $d$, of the antibiotic drug vancomycin is directly proportional to a person's weight, $w$. If Carmen Brown, who is 132 pounds, is given 182 milligrams, find the recommended dosage for Bill Glenn, who is 198 pounds.

**89. Runner's Speed** The time, $t$, it takes a runner to cover a specified distance is inversely proportional to the runner's speed. If Nhat Chung runs at an average of 6 miles per hour, he will finish a race in 1.4 hours. How long will it take Leif Lundgren, who runs at 5 miles per hour, to finish the race?

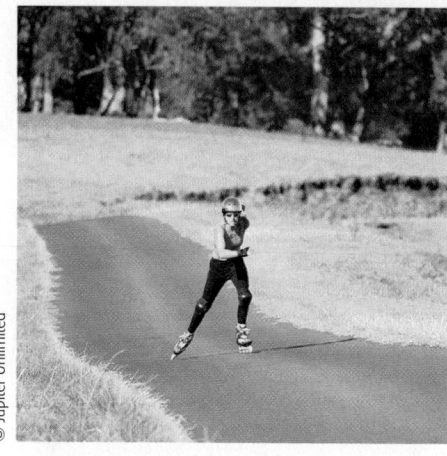

© Jupiter Unlimited

See Exercise 85.

## Chapter 7 Practice Test

*Chapter Test Prep Videos provide fully worked-out solutions to any of the exercises you want to review. Chapter Test Prep Videos are available via* MyMathLab®, *or on* You Tube *(search "Angel Elementary and Intermidiate Algebra" and click on "Channels").*

*Simplify.*

**1.** $\dfrac{-8 + x}{x - 8}$

**2.** $\dfrac{x^3 - 1}{x^2 - 1}$

*Perform each indicated operation.*

**3.** $\dfrac{20x^2 y^3}{4z^2} \cdot \dfrac{8xz^3}{5xy^4}$

**4.** $\dfrac{a^2 - 9a + 14}{a - 2} \cdot \dfrac{a^2 - 4a - 21}{(a - 7)^2}$

**5.** $\dfrac{x^2 - x - 6}{x^2 - 9} \cdot \dfrac{x^2 - 6x + 9}{x^2 + 4x + 4}$

**6.** $\dfrac{x^2 - 1}{x + 2} \cdot \dfrac{x + 2}{1 - x^2}$

**7.** $\dfrac{x^2 - 4y^2}{5x + 20y} \div \dfrac{x + 2y}{x + 4y}$

**8.** $\dfrac{15}{y^2 + 2y - 15} \div \dfrac{5}{y - 3}$

**9.** $\dfrac{m^2 + 3m - 18}{m - 3} \div \dfrac{m^2 - 8m + 15}{3 - m}$

**10.** $\dfrac{4x + 3}{8y} + \dfrac{2x - 5}{8y}$

**11.** $\dfrac{7x^2 - 4}{x + 3} - \dfrac{6x + 9}{x + 3}$

**12.** $\dfrac{2}{xy} - \dfrac{8}{xy^3}$

**13.** $3 - \dfrac{5z}{z - 5}$

**14.** $\dfrac{x - 5}{x^2 - 16} - \dfrac{x - 2}{x^2 + 2x - 8}$

*Simplify.*

**15.** $\dfrac{2 + \dfrac{1}{2}}{3 - \dfrac{1}{5}}$

**16.** $\dfrac{x + \dfrac{x}{y}}{\dfrac{7}{x}}$

**17.** $\dfrac{4 + \dfrac{3}{x}}{\dfrac{9}{x} - 5}$

*Solve.*

**18.** $2 + \dfrac{8}{x} = 6$

**19.** $\dfrac{2x}{3} - \dfrac{x}{4} = x + 1$

**20.** $\dfrac{x}{x - 8} + \dfrac{6}{x - 2} = \dfrac{x^2}{x^2 - 10x + 16}$

*Solve.*

**21. Working Together** Mr. Jackson, on his tractor, can clear a 1-acre field in 10 hours. Mr. Hackett, on his tractor, can clear a 1-acre field in 15 hours. If they work together, how long will it take them to clear a 1-acre field?

**22. Determine a Number** The sum of a positive number and its reciprocal is 2. Determine the number.

**23. Area of Triangle** The area of a triangle is 30 square inches. If the height is 2 inches less than 2 times the base, determine the height and base of the triangle.

**24. Exercising** LaConya Bertrell exercises for $1\frac{1}{2}$ hours each day. During the first part of her routine, she rides a bicycle and averages 10 miles per hour. For the remainder of the time, she rollerblades and averages 4 miles per hour. If the total distance she travels is 12 miles, how far did she travel on the rollerblades?

**25. Making Music** The wavelength of sound waves, $w$, is inversely proportional to the frequency, $f$ (or pitch). If a frequency of 263 cycles per second (middle C on a piano) produces a wavelength of about 4.3 feet, determine the length of a wavelength of a frequency of 1000 cycles per second.

© Losevsky Pavel/Shutterstock

## Cumulative Review Test

*Take the following test and check your answers with those given in the back of the book. Review any questions that you answered incorrectly. The section where the material was covered is indicated after the answer.*

**1.** Evaluate $3x^2 - 5xy^2 - 7$, when $x = -4$ and $y = -2$.

**2.** Evaluate $[6 - [3(8 \div 4)^2 + 9 \cdot 4]^2$.

**3.** Solve $5z + 4 = -3(z - 7)$.

**4.** The cost of a 2009 Ford Mustang increased by 1.8% over the cost of the 2008 Mustang. Write an expression for the sum of the costs of a 2009 and a 2008 Mustang.

**5.** Find the slope of the line through the points $(-7, 8)$ and $(3, 8)$.

**6.** Write the equation $3x - 4y = 12$ in slope-intercept form.

**7.** Simplify $(6x^2 - 3x - 5) - (-2x^2 - 8x - 19)$.

**8.** Multiply $(3n^2 - 4n + 3)(2n - 5)$.

**9.** Divide $\dfrac{4x - 38}{8}$.

**10.** Factor $8a^2 - 8a - 5a + 5$.

**11.** Factor $13x^2 + 26x - 39$.

**12.** Solve $2x^2 = 11x - 12$.

**13.** Multiply $\dfrac{x^2 + x - 12}{x^2 - x - 6} \cdot \dfrac{x^2 - 2x - 8}{2x^2 - 7x - 4}$.

**14.** Subtract $\dfrac{r}{r + 2} - \dfrac{3}{r - 5}$.

**15.** Add $\dfrac{4}{x^2 - 3x - 10} + \dfrac{6}{x^2 + 5x + 6}$.

**16.** Solve the equation $\dfrac{x}{9} - \dfrac{x}{6} = \dfrac{1}{12}$.

**17.** Solve the equation $\dfrac{7}{x + 3} + \dfrac{5}{x + 2} = \dfrac{5}{x^2 + 5x + 6}$.

**18. Medical Plans** A school district allows its employees to choose from two medical plans. With plan 1, the employee pays 10% of all medical bills (the school district pays the balance). With plan 2, the employee pays the school district a one-time payment of $150, then the employee pays 5% of all medical bills. What total medical bills would result in the employee paying the same amount with the two plans?

**19. Bird Seed** A feed store owner wishes to make his own store-brand mixture of bird seed by mixing sunflower seed that costs $0.50 per pound with a premixed assorted seed that costs $0.20 per pound. How many pounds of each will he have to use to make a 50-pound mixture that will cost $16.00?

**20. Sailing** During the first leg of a race, the sailboat *Thumper* sailed at an average speed of 6.5 miles per hour. During the second leg of the race, the winds increased and *Thumper* sailed at an average speed of 9.5 miles per hour. If the total distance sailed by *Thumper* was 12.75 miles, and the total time spent racing was 1.5 hours, determine the distance traveled by *Thumper* on each leg of the race.

© Allen R. Angel

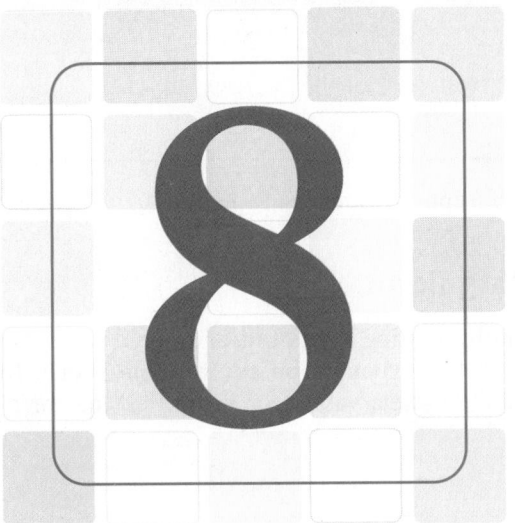

# 8

# Functions and Their Graphs

## Goals of This Chapter

The two primary goals of this chapter are to provide you with a good understanding of graphing and functions. Graphing is a key element to this and many mathematics courses. Functions are closely related to graphing and are a unifying concept throughout mathematics. We will use both graphing and functions throughout the rest of this book.

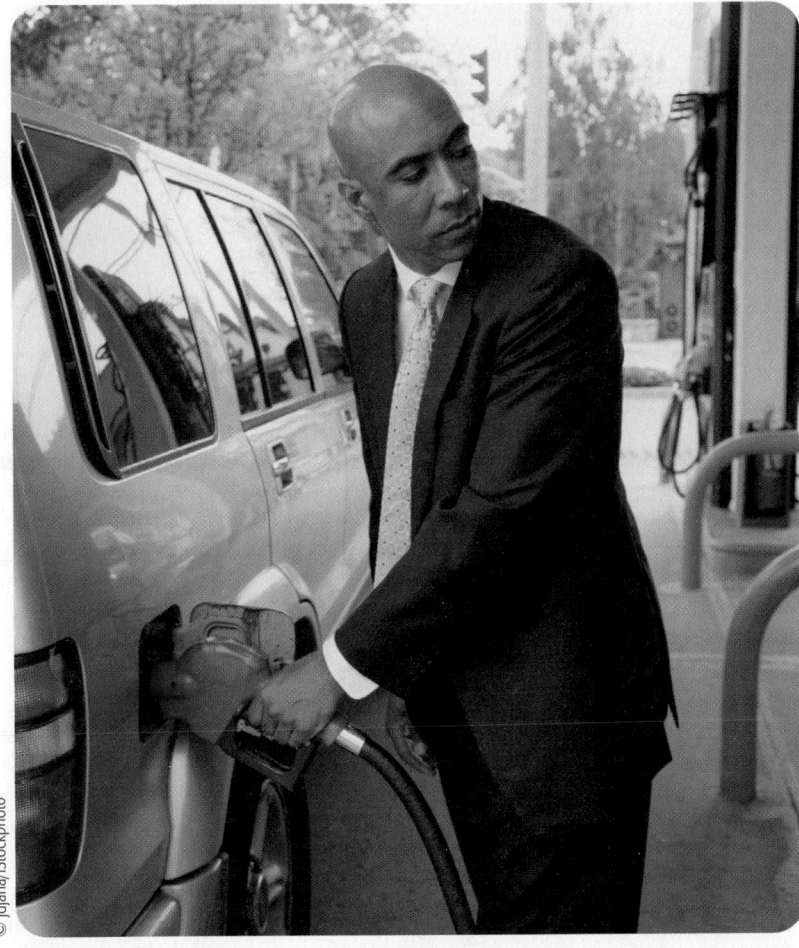

© jdjana/iStockphoto

**We see graphs daily** in newspapers, magazines, and on the Internet. You will see many such graphs in this chapter. For example, in Exercise 106 on page 474, a graph is used to show inflation-adjusted gasoline prices.

# 8.1 More on Graphs

1 Graph equations by plotting points.

2 Interpret graphs.

In this section we will expand upon the concepts introduced in Chapter 4.

## 1 Graph Equations by Plotting Points

### Understanding Algebra

A *linear equation in two variables* is an equation that can be put into the form $ax + by = c$ where $a$, $b$, and $c$ are real numbers.

In section 4.2, we graphed linear equations by plotting points. Linear equations are also called **first-degree** equations since the greatest exponent on any variable is one. To graph a linear equation by plotting points we select values for $x$ and find the corresponding values of $y$.

**EXAMPLE 1**  Graph $y = -\dfrac{1}{3}x + 1$.

*Solution*   When we select values for $x$, we will select some positive values, some negative values, and 0. We will select multiples of 3 so the values of $y$ are integer values. The graph is illustrated in **Figure 8.1**.

### Understanding Algebra

When graphing linear equations in which the coefficient on $x$ is a fraction, choose values of $x$ that are multiples of the denominator. This may result in integer values for $y$.

| $x$ | $y$ |
|-----|-----|
| $-6$ | $3$ |
| $-3$ | $2$ |
| $0$ | $1$ |
| $3$ | $0$ |
| $6$ | $-1$ |

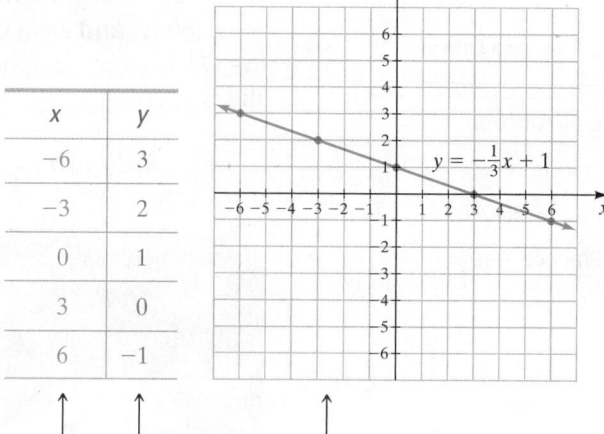

FIGURE 8.1

1. Select values for $x$.
2. Compute $y$.
3. Plot the points and draw the graph.

Now Try Exercise 13

If we are asked to graph a linear equation not solved for $y$, such as $x + 3y = 3$, our first step will be to solve the equation for $y$. For example, if we solve $x + 3y = 3$ for $y$ we obtain

$$x + 3y = 3$$

$$3y = -x + 3 \qquad \text{Subtract } x \text{ from both sides.}$$

$$y = \frac{-x + 3}{3} \qquad \text{Divide both sides by 3.}$$

$$y = \frac{-x}{3} + \frac{3}{3} = -\frac{1}{3}x + 1$$

The resulting equation, $y = -\dfrac{1}{3}x + 1$, is the same equation we graphed in Example 1.

Therefore, the graph of $x + 3y = 3$ is also illustrated in **Figure 8.1** above.

Equations whose graphs are not straight lines are called **nonlinear equations**. To graph nonlinear equations by plotting points, we follow the same procedure used to graph linear equations. However, since the graphs are not straight lines, we may need to plot more points to draw the graphs.

**EXAMPLE 2** Graph $y = x^2 - 4$.

***Solution***   We select some values for $x$ and find the corresponding values of $y$. Then we plot the points and connect them with a smooth curve. The graph is shown in **Figure 8.2**.

| $x$ | $y$ |
|-----|-----|
| $-3$ | $5$ |
| $-2$ | $0$ |
| $-1$ | $-3$ |
| $0$ | $-4$ |
| $1$ | $-3$ |
| $2$ | $0$ |
| $3$ | $5$ |

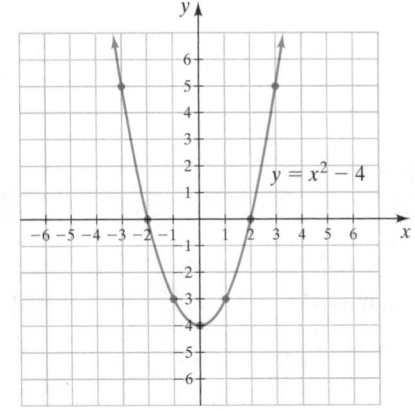

FIGURE 8.2

Now Try Exercise 19

**EXAMPLE 3** Graph $y = \dfrac{1}{x}$.

***Solution***   When selecting values for $x$, notice that $x = 0$ gives us $y = \dfrac{1}{0}$, which is undefined. Therefore, there will be no part of the graph at $x = 0$. Also notice that when $x = \dfrac{1}{2}$, we get $\dfrac{1}{\frac{1}{2}} = 1 \div \dfrac{1}{2} = 1 \cdot \dfrac{2}{1} = 2$. This graph has two branches, one to the left of the $y$-axis and one to the right of the $y$-axis, as shown in **Figure 8.3** below.

| $x$ | $y$ |
|-----|-----|
| $-3$ | $-\dfrac{1}{3}$ |
| $-2$ | $-\dfrac{1}{2}$ |
| $-1$ | $-1$ |
| $-\dfrac{1}{2}$ | $-2$ |
| $\dfrac{1}{2}$ | $2$ |
| $1$ | $1$ |
| $2$ | $\dfrac{1}{2}$ |
| $3$ | $\dfrac{1}{3}$ |

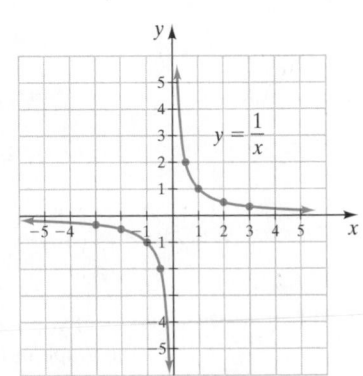

FIGURE 8.3

Now Try Exercise 29

   In the graph for Example 3, notice that for values of $x$ far to the right of 0, or far to the left of 0, the graph approaches the $x$-axis but does not touch it. For example when $x = 1000$, $y = 0.001$ and when $x = -1000$, $y = -0.001$.

EXAMPLE 4  Graph $y = |x|$. ——————

*Solution*    To graph this absolute value equation, we select some values for $x$ and find the corresponding values of $y$. Then we plot the points and draw the graph. Notice that this graph is V-shaped, as shown in **Figure 8.4**.

| $x$ | $y$ |
|-----|-----|
| $-4$ | $4$ |
| $-3$ | $3$ |
| $-2$ | $2$ |
| $-1$ | $1$ |
| $0$ | $0$ |
| $1$ | $1$ |
| $2$ | $2$ |
| $3$ | $3$ |
| $4$ | $4$ |

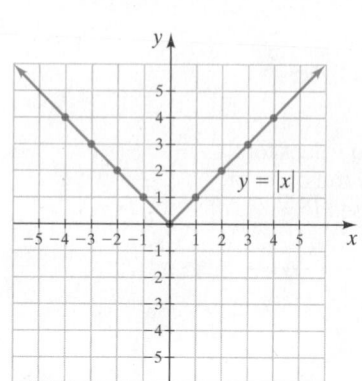

FIGURE 8.4

Now Try Exercise 23

> ### Understanding Algebra
>
> The absolute value of a number $x$, written $|x|$, is the distance that $x$ is from 0 on the number line.

**Avoiding Common Errors**

When graphing nonlinear equations, be sure to plot enough points to get a true picture of the graph. For example, when graphing $y = \dfrac{1}{x}$ many students consider only integer values of $x$. Following is a table of values for the equation and two graphs that contain the points indicated in the table.

| $x$ | $-3$ | $-2$ | $-1$ | $1$ | $2$ | $3$ |
|-----|------|------|------|-----|-----|-----|
| $y$ | $-\dfrac{1}{3}$ | $-\dfrac{1}{2}$ | $-1$ | $1$ | $\dfrac{1}{2}$ | $\dfrac{1}{3}$ |

| CORRECT | INCORRECT |
|---------|-----------|
|  | 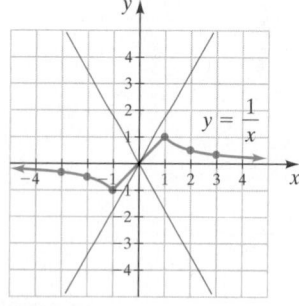 |
| FIGURE 8.5 | FIGURE 8.6 |

If you select and plot fractional values of $x$ near 0, as was done in Example 3, you get the graph in **Figure 8.5**. The graph in **Figure 8.6** cannot be correct because the equation is not defined when $x$ is 0 and therefore the graph cannot cross the $y$-axis. Whenever you plot a graph that contains a variable in the denominator, select values for the variable that are very close to the value that makes the denominator 0 and observe what happens. For example, when graphing $y = \dfrac{1}{x - 3}$ you should use values of $x$ close to 3, such as 2.9 and 3.1 or 2.99 and 3.01, and see what values you obtain for $y$.

*(continued next page)*

Also, when graphing nonlinear equations, it is a good idea to consider both positive and negative values. For example, if you used only positive values of $x$ when graphing $y = |x|$, the graph would appear to be a straight line going through the origin, instead of the V-shaped graph shown in **Figure 8.4** on page 452.

## 2  Interpret Graphs

We see many different types of graphs daily in newspapers, in magazines, on the Internet and so on. Throughout this book, we present a variety of graphs. Since being able to draw and interpret graphs is very important, we will study this further in Section 8.2. In Example 5 you must understand and interpret graphs to answer the question.

EXAMPLE 5   When Jim Herring went to see his mother in Cincinnati, he boarded a Southwest Airlines plane. The plane sat on the runway for 20 minutes and then took off. The plane flew at about 600 miles per hour for about 2 hours. It then reduced its speed to about 300 miles per hour and circled the Cincinnati Airport for about 15 minutes before it came in for a landing. After landing, the plane taxied to the gate and stopped. Which graph in **Figures 8.7a–8.7d** best illustrates this situation?

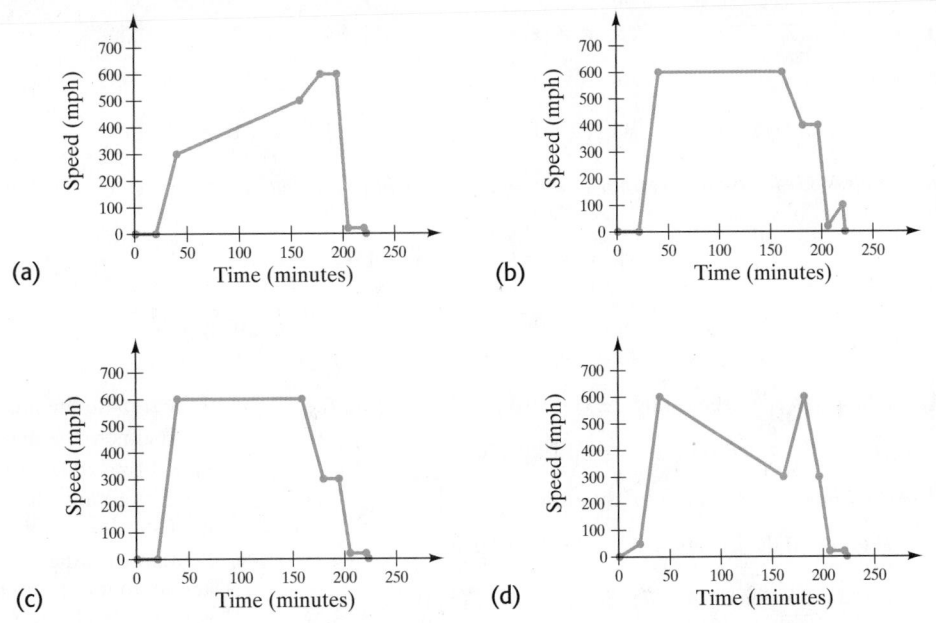

FIGURE 8.7

*Solution*   The graph that depicts the situation described is **(c)**, reproduced with annotations in **Figure 8.8**. The graph shows speed versus time, with time on the horizontal axis.

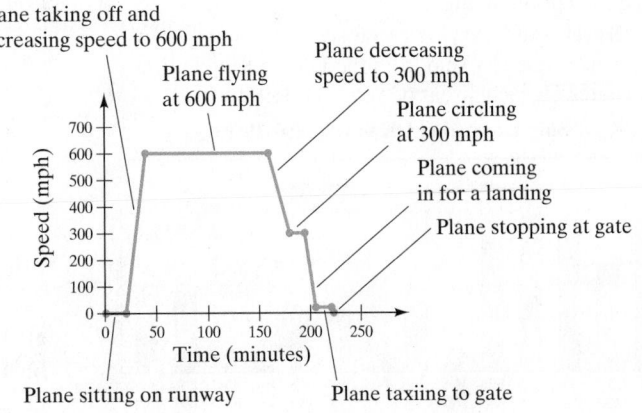

FIGURE 8.8

Now Try Exercise 53

# EXERCISE SET 8.1

MathXL®    MyMathLab®
MathXL®     MyMathLab

## Warm-Up Exercises

*Fill in the blanks with the appropriate word, phrase, or symbol(s) from the following list.*

linear         distance         $ax + by = c$         nonlinear         first-degree         value

1. A linear equation in two variables is an equation that can be put into the form _____.

2. Linear equations are also called _____ equations since the greatest exponent on any variable is one.

3. Equations whose graphs are not straight lines are _____ equations.

4. The absolute value of a number is the _____ that number is from 0 on the number line.

## Practice the Skills

*Graph each equation.*

5. $y = x + 1$

6. $y = 3x$

7. $y = -3x - 5$

8. $y = -2x + 2$

9. $y = 2x + 4$

10. $y = x + 2$

11. $y = \dfrac{1}{2}x$

12. $y = -\dfrac{1}{3}x$

13. $y = \dfrac{1}{2}x - 1$

14. $y = -\dfrac{1}{2}x - 3$

15. $y = -\dfrac{1}{3}x + 2$

16. $y = -\dfrac{1}{3}x + 4$

17. $y = x^2$

18. $y = x^2 - 2$

19. $y = -x^2$

20. $y = -x^2 + 4$

21. $y = |x| + 1$

22. $y = |x| + 2$

23. $y = -|x|$

24. $y = -|x| - 3$

25. $y = x^3$

26. $y = -x^3$

27. $y = x^3 + 1$

28. $y = \dfrac{1}{x}$

29. $y = -\dfrac{1}{x}$

30. $x^2 = 1 + y$

31. $x = |y|$

32. $x = y^2$

*In Exercises 33–40, use a calculator to obtain at least eight points that are solutions to the equation. Then graph the equation by plotting the points.*

33. $y = x^3 - x^2 - x + 1$

34. $y = -x^3 + x^2 + x - 1$

35. $y = \dfrac{1}{x + 1}$

36. $y = \dfrac{1}{x} + 1$

37. $y = \sqrt{x}$

38. $y = \sqrt{x + 4}$

39. $y = \dfrac{1}{x^2}$

40. $y = \dfrac{|x^2|}{2}$

41. Is the point represented by the ordered pair $\left(\dfrac{1}{3}, \dfrac{1}{12}\right)$ on the graph of the equation $y = \dfrac{x^2}{x + 1}$? Explain.

42. Is the point represented by the ordered pair $\left(-\dfrac{1}{2}, -\dfrac{3}{5}\right)$ on the graph of the equation $y = \dfrac{x^2 + 1}{x^2 - 1}$? Explain.

43. a) Plot the points $A(2, 7)$, $B(2, 3)$, and $C(6, 3)$, and then draw $\overline{AB}$, $\overline{AC}$, and $\overline{BC}$. ($\overline{AB}$ represents the line segment from $A$ to $B$.)

   b) Find the area of the figure.

44. a) Plot the points $A(-4, 5)$, $B(2, 5)$, $C(2, -3)$, and $D(-4, -3)$, and then draw $\overline{AB}$, $\overline{BC}$, $\overline{CD}$, and $\overline{DA}$.

   b) Find the area of the figure.

45. **UK to US Travel** The following graph shows the number of passengers that traveled by ship from the United Kingdom to the United States for each 5-year period from 1890 to 1959.

**Ship Travel from UK to US, 1890–1959**

*Source:* BT27 Passenger Lists on findmypast.com

a) Estimate the number of passengers during 1895–1899.

b) Estimate the number of passengers during 1955–1959.

c) During which 5-year periods was the number of passengers greater than 1,000,000 passengers?

d) Does this graph appear to be linear?

46. **Scotland Temperature** The following graph shows the average temperature in degrees Celsius for each month in Glasgow, Scotland.

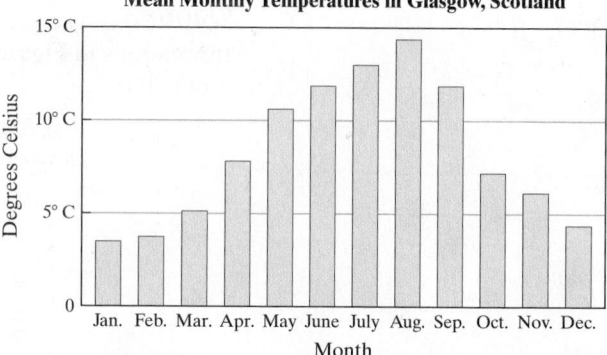

**Mean Monthly Temperatures in Glasgow, Scotland**

*Source:* www.wunderground.com

a) Estimate the average temperature in February.

b) Estimate the average temperature in July.

c) During which months was the average temperature greater than 10°C?

d) Does this graph appear to be linear?

*Match Exercises 47–50 with the corresponding graph of elevation above sea level versus time, labeled a–d, below.*

**47.** Nancy Johnson started out by walking up a steep hill for 5 minutes. For the next 5 minutes she walked down a steep hill to an elevation lower than her starting point. For the next 10 minutes she walked on level ground. For the next 10 minutes she walked up a slight hill, at which time she reached her starting elevation.

**48.** James Condor started out by walking up a hill for 5 minutes. For the next 10 minutes he walked down a hill to an elevation equal to his starting elevation. For the next 10 minutes he walked on level ground. For the next 5 minutes he walked downhill.

**49.** Mary Beth Headlee walked for 5 minutes on level ground. Then for 5 minutes she climbed a slight hill. Then she walked on level ground for 5 minutes. Then for the next 5 minutes she climbed a steep hill. During the next 10 minutes she descended uniformly until she reached the height at which she had started.

**50.** Don Ransford walked on level ground for 5 minutes. Then he walked down a steep hill for 10 minutes. For the next 5 minutes he walked on level ground. For the next 5 minutes he walked back up to his starting height. For the next 5 minutes he walked on level ground.

*Match Exercises 51–54 with the corresponding graph of speed versus time, labeled a–d, below.*

**51.** To go to work, Donna Clark drove on a country road for 10 minutes, then drove on a highway for 12 minutes, then drove in stop-and-go traffic for 8 minutes.

**52.** To go to work, Bob Plough drove in stop-and-go traffic for 5 minutes, then drove on the expressway for 20 minutes, then drove in stop-and-go traffic for 5 minutes.

**53.** To go to work, Ron Breitfelder walked for 3 minutes, waited for the train for 5 minutes, rode the train for 15 minutes, then walked for 7 minutes.

**54.** To go to work, Kim Ghiselin rode her bike uphill for 10 minutes, then rode downhill for 15 minutes, then rode on a level street for 5 minutes.

(a)

(b)

(c)

(d)

*Match Exercises 55–58 with the corresponding graph of speed versus time, labeled a–d below.*

**55.** Christina Dwyer walked for 5 minutes to warm up, jogged for 20 minutes, and then walked for 5 minutes to cool down.

**56.** Annie Droullard went for a leisurely bike ride at a constant speed for 30 minutes.

**57.** Michael Odu took a 30-minute walk through his neighborhood. He stopped very briefly on 7 occasions to pick up trash.

**58.** Richard Dai walked through his neighborhood and stopped 3 times to chat with his neighbors. He was gone from his house a total of 30 minutes.

(a)

(b)

(c)

(d)

*Match Exercises 59–62 with the corresponding graph of distance traveled versus time, labeled a–d. Recall from Chapter 2 that distance = rate × time. Selected distances are indicated on the graphs.*

**59.** Train A traveled at a speed of 40 mph for 1 hour, then 80 mph for 2 hours, and then 60 mph for 3 hours.

**60.** Train C traveled at a speed of 80 mph for 2 hours, then stayed in a station for 1 hour, and then traveled 40 mph for 3 hours.

**61.** Train B traveled at a speed of 20 mph for 2 hours, then 60 mph for 3 hours, and then 80 mph for 1 hour.

**62.** Train D traveled at 30 mph for 1 hour, then 65 mph for 2 hours, and then 30 mph for 3 hours.

(a)

(b)

(c)

(d)

Use a graphing calculator to graph each function. Make sure you select values for the window that will show the curvature of the graph. Then, if your calculator can display tables, display a table of values in which the x-values extend by units, from 0 to 6.

**63.** $y = 2x - 3$

**64.** $y = \frac{1}{3}x + 2$

**65.** $y = x^2 - 2x - 8$

**66.** $y = -x^2 + 16$

**67.** $y = x^3 - 2x + 4$

**68.** $y = 2x^3 - 6x^2 - 1$

## Challenge Problems

We will discuss many of the concepts introduced in Exercises 69–76 in Section 8.4.

**69.** Graph $y = x + 1$, $y = x + 3$, and $y = x - 1$ on the same axes.

a) What do you notice about the graphs of the equations and the values where the graphs intersect the y-axis?

b) Do all the graphs seem to have the same slant (or slope)?

**70.** Graph $y = \frac{1}{2}x$, $y = \frac{1}{2}x + 3$, and $y = \frac{1}{2}x - 4$ on the same axes.

a) What do you notice about the graphs of equations and the values where the graphs intersect the y-axis?

b) Do all of these graphs seem to have the same slant (or slope)?

**71.** Graph $y = 2x$. Determine the *rate of change* of y with respect to x. That is, by how many units does y change compared to each unit change in x?

**72.** Graph $y = 4x$. Determine the rate of change of y with respect to x.

**73.** Graph $y = 3x + 2$. Determine the rate of change of y with respect to x.

**74.** Graph $y = \frac{1}{2}x$. Determine the rate of change of y with respect to x.

**75.** The ordered pair $(3, -7)$ represents one point on the graph of a linear equation. If y increases 4 units for each unit increase in x on the graph, find two other solutions to the equation.

**76.** The ordered pair $(1, -4)$ represents one point on the graph of a linear equation. If y increases 3 units for each unit increase in x on the graph, find two other solutions to the equation.

Graph each equation.

**77.** $y = |x - 2|$

**78.** $x = y^2 + 2$

## Group Activity

Discuss and work Exercises 79–80 as a group.

**79.** **a)** Group member 1: Plot the points $(-2, 4)$ and $(6, 8)$. Determine the *midpoint* of the line segment connecting these points.

Group member 2: Follow the above instructions for the points $(-3, -2)$ and $(5, 6)$.

Group member 3: Follow the above instructions for the points $(4, 1)$ and $(-2, 4)$.

**b)** As a group, determine a formula for the midpoint of the line segment connecting the points $(x_1, y_1)$ and $(x_2, y_2)$. (*Note*: We will discuss the midpoint formula further in Chapter 14.)

**80.** Three points on a parallelogram are $A(3, 5)$, $B(8, 5)$, and $C(-1, -3)$.

a) Individually determine a fourth point D that completes the parallelogram.

b) Individually compute the area of your parallelogram.

c) Compare your answers. Did you all get the same answers? If not, why not?

d) Is there more than one point that can be used to complete the parallelogram? If so, give the points and find the corresponding areas of each parallelogram.

## Cumulative Review Exercises

[1.9] **81.** Evaluate $x^2 - 4x + 5$ when $x = -3$.

[4.3] **82.** Determine the slope of the line through the points $(-5, 3)$ and $(6, 7)$.

[7.6] *Solve.*

**83.** $\dfrac{2x}{x^2 - 4} + \dfrac{1}{x - 2} = \dfrac{2}{x + 2}$

**84.** $\dfrac{4x}{x^2 + 6x + 9} - \dfrac{2x}{x + 3} = \dfrac{x + 1}{x + 3}$

# 8.2 Functions

**1** Understand set builder notation and interval notation.

**2** Understand relations.

**3** Recognize functions.

**4** Use the vertical line test.

**5** Understand function notation.

**6** Study applications of functions in daily life.

## **1** Understand Set Builder Notation and Interval Notation

We begin with a discussion of methods used to represent sets of numbers. When we list the elements in a set separated by commas, the set is said to be in **roster form**. A set of numbers that cannot be listed in roster form generally can be written using either *set builder notation* or *interval notation*. We begin with set builder notation.

When using set builder notation, symbols are often used to represent sets of numbers. The following chart gives some of the symbols we will use to represent sets of numbers.

### Sets of Numbers

| Sets of Numbers | Symbol |
|---|---|
| Set of natural or counting numbers | $N$ |
| Set of whole numbers | $W$ |
| Set of integers | $I$ |
| Set of real numbers | $\mathbb{R}$ |

The symbol $\in$ is read "is an element of." For example, we read the symbols $x \in N$

$$x \qquad \in \qquad\qquad N$$
$$\uparrow \qquad\qquad\quad \uparrow$$
is an element of   the set of natural numbers

Set builder notation also generally uses the following inequality symbols.

### Inequality Symbols

> is read "is greater than."

$\geq$ is read "is greater than or equal to."

< is read "is less than."

$\leq$ is read "is less than or equal to."

$\neq$ is read "is not equal to."

Inequalities can be explained using the real number line (**Fig. 8.9**).

FIGURE 8.9

The number $a$ is greater than the number $b$, $a > b$, when $a$ is to the right of $b$ on the number line (**Fig. 8.10**). We can also state that the number $b$ is less than $a$, $b < a$, when $b$ is to the left of $a$ on the number line. The inequality $a \neq b$ means either $a < b$ or $a > b$.

FIGURE 8.10

Remember that the symbol used in an inequality, if it is true, always points to the smaller of the two numbers.

| Notation | Is read as |
|---|---|
| $x > 2$ | $x$ is any real number greater than 2. |
| $x \leq -3$ | $x$ is any real number less than or equal to $-3$. |
| $-4 \leq x < 3$ | $x$ is any real number greater than or equal to $-4$ and less than 3. |

In the inequalities $x > 2$ and $x \leq -3$, the 2 and the $-3$ are called **endpoints**. In the inequality $-4 \leq x < 3$, the $-4$ and 3 are the endpoints. The solutions to inequalities that use either $<$ or $>$ do not include the endpoints, but the solutions to inequalities that use either $\leq$ or $\geq$ do include the endpoints. This is shown as follows:

Endpoint not          Endpoint
included               included

Below are three illustrations of inequalities indicated on the number line.

| Inequality | Inequality Indicated on the Number Line |
|---|---|
| $x > 2$ | $-6\ -5\ -4\ -3\ -2\ -1\ \ 0\ \ 1\ \ 2\ \ 3\ \ 4\ \ 5\ \ 6$ |
| $x \leq -1$ | $-6\ -5\ -4\ -3\ -2\ -1\ \ 0\ \ 1\ \ 2\ \ 3\ \ 4\ \ 5\ \ 6$ |
| $-4 \leq x < 3$ | $-6\ -5\ -4\ -3\ -2\ -1\ \ 0\ \ 1\ \ 2\ \ 3\ \ 4\ \ 5\ \ 6$ |

The word *between* indicates that the endpoints are not included in the answer. For example, the set of natural numbers between 2 and 6 is {3, 4, 5}. If we wish to include the endpoints, we can use the word *inclusive*. For example, the set of natural numbers between 2 and 6 inclusive is {2, 3, 4, 5, 6}.

We are now ready to describe a set using **set builder notation**. An example of set builder notation is

$$E = \{x \mid x \text{ is a natural number greater than 7}\}$$

This is read "Set $E$ is the set of all elements $x$, such that $x$ is a natural number greater than 7." In roster form, this set is written

$$E = \{8, 9, 10, 11, 12, \ldots\}$$

The general form of set builder notation is

We often will use the variable $x$ when using set builder notation, although any variable can be used.

Two condensed ways of writing set $E = \{x \mid x \text{ is a natural number greater than 7}\}$ in set builder notation follow.

$$E = \{x \mid x > 7 \text{ and } x \in N\} \quad \text{or} \quad E = \{x \mid x \geq 8 \text{ and } x \in N\}$$

The set $A = \{x\,|\,{-3} < x \le 4 \text{ and } x \in I\}$ is the set of integers greater than $-3$ and less than or equal to 4. The set written in roster form is $\{-2, -1, 0, 1, 2, 3, 4\}$. Notice that the endpoint $-3$ is not included in the set but the endpoint 4 is included.

How do the sets $B = \{x\,|\,x > 2 \text{ and } x \in N\}$ and $C = \{x\,|\,x > 2\}$ differ? Set $B$ contains only the natural numbers greater than 2, that is, $\{3, 4, 5, 6, \ldots\}$. Set $C$ contains not only the natural numbers greater than 2 but also fractions and decimal numbers greater than 2. Since there is no smallest number greater than 2, this set cannot be written in roster form. Below we illustrate these two sets on the number line. We have also illustrated two other sets.

| Set | Set Indicated on the Number Line |
|---|---|
| $\{x\,|\,x > 2 \text{ and } x \in N\}$ | |
| $\{x\,|\,x > 2\}$ | |
| $\{x\,|\,{-1} \le x < 4 \text{ and } x \in I\}$ | |
| $\{x\,|\,{-1} \le x < 4\}$ | |

In addition to set builder notation, we will also use **interval notation** to describe sets of numbers. The following chart shows the correspondence between inequality symbols, number line endpoints, and interval notation symbols.

| Inequality Symbol | Number Line Endpoints | Interval Notation Symbols | Is endpoint included in solution? |
|---|---|---|---|
| $<$ or $>$ | ○ | ( or ) | No |
| $\le$ or $\ge$ | ● | [ or ] | Yes |

The symbol $\infty$ is read "infinity"; it indicates that the solution set continues indefinitely.

## Understanding Algebra

Whenever $\infty$ is used in interval notation, a *parenthesis* must be used on the corresponding side of the interval notation.

| Solution of Inequality | Solution Set Indicated on Number Line | Solution Set Represented in Interval Notation |
|---|---|---|
| $x \ge 5$ | | $[5, \infty)$ |
| $x < 3$ | | $(-\infty, 3)$ |
| $2 < x \le 6$ | | $(2, 6]$ |
| $-6 \le x \le -1$ | | $[-6, -1]$ |
| $x > a$ | | $(a, \infty)$ |
| $x \ge a$ | | $[a, \infty)$ |
| $x < a$ | | $(-\infty, a)$ |
| $x \le a$ | | $(-\infty, a]$ |
| $a < x < b$ | | $(a, b)$ |
| $a \le x \le b$ | | $[a, b]$ |
| $a < x \le b$ | | $(a, b]$ |
| $a \le x < b$ | | $[a, b)$ |

The set of real numbers, $\mathbb{R}$, is represented in interval notation as $(-\infty, \infty)$.

## 2 Understand Relations

We often find that one quantity is related to a second quantity. For example, the amount you spend for oranges is related to the number of oranges you purchase.

Suppose oranges cost 30 cents apiece. Then one orange costs 30 cents, two oranges cost 60 cents, three oranges cost 90 cents, and so on. The ordered pairs that represent this situation are (1, 30), (2, 60), (3, 90), and so on. An equation that represents this situation is

$$C = 30n$$

The cost of the oranges *depends* on the number of oranges purchased. Thus, cost is the *dependent* variable.

The number of oranges is the independent variable.

Now consider the equation $y = 2x + 3$. Some ordered pairs that satisfy this equation are $(-2, -1), (-1, 1), (0, 3), (1, 5), (2, 7)$, and so on.

$$y = 2x + 3$$

The value of y *depends* on the value of $x$. Thus, $y$ is the *dependent* variable.

$x$ is the independent variable.

### Dependent and Independent Variables

For an equation in variables $x$ and $y$, if the value of $y$ depends on the value of $x$, then $y$ is the **dependent variable** and $x$ is the **independent variable**.

Since related quantities can be represented as ordered pairs, the concept of a *relation* can be defined as follows.

### Relation, Domain, Range

A **relation** is any set of ordered pairs of the form $(x, y)$. The set of $x$-coordinates is called the **domain** of the relation. The set of $y$-coordinates is called the **range** of the relation.

Since the equation $y = 2x + 3$ can be represented as a set of ordered pairs, it is a relation.

## 3 Recognize Functions

We now discuss functions—one of the most important concepts in mathematics.

### Function

A **function** is a relation in which each element in the domain corresponds to exactly one element in the range.

Consider the oranges that cost 30 cents apiece that we just discussed. We can illustrate the number of oranges and the cost of the oranges using **Figure 8.11** on page 462.

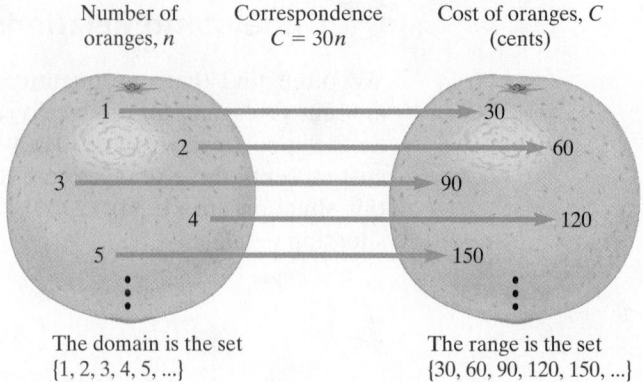

FIGURE 8.11

The domain is the set
{1, 2, 3, 4, 5, ...}

The range is the set
{30, 60, 90, 120, 150, ...}

## Understanding Algebra

In a relation (or function), the set of values for the independent variable is called the *domain*. The set of values for the dependent variable is called the *range*.

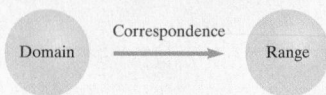

Notice that each number in the set of numbers of oranges, *n*, corresponds to (or is paired with) exactly one number in the set of cost of oranges, *C*. Therefore, this correspondence is a function.

EXAMPLE 1  Determine whether each correspondence is a function.

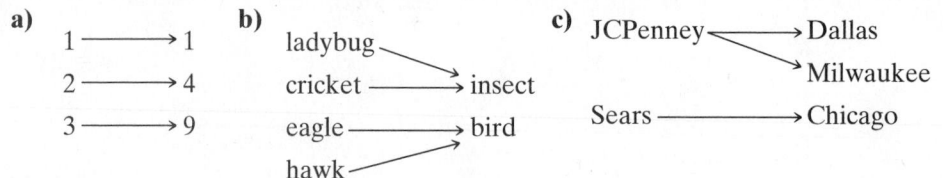

### Solution

**a)** For a correspondence to be a function, each element in the domain must correspond with exactly one element in the range. Here the domain is {1, 2, 3} and the range is {1, 4, 9}. Since each element in the domain corresponds to exactly one element in the range, this correspondence is a function.

**b)** Here the domain is {ladybug, cricket, eagle, hawk} and the range is {insect, bird}. Even though the domain has four elements and the range has two elements, each element in the domain corresponds with exactly one element in the range. Thus, this correspondence is a function.

**c)** Here the domain is {JCPenney, Sears} and the range is {Dallas, Milwaukee, Chicago}. Notice that JCPenney corresponds to both Dallas and Milwaukee. Therefore each element in the domain *does not* correspond to exactly one element in the range. Thus, this correspondence is a relation but *not* a function.

Now Try Exercise 17

EXAMPLE 2  Give the domain and range, then determine whether the relation is a function.

**a)** {(1, 4), (2, 3), (3, 5), (−1, 3), (0, 6)}    **b)** {(−1, 3), (4, 2), (3, 1), (2, 6), (3, 5)}

### Solution

**a)** The domain is {1, 2, 3, −1, 0} and the range is {4, 3, 5, 6}. Notice that when listing the range, we only include the number 3 once, even though it appears in both (2, 3) and (−1, 3). Each number in the domain corresponds with exactly one number in the range. For example, the 1 in the domain corresponds with only the 4 in the range, and so on. Since no *x*-value corresponds to more than one *y*-value, this relation *is a function*.

**b)** The domain is {−1, 4, 3, 2} and the range is {3, 2, 1, 6, 5}. Since the ordered pairs (3, 1) and (3, 5) have *the same first coordinate* and a different second coordinate, each value in the domain does not correspond to exactly one value in the range. Therefore, this relation is *not a function*.

Now Try Exercise 23

Function

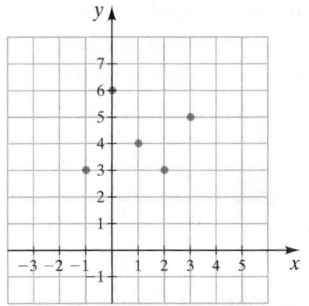

(a) First set of ordered pairs

Not a function

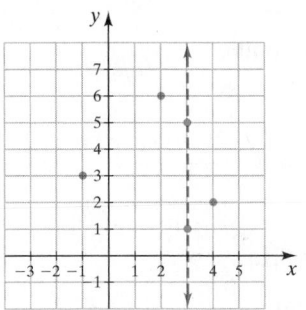

(b) Second set of ordered pairs

FIGURE 8.12

Example 2 leads to an alternate definition of function.

### Function

A **function** is a set of ordered pairs in which no *first* coordinate is repeated.

Recall the equation $y = 2x + 3$, introduced on page 461. Some ordered pairs that satisfy this equation are $(-2, -1), (-1, 1), (0, 3), (1, 5),$ and $(2, 7)$. Notice that each value of $x$ gives a unique value of $y$. Therefore, the equation $y = 2x + 3$ is not only a relation, it is also a function.

## 4  Use the Vertical Line Test

### Graph of Function or Relation

The **graph of a function or relation** is the graph of its set of ordered pairs.

The two sets of ordered pairs in Example 2 **a)** and **b)** are graphed in **Figures 8.12a** and **8.12b**, respectively. Notice that in the function in **Figure 8.12a** it is not possible to draw a vertical line that intersects two points. In **Figure 8.12b** we *can* draw a vertical line through the points $(3, 1)$ and $(3, 5)$. This shows that each $x$-value does not correspond to exactly one $y$-value, and the graph does not represent a function.

This method of determining whether a graph represents a function is called the **vertical line test**.

### Vertical Line Test

If a vertical line can be drawn so that it intersects a graph at more than one point, then the graph does not represent a function. If a vertical line cannot be drawn so that it intersects a graph at more than one point, then the graph represents a function.

The vertical line test shows that **Figure 8.13b** represents a function and **Figures 8.13a** and **8.13c** do not represent functions.

Not a function

(a)

Function

(b)

Not a function

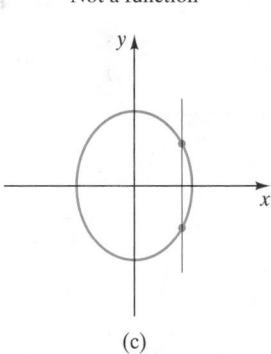

(c)

FIGURE 8.13

**EXAMPLE 3** Use the vertical line test to determine whether the following graphs represent functions. Also determine the domain and range of each function or relation.

a)

FIGURE 8.14

b)

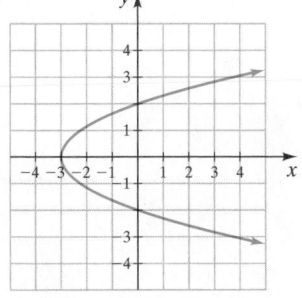

FIGURE 8.15

*Solution*

**a)** A vertical line cannot be drawn to intersect the graph in **Figure 8.14** on page 463 at more than one point. Thus this is the graph of a function. Since the line extends indefinitely in both directions, every value of $x$ will be included in the domain. The domain is the set of real numbers.

$$\text{Domain:} \quad \mathbb{R} \quad \text{or} \quad (-\infty, \infty)$$

Since all values of $y$ are included on the graph, the range is also the set of real numbers.

$$\text{Range:} \quad \mathbb{R} \quad \text{or} \quad (-\infty, \infty)$$

**b)** Since a vertical line can be drawn to intersect the graph in **Figure 8.15** on page 463 at more than one point, this is *not* the graph of a function. The domain of this relation is the set of values greater than or equal to $-3$.

$$\text{Domain:} \quad \{x | x \geq -3\} \quad \text{or} \quad [-3, \infty)$$

The range is the set of $y$-values, which can be any real number.

$$\text{Range:} \quad \mathbb{R} \quad \text{or} \quad (-\infty, \infty)$$

*Now Try Exercise 65*

**FIGURE 8.16**

**EXAMPLE 4** Consider the function graphed in **Figure 8.16**.

**a)** What element of the range is paired with 4 in the domain?

**b)** What elements of the domain are paired with $-2$ in the range?

**c)** What is the domain of the function?

**d)** What is the range of the function?

*Solution*

**a)** The range is the set of $y$-values. The $y$-value paired with the $x$-value 4 is 3.

**b)** The domain is the set of $x$-values. The $x$-values paired with the $y$-value $-2$ are 2 and 6.

**c)** The domain is the set of $x$-values, 0 through 8. Thus the domain is

$$\{x | 0 \leq x \leq 8\} \quad \text{or} \quad [0, 8]$$

**d)** The range is the set of $y$-values, $-2$ through 3. Thus, the range is

$$\{y | -2 \leq y \leq 3\} \quad \text{or} \quad [-2, 3]$$

*Now Try Exercise 71*

**EXAMPLE 5** **Figure 8.17** illustrates a graph of speed versus time of a man out for a walk and run. Write a story about the man's outing that corresponds to this function.

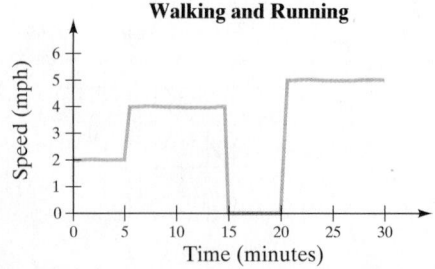

**FIGURE 8.17**

*Solution*   Understand   The horizontal axis is time and the vertical axis is speed. When the graph is horizontal it means the person is traveling at the constant speed indicated on the vertical axis. The steep portions of the graph that increase with time indicate an increase in speed, whereas the steep portions of the graph that decrease with time indicate a decrease in speed.

Answer   Here is one possible interpretation of the graph. The man walks for about 5 minutes at a speed of about 2 miles per hour. Then the man speeds up to about 4 miles per hour and walks fast or runs at about this speed for about 10 minutes. Then the man slows down and stops, and then rests for about 5 minutes. Finally, the man speeds up to about 5 miles per hour and runs at this speed for about 10 minutes.

Now Try Exercise 99

## 5 Understand Function Notation

In Section 8.1 we graphed the equations shown in **Table 8.1** below. Notice that each of these graphs passes the vertical line test. Therefore, the *graphs* represent functions and we say the *equations* define functions. We will use specific notation to name functions that are defined by equations.

**TABLE 8.1   Graphs of Functions**

| Section 8.1 example | Equation graphed | Graph | Does the graph represent a function? | Domain | Range |
|---|---|---|---|---|---|
| 1 | $y = -\dfrac{1}{3}x + 1$ | | Yes | $(-\infty, \infty)$ | $(-\infty, \infty)$ |
| 2 | $y = x^2 - 4$ | | Yes | $(-\infty, \infty)$ | $[-4, \infty)$ |
| 3 | $y = \dfrac{1}{x}$ | | Yes | $(-\infty, 0) \cup (0, \infty)$ | $(-\infty, 0) \cup (0, \infty)$ |
| 4 | $y = |x|$ | | Yes | $(-\infty, \infty)$ | $[0, \infty)$ |

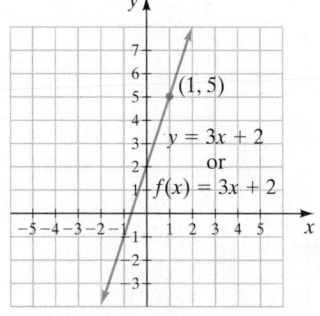

FIGURE 8.18

### Function Notation

If an equation involving $x$ as the independent variable and $y$ as the dependent variable defines a function, we say **$y$ is a function of $x$** and we write **$y = f(x)$** (pronounced "f of x.")

Consider the equation  as seen in **Figure 8.18**. Since this graph passes the vertical line test, this equation defines a function and we can write the equation using function notation as $f(x) = 3x + 2$.

$$f(x) = \underline{3x + 2}$$

$f$ is the ———— name of the function

$x$ is the ———— independent variable

substitute values of $x$ to get the corresponding values of $y$ or $f(x)$

For example, if $f(x) = 3x + 2$, then $f(1)$, read "f of one" is found as follows:

$$f(x) = 3x + 2$$
$$f(1) = 3(1) + 2 = 5$$

Therefore, when $x$ is $1$, $y$ is $5$. The ordered pair $(1, 5)$ appears on the graph of $y = 3x + 2$ in **Figure 8.18** on page 465.

> **Helpful Hint**
>
> Linear equations that are not solved for $y$ can be written using function notation by solving the equation for $y$, then replacing $y$ with $f(x)$. For example, the equation $-9x + 3y = 6$ becomes $y = 3x + 2$ and we can write $f(x) = 3x + 2$.

**EXAMPLE 6** If $f(x) = -4x^2 + 3x - 2$, find

**a)** $f(2)$        **b)** $f(-1)$        **c)** $f(a)$

*Solution*

**a)** $f(x) = -4x^2 + 3x - 2$
$f(2) = -4(2)^2 + 3(2) - 2 = -4(4) + 6 - 2 = -16 + 6 - 2 = -12$

**b)** $f(-1) = -4(-1)^2 + 3(-1) - 2 = -4(1) - 3 - 2 = -4 - 3 - 2 = -9$

**c)** To evaluate the function at $a$, we replace each $x$ in the function with an $a$.

$$f(x) = -4x^2 + 3x - 2$$
$$f(a) = -4a^2 + 3a - 2$$

Now Try Exercise 77

**EXAMPLE 7** Determine each indicated function value.

**a)** $g(-2)$ for $g(t) = \dfrac{1}{t + 8}$

**b)** $h(5)$ for $h(s) = 2|s - 6|$

**c)** $j(-3)$ for $j(r) = \sqrt{22 - r}$

*Solution*    In each part, substitute the indicated value into the function and evaluate.

**a)** $g(-2) = \dfrac{1}{-2 + 8} = \dfrac{1}{6}$

**b)** $h(5) = 2|5 - 6| = 2|-1| = 2(1) = 2$

**c)** $j(-3) = \sqrt{22 - (-3)} = \sqrt{22 + 3} = \sqrt{25} = 5$

Now Try Exercise 81

**6  Study Applications of Functions in Daily Life**

Now we examine additional applications of functions.

**EXAMPLE 8  Business Jets**  The graph in **Figure 8.19** shows the number of business jets manufactured for the years from 1994 through 2008, projected through 2013.

**Business-Jet Market**

FIGURE 8.19        *Source:* Forecast International

**a)** Explain why the graph in **Figure 8.19** represents a function.

**b)** Determine the number of business jets projected to be manufactured in 2010.

**c)** Determine the projected percent increase in the number of business jets to be manufactured from 2003 to 2011.

**d)** Determine the percent decrease in the number of business jets manufactured from 2001 to 2003.

*Solution*

**a)** The graph represents a function since each year corresponds to a specific number of business jets manufactured. Notice that the graph passes the vertical line test.

**b)** In 2010, the graph shows that 1325 business jets should be manufactured. If we let the function be represented by $J$, then $J(2010) = 1325$.

**c)** We will follow the problem-solving procedure to solve this problem.

Understand and Translate   We need to determine the percent increase in the number of business jets to be manufactured from 2003 to 2011. To do this, use the formula

$$\text{percent change (increase or decrease)} = \frac{\left(\begin{array}{c}\text{value in}\\\text{latest period}\end{array}\right) - \left(\begin{array}{c}\text{value in}\\\text{previous period}\end{array}\right)}{\text{value in previous period}}$$

The latest period is 2011 and the previous period is 2003. Substituting the values, we get

$$\text{percent change} = \frac{1400 - 525}{525}$$

Carry Out

$$= \frac{875}{525} \approx 1.667 = 166.7\%$$

Check and Answer   Our calculations appear correct. There is projected to be about a 166.7% increase in the number of business jets manufactured from 2003 to 2011.

**d)** To find the percent decrease from 2001 to 2003, we follow the same procedure as in part **c)**. The latest period is 2003 and the previous period is 2001.

$$\text{percent change (increase or decrease)} = \frac{\left(\begin{array}{c}\text{value in}\\\text{latest period}\end{array}\right) - \left(\begin{array}{c}\text{value in}\\\text{previous period}\end{array}\right)}{\text{value in previous period}}$$

$$= \frac{525 - 785}{785} = \frac{-260}{785} \approx -0.331 = -33.1\%$$

The negative sign preceding the 33.1% indicates a percent decrease. Thus, there was about a 33.1% decrease in the manufacture of business jets from 2001 to 2003.

Now Try Exercise 105

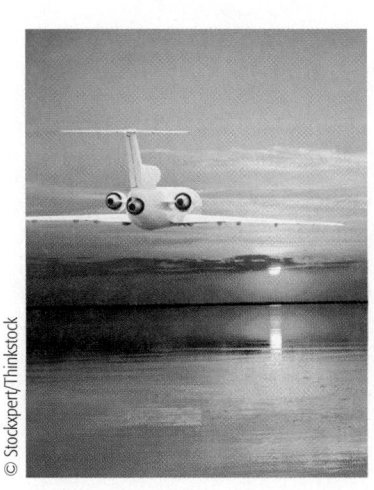

**EXAMPLE 9** **30-Year Mortgage Rates** The graph in **Figure 8.20** shows the 30-year mortgage rate from 1971 to 2007.

**a)** Using the graph in **Figure 8.20**, explain why this set of points represents a function.

**b)** Using the graph in **Figure 8.21**, estimate the 30-year mortgage rate in 2006.

FIGURE 8.20

FIGURE 8.21

*Solution*

**a)** Since each year corresponds with exactly one 30-year mortgage rate, the set of points represents a function. Notice that this graph passes the vertical line test.

**b)** If we connect the points with line segments as in **Figure 8.21**, we can estimate from the graph that the 30-year mortgage rate in 2006 was about 6%. If we call this function $f$, then $f(2006) \approx 6\%$.

Now Try Exercise 107

Often formulas are written using function notation as shown next.

**EXAMPLE 10** The Celsius temperature, $C$, is a function of the Fahrenheit temperature, $F$.

$$C(F) = \frac{5}{9}(F - 32)$$

Determine the Celsius temperature that corresponds to 50°F.

*Solution* We need to find $C(50)$. We do so by substitution.

$$C(F) = \frac{5}{9}(F - 32)$$

$$C(50) = \frac{5}{9}(50 - 32)$$

$$= \frac{5}{9}(18) = 10$$

Therefore, 50°F = 10°C.

Now Try Exercise 87

# EXERCISE SET 8.2

MathXL®   MyMathLab®
MathXL®        MyMathLab

## Warm-Up Exercises

*Fill in the blanks with the appropriate word, phrase, or symbol(s) from the following list.*

relation              function           domain            graph              range

x-coordinate          y-coordinates      $f(x)$            vertical line

vertical line test    dependent          independent       ordered pairs

**1.** The set of x-coordinates of a relation is called the _____ of the relation.

**2.** The set of _____ of a relation is called the range of the relation.

**3.** A relation is any set of _____ .

**4.** A function is a _____ in which each element of the domain corresponds to exactly one element of the range.

**5.** If a _____ intersects a graph in more than one point, then this graph is not the graph of a function.

**6.** In a function, every _____ corresponds to exactly one y-coordinate.

**7.** In a function, every element in the domain must correspond to exactly one element in the _____ .

**8.** The notation _____ is pronounced "f of x."

**9.** The notation $y = f(x)$, means that y is a _____ of x.

**10.** In the notation $y = f(x)$, y is the _____ variable.

**11.** In the notation $y = f(x)$, x is the _____ variable.

**12.** The _____ of a function or relation is the graph of its set of ordered pairs.

## Practice the Skills

*In Exercises 13–20,* **a)** *determine if the relation illustrated is a function.* **b)** *Give the domain and range of each function or relation.*

**13.**  minimum driving age

Idaho ———→ 15

Texas ———→ 16

Georgia ———→ 18

**14.**  minimum driving age

New York
Florida ———→ 16
Michigan

**15.**  twice a number

3 ———→ 6

5 ———→ 10

11 ———→ 22

**16.**  nicknames

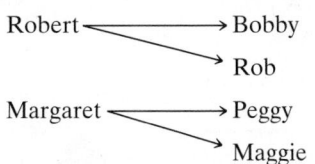

Robert ←———→ Bobby
         ———→ Rob

Margaret ←———→ Peggy
          ———→ Maggie

**17.**  number of siblings

Cameron ———→ 3

Tyrone ———→ 6

Vishnu

**18.**  a number squared

4 ———→ 16

5 ———→ 25

7 ———→ 49

**19.**  cost of a stamp

1990 ———→ 20

2001 ———→ 34

2002 ———→ 37

**20.**  absolute value

$|-8|$ ———→ 8

$|8|$

$|0|$ ———→ 0

*In Exercises 21–28,* **a)** *determine which of the following relations are also functions.* **b)** *Give the domain and range of each relation or function.*

**21.** $\{(1,4), (2,2), (3,5), (4,3), (5,1)\}$

**22.** $\{(1,0), (4,2), (9,3), (1,-1), (4,-2), (9,-3)\}$

**23.** $\{(3,-1), (5,0), (1,2), (4,4), (2,2), (7,9)\}$

**24.** $\{(-1,1), (0,-3), (3,4), (4,5), (-2,-2)\}$

**25.** $\{(1,4), (2,5), (3,6), (2,2), (1,1)\}$

**26.** $\{(6,3), (-3,4), (0,3), (5,2), (3,5), (2,8)\}$

**27.** $\{(0,3), (1,3), (2,2), (1,-1), (2,-7)\}$

**28.** $\{(3,5), (2,5), (1,5), (0,5), (-1,5)\}$

*In Exercises 29–38, list each set in roster form.*

**29.** $A = \{x | -1 < x < 1 \text{ and } x \in I\}$

**30.** $B = \{y | y \text{ is an odd natural number less than 6}\}$

**31.** $C = \{z | z \text{ is an even integer greater than 16 and less than or equal to 20}\}$

**32.** $D = \{x | x \geq -3 \text{ and } x \in I\}$

**33.** $E = \{x | x < 3 \text{ and } x \in W\}$

**34.** $F = \left\{x \,\middle|\, -\dfrac{6}{5} \leq x < \dfrac{15}{4} \text{ and } x \in N\right\}$

**35.** $H = \{x | x \text{ is a whole number multiple of 7}\}$

**36.** $L = \{x | x \text{ is an integer greater than } -5\}$

**37.** $J = \{x | x > 0 \text{ and } x \in I\}$

**38.** $K = \{x | x \text{ is a whole number between 9 and 10}\}$

*In Exercises 39 and 40,* **a)** *write out how you would read each set;* **b)** *write the set in roster form.*

**39.** $A = \{x | x < 7 \text{ and } x \in N\}$

**40.** $B = \{x | x \text{ is one of the last five capital letters in the English alphabet}\}$

*Illustrate each set on a number line.*

**41.** $\{x | x \geq 0\}$

**42.** $\{w | w > -5\}$

**43.** $\{z | z \leq 2\}$

**44.** $\{y | y < 4\}$

**45.** $\{p | -6 \leq p < 3\}$

**46.** $\{x | -1.67 \leq x < 5.02\}$

**47.** $\{q | q > -3 \text{ and } q \in N\}$

**48.** $\{x | -1.93 \leq x \leq 2 \text{ and } x \in I\}$

**49.** $\{r | r \leq \pi \text{ and } r \in W\}$

**50.** $\left\{x \,\middle|\, \dfrac{5}{12} < x \leq \dfrac{7}{12} \text{ and } x \in N\right\}$

*Express in set builder notation each set of numbers that are indicated on the number line.*

**51.**

**52.**

**53.**

**54.**

**55.**

**56.**

**57.**

**58.**

**59.**

**60.**

*In Exercises 61–72,* **a)** *determine whether the graph illustrated represents a function.* **b)** *Give the domain and range of each function or relation.* **c)** *Approximate the value or values of x where y = 2.*

**61.**

**62.**

**63.**

**64.**

**65.**

**66.**

**67.**

**68.**

**69.**

**70.**

**71.**

**72.**

*Evaluate each function at the indicated values.*

**73.** $f(x) = 4x - 3$; find
   **a)** $f(2)$.
   **b)** $f(-2)$.

**74.** $f(a) = \dfrac{1}{3}a + 4$; find
   **a)** $f(0)$.
   **b)** $f(-12)$.

**75.** $h(x) = x^2 - x - 6$; find
   **a)** $h(0)$.
   **b)** $h(-1)$.

**76.** $g(x) = -2x^2 + 7x - 11$; find
   **a)** $g(2)$.
   **b)** $g\left(\dfrac{1}{2}\right)$.

**77.** $r(t) = -t^3 - 2t^2 + t + 4$; find
   **a)** $r(1)$.
   **b)** $r(-2)$.

**78.** $g(t) = 4 - 3t + 16t^2 - 2t^3$; find
   **a)** $g(0)$.
   **b)** $g(3)$.

**79.** $h(z) = |5 - 2z|$; find
   **a)** $h(6)$.
   **b)** $h\left(\dfrac{5}{2}\right)$.

**80.** $q(x) = -2|x + 8| + 13$; find
   **a)** $q(0)$.
   **b)** $q(-4)$.

**81.** $s(t) = \sqrt{t + 3}$; find
   **a)** $s(-3)$.
   **b)** $s(6)$.

**82.** $f(t) = \sqrt{5 - 2t}$; find
   **a)** $f(-2)$.
   **b)** $f(2)$.

**83.** $g(x) = \dfrac{x^3 - 2}{x - 2}$; find
   **a)** $g(0)$.
   **b)** $g(2)$.

**84.** $h(x) = \dfrac{x^2 + 4x}{x + 6}$; find
   **a)** $h(-3)$.
   **b)** $h\left(\dfrac{2}{5}\right)$.

## Problem Solving

**85. Area of a Rectangle**  The formula for the area of a rectangle is $A = lw$. If the length of a rectangle is 4 inches, then the area is a function of its width, $A(w) = 4w$. Find the area when the width is

   **a)** 4 feet.

   **b)** 6.5 feet.

**86. Simple Interest**  The formula for the simple interest earned for a period of 1 year is $i = pr$, where $p$ is the principal invested and $r$ is the simple interest rate. If $1000 is invested, the simple interest earned in 1 year is a function of the simple interest rate, $i(r) = 1000r$. Determine the simple interest earned in 1 year if the interest rate is

   **a)** 2.5%.

   **b)** 4.25%.

**87. Area of a Circle**  The formula for the area of a circle is $A = \pi r^2$. The area is a function of the radius.

   **a)** Write this function using function notation.

   **b)** Determine the area when the radius is 12 yards.

**88. Perimeter of a Square**  The formula for the perimeter of a square is $P = 4s$ where $s$ represents the length of any one of the sides of the square.

   **a)** Write this function using function notation.

   **b)** Determine the perimeter of a square with sides of length 7 meters.

**89. Temperature**  The formula for changing Fahrenheit temperature into Celsius temperature is $C = \dfrac{5}{9}(F - 32)$. The Celsius temperature is a function of Fahrenheit temperature.

   **a)** Write this function using function notation.

   **b)** Find the Celsius temperature that corresponds to $-31°F$

**90. Volume of a Cylinder**  The formula for the volume of a right circular cylinder is $V = \pi r^2 h$. If the height, $h$, is 3 feet, then the volume is a function of the radius, $r$.

   **a)** Write this formula in function notation, where the height is 3 feet.

   **b)** Find the volume if the radius is 2 feet.

**91. Sauna Temperature**  The temperature, $T$, in degrees Celsius, in a sauna $n$ minutes after being turned on is given by the function $T(n) = -0.03n^2 + 1.5n + 14$. Find the sauna's temperature after

   **a)** 3 minutes.

   **b)** 12 minutes.

**92. Stopping Distance**  The stopping distance, $d$, in meters for a car traveling $v$ kilometers per hour is given by the function $d(v) = 0.18v + 0.01v^2$. Find the stopping distance for the following speeds:

   **a)** 60 km/hr

   **b)** 25 km/hr

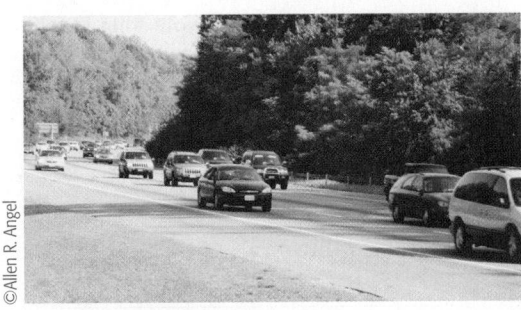

See Exercise 92 on page 472.

**93. Air Conditioning** When an air conditioner is turned on maximum in a bedroom at $80°$, the temperature, $T$, in the room after $A$ minutes can be approximated by the function $T(A) = -0.02A^2 - 0.34A + 80, 0 \leq A \leq 15$.

  **a)** Estimate the room temperature 4 minutes after the air conditioner is turned on.

  **b)** Estimate the room temperature 12 minutes after the air conditioner is turned on.

**94. Accidents** The number of accidents, $n$, in 1 month involving drivers $x$ years of age can be approximated by the function $n(x) = 2x^2 - 150x + 4000$. Find the approximate number of accidents in 1 month that involved

  **a)** 18-year-olds.       **b)** 25-year-olds.

**95. Oranges** The total number of oranges, $T$, in a square pyramid whose base is $n$ by $n$ oranges is given by the function

$$T(n) = \frac{1}{3}n^3 + \frac{1}{2}n^2 + \frac{1}{6}n$$

Find the number of oranges if the base is

  **a)** 6 by 6 oranges.       **b)** 8 by 8 oranges.

**96. Rock Concert** If the cost of a ticket to a rock concert is increased by $x$ dollars, the estimated increase in revenue, $R$, in thousands of dollars is given by the function $R(x) = 24 + 5x - x^2, x < 8$. Find the increase in revenue if the cost of the ticket is increased by

  **a)** \$1.

  **b)** \$4.

**97. Supply and Demand** The price of a bushel of soybeans can be estimated by the function

$$f(Q) = -0.00004Q + 4.25, \quad 10,000 \leq Q \leq 60,000$$

where $f(Q)$ is the price of a bushel of soybeans and $Q$ is the annual number of bushels of soybeans produced.

  **a)** Construct a graph showing the relationship between the number of bushels of soybeans produced and the price of a bushel of soybeans.

  **b)** Estimate the cost of a bushel of soybeans if 40,000 bushels of soybeans are produced in a given year.

**98. Household Expenditures** The average annual household expenditure is a function of the average annual household income. The average expenditure can be estimated by the function

$$f(i) = 0.6i + 5000, \quad \$3500 \leq i \leq \$50,000$$

where $f(i)$ is the average household expenditure and $i$ is the average household income.

  **a)** Draw a graph showing the relationship between average household income and the average household expenditure.

  **b)** Estimate the average household expenditure for a family whose average household income is \$30,000.

## Concept/Writing Exercises

*Review Example 5 before working Exercises 99–104.*

**99. Heart Rate** The following graph shows a person's heart rate while doing exercise. Write a story that this graph may represent.

**100. Water Level** The following graph shows the water level at a certain point during a flood. Write a story that this graph may represent.

**101. Height above Sea Level** The following graph shows height above sea level versus time when a man leaves his house and goes for a walk. Write a story that this graph may represent.

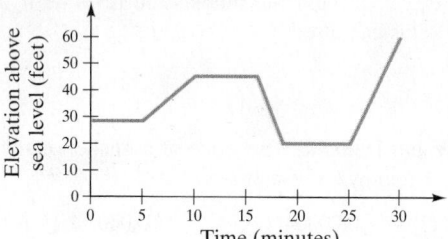

**102. Water Level in a Bathtub** The following graph shows the level of water in a bathtub versus time. Write a story that this graph may represent.

**103. Speed of a Car** The following graph shows the speed of a car versus time. Write a story that this graph may represent.

**104. Distance Traveled** The following graph shows the distance traveled by a person in a car versus time. Write a story that this graph may represent.

**105. College Costs** The following graph compares the average cost (in 2006 dollars) of attending college for one year at private colleges and public colleges.

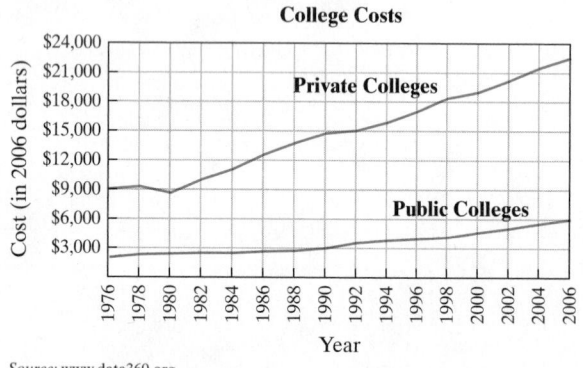

*Source:* www.data360.org

a) Does each line shown represent a function? Explain.

b) In these graphs, which variable is the independent variable?

c) If $f$ represents the average cost of attending private college for one year, determine $f(2006)$.

d) If $g$ represents the average cost of attending public college for one year, determine $g(2006)$.

e) Determine the percent increase from 1990 to 2006 in the average cost of one year of public college.

**106. Price of Gasoline** The following graph shows the inflation-adjusted average price of a gallon (in 2008 dollars) of gasoline for the years from 1918 to 2008.

*Source:* US Energy Information Administration

a) Does this graph represent a function?

b) In this graph, what is the dependent variable?

c) If $p$ represents this function, determine $p(2006)$.

d) Determine the percent decrease from 1981 to 2008 (round your answer to the nearest tenth of a percent)

**107. Super Bowl Commercials** The average price of the cost of a 30-second commercial during the Super Bowl has been increasing over the years. The following chart gives the approximate cost of a 30-second commercial for selected years from 1981 through 2009.

| Year | Cost ($1000s) |
|------|---------------|
| 1981 | 280 |
| 1985 | 500 |
| 1989 | 740 |
| 1993 | 970 |
| 1997 | 1200 |
| 2001 | 2000 |
| 2005 | 2400 |
| 2009 | 3000 |

**a)** Draw a line graph that displays this information.

**b)** Does the graph appear to be approximately linear? Explain.

**c)** From the graph, estimate the cost of a 30-second commercial in 2004.

**108. Shipments of LCD Monitors** The following graph shows the shipments of LCD monitors, in millions of units, for the years from 2002 to 2008.

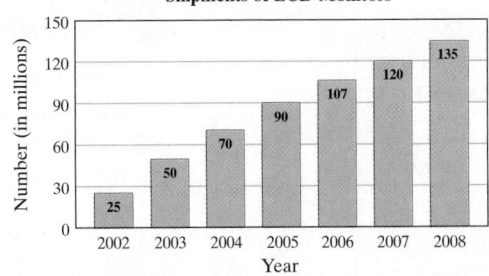

Shipments of LCD Monitors

*Source:* DisplaySearch, Market Intelligence Center, *Wall Street Journal*

**a)** Draw a line graph that displays this information.

**b)** Does the graph you drew in part **a)** appear to be approximately linear? Explain.

**c)** Assuming this trend continues, from the line graph you drew, estimate the number of LCD monitors to be shipped in 2009.

**d)** Does the bar graph represent a function?

**e)** Does the line graph you drew in part **a)** represent a function?

---

## Group Activity

*In many real-life situations, more than one function may be needed to represent a problem. This often occurs where two or more different rates are involved. For example, when discussing federal income taxes, there are different tax rates. When two or more functions are used to represent a problem, the function is called a **piecewise function**. Following are two examples of piecewise functions and their graphs.*

$$f(x) = \begin{cases} -x + 2, & 0 \le x < 4 \\ 2x - 10, & 4 \le x < 8 \end{cases}$$

$$f(x) = \begin{cases} 2x - 1, & -2 \le x < 2 \\ x - 2, & 2 \le x < 4 \end{cases}$$

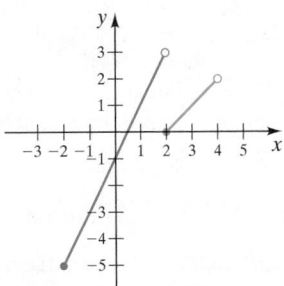

*As a group, graph the following piecewise functions.*

**109.** $f(x) = \begin{cases} x + 3, & -1 \le x < 2 \\ 7 - x, & 2 \le x < 4 \end{cases}$

**110.** $g(x) = \begin{cases} 2x + 3, & -3 < x < 0 \\ -3x + 1, & 0 \le x < 2 \end{cases}$

---

## Cumulative Review Exercises

[2.5] **111.** Solve $3x - 2 = \dfrac{1}{3}(3x - 3)$.

[2.6] **112.** Solve the following formula for $p_2$.

$$E = a_1 p_1 + a_2 p_2 + a_3 p_3$$

[4.4] **113.** Write the equation $3x + 6y = 9$ in slope-intercept form and indicate the slope and the $y$-intercept.

[7.2] **114.** Divide $\dfrac{3x^2 - 16x - 12}{3x^2 - 10x - 8} \div \dfrac{x^2 - 7x + 6}{3x^2 - 11x - 4}$.

# 8.3 Linear Functions

1  Graph linear functions.

2  Graph linear functions using intercepts.

3  Study applications of functions.

## 1  Graph Linear Functions

In Section 8.1 we graphed linear equations. To graph the linear equation $y = 2x + 4$, we can make a table of values, plot the points, and draw the graph, as shown in **Figure 8.22**. Since this graph passes the vertical line test, the equation defines a function and we may write $f(x) = 2x + 4$.

| x | y |
|---|---|
| -2 | 0 |
| 0 | 4 |
| 1 | 6 |

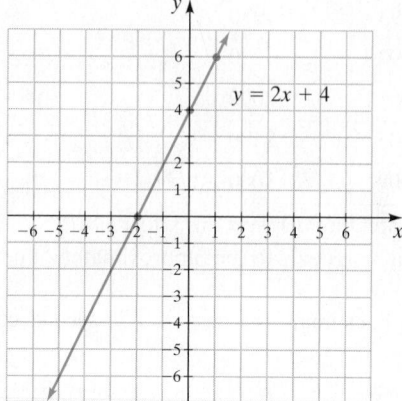

**FIGURE 8.22**

This is an example of a linear function.

---

**Linear Function**

A linear function is a function of the form $f(x) = ax + b$.

- The graph of any linear equation is a straight line.
- The domain of any linear function is all real numbers, $\mathbb{R}$.
- If $a \neq 0$, then the range of any linear function is all real numbers, $\mathbb{R}$.

---

**Helpful Hint**

To graph linear functions, we treat $f(x)$ as $y$ and follow the same procedure used to graph linear equations discussed in Section 8.1.

---

**EXAMPLE 1**  Graph $f(x) = \frac{1}{2}x - 1$.

*Solution*    We construct a table of values by substituting values for $x$ and finding corresponding values of $f(x)$ or $y$. Then we plot the points and draw the graph, as illustrated in **Figure 8.23**.

| x | f(x) |
|---|---|
| -2 | -2 |
| 0 | -1 |
| 2 | 0 |

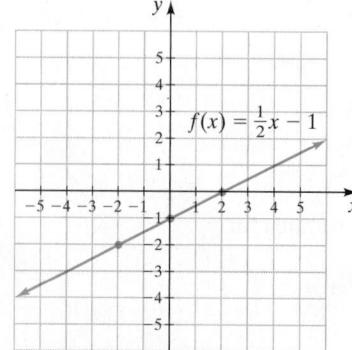

**FIGURE 8.23**

Now Try Exercise 9

Note that the vertical axis in **Figure 8.23** may also be labeled as $f(x)$ instead of $y$. In this book we will continue to label it $y$.

## 2 Graph Linear Functions Using Intercepts

In Section 4.2 we discussed graphing linear equations using their intercepts. In the box below we refresh your memory on how to find the intercepts.

> **To Find the *x*- and *y*-Intercepts**
>
> To find the *y*-intercept, set $x = 0$ and solve for *y*.
> To find the *x*-intercept, set $y = 0$ and solve for *x*.

To graph linear functions using intercepts just remember that $f(x)$ is the same as *y*. Example 2 explains how to graph a linear function using its intercepts.

**EXAMPLE 2**  Graph $f(x) = -\dfrac{1}{3}x - 1$ using the *x*- and *y*-intercepts.

*Solution*   Treat $f(x)$ the same as *y*. To find the *y*-intercept, set $x = 0$ and solve for $f(x)$.

$$f(x) = -\frac{1}{3}x - 1$$

$$f(x) = -\frac{1}{3}(0) - 1 = -1$$

The *y*-intercept is $(0, -1)$.
To find the *x*-intercept, set $f(x) = 0$ and solve for *x*.

FIGURE 8.24

$$f(x) = -\frac{1}{3}x - 1$$

$$0 = -\frac{1}{3}x - 1$$

$$3(0) = 3\left(-\frac{1}{3}x - 1\right) \qquad \text{Multiply both sides by 3.}$$

$$0 = -x - 3 \qquad \text{Distributive property}$$

$$x = -3 \qquad \text{Add } x \text{ to both sides.}$$

The *x*-intercept is $(-3, 0)$. The graph is shown in **Figure 8.24**.

Now Try Exercise 15

> **Helpful Hint**
>
> When graphing a linear equation by using the *x*-intercept and *y*-intercept, it is helpful to plot a checkpoint. For example, in Example 2, if we substitute $x = 3$ into the equation and solve for *y*, we get $y = -2$. Thus, a third point on the graph should be $(3, -2)$. Looking at **Figure 8.24**, we see that this point is on the graph and we are reassured that our graph is correct.

### Using Your Graphing Calculator ◼◼◼◼◼

In Section 4.2 we used the graphing calculator to find intercepts. Here we will use this information to find the intercepts of the graph of $f(x) = -\dfrac{1}{3}x - 1$ from example 2. The graph of the function is obtained by pressing the keys shown below. The graph is shown in **Figure 8.25a**.

$$\boxed{Y =}\ \boxed{(}\ \boxed{(-)}\ \boxed{1}\ \boxed{\div}\ \boxed{3}\ \boxed{)}\ \boxed{\text{X,T,}\Theta\text{,n}}\ \boxed{-}\ \boxed{1}\ \boxed{\text{GRAPH}}$$

FIGURE 8.25

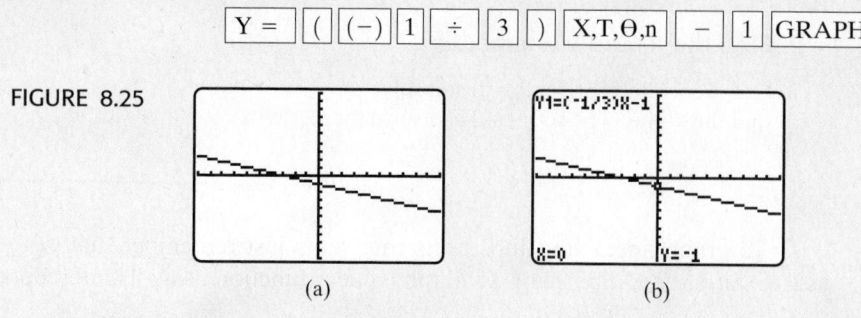

           (a)                      (b)

To get the $y$-intercept press $\boxed{\text{TRACE}}$. The $y$-intercept is $(0, -1)$ as shown in **Figure 8.25b**. To get the $x$-intercept, we use the "zero" feature by first pressing the keys

$$\boxed{\text{2nd}}\ \boxed{\text{TRACE}}\ \boxed{2}.$$

To enter a left-bound, a right-bound, and a guess, move the flashing cursor left or right by using the keys $\boxed{<}$ or $\boxed{>}$ followed by pressing $\boxed{\text{ENTER}}$. The screen is shown in **Figure 8.26**. The $x$-intercept is $(-3, 0)$.

FIGURE 8.26

### 3 Study Applications of Functions

Graphs are often used to show the relationship between variables.

**EXAMPLE 3**  **Tire Store Profit**  The yearly profit, $p$, of a tire store can be estimated by the function $p(n) = 20n - 30{,}000$, where $n$ is the number of tires sold per year.

**a)** Draw a graph of profit versus tires sold for up to and including 6000 tires.

**b)** Estimate the number of tires that must be sold for the company to break even.

**c)** Estimate the number of tires sold if the company has a $70,000 profit.

---

> **Understanding Algebra**
>
> A company's break-even point is the point where the profit is equal to 0. In other words, the company neither makes money nor loses money.

---

*Solution*

**a)** Understand

$$p \qquad \text{is a function of} \qquad n$$

$p$ is the profit.⤶                           ⤶ $n$ is the number of tires sold.
$p$ is the dependent variable.                 $n$ is the independent variable.
The vertical axis will be labeled $p$.      The horizontal axis will be labeled $n$.

Since the minimum number of tires that can be sold is 0, the horizontal axis will go from 0 to 6000 tires. We will graph this equation by determining and plotting the intercepts.

Translate and Carry Out     To find the $p$-intercept, we set $n = 0$ and solve for $p(n)$.

$$p(n) = 20n - 30{,}000$$
$$p(n) = 20(0) - 30{,}000 = -30{,}000$$

Thus, the $p$-intercept is $(0, -30{,}000)$.

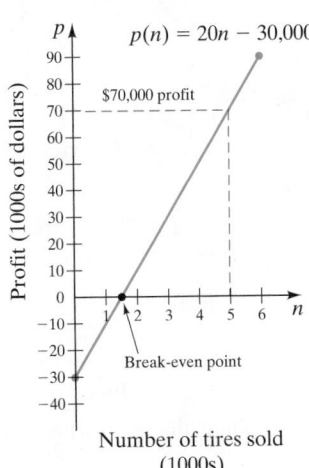

$p(n) = 20n - 30,000$

$70,000 profit

Break-even point

Number of tires sold
(1000s)

FIGURE 8.27

To find the *n*-intercept, we set $p(n) = 0$ and solve for *n*.

$$p(n) = 20n - 30,000$$
$$0 = 20n - 30,000$$
$$30,000 = 20n$$
$$1500 = n$$

Thus the *n*-intercept is $(1500, 0)$.

Answer   Now we use the *p*- and *n*-intercepts to draw the graph (see **Fig. 8.27**).

**b)** The break-even point is where the graph intersects the *n*-axis, for this is where the profit, *p*, is 0. To break even, approximately 1500 tires must be sold.

**c)** To make $70,000, approximately 5000 tires must be sold (shown by the dashed red line in **Fig. 8.27**).

Now Try Exercise 31

> ### Helpful Hint
>
> Sometimes it is difficult to read an exact answer from a graph. To determine the exact number of tires needed to break even in Example 3, substitute 0 for $p(n)$ in the function $p(n) = 20n - 30,000$ and solve for *n*. To determine the exact number of tires needed to obtain a $70,000 profit, substitute 70,000 for $p(n)$ and solve the equation for *n*.

**EXAMPLE 4  Toy Store Sales** Rob Kimball is the owner of a toy store. His monthly salary consists of $200 plus 10% of the store's sales for that month.

**a)** Write a function expressing his monthly salary, *m*, in terms of the store's sales, *s*.

**b)** Draw a graph of his monthly salary for sales up to and including $20,000.

**c)** If the store's sales for the month of April are $15,000, what will Rob's salary be for April?

### Solution

**a)** Since Rob's monthly salary depends on the store's sales, *m* depends on *s*, and *m* is a function of *s*.

Rob's monthly salary consists of $200 plus 10% of the store's sales.

$$m(s) \quad = \quad 200 + 0.10 \cdot s$$

or  $m(s) = 200 + 0.10s$

| *s* | *m* |
|---|---|
| 0 | 200 |
| 10,000 | 1200 |
| 20,000 | 2200 |

**b)** Since *m* is a function of *s*, *s* will be on the horizontal axis and *m* will be on the vertical axis. We will graph this function by plotting points. We select values for *s* and find the corresponding values for *m*. The table in the margin shows three ordered pairs which we will plot and then draw the graph shown in **Figure 8.28**.

**c)** By reading our graph carefully, we can estimate that when the store's sales are $15,000, Rob's monthly salary is about $1700.

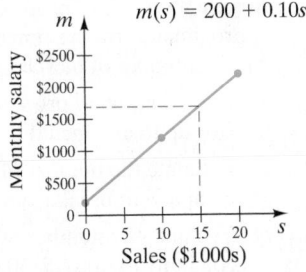

$m(s) = 200 + 0.10s$

FIGURE 8.28

Now Try Exercise 33

# EXERCISE SET 8.3    MathXL®   MyMathLab®
MathXL®        MyMathLab

## Warm-Up Exercises

*Fill in the blanks with the appropriate word, phrase, or symbol(s) from the following list.*

linear            domain           range              x-intercept          y-intercept

integers          break-even       profit             real numbers

1. The domain, or set of x-coordinates, of every linear function is all _____, symbolized $\mathbb{R}$.

2. The _____ or set of y-coordinates, of every linear function $f(x) = ax + b, a \neq 0$, is all real numbers, symbolized $\mathbb{R}$.

3. The _____ is the point where the graph crosses the x-axis.

4. The _____ is the point where the graph crosses the y-axis.

5. A function of the form $f(x) = ax + b$, where a and b are real numbers, is known as a _____ function.

6. A company's _____ point is the point where the profit is equal to 0.

## Practice the Skills

*Graph each function by plotting points.*

7. $f(x) = 3x - 2$

8. $g(x) = -2x + 3$

9. $g(x) = -\frac{2}{3}x + 4$

10. $f(x) = \frac{1}{2}x - 6$

11. $f(x) = \frac{3}{4}x + 1$

12. $g(x) = 6x - 6$

*Graph each function using the x- and y-intercepts.*

13. $f(x) = 3x - 6$

14. $g(x) = -2x + 6$

15. $f(x) = 2x + 3$

16. $f(x) = -6x + 5$

17. $g(x) = 4x - 8$

18. $p(x) = -\frac{3}{4}x + 3$

19. $s(x) = \frac{4}{3}x + 3$

20. $g(x) = -\frac{1}{4}x + 2$

21. $h(x) = -\frac{6}{5}x + 2$

22. $f(x) = \frac{1}{2}x - 2$

23. $g(x) = -\frac{1}{2}x + 2$

24. $p(x) = -\frac{1}{4}x - 6$

25. $w(x) = \frac{1}{3}x - 2$

26. $h(x) = -\frac{1}{3}x + 4$

27. $s(x) = -\frac{4}{3}x + 48$

28. $d(x) = -\frac{1}{3}x - 2$

## Problem Solving

29. **Distance** Using the distance formula

$$\text{distance} = \text{rate} \cdot \text{time, or } d = rt$$

draw a graph of distance versus time for a constant rate of 30 miles per hour.

30. **Simple Interest** Using the simple interest formula

$$\text{interest} = \text{principal} \cdot \text{rate} \cdot \text{time, or } i = prt$$

draw a graph of interest versus time for a principal of $1000 and a rate of 3%.

31. **Bicycle Profit** The profit of a bicycle manufacturer can be approximated by the function $p(x) = 60x - 80,000$, where x is the number of bicycles produced and sold.

a) Draw a graph of profit versus the number of bicycles sold (for up to and including 5000 bicycles).

b) Estimate the number of bicycles that must be sold for the company to break even.

c) Estimate the number of bicycles that must be sold for the company to make $150,000 profit.

32. **Taxi Operating Costs** Raul Lopez's weekly cost of operating a taxi is $75 plus 15¢ per mile.

a) Write a function expressing Raul's weekly cost, c, in terms of the number of miles, m.

b) Draw a graph illustrating weekly cost versus the number of miles, for up to and including 200, driven per week.

c) If during 1 week, Raul drove the taxi 150 miles, what would be the cost?

d) How many miles would Raul have to drive for the weekly cost to be $135?

© Jerryb 8/Dreamstime.com

33. **Salary Plus Commission** Jayne Haydack's weekly salary at Charter Network is $500 plus 15% commission on her weekly sales.

a) Write a function expressing Jayne's weekly salary, s, in terms of her weekly sales, x.

b) Draw a graph of Jayne's weekly salary versus her weekly sales, for up to and including $5000 in sales.

c) What is Jayne's weekly salary if her sales were $3000?

d) If Jayne's weekly salary for the week was $1100, what were her weekly sales?

**34. Salary Plus Commission** Christina Miller, a real estate agent, makes $100 per week plus a 3% sales commission on each property she sells.

**a)** Write a function expressing her weekly salary, $s$, in terms of sales, $x$.

**b)** Draw a graph of her salary versus her weekly sales, for sales up to and including $100,000.

**c)** If she sells one house per week for $75,000, what will her weekly salary be?

© Morgan Lane Photography/Shutterstock

## Concept/Writing Exercises

**35. Weight of Girls** The following graph shows weight, in kilograms, for girls (up to 36 months of age) versus length (or height), in centimeters. The red line is the average weight for all girls of the given length, and the green lines represent the upper and lower limits of the normal range.

**Girls: Birth to 36 Months Physical Growth**

*Source:* National Center for Health Statistics

**a)** Explain why the red line represents a function.

**b)** What is the independent variable? What is the dependent variable?

**c)** Is the graph of weight versus length approximately linear?

**d)** What is the weight in kilograms of the average girl who is 85 centimeters long?

**e)** What is the average length in centimeters of the average girl with a weight of 7 kilograms?

**f)** What weights are considered normal for a girl 95 centimeters long?

**g)** What is happening to the normal range as the lengths increase? Is this what you would expect to happen? Explain.

**36. Compound Interest** The following graph shows the effect of compound interest.

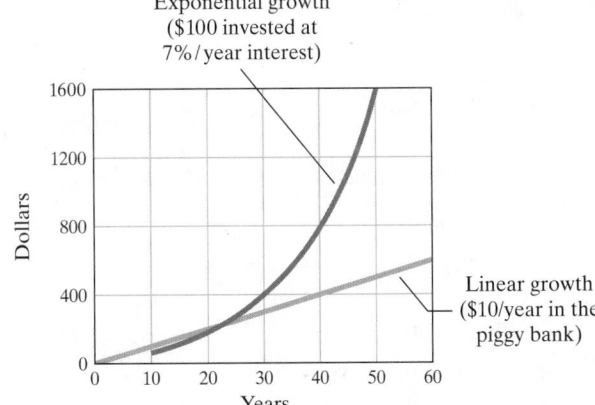

If a child puts $10 each year in a piggy bank, the savings will grow linearly, as shown by the lower curve. If, at age 10, the child invests $100 at 7% interest compounded annually, that $100 will grow exponentially.

**a)** Explain why both graphs represent functions.

**b)** What is the independent variable? What is the dependent variable?

**c)** Using the linear growth curve, determine how long it would take to save $600.

**d)** Using the exponential growth curve, which begins at year 10, how long after the account is opened would the amount reach $600?

**e)** Starting at year 20, how long would it take for the money growing at a linear rate to double?

**f)** Starting at year 20, how long would it take for the money growing exponentially to double? (Exponential growth will be discussed at length in Chapter 13.)

**37.** When, if ever, will the $x$- and $y$-intercepts of a graph be the same? Explain.

**38.** Write two linear functions whose $x$- and $y$-intercepts are both $(0, 0)$.

**39.** Write a function whose graph will have no $x$-intercept but will have a $y$-intercept at $(0, 4)$.

**40.** Write an equation whose graph will have no $y$-intercept but will have an $x$-intercept at $-5$.

## Challenge Problems

**41.** If the $x$- and $y$-intercepts of a linear function are at 1 and $-3$, respectively, what will be the new $x$- and $y$-intercepts if the graph is moved (or translated) up 3 units?

**42.** If the $x$- and $y$-intercepts of a linear function are $-1$ and 3, respectively, what will be the new $x$- and $y$-intercepts if the graph is moved (or translated) down 4 units?

 *Find the x- and y-intercepts of the graph of each equation using your graphing calculator.*

**43.** $y = 2(x + 3.2)$      **44.** $5x - 2y = 7$      **45.** $-4x - 3.2y = 8$      **46.** $y = \dfrac{3}{5}x - \dfrac{1}{2}$

## Group Activity

*In Exercises 47 and 48, we give two ordered pairs, which are on a graph.* **a)** *Plot the points and draw the line through the points.* **b)** *Find the change in y, or the vertical change, between the points.* **c)** *Find the change in x, or the horizontal change, between the points.* **d)** *Find the ratio of the vertical change to the horizontal change between these two points. Do you know what this ratio represents? (We will discuss this further in Section 8.4.)*

**47.** $(0, 2)$ and $(-4, 0)$

**48.** $(3, 5)$ and $(-1, -1)$

## Cumulative Review Exercises

[5.4] **49.** Add $7x^2 - 3x - 100$ and $-4x^2 - 9x + 12$.

[6.4] **50.** Factor $3x^2 - 12x - 96$.

[7.1] **51.** Simplify $\dfrac{x - 2}{x^2 - 4}$.

[7.6] **52.** Solve $x + \dfrac{24}{x} = 10$.

## Mid-Chapter Test: 8.1–8.3

*To find out how well you understand the chapter material to this point, take this brief test. The answers, and the section where the material was initially discussed, are given in the back of the book. Review any questions you answered incorrectly.*

*Graph each equation.*

**1.** $y = 3x + 2$      **2.** $y = -x^2 + 3$

**3.** $y = |x| - 4$      **4.** $y = \sqrt{x - 4}$

**5. a)** What is a relation?

   **b)** What is a function?

   **c)** Is every relation a function? Explain.

   **d)** Is every function a relation? Explain.

*In Exercises 6–8, determine which of the following relations are also functions. Give the domain and range of each relation or function.*

**6.** $\{(1, 5), (2, -3), (7, -1), (-5, 6)\}$

**7.**

**8.**

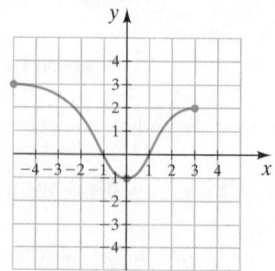

**9.** List $E = \{x \mid x \text{ is a whole multiple of } 9\}$ in roster form.

**10.** Illustrate $\{z \mid z \leq -5\}$ on a number line.

**11.** If $g(x) = 2x^2 + 8x - 13$, find $g(-2)$.

**12.** The height, $h$, in feet of an apple thrown from the top of a building is

$$h(t) = -6t^2 + 3t + 150$$

where $t$ is time in seconds. Find the height of the apple 3 seconds after it is thrown.

*Graph each function.*

**13.** $f(x) = 2x + 3$      **14.** $g(x) = \dfrac{1}{2}x - 4$

**15.** Graph $m(x) = -\dfrac{1}{3}x + 2$ using the $x$- and $y$- intercepts.

**16. Profit** The daily profit, in dollars, for a shoe company is $p(x) = 30x - 660$, where $x$ is the number of pairs of shoes manufactured and sold.

   **a)** Draw a graph of profit versus the number of pairs of shoes sold (for up to and including 40 pairs).

   **b)** Determine the number of pairs of shoes that must be sold for the company to break even.

   **c)** Determine the number of pairs of shoes that must be sold for the company to make a daily profit of $360.

# 8.4 Slope, Modeling, and Linear Relationships

1 Recognize slope as a rate of change.

2 Use the slope-intercept form to construct models from graphs.

3 Use the point-slope form to construct models from graphs.

4 Recognize vertical translations.

5 Use slope to identify and construct perpendicular lines.

## Understanding Algebra

The Greek letter delta, $\Delta$, is often used to represent the phrase "the change in." So, $\Delta y$ represents "the change in $y$" and $\Delta x$ represents "the change in $x$" and the formula for slope can be given as

$$m = \frac{\Delta y}{\Delta x} = \frac{y_2 - y_1}{x_2 - x_1}$$

We introduced slope, the slope-intercept form of a linear equation, and the point-slope form of a linear equation in Sections 4.3 and 4.4. In this section, we briefly review and expand upon these topics.

## 1 Recognize Slope as a Rate of Change

Recall from Section 4.3 that the slope of a line is the ratio "rise over run" or the ratio of the vertical change to the horizontal change between any two selected points on the line.

### Slope

The **slope** of the line through the distinct points $(x_1, y_1)$ and $(x_2, y_2)$ is

$$m = \text{slope} = \frac{\text{change in } y \text{ (vertical change)}}{\text{change in } x \text{ (horizontal change)}} = \frac{\text{rise}}{\text{run}} = \frac{y_2 - y_1}{x_2 - x_1}$$

provided that $x_1 \neq x_2$.

### Helpful Hint

It makes no difference which two points on the line are selected when finding the slope of a line. It also makes no difference which point you label $(x_1, y_1)$ or $(x_2, y_2)$.

EXAMPLE 1  Find the slope of the line in **Figure 8.29.**

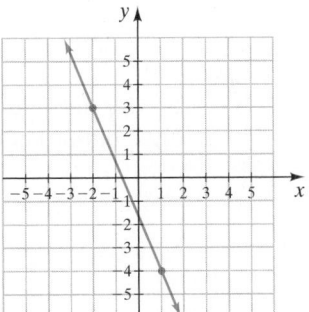

FIGURE 8.29

*Solution*  Two points on the line are $(-2, 3)$ and $(1, -4)$. Let $(x_2, y_2) = (-2, 3)$ and $(x_1, y_1) = (1, -4)$. Then

$$m = \frac{y_2 - y_1}{x_2 - x_1} = \frac{3 - (-4)}{-2 - 1} = \frac{3 + 4}{-3} = -\frac{7}{3}$$

The slope of the line is $-\dfrac{7}{3}$. Note that if we had let $(x_1, y_1) = (-2, 3)$ and $(x_2, y_2) = (1, -4)$, the slope would still be $-\dfrac{7}{3}$. Try it and see.

Now Try Exercise 11

Sometimes it is helpful to describe slope as a *rate of change*. Consider a slope of $\dfrac{3}{4}$. This means that the $y$-value increases 3 units for each 4-unit increase in $x$. Equivalently, we can say that the $y$-value increases $\dfrac{3}{4}$ units, or 0.75 units, for each 1-unit increase in $x$.

When we give the change in $y$ per unit change in $x$ we are giving the slope as a **rate of change**. When discussing real-life situations or when creating mathematical models, it is often useful to discuss slope as a rate of change. In these models, the independent variable is often *time*.

**EXAMPLE 2  Public Debt** The following table and corresponding graph illustrate the U.S. public debt in trillions of dollars from 1980 through 2008.

| Year | U.S. Public Debt (trillions of dollars) |
|------|------|
| 1980 | 0.91 |
| 1984 | 1.57 |
| 1988 | 2.60 |
| 1992 | 4.06 |
| 1996 | 5.22 |
| 2000 | 5.67 |
| 2004 | 7.38 |
| 2008 | 9.67 |

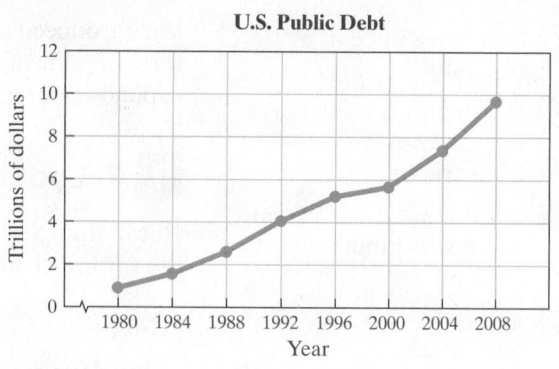

FIGURE 8.30

*Source:* U.S. Department of the Treasury

**a)** Determine the slope of the line segments between 1996 and 2000 and between 2004 and 2008.

**b)** Compare the two slopes found in part **a)** and explain what this means in terms of the U.S. public debt.

*Solution*

**a)** Understand   To find the slope between any two years, find the ratio of the change in debt to the change in time.

Slope from 1996 to 2000

$$m = \frac{5.67 - 5.22}{2000 - 1996} = \frac{0.45}{4} = 0.1125$$

The U.S. public debt from 1996 to 2000 increased at a rate of $0.1125 trillion (or $112.5 billion) per year.

Slope from 2004 to 2008

$$m = \frac{9.67 - 7.38}{2008 - 2004} = \frac{2.29}{4} = 0.5725$$

The U.S. public debt from 2004 to 2008 increased at a rate of $0.5725 trillion (or $572.5 billion) per year.

**b)** Slope measures a rate of change. There was a much greater (over 5 times greater) increase in the average rate of change in the public debt from 2004 to 2008 than from 1996 to 2000.

Now Try Exercise 57

Since the slope of a line can be a positive number, a negative number, or zero, the rate of change can be increasing, decreasing, or zero. We will now consider a function that has a zero rate of change.

Many amusement parks charge a flat fee to enter the park for the day. The cost for a customer is the same regardless of the number of rides the customer takes. If the daily admission is $34.00, the cost, $c$, as a function of the number of rides can be represented as $c(x) = 34$. The graph of this function is illustrated in **Figure 8.31**. Let us select two points on the graph, $(10, 34)$ and $(30, 34)$. If we determine the slope by the formula, we find the slope is indeed 0. Let $(x_2, y_2) = (30, 34)$ and $(x_1, y_1) = (10, 34)$.

$$m = \frac{y_2 - y_1}{x_2 - x_1} = \frac{34 - 34}{30 - 10} = \frac{0}{20} = 0$$

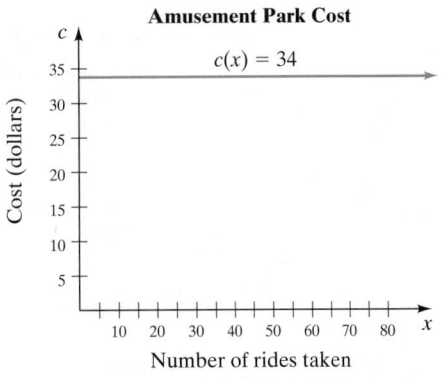

**FIGURE 8.31**

Notice that this function, $c(x) = 34$, is defined using only a constant. It is called a **constant function**. Any constant function will have the form $f(x) = b$ and its graph will be a horizontal line whose slope is 0.

## 2 Use the Slope-Intercept Form to Construct Models from Graphs

**Slope-Intercept Form**

The **slope-intercept form of a linear equation** is

$$y = mx + b$$

where *m* **is the slope** of the line and **(0, b) is the y-intercept** of the line.

### Examples of Equations in Slope-Intercept Form

$$y = 3x - 6 \qquad y = \frac{1}{2}x + \frac{3}{2}$$

Slope → $y = mx + b$ ← *y*-intercept is $(0, b)$

| Equation | Slope | y-Intercept |
|---|---|---|
| $y = 3x - 6$ | 3 | $(0, -6)$ |
| $y = \frac{1}{2}x + \frac{3}{2}$ | $\frac{1}{2}$ | $\left(0, \frac{3}{2}\right)$ |

**Writing an Equation in Slope-Intercept Form**

**To write an equation in slope-intercept form,** solve the equation for *y*.

We can often use the slope-intercept form of a linear equation to determine a function that models a real-life situation.

EXAMPLE 3 **Newspapers** Consider the green graph in **Figure 8.32** on page 486, which shows the declining number of adults who read a daily newspaper. Notice that the graph is somewhat linear. The dashed red line is a linear function which was drawn to approximate the green graph.

  **a)** Write a linear function to represent the dashed red line.

  **b)** Assuming this trend continues, use the function determined in part **a)** to estimate the percent of adults who will read a newspaper in 2015.

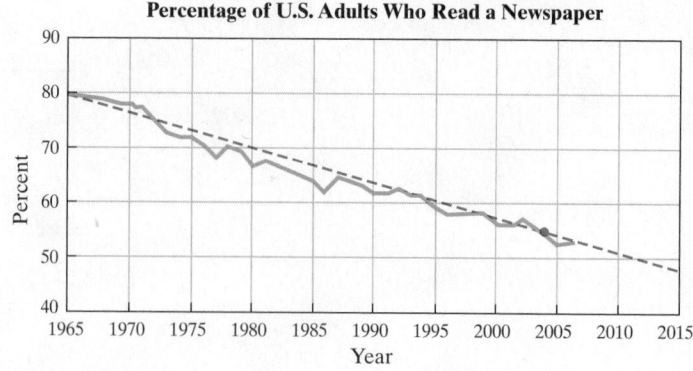

**FIGURE 8.32**    *Source:* NAA Market & Business Analysis

## Solution

**a)** We will let $x$ = the number of years since 1965. Then on the $x$-axis we can replace 1965 with 0, 1966 with 1, 1967 with 2, and so on. Then 2004 would be 39 and 2005 would be 40 (see **Fig. 8.33**). We will let $y$ = percent.

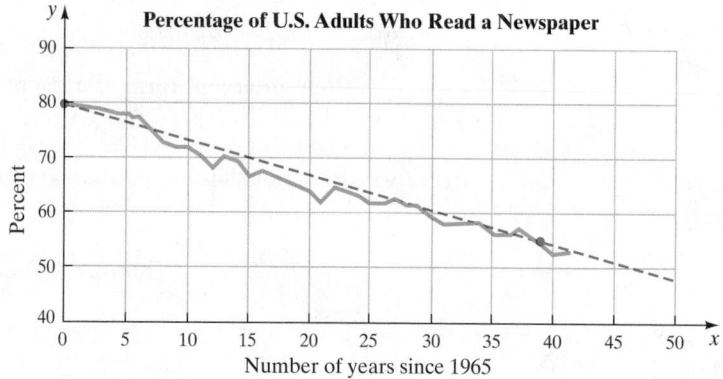

**FIGURE 8.33**    *Source:* NAA Market & Business Analysis

We will select two points on the graph that will allow us to find the slope of the graph. The $y$-intercept is at 80. Thus, one point on the graph is $(0, 80)$. In 2004, or year 39 in **Figure 8.33**, it appears that about 55% of the adult population read a daily newspaper. Let's select $(39, 55)$ as a second point.

$$\text{slope} = \frac{y_2 - y_1}{x_2 - x_1} = \frac{55 - 80}{39 - 0} = \frac{-25}{39} \approx -0.641$$

Since the slope is approximately $-0.641$ and the $y$-intercept is $(0, 80)$, the equation of the straight line is $y = -0.641x + 80$. This equation in function notation is $f(x) = -0.641x + 80$. To use this function remember that $x = 0$ represents 1965, $x = 1$ represents 1966, and so on. Note that $f(x)$, the percent, is a function of $x$, the number of years since 1965.

**b)** To determine the approximate percent of readers in 2015, we substitute $2015 - 1965$, or 50, for $x$ in the function.

$$f(x) = -0.641x + 80$$
$$f(50) = -0.641(50) + 80$$
$$= -32.05 + 80$$
$$= 47.95$$

Thus, if the current trend continues, about 47.95% of adults will read a daily newspaper in 2015.

Now Try Exercise 61

# **3** Use the Point-Slope Form to Construct Models from Graphs

Earlier we introduced the point-slope form of a linear equation. We repeat it here for your convenience.

---

**Point-Slope Form**

The **point-slope form of a linear equation** is

$$y - y_1 = m(x - x_1)$$

where **$m$ is the slope** of the line and $(x_1, y_1)$ is a point on the line.

---

Now let's look at an application where we use the point-slope form to determine a function that models a given situation.

**EXAMPLE 4   Burning Calories**   The number of calories burned in 1 hour of bicycle riding is a linear function of the speed of the bicycle. A person riding at 12 mph will burn about 564 calories in 1 hour and while riding at 18 mph will burn about 846 calories in 1 hour. This information is shown in **Figure 8.34**.

**Calories Burned While Bike Riding**

**FIGURE 8.34**     *Source:* American Heart Association

**a)** Determine a linear function that can be used to estimate the number of calories, $C$, burned in 1 hour when a bicycle is ridden at $r$ mph, for $6 \leq r \leq 24$.

**b)** Use the function determined in part **a)** to estimate the number of calories burned in 1 hour when a bicycle is ridden at 20 mph.

**c)** Use the function determined in part **a)** to estimate the speed at which a bicycle should be ridden to burn 800 calories in 1 hour.

*Solution*

**a)** Understand and Translate   We will use the variables $r$ (for rate) and $C$ (for calories) instead of $x$ and $y$, respectively. To find the necessary function, we will use the points $(12, 564)$ and $(18, 846)$ and proceed as we did in Example 3. We will first calculate the slope and then use the point-slope form to determine the equation of the line.

Carry Out
$$m = \frac{C_2 - C_1}{r_2 - r_1}$$

$$= \frac{846 - 564}{18 - 12} = \frac{282}{6} = 47$$

Now we write the equation using the point-slope form. We will choose the point $(12, 564)$ for $(r_1, C_1)$.

$$C - C_1 = m(r - r_1)$$
$$C - 564 = 47(r - 12) \qquad \text{Point-slope form}$$
$$C - 564 = 47r - 564$$
$$C = 47r \qquad \text{Slope-intercept form}$$

**Answer**  Since the number of calories burned, $C$, is a function of the rate, $r$, the function we are seeking is

$$C(r) = 47r$$

**b)** To estimate the number of calories burned in 1 hour while riding at 20 mph, we substitute 20 for $r$ in the function.

$$C(r) = 47r$$
$$C(20) = 47(20) = 940$$

Therefore, 940 calories are burned while riding at 20 mph for 1 hour.

**c)** To estimate the speed at which a bicycle should be ridden to burn 800 calories in 1 hour, we substitute 800 for $C(r)$ in the function.

$$C(r) = 47r$$
$$800 = 47r$$
$$\frac{800}{47} = r$$
$$r \approx 17.02$$

Thus the person would need to ride the bicycle at about 17.02 mph to burn 800 calories in 1 hour.

Now Try Exercise 75

In Example 4, the function was $C(r) = 47r$. The graph of this function has a slope of 47 and a $y$-intercept at $(0,0)$. If the graph in **Figure 8.34** on page 487 was extended to the left, it would intersect the origin. This makes sense since a rate of 0 miles per hour would result in 0 calories being burned.

## 4  Recognize Vertical Translations

**Figure 8.35** illustrates two parallel lines. In Chapter 4, we defined parallel lines as lines that never intersect. We found that lines that do not intersect have the same slope.

Consider the three equations

$$y = 2x + 3$$
$$y = 2x$$
$$y = 2x - 3$$

Observe the graphs shown in **Figure 8.36a** and **8.36b**. The graph of $y = 2x$ is shown in blue in both figures and has a $y$-intercept of $(0,0)$. Notice that the graph of $y = 2x + 3$ has $y$-intercept $(0, 3)$ and that the graph of $y = 2x - 3$ has $y$-intercept $(0, -3)$. Notice also that the lines are **parallel**—that is, the lines do not intersect.

Parallel lines

FIGURE 8.35

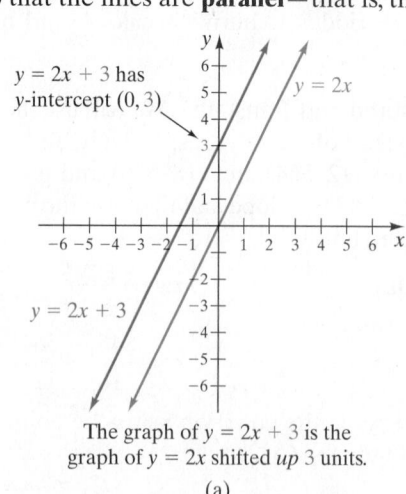

The graph of $y = 2x + 3$ is the graph of $y = 2x$ shifted *up* 3 units.

(a)

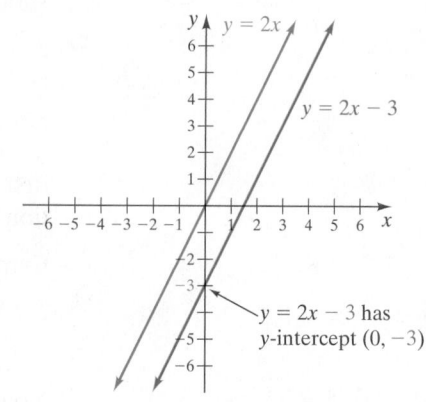

The graph of $y = 2x - 3$ is the graph of $y = 2x$ shifted *down* 3 units.

(b)

FIGURE 8.36

The graphs of $y = 2x + 3$ and $y = 2x - 3$ are identical to $y = 2x$ except for the $y$-intercept. We say that the graphs of $y = 2x + 3$ and $y = 2x - 3$ are **vertical translations** of the graph of $y = 2x$.

After viewing the graphs shown in **Figure 8.36**, can you predict what the graph of $y = 2x + 4$ would look like? The graph of $y = 2x + 4$ is parallel to $y = 2x$ and has y-intercept $(0, 4)$. We say that the graph of $y = 2x + 4$ is the graph of $y = 2x$ shifted or **translated** up 4 units (see **Figure 8.37**)

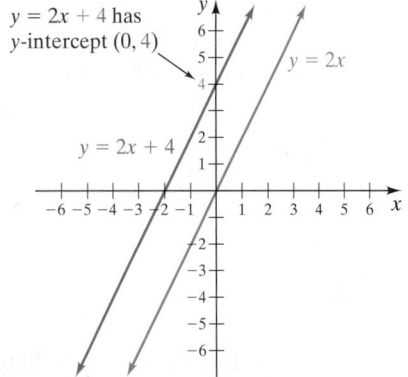

**FIGURE 8.37**

Notice in the equations $y = 2x$, $y = 2x + 3$, $y = 2x - 3$, and $y = 2x + 4$ that the coefficient on $x$ is 2 and that all four lines are parallel. When lines are parallel they have the same *steepness* or *slope*.

### 5 Use Slope to Identify and Construct Perpendicular Lines

When two lines intersect to form a right (or 90°) angle, we say the lines are **perpendicular lines**. **Figure 8.38** illustrates perpendicular lines.

Perpendicular lines

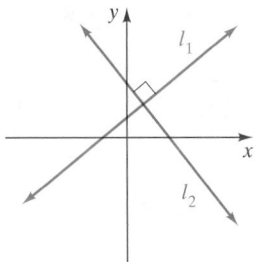

**FIGURE 8.38**

> **Perpendicular Lines**
>
> Two lines are **perpendicular** when their slopes are *negative reciprocals*.

For any nonzero number $a$, its **negative reciprocal** is $\dfrac{-1}{a}$ or $-\dfrac{1}{a}$. For example, the negative reciprocal of 2 is $\dfrac{-1}{2}$ or $-\dfrac{1}{2}$. The product of any nonzero number and its negative reciprocal is $-1$.

**EXAMPLE 5** Two points on line $l_1$ are $(0, 2)$ and $(3, 4)$. Two points on line $l_2$ are $(-2, 4)$ and $(2, -2)$. By comparing their slopes, determine whether $l_1$ and $l_2$ are parallel lines, perpendicular lines, or neither parallel nor perpendicular lines.

*Solution*   First, determine the slope $m_1$ of $l_1$.

$$m_1 = \frac{4 - 2}{3 - 0} = \frac{2}{3}$$

Thus, the slope of the first line is $\dfrac{2}{3}$. Next, determine the slope $m_2$ of $l_2$.

$$m_2 = \frac{-2 - 4}{2 - (-2)} = \frac{-6}{4} = -\frac{3}{2}$$

Thus, the slope of the second line is $-\dfrac{3}{2}$. Since the slopes are not equal, the lines are not parallel. The slopes are negative reciprocals of each other. Note that

$$m_1 \cdot m_2 = \frac{2}{3} \cdot \left(-\frac{3}{2}\right) = -1$$

Since the slopes are negative reciprocals of each other, the lines are perpendicular.

Now Try Exercise 19

**EXAMPLE 6** Consider the equation $2x + 4y = 8$. Determine the equation of the line that has a $y$-intercept of 5 and is **a)** parallel to the given line and **b)** perpendicular to the given line.

*Solution*

**a)** If we know the slope, $m$, of a line and its $y$-intercept, $(0, b)$, we can use the slope-intercept form, $y = mx + b$, to write the equation. We begin by solving the given equation for $y$.

$$2x + 4y = 8$$
$$4y = -2x + 8$$
$$y = \frac{-2x + 8}{4}$$
$$y = -\frac{1}{2}x + 2$$

Two lines are parallel when they have the same slope. Therefore, the slope of the line parallel to the given line must be $-\frac{1}{2}$. Since its slope is $-\frac{1}{2}$ and its $y$-intercept is 5, its equation must be

$$y = -\frac{1}{2}x + 5.$$

The graphs of $2x + 4y = 8$ (in blue) and $y = -\frac{1}{2}x + 5$ (in green) are shown in **Figure 8.39**.

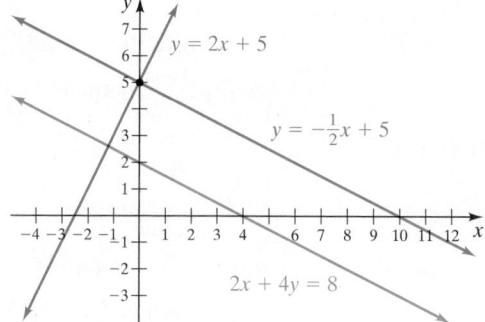

FIGURE 8.39

**b)** Two lines are perpendicular when their slopes are negative reciprocals. We know that the slope of the given line is $-\frac{1}{2}$. Therefore, the slope of the perpendicular line must be $\frac{-1}{-\frac{1}{2}}$ or 2. The line perpendicular to the given line has a $y$-intercept of 5. Thus the equation is

$$y = 2x + 5.$$

**Figure 8.39** also shows the graph of $y = 2x + 5$ (in red).

Now Try Exercise 39

**EXAMPLE 7** Consider the equation $5y = -10x + 7$.

**a)** Determine the equation of a line that passes through $\left(4, \frac{1}{3}\right)$ that is perpendicular to the graph of the given equation. Write the equation in standard form.

**b)** Write the equation determined in part **a)** using function notation.

*Solution*

**a)** Determine the slope of the given line by solving the equation for $y$.

$$5y = -10x + 7$$
$$y = \frac{-10x + 7}{5}$$
$$y = -2x + \frac{7}{5}$$

Since the slope of the given line is $-2$, the slope of a line perpendicular to it must be the negative reciprocal of $-2$, which is $\frac{1}{2}$. Since the line whose equation we are seeking contains point $\left(4, \frac{1}{3}\right)$ and has slope $\frac{1}{2}$, we will use the point-slope form to obtain

$$y - y_1 = m(x - x_1)$$
$$y - \frac{1}{3} = \frac{1}{2}(x - 4) \qquad \text{Point-slope form}$$

Now multiply both sides of the equation by the least common denominator, 6, to eliminate fractions.

$$6\left(y - \frac{1}{3}\right) = 6\left[\frac{1}{2}(x - 4)\right]$$
$$6y - 2 = 3(x - 4)$$
$$6y - 2 = 3x - 12$$

Now write the equation in standard form.

$$-3x + 6y - 2 = -12$$
$$-3x + 6y = -10 \qquad \text{Standard form}$$

Note that $3x - 6y = 10$ is also an acceptable answer. **Figure 8.40** shows the graph of the given equation, $5y = -10x + 7$ (in blue), and the graph of the equation of the perpendicular line, $-3x + 6y = -10$ (in red).

**b)** To write the equation using function notation, we solve the equation determined in part **a)** for $y$, and then replace $y$ with $f(x)$.

We will leave it to you to show that the function is $f(x) = \frac{1}{2}x - \frac{5}{3}$.

Now Try Exercise 43

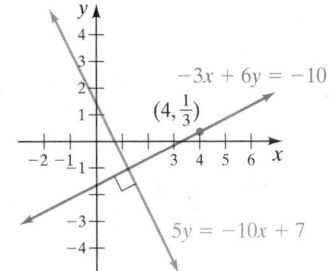

**FIGURE 8.40**

---

# EXERCISE SET 8.4    MathXL®    MyMathLab®
MathXL®    MyMathLab

## Warm-Up Exercises

*Fill in the blanks with the appropriate word, phrase, or symbol(s) from the following list.*

| | | | |
|---|---|---|---|
| constant | linear | translated | perpendicular |
| parallel | rate | reflected | slope |

**1.** The _____ of a line is the ratio of the vertical change to the horizontal change between any two selected points on the line.

**2.** When we give the change in $y$ per unit change in $x$ we are giving the slope as a _____ of change.

**3.** A function of the form $f(x) = b$ where $b$ is a real number is a _____ function.

**4.** When two lines are _____ they have the same slope.

**5.** When two lines are _____ they have slopes that are negative reciprocals.

**6.** The graph of $y = 2x + 4$ is the graph of $y = 2x$ _____ up 4 units.

## Practice the Skills

*Find the slope of the line in each of the figures. If the slope of the line is undefined, so state. Then write an equation of the given line using function notation when possible.*

**7.**

**8.**

**9.**

**10.**

**11.**

**12.**

**13.**

**14.**
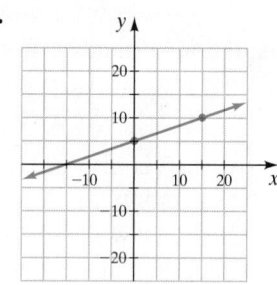

*Graph each constant function.*

**15.** $f(x) = -5$        **16.** $g(x) = 2$        **17.** $g(x) = -15$        **18.** $h(x) = 75$

*Two points on $l_1$ and two points on $l_2$ are given. Determine whether $l_1$ is parallel to $l_2$, $l_1$ is perpendicular to $l_2$, or neither.*

**19.** $l_1$: $(2, 0)$ and $(0, 2)$; $l_2$: $(3, 0)$ and $(0, 3)$

**20.** $l_1$: $(7, 6)$ and $(3, 9)$; $l_2$: $(5, -1)$ and $(9, -4)$

**21.** $l_1$: $(4, 6)$ and $(5, 7)$; $l_2$: $(-1, -1)$ and $(1, 4)$

**22.** $l_1$: $(-3, 4)$ and $(4, -3)$; $l_2$: $(-5, -6)$ and $(6, -5)$

**23.** $l_1$: $(3, 2)$ and $(-1, -2)$; $l_2$: $(2, 0)$ and $(3, -1)$

**24.** $l_1$: $(3, 5)$ and $(9, 1)$; $l_2$: $(4, 0)$ and $(6, 3)$

*Determine whether the two equations represent lines that are parallel, perpendicular, or neither.*

**25.** $y = \dfrac{1}{5}x + 9$

  $y = -5x + 2$

**26.** $2x + 3y = 11$

  $y = -\dfrac{2}{3}x + 4$

**27.** $4x + 2y = 8$

  $8x = 4 - 4y$

**28.** $2x - y = 4$

  $3x + 6y = 18$

**29.** $2x - y = 4$

  $-x + 4y = 4$

**30.** $6x + 2y = 8$

  $4x - 5 = -y$

**31.** $y = \dfrac{1}{2}x - 6$

  $-4y = 8x + 15$

**32.** $2y - 8 = -5x$

  $y = -\dfrac{5}{2}x - 2$

**33.** $y = \dfrac{1}{2}x + 6$

  $-2x + 4y = 8$

**34.** $-4x + 6y = 11$

  $2x - 3y = 5$

**35.** $x - 2y = -9$

  $y = x + 6$

**36.** $\dfrac{1}{2}x - \dfrac{3}{4}y = 1$

  $\dfrac{3}{5}x + \dfrac{2}{5}y = -1$

*Find the equation of a line with the properties given. Write the equation in the form indicated.*

**37.** Through $(2, 5)$ and parallel to the graph of $y = 2x + 4$ (slope-intercept form)

**38.** Through $(-1, 6)$ and parallel to the graph of $4x - 2y = 6$ (slope-intercept form)

**39.** Through $(-3, -5)$ and parallel to the graph of $2x - 5y = 7$ (standard form)

**40.** Through $(-1, 4)$ and perpendicular to the graph of $y = -2x - 1$ (standard form)

**41.** With $x$-intercept $(3, 0)$ and $y$-intercept $(0, 5)$ (slope-intercept form)

**42.** Through $(-2, -1)$ and perpendicular to the graph of $f(x) = -\dfrac{1}{5}x + 1$ (function notation)

**43.** Through $(5, -2)$ and perpendicular to the graph of $y = \dfrac{1}{3}x + 1$ (function notation)

**44.** Through $(-3, 5)$ and perpendicular to the line with $x$-intercept $(2, 0)$ and $y$-intercept $(0, 2)$ (standard form)

**45.** Through $(6, 2)$ and perpendicular to the line with $x$-intercept $(2, 0)$ and $y$-intercept $(0, -3)$ (slope-intercept form)

**46.** Through the point $(1, 2)$ and parallel to the line through the points $(3, 5)$ and $(-2, 3)$ (function notation)

## Problem Solving

**47.** In the following figure, the green line is a translation of the blue line.

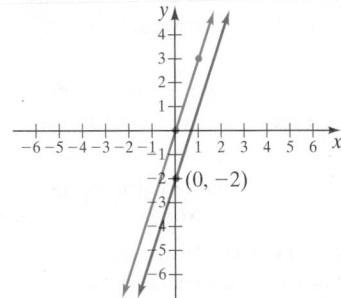

a) Determine the equation of the blue line.

b) Use the equation of the blue line to determine the equation of the green line.

**48.** In the following figure, the green line is a translation of the blue line.

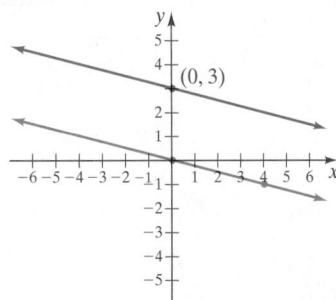

a) Determine the equation of the blue line.

b) Use the equation of the blue line to determine the equation of the green line.

**49.** The graph of $y = x$ is translated up 4 units. Determine

a) the slope of the translated graph.

b) the $y$-intercept of the translated graph.

c) the equation of the translated graph.

**50.** The graph of $y = -\dfrac{3}{2}x$ is translated down 3 units. Determine

a) the slope of the translated graph.

b) the $y$-intercept of the translated graph.

c) the equation of the translated graph.

**51.** The graph of $3x - 2y = 6$ is translated down 4 units. Find the equation of the translated graph.

**52.** The graph of $-3x - 5y = 15$ is translated up 3 units. Find the equation of the translated graph.

**53.** If a line passes through the points $(6, 4)$ and $(-4, 2)$, find the change of $y$ with respect to a 1-unit change in $x$.

**54.** If a line passes through the points $(-3, -4)$ and $(5, 2)$, find the change of $y$ with respect to a 1-unit change in $x$.

**TV Sales** *For Exercises 55 and 56, use the graphs below. The graph on the left shows digital TV sales (in millions) and the graph on the right shows analog TV sales (in millions) for the years from 2004 to 2008.*

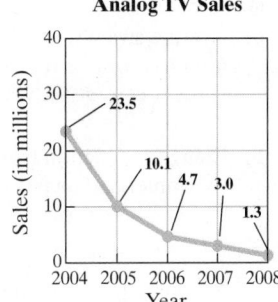

*Source:* Consumer Electronics Association

**55. a)** For the graph of digital TV sales, determine the slope of the line segment from 2005 to 2006.

**b)** Is the slope of the line segment positive or negative?

**c)** Find the average rate of change from 2004 to 2008.

**56. a)** For the graph of analog TV sales, determine the slope of the line segment from 2005 to 2006.

**b)** Is the slope of the line segment positive or negative?

**c)** Find the average rate of change from 2004 to 2008.

**57. Amtrak Expenses** The following table gives the expenses, in millions of dollars, of Amtrak for selected years.

| Year | Amtrak Expenses (in millions of dollars) |
|------|------------------------------------------|
| 1995 | $ 2257 |
| 2000 | $ 2876 |
| 2004 | $ 3133 |
| 2008 | $ 3260 |

*Source:* Amtrak

**a)** Plot these points on a graph.

**b)** Connect these points using line segments.

**c)** Determine the slopes of each of the three line segments.

**d)** During which period was there the greatest average rate of change? Explain.

**58. Demand for Steel**  The table above and to the right gives the world demand for steel, in millions of metric tons, for the years from 2004 to 2007.

**a)** Plot these points on a graph.

**b)** Determine the slope of each line segment.

**c)** Is this graph an example of a linear function? Explain.

**d)** During which period was rate of change highest? Explain.

| Year | World Demand for Steel (in millions of metric tons) |
|------|-----------------------------------------------------|
| 2004 | 950 |
| 2005 | 1029 |
| 2006 | 1121 |
| 2007 | 1179 |

*Source:* International Iron and Steel Institute

**59. Heart Rate**  The following bar graph shows the maximum recommended heart rate, in beats per minute, under stress for men of different ages. The bars are connected by a straight line.

**a)** Use the straight line to determine a function that can be used to estimate the maximum recommended heart rate, $h$, for $0 \le x \le 50$, where $x$ is the number of years after age 20.

**b)** Using the function from part **a)**, determine the maximum recommended heart rate for a 34-year-old man.

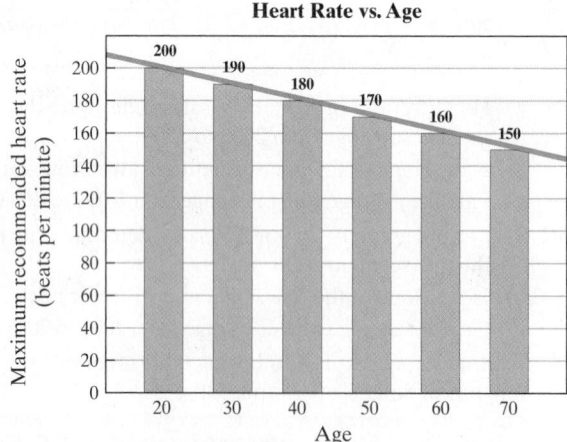

**Heart Rate vs. Age**

*Source:* The American Geriatric Society

**60. Poverty Threshold**  The poverty threshold is an estimate of the annual family income necessary to have a minimally acceptable standard of living. The following bar graph shows the poverty threshold for a family of four for the years 2003 through 2007.

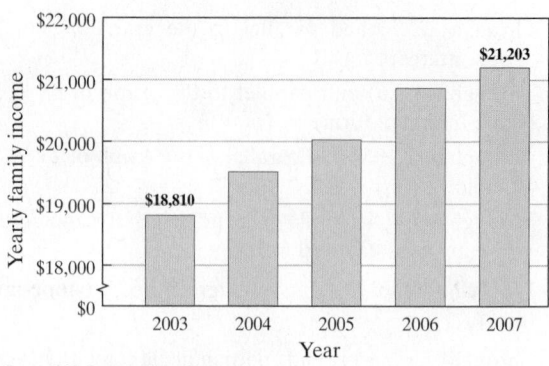

**U.S. Poverty Threshold for a Family of Four**

*Source:* U.S. Bureau of the Census

**a)** Determine a linear function that can be used to estimate the poverty threshold for a family of four, $P$, from 2003 through 2007. Let $t$ represent the number of years since 2003.

**b)** Using the function from part **a)**, determine the poverty threshold in 2004. Compare your answer with the graph to see whether the graph supports your answer.

**c)** Assuming this trend continues, determine the poverty threshold for a family of four in the year 2015.

**d)** Assuming this trend continues, in which year will the poverty threshold for a family of four reach $22,997.75?

**61. Teacher Salaries**  The following graph shows teacher salaries for the 2008–2009 school year in the Manatee County, Florida, school system for teachers whose highest degree is a bachelor's degree. Teachers with 0 years of experience earn $37,550 per year and teachers with 5 years of experience earn $38,600. Let $S$ represent the annual teacher salary and let $t$ represent years of experience.

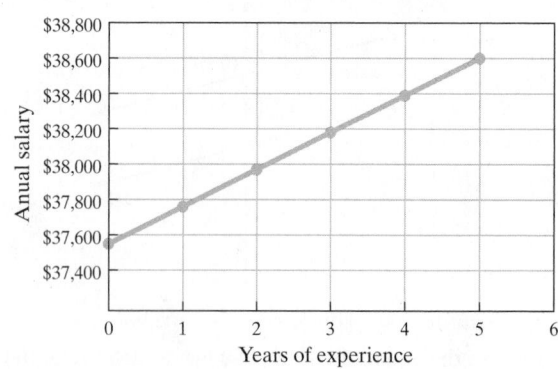

**Teacher Salary Schedule**

*Source:* Manatee County School Board

**a)** Determine a linear function $S(t)$ that fits this data.

**b)** Using the function from part **a)**, estimate the annual salary for a teacher with 3 years of experience. Compare your answer with the graph to see whether the graph supports your answer.

**c)** Assuming this trend continues, what will be the annual salary for a teacher with 10 years of experience?

**d)** Assuming this trend continues, how many years of experience must a teacher have to earn $40,070 per year?

**62. Firefighter Salaries** In Livonia, Michigan, firefighters with 0 years of experience earn an annual salary of $33,259 and firefighters with 5 years of experience earn an annual salary of $47,091. Let $S$ represent the annual firefighters salary and let $t$ represent years of experience.

*Source:* www.firehouse.com

**a)** Determine a linear function $S(t)$ that fits this data.

**b)** Use the function from part **a)** to estimate the annual salary for a firefighter with 3 years of experience.

**c)** Assuming this trend continues, what will be the annual salary for a firefighter with 10 years of experience?

**d)** Assuming this trend continues, how many years of experience must a firefighter have to earn an annual salary of $52,623.80?

**63. Park Ranger Salaries** In Maryland, state park rangers with 0 years of experience earn an annual salary of $37,855 and park rangers with 5 years of experience earn an annual salary of $47,123. Let $S$ represent the annual park ranger salary and let $t$ represent years of experience.

*Source:* www.dbm.maryland.gov

**a)** Determine a linear function $S(t)$ that fits this data.

**b)** Use the function from part **a)** to estimate the annual salary for a park ranger with 3 years of experience.

**c)** Assuming this trend continues, what will be the annual salary for a park ranger with 10 years of experience?

**d)** Assuming this trend continues, how many years of experience must a park ranger have to earn an annual salary of $52,683.80?

**64. Social Security** The number of workers per Social Security beneficiary has been declining approximately linearly since 1970. In 1970 there were 3.7 workers per beneficiary. In 2050 it is projected there will be 2.0 workers per beneficiary. Let $W$ be the workers per Social Security beneficiary and $t$ be the number of years since 1970.

**a)** Find a function $W(t)$ that fits the data.

**b)** Estimate the number of workers per beneficiary in 2020.

**65. Treadmill** The number of calories burned in 1 hour on a treadmill is a function of the speed of the treadmill. A person walking on a treadmill at a speed of 2.5 miles per hour will burn about 210 calories. At 6 miles per hour the person will burn about 370 calories. Let $C$ be the calories burned in 1 hour and $s$ be the speed of the treadmill.

**a)** Determine a linear function $C(s)$ that fits the data.

**b)** Estimate the calories burned by the person on a treadmill in 1 hour at a speed of 5 miles per hour.

**66. Inclined Treadmill** The number of calories burned for 1 hour on a treadmill going at a constant speed is a function of the incline of the treadmill. At 4 miles per hour a person on a 5° incline will burn 525 calories. At 4 mph on a 15° incline the person will burn 880 calories. Let $C$ be the calories burned and $d$ be the degrees of incline of the treadmill.

**a)** Determine a linear function $C(d)$ that fits the data.

**b)** Determine the number of calories burned by the person in 1 hour on a treadmill going 4 miles per hour and at a 9° incline.

**67. Demand for iPods** The *demand* for a product is the number of items the public is willing to buy at a given price. Suppose the demand, $d$, for iPods sold in 1 month is a linear function of the price, $p$, for $150 \le p \le $400. If the price is $200, then 50 iPods will be sold each month. If the price is $300, only 30 iPods will be sold.

**a)** Using ordered pairs of the form $(p, d)$, write an equation for the demand, $d$, as a function of price, $p$.

**b)** Using the function from part **a)**, determine the demand when the price of the iPods is $260.

**c)** Using the function from part **a)**, determine the price charged if the demand for iPods is 45.

**68. Demand for New Sandwiches** The marketing manager of Arby's restaurants determines that the demand, $d$, for a new sandwich is a linear function of the price, $p$, for $0.80 \le p \le $4.00. If the price is $1.00, then 530 sandwiches will be sold each month. If the price is $2.00, only 400 sandwiches will be sold each month.

**a)** Using ordered pairs of the form $(p, d)$, write an equation for the demand, $d$, as a function of price, $p$.

**b)** Using the function from part **a)**, determine the demand when the price of the sandwich is $2.60.

**c)** Using the function from part **a)**, determine the price charged if the demand for sandwiches is 244 sandwiches.

**69. Supply of Kites** The *supply* of a product is the number of items a seller is willing to sell at a given price. The maker of a new kite for children determines that the number of kites she is willing to supply, $s$, is a linear function of the selling price $p$ for $2.00 \le p \le $4.00. If a kite sells for $2.00, then 130 per month will be supplied. If a kite sells for $4.00, then 320 per month will be supplied.

**a)** Using ordered pairs of the form $(p, s)$, write an equation for the supply, $s$, as a function of price, $p$.

**b)** Using the function from part **a)**, determine the supply when the price of a kite is $2.80.

**c)** Using the function from part **a)**, determine the price paid if the supply is 225 kites.

**70. Supply of Baby Strollers** The manufacturer of baby strollers determines that the supply, $s$, is a linear function of the selling price, $p$, for $\$200 \leq p \leq \$300$. If a stroller sells for $210.00, then 20 strollers will be supplied per month. If a stroller sells for $230.00, then 30 strollers will be supplied per month.

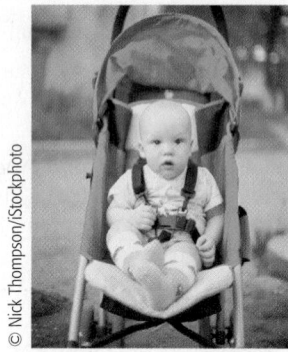

© Nick Thompson/iStockphoto

**a)** Using ordered pairs of the form $(p, s)$, write an equation for the supply, $s$, as a function of price, $p$.

**b)** Using the function from part **a)**, determine the supply when the price of a stroller is $220.00.

**c)** Using the function from part **a)**, determine the selling price if the supply is 35 strollers.

**71. High School Play** The income, $i$, from a high school play is a linear function of the number of tickets sold, $t$. When 80 tickets are sold, the income is $1000. When 200 tickets are sold, the income is $2500.

**a)** Use these data to write the income, $i$, as a function of the number of tickets sold, $t$.

**b)** Using the function from part **a)**, determine the income if 120 tickets are sold.

**c)** If the income is $2200, how many tickets were sold?

**72. Gas Mileage of a Car** The gas mileage, $m$, of a specific car is a linear function of the speed, $s$, at which the car is driven, for $30 \leq s \leq 60$. If the car is driven at a rate of 30 mph, the car's gas mileage is 35 miles per gallon. If the car is driven at 60 mph, the car's gas mileage is 20 miles per gallon.

**a)** Use this data to write the gas mileage, $m$, as a function of speed, $s$.

**b)** Using the function from part **a)**, determine the gas mileage if the car is driven at a speed of 48 mph.

**c)** Using the function from part **a)**, determine the speed at which the car must be driven to get gas mileage of 40 miles per gallon.

**73. Auto Registration** The registration fee, $r$, for a vehicle in a certain region is a linear function of the weight of the vehicle, $w$, for $1000 \leq w \leq 6000$ pounds. When the weight is 2000 pounds, the registration fee is $30. When the weight is 4000 pounds, the registration fee is $50.

**a)** Use these data to write the registration fee, $r$, as a function of the weight of the vehicle, $w$.

**b)** Using the function from part **a)**, determine the registration fee for a 2006 Ford Mustang if the weight of the vehicle is 3613 pounds.

**c)** If the cost of registering a vehicle is $60, determine the weight of the vehicle.

© Allen R. Angel

**74. Lecturer Salary** Suppose the annual salary of a lecturer at Chaumont University is a linear function of the number of years of teaching experience. A lecturer with 9 years of teaching experience is paid $41,350. A lecturer with 15 years of teaching experience is paid $46,687.

**a)** Use this data to write the annual salary of a lecturer, $s$, as a function of the number of years of teaching experience, $n$.

**b)** Using the function from part **a)**, determine the annual salary of a lecturer with 10 years of teaching experience.

**c)** Using the function from part **a)**, estimate the number of years of teaching experience a lecturer must have to obtain an annual salary of $44,908.

**75. Life Expectancy** As seen in the following graph, the expected number of remaining years of life of a person, $y$, *approximates* a linear function. The expected number of remaining years is a function of the person's current age, $a$, for $30 \leq a \leq 80$.

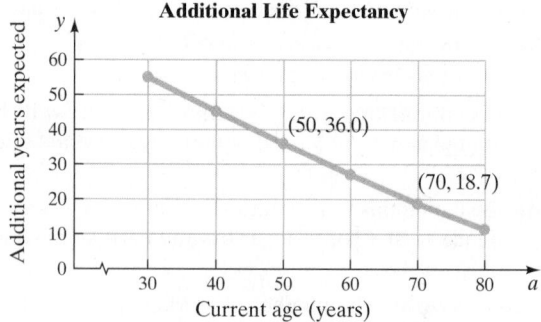

Source: TIAA/CREF

**a)** Using the two points on the graph, determine the function $y(a)$ that can be used to approximate the graph.

**b)** Using the function from part **a)**, estimate the additional life expectancy of a person who is currently 37 years old.

**c)** Using the function from part **a)**, estimate the current age of a person who has an additional life expectancy of 25 years.

**76. Guarneri del Gesù Violin** The graph on page 497 shows that the projected value, $v$, of a Guarneri del Gesù violin is a linear function of the age, $a$, in years, of the violin, for $261 \leq a \leq 290$.

**Value of a del Gesù Violin**

(Graph: Projected value (millions) on vertical axis *v*, Age of violin (years) on horizontal axis *a*. Points shown: (261, 3.5), (275, 10.5), and a point near (290, 18).)

*Source:* Machold Rare Violins, LTD

**a)** Determine the function $v(a)$ represented by this line.

**b)** Using the function from part **a)**, determine the projected value of a 265-year-old Guarneri del Gesù violin.

**c)** Using the function from part **a)**, determine the age of a Guarneri del Gesù violin with a projected value of $15 million.

Guarneri del Gesù, "Sainton," 1741

© Sibrikov Valery/Shutterstock

**77. Boys' Weights** The diagram below shows percentiles for boys' lengths (heights) and weights from birth to age 36 months. Certain portions of the graphs can be approximated with a linear function. For example, the graph representing the 95th percentile of boys' weights (the top red line) from age 18 months to age 36 months is approximately linear.

**Boys: Birth to 36 months**
Length-for-Age and Weight-for-Age Percentiles

*Source:* National Center for Health Statistics

**a)** Use the points shown on the graph of the 95th percentile to write weight, $w$, as a linear function of age, $a$, for boys between 18 and 36 months old.

**b)** Using the function from part **a)**, estimate the weight of a 22-month-old boy who is in the 95th percentile for weight. Compare your answer with the graph to see whether the graph supports your answer.

**78. Boys' Lengths** The diagram in Exercise 77 shows that the graph representing the 95th percentile of boys' lengths (the top yellow line) from age 18 months to age 36 months is approximately linear.

**a)** Use the points shown on the graph of the 95th percentile to write length, $l$, as a linear function of age, $a$, for boys between ages 18 and 36 months.

**b)** Using the function from part **a)**, estimate the length of a 21-month-old boy who is in the 95th percentile. Compare your answer with the graph to see whether the graph supports your answer.

## Concept/Writing Exercises

*Suppose you are attempting to graph the equations shown and you get the screens shown. Explain how you know that you have made a mistake in entering each equation. The standard window setting is used on each graph.*

**79.** $y = 3x + 6$

**80.** $y = -2x - 4$

**81.** $y = \dfrac{1}{2}x + 4$

**82.** $y = -4x - 1$

## Challenge Problems

**83. Castle** The photo below is the Castle at Chichén Itzá, Mexico. Each side of the castle has a stairway consisting of 91 steps. The steps of the castle are quite narrow and steep, which makes them hard to climb. The total vertical distance of the 91 steps is 1292.2 inches. If a straight line were to be drawn connecting the tips of the steps, the absolute value of the slope of this line would be 2.21875. Find the average height and width of a step.

© Allen R. Angel

**84.** A **tangent line** is a straight line that touches a curve at a single point (the tangent line may cross the curve at a different point if extended). **Figure 8.41 a)** shows three tangent lines to the curve at points *a*, *b*, and *c*. Note that the tangent line at point *a* has a positive slope, the tangent line at point *b* has a slope of 0, and the tangent line at point *c* has a negative slope. Now consider the curve in **Figure 8.41 b)**. Imagine that tangent lines are drawn at all points on the curve except at endpoints *a* and *e*. Where on the curve in **Figure 8.41 b)** would the tangent lines have a positive slope, a slope of 0, and a negative slope?

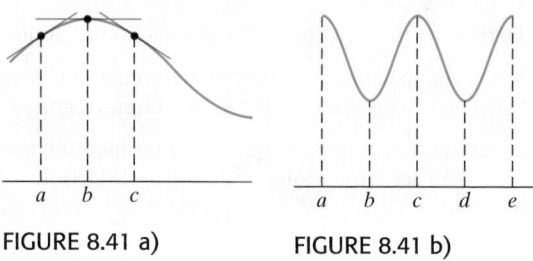

FIGURE 8.41 a)          FIGURE 8.41 b)

## Group Activity

**85.** The following graph from *Consumer Reports* shows the depreciation on a typical car. The initial purchase price is represented as 100%.

**a)** Group member 1: Determine the 1-year period in which a car depreciates most. Estimate from the graph the percent a car depreciates during this period.

**b)** Group member 2: Determine between which years the depreciation appears linear or nearly linear.

**c)** Group member 3: Determine between which 2 years the depreciation is the lowest.

**d)** As a group, estimate the slope of the line segment from year 0 to year 1. Explain what this means in terms of rate of change.

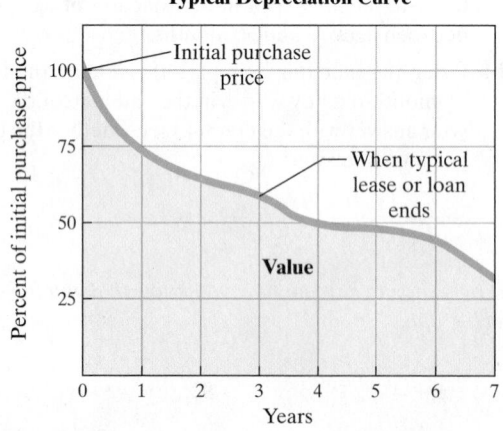

**86.** The graph on the right shows the growth of the circumference of a girl's head. The red line is the average head circumference of all girls for the given age while the green lines represent the

upper and lower limits of the normal range. Discuss and answer the following questions as a group.

**a)** Explain why the graph of the average head circumference represents a function.

**b)** What is the independent variable? What is the dependent variable?

**c)** What is the domain of the graph of the average head circumference? What is the range of the average head circumference graph?

**d)** What interval is considered normal for girls of age 18?

**e)** For this graph, is head circumference a function of age or is age a function of head circumference? Explain your answer.

**f)** Estimate the average girl's head circumference at age 10 and at age 14.

**g)** This graph appears to be nearly linear. Determine an equation or function that can be used to estimate the orange line between (2, 48) and (18, 55).

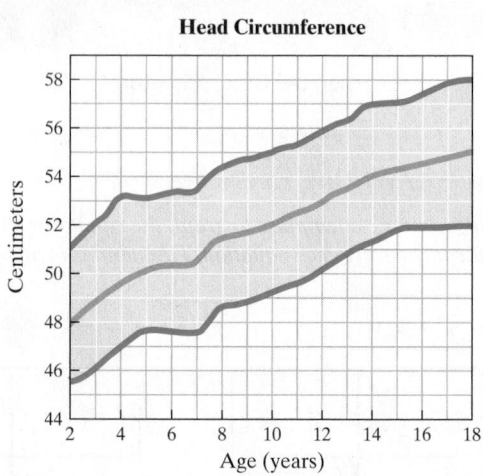

*Source:* National Center for Health Statistics

## Cumulative Review Exercises

[1.9]   **87.** Evaluate $\dfrac{-6^2 - 16 \div 2 \div |-4|}{5 - 3 \cdot 2 - 4 \div 2^2}$.

*Solve each equation.*

[2.4]   **88.** $2.6x - (-1.4x + 3.4) = 6.2$

[2.5]   **89.** $\dfrac{3}{4}x + \dfrac{1}{5} = \dfrac{2}{3}(x - 2).$

[8.2]   **90.  a)** What is a relation?

**b)** What is a function?

**c)** Draw a graph that is a relation but not a function.

**91.** Find the domain and range of the function $\{(4, 3), (5, -2), (3, 2), (6, -1)\}$.

# 8.5 The Algebra of Functions

**1** Find the sum, difference, product, and quotient of functions.

**2** Graph the sum of functions.

## 1 Find the Sum, Difference, Product, and Quotient of Functions

In this section we introduce some ways that functions can be combined. Consider the functions $f(x) = x - 3$ and $g(x) = x^2 + 2x$. We find $f(5)$ and $g(5)$ as follows.

$$f(x) = x - 3 \qquad\qquad g(x) = x^2 + 2x$$
$$f(5) = 5 - 3 = 2 \qquad g(5) = 5^2 + 2(5) = 35$$

Next, if we look at the sum of the two functions, we get

$$f(x) \quad + \quad g(x)$$
$$\downarrow \qquad \downarrow \qquad \downarrow$$
$$(x - 3) + (x^2 + 2x) = x^2 + 3x - 3$$

This new function is designated as $(f + g)(x)$ and we write

$$(f + g)(x) = x^2 + 3x - 3$$

We find $(f + g)(5)$ as follows.

$$(f + g)(\ ) = 5^2 + 3(5) - 3$$
$$= 25 + 15 - 3 = 37$$

Notice that

$$f(5) + g(5) = (f + g)(5)$$
$$2 + 35 = 37 \qquad \text{True}$$

In fact, for any real number substituted for $x$ you will find that

$$f(x) + g(x) = (f + g)(x)$$

Similar notation exists for subtraction, multiplication, and division of functions.

### Understanding Algebra

The domain of a function is the set of values that can be used for the independent variable. For example, the domain of

- $f(x) = 2x^2 - 6x + 5$ is all real numbers since any real number can be substituted for $x$.

- $g(x) = \dfrac{1}{x - 8}$ is all real numbers except 8, since $x = 8$ leads to $\dfrac{1}{0}$, which is undefined.

---

**Operations on Functions**

If $f(x)$ represents one function, $g(x)$ represents a second function, and $x$ is in the domain of both functions, then the following operations on functions may be performed:

**Sum of functions:** $(f + g)(x) = f(x) + g(x)$

**Difference of functions:** $(f - g)(x) = f(x) - g(x)$

**Product of functions:** $(f \cdot g)(x) = f(x) \cdot g(x)$

**Quotient of functions:** $(f/g)(x) = \dfrac{f(x)}{g(x)}$, provided that $g(x) \neq 0$

EXAMPLE 1  If $f(x) = x^2 + x - 6$ and $g(x) = x - 3$, find

**a)** $(f + g)(x)$             **b)** $(f - g)(x)$

**c)** $(g - f)(x)$           **d)** Does $(f - g)(x) = (g - f)(x)$?

*Solution*    To answer parts **a)–c)**, we perform the indicated operation.

**a)** $(f + g)(x) = f(x) + g(x)$

$= (x^2 + x - 6) + (x - 3)$

$= x^2 + x - 6 + x - 3$

$= x^2 + 2x - 9$

**b)** $(f - g)(x) = f(x) - g(x)$

$= (x^2 + x - 6) - (x - 3)$

$= x^2 + x - 6 - x + 3$

$= x^2 - 3$

**c)** $(g - f)(x) = g(x) - f(x)$

$= (x - 3) - (x^2 + x - 6)$

$= x - 3 - x^2 - x + 6$

$= -x^2 + 3$

**d)** By comparing the answers to parts **b)** and **c)**, we see that

$$(f - g)(x) \neq (g - f)(x)$$

Now Try Exercise 11

EXAMPLE 2  If $f(x) = x^2 - 4$ and $g(x) = x - 2$, find

**a)** $(f - g)(6)$      **b)** $(f \cdot g)(5)$      **c)** $(f/g)(8)$

*Solution*

**a)** $(f - g)(x) = f(x) - g(x)$

$= (x^2 - 4) - (x - 2)$

$= x^2 - x - 2$

$(f - g)(6) = 6^2 - 6 - 2$

$= 36 - 6 - 2$

$= 28$

We could have also found the solution as follows:

$$f(x) = x^2 - 4 \qquad\qquad g(x) = x - 2$$
$$f(6) = 6^2 - 4 = 32 \qquad\qquad g(6) = 6 - 2 = 4$$
$$(f - g)(6) = f(6) - g(6)$$
$$= 32 - 4 = 28$$

**b)** We will find $(f \cdot g)(5)$ using the fact that

$$(f \cdot g)(5) = f(5) \cdot g(5)$$
$$f(x) = x^2 - 4 \qquad\qquad g(x) = x - 2$$
$$f(5) = 5^2 - 4 = 21 \qquad\qquad g(5) = 5 - 2 = 3$$

Thus $f(5) \cdot g(5) = 21 \cdot 3 = 63$. Therefore, $(f \cdot g)(5) = 63$.

**c)** We will find $(f/g)(8)$ by using the fact that

$$(f/g)(8) = f(8)/g(8)$$
$$f(x) = x^2 - 4 \qquad\qquad g(x) = x - 2$$
$$f(8) = 8^2 - 4 = 60 \qquad\qquad g(8) = 8 - 2 = 6$$

Therefore, $(f/g)(8) = f(8)/g(8) = 60/6 = 10$. Therefore, $(f/g)(8) = 10$.

Now Try Exercise 31

**FIGURE 8.42**

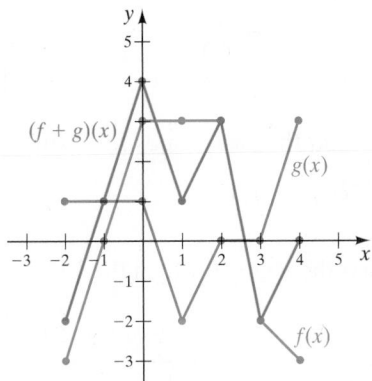

**FIGURE 8.43**

## 2 Graph the Sum of Functions

Now we will explain how we can graph the sum, difference, product, or quotient of two functions. **Figure 8.42** shows two functions, $f(x)$, illustrated in blue, and $g(x)$, illustrated in red.

The table below gives the integer values of $x$ from $-2$ to $4$, the values of $f(-2)$ through $f(4)$, and the values of $g(-2)$ through $g(4)$. These values are taken directly from **Figure 8.42**. The values of $(f+g)(-2)$ through $(f+g)(4)$ are determined by adding the values of $f(x)$ and $g(x)$. The graph of $(f+g)(x) = f(x) + g(x)$ is illustrated in green in **Figure 8.43**.

| $x$ | $f(x)$ | $g(x)$ | $(f+g)(x)$ |
|---|---|---|---|
| $-2$ | $-3$ | $1$ | $-3 + 1 = -2$ |
| $-1$ | $0$ | $1$ | $0 + 1 = 1$ |
| $0$ | $3$ | $1$ | $3 + 1 = 4$ |
| $1$ | $3$ | $-2$ | $3 + (-2) = 1$ |
| $2$ | $3$ | $0$ | $3 + 0 = 3$ |
| $3$ | $-2$ | $0$ | $-2 + 0 = -2$ |
| $4$ | $-3$ | $3$ | $-3 + 3 = 0$ |

We could graph the difference, product, or quotient of the two functions using a similar technique. For example, to graph the product function $(f \cdot g)(x)$, we would evaluate $(f \cdot g)(-2)$ as follows:

$$(f \cdot g)(-2) = f(-2) \cdot g(-2)$$
$$= (-3)(1) = -3$$

Thus, the graph of $(f \cdot g)(x)$ would have an ordered pair at $(-2, -3)$. Other ordered pairs would be determined by the same procedure.

In newspapers, magazines, and on the Internet we often find graphs that show the sum of two or more functions. Graphs that show the sum of functions are generally indicated in one of three ways: line graphs, bar graphs, or stacked (or cumulative) line graphs. Examples 3 through 5 show the three general methods. Each of these examples will use the same data pertaining to cholesterol.

**EXAMPLE 3**  **Line Graph** Ray Hundley has kept a record of his bad cholesterol (low-density lipoprotein, or LDL) and his good cholesterol (high-density lipoprotein or HDL) from 2006 through 2010. **Table 8.2** shows his LDL and his HDL for these years.

**TABLE 8.2    Cholesterol**

|  | 2006 | 2007 | 2008 | 2009 | 2010 |
|---|---|---|---|---|---|
| **LDL** | 220 | 240 | 140 | 235 | 130 |
| **HDL** | 30 | 40 | 70 | 35 | 40 |

**a)** Explain why the data consisting of the years and the LDL values are a function, and the data consisting of the years and the HDL values are also a function.

**b)** Draw a line graph that shows the LDL, the HDL, and the total cholesterol from 2006 through 2010. The total cholesterol is the sum of the LDL and the HDL.

**c)** If $L$ represents the amount of LDL and $H$ represents the amount of HDL, show that $(L + H)(2010) = 170$.

**d)** By looking at the graph drawn in part **b)**, determine the years in which the LDL was less than 180.

### Solution

**a)** The data consisting of the years and the LDL values are a function because for each year there is exactly one LDL value. Note that the year is the independent variable, and the LDL value is the dependent variable. The data consisting of the years and the HDL values are a function for the same reason.

**b)** For any given year, the total cholesterol is the sum of the LDL and HDL for that year. For example, for 2009, to find the cholesterol, we add $235 + 35 = 270$. The graph in **Figure 8.44** shows LDL, HDL, and total cholesterol for the years 2006 through 2010.

**LDL, HDL, and Total Cholesterol**

FIGURE 8.44

**c)** To find LDL + HDL, or the total cholesterol, we add the two values for 2010.

$$(L + H)(2010) = L(2010) + H(2010)$$
$$= 130 + 40 = 170$$

**d)** By looking at the graph drawn in part **b)**, we see that the years in which the LDL was less than 180 are 2008 and 2010.

Now Try Exercise 63a

## EXAMPLE 4  Bar Graph

**a)** Using the data given in **Table 8.2** on page 501, draw a bar graph that shows the LDL, HDL, and total cholesterol for the years 2006 through 2010.

**b)** If $L$ represents the amount of LDL and $H$ represents the amount of HDL, use the graph drawn in part **a)** to determine $(L + H)(2007)$.

**c)** By observing the graph drawn in part **a)**, determine in which years the total cholesterol was less than 220.

**d)** By observing the graph drawn in part **a)**, estimate the HDL in 2008.

*Solution*

**a)** To obtain a bar graph showing the total cholesterol, we add the HDL to the LDL for each given year. For example, for 2006, we start by drawing a bar up to 220 to represent the LDL. Directly on top of that bar we add a second bar of 30 units to represent the HDL. This brings the total bar to $220 + 30$ or 250 units. We use the same procedure for each year from 2006 to 2010. The bar graph is shown in **Figure 8.45**.

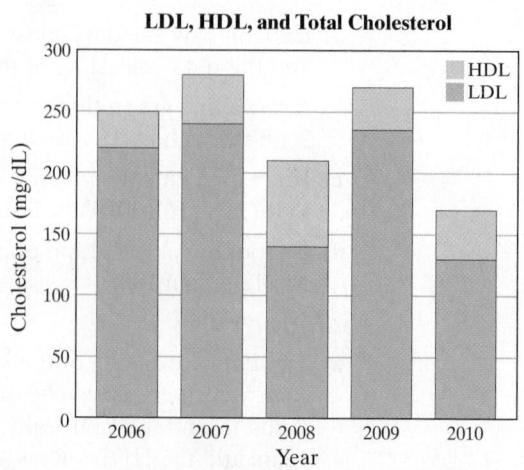

**LDL, HDL, and Total Cholesterol**

FIGURE 8.45

**b)** By observing the graph in **Figure 8.45**, we see that the $(L + H)(2007)$, or the total cholesterol for 2007, is about 280.

**c)** By observing the graph, we see that the total cholesterol was less than 220 in 2008 and 2010.

**d)** For 2008, the HDL bar begins at about 140 and ends at about 210. The difference in these amounts, $210 - 140 = 70$, represents the amount of HDL in 2008. Therefore, the HDL in 2008 was about 70.

Now Try Exercise 63b

## EXAMPLE 5  Stacked Line Graph

**a)** Using the data from **Table 8.2** on page 501, draw a stacked (or cumulative) line graph that shows the LDL, HDL, and total cholesterol for the years 2006–2010.

**b)** Using the graph drawn in part **a)**, determine which years the total cholesterol was greater than or equal to 200.

**c)** Using the graph drawn in part **a)**, estimate the amount of HDL in 2010.

**d)** Using the graph from part **a)**, determine the years in which the LDL was greater than or equal to 180 and the total cholesterol was less than or equal to 250.

### Solution

**a)** To obtain a stacked line graph, begin by drawing a line graph to represent the LDL. This is the same line graph used in **Figure 8.44** on page 502. The area under this line graph (the blue area in **Figure 8.46**) represents the LDL. Next, going year by year, we add the HDL to the LDL to create a second line graph. For example, in year 2006, the second line graph has a value at $220 + 30$ or 250. The area between the two line graphs (the green area in **Figure 8.46**) represents the HDL and the total area under the top line graph represents the total cholesterol.

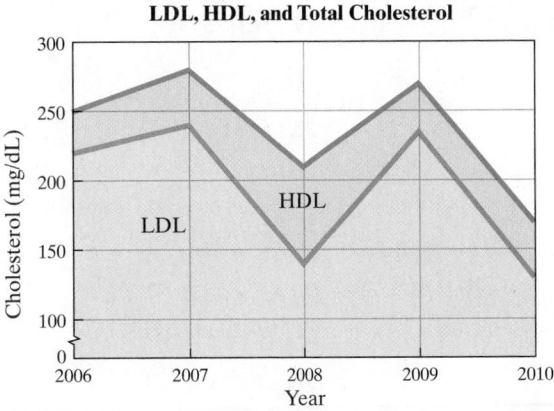

**FIGURE 8.46**

**b)** By looking at the graph, we see that the total cholesterol, indicated by the dark green line, was greater than or equal to 200 in 2006, 2007, 2008, and 2009.

**c)** By looking at the HDL area of the graph, we can see that in 2010 the HDL starts at around 130 and ends at about 170. If we subtract, we obtain $170 - 130 = 40$. Therefore, the HDL in 2010 is about 40. If we let $H$ represent the amount of HDL, then $H(2010) \approx 40$.

**d)** By looking at the graph, we can determine that the only year in which the LDL was greater than or equal to 180 and the total cholesterol was less than or equal to 250 was 2006.

Now Try Exercise 63c

# EXERCISE SET 8.5

MathXL®    MyMathLab®

## Warm-Up Exercises

*Fill in the blanks with the appropriate word, phrase, or symbol(s) from the following list.*

| domain | range | $\dfrac{x^2}{2x-5}$ | $x^2 + 2x - 5$ | $x^2 - 2x + 5$ | $x^2 - 2x - 5$ |
|---|---|---|---|---|---|
| $2x^3 - 5x^2$ | all real numbers | 3 | 5 | 7 | all real numbers except 3 |

*For Exercises 1–6, let $f(x) = x^2$ and $g(x) = 2x - 5$.*

**1.** $(f - g)(x) = $ _____.

**2.** $(f - g)(2) = $ _____.

**3.** $(f + g)(x) = $ _____.

**4.** $(f + g)(2) = $ _____.

**5.** $(f \cdot g)(x) = $ _____.

**6.** $(f/g)(x) = $ _____.

**7.** The domain of the function $f(x) = x - 3$ is _____.

**8.** The domain of the function $f(x) = \dfrac{1}{x-3}$ is _____.

## Practice the Skills

*For each pair of functions, find **a)** $(f + g)(x)$, **b)** $(f + g)(a)$, and **c)** $(f + g)(2)$.*

**9.** $f(x) = x + 5, g(x) = 3x - 2$

**10.** $f(x) = x^2 - x - 8, g(x) = x^2 + 1$

**11.** $f(x) = -3x^2 + x - 4, g(x) = x^3 + 3x^2$

**12.** $f(x) = 4x^3 + 2x^2 - x - 1, g(x) = x^3 - x^2 + 2x + 6$

**13.** $f(x) = 4x^3 - 3x^2 - x, g(x) = 3x^2 + 4$

**14.** $f(x) = 3x^2 - x + 2, g(x) = 6 - 4x^2$

*Let $f(x) = x^2 - 4$ and $g(x) = -5x + 3$. Find the following.*

**15.** $f(3) + g(3)$

**16.** $f(5) + g(5)$

**17.** $f(4) - g(4)$

**18.** $f\left(\dfrac{1}{4}\right) - g\left(\dfrac{1}{4}\right)$

**19.** $f(3) \cdot g(3)$

**20.** $f(-1) \cdot g(-1)$

**21.** $\dfrac{f\left(\dfrac{3}{5}\right)}{g\left(\dfrac{3}{5}\right)}$

**22.** $f(-1)/g(-1)$

**23.** $g(-3) - f(-3)$

**24.** $g(6) \cdot f(6)$

**25.** $g(0)/f(0)$

**26.** $f(2)/g(2)$

*Let $f(x) = 2x^2 - x$ and $g(x) = x - 6$. Find the following.*

**27.** $(f + g)(x)$

**28.** $(f + g)(a)$

**29.** $(f + g)(1)$

**30.** $(f + g)(-3)$

**31.** $(f - g)(-2)$

**32.** $(f - g)(1)$

**33.** $(f \cdot g)(0)$

**34.** $(f \cdot g)(3)$

**35.** $(f/g)(-1)$

**36.** $(f/g)(6)$

**37.** $(g/f)(5)$

**38.** $(g - f)(4)$

**39.** $(g - f)(x)$

**40.** $(g - f)(r)$

## Problem Solving

*Using the graph on the right, find the value of the following.*

**41.** $(f + g)(0)$

**42.** $(f - g)(0)$

**43.** $(f \cdot g)(2)$

**44.** $(f/g)(1)$

**45.** $(g - f)(-1)$

**46.** $(g + f)(-3)$

**47.** $(g/f)(4)$

**48.** $(g \cdot f)(-1)$

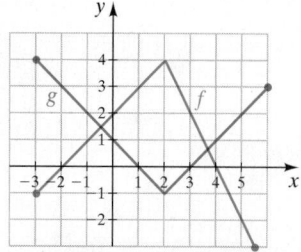

*Using the graph on the right, find the value of the following.*

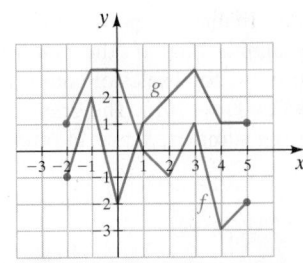

**49.** $(f + g)(-2)$

**50.** $(f - g)(-1)$

**51.** $(f \cdot g)(1)$

**52.** $(g - f)(3)$

**53.** $(f/g)(4)$

**54.** $(g/f)(5)$

**55.** $(g/f)(2)$

**56.** $(g \cdot f)(0)$

**57. Retirement Account** The following graph shows the amount of money Sharon and Frank Dangman have contributed to a joint retirement account for the years 2006 to 2010.

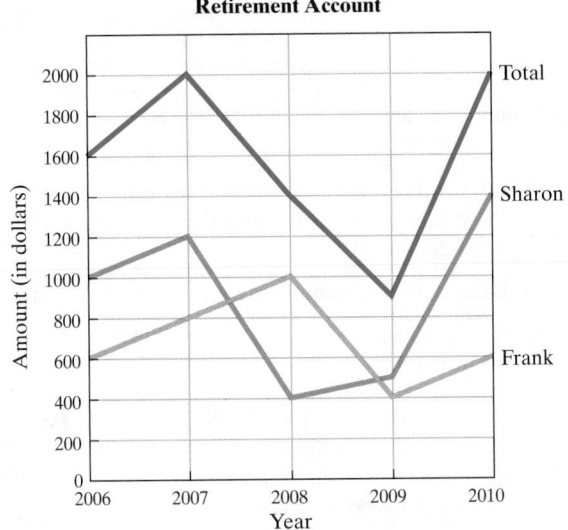

**a)** In which year did Frank contribute $1000?

**b)** In 2010, estimate how much more Sharon contributed to the retirement account than Frank contributed.

**c)** For this five-year period, estimate the total amount Sharon and Frank contributed to the joint retirement account.

**d)** If Frank's income is a function $F(x)$ and Sharon's income is a function $S(x)$, estimate $(F + S)(2009)$.

**58. College Employment** Kelly Housman is a college student who works as a waitress on weekends and as a tutor during the week. The following graph shows her income for the months of August, September, October, November, and December.

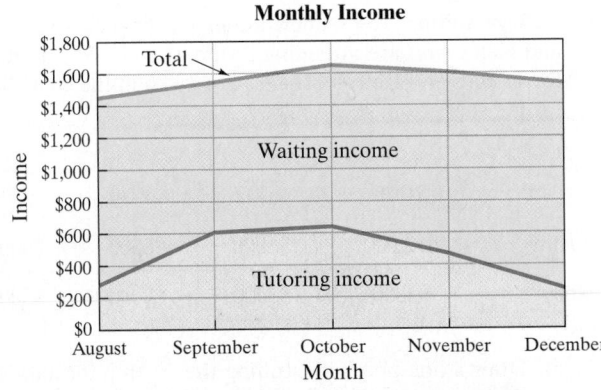

**a)** In which month did Kelly earn a total of over $1600?

**b)** In September, estimate Kelly's income from waiting.

**c)** For this 5-month period, estimate Kelly's total income.

**d)** If the tutoring income is a function $T(x)$ and waiting income is a function $W(x)$, estimate $(W + T)(\text{October})$.

**59. Gender Makeup of Classes** The following bar graph shows the total number of students in Dr. James Condor's Intermediate Algebra class for each semester shown. The blue bars on the bottom of the graph represent the number of female students, $F$, and the red bars on the top of the graph represent the number of male students, $M$.

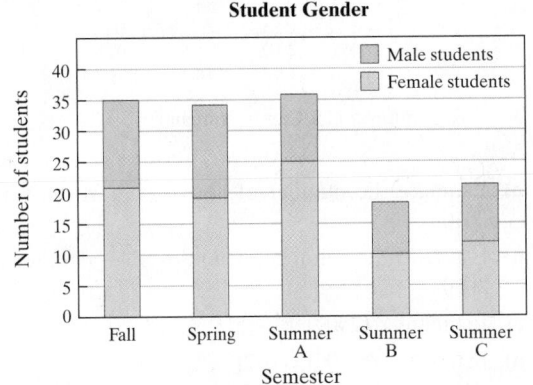

**a)** In what semester did Dr. Condor have the greatest number of Intermediate Algebra students? How many students were in that class?

**b)** In what semester did Dr. Condor have the fewest number of female students? How many female students did he have in class that semester?

**c)** Estimate $M(\text{Summer A})$.

**d)** Estimate $F(\text{Spring})$.

**60. Global Population** The following graph shows the projected total global population and the projected population of children 0–14 years of age from 2002 to 2050.

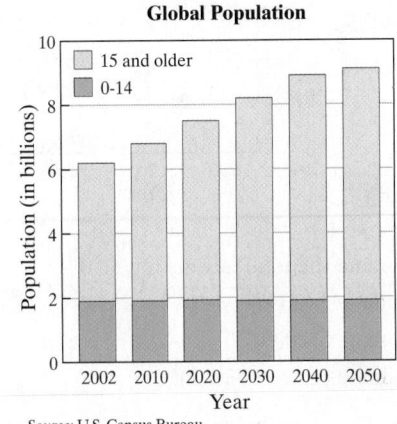

*Source:* U.S. Census Bureau

**a)** Estimate the projected global population in 2050.

**b)** Estimate the projected number of children 0–14 years of age in 2050.

**c)** Estimate the projected number of people 15 years of age and older in 2050.

**d)** Estimate the projected difference in the total global population between 2002 and 2050.

**61. House Sales** In many regions of the country, houses sell better in the summer than at other times of the year. The graph below shows the total sales of houses in the town of Mineral Point from 2006 to 2010. The graph also shows the sale of houses in the summer, $S$, and in other times of the year, $Y$.

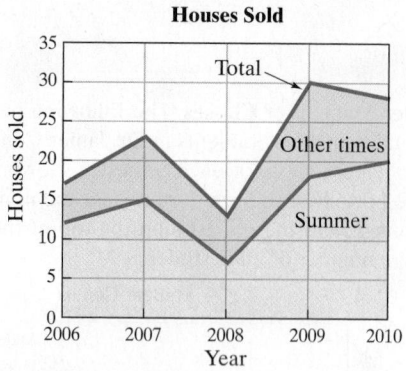

**Houses Sold**

*Source:* U.S. Census Bureau

**a)** Estimate the number of houses sold in the summer of 2010.

**b)** Estimate the number of houses sold at other times in 2010.

**c)** Estimate $Y(2009)$.

**d)** Estimate $(S + Y)(2007)$.

**62. Income** Rod Sac deCrasse owns a business where he does landscaping in the summer and snow removal in the winter. The graph below shows the total income, $T$, for the years 2006–2010 broken down into his landscaping income, $L$, and his snow removal income, $S$.

**Income**

**a)** Estimate the total income for 2010.

**b)** Estimate $L(2006)$.

**c)** Estimate $S(2009)$.

**d)** Estimate $(L + S)(2007)$.

**63. Income** The chart below shows Mr. and Mrs. Abrams's income for the years from 2006 to 2010.

|  | 2006 | 2007 | 2008 | 2009 | 2010 |
|---|---|---|---|---|---|
| Mr. Abrams | $15,500 | $17,000 | $8,000 | $25,000 | $20,000 |
| Mrs. Abrams | $4,500 | $18,000 | $28,000 | $7,000 | $22,500 |

**a)** Draw a line graph illustrating Mr. Abram's income, Mrs. Abram's income, and their total income for the years 2006–2010. See Example 3.

**b)** Draw a bar graph illustrating the given information. See Example 4.

**c)** Draw a stacked line graph illustrating the given information. See Example 5.

**64. Telephone Bills** The chart below shows Kelly Lopez's home telephone bills and cellular telephone bills (rounded to the nearest $10) for the years from 2006 to 2010.

|  | 2006 | 2007 | 2008 | 2009 | 2010 |
|---|---|---|---|---|---|
| Home | $40 | $50 | $60 | $50 | $0 |
| Cellular | $80 | $50 | $20 | $50 | $60 |

**a)** Draw a line graph illustrating the home telephone bills, the cellular phone bills, and the total phone bills for the years 2006–2010.

**b)** Draw a bar graph illustrating the given information.

**c)** Draw a stacked line graph illustrating the given information.

**65. Taxes** Maria Cisneros pays both federal and state income taxes. The chart shows the amount of income taxes she paid to the federal government and to her state government for the years from 2006 to 2010.

|  | 2006 | 2007 | 2008 | 2009 | 2010 |
|---|---|---|---|---|---|
| Federal | $4000 | $5000 | $3000 | $6000 | $6500 |
| State | $1600 | $2000 | $0 | $1700 | $1200 |

**a)** Draw a line graph illustrating the amount spent on federal taxes, the amount spent on state taxes, and the total amount spent on these two taxes for the years 2006–2010.

**b)** Draw a bar graph illustrating the given information.

**c)** Draw a stacked line graph illustrating the given information.

**66. College Tuition** The Olmert family has twin children, Justin and Kelly, who are attending different colleges. The tuition for Justin's and Kelly's colleges are given in the chart below for the years from 2006 to 2009.

|  | 2006 | 2007 | 2008 | 2009 |
|---|---|---|---|---|
| Justin | $12,000 | $6000 | $8000 | $9000 |
| Kelly | $2000 | $8000 | $8000 | $5000 |

**a)** Draw a line graph illustrating the given information, including the total tuition spent on college for both Justin and Kelly for the years 2006–2009.

**b)** Draw a bar graph illustrating the given information.

**c)** Draw a stacked line graph illustrating the given information.

## Concept/Writing Exercises

*For Exercises 67–76, let f and g represent two functions that are graphed on the same axes.*

**67.** What restriction is placed on the property $f(x)/g(x) = (f/g)(x)$? Explain.

**68.** Does $f(x) - g(x) = (f - g)(x)$ for all values of $x$? Explain.

**69.** Does $(f - g)(x) = (g - f)(x)$ for all values of $x$? Explain and give an example to support your answer.

**70.** Does $(f + g)(x) = (g + f)(x)$ for all values of $x$? Explain and give an example to support your answer.

**71.** If, at $a$, $(f + g)(a) = 0$, what must be true about $f(a)$ and $g(a)$?

**72.** If, at $a$, $(f \cdot g)(a) = 0$, what must be true about $f(a)$ and $g(a)$?

**73.** If, at $a$, $(f - g)(a) = 0$, what must be true about $f(a)$ and $g(a)$?

**74.** If, at $a$, $(f - g)(a) < 0$, what must be true about $f(a)$ and $g(a)$?

**75.** If, at $a$, $(f/g)(a) < 0$, what must be true about $f(a)$ and $g(a)$?

**76.** If, at $a$, $(f \cdot g)(a) < 0$, what must be true about $f(a)$ and $g(a)$?

## Group Activity

**77. SAT Scores** The graph shows the average math and reading/verbal scores of entering college classes on the SAT college entrance exam for the years 2000 through 2006. Let $f$ represent the math scores, and $g$ represent the reading/verbal scores, and let $t$ represent the year. As a group, draw a graph that represents $(f + g)(t)$.

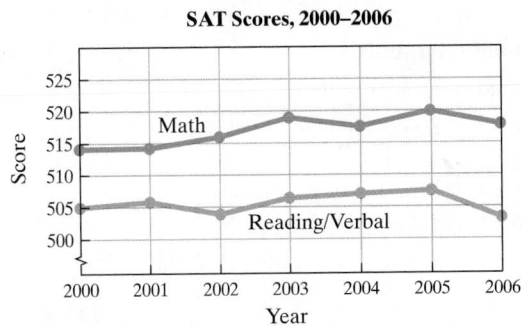

SAT Scores, 2000–2006

## Cumulative Review Exercises

[1.9] **78.** Evaluate $(-4)^{-3}$.

[2.6] **79.** Solve the formula $A = \frac{1}{2}bh$ for $h$.

[3.2] **80. Washing Machine** The cost of a washing machine, including a 6% sales tax, is $477. Determine the pre-tax cost of the washing machine.

[4.2] **81.** Graph $3x - 4y = 12$.

[5.3] **82.** Express 2,960,000 in scientific notation.

[8.1] **83.** Graph $y = |x| - 2$.

# Chapter 8 Summary

| IMPORTANT FACTS AND CONCEPTS | EXAMPLES |
|---|---|
| | Section 8.1 |
| A **nonlinear equation** is an equation whose graph is not a straight line. | $y = x^2 + 2$ is a nonlinear equation whose graph is illustrated below. |

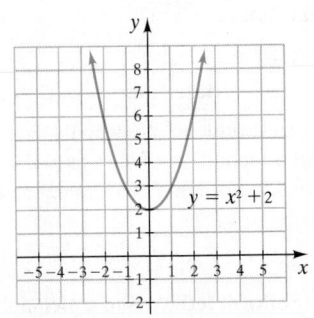

| IMPORTANT FACTS AND CONCEPTS | EXAMPLES |
|---|---|

### Section 8.1 (cont.)

When graphing an equation where a variable appears in a denominator, the equation is not defined at the value that makes the denominator 0.

The graph of $y = \dfrac{1}{x}$ is not defined at $x = 0$.

### Section 8.2

**Inequality Symbols**

$>$ is read "is greater than."

$\geq$ is read "is greater than or equal to."

$<$ is read "is less than."

$\leq$ is read "is less than or equal to."

$\neq$ is read "is not equal to."

Inequalities can be graphed on a real number line.

$6 > 2$ is read 6 is greater than 2

$5 \geq 5$ is read 5 is greater than or equal to 5

$-4 < 3$ is read $-4$ is less than 3

$-10 \leq -1$ is read $-10$ is less than or equal to $-1$

$-5 \neq 17$ is read $-5$ is not equal to 17

$x > 3$

**Set builder notation** has the form

$$\{\ x\ |\ x \text{ has property } p\ \}$$

The set of — all elements $x$ — such that — $x$ has the given property

$\{x \mid -1 \leq x < 2\}$

$\{x \mid x > 4 \text{ and } x \in N\}$

**Interval Notation**

When writing sets of numbers in interval notation, a parenthesis is used to indicate that an endpoint is not included in the interval and a bracket is used to indicate that an endpoint is included.

| Set Builder Notation | Interval Notation |
|---|---|
| $\{x \mid -1 < x < 3\}$ | $(-1, 3)$ |
| $\{x \mid -1 \leq x < 3\}$ | $[-1, 3)$ |
| $\{x \mid -1 < x \leq 3\}$ | $(-1, 3]$ |
| $\{x \mid -1 \leq x \leq 3\}$ | $[-1, 3]$ |
| $\{x \mid -4 < x\}$ | $(-4, \infty)$ |
| $\{x \mid x \leq 5\}$ | $(-\infty, 5]$ |

For an equation in variables $x$ and $y$, if the value of $y$ depends on the value of $x$, then $y$ is the **dependent variable** and $x$ is the **independent variable**.

In the equation $y = 2x^2 + 3x - 4$, $x$ is the independent variable and $y$ is the dependent variable.

A **relation** is any set of ordered pairs.

A **function** is a correspondence between a first set of elements, the **domain**, and a second set of elements, the **range**, such that each element of the domain corresponds to exactly one element of the range.

**Alternate Definition**

A **function** is a set of ordered pairs in which no first coordinate is repeated.

$\{(1, 2), (2, 3), (1, 4)\}$ is a relation, but not a function.

$\{(1, 6), (2, 7), (3, 10)\}$ is a relation. It is also a function since each element in the domain corresponds to exactly one element in the range.

domain: $\{1, 2, 3\}$, range $= \{6, 7, 10\}$

| IMPORTANT FACTS AND CONCEPTS | EXAMPLES |
|---|---|

**Section 8.2 (cont.)**

| | |
|---|---|
| The **vertical line test** can be used to determine if a graph represents a function.<br>If a vertical line can be drawn through any part of the graph and the line intersects another part of the graph, the graph does not represent a function. If a vertical line cannot be drawn to intersect the graph at more than one point, the graph represents a function. | 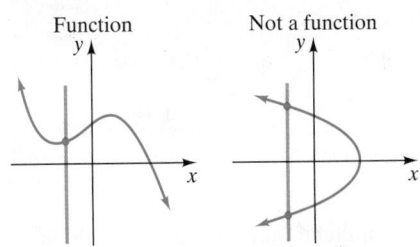 |
| **Function notation** can be used to write an equation when $y$ is a function of $x$. For function notation, replace $y$ with $f(x)$, $g(x)$, $h(x)$, and so on. | $y = 7x - 9$ can be written as $f(x) = 7x - 9$ |
| Given $y = f(x)$, to find $f(a)$, replace each $x$ with $a$. | Let $\qquad f(x) = x^2 + 2x - 8.$<br>Then $\qquad f(1) = 1^2 + 2(1) - 8 = -5$<br>$\qquad\qquad f(a) = a^2 + 2a - 8.$ |

**Section 8.3**

| | |
|---|---|
| A **linear function** is a function of the form $f(x) = ax + b$. The graph of a linear function is a straight line. | Graph $f(x) = \dfrac{1}{3}x - 2$.<br>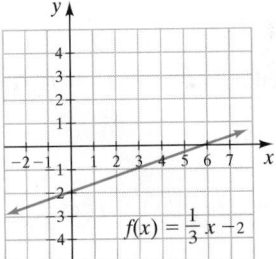 |
| Many real-life applications can be represented as linear functions. | John earns a weekly salary of \$300 plus 10% of all dollar sales he makes. Represent his weekly salary, $w$, as a function of his weekly dollar sales, $s$.<br>$$w(s) = 300 + 0.10s$$ |

**Section 8.4**

| | |
|---|---|
| Slope can be represented as a **rate of change**. When we give the change in $y$ per unit change in $x$, we are giving slope as a rate of change. | In 2005 the U.S. public debt was \$7832.6 billion. In 2002 the U.S. public debt was \$5957.2 billion. Determine the rate of change in the U.S. public debt from 2002 to 2005.<br>$$m = \frac{7832.6 - 5957.2}{2005 - 2002}$$<br>$$= \frac{1875.4}{3} \approx 625.13$$<br>The rate of change is \$625.13 billion per year. |
| **Vertical Translations**<br>When a graph is shifted or translated vertically, we refer to it as a vertical translation. | 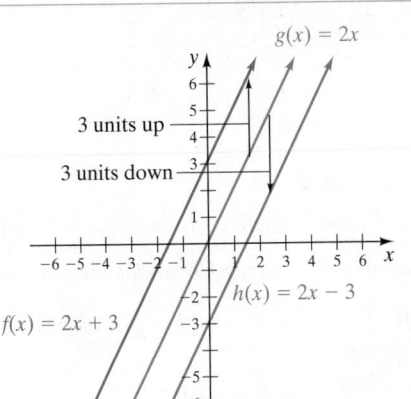 |

| IMPORTANT FACTS AND CONCEPTS | EXAMPLES |
|---|---|

### Section 8.4 (cont.)

**Negative Reciprocals**

The product of a number and its negative reciprocal is $-1$.

$$a\left(-\frac{1}{a}\right) = -1$$

The negative reciprocal of 9 is $-\frac{1}{9}$ since

$$9\left(-\frac{1}{9}\right) = -1.$$

---

Two lines are **parallel** if they have the same slope.

Two lines are **perpendicular** if their slopes are negative reciprocals. For any real number $a \neq 0$, its negative reciprocal is $-\frac{1}{a}$.

The graphs of $y = 2x + 4$ and $y = 2x + 7$ are parallel since both graphs have the same slope of 2, but different $y$-intercepts.

The graphs of $y = 3x - 5$ and $y = -\frac{1}{3}x + 8$ are perpendicular since one graph has a slope of 3 and the other graph has a slope of $-\frac{1}{3}$. The number $-\frac{1}{3}$ is the negative reciprocal of 3.

### Section 8.5

**Operations on Functions**

**Sum of functions:** $(f + g)(x) = f(x) + g(x)$

**Difference of functions:** $(f - g)(x) = f(x) - g(x)$

**Product of functions:** $(f \cdot g)(x) = f(x) \cdot g(x)$

**Quotient of functions:** $(f/g)(x) = \dfrac{f(x)}{g(x)}, \ g(x) \neq 0$

If $f(x) = x^2 + 2x - 5$ and $g(x) = x - 3$, then

$$(f + g)(x) = f(x) + g(x) = (x^2 + 2x - 5) + (x - 3)$$
$$= x^2 + 3x - 8$$

$$(f - g)(x) = f(x) - g(x) = (x^2 + 2x - 5) - (x - 3)$$
$$= x^2 + x - 2$$

$$(f \cdot g)(x) = f(x) \cdot g(x)$$
$$= (x^2 + 2x - 5)(x - 3)$$
$$= x^3 - x^2 - 11x + 15$$

$$(f/g)(x) = \frac{f(x)}{g(x)} = \frac{x^2 + 2x - 5}{x - 3}, \quad x \neq 3$$

## Chapter 8 Review Exercises

[8.1] *Graph each equation.*

**1.** $y = -2x - 1$  
**2.** $y = \dfrac{1}{2}x$  
**3.** $y = \dfrac{1}{2}x + 3$  
**4.** $y = -\dfrac{3}{2}x + 1$  
**5.** $y = x^2$

**6.** $y = x^2 - 1$  
**7.** $y = |x|$  
**8.** $y = |x| - 1$  
**9.** $y = x^3$  
**10.** $y = x^3 + 4$

[8.2]

**11.** Define function.

**12.** Is every relation a function? Is every function a relation? Explain.

*Determine whether the following relations are functions. Explain your answers.*

**13.**
$$
\begin{array}{l}
a \searrow 0 \\
b \nearrow 2 \\
c \longrightarrow 4
\end{array}
$$

**14.** $\{(1, 8), (2, -4), (7, -9), (-6, -1), (2, -2)\}$

*List each set in roster form.*

**15.** $A = \{x | 3 \le x \le 6 \text{ and } x \in N\}$

**16.** $B = \{x | x \ge 6 \text{ and } x \in I\}$

*Represent each set using interval notation.*

**17.** $\{x | x > 4\}$

**18.** $\{x | x \le -2\}$

**19.** $\{x | -1.5 < x \le 2.7\}$

*For Exercises 20–23,* **a)** *determine whether the following graphs represent functions;* **b)** *determine the domain and range of each relation or function.*

**20.**

**21.**

**22.**

**23.**

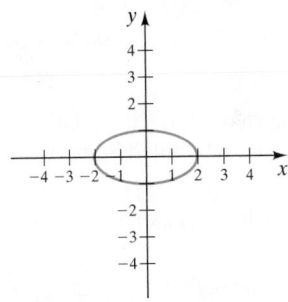

**24.** If $f(x) = -x^2 + 3x - 4$, find

   **a)** $f(2)$ and

   **b)** $f(h)$.

**25.** If $g(t) = 2t^3 - 3t^2 + 6$, find

   **a)** $g(-1)$ and

   **b)** $g(a)$.

**26.** **Speed of Car** Jane Covillion goes for a ride in a car. The following graph shows the car's speed as a function of time. Make up a story that corresponds to this graph.

**27.** **Apple Orchard** The number of baskets of apples, $N$, that are produced by $x$ trees in a small orchard ($x \le 100$) is given by the function $N(x) = 40x - 0.2x^2$. How many baskets of apples are produced by

   **a)** 30 trees?

   **b)** 50 trees?

**28.** **Falling Ball** If a ball is dropped from the top of a 196-foot building, its height above the ground, $h$, at any time, $t$, can be found by the function $h(t) = -16t^2 + 196, 0 \le t \le 3.5$. Find the height of the ball at

   **a)** 1 second.

   **b)** 3 seconds.

[8.3] *Graph each function.*

**29.** $f(x) = \dfrac{1}{2}x - 4$

**30.** $f(x) = \dfrac{8}{3}x - 80$

**31.** $f(x) = 4$

**32.** **Bagel Company** The yearly profit, $p$, of a bagel company can be estimated by the function $p(x) = 0.1x - 5000$, where $x$ is the number of bagels sold per year.

   **a)** Draw a graph of profits versus bagels sold for up to and including 250,000 bagels.

   **b)** Estimate the number of bagels that must be sold for the company to break even.

   **c)** Estimate the number of bagels sold if the company has $22,000 profit.

**33.** **Interest** Draw a graph illustrating the interest on a $12,000 loan for a 1-year period for various interest rates up to and including 20%. Use interest = principal · rate · time.

[8.4] *In Exercises 34–36,* **a)** *indicate if the graph represents a function.* **b)** *If so, express the equation of the graph using function notation.*

**34.**

**35.**

**36.**

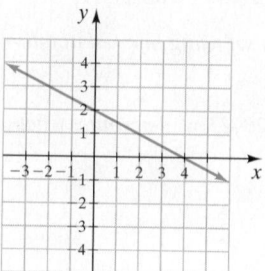

**37.** If the graph of $y = -2x$ is translated up 1 unit, determine
   **a)** the slope of the translated graph.
   **b)** the $y$-intercept of the translated graph.
   **c)** the equation of the translated graph.

**38.** If one point on a graph is $(-6, -4)$ and the slope is $\frac{2}{3}$, find the $y$-intercept of the graph.

**39.** **Typhoid Fever** The following chart shows the number of reported cases of typhoid fever in the United States for select years from 1970 through 2000.
   **a)** Plot each point and draw line segments from point to point.
   **b)** Compute the slope of the line segments.
   **c)** During which 10-year period did the number of reported cases of typhoid fever increase the most?

| Year | Number of reported typhoid fever cases |
|------|----------------------------------------|
| 1970 | 346 |
| 1980 | 510 |
| 1990 | 552 |
| 2000 | 317 |

*Source:* U.S. Dept. of Health and Human Services

**40.** **Social Security** The following graph shows the number of Social Security beneficiaries from 1980 projected through 2070. Use the slope-intercept form to find the function $n(t)$ (represented by the straight line) that can be used to represent this data.

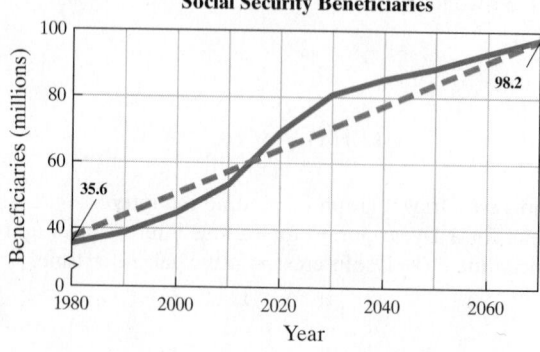

**Social Security Beneficiaries**

*Determine whether the two given lines are parallel, perpendicular, or neither.*

**41.** $3x - 7y = 10$
$$y = \frac{3}{7}x - 5$$

**42.** $2x - 3y = 7$
$$-3x - 2y = 8$$

**43.** $4x - 2y = 13$
$$-2x + 4y = -9$$

*Find the equation of the line with the properties given. Write each answer in slope-intercept form.*

**44.** Slope $= \frac{1}{2}$, through $(2, 8)$

**45.** Through $(-3, 1)$ and $(4, -6)$

**46.** Through $(0, 6)$ and parallel to the graph of $y = -\frac{2}{3}x + 1$

**47.** Through $(2, 8)$ and parallel to the graph whose equation is $5x - 2y = 7$

**48.** Through $(-3, 1)$ and perpendicular to the graph whose equation is $y = \frac{3}{5}x + 5$

**49.** Through $(4, 5)$ and perpendicular to the graph whose equation is $4x - 2y = 8$

*Two points on $l_1$ and two points on $l_2$ are given. Determine whether $l_1$ is parallel to $l_2$, $l_1$ is perpendicular to $l_2$, or neither.*

**50.** $l_1$: $(3, 7)$ and $(0, -3)$; $l_2$: $(1, -1)$ and $(2, -2)$

**51.** $l_1$: $(3, 2)$ and $(2, 3)$; $l_2$: $(4, 1)$ and $(1, 4)$

**52.** $l_1$: $(7, 3)$ and $(4, 6)$; $l_2$: $(5, 2)$ and $(6, 3)$

**53.** $l_1$: $(-3, 5)$ and $(2, 3)$; $l_2$: $(-4, -2)$ and $(-1, 2)$

**54. Insurance Rates** The monthly rates for $100,000 of life insurance from the General Financial Group for men increases approximately linearly from age 35 through age 50. The rate for a 35-year-old man is $10.76 per month and the rate for a 50-year-old man is $19.91 per month. Let *r* be the rate and let *a* be the age of a man between 35 and 50 years of age.

a) Determine a linear function $r(a)$ that fits these data.

b) Using the function in part **a)**, estimate the monthly rate for a 40-year-old man.

**55. Burning Calories** The number of calories burned in 1 hour of swimming, when swimming between 20 and 50 yards per minute, is a linear function of the speed of the swimmer. A person swimming at 30 yards per minute will burn about 489 calories in 1 hour. While swimming at 50 yards per minute a person will burn about 525 calories in 1 hour. This information is shown in the following graph.

**Calories Burned while Swimming**

*Source: Health Magazine* Web Site, www.health.com

a) Determine a linear function that can be used to estimate the number of calories, *C*, burned in 1 hour when a person swims at *r* yards per minute.

b) Use the function determined in part **a)** to determine the number of calories burned in 1 hour when a person swims at 40 yards per minute.

c) Use the function determined in part **a)** to estimate the speed at which a person needs to swim to burn 600 calories in 1 hour.

[8.5] *Given* $f(x) = x^2 - 3x + 4$ *and* $g(x) = 2x - 5$, *find the following.*

**56.** $(f + g)(x)$

**57.** $(f + g)(4)$

**58.** $(g - f)(x)$

**59.** $(g - f)(-1)$

**60.** $(f \cdot g)(-1)$

**61.** $(f \cdot g)(3)$

**62.** $(f/g)(1)$

**63.** $(f/g)(2)$

**64. Female Population** According to the U.S. Census, the female population is expected to grow worldwide. The following graph shows the female population worldwide for selected years from 2002 to 2050.

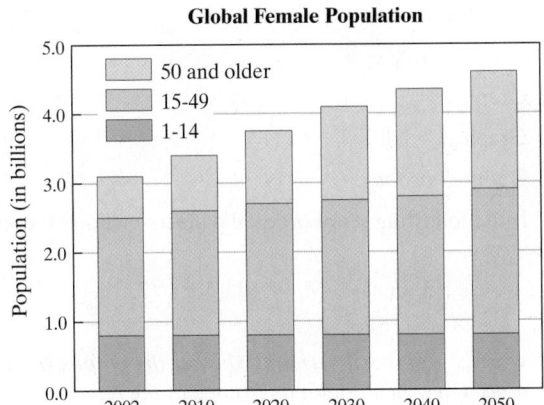

**Global Female Population**

*Source:* U.S. Census Bureau

a) Estimate the projected female population worldwide in 2050.

b) Estimate the projected number of women 15–49 years of age in 2050.

c) Estimate the number of women who are projected to be in the 50 years and older age group in 2010.

d) Estimate the projected percent increase in the number of women 50 years and older from 2002 to 2010.

**65. Retirement Income** Joni Burnette recently retired from her full-time job. The following graph shows her retirement income for the years 2007–2010.

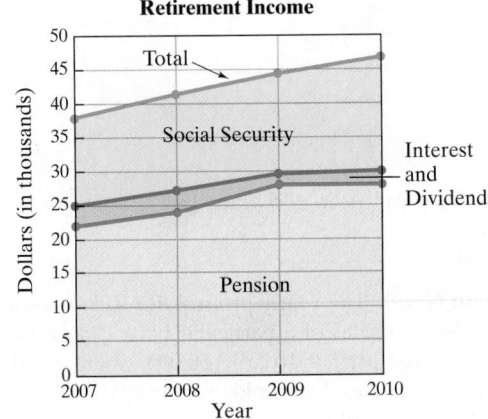

**Retirement Income**

a) Estimate Joni's total retirement income in 2010.

b) Estimate Joni's pension income in 2009.

c) Estimate Joni's interest and dividend income in 2007.

## Chapter 8 Practice Test

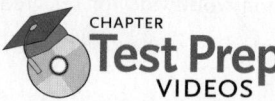

*Chapter Test Prep Videos provide fully worked-out solutions to any of the exercises you want to review. Chapter Test Prep Videos are available via MyMathLab®, or on You Tube (search "Angel Elementary and Intermediate Algebra" and click on "Channels").*

1. Graph $y = -2x + 1$.

2. Graph $y = \sqrt{x}$.

3. Graph $y = x^2 - 4$.

4. Graph $y = |x|$.

5. Define *function*.

6. Is the following set of ordered pairs a function? Explain your answer.

$$\{(3, 1), (-2, 6), (4, 6), (5, 2), (7, 3)\}$$

*In Exercises 7 and 8, determine whether the graphs represent functions. Give the domain and range of the relation or function.*

7.

8.

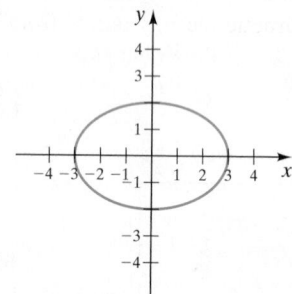

9. If $f(x) = 3x^2 - 6x + 5$, find $f(-2)$.

10. Graph $f(x) = -3$

11. **Profit Graph** The yearly profit, $p$, for Zico Publishing Company on the sales of a particular book can be estimated by the function $p(x) = 10.2x - 50,000$, where $x$ is the number of books produced and sold.

a) Draw a graph of profit versus books sold for up to and including 30,000 books.

b) Use function $p(x)$ to estimate the number of books that must be sold for the company to break even.

c) Use function $p(x)$ to estimate the number of books that the company must sell to make a $100,000 profit.

12. **U.S. Population** Determine the function represented by the red line on the graph that can be used to estimate the projected U.S. population, $p$, from 2000 to 2050. Let 2000 be the reference year so that 2000 corresponds to $t = 0$.

**U.S. Population Projections 2000–2050**

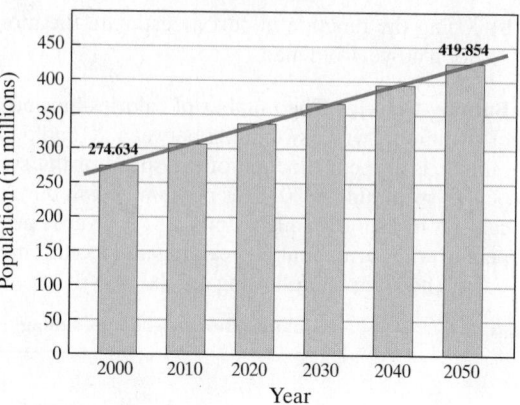

*Source:* U.S. Bureau of the Census

13. Determine whether the graphs of the two equations are parallel, perpendicular, or neither. Explain your answer.

$$2x - 3y = 12$$
$$4x + 10 = 6y$$

14. Determine the equation, in slope-intercept form, of the line that goes through the point $(6, -5)$ and is perpendicular to the graph of $y = \dfrac{1}{2}x + 1$.

15. **Heart Disease** Deaths due to heart disease has been declining approximately linearly since the year 2000. The bar graph below shows the number of deaths, per 100,000 deaths, due to heart disease in selected years, projected for 2006–2010.

**Heart Disease Death Rate**

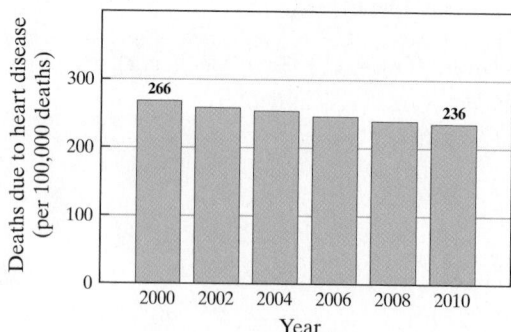

*Source:* U.S. Dept. of Health and Human Services

a) Let $r$ be the number of deaths due to heart disease per 100,000 deaths, and let $t$ represent the years since 2000. Write the linear function $r(t)$ that can be used to approximate the data.

b) Use the function from part **a)** to estimate the death rate due to heart disease in 2006.

c) Assuming this trend continues until the year 2020, estimate the death rate due to heart disease in 2020.

*In Exercises 16–18, if* $f(x) = 2x^2 - x$ *and* $g(x) = x - 6$, *find*

**16.** $(f + g)(3)$

**17.** $(f/g)(-1)$

**18.** $f(a)$

**19. Paper Use** The following graph shows paper use in 1995 and projected paper use from 1995 through 2015.

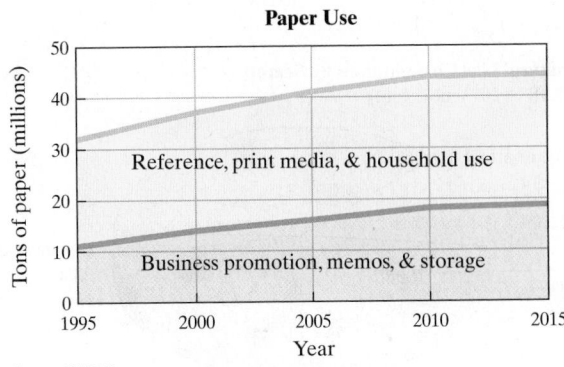

Paper Use

*Source:* CAP Ventures

**a)** Estimate the total number of tons of paper to be used in 2010.

**b)** Estimate the number of tons of paper to be used by businesses in 2010.

**c)** Estimate the number of tons of paper to be used for reference, print media, and household use in 2010.

## Cumulative Review Test

*Take the following test and check your answers with those given in the back of the book. Review any questions that you answered incorrectly. The section where the material was covered is indicated after the answer.*

**1.** Evaluate $4x^2 - 7xy^2 + 6$ when $x = -3$ and $y = -2$.

**2.** Evaluate $2 - \{3[6 - 4(6^2 \div 4)]\}$.

**3.** Solve $2(x + 4) - 5 = -3[x - (2x + 1)]$.

**4.** Solve the formula $P = 2E + 3R$ for $R$.

**5.** Write the equation $5x - 2y = 12$ in slope-intercept form.

**6.** Simplify $\left(\dfrac{6x^2y^3}{2x^5y}\right)^3$.

**7.** Simplify $(6x^2 - 3x - 2) - (-2x^2 - 8x - 1)$.

**8.** Multiply $(4x^2 - 6x + 3)(3x - 5)$.

**9.** Factor $6a^2 - 6a - 5a + 5$.

**10.** Solve $2x^2 = x + 15$.

**11.** Multiply $\dfrac{x^2 - 9}{x^2 - x - 6} \cdot \dfrac{x^2 - 2x - 8}{2x^2 - 7x - 4}$.

**12.** Subtract $\dfrac{x}{x + 4} - \dfrac{3}{x - 5}$.

**13.** Add $\dfrac{4}{x^2 - 3x - 10} + \dfrac{2}{x^2 + 5x + 6}$.

**14.** Solve $\dfrac{x}{9} - \dfrac{x}{6} = \dfrac{1}{12}$.

**15.** Solve $\dfrac{7}{x + 3} + \dfrac{5}{x + 2} = \dfrac{5}{x^2 + 5x + 6}$.

**16. a)** Determine whether the following graph represents a function.

   **b)** Find the domain and range of the graph.

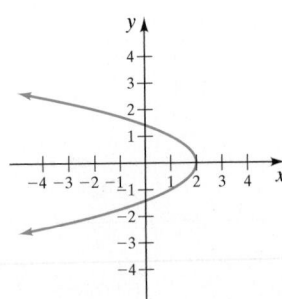

**17.** Determine whether the graphs of the two given equations are parallel, perpendicular, or neither.

$$2x - 5y = 6$$
$$5x - 2y = 9$$

*In Exercises 18 and 19, if* $f(x) = x^2 + 3x - 2$ *and* $g(x) = 4x - 6$, *find*

**18.** $(f + g)(x)$.

**19.** $(f \cdot g)(4)$.

**20. Natural Gas Consumption**  The total consumption of natural gas in 2003 was 21.8 trillion cubic feet ($2.18 \times 10^{13}$). The following pie chart shows the breakdown of consumption by sector.

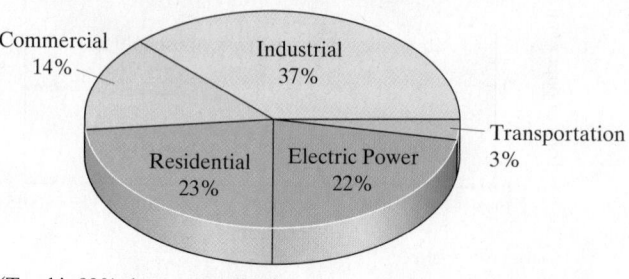

**Natural Gas Consumption by Sector**
**($2.18 \times 10^{13}$ cubic feet)**

Commercial 14%

Industrial 37%

Residential 23%

Electric Power 22%

Transportation 3%

(Total is 99% due to rounding.)

*Source:* Energy Information Administration

Answer the following questions using scientific notation.

**a)** What was the amount of natural gas consumption by the commercial sector in 2003?

**b)** How much more natural gas was consumed by the industrial sector than by the transportation sector in 2003?

**c)** If the consumption of natural gas is expected to increase by a total of 10% from 2003 to 2006, what will be the consumption of natural gas in 2006?

# 9

# Systems of Linear Equations

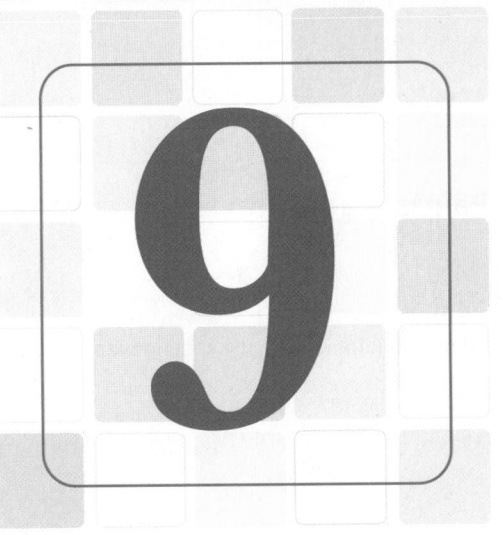

## Goals of This Chapter

In this chapter we solve systems of linear equations using the following methods: by graphing, by substitution, by the addition method, by using matrices, and by using determinants and Cramer's rule. Throughout the chapter there are many real-life applications. The chapter covers essential topics used by businesses to consider the relationships among variables involved in the day-to-day operations of a business.

© Allen R. Angel

**Systems of equations are** frequently used to solve real-life problems. For example, in Example 6 on page 553 we use a system of equations to study business loans taken out by a toy store.

# 9.1 Solving Systems of Equations Graphically

1 Determine if an ordered pair is a solution to a system of equations.

2 Determine if a system of equations is consistent, inconsistent, or dependent.

3 Solve a system of equations graphically.

## 1 Determine If an Ordered Pair Is a Solution to a System of Equations

### System of Linear Equations

When two or more linear equations are considered simultaneously, the equations are called a **system of linear equations**.

For example,

$$\left.\begin{array}{l}(1)\ y = x + 5 \\ (2)\ y = 2x + 4\end{array}\right\} \quad \text{System of linear equations}$$

### Solution

A **solution to a system of linear equations in two variables** is an ordered pair that satisfies each equation in the system.

---

**Understanding Algebra**

A solution to a system of two linear equations with variables $x$ and $y$ is an ordered pair $(x, y)$ that satisfies *both* equations.

---

The solution to the system above is the ordered pair $(1, 6)$.

Check | In Equation (1) | In Equation (2)
| $(1, 6)$ | $(1, 6)$
| $y = x + 5$ | $y = 2x + 4$
| $6 \overset{?}{=} 1 + 5$ | $6 \overset{?}{=} 2(1) + 4$
| $6 = 6$ True | $6 = 6$ True

Because the ordered pair $(1, 6)$ satisfies *both* equations, it is a solution to the system of equations. Notice that the ordered pair $(3, 8)$ satisfies the first equation but does not satisfy the second equation.

Check | In Equation (1) | In Equation (2)
| $(3, 8)$ | $(3, 8)$
| $y = x + 5$ | $y = 2x + 4$
| $8 \overset{?}{=} 3 + 5$ | $8 \overset{?}{=} 2(3) + 4$
| $8 = 8$ True | $8 = 10$ False

Since the ordered pair $(3, 8)$ does not satisfy *both* equations, it is *not* a solution to the system of equations.

EXAMPLE 1  Determine which of the following ordered pairs satisfy the system of equations.

$$y = 2x - 8$$
$$2x + y = 4$$

**a)** $(1, -6)$     **b)** $(3, -2)$

*Solution*

**a)** Substitute 1 for $x$ and $-6$ for $y$ in each equation.

| $y = 2x - 8$ | $2x + y = 4$ |
| $-6 \overset{?}{=} 2(1) - 8$ | $2(1) + (-6) \overset{?}{=} 4$ |
| $-6 \overset{?}{=} 2 - 8$ | $2 - 6 \overset{?}{=} 4$ |
| $-6 \overset{?}{=} -6$ True | $-4 = 4$ False |

Since $(1, -6)$ does not satisfy both equations, it is not a solution to the system of equations.

**b)** Substitute $3$ for $x$ and $-2$ for $y$ in each equation.

$$y = 2x - 8 \qquad\qquad\qquad 2x + y = 4$$
$$-2 \overset{?}{=} 2(3) - 8 \qquad\qquad 2(3) + (-2) \overset{?}{=} 4$$
$$-2 \overset{?}{=} 6 - 8 \qquad\qquad\qquad 6 - 2 \overset{?}{=} 4$$
$$-2 = -2 \quad \text{True} \qquad\qquad\quad 4 = 4 \quad \text{True}$$

Since $(3, -2)$ satisfies both equations, it is a solution to the system of linear equations.

Now Try Exercise 13

In this chapter we discuss three methods for finding the solution to a system of equations: the *graphical method*, the *substitution method*, and the *addition method*. In this section we discuss the graphical method.

## 2 Determine If a System of Equations Is Consistent, Inconsistent, or Dependent

Recall from Chapter 4 that the graph of a linear equation is the set of points whose ordered pairs are solutions to that equation. Thus, the solution to a system of linear equations is represented by the point or points that are on the graphs of *each* of the equations in the system. When two linear equations are graphed, three situations are possible, as illustrated in **Figure 9.1**.

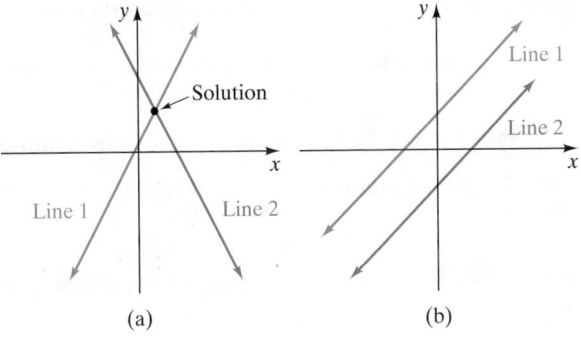

FIGURE 9.1

| (a) | (b) | (c) |
|---|---|---|
| Line 1 *Intersects* Line 2 | Line 1 is *Parallel* to Line 2 | Line 1 is the *Same Line* as Line 2 |
| • Exactly one solution | • No solution | • Infinite number of solutions |
| • The solution is at the point of intersection. | • Since parallel lines do not intersect, there is no solution. | • Every point on the common line is a solution. |
| • **Consistent system** | • **Inconsistent system** | • **Dependent system** |
| • Lines have different slopes | • Lines have same slopes and different *y*-intercepts | • Lines have same slope and the same *y*-intercept |

We can determine if a system of linear equations is consistent, inconsistent, or dependent by writing each equation in slope-intercept form and comparing the slopes and *y*-intercepts.

**EXAMPLE 2** Determine whether the following system of equations has exactly one solution, no solution, or an infinite number of solutions.

$$y = \frac{1}{2}x + 4$$
$$x - 2y = 6$$

*Solution*   The first equation is given in slope-intercept form:

$$y = \frac{1}{2}x \;+\; 4$$

the slope      the *y*-intercept

is $\frac{1}{2}$        is $(0, 4)$

Next, write the second equation in slope-intercept form by solving for *y*.

$$x - 2y = 6$$
$$-2y = -x + 6$$
$$y = \frac{1}{2}x \;-\; 3$$

the slope      the *y*-intercept

is $\frac{1}{2}$        is $(0, -3)$

The lines that represent the two equations in the system have the same slope and different *y*-intercepts. Thus, the two lines are parallel. The system is inconsistent and has no solution.

Now Try Exercise 29

## 3 Solve a System of Equations Graphically

Now we will see how to solve systems of equations graphically.

> **To Obtain the Solution to a System of Equations Graphically**
>
> Graph each equation and determine the point or points of intersection.

EXAMPLE 3   Solve the following system of equations graphically.

$$2x + y = 11$$
$$x + 3y = 18$$

*Solution*   Find the *x*-intercept and the *y*-intercept of the first equation.

$2x + y = 11$       **Ordered Pair**

Let $x = 0$; then $y = 11$.      $(0, 11)$

Let $y = 0$; then $x = \frac{11}{2}$.      $\left(\frac{11}{2}, 0\right)$

Find the *x*-intercept and the *y*-intercept of the second equation.

$x + 3y = 18$       **Ordered Pair**

Let $x = 0$; then $y = 6$.      $(0, 6)$

Let $y = 0$; then $x = 18$.      $(18, 0)$

Now use the intercepts to graph the equations on the same set of axes.

The two graphs (**Fig. 9.2** on page 521) appear to intersect at the point $(3, 5)$. The point $(3, 5)$ may be the solution to the system of equations. To be sure, however, we must check to see that $(3, 5)$ satisfies *both* equations.

**FIGURE 9.2**

Check

$$2x + y = 11 \qquad\qquad x + 3y = 18$$
$$2(3) + 5 \overset{?}{=} 11 \qquad\qquad 3 + 3(5) \overset{?}{=} 18$$
$$11 = 11 \quad \text{True} \qquad\qquad 18 = 18 \quad \text{True}$$

Since the ordered pair $(3, 5)$ checks in both equations, it is the solution to the system of equations. This system of equations is consistent.

Now Try Exercise 43

**EXAMPLE 4** Solve the following system of equations graphically.

$$2x + y = 3$$
$$4x + 2y = 12$$

*Solution*    Find the $x$- and $y$-intercepts of each graph; then draw the graphs.

|  | $2x + y = 3$ | Ordered Pair |
|--|--|--|
| | Let $x = 0$; then $y = 3$. | $(0, 3)$ |
| | Let $y = 0$; then $x = \dfrac{3}{2}$. | $\left(\dfrac{3}{2}, 0\right)$ |

|  | $4x + 2y = 12$ | Ordered Pair |
|--|--|--|
| | Let $x = 0$; then $y = 6$. | $(0, 6)$ |
| | Let $y = 0$; then $x = 3$. | $(3, 0)$ |

The two lines (**Fig. 9.3**) appear to be parallel.

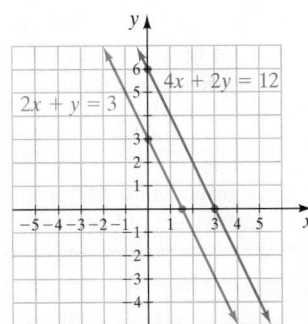

**FIGURE 9.3**

To show that the two lines are indeed parallel, write each equation in slope-intercept form.

$$2x + y = 3 \qquad\qquad 4x + 2y = 12$$
$$y = -2x + 3 \qquad\qquad 2y = -4x + 12$$
$$y = -2x + 6$$

Both equations have the same slope, $-2$, and different $y$-intercepts. Thus the lines are parallel. The system is inconsistent, and has no solution.

Now Try Exercise 55

EXAMPLE 5  Solve the following system of equations graphically.

$$x - \frac{1}{2}y = 2$$
$$y = 2x - 4$$

*Solution*  Find the $x$- and $y$-intercepts of each graph; then draw the graphs.

| $x - \frac{1}{2}y = 2$ | Ordered Pair |
|---|---|
| Let $x = 0$; then $y = -4$. | $(0, -4)$ |
| Let $y = 0$; then $x = 2$. | $(2, 0)$ |

| $y = 2x - 4$ | Ordered Pair |
|---|---|
| Let $x = 0$; then $y = -4$. | $(0, -4)$ |
| Let $y = 0$; then $x = 2$. | $(2, 0)$ |

Because the lines have the same $x$- and $y$-intercepts, both equations represent the same line (**Fig. 9.4**). When the equations are written in slope-intercept form, it becomes clear that the equations are represented by the same line. The system is dependent, and there are an infinite number of solutions.

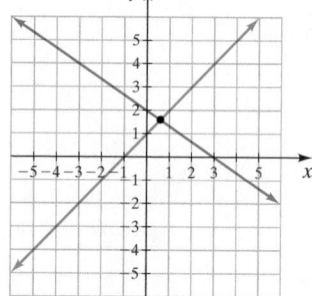

FIGURE 9.4

$$x - \frac{1}{2}y = 2 \qquad\qquad y = 2x - 4$$
$$2\left(x - \frac{1}{2}y\right) = 2(2)$$
$$2x - y = 4$$
$$-y = -2x + 4$$
$$y = 2x - 4$$

The solution to this system of equations is all the points on the line.

Now Try Exercise 53

Can you determine the solution to the system of equations shown in **Figure 9.5**? You may estimate the solution to be $\left(\frac{7}{10}, \frac{3}{2}\right)$ when it may actually be $\left(\frac{4}{5}, \frac{8}{5}\right)$. The accuracy of your answer will depend on how carefully you draw the graphs and on the scale of the graph paper used. In Section 9.2 we present an algebraic method that gives exact solutions to systems of equations.

Next, we work an application problem using two variables and illustrate the solution in the form of a graph. A graph may help you better visualize the total picture in application problems.

FIGURE 9.5

## Using Your Graphing Calculator

In Chapter 4 we learned how to use a graphing calculator to graph the equations of lines. Here we will use a graphing calculator to solve a system of linear equations graphically.

EXAMPLE  Use your graphing calculator to solve the systems of equations.

$$y = 2x - 3$$
$$y = -\frac{3}{2}x + 4$$

FIGURE 9.6

*Solution*  Press $\boxed{Y=}$ and enter these equations as $Y_1$ and $Y_2$, respectively, and press $\boxed{GRAPH}$. The graphs of the equations are shown in **Figure 9.6**.

The solution is at the point of intersection of the two lines. To determine the point, press $\boxed{2nd}$ $\boxed{TRACE}$ $\boxed{5}$ to activate the "intersect" feature. The calculator will ask you to choose the curves. Do so by pressing $\boxed{ENTER}$ $\boxed{ENTER}$. Now you are asked to make a guess. Press $\boxed{\blacktriangleleft}$ or $\boxed{\blacktriangleright}$ until you see the flashing cursor near the intersection point and then press $\boxed{ENTER}$. **Figure 9.7** shows that the intersection of the two lines is at the point $(2, 1)$. Substitute $x = 2$ and $y = 1$ into each of the original equations to confirm that $(2, 1)$ is a solution to the system of equations.

FIGURE 9.7

EXAMPLE 6 **Security Systems** Meghan O'Donnell plans to install a security system in her house. She has narrowed her choices to two security dealers: Moneywell and Doile. Moneywell's system costs $3580 to install and the monitoring fee is $20 per month. Doile's equivalent system costs only $2620 to install, but the monitoring fee is $32 per month.

**a)** Assuming that the monthly monitoring fees do not change, in how many months would the total cost of Moneywell's and Doile's systems be the same?

**b)** If both dealers guarantee not to raise monthly fees for 10 years, and if Meghan plans to use the system for 10 years, which system would be the least expensive?

*Solution*

**a) Understand and Translate**   We need to determine the number of months for which both systems will have the same total cost.

$$\text{Let } n = \text{number of months}$$
$$c = \text{total cost of the security system over } n \text{ months.}$$

Now we can write an equation to represent the cost of each system using the two variables $c$ and $n$.

| Moneywell | Doile |
|---|---|
| $\text{Total cost} = \begin{pmatrix}\text{initial}\\\text{cost}\end{pmatrix} + \begin{pmatrix}\text{fees over}\\n \text{ months}\end{pmatrix}$ | $\text{Total cost} = \begin{pmatrix}\text{initial}\\\text{cost}\end{pmatrix} + \begin{pmatrix}\text{fees over}\\n \text{ months}\end{pmatrix}$ |
| $c = 3580 + 20n$ | $c = 2620 + 32n$ |

Thus, our system of equations is

$$c = 3580 + 20n$$
$$c = 2620 + 32n$$

**Carry Out**   Now let's graph each equation. Following are tables of values.

$c = 3580 + 20n$

| | | $n$ | $c$ |
|---|---|---|---|
| Let $n = 0$. | $c = 3580 + 20(0) = 3580$ | 0 | 3580 |
| Let $n = 100$. | $c = 3580 + 20(100) = 5580$ | 100 | 5580 |
| Let $n = 160$. | $c = 3580 + 20(160) = 6780$ | 160 | 6780 |

$c = 2620 + 32n$

| | | $n$ | $c$ |
|---|---|---|---|
| Let $n = 0$. | $c = 2620 + 32(0) = 2620$ | 0 | 2620 |
| Let $n = 100$. | $c = 2620 + 32(100) = 5820$ | 100 | 5820 |
| Let $n = 160$. | $c = 2620 + 32(160) = 7740$ | 160 | 7740 |

**Figure 9.8** shows the graphs of the equations.

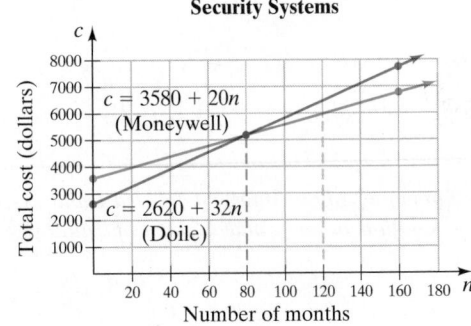

FIGURE 9.8

**Check and Answer**   The graph (**Fig. 9.8**) shows that the total cost of the two security systems would be the same in 80 months.

**b)** Since 10 years is 120 months, we draw a dashed vertical line at $n = 120$ months and see where it intersects the two lines. Since at 120 months the Doile line is higher than the Moneywell line, the cost for the Doile system for 120 months is more than the cost of the Moneywell system. Therefore, the cost of the Moneywell system would be less expensive for 10 years.

Now Try Exercise 63

# EXERCISE SET 9.1

MathXL®   MyMathLab®

MathXL®   MyMathLab

## Warm-Up Exercises

*Fill in the blanks with the appropriate word, phrase, or symbol(s) from the following list.*

unique solution                 dependent                consistent               inconsistent

intersecting lines              parallel lines            the same line            ordered pair

1. For a system with two equations and two variables, a solution is an _____.

2. The lines that represent the equations in a consistent system of linear equations are _____.

3. A system of equations that has at least one solution is a _____ system.

4. The _____ to a system of two linear equations is located at the intersection point of the graphs of the equations.

5. The lines that represent the equations in an inconsistent system of linear equations are _____.

6. The lines that represent the equations in a dependent system of linear equations are _____.

## Practice the Skills

*Determine which, if any, of the following ordered pairs satisfy each system of linear equations.*

7. $y = 5x - 8$
   $y = -3x$
   **a)** $(2, 2)$   **b)** $(0, 0)$   **c)** $(1, -3)$

8. $y = -4x$
   $y = -2x + 8$
   **a)** $(0, 8)$   **b)** $(-4, 16)$   **c)** $(3, -12)$

9. $y = 2x - 3$
   $y = x + 5$
   **a)** $(8, 13)$   **b)** $(4, 5)$   **c)** $(5, 7)$

10. $x + 2y = 4$
    $y = 3x - 5$
    **a)** $(0, 2)$   **b)** $(2, 1)$   **c)** $(4, 0)$

11. $4x + y = 15$
    $5x + y = 10$
    **a)** $(3, 3)$   **b)** $(2, 0)$   **c)** $(-1, 19)$

12. $y = 2x + 6$
    $y = 2x - 1$
    **a)** $(0, 6)$   **b)** $(3, 5)$   **c)** $(-2, 0)$

13. $4x - 6y = 12$
    $y = \dfrac{2}{3}x - 2$
    **a)** $(3, 0)$   **b)** $(9, 4)$   **c)** $(6, 1)$

14. $y = -x + 5$
    $2y = -2x + 10$
    **a)** $(6, -1)$   **b)** $(0, 5)$   **c)** $(-2, 3)$

15. $3x - 4y = 8$
    $2y = \dfrac{2}{3}x - 4$
    **a)** $(0, -2)$   **b)** $(1, -6)$   **c)** $\left(-\dfrac{1}{3}, -\dfrac{9}{4}\right)$

16. $2x + 3y = 6$
    $-x + \dfrac{5}{2} = \dfrac{1}{2}y$
    **a)** $\left(\dfrac{1}{2}, \dfrac{5}{3}\right)$   **b)** $\left(-2, \dfrac{10}{3}\right)$   **c)** $\left(\dfrac{9}{4}, \dfrac{1}{2}\right)$

*Identify each system of linear equations (lines are labeled 1 and 2) as consistent, inconsistent, or dependent. State whether the system has exactly one solution, no solution, or an infinite number of solutions.*

17.

18.

19.

20.

**21.**

**22.**

**23.**

**24.**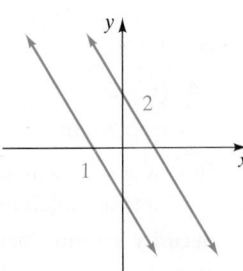

*Express each equation in slope-intercept form. Without graphing the equations, state whether the system of equations has exactly one solution, no solution, or an infinite number of solutions.*

**25.** $y = \dfrac{1}{2}x - 5$

$3x - 4y = 8$

**26.** $x + y = 8$

$x - y = 8$

**27.** $2y = 3x + 3$

$y = \dfrac{3}{2}x - 2$

**28.** $y = \dfrac{1}{2}x + 3$

$2y = x + 6$

**29.** $2x = y - 6$

$3x = 3y + 5$

**30.** $x + 3y = 6$

$4x + y = 4$

**31.** $3x + 5y = -7$

$-3x - 5y = -10$

**32.** $x - y = 2$

$2x - 2y = -6$

**33.** $x = 3y + 5$

$2x - 6y = 10$

**34.** $x - y = 3$

$\dfrac{1}{2}x - 2y = -8$

**35.** $y = \dfrac{3}{2}x + \dfrac{1}{2}$

$3x - 2y = \dfrac{5}{2}$

**36.** $2y = \dfrac{7}{3}x - 9$

$4y = 8x + 9$

*Determine the solution to each system of equations graphically. If the system is dependent or inconsistent, so state.*

**37.** $y = x + 3$

$y = -x + 3$

**38.** $y = -3x + 9$

$y = 2x - 6$

**39.** $y = 3x - 6$

$y = -x + 6$

**40.** $y = 3x - 4$

$y = -x$

**41.** $4x = 8$

$y = -3$

**42.** $2x - y = 7$

$2y = 2x - 6$

**43.** $x + y = 5$

$-x + y = 1$

**44.** $-x + 2y = 7$

$2x - y = -2$

**45.** $y = -\dfrac{1}{2}x + 4$

$x + 2y = 6$

**46.** $-x + 2y = 0$

$2x - y = -3$

**47.** $x + 2y = 8$

$5x + 2y = 0$

**48.** $3x + y = -6$

$2x = 1 + y$

**49.** $2x + 3y = 6$

$4x = -6y + 12$

**50.** $2x + 5y - 6 = 0$

$2x - 3y = 6$

**51.** $y = 3$

$y = 2x - 3$

**52.** $x = 5$

$y = 2x - 8$

**53.** $x - 2y = 4$

$2x - 4y = 8$

**54.** $4x - y = 6$

$2y = 8x - 12$

**55.** $2x + y = -2$

$6x + 3y = 6$

**56.** $y = 2x - 1$

$2y = 4x + 5$

**57.** $4x - 3y = 6$

$2x + 4y = 14$

**58.** $2x + 6y = 12$

$y = -\dfrac{1}{3}x + 2$

**59.** $2x - 3y = 0$

$x + 2y = 0$

**60.** $4x = 4y - 12$

$-8x + 8y = 8$

## Problem Solving

*In Exercises 61–64, find each solution by graphing the system of equations.*

**61. Furnace Repair** Edith Hall's furnace is 10 years old and has a problem. The furnace repair man indicates that it will cost Edith $600 to repair her furnace. She can purchase a new, more efficient furnace for $1800. Her present furnace aver- ages about $650 per year for energy cost and the new furnace would average about $450 per year.

We can represent the total cost, $c$, of repair or replacement, plus energy cost over $n$ years by the following system of equations.

(repair)                        $c = 600 + 650n$

(replacement)          $c = 1800 + 450n$

In how many years would the total cost of repair equal the total cost of replacement?

**62. Security Systems** Juan Varges is considering the two security systems discussed in Example 6. If Moneywell's system costs $4400 plus $15 per month and Doile's system costs $3400 plus $25 per month, after how many months would the total cost of the two systems be the same?

**63. Boat Ride** Rudy has visitors at his home and wants to take them out on a pontoon boat for a day. There are two pontoon boat rental agencies on the lake. Bob's Boat Rental charges $25 per hour for the boat rental, which includes all the gasoline used. Hopper's Rental charges $21 per hour plus a flat charge of $28 for the gasoline used. The equations that represent the total cost, $c$, follow. In the equations $h$ represents the number of hours the boats are rented.

$$c = 25h$$
$$c = 21h + 28$$

Determine the number of hours the boats must be rented for the total cost to be the same.

**64. Landscaping** The Evergreen Landscape Service charges a consultation fee of $200 plus $50 per hour for labor. The Out of Sight Landscape Service charges a consultation fee of $300 plus $40 per hour for labor. We can represent this situation with the system of equations

$$c = 200 + 50h$$
$$c = 300 + 40h$$

where $c$ is the total cost and $h$ is the number of hours of labor. Find the number of hours of labor for the two services to have the same total cost.

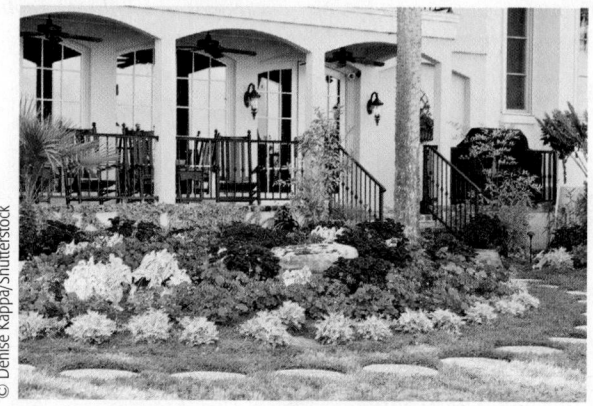
© Denise Kappa/Shutterstock

## Concept/Writing Exercises

**65.** Given the system of equations $6x - 4y = 12$ and $12y = 18x - 24$, determine without graphing whether the graphs of the two equations will be parallel lines. Explain how you determined your answer.

**66.** Given the system of equations $4x - 8y = 12$ and $2x - 8 = 4y$, determine without graphing whether the graphs of the two equations will be parallel lines. Explain how you determined your answer.

**67.** If a system of linear equations has solutions $(4, 3)$ and $(6, 5)$, how many solutions does the system have? Explain.

**68.** If the slope of one line in a system of linear equations is 2 and the slope of the second line in the system is 3, how many solutions does the system have? Explain.

**69.** If two distinct lines are parallel, how many solutions does the system have? Explain.

**70.** If two different lines in a linear system of equations pass through the origin, must the solution to the system be $(0, 0)$? Explain.

**71.** Consider the system $x = 5$ and $y = 3$. How many solutions does the system have? What is the solution?

**72.** A system of linear equations has $(3, -1)$ as its solution. If one line in the system is vertical and the other line is horizontal, determine the equations in the system.

## Group Activity

*Discuss and answer Exercises 73–78 as a group. Suppose that a system of three linear equations in two variables is graphed on the same axes. Find the maximum number of points where two or more of the lines can intersect if*

**73.** the three lines have the same slope but different $y$-intercepts.

**74.** the three lines have the same slope and the same $y$-intercept.

**75.** two lines have the same slope but different $y$-intercepts and the third line has a different slope.

**76.** the three lines have different slopes but the same $y$-intercept.

**77.** the three lines have different slopes but two have the same $y$-intercept.

**78.** the three lines have different slopes and different $y$-intercepts.

## Cumulative Review Exercises

[2.1] **79.** Simplify $3x - (x - 6) + 4(3 - x)$.

[2.5] **80.** Solve $2(x + 3) - x = 5x + 2$.

[2.6] **81.** If $A = p(1 + rt)$, find $r$ when $A = 1000$, $t = 20$, and $p = 500$.

[4.2] **82. a)** Find the $x$- and $y$-intercepts of $2x + 3y = 12$.
         **b)** Use the intercepts to graph the equation.

[7.1] **83.** Simplify $\dfrac{x^2 - 9x + 14}{2 - x}$.

[7.6] **84.** Solve $\dfrac{4}{b} + 2b = \dfrac{38}{3}$.

# 9.2 Solving Systems of Equations by Substitution

**1** Solve systems of equations by substitution.

Although solving systems of equations graphically helps us visualize a system of equations and its solution, an exact solution is sometimes difficult to determine from the graph. For this reason it may be necessary to use algebraic methods to find the solution to a system of linear equations in two variables. The first of these methods is **substitution**.

## **1** Solve Systems of Equations by Substitution

### To Solve a System of Equations by Substitution

1. Solve for a variable in either equation. (If possible, solve for a variable with a numerical coefficient of 1 or −1 to avoid working with fractions.)
2. Substitute the expression found for the variable in step 1 into the other equation.
3. Solve the equation determined in step 2 to find the value of one variable.
4. Substitute the value found in step 3 into the equation obtained in step 1 to find the value of the other variable.
5. Check by substituting both values in both original equations.

**EXAMPLE 1** Solve the following system of equations by substitution.

$$x + 3y = 18$$
$$2x + y = 11$$

*Solution*    We begin by solving for a variable in either equation. Since both equations have a variable with a numerical coefficient of 1, we can solve either equation without introducing fractions. We will solve for $x$ in the first equation, $x + 3y = 18$.

Step 1                     $x + 3y = 18$           Solve first equation for $x$.
                            $x = -3y + 18$

Next, substitute $-3y + 18$ for $x$ in the *other equation*, $2x + y = 11$, and solve for the remaining variable, $y$.

Step 2                     $2x + y = 11$           Second equation
                    $2(\overbrace{-3y + 18}) + y = 11$      Substitution; this is now an equation in only one variable, $y$.

Step 3                     $-6y + 36 + y = 11$
                            $-5y + 36 = 11$
                            $-5y = -25$
                            $y = 5$

Finally, substitute $y = 5$ in the equation found in Step 1 and find the value of $x$.

                            $x = -3y + 18$
Step 4                      $x = -3(5) + 18$
                            $x = -15 + 18$
                            $x = 3$

### Understanding Algebra

In Example 1, we also could have solved the system by solving the second equation for $y$, and substituting this expression into the first equation. Try this now. You should get the same answer.

Check the solution by substituting $x = 3$ and $y = 5$ into both equations.

Step 5          $x + 3y = 18$              $2x + y = 11$
                $3 + 3(5) = 18$            $2(3) + 5 = 11$
                $3 + 15 = 18$              $6 + 5 = 11$
                $18 = 18$   True           $11 = 11$   True

Since $(3, 5)$ satisfies both equations, it is the solution to the system of equations.

Now Try Exercise 5

The system from Example 1 was solved graphically in Example 3 of Section 9.1. **Figure 9.2** on page 521 shows that the solution $(3, 5)$ is the point of intersection of the graphs of the two lines.

**EXAMPLE 2** Solve the following system of equations by substitution.

$$2x + y = 3$$
$$4x + 2y = 12$$

*Solution*  First solve for $y$ in $2x + y = 3$.

Step 1
$$2x + y = 3$$
$$y = -2x + 3$$

Now substitute the expression $-2x + 3$ for $y$ in the *other equation*, $4x + 2y = 12$, and solve for $x$.

$$4x + 2y = 12$$

Step 2
$$4x + 2(\overbrace{-2x + 3}) = 12 \qquad \text{Substitution; this is now an equation in only one variable, } x.$$
$$4x - 4x + 6 = 12$$
$$6 = 12 \quad \text{False}$$

Since the statement $6 = 12$ is false, the system has no solution. Therefore, the graphs of the equations will be parallel lines and the system is inconsistent because it has no solution.

Now Try Exercise 9

**Understanding Algebra**

If, when solving a system of equations by the substitution method, you obtain an equation that is always

• *false*, such as $6 = 12$ or $0 = 6$, the system is inconsistent and has no solution.

• *true*, such as $2 = 2$ or $0 = 0$, the system is dependent and has an infinite number of solutions.

Note that the solution in Example 2 is identical to the graphical solution obtained in Example 4 of Section 9.1. **Figure 9.3** on page 521 shows the parallel lines.

**EXAMPLE 3** Solve the following system of equations by substitution.

$$x - \frac{1}{2}y = 2$$
$$y = 2x - 4$$

*Solution*  The equation $y = 2x - 4$ is already solved for $y$. Substitute $2x - 4$ for $y$ in the other equation, $x - \frac{1}{2}y = 2$, and solve for $x$.

$$x - \frac{1}{2}y = 2$$

Step 2
$$x - \frac{1}{2}(\overbrace{2x - 4}) = 2$$
$$x - x + 2 = 2$$
$$2 = 2 \quad \text{True}$$

Since the statement $2 = 2$ is true, this system has an infinite number of solutions. Therefore, the graphs of the equations represent the same line and the system is dependent.

Now Try Exercise 17

Note that the solution in Example 3 is identical to the solution obtained graphically in Example 5 of Section 9.1. **Figure 9.4** on page 522 shows that the graphs of both equations are the same line.

**EXAMPLE 4** Solve the following system of equations by substitution.

$$3x + 6y = 9$$
$$2x - 3y = 6$$

*Solution*  None of the variables in either equation has a numerical coefficient of 1. However, since the numbers 3, 6, and 9 are all divisible by 3, if you solve the first equation for $x$, you will avoid having to work with fractions.

$$3x + 6y = 9$$
$$3x = -6y + 9$$
$$\frac{3x}{3} = \frac{-6y + 9}{3}$$
$$x = -\frac{6}{3}y + \frac{9}{3}$$
$$x = -2y + 3$$

Now substitute $-2y + 3$ for $x$ in the other equation, $2x - 3y = 6$, and solve for the remaining variable, $y$.

$$2x - 3y = 6$$
$$2(\overbrace{-2y + 3}) - 3y = 6$$
$$-4y + 6 - 3y = 6$$
$$-7y + 6 = 6$$
$$-7y = 0$$
$$y = 0$$

Finally, solve for $x$ by substituting $y = 0$ in the equation previously solved for $x$.

$$x = -2y + 3$$
$$x = -2(0) + 3 = 0 + 3 = 3$$

The solution is $(3, 0)$.

Now Try Exercise 23

**Helpful Hint**

Remember that a solution to a system of linear equations must contain both an $x$- and a $y$-value. Write the solution as an ordered pair.

**EXAMPLE 5** Solve the following system of equations by substitution.

$$6x + 8y = 3$$
$$3x = 3y + 5$$

*Solution*  We will solve for $x$ in the second equation.

$$3x = 3y + 5$$
$$x = \frac{3y + 5}{3}$$
$$x = y + \frac{5}{3}$$

Now substitute $y + \dfrac{5}{3}$ for $x$ in the other equation.

$$6x + 8y = 3$$

$$6\left(\overbrace{y + \dfrac{5}{3}}\right) + 8y = 3$$

$$6y + 10 + 8y = 3$$

$$14y + 10 = 3$$

$$14y = -7$$

$$y = \dfrac{-7}{14} = -\dfrac{1}{2}$$

Finally, find the value of $x$.

$$x = y + \dfrac{5}{3}$$

$$x = -\dfrac{1}{2} + \dfrac{5}{3} = -\dfrac{3}{6} + \dfrac{10}{6} = \dfrac{7}{6}$$

The solution is the ordered pair $\left(\dfrac{7}{6}, -\dfrac{1}{2}\right)$.

Now Try Exercise 25

**EXAMPLE 6  Wisconsin Towns**  The total population of the towns of Dickeyville and Muscoda, Wisconsin, is 2462. If the population of Muscoda is 372 more than the population of Dickeyville, find the population of each town.

*Solution*    Understand and Translate    We will use two variables to solve this problem.

Let $x$ = population of Dickeyville.

Let $y$ = population of Muscoda.

Since we use two variables we must write two equations. The first equation comes from the phrase

*the total population of the towns . . . is 2462*
$$x + y = 2462$$

The second equation comes from the phrase

*the population of Muscoda is 372 more than the population of Dickeyville*

$$y = x + 372$$

Thus we have the system of equations

$$x + y = 2462$$
$$y = x + 372$$

Carry Out    Because the second equation, $y = x + 372$, is already solved for $y$, we will substitute $x + 372$ for $y$ in the first equation.

$$x + y = 2462$$

$$x + \overbrace{x + 372} = 2462$$

$$2x + 372 = 2462$$

$$2x = 2090$$

$$x = 1045$$

Check and Answer    The population of Dickeyville is 1045. The population of Muscoda is therefore $x + 372 = 1045 + 372 = 1417$. We check the solution by noting the total population of the two towns is $1045 + 1417 = 2462$.

Now Try Exercise 35

# EXERCISE SET 9.2

MathXL® MyMathLab®
MathXL® MyMathLab

## Warm-Up Exercises

*Fill in the blanks with the appropriate word, phrase, or symbol(s) from the following list.*

| factor | $x$-intercept | unique solution | an infinite number of solutions |
|---|---|---|---|
| point of intersection | no solution | coefficient | constant |

**1.** When solving a system of linear equations by substitution, if possible, solve for a variable with a _____ of 1 or $-1$ to avoid working with fractions.

**2.** The solution to a system of linear equations occurs at the _____ of graphs of the equations in the system.

**3.** If, when solving a system of equations by the substitution method, you obtain an equation that is always false, the system is inconsistent and has _____ .

**4.** If, when solving a system of equations by the substitution method, you obtain an equation that is always true, the system is dependent and has _____ .

## Practice the Skills

*Find the solution to each system of equations by substitution.*

**5.** $x - 3y = -1$
$2x + 4y = 8$

**6.** $y = 2x + 7$
$y = -x - 5$

**7.** $x + y = -2$
$x - y = 0$

**8.** $x + 2y = 6$
$4y = 12 - 2x$

**9.** $3x + y = 3$
$3x + y + 5 = 0$

**10.** $3x - y = 8$
$6x - 2y = 10$

**11.** $x = 3$
$x + y + 5 = 0$

**12.** $y = 2x + 4$
$y = -2$

**13.** $x = y + 1$
$4x + 2y = -14$

**14.** $2x + 3y = 7$
$6x - y = 1$

**15.** $2x + y = 11$
$y = 3x - 4$

**16.** $y = -2x + 7$
$x + 4y = 0$

**17.** $y = \dfrac{1}{3}x - 2$
$x - 3y = 6$

**18.** $x - \dfrac{1}{2}y = 7$
$y = 2x - 14$

**19.** $2x + 5y = 9$
$6x - 2y = 10$

**20.** $3x - 3y = 4$
$2x + 3y = 5$

**21.** $y = 2x - 13$
$-4x - 7 = 9y$

**22.** $5x - 2y = -7$
$8 = y - 4x$

**23.** $4x - 5y = -4$
$3x = 2y - 3$

**24.** $2x - 3y = 9$
$5x + 2y = -6$

**25.** $4x + 5y = -6$
$2x - \dfrac{10}{3}y = -4$

**26.** $4x + y = 1$
$10x - 5y = -2$

**27.** $3x - 4y = 15$
$-6x + 8y = -14$

**28.** $\dfrac{1}{2}x + 2y = 5$
$3x - \dfrac{1}{2}y = 5$

**29.** $4x - y = 1$
$10x + \dfrac{1}{2}y = 1$

**30.** $-\dfrac{1}{5}x + \dfrac{3}{7}y = 2$
$\dfrac{3}{5}x - \dfrac{2}{7}y = 1$

## Problem Solving

**31. Positive Integers** The sum of two positive integers is 63. The first number is twice the second number. Find the two numbers.

**32. Positive Integers** The difference of two positive integers is 44. Find the integers if the larger number is twice the smaller number.

**33. Rectangle** The perimeter of a rectangle is 50 feet. Find the dimensions of the rectangle if the length is 9 feet greater than the width.

**34. Rectangle** The perimeter of a rectangle is 60 feet. Find the dimensions of the rectangle if its length is 5 times its width.

**35. Rodeo Attendance** At a rodeo the total paid attendance was 2500. If the number of people who received a discount on the admission fee was 622 less than the number who did not, determine the number of paid attendees that paid the full fee.

See Exercise 35.

36. **Wooden Horse** To buy a statue of a wooden horse Billy and Jean combined their money. Together they had $530. If Jean had $130 more than Billy, how much money did each have?

37. **Legal Settlement** After a legal settlement, the client's portion of the award was three times as much money as the attorney's portion. If the total award was $40,000, how much did the client get?

38. **Jelly Beans** A candy store mixed green and red jelly beans in a barrel. The barrel contains 42 pounds of the mixture. If there are 3 times as many pounds of the green jelly beans as red jelly beans, find the number of pounds of green jelly beans and the number of pounds of red jelly beans in the barrel.

39. **Refinancing** Dona Lee is considering refinancing her house. The cost of refinancing is a one-time charge of $1280. With her reduced mortgage rate, her monthly interest and principal payments would be $794 per month. Her total cost, $c$, for $n$ months could be represented by $c = 1280 + 794n$. At her current rate her mortgage payments are $874 per month and the total cost for $n$ months can be represented by $c = 874n$.

   a) In how many months would both mortgage plans have the same total cost?

   b) If Dona plans to remain in her house for 12 years, should she refinance?

40. **Temperatures** In Seattle the temperature is 86°F, but it is decreasing by 2 degrees per hour. The temperature, $T$, at time, $t$, in hours, is represented by $T = 86 - 2t$. In Spokane the temperature is 59°F, but it is increasing by 2.5 degrees per hour. The temperature, $T$, can be represented by $T = 59 + 2.5t$.

a) If the temperature continues decreasing and increasing at the same rate in these cities, how long will it be before both cities have the same temperature?

b) When both cities have the same temperature, what will that temperature be?

Seattle, Washington

41. **Traveling by Car** Jean Woody's car is at the 80 mile marker on a highway. Roberta Kieronski's car is 15 miles behind Jean's car. Jean's car is traveling at 60 miles per hour. The mile marker that Jean's car will be at in $t$ hours can be found by the equation $m = 80 + 60t$. Roberta's car is traveling at 72 miles per hour. The mile marker that Roberta's car will be at in $t$ hours can be found by the equation $m = 65 + 72t$.

   a) Determine the time it will take for Roberta's car to catch up with Jean's car.

   b) At which mile marker will they be when they meet?

42. **Computer Store** Will Worthy's present salary consists of a fixed weekly salary of $300 plus a $20 bonus for each computer system he sells. His weekly salary can be represented by $s = 300 + 20n$, where $n$ is the number of computer systems he sells. He is considering another position where his weekly salary would be $400 plus a $10 bonus for each computer system he sells. The other position's weekly salary can be represented by $s = 400 + 10n$. How many computer systems would Will need to sell in a week for his salary to be the same with both employers?

## Challenge Problems

*Answer parts **a**) through **d**) on your own.*

43. **Heat Transfer** In a laboratory during an experiment on heat transfer, a large metal ball is heated to a temperature of 180°F. This metal ball is then placed in a gallon of oil at a temperature of 20°F. Assume that when the ball is placed in the oil it loses temperature at the rate of 10 degrees per minute while the oil's temperature rises at a rate of 6 degrees per minute.

   a) Write an equation that can be used to determine the ball's temperature $t$ minutes after being placed in the oil.

   b) Write an equation that can be used to determine the oil's temperature $t$ minutes after the ball is placed in it.

   c) Determine how long it will take for the ball and oil to reach the same temperature.

   d) When the ball and oil reach the same temperature, what will the temperature be?

## Group Activity

*Discuss and work Exercise 44 as a group.*

44. In intermediate algebra you may solve systems containing three equations with three variables. As a group, solve the system of equations on the right. Your answer will be in the form of an **ordered triple** $(x, y, z)$.

$$x = 4$$
$$2x - y = 6$$
$$-x + y + z = -3$$

## Cumulative Review Exercises

[2.6]   **45. Willow Tree** The diameter of a willow tree grows about 1.2 inches per year. What is the approximate age of a willow tree whose diameter is 27.6 inches?

[4.2]   **46.** Find the missing coordinate in the given solution $(3, ?)$ for $4x + 5y = 22$.

**47.** Graph $4x - 8y = 16$ using the intercepts.

[4.4]   **48.** Find the slope and $y$-intercept of the graph of the equation $3x - 5y = 25$.

**49.** Determine the equation of the following line.

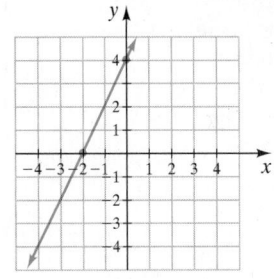

[5.5]   **50.** Multiply $(6x + 7)(3x - 2)$.

# 9.3 Solving Systems of Equations by the Addition Method

1 Solve systems of equations by the addition method.

## Understanding Algebra

The goal of both the substitution method and the addition method is to obtain a single equation with only one variable. With the addition method, we add two equations together so that one of the variables "drops out" and we are able to solve for the other variable.

## 1 Solve Systems of Equations by the Addition Method

A third method of solving a system of linear equations, and the second algebraic method, is the **addition (or elimination) method**.

In the addition method, we use the fact that if $a = b$ and $c = d$, then $a + c = b + d$. Suppose we have the following two equations.

$$x - 2y = 9$$
$$3x + 2y = 11$$

Notice the like terms of the two equations are aligned. Also notice that $-2y$ and $+2y$ are opposites of each other. If we *add* the like terms of the two equations together we are able to *eliminate* the $y$ variable.

$$\begin{aligned} x - 2y &= 9 \\ 3x + 2y &= 11 \\ \hline 4x &= 20 \end{aligned}$$   Add the like terms of the two equations.

Now we are able to solve this single equation for $x$.

$$4x = 20$$
$$x = 5$$

We can now substitute $5$ for $x$ in either of the original equations to find $y$. We will substitute $5$ for $x$ in $x - 2y = 9$.

$$x - 2y = 9$$
$$5 - 2y = 9$$   Substitute $x = 5$ into first equation.
$$-2y = 4$$   Solve for $y$.
$$y = -2$$

The solution to the system is $(5, -2)$. This ordered pair will check in both equations. Check this solution now.

### To Solve a System of Equations Using the Addition (or Elimination) Method

**1.** If necessary, rewrite each equation so that the terms containing variables appear on the left side of the equals sign and any constants appear on the right side of the equals sign.

*(continued on p. 534)*

2. If necessary, multiply one or both equations by a constant(s) so that when the equations are added the resulting sum will contain only one variable.

3. Add the equations. This will result in a single equation containing only one variable.

4. Solve for the variable in the equation from step 3.

5. **a)** Substitute the value found in step 4 into either of the original equations. Solve that equation to find the value of the remaining variable.

   or

   **b)** Repeat steps 2–4 to eliminate the other variable.

6. Check the values obtained in all original equations.

**EXAMPLE 1** Solve the following system of equations using the addition method.

$$x + y = 6$$
$$2x - y = 3$$

*Solution*   Note that the first equation contains $+y$ and the second equation contains $-y$. By *adding* the equations, we can *eliminate* the variable $y$ and obtain one equation containing only one variable, $x$.

$$
\begin{array}{ll}
x + y = 6 & \\
\underline{2x - y = 3} & \text{Add the like terms of the two equations.} \\
3x \quad\;\; = 9 &
\end{array}
$$

Now we solve for the remaining variable, $x$.

$$\frac{3x}{3} = \frac{9}{3} \qquad \text{Solve for } x.$$
$$x = 3$$

Finally, we solve for $y$ by substituting $x = 3$ in either of the original equations.

$$x + y = 6$$
$$3 + y = 6$$
$$y = 3$$

The solution is $(3, 3)$.

Check   We check the ordered pair $(3, 3)$ in *both* equations.

$$
\begin{array}{ll}
x + y = 6 & \qquad\qquad 2x - y = 3 \\
3 + 3 \stackrel{?}{=} 6 & \qquad\qquad 2(3) - 3 \stackrel{?}{=} 3 \\
\quad\;\; 6 = 6 \quad \text{True} & \qquad\qquad\quad 6 - 3 \stackrel{?}{=} 3 \\
& \qquad\qquad\qquad\quad\; 3 = 3 \quad \text{True}
\end{array}
$$

Now Try Exercise 7

When solving systems of equations with the addition method, it is often necessary to multiply one or both equations in the system by a constant. This is done to get opposite terms that will be eliminated when the equations are added.

For example, consider the system of equations below.

$$3x + 2y = 6 \qquad eq.\,1$$
$$x - y = 8 \qquad eq.\,2$$

To get the $x$-terms to be opposites of each other, we can multiply *eq.* 2 by $-3$.

$$
\begin{array}{lll}
3x + 2y = 6 & \xrightarrow{\;eq.\,1 \text{ stays the same}\;} & 3x + 2y = 6 \\
-3(x - y) = -3(8) & \xrightarrow{\;eq.\,2 \text{ multiplied by } -3\;} & -3x + 3y = -24
\end{array}
$$

After this step, we can continue to solve the system in a manner similar to what we did in Example 1. We could have also multiplied *eq.* 2 by 2 so that the *y*-terms would be opposite and would drop out when the equations are added.

**EXAMPLE 2** Solve the following system of equations using the addition method.

$$x + 3y = 13 \qquad eq. 1$$
$$x + 4y = 18 \qquad eq. 2$$

*Solution*   Our goal is to rewrite the system so that we have opposite terms that are eliminated when the two equations are added together. We will multiply *eq.* 1 by $-1$.

$$-1(x + 3y) = -1 \cdot 13 \xrightarrow{\;eq.\ 1\ multiplied\ by\ -1\;} -x - 3y = -13$$
$$x + 4y = 18 \xrightarrow{\;eq.\ 2\ stays\ the\ same\;} x + 4y = 18$$

Now add the two equations.

$$\begin{array}{r} -x - 3y = -13 \\ x + 4y = \phantom{-}18 \\ \hline y = \phantom{-}5 \end{array}$$

**Understanding Algebra**

When using the addition method, if you need to multiply an equation by a constant, remember that *both sides of the equation must be multiplied* by the constant. This will produce an equivalent equation and will not change the solution to the system of equations.

Next, we solve for *x* in either of the original equations. We will substitute $y = 5$ into *eq.* 1.

$$x + 3y = 13$$
$$x + 3(5) = 13$$
$$x + 15 = 13$$
$$x = -2$$

A check will show that the solution is $(-2, 5)$.

*Now Try Exercise 13*

**EXAMPLE 3** Solve the following system of equations using the addition method.

$$2x + y = 11 \qquad eq. 1$$
$$x + 3y = 18 \qquad eq. 2$$

*Solution*   To eliminate the variable *x*, multiply *eq.* 2 by $-2$ and add the two equations.

$$2x + y = 11 \xrightarrow{\;eq.\ 1\ stays\ the\ same\;} 2x + y = 11$$
$$-2(x + 3y) = -2 \cdot 18 \xrightarrow{\;eq.\ 2\ multiplied\ by\ -2\;} -2x - 6y = -36$$

Now add:

$$\begin{array}{r} 2x + \phantom{6}y = \phantom{-}11 \\ -2x - 6y = -36 \\ \hline -5y = -25 \\ y = \phantom{-}5 \end{array}$$

Solve for *x* by substituting $y = 5$ into *eq.* 1.

$$2x + y = 11$$
$$2x + 5 = 11$$
$$2x = 6$$
$$x = 3$$

**Understanding Algebra**

In Example 3, we could have solved the system by first multiplying *eq.* 1 by $-3$ to eliminate the *y*-terms. Try this now. You should get the same answer.

The solution is $(3, 5)$.

*Now Try Exercise 23*

*Note that the solution in Example 3 is the same as the solution obtained graphically in Example 3 of Section 9.1 (see* **Fig. 9.2** *on page 521) and by substitution in Example 1 of Section 9.2 on page 527.*

**EXAMPLE 4** Solve the system of equations using the addition method.

$$3x = 2y + 5$$
$$6y = -5x + 27$$

*Solution* Our first step is to rewrite each equation so the variable terms are on the left side and the constants are on the right side of the equals sign.

$3x = 2y + 5$ $\xrightarrow{\text{Subtracted } 2y \text{ from both sides}}$ $3x - 2y = 5$   *eq. 1*

$6y = -5x + 27$ $\xrightarrow{\text{Added } 5x \text{ to both sides}}$ $5x + 6y = 27$   *eq. 2*

Now, continue to solve the system. To eliminate the *y*-terms, multiply *eq.* 1 by 3.

$3(3x - 2y) = 3 \cdot 5$ $\xrightarrow{\textit{eq. } 1 \text{ multiplied by } 3}$ $9x - 6y = 15$

$5x + 6y = 27$ $\xrightarrow{\textit{eq. } 2 \text{ stays the same}}$ $5x + 6y = 27$

Next, add the two equations and solve for *x*.

$$9x - 6y = 15$$
$$\underline{5x + 6y = 27}$$
$$14x \quad\quad = 42$$
$$x = 3$$

Solve for *y* by substituting $x = 3$ into *eq.* 2.

$$6y = -5x + 27$$
$$6y = -5(3) + 27$$
$$6y = -15 + 27$$
$$6y = 12$$
$$y = 2$$

The solution is $(3, 2)$. Check the solution by substituting $x = 3$ and $y = 2$ into both of the *original* equations.

Now Try Exercise 31

> **Helpful Hint**
>
> When solving a system of equations, if a mistake is made midway through the solution process, you will discover the error by substituting the values for the variables back into the *original* equations during the *check* phase.

**EXAMPLE 5** Solve the following system of equations using the addition method.

$$2x + 3y = 6 \quad\quad \textit{eq. } 1$$
$$5x - 4y = -8 \quad\quad \textit{eq. } 2$$

*Solution* The variable *x* can be eliminated by multiplying *eq.* 1 by −5 and *eq.* 2 by 2 and then adding the equations.

$-5(2x + 3y) = -5 \cdot 6$ $\xrightarrow{\textit{eq. } 1 \text{ multiplied by } -5}$ $-10x - 15y = -30$

$2(5x - 4y) = 2 \cdot (-8)$ $\xrightarrow{\textit{eq. } 2 \text{ multiplied by } 2}$ $10x - 8y = -16$

$$-10x - 15y = -30$$
$$\underline{10x - \;\;8y = -16}$$
$$-23y = -46$$
$$y = \quad 2$$

**Understanding Algebra**

In Example 5, we also could have solved the system by first multiplying *eq.* 1 by 4 and *eq.* 2 by 3 to eliminate the *y*-terms. Try this now. You should get the same answer.

Solve for *x* by substituting $y = 2$ into *eq.* 1.

$$2x + 3y = 6$$
$$2x + 3(2) = 6$$
$$2x + 6 = 6$$
$$2x = 0$$
$$x = 0$$

The solution is $(0, 2)$.

*Now Try Exercise 19*

**EXAMPLE 6**  Solve the following system of equations using the addition method.

$$2x + y = 3 \qquad \textit{eq. 1}$$
$$4x + 2y = 12 \qquad \textit{eq. 2}$$

*Solution*    The variable *y* can be eliminated by multiplying *eq.* 1 by $-2$ and then adding the two equations.

$$-2(2x + y) = -2 \cdot 3 \xrightarrow{\textit{eq. 1 multiplied by } -2} -4x - 2y = -6$$
$$4x + 2y = 12 \xrightarrow{\textit{eq. 2 stays the same}} 4x + 2y = 12$$

$$\begin{array}{r} -4x - 2y = -6 \\ 4x + 2y = 12 \\ \hline 0 = 6 \quad \text{False} \end{array}$$

Since $0 = 6$ is a false statement, this system has no solution. The system is inconsistent. The graphs of the equations will be parallel lines.

*Now Try Exercise 25*

*Note that the solution in Example 6 is identical to the solutions obtained by graphing in Example 4 of Section 9.1 (see* **Fig. 9.3** *on page 521) and by substitution in Example 2 of Section 9.2 on page 528.*

**EXAMPLE 7**  Solve the following system of equations using the addition method.

$$x - \frac{1}{2}y = 2$$
$$y = 2x - 4$$

*Solution*    First align the *x*- and *y*-terms on the left side of the equals sign by subtracting $2x$ from both sides of the second equation.

$$x - \frac{1}{2}y = 2 \qquad \textit{eq. 1}$$
$$-2x + y = -4 \qquad \textit{eq. 2}$$

**Understanding Algebra**

If, when solving a system of equations by the addition method, you obtain an equation that is always

- *false*, such as $0 = 6$, the system is inconsistent and has no solution.
- *true*, such as $0 = 0$, the system is dependent and has an infinite number of solutions.

Now proceed as in the previous examples. Begin by multiplying *eq.* 1 by 2.

$$2\left(x - \frac{1}{2}y\right) = 2 \cdot 2 \xrightarrow{\textit{eq. 1 multiplied by 2}} 2x - y = 4$$
$$-2x + y = -4 \xrightarrow{\textit{eq. 2 stays the same}} -2x + y = -4$$

$$\begin{array}{r} 2x - y = 4 \\ -2x + y = -4 \\ \hline 0 = 0 \quad \text{True} \end{array}$$

Since $0 = 0$ is a true statement, the system is dependent and has an infinite number of solutions. When graphed, both equations will be the same line.

*Now Try Exercise 21*

*The solution in Example 7 is the same as the solutions obtained by graphing in Example 5 of Section 9.1 (see* **Fig. 9.4** *on page 522) and by substitution in Example 3 of Section 9.2 on page 528.*

**EXAMPLE 8** Solve the following system of equations using the addition method.

$$2x + 3y = 7 \qquad eq. 1$$
$$5x - 7y = -3 \qquad eq. 2$$

*Solution*   We can eliminate the variable $x$ by multiplying *eq.* 1 by $-5$ and *eq.* 2 by 2.

$$-5(2x + 3y) = -5 \cdot 7 \xrightarrow{\ eq.\ 1\ multiplied\ by\ -5\ } -10x - 15y = -35$$
$$2(5x - 7y) = 2(-3) \xrightarrow{\ eq.\ 2\ multiplied\ by\ 2\ } 10x - 14y = -6$$

$$
\begin{array}{r}
-10x - 15y = -35 \\
10x - 14y = -6 \\
\hline
-29y = -41 \\
y = \dfrac{41}{29}
\end{array}
$$

We can now find $x$ by substituting $y = \dfrac{41}{29}$ into one of the original equations and solving for $x$. An easier method of solving for $x$ is to go back to the original equations and eliminate the variable $y$. We can do this by multiplying *eq.* 1 by 7 and *eq.* 2 by 3.

$$7(2x + 3y) = 7 \cdot 7 \quad \text{or} \quad 14x + 21y = 49$$
$$3(5x - 7y) = 3(-3) \quad \text{or} \quad 15x - 21y = -9$$

$$
\begin{array}{r}
14x + 21y = 49 \\
15x - 21y = -9 \\
\hline
29x \qquad\quad = 40 \\
x = \dfrac{40}{29}
\end{array}
$$

The solution is $\left( \dfrac{40}{29}, \dfrac{41}{29} \right)$.

Now Try Exercise 33

---

**Helpful Hint**

In Sections 9.1, 9.2, and 9.3 we have introduced three different methods for solving linear systems of equations in two variables. Which method should you use? The table below can help you decide.

| Method | Comments |
|---|---|
| Graphing | • Graphing is used when you wish to see a visual representation of the system of equations and the solution.<br>• Graphing may not be the best method to use when an exact answer is needed—especially if the solution includes fractions. |
| Substitution | • Substitution will give you an exact answer.<br>• Substitution works well if at least one of the variable terms in at least one of the equations has a coefficient of 1 or −1. |
| Addition (or Elimination) | • Addition will give you an exact answer.<br>• Addition works well when neither of the variable terms in either equation has a coefficient of 1 or −1. |

# EXERCISE SET 9.3  MathXL® MyMathLab®

## Warm-Up Exercises

*Fill in the blanks with the appropriate word, phrase, or symbol(s) from the following list.*

| factor | eliminate | unique solution | an infinite number of solutions |
|---|---|---|---|
| point of intersection | no solution | constant | |

1. When solving a system of equations using the addition method, our goal is to _____ one of the variables when we add the two equations together.

2. The solution to a system of linear equations occurs at the _____ of the graphs of the equations in the system.

3. If, when solving a system of equations by the addition method, you obtain an equation that is always true, the system is dependent and has _____ .

4. If, when solving a system of equations by the addition method, you obtain an equation that is always false, the system is inconsistent and has _____ .

## Practice the Skills

*Solve each system of equations using the addition method.*

5. $x + y = 6$
   $x - y = 4$

6. $x - y = 3$
   $x + y = 5$

7. $x - 3y = 1$
   $x + 3y = -5$

8. $5x + y = 14$
   $4x - y = 4$

9. $x + 2y = 21$
   $2x - 2y = -6$

10. $x - y = 3$
    $x + y = -3$

11. $4x + y = 6$
    $-8x - 2y = 20$

12. $x - y = 2$
    $3x - 3y = 1$

13. $-5x + y = 14$
    $-3x + y = -2$

14. $6x + 3y = 30$
    $2x + 3y = 18$

15. $2x + y = -6$
    $2x - 2y = 3$

16. $-4x + 3y = 0$
    $7x - 6y = 3$

17. $2y = 6x + 16$
    $y = -3x - 4$

18. $2x - 3y = 4$
    $2x + y = -4$

19. $5x + 3y = 12$
    $3x - 6y = 15$

20. $8x - 4y = 12$
    $2x - 8y = 3$

21. $-2y = -4x + 12$
    $y = 2x - 6$

22. $4x - 2y = 12$
    $4y = 8x - 24$

23. $5x - 4y = -3$
    $7y = 2x + 12$

24. $2x + 3y = -3$
    $-3x - 5y = 7$

25. $5x - 4y = 1$
    $-10x + 8y = -3$

26. $2x - 3y = 11$
    $5y = 3x - 17$

27. $5x - 6y = 0$
    $3x + 4y = 0$

28. $4x - 2y = -4$
    $5x - 6y = -26$

29. $-5x + 4y = -20$
    $3x - 2y = 15$

30. $4x - 3y = -4$
    $3x - 5y = 10$

31. $6x = 4y + 12$
    $3y - 5x = -6$

32. $5x = 2y - 4$
    $3x - 5y = 6$

33. $4x + 5y = 0$
    $3x = 6y + 4$

34. $4x - 3y = 8$
    $-3x + 4y = 9$

35. $x - \dfrac{1}{2}y = 4$
    $3x + y = 6$

36. $2x - \dfrac{1}{3}y = 6$
    $3x - \dfrac{5}{6}y = -13$

37. $3x - y = 4$
    $2x - \dfrac{2}{3}y = 8$

38. $-5x + 6y = -12$
    $\dfrac{5}{3}x - 4 = 2y$

## Problem Solving

39. **Sum of Numbers** The sum of two numbers is 60. When the second number is subtracted from the first number, the difference is 38. Find the two numbers.

40. **Sum of Numbers** The sum of two numbers is 46. When the first number is subtracted from the second number, the difference is 6. Find the two numbers.

41. **Sum of Numbers** The sum of a number and twice a second number is 14. When the second number is subtracted from the first number, the difference is 2. Find the two numbers.

42. **Sum of Numbers** The sum of two numbers is 9. Twice the first number subtracted from three times the second number is 7. Find the two numbers.

**43. Rectangles** When the length of a rectangle is $x$ inches and the width is $y$ inches, the perimeter is 18 inches. If the length is doubled and the width is tripled, the perimeter becomes 42 inches. Find the length and width of the original rectangle.

**44. Perimeter of a Rectangle** When the length of a rectangle is $x$ inches and the width is $y$ inches, the perimeter is 28 inches. If the length is doubled and the width is tripled, the perimeter becomes 66 inches. Find the length and width of the original rectangle.

**45. Photograph** A photograph has a perimeter of 36 inches. The difference between the photograph's length and width is 2 inches. Find the length and width of the photograph.

**46. Rectangular Garden** John has a large rectangular garden with a perimeter of 82 feet. The difference between the garden's length and width is 11 feet. Determine the length and width of his garden.

## Concept/Writing Exercises

**47.** Construct a system of two equations that has no solution. Explain how you know the system has no solution.

**48.** Construct a system of two equations that has an infinite number of solutions. Explain how you know the system has an infinite number of solutions.

**49. a)** Solve the system of equations

$$4x + 2y = 1000$$
$$2x + 4y = 800$$

**b)** If we divide all the terms in the top equation by 2, we get the system shown at right:

$$2x + y = 500$$
$$2x + 4y = 800$$

How will the solution to this system compare to the solution in part **a)**? Explain and then check your explanation by solving this system.

**50.** Suppose we divided all the terms in both equations given in Exercise 49 **a)** by 2, and then solved the system. How will the solution to this system compare to the solution in part **a)**? Explain and then check your explanation by solving each system.

## Challenge Problems

*In Exercises 51 and 52, solve each system of equations using the addition method. (Hint: First remove all fractions by multiplying both sides of the equation by the LCD.)*

**51.** 
$$\frac{x+2}{2} - \frac{y+4}{3} = 4$$
$$\frac{x+y}{2} = \frac{1}{2} + \frac{x-y}{3}$$

**52.** 
$$\frac{5}{2}x + 3y = \frac{9}{2} + y$$
$$\frac{1}{4}x - \frac{1}{2}y = 6x + 12$$

*In intermediate algebra you may solve systems of three equations with three unknowns. Solve the following system.*

**53.**
$$x + 2y - z = 2$$
$$2x - y + z = 3$$
$$3x + y + z = 8$$

*Hint:* Work with *one pair* of equations to get one equation in two unknowns. Then work with *a different pair* of the original equations to get another equation in the same two unknowns. Then solve the system of two equations in two unknowns. List your answer as an *ordered triple* of the form $(x, y, z)$.

## Group Activity

*Work parts* **a)** *and* **b)** *of Exercise 54 on your own. Then discuss and work parts* **c)** *and* **d)** *as a group.*

**54.** How difficult is it to construct a system of linear equations that has a specific solution? It is really not too difficult to do. Consider:

$$2(3) + 4(5) = 26$$
$$4(3) - 7(5) = -23$$

The system of equations

$$2x + 4y = 26$$
$$4x - 7y = -23$$

has solution $(3, 5)$.

**a)** Using the information provided, determine another system of equations that has $(3, 5)$ as a solution.

**b)** Determine a system of linear equations that has $(2, 3)$ as a solution.

**c)** Compare your answer with the answers of the other members of your group.

**d)** As a group, determine the number of systems of equations that have $(2, 3)$ as a solution.

## Cumulative Review Exercises

[1.9]  **55.** Evaluate $5^3$.

[2.5]  **56.** Solve the equation $2(2x - 3) = 2x + 8$.

[5.4]  **57.** Simplify $(4x^2y - 3xy + y) - (2x^2y + 6xy - 3y)$.

[5.5]  **58.** Multiply $(8a^4b^2c)(4a^2b^7c^4)$.

[6.2]  **59.** Factor $xy + xc - ay - ac$ by grouping.

[8.2]  **60.** If $f(x) = 2x^2 - 4$, find $f(-3)$.

# 9.4 Solving Systems of Linear Equations in Three Variables

1. Solve systems of linear equations in three variables.

2. Learn the geometric interpretation of a system of linear equations in three variables.

3. Recognize inconsistent and dependent systems.

## 1 Solve Systems of Linear Equations in Three Variables

The equation $2x - 3y + 4z = 8$ is an example of a **linear equation in three variables**. The solution to a linear equation in three variables is an *ordered triple* of the form $(x, y, z)$. One solution to the equation given is $(1, 2, 3)$. Check now to verify that $(1, 2, 3)$ is a solution to the equation.

To solve systems of linear equations with three variables, we can use either substitution or the addition method, as discussed in Sections 9.2 and 9.3.

**EXAMPLE 1** Solve the following system by substitution.

$$x = -3$$
$$3x + 4y = 7$$
$$-2x - 3y + 5z = 19$$

*Solution*   Since we know that $x = -3$, we substitute $-3$ for $x$ in the equation $3x + 4y = 7$ and solve for $y$.

$$3x + 4y = 7$$
$$3(-3) + 4y = 7$$
$$-9 + 4y = 7$$
$$4y = 16$$
$$y = 4$$

Now we substitute $x = -3$ and $y = 4$ into the last equation and solve for $z$.

$$-2x - 3y + 5z = 19$$
$$-2(-3) - 3(4) + 5z = 19$$
$$6 - 12 + 5z = 19$$
$$-6 + 5z = 19$$
$$5z = 25$$
$$z = 5$$

Check   $x = -3, y = 4, z = 5$. The solution must be checked in *all three* original equations.

| $x = -3$ | $3x + 4y = 7$ | $-2x - 3y + 5z = 19$ |
|---|---|---|
| $-3 = -3$   True | $3(-3) + 4(4) \overset{?}{=} 7$ | $-2(-3) - 3(4) + 5(5) \overset{?}{=} 19$ |
| | $7 = 7$   True | $19 = 19$   True |

The solution is the ordered triple $(-3, 4, 5)$. Remember that the ordered triple lists the $x$-value first, the $y$-value second, and the $z$-value third.

Now Try Exercise 5

Not every system of linear equations in three variables can be solved easily by substitution. We can also find the solution by the addition method, as illustrated in Example 2.

**EXAMPLE 2** Solve the following system of equations using the addition method.

$$3x + 2y + z = 4 \quad (eq.\ 1)$$
$$2x - 3y + 2z = -7 \quad (eq.\ 2)$$
$$x + 4y - z = 10 \quad (eq.\ 3)$$

*Solution*    Our first step is to use the addition method to write two new equations where each has the same two variables. We do this by eliminating one of the three variables $x$, $y$, or $z$. Here we will eliminate the variable $z$. Use (*eq.* 1) and (*eq.* 3) to eliminate $z$ and call this new equation (*eq.* 4).

$$
\begin{array}{ll}
3x + 2y + z = 4 & (eq.\ 1) \\
\underline{x + 4y - z = 10} & (eq.\ 3) \\
4x + 6y \quad\ \ = 14 & \text{Sum of equations, } (eq.\ 4)
\end{array}
$$

Now, use (*eq.* 1) and (*eq.* 2) to eliminate $z$ and call this new equation (*eq.* 5). Notice that we multiplied (*eq.* 1) by $-2$ so that after adding (*eq.* 1) and (*eq.* 2), $z$ would be eliminated from (*eq.* 5).

$$
\begin{array}{ll}
-6x - 4y - 2z = -8 & (eq.\ 1) \text{ Multiplied by } -2 \\
\underline{2x - 3y + 2z = -7} & (eq.\ 2) \\
-4x - 7y \qquad = -15 & \text{Sum of equations, } (eq.\ 5)
\end{array}
$$

We now have a system of two equations with two variables, (*eq.* 4), and (*eq.* 5). We can solve this system using the addition method.

$$
\begin{array}{ll}
4x + 6y = 14 & (eq.\ 4) \\
\underline{-4x - 7y = -15} & (eq.\ 5) \\
-y = -1 & \text{Sum of equations} \\
y = 1
\end{array}
$$

Next we substitute $y = 1$ into either one of the two equations containing only two variables [(*eq.* 4) or (*eq.* 5)] and solve for $x$.

$$
\begin{array}{ll}
4x + 6y = 14 & (eq.\ 4) \\
4x + 6(1) = 14 & \text{Substitute 1 for } y \text{ in } (eq.\ 4). \\
4x + 6 = 14 \\
4x = 8 \\
x = 2
\end{array}
$$

Finally, we substitute $x = 2$ and $y = 1$ into any of the original equations and solve for $z$.

$$
\begin{array}{ll}
3x + 2y + z = 4 & (eq.\ 1) \\
3(2) + 2(1) + z = 4 & \text{Substitute 2 for } x \text{ and 1 for } y \text{ in } (eq.\ 1). \\
6 + 2 + z = 4 \\
8 + z = 4 \\
z = -4
\end{array}
$$

The solution is the ordered triple $(2, 1, -4)$. Check this solution in *all three* original equations.

Now Try Exercise 17

EXAMPLE 3    Solve the following system of equations.

$$
\begin{array}{ll}
2x - 3y + 2z = -1 & (eq.\ 1) \\
x + 2y \qquad\ = 14 & (eq.\ 2) \\
x \qquad\ - 5z = -11 & (eq.\ 3)
\end{array}
$$

*Solution*   The third equation only has variables $x$ and $z$. Therefore, use the addition method to write a new equation that also only has variables $x$ and $z$. We use (*eq.* 1) and (*eq.* 2). We will multiply (*eq.* 1) by 2 and (*eq.* 2) by 3 so that we can eliminate the variable $y$ when we add the equations. We will call this new equation (*eq.* 4).

$$
\begin{array}{ll}
4x - 6y + 4z = -2 & (eq.\,1)\text{ Multiplied by 2} \\
\underline{3x + 6y \qquad\;\; = \;\; 42} & (eq.\,2)\text{ Multiplied by 3} \\
7x \qquad\;\; + 4z = \;\; 40 & \text{Sum of equations, } (eq.\,4)
\end{array}
$$

We now have two equations containing only the variables $x$ and $z$.

$$
\begin{array}{ll}
7x + 4z = \;\;\;\, 40 & (eq.\,4) \\
x - 5z = -11 & (eq.\,3)
\end{array}
$$

Let's now eliminate the variable $x$.

$$
\begin{array}{ll}
7x + \;\;\, 4z = \;\;\, 40 & (eq.\,4) \\
\underline{-7x + 35z = \;\;\, 77} & (eq.\,3)\;\; \text{Multiplied by } -7 \\
39z = 117 & \text{Sum of equations} \\
z = 3 &
\end{array}
$$

Now we solve for $x$ by using one of the equations containing only the variables $x$ and $z$. We substitute 3 for $z$ in (*eq.* 3).

$$
\begin{array}{ll}
x - 5z = -11 & (eq.\,3) \\
x - 5(3) = -11 & \text{Substitute 3 for } z \text{ in } (eq.\,3). \\
x - 15 = -11 & \\
x = 4 &
\end{array}
$$

Finally, we solve for $y$ using any of the original equations that contains $y$.

$$
\begin{array}{ll}
x + 2y = 14 & (eq.\,2) \\
4 + 2y = 14 & \text{Substitute 4 for } x \text{ in } (eq.\,2). \\
2y = 10 & \\
y = 5 &
\end{array}
$$

The solution is the ordered triple $(4, 5, 3)$.

Check    (eq. 1)

$$
\begin{array}{l}
2x - 3y + 2z = -1 \\
2(4) - 3(5) + 2(3) \overset{?}{=} -1 \\
8 - 15 + 6 \overset{?}{=} -1 \\
-1 = -1 \\
\qquad\qquad \text{True}
\end{array}
$$

(eq. 2)

$$
\begin{array}{l}
x + 2y = 14 \\
4 + 2(5) \overset{?}{=} 14 \\
4 + 10 \overset{?}{=} 14 \\
14 = 14 \\
\text{True}
\end{array}
$$

(eq. 3)

$$
\begin{array}{l}
x - 5z = -11 \\
4 - 5(3) \overset{?}{=} -11 \\
4 - 15 \overset{?}{=} -11 \\
-11 = -11 \\
\text{True}
\end{array}
$$

Now Try Exercise 13

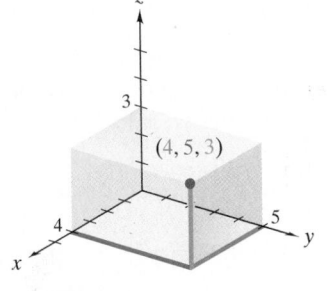

FIGURE 9.9

## 2 Learn the Geometric Interpretation of a System of Linear Equations in Three Variables

The solution to a system of linear equations in *two* variables occurs at the intersection point of the two *lines* that represent the equations in the system. The lines are drawn in the 2-dimensional Cartesian coordinate system.

The solution to a system of linear equations in *three* variables occurs at the intersection point of three *planes* that represent the equations in the system. (See Exercise 47 on page 547.) The planes are drawn in a 3-dimensional coordinate system in which an $x$-axis, a $y$-axis, and $z$-axis are all perpendicular to each other (see **Figure 9.9**.)

In Example 3, the solution to the system was the ordered triple $(4, 5, 3)$. The point represented by this ordered triple is shown in **Figure 9.9**. Graphing planes in a 3-dimensional coordinate system is beyond the scope of this book.

## 3 Recognize Inconsistent and Dependent Systems

Recall that *an inconsistent system of equations is one that has no solution.* If, while solving a system of linear equations in three variables, you obtain a statement that is always false, such as $3 = 0$, the system is inconsistent and has no solution. In such a system, at least two of the planes that represent the equations are parallel and therefore a single point of intersection of the three planes does not exist. (See Exercises 45 and 46 on page 546).

Recall that *a dependent system of equations is one that has an infinite number of solutions.* If, while solving a system of linear equations in three variables, you obtain a statement that is always true, such as $0 = 0$, the system is dependent and has an infinite number of solutions. In such a system, either the planes that represent the equations are all the same plane or all three planes intersect at a line. (See Exercise 48 on page 547).

**EXAMPLE 4** Solve the following system of equations.

$$\begin{aligned} -3x + 5y + z &= -3 \quad (eq.\,1) \\ 6x - 10y - 2z &= 1 \quad (eq.\,2) \\ 7x - 4y + 11z &= -6 \quad (eq.\,3) \end{aligned}$$

**Solution**    We will begin by eliminating the variable $x$ from ($eq.\,1$) and ($eq.\,2$).

$$\begin{array}{ll} -6x + 10y + 2z = -6 & (eq.\,1) \quad \text{Multiplied by 2} \\ \underline{6x - 10y - 2z = \phantom{-}1} & (eq.\,2) \\ 0 = -5 & \text{False} \end{array}$$

Since we obtained the false statement $0 = -5$, this system is inconsistent and has no solution.

Now Try Exercise 33

**EXAMPLE 5** Solve the following system of equations.

$$\begin{aligned} x - y + z &= 1 \quad (eq.\,1) \\ x + 2y - z &= 1 \quad (eq.\,2) \\ x - 4y + 3z &= 1 \quad (eq.\,3) \end{aligned}$$

**Solution**    We will begin by eliminating the variable $x$ from ($eq.\,1$) and ($eq.\,2$) and then from ($eq.\,1$) and ($eq.\,3$).

$$\begin{array}{ll} -x + y - z = -1 & (eq.\,1) \quad \text{Multiplied by } -1 \\ \underline{x + 2y - z = \phantom{-}1} & (eq.\,2) \\ 3y - 2z = \phantom{-}0 & \text{Sum of equations, } (eq.\,4) \end{array}$$

$$\begin{array}{ll} x - y + z = \phantom{-}1 & (eq.\,1) \\ \underline{-x + 4y - 3z = -1} & (eq.\,3) \quad \text{Multiplied by } -1 \\ 3y - 2z = \phantom{-}0 & \text{Sum of equations, } (eq.\,5) \end{array}$$

Now we eliminate the variable $y$ using ($eq.\,4$) and ($eq.\,5$).

$$\begin{array}{ll} -3y + 2z = 0 & (eq.\,4) \quad \text{Multiplied by } -1 \\ \underline{3y - 2z = 0} & (eq.\,5) \\ 0 = 0 & \text{True} \end{array}$$

Since we obtained the true statement $0 = 0$, this system is dependent and has an infinite number of solutions.

Recall from Section 9.1 that systems of equations that are dependent are also consistent since they have a solution.

Now Try Exercise 35

> ## Helpful Hint
>
> If an equation in a system contains fractions, eliminate the fractions by multiplying each term in the equation by the least common denominator. Then continue to solve the system. If, for example, one equation in the system is $\frac{3}{4}x - \frac{5}{8}y + z = \frac{1}{2}$, multiply both sides of the equation by 8 to obtain the equivalent equation $6x - 5y + 8z = 4$.

# EXERCISE SET 9.4

MathXL®   MyMathLab®

## Warm-Up Exercises

*Fill in the blanks with the appropriate word, phrase, or symbol(s) from the following list.*

ordered pair          ordered triple          plane          line          dependent          inconsistent

**1.** The solution to a system of linear equations in three variables is an _____ .

**2.** A linear equation in three variables is represented by a _____ in a 3-dimensional coordinate system.

**3.** If, while solving a system of linear equations in three variables, you obtain a statement that is always false, the system is _____ and has no solution.

**4.** If, while solving a system of linear equations in three variables, you obtain a statement that is always true, the system is _____ and has an infinite number of solutions.

## Practice the Skills

*Solve by substitution.*

**5.** $x = 2$
$2x - y = 4$
$3x + 2y - 2z = 4$

**6.** $-x + 3y - 5z = -7$
$2y - z = -1$
$z = 3$

**7.** $5x - 6z = -17$
$3x - 4y + 5z = -1$
$2z = -6$

**8.** $2x - 5y = 12$
$-3y = -9$
$2x - 3y + 4z = 8$

**9.** $x + 2y = 6$
$3y = 9$
$x + 2z = 12$

**10.** $x - y + 5z = -4$
$3x - 2z = 6$
$4z = 2$

*Solve using the addition method.*

**11.** $x - 2y = -1$
$3x + 2y = 5$
$2x - 4y + z = -1$

**12.** $x - y + 2z = 1$
$y - 4z = 2$
$-2x + 2y - 5z = 2$

**13.** $2y + 4z = 2$
$x + y + 2z = -2$
$2x + y + z = 2$

**14.** $2x + y - 8 = 0$
$3x - 4z = -3$
$2x - 3z = 1$

**15.** $3p + 2q = 11$
$4q - r = 6$
$6p + 7r = 4$

**16.** $3s + 5t = -12$
$2t - 2u = 2$
$-s + 6u = -2$

**17.** $p + q + r = 4$
$p - 2q - r = 1$
$2p - q - 2r = -1$

**18.** $x - 2y + 3z = -7$
$2x - y - z = 7$
$-4x + 3y + 2z = -14$

**19.** $2x - 2y + 3z = 5$
$2x + y - 2z = -1$
$4x - y - 3z = 0$

**20.** $2x - y - 2z = 3$
$x - 3y - 4z = 2$
$x + y + 2z = -1$

**21.** $r - 2s + t = 2$
$2r + 3s - t = -3$
$2r - s - 2t = 1$

**22.** $3a - 3b + 4c = -1$
$a - 2b + 2c = 2$
$2a - 2b - c = 3$

**23.** $2a + 2b - c = 2$
$3a + 4b + c = -4$
$5a - 2b - 3c = 5$

**24.** $x - 2y + 2z = 3$
$2x - 3y + 2z = 5$
$x + y + 6z = -2$

**25.** $-x + 3y + z = 0$
$-2x + 4y - z = 0$
$3x - y + 2z = 0$

**26.** $x + y + z = 0$
$-x - y + z = 0$
$-x + y + z = 0$

**27.** $-\dfrac{1}{4}x + \dfrac{1}{2}y - \dfrac{1}{2}z = -2$
$\dfrac{1}{2}x + \dfrac{1}{3}y - \dfrac{1}{4}z = 2$
$\dfrac{1}{2}x - \dfrac{1}{2}y + \dfrac{1}{4}z = 1$

**28.** $\dfrac{2}{3}x + y - \dfrac{1}{3}z = \dfrac{1}{3}$
$\dfrac{1}{2}x + y + z = \dfrac{5}{2}$
$\dfrac{1}{4}x - \dfrac{1}{4}y + \dfrac{1}{4}z = \dfrac{3}{2}$

**29.** $x - \dfrac{2}{3}y - \dfrac{2}{3}z = -2$
$\dfrac{2}{3}x + y - \dfrac{2}{3}z = \dfrac{1}{3}$
$-\dfrac{1}{4}x + y - \dfrac{1}{4}z = \dfrac{3}{4}$

**30.** $\dfrac{1}{8}x + \dfrac{1}{4}y + z = 2$
$\dfrac{1}{3}x + \dfrac{1}{4}y + z = \dfrac{17}{6}$
$-\dfrac{1}{4}x + \dfrac{1}{3}y - \dfrac{1}{2}z = -\dfrac{5}{6}$

**31.** $0.2x + 0.3y + 0.3z = 1.1$
$0.4x - 0.2y + 0.1z = 0.4$
$-0.1x - 0.1y + 0.3z = 0.4$

**32.** $0.6x - 0.4y + 0.2z = 2.2$
$-0.1x - 0.2y + 0.3z = 0.9$
$-0.2x - 0.1y - 0.3z = -1.2$

*Determine whether the following systems are inconsistent, dependent, or neither.*

**33.** $2x + y + 2z = 1$
$x - 2y - z = 0$
$3x - y + z = 2$

**34.** $2p - 4q + 6r = 5$
$-p + 2q - 3r = -3$
$3p + 4q + 5r = 8$

**35.** $x - 4y - 3z = -1$
$-3x + 12y + 9z = 3$
$2x - 10y - 7z = 5$

**36.** $5a - 4b + 2c = 5$
$-10a + 8b - 4c = -10$
$-7a - 4b + c = 7$

**37.** $x + 3y + 2z = 6$
$x - 2y - z = 8$
$-3x - 9y - 6z = -7$

**38.** $2x - 2y + 4z = 2$
$-3x + y = -9$
$2x - y + z = 5$

## Problem Solving

**39.** Three solutions to the equation $Ax + By + Cz = 1$ are $(-1, 2, -1)$, $(-1, 1, 2)$, and $(1, -2, 2)$. Determine the values of $A$, $B$, and $C$ and write the equation using the numerical values found.

**40.** Three solutions to the equation $Ax + By + Cz = 14$ are $(3, -1, 2)$, $(2, -2, 1)$, and $(-5, 3, -24)$. Find the values of $A$, $B$, and $C$ and write the equation using the numerical values found.

*In Exercises 41 and 42, write a system of linear equations in three variables that has the given solution.*

**41.** $(3, 1, 6)$

**42.** $(-2, 5, 3)$

**43. a)** Find the values of $a$, $b$, and $c$ such that the points $(1, -1), (-1, -5)$, and $(3, 11)$ lie on the graph of $y = ax^2 + bx + c$.

**b)** Find the quadratic equation whose graph passes through the three points indicated. Explain how you determined your answer.

**44. a)** Find the values of $a$, $b$, and $c$ such that the points $(1, 7)$, $(-2, -5)$, and $(3, 5)$ lie on the graph of $y = ax^2 + bx + c$.

**b)** Find the quadratic equation whose graph passes through the three points indicated. Explain how you determined your answer.

## Concept/Writing Exercises

*An equation in three variables represents a plane. Consider a system of equations consisting of three equations in three variables. Answer the following questions.*

**45.** If the three planes are parallel to one another as illustrated in the figure, how many points will be common to all three planes? Is the system consistent or inconsistent? Explain your answer.

**46.** If two of the planes are parallel to each other and the third plane intersects each of the other two planes, how many points will be common to all three planes? Is the system consistent or inconsistent? Explain your answer.

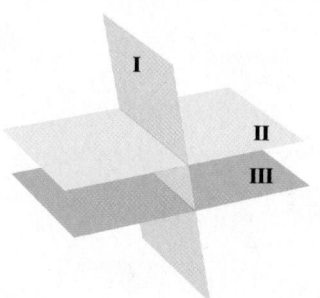

**47.** If the three planes are as illustrated in the figure, how many points will be common to all three planes? Is the system consistent or inconsistent? Explain your answer.

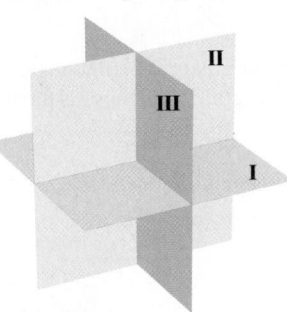

**48.** If the three planes are as illustrated in the figure, how many points will be common to all three planes? Is the system dependent? Explain your answer.

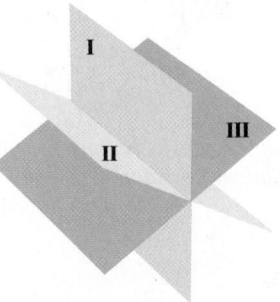

**49.** Is it possible for a system of linear equations in three variables to have exactly

   **a)** no solution,

   **b)** one solution,

   **c)** two solutions? Explain your answer.

**50.** In a system of linear equations in three variables, if the graphs of two equations are parallel planes, is it possible for the system to be

   **a)** consistent,          **b)** dependent,

   **c)** inconsistent? Explain your answer.

## Challenge Problems

*Find the solution to the following systems of equations.*

**51.** $3p + 4q = 11$

   $2p + r + s = 9$

   $q - s = -2$

   $p + 2q - r = 2$

**52.** $3a + 2b - c = 0$

   $2a + 2c + d = 5$

   $a + 2b - d = -2$

   $2a - b + c + d = 2$

## Cumulative Review Exercises

[1.10] **53.** Name the properties illustrated.

   **a)** $x + 4 = 4 + x$

   **b)** $(3x)y = 3(xy)$

   **c)** $4(x + 2) = 4x + 8$

[2.5] **54.** Solve $3x + 4 = -(x - 6)$.

[3.2] **55. Cross-Country Skiing** Margie Steiner begins skiing along a trail at 3 miles per hour. Ten minutes $\left(\frac{1}{6} \text{ hour}\right)$ later, her husband, David, begins skiing along the same trail at 5 miles per hour.

   **a)** How long after David leaves will he catch up to Margie?

   **b)** How far from the starting point will they be when they meet?

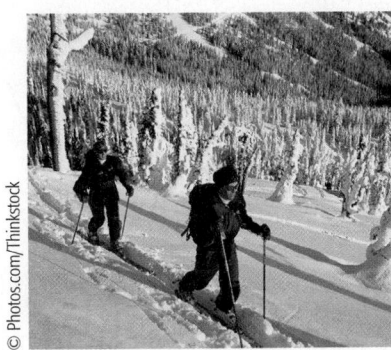

© Photos.com/Thinkstock

[3.3] **56. Perimeter** The perimeter of a rectangle is 22 feet. Find the dimensions of the rectangle if the length is two feet more than twice the width.

## Mid-Chapter Test: 9.1–9.4

*To find out how well you understand the chapter material to this point, take this brief test. The answers, and the section where the material was initially discussed, are given in the back of the book. Review any questions that you answered incorrectly.*

*Determine which of the following ordered pairs satisfy each system of equations.*

**1.** $4x + 3y = -1$

   $x - 2y = 8$

   **a)** $(-1, 1)$

   **b)** $(2, -3)$

**2.** $6x - y = -2$

$7x + \dfrac{1}{2}y = 6$

**a)** $\left(\dfrac{1}{2}, 5\right)$   **b)** $\left(\dfrac{1}{3}, 4\right)$

*Without graphing, state whether the system of equations has exactly one solution, no solution, or an infinite number of solutions.*

**3.** $2x + y = 8$

$3x - 4y = 1$

**4.** $\dfrac{1}{2}x - 3y = 5$

$-2x + 12y = -20$

**5.** $y = \dfrac{3}{2}x + \dfrac{5}{2}$

$3x - 2y = 7$

*Determine the solution to each system of equations graphically. If the system is dependent or inconsistent, so state.*

**6.** $y = 2x + 1$

$y = -x + 4$

**7.** $x = 5$

$y = -3$

*Solve each system of equations by substitution.*

**8.** $3x + y = -2$

$2x - 3y = -16$

**9.** $x - 3y = 2$

$4x + 9y = 1$

**10.** $3x - y = 5$

$x - \dfrac{1}{3}y = 2$

**11. Rectangle** The perimeter of a rectangle is 44 feet. Find the length and the width if the length is 8 feet greater than the width.

*Solve each system of equations using the addition method.*

**12.** $x + 3y = 1$

$2x - 3y = 11$

**13.** $4x + 3y = 4$

$-8x + 5y = 14$

**14.** $5x - 2y = 1$

$-10x + 4y = -2$

**15.** In solving the system of equations

$$3x - 5y = -16$$
$$2x + 3y = 21,$$

Hugo Platt stated that the solution was $x = 3$. This is incorrect. Why? Explain your answer. Give the correct solution to the system.

*Solve each system of equations.*

**16.** $x + y + z = 2$

$2x - y + 2z = -2$

$3x + 2y + 6z = 1$

**17.** $2x - y - z = 1$

$3x + 5y + 2z = 12$

$-6x - 4y + 5z = 3$

# 9.5 Systems of Linear Equations: Applications and Problem Solving

**1** Use systems of equations to solve applications.

**2** Use linear systems in three variables to solve applications.

## **1** Use Systems of Equations to Solve Applications

Many of the applications solved in earlier chapters using only one variable can now be solved using two variables. Whenever we use two variables to solve an application problem, we must write a system of two equations.

**EXAMPLE 1 Land Area of the Carolinas** The combined land area of North Carolina and South Carolina is 78,820 square miles. The difference between the two states' land areas is 18,602 square miles. If North Carolina has the larger land area, find the land area of each state.

*Solution* **Understand** We need to determine the land area of North Carolina and the land area of South Carolina. We will use two variables and therefore will need to determine two equations.

### Understanding Algebra

Whenever we use *two* variables to solve an application problem, we must determine *two* equations to solve the problem.

Translate          Let $N$ = the land area of North Carolina.
                   Let $S$ = the land area of South Carolina.

Since the total area of the two states is 78,820 square miles, the first equation is

$$N + S = 78{,}820$$

Since North Carolina has a larger land area and since the difference in the land area is 18,602 square miles, the second equation is

$$N - S = 18{,}602$$

The system of two equations is

$$N + S = 78{,}820 \quad (eq.\,1)$$
$$N - S = 18{,}602 \quad (eq.\,2)$$

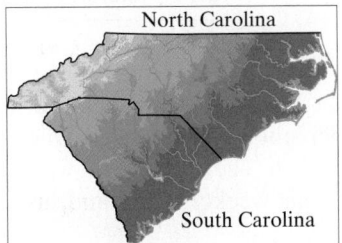

Carry Out    We will use the addition method to solve this system of equations.

$$
\begin{aligned}
N + S &= 78{,}820 \\
\underline{N - S} &= \underline{18{,}602} \\
2N \phantom{{}+S} &= 97{,}422 \quad \text{Sum of the two equations} \\
N &= 48{,}711 \quad \text{Divided both sides by 2.}
\end{aligned}
$$

Thus, $N = 48{,}711$. To determine the value for $S$, substitute 48,711 into $(eq.\,1)$.

$$
\begin{aligned}
N + S &= 78{,}820 \\
48{,}711 + S &= 78{,}820 \\
S &= 30{,}109 \quad \text{Subtracted 48,711 from both sides.}
\end{aligned}
$$

Answer    The land area of North Carolina is 48,711 square miles and the land area of South Carolina is 30,109 square miles.

NowTry Exercise 1

EXAMPLE 2 **Canoe Speed** The Burnhams are canoeing on the Suwannee River. They travel at an average speed of 4.75 miles per hour when paddling with the current and 2.25 miles per hour when paddling against the current. Determine the speed of the canoe in still water and the speed of the current.

*Solution*    Understand    When they are traveling with the current, the canoe's speed is the canoe's speed in still water *plus* the current's speed. When traveling against the current, the canoe's speed is the canoe's speed in still water *minus* the current's speed.

Translate          Let $s$ = speed of the canoe in still water.
                   Let $c$ = speed of the current.

The system of equations is:

speed of the canoe traveling with the current:    $s + c = 4.75$
speed of the canoe traveling against the current:    $s - c = 2.25$

Carry Out    We will use the addition method, as we discussed in Section 9.3, to solve this system of equations.

$$
\begin{aligned}
s + c &= 4.75 \\
\underline{s - c} &= \underline{2.25} \\
2s \phantom{{}+c} &= 7.00 \\
s &= 3.5
\end{aligned}
$$

The speed of the canoe in still water is 3.5 miles per hour. We now determine the speed of the current.

$$s + c = 4.75$$
$$3.5 + c = 4.75$$
$$c = 1.25$$

**Answer**   The speed of the current is 1.25 miles per hour, and the speed of the canoe in still water is 3.5 miles per hour.

Now Try Exercise 13

**EXAMPLE 3  Salary** Yamil Bermudez, a salesman at Hancock Appliances, receives a weekly salary plus a commission, which is a percentage of his sales. One week, with sales of $3000, his total take-home pay was $850. The next week, with sales of $4000, his total take-home pay was $1000. Find his weekly salary and his commission rate.

*Solution*   Understand   Yamil's take-home pay consists of his weekly salary plus commission. We are given information about two specific weeks that we can use to find his weekly salary and his commission rate.

Translate                        Let $s$ = his weekly salary.

                        Let $r$ = his commission rate.

In week 1, his commission on $3000 is $3000r$, and in week 2 his commission on $4000 is $4000r$. Thus the system of equations is

salary + commission = take-home salary

| | |
|---|---|
| 1st week | $s + 3000r = 850$ |
| 2nd week | $s + 4000r = 1000$ |

System of equations

Carry Out

$$-s - 3000r = -850 \quad \text{1st week multiplied by } -1$$
$$\underline{s + 4000r = 1000} \quad \text{2nd week}$$
$$1000r = 150 \quad \text{Sum of equations}$$
$$r = \frac{150}{1000}$$
$$r = 0.15$$

Yamil's commission rate is 15%. Now we find his weekly salary by substituting 0.15 for $r$ in either equation.

$$s + 3000r = 850$$
$$s + 3000(0.15) = 850 \quad \text{Substitute 0.15 for } r \text{ in the 1st-week}$$
$$s + 450 = 850 \quad \text{equation.}$$
$$s = 400$$

**Answer**   Yamil's weekly salary is $400 and his commission rate is 15%.

Now Try Exercise 15

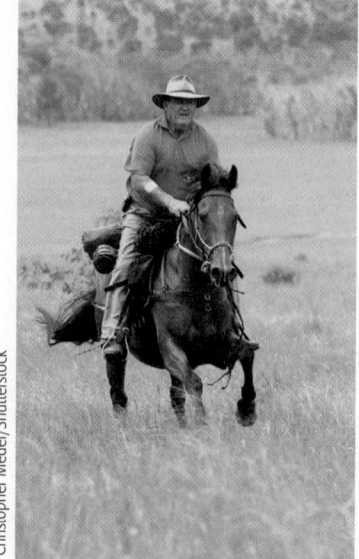

**EXAMPLE 4  Riding Horses** Ben Campbell leaves his ranch riding on his horse at 5 miles per hour. One-half hour later, Joe Campbell leaves the same ranch and heads along the same route on his horse at 8 miles per hour.

**a)** How long after Joe leaves the ranch will he catch up to Ben?

**b)** When Joe catches up to Ben, how far from the ranch will they be?

*Solution*

**a)** Understand   When Joe catches up to Ben, they both will have traveled the same distance. Joe will have traveled the distance in $\frac{1}{2}$ hour less time since he left $\frac{1}{2}$ hour after Ben. We will use the formula distance = rate · time to solve this problem.

**Translate**
$$\text{Let } b = \text{time traveled by Ben.}$$
$$\text{Let } j = \text{time traveled by Joe.}$$

We will set up a table to organize the given information.

|  | Rate | Time | Distance |
|---|---|---|---|
| Ben | 5 | $b$ | $5b$ |
| Joe | 8 | $j$ | $8j$ |

Since both Ben and Joe cover the same distance, we write

$$\text{Ben's distance} = \text{Joe's distance}$$
$$5b = 8j$$

Our second equation comes from the fact that Joe is traveling for $\frac{1}{2}$ hour less time than Ben. Therefore, $j = b - \frac{1}{2}$. Thus our system of equations is:

$$5b = 8j$$
$$j = b - \frac{1}{2}$$

**Carry Out** We will solve this system of equations using substitution. Since $j = b - \frac{1}{2}$, substitute $b - \frac{1}{2}$ for $j$ in the first equation and solve for $b$.

$$5b = 8j$$
$$5b = 8\left(b - \frac{1}{2}\right)$$
$$5b = 8b - 4$$
$$-3b = -4$$
$$b = \frac{-4}{-3} = 1\frac{1}{3}$$

Therefore, the time Ben has been traveling is $1\frac{1}{3}$ hours. To get the time Joe has been traveling, we will subtract $\frac{1}{2}$ hour from Ben's time.

$$j = b - \frac{1}{2}$$
$$j = 1\frac{1}{3} - \frac{1}{2}$$
$$j = \frac{4}{3} - \frac{1}{2} = \frac{8}{6} - \frac{3}{6} = \frac{5}{6}$$

**Answer** Joe will catch up to Ben $\frac{5}{6}$ of an hour (or 50 minutes) after Joe leaves the ranch.

**b)** We can use either Ben's or Joe's distance to determine the distance traveled from the ranch. We will use Joe's distance.

$$d = 8j = 8\left(\frac{5}{6}\right) = \frac{\overset{4}{\cancel{8}}}{1} \cdot \frac{5}{\underset{3}{\cancel{6}}} = \frac{20}{3} = 6\frac{2}{3}$$

**Answer** Thus, Joe will catch up to Ben when they are $6\frac{2}{3}$ miles from the ranch.

Now Try Exercise 33

EXAMPLE 5 **Mixing Solutions** Chung Song, a chemist with Johnson and Johnson, wishes to create a new household cleaner containing 30% trisodium phosphate (TSP). Chung needs to mix a 16% TSP solution with a 72% TSP solution to get 6 liters of a 30% TSP solution. How many liters of the 16% solution and of the 72% solution will he need to mix?

*Solution*    Understand    To solve this problem we use the fact that the amount of TSP in a solution is found by multiplying the percent strength of the solution by the number of liters (the volume) of the solution. Chung needs to mix a 16% solution and a 72% solution to obtain 6 liters of a solution whose strength, 30%, is between the strengths of the two solutions being mixed.

Translate         Let $x$ = number of liters of the 16% solution.
                  Let $y$ = number of liters of the 72% solution.

We will draw a sketch (**Fig. 9.10**) and then use a table to help analyze the problem.

FIGURE 9.10

| Solution | Strength of Solution | Number of Liters | Amount of TSP |
|----------|----------------------|------------------|---------------|
| 16% solution | 0.16 | $x$ | $0.16x$ |
| 72% solution | 0.72 | $y$ | $0.72y$ |
| Mixture | 0.30 | 6 | $0.30(6)$ |

Since the sum of the volumes of the 16% solution and the 72% solution is 6 liters, our first equation is

$$x + y = 6$$

The second equation comes from the fact that the solutions are mixed.

$$\left(\begin{array}{c}\text{amount of TSP}\\\text{in 16\% solution}\end{array}\right) + \left(\begin{array}{c}\text{amount of TSP}\\\text{in 72\% solution}\end{array}\right) = \left(\begin{array}{c}\text{amount of TSP}\\\text{in mixture}\end{array}\right)$$

$$0.16x \qquad + \qquad 0.72y \qquad = \qquad 0.30(6)$$

Therefore, the system of equations is

$$x + y = 6$$
$$0.16x + 0.72y = 0.30(6)$$

Carry Out   Solving $x + y = 6$ for $y$, we get $y = -x + 6$. Substituting $-x + 6$ for $y$ in the second equation gives us

$$0.16x + 0.72y = 0.30(6)$$
$$0.16x + 0.72(-x + 6) = 0.30(6)$$
$$0.16x - 0.72x + 4.32 = 1.8$$
$$-0.56x + 4.32 = 1.8$$
$$-0.56x = -2.52$$
$$x = \frac{-2.52}{-0.56} = 4.5$$

Therefore, Chung must use 4.5 liters of the 16% solution. Since the two solutions must total 6 liters, he must use 6 − 4.5 or 1.5 liters of the 72% solution.

Now Try Exercise 17

> **Helpful Hint**
>
> In Example 5, the equation $0.16x + 0.72y = 0.30(6)$ could have been simplified by multiplying both sides of the equation by 100. This would give the equation $16x + 72y = 30(6)$ or $16x + 72y = 180$. Then the system of equations would be $x + y = 6$ and $16x + 72y = 180$. If you solve this system, you should obtain the same solution. Try it and see.

## 2 Use Linear Systems in Three Variables to Solve Applications

Now let us look at some applications that involve three equations and three variables.

**EXAMPLE 6 Bank Loans** Tiny Tots Toys must borrow $25,000 to pay for an expansion. It is not able to obtain a loan for the total amount from a single bank, so it takes out loans from three different banks. It borrows some of the money at a bank that charges it 8% interest. At the second bank, it borrows $2000 more than one-half the amount borrowed from the first bank. The interest rate at the second bank is 10%. The balance of the $25,000 is borrowed from a third bank, where Tiny Tots pays 9% interest. The total annual interest Tiny Tots Toys pays for the three loans is $2220. How much does it borrow at each rate?

*Solution* **Understand** We are asked to determine how much is borrowed at each of the three different rates. Therefore, this problem will contain three variables, one for each amount borrowed. Since the problem will contain three variables, we will need to determine three equations to use in our system of equations.

**Translate**    Let $x =$ amount borrowed at first bank.

Let $y =$ amount borrowed at second bank.

Let $z =$ amount borrowed at third bank.

Since the total amount borrowed is $25,000 we know that

$$x + y + z = 25{,}000 \qquad \text{Total amount borrowed is \$25,000.}$$

At the second bank, Tiny Tots Toys borrows $2000 more than one-half the amount borrowed from the first bank. Therefore, our second equation is

$$y = \frac{1}{2}x + 2000 \qquad \text{Second, } y, \text{ is \$2000 more than } \frac{1}{2} \text{ of first, } x.$$

Our last equation comes from the fact that the total annual interest charged by the three banks is $2220. The interest at each bank is found by multiplying the interest rate by the amount borrowed.

$$0.08x + 0.10y + 0.09z = 2220 \qquad \text{Total interest is \$2220.}$$

Thus, our system of equations is

$$x + y + z = 25{,}000 \qquad (1)$$

$$y = \frac{1}{2}x + 2000 \qquad (2)$$

$$0.08x + 0.10y + 0.09z = 2220 \qquad (3)$$

Both sides of equation (2) can be multiplied by 2 to remove fractions.

$$2(y) = 2\left(\frac{1}{2}x + 2000\right)$$

$$2y = x + 4000 \qquad \text{Distributive property}$$

$$-x + 2y = 4000 \qquad \text{Subtract } x \text{ from both sides.}$$

The decimals in equation (3) can be removed by multiplying both sides of the equation by 100. This gives

$$8x + 10y + 9z = 222{,}000$$

### Understanding Algebra

Whenever we use *three* variables to solve an application problem, we must determine *three* equations to solve the problem.

Our simplified system of equations is therefore

$$
\begin{aligned}
x + y + z &= 25{,}000 \quad (eq.\,1)\\
-x + 2y \phantom{+ z} &= 4000 \quad (eq.\,2)\\
8x + 10y + 9z &= 222{,}000 \quad (eq.\,3)
\end{aligned}
$$

**Carry Out**   There are various ways of solving this system. Let's use $(eq.\,1)$ and $(eq.\,3)$ to eliminate the variable $z$.

$$
\begin{array}{ll}
-9x - 9y - 9z = -225{,}000 & (eq.\,1) \quad \text{Multiplied by } -9\\
\underline{\phantom{-}8x + 10y + 9z = \phantom{-}222{,}000} & (eq.\,3)\\
-x + \phantom{10}y \phantom{+ 9z} = \phantom{-}{-3{,}000} & \text{Sum of equations, } (eq.\,4)
\end{array}
$$

Now we use $(eq.\,2)$ and $(eq.\,4)$ to eliminate the variable $x$ and solve for $y$.

$$
\begin{array}{ll}
x - 2y = -4000 & (eq.\,2) \quad \text{Multiplied by } -1\\
\underline{-x + \phantom{2}y = -3000} & (eq.\,4)\\
\phantom{-x +}{-y} = -7000 & \text{Sum of equations}\\
\phantom{-x +}y = \phantom{-}7000 &
\end{array}
$$

Now that we know the value of $y$ we can solve for $x$.

$$
\begin{array}{ll}
-x + 2y = 4000 & (eq.\,2)\\
-x + 2(7000) = 4000 & \text{Substitute 7000 for } y \text{ in } (eq.\,2).\\
-x + 14{,}000 = 4000 &\\
-x = -10{,}000 &\\
x = \phantom{-}10{,}000 &
\end{array}
$$

Finally, we solve for $z$.

$$
\begin{aligned}
x + y + z &= 25{,}000 \quad (eq.\,1)\\
10{,}000 + 7000 + z &= 25{,}000\\
17{,}000 + z &= 25{,}000\\
z &= 8000
\end{aligned}
$$

**Answer**   Tiny Tots Toys borrows \$10,000 at 8%, \$7000 at 10%, and \$8000 at 9% interest.

Now Try Exercise 55

### EXAMPLE 7   **Inflatable Boats**   Hobson, Inc., has a small manufacturing plant that makes three types of inflatable boats: one-person, two-person, and four-person models. Each boat requires the service of three departments: cutting, assembly, and packaging. The cutting, assembly, and packaging departments are allowed to use a total of 380, 330, and 120 person-hours per week, respectively. The time requirements for each boat and department are specified in the following table. Determine how many of each type of boat Hobson must produce each week for its plant to operate at full capacity.

| | Time (person-hr) per Boat | | |
|---|---|---|---|
| Department | One-Person Boat | Two-Person Boat | Four-Person Boat |
| Cutting | 0.6 | 1.0 | 1.5 |
| Assembly | 0.6 | 0.9 | 1.2 |
| Packaging | 0.2 | 0.3 | 0.5 |

*Solution*   **Understand**   We are told that three different types of boats are produced and we are asked to determine the number of each type produced. Since this problem involves three amounts to be found, the system will contain three equations in three variables.

**Translate**   We will use the information given in the table.

Let $x$ = number of one-person boats.

Let $y$ = number of two-person boats.

Let $z$ = number of four-person boats.

The total number of cutting hours for the three types of boats must equal 380 person-hours.

$$0.6x + 1.0y + 1.5z = 380$$

The total number of assembly hours must equal 330 person-hours.

$$0.6x + 0.9y + 1.2z = 330$$

The total number of packaging hours must equal 120 person-hours.

$$0.2x + 0.3y + 0.5z = 120$$

Therefore, the system of equations is

$$0.6x + 1.0y + 1.5z = 380$$
$$0.6x + 0.9y + 1.2z = 330$$
$$0.2x + 0.3y + 0.5z = 120$$

Multiplying each equation in the system by 10 will eliminate the decimal numbers and give a simplified system of equations.

$$6x + 10y + 15z = 3800 \quad (eq.\,1)$$
$$6x + 9y + 12z = 3300 \quad (eq.\,2)$$
$$2x + 3y + 5z = 1200 \quad (eq.\,3)$$

**Carry Out**   Let's first eliminate the variable $x$ using $(eq.\,1)$ and $(eq.\,2)$, and then $(eq.\,1)$ and $(eq.\,3)$.

$$
\begin{array}{ll}
6x + 10y + 15z = \phantom{-}3800 & (eq.\,1) \\
\underline{-6x - 9y - 12z = -3300} & (eq.\,2) \ \text{Multiplied by } -1 \\
\phantom{6x +} y + 3z = \phantom{-}500 & \text{Sum of equations, } (eq.\,4)
\end{array}
$$

$$
\begin{array}{ll}
6x + 10y + 15z = \phantom{-}3800 & (eq.\,1) \\
\underline{-6x - 9y - 15z = -3600} & (eq.\,3) \ \text{Multiplied by } -3 \\
\phantom{6x +} y \phantom{+ 3z} = \phantom{-}200 & \text{Sum of equations, } (eq.\,5)
\end{array}
$$

Note that when we added the last two equations, both variables $x$ and $z$ were eliminated at the same time. Now we know the value of $y$ and can solve for $z$.

$$y + 3z = 500 \qquad (eq.4)$$
$$200 + 3z = 500 \qquad \text{Substitute 200 for } y.$$
$$3z = 300$$
$$z = 100$$

Finally, we find $x$.

$$6x + 10y + 15z = 3800 \qquad (eq.\,1)$$
$$6x + 10(200) + 15(100) = 3800$$
$$6x + 2000 + 1500 = 3800$$
$$6x + 3500 = 3800$$
$$6x = 300$$
$$x = 50$$

**Answer**   Hobson should produce 50 one-person boats, 200 two-person boats, and 100 four-person boats per week.

Now Try Exercise 59

# EXERCISE SET 9.5

MathXL®    MyMathLab®
MathXL®     MyMathLab

## Practice the Skills/Problem Solving

1. **Land Area** The combined land area of the countries of Georgia and Ireland is 139,973 square kilometers. The difference between the two countries' land areas is 573 square kilometers. If Ireland has the larger land area, determine the land area of each country.

Cliffs of Moher, Ireland

2. **Daytona 500 Wins** As of this writing, Richard Petty has won the Daytona 500 race the greatest number of times and Cale Yarborough has won the second greatest number of Daytona 500 races. Petty's number of wins is one less than twice Yarborough's number of wins. The total number of wins by the two drivers is 11. Determine the number of wins by Petty and by Yarborough.

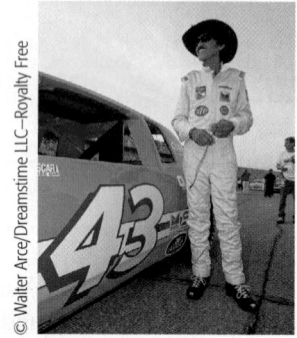

Richard Petty

3. **Fat Content** A nutritionist finds that a large order of fries at McDonald's has more fat than a McDonald's quarter-pound hamburger. The fries have 4 grams more than three times the amount of fat that the hamburger has. The difference in the fat content between the fries and the hamburger is 46 grams. Find the fat content of the hamburger and of the fries.

4. **Theme Parks** The two most visited theme parks in the United States in 2007 were Walt Disney's Magic Kingdom in Florida and Disneyland in California. The total number of visitors to these parks was 31.8 million people. The number of people who visited the Magic Kingdom was 2.2 million more than the number of people who visited Disneyland. How many people visited each of these parks in 2004? *Source:* www.coastergrotto.com

5. **Hot Dog Stand** At Big Al's hot dog stand, 2 hot dogs and 3 sodas cost $7. The cost of 4 hot dogs and 2 sodas is $10. Determine the cost of a hot dog and the cost of a soda.

6. **Water and Pretzel** At a professional football game, the cost of 2 bottles of water and 3 pretzels is $16.50. The cost of 4 bottles of water and 1 pretzel is $15.50. Determine the cost of a bottle of water and the cost of a pretzel.

7. **Memory Cards** For her digital camera, Judy Ackerman purchased a 512-MB memory card and a 4-GB memory card. Together the two memory cards can store a total of 996 photos. The 4-GB memory card can store 15 more than eight times the number of photos the 512-MB memory card can store. How many photos can each memory card store?

8. **Washer and Dryer** Beth Rechsteiner purchased a new washer and dryer for a total price of $1650. The dryer cost $150 less than the washer. Determine the price of the washer and the price of the dryer.

9. **Complementary Angles** Two angles are **complementary angles** if the sum of their measures is 90°. (See Section 2.3.) If the measure of the larger of two complementary angles is 15° more than two times the measure of the smaller angle, find the measures of the two angles.

10. **Complementary Angles** The difference between the measures of two complementary angles is 46°. Determine the measures of the two angles.

11. **Supplementary Angles** Two angles are **supplementary angles** if the sum of their measures is 180°. (See Section 2.3.) Find the measures of two supplementary angles if the measure of one angle is 28° less than three times the measure of the other.

12. **Supplementary Angles** Determine the measures of two supplementary angles if the measure of one angle is three and one half times larger than the measure of the other angle.

13. **Rowing Speed** The Heart O'Texas Rowing Team, while practicing in Austin, Texas, rowed an average of 15.6 miles per hour with the current and 8.8 miles per hour against the current. Determine the team's rowing speed in still water and the speed of the current.

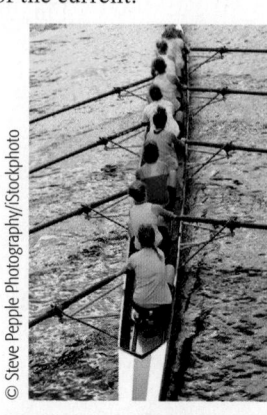

14. **Flying Speed** Jung Lee, in his Piper Cub airplane, flew an average of 121 miles per hour with the wind and 87 miles per hour against the wind. Determine the speed of the airplane in still air and the speed of the wind.

15. **Salary Plus Commission** Don Lavigne, an office equipment sales representative, earns a weekly salary plus a commission on his sales. One week his total compensation on sales of $4000 was $660. The next week his total compensation on sales of $6000 was $740. Find Don's weekly salary and his commission rate.

16. **Truck Rental** A truck rental agency charges a daily fee plus a mileage fee. Hugo was charged $85 for 2 days and 100 miles and Christina was charged $165 for 3 days and 400 miles. What is the agency's daily fee, and what is the mileage fee?

17. **Lavender Oil** Pola Sommers, a massage therapist, needs 3 ounces of a 20% lavender oil solution. She has only 5% and 30% lavender oil solutions available. How many ounces of each should Pola mix to obtain the desired solution?

18. **Fertilizer Solutions** Frank Ditlman needs to apply a 10% liquid nitrogen solution to his rose garden, but he only has a 4% liquid nitrogen solution and a 20% liquid nitrogen solution available. How much of the 4% solution and how much of the 20% solution should Frank mix together to get 10 gallons of the 10% solution?

19. **Weed Killer** Round-Up Concentrate Grass and Weed Killer consists of an 18% active ingredient glyphosate (and 82% inactive ingredients). The concentrate is to be mixed with water and the mixture applied to weeds. If the final mixture is to contain 0.9% active ingredient, how much concentrate and how much water should be mixed to make 200 gallons of the final mixture?

20. **Lawn Fertilizer** Scott's Winterizer Lawn Fertilizer is 22% nitrogen. Schultz's Lime with Lawn Fertilizer is 4% nitrogen. William Weaver, owner of Weaver's Nursery, wishes to mix these two fertilizers to make 400 pounds of a special 10% nitrogen mixture for midseason lawn feeding. How much of each fertilizer should he mix?

21. **Birdseed** Birdseed costs $0.59 a pound and sunflower seeds cost $0.89 a pound. Angela Leinenbachs' pet store wishes to make a 40-pound mixture of birdseed and sunflower seeds that sells for $0.76 per pound. How many pounds of each type of seed should she use?

22. **Coffee** Franco Manue runs a grocery store. He wishes to mix 30 pounds of coffee to sell for a total cost of $170. To obtain the mixture, he will mix coffee that sells for $5.20 per pound with coffee that sells for $6.30 per pound. How many pounds of each coffee should he use?

23. **Amtrak** Ann Marie Whittle has been pricing Amtrak fares for a group to visit New York. Three adults and four children would cost a total of $159. Two adults and three children would cost a total of $112. Determine the price of an adult's ticket and a child's ticket.

24. **Buffalo Wings** The Wing House sells both regular size and jumbo size orders of Buffalo chicken wings. Three regular orders and five jumbo orders of wings cost $67. Four regular and four jumbo orders of wings cost $64. Determine the cost of a regular order of wings and a jumbo order of wings.

25. **Savings Accounts** Mr. and Mrs. Gamton invest a total of $10,000 in two savings accounts. One account pays 5% interest and the other 6%. Find the amount placed in each account if the accounts receive a total of $540 in interest after 1 year. Use interest = principal · rate · time.

26. **Investments** Louis Okonkwo invested $30,000, part at 9% and part at 5%. If he had invested the entire amount at 6.5%, his total annual interest would be the same as the sum of the annual interest received from the two other accounts. How much was invested at each interest rate?

27. **Milk** Becky Slaats is a plant engineer at Velda Farms Dairy Cooperative. She wishes to mix whole milk, which is 3.25% fat, and skim milk, which has no fat, to obtain 260 gallons of a mixture of milk that contains 2% fat. How many gallons of whole milk and how many gallons of skim milk should Becky mix to obtain the desired mixture?

28. **Quiche Lorraine** Lambert Marx's recipe for quiche lorraine calls for 2 cups (16 ounces) of light cream that is 20% butterfat. It is often difficult to find light cream with 20% butterfat at the supermarket. What is commonly found is heavy cream, which is 36% butterfat, and half-and-half, which is 10.5% butterfat. How much of the heavy cream and how much of the half-and-half should Lambert mix to obtain the mixture necessary for the recipe?

29. **Birdseed** By ordering directly through www.birdseed.com, the Carters can purchase Season's Choice birdseed for $1.79 per pound and Garden Mix birdseed for $1.19 per pound. If they wish to purchase 20 pounds and spend $28 on birdseed, how many pounds of each type should they buy?

30. **Juice** The Healthy Favorites Juice Company sells apple juice for 8.3¢ an ounce and raspberry juice for 9.3¢ an ounce. The company wishes to market and sell 8-ounce cans of apple-raspberry juice for 8.7¢ an ounce. How many ounces of each should be mixed?

31. **Car Travel** Two cars start at the same point in Alexandria, Virginia, and travel in opposite directions. One car travels 5 miles per hour faster than the other car. After 4 hours, the two cars are 420 miles apart. Find the speed of each car.

32. **Road Construction** Kip Ortiz drives from Atlanta to Louisville, a distance of 430 miles. Due to road construction and heavy traffic, during the first part of his trip, Kip drives at an average rate of 50 miles per hour. During the rest of his trip he drives at an average rate of 70 miles per hour. If his total trip takes 7 hours, how many hours does he drive at each speed?

**33. Avon Conference** Cabrina Wilson and Dabney Jefferson are Avon representatives who are attending a conference in Seattle. After the conference, Cabrina drives home to Boise at an average speed of 65 miles per hour and Dabney drives home to Portland at an average speed of 50 miles per hour. If the sum of their driving times is 11.4 hours and if the sum of the distances driven is 690 miles, determine the time each representative spent driving home.

**34. Exercise** For her exercise routine, Cynthia Harrison rides a bicycle for half an hour and then rollerblades for half an hour. Cynthia rides the bicycle at a speed that is twice the speed at which she rollerblades. If the total distance covered is 12 miles, determine the speed at which she bikes and rollerblades.

**35. Dog Diet** Bill Lutes's dog is on a strict diet. The dog is to receive, among other nutrients, 20 grams of protein and 6 grams of carbohydrates. Bill has only two food mixes available of the following compositions. How many grams of each mix should be used to obtain the right diet for his dog?

| Mix | Protein (%) | Carbohydrate (%) |
|-----|-------------|------------------|
| Mix $A$ | 10 | 6 |
| Mix $B$ | 20 | 2 |

© Micmakin/Shutterstock

**36. Chair Manufacturing** A company makes two models of chairs. Information about the construction of the chairs is given in the table shown. On a particular day the company allocated 46.4 person-hours for assembling and 8.8 person-hours for painting. How many of each chair can be made?

| Model | Time to Assemble | Time to Paint |
|-------|------------------|---------------|
| Model $A$ | 1 hr | 0.5 hr |
| Model $B$ | 3.2 hr | 0.4 hr |

**37. Brass Alloy** By weight, one alloy of brass is 70% copper and 30% zinc. Another alloy of brass is 40% copper and 60% zinc. How many grams of each of these alloys need to be melted and combined to obtain 300 grams of a brass alloy that is 60% copper and 40% zinc?

**38. Silver Alloy** Sterling silver is 92.5% pure silver. How many grams of pure (100%) silver and how many grams of sterling silver must be mixed to obtain 250 g of a 94% silver alloy?

**39. Internal Revenue Service** The following graph shows the number of paper tax returns and the number of online tax returns filed with the IRS in the years from 2002 to 2010. If $t$ represents the number of years since 2002, the number of paper tax returns, in millions, filed with the IRS can be estimated by the function $P(t) = -2.73t + 58.37$ and the number of online tax returns, in millions, filed with the IRS can be estimated by the function $o(t) = 1.95t + 10.58$. Assuming this trend continues, solve this system of equations to determine the year that the number of paper tax returns will be the same as the number of online tax returns.

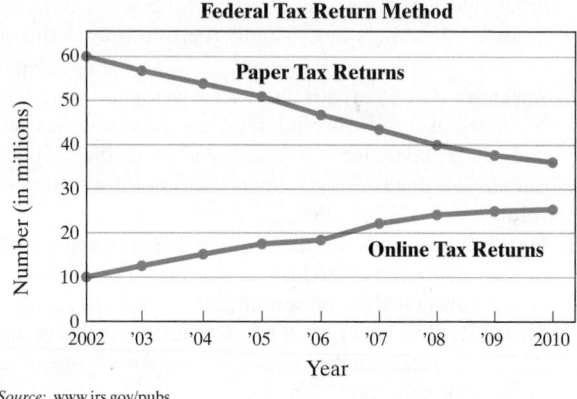

*Source:* www.irs.gov/pubs

**40. Walking and Jogging** Cuong Tham tries to exercise every day. He walks at 3 miles per hour and then jogs at 5 miles per hour. If it takes him 0.9 hours to travel a total of 3.5 miles, how long does he jog?

**41. Texas Driving** Tom Johnson and Melissa Acino started driving at the same time in different cars from Oklahoma City. They both traveled south on Route 35. When Melissa reached the Dallas/Ft. Worth area, a distance of 150 miles, Tom had only reached Denton, Texas, a distance of 120 miles. If Melissa averaged 15 miles per hour faster than Tom, find the average speed of each car.

**42. Photocopy Costs** At a local copy center two plans are available.
   Plan 1: 10¢ per copy
   Plan 2: an annual charge of $120 plus 4¢ per copy

   **a)** Represent this information as a system of equations.

   **b)** Graph the system of equations for up to 4000 copies made.

   **c)** From the graph, estimate the number of copies a person would have to make in a year for the two plans to have the same total cost.

   **d)** Solve the system algebraically. If your answer does not agree with your answer in part **c)**, explain why.

*In Exercises 43–62, solve each problem using a system of three equations in three unknowns.*

**43. Mail Volume** The average American household receives 24 pieces of mail each week. The number of bills and statements is two less than twice the number of pieces of personal mail. The number of advertisements is two more than five times the number of pieces of personal mail. How many pieces of personal mail, bills and statements, and advertisements does the average family get each week?
*Source:* Arthur D. Little, Inc.

**44. Submarine Personnel** A 141-person crew is standard on a Los Angeles class submarine. The number of chief petty officers (enlisted) is four more than the number of commissioned officers. The number of other enlisted men is three less than eight times the number of commissioned officers. Determine the number of commissioned officers, chief petty officers, and other enlisted people on the submarine.

USS *Greenville*

**45. College Football Bowl Games** Through 2007, the Universities of Alabama, Tennessee, and Texas have had the most appearances in college football bowl games. These three schools have had a total of 147 bowl appearances. Alabama has had 8 more appearances than Texas. Together, the number of appearances by Tennessee and Texas is 39 more than the number of appearances by Alabama. Determine the number of bowl appearances for each school.

**46. Summer Olympics** In the 2008 Summer Olympics in Beijing, China, the countries earning the most medals were the United States, China, and Russia. Together, these three countries earned a total of 282 medals. The United States earned 10 more medals than China. Together the number of medals earned by the United States and Russia is 18 less than twice the number of medals earned by China. Determine the number of medals each country earned.
*Source:* en.beijing2008.cn

**47. NPF Softball** The three teams with the most wins in the 2008 National Pro Fastpitch (NPF) softball league regular season were the Chicago Bandits, the Philadelphia Force, and the Washington Glory. These three teams had a total of 93 wins. The Bandits had two more wins than the Glory and the Force had one less win than the Bandits. Determine the number of wins each team had.
*Source:* www.profastpitch.com

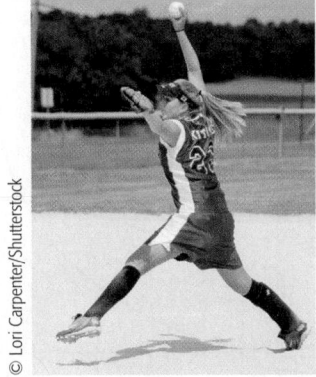

**48. Wettest Cities** The three cities within the continental United States that receive the highest average annual rainfall are Mobile, Alabama, Pensacola, Florida, and New Orleans, Louisiana. The total average annual rainfall for all three cities is 196 inches. Mobile has an average rainfall that is 3 inches more than that of New Orleans. New Orleans has an average rainfall that is 1 inch less than that of Pensacola. Determine the average annual rainfall for each of the three cities.
*Source:* www.livescience.com

**49. Top-Selling U.S. Albums** As of 2008, the best-selling music albums based on sales in the United States are The Eagles—*Their Greatest Hits 1971–1975,* Michael Jackson—*Thriller,* and Pink Floyd—*The Wall.* The total number of the three albums sold in the United States is 79.5 million. The number of *Thriller* albums sold was 2 million less than the number of *Their Greatest Hits 1971–1975* albums sold. The number of *The Wall* albums sold was 3.5 million less than the number of *Thriller* albums sold. Determine the number of each album sold.
*Source:* Recording Industry Association of America

The Eagles—*Their Greatest Hits 1971–1975*

**50. Best-Selling Movies** As of 2008, The best-selling movies based on box-offices sales of all time are *Titanic, The Lord of the Rings: The Return of the King,* and *Pirates of the Caribbean: Dead Man's Chest.* The total sales of all three movies are $4024 million. The combined sales of *The Return of the King* and *Dead Man's Chest* are $354 million more than sales of *Titanic.* Sales of *Return of the King* are $69 million more than sales of *Dead Man's Chest.* Determine the sales for each of the movies.
*Source:* All-Time Worldwide Box Office

**51. Super Bowls** As of 2010, the states of Florida, California, and Louisiana have hosted the most Super Bowls. These three states hosted a total of 35 Super Bowls. Florida hosted 6 more Super Bowls than Louisiana. Together, Florida and Louisiana hosted two more than twice the number California hosted. Determine the number of Super Bowls hosted by each of these three states.
*Source:* www.nfl.com

**52. Concert Tickets** Three kinds of tickets for a Soggy Bottom Boys concert are available: up-front, main-floor, and balcony. The most expensive tickets, the up-front tickets, are twice as expensive as balcony tickets. Balcony tickets are $10 less than main-floor tickets and $30 less than up-front tickets. Determine the price of each kind of ticket.

**53. Triangle** The sum of the measures of the angles of a triangle is 180°. The smallest angle of the triangle has a measure $\frac{2}{3}$ the measure of the second smallest angle. The largest angle has a measure that is 30° less than three times the measure of the second smallest angle. Determine the measure of each angle.

**54. Triangle** The largest angle of a triangle has a measure that is 10° less than three times the measure of the second smallest angle. The measure of the smallest angle is equal to the difference between the measure of the largest angle and twice the measure of the second smallest angle. Determine the measures of the three angles of the triangle.

**55. Investments** Tam Phan received a check for $10,000. She decided to divide the money (not equally) into three different investments. She placed part of her money in a savings account paying 3% interest. The second amount, which was twice the first amount, was placed in a certificate of deposit paying 5% interest. She placed the balance in a money market fund paying 6% interest. If Tam's total interest over the period of 1 year was $525.00, how much was placed in each account?

**56. Bonus** Nick Pfaff, an attorney, divided his $15,000 holiday bonus check among three different investments. With some of the money, he purchased a municipal bond paying 5.5% simple interest. He invested twice the amount of money that he paid for the municipal bond in a certificate of deposit paying 4.5% simple interest. Nick placed the balance of the money in a money market account paying 3.75% simple interest. If Nick's total interest for 1 year was $692.50, how much was placed in each account?

**57. Hydrogen Peroxide** A 10% solution, a 12% solution, and a 20% solution of hydrogen peroxide are to be mixed to get 8 liters of a 13% solution. How many liters of each must be mixed if the volume of the 20% solution must be 2 liters less than the volume of the 10% solution?

**58. Sulfuric Acid** An 8% solution, a 10% solution, and a 20% solution of sulfuric acid are to be mixed to get 100 milliliters of a 12% solution. If the *volume of acid* from the 8% solution is to equal half the *volume of acid* from the other two solutions, how much of each solution is needed?

**59. Furniture Manufacturing** Donaldson Furniture Company produces three types of rocking chairs: the children's model, the standard model, and the executive model. Each chair is made in three stages: cutting, construction, and finishing. The time needed for each stage of each chair is given in the following chart. During a specific week the company has available a maximum of 154 hours for cutting, 94 hours for construction, and 76 hours for finishing. Determine how many of each type of chair the company should make to be operating at full capacity.

| Stage | Children's | Standard | Executive |
|---|---|---|---|
| Cutting | 5 hr | 4 hr | 7 hr |
| Construction | 3 hr | 2 hr | 5 hr |
| Finishing | 2 hr | 2 hr | 4 hr |

**60. Bicycle Manufacturing** The Jamis Bicycle Company produces three models of bicycles: Dakar, Komodo, and Aragon. Each bicycle is made in three stages: welding, painting, and assembling. The time needed for each stage of each bicycle is given in the chart below. During a specific week, the company has available a maximum of 133 hours for welding, 78 hours for painting, and 96 hours for assembling. Determine how many of each type of bicycle the company should make to be operating at full capacity.

| Stage | Dakar | Komodo | Aragon |
|---|---|---|---|
| Welding | 2 | 3 | 4 |
| Painting | 1 | 2 | 2.5 |
| Assembling | 1.5 | 2 | 3 |

**61. Current Flow** In electronics it is necessary to analyze current flow through paths of a circuit. In three paths ($A$, $B$, and $C$) of a circuit, the relationships are the following:

$$I_A + I_B + I_C = 0$$
$$-8I_B + 10I_C = 0$$
$$4I_A - 8I_B = 6$$

where $I_A$, $I_B$, and $I_C$ represent the current in paths $A$, $B$, and $C$, respectively. Determine the current in each path of the circuit.

**62. Forces on a Beam** In physics we often study the forces acting on an object. For three forces, $F_1$, $F_2$, and $F_3$, acting on a beam, the following equations were obtained.

$$3F_1 + F_2 - F_3 = 2$$
$$F_1 - 2F_2 + F_3 = 0$$
$$4F_1 - F_2 + F_3 = 3$$

Find the three forces.

## Group Activity

*Discuss and answer Exercise 63 as a group.*

**63. Two Cars** A *nonlinear system of equations* is a system of equations containing at least one equation that is not linear. (Nonlinear systems of equations will be discussed in Chapter 14.) The graph shows a nonlinear system of equations. The curves represent speed versus time for two cars.

   **a)** Are the two curves functions? Explain.

   **b)** Discuss the meaning of this graph.

   **c)** At time $t = 0.5$ hr, which car is traveling at a greater speed? Explain your answer.

   **d)** Assume the two cars start at the same position and are traveling in the same direction. Which car, $A$ or $B$, traveled farther in 1 hour? Explain your answer.

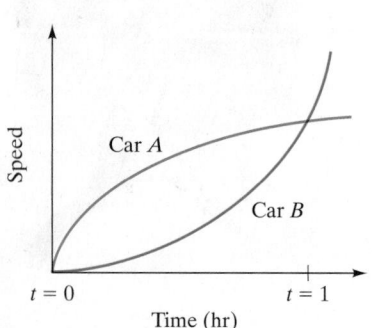

## Cumulative Review Exercises

[1.9]   **64.** Evaluate $\frac{1}{2}x + \frac{2}{5}xy + \frac{1}{8}y$ when $x = -2$, $y = 5$.

[2.5]   **65.** Solve $4 - 2[(x - 5) + 2x] = -(x + 6)$.

[4.4]   **66.** Write an equation of the line that passes through points $(6, -4)$ and $(2, -8)$.

[8.2]   **67.** Explain how to determine whether a graph represents a function.

# 9.6 Solving Systems of Equations Using Matrices

**1** Write an augmented matrix.

**2** Solve systems of linear equations.

**3** Solve systems of linear equations in three variables.

**4** Recognize inconsistent and dependent systems.

## **1** Write an Augmented Matrix

### Matrix

- A **matrix** is a rectangular array of numbers written within brackets. The plural of *matrix* is **matrices**.
- The numbers in a matrix are called the **elements** of the matrix.
- The **dimensions** of a matrix are the number of rows by the number of columns.

An example of a matrix is

$$2 \text{ rows} \xrightarrow{\overrightarrow{\phantom{xx}}} \begin{bmatrix} 5 & 7 & 2 \\ -1 & 3 & 4 \end{bmatrix} \quad \overset{\displaystyle 3 \text{ columns}}{\downarrow \ \downarrow \ \downarrow}$$

The dimensions of this matrix are 2 rows by 3 columns. We write $2 \times 3$ and we say "2 by 3."

A **square matrix** is a matrix with the same number of rows and columns. An example of a $2 \times 2$ square matrix is

$$2 \text{ rows} \xrightarrow{\overrightarrow{\phantom{xx}}} \begin{bmatrix} 4 & 6 \\ 9 & -2 \end{bmatrix} \quad \overset{\displaystyle 2 \text{ columns}}{\downarrow \ \downarrow}$$

We will solve systems of linear equations using an *augmented matrix*. An **augmented matrix** is a matrix made up of two smaller matrices separated by a vertical line. An augmented matrix that represents a system of linear equations will have the numerical coefficients to the left of the vertical line and the constants to the right of the vertical line as illustrated below. The system of equations

$$2x - 3y = 10$$
$$4x - 5y = 9$$

is represented with the augmented matrix

$$\begin{bmatrix} 2 & -3 & \vert & 10 \\ 4 & -5 & \vert & 9 \end{bmatrix}$$

coefficients on $x$   constants

coefficients on $y$

The $2 \times 2$ matrix $\begin{bmatrix} 2 & -3 \\ 4 & -5 \end{bmatrix}$ is the **coefficient matrix**. The $2 \times 1$ matrix $\begin{bmatrix} 10 \\ 9 \end{bmatrix}$ is the **constant matrix**. So, the matrix that represents the system of equations $\begin{bmatrix} 2 & -3 & | & 10 \\ 4 & -5 & | & 9 \end{bmatrix}$ is the coefficient matrix *augmented* by the constant matrix.

When we represent a system of linear equations with an augmented matrix, each equation needs to be in the form $ax + by = c$. For example, the system of equations

$$\begin{array}{l} y = -2x + 5 \\ x = 7y - 4 \end{array} \quad \text{would be rewritten as} \quad \begin{array}{l} 2x + y = 5 \\ x - 7y = -4 \end{array}$$

This system can be represented with the augmented matrix

$$\begin{bmatrix} 2 & 1 & | & 5 \\ 1 & -7 & | & -4 \end{bmatrix}$$

When using an augmented matrix to solve a system of linear equations, we will use a method very similar to the addition method discussed in Section 9.3.

## 2 Solve Systems of Linear Equations

To solve a system of two linear equations using matrices, we perform certain procedures discussed below so that we can rewrite the augmented matrix in **row echelon** (or **triangular**) **form**:

$$\begin{bmatrix} 1 & a & | & p \\ 0 & 1 & | & q \end{bmatrix}$$

where the $a$, $p$, and $q$ represent numbers. From this type of augmented matrix we can write an equivalent system of equations. This matrix represents the linear system

$$\begin{array}{l} 1x + ay = p \\ 0x + 1y = q \end{array} \quad \text{or} \quad \begin{array}{l} x + ay = p \\ y = q \end{array}$$

For example,

$$\begin{bmatrix} 1 & 2 & | & 4 \\ 0 & 1 & | & 5 \end{bmatrix} \quad \text{represents} \quad \begin{array}{l} x + 2y = 4 \\ y = 5 \end{array}$$

Note that the system above on the right can be easily solved by substitution. Its solution is $(-6, 5)$.

We use **row transformations** to rewrite the augmented matrix in row echelon form. We will use three row transformation procedures. Row transformations on a matrix are equivalent to operations you would perform on the equations in the system.

### Procedures for Row Transformations

1. All the numbers in a row may be multiplied (or divided) by any nonzero real number. (This is the same as multiplying both sides of an equation by a nonzero real number.)

2. All the numbers in a row may be multiplied by any nonzero real number. These products may then be added to the corresponding numbers in any other row. (This is equivalent to eliminating a variable from a system of equations using the addition method.)

3. The order of the rows may be switched. (This is equivalent to switching the order of the equations in a system of equations.)

Generally, when obtaining a 1 in the augmented matrix we use row transformation procedure 1, and when obtaining a 0 we use row transformation procedure 2. *Work by columns starting from the left.* Start with the first column, first row.

**EXAMPLE 1** Solve the following system of equations using matrices.

$$2x + 4y = -2$$
$$3x - 2y = 5$$

*Solution*  First we write the augmented matrix.

$$\begin{bmatrix} 2 & 4 & | & -2 \\ 3 & -2 & | & 5 \end{bmatrix}$$

Our goal is to obtain a matrix of the form $\begin{bmatrix} 1 & a & | & p \\ 0 & 1 & | & q \end{bmatrix}$. We begin by using row transformation procedure 1 to replace the 2 in the first column, first row, with 1. To do so, we multiply the first row of numbers by $\frac{1}{2}$. We abbreviate this multiplication as $\frac{1}{2}R_1$ and place it to the right of the matrix in the same row where the operation was performed.

$$\begin{bmatrix} \left(\frac{1}{2}\right)2 & \left(\frac{1}{2}\right)4 & | & \left(\frac{1}{2}\right)(-2) \\ 3 & -2 & | & 5 \end{bmatrix} \quad \frac{1}{2}R_1$$

or

$$\begin{bmatrix} 1 & 2 & | & -1 \\ 3 & -2 & | & 5 \end{bmatrix}$$

We next use the row transformations to produce a 0 in the first column, second row where currently a 3 is in this position. We will multiply the elements in row 1 by $-3$ and add the products to row 2. We abbreviate this row transformation as $-3R_1 + R_2$.

The elements in the first row multiplied by $-3$ are:

$$-3(1) \qquad -3(2) \qquad -3(-1)$$

or

$$-3 \qquad -6 \qquad 3$$

Now add these products to their respective elements in the second row to get

$$\begin{bmatrix} 1 & 2 & | & -1 \\ -3 + 3 & -6 + (-2) & | & 3 + 5 \end{bmatrix} \quad -3R_1 + R_2$$

or

$$\begin{bmatrix} 1 & 2 & | & -1 \\ 0 & -8 & | & 8 \end{bmatrix}$$

We next use row transformations to produce a 1 in the second row, second column where currently a $-8$ is in this position. We do this by multiplying the second row by $-\frac{1}{8}$.

$$\begin{bmatrix} 1 & 2 & | & -1 \\ \left(-\frac{1}{8}\right)0 & \left(-\frac{1}{8}\right)(-8) & | & \left(-\frac{1}{8}\right)8 \end{bmatrix} \quad -\frac{1}{8}R_2$$

or

$$\begin{bmatrix} 1 & 2 & | & -1 \\ 0 & 1 & | & -1 \end{bmatrix}$$

The matrix is now in row echelon form and the equivalent system of equations is

$$x + 2y = -1$$
$$y = -1$$

Now we can solve for $x$ using substitution.

$$x + 2y = -1$$
$$x + 2(-1) = -1$$
$$x - 2 = -1$$
$$x = 1$$

A check will shown that $(1, -1)$ is the solution to the original system.

Now Try Exercise 19

## Understanding Algebra

When rewriting the augmented matrix in row echelon form, work by columns, from the left-hand column to the right-hand column. Always complete one column before moving to the next column. In each column, first obtain the 1 in the appropriate position, and then obtain the 0's.

## 3 Solve Systems of Linear Equations in Three Variables

Now we will use matrices to solve a system of three linear equations in three variables. We use the same row transformation procedures used when solving a system of two linear equations. Our goal is to produce an augmented matrix in the row echelon form

$$\begin{bmatrix} 1 & a & b & p \\ 0 & 1 & c & q \\ 0 & 0 & 1 & r \end{bmatrix}$$

where $a, b, c, p, q$ and $r$ represent numbers. This matrix represents the following system of equations.

$$1x + ay + bz = p \qquad x + ay + bz = p$$
$$0x + 1y + cz = q \quad \text{or} \quad y + cz = q$$
$$0x + 0y + 1z = r \qquad z = r$$

> ### Helpful Hint
>
> **Study Tip**
>
> When using matrices, be careful to keep all the numbers lined up neatly in rows and columns. One slight mistake in copying numbers from one matrix to another will lead to an incorrect attempt at solving a system of equations.
>
> $$x - 3y + z = 3$$
> For example, the system of equations, $4x + 2y - 5z = 20$, when correctly represented
> $$-5x - y - 4z = 13$$
>
> with the augmented matrix, $\begin{bmatrix} 1 & -3 & 1 & 3 \\ 4 & 2 & -5 & 20 \\ -5 & -1 & -4 & 13 \end{bmatrix}$, leads to the solution $(1, -2, -4)$.
>
> However, a matrix that looks quite similar, $\begin{bmatrix} 1 & -3 & 1 & 3 \\ 4 & -1 & -5 & 20 \\ -5 & 2 & -4 & 13 \end{bmatrix}$, leads to the incorrect ordered triple of $\left( -\dfrac{25}{53}, -\dfrac{130}{53}, -\dfrac{206}{53} \right)$.

EXAMPLE 2 Solve the following system of equations using matrices.

$$x - 2y + 3z = -7$$
$$2x - y - z = 7$$
$$-x + 3y + 2z = -8$$

*Solution* First write the augmented matrix.

$$\begin{bmatrix} 1 & -2 & 3 & -7 \\ 2 & -1 & -1 & 7 \\ -1 & 3 & 2 & -8 \end{bmatrix}$$

Next we use row transformations to produce a first column of $\begin{smallmatrix} 1 \\ 0 \\ 0 \end{smallmatrix}$. Since the number in the first column, first row, is already a 1, we will work with the 2 in the first column, second row. Multiplying the numbers in the first row by $-2$ and adding those products to the respective numbers in the second row will replace 2 with 0. The matrix is now

$$\left[\begin{array}{ccc|c} 1 & -2 & 3 & -7 \\ 0 & 3 & -7 & 21 \\ -1 & 3 & 2 & -8 \end{array}\right] \quad -2R_1 + R_2$$

Continuing down the first column, we wish to replace the $-1$ with 0. By multiplying the numbers in the first row by 1, and then adding the products to the third row, we get

$$\left[\begin{array}{ccc|c} 1 & -2 & 3 & -7 \\ 0 & 3 & -7 & 21 \\ 0 & 1 & 5 & -15 \end{array}\right] \quad 1R_1 + R_3$$

Now we work with the second column. We wish to produce a second column of the form $\begin{smallmatrix} a \\ 1 \\ 0 \end{smallmatrix}$ where $a$ represents a number. Since there is presently a 1 in the third row, second column, and we want a 1 in the second row, second column, we switch the second and third rows of the matrix. This gives

$$\left[\begin{array}{ccc|c} 1 & -2 & 3 & -7 \\ 0 & 1 & 5 & -15 \\ 0 & 3 & -7 & 21 \end{array}\right] \quad \text{Switch } R_2 \text{ and } R_3.$$

Continuing down the second column, we wish to replace the 3 in the third row with 0 by multiplying the numbers in the second row by $-3$ and adding those products to the third row. This gives

$$\left[\begin{array}{ccc|c} 1 & -2 & 3 & -7 \\ 0 & 1 & 5 & -15 \\ 0 & 0 & -22 & 66 \end{array}\right] \quad -3R_2 + R_3$$

Now we work with the third column. We wish to produce a third column of the form $\begin{smallmatrix} b \\ c \\ 1 \end{smallmatrix}$ where $b$ and $c$ represent numbers. We can multiply the numbers in the third row by $-\dfrac{1}{22}$ to replace $-22$ with 1.

$$\left[\begin{array}{ccc|c} 1 & -2 & 3 & -7 \\ 0 & 1 & 5 & -15 \\ 0 & 0 & 1 & -3 \end{array}\right] \quad -\tfrac{1}{22}R_3$$

This matrix is now in row echelon form. From this matrix we obtain the system of equations

$$x - 2y + 3z = -7$$
$$y + 5z = -15$$
$$z = -3$$

The third equation gives us the value of $z$ in the solution. Now we can solve for $y$ by substituting $-3$ for $z$ in the second equation.

$$y + 5z = -15$$
$$y + 5(-3) = -15$$
$$y - 15 = -15$$
$$y = 0$$

Now we solve for $x$ by substituting 0 for $y$ and $-3$ for $z$ in the first equation.

$$x - 2y + 3z = -7$$
$$x - 2(0) + 3(-3) = -7$$
$$x - 0 - 9 = -7$$
$$x - 9 = -7$$
$$x = 2$$

The solution is $(2, 0, -3)$. Check this now by substituting the appropriate values into each of the original equations.

**Now Try Exercise 33**

## Understanding Algebra

When working with augmented matrices to solve a system of linear equations, if you produce

- A row with 0's to the left and a non-zero number to the right of the vertical bar, the system is inconsistent and has no solution.

- A row with all 0's, the system is dependent and has an infinite number of solutions.

## 4 Recognize Inconsistent and Dependent Systems

When solving a system of two equations, if you obtain an augmented matrix in which one row of numbers on the left side of the vertical line is all zeros but a zero does not appear in the same row on the right side of the vertical line, the system is inconsistent and has no solution. For example, a system of equations that yields the following augmented matrix is an inconsistent system.

$$\begin{bmatrix} 1 & 2 & | & 5 \\ 0 & 0 & | & 3 \end{bmatrix} \longleftarrow \text{Inconsistent system}$$

The second row of the matrix represents the equation

$$0x + 0y = 3$$

which is never true.

If you obtain a matrix in which a 0 appears across an entire row, the system of equations is dependent. For example, a system of equations that yields the following augmented matrix is a dependent system.

$$\begin{bmatrix} 1 & -3 & | & -4 \\ 0 & 0 & | & 0 \end{bmatrix} \longleftarrow \text{Dependent system}$$

The second row of the matrix represents the equation

$$0x + 0y = 0$$

which is always true.

Similar rules hold for systems with three equations.

$$\begin{bmatrix} 1 & 3 & 7 & | & 5 \\ 0 & 0 & 0 & | & -1 \\ 0 & 1 & -2 & | & 3 \end{bmatrix} \longleftarrow \text{Inconsistent system}$$

$$\begin{bmatrix} 1 & 3 & -1 & | & 2 \\ 0 & 0 & 0 & | & 0 \\ 0 & 5 & 6 & | & -4 \end{bmatrix} \longleftarrow \text{Dependent system}$$

# EXERCISE SET 9.6    MathXL®    MyMathLab®

MathXL®    MyMathLab

## Warm-Up Exercises

*Fill in the blanks with the appropriate word, phrase, or symbol(s) from the following list.*

| row transformations | augmented matrix | square matrix | dimensions | inconsistent | elements |
| row echelon | functions | rectangular form | column transformations | dependent | |

**1.** To solve a system of linear equations using matrices, we use _____ to rewrite the augmented matrix in row echelon form.

**2.** The numbers inside the matrix are called the _____.

**3.** The number of rows by the number of columns refer to the _____ of the matrix.

**4.** When a matrix consists of two smaller matrices separated by a vertical line it is called a(n) _____.

**5.** If, while solving a system of equations using an augmented matrix, you produce the matrix $\begin{bmatrix} -2 & 7 & | & 1 \\ 0 & 0 & | & 6 \end{bmatrix}$, you can conclude (assuming that all of your calculations are correct) that the system is _____ and has no solution.

**6.** If, while solving a system of equations using an augmented matrix, you produce the matrix $\begin{bmatrix} -2 & 7 & | & 1 \\ 0 & 0 & | & 0 \end{bmatrix}$, you can conclude (assuming all of your calculations are correct) that the system is _____ and has an infinite number of solutions.

**7.** The matrix $\begin{bmatrix} 2 & 4 & 0 \\ -1 & 1 & 5 \\ 3 & 6 & -2 \end{bmatrix}$ is an example of a _____.

**8.** The augmented matrix $\begin{bmatrix} 1 & 4 & 6 & | & 1 \\ 0 & 1 & 3 & | & -2 \\ 0 & 0 & 1 & | & 7 \end{bmatrix}$ is in _____ form.

## Practice the Skills

*Perform each row transformation indicated and write the new matrix.*

**9.** $\begin{bmatrix} 7 & -14 & | & -35 \\ 3 & -7 & | & -4 \end{bmatrix}$ Multiply numbers in the first row by $\frac{1}{7}$.

**10.** $\begin{bmatrix} 1 & 8 & | & 3 \\ 0 & 4 & | & -3 \end{bmatrix}$ Multiply numbers in the second row by $\frac{1}{4}$.

**11.** $\begin{bmatrix} 4 & 7 & 2 & | & -1 \\ 3 & 2 & 1 & | & -5 \\ 1 & 1 & 3 & | & -8 \end{bmatrix}$ Switch row 1 and row 3.

**12.** $\begin{bmatrix} 1 & 5 & 7 & | & 2 \\ 0 & 8 & -1 & | & -6 \\ 0 & 1 & 3 & | & -4 \end{bmatrix}$ Switch row 2 and row 3.

**13.** $\begin{bmatrix} 1 & 3 & | & 12 \\ -4 & 11 & | & -6 \end{bmatrix}$ Multiply numbers in the first row by 4 and add the products to the second row.

**14.** $\begin{bmatrix} 1 & 5 & | & 6 \\ \frac{1}{2} & 10 & | & -4 \end{bmatrix}$ Multiply numbers in the first row by $-\frac{1}{2}$ and add the products to the second row.

**15.** $\begin{bmatrix} 1 & 0 & 8 & | & \frac{1}{4} \\ 5 & 2 & 2 & | & -2 \\ 6 & -3 & 1 & | & 0 \end{bmatrix}$ Multiply numbers in the first row by $-5$ and add the products to the second row.

**16.** $\begin{bmatrix} 1 & 2 & -1 & | & 6 \\ 0 & 1 & 5 & | & 0 \\ 0 & 0 & 3 & | & 12 \end{bmatrix}$ Multiply numbers in the third row by $\frac{1}{3}$.

*Solve each system using matrices.*

**17.** $x + 3y = 3$
$-x + y = -3$

**18.** $x - 2y = 0$
$2x + 5y = 9$

**19.** $x + 3y = -2$
$-2x - 7y = 3$

**20.** $2x + 4y = -8$
$3x - 5y = -1$

**21.** $5a - 10b = -10$
$2a + b = 1$

**22.** $3s - 2t = 1$
$-2s + 4t = -6$

**23.** $2x - 5y = -6$
$-4x + 10y = 12$

**24.** $-2m - 4n = 7$
$3m + 6n = -8$

**25.** $12x + 2y = 2$
$6x - 3y = -11$

**26.** $4r + 2s = -10$
$-2r + s = -7$

**27.** $-3x + 6y = 5$
$2x - 4y = 7$

**28.** $8x = 4y + 12$
$-2x + y = -3$

**29.** $12x - 8y = 6$
$\quad\;\; -3x + 4y = -1$

**30.** $2x - 3y = 3$
$\quad\;\; -5x + 9y = -7$

**31.** $10m = 8n + 15$
$\quad\;\; 16n = -15m - 2$

**32.** $8x = 9y + 4$
$\quad\;\; 16x - 27y = 11$

*Solve each system using matrices.*

**33.** $x + 2y - 4z = 5$
$\quad\;\; 3x - y + z = 1$
$\quad\;\; 2x + 2y - 3z = 6$

**34.** $a - 3b + 4c = 7$
$\quad\;\; 4a + b + c = -2$
$\quad\;\; -2a - 3b + 5c = 12$

**35.** $x + 2y = 5$
$\quad\;\; y - z = -1$
$\quad\;\; 2x - 3z = 0$

**36.** $3a - 5c = 3$
$\quad\;\; a + 2b = -6$
$\quad\;\; 7b - 4c = 5$

**37.** $x - 2y + 4z = 5$
$\quad\;\; -3x + 4y - 2z = -8$
$\quad\;\; 4x + 5y - 4z = -3$

**38.** $3x + 5y + 2z = 3$
$\quad\;\; -x - y - z = -2$
$\quad\;\; 2x - 2y + 5z = 11$

**39.** $2x - 5y + z = 1$
$\quad\;\; 3x - 5y + z = 3$
$\quad\;\; -4x + 10y - 2z = -2$

**40.** $x + 2y + 3z = 1$
$\quad\;\; 4x + 5y + 6z = -3$
$\quad\;\; 7x + 8y + 9z = 0$

**41.** $4p - q + r = 4$
$\quad\;\; -6p + 3q - 2r = -5$
$\quad\;\; 2p + 5q - r = 7$

**42.** $-4r + 3s - 6t = 14$
$\quad\;\; 4r + 2s - 2t = -3$
$\quad\;\; 2r - 5s - 8t = -23$

**43.** $2x - 4y + 3z = -12$
$\quad\;\; 3x - y + 2z = -3$
$\quad\;\; -4x + 8y - 6z = 10$

**44.** $3x - 2y + 4z = -1$
$\quad\;\; 5x + 2y - 4z = 9$
$\quad\;\; -6x + 4y - 8z = 2$

**45.** $5x - 3y + 4z = 22$
$\quad\;\; -x - 15y + 10z = -15$
$\quad\;\; -3x + 9y - 12z = -6$

**46.** $9x - 4y + 5z = -2$
$\quad\;\; -9x + 5y - 10z = -1$
$\quad\;\; 9x + 3y + 10z = 1$

## Problem Solving

*Solve using matrices.*

**47. Angles of a Roof** In a triangular cross section of a roof, the largest angle is 55° greater than the smallest angle. The largest angle is 20° greater than the remaining angle. Find the measure of each angle.

**48. Right Angle** A right angle is divided into three smaller angles. The largest of the three angles is twice the smallest. The remaining angle is 10° greater than the smallest. Find the measure of each angle.

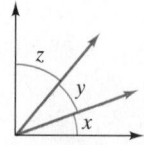

**49. Most Valuable Sports Franchises** As of 2008, the three most valuable National Football League franchises are in Washington, D.C., Dallas, and Houston, respectively. The total value of these three franchises is $2928 million. The Washington franchise is worth $177 million more than the Dallas franchise.

The Houston franchise is worth $18 million less than the Dallas franchise. Determine the value of each franchise.

*Source*: www.espn.com

**50. Baseball Player Autograph Collection** Alex Runde has a large collection of baseball players' autographs from players from the Tampa Bay Rays, the Milwaukee Brewers, and the Colorado Rockies. The total number of autographs from players from all three teams is 42. The number of autographs from the Rays is 5 more than twice the number of autographs from the Rockies. The number of autographs from the Brewers is 8 less than the number of autographs from the Rays. Determine the number of autographs Alex has from players on each team.

## Concept/Writing Exercises

**51.** When solving a system of linear equations by matrices, if two rows are identical, will the system be consistent, dependent, or inconsistent?

**52.** When solving a system of equations using matrices, how will you know if the system is

**a)** dependent,

**b)** inconsistent?

**53.** When solving a system of linear equations using matrices, if two rows of a matrix are switched, will the solution to the system change? Explain.

**54.** You can tell whether a system of two equations in two variables is consistent, dependent, or inconsistent by comparing the slopes and *y*-intercepts of the graphs of the equations. Can you tell, without solving, if a system of three equations in three variables is consistent, dependent, or inconsistent? Explain.

## Cumulative Review Exercises

[5.6]  **55.** Divide $\dfrac{3x^2 + 4x - 23}{x + 4}$.

[6.6]  **56.** Solve $2x^2 - x - 36 = 0$.

[7.4]  **57.** Subtract $\dfrac{1}{x^2 - 4} - \dfrac{2}{x - 2}$.

[7.7]  **58. Stacking Wood** Sam Davidson can stack a cord of wood in 20 minutes. With his wife's help, they can stack the wood in 12 minutes. How long would it take his wife, Terry, to stack the wood by herself?

# 9.7 Solving Systems of Equations Using Determinants and Cramer's Rule

1 Evaluate the determinant of a 2 × 2 matrix.

2 Use Cramer's rule.

3 Evaluate the determinant of a 3 × 3 matrix.

4 Use Cramer's rule with systems in three variables.

## Understanding Algebra

To evaluate a 2 × 2 determinant, subtract the product of the two diagonals as shown below:

$$\begin{vmatrix} a_1 & b_1 \\ a_2 & b_2 \end{vmatrix} = \begin{vmatrix} a_1 & b_1 \\ a_2 & b_2 \end{vmatrix}$$

$$= a_1 b_2 - a_2 b_1$$

## 1 Evaluate the Determinant of a 2 × 2 Matrix

Associated with every square matrix is a number called its **determinant**. For a 2 × 2 matrix, its determinant is defined as follows.

### Determinant

The **determinant** of a 2 × 2 matrix $\begin{bmatrix} a_1 & b_1 \\ a_2 & b_2 \end{bmatrix}$ is denoted $\begin{vmatrix} a_1 & b_1 \\ a_2 & b_2 \end{vmatrix}$ and is evaluated as

$$\begin{vmatrix} a_1 & b_1 \\ a_2 & b_2 \end{vmatrix} = a_1 b_2 - a_2 b_1$$

A system of linear equations may also be solved using determinants.

EXAMPLE 1  Evaluate each determinant.

**a)** $\begin{vmatrix} 2 & -1 \\ 3 & -5 \end{vmatrix}$   **b)** $\begin{vmatrix} 2 & 3 \\ -1 & 4 \end{vmatrix}$

*Solution*

**a)** $a_1 = 2, a_2 = 3, b_1 = -1, b_2 = -5$

$$\begin{vmatrix} 2 & -1 \\ 3 & -5 \end{vmatrix} = 2(-5) - (3)(-1) = -10 + 3 = -7$$

**b)** $\begin{vmatrix} 2 & 3 \\ -1 & 4 \end{vmatrix} = (2)(4) - (-1)(3) = 8 + 3 = 11$

Now Try Exercise 7

## 2 Use Cramer's Rule

If we begin with the equations

$$a_1x + b_1y = c_1$$
$$a_2x + b_2y = c_2$$

we can use the addition method to show that

$$x = \frac{c_1b_2 - c_2b_1}{a_1b_2 - a_2b_1} \quad \text{and} \quad y = \frac{a_1c_2 - a_2c_1}{a_1b_2 - a_2b_1}$$

(see Challenge Problem 67 on page 576). Notice that the *denominators* of $x$ and $y$ are both $a_1b_2 - a_2b_1$. Following is the determinant that yields this denominator. We have labeled this denominator $D$.

$$D = \begin{vmatrix} a_1 & b_1 \\ a_2 & b_2 \end{vmatrix} = a_1b_2 - a_2b_1$$

The *numerators* of $x$ and $y$ are different. Following are two determinants, labeled $D_x$ and $D_y$, that yield the numerators of $x$ and $y$.

$$D_x = \begin{vmatrix} c_1 & b_1 \\ c_2 & b_2 \end{vmatrix} = c_1b_2 - c_2b_1 \qquad D_y = \begin{vmatrix} a_1 & c_1 \\ a_2 & c_2 \end{vmatrix} = a_1c_2 - a_2c_1$$

We use determinants $D$, $D_x$, and $D_y$ in *Cramer's rule*. **Cramer's rule** can be used to solve systems of linear equations.

### Cramer's Rule for System of Linear Equations

For a system of linear equations of the form

$$a_1x + b_1y = c_1$$
$$a_2x + b_2y = c_2$$

with

$$D = \begin{vmatrix} a_1 & b_1 \\ a_2 & b_2 \end{vmatrix}, D_x = \begin{vmatrix} c_1 & b_1 \\ c_2 & b_2 \end{vmatrix}, \text{ and } D_y = \begin{vmatrix} a_1 & c_1 \\ a_2 & c_2 \end{vmatrix}$$

then

$$x = \frac{D_x}{D} \quad \text{and} \quad y = \frac{D_y}{D}, D \neq 0.$$

### Helpful Hint

The elements in determinant $D$ are the coefficients of the $x$ and $y$ terms.

$$a_1x + b_1y = c_1$$
$$a_2x + b_2y = c_2$$
$$\downarrow \qquad \downarrow$$
$$D = \begin{vmatrix} a_1 & b_1 \\ a_2 & b_2 \end{vmatrix}$$

To obtain $D_x$ replace the values in the first column with the constants.

$$a_1x + b_1y = c_1$$
$$a_2x + b_2y = c_2$$
$$\downarrow \qquad \downarrow$$
$$D_x = \begin{vmatrix} c_1 & b_1 \\ c_2 & b_2 \end{vmatrix}$$

To obtain $D_y$ replace the values in the second column with the constants.

$$a_1x + b_1y = c_1$$
$$a_2x + b_2y = c_2$$

$$D_y = \begin{vmatrix} a_1 & c_1 \\ a_2 & c_2 \end{vmatrix}$$

**EXAMPLE 2**  Solve the following system of equations using Cramer's rule.

$$3x + 5y = 7$$
$$4x - y = -6$$

*Solution*    Both equations are given in the desired form, $ax + by = c$. When labeling $a$, $b$, and $c$, we will refer to $3x + 5y = 7$ as equation 1 and $4x - y = -6$ as equation 2 for the subscripts.

$$\begin{array}{ccc} a_1 & b_1 & c_1 \\ \downarrow & \downarrow & \downarrow \end{array}$$
$$3x + 5y = 7$$
$$4x - 1y = -6$$
$$\begin{array}{ccc} \uparrow & \uparrow & \uparrow \\ a_2 & b_2 & c_2 \end{array}$$

We now determine $D$, $D_x$, $D_y$.

$$D = \begin{vmatrix} a_1 & b_1 \\ a_2 & b_2 \end{vmatrix} = \begin{vmatrix} 3 & 5 \\ 4 & -1 \end{vmatrix} = 3(-1) - 4(5) = -3 - 20 = -23$$

$$D_x = \begin{vmatrix} c_1 & b_1 \\ c_2 & b_2 \end{vmatrix} = \begin{vmatrix} 7 & 5 \\ -6 & -1 \end{vmatrix} = 7(-1) - (-6)(5) = -7 + 30 = 23$$

$$D_y = \begin{vmatrix} a_1 & c_1 \\ a_2 & c_2 \end{vmatrix} = \begin{vmatrix} 3 & 7 \\ 4 & -6 \end{vmatrix} = 3(-6) - 4(7) = -18 - 28 = -46$$

Now we find the values of $x$ and $y$.

$$x = \frac{D_x}{D} = \frac{23}{-23} = -1$$

$$y = \frac{D_y}{D} = \frac{-46}{-23} = 2$$

Thus the solution is $x = -1$, $y = 2$ or the ordered pair $(-1, 2)$. A check will show that this ordered pair satisfies both equations.

Now Try Exercise 15

When the determinant $D = 0$, Cramer's rule cannot be used since division by 0 is undefined. You must then use a different method to solve the system. However, you may evaluate $D_x$ and $D_y$ to determine whether the system is dependent or inconsistent.

---

**When $D = 0$**

For a system of two linear equations with two variables,

if $D = 0$, $D_x = 0$, and $D_y = 0$, then the system is dependent.

if $D = 0$ and either $D_x \neq 0$ or $D_y \neq 0$, then the system is inconsistent.

### 3 Evaluate the Determinant of a 3 × 3 Matrix

For the determinant

$$\begin{vmatrix} a_1 & b_1 & c_1 \\ a_2 & b_2 & c_2 \\ a_3 & b_3 & c_3 \end{vmatrix}$$

the **minor determinant** of $a_1$ is found by crossing out the elements in the same row and column in which the element $a_1$ appears. The remaining elements form the minor determinant of $a_1$. The minor determinants of other elements are found similarly.

$$\begin{vmatrix} a_1 & b_1 & c_1 \\ a_2 & b_2 & c_2 \\ a_3 & b_3 & c_3 \end{vmatrix} \qquad \begin{vmatrix} b_2 & c_2 \\ b_3 & c_3 \end{vmatrix} \qquad \text{Minor determinant of } a_1$$

$$\begin{vmatrix} a_1 & b_1 & c_1 \\ a_2 & b_2 & c_2 \\ a_3 & b_3 & c_3 \end{vmatrix} \qquad \begin{vmatrix} b_1 & c_1 \\ b_3 & c_3 \end{vmatrix} \qquad \text{Minor determinant of } a_2$$

$$\begin{vmatrix} a_1 & b_1 & c_1 \\ a_2 & b_2 & c_2 \\ a_3 & b_3 & c_3 \end{vmatrix} \qquad \begin{vmatrix} b_1 & c_1 \\ b_2 & c_2 \end{vmatrix} \qquad \text{Minor determinant of } a_3$$

To evaluate determinants of a 3 × 3 matrix, we use minor determinants. The following box shows how such a determinant may be evaluated by **expansion by the minors of the first column**.

---

**Expansion of the Determinant by the Minors of the First Column**

$$\begin{vmatrix} a_1 & b_1 & c_1 \\ a_2 & b_2 & c_2 \\ a_3 & b_3 & c_3 \end{vmatrix} = a_1 \begin{vmatrix} b_2 & c_2 \\ b_3 & c_3 \end{vmatrix} - a_2 \begin{vmatrix} b_1 & c_1 \\ b_3 & c_3 \end{vmatrix} + a_3 \begin{vmatrix} b_1 & c_1 \\ b_2 & c_2 \end{vmatrix}$$

where the terms correspond to the Minor determinant of $a_1$, Minor determinant of $a_2$, and Minor determinant of $a_3$.

---

**EXAMPLE 3** Evaluate $\begin{vmatrix} 4 & -2 & 6 \\ 3 & 5 & 0 \\ 1 & -3 & -1 \end{vmatrix}$ using expansion by the minors of the first column.

*Solution*    We will follow the procedure given in the box.

$$\begin{vmatrix} 4 & -2 & 6 \\ 3 & 5 & 0 \\ 1 & -3 & -1 \end{vmatrix} = 4 \begin{vmatrix} 5 & 0 \\ -3 & -1 \end{vmatrix} - 3 \begin{vmatrix} -2 & 6 \\ -3 & -1 \end{vmatrix} + 1 \begin{vmatrix} -2 & 6 \\ 5 & 0 \end{vmatrix}$$

$$= 4[5(-1) - (-3)0] - 3[(-2)(-1) - (-3)6] + 1[(-2)0 - 5(6)]$$

$$= 4(-5 + 0) - 3(2 + 18) + 1(0 - 30)$$

$$= 4(-5) - 3(20) + 1(-30)$$

$$= -20 - 60 - 30$$

$$= -110$$

The determinant has a value of $-110$.

Now Try Exercise 13

## 4  Use Cramer's Rule with Systems in Three Variables

Cramer's rule can also be used to solve systems of equations in three variables as follows.

### Cramer's Rule for a System of Equations in Three Variables

To solve the system

$$a_1x + b_1y + c_1z = d_1$$
$$a_2x + b_2y + c_2z = d_2$$
$$a_3x + b_3y + c_3z = d_3$$

with

$$D = \begin{vmatrix} a_1 & b_1 & c_1 \\ a_2 & b_2 & c_2 \\ a_3 & b_3 & c_3 \end{vmatrix} \qquad D_x = \begin{vmatrix} d_1 & b_1 & c_1 \\ d_2 & b_2 & c_2 \\ d_3 & b_3 & c_3 \end{vmatrix}$$

$$D_y = \begin{vmatrix} a_1 & d_1 & c_1 \\ a_2 & d_2 & c_2 \\ a_3 & d_3 & c_3 \end{vmatrix} \qquad D_z = \begin{vmatrix} a_1 & b_1 & d_1 \\ a_2 & b_2 & d_2 \\ a_3 & b_3 & d_3 \end{vmatrix}$$

then

$$x = \frac{D_x}{D} \qquad y = \frac{D_y}{D} \qquad z = \frac{D_z}{D}, \quad D \neq 0$$

### Helpful Hint

Note that the denominators of the expressions for $x, y,$ and $z$ are all the same determinant, $D$. Note that the $d$'s replace the $a$'s, the numerical coefficients of the $x$-terms, in $D_x$. The $d$'s replace the $b$'s, the numerical coefficients of the $y$-terms, in $D_y$. And the $d$'s replace the $c$'s, the numerical coefficients of the $z$-terms, in $D_z$.

**EXAMPLE 4**  Solve the following system of equations using determinants.

$$3x - 2y - z = -6$$
$$2x + 3y - 2z = 1$$
$$x - 4y + z = -3$$

*Solution*

$$a_1 = 3 \quad b_1 = -2 \quad c_1 = -1 \quad d_1 = -6$$
$$a_2 = 2 \quad b_2 = 3 \quad c_2 = -2 \quad d_2 = 1$$
$$a_3 = 1 \quad b_3 = -4 \quad c_3 = 1 \quad d_3 = -3$$

We will use expansion by the minor determinants of the first column to evaluate $D$, $D_x$, $D_y$, and $D_z$.

$$D = \begin{vmatrix} 3 & -2 & -1 \\ 2 & 3 & -2 \\ 1 & -4 & 1 \end{vmatrix} = 3\begin{vmatrix} 3 & -2 \\ -4 & 1 \end{vmatrix} - 2\begin{vmatrix} -2 & -1 \\ -4 & 1 \end{vmatrix} + 1\begin{vmatrix} -2 & -1 \\ 3 & -2 \end{vmatrix}$$

$$= 3(-5) - 2(-6) + 1(7)$$

$$= -15 + 12 + 7 = 4$$

$$D_x = \begin{vmatrix} -6 & -2 & -1 \\ 1 & 3 & -2 \\ -3 & -4 & 1 \end{vmatrix} = -6\begin{vmatrix} 3 & -2 \\ -4 & 1 \end{vmatrix} - 1\begin{vmatrix} -2 & -1 \\ -4 & 1 \end{vmatrix} + (-3)\begin{vmatrix} -2 & -1 \\ 3 & -2 \end{vmatrix}$$

$$= -6(-5) - 1(-6) - 3(7)$$

$$= 30 + 6 - 21 = 15$$

$$D_y = \begin{vmatrix} 3 & -6 & -1 \\ 2 & 1 & -2 \\ 1 & -3 & 1 \end{vmatrix} = 3\begin{vmatrix} 1 & -2 \\ -3 & 1 \end{vmatrix} - 2\begin{vmatrix} -6 & -1 \\ -3 & 1 \end{vmatrix} + 1\begin{vmatrix} -6 & -1 \\ 1 & -2 \end{vmatrix}$$

$$= 3(-5) - 2(-9) + 1(13)$$

$$= -15 + 18 + 13 = 16$$

## Understanding Algebra

When we have a system of equations in three variables in which one or more equations are missing a variable, we represent the missing variable with a coefficient of 0. Thus,

$$2x - 3y + 2z = -1$$
$$x + 2y \qquad = 14$$
$$x \qquad -3z = -5$$

is written

$$2x - 3y + 2z = -1$$
$$x + 2y + 0z = 14$$
$$x + 0y - 3z = -5$$

$$D_z = \begin{vmatrix} 3 & -2 & -6 \\ 2 & 3 & 1 \\ 1 & -4 & -3 \end{vmatrix} = 3\begin{vmatrix} 3 & 1 \\ -4 & -3 \end{vmatrix} - 2\begin{vmatrix} -2 & -6 \\ -4 & -3 \end{vmatrix} + 1\begin{vmatrix} -2 & -6 \\ 3 & 1 \end{vmatrix}$$

$$= 3(-5) - 2(-18) + 1(16)$$

$$= -15 + 36 + 16 = 37$$

We found that $D = 4$, $D_x = 15$, $D_y = 16$, and $D_z = 37$. Therefore,

$$x = \frac{D_x}{D} = \frac{15}{4} \qquad y = \frac{D_y}{D} = \frac{16}{4} = 4 \qquad z = \frac{D_z}{D} = \frac{37}{4}$$

The solution to the system is $\left(\frac{15}{4}, 4, \frac{37}{4}\right)$. Note the ordered triple lists $x$, $y$, and $z$ in this order.

Now Try Exercise 33

### Helpful Hint

When evaluating determinants, if any two rows (or columns) are identical, or identical except for opposite signs, the determinant has a value of 0. For example,

$$\begin{vmatrix} 5 & -2 \\ 5 & -2 \end{vmatrix} = 0 \quad \text{and} \quad \begin{vmatrix} 5 & -2 \\ -5 & 2 \end{vmatrix} = 0$$

$$\begin{vmatrix} 5 & -3 & 4 \\ 2 & 6 & 5 \\ 5 & -3 & 4 \end{vmatrix} = 0 \quad \text{and} \quad \begin{vmatrix} 5 & -3 & 4 \\ -5 & 3 & -4 \\ 6 & 8 & 2 \end{vmatrix} = 0$$

As with determinants of a $2 \times 2$ matrix, when the determinant $D = 0$, Cramer's rule cannot be used since division by 0 is undefined. You must then use a different method to solve the system. However, you may evaluate $D_x$, $D_y$, and $D_z$ to determine whether the system is dependent or inconsistent.

### When $D = 0$

For a system of three linear equations with three variables,

if $D = 0$, $D_x = 0$, $D_y = 0$, and $D_z = 0$, then the system is dependent.

if $D = 0$ and $D_x \neq 0$, $D_y \neq 0$, or $D_z \neq 0$, then the system is inconsistent.

# EXERCISE SET 9.7    MathXL®    MyMathLab®

MathXL®    MyMathLab

## Warm-Up Exercises

*Fill in the blanks with the appropriate word, phrase, or symbol(s) from the following list.*

$D \qquad D_x \qquad D_y \qquad D_z \qquad$ determinant $\qquad$ dependent $\qquad$ inconsistent $\qquad$ unique

**1.** Associated with every square matrix is a number called its _____.

*For Exercises 2–4, refer to the system of equations*

$$2x - 3y = 2$$
$$x - 7y = 6$$

**2.** The determinant _____ $= \begin{vmatrix} 2 & -3 \\ 1 & -7 \end{vmatrix}$.

**3.** The determinant _____ $= \begin{vmatrix} 2 & 2 \\ 1 & 6 \end{vmatrix}$.

**4.** The determinant _____ $= \begin{vmatrix} 2 & -3 \\ 6 & -7 \end{vmatrix}$.

**5.** In a system of two linear equations with two variables, if $D = 0$, $D_x = 0$, and $D_y = 0$, then the system is _____ and the system has an infinite number of solutions.

**6.** In a system of two linear equations with two variables, if $D = 0$, and either $D_x \neq 0$ or $D_y \neq 0$, then the system is _____ and the system has no solutions.

## Practice the Skills

*Evaluate each determinant.*

**7.** $\begin{vmatrix} 3 & 2 \\ 5 & 7 \end{vmatrix}$

**8.** $\begin{vmatrix} 3 & 5 \\ -1 & -2 \end{vmatrix}$

**9.** $\begin{vmatrix} \frac{1}{2} & 3 \\ 2 & -4 \end{vmatrix}$

**10.** $\begin{vmatrix} 13 & -\frac{2}{3} \\ -1 & 0 \end{vmatrix}$

**11.** $\begin{vmatrix} 3 & 2 & 0 \\ 0 & 5 & 3 \\ -1 & 4 & 2 \end{vmatrix}$

**12.** $\begin{vmatrix} 4 & 1 & 1 \\ 0 & 0 & 3 \\ 2 & 2 & 9 \end{vmatrix}$

**13.** $\begin{vmatrix} 2 & 3 & 1 \\ 1 & -3 & -6 \\ -4 & 5 & 9 \end{vmatrix}$

**14.** $\begin{vmatrix} 5 & -8 & 6 \\ 3 & 0 & 4 \\ -5 & -2 & 1 \end{vmatrix}$

*Solve each system of equations using determinants.*

**15.** $3x + 4y = 10$
$x + 3y = 5$

**16.** $2x + 4y = -2$
$-5x - 2y = 13$

**17.** $-x - 2y = 2$
$x + 3y = -6$

**18.** $2r + 3s = -9$
$3r + 5s = -16$

**19.** $6x = 4y + 7$
$8x - 1 = -3y$

**20.** $6x + 3y = -4$
$9x + 5y = -6$

**21.** $5p - 7q = -21$
$-4p + 3q = 22$

**22.** $4x = -5y - 2$
$-2x = y + 4$

**23.** $x + 5y = 3$
$2x - 6 = -10y$

**24.** $9x + 6y = -3$
$6x + 4y = -2$

**25.** $3r = -4s - 6$
$3s = -5r + 1$

**26.** $x = y - 1$
$3y = 2x + 9$

**27.** $5x - 5y = 3$
$-x + y = -4$

**28.** $2x - 5y = -3$
$-4x + 10y = 7$

**29.** $6.3x - 4.5y = -9.9$
$-9.1x + 3.2y = -2.2$

**30.** $-1.1x + 8.3y = 36.5$
$3.5x + 1.6y = -4.1$

*Solve each system using determinants.*

**31.** $x + y - z = 2$
$2x + 3y - 2z = 6$
$5x - 2y + 3z = 4$

**32.** $2x + 3y = 4$
$3x + 7y - 4z = -3$
$x - y + 2z = 9$

**33.** $3x - 5y - 4z = -4$
$4x + 2y = 1$
$6y - 4z = -11$

**34.** $2x + 5y + 3z = 2$
$6x - 9y = 5$
$3y + 2z = 1$

**35.** $x + 4y - 3z = -6$
$2x - 8y + 5z = 12$
$3x + 4y - 2z = -3$

**36.** $2x + y - 2z = 4$
$2x + 2y - 4z = 1$
$-6x + 8y - 4z = 1$

**37.** $a - b + 2c = 3$
$a - b + c = 1$
$2a + b + 2c = 2$

**38.** $-2x + y + 8 = -2$
$3x + 2y + z = 3$
$x - 3y - 5z = 5$

**39.** $a + 2b + c = 1$
$a - b + 3c = 2$
$2a + b + 4c = 3$

**40.** $4x - 2y + 6z = 2$
$-6x + 3y - 9z = -3$
$2x - 7y + 11z = -5$

**41.** $1.1x + 2.3y - 4.0z = -9.2$
$-2.3x + 4.6z = 6.9$
$-8.2y - 7.5z = -6.8$

**42.** $4.6y - 2.1z = 24.3$
$-5.6x + 1.8y = -5.8$
$2.8x - 4.7y - 3.1z = 7.0$

**43.** $-6x + 3y - 12z = -13$
$5x + 2y - 3z = 1$
$2x - y + 4z = -5$

**44.** $x - 2y + z = 2$
$4x - 6y + 2z = 3$
$2x - 3y + z = 0$

**45.** $2x + \frac{1}{2}y - 3z = 5$
$-3x + 2y + 2z = 1$
$4x - \frac{1}{4}y - 7z = 4$

**46.** $\frac{1}{4}x - \frac{1}{2}y + 3z = -3$
$2x - 3y + 2z = -1$
$\frac{1}{6}x + \frac{1}{3}y - \frac{1}{3}z = 1$

**47.** $0.3x - 0.1y - 0.3z = -0.2$
$0.2x - 0.1y + 0.1z = -0.9$
$0.1x + 0.2y - 0.4z = 1.7$

**48.** $0.6u - 0.4v + 0.5w = 3.1$
$0.5u + 0.2v + 0.2w = 1.3$
$0.1u + 0.1v + 0.1w = 0.2$

## Problem Solving

*Solve for the given letter.*

**49.** $\begin{vmatrix} 4 & 6 \\ -2 & y \end{vmatrix} = 32$

**50.** $\begin{vmatrix} b-3 & -4 \\ b+2 & -6 \end{vmatrix} = 14$

**51.** $\begin{vmatrix} 4 & 7 & y \\ 3 & -1 & 2 \\ 4 & 1 & 5 \end{vmatrix} = -35$

**52.** $\begin{vmatrix} 3 & 2 & -2 \\ 0 & 5 & -6 \\ -1 & x & -7 \end{vmatrix} = -31$

*Solve using determinants.*

**53. Ticket Prices** Dayton Sinkia purchased 5 adult tickets and 8 student tickets to see the play *A Servant of Two Masters* at State College of Florida for $90. Danielle Zoller purchased 3 adult tickets and 7 student tickets for $65. Determine the price of an adult ticket and the price of a student ticket.

**54. Little League Concession Stand Prices** At Braden River Little League, Beth Van Vranken purchased 3 hot dogs, 4 bottles of water, and 2 bags of sunflower seeds for $11. Jeff Parrill purchased 5 hot dogs, 3 bottles of water, and 4 bags of sunflower seeds for $14.25. Dave Hauck purchased 1 hot dog, 2 bottles of water, and 5 bags of sunflower seeds for $7.75. Determine the price of a hot dog, a bottle of water, and a bag of sunflower seeds.

© RBFried/iStockphoto

## Concept/Writing Exercises

**55.** Given a determinant of the form $\begin{vmatrix} a_1 & b_1 \\ a_2 & b_2 \end{vmatrix}$, how will the value of the determinant change if the $a$'s are switched with each other and the $b$'s are switched with each other, $\begin{vmatrix} a_2 & b_2 \\ a_1 & b_1 \end{vmatrix}$? Explain your answer.

**56.** Given a determinant of the form $\begin{vmatrix} a_1 & b_1 \\ a_2 & b_2 \end{vmatrix}$, how will the value of the determinant change if the $a$'s are switched with the $b$'s, $\begin{vmatrix} b_1 & a_1 \\ b_2 & a_2 \end{vmatrix}$? Explain your answer.

**57.** In a $2 \times 2$ determinant, if the rows are the same, what is the value of the determinant?

**58.** If all the numbers in one row or one column of a $2 \times 2$ determinant are 0, what is the value of the determinant?

**59.** If all the numbers in one row or one column of a $3 \times 3$ determinant are 0, what is the value of the determinant?

**60.** Given a $3 \times 3$ determinant, if all the numbers in one row are multiplied by $-1$, will the value of the new determinant change? Explain.

**61.** Given a $3 \times 3$ determinant, if the first and second rows are switched, will the value of the new determinant change? Explain.

**62.** In a $3 \times 3$ determinant, if any two rows are the same, can you make a generalization about the value of the determinant?

**63.** In a $3 \times 3$ determinant, if the numbers in the first row are multiplied by $-1$ and the numbers in the second row are multiplied by $-1$, will the value of the new determinant change? Explain.

**64.** In a $3 \times 3$ determinant, if the numbers in the second row are multiplied by $-1$ and the numbers in the third row are multiplied by $-1$, will the value of the new determinant change? Explain.

**65.** In a $3 \times 3$ determinant, if the numbers in the second row are multiplied by 2, will the value of the new determinant change? Explain.

**66.** In a $3 \times 3$ determinant, if the numbers in the first row are multiplied by 3 and the numbers in the third row are multiplied by 4, will the value of the new determinant change? Explain.

## Challenge Problems

**67.** Use the addition method to solve the following system for **a)** $x$ and **b)** $y$.

$$a_1x + b_1y = c_1$$

$$a_2x + b_2y = c_2$$

## Cumulative Review Exercises

[2.6]   **68.** Solve the equation $2x - 5y = 6$ for $y$.

*Graph $3x + 4y = 8$ using the indicated method.*

[4.2]   **69.** By plotting points

**70.** Using the $x$- and $y$-intercepts

[4.4]   **71.** Using the slope and $y$-intercept

# Chapter 9 Summary

| IMPORTANT FACTS AND CONCEPTS | EXAMPLES |
|---|---|

### Section 9.1

A **system of linear equations** is a system having two or more linear equations.

The **solution to a system of equations** is the ordered pair or pairs that satisfy all the equations in the system.

$$y = 2x - 5$$
$$y = -4x + 7$$

The solution to the above system is $(2, -1)$.

---

A linear system of equations may have exactly one solution (intersecting lines) no solution (parallel lines) or an infinite number of solutions (same line).

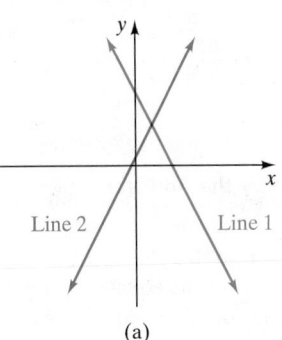

Consistent, exactly 1 solution

(a)

Inconsistent, no solution

(b)

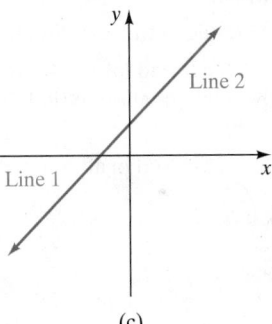

Dependent, infinite number of solutions

(c)

---

To **solve a system of equations graphically**, graph each equation and determine the point or points of intersection.

Solve the system of equations graphically.

$$y = -x + 5$$
$$y = x + 3$$

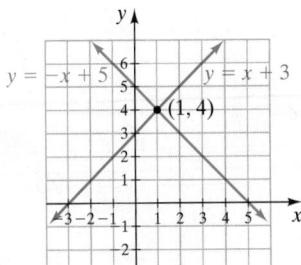

The solution is $(1, 4)$.

### Section 9.2

**To Solve a System of Equations by Substitution**

1. Solve for a variable in either equation.

2. Substitute the expression found for the variable in step 1 into the other equation.

Solve the system of equations by substitution.

$$y = -x + 5$$
$$y = x + 3$$

| IMPORTANT FACTS AND CONCEPTS | EXAMPLES |
|---|---|

### Section 9.2 (cont.)

**3.** Solve the equation determined in step 2 to find the value of one variable.

**4.** Substitute the value found in step 3 into the equation obtained in step 1 to find the value of the other variable.

**5.** Check by substituting both values in both original equations.

Substitute $x + 3$ for $y$ in the first equation.

$$x + 3 = -x + 5$$
$$2x + 3 = 5$$
$$2x = 2$$
$$x = 1$$

Now, solve for $y$ by substituting $x = 1$ into the first equation.

$$y = -x + 5$$
$$y = -1 + 5$$
$$= 4$$

The solution is $(1, 4)$.

### Section 9.3

**To Solve a System of Equations by the Addition (or Elimination) Method**

**1.** If necessary, rewrite each equation so that the terms containing variables appear on the left side of the equals sign and any constants appear on the right side of the equals sign.

**2.** If necessary, multiply one or both equations by a constant(s) so that when the equations are added the resulting sum will contain only one variable.

**3.** Add the equations.

**4.** Solve for the variable in the equation from step 3.

**5.** Substitute the value found in step 4 into either of the original equations. Solve that equation to find the value of the remaining variable.

**6.** Check the values obtained in all original equations.

Solve the system of equations by the addition method.

$$2x + 3y = -4 \qquad eq.\,1$$
$$3x - y = -17. \qquad eq.\,2$$

Multiply *eq.* 2 by 3, then add:

$$\begin{array}{ll} 2x + 3y = \phantom{-}-4 & eq.\,1 \text{ stays the same} \\ \underline{9x - 3y = -51} & eq.\,2 \text{ multiplied by 3} \\ 11x \phantom{- 3y} = -55 & \\ x = \phantom{-}-5 & \end{array}$$

Substitute $-5$ for $x$ into *eq.* 2.

$$3x - y = -17$$
$$3(-5) - y = -17$$
$$-15 - y = -17$$
$$-y = -2$$
$$y = 2$$

The solution is $(-5, 2)$.

### Section 9.4

**To Solve Three Linear Equations in Three Variables**

To solve a system of three linear equations, use the substitution method or addition method.

Solve the system of equations.

$$x - y + 3z = -1 \qquad (eq.\,1)$$
$$4y - 7z = 2 \qquad (eq.\,2)$$
$$z = 2 \qquad (eq.\,3)$$

Substitute 2 for $z$ in $(eq.\,2)$ to obtain the value for $y$.

$$4y - 7z = 2$$
$$4y - 7(2) = 2$$
$$4y = 16$$
$$y = 4.$$

Substitute 4 for $y$ and 2 for $z$ in $(eq.\,1)$ to obtain the value for $x$.

$$x - y + 3z = -1$$
$$x - 4 + 3(2) = -1$$
$$x = -3.$$

A check shows that $(-3, 4, 2)$ is a solution to the system of equations.

| IMPORTANT FACTS AND CONCEPTS | EXAMPLES |
|---|---|

### Section 9.5

**Applications**

Systems of two linear equations in two unknowns.

The sum of the areas of two circles is 180 square meters. The difference of their areas is 20 square meters. Determine the area of each circle.

**Solution**

Let $x$ be the area of the larger circle and $y$ be the area of the smaller circle.

The two equations for this systems are

$$
\begin{array}{ll}
x + y = 180 & \text{Sum of areas} \\
\underline{x - y = \phantom{0}20} & \text{Difference of areas} \\
2x \phantom{+ y} = 200 & \\
\phantom{2}x = 100 &
\end{array}
$$

Substitute 100 for $x$ in the first equation to get

$$
\begin{aligned}
x + y &= 180 \\
100 + y &= 180 \\
y &= 80
\end{aligned}
$$

The area of the larger circle is 100 square meters and the area of the smaller circle is 80 square meters.

### Section 9.6

A **matrix** is a rectangular array of numbers within brackets. The numbers inside the brackets are called **elements**.

A **square matrix** has the same number of rows and columns.

$$
\begin{bmatrix} 8 & 1 & 4 \\ -3 & 0 & 2 \end{bmatrix}, \begin{bmatrix} 5 & 0 \\ -2 & 8 \\ 6 & -11 \end{bmatrix}
$$

$$
\begin{bmatrix} 5 & -1 \\ 8 & 2 \end{bmatrix}, \begin{bmatrix} 3 & 0 & -6 \\ -1 & 5 & 2 \\ 9 & 10 & -7 \end{bmatrix}
$$

An **augmented matrix** is a matrix separated by a vertical line. For a system of equations, the coefficients of the variables are placed on the left side of the vertical line and the constants are placed on the right side in the augmented matrix.

The **row echelon** (or **triangular**) **form** of an augmented matrix is

$$
\begin{bmatrix} 1 & a & | & p \\ 0 & 1 & | & q \end{bmatrix}
$$

|  System  | Augmented Matrix |
|---|---|

$$
\begin{aligned}
2x - 3y &= 8 \\
5x + 7y &= -4
\end{aligned}
\qquad
\left[ \begin{array}{cc|c} 2 & -3 & 8 \\ 5 & 7 & -4 \end{array} \right]
$$

row echelon form $\qquad \left[ \begin{array}{cc|c} 1 & -6 & 2 \\ 0 & 1 & 9 \end{array} \right]$

**Row transformations** can be used to rewrite a matrix into row echelon form.

**Procedures for Row Transformations**

1. All the numbers in a row may be multiplied (or divided) by any nonzero real number.

2. All the numbers in a row may be multiplied by any nonzero real number. These products may then be added to the corresponding numbers in any other row.

3. The order of the rows may be switched.

Solve the system of equations

$$
\begin{aligned}
x + 4y &= -7 \\
6x - 5y &= 16
\end{aligned}
$$

The augmented matrix is

$$
\left[ \begin{array}{cc|c} 1 & 4 & -7 \\ 6 & -5 & 16 \end{array} \right] = \left[ \begin{array}{cc|c} 1 & 4 & -7 \\ 0 & -29 & 58 \end{array} \right] \begin{array}{l} \\ -6R_1 + R_2 \end{array}
$$

$$
= \left[ \begin{array}{cc|c} 1 & 4 & -7 \\ 0 & 1 & -2 \end{array} \right] -\frac{1}{29} R_2
$$

The equivalent system of equations is

$$
\begin{aligned}
x + 4y &= -7 \\
y &= -2
\end{aligned}
$$

| IMPORTANT FACTS AND CONCEPTS | EXAMPLES |
|---|---|

### Section 9.6 (cont.)

| | Substitute $-2$ for $y$ into the first equation. |
|---|---|
| | $$x + 4(-2) = -7$$ $$x - 8 = -7$$ $$x = 1.$$ The solution is $(1, -2)$. |
| A system of equations is **inconsistent** and has **no solution** if you obtain an augmented matrix in which one row of numbers has zeros on the left side of the vertical line and a nonzero number on the right side of the vertical line. | $$\left[\begin{array}{ccc|c} 1 & 2 & -3 & 23 \\ 0 & 0 & 0 & 8 \\ -1 & 7 & 6 & 9 \end{array}\right]$$ |
| A system of equations is **dependent** and has an **infinite number of solutions** if you obtain an augmented matrix in which a 0 appears across an entire row. | The second row shows this system is inconsistent and has no solution. $$\left[\begin{array}{ccc|c} 1 & 6 & -1 & 15 \\ 0 & 0 & 0 & 0 \\ 3 & 5 & 8 & -12 \end{array}\right]$$ The second row shows this system is dependent and has an infinite number of solutions. |

### Section 9.7

| The **determinant** of a $2 \times 2$ matrix $\begin{bmatrix} a_1 & b_1 \\ a_2 & b_2 \end{bmatrix}$ is denoted $\begin{vmatrix} a_1 & b_1 \\ a_2 & b_2 \end{vmatrix}$ and is evaluated as $$\begin{vmatrix} a_1 & b_1 \\ a_2 & b_2 \end{vmatrix} = a_1 b_2 - a_2 b_1$$ | $$\begin{vmatrix} 3 & -2 \\ 5 & 1 \end{vmatrix} = (3)(1) - (5)(-2) = 3 + 10 = 13$$ |
|---|---|
| **Cramer's Rule for Systems of Linear Equations** For a system of linear equations of the form $$a_1 x + b_1 y = c_1$$ $$a_2 x + b_2 y = c_2$$ with $$D = \begin{vmatrix} a_1 & b_1 \\ a_2 & b_2 \end{vmatrix}, D_x = \begin{vmatrix} c_1 & b_1 \\ c_2 & b_2 \end{vmatrix}, \text{ and } D_y = \begin{vmatrix} a_1 & c_1 \\ a_2 & c_2 \end{vmatrix}$$ then $$x = \frac{D_x}{D} \text{ and } y = \frac{D_y}{D}, D \neq 0$$ | Solve the system of equations. $$2x + y = 6$$ $$4x - 3y = -13$$ $$D = \begin{vmatrix} 2 & 1 \\ 4 & -3 \end{vmatrix} = -10$$ $$D_x = \begin{vmatrix} 6 & 1 \\ -13 & -3 \end{vmatrix} = -5 \qquad D_y = \begin{vmatrix} 2 & 6 \\ 4 & -13 \end{vmatrix} = -50$$ Then $$x = \frac{D_x}{D} = \frac{-5}{-10} = \frac{1}{2}, \quad y = \frac{D_y}{D} = \frac{-50}{-10} = 5$$ The solution is $\left(\frac{1}{2}, 5\right)$. |
| For the determinant $$\begin{vmatrix} a_1 & b_1 & c_1 \\ a_2 & b_2 & c_2 \\ a_3 & b_3 & c_3 \end{vmatrix}$$ The **minor determinant of $a_1$** is found by crossing out the elements in the same row and column containing the element $a_1$. | For $\begin{vmatrix} 6 & 2 & -1 \\ 0 & 3 & 5 \\ 7 & 1 & 9 \end{vmatrix}$, the minor determinant of $a_1$ is $\begin{vmatrix} 3 & 5 \\ 1 & 9 \end{vmatrix}$. Evaluate $\begin{vmatrix} 2 & 0 & 3 \\ -1 & -5 & 2 \\ 3 & 6 & -4 \end{vmatrix}$ using expansion by minors of the first column. |

| IMPORTANT FACTS AND CONCEPTS | EXAMPLES |
| --- | --- |

**Section 9.7 (cont.)**

**Expansion of the Determinant by the Minors of the First Column**

$$
\begin{vmatrix} a_1 & b_1 & c_1 \\ a_2 & b_2 & c_2 \\ a_3 & b_3 & c_3 \end{vmatrix} = a_1 \begin{vmatrix} b_2 & c_2 \\ b_3 & c_3 \end{vmatrix} - a_2 \begin{vmatrix} b_1 & c_1 \\ b_3 & c_3 \end{vmatrix} + a_3 \begin{vmatrix} b_1 & c_1 \\ b_2 & c_2 \end{vmatrix}
$$

Minor determinant of $a_1$, Minor determinant of $a_2$, Minor determinant of $a_3$

$$
\begin{vmatrix} 2 & 0 & 3 \\ -1 & -5 & 2 \\ 3 & 6 & -4 \end{vmatrix} = 2 \begin{vmatrix} -5 & 2 \\ 6 & -4 \end{vmatrix} - (-1) \begin{vmatrix} 0 & 3 \\ 6 & -4 \end{vmatrix} + 3 \begin{vmatrix} 0 & 3 \\ -5 & 2 \end{vmatrix}
$$

$$
= 2(8) + 1(-18) + 3(15)
$$
$$
= 16 - 18 + 45
$$
$$
= 43
$$

**Cramer's Rule for a System of Equations in Three Variables**

To solve the system

$$
a_1x + b_1y + c_1z = d_1
$$
$$
a_2x + b_2y + c_2z = d_2
$$
$$
a_3x + b_3y + c_3z = d_3
$$

with

$$
D = \begin{vmatrix} a_1 & b_1 & c_1 \\ a_2 & b_2 & c_2 \\ a_3 & b_3 & c_3 \end{vmatrix} \qquad D_x = \begin{vmatrix} d_1 & b_1 & c_1 \\ d_2 & b_2 & c_2 \\ d_3 & b_3 & c_3 \end{vmatrix}
$$

$$
D_y = \begin{vmatrix} a_1 & d_1 & c_1 \\ a_2 & d_2 & c_2 \\ a_3 & d_3 & c_3 \end{vmatrix} \qquad D_z = \begin{vmatrix} a_1 & b_1 & d_1 \\ a_2 & b_2 & d_2 \\ a_3 & b_3 & d_3 \end{vmatrix}
$$

then

$$
x = \frac{D_x}{D} \quad y = \frac{D_y}{D} \quad z = \frac{D_z}{D}, \quad D \neq 0
$$

Solve the system of equations.

$$
2x + y + z = 0
$$
$$
4x - y + 3z = -9
$$
$$
6x + 2y + 5z = -8
$$

$$
D = \begin{vmatrix} 2 & 1 & 1 \\ 4 & -1 & 3 \\ 6 & 2 & 5 \end{vmatrix} = -10 \qquad D_x = \begin{vmatrix} 0 & 1 & 1 \\ -9 & -1 & 3 \\ -8 & 2 & 5 \end{vmatrix} = -5
$$

$$
D_y = \begin{vmatrix} 2 & 0 & 1 \\ 4 & -9 & 3 \\ 6 & -8 & 5 \end{vmatrix} = -20 \qquad D_z = \begin{vmatrix} 2 & 1 & 0 \\ 4 & -1 & -9 \\ 6 & 2 & -8 \end{vmatrix} = 30
$$

Then

$$
x = \frac{D_x}{D} = \frac{-5}{-10} = \frac{1}{2}, \quad y = \frac{D_y}{D} = \frac{-20}{-10} = 2 \quad z = \frac{D_z}{D} = \frac{30}{-10} = -3
$$

The solution is $\left(\frac{1}{2}, 2, -3\right)$.

## Chapter 9 Review Exercises

**[9.1]** *Determine which, if any, of the ordered pairs satisfy each system of equations.*

**1.** $y = 4x - 2$
$2x + 3y = 8$
**a)** $(0, -2)$     **b)** $(4, 0)$     **c)** $(1, 2)$

**2.** $y = -x + 4$
$3x + 5y = 15$
**a)** $\left(\frac{5}{2}, \frac{3}{2}\right)$     **b)** $(-1, 5)$     **c)** $\left(\frac{1}{2}, \frac{3}{5}\right)$

*Identify each system of linear equations as consistent, inconsistent, or dependent. State whether the system has exactly one solution, no solution, or an infinite number of solutions.*

**3.**      **4.**      **5.**      **6.**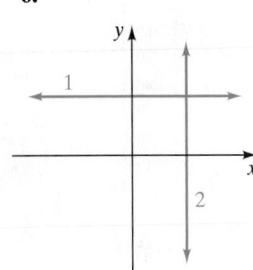

*Write each equation in slope-intercept form. Without graphing or solving the system of equations, state whether the system of linear equations has exactly one solution, no solution, or an infinite number of solutions.*

**7.** $x + 2y = 7$
$4x = -8y + 15$

**8.** $y = 4x - 3$
$3x - y = 7$

**9.** $y = \frac{1}{3}x + \frac{2}{3}$
$6y - 2x = 4$

**10.** $6x = 4y - 20$
$4x = 6y + 20$

*Determine the solution to each system of equations graphically.*

**11.** $y = x - 4$
  $y = 2x - 7$

**12.** $x = -2$
  $y = 5$

**13.** $x + 2y = 8$
  $2x - y = -4$

**14.** $x + 4y = 8$
  $y = 2$

**15.** $y = 3$
  $y = -2x + 5$

**16.** $y = x - 3$
  $3x - 3y = 9$

**17.** $3x + y = 0$
  $3x - 3y = 12$

**18.** $x + 5y = 10$
  $\dfrac{1}{5}x + y = -1$

[9.2]  *Find the solution to each system of equations by substitution.*

**19.** $y = 2x - 8$
  $2x - 5y = 0$

**20.** $x = 3y - 9$
  $x + 2y = 1$

**21.** $2x - y = 7$
  $x + 2y = 6$

**22.** $x = -3y$
  $x + 4y = 5$

**23.** $4x - 2y = 7$
  $y = 2x + 3$

**24.** $2x - 4y = 7$
  $-4x + 8y = -14$

**25.** $2x - 3y = 8$
  $6x - 5y = 20$

**26.** $3x - y = -5$
  $x + 2y = 8$

[9.3]  *Find the solution to each system of equations using the addition method.*

**27.** $x + y = -8$
  $-x + 6y = -6$

**28.** $x + 2y = -3$
  $5x - 2y = 9$

**29.** $x + y = 12$
  $2x + y = 5$

**30.** $4x - 3y = 8$
  $2x + 5y = 8$

**31.** $-2x + 3y = 15$
  $7x + 3y = 6$

**32.** $2x + y = 3$
  $-4x - 2y = 5$

**33.** $3x = -4y + 15$
  $8y = -6x + 30$

**34.** $2x - 5y = 12$
  $3x - 4y = -6$

[9.4]  *Determine the solution to each system of equations using substitution or the addition method.*

**35.** $x - 2y - 4z = 13$
  $3y + 2z = -2$
  $5z = -20$

**36.** $2a + b - 2c = 5$
  $3b + 4c = 1$
  $3c = -6$

**37.** $x + 2y + 3z = 3$
  $-2x - 3y - z = 5$
  $3x + 3y + 7z = 2$

**38.** $-x - 4y + 2z = 1$
  $2x + 2y + z = 0$
  $-3x - 2y - 5z = 5$

**39.** $3y - 2z = -4$
  $3x - 5z = -7$
  $2x + y = 6$

**40.**  $a + 2b - 5c = 19$
  $2a - 3b + 3c = -15$
  $5a - 4b - 2c = -2$

**41.**  $x - y + 3z = 1$
  $-x + 2y - 2z = 1$
  $x - 3y + z = 2$

**42.** $-2x + 2y - 3z = 6$
  $4x - y + 2z = -2$
  $2x + y - z = 4$

[9.5]  *Express each problem as a system of linear equations and use the method of your choice to find the solution to the problem.*

**43. Ages** Luan Baker is 10 years older than his niece, Jennifer Miesen. If the sum of their ages is 66, find Luan's age and Jennifer's age.

**44. Air Speed** An airplane can travel 560 miles per hour with the wind and 480 miles per hour against the wind. Determine the speed of the plane in still air and the speed of the wind.

**45. Mixing Solutions** Sally Dove has two acid solutions, as illustrated. How much of each must she mix to get 6 liters of a 40% acid solution?

**46. Ice Hockey** The admission at an ice hockey game is $15 for adults and $11 for children. A total of 650 tickets were sold. Determine how many children's tickets and how many adults' tickets were sold if a total of $8790 was collected.

**47. Return to Space** John Glenn was the first American astronaut to go into orbit around Earth. Many years later he returned to space. The second time he returned to space he was 5 years younger than twice his age when he went into space for the first time. The sum of his ages for both times he was in space is 118. Find his age each time he was in space.

**48. Savings Accounts** Jorge Minez has a total of $40,000 invested in three different savings accounts. He has some money invested in one account that gives 7% interest. The second account has $5000 less than the first account and gives 5% interest. The third account gives 3% interest. If the total annual interest that Jorge receives in a year is $2300, find the amount in each account.

[9.6]  *Solve each system of equations using matrices.*

**49.** $x + 5y = 1$
  $-2x - 8y = -6$

**50.** $-2x + 4y = -2$
  $5x - 8y = 7$

**51.** $3y = 6x - 12$
  $4x = 2y + 8$

**52.** $2x - y - z = 5$
  $x + 2y + 3z = -2$
  $3x - 2y + z = 2$

**53.** $3a - b + c = 2$
  $2a - 3b + 4c = 4$
  $a + 2b - 3c = -6$

**54.** $x + y + z = 3$
  $3x + 4y = -1$
  $y - 3z = -10$

[9.7]  *Solve each system of equations using determinants.*

**55.** $7x - 8y = -10$
$-5x + 4y = 2$

**56.** $x + 4y = 5$
$5x + 3y = -9$

**57.** $9m + 4n = -1$
$7m - 2n = -11$

**58.** $p + q + r = 5$
$2p + q - r = -5$
$3p + 2q - 3r = -12$

**59.** $-2a + 3b - 4c = -7$
$2a + b + c = 5$
$-2a - 3b + 4c = 3$

**60.** $y + 3z = 4$
$-x - y + 2z = 0$
$x + 2y + z = 1$

## Chapter 9 Practice Test

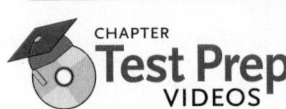 *Chapter Test Prep Videos provide fully worked-out solutions to any of the exercises you want to review. Chapter Test Prep Videos are available via MyMathLab®, or on YouTube (search "Angel Elementary and Intermediate Algebra" and click on "Channels").*

**1.** Determine which, if any, of the ordered pairs satisfy the system of equations.

$$x + 2y = -6$$
$$3x + 2y = -12$$

**a)** $(-6, 0)$    **b)** $\left(-3, -\frac{3}{2}\right)$    **c)** $(2, -4)$

*Identify each system as consistent, inconsistent, or dependent. State whether the system has exactly one solution, no solution, or an infinite number of solutions.*

**2.**

**3.**

**4.**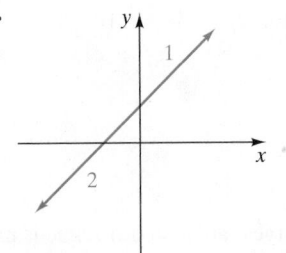

*Determine, without solving the system, whether the system of equations is consistent, inconsistent, or dependent. State whether the system has exactly one solution, no solution, or an infinite number of solutions.*

**5.** $5x + 2y = 4$
$6x = 3y - 7$

**6.** $5x + 3y = 9$
$2y = -\frac{10}{3}x + 6$

**7.** $5x - 4y = 6$
$-10x + 8y = -10$

*Solve each system of equations by the method indicated.*

**8.** $y = 3x - 2$
$y = -2x + 8$
graphically

**9.** $y = -x + 6$
$y = 2x + 3$
graphically

**10.** $y = 4x - 3$
$y = 5x - 4$
substitution

**11.** $4a + 7b = 2$
$5a + b = -13$
substitution

**12.** $0.3x = 0.2y + 0.4$
$-1.2x + 0.8y = -1.6$
addition

**13.** $\frac{3}{2}a + b = 6$
$a - \frac{5}{2}b = -4$
addition

**14.** $x + y + z = 2$
$-2x - y + z = 1$
$x - 2y - z = 1$
addition

**15.** Write the augmented matrix for the following system of equations.

$$-2x + 3y + 7z = 5$$
$$3x - 2y + z = -2$$
$$x - 6y + 9z = -13$$

**16.** Consider the following augmented matrix.

$$\begin{bmatrix} 6 & -2 & 4 & | & 4 \\ 4 & 3 & 5 & | & 6 \\ 2 & -1 & 4 & | & -3 \end{bmatrix}$$

Show the results obtained by multiplying the elements in the third row by $-2$ and adding the products to their corresponding elements in the second row.

*Solve each system of equations using matrices.*

**17.** $2x + 7y = 1$
$3x + 5y = 7$

**18.** $x - 2y + z = 7$
$-2x - y - z = -7$
$4x + 5y - 2z = 3$

*Evaluate each determinant.*

**19.** $\begin{vmatrix} 3 & -1 \\ 5 & -2 \end{vmatrix}$

**20.** $\begin{vmatrix} 8 & 2 & -1 \\ 3 & 0 & 5 \\ 6 & -3 & 4 \end{vmatrix}$

*Solve each system of equations using determinants and Cramer's rule.*

**21.** $4x + 3y = -6$
$-2x + 5y = 16$

**22.** $2r - 4s + 3t = -1$
$-3r + 5s - 4t = 0$
$-2r + s - 3t = -2$

*Use the method of your choice to find the solution to each problem.*

**23. Sunflower Seed Mixture** Agway Gardens has sunflower seeds in a barrel that sell for $0.49 per pound and gourmet bird seed mix that sells for $0.89 per pound. How much of each must be mixed to get a 20-pound mixture that sells for $0.73 per pound?

**24. Mixing Solutions** Tyesha Blackwell, a chemist, has 6% and 15% solutions of sulfuric acid. How much of each solution should she mix to get 10 liters of a 9% solution?

**25. Sum of Numbers** The sum of three numbers is 29. The greatest number is four times the smallest number. The remaining number is 1 more than twice the smallest number. Find the three numbers.

Exercise 23.

## Cumulative Review Test

*Take the following test and check your answers with those given in the back of the book. Review any questions that you answered incorrectly. The section where the material was covered is indicated after the answer.*

**1.** Consider the following set of numbers.
$$\left\{ \frac{1}{2}, -4, 9, 0, \sqrt{3}, -4.63, 1 \right\}$$
List the elements of the set that are

a) natural numbers;

b) rational numbers;

c) real numbers.

**2.** Evaluate $16 \div \left\{ 4\left[ 3 + \left( \frac{5+10}{5} \right)^2 \right] - 32 \right\}$.

**3.** Solve $-7(3 - x) = 4(x + 2) - 3x$.

**4.** Graph $4x - 8y = 16$.

**5.** Find the equation of the line that has a slope of $\frac{2}{5}$ and goes through the point $(-3, 1)$.

**6.** Write the equation of the graph in the accompanying figure.

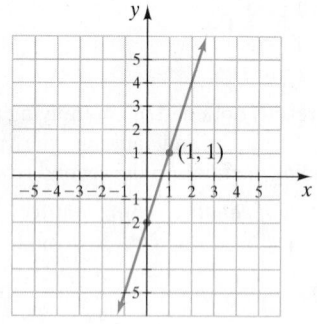

**7.** Simplify $\frac{4a^3b^{-5}}{28a^8b}$.

**8.** Factor $3x^3 + 4x^2 + 6x + 8$.

**9.** Factor $x^2 - 16x + 28$.

**10.** Use factoring to solve $x^2 - 5x = 0$.

**11.** Simplify $\frac{y-5}{8} + \frac{2y+7}{8}$.

**12.** Solve $\frac{5}{y+2} = \frac{3}{2y-7}$.

**13.** Determine which of the following graphs represent functions. Explain.

a)

b)

c)

**14.** If $f(x) = \frac{x+3}{x^2-9}$, find

a) $f(-4)$.         b) $f(h)$.         c) $f(3)$.

*Solve each system of equations.*

**15.** $3x + y = 6$
$y = 4x - 1$

**16.** $2p + 3q = 11$
$-3p - 5q = -16$

**17.** $x - 2y = 0$
$2x + z = 7$
$y - 2z = -5$

**18. Angles of a Triangle** If the largest angle of a triangle is nine times the measure of the smallest angle, and the middle-sized angle is 70° greater than the measure of the smallest angle, find the measure of the three angles.

**19. Walking and Jogging** Mark Simmons walks at 4 miles per hour and Judy Bolin jogs at 6 miles per hour. Mark begins walking $\frac{1}{2}$ hour before Judy starts jogging. If Judy jogs on the same path that Mark walks, how long after Judy begins jogging will she catch up to Mark?

**20. Rock Concert** There are two different prices of seats at a rock concert. The higher priced seats sell for $20 and the less expensive seats sell for $16. If a total of 1000 tickets are sold and the total ticket sales are $18,400, how many of each type of ticket are sold?

# 10

# Inequalities in One and Two Variables

## Goals of This Chapter

In this chapter we discuss inequalities. Inequalities are similar to equations in that they state a relationship between two expressions. We will do many of the same things to inequalities that we have done to equations. We will solve them, graph them, and work with systems involving them.

　In Section 10.1, we solve linear inequalities in one variable. In Section 10.2 we expand on earlier material by solving equations and inequalities involving absolute value. In Section 10.3, we graph linear inequalities in two variables and then use this techinque to graph systems of linear inequalities.

© Nruboc/Dreamstime.com

**When mailing a package,** the size of the package can come into play with respect to postage or even whether the package can be sent in the method you want. The restrictions on the size of a package can be expressed using inequalities. In Exercise 73 on page 597 you will work with package restrictions set by the United Parcel Service (UPS).

# 10.1 Solving Linear Inequalities in One Variable

1 Solve inequalities.

2 Show a solution on a number line, in interval notation, or as a set.

3 Find the union and intersection of sets.

4 Solve compound inequalities involving *and*.

5 Solve compound inequalities involving *or*.

## 1 Solve Inequalities

Inequalities and set builder notation were introduced in Section 8.2. You may wish to review that section now. The inequality symbols follow.*

| Inequality Symbols | |
|---|---|
| > | is greater than |
| ≥ | is greater than or equal to |
| < | is less than |
| ≤ | is less than or equal to |

A mathematical expression containing one or more of these symbols is called an **inequality**.

Examples of Inequalities in One Variable

$$2x + 3 \le 5 \qquad 4x > 3x - 5 \qquad 1.5 \le -2.3x + 4.5 \qquad \frac{1}{2}x + 3 \ge 0$$

*To solve an inequality, we must isolate the variable on one side of the inequality symbol.* To isolate the variable, we use the same basic techniques used in solving equations.

### Properties Used to Solve Linear Inequalities

**1.** If $a > b$, then $a + c > b + c$.

**2.** If $a > b$, then $a - c > b - c$.

**3.** If $a > b$, and $c > 0$, then $ac > bc$.

**4.** If $a > b$, and $c > 0$, then $\dfrac{a}{c} > \dfrac{b}{c}$.

**5.** If $a > b$, and $c < 0$, then $ac < bc$.

**6.** If $a > b$, and $c < 0$, then $\dfrac{a}{c} < \dfrac{b}{c}$.

**Helpful Hint**

The properties used to solve linear inequalities can be summarized in three sentences:

**1.** The same number can be added to or subtracted from both sides of an inequality.

**2.** Both sides of an inequality can be multiplied or divided by any positive number.

**3.** *When both sides of an inequality are multiplied or divided by a negative number, the direction of the inequality symbol reverses.*

---

\*≠, is not equal to, is also an inequality. ≠ means < or >. Thus, 2 ≠ 3 means 2 < 3 or 2 > 3.

| Example of Multiplication by a Negative Number | | Example of Division by a Negative Number | |
|---|---|---|---|
| Multiply both sides of the inequality by $-1$ and reverse the direction of the inequality symbol. | $4 > -2$ $-1(4) < -1(-2)$ $-4 < 2$ | $10 \geq -4$ $\dfrac{10}{-2} \leq \dfrac{-4}{-2}$ $-5 \leq 2$ | Divide both sides of the inequality by $-2$ and reverse the direction of the inequality symbol. |

---

**Helpful Hint**

Do not forget to reverse the direction of the inequality symbol when multiplying or dividing both sides of the inequality by a negative number.

| Inequality | Direction of Inequality Symbol |
|---|---|
| $-3x < 6$ | $\dfrac{-3x}{-3} > \dfrac{6}{-3}$ |
| $-\dfrac{x}{2} > 5$ | $(-2)\left(-\dfrac{x}{2}\right) < (-2)(5)$ |

---

**EXAMPLE 1** Solve the inequalities. **a)** $5x - 7 \geq -17$   **b)** $-6x + 4 < -14$

*Solution*

**a)**

$$5x - 7 \geq -17$$
$$5x - 7 + 7 \geq -17 + 7 \quad \text{Add 7 to both sides.}$$
$$5x \geq -10$$
$$\frac{5x}{5} \geq \frac{-10}{5} \quad \text{Divide both sides by 5.}$$
$$x \geq -2$$

The solution set is $\{x | x \geq -2\}$. Any real number greater than or equal to $-2$ will satisfy the inequality.

**b)**

$$-6x + 4 < -14$$
$$-6x + 4 - 4 < -14 - 4 \quad \text{Subtract 4 from both sides.}$$
$$-6x < -18$$
$$\frac{-6x}{-6} > \frac{-18}{-6} \quad \text{Divide both sides by } -6 \text{ and reverse the direction of the inequality.}$$
$$x > 3$$

The solution set is $\{x | x > 3\}$. Any number greater than 3 will satisfy the inequality.

Now Try Exercise 17

## 2 Show a Solution on a Number Line, in Interval Notation, or as a Set

The solution to an inequality can be shown on a number line, written as a solution set, or written in interval notation. We used interval notation previously in Section 8.2 and review it here.

The following symbols will be used when representing solutions to inequalities.

| Inequality Symbol | Number Line Endpoints | Interval Notation Symbols | Is endpoint included in solution? |
|---|---|---|---|
| $<$ or $>$ | ○ | ( or ) | No |
| $\leq$ or $\geq$ | ● | [ or ] | Yes |

The symbol $\infty$ is read "infinity"; it indicates that the solution set continues indefinitely.

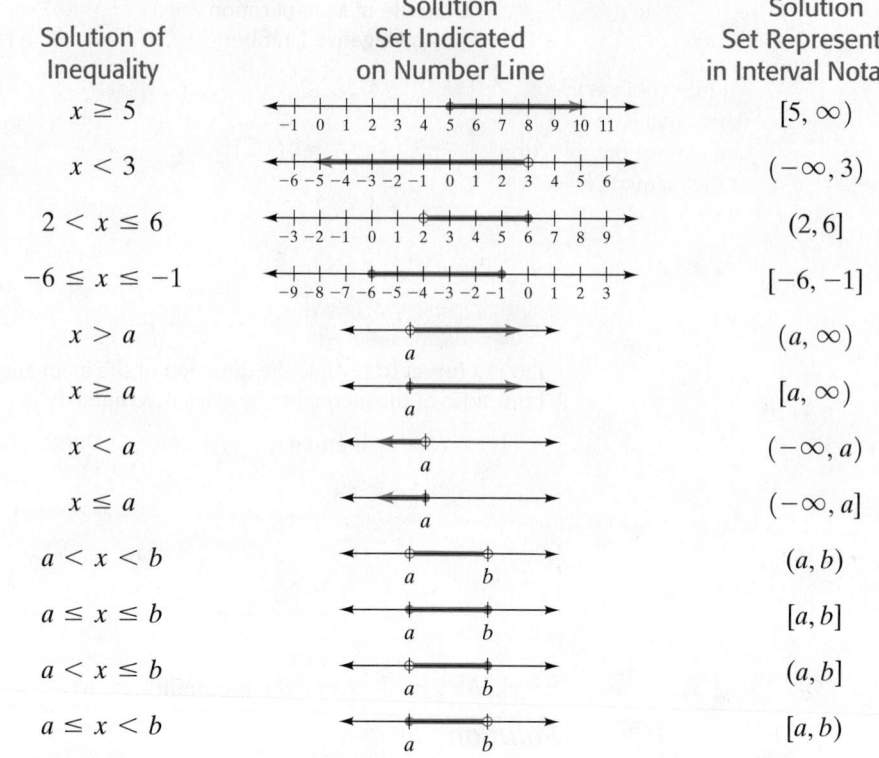

| Solution of Inequality | Solution Set Indicated on Number Line | Solution Set Represented in Interval Notation |
|---|---|---|
| $x \geq 5$ | | $[5, \infty)$ |
| $x < 3$ | | $(-\infty, 3)$ |
| $2 < x \leq 6$ | | $(2, 6]$ |
| $-6 \leq x \leq -1$ | | $[-6, -1]$ |
| $x > a$ | | $(a, \infty)$ |
| $x \geq a$ | | $[a, \infty)$ |
| $x < a$ | | $(-\infty, a)$ |
| $x \leq a$ | | $(-\infty, a]$ |
| $a < x < b$ | | $(a, b)$ |
| $a \leq x \leq b$ | | $[a, b]$ |
| $a < x \leq b$ | | $(a, b]$ |
| $a \leq x < b$ | | $[a, b)$ |

**EXAMPLE 2** Solve the following inequality and give the solution both on a number line and in interval notation.

$$\frac{1}{4}z - \frac{1}{2} < \frac{2z}{3} + 2$$

*Solution*   We can eliminate fractions from an inequality by multiplying both sides of the inequality by the least common denominator, LCD, of the fractions.

$$\frac{1}{4}z - \frac{1}{2} < \frac{2z}{3} + 2$$

$$12\left(\frac{1}{4}z - \frac{1}{2}\right) < 12\left(\frac{2z}{3} + 2\right) \qquad \text{Multiply both sides by the LCD, 12.}$$

$$3z - 6 < 8z + 24 \qquad \text{Distributive property}$$

$$3z - 8z - 6 < 8z - 8z + 24 \qquad \text{Subtract } 8z \text{ from both sides.}$$

$$-5z - 6 < 24$$

$$-5z - 6 + 6 < 24 + 6 \qquad \text{Add 6 to both sides.}$$

$$-5z < 30$$

$$\frac{-5z}{-5} > \frac{30}{-5} \qquad \text{Divide both sides by } -5 \text{ and reverse the direction of inequality symbol.}$$

$$z > -6$$

**Number Line**

**Interval Notation**

$$(-6, \infty)$$

The solution set is $\{z | z > -6\}$.

Now Try Exercise 31

In Example 2 we illustrated the solution on a number line, in interval notation, and as a solution set.

**EXAMPLE 3** Solve the inequality $2(3p - 5) + 9 \leq 8(p + 1) - 2(p - 3)$.

*Solution*
$$2(3p - 5) + 9 \leq 8(p + 1) - 2(p - 3)$$
$$6p - 10 + 9 \leq 8p + 8 - 2p + 6$$
$$6p - 1 \leq 6p + 14$$
$$6p - 6p - 1 \leq 6p - 6p + 14$$
$$-1 \leq 14$$

Since $-1$ is always less than or equal to 14, the inequality is true for all real numbers. When an inequality is true for all real numbers, the solution set is *the set of all real numbers*, $\mathbb{R}$. The solution set to this example can also be indicated on a number line or given in interval notation.

$$\text{or } (-\infty, \infty)$$

NowTry Exercise 23

## Understanding Algebra

If, while solving an inequality, you get an inequality with no variables that is always

- true, such as $-1 \leq 14$, then the solution set is the set of *all real numbers*, $\mathbb{R}$.

- false, such as $5 < -3$, then the solution set is the empty set, $\varnothing$.

If Example 3 had resulted in the expression $-1 \geq 14$, the inequality would never have been true, since $-1$ is never greater than or equal to 14. When an inequality is never true, it has no solution. The solution set of an inequality that has no solution is the *empty* or *null set*, { } or $\varnothing$. We will represent the empty set on the number line as follows, $\xleftarrow{\hspace{1cm}}_{0}\xrightarrow{\hspace{1cm}}$.

> **Helpful Hint**
>
> Generally, when writing a solution to an inequality, we write the variable on the left. For example, when solving an inequality, if we obtain $5 \geq y$ we would write the solution as $y \leq 5$. For example,
>
> $-6 < x$ means $x > -6$   (inequality symbol points to $-6$ in both cases)
> $4 > x$ means $x < 4$    (inequality symbol points to $x$ in both cases)
> $a < x$ means $x > a$    (inequality symbol points to $a$ in both cases)
> $a > x$ means $x < a$    (inequality symbol points to $x$ in both cases)

**EXAMPLE 4  Packages on a Boat** A small boat has a maximum weight load of 750 pounds. Millie Harrison has to transport packages weighing 42.5 pounds each.

a) Write an inequality that can be used to determine the maximum number of packages that Millie can safely place on the boat if she weighs 128 pounds.

b) Find the maximum number of packages that Millie can transport.

*Solution*

a) Understand and Translate   Let $n$ = number of packages.

$$\text{Millie's weight} \quad + \quad \text{weight of } n \text{ packages} \leq 750$$
$$128 \quad\quad + \quad\quad 42.5n \quad\quad\quad \leq 750$$

b) Carry Out
$$128 + 42.5n \leq 750$$
$$42.5n \leq 622$$
$$n \leq 14.6$$

Answer   Therefore, Millie can transport up to 14 packages on the boat.

NowTry Exercise 75

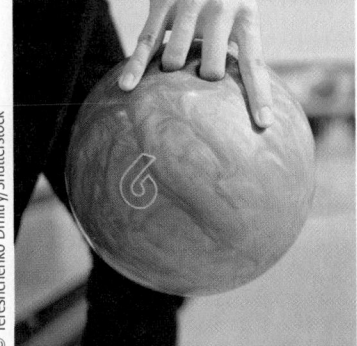

**EXAMPLE 5  Bowling Alley Rates** At the Corbin Bowl bowling alley in Tarzana, California, it costs $2.50 to rent bowling shoes and it costs $4.00 per game bowled.

a) Write an inequality that can be used to determine the maximum number of games that Ricky Olson can bowl if he has only $20.

b) Find the maximum number of games that Ricky can bowl.

*Solution*

**a)** Understand and Translate

$$\text{Let } g = \text{number of games bowled.}$$
$$\text{Then } 4.00g = \text{cost of bowling } g \text{ games.}$$

$$\text{cost of shoe rental } + \text{ cost of bowling } g \text{ games } \leq \text{ money Ricky has}$$
$$2.50 \qquad + \qquad 4.00g \qquad\qquad \leq \qquad 20$$

**b)** Carry Out

$$2.50 + 4.00g \leq 20$$
$$4.00g \leq 17.50$$
$$\frac{4.00g}{4.00} \leq \frac{17.50}{4.00}$$
$$g \leq 4.375$$

**Answer and Check**   Since he can't play a portion of a game, the maximum number of games that he can afford to bowl is 4. If Ricky were to bowl 5 games, he would owe $2.50 + 5($4.00) = $22.50$, which is more money than the $20 that he has.

Now Try Exercise 77

**EXAMPLE 6   Profit**  For a business to realize a profit, its revenue (or income), $R$, must be greater than its cost, $C$. That is, a profit will be obtained when $R > C$ (the company breaks even when $R = C$). A company that produces playing cards has a weekly cost equation of $C = 1525 + 1.7x$ and a weekly revenue equation of $R = 4.2x$, where $x$ is the number of decks of playing cards produced and sold in a week. How many decks of cards must be produced and sold in a week for the company to make a profit?

*Solution*   Understand and Translate   The company will make a profit when $R > C$, or

$$4.2x > 1525 + 1.7x$$

Carry Out

$$2.5x > 1525$$
$$x > \frac{1525}{2.5}$$
$$x > 610$$

**Answer**   The company will make a profit when more than 610 decks are produced and sold in a week.

Now Try Exercise 79

**EXAMPLE 7   Tax Tables**  The 2008 tax rate schedule for married couples who file a joint tax return is shown below.

**Schedule Y-1** Use if your filing status is **Married filing jointly or Qualifying widow(er)**

| If the taxable income is over- | But not over- | The tax is: | Of the amount over- |
|---|---|---|---|
| $0 | $16,050 | 10% | $0 |
| $16,050 | $65,100 | $1,605.00 + 15% | $16,050 |
| $65,100 | $131,450 | $8,962.50 + 25% | $65,100 |
| $131,450 | $200,300 | $25,550.00 + 28% | $131,450 |
| $200,300 | $357,700 | $44,828.00 + 33% | $200,300 |
| $357,700 | ∞ | $96,770.00 + 35% | $357,700 |

**a)** Write, in interval notation, the amounts of taxable income that make up each of the six listed tax brackets, that is, the 10%, 15%, 25%, 28%, 33%, and 35% tax brackets.

**b)** Determine the tax for a married couple filing jointly if their taxable income is $13,500.

**c)** Determine the tax for a married couple filing jointly if their taxable income is $136,000.

*Solution*

**a)** The words *But Not Over* mean "less than or equal to." The taxable incomes that make up the six tax brackets are

(0, 16,050] for the 10% tax bracket

(16,050, 65,100] for the 15% tax bracket

(65,100, 131,450] for the 25% tax bracket

(131,450, 200,300] for the 28% tax bracket

(200,300, 357,700] for the 33% tax bracket

$(357,700, \infty)$ for the 35% tax bracket

**b)** The tax for a married couple filing jointly with taxable income of $13,500 is 10% of $13,500. Therefore,

$$\text{tax} = 0.10(13{,}500) = 1350$$

The tax is $1350.

**c)** A taxable income of $136,000 places a married couple filing jointly in the 28% tax bracket. The tax is $25,500 + 28% of the taxable income over $131,450. The taxable income over $131,450 is $136,000 − $131,450 = $4550. Therefore,

$$\text{tax} = 25{,}550.00 + 0.28(4550) = 25{,}550 + 1274 = 26{,}824$$

The tax is $26,824.

Now Try Exercise 89

## 3 Find the Union and Intersection of Sets

After finding the solution set to an inequality, it is sometimes necessary to combine the solution set with another solution set. Just as *operations* such as addition and multiplication are performed on numbers, operations can be performed on sets. Two set operations are *union* and *intersection*.

### Union of Two Sets

The **union** of set $A$ and set $B$, written $A \cup B$, is the set of elements that belong to either set $A$ *or* set $B$.

Because the word *or*, as used in this context, means belonging to set $A$ or set $B$ or both sets, the union is formed by combining, or joining together, the elements in set $A$ with those in set $B$. If an item is an element in either set $A$, or set $B$, or in both sets, then it is an element in the union of the sets. If an element appears in both sets, we only list it once when we write the union of two sets.

#### Examples of Union of Sets

$A = \{1, 2, 3, 4, 5\}, \quad B = \{3, 4, 5, 6, 7\}, \quad A \cup B = \{1, 2, 3, 4, 5, 6, 7\}$

$A = \{a, b, c, d, e\}, \quad B = \{x, y, z\}, \quad A \cup B = \{a, b, c, d, e, x, y, z\}$

In set builder notation we can express $A \cup B$ as

### Union

$$A \cup B = \{x \mid x \in A \text{ or } x \in B\}$$

### Intersection of Two Sets

The **intersection** of set $A$ and set $B$, written $A \cap B$, is the set of all elements that are common to both set $A$ *and* set $B$.

Because the word *and*, as used in this context, means belonging to *both* set $A$ and set $B$, the intersection is formed by using only those elements that are in both set $A$ and set $B$. If an item is an element in only one of the two sets, then it is not an element in the intersection of the sets.

Examples of Intersection of Sets

$$A = \{1, 2, 3, 4, 5\}, \qquad B = \{3, 4, 5, 6, 7\}, \qquad A \cap B = \{3, 4, 5\}$$
$$A = \{a, b, c, d, e\}, \qquad B = \{x, y, z\}, \qquad A \cap B = \{\ \}$$

Note that in the last example, sets $A$ and $B$ have no elements in common. Therefore, their intersection is the empty set. In set builder notation we can express $A \cap B$ as

### Intersection

$$A \cap B = \{x | x \in A \text{ and } x \in B\}$$

## 4  Solve Compound Inequalities Involving *And*

A **compound inequality** is formed by joining two inequalities with the word *and* or *or*. Sometimes the word *and* is implied without being written.

Examples of Compound Inequalities

$$3 < x \quad \text{and} \quad x < 5$$
$$x + 4 > 2 \quad \text{or} \quad 2x - 3 < 6$$
$$4x - 6 \geq -3 \quad \text{and} \quad x - 6 < 17$$

The solution of a compound inequality using the word *and* is all the numbers that make *both* parts of the inequality true. Consider

$$3 < x \quad \text{and} \quad x < 5$$

The numbers that satisfy both inequalities may be easier to see if we graph the solution to each inequality on a number line (see **Fig. 10.1**). Now we can see that the numbers that satisfy both inequalities are the numbers between 3 and 5. The solution set is $\{x | 3 < x < 5\}$.

**FIGURE 10.1**

Recall that the intersection of two sets is the set of elements common to both sets. *To find the solution set of an inequality containing the word* **and**, *take the* **intersection** *of the solution sets of the two inequalities.*

**EXAMPLE 8**  Solve $x + 5 \leq 8$ and $2x - 9 > -7$

*Solution*    Begin by solving each inequality separately.

$$x + 5 \leq 8 \quad \text{and} \quad 2x - 9 > -7$$
$$x \leq 3 \qquad\qquad\qquad 2x > 2$$
$$\qquad\qquad\qquad\qquad x > 1$$

Now take the intersection of the sets $\{x | x \leq 3\}$ and $\{x | x > 1\}$. When we find $\{x | x \leq 3\} \cap \{x | x > 1\}$, we are finding the values of $x$ common to both sets. **Figure 10.2** illustrates that the solution set is $\{x | 1 < x \leq 3\}$. In interval notation, the solution is $(1, 3]$.

**FIGURE 10.2**

Now Try Exercise 67

Sometimes a compound inequality using the word *and* can be written in a shorter form.

$$3 < x \text{ and } x < 5 \quad \text{is written} \quad 3 < x < 5,$$
$$-1 < x + 3 \text{ and } x + 3 \le 5 \quad \text{is written} \quad -1 < x + 3 \le 5$$

**EXAMPLE 9** Solve $-1 < x + 3 \le 5$

*Solution*    $-1 < x + 3 \le 5$ means $-1 < x + 3$ and $x + 3 \le 5$. Solve each inequality separately.

$$-1 < x + 3 \quad \text{and} \quad x + 3 \le 5$$
$$-4 < x \qquad\qquad\qquad x \le 2$$

Remember that $-4 < x$ means $x > -4$. **Figure 10.3** illustrates that the solution set is $\{x | -4 < x \le 2\}$. In interval notation, the solution is $(-4, 2]$.

**FIGURE 10.3**

Now Try Exercise 45

The inequality in Example 9, $-1 < x + 3 \le 5$, can be solved in another way. We could have subtracted 3 from all three parts to isolate the variable in the middle and solve the inequality.

$$-1 < x + 3 \le 5$$
$$-1 \boxed{-3} < x + \boxed{-3} \le 5 \boxed{-3}$$
$$-4 < x \le 2$$

Note that this is the same solution as obtained in Example 9.

**Understanding Algebra**

When solving an inequality with more than two *parts*, whatever we do to one part, we must do to *all* parts.

**EXAMPLE 10** Solve the inequality $-3 \le 2t - 7 < 8$.

*Solution*    We wish to isolate the variable $t$. We begin by adding 7 to all three parts of the inequality.

$$-3 \le 2t - 7 < 8$$
$$-3 \boxed{+7} \le 2t - 7 \boxed{+7} < 8 \boxed{+7}$$
$$4 \le 2t < 15$$

Now divide all three parts of the inequality by 2.

$$\frac{4}{2} \le \frac{2t}{2} < \frac{15}{2}$$
$$2 \le t < \frac{15}{2}$$

The solution may also be illustrated on a number line, written in interval notation, or written as a solution set. Below we show each form.

The answer in interval notation is $\left[2, \frac{15}{2}\right)$. The solution set is $\left\{ t \Big| 2 \le t < \frac{15}{2} \right\}$.

Now Try Exercise 49

**EXAMPLE 11** Solve the inequality $-2 < \dfrac{4 - 3x}{5} < 8$.

*Solution*    Multiply all three parts by 5 to eliminate the denominator.

$$-2 < \frac{4 - 3x}{5} < 8$$

$$-2(5) < 5\left(\frac{4 - 3x}{5}\right) < 8(5)$$

$$-10 < 4 - 3x < 40$$

$$-10 - 4 < 4 - 4 - 3x < 40 - 4$$

$$-14 < -3x < 36$$

Now divide all three parts of the inequality by $-3$. Remember that when we multiply or divide an inequality by a negative number, the direction of the inequality symbols reverse.

$$\frac{-14}{-3} > \frac{-3x}{-3} > \frac{36}{-3}$$

$$\frac{14}{3} > x > -12$$

Although $\dfrac{14}{3} > x > -12$ is correct, we generally write compound inequalities with the smaller value on the left. We will, therefore, rewrite the solution as

$$-12 < x < \frac{14}{3}$$

The solution may also be illustrated on a number line, written in interval notation, or written as a solution set.

The solution in interval notation is $\left(-12, \dfrac{14}{3}\right)$. The solution set is $\left\{x \,\middle|\, -12 < x < \dfrac{14}{3}\right\}$.

Now Try Exercise 53

**Avoiding Common Errors**

You must be very careful when writing the solution to a compound inequality. In Example 11 we can change the solution from

$$\frac{14}{3} > x > -12 \quad \text{to} \quad -12 < x < \frac{14}{3}$$

This is correct since both say that $x$ is greater than $-12$ and less than $\dfrac{14}{3}$. Notice that the inequality symbol in both cases is pointing to the smaller number.

In Example 11, had we written the answer $\dfrac{14}{3} < x < -12$, we would have given the incorrect solution. Remember that the inequality $\dfrac{14}{3} < x < -12$ means that $\dfrac{14}{3} < x$ and $x < -12$. There is no number that is both greater than $\dfrac{14}{3}$ and less than $-12$. Also, by examining the inequality $\dfrac{14}{3} < x < -12$, it appears as if we are saying that $-12$ is a greater number than $\dfrac{14}{3}$, which is obviously incorrect.

It would also be incorrect to write the answer as

$$-12 < x > \frac{14}{3} \quad \text{or} \quad \frac{14}{3} < x > -12$$

**EXAMPLE 12 Calculating Grades** In an anatomy and physiology course, an average score greater than or equal to 80 and less than 90 will result in a final grade of B. Steve Reinquist received scores of 85, 90, 68, and 70 on his first four exams. For Steve to receive a final grade of B in the course, between which two scores must his fifth (and last) exam fall?

*Solution* Let $x$ = Steve's last exam score.

$$80 \leq \text{average of five exams} < 90$$
$$80 \leq \frac{85 + 90 + 68 + 70 + x}{5} < 90$$
$$80 \leq \frac{313 + x}{5} < 90$$
$$400 \leq 313 + x < 450$$
$$400 - 313 \leq 313 - 313 + x < 450 - 313$$
$$87 \leq x < 137$$

Steve would need a minimum score of 87 on his last exam to obtain a final grade of B. If the highest score he could receive on the test is 100, is it possible for him to obtain a final grade of A (90 average or higher)? Explain.

*Now Try Exercise 85*

### 5 Solve Compound Inequalities Involving *Or*

The solution to a compound inequality using the word *or* is all the numbers that make *either* of the inequalities a true statement. Consider the compound inequality

$$x > 3 \quad \text{or} \quad x < 5$$

What are the numbers that satisfy the compound inequality? Let's graph the solution to each inequality on the number line (see **Fig. 10.4**). Note that every real number satisfies at least one of the two inequalities. Therefore, the solution set to the compound inequality is the set of all real numbers, $\mathbb{R}$.

**FIGURE 10.4**

The *union* of two sets is the set of elements that belong to *either* of the sets. *To find the solution set of an inequality containing the word* **or**, *take the* **union** *of the solution sets of the two inequalities.*

**EXAMPLE 13** Solve $r - 2 \leq -6$ or $-4r + 3 < -5$.

*Solution* Solve each inequality separately.

$$r - 2 \leq -6 \quad \text{or} \quad -4r + 3 < -5$$
$$r \leq -4 \qquad\qquad -4r < -8$$
$$\qquad\qquad\qquad r > 2$$

Now graph each solution on its own number line and then find the union (see **Fig. 10.5**). The union is $r \leq -4$ or $r > 2$.

**FIGURE 10.5**

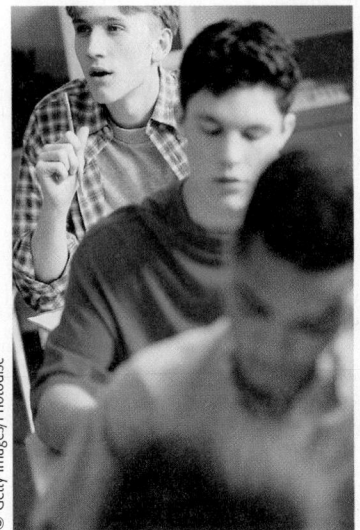

© Getty Images/Photodisc

The solution set is $\{r|r \le -4\} \cup \{r|r > 2\}$, which can be written as $\{r|r \le -4$ or $r > 2\}$. In interval notation, the solution is $(-\infty, -4] \cup (2, \infty)$.

NowTry Exercise 69

> **Helpful Hint**
>
> There are various ways to write the solution to an inequality problem. Be sure to indicate the solution to an inequality problem in the form requested by your professor. Examples of various forms follow.
>
> | Inequality | Number Line | Interval Notation | Solution Set |
> |---|---|---|---|
> | $x < \dfrac{5}{3}$ | | $\left(-\infty, \dfrac{5}{3}\right)$ | $\left\{x \middle| x < \dfrac{5}{3}\right\}$ |
> | $-4 < t \le \dfrac{5}{3}$ | | $\left(-4, \dfrac{5}{3}\right]$ | $\left\{t \middle| -4 < t \le \dfrac{5}{3}\right\}$ |

---

# EXERCISE SET 10.1   MathXL®   MyMathLab®

## Warm-Up Exercises

*Fill in the blanks with the appropriate word, phrase, or symbol(s) from the following list.*

direction          open          closed          intersection          union          compound          simple

**1.** A _____ inequality is formed by joining two in-equalities with the word *and* or *or*.

**2.** A(n) _____ circle on the number line indicates that the endpoint is not part of the solution.

**3.** A(n) _____ circle on the number line indicates that the endpoint is part of the solution.

**4.** To find the solution set of an inequality containing the word *and*, take the _____ of the solution sets of the two inequalities.

**5.** To find the solution set of an inequality containing the word *or*, take the _____ of the solution sets of the two inequalities.

**6.** Whenever you multiply or divide both sides of an inequality by a negative number, you must change the _____ of the inequality symbol.

## Practice the Skills

*Express each inequality* **a)** *using a number line,* **b)** *in interval notation, and* **c)** *as a solution set (use set builder notation).*

**7.** $x > -3$

**8.** $t > \dfrac{3}{4}$

**9.** $w \le \pi$

**10.** $-4 < x < 3$

**11.** $-3 < q \le \dfrac{4}{5}$

**12.** $x \ge -\dfrac{6}{5}$

**13.** $-7 < x \le -4$

**14.** $-2\dfrac{7}{8} \le k < -1\dfrac{2}{3}$

*Solve each inequality and graph the solution on the number line.*

**15.** $x + 8 > 10$

**16.** $2x + 3 > 4$

**17.** $3 - x < -4$

**18.** $12b - 5 \le 8b + 7$

**19.** $4.7x - 5.48 \ge 11.44$

**20.** $1.4x + 2.2 < 2.6x - 0.2$

**21.** $4(x + 2) \le 4x + 8$

**22.** $15.3 > 3(a - 1.4)$

**23.** $5b - 6 \ge 3(b + 3) + 2b$

**24.** $-6(d + 2) < -9d + 3(d - 1)$

**25.** $2y - 6y + 8 \le 2(-2y + 9)$

**26.** $\dfrac{y}{2} + \dfrac{4}{5} \le 3$

*Solve each inequality and give the solution in interval notation.*

**27.** $4 + \dfrac{4x}{3} < 6$

**28.** $4 - 3x < 5 + 2x + 17$

**29.** $\dfrac{v - 5}{3} - v \ge -3(v - 1)$

**30.** $\dfrac{h}{2} - \dfrac{5}{6} < \dfrac{7}{8} + h$

**31.** $\dfrac{t}{3} - t + 7 \le -\dfrac{4t}{3} + 8$

**32.** $\dfrac{6(x - 2)}{5} > \dfrac{10(2 - x)}{3}$

**33.** $-3x + 1 < 3[(x + 2) - 2x] - 1$

**34.** $4[x - (3x - 2)] > 3(x + 5) - 15$

*Find $A \cup B$ and $A \cap B$ for each set A and B.*

**35.** $A = \{5, 6, 7\}, B = \{6, 7, 8\}$

**36.** $A = \{2, 4, 6, 8\}, B = \{1, 3, 5, 7\}$

**37.** $A = \{-2, -4, -5\}, B = \{-1, -2, -4, -6\}$

**38.** $A = \{-1, 0, 1\}, B = \{0, 2, 4, 6\}$

**39.** $A = \{\ \}, B = \{0, 1, 2, 3\}$

**40.** $A = \{2, 4, 6\}, B = \{2, 4, 6, 8, \ldots\}$

**41.** $A = \{0, 2, 4, 6, 8\}, B = \{1, 3, 5, 7\}$

**42.** $A = \{1, 3, 5\}, B = \{1, 3, 5, 7, \ldots\}$

**43.** $A = \{0.1, 0.2, 0.3\}, B = \{0.2, 0.3, 0.4, 0.5, \ldots\}$

**44.** $A = \left\{1, \dfrac{1}{2}, \dfrac{1}{4}, \dfrac{1}{6}, \ldots\right\}, B = \left\{\dfrac{1}{4}, \dfrac{1}{6}, \dfrac{1}{8}\right\}$

*Solve each inequality and give the solution in interval notation.*

**45.** $-2 \leq t + 3 < 4$

**46.** $-7 < p - 6 \leq -5$

**47.** $-15 \leq -3z \leq 12$

**48.** $-16 < 5 - 3n \leq 13$

**49.** $4 \leq 2x - 4 < 7$

**50.** $-12 < 3x - 5 \leq -1$

**51.** $14 \leq 2 - 3g < 15$

**52.** $\dfrac{1}{2} < 3x + 4 < 13$

*Solve each inequality and give the solution set.*

**53.** $5 \leq \dfrac{3x + 1}{2} < 11$

**54.** $\dfrac{3}{5} < \dfrac{-x - 5}{3} < 2$

**55.** $-6 \leq -3(2x - 4) < 12$

**56.** $-6 < \dfrac{4 - 3x}{2} < \dfrac{2}{3}$

**57.** $0 \leq \dfrac{3(u - 4)}{7} \leq 1$

**58.** $-15 < \dfrac{3(x - 2)}{5} \leq 0$

*Solve each inequality and indicate the solution set.*

**59.** $c \leq 1$ and $c > -3$

**60.** $d > 0$ or $d \leq 8$

**61.** $x < 2$ and $x > 4$

**62.** $w \leq -1$ or $w > 6$

**63.** $x + 1 < 3$ and $x + 1 > -4$

**64.** $5x - 3 \leq 7$ or $-2x + 5 < -3$

*Solve each inequality and give the solution in interval notation.*

**65.** $2s + 3 < 7$ or $-3s + 4 \leq -17$

**66.** $4a + 7 \geq 9$ and $-3a + 4 \leq -17$

**67.** $4x + 5 \geq 5$ and $3x - 7 \leq -1$

**68.** $5 - 3x < -3$ and $5x - 3 > 10$

**69.** $4 - r < -2$ or $3r - 1 < -1$

**70.** $-x + 3 < 0$ or $2x - 5 \geq 3$

**71.** $2k + 5 > -1$ and $7 - 3k \leq 7$

**72.** $2q - 11 \leq -7$ or $2 - 3q < 11$

## Problem Solving

**73. UPS Packages** To avoid a surcharge, the length plus the girth of a package to be shipped by United Parcel Service (UPS) can be no larger than 130 inches.

  **a)** Write an inequality that expresses this information, using *l* for the length and *g* for the girth.

  **b)** UPS has defined girth as twice the width plus twice the depth. Write an inequality using length, *l*, width, *w*, and depth, *d*, to indicate the maximum allowable dimensions of a package that may be shipped without paying the surcharge.

  **c)** If the length of a package is 40 inches and the width of a package is 20.5 inches, find the maximum allowable depth of the package.

**74. Carry-On Luggage** Many airlines have limited the size of luggage passengers may carry onboard on domestic flights. The length, *l*, plus the width, *w*, plus the depth, *d*, of the carry-on luggage must not exceed 45 inches.

  **a)** Write an inequality that describes this restriction, using *l*, *w*, and *d* as described above.

  **b)** If Ryan McHenry's luggage is 23 inches long and 12 inches wide, what is the maximum depth it can have and still be carried on the plane?

© Allen R. Angel

*In Exercises 75–88, set up an inequality that can be used to solve the problem. Solve the problem and find the desired value.*

**75. Weight Limit** Cal Worth, a janitor, must move a large shipment of books from the first floor to the fifth floor. The sign on the elevator reads "maximum weight 800 pounds." If each box of books weighs 70 pounds, find the maximum number of boxes that Cal should place in the elevator, if he does not ride.

**76. Elevator Limit** If the janitor in Exercise 75, weighing 195 pounds, must ride up with the boxes of books, find the maximum number of boxes that can be placed into the elevator.

**77. Text Messages** The Verizon Wireless Message Pack 200 plan includes 200 text messages per month for $5.00. Additional text messages are $0.15 each. If Berma Williams uses this plan, how many text messages can she purchase for $20?

© iStockphoto

**78. Parking Garage** A downtown parking garage in Austin, Texas, charges $1.25 for the first hour and $0.75 for each additional hour or part thereof. What is the maximum length of time you can park in the garage if you wish to pay no more than $3.75?

**79. Book Profit** April Lemons is considering writing and publishing her own book. She estimates her revenue equation as $R = 6.42x$, and her cost equation as $C = 10{,}025 + 1.09x$, where $x$ is the number of books she sells. Find the minimum number of books she must sell to make a profit. See Example 6.

**80. Dry Cleaning Profit** Peter Collinge is opening a dry cleaning store. He estimates his cost equation as $C = 8000 + 0.08x$ and his revenue equation as $R = 1.85x$, where $x$ is the number of garments dry cleaned in a year. Find the minimum number of garments that must be dry cleaned in a year for Peter to make a profit.

© Allen R. Angel

**81. Mailing Large Envelopes** The cost to mail a large envelope is $0.83 for the first ounce and $0.17 for each additional ounce. What is the maximum weight of a large envelope that Joni Burnette can mail for $2.70?

**82. Presorted First-Class Mail** Companies can send pieces of mail weighing up to 1 ounce by using *presorted first-class* mail. The company must first purchase a bulk permit for $180 per year, and then pay $0.394 per piece sent. Without the permit, each piece would cost $0.42. Determine the minimum number of pieces of mail that would have to be mailed for it to be financially worthwhile for a company to use presorted first-class mail.

**83. Comparing Payment Plans** Melissa Pfistner recently accepted a sales position in Ohio. She can select between two payment plans. Plan 1 is a salary of $300 per week plus a 10% commission on sales. Plan 2 is a salary of $400 per week plus an 8% commission on sales. For what amount of weekly sales would Melissa earn more by plan 1?

**84. College Employment** To be eligible to continue her financial assistance for college, Katie Hanenberg can earn no more than $2000 during her 8-week summer employment. She already earns $90 per week as a day-care assistant. She is considering adding an evening job at a fast-food restaurant, where she will earn $6.25 per hour. What is the maximum number of hours she can work at the restaurant without jeopardizing her financial assistance?

**85. A Passing Grade** To pass a course, Corrina Schultz needs an average score of 60 or more. If Corrina's scores are 66, 72, 90, 49, and 59, find the minimum score that she can get on her sixth and last exam and pass the course.

**86. Minimum Grade** To receive an A in a course, Stephen Heasley must obtain an average score of 90 or higher on five exams. If Stephen's first four exam scores are 92, 87, 96, and 77, what is the minimum score that Stephen can receive on the fifth exam to get an A in the course?

**87. Averaging Grades** Calisha Mahoney's grades on her first four exams are 85, 92, 72, and 75. An average greater than or equal to 80 and less than 90 will result in a final grade of B. What range of grades on Calisha's fifth and last exam will result in a final grade of B? Assume a maximum grade of 100.

**88. Clean Air** For air to be considered "clean," the average of three pollutants must be less than 3.2 parts per million. If the first two pollutants are 2.7 and 3.42 ppm, what values of the third pollutant will result in clean air?

**89. Income Taxes** Refer to Example 7 on page 590. Su-hua and Ting-Fang Zheng file a joint tax return. Determine the 2008 income tax Su-hua and Ting-Fang will owe if their taxable income is

**a)** $78,221.          **b)** $301,233.

**90. Income Taxes** Refer to Example 7 on page 590. Jose and Mildred Battiste file a joint tax return. Determine the 2008 income tax Jose and Mildred will owe if their taxable income is

**a)** $128,479.          **b)** $275,248.

**Velocity** *In physics, an object that is traveling upwards has a positive velocity ($v > 0$) and an object that is traveling downwards has a negative velocity ($v < 0$). In Exercises 91–96, the velocity, v, is given for an object t seconds after it is projected upward. Using interval notation, determine the intervals of time when the object is traveling* **a)** *upward or* **b)** *downward.*

**91.** $v = -32t + 96$,          $0 \le t \le 10$

**92.** $v = -32t + 172.8$,          $0 \le t \le 12$

**93.** $v = -9.8t + 49$,          $0 \le t \le 13$

**94.** $v = -9.8t + 31.36$,          $0 \le t \le 6$

**95.** $v = -32t + 320$,          $0 \le t \le 8$

**96.** $v = -9.8t + 68.6$,          $0 \le t \le 5$

**97. Water Acidity** Thomas Hayward is checking the water acidity in a swimming pool. The water acidity is considered normal when the average pH reading of three daily measurements is greater than 7.2 and less than 7.8. If the first two pH readings are 7.48 and 7.15, find the range of pH values for the third reading that will result in the acidity level being normal.

© Allen R. Angel

See Exercise 97 on page 598.

**98.** If $a > b$, will $a^2$ always be greater than $b^2$? Explain and give an example to support your answer.

**99. Insurance Policy** A Blue Cross/Blue Shield insurance policy has a $100 deductible, after which it pays 80% of the total

medical cost, $c$. The customer pays 20% until the customer has paid a total of $500, after which the policy pays 100% of the medical cost. We can describe this policy as follows:

Blue Cross Pays

$$
\begin{array}{ll}
0, & \text{if } c \le \$100 \\
0.80(c - 100), & \text{if } \$100 < c \le \$2100 \\
c - 500, & \text{if } c > \$2100
\end{array}
$$

Explain why this set of inequalities describes Blue Cross/Blue Shield's payment plan.

**100.** Explain why the inequality $a < bx + c < d$ cannot be solved for $x$ unless additional information is given.

**Growth Charts** *Exercises 101 and 102 show growth charts for children from birth to age 36 months. In general, the nth percentile represents that value that is above n% and below $(100 - n)$% of the items being measured. For example, a 24-month-old boy at the 60th percentile for weight weighs more than 60% and less than 40% of 24-month-old boys.*

**101.** The following chart shows the weight-for-age percentiles for boys from birth to age 36 months. The red curve is the 50th percentile. The orange region is between the 10th percentile (blue curve) and the 90th percentile (green curve). That is, 80% of the weights are between the values represented by the blue curve and the green curve. Use this graph to determine, in interval notation, where 80% of the weights occur for boys of age

a) 9 months.          b)  21 months.

c)  36 months.

**102.** The following chart shows the weight-for-age percentiles for girls from birth to age 36 months. The orange region is between the 10th percentile (blue curve) and the 90th percentile (green curve) and 80% of the weights are in this region. Use this graph to determine, in interval notation, where 80% of the weights occur for girls of age

a)  9 months.          b) 21 months.

c)  36 months.

**Weight-for-age percentiles:**
Boys, birth to 36 months

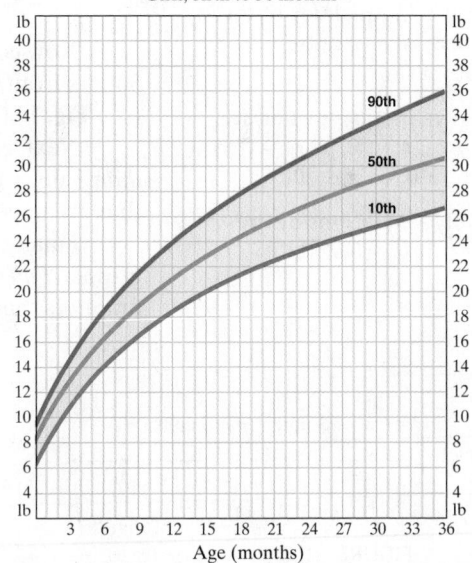

**Weight-for-age percentiles:**
Girls, birth to 36 months

*Source:* National Center for Health Statistics

## Challenge Problems

**103. Calculating Grades** Stephen Heasley's first five scores in European history were 82, 90, 74, 76, and 68. The final exam for the course is to count one-third in computing the final average. A final average greater than or equal to 80 and less

than 90 will result in a final grade of B. What range of final exam scores will result in Stephen's receiving a final grade of B in the course? Assume that a maximum score of 100 is possible.

*In Exercises 104–106,* **a)** *explain how to solve the inequality, and* **b)** *solve the inequality and give the solution in interval notation.*

**104.** $x < 3x - 10 < 2x$

**105.** $x < 2x + 3 < 2x + 5$

**106.** $x + 5 < x + 3 < 2x + 2$

## Cumulative Review Exercises

[1.4]    **107.** For $A = \left\{-3, 4, \dfrac{5}{2}, \sqrt{7}, 0, -\dfrac{13}{29}\right\}$, list the elements that are

          **a)** counting numbers

          **b)** whole numbers

          **c)** rational numbers

          **d)** real numbers

[1.10]    *For Exercises 108 and 109, name each illustrated property.*

    **108.** $(3x + 8) + 4y = 3x + (8 + 4y)$

    **109.** $5x + y = y + 5x$

[2.6]    **110.** Solve the formula $R = L + (V - D)r$ for $V$.

# 10.2 Solving Equations and Inequalities Containing Absolute Values

1　Understand absolute value on the number line.

2　Solve equations of the form $|x| = a, a > 0$.

3　Solve inequalities of the form $|x| < a, a > 0$.

4　Solve inequalities of the form $|x| > a, a > 0$.

5　Solve inequalities of the form $|x| < a$ or $|x| > a$, $a < 0$.

6　Solve inequalities of the form $|x| < 0, |x| \le 0$, $|x| > 0$, or $|x| \ge 0$.

7　Solve equations of the form $|x| = |y|$.

## 1　Understand Absolute Value on the Number Line

**Absolute value**

The **absolute value** of a number $x$, symbolized $|x|$, is the distance $x$ is from 0 on the number line.

$|3| = 3$ because the number 3 is 3 units from 0 on the number line.

$|-3| = 3$ because the number $-3$ is 3 units from 0 on the number line.

Now consider the equation $|x| = 3$. We are looking for values of $x$ that *are exactly* 3 units from 0 on the number line. Thus, the solutions to $|x| = 3$ are $x = 3$ and $x = -3$ (see **Fig. 10.6a**).

Next, consider the inequality $|x| < 3$. We are looking for values of $x$ that *are less than* 3 units from 0 on the number line. Thus, the solutions to $|x| < 3$ are the numbers that are between $-3$ and 3 on the number line (see **Fig. 10.6b**).

Finally, consider the inequality $|x| > 3$. We are looking for values of $x$ that *are greater than* 3 units from 0 on the number line. Thus, the solutions to $|x| > 3$ are the numbers that are less than $-3$ or greater than 3 on the number line (see **Fig. 10.6c**).

(a)

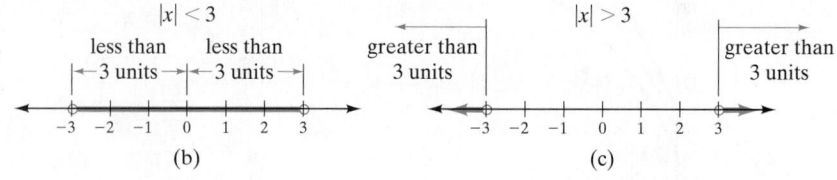

(b)　　　　　　　　(c)

FIGURE 10.6

We will use these examples and their illustrations on the number line to develop methods for solving equations and inequalities involving absolute value.

## 2　Solve Equations of the Form $|x| = a, a > 0$

When solving an equation of the form $|x| = a, a > 0$, we are finding the values that are exactly $a$ units from 0 on the number line.

**To Solve Equations of the Form $|x| = a$**

If $|x| = a$ and $a > 0$, then $x = a$ or $x = -a$.

**EXAMPLE 1** Solve each equation.

**a)** $|x| = 2$          **b)** $|x| = 0$          **c)** $|x| = -2$

*Solution*

**a)** Using the procedure, we get $x = 2$ or $x = -2$. The solution set is $\{-2, 2\}$.

**b)** The only real number whose absolute value equals 0 is 0. Thus, the solution set for $|x| = 0$ is $\{0\}$.

**c)** The absolute value of a number is never negative, so there are no solutions to this equation. The solution set is $\varnothing$.

Now Try Exercise 13

> **Understanding Algebra**
>
> The absolute value of a number is never negative.

**EXAMPLE 2** Solve the equation $|2w - 1| = 5$.

*Solution*   We are looking for the values of $w$ such that $2w - 1$ is exactly 5 units from 0 on a number line. Thus, the quantity $2w - 1$ must be equal to 5 or $-5$.

$$2w - 1 = 5 \quad \text{or} \quad 2w - 1 = -5$$
$$2w = 6 \qquad\qquad 2w = -4$$
$$w = 3 \qquad\qquad w = -2$$

Check          $w = 3$     $|2w - 1| = 5$          $w = -2$     $|2w - 1| = 5$

$$|2(3) - 1| \overset{?}{=} 5 \qquad\qquad |2(-2) - 1| \overset{?}{=} 5$$
$$|6 - 1| \overset{?}{=} 5 \qquad\qquad |-4 - 1| \overset{?}{=} 5$$
$$|5| \overset{?}{=} 5 \qquad\qquad |-5| \overset{?}{=} 5$$
$$5 = 5 \quad \text{True} \qquad\qquad 5 = 5 \quad \text{True}$$

The solutions 3 and $-2$ each result in $2w - 1$ being 5 units from 0 on the number line. The solution set is $\{-2, 3\}$.

Now Try Exercise 19

Consider the equation $|2w - 1| - 3 = 2$. The first step in solving this equation is to isolate the absolute value term. We do this by adding 3 to both sides of the equation. This results in the equation $|2w - 1| = 5$, which we solved in Example 2.

## 3  Solve Inequalities of the Form $|x| < a$, $a > 0$

Recall our discussion earlier regarding $|x| < 3$. The solutions to $|x| < 3$ are the numbers that are between $-3$ and $3$ on the number line (See **Fig. 10.6b** on page 600). Similarly, the solutions to $|x| < a$ are the numbers that are between $-a$ and $a$ on the number line.

To solve inequalities of the form $|x| < a$, we can use the following procedure.

> **To Solve Inequalities of the Form $|x| < a$**
>
> If $|x| < a$ and $a > 0$, then $-a < x < a$.

**EXAMPLE 3** Solve the inequality $|2x - 3| < 5$.

*Solution*   The solution to this inequality will be the set of values such that the distance between $2x - 3$ and 0 on a number line will be less than 5 units (see **Fig. 10.7**). Using **Figure 10.7**, we can see that $-5 < 2x - 3 < 5$.

**FIGURE 10.7**

Solving, we get

$$-5 < 2x - 3 < 5$$
$$-2 < 2x < 8$$
$$-1 < x < 4$$

The solution set is $\{x|-1 < x < 4\}$.

Now Try Exercise 33

**EXAMPLE 4** Solve the inequality $|2x + 1| \leq 9$ and graph the solution on a number line.

*Solution*    Since this inequality is of the form $|x| \leq a$, we write

$$-9 \leq 2x + 1 \leq 9$$
$$-10 \leq 2x \leq 8$$
$$-5 \leq x \leq 4$$

Now Try Exercise 75

**EXAMPLE 5** Solve the inequality $|7.8 - 4x| - 5.3 < 14.1$ and graph the solution on a number line.

*Solution*    First isolate the absolute value by adding 5.3 to both sides of the inequality. Then solve as in the previous examples.

$$|7.8 - 4x| - 5.3 < 14.1$$
$$|7.8 - 4x| < 19.4$$
$$-19.4 < 7.8 - 4x < 19.4$$
$$-27.2 < -4x < 11.6$$
$$\frac{-27.2}{-4} > \frac{-4x}{-4} > \frac{11.6}{-4}$$
$$6.8 > x > -2.9 \quad \text{or} \quad -2.9 < x < 6.8$$

The solution set is $\{x|-2.9 < x < 6.8\}$.

Now Try Exercise 43

**Understanding Algebra**

In Example 5, the solution

$$-2.9 < x < 6.8$$

written in interval notation is

$$(-2.9, 6.8).$$

### 4 Solve Inequalities of the Form $|x| > a, a > 0$

Recall our discussion earlier regarding $|x| > 3$. The solutions to $|x| > 3$ are the numbers that are less than $-3$ or greater than 3 on the number line (See **Fig. 10.6c** on page 600). Similarly, the solutions to $|x| > a$ are the numbers that are less than $-a$ or greater than $a$ on the number line.

To solve inequalities of the form $|x| > a$, we can use the following procedure.

**To Solve Inequalities of the Form $|x| > a$**

If $|x| > a$ and $a > 0$, then $x < -a$ or $x > a$.

**EXAMPLE 6** Solve the inequality $|2x - 3| > 5$ and graph the solution on a number line.

*Solution*    The solution to $|2x - 3| > 5$ is the set of values such that the distance between $2x - 3$ and 0 on a number line will be greater than 5. The quantity $2x - 3$ must be either less than $-5$ or greater than 5 (see **Fig. 10.8**).

**FIGURE 10.8**

Since $2x - 3$ must be either less than $-5$ or greater than 5, we set up and solve the following compound inequality:

$$2x - 3 < -5 \quad \text{or} \quad 2x - 3 > 5$$
$$2x < -2 \qquad\qquad\quad 2x > 8$$
$$x < -1 \qquad\qquad\quad x > 4$$

The solution set to $|2x - 3| > 5$ is $\{x | x < -1 \text{ or } x > 4\}$.

Now Try Exercise 51

**EXAMPLE 7** Solve the inequality $|2x - 1| \geq 7$ and graph the solution on a number line.

*Solution*  Since this inequality is of the form $|x| \geq a$, we use the procedure given above.

$$2x - 1 \leq -7 \quad \text{or} \quad 2x - 1 \geq 7$$
$$2x \leq -6 \qquad\qquad\quad 2x \geq 8$$
$$x \leq -3 \qquad\qquad\quad x \geq 4$$

Now Try Exercise 53

**EXAMPLE 8** Solve the inequality $\left|\dfrac{3x - 4}{2}\right| \geq 9$ and graph the solution on a number line.

*Solution*  Since this inequality is of the form $|x| \geq a$, we write

$$\frac{3x - 4}{2} \leq -9 \quad \text{or} \quad \frac{3x - 4}{2} \geq 9$$

Now multiply both sides of each inequality by the least common denominator, 2. Then solve each inequality.

$$2\left(\frac{3x - 4}{2}\right) \leq -9 \cdot 2 \quad \text{or} \quad 2\left(\frac{3x - 4}{2}\right) \geq 9 \cdot 2$$
$$3x - 4 \leq -18 \qquad\qquad\quad 3x - 4 \geq 18$$
$$3x \leq -14 \qquad\qquad\qquad 3x \geq 22$$
$$x \leq -\frac{14}{3} \qquad\qquad\qquad x \geq \frac{22}{3}$$

Now Try Exercise 57

### Understanding Algebra

In Example 7, any value of $x$ less than or equal to $-3$, or greater than or equal to 4, would result in $2x - 1$ representing a number that is greater than or equal to 7 units from 0 on a number line. The solution set is $\{x | x \leq -3 \text{ or } x \geq 4\}$. In interval notation, the solution is $(-\infty, -3] \cup [4, \infty)$.

---

### Helpful Hint

Some general information about equations and inequalities containing absolute value follows. For real numbers $a, b,$ and $c$, where $a \neq 0$ and $c > 0$:

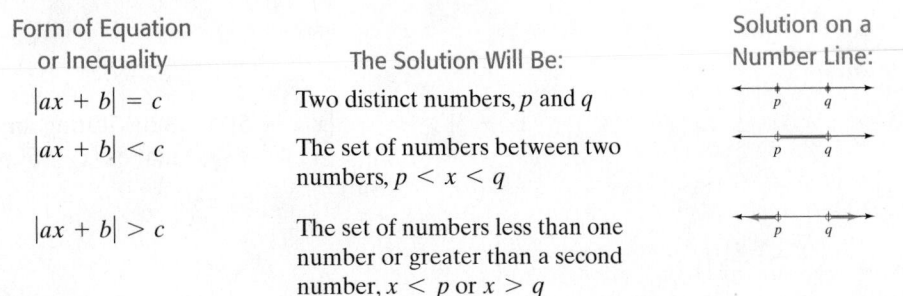

| Form of Equation or Inequality | The Solution Will Be: | Solution on a Number Line: |
|---|---|---|
| $\|ax + b\| = c$ | Two distinct numbers, $p$ and $q$ | |
| $\|ax + b\| < c$ | The set of numbers between two numbers, $p < x < q$ | |
| $\|ax + b\| > c$ | The set of numbers less than one number or greater than a second number, $x < p$ or $x > q$ | |

## 5  Solve Inequalities of the Form $|x| < a$ or $|x| > a, a < 0$

Consider the inequality $|x| < -3$. Since $|x|$ will always have a value greater than or equal to 0 for any real number $x$, this inequality can never be true, and the solution is the empty set, $\varnothing$.

**EXAMPLE 9**  Solve the inequality $|6x - 8| + 5 < 3$.

*Solution*    Begin by subtracting 5 from both sides of the inequality.

$$|6x - 8| + 5 < 3$$
$$|6x - 8| < -2$$

Since $|6x - 8|$ will always be greater than or equal to 0 for any real number $x$, this inequality can never be true. Therefore, the solution is the empty set, $\varnothing$.

Now Try Exercise 41

Now consider the inequality $|x| > -3$. Since $|x|$ will always have a value greater than or equal to 0 for any real number $x$, this inequality will always be true and the solution is the set of all real numbers, $\mathbb{R}$.

**EXAMPLE 10**  Solve the inequality $|5x + 3| + 4 \geq -9$.

*Solution*    Begin by subtracting 4 from both sides of the inequality.

$$|5x + 3| + 4 \geq -9$$
$$|5x + 3| \geq -13$$

Since $|5x + 3|$ will always be greater than or equal to 0 for any real number $x$, this inequality is true for all real numbers. Thus, the solution is the set of all real numbers, $\mathbb{R}$.

Now Try Exercise 59

## 6  Solve Inequalities of the Form $|x| < 0$, $|x| \leq 0$, $|x| > 0$, or $|x| \geq 0$

Each of the examples below requires you to recall that the absolute value of a number can never be negative.

| Inequality | Solution Set | Explanation |
|---|---|---|
| $|x - 5| < 0$ | $\varnothing$ | The absolute value of a number can never be $< 0$. |
| $|x - 5| \leq 0$ | $\{5\}$ | The absolute value of a number can never be $< 0$ but it can be $= 0$. When $x = 5$, we get $$|x - 5| \leq 0$$ $$|5 - 5| \leq 0$$ $$0 \leq 0 \quad \text{True}$$ |
| $|x - 5| > 0$ | $\{x \mid x \neq 5\}$ | Substituting any real number for $x$ except for 5 makes $|x - 5|$ positive. When $x = 5$, we get $$|x - 5| > 0$$ $$|5 - 5| > 0$$ $$0 > 0 \quad \text{False}$$ |
| $|x - 5| \geq 0$ | $\mathbb{R}$ | The absolute value of any number is always $\geq 0$. |

**EXAMPLE 11** Solve each inequality.   **a)** $|x + 2| > 0$   **b)** $|3x - 8| \le 0$

*Solution*

**a)** The inequality will be true for every value of $x$ except $-2$. The solution set is $\{x \mid x < -2 \text{ or } x > -2\}$.

**b)** Determine the number that makes the absolute value equal to 0 by setting the expression within the absolute value sign equal to 0 and solving for $x$.

$$3x - 8 = 0$$
$$3x = 8$$
$$x = \frac{8}{3}$$

The inequality will be true only when $x = \dfrac{8}{3}$. The solution set is $\left\{ \dfrac{8}{3} \right\}$.

Now Try Exercise 61

## 7   Solve Equations of the Form $|x| = |y|$

Now we will discuss absolute value equations where an absolute value appears on both sides of the equation.

When solving an absolute value equation with an absolute value expression on each side of the equals sign, the two expressions must have the same absolute value. Therefore, *the expressions must be equal to each other or be opposites of each other*.

> **To Solve Equations of the Form $|x| = |y|$**
>
> If $|x| = |y|$, then $x = y$ or $x = -y$.

**EXAMPLE 12** Solve the equation $|z + 3| = |2z - 7|$.

*Solution*   If we let $z + 3$ be $x$ and $2z - 7$ be $y$, this equation is of the form $|x| = |y|$. Using the procedure given above, we obtain the two equations

$$z + 3 = 2z - 7 \quad \text{or} \quad z + 3 = -(2z - 7)$$

Now solve each equation.

$$
\begin{array}{ll}
z + 3 = 2z - 7 & \text{or} \quad z + 3 = -(2z - 7) \\
3 = z - 7 & \qquad z + 3 = -2z + 7 \\
10 = z & \qquad 3z + 3 = 7 \\
 & \qquad 3z = 4 \\
 & \qquad z = \dfrac{4}{3}
\end{array}
$$

**Understanding Algebra**

If $|x| = |y|$, then $x = y$ or $x = -y$.

Check      $z = 10 \quad |z + 3| = |2z - 7| \qquad\qquad z = \dfrac{4}{3} \quad |z + 3| = |2z - 7|$

$$|10 + 3| \stackrel{?}{=} |2(10) - 7| \qquad\qquad \left|\frac{4}{3} + 3\right| \stackrel{?}{=} \left|2\left(\frac{4}{3}\right) - 7\right|$$

$$|13| \stackrel{?}{=} |20 - 7| \qquad\qquad\qquad \left|\frac{13}{3}\right| \stackrel{?}{=} \left|\frac{8}{3} - \frac{21}{3}\right|$$

$$|13| \stackrel{?}{=} |13| \qquad\qquad\qquad\quad \left|\frac{13}{3}\right| \stackrel{?}{=} \left|-\frac{13}{3}\right|$$

$$13 = 13 \quad \text{True} \qquad\qquad\qquad \frac{13}{3} = \frac{13}{3} \quad \text{True}$$

The solution set is $\left\{ 10, \dfrac{4}{3} \right\}$.

Now Try Exercise 63

**EXAMPLE 13** Solve the equation $|4x - 7| = |6 - 4x|$.

*Solution*

$$4x - 7 = 6 - 4x \qquad \text{or} \qquad 4x - 7 = -(6 - 4x)$$
$$8x - 7 = 6 \qquad\qquad\qquad 4x - 7 = -6 + 4x$$
$$8x = 13 \qquad\qquad\qquad\qquad -7 = -6 \qquad \text{False}$$
$$x = \frac{13}{8}$$

Since the equation $4x - 7 = -(6 - 4x)$ results in a false statement, the absolute value equation has only one solution. A check will show that the solution set is $\left\{ \dfrac{13}{8} \right\}$.

Now Try Exercise 69

---

### Summary of Procedures for Solving Equations and Inequalities Containing Absolute Value

For $a > 0$,

$$\text{If } |x| = a, \text{ then } x = a \text{ or } x = -a.$$
$$\text{If } |x| < a, \text{ then } -a < x < a.$$
$$\text{If } |x| > a, \text{ then } x < -a \text{ or } x > a.$$
$$\text{If } |x| = |y|, \text{ then } x = y \text{ or } x = -y.$$

---

# EXERCISE SET 10.2   MathXL®   MyMathLab®

MathXL®          MyMathLab

## Warm-Up Exercises

*Fill in the blanks with the appropriate word, phrase, or symbol(s) from the following list.*

| | | | | | |
|---|---|---|---|---|---|
| $\lvert x \rvert = 4$ | $\lvert x \rvert < 4$ | $\lvert x \rvert \le 4$ | $\lvert x \rvert > 4$ | $\lvert x \rvert \ge 4$ | $\lvert x \rvert = 5$ |
| $\lvert x \rvert < 5$ | $\lvert x \rvert \le 5$ | $\lvert x \rvert > 5$ | $\lvert x \rvert \ge 5$ | $\lvert x \rvert < -6$ | $\lvert x \rvert > -6$ |

**1.** The graph of the solution to _____ on the number line is .

**2.** The graph of the solution to _____ on the number line is .

**3.** The graph of the solution to _____ on the number line is .

**4.** The graph of the solution to _____ on the number line is .

**5.** The graph of the solution to _____ on the number line is .

**6.** The solution set of _____ is $\{x \mid -5 \le x \le 5\}$.

**7.** The solution set of _____ is $\{x \mid x \le -5 \text{ or } x \ge 5\}$.

**8.** The solution set of _____ is $\{-5, 5\}$.

**9.** The solution set of _____ is $\{x \mid -5 < x < 5\}$.

**10.** The solution set of _____ is $\{x \mid x < -5 \text{ or } x > 5\}$.

**11.** The solution set of _____ is $\mathbb{R}$.

**12.** The solution set of _____ is $\varnothing$.

## Practice the Skills

*Find the solution set for each equation.*

**13.** $|a| = 7$

**14.** $|b| = 17$

**15.** $|c| = \dfrac{1}{2}$

**16.** $|x| = 0$

**17.** $|d| = -\dfrac{5}{6}$

**18.** $|l + 4| = 6$

**19.** $|x + 5| = 8$

**20.** $|3 + y| = \dfrac{3}{5}$

**21.** $|4.5q + 31.5| = 0$

**22.** $|4.7 - 1.6z| = 14.3$

**23.** $|5 - 3x| = \dfrac{1}{2}$

**24.** $|6(y + 4)| = 24$

**25.** $\left|\dfrac{x-3}{4}\right| = 5$

**26.** $\left|\dfrac{3z+5}{6}\right| - 2 = 7$

**27.** $\left|\dfrac{x-3}{4}\right| + 8 = 8$

**28.** $\left|\dfrac{5x-3}{2}\right| + 5 = 9$

**29.** $|x-5| + 4 = 3$

**30.** $|2x+3| - 5 = -8$

*Find the solution set for each inequality.*

**31.** $|w| < 1$

**32.** $|p| \le 9$

**33.** $|q+5| \le 8$

**34.** $|7-x| < 6$

**35.** $|5b-15| < 10$

**36.** $|x-3| - 7 < -2$

**37.** $|2x+3| - 5 \le 10$

**38.** $|4-3x| - 4 < 11$

**39.** $|3x-7| + 8 < 14$

**40.** $\left|\dfrac{2x-1}{9}\right| \le \dfrac{5}{9}$

**41.** $|2x-6| + 5 \le 1$

**42.** $|2x-3| < -10$

**43.** $\left|\dfrac{1}{2}j + 4\right| < 7$

**44.** $\left|\dfrac{k}{4} - \dfrac{3}{8}\right| < \dfrac{7}{16}$

**45.** $\left|\dfrac{x-3}{2}\right| - 4 \le -2$

**46.** $\left|7x - \dfrac{1}{2}\right| < 0$

*Find the solution set for each inequality.*

**47.** $|y| > 8$

**48.** $|a| \ge 13$

**49.** $|x+4| > 5$

**50.** $|2b-7| > 3$

**51.** $|7-3b| > 5$

**52.** $\left|\dfrac{6+2z}{3}\right| > 2$

**53.** $|2h-5| > 3$

**54.** $|2x-1| \ge 12$

**55.** $|0.1x - 0.4| + 0.4 > 0.6$

**56.** $|3.7d + 6.9| - 2.1 > -5.4$

**57.** $\left|\dfrac{x}{2} + 4\right| \ge 5$

**58.** $\left|4 - \dfrac{3x}{5}\right| \ge 9$

**59.** $|7w+3| - 12 \ge -12$

**60.** $|2.6 - x| \ge 0$

**61.** $|4-2x| > 0$

**62.** $|2c-8| > 0$

*Find the solution set for each equation.*

**63.** $|3p-5| = |2p+10|$

**64.** $|6n+3| = |4n-13|$

**65.** $|6x| = |3x-9|$

**66.** $|5t-10| = |10-5t|$

**67.** $\left|\dfrac{2r}{3} + \dfrac{5}{6}\right| = \left|\dfrac{r}{2} - 3\right|$

**68.** $|3x-8| = |3x+8|$

**69.** $\left|-\dfrac{3}{4}m + 8\right| = \left|7 - \dfrac{3}{4}m\right|$

**70.** $\left|\dfrac{3}{2}r + 2\right| = \left|8 - \dfrac{3}{2}r\right|$

*Find the solution set for each equation or inequality.*

**71.** $|h| = 9$

**72.** $|y| \le 8$

**73.** $|q+6| > 2$

**74.** $|9d+7| \le -9$

**75.** $|2w-7| \le 9$

**76.** $|2z-7| + 5 > 8$

**77.** $|5a-1| = 9$

**78.** $|2x-4| + 5 = 13$

**79.** $|5+2x| > 0$

**80.** $|7-3b| = |5b+15|$

**81.** $|4+3x| \le 9$

**82.** $|2.4x + 4| + 4.9 > 3.9$

**83.** $|3n+8| - 4 = -10$

**84.** $|4-2x| - 3 = 7$

**85.** $\left|\dfrac{w+4}{3}\right| + 5 < 9$

**86.** $\left|\dfrac{5t-10}{6}\right| > \dfrac{5}{3}$

**87.** $\left|\dfrac{3x-2}{4}\right| - \dfrac{1}{3} \ge -\dfrac{1}{3}$

**88.** $\left|\dfrac{2x-4}{5}\right| = 14$

**89.** $|2x-8| = \left|\dfrac{1}{2}x + 3\right|$

**90.** $\left|\dfrac{1}{3}y + 3\right| = \left|\dfrac{2}{3}y - 1\right|$

**91.** $|2-3x| = \left|4 - \dfrac{5}{3}x\right|$

**92.** $\left|\dfrac{-2u+3}{7}\right| \le 5$

## Problem Solving

**93. Glass Thickness** Certain types of glass manufactured by PPG Industries ideally will have a thickness of 0.089 inches. However, due to limitations in the manufacturing process, the thickness is allowed to vary from the ideal thickness by up to 0.004 inch. If $t$ represents the actual thickness of the glass, then the allowable range of thicknesses can be represented using the inequality $|t - 0.089| \le 0.004$.
*Source:* www.ppg.com

**a)** Solve this inequality for $t$ (use interval notation).
**b)** What is the smallest thickness the glass is allowed to be?
**c)** What is the largest thickness the glass is allowed to be?

**94. Plywood Guarantee** Certain plywood manufactured by Lafor International is guaranteed to be $\frac{5}{8}$ inch thick with a tolerance of plus or minus $\frac{1}{56}$ of an inch. If $t$ represents the actual thickness of the plywood, then the allowable range of thicknesses can be represented using the inequality $\left| t - \frac{5}{8} \right| \le \frac{1}{56}$.

*Source:* www.sticktrade.com

a) Solve this inequality for $t$ (use interval notation).

b) What is the smallest thickness the plywood is allowed to be?

c) What is the largest thickness the plywood is allowed to be?

**95. Submarine Depth** A submarine is 160 feet below sea level. It has rock formations above and below it, and should not change its depth by more than 28 feet. Its distance below sea level, $d$, can be described by the inequality $|d - 160| \le 28$.

a) Solve this inequality for $d$. Write your answer in interval notation.

b) Between what vertical distances, measured from sea level, may the submarine move?

**96. A Bouncing Spring** A spring hanging from a ceiling is bouncing up and down so that its distance, $d$, above the ground satisfies the inequality $|d - 4| \le \frac{1}{2}$ foot (see the figure).

a) Solve this inequality for $d$. Write your answer in interval notation.

b) Between what distances, measured from the ground, will the spring oscillate?

**97.** How many solutions are there to the following equations or inequalities if $a \ne 0$ and $k > 0$?

a) $|ax + b| = k$

b) $|ax + b| < k$

c) $|ax + b| > k$

**98.** Suppose $|x| < |y|$ and $x < 0$ and $y < 0$.

a) Which of the following must be true: $x < y$, $x > y$, or $x = y$?

b) Give an example to support your answer to part **a)**.

**99.** How many solutions will $|ax + b| = k$, $a \ne 0$ have if

a) $k < 0$,

b) $k = 0$,

c) $k > 0$?

**100.** Suppose $m$ and $n$ $(m < n)$ are two distinct solutions to the equation $|ax + b| = c$. Indicate the solutions, using both inequality symbols and the number line, to each inequality. (See the Helpful Hint on page 603.)

a) $|ax + b| < c$

b) $|ax + b| > c$

## Concept/Writing Exercises

**101.** For what value of $x$ will the inequality $|ax + b| \le 0$ be true? Explain.

**102.** For what value of $x$ will the inequality $|ax + b| > 0$ *not* be true? Explain.

**103. a)** Explain how to find the solution to the equation $|ax + b| = c$. (Assume that $c > 0$ and $a \ne 0$.)

b) Solve this equation for $x$.

**104. a)** Explain how to find the solution to the inequality $|ax + b| < c$. (Assume that $a > 0$ and $c > 0$.)

b) Solve this inequality for $x$.

**105. a)** Explain how to find the solution to the inequality $|ax + b| > c$. (Assume that $a > 0$ and $c > 0$.)

b) Solve this inequality for $x$.

**106. a)** What is the first step in solving the inequality $-4|3x - 5| \le -12$?

b) Solve this inequality and give the solution in interval notation.

*Determine what values of x will make each equation true. Explain your answer.*

**107.** $|x - 4| = |4 - x|$        **108.** $|x - 4| = -|x - 4|$        **109.** $|x| = x$        **110.** $|x + 2| = x + 2$

*Solve. Explain how you determined your answer.*

**111.** $|x + 1| = 2x - 1$             **112.** $|3x + 1| = x - 3$          **113.** $|x - 4| = -(x - 4)$

## Challenge Problems

*Solve by considering the possible signs for x.*

**114.** $|x| + x = 8$        **115.** $x + |-x| = 8$        **116.** $|x| - x = 8$        **117.** $x - |x| = 8$

## Group Activity

*Discuss and answer Exercise 118 as a group.*

**118.** Consider the equation $|x + y| = |y + x|$.

    **a)** Have each group member select an *x* value and a *y* value and determine whether the equation holds. Repeat for two other pairs of *x*- and *y*-values.

    **b)** As a group, determine for what values of *x* and *y* the equation is true. Explain your answer.

    **c)** Now consider $|x - y| = -|y - x|$. Under what conditions will this equation be true?

## Cumulative Review Exercises

*Evaluate.*

[1.9] **119.** $\dfrac{1}{3} + \dfrac{1}{4} \div \dfrac{2}{5}\left(\dfrac{1}{3}\right)^2$

    **120.** $4(x + 3y) - 5xy$ when $x = 1$, $y = 3$

[3.4] **121.** **Swimming** Terry Chong swims across a lake averaging 2 miles an hour. Then he turns around and swims back across the lake, averaging 1.6 miles per hour. If his total swimming time is 1.5 hours, what is the width of the lake?

[10.1] **122.** Find the solution set to the inequality

$$-7(x - 3) + 5(x + 1) \geq 20$$

---

## Mid-Chapter Test: 10.1–10.2

*To find out how well you understand the chapter material to this point, take this brief test. The answers, and the section where the material was initially discussed, are given in the back of the book. Review any questions that you answered incorrectly.*

**1.** Express $x > \dfrac{5}{2}$ on a number line.

**2.** Express $-4 \leq w < 5$ in interval notation.

*Solve each inequality and graph the solution on a number line.*

**3.** $6x - 5 \leq 4x + 61$

**4.** $5y - 9y + 8 \leq 2(-2y + 1)$

**5.** $2.4w - 3.2 < 3.6w - 0.8$

*Solve each inequality and give the solution in interval notation.*

**6.** $3 - 2x < 6 + 2x + 1$

**7.** $-8 < p - 3 \leq 4$

**8.** $\dfrac{1}{3} \leq 4x + 1 < 5$

*Solve each inequality and indicate the solution set.*

**9.** $-3 \leq \dfrac{2x + 1}{5} \leq 1$

**10.** $m \leq 7$ and $m > -6$

**11.** $9 - x < 3$ or $7x - 5 \leq -5$

**12.** Find $A \cup B$ and $A \cap B$ for $A = \{7, 8, 9, 10\}$ and $B = \{8, 10, 12, 15\}$.

*Find the solution set for each equation or inequality.*

**13.** $|t - 4| = 6$

**14.** $\left|\dfrac{w - 5}{3}\right| \leq 7$

**15.** $|z| \geq 8$

**16.** $|5 - 2x| \geq 0$

**17.** $|4 - 2x| = 5$

**18.** $|-3m + 4| < 7$

**19.** $|6y + 9| = |4y - 1|$

**20.** $\left|\dfrac{1}{2}z - 3\right| = \left|\dfrac{3}{2}z + 2\right|$

# 10.3 Graphing Linear Inequalities in Two Variables and Systems of Linear Inequalities

**1** Graph linear inequalities in two variables.

**2** Solve systems of linear inequalities.

**3** Solve linear programming problems.

**4** Solve systems of linear inequalities containing absolute value.

## Understanding Algebra

The following symbols are used in linear inequalities:

- $<$ is less than
- $>$ is greater than
- $\leq$ is less than or equal to
- $\geq$ is greater than or equal to

## 1 Graph Linear Inequalities in Two Variables

### Linear Inequality in Two Variables

A **linear inequality in two variables** can be written in one of the following forms:

$$ax + by < c, \quad ax + by > c, \quad ax + by \leq c, \quad ax + by \geq c$$

where $a, b,$ and $c$ are real numbers and $a$ and $b$ are not both 0.

### Examples of Linear Inequalities in Two Variables

$$2x + 3y > 2 \qquad\qquad 3y < 4x - 9$$

$$-x - 2y \leq 3 \qquad\qquad 5x \geq 2y - 7$$

Consider the graph of the equation $x + y = 3$ shown in **Figure 10.9**. The line acts as a **boundary** between two *half-planes* and divides the plane into three distinct sets of points: the points on the line itself, the points in the half-plane on one side of the line, and the points in the half-plane on the other side of the line.

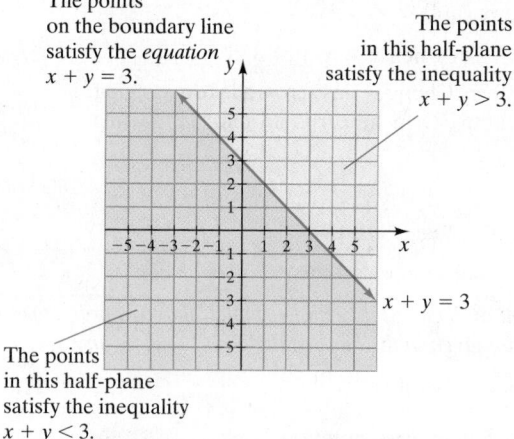

The points on the boundary line satisfy the *equation* $x + y = 3$.

The points in this half-plane satisfy the inequality $x + y > 3$.

The points in this half-plane satisfy the inequality $x + y < 3$.

**FIGURE 10.9**

## Understanding Algebra

If a linear inequality has

- $<$ or $>$, draw a *dashed* boundary line.
- $\leq$ or $\geq$, draw a *solid* boundary line.

When we graph linear inequalities, we generally will shade only one of the two half-planes established by the boundary line. If the inequality is written using $<$ or $>$, we draw a dashed line to indicate that the boundary line is not part of the solution set. If the inequality is written using $\leq$ or $\geq$, we will draw a solid line to indicate that the boundary line is part of the solution set.

### To Graph a Linear Inequality in Two Variables

1. To get the equation of the boundary line, replace the inequality symbol with an equals sign.

2. Draw the graph of the equation in step 1. If the original inequality contains a $\geq$ or $\leq$ symbol, draw the boundary line using a solid line. If the original inequality contains a $>$ or $<$ symbol, draw the boundary line using a dashed line.

3. Select any point not on the boundary line and determine if this point is a solution to the original inequality. If the point selected is a solution, shade the half-plane on the side of the line containing this point. If the selected point does not satisfy the inequality, shade the half-plane on the side of the line not containing this point.

In step 3 we are deciding which half-plane contains points that satisfy the given inequality.

**EXAMPLE 1** Graph the inequality $y < \frac{2}{3}x - 3$.

*Solution*   First graph the equation $y = \frac{2}{3}x - 3$. Since the original inequality contains a less than sign, $<$, use a dashed line when drawing the graph (**Fig. 10.10**). The dashed line indicates that the points on this line are not solutions to the inequality $y < \frac{2}{3}x - 3$. Select a point not on the line and determine if this point satisfies the inequality. Often the easiest point to use is the origin, $(0, 0)$.

Checkpoint $(0, 0)$

$$y < \frac{2}{3}x - 3$$

$$0 \overset{?}{<} \frac{2}{3}(0) - 3$$

$$0 \overset{?}{<} 0 - 3$$

$$0 < -3 \qquad \text{False}$$

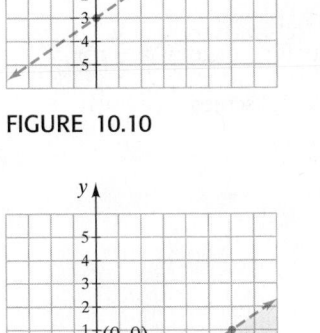

FIGURE 10.10

Since 0 is not less than $-3$, the point $(0, 0)$ does not satisfy the inequality. The solution will be all points in the half-plane that does *not* contain point $(0, 0)$. Shade in this half-plane (**Fig. 10.11**). Every point in the shaded half-plane satisfies the given inequality. Let's check a few selected points $A$, $B$, and $C$.

| Point $A$ | Point $B$ | Point $C$ |
|:---:|:---:|:---:|
| $(6, 0)$ | $(3, -3)$ | $(0, -4)$ |
| $y < \frac{2}{3}x - 3$ | $y < \frac{2}{3}x - 3$ | $y < \frac{2}{3}x - 3$ |
| $0 \overset{?}{<} \frac{2}{3}(6) - 3$ | $-3 \overset{?}{<} \frac{2}{3}(3) - 3$ | $-4 \overset{?}{<} \frac{2}{3}(0) - 3$ |
| $0 \overset{?}{<} 4 - 3$ | $-3 \overset{?}{<} 2 - 3$ | $-4 \overset{?}{<} 0 - 3$ |
| $0 < 1 \quad$ True | $-3 < -1 \quad$ True | $-4 < -3 \quad$ True |

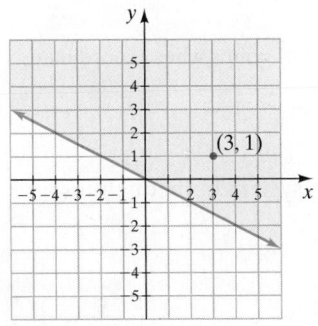

FIGURE 10.11

Now Try Exercise 13

**EXAMPLE 2** Graph the inequality $y \geq -\frac{1}{2}x$.

*Solution*   First, we graph the equation $y = -\frac{1}{2}x$. Since the inequality is $\geq$, we use a solid boundary line to indicate that the points on the line are solutions to the inequality (**Fig. 10.12**). Since the point $(0, 0)$ is on the line, we cannot select that point as our checkpoint. Let's select the point $(3, 1)$.

Checkpoint $(3, 1)$

$$y \geq -\frac{1}{2}x$$

$$1 \overset{?}{\geq} -\frac{1}{2}(3)$$

$$1 \geq -\frac{3}{2} \qquad \text{True}$$

Since the point $(3, 1)$ satisfies the inequality, every point in the same half-plane as $(3, 1)$ will also satisfy the inequality $y \geq -\frac{1}{2}x$. Shade this half-plane as indicated.

Every point in the shaded half-plane, as well as every point on the boundary line, satisfies the inequality.

Now Try Exercise 23

FIGURE 10.12

**EXAMPLE 3** Graph the inequality $3x - 2y < -6$.

*Solution* First, we graph the equation $3x - 2y = -6$. Since the inequality is $<$, we use a dashed boundary line (**Fig. 10.13**). Substituting the checkpoint $(0, 0)$ into the inequality results in a false statement.

$$\text{Checkpoint } (0, 0)$$
$$3x - 2y < -6$$
$$3(0) - 2(0) \overset{?}{<} -6$$
$$0 < -6 \qquad \text{False}$$

The solution is, therefore, the half-plane that does not contain the origin.

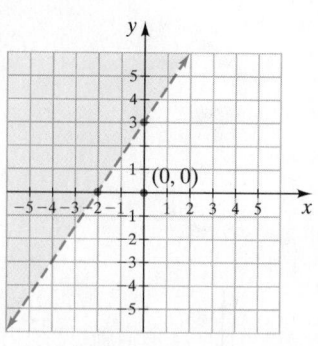

**FIGURE 10.13**

Now Try Exercise 21

## Using Your Graphing Calculator

We will graph the inequality $3x - 2y < -6$ that we graphed in Example 3. First, solve the inequality for $y$ to get $y > \frac{3}{2}x + 3$.

We begin by entering the equation of the boundary line $y = \frac{3}{2}x + 3$. **Figure 10.14a** shows the TI-84 Plus screen.

(a)

(b)

**FIGURE 10.14**

Notice the symbol to the left of $Y_1$. The symbol indicates the shading will go above the boundary line since the inequality used is $>$. To get this symbol, use the left arrow key until the cursor is in this position on the screen. Then press $\boxed{\text{ENTER}}$ until the symbol appears. After pressing $\boxed{\text{GRAPH}}$ the window shows the graph in **Figure 10.14b**. Compare **Figure 10.14b** to **Figure 10.13** above. A word of caution: notice that the window does not show a dashed boundary line.

## 2 Solve Systems of Linear Inequalities

In Section 9.1 we learned how to solve systems of equations graphically. In this section we show how to solve **systems of linear inequalities** graphically.

### To Solve a System of Linear Inequalities

Graph each inequality on the same axes. The solution is the set of points whose coordinates satisfy all the inequalities in the system.

**EXAMPLE 4** Determine the solution to the following system of inequalities.

$$y < -\frac{1}{2}x + 2$$
$$x - y \le 4$$

*Solution* First graph the inequality $y < -\dfrac{1}{2}x + 2$ (**Fig. 10.15** below). Now on the same axes graph the inequality $x - y \le 4$ (**Fig. 10.16** below). The solution is the set of points common to the graphs of both inequalities. It is the part of the graph that contains both shadings. The dashed line is not part of the solution, but the part of the solid line that satisfies both inequalities is part of the solution.

FIGURE 10.15

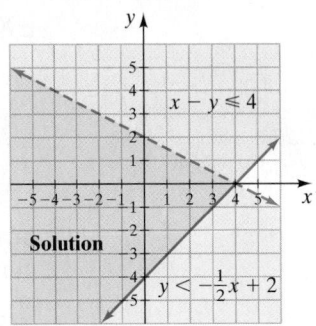

FIGURE 10.16

Now Try Exercise 29

**EXAMPLE 5** Determine the solution to the following system of inequalities.

$$3x - y < 6$$
$$2x + 2y \ge 5$$

*Solution* Graph $3x - y < 6$ (see **Fig. 10.17**). Graph $2x + 2y \ge 5$ on the same axes (**Fig. 10.18**). The solution is the part of the graph that contains both shadings and the part of the solid line that satisfies both inequalities.

FIGURE 10.17

FIGURE 10.18

Now Try Exercise 35

**EXAMPLE 6** Determine the solution to the following system of inequalities.

$$y > -1$$
$$x \le 4$$

*Solution* The solution is illustrated in **Figure 10.19**.

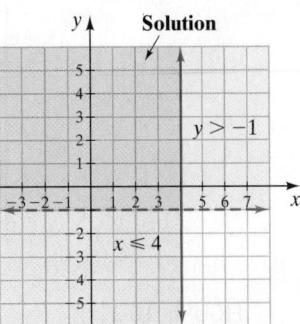

FIGURE 10.19

Now Try Exercise 39

### 3 Solve Linear Programming Problems

**Linear programming** is a mathematical process that includes graphing more than two linear inequalities on the same axes. The inequalities involved in linear programming are called the **constraints**.

**EXAMPLE 7** Determine the solution to the following system of inequalities.

$$x \geq 0$$
$$y \geq 0$$
$$2x + 3y \leq 12$$
$$2x + y \leq 8$$

*Solution*  The first two inequalities, $x \geq 0$ and $y \geq 0$, indicate that the solution must be in the first quadrant because that is the only quadrant where both $x$ and $y$ are positive. **Figure 10.20** illustrates the graphs of the four inequalities.

**FIGURE 10.20**

**Figure 10.21** illustrates the graphs on the same axes and the solution to the system of inequalities. Note that every point in the shaded area and every point on the lines that form the polygonal region is part of the answer.

Now Try Exercise 47

**FIGURE 10.21**

**EXAMPLE 8** Determine the solution to the following system of inequalities.

$$x \geq 0$$
$$y \geq 0$$
$$x \leq 15$$
$$8x + 8y \leq 160$$
$$4x + 12y \leq 180$$

*Solution*  The first two inequalities indicate that the solution must be in the first quadrant. The third inequality indicates that $x$ must be a value less than or equal to 15. **Figure 10.22a** indicates the graphs of corresponding equations and shows the region that satisfies all the inequalities in the system. **Figure 10.22b** indicates the solution to the system of inequalities.

**FIGURE 10.22**　　　　　(a)　　　　　　　　　　　(b)

Now Try Exercise 53

## **4** Solve Systems of Linear Inequalities Containing Absolute Value

Now we will graph systems of linear inequalities containing absolute value in the Cartesian coordinate system. Before we do some examples, let us recall the rules for absolute value inequalities that we learned in Section 10.2.

---

### Solving Absoute Value Inequalities

If $|x| < a$ and $a > 0$, then $-a < x < a$.

If $|x| > a$ and $a > 0$, then $x < -a$ or $x > a$.

---

**EXAMPLE 9** Graph $|x| < 3$ in the Cartesian coordinate system.

*Solution*    From the rules given, we know that $|x| < 3$ means $-3 < x < 3$. We draw dashed vertical lines through $-3$ and $3$ and shade the area between the two (**Fig. 10.23**).

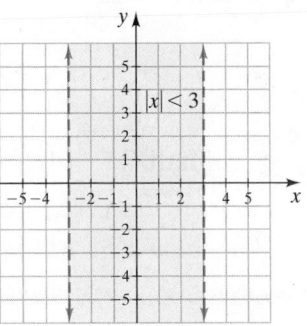

**FIGURE 10.23**

Now Try Exercise 57

**EXAMPLE 10** Graph $|y + 1| > 3$ in the Cartesian coordinate system.

*Solution*    From the rules given, we know that $|y + 1| > 3$ means $y + 1 < -3$ or $y + 1 > 3$. First we solve each inequality.

$$y + 1 < -3 \quad \text{or} \quad y + 1 > 3$$
$$y < -4 \qquad\qquad y > 2$$

Now we graph both inequalities and take the *union* of the two graphs. The solution is the shaded area in **Fig. 10.24**.

**FIGURE 10.24**

Now Try Exercise 59

EXAMPLE 11 Graph the following system of inequalities.

$$|x| < 3$$
$$|y + 1| > 3$$

**Solution**   We draw both inequalities on the same axes. Therefore, we combine the graph drawn in Example 9 with the graph drawn in Example 10 (see **Fig. 10.25**). The points common to both inequalities form the solution to the system.

FIGURE 10.25

Now Try Exercise 65

# EXERCISE SET 10.3   MathXL®   MyMathLab®

## Warm-Up Exercises

*Fill in the blanks with the appropriate word, phrase, or symbol(s) from the following list.*

| solid | checkpoint | boundary line | dashed | half-planes |
| constraints | constants | linear programming | satisfy | geometry |

1. When graphing a linear inequality, the _____ divides the plane into two half-planes.

2. If the linear inequality contains < or > draw a _____ boundary line.

3. If the linear inequality contains ≤ or ≥ draw a _____ boundary line.

4. When graphing linear inequalities, we generally only shade one of the two _____ established by the boundary line.

5. To determine which half-plane to shade, pick a _____ that is not on the boundary line.

6. The solution to a system of linear inequalities is the set of points whose coordinates _____ all the inequalities in the system.

7. A mathematical process for which you often must graph more than two linear inequalities on the same axes is called _____ .

8. The inequalities in a linear programming problem are called the _____ .

## Practice the Skills

*Graph each inequality.*

9. $y < 2x + 1$

10. $y \geq 3x - 1$

11. $y > 2x - 1$

12. $y \leq -x + 4$

13. $y \geq \dfrac{1}{2}x - 3$

14. $y < 3x + 2$

15. $2x + 3y > 6$

16. $2x - 3y \geq 12$

17. $y \leq -3x + 5$

18. $y \leq \dfrac{2}{3}x + 3$

19. $2x + y < 4$

20. $3x - 4y \leq 12$

21. $10 \geq 5x - 2y$

22. $-x - 2y > 4$

23. $y \geq -\dfrac{1}{2}x$

24. $y < \dfrac{1}{2}x$

25. $x > 1$

26. $x \geq 4$

27. $y < -2$

28. $y < x$

*Determine the solution to each system of inequalities.*

29. $2x - y < 4$
    $y \geq -x + 2$

30. $y \leq -2x + 1$
    $y > -3x$

31. $y < 3x - 2$
    $y \leq -2x + 3$

32. $y \geq 2x - 5$
    $y > -3x + 5$

33. $y < x$
    $y \geq 3x + 2$

34. $-3x + 2y \geq -5$
    $y \leq -4x + 7$

35. $-2x + 3y < -5$
    $3x - 8y > 4$

36. $-4x + 3y \geq -4$
    $y > -3x + 3$

**37.** $-4x + 5y < 20$
   $x \geq -3$

**38.** $y \geq -\dfrac{2}{3}x + 1$
   $y > -4$

**39.** $x \leq 4$
   $y \geq -2$

**40.** $x \geq 0$
   $x - 3y < 6$

**41.** $5x + 2y > 10$
   $3x - y > 3$

**42.** $3x + 2y > 8$
   $x - 5y < 5$

**43.** $-2x > y + 4$
   $-x < \dfrac{1}{2}y - 1$

**44.** $y \leq 3x - 2$
   $\dfrac{1}{3}y < x + 1$

**45.** $y < 3x - 4$
   $6x \geq 2y + 8$

**46.** $\dfrac{1}{2}x + \dfrac{1}{2}y \geq 2$
   $2x - 3y \leq -6$

*Determine the solution to each system of inequalities. Use the method discussed in Examples 7 and 8.*

**47.** $x \geq 0$
   $y \geq 0$
   $2x + 3y \leq 6$
   $4x + y \leq 4$

**48.** $x \geq 0$
   $y \geq 0$
   $x + y \leq 6$
   $7x + 4y \leq 28$

**49.** $x \geq 0$
   $y \geq 0$
   $2x + 3y \leq 8$
   $4x + 2y \leq 8$

**50.** $x \geq 0$
   $y \geq 0$
   $3x + 2y \leq 18$
   $2x + 4y \leq 20$

**51.** $x \geq 0$
   $y \geq 0$
   $3x + y \leq 9$
   $2x + 5y \leq 10$

**52.** $x \geq 0$
   $y \geq 0$
   $5x + 4y \leq 16$
   $x + 6y \leq 18$

**53.** $x \geq 0$
   $y \geq 0$
   $x \leq 4$
   $x + y \leq 6$
   $x + 2y \leq 8$

**54.** $x \geq 0$
   $y \geq 0$
   $x \leq 4$
   $2x + 3y \leq 18$
   $4x + 2y \leq 20$

**55.** $x \geq 0$
   $y \geq 0$
   $x \leq 15$
   $30x + 25y \leq 750$
   $10x + 40y \leq 800$

**56.** $x \geq 0$
   $y \geq 0$
   $x \leq 15$
   $40x + 25y \leq 1000$
   $5x + 30y \leq 900$

*Determine the solution to each inequality.*

**57.** $|x| < 2$

**58.** $|x| > 1$

**59.** $|y - 2| \leq 4$

**60.** $|y| \geq 2$

*Determine the solution to each system of inequalities.*

**61.** $|y| > 2$
   $y \leq x + 3$

**62.** $|x| > 1$
   $y \leq 3x + 2$

**63.** $|y| < 4$
   $y \geq -2x + 2$

**64.** $|x - 2| \leq 3$
   $x - y > 2$

**65.** $|x + 2| < 3$
   $|y| > 4$

**66.** $|x - 2| > 1$
   $y > -2$

**67.** $|x - 3| \leq 4$
   $|y + 2| \leq 1$

**68.** $|x + 1| \leq 2$
   $|y - 3| \leq 1$

## Concept/Writing Exercises

**69.** If in a system of two inequalities, one inequality contains $<$ and the other inequality contains $\geq$, is the point of intersection of the two boundary lines of the inequalities in the solution set? Explain.

**70.** If in a system of two inequalities, one inequality contains $\leq$ and the other inequality contains $\geq$, is the point of intersection of the two boundary lines of the inequalities in the solution set? Explain.

**71.** If in a system of two inequalities, one inequality contains $<$ and the other inequality contains $>$, is the point of intersec-

tion of the two boundary lines of the inequalities in the solution set? Explain.

**72. a)** Is it possible for a system of linear inequalities to have no solution? Explain. Make up an example to support your answer.

**b)** Is it possible for a system of two linear inequalities to have exactly one solution? Explain. If you answer yes, make up an example to support your answer.

*Without graphing, determine the number of solutions in each indicated system of inequalities. Explain your answers.*

**73.** $3x - y \leq 4$
   $3x - y > 4$

**74.** $2x + y < 6$
   $2x + y > 6$

**75.** $5x - 2y \leq 3$
   $5x - 2y \geq 3$

**76.** $5x - 3y > 5$
   $5x - 3y > -1$

**77.** $2x - y < 7$
   $3x - y < -2$

**78.** $x + y \leq 0$
   $x - y \geq 0$

## Challenge Problems

*Determine the solution to each system of inequalities.*

**79.** $y \geq x^2$
$\quad y \leq 4$

**80.** $y < 4 - x^2$
$\quad y > -5$

**81.** $y < |x|$
$\quad y < 4$

**82.** $y \geq |x - 2|$
$\quad y \leq -|x - 2|$

## Cumulative Review Exercises

[2.6]  **83.** A formula for levers in physics is $f_1 d_1 + f_2 d_2 = f_3 d_3$. Solve this formula for $f_2$.

[8.2]  *State the domain and range of each function.*

**84.** $\{(4, 3), (5, -2), (-1, 2), (0, -5)\}$

**85.** $f(x) = \dfrac{2}{3}x - 4$

**86.**

(2, -1)

# Chapter 10 Summary

| IMPORTANT FACTS AND CONCEPTS | EXAMPLES |
|---|---|

### Section 10.1

**Properties Used to Solve Inequalities**

| | |
|---|---|
| **1.** If $a > b$, then $a + c > b + c$. | **1.** If $6 > 5$, then $6 + 3 > 5 + 3$. |
| **2.** If $a > b$, then $a - c > b - c$. | **2.** If $6 > 5$, then $6 - 3 > 5 - 3$. |
| **3.** If $a > b$, and $c > 0$, then $ac > bc$. | **3.** If $7 > 3$, then $7 \cdot 4 > 3 \cdot 4$. |
| **4.** If $a > b$, and $c > 0$, then $\dfrac{a}{c} > \dfrac{b}{c}$. | **4.** If $7 > 3$, then $\dfrac{7}{4} > \dfrac{3}{4}$. |
| **5.** If $a > b$, and $c < 0$, then $ac < bc$. | **5.** If $9 > 2$, then $9(-3) < 2(-3)$. |
| **6.** If $a > b$, and $c < 0$, then $\dfrac{a}{c} < \dfrac{b}{c}$. | **6.** If $9 > 2$, then $\dfrac{9}{-3} < \dfrac{2}{-3}$. |

| | |
|---|---|
| A **compound inequality** is formed by joining two inequalities with the word *and* or *or*. | $x \leq 7$ and $x > 5$ is a compound inequality.<br>$x < -1$ or $x \geq 4$ is a compound inequality. |
| To find the solution set of an inequality containing the word *and* take the **intersection** of the solution sets of the two inequalities. | Solve $x \leq 7$ and $x > 5$.<br>The intersection of $\{x \mid x \leq 7\}$ and $\{x \mid x > 5\}$ is $\{x \mid 5 < x \leq 7\}$ or $(5, 7]$. |
| To find the solution set of an inequality containing the word *or*, take the **union** of the solution sets of the two inequalities. | Solve $x < -1$ or $x \geq 4$.<br>The union of $\{x \mid x < -1\}$ or $\{x \mid x \geq 4\}$ is $\{x \mid x < -1 \text{ or } x \geq 4\}$ or $(-\infty, -1) \cup [4, \infty)$. |

### Section 10.2

| | |
|---|---|
| **To Solve Equations of the Form $\lvert x \rvert = a$**<br>If $\lvert x \rvert = a$ and $a > 0$, then $x = a$ or $x = -a$. | Solve $\lvert x \rvert = 6$.<br>$\quad \lvert x \rvert = 6$ gives $x = 6$ or $x = -6$. |
| **To Solve Inequalities of the Form $\lvert x \rvert < a$**<br>If $\lvert x \rvert < a$ and $a > 0$, then $-a < x < a$. | Solve $\lvert 3x + 1 \rvert < 13$.<br>$$-13 < 3x + 1 < 13$$<br>$$-14 < 3x < 12$$<br>$$-\dfrac{14}{3} < x < 4$$<br>$$\left\{ x \mid -\dfrac{14}{3} < x < 4 \right\} \text{ or } \left( -\dfrac{14}{3}, 4 \right)$$ |

| IMPORTANT FACTS AND CONCEPTS | EXAMPLES |
|---|---|

**Section 10.2 (cont.)**

| | |
|---|---|
| **To Solve Inequalities of the Form $\lvert x \rvert > a$** | Solve $\lvert 2x - 3 \rvert \geq 5$. |
| If $\lvert x \rvert > a$ and $a > 0$, then $x < -a$ or $x > a$. | $\begin{array}{lll} 2x - 3 \leq -5 & \text{or} & 2x - 3 \geq 5 \\ 2x \leq -2 & & 2x \geq 8 \\ x \leq -1 & & x \geq 4 \end{array}$ |
| | $\{x \mid x \leq -1 \text{ or } x \geq 4\}$ or $(-\infty, -1] \cup [4, \infty)$ |
| If $\lvert x \rvert > a$ and $a < 0$, the solution set is $\mathbb{R}$. | $\lvert x \rvert > -7$, the solution set is $\mathbb{R}$. |
| If $\lvert x \rvert < a$ and $a < 0$, the solution set is $\varnothing$. | $\lvert x \rvert < -7$, the solution set is $\varnothing$. |
| **To Solve Equations of the Form $\lvert x \rvert = \lvert y \rvert$.** | Solve $\lvert x \rvert = \lvert 3 \rvert$. |
| If $\lvert x \rvert = \lvert y \rvert$, then $x = y$ or $x = -y$. | $x = 3$ or $x = -3$. |

**Section 10.3**

| | |
|---|---|
| A **linear inequality** results when the equals sign of a linear equation is replaced with an inequality sign. | $3x - 4y > 1$ and $2x + 5y \leq -4$ are both linear inequalities in two variables. |
| **To Graph a Linear Inequality in Two Variables** | Graph $y > -x + 1$. |
| 1. Replace the inequality symbol with an equals sign. | 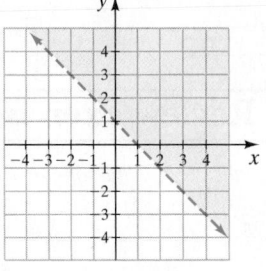 |
| 2. Draw the graph of the equation in step 1. If the original inequality is $\geq$ or $\leq$ draw a solid line. If the original inequality is a $>$ or $<$ draw a dashed line. | |
| 3. Select any point not on the line. If the point selected is a solution, shade the half-plane on the side of the line containing this point. If the selected point does not satisfy the inequality, shade the half-plane on the side of the line not containing this point. | |
| To solve a system of linear inequalities, graph each inequality on the same axes. The solution is the set of points whose coordinates satisfy all the inequalities in the system. | Determine the solution to the system of inequalities. $$y < -\frac{1}{3}x + 1$$ $$x - y \leq 2$$ 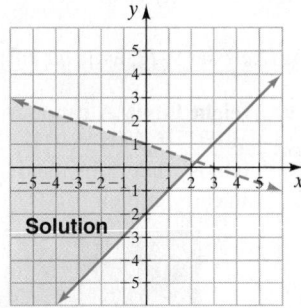 |
| **Linear programming** is a process where more than two linear inequalities are graphed on the same axes. | Determine the solution to the system of inequalities. $$\begin{array}{c} x \geq 0 \\ y \geq 0 \\ x \leq 8 \\ x + y \leq 10 \\ x + 2y \leq 16 \end{array}$$ 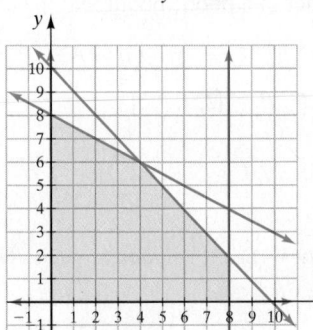 |

| IMPORTANT FACTS AND CONCEPTS | EXAMPLES |
|---|---|

**Section 10.3 (cont.)**

For systems of linear inequalities with absolute values:

If $|x| < a$ and $a > 0$, then $-a < x < a$.

If $|x| > a$ and $a > 0$, then $x < -a$ or $x > a$.

Determine the solution to the system of inequalities.

$$|x| < 2$$
$$|y - 1| > 3$$

## Chapter 10 Review Exercises

*[10.1]  Solve the inequality. Graph the solution on a real number line.*

**1.** $2x + 3 \le 7$

**2.** $8 - 2w > -4$

**3.** $2x + 1 > 6$

**4.** $26 \le 4x + 5$

**5.** $\dfrac{4x + 3}{3} > -5$

**6.** $2(x - 1) > 3x + 8$

**7.** $-4(x - 2) \ge 6x + 8 - 10x$

**8.** $\dfrac{x}{2} + \dfrac{3}{4} > x - \dfrac{x}{2} + 1$

*In Exercises 9–12, write an inequality that can be used to solve each problem. Solve the inequality and answer the question.*

**9. Weight Limit** A canoe can safely carry a total weight of 560 pounds. If Bob and Kathie together weigh a total of 300 pounds, what is the maximum number of 40-pound boxes they can carry in their canoe?

© Allen R. Angel

**10. Phone Booth Call** Michael Lamb, a telephone operator, informs a customer in a phone booth that the charge for calling

Omaha, Nebraska, is \$4.50 for the first 3 minutes and 95¢ for each additional minute or any part thereof. How long can the customer talk if he has \$8.65?

**11. Fitness Center** A fitness center guarantees that customers will lose a minimum of 5 pounds the first week and $1\dfrac{1}{2}$ pounds each additional week. Find the maximum amount of time needed to lose 27 pounds.

**12. Exam Scores** Patrice Lee's first four exam scores are 94, 73, 72, and 80. If a final average greater than or equal to 80 and less than 90 is needed to receive a final grade of B in the course, what range of scores on the fifth and last exam will result in Patrice's receiving a B in the course? Assume a maximum score of 100.

*Find $A \cup B$ and $A \cap B$ for each set A and B.*

**13.** $A = \{1, 2, 3, 4, 5\}, B = \{2, 3, 4, 5\}$

**14.** $A = \{3, 5, 7, 9\}, B = \{2, 4, 6, 8\}$

**15.** $A = \{1, 2, 3, 4, \dots\}, B = \{2, 4, 6, \dots\}$

**16.** $A = \{4, 6, 9, 10, 11\}, B = \{3, 5, 9, 10, 12\}$

*Solve each inequality. Write the solution in interval notation.*

**17.** $0 < x - 5 < 6$

**18.** $8 < p + 11 \le 16$

**19.** $3 < 2x - 4 < 12$

**20.** $-12 < 6 - 3x < -2$

**21.** $-1 < \dfrac{5}{9}x + \dfrac{2}{3} \le \dfrac{11}{9}$

**22.** $-8 < \dfrac{4 - 2x}{3} < 0$

*Find the solution set to each compound inequality.*

**23.** $h \le 1$ and $7h - 4 > -25$

**24.** $2x - 1 > 5$ or $3x - 2 \le 10$

**25.** $4x - 5 < 11$ and $-3x - 4 \ge 8$

**26.** $\dfrac{7 - 2g}{3} \le -5$ or $\dfrac{3 - g}{9} > 1$

*[10.2]  Find the solution set to each equation or inequality.*

**27.** $|a| = 2$

**28.** $|x| < 8$

**29.** $|x| \ge 9$

**30.** $|l + 5| = 13$

**31.** $|x - 2| \ge 5$

**32.** $|4 - 2x| = 5$

**33.** $|-2q + 9| < 7$

**34.** $\left|\dfrac{2x - 3}{5}\right| = 1$

**35.** $\left|\dfrac{x - 4}{3}\right| < 6$

**36.** $|4d - 1| = |6d + 9|$

**37.** $|2x - 3| + 4 \ge -17$

*[10.1, 10.2]  Solve each inequality. Give the solution in interval notation.*

**38.** $|3c + 8| - 6 \le 1$

**39.** $3 < 2x - 5 \le 11$

**40.** $-2 \le \dfrac{5 - 2x}{11} < 2$

**41.** $2p - 5 < 7$ and $9 - 3p \le 15$

**42.** $x - 3 \le 4$ or $2x - 5 > 7$

**43.** $-10 < 3(x - 4) \le 18$

*[10.3]  Graph each inequality.*

**44.** $y \ge -5$

**45.** $x < 4$

**46.** $y \le 4x - 3$

**47.** $y < \dfrac{1}{3}x - 2$

*Determine the solution to each system of inequalities.*

**48.** $-x + 3y > 6$
$2x - y \le 2$

**49.** $5x - 2y \le 10$
$3x + 2y > 6$

**50.** $y > 2x + 3$
$y < -x + 4$

**51.** $x > -2y + 4$
$y < -\dfrac{1}{2}x - \dfrac{3}{2}$

*Determine the solution to the system of inequalities.*

**52.** $x \ge 0$
$y \ge 0$
$x + y \le 6$
$4x + y \le 8$

**53.** $x \ge 0$
$y \ge 0$
$2x + y \le 6$
$4x + 5y \le 20$

**54.** $|x| \le 3$
$|y| > 2$

**55.** $|x| > 4$
$|y - 2| \le 3$

# Chapter 10 Practice Test

CHAPTER
**Test Prep**
VIDEOS

*Chapter Test Prep Videos provide fully worked-out solutions to any of the exercises you want to review. Chapter Test Prep Videos are available via MyMathLab®, or on You Tube (search "Angel Elementary and Intermediate Algebra" and click on "Channels").*

*Find $A \cup B$ and $A \cap B$ for sets A and B.*

**1.** $A = \{8, 10, 11, 14\}$, $B = \{5, 7, 8, 9, 10\}$

**2.** $A = \{1, 3, 5, 7, \ldots\}$, $B = \{3, 5, 7, 9, 11\}$

*Solve each inequality and graph the solution on a number line.*

**3.** $3(2q + 4) < 5(q - 1) + 7$

**4.** $\dfrac{6 - 2x}{5} \ge -12$

*Solve each inequality and write the solution in interval notation.*

**5.** $x - 3 \le 4$ and $2x + 1 > 10$

**6.** $7 \le \dfrac{2u - 5}{3} < 9$

*Find the solution set to the following equations.*

**7.** $|2b + 5| = 9$

**8.** $|2x - 3| = \left|\dfrac{1}{2}x - 10\right|$

**9.** $|4z + 12| = 0$

*Find the solution set to the following inequalities.*

**10.** $|2x - 3| + 6 > 11$

**11.** $\left|\dfrac{2x - 3}{8}\right| \le \dfrac{1}{4}$

**12.** Graph $y < 3x - 2$.

*Determine the solution to each system of inequalities.*

**13.** $3x + 2y < 9$
$-2x + 5y \le 10$

**14.** $|x| > 3$
$|y| \le 1$

## Cumulative Review Test

*Take the following test and check your answers with those given in the back of the book. Review any questions that you answered incorrectly. The section where the material was covered is indicated after the answer.*

1. Evaluate $-40 - 3(4 - 8)^2$.

2. Evaluate $2(x + 2y) + 4x - 5y$ when $x = 2$ and $y = 3$.

3. Simplify $\dfrac{1}{2}(x + 3) + \dfrac{1}{3}(3x + 6)$.

4. Solve $\dfrac{a - 7}{3} = \dfrac{a + 5}{2} - \dfrac{7a - 1}{6}$.

5. **Spreading Fertilizer** If a 40-pound bag of fertilizer covers 5000 square feet, how many pounds of fertilizer are needed to cover an area of 26,000 square feet?

6. **Financial Planning** Belen Poltorade, a financial planner, is offering her customers two financial plans for managing their assets. With plan 1 she charges a planning fee of $1000 plus 1% of the assets she will manage for the customers. With plan 2 she charges a planning fee of $500 plus 2% of the assets she will manage. How much in customer assets would result in both plans having the same total fee?

7. **Bird Food** At Agway Gardens, bird food is sold in bulk. In one barrel are sunflower seeds that sell for $1.80 per pound. In a second barrel is cracked corn that sells for $1.40 per pound. If the store makes bags of a mixture of the two by mixing 2.5 pounds of the sunflower seeds with 1 pound of the cracked corn, what should be the cost per pound of the mixture?

8. Simplify $\left(\dfrac{4a^3 b^{-2}}{2a^{-2} b^{-3}}\right)^{-2}$.

9. Multiply $(0.003)(0.00015)$ and write the answer in scientific notation.

10. Factor $24p^3 q + 16p^2 q - 30pq$.

11. Solve $\dfrac{2b}{b + 1} = 2 - \dfrac{5}{2b}$.

12. Solve $\dfrac{4}{r + 5} + \dfrac{1}{r + 3} = \dfrac{2}{r^2 + 8r + 15}$.

13. Graph $y = x^2 - 2$.

14. Indicate which of the following relations are functions.

   a) $\{(4, 2), (-5, 3), (6, 3), (5, 3)\}$

   b)

   c)

15. Find the equation of the line whose graph has a $y$-intercept of $(0, -2)$ and is parallel to the graph of $3x - 5y = 7$.

16. Find the equation of the line whose graph passes through $(3, -2)$ and is perpendicular to the graph of $4x - 3y = 12$.

17. Given $f(x) = x^2 - 11x + 30$ and $g(x) = x - 5$, find

   a) $(f \cdot g)(x)$.

   b) $(f/g)(x)$.

   c) the domain of $(f/g)(x)$.

18. Solve the system of equations.

$$2a - b - 2c = -1$$
$$a - 2b - c = 1$$
$$a + b + c = 4$$

19. Solve $\dfrac{1}{2} < 3x + 4 < 6$ and give the solution in interval notation.

20. Graph $y < 2x - 3$.

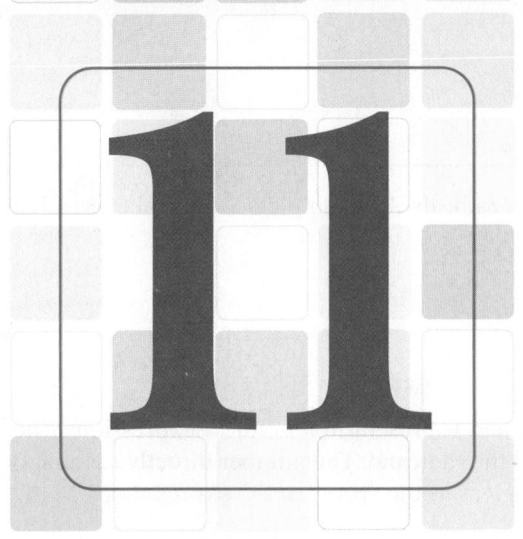

# 11 Roots, Radicals, and Complex Numbers

## Goals of This Chapter

In this chapter, we explain how to add, subtract, multiply, and divide radical expressions. We also graph radical functions, solve radical equations, and introduce imaginary numbers and complex numbers.

Make sure that you understand the three requirements for simplifying radical expressions as discussed in Section 11.5.

**Many scientific formulas, including** many that pertain to real-life situations, contain radical expressions. In Exercise 106 on page 630, we will see how a radical is used to determine the relationship between wind speed and the height of waves in certain areas of the ocean.

© Epic Stock /Shutterstock

# 11.1 Roots and Radicals

1 Find square roots.

2 Find cube roots.

3 Understand odd and even roots.

4 Evaluate radicals using absolute value.

In this section we introduce the concept of radicals. An example of a **radical expression** is $\sqrt[3]{x}$

The symbol $\sqrt{\ }$ is called the **radical sign** and is present in all radical expressions. The expression under the radical sign is called the **radicand**. The number directly to the left of the radical sign is called the **index** and gives us the "root" of the expression.

## 1 Find Square Roots

A **square root** is a radical expression that has an index of 2. The index of a square root is generally not written. Thus,

$$\sqrt{x} \text{ means } \sqrt[2]{x}$$

Every positive real number has two square roots—a principal or positive square root and a negative square root.

> **Square Roots**
>
> For any positive real number $a$,
>
> • The **principal** or **positive square root** of $a$, written $\sqrt{a}$, is the *positive* number $b$ such that $b^2 = a$.
> • The **negative square root** of $a$, written $-\sqrt{a}$, is the *opposite of the principal square root of a*.

For example, the real number 25 has two square roots:

• the principal or positive square root of 25, written $\sqrt{25}$, is 5 since 5 is a positive number such that $5^2 = 25$.
• The negative square root of 25, written $-\sqrt{25}$, is $-5$ since it is the opposite of the principal square root, 5. Also note that $(-5)^2 = (-5)(-5) = 25$.

*In this book, whenever we use the words "square root," we are referring to the principal or positive square root.* If you are asked to find the *square root* of 36, you are being asked to find the positive square root of 36, $\sqrt{36}$, and your answer will be 6.

When we evaluate square roots using our calculator, if we get a decimal number we can usually see if the decimal number terminates or repeats. In Chapter 1, we learned that a rational number can be written as either a terminating or a repeating decimal number. If a decimal number does not terminate and does not repeat, then it is an irrational number. The table below gives some examples.

| Radical | Calculator Result | Terminating, Repeating, or Neither? | Rational or Irrational? |
|---|---|---|---|
| $\sqrt{49}$ | 7 | Terminating | Rational |
| $\sqrt{0.01}$ | .1 | Terminating | Rational |
| $\sqrt{\dfrac{25}{4}}$ | 2.5 | Terminating | Rational |
| $\sqrt{\dfrac{4}{9}}$ | .6666666666 | Repeating | Rational |
| $\sqrt{2}$ | 1.414213562 | Neither | Irrational |
| $\sqrt{2.5}$ | 1.58113883 | Neither | Irrational |
| $\sqrt{\dfrac{1}{2}}$ | .7071067812 | Neither | Irrational |

---

**Understanding Algebra**

When we say "square root" we refer to a radical expression with an index of 2. Furthermore, we usually do not write the index of 2. Thus, "*square root of x*" means

$$\sqrt[2]{x}$$

but we write

$$\sqrt{x}$$

Now let's consider $\sqrt{-25}$. Since the square of any real number will always be greater than or equal to 0, there is no real number that when squared equals $-25$. For this reason, $\sqrt{-25}$ is *not a real number*. Since the square of any real number cannot be negative, *the square root of a negative number is not a real number*. If you evaluate $\sqrt{-25}$ on most calculators, you will get an error message. We will discuss numbers like $\sqrt{-25}$ later in this chapter.

---

**Helpful Hint**

Do not confuse $-\sqrt{36}$ with $\sqrt{-36}$. Because $\sqrt{36} = 6$, $-\sqrt{36} = -6$. However, $\sqrt{-36}$ is not a real number. The square root of a negative number is not a real number.

$$\sqrt{36} = 6$$
$$-\sqrt{36} = -6$$
$$\sqrt{-36} \text{ is not a real number.}$$

---

**The Square Root Function**   When graphing square root functions, functions of the form $f(x) = \sqrt{x}$, we must always remember that the radicand, $x$, cannot be negative. Thus, the domain of $f(x) = \sqrt{x}$ is $\{x | x \geq 0\}$, or $[0, \infty)$ in interval notation. To graph $f(x) = \sqrt{x}$, we can select some convenient values of $x$ and find the corresponding values of $f(x)$ or $y$ and then plot the ordered pairs, as shown in **Figure 11.1**.

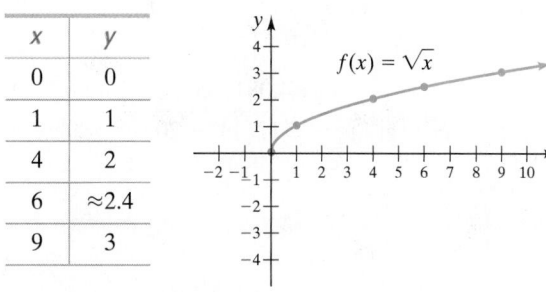

| x | y |
|---|---|
| 0 | 0 |
| 1 | 1 |
| 4 | 2 |
| 6 | ≈2.4 |
| 9 | 3 |

FIGURE 11.1

Since the value of $f(x)$ can never be negative, the range of $f(x) = \sqrt{x}$ is $\{y | y \geq 0\}$, or $[0, \infty)$ in interval notation.

EXAMPLE 1   For each function, find the indicated value(s).

**a)** $f(x) = \sqrt{11x - 2}$, $f(6)$        **b)** $g(r) = -\sqrt{-3r + 1}$, $g(-5)$ and $g(7)$

*Solution*

**a)** $f(6) = \sqrt{11(6) - 2}$        Substitute 6 for x.

$\quad\quad = \sqrt{64}$

$\quad\quad = 8$

**b)** $g(-5) = -\sqrt{-3(-5) + 1}$        Substitute −5 for r.

$\quad\quad\quad = -\sqrt{16}$

$\quad\quad\quad = -4$

$\quad g(7) = -\sqrt{-3(7) + 1}$        Substitute 7 for r.

$\quad\quad\quad = -\sqrt{-20}$        Not a real number.

Thus, $g(7)$ is not a real number.

Now Try Exercise 77

## 2 Find Cube Roots

The concept used to explain square roots can be expanded to explain cube roots. The index of a cube root is 3. The cube root of a number $a$ is written $\sqrt[3]{a}$.

---

**Cube Root**

The **cube root** of a number $a$, written $\sqrt[3]{a}$, is the number $b$ such that $b^3 = a$.

---

### Examples

$$\sqrt[3]{8} = 2 \qquad\qquad \text{since } 2^3 = 8$$
$$\sqrt[3]{-27} = -3 \qquad \text{since } (-3)^3 = -27$$

*For each real number, there is only one cube root. The cube root of a positive number is positive and the cube root of a negative number is negative.*

**EXAMPLE 2** For each function, find the indicated value(s).

**a)** $f(x) = \sqrt[3]{10x + 34}, f(3)$      **b)** $g(r) = \sqrt[3]{12r - 20}, g(-4)$ and $g(1)$

*Solution*

**a)** $f(3) = \sqrt[3]{10(3) + 34}$     Substitute 3 for $x$.

       $= \sqrt[3]{64} = 4$

**b)** $g(-4) = \sqrt[3]{12(-4) - 20}$     Substitute $-4$ for $r$.

        $= \sqrt[3]{-68}$

        $\approx -4.081655102$     From a calculator

    $g(1) = \sqrt[3]{12(1) - 20}$     Substitute 1 for $r$.

        $= \sqrt[3]{-8}$

        $= -2$

Now Try Exercise 83

*The Cube Root Function* **Figure 11.2** shows the graph of $y = \sqrt[3]{x}$. To obtain this graph, we substituted values for $x$ and found the corresponding values of $f(x)$ or $y$.

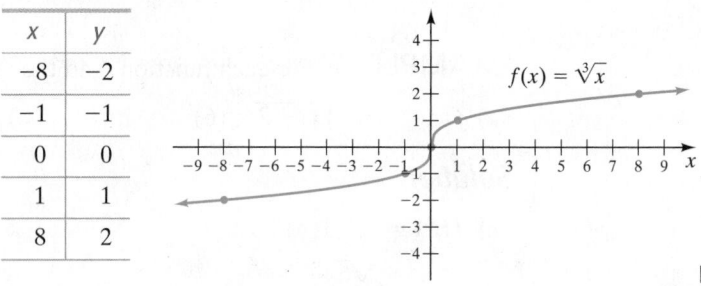

| $x$ | $y$ |
|-----|-----|
| $-8$ | $-2$ |
| $-1$ | $-1$ |
| $0$ | $0$ |
| $1$ | $1$ |
| $8$ | $2$ |

$f(x) = \sqrt[3]{x}$

FIGURE 11.2

Notice that both the domain and range are all real numbers, $\mathbb{R}$.

## 3 Understand Odd and Even Roots

Up to this point we have discussed square and cube roots. Other radical expressions have different indices. For example, in the expression $\sqrt[5]{xy}$, (read "the fifth root of $xy$") the index is 5 and the radicand is $xy$.

Radical expressions that have an even-number index such as $2, 4, 6, \ldots$, are **even roots**. Square roots are even roots since their index is 2. Radical expressions that have an odd-number index such as $1, 3, 5, \ldots$, are **odd roots**.

**Even Root**

The $n$th root of $a$, $\sqrt[n]{a}$, where $n$ is an *even index* and $a$ is a nonnegative real number, is called an **even root** and is the nonnegative real number $b$ such that $b^n = a$.

### Examples of Even Roots

$$\sqrt{9} = 3 \qquad \text{since } 3^2 = 3 \cdot 3 = 9$$
$$\sqrt[4]{16} = 2 \qquad \text{since } 2^4 = 2 \cdot 2 \cdot 2 \cdot 2 = 16$$
$$\sqrt[6]{729} = 3 \qquad \text{since } 3^6 = 3 \cdot 3 \cdot 3 \cdot 3 \cdot 3 \cdot 3 = 729$$
$$\sqrt[4]{\frac{1}{256}} = \frac{1}{4} \qquad \text{since } \left(\frac{1}{4}\right)^4 = \left(\frac{1}{4}\right)\left(\frac{1}{4}\right)\left(\frac{1}{4}\right)\left(\frac{1}{4}\right) = \frac{1}{256}$$

Any real number when raised to an even power results in a positive real number. Thus, *the even root of a negative number is not a real number.*

**Helpful Hint**

There is an important difference between $-\sqrt[4]{16}$ and $\sqrt[4]{-16}$. The number $-\sqrt[4]{16}$ is the opposite of $\sqrt[4]{16}$. Because $\sqrt[4]{16} = 2$, $-\sqrt[4]{16} = -2$. However, $\sqrt[4]{-16}$ is not a real number since no real number when raised to the fourth power equals $-16$.

$$-\sqrt[4]{16} = -(\sqrt[4]{16}) = -2$$
$$\sqrt[4]{-16} \text{ is not a real number.}$$

## Understanding Algebra

- A radical with an even index must have a nonnegative radicand if it is to be a real number.
- A radical with an odd index will be a real number with any real number as its radicand.
- Note that $\sqrt[n]{0} = 0$ regardless of whether $n$ is an odd or even index.

**Odd Root**

The $n$th root of $a$, $\sqrt[n]{a}$, where $n$ is an *odd index* and $a$ is *any real number*, is called an **odd root** and is the real number $b$ such that $b^n = a$.

### Examples of Odd Roots

$$\sqrt[3]{8} = 2 \qquad \text{since } 2^3 = 2 \cdot 2 \cdot 2 = 8$$
$$\sqrt[3]{-8} = -2 \qquad \text{since } (-2)^3 = (-2)(-2)(-2) = -8$$
$$\sqrt[5]{243} = 3 \qquad \text{since } 3^5 = 3 \cdot 3 \cdot 3 \cdot 3 \cdot 3 = 243$$
$$\sqrt[5]{-243} = -3 \qquad \text{since } (-3)^5 = (-3)(-3)(-3)(-3)(-3) = -243$$

*An odd root of a positive number is a positive number, and an odd root of a negative number is a negative number.*

EXAMPLE 3  Indicate whether or not each radical expression is a real number. If the expression is a real number, find its value.

a) $\sqrt[4]{-81}$         b) $-\sqrt[4]{81}$         c) $\sqrt[5]{-32}$         d) $-\sqrt[5]{-32}$

*Solution*

a) Not a real number. Even roots of negative numbers are not real numbers.
b) Real number, $-\sqrt[4]{81} = -(\sqrt[4]{81}) = -(3) = -3$
c) Real number, $\sqrt[5]{-32} = -2$ since $(-2)^5 = -32$
d) Real number, $-\sqrt[5]{-32} = -(-2) = 2$

Now Try Exercise 21

**Table 11.1** summarizes the information about even and odd roots.

**TABLE 11.1**

|  | $n$ **Is Even** | $n$ **Is Odd** |
|---|---|---|
| $a > 0$ | $\sqrt[n]{a}$ is a positive real number. | $\sqrt[n]{a}$ is a positive real number. |
| $a < 0$ | $\sqrt[n]{a}$ is not a real number. | $\sqrt[n]{a}$ is a negative real number. |
| $a = 0$ | $\sqrt[n]{0} = 0$ | $\sqrt[n]{0} = 0$ |

## 4 Evaluate Radicals Using Absolute Value

You may think that $\sqrt{a^2} = a$, but this is not necessarily true. Below we evaluate $\sqrt{a^2}$ for $a = 2$ and $a = -2$. You will see that when $a = -2$, $\sqrt{a^2} \neq a$.

$$a = 2: \qquad \sqrt{a^2} = \sqrt{2^2} = \sqrt{4} = 2 \qquad \text{Note that } \sqrt{2^2} = 2.$$
$$a = -2: \qquad \sqrt{a^2} = \sqrt{(-2)^2} = \sqrt{4} = 2 \qquad \text{Note that } \sqrt{(-2)^2} \neq -2.$$

By examining these examples, we can reason that $\sqrt{a^2}$ *will always be a positive real number* for any nonzero real number $a$. Recall from Section 1.5 that the *absolute value* of any real number $a$, or $|a|$, is also a positive number for any nonzero number. We use these facts to reason that

### Radicals and Absolute Value

For any real number $a$,

$$\sqrt{a^2} = |a|$$

This indicates that the principal square root of $a^2$ is the absolute value of $a$.

EXAMPLE 4 Use absolute value to evaluate. ───────

**a)** $\sqrt{9^2}$       **b)** $\sqrt{0^2}$       **c)** $\sqrt{(-15.7)^2}$

*Solution*

**a)** $\sqrt{9^2} = |9| = 9$    **b)** $\sqrt{0^2} = |0| = 0$    **c)** $\sqrt{(-15.7)^2} = |-15.7| = 15.7$

Now Try Exercise 41

When simplifying a square root, if the radicand contains a variable and we are not sure that the radicand is positive, we need to use absolute value signs when simplifying.

EXAMPLE 5 Simplify. ───────

**a)** $\sqrt{(x + 8)^2}$    **b)** $\sqrt{16x^2}$    **c)** $\sqrt{25y^6}$    **d)** $\sqrt{a^2 - 6a + 9}$

*Solution*  Each square root has a radicand that contains a variable. Since we do not know the value of the variable, we do not know whether it is positive or negative. Therefore, we must use absolute value signs when simplifying.

**a)** $\sqrt{(x + 8)^2} = |x + 8|$

**b)** Write $16x^2$ as $(4x)^2$, then simplify.

$$\sqrt{16x^2} = \sqrt{(4x)^2} = |4x|$$

**c)** Write $25y^6$ as $(5y^3)^2$, then simplify.

$$\sqrt{25y^6} = \sqrt{(5y^3)^2} = |5y^3|$$

**d)** Notice that $a^2 - 6a + 9$ is a perfect square trinomial. Write the trinomial as the square of a binomial, then simplify.

$$\sqrt{a^2 - 6a + 9} = \sqrt{(a - 3)^2} = |a - 3|$$

Now Try Exercise 63

If you have a square root whose radicand contains a variable and are given instructions like "Assume all variables represent positive values and the radicand is nonnegative," then it is not necessary to use the absolute value sign when simplifying.

EXAMPLE 6 Simplify. Assume all variables represent positive values and the radicand is nonnegative.

**a)** $\sqrt{64x^2}$    **b)** $\sqrt{81p^4}$    **c)** $\sqrt{49x^6}$    **d)** $\sqrt{4x^2 - 12xy + 9y^2}$

*Solution*

**a)** $\sqrt{64x^2} = \sqrt{(8x)^2} = 8x$      Write $64x^2$ as $(8x)^2$.

**b)** $\sqrt{81p^4} = \sqrt{(9p^2)^2} = 9p^2$      Write $81p^4$ as $(9p^2)^2$.

**c)** $\sqrt{49x^6} = \sqrt{(7x^3)^2} = 7x^3$      Write $49x^6$ as $(7x^3)^2$.

**d)** $\sqrt{4x^2 - 12xy + 9y^2} = \sqrt{(2x - 3y)^2}$      Write $4x^2 - 12xy + 9y^2$ as $(2x - 3y)^2$.

$\qquad\qquad\qquad\qquad\qquad = 2x - 3y$

Now Try Exercise 67

We only need to be concerned about using absolute value signs when discussing square (and other even) roots. We do not need to use absolute value signs when the index is odd.

# EXERCISE SET 11.1

MathXL®   MyMathLab®

## Warm-Up Exercises

*Fill in the blanks with the appropriate word, phrase, or symbol(s) from the following list.*

| | | | | | |
|---|---|---|---|---|---|
| odd | index | principal | cube | rational | radical |
| square | radicand | even | irrational | negative | |

**1.** The symbol $\sqrt{\ }$ is called the _____ sign.

**2.** In the radical expression $\sqrt[3]{5}$ the 3 is the _____.

**3.** In the radical expression $\sqrt[3]{5}$ the 5 is the _____.

**4.** When we say "_____ root" we refer to a radical expression with an index of 2.

**5.** When we say "_____ root" we refer to a radical expression with an index of 3.

**6.** The _____ square root of $a$, written $\sqrt{a}$, is the positive number $b$ such that $b^2 = a$.

**7.** The _____ square root of $a$, written $-\sqrt{a}$, is the opposite of the principal square root of $a$.

**8.** A _____ number can be written as either a terminating or a repeating decimal number.

**9.** The _____ root of a negative number is not a real number.

**10.** The _____ root of a negative number is a negative number.

## Practice the Skills

*Evaluate each radical expression if it is a real number. Use a calculator to approximate irrational numbers to the nearest hundredth. If the expression is not a real number, so state.*

**11. a)** $\sqrt{9}$    **b)** $-\sqrt{9}$    **c)** $\sqrt{-9}$    **d)** $-\sqrt{-9}$

**12. a)** $\sqrt{16}$    **b)** $-\sqrt{16}$    **c)** $\sqrt{-16}$    **d)** $-\sqrt{-16}$

**13.** $\sqrt[3]{-64}$      **14.** $\sqrt[3]{125}$      **15.** $\sqrt[3]{-125}$      **16.** $-\sqrt[3]{-125}$

**17.** $\sqrt[3]{-1}$      **18.** $-\sqrt[5]{-1}$      **19.** $\sqrt[5]{1}$      **20.** $\sqrt[6]{64}$

**21.** $\sqrt[6]{-64}$      **22.** $\sqrt[4]{-81}$      **23.** $\sqrt[3]{-343}$      **24.** $\sqrt{121}$

**25.** $\sqrt{-36}$      **26.** $\sqrt{45.3}$      **27.** $\sqrt{-45.3}$      **28.** $\sqrt{53.9}$

**29.** $\sqrt{\dfrac{1}{25}}$      **30.** $\sqrt{-\dfrac{1}{25}}$      **31.** $\sqrt[3]{\dfrac{1}{8}}$      **32.** $\sqrt[3]{-\dfrac{1}{8}}$

**33.** $\sqrt{\dfrac{4}{49}}$      **34.** $\sqrt[3]{\dfrac{8}{27}}$      **35.** $\sqrt[3]{-\dfrac{8}{27}}$      **36.** $\sqrt[4]{-8.9}$

**37.** $-\sqrt[4]{18.2}$      **38.** $\sqrt[5]{93}$

*Use absolute value to evaluate.*

**39.** $\sqrt{7^2}$      **40.** $\sqrt{(-7)^2}$      **41.** $\sqrt{(-3)^2}$      **42.** $\sqrt{3^2}$

**43.** $\sqrt{119^2}$      **44.** $\sqrt{(-119)^2}$      **45.** $\sqrt{(235.23)^2}$      **46.** $\sqrt{(-201.5)^2}$

**47.** $\sqrt{(0.06)^2}$      **48.** $\sqrt{(-0.19)^2}$      **49.** $\sqrt{\left(\dfrac{12}{13}\right)^2}$      **50.** $\sqrt{\left(-\dfrac{101}{319}\right)^2}$

*Write as an absolute value.*

**51.** $\sqrt{(x-8)^2}$  **52.** $\sqrt{(a+10)^2}$  **53.** $\sqrt{(x-3)^2}$  **54.** $\sqrt{(7a-11b)^2}$

**55.** $\sqrt{(3x^2-1)^2}$  **56.** $\sqrt{(7y^2-3y)^2}$  **57.** $\sqrt{(6a^3-5b^4)^2}$  **58.** $\sqrt{(9y^4-2z^3)^2}$

*Use absolute value to simplify. You may need to factor first.*

**59.** $\sqrt{x^{10}}$  **60.** $\sqrt{y^{22}}$  **61.** $\sqrt{z^{32}}$  **62.** $\sqrt{x^{200}}$

**63.** $\sqrt{a^2-8a+16}$  **64.** $\sqrt{x^2-12x+36}$  **65.** $\sqrt{9a^2+12ab+4b^2}$  **66.** $\sqrt{4x^2+20xy+25y^2}$

*Simplify. Assume that all variables represent positive values and that the radicand is nonnegative.*

**67.** $\sqrt{25a^2}$  **68.** $\sqrt{100a^4}$  **69.** $\sqrt{16c^6}$  **70.** $\sqrt{121z^8}$

**71.** $\sqrt{x^2+4x+4}$  **72.** $\sqrt{9a^2-6a+1}$  **73.** $\sqrt{4x^2+4xy+y^2}$  **74.** $\sqrt{16b^2-40bc+25c^2}$

*Find the indicated value of each function. Use your calculator to approximate irrational numbers. Round irrational numbers to the nearest thousandth.*

**75.** $f(x)=\sqrt{5x-6}, f(2)$  **76.** $f(c)=\sqrt{7c+1}, f(5)$  **77.** $q(x)=\sqrt{76-3x}, q(4)$

**78.** $q(b)=\sqrt{9b+34}, q(-1)$  **79.** $t(a)=\sqrt{-15a-9}, t(-6)$  **80.** $f(a)=\sqrt{14a-36}, f(4)$

**81.** $g(x)=\sqrt{64-8x}, g(-3)$  **82.** $p(x)=\sqrt[3]{8x+9}, p(2)$  **83.** $h(x)=\sqrt[3]{9x^2+4}, h(4)$

**84.** $k(c)=\sqrt[4]{16c-5}, k(6)$  **85.** $f(x)=\sqrt[3]{-2x^2+x-6}, f(-3)$  **86.** $t(x)=\sqrt[4]{2x^3-3x^2+6x}, t(2)$

## Problem Solving

**87.** Find $f(81)$ if $f(x)=x+\sqrt{x}+7$.

**88.** Find $g(25)$ if $g(x)=x^2+\sqrt{x}-13$.

**89.** Find $t(18)$ if $t(x)=\dfrac{x}{2}+\sqrt{2x}-4$.

**90.** Find $m(36)$ if $m(x)=\dfrac{x}{3}+\sqrt{4x}+10$.

**91.** Find $k(8)$ if $k(x)=x^2+\sqrt{\dfrac{x}{2}}-21$.

**92.** Find $r(45)$ if $r(x)=\dfrac{x}{9}+\sqrt{\dfrac{x}{5}}+13$.

**93.** Select a value for $x$ for which $\sqrt{(2x+1)^2}\neq 2x+1$.

**94.** Select a value for $x$ for which $\sqrt{(5x-3)^2}\neq 5x-3$.

*By considering the domains of the functions in Exercises 95 through 98, match each function with its graph, labeled **a)** through **d)**.*

**95.** $f(x)=\sqrt{x}$  **96.** $f(x)=\sqrt{x^2}$  **97.** $f(x)=\sqrt{x-5}$  **98.** $f(x)=\sqrt{x+5}$

**a)**   **b)**   **c)**   **d)**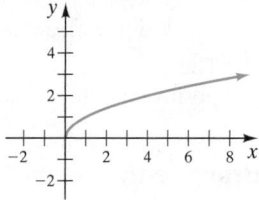

**99.** Graph $f(x)=\sqrt{x}+1$.  **100.** Graph $g(x)=-\sqrt{x}$.

**101.** Graph $g(x)=\sqrt{x+1}$.  **102.** Graph $f(x)=\sqrt{x-2}$.

**103.** Give a radical function whose domain is $\{x|x\geq 8\}$.

**104.** Give a radical function whose domain is $\{x|x\leq 5\}$.

**105. Velocity of an Object** The velocity, $V$, of an object, in feet per second, after it has fallen a distance, $h$, in feet, can be found by the formula $V=\sqrt{64.4h}$. A pile driver is a large mass that is used as a hammer to drive pilings into soft earth to support a building or other structure.

With what velocity will the hammer hit the piling if it falls from

**a)** 20 feet above the top of the piling?

**b)** 40 feet above the top of the piling?

**106. Wave Action** Scripps Institute of Oceanography in La Jolla, California, developed the formula for relating wind speed, $u$, in knots, with the height, $H$, in feet, of the waves the wind produces in certain areas of the ocean. This formula is

$$u=\sqrt{\dfrac{H}{0.026}}$$

If waves produced by a storm have a height of 15 feet, what is the wind speed producing the waves?

## Concept/Writing Exercises

**107.** Explain why $\sqrt{-81}$ is not a real number.

**108.** Will a radical expression with an odd index and a real number as the radicand always be a real number? Explain your answer.

**109.** Will a radical expression with an even index and a real number as the radicand always be a real number? Explain your answer.

**110. a)** To what is $\sqrt{a^2}$ equal?

 **b)** To what is $\sqrt{a^2}$ equal if we know $a \geq 0$?

**111. a)** Evaluate $\sqrt{a^2}$ for $a = 1.3$.

 **b)** Evaluate $\sqrt{a^2}$ for $a = -1.3$.

**112. a)** Evaluate $\sqrt[4]{16}$

 **b)** Evaluate $-\sqrt[4]{16}$

 **c)** Evaluate $\sqrt[4]{-16}$.

**113.** For what values of $x$ will $\sqrt{(x-1)^2} = x - 1$? Explain how you determined your answer.

**114.** For what values of $x$ will $\sqrt{(x+3)^2} = x + 3$? Explain how you determined your answer.

**115.** For what values of $x$ will $\sqrt{(2x-6)^2} = 2x - 6$? Explain how you determined your answer.

**116.** For what values of $x$ will $\sqrt{(3x-8)^2} = 3x - 8$? Explain how you determined your answer.

**117. a)** For what values of $a$ is $\sqrt{a^2} = |a|$?

 **b)** For what values of $a$ is $\sqrt{a^2} = a$?

 **c)** For what values of $a$ is $\sqrt[3]{a^3} = a$?

**118.** Under what circumstances is the expression $\sqrt[n]{x}$ not a real number?

**119.** Explain why the expression $\sqrt[n]{x^n}$ is a real number for any real number $x$.

**120.** Under what circumstances is the expression $\sqrt[n]{x^m}$ not a real number?

**121.** Find the domain of $\dfrac{\sqrt{x+5}}{\sqrt[3]{x+5}}$. Explain how you determined your answer.

**122.** Find the domain of $\dfrac{\sqrt[3]{x-2}}{\sqrt[6]{x+1}}$. Explain how you determined your answer.

**123.** If $f(x) = -\sqrt{x}$, can $f(x)$ ever be

 **a)** greater than 0,

 **b)** equal to 0,

 **c)** less than 0?

Explain your answers.

**124.** If $f(x) = \sqrt{x+5}$, can $f(x)$ ever be

 **a)** less than 0,

 **b)** equal to 0,

 **c)** greater than 0?

Explain your answers.

## Group Activity

*In this activity, you will determine the conditions under which certain properties of radicals are true. We will discuss these properties later in this chapter. Discuss and answer these exercises as a group.*

**125.** The property $\sqrt[n]{a} \cdot \sqrt[n]{b} = \sqrt[n]{ab}$, called the *multiplication property for radicals*, is true for certain real numbers $a$ and $b$. By substituting values for $a$ and $b$, determine under what conditions this property is true.

**126.** The property $\dfrac{\sqrt[n]{a}}{\sqrt[n]{b}} = \sqrt[n]{\dfrac{a}{b}}$, called the *division property for radicals*, is true for certain real numbers $a$ and $b$. By substituting values for $a$ and $b$, determine under what conditions this property is true.

## Cumulative Review Exercises

*Factor.*

[6.2] **127.** $9ax - 3bx + 12ay - 4by$

[6.4] **128.** $3x^3 - 18x^2 + 24x$

 **129.** $8x^4 + 10x^2 - 3$

[6.5] **130.** $x^3 - \dfrac{8}{27}y^3$

# 11.2 Rational Exponents

**1** Change a radical expression to an exponential expression.

**2** Simplify radical expressions.

**3** Apply the rules of exponents to rational and negative exponents.

**4** Factor expressions with rational exponents.

## **1** Change a Radical Expression to an Exponential Expression

So far, we have not discussed exponential expressions with rational exponents such as $5^{1/3}$, $x^{3/4}$, and $-27^{-4/3}$. In this section, we discuss the relationship between such expressions and radical expressions.

Consider $x = 5^{1/3}$. Now cube both sides of this equation and simplify using the rules of exponents.

$$x = 5^{1/3}$$
$$x^3 = (5^{1/3})^3 = 5^{(1/3)\cdot 3} = 5^1 = 5$$

Thus, $5^{1/3}$ is a number whose cube is 5. Recall that $\sqrt[3]{5}$ is also a number whose cube is 5. Thus, we can conclude that

$$5^{1/3} = \sqrt[3]{5}$$

The following rule shows that a radical expression can be rewritten as an expression with a rational exponent.

---

### Exponential Form of $\sqrt[n]{a}$

$$\sqrt[n]{a} = a^{1/n}$$

When $a$ is nonnegative, $n$ can be any index.
When $a$ is negative, $n$ must be odd.

---

*For the remainder of this chapter, unless you are instructed otherwise, assume that all variables in radicands represent nonnegative real numbers and that the radicand is nonnegative. This assumption will allow us to write many answers without absolute value signs.*

**EXAMPLE 1** Write each expression in exponential form (with rational exponents).

**a)** $\sqrt{7}$      **b)** $\sqrt[3]{15ab}$      **c)** $\sqrt[7]{-4x^2y^5}$      **d)** $\sqrt[8]{\dfrac{5x^7}{2z^{11}}}$

*Solution*

**a)** $\sqrt{7} = 7^{1/2}$     Recall that the index of a square root is 2.

**b)** $\sqrt[3]{15ab} = (15ab)^{1/3}$     **c)** $\sqrt[7]{-4x^2y^5} = (-4x^2y^5)^{1/7}$     **d)** $\sqrt[8]{\dfrac{5x^7}{2z^{11}}} = \left(\dfrac{5x^7}{2z^{11}}\right)^{1/8}$

Now Try Exercise 19

Exponential expressions can also be converted to radical expressions.

**EXAMPLE 2** Write each expression in radical form (without rational exponents).

**a)** $9^{1/2}$      **b)** $(-8)^{1/3}$      **c)** $y^{1/4}$      **d)** $(10x^2y)^{1/7}$      **e)** $5rs^{1/2}$

*Solution*

**a)** $9^{1/2} = \sqrt{9} = 3$      **b)** $(-8)^{1/3} = \sqrt[3]{-8} = -2$      **c)** $y^{1/4} = \sqrt[4]{y}$

**d)** $(10x^2y)^{1/7} = \sqrt[7]{10x^2y}$      **e)** $5rs^{1/2} = 5r\sqrt{s}$

Now Try Exercise 33

## **2** Simplify Radical Expressions

We can expand the preceding rule so that radicals of the form $\sqrt[n]{a^m}$ can be written as exponential expressions. Consider $a^{2/3}$. We can write $a^{2/3}$ as $(a^{1/3})^2$ or $(a^2)^{1/3}$. This suggests $a^{2/3} = (\sqrt[3]{a})^2 = \sqrt[3]{a^2}$.

## Exponential Form of $\sqrt[n]{a^m}$

For any nonnegative number $a$, and integers $m$ and $n$,

$$\sqrt[n]{a^m} = \left(\sqrt[n]{a}\right)^m = a^{m/n}$$

Power ↗    Index ←

This rule shows the relationship between radical expressions and exponential expressions with rational exponents.

When changing a radical expression to an exponential expression

The power of the radicand becomes the numerator of the rational exponent.

$$\sqrt[3]{x^4} = x^{4/3}$$

The index of the radical becomes the denominator of the rational exponent.

### Examples

$$\sqrt{y^3} = y^{3/2} \qquad \sqrt[3]{z^2} = z^{2/3} \qquad \sqrt[5]{2^8} = 2^{8/5}$$

$$(\sqrt{p})^3 = p^{3/2} \qquad (\sqrt[4]{x})^3 = x^{3/4} \qquad (\sqrt[4]{7})^3 = 7^{3/4}$$

By this rule, for nonnegative values of the variable we can write

$$\sqrt{x^5} = (\sqrt{x})^5 \qquad (\sqrt[4]{p})^3 = \sqrt[4]{p^3}$$

**EXAMPLE 3** Write each expression in exponential form (with rational exponents) and then simplify.

**a)** $\sqrt[4]{x^{12}}$ **b)** $(\sqrt[3]{y})^{15}$ **c)** $(\sqrt[6]{x})^{12}$

*Solution*

**a)** $\sqrt[4]{x^{12}} = x^{12/4} = x^3$ **b)** $(\sqrt[3]{y})^{15} = y^{15/3} = y^5$ **c)** $(\sqrt[6]{x})^{12} = x^{12/6} = x^2$

Now Try Exercise 45

When changing an exponential expression with a rational exponent to a radical expression

The numerator of the exponent becomes the power of the radicand.

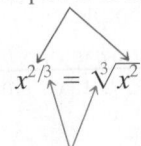

$$x^{2/3} = \sqrt[3]{x^2}$$

The denominator of the exponent becomes the index of the radical.

### Examples

$$x^{1/2} = \sqrt{x} \qquad\qquad 5^{1/3} = \sqrt[3]{5}$$

$$7^{2/3} = \sqrt[3]{7^2} \text{ or } (\sqrt[3]{7})^2 \qquad\qquad y^{3/10} = \sqrt[10]{y^3} \text{ or } (\sqrt[10]{y})^3$$

$$x^{9/5} = \sqrt[5]{x^9} \text{ or } (\sqrt[5]{x})^9 \qquad\qquad z^{10/3} = \sqrt[3]{z^{10}} \text{ or } (\sqrt[3]{z})^{10}$$

Notice that you may choose, for example, to write $6^{2/3}$ as either $\sqrt[3]{6^2}$ or $(\sqrt[3]{6})^2$.

**EXAMPLE 4** Write each expression in radical form (without rational exponents).

**a)** $x^{2/5}$    **b)** $(3ab)^{5/4}$

*Solution*

**a)** $x^{2/5} = \sqrt[5]{x^2}$ or $(\sqrt[5]{x})^2$    **b)** $(3ab)^{5/4} = \sqrt[4]{(3ab)^5}$ or $(\sqrt[4]{3ab})^5$

Now Try Exercise 35

**EXAMPLE 5** Simplify.

**a)** $4^{3/2}$    **b)** $\sqrt[6]{(49)^3}$    **c)** $\sqrt[4]{(xy)^{20}}$    **d)** $(\sqrt[15]{z})^5$

*Solution*

**a)** Sometimes an expression with a rational exponent can be simplified more easily by writing the expression as a radical, as illustrated.

$$4^{3/2} = (\sqrt{4})^3 \quad \text{Write as a radical.}$$
$$= (2)^3$$
$$= 8$$

**b)** Sometimes a radical expression can be simplified more easily by writing the expression with rational exponents, as illustrated in parts **b)** through **d)**.

$$\sqrt[6]{(49)^3} = 49^{3/6} \quad \text{Write with a rational exponent.}$$
$$= 49^{1/2} \quad \text{Reduce exponent.}$$
$$= \sqrt{49} \quad \text{Write as a radical.}$$
$$= 7 \quad \text{Simplify.}$$

**c)** $\sqrt[4]{(xy)^{20}} = (xy)^{20/4} = (xy)^5$

**d)** $(\sqrt[15]{z})^5 = z^{5/15} = z^{1/3}$ or $\sqrt[3]{z}$

Now Try Exercise 51

Now let's consider $\sqrt[5]{x^5}$. When written in exponential form, this is $x^{5/5} = x^1 = x$. This leads to the following rule.

---

**Exponential form of $\sqrt[n]{a^n}$**

For any nonnegative real number $a$,

$$\sqrt[n]{a^n} = (\sqrt[n]{a})^n = a^{n/n} = a$$

---

If $n$ is an even index and $a$ is a negative real number, $\sqrt[n]{a^n} = |a|$ and not $a$. For example, $\sqrt[6]{(-5)^6} = |-5| = 5$. *Since we are assuming, except where noted otherwise, that variables in radicands represent nonnegative real numbers*, we may write $\sqrt[6]{x^6} = x$ and not $|x|$. This assumption also lets us write $\sqrt{x^2} = x$ and $(\sqrt[4]{z})^4 = z$.

Examples

$$\sqrt{3^2} = 3 \qquad\qquad \sqrt[4]{y^4} = y$$
$$\sqrt[6]{(xy)^6} = xy \qquad\qquad (\sqrt[5]{z})^5 = z$$

## 3 Apply the Rules of Exponents to Rational and Negative Exponents

In Section 5.1, we introduced and discussed the rules of exponents where we only used exponents that were whole numbers. The rules still apply when the exponents are rational numbers. Let's review those rules now.

### Rules of Exponents

For all real numbers $a$ and $b$ and all rational numbers $m$ and $n$,

Product rule $\qquad\qquad\qquad\qquad\quad a^m \cdot a^n = a^{m+n}$

Quotient rule $\qquad\qquad\qquad\qquad\quad \dfrac{a^m}{a^n} = a^{m-n}, \quad a \neq 0$

Negative exponent rule $\qquad\qquad\quad a^{-m} = \dfrac{1}{a^m}, \quad a \neq 0$

Zero exponent rule $\qquad\qquad\qquad a^0 = 1, \quad a \neq 0$

Raising a power to a power $\qquad\quad (a^m)^n = a^{m\cdot n}$

Raising a product to a power $\qquad (ab)^m = a^m b^m$

Raising a quotient to a power $\qquad \left(\dfrac{a}{b}\right)^m = \dfrac{a^m}{b^m}, \quad b \neq 0$

Using these rules, we will now work some problems in which the exponents are rational numbers.

**EXAMPLE 6**  Evaluate.  **a)** $8^{-2/3}$  **b)** $(-27)^{-5/3}$  **c)** $(-32)^{-6/5}$

*Solution*

**a)** Begin by using the negative exponent rule.

$$8^{-2/3} = \frac{1}{8^{2/3}} \qquad \text{Negative exponent rule}$$

$$= \frac{1}{(\sqrt[3]{8})^2} \qquad \text{Write the denominator as a radical.}$$

$$= \frac{1}{2^2} \qquad \text{Simplify the denominator.}$$

$$= \frac{1}{4}$$

**b)** $(-27)^{-5/3} = \dfrac{1}{(-27)^{5/3}} = \dfrac{1}{(\sqrt[3]{-27})^5} = \dfrac{1}{(-3)^5} = -\dfrac{1}{243}$

**c)** $(-32)^{-6/5} = \dfrac{1}{(-32)^{6/5}} = \dfrac{1}{(\sqrt[5]{-32})^6} = \dfrac{1}{(-2)^6} = \dfrac{1}{64}$

*Now Try Exercise 81*

Note that Example 6 **a)** could have been evaluated as follows:

$$8^{-2/3} = \frac{1}{8^{2/3}} = \frac{1}{\sqrt[3]{8^2}} = \frac{1}{\sqrt[3]{64}} = \frac{1}{4}$$

However, it is generally easier to evaluate the root before applying the power.

Consider the expression $(-16)^{3/4}$. This can be rewritten as $(\sqrt[4]{-16})^3$. Since $(\sqrt[4]{-16})^3$ is not a real number, the expression $(-16)^{3/4}$ is not a real number.

In Chapter 5, we indicated that

$$\left(\frac{a}{b}\right)^{-n} = \left(\frac{b}{a}\right)^n$$

We use this fact in the following example.

**EXAMPLE 7**  Evaluate.  **a)** $\left(\dfrac{9}{25}\right)^{-1/2}$  **b)** $\left(\dfrac{27}{8}\right)^{-1/3}$

*Solution*

**a)** $\left(\dfrac{9}{25}\right)^{-1/2} = \left(\dfrac{25}{9}\right)^{1/2} = \sqrt{\dfrac{25}{9}} = \dfrac{5}{3}$  **b)** $\left(\dfrac{27}{8}\right)^{-1/3} = \left(\dfrac{8}{27}\right)^{1/3} = \sqrt[3]{\dfrac{8}{27}} = \dfrac{2}{3}$

*Now Try Exercise 83*

**EXAMPLE 8** Simplify each expression and write the answer without negative exponents.

**a)** $a^{1/2} \cdot a^{-2/3}$    **b)** $(6x^2y^{-4})^{-1/2}$    **c)** $3.2x^{1/3}(2.4x^{1/2} + x^{-1/4})$    **d)** $\left(\dfrac{9x^{-4}z^{2/5}}{z^{-3/5}}\right)^{1/8}$

*Solution*

**a)** $a^{1/2} \cdot a^{-2/3} = a^{(1/2)-(2/3)}$          Product rule

$\qquad\qquad\qquad = a^{-1/6}$          Find the LCD and subtract the exponents.

$\qquad\qquad\qquad = \dfrac{1}{a^{1/6}}$          Negative exponent rule

**b)** $(6x^2y^{-4})^{-1/2} = 6^{-1/2}x^{2(-1/2)}y^{-4(-1/2)}$          Raise the product to a power.

$\qquad\qquad\qquad = 6^{-1/2}x^{-1}y^2$          Multiply the exponents.

$\qquad\qquad\qquad = \dfrac{y^2}{6^{1/2}x} \left(\text{or } \dfrac{y^2}{x\sqrt{6}}\right)$          Negative exponent rule

**c)** Begin by using the distributive property.

$3.2x^{1/3}(2.4x^{1/2} + x^{-1/4}) = (3.2x^{1/3})(2.4x^{1/2}) + (3.2x^{1/3})(x^{-1/4})$   Distributive property

$\qquad\qquad\qquad\qquad\qquad = (3.2)(2.4)(x^{(1/3)+(1/2)}) + 3.2x^{(1/3)-(1/4)}$   Product rule

$\qquad\qquad\qquad\qquad\qquad = 7.68x^{5/6} + 3.2x^{1/12}$

**d)** $\left(\dfrac{9x^{-4}z^{2/5}}{z^{-3/5}}\right)^{1/8} = (9x^{-4}z^{(2/5)-(-3/5)})^{1/8}$          Quotient rule

$\qquad\qquad\qquad = (9x^{-4}z)^{1/8}$          Subtract the exponents.

$\qquad\qquad\qquad = 9^{1/8}x^{-4(1/8)}z^{1/8}$          Raise the product to a power.

$\qquad\qquad\qquad = 9^{1/8}x^{-4/8}z^{1/8}$          Multiply the exponents.

$\qquad\qquad\qquad = 9^{1/8}x^{-1/2}z^{1/8}$          Simplify exponent.

$\qquad\qquad\qquad = \dfrac{9^{1/8}z^{1/8}}{x^{1/2}}$          Negative exponent rule

Now Try Exercise 105

**Understanding Algebra**

When using the rules of exponents, there are many different ways to simplify exponential expressions. As long as you use the rules properly, you should arrive at the correct simplified expression.

**EXAMPLE 9** Simplify.   **a)** $\sqrt[15]{(7y)^5}$   **b)** $\left(\sqrt[4]{a^2b^3c}\right)^{20}$   **c)** $\sqrt[4]{\sqrt[3]{x}}$

*Solution*

**a)** $\sqrt[15]{(7y)^5} = (7y)^{5/15}$          Write with a rational exponent.

$\qquad\qquad = (7y)^{1/3}$          Simplify the exponent.

$\qquad\qquad = \sqrt[3]{7y}$          Write as a radical.

**b)** $\left(\sqrt[4]{a^2b^3c}\right)^{20} = (a^2b^3c)^{20/4}$          Write with a rational exponent.

$\qquad\qquad\qquad = (a^2b^3c)^5$

$\qquad\qquad\qquad = a^{10}b^{15}c^5$          Raise the product to a power.

**c)** $\sqrt[4]{\sqrt[3]{x}} = \sqrt[4]{x^{1/3}}$          Write $\sqrt[3]{x}$ as $x^{1/3}$.

$\qquad\qquad = (x^{1/3})^{1/4}$          Write with a rational exponent.

$\qquad\qquad = x^{1/12}$          Raise the power to a power.

$\qquad\qquad = \sqrt[12]{x}$          Write as a radical.

Now Try Exercise 53

## Using Your Calculator

**Finding Roots or Expressions with Rational Exponents on a Scientific or Graphing Calculator**

There are several ways to evaluate an expression like $845^{-3/5}$ on a calculator.*

**Scientific Calculator**

To evaluate $845^{-3/5}$, press

Answer displayed

845 $\boxed{y^x}$ $\boxed{(}$ 3 $\boxed{+/_-}$ $\boxed{\div}$ 5 $\boxed{)}$ $\boxed{=}$ 0.017534201

**Graphing Calculator**

To evaluate $845^{-3/5}$, press the following keys.

Answer displayed

845 $\boxed{\wedge}$ $\boxed{(}$ $\boxed{(-)}$ 3 $\boxed{\div}$ 5 $\boxed{)}$ $\boxed{\text{ENTER}}$ .0175342008

*Keystrokes to use vary from calculator to calculator. For calculators with REAL and COMPLEX mode, these results shown assume the calculator is in REAL mode. Read your calculator manual to learn more on how to evaluate exponential expressions.

## 4 Factor Expressions with Rational Exponents

To factor an expression with rational exponents, factor out the term with the smallest exponent.

**EXAMPLE 10** Factor $x^{2/5} + x^{-3/5}$.

**Solution** The smallest of the two exponents is $-3/5$. Therefore, we will factor $x^{-3/5}$ from both terms. To find the new exponent on the variable that originally had the greater exponent, we subtract the exponent that was factored out from the original exponent.

Original exponent

Exponent factored out

$$x^{2/5} + x^{-3/5} = x^{-3/5}\left(x^{2/5-(-3/5)} + 1\right)$$
$$= x^{-3/5}(x^1 + 1)$$
$$= x^{-3/5}(x + 1)$$
$$= \frac{x + 1}{x^{3/5}}$$

We can check our factoring by multiplying.

$$x^{-3/5}(x + 1) = x^{-3/5} \cdot x + x^{-3/5} \cdot 1$$
$$= x^{(-3/5)+1} + x^{-3/5}$$
$$= x^{2/5} + x^{-3/5}$$

Since we obtained the original expression, the factoring is correct.

Now Try Exercise 135

# EXERCISE SET 11.2    MathXL®    MyMathLab®
MathXL®        MyMathLab

## Warm-Up Exercises

*Fill in the blanks with the appropriate word, phrase, or symbol(s) from the following list.*

multiply        denominator        numerator        add        index        power        divide        function

**1.** When changing a radical expression, $\sqrt[n]{x^m}$, to an exponential expression, the exponent of the radicand, $m$, becomes the _____ of the rational exponent.

**2.** When changing a radical expression, $\sqrt[n]{x^m}$, to an exponential expression, the index of the radical, $n$, becomes the _____ of the rational exponent.

**3.** When changing an exponential expression with a rational exponent, $x^{m/n}$, to a radical expression, the denominator, $n$, becomes the _____ of the radical.

**4.** When changing an exponential expression with a rational exponent, $x^{m/n}$, to a radical expression, the numerator, $m$, becomes the _____ of the radicand.

**5.** To simplify the expression $x^{2/3} \cdot x^{1/4}$, _____ the exponents.

**6.** To simplify the expression $(x^{2/3})^{1/4}$, _____ the exponents.

## Practice the Skills

*Write each expression in exponential form. Assume that all variables represent positive real numbers.*

**7.** $\sqrt{x^5}$

**8.** $\sqrt{b^3}$

**9.** $\sqrt{9^5}$

**10.** $\sqrt[3]{y}$

**11.** $\sqrt[3]{z^5}$

**12.** $\sqrt[3]{x^{11}}$

**13.** $\sqrt[3]{7^{10}}$

**14.** $\sqrt[5]{9^{11}}$

**15.** $\sqrt[4]{9^7}$

**16.** $(\sqrt{x})^9$

**17.** $(\sqrt[3]{y})^{14}$

**18.** $\sqrt{ab^5}$

**19.** $\sqrt[4]{a^3b}$

**20.** $\sqrt[3]{x^4y}$

**21.** $\sqrt[4]{x^9z^5}$

**22.** $\sqrt[6]{y^{11}z}$

**23.** $\sqrt[6]{3a + 8b}$

**24.** $\sqrt[9]{3x + 5z^4}$

**25.** $\sqrt[5]{\dfrac{2x^6}{11y^7}}$

**26.** $\sqrt[4]{\dfrac{3a^8}{11b^5}}$

*Write each expression in radical form. Assume that all variables represent positive real numbers.*

**27.** $a^{1/2}$

**28.** $a^{1/3}$

**29.** $a^{2/3}$

**30.** $19^{1/2}$

**31.** $18^{5/3}$

**32.** $y^{17/6}$

**33.** $(24x^3)^{1/2}$

**34.** $(85a^3)^{5/2}$

**35.** $(11b^2c)^{3/5}$

**36.** $(8x^3y^2)^{7/4}$

**37.** $(6a + 5b)^{1/5}$

**38.** $(8x^2 + 9y)^{7/3}$

**39.** $(b^3 - d)^{-1/3}$

**40.** $(7x^2 - 2y^3)^{-1/6}$

*Simplify each radical expression by changing the expression to exponential form. Write the answer in radical form when appropriate. Assume that all variables represent positive real numbers.*

**41.** $\sqrt{a^2}$

**42.** $\sqrt[3]{b^6}$

**43.** $\sqrt[3]{x^9}$

**44.** $\sqrt[4]{x^{12}}$

**45.** $\sqrt[6]{y^2}$

**46.** $\sqrt[8]{b^4}$

**47.** $\sqrt[6]{y^3}$

**48.** $\sqrt[12]{z^4}$

**49.** $(\sqrt{19.3})^2$

**50.** $\sqrt[4]{(6.83)^4}$

**51.** $(\sqrt[3]{xy^2})^{15}$

**52.** $(\sqrt[4]{a^4bc^3})^{40}$

**53.** $(\sqrt[8]{xyz})^4$

**54.** $(\sqrt[9]{a^2bc^4})^3$

**55.** $\sqrt{\sqrt{x}}$

**56.** $\sqrt{\sqrt[3]{a}}$

**57.** $\sqrt{\sqrt[4]{y}}$

**58.** $\sqrt[3]{\sqrt[4]{b}}$

**59.** $\sqrt[3]{\sqrt[3]{x^2y}}$

**60.** $\sqrt[4]{\sqrt[3]{7y}}$

**61.** $\sqrt{\sqrt[5]{a^9}}$

**62.** $\sqrt[5]{\sqrt[4]{ab}}$

*Evaluate if possible. If the expression is not a real number, so state.*

**63.** $9^{1/2}$

**64.** $100^{1/2}$

**65.** $64^{1/3}$

**66.** $81^{1/4}$

**67.** $64^{2/3}$

**68.** $27^{2/3}$

**69.** $(-49)^{1/2}$

**70.** $(-64)^{1/4}$

**71.** $\left(\dfrac{25}{9}\right)^{1/2}$

**72.** $\left(\dfrac{100}{49}\right)^{1/2}$

**73.** $\left(\dfrac{1}{8}\right)^{1/3}$

**74.** $\left(\dfrac{1}{32}\right)^{1/5}$

**75.** $-81^{1/2}$

**76.** $(-81)^{1/2}$

**77.** $-64^{1/3}$

**78.** $(-64)^{1/3}$

**79.** $64^{-1/3}$

**80.** $49^{-1/2}$

**81.** $16^{-3/2}$

**82.** $64^{-2/3}$

**83.** $\left(\dfrac{64}{27}\right)^{-1/3}$

**84.** $(-81)^{3/4}$

**85.** $(-100)^{3/2}$

**86.** $-\left(\dfrac{25}{49}\right)^{-1/2}$

**87.** $121^{1/2} + 169^{1/2}$

**88.** $49^{-1/2} + 36^{-1/2}$

**89.** $343^{-1/3} + 16^{-1/2}$

**90.** $16^{-1/2} - 256^{-3/4}$

*Simplify. Write the answer in exponential form without negative exponents. Assume that all variables represent positive real numbers.*

**91.** $x^4 \cdot x^{1/2}$

**92.** $x^6 \cdot x^{1/2}$

**93.** $\dfrac{x^{1/2}}{x^{1/3}}$

**94.** $x^{-6/5}$

**95.** $\left(x^{1/2}\right)^{-2}$

**96.** $\left(a^{-1/3}\right)^{-1/2}$

**97.** $\left(9^{-1/3}\right)^{0}$

**98.** $\dfrac{x^4}{x^{-1/2}}$

**99.** $\dfrac{5y^{-1/3}}{60y^{-2}}$

**100.** $x^{-1/2}x^{-2/5}$

**101.** $4x^{5/3}3x^{-7/2}$

**102.** $\left(x^{-4/5}\right)^{1/3}$

**103.** $\left(\dfrac{3}{24x}\right)^{1/3}$

**104.** $\left(\dfrac{54}{2x^4}\right)^{1/3}$

**105.** $\left(\dfrac{22x^{3/7}}{2x^{1/2}}\right)^{2}$

**106.** $\left(\dfrac{x^{-1/3}}{x^{-2}}\right)^{2}$

**107.** $\left(\dfrac{a^4}{4a^{-2/5}}\right)^{-3}$

**108.** $\left(\dfrac{27z^{1/4}y^3}{3z^{1/4}}\right)^{1/2}$

**109.** $\left(\dfrac{x^{3/4}y^{-3}}{x^{1/2}y^2}\right)^{4}$

**110.** $\left(\dfrac{250a^{-3/4}b^5}{2a^{-2}b^2}\right)^{2/3}$

*Multiply. Assume that all variables represent positive real numbers.*

**111.** $4z^{-1/2}(2z^4 - z^{1/2})$

**112.** $-3a^{-4/9}(5a^{1/9} - a^2)$

**113.** $5x^{-1}(x^{-4} + 4x^{-1/2})$

**114.** $-9z^{3/2}(z^{3/2} - z^{-3/2})$

**115.** $-6x^{5/3}(-2x^{1/2} + 3x^{1/3})$

**116.** $\dfrac{1}{2}x^{-2}(10x^{4/3} - 38x^{-1/2})$

*Use a calculator to evaluate each expression. Give the answer to the nearest hundredth.*

**117.** $\sqrt{180}$

**118.** $\sqrt[3]{168}$

**119.** $\sqrt[5]{402.83}$

**120.** $\sqrt[4]{1096}$

**121.** $93^{2/3}$

**122.** $38.2^{3/2}$

**123.** $1000^{-1/2}$

**124.** $8060^{-3/2}$

## Problem Solving

**125.** Under what conditions will $\sqrt[n]{a^n} = \left(\sqrt[n]{a}\right)^n = a$?

**126.** By selecting values for $a$ and $b$, show that $\left(a^2 + b^2\right)^{1/2}$ is not equal to $a + b$.

**127.** By selecting values for $a$ and $b$, show that $\left(a^{1/2} + b^{1/2}\right)^2$ is not equal to $a + b$.

**128.** By selecting values for $a$ and $b$, show that $\left(a^3 + b^3\right)^{1/3}$ is not equal to $a + b$.

**129.** By selecting values for $a$ and $b$, show that $\left(a^{1/3} + b^{1/3}\right)^3$ is not equal to $a + b$.

**130.** Determine whether $\sqrt[3]{\sqrt{x}} = \sqrt{\sqrt[3]{x}}$, $x \ge 0$.

*Factor. Write the answer without negative exponents. Assume that all variables represent positive real numbers.*

**131.** $x^{3/2} + x^{1/2}$

**132.** $x^{1/4} - x^{5/4}$

**133.** $y^{1/3} - y^{7/3}$

**134.** $x^{-1/2} + x^{1/2}$

**135.** $y^{-2/5} + y^{8/5}$

**136.** $a^{6/5} + a^{-4/5}$

*In Exercises 137 through 142, use a calculator where appropriate.*

**137. Growing Bacteria** The function $B(t) = 2^{10} \cdot 2^t$ approximates the number of bacteria in a certain culture after $t$ hours.

  **a)** The initial number of bacteria is determined when $t = 0$. What is the initial number of bacteria?

  **b)** How many bacteria are there after $\dfrac{1}{2}$ hour?

**138. Carbon Dating** Carbon dating is used by scientists to find the age of fossils, bones, and other items. The formula used in carbon dating is $P = P_0 2^{-t/5600}$, where $P_0$ represents the original amount of carbon 14 ($C_{14}$) present and $P$ represents the amount of $C_{14}$ present after $t$ years. If 10 milligrams (mg) of $C_{14}$ is present in an animal bone recently excavated, how many milligrams will be present in 5000 years?

**139. Resting Metabolic Rate** A person's resting metabolic rate (RMR) is the number of calories a person burns in one day of resting. A person's RMR can be estimated using the function $R(x) = 70x^{3/4}$ where $R(x)$ is measured in calories per day and $x$ is the person's mass in kilo-

grams. If Ken Machol has a mass of 102 kg, estimate his RMR to the nearest calorie.

*Source*: www.bodybuilding.com

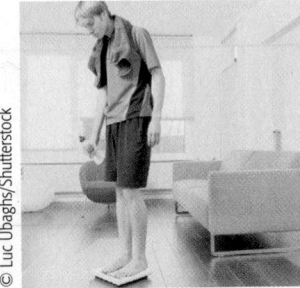

**140. Resting Metabolic Rate** If Martin Alexander has a mass of 82 kg, estimate his resting metabolic rate (see Exercise 139) to the nearest calorie.

**141.** Find the domain of $f(x) = (x - 7)^{1/2}(x + 3)^{-1/2}$.

**142.** Find the domain of $f(x) = (x + 4)^{1/2}(x - 3)^{-1/2}$.

*Determine the index to be placed in the shaded area to make the statement true.*

**143.** $\sqrt[4]{\sqrt[5]{\sqrt{\sqrt[3]{z}}}} = z^{1/120}$

**144.** $\sqrt[4]{\sqrt{\sqrt{\sqrt{x}}}} = x^{1/24}$

## Concept/Writing Exercises

**145. a)** Under what conditions is $\sqrt[n]{a}$ a real number?

   **b)** When $\sqrt[n]{a}$ is a real number, how can it be expressed with rational exponents?

**146. a)** Under what conditions is $\sqrt[n]{a^m}$ a real number?

   **b)** Under what conditions is $(\sqrt[n]{a})^m$ a real number?

   **c)** When $\sqrt[n]{a^m}$ is a real number, how can it be expressed with rational exponents?

**147. a)** Under what conditions is $\sqrt[n]{a^n}$ a real number?

   **b)** When $n$ is even and $a \geq 0$, what is $\sqrt[n]{a^n}$ equal to?

   **c)** When $n$ is odd, what is $\sqrt[n]{a^n}$ equal to?

   **d)** When $n$ is even and $a$ may be any real number, what is $\sqrt[n]{a^n}$ equal to?

**148. a)** Explain the difference between $-16^{1/2}$ and $(-16)^{1/2}$.

   **b)** Evaluate each expression in part **a)** if possible.

**149. a)** Is $(xy)^{1/2} = xy^{1/2}$? Explain.

   **b)** Is $(xy)^{-1/2} = \dfrac{x^{1/2}}{y^{-1/2}}$? Explain.

**150. a)** Is $\sqrt[9]{(3y)^3} = (3y)^{6/3}$? Explain.

   **b)** Is $\sqrt{(ab)^4} = (ab)^2$? Explain.

**151.** Evaluate $(3^{\sqrt{2}})^{\sqrt{2}}$. Explain how you determined your answer.

**152. a)** On your calculator, evaluate $3^{\pi}$.

   **b)** Explain why your value from part **a)** does or does not make sense.

## Cumulative Review Exercises

[7.5] **153.** Simplify $\dfrac{a^{-2} + ab^{-1}}{ab^{-2} - a^{-2}b^{-1}}$.

[7.6] **154.** Solve $\dfrac{3x - 2}{x + 4} = \dfrac{2x + 1}{3x - 2}$.

[7.7] **155. Flying a Plane** Amy Mayfield can fly her plane 500 miles against the wind in the same time it takes her to fly 560 miles with the wind. If the wind blows at 25 miles per hour, find the speed of the plane in still air.

[8.2] **156.** Determine which of the following relations are also functions.

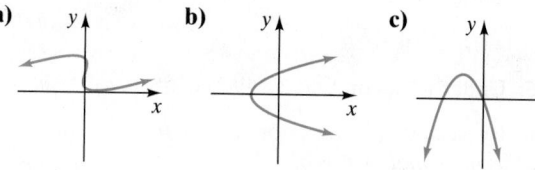

# 11.3 Simplifying Radicals

1   Understand perfect powers.

2   Simplify radicals using the product rule for radicals.

3   Simplify radicals using the quotient rule for radicals.

## 1   Understand Perfect Powers

In this section, we will simplify radical expressions. We begin with a discussion of perfect powers.

> **Perfect Power, Perfect Square, Perfect Cube**
>
> • A **perfect power** is a number or expression that can be written as an expression raised to a power that is a whole number greater than 1.
> • A **perfect square** is a number or expression that can be written as a square of an expression. A perfect square is a perfect *second* power.
> • A **perfect cube** is a number or expression that can be written as a cube of an expression. A perfect cube is a perfect *third* power.

Examples of perfect squares are illustrated below.

| Perfect squares | 1, | 4, | 9, | 16, | 25, | 36,... |
|---|---|---|---|---|---|---|
| | ↓ | ↓ | ↓ | ↓ | ↓ | ↓ |
| Square of a number | $1^2$, | $2^2$, | $3^2$, | $4^2$, | $5^2$, | $6^2$,... |

Variables with exponents may also be perfect squares, as illustrated below.

| Perfect squares | $x^2$, | $x^4$, | $x^6$, | $x^8$, | $x^{10}$,... |
|---|---|---|---|---|---|
| | ↓ | ↓ | ↓ | ↓ | ↓ |
| Square of an expression | $(x)^2$, | $(x^2)^2$, | $(x^3)^2$, | $(x^4)^2$, | $(x^5)^2$,... |

*Notice that the exponents on the variables in the perfect squares are all multiples of 2.*

Examples of perfect cubes are illustrated below.

| Perfect cubes | 1, | 8, | 27, | 64, | 125, | 216,... |
|---|---|---|---|---|---|---|
| | ↓ | ↓ | ↓ | ↓ | ↓ | ↓ |
| Cube of a number | $1^3$, | $2^3$, | $3^3$, | $4^3$, | $5^3$, | $6^3$,... |
| Perfect cubes | $x^3$, | $x^6$, | $x^9$, | $x^{12}$, | $x^{15}$,... | |
| | ↓ | ↓ | ↓ | ↓ | ↓ | |
| Cube of an expression | $(x)^3$, | $(x^2)^3$, | $(x^3)^3$, | $(x^4)^3$, | $(x^5)^3$,... | |

*Notice that the exponents on the variables in the perfect cubes are all multiples of 3.*

When simplifying radicals, we will be looking for perfect powers in the radicand. For example, if we are simplifying a square root, then we are interested in finding perfect squares. If we are simplifying a cube root, then we are interested in finding perfect cubes. If we are simplifying a fourth root, then we are interested in finding perfect fourth powers, and so on.

## Understanding Algebra

A variable with an exponent is a

- Perfect square if the exponent is divisible by 2. For example, $x^2, x^4, x^6, \ldots$ are perfect squares.
- Perfect cube if the exponent is divisible by 3. For example, $x^3, x^6, x^9, \ldots$ are perfect cubes.
- Perfect fourth power if the exponent is divisible by 4. For example, $x^4, x^8, x^{12}, \ldots$ are perfect fourth powers.

This pattern continues for higher perfect powers.

### Helpful Hint

A quick way to determine if a radicand $x^m$ is a perfect power for an index is to determine if the exponent $m$ is divisible by the index of the radical. For example, consider $\sqrt[5]{x^{20}}$. Since the exponent, 20, is divisible by the index, 5, $x^{20}$ is a perfect fifth power. Now consider $\sqrt[6]{x^{20}}$. Since the exponent, 20, is not divisible by the index, 6, $x^{20}$ is not a perfect sixth power. However, $x^{18}$ and $x^{24}$ are both perfect sixth powers since 6 divides both 18 and 24.

Notice that the square root of a perfect square simplifies to an expression without a radical sign, the cube root of a perfect cube simplifies to an expression without a radical sign, and so on.

### Examples

$$\sqrt{36} = \sqrt{6^2} = 6^{2/2} = 6$$
$$\sqrt[3]{27} = \sqrt[3]{3^3} = 3^{3/3} = 3$$
$$\sqrt{x^6} = x^{6/2} = x^3$$
$$\sqrt[3]{z^{12}} = z^{12/3} = z^4$$
$$\sqrt[5]{n^{35}} = n^{35/5} = n^7$$

Now we are ready to discuss the product rule for radicals.

## 2   Simplify Radicals Using the Product Rule for Radicals

To introduce the **product rule for radicals**, observe that $\sqrt{4} \cdot \sqrt{9} = 2 \cdot 3 = 6$. Also, $\sqrt{4 \cdot 9} = \sqrt{36} = 6$. We see that $\sqrt{4} \cdot \sqrt{9} = \sqrt{4 \cdot 9}$. This is one example of the product rule for radicals.

### Product Rule for Radicals

For nonnegative real numbers $a$ and $b$,
$$\sqrt[n]{a} \cdot \sqrt[n]{b} = \sqrt[n]{ab}$$

### Examples of the Product Rule for Radicals

$$\sqrt{20} = \begin{cases} \sqrt{1} \cdot \sqrt{20} \\ \sqrt{2} \cdot \sqrt{10} \\ \sqrt{4} \cdot \sqrt{5} \end{cases} \qquad \sqrt[3]{20} = \begin{cases} \sqrt[3]{1} \cdot \sqrt[3]{20} \\ \sqrt[3]{2} \cdot \sqrt[3]{10} \\ \sqrt[3]{4} \cdot \sqrt[3]{5} \end{cases}$$

$\sqrt{20}$ can be factored into any of these forms.    $\sqrt[3]{20}$ can be factored into any of these forms.

$$\sqrt{x^7} = \begin{cases} \sqrt{x} \cdot \sqrt{x^6} \\ \sqrt{x^2} \cdot \sqrt{x^5} \\ \sqrt{x^3} \cdot \sqrt{x^4} \end{cases} \qquad \sqrt[3]{x^7} = \begin{cases} \sqrt[3]{x} \cdot \sqrt[3]{x^6} \\ \sqrt[3]{x^2} \cdot \sqrt[3]{x^5} \\ \sqrt[3]{x^3} \cdot \sqrt[3]{x^4} \end{cases}$$

$\sqrt{x^7}$ can be factored into any of these forms.    $\sqrt[3]{x^7}$ can be factored into any of these forms.

Here is a general procedure that can be used to simplify radicals using the product rule.

> **To Simplify Radicals Using the Product Rule**
>
> 1. If the radicand contains a coefficient other than 1, write it as a product of two numbers, one of which is the largest perfect power for the index.
> 2. Write each variable factor as a product of two factors, one of which is the largest perfect power of the variable for the index.
> 3. Use the product rule to write the radical expression as a product of radicals. Place all the perfect powers (numbers and variables) under the same radical.
> 4. Simplify the radical containing the perfect powers.

If we are simplifying a *square* root, we will write the radicand as the product of the largest *perfect square* and another expression. If we are simplifying a *cube* root, we will write the radicand as the product of the largest *perfect cube* and another expression, and so on.

**EXAMPLE 1** Simplify.  **a)** $\sqrt{32}$   **b)** $\sqrt{60}$   **c)** $\sqrt[3]{54}$   **d)** $\sqrt[4]{96}$

*Solution*   The radicands in this example contain no variables. We will follow step 1 of the procedure.

**a)** Since we are evaluating a square root, we look for the largest perfect square that divides 32. The largest perfect square that divides, or is a factor of, 32 is 16.

$$\sqrt{32} = \sqrt{16 \cdot 2} = \sqrt{16}\,\sqrt{2} = 4\sqrt{2}$$

**b)** The largest perfect square that is a factor of 60 is 4.

$$\sqrt{60} = \sqrt{4 \cdot 15} = \sqrt{4}\,\sqrt{15} = 2\sqrt{15}$$

**c)** The largest perfect cube that is a factor of 54 is 27.

$$\sqrt[3]{54} = \sqrt[3]{27 \cdot 2} = \sqrt[3]{27}\sqrt[3]{2} = 3\sqrt[3]{2}$$

**d)** The largest perfect fourth power that is a factor of 96 is 16.

$$\sqrt[4]{96} = \sqrt[4]{16 \cdot 6} = \sqrt[4]{16}\,\sqrt[4]{6} = 2\sqrt[4]{6}$$

Now Try Exercise 19

**Understanding Algebra**

To simplify $\sqrt{20}$ we write it as $\sqrt{4} \cdot \sqrt{5}$ because 4 is the largest perfect square that divides 20:

$$\sqrt{20} = \sqrt{4} \cdot \sqrt{5} = 2\sqrt{5}$$

Similarly, to simplify $\sqrt[3]{24}$ we write it as $\sqrt[3]{8} \cdot \sqrt[3]{3}$ because 8 is the largest perfect cube that divides 24:

$$\sqrt[3]{24} = \sqrt[3]{8} \cdot \sqrt[3]{3} = 2\sqrt[3]{3}$$

**Helpful Hint**

In Example 1 **a)**, if you first thought that 4 was the largest perfect square that divided 32, you could proceed as follows:

$$\sqrt{32} = \sqrt{4 \cdot 8} = \sqrt{4}\,\sqrt{8} = 2\sqrt{8}$$
$$= 2\sqrt{4 \cdot 2} = 2\sqrt{4}\,\sqrt{2} = 2 \cdot 2\sqrt{2} = 4\sqrt{2}$$

Note that the final result is the same, but you must perform more steps. The lists of perfect squares and perfect cubes on pages 640–641 can help you determine the largest perfect square or perfect cube that is a factor of a radicand.

In Example 1 **b)**, $\sqrt{15}$ can be factored as $\sqrt{5 \cdot 3}$; however, since neither 5 nor 3 is a perfect square, $\sqrt{15}$ cannot be simplified.

When the radicand is a perfect power for the index, the radical can be simplified by writing it in exponential form, as in Example 2.

**EXAMPLE 2** Simplify.   **a)** $\sqrt{x^4}$   **b)** $\sqrt[3]{x^{12}}$   **c)** $\sqrt[5]{z^{40}}$

*Solution*

**a)** $\sqrt{x^4} = x^{4/2} = x^2$   **b)** $\sqrt[3]{x^{12}} = x^{12/3} = x^4$   **c)** $\sqrt[5]{z^{40}} = z^{40/5} = z^8$

Now Try Exercise 33

EXAMPLE 3  Simplify.  **a)** $\sqrt{x^9}$  **b)** $\sqrt[5]{x^{23}}$  **c)** $\sqrt[4]{y^{33}}$

*Solution*   Because the radicands have coefficients of 1, we start with step 2 of the procedure.

**a)** The largest perfect square less than or equal to $x^9$ is $x^8$.

$$\sqrt{x^9} = \sqrt{x^8 \cdot x} = \sqrt{x^8} \cdot \sqrt{x} = x^{8/2}\sqrt{x} = x^4\sqrt{x}$$

**b)** The largest perfect fifth power less than or equal to $x^{23}$ is $x^{20}$.

$$\sqrt[5]{x^{23}} = \sqrt[5]{x^{20} \cdot x^3} = \sqrt[5]{x^{20}}\,\sqrt[5]{x^3} = x^{20/5}\sqrt[5]{x^3} = x^4\sqrt[5]{x^3}$$

**c)** The largest perfect fourth power less than or equal to $y^{33}$ is $y^{32}$.

$$\sqrt[4]{y^{33}} = \sqrt[4]{y^{32} \cdot y} = \sqrt[4]{y^{32}}\,\sqrt[4]{y} = y^{32/4}\sqrt[4]{y} = y^8\sqrt[4]{y}$$

Now Try Exercise 39

If you observe the answers to Example 3, you will see that the exponent on the variable in the radicand is always less than the index. *When a radical is simplified, the exponents on the variables in the radicand are less than the index.*

In Example 3 **b)**, we simplified $\sqrt[5]{x^{23}}$. If we divide 23, the exponent in the radicand, by 5, the index, we obtain

$$\begin{array}{r} 4 \quad \longleftarrow \quad \text{Quotient} \\ 5\overline{)23} \\ \underline{20} \\ 3 \quad \longleftarrow \quad \text{Remainder} \end{array}$$

Notice that $\sqrt[5]{x^{23}}$ simplifies to $x^4\sqrt[5]{x^3}$ and

$$\text{Quotient} \longrightarrow x^4\sqrt[5]{x^3} \longleftarrow \text{Remainder}$$

When simplifying a radical, if you divide the exponent within the radical by the index, the quotient will be the exponent on the variable outside the radical sign and the remainder will be the exponent on the variable within the radical sign. Simplify Example 3 **c)** using this technique now.

EXAMPLE 4  Simplify.  **a)** $\sqrt{x^{12}y^{17}}$  **b)** $\sqrt[4]{x^6 y^{23}}$

*Solution*

**a)** $x^{12}$ is a perfect square. The largest perfect square that is a factor of $y^{17}$ is $y^{16}$. Write $y^{17}$ as $y^{16} \cdot y$.

$$\sqrt{x^{12}y^{17}} = \sqrt{x^{12} \cdot y^{16} \cdot y} = \sqrt{x^{12}y^{16}}\,\sqrt{y}$$
$$= \sqrt{x^{12}}\sqrt{y^{16}}\,\sqrt{y}$$
$$= x^6 y^8 \sqrt{y}$$

**Understanding Algebra**

Often the steps when simplifying a radical are done mentally. For example, in Example 4 **a)** we do not show $\sqrt{x^{12}} = x^{12/2} = x^6$ and $\sqrt{y^{16}} = y^{16/2} = y^8$.

**b)** We begin by finding the largest perfect fourth power factors of $x^6$ and $y^{23}$. For an index of 4, the largest perfect power that is a factor of $x^6$ is $x^4$. The largest perfect power that is a factor of $y^{23}$ is $y^{20}$.

$$\sqrt[4]{x^6 y^{23}} = \sqrt[4]{x^4 \cdot x^2 \cdot y^{20} \cdot y^3}$$
$$= \sqrt[4]{x^4 y^{20} \cdot x^2 y^3}$$
$$= \sqrt[4]{x^4 y^{20}}\,\sqrt[4]{x^2 y^3}$$
$$= xy^5 \sqrt[4]{x^2 y^3}$$

Now Try Exercise 51

Often the steps where we change the radical expression to exponential form are done mentally, and those steps are not illustrated.

> **Helpful Hint**
>
> In Example 4 **b)**, we showed that $\sqrt[4]{x^6y^{23}} = xy^5\sqrt[4]{x^2y^3}$. We can also simplify this radical by dividing the exponents on the variables in the radicand, 6 and 23, by the index, 4. Note the location of the quotients and the remainders.
>
> 6 ÷ 4 gives a quotient of 1 and a remainder of 2
>
> $$\sqrt[4]{x^6y^{23}} = x^1y^5\sqrt[4]{x^2y^3}$$
>
> 23 ÷ 4 gives a quotient of 5 and a remainder of 3

**EXAMPLE 5** Simplify.   **a)** $\sqrt{80x^5y^{12}z^3}$    **b)** $\sqrt[3]{54x^{17}y^{25}}$

*Solution*

**a)** The largest perfect square that is a factor of 80 is 16. The largest perfect square that is a factor of $x^5$ is $x^4$. The expression $y^{12}$ is a perfect square. The largest perfect square that is a factor of $z^3$ is $z^2$. Place all the perfect squares under the same radical, and then simplify.

$$\sqrt{80x^5y^{12}z^3} = \sqrt{16\cdot5\cdot x^4\cdot x\cdot y^{12}\cdot z^2\cdot z}$$
$$= \sqrt{16x^4y^{12}z^2\cdot 5xz}$$
$$= \sqrt{16x^4y^{12}z^2}\cdot\sqrt{5xz}$$
$$= 4x^2y^6z\sqrt{5xz}$$

**b)** The largest perfect cube that is a factor of 54 is 27. The largest perfect cube that is a factor of $x^{17}$ is $x^{15}$. The largest perfect cube that is a factor of $y^{25}$ is $y^{24}$.

$$\sqrt[3]{54x^{17}y^{25}} = \sqrt[3]{27\cdot2\cdot x^{15}\cdot x^2\cdot y^{24}\cdot y}$$
$$= \sqrt[3]{27x^{15}y^{24}\cdot 2x^2y}$$
$$= \sqrt[3]{27x^{15}y^{24}}\cdot\sqrt[3]{2x^2y}$$
$$= 3x^5y^8\sqrt[3]{2x^2y}$$

Now Try Exercise 57

### 3  Simplify Radicals Using the Quotient Rule for Radicals

In mathematics we sometimes need to simplify a quotient of two radicals. To do so we use the **quotient rule for radicals**.

> **Quotient Rule for Radicals**
>
> For nonnegative real numbers $a$ and $b$,
>
> $$\frac{\sqrt[n]{a}}{\sqrt[n]{b}} = \sqrt[n]{\frac{a}{b}}, \qquad b \neq 0$$

Examples of the Quotient Rule for Radicals

$$\frac{\sqrt{18}}{\sqrt{3}} = \sqrt{\frac{18}{3}} \qquad \sqrt{\frac{9}{25}} = \frac{\sqrt{9}}{\sqrt{25}}$$

$$\frac{\sqrt{x^3}}{\sqrt{x}} = \sqrt{\frac{x^3}{x}} \qquad \sqrt{\frac{x^4}{y^2}} = \frac{\sqrt{x^4}}{\sqrt{y^2}}$$

$$\frac{\sqrt[3]{y^5}}{\sqrt[3]{y^2}} = \sqrt[3]{\frac{y^5}{y^2}} \qquad \sqrt[3]{\frac{z^9}{27}} = \frac{\sqrt[3]{z^9}}{\sqrt[3]{27}}$$

Examples 6 and 7 illustrate how to use the quotient rule to simplify radical expressions.

**EXAMPLE 6** Simplify.   **a)** $\dfrac{\sqrt{75}}{\sqrt{3}}$   **b)** $\dfrac{\sqrt[3]{24x}}{\sqrt[3]{3x}}$   **c)** $\dfrac{\sqrt[3]{x^4 y^7}}{\sqrt[3]{xy^{-5}}}$

*Solution*   In each part we use the quotient rule to write the quotient of radicals as a single radical. Then we simplify.

**a)** $\dfrac{\sqrt{75}}{\sqrt{3}} = \sqrt{\dfrac{75}{3}} = \sqrt{25} = 5$

**b)** $\dfrac{\sqrt[3]{24x}}{\sqrt[3]{3x}} = \sqrt[3]{\dfrac{24x}{3x}} = \sqrt[3]{8} = 2$

**c)** $\dfrac{\sqrt[3]{x^4 y^7}}{\sqrt[3]{xy^{-5}}} = \sqrt[3]{\dfrac{x^4 y^7}{xy^{-5}}}$   Quotient rule for radicals

$\qquad\qquad = \sqrt[3]{x^3 y^{12}}$   Simplify the radicand.

$\qquad\qquad = xy^4$

Now Try Exercise 93

In Section 11.1 when we introduced radicals, we indicated that $\sqrt{\dfrac{4}{9}} = \dfrac{2}{3}$ since $\dfrac{2}{3} \cdot \dfrac{2}{3} = \dfrac{4}{9}$.

The quotient rule may be helpful in evaluating square roots containing fractions as illustrated in Example 7 **a)**.

**EXAMPLE 7** Simplify.   **a)** $\sqrt{\dfrac{121}{25}}$   **b)** $\sqrt[3]{\dfrac{8x^4 y}{27xy^{10}}}$   **c)** $\sqrt[4]{\dfrac{18xy^5}{3x^9 y}}$

*Solution*   In each part we first simplify the radicand, if possible. Then we use the quotient rule to write the given radical as a quotient of radicals.

**a)** $\sqrt{\dfrac{121}{25}} = \dfrac{\sqrt{121}}{\sqrt{25}} = \dfrac{11}{5}$

**b)** $\sqrt[3]{\dfrac{8x^4 y}{27xy^{10}}} = \sqrt[3]{\dfrac{8x^3}{27y^9}} = \dfrac{\sqrt[3]{8x^3}}{\sqrt[3]{27y^9}} = \dfrac{2x}{3y^3}$

**c)** $\sqrt[4]{\dfrac{18xy^5}{3x^9 y}} = \sqrt[4]{\dfrac{6y^4}{x^8}} = \dfrac{\sqrt[4]{6y^4}}{\sqrt[4]{x^8}} = \dfrac{\sqrt[4]{y^4}\,\sqrt[4]{6}}{x^2} = \dfrac{y\sqrt[4]{6}}{x^2}$

Now Try Exercise 97

---

**Avoiding Common Errors**

The following simplifications are correct because the numbers and variables divided out are not within square roots.

| CORRECT | CORRECT |
|---|---|
| $\dfrac{\overset{2}{\cancel{6}}\,\sqrt{2}}{\underset{1}{\cancel{3}}} = 2\sqrt{2}$ | $\dfrac{\cancel{x}\sqrt{2}}{\cancel{x}} = \sqrt{2}$ |

An expression within a square root cannot be divided by an expression not within the square root.

| CORRECT | INCORRECT |
|---|---|
| $\dfrac{\sqrt{2}}{2}$   Cannot be simplified further | $\dfrac{\sqrt{\overset{1}{\cancel{2}}}}{\underset{1}{\cancel{2}}} = \sqrt{1} = 1$ |
| $\dfrac{\sqrt{x^3}}{x} = \dfrac{\sqrt{x^2}\,\sqrt{x}}{x} = \dfrac{\cancel{x}\sqrt{x}}{\cancel{x}} = \sqrt{x}$ | $\dfrac{\sqrt{x^{\overset{2}{\cancel{3}}}}}{\cancel{x}} = \sqrt{x^2} = x$ |

# EXERCISE SET 11.3    MathXL®    MyMathLab®
<small>MathXL®        MyMathLab</small>

## Warm-Up Exercises

*Fill in the blanks with the appropriate word, phrase, or symbol(s) from the following list.*

| quotient | square | less | 3 | cube | greater | 2 | product | 4 |

**1.** A perfect _____ is a number or expression that can be written as a square of an expression.

**2.** A perfect _____ is a number or expression that can be written as a cube of an expression.

**3.** A variable with an exponent is a perfect cube if the exponent is divisible by _____ .

**4.** A variable with an exponent is a perfect square if the exponent is divisible by _____ .

**5.** A variable with an exponent is a perfect fourth power if the exponent is divisible by _____ .

**6.** When a radical is simplified, the exponents on the variables in the radicand are _____ than the index.

**7.** The _____ rule for radicals states that for nonnegative real numbers, $\sqrt[n]{a} \cdot \sqrt[n]{b} = \sqrt[n]{ab}$.

**8.** The _____ rule for radicals states that for nonnegative real numbers, $\dfrac{\sqrt[n]{a}}{\sqrt[n]{b}} = \sqrt[n]{\dfrac{a}{b}}, b \neq 0$.

## Practice the Skills

*Simplify. Assume that all variables represent positive real numbers.*

**9.** $\sqrt{49}$     **10.** $\sqrt{100}$     **11.** $\sqrt{24}$     **12.** $\sqrt{18}$

**13.** $\sqrt{32}$     **14.** $\sqrt{12}$     **15.** $\sqrt{50}$     **16.** $\sqrt{72}$

**17.** $\sqrt{75}$     **18.** $\sqrt{300}$     **19.** $\sqrt{40}$     **20.** $\sqrt{600}$

**21.** $\sqrt[3]{16}$     **22.** $\sqrt[3]{24}$     **23.** $\sqrt[3]{54}$     **24.** $\sqrt[3]{81}$

**25.** $\sqrt[3]{32}$     **26.** $\sqrt[3]{108}$     **27.** $\sqrt[3]{40}$     **28.** $\sqrt[4]{80}$

**29.** $\sqrt[4]{48}$     **30.** $\sqrt[4]{162}$     **31.** $-\sqrt[5]{64}$     **32.** $-\sqrt[5]{243}$

**33.** $\sqrt[3]{b^9}$     **34.** $6\sqrt{y^{12}}$     **35.** $\sqrt[3]{x^6}$     **36.** $\sqrt[5]{y^{20}}$

**37.** $\sqrt{x^3}$     **38.** $-\sqrt{x^5}$     **39.** $\sqrt{a^5}$     **40.** $\sqrt{b^7}$

**41.** $8\sqrt[3]{z^{32}}$     **42.** $\sqrt[3]{a^7}$     **43.** $\sqrt[4]{b^{23}}$     **44.** $\sqrt[5]{z^7}$

**45.** $\sqrt[6]{x^9}$     **46.** $\sqrt[7]{y^{15}}$     **47.** $3\sqrt[5]{y^{23}}$     **48.** $\sqrt{24x^3}$

**49.** $2\sqrt{50y^9}$     **50.** $\sqrt{75a^7b^{11}}$     **51.** $\sqrt[3]{x^3y^7}$     **52.** $\sqrt{x^5y^9}$

**53.** $\sqrt[5]{a^6b^{23}}$     **54.** $-\sqrt{20x^6y^7z^{12}}$     **55.** $\sqrt{24x^{15}y^{20}z^{27}}$     **56.** $\sqrt[3]{16x^3y^6}$

**57.** $\sqrt[3]{81a^6b^8}$     **58.** $\sqrt[3]{128a^{10}b^{11}c^{12}}$     **59.** $\sqrt[4]{32x^8y^9z^{19}}$     **60.** $\sqrt[4]{48x^{11}y^{21}}$

**61.** $\sqrt[4]{81a^8b^9}$     **62.** $-\sqrt[4]{32x^{18}y^{31}}$     **63.** $\sqrt[5]{32a^{10}b^{12}}$     **64.** $\sqrt[6]{64x^{12}y^{23}z^{50}}$

*Simplify. Assume that all variables represent positive real numbers.*

**65.** $\sqrt{\dfrac{18}{2}}$     **66.** $\sqrt{\dfrac{45}{5}}$     **67.** $\sqrt{\dfrac{81}{100}}$     **68.** $\sqrt{\dfrac{8}{50}}$

**69.** $\dfrac{\sqrt{27}}{\sqrt{3}}$     **70.** $\dfrac{\sqrt{72}}{\sqrt{2}}$     **71.** $\dfrac{\sqrt{3}}{\sqrt{48}}$     **72.** $\dfrac{\sqrt{15}}{\sqrt{60}}$

**73.** $\sqrt[3]{\dfrac{3}{24}}$     **74.** $\sqrt[3]{\dfrac{2}{54}}$     **75.** $\dfrac{\sqrt[3]{3}}{\sqrt[3]{81}}$     **76.** $\dfrac{\sqrt[3]{32}}{\sqrt[3]{4}}$

**77.** $\sqrt[4]{\dfrac{3}{48}}$     **78.** $\dfrac{\sqrt[4]{243}}{\sqrt[4]{3}}$     **79.** $\sqrt[5]{\dfrac{96}{3}}$     **80.** $\dfrac{\sqrt[5]{2}}{\sqrt[5]{64}}$

**81.** $\sqrt{\dfrac{x^2}{9}}$     **82.** $\sqrt{\dfrac{9y^4}{z^2}}$     **83.** $\sqrt{\dfrac{16x^4}{25y^{10}}}$     **84.** $\sqrt{\dfrac{49a^8b^{10}}{121c^{14}}}$

**85.** $\sqrt[3]{\dfrac{c^6}{64}}$     **86.** $\sqrt[3]{\dfrac{27x^6}{y^{12}}}$     **87.** $\sqrt[3]{\dfrac{a^8b^{12}}{b^{-8}}}$     **88.** $\sqrt[4]{\dfrac{16x^{16}y^{32}}{81x^{-4}}}$

**89.** $\dfrac{\sqrt{24}}{\sqrt{3}}$     **90.** $\dfrac{\sqrt{64x^5}}{\sqrt{2x^3}}$     **91.** $\dfrac{\sqrt{27x^6}}{\sqrt{3x^2}}$     **92.** $\dfrac{\sqrt{72x^3y^5}}{\sqrt{8x^3y^7}}$

**93.** $\dfrac{\sqrt{48x^6y^9}}{\sqrt{6x^2y^6}}$     **94.** $\dfrac{\sqrt{300a^{10}b^{11}}}{\sqrt{2ab^4}}$     **95.** $\sqrt[3]{\dfrac{5xy}{8x^{13}}}$     **96.** $\sqrt[3]{\dfrac{64a^5b^{12}}{27a^{14}b^5}}$

**97.** $\sqrt[3]{\dfrac{25x^2y^9}{5x^8y^2}}$     **98.** $\sqrt[3]{\dfrac{54xy^4z^{17}}{18x^{13}z^4}}$     **99.** $\sqrt[4]{\dfrac{10x^4y}{81x^{-8}}}$     **100.** $\sqrt[4]{\dfrac{3a^6b^5}{16a^{-6}b^{13}}}$

## Concept/Writing Exercises

**101. a)** How do you obtain the numbers that are perfect squares?

**b)** List the first six perfect squares.

**102. a)** How do you obtain the numbers that are perfect cubes?

**b)** List the first six perfect cube numbers.

**103.** When we gave the product rule, we stated that for nonnegative real numbers $a$ and $b$, $\sqrt[n]{a} \cdot \sqrt[n]{b} = \sqrt[n]{ab}$. Why is it necessary to specify that both $a$ and $b$ are nonnegative real numbers?

**104.** When we gave the quotient rule, we stated that for nonnegative real numbers $a$ and $b$, $\dfrac{\sqrt[n]{a}}{\sqrt[n]{b}} = \sqrt[n]{\dfrac{a}{b}}$, $b \neq 0$. Why is it necessary to state that both $a$ and $b$ are nonnegative real numbers?

**105.** Prove $\sqrt{a \cdot b} = \sqrt{a} \sqrt{b}$ by converting $\sqrt{a \cdot b}$ to exponential form.

**106.** Will the product of two radicals always be a radical? Give an example to support your answer.

**107.** Will the quotient of two radicals always be a radical? Give an example to support your answer.

**108.** Prove $\sqrt[n]{\dfrac{a}{b}} = \dfrac{\sqrt[n]{a}}{\sqrt[n]{b}}$ by converting $\sqrt[n]{\dfrac{a}{b}}$ to exponential form.

**109. a)** Will $\dfrac{\sqrt[n]{x}}{\sqrt[n]{x}}$ always equal 1?

**b)** If your answer to part **a)** was no, under what conditions does $\dfrac{\sqrt[n]{x}}{\sqrt[n]{x}}$ equal 1?

## Cumulative Review Exercises

[2.6] **110.** Solve the formula $F = \dfrac{9}{5}C + 32$ for $C$.

[5.6] **111.** Divide $\dfrac{15x^{12} - 5x^9 + 20x^6}{5x^6}$.

[6.5] **112.** Factor $(x - 3)^3 + 8$.

[6.6] **113.** Solve $(2x - 3)(x - 2) = 4x - 6$.

[10.2] **114.** Solve $\left| \dfrac{2x - 4}{5} \right| = 12$.

# 11.4 Adding, Subtracting, and Multiplying Radicals

**1** Add and subtract radicals.

**2** Multiply radicals.

## 1 Add and Subtract Radicals

**Like Radicals and Unlike Radicals**

- **Like radicals** are radicals having the same radicand and index.

- **Unlike radicals** are radicals differing in either the radicand or the index.

| Examples of Like Radicals | Examples of Unlike Radicals | |
|---|---|---|
| $\sqrt{5}, 3\sqrt{5}$ | $\sqrt{5}, \sqrt[3]{5}$ | Indices differ. |
| $6\sqrt{7}, -2\sqrt{7}$ | $\sqrt{6}, \sqrt{7}$ | Radicands differ. |
| $\sqrt{x}, 5\sqrt{x}$ | $\sqrt{x}, \sqrt{2x}$ | Radicands differ. |
| $\sqrt[3]{2x}, -4\sqrt[3]{2x}$ | $\sqrt{x}, \sqrt[3]{x}$ | Indices differ. |
| $\sqrt[4]{x^2y^5}, -\sqrt[4]{x^2y^5}$ | $\sqrt[3]{xy}, \sqrt[3]{x^2y}$ | Radicands differ. |

The distributive property is used to add or subtract like radicals in the same way that like terms are added or subtracted.

**Examples of Adding and Subtracting Like Radicals**

$$3\sqrt{6} + 2\sqrt{6} = (3 + 2)\sqrt{6} = 5\sqrt{6}$$
$$5\sqrt{x} - 7\sqrt{x} = (5 - 7)\sqrt{x} = -2\sqrt{x}$$
$$\sqrt[3]{4x^2} + 5\sqrt[3]{4x^2} = (1 + 5)\sqrt[3]{4x^2} = 6\sqrt[3]{4x^2}$$
$$4\sqrt{5x} - y\sqrt{5x} = (4 - y)\sqrt{5x}$$

### Understanding Algebra

We add like radicals using the distributive property in the same way we add like terms. For example, we add like terms $3x$ and $4x$ as follows:

$$3x + 4x = (3 + 4)x = 7x$$

We add like radicals $3\sqrt{2}$ and $4\sqrt{2}$ as follows:

$$3\sqrt{2} + 4\sqrt{2} =$$
$$(3 + 4)\sqrt{2} = 7\sqrt{2}$$

**EXAMPLE 1** Simplify.

**a)** $6 + 4\sqrt{2} - \sqrt{2} + 7$          **b)** $2\sqrt[3]{x} + 8x + 4\sqrt[3]{x} - 3$

*Solution*

**a)** $6 + 4\sqrt{2} - \sqrt{2} + 7 = 6 + 7 + 4\sqrt{2} - \sqrt{2}$     Place like terms together.

$$= 13 + (4 - 1)\sqrt{2}$$

$$= 13 + 3\sqrt{2} \quad (\text{or } 3\sqrt{2} + 13)$$

**b)** $2\sqrt[3]{x} + 8x + 4\sqrt[3]{x} - 3 = 6\sqrt[3]{x} + 8x - 3$

Now Try Exercise 15

It is sometimes possible to convert unlike radicals into like radicals by simplifying one or more of the radicals, as was discussed in Section 11.3.

**EXAMPLE 2** Simplify $\sqrt{3} + \sqrt{27}$.

*Solution*   Since $\sqrt{3}$ and $\sqrt{27}$ are unlike radicals, they cannot be added in their present form. We can simplify $\sqrt{27}$ to obtain like radicals.

$$\sqrt{3} + \sqrt{27} = \sqrt{3} + \sqrt{9}\sqrt{3}$$
$$= \sqrt{3} + 3\sqrt{3} = 4\sqrt{3}$$

Now Try Exercise 19

### To Add or Subtract Radicals

1. Simplify each radical expression.
2. Combine like radicals (if there are any).

**EXAMPLE 3** Simplify.

**a)** $5\sqrt{24} + \sqrt{54}$    **b)** $2\sqrt{45} - \sqrt{80} + \sqrt{20}$    **c)** $\sqrt[3]{27} + \sqrt[3]{81} - 7\sqrt[3]{3}$

*Solution*

**a)** $5\sqrt{24} + \sqrt{54} = 5 \cdot \sqrt{4} \cdot \sqrt{6} + \sqrt{9} \cdot \sqrt{6}$

$$= 5 \cdot 2\sqrt{6} + 3\sqrt{6}$$

$$= 10\sqrt{6} + 3\sqrt{6} = 13\sqrt{6}$$

**b)** $2\sqrt{45} - \sqrt{80} + \sqrt{20} = 2 \cdot \sqrt{9} \cdot \sqrt{5} - \sqrt{16} \cdot \sqrt{5} + \sqrt{4} \cdot \sqrt{5}$

$$= 2 \cdot 3\sqrt{5} - 4\sqrt{5} + 2\sqrt{5}$$

$$= 6\sqrt{5} - 4\sqrt{5} + 2\sqrt{5} = 4\sqrt{5}$$

**c)** $\sqrt[3]{27} + \sqrt[3]{81} - 7\sqrt[3]{3} = 3 + \sqrt[3]{27} \cdot \sqrt[3]{3} - 7\sqrt[3]{3}$

$$= 3 + 3\sqrt[3]{3} - 7\sqrt[3]{3} = 3 - 4\sqrt[3]{3}$$

Now Try Exercise 23

**EXAMPLE 4** Simplify.    **a)** $\sqrt{x^2} - \sqrt{x^2 y} + x\sqrt{y}$    **b)** $\sqrt[3]{x^{13}y^2} - \sqrt[3]{x^4 y^8}$

*Solution*

**a)** $\sqrt{x^2} - \sqrt{x^2 y} + x\sqrt{y} = x - \sqrt{x^2} \cdot \sqrt{y} + x\sqrt{y}$

$$= x - x\sqrt{y} + x\sqrt{y}$$

$$= x$$

**b)** $\sqrt[3]{x^{13}y^2} - \sqrt[3]{x^4 y^8} = \sqrt[3]{x^{12}} \cdot \sqrt[3]{xy^2} - \sqrt[3]{x^3 y^6} \cdot \sqrt[3]{xy^2}$

$$= x^4 \sqrt[3]{xy^2} - xy^2 \sqrt[3]{xy^2}$$

Now factor out the common factor, $\sqrt[3]{xy^2}$.

$$= (x^4 - xy^2)\sqrt[3]{xy^2}$$

Now Try Exercise 35

> **Helpful Hint**
>
> The product rule and quotient rule for radicals presented in Section 11.3 are
>
> $$\sqrt[n]{a} \cdot \sqrt[n]{b} = \sqrt[n]{ab} \qquad \frac{\sqrt[n]{a}}{\sqrt[n]{b}} = \sqrt[n]{\frac{a}{b}}$$
>
> Students often incorrectly assume similar properties exist for addition and subtraction. They do not. To illustrate this, let $n$ be a square root (index 2), $a = 9$, and $b = 16$.
>
> $$\sqrt[n]{a} + \sqrt[n]{b} \neq \sqrt[n]{a+b}$$
> $$\sqrt{9} + \sqrt{16} \neq \sqrt{9 + 16}$$
> $$3 + 4 \neq \sqrt{25}$$
> $$7 \neq 5$$

## 2  Multiply Radicals

To multiply radicals, we use the product rule given earlier. After multiplying, we can often simplify the new radical (see Examples 5 and 6).

**EXAMPLE 5**  Multiply and simplify.

**a)** $\sqrt{6x^3}\sqrt{8x^6}$     **b)** $\sqrt[3]{2x}\sqrt[3]{4x^2}$     **c)** $\sqrt[4]{4x^{11}y}\sqrt[4]{16x^6y^{22}}$

*Solution*

**a)** $\sqrt{6x^3}\sqrt{8x^6} = \sqrt{6x^3 \cdot 8x^6}$     Product rule for radicals

$= \sqrt{48x^9}$

$= \sqrt{16x^8}\sqrt{3x}$     $16x^8$ is a perfect square.

$= 4x^4\sqrt{3x}$

**b)** $\sqrt[3]{2x}\sqrt[3]{4x^2} = \sqrt[3]{2x \cdot 4x^2}$     Product rule for radicals

$= \sqrt[3]{8x^3}$     $8x^3$ is a perfect cube.

$= 2x$

**c)** $\sqrt[4]{4x^{11}y}\sqrt[4]{16x^6y^{22}} = \sqrt[4]{4x^{11}y \cdot 16x^6y^{22}}$     Product rule for radicals

$= \sqrt[4]{64x^{17}y^{23}}$

$= \sqrt[4]{16x^{16}y^{20}}\sqrt[4]{4xy^3}$     The largest perfect fourth root factors are $16$, $x^{16}$, and $y^{20}$.

$= 2x^4y^5\sqrt[4]{4xy^3}$

Now Try Exercise 47

*As stated earlier, when a radical is simplified, the exponents on the variables in the radicand are less than the index.*

**EXAMPLE 6**  Multiply and simplify $\sqrt{2x}(\sqrt{8x} - \sqrt{50})$.

*Solution*   Begin by using the distributive property.

$$\sqrt{2x}(\sqrt{8x} - \sqrt{50}) = (\sqrt{2x})(\sqrt{8x}) + (\sqrt{2x})(-\sqrt{50})$$
$$= \sqrt{16x^2} - \sqrt{100x}$$
$$= 4x - \sqrt{100}\sqrt{x}$$
$$= 4x - 10\sqrt{x}$$

Now Try Exercise 53

Note in Example 6 that the same result could be obtained by first simplifying $\sqrt{8x}$ and $\sqrt{50}$ and then multiplying. You may wish to try this now.

We next will multiply radical expressions involving sums or differences of radicals such as $(\sqrt{a} + \sqrt{b})(\sqrt{c} + \sqrt{d})$. We will use the FOIL method just as we did when multiplying two binomials.

**EXAMPLE 7** Multiply $(\sqrt{x} - \sqrt{y})(\sqrt{x} - y)$.

*Solution*   We will multiply using the FOIL method.

$$\begin{array}{cccc} F & O & I & L \\ \downarrow & \downarrow & \downarrow & \downarrow \\ (\sqrt{x})(\sqrt{x}) & + (\sqrt{x})(-y) & + (-\sqrt{y})(\sqrt{x}) & + (-\sqrt{y})(-y) \end{array}$$

$$= \sqrt{x^2} \quad - \quad y\sqrt{x} \quad - \quad \sqrt{xy} \quad + \quad y\sqrt{y}$$

$$= x - y\sqrt{x} - \sqrt{xy} + y\sqrt{y}$$

*Now Try Exercise 63*

**EXAMPLE 8** Simplify.  **a)** $(2\sqrt{6} - \sqrt{3})^2$  **b)** $(\sqrt[3]{x} - \sqrt[3]{2y^2})(\sqrt[3]{x^2} - \sqrt[3]{8y})$

*Solution*

**a)** $(2\sqrt{6} - \sqrt{3})^2 = (2\sqrt{6} - \sqrt{3})(2\sqrt{6} - \sqrt{3})$

Now multiply using the FOIL method.

$$\begin{array}{cccc} F & O & I & L \end{array}$$
$$(2\sqrt{6})(2\sqrt{6}) + (2\sqrt{6})(-\sqrt{3}) + (-\sqrt{3})(2\sqrt{6}) + (-\sqrt{3})(-\sqrt{3})$$
$$= 4(6) - 2\sqrt{18} - 2\sqrt{18} + 3$$
$$= 24 - 2\sqrt{18} - 2\sqrt{18} + 3$$
$$= 27 - 4\sqrt{18}$$
$$= 27 - 4\sqrt{9}\sqrt{2}$$
$$= 27 - 12\sqrt{2}$$

**b)** Multiply using the FOIL method.

$$\begin{array}{cccc} F & O & I & L \end{array}$$
$$(\sqrt[3]{x} - \sqrt[3]{2y^2})(\sqrt[3]{x^2} - \sqrt[3]{8y}) = (\sqrt[3]{x})(\sqrt[3]{x^2}) + (\sqrt[3]{x})(-\sqrt[3]{8y}) + (-\sqrt[3]{2y^2})(\sqrt[3]{x^2}) + (-\sqrt[3]{2y^2})(-\sqrt[3]{8y})$$
$$= \sqrt[3]{x^3} - \sqrt[3]{8xy} - \sqrt[3]{2x^2y^2} + \sqrt[3]{16y^3}$$
$$= \sqrt[3]{x^3} - \sqrt[3]{8}\sqrt[3]{xy} - \sqrt[3]{2x^2y^2} + \sqrt[3]{8y^3}\sqrt[3]{2}$$
$$= x - 2\sqrt[3]{xy} - \sqrt[3]{2x^2y^2} + 2y\sqrt[3]{2}$$

*Now Try Exercise 99*

**EXAMPLE 9** Multiply $(3 + \sqrt{6})(3 - \sqrt{6})$.

*Solution*   We can multiply using the FOIL method.

$$\begin{array}{cccc} F & O & I & L \end{array}$$
$$(3 + \sqrt{6})(3 - \sqrt{6}) = 3(3) + 3(-\sqrt{6}) + (\sqrt{6})(3) + (\sqrt{6})(-\sqrt{6}).$$
$$= 9 \quad - \quad 3\sqrt{6} \quad + \quad 3\sqrt{6} \quad - \quad \sqrt{36}$$
$$= 9 - \sqrt{36}$$
$$= 9 - 6 = 3$$

*Now Try Exercise 59*

Note that in Example 9, we multiplied *the sum and difference of the same two radical expressions*. Recall from Section 6.5 that $(a + b)(a - b) = a^2 - b^2$. If we let $a = 3$ and $b = \sqrt{6}$, then we can multiply as follows.

$$(a + b)(a - b) = a^2 - b^2$$
$$(3 + \sqrt{6})(3 - \sqrt{6}) = 3^2 - (\sqrt{6})^2$$
$$= 9 - 6$$
$$= 3$$

When multiplying the sum and difference of the same two radical expressions, you may obtain the answer using the difference of the squares of the two radical expressions.

**EXAMPLE 10** If $f(x) = \sqrt[3]{x^2}$ and $g(x) = \sqrt[3]{x^4} + \sqrt[3]{x^2}$, find **a)** $(f \cdot g)(x)$ and **b)** $(f \cdot g)(6)$.

*Solution*

**a)** From Section 8.5, we know that $(f \cdot g)(x) = f(x) \cdot g(x)$.

$$
\begin{aligned}
(f \cdot g)(x) &= f(x) \cdot g(x) \\
&= \sqrt[3]{x^2}\left(\sqrt[3]{x^4} + \sqrt[3]{x^2}\right) && \text{Substitute given values.} \\
&= \sqrt[3]{x^2}\,\sqrt[3]{x^4} + \sqrt[3]{x^2}\,\sqrt[3]{x^2} && \text{Distributive property} \\
&= \sqrt[3]{x^6} + \sqrt[3]{x^4} && \text{Product rule for radicals} \\
&= x^2 + x\sqrt[3]{x} && \text{Simplify radicals.}
\end{aligned}
$$

**b)** To compute $(f \cdot g)(6)$, substitute 6 for $x$ in the answer obtained in part **a)**.

$$
\begin{aligned}
(f \cdot g)(x) &= x^2 + x\sqrt[3]{x} \\
(f \cdot g)(6) &= 6^2 + 6\sqrt[3]{6} && \text{Substitute 6 for } x. \\
&= 36 + 6\sqrt[3]{6}
\end{aligned}
$$

Now Try Exercise 77

As stated earlier in this chapter, unless we are instructed otherwise, we assume that variable expressions in radicands represent nonnegative real numbers. Example 11 demonstrates how we must use absolute value in the cases when the radicand can represent any real number.

**EXAMPLE 11** Simplify $f(x)$ if

**a)** $f(x) = \sqrt{x^2 + 6x + 9}$ assuming $x \geq -3$.

**b)** $f(x) = \sqrt{x^2 + 6x + 9}$ assuming $x$ can be any real number.

*Solution*

**a)**
$$
\begin{aligned}
f(x) &= \sqrt{x^2 + 6x + 9} \\
&= \sqrt{(x + 3)^2} && x^2 + 6x + 9 \text{ was factored as } (x + 3)^2. \\
&= x + 3 && \text{Since } x \geq -3, x + 3 \geq 0 \text{ and no absolute} \\
& && \text{value bars are needed.}
\end{aligned}
$$

**b)**
$$
\begin{aligned}
f(x) &= \sqrt{x^2 + 6x + 9} \\
&= \sqrt{(x + 3)^2} && x^2 + 6x + 9 \text{ was factored as } (x + 3)^2. \\
&= |x + 3| && \text{Since } x \text{ can be any real number, } x + 3 \text{ may be} \\
& && \text{negative and absolute value bars are required.}
\end{aligned}
$$

Now Try Exercise 105

# EXERCISE SET 11.4    MathXL®    MyMathLab®
MathXL®    MyMathLab

## Warm-Up Exercises

*Fill in the blanks with the appropriate word, phrase, or symbol(s) from the following list.*

commutative        distributive        less        like        FOIL        first        greater        unlike

1. Radicals having the same radicand and index are _____ radicals.

2. Radicals differing in either the radicand or the index are _____ radicals.

3. The _____ property is used to add or subtract like radicals in the same way that like terms are added or subtracted.

4. When a radical is simplified, the exponents on the variables in the radicand are _____ than the index.

5. When multiplying radical expressions involving sums or differences of radicals, such as $(\sqrt{a} + \sqrt{b})(\sqrt{c} + \sqrt{d})$, use the _____ method.

6. FOIL is an acronym for _____, outer, inner, last.

## Practice the Skills

*In this exercise set, assume that all variables represent positive real numbers.*

*Simplify.*

7. $\sqrt{2} - \sqrt{2}$

8. $4\sqrt{3} - \sqrt{3}$

9. $6\sqrt{5} - 2\sqrt{5}$

10. $3\sqrt{2} + 7\sqrt{2} - 11$

11. $2\sqrt{3} - 2\sqrt{3} - 4\sqrt{3} + 5$

12. $6\sqrt[3]{7} - 8\sqrt[3]{7}$

13. $2\sqrt[4]{y} - 9\sqrt[4]{y}$

14. $3\sqrt[5]{a} + 7 + 5\sqrt[5]{a} - 2$

15. $3\sqrt{5} - \sqrt[3]{x} + 6\sqrt{5} + 3\sqrt[3]{x}$

16. $9 + 4\sqrt[4]{a} - 7\sqrt[4]{a} + 5$

17. $5\sqrt{x} - 8\sqrt{y} + 3\sqrt{x} + 2\sqrt{y} - \sqrt{x}$

18. $8\sqrt{a} + 4\sqrt[3]{b} + 7\sqrt{a} - 12\sqrt[3]{b}$

*Simplify.*

19. $\sqrt{3} + \sqrt{12}$

20. $\sqrt{8} + \sqrt{18}$

21. $-6\sqrt{75} + 5\sqrt{125}$

22. $3\sqrt{250} + 4\sqrt{160}$

23. $-4\sqrt{90} + 3\sqrt{40} + 2\sqrt{10}$

24. $3\sqrt{40x^2y} + 2x\sqrt{490y}$

25. $\sqrt{500xy^2} + y\sqrt{320x}$

26. $5\sqrt{8} + 2\sqrt{50} - 3\sqrt{72}$

27. $2\sqrt{5x} - 3\sqrt{20x} - 4\sqrt{45x}$

28. $3\sqrt{27c^2} - 2\sqrt{108c^2} - \sqrt{48c^2}$

29. $3\sqrt{50a^2} - 3\sqrt{72a^2} - 8a\sqrt{18}$

30. $4\sqrt[3]{5} - 5\sqrt[3]{40}$

31. $\sqrt[3]{108} + \sqrt[3]{32}$

32. $3\sqrt[3]{16} + \sqrt[3]{54}$

33. $\sqrt[3]{27} - 5\sqrt[3]{8}$

34. $3\sqrt{45x^3} + \sqrt{5x}$

35. $2\sqrt[3]{a^4b^2} + 4a\sqrt[3]{ab^2}$

36. $5y\sqrt[4]{48x^5} - x\sqrt[4]{3x^5y^4}$

37. $\sqrt{4r^7s^5} + 3r^2\sqrt{r^3s^5} - 2rs\sqrt{r^5s^3}$

38. $x\sqrt[3]{27x^5y^2} - x^2\sqrt[3]{x^2y^2} + 4\sqrt[3]{x^8y^2}$

39. $\sqrt[3]{128x^8y^{10}} - 2x^2y\sqrt[3]{16x^2y^7}$

40. $5\sqrt[3]{320x^5y^8} + 3x\sqrt[3]{135x^2y^8}$

*Simplify.*

41. $\sqrt{2}\sqrt{8}$

42. $\sqrt{3}\,\sqrt{27}$

43. $\sqrt[3]{4}\,\sqrt[3]{14}$

44. $\sqrt[3]{3}\,\sqrt[3]{54}$

45. $\sqrt{9m^3n^7}\,\sqrt{3mn^4}$

46. $\sqrt[3]{5ab^2}\,\sqrt[3]{25a^4b^{12}}$

47. $\sqrt[3]{9x^7y^{10}}\,\sqrt[3]{6x^4y^3}$

48. $\sqrt[4]{3x^9y^{12}}\,\sqrt[4]{54x^4y^7}$

49. $\sqrt[5]{x^{24}y^{30}z^9}\,\sqrt[5]{x^{13}y^8z^7}$

50. $\sqrt[4]{8x^4yz^3}\,\sqrt[4]{2x^2y^3z^7}$

51. $\left(\sqrt[3]{2x^3y^4}\right)^2$

52. $\sqrt{2}\left(\sqrt{6} + \sqrt{18}\right)$

53. $\sqrt{5}\left(\sqrt{5} - \sqrt{3}\right)$

54. $\sqrt{3}\left(\sqrt{12} + \sqrt{8}\right)$

55. $\sqrt[3]{y}\left(2\sqrt[3]{y} - \sqrt[3]{y^8}\right)$

56. $\sqrt{3y}\left(\sqrt{27y^2} - \sqrt{y}\right)$

57. $2\sqrt[3]{x^4y^5}\left(\sqrt[3]{8x^{12}y^4} + \sqrt[3]{16xy^9}\right)$

58. $\sqrt[5]{16x^7y^6}\left(\sqrt[5]{2x^6y^9} - \sqrt[5]{10x^3y^7}\right)$

59. $(8 + \sqrt{5})(8 - \sqrt{5})$

60. $(9 - \sqrt{5})(9 + \sqrt{5})$

61. $(\sqrt{6} + x)(\sqrt{6} - x)$

62. $(\sqrt{x} + y)(\sqrt{x} - y)$

63. $(\sqrt{7} - \sqrt{z})(\sqrt{7} + \sqrt{z})$

64. $(3\sqrt{a} - 5\sqrt{b})(3\sqrt{a} + 5\sqrt{b})$

65. $(\sqrt{3} + 4)(\sqrt{3} + 5)$

66. $(1 + \sqrt{5})(8 + \sqrt{5})$

67. $(3 - \sqrt{2})(4 - \sqrt{8})$

68. $(5\sqrt{6} + 3)(4\sqrt{6} - 1)$

69. $(4\sqrt{3} + \sqrt{2})(\sqrt{3} - \sqrt{2})$

70. $(\sqrt{3} + 7)^2$

71. $(2\sqrt{5} - 3)^2$

72. $(\sqrt{y} + \sqrt{6z})(\sqrt{2z} - \sqrt{8y})$

73. $(2\sqrt{3x} - \sqrt{y})(3\sqrt{3x} + \sqrt{y})$

74. $(\sqrt[3]{9} + \sqrt[3]{2})(\sqrt[3]{3} + \sqrt[3]{4})$

75. $(\sqrt[3]{4} - \sqrt[3]{6})(\sqrt[3]{2} - \sqrt[3]{36})$

76. $(\sqrt[3]{4x} - \sqrt[3]{2y})(\sqrt[3]{4x} + \sqrt[3]{10})$

*In Exercises 77–82, f(x) and g(x) are given. Find $(f \cdot g)(x)$.*

77. $f(x) = \sqrt{x}, g(x) = \sqrt{x} - \sqrt{3}$

78. $f(x) = \sqrt{6x}, g(x) = \sqrt{6x} - \sqrt{10x}$

79. $f(x) = \sqrt[3]{x}, g(x) = \sqrt[3]{x^5} + \sqrt[3]{x^4}$

80. $f(x) = \sqrt[3]{2x^2}, g(x) = \sqrt[3]{4x} + \sqrt[3]{32x^2}$

81. $f(x) = \sqrt[4]{3x^2}, g(x) = \sqrt[4]{9x^4} - \sqrt[4]{x^7}$

82. $f(x) = \sqrt[4]{2x^3}, g(x) = \sqrt[4]{8x^5} - \sqrt[4]{5x^6}$

*Simplify. These exercises are a combination of the types of exercises presented earlier in this exercise set.*

**83.** $\sqrt{18}$

**84.** $\sqrt{300}$

**85.** $\sqrt{125} - \sqrt{20}$

**86.** $4\sqrt{7} + 2\sqrt{63} - 2\sqrt{28}$

**87.** $(3\sqrt{2} - 4)(\sqrt{2} + 5)$

**88.** $(\sqrt{5} + \sqrt{2})(\sqrt{2} + \sqrt{20})$

**89.** $\sqrt{6}(5 - \sqrt{2})$

**90.** $3\sqrt[3]{81} + 4\sqrt[3]{24}$

**91.** $\sqrt{150}\sqrt{3}$

**92.** $\sqrt[4]{2}\,\sqrt[4]{40}$

**93.** $\sqrt[3]{80x^{11}}$

**94.** $\sqrt[3]{x^9 y^{11} z}$

**95.** $\sqrt[6]{128ab^{17}c^9}$

**96.** $\sqrt[5]{14x^4y^2}\,\sqrt[5]{3x^4y^3}$

**97.** $2b\sqrt[4]{a^4b} + ab\sqrt[4]{16b}$

**98.** $2\sqrt[3]{24a^3y^4} + 4a\sqrt[3]{81y^4}$

**99.** $\left(\sqrt[3]{x^2} - \sqrt[3]{y}\right)\left(\sqrt[3]{x} - 2\sqrt[3]{y^2}\right)$

**100.** $\left(\sqrt[3]{a} + 5\right)\left(\sqrt[3]{a^2} - 6\right)$

**101.** $\sqrt[3]{3ab^2}\left(\sqrt[3]{4a^4b^3} - \sqrt[3]{8a^5b^4}\right)$

**102.** $\sqrt[4]{4st^2}\left(\sqrt[4]{2s^5t^6} + \sqrt[4]{5s^9t^2}\right)$

*Simplify the following. In Exercises 105 and 106, assume the variable can be any real number. See Example 11.*

**103.** $f(x) = \sqrt{2x - 5}\sqrt{2x - 5}, x \geq \dfrac{5}{2}$

**104.** $g(a) = \sqrt{3a + 7}\,\sqrt{3a + 7}, a \geq -\dfrac{7}{3}$

**105.** $h(r) = \sqrt{4r^2 - 32r + 64}$

**106.** $f(b) = \sqrt{20b^2 + 60b + 45}$

## Problem Solving

*Find the perimeter and area of the following figures. Write the perimeter and area in radical form with the radicals simplified. (Hint: See page 134 for the needed formulas.)*

**107.**

**108.**

**109.**

**110.**

**111. Skid Marks** Law enforcement officials sometimes use the formula $s = \sqrt{30FB}$ to determine a car's speed, $s$, in miles per hour, from a car's skid marks. The $F$ in the formula represents the "road factor," which is determined by the road's surface, and the $B$ represents the braking distance, in feet. Officer Jenkins is investigating an accident. Find the car's speed if the skid marks are 80 feet long and **a)** the road was dry asphalt, whose road factor is 0.85, **b)** the road was wet gravel, whose road factor is 0.52.

**112. Water through a Fire Hose** The rate at which water flows through a particular fire hose, $R$, in gallons per minute, can be approximated by the formula $R = 28d^2\sqrt{P}$, where $d$ is the diameter of the nozzle, in inches, and $P$ is the nozzle pressure, in pounds per square inch. If a nozzle has a diameter of 2.5 inches and the nozzle pressure is 80 pounds per square inch, find the flow rate.

**113. Height of Girls** The function $f(t) = 3\sqrt{t} + 19$ can be used to approximate the median height, $f(t)$, in inches, for U.S. girls of age $t$, in months, where $1 \leq t \leq 60$. Estimate the median height of girls at age **a)** 36 months and **b)** 40 months.

**114. Standard Deviation** In statistics, the standard deviation of the population, $\sigma$, read "sigma," is a measure of the spread of a set of data about the mean of the data. One formula used to determine sigma is $\sigma = \sqrt{npq}$, where $n$ represents the sample size, $p$ represents the percent chance that something specific happens, and $q$ represents the percent chance that the specific thing does not happen. In a sample of 600 people who purchase airline tickets, the percent that showed up for their flight, $p$, was 0.93, and the percent that did not show up for their flight, $q$, was 0.07. Use this information to find $\sigma$.

**115.** The graph of $f(x) = \sqrt{x}$ is shown.

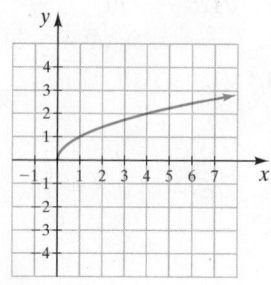

**a)** If $g(x) = 2$, sketch the graph of $(f + g)(x)$.

**b)** What effect does adding 2 have to the graph of $f(x)$?

**117.** You are given that $f(x) = \sqrt{x}$ and $g(x) = \sqrt{x} - 2$.

**a)** Sketch the graph of $(f - g)(x)$.

**b)** What is the domain of $(f - g)(x)$?

**119.** Graph the function $f(x) = \sqrt{x^2}$.

**116.** The graph of $f(x) = -\sqrt{x}$ is shown.

**a)** If $g(x) = 3$, sketch the graph of $(f + g)(x)$.

**b)** What effect does adding 3 have to the graph of $f(x)$?

**118.** You are given that $f(x) = \sqrt{x}$ and $g(x) = -\sqrt{x} - 3$.

**a)** Sketch the graph of $(f + g)(x)$.

**b)** What is the domain of $(f + g)(x)$?

**120.** Graph the function $f(x) = \sqrt{x^2} - 4$.

## Concept/Writing Exercises

**121.** Does $\sqrt{a} + \sqrt{b} = \sqrt{a + b}$? Explain your answer and give an example supporting your answer.

**122.** Since $64 + 36 = 100$, does $\sqrt{64} + \sqrt{36} = \sqrt{100}$? Explain your answer.

**123.** Will the sum of two radicals always be a radical? Give an example to support your answer.

**124.** Will the difference of two radicals always be a radical? Give an example to support your answer.

## Cumulative Review Exercises

[1.4] **125.** What is a rational number?

**126.** What is a real number?

**127.** What is an irrational number?

[1.5] **128.** What is the definition of $|a|$?

[2.6] **129.** Solve the formula $E = \frac{1}{2}mv^2$ for $m$.

[10.1] **130.** Solve the inequality $-4 < 2x - 3 \le 7$ and indicate the solution **a)** on the number line; **b)** in interval notation; **c)** in set builder notation.

## Mid-Chapter Test: 11.1–11.4

*To find out how well you understand the chapter material to this point, take this brief test. The answers, and the section where the material was initially discussed, are given in the back of the book. Review any questions that you answered incorrectly.*

*Find the indicated root.*

**1.** $\sqrt{121}$

**2.** $\sqrt[3]{-\dfrac{27}{64}}$

*Use absolute value to evaluate.*

**3.** $\sqrt{(-16.3)^2}$

**4.** $\sqrt{(3a^2 - 4b^3)^2}$

**5.** Find $g(16)$ if $g(x) = \dfrac{x}{8} + \sqrt{4x} - 7$.

**6.** Write $\sqrt[5]{7a^4b^3}$ in exponential form.

**7.** Evaluate $-49^{1/2} + 81^{3/4}$.

*Simplify each expression.*

**8.** $\left(\sqrt[4]{a^2b^3c}\right)^{20}$

**9.** $7x^{-5/2} \cdot 2x^{3/2}$

**10.** Multiply $8x^{-2}(x^3 + 2x^{-1/2})$.

*Simplify each radical.*

**11.** $\sqrt{32x^4y^9}$

**12.** $\sqrt[6]{64a^{13}b^{23}c^{15}}$

**13.** $\dfrac{\sqrt[3]{3}}{\sqrt[3]{81}}$

**14.** $\dfrac{\sqrt{20x^5y^{12}}}{\sqrt{180x^{15}y^7}}$

*Simplify. Assume that all variables represent positive real numbers.*

**15.** $2\sqrt{x} - 3\sqrt{y} + 9\sqrt{x} + 15\sqrt{y}$

**16.** $2\sqrt{90x^2y} + 3x\sqrt{490y}$

**17.** $(x + \sqrt{5})(2x - 3\sqrt{5})$

**18.** $2\sqrt{3a}\left(\sqrt{27a^2} - 5\sqrt{4a}\right)$

**19.** $3b\sqrt[4]{a^5b} + 2ab\sqrt[4]{16ab}$

**20.** When simplifying the following square roots, in which parts will the answer contain an absolute value? Explain your answer and simplify parts **a)** and **b)**.

**a)** $\sqrt{(x - 3)^2}$

**b)** $\sqrt{64x^2}, x \ge 0$

# 11.5 Dividing Radicals

1 Rationalize denominators.

2 Rationalize a denominator using the conjugate.

3 Understand when a radical is simplified.

4 Use rationalizing the denominator in an addition problem.

5 Divide radical expressions with different indices.

## 1 Rationalize Denominators

When the denominator of a fraction contains radicals, it is customary to rewrite the fraction as an equivalent fraction in which the denominator does not contain radicals. The process we use to rewrite the fraction is known as **rationalizing the denominator**. In this section we will use the quotient rule for radicals, introduced in Section 11.3, to rationalize denominators.

> **To Rationalize a Denominator**
>
> Multiply both the numerator and the denominator of the fraction by a radical that will produce a radicand in the denominator that is a perfect power for the index.

When both the numerator and denominator are multiplied by the same radical expression, you are in effect multiplying the fraction by 1, which does not change its value. Also recall that we will assume that all variables in the radicands will represent nonnegative real numbers.

**EXAMPLE 1** Simplify. **a)** $\dfrac{1}{\sqrt{5}}$ **b)** $\dfrac{x}{4\sqrt{3}}$ **c)** $\dfrac{11}{\sqrt{2x}}$ **d)** $\dfrac{\sqrt[3]{16a^4}}{\sqrt[3]{b}}$

*Solution*   To simplify each expression, we must rationalize the denominator. We do so by multiplying both the numerator and denominator by a radical that will produce a denominator that is a perfect power for the given index.

**a)** $\dfrac{1}{\sqrt{5}} = \dfrac{1}{\sqrt{5}} \cdot \dfrac{\sqrt{5}}{\sqrt{5}} = \dfrac{\sqrt{5}}{\sqrt{25}} = \dfrac{\sqrt{5}}{5}$   **b)** $\dfrac{x}{4\sqrt{3}} = \dfrac{x}{4\sqrt{3}} \cdot \dfrac{\sqrt{3}}{\sqrt{3}} = \dfrac{x\sqrt{3}}{4 \cdot 3} = \dfrac{x\sqrt{3}}{12}$

**c)** There are two factors in the radicand, 2 and $x$. Since $2^2$ or 4 is a perfect square and $x^2$ is a perfect square, we multiply both numerator and denominator by $\sqrt{2x}$.

$$\frac{11}{\sqrt{2x}} = \frac{11}{\sqrt{2x}} \cdot \frac{\sqrt{2x}}{\sqrt{2x}}$$

$$= \frac{11\sqrt{2x}}{\sqrt{4x^2}}$$

$$= \frac{11\sqrt{2x}}{2x}$$

**d)** There are no common factors in the numerator and denominator. Before we rationalize the denominator, we will simplify the numerator.

$$\frac{\sqrt[3]{16a^4}}{\sqrt[3]{b}} = \frac{\sqrt[3]{8a^3}\,\sqrt[3]{2a}}{\sqrt[3]{b}} \qquad \text{Product rule for radicals}$$

$$= \frac{2a\sqrt[3]{2a}}{\sqrt[3]{b}} \qquad \text{Simplify the numerator.}$$

Now we rationalize the denominator. Since the denominator is a cube root, we need to produce a radicand that is a perfect cube. Since the denominator contains $b$ and we want $b^3$, we need two more factors of $b$, or $b^2$. We therefore multiply both numerator and denominator by $\sqrt[3]{b^2}$.

$$= \frac{2a\sqrt[3]{2a}}{\sqrt[3]{b}} \cdot \frac{\sqrt[3]{b^2}}{\sqrt[3]{b^2}}$$

$$= \frac{2a\sqrt[3]{2ab^2}}{\sqrt[3]{b^3}}$$

$$= \frac{2a\sqrt[3]{2ab^2}}{b}$$

Now Try Exercise 15

## Understanding Algebra

Historically, denominators were rationalized so that obtaining a decimal approximation would be easier. Example 1 **a)** showed that $\dfrac{1}{\sqrt{5}} = \dfrac{\sqrt{5}}{5}$.

Approximating $\dfrac{1}{\sqrt{5}}$ using long division requires you to divide 1 by an approximation of $\sqrt{5}$ such as 2.236.

However, approximating $\dfrac{\sqrt{5}}{5}$ requires you to divide 2.236 by 5—a much easier task!

**EXAMPLE 2** Simplify.   **a)** $\sqrt{\dfrac{5}{7}}$   **b)** $\sqrt[3]{\dfrac{x}{2y^2}}$   **c)** $\sqrt[4]{\dfrac{32x^9y^6}{3z^2}}$

*Solution*   In each part, we will use the quotient rule to rewrite the radical as a quotient of two radicals.

**a)** $\sqrt{\dfrac{5}{7}} = \dfrac{\sqrt{5}}{\sqrt{7}} \cdot \dfrac{\sqrt{7}}{\sqrt{7}} = \dfrac{\sqrt{35}}{\sqrt{49}} = \dfrac{\sqrt{35}}{7}$

**b)** $\sqrt[3]{\dfrac{x}{2y^2}} = \dfrac{\sqrt[3]{x}}{\sqrt[3]{2y^2}}$

The denominator is $\sqrt[3]{2y^2}$. We will multiply both the numerator and denominator by the cube root of an expression that will produce the denominator $\sqrt[3]{2^3y^3}$. Since $2 \cdot 2^2 = 2^3$ and $y^2 \cdot y = y^3$, we multiply both numerator and denominator by $\sqrt[3]{2^2y}$.

$$\dfrac{\sqrt[3]{x}}{\sqrt[3]{2y^2}} = \dfrac{\sqrt[3]{x}}{\sqrt[3]{2y^2}} \cdot \dfrac{\sqrt[3]{2^2y}}{\sqrt[3]{2^2y}}$$

$$= \dfrac{\sqrt[3]{x}\,\sqrt[3]{4y}}{\sqrt[3]{2^3y^3}}$$

$$= \dfrac{\sqrt[3]{4xy}}{2y}$$

**c)** After using the quotient rule, we simplify the numerator.

$$\sqrt[4]{\dfrac{32x^9y^6}{3z^2}} = \dfrac{\sqrt[4]{32x^9y^6}}{\sqrt[4]{3z^2}} \qquad \text{Quotient rule for radicals}$$

$$= \dfrac{\sqrt[4]{16x^8y^4}\,\sqrt[4]{2xy^2}}{\sqrt[4]{3z^2}} \qquad \text{Product rule for radicals}$$

$$= \dfrac{2x^2y\sqrt[4]{2xy^2}}{\sqrt[4]{3z^2}} \qquad \text{Simplify the numerator.}$$

Now we rationalize the denominator. We need to multiply both the numerator and the denominator by a fourth root to produce a radicand that is a perfect fourth power. Since the denominator contains one factor of 3, we need 3 more factors of 3, or $3^3$. Since there are two factors of $z$, we need 2 more factors of $z$, or $z^2$. Thus we will multiply both numerator and denominator by $\sqrt[4]{3^3z^2}$.

$$= \dfrac{2x^2y\sqrt[4]{2xy^2}}{\sqrt[4]{3z^2}} \cdot \dfrac{\sqrt[4]{3^3z^2}}{\sqrt[4]{3^3z^2}}$$

$$= \dfrac{2x^2y\sqrt[4]{2xy^2}\,\sqrt[4]{27z^2}}{\sqrt[4]{3z^2}\,\sqrt[4]{3^3z^2}}$$

$$= \dfrac{2x^2y\sqrt[4]{54xy^2z^2}}{\sqrt[4]{3^4z^4}} \qquad \text{Product rule for radicals}$$

$$= \dfrac{2x^2y\sqrt[4]{54xy^2z^2}}{3z}$$

*Note:* There are no perfect fourth power factors of 54, and each exponent in the radicand is less than the index.

Now Try Exercise 53

## 2 Rationalize a Denominator Using the Conjugate

The denominator of some fractions is a sum or difference involving radicals that cannot be simplified. One such fraction is $\dfrac{13}{4 + \sqrt{3}}$. To rationalize the denominator of this fraction, we multiply the numerator and the denominator by $4 - \sqrt{3}$. The expressions $4 + \sqrt{3}$ and $4 - \sqrt{3}$ are **conjugates** of each other. Other examples of conjugates are shown below.

| Radical Expression | Conjugate |
|---|---|
| $9 + \sqrt{2}$ | $9 - \sqrt{2}$ |
| $8\sqrt{3} - \sqrt{5}$ | $8\sqrt{3} + \sqrt{5}$ |
| $\sqrt{x} + \sqrt{y}$ | $\sqrt{x} - \sqrt{y}$ |
| $-6a - \sqrt{b}$ | $-6a + \sqrt{b}$ |

When a radical expression is multiplied by its conjugate using the FOIL method, the product simplifies to an expression that contains no square roots. Our next example shows the product of a radical expression and its conjugate.

EXAMPLE 3  Multiply $(6 + \sqrt{3})(6 - \sqrt{3})$.

*Solution*   Multiply using the FOIL method.

$$\overset{\text{F}\qquad\text{O}\qquad\text{I}\qquad\text{L}}{(6 + \sqrt{3})(6 - \sqrt{3}) = 6(6) + 6(-\sqrt{3}) + 6(\sqrt{3}) + \sqrt{3}(-\sqrt{3})}$$

$$= 36 \ \boxed{-6\sqrt{3} + 6\sqrt{3}} \ - \sqrt{9}$$

$$= 36 - \sqrt{9}$$

$$= 36 - 3$$

$$= 33$$

Now Try Exercise 57

### Understanding Algebra

In Chapter 5 we discussed the product

$$(a + b)(a - b) = a^2 - b^2$$

Notice how the sum of the outer and inner products, $-ab + ab$, is always 0. This same procedure can be used when multiplying a radical expression by its conjugate. For example,

$$(3 + \sqrt{2})(3 - \sqrt{2}) = 3^2 - (\sqrt{2})^2$$
$$= 9 - 2 = 7$$

In Example 3, we would get the same result using the formula for the product of the sum and difference of the same two terms. The product results in the difference of two squares, $(a + b)(a - b) = a^2 - b^2$. In Example 3, if we let $a = 6$ and $b = \sqrt{3}$, then using the formula we get the following.

$$(a + b)(a - b) = a^2 - b^2$$
$$(6 + \sqrt{3})(6 - \sqrt{3}) = 6^2 - (\sqrt{3})^2$$
$$= 36 - 3$$
$$= 33$$

### To Rationalize the Denominator Using the Conjugate

Multiply the numerator and the denominator by the conjugate of the denominator. Simplify using the distributive property, FOIL, or the special product $(a + b)(a - b) = a^2 - b^2$.

Now let's work an example where we rationalize a denominator using the conjugate.

**EXAMPLE 4** Simplify.    **a)** $\dfrac{13}{4 + \sqrt{3}}$    **b)** $\dfrac{6}{\sqrt{5} - \sqrt{2}}$    **c)** $\dfrac{a - \sqrt{b}}{a + \sqrt{b}}$

*Solution*    In each part, we rationalize the denominator by multiplying the numerator and the denominator by the conjugate of the denominator.

**a)**
$$\frac{13}{4 + \sqrt{3}} = \frac{13}{4 + \sqrt{3}} \cdot \frac{4 - \sqrt{3}}{4 - \sqrt{3}}$$

$$= \frac{13(4 - \sqrt{3})}{(4 + \sqrt{3})(4 - \sqrt{3})}$$

$$= \frac{13(4 - \sqrt{3})}{16 - 3}$$

$$= \frac{\overset{1}{\cancel{13}}(4 - \sqrt{3})}{\underset{1}{\cancel{13}}} \text{ or } 4 - \sqrt{3}$$

**b)**
$$\frac{6}{\sqrt{5} - \sqrt{2}} = \frac{6}{\sqrt{5} - \sqrt{2}} \cdot \frac{\sqrt{5} + \sqrt{2}}{\sqrt{5} + \sqrt{2}}$$

$$= \frac{6(\sqrt{5} + \sqrt{2})}{5 - 2}$$

$$= \frac{\overset{2}{\cancel{6}}(\sqrt{5} + \sqrt{2})}{\underset{1}{\cancel{3}}}$$

$$= 2(\sqrt{5} + \sqrt{2}) \quad \text{or} \quad 2\sqrt{5} + 2\sqrt{2}$$

**c)**
$$\frac{a - \sqrt{b}}{a + \sqrt{b}} = \frac{a - \sqrt{b}}{a + \sqrt{b}} \cdot \frac{a - \sqrt{b}}{a - \sqrt{b}}$$

$$= \frac{a^2 - a\sqrt{b} - a\sqrt{b} + \sqrt{b^2}}{a^2 - b}$$

$$= \frac{a^2 - 2a\sqrt{b} + b}{a^2 - b}$$

Remember that you cannot divide out $a^2$ or $b$ because they are terms, not factors.

*Now Try Exercise 75*

Now that we have illustrated how to rationalize denominators, let's discuss the criteria a radical must meet to be considered simplified.

## 3  Understand When a Radical Is Simplified

After you have simplified a radical expression, you should check it to make sure that it is simplified as far as possible.

---

**A Radical Expression Is Simplified When the Following Are All True**

1. No perfect powers are factors of the radicand and all exponents in the radicand are less than the index.

2. No radicand contains a fraction.

3. No denominator contains a radical.

**EXAMPLE 5**  Determine whether the following expressions are simplified. If not, explain why. Simplify the expressions if not simplified.

**a)** $\sqrt{48x^5}$ 　　　　　　　 **b)** $\sqrt{\dfrac{1}{2}}$ 　　　　　　　 **c)** $\dfrac{1}{\sqrt{6}}$

*Solution*

**a)** This expression is not simplified because 16 is a perfect square factor of 48 and because $x^4$ is a perfect square factor of $x^5$. Let's simplify the radical.

$$\sqrt{48x^5} = \sqrt{16x^4 \cdot 3x} = \sqrt{16x^4} \cdot \sqrt{3x} = 4x^2\sqrt{3x}$$

**b)** This expression is not simplified since the radicand contains the fraction $\dfrac{1}{2}$. We simplify by first using the quotient rule, and then we rationalize the denominator as follows.

$$\sqrt{\dfrac{1}{2}} = \dfrac{\sqrt{1}}{\sqrt{2}} \cdot \dfrac{\sqrt{2}}{\sqrt{2}} = \dfrac{\sqrt{2}}{2}$$

**c)** This expression is not simplified since the denominator, $\sqrt{6}$, contains a radical. We simplify by rationalizing the denominator as follows.

$$\dfrac{1}{\sqrt{6}} = \dfrac{1}{\sqrt{6}} \cdot \dfrac{\sqrt{6}}{\sqrt{6}} = \dfrac{\sqrt{6}}{6}$$

Now Try Exercise 7

### Understanding Algebra

When written in simplified form, the following must be true about the radicand:

- The coefficient of the radicand has no perfect power factors.
- The exponents on the variables are all less than the index of the radical.

## 4 Use Rationalizing the Denominator in an Addition Problem

Now let's work an addition problem that requires rationalizing the denominator. This example makes use of the methods discussed in Sections 11.3 and 11.4 to add and subtract radicals.

**EXAMPLE 6**  Simplify $4\sqrt{2} - \dfrac{3}{\sqrt{8}} + \sqrt{32}$.

*Solution*　Begin by rationalizing the denominator and by simplifying $\sqrt{32}$.

$$4\sqrt{2} - \dfrac{3}{\sqrt{8}} + \sqrt{32} = 4\sqrt{2} - \dfrac{3}{\sqrt{8}} \cdot \dfrac{\sqrt{2}}{\sqrt{2}} + \sqrt{16}\sqrt{2} \qquad \text{Rationalize denominator.}$$

$$= 4\sqrt{2} - \dfrac{3\sqrt{2}}{\sqrt{16}} + 4\sqrt{2} \qquad \text{Product rule}$$

$$= 4\sqrt{2} - \dfrac{3}{4}\sqrt{2} + 4\sqrt{2} \qquad \text{Write } \dfrac{3\sqrt{2}}{\sqrt{16}} \text{ as } \dfrac{3}{4}\sqrt{2}.$$

$$= \left(4 - \dfrac{3}{4} + 4\right)\sqrt{2} \qquad \text{Simplify.}$$

$$= \dfrac{29\sqrt{2}}{4}$$

Now Try Exercise 115

## 5 Divide Radical Expressions with Different Indices

Now we will divide radical expressions where the radicals have different indices. To divide such problems, write each radical in exponential form. Then use the rules of exponents with the rational exponents, as was discussed in Section 11.2, to simplify the expression. Example 7 illustrates this procedure.

EXAMPLE 7  Simplify.    **a)** $\dfrac{\sqrt[5]{(m+n)^7}}{\sqrt[3]{(m+n)^4}}$    **b)** $\dfrac{\sqrt[3]{a^5b^4}}{\sqrt{a^2b}}$

*Solution*  Begin by writing the numerator and denominator with rational exponents.

**a)** $\dfrac{\sqrt[5]{(m+n)^7}}{\sqrt[3]{(m+n)^4}} = \dfrac{(m+n)^{7/5}}{(m+n)^{4/3}}$    Write with rational exponents.

$= (m+n)^{(7/5)-(4/3)}$    Quotient rule for exponents

$= (m+n)^{1/15}$

$= \sqrt[15]{m+n}$    Write as a radical.

**b)** $\dfrac{\sqrt[3]{a^5b^4}}{\sqrt{a^2b}} = \dfrac{(a^5b^4)^{1/3}}{(a^2b)^{1/2}}$    Write with rational exponents.

$= \dfrac{a^{5/3}b^{4/3}}{ab^{1/2}}$    Raise the product to a power.

$= a^{(5/3)-1}b^{(4/3)-(1/2)}$    Quotient rule for exponents

$= a^{2/3}b^{5/6}$

$= a^{4/6}b^{5/6}$    Write the fractions with denominator 6.

$= (a^4b^5)^{1/6}$    Rewrite using the laws of exponents.

$= \sqrt[6]{a^4b^5}$    Write as a radical.

Now Try Exercise 133

---

# EXERCISE SET 11.5    MathXL®    MyMathLab®
MathXL®        MyMathLab

## Warm-Up Exercises

*Fill in the blanks with the appropriate word, phrase, or symbol(s) from the following list.*

conjugate      sum      power      product      rationalizing      radicand      denominator      numerator

**1.** The process we use to rewrite the fraction without any radicals in the denominator is known as _____ the denominator.

**2.** When using the FOIL method on the product $(a+b)(a-b)$ the _____ of the outer and inner products is always 0.

**3.** To rationalize the denominator of a fraction containing a radical expression, multiply both the numerator and the denominator of the fraction by a radical that will produce a radicand in the denominator that is a perfect _____ for the index.

**4.** To rationalize the denominator using the conjugate, multiply the numerator and the denominator by the _____ of the denominator.

**5.** The expression $\sqrt{\dfrac{1}{2}}$ is not considered simplified since the _____ contains a fraction.

**6.** The expression $\dfrac{\sqrt{2}}{\sqrt{5}}$ is not considered simplified since the _____ contains a radical.

## Practice the Skills

*Simplify. In this exercise set, assume all variables represent positive real numbers.*

**7.** $\dfrac{1}{\sqrt{2}}$

**8.** $\dfrac{1}{\sqrt{11}}$

**9.** $\dfrac{4}{\sqrt{5}}$

**10.** $\dfrac{3}{\sqrt{7}}$

**11.** $\dfrac{6}{\sqrt{6}}$

**12.** $\dfrac{17}{\sqrt{17}}$

**13.** $\dfrac{1}{\sqrt{z}}$

**14.** $\dfrac{y}{\sqrt{y}}$

**15.** $\dfrac{p}{\sqrt{2}}$

**16.** $\dfrac{m}{\sqrt{13}}$

**17.** $\dfrac{\sqrt{y}}{\sqrt{7}}$

**18.** $\dfrac{\sqrt{19}}{\sqrt{q}}$

**19.** $\dfrac{6\sqrt{3}}{\sqrt{6}}$

**20.** $\dfrac{15x}{\sqrt{x}}$

**21.** $\dfrac{\sqrt{x}}{\sqrt{y}}$

**22.** $\dfrac{2\sqrt{3}}{\sqrt{a}}$

**23.** $\sqrt{\dfrac{5m}{8}}$

**24.** $\dfrac{9\sqrt{3}}{\sqrt{y^3}}$

**25.** $\dfrac{2n}{\sqrt{18n}}$

**26.** $\sqrt{\dfrac{120x}{4y^3}}$

**27.** $\sqrt{\dfrac{18x^4y^3}{2z^3}}$

**28.** $\sqrt{\dfrac{7pq^4}{2r}}$

**29.** $\sqrt{\dfrac{20y^4z^3}{3xy^{-4}}}$

**30.** $\sqrt{\dfrac{5xy^6}{3z}}$

**31.** $\sqrt{\dfrac{48x^6y^5}{3z^3}}$

**32.** $\sqrt{\dfrac{45y^{12}z^{10}}{2x}}$

*Simplify.*

**33.** $\dfrac{1}{\sqrt[3]{2}}$

**34.** $\dfrac{1}{\sqrt[3]{4}}$

**35.** $\dfrac{8}{\sqrt[3]{y}}$

**36.** $\dfrac{2}{\sqrt[3]{a^2}}$

**37.** $\dfrac{1}{\sqrt[4]{2}}$

**38.** $\dfrac{1}{\sqrt[4]{4}}$

**39.** $\dfrac{a}{\sqrt[4]{8}}$

**40.** $\dfrac{8}{\sqrt[4]{z}}$

**41.** $\dfrac{5}{\sqrt[4]{z^2}}$

**42.** $\dfrac{13}{\sqrt[4]{z^3}}$

**43.** $\dfrac{10}{\sqrt[5]{y^3}}$

**44.** $\dfrac{x}{\sqrt[5]{y^4}}$

**45.** $\dfrac{2}{\sqrt[7]{a^4}}$

**46.** $\sqrt[3]{\dfrac{4x}{y}}$

**47.** $\sqrt[3]{\dfrac{1}{2x}}$

**48.** $\sqrt[3]{\dfrac{7c}{9y^2}}$

**49.** $\dfrac{5m}{\sqrt[4]{2}}$

**50.** $\dfrac{3}{\sqrt[4]{a}}$

**51.** $\sqrt[4]{\dfrac{5}{3x^3}}$

**52.** $\sqrt[4]{\dfrac{2x^3}{4y^2}}$

**53.** $\sqrt[3]{\dfrac{3x^2}{2y^2}}$

**54.** $\sqrt[3]{\dfrac{15x^6y^7}{2z^2}}$

**55.** $\sqrt[3]{\dfrac{14xy^2}{2z^2}}$

**56.** $\sqrt[6]{\dfrac{r^4s^9}{2r^5}}$

*Multiply.*

**57.** $(4 + \sqrt{5})(4 - \sqrt{5})$

**58.** $(2 + \sqrt{7})(2 - \sqrt{7})$

**59.** $(8 + \sqrt{2})(8 - \sqrt{2})$

**60.** $(6 - \sqrt{7})(6 + \sqrt{7})$

**61.** $(2 - \sqrt{10})(2 + \sqrt{10})$

**62.** $(3 + \sqrt{17})(3 - \sqrt{17})$

**63.** $(\sqrt{a} - \sqrt{b})(\sqrt{a} + \sqrt{b})$

**64.** $(\sqrt{x} - \sqrt{y})(\sqrt{x} + \sqrt{y})$

**65.** $(2\sqrt{x} - 3\sqrt{y})(2\sqrt{x} + 3\sqrt{y})$

**66.** $(5\sqrt{c} - 4\sqrt{d})(5\sqrt{c} + 4\sqrt{d})$

*Simplify by rationalizing the denominator.*

**67.** $\dfrac{1}{\sqrt{2} + 1}$

**68.** $\dfrac{4}{\sqrt{3} - 1}$

**69.** $\dfrac{1}{2 + \sqrt{3}}$

**70.** $\dfrac{3}{5 - \sqrt{7}}$

**71.** $\dfrac{5}{\sqrt{2} - 7}$

**72.** $\dfrac{6}{\sqrt{2} + \sqrt{3}}$

**73.** $\dfrac{\sqrt{5}}{2\sqrt{5} - \sqrt{6}}$

**74.** $\dfrac{1}{\sqrt{17} - \sqrt{8}}$

**75.** $\dfrac{3}{6 + \sqrt{x}}$

**76.** $\dfrac{4\sqrt{5}}{\sqrt{a} - 3}$

**77.** $\dfrac{4\sqrt{x}}{\sqrt{x} - y}$

**78.** $\dfrac{\sqrt{8x}}{x + \sqrt{y}}$

**79.** $\dfrac{\sqrt{2} - 2\sqrt{3}}{\sqrt{2} + 4\sqrt{3}}$

**80.** $\dfrac{\sqrt{c} - \sqrt{2d}}{\sqrt{c} - \sqrt{d}}$

**81.** $\dfrac{\sqrt{a^3} + \sqrt{a^7}}{\sqrt{a}}$

**82.** $\dfrac{2\sqrt{xy} - \sqrt{xy}}{\sqrt{x} + \sqrt{y}}$

**83.** $\dfrac{4}{\sqrt{x + 2} - 3}$

**84.** $\dfrac{8}{\sqrt{y - 3} + 6}$

*Simplify. These exercises are a combination of the types of exercises presented earlier in this exercise set.*

**85.** $\sqrt{\dfrac{a^2}{25}}$

**86.** $\sqrt[3]{\dfrac{a^3}{8}}$

**87.** $\sqrt{\dfrac{2}{9}}$

**88.** $\sqrt{\dfrac{a}{b}}$

**89.** $(\sqrt{7} + \sqrt{6})(\sqrt{7} - \sqrt{6})$

**90.** $\sqrt[3]{\dfrac{1}{16}}$

**91.** $\sqrt{\dfrac{24x^3y^6}{5z}}$

**92.** $\dfrac{5}{4 - \sqrt{y}}$

**93.** $\sqrt{\dfrac{28xy^4}{2x^3y^4}}$

**94.** $\dfrac{8x}{\sqrt[3]{5y}}$

**95.** $\dfrac{1}{\sqrt{a} + 7}$

**96.** $\dfrac{\sqrt{x}}{\sqrt{x} + 6\sqrt{y}}$

**97.** $-\dfrac{7\sqrt{x}}{\sqrt{98}}$

**98.** $\sqrt{\dfrac{2xy^4}{50xy^2}}$

**99.** $\sqrt[4]{\dfrac{3y^2}{2x}}$

**100.** $\sqrt{\dfrac{49x^2y^5}{3z}}$

**101.** $\sqrt[3]{\dfrac{32y^{12}z^{10}}{2x}}$

**102.** $\dfrac{\sqrt{3} + 2}{\sqrt{2} + \sqrt{3}}$

**103.** $\dfrac{\sqrt{ar}}{\sqrt{a} - 2\sqrt{r}}$

**104.** $\sqrt[4]{\dfrac{2}{9x}}$

**105.** $\dfrac{\sqrt[3]{6x}}{\sqrt[3]{5xy}}$

**106.** $\dfrac{\sqrt[3]{16m^2n}}{\sqrt[3]{2mn^2}}$

**107.** $\sqrt[4]{\dfrac{2x^7y^{12}z^4}{3x^9}}$

**108.** $\dfrac{9}{\sqrt{y + 9} - \sqrt{y}}$

*Simplify.*

**109.** $\dfrac{1}{\sqrt{2}} + \dfrac{\sqrt{2}}{2}$

**110.** $\dfrac{1}{\sqrt{3}} + \dfrac{\sqrt{3}}{3}$

**111.** $\sqrt{5} - \dfrac{2}{\sqrt{5}}$

**112.** $\dfrac{\sqrt{6}}{2} - \dfrac{2}{\sqrt{6}}$

**113.** $4\sqrt{\dfrac{1}{6}} + \sqrt{24}$

**114.** $5\sqrt{3} - \dfrac{3}{\sqrt{3}} + 2\sqrt{18}$

**115.** $5\sqrt{2} - \dfrac{2}{\sqrt{8}} + \sqrt{50}$

**116.** $\dfrac{2}{3} + \dfrac{1}{\sqrt{3}} + \sqrt{75}$

**117.** $\sqrt{\dfrac{1}{2}} + 7\sqrt{2} + \sqrt{18}$

**118.** $\dfrac{1}{2}\sqrt{18} - \dfrac{3}{\sqrt{2}} - 9\sqrt{50}$

**119.** $\dfrac{2}{\sqrt{50}} - 3\sqrt{50} - \dfrac{1}{\sqrt{8}}$

**120.** $\dfrac{\sqrt{3}}{3} + \dfrac{5}{\sqrt{3}} + \sqrt{12}$

**121.** $\sqrt{\dfrac{3}{8}} + \sqrt{\dfrac{3}{2}}$

**122.** $2\sqrt{\dfrac{8}{3}} - 4\sqrt{\dfrac{100}{6}}$

**123.** $-2\sqrt{\dfrac{x}{y}} + 3\sqrt{\dfrac{y}{x}}$

**124.** $-5x\sqrt{\dfrac{y}{y^2}} + 9x\sqrt{\dfrac{1}{y}}$

**125.** $\dfrac{3}{\sqrt{a}} - \sqrt{\dfrac{9}{a}} + 2\sqrt{a}$

**126.** $6\sqrt{x} + \dfrac{1}{\sqrt{x}} + \sqrt{\dfrac{1}{x}}$

*Simplify.*

**127.** $\dfrac{\sqrt{(a+b)^4}}{\sqrt[3]{a+b}}$

**128.** $\dfrac{\sqrt[3]{c+2}}{\sqrt[4]{(c+2)^3}}$

**129.** $\dfrac{\sqrt[5]{(a+2b)^4}}{\sqrt[3]{(a+2b)^2}}$

**130.** $\dfrac{\sqrt[6]{(r+3)^5}}{\sqrt[3]{(r+3)^5}}$

**131.** $\dfrac{\sqrt[3]{r^2 s^4}}{\sqrt{rs}}$

**132.** $\dfrac{\sqrt{a^2 b^4}}{\sqrt[3]{ab^2}}$

**133.** $\dfrac{\sqrt[5]{x^4 y^6}}{\sqrt[3]{(xy)^2}}$

**134.** $\dfrac{\sqrt[6]{4m^8 n^4}}{\sqrt[4]{m^4 n^2}}$

## Problem Solving

**135. Illumination of a Light** Under certain conditions the formula

$$d = \sqrt{\dfrac{72}{I}}$$

is used to show the relationship between the illumination on an object, $I$, in lumens per meter, and the distance, $d$, in meters, the object is from the light source. If the illumination on a person standing near a light source is 5.3 lumens per meter, how far is the person from the light source?

**136. Strength of a Board** When sufficient pressure is applied to a particular particle board, the particle board will break (or rupture). The thicker the particle board the greater will be the pressure that will need to be applied before the board breaks. The formula

$$T = \sqrt{\dfrac{0.05\,LB}{M}}$$

relates the thickness of a specific particle board, $T$, in inches, the board's length, $L$, in inches, the board's load that will cause the board to rupture, $B$, in pounds, and the modulus of rupture, $M$, in pounds per square inch. The modulus of rupture is a constant determined by sample tests on the specific type of particle board.

Find the thickness of a 36-inch-long particle board if the modulus of rupture is 2560 pounds per square inch and the board ruptures when 800 pounds are applied.

**137. Volume of a Fish Tank** A new restaurant wants to have a spherical fish tank in its lobby. The radius, $r$, in inches, of a spherical tank can be found by the formula

$$r = \sqrt[3]{\dfrac{3V}{4\pi}}$$

where $V$ is the volume of the tank in cubic inches. Find the radius of a spherical tank whose volume is 7238.23 cubic inches.

**138. Consecutive Numbers** If we consider the set of consecutive natural numbers $1, 2, 3, 4, \ldots, n$ to be the population, the standard deviation, $\sigma$, which is a measure of the spread of the data from the mean, can be calculated by the formula

$$\sigma = \sqrt{\dfrac{n^2 - 1}{12}}$$

where $n$ represents the number of natural numbers in the population. Find the standard deviation for the first 100 consecutive natural numbers.

**139. U.S. Farms** The number of farms in the United States is declining annually (however, the size of the remaining farms is increasing). A function that can be used to estimate the number of farms, $N(t)$, in millions, is

$$N(t) = \dfrac{6.21}{\sqrt[4]{t}}$$

where $t$ is years since 1959 and $1 \le t \le 50$. Estimate the number of farms in the United States in **a)** 1960 and **b)** 2008.

**140. Infant Mortality Rate** The U.S. infant mortality rate has been declining steadily. The infant mortality rate, $N(t)$, defined as deaths per 1000 live births, can be estimated by the function

$$N(t) = \dfrac{28.46}{\sqrt[3]{t^2}}$$

where $t$ is years since 1969 and $1 \le t \le 37$. Estimate the infant mortality rate in **a)** 1970 and **b)** 2006.

*In higher math courses, it may be necessary to rationalize the numerators of radical expressions. Rationalize the numerators of the following expressions. (Your answers will contain radicals in the denominators.)*

**141.** $\dfrac{5 - \sqrt{5}}{6}$

**142.** $\dfrac{\sqrt{7}}{3}$

**143.** $\dfrac{\sqrt{x + h} - \sqrt{x}}{h}$

**144.** $\dfrac{6\sqrt{x} - \sqrt{3}}{x}$

## Concept/Writing Exercises

**145.** Which is greater, $\dfrac{2}{\sqrt{2}}$ or $\dfrac{3}{\sqrt{3}}$? Explain.

**146.** Which is greater, $\dfrac{\sqrt{3}}{2}$ or $\dfrac{2}{\sqrt{3}}$? Explain.

**147.** Which is greater, $\dfrac{1}{\sqrt{3} + 2}$ or $2 + \sqrt{3}$? (Do not use a calculator.) Explain how you determined your answer.

**148.** Which is greater, $\dfrac{1}{\sqrt{3}} + \sqrt{75}$ or $\dfrac{2}{\sqrt{12}} + \sqrt{48} + 2\sqrt{3}$? (Do not use a calculator.) Explain how you determined your answer.

**149.** Consider the functions $f(x) = x^{a/2}$ and $g(x) = x^{b/3}$.
  **a)** List three values for $a$ that will result in $x^{a/2}$ being a perfect square.
  **b)** List three values for $b$ that will result in $x^{b/3}$ being a perfect cube.
  **c)** If $x \geq 0$, find $(f \cdot g)(x)$.
  **d)** If $x \geq 0$, find $(f/g)(x)$.

## Group Activity

**Similar Figures** *The following two exercises will reinforce many of the concepts presented in this chapter. Work each problem as a group. Make sure each member of the group understands each step in obtaining the solution. The figures in each exercise are similar. For each exercise, use a proportion to find the length of side x. Write the answer in radical form with a rationalized denominator.*

**150.**

**151.**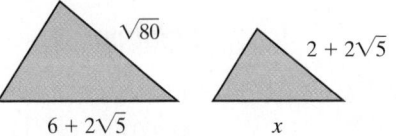

## Cumulative Review Exercises

[2.6]   **152.** Solve the equation $A = \dfrac{1}{2}h(b_1 + b_2)$ for $b_2$.

[3.4]   **153. Moving Vehicles** Two cars leave from West Point at the same time traveling in opposite directions. One travels 10 miles per hour faster than the other. If the two cars are 270 miles apart after 3 hours, find the speed of each car.

[5.5]   **154.** Multiply $(x - 2)(4x^2 + 9x - 2)$.

[7.6]   **155.** Solve $\dfrac{x}{2} - \dfrac{4}{x} = -\dfrac{7}{2}$.

# 11.6 Solving Radical Equations

1  Solve equations containing one radical.

2  Solve equations containing two radicals.

3  Solve equations containing two radical terms and a nonradical term.

4  Solve applications using radical equations.

5  Solve for a variable in a radicand.

## 1  Solve Equations Containing One Radical

**Radical Equation**

A **radical equation** is an equation that contains a variable in a radicand.

Examples of Radical Equations

$$\sqrt{x} = 5, \qquad \sqrt[3]{y + 4} = 9, \qquad \sqrt{x - 2} = 7 + \sqrt{x + 8}$$

**To Solve Radical Equations**

1. Rewrite the equation so that one radical containing a variable is by itself (isolated) on one side of the equation.
2. Raise each side of the equation to a power equal to the index of the radical.
3. Combine like terms.
4. If the equation still contains a term with a variable in a radicand, repeat steps 1 through 3.
5. Solve the resulting equation for the variable.
6. Check all solutions in the original equation for extraneous solutions.

*Whenever we raise both sides of an equation to an even power, we could be introducing an extraneous, or false, solution. Therefore, we must check the results in the original equation.*

**EXAMPLE 1** Solve the equation $\sqrt{x} = 5$. ———————————————

**Solution**    The square root containing the variable is already by itself on one side of the equation. Square both sides of the equation.

$$\sqrt{x} = 5$$
$$(\sqrt{x})^2 = (5)^2$$
$$x = 25$$

Check    $\sqrt{x} = 5$
$$\sqrt{25} \overset{?}{=} 5$$
$$5 = 5 \quad \text{True}$$

Now Try Exercise 11

**EXAMPLE 2** Solve. ———————————————————

**a)** $\sqrt{x-4} - 6 = 0$    **b)** $\sqrt[3]{x} + 10 = 8$    **c)** $\sqrt{x} + 3 = 0$

**Solution**    The first step in each part will be to isolate the term containing the radical.

**a)**
$$\sqrt{x-4} - 6 = 0$$
$$\sqrt{x-4} = 6 \qquad \text{Isolate the radical containing the variable.}$$
$$(\sqrt{x-4})^2 = 6^2 \qquad \text{Square both sides.}$$
$$x - 4 = 36 \qquad \text{Solve for the variable.}$$
$$x = 40$$

A check will show that 40 is the solution.

**b)**
$$\sqrt[3]{x} + 10 = 8$$
$$\sqrt[3]{x} = -2 \qquad \text{Isolate the radical containing the variable.}$$
$$(\sqrt[3]{x})^3 = (-2)^3 \qquad \text{Cube both sides.}$$
$$x = -8$$

A check will show that $-8$ is the solution.

> ## Understanding Algebra
>
> Whenever we raise both sides of an equation to an *even* power, we could be introducing an extraneous or false solution. Therefore we must check our answers. Example 2 **c)** shows that when we square both sides of the equation $\sqrt{x} = -3$ we obtain the result $x = 9$. By checking this result in the original equation we are able to see that 9 is an extraneous solution.

**c)**
$$\sqrt{x} + 3 = 0$$
$$\sqrt{x} = -3 \qquad \text{Isolate the radical containing the variable.}$$
$$(\sqrt{x})^2 = (-3)^2 \qquad \text{Square both sides.}$$
$$x = 9$$

Check    $\sqrt{x} + 3 = 0$
$$\sqrt{9} + 3 \overset{?}{=} 0$$
$$3 + 3 \overset{?}{=} 0$$
$$6 = 0 \quad \text{False}$$

A check shows that 9 is not a solution. The answer to part **c)** is "no real solution." You may have realized there was no real solution to the problem when you obtained the equation $\sqrt{x} = -3$, because $\sqrt{x}$ cannot equal a negative real number.

Now Try Exercise 17

EXAMPLE 3   Solve $\sqrt{2x-3} = x-3$.

*Solution*   Since the radical is already isolated, we square both sides of the equation, resulting in a quadratic equation.

$$(\sqrt{2x-3})^2 = (x-3)^2$$
$$2x-3 = (x-3)(x-3)$$
$$2x-3 = x^2 - 6x + 9$$
$$0 = x^2 - 8x + 12$$

Now we factor and use the zero-factor property to solve the quadratic equation.

$$x^2 - 8x + 12 = 0$$
$$(x-6)(x-2) = 0$$
$$x - 6 = 0 \quad \text{or} \quad x - 2 = 0$$
$$x = 6 \qquad\qquad x = 2$$

Check

$$
\begin{array}{cc}
x = 6 & x = 2 \\
\sqrt{2x-3} = x-3 & \sqrt{2x-3} = x-3 \\
\sqrt{2(6)-3} \overset{?}{=} 6-3 & \sqrt{2(2)-3} \overset{?}{=} 2-3 \\
\sqrt{9} \overset{?}{=} 3 & \sqrt{1} \overset{?}{=} -1 \\
3 = 3 \quad \text{True} & 1 = -1 \quad \text{False}
\end{array}
$$

Thus, 6 is a solution, but 2 is not a solution to the equation. The 2 is an extraneous solution because 2 satisfies the equation $(\sqrt{2x-3})^2 = (x-3)^2$, but not the original equation, $\sqrt{2x-3} = x-3$.

Now Try Exercise 43

## Using Your Graphing Calculator ▪▪▪▪▪

We can use a graphing calculator to solve or check radical equations. In Example 3, we found the solution to $\sqrt{2x-3} = x-3$ to be 6. By graphing both

$$Y_1 = \sqrt{2x-3} \quad \text{and} \quad Y_2 = x-3$$

on a graphing calculator, we can see from the graph in **Figure 11.3** that the two functions appear to intersect at $x = 6$.

FIGURE 11.3

FIGURE 11.4

We can also use the *intersect* feature to determine the solutions to radical equations. **Figure 11.4** shows that the two graphs intersect when $x = 6$ and $y = 3$.

EXAMPLE 4   Solve $x - 2\sqrt{x} - 3 = 0$.

*Solution*   First, isolate the radical term by writing the radical term by itself on one side of the equation.

$$x - 2\sqrt{x} - 3 = 0$$
$$-2\sqrt{x} = -x + 3$$
$$2\sqrt{x} = x - 3$$

Now square both sides of the equation.

$$(2\sqrt{x})^2 = (x - 3)^2$$
$$4x = x^2 - 6x + 9$$
$$0 = x^2 - 10x + 9$$
$$0 = (x - 1)(x - 9)$$
$$x - 1 = 0 \quad \text{or} \quad x - 9 = 0$$
$$x = 1 \qquad\qquad x = 9$$

Check

$$x = 1 \qquad\qquad\qquad\qquad x = 9$$
$$x - 2\sqrt{x} - 3 = 0 \qquad\qquad x - 2\sqrt{x} - 3 = 0$$
$$1 - 2\sqrt{1} - 3 \overset{?}{=} 0 \qquad\qquad 9 - 2\sqrt{9} - 3 \overset{?}{=} 0$$
$$1 - 2(1) - 3 \overset{?}{=} 0 \qquad\qquad 9 - 2(3) - 3 \overset{?}{=} 0$$
$$1 - 2 - 3 \overset{?}{=} 0 \qquad\qquad 9 - 6 - 3 \overset{?}{=} 0$$
$$-4 = 0 \quad \text{False} \qquad\qquad 3 - 3 \overset{?}{=} 0$$
$$0 = 0 \quad \text{True}$$

The solution is 9. The value 1 is an extraneous solution.

Now Try Exercise 41

## 2 Solve Equations Containing Two Radicals

Now we will look at some equations that contain two radicals.

**EXAMPLE 5** Solve $\sqrt{9x^2 + 6} = 3\sqrt{x^2 + x - 2}$.

*Solution* Since the two radicals appear on different sides of the equation, we square both sides of the equation.

$$\left(\sqrt{9x^2 + 6}\right)^2 = \left(3\sqrt{x^2 + x - 2}\right)^2 \quad \text{Square both sides.}$$
$$9x^2 + 6 = 9(x^2 + x - 2)$$
$$9x^2 + 6 = 9x^2 + 9x - 18 \qquad \text{Distributive property}$$
$$6 = 9x - 18 \qquad\qquad 9x^2 \text{ was subtracted from both sides.}$$
$$24 = 9x$$
$$\frac{8}{3} = x$$

A check will show that $\frac{8}{3}$ is the solution.

Now Try Exercise 27

Equations are sometimes given using exponents rather than radicals. Example 6 illustrates such an equation.

**EXAMPLE 6** For $f(x) = 3(x - 2)^{1/3}$ and $g(x) = (17x - 14)^{1/3}$, find all values of $x$ for which $f(x) = g(x)$.

*Solution* We set the two functions equal to each other and solve for $x$.

$$f(x) = g(x)$$
$$3(x - 2)^{1/3} = (17x - 14)^{1/3}$$
$$[3(x - 2)^{1/3}]^3 = [(17x - 14)^{1/3}]^3 \quad \text{Cube both sides.}$$
$$3^3(x - 2) = 17x - 14$$
$$27(x - 2) = 17x - 14$$
$$27x - 54 = 17x - 14$$
$$10x - 54 = -14$$
$$10x = 40$$
$$x = 4$$

**Understanding Algebra**

The functions in Example 6 could also be written using radicals. Thus, we could also answer the question by solving the equation

$$3\sqrt[3]{x - 2} = \sqrt[3]{17x - 14}$$

A check will show that the solution is 4. If you substitute 4 into both $f(x)$ and $g(x)$, you will find they both simplify to $3\sqrt[3]{2}$. Check this now.

Now Try Exercise 69

### 3  Solve Equations Containing Two Radical Terms and a Nonradical Term

When a radical equation contains two radical terms and a third nonradical term, you will sometimes need to raise both sides of the equation to a given power twice to obtain the solution. This procedure is illustrated in Example 7.

**EXAMPLE 7**  Solve $\sqrt{5x - 1} - \sqrt{3x - 2} = 1$.

*Solution*  We must isolate one radical term on one side of the equation. We will begin by adding $\sqrt{3x - 2}$ to both sides of the equation to isolate $\sqrt{5x - 1}$. Then we will square both sides of the equation and combine like terms.

$$\sqrt{5x - 1} = 1 + \sqrt{3x - 2} \qquad \text{Isolate } \sqrt{5x - 1}.$$
$$(\sqrt{5x - 1})^2 = (1 + \sqrt{3x - 2})^2 \qquad \text{Square both sides.}$$
$$5x - 1 = (1 + \sqrt{3x - 2})(1 + \sqrt{3x - 2}) \qquad \text{Write as a product.}$$
$$5x - 1 = 1 + \sqrt{3x - 2} + \sqrt{3x - 2} + (\sqrt{3x - 2})^2 \quad \text{Multiply.}$$
$$5x - 1 = 1 + 2\sqrt{3x - 2} + 3x - 2 \qquad \text{Combine like terms; simplify.}$$
$$5x - 1 = 3x - 1 + 2\sqrt{3x - 2} \qquad \text{Combine like terms.}$$
$$2x = 2\sqrt{3x - 2} \qquad \text{Isolate the radical term.}$$
$$x = \sqrt{3x - 2} \qquad \text{Both sides were divided by 2.}$$

We have isolated the remaining radical term. We now square both sides of the equation again and solve for $x$.

$$x = \sqrt{3x - 2}$$
$$x^2 = (\sqrt{3x - 2})^2 \quad \text{Square both sides.}$$
$$x^2 = 3x - 2$$
$$x^2 - 3x + 2 = 0$$
$$(x - 2)(x - 1) = 0$$
$$x - 2 = 0 \quad \text{or} \quad x - 1 = 0$$
$$x = 2 \qquad\qquad x = 1$$

A check will show that both 2 and 1 are solutions of the equation.

Now Try Exercise 61

**EXAMPLE 8**  For $f(x) = \sqrt{5x - 1} - \sqrt{3x - 2}$, find all values of $x$ for which $f(x) = 1$.

*Solution*  Substitute 1 for $f(x)$. This gives

$$1 = \sqrt{5x - 1} - \sqrt{3x - 2}$$

Since this is the same equation we solved in Example 7, the answers are $x = 2$ and $x = 1$. Verify for yourself that $f(2) = 1$ and $f(1) = 1$.

Now Try Exercise 121

> **Avoiding Common Errors**
>
> In Chapter 5, we stated that $(a + b)^2 \neq a^2 + b^2$. Be careful when squaring a binomial like $1 + \sqrt{x}$. Look at the following computations carefully so that you do not make the mistake shown on the right.
>
> | CORRECT | INCORRECT |
> |---|---|
> | $(1 + \sqrt{x})^2 = (1 + \sqrt{x})(1 + \sqrt{x})$ | $(1 + \sqrt{x})^2 = 1^2 + (\sqrt{x})^2$ |
> | $\phantom{(1 + \sqrt{x})^2 =}$ F    O    I    L | $\phantom{(1 + \sqrt{x})^2 =} = 1 + x$ |
> | $\phantom{(1 + \sqrt{x})^2} = 1 + \sqrt{x} + \sqrt{x} + \sqrt{x}\sqrt{x}$ | |
> | $\phantom{(1 + \sqrt{x})^2} = 1 + 2\sqrt{x} + x$ | |

## 4   Solve Applications Using Radical Equations

Now we will look at a few of the many applications of radicals.

**EXAMPLE 9   The Green Monster**  In Fenway Park, where the Boston Red Sox play baseball, the distance from home plate down the third base line to the bottom of the wall in left field is 310 feet. In left field at the end of the baseline there is a green wall perpendicular to the ground that is 37 feet tall. This green wall is commonly known as *the Green Monster* (see photo). Determine the distance from home plate to the top of the Green Monster along the third base line.

*Solution*   Understand   **Figure 11.5** illustrates the problem. We need to find the distance from home plate to the top of the wall in left field.

FIGURE 11.5

Translate   To solve the problem we use the Pythagorean Theorem, which was discussed earlier, $\text{leg}^2 + \text{leg}^2 = \text{hyp}^2$, or $a^2 + b^2 = c^2$.

$$310^2 + 37^2 = c^2 \qquad \text{Substitute known values.}$$

Carry Out

$$96{,}100 + 1369 = c^2$$

$$97{,}469 = c^2$$

$$\sqrt{97{,}469} = \sqrt{c^2} \qquad \text{Take the square root of both sides.}$$

$$\sqrt{97{,}469} = c \qquad \text{* See footnote.}$$

$$312.20 \approx c$$

Answer   The distance from home plate to the top of the wall is about 312.20 feet.

Now Try Exercise 99

**EXAMPLE 10   Period of a Pendulum**  The length of time it takes for a pendulum to make one complete swing back and forth is called the *period* of the pendulum. See **Figure 11.6.** The period of a pendulum, $T$, in seconds, can be found by the formula $T = 2\pi\sqrt{\dfrac{L}{32}}$, where $L$ is the length of the pendulum in feet. Find the period of a pendulum if its length is 5 feet.

---

*$c^2 = 97{,}469$ has two solutions: $c = \sqrt{97{,}469}$ and $c = -\sqrt{97{,}469}$. Since we are solving for a length, which must be a positive quantity, we use the positive root.

**FIGURE 11.6**

*Solution*   Substitute 5 for $L$ and 3.14 for $\pi$ in the formula. If you have a calculator that has a $\boxed{\pi}$ key, use it to enter $\pi$.

$$T = 2\pi\sqrt{\frac{L}{32}}$$

$$\approx 2(3.14)\sqrt{\frac{5}{32}}$$

$$\approx 2(3.14)\sqrt{0.15625} \approx 2.48$$

Thus, the period is about 2.48 seconds. If you have a grandfather clock with a 5-foot pendulum, it will take about 2.48 seconds for it to swing back and forth.

Now Try Exercise 103

## 5  Solve for a Variable in a Radicand

Many formulas involve radicals that contain variables. It is sometimes necessary to solve for a variable that is part of a radicand. To do so, follow the same general procedure we have used thus far.

**EXAMPLE  11  Margin of Error**  A formula in statistics for finding the margin of error is $E = t\dfrac{s}{\sqrt{n}}$.

a) Find $E$ if $t = 2.064$, $s = 15$, and $n = 25$

b) Solve this equation for $n$.

*Solution*

a) $E = t\dfrac{s}{\sqrt{n}} = 2.064\left(\dfrac{15}{\sqrt{25}}\right) = 2.064\left(\dfrac{15}{5}\right) = 6.192$

b) First eliminate the fraction by multiplying both sides of the equation by $\sqrt{n}$.

$$E = t\frac{s}{\sqrt{n}}$$

$$E\sqrt{n} = \left(t\frac{s}{\sqrt{n}}\right)\sqrt{n} \qquad \text{Eliminate the fraction.}$$

$$E\sqrt{n} = ts$$

$$\sqrt{n} = \frac{ts}{E} \qquad \text{Isolate the radical by dividing both sides by } E.$$

$$(\sqrt{n})^2 = \left(\frac{ts}{E}\right)^2 \qquad \text{Square both sides.}$$

$$n = \left(\frac{ts}{E}\right)^2 \quad \text{or} \quad n = \frac{t^2s^2}{E^2}$$

Now Try Exercise 75

# EXERCISE SET 11.6    MathXL®   MyMathLab®
MathXL®        MyMathLab

## Warm-Up Exercises

*Fill in the blanks with the appropriate word, phrase, or symbol(s) from the following list.*

original     radical     isolated     raise     square     index     cube     extraneous     root     radicand

**1.** A _____ equation is an equation that contains a variable in the radicand.

**2.** Whenever we raise both sides of an equation to an even power, we could be introducing an _____, or false, solution.

**3.** To determine whether the result of solving an equation is an extraneous solution, we must check the result in the _____ equation.

**4.** When solving a radical equation, we rewrite the equation so one radical containing a variable is _____.

**5.** To eliminate an isolated radical from an equation, raise each side of the equation to the _____ of the radical.

**6.** To eliminate the radical in the equation $\sqrt{2x - 5} = 7$, _____ both sides of the equation.

**7.** To eliminate the radical in the equation $\sqrt[3]{x + 1} = 2$, _____ both sides of the equation.

**8.** To eliminate the radical in the equation $\sqrt[4]{x - 1} = 3$, _____ both sides of the equation to the fourth power.

## Practice the Skills

*Solve and check your solution(s). If the equation has no real solution, so state.*

**9.** $\sqrt{x} = 4$

**10.** $\sqrt{x} = 2$

**11.** $\sqrt{x} = -2$

**12.** $\sqrt[3]{x} = 2$

**13.** $\sqrt[3]{x} = -4$

**14.** $\sqrt{a} + 5 = 0$

**15.** $\sqrt{2x + 3} = 5$

**16.** $\sqrt[3]{7x - 6} = 4$

**17.** $\sqrt[3]{3x} + 4 = 7$

**18.** $2\sqrt{4x + 5} = 14$

**19.** $\sqrt[3]{2x + 29} = 3$

**20.** $\sqrt[3]{6x + 2} = -4$

**21.** $\sqrt[4]{x} = 3$

**22.** $\sqrt[4]{x} = -3$

**23.** $\sqrt[4]{x + 10} = 3$

**24.** $\sqrt[4]{3x - 2} = 2$

**25.** $\sqrt{x - 1} + 3 = 8$

**26.** $\sqrt{x + 3} + 4 = 5$

**27.** $\sqrt{x + 8} = \sqrt{x - 8}$

**28.** $\sqrt{r + 5} + 7 = 10$

**29.** $2\sqrt[3]{x - 1} = \sqrt[3]{x^2 + 2x}$

**30.** $\sqrt[3]{6t - 1} = \sqrt[3]{2t + 3}$

**31.** $\sqrt[4]{x + 8} = \sqrt[4]{2x}$

**32.** $\sqrt[4]{3x - 1} + 4 = 0$

**33.** $\sqrt{5x + 1} - 6 = 0$

**34.** $\sqrt{x^2 + 12x + 3} = -x$

**35.** $\sqrt{m^2 + 6m - 4} = m$

**36.** $\sqrt{x^2 + 3x + 12} = x$

**37.** $\sqrt{5c + 1} - 9 = 0$

**38.** $\sqrt{b^2 - 2} = b + 4$

**39.** $\sqrt{z^2 + 5} = z + 1$

**40.** $\sqrt{x} + 6x = 1$

**41.** $\sqrt{2y + 5} + 5 - y = 0$

**42.** $\sqrt{4x + 1} = \dfrac{1}{2}x + 2$

**43.** $\sqrt{5x + 6} = 2x - 6$

**44.** $\sqrt{4b + 5} + b = 10$

**45.** $(2a + 9)^{1/2} - a + 3 = 0$

**46.** $(3x + 4)^{1/2} - x = -2$

**47.** $(2x^2 + 4x + 9)^{1/2} = (2x^2 + 9)^{1/2}$

**48.** $(2x + 1)^{1/2} + 7 = x$

**49.** $(r + 4)^{1/3} = (3r + 10)^{1/3}$

**50.** $(7x + 6)^{1/3} + 4 = 0$

**51.** $(5x + 7)^{1/4} = (9x + 1)^{1/4}$

**52.** $(5b + 3)^{1/4} = (2b + 17)^{1/4}$

**53.** $\sqrt[4]{x + 5} = -2$

**54.** $\sqrt{x^2 + x - 1} = -\sqrt{x + 3}$

*Solve. You will have to square both sides of the equation twice to eliminate all radicals.*

**55.** $\sqrt{x} + 1 = \sqrt{x + 1}$

**56.** $\sqrt{x} - 1 = \sqrt{x - 1}$

**57.** $\sqrt{3a + 1} = \sqrt{a - 4} + 3$

**58.** $\sqrt{x + 1} = 2 - \sqrt{x}$

**59.** $\sqrt{x + 3} = \sqrt{x} - 3$

**60.** $\sqrt{y + 1} = 2 + \sqrt{y - 7}$

**61.** $\sqrt{x + 7} = 6 - \sqrt{x - 5}$

**62.** $\sqrt{b - 3} = 4 - \sqrt{b + 5}$

**63.** $\sqrt{4x - 3} = 2 + \sqrt{2x - 5}$

**64.** $\sqrt{r + 10} + 2 + \sqrt{r - 5} = 0$

**65.** $\sqrt{y + 1} = \sqrt{y + 10} - 3$

**66.** $3 + \sqrt{x + 1} = \sqrt{3x + 12}$

*For each pair of functions, find all real values of x where $f(x) = g(x)$.*

**67.** $f(x) = \sqrt{2x + 3}, g(x) = \sqrt{x + 5}$

**68.** $f(x) = \sqrt{x^2 - 2x}, g(x) = \sqrt{x - 2}$

**69.** $f(x) = \sqrt[3]{5x - 19}, g(x) = \sqrt[3]{6x - 23}$

**70.** $f(x) = (14x - 8)^{1/2}, g(x) = 2(3x + 2)^{1/2}$

**71.** $f(x) = 2(8x + 24)^{1/3}, g(x) = 4(2x - 2)^{1/3}$

**72.** $f(x) = 2\sqrt{x + 2}, g(x) = 8 - \sqrt{x + 14}$

*Solve each formula for the indicated variable.*

**73.** $p = \sqrt{2v}$, for $v$

**74.** $l = \sqrt{4r}$, for $r$

**75.** $v = \sqrt{2gh}$, for $g$

**76.** $v = \sqrt{\dfrac{2E}{m}}$, for $E$

**77.** $v = \sqrt{\dfrac{FR}{M}}$, for $F$

**78.** $\omega = \sqrt{\dfrac{a_0}{b_0}}$, for $b_0$

**79.** $x = \sqrt{\dfrac{m}{k}} V_0$, for $m$

**80.** $T = 2\pi\sqrt{\dfrac{L}{32}}$, for $L$

**81.** $r = \sqrt{\dfrac{A}{\pi}}$, for $A$

**82.** $r = \sqrt[3]{\dfrac{3V}{4\pi}}$, for $V$

## Problem Solving

*Use the Pythagorean Theorem to find the length of the unknown side of each triangle. Write the answer as a radical in simplified form.*

**83.**

$x$   9

$\sqrt{6}$

**84.**

$\sqrt{70}$   $x$

$\sqrt{10}$

**85.**

$\sqrt{65}$   5

$x$

**86.**

$x$   6

4

*Solve. You will need to square both sides of the equation twice.*

**87.** $\sqrt{x + 3} + \sqrt{x} = \sqrt{x + 8}$

**89.** $\sqrt{4y + 6} + \sqrt{y + 5} = \sqrt{y + 1}$

**91.** $\sqrt{c + 1} + \sqrt{c - 2} = \sqrt{3c}$

**93.** $\sqrt{a + 2} - \sqrt{a - 3} = \sqrt{a - 6}$

**88.** $\sqrt{2x} - \sqrt{x - 4} = \sqrt{12 - x}$

**90.** $\sqrt{2b - 2} + \sqrt{b - 5} = \sqrt{4b}$

**92.** $\sqrt{2t - 1} + \sqrt{t - 4} = \sqrt{3t + 1}$

**94.** $\sqrt{r - 1} - \sqrt{r + 6} = \sqrt{r - 9}$

*Solve. You will need to square both sides of the equation twice.*

**95.** $\sqrt{2 - \sqrt{x}} = \sqrt{x}$

**97.** $\sqrt{2 + \sqrt{x + 1}} = \sqrt{7 - x}$

**96.** $\sqrt{6 + \sqrt{x + 4}} = \sqrt{2x - 1}$

**98.** $\sqrt{1 + \sqrt{x - 1}} = \sqrt{x - 6}$

**99. Baseball Diamond** A regulation baseball diamond is a square with 90 feet between bases. How far is second base from home plate?

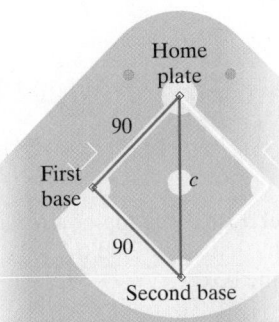

Home plate

90

First base   $c$

90

Second base

**100. Little League Baseball Diamond** A Little League baseball diamond is a square with 60 feet between bases (see Exercise 99). How far is second base from home plate?

**101. Wire from Telephone Pole** A telephone pole is at a right, or 90°, angle with the ground as shown in the figure. Find the length of the wire that connects to the pole 40 feet above the ground and is anchored to the ground 20 feet from the base of the pole.

40 ft   $l$

20 ft

**102. Radius of Basketball Hoop** When you are given the area of a circle, its radius can be found by the formula $r = \sqrt{A/\pi}$.

© Allen R. Angel

**a)** Find the radius of a basketball hoop if the area enclosed by the hoop is 254.47 square inches.

**b)** If the diameter of a basketball is 9 inches, what is the minimum distance possible between the hoop and the ball when the center of the ball is in the center of the hoop?

**103. Period of a Pendulum** The formula for the period of a pendulum is

$$T = 2\pi\sqrt{\frac{l}{g}}$$

where $T$ is the period in seconds, $l$ is its length in feet, and $g$ is the acceleration of gravity. On Earth, gravity is 32 ft/sec². The formula when used on Earth becomes

$$T = 2\pi\sqrt{\frac{l}{32}}$$

**a)** Find the period of a pendulum whose length is 8 feet.

**b)** If the length of a pendulum is doubled, what effect will this have on the period?

**c)** The gravity on the Moon is 1/6 that on Earth. If a pendulum has a period of 2 seconds on Earth, what will be the period of the same pendulum on the Moon?

**104. Diagonal of a Suitcase** A formula for the length of a diagonal from the upper corner of a box to the opposite lower corner is $d = \sqrt{L^2 + W^2 + H^2}$, where $L$, $W$, and $H$ are the length, width, and height, respectively.

a) Find the length of the diagonal of a suitcase of length 22 inches, width 15 inches, and height 12 inches.

b) If the length, width, and height are all doubled, how will the diagonal change?

c) Solve the formula for $W$.

**105. Blood Flowing in an Artery** The formula

$$r = \sqrt[4]{\frac{8\mu l}{\pi R}}$$

is used in determining movement of blood through arteries. In the formula, $R$ represents the resistance to blood flow, $\mu$ is the viscosity of blood, $l$ is the length of the artery, and $r$ is the radius of the artery. Solve this equation for $R$.

**106. Falling Object** The formula

$$t = \frac{\sqrt{19.6s}}{9.8}$$

can be used to tell the time, $t$, in seconds, that an object has been falling if it has fallen $s$ meters. Suppose an object has been dropped from a helicopter and has fallen 100 meters. How long has it been in free fall?

**107. Earth Days** For any planet in our solar system, its "year" is the time it takes for the planet to revolve once around the Sun. The number of Earth days in a given planet's year, $N$, is approximated by the formula $N = 0.2(\sqrt{R})^3$, where $R$ is the mean distance of the planet to the Sun in millions of kilometers. Find the number of Earth days in the year of the planet Earth, whose mean distance to the Sun is 149.4 million kilometers.

**108. Earth Days** Find the number of Earth days in the year of the planet Mercury, whose mean distance to the Sun is 58 million kilometers. See Exercise 107.

**109. Forces on a Car** When two forces, $F_1$ and $F_2$, pull at right angles to each other as illustrated below, the resultant, or the effective force, $R$, can be found by the formula $R = \sqrt{F_1^2 + F_2^2}$. Two cars are trying to pull a third out of the mud, as illustrated. If car $A$ is exerting a force of 60 pounds and car $B$ is exerting a force of 80 pounds, find the resulting force on the car stuck in the mud.

 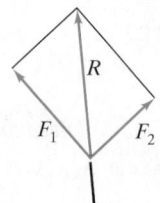

**110. Escape Velocity** The escape velocity, or the velocity needed for a spacecraft to escape a planet's gravitational field, is found by the formula $v_e = \sqrt{2gR}$, where $g$ is the force of gravity of the planet and $R$ is the radius of the planet. Find the escape velocity for Earth, in meters per second, where $g = 9.75$ meters per second squared and $R = 6,370,000$ meters.

**111. Motion of a Wave** A formula used in the study of shallow water wave motion is $c = \sqrt{gH}$, in which $c$ is wave velocity, $H$ is water depth, and $g$ is the acceleration due to gravity. Find the wave velocity if the water's depth is 10 feet. (Use $g = 32$ ft/sec².)

**112. Diagonal of a Box** The top of a rectangular box measures 20 inches by 32 inches. Find the length of the diagonal for the top of the box.

**113. Flower Garden** A rectangular flower garden measures 25 meters by 32 meters. Find the length of the diagonal for the garden.

**114. Speed of Sound** When sound travels through air (or any gas), the velocity of the sound wave is dependent on the air (or gas) temperature. The velocity, $v$, in meters per second, at air temperature, $t$, in degrees Celsius, can be found by the formula

$$v = 331.3\sqrt{1 + \frac{t}{273}}$$

Find the speed of sound in air whose temperature is 20°C (equivalent to 68°F).

*A formula that we have already mentioned and that we will be discussing in more detail shortly is the quadratic formula*

$$x = \frac{-b \pm \sqrt{b^2 - 4ac}}{2a}$$

**115.** Find $x$ when $a = 1$, $b = 0$, $c = -4$.

**116.** Find $x$ when $a = 1$, $b = 1$, $c = -12$.

**117.** Find $x$ when $a = -1$, $b = 4$, $c = 5$.

**118.** Find $x$ when $a = 2$, $b = 5$, $c = -12$.

*Given $f(x)$, find all values of x for which $f(x)$ is the indicated value.*

**119.** $f(x) = \sqrt{x - 5}, f(x) = 5$

**120.** $f(x) = \sqrt[3]{2x + 3}, f(x) = 3$

**121.** $f(x) = \sqrt{3x^2 - 11} + 7, f(x) = 15$

**122.** $f(x) = 8 + \sqrt[3]{x^2 + 152}, f(x) = 14$

*In Exercises 123 and 124, solve the equation.*

**123.** $\sqrt{x^2 + 49} = (x^2 + 49)^{1/2}$

**124.** $\sqrt{x^2 - 16} = (x^2 - 16)^{1/2}$

## Concept/Writing Exercises

**125.** Consider the equation $\sqrt{x + 3} = -\sqrt{2x - 1}$. Explain why this equation can have no real solution.

**126.** Consider the equation $-\sqrt{x^2} = \sqrt{(-x)^2}$. By studying the equation, can you determine its solution? Explain.

**127.** Consider the equation $\sqrt[3]{x^2} = -\sqrt[3]{x^2}$. By studying the equation, can you determine its solution? Explain.

**128.** Explain without solving the equation how you can tell that $\sqrt{x - 3} + 4 = 0$ has no solution.

**129.** Does the equation $\sqrt{x} = 5$ have one or two solutions? Explain.

**130.** Does the equation $x^2 = 9$ have one or two solutions? Explain

**131. a)** Consider the equation $\sqrt{4x - 12} = x - 3$. Setting each side of the equation equal to $y$ yields the following system of equations.

$$y = \sqrt{4x - 12}$$

$$y = x - 3$$

The graphs of the equations in the system are illustrated in the figure.

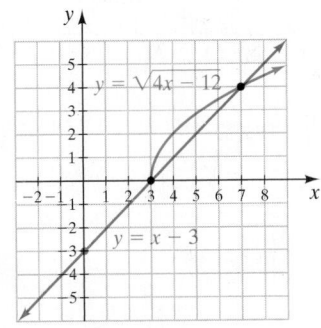

From the graphs, determine the values that appear to be solutions to the equation $\sqrt{4x - 12} = x - 3$. Explain how you determined your answer.

**b)** Substitute the values found in part **a)** into the original equation and determine whether they are the solutions to the equation.

**c)** Solve the equation $\sqrt{4x - 12} = x - 3$ algebraically and see if your solution agrees with the values obtained in part **a)**.

**132.** If the graph of a radical function, $f(x)$, does not intersect the x-axis, then the equation $f(x) = 0$ has no real solutions. Explain why.

**133.** Suppose we are given a rational function $g(x)$. If $g(4) = 0$, then the graph of $g(x)$ must intersect the x-axis at 4. Explain why.

**134.** The graph of the equation $y = \sqrt{x - 3} + 2$ is illustrated in the figure.

**a)** What is the domain of the function?

**b)** How many real solutions does the equation $\sqrt{x - 3} + 2 = 0$ have? List all the real solutions. Explain how you determined your answer.

## Challenge Problems

*Solve.*

**135.** $\sqrt{\sqrt{x + 25} - \sqrt{x}} = 5$

**136.** $\sqrt{\sqrt{x + 9} + \sqrt{x}} = 3$

*Solve each equation for n.*

**137.** $z = \dfrac{\overline{x} - \mu}{\dfrac{\sigma}{\sqrt{n}}}$

**138.** $z = \dfrac{p' - p}{\sqrt{\dfrac{pq}{n}}}$

**139. Confidence Interval** In statistics, a "confidence interval" is a range of values that is likely to contain the true value of the population. For a "95% confidence interval," the lower limit of the range, $L_1$, and the upper limit of the range, $L_2$, can be found by the formulas (above and to the right)

$$L_1 = p - 1.96\sqrt{\dfrac{p(1 - p)}{n}}$$

$$L_2 = p + 1.96\sqrt{\dfrac{p(1 - p)}{n}}$$

where $p$ represents the percent obtained from a sample and $n$ is the size of the sample. Francesco, a statistician, takes a sample of 36 families and finds that 60% of those surveyed use an answering machine in their home. He can be 95% certain that the true percent of families that use an answering machine in their home is between $L_1$ and $L_2$. Find the values of $L_1$ and $L_2$. Use $p = 0.60$ and $n = 36$ in the formulas.

**140. Quadratic Mean** The *quadratic mean* (or *root mean square, RMS*) is often used in physical applications. In power distribution systems, for example, voltages and currents are usually referred to in terms of their RMS values. The quadratic mean of a set of scores is obtained by squaring each score and adding the results (signified by $\Sigma x^2$), then dividing the value obtained by the number of scores, $n$, and then taking the square root of this value. We may express the formula as

$$\text{quadratic mean} = \sqrt{\frac{\Sigma x^2}{n}}$$

Find the quadratic mean of the numbers 2, 4, and 10.

## Group Activity

*Discuss and answer Exercise 141 as a group.*

**141. Heron's Formula** The area of a triangle is $A = \frac{1}{2}bh$. If the height is not known but we know the lengths of the three sides, we can use Heron's formula to find the area, $A$. Heron's formula is

$$A = \sqrt{S(S - a)(S - b)(S - c)}$$

where $a, b$, and $c$ are the lengths of the three sides and

$$S = \frac{a + b + c}{2}$$

**a)** Have each group member use Heron's formula to find the area of a triangle whose sides are 3 inches, 4 inches, and 5 inches.

**b)** Compare your answers for part **a)**. If any member of the group did not get the correct answer, make sure they understand their error.

**c)** Have each member of the group do the following.

    **1.** Draw a triangle on the following grid. Place each vertex of the triangle at the intersection of two grid lines.

    **2.** Measure with a ruler the length of each side of your triangle.

    **3.** Use Heron's formula to find the area of your triangle.

**d)** Compare and discuss your work from part **c)**.

## Cumulative Review Exercises

[2.6]   **142.** Solve the formula $P_1P_2 - P_1P_3 = P_2P_3$ for $P_2$.

[7.1]   **143.** Simplify $\dfrac{x(x - 5) + x(x - 2)}{2x - 7}$.

*Perform each indicated operation.*

[7.2]   **144.** $\dfrac{4a^2 - 9b^2}{4a^2 + 12ab + 9b^2} \cdot \dfrac{6a^2b}{8a^2b^2 - 12ab^3}$

[7.4]   **145.** $(t^2 - 2t - 15) \div \dfrac{t^2 - 9}{t^2 - 3t}$

[7.4]   **146.** $\dfrac{2}{x + 3} - \dfrac{1}{x - 3} + \dfrac{2x}{x^2 - 9}$

[7.6]   **147.** Solve $2 + \dfrac{3x}{x - 1} = \dfrac{8}{x - 1}$.

# 11.7 Complex Numbers

1   Recognize a complex number.

2   Add and subtract complex numbers.

3   Multiply complex numbers.

4   Divide complex numbers.

5   Find powers of *i*.

## 1   Recognize a Complex Number

In Section 11.1 we mentioned that the square roots of negative numbers, such as $\sqrt{-4}$, are not real numbers. Numbers like $\sqrt{-4}$ are called **imaginary numbers**. Although they do not belong to the set of real numbers, the imaginary numbers do exist and are very useful in mathematics and science.

    Every imaginary number has $\sqrt{-1}$ as a factor. The $\sqrt{-1}$, called the **imaginary unit**, is denoted by the letter *i*.

**Imaginary Unit**

$$i = \sqrt{-1}$$

## Understanding Algebra

The square root of any negative number is an imaginary number. For example, $\sqrt{-25}$ is an imaginary number that we write as follows:

$$\sqrt{-25} = \sqrt{-1 \cdot 25}$$
$$= \sqrt{-1} \cdot \sqrt{25}$$
$$= i \cdot 5 \text{ or } 5i$$

$\sqrt{-3}$ is an imaginary number that we write as follows:

$$\sqrt{-3} = \sqrt{-1 \cdot 3}$$
$$= \sqrt{-1} \cdot \sqrt{3}$$
$$= i\sqrt{3}$$

To write the square root of a negative number in terms of $i$, use the following property.

### Square Root of a Negative Number

For any positive real number $n$,

$$\sqrt{-n} = \sqrt{-1 \cdot n} = \sqrt{-1}\,\sqrt{n} = i\sqrt{n}$$

Therefore, we can write

$$\sqrt{-4} = \sqrt{-1}\,\sqrt{4} = i2 \quad \text{or} \quad 2i$$
$$\sqrt{-9} = \sqrt{-1}\,\sqrt{9} = i3 \quad \text{or} \quad 3i$$
$$\sqrt{-7} = \sqrt{-1}\,\sqrt{7} = i\sqrt{7}$$

In this book we will generally write $i\sqrt{7}$ rather than $\sqrt{7}i$ to avoid confusion with $\sqrt{7i}$. Also, $3\sqrt{5}i$ is written as $3i\sqrt{5}$.

### Examples

$$\sqrt{-81} = 9i \qquad\qquad \sqrt{-6} = i\sqrt{6}$$
$$\sqrt{-49} = 7i \qquad\qquad \sqrt{-10} = i\sqrt{10}$$

The real number system is a part of a larger number system, called the *complex number system*. Now we will discuss **complex numbers**.

### Complex Number

Every number of the form

$$a + bi$$

where $a$ and $b$ are real numbers and $i$ is the imaginary unit, is a **complex number**.

*Every real number and every imaginary number are also complex numbers.* A complex number has two parts: a real part, $a$, and an imaginary part, $b$.

Real part —⟶ $a$ + $b\,i$ ⟵— Imaginary part

If $b = 0$, the complex number is a real number. If $a = 0$, the complex number is a **pure imaginary number**.

### Examples of Complex Numbers

$$3 + 2i \qquad a = 3, b = 2$$
$$5 - i\sqrt{6} \qquad a = 5, b = -\sqrt{6}$$
$$4 \qquad a = 4, b = 0 \qquad \text{(real number, } b = 0)$$
$$8i \qquad a = 0, b = 8 \qquad \text{(imaginary number, } a = 0)$$
$$-i\sqrt{7} \qquad a = 0, b = -\sqrt{7} \qquad \text{(imaginary number, } a = 0)$$

We stated that all real numbers and imaginary numbers are also complex numbers. The relationship between the various sets of numbers is illustrated in **Figure 11.7**.

| Complex Numbers | | |
|---|---|---|
| Real Numbers | | Nonreal Numbers |
| Rational numbers $\frac{1}{2}, -\frac{3}{5}, \frac{9}{4}$ <br> Integers $-4, -9$ <br> Whole numbers $0, 4, 12$ | Irrational numbers $\sqrt{2}, \sqrt{3}$ <br> $-\sqrt{7}, \pi$ | $2 + 3i$ <br> $6 - 4i$ <br> $\sqrt{2} + i\sqrt{3}$ <br> Pure imaginary numbers $i\sqrt{5}$ <br> $6i$ <br> $-3i$ |

FIGURE 11.7

**EXAMPLE 1** Write each complex number in the form $a + bi$.

**a)** $7 + \sqrt{-36}$    **b)** $4 - \sqrt{-12}$    **c)** $19$    **d)** $\sqrt{-50}$    **e)** $6 + \sqrt{10}$

*Solution*

**a)** $\begin{aligned} 7 + \sqrt{-36} &= 7 + \sqrt{-1}\sqrt{36} \\ &= 7 + i6 \quad \text{or} \quad 7 + 6i \end{aligned}$

**b)** $\begin{aligned} 4 - \sqrt{-12} &= 4 - \sqrt{-1}\sqrt{12} \\ &= 4 - \sqrt{-1}\sqrt{4}\sqrt{3} \\ &= 4 - i(2)\sqrt{3} \quad \text{or} \quad 4 - 2i\sqrt{3} \end{aligned}$

**c)** $19 = 19 + 0i$

**d)** $\begin{aligned} \sqrt{-50} &= 0 + \sqrt{-50} \\ &= 0 + \sqrt{-1}\sqrt{25}\sqrt{2} \\ &= 0 + i(5)\sqrt{2} \quad \text{or} \quad 0 + 5i\sqrt{2} \end{aligned}$

**e)** Both 6 and $\sqrt{10}$ are real numbers. Written as a complex number, the answer is $(6 + \sqrt{10}) + 0i$.

*Now Try Exercise 23*

Complex numbers can be added, subtracted, multiplied, and divided. To perform these operations, we use the definitions that $i = \sqrt{-1}$ and $i^2 = -1$.

---

**Definition of $i^2$**

If $i = \sqrt{-1}$, then

$$i^2 = -1$$

---

## 2  Add and Subtract Complex Numbers

We now explain how to add or subtract complex numbers.

---

**To Add or Subtract Complex Numbers**

1. Change all imaginary numbers to $bi$ form.
2. Add (or subtract) the real parts of the complex numbers.
3. Add (or subtract) the imaginary parts of the complex numbers.
4. Write the answer in the form $a + bi$.

---

**EXAMPLE 2** Add $(9 + 15i) + (-6 - 2i) + 18$.

*Solution*   $\begin{aligned} (9 + 15i) + (-6 - 2i) + 18 &= 9 + 15i - 6 - 2i + 18 \\ &= 9 - 6 + 18 + 15i - 2i \\ &= 21 + 13i \end{aligned}$

*Now Try Exercise 27*

**EXAMPLE 3** Subtract $(8 - \sqrt{-27}) - (-3 + \sqrt{-48})$.

*Solution*

$\begin{aligned} (8 - \sqrt{-27}) - (-3 + \sqrt{-48}) &= (8 - \sqrt{-1}\sqrt{27}) - (-3 + \sqrt{-1}\sqrt{48}) \\ &= (8 - \sqrt{-1}\sqrt{9}\sqrt{3}) - (-3 + \sqrt{-1}\sqrt{16}\sqrt{3}) \\ &= (8 - 3i\sqrt{3}) - (-3 + 4i\sqrt{3}) \\ &= 8 - 3i\sqrt{3} + 3 - 4i\sqrt{3} \\ &= 8 + 3 - 3i\sqrt{3} - 4i\sqrt{3} \\ &= 11 - 7i\sqrt{3} \end{aligned}$

*Now Try Exercise 35*

## 3 Multiply Complex Numbers

Now let's discuss how to multiply complex numbers.

### To Multiply Complex Numbers

1. Change all imaginary numbers to $bi$ form.
2. Multiply the complex numbers as you would multiply polynomials.
3. Substitute $-1$ for each $i^2$.
4. Combine the real parts and the imaginary parts. Write the answer in $a + bi$ form.

**EXAMPLE 4** Multiply.

**a)** $5i(6 - 2i)$     **b)** $\sqrt{-9}(\sqrt{-3} + 8)$     **c)** $(2 - \sqrt{-18})(\sqrt{-2} + 5)$

*Solution*

**a)** $5i(6 - 2i) = 5i(6) + 5i(-2i)$      Distributive property

$= 30i - 10i^2$

$= 30i - 10(-1)$      Replace $i^2$ with $-1$.

$= 30i + 10$    or    $10 + 30i$

**b)** $\sqrt{-9}(\sqrt{-3} + 8) = 3i(i\sqrt{3} + 8)$      Change imaginary numbers to $bi$ form.

$= 3i(i\sqrt{3}) + 3i(8)$      Distributive property

$= 3i^2\sqrt{3} + 24i$

$= 3(-1)\sqrt{3} + 24i$      Replace $i^2$ with $-1$.

$= -3\sqrt{3} + 24i$

**c)** $(2 - \sqrt{-18})(\sqrt{-2} + 5) = (2 - \sqrt{-1}\sqrt{18})(\sqrt{-1}\sqrt{2} + 5)$

$= (2 - \sqrt{-1}\sqrt{9}\sqrt{2})(\sqrt{-1}\sqrt{2} + 5)$

$= (2 - 3i\sqrt{2})(i\sqrt{2} + 5)$

Now use the FOIL method to multiply.

$(2 - 3i\sqrt{2})(i\sqrt{2} + 5) = (2)(i\sqrt{2}) + (2)(5) + (-3i\sqrt{2})(i\sqrt{2}) + (-3i\sqrt{2})(5)$

$= 2i\sqrt{2} + 10 - 3i^2(2) - 15i\sqrt{2}$

$= 2i\sqrt{2} + 10 - 3(-1)(2) - 15i\sqrt{2}$

$= 2i\sqrt{2} + 10 + 6 - 15i\sqrt{2}$

$= 16 - 13i\sqrt{2}$

Now Try Exercise 45

### Avoiding Common Errors

What is $\sqrt{-4} \cdot \sqrt{-2}$?

**CORRECT**

$\sqrt{-4} \cdot \sqrt{-2} = 2i \cdot i\sqrt{2}$

$= 2i^2\sqrt{2}$

$= 2(-1)\sqrt{2}$

$= -2\sqrt{2}$

**INCORRECT**

$\sqrt{-4} \cdot \sqrt{-2} = \sqrt{8}$

$= \sqrt{4} \cdot \sqrt{2}$

$= 2\sqrt{2}$

Recall that $\sqrt{a} \cdot \sqrt{b} = \sqrt{ab}$ only for *nonnegative* real numbers $a$ and $b$.

## 4 Divide Complex Numbers

The **conjugate of a complex number** $a + bi$ is $a - bi$. For example,

| Complex Number | Conjugate |
|---|---|
| $3 + 7i$ | $3 - 7i$ |
| $1 - i\sqrt{3}$ | $1 + i\sqrt{3}$ |
| $2i$ (or $0 + 2i$) | $-2i$ (or $0 - 2i$) |

When a complex number is multiplied by its conjugate using the FOIL method, the inner and outer products have a sum of 0, and the result is a real number. For example,

$$(5 + 3i)(5 - 3i) = 25 - 15i + 15i - 9i^2$$
$$= 25 - 9i^2$$
$$= 25 - 9(-1)$$
$$= 25 + 9 = 34$$

Now we explain how to divide complex numbers.

> **To Divide Complex Numbers**
>
> 1. Change all imaginary numbers to $bi$ form.
> 2. Multiply both the numerator and denominator by the conjugate of the denominator.
> 3. Write the answer in $a + bi$ form.

**EXAMPLE 5**  Divide $\dfrac{9 + i}{i}$.

*Solution*  Begin by multiplying both numerator and denominator by $-i$, the conjugate of $i$.

$$\frac{9 + i}{i} \cdot \frac{-i}{-i} = \frac{(9 + i)(-i)}{-i^2}$$

$$= \frac{-9i - i^2}{-i^2} \qquad \text{Distributive property}$$

$$= \frac{-9i - (-1)}{-(-1)} \qquad \text{Replace } i^2 \text{ with } -1.$$

$$= \frac{-9i + 1}{1}$$

$$= 1 - 9i$$

Now Try Exercise 59

**EXAMPLE 6**  Divide $\dfrac{3 + 2i}{4 - i}$.

*Solution*  Multiply both numerator and denominator by $4 + i$, the conjugate of $4 - i$.

$$\frac{3 + 2i}{4 - i} \cdot \frac{4 + i}{4 + i} = \frac{12 + 3i + 8i + 2i^2}{16 - i^2}$$

$$= \frac{12 + 11i + 2(-1)}{16 - (-1)}$$

$$= \frac{10 + 11i}{17} = \frac{10}{17} + \frac{11}{17}i$$

Now Try Exercise 65

**EXAMPLE 7**  **Impedance**  A concept in the study of electronics is *impedance*. Impedance affects the current in a circuit. The impedance, $Z$, in a circuit is found by the formula $Z = \dfrac{V}{I}$, where $V$ is voltage and $I$ is current. Find $Z$ when $V = 1.6 - 0.3i$ and $I = -0.2i$, where $i = \sqrt{-1}$.

*Solution* $Z = \dfrac{V}{I} = \dfrac{1.6 - 0.3i}{-0.2i}$. Now multiply both numerator and denominator by the conjugate of the denominator, $0.2i$.

$$Z = \frac{1.6 - 0.3i}{-0.2i} \cdot \frac{0.2i}{0.2i} = \frac{0.32i - 0.06i^2}{-0.04i^2}$$

$$= \frac{0.32i + 0.06}{0.04}$$

$$= \frac{0.32i}{0.04} + \frac{0.06}{0.04}$$

$$= 8i + 1.5 = 1.5 + 8i$$

Now Try Exercise 121

## 5 Find Powers of *i*

Using $i = \sqrt{-1}$ and $i^2 = -1$, we can find other **powers of *i***. For example,

$$i^3 = i^2 \cdot i = -1 \cdot i = -i \qquad i^6 = i^4 \cdot i^2 = 1(-1) = -1$$
$$i^4 = i^2 \cdot i^2 = (-1)(-1) = 1 \qquad i^7 = i^4 \cdot i^3 = 1(-i) = -i$$
$$i^5 = i^4 \cdot i^1 = 1 \cdot i = i \qquad i^8 = i^4 \cdot i^4 = (1)(1) = 1$$

Note that successive powers of *i* rotate through the four values $i$, $-1$, $-i$, and $1$ (see **Fig. 11.8**).

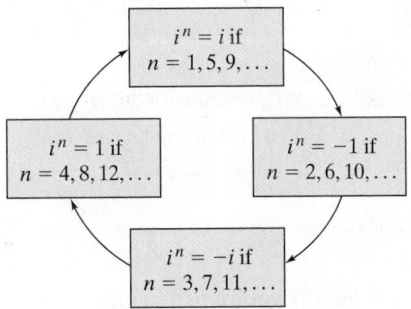

**FIGURE 11.8**

## EXAMPLE 8 Evaluate. **a)** $i^{35}$ **b)** $i^{101}$

*Solution* Write each expression as a product of factors such that the exponent of one factor is the largest multiple of 4 less than or equal to the given exponent. Then write this factor as $i^4$ raised to some power. Since $i^4$ has a value of 1, the expression $i^4$ raised to a power will also have a value of 1.

**a)** $i^{35} = i^{32} \cdot i^3 = (i^4)^8 \cdot i^3 = 1 \cdot i^3 = 1(-i) = -i$

**b)** $i^{101} = i^{100} \cdot i^1 = (i^4)^{25} \cdot i = 1 \cdot i = i$

Now Try Exercise 101

### Helpful Hint

A quick way of evaluating $i^n$ is to divide the exponent by 4 and observe the remainder.

If the remainder is 0, the value is 1.      If the remainder is 2, the value is $-1$.
If the remainder is 1, the value is $i$.      If the remainder is 3, the value is $-i$.

For Example 8 **a)**
$$\begin{array}{r} 8 \\ 4\overline{)35} \\ \underline{32} \\ 3 \end{array}$$
Answer is $-i$.

For 8 **b)**
$$\begin{array}{r} 25 \\ 4\overline{)101} \\ \underline{8} \\ 21 \\ \underline{20} \\ 1 \end{array}$$
Answer is $i$.

$i^{35} = (i^4)^8 \cdot i^3 = (1)^8 \cdot i^3 = 1 \cdot i^3 = i^3 = -i$

EXAMPLE 9  Let $f(x) = x^2$. Find   **a)** $f(6i)$      **b)** $f(3 + 7i)$.

*Solution*

**a)** $f(x) = x^2$

$f(6i) = (6i)^2 = 36i^2 = 36(-1) = -36$

**b)** $f(x) = x^2$

$$f(3 + 7i) = (3 + 7i)^2 = (3)^2 + 2(3)(7i) + (7i)^2$$
$$= 9 + 42i + 49i^2$$
$$= 9 + 42i + 49(-1)$$
$$= 9 + 42i - 49$$
$$= -40 + 42i$$

Now Try Exercise 111

# EXERCISE SET 11.7    MathXL®    MyMathLab®

## Warm-Up Exercises

*Fill in the blanks with the appropriate word, phrase, or symbol(s) from the following list.*

| | | | | | |
|---|---|---|---|---|---|
| −1 | pure | real | complex | number | −i |
| unit | part | 1 | conjugate | imaginary | rational |

**1.** The square root of any negative number is an _____ number.

**2.** The imaginary _____ is denoted with the letter $i$.

**3.** Every number of the form $a + bi$ where $a$ and $b$ are real numbers and $i$ is the imaginary unit, is a _____ number.

**4.** In a complex number, $a + bi$, $a$ is known as the _____ part.

**5.** In a complex number, $a + bi$, $b$ is known as the imaginary _____.

**6.** If $a = 0$, the complex number $a + bi$ is a _____ imaginary number.

**7.** If $b = 0$, the complex number $a + bi$ is a real _____.

**8.** For a complex number $a + bi$ the _____ is $a - bi$.

**9.** If $i$ is the imaginary unit, then $i^2 =$ _____.

**10.** If $i$ is the imaginary unit, then $i^3 =$ _____.

## Practice the Skills

*Write each expression as a complex number in the form $a + bi$.*

**11.** 4

**12.** $2i$

**13.** $\sqrt{25}$

**14.** $\sqrt{-100}$

**15.** $21 - \sqrt{-36}$

**16.** $\sqrt{3} + \sqrt{-3}$

**17.** $\sqrt{-24}$

**18.** $\sqrt{49} - \sqrt{-49}$

**19.** $8 - \sqrt{-12}$

**20.** $\sqrt{-9} + \sqrt{-81}$

**21.** $3 + \sqrt{-98}$

**22.** $\sqrt{-9} + 7i$

**23.** $12 - \sqrt{-25}$

**24.** $10 + \sqrt{-32}$

**25.** $7i - \sqrt{-45}$

**26.** $\sqrt{144} + \sqrt{-96}$

*Add or subtract.*

**27.** $(3 + 7i) + (1 - 3i)$

**28.** $(-4 + 5i) + (2 - 4i)$

**29.** $(3 + 7i) - (1 - 3i)$

**30.** $(7 - \sqrt{-4}) - (-1 - \sqrt{-16})$

**31.** $(1 + \sqrt{-1}) + (-18 - \sqrt{-169})$

**32.** $(16 - i\sqrt{3}) + (17 - \sqrt{-3})$

**33.** $(\sqrt{3} + \sqrt{2}) + (3\sqrt{2} - \sqrt{-8})$

**34.** $(8 - \sqrt{2}) - (5 + \sqrt{-15})$

**35.** $(5 - \sqrt{-72}) + (6 + \sqrt{-8})$

**36.** $(29 + \sqrt{-75}) + (\sqrt{-147})$

**37.** $(\sqrt{4} - \sqrt{-45}) + (-\sqrt{25} + \sqrt{-5})$

**38.** $(\sqrt{20} - \sqrt{-12}) + (2\sqrt{5} + \sqrt{-75})$

*Multiply.*

**39.** $3(2 - 4i)$

**40.** $-2(-3 + i)$

**41.** $i(4 + 9i)$

**42.** $3i(6 - i)$

**43.** $\sqrt{-9}(6 + 11i)$

**44.** $\dfrac{1}{2}i\left(\dfrac{1}{3} - 18i\right)$

**45.** $\sqrt{-16}(\sqrt{3} - 7i)$

**46.** $-\sqrt{-24}(\sqrt{6} - \sqrt{-3})$

**47.** $\sqrt{-27}(\sqrt{3} - \sqrt{-3})$

**48.** $\sqrt{-32}(\sqrt{2} + \sqrt{-8})$

**49.** $(3 + 2i)(1 + i)$

**50.** $(6 - 2i)(3 + i)$

**51.** $(10 - 3i)(10 + 3i)$

**52.** $(-4 + 3i)(2 - 5i)$

**53.** $(7 + \sqrt{-2})(5 - \sqrt{-8})$

**54.** $(\sqrt{4} - 3i)(4 + \sqrt{-4})$

**55.** $\left(\dfrac{1}{2} - \dfrac{1}{3}i\right)\left(\dfrac{1}{4} + \dfrac{2}{3}i\right)$

**56.** $\left(\dfrac{3}{5} - \dfrac{1}{4}i\right)\left(\dfrac{2}{3} + \dfrac{2}{5}i\right)$

*Divide.*

**57.** $\dfrac{2}{3i}$

**58.** $\dfrac{-3}{4i}$

**59.** $\dfrac{2 + 3i}{2i}$

**60.** $\dfrac{7 - 3i}{2i}$

**61.** $\dfrac{6}{2 - i}$

**62.** $\dfrac{9}{5 + i}$

**63.** $\dfrac{3}{1 - 2i}$

**64.** $\dfrac{13}{-3 - 4i}$

**65.** $\dfrac{6 - 3i}{4 + 2i}$

**66.** $\dfrac{4 - 3i}{4 + 3i}$

**67.** $\dfrac{4}{6 - \sqrt{-4}}$

**68.** $\dfrac{2}{3 + \sqrt{-5}}$

**69.** $\dfrac{\sqrt{2}}{5 + \sqrt{-12}}$

**70.** $\dfrac{\sqrt{6}}{\sqrt{3} - \sqrt{-9}}$

**71.** $\dfrac{5 + \sqrt{-4}}{4 - \sqrt{-16}}$

**72.** $\dfrac{2 + \sqrt{-25}}{4 - \sqrt{-9}}$

**73.** $\dfrac{\sqrt{-75}}{\sqrt{-3}}$

**74.** $\dfrac{\sqrt{-30}}{\sqrt{-2}}$

**75.** $\dfrac{\sqrt{-32}}{\sqrt{-18}\sqrt{8}}$

**76.** $\dfrac{\sqrt{-40}\sqrt{-20}}{\sqrt{-4}}$

*Perform each indicated operation. These exercises are a combination of the types of exercises presented earlier in this exercise set.*

**77.** $(-5 + 7i) + (5 - 7i)$

**78.** $\left(\dfrac{4}{3} + \dfrac{1}{2}i\right) - \left(\dfrac{1}{3} - \dfrac{1}{2}i\right)$

**79.** $(\sqrt{50} - \sqrt{2}) - (\sqrt{-12} - \sqrt{-48})$

**80.** $(8 - \sqrt{-6}) - (2 - \sqrt{-24})$

**81.** $5.2(4 - 3.2i)$

**82.** $\sqrt{-6}(\sqrt{3} - \sqrt{-10})$

**83.** $(9 + 2i)(3 - 5i)$

**84.** $(\sqrt{3} + 2i)(\sqrt{6} - \sqrt{-8})$

**85.** $\dfrac{11 + 4i}{2i}$

**86.** $\dfrac{1}{4 + 3i}$

**87.** $\dfrac{6}{\sqrt{3} - \sqrt{-4}}$

**88.** $\dfrac{5 - 2i}{3 + 2i}$

**89.** $\left(11 - \dfrac{5}{9}i\right) - \left(4 - \dfrac{3}{5}i\right)$

**90.** $\dfrac{8}{7}\left(4 - \dfrac{2}{5}i\right)$

**91.** $\left(\dfrac{2}{3} - \dfrac{1}{5}i\right)\left(\dfrac{3}{5} - \dfrac{3}{4}i\right)$

**92.** $\sqrt{\dfrac{4}{9}}\left(\sqrt{\dfrac{25}{36}} - \sqrt{-\dfrac{4}{25}}\right)$

**93.** $\dfrac{\sqrt{-48}}{\sqrt{-12}}$

**94.** $\dfrac{2 - \sqrt{-9}}{4 + \sqrt{-49}}$

**95.** $(5.23 - 6.41i) - (9.56 + 4.5i)$

**96.** $(7 + \sqrt{-9})(4 - \sqrt{-4})$

*For each imaginary number, indicate whether its value is $i$, $-1$, $-i$, or $1$.*

**97.** $i^6$

**98.** $i^{63}$

**99.** $i^{160}$

**100.** $i^{231}$

**101.** $i^{93}$

**102.** $i^{103}$

**103.** $i^{811}$

**104.** $i^{1213}$

## Problem Solving

**105.** Consider the complex number $2 + 3i$.
  **a)** Find the additive inverse.
  **b)** Find the multiplicative inverse. Write the answer in simplified form.

**106.** Consider the complex number $4 - 5i$.
  **a)** Find the additive inverse.
  **b)** Find the multiplicative inverse. Write the answer in simplified form.

**107.** If $f(x) = x^2$, find $f(2i)$.

**108.** If $f(x) = x^2$, find $f(4i)$.

**109.** If $f(x) = x^4 - 2x$, find $f(2i)$.

**110.** If $f(x) = x^3 - 4x^2$, find $f(5i.)$

**111.** If $f(x) = x^2 + 2x$, find $f(3 + i)$.

**112.** If $f(x) = \dfrac{x^2}{x - 2}$, find $f(4 - i)$.

*Evaluate each expression for the given value of x.*

**113.** $x^2 - 2x + 5, x = 1 + 2i$

**114.** $x^2 - 2x + 5, x = 1 - 2i$

**115.** $x^2 + 2x + 7, x = -1 + i\sqrt{5}$

**116.** $x^2 + 2x + 9, x = -1 - i\sqrt{5}$

*In Exercises 117–120, determine whether the given value of x is a solution to the equation.*

**117.** $x^2 - 4x + 5 = 0, x = 2 - i$

**118.** $x^2 - 4x + 5 = 0, x = 2 + i$

**119.** $x^2 - 6x + 11 = 0, x = -3 + i\sqrt{3}$

**120.** $x^2 - 6x + 15 = 0, x = 3 - i\sqrt{3}$

**121. Impedance** Find the impedance, $Z$, using the formula $Z = \dfrac{V}{I}$ when $V = 1.8 + 0.5i$ and $I = 0.6i$. See Example 7.

**122. Impedance** Refer to Exercise 121. Find the impedance when $V = 2.4 - 0.6i$ and $I = -0.4i$.

**123. Impedance** Under certain conditions, the total impedance, $Z_T$, of a circuit is given by the formula

$$Z_T = \frac{Z_1 Z_2}{Z_1 + Z_2}$$

Find $Z_T$ when $Z_1 = 2 - i$ and $Z_2 = 4 + i$.

**124. Impedance** Refer to Exercise 123. Find $Z_T$ when $Z_1 = 3 - i$ and $Z_2 = 5 + i$.

**125.** Determine whether $i^{-1}$ is equal to $i, -1, -i$, or 1. Show your work.

**126.** Determine whether $i^{-5}$ is equal to $i, -1, -i$, or 1. Show your work.

*In Chapter 12, we will use the quadratic formula* $x = \dfrac{-b \pm \sqrt{b^2 - 4ac}}{2a}$ *to solve equations of the form* $ax^2 + bx + c = 0$. **a)** *Use the quadratic formula to solve the following quadratic equations.* **b)** *Check each of the two solutions by substituting the values found for x (one at a time) back into the original equation. In these exercises, the* ± *(read "plus or minus") results in two distinct complex answers.*

**127.** $x^2 - 2x + 6 = 0$

**128.** $x^2 - 4x + 6 = 0$

*Given the complex numbers* $a = 5 + 2i\sqrt{3}, b = 1 + i\sqrt{3}$, *evaluate each expression.*

**129.** $a + b$

**130.** $a - b$

**131.** $ab$

**132.** $\dfrac{a}{b}$

## Concept/Writing Exercises

**133.** Are all of the following complex numbers? If any are not complex numbers, explain why.

**a)** 9     **b)** $-\dfrac{1}{2}$     **c)** $4 - \sqrt{-2}$

**d)** $7 - 3i$     **e)** $4.2i$     **f)** $11 + \sqrt{3}$

**134.** Is every real and every imaginary number a complex number?

**135.** Is every complex number a real number?

**136. a)** Is $i \cdot i$ a real number? Explain.

**b)** Is $i \cdot i \cdot i$ a real number? Explain.

*In Exercises 137–140, answer true or false. Support your answer with an example.*

**137.** The product of two pure imaginary numbers is always a real number.

**138.** The sum of two pure imaginary numbers is always an imaginary number.

**139.** The product of two complex numbers is always a real number.

**140.** The sum of two complex numbers is always a complex number.

**141.** What values of $n$ will result in $i^n$ being a real number? Explain.

**142.** What values of $n$ will result in $i^{2n}$ being a real number? Explain.

## Cumulative Review Exercises

[5.6] **143.** Divide $\dfrac{8c^2 + 6c - 35}{4c + 9}$.

[7.4] **144.** Add $\dfrac{b}{a - b} + \dfrac{a + b}{b}$.

[7.6] **145.** Solve $\dfrac{x}{4} + \dfrac{1}{2} = \dfrac{x - 1}{2}$.

[9.5] **146. Mixture** Berreda Coughlin, a grocer in Dallas, has two coffees, one selling for $5.50 per pound and the other for $6.30 per pound. How many pounds of each type of coffee should he mix to make 40 pounds of coffee to sell for $6.00 per pound?

© Andrey Armyagov/Shutterstock

# Chapter 11 Summary

| IMPORTANT FACTS AND CONCEPTS | EXAMPLES |
|---|---|
| **Section 11.1** ||
| A **radical expression** has the form $\sqrt[n]{x}$, where $n$ is the index and $x$ is the radicand. | In the radical expression $\sqrt[3]{x}$, 3 is the index and $x$ is the radicand. |
| The **principal square root** of a positive number $a$, written $\sqrt{a}$, is the positive number $b$ such that $b^2 = a$. | $\sqrt{81} = 9$, since $9^2 = 81$<br>$\sqrt{0.36} = 0.6$ since $(0.6)^2 = 0.36$ |
| The **square root function** is $f(x) = \sqrt{x}$. Its domain is $[0, \infty)$ and its range is $[0, \infty)$. | <br><br>$f(x) = \sqrt{x}$<br><br><br> |
| The **cube root** of a number $a$, written $\sqrt[3]{a}$, is the number $b$ such that $b^3 = a$. | $\sqrt[3]{27} = 3$ since $3^3 = 27$<br>$\sqrt[3]{-125} = -5$ since $(-5)^3 = -125$ |
| The **cube root function** is $f(x) = \sqrt[3]{x}$. Its domain is $(-\infty, \infty)$ or $\mathbb{R}$ and its range is $(-\infty, \infty)$ or $\mathbb{R}$. | <br><br>$f(x) = \sqrt[3]{x}$<br><br><br> |
| The $n$th root of $a$, $\sqrt[n]{a}$, where $n$ is an *even index* and $a$ is a nonnegative real number, is called an **even root** and is the nonnegative number $b$ such that $b^n = a$. | $\sqrt{4} = 2$ since $2^2 = 2 \cdot 2 = 4$<br>$\sqrt[4]{81} = 3$ since $3^4 = 3 \cdot 3 \cdot 3 \cdot 3 = 81$ |
| The $n$th root of $a$, $\sqrt[n]{a}$, where $n$ is an *odd index* and $a$ is any real number, is called an **odd root** and is the real number $b$ such that $b^n = a$. | $\sqrt[3]{27} = 3$ since $3^3 = 3 \cdot 3 \cdot 3 = 27$<br>$\sqrt[5]{-32} = -2$ since $(-2)^5 = (-2)(-2)(-2)(-2)(-2) = -32$ |
| For any real number $a$, $\sqrt{a^2} = |a|$. | $\sqrt{(-6)^2} = |-6| = 6$<br>$\sqrt{(y + 8)^2} = |y + 8|$ |
| **Section 11.2** ||
| **Rational Exponent**<br>$$\sqrt[n]{a} = a^{1/n}$$<br>When $a$ is nonnegative, $n$ can be any index.<br>When $a$ is negative, $n$ must be odd. | $\sqrt{17} = 17^{1/2}$<br>$\sqrt[4]{21x^3y^2} = (21x^3y^2)^{1/4}$ |
| For any nonnegative number $a$, and integers $m$ and $n$,<br> | $\sqrt[4]{z^9} = (\sqrt[4]{z})^9 = z^{9/4}$ |
| For any nonnegative real number $a$,<br>$$\sqrt[n]{a^n} = (\sqrt[n]{a})^n = a^{n/n} = a$$ | $\sqrt[4]{y^4} = y, \qquad \sqrt[8]{14^8} = 14$ |

| IMPORTANT FACTS AND CONCEPTS | EXAMPLES |
|---|---|

### Section 11.2 (cont.)

**Rules of Exponents**

For all real numbers $a$ and $b$ and all rational numbers $m$ and $n$,

Product rule $\qquad\qquad a^m \cdot a^n = a^{m+n}$

$$x^{1/3} \cdot x^{4/3} = x^{(1/3)+(4/3)} = x^{5/3}$$

Quotient rule $\qquad\qquad \dfrac{a^m}{a^n} = a^{m-n}, \quad a \neq 0$

$$\dfrac{x^{4/5}}{x^{1/2}} = x^{(4/5)-(1/2)} = x^{(8/10)-(5/10)} = x^{3/10}$$

Negative exponent rule $\qquad a^{-m} = \dfrac{1}{a^m}, \quad a \neq 0$

$$x^{-1/7} = \dfrac{1}{x^{1/7}}$$

Zero exponent rule $\qquad a^0 = 1, \quad a \neq 0$

$$m^0 = 1$$

Raising a power to a power $\quad (a^m)^n = a^{m\cdot n}$

$$(c^{1/8})^{16} = c^{(1/8)\cdot 16} = c^2$$

Raising a product to a power $\quad (ab)^m = a^m b^m$

$$(p^3 q^4)^{1/8} = p^{3/8} q^{1/2}$$

Raising a quotient to a power $\quad \left(\dfrac{a}{b}\right)^m = \dfrac{a^m}{b^m}, \quad b \neq 0$

$$\left(\dfrac{a}{b}\right)^{-n} = \left(\dfrac{b}{a}\right)^n = \dfrac{b^n}{a^n}, \quad a \neq 0, b \neq 0$$

$$\left(\dfrac{81}{49}\right)^{-1/2} = \left(\dfrac{49}{81}\right)^{1/2} = \dfrac{49^{1/2}}{81^{1/2}} = \dfrac{7}{9}$$

### Section 11.3

A number or expression is a **perfect square** if it is the square of an expression.

A number or expression is a **perfect cube** if it is the cube of an expression.

| | | | | |
|---|---|---|---|---|
| Perfect squares: | 49 | 81 | $x^{12}$ | $y^{50}$ |
| | ↓ | ↓ | ↓ | ↓ |
| Square of a number or expression: | $7^2$ | $9^2$ | $(x^6)^2$ | $(y^{25})^2$ |
| Perfect cubes: | 27 | −27 | $y^{18}$ | $z^{30}$ |
| | ↓ | ↓ | ↓ | ↓ |
| Cube of a number or expression: | $3^3$ | $(-3)^3$ | $(y^6)^3$ | $(z^{10})^3$ |

**Product Rule for Radicals**

For nonnegative real numbers $a$ and $b$,

$$\sqrt[n]{a} \cdot \sqrt[n]{b} = \sqrt[n]{ab}$$

$$\sqrt{2} \cdot \sqrt{8} = \sqrt{16} = 4, \quad \sqrt[3]{2x^3} = \sqrt[3]{x^3} \cdot \sqrt[3]{2} = x\sqrt[3]{2}$$

**To Simplify Radicals Using the Product Rule**

1. If the radicand contains a coefficient other than 1, write it as a product of two numbers, one of which is the largest perfect power for the index.

2. Write each variable factor as a product of two factors, one of which is the largest perfect power of the variable for the index.

3. Use the product rule to write the radical expression as a product of radicals. Place all the perfect powers (numbers and variables) under the same radical.

4. Simplify the radical containing the perfect powers.

$$\sqrt{24} = \sqrt{4 \cdot 6} = \sqrt{4}\sqrt{6} = 2\sqrt{6}$$

$$\sqrt[3]{16x^5 y^9} = \sqrt[3]{8x^3 y^9 \cdot 2x^2}$$

$$= \sqrt[3]{8x^3 y^9} \sqrt[3]{2x^2}$$

$$= 2xy^3 \sqrt[3]{2x^2}$$

**Quotient Rule for Radicals**

For nonnegative real numbers $a$ and $b$,

$$\dfrac{\sqrt[n]{a}}{\sqrt[n]{b}} = \sqrt[n]{\dfrac{a}{b}}, \quad b \neq 0$$

$$\dfrac{\sqrt{32}}{\sqrt{2}} = \sqrt{\dfrac{32}{2}} = \sqrt{16} = 4, \quad \sqrt[3]{\dfrac{x^6}{y^{12}}} = \dfrac{\sqrt[3]{x^6}}{\sqrt[3]{y^{12}}} = \dfrac{x^2}{y^4}$$

### Section 11.4

**Like radicals** are radicals with the same radicand and index.
**Unlike radicals** are radicals with a different radicand or index.

| Like Radicals | Unlike Radicals |
|---|---|
| $\sqrt{3}, \quad 12\sqrt{3}$ | $\sqrt{3}, \quad 7\sqrt[3]{3}$ |
| $2\sqrt[4]{xy^3}, \quad -3\sqrt[4]{xy^3}$ | $\sqrt[5]{xy^3}, \quad x\sqrt[5]{y^3}$ |

**To Add or Subtract Radicals**

1. Simplify each radical expression.

2. Combine like radicals (if there are any).

$$\sqrt{27} + \sqrt{48} - 2\sqrt{75} = \sqrt{9} \cdot \sqrt{3} + \sqrt{16} \cdot \sqrt{3} - 2 \cdot \sqrt{25} \cdot \sqrt{3}$$

$$= 3\sqrt{3} + 4\sqrt{3} - 10\sqrt{3}$$

$$= -3\sqrt{3}$$

| IMPORTANT FACTS AND CONCEPTS | EXAMPLES |
|---|---|

**Section 11.4 (cont.)**

**To Multiply Radicals**

Use the product rule.

$$\sqrt[n]{a} \cdot \sqrt[n]{b} = \sqrt[n]{ab}$$

$$\sqrt[4]{8c^2}\,\sqrt[4]{4c^3} = \sqrt[4]{32c^5} = \sqrt[4]{16c^4}\,\sqrt[4]{2c}$$
$$= 2c\,\sqrt[4]{2c}$$

**Section 11.5**

To **rationalize a denominator** multiply both the numerator and the denominator of the fraction by a radical that will result in the radicand in the denominator becoming a perfect power.

$$\frac{6}{\sqrt{3x}} \cdot \frac{\sqrt{3x}}{\sqrt{3x}} = \frac{6\sqrt{3x}}{\sqrt{9x^2}} = \frac{6\sqrt{3x}}{3x} = \frac{2\sqrt{3x}}{x}$$

**A Radical Expression Is Simplified When the Following Are All True**

1. No perfect powers are factors of the radicand and all exponents in the radicand are less than the index.

2. No radicand contains a fraction.

3. No denominator contains a radical.

|  | Not Simplified | Simplified |
|---|---|---|
| **1.** | $\sqrt{x^3}$ | $x\sqrt{x}$ |
| **2.** | $\sqrt{\dfrac{1}{2}}$ | $\dfrac{\sqrt{2}}{2}$ |
| **3.** | $\dfrac{1}{\sqrt{2}}$ | $\dfrac{\sqrt{2}}{2}$ |

**Section 11.6**

**To Solve Radical Equations**

1. Rewrite the equation so that one radical containing a variable is by itself (isolated) on one side of the equation.

2. Raise each side of the equation to a power equal to the index of the radical.

3. Combine like terms.

4. If the equation still contains a term with a variable in a radicand, repeat steps 1 through 3.

5. Solve the resulting equation for the variable.

6. Check all solutions in the original equation for extraneous solutions.

Solve $\sqrt{x} - 8 = 0$.

$$\sqrt{x} - 8 = 0$$
$$\sqrt{x} = 8$$
$$(\sqrt{x})^2 = 8^2$$
$$x = 64$$

A check shows that 64 is the solution.

**Section 11.7**

The **imaginary unit**, $i$ is defined as $i = \sqrt{-1}$. (Also, $i^2 = -1$.)

$$\sqrt{-25} = \sqrt{25}\,\sqrt{-1} = 5i$$

**Imaginary Number**

For any positive number $n$,
$$\sqrt{-n} = i\sqrt{n}.$$

$$\sqrt{-19} = i\sqrt{19}$$

A **complex number** is a number of the form $a + bi$, where $a$ and $b$ are real numbers.

$3 + 2i$ and $26 - 15i$ are complex numbers.

**To Add or Subtract Complex Numbers**

1. Change all imaginary numbers to $bi$ form.

2. Add (or subtract) the real parts of the complex numbers.

3. Add (or subtract) the imaginary parts of the complex numbers.

4. Write the answer in the form $a + bi$.

Add $(8 - 3i) + (12 + 5i)$.

$$(8 - 3i) + (12 + 5i)$$
$$= 8 + 12 - 3i + 5i$$
$$= 20 + 2i$$

**To Multiply Complex Numbers**

1. Change all imaginary numbers to $bi$ form.

2. Multiply the complex numbers as you would multiply polynomials.

3. Substitute $-1$ for each $i^2$.

4. Combine the real parts and the imaginary parts. Write the answer in $a + bi$ form.

Multiply $(7 + 2i\sqrt{3})(5 - 4i\sqrt{3})$.

$$(7 + 2i\sqrt{3})(5 - 4i\sqrt{3})$$
$$= 35 - 28i\sqrt{3} + 10i\sqrt{3} - 8(i^2)(3)$$
$$= 35 - 28i\sqrt{3} + 10i\sqrt{3} + 24$$
$$= 59 - 18i\sqrt{3}$$

| IMPORTANT FACTS AND CONCEPTS | EXAMPLES |
|---|---|

**Section 11.7 (cont.)**

| The **conjugate of a complex number** $a + bi$ is $a - bi$. | Complex Number | Conjugate |
|---|---|---|
| | $14 + 2i$ | $14 - 2i$ |
| | $-17 - 8i$ | $-17 + 8i$ |

**To Divide Complex Numbers**

1. Change all imaginary numbers to $bi$ form.
2. Rationalize the denominator by multiplying both the numerator and denominator by the conjugate of the denominator.
3. Write the answer in $a + bi$ form.

Divide $\dfrac{2 - i}{5 + 3i}$.

$$\frac{2 - i}{5 + 3i} \cdot \frac{5 - 3i}{5 - 3i} = \frac{10 - 6i - 5i + 3i^2}{25 - 9i^2} = \frac{7 - 11i}{34}$$

**Powers of $i$**
$$i^2 = -1, i^3 = -i, i^4 = 1, i^5 = i$$

$$i^{38} = i^{36} \cdot i^2 = (i^4)^9 \cdot i^2 = 1^9 \cdot (-1) = -1$$
$$i^{63} = i^{60} \cdot i^3 = (i^4)^{15} \cdot i^3 = 1^{15}(-i) = -i$$

## Chapter 11 Review Exercises

[11.1]   *Evaluate.*

**1.** $\sqrt{9}$    **2.** $\sqrt[3]{-27}$    **3.** $\sqrt[3]{-125}$    **4.** $\sqrt[4]{256}$

*Use absolute value to evaluate.*

**5.** $\sqrt{(-5)^2}$    **6.** $\sqrt{(38.2)^2}$

*Write as an absolute value.*

**7.** $\sqrt{x^2}$    **8.** $\sqrt{(x + 7)^2}$    **9.** $\sqrt{(x - y)^2}$    **10.** $\sqrt{(x^2 - 4x + 12)^2}$

**11.** Let $f(x) = \sqrt{10x + 9}$. Find $f(4)$.    **12.** Let $k(x) = 2x + \sqrt{\dfrac{x}{3}}$. Find $k(27)$.

**13.** Let $g(x) = \sqrt[3]{2x + 3}$. Find $g(4)$ and round the answer to the nearest tenth.    **14. Area** The area of a square is 144 square meters. Find the length of its side.

*For the remainder of these review exercises, assume that all variables represent positive real numbers.*

[11.2]   *Write in exponential form.*

**15.** $\sqrt{x^3}$    **16.** $\sqrt[3]{x^5}$    **17.** $(\sqrt[4]{y})^{13}$    **18.** $\sqrt[7]{6^{-2}}$

*Write in radical form.*

**19.** $x^{1/2}$    **20.** $a^{2/3}$    **21.** $(8m^2n)^{7/4}$    **22.** $(x + y)^{-5/3}$

*Simplify each radical expression by changing the expression to exponential form. Write the answer in radical form when appropriate.*

**23.** $\sqrt[3]{4^6}$    **24.** $\sqrt{x^{12}}$    **25.** $(\sqrt[4]{9})^8$    **26.** $\sqrt[20]{a^5}$

*Evaluate if possible. If the expression is not a real number, so state.*

**27.** $-36^{1/2}$    **28.** $(-36)^{1/2}$    **29.** $\left(\dfrac{64}{27}\right)^{-1/3}$    **30.** $64^{-1/2} + 8^{-2/3}$

*Simplify. Write the answer without negative exponents.*

**31.** $x^{3/5} \cdot x^{-1/3}$    **32.** $\left(\dfrac{64}{y^9}\right)^{1/3}$    **33.** $\left(\dfrac{a^{-6/5}}{a^{2/5}}\right)^{2/3}$    **34.** $\left(\dfrac{20x^5y^{-3}}{4y^{1/2}}\right)^2$

*Multiply.*

**35.** $a^{1/2}(5a^{3/2} - 3a^2)$    **36.** $4x^{-2/3}\left(x^{-1/2} + \dfrac{11}{4}x^{2/3}\right)$

*Factor each expression. Write the answer without negative exponents.*

**37.** $x^{2/5} + x^{7/5}$    **38.** $a^{-1/2} + a^{3/2}$

*For each function, find the indicated value of the function. Use your calculator to evaluate irrational numbers. Round irrational numbers to the nearest thousandth.*

**39.** If $f(x) = \sqrt{6x - 11}$, find $f(6)$.    **40.** If $g(x) = \sqrt[3]{9x - 17}$, find $g(4)$.

[11.1]   *Graph the following functions.*

**41.** $f(x) = \sqrt{x}$    **42.** $f(x) = \sqrt{x} - 4$

[11.2–11.5] *Simplify.*

**43.** $\sqrt{18}$

**44.** $\sqrt[3]{16}$

**45.** $\sqrt{\dfrac{49}{9}}$

**46.** $\sqrt[3]{\dfrac{8}{125}}$

**47.** $-\sqrt{\dfrac{81}{49}}$

**48.** $\sqrt[3]{-\dfrac{27}{125}}$

**49.** $\sqrt{32}\,\sqrt{2}$

**50.** $\sqrt[3]{32}\,\sqrt[3]{2}$

**51.** $\sqrt{18x^2y^3z^4}$

**52.** $\sqrt{75x^3y^7}$

**53.** $\sqrt[3]{54a^7b^{10}}$

**54.** $\sqrt[3]{125x^8y^9z^{16}}$

**55.** $\left(\sqrt[6]{x^2y^3z^5}\right)^{42}$

**56.** $\left(\sqrt[5]{2ab^4c^6}\right)^{15}$

**57.** $\sqrt{5x}\,\sqrt{8x^5}$

**58.** $\sqrt[3]{2x^2y}\,\sqrt[3]{4x^9y^4}$

**59.** $\sqrt[3]{2x^4y^5}\,\sqrt[3]{16x^4y^4}$

**60.** $\sqrt[4]{4x^4y^7}\,\sqrt[4]{4x^5y^9}$

**61.** $\sqrt{3x}\left(\sqrt{12x}-\sqrt{20}\right)$

**62.** $\sqrt[3]{2x^2y}\left(\sqrt[3]{4x^4y^7}+\sqrt[3]{9x}\right)$

**63.** $\sqrt{\sqrt{a^3b^2}}$

**64.** $\sqrt{\sqrt[3]{x^5y^2}}$

**65.** $\left(\dfrac{4r^2p^{1/3}}{r^{1/2}p^{4/3}}\right)^3$

**66.** $\left(\dfrac{6y^{2/5}z^{1/3}}{x^{-1}y^{3/5}}\right)^{-1}$

**67.** $\sqrt{\dfrac{3}{5}}$

**68.** $\sqrt[3]{\dfrac{7}{9}}$

**69.** $\sqrt[4]{\dfrac{5}{4}}$

**70.** $\dfrac{x}{\sqrt{10}}$

**71.** $\dfrac{8}{\sqrt{x}}$

**72.** $\dfrac{m}{\sqrt[3]{25}}$

**73.** $\dfrac{10}{\sqrt[3]{y^2}}$

**74.** $\dfrac{9}{\sqrt[4]{z}}$

**75.** $\sqrt[3]{\dfrac{x^3}{27}}$

**76.** $\dfrac{\sqrt[3]{2x^{10}}}{\sqrt[3]{16x^7}}$

**77.** $\sqrt{\dfrac{32x^2y^5}{2x^8y}}$

**78.** $\sqrt[4]{\dfrac{48x^9y^{15}}{3xy^3}}$

**79.** $\sqrt{\dfrac{6x^4}{y}}$

**80.** $\sqrt{\dfrac{12a}{7b}}$

**81.** $\sqrt{\dfrac{18x^4y^5}{3z}}$

**82.** $\sqrt{\dfrac{125x^2y^5}{3z}}$

**83.** $\sqrt[3]{\dfrac{108x^3y^7}{2y^3}}$

**84.** $\sqrt[3]{\dfrac{3x}{5y}}$

**85.** $\sqrt[3]{\dfrac{9x^5y^3}{x^6}}$

**86.** $\sqrt[3]{\dfrac{y^6}{5x^2}}$

**87.** $\sqrt[4]{\dfrac{2a^2b^{11}}{a^5b}}$

**88.** $\sqrt[4]{\dfrac{3x^2y^6}{8x^3}}$

**89.** $(3-\sqrt{2})(3+\sqrt{2})$

**90.** $(\sqrt{3}+\sqrt{5})(\sqrt{3}-\sqrt{5})$

**91.** $(x-\sqrt{y})(x+\sqrt{y})$

**92.** $(\sqrt{3}+2)^2$

**93.** $(\sqrt{x}-\sqrt{3y})(\sqrt{x}+\sqrt{5y})$

**94.** $(\sqrt[3]{2x}-\sqrt[3]{3y})(\sqrt[3]{3x}-\sqrt[3]{2y})$

**95.** $\dfrac{6}{2+\sqrt{5}}$

**96.** $\dfrac{x}{4+\sqrt{x}}$

**97.** $\dfrac{a}{4-\sqrt{b}}$

**98.** $\dfrac{x}{\sqrt{y}-7}$

**99.** $\dfrac{\sqrt{x}}{\sqrt{x}+\sqrt{y}}$

**100.** $\dfrac{\sqrt{x}-3\sqrt{y}}{\sqrt{x}-\sqrt{y}}$

**101.** $\dfrac{2}{\sqrt{a-1}-2}$

**102.** $\dfrac{5}{\sqrt{y+2}-3}$

**103.** $\sqrt[3]{x}+10\sqrt[3]{x}-2\sqrt[3]{x}$

**104.** $\sqrt{3}+\sqrt{27}-\sqrt{192}$

**105.** $\sqrt[3]{16}-5\sqrt[3]{54}+3\sqrt[3]{64}$

**106.** $\sqrt{2}-\dfrac{3}{\sqrt{32}}+\sqrt{50}$

**107.** $9\sqrt{x^5y^6}-\sqrt{16x^7y^8}$

**108.** $8\sqrt[3]{x^7y^8}-\sqrt[3]{x^4y^2}+3\sqrt[3]{x^{10}y^2}$

*In Exercises 109 and 110, $f(x)$ and $g(x)$ are given. Find $(f \cdot g)(x)$.*

**109.** $f(x)=\sqrt{3x},\ g(x)=\sqrt{6x}-\sqrt{15}$

**110.** $f(x)=\sqrt[3]{2x^2},\ g(x)=\sqrt[3]{4x^4}+\sqrt[3]{16x^5}$

*Simplify. In Exercise 112, assume that the variable can be any real number.*

**111.** $f(x)=\sqrt{2x+7}\,\sqrt{2x+7},\ x\geq-\dfrac{7}{2}$

**112.** $g(a)=\sqrt{20a^2+100a+125}$

*Simplify.*

**113.** $\dfrac{\sqrt[3]{(x+5)^5}}{\sqrt{(x+5)^3}}$

**114.** $\dfrac{\sqrt[3]{a^3b^2}}{\sqrt[4]{a^4b}}$

**Perimeter and Area** *For each figure, find* **a)** *the perimeter and* **b)** *the area. Write the perimeter and area in radical form with the radicals simplified.*

**115.**

Rectangle

**116.**

**117.** The graph of $f(x) = \sqrt{x} + 2$ is given.

**a)** For $g(x) = -3$, sketch the graph of $(f + g)(x)$.
**b)** What is the domain of $(f + g)(x)$?

**118.** The graph of $f(x) = -\sqrt{x}$ is given.

**a)** For $g(x) = \sqrt{x} + 2$, sketch the graph of $(f + g)(x)$.
**b)** What is the domain of $(f + g)(x)$?

[11.6] *Solve each equation and check your solutions.*

**119.** $\sqrt{x} = 4$  **120.** $\sqrt{x} = -4$  **121.** $\sqrt[3]{x} = 4$
**122.** $\sqrt[3]{x} = -5$  **123.** $7 + \sqrt{x} = 10$  **124.** $7 + \sqrt[3]{x} = 12$
**125.** $\sqrt{3x + 4} = \sqrt{5x + 14}$  **126.** $\sqrt{x^2 + 2x - 8} = x$  **127.** $\sqrt[3]{x - 9} = \sqrt[3]{5x + 3}$
**128.** $(x^2 + 7)^{1/2} = x + 1$  **129.** $\sqrt{x} + 3 = \sqrt{3x + 9}$  **130.** $\sqrt{6x - 5} - \sqrt{2x + 6} - 1 = 0$

*For each pair of functions, find all values of x for which $f(x) = g(x)$.*

**131.** $f(x) = \sqrt{3x + 4}, g(x) = 2\sqrt{2x - 4}$

**132.** $f(x) = (4x + 5)^{1/3}, g(x) = (6x - 7)^{1/3}$

*Solve the following for the indicated variable.*

**133.** $V = \sqrt{\dfrac{2L}{w}}$, for $L$

**134.** $r = \sqrt{\dfrac{A}{\pi}}$, for $A$

*Find the length of the unknown side of each right triangle. Write the answer as a radical in simplified form.*

**135.**

**136.**

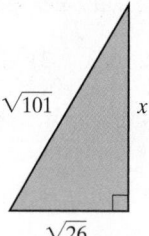

*Solve.*

**137. Telephone Pole** How long a wire does a phone company need to reach the top of a 5-meter telephone pole from a point on the ground 2 meters from the base of the pole?

**138. Velocity** Use the formula $v = \sqrt{2gh}$ to find the velocity of an object after it has fallen 20 feet ($g = 32$ ft/s$^2$).

**139. Pendulum** Use the formula

$$T = 2\pi\sqrt{\dfrac{L}{32}}$$

to find the period of a pendulum, $T$, if its length, $L$, is 64 feet.

**140. Kinetic Energy** Kinetic energy is the energy of an object in motion. The formula

$$V = \sqrt{\dfrac{2K}{m}}$$

can be used to determine the velocity, $V$, in meters per second, when a mass, $m$, in kilograms, has a kinetic energy, $K$, in joules. A 0.145-kg baseball is thrown. If the kinetic energy of the moving ball is 45 joules, at what speed is the ball moving?

**141. Speed of Light** Albert Einstein found that if an object at rest, with mass $m_0$, is made to travel close to the speed of light, its mass increases to $m$, where

$$m = \dfrac{m_0}{\sqrt{1 - \dfrac{v^2}{c^2}}}$$

In the formula, $v$ is the velocity of the moving object and $c$ is the speed of light.* In an accelerator used for cancer therapy, particles travel at speeds of $0.98c$, that is, at 98% the speed of light. At a speed of $0.98c$, determine a particle's mass, $m$, in terms of its resting mass, $m_0$. Use $v = 0.98c$ in the above formula.

Albert Einstein

*The speed of light is $3.00 \times 10^8$ meters per second. However, you do not need this information to solve this problem.

[11.7]  *Write each expression as a complex number in the form a + bi.*

**142.** 1              **143.** −8              **144.** $7 - \sqrt{-256}$              **145.** $9 + \sqrt{-16}$

*Perform each indicated operation.*

**146.** $(3 + 2i) + (10 - i)$          **147.** $(4 + i) - (7 - 2i)$          **148.** $(\sqrt{3} + \sqrt{-5}) + (11\sqrt{3} - \sqrt{-7})$

**149.** $\sqrt{-6}(\sqrt{6} + \sqrt{-6})$          **150.** $(4 + 3i)(2 - 3i)$          **151.** $(6 + \sqrt{-3})(4 - \sqrt{-15})$

**152.** $\dfrac{8}{3i}$          **153.** $\dfrac{2 + \sqrt{3}}{2i}$          **154.** $\dfrac{4}{3 + 2i}$

**155.** $\dfrac{\sqrt{3}}{5 - \sqrt{-6}}$

*Evaluate each expression for the given value of x.*

**156.** $x^2 - 2x + 9, x = 1 + 2i\sqrt{2}$          **157.** $x^2 - 2x + 12, x = 1 - 2i$

*For each imaginary number, indicate whether its value is i, −1, −i, or 1.*

**158.** $i^{33}$          **159.** $i^{59}$          **160.** $i^{404}$          **161.** $i^{802}$

## Chapter 11 Practice Test

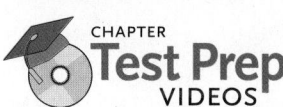

*Chapter Test Prep Videos provide fully worked-out solutions to any of the exercises you want to review. Chapter Test Prep Videos are available via* MyMathLab®, *or on* YouTube *(search "Angel Elementary and Intermediate Algebra" and click on "Channels")*

**1.** Write $\sqrt{(5x - 3)^2}$ as an absolute value.

**2.** Simplify $\left(\dfrac{x^{2/5} \cdot x^{-1}}{x^{3/5}}\right)^2$.

**3.** Factor $x^{-2/3} + x^{4/3}$.

**4.** Graph $g(x) = \sqrt{x} + 1$.

*In Exercises 5–14, simplify. Assume that all variables represent positive real numbers.*

**5.** $\sqrt{54x^7y^{10}}$

**6.** $\sqrt[3]{25x^5y^2} \sqrt[3]{10x^6y^8}$

**7.** $\sqrt{\dfrac{7x^6y^3}{8z}}$

**8.** $\dfrac{9}{\sqrt[3]{x}}$

**9.** $\dfrac{\sqrt{3}}{3 + \sqrt{27}}$

**10.** $2\sqrt{24} - 6\sqrt{6} + 3\sqrt{54}$

**11.** $\sqrt[3]{8x^3y^5} + 4\sqrt[3]{x^6y^8}$

**12.** $(\sqrt{3} - 2)(6 - \sqrt{8})$

**13.** $\sqrt[4]{\sqrt{x^5y^3}}$

**14.** $\dfrac{\sqrt[4]{(7x + 2)^5}}{\sqrt[3]{(7x + 2)^2}}$

*In Exercises 15–17, solve the equation.*

**15.** $\sqrt{2x + 19} = 3$

**16.** $\sqrt{x^2 - x - 12} = x + 3$

**17.** $\sqrt{a - 8} = \sqrt{a} - 2$

**18.** For $f(x) = (9x + 37)^{1/3}$ and $g(x) = 2(2x + 2)^{1/3}$, find all values of x such that $f(x) = g(x)$.

**19.** Solve the formula $w = \dfrac{\sqrt{2gh}}{4}$ for g.

**20.** **Falling Object** The velocity, V, in feet per second, after an object has fallen a distance, h, in feet, can be found by the formula $V = \sqrt{64.4h}$. Find the velocity of a pen after it has fallen 200 feet.

**21.** **Ladder** A ladder is placed against a house. If the base of the ladder is 5 feet from the house and the ladder rests on the house 12 feet above the ground, find the length of the ladder.

**22.** **Springs** A formula used in the study of springs is

$$T = 2\pi\sqrt{\dfrac{m}{k}}$$

where T is the period of a spring (the time for the spring to stretch and return to its resting point), m is the mass on the spring, in kilograms, and k is the spring's constant, in newtons/meter. A mass of 1400 kilograms rests on a spring. Find the period of the spring if the spring's constant is 65,000 newtons/meter.

**23.** Multiply $(6 - \sqrt{-4})(2 + \sqrt{-16})$.

**24.** Divide $\dfrac{5 - i}{7 + 2i}$.

**25.** Evaluate $x^2 + 6x + 12$ for $x = -3 + i$.

## Cumulative Review Test

*Take the following test and check your answers with those given in the back of the book. Review any questions that you answered incorrectly. The section where the material was covered is indicated after the answer.*

**1.** Solve $\frac{1}{5}(x - 3) = \frac{3}{4}(x + 3) - x$.

**2.** Solve $3(x - 4) = 6x - (4 - 5x)$.

**3. Sweater** When the price of a sweater is decreased by 60%, it costs $16. Find the original price of the sweater.

**4.** Graph $y = \frac{3}{2}x - 3$.

**5.** Multiply $(5xy - 3)(5xy + 3)$.

**6. Volume** The volume of the box that follows is $6r^3 + 5r^2 + r$. Find $w$ in terms of $r$.

**7.** Factor $4x^3 - 9x^2 + 5x$.

**8.** Factor $(x + 1)^3 - 27$.

**9.** Solve $8x^2 - 3 = -10x$.

**10.** Multiply $\dfrac{4x + 4y}{x^2y} \cdot \dfrac{y^3}{12x}$.

**11.** Add $\dfrac{x - 4}{x - 5} - \dfrac{3}{x + 5} - \dfrac{10}{x^2 - 25}$.

**12.** Solve $\dfrac{4}{x} - \dfrac{1}{6} = \dfrac{1}{x}$.

**13.** Determine whether the graphs of the given equations are parallel lines, perpendicular lines, or neither.

$$y = 3x - 8$$
$$6y = 18x + 12$$

**14.** Find the equation of the line through $(1, -4)$ that is perpendicular to the graph of $3x - 2y = 6$.

**15.** Given $f(x) = x^2 - 3x + 4$ and $g(x) = 2x - 9$, find $(g - f)(x)$.

**16.** Solve the system of equations.

$$x + 2y = 12$$
$$4x = 8$$
$$3x - 4y + 5z = 20$$

**17.** Evaluate the determinant.

$$\begin{vmatrix} 3 & -6 & -1 \\ 2 & 1 & -2 \\ 1 & 3 & 1 \end{vmatrix}$$

**18.** Find the solution set of $|3 - 2x| < 5$.

**19.** Solve $\sqrt{2x^2 + 7} + 3 = 8$.

**20. Falling Object** The distance, $d$, an object drops in free fall is directly proportional to the square of the time, $t$. If an object falls 16 feet in 1 second, how far will the object fall in 5 seconds?

# 12 Quadratic Functions

## Goals of This Chapter

We introduced quadratic equations in Section 6.6 and functions in Section 8.2. Now we expand on the concepts. We introduce completing the square and the quadratic formula. After studying these sections, you will know three techniques for solving quadratic equations: factoring (when possible), completing the square, and the quadratic formula. We will also discuss techniques for graphing quadratic functions and study nonlinear inequalities in one variable.

© NASA/Johnson Space Center

**There are many real-life** situations that can be represented or approximated with quadratic equations. As you read through this chapter, you will see several interesting applications of quadratic equations and quadratic functions. For instance, in Exercises 99 and 100 on page 712, we will use quadratic equations and the quadratic formula to discuss gravity on the moon.

# 12.1 Solving Quadratic Equations by Completing the Square

1 Use the square root property to solve equations.

2 Understand perfect square trinomials.

3 Solve quadratic equations by completing the square.

In Chapter 6 we solved quadratic equations by factoring. However, not every quadratic equation can be solved by factoring. In this section we introduce two additional procedures used to solve quadratic equations: the square root property and completing the square.

## 1 Use the Square Root Property to Solve Quadratic Equations

In Chapter 11 we learned that every positive number has two square roots—a positive square root and a negative square root. For example, 25 has two square roots:

| Positive Square Root of 25 | Negative Square Root of 25 |
|---|---|
| $\sqrt{25} = 5$ | $-\sqrt{25} = -5$ |

We can refer to both square roots of 25 using the symbol $\pm$, which is read "plus or minus." Thus, $\pm 5$ refers to both 5 and $-5$.

Next, consider the quadratic equation $x^2 = 25$. Since $5^2 = 25$ and $(-5)^2 = 25$, both $x = 5$ and $x = -5$ are solutions to the equation. We can summarize the solution process as follows

$$x^2 = 25$$
$$x = \pm\sqrt{25} \qquad \pm \text{ means take the positive and the negative square roots.}$$
$$x = \pm 5 \qquad \text{The solutions are 5 and } -5.$$

In general, when solving equations of the form $x^2 = a$, we can use the *square root property*.

> ### Understanding Algebra
>
> Recall that a quadratic equation is an equation of the form $ax^2 + bx + c = 0$, where $a$, $b$, and $c$ are real numbers and $a \neq 0$.

> ### Understanding Algebra
>
> A convenient way to write both the positive and the negative square roots is to use the symbol $\pm$, which is read "plus or minus." The solutions to the equation $x^2 = 25$ can be written as $\pm\sqrt{25} = \pm 5$. Thus the solutions are 5 and $-5$.

> **Square Root Property**
>
> If $x^2 = a$, where $a$ is a real number, then $x = \pm\sqrt{a}$.

EXAMPLE 1 Solve the following equations.

**a)** $x^2 = 49$      **b)** $x^2 - 9 = 0$      **c)** $x^2 + 10 = 85$

*Solution*

**a)** Use the square root property to solve the equation.

$$x^2 = 49 \qquad \text{Use the square root property.}$$
$$x = \pm\sqrt{49} \qquad \text{The positive and the negative square roots of 49.}$$
$$x = \pm 7 \qquad \text{The solutions are 7 and } -7.$$

Check the solutions in the original equation.

| $x = 7$ | $x = -7$ |
|---|---|
| $x^2 = 49$ | $x^2 = 49$ |
| $(7)^2 = 49$ | $(-7)^2 = 49$ |
| $49 = 49$   True | $49 = 49$   True |

In both cases we have true statements. Therefore, both 7 and $-7$ are solutions to the equation.

**b)**

$$x^2 - 9 = 0 \qquad \text{Add 9 to both sides.}$$
$$x^2 = 9 \qquad \text{Use the square root property.}$$
$$x = \pm\sqrt{9}$$
$$= \pm 3 \qquad \text{The solutions are 3 and } -3.$$

Check the solutions in the original equation.

$$x = 3 \qquad\qquad x = -3$$
$$x^2 - 9 = 0 \qquad\qquad x^2 - 9 = 0$$
$$3^2 - 9 \stackrel{?}{=} 0 \qquad\qquad (-3)^2 - 9 \stackrel{?}{=} 0$$
$$0 = 0 \quad \text{True} \qquad\qquad 0 = 0 \quad \text{True}$$

In both cases the check is true, which means that both 3 and $-3$ are solutions to the equation.

**c)**

$$x^2 + 10 = 85 \qquad \text{Subtract 10 from both sides.}$$
$$x^2 = 75 \qquad \text{Use the square root property.}$$
$$x = \pm\sqrt{75} \qquad \text{Simplify } \sqrt{75}.$$
$$= \pm\sqrt{25}\sqrt{3}$$
$$= \pm 5\sqrt{3}$$

The solutions are $5\sqrt{3}$ and $-5\sqrt{3}$.

*Now Try Exercise 13*

Not all quadratic equations have real number solutions, as is illustrated in Example 2.

**EXAMPLE 2**  Solve the equation $x^2 + 7 = 0$.

*Solution*

$$x^2 + 7 = 0 \qquad \text{Subtract 7 from both sides.}$$
$$x^2 = -7 \qquad \text{Use the square root property.}$$
$$x = \pm\sqrt{-7} \qquad \text{Simplify the imaginary solutions.}$$
$$= \pm i\sqrt{7}$$

The solutions are $i\sqrt{7}$ and $-i\sqrt{7}$, both of which are imaginary numbers.

*Now Try Exercise 15*

## Understanding Algebra

Recall from Chapter 11 that the square root of a negative number is an *imaginary number*. In general, for any positive real number, $n$,

$$\sqrt{-n} = \sqrt{-1}\sqrt{n} = i\sqrt{n}$$

where $i = \sqrt{-1}$.

**EXAMPLE 3**  Solve    **a)** $(a - 5)^2 = 32$    **b)** $(z + 3)^2 + 28 = 0$.

*Solution*

**a)** Begin by using the square root property.

$$(a - 5)^2 = 32 \qquad \text{Use the square root property.}$$
$$a - 5 = \pm\sqrt{32} \qquad \text{Add 5 to both sides.}$$
$$a = 5 \pm \sqrt{32} \qquad \text{Simplify } \sqrt{32}.$$
$$= 5 \pm \sqrt{16}\sqrt{2}$$
$$= 5 \pm 4\sqrt{2}$$

The solutions are $5 + 4\sqrt{2}$ and $5 - 4\sqrt{2}$.

**b)** Begin by subtracting 28 from both sides of the equation to isolate the term containing the variable.

$$(z + 3)^2 + 28 = 0 \qquad \text{Subtract 28 from both sides.}$$
$$(z + 3)^2 = -28 \qquad \text{Use the square root property.}$$
$$z + 3 = \pm\sqrt{-28} \qquad \text{Subtract 3 from both sides.}$$
$$z = -3 \pm \sqrt{-28} \qquad \text{Simplify } \sqrt{-28}.$$
$$= -3 \pm \sqrt{28}\sqrt{-1}$$
$$= -3 \pm i\sqrt{4}\sqrt{7}$$
$$= -3 \pm 2i\sqrt{7}$$

## Understanding Algebra

Recall that *complex numbers* are numbers of the form

$$a + bi$$

where $a$ and $b$ are real numbers and

$$i = \sqrt{-1}.$$

The solutions are $-3 + 2i\sqrt{7}$ and $-3 - 2i\sqrt{7}$. Note that the solutions to the equation $(z + 3)^2 + 28 = 0$ are complex numbers.

*Now Try Exercise 23*

## 2 Understand Perfect Square Trinomials

Recall that a **perfect square trinomial** is a trinomial that can be expressed as the square of a binomial. Some examples follow.

| Perfect Square Trinomials | | Factors | | Square of a Binomial |
|---|---|---|---|---|
| $x^2 + 8x + 16$ | $=$ | $(x + 4)(x + 4)$ | $=$ | $(x + 4)^2$ |
| $x^2 - 8x + 16$ | $=$ | $(x - 4)(x - 4)$ | $=$ | $(x - 4)^2$ |
| $x^2 + 10x + 25$ | $=$ | $(x + 5)(x + 5)$ | $=$ | $(x + 5)^2$ |
| $x^2 - 10x + 25$ | $=$ | $(x - 5)(x - 5)$ | $=$ | $(x - 5)^2$ |

In a perfect square trinomial with a leading coefficient of 1, the constant term is the square of one-half the coefficient of the first-degree term.

Let's examine some perfect square trinomials for which the leading coefficient is 1.

$$x^2 + 8x + 16 = (x + 4)^2$$
$$\left[\tfrac{1}{2}(8)\right]^2 = (4)^2$$

$$x^2 - 10x + 25 = (x - 5)^2$$
$$\left[\tfrac{1}{2}(-10)\right]^2 = (-5)^2$$

When a perfect square trinomial with a leading coefficient of 1 is written as the square of a binomial, the constant in the binomial is one-half the coefficient of the first-degree term in the trinomial. For example,

$$x^2 + 8x + 16 = (x + 4)^2$$
$$\tfrac{1}{2}(8)$$

$$x^2 - 10x + 25 = (x - 5)^2$$
$$\tfrac{1}{2}(-10)$$

## 3 Solve Quadratic Equations by Completing the Square

To solve a quadratic equation by **completing the square**, we add a constant to both sides of the equation so the resulting trinomial is a perfect square trinomial. Then we use the square root property to solve the resulting equation.

### To Solve a Quadratic Equation by Completing the Square

1. Use the multiplication (or division) property of equality if necessary to make the leading coefficient 1.
2. Rewrite the equation with the constant by itself on the right side of the equation.
3. Take one-half the numerical coefficient of the first-degree term, square it, and add this quantity to both sides of the equation.
4. Factor the trinomial as the square of a binomial.
5. Use the square root property to take the square root of both sides of the equation.
6. Solve for the variable.
7. Check your solutions in the *original* equation.

**EXAMPLE 4** Solve the equation $x^2 + 6x + 5 = 0$ by completing the square.

*Solution*   Since the leading coefficient is 1, step 1 is not necessary.

**Step 2**  Subtract 5 from both sides of the equation.

$$x^2 + 6x + 5 = 0$$
$$x^2 + 6x = -5$$

**Step 3**  Determine the square of one-half the numerical coefficient of the first-degree term, 6.

$$\frac{1}{2}(6) = 3, \qquad 3^2 = 9$$

Add this value to both sides of the equation.

$$x^2 + 6x \;+\,9 = -5 \;+\,9$$
$$x^2 + 6x + 9 = 4$$

**Step 4**  By following this procedure, we produce a perfect square trinomial on the left side of the equation. The expression $x^2 + 6x + 9$ is a perfect square trinomial that can be factored as $(x + 3)^2$.

$$\frac{1}{2} \text{ the numerical coefficient of the}$$
$$\text{first-degree term is } \tfrac{1}{2}(6) = +3$$

$$(x + 3)^2 = 4$$

**Step 5**  Use the square root property.

$$x + 3 = \pm\sqrt{4}$$
$$x + 3 = \pm 2$$

**Step 6**  Finally, solve for $x$ by subtracting 3 from both sides of the equation.

$$x + 3 \;-\; 3 = \;-3\; \pm 2$$
$$x = -3 \pm 2$$
$$x = -3 + 2 \quad \text{or} \quad x = -3 - 2$$
$$x = -1 \qquad\qquad x = -5$$

**Step 7**  Check both solutions in the original equation.

$$x = -1 \qquad\qquad\qquad\qquad x = -5$$
$$x^2 + 6x + 5 = 0 \qquad\qquad\qquad x^2 + 6x + 5 = 0$$
$$(-1)^2 + 6(-1) + 5 \stackrel{?}{=} 0 \qquad\qquad (-5)^2 + 6(-5) + 5 \stackrel{?}{=} 0$$
$$1 - 6 + 5 \stackrel{?}{=} 0 \qquad\qquad\qquad 25 - 30 + 5 \stackrel{?}{=} 0$$
$$0 = 0 \quad \text{True} \qquad\qquad\qquad\qquad 0 = 0 \quad \text{True}$$

Since each number checks, both $-1$ and $-5$ are solutions to the original equation.

Now Try Exercise 49

**EXAMPLE 5** Solve the equation $-x^2 = -3x - 18$ by completing the square.

*Solution*   Begin by multiplying both sides of the equation by $-1$ to make the coefficient of the squared term equal to 1.

$$-x^2 = -3x - 18$$
$$-1(-x^2) = -1(-3x - 18) \qquad \text{Multiply both sides by } -1.$$
$$x^2 = 3x + 18 \qquad\qquad \text{Subtract } 3x \text{ from both sides.}$$
$$x^2 - 3x = 18$$

Take half the numerical coefficient of the first-degree term, square it, and add this product to both sides of the equation.

$$\frac{1}{2}(-3) = -\frac{3}{2} \qquad \left(-\frac{3}{2}\right)^2 = \frac{9}{4}$$

$$x^2 - 3x + \frac{9}{4} = 18 + \frac{9}{4} \qquad \text{Complete the square.}$$

$$\left(x - \frac{3}{2}\right)^2 = 18 + \frac{9}{4} \qquad \text{Factor the trinomial as the square of a binomial.}$$

$$\left(x - \frac{3}{2}\right)^2 = \frac{72}{4} + \frac{9}{4}$$

$$\left(x - \frac{3}{2}\right)^2 = \frac{81}{4}$$

$$x - \frac{3}{2} = \pm\sqrt{\frac{81}{4}} \qquad \text{Square root property}$$

$$x - \frac{3}{2} = \pm\frac{9}{2} \qquad \text{Simplify.}$$

$$x = \frac{3}{2} \pm \frac{9}{2} \qquad \text{Added } \frac{3}{2} \text{ to both sides.}$$

$$x = \frac{3}{2} + \frac{9}{2} \qquad \text{or} \qquad x = \frac{3}{2} - \frac{9}{2}$$

$$x = \frac{12}{2} = 6 \qquad\qquad x = -\frac{6}{2} = -3$$

The solutions are 6 and −3.

Now Try Exercise 53

### Understanding Algebra

When a quadratic equation has a leading coefficient that is not 1, you can multiply each term on both sides of the equation by the reciprocal of the leading coefficient. This will result in an equivalent equation with a leading coefficient of 1. For example, the equation

$$-\frac{1}{4}x^2 + 2x - 8 = 0$$

can be multiplied by −4 to obtain the equivalent equation

$$x^2 - 8x + 32 = 0$$

which has a leading coefficient of 1.

In the following examples we will not show some of the intermediate steps.

**EXAMPLE 6** Solve $x^2 - 8x + 34 = 0$ by completing the square.

*Solution*
$$x^2 - 8x + 34 = 0 \qquad \text{Subtract 34 from both sides.}$$
$$x^2 - 8x = -34$$
$$x^2 - 8x + 16 = -34 + 16 \qquad \text{Complete the square.}$$
$$(x - 4)^2 = -18 \qquad \text{Factor the trinomial as the square of a binomial.}$$
$$x - 4 = \pm\sqrt{-18} \qquad \text{Square root property}$$
$$x - 4 = \pm 3i\sqrt{2} \qquad \text{Simplify.}$$
$$x = 4 \pm 3i\sqrt{2} \qquad \text{Solve for } x.$$

The solutions are $4 + 3i\sqrt{2}$ and $4 - 3i\sqrt{2}$.

Now Try Exercise 61

**EXAMPLE 7** Solve the equation $-4x^2 + 8x + 32 = 0$ by completing the square.

*Solution*
$$-4x^2 + 8x + 32 = 0$$
$$-\frac{1}{4}(-4x^2 + 8x + 32) = -\frac{1}{4}(0) \qquad \begin{array}{l}\text{Multiply by } -1/4 \text{ to obtain} \\ \text{a leading coefficient of 1.}\end{array}$$
$$x^2 - 2x - 8 = 0$$
$$x^2 - 2x = 8 \qquad \text{Added 8 to both sides.}$$
$$x^2 - 2x + 1 = 8 + 1 \qquad \text{Complete the square.}$$
$$(x - 1)^2 = 9 \qquad \begin{array}{l}\text{Factor the trinomial as the} \\ \text{square of a binomial.}\end{array}$$
$$x - 1 = \pm 3 \qquad \text{Square root property}$$
$$x = 1 \pm 3 \qquad \text{Solve for } x.$$
$$x = 1 + 3 \qquad \text{or} \qquad x = 1 - 3$$
$$x = 4 \qquad\qquad x = -2$$

Now Try Exercise 75

## Using Your Graphing Calculator ▪▪▪▪▪

We can verify the *real* solutions to quadratic equations using a graphing calculator. Recall that the *x*-intercepts of a function occur when $y = 0$. Thus, to verify the solutions to the equation in Example 7, $-4x^2 + 8x + 32 = 0$, we will observe the *x*-intercepts of the function $y = -4x^2 + 8x + 32$.

The figure above verifies that the solutions to $-4x^2 + 8x + 32 = 0$ are $-2$ and $4$. Remember that the *x*-intercepts only represent *real* solutions to an equation. Thus, we cannot use a graphing calculator to verify the solutions to the equation in Example 6 since the solutions are complex numbers.

Generally, quadratic equations that cannot be easily solved by factoring will be solved by the *quadratic formula*, which we present in the next section.

EXAMPLE 8  **Compound Interest**  The compound interest formula $A = p\left(1 + \dfrac{r}{n}\right)^{nt}$ can be used to find the amount, $A$, when an initial principal, $p$, is invested at an annual interest rate, $r$, compounded $n$ times a year for $t$ years.

a) Josh Adams initially invested \$1000 in a savings account where interest is compounded annually (once a year). If after 2 years the amount, or balance, in the account is \$1102.50, find the annual interest rate, $r$.

b) Trisha McDowell initially invested \$1000 in a savings account where interest is compounded quarterly (four times per year). If after 3 years the amount in the account is \$1195.62, find the annual interest rate, $r$.

*Solution*

a) Understand   We are given the following information:

$$p = \$1000, \qquad A = \$1102.50, \qquad n = 1, \qquad t = 2$$

We are asked to find the annual interest rate, $r$. To do so, we substitute the appropriate values into the formula and solve for $r$.

Translate
$$A = p\left(1 + \frac{r}{n}\right)^{nt}$$

$$1102.50 = 1000\left(1 + \frac{r}{1}\right)^{1(2)}$$

Carry Out

$1102.50 = 1000(1 + r)^2$     Divide both sides by 1000.

$1.10250 = (1 + r)^2$     Square root property

$\pm\sqrt{1.10250} = 1 + r$     We will only use the positive square root since $r$ must be positive.

$1.05 = 1 + r$     Subtract 1 from both sides.

$0.05 = r$

Answer   The annual interest rate is 0.05 or 5%.

## Understanding Algebra

Recall from Chapter 11 that
$$\sqrt[n]{a} = a^{1/n}.$$
Therefore, from Example 8,
$$\sqrt[12]{1.19562} = 1.19562^{1/12}$$

**b) Understand** We are given
$$p = \$1000, \qquad A = \$1195.62, \qquad n = 4, \qquad t = 3$$
To find $r$, we substitute the appropriate values into the formula and solve for $r$.

Translate
$$A = p\left(1 + \frac{r}{n}\right)^{nt}$$

$$1195.62 = 1000\left(1 + \frac{r}{4}\right)^{4(3)}$$

Carry Out
$$1.19562 = \left(1 + \frac{r}{4}\right)^{12} \qquad \text{Divided both sides by 1000.}$$

$$\sqrt[12]{1.19562} = 1 + \frac{r}{4} \qquad \begin{array}{l}\text{Take the 12th root of both} \\ \text{sides (or raise both sides to} \\ \text{the 1/12 power).}\end{array}$$

$$1.015 \approx 1 + \frac{r}{4} \qquad \begin{array}{l}\text{Approximated } 1.19562^{1/12} \\ \text{on a calculator.}\end{array}$$

$$0.015 \approx \frac{r}{4} \qquad \text{Subtracted 1 from both sides.}$$

$$0.06 \approx r \qquad \text{Multiplied both sides by 4.}$$

**Answer** The annual interest rate is approximately 0.06 or 6%.

Now Try Exercise 103

### Helpful Hint

*Study Tip*

In this chapter, you will be working with roots and radicals. This material was discussed in Chapter 11. If you do not remember how to evaluate or simplify radicals, review Chapter 11 now.

## EXERCISE SET 12.1　MathXL®　MyMathLab®

### Warm-Up Exercises

*Fill in the blanks with the appropriate word, phrase, or symbol(s) from the following list.*

| reciprocal | imaginary | quadratic | square root property | perfect square trinomial | $\frac{2}{3}$ | $\frac{3}{2}$ |
| 25 | 10 | 100 | one-half | complex | binomial | |

1. A _____ equation is an equation of the form $ax^2 + bx + c = 0$, where $a, b,$ and $c$ are real numbers and $a \neq 0$.

2. The square root of a negative number is an _____ number.

3. A number in the form $a + bi$ is a _____ number.

4. A perfect square trinomial is a trinomial that can be expressed as the square of a _____.

5. To solve a quadratic equation by completing the square, we add a constant to both sides of the equation so the resulting trinomial is a _____.

6. Once one side of a quadratic equation is rewritten as a perfect square trinomial, the equation can be solved by using the _____.

7. When a quadratic equation has a leading coefficient that is not 1, you can multiply each term on both sides of the equation by the _____ of the leading coefficient.

8. In a perfect square trinomial with a leading coefficient of 1, the constant term is the square of _____ the coefficient of the first-degree term.

9. To solve the quadratic equation $x^2 - 10x = 24$ by completing the square, you would add _____ to both sides of the equation.

10. To obtain an equivalent equation with a leading coefficient of 1, multiply both sides of the equation $\frac{2}{3}x^2 + 3x - 5 = 0$ by _____.

## Practice the Skills

*For Exercises 11–36, use the square root property to solve each equation.*

**11.** $x^2 = 81$

**12.** $x^2 = 100$

**13.** $x^2 - 25 = 0$

**14.** $x^2 - 49 = 0$

**15.** $x^2 + 49 = 0$

**16.** $x^2 - 24 = 0$

**17.** $x^2 + 24 = 0$

**18.** $y^2 - 10 = 51$

**19.** $y^2 + 10 = -51$

**20.** $(x - 3)^2 = 49$

**21.** $(p - 4)^2 = 16$

**22.** $(x + 3)^2 = 49$

**23.** $(x + 3)^2 + 25 = 0$

**24.** $(a - 3)^2 = 45$

**25.** $(a - 2)^2 + 45 = 0$

**26.** $(a + 2)^2 + 45 = 0$

**27.** $\left(b + \dfrac{1}{3}\right)^2 = \dfrac{4}{9}$

**28.** $\left(b - \dfrac{1}{3}\right)^2 = \dfrac{4}{9}$

**29.** $\left(b - \dfrac{2}{3}\right)^2 + \dfrac{4}{9} = 0$

**30.** $(x - 0.2)^2 = 0.64$

**31.** $(x + 0.8)^2 = 0.81$

**32.** $\left(x + \dfrac{1}{2}\right)^2 = \dfrac{16}{9}$

**33.** $(2a - 5)^2 = 18$

**34.** $(4y + 1)^2 = 12$

**35.** $\left(2y + \dfrac{1}{2}\right)^2 = \dfrac{4}{25}$

**36.** $\left(3x - \dfrac{1}{4}\right)^2 = \dfrac{9}{25}$

*Solve each equation by completing the square.*

**37.** $x^2 + 6x + 5 = 0$

**38.** $x^2 - 2x - 3 = 0$

**39.** $x^2 + 8x + 15 = 0$

**40.** $x^2 - 8x + 15 = 0$

**41.** $x^2 + 6x + 8 = 0$

**42.** $x^2 - 6x + 8 = 0$

**43.** $x^2 - 7x + 6 = 0$

**44.** $x^2 + 9x + 18 = 0$

**45.** $2x^2 + x - 1 = 0$

**46.** $3c^2 - 4c - 4 = 0$

**47.** $2z^2 - 7z - 4 = 0$

**48.** $4a^2 + 9a = 9$

**49.** $x^2 - 13x + 40 = 0$

**50.** $x^2 + x - 12 = 0$

**51.** $-x^2 + 6x + 7 = 0$

**52.** $-a^2 - 5a + 14 = 0$

**53.** $-z^2 + 9z - 20 = 0$

**54.** $-z^2 - 4z + 12 = 0$

**55.** $b^2 = 3b + 28$

**56.** $-x^2 = 6x - 27$

**57.** $x^2 + 10x = 11$

**58.** $-x^2 + 40 = -3x$

**59.** $x^2 - 4x - 10 = 0$

**60.** $x^2 - 6x + 2 = 0$

**61.** $r^2 + 8r + 5 = 0$

**62.** $a^2 + 4a - 8 = 0$

**63.** $c^2 - c - 3 = 0$

**64.** $p^2 - 5p = 4$

**65.** $x^2 + 3x + 6 = 0$

**66.** $z^2 - 5z + 7 = 0$

**67.** $9x^2 - 9x = 0$

**68.** $4y^2 + 12y = 0$

**69.** $-\dfrac{3}{4}b^2 - \dfrac{1}{2}b = 0$

**70.** $\dfrac{1}{3}a^2 - \dfrac{5}{3}a = 0$

**71.** $36z^2 - 6z = 0$

**72.** $x^2 = \dfrac{9}{2}x$

**73.** $-\dfrac{1}{2}p^2 - p + \dfrac{3}{2} = 0$

**74.** $2x^2 + 6x = 20$

**75.** $2x^2 = 8x + 64$

**76.** $3x^2 + 33x + 72 = 0$

**77.** $2x^2 + 18x + 4 = 0$

**78.** $\dfrac{2}{3}x^2 + \dfrac{4}{3}x + 1 = 0$

**79.** $\dfrac{3}{4}w^2 + \dfrac{1}{2}w - \dfrac{1}{4} = 0$

**80.** $\dfrac{3}{4}c^2 - 2c + 1 = 0$

**81.** $2x^2 - x = -5$

**82.** $\dfrac{5}{2}x^2 + \dfrac{3}{2}x - \dfrac{5}{4} = 0$

**83.** $-3x^2 + 6x = 6$

**84.** $x^2 + 2x = -5$

## Problem Solving

**Area**   *In Exercises 85–88, the area, A, of each rectangle is given.* **a)** *Write an equation for the area.* **b)** *Solve the equation for x.*

**85.**

$A = 21$   $x - 2$
$x + 2$

**86.**

$A = 35$   $x + 3$
$x + 5$

**87.**

$A = 18$   $x + 2$
$x + 4$

**88.**

$A = 23$   $x - 3$
$x - 1$

**89. Stopping Distance on Snow**  The formula for approximating the stopping distance, $d$, in feet, for a specific car on snow is $d = \dfrac{1}{6}x^2$, where $x$ is the speed of the car, in miles per hour, before the brakes are applied. If the car's stopping distance was 24 feet, what was the car's speed before the brakes were applied?

**90. Stopping Distance on Dry Pavement** The formula for approximating the stopping distance, $d$, in feet, for a specific car on dry pavement is $d = \frac{1}{10}x^2$, where $x$ is the speed of the car, in miles per hour, before the brakes are applied. If the car's stopping distance was 90 feet, what was the car's speed before the brakes were applied?

**91. Integers** The product of two consecutive positive odd integers is 35. Find the two odd integers.

**92. Integers** The larger of two integers is 2 more than twice the smaller. Find the two numbers if their product is 12.

**93. Rectangular Garden** Donna Simm has marked off an area in her yard where she will plant tomato plants. Find the dimensions of the rectangular area if the length is 2 feet more than twice the width and the area is 60 square feet.

**94. Driveway** Manuel Cortez is planning to blacktop his driveway. Find the dimensions of the rectangular driveway if its area is 381.25 square feet and its length is 18 feet greater than its width.

**95. Patio** Bill Justice is designing a square patio whose diagonal is 6 feet longer than the length of a side. Find the dimensions of the patio.

**96. Wading Pool** The Lakeside Hotel is planning to build a shallow wading pool for children. If the pool is to be square and the diagonal of the square is 7 feet longer than a side, find the dimensions of the pool.

**97. Inscribed Triangle** When a triangle is inscribed in a semicircle where a diameter of the circle is a side of the triangle, the triangle formed is always a right triangle. If an isosceles triangle (two equal sides) is inscribed in a semicircle of radius 10 inches, find the length of the other two sides of the triangle.

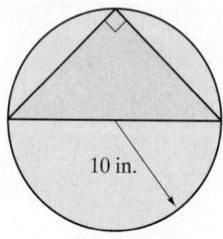

10 in.

**98. Inscribed Triangle** Refer to Exercise 97. Suppose a triangle is inscribed in a semicircle whose diameter is 12 meters. If one side of the inscribed triangle is 6 meters, find the third side.

**99. Area of Circle** The area of a circle is $24\pi$ square feet. Use the formula $A = \pi r^2$ to find the radius of the circle.

**100. Area of Circle** The area of a circle is $16.4\pi$ square meters. Find the radius of the circle.

*Use the formula $A = p\left(1 + \dfrac{r}{n}\right)^{nt}$ to answer Exercises 101–104.*

**101. Savings Account** Frank Dipalo initially invested $500 in a savings account where interest is compounded annually. If after 2 years the amount in the account is $540.80, find the annual interest rate.

**102. Savings Account** Margret Chang initially invested $1500 in a savings account where interest is compounded annually. If after 3 years the amount in the account is $1687.30, find the annual interest rate.

**103. Savings Account** Steve Rodi initially invested $1200 in a savings account where interest is compounded semiannually. If after 3 years the amount in the account is $1432.86, find the annual interest rate.

**104. Savings Account** Angela Reyes initially invested $1500 in a savings account where interest is compounded semiannually. If after 4 years the amount in the account is $2052.85, find the annual interest rate.

**105. Surface Area and Volume** The surface area, $S$, and volume, $V$, of a right circular cylinder of radius, $r$, and height, $h$, are given by the formulas

$$S = 2\pi r^2 + 2\pi rh, \qquad V = \pi r^2 h$$

**a)** Find the surface area of the cylinder if its height is 10 inches and its volume is 160 cubic inches.

**b)** Find the radius if the height is 10 inches and the volume is 160 cubic inches.

**c)** Find the radius if the height is 10 inches and the surface area is 160 square inches.

## Group Activity

*Discuss and answer Exercise 106 as a group.*

**106.** On the following grid, the points $(x_1, y_1)$, $(x_2, y_2)$, and $(x_1, y_2)$ are plotted.

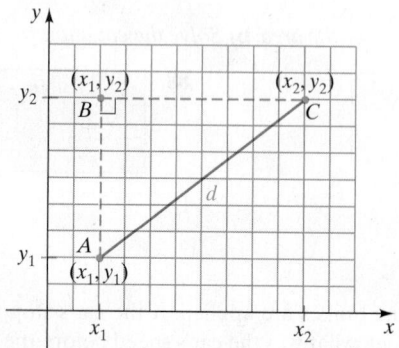

**a)** Explain why $(x_1, y_2)$ is placed where it is and not somewhere else on the graph.

**b)** Express the length of the orange dashed line in terms of $y_2$ and $y_1$. Explain how you determined your answer.

**c)** Express the length of the green dashed line in terms of $x_2$ and $x_1$.

**d)** Using the Pythagorean Theorem and the right triangle $ABC$, derive a formula for the distance, $d$, between points $(x_1, y_1)$ and $(x_2, y_2)$.* Explain how you determined the formula.

**e)** Use the formula you determined in part **d)** to find the distance of the line segment between the points $(1, 4)$ and $(3, 7)$.

_____

*The distance formula will be discussed in a later chapter.

## Cumulative Review Exercises

[2.5] **107.** Solve $-4(2z - 6) = -3(z - 4) + z$.

[3.4] **108. Investment** Thea Prettyman invested $10,000 for 1 year, part at 7% and part at $6\frac{1}{4}$%. If she earned a total interest of $656.50, how much was invested at each rate?

[4.3] **109.** Find the slope of the line through $(-2, 5)$ and $(0, 5)$.

[5.5] **110.** Multiply $(x - 2)(4x^2 + 9x - 3)$.

[10.2] **111.** Solve $|x + 3| = |2x - 7|$.

# 12.2 Solving Quadratic Equations by the Quadratic Formula

1 Derive the quadratic formula.

2 Use the quadratic formula to solve equations.

3 Determine a quadratic equation given its solutions.

4 Use the discriminant to determine the number of real solutions to a quadratic equation.

5 Study applications that use quadratic equations.

## Understanding Algebra

Although any quadratic equation can also be solved by completing the square, we generally prefer to use the quadratic formula because it is easier and more efficient to use. However, completing the square is a useful skill that can be used in other areas of algebra including the study of conic sections in Chapter 14.

## 1 Derive the Quadratic Formula

The quadratic formula can be used to solve any quadratic equation. *It is the most useful and most versatile method of solving quadratic equations.*

The standard form of a quadratic equation is $ax^2 + bx + c = 0$, where $a$ is the coefficient of the squared term, $b$ is the coefficient of the first-degree term, and $c$ is the constant.

| Quadratic Equation in Standard Form | Values of Coefficients |
|---|---|
| $x^2 - 3x + 4 = 0$ | $a = 1, \quad b = -3, \quad c = 4$ |
| $1.3x^2 - 7.9 = 0$ | $a = 1.3, \quad b = 0, \quad c = -7.9$ |
| $-\frac{5}{6}x^2 + \frac{3}{8}x = 0$ | $a = -\frac{5}{6}, \quad b = \frac{3}{8}, \quad c = 0$ |

We can derive the quadratic formula by starting with a quadratic equation in standard form and completing the square, as discussed in the preceding section.

$$ax^2 + bx + c = 0$$

$$\frac{ax^2}{a} + \frac{b}{a}x + \frac{c}{a} = 0 \qquad \text{Divide both sides by } a.$$

$$x^2 + \frac{b}{a}x = -\frac{c}{a} \qquad \text{Subtract } c/a \text{ from both sides.}$$

$$x^2 + \frac{b}{a}x + \boxed{\frac{b^2}{4a^2}} = -\frac{c}{a} + \boxed{\frac{b^2}{4a^2}} \qquad \text{Take } 1/2 \text{ of } b/a \text{ (that is, } b/2a\text{), and square it to get } b^2/4a^2. \text{ Then add this expression to both sides.}$$

$$\left(x + \frac{b}{2a}\right)^2 = \frac{b^2}{4a^2} - \frac{c}{a} \qquad \text{Rewrite the left side of the equation as the square of a binomial.}$$

$$\left(x + \frac{b}{2a}\right)^2 = \frac{b^2 - 4ac}{4a^2} \qquad \text{Write the right side with a common denominator.}$$

$$x + \frac{b}{2a} = \pm\sqrt{\frac{b^2 - 4ac}{4a^2}} \qquad \text{Square root property}$$

$$x + \frac{b}{2a} = \pm\frac{\sqrt{b^2 - 4ac}}{2a} \qquad \text{Quotient rule for radicals}$$

$$x = -\frac{b}{2a} \pm \frac{\sqrt{b^2 - 4ac}}{2a} \qquad \text{Subtract } b/2a \text{ from both sides.}$$

$$x = \frac{-b \pm \sqrt{b^2 - 4ac}}{2a} \qquad \text{Write with a common denominator to get the quadratic formula.}$$

## 2 Use the Quadratic Formula to Solve Equations

Now that we have derived the quadratic formula, we will use it to solve quadratic equations.

---

**To Solve a Quadratic Equation by the Quadratic Formula**

1. Write the quadratic equation in standard form, $ax^2 + bx + c = 0$, and determine the numerical values for $a, b$, and $c$.

2. Substitute the values for $a, b$, and $c$ into the quadratic formula and then evaluate the formula to obtain the solution.

The Quadratic Formula

$$x = \frac{-b \pm \sqrt{b^2 - 4ac}}{2a}$$

---

**EXAMPLE 1**  Solve $x^2 + 2x - 8 = 0$ by using the quadratic formula.

*Solution*   In this equation, $a = 1, b = 2$, and $c = -8$.

$$x = \frac{-b \pm \sqrt{b^2 - 4ac}}{2a}$$

$$x = \frac{-2 \pm \sqrt{2^2 - 4(1)(-8)}}{2(1)}$$

$$= \frac{-2 \pm \sqrt{4 + 32}}{2}$$

$$= \frac{-2 \pm \sqrt{36}}{2}$$

$$= \frac{-2 \pm 6}{2}$$

$$x = \frac{-2 + 6}{2} \qquad \text{or} \qquad x = \frac{-2 - 6}{2}$$

$$x = \frac{4}{2} = 2 \qquad\qquad\qquad x = \frac{-8}{2} = -4$$

A check will show that both 2 and $-4$ are solutions to the equation. Note that the solutions to the equation $x^2 + 2x - 8 = 0$ are two real numbers.

Now Try Exercise 23

---

**Helpful Hint**

The solution to Example 1 could also be obtained by factoring, as follows:

$$x^2 + 2x - 8 = 0$$
$$(x + 4)(x - 2) = 0$$
$$x + 4 = 0 \qquad \text{or} \qquad x - 2 = 0$$
$$x = -4 \qquad\qquad\qquad x = 2$$

When you are given a quadratic equation to solve and the method to solve it has not been specified, you may try solving by factoring first (as we discussed in Section 6.6). If the equation cannot be easily factored, use the quadratic formula.

---

**EXAMPLE 2**  Solve $-9x^2 = -6x + 1$ by the quadratic formula.

*Solution*   Begin by adding $9x^2$ to both sides of the equation to obtain

$$0 = 9x^2 - 6x + 1$$
$$\text{or} \quad 9x^2 - 6x + 1 = 0$$

$$a = 9, \qquad b = -6, \qquad c = 1$$

$$x = \frac{-b \pm \sqrt{b^2 - 4ac}}{2a}$$

$$= \frac{-(-6) \pm \sqrt{(-6)^2 - 4(9)(1)}}{2(9)}$$

$$= \frac{6 \pm \sqrt{36 - 36}}{18} = \frac{6 \pm \sqrt{0}}{18} = \frac{6}{18} = \frac{1}{3}$$

Note that the solution to the equation $-9x^2 = -6x + 1$ is $\frac{1}{3}$, a single value. Some quadratic equations have just one value as the solution.

Now Try Exercise 39

**Avoiding Common Errors**

The entire numerator of the quadratic formula must be divided by $2a$.

CORRECT

$$x = \frac{-b \pm \sqrt{b^2 - 4ac}}{2a}$$

INCORRECT

$$x = -b \pm \frac{\sqrt{b^2 - 4ac}}{2a}$$

$$x = \frac{-b}{2a} \pm \sqrt{b^2 - 4ac}$$

EXAMPLE 3  Solve $p^2 + \frac{1}{3}p + \frac{5}{6} = 0$ by using the quadratic formula.

*Solution*  Begin by clearing the fractions from the equation by multiplying both sides of the equation by the least common denominator, 6.

$$6\left(p^2 + \frac{1}{3}p + \frac{5}{6}\right) = 6(0)$$

$$6p^2 + 2p + 5 = 0$$

Now we can use the quadratic formula with $a = 6, b = 2$, and $c = 5$.

$$p = \frac{-b \pm \sqrt{b^2 - 4ac}}{2a}$$

$$= \frac{-2 \pm \sqrt{2^2 - 4(6)(5)}}{2(6)}$$

$$= \frac{-2 \pm \sqrt{-116}}{12}$$

$$= \frac{-2 \pm \sqrt{-4}\sqrt{29}}{12}$$

$$= \frac{-2 \pm 2i\sqrt{29}}{12}$$

$$= \frac{\overset{1}{2}(-1 \pm i\sqrt{29})}{\underset{6}{12}}$$

$$= \frac{-1 \pm i\sqrt{29}}{6}$$

The solutions are $\dfrac{-1 + i\sqrt{29}}{6}$ and $\dfrac{-1 - i\sqrt{29}}{6}$. Note that neither solution is a real number. Both solutions are complex numbers.

Now Try Exercise 53

**Avoiding Common Errors**

Some students use the quadratic formula correctly until the last step, where they make an error. Below are illustrated both the correct and incorrect procedures for simplifying an answer.

When *both* terms in the numerator *and* the denominator have a common factor, that common factor may be divided out, as follows:

CORRECT

$$\frac{2 + 4\sqrt{3}}{2} = \frac{\overset{1}{\cancel{2}}(1 + 2\sqrt{3})}{\underset{1}{\cancel{2}}} = 1 + 2\sqrt{3}$$

$$\frac{6 + 3\sqrt{3}}{6} = \frac{\overset{1}{\cancel{3}}(2 + \sqrt{3})}{\underset{2}{\cancel{6}}} = \frac{2 + \sqrt{3}}{2}$$

Below are some common errors. Study them carefully so you will not make them. Can you explain why each of the following procedures is incorrect?

INCORRECT

$$\frac{2 + 3}{2} = \frac{\overset{1}{\cancel{2}} + 3}{\underset{1}{\cancel{2}}} \qquad\qquad \frac{3 + 2\sqrt{5}}{2} = \frac{3 + \overset{1}{\cancel{2}}\sqrt{5}}{\underset{1}{\cancel{2}}}$$

$$\frac{3 + \cancel{\sqrt{6}}}{2} = \frac{3 + \sqrt{6^3}}{\underset{1}{\cancel{2}}} \qquad\qquad \frac{4 + 3\sqrt{5}}{2} = \frac{\overset{2}{\cancel{4}} + 3\sqrt{5}}{\underset{1}{\cancel{2}}}$$

Note that $\frac{2 + 3}{2}$ simplifies to $\frac{5}{2}$. However, $\frac{3 + 2\sqrt{5}}{2}$, $\frac{3 + \sqrt{6}}{2}$, and $\frac{4 + 3\sqrt{5}}{2}$ cannot be simplified any further.

**EXAMPLE 4**  Given $f(x) = 2x^2 + 4x$, find all real values of $x$ for which $f(x) = 5$.

*Solution*    We wish to determine all real values of $x$ for which $2x^2 + 4x = 5$

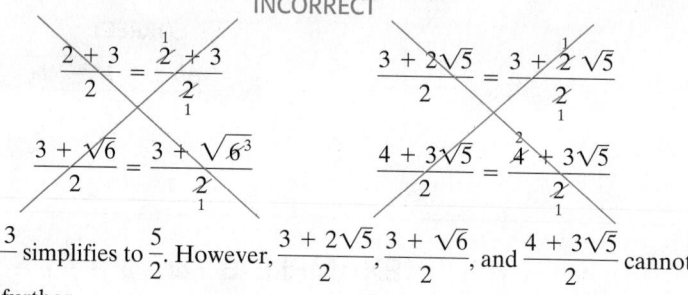

$$\overbrace{2x^2 + 4x} = 5 \qquad\text{Subtract 5 from both sides.}$$
$$2x^2 + 4x - 5 = 0 \qquad\text{Use quadratic formula with } a = 2, b = 4, c = -5.$$

$$x = \frac{-b \pm \sqrt{b^2 - 4ac}}{2a}$$

$$= \frac{-4 \pm \sqrt{4^2 - 4(2)(-5)}}{2(2)} = \frac{-4 \pm \sqrt{56}}{4} = \frac{-4 \pm 2\sqrt{14}}{4}$$

Next, factor out 2 from both terms in the numerator, and then divide out the common factor.

$$x = \frac{\overset{1}{\cancel{2}}(-2 \pm \sqrt{14})}{\underset{2}{\cancel{4}}} = \frac{-2 \pm \sqrt{14}}{2}\text{ *}$$

Thus, the solutions are $\dfrac{-2 + \sqrt{14}}{2}$ and $\dfrac{-2 - \sqrt{14}}{2}$.

Note that the expression in Example 4, $2x^2 + 4x - 5$, is not factorable. Therefore, Example 4 could not be solved by factoring.

Now Try Exercise 69

**Understanding Algebra**

If the numerical coefficients of a quadratic equation have a common factor, divide each term by the common factor. For example, for the equation $3x^2 + 12x + 3 = 0$, first divide each term by the common factor, 3, and simplify:

$$\frac{3x^2}{3} + \frac{12x}{3} + \frac{3}{3} = \frac{0}{3}$$
$$x^2 + 4x + 1 = 0$$

This new equation is equivalent to the original equation and is easier to solve.

*Solutions will be given in this form in the Answer Section.

## 3 Determine a Quadratic Equation Given Its Solutions

If we are given the solutions of an equation, we can find the equation by working backward. This procedure is illustrated in Example 5.

**EXAMPLE 5** Determine an equation that has the following solutions:

**a)** $-5$ and $1$         **b)** $3 + 2i$ and $3 - 2i$

*Solution*

**a)** If the solutions are $-5$ and $1$ we write

$$x = -5 \quad \text{or} \quad x = 1$$
$$x + 5 = 0 \qquad\qquad x - 1 = 0 \qquad \text{Set equations equal to 0.}$$
$$(x + 5)(x - 1) = 0 \qquad \text{Zero-factor property}$$
$$x^2 - x + 5x - 5 = 0 \qquad \text{Multiply factors.}$$
$$x^2 + 4x - 5 = 0 \qquad \text{Combine like terms.}$$

Thus, the equation is $x^2 + 4x - 5 = 0$. Many other equations have solutions $-5$ and $1$. In fact, any equation of the form $k(x^2 + 4x - 5) = 0$, where $k$ is a nonzero constant, has those solutions.

**b)**
$$x = 3 + 2i \quad \text{or} \quad x = 3 - 2i$$
$$x - (3 + 2i) = 0 \qquad x - (3 - 2i) = 0 \qquad \text{Set equations equal to 0.}$$
$$[x - (3 + 2i)][x - (3 - 2i)] = 0 \qquad \text{Zero-factor property}$$
$$x \cdot x - x(3 - 2i) - x(3 + 2i) + (3 + 2i)(3 - 2i) = 0 \qquad \text{Multiply.}$$
$$x^2 - 3x + 2xi - 3x - 2xi + (9 - 4i^2) = 0 \qquad \text{Distributive property; multiply}$$
$$x^2 - 6x + 9 - 4i^2 = 0 \qquad \text{Combine like terms.}$$
$$x^2 - 6x + 9 - 4(-1) = 0 \qquad \text{Substitute: } i^2 = -1.$$
$$x^2 - 6x + 13 = 0 \qquad \text{Simplify.}$$

The equation $x^2 - 6x + 13 = 0$ has the complex solutions $3 + 2i$ and $3 - 2i$.

Now Try Exercise 75

In Example 5 **a)**, the equation $x^2 + 4x - 5 = 0$ had the real number solutions $-5$ and $1$. The solutions correspond to the $x$-intercepts $(-5, 0)$ and $(1, 0)$ of the graph of the function $f(x) = x^2 + 4x - 5$, as shown in **Figure 12.1**.

FIGURE 12.1

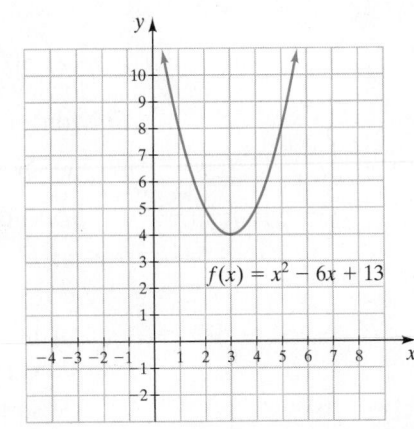

FIGURE 12.2

### Understanding Algebra

Recall from Chapter 4 that an *x-intercept* of a graph is a point $(x, 0)$ where a graph crosses the *x*-axis. To find the *x*-intercept of a graph, we set $y = 0$ or set $f(x) = 0$ and solve for *x*. If *x* is a real number then the point $(x, 0)$ is an *x*-intercept.

In Example 5 **b)**, the equation $x^2 - 6x + 13 = 0$ had only complex solutions and no real number solutions. Thus the graph of the function $f(x) = x^2 - 6x + 13$ has no *x*-intercepts, as shown in **Figure 12.2**.

## 4 Use the Discriminant to Determine the Number of Real Solutions to a Quadratic Equation

### Discriminant

The **discriminant** of a quadratic equation is the expression under the radical sign in the quadratic formula.

$$\underbrace{b^2 - 4ac}_{\text{Discriminant}}$$

The discriminant provides information to determine the number and kinds of solutions of a quadratic equation.

### Solutions of a Quadratic Equation

For a quadratic equation of the form $ax^2 + bx + c = 0, a \neq 0$:

If $b^2 - 4ac > 0$, the quadratic equation has two distinct real number solutions.

If $b^2 - 4ac = 0$, the quadratic equation has a single real number solution.

If $b^2 - 4ac < 0$, the quadratic equation has no real number solution.

### EXAMPLE 6

**a)** Find the discriminant of the equation $x^2 - 8x + 16 = 0$.

**b)** How many real number solutions does the given equation have?

**c)** Use the quadratic formula to find the solution(s).

*Solution*

**a)** $a = 1,$     $b = -8,$     $c = 16$

$$b^2 - 4ac = (-8)^2 - 4(1)(16)$$
$$= 64 - 64 = 0$$

**b)** Since the discriminant equals 0, there is a single real number solution.

**c)**
$$x = \frac{-b \pm \sqrt{b^2 - 4ac}}{2a}$$
$$= \frac{-(-8) \pm \sqrt{0}}{2(1)} = \frac{8 \pm 0}{2} = \frac{8}{2} = 4$$

The only solution is 4.

Now Try Exercise 9

### EXAMPLE 7

Without actually finding the solutions, determine whether the following equations have two distinct real number solutions, a single real number solution, or no real number solution.

**a)** $2x^2 - 4x + 6 = 0$     **b)** $x^2 - 5x - 3 = 0$     **c)** $4x^2 - 12x = -9$

*Solution*   We use the discriminant of the quadratic formula to answer these questions.

**a)** $b^2 - 4ac = (-4)^2 - 4(2)(6) = 16 - 48 = -32$

Since the discriminant is negative, this equation has no real number solution.

**b)** $b^2 - 4ac = (-5)^2 - 4(1)(-3) = 25 + 12 = 37$

Since the discriminant is positive, this equation has two distinct real number solutions.

**c)** First, rewrite $4x^2 - 12x = -9$ as $4x^2 - 12x + 9 = 0$.

$$b^2 - 4ac = (-12)^2 - 4(4)(9) = 144 - 144 = 0$$

Since the discriminant is 0, this equation has a single real number solution.

Now Try Exercise 15

The discriminant can be used to find the number of real solutions to an equation of the form $ax^2 + bx + c = 0$. Since the $x$-intercepts of a quadratic function, $f(x) = ax^2 + bx + c$, occur where $f(x) = 0$, the discriminant can also be used to find the number of $x$-intercepts of a quadratic function. **Figure 12.3** shows the relationship between the discriminant and the number of $x$-intercepts for a function of the form $f(x) = ax^2 + bx + c$.

### Graphs of $f(x) = ax^2 + bx + c$

If $b^2 - 4ac > 0$, $f(x)$ has two distinct $x$-intercepts.

If $b^2 - 4ac = 0$, $f(x)$ has a single $x$-intercept.

If $b^2 - 4ac < 0$, $f(x)$ has no $x$-intercepts.

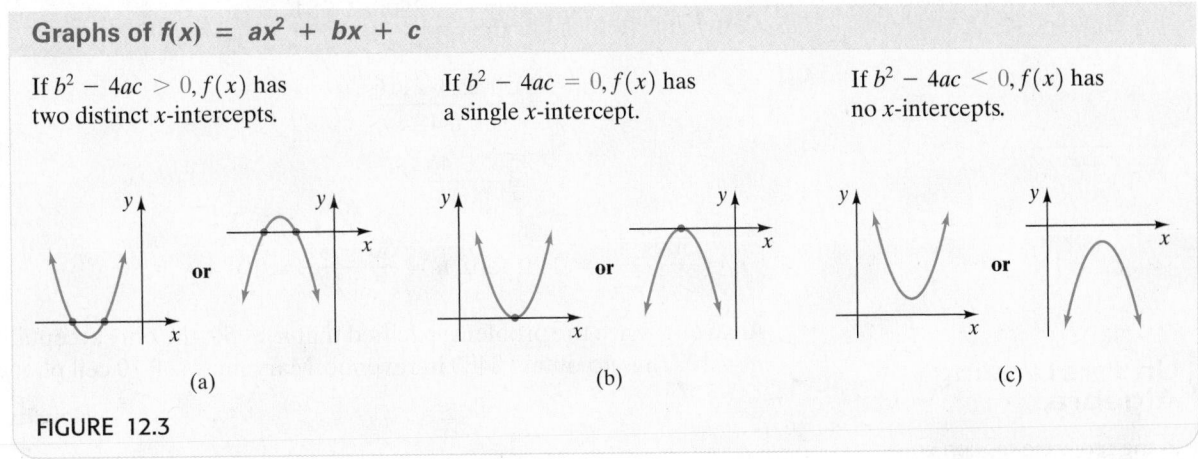

(a)                                   (b)                                   (c)

FIGURE 12.3

## 5  Study Applications That Use Quadratic Equations

We will now look at some applications of quadratic equations.

**EXAMPLE 8  Cell Phones**  Mary Olson owns a business that sells cell phones. The revenue, $R(n)$, from selling the cell phones is determined by multiplying the number of cell phones by the selling price per phone. Suppose the revenue from selling $n$ cell phones, $n \leq 50$, is

$$R(n) = n(50 - 0.2n)$$

where $(50 - 0.2n)$ is the price per cell phone, in dollars.

**a)** Find the revenue when 30 cell phones are sold.

**b)** How many cell phones must be sold to have a revenue of $480?

*Solution*

**a)** To find the revenue when 30 cell phones are sold, we evaluate the revenue function for $n = 30$.

$$R(n) = n(50 - 0.2n)$$
$$R(30) = 30[50 - 0.2(30)]$$
$$= 30(50 - 6)$$
$$= 30(44)$$
$$= 1320$$

The revenue from selling 30 cell phones is $1320.

**b)** Understand   We want to find the number of cell phones that need to be sold to have $480 in revenue. Thus, we need to let $R(n) = 480$ and solve for $n$.

$$R(n) = n(50 - 0.2n)$$
$$480 = n(50 - 0.2n)$$
$$480 = 50n - 0.2n^2$$
$$0.2n^2 - 50n + 480 = 0$$

Now we can use the quadratic formula to solve the equation.

Translate    $a = 0.2,$      $b = -50,$      $c = 480$

$$n = \frac{-b \pm \sqrt{b^2 - 4ac}}{2a}$$

$$= \frac{-(-50) \pm \sqrt{(-50)^2 - 4(0.2)(480)}}{2(0.2)}$$

Carry Out    $= \frac{50 \pm \sqrt{2500 - 384}}{0.4}$

$$= \frac{50 \pm \sqrt{2116}}{0.4}$$

$$= \frac{50 \pm 46}{0.4}$$

$$n = \frac{50 + 46}{0.4} = 240 \quad \text{or} \quad n = \frac{50 - 46}{0.4} = 10$$

**Answer**   Since the problem specified that $n \leq 50$, the only acceptable solution is $n = 10$. Thus, to obtain $480 in revenue, Mary must sell 10 cell phones.

Now Try Exercise 87

## Understanding Algebra

In the projectile motion equation, the variables $v_0$ and $h_0$ have *subscripts* of 0. Subscripts of 0 usually refer to the *initial* value for a variable. Thus, $v_0$ refers to the *initial* velocity with which the object is projected upward and $h_0$ refers to the *initial* height from which the object is projected.

An important equation in physics relates the height of an object to the time after the object is projected upward.

### Projectile Motion Equation

The height, $h$, of an object $t$ seconds after it is projected upward can be found by solving the equation

$$h = \frac{1}{2}gt^2 + v_0t + h_0, \text{ where}$$

- $g$ is the acceleration due to gravity,
- $v_0$ is the initial velocity of the object, and
- $h_0$ is the initial height of the object.

Before we use the equation, a few notes about $g$, the acceleration due to gravity.

- When we measure the height of the object in feet, the acceleration due to gravity on Earth is $-32$ ft/sec$^2$ or $g = -32$.
- When we measure the height of an object in meters, the acceleration due to gravity on Earth is $-9.8$ m/sec$^2$ or $g = -9.8$.
- The value of $g$ will be different on the moon or on a planet other than Earth, but we can still use the projectile motion formula.

EXAMPLE 9 **Throwing a Ball** Betty Heller is on top of a building and throws a baseball upward from an initial height of 60 feet with an initial velocity of 30 feet per second. Use the projectile motion equation to answer the following questions. Round your answers to the nearest tenth of a second.

a) How long after Betty throws the ball will the ball be 25 feet above the ground?

b) How long after Betty throws the ball will the ball hit the ground?

c) What is the height of the ball after 2 seconds?

*Solution*

a) Understand    The problem is illustrated in **Figure 12.4**. We are asked to find the time, $t$, it takes for the object to reach a height, $h$, of 25 feet. Thus we have the following values to substitute into the projectile motion equation:   $h = 25$, $g = -32$, $v_0 = 30$, and $h_0 = 60$.

FIGURE 12.4

Translate        $h = \dfrac{1}{2}gt^2 + v_0t + h_0,$

$25 = \dfrac{1}{2}(-32)t^2 + (30)t + 60$

Carry Out    $25 = -16t^2 + 30t + 60$        Add $16t^2$, subtract $30t$, and subtract 60 from
both sides to get a quadratic equation in stan-
dard form with a positive leading coefficient.

$16t^2 - 30t - 35 = 0$

$t = \dfrac{-b \pm \sqrt{b^2 - 4ac}}{2a}$        Solve using the quadratic formula with
$a = 16, b = -30,$ and $c = -35.$

$= \dfrac{-(-30) \pm \sqrt{(-30)^2 - 4(16)(-35)}}{2(16)}$

$= \dfrac{30 \pm \sqrt{3140}}{32}$

$t = \dfrac{30 + \sqrt{3140}}{32}$    or    $t = \dfrac{30 - \sqrt{3140}}{32}$

$\approx 2.7$                    $\approx -0.8$

Answer    Since time cannot be negative, the only reasonable solution is 2.7 seconds. Thus, about 2.7 seconds after the ball is thrown upward, it will be 25 feet
above the ground.

b) Understand    When the ball strikes the ground, its distance above the ground is
0. We substitute $h = 0$ into the formula and solve for $t$.

Translate                            $h = \dfrac{1}{2}gt^2 + v_0t + h_0$

$0 = \dfrac{1}{2}(-32)t^2 + 30t + 60$

Carry Out    $0 = -16t^2 + 30t + 60$

$16t^2 - 30t - 60 = 0$        Multiply both sides of the equation by $-1$
to get a positive leading coefficient.

$t = \dfrac{-b \pm \sqrt{b^2 - 4ac}}{2a}$        Solve using the quadratic formula with $a = 16, b = -30,$
and $c = -60$

$= \dfrac{-(-30) \pm \sqrt{(-30)^2 - 4(16)(-60)}}{2(16)}$

$= \dfrac{30 \pm \sqrt{4740}}{32}$

$t = \dfrac{30 + \sqrt{4740}}{32}$    or    $t = \dfrac{30 - \sqrt{4740}}{32}$

$\approx 3.1$                    $\approx -1.2$

Answer    Since time cannot be negative, the only reasonable solution is 3.1 seconds. Thus, the ball will strike the ground after about 3.1 seconds.

c) Understand    We are asked to find the height, $h$, after 2 seconds. We will substitute $t = 2$ into the projectile motion equation and solve for $h$.

$h = \dfrac{1}{2}gt^2 + v_0t + h_0$

Translate and Carry Out    $h = \dfrac{1}{2}(-32)(2)^2 + 30(2) + 60$

$= -64 + 60 + 60 = 56$

Answer    Thus, the ball will be at a height of 56 feet after 2 seconds.

Now Try Exercise 97

# EXERCISE SET 12.2   MathXL® MyMathLab®
MathXL®    MyMathLab

## Warm-Up Exercises

*Fill in the blanks with the appropriate word, phrase, or symbol(s) from the following list.*

| no | one | two | completing the square | discriminant | positive integer |
|---|---|---|---|---|---|
| projectile motion | divide | multiply | factoring | linear | |

1. Although any quadratic equation can be solved by _____ we generally prefer to use the quadratic formula because it is easier and more efficient to use.

2. When using the quadratic formula, the calculations are generally easier if the leading coefficient, $a$, is a _____.

3. If the numerical coefficients of all of the terms in a quadratic equation have a common factor, _____ each term by the common factor.

4. The _____ of a quadratic equation is the expression under the radical sign in the quadratic formula: $b^2 - 4ac$.

5. If the discriminant of a quadratic equation is positive, the equation has _____ distinct real number solution(s).

6. If the discriminant of a quadratic equation is equal to zero, the equation has _____ real number solution(s).

7. If the discriminant of a quadratic equation is negative, the equation has _____ real number solution(s).

8. The height, $h$, of an object $t$ seconds after it is projected upward can be found using the _____ equation.

## Practice the Skills

*Use the discriminant to determine whether each equation has two distinct real solutions, a single real solution, or no real solution.*

9. $x^2 + 6x + 2 = 0$

10. $2x^2 + x + 3 = 0$

11. $4z^2 + 6z + 5 = 0$

12. $-a^2 + 3a - 6 = 0$

13. $5p^2 + 3p - 7 = 0$

14. $2x^2 = 16x - 32$

15. $-5x^2 + 5x - 8 = 0$

16. $4.1x^2 - 3.1x - 2.8 = 0$

17. $x^2 + 10.2x + 26.01 = 0$

18. $\frac{1}{2}x^2 + \frac{2}{3}x + 10 = 0$

19. $b^2 = -3b - \frac{9}{4}$

20. $\frac{x^2}{3} = \frac{2x}{7}$

*Solve each equation by using the quadratic formula.*

21. $x^2 + 7x + 10 = 0$

22. $x^2 + 9x + 18 = 0$

23. $a^2 - 6a + 8 = 0$

24. $a^2 + 6a + 8 = 0$

25. $x^2 = -6x + 7$

26. $-a^2 - 9a + 10 = 0$

27. $-b^2 = 4b - 20$

28. $a^2 - 16 = 0$

29. $b^2 - 64 = 0$

30. $2x^2 = 4x + 1$

31. $3w^2 - 4w + 5 = 0$

32. $x^2 - 6x = 0$

33. $c^2 - 5c = 0$

34. $-t^2 - t - 1 = 0$

35. $4s^2 - 8s + 6 = 0$

36. $-3r^2 = 9r + 6$

37. $a^2 + 2a + 1 = 0$

38. $y^2 + 16y + 64 = 0$

39. $16x^2 - 8x + 1 = 0$

40. $100m^2 + 20m + 1 = 0$

41. $x^2 - 2x - 1 = 0$

42. $2 - 3r^2 = -4r$

43. $-n^2 = 3n + 6$

44. $-9d - 3d^2 = 5$

45. $2x^2 + 5x - 3 = 0$

46. $(r - 3)(3r + 4) = -10$

47. $(2a + 3)(3a - 1) = 2$

48. $6x^2 = 21x + 27$

49. $\frac{1}{2}t^2 + t - 12 = 0$

50. $\frac{2}{3}x^2 = 8x - 18$

51. $9r^2 + 3r - 2 = 0$

52. $2x^2 - 4x - 2 = 0$

53. $\frac{1}{2}x^2 + 2x + \frac{2}{3} = 0$

54. $x^2 - \frac{11}{3}x = \frac{10}{3}$

55. $a^2 - \frac{a}{5} - \frac{1}{3} = 0$

56. $b^2 = -\frac{b}{2} + \frac{2}{3}$

57. $c = \frac{c - 6}{4 - c}$

58. $3y = \frac{5y + 6}{2y + 3}$

59. $2x^2 - 4x + 5 = 0$

60. $3a^2 - 4a = -5$

61. $y^2 + \frac{y}{2} = -\frac{3}{2}$

62. $2b^2 - \frac{7}{3}b + \frac{4}{3} = 0$

63. $0.1x^2 + 0.6x - 1.2 = 0$

64. $2.3x^2 - 5.6x - 0.4 = 0$

*For each function, determine all real values of the variable for which the function has the value indicated.*

**65.** $f(x) = 2x^2 - 3x + 7, f(x) = 7$

**66.** $g(x) = x^2 + 3x + 8, g(x) = 8$

**67.** $k(x) = x^2 - x - 15, k(x) = 15$

**68.** $p(r) = r^2 + 17r + 81, p(r) = 9$

**69.** $h(t) = 2t^2 - 7t + 6, h(t) = 2$

**70.** $t(x) = x^2 + 5x - 4, t(x) = 3$

**71.** $g(a) = 2a^2 - 3a + 16, g(a) = 14$

**72.** $h(x) = 6x^2 + 3x + 1, h(x) = -7$

*Determine a quadratic equation that has the given solutions.*

**73.** $1, 6$

**74.** $-3, 4$

**75.** $1, -9$

**76.** $-2, -6$

**77.** $-\dfrac{3}{5}, \dfrac{2}{3}$

**78.** $-\dfrac{1}{3}, -\dfrac{3}{4}$

**79.** $\sqrt{2}, -\sqrt{2}$

**80.** $\sqrt{5}, -\sqrt{5}$

**81.** $3i, -3i$

**82.** $8i, -8i$

**83.** $3 + \sqrt{2}, 3 - \sqrt{2}$

**84.** $5 - \sqrt{3}, 5 + \sqrt{3}$

**85.** $2 + 3i, 2 - 3i$

**86.** $5 - 4i, 5 + 4i$

## Problem Solving

*In Exercises 87–90,* **a)** *set up a revenue function, R(n), that can be used to solve the problem, and then,* **b)** *solve the problem. See Example 8.*

**87. Selling Lamps** A business sells $n$ lamps, $n \le 65$, at a price of $(10 - 0.02n)$ dollars per lamp. How many lamps must be sold to have a revenue of $450?

**88. Selling Batteries** A business sells $n$ batteries, $n \le 26$, at a price of $(25 - 0.1n)$ dollars per battery. How many batteries must be sold to have a revenue of $460?

**89. Selling Chairs** A business sells $n$ chairs, $n \le 50$, at a price of $(50 - 0.4n)$ dollars per chair. How many chairs must be sold to have a revenue of $660?

**90. Selling Watches** A business sells $n$ watches, $n \le 75$, at a price of $(30 - 0.15n)$ dollars per watch. How many watches must be sold to have a revenue of $1260?

© Tatiana Popova/Shutterstock

*In Exercises 91–108, use a calculator as needed to give the solution in decimal form. Round irrational numbers to the nearest hundredth.*

**91. Numbers** Twice the square of a positive number increased by three times the number is 27. Find the number.

**92. Numbers** Three times the square of a positive number decreased by twice the number is 21. Find the number.

**93. Rectangular Garden** The length of a rectangular garden is 1 foot less than 3 times its width. Find the length and width if the area of the garden is 24 square feet.

**94. Rectangular Region** Lora Wallman wishes to fence in a rectangular region along a riverbank by constructing fencing as il-

lustrated in the diagram. If she has only 400 feet of fencing and wishes to enclose an area of 15,000 square feet, find the dimensions of the rectangular region.

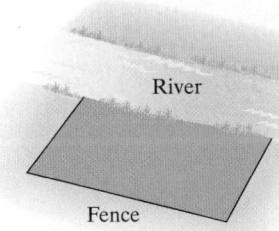

**95. Photo** John Williams, a professional photographer, has a 6-inch-by-8-inch photo. He wishes to reduce the photo by the same amount on each side so that the resulting photo will have half the area of the original photo. By how much will he have to reduce the length of each side?

**96. Rectangular Garden** Bart Simmons has a 12-meter-by-9-meter flower garden. He wants to build a gravel path of uniform width along the inside of the garden on each side so that the resulting garden will have half the area of the original garden. What will be the width of the gravel path?

*In Exercises 97–100, use the projectile motion equation* $h = \dfrac{1}{2}gt^2 + v_0t + h_0$. *If necessary, round your answers to the nearest tenth of a second. See Example 9.*

**97. Hitting a Tennis Ball** A tennis ball is hit upward from an initial height of 4 feet with an initial velocity of 40 feet per second.

   **a)** How long after the ball is hit will it be 20 feet above the ground?

   **b)** How long after the ball is hit will it hit the ground?

   **c)** What is the height of the ball after 1 second?

**98. Throwing a Horseshoe** A horseshoe is thrown upward from an initial height of 3 feet with an initial velocity of 25 feet per second.

   **a)** How long after the horseshoe is thrown will it be 10 feet above the ground?

   **b)** How long after the horseshoe is thrown will it hit the ground?

   **c)** What is the height of the horseshoe after 1 second?

**99. Golfing on the Moon** On February 6, 1971, astronaut Alan Shepard hit a golf ball while standing on the moon. The acceleration due to gravity, $g$, on the moon is about $-5.3$ ft/sec$^2$. Assume the ball was hit upward from an initial height of 0 feet and with an initial velocity of 26.5 ft/sec.

**a)** How long after the golf ball was hit would it reach a height of 23.85 feet?

(*Hint:* Substitute the appropriate values into the projectile motion equation and then divide each term by the leading coefficient.)

**b)** How long after the golf ball was hit would it return to the surface of the moon?

**c)** What would be the height of the golf ball after 5 seconds?

**100. Baseball on the Moon** Suppose Neil Armstrong had hit a baseball when he first set foot on the moon on July 20, 1969. Assume the ball was hit upward from an initial height of 2.65 feet and with an initial velocity of 132.5 ft/sec. See Exercise 99.

**a)** How long after the baseball was hit would it reach a height of 132.5 feet?

**b)** How long after the baseball was hit would it hit the ground?

**c)** What would be the height of the baseball after 25 seconds?

Neil Armstrong on the moon

## Concept/Writing Exercises

**101.** Consider the two equations $-6x^2 + \frac{1}{2}x - 5 = 0$ and $6x^2 - \frac{1}{2}x + 5 = 0$. Must the solutions to these two equations be the same? Explain your answer.

**102.** Consider $12x^2 - 15x - 6 = 0$ and $3(4x^2 - 5x - 2) = 0$.

**a)** Will the solution to the two equations be the same? Explain.

**b)** Solve $12x^2 - 15x - 6 = 0$.

**c)** Solve $3(4x^2 - 5x - 2) = 0$.

**103. a)** Explain how to find the discriminant.

**b)** What is the discriminant for the equation $3x^2 - 6x + 10 = 0$?

**c)** Write a paragraph or two explaining the relationship between the value of the discriminant and the number of real solutions to a quadratic equation. In your paragraph, explain *why* the value of the discriminant determines the number of real solutions.

**104.** Write a paragraph or two explaining the relationship between the value of the discriminant and the number of $x$-intercepts of $f(x) = ax^2 + bx + c$. In your paragraph, explain when the function will have no, one, and two $x$-intercepts.

**105.** Give your own example of a quadratic equation that can be solved by the quadratic formula but not by factoring over the set of integers.

**106.** Are there any quadratic equations that **a)** can be solved by the quadratic formula that cannot be solved by completing the square? **b)** can be solved by completing the square that cannot be solved by factoring over the set of integers?

**107.** When solving a quadratic equation by the quadratic formula, if the discriminant is a perfect square, must the equation be factorable over the set of integers?

**108.** When solving a quadratic equation by the quadratic formula, if the discriminant is a natural number, must the equation be factorable over the set of integers?

## Challenge Problems

**109. Heating a Metal Cube** A metal cube expands when heated. If each edge increases 0.20 millimeter after being heated and the total volume increases by 6 cubic millimeters, find the original length of a side of the cube.

**110. Six Solutions** The equation $x^n = 1$ has $n$ solutions (including the complex solutions). Find the six solutions to $x^6 = 1$. (*Hint:* Rewrite the equation as $x^6 - 1 = 0$, then factor using the formula for the difference of two squares.)

**111. Throwing a Rock** Travis Hawley is on the fourth floor of an eight-story building and Courtney Prenzlow is on the roof. Travis is 60 feet above the ground while Courtney is 120 feet above the ground.

**a)** If Travis drops a rock out of a window, determine the time it takes for the rock to strike the ground.

**b)** If Courtney drops a rock off the roof, determine the time it takes for the rock to strike the ground.

**c)** If Travis throws a rock upward with an initial velocity of 100 feet per second at the same time that Courtney throws a rock upward at 60 feet per second, whose rock will strike the ground first? Explain.

**d)** Will the rocks ever be at the same distance above the ground? If so, at what time?

[5.3]   **112.** Evaluate $\dfrac{5.55 \times 10^3}{1.11 \times 10^1}$.

[8.2]   **113.** If $f(x) = x^2 + 2x - 8$, find $f(3)$.

[9.3]   **114.** Solve the system of equations.
$$3x + 4y = 2$$
$$2x = -5y - 1$$

[11.5]   **115.** Simplify $\dfrac{x + \sqrt{y}}{x - \sqrt{y}}$.

[11.6]   **116.** Solve $\sqrt{x^2 - 6x - 4} = x$.

# 12.3 Quadratic Equations: Applications and Problem Solving

**1** Solve additional applications.

**2** Solve for a variable in a formula.

## **1** Solve Additional Applications

In this section, we will explore several more applications of quadratic equations. We start by investigating the profit of a new company.

**EXAMPLE 1   Company Profit**  Laserox, a start-up company, projects that its annual profit, $p(t)$, in thousands of dollars, over the first 6 years of operation can be approximated by the function $p(t) = 1.2t^2 + 4t - 8$, where $t$ is the number of years completed.

**a)** Estimate the profit (or loss) of the company after the first year.

**b)** Estimate the profit (or loss) of the company after 6 years.

**c)** Estimate the time needed for the company to break even.

### Solution

**a)** To estimate the profit after 1 year, we evaluate the function at 1.
$$p(t) = 1.2t^2 + 4t - 8$$
$$p(1) = 1.2(1)^2 + 4(1) - 8 = -2.8$$

Thus, at the end of the first year the company projects a loss of $2.8 thousand or a loss of $2800.

**b)** $p(6) = 1.2(6)^2 + 4(6) - 8 = 59.2$

Thus, at the end of the sixth year the company's projected profit is $59.2 thousand, or a profit of $59,200.

**c)** Understand   The company will break even when the profit is 0. Thus, to find the break-even point (no profit or loss) we solve the equation
$$1.2t^2 + 4t - 8 = 0$$

We can use the quadratic formula to solve this equation.

Translate                         $a = 1.2, \qquad b = 4, \qquad c = -8$
$$t = \frac{-b \pm \sqrt{b^2 - 4ac}}{2a}$$
$$= \frac{-4 \pm \sqrt{4^2 - 4(1.2)(-8)}}{2(1.2)}$$

Carry Out
$$= \frac{-4 \pm \sqrt{16 + 38.4}}{2.4}$$
$$= \frac{-4 \pm \sqrt{54.4}}{2.4}$$
$$\approx \frac{-4 \pm 7.376}{2.4}$$

$$t \approx \frac{-4 + 7.376}{2.4} \approx 1.4 \qquad \text{or} \qquad t \approx \frac{-4 - 7.376}{2.4} \approx -4.74$$

Answer   Since time cannot be negative, the break-even time is about 1.4 years.

Now Try Exercise 29

EXAMPLE 2 **Life Expectancy** The function $N(t) = 0.0054t^2 - 1.46t + 95.11$ can be used to estimate the average number of years of life expectancy remaining for a person of age $t$ years where $30 \le t \le 100$.

**a)** Estimate the remaining life expectancy of a person of age 40.

**b)** If a person has a remaining life expectancy of 14.3 years, estimate the age of the person.

*Solution*

**a)** Understand    To determine the remaining life expectancy for a 40-year-old, we substitute 40 for $t$ in the function and evaluate.

Translate

$$N(t) = 0.0054t^2 - 1.46t + 95.11$$
$$N(40) = 0.0054(40)^2 - 1.46(40) + 95.11$$

Carry Out

$$= 0.0054(1600) - 58.4 + 95.11$$
$$= 8.64 - 58.4 + 95.11$$
$$= 45.35$$

Answer and Check    The answer appears reasonable. Thus, on the average, a 40-year-old can expect to live another 45.35 years to an age of 85.35 years.

**b)** Understand    Here we are given the remaining life expectancy, $N(t)$, and asked to find the age of the person, $t$. To solve this problem, we substitute 14.3 for $N(t)$ and solve for $t$. We will use the quadratic formula.

Translate

$$N(t) = 0.0054t^2 - 1.46t + 95.11$$
$$14.3 = 0.0054t^2 - 1.46t + 95.11$$

Carry Out

$$0 = 0.0054t^2 - 1.46t + 80.81$$
$$a = 0.0054, \quad b = -1.46, \quad c = 80.81$$

$$t = \frac{-b \pm \sqrt{b^2 - 4ac}}{2a}$$

$$= \frac{-(-1.46) \pm \sqrt{(-1.46)^2 - 4(0.0054)(80.81)}}{2(0.0054)}$$

$$= \frac{1.46 \pm \sqrt{2.1316 - 1.745496}}{0.0108}$$

$$= \frac{1.46 \pm \sqrt{0.386104}}{0.0108}$$

$$\approx \frac{1.46 \pm 0.6214}{0.0108}$$

$$t \approx \frac{1.46 + 0.6214}{0.0108} \quad \text{or} \quad t \approx \frac{1.46 - 0.6214}{0.0108}$$

$$\approx 192.72 \qquad\qquad\qquad \approx 77.65$$

Answer    Since 192.72 is not a reasonable age, we can exclude that as a possibility. Thus, the average person who has a life expectancy of 14.3 years is about 77.65 years old.

Now Try Exercise 33

**Motion Problems** We first discussed motion problems in Section 3.4.

EXAMPLE 3 **Motorboat Ride** Charles Curtis travels in his motorboat downstream for 12 miles with the current. He then turns around and heads back to the starting point going upstream against the current. The total time of his trip is 5 hours and the river current is 2 miles per hour. If during the entire trip he did not touch the throttle to change the speed, find the speed the boat would have traveled in still water.

*Solution*   Understand   We are asked to find the rate of the boat in still water. Let $r$ = the rate of the boat in still water. We know that the total time of the trip is 5 hours. Thus, the time downriver plus the time upriver must total 5 hours. Since distance = rate · time, we can find the time by dividing the distance by the rate.

| Direction | Distance | Rate | Time |
|---|---|---|---|
| Downriver (with current) | 12 | $r + 2$ | $\dfrac{12}{r + 2}$ |
| Upriver (against current) | 12 | $r - 2$ | $\dfrac{12}{r - 2}$ |

### Understanding Algebra

The distance formula is usually written as

$$\text{distance} = \text{rate} \cdot \text{time}$$

However, it is sometimes convenient to solve the formula for *time*:

$$\frac{\text{distance}}{\text{rate}} = \frac{\text{rate} \cdot \text{time}}{\text{rate}}$$

$$\frac{\text{distance}}{\text{rate}} = \text{time}$$

or    $\text{time} = \dfrac{\text{distance}}{\text{rate}}$

Translate        time downriver + time upriver = total time

$$\frac{12}{r + 2} + \frac{12}{r - 2} = 5$$

Carry Out   $(r + 2)(r - 2)\left(\dfrac{12}{r + 2} + \dfrac{12}{r - 2}\right) = (r + 2)(r - 2)\,(5)$   Multiply by the LCD.

$(r + 2)(r - 2)\left(\dfrac{12}{r + 2}\right) + (r + 2)(r - 2)\left(\dfrac{12}{r - 2}\right) = (r + 2)(r - 2)(5)$   Distributive property

$$12(r - 2) + 12(r + 2) = 5(r^2 - 4)$$

$$12r - 24 + 12r + 24 = 5r^2 - 20$$   Distributive property

$$24r = 5r^2 - 20$$   Simplify.

or   $5r^2 - 24r - 20 = 0$

Using the quadratic formula with $a = 5, b = -24$, and $c = -20$, we obtain

$$r = \frac{24 \pm \sqrt{976}}{10}$$

$$r \approx 5.5 \quad \text{or} \quad r \approx -0.7$$

Answer   Since the rate cannot be negative, the rate or speed of the boat in still water is about 5.5 miles per hour.

Now Try Exercise 43

**Work Problems**   Let's do an example involving a work problem, first discussed in Section 7.7. You may wish to review that section before studying the next example.

**EXAMPLE 4   Pumping Water**   The Durals need to pump water from their flooded basement. They have a pump and they borrow another from their neighbors, the Sullivans. With both pumps working together, the job can be done in 6 hours. The Sullivans' pump working alone would do the job in 2 hours less than the Durals' pump working alone. How long would it take for each pump to do the job working alone?

*Solution*   Understand   Recall from Section 7.7 that the rate of work multiplied by the time worked gives the part of the task completed. Let $t$ = number of hours for the Durals' (slower) pump to complete the job by itself, then $t - 2$ = number of hours for the Sullivans' pump to complete the job by itself.

| Pump | Rate of Work | Time Worked | Part of Task Completed |
|---|---|---|---|
| Durals' pump | $\dfrac{1}{t}$ | 6 | $\dfrac{6}{t}$ |
| Sullivans' pump | $\dfrac{1}{t - 2}$ | 6 | $\dfrac{6}{t - 2}$ |

Translate $\left( \begin{array}{c} \text{part of task} \\ \text{by Durals' pump} \end{array} \right) + \left( \begin{array}{c} \text{part of task} \\ \text{by Sullivans' pump} \end{array} \right) = 1$

$$\frac{6}{t} + \frac{6}{t-2} = 1$$

Carry Out $\quad t(t-2)\left( \dfrac{6}{t} + \dfrac{6}{t-2} \right) = t(t-2)(1)$  Multiply both sides by the LCD, $t(t-2)$.

$$t(t-2)\left( \frac{6}{t} \right) + t(t-2)\left( \frac{6}{t-2} \right) = t^2 - 2t \qquad \text{Distributive property}$$

$$6(t-2) + 6t = t^2 - 2t$$
$$6t - 12 + 6t = t^2 - 2t$$
$$t^2 - 14t + 12 = 0$$

Using the quadratic formula, we obtain

$$t = \frac{14 \pm \sqrt{148}}{2}$$

$$t \approx 13.1 \quad \text{or} \quad t \approx 0.9$$

**Answer** Both $t = 13.1$ and $t = 0.9$ satisfy the original equation. However, if $t = 0.9$ then the Durals' pump could complete the job in 0.9 hours and the Sullivans' pump could complete the job in $0.9 - 2$ or $-1.1$ hours, which is not possible. Therefore, the only acceptable solution is $t = 13.1$. Thus, the Durals' pump takes about 13.1 hours by itself and the Sullivans' pump takes about $13.1 - 2$ or 11.1 hours by itself to complete the job.

Now Try Exercise 45

## 2 Solve for a Variable in a Formula

When solving for a variable in a formula, it may be necessary to use the square root property to isolate the variable. However, often the variable can only represent a nonnegative number. Therefore, *when using the square root property to solve for a variable in most formulas, we will use only the positive square root.* This will be demonstrated in our next example.

## EXAMPLE 5

**a)** The formula for the area of a circle is $A = \pi r^2$. Solve this equation for the radius, $r$.

**b)** *Newton's law of universal gravity* is

$$F = G\frac{m_1 m_2}{r^2}$$

Solve the equation for $r$, which measures distance.

*Solution*

**a)**
$$A = \pi r^2$$
$$\frac{A}{\pi} = r^2 \qquad \text{Isolate } r^2 \text{ by dividing both sides by } \pi.$$
$$\sqrt{\frac{A}{\pi}} = r \qquad \text{Square root property; } r \text{ must be positive.}$$

**b)**
$$F = G\frac{m_1 m_2}{r^2}$$
$$r^2 F = Gm_1 m_2 \qquad \text{Multiply both sides of formula by } r^2.$$
$$r^2 = \frac{Gm_1 m_2}{F} \qquad \text{Isolate } r^2 \text{ by dividing both sides by } F.$$
$$r = \sqrt{\frac{Gm_1 m_2}{F}} \qquad \text{Square root property; } r \text{ must be positive.}$$

Now Try Exercise 23

In both parts of Example 5, since $r$ must be greater than 0, when we used the square root property, we listed only the positive square root.

EXAMPLE 6  **Diagonal of a Suitcase**  The diagonal of a box can be calculated by the formula

$$d = \sqrt{L^2 + W^2 + H^2}$$

where $L$ is the length, $W$ is the width, and $H$ is the height of the box. See **Figure 12.5**.

FIGURE 12.5

a) Find the diagonal of a suitcase of length 30 inches, width 15 inches, and height 10 inches.

b) Solve the equation for the width, $W$.

### Understanding Algebra

Recall from Section 11.6 that when we need to eliminate a square root from an equation, we can isolate the square root and then square both sides of the equation. We simplify using the property

$$(\sqrt{a})^2 = a, a \geq 0$$

*Solution*

a) Understand   To find the diagonal, we need to substitute the appropriate values into the formula and solve for the diagonal, $d$.

Translate
$$d = \sqrt{L^2 + W^2 + H^2}$$
$$d = \sqrt{(30)^2 + (15)^2 + (10)^2}$$

Carry Out
$$= \sqrt{900 + 225 + 100}$$
$$= \sqrt{1225}$$
$$= 35$$

Answer   Thus, the diagonal of the suitcase is 35 inches.

b) Our first step in solving for $W$ is to square both sides of the formula.

$$d = \sqrt{L^2 + W^2 + H^2}$$
$$d^2 = \left(\sqrt{L^2 + W^2 + H^2}\right)^2 \qquad \text{Square both sides.}$$
$$d^2 = L^2 + W^2 + H^2 \qquad \text{Use } (\sqrt{a})^2 = a, a \geq 0.$$
$$d^2 - L^2 - H^2 = W^2 \qquad \text{Isolate } W^2.$$
$$\sqrt{d^2 - L^2 - H^2} = W \qquad \text{Square root property}$$

Now Try Exercise 15

EXAMPLE 7  **Traffic Cones**  The surface area of a right circular cone is

$$s = \pi r \sqrt{r^2 + h^2}$$

a) An orange traffic cone used on roads is 18 inches high with a radius of 12 inches. Find the surface area of the cone.

b) Solve the formula for $h$.

*Solution*

**a) Understand and Translate**    To find the surface area, we substitute the appropriate values into the formula.

$$s = \pi r \sqrt{r^2 + h^2}$$
$$= \pi(12)\sqrt{(12)^2 + (18)^2}$$

Carry Out

$$= 12\pi\sqrt{144 + 324}$$
$$= 12\pi\sqrt{468}$$
$$\approx 815.56$$

**Answer**    The surface area is about 815.56 square inches.

**b)** To solve for $h$ we need to isolate $h$ on one side of the equation. There are various ways to solve the equation for $h$.

$$s = \pi r \sqrt{r^2 + h^2}$$

$$\frac{s}{\pi r} = \sqrt{r^2 + h^2} \qquad \text{Divide both sides by } \pi r.$$

$$\left(\frac{s}{\pi r}\right)^2 = \left(\sqrt{r^2 + h^2}\right)^2 \qquad \text{Square both sides.}$$

$$\frac{s^2}{\pi^2 r^2} = r^2 + h^2 \qquad \text{Use } (\sqrt{a})^2 = a, a \geq 0.$$

$$\frac{s^2}{\pi^2 r^2} - r^2 = h^2 \qquad \text{Subtract } r^2 \text{ from both sides.}$$

$$\sqrt{\frac{s^2}{\pi^2 r^2} - r^2} = h \qquad \text{Square root property}$$

Other acceptable answers are $h = \sqrt{\dfrac{s^2 - \pi^2 r^4}{\pi^2 r^2}}$ and $h = \dfrac{\sqrt{s^2 - \pi^2 r^4}}{\pi r}$.

*Now Try Exercise 27*

---

# EXERCISE SET 12.3    MathXL®    MyMathLab®
MathXL®    MyMathLab

## Warm-Up Exercises

*Fill in the blanks with the appropriate word, phrase, or symbol(s) from the following list.*

| square root | profit | square | $\dfrac{\text{rate}}{\text{distance}}$ | $\dfrac{\text{distance}}{\text{rate}}$ | revenue | one whole task | part of the task |

**1.** A company will break even when it has a _____ of 0.

**2.** If we solve the distance formula for *time*, we get time = _____.

**3.** To solve work problems we use the fact that the part of the task done by the first person (or machine) plus the part of the task done by the second person (or machine) is equal to _____ completed.

**4.** When we need to eliminate a square root from an equation, we can isolate the square root and then _____ both sides of the equation.

---

## Practice the Skills

*Solve for the indicated variable. Assume the variable you are solving for must be greater than 0.*

**5.** $A = s^2$, for $s$ (area of a square)

**6.** $A = (s + 1)^2$, for $s$ (area of a square)

**7.** $E = i^2 r$, for $i$ (current in electronics)

**8.** $A = 4\pi r^2$, for $r$ (surface area of a sphere)

**9.** $d = 16t^2$, for $t$ (distance of a falling object)

**10.** $d = \dfrac{1}{9}x^2$, for $x$ (stopping distance on pavement)

11. $E = mc^2$, for $c$ (Einstein's famous energy formula)

12. $V = \pi r^2 h$, for $r$ (volume of a right circular cylinder)

13. $V = \dfrac{1}{3}\pi r^2 h$, for $r$ (volume of a right circular cone)

14. $d = \sqrt{L^2 + W^2}$, for $L$ (diagonal of a rectangle)

15. $d = \sqrt{L^2 + W^2}$, for $W$ (diagonal of a rectangle)

16. $a^2 + b^2 = c^2$, for $a$ (Pythagorean Theorem)

17. $a^2 + b^2 = c^2$, for $b$ (Pythagorean Theorem)

18. $d = \sqrt{L^2 + W^2 + H^2}$, for $L$ (diagonal of a box)

19. $d = \sqrt{L^2 + W^2 + H^2}$, for $H$ (diagonal of a box)

20. $A = P(1 + r)^2$, for $r$ (compound interest formula)

21. $h = -16t^2 + s_0$, for $t$ (height of an object)

22. $h = -4.9t^2 + s_0$, for $t$ (height of an object)

23. $E = \dfrac{1}{2}mv^2$, for $v$ (kinetic energy)

24. $f_x^2 + f_y^2 = f^2$, for $f_x$ (forces acting on an object)

25. $a = \dfrac{v_2^2 - v_1^2}{2d}$, for $v_1$ (acceleration of a vehicle)

26. $A = 4\pi(R^2 - r^2)$, for $R$ (surface area of two spheres)

27. $v' = \sqrt{c^2 - v^2}$, for $c$ (relativity; $v'$ is read "$v$ prime")

28. $L = L_0\sqrt{1 - \dfrac{v^2}{c^2}}$, for $v$ (art, a painting's contraction)

## Problem Solving

29. **Coffee Shop Profit** Benny's Beans, a new coffee shop, projects that its annual profit, $p(t)$, in thousands of dollars, over the first 6 years in business can be approximated by the function $p(t) = 0.2t^2 + 5.4t - 8$, where $t$ is measured in years.

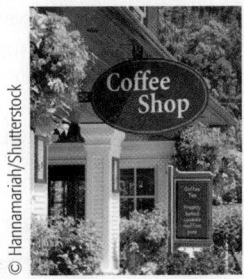

Coffee Shop

© Hannamariah/Shutterstock

a) Estimate the profit (or loss) of the coffee shop after the first year.

b) Estimate the profit (or loss) of the coffee shop after the fifth year.

c) Estimate the time needed for the coffee shop to break even.

30. **Organic Food Store Profit** Hatty's Health Foods, an organic food store, projects that its annual profit, $p(t)$, in thousands of dollars, over the first 6 years in business can be approximated by the function $p(t) = 0.3t^2 + 4.8t - 9$, where $t$ is measured in years.

a) Estimate the profit (or loss) of the food store after the first year.

b) Estimate the profit (or loss) of the food store after the fifth year.

c) Estimate the time needed for the food store to break even.

31. **Souvenir Shop Profit** Bull Snort, a University of South Florida souvenir shop, projects that its annual profit, $p(t)$, in thousands of dollars, over the first 7 years in business can be approximated by the function $p(t) = 0.1t^2 + 3.9t - 6$, where $t$ is measured in years.

a) Estimate the profit (or loss) of the souvenir shop after the first year.

b) Estimate the profit (or loss) of the souvenir shop after the sixth year.

c) Estimate the time needed for the souvenir shop to break even.

32. **Watch Repair Shop Profit** Repair Time, a watch and clock repair shop, projects that its annual profit, $p(t)$, in thousands of dollars, over the first 7 years in business can be approximated by the function $p(t) = 0.4t^2 + 4.4t - 8$, where $t$ is measured in years.

a) Estimate the profit (or loss) of the repair shop after the first year.

b) Estimate the profit (or loss) of the repair shop after the sixth year.

c) Estimate the time needed for the repair shop to break even.

33. **Temperature** The temperature, $T$, in degrees Fahrenheit, in a car's radiator during the first 4 minutes of driving is a function of time, $t$. The temperature can be found by the formula $T = 6.2t^2 + 12t + 32, 0 \le t \le 4$.

a) What is the car radiator's temperature at the instant the car is turned on?

b) What is the car radiator's temperature after the car has been driven for 2 minutes?

c) How long after the car has begun operating will the car radiator's temperature reach 120°F?

34. **Grade Point Average** At a college, records show that the average person's grade point average, $G$, is a function of the number of hours he or she studies and does homework per week, $h$. The grade point average can be estimated by the equation $G = 0.01h^2 + 0.2h + 1.2, 0 \le h \le 8$.

a) What is the GPA of the average student who studies for 0 hours a week?

b) What is the GPA of the average student who studies 3 hours per week?

c) To obtain a 3.2 GPA, how many hours per week would the average student need to study?

**35. Age of Drivers and Accidents** The number of fatal automobile crashes in the United States per 100 million miles driven, $A(t)$, for drivers age $t$ years old, can be estimated by the function $A(t) = 0.013t^2 - 1.19t + 28.24$.

*Source:* Insurance Institute for Highway Safety

a) Estimate the number of fatal crashes per 100 million miles driven for drivers age $t = 20$ years old.

b) Estimate the number of fatal crashes per 100 million miles driven for drivers age $t = 75$ years old.

c) Estimate the ages at which about 10 fatal crashes per 100 million miles driven occur.

**36. Drug-Free Schools** The following graph summarizes data on the percent of students at various ages who say their school is not drug free.

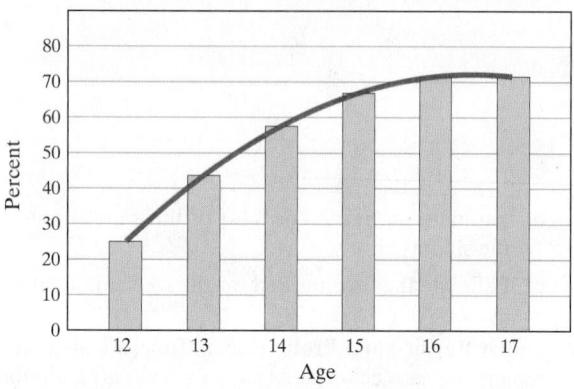

Students Who Say Their School Is Not Drug Free

*Source:* National Center on Addiction and Substance Abuse

The function $f(a) = -2.32a^2 + 76.58a - 559.87$ can be used to estimate the percent of students who say their school is not drug free. In the function, $a$ represents the student's age, where $12 \le a \le 17$. Use the function to answer the following questions.

a) Estimate the percent of 14-year-olds who say their school is not drug free.

b) At what age do 70% of the students say their school is not drug free?

**37. Social Security Assets** The projected assets for Social Security can be estimated by the function $f(t) = -20.57t^2 + 758.9t - 3140$, where $f(t)$ is the total amount of assets in billions of dollars and $t$ represents the number of years since 2000.

*Source:* Social Security Administration

a) Estimate the projected assets for Social Security in 2015.

b) Estimate the projected assets for Social Security in 2030.

c) During what years will the projected assets for Social Security be about $2,000 billion?

**38. Profit** A video store's weekly profit, $P$, in thousands of dollars, is a function of the rental price of the tapes, $t$. The profit equation is $P = 0.2t^2 + 1.5t - 1.2, 0 \le t \le 5$.

a) What is the store's weekly profit or loss if they charge $3 per tape?

b) What is the weekly profit if they charge $5 per tape?

c) At what tape rental price will their weekly profit be $1.4 thousand?

**39. Playground** The area of a children's rectangular playground is 600 square meters. The length is 10 meters longer than the width. Find the length and width of the playground.

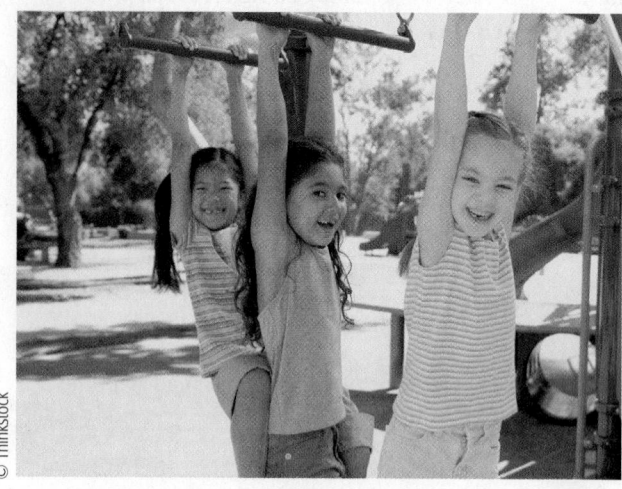

© Thinkstock

**40. Travel** Hana Juarez drove for 80 miles in heavy traffic. She then reached the highway where she drove 260 miles at an average speed that was 25 miles per hour greater than the average speed she drove in heavy traffic. If the total trip took 6 hours, determine her average speed in heavy traffic and on the highway.

**41. Drilling a Well** Paul and Rima Jones hired the Ruth Cardiff Drilling Company to drill a well. The company had to drill 64 feet to hit water. The company informed the Joneses that they had just ordered new drilling equipment that drills at an average of 1 foot per hour faster, and that with their new equipment, they would have hit water in 3.2 hours less time. Find the rate at which their present equipment drills.

**42. Car Carrier** Frank Sims, a truck driver, was transporting a heavy load of new cars on a car carrier from Detroit, Michigan, to Indianapolis, Indiana. On his return trip to Detroit, he averaged 10 miles per hour faster than on his trip to Indianapolis. If the total distance traveled each way was 300 miles and the total time he spent driving was 11 hours, find his average speed going and returning.

**43. Runner** Latoya Williams, a long-distance runner, starts jogging at her house. She jogs 6 miles and then turns around and jogs back to her house. The first part of her jog is mostly uphill, so her speed averages 2 miles per hour less than her returning speed. If the total time she spends jogging is $1\frac{3}{4}$ hours, find her speed going and her speed returning.

**44. Red Rock Canyon** Kathy Nickell traveled from the Red Rock Canyon Conservation Area, just outside Las Vegas, to Phoenix, Arizona. The total distance she traveled was 300 miles. After she got to Phoenix, she figured out that had she averaged 10 miles per hour faster, she would have arrived 1 hour earlier. Find the average speed that Kathy drove.

Red Rock Canyon

**45. Build an Engine** Two mechanics, Bonita Rich and Pamela Pearson, take 6 hours to rebuild an engine when they work together. If each worked alone, Bonita, the more experienced mechanic, could complete the job 1 hour faster than Pamela. How long would it take each of them to rebuild the engine working alone?

**46. Riding a Bike** Ricky Bullock enjoys riding his bike from Washington, D.C., to Bethesda, Maryland, and back, a total of 30 miles on the Capital Crescent path. The trip to Bethesda is uphill most of the distance. The bike's average speed going to Bethesda is 5 miles per hour slower than the average speed returning to D.C. If the round trip takes 4.5 hours, find the average speed in each direction.

**47. Flying a Plane** Dole Rohm flew his single-engine Cessna airplane 80 miles with the wind from Jackson Hole, Wyoming, to above Blackfoot, Idaho. He then turned around and flew back to Jackson Hole against the wind. If the wind was a constant 30 miles per hour, and the total time going and returning was 1.3 hours, find the speed of the plane in still air.

**48. Ships** After a small oil spill, two cleanup ships are sent to siphon off the oil floating in Baffin Bay. The newer ship can clean up the entire spill by itself in 3 hours less time than the older ship takes by itself. Working together the two ships can clean up the oil spill in 8 hours. How long will it take the newer ship by itself to clean up the spill?

**49. Janitorial Service** The O'Connors own a small janitorial service. John requires $\frac{1}{2}$ hour more time to clean the Moose Club by himself than Chris does working by herself. If together they can clean the club in 6 hours, find the time required by each to clean the club.

**50. Electric Heater** A small electric heater requires 6 minutes longer to raise the temperature in an unheated garage to a comfortable level than does a larger electric heater. Together the two heaters can raise the garage temperature to a comfortable level in 42 minutes. How long would it take each heater by itself to raise the temperature in the garage to a comfortable level?

**51. Travel** Shywanda Moore drove from San Antonio, Texas, to Austin, Texas, a distance of 75 miles. She then stopped for 2 hours to see a friend in Austin before continuing her journey from Austin to Dallas, Texas, a distance of 195 miles. If she drove 10 miles per hour faster from San Antonio to Austin and the total time of the trip was 6 hours, find her average speed from San Antonio to Austin.

River Walk, San Antonio, Texas

**52. Travel** Lewis and his friend George are traveling from Nashville to Baltimore. Lewis travels by car while George travels by train. The train and car leave Nashville at the same time from the same point. During the trip Lewis and George speak by cellular phone, and Lewis informs George that he has just stopped for the evening after traveling for 500 miles. One and two-thirds hours later, George calls Lewis and informs him that the train had just reached Baltimore, a distance of 800 miles from Nashville. Assuming the train averaged 20 miles per hour faster than the car, find the average speed of the car and the train.

**53. Widescreen TVs** A widescreen television (see figure) has an aspect ratio of 16 : 9. This means that the ratio of the length to the height of the screen is 16 to 9. The figure drawn on the photo illustrates how the length and the height of the screen of a 40-inch widescreen television can be found. Determine the length and height of a 40-inch widescreen television.

**54. Standard TVs** Many picture tube televisions have a screen aspect ratio of 4:3. Determine the length and height of a television screen that has an aspect ratio of 4:3 and whose diagonal is 36 inches. See Exercise 53.

## Concept/Writing Exercises

**55.** Write your own motion problem and solve it.

**56.** Write your own work problem and solve it.

**57.** In general, when solving for a variable in a formula, whether you use the square root property or the quadratic formula, you use only the positive square root. Explain why.

**58.** Suppose $P = ☻^2 + □^2$ is a real formula. Solving for ☻ gives ☻ $= \sqrt{P - □^2}$. If ☻ is to be a real number, what relationship must exist between $P$ and $□$?

## Challenge Problems

**59. Area** The area of a rectangle is 18 square meters. When the length is increased by 2 meters and the width by 3 meters, the area becomes 48 square meters. Find the dimensions of the smaller rectangle.

**60. Area** The area of a rectangle is 35 square inches. When the length is decreased by 1 inch and the width is increased by 1 inch, the area of the new rectangle is 36 square inches. Find the dimensions of the original rectangle.

## Cumulative Review Exercises

[1.9] **61.** Evaluate $-[4(5 - 3)^3] + 2^4$.

[2.6] **62.** Solve $IR + Ir = E$ for $R$.

[7.4] **63.** Add $\dfrac{r}{r - 4} - \dfrac{r}{r + 4} + \dfrac{32}{r^2 - 16}$.

[11.2] **64.** Simplify $\left( \dfrac{x^{3/4} y^{-2}}{x^{1/2} y^2} \right)^8$.

[11.6] **65.** Solve $\sqrt{x^2 + 3x + 12} = x$.

## Mid-Chapter Test: 12.1–12.3

*To find out how well you understand the chapter material to this point, take this brief test. The answers, and the section where the material was initially discussed, are given in the back of the book. Review any questions that you answered incorrectly.*

*Use the square root property to solve each equation.*

**1.** $x^2 - 12 = 86$

**2.** $(a - 3)^2 + 20 = 0$

**3.** $(2m + 7)^2 = 36$

*Solve each equation by completing the square.*

**4.** $y^2 + 4y - 12 = 0$

**5.** $3a^2 - 12a - 30 = 0$

**6.** $4c^2 + c = -9$

**7. Patio** The patio of a house is a square where the diagonal is 6 meters longer than a side. Find the length of one side of the patio.

**8. a)** Give the formula for the discriminant of a quadratic equation.

   **b)** Explain how to determine if a quadratic equation has two distinct real solutions, a single real solution, or no real solution.

**9.** Use the discriminant to determine if the equation $2b^2 - 6b - 11 = 0$ has two distinct real solutions, a single real solution, or no real solution.

*Solve each equation by the quadratic formula.*

**10.** $6n^2 + n = 15$

**11.** $p^2 = -4p + 8$

**12.** $3d^2 - 2d + 5 = 0$

*In Exercises 13 and 14, determine an equation that has the given solutions.*

**13.** $7, -2$

**14.** $2 + \sqrt{5}, 2 - \sqrt{5}$

**15. Lamps** A business sells $n$ lamps, $n \leq 20$, at a price of $(60 - 0.5n)$ dollars per lamp. How many lamps must be sold to have revenue of $550?

*In Exercises 16–18, solve for the indicated variable. Assume all variables are positive.*

**16.** $y = x^2 - r^2$, for $r$

**17.** $A = \dfrac{1}{3} kx^2$, for $x$

**18.** $D = \sqrt{x^2 + y^2}$, for $y$

**19. Area** The length of a rectangle is two feet more than twice the width. Find the dimensions if its area is 60 square feet.

**20. Clocks** The profit from a company selling $n$ clocks is $p(n) = 2n^2 + n - 35$, where $p(n)$ is hundreds of dollars. How many clocks must be sold to have a profit of $2000?

# 12.4 Factoring Expressions and Solving Equations That Are Quadratic in Form

**1** Factor expressions that are quadratic in form.

**2** Solve equations that are quadratic in form.

**3** Solve equations with rational exponents.

## Understanding Algebra

In general, when rewriting an expression that is quadratic in form, we will let $u$ equal the "middle" variable expression without the numerical coefficient. For example, in the expression $x^6 + 7x^3 - 8$, we will let $u = x^3$. Since $u^2 = (x^3)^2 = x^6$, the expression becomes: $u^2 + 7u - 8$.

In this section we will work with expressions and equations that are not quadratic, but can be rewritten as quadratic expressions and quadratic equations respectively.

## **1** Factor Expressions That Are Quadratic in Form

We begin by defining expressions that are quadratic in form.

### Expression Quadratic in Form

An expression that can be rewritten in the form $au^2 + bu + c$ for $a \neq 0$, where $u$ is an algebraic expression, is called an **expression that is quadratic in form**.

In Chapter 6 we factored polynomials. Our next three examples will illustrate factoring polynomials that are quadratic in form by performing a substitution.

**EXAMPLE 1** Factor $y^4 - y^2 - 6$.

**Solution**   If we can rewrite this expression in the form $ax^2 + bx + c$, it will be easier to factor. Since $(y^2)^2 = y^4$, if we substitute $u$ for $y^2$, the trinomial becomes

$$y^4 - y^2 - 6 = (y^2)^2 - y^2 - 6$$
$$= u^2 - u - 6 \qquad \text{Substitute } u \text{ for } y^2.$$

Now factor $u^2 - u - 6$.

$$= (u + 2)(u - 3)$$

Finally, substitute $y^2$ in place of $u$ to obtain

$$= (y^2 + 2)(y^2 - 3) \quad \text{Substitute } y^2 \text{ for } u.$$

Thus, $y^4 - y^2 - 6 = (y^2 + 2)(y^2 - 3)$. Note that $u$ was substituted for $y^2$, and then $y^2$ was substituted back for $u$.

Now Try Exercise 7

**EXAMPLE 2** Factor $3z^4 - 17z^2 - 28$.

**Solution**   Let $u = z^2$. Then the trinomial can be rewritten

$$3z^4 - 17z^2 - 28 = 3(z^2)^2 - 17z^2 - 28$$
$$= 3u^2 - 17u - 28 \qquad \text{Substitute } u \text{ for } z^2.$$
$$= (3u + 4)(u - 7) \qquad \text{Factor.}$$

Now substitute $z^2$ for $u$.

$$= (3z^2 + 4)(z^2 - 7) \quad \text{Substitute } z^2 \text{ for } u.$$

Thus, $3z^4 - 17z^2 - 28 = (3z^2 + 4)(z^2 - 7)$.

Now Try Exercise 11

**EXAMPLE 3** Factor $2(x + 5)^2 - 5(x + 5) - 12$.

**Solution**   Again use a substitution, as in Examples 1 and 2. Substitute $u = x + 5$ in the equation, to obtain

$$2(x + 5)^2 - 5(x + 5) - 12$$
$$= 2u^2 - 5u - 12 \qquad \text{Substitute } u \text{ for } (x + 5).$$

Now factor $2u^2 - 5u - 12$.

$$= (2u + 3)(u - 4)$$

Finally, replace $u$ with $x + 5$ to obtain

$$= [2(x + 5) + 3][(x + 5) - 4] \quad \text{Substitute } (x + 5) \text{ for } u.$$
$$= [2x + 10 + 3][x + 1]$$
$$= (2x + 13)(x + 1)$$

Thus, $2(x + 5)^2 - 5(x + 5) - 12 = (2x + 13)(x + 1)$. Note that $u$ was substituted for $x + 5$, and then $x + 5$ was substituted back for $u$.

Now Try Exercise 15

## 2 Solve Equations That Are Quadratic in Form

We now solve equations that are not quadratic, but can be rewritten in the form of a quadratic equation.

### Equations Quadratic in Form

An equation that can be rewritten in the form $au^2 + bu + c = 0$ for $a \neq 0$, where $u$ is an algebraic expression, is called an **equation that is quadratic in form**.

When given an equation quadratic in form, we will make a substitution to rewrite the equation in the form $au^2 + bu + c = 0$. For example, consider the equation $2(x - 3)^2 + 5(x - 3) - 7 = 0$.

We will let $u = x - 3$. Then $u^2 = (x - 3)^2$ and the equation can then be rewritten as follows

$$2\underbrace{(x - 3)^2} + 5\underbrace{(x - 3)} - 7 = 0$$
$$2 \quad u^2 \quad + 5 \quad u \quad - 7 = 0$$

The equation $2u^2 + 5u - 7 = 0$ can be solved by factoring, completing the square, or by using the quadratic formula. Other examples are shown in the table below.

> **Understanding Algebra**
>
> In general, when rewriting an equation that is quadratic in form, we will let $u$ equal the "middle" variable expression without the numerical coefficient. For example, in the equation $4a - 5\sqrt{a} + 1 = 0$, we will let $u = \sqrt{a}$. Since $u^2 = (\sqrt{a})^2 = a$, the equation becomes $4u^2 - 5u + 1 = 0$.

| Equation Quadratic in Form | Substitution | Equation with Substitution |
|---|---|---|
| $y^4 - y^2 - 6 = 0$ | $u = y^2$ | $u^2 - u - 6 = 0$ |
| $2(x + 5)^2 - 5(x + 5) - 12 = 0$ | $u = x + 5$ | $2u^2 - 5u - 12 = 0$ |
| $x^{2/3} + 4x^{1/3} - 3 = 0$ | $u = x^{1/3}$ | $u^2 + 4u - 3 = 0$ |

To solve equations quadratic in form, we use the following procedure.

### To Solve Equations Quadratic in Form

1. Make a substitution that will result in an equation of the form $au^2 + bu + c = 0$, $a \neq 0$, where $u$ is a function of the original variable.
2. Solve the equation $au^2 + bu + c = 0$ for $u$.
3. Replace $u$ with the function of the original variable from step 1 and solve the resulting equation for the original variable.
4. Check for extraneous solutions by substituting the apparent solutions into the original equation.

## EXAMPLE 4

a) Solve $x^4 - 5x^2 + 4 = 0$.

b) Find the $x$-intercepts of the graph of the function $f(x) = x^4 - 5x^2 + 4$.

*Solution*

**a)** To obtain an equation quadratic in form, we will let $u = x^2$. Then $u^2 = (x^2)^2 = x^4$.

$$x^4 - 5x^2 + 4 = 0$$

$$\downarrow$$

$$(x^2)^2 - 5x^2 + 4 = 0 \quad x^4 \text{ was replaced with } (x^2)^2.$$

$$\downarrow \qquad \downarrow$$

$$u^2 - 5u + 4 = 0 \quad u \text{ was substituted for } x^2.$$

We now have a quadratic equation we can solve by factoring

$$u^2 - 5u + 4 = 0$$
$$(u - 4)(u - 1) = 0$$
$$u - 4 = 0 \quad \text{or} \quad u - 1 = 0$$
$$u = 4 \qquad\qquad u = 1$$

Next, we replace $u$ with $x^2$ and solve for $x$.

$$u = 4 \qquad\qquad u = 1$$
$$\downarrow \qquad\qquad \downarrow$$
$$x^2 = 4 \qquad\qquad x^2 = 1 \qquad u \text{ was replaced with } x^2.$$
$$x = \pm\sqrt{4} \qquad x = \pm\sqrt{1} \quad \text{Square root property}$$
$$x = \pm 2 \qquad\qquad x = \pm 1$$

Check the four possible solutions in the original equation.

| $x = 2$ | $x = -2$ | $x = 1$ | $x = -1$ |
|---|---|---|---|
| $x^4 - 5x^2 + 4 = 0$ | $x^4 - 5x^2 + 4 = 0$ | $x^4 - 5x^2 + 4 = 0$ | $x^4 - 5x^2 + 4 = 0$ |
| $2^4 - 5(2)^2 + 4 \overset{?}{=} 0$ | $(-2)^4 - 5(-2)^2 + 4 \overset{?}{=} 0$ | $1^4 - 5(1)^2 + 4 \overset{?}{=} 0$ | $(-1)^4 - 5(-1)^2 + 4 \overset{?}{=} 0$ |
| $16 - 20 + 4 \overset{?}{=} 0$ | $16 - 20 + 4 \overset{?}{=} 0$ | $1 - 5 + 4 \overset{?}{=} 0$ | $1 - 5 + 4 \overset{?}{=} 0$ |
| $0 = 0$ | $0 = 0$ | $0 = 0$ | $0 = 0$ |
| True | True | True | True |

Thus, the solutions are $2, -2, 1,$ and $-1$.

**b)** The $x$-intercepts occur where $f(x) = 0$. Therefore, the graph will cross the $x$-axis at the solutions to the equation $x^4 - 5x^2 + 4 = 0$.

From part **a)**, we know the solutions are $2, -2, 1,$ and $-1$. Thus, the $x$-intercepts are $(2, 0), (-2, 0), (1, 0),$ and $(-1, 0)$. **Figure 12.6** is the graph of $f(x) = x^4 - 5x^2 + 4 = 0$ as illustrated on a graphing calculator. Notice that the graph crosses the $x$-axis at $x = 2, x = -2, x = 1,$ and $x = -1$.

Now Try Exercise 31

## Understanding Algebra

Recall that the $x$-intercepts of a function are the points where the graph crosses the $x$-axis. The $x$-intercepts always have a $y$-coordinate of 0. To determine the $x$-intercepts of a function, set $y$ or $f(x) = 0$ and solve for $x$.

$-3, 3, 1, -3, 6, 1$

**FIGURE 12.6**

**EXAMPLE 5** Solve $p^4 + 2p^2 = 8$.

*Solution*    We will let $u = p^2$, then $u^2 = (p^2)^2 = p^4$.

$$p^4 + 2p^2 - 8 = 0 \qquad \text{Equation was set equal to 0.}$$

$$(p^2)^2 + 2p^2 - 8 = 0 \qquad p^4 \text{ was written as } (p^2)^2.$$

$$\downarrow \qquad \downarrow$$

$$u^2 + 2u - 8 = 0 \qquad u \text{ was substituted for } p^2.$$

$$(u + 4)(u - 2) = 0$$

$$u + 4 = 0 \quad \text{or} \quad u - 2 = 0$$

$$u = -4 \qquad\qquad u = 2$$

Next, we substitute back $p^2$ for $u$ and solve for $p$.

$$p^2 = -4 \qquad\qquad p^2 = 2 \qquad\qquad \text{$u$ was replaced with $p^2$.}$$
$$p = \pm\sqrt{-4} \qquad\qquad p = \pm\sqrt{2} \qquad\qquad \text{Square root property}$$
$$p = \pm 2i$$

Check the four possible solutions in the *original* equation.

$$
\begin{array}{cccc}
p = 2i & p = -2i & p = \sqrt{2} & p = -\sqrt{2} \\
p^4 + 2p^2 = 8 & p^4 + 2p^2 = 8 & p^4 + 2p^2 = 8 & p^4 + 2p^2 = 8 \\
(2i)^4 + 2(2i)^2 \overset{?}{=} 8 & (-2i)^4 + 2(-2i)^2 \overset{?}{=} 8 & (\sqrt{2})^4 + 2(\sqrt{2})^2 \overset{?}{=} 8 & (-\sqrt{2})^4 + 2(-\sqrt{2})^2 \overset{?}{=} 8 \\
2^4 i^4 + 2(2^2)(i^2) \overset{?}{=} 8 & (-2)^4 i^4 + 2(-2)^2 i^2 \overset{?}{=} 8 & 4 + 2(2) \overset{?}{=} 8 & 4 + 2(2) \overset{?}{=} 8 \\
16(1) + 8(-1) \overset{?}{=} 8 & 16(1) + 8(-1) \overset{?}{=} 8 & 8 = 8 & 8 = 8 \\
16 - 8 = 8 & 16 - 8 = 8 & \text{True} & \text{True} \\
\text{True} & \text{True} & &
\end{array}
$$

Thus, the solutions are $2i$, $-2i$, $\sqrt{2}$, and $-\sqrt{2}$.

Now Try Exercise 41

**Helpful Hint**

Students sometimes solve the equation for $u$ but then forget to complete the problem by solving for the original variable. Remember that if the original equation is in $x$ you must obtain values for $x$.

**EXAMPLE 6**  Solve $4(2w + 1)^2 - 16(2w + 1) + 15 = 0$.

**Solution**  We will let $u = 2w + 1$, then $u^2 = (2w + 1)^2$ and the equation becomes

$$4(2w + 1)^2 - 16(2w + 1) + 15 = 0$$
$$4u^2 - 16u + 15 = 0 \qquad \text{$u$ was substituted for $2w + 1$.}$$

Now we can factor and solve.

$$(2u - 3)(2u - 5) = 0$$
$$2u - 3 = 0 \qquad \text{or} \qquad 2u - 5 = 0$$
$$2u = 3 \qquad\qquad\qquad 2u = 5$$
$$u = \frac{3}{2} \qquad\qquad\qquad u = \frac{5}{2}$$

Next, we substitute back $2w + 1$ for $u$ and solve for $w$.

$$2w + 1 = \frac{3}{2} \qquad\qquad 2w + 1 = \frac{5}{2} \qquad \text{$2w + 1$ was substituted for $u$.}$$
$$2w = \frac{1}{2} \qquad\qquad\qquad 2w = \frac{3}{2}$$
$$w = \frac{1}{4} \qquad\qquad\qquad w = \frac{3}{4}$$

A check will show that both $\frac{1}{4}$ and $\frac{3}{4}$ are solutions to the original equation.

Now Try Exercise 53

**EXAMPLE 7**  Find the $x$-intercepts of the graph of the function $f(x) = 2x^{-2} + x^{-1} - 1$.

**Solution**  The $x$-intercepts occur where $f(x) = 0$. Therefore, to find the $x$-intercepts we must solve the equation

$$2x^{-2} + x^{-1} - 1 = 0$$

We will let $u = x^{-1}$, then $u^2 = (x^{-1})^2 = x^{-2}$.

$$2(x^{-1})^2 + x^{-1} - 1 = 0$$
$$2u^2 + u - 1 = 0 \qquad u \text{ was substituted for } x^{-1}.$$
$$(2u - 1)(u + 1) = 0$$
$$2u - 1 = 0 \qquad \text{or} \qquad u + 1 = 0$$
$$u = \frac{1}{2} \qquad\qquad\qquad u = -1$$

Now we substitute $x^{-1}$ for $u$.

$$x^{-1} = \frac{1}{2} \qquad \text{or} \qquad x^{-1} = -1$$
$$\frac{1}{x} = \frac{1}{2} \qquad\qquad \frac{1}{x} = -1$$
$$x = 2 \qquad\qquad\qquad x = -1$$

A check will show that both 2 and −1 are solutions to the original equation. Thus, the $x$-intercepts are $(2, 0)$ and $(-1, 0)$.

Now Try Exercise 85

### Understanding Algebra

Recall from Chapter 5 that an expression raised to a negative exponent can be written as follows:

$$a^{-m} = \frac{1}{a^m}, \, a \neq 0$$

## 3 Solve Equations with Rational Exponents

In our next two examples, while solving the equations that are quadratic in form, we will raise both sides of the equation to a power to eliminate rational exponents (or radicals). Recall from Chapter 11 that whenever we raise both sides of an equation to a power, it is possible to introduce extraneous solutions. *Therefore, whenever you raise both sides of an equation with rational exponents to a power, you must check all apparent solutions in the original equation.*

**EXAMPLE 8**   Solve $x^{2/5} + x^{1/5} - 6 = 0$. ───────────

*Solution*   We will let $u = x^{1/5}$, then $u^2 = (x^{1/5})^2 = x^{2/5}$. The equation becomes

$$(x^{1/5})^2 + x^{1/5} - 6 = 0$$
$$u^2 + u - 6 = 0 \qquad u \text{ was substituted for } x^{1/5}.$$
$$(u + 3)(u - 2) = 0$$
$$u + 3 = 0 \qquad \text{or} \qquad u - 2 = 0$$
$$u = -3 \qquad\qquad\qquad u = 2$$

Now substitute $x^{1/5}$ for $u$ and raise both sides of the equation to the fifth power to remove the rational exponents.

$$x^{1/5} = -3 \qquad \text{or} \qquad x^{1/5} = 2$$
$$(x^{1/5})^5 = (-3)^5 \qquad\qquad (x^{1/5})^5 = 2^5$$
$$x = -243 \qquad\qquad\qquad x = 32$$

The two *possible* solutions are −243 and 32. Remember that whenever you raise both sides of an equation to a power, as you did here, you need to check for extraneous solutions.

### Understanding Algebra

Recall from Chapter 11 that an expression with rational exponents can be rewritten as a radical expression using the following rule:

$$a^{m/n} = (\sqrt[n]{a})^m$$

Check

$$x = -243 \qquad\qquad\qquad\qquad x = 32$$
$$x^{2/5} + x^{1/5} - 6 = 0 \qquad\qquad\qquad x^{2/5} + x^{1/5} - 6 = 0$$
$$(-243)^{2/5} + (-243)^{1/5} - 6 \stackrel{?}{=} 0 \qquad (32)^{2/5} + (32)^{1/5} - 6 \stackrel{?}{=} 0$$
$$(\sqrt[5]{-243})^2 + \sqrt[5]{-243} - 6 \stackrel{?}{=} 0 \qquad (\sqrt[5]{32})^2 + \sqrt[5]{32} - 6 \stackrel{?}{=} 0$$
$$(-3)^2 - 3 - 6 \stackrel{?}{=} 0 \qquad\qquad\qquad 2^2 + 2 - 6 \stackrel{?}{=} 0$$
$$9 - 3 - 6 \stackrel{?}{=} 0 \qquad\qquad\qquad\qquad 4 + 2 - 6 \stackrel{?}{=} 0$$
$$0 = 0 \quad \text{True} \qquad\qquad\qquad\qquad 0 = 0 \quad \text{True}$$

Since both values check, the solutions are −243 and 32.

Now Try Exercise 75

**EXAMPLE 9** Solve $2p - \sqrt{p} - 10 = 0$.

*Solution*    Since $\sqrt{p} = p^{1/2}$ we can express the equation as

$$2p - p^{1/2} - 10 = 0$$

We will let $u = p^{1/2}$, then $u^2 = (p^{1/2})^2 = p$ and the equation becomes

$$2p - p^{1/2} - 10 = 0$$
$$2(p^{1/2})^2 - p^{1/2} - 10 = 0$$

If we let $u = p^{1/2}$, this equation is quadratic in form.

$$2u^2 - u - 10 = 0$$
$$(2u - 5)(u + 2) = 0$$

$$2u - 5 = 0 \quad \text{or} \quad u + 2 = 0$$
$$2u = 5 \qquad\qquad u = -2$$
$$u = \frac{5}{2}$$

Next, we substitute $p^{1/2}$ for $u$.

$$p^{1/2} = \frac{5}{2} \qquad\qquad p^{1/2} = -2$$

Now we square both sides of the equation.

$$(p^{1/2})^2 = \left(\frac{5}{2}\right)^2 \qquad (p^{1/2})^2 = (-2)^2$$

$$p = \frac{25}{4} \qquad\qquad p = 4$$

## Understanding Algebra

When solving an equation involving variables with rational exponents or radicals, we will often raise both sides of the equation to a power in order to eliminate the rational exponent or radical. Whenever we do this, it is possible to introduce an extraneous or false solution. Therefore, we must be careful to check our answers in the original equation.

We must now check both apparent solutions in the original equation.

Check
$$p = \frac{25}{4} \qquad\qquad\qquad p = 4$$
$$2p - \sqrt{p} - 10 = 0 \qquad\qquad 2p - \sqrt{p} - 10 = 0$$
$$2\left(\frac{25}{4}\right) - \sqrt{\frac{25}{4}} - 10 \overset{?}{=} 0 \qquad 2(4) - \sqrt{4} - 10 \overset{?}{=} 0$$
$$\frac{25}{2} - \frac{5}{2} - 10 \overset{?}{=} 0 \qquad\qquad 8 - 2 - 10 \overset{?}{=} 0$$
$$0 = 0 \quad \text{True} \qquad\qquad -4 = 0 \quad \text{False}$$

Since 4 does not check, it is an extraneous solution. The only solution is $\frac{25}{4}$.

Now Try Exercise 49

# EXERCISE SET 12.4    MathXL®    MyMathLab®

MathXL®    MyMathLab

## Warm-Up Exercises

*Fill in the blanks with the appropriate word, phrase, or symbol(s) from the following list.*

| | | | |
|---|---|---|---|
| $h - 2$ | extraneous solution | $v$ | quadratic in form | $x^3$ |
| $\sqrt{v}$ | $(h - 2)^2$ | $c^2$ | $x^6$ | |

1. An expression that can be rewritten in the form $au^2 + bu + c = 0$ for $a \neq 0$, where $u$ is an algebraic expression, is called an expression that is _____.

2. In the expression $x^6 + 7x^3 - 8$, if we let $u =$ _____ the expression can be rewritten as $u^2 + 7u - 8$.

3. In the expression $(h - 2)^2 + (h - 2) - 42$, if we let $u =$ _____ the expression can be rewritten as $u^2 + u - 42$.

4. To solve the equation $c^4 + c^2 - 2 = 0$, the best choice for $u$ to obtain an equation quadratic in form is _____.

5. To solve the equation $v - 3\sqrt{v} - 28 = 0$, the best choice for $u$ to obtain an equation quadratic in form is _____.

6. Whenever we solve an equation by raising both sides of the equation to a power we may introduce an _____.

## Practice the Skills

*Factor each trinomial completely.*

**7.** $x^4 + x^2 - 6$

**8.** $x^4 - 3x^2 - 10$

**9.** $x^4 + 5x^2 + 6$

**10.** $x^4 + 2x^2 - 35$

**11.** $6a^4 + 5a^2 - 25$

**12.** $(2x + 1)^2 + 2(2x + 1) - 15$

**13.** $4(x + 1)^2 + 8(x + 1) + 3$

**14.** $(2x + 3)^2 - (2x + 3) - 6$

**15.** $6(a + 2)^2 - 7(a + 2) - 5$

**16.** $6(p - 5)^2 + 11(p - 5) + 3$

**17.** $a^2b^2 + 8ab + 15$

**18.** $x^2y^2 - 10xy + 24$

**19.** $3x^2y^2 - 2xy - 5$

**20.** $3p^2q^2 + 11pq + 6$

**21.** $2a^2(5 - a) - 7a(5 - a) + 5(5 - a)$

**22.** $2y^2(y + 2) + 13y(y + 2) + 15(y + 2)$

**23.** $2x^2(x - 3) + 7x(x - 3) + 6(x - 3)$

**24.** $3x^2(x - 2) + 5x(x - 2) - 2(x - 2)$

**25.** $y^4 + 13y^2 + 30$

**26.** $3z^4 - 16z^2 + 5$

**27.** $x^2(x + 3) + 3x(x + 3) + 2(x + 3)$

**28.** $x^2(x - 1) - x(x - 1) - 30(x - 1)$

**29.** $5a^5b^2 - 8a^4b^3 + 3a^3b^4$

**30.** $2x^2y^6 + 3xy^5 - 9y^4$

*Solve each equation.*

**31.** $x^4 - 10x^2 + 9 = 0$

**32.** $x^4 - 37x^2 + 36 = 0$

**33.** $x^4 + 17x^2 + 16 = 0$

**34.** $x^4 + 50x^2 + 49 = 0$

**35.** $x^4 - 13x^2 + 36 = 0$

**36.** $x^4 + 13x^2 + 36 = 0$

**37.** $a^4 - 7a^2 + 12 = 0$

**38.** $b^4 + 7b^2 + 12 = 0$

**39.** $4x^4 - 17x^2 + 4 = 0$

**40.** $9d^4 - 13d^2 + 4 = 0$

**41.** $r^4 - 8r^2 = -15$

**42.** $p^4 - 8p^2 = -12$

**43.** $z^4 - 7z^2 = 18$

**44.** $a^4 + a^2 = 42$

**45.** $-c^4 = 4c^2 - 5$

**46.** $9b^4 = 57b^2 - 18$

**47.** $\sqrt{x} = 2x - 6$

**48.** $x - 2\sqrt{x} = 8$

**49.** $x - \sqrt{x} = 6$

**50.** $x - 4 = -3\sqrt{x}$

**51.** $9x + 3\sqrt{x} = 2$

**52.** $8x + 2\sqrt{x} = 1$

**53.** $(x + 3)^2 + 2(x + 3) = 24$

**54.** $(x + 1)^2 + 4(x + 1) + 3 = 0$

**55.** $6(a - 2)^2 = -19(a - 2) - 10$

**56.** $10(z + 2)^2 = 3(z + 2) + 1$

**57.** $(x^2 - 3)^2 - (x^2 - 3) - 6 = 0$

**58.** $(a^2 - 1)^2 - 5(a^2 - 1) - 14 = 0$

**59.** $2(b + 3)^2 + 5(b + 3) - 3 = 0$

**60.** $(z^2 - 6)^2 + 2(z^2 - 6) - 24 = 0$

**61.** $18(x^2 - 5)^2 + 27(x^2 - 5) + 10 = 0$

**62.** $28(x^2 - 8)^2 - 23(x^2 - 8) - 15 = 0$

**63.** $a^{-2} + 4a^{-1} + 4 = 0$

**64.** $x^{-2} + 10x^{-1} + 25 = 0$

**65.** $12b^{-2} - 7b^{-1} + 1 = 0$

**66.** $5x^{-2} + 4x^{-1} - 1 = 0$

**67.** $2b^{-2} = 7b^{-1} - 3$

**68.** $10z^{-2} - 3z^{-1} - 1 = 0$

**69.** $x^{-2} + 9x^{-1} = 10$

**70.** $6a^{-2} = a^{-1} + 12$

**71.** $x^{-2} = 4x^{-1} + 12$

**72.** $x^{2/3} - 5x^{1/3} + 6 = 0$

**73.** $x^{2/3} - 4x^{1/3} = -3$

**74.** $x^{2/3} = 3x^{1/3} + 4$

**75.** $b^{2/3} - 9b^{1/3} + 18 = 0$

**76.** $c^{2/3} - 4 = 0$

**77.** $-2a - 5a^{1/2} + 3 = 0$

**78.** $r^{2/3} - 7r^{1/3} + 10 = 0$

**79.** $c^{2/5} + 3c^{1/5} + 2 = 0$

**80.** $x^{2/5} - 5x^{1/5} + 6 = 0$

*Find all x-intercepts of each function.*

**81.** $f(x) = x - 5\sqrt{x} + 4$

**82.** $g(x) = x - 15\sqrt{x} + 56$

**83.** $h(x) = x + 14\sqrt{x} + 45$

**84.** $k(x) = x + 7\sqrt{x} + 12$

**85.** $p(x) = 4x^{-2} - 19x^{-1} - 5$

**86.** $g(x) = 4x^{-2} + 12x^{-1} + 9$

**87.** $f(x) = x^{2/3} - x^{1/3} - 6$

**88.** $f(x) = x^{1/2} + 6x^{1/4} - 7$

**89.** $g(x) = (x^2 - 3x)^2 + 2(x^2 - 3x) - 24$

**90.** $g(x) = (x^2 - 6x)^2 - 5(x^2 - 6x) - 24$

**91.** $f(x) = x^4 - 29x^2 + 100$

**92.** $h(x) = x^4 - 4x^2 + 3$

## Problem Solving

**93.** Solve the equation $\dfrac{3}{x^2} - \dfrac{3}{x} = 60$ by

    **a)** multiplying both sides of the equation by the LCD.

    **b)** writing the equation with negative exponents.

**94.** Solve the equation $1 = \dfrac{2}{x} - \dfrac{2}{x^2}$ by

    **a)** multiplying both sides of the equation by the LCD.

    **b)** writing the equation with negative exponents.

*Find all real solutions to each equation.*

**95.** $15(r + 2) + 22 = -\dfrac{8}{r + 2}$

**96.** $2(p + 3) + 5 = \dfrac{3}{p + 3}$

**97.** $4 - (x - 1)^{-1} = 3(x - 1)^{-2}$

**98.** $3(x - 4)^{-2} = 16(x - 4)^{-1} + 12$

**99.** $x^6 - 9x^3 + 8 = 0$

**100.** $x^6 - 28x^3 + 27 = 0$

**101.** $(x^2 + 2x - 2)^2 - 7(x^2 + 2x - 2) + 6 = 0$

**102** $(x^2 + 3x - 2)^2 - 10(x^2 + 3x - 2) + 16 = 0$

*Find all solutions to each equation.*

**103.** $2n^4 - 6n^2 - 3 = 0$

**104.** $3x^4 + 8x^2 - 1 = 0$

## Concept/Writing Exercises

**105.** Give a general procedure for solving an equation of the form $ax^4 + bx^2 + c = 0$.

**106.** Give a general procedure for solving an equation of the form $ax^{2n} + bx^n + c = 0$.

**107.** Give a general procedure for solving an equation of the form $ax^{-2} + bx^{-1} + c = 0$.

**108.** Give a general procedure for solving an equation of the form $a(x - r)^2 + b(x - r) - c = 0$.

**109.** Determine an equation of the form $ax^4 + bx^2 + c = 0$ that has solutions $\pm 2$ and $\pm 1$. Explain how you obtained your answer.

**110.** Determine an equation of the form $ax^4 + bx^2 + c = 0$ that has solutions $\pm 3$ and $\pm 2i$. Explain how you obtained your answer.

**111.** Determine an equation of the form $ax^4 + bx^2 + c = 0$ that has solutions $\pm \sqrt{2}$ and $\pm \sqrt{5}$. Explain how you obtained your answer.

**112.** Determine an equation of the form $ax^4 + bx^2 + c = 0$ that has solutions $\pm 2i$ and $\pm 5i$. Explain how you obtained your answer.

**113.** Is it possible for an equation of the form $ax^4 + bx^2 + c = 0$ to have exactly one imaginary solution? Explain.

**114.** Is it possible for an equation of the form $ax^4 + bx^2 + c = 0$ (where $a$, $b$, and $c \neq 0$) to have exactly one real solution? Explain.

## Cumulative Review Exercises

[1.9] **115.** Evaluate $\dfrac{4}{5} - \left(\dfrac{3}{4} - \dfrac{2}{3}\right)$.

[2.4] **116.** Solve $3(x + 2) - 2(3x + 3) = -3$

[8.2] **117.** State the domain and range for $y = (x - 3)^2$.

[11.3] **118.** Simplify $\sqrt[3]{16x^3y^6}$.

[11.4] **119.** Add $\sqrt{75} + \sqrt{48}$.

# 12.5 Graphing Quadratic Functions

1  Identify some key characteristics of graphs of polynomial functions.

2  Find the axis of symmetry, vertex, and x-intercepts of a parabola.

3  Graph quadratic functions using the axis of symmetry, vertex, and intercepts.

4  Solve maximum and minimum problems.

5  Understand translations of parabolas.

6  Write functions in the form $f(x) = a(x - h)^2 + k$.

## 1  Identify Some Key Characteristics of Graphs of Polynomial Functions

Quadratic functions are part of a larger family of functions called *polynomial functions*. In Section 5.4 we discussed polynomials and in Section 8.2 we discussed functions.

**Polynomial and Polynomial Function**

A **polynomial in $x$** is an expression containing a finite sum of terms of the form $ax^n$ for any real number $a$ and any whole number $n$. A **polynomial function** is a function in which the expression used to describe the function is a polynomial.

The following are some examples of polynomial functions.

$$p(x) = x + 3$$

$$p(x) = 2x^2 - 5x + 6$$

$$p(x) = 5x^3 - \frac{1}{2}x^2 + 6x - 4$$

We'll now discuss some key characteristics of the graphs of polynomial functions. *Like any function, we can plot points to help us understand the shape of the graph of a polynomial function.* In the following example, we will evaluate a polynomial function at given values and identify the corresponding point on the graph of the polynomial function.

EXAMPLE 1   For the polynomial function $p(x) = x^4 - x^3 - 2x^2 - x + 3$, evaluate the function at the given values and identify the corresponding point on the graph of the function.

**a)** $p(0)$          **b)** $p(2)$          **c)** $p(-1)$

*Solution*

**a)** $p(x) = x^4 - x^3 - 2x^2 - x + 3$
$p(0) = 0^4 - 0^3 - 2 \cdot 0^2 - 0 + 3$
$\quad = 0 - 0 - 2 \cdot 0 - 0 + 3 = 3$

Thus, when $x = 0$, $p(x) = 3$ and the point $(0, 3)$ is a point on the graph of $p(x) = x^4 - x^3 - 2x^2 - x + 3$.

**b)** $p(2) = 2^4 - 2^3 - 2 \cdot 2^2 - 2 + 3$
$\quad = 16 - 8 - 2 \cdot 4 - 2 + 3 = 1$

The point $(2, 1)$ is a point on the graph of $p(x) = x^4 - x^3 - 2x^2 - x + 3$.

**c)** $p(-1) = (-1)^4 - (-1)^3 - 2(-1)^2 - (-1) + 3$
$\quad = 1 - (-1) - 2 \cdot 1 + 1 + 3$
$\quad = 1 + 1 - 2 + 1 + 3 = 4$

The point $(-1, 4)$ is a point on the graph of $p(x) = x^4 - x^3 - 2x^2 - x + 3$.

**Figure 12.7** shows the graph of $p(x) = x^4 - x^3 - 2x^2 - x + 3$. Note the three points found are indicated on the graph.

Now Try Exercise 15

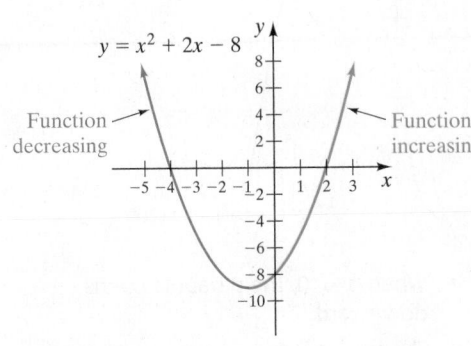

$p(x) = x^4 - x^3 - 2x^2 - x + 3$

**FIGURE 12.7**

Although plotting points can help us determine the graphs of polynomial functions, we need some additional guidelines as to what polynomial functions look like. *The graphs of all polynomial functions are smooth, continuous curves.* The graph shown in **Figure 12.7** and the graphs shown in **Figures 12.8–12.13** are all smooth, continuous curves.

We now will focus on the general shape of quadratic functions and cubic functions. The graphs of all quadratic functions with a *positive leading coefficient* will have the shape of the graph in **Figure 12.8.**

The graph of a cubic polynomial function with a *positive leading coefficient* may have the shape of the graph in either **Figure 12.9** or **Figure 12.10**. Notice that *whenever the leading coefficient of a polynomial function is positive, the polynomial function will increase (or move upward as x increases—the green part of the curve) to the right of some value of x.*

**FIGURE 12.8**

**FIGURE 12.9**

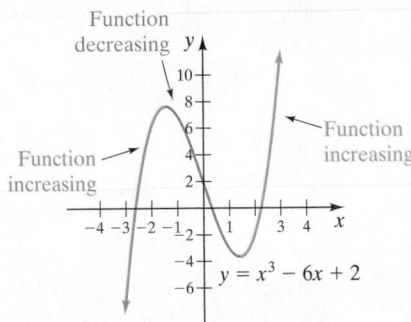

**FIGURE 12.10**

*Polynomial functions with a negative leading coefficient will decrease (or move downward as x increases—the red part of the curve) to the right of some value of x.* A quadratic polynomial function with a negative leading coefficient is shown in

**Figure 12.11**, and cubic polynomial functions with negative leading coefficients are shown in **Figure 12.12** and **Figure 12.13**.

FIGURE 12.11

FIGURE 12.12

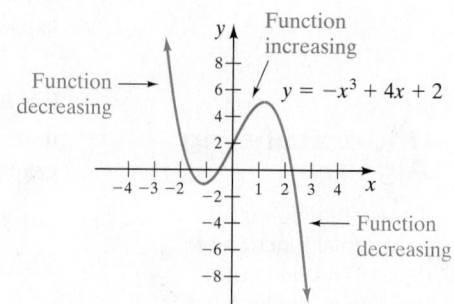

FIGURE 12.13

The leading coefficient is the coefficient of the term with the greatest exponent on the variable. As *x* increases, this term will eventually dominate all the other terms in the function. So if the coefficient of this term is positive, the function will *eventually* increase as *x* increases. If the leading coefficient is negative, the function will *eventually* decrease as *x* increases.

## 2 Find the Axis of Symmetry, Vertex, and *x*-Intercepts of a Parabola

Now let's look at the graphs of quadratic functions a bit more closely. We begin with the definition of a quadratic function.

**Quadratic Function**

A **quadratic function** is a function that can be written in the form

$$f(x) = ax^2 + bx + c \qquad \text{for real numbers } a, b, \text{ and } c, \text{ with } a \neq 0.$$

The graph of quadratic function is a **parabola**. A parabola has a shape that resembles, but is not the same as, the letter *U*. For a quadratic function, the sign of the leading coefficient, *a*, determines whether the parabola opens upward (see **Fig. 12.14a**) or downward (see **Fig. 12.14b**).

$a > 0$,
Parabola
opens upward

$f(x) = ax^2 + bx + c$

$a < 0$,
Parabola
opens downward

(a)

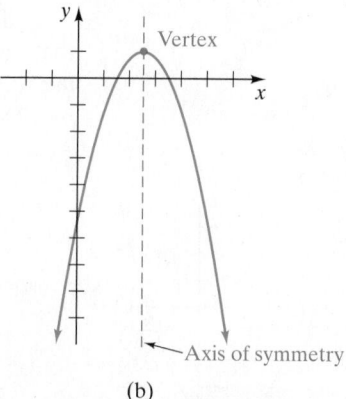

(b)

FIGURE 12.14

• When $a > 0$, the parabola opens upward.
• The **vertex** is the lowest point on the curve.
• The **minimum value of the function** is the *y*-coordinate of the vertex.

• When $a < 0$, the parabola opens downward.
• The **vertex** is the highest point on the curve.
• The **maximum value of the function** is the *y*-coordinate of the vertex.

Graphs of quadratic functions will have **symmetry** about a vertical line, called the **axis of symmetry**, which goes through the vertex. This means that if we fold the paper along this imaginary line, the right and left sides of the parabola will coincide (see both graphs in **Fig. 12.14** on p. 732).

Now we will derive the formula for the axis of symmetry, and find the coordinates of the vertex of a parabola, by beginning with a quadratic function of the form $f(x) = ax^2 + bx + c$ and completing the square on the first two terms.

$$f(x) = ax^2 + bx + c$$

$$= a\left(x^2 + \frac{b}{a}x\right) + c \quad \text{Factor out } a.$$

One half the coefficient of $x$ is $\frac{b}{2a}$. Its square is $\frac{b^2}{4a^2}$. Add and subtract this term inside the parentheses. The sum of these two terms is zero.

$$f(x) = a\left(x^2 + \frac{b}{a}x + \frac{b^2}{4a^2} - \frac{b^2}{4a^2}\right) + c$$

Now rewrite the function as follows.

$$f(x) = a\left(x^2 + \frac{b}{a}x + \frac{b^2}{4a^2}\right) - a\left(\frac{b^2}{4a^2}\right) + c$$

$$= a\left(x + \frac{b}{2a}\right)^2 - \frac{b^2}{4a} + c \quad \begin{array}{l}\text{Replace the trinomial with the}\\ \text{square of a binomial.}\end{array}$$

$$= a\left(x + \frac{b}{2a}\right)^2 - \frac{b^2}{4a} + \frac{4ac}{4a} \quad \begin{array}{l}\text{Write fractions with a common}\\ \text{denominator.}\end{array}$$

$$= a\left(x + \frac{b}{2a}\right)^2 + \frac{4ac - b^2}{4a} \quad \begin{array}{l}\text{Combine the last two terms;}\\ \text{write with the variable } a \text{ first.}\end{array}$$

$$= a\left(x - \left(-\frac{b}{2a}\right)\right)^2 + \frac{4ac - b^2}{4a}$$

Now consider the following.

- The expression $\left(x - \left(-\frac{b}{2a}\right)\right)^2$ will always be greater than or equal to 0.

- If $a > 0$, the parabola opens upward and the function has a minimum value. This minimum value will occur when $x = -\frac{b}{2a}$.

- If $a < 0$, the parabola opens downward and the function has a maximum value. This maximum value will occur when $x = -\frac{b}{2a}$.

- Therefore, the $x$-coordinate of the vertex can be found using the formula $x = -\frac{b}{2a}$.

  To determine the $y$-coordinate of the vertex, evaluate $f\left(-\frac{b}{2a}\right)$. We find that the $y$-coordinate of the vertex is $y = \frac{4ac - b^2}{4a}$.

## Vertex of a Parabola

The parabola represented by the function $f(x) = ax^2 + bx + c$ will have vertex

$$\left(-\frac{b}{2a}, \frac{4ac - b^2}{4a}\right)$$

Since we often find the $y$-coordinate of the vertex by substituting the $x$-coordinate of the vertex into $f(x)$, the vertex may also be designated as

$$\left(-\frac{b}{2a}, f\left(-\frac{b}{2a}\right)\right)$$

Since the axis of symmetry is the vertical line through the vertex, its equation is found using the same formula used to find the $x$-coordinate of the vertex.

## Axis of Symmetry of a Parabola

For a quadratic function of the form $f(x) = ax^2 + bx + c$, the equation of the **axis of symmetry** of the parabola is

$$x = -\frac{b}{2a}$$

Recall that to find the $x$-intercepts of a function, we set $y$ or $f(x) = 0$ and solve for $x$.

## $x$-Intercepts of a Parabola

To find the $x$-intercepts (if there are any) of a quadratic function, solve the equation $ax^2 + bx + c = 0$ for $x$.

This equation may be solved by factoring, by using the quadratic formula, or by completing the square.

As we mentioned in Section 12.2, the discriminant, $b^2 - 4ac$, may be used to determine the *number of x-intercepts*. The following table summarizes information about the discriminant.

| Discriminant, $b^2 - 4ac$ | Number of $x$-Intercepts | Possible Graphs of $f(x) = ax^2 + bx + c$ |
|:---:|:---:|:---:|
| $> 0$ | Two | |
| $= 0$ | One | |
| $< 0$ | None | |

### 3 Graph Quadratic Functions Using the Axis of Symmetry, Vertex, and Intercepts

Now we will draw graphs of quadratic functions.

EXAMPLE 2  Consider the quadratic function $y = -x^2 + 8x - 12$.

a) Determine whether the parabola opens upward or downward.

b) Find the $y$-intercept.

c) Find the vertex.

d) Find the $x$-intercepts, if any.

e) Draw the graph.

*Solution*

a) Since $a$ is $-1$, which is less than 0, the parabola opens downward.

b) To find the $y$-intercept, set $x = 0$ and solve for $y$.

$$y = -(0)^2 + 8(0) - 12 = -12$$

The $y$-intercept is $(0, -12)$.

c) First, find the $x$-coordinate, then find the $y$-coordinate of the vertex. From the function, $a = -1$, $b = 8$, and $c = -12$.

$$x = -\frac{b}{2a} = -\frac{8}{2(-1)} = 4$$

$$y = \frac{4ac - b^2}{4a} = \frac{4(-1)(-12) - 8^2}{4(-1)} = \frac{48 - 64}{-4} = 4$$

The vertex is at $(4, 4)$. The $y$-coordinate of the vertex could also be found by substituting 4 for $x$ in the function and finding the corresponding value of $y$, which is 4.

d) To find the $x$-intercepts, set $y = 0$.

$$0 = -x^2 + 8x - 12$$

$$\text{or} \quad x^2 - 8x + 12 = 0$$

$$(x - 6)(x - 2) = 0$$

$$x - 6 = 0 \quad \text{or} \quad x - 2 = 0$$

$$x = 6 \quad \text{or} \quad x = 2$$

Thus, the $x$-intercepts are $(2, 0)$ and $(6, 0)$. These values could also be found by the quadratic formula (or by completing the square).

e) Use all this information to draw the graph (**Fig. 12.15**).

Now Try Exercise 21

---

**Understanding Algebra**

Recall from Chapter 8, $y$ is a function of $x$, and we write $y = f(x)$. Therefore, when graphing the function in Example 2, we can write $f(x) = -x^2 + 8x - 12$, or equivalently $y = -x^2 + 8x - 12$.

FIGURE 12.15

Notice that in Example 2, the equation is $y = -x^2 + 8x - 12$ and the $y$-intercept is $(0, -12)$. In general, for any equation of the form $y = ax^2 + bx + c$, the $y$-intercept will be $(0, c)$.

If you obtain irrational values when finding $x$-intercepts by the quadratic formula, use your calculator to estimate these values, and then plot these decimal values. For example, if you obtain $x = \frac{2 \pm \sqrt{10}}{2}$, you would evaluate $\frac{2 + \sqrt{10}}{2}$ and $\frac{2 - \sqrt{10}}{2}$ on your calculator and obtain 2.58 and $-0.58$, respectively, to the nearest hundredth. The $x$-intercepts would therefore be $(2.58, 0)$ and $(-0.58, 0)$.

**EXAMPLE 3** Consider the quadratic function $f(x) = 2x^2 + 6x + 5$.

**a)** Determine whether the parabola opens upward or downward.

**b)** Find the $y$-intercept.       **c)** Find the vertex.

**d)** Find the $x$-intercepts, if any.       **e)** Draw the graph.

*Solution*

**a)** Since $a$ is 2, which is greater than 0, the parabola opens upward.

**b)** Since $f(x)$ is the same as $y$, to find the $y$-intercept, set $x = 0$ and solve for $f(x)$, or $y$.

$$f(0) = 2(0)^2 + 6(0) + 5 = 5$$

The $y$-intercept is $(0, 5)$.

**c)** Here $a = 2, b = 6$, and $c = 5$.

$$x = -\frac{b}{2a} = -\frac{6}{2(2)} = -\frac{6}{4} = -\frac{3}{2}$$

$$y = \frac{4ac - b^2}{4a} = \frac{4(2)(5) - 6^2}{4(2)} = \frac{40 - 36}{8} = \frac{4}{8} = \frac{1}{2}$$

The vertex is $\left(-\frac{3}{2}, \frac{1}{2}\right)$. The $y$-coordinate of the vertex can also be found by evaluating $f\left(-\frac{3}{2}\right)$.

**d)** To find the $x$-intercepts, set $f(x) = 0$.

$$0 = 2x^2 + 6x + 5$$

This trinomial cannot be factored. To determine whether this equation has any real solutions, evaluate the discriminant.

$$b^2 - 4ac = 6^2 - 4(2)(5) = 36 - 40 = -4$$

Since the discriminant is less than 0, this equation has no real solutions and the graph has no $x$-intercepts.

**e)** The graph is given in **Figure 12.16**.

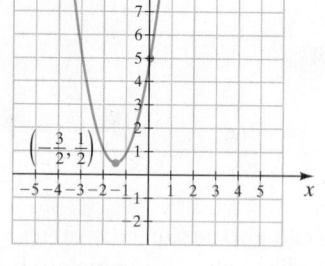

FIGURE 12.16

Now Try Exercise 45

## 4 Solve Maximum and Minimum Problems

A parabola that opens upward has a **minimum value** at its vertex, as illustrated in **Figure 12.17a**. A parabola that opens downward has a **maximum value** at its vertex, as shown in **Figure 12.17b**. If you are given a function of the form $f(x) = ax^2 + bx + c$, the maximum or minimum value will occur at $-\dfrac{b}{2a}$, and the value will be $\dfrac{4ac - b^2}{4a}$.

There are many real-life problems that require finding maximum and minimum values.

$$y = ax^2 + bx + c$$

FIGURE 12.17       (a)       (b)

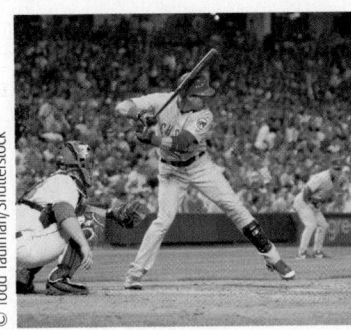

© Todd Taulman/Shutterstock

EXAMPLE 4  **Baseball**  Mark DeRosa hits a baseball at a height of 3 feet above the ground. The height of the ball above the ground, $f(t)$, in feet, at time, $t$, in seconds, can be estimated by the function

$$f(t) = -16t^2 + 52t + 3$$

**a)** Find the maximum height attained by the baseball.

**b)** Find the time it takes for the baseball to reach its maximum height.

**c)** Find the time it takes for the baseball to strike the ground.

*Solution*

**a)** Understand   The baseball will follow the path of a parabola that opens downward ($a < 0$). The baseball will rise to a maximum height, then begin its fall back to the ground due to gravity. To find the maximum height, we use the formula $y = \dfrac{4ac - b^2}{4a}$.

Translate                $a = -16, \qquad b = 52, \qquad c = 3$

$$y = \frac{4ac - b^2}{4a}$$

Carry Out

$$= \frac{4(-16)(3) - (52)^2}{4(-16)}$$

$$= \frac{-192 - 2704}{-64}$$

$$= \frac{-2896}{-64}$$

$$= 45.25$$

Answer   The maximum height attained by the baseball is 45.25 feet.

**b)** The baseball reaches its maximum height at

$$t = -\frac{b}{2a} = -\frac{52}{2(-16)} = -\frac{52}{-32} = \frac{13}{8} \quad \text{or} \quad 1\frac{5}{8} \quad \text{or} \quad 1.625 \text{ seconds}$$

**c)** Understand and Translate   When the baseball strikes the ground, its height, $y$, above the ground is 0. Thus, to determine when the baseball strikes the ground, we solve the equation

$$-16t^2 + 52t + 3 = 0$$

We will use the quadratic formula to solve the equation.

$$t = \frac{-b \pm \sqrt{b^2 - 4ac}}{2a}$$

$$= \frac{-52 \pm \sqrt{(52)^2 - 4(-16)(3)}}{2(-16)}$$

Carry Out

$$= \frac{-52 \pm \sqrt{2704 + 192}}{-32}$$

$$= \frac{-52 \pm \sqrt{2896}}{-32}$$

$$\approx \frac{-52 \pm 53.81}{-32}$$

$$t \approx \frac{-52 + 53.81}{-32} \quad \text{or} \quad t \approx \frac{-52 - 53.81}{-32}$$

$$\approx -0.06 \text{ second} \qquad\qquad \approx 3.31 \text{ seconds}$$

Now Try Exercise 103

**Answer**  The only acceptable value is 3.31 seconds. The baseball strikes the ground in about 3.31 seconds. Notice in part **b)** that the time it takes the baseball to reach its maximum height, 1.625 seconds, is not quite half the total time the baseball was in flight, 3.31 seconds. The reason for this is that the baseball was hit from a height of 3 feet and not at ground level.

EXAMPLE 5  **Area of a Rectangle**  Consider the rectangle below where the length is $x + 3$ and the width is $10 - x$.

$10 - x$

$x + 3$

**a)** Find an equation for the area, $A(x)$.

**b)** Find the value for $x$ that gives the largest (maximum) area.

**c)** Find the maximum area.

*Solution*

**a)** Area is length times width. The area function is

$$A(x) = (x + 3)(10 - x)$$
$$= -x^2 + 7x + 30$$

**b)** Understand and Translate  The graph of the function is a parabola that opens downward. Thus, the maximum value occurs at the vertex. Therefore, the maximum area occurs at $x = -\dfrac{b}{2a}$. Here $a = -1$ and $b = 7$.

Carry Out

$$x = -\frac{b}{2a} = -\frac{7}{2(-1)} = \frac{7}{2} = 3.5$$

**Answer**   The maximum area occurs when $x$ is 3.5 units.

**c)** To find the maximum area, substitute 3.5 for each $x$ in the equation determined in part **a)**.

$$A(x) = -x^2 + 7x + 30$$

$$A(3.5) = -(3.5)^2 + 7(3.5) + 30$$

$$= -12.25 + 24.5 + 30$$

$$= 42.25$$

Observe that for this rectangle, the length is $x + 3 = 3.5 + 3 = 6.5$ units and the width is $10 - x = 10 - 3.5 = 6.5$ units. The rectangle is actually a square, and its area is $(6.5)(6.5) = 42.25$ square units. Therefore, the maximum area is 42.25 square units.

Now Try Exercise 83

In Example 5 **c)**, the maximum area could have been determined by using the formula $y = \dfrac{4ac - b^2}{4a}$. Determine the maximum area now using this formula. You should obtain the same answer, 42.25 square units.

**FIGURE 12.18**

**EXAMPLE 6 Rectangular Corral** Greg Fierro is building a corral for newborn calves in the shape of a rectangle (see **Fig. 12.18**). If he plans to use 160 meters of fencing, find the dimensions of the corral that will give the greatest area.

*Solution*   Understand   We are given the perimeter of the corral, 160 meters. The formula for the perimeter of a rectangle is $P = 2l + 2w$. For this problem, $160 = 2l + 2w$. We are asked to maximize the area, $A$, where

$$A = lw$$

We need to express the area in terms of one variable, not two. To express the area in terms of $l$, we solve the perimeter formula, $160 = 2l + 2w$, for $w$, then make a substitution.

Translate
$$160 = 2l + 2w$$
$$160 - 2l = 2w$$
$$80 - l = w$$

Carry Out   Now we substitute $80 - l$ for $w$ into $A = lw$. This gives

$$A = lw$$
$$A = l(80 - l)$$
$$A = -l^2 + 80l$$

In this quadratic equation, $a = -1$, $b = 80$, and $c = 0$. The maximum area will occur at

$$l = -\frac{b}{2a} = -\frac{80}{2(-1)} = 40$$

Answer   The length that will give the largest area is 40 meters. The width, $w = 80 - l$, will also be 40 meters. Thus, a square with dimensions 40 by 40 meters will give the largest area.

The largest area can be found by substituting $l = 40$ into the formula $A = l(80 - l)$ or by using $A = \frac{4ac - b^2}{4a}$. In either case, we obtain an area of 1600 square meters.

Now Try Exercise 105

## 5  Understand Translations of Parabolas

Our next method of graphing parabolas begins with the graphs of quadratic functions of the form $f(x) = ax^2$. Consider the functions $f(x) = x^2$, $g(x) = 2x^2$, and $h(x) = \frac{1}{2}x^2$ shown in **Figure 12.19**. Notice that the *value* of $a$ determines the width of the parabola.

Next, consider the functions $f(x) = -x^2$, $g(x) = -2x^2$, and $h(x) = -\frac{1}{2}x^2$ shown in **Figure 12.20**. Although the parabolas open downward, the value of $a$ still determines the width of the parabola.

**Understanding Algebra**

In general, when graphing the parabola that corresponds to a quadratic function of the form $f(x) = ax^2$, as $|a|$ gets larger, the parabola gets narrower. As $|a|$ gets smaller, the parabola gets wider.

**FIGURE 12.19**

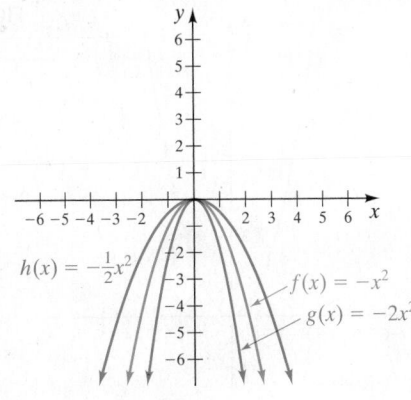

**FIGURE 12.20**

From **Figure 12.19** and **Figure 12.20** we can see that, *in general, as |a| gets larger, the parabola gets narrower. As |a| gets smaller, the parabola gets wider.*

We now will **translate**, or shift, the position of graphs of the form $f(x) = ax^2$ to obtain the graphs of other quadratic functions. For example, consider the three functions $f(x) = x^2, g(x) = (x - 2)^2$, and $h(x) = (x + 2)^2$ shown in **Figure 12.21**. Notice that the three graphs are identical in *shape* but are in different positions.

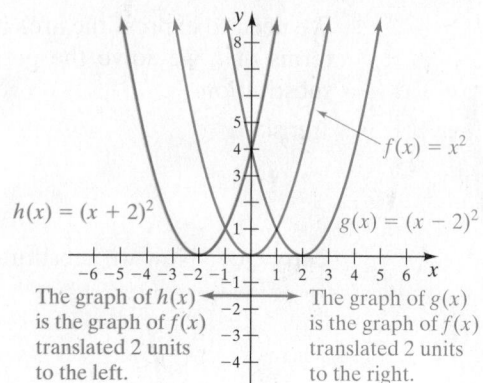

FIGURE 12.21

Note also that the graph of $g(x)$ is translated, or shifted, 2 units to the right of the graph of $f(x)$. The graph of $h(x)$ is translated 2 units to the left of the graph of $f(x)$.

*In general, the graph of $g(x) = a(x - h)^2$ will have the same shape as the graph of $f(x) = ax^2$. If h is positive, then the graph of $g(x)$ will be shifted h units to the right of the graph of $f(x)$. If h is negative, then the graph of $g(x)$ will be shifted |h| units to the left of the graph of $f(x)$.*

Next, consider the graphs of $f(x) = x^2, g(x) = x^2 + 3$, and $h(x) = x^2 - 3$ shown in **Figure 12.22**. Notice again that the three graphs are identical in *shape* but are in different positions.

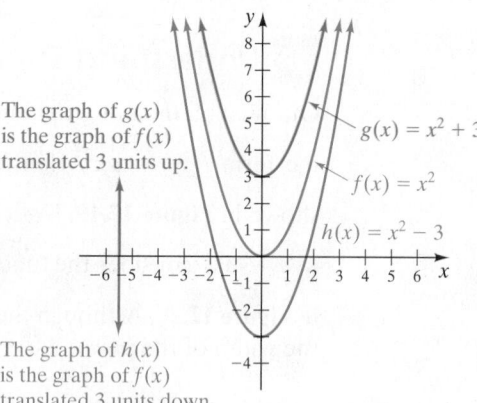

FIGURE 12.22

Note also that the graph of $g(x)$ is translated up from the graph of $f(x)$. The graph of $h(x)$ is translated 3 units down from the graph of $f(x)$.

*In general, the graph of $g(x) = ax^2 + k$ will have the same shape as the graph of $f(x) = ax^2$. If k is positive, then the graph of $g(x)$ will be shifted k units up from the graph of $f(x)$. If k is negative, then the graph of $g(x)$ will be shifted |k| units down from the graph of $f(x)$.*

Now consider the graphs of $f(x) = x^2$ and $g(x) = (x - 2)^2 + 3$, shown in **Figure 12.23**. Notice that the graph of $g(x)$ has the same shape as that of $f(x)$. The graph of $g(x)$ is the graph of $f(x)$ translated 2 units to the right and 3 units up. This graph and the discussion preceding it lead to the following important facts.

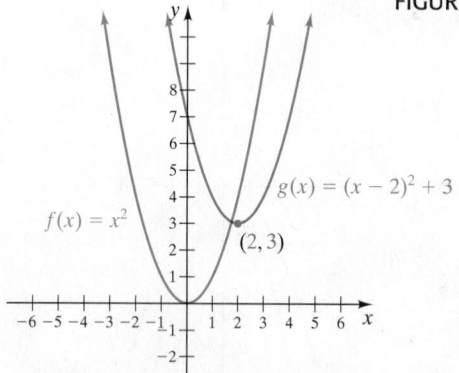

FIGURE 12.23

### Parabola Shifts

For any function $f(x) = ax^2$, the graph of $g(x) = a(x - h)^2 + k$ will have the same shape as the graph of $f(x)$. The graph of $g(x)$ will be the graph of $f(x)$ shifted as follows:

- If $h$ is a positive real number, the graph will be shifted $h$ units to the right.
- If $h$ is a negative real number, the graph will be shifted $|h|$ units to the left.
- If $k$ is a positive real number, the graph will be shifted $k$ units up.
- If $k$ is a negative real number, the graph will be shifted $|k|$ units down.

Examine the graph of $g(x) = (x - 2)^2 + 3$ in **Figure 12.23** on page 740. Notice that its axis of symmetry is $x = 2$ and its vertex is $(2, 3)$.

### Axis of Symmetry and Vertex of a Parabola

The graph of any function of the form

$$f(x) = a(x - h)^2 + k$$

will be a parabola with axis of symmetry $x = h$ and vertex at $(h, k)$.

| Example | Axis of Symmetry | Vertex | Parabola Opens |
|---------|------------------|--------|----------------|
| $f(x) = 2(x - 5)^2 + 7$ | $x = 5$ | $(5, 7)$ | upward, $a > 0$ |
| $f(x) = -\dfrac{1}{2}(x - 6)^2 - 3$ | $x = 6$ | $(6, -3)$ | downward, $a < 0$ |

Now consider $f(x) = 2(x + 5)^2 + 3$. We can rewrite this as $f(x) = 2[x - (-5)]^2 + 3$. Therefore, $h$ has a value of $-5$ and $k$ has a value of $3$. The graph of this function has axis of symmetry $x = -5$ and vertex at $(-5, 3)$.

| Example | Axis of Symmetry | Vertex | Parabola Opens |
|---------|------------------|--------|----------------|
| $f(x) = 3(x + 4)^2 - 2$ | $x = -4$ | $(-4, -2)$ | upward, $a > 0$ |
| $f(x) = -\dfrac{1}{2}\left(x + \dfrac{1}{3}\right)^2 + \dfrac{1}{4}$ | $x = -\dfrac{1}{3}$ | $\left(-\dfrac{1}{3}, \dfrac{1}{4}\right)$ | downward, $a < 0$ |

Now we are ready to graph parabolas using translations.

**EXAMPLE 7**  The graph of $f(x) = -2x^2$ is illustrated in **Figure 12.24**. Using this graph as a guide, graph $g(x) = -2(x + 3)^2 - 4$.

*Solution*    The function $g(x)$ may be written $g(x) = -2[x - (-3)]^2 - 4$. Therefore, in the function, $h$ has a value of $-3$ and $k$ has a value of $-4$. The graph of $g(x)$ will therefore be the graph of $f(x)$ translated 3 units to the left (because $h = -3$) and 4 units down (because $k = -4$). The graphs of $f(x)$ and $g(x)$ are illustrated in **Figure 12.25**.

FIGURE 12.24

FIGURE 12.25

Now Try Exercise 55

In objective 2, we started with a function of the form $f(x) = ax^2 + bx + c$ and completed the square to obtain

$$f(x) = a\left[x - \left(-\frac{b}{2a}\right)\right]^2 + \frac{4ac - b^2}{4a}$$

We stated that the vertex of the parabola of this function is $\left(-\dfrac{b}{2a}, \dfrac{4ac - b^2}{4a}\right)$.

Suppose we substitute $h$ for $-\dfrac{b}{2a}$ and $k$ for $\dfrac{4ac - b^2}{4a}$ in the function. We then get

$$f(x) = a(x - h)^2 + k$$

which we know is a parabola with vertex at $(h, k)$. Therefore, both functions $f(x) = ax^2 + bx + c$ and $f(x) = a(x - h)^2 + k$ yield the same vertex and axis of symmetry for any given function.

## 6 Write Functions in the Form $f(x) = a(x - h)^2 + k$

If we wish to graph parabolas using translations, we need to change the form of a function from $f(x) = ax^2 + bx + c$ to $f(x) = a(x - h)^2 + k$. To do this, we *complete the square* as was discussed in Section 12.1.

EXAMPLE 8   Given $f(x) = x^2 - 6x + 10$, ─────────────

**a)** Write $f(x)$ in the form $f(x) = a(x - h)^2 + k$.
**b)** Graph $f(x)$.

*Solution*

**a)** We use the $x^2$ and $-6x$ terms to obtain a perfect square trinomial.

$$f(x) = (x^2 - 6x) + 10$$

Now we take half the coefficient of the $x$-term and square it.

$$\left[\frac{1}{2}(-6)\right]^2 = \boxed{9}$$

We then add this value, 9, within the parentheses. Since we are adding 9 within parentheses, we add $-9$ outside the parentheses. Adding 9 and $-9$ to an expression is the same as adding 0, which does not change the value of the expression.

$$f(x) = (x^2 - 6x + 9) - 9 + 10$$

By doing this we have created a perfect square trinomial within the parentheses, plus a constant outside the parentheses. We express the perfect square trinomial as the square of a binomial.

$$f(x) = (x - 3)^2 + 1$$

The function is now in the form we are seeking.

**b)** Since $a = 1$, which is greater than 0, the parabola opens upward. The axis of symmetry of the parabola is $x = 3$ and the vertex is at $(3, 1)$. The $y$-intercept can be obtained by substituting $x = 0$ and finding $f(x)$. When $x = 0$, $f(x) = (-3)^2 + 1 = 10$. Thus, the $y$-intercept is at 10. By plotting the vertex, $y$-intercept, and a few other points, we obtain the graph in **Figure 12.26**. The figure also shows the graph of $y = x^2$ for comparison.

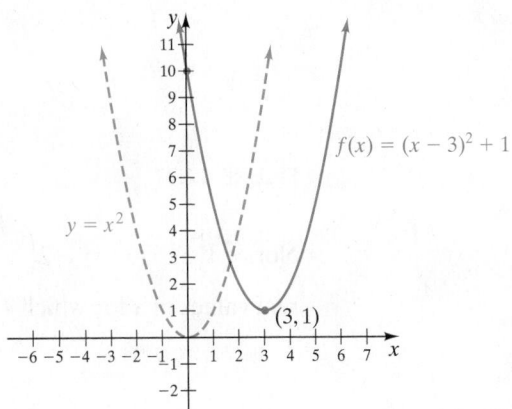

FIGURE 12.26

Now Try Exercise 65

EXAMPLE 9    Given $f(x) = -2x^2 - 10x - 13$,

**a)** Write $f(x)$ in the form $f(x) = a(x - h)^2 + k$.

**b)** Graph $f(x)$.

*Solution*

**a)** When the leading coefficient is not 1, we factor out the leading coefficient from the terms containing the variable.

$$f(x) = -2(x^2 + 5x) - 13$$

Now we complete the square.

Half of coefficient of
first-degree term squared

$$\left[\frac{1}{2}(5)\right]^2 = \frac{25}{4}$$

If we add $\dfrac{25}{4}$ within the parentheses, we are actually adding $-2\left(\dfrac{25}{4}\right)$ or $-\dfrac{25}{2}$, since each term in parentheses is multiplied by $-2$. Therefore, to compensate, we must add $\dfrac{25}{2}$ outside the parentheses.

$$f(x) = -2\left(x^2 + 5x + \frac{25}{4}\right) + \frac{25}{2} - 13$$

$$= -2\left(x + \frac{5}{2}\right)^2 - \frac{1}{2}$$

**b)** Since $a = -2$, the parabola opens downward. The axis of symmetry is $x = -\dfrac{5}{2}$ and the vertex is $\left(-\dfrac{5}{2}, -\dfrac{1}{2}\right)$. The $y$-intercept is at $f(0) = -13$. We plot a few points and draw the graph in **Figure 12.27** on page 744. In the figure, we also show the graph of $y = -2x^2$ for comparison.

**FIGURE 12.27**

Notice that $f(x) = -2\left(x + \dfrac{5}{2}\right)^2 - \dfrac{1}{2}$ has no $x$-intercepts. Therefore, there are no real values of $x$ for which $f(x) = 0$.

*Now Try Exercise 69*

A second way to change the equation from $f(x) = ax^2 + bx + c$ to $f(x) = a(x - h)^2 + k$ form is to let $h = -\dfrac{b}{2a}$ and $k = \dfrac{4ac - b^2}{4a}$. Find the values for $h$ and $k$ and then substitute the values obtained into $f(x) = a(x - h)^2 + k$. For example, for the function $f(x) = -2x^2 - 10x - 13$ in Example 9, $a = -2, b = -10$, and $c = -13$. Then

$$h = -\frac{b}{2a} = -\frac{-10}{2(-2)} = -\frac{5}{2}$$

$$k = \frac{4ac - b^2}{4a} = \frac{4(-2)(-13) - (-10)^2}{4(-2)} = -\frac{1}{2}$$

Therefore,

$$f(x) = a(x - h)^2 + k$$

$$= -2\left[x - \left(-\frac{5}{2}\right)\right]^2 - \frac{1}{2}$$

$$= -2\left(x + \frac{5}{2}\right)^2 - \frac{1}{2}$$

This answer checks with that obtained in Example 9.

---

# EXERCISE SET 12.5   MathXL®   MyMathLab®

MathXL®   MyMathLab

## Warm-Up Exercises

*Fill in the blanks with the appropriate word, phrase, or symbol(s) from the following list.*

| down | axis of symmetry | $y$-intercept | $-\dfrac{b}{2a}$ | opens downward |
|---|---|---|---|---|
| vertex | narrower | right | left | $x$-intercepts |
| $\dfrac{4ac - b^2}{4a}$ | decrease | up | increase | |

**1.** If the leading coefficient of a polynomial function is positive, the graph will eventually _____ as $x$ increases.

**2.** If the leading coefficient of a polynomial function is negative, the graph will eventually _____ as $x$ increases.

**3.** When $a < 0$, the graph of $f(x) = ax^2 + bx + c$ is a parabola that _____ and the vertex is the highest point on the curve.

**4.** When $a > 0$, the graph of $f(x) = ax^2 + bx + c$ is a parabola that opens upward and the _____ is the lowest point on the curve.

**5.** The equation $x =$ _____ gives the equation of the axis of symmetry and the $x$-coordinate of the vertex.

**6.** To find the _____ of a quadratic function (if there are any), set $y$ or $f(x)$ equal to 0 and solve for $x$.

**7.** To find the _____ of a quadratic function, set $x = 0$ and solve for $y$ or $f(x)$.

**8.** In general, when graphing the parabola that corresponds to a quadratic function of the form $f(x) = ax^2$, as $|a|$ gets larger, the parabola gets _____.

**9.** If $h$ is positive, then the graph of $g(x) = a(x - h)^2$ will have the same shape as the graph of $f(x) = ax^2$ but will be shifted $h$ units to the _____ of the graph of $f(x)$.

**10.** If $h$ is negative, then the graph of $g(x) = a(x - h)^2$ will have the same shape as the graph of $f(x) = ax^2$ but will be shifted $|h|$ units to the _____ of the graph of $f(x)$.

**11.** If $k$ is positive, then the graph of $g(x) = ax^2 + k$ will have the same shape as the graph of $f(x) = ax^2$ but will be shifted $k$ units _____ from the graph of $f(x)$.

**12.** If $k$ is negative, then the graph of $g(x) = ax^2 + k$ will have the same shape as the graph of $f(x) = ax^2$ but will be shifted $|k|$ units _____ from the graph of $f(x)$.

## Practice the Skills

*Evaluate each polynomial function at the given value.*

**13.** Find $p(0)$ if $p(x) = x^4 - 6x^3 + 7x^2 - x + 3$

**14.** Find $p(1)$ if $p(x) = x^4 - 2x^3 + x^2 - 5x + 1$

**15.** Find $p(2)$ if $p(x) = x^2 - 6x + 4$.

**16.** Find $p(-1)$ if $p(x) = 4x^2 - 6x + 22$.

**17.** Find $p\left(\dfrac{1}{2}\right)$ if $p(x) = 2x^2 - 3x - 6$.

**18.** Find $p\left(\dfrac{1}{3}\right)$ if $p(x) = \dfrac{1}{2}x^3 - x^2 + 6$.

**19.** Find $p(0.4)$ if $p(x) = 0.2x^3 + 1.6x^2 - 2.3$.

**20.** Find $p(-1.2)$ if $p(x) = -1.6x^3 - 4.6x^2 - 0.1x$.

*In Exercises 21–46,* **a)** *determine whether the parabola opens upward or downward,* **b)** *find the y-intercept,* **c)** *find the vertex,* **d)** *find the x-intercepts (if any), and* **e)** *draw the graph.*

**21.** $f(x) = x^2 + 8x + 15$

**22.** $g(x) = x^2 + 2x - 3$

**23.** $f(x) = x^2 - 4x + 3$

**24.** $h(x) = x^2 - 2x - 8$

**25.** $f(x) = -x^2 - 2x + 8$

**26.** $p(x) = -x^2 + 8x - 15$

**27.** $g(x) = -x^2 + 4x + 5$

**28.** $n(x) = -x^2 - 2x + 24$

**29.** $t(x) = -x^2 + 4x - 5$

**30.** $g(x) = x^2 + 6x + 13$

**31.** $f(x) = x^2 - 4x + 4$

**32.** $r(x) = -x^2 + 10x - 25$

**33.** $r(x) = x^2 + 2$

**34.** $f(x) = x^2 + 4x$

**35.** $l(x) = -x^2 + 5$

**36.** $g(x) = -x^2 + 6x$

**37.** $f(x) = -2x^2 + 4x - 8$

**38.** $g(x) = -2x^2 - 6x + 4$

**39.** $m(x) = 3x^2 + 4x + 3$

**40.** $p(x) = -2x^2 + 5x + 4$

**41.** $y = 3x^2 + 4x - 6$

**42.** $y = x^2 - 6x + 4$

**43.** $y = 2x^2 - x - 6$

**44.** $g(x) = -4x^2 + 6x - 9$

**45.** $f(x) = -x^2 + 3x - 5$

**46.** $h(x) = -2x^2 + 4x - 5$

*Using the graphs in* **Figures 12.19** *through* **12.23** *as a guide, graph each function and label the vertex.*

**47.** $f(x) = (x - 3)^2$
**48.** $f(x) = (x - 4)^2$
**49.** $f(x) = (x + 1)^2$
**50.** $f(x) = (x + 2)^2$

**51.** $f(x) = x^2 + 3$
**52.** $f(x) = x^2 + 5$
**53.** $f(x) = x^2 - 1$
**54.** $f(x) = x^2 - 4$

**55.** $f(x) = (x - 2)^2 + 3$
**56.** $f(x) = (x - 3)^2 - 4$
**57.** $f(x) = (x + 4)^2 + 4$
**58.** $h(x) = (x + 4)^2 - 1$

**59.** $g(x) = -(x + 3)^2 - 2$
**60.** $g(x) = (x - 1)^2 + 4$
**61.** $y = -2(x - 2)^2 + 2$
**62.** $y = -2(x - 3)^2 + 1$

**63.** $h(x) = -2(x + 1)^2 - 3$
**64.** $f(x) = -(x - 5)^2 + 2$

*In Exercises 65–74,* **a)** *express each function in the form* $f(x) = a(x - h)^2 + k$ *and* **b)** *draw the graph of each function and label the vertex.*

**65.** $f(x) = x^2 - 6x + 8$
**66.** $g(x) = x^2 + 6x + 2$

**67.** $g(x) = x^2 - x - 3$
**68.** $f(x) = x^2 - x + 1$

**69.** $f(x) = -x^2 - 4x - 6$
**70.** $h(x) = -x^2 + 6x + 1$

**71.** $g(x) = x^2 - 4x - 1$
**72.** $p(x) = x^2 - 2x - 6$

**73.** $f(x) = 2x^2 + 5x - 3$
**74.** $k(x) = 2x^2 + 7x - 4$

## Problem Solving

*In Exercises 75–78, determine which of the graphs—* **a)**, **b)**, *or* **c)** *—is the graph of the given equation. Explain how you determined your answer.*

**75.** $y = x^2 + 3x - 4$

**a)**

**b)**

**c)**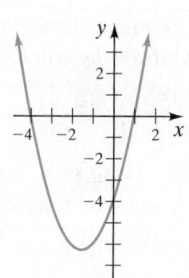

**76.** $y = x^3 + 2x^2 - 4$

**a)**

**b)**

**c)**

**77.** $y = -x^3 + 2x - 6$

**a)**

**b)**

**c)**

**78.** $y = x^3 + 4x^2 - 5$

**a)**

**b)**

**c)**

*Match the functions in Exercises 79–82 with the appropriate graphs labeled* **a)** *through* **d)**.

**a)**   **b)**   **c)**   **d)**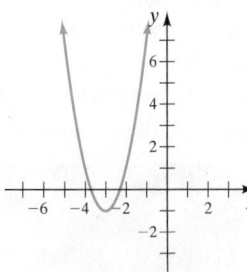

**79.** $f(x) = 2(x + 3)^2 - 1$  **80.** $f(x) = -2(x + 3)^2 - 1$  **81.** $f(x) = 2(x - 1)^2 + 3$  **82.** $f(x) = -2(x - 1)^2 + 3$

**Area** *For each rectangle,* **a)** *find the value for x that gives the maximum area and* **b)** *find the maximum area.*

**83.**

$18 - x$

$x + 4$

**84.**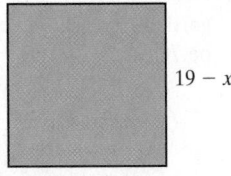

$19 - x$

$x + 7$

**85.**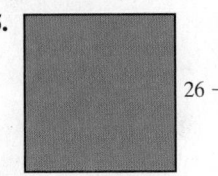

$26 - x$

$x + 5$

**86.**

$19 - x$

$x + 2$

**87. Selling Batteries** The revenue function for selling $n$ batteries is $R(n) = n(8 - 0.02n) = -0.02n^2 + 8n$. Find **a)** the number of batteries that must be sold to obtain the maximum revenue and **b)** the maximum revenue.

**88. Selling Watches** The revenue function for selling $n$ watches is $R(n) = n(25 - 0.1n) = -0.1n^2 + 25n$. Find **a)** the number of watches that must be sold to obtain the maximum revenue and **b)** the maximum revenue.

**89. Enrollment** The enrollment in a high school in the Naplewood School District can be approximated by the function

$$N(t) = -0.043t^2 + 1.82t + 46.0$$

where $t$ is the number of years since 1989 and $1 \le t \le 22$. In what year will the maximum enrollment be obtained?

**90. Drug-Free Schools** The percent of students in schools in the United States who say their school is not drug free can be approximated by the function

$$f(a) = -2.32a^2 + 76.58a - 559.87$$

where $a$ is the student's age and $12 < a < 20$. What age has the highest percent of students who say their school is not drug free?

**91.** What is the distance between the vertices of the graphs of $f(x) = (x - 2)^2 + \dfrac{5}{2}$ and $g(x) = (x - 2)^2 - \dfrac{3}{2}$?

**92.** What is the distance between the vertices of the graphs of $f(x) = 2(x - 4)^2 - 3$ and $g(x) = -3(x - 4)^2 + 2$?

**93.** What is the distance between the vertices of the graphs of $f(x) = 2(x + 4)^2 - 3$ and $g(x) = -(x + 1)^2 - 3$?

**94.** What is the distance between the vertices of the graphs of $f(x) = -\dfrac{1}{3}(x - 3)^2 - 2$ and $g(x) = 2(x + 5)^2 - 2$?

**95.** Write the function whose graph has the shape of the graph of $f(x) = 2x^2$ and has a vertex at $(3, -2)$.

**96.** Write the function whose graph has the shape of the graph of $f(x) = -\dfrac{1}{2}x^2$ and has a vertex at $\left(\dfrac{2}{3}, -5\right)$.

**97.** Write the function whose graph has the shape of the graph of $f(x) = -4x^2$ and has a vertex at $\left(-\dfrac{3}{5}, -\sqrt{2}\right)$.

**98.** Write the function whose graph has the shape of the graph of $f(x) = \dfrac{3}{5}x^2$ and has a vertex at $(-\sqrt{3}, \sqrt{5})$.

**99. Selling Tickets** The Johnson High School Theater Club is trying to set the price of tickets for a play. If the price is too low, they will not make enough money to cover expenses, and if the price is too high, not enough people will pay the price of a ticket. They estimate that their total income per play, $I$, in hundreds of dollars, can be approximated by the formula

$$I = -x^2 + 24x - 44, 0 \le x \le 24$$

where $x$ is the cost of a ticket.

**a)** Draw a graph of income versus the cost of a ticket.

**b)** Determine the minimum cost of a ticket for the theater club to break even.

**c)** Determine the maximum cost of a ticket that the theater club can charge and break even.

**d)** How much should they charge to receive the maximum income?

**e)** Find the maximum income.

**100. Throwing an Object** An object is projected upward with an initial velocity of 192 feet per second. The object's distance above the ground, $d$, after $t$ seconds may be found by the formula $d = -16t^2 + 192t$.

**a)** Find the object's distance above the ground after 3 seconds.

**b)** Draw a graph of distance versus time.

**c)** What is the maximum height the object will reach?

**d)** At what time will it reach its maximum height?

**e)** At what time will the object strike the ground?

**101. Profit** The Fulton Bird House Company earns a weekly profit according to the function $f(x) = -0.4x^2 + 80x - 200$, where $x$ is the number of bird feeders built and sold.

**a)** Find the number of bird feeders that the company must sell in a week to obtain the maximum profit.

**b)** Find the maximum profit.

**102. Profit** The A. B. Bronson Company earns a weekly profit according to the function $f(x) = -1.2x^2 + 180x - 280$, where $x$ is the number of rocking chairs built and sold.

**a)** Find the number of rocking chairs that the company must sell in a week to obtain the maximum profit.

**b)** Find the maximum profit.

**103. Firing a Cannon** If a certain cannon is fired from a height of 9.8 meters above the ground, at a certain angle, the height of the cannonball above the ground, $h$, in meters, at time, $t$, in seconds, is found by the function

$$h(t) = -4.9t^2 + 24.5t + 9.8$$

**a)** Find the maximum height attained by the cannonball.

**b)** Find the time it takes for the cannonball to reach its maximum height.

**c)** Find the time it takes for the cannonball to strike the ground.

**104. Throwing a Ball** Ramon Loomis throws a ball into the air with an initial velocity of 32 feet per second. The height of the ball at any time $t$ is given by the formula $h = 96t - 16t^2$. At what time does the ball reach its maximum height? What is the maximum height?

**105. Room in a House** Jake Kishner is designing plans for his house. What is the maximum possible area of a room if its perimeter is to be 80 feet?

**106. Greatest Area** What are the dimensions of a rectangular garden that will have the greatest area if its perimeter is to be 70 feet?

**107. Minimum Product** What is the minimum product of two numbers that differ by 8? What are the numbers?

**108. Minimum Product** What is the minimum product of two numbers that differ by 10? What are the numbers?

**109. Maximum Product** What is the maximum product of two numbers that add to 60? What are the numbers?

**110. Maximum Product** What is the maximum product of two numbers that add to 5? What are the numbers?

*The profit of a company, in dollars, is the difference between the company's revenue and cost. Exercises 111 and 112 give cost, C(x), and revenue, R(x), functions for a particular company. The x represents the number of items produced and sold to distributors. Determine* **a)** *the maximum profit of the company and* **b)** *the number of items that must be produced and sold to obtain the maximum profit.*

**111.** $C(x) = 2000 + 40x$
   $R(x) = 800x - x^2$

**112.** $C(x) = 5000 + 12x$
   $R(x) = 2000x - x^2$

## Concept/Writing Exercises

**113.** Consider the graph of $f(x) = ax^2$. What is the general shape of $f(x)$ if **a)** $a > 0$, **b)** $a < 0$?

**114.** Consider the graph of $f(x) = ax^2$. Explain how the shape of $f(x)$ changes as $|a|$ increases and as $|a|$ decreases.

**115.** Does the function $f(x) = 3x^2 - 4x + 2$ have a maximum or a minimum value? Explain.

**116.** Does the function $g(x) = -\frac{1}{2}x^2 + 2x - 7$ have a maximum or a minimum value? Explain.

**117.** Consider $f(x) = x^2 - 8x + 12$ and $g(x) = -x^2 + 8x - 12$.

**a)** Without graphing, can you explain how the graphs of the two functions compare?

**b)** Will the graphs have the same $x$-intercepts? Explain.

**c)** Will the graphs have the same vertex? Explain.

**d)** Graph both functions on the same axes.

**118.** By observing the leading coefficient in a quadratic function and by determining the coordinates of the vertex of its graph, explain how you can determine the number of $x$-intercepts the parabola has.

## Challenge Problems

In Exercises 119 and 120, determine which of the graphs—**a)**, **b)**, or **c)**—is the graph of the given equation. Explain how you determined your answer.

**119.** $y = -x^4 + 3x^3 - 5$

**a)**

**b)**

**c)**

**120.** $y = 2x^4 + 9x^2 - 5$

**a)**

**b)**

**c)**

**121. Baseball** In Example 4 of this section, we used the function $f(t) = -16t^2 + 52t + 3$ to find that the maximum height, $f$, attained by a baseball hit by Mark DeRosa was 45.25 feet. The ball reached this height at 1.625 seconds after the baseball was hit.
Review Example 4 now.

**a)** Write $f(t)$ in the form $f(t) = a(t - h)^2 + k$ by completing the square.

**b)** Using the function you obtained in part **a)**, determine the maximum height attained by the baseball, and the time after it was hit that the baseball attained its maximum value.

**c)** Are the answers you obtained in part **b)** the same answers obtained in Example 4? If not, explain why not.

## Group Activity

*Discuss and answer Exercises 122–123 as a group.*

**122.** If the leading term of a polynomial function is $3x^3$, which of the following could possibly be the graph of the polynomial? Explain. Consider what happens for large positive values of $x$ and for large negative values of $x$.

**a)**

**b)**

**c)**

**123.** If the leading term of a polynomial is $-2x^4$, which of the following could possibly be the graph of the polynomial? Explain.

**a)**

**b)**

**c)**

*Discuss and answer Exercise 124 as a group.*

**124. a)** Group member 1: Write two quadratic functions $f(x)$ and $g(x)$ so that the functions will not intersect.

**b)** Group member 2: Write two quadratic functions $f(x)$ and $g(x)$ so that neither function will have $x$-intercepts, and the vertices of the functions are on opposite sides of the $x$-axis.

**c)** Group member 3: Write two quadratic functions $f(x)$ and $g(x)$ so that both functions have the same vertex but one function opens upward and the other opens downward.

**d)** As a group, review each answer in parts **a)**–**c)** and decide whether each answer is correct. Correct any answer that is incorrect.

**Cumulative Review Exercises**

[3.3]    **125.** Find the area shaded blue in the figure.

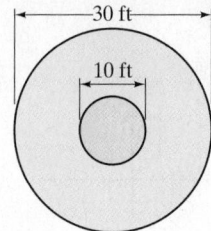

[7.2]    **126.** Divide $(x - 3) \div \dfrac{x^2 + 3x - 18}{x}$.

[9.4]    **127.** Solve the system of equations.

$$x - y = -5$$
$$2x + 2y - z = 0$$
$$x + y + z = 3$$

[9.7]    **128.** Evaluate the determinant.

$$\begin{vmatrix} \dfrac{1}{2} & 3 \\ 2 & -4 \end{vmatrix}$$

[10.3]    **129.** Graph $y \le \dfrac{2}{3}x + 3$.

# 12.6 Quadratic and Other Inequalities in One Variable

**1** Solve quadratic inequalities.

**2** Solve other polynomial inequalities.

**3** Solve rational inequalities.

In Section 10.1, we discussed linear inequalities in one variable. Now we discuss quadratic inequalities in one variable.

### Quadratic Inequality

A **quadratic inequality*** is an inequality that can be written in one of the following forms

$$ax^2 + bx + c < 0 \qquad ax^2 + bx + c > 0$$
$$ax^2 + bx + c \le 0 \qquad ax^2 + bx + c \ge 0$$

where $a$, $b$, and $c$ are real numbers, with $a \ne 0$.

Examples of Quadratic Inequalities
$$x^2 + x - 12 > 0, \qquad 2x^2 - 9x - 5 \le 0$$

### Solution to a Quadratic Inequality

The **solution to a quadratic inequality** is the set of all values that make the inequality a true statement.

For example, substitute 5 for $x$ in $x^2 + x - 12 > 0$.

$$x^2 + x - 12 > 0$$
$$5^2 + 5 - 12 \overset{?}{>} 0$$
$$18 > 0 \quad \text{True}$$

The inequality is true when $x$ is 5, so 5 is a solution to (or satisfies) the inequality. However, 5 is not the only solution. There are other values that are solutions to the inequality.

---

*An inequality of the form $ax^2 + bx + c \ne 0$ is also a quadratic inequality that we will not be discussing in this book.

## 1 Solve Quadratic Inequalities

We will begin our study of quadratic inequalities by studying the graphs of quadratic functions. For example, consider once again the inequality $x^2 + x - 12 > 0$. **Figure 12.28a** shows the graph of the function $f(x) = x^2 + x - 12$. Notice that when $x < -4$ or when $x > 3$, $f(x) > 0$ (shown in red in **Figure 12.28b**). Thus, when $x < -4$ or when $x > 3$, $x^2 + x - 12 > 0$.

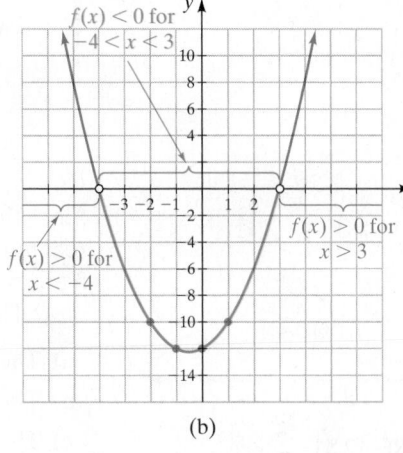

**FIGURE 12.28**                    (a)                              (b)

Notice also that when $-4 < x < 3$, $f(x) < 0$ (shown in green in **Figure 12.28b**). Thus, when $-4 < x < 3$, $x^2 + x - 12 < 0$.

Although the method we just described to solve quadratic inequalities will work for any quadratic inequality, for many inequalities it may be inconvenient or take too much time to graph the function. In our next example, we will outline a more efficient procedure to find the solution to a quadratic inequality.

EXAMPLE 1   Solve the inequality $x^2 + x - 12 > 0$. Give the solution **a)** on a number line, **b)** in interval notation, and **c)** in set builder notation.

*Solution*     First, set the inequality equal to 0 and solve the equation.

$$x^2 + x - 12 = 0$$

$$(x + 4)(x - 3) = 0$$

$$x + 4 = 0 \quad \text{or} \quad x - 3 = 0$$

$$x = -4 \qquad\qquad x = 3$$

The values $x = -4$ and $x = 3$ are called **boundary values** since they will be on the boundaries of the intervals that make up the solution to the inequality. The boundary values are used to break up a number line into intervals. Whenever the original inequality symbol is $<$ or $>$, we will indicate the boundary values on the number line using open circles, $\circ$. This indicates the boundary values are not part of the solution. Whenever the original inequality symbol is $\leq$ or $\geq$, we will indicate the boundary values on the number line using closed circles, $\bullet$. This indicates the boundary values are part of the solution.

$$
\begin{array}{ccccc}
A & | & B & | & C \\
(-\infty, -4) & | & (-4, 3) & | & (3, \infty)
\end{array}
$$

$-4 \qquad 3$

boundary values indicated
with open circles

**FIGURE 12.29**

In **Figure 12.29**, we have labeled the intervals $A$, $B$, and $C$. Next, we select one test value in *each* interval. Then we substitute each of those numbers, one at a time, into the *inequality* $x^2 + x - 12 > 0$. If the test value results in a true statement, all

**Understanding Algebra**

While solving quadratic inequalities, use the following when marking your boundary values on the number line:

• If the inequality symbol is $<$ or $>$, use an open circle, $\circ$.

• If the inequality symbol is $\leq$ or $\geq$, use a closed circle, $\bullet$.

$x = -5, x = 0, x = 4$
are test values

**FIGURE 12.30**

values in that interval will also satisfy the inequality. If the test value results in a false statement, no numbers in that interval will satisfy the inequality.

In this example, we will use the test values of $-5$ in interval $A$, 0 in interval $B$, and 4 in interval $C$ (see **Fig. 12.30**).

| Interval $A$ | Interval $B$ | Interval $C$ |
|:---:|:---:|:---:|
| $(-\infty, -4)$ | $(-4, 3)$ | $(3, \infty)$ |
| Test value, $-5$ | Test value, 0 | Test value, 4 |
| Is $x^2 + x - 12 > 0$? | Is $x^2 + x - 12 > 0$? | Is $x^2 + x - 12 > 0$? |
| $(-5)^2 - 5 - 12 \overset{?}{>} 0$ | $0^2 + 0 - 12 \overset{?}{>} 0$ | $4^2 + 4 - 12 \overset{?}{>} 0$ |
| $8 > 0$ | $-12 > 0$ | $8 > 0$ |
| True | False | True |

Since the test values in both intervals $A$ and $C$ satisfy the inequality, the solution is all real numbers in intervals $A$ or $C$. Since the inequality symbol is $>$, the values $-4$ and 3 are not included in the solution because they make the inequality equal to 0.

The answers to parts **a)**, **b)**, and **c)** follow.

**FIGURE 12.31**

**a)** The solution is illustrated on a number line in **Figure 12.31**.

**b)** The solution in interval notation is $(-\infty, -4) \cup (3, \infty)$.

**c)** The solution in set builder notation is $\{x \mid x < -4 \text{ or } x > 3\}$.

Note that the solution, in any form, is consistent with the red portion of the graph in **Figure 12.28b** on page 751.

Now Try Exercise 15

---

**To Solve Quadratic and Other Inequalities**

1. Write the inequality as an equation and solve the equation. The solutions are the boundary values.

2. Construct a number line and mark each boundary value from step 1 as follows:
   • If the inequality symbol is $<$ or $>$, use an open circle, $\circ$.
   • If the inequality symbol is $\leq$ or $\geq$, use a closed circle, $\bullet$.

3. If solving a rational inequality, determine the values that make any denominator 0. These values are also boundary values. Indicate these boundary values on your number line with an open circle, $\circ$.

4. Select a test value in each interval and determine whether it satisfies the inequality.

5. The solution is the set of points that satisfy the inequality.

6. Write the solution in the form instructed.

---

**EXAMPLE 2** Solve the inequality $x^2 - 4x \geq -4$. Give the solution **a)** on a number line, **b)** in interval notation, and **c)** in set builder notation.

*Solution*    Write the inequality as an equation, then solve the equation.

$$x^2 - 4x = -4$$
$$x^2 - 4x + 4 = 0$$
$$(x - 2)(x - 2) = 0$$
$$x - 2 = 0 \quad \text{or} \quad x - 2 = 0$$
$$x = 2 \qquad\qquad x = 2$$

**FIGURE 12.32**

Since both factors are the same there is only one boundary value, 2, indicated in **Figure 12.32** with a closed circle. Both test values, 1 and 3, result in true statements.

| Interval $A$ | Interval $B$ |
|---|---|
| $(-\infty, 2]$ | $[2, \infty)$ |
| Test value, 1 | Test value, 3 |
| $x^2 - 4x \geq -4$ | $x^2 - 4x \geq -4$ |
| $1^2 - 4(1) \overset{?}{\geq} -4$ | $3^2 - 4(3) \overset{?}{\geq} -4$ |
| $1 - 4 \overset{?}{\geq} -4$ | $9 - 12 \overset{?}{\geq} -4$ |
| $-3 \geq -4$   True | $-3 \geq -4$   True |

The solution set includes both intervals and the boundary value, 2. The solution set is the set of real numbers, $\mathbb{R}$. The answers to parts **a)**, **b)**, and **c)** follow.

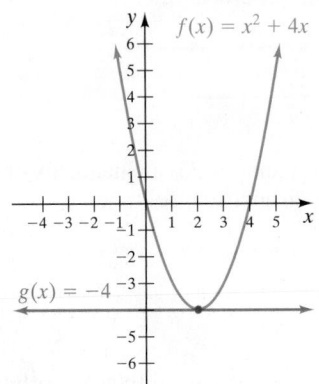

$f(x) = x^2 + 4x$

$g(x) = -4$

**FIGURE 12.33**

**a)** ⟷ (number line marked at 2)   **b)** $(-\infty, \infty)$   **c)** $\{x \mid -\infty < x < \infty\}$

Now Try Exercise 11

We can check the solution to Example 2 using graphing. Let $f(x) = x^2 - 4x$ and $g(x) = -4$. For $x^2 - 4x \geq -4$ to be true, we want $f(x) \geq g(x)$. The graphs of $f(x)$ and $g(x)$ are given in **Figure 12.33**.

Observe that $f(x) = g(x)$ at $x = 2$ and $f(x) > g(x)$ for all other values of $x$. Thus, $f(x) \geq g(x)$ for all values of $x$, and the solution set is the set of real numbers.

In Example 2, if we rewrite the inequality $x^2 - 4x \geq -4$ as $x^2 - 4x + 4 \geq 0$ and then as $(x - 2)^2 \geq 0$ we can see that the solution must be the set of real numbers, since $(x - 2)^2$ must be greater than or equal to 0 for any real number $x$.

**EXAMPLE 3** Solve the inequality $x^2 - 2x - 4 \leq 0$. Express the solution in interval notation.

*Solution*   First we need to solve the equation $x^2 - 2x - 4 = 0$. Since this equation is not factorable, we use the quadratic formula to solve. Here $a = 1, b = -2$, and $c = -4$.

$$x = \frac{-b \pm \sqrt{b^2 - 4ac}}{2a}$$

$$= \frac{-(-2) \pm \sqrt{(-2)^2 - 4(1)(-4)}}{2(1)} = \frac{2 \pm \sqrt{20}}{2} = \frac{2 \pm 2\sqrt{5}}{2} = 1 \pm \sqrt{5}$$

| $A$ | $B$ | $C$ |
|---|---|---|
| $-2$ | $0$ | $4$ |
| $1 - \sqrt{5}$ | | $1 + \sqrt{5}$ |

**FIGURE 12.34**

The boundary values are $1 - \sqrt{5}$ and $1 + \sqrt{5}$. The value of $1 - \sqrt{5}$ is about $-1.24$ and the value of $1 + \sqrt{5}$ is about $3.24$. We indicate the boundary values on the number line with closed circles and we will select test values of $-2$, 0, and 4 (see **Fig. 12.34**).

| Interval $A$ | Interval $B$ | Interval $C$ |
|---|---|---|
| $(-\infty, 1 - \sqrt{5}]$ | $[1 - \sqrt{5}, 1 + \sqrt{5}]$ | $[1 + \sqrt{5}, \infty)$ |
| Test value, $-2$ | Test value, 0 | Test value, 4 |
| $x^2 - 2x - 4 \leq 0$ | $x^2 - 2x - 4 \leq 0$ | $x^2 - 2x - 4 \leq 0$ |
| $(-2)^2 - 2(-2) - 4 \overset{?}{\leq} 0$ | $0^2 - 2(0) - 4 \overset{?}{\leq} 0$ | $4^2 - 2(4) - 4 \overset{?}{\leq} 0$ |
| $4 + 4 - 4 \overset{?}{\leq} 0$ | $0 - 0 - 4 \overset{?}{\leq} 0$ | $16 - 8 - 4 \overset{?}{\leq} 0$ |
| $4 \leq 0$ | $-4 \leq 0$ | $4 \leq 0$ |
| False | True | False |

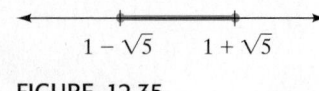

FIGURE 12.35

The boundary values are part of the solution because the inequality symbol is $\leq$ and the boundary values make the inequality equal to 0. Thus, the solution in interval notation is $[1 - \sqrt{5}, 1 + \sqrt{5}]$. The solution is illustrated on the number line in **Figure 12.35**.

*Now Try Exercise 19*

> **Helpful Hint**
>
> If $ax^2 + bx + c = 0$, with $a > 0$, has two distinct real solutions, then:
>
> | Inequality of Form | Solution Is | Solution on Number Line |
> |---|---|---|
> | $ax^2 + bx + c \geq 0$ | End intervals | ← ——+    +—— → |
> | $ax^2 + bx + c \leq 0$ | Center interval | ← ——+——+—— → |
>
> Example 1 is an inequality of the form $ax^2 + bx + c > 0$, and Example 3 is an inequality of the form $ax^2 + bx + c \leq 0$. Example 2 does not have two distinct real solutions, so this Helpful Hint does not apply.

## 2  Solve Other Polynomial Inequalities

The same procedure we used to solve quadratic inequalities can be used to solve other **polynomial inequalities**, as illustrated in the following examples.

**EXAMPLE 4** Solve the polynomial inequality $(3x - 2)(x + 3)(x + 5) < 0$. Illustrate the solution on a number line and write the solution in both interval notation and set builder notation.

*Solution* We use the zero-factor property to solve the equation $(3x - 2)(x + 3)(x + 5) = 0$.

$$3x - 2 = 0 \quad \text{or} \quad x + 3 = 0 \quad \text{or} \quad x + 5 = 0$$
$$x = \frac{2}{3} \qquad\qquad x = -3 \qquad\qquad x = -5$$

The solutions $-5, -3$, and $\frac{2}{3}$ are indicated with open circles and break the number line into four intervals (see **Fig. 12.36**). The test values we will use are $-6, -4, 0$, and 1. We show the results in the following table.

FIGURE 12.36

| Interval | Test value | $(3x - 2)(x + 3)(x + 5)$ | $< 0$ |
|---|---|---|---|
| $A: (-\infty, -5)$ | $-6$ | $-60$ | True |
| $B: (-5, -3)$ | $-4$ | $14$ | False |
| $C: \left(-3, \dfrac{2}{3}\right)$ | $0$ | $-30$ | True |
| $D: \left(\dfrac{2}{3}, \infty\right)$ | $1$ | $24$ | False |

FIGURE 12.37

Since the original inequality symbol is $<$, the boundary values are not part of the solution. The solution, intervals $A$ and $C$, is illustrated on the number line in **Figure 12.37**. The solution in interval notation is $(-\infty, -5) \cup \left(-3, \frac{2}{3}\right)$. The solution in set builder notation is $\left\{ x \mid x < -5 \text{ or } -3 < x < \frac{2}{3} \right\}$.

*Now Try Exercise 27*

**EXAMPLE 5** Given $f(x) = 3x^3 - 3x^2 - 6x$, find all values of $x$ for which $f(x) \geq 0$. Illustrate the solution on a number line and give the solution in interval notation.

*Solution*   We need to solve the inequality

$$3x^3 - 3x^2 - 6x \geq 0$$

We start by solving the equation $3x^3 - 3x^2 - 6x = 0$.

$$3x(x^2 - x - 2) = 0$$
$$3x(x - 2)(x + 1) = 0$$
$$3x = 0 \quad \text{or} \quad x - 2 = 0 \quad \text{or} \quad x + 1 = 0$$
$$x = 0 \qquad\qquad x = 2 \qquad\qquad x = -1$$

The solutions $-1, 0$, and $2$ are indicated with closed circles and break the number line into four intervals (see **Fig. 12.38**). The test values that we will use are $-2$, $-\dfrac{1}{2}$, $1$, and $3$.

| Interval | Test Value | $3x^3 - 3x^2 - 6x$ | $\geq 0$ |
|---|---|---|---|
| $A: (-\infty, -1]$ | $-2$ | $-24$ | False |
| $B: [-1, 0]$ | $-\dfrac{1}{2}$ | $\dfrac{15}{8}$ | True |
| $C: [0, 2]$ | $1$ | $-6$ | False |
| $D: [2, \infty)$ | $3$ | $36$ | True |

Since the original inequality is $\geq$, the boundary values are part of the solution. The solution, intervals $B$ and $D$, is illustrated on the number line in **Figure 12.39a**. The solution in interval notation is $[-1, 0] \cup [2, \infty)$. **Figure 12.39b** shows the graph of $f(x) = 3x^3 - 3x^2 - 6x$. Notice $f(x) \geq 0$ for $-1 \leq x \leq 0$ and for $x \geq 2$, which agrees with our solution.

FIGURE 12.38

(On the left margin, above Figure 12.38:)

A  B  C  D

$-2 \quad -\dfrac{1}{2} \quad 1 \quad 3$

$-1 \quad 0 \quad 2$

---

## Understanding Algebra

Consider the inequality

$$-3x^3 + 3x^2 + 6x \leq 0.$$

It is generally easier to solve a polynomial inequality with a positive leading coefficient. We can change the leading coefficient to a positive number by multiplying both sides of the inequality by $-1$. When doing this, we must change the direction of the inequality symbol.

$$-3x^3 + 3x^2 + 6x \leq 0$$
$$-1(-3x^3 + 3x^2 + 6x) \geq -1(0)$$
$$3x^3 - 3x^2 - 6x \geq 0$$

This inequality was solved in Example 5.

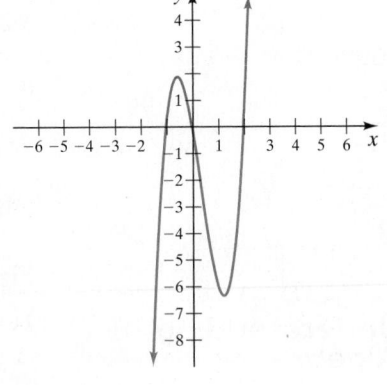

(a)

(b)

FIGURE 12.39

Now Try Exercise 41

---

**3  Solve Rational Inequalities**

In Examples 6 and 7, we solve **rational inequalities**, which are inequalities that contain at least one rational expression.

**EXAMPLE 6** Solve the inequality $\dfrac{x-1}{x+3} \geq 2$ and graph the solution on a number line.

**Solution**   Change the $\geq$ to $=$ and solve the resulting equation.

$$\frac{x-1}{x+3} = 2$$

$$\cancel{x+3} \cdot \frac{x-1}{\cancel{x+3}} = 2(x+3) \qquad \text{Multiply both sides by } x+3.$$

$$x - 1 = 2x + 6$$

$$-1 = x + 6$$

$$-7 = x$$

The solution $-7$ is a boundary value and is indicated with a closed circle on the number line (see **Fig. 12.40**).

When solving rational inequalities, we also need to determine the value or values that make any denominator 0. We set the denominator equal to 0 and solve.

$$x + 3 = 0$$

$$x = -3$$

Since $-3$ cannot be a solution, we indicate it on the number line with an open circle (see **Fig. 12.40**).

We use the solution to the equation, $-7$, and the value that makes the denominator 0, $-3$, to determine the intervals, shown in **Figure 12.40**. We will use $-8$, $-5$, and 0 as our test values.

FIGURE 12.40

| Interval $A$ | Interval $B$ | Interval $C$ |
|:---:|:---:|:---:|
| $(-\infty, -7]$ | $[-7, -3)$ | $(-3, \infty)$ |
| Test value, $-8$ | Test value, $-5$ | Test value, 0 |
| $\dfrac{x-1}{x+3} \geq 2$ | $\dfrac{x-1}{x+3} \geq 2$ | $\dfrac{x-1}{x+3} \geq 2$ |
| $\dfrac{-8-1}{-8+3} \overset{?}{\geq} 2$ | $\dfrac{-5-1}{-5+3} \overset{?}{\geq} 2$ | $\dfrac{0-1}{0+3} \overset{?}{\geq} 2$ |
| $\dfrac{9}{5} \geq 2$  False | $3 \geq 2$  True | $-\dfrac{1}{3} \geq 2$  False |

Now check the boundary values $-7$ and $-3$. Since $-7$ results in the inequality $-2 \geq -2$, which is true, $-7$ is a solution. Since division by 0 is not permitted, $-3$ is not a solution. The solution is illustrated on the number line in **Figure 12.41**, and the solution is $[-7, -3)$.

FIGURE 12.41

Now Try Exercise 81

In Example 6, we solved $\dfrac{x-1}{x+3} \geq 2$. Suppose we graphed $f(x) = \dfrac{x-1}{x+3}$.

**Figure 12.42** shows the graph of $f(x) = \dfrac{x-1}{x+3}$ and the graph of $y = 2$. Notice $f(x) \geq 2$ when $-7 \leq x < -3$.

## Understanding Algebra

When solving a rational inequality, the values at which any denominator is equal to 0 are boundary values. However, since a value that makes the denominator 0 can never be a solution, we will *always* indicate these values with an open circle. This shows they are not included in the solution set.

FIGURE 12.42

**EXAMPLE 7** Solve the inequality $\dfrac{(x-3)(x+4)}{x+1} \geq 0$. Graph the solution on a number line and give the solution in interval notation.

*Solution*   The solutions to the equation $\dfrac{(x-3)(x+4)}{x+1} = 0$ are 3 and $-4$. We indicate the solutions on the number line with closed circles (see **Fig. 12.43**) since the inequality symbol is $\geq$. The inequality is not defined at $-1$. We indicate this value on the number line with an open circle (see **Fig. 12.43**) since $-1$ cannot be a solution to the inequality.

We therefore use the values $-4$, $-1$, and 3 to determine the intervals on the number line (see **Fig. 12.43**). Checking test values at $-5$, $-2$, 0, and 4, we find that the values in intervals $B$ and $D$, $-4 \leq x < -1$ and $x \geq 3$, satisfy the inequality. Check the test values yourself to verify this. The values 3 and $-4$ make the inequality equal to 0 and are part of the solution. The inequality is not defined at $-1$, so $-1$ is not part of the solution. The solution is $[-4, -1) \cup [3, \infty)$. The solution is illustrated on the number line in **Figure 12.44**.

Now Try Exercise 71

FIGURE 12.43

FIGURE 12.44

# EXERCISE SET 12.6    MathXL®   MyMathLab®
MathXL®   MyMathLab

## Warm-Up Exercises

*Fill in the blanks with the appropriate word, phrase, or symbol(s) from the following list.*

not included        boundary values        closed        included        solution        open        vertices        divisor

**1.** The _____ to an inequality is the set of all values that make the inequality a true statement.

**2.** Consider the inequality $x^2 - 3x - 4 \leq 0$. The solutions to the equation $x^2 - 3x - 4 = 0$ are called _____.

**3.** In solving the inequality $(x-5)(x+3) \geq 0$ the boundary values 5 and $-3$ are _____ in the solution set.

**4.** In solving the inequality $(x-2)(x+4) < 0$ the boundary values 2 and $-4$ are _____ in the solution set.

**5.** In solving the inequality $\dfrac{1}{x-3} \leq 0$ the boundary value of 3 is indicated on the number line with a(n) _____ circle.

**6.** In solving the inequality $\dfrac{x+1}{x-3} \leq 0$ the boundary value of $-1$ is indicated on the number line with a(n) _____ circle.

## Practice the Skills

*Solve each quadratic inequality and graph the solution on a number line.*

**7.** $x^2 - 2x - 3 \geq 0$

**8.** $x^2 - 2x - 3 < 0$

**9.** $x^2 + 7x + 6 > 0$

**10.** $x^2 + 8x + 7 < 0$

**11.** $n^2 - 6n + 9 \geq 0$

**12.** $x^2 - 8x \geq 0$

**13.** $x^2 - 16 < 0$

**14.** $r^2 - 5r < 0$

**15.** $2x^2 + 5x - 3 \geq 0$

**16.** $3n^2 - 7n \leq 6$

**17.** $5x^2 + 6x \leq 8$

**18.** $3x^2 + 5x - 3 \leq 0$

**19.** $2x^2 - 12x + 9 \leq 0$

**20.** $5x^2 \leq -20x - 4$

*Solve each polynomial inequality and give the solution in interval notation.*

**21.** $(x-1)(x+2)(x-3) \leq 0$

**22.** $(x-2)(x+2)(x+5) \leq 0$

**23.** $(a-3)(a+2)(a+4) < 0$

**24.** $(r-1)(r+2)(r+7) < 0$

**25.** $(2c + 5)(3c - 6)(c + 6) > 0$

**26.** $(a - 4)(a - 2)(a + 8) > 0$

**27.** $(3x + 5)(x - 3)(x + 1) > 0$

**28.** $(3c - 1)(c + 4)(3c + 6) \leq 0$

**29.** $(x + 2)(x + 2)(3x - 8) \geq 0$

**30.** $(x + 3)^2(4x - 7) \leq 0$

**31.** $x^3 - 6x^2 + 9x < 0$

**32.** $x^3 + 3x^2 - 40x > 0$

*For each function provided, find all values of x for which $f(x)$ satisfies the indicated conditions. Graph the solution on a number line.*

**33.** $f(x) = x^2 - 2x, f(x) \geq 0$

**34.** $f(x) = x^2 - 7x, f(x) > 0$

**35.** $f(x) = x^2 + 4x, f(x) > 0$

**36.** $f(x) = x^2 + 8x, f(x) \leq 0$

**37.** $f(x) = x^2 - 14x + 48, f(x) < 0$

**38.** $f(x) = x^2 - 2x - 15, f(x) < 0$

**39.** $f(x) = 2x^2 + 9x - 1, f(x) \leq 5$

**40.** $f(x) = x^2 + 5x - 3, f(x) \leq 4$

**41.** $f(x) = 2x^3 + 9x^2 - 35x, f(x) \geq 0$

**42.** $f(x) = x^3 - 9x, f(x) \leq 0$

*Solve each rational inequality and give the solution in set builder notation.*

**43.** $\dfrac{x + 1}{x - 3} \leq 0$

**44.** $\dfrac{x + 2}{x - 4} \geq 0$

**45.** $\dfrac{x - 1}{x + 5} < 0$

**46.** $\dfrac{x - 1}{x + 5} \leq 0$

**47.** $\dfrac{x + 3}{x - 2} \geq 0$

**48.** $\dfrac{x - 4}{x + 6} > 0$

**49.** $\dfrac{a - 9}{a + 5} < 0$

**50.** $\dfrac{b + 7}{b + 1} \leq 0$

**51.** $\dfrac{c - 10}{c - 4} > 0$

**52.** $\dfrac{2d - 6}{d - 1} < 0$

**53.** $\dfrac{3y + 6}{y + 4} \leq 0$

**54.** $\dfrac{4z - 8}{z - 9} \geq 0$

**55.** $\dfrac{5a + 10}{3a - 1} \geq 0$

**56.** $\dfrac{x + 4}{x - 4} \leq 0$

**57.** $\dfrac{3x + 4}{2x - 1} < 0$

**58.** $\dfrac{k + 3}{k} \geq 0$

**59.** $\dfrac{3x + 8}{x - 2} \leq 0$

**60.** $\dfrac{4x - 2}{2x - 8} > 0$

*Solve each rational inequality and give the solution in interval notation.*

**61.** $\dfrac{(x - 2)(x - 4)}{x + 1} < 0$

**62.** $\dfrac{(x + 1)(x - 6)}{x + 3} \leq 0$

**63.** $\dfrac{(x - 2)(x + 3)}{x - 5} > 0$

**64.** $\dfrac{(x - 2)(x + 3)}{x - 5} \geq 0$

**65.** $\dfrac{(a - 1)(a - 7)}{a + 2} \geq 0$

**66.** $\dfrac{(b - 2)(b + 4)}{b} < 0$

**67.** $\dfrac{c}{(c - 3)(c + 8)} \leq 0$

**68.** $\dfrac{z - 5}{(z + 6)(z - 9)} \geq 0$

**69.** $\dfrac{x - 6}{(x + 4)(x - 1)} \leq 0$

**70.** $\dfrac{x + 9}{(x - 2)(x + 4)} > 0$

**71.** $\dfrac{(x - 3)(2x + 5)}{x - 4} \geq 0$

**72.** $\dfrac{r(r - 8)}{2r + 6} < 0$

*Solve each rational inequality and graph the solution on a number line.*

**73.** $\dfrac{2}{x - 4} \geq 1$

**74.** $\dfrac{2}{x - 4} > 1$

**75.** $\dfrac{3}{x - 1} > -1$

**76.** $\dfrac{3}{x + 1} \geq -1$

**77.** $\dfrac{5}{x + 2} \leq 1$

**78.** $\dfrac{5}{x + 2} < 1$

**79.** $\dfrac{2p-5}{p-4} \leq 1$

**80.** $\dfrac{2}{2a-1} > 2$

**81.** $\dfrac{4}{x+2} \geq 2$

**82.** $\dfrac{x+6}{x+2} > 1$

**83.** $\dfrac{w}{3w-2} > -2$

**84.** $\dfrac{x-1}{2x+6} \leq -3$

## Concept/Writing Exercises

**85.** The graph of $f(x) = x^2 - 7x + 10$ is given. Find the solution to **a)** $f(x) > 0$ and **b)** $f(x) < 0$.

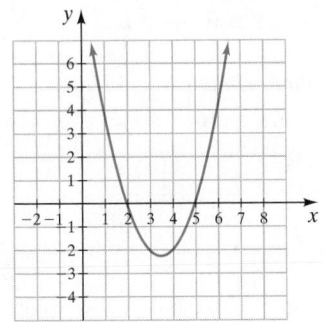

**86.** The graph of $f(x) = -x^2 - 4x + 5$ is given. Find the solution to **a)** $f(x) \geq 0$ and **b)** $f(x) \leq 0$.

**87.** The graph of $y = \dfrac{x^2 - 4x + 4}{x - 4}$ is illustrated. Determine the solution to the following inequalities.

**a)** $\dfrac{x^2 - 4x + 4}{x - 4} > 0$

**b)** $\dfrac{x^2 - 4x + 4}{x - 4} < 0$

Explain how you determined your answer.

**88.** The graph of $y = \dfrac{x^2 + x - 6}{x - 4}$ is illustrated. Determine the solution to the following inequalities.

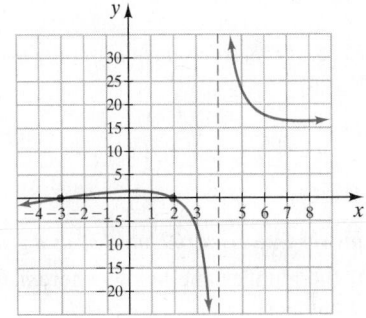

**a)** $\dfrac{x^2 + x - 6}{x - 4} \geq 0$

**b)** $\dfrac{x^2 + x - 6}{x - 4} < 0$

Explain how you determined your answer.

**89.** Write a quadratic inequality whose solution is

**90.** Write a quadratic inequality whose solution is

**91.** Write a rational inequality whose solution is

**92.** Write a rational inequality whose solution is

**93.** What is the solution to the inequality $(x + 3)^2(x - 1)^2 \geq 0$? Explain your answer.

**94.** What is the solution to the inequality $x^2(x - 3)^2 (x + 4)^2 < 0$? Explain your answer.

**95.** What is the solution to the inequality $\dfrac{x^2}{(x + 2)^2} \geq 0$? Explain your answer.

**96.** What is the solution to the inequality $\dfrac{x^2}{(x - 3)^2} > 0$? Explain your answer.

**97.** If $f(x) = ax^2 + bx + c$ where $a > 0$ and the discriminant is negative, what is the solution to $f(x) < 0$? Explain.

**98.** If $f(x) = ax^2 + bx + c$ where $a < 0$ and the discriminant is negative, what is the solution to $f(x) > 2$? Explain.

## Challenge Problems

*Solve each inequality and graph the solution on the number line.*

**99.** $(x + 1)(x - 3)(x + 5)(x + 8) \geq 0$

**100.** $\dfrac{(x - 4)(x + 2)}{x(x + 9)} \geq 0$

*Write a quadratic inequality with the following solutions. Many answers are possible. Explain how you determined your answers.*

**101.** $(-\infty, 0) \cup (3, \infty)$     **102.** $\{2\}$     **103.** $\varnothing$     **104.** $\mathbb{R}$

*In Exercises 105 and 106, solve each inequality and give the solution in interval notation. Use techniques from Section 12.5 to help you find the solution.*

**105.** $x^4 - 10x^2 + 9 > 0$     **106.** $x^4 - 26x^2 + 25 \leq 0$

*In Exercises 107 and 108, solve each inequality using factoring by grouping. Give the solution in interval notation.*

**107.** $x^3 + x^2 - 4x - 4 \geq 0$     **108.** $2x^3 + x^2 - 32x - 16 < 0$

## Group Activity

*Discuss and answer Exercises 109 and 110 as a group.*

**109.** Consider the number line below, where $a, b$, and $c$ are distinct real numbers.

  **a)** In which intervals will the real numbers satisfy the inequality $(x - a)(x - b)(x - c) > 0$? Explain.

  **b)** In which intervals will the real numbers satisfy the inequality $(x - a)(x - b)(x - c) < 0$? Explain.

Interval  Interval  Interval  Interval
  1      2      3      4
       $a$      $b$      $c$

**110.** Consider the number line below where $a, b, c$, and $d$ are distinct real numbers.

Interval  Interval  Interval  Interval  Interval
  1      2      3      4      5
       $a$      $b$      $c$      $d$

  **a)** In which intervals do the real numbers satisfy the inequality $(x - a)(x - b)(x - c)(x - d) > 0$? Explain.

  **b)** In which interval do the real numbers satisfy the inequality $(x - a)(x - b)(x - c)(x - d) < 0$? Explain.

## Cumulative Review Exercises

[3.4]  **111. Antifreeze** How many quarts of a 100% antifreeze solution should Paul Simmons add to 10 quarts of a 20% antifreeze solution to make a 50% antifreeze solution?

[5.4]  **112.** Add $(6r + 5s - t) + (-3r - 2s - 8t)$.

[7.5]  **113.** Simplify $\dfrac{1 + \dfrac{x}{x + 1}}{\dfrac{2x + 1}{x - 3}}$.

[8.2]  **114.** If $h(x) = \dfrac{x^2 + 4x}{x + 9}$, find $h(-3)$.

[11.7]  **115.** Multiply $(3 - 4i)(6 + 5i)$.

# Chapter 12 Summary

| IMPORTANT FACTS AND CONCEPTS | EXAMPLES |
|---|---|

### Section 12.1

**Square Root Property**

If $x^2 = a$, where $a$ is a real number, then $x = \pm\sqrt{a}$.

Solve $x^2 - 36 = 0$.

$$x^2 - 36 = 0$$
$$x^2 = 36$$
$$x = \pm\sqrt{36} = \pm 6$$

The solutions are $-6$ and $6$.

---

A **perfect square trinomial** is a trinomial that can be expressed as the square of a binomial.

$$x^2 - 10x + 25 = (x - 5)^2$$

---

**To Solve a Quadratic Equation by Completing the Square**

1. Use the multiplication (or division) property of equality if necessary to make the leading coefficient 1.
2. Rewrite the equation with the constant by itself on the right side of the equation.
3. Take one-half the numerical coefficient of the first-degree term, square it, and add this quantity to both sides of the equation.
4. Factor the trinomial as the square of a binomial.
5. Use the square root property to take the square root of both sides of the equation.
6. Solve for the variable.

Solve $x^2 + 4x - 12 = 0$ by completing the square.

$$x^2 + 4x - 12 = 0$$
$$x^2 + 4x = 12$$
$$x^2 + 4x + 4 = 12 + 4$$
$$(x + 2)^2 = 16$$
$$x + 2 = \pm\sqrt{16}$$
$$x + 2 = \pm 4$$
$$x = -2 \pm 4$$
$$x = -2 - 4 = -6 \quad \text{or} \quad x = -2 + 4 = 2$$

The solutions are $-6$ and $2$.

### Section 12.2

The **standard form of a quadratic equation** is $ax^2 + bx + c = 0$, $a \neq 0$.

$$x^2 - 5x + 17 = 0$$

---

**To Solve a Quadratic Equation by the Quadratic Formula**

1. Write the quadratic equation in standard form, $ax^2 + bx + c = 0$, and determine the numerical values for $a, b,$ and $c$.
2. Substitute the values for $a, b,$ and $c$ into the quadratic formula and then evaluate the formula to obtain the solution.

The Quadratic Formula
$$x = \frac{-b \pm \sqrt{b^2 - 4ac}}{2a}$$

Solve $x^2 - 2x - 15 = 0$ by the quadratic formula.

$$a = 1, \quad b = -2, \quad c = -15$$
$$x = \frac{-b \pm \sqrt{b^2 - 4ac}}{2a}$$
$$= \frac{-(-2) \pm \sqrt{(-2)^2 - 4(1)(-15)}}{2(1)}$$
$$= \frac{2 \pm \sqrt{64}}{2}$$
$$= \frac{2 \pm 8}{2}$$

$$x = \frac{2 + 8}{2} = \frac{10}{2} = 5 \quad \text{or} \quad x = \frac{2 - 8}{2} = \frac{-6}{2} = -3$$

The solutions are $5$ and $-3$.

---

**Solutions of a Quadratic Equation**

For a quadratic equation of the form $ax^2 + bx + c = 0, a \neq 0$, the **discriminant** is $b^2 - 4ac$.

If $b^2 - 4ac > 0$, the quadratic equation has two distinct real number solutions.

If $b^2 - 4ac = 0$, the quadratic equation has a single real number solution.

If $b^2 - 4ac < 0$, the quadratic equation has no real number solution.

Determine the number of solutions of $3x^2 - x + 7 = 0$.

$$a = 3, b = -1, c = 7$$
$$b^2 - 4ac = (-1)^2 - 4(3)(7)$$
$$= 1 - 84$$
$$= -83$$

Since the discriminant is negative, the equation has no real number solution.

| IMPORTANT FACTS AND CONCEPTS | EXAMPLES |
|---|---|

### Section 12.3

**Solve Applications That Involve Quadratic Equations**

The function $N(t) = 0.0054t^2 - 1.46t + 95.11$ can be used to estimate the average number of years of life expectancy remaining for a person of age $t$ years where $30 \leq t \leq 100$.

a)  Estimate the remaining life expectancy of a person of age 40.

b)  If a person has a remaining life expectancy of 14.3 years, estimate the age of the person.

a)
$$N(t) = 0.0054t^2 - 1.46t + 95.11$$
$$N(40) = 0.0054(40)^2 - 1.46(40) + 95.11$$
$$= 0.0054(1600) - 58.4 + 95.11$$
$$= 8.64 - 58.4 + 95.11$$
$$= 45.35 \text{ years}$$

b)
$$N(t) = 0.0054t^2 - 1.46t + 95.11$$
$$14.3 = 0.0054t^2 - 1.46t + 95.11$$
$$0 = 0.0054t^2 - 1.46t + 80.81$$
$$a = 0.0054, \quad b = -1.46, \quad c = 80.81$$
$$t = \frac{-b \pm \sqrt{b^2 - 4ac}}{2a}$$
$$= \frac{-(-1.46) \pm \sqrt{(-1.46)^2 - 4(0.0054)(80.81)}}{2(0.0054)}$$
$$= \frac{1.46 \pm \sqrt{2.1316 - 1.745496}}{0.0108}$$
$$= \frac{1.46 \pm \sqrt{0.386104}}{0.0108}$$
$$\approx \frac{1.46 \pm 0.6214}{0.0108}$$
$$t \approx \frac{1.46 + 0.6214}{0.0108} \quad \text{or} \quad t \approx \frac{1.46 - 0.6214}{0.0108}$$
$$\approx \cancel{192.72} \qquad\qquad\qquad \approx 77.65 \text{ years}$$

**Solve for a Variable in a Formula**

The formula for the area of a circle is $A = \pi r^2$. Solve this equation for the radius, $r$.

$$A = \pi r^2$$
$$\frac{A}{\pi} = r^2 \quad \text{Isolate } r^2 \text{ by dividing both sides by } \pi.$$
$$\sqrt{\frac{A}{\pi}} = r \quad \text{Square root property}$$

### Section 12.4

An equation that can be rewritten in the form $au^2 + bu + c = 0$ for $a \neq 0$, where $u$ is an algebraic expression, is called **quadratic in form**.

**To Solve Equations Quadratic in Form**

1.  Make a substitution that will result in an equation of the form $au^2 + bu + c = 0, a \neq 0$, where $u$ is a function of the original variable.

2.  Solve the equation $au^2 + bu + c = 0$ for $u$.

3.  Replace $u$ with the function of the original variable from step 1 and solve the resulting equation for the original variable.

4.  Check for extraneous solutions by substituting the apparent solutions into the original equation.

Solve $x^4 - 17x^2 + 16 = 0$.

$$\text{Let } u = x^2.$$

Then,
$$u^2 - 17u + 16 = 0$$
$$(u - 16)(u - 1) = 0$$
$$u - 16 = 0 \quad \text{or} \quad u - 1 = 0$$
$$u = 16 \qquad\qquad u = 1$$
$$x^2 = 16 \qquad\qquad x^2 = 1$$
$$x = \pm 4 \qquad\qquad x = \pm 1$$

A check will show the solutions are $4, -4, 1,$ and $-1$.

| IMPORTANT FACTS AND CONCEPTS | EXAMPLES |
|---|---|
| **Section 12.5** | |

**Polynomial and Polynomial Functions**

A **polynomial in $x$** is an expression containing a finite sum of terms of the form $ax^n$ for any real number $a$ and any whole number $n$. A **polynomial function** is a function in which the expression used to describe the function is a polynomial.

$$p(x) = x + 3$$
$$p(x) = 2x^2 - 5x + 6$$
$$p(x) = 5x^3 - \frac{1}{2}x^2 + 6x - 4$$

---

Polynomial functions have graphs that are smooth continuous curves.

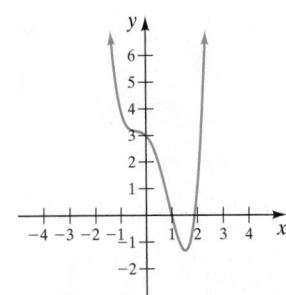

---

Polynomial functions with a positive leading coefficient will increase to the right of some value of $x$.

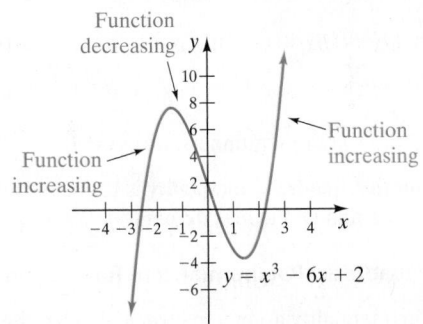

---

Polynomial functions with a negative leading coefficient will decrease to the right of some value of $x$.

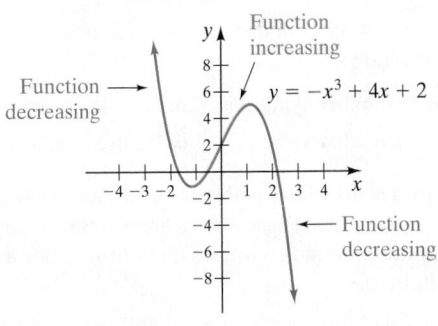

---

**Quadratic Functions**

Functions of the form $f(x) = ax^2 + bx + c$ are called **quadratic functions** and have graphs that are parabolas.

a) The parabola opens upward when $a > 0$ and downward when $a < 0$.

b) The axis of symmetry is the line $x = -\dfrac{b}{2a}$.

c) The vertex is the point $\left( -\dfrac{b}{2a}, \dfrac{4ac - b^2}{4a} \right)$ or $\left[ -\dfrac{b}{2a}, f\left( -\dfrac{b}{2a} \right) \right]$.

d) The $y$-intercept is the point $(0, c)$.

e) To obtain the $x$-intercept(s) (if there are any), solve the equation $ax^2 + bx + c = 0$ for $x$.

The graph of the quadratic function $f(x) = x^2 - 2x - 3$ is a parabola.

a) It opens upward since $a > 0$.

b) The axis of symmetry is $x = -\dfrac{-2}{2(1)} = 1$.

c) The vertex is $(1, -4)$.

d) The $y$-intercept is $(0, -3)$.

e)
$$x^2 - 2x - 3 = 0$$
$$(x - 3)(x + 1) = 0$$
$$x - 3 = 0 \quad \text{or} \quad x + 1 = 0$$
$$x = 3 \qquad\qquad x = -1$$

The $x$-intercepts are $(3, 0)$ and $(-1, 0)$.

| IMPORTANT FACTS AND CONCEPTS | EXAMPLES |
|---|---|

## Section 12.5 (cont.)

The graph of $f(x) = x^2 - 2x - 3$ is shown below.

## Section 12.6

A **quadratic inequality** is an inequality that can be written in one of the following forms

$$ax^2 + bx + c < 0 \qquad ax^2 + bx + c > 0$$
$$ax^2 + bx + c \leq 0 \qquad ax^2 + bx + c \geq 0$$

where $a$, $b$, and $c$ are real numbers, with $a \neq 0$.

The **solution to a quadratic inequality** is the set of all values that make the inequality a true statement.

$$x^2 - 5x + 7 > 0$$

### To Solve Quadratic, Polynomial, and Rational Inequalities

1. Write the inequality as an equation and solve the equation. The solutions are the boundary values.
2. Construct a number line and mark each boundary value from step 1 as follows:
   - If the inequality symbol is $<$ or $>$, use an open circle, ○.
   - If the inequality symbol is $\leq$ or $\geq$, use a closed circle, ●.
3. If solving a rational inequality, determine the values that make any denominator 0. These values are also boundary values. Indicate these boundary values on your number line with an open circle, ○.
4. Select a test value in each interval and determine whether it satisfies the inequality.
5. The solution is the set of points that satisfy the inequality.
6. Write the solution in the form instructed.

Solve $(2x - 1)(x - 3)(x + 1) < 0$.

$$(2x - 1)(x - 3)(x + 1) < 0$$
$$(2x - 1)(x - 3)(x + 1) = 0$$

$$2x - 1 = 0 \quad \text{or} \quad x - 3 = 0 \quad \text{or} \quad x + 1 = 0$$

$$x = \frac{1}{2} \qquad\qquad x = 3 \qquad\qquad x = -1$$

The intervals and test values selected are shown below.

| Interval | Test value | $(2x - 1)(x - 3)(x + 1) < 0$ | |
|---|---|---|---|
| $(-\infty, -1)$ | $-2$ | $-25$ | True |
| $\left(-1, \frac{1}{2}\right)$ | $0$ | $3$ | False |
| $\left(\frac{1}{2}, 3\right)$ | $1$ | $-4$ | True |
| $(3, \infty)$ | $5$ | $108$ | False |

Solution on a number line:

Solution in interval notation: $(-\infty, -1) \cup \left(\frac{1}{2}, 3\right)$

Solution in set builder notation:

$$\left\{ x \,\middle|\, x < -1 \quad \text{or} \quad \frac{1}{2} < x < 3 \right\}$$

## Chapter 12 Review Exercises

**[12.1]** *Use the square root property to solve each equation.*

**1.** $(x - 5)^2 = 16$       **2.** $(2x + 1)^2 = 60$       **3.** $\left(x - \frac{1}{3}\right)^2 = \frac{4}{9}$       **4.** $\left(2x - \frac{1}{2}\right)^2 = 4$

*Solve each equation by completing the square.*

**5.** $x^2 - 8x + 12 = 0$          **6.** $x^2 + 4x - 32 = 0$          **7.** $a^2 + 2a - 9 = 0$

**8.** $z^2 + 6z = 12$          **9.** $x^2 - 2x + 10 = 0$          **10.** $2r^2 - 8r = -64$

**Area**   *In Exercises 11 and 12, the area, A, of each rectangle is given.* **a)** *Write an equation for the area.* **b)** *Solve the equation for x.*

**11.**

$A = 32$   $x + 1$

$x + 5$

**12.**

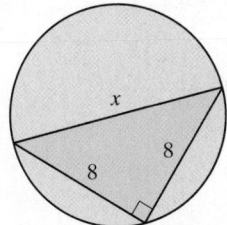

$A = 63$   $x + 2$

$x + 4$

**13. Consecutive Integers** The product of two consecutive positive integers is 56. Find the two integers.

**14. Living Room** Nedal Williams just moved into a new house where the living room is a square whose diagonal is 7 feet longer than the length of a side. Find the dimensions of the living room.

**[12.2]** *Determine whether each equation has two distinct real solutions, a single real solution, or no real solution.*

**15.** $3x^2 - 7x + 1 = 0$          **16.** $3x^2 + 2x = -6$          **17.** $r^2 + 16r = -64$

**18.** $5x^2 - x + 2 = 0$          **19.** $a^2 - 14a = -49$          **20.** $\frac{1}{2}x^2 - 3x = 8$

*Solve each equation by using the quadratic formula.*

**21.** $6x^2 - 5x - 50 = 0$          **22.** $x^2 - 11x = -18$          **23.** $r^2 = 3r + 40$

**24.** $7x^2 = 9x$          **25.** $6a^2 + a - 15 = 0$          **26.** $4x^2 + 11x = 3$

**27.** $x^2 + 8x + 5 = 0$          **28.** $b^2 + 4b = 8$          **29.** $2x^2 + 4x - 3 = 0$

**30.** $3y^2 - 6y = 8$          **31.** $x^2 - x + 13 = 0$          **32.** $x^2 - 2x + 11 = 0$

**33.** $2x^2 - \frac{5}{3}x = \frac{25}{3}$          **34.** $4x^2 + 5x - \frac{3}{2} = 0$

*For the given function, determine all real values of the variable for which the function has the value indicated.*

**35.** $f(x) = x^2 - 4x - 35, f(x) = 25$          **36.** $g(x) = 6x^2 + 5x, g(x) = 6$

**37.** $h(r) = 5r^2 - 7r - 10, h(r) = -8$          **38.** $f(x) = -2x^2 + 6x + 7, f(x) = -2$

*Determine an equation that has the given solutions.*

**39.** $3, -1$          **40.** $\frac{2}{3}, -2$          **41.** $-\sqrt{11}, \sqrt{11}$          **42.** $3 - 2i, 3 + 2i$

**[12.1–12.3]**

**43. Rectangular Garden** Sophia Yang is designing a rectangular flower garden. If the area is to be 96 square feet and the length is to be 4 feet greater than the width, find the dimensions of the garden.

**44. Triangle and Circle** Find the length of side $x$ in the figure.

**45. Savings Account** Samuel Rivera invested $1000 in a savings account where the interest is compounded annually. If after 2 years the amount in the account is $1081.60, find the annual interest rate.

**46. Numbers** The larger of two positive numbers is 4 greater than the smaller. Find the two numbers if their product is 77.

**47. Rectangle** The length of a rectangle is 4 inches less than twice its width. Find the dimensions if its area is 96 square inches.

**48. Wheat Crop** The value, $V$, in dollars per acre of a wheat crop $d$ days after planting is given by the formula $V = 12d - 0.05d^2, 20 < d < 80$. Find the value of an acre of wheat 60 days after it has been planted.

**49. Redwood Growth** The growth, $f(x)$, in inches per year of a redwood tree less than 30 years old is given by the function $f(x) = -0.02x^2 + x + 1$, where $x$ represents the number of inches of rainfall per year.

*Source:* www.humboldtredwoods.org

**a)** Find the growth per year of a redwood tree in a year in which there is 12 inches of rainfall.

**b)** If the growth of a redwood tree in one year is 10 inches, find the number of inches of rainfall for that year (round your answer to 2 decimal places).

**50. Falling Object** The distance, $d$, in feet, that an object is from the ground $t$ seconds after being dropped from an airplane is given by the formula $d = -16t^2 + 784$.

**a)** Find the distance the object is from the ground 2 seconds after it has been dropped.

**b)** When will the object hit the ground?

**51. Oil Leak** A tractor has an oil leak. The amount of oil, $L(t)$, in milliliters per hour that leaks out is a function of the tractor's operating temperature, $t$, in degrees Celsius. The function is

$$L(t) = 0.0004t^2 + 0.16t + 20, 100°C \leq t \leq 160°C$$

**a)** How many milliliters of oil will leak out in 1 hour if the operating temperature of the tractor is 100°C?

**b)** If oil is leaking out at 53 milliliters per hour, what is the operating temperature of the tractor?

© Ilja Masik/Shutterstock

**52. Molding Machines** Two molding machines can complete an order in 12 hours. The larger machine can complete the order

by itself in 1 hour less time than the smaller machine can by itself. How long will it take each machine to complete the order working by itself?

**53. Travel Time** Steve Forrester drove 25 miles at a constant speed, and then he increased his speed by 15 miles per hour for the next 65 miles. If the time required to travel 90 miles was 1.5 hours, find the speed he drove during the first 25 miles.

**54. Canoe Trip** Joan Banker canoed downstream going with the current for 3 miles, then turned around and canoed upstream against the current to her starting point. If the total time she spent canoeing was 4 hours and the current was 0.4 mile per hour, what is the speed she canoes in still water?

**55. Area** The area of a rectangle is 80 square units. If the length is $x$ units and the width is $x - 2$ units, find the length and the width. Round your answer to the nearest tenth of a unit.

$$\boxed{A = 80} \quad x - 2$$

$$x$$

**56. Selling Tables** A business sells $n$ tables, $n \leq 40$, at a price of $(60 - 0.3n)$ dollars per table. How many tables must be sold to have a revenue of $1080?

*In Exercises 57–60, solve each equation for the variable indicated.*

**57.** $a^2 + b^2 = c^2$, for $a$ (Pythagorean Theorem)

**58.** $h = -4.9t^2 + c$, for $t$ (height of an object)

**59.** $v_x^2 + v_y^2 = v^2$, for $v_y$ (vectors)

**60.** $a = \dfrac{v_2^2 - v_1^2}{2d}$, for $v_2$

[12.4] *Factor.*

**61.** $x^4 - x^2 - 20$

**62.** $4x^4 + 4x^2 - 3$

**63.** $(x + 5)^2 + 10(x + 5) + 24$

**64.** $4(2x + 3)^2 - 12(2x + 3) + 5$

*Solve each equation.*

**65.** $x^4 - 13x^2 + 36 = 0$

**66.** $x^4 - 21x^2 + 80 = 0$

**67.** $a^4 = 5a^2 + 24$

**68.** $3y^{-2} + 16y^{-1} = 12$

**69.** $3r + 11\sqrt{r} - 4 = 0$

**70.** $2p^{2/3} - 7p^{1/3} + 6 = 0$

**71.** $6(x - 2)^{-2} = -13(x - 2)^{-1} + 8$

**72.** $10(r + 1) = \dfrac{12}{r + 1} - 7$

*Find all x-intercepts of the given function.*

**73.** $f(x) = x^4 - 82x^2 + 81$

**74.** $f(x) = 30x + 13\sqrt{x} - 10$

**75.** $f(x) = x - 6\sqrt{x} + 12.$

**76.** $f(x) = (x^2 - 6x)^2 - 5(x^2 - 6x) - 24$

[12.5] **a)** *Determine whether the parabola opens upward or downward.* **b)** *Find the y-intercept.* **c)** *Find the vertex.* **d)** *Find the x-intercepts if they exist.* **e)** *Draw the graph.*

**77.** $f(x) = x^2 + 5x$

**78.** $f(x) = x^2 - 2x - 8$

**79.** $g(x) = -x^2 - 2$

**80.** $g(x) = -2x^2 - x + 15$

**81. Selling Tickets** The Hamilton Outdoor Theater estimates that its total income, $I$, in hundreds of dollars, for its production of a play, can be approximated by the formula $I = -x^2 + 22x - 45, 2 \le x \le 20$, where $x$ is the cost of a ticket.

**a)** How much should the theater charge to maximize its income?

**b)** What is the maximum income?

**82. Tossing a Ball** Josh Vincent tosses a ball upward from the top of a 75-foot building. The height, $s(t)$, of the ball at any time $t$ can be determined by the function $s(t) = -16t^2 + 80t + 75$.

**a)** At what time will the ball attain its maximum height?

**b)** What is the maximum height?

*Graph each function.*

**83.** $f(x) = (x - 3)^2$

**84.** $f(x) = -(x + 2)^2 - 3$

**85.** $g(x) = -2(x + 4)^2 - 1$   **86.** $h(x) = \dfrac{1}{2}(x - 1)^2 + 3$

[12.6] *Graph the solution to each inequality on a number line.*

**87.** $x^2 + 4x + 3 \ge 0$

**88.** $x^2 + 3x - 10 \le 0$

**89.** $x^2 \le 11x - 20$

**90.** $3x^2 + 8x > 16$

**91.** $4x^2 - 9 \le 0$

**92.** $6x^2 - 30 > 0$

*Solve each inequality and give the solution in set builder notation.*

**93.** $\dfrac{x + 1}{x - 5} > 0$

**94.** $\dfrac{x - 3}{x + 2} \le 0$

**95.** $\dfrac{2x - 4}{x + 3} \ge 0$

**96.** $\dfrac{3x + 5}{x - 6} < 0$

**97.** $(x + 4)(x + 1)(x - 2) > 0$

**98.** $x(x - 3)(x - 6) \le 0$

*Solve each inequality and give the solution in interval notation.*

**99.** $(3x + 4)(x - 1)(x - 3) \ge 0$

**100.** $2x(x + 2)(x + 4) < 0$

**101.** $\dfrac{x(x - 4)}{x + 2} > 0$

**102.** $\dfrac{(x - 2)(x - 8)}{x + 3} < 0$

**103.** $\dfrac{x - 3}{(x + 2)(x - 7)} \ge 0$

**104.** $\dfrac{x(x - 6)}{x + 3} \le 0$

*Solve each inequality and graph the solution on a number line.*

**105.** $\dfrac{5}{x + 4} \ge -1$

**106.** $\dfrac{2x}{x - 2} \le 1$

**107.** $\dfrac{2x + 3}{3x - 5} < 4$

## Chapter 12 Practice Test

*Chapter Test Prep Videos provide fully worked-out solutions to any of the exercises you want to review. Chapter Test Prep Videos are available via MyMathLab®, or on You[Tube] (search "Angel Elementary and Intermediate Algebra" and click on "Channels")*

*Solve by completing the square.*

**1.** $x^2 + 2x - 15 = 0$

**2.** $a^2 + 7 = 6a$

*Solve by the quadratic formula.*

**3.** $x^2 - 6x - 16 = 0$

**4.** $x^2 - 4x = -11$

*Solve by the method of your choice.*

**5.** $3r^2 + r = 2$

**6.** $p^2 + 4 = -7p$

**7.** Write an equation that has *x*-intercepts 4, $-\dfrac{2}{5}$.

**8.** Solve the formula $K = \dfrac{1}{2}mv^2$ for *v*.

**9. Cost** The cost, *c*, of a house in Duquoin, Illinois, is a function of the number of square feet, *s*, of the house. The cost of the house can be approximated by

$$c(s) = -0.01s^2 + 78s + 22{,}000, \quad 1300 \le s \le 3900$$

   **a)** Estimate the cost of a 1600-square-foot house.

   **b)** How large a house can Clarissa Skocy purchase if she wishes to spend $160,000 on a house?

**10. Trip in a Park** Tom Ficks drove his 4-wheel-drive Jeep from Anchorage, Alaska, to the Chena River State Recreation Park, a distance of 520 miles. Had he averaged 15 miles per hour faster, the trip would have taken 2.4 hours less. Find the average speed that Tom drove.

Chena River State Recreation Park

*Solve.*

**11.** $2x^4 + 15x^2 - 50 = 0$

**12.** $3r^{2/3} + 11r^{1/3} - 42 = 0$

**13.** Find all *x*-intercepts of $f(x) = 16x - 24\sqrt{x} + 9$.

*Graph each function.*

**14.** $f(x) = (x - 3)^2 + 2$

**15.** $h(x) = -\dfrac{1}{2}(x - 2)^2 - 2$

**16.** Determine whether $6x^2 = 2x + 3$ has two distinct real solutions, a single real solution, or no real solution. Explain your answer.

**17.** Consider the quadratic equation $y = x^2 + 2x - 8$.

   **a)** Determine whether the parabola opens upward or downward.

   **b)** Find the *y*-intercept.

   **c)** Find the vertex.

   **d)** Find the *x*-intercepts (if they exist).

   **e)** Draw the graph.

**18.** Write a quadratic equation whose *x*-intercepts are $(-7, 0)$, $\left(\dfrac{1}{2}, 0\right)$.

*Solve each inequality and graph the solution on a number line.*

**19.** $x^2 - x \ge 42$

**20.** $\dfrac{(x + 5)(x - 4)}{x + 1} \ge 0$

© Michael Klenetsky /Shutterstock

*Solve the following inequality. Write the answer in* **a)** *interval notation and* **b)** *set builder notation.*

**21.** $\dfrac{x + 3}{x + 2} \le -1$

**22. Carpet** The length of a rectangular Persian carpet is 3 feet greater than twice its width. Find the length and width of the carpet if its area is 65 square feet.

**23. Throwing a Ball** Jose Ramirez throws a ball upward from the top of a building. The distance, $d$, of the ball from the ground at any time, $t$, is $d = -16t^2 + 80t + 96$. How long will it take for the ball to strike the ground?

**24. Profit** The Leigh Ann Sims Company earns a weekly profit according to the function $f(x) = -1.4x^2 + 56x - 70$, where $x$ is the number of wood carvings made and sold each week.

    **a)** Find the number of carvings the company must sell in a week to maximize its profit.

    **b)** What is its maximum weekly profit?

**25. Selling Brooms** A business sells $n$ brooms, $n \le 32$, at a price of $(10 - 0.1n)$ dollars per broom. How many brooms must be sold to have a revenue of $210?

## Cumulative Review Test

*Take the following test and check your answers with those given in the back of the book. Review any questions that you answered incorrectly. The section where the material was covered is indicated after the answer.*

**1.** Evaluate $-4 \div (-2) + 18 - \sqrt{49}$.

**2.** Evaluate $2x^2 + 3x + 4$ when $x = 2$.

**3.** Simplify $6x - \{3 - [2(x - 2) - 5x]\}$.

**4.** Solve the equation $-\dfrac{1}{2}(4x - 6) = \dfrac{1}{3}(3 - 6x) + 2$.

**5.** Graph each equation.

    **a)** $x = -4$

    **b)** $y = 2$

**6.** Write the equation in slope-intercept form of a line passing through the points $(6, 5)$ and $(4, 3)$.

**7.** Express 2,540,000 in scientific notation.

**8.** Add $\dfrac{x + 2}{x^2 - x - 6} + \dfrac{x - 3}{x^2 - 8x + 15}$.

**9.** Solve the equation

$$\frac{1}{a - 2} = \frac{4a - 1}{a^2 + 5a - 14} + \frac{2}{a + 7}.$$

**10. Wattage Rating** The wattage rating of an appliance, $w$, varies jointly as the square of the current, $I$, and the resistance, $R$. If the wattage is 12 when the current is 2 amperes and the resistance is 100 ohms, find the wattage when the current is 0.8 ampere and the resistance is 600 ohms.

**11. Small Orchard** The number of baskets of apples, $N$, that are produced by $x$ trees in a small orchard is given by the function $N(x) = -0.2x^2 + 40x$. How many baskets of apples are produced by 50 trees?

**12. a)** Determine whether the following graph represents a function. Explain your answer.

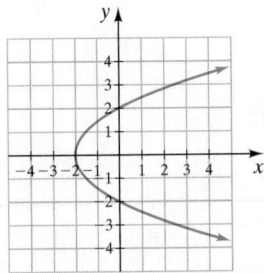

    **b)** Determine the domain and range of the function or relation.

**13.** Solve the system of equations.

$$4x - 3y = 10$$
$$2x + y = 5$$

**14.** Evaluate the following determinant.

$$\begin{vmatrix} 4 & 0 & -2 \\ 3 & 5 & 1 \\ 1 & -1 & 7 \end{vmatrix}$$

**15.** Solve the inequality $-4 < \dfrac{x+4}{2} < 6$. Write the solution in interval notation.

**16.** Find the solution set for the equation $|4 - 2x| = 5$.

**17.** Simplify $\dfrac{3 - 4i}{2 + 5i}$.

**18.** Solve $4x^2 = -3x - 12$.

**19.** Solve the formula $V = \dfrac{1}{3}\pi r^2 h$ for $r$.

**20.** Factor $(x + 3)^2 + 10(x + 3) + 24$.

# 13 Exponential and Logarithmic Functions

## Goals of This Chapter

Exponential and logarithmic functions have a wide variety of uses, some of which you will see as you read through this chapter. You often read in newspaper and magazine articles that health care spending, use of the Internet, and the world population, to list just a few, are growing exponentially. By the time you finish this chapter you should have a clear understanding of just what this means.

We also introduce two special functions, the natural exponential function and the natural logarithmic function. Many natural phenomena, such as carbon dating, radioactive decay, and the growth of savings invested in an account compounding interest continuously, can be described by natural exponential functions.

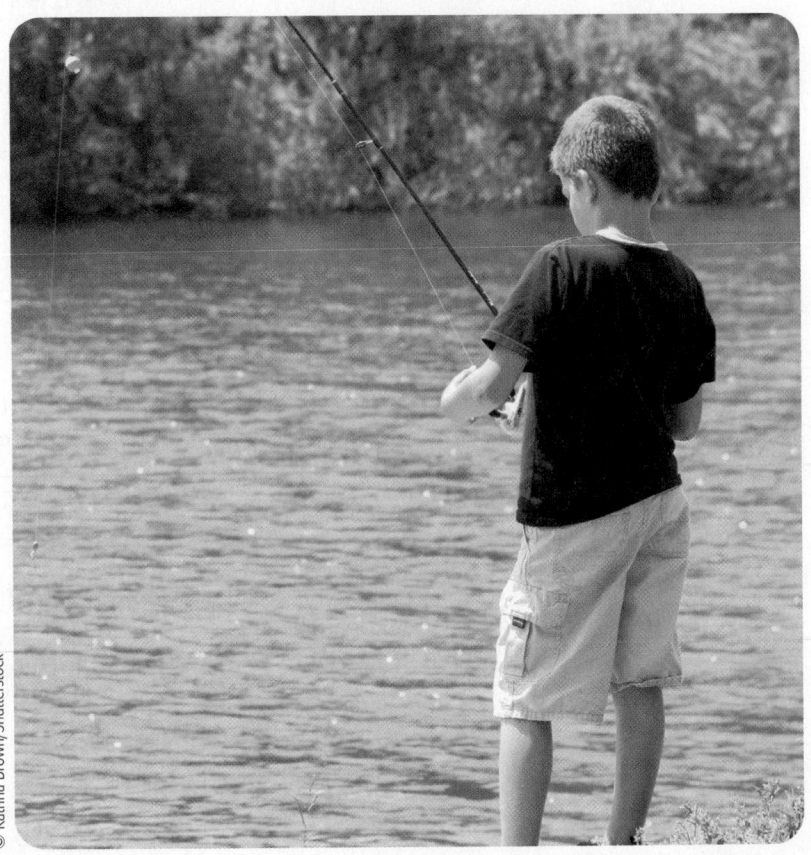

© Katrina Brown/Shutterstock

**The populations of many species** can be modeled using either exponential or logarithmic functions. Specific examples include modeling the human populations of cities, countries, and of the entire world. Other examples involve the population of bears in a forest, manatees in an estuary, and bald eagles in a national park. In Exercise 74 on page 826, we use an exponential function to model the population of trout in a lake.

# 13.1 Composite and Inverse Functions

1 Find composite functions.

2 Understand one-to-one functions.

3 Find inverse functions.

4 Find the composition of a function and its inverse.

We begin with composite functions, one-to-one functions, and inverse functions. These concepts will be used in our discussion of logarithmic and exponential functions.

## 1 Find Composite Functions

Often one variable is a function of another variable which in turn is a function of a third variable. We describe the relationship among such functions as a **composition of functions**. For example, suppose that 1 U.S. dollar can be converted into 1.20 Canadian dollars, and 1 Canadian dollar can be converted into 11.40 Mexican pesos. Using this information, we can convert 20 U.S. dollars into Mexican pesos. We have the following functions.

$$g(x) = 1.20x \qquad (x \text{ U.S. dollars to Canadian dollars})$$

$$f(x) = 11.40x \qquad (x \text{ Canadian dollars to Mexican pesos})$$

If we let $x = 20$, for \$20 U.S., then it can be converted into \$24 Canadian using function $g$:

$$g(x) = 1.20x$$

$$g(20) = 1.20(20) = \$24 \text{ Canadian}$$

The \$24 Canadian can, in turn, be converted into 273.60 Mexican pesos using function $f$:

$$f(x) = 11.40x$$

$$f(24) = 11.40(24) = 273.60 \text{ Mexican pesos}$$

This conversion can be found directly without performing this string of calculations. One U.S. dollar can be converted into Mexican pesos by substituting the $1.20x$ found in function $g(x)$ for the $x$ in $f(x)$. This gives a new function, $h$, which converts U.S. dollars directly into Mexican pesos.

$$g(x) = 1.20x \qquad f(x) = 11.40x$$

$$h(x) = f[g(x)]$$

$$= 11.40(\boxed{1.20x}) \qquad \text{Substitute } g(x) \text{ for } x \text{ in } f(x).$$

$$= 13.68x$$

Thus, each U.S. dollar, $x$, is worth 13.68 Mexican pesos. If we substitute \$20 for $x$, we get 273.60 pesos, which is what we expected.

$$h(x) = 13.68x$$

$$h(20) = 13.68(20) = 273.60$$

Function $h$ is called a **composition of $f$ with $g$** or the **composite function of $f$ with $g$**. We denote the composite function $(f \circ g)(x)$ and is read "$f$ of $g$ of $x$" or "$f$ composed with $g$ of $x$." **Figure 13.1** shows how the composite function $h$ relates to functions $f$ and $g$.

**FIGURE 13.1**

We now define the **composite function**.

### Composite Function

The **composite function** $(f \circ g)(x)$ is defined as

$$(f \circ g)(x) = f[g(x)]$$

When we are given $f(x)$ and $g(x)$, to find $(f \circ g)(x)$ we substitute $g(x)$ for $x$ in $f(x)$ to get $f[g(x)]$.

EXAMPLE 1   Given $f(x) = x^2 - 2x + 3$ and $g(x) = x - 5$, find

a) $f(4)$            b) $f(a)$            c) $(f \circ g)(x)$            d) $(f \circ g)(3)$

*Solution*

a) To find $f(4)$, we substitute 4 for each $x$ in $f(x)$.

$$f(x) = x^2 - 2x + 3$$
$$f(4) = 4^2 - 2 \cdot 4 + 3 = 16 - 8 + 3 = 11$$

b) To find $f(a)$, we substitute $a$ for each $x$ in $f(x)$.

$$f(x) = x^2 - 2x + 3$$
$$f(a) = a^2 - 2a + 3$$

c) $(f \circ g)(x) = f[g(x)]$. To find $(f \circ g)(x)$, we substitute $g(x)$, which is $x - 5$, for each $x$ in $f(x)$.

$$f(\,x\,) = x^2 - 2x + 3$$
$$f[g(x)] = [g(x)]^2 - 2[g(x)] + 3$$

Since $g(x) = x - 5$, we substitute as follows

$$f[g(x)] = (\,x - 5\,)^2 - 2(\,x - 5\,) + 3$$
$$= (x - 5)(x - 5) - 2x + 10 + 3$$
$$= x^2 - 10x + 25 - 2x + 13$$
$$= x^2 - 12x + 38$$

Therefore, the composite function of $f$ with $g$ is $x^2 - 12x + 38$.

$$(f \circ g)(x) = f[g(x)] = x^2 - 12x + 38$$

d) To find $(f \circ g)(3)$, we substitute 3 for $x$ in $(f \circ g)(x)$.

$$(f \circ g)(x) = x^2 - 12x + 38$$
$$(f \circ g)(3) = 3^2 - 12(3) + 38 = 11$$

Now Try Exercise 9

### Understanding Algebra

Several examples and exercises in this section show us that composition of functions is not a commutative operation. That is, in general,

$$(f \circ g)(x) \neq (g \circ f)(x)$$

To determine $(g \circ f)(x)$ or $g[f(x)]$, substitute $f(x)$ for each $x$ in $g(x)$. Using $f(x)$ and $g(x)$ as given in Example 1, we find $(g \circ f)(x)$ as follows.

$$g(x) = x - 5, \qquad f(x) = x^2 - 2x + 3$$
$$g[f(x)] = f(x) - 5$$
$$g[f(x)] = (x^2 - 2x + 3) - 5$$
$$= x^2 - 2x + 3 - 5$$
$$= x^2 - 2x - 2$$

Therefore, the composite function of $g$ with $f$ is $x^2 - 2x - 2$.

$$(g \circ f)(x) = g[f(x)] = x^2 - 2x - 2$$

*By comparing the illustrations above, we see that in this example, $f[g(x)] \neq g[f(x)]$.*

EXAMPLE 2  Given $f(x) = x^2 + 4$ and $g(x) = \sqrt{x - 1}$, find

**a)** $(f \circ g)(x)$           **b)** $(g \circ f)(x)$

*Solution*

**a)** To find $(f \circ g)(x)$, we substitute $g(x)$, which is $\sqrt{x - 1}$, for each $x$ in $f(x)$. You should realize that $\sqrt{x - 1}$ is a real number only when $x \geq 1$.

$$f(x) = \boxed{x}^2 + 4$$

$$(f \circ g)(x) = f[g(x)] = (\boxed{\sqrt{x - 1}})^2 + 4 = x - 1 + 4 = x + 3, x \geq 1$$

Since values of $x < 1$ are not in the domain of $g(x)$, values of $x < 1$ are not in the domain of $(f \circ g)(x)$.

**b)** To find $(g \circ f)(x)$, we substitute $f(x)$, which is $x^2 + 4$, for each $x$ in $g(x)$.

$$g(x) = \sqrt{\boxed{x} - 1}$$

$$(g \circ f)(x) = g[f(x^2 + 4)] = \sqrt{\boxed{(x^2 + 4)} - 1} = \sqrt{x^2 + 3}$$

*Now Try Exercise 19*

EXAMPLE 3  Given $f(x) = x - 3$ and $g(x) = x + 7$, find

**a)** $(f \circ g)(x)$     **b)** $(f \circ g)(2)$     **c)** $(g \circ f)(x)$     **d)** $(g \circ f)(2)$

*Solution*

**a)**
$$f(x) = \boxed{x} - 3$$
$$(f \circ g)(x) = f[g(x)] = (\boxed{x + 7}) - 3 = x + 4$$

**b)** We find $(f \circ g)(2)$ by substituting 2 for each $x$ in $(f \circ g)(x)$.
$$(f \circ g)(x) = x + 4$$
$$(f \circ g)(2) = 2 + 4 = 6$$

**c)**
$$g(x) = \boxed{x} + 7$$
$$(g \circ f)(x) = g[f(x)] = (\boxed{x - 3}) + 7 = x + 4$$

**d)** Since $(g \circ f)(x) = x + 4$, $(g \circ f)(2) = 2 + 4 = 6$.

*Now Try Exercise 11*

In general, $(f \circ g)(x) \neq (g \circ f)(x)$ as we saw at the end of Example 1. In Example 3, $(f \circ g)(x) = (g \circ f)(x)$, but this is only due to the specific functions used.

---

**Helpful Hint**

Do not confuse finding the product of two functions with finding a composite function.

Product of functions $f$ and $g$:     $(fg)(x) = (f \cdot g)(x) = f(x) \cdot g(x)$

Composite function of $f$ with $g$:     $(f \circ g)(x) = f[g(x)]$

When multiplying functions $f$ and $g$, we can use a dot between the $f$ and $g$. When finding the composite function of $f$ with $g$, we use a small *open* circle.

## 2  Understand One-to-One Fuctions

Before discussing one-to-one functions, take a moment to review the definitions of relation, domain, range, and function shown in the Understanding Algebra box.

### One-to-One Function

A function is a **one-to-one function** if each element in the range corresponds to exactly one element in the domain.

Consider the following two sets of ordered pairs.

$$A = \{(1, 2), (3, 5), (4, 6), (-2, 1)\}$$

$$B = \{(1, 2), (3, 5), (4, 6), (-2, 5)\}$$

- Both sets $A$ and $B$ represent functions since each $x$-coordinate corresponds to exactly one $y$-coordinate.
- Set $A$ is also a one-to-one function since each $y$-coordinate also corresponds to exactly one $x$-coordinate.
- Set $B$ is not a one-to-one function since the $y$-coordinate 5 corresponds to two $x$-coordinates, 3 and $-2$.

**Figure 13.2** shows the graph of function $A$ and **Figure 13.3** shows the graph of function $B$.

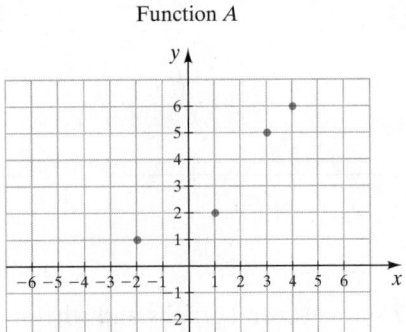

Function A

Function B

- Each *x*-coordinate corresponds to exactly one *y*-coordinate.

  *Therefore, set A is a function.*

- Each *y*-coordinate also corresponds to exactly one *x*-coordinate. *Therefore, set A is a one-to-one function.*

FIGURE 13.2

- Each *x*-coordinate corresponds to exactly one *y*-coordinate.

  *Therefore, set B is a function.*

- The *y*-coordinate 5 corresponds to two *x*-coordinates, 3 and $-2$. *Therefore, set B is not a one-to-one function.*

FIGURE 13.3

For a graph to represent a function, it must pass the *vertical* line test (see Understanding Algebra box). For a graph to represent a one-to-one function it must also pass the *horizontal* line test.

### Horizontal Line Test

If a horizontal line can be drawn so that it intersects the graph of a function at more than one point, the graph is not a one-to-one function.

Observe the graph of the function $f(x) = x^2$ shown in **Figure 13.4** First, it is a function since it passes the *vertical* line test. However, it is not a one-to-one function since it fails the *horizontal* line test. **Figure 13.5** shows that $f(x) = x^2$ has a $y$-coordinate that corresponds to two different $x$-coordinates.

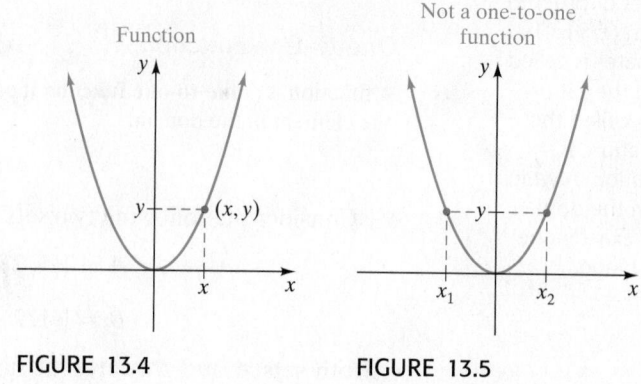

FIGURE 13.4                    FIGURE 13.5

Often we can restrict the domain of a function that is not a one-to-one function so that the resulting function is a one-to-one function. For example, both $f(x) = x^2, x \geq 0$ **(Fig. 13.6a)** and $f(x) = x^2, x \leq 0$ **(Figure 13.6b)** are examples of one-to-one functions. Note that both graphs pass the horizontal line test.

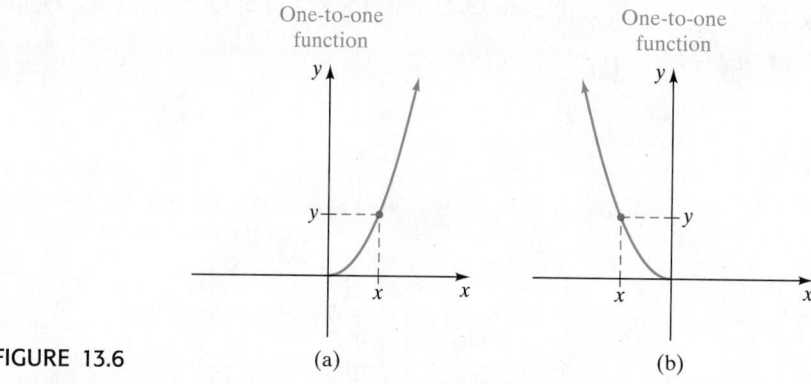

FIGURE 13.6              (a)                              (b)

**Figure 13.7** shows several graphs and whether or not they represent one-to-one functions. Notice that the graph in part $(f)$ does not represent a function since it fails the vertical line test. Therefore this graph cannot represent a one-to-one function even though the graph passes the horizontal line test.

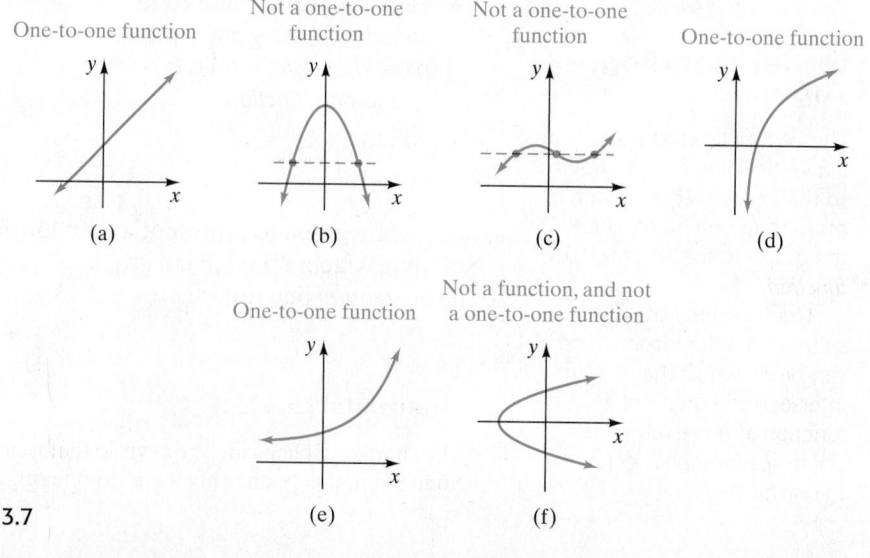

FIGURE 13.7              (e)                              (f)

## 3 Find Inverse Functions

One-to-one functions consist of ordered pairs in which each $x$-coordinate corresponds to exactly one $y$-coordinate *and* each $y$-coordinate corresponds to exactly one $x$-coordinate. Such a relationship allows us to create a new function called the *inverse* function. *Only one-to-one functions have inverses.*

> **Inverse Function**
>
> If $f(x)$ is a one-to-one function with ordered pairs of the form $(x, y)$, the **inverse function**, $f^{-1}(x)$, is a one-to-one function with ordered pairs of the form $(y, x)$.

We begin with an example.

Function, $f(x)$:         $\{(1, 4), (2, 0), (3, 7), (-2, 1), (-1, -5)\}$

Inverse function, $f^{-1}(x)$: $\{(4, 1), (0, 2), (7, 3), (1, -2), (-5, -1)\}$

If we graph the points in the function and the points in the inverse function **(Fig. 13.8)**, we see that the points are symmetric with respect to the line $y = x$.

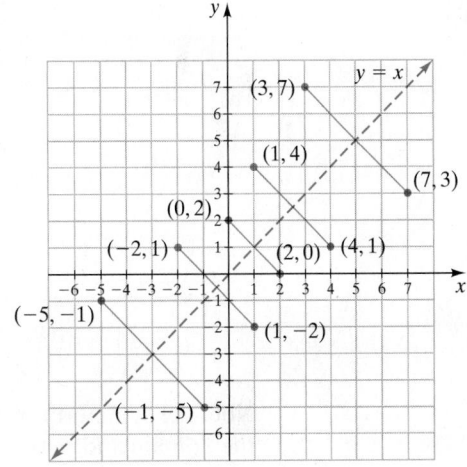

**FIGURE 13.8**

- Ordered pair in function, $f(x)$
- Ordered pair in inverse function, $f^{-1}(x)$

Some important information about inverse functions:

- Only one-to-one functions have inverse functions.
- The domain of $f(x)$ is the range of $f^{-1}(x)$.
- The range of $f(x)$ is the domain of $f^{-1}(x)$.
- When the graphs of $f(x)$ and $f^{-1}(x)$ are graphed on the same axes, the graphs of $f(x)$ and $f^{-1}(x)$ are symmetric about the line $y = x$.

When a one-to-one function is given as an equation, its inverse function can be found by the following procedure.

> **To Find the Inverse Function of a One-to-One Function**
>
> 1. Replace $f(x)$ with $y$.
> 2. Interchange the two variables $x$ and $y$.
> 3. Solve the equation for $y$.
> 4. Replace $y$ with $f^{-1}(x)$ (this gives the inverse function using inverse function notation).

EXAMPLE 4

**a)** Find the inverse function of $f(x) = 4x + 2$.

**b)** On the same axes, graph both $f(x)$ and $f^{-1}(x)$.

*Solution*

**a)** This function is one-to-one, therefore we will follow the four-step procedure.

$$f(x) = 4x + 2 \quad \text{Original function}$$

Step 1 $\qquad\qquad\qquad y = 4x + 2 \quad$ Replace $f(x)$ with $y$.

Step 2 $\qquad\qquad\qquad x = 4y + 2 \quad$ Interchange $x$ and $y$.

Step 3 $\qquad\qquad\qquad x - 2 = 4y \qquad$ Solve for $y$.

$$\frac{x - 2}{4} = y$$

$$\text{or} \quad y = \frac{x - 2}{4}$$

Step 4 $\qquad\qquad\qquad f^{-1}(x) = \dfrac{x - 2}{4} \quad$ Replace $y$ with $f^{-1}(x)$.

**b)** Below we show tables of values for $f(x)$ and $f^{-1}(x)$. The graphs of $f(x)$ and $f^{-1}(x)$ are shown in **Figure 13.9**.

| $x$ | $y = f(x)$ |
|---|---|
| 0 | 2 |
| 1 | 6 |

| $x$ | $y = f^{-1}(x)$ |
|---|---|
| 2 | 0 |
| 6 | 1 |

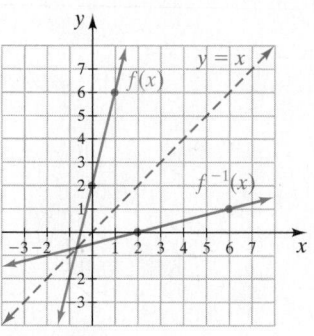

FIGURE 13.9

Note the symmetry of $f(x)$ and $f^{-1}(x)$ about the line $y = x$. Also note that both the domain and range of both $f(x)$ and $f^{-1}(x)$ are the set of real numbers, $\mathbb{R}$.

Now Try Exercise 67

EXAMPLE 5

**a)** Find the inverse function of $f(x) = x^3 + 2$.

**b)** On the same axes, graph both $f(x)$ and $f^{-1}(x)$.

*Solution*

**a)** This function is one-to-one; therefore we will follow the four-step procedure to find its inverse.

$$f(x) = x^3 + 2 \quad \text{Original function}$$

Step 1 $\qquad\qquad\qquad y = x^3 + 2 \quad$ Replace $f(x)$ with $y$.

Step 2 $\qquad\qquad\qquad x = y^3 + 2 \quad$ Interchange $x$ and $y$.

Step 3 $\qquad\qquad\qquad x - 2 = y^3 \qquad$ Solve for $y$.

$$\sqrt[3]{x - 2} = \sqrt[3]{y^3} \quad \text{Take the cube root of both sides.}$$

$$\sqrt[3]{x - 2} = y$$

$$\text{or} \qquad y = \sqrt[3]{x - 2}$$

Step 4 $\qquad\qquad\qquad f^{-1}(x) = \sqrt[3]{x - 2} \quad$ Replace $y$ with $f^{-1}(x)$.

**b)** Below we show tables of values for $f(x)$ and $f^{-1}(x)$.

| $x$ | $y = f(x)$ | $x$ | $y = f^{-1}(x)$ |
|---|---|---|---|
| $-2$ | $-6$ | $-6$ | $-2$ |
| $-1$ | $1$ | $1$ | $-1$ |
| $0$ | $2$ | $2$ | $0$ |
| $1$ | $3$ | $3$ | $1$ |
| $2$ | $10$ | $10$ | $2$ |

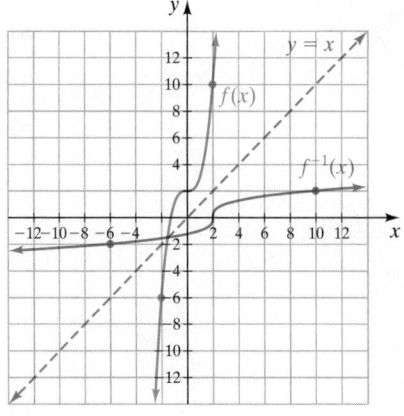

**FIGURE 13.10**

The graphs of $f(x)$ and $f^{-1}(x)$ are shown in **Figure 13.10**. Notice for each point $(a, b)$ on the graph of $f(x)$, the point $(b, a)$ appears on the graph of $f^{-1}(x)$. For example, the points $(2, 10)$ and $(-2, -6)$, indicated in blue, appear on the graph of $f(x)$, and the points $(10, 2)$ and $(-6, -2)$, indicated in red, appear on the graph of $f^{-1}(x)$.

Now Try Exercise 61

## 4 Find the Composition of a Function and Its Inverse

To help reinforce the relationship between a function and its inverse, we will evaluate the composition of $f^{-1}(x)$ and $f(x)$ from Example 5 for the values of $x = -2, x = 0$, and $x = 2$.

$$(f^{-1} \circ f)(-2) = f^{-1}[f(-2)] = f^{-1}(-6) = -2$$

$$(f^{-1} \circ f)(0) = f^{-1}[f(0)] = f^{-1}(2) = 0$$

$$(f^{-1} \circ f)(2) = f^{-1}[f(2)] = f^{-1}(10) = 2$$

Notice that the output of the composition of a function and its inverse is always equal to the input. Similarly, we will evaluate the composition of $f(x)$ and $f^{-1}(x)$ from Example 5 for the values of $x = -6, x = 2$, and $x = 10$.

$$(f \circ f^{-1})(-6) = f[f^{-1}(-6)] = f(-2) = -6$$

$$(f \circ f^{-1})(2) = f[f^{-1}(2)] = f(0) = 2$$

$$(f \circ f^{-1})(10) = f[f^{-1}(10)] = f(2) = 10$$

Notice again that the output is always equal to the input. This relationship is summarized as follows.

### The Composition of a Function and Its Inverse

For any one-to-one function $f(x)$ and its inverse $f^{-1}(x)$,

$$(f \circ f^{-1})(x) = x \quad \text{and} \quad (f^{-1} \circ f)(x) = x.$$

**EXAMPLE 6** In Example 4, we determined that for $f(x) = 4x + 2$, $f^{-1}(x) = \dfrac{x-2}{4}$. Show that

**a)** $(f \circ f^{-1})(x) = x$           **b)** $(f^{-1} \circ f)(x) = x$

*Solution*

**a)** To determine $(f \circ f^{-1})(x)$, substitute $f^{-1}(x)$ for each $x$ in $f(x)$.

$$f(x) = 4\,\boxed{x}\, + 2$$

$$(f \circ f^{-1})(x) = 4\left(\boxed{\dfrac{x-2}{4}}\right) + 2$$

$$= x - 2 + 2 = x$$

**b)** To determine $(f^{-1} \circ f)(x)$, substitute $f(x)$ for each $x$ in $f^{-1}(x)$.

$$f^{-1}(x) = \dfrac{\boxed{x}\,-2}{4}$$

$$(f^{-1} \circ f)(x) = \dfrac{4x + 2 - 2}{4}$$

$$= \dfrac{4x}{4} = x$$

Thus, $(f \circ f^{-1})(x) = (f^{-1} \circ f)(x) = x$.

Now Try Exercise 77

**EXAMPLE 7** In Example 5, we determined that $f(x) = x^3 + 2$ and $f^{-1}(x) = \sqrt[3]{x-2}$ are inverse functions. Show that

**a)** $(f \circ f^{-1})(x) = x$           **b)** $(f^{-1} \circ f)(x) = x$

*Solution*

**a)** To determine $(f \circ f^{-1})(x)$, substitute $f^{-1}(x)$ for each $x$ in $f(x)$.

$$f(x) = \boxed{x}^{\,3} + 2$$

$$(f \circ f^{-1})(x) = \left(\boxed{\sqrt[3]{x-2}}\right)^3 + 2$$

$$= x - 2 + 2 = x$$

**b)** To determine $(f^{-1} \circ f)(x)$, substitute $f(x)$ for each $x$ in $f^{-1}(x)$.

$$f^{-1}(x) = \sqrt[3]{\boxed{x}-2}$$

$$(f^{-1} \circ f)(x) = \sqrt[3]{(x^3 + 2) - 2}$$

$$= \sqrt[3]{x^3} = x$$

Thus, $(f \circ f^{-1})(x) = (f^{-1} \circ f)(x) = x$.

Now Try Exercise 79

Because a function and its inverse "undo" each other, the composite of a function with its inverse results in the given value from the domain. For example, for any function $f(x)$ and its inverse $f^{-1}(x)$, $(f^{-1} \circ f)(3) = 3$, and $(f \circ f^{-1})\left(-\dfrac{1}{2}\right) = -\dfrac{1}{2}$.

# EXERCISE SET 13.1   MathXL®   MyMathLab®
<sub>MathXL®   MyMathLab</sub>

## Warm-Up Exercises

*Fill in the blanks with the appropriate word, phrase, or symbol(s) from the following list.*

| | | | | |
|---|---|---|---|---|
| inverse | domain | horizontal | $x$ | $f(x)$ |
| vertical | range | one-to-one | composition | $y$ |

1. When one variable is a function of another variable which in turn is a function of a third variable, we describe the relationship among such functions as a _____ of functions.

2. A function is a _____ function if each element in the range corresponds to exactly one element in the domain.

3. The _____ line test states that, if a _____ line can be drawn so that it intersects a graph at more than one point, then the graph is not the graph of a function.

4. The _____ line test states that, if a _____ line can be drawn so that it intersects the

graph of a function at more than one point, the function is not a one-to-one function.

5. If a one-to-one function has ordered pairs of the form $(x, y)$, the _____ function is a one-to-one function with ordered pairs of the form $(y, x)$.

6. For a one-to-one function $f(x)$ and its inverse function $f^{-1}(x)$, the domain of $f(x)$ is the _____ of $f^{-1}(x)$.

7. For a one-to-one function $f(x)$ and its inverse function $f^{-1}(x)$, the range of $f(x)$ is the _____ of $f^{-1}(x)$.

8. For any one-to-one function $f(x)$ and its inverse $f^{-1}(x)$, $(f \circ f^{-1})(x) = $ _____ and $(f^{-1} \circ f)(x) = $ _____.

## Practice the Skills

*For each pair of functions, find **a)** $(f \circ g)(x)$, **b)** $(f \circ g)(4)$, **c)** $(g \circ f)(x)$, and **d)** $(g \circ f)(4)$.*

9. $f(x) = x + 4, g(x) = 2x - 3$

10. $f(x) = 3x - 2, g(x) = x + 1$

11. $f(x) = x + 3, g(x) = x^2 + x - 4$

12. $f(x) = x + 2, g(x) = x^2 + 4x - 2$

13. $f(x) = \dfrac{1}{x}, g(x) = 2x + 3$

14. $f(x) = \dfrac{2}{x}, g(x) = x^2 + 1$

15. $f(x) = 3x + 1, g(x) = \dfrac{3}{x}$

16. $f(x) = x^2 - 5, g(x) = \dfrac{4}{x}$

17. $f(x) = x^2 + 1, g(x) = x^2 + 5$

18. $f(x) = x^2 - 4, g(x) = x^2 + 3$

19. $f(x) = x - 4, g(x) = \sqrt{x + 5}, x \geq -5$

20. $f(x) = \sqrt{x + 6}, x \geq -6, g(x) = x + 7$

*In Exercises 21–42, determine whether each function is a one-to-one function.*

21.

22.

23.

24.
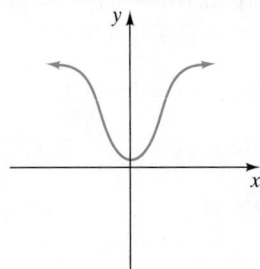

25. $\{(1, 2), (2, 3), (3, 4), (4, 5)\}$

26. $\{(1, 1), (2, 2), (3, 3), (4, 4)\}$

27. $\{(-4, 2), (5, 3), (0, 2), (4, 8)\}$

28. $\{(0, 5), (1, 4), (-3, 5), (4, 2)\}$

29. $y = 2x + 5$

30. $y = 3x - 8$

31. $y = x^2 - 1$

32. $y = -x^2 + 3$

33. $y = x^2 - 2x + 5$

34. $y = x^2 - 2x + 6, x \geq 1$

35. $y = x^2 - 9, x \geq 0$

36. $y = x^2 - 9, x \leq 0$

37. $y = \sqrt{x}$

38. $y = -\sqrt{x}$

39. $y = |x|$

40. $y = -|x|$

41. $y = \sqrt[3]{x}$

42. $y = x^3$

*In Exercises 43–48, for the given function, find the domain and range of both $f(x)$ and $f^{-1}(x)$.*

43. $\{(4, 0), (8, 9), (2, 7), (-1, 6), (-2, 4)\}$

44. $\left\{(-2, -3), (-4, 0), (5, 3), (6, 2), \left(2, \dfrac{1}{2}\right)\right\}$

**45.**

**46.**

**47.**

**48.**

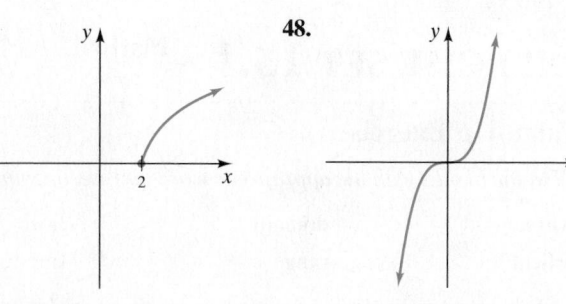

*For each function,* **a)** *determine whether it is one-to-one;* **b)** *if it is one-to-one, find its inverse function.*

**49.** $f(x) = x + 3$

**50.** $f(x) = x - 4$

**51.** $h(x) = 4x$

**52.** $k(x) = 2x - 7$

**53.** $p(x) = 3x^2$

**54.** $r(x) = |x|$

**55.** $t(x) = x^2 + 3$

**56.** $m(x) = -x^2 + x + 8$

**57.** $g(x) = \dfrac{1}{x}$

**58.** $h(x) = \dfrac{5}{x}$

**59.** $f(x) = x^2 + 10$

**60.** $g(x) = x^3 + 9$

**61.** $g(x) = x^3 - 6$

**62.** $f(x) = \sqrt{x}, x \geq 0$

**63.** $g(x) = \sqrt{x + 2}, x \geq -2$

**64.** $f(x) = x^2 - 3, x \geq 0$

**65.** $h(x) = x^2 - 4, x \geq 0$

**66.** $h(x) = |x|$

*For each one-to-one function,* **a)** *find* $f^{-1}(x)$ *and* **b)** *graph* $f(x)$ *and* $f^{-1}(x)$ *on the same axes.*

**67.** $f(x) = 2x + 8$

**68.** $f(x) = -3x + 6$

**69.** $f(x) = \sqrt{x}, x \geq 0$

**70.** $f(x) = -\sqrt{x}, x \geq 0$

**71.** $f(x) = \sqrt{x - 1}, x \geq 1$

**72.** $f(x) = \sqrt{x + 4}, x \geq -4$

**73.** $f(x) = \sqrt[3]{x}$

**74.** $f(x) = \sqrt[3]{x + 3}$

**75.** $f(x) = \dfrac{1}{x}, x > 0$

**76.** $f(x) = \dfrac{1}{x}$

*For each pair of inverse functions, show that* $(f \circ f^{-1})(x) = x$ *and* $(f^{-1} \circ f)(x) = x$.

**77.** $f(x) = x + 5, f^{-1}(x) = x - 5$

**78.** $f(x) = 3x, f^{-1}(x) = \dfrac{x}{3}$

**79.** $f(x) = \dfrac{1}{2}x + 3, f^{-1}(x) = 2x - 6$

**80.** $f(x) = -\dfrac{1}{3}x + 2, f^{-1}(x) = -3x + 6$

**81.** $f(x) = \sqrt[3]{x - 2}, f^{-1}(x) = x^3 + 2$

**82.** $f(x) = \sqrt[3]{x + 9}, f^{-1}(x) = x^3 - 9$

**83.** $f(x) = \dfrac{3}{x}, f^{-1}(x) = \dfrac{3}{x}$

**84.** $f(x) = \sqrt{x + 5}, f^{-1}(x) = x^2 - 5, x \geq 0$

## Problem Solving

**85.** The function $f(x) = 3x$ converts yards, $x$, into feet. Find the inverse function that converts feet into yards. In the inverse function, what do $x$ and $f^{-1}(x)$ represent?

**86.** The function $f(x) = 12x$ converts feet, $x$, into inches. Find the inverse function that converts inches into feet. In the inverse function, what do $x$ and $f^{-1}(x)$ represent?

**87.** The function $f(x) = \dfrac{5}{9}(x - 32)$ converts degrees Fahrenheit, $x$, to degrees Celsius. Find the inverse function that changes degrees Celsius into degrees Fahrenheit.

**88.** The function $f(x) = \dfrac{22}{15}x$ converts miles per hour, $x$, into feet per second. Find the inverse function that converts feet per second into miles per hour.

See Exercise 87.

*In Exercises 89–92, the functions $f(x)$ and $g(x)$ are given. Determine the composition $(f \circ g)(x)$. For the composition function, what does $x$ represent and what does $(f \circ g)(x)$ represent?*

**89.** $f(x) = 16x$ converts pounds, $x$, to ounces. $g(x) = 28.35x$ converts ounces, $x$ to grams.

**90.** $f(x) = 2000x$ converts tons, $x$, to pounds. $g(x) = 16x$ converts pounds, $x$, to ounces.

**91.** $f(x) = 3x$ converts yards, $x$, to feet. $g(x) = 0.305x$ converts feet, $x$, to meters.

**92.** $f(x) = 1760x$ converts miles, $x$, to yards. $g(x) = 0.915x$ converts yards, $x$, to meters.

## Concept/Writing Exercises

**93.** Is $(f \circ g)(x) = (g \circ f)(x)$ for all values of $x$? Explain and give an example to support your answer.

**94.** Consider the functions $f(x) = \sqrt{x + 5}$, $x \geq -5$, and $g(x) = x^2 - 5$, $x \geq 0$.

 **a)** Show that $(f \circ g)(x) = (g \circ f)(x)$ for $x \geq 0$.

 **b)** Explain why we need to stipulate that $x \geq 0$ for part **a)** to be true.

**95.** Consider the functions $f(x) = x^3 + 2$ and $g(x) = \sqrt[3]{x - 2}$.

 **a)** Show that $(f \circ g)(x) = (g \circ f)(x)$.

 **b)** What are the domains of $f(x)$, $g(x)$, $(f \circ g)(x)$, and $(g \circ f)(x)$? Explain.

**96.** For the function $f(x) = x^3$, $f(2) = 2^3 = 8$. Explain why $f^{-1}(8) = 2$.

**97.** For the function $f(x) = x^4$, $x > 0$, $f(2) = 16$. Explain why $f^{-1}(16) = 2$.

**98. a)** Does the function $f(x) = |x|$ have an inverse? Explain.

 **b)** If the domain is limited to $x \geq 0$, does the function have an inverse? Explain.

 **c)** Find the inverse function of $f(x) = |x|$, $x \geq 0$.

## Challenge Problems

**99. Area** When a pebble is thrown into a pond, the circle formed by the pebble hitting the water expands with time. The area of the expanding circle may be found by the formula $A = \pi r^2$. The radius, $r$, of the circle, in feet, is a function of time, $t$, in seconds. Suppose that the function is $r(t) = 2t$.

© Daniel Rybkin/Shutterstock

 **a)** Find the radius of the circle at 3 seconds.

 **b)** Find the area of the circle at 3 seconds.

 **c)** Express the area as a function of time by finding $A \circ r$.

 **d)** Using the function found in part **c)**, find the area of the circle at 3 seconds.

 **e)** Do your answers in parts **b)** and **d)** agree? If not, explain.

**100. Surface Area** The surface area, $S$, of a spherical balloon of radius $r$, in inches, is found by $S(r) = 4\pi r^2$. If the balloon is being blown up at a constant rate by a machine, then the radius of the balloon is a function of time. Suppose that this function is $r(t) = 1.2t$, where $t$ is in seconds.

 **a)** Find the radius of the balloon at 2 seconds.

 **b)** Find the surface area at 2 seconds.

 **c)** Express the surface area as a function of time by finding $S \circ r$.

 **d)** Using the function found in part **c)**, find the surface area after 2 seconds.

 **e)** Do your answers in parts **b)** and **d)** agree? If not, explain why not.

## Group Activity

*Discuss and answer Exercise 101 as a group.*

**101.** Consider the function $f(x) = 2^x$. This is an example of an *exponential function*, which we will discuss in the next section.

 **a)** Graph this function by substituting values for $x$ and finding the corresponding values of $f(x)$.

 **b)** Do you think this function has an inverse? Explain your answer.

 **c)** Using the graph in part **a)**, draw the inverse function, $f^{-1}(x)$ on the same axes.

 **d)** Explain how you obtained the graph of $f^{-1}(x)$.

## Cumulative Review Exercises

[4.4] **102.** Determine the equation of a line in standard form that passes through $\left(\dfrac{1}{2}, 3\right)$ and is parallel to the graph of $2x + 3y - 9 = 0$.

[5.6] **103.** Divide using synthetic division
$(x^3 + 6x^2 + 6x - 8) \div (x + 2)$

[7.5] **104.** Simplify $\dfrac{\dfrac{3}{x^2} - \dfrac{2}{x}}{\dfrac{x}{6}}$.

[11.5] **105.** Assuming all variables represent positive real numbers, simplify $\sqrt{\dfrac{24x^3 y^2}{3xy^3}}$.

[12.1] **106.** Solve $x^2 + 2x - 10 = 0$ by completing the square.

# 13.2 Exponential Functions

## 1  Graph Exponential Functions

There are many applications of exponential functions. Some examples include the growth of populations, the doubling of bacteria in a biology experiment, the value of money in a bank account with compound interest, the declining amount of carbon 14 remaining in a fossil, and many others. The graphs shown in **Figure 13.11** and **Figure 13.12** show two examples of exponential functions.

**FIGURE 13.11**        **FIGURE 13.12**

As seen in the definition below, an exponential function will always have a variable exponent.

### Understanding Algebra

What is the difference between the two functions

$$f(x) = 2^x \text{ and } g(x) = x^2?$$

Exponential Function

$$f(x) = 2^x$$
        variable
        *exponent*

Notice the location of the variable *x*. In an *exponential* function, the variable is in the *exponent* position.

Polynomial (Quadratic) Function

$$g(x) = x^2$$
        variable
        *base*

In a polynomial (in this case quadratic) function, the variable is in the *base* position.

### Exponential Function

For any real number $a > 0$ and $a \neq 1$,

$$f(x) = a^x \quad \text{or} \quad y = a^x$$

is an **exponential function**.

An exponential function is a function of the form $f(x) = a^x$, or $y = a^x$, where $a$ is a positive real number not equal to 1. Notice the variable is in the exponent.

### Examples of Exponential Functions

$$f(x) = 2^x, \qquad y = 5^x, \qquad g(x) = \left(\frac{1}{2}\right)^x$$

Exponential functions can be graphed by selecting values for $x$, finding the corresponding values of $y$ [or $f(x)$], and plotting the points.

Before we graph exponential functions, let's discuss some characteristics of the graphs of exponential functions.

### Graphs of Exponential Functions

For all exponential functions of the form $y = a^x$ or $f(x) = a^x$, where $a > 0$ and $a \neq 1$,

1. The domain of the function is $(-\infty, \infty)$.

2. The range of the function is $(0, \infty)$.

3. The graph of the function passes through the points $\left(-1, \dfrac{1}{a}\right)$, $(0, 1)$, and $(1, a)$.

**EXAMPLE 1** Graph the exponential function $y = 2^x$. State the domain and range of the function.

*Solution* The function is of the form $y = a^x$, where $a = 2$. First, construct a table of values. In the table, the three points listed in item 3 of the box on page 784 are shown in red.

| $x$ | $-4$ | $-3$ | $-2$ | $-1$ | 0 | 1 | 2 | 3 | 4 |
|---|---|---|---|---|---|---|---|---|---|
| $y$ | $\frac{1}{16}$ | $\frac{1}{8}$ | $\frac{1}{4}$ | $\frac{1}{2}$ | 1 | 2 | 4 | 8 | 16 |

Now plot these points and connect them with a smooth curve (**Fig. 13.13**). The three ordered pairs in red in the table are marked in red on the graph.

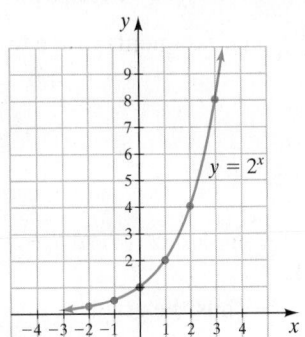

**FIGURE 13.13**

The domain of this function is the set of real numbers, $\mathbb{R}$. The range is $\{y | y > 0\}$.

Now Try Exercise 7

**EXAMPLE 2** Graph $y = \left(\frac{1}{2}\right)^x$. State the domain and range of the function.

*Solution* This function is of the form $y = a^x$, where $a = \frac{1}{2}$. Construct a table of values and plot the curve (**Fig. 13.14**).

| $x$ | $-4$ | $-3$ | $-2$ | $-1$ | 0 | 1 | 2 | 3 | 4 |
|---|---|---|---|---|---|---|---|---|---|
| $y$ | 16 | 8 | 4 | 2 | 1 | $\frac{1}{2}$ | $\frac{1}{4}$ | $\frac{1}{8}$ | $\frac{1}{16}$ |

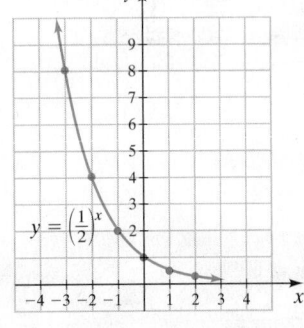

**FIGURE 13.14**

The domain is the set of real numbers, $\mathbb{R}$. The range is $\{y | y > 0\}$.

Now Try Exercise 13

Note that the graphs in **Figures 13.13** and **13.14** represent one-to-one functions since each graph passes the horizontal line test.

## Understanding Algebra

When graphing functions of the form $y = a^x$ or $f(x) = a^x$, we can predict the shape of the graph by observing the three points

$$\left(-1, \frac{1}{a}\right), (0, 1) \text{ and } (1, a)$$

- When $a > 1$, the graph becomes almost horizontal to the left of $\left(-1, \frac{1}{a}\right)$ and somewhat vertical to the right of $(1, a)$; see Example 1.

- When $0 < a < 1$, the graph becomes almost horizontal to the right of $(1, a)$ and somewhat vertical to the left of $\left(-1, \frac{1}{a}\right)$; see Example 2.

---

**Helpful Hint**

When graphing exponential functions of the form $y = a^x$ where $x > 0$, if

- $a > 1$, the graph will rise from left to right. See the graph of $y = 2^x$ in **Figure 13.13**.

- $0 < a < 1$, the graph will fall from left to right. See the graph of $y = \left(\frac{1}{2}\right)^x$ in **Figure 13.14**.

Whenever we encounter an exponential function with a negative exponent such as $y = 2^{-x}$, we can recall our rules of exponents to see that

$$y = 2^{-x}$$
$$= \frac{1}{2^x}$$
$$= \left(\frac{1}{2}\right)^x$$

Thus, the graph of $y = 2^{-x}$ is the graph of $y = \left(\frac{1}{2}\right)^x$ shown in **Figure 13.14** on page 785.

Similarly, when we encounter a function such as $y = \left(\frac{1}{2}\right)^{-x}$, we can use the rules of exponents to see that

$$y = \left(\frac{1}{2}\right)^{-x}$$
$$= \left(\frac{2}{1}\right)^x$$
$$= 2^x$$

Thus, the graph of $y = \left(\frac{1}{2}\right)^{-x}$ is the graph of $y = 2^x$ shown in **Figure 13.13** on page 785.

## 2 Solve Applications of Exponential Functions

Exponential functions are often used to describe the growth and decay of certain quantities. We illustrate exponential functions in the next five examples.

EXAMPLE 3  **Pennies Add Up**  Jennifer Hewlett told her young son that if he did his chores, she would give him 2 cents the first week and double the amount each week for the next 10 weeks. The number of cents her son would receive in any given week, $w$, can be determined by the function $n(w) = 2^w$. Determine the number of cents Jennifer would give her son in week 8.

*Solution*    By evaluating $2^8$, we determine that in week 8 Jennifer would give her son 256 cents, or $2.56.

Now Try Exercise 29

EXAMPLE 4  **Value of a Jeep**  Ronald Yates just bought a new Jeep Compass for $22,000. Assume the value of the Jeep depreciates at a rate of 20% per year. Therefore, the value of the Jeep is 80% of the previous year's value. One year from now, its value will be $22,000(0.80). Two years from now, its value will be $22,000(0.80)(0.80) = $22,000(0.80)^2$ and so on. Therefore, the formula for the value of the Jeep is

$$v(t) = 22,000(0.80)^t$$

where $t$ is time in years. Find the value of the Jeep **a)** 1 year from now and **b)** 5 years from now.

*Solution*

**a)** To find the value 1 year from now, substitute 1 for $t$.

$$v(t) = 22,000(0.80)^t$$
$$v(1) = 22,000(0.80)^1 \quad \text{Substitute 1 for } t.$$
$$= 17,600$$

One year from now, the value of the Jeep will be $17,600.

**b)** To find the value 5 years from now, substitute 5 for $t$.

$$v(t) = 22{,}000(0.80)^t$$

$$v(5) = 22{,}000(0.80)^5 \qquad \text{Substitute 5 for } t.$$

$$= 22{,}000(0.32768)$$

$$= 7208.96$$

Five years from now, the value of the Jeep will be $7208.96.

Now Try Exercise 45

Previously, we have used the compound interest formula to determine the amount of money we accumulate in a savings or other investment account.

---

**Compound Interest Formula**

The accumulated amount, $A$, in a compound interest account can be found using the formula

$$A = p\left(1 + \frac{r}{n}\right)^{nt}$$

where $p$ is the principal or the initial investment amount, $r$ is the interest rate as a decimal, $n$ is the number of compounding periods per year, and $t$ is the time in years.

---

**EXAMPLE 5  Compound Interest**  Nancy Johnson invests $10,000 in a certificate of deposit (CD) with 5% interest compounded quarterly for 6 years. Determine the value of the CD after 6 years.

*Solution*   Understand   We are given that the principal, $p$, is $10,000. We are also given that the interest rate, $r$, is 5%. Because the interest is compounded quarterly, the number of compounding periods, $n$, is 4. The money is invested for 6 years. Therefore, $t$ is 6.

Translate   Now we substitute these values into the formula

$$A = p\left(1 + \frac{r}{n}\right)^{nt}$$

$$= 10{,}000\left(1 + \frac{0.05}{4}\right)^{4(6)}$$

Carry Out

$$= 10{,}000(1 + 0.0125)^{24}$$

$$= 10{,}000(1.0125)^{24}$$

$$\approx 10{,}000(1.347351) \qquad \text{From a calculator}$$

$$\approx 13{,}473.51$$

Answer   The original $10,000 has grown to about $13,473.51 after 6 years.

Now Try Exercise 33

## Understanding Algebra

When using the compound interest formula you need to be careful to write the interest rate as a decimal. For example, an interest rate of 5% means that $r = 0.05$. An interest rate of 2.75% means that $r = 0.0275$, and so on.

When writing $n$, the number of compounding periods per year, here are the most commonly used values:

semiannually: $n = 2$

quarterly: $n = 4$

monthly: $n = 12$

---

**EXAMPLE 6  Carbon 14 Dating**  Carbon 14 dating is used by scientists to find the age of fossils and artifacts. The formula used in carbon dating is

$$A = A_0 \cdot 2^{-t/5600}$$

where $A_0$ represents the amount of carbon 14 present when the organism died and $A$ represents the amount of carbon 14 present after $t$ years. If 500 grams of carbon 14 were present when an organism died, how many grams will be found in the fossil 2000 years later?

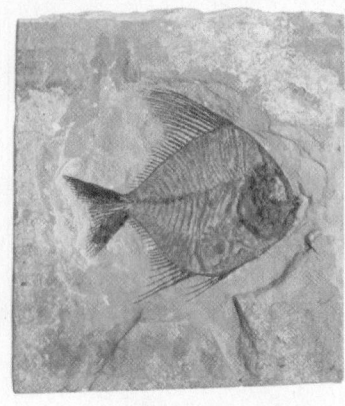

*Solution*   Understand   When the organism died, 500 grams of carbon 14 were present. Therefore, $A_0 = 500$. To find out how many grams of carbon 14 will be present 2000 years later, we substitute 2000 for $t$ in the formula.

Translate

$$A = A_0 \cdot 2^{-t/5600}$$
$$= 500(2)^{-2000/5600}$$

Carry Out

$$\approx 500(0.7807092) \qquad \text{From a calculator}$$
$$\approx 390.35 \text{ grams}$$

Answer   After 2000 years, about 390.35 of the original 500 grams of carbon 14 are still present.

Now Try Exercise 39

**EXAMPLE 7   Exports to China.   Figure 13.15** shows the annual amount of exports from the United States to China for the years 2000 through 2007 in billions of dollars.

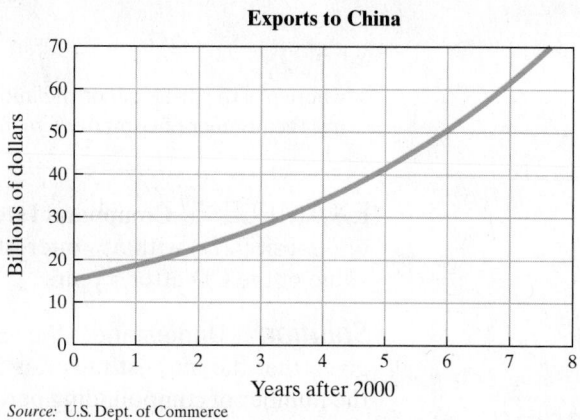

**Exports to China**

*Source:* U.S. Dept. of Commerce

**FIGURE 13.15**

An exponential function that closely approximates this curve is $f(t) = 15.37(1.22)^t$. In this function, $f(t)$ is the total value of exports from the United States to China and $t$ is the number of years since 2000. Assume that this trend continues. Use this function to estimate the value of the exports to China in **a)** 2010 and **b)** 2015. Round your answers to the nearest billion dollars.

*Solution*

**a)** Understand   In this function, $t$ is years since 2000. Thus, the year 2010 is represented by $t = 10$. To estimate the value of the exports to China in 2010, we need to evaluate this function for $t = 10$.

Translate and Carry Out     $f(t) = 15.37(1.22)^t$

$$f(10) = 15.37(1.22)^{10} \approx 112.27$$

Answer   Therefore, if this trend continues, the exports to China in 2010 will be about $112 billion.

**b)** The year 2015 is represented by $t = 15$ and we need to evaluate $f(15)$.

$$f(t) = 15.37(1.22)^t$$

$$f(15) = 15.37(1.22)^{15} \approx 303.44$$

Answer   Therefore, if this trend continues, the exports to China in 2015 will be about $303 billion.

Now Try Exercise 47

# EXERCISE SET 13.2   MathXL®  MyMathLab®
MathXL®   MyMathLab

## Warm-Up Exercises

*Fill in the blanks with the appropriate word, phrase, or symbol(s) from the following list.*

periods     base     fall     exponent     principal     rise     rate     time

**1.** In an exponential function, the variable is in the _____ position.

**2.** In a quadratic function, the variable is in the _____ position.

**3.** When graphing exponential functions of the form $y = a^x$, if $a > 1$, the graph will _____ from left to right.

**4.** When graphing exponential functions of the form $y = a^x$, if $0 < a < 1$, the graph will _____ from left to right.

**5.** In the compound interest formula $A = p\left(1 + \dfrac{r}{n}\right)^{nt}$, $p$ is the _____ or the initial investment amount.

**6.** In the compound interest formula $A = p\left(1 + \dfrac{r}{n}\right)^{nt}$, $n$ is the number of compounding _____ per year.

## Practice the Skills

*Graph each exponential function.*

**7.** $y = 2^x$

**8.** $y = 3^x$

**9.** $y = \left(\dfrac{1}{2}\right)^x$

**10.** $y = \left(\dfrac{1}{3}\right)^x$

**11.** $y = 4^x$

**12.** $y = 5^x$

**13.** $y = \left(\dfrac{1}{4}\right)^x$

**14.** $y = \left(\dfrac{1}{5}\right)^x$

**15.** $y = 3^{-x}$

**16.** $y = 4^{-x}$

**17.** $y = \left(\dfrac{1}{3}\right)^{-x}$

**18.** $y = \left(\dfrac{1}{4}\right)^{-x}$

**19.** $y = 2^{x-1}$

**20.** $y = 2^{x+1}$

**21.** $y = \left(\dfrac{1}{3}\right)^{x+1}$

**22.** $y = \left(\dfrac{1}{3}\right)^{x-1}$

**23.** $y = 2^x + 1$

**24.** $y = 2^x - 1$

**25.** $y = 3^x - 1$

**26.** $y = 3^x + 2$

## Problem Solving

**27. U.S. Population** The following graph shows the growth in the U.S. population of people age 85 and older for the years from 1960 to 2000 and projected to 2050. The exponential function that closely approximates this graph is

$$f(t) = 0.592(1.042)^t$$

In this function, $f(t)$ is the population, in millions, of people age 85 and older and $t$ is the number of years since 1960. Assuming this trend continues, use this function to estimate the number of U.S. people age 85 and older in **a)** 2060. **b)** 2100.

**U.S. Population 85 and Older**

Number (millions) vs Year

85 and older

1960 1970 1980 1990 2000 2010 2020 2030 2040 2050

Projected

Year

*Source:* U.S. Census Bureau

**28. World Population** Since about 1650 the world population has been growing exponentially. The exponential function that closely approximates the world population from 1650 and projected to 2015 is

$$f(t) = \frac{1}{2}(2.718)^{0.0072t}$$

In this function, $f(t)$ is the world population, in billions of people, and $t$ is the number of years since 1650. If this trend continues, estimate the world population in **a)** 2010. **b)** 2015.

**29. Doubling** If $2 is doubled each day for 9 days, determine the amount on day 9.

**30. Doubling** If $2 is doubled each day for 12 days, determine the amount on day 12.

**31. Bacteria in a Petri Dish** Five bacteria are placed in a petri dish. The population will triple every day. The formula for the number of bacteria in the dish on day $t$ is

$$N(t) = 5(3)^t$$

where $t$ is the number of days after the five bacteria are placed in the dish. How many bacteria are in the dish 2 days after the five bacteria are placed in the dish?

**32. Bacteria in a Petri Dish** Refer to Exercise 31. How many bacteria are in the dish 6 days after the five bacteria are placed in the dish?

33. **Compound Interest** If Don Gecewicz invests $5000 at 6% interest compounded quarterly, find the amount after 4 years (see Example 5).

34. **Compound Interest** If Don Treadwell invests $8000 at 4% interest compounded quarterly, find the amount after 5 years.

35. **Certificate of Deposit** Joni Burnette receives a bonus of $5000 for meeting her annual sales quota. She invests the bonus in a certificate of deposit (CD) that pays 4.2% interest compounded monthly. Determine the value of the CD after 5 years.

36. **Money Market Account** Martha Goshaw invests $2500 in a money market account that pays 3.6% interest compounded quarterly. Determine the accumulated amount after 2 years.

37. **Savings Account** Byron Dyce deposits $3000 in a savings account that pays 2.4% interest compounded quarterly. Determine the accumulated amount after 2 years.

38. **Retirement Account** To invest for his retirement, John Salak invests $10,000 in an account paying 6% interest compounded semiannually. Determine the accumulated amount after 25 years.

39. **Carbon 14 Dating** If 12 grams of carbon 14 are originally present in a certain animal bone, how much will remain at the end of 1000 years? Use $A = A_0 \cdot 2^{-t/5600}$ (see Example 6).

40. **Carbon 14 Dating** If 60 grams of carbon 14 are originally present in the fossil Tim Jonas found at an archeological site, how much will remain after 10,000 years?

41. **Radioactive Substance** The amount of a radioactive substance present, in grams, at time $t$, in years, is given by the formula $y = 80(2)^{-0.4t}$. Find the number of grams present in **a)** 10 years. **b)** 100 years.

42. **Radioactive Substance** The amount of a radioactive substance present, in grams, at time $t$, in years, is given by the formula $y = 20(3)^{-0.6t}$. Find the number of grams present in 4 years.

43. **Population** The expected future population of Ackworth, which now has 2000 residents, can be approximated by the formula $y = 2000(1.2)^{0.1t}$, where $t$ is the number of years in the future. Find the expected population of the town in **a)** 10 years. **b)** 50 years.

44. **Population** The expected future population of Antwerp, which currently has 6800 residents, can be approximated by the formula $y = 6800(1.4)^{-0.2t}$, where $t$ is the number of years in the future. Find the expected population of the town 30 years in the future.

45. **Value of an SUV** The cost of a new SUV is $24,000. If it depreciates at a rate of 18% per year, the value of the SUV in $t$ years can be approximated by the formula

$$V(t) = 24,000(0.82)^t$$

Find the value of the SUV in 4 years.

46. **Value of an ATV** The cost of a new all-terrain vehicle is $6200. If it depreciates at a rate of 15% per year, the value of the ATV in $t$ years can be approximated by the formula

$$V(t) = 6200(0.85)^t$$

Find the value of the ATV in 10 years.

© Allen R. Angel

See Exercise 46.

47. **Atmospheric Pressure** Atmospheric pressure varies with altitude. The greater the altitude, the lower the pressure, as shown in the following graph.

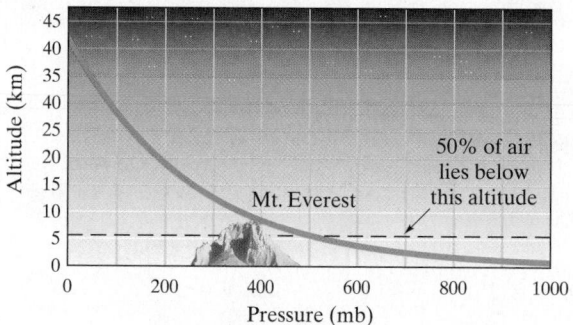

The equation $A = 41.97(0.996)^x$ can be used to estimate the altitude, $A$, in kilometers, for a given pressure, $x$, in millibars (mb). If the atmospheric pressure on top of Mt. Everest is about 389 mb, estimate the altitude of the top of Mt. Everest.

48. **Centenarians** Based on projections of the U.S. Census Bureau, the number of centenarians (people age 100 or older) will grow exponentially beyond 1995 (see the graph below). The function

$$f(t) = 71.24(1.045)^t$$

can be used to approximate the number of centenarians, in thousands, in the United States where $t$ is time in years since 1995. Use this function to estimate the number of centenarians in **a)** 2060. **b)** 2070.

**Number of Centenarians in the United States**

*Source:* U.S. Bureau of the Census

**49. Bike Shop Sales** Spokes for Folks, a bike shop, has annual sales for years 2006–2010 (in thousands of dollars) as shown in the graph below.

**Spokes for Folks Sales**

*Source:* U.S. Dept. of Commerce

The annual sales can be estimated by the function $S(t) = 51.4(1.85)^t$, where $S(t)$ is annual sales in thousands of dollars and $t$ is the number of years after 2006. Assuming this trend continues, determine the annual sales for the following years. Round your answers to the nearest thousand dollars.

**a)** 2015

**b)** 2020

**50. Sign Sales** Signs 2 Go, a sign-making print shop, has annual sales for years 2006–2010 (in thousands of dollars) as shown in the graph below.

**Signs 2 Go Sales**

The annual sales can be estimated by the function $S(t) = 23.1(1.19)^t$, where $S(t)$ is annual sales in thousands of dollars and $t$ is the number of years after 2006. Assuming this trend

continues, determine the annual sales for the following years. Round your answers to the nearest thousand dollars.

**a)** 2015

**b)** 2020

**51. Simple and Compound Interest** The following graph indicates linear growth of $100 invested at 7% simple interest and exponential growth at 7% interest compounded annually. In the formulas, $A$ represents the amount in dollars and $t$ represents the time in years.

**a)** Use the graph to estimate the doubling time for $100 invested at 7% simple interest.

**b)** Estimate the doubling time for $100 invested at 7% interest compounded annually.

**c)** Estimate the difference in amounts after 10 years for $100 invested by each method.

**d)** Most banks compound interest daily instead of annually. What effect does this have on the total amount? Explain.

**52.** In Exercise 51, we graphed the amount for various years when $100 is invested at 7% simple interest and at 7% interest compounded annually.

**a)** Use the compound interest formula to determine the amount if $100 is compounded daily at 7% for 10 years (assume 365 days per year).

**b)** Estimate the difference in the amount in 10 years for the $100 invested at 7% simple interest versus the 7% interest compounded daily.

## Concept/Writing Exercises

**53.** Consider the exponential function $y = \left(\frac{1}{2}\right)^x$.

**a)** As $x$ increases, what happens to $y$?

**b)** Can $y$ ever be 0? Explain.

**c)** Can $y$ ever be negative? Explain.

**54.** Consider the exponential function $y = 2^x$.

**a)** As $x$ increases, what happens to $y$?

**b)** Can $y$ ever be 0? Explain.

**c)** Can $y$ ever be negative? Explain.

**55.** Consider the equations $y = 2^x$ and $y = 3^x$.

**a)** Will both graphs have the same or different $y$-intercepts? Determine their $y$-intercepts.

**b)** How will the graphs of the two functions compare?

**56.** Consider the equations $y = \left(\frac{1}{2}\right)^x$ and $y = \left(\frac{1}{3}\right)^x$.

**a)** Will both graphs have the same or different $y$-intercepts? Determine their $y$-intercepts.

**b)** How will the graphs of the two functions compare?

**57.** We stated earlier that, for exponential functions $f(x) = a^x$, the value of $a$ cannot equal 1.

**a)** What does the graph of $f(x) = a^x$ look like when $a = 1$?

**b)** Is $f(x) = a^x$ a function when $a = 1$?

**c)** Does $f(x) = a^x$ have an inverse function when $a = 1$? Explain your answer.

**58.** How will the graphs of $y = a^x$ and $y = a^x + k, k > 0$, compare?

**59.** How will the graphs of $y = a^x$ and $y = a^x - k, k > 0$, compare?

**60.** For $a > 1$, how will the graphs of $y = a^x$ and $y = a^{x+1}$ compare?

**61.** For $a > 1$, how will the graphs of $y = a^x$ and $y = a^{x+2}$ compare?

**62. a)** Is $y = x^{\pi}$ an exponential function? Explain.

 **b)** Is $y = \pi^x$ an exponential function? Explain.

## Challenge Problem

**63.** Suppose Bob Jenkins gives Carol Dantuma $1 on day 1, $2 on day 2, $4 on day 3, $8 on day 4, and continues this doubling process for 30 days.

 **a)** Determine how much Bob will give Carol on day 15.

 **b)** Determine how much Bob will give Carol on day 20.

 **c)** Express the amount, using exponential form, that Bob gives Carol on day $n$.

 **d)** How much, in dollars, will Bob give Carol on day 30? Write the amount in exponential form. Then use a calculator to evaluate.

 **e)** Express the total amount Bob gives Carol over the 30 days as a sum of exponential terms. (Do not find the actual value.)

## Group Activity

**64.** Functions that are exponential or are approximately exponential are commonly seen.

 **a)** Have each member of the group individually determine a function not given in this section that may approximate an exponential function. You may use newspapers, books, or other sources.

 **b)** As a group, discuss one another's functions. Determine whether each function presented is an exponential function.

 **c)** As a group, write a paper that discusses each of the exponential functions and state why you believe each function is exponential.

## Cumulative Review Exercises

[5.4]  **65.** Consider the polynomial

$$2.3x^4y - 6.2x^6y^2 + 9.2x^5y^2$$

 **a)** Write the polynomial in descending order of the variable $x$.

 **b)** What is the degree of the polynomial?

 **c)** What is the leading coefficient?

[8.5]   **66.** If $f(x) = x + 5$ and $g(x) = x^2 - 2x + 4$, find $(f \cdot g)(x)$.

[11.1]  **67.** Write $\sqrt{a^2 - 8a + 16}$ as an absolute value.

[11.3]  **68.** Simplify $\sqrt[4]{\dfrac{32x^5y^9}{2y^3z}}$.

# 13.3 Logarithmic Functions

1 Define a logarithm.

2 Convert from exponential form to logarithmic form.

3 Graph logarithmic functions.

4 Compare the graphs of exponential and logarithmic functions.

5 Solve applications of logarithmic functions.

## 1 Define a Logarithm

Consider the exponential function $y = 2^x$. In **Figure 13.13** on page 785, we see that the graph of this function passes the horizontal line test and therefore this function is a one-to-one function and has an inverse. To find the inverse of $y = 2^x$ we interchange $x$ and $y$ to get the equation $x = 2^y$. To solve this equation for $y$, we introduce a new definition.

**Logarithm**

For $x > 0$ and $a > 0, a \neq 1$,

$$y = \log_a x \text{ means } x = a^y$$

The expression $\log_a x$ is read "the logarithm of $x$ to the base $a$" or simply "log, base $a$, of $x$."

Using this definition, we can solve the equation $x = 2^y$ for $y$ by rewriting the exponential equation as an equation involving a logarithm.

$$\underbrace{x = 2^y}_{\substack{\text{Exponential} \\ \text{form}}} \text{ is an equivalent equation to } \underbrace{y = \log_2 x}_{\substack{\text{Logarithmic} \\ \text{form}}}$$

Thus, $y = 2^x$ and $y = \log_2 x$ are inverse functions. In general, $y = a^x$ and $y = \log_a x$ are inverse functions.

## 2  Convert from Exponential Form to Logarithmic Form

To convert from an exponential equation to a logarithmic equation consider the following diagram:

For example:

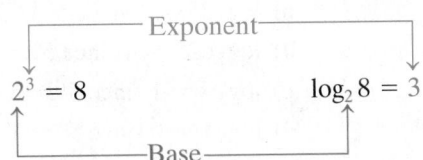

Thus the exponential equation $2^3 = 8$ is equivalent to the logarithmic equation $\log_2 8 = 3$ or $3 = \log_2 8$.

Following are some examples of how an exponential expression can be converted to a logarithmic expression.

| Exponential Form | Logarithmic Form |
|---|---|
| $10^0 = 1$ | $\log_{10} 1 = 0$ |
| $4^2 = 16$ | $\log_4 16 = 2$ |
| $\left(\dfrac{1}{2}\right)^5 = \dfrac{1}{32}$ | $\log_{1/2} \dfrac{1}{32} = 5$ |
| $5^{-2} = \dfrac{1}{25}$ | $\log_5 \dfrac{1}{25} = -2$ |

Now let's do a few examples involving conversion from exponential form into logarithmic form, and vice versa.

**EXAMPLE 1**  Write each equation in logarithmic form.

**a)** $3^4 = 81$        **b)** $\left(\dfrac{1}{5}\right)^3 = \dfrac{1}{125}$        **c)** $2^{-5} = \dfrac{1}{32}$

*Solution*

**a)** $\log_3 81 = 4$        **b)** $\log_{1/5} \dfrac{1}{125} = 3$        **c)** $\log_2 \dfrac{1}{32} = -5$

Now Try Exercise 19

**EXAMPLE 2**  Write each equation in exponential form.

**a)** $\log_7 49 = 2$        **b)** $\log_4 64 = 3$        **c)** $\log_{1/3} \dfrac{1}{81} = 4$

*Solution*

**a)** $7^2 = 49$        **b)** $4^3 = 64$        **c)** $\left(\dfrac{1}{3}\right)^4 = \dfrac{1}{81}$

Now Try Exercise 35

**EXAMPLE 3** Write each equation in exponential form; then find the unknown value.

**a)** $y = \log_5 25$        **b)** $2 = \log_a 16$        **c)** $3 = \log_{1/2} x$

*Solution*

**a)** $5^y = 25$. Since $5^2 = 25$, $y = 2$.

**b)** $a^2 = 16$. Since $4^2 = 16$, $a = 4$. Note that $a$ must be greater than 0, so $-4$ is not a possible answer for $a$.

**c)** $\left(\frac{1}{2}\right)^3 = x$. Since $\left(\frac{1}{2}\right)^3 = \frac{1}{8}$, $x = \frac{1}{8}$.

Now Try Exercise 53

**EXAMPLE 4** Evaluate the following

**a)** $\log_5 25$        **b)** $\log_5 625$        **c)** $\log_5 5$

**d)** $\log_5 1$        **e)** $\log_5 \left(\frac{1}{5}\right)$        **f)** $\log_5 \sqrt{5}$

*Solution*

**a)** $\log_5 25 = 2$ since $5^2 = 25$.

**b)** $\log_5 625 = 4$ since $5^4 = 625$.

**c)** $\log_5 5 = 1$ since $5^1 = 5$.

**d)** $\log_5 1 = 0$ since $5^0 = 1$.

**e)** $\log_5 \left(\frac{1}{5}\right) = -1$ since $5^{-1} = \frac{1}{5}$.

**f)** $\log_5 \sqrt{5} = \frac{1}{2}$ since $5^{1/2} = \sqrt{5}$.

Now Try Exercise 75

**Helpful Hint**

Since a logarithm is an exponent, it is very important to know the rules of exponents when evaluating logarithmic expressions. It would be helpful to review the rules of exponents from Sections 5.1 and 5.2 as well as the relationship between radicals and rational exponents from Section 11.2.

## 3   Graph Logarithmic Functions

We are now ready to introduce logarithmic functions.

**Logarithmic Function**

For any real number $a > 0$, $a \neq 1$, and $x > 0$,

$$f(x) = \log_a x \text{ or } y = \log_a x$$

is a **logarithmic function**.

### Examples of Logarithmic Functions

$$f(x) = \log_5 x \qquad y = \log_{1/2} x \qquad g(x) = \log_5 x$$

Logarithmic functions can be graphed by converting the logarithmic equation to an exponential equation and then plotting points.

Before we graph logarithmic functions, let's discuss some characteristics of the graphs of logarithmic functions.

### Graphs of Logarithmic Functions

For all logarithmic functions of the form $y = \log_a x$ or $f(x) = \log_a x$, where $a > 0$, $a \neq 1$, and $x > 0$

1. The domain of the function is $(0, \infty)$.
2. The range of the function is $(-\infty, \infty)$.
3. The graph passes through the points $\left(\dfrac{1}{a}, -1\right)$, $(1, 0)$, and $(a, 1)$.

**EXAMPLE 5**  Graph $y = \log_2 x$. State the domain and range of the function.

**Solution**  This is an equation of the form $y = \log_a x$, where $a = 2$. $y = \log_2 x$ means $x = 2^y$. Using $x = 2^y$, construct a table of values. The table will be easier to develop by selecting values for $y$ and finding the corresponding values for $x$. In the table, the three points listed in item 3 in the box are shown in blue.

| $x$ | $\dfrac{1}{16}$ | $\dfrac{1}{8}$ | $\dfrac{1}{4}$ | $\dfrac{1}{2}$ | 1 | 2 | 4 | 8 | 16 |
|---|---|---|---|---|---|---|---|---|---|
| $y$ | $-4$ | $-3$ | $-2$ | $-1$ | 0 | 1 | 2 | 3 | 4 |

**FIGURE 13.16**

Now draw the graph (**Fig. 13.16**). The three ordered pairs in blue in the table are marked in blue on the graph. The domain, the set of $x$-values, is $\{x \mid x > 0\}$. The range, the set of $y$-values, is the set of all real numbers, $\mathbb{R}$.

Now Try Exercise 91

**EXAMPLE 6**  Graph $y = \log_{1/2} x$. State the domain and range of the function.

**Solution**  This is an equation of the form $y = \log_a x$, where $a = \dfrac{1}{2}$. $y = \log_{1/2} x$ means $x = \left(\dfrac{1}{2}\right)^y$. Construct a table of values by selecting values for $y$ and finding the corresponding values of $x$.

| $x$ | 16 | 8 | 4 | 2 | 1 | $\dfrac{1}{2}$ | $\dfrac{1}{4}$ | $\dfrac{1}{8}$ | $\dfrac{1}{16}$ |
|---|---|---|---|---|---|---|---|---|---|
| $y$ | $-4$ | $-3$ | $-2$ | $-1$ | 0 | 1 | 2 | 3 | 4 |

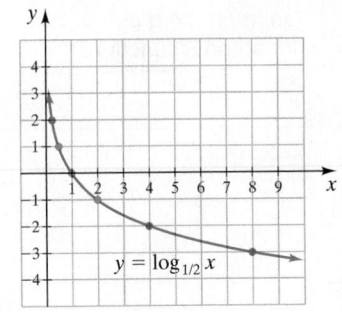

**FIGURE 13.17**

The graph is illustrated in **Figure 13.17**. The domain is $\{x \mid x > 0\}$. The range is the set of real numbers, $\mathbb{R}$.

Now Try Exercise 93

If we study the domains in Examples 5 and 6, we see that the domains of both $y = \log_2 x$ and $y = \log_{1/2} x$ are $\{x \mid x > 0\}$. In fact, *for any logarithmic function* $y = \log_a x$, *the domain is* $\{x \mid x > 0\}$. Also note that the graphs in Examples 5 and 6 are both graphs of one-to-one functions.

## Understanding Algebra

When graphing functions of the form $y = \log_a x$ or $f(x) = \log_a x$, we can predict the shape of the graph by observing three points:

$$\left(\dfrac{1}{a}, -1\right), \ (1, 0), \text{ and } (a, 1).$$

- When $a > 1$, the graph becomes almost vertical to left of $\left(\dfrac{1}{a}, -1\right)$ and somewhat horizontal to the right of $(a, 1)$; see Example 5.
- When $0 < a < 1$, the graph becomes almost vertical to the left of $(a, 1)$ and somewhat horizontal to the right of $\left(\dfrac{1}{a}, -1\right)$; see Example 6.

## 4  Compare the Graphs of Exponential and Logarithmic Functions

Recall that $y = a^x$ and $y = \log_a x$ are *inverse functions*. We may therefore write: if $f(x) = a^x$ then $f^{-1}(x) = \log_a x$. In the box below we highlight some of the characteristics of the graphs of the general exponential function $y = a^x$ and the general logarithmic function $y = \log_a x$.

**Graph Characteristics**

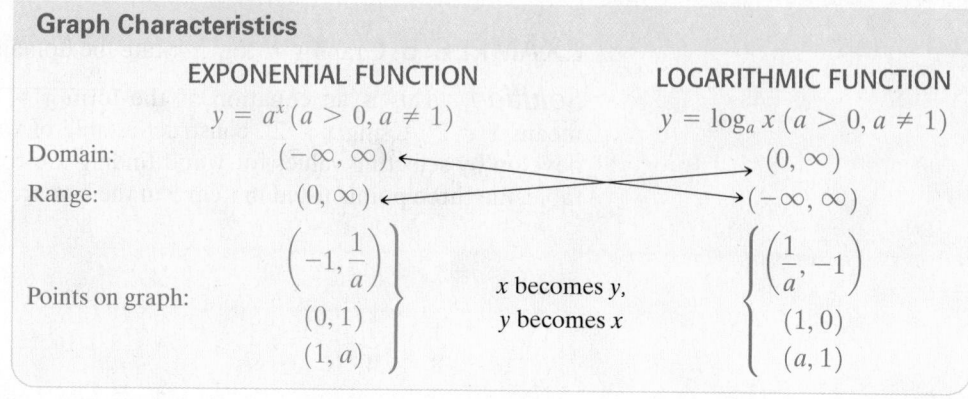

|  | EXPONENTIAL FUNCTION $y = a^x \,(a > 0, a \neq 1)$ |  | LOGARITHMIC FUNCTION $y = \log_a x \,(a > 0, a \neq 1)$ |
|---|---|---|---|
| Domain: | $(-\infty, \infty)$ |  | $(0, \infty)$ |
| Range: | $(0, \infty)$ |  | $(-\infty, \infty)$ |
| Points on graph: | $\left(-1, \dfrac{1}{a}\right)$ <br> $(0, 1)$ <br> $(1, a)$ | $x$ becomes $y$, <br> $y$ becomes $x$ | $\left(\dfrac{1}{a}, -1\right)$ <br> $(1, 0)$ <br> $(a, 1)$ |

In **Figure 13.18**, we show the graphs of $y = a^x$ and $y = \log_a x$ for $a > 1$. Notice the graphs are symmetric about the line $y = x$.

FIGURE 13.18

The graphs of $y = 2^x$ and $y = \log_2 x$ are illustrated in **Figure 13.19**. The graphs of $y = \left(\dfrac{1}{2}\right)^x$ and $y = \log_{1/2} x$ are illustrated in **Figure 13.20**. In each figure, the graphs are inverses of each other and are symmetric with respect to the line $y = x$.

FIGURE 13.19

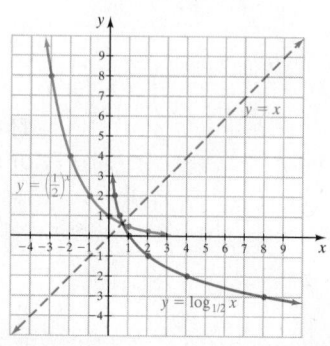

FIGURE 13.20

## 5 Solve Applications of Logarithmic Functions

We will see many applications of logarithms later, but let's look at one important application now.

EXAMPLE 7   **Earthquakes**   Logarithms are used to measure the magnitude of earthquakes. The Richter scale for measuring earthquakes was developed by Charles R. Richter. The magnitude, $R$, of an earthquake on the Richter scale is given by the formula

$$R = \log_{10} I$$

where $I$ represents the number of times greater (or more intense) the earthquake is than the smallest measurable activity that can be measured on a seismograph.

a) If an earthquake measures 4 on the Richter scale, how many times more intense is it than the smallest measurable activity?

b) How many times more intense is an earthquake that measures 5 on the Richter scale than an earthquake that measures 4?

*Solution*

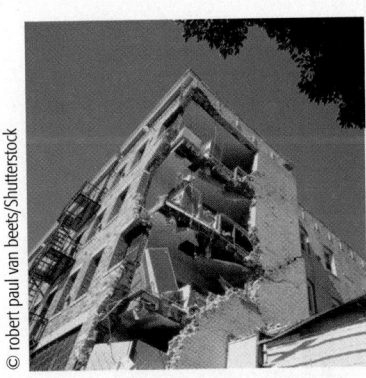

© robert paul van beets/Shutterstock

a) Understand   The Richter number, $R$, is 4. To find how many times more intense the earthquake is than the smallest measurable activity, $I$, we substitute $R = 4$ into the formula and solve for $I$.

Translate                    $R = \log_{10} I$

                             $4 = \log_{10} I$

Carry Out                    $10^4 = I$        Change to exponential form.

                        $10,000 = I$

Answer   Therefore, an earthquake that measures 4 on the Richter scale is 10,000 times more intense than the smallest measurable activity.

b)                           $5 = \log_{10} I$

                             $10^5 = I$        Change to exponential form.

                       $100,000 = I$

Since $(10,000)(10) = 100,000$, an earthquake that measures 5 on the Richter scale is 10 times more intense than an earthquake that measures 4.

Now Try Exercise 99

---

# EXERCISE SET 13.3    MathXL®   MyMathLab®
                        MathXL®      MyMathLab

## Warm-Up Exercises

*Fill in the blanks with the appropriate word, phrase, or symbol(s) from the following list.*

logarithmic        exponent        symmetric        base        inverse        domain        range        composite

**1.** The expression $\log_a x$ represents the _____ to which the base $a$ must be raised to obtain $x$.

**2.** In general, $y = a^x$ and $y = \log_a x$ are _____ functions.

**3.** In both of the equations $a^y = x$ and $\log_a x = y$, $a$ is called the _____.

**4.** $f(x) = \log_5 x$ is an example of a _____ function.

**5.** For any logarithmic function, the _____ is $\{x \mid x > 0\}$.

**6.** The graphs of $y = a^x$ and $y = \log_a x$ are _____ with respect to the line $y = x$.

## Practice the Skills

*Write each equation in logarithmic form.*

**7.** $5^2 = 25$

**8.** $4^2 = 16$

**9.** $3^2 = 9$

**10.** $2^6 = 64$

**11.** $16^{1/2} = 4$

**12.** $49^{1/2} = 7$

**13.** $8^{1/3} = 2$

**14.** $16^{1/4} = 2$

**15.** $\left(\dfrac{1}{2}\right)^5 = \dfrac{1}{32}$

**16.** $\left(\dfrac{1}{3}\right)^4 = \dfrac{1}{81}$

**17.** $2^{-3} = \dfrac{1}{8}$

**18.** $6^{-3} = \dfrac{1}{216}$

**19.** $4^{-3} = \dfrac{1}{64}$

**20.** $81^{1/2} = 9$

**21.** $64^{1/3} = 4$

**22.** $5^{-4} = \dfrac{1}{625}$

**23.** $8^{-1/3} = \dfrac{1}{2}$

**24.** $16^{-1/2} = \dfrac{1}{4}$

**25.** $81^{-1/4} = \dfrac{1}{3}$

**26.** $32^{-1/5} = \dfrac{1}{2}$

**27.** $10^{0.8451} = 7$

**28.** $10^{1.0792} = 12$

**29.** $e^2 = 7.3891$

**30.** $e^{-1/2} = 0.6065$

**31.** $a^n = b$

**32.** $c^b = w$

*Write each equation in exponential form.*

**33.** $\log_3 9 = 2$

**34.** $\log_4 64 = 3$

**35.** $\log_{1/3} \dfrac{1}{27} = 3$

**36.** $\log_{1/2} \dfrac{1}{64} = 6$

**37.** $\log_5 \dfrac{1}{25} = -2$

**38.** $\log_5 \dfrac{1}{625} = -4$

**39.** $\log_{49} 7 = \dfrac{1}{2}$

**40.** $\log_{64} 4 = \dfrac{1}{3}$

**41.** $\log_9 \dfrac{1}{81} = -2$

**42.** $\log_{10} \dfrac{1}{100} = -2$

**43.** $\log_{10} \dfrac{1}{1000} = -3$

**44.** $\log_{10} 1000 = 3$

**45.** $\log_6 216 = 3$

**46.** $\log_4 1024 = 5$

**47.** $\log_{10} 0.62 = -0.2076$

**48.** $\log_{10} 8 = 0.9031$

**49.** $\log_e 6.52 = 1.8749$

**50.** $\log_e 30 = 3.4012$

**51.** $\log_w s = -p$

**52.** $\log_r c = -a$

*Write each equation in exponential form; then find the unknown value.*

**53.** $\log_4 64 = y$

**54.** $\log_5 25 = y$

**55.** $\log_a 125 = 3$

**56.** $\log_a 81 = 4$

**57.** $\log_3 x = 3$

**58.** $\log_2 x = 5$

**59.** $\log_2 \dfrac{1}{16} = y$

**60.** $\log_8 \dfrac{1}{64} = y$

**61.** $\log_{1/2} x = 6$

**62.** $\log_{1/3} x = 4$

**63.** $\log_a \dfrac{1}{27} = -3$

**64.** $\log_9 \dfrac{1}{81} = y$

**65.** $\log_{25} 5 = y$

**66.** $\log_{36} 6 = y$

*Evaluate the following.*

**67.** $\log_{10} 1$

**68.** $\log_{10} 10$

**69.** $\log_{10} 100$

**70.** $\log_{10} 1000$

**71.** $\log_{10} \dfrac{1}{100}$

**72.** $\log_{10} \dfrac{1}{1000}$

**73.** $\log_{10} 10{,}000$

**74.** $\log_{10} 100{,}000$

**75.** $\log_4 256$

**76.** $\log_{13} 169$

**77.** $\log_3 \dfrac{1}{81}$

**78.** $\log_5 \dfrac{1}{125}$

**79.** $\log_8 \dfrac{1}{64}$

**80.** $\log_{14} \dfrac{1}{14}$

**81.** $\log_7 \sqrt{7}$

**82.** $\log_7 \sqrt[3]{7}$

**83.** $\log_9 9$

**84.** $\log_{12} 12$

**85.** $\log_{100} 10$

**86.** $\log_{1000} 10$

*Graph the logarithmic function.*

**87.** $y = \log_2 x$

**88.** $y = \log_3 x$

**89.** $y = \log_{1/2} x$

**90.** $y = \log_{1/3} x$

**91.** $y = \log_5 x$

**92.** $y = \log_4 x$

**93.** $y = \log_{1/5} x$

**94.** $y = \log_{1/4} x$

*Graph each pair of functions on the same axes.*

**95.** $y = 2^x$, $y = \log_{1/2} x$

**96.** $y = \left(\dfrac{1}{2}\right)^x$, $y = \log_2 x$

**97.** $y = 2^x$, $y = \log_2 x$

**98.** $y = \left(\dfrac{1}{2}\right)^x$, $y = \log_{1/2} x$

## Problem Solving

99. **Earthquake** If the magnitude of an earthquake is 7 on the Richter scale, how many times more intense is the earthquake than the smallest measurable activity? Use $R = \log_{10} I$ (see Example 7).

100. **Earthquake** If the magnitude of an earthquake is 5 on the Richter scale, how many times more intense is the earthquake than the smallest measurable activity? Use $R = \log_{10} I$.

101. **Earthquake** How many times more intense is an earthquake that measures 6 on the Richter scale than an earthquake that measures 2?

102. **Earthquake** How many times more intense is an earthquake that measures 4 on the Richter scale than an earthquake that measures 1?

103. Graph $y = \log_2 (x - 1)$.

104. Graph $y = \log_3 (x - 2)$.

## Concept/Writing Exercises

105. Consider the logarithmic function $y = \log_a x$.

    **a)** What are the restrictions on $a$?

    **b)** What is the domain of the function?

    **c)** What is the range of the function?

106. For the logarithmic function $y = \log_a(x - 3)$, what must be true about $x$? Explain.

107. If some points on the graph of the exponential function $f(x) = a^x$ are $\left(-3, \frac{1}{27}\right)$, $\left(-2, \frac{1}{9}\right)$, $\left(-1, \frac{1}{3}\right)$, $(0, 1)$, $(1, 3)$, $(2, 9)$, and $(3, 27)$, list some points on the graph of the logarithmic function $g(x) = \log_a x$. Explain how you determined your answer.

108. What is the $x$-intercept of the graph of an equation of the form $y = \log_a x$?

109. If $f(x) = 5^x$, what is $f^{-1}(x)$?

110. If $f(x) = \log_6 x$, what is $f^{-1}(x)$?

111. Between which two integers must $\log_3 62$ lie? Explain.

112. Between which two integers must $\log_{10} 0.672$ lie? Explain.

113. Between which two integers must $\log_{10} 425$ lie? Explain.

114. Between which two integers must $\log_5 0.3256$ lie? Explain.

115. For $x > 1$, which will grow faster as $x$ increases, $2^x$ or $\log_{10} x$? Explain.

116. For $x > 1$, which will grow faster as $x$ increases, $x$ or $\log_{10} x$? Explain.

## Cumulative Review Exercises

[6.3–6.5]   *Factor.*

117. $2x^3 - 6x^2 - 36x$

118. $x^4 - 16$

119. $40x^2 + 52x - 12$

120. $6r^2s^2 + rs - 1$

# 13.4 Properties of Logarithms

**1** Use the product rule for logarithms.

**2** Use the quotient rule for logarithms.

**3** Use the power rule for logarithms.

**4** Use additional properties of logarithms.

In this section we will study several properties of logarithms. We begin with an important definition.

> **Argument**
>
> In the logarithmic expression $\log_a x$, $x$ is called the **argument** of the logarithm.

| Logarithmic Expression | Argument |
|:---:|:---:|
| $\log_{10} 3$ | 3 |
| $\log_2 (x - 5)$ | $x - 5$ |
| $\log_7 (x^2 - 4x + 2)$ | $x^2 - 4x + 2$ |

*Since we can only take the logarithm of positive numbers, when an argument contains a variable, we will assume that the argument represents a positive number.*

## 1 Use the Product Rule for Logarithms

### Product Rule for Logarithms

For positive real numbers $x, y$, and $a, a \neq 1$,

$$\log_a xy = \log_a x + \log_a y \qquad \text{Property 1}$$

The logarithm of a product equals the sum of the logarithms of the factors.

### Examples of Property 1

$$\log_3 (6 \cdot 7) = \log_3 6 + \log_3 7$$

$$\log_4 3z = \log_4 3 + \log_4 z$$

$$\log_8 x^2 = \log_8 (x \cdot x) = \log_8 x + \log_8 x \text{ or } 2 \log_8 x$$

Property 1, the product rule, can be expanded to three or more factors, for example, $\log_a xyz = \log_a x + \log_a y + \log_a z$.

## 2 Use the Quotient Rule for Logarithms

### Quotient Rule for Logarithms

For positive real numbers $x, y$, and, $a, a \neq 1$,

$$\log_a \frac{x}{y} = \log_a x - \log_a y \qquad \text{Property 2}$$

The logarithm of a quotient equals the logarithm of the numerator minus the logarithm of the denominator.

### Examples of Property 2

$$\log_3 \frac{19}{4} = \log_3 19 - \log_3 4$$

$$\log_6 \frac{x}{3} = \log_6 x - \log_6 3$$

$$\log_5 \frac{z}{z + 2} = \log_5 z - \log_5 (z + 2)$$

## 3 Use the Power Rule for Logarithms

### Power Rule for Logarithms

If $x$ and $a$ are positive real numbers, $a \neq 1$, and $n$ is any real number, then

$$\log_a x^n = n \log_a x \qquad \text{Property 3}$$

The logarithm of a number raised to an exponent equals the exponent times the logarithm of the number.

Examples of Property 3

$$\log_2 4^3 = 3 \log_2 4$$

$$\log_3 x^2 = 2 \log_3 x$$

$$\log_5 \sqrt{12} = \log_5 (12)^{1/2} = \frac{1}{2} \log_5 12$$

$$\log_8 \sqrt[5]{z + 3} = \log_8 (z + 3)^{1/5} = \frac{1}{5} \log_8 (z + 3)$$

**EXAMPLE 1** Use properties 1 through 3 to expand.

**a)** $\log_8 \dfrac{29}{43}$      **b)** $\log_4 (64 \cdot 180)$      **c)** $\log_{10} (22)^{1/5}$

*Solution*

**a)** $\log_8 \dfrac{29}{43} = \log_8 29 - \log_8 43$          Quotient rule

**b)** $\log_4 (64 \cdot 180) = \log_4 64 + \log_4 180$          Product rule

**c)** $\log_{10} (22)^{1/5} = \dfrac{1}{5} \log_{10} 22$          Power rule

Now Try Exercise 11

Often we will have to use two or more of these properties in the same problem.

**EXAMPLE 2** Expand.

**a)** $\log_{10} 4(x + 2)^3$          **b)** $\log_5 \dfrac{(4 - a)^2}{3}$

**c)** $\log_5 \left( \dfrac{4 - a}{3} \right)^2$          **d)** $\log_5 \dfrac{[x(x + 4)]^3}{8}$

*Solution*

**a)**  $\log_{10} 4(x + 2)^3 = \log_{10} 4 + \log_{10} (x + 2)^3$          Product rule

$= \log_{10} 4 + 3 \log_{10} (x + 2)$          Power rule

**b)**  $\log_5 \dfrac{(4 - a)^2}{3} = \log_5 (4 - a)^2 - \log_5 3$          Quotient rule

$= 2 \log_5 (4 - a) - \log_5 3$          Power rule

**c)**  $\log_5 \left( \dfrac{4 - a}{3} \right)^2 = 2 \log_5 \left( \dfrac{4 - a}{3} \right)$          Power rule

$= 2[\log_5 (4 - a) - \log_5 3]$          Quotient rule

$= 2 \log_5 (4 - a) - 2 \log_5 3$          Distributive property

**d)** $\log_5 \dfrac{[x(x + 4)]^3}{8} = \log_5 [x(x + 4)]^3 - \log_5 8$          Quotient rule

$= 3 \log_5 x(x + 4) - \log_5 8$          Power rule

$= 3[\log_5 x + \log_5 (x + 4)] - \log_5 8$          Product rule

$= 3 \log_5 x + 3 \log_5 (x + 4) - \log_5 8$          Distributive property

Now Try Exercise 21

> **Helpful Hint**
>
> In Example 2 **b)**, when we expanded $\log_5 \dfrac{(4-a)^2}{3}$, we first used the quotient rule. In Example 2 **c)**, when we expanded $\log_5 \left(\dfrac{4-a}{3}\right)^2$, we first used the power rule. Do you see the difference in the two problems? In $\log_5 \dfrac{(4-a)^2}{3}$, just the numerator of the argument is squared; therefore, we use the quotient rule first. In $\log_5 \left(\dfrac{4-a}{3}\right)^2$, the entire argument is squared, so we use the power rule first.

EXAMPLE 3  Write each of the following as the logarithm of a single expression.

**a)** $3 \log_8 (z + 2) - \log_8 z$

**b)** $\log_7 (x + 1) + 2 \log_7 (x + 4) - 3 \log_7 (x - 5)$

*Solution*

**a)** $3 \log_8 (z + 2) - \log_8 z = \log_8 (z + 2)^3 - \log_8 z$      Power rule

$$= \log_8 \frac{(z + 2)^3}{z} \qquad \text{Quotient rule}$$

**b)**    $\log_7 (x + 1) + 2 \log_7 (x + 4) - 3 \log_7 (x - 5)$

$= \log_7 (x + 1) + \log_7 (x + 4)^2 - \log_7 (x - 5)^3$      Power rule

$= \log_7 (x + 1)(x + 4)^2 - \log_7 (x - 5)^3$      Product rule

$$= \log_7 \frac{(x + 1)(x + 4)^2}{(x - 5)^3} \qquad \text{Quotient rule}$$

Now Try Exercise 39

> **Avoiding Common Errors**
>
> THE CORRECT RULES ARE
>
> $$\log_a xy = \log_a x + \log_a y$$
>
> $$\log_a \frac{x}{y} = \log_a x - \log_a y$$
>
> Note that
>
> $$\log_a (x + y) \neq \log_a x + \log_a y \qquad \log_a xy \neq (\log_a x)(\log_a y)$$
>
> $$\log_a (x - y) \neq \log_a x - \log_a y \qquad \log_a \frac{x}{y} \neq \frac{\log_a x}{\log_a y}$$

## Understanding Algebra

Our last two logarithm properties are related to the properties of inverse functions. Earlier in this chapter we found that if two functions $f(x)$ and $f^{-1}(x)$ are inverses of each other, then

$$(f^{-1} \circ f)(x) = x$$

and

$$(f \circ f^{-1})(x) = x$$

We also discussed that the functions

$$f(x) = a^x \text{ and } f^{-1}(x) = \log_a x$$

are inverse functions. Therefore, we have

$$(f^{-1} \circ f)(x) = f^{-1}[f(x)]$$
$$= f^{-1}(a^x)$$
$$= \log_a a^x = x$$

and

$$(f \circ f^{-1})(x) = f[f^{-1}(x)]$$
$$= f(\log_a x)$$
$$= a^{\log_a x} = x$$

## 4  Use Additional Properties of Logarithms

The last properties we discuss in this section will be used to solve equations in Section 13.6.

**Additional Properties of Logarithms**

If $a > 0$, and $a \neq 1$, then

$$\log_a a^x = x \qquad \qquad \text{Property 4}$$

and      $a^{\log_a x} = x \ (x > 0)$      Property 5

| Examples of Property 4 | Examples of Property 5 |
|---|---|
| $\log_6 6^5 = 5$ | $3^{\log_3 7} = 7$ |
| $\log_9 9^x = x$ | $5^{\log_5 x} = x \ (x > 0)$ |

EXAMPLE 4  Evaluate.    **a)** $\log_5 25$    **b)** $\sqrt{16}^{\,\log_4 9}$

*Solution*

**a)** $\log_5 25$ may be written as $\log_5 5^2$. By property 4,

$$\log_5 25 = \log_5 5^2 = 2$$

**b)** $\sqrt{16}^{\,\log_4 9}$ may be written $4^{\log_4 9}$. By property 5,

$$\sqrt{16}^{\,\log_4 9} = 4^{\log_4 9} = 9$$

Now Try Exercise 55

## EXERCISE SET 13.4   MathXL®   MyMathLab®

### Warm-Up Exercises

*Fill in the blanks with the appropriate word, phrase, or symbol(s) from the following list.*

| times | plus | minus | positive | inverses | argument | sum | composites |
|---|---|---|---|---|---|---|---|

**1.** We can only take the logarithm of _____ numbers.

**2.** In the logarithmic expression $\log_a x$, the $x$ is called the _____ of the logarithm.

**3.** The logarithm of a product equals the _____ of the logarithms of the factors.

**4.** The logarithm of a quotient equals the logarithm of the numerator _____ the logarithm of the denominator.

**5.** The logarithm of a number raised to an exponent equals the exponent _____ the logarithm of the number.

**6.** If two functions $f(x)$ and $f^{-1}(x)$ are _____ of each other, then $(f^{-1} \circ f)(x) = x$ and $(f \circ f^{-1})(x) = x$.

### Practice the Skills

*Use properties 1–3 to expand.*

**7.** $\log_2 (3 \cdot 5)$

**8.** $\log_3 (2 \cdot 11)$

**9.** $\log_8 7(x + 3)$

**10.** $\log_9 x(x + 2)$

**11.** $\log_2 \dfrac{27}{11}$

**12.** $\log_5 (41 \cdot 9)$

**13.** $\log_{10} \dfrac{\sqrt{x}}{x - 9}$

**14.** $\log_5 3^8$

**15.** $\log_6 x^7$

**16.** $\log_9 12(4)^6$

**17.** $\log_4 (r + 7)^5$

**18.** $\log_8 b^3(b - 2)$

**19.** $\log_4 \sqrt{\dfrac{a^3}{a + 2}}$

**20.** $\log_9 (x - 6)^3 x^2$

**21.** $\log_3 \dfrac{d^6}{(a - 8)^4}$

**22.** $\log_7 x^2(x - 13)$

**23.** $\log_8 \dfrac{y(y + 4)}{y^3}$

**24.** $\log_{10} \left(\dfrac{z}{6}\right)^2$

**25.** $\log_{10} \dfrac{9m}{8n}$

**26.** $\log_5 \dfrac{\sqrt{a}\,\sqrt[3]{b}}{\sqrt[4]{c}}$

*Write as a logarithm of a single expression.*

**27.** $\log_2 3 + \log_2 7$

**28.** $\log_7 4 + \log_7 3$

**29.** $\log_2 9 - \log_2 5$

**30.** $\log_7 17 - \log_7 3$

**31.** $6 \log_4 2$

**32.** $\dfrac{1}{3} \log_8 7$

**33.** $\log_{10} x + \log_{10} (x + 3)$

**34.** $\log_5 (a + 1) - \log_5 (a + 10)$

**35.** $2 \log_9 z - \log_9 (z - 2)$

**36.** $3 \log_8 y + 2 \log_8 (y - 9)$

**37.** $4(\log_5 p - \log_5 3)$

**38.** $\dfrac{1}{2} [\log_6 (r - 1) - \log_6 r]$

**39.** $\log_2 n + \log_2 (n + 4) - \log_2 (n - 3)$

**40.** $2 \log_5 t + 5 \log_5 (t - 6) + \log_5 (3t + 7)$

**41.** $\dfrac{1}{2}[\log_5 (x - 8) - \log_5 x]$

**42.** $6 \log_7 (a + 3) + 2 \log_7 (a - 1) - \dfrac{1}{2} \log_7 a$

**43.** $2 \log_9 4 + \dfrac{1}{3} \log_9 (r - 6) - \dfrac{1}{2} \log_9 r$

**44.** $5 \log_6 (x + 3) - [2 \log_6 (7x + 1) + 3 \log_6 x]$

**45.** $4 \log_6 3 - [2 \log_6 (x + 3) + 4 \log_6 x]$

**46.** $2 \log_7 (m - 4) + 3 \log_7 (m + 3) - [5 \log_7 2 + 3 \log_7 (m - 2)]$

*Find the value by writing each argument using the numbers 2 and/or 5 and using the values $\log_a 2 = 0.3010$ and $\log_a 5 = 0.6990$.*

**47.** $\log_a 10$

**48.** $\log_a 2.5$

**49.** $\log_a 0.4$

**50.** $\log_a \dfrac{1}{8}$

**51.** $\log_a 25$

**52.** $\log_a \sqrt[3]{5}$

*Evaluate (see Example 4).*

**53.** $7^{\log_7 2}$

**54.** $\log_4 4$

**55.** $(2^3)^{\log_8 7}$

**56.** $\log_8 64$

**57.** $\log_3 27$

**58.** $2 \log_9 \sqrt{9}$

**59.** $5(\sqrt[3]{27})^{\log_3 5}$

**60.** $\dfrac{1}{2} \log_6 \sqrt[3]{6}$

## Problem Solving

**61.** Express $\log_a (x^2 - 4) - \log_a (x + 2)$ as a single logarithm and simplify.

**62.** Express $\log_a (x - 3) - \log_a (x^2 + 5x - 24)$ as a single logarithm and simplify.

*Use properties 1–3 to expand.*

**63.** $\log_2 \dfrac{\sqrt[4]{xy} \; \sqrt[3]{a}}{\sqrt[5]{a - b}}$

**64.** $\log_3 \left[ \dfrac{(a^2 + b^2)(c^2)}{(a - b)(b + c)(c + d)} \right]^2$

*If $\log_{10} x = 0.4320$, find the following.*

**65.** $\log_{10} x^2$

**66.** $\log_{10} \sqrt[3]{x}$

**67.** $\log_{10} \sqrt[4]{x}$

**68.** $\log_{10} x^{11}$

*If $\log_{10} x = 0.5000$ and $\log_{10} y = 0.2000$, find the following.*

**69.** $\log_{10} xy$

**70.** $\log_{10} \left( \dfrac{x}{y} \right)$

**71.** Using the information given in the instructions for Exercises 69 and 70, is it possible to find $\log_{10} (x + y)$? Explain.

**72.** Are the graphs of $y = \log_b x^2$ and $y = 2 \log_b x$ the same? Explain your answer by discussing the domain of each equation.

## Concept/Writing Exercises

**73.** Is $\log_a(xyz) = \log_a x + \log_a y + \log_a z$? Explain.

**74.** Is $\log_b(x + y + z) = \log_b x + \log_b y + \log_b z$? Explain.

**75.** Is $\log_a (x^2 + 8x + 16) = 2 \log_a (x + 4)$? Explain.

**76.** Is $\log_a (4x^2 - 20x + 25) = 2 \log_a (2x - 5)$? Explain.

## Challenge Problems

**77.** For $x > 0$ and $y > 0$, is $\log_a \dfrac{x}{y} = \log_a xy^{-1}$

$= \log_a x + \log_a y^{-1} = \log_a x + \log_a \dfrac{1}{y}$?

**78.** Read Exercise 77. By the quotient rule, $\log_a \dfrac{x}{y} = \log_a x - \log_a y$. Can we therefore conclude that $\log_a x - \log_a y = \log_a x + \log_a \dfrac{1}{y}$?

**79.** Use the product rule to show that

$$\log_a \dfrac{x}{y} = \log_a x + \log_a \dfrac{1}{y}$$

**80. a)** Explain why

$$\log_a \dfrac{3}{xy} \neq \log_a 3 - \log_a x + \log_a y$$

**b)** Expand $\log_a \dfrac{3}{xy}$ correctly.

## Group Activity

*Discuss and answer Exercise 81 as a group.*

**81.** Consider $\log_a \dfrac{\sqrt{x^4 y}}{\sqrt{xy^3}}$, where $x > 0$ and $y > 0$.

**a)** Group member 1: Expand the expression using the quotient rule.

**b)** Group member 2: Expand the expression using the product rule.

**c)** Group member 3: First simplify $\dfrac{\sqrt{x^4 y}}{\sqrt{xy^3}}$, then expand the resulting logarithm.

**d)** Check each other's work and make sure all answers are correct. Can this expression be simplified by all three methods?

## Cumulative Review Exercises

*For Exercises 82 and 83, perform each indicated operation.*

[7.2]  **82.** $\dfrac{2x + 5}{x^2 - 7x + 12} \div \dfrac{x - 4}{2x^2 - x + 15}$

[7.4]  **83.** $\dfrac{2x + 5}{x^2 - 7x + 12} - \dfrac{x - 4}{2x^2 - x - 15}$

[7.7]  **84. Paint a House.** Mike Eisen can paint a house by himself in 4 days and Jill McGhee can paint the same house by herself in 5 days. How long would it take them to paint the house together?

[11.4]  **85.** Multiply and then simplify $\sqrt[3]{4x^4y^7} \cdot \sqrt[3]{12x^7y^{10}}$.

[11.6]  **86.** Solve $2a - 7\sqrt{a} = 30$.

---

## Mid-Chapter Test: 13.1–13.4

*To find out how well you understand the chapter material to this point, take this brief test. The answers, and the section where the material was initially discussed, are given in the back of the book. Review any questions you answered incorrectly.*

**1. a)** Explain how to find $(f \circ g)(x)$.

 **b)** If $f(x) = 3x + 3$ and $g(x) = 2x + 5$, find $(f \circ g)(x)$.

**2.** Let $f(x) = x^2 + 5$ and $g(x) = \dfrac{6}{x}$; find

 **a)** $(f \circ g)(x)$

 **b)** $(f \circ g)(3)$

 **c)** $(g \circ f)(x)$

 **d)** $(g \circ f)(3)$

**3. a)** Explain what it means when a function is a one-to-one function.

 **b)** Is the function represented by the following graph a one-to-one function? Explain.

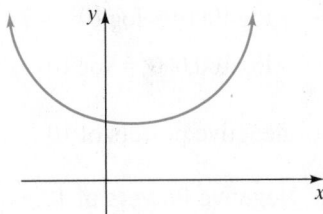

*In Exercises 4–6, for each function, **a)** determine whether it is a one-to-one function; **b)** if it is a one-to-one function, find its inverse function.*

 **4.** $\{(-3, 2), (2, 3), (5, 1), (6, 8)\}$

 **5.** $p(x) = \dfrac{1}{3}x - 5$

 **6.** $k(x) = \sqrt{x - 4}, \quad x \ge 4$

**7.** Let $m(x) = -2x + 4$. Find $m^{-1}(x)$ and then graph $m(x)$ and $m^{-1}(x)$ on the same axes.

*Graph each exponential function.*

**8.** $y = 2^x$

**9.** $y = 3^{-x}$

**10.** Graph the logarithmic function $y = \log_2 x$.

**11. Bacteria** The number of bacteria in a petri dish is $N(t) = 5(2)^t$, where $t$ is the number of hours after the 5 original bacteria are placed in the dish. How many bacteria are in the dish

 **a)** 1 hour later?

 **b)** 6 hours later?

**12.** Write $27^{2/3} = 9$ in logarithmic form.

**13.** Write $\log_2 \dfrac{1}{64} = -6$ in exponential form.

**14.** Evaluate $\log_5 125$.

**15.** Solve the equation $\log_{1/4} \dfrac{1}{16} = x$ for $x$.

**16.** Solve the equation $\log_x 64 = 3$ for $x$.

*Use properties 1–3 to write as a sum or difference of logarithms.*

**17.** $\log_9 x^2(x - 5)$

**18.** $\log_5 \dfrac{7m}{\sqrt{n}}$

*Write as a single logarithm.*

**19.** $3 \log_2 x + \log_2 (x + 7) - 4 \log_2 (x + 1)$

**20.** $\dfrac{1}{2}[\log_7 (x + 2) - \log_7 x]$

# 13.5 Common Logarithms

1 Find common logarithms of powers of 10.

2 Approximate common logarithms.

3 Approximate powers of 10.

The properties of logarithms discussed in Section 13.4 apply to any logarithms with a real number base $a$ with $a > 0$ and $a \neq 1$. Since our number system is based on the number 10, we often use logarithms with a base of 10, which are called *common logarithms*.

---

**Common Logarithm**

A **common logarithm** is a logarithm with a base of 10. When the base of a logarithm is not indicated, we assume the base is 10. Thus

$$\log x = \log_{10} x \qquad (x > 0)$$

---

The five properties we studied in Section 13.4 can be rewritten as properties of common logarithms.

---

**Common Logarithm Properties**

1. $\log xy = \log x + \log y$

2. $\log \dfrac{x}{y} = \log x - \log y$

3. $\log x^n = n \log x$

4. $\log 10^x = x$

5. $10^{\log x} = x$

---

## Understanding Algebra

Remember that *a logarithm is an exponent*. A common logarithm is the exponent to which you would raise 10 in order to get the argument. For example, log 100 is the exponent to which you would raise 10 in order to get 100. Thus, log 100 = 2.

## Understanding Algebra

To find common logarithms of 10 raised to a negative power, we need to remember the following rule of exponents:

$$a^{-m} = \frac{1}{a^m}.$$

## 1 Find Common Logarithms of Powers of 10

We can use the fourth property of common logarithms to evaluate common logarithms of numbers that are powers of 10. We begin with some examples of common logarithms of non-negative powers of 10.

**Common Logarithms of Non-Negative Powers of 10**

$$\log 1 = \log 10^0 = 0 \qquad\qquad \log 10 = \log 10^1 = 1$$

$$\log 100 = \log 10^2 = 2 \qquad\qquad \log 1000 = \log 10^3 = 3$$

$$\log 10{,}000 = \log 10^4 = 4 \qquad\qquad \log 100{,}000 = \log 10^5 = 5$$

We can also evaluate common logarithms of negative powers of 10.

**Common Logarithms of Negative Powers of 10**

$$\log 0.1 = \log \frac{1}{10} = \log 10^{-1} = -1$$

$$\log 0.01 = \log \frac{1}{100} = \log 10^{-2} = -2$$

$$\log 0.001 = \log \frac{1}{1000} = \log 10^{-3} = -3$$

$$\log 0.0001 = \log \frac{1}{10{,}000} = \log 10^{-4} = -4$$

$$\log 0.00001 = \log \frac{1}{100{,}000} = \log 10^{-5} = -5$$

## 2 Approximate Common Logarithms

We will use a scientific or graphing calculator to approximate most common loga-rithms. Before we do this, we introduce a method to estimate the value of a common logarithm between two integers. For example, suppose we wish to estimate the value of log 5. Since 5 is between 1 and 10, we can conclude that log 5 is between log 1 and log 10 and we see the following:

$$1 < 5 < 10$$

$$\log 1 < \log 5 < \log 10$$

$$0 < \log 5 < 1$$

Thus we can conclude that log 5 is a number between 0 and 1.

EXAMPLE 1   Without using a calculator, estimate between which two integers each common logarithm will be.

**a)** log 82                          **b)** log 5091
**c)** log 0.7                         **d)** log 0.03

*Solution*

**a)** Since 82 is between 10 and 100, we have the following

$$10 < 82 < 100$$

$$\log 10 < \log 82 < \log 100$$

$$1 < \log 82 < 2$$

Thus, log 82 is a number between 1 and 2.

**b)** Since 5091 is between 1000 and 10,000, we have the following

$$1000 < 5091 < 10{,}000$$

$$\log 1000 < \log 5091 < \log 10{,}000$$

$$3 < \log 5091 < 4$$

Thus, log 5091 is a number between 3 and 4.

**c)** Since 0.7 is between 0.1 and 1, we have the following

$$0.1 < 0.7 < 1$$

$$\log 0.1 < \log 0.7 < \log 1$$

$$-1 < \log 0.7 < 0$$

Thus, log 0.7 is a number between −1 and 0.

**d)** Since 0.03 is between 0.01 and 0.1, we have the following

$$0.01 < 0.03 < 0.1$$

$$\log 0.01 < \log 0.03 < \log 0.1$$

$$-2 < \log 0.03 < -1$$

Thus, log 0.03 is a number between −2 and −1.

Now Try Exercise 15

We now will approximate common logarithms using a scientific or graphing calculator using the key $\boxed{\text{LOG}}$ as shown on page 808.

*Using Your Calculator*

**Approximating Common Logarithms**

**Scientific Calculator**

To approximate common logarithms on many scientific calculators, enter the argument, then press the logarithm key.

| EXAMPLE | KEYS TO PRESS | ANSWER DISPLAYED |
|---|---|---|
| Approximate log 400. | 400 LOG | 2.60206 |

**Graphing Calculator**

On graphing calculators and on many scientific calculators, you first press the LOG key and then you enter the number. For example, on the TI-84 Plus, you would do the following:

| EXAMPLE | KEYS TO PRESS | ANSWER DISPLAYED |
|---|---|---|
| Approximate log 400. | LOG (400 ) ENTER | 2.602059991 |

↑
Generated by calculator

**EXAMPLE 2**  Use a calculator to approximate the following common logarithms. Round your answers to four decimal places. Compare your answers to those from Example 1 on page 807.

**a)** log 82            **b)** log 5091

**c)** log 0.7           **d)** log 0.03

*Solution*    Using a calculator as shown above, we obtain the following.

**a)** $\log 82 \approx 1.9138$. Note in Example 1 **a)** we correctly estimated that log 82 is a number between 1 and 2.

**b)** $\log 5091 \approx 3.7068$. Note in Example 1 **b)** we correctly estimated that log 5091 is a number between 3 and 4.

**c)** $\log 0.7 \approx -0.1549$. Note in Example 1 **c)** we correctly estimated that log 0.7 is a number between $-1$ and 0.

**d)** $\log 0.03 \approx -1.5229$. Note in Example 1 **d)** we correctly estimated that log 0.03 is a number between $-2$ and $-1$.

Now Try Exercise 27

Recall from our definition of logarithm from Section 13.3 that $y = \log_a x$ means that $x = a^y$. We restate this definition for common logarithms.

---

**Definition of a Common Logarithm**

For all positive numbers $x$

$$y = \log x \quad \text{means} \quad x = 10^y$$

The **common logarithm** of a positive number $x$ is the exponent to which the base 10 must be raised to obtain the number $x$.

---

**EXAMPLE 3**  Find the exponent to which 10 must be raised to obtain each of the following numbers. Round your answer to four decimal places.

**a)** 75         **b)** 3594         **c)** 0.00324

*Solution*    We are asked to find the *exponent* of 10. Using the definition of a common logarithm, we can see that we are asked to find the common logarithm of each of these numbers.

**a)** $\log 75 \approx 1.8751$          Note: $10^{1.8751} \approx 75$

**b)** $\log 3594 \approx 3.5556$          Note: $10^{3.5556} \approx 3594$

**c)** $\log 0.00324 \approx -2.4895$          Note: $10^{-2.4895} \approx 0.00324$

*Now Try Exercise 37*

## 3  Approximate Powers of 10

While solving equations that involve common logarithms, we will often need to evaluate a power of 10. For example, if $\log x = 3$ then, using the definition of a common logarithm, we see that $x = 10^3$ or 1000. In many equations the power of 10 will not be an integer. We will use a scientific or graphing calculator to approximate such powers of 10.

### Using Your Calculator

**Approximating Powers of 10**

To approximate powers of 10 on your calculator, use the $10^x$ function which is located directly above the $\boxed{\text{LOG}}$ key. To access this function, press the $\boxed{\text{2ND}}$, $\boxed{\text{INV}}$, or $\boxed{\text{Shift}}$ key prior to pressing the $\boxed{\text{LOG}}$ key.

**EXAMPLE 4**  Approximate the following powers of 10. Round your answers to four decimal places.

**a)** $10^{1.394}$                    **b)** $10^{2.827}$                    **c)** $10^{-0.356}$

*Solution*    Use a calculator to approximate each power of 10.

**a)** $10^{1.394} \approx 24.7742$

**b)** $10^{2.827} \approx 671.4289$

**c)** $10^{-0.356} \approx 0.4406$

*Now Try Exercise 49*

When we evaluate a power of 10, the number obtained may be referred to as an **antilogarithm** or an **antilog**. For example, in Example 4 **a)**, we found that $10^{1.394} \approx 24.7742$. Thus, we may write antilog $1.394 \approx 24.7742$. Note that $\log 24.7742 \approx 1.394$.

**EXAMPLE 5**  Solve for $x$ in each of the following equations. Round your answers to four decimal places.

**a)** $\log x = 0.132$          **b)** $\log x = -1.203$

*Solution*

**a)** Using the definition of a common logarithm, we know that

$$\log x = 0.132 \quad \text{means} \quad x = 10^{0.132} \approx 1.3552$$

**b)**
$$\log x = -1.203 \quad \text{means} \quad x = 10^{-1.203} \approx 0.0627$$

*Now Try Exercise 65*

In Example 5 **a)**, we solved the equation $\log x = 0.132$ to get the solution $x = 10^{0.132} \approx 1.3552$. We may also have written $x = $ antilog $0.132 \approx 1.3552$. Note that $\log 1.3552 \approx 0.132$.

**EXAMPLE 6  Earthquake**  The magnitude of an earthquake on the Richter scale is given by the formula $R = \log I$, where $I$ is the number of times more intense the quake is than the smallest measurable activity. How many times more intense is an earthquake measuring 6.2 on the Richter scale than the smallest measurable activity?

*Solution*    We want to find the value for $I$. We are given that $R = 6.2$. Substitute 6.2 for $R$ in the formula $R = \log I$, and then solve for $I$.

$$R = \log I$$
$$6.2 = \log I \quad \text{Substitute 6.2 for } R.$$

To find $I$, we will rewrite the logarithmic equation as an exponential equation using the definition of a common logarithm.

$$6.2 = \log I \quad \text{means} \quad I = 10^{6.2}$$
$$\text{and we get} \quad I \approx 1{,}584{,}893.$$

Thus, this earthquake is about 1,584,893 times more intense than the smallest measurable activity.

Now Try Exercise 89

# EXERCISE SET 13.5    MathXL®   MyMathLab®

## Warm-Up Exercises

*Fill in the blanks with the appropriate word, phrase, or symbol(s) from the following list.*

| $10^y$ | common | 4 | $y^{10}$ | exponent | 2 | $10^x$ |

**1.** A _____ logarithm is a logarithm with a base of 10.

**2.** A common logarithm is the _____ to which you would raise 10 to get the argument.

**3.** Using the definition of a common logarithm, $y = \log x$ means $x =$ _____.

**4.** Since 2010 is between 1000 and 10,000, log 2010 is between 3 and _____.

## Practice the Skills

*Evaluate the common logarithm of each power of 10 without the use of a calculator.*

**5.** log 1              **6.** log 100            **7.** log 0.1            **8.** log 1000
**9.** log 0.01          **10.** log 10            **11.** log 0.001         **12.** 0.0001

*Without using a calculator, estimate between which two integers each logarithm will be. See Example 1.*

**13.** 86               **14.** 352               **15.** 19,200            **16.** 1001
**17.** 0.0613           **18.** 941,000           **19.** 101               **20.** 0.000835
**21.** 3.75             **22.** 0.375             **23.** 0.0173            **24.** 0.00872

*Use a calculator to approximate the following common logarithms. Round your answers to four decimal places. Compare your answers to those from Exercises 13–24. See Example 2.*

**25.** 86               **26.** 352               **27.** 19,200            **28.** 1001
**29.** 0.0613           **30.** 941,000           **31.** 101               **32.** 0.000835
**33.** 3.75             **34.** 0.375             **35.** 0.0173            **36.** 0.00872

*Find the exponent to which 10 must be raised to obtain each of the following numbers. Round your answers to four decimal places. See Example 3.*

**37.** 3560             **38.** 817,000           **39.** 0.0727            **40.** 0.00612
**41.** 243              **42.** 8.16              **43.** 0.00592           **44.** 73,700,000
**45.** 0.0098           **46.** 0.0037            **47.** 15.491            **48.** 10.892

*Find the following powers of 10. Round your answers to four decimal places. See Example 4.*

**49.** $10^{0.2137}$    **50.** $10^{1.3845}$     **51.** $10^{4.6283}$     **52.** $10^{3.5527}$
**53.** $10^{-1.7086}$   **54.** $10^{-2.7431}$    **55.** $10^{0.001}$      **56.** $10^{-0.001}$
**57.** $10^{2.7625}$    **58.** $10^{-0.1543}$    **59.** $10^{-2.014}$     **60.** $10^{5.5922}$

*Solve for x in each of the following equations. If necessary, round your answers to four decimal places. See Example 5.*

**61.** $\log x = 2.0000$   **62.** $\log x = 1.4612$   **63.** $\log x = 3.3817$   **64.** $\log x = 1.9330$
**65.** $\log x = 4.1409$   **66.** $\log x = -2.103$   **67.** $\log x = -1.06$    **68.** $\log x = -3.1469$
**69.** $\log x = -0.6218$  **70.** $\log x = 1.5177$   **71.** $\log x = -0.1256$  **72.** $\log x = -1.3206$

*Use the common logarithm properties 4 and 5 on page 806 to evaluate the following.*

**73.** $\log 10^7$

**74.** $\log 10^{3.4}$

**75.** $10^{\log 7}$

**76.** $10^{\log 3.4}$

**77.** $4 \log 10^{5.2}$

**78.** $8 \log 10^{1.2}$

**79.** $5(10^{\log 8.3})$

**80.** $2.3(10^{\log 5.2})$

## Problem Solving

*Solve Exercises 81–84 using $R = \log I$ (see Example 6). Round your answers to four decimal places.*

**81.** Find $I$ if $R = 3.4$

**82.** Find $I$ if $R = 4.9$

**83.** Find $I$ if $R = 5.7$

**84.** Find $I$ if $R = 0.1$

**85. Astronomy** In astronomy, a formula used to find the diameter, in kilometers, of minor planets (also called asteroids) is $\log d = 3.7 - 0.2g$, where $g$ is a quantity called the absolute magnitude of the minor planet. Find the diameter of a minor planet if its absolute magnitude is **a)** 11 and **b)** 20. **c)** Find the absolute magnitude of the minor planet whose diameter is 5.8 kilometers.

© serjoe/Shutterstock

**86. Standardized Test** The average score on a standardized test is a function of the number of hours studied for the test. The average score, $f(x)$, in points, can be approximated by $f(x) = \log 0.3x + 1.8$, where $x$ is the number of hours studied for the test. The maximum possible score on the test is 4.0. Find the score received by the average person who studied for **a)** 15 hours. **b)** 55 hours.

**87. Learning Retention** Sammy Barcia just finished a course in physics. The percent of the course he will remember $t$ months later can be approximated by the function

$$R(t) = 94 - 46.8 \log (t + 1)$$

for $0 \le t \le 48$. Find the percent of the course Sammy will remember **a)** 2 months later. **b)** 48 months later.

**88. Learning Retention** Karen Frye just finished a course in psychology. The percent of the course she will remember $t$ months later can be approximated by the function

$$R(t) = 85 - 41.9 \log (t + 1)$$

for $0 \le t \le 48$. Find the percent of the course she will remember **a)** 10 months later. **b)** 25 months later.

**89. Earthquake** How many times more intense is an earthquake having a Richter scale number of 3.8 than the smallest measurable activity? See Example 6.

**90. Earthquake** The strongest earthquake ever recorded took place in Chile on May 22, 1960. It measured 9.5 on the Richter scale. How many times more intense was this earthquake than the smallest measurable activity?

**91. Energy of an Earthquake** A formula sometimes used to estimate the seismic energy released by an earthquake is $\log E = 11.8 + 1.5m_s$, where $E$ is the seismic energy and $m_s$ is the surface wave magnitude.

**a)** Find the energy released in an earthquake whose surface wave magnitude is 6.

**b)** If the energy released during an earthquake is $1.2 \times 10^{15}$, what is the magnitude of the surface wave?

**92. Sound Pressure** The sound pressure level, $s_p$, is given by the formula $s_p = 20 \log \dfrac{p_r}{0.0002}$, where $p_r$ is the sound pressure in dynes/cm$^2$.

**a)** Find the sound pressure level if the sound pressure is 0.0036 dynes/cm$^2$

**b)** If the sound pressure level is 10.0, find the sound pressure.

**93. Earthquake** The Richter scale, used to measure the strength of earthquakes, relates the magnitude, $M$, of the earthquake to the release of energy, $E$, in ergs, by the formula

$$M = \frac{\log E - 11.8}{1.5}$$

An earthquake releases $1.259 \times 10^{21}$ ergs of energy. What is the magnitude of such an earthquake on the Richter scale?

**94. pH of a Solution** The pH is a measure of the acidity or alkalinity of a solution. The pH of water, for example, is 7. In general, acids have pH numbers less than 7 and alkaline solutions have pH numbers greater than 7. The pH of a solution is defined as pH $= -\log[H_3O^+]$, where $H_3O^+$ represents the hydronium ion concentration of the solution. Find the pH of a solution whose hydronium ion concentration is $2.8 \times 10^{-3}$.

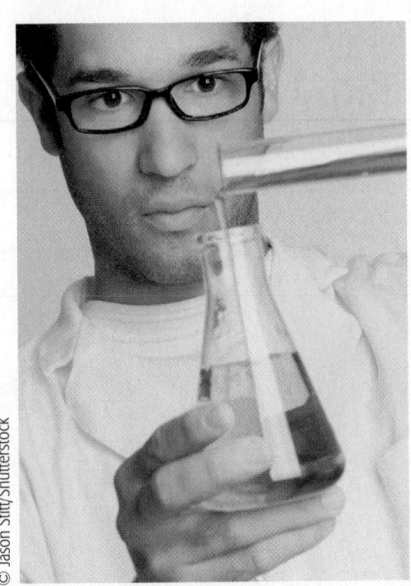

© Jason Stitt/Shutterstock

## Concept/Writing Exercises

**95.** On your calculator, you find log 462 and obtain the value 1.6646. Can this value be correct? Explain.

**96.** On your calculator, you find log 6250 and obtain the value 2.7589. Can this value be correct? Explain.

**97.** On your calculator, you find log 0.163 and obtain the value $-2.7878$. Can this value be correct? Explain.

**98.** On your calculator, you find $\log(-1.23)$ and obtain the value 0.08991. Can this value be correct? Explain.

**99.** Is $\log \dfrac{y}{4x} = \log y - \log 4 + \log x$? Explain.

**100.** Is $\log \dfrac{5x^2}{3} = 2(\log 5 + \log x) - \log 3$? Explain.

## Challenge Problems

**101.** Solve the formula $R = \log I$ for $I$.

**102.** Solve the formula $\log E = 11.8 + 1.5m$ for $E$.

**103.** Solve the formula $R = 26 - 41.9 \log(t + 1)$ for $t$.

**104.** Solve the formula $f = 76 - \log x$ for $x$.

## Group Activity

**105.** In Section 13.7, we introduce the *change of base formula*, $\log_a x = \dfrac{\log_b x}{\log_b a}$, where $a$ and $b$ are bases and $x$ is a positive number.

    **a)** Group member 1: Use the change of base formula to evaluate $\log_3 45$. (*Hint*: Let $b = 10$.)

    **b)** Group member 2: Repeat part **a)** for $\log_5 30$.

    **c)** Group member 3: Repeat part **a)** for $\log_6 40$.

    **d)** As a group, use the fact that $\log_a x = \dfrac{\log_b x}{\log_b a}$, where $b = 10$, to graph the equation $y = \log_2 x$ for $x > 0$. Use a graphing calculator if available.

## Cumulative Review Exercises

[9.3]  **106.** Solve the system of equations.

$$3r = -4s - 6$$
$$3s = -5r + 1$$

[9.5]  **107.** **Cars** Two cars start at the same point in Alexandria, Virginia, and travel in opposite directions. One car travels 5 miles per hour faster than the other car. After 4 hours, the two cars are 420 miles apart. Find the speed of each car.

[12.3]  **108.** **Motorboat** In 4 hours the Simpsons traveled 15 miles downriver in their motorboat, and then turned around and returned home. If the river current is 5 miles per hour, find the speed of their boat in still water.

[12.5]  **109.** Draw the graph of $y = (x - 2)^2 + 1$.

[12.6]  **110.** Graph the solution to $\dfrac{2x - 3}{5x + 10} < 0$ on a number line.

# 13.6 Exponential and Logarithmic Equations

**1** Solve exponential and logarithmic equations.

**2** Solve applications.

## 1 Solve Exponential and Logarithmic Equations

In this section, we will further study *exponential equations* and *logarithmic equations*. Below we list some of the properties we will use when solving such equations.

**Properties for Solving Exponential and Logarithmic Equations**

    **a.** If $x = y$, then $a^x = a^y$.

    **b.** If $a^x = a^y$, then $x = y$.

    **c.** If $x = y$, then $\log_b x = \log_b y$  $(x > 0, y > 0)$.

    **d.** If $\log_b x = \log_b y$, then $x = y$  $(x > 0, y > 0)$.    Properties 6a–6d

**EXAMPLE 1** Solve the equation $8^x = \dfrac{1}{2}$.

**Solution**   To solve this equation, we will write both sides of the equation with the same base, 2, and then use property 6b.

$$8^x = \frac{1}{2}$$

$$(2^3)^x = \frac{1}{2} \qquad \text{Write 8 as } 2^3.$$

$$2^{3x} = 2^{-1} \qquad \text{Write } \frac{1}{2} \text{ as } 2^{-1}.$$

Using property 6b, we can write

$$3x = -1$$

$$x = -\frac{1}{3}$$

*Now Try Exercise 7*

When both sides of the exponential equation cannot be written as a power of the same base, we often begin by taking the logarithm of both sides of the equation, as in Example 2. In the following examples, we will round logarithms to four decimal places.

**EXAMPLE 2** Solve the equation $5^n = 28$.

**Solution**   Take the logarithm of both sides of the equation and solve for $n$.

$$\log 5^n = \log 28$$

$$n \log 5 = \log 28 \qquad\qquad \text{Power rule}$$

$$n = \frac{\log 28}{\log 5} \qquad\qquad \text{Divide both sides by } \log 5.$$

$$\approx \frac{1.4472}{0.6990} \approx 2.0704$$

*Now Try Exercise 23*

Some logarithmic equations can be solved by expressing the equation in exponential form. *It is necessary to check logarithmic equations for extraneous solutions.* When checking a solution, if you obtain the logarithm of a nonpositive number, the solution is extraneous.

**EXAMPLE 3** Solve the equation $\log_2 (x + 3)^3 = 4$.

**Solution**   Write the equation in exponential form.

$$(x + 3)^3 = 2^4 \qquad\qquad \text{Write in exponential form.}$$

$$(x + 3)^3 = 16$$

$$x + 3 = \sqrt[3]{16} \qquad\qquad \text{Take the cube root of both sides.}$$

$$x = -3 + \sqrt[3]{16} \qquad\qquad \text{Solve for } x.$$

Check

$$\log_2 (x + 3)^3 = 4$$

$$\log_2 [(-3 + \sqrt[3]{16}) + 3]^3 \overset{?}{=} 4$$

$$\log_2 (\sqrt[3]{16})^3 \overset{?}{=} 4$$

$$\log_2 16 \overset{?}{=} 4 \qquad (\sqrt[3]{16})^3 = 16$$

$$2^4 \overset{?}{=} 16 \qquad \text{Write in exponential form.}$$

$$16 = 16 \qquad \text{True}$$

*Now Try Exercise 43*

Other logarithmic equations can be solved using the properties of logarithms given in earlier sections.

**EXAMPLE 4** Solve the equation $\log (3x + 2) + \log 9 = \log (x + 5)$.

**Solution**

$$\log (3x + 2) + \log 9 = \log (x + 5)$$
$$\log [(3x + 2)(9)] = \log (x + 5) \qquad \text{Product rule}$$
$$(3x + 2)(9) = (x + 5) \qquad \text{Property 6d}$$
$$27x + 18 = x + 5$$
$$26x + 18 = 5$$
$$26x = -13$$
$$x = -\frac{1}{2}$$

Check for yourself that the solution is $-\frac{1}{2}$.

Now Try Exercise 51

**EXAMPLE 5** Solve the equation $\log x + \log (x + 1) = \log 12$.

**Solution**

$$\log x + \log (x + 1) = \log 12$$
$$\log x(x + 1) = \log 12 \qquad \text{Product rule}$$
$$x(x + 1) = 12 \qquad \text{Property 6d}$$
$$x^2 + x = 12$$
$$x^2 + x - 12 = 0$$
$$(x + 4)(x - 3) = 0$$
$$x + 4 = 0 \qquad \text{or} \qquad x - 3 = 0$$
$$x = -4 \qquad\qquad x = 3$$

> **Understanding Algebra**
>
> Recall from Section 13.3 that the domain of logarithmic functions is $(0, \infty)$. In other words, *we cannot take the logarithm of zero or a negative number.* Thus, we must check our results from solving logarithmic equations to make sure the argument of the logarithm is positive.

Check

$\underline{x = -4}$

$$\log x + \log (x + 1) = \log 12$$
$$\log (-4) + \log (-4 + 1) \stackrel{?}{=} \log 12$$
$$\log (-4) + \log (-3) \stackrel{?}{=} \log 12$$

Stop.        ↑           ↑
Logarithms of negative numbers are not real numbers.

$\underline{x = 3}$

$$\log x + \log (x + 1) = \log 12$$
$$\log 3 + \log (3 + 1) \stackrel{?}{=} \log 12$$
$$\log 3 + \log 4 \stackrel{?}{=} \log 12$$
$$\log [(3)(4)] \stackrel{?}{=} \log 12$$
$$\log 12 = \log 12 \qquad \text{True}$$

Thus, $-4$ is an extraneous solution. The only solution is 3.

Now Try Exercise 65

## Using Your Graphing Calculator ▪▪▪▪▪

Previously, we have shown how to solve equations in one variable using a graphing calculator. To solve the equation $\log x + \log (x + 1) = \log 12$ (see Example 5) using a TI-84 Plus, we can graph

$Y_1 = \log (x) + \log (x + 1)$
$Y_2 = \log (12)$

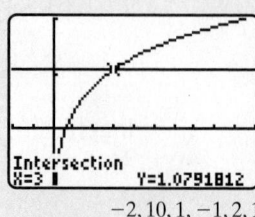

$-2, 10, 1, -1, 2, 1$

**FIGURE 13.21**

The point of intersection of these graphs can be found using the *intersect* function from the CALC menu.
**Figure 13.21** shows that the x-coordinate of the point of intersection, 3, is the solution to the equation.

**EXAMPLE 6** Solve the equation $\log(5x - 3) - \log(2x) = 1$.

*Solution*

$$\log(5x - 3) - \log(2x) = 1$$

$$\log\left(\frac{5x - 3}{2x}\right) = 1 \qquad \text{Quotient rule}$$

$$\left(\frac{5x - 3}{2x}\right) = 10^1 \qquad \text{Write in exponential form.}$$

$$\frac{5x - 3}{2x} = 10$$

$$5x - 3 = 20x \qquad \text{Multiply both sides by } 2x.$$

$$-3 = 15x \qquad \text{Subtract } 5x \text{ from both sides.}$$

$$x = -\frac{3}{15} = -\frac{1}{5} = -0.2$$

Check
$$\log(5x - 3) - \log(2x) = 1$$
$$\log[5(-0.2) - 3] - \log[2(-0.2)] = 1$$
$$\log(-4) - \log(-0.4) = 1$$

Since we get the logarithms of negative numbers, $-0.2$ is an extraneous solution. Thus, the equation has no solution. Its solution set is the empty set, $\varnothing$.

Now Try Exercise 57

## 2 Solve Applications

Now we will look at an application that involves an exponential equation.

**EXAMPLE 7** **Bacteria** If there are initially 1000 bacteria in a culture, and the number of bacteria doubles each hour, the number of bacteria after $t$ hours can be found by the formula

$$N = 1000(2)^t$$

How long will it take for the culture to grow to 30,000 bacteria?

*Solution*

$$N = 1000(2)^t$$
$$30{,}000 = 1000(2)^t \qquad \text{Substitute 30,000 for } N.$$
$$30 = (2)^t \qquad \text{Divide both sides by 1000.}$$

We want to find the value for $t$. To accomplish this we will use logarithms. Begin by taking the logarithm of both sides of the equation.

$$\log 30 = \log(2)^t$$
$$\log 30 = t \log 2 \qquad \text{Power rule}$$
$$\frac{\log 30}{\log 2} = t \qquad \text{Divide both sides by log 2.}$$
$$\frac{1.4771}{0.3010} \approx t$$
$$4.91 \approx t$$

© Olivier Le Queinec/Shutterstock

It will take about 4.91 hours for the culture to grow to 30,000 bacteria.

Now Try Exercise 69

# EXERCISE SET 13.6   MathXL®   MyMathLab®

## Warm-Up Exercises

*Fill in the blanks with the appropriate word, phrase, or symbol(s) from the following list.*

extraneous        exponential        base        logarithm        quotient        sum        product        difference

**1.** To solve the equation $9^x = \dfrac{1}{3}$, write both sides with the same _____, 3, and then use property 6b.

**2.** To solve the equation $3^x = 41$, take the common _____ of both sides.

**3.** It is necessary to check logarithmic equations for _____ solutions.

**4.** To solve the equation $\log_3(x + 1) = 2$, first write the equation in _____ form.

**5.** To solve the equation $\log_3 x + \log_3(x - 2) = 1$, first rewrite the left side of the equation as a single logarithm using the _____ rule for logarithms.

**6.** To solve the equation $\log_3 x - \log_3(x - 2) = 1$, first rewrite the left side of the equation as a single logarithm using the _____ rule for logarithms.

## Practice the Skills

*Solve each exponential equation without using a calculator.*

**7.** $5^x = 125$

**8.** $2^x = 128$

**9.** $3^x = 81$

**10.** $4^x = 256$

**11.** $64^x = 8$

**12.** $81^x = 3$

**13.** $7^{-x} = \dfrac{1}{49}$

**14.** $6^{-x} = \dfrac{1}{216}$

**15.** $27^x = \dfrac{1}{3}$

**16.** $25^x = \dfrac{1}{5}$

**17.** $2^{x+2} = 64$

**18.** $3^{x-6} = 81$

**19.** $2^{3x-2} = 128$

**20.** $64^x = 4^{4x+1}$

**21.** $27^x = 3^{2x+3}$

**22.** $\left(\dfrac{1}{2}\right)^x = 16$

*Solve each exponential equation. Use a calculator and round your answers to the nearest hundredth.*

**23.** $7^x = 50$

**24.** $1.05^x = 23$

**25.** $4^{x-1} = 35$

**26.** $2.3^{x-1} = 26.2$

**27.** $1.63^{x+1} = 25$

**28.** $4^x = 9^{x-2}$

**29.** $3^{x+4} = 6^x$

**30.** $5^x = 2^{x+5}$

*Solve each logarithmic equation. Use a calculator where appropriate. If the answer is irrational, round the answer to the nearest hundredth.*

**31.** $\log_{36} x = \dfrac{1}{2}$

**32.** $\log_{81} x = \dfrac{1}{2}$

**33.** $\log_{125} x = \dfrac{1}{3}$

**34.** $\log_{81} x = \dfrac{1}{4}$

**35.** $\log_2 x = -4$

**36.** $\log_7 x = -2$

**37.** $\log x = 2$

**38.** $\log x = 4$

**39.** $\log_2(5 - 3x) = 3$

**40.** $\log_4(3x + 7) = 3$

**41.** $\log_5(x + 1)^2 = 2$

**42.** $\log_3(a - 2)^2 = 2$

**43.** $\log_2(r + 4)^2 = 4$

**44.** $\log_2(p - 3)^2 = 6$

**45.** $\log(x + 8) = 2$

**46.** $\log(3x - 8) = 1$

**47.** $\log_2 x + \log_2 5 = 2$

**48.** $\log_3 2x + \log_3 x = 4$

**49.** $\log(r + 2) = \log(3r - 1)$

**50.** $\log 2a = \log(1 - a)$

**51.** $\log(2x + 1) + \log 4 = \log(7x + 8)$

**52.** $\log(x - 5) + \log 3 = \log(2x)$

**53.** $\log n + \log(3n - 5) = \log 2$

**54.** $\log(x + 4) - \log x = \log(x + 1)$

**55.** $\log 6 + \log y = 0.72$

**56.** $\log(x + 4) - \log x = 1.22$

**57.** $2 \log x - \log 9 = 2$

**58.** $\log 6000 - \log(x + 2) = 3.15$

**59.** $\log x + \log(x - 3) = 1$

**60.** $2 \log_2 x = 4$

**61.** $\log x = \dfrac{1}{3} \log 64$

**62.** $\log_7 x = \dfrac{3}{2} \log_7 9$

**63.** $\log_8 x = 4 \log_8 2 - \log_8 8$

**64.** $\log_4 x + \log_4(6x - 7) = \log_4 5$

**65.** $\log_5(x + 3) + \log_5(x - 2) = \log_5 6$

**66.** $\log_7(x + 6) - \log_7(x - 3) = \log_7 4$

**67.** $\log_2(x + 3) - \log_2(x - 6) = \log_2 4$

**68.** $\log(x - 7) - \log(x + 3) = \log 6$

## Problem Solving

*Solve each problem. Round your answers to the nearest hundredth.*

**69. Bacteria** If the initial number of bacteria in the culture in Example 7 is 4500, when will the number of bacteria in the culture reach 50,000? Use $N = 4500(2)^t$.

**70. Bacteria** If after 4 hours the culture in Example 7 contains 2224 bacteria, how many bacteria were present initially?

**71. Radioactive Decay** The amount, $A$, of 200 grams of a certain radioactive material remaining after $t$ years can be found by the equation $A = 200(0.75)^t$. When will 80 grams remain?

**72. Radioactive Decay** The amount, $A$, of 70 grams of a certain radioactive material remaining after $t$ years can be found by the equation $A = 70(0.62)^t$. When will 10 grams remain?

**73. Savings Account** Paul Trapper invests $2000 in a savings account earning interest at a rate of 5% compounded annually. How long will it take for the $2000 to grow to $4600? Use the compound interest formula, $A = p\left(1 + \dfrac{r}{n}\right)^{nt}$, which was discussed on page 787.

**74. Savings Account** If Tekar Werner invests $600 in a savings account earning interest at a rate of 6% compounded semi-annually, how long will it take for the $600 to grow to $1800?

**75. Money Market Account** If Jacci White invests $2500 in a money market account earning interest at a rate of 4% compounded quarterly, how long will it take for the $2500 to grow to $4000?

**76. Credit Union Account** Charlotte Newsome invests $1000 in a share savings account at her credit union. If her money earns interest at a rate of 3% compounded monthly, how long will it take for the $1000 to grow to $1500?

**77. Depreciation** A machine purchased for business use can be depreciated to reduce income tax. The value of the machine at the end of its useful life is called its *scrap value*. When the machine depreciates by a constant percentage annually, its scrap value, $S$, is $S = c(1 - r)^n$, where $c$ is the original cost, $r$ is the annual rate of depreciation as a decimal, and $n$ is the useful life in years. Find the scrap value of a machine that costs $50,000, has a useful life of 12 years, and has an annual depreciation rate of 15%.

**78. Depreciation** If the machine in Exercise 77 costs $100,000, has a useful life of 15 years, and has an annual depreciation rate of 8%, find its scrap value.

**79. Power Gain of an Amplifier** The power gain, $P$, of an amplifier is defined as

$$P = 10 \log\left(\frac{P_{out}}{P_{in}}\right)$$

where $P_{out}$ is the output power in watts and $P_{in}$ is the input power in watts. If an amplifier has an output power of 12.6 watts and an input power of 0.146 watts, find the power gain.

**80. Earthquake** Measured on the Richter scale, the magnitude, $R$, of an earthquake of intensity $I$ is defined by $R = \log I$, where $I$ is the number of times more intense the earthquake is than the minimum level for comparison.

a) How many times more intense was the 1906 San Francisco earthquake, which measured 8.25 on the Richter scale, than the minimum level for comparison?

b) How many times more intense is an earthquake that measures 8.3 on the Richter scale than one that measures 4.7?

**81. Magnitude of Sound** The decibel scale is used to measure the magnitude of sound. The magnitude $d$, in decibels, of a sound is defined to be $d = 10 \log I$, where $I$ is the number of times greater (or more intense) the sound is than the minimum intensity of audible sound.

a) An airplane engine (nearby) measures 120 decibels. How many times greater than the minimum level of audible sound is the airplane engine?

b) The intensity of the noise in a busy city street is 50 decibels. How many times greater is the intensity of the sound of the airplane engine than the sound of the city street?

**82.** In the following procedure, we begin with a true statement and end with a false statement. Can you find the error?

| | |
|---|---|
| $2 < 3$ | True |
| $2 \log (0.1) < 3 \log (0.1)$ | Multiply both sides by log (0.1). |
| $\log (0.1)^2 < \log (0.1)^3$ | Property 3 |
| $(0.1)^2 < (0.1)^3$ | Property 6d |
| $0.01 < 0.001$ | False |

**83.** Solve $8^x = 16^{x-2}$.

**84.** Solve $27^x = 81^{x-3}$.

**85.** Use equations that are quadratic in form to solve the equation $2^{2x} - 6(2^x) + 8 = 0$.

**86.** Use equations that are quadratic in form to solve the equation $2^{2x} - 18(2^x) + 32 = 0$.

*Change the exponential or logarithmic equation to the form ax + by = c, and then solve the system of equations.*

**87.** $2^x = 8^y$
$x + y = 4$

**88.** $3^{2x} = 9^{y+1}$
$x - 2y = -3$

**89.** $\log (x + y) = 2$
$x - y = 8$

**90.** $\log (x + y) = 3$
$2x - y = 5$

 *Use your calculator to estimate the solutions to the nearest tenth. If a real solution does not exist, so state.*

**91.** $\log (x + 3) + \log x = \log 16$

**92.** $\log (3x + 5) = 2.3x - 6.4$

**93.** $5.6 \log (5x - 12) = 2.3 \log (x - 5.4)$

**94.** $5.6 \log (x + 12.2) - 1.6 \log (x - 4) = 20.3 \log (2x - 6)$

### Concept/Writing Exercises

**95.** How can you tell quickly that $\log (x + 4) = \log (-2)$ has no real solution?

**96.** In properties 6c and 6d, we specify that both $x$ and $y$ must be positive. Explain why.

### Cumulative Review Exercises

[2.6] **97.** Consider the following two figures. Which has a greater volume, and by how much?

[8.5] **98.** Let $f(x) = x^2 - x$ and $g(x) = x - 1$. Find $(g - f)(3)$.

[10.3] **99.** Determine the solution set to the system of inequalities
$$3x - 4y \le 6$$
$$y > -x + 4$$

[11.5] **100.** Simplify $\dfrac{2\sqrt{xy} - \sqrt{xy}}{\sqrt{x} + \sqrt{y}}$.

[12.3] **101.** Solve $E = mc^2$ for $c$.

[12.5] **102.** Determine the function for the parabola that has the shape of $f(x) = 2x^2$ and has its vertex at $(3, -5)$.

# 13.7 Natural Exponential and Natural Logarithmic Functions

1 Identify the natural exponential function.

2 Identify the natural logarithmic function.

3 Approximate natural logarithms and powers of $e$ on a calculator.

4 Use the change of base formula.

5 Solve natural logarithmic and natural exponential equations.

6 Solve applications.

In this section we discuss the *natural exponential function*, and its inverse, *the natural logarithmic function*. These functions are frequently used to describe naturally occurring events, thus the adjective *natural*. Both of these functions rely on an irrational number designated by the letter $e$.

> **The Natural Base, $e$**
>
> The **natural base, $e$,** is an irrational number that serves as the base for the *natural exponential function* and the *natural logarithmic function*.
>
> $$e \approx 2.7183$$

## 1 Identify the Natural Exponential Function

In Section 13.2, we discussed exponential functions. Recall that exponential functions are of the form $f(x) = a^x, a > 0$ and $a \ne 1$. We now introduce the *natural exponential function*.

**Understanding Algebra**

There are two very important types of logarithms whose notations differ from other logarithms.

- *Common* logarithms have a base of 10 and are written using the notation log *x*. Thus,

$$\log_{10} x = \log x$$

- *Natural* logarithms have a base of *e*, the natural base, and are written using the notation ln *x*. Thus,

$$\log_e x = \ln x$$

**The Natural Exponential Function**

The **natural exponential function** is

$$f(x) = e^x$$

where *e* is the natural base.

## 2 Identify the Natural Logarithmic Function

We begin by defining natural logarithms.

**Natural Logarithms**

**Natural logarithms** are logarithms with a base of *e*, the natural base. We indicate natural logarithms with the notation ln.

$$\log_e x = \ln x \qquad (x > 0)$$

ln *x* is read "the natural logarithm of *x*."

In Section 13.3, we discussed logarithmic functions. Recall that logarithmic functions are of the form $f(x) = \log_a x$, $a > 0$, $a \neq 1$, $x > 0$. We now introduce the *natural logarithmic function*.

**The Natural Logarithmic Function**

The **natural logarithmic function** is

$$f(x) = \ln x \quad (x > 0)$$

where $\ln x = \log_e x$ and *e* is the natural base.

When we change a natural logarithm to exponential form, the base of the exponential expression is the natural base, *e*.

**Natural Logarithm in Exponential Form**

$$\text{For } x > 0, \text{ if } y = \ln x, \text{ then } e^y = x.$$

**EXAMPLE 1** Find the value of the expression by changing the natural logarithm to exponential form.

**a)** ln 1                    **b)** ln *e*

*Solution*

**a)** Let $y = \ln 1$; then $e^y = 1$. Since any nonzero value to the 0th power equals 1, *y* must equal 0. Thus, ln 1 = 0.

**b)** Let $y = \ln e$; then $e^y = e$. For $e^y$ to equal *e*, *y* must equal 1. Thus, ln *e* = 1.

Now Try Exercise 11

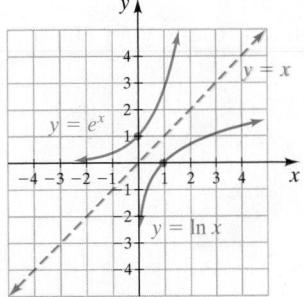

FIGURE 13.22

Recall from Section 13.3 that the functions $y = a^x$ and $y = \log_a x$ are inverse functions. Thus, $y = e^x$ and $y = \ln x$ are inverse functions. The graphs are shown in **Figure 13.22**. Notice that the graphs are symmetric about the line $y = x$. Notice also that the graph of $y = e^x$ is similar to the graph of $y = a^x$ where $a > 1$ and the graph of $y = \ln x$ is similar to the graph of $y = \log_a x$ where $a > 1$.

### 3 Approximate Natural Logarithms and Powers of $e$ on a Calculator

In a manner similar to how we approximated common logarithms and powers of 10 in Section 13.5, we can use a scientific or graphing calculator to approximate natural logarithms and powers of $e$.

 *Using Your Calculator* ▪▪▪▪▪

**Approximating Natural Logarithms**

**Scientific Calculator**

To approximate natural logarithms on many scientific calculators, enter the argument, then press the $\boxed{\text{LN}}$ key.

| EXAMPLE | KEYS TO PRESS | ANSWER DISPLAYED |
|---|---|---|
| Approximate ln 31 | 31 $\boxed{\text{LN}}$ | 3.433987204 |

**Graphing Calculator**

To approximate natural logarithms on a graphing calculator and on many scientific calculators, first press $\boxed{\text{LN}}$ and then enter the number. For example, on the TI-84 Plus, you would do the following:

| EXAMPLE | KEYS TO PRESS | ANSWER DISPLAYED |
|---|---|---|
| Approximate ln 31 | $\boxed{\text{LN}}$ (31) $\boxed{\text{ENTER}}$ | 3.433987204 |

↑
Generated by calculator

Remember that *a logarithm is an exponent*. Thus, the natural logarithm of a number is the exponent to which the natural base $e$ must be raised to obtain that number. For example:

$$\ln 242 \approx 5.488937726 \qquad \text{therefore} \qquad e^{5.488937726} \approx 242$$
$$\ln 0.85 \approx -0.1625189295 \qquad \text{therefore} \qquad e^{-0.1625189295} \approx 0.85$$

*Using Your Calculator* ▪▪▪▪▪

**Approximating Powers of $e$**

To evaluate powers of the natural base $e$ on your calculator, use the $e^x$ function, which is located directly above the $\boxed{\text{LN}}$ key. To access this function, press the $\boxed{\text{2ND}}$, $\boxed{\text{INV}}$, or $\boxed{\text{SHIFT}}$ key prior to pressing the $\boxed{\text{LN}}$ key.

**EXAMPLE 2** Approximate the following powers of the natural base $e$. Round your answers to four decimal places

**a)** $e^{1.394}$     **b)** $e^{2.827}$     **c)** $e^{-0.356}$

*Solution*    Use a calculator to approximate each power of $e$.

**a)** $e^{1.394} \approx 4.0309$    **b)** $e^{2.827} \approx 16.8947$    **c)** $e^{-0.356} \approx 0.7005$

Now Try Exercise 15

**EXAMPLE 3** Solve for $x$ in each of the following equations. Round your answers to four decimal places.

**a)** $\ln x = 0.132$     **b)** $\ln x = -1.203$

*Solution*

**a)** Using the definition of a logarithm, we know that
$$\ln x = 0.132 \quad \text{means} \quad x = e^{0.132} \approx 1.1411$$

**b)** $\ln x = -1.203 \quad \text{means} \quad x = e^{-1.203} \approx 0.3002$

Now Try Exercise 21

## 4 Use the Change of Base Formula

If you are given a logarithm in a base other than 10 or $e$, you will not be able to evaluate it on your calculator directly. When this occurs, you can use the **change of base formula**.

### Change of Base Formula

For any logarithm bases $a$ and $b$, and positive number $x$,

$$\log_a x = \frac{\log_b x}{\log_b a}$$

In the change of base formula, 10 is often used in place of base $b$ because we can approximate common logarithms on a calculator. Replacing base $b$ with 10, we get

$$\log_a x = \frac{\log_{10} x}{\log_{10} a} \quad \text{or} \quad \log_a x = \frac{\log x}{\log a}$$

EXAMPLE 4   Use the change of base formula to approximate $\log_3 24$.

*Solution*   If we substitute 3 for $a$ and 24 for $x$ in $\log_a x = \dfrac{\log x}{\log a}$, we obtain

$$\log_3 24 = \frac{\log 24}{\log 3} \approx 2.8928$$

Note that $3^{2.8928} \approx 24$.

Now Try Exercise 27

## 5 Solve Natural Logarithmic and Natural Exponential Equations

The properties of logarithms discussed in Section 13.4 also hold true for natural logarithms.

### Properties for Natural Logarithms

| | | |
|---|---|---|
| $\ln xy = \ln x + \ln y$ | ($x > 0$ and $y > 0$) | Product rule |
| $\ln \dfrac{x}{y} = \ln x - \ln y$ | ($x > 0$ and $y > 0$) | Quotient rule |
| $\ln x^n = n \ln x$ | ($x > 0$) | Power rule |

Properties 4 and 5 from page 802 can also be written using natural logarithms. Thus, $\ln e^x = x$ and $e^{\ln x} = x$. We will refer to these properties as properties 7 and 8, respectively.

### Additional Properties for Natural Logarithms and Natural Exponential Expressions

| | | |
|---|---|---|
| $\ln e^x = x$ | | Property 7 |
| $e^{\ln x} = x,$ | $x > 0$ | Property 8 |

Using property 7, $\ln e^x = x$, we can state, for example, that $\ln e^{kt} = kt$, and $\ln e^{-2.06t} = -2.06t$. Using property 8, $e^{\ln x} = x$, we can state, for example, that $e^{\ln (t+2)} = t + 2$ and $e^{\ln kt} = kt$.

**EXAMPLE 5** Solve the equation $\ln y - \ln (x + 9) = t$ for $y$.

*Solution*

$$\ln y - \ln (x + 9) = t$$

$$\ln \frac{y}{x + 9} = t \qquad \text{Quotient rule}$$

$$\frac{y}{x + 9} = e^t \qquad \text{Write in exponential form.}$$

$$y = e^t(x + 9) \qquad \text{Solve for } y.$$

Now Try Exercise 63

**EXAMPLE 6** Solve the equation $225 = 450e^{-0.4t}$ for $t$.

*Solution*   Begin by dividing both sides of the equation by 450 to isolate $e^{-0.4t}$.

$$\frac{225}{450} = \frac{450e^{-0.4t}}{450}$$

$$0.5 = e^{-0.4t}$$

Now take the natural logarithm of both sides of the equation to eliminate the exponential expression on the right side of the equation.

$$\ln 0.5 = \ln e^{-0.4t}$$

$$\ln 0.5 = -0.4t \qquad \text{Property 7}$$

$$-0.6931472 = -0.4t$$

$$\frac{-0.6931472}{-0.4} = t$$

$$1.732868 = t$$

Now Try Exercise 49

**EXAMPLE 7** Solve the equation $P = P_0 e^{kt}$ for $t$.

*Solution*   We can follow the same procedure as used in Example 6.

$$P = P_0 e^{kt}$$

$$\frac{P}{P_0} = \frac{P_0 e^{kt}}{P_0} \qquad \text{Divide both sides by } P_0.$$

$$\frac{P}{P_0} = e^{kt}$$

$$\ln \frac{P}{P_0} = \ln e^{kt} \qquad \text{Take natural log of both sides.}$$

$$\ln P - \ln P_0 = \ln e^{kt} \qquad \text{Quotient rule}$$

$$\ln P - \ln P_0 = kt \qquad \text{Property 7}$$

$$\frac{\ln P - \ln P_0}{k} = t \qquad \text{Solve for } t.$$

Now Try Exercise 59

## 6  Solve Applications

We now introduce some applications involving the natural base and natural logarithms. We begin with a formula used when a quantity increases or decreases at an *exponential rate*.

### Exponential Growth or Decay Formula

When a quantity $P$ increases (grows) or decreases (decays) at an exponential rate, the value of $P$ after time $t$ can be found using the formula

$$P = P_0 e^{kt},$$

where $P_0$ is the initial or starting value of the quantity $P$, and $k$ is the constant growth rate or decay rate.

When $k > 0$, $P$ increases as $t$ increases.

When $k < 0$, $P$ decreases and gets closer to 0 as $t$ increases.

**EXAMPLE 8  Interest Compounded Continuously** When interest is compounded continuously, the balance, $P$, in the account at any time, $t$, can be calculated by the exponential growth formula $P = P_0 e^{kt}$, where $P_0$ is the principal initially invested and $k$ is the interest rate.

**a)** Suppose the interest rate is 6% compounded continuously and $1000 is initially invested. Determine the balance in the account after 3 years.

**b)** How long will it take the account to double in value?

*Solution*

**a) Understand and Translate**  We are told that the principal initially invested, $P_0$, is $1000. We are also given that the time, $t$, is 3 years and that the interest rate, $k$, is 6% or 0.06. We substitute these values into the given formula and solve for $P$.

$$P = P_0 e^{kt}$$
$$P = 1000\, e^{(0.06)(3)}$$

Carry Out
$$= 1000 e^{0.18} = 1000(1.1972174) \qquad \text{From a calculator}$$
$$\approx 1197.22$$

**Answer**  After 3 years, the balance in the account is $\approx$$1197.22.

**b) Understand and Translate**  For the value of the account to double, the balance in the account would have to reach $2000. Therefore, we substitute 2000 for $P$ and solve for $t$.

$$P = P_0 e^{kt}$$
$$2000 = 1000\, e^{0.06t}$$

Carry Out
$$2 = e^{0.06t} \qquad \text{Divide both sides by 1000.}$$
$$\ln 2 = \ln e^{0.06t} \qquad \text{Take natural log of both sides.}$$
$$\ln 2 = 0.06t \qquad \text{Property 7}$$
$$\frac{\ln 2}{0.06} = t$$
$$\frac{0.6931472}{0.06} = t$$
$$11.552453 \approx t$$

**Answer**  Thus, with an interest rate of 6% compounded continuously, the account will double in about 11.6 years.

Now Try Exercise 69

**EXAMPLE 9  Radioactive Decay**  Strontium 90 is a radioactive isotope that decays exponentially at 2.8% per year. Suppose there are initially 1000 grams of strontium 90 in a substance.

**a)** Find the number of grams of strontium 90 left after 50 years.

**b)** Find the half-life of strontium 90.

*Solution*

**a) Understand**   Since the strontium 90 is decaying over time, the value of $k$ in the formula $P = P_0 e^{kt}$ is negative. Since the rate of decay is 2.8% per year, we use $k = -0.028$. Therefore, the formula we use is $P = P_0 e^{-0.028t}$.

Translate    $P = P_0 e^{-0.028t}$

$= 1000 e^{-0.028(50)}$

Carry Out    $= 1000 e^{-1.4} = 1000(0.246597) = 246.597$

Answer    Thus, after 50 years, 246.597 grams of strontium 90 remain.

**b)** To find the half-life, we need to determine when 500 grams of strontium 90 are left.

$$P = P_0 e^{-0.028t}$$

$$500 = 1000 e^{-0.028t}$$

$$0.5 = e^{-0.028t} \qquad \text{Divide both sides by 1000.}$$

$$\ln 0.5 = \ln e^{-0.028t} \qquad \text{Take natural log of both sides.}$$

$$-0.6931472 = -0.028t \qquad \text{Property 7}$$

$$\frac{-0.6931472}{-0.028} = t$$

$$24.755257 \approx t$$

Thus, the half-life of strontium 90 is about 24.8 years.

Now Try Exercise 71

**EXAMPLE 10 Selling Toys** The formula for estimating the amount of money, $A$, spent on advertising a certain toy is $A = 350 + 650 \ln n$ where $n$ is the expected number of toys to be sold.

**a)** If the company wishes to sell 2200 toys, how much money should the company expect to spend on advertising?

**b)** How many toys can be expected to be sold if $6000 is spent on advertising?

*Solution*

**a)**   $A = 350 + 650 \ln n$

$= 350 + 650 \ln 2200 \qquad \text{Substitute 2200 for } n.$

$= 350 + 650(7.6962126)$

$= 5352.54$

Thus, $5352.54 should be expected to be spent on advertising.

**b) Understand and Translate**   We are asked to find the number of toys expected to be sold, $n$, if $6000 is spent on advertising. We substitute the given values into the equation and solve for $n$.

$$A = 350 + 650 \ln n$$

Carry Out    $6000 = 350 + 650 \ln n \qquad \text{Substitute 6000 for } A.$

$$5650 = 650 \ln n \qquad \text{Subtract 350 from both sides.}$$

$$\frac{5650}{650} = \ln n \qquad \text{Divide both sides by 650.}$$

$$8.69231 \approx \ln n$$

$$e^{8.69231} \approx n \qquad \text{Change to exponential form.}$$

$$5957 \approx n \qquad \text{Obtain answer from a calculator.}$$

Answer    Thus, about 5957 toys can be expected to be sold if $6000 is spent on advertising.

Now Try Exercise 75

## Using Your Graphing Calculator ▪▪▪▪

To approximate the solutions to the equation $4e^{0.3x} - 5 = x + 3$ using a TI-84 Plus, we can graph

$$Y_1 = 4e^{\wedge}(0.3x) - 5$$
$$Y_2 = x + 3$$

FIGURE 13.23                    FIGURE 13.24

The points of intersection of these graphs can be found using the *intersect* function from the CALC menu. **Figure 13.23** shows that the x-coordinate of one point of intersection, $x \approx -7.5896$, is one solution. **Figure 13.24** shows that the x-coordinate of the other point of intersection, $x \approx 3.5284$, is the other solution.

# EXERCISE SET 13.7   MathXL®  MyMathLab®

## Warm-Up Exercises

*Fill in the blanks with the appropriate word, phrase, or symbol(s) from the following list.*

write          natural          inverse          quarterly          exponential          composite

logarithmic    continuously     base             change             common               growth

1. The natural _____, $e$, is an irrational number that serves as the base for the natural exponential function and the natural logarithmic function.

2. The natural exponential function and the natural logarithmic function are _____ functions.

3. The natural _____ function is $f(x) = e^x$, where $e$ is the natural base.

4. The natural _____ function is $f(x) = \ln x$, $x > 0$, where $e$ is the natural base.

5. The formula $\log_a x = \dfrac{\log_b x}{\log_b a}$ is the _____ of base formula.

6. In the change of base formula, 10 is often used in place of $b$ because we can approximate _____ logarithms on a calculator.

7. To solve the equation $5.7 = e^x$, take the _____ logarithm of both sides of the equation.

8. To solve the equation $\ln x = 1.239$, _____ the equation in exponential form.

9. The exponential _____ or decay formula states that when a quantity $P$ increases or decreases at an exponential rate, the value of $P$ after time $t$ can be found using the formula $P = P_0 e^{kt}$.

10. When interest is compounded _____, the balance in the account can be calculated using the exponential growth formula.

## Practice the Skills

*Approximate the following values. Round your answers to four decimal places.*

**11.** $\ln 27$       **12.** $\ln 810$       **13.** $\ln 0.415$       **14.** $\ln 0.000176$

*Approximate the following values. Round your answers to four decimal places.*

**15.** $e^{1.2}$       **16.** $e^{4.8}$       **17.** $e^{-0.56}$       **18.** $e^{-2.6}$

*Approximate the value of x. Round your answers to four decimal places.*

**19.** $\ln x = 1.6$       **20.** $\ln x = 5.2$       **21.** $\ln x = -2.85$       **22.** $\ln x = -0.674$

*Approximate the value of x. Round your answers to four decimal places.*

**23.** $e^x = 98$       **24.** $e^x = 2010$       **25.** $e^x = 0.0574$       **26.** $e^x = 0.000231$

*Use the change of base formula to approximate the value of the following logarithms. Round your answers to four decimal places.*

**27.** $\log_2 21$       **28.** $\log_2 89$       **29.** $\log_4 11$       **30.** $\log_4 316$

**31.** $\log_5 82$

**32.** $\log_5 1893$

**33.** $\log_6 185$

**34.** $\log_6 806$

**35.** $\log_5 0.463$

**36.** $\log_3 0.0365$

*Solve the following logarithmic equations.*

**37.** $\ln x + \ln (x - 1) = \ln 12$

**38.** $\ln (x + 4) + \ln (x - 2) = \ln 16$

**39.** $\ln x + \ln (x + 4) = \ln 5$

**40.** $\ln (x + 3) + \ln (x - 3) = \ln 40$

**41.** $\ln x = 5 \ln 2 - \ln 8$

**42.** $\ln x = \dfrac{3}{2} \ln 16$

**43.** $\ln (x^2 - 4) - \ln (x + 2) = \ln 4$

**44.** $\ln (x + 12) - \ln (x - 4) = \ln 5$

*Each of the following equations is in the form $P = P_0 e^{kt}$. Solve each equation for the remaining variable. Remember, e is a constant. Round your answers to four decimal places.*

**45.** $P = 120 e^{2.3(1.6)}$

**46.** $900 = P_0 e^{(0.4)(3)}$

**47.** $50 = P_0 e^{-0.5(3)}$

**48.** $18 = 9 e^{2t}$

**49.** $60 = 20 e^{1.4t}$

**50.** $29 = 58 e^{-0.5t}$

**51.** $86 = 43 e^{k(3)}$

**52.** $15 = 75 e^{k(4)}$

**53.** $20 = 40 e^{k(2.4)}$

**54.** $100 = A_0 e^{-0.02(3)}$

**55.** $A = 6000 e^{-0.08(3)}$

**56.** $51 = 68 e^{-0.04t}$

*Solve for the indicated variable.*

**57.** $V = V_0 e^{kt}$, for $V_0$

**58.** $P = P_0 e^{kt}$, for $P_0$

**59.** $P = 150 e^{7t}$, for $t$

**60.** $361 = P_0 e^{kt}$, for $t$

**61.** $A = A_0 e^{kt}$, for $k$

**62.** $167 = R_0 e^{kt}$, for $k$

**63.** $\ln y - \ln x = 2.3$, for $y$

**64.** $\ln y + 9 \ln x = \ln 2$, for $y$

**65.** $\ln y - \ln (x + 6) = 5$ for $y$

**66.** $\ln (x + 2) - \ln (y - 1) = \ln 5$, for $y$

## Problem Solving

 *Use a calculator to solve.*

**67. Interest Compounded Continuously** If $5000 is invested at 6% compounded continuously,

   **a)** determine the balance in the account after 2 years.

   **b)** How long would it take the value of the account to double? (See Example 8.)

**68. Interest Compounded Continuously** If $3000 is invested at 3% compounded continuously,

   **a)** determine the balance in the account after 30 years.

   **b)** How long would it take the value of the account to double?

**69. Credit Union Account** Lucy Alfonso invested $2500 in her credit union savings account earning 4% interest compounded continuously.

   **a)** Determine the balance in Lucy's account after 3 years.

   **b)** How long will it take for the value of Lucy's account to double?

**70. Retirement Account** Kristie Paulson invested $10,000 in a retirement savings account earning 5% interest compounded continuously.

   **a)** Determine the balance in Kristie's account after 25 years.

   **b)** How long will it take for the value of Kristie's account to triple?

**71. Radioactive Decay** Refer to Example 9. Determine the amount of strontium 90 remaining after 20 years if there were originally 70 grams.

**72. Strontium 90** Refer to Example 9. Determine the amount of strontium 90 remaining after 40 years if there were originally 200 grams.

**73. Soft Drinks** For a certain soft drink, the percent of a target market, $f(t)$, that buys the soft drink is a function of the

number of days, $t$, that the soft drink is advertised. The function that describes this relationship is $f(t) = 1 - e^{-0.04t}$.

   **a)** What percent of the target market buys the soft drink after 50 days of advertising?

   **b)** How many days of advertising are needed if 75% of the target market is to buy the soft drink?

© Allen R. Angel

**74. Trout in a Lake** In 2010, Thomas Pearse Lake had 300 trout. The growth in the number of trout is estimated by the function $g(t) = 300 e^{0.07t}$, where $t$ is the number of years after 2010. How many trout will be in the lake in **a)** 2015? **b)** 2020?

© Allen R. Angel

**75. Walking Speed** It was found in a psychological study that the average walking speed, $f(P)$, of a person living in a city is a function of the population of the city. For a city of population $P$, the average walking speed in feet per second is given by $f(P) = 0.37 \ln P + 0.05$. The population of Nashville, Tennessee, is 972,000.

   a) What is the average walking speed of a person living in Nashville?

   b) What is the average walking speed of a person living in New York City, population 8,567,000?

   c) If the average walking speed of the people in a certain city is 5.0 feet per second, what is the population of the city?

**76. Advertising** For a certain type of tie, the number of ties sold, $N(a)$, is a function of the dollar amount spent on advertising, $a$ (in thousands of dollars). The function that describes this relationship is $N(a) = 800 + 300 \ln a$.

   a) How many ties were sold after $1500 (or $1.5 thousand) was spent on advertising?

   b) How much money must be spent on advertising to sell 1000 ties?

**77.** Assume that the value of the island of Manhattan has grown at an exponential rate of 8% per year since 1626 when Peter Minuet of the Dutch West India Company purchased it for $24. Then the value of Manhattan can be determined by the equation $V = 24e^{0.08t}$, where $t$ is the number of years since 1626. Determine the value of the island of Manhattan in 2010, that is when $t = 384$ years.

Times Square in Manhattan

**78. Prescribing a Drug** The percent of doctors who accept and prescribe a new drug is given by the function $P(t) = 1 - e^{-0.22t}$, where $t$ is the time in months since the drug was placed on the market. What percent of doctors accept a new drug 2 months after it is placed on the market?

**79. World Population** The word population in February 2009 was about 6.76 billion people. Assume that the world population continues to grow exponentially at the growth rate of about 1.2% per year. Thus, the expected world population, in billions of people, in $t$ years, is given by the function $P(t) = 6.76e^{0.012t}$, where $t$ is the number of years since 2009.

   a) Use this function to estimate the world population in 2015.

   b) At the current rate, how long will it take for the world population to double?

**80. United States Population** The United States population in February 2009 was about 305.8 million people. Assume that the United States population continues to grow exponentially at the growth rate of about 1.0% per year. Thus, the expected United States population, in millions of people, in $t$ years, is given by the function $P(t) = 305.8e^{0.010t}$, where $t$ is the number of years since 2009.

   a) Use this function to estimate the United States population in 2015.

   b) At the current rate, how long will it take for the United States population to double?

**81. China Population** The population of China in February 2009 was about 1.34 billion people. Assume that the population of China continues to grow exponentially at the growth rate of about 0.6% per year. Thus, the expected population of China, in billions of people, in $t$ years, is given by the function $P(t) = 1.34e^{0.006t}$, where $t$ is the number of years since 2009.

   a) Use this function to estimate the population of China in 2015.

   b) At the current rate, how long will it take for the China population to double?

**82. Mexico Population** The population of Mexico in February 2009 was about 111.2 million people. Assume that the population of Mexico continues to grow exponentially at the growth rate of about 1.1% per year. Thus, the expected population of Mexico, in millions of people, in $t$ years, is given by the function $P(t) = 111.2e^{0.011t}$, where $t$ is the number of years since 2009.

   a) Use this function to estimate the population of Mexico in 2015.

   b) At the current rate, how long will it take for the population of Mexico to double?

Mexico City

**83. Boys' Heights** The shaded area of the following graph shows the normal range (from the 5th to the 95th percentile) of heights for boys from birth up to 36 months. The median, or 50th percentile, of heights is indicated by the green line. The function $y = 15.29 + 5.93 \ln x$ can be used to estimate the median height of boys from 3 months to 36 months. Use this function to estimate the median height for boys at age **a)** 18 months. **b)** 30 months.

**Boys**

*Source: Newsweek*

**84. Girls' Weights** The shaded area of the following graph shows the normal range (from the 5th to the 95th percentile) of weights for girls from birth up to 36 months. The median, or 50th percentile, of weights is indicated by the orange line. The function $y = 3.17 + 7.32 \ln x$ can be used to estimate the median weight of girls from 3 months to 36 months. Use this function to estimate the median weight for girls at **a)** 18 months. **b)** 30 months.

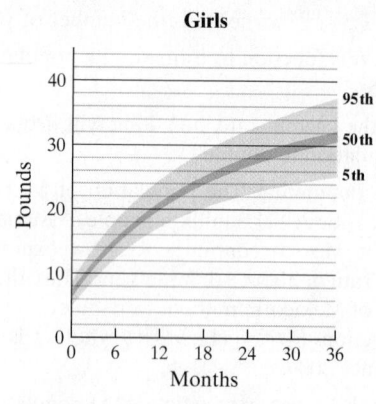

**Girls**

*Source: Newsweek*

**85. Radioactive Decay** Uranium 234 (U-234) decays exponentially at a rate of 0.0003% per year. Thus, the formula $A = A_0 e^{-0.000003t}$ can be used to determine the amount of U-234 remaining from an initial amount $A_0$, after $t$ years.

**a)** If 1000 grams of U-234 are present in 2010, how many grams will remain in 2110, 100 years later?

**b)** Find the half-life of U-234.

**86. Radioactive Decay** Plutonium, which is commonly used in nuclear reactors, decays exponentially at a rate of 0.003% per year. The formula $A = A_0 e^{kt}$ can be used to find the

amount of plutonium remaining from an initial amount, $A_0$, after $t$ years. In the formula, the $k$ is replaced with $-0.00003$.

**a)** If 1000 grams of plutonium are present in 2009, how many grams of plutonium will remain in the year 2109, 100 years later?

**b)** Find the half-life of plutonium.

**87. Carbon Dating** Carbon dating is used to estimate the age of ancient plants and objects. The radioactive element, carbon 14, is most often used for this purpose. Carbon 14 decays exponentially at a rate of 0.01205% per year. The amount of carbon 14 remaining in an object after $t$ years can be found by the function $f(t) = v_0 e^{-0.0001205t}$, where $v_0$ is the initial amount present.

**a)** If an ancient animal bone originally had 20 grams of carbon 14, and when found it had 9 grams of carbon 14, how old is the bone?

**b)** How old is an item that has 50% of its original carbon 14 remaining?

**88. Compound Interest** At what rate, compounded continuously, must a sum of money be invested if it is to double in 13 years?

**89. Compound Interest** How much money must be deposited today to become $20,000 in 18 years if invested at 6% compounded continuously?

**90. Radioisotope** The power supply of a satellite is a radioisotope. The power $P$, in watts, remaining in the power supply is a function of the time the satellite is in space.

**a)** If there are 50 grams of the isotope originally, the power remaining after $t$ days is $P = 50e^{-0.002t}$. Find the power remaining after 50 days.

**b)** When will the power remaining drop to 10 watts?

**91. Radioactive Decay** During the nuclear accident at Chernobyl in Ukraine in 1986, two of the radioactive materials that escaped into the atmosphere were cesium 137, with a decay rate of 2.3% and strontium 90, with a decay rate of 2.8%.

**a)** Which material will decompose more quickly? Explain.

**b)** What percentage of the cesium will remain in 2036, 50 years after the accident?

**92. Radiometric Dating** In the study of radiometric dating (using radioactive isotopes to determine the age of items), the formula

$$t = \frac{t_h}{0.693} \ln\left(\frac{N_0}{N}\right)$$

is often used. In the formula, $t$ is the age of the item, $t_h$ is the half-life of the radioactive isotope used, $N_0$ is the original number of radioactive atoms present, and $N$ is the number remaining at time $t$. Suppose a rock originally contained $5 \times 10^{12}$ atoms of uranium 238. Uranium 238 has a half-life of $4.5 \times 10^9$ years. If at present there are $4 \times 10^{12}$ atoms, how old is the rock?

## Challenge Exercises

*In Exercises 93–96, when you solve for the given variable, write the answer without using the natural logarithm.*

**93. Velocity** The distance traveled by a train initially moving at velocity $v_0$ after the engine is shut off can be calculated by the formula $x = \frac{1}{k} \ln (k v_0 t + 1)$. Solve this equation for $v_0$.

**94. Intensity of Light** The intensity of light as it passes through a certain medium is found by the formula $x = k(\ln I_0 - \ln I)$. Solve this equation for $I_0$.

**95. Electric Circuit** An equation relating the current and time in an electric circuit is $\ln i - \ln I = \frac{-t}{RC}$. Solve this equation for $i$.

**96. Molecule** A formula used in studying the action of a protein molecule is $\ln M = \ln Q - \ln (1 - Q)$. Solve this equation for $Q$.

## Cumulative Review Exercises

[5.5]   **97.** Multiply $(3xy^2 + y)(4x - 3xy)$.

[6.5]   **98.** Find two values of $b$ that will make $4x^2 + bx + 25$ a perfect square trinomial.

[8.2]   **99.** Let $h(x) = \dfrac{x^2 + 4x}{x + 6}$. Find **a)** $h(-4)$. **b)** $h\left(\dfrac{2}{5}\right)$.

[9.5]   **100. Tickets** The admission at an ice hockey game is $15 for adults and $11 for children. A total of 550 tickets were sold. Determine how many children's tickets and how many adults' tickets were sold if a total of $7290 was collected.

[11.4]   **101.** Multiply $\sqrt[3]{x}\left(\sqrt[3]{x^2} + \sqrt[3]{x^5}\right)$.

# Chapter 13 Summary

| IMPORTANT FACTS AND CONCEPTS | EXAMPLES |
|---|---|

### Section 13.1

The **composite function** $(f \circ g)(x)$ is defined as

$$(f \circ g)(x) = f[g(x)]$$

Given $f(x) = x^2 + 3x - 1$ and $g(x) = x - 4$, then

$$(f \circ g)(x) = f[g(x)] = (x - 4)^2 + 3(x - 4) - 1$$
$$= x^2 - 8x + 16 + 3x - 12 - 1$$
$$= x^2 - 5x + 3$$
$$(g \circ f)(x) = g[f(x)] = (x^2 + 3x - 1) - 4$$
$$= x^2 + 3x - 5$$

---

A function is a **one-to-one function** if each element in the range corresponds with exactly one element in the domain.

The set $\{(1, 3), (-2, 5), (6, 2), (4, -1)\}$ is a one-to-one function since each element in the range corresponds with exactly one element in the domain.

---

For a function to be a one-to-one function, its graph must pass the **vertical line test** (the test to ensure it is a function) and the **horizontal line test** (to test the one-to-one criteria).

Function, but not a one-to-one function

One-to-one function

---

If $f(x)$ is a one-to-one function with ordered pairs of the form $(x, y)$, its **inverse function**, $f^{-1}(x)$, is a one-to-one function with ordered pairs of the form $(y, x)$. Only one-to-one functions have inverse functions.

**To Find the Inverse Function of a One-to-One Function**

**1.** Replace $f(x)$ with $y$.

**2.** Interchange the two variables $x$ and $y$.

**3.** Solve the equation for $y$.

**4.** Replace $y$ with $f^{-1}(x)$ (this gives the inverse function using inverse function notation).

Find the inverse function for $f(x) = 2x + 5$. Graph $f(x)$ and $f^{-1}(x)$ on the same set of axes.

Solution:

$$f(x) = 2x + 5$$
$$y = 2x + 5$$
$$x = 2y + 5$$
$$x - 5 = 2y$$
$$\frac{1}{2}x - \frac{5}{2} = y$$

or $\quad f^{-1}(x) = \dfrac{1}{2}x - \dfrac{5}{2}$

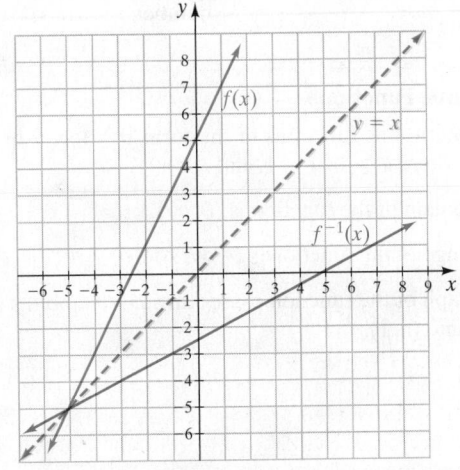

| IMPORTANT FACTS AND CONCEPTS | EXAMPLES |
|---|---|

**Section 13.1 (cont.)**

If two functions $f(x)$ and $f^{-1}(x)$ are inverses of each other, $(f \circ f^{-1})(x) = x$ and $(f^{-1} \circ f)(x) = x$.

For the previous example with $f(x) = 2x + 5$ and $f^{-1}(x) = \frac{1}{2}x - \frac{5}{2}$, then

$$(f \circ f^{-1})(x) = f[f^{-1}(x)] = 2\left(\frac{1}{2}x - \frac{5}{2}\right) + 5$$

$$= x - 5 + 5 = x$$

and

$$(f^{-1} \circ f)(x) = f^{-1}[f(x)] = \frac{1}{2}(2x + 5) - \frac{5}{2}$$

$$= x + \frac{5}{2} - \frac{5}{2} = x$$

**Section 13.2**

For any real number $a > 0$ and $a \neq 1$,

$$f(x) = a^x \quad \text{or} \quad y = a^x$$

is an **exponential function**.

For all exponential functions of the form $y = a^x$ or $f(x) = a^x$, where $a > 0$ and $a \neq 1$,

1. The domain of the function is $(-\infty, \infty)$.

2. The range of the function is $(0, \infty)$.

3. The graph of the function passes through the points $\left(-1, \frac{1}{a}\right)$, $(0, 1)$, and $(1, a)$.

Graph $y = 3^x$.

| x | y |
|---|---|
| −2 | 1/9 |
| −1 | 1/3 |
| 0 | 1 |
| 1 | 3 |
| 2 | 9 |

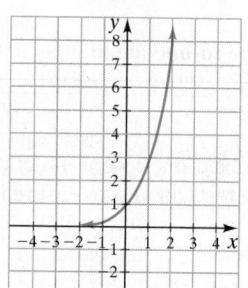

**Section 13.3**

**Logarithms**

For $x > 0$ and $a > 0, a \neq 1$

$$y = \log_a x \quad \text{means} \quad x = a^y$$

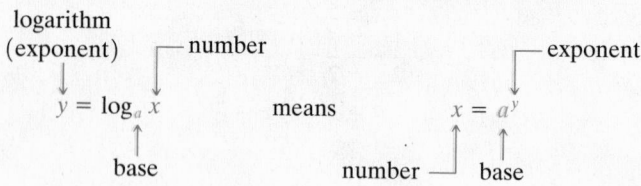

| Exponential Form | Logarithmic Form |
|---|---|
| $9^2 = 81$ | $\log_9 81 = 2$ |
| $\left(\frac{1}{4}\right)^3 = \frac{1}{64}$ | $\log_{1/4} \frac{1}{64} = 3$ |

**Logarithmic Functions**

For all logarithmic functions of the form $y = \log_a x$ or $f(x) = \log_a x$, where $a > 0$, $a \neq 1$, and $x > 0$,

1. The domain of the function is $(0, \infty)$.

2. The range of the function is $(-\infty, \infty)$.

3. The graph of the function passes through the points $\left(\frac{1}{a}, -1\right)$, $(1, 0)$, and $(a, 1)$.

Graph $y = \log_4 x$.

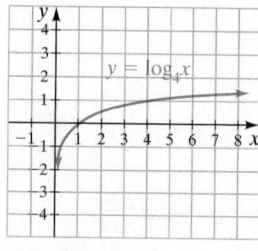

| IMPORTANT FACTS AND CONCEPTS | EXAMPLES |
|---|---|

### Section 13.3 (cont.)

**Characteristics of Exponential and Logarithmic Functions**

$y = a^x$ and $y = \log_a x$ are inverse functions.

| Exponential Function | Logarithmic Function |
|---|---|
| $y = a^x \ (a > 0, a \neq 1)$ | $y = \log_a x \ (a > 0, a \neq 1)$ |

Domain:  $(-\infty, \infty)$  ⟷  $(0, \infty)$
Range:  $(0, \infty)$  ⟷  $(-\infty, \infty)$

Points on graph:  $\left(-1, \dfrac{1}{a}\right)$ ⎫
$(0, 1)$  ⎬ *x* becomes *y*
$(1, a)$ ⎭ *y* becomes *x*

$\left(\dfrac{1}{a}, -1\right)$
$(1, 0)$
$(a, 1)$

Graph $y = 3^x$ and $y = \log_3 x$ on the same set of axes.

### Section 13.4

**Product Rule for Logarithms**

For positive real numbers $x, y,$ and $a, a \neq 1$,

$$\log_a xy = \log_a x + \log_a y \qquad \text{Property 1}$$

$$\log_5 (9 \cdot 13) = \log_5 9 + \log_5 13$$
$$\log_7 mn = \log_7 m + \log_7 n$$

**Quotient Rule for Logarithms**

For positive real numbers $x, y,$ and $a, a \neq 1$,

$$\log_a \frac{x}{y} = \log_a x - \log_a y \qquad \text{Property 2}$$

$$\log_3 \frac{15}{4} = \log_3 15 - \log_3 4$$

$$\log_8 \frac{z + 1}{z + 3} = \log_8 (z + 1) - \log_8 (z + 3)$$

**Power Rule for Logarithms**

If $x$ and $a$ are positive real numbers, $a \neq 1$, and $n$ is any real number, then

$$\log_a x^n = n \log_a x \qquad \text{Property 3}$$

$$\log_9 23^5 = 5 \log_9 23$$

$$\log_6 \sqrt[3]{x + 4} = \log_6 (x + 4)^{1/3} = \frac{1}{3} \log_6 (x + 4)$$

**Additional Properties of Logarithms**

If $a > 0$, and $a \neq 1$, then

$$\log_a a^x = x \qquad \text{Property 4}$$

and $\quad a^{\log_a x} = x \ (x > 0) \qquad \text{Property 5}$

$$\log_4 16 = \log_4 4^2 = 2$$

$$7^{\log_7 3} = 3$$

### Section 13.5

**Common Logarithm**

Base 10 logarithms are called common logarithms.

$$\log x \text{ means } \log_{10} x$$

The **common logarithm** of a positive real number is the *exponent* to which the base 10 is raised to obtain the number.

$$\text{If } \log N = L, \quad \text{then} \quad 10^L = N.$$

To find a common logarithm, use a scientific or graphing calculator. Here we round the answer to four decimal places.

$\log 17$ means $\log_{10} 17$
$\log (b + c)$ means $\log_{10} (b + c)$

If $\log 14 \approx 1.1461$, then $10^{1.1461} \approx 14$.
If $\log 0.6 \approx -0.2218$, then $10^{-0.2218} \approx 0.6$.

$\log 183 \approx 2.2625$

$\log 0.42 \approx -0.3768$

**Powers of Ten**

$$\text{If } \log x = y \quad \text{then} \quad x = 10^y$$

To find powers of 10, use a scientific or graphing calculator.

If $\log 1890.1662 \approx 3.2765$, then $10^{3.2765} \approx 1890.1662$.
If $\log 0.0143 \approx -1.8447$, then $10^{-1.8447} \approx 0.0143$.

| IMPORTANT FACTS AND CONCEPTS | EXAMPLES |
|---|---|

### Section 13.6

**Properties for Solving Exponential and Logarithmic Equations**

a) If $x = y$, then $a^x = a^y$.

b) If $a^x = a^y$, then $x = y$.

c) If $x = y$, then $\log_b x = \log_b y$   $(x > 0, y > 0)$.

d) If $\log_b x = \log_b y$, then $x = y$   $(x > 0, y > 0)$.

Properties 6a–6d

a) If $x = 5$, then $3^x = 3^5$.

b) If $3^x = 3^5$, then $x = 5$.

c) If $x = 2$, then $\log x = \log 2$.

d) If $\log x = \log 2$, then $x = 2$.

### Section 13.7

The **natural exponential function** is
$$f(x) = e^x$$
where $e \approx 2.7183$.

**Natural logarithms** are logarithms to the base $e$. Natural logarithms are indicated by the notation ln.
$$\log_e x = \ln x$$
For $x > 0$, if $y = \ln x$, then $e^y = x$.

The **natural logarithmic function** is
$$g(x) = \ln x$$
where the base $e \approx 2.7183$.

To approximate natural exponential and natural logarithmic values, use a scientific or graphing calculator.

The natural exponential function, $f(x) = e^x$, and the natural logarithmic function, $g(x) = \ln x$, are inverses of each other.

Graph $f(x) = e^x$ and $g(x) = \ln x$ on the same set of axes.

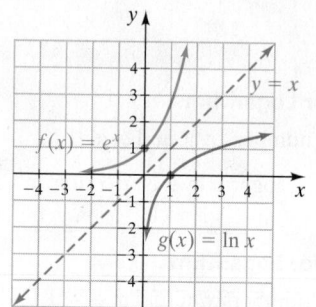

$\ln 5.83 \approx 1.7630$

If $\ln x = -2.09$, then $x = e^{-2.09} = 0.1237$.

**Change of Base Formula**

For any logarithm bases $a$ and $b$, and positive number $x$,
$$\log_a x = \frac{\log_b x}{\log_b a}$$

$$\log_5 98 = \frac{\log 98}{\log 5} \approx 2.8488$$

**Properties for Natural Logarithms**

| | | |
|---|---|---|
| $\ln xy = \ln x + \ln y$ | $(x > 0 \text{ and } y > 0)$ | Product rule |
| $\ln \dfrac{x}{y} = \ln x - \ln y$ | $(x > 0 \text{ and } y > 0)$ | Quotient rule |
| $\ln x^n = n \ln x$ | $(x > 0)$ | Power rule |

$$\ln 7 \cdot 30 = \ln 7 + \ln 30$$
$$\ln \frac{x+1}{x+8} = \ln(x+1) - \ln(x+8)$$
$$\ln m^5 = 5 \ln m$$

**Additional Properties for Natural Logarithms and Natural Exponential Expressions**

| | |
|---|---|
| $\ln e^x = x$ | Property 7 |
| $e^{\ln x} = x, \quad x > 0$ | Property 8 |

$$\ln e^{19} = 19$$
$$e^{\ln 2} = 2$$

## Chapter 13 Review Exercises

[13.1]   *Given $f(x) = x^2 - 3x + 4$ and $g(x) = 2x - 5$, find the following.*

**1.** $(f \circ g)(x)$          **2.** $(f \circ g)(3)$          **3.** $(g \circ f)(x)$          **4.** $(g \circ f)(-3)$

*Given $f(x) = 6x + 7$ and $g(x) = \sqrt{x - 3}, x \geq 3$, find the following.*

**5.** $(f \circ g)(x)$                              **6.** $(g \circ f)(x)$

*Determine whether each function is a one-to-one function.*

**7.**

**8.**

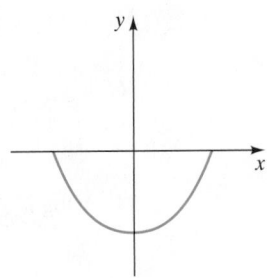

**9.** $\{(6, 2), (4, 0), (-5, 7), (3, 8)\}$

**10.** $\left\{(0, -2), (6, 1), (3, -2), \left(\dfrac{1}{2}, 4\right)\right\}$

**11.** $y = \sqrt{x + 8},\ x \geq -8$

**12.** $y = x^2 - 9$

*In Exercises 13 and 14, for each function, find the domain and range of both $f(x)$ and $f^{-1}(x)$.*

**13.** $\{(5, 3), (6, 2), (-4, -3), (-1, 8)\}$

**14.**

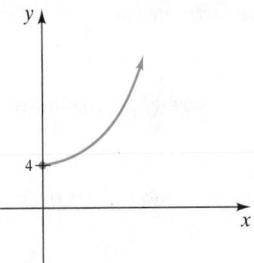

*In Exercises 15 and 16, find $f^{-1}(x)$ and graph $f(x)$ and $f^{-1}(x)$ on the same axes.*

**15.** $y = f(x) = 4x - 2$

**16.** $y = f(x) = \sqrt[3]{x - 1}$

**17. Yards to Feet** The function $f(x) = 36x$ converts yards, $x$, into inches. Find the inverse function that converts inches into yards. In the inverse function, what do $x$ and $f^{-1}(x)$ represent?

**18. Gallons to Quarts** The function $f(x) = 4x$ converts gallons, $x$, into quarts. Find the inverse function that converts quarts into gallons. In the inverse function, what do $x$ and $f^{-1}(x)$ represent?

[13.2]   *Graph the following functions.*

**19.** $y = 2^x$

**20.** $y = \left(\dfrac{1}{2}\right)^x$

**21. Compound Interest** Jim Marino invested $1500 in a savings account that earns 4% interest compounded quarterly. Use the compound interest formula $A = p\left(1 + \dfrac{r}{n}\right)^{nt}$ to determine the amount in Jim's account after 5 years.

© Thinkstock

[13.3]   *Write each equation in logarithmic form.*

**22.** $8^2 = 64$

**23.** $81^{1/4} = 3$

**24.** $5^{-3} = \dfrac{1}{125}$

*Write each equation in exponential form.*

**25.** $\log_2 32 = 5$

**26.** $\log_{1/4} \dfrac{1}{16} = 2$

**27.** $\log_6 \dfrac{1}{36} = -2$

*Write each equation in exponential form and find the missing value.*

**28.** $3 = \log_4 x$

**29.** $4 = \log_a 81$

**30.** $-3 = \log_{1/5} x$

*Graph the following functions.*

**31.** $y = \log_3 x$

**32.** $y = \log_{1/2} x$

[13.4]   *Use the properties of logarithms to expand each expression.*

**33.** $\log_5 17^8$

**34.** $\log_3 \sqrt{x-9}$

**35.** $\log \dfrac{6(a+1)}{19}$

**36.** $\log \dfrac{x^4}{7(2x+3)^5}$

*Write the following as the logarithm of a single expression.*

**37.** $5\log x - 3\log(x+1)$

**38.** $4(\log 2 + \log x) - \log y$

**39.** $\dfrac{1}{3}[\ln x - \ln(x+2)] - \ln 2$

**40.** $3\ln x + \dfrac{1}{2}\ln(x+1) - 6\ln(x+4)$

*Evaluate.*

**41.** $8^{\log_8 10}$

**42.** $\log_4 4^5$

**43.** $11^{\log_9 81}$

**44.** $9^{\log_9 \sqrt{8}}$

[13.5, 13.7]   *Use a calculator to approximate each logarithm. Round your answers to four decimal places.*

**45.** $\log 819$

**46.** $\ln 0.0281$

*Use a calculator to approximate each power of 10. Round your answers to four decimal places.*

**47.** $10^{3.159}$

**48.** $10^{-3.157}$

*Use a calculator to find x. Round your answers to four decimal places.*

**49.** $\log x = 4.063$

**50.** $\log x = -1.2262$

*Evaluate.*

**51.** $\log 10^5$

**52.** $10^{\log 9}$

**53.** $7\log 10^{3.2}$

**54.** $2(10^{\log 4.7})$

[13.6]   *Solve without using a calculator.*

**55.** $625 = 5^x$

**56.** $49^x = \dfrac{1}{7}$

**57.** $2^{3x-1} = 32$

**58.** $27^x = 3^{2x+5}$

*Solve using a calculator. Round your answers to four decimal places.*

**59.** $7^x = 152$

**60.** $3.1^x = 856$

**61.** $12.5^{x+1} = 381$

**62.** $3^{x+2} = 8^x$

*Solve the logarithmic equation.*

**63.** $\log_7(2x-3) = 2$

**64.** $\log x + \log(4x-19) = \log 5$

**65.** $\log_3 x + \log_3(2x+1) = 1$

**66.** $\ln(x+1) - \ln(x-2) = \ln 4$

[13.7]   *Solve each exponential equation for the remaining variable. Round your answers to four decimal places.*

**67.** $50 = 25e^{0.6t}$

**68.** $100 = A_0 e^{-0.42(3)}$

*Solve for the indicated variable.*

**69.** $A = A_0 e^{kt}$, for $t$

**70.** $200 = 800e^{kt}$, for $k$

**71.** $\ln y - \ln x = 6$, for $y$

**72.** $\ln(y+1) - \ln(x+8) = \ln 3$, for $y$

*Use the change of base formula to evaluate. Write the answers rounded to four decimal places.*

**73.** $\log_2 196$

**74.** $\log_3 47$

[13.2–13.7]

**75. Compound Interest** Find the amount of money accumulated if Justine Elwood puts $12,000 in a savings account yielding 6% interest compounded annually for 8 years. Use

$$A = p\left(1 + \dfrac{r}{n}\right)^{nt}.$$

**76. Interest Compounded Continuously** If $6000 is placed in a savings account paying 4% interest compounded continuously, find the time needed for the account to double in value.

**77. Bacteria** The bacteria *Escherichia coli* are commonly found in the bladders of humans. Suppose that 2000 bacteria are present at time 0. Then the number of bacteria present $t$ minutes later may be found by the function $N(t) = 2000(2)^{0.05t}$.

a) When will 50,000 bacteria be present?

b) Suppose that a human bladder infection is classified as a condition with 120,000 bacteria. When would a person develop a bladder infection if he or she started with 2000 bacteria?

**78. Atmospheric Pressure** The atmospheric pressure, $P$, in pounds per square inch at an elevation of $x$ feet above sea level can be found by the formula $P = 14.7e^{-0.00004x}$. Find the atmospheric pressure at the top of Half Dome in Yosemite National Park, an elevation of 8842 feet.

Yosemite National Park

© Allen R. Angel

**79. Retention** A class of history students is given a final exam at the end of the course. As part of a research project, the students are also given equivalent forms of the exam each month for $n$ months. The average grade of the class after $n$ months may be found by the function $A(n) = 72 - 18 \log(n + 1), n \geq 0$.

a) What was the class average when the students took the original exam ($n = 0$)?

b) What was the class average for the exam given 3 months later?

c) After how many months was the class average 58.0?

## Chapter 13 Practice Test

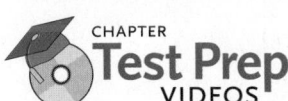

CHAPTER
**Test Prep**
VIDEOS

*Chapter Test Prep Videos provide fully worked-out solutions to any of the exercises you want to review. Chapter Test Prep Videos are available via MyMathLab®, or on YouTube (search "Angel Elementary and Intermediate Algebra" and click on "Channels")*

**1. a)** Determine whether the following function is a one-to-one function.

$$\{(4, 2), (-3, 8), (-1, 3), (6, -7)\}$$

**b)** List the set of ordered pairs in the inverse function.

**2.** Given $f(x) = x^2 - 3$ and $g(x) = x + 2$, find **a)** $(f \circ g)(x)$. **b)** $(f \circ g)(6)$.

**3.** Given $f(x) = x^2 + 8$ and $g(x) = \sqrt{x - 5}, x \geq 5$, find **a)** $(g \circ f)(x)$. **b)** $(g \circ f)(7)$.

*In Exercises 4 and 5, **a)** find $f^{-1}(x)$ and **b)** graph $f(x)$ and $f^{-1}(x)$ on the same axes.*

**4.** $y = f(x) = -3x - 5$

**5.** $y = f(x) = \sqrt{x - 1}, x \geq 1$

**6.** What is the domain of $y = \log_5 x$?

**7.** Evaluate $\log_4 \dfrac{1}{256}$.

**8.** Graph $y = 3^x$.

**9.** Graph $y = \log_2 x$.

**10.** Write $2^{-5} = \dfrac{1}{32}$ in logarithmic form.

**11.** Write $\log_5 125 = 3$ in exponential form.

*Write Exercises 12 and 13 in exponential form and find the missing value.*

**12.** $4 = \log_2 (x + 3)$

**13.** $y = \log_{64} 16$

**14.** Expand $\log_2 \dfrac{x^3(x - 4)}{x + 2}$.

**15.** Write as the logarithm of a single expression.

$$7 \log_6 (x - 4) + 2 \log_6 (x + 3) - \frac{1}{2} \log_6 x.$$

**16.** Evaluate $10 \log_9 \sqrt{9}$.

**17. a)** Find $\log 4620$ rounded to 4 decimal places.

**b)** Find $\ln 0.0692$ rounded to 4 decimal places.

**18.** Solve $3^x = 19$ for $x$.

**19.** Solve $\log 4x = \log (x + 3) + \log 2$ for $x$.

**20.** Solve $\log (x + 5) - \log (x - 2) = \log 6$ for $x$.

**21.** Find $N$, rounded to 4 decimal places, if $\ln N = 2.79$.

**22.** Evaluate $\log_6 40$, rounded to 4 decimal places, using the change of base formula.

**23.** Solve $100 = 250e^{-0.03t}$ for $t$, rounded to 4 decimal places.

**24. Savings Account** What amount of money accumulates if Kim Lee puts $3500 in a savings account yielding 4% interest compounded quarterly for 10 years?

**25. Carbon 14** The amount of carbon 14 remaining after $t$ years is found by the formula $v = v_0 e^{-0.0001205t}$, where $v_0$ is the original amount of carbon 14. If a fossil originally contained 60 grams of carbon 14, and now contains 40 grams of carbon 14, how old is the fossil?

## Cumulative Review Test

*Take the following test and check your answers with those given in the back of the book. Review any questions that you answered incorrectly. The section where the material is covered is indicated after the answer.*

**1.** Evaluate $\dfrac{6 - |-18| \div 3^2 - 6}{4 - |-8| \div 2^2}$.

**2.** Solve $4 - (6x + 6) = -(-2x + 10)$.

**3.** Solve the equation $2x - 3y = 5$ for $y$.

**4. Dinner** Thomas Furgeson took his wife out to dinner at a nice restaurant. The cost of the meal before tax was $92. If the total price, including tax, was $98.90, find the tax rate.

© Angel R. Angel

**5. Jogging** Jason Sykora jogs at 4 mph while Kendra Rathbun jogs at 5 mph. Kendra starts jogging $\dfrac{1}{2}$ hour after Jason and jogs along the same path.

 **a)** How long after Kendra starts will the two meet?

 **b)** How far from the starting point will they be when they meet?

**6.** Find the slope of the line in the figure below. Then write the equation of the given line.

**7.** Simplify $\left(\dfrac{3x^4 y^{-3}}{6xy^4 z^2}\right)^{-3}$.

**8.** Divide $\dfrac{x^3 + 3x^2 + 5x + 4}{x + 1}$.

**9.** Factor $12x^2 - 5xy - 3y^2$.

**10.** Factor $x^2 - 2xy + y^2 - 25$.

**11.** Solve $\dfrac{x + 1}{x + 2} + \dfrac{x - 2}{x - 3} = \dfrac{x^2 - 4}{x^2 - x - 6}$.

**12.** $L$ varies inversely as the square of $P$. Find $L$ when $P = 4$ and $k = 100$.

**13.** Let $h(x) = \dfrac{x^2 + 4x}{x + 6}$. Find $h(-3)$.

**14.** Solve the system of equations.

$$0.4x + 0.6y = 3.2$$
$$1.4x - 0.3y = 1.6$$

**15.** Solve the system of equations using matrices.

$$x + y = 6$$
$$-2x + y = 3$$

**16.** Evaluate the following determinant.

$$\begin{vmatrix} 3 & 0 & -1 \\ 2 & 5 & 3 \\ -1 & 4 & 6 \end{vmatrix}$$

**17.** Graph $y \le \dfrac{1}{3}x + 6$.

**18.** Let $g(x) = x^2 - 4x - 5$.

 **a)** Express $g(x)$ in the form

$$g(x) = a(x - h)^2 + k.$$

 **b)** Draw the graph and label the vertex.

**19.** Given $f(x) = x^3 - 6x^2 + 5x$, find all values of $x$ for which $f(x) \ge 0$. Write the answer in interval notation.

**20. Radioactive Decay** Strontium 90 is a radioactive substance that decays exponentially at 2.8% per year. Suppose there are originally 600 grams of strontium 90. **a)** Find the number of grams left after 60 years. **b)** What is the half-life of strontium 90?

# 14 Conic Sections

## Goals of This Chapter

The focus of this chapter is on graphing conic sections. These include the circle, ellipse, parabola, and hyperbola. We have already discussed parabolas. In this chapter, we will learn more about them. We will also solve nonlinear systems of equations both algebraically and graphically.

**The shape of an** ellipse gives it an unusual feature. Anything bounced off the wall of an elliptical shape from one focal point will ricochet to the other focal point. This feature has been used in architecture and medicine. One example is the National Statuary Hall in the Capitol Building, which has an elliptically shaped domed ceiling. If you whisper at one focal point, your whisper can be heard at the other focal point. Similarly, a ball hit from one focus of an elliptical billiard table will rebound to the other focal point. In Exercise 54 on page 853, you will determine the location of the foci of an elliptical billiard table.

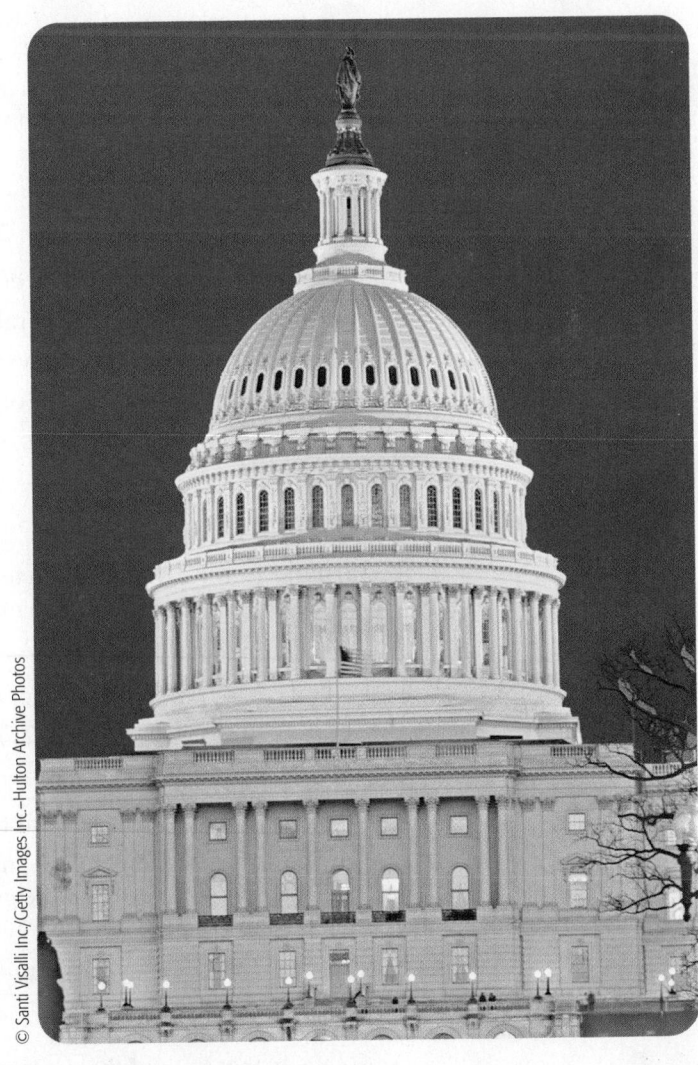

© Santi Visalli Inc./Getty Images Inc.–Hulton Archive Photos

# 4.1 The Parabola and the Circle

1  Identify and describe the conic sections.

2  Review parabolas.

3  Graph parabolas of the form $x = a(y - k)^2 + h$.

4  Learn the distance and midpoint formulas.

5  Graph circles with centers at the origin.

6  Graph circles with centers at $(h, k)$.

## 1 Identify and Describe the Conic Sections

A *parabola* is one type of conic section. Other conic sections are circles, ellipses, and hyperbolas. Each of these shapes is called a conic section because each can be made by slicing a cone and observing the shape of the slice. The methods used to slice the cone to obtain each conic section are illustrated in **Figure 14.1**.

FIGURE 14.1        Parabola          Circle          Ellipse        Hyperbola

## 2 Review Parabolas

Parabolas were first discussed in Section 12.5. They will be discussed further in this chapter. Example 1 will refresh your memory on how to graph parabolas in the forms $y = ax^2 + bx + c$ and $y = a(x - h)^2 + k$.

### EXAMPLE 1  Consider $y = 2x^2 + 4x - 6$.

**a)** Write the equation in $y = a(x - h)^2 + k$ form.

**b)** Determine whether the parabola opens upward or downward.

**c)** Determine the vertex of the parabola.

**d)** Determine the $y$-intercept of the parabola.

**e)** Determine the $x$-intercepts of the parabola.

**f)** Graph the parabola.

*Solution*

**a)** First, factor 2 from the two terms containing the variable to make the coefficient of the squared term 1. (Do not factor 2 from the constant, $-6$.)

$$y = 2x^2 + 4x - 6$$
$$= 2(x^2 + 2x) - 6 \qquad \text{2 was factored from the first two terms.}$$
$$= 2(x^2 + 2x + 1) - 2 - 6 \quad \text{Complete the square.}$$
$$= 2(x + 1)^2 - 8 \qquad \text{Simplify.}$$

**b)** The parabola opens upward because $a = 2$, which is greater than 0.

**c)** The vertex of the graph of an equation in the form $y = a(x - h)^2 + k$ is $(h, k)$. Therefore, the vertex of the graph of $y = 2(x + 1)^2 - 8$ is $(-1, -8)$. The vertex of a parabola can also be found using

$$\left(-\frac{b}{2a}, \frac{4ac - b^2}{4a}\right) \quad \text{or} \quad \left(-\frac{b}{2a}, f\left(-\frac{b}{2a}\right)\right)$$

Show that both of these procedures give $(-1, -8)$ as the vertex of the parabola now.

**d)** To determine the *y*-intercept, let $x = 0$ and solve for *y*.

$$y = 2(x + 1)^2 - 8$$
$$= 2(0 + 1)^2 - 8$$
$$= -6$$

The *y*-intercept is $(0, -6)$.

**e)** To determine the *x*-intercepts, let $y = 0$ and solve for *x*.

$$y = 2(x + 1)^2 - 8$$

$0 = 2(x + 1)^2 - 8$     0 was substituted for *y*.

$8 = 2(x + 1)^2$     8 was added to both sides.

$4 = (x + 1)^2$     Both sides were divided by 2.

$\pm 2 = x + 1$     Square root property

$-1 \pm 2 = x$     1 was subtracted from both sides.

| | | |
|---|---|---|
| $x = -1 - 2$ | or | $x = -1 + 2$ |
| $x = -3$ | | $x = 1$ |

FIGURE 14.2

The *x*-intercepts are $(-3, 0)$ and $(1, 0)$. The *x*-intercepts could also be found by substituting 0 for *y* in $y = 2x^2 + 4x - 6$ and solving for *x* using factoring or the quadratic formula. Do this now and show that you get the same *x*-intercepts.

**f)** We use the vertex and the *x*- and *y*-intercepts to draw the graph, which is shown in **Figure 14.2**.

*Now Try Exercise 9*

**3** Graph Parabolas of the Form $x = a(y - k)^2 + h$

Parabolas can also open to the right or left. The graph of an equation of the form $x = a(y - k)^2 + h$ will be a parabola whose vertex is at the point $(h, k)$. If *a* is a positive number, the parabola will open to the right, and if *a* is a negative number, the parabola will open to the left.

The four different forms of a parabola are shown in **Figure 14.3**.

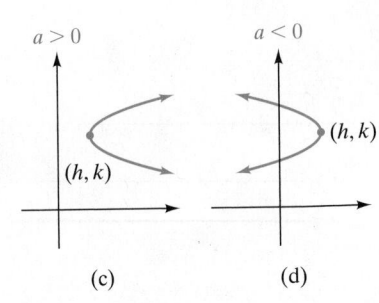

FIGURE 14.3    (a)       (b)       (c)       (d)

**Understanding Algebra**

With a parabola of the form $x = a(y - k)^2 + h$

- If $a > 0$, the parabola opens to the right.
- If $a < 0$, the parabola opens to the left.

**Parabola with Vertex at (*h*, *k*)**

1. $y = a(x - h)^2 + k, a > 0$ (opens upward)    Figure 14.3a
2. $y = a(x - h)^2 + k, a < 0$ (opens downward)    Figure 14.3b
3. $x = a(y - k)^2 + h, a > 0$ (opens to the right)    Figure 14.3c
4. $x = a(y - k)^2 + h, a < 0$ (opens to the left)    Figure 14.3d

Note that equations of the form $y = a(x - h)^2 + k$ are functions since their graphs pass the vertical line test. However, equations of the form $x = a(y - k)^2 + h$ are not functions since their graphs do not pass the vertical line test.

**EXAMPLE 2** Sketch the graph of $x = -2(y + 4)^2 - 1$.

*Solution* The graph opens to the left since the equation is of the form $x = a(y - k)^2 + h$ and $a = -2$, which is less than 0. The equation can be expressed as $x = -2[y - (-4)]^2 - 1$. Thus, $h = -1$ and $k = -4$. The vertex of the graph is $(-1, -4)$. See **Figure 14.4**.

**FIGURE 14.4**

If we set $y = 0$, we see that the $x$-intercept is at $-2(0 + 4)^2 - 1 = -2(16) - 1$ or $-33$. By substituting values for $y$ you can find the corresponding values of $x$. When $y = -2$, $x = -9$, and when $y = -6$, $x = -9$. These points are marked on the graph. Notice that this graph has no $y$-intercept.

Now Try Exercise 21

**EXAMPLE 3**

**a)** Write the equation $x = 2y^2 + 12y + 13$ in the form $x = a(y - k)^2 + h$.

**b)** Graph $x = 2y^2 + 12y + 13$.

*Solution*

**a)** First factor 2 from the first two terms. Then, complete the square on the expression within the parentheses.

$$x = 2y^2 + 12y + 13$$
$$= \boxed{2}\,(y^2 + 6y) + 13 \qquad \text{2 was factored from the first two terms.}$$
$$= \boxed{2}\,(y^2 + 6y + 9) + (\,\boxed{2}\,)(-9) + 13 \quad \text{Complete the square.}$$
$$= 2(y^2 + 6y + 9) - 18 + 13 \qquad \text{Simplify.}$$
$$= 2(y + 3)^2 - 5$$

**b)** Since $a > 0$, the parabola opens to the right. Note that when $y = 0$, $x = 2(0)^2 + 12(0) + 13 = 13$. Thus, the $x$-intercept is $(13, 0)$.

The vertex of the parabola is $(-5, -3)$. When $y = -6$, we find that $x = 13$. Thus, another point on the graph is $(13, -6)$.

Using the quadratic formula, we can determine that the $y$-intercepts are about $(0, -4.6)$ and $(0, -1.4)$. The graph is shown in **Figure 14.5**.

**FIGURE 14.5**

Now Try Exercise 35

## 4 Learn the Distance and Midpoint Formulas

Now we will derive a formula to find the *distance* between two points on a line. Consider **Figure 14.6**.

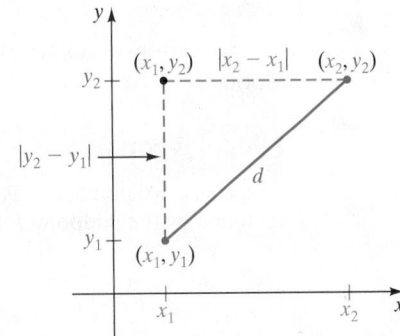

**FIGURE 14.6**

The horizontal distance between the two points $(x_1, y_2)$ and $(x_2, y_2)$, indicated by the blue dashed line, is $|x_2 - x_1|$.

The vertical distance between the points $(x_1, y_1)$ and $(x_1, y_2)$, indicated by the red dashed line, is $|y_2 - y_1|$.

Using the Pythagorean Theorem where $d$ is the distance between the two points $(x_1, y_1)$ and $(x_2, y_2)$, we get

$$d^2 = |x_2 - x_1|^2 + |y_2 - y_1|^2$$

Since any nonzero number squared is positive, we do not need absolute value signs. We can therefore write

$$d^2 = (x_2 - x_1)^2 + (y_2 - y_1)^2$$

Using the square root property, with the principal square root, we get the distance between the points $(x_1, y_1)$ and $(x_2, y_2)$, which is $d = \sqrt{(x_2 - x_1)^2 + (y_2 - y_1)^2}$.

**Distance Formula**

The distance, $d$, between any two points $(x_1, y_1)$ and $(x_2, y_2)$ can be found by the distance formula:

$$d = \sqrt{(x_2 - x_1)^2 + (y_2 - y_1)^2}$$

**Helpful Hint**

The distance between any two points will always be a positive number. When finding the distance, it makes no difference which point we designate as point 1, $(x_1, y_1)$, or point 2, $(x_2, y_2)$. Note that the square of any real number will always be greater than or equal to 0. For example, $(5 - 2)^2 = (2 - 5)^2 = 9$.

**EXAMPLE 4**  Determine the distance between the points $(4, 5)$ and $(-2, 3)$.

*Solution*  We plot the points (**Fig. 14.7**). Label $(4, 5)$ point 1 and $(-2, 3)$ point 2. Thus, $(x_2, y_2)$ represents $(-2, 3)$ and $(x_1, y_1)$ represents $(4, 5)$. Now use the distance formula to find the distance, $d$.

$$\begin{aligned}
d &= \sqrt{(x_2 - x_1)^2 + (y_2 - y_1)^2} \\
&= \sqrt{(-2 - 4)^2 + (3 - 5)^2} \\
&= \sqrt{(-6)^2 + (-2)^2} \\
&= \sqrt{36 + 4} \\
&= \sqrt{40} \quad \text{or} \quad \approx 6.32
\end{aligned}$$

**FIGURE 14.7**

Thus, the distance between the points $(4, 5)$ and $(-2, 3)$ is $\sqrt{40}$ or about 6.32 units.

Now Try Exercise 47

> **Avoiding Common Errors**
>
> Students will sometimes begin finding the distance correctly using the distance formula but will forget to take the square root of the sum $(x_2 - x_1)^2 + (y_2 - y_1)^2$ to obtain the correct answer. When taking the square root, remember that $\sqrt{a^2 + b^2} \neq a + b$.

It is often necessary to find the **midpoint** of a line segment between two given endpoints.

### Midpoint Formula

Given any two points $(x_1, y_1)$ and $(x_2, y_2)$, the point halfway between the given points can be found by the midpoint formula:

$$\text{midpoint} = \left(\frac{x_1 + x_2}{2}, \frac{y_1 + y_2}{2}\right)$$

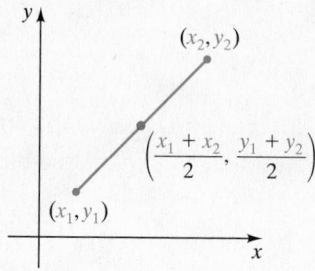

To find the midpoint, we take the average (the mean) of the $x$-coordinates and of the $y$-coordinates.

**EXAMPLE 5** A line segment through the center of a circle intersects the circle at the points $(-3, 6)$ and $(4, 1)$. Find the center of the circle.

*Solution* To find the center of the circle, we find the midpoint of the line segment between $(-3, 6)$ and $(4, 1)$. It makes no difference which points we label $(x_1, y_1)$ and $(x_2, y_2)$. We will let $(-3, 6)$ be $(x_1, y_1)$ and $(4, 1)$ be $(x_2, y_2)$. See **Figure 14.8**.

$$\text{midpoint} = \left(\frac{x_1 + x_2}{2}, \frac{y_1 + y_2}{2}\right)$$
$$= \left(\frac{-3 + 4}{2}, \frac{6 + 1}{2}\right) = \left(\frac{1}{2}, \frac{7}{2}\right)$$

The point $\left(\frac{1}{2}, \frac{7}{2}\right)$ is halfway between the points $(-3, 6)$ and $(4, 1)$. It is also the center of the circle.

Now Try Exercise 59

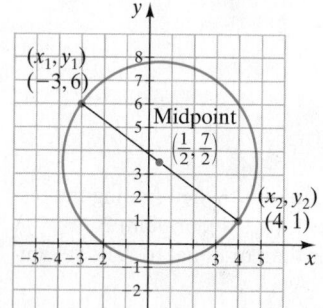

**FIGURE 14.8**

## 5 Graph Circles with Centers at the Origin

### Circle

A **circle** is the set of points in a plane that are the same distance, called the **radius**, from a fixed point, called the **center**.

The *standard form* of the equation of a circle whose center is at the origin may be derived using the distance formula. Let $(x, y)$ be a point on a circle of radius $r$ with center at $(0, 0)$. See **Figure 14.9**. Using the distance formula, we have

$$d = \sqrt{(x_2 - x_1)^2 + (y_2 - y_1)^2} \qquad \text{Distance formula}$$
$$\text{or} \quad r = \sqrt{(x - 0)^2 + (y - 0)^2} \qquad \begin{array}{l}\text{Substitute } r \text{ for } d, (x, y) \text{ for}\\ (x_2, y_2), \text{ and } (0, 0) \text{ for } (x_1, y_1).\end{array}$$
$$r = \sqrt{x^2 + y^2} \qquad \text{Simplify the radicand.}$$
$$r^2 = x^2 + y^2 \qquad \text{Square both sides.}$$

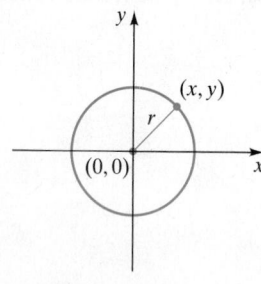

**FIGURE 14.9**

## Circle with Its Center at the Origin and Radius *r*

$$x^2 + y^2 = r^2$$

For example, $x^2 + y^2 = 16$ is a circle with its center at the origin and radius 4, and $x^2 + y^2 = 10$ is a circle with its center at the origin and radius $\sqrt{10}$. Note that $4^2 = 16$ and $(\sqrt{10})^2 = 10$. Both circles are illustrated below.

$x^2 + y^2 = 16$

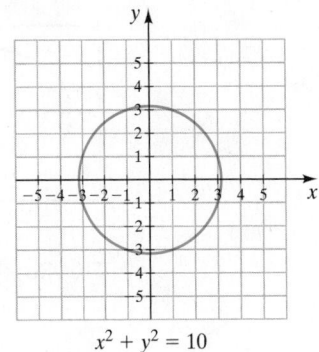
$x^2 + y^2 = 10$

**EXAMPLE 6**  Graph the following equations.

**a)** $x^2 + y^2 = 64$    **b)** $y = \sqrt{64 - x^2}$    **c)** $y = -\sqrt{64 - x^2}$

*Solution*

**a)** If we rewrite the equation as

$$x^2 + y^2 = 8^2$$

we see that this is the equation of a circle with center at the origin and radius 8. The graph is illustrated is **Figure 14.10**.

**b)** If we solve the equation $x^2 + y^2 = 64$ for *y*, we obtain

$$y^2 = 64 - x^2$$
$$y = \pm\sqrt{64 - x^2}$$

In the equation $y = \pm\sqrt{64 - x^2}$, the equation $y = +\sqrt{64 - x^2}$ or, simply, $y = \sqrt{64 - x^2}$, represents the top half of the circle, while the equation $y = -\sqrt{64 - x^2}$ represents the bottom half of the circle. Thus, the graph of $y = \sqrt{64 - x^2}$, where *y* is the principal square root, lies above and on the *x*-axis. The graph is the semicircle shown in **Figure 14.11**.

**c)** The graph of $y = -\sqrt{64 - x^2}$ is also a semicircle. However, this graph lies below and on the *x*-axis. The graph is shown in **Figure 14.12**.

**FIGURE 14.10**

**FIGURE 14.11**

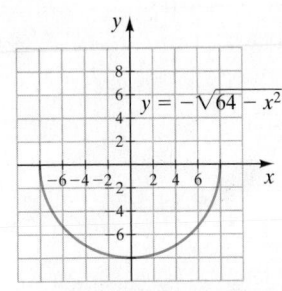

**FIGURE 14.12**

Now Try Exercise 91

Consider the equations $y = \sqrt{64 - x^2}$ and $y = -\sqrt{64 - x^2}$ in Example **6 b)** and **6 c)**. If you square both sides of the equations and rearrange the terms, you will obtain $x^2 + y^2 = 64$.

## Using Your Graphing Calculator ■■■■■

When using your calculator, you insert the function you wish to graph to the right of $y =$. Circles are not functions since they do not pass the vertical line test. To graph the equation $x^2 + y^2 = 64$, which is a circle of radius 8, we solve the equation for $y$ to obtain $y = \pm\sqrt{64 - x^2}$. We then graph the two functions $Y_1 = \sqrt{64 - x^2}$ and $Y_2 = -\sqrt{64 - x^2}$ on the same axes to obtain the circle. These graphs are illustrated in **Figure 14.13**.

Because of the distortion due to unequal scales on the axes, the graph does not appear to be a circle. When you use the $Z$ Square feature of the calculator found in the ZOOM menu, the figure appears as a circle (see **Fig. 14.14**).

$-10, 10, 1, -10, 10, 1$

**FIGURE 14.13**

$\approx-15.2, \approx15.2, 1, -10, 10, 1$

**FIGURE 14.14**

## 6  Graph Circles with Centers at $(h, k)$

The standard form of a circle with center at $(h, k)$ and radius $r$ can be derived using the distance formula. Let $(h, k)$ be the center of the circle and let $(x, y)$ be any point on the circle (see **Fig. 14.15**). If the radius, $r$, represents the distance between a point, $(x, y)$, on the circle and its center, $(h, k)$, then by the distance formula

$$r = \sqrt{(x - h)^2 + (y - k)^2}$$

We now square both sides of the equation to obtain the standard form of a circle with center at $(h, k)$ and radius $r$.

$$r^2 = (x - h)^2 + (y - k)^2$$

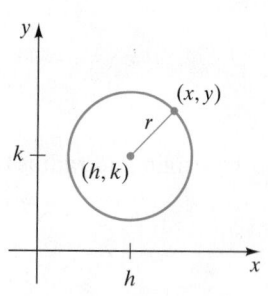

**FIGURE 14.15**

### Circle with Its Center at $(h, k)$ and Radius $r$

$$(x - h)^2 + (y - k)^2 = r^2$$

**EXAMPLE 7**  Determine the equation of the circle shown in **Figure 14.16**.

*Solution*   The center is $(-3, 2)$ and the radius is 3.

$$(x - h)^2 + (y - k)^2 = r^2$$
$$[x - (-3)]^2 + (y - 2)^2 = 3^2$$
$$(x + 3)^2 + (y - 2)^2 = 9$$

**FIGURE 14.16**

Now Try Exercise 77

**EXAMPLE 8**

a) Show that the graph of the equation $x^2 + y^2 + 6x - 2y - 6 = 0$ is a circle.

b) Determine the center and radius of the circle and then draw the circle.

c) Find the area of the circle.

*Solution*

a) We will write this equation in standard form by completing the square. First we rewrite the equation, placing all the terms containing $x$'s together and the terms containing $y$'s together.

$$x^2 + 6x + y^2 - 2y - 6 = 0$$

Understanding Algebra

The area of a circle is

$$A = \pi r^2.$$

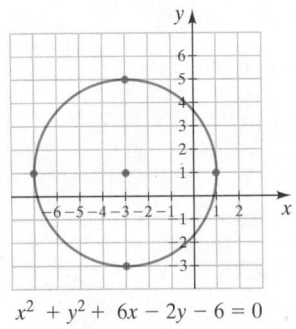

$x^2 + y^2 + 6x - 2y - 6 = 0$

**FIGURE 14.17**

Then we rewrite the equation with the constant on the right side of the equation.

$$x^2 + 6x + y^2 - 2y = 6 \qquad \text{Rewrite equation.}$$

Now we complete the square twice, once for each variable.

$$x^2 + 6x \boxed{+\ 9} + y^2 - 2y = 6 \boxed{+\ 9} \qquad \text{Complete the square on the } x\text{-terms.}$$

$$x^2 + 6x + 9 + y^2 - 2y \boxed{+\ 1} = 6 + 9 \boxed{+\ 1} \qquad \text{Complete the square on the } y\text{-terms.}$$

or

$$\underbrace{x^2 + 6x + 9}_{} \ + \ \underbrace{y^2 - 2y + 1}_{} \ = 16$$

$$(x + 3)^2 \ + \ (y - 1)^2 \ = 16$$

$$(x + 3)^2 \ + \ (y - 1)^2 \ = 4^2$$

**b)** The center of the circle is at $(-3, 1)$ and the radius is 4. The circle is sketched in **Figure 14.17**.

**c)** The area is

$$A = \pi r^2 = \pi (4)^2 = 16\pi \approx 50.3 \text{ square units}$$

Now Try Exercise 101

# EXERCISE SET 14.1   MathXL®   MyMathLab®

## Warm-Up Exercises

*Fill in the blanks with the appropriate word, phrase, or symbol(s) from the following list.*

distance          $(0, 0)$          parabola          circle          $(k, h)$          $(h, k)$          $(x, y)$          midpoint

**1.** The formula, midpoint $= \left( \dfrac{x_1 + x_2}{2}, \dfrac{y_1 + y_2}{2} \right)$, is called the _____ formula.

**2.** The formula, $d = \sqrt{(x_2 - x_1)^2 + (y_2 - y_1)^2}$, is called the _____ formula.

**3.** The equation $y = a(x - h)^2 + k$ is the equation of a _____.

**4.** The equation $x^2 + y^2 = r^2$ is the equation of a circle with center at _____ and radius $r$.

**5.** The equation $(x - h)^2 + (y - k)^2 = r$ is the equation of a circle with center at _____ and radius $r$.

**6.** The four conic sections are the parabola, _____, ellipse, and hyperbola.

## Practice the Skills

*Graph each equation.*

**7.** $y = (x - 2)^2 + 3$       **8.** $y = (x - 2)^2 - 3$       **9.** $y = (x + 3)^2 + 2$       **10.** $y = (x + 3)^2 - 4$

**11.** $y = (x - 2)^2 - 1$       **12.** $y = (x + 2)^2 + 1$       **13.** $y = -(x - 1)^2 + 1$       **14.** $y = -(x + 4)^2 - 5$

**15.** $y = -(x + 3)^2 + 4$       **16.** $y = 2(x + 1)^2 - 3$       **17.** $y = -3(x - 5)^2 + 3$       **18.** $x = (y - 1)^2 + 1$

**19.** $x = (y - 4)^2 - 3$       **20.** $x = -(y - 2)^2 + 1$       **21.** $x = -(y - 5)^2 + 4$       **22.** $x = -2(y + 4)^2 - 3$

**23.** $x = -5(y + 3)^2 - 6$       **24.** $x = 3(y + 1)^2 + 5$       **25.** $y = -2\left(x + \dfrac{1}{2}\right)^2 + 6$       **26.** $y = -\left(x - \dfrac{5}{2}\right)^2 + \dfrac{1}{2}$

*In Exercises 27–40,* **a)** *Write the equation in the form* $y = a(x - h)^2 + k$ *or* $x = a(y - k)^2 + h$. **b)** *Graph the equation.*

**27.** $y = x^2 + 2x$                **28.** $y = x^2 - 2x$                **29.** $y = x^2 + 6x$

**30.** $y = x^2 - 4x$                **31.** $x = y^2 + 4y$                **32.** $x = y^2 - 6y$

**33.** $y = x^2 + 7x + 10$          **34.** $y = x^2 + 2x - 7$          **35.** $x = -y^2 + 6y - 9$

**36.** $x = -y^2 - 2y + 5$          **37.** $y = -x^2 + 4x - 4$          **38.** $y = 2x^2 - 4x - 4$

**39.** $x = -y^2 + 3y - 4$          **40.** $x = 3y^2 - 12y - 36$

*Determine the distance between each pair of points. Use a calculator where appropriate and round your answers to the nearest hundredth.*

**41.** $(5, -1)$ and $(5, -6)$          **42.** $(-7, 2)$ and $(-3, 2)$          **43.** $(-1, 6)$ and $(8, 6)$

**44.** $(1, 11)$ and $(4, 15)$          **45.** $(-1, -3)$ and $(4, 9)$          **46.** $(-4, -7)$ and $(2, 1)$

**47.** $(-4, -5)$ and $(5, -2)$

**48.** $(6, 7)$ and $(11, 0)$

**49.** $(3, -1)$ and $\left(\frac{1}{2}, 4\right)$

**50.** $\left(-\frac{1}{4}, 2\right)$ and $\left(-\frac{3}{2}, 6\right)$

**51.** $(-1.6, 3.5)$ and $(-4.3, -1.7)$

**52.** $(5.2, -3.6)$ and $(-1.6, 2.3)$

**53.** $(\sqrt{7}, \sqrt{3})$ and $(0, 0)$

**54.** $(-\sqrt{2}, -\sqrt{5})$ and $(0, 0)$

*Determine the midpoint of the line segment between each pair of points.*

**55.** $(1, 3)$ and $(5, 7)$

**56.** $(0, 8)$ and $(4, -6)$

**57.** $(-7, 3)$ and $(7, -3)$

**58.** $(4, 9)$ and $(1, -5)$

**59.** $(-1, 4)$ and $(4, 6)$

**60.** $(-8, -9)$ and $(-6, -3)$

**61.** $\left(3, \frac{1}{2}\right)$ and $(2, -4)$

**62.** $\left(\frac{5}{2}, 1\right)$ and $\left(2, \frac{9}{2}\right)$

**63.** $(\sqrt{3}, 2)$ and $(\sqrt{2}, 5)$

**64.** $(-\sqrt{7}, 8)$ and $(\sqrt{5}, \sqrt{3})$

*Write the equation of each circle with the given center and radius.*

**65.** Center $(0, 0)$, radius 5

**66.** Center $(0, 0)$, radius 9

**67.** Center $(2, 0)$, radius 8

**68.** Center $(-3, 0)$, radius 7

**69.** Center $(0, 5)$, radius 2

**70.** Center $(0, -6)$, radius 10

**71.** Center $(3, 4)$, radius 4

**72.** Center $(-5, 2)$, radius 3

**73.** Center $(7, -6)$, radius 12

**74.** Center $(-6, -1)$, radius 15

**75.** Center $(1, 2)$, radius $\sqrt{5}$

**76.** Center $(-7, -2)$, radius $\sqrt{13}$

*Write the equation of each circle. Assume the radius is a whole number.*

**77.**

**78.**

**79.**

**80.**

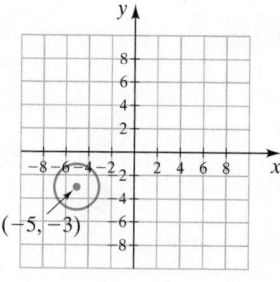

*Graph each equation.*

**81.** $x^2 + y^2 = 16$

**82.** $x^2 + y^2 = 5$

**83.** $x^2 + y^2 = 10$

**84.** $(x - 1)^2 + y^2 = 7$

**85.** $(x + 4)^2 + y^2 = 25$

**86.** $x^2 + (y + 1)^2 = 9$

**87.** $x^2 + (y - 3)^2 = 4$

**88.** $(x - 2)^2 + (y + 3)^2 = 16$

**89.** $(x + 8)^2 + (y + 2)^2 = 9$

**90.** $(x + 3)^2 + (y - 4)^2 = 36$

**91.** $y = \sqrt{25 - x^2}$

**92.** $y = \sqrt{16 - x^2}$

**93.** $y = -\sqrt{4 - x^2}$

**94.** $y = -\sqrt{49 - x^2}$

*In Exercises 95–102,* **a)** *use the method of completing the square to write each equation in standard form.* **b)** *Draw the graph.*

**95.** $x^2 + y^2 + 8x + 15 = 0$

**96.** $x^2 + y^2 + 4y = 0$

**97.** $x^2 + y^2 + 6x - 4y + 4 = 0$

**98.** $x^2 + y^2 + 2x - 4y - 4 = 0$

**99.** $x^2 + y^2 + 6x - 2y + 6 = 0$

**100.** $x^2 + y^2 + 4x - 6y - 3 = 0$

**101.** $x^2 + y^2 - 8x + 2y + 13 = 0$

**102.** $x^2 + y^2 - x + 3y - \frac{3}{2} = 0$

## Problem Solving

**103.** Find the area of the circle in Exercise 85.

**104.** Find the area of the circle in Exercise 87.

In Exercises 105–108, find the x- and y-intercepts, if they exist, of the graph of each equation.

**105.** $x = y^2 - 6y - 7$

**106.** $x = -y^2 + 8y - 12$

**107.** $x = 2(y - 3)^2 + 6$

**108.** $x = -(y + 2)^2 - 8$

**109.** If you know the midpoint of a line segment, is it possible to determine the length of the line segment? Explain.

**110.** If you know one endpoint of a line segment and the length of the line segment, can you determine the other endpoint? Explain.

**111.** Find the length of the line segment whose midpoint is $(4, -6)$ with one endpoint at $(7, -2)$.

**112.** Find the length of the line segment whose midpoint is $(-2, 4)$ with one endpoint at $(3, 6)$.

**113.** Find the equation of a circle with center at $(-6, 2)$ that is tangent to the x-axis (that is, the circle touches the x-axis at only one point).

**114.** Find the equation of a circle with center at $(-3, 5)$ that is tangent to the y-axis.

In Exercises 115 and 116, find **a)** the radius of the circle whose diameter is along the line shown, **b)** the center of the circle, and **c)** the equation of the circle.

**115.**

**116.**

**117. Points of Intersection** What is the maximum number and the minimum number of points of intersection possible for the graphs of $y = a(x - h_1)^2 + k_1$ and $x = a(y - k_2)^2 + h_2$? Explain.

**118. Triangle** Consider the figure below.

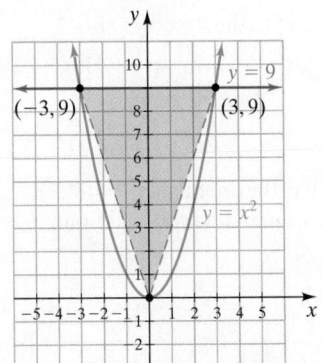

**a)** Find the area of the triangle outlined in green.

**b)** When a triangle is inscribed within a parabola, as in the figure, the area within the parabola from the base of the triangle is $\frac{4}{3}$ the area of the triangle. Find the area within the parabola from $x = -3$ to $x = 3$.

**119. Ferris Wheel** The Ferris wheel at Navy Pier in Chicago is 150 feet tall. The radius of the wheel itself is 68.2 feet.

**a)** What is the clearance below the wheel?

**b)** How high is the center of the wheel from the ground?

**c)** Find the equation of the wheel. Assume the origin is on the ground directly below the center of the wheel.

**120. Shaded Area** Find the shaded area of the square in the figure. The equation of the circle is $x^2 + y^2 = 9$.

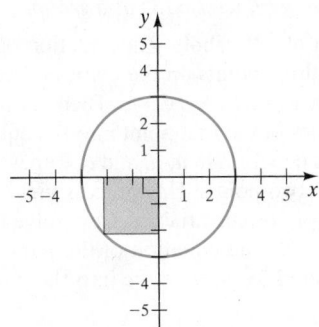

**121. Shaded Area** Consider the figure below. Write an equation for

**a)** the blue circle,

**b)** the red circle, and

**c)** the green circle.

**d)** Find the shaded area.

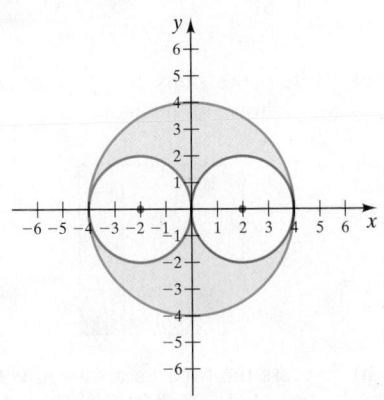

**122. Points of Intersection** Consider the equations $x^2 + y^2 = 16$ and $(x - 2)^2 + (y - 2)^2 = 16$. By considering the center and radius of each circle, determine the number of points of intersection of the two circles.

**123. Concentric Circles** Find the area between the two concentric circles whose equations are $(x - 2)^2 + (y + 4)^2 = 16$ and $(x - 2)^2 + (y + 4)^2 = 64$. *Concentric circles* are circles that have the same center.

**124. Tunnel** A highway department is planning to construct a semicircular one-way tunnel through a mountain. The tunnel is to be large enough so that a truck 8 feet wide and 10 feet tall will pass through the center of the tunnel with 1 foot to spare directly above the corner of the truck when it

is driving down the center of the tunnel (as shown in the figure below). Determine the minimum radius of the tunnel.

## Concept/Writing Exercises

**125.** Will all parabolas of the form $y = a(x - h)^2 + k, a > 0$ be functions? Explain. What will be the domain and range of $y = a(x - h)^2 + k, a > 0$?

**126.** Will all parabolas of the form $x = a(y - k)^2 + h, a > 0$ be functions? Explain. What will be the domain and range of $x = a(y - k)^2 + h, a > 0$?

**127.** How will the graphs of $y = 2(x - 3)^2 + 4$ and $y = -2(x - 3)^2 + 4$ compare?

**128.** What is the definition of a circle?

**129.** Is $x^2 - y^2 = 9$ an equation for a circle? Explain.

**130.** Is $-x^2 + y^2 = 25$ an equation for a circle? Explain.

**131.** Is $2x^2 + 3y^2 = 6$ an equation for a circle? Explain.

**132.** Is $x = y^2 - 6y + 3$ an equation for a parabola? Explain.

**133.** Is $x^2 = y^2 - 6y + 3$ an equation for a parabola? Explain.

**134.** Is $x = y + 2$ an equation for a parabola? Explain.

## Group Activity

*Discuss and answer Exercise 135 as a group.*

**135. Equation of a Parabola** The equation of a parabola can be found if three points on the parabola are known. To do so, start with $y = ax^2 + bx + c$. Then substitute the $x$- and $y$-coordinates of the first point into the equation. This will result in an equation in $a$, $b$, and $c$. Repeat the procedure for the other two points. This process yields a system of three equations in three variables. Next solve the system for $a$, $b$, and $c$. To find the equation of the parabola, substitute the values found for $a$, $b$, and $c$ into the equation $y = ax^2 + bx + c$.

Three points on a parabola are $(0, 12)$, $(3, -3)$, and $(-2, 32)$.

**a)** Individually, find a system of equations in three variables that can be used to find the equation of the parabola. Then compare your answers. If each member

of the group does not have the same system, determine why.

**b)** Individually, solve the system and determine the values of $a$, $b$, and $c$. Then compare your answers.

**c)** Individually, write the equation of the parabola passing through $(0, 12)$, $(3, -3)$, and $(-2, 32)$. Then compare your answers.

**d)** Individually, write the equation in

$$y = a(x - h)^2 + k$$

form. Then compare your answers.

**e)** Individually, graph the equation in part **d)**. Then compare your answers.

## Cumulative Review Exercises

[5.2] **136.** Simplify $\dfrac{6x^{-3}y^4}{18x^{-2}y^3}$.

[5.5] **137. a)** Write expressions to represent each of the four areas shown in the figure.

|   | $a$ | $b$ |
|---|-----|-----|
| $a$ | 1 | 2 |
| $b$ | 3 | 4 |

**b)** Express the total area shown as the square of a binomial.

[9.7] **138.** Evaluate the determinant.

$$\begin{vmatrix} 4 & 0 & 3 \\ 5 & 2 & -1 \\ 3 & 6 & 4 \end{vmatrix}$$

[10.1] **139.** Solve the inequality $-4 < 3x - 4 < 17$. Write the solution in interval notation.

[14.1] **140.** Graph $y = (x - 4)^2 + 1$.

# 14.2 The Ellipse

1 Graph ellipses.

2 Graph ellipses with centers at $(h, k)$.

FIGURE 14.19

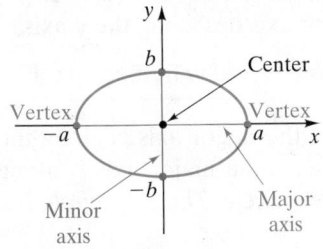

FIGURE 14.20

## 1 Graph Ellipses

### Ellipse

An **ellipse** is a set of points in a plane, the sum of whose distances from two fixed points is a constant. The two fixed points are called the **foci** (each is a **focus**) of the ellipse.

In **Figure 14.18**, $F_1$ and $F_2$ represent the two foci of an ellipse.

FIGURE 14.18

We can construct an ellipse using a length of string and two thumbtacks. Place the two thumbtacks fairly close together (**Fig. 14.19**). Then tie the ends of the string to the thumbtacks. With a pencil or pen pull the string taut, and, while keeping the string taut, draw the ellipse by moving the pencil around the thumbtacks.

In **Figure 14.20**, the line segment from $-a$ to $a$ on the $x$-axis is the *longer* or **major axis** and the line segment from $-b$ to $b$ is the *shorter* or **minor axis** of the ellipse. The major axis of an ellipse may also be on the $y$-axis. **Figure 14.20** also shows the *center* of the ellipse and the two vertices (the red dots). The vertices are the endpoints of the major axis.

### Ellipse with Its Center at the Origin

The standard form of an ellipse with its center at the origin is

$$\frac{x^2}{a^2} + \frac{y^2}{b^2} = 1$$

where $(a, 0)$ and $(-a, 0)$ are the $x$-intercepts and $(0, b)$ and $(0, -b)$ are the $y$-intercepts.

Observe that:

• The $x$-intercepts are obtained from the constant in the denominator of the $x$-term.
• The $y$-intercepts are obtained from the constant in the denominator of the $y$-term.
• If $a^2 > b^2$, the major axis is along the $x$-axis.
• If $b^2 > a^2$, the major axis is along the $y$-axis.

In Example 1, the major axis of the ellipse is along the $x$-axis.

EXAMPLE 1  Graph $\dfrac{x^2}{9} + \dfrac{y^2}{4} = 1$.

*Solution*   We can rewrite the equation as

$$\frac{x^2}{3^2} + \frac{y^2}{2^2} = 1$$

Thus, $a = 3$ and the $x$-intercepts are $(3, 0)$ and $(-3, 0)$.
Since $b = 2$, the $y$-intercepts are $(0, 2)$ and $(0, -2)$.
The ellipse is illustrated in **Figure 14.21**.

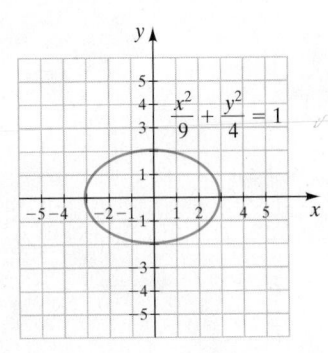

FIGURE 14.21

Now Try Exercise 15

An equation may be written so that it may not be obvious that its graph is an ellipse, as illustrated in Example 2.

**EXAMPLE 2** Graph $20x^2 + 9y^2 = 180$.

*Solution*

$$20x^2 + 9y^2 = 180$$

$$\frac{20x^2 + 9y^2}{180} = \frac{180}{180} \qquad \text{Both sides were divided by 180.}$$

$$\frac{20x^2}{180} + \frac{9y^2}{180} = 1 \qquad \text{Simplify}$$

$$\frac{x^2}{9} + \frac{y^2}{20} = 1 \qquad \text{Equation of ellipse}$$

The equation can now be recognized as an ellipse in standard form.

$$\frac{x^2}{a^2} + \frac{y^2}{b^2} = 1$$

Since $a^2 = 9$, $a = 3$. We know that $b^2 = 20$; thus $b = \sqrt{20}$ (or approximately 4.47).

$$\frac{x^2}{3^2} + \frac{y^2}{(\sqrt{20})^2} = 1$$

The $x$-intercepts are $(3, 0)$ and $(-3, 0)$. The $y$-intercepts are $(0, -\sqrt{20})$ and $(0, \sqrt{20})$. The graph is illustrated in **Figure 14.22**. Note that the major axis lies along the $y$-axis.

*Now Try Exercise 19*

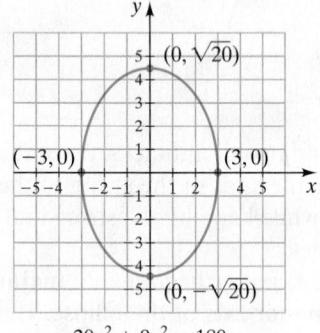

$20x^2 + 9y^2 = 180$

**FIGURE 14.22**

In Example 1, since $a^2 = 9$ and $b^2 = 4$ and $a^2 > b^2$, the major axis is along the $x$-axis. In Example 2, since $a^2 = 9$ and $b^2 = 20$ and $b^2 > a^2$, the major axis is along the $y$-axis. In the specific case where $a^2 = b^2$, the figure is a circle. *Thus, the circle is a special case of an ellipse.*

**EXAMPLE 3** Write the equation of the ellipse shown in **Figure 14.23**.

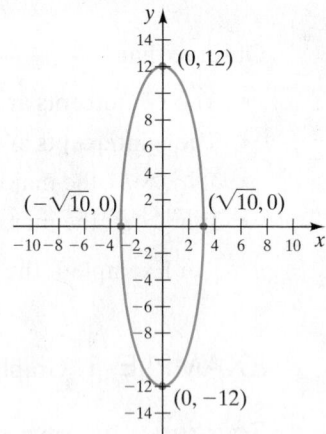

**FIGURE 14.23**

*Solution* The $x$-intercepts are $(-\sqrt{10}, 0)$ and $(\sqrt{10}, 0)$; thus, $a = \sqrt{10}$ and $a^2 = 10$. The $y$-intercepts are $(0, -12)$ and $(0, 12)$; thus, $b = 12$ and $b^2 = 144$.

$$\frac{x^2}{a^2} + \frac{y^2}{b^2} = 1$$

$$\frac{x^2}{10} + \frac{y^2}{144} = 1$$

*Now Try Exercise 45*

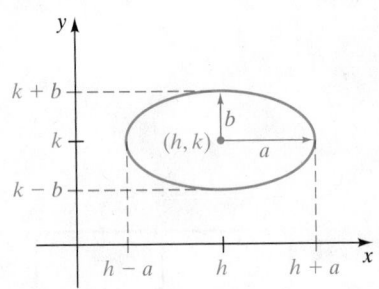

FIGURE 14.24

The formula for the **area of an ellipse** is $A = \pi ab$. In Example 1, where $a = 3$ and $b = 2$, the area is $A = \pi(3)(2) = 6\pi \approx 18.8$ square units.

In Example 2, where $a = 3$ and $b = \sqrt{20}$, the area is $A = \pi(3)(\sqrt{20}) = \pi(3)(2\sqrt{5}) = 6\pi\sqrt{5} \approx 42.1$ square units.

## 2  Graph Ellipses with Centers at (h, k)

Horizontal and vertical translations, similar to those used in Chapter 12, may be used to obtain the equation of an ellipse with center at $(h, k)$.

> **Ellipse with Its Center at (h, k)**
>
> $$\frac{(x-h)^2}{a^2} + \frac{(y-k)^2}{b^2} = 1$$

In the formula, the $h$ shifts the graph left or right from the origin and $k$ shifts the graph up or down from the origin, as shown in **Figure 14.24**.

**EXAMPLE 4**   Graph $\dfrac{(x-2)^2}{25} + \dfrac{(y+3)^2}{16} = 1$. 

*Solution*   This is the graph of $\dfrac{x^2}{25} + \dfrac{y^2}{16} = 1$ or $\dfrac{x^2}{5^2} + \dfrac{y^2}{4^2} = 1$ translated so that its center is at $(2, -3)$. Note that $a = 5$ and $b = 4$. The graph is shown in **Figure 14.25**.

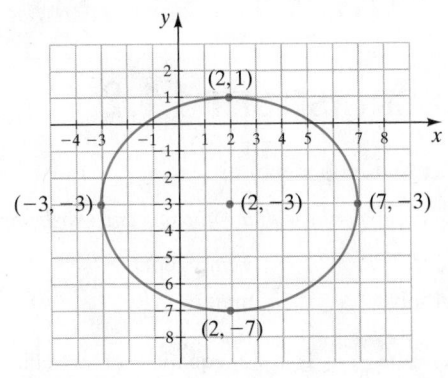

FIGURE 14.25

Now Try Exercise 33

An understanding of ellipses is useful in many areas. Astronomers know that planets revolve in elliptical orbits around the Sun. Communications satellites move in elliptical orbits around Earth (see **Fig. 14.26**).

Ellipses are used in medicine to smash kidney stones. When a signal emerges from one focus of an ellipse, the signal is reflected to the other focus. In kidney stone machines, the person is situated so that the stone to be smashed is at one focus of an elliptically shaped chamber called a lithotripter (see **Fig. 14.27** and Exercises 55 and 56).

In certain buildings with ellipsoidal ceilings, a person standing at one focus can whisper something and a person standing at the other focus can clearly hear what the person whispered. There are many other uses for ellipses, including lamps that are made to concentrate light at a specific point.

FIGURE 14.26

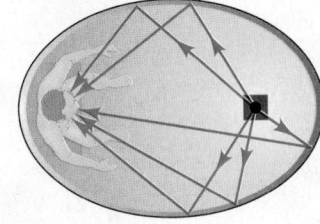

FIGURE 14.27

## Using Your Graphing Calculator ▪▪▪▪▪

Ellipses are not functions. To graph ellipses on a graphing calculator, we solve the equation for $y$. This will give the two equations that we use to graph the ellipse.

In Example 1, we graphed $\dfrac{x^2}{9} + \dfrac{y^2}{4} = 1$. Solving this equation for $y$, we get

$$\frac{x^2}{9} + \frac{y^2}{4} = 1$$

$$\boxed{36} \cdot \frac{x^2}{9} + \boxed{36} \cdot \frac{y^2}{4} = 1 \cdot \boxed{36} \qquad \text{Multiply by the LCD.}$$

$$4x^2 + 9y^2 = 36$$

$$9y^2 = 36 - 4x^2$$

$$y^2 = \frac{36 - 4x^2}{9}$$

$$y^2 = \frac{4(9 - x^2)}{9} \qquad \text{Factor 4 from the numerator.}$$

$$y = \pm\frac{2}{3}\sqrt{9 - x^2} \qquad \text{Square root property}$$

To graph the ellipse, we let $Y_1 = \dfrac{2}{3}\sqrt{9 - x^2}$ and $Y_2 = -\dfrac{2}{3}\sqrt{9 - x^2}$ and graph both equations. The graphs of $Y_1$ and $Y_2$ are illustrated in **Figure 14.28**.

FIGURE 14.28

# EXERCISE SET 14.2    MathXL®    MyMathLab®

## Warm-Up Exercises

*Fill in the blanks with the appropriate word, phrase, or symbol(s) from the following list.*

| major | $y$-intercepts | foci | divide | subtract | $(0,0)$ |
| multiply | minor | $x$-intercepts | $(k,h)$ | ellipse | $(h,k)$ |

1. $\dfrac{x^2}{a^2} + \dfrac{y^2}{b^2} = 1$ is an equation of an ellipse with center at _____.

2. $\dfrac{(x-h)^2}{a^2} + \dfrac{(y-k)^2}{b^2} = 1$ is an equation of an ellipse with center at _____.

3. A set of points in a plane the sum of whose distances from two fixed points is a constant is an _____.

4. An ellipse is the set of points in a plane whose distance from two fixed points, called the _____, is a constant.

5. For the ellipse $\dfrac{x^2}{81} + \dfrac{y^2}{16} = 1$, the points $(-9, 0)$ and $(9, 0)$ are the _____.

6. For the ellipse $\dfrac{x^2}{81} + \dfrac{y^2}{16} = 1$, the points $(0,-4)$ and $(0, 4)$ are the _____.

7. For the ellipse $\dfrac{x^2}{49} + \dfrac{y^2}{25} = 1$, the line from $-7$ to $7$ on the $x$-axis is the _____ axis.

8. For the ellipse $\dfrac{x^2}{49} + \dfrac{y^2}{25} = 1$, the line from $-5$ to $5$ on the $y$-axis is the _____ axis.

9. The first step to graph the ellipse $25x^2 + 50y^2 = 100$ is to _____ both sides by 100.

10. The first step to graph the ellipse $-\dfrac{x^2}{4} - \dfrac{y^2}{9} = -1$ is to _____ both sides by $-1$.

## Practice the Skills

*Graph each equation.*

11. $\dfrac{x^2}{4} + \dfrac{y^2}{1} = 1$

12. $\dfrac{x^2}{1} + \dfrac{y^2}{4} = 1$

13. $\dfrac{x^2}{4} + \dfrac{y^2}{9} = 1$

14. $\dfrac{x^2}{9} + \dfrac{y^2}{4} = 1$

15. $\dfrac{x^2}{25} + \dfrac{y^2}{9} = 1$

16. $\dfrac{x^2}{100} + \dfrac{y^2}{16} = 1$

17. $\dfrac{x^2}{16} + \dfrac{y^2}{25} = 1$

18. $\dfrac{x^2}{36} + \dfrac{y^2}{64} = 1$

19. $x^2 + 16y^2 = 16$

20. $x^2 + 25y^2 = 25$

21. $49x^2 + y^2 = 49$

22. $9x^2 + 25y^2 = 225$

23. $4x^2 + 9y^2 = 36$

24. $9x^2 + 16y^2 = 144$

25. $25x^2 + 100y^2 = 400$

26. $100x^2 + 25y^2 = 400$

27. $x^2 + 2y^2 = 8$

28. $x^2 + 36y^2 = 36$

29. $\dfrac{x^2}{16} + \dfrac{(y-2)^2}{9} = 1$

30. $\dfrac{(x-1)^2}{16} + \dfrac{y^2}{1} = 1$

**31.** $\dfrac{(x+3)^2}{9} + \dfrac{y^2}{25} = 1$   **32.** $\dfrac{(x-3)^2}{25} + \dfrac{(y+2)^2}{49} = 1$   **33.** $\dfrac{(x+1)^2}{9} + \dfrac{(y-2)^2}{4} = 1$   **34.** $\dfrac{(x-3)^2}{16} + \dfrac{(y-4)^2}{25} = 1$

**35.** $(x+3)^2 + 9(y+1)^2 = 81$   **36.** $18(x-1)^2 + 2(y+3)^2 = 72$   **37.** $(x-5)^2 + 4(y+4)^2 = 4$

**38.** $4(x-2)^2 + 9(y+2)^2 = 36$   **39.** $12(x+4)^2 + 3(y-1)^2 = 48$   **40.** $25(x-2)^2 + 9(y-1)^2 = 225$

## Problem Solving

**41.** Find the area of the ellipse in Exercise 11.

**42.** Find the area of the ellipse in Exercise 15.

**43.** Find the area of the ellipse in Exercise 17.

**44.** Find the area of the ellipse in Exercise 29.

*In Exercises 45–48, find the equation of the ellipse that has the four points as endpoints of the major and minor axes.*

**45.** $(5,0), (-5,0), (0,4), (0,-4)$

**46.** $(6,0), (-6,0), (0,5), (0,-5)$

**47.** $(2,0), (-2,0), (0,3), (0,-3)$

**48.** $(1,0), (-1,0), (0,9), (0,-9)$

*In Exercises 49 and 50, write the equation in standard form. Determine the center of each ellipse.*

**49.** $x^2 + 4y^2 + 6x + 16y - 11 = 0$

**50.** $x^2 + 4y^2 - 4x - 8y - 92 = 0$

**51. Art Gallery** An art gallery has an elliptical hall. The maximum distance from one focus to the wall is 90.2 feet and the minimum distance is 20.7 feet. Find the distance between the foci.

**52. Communications Satellite** A space shuttle transported a communications satellite to space. The satellite travels in an elliptical orbit around Earth. The maximum distance of the satellite from Earth is 23,200 miles and the minimum distance is 22,800 miles. Earth is at one focus of the ellipse. Find the distance from Earth to the other focus.

**53. Tunnel through a Mountain** The tunnel in the photo is the top half of an ellipse. The width of the tunnel is 20 feet and the height is 24 feet.

© Allen R. Angel

**a)** If you pictured a completed ellipse with the center of the ellipse being at the center of the road, determine the equation of the ellipse.

**b)** Find the area of the ellipse found in part **a)**.

**c)** Find the area of the opening of the tunnel.

**54. Billiard Table** An elliptical billiard table is 8 feet long by 5 feet wide. Determine the location of the foci. On such a table, if a ball is put at each focus and one ball is hit with enough force, it would hit the ball at the other focus no matter where it banks on the table.

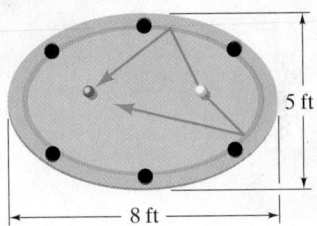

5 ft

8 ft

**55. Lithotripter Machine** Suppose the lithotripter machine described on page 851 is 6 feet long and 4 feet wide. Describe the location of the foci.

**56. Lithotripter** On page 851 we gave a brief introduction to the lithotripter, which uses ultrasound waves to shatter kidney stones. Do research and write a detailed report describing the procedure used to shatter kidney stones. Make sure that you explain how the waves are directed onto the stone.

**57. Whispering Gallery** The National Statuary Hall in the Capitol Building in Washington, D.C., is a "whispering gallery." Do research and explain why one person standing at a certain point can whisper something and someone standing a considerable distance away can hear it.

**58.** Check your answer to Exercise 11 on your graphing calculator.

**59.** Check your answer to Exercise 17 on your graphing calculator.

## Concept/Writing Exercises

**60.** Discuss the graphs of $\dfrac{x^2}{a^2} + \dfrac{y^2}{b^2} = 1$ when $a > b$, $a < b$, and $a = b$.

**61.** Explain why the circle is a special case of the ellipse.

**62.** Is $\dfrac{x^2}{36} - \dfrac{y^2}{49} = 1$ an equation for an ellipse? Explain.

**63.** Is $-\dfrac{x^2}{49} + \dfrac{y^2}{81} = 1$ an equation for an ellipse? Explain.

**64.** How many points are on the graph of $16x^2 + 25y^2 = 0$? Explain.

**65.** Consider the graph of the equation $\dfrac{x^2}{a^2} + \dfrac{y^2}{b^2} = 1$. Explain what will happen to the shape of the graph as the value of $b$ gets closer to the value of $a$. What is the shape of the graph when $a = b$?

**66.** How many points of intersection will the graphs of the equations $x^2 + y^2 = 49$ and $\dfrac{x^2}{16} + \dfrac{y^2}{25} = 1$ have? Explain.

**67.** How many points of intersection will the graphs of the equations $y = 2(x-2)^2 - 3$ and $\dfrac{(x-2)^2}{4} + \dfrac{(y+3)^2}{9} = 1$ have? Explain.

## Challenge Problems

*Determine the equation of the ellipse that has the following four points as endpoints of the major and minor axes.*

**68.** $(-7, 3), (5, 3), (-1, 5), (-1, 1)$

**69.** $(-3, 2), (11, 2), (4, 5), (4, -1)$

## Group Activity

*Work Exercise 70 individually. Then compare your answers.*

**70. Tunnel** The photo shows an elliptical tunnel (with the bottom part of the ellipse not shown) near Rockefeller Center in New York City. The maximum width of the tunnel is 18 feet and the maximum height *from the ground to the top* is 10.5 feet.

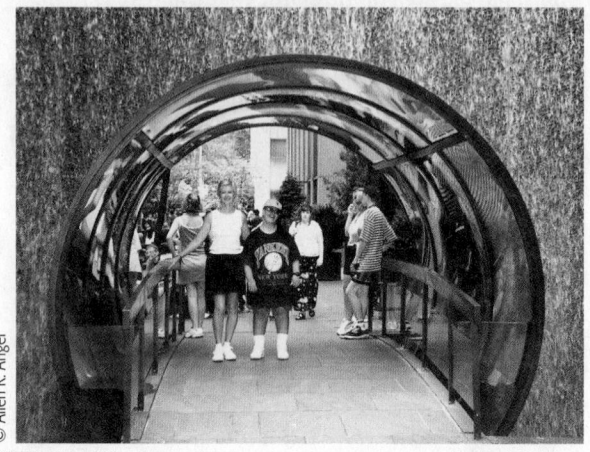

© Allen R. Angel

**a)** If the *completed ellipse* would have a maximum height of 15 feet, how high from the ground is the center of the elliptical tunnel?

**b)** Consider the following graph, which could be used to represent the tunnel.

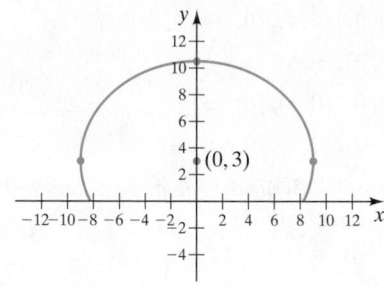

If the ellipse were continued, what would be the other *y*-intercept of the graph?

**c)** Write the equation of the ellipse, if completed, in part **b)**.

## Cumulative Review Exercises

[2.6]  **71.** Solve the formula $S = \dfrac{n}{2}(f + l)$ for $l$.

[5.6]  **72.** Divide $\dfrac{2x^2 + 2x - 7}{2x - 3}$.

[11.6]  **73.** Solve $\sqrt{3b - 2} = 10 - b$.

[12.6]  **74.** Solve $\dfrac{3x + 5}{x - 4} \le 0$, and give the solution in interval notation.

[13.7]  **75.** Find $\log_8 321$.

---

### Mid-Chapter Test: 14.1–14.2

*To find out how well you understand the chapter material to this point, take this brief test. The answers, and the section where the material was initially discussed, are given in the back of the book. Review any questions you answered incorrectly.*

*Graph each equation.*

**1.** $y = (x - 2)^2 - 1$

**2.** $y = -(x + 1)^2 + 3$

**3.** $x = -(y - 4)^2 + 1$

**4.** $x = 2(y + 3)^2 - 2$

**5.** $y = x^2 + 6x + 10$

*Find the distance between each pair of points. Where appropriate, round your answers to the nearest hundredth.*

**6.** $(-7, 4)$ and $(-2, -8)$

**7.** $(5, -3)$ and $(2, 9)$

*Find the midpoint of the line segment between each pair of points.*

**8.** $(9, -1)$ and $(-11, 6)$

**9.** $\left(-\dfrac{5}{2}, 7\right)$ and $\left(8, \dfrac{1}{2}\right)$

**10.** Write the equation of the circle with center at $(-3, 2)$ and a radius of 5 units.

*Graph each equation.*

**11.** $x^2 + (y - 1)^2 = 16$

**12.** $y = \sqrt{36 - x^2}$

**13.** $x^2 + y^2 - 2x + 4y - 4 = 0$

**14.** What is the definition of a circle?

*Graph each equation.*

**15.** $\dfrac{x^2}{4} + \dfrac{y^2}{9} = 1$

**16.** $\dfrac{x^2}{81} + \dfrac{y^2}{25} = 1$

**17.** $\dfrac{(x - 1)^2}{49} + \dfrac{(y + 2)^2}{4} = 1$

**18.** $36(x + 3)^2 + (y - 4)^2 = 36$.

**19.** Find the area of the ellipse in Exercise 15.

**20.** Find the equation of the ellipse that has the four points $(8, 0), (-8, 0), (0, 5),$ and $(0, -5)$ as the endpoints of the major and minor axes.

# 14.3 The Hyperbola

1 Graph hyperbolas.

2 Review conic sections.

## 1 Graph Hyperbolas

### Hyperbola

A **hyperbola** is the set of points in a plane, the difference of whose distances from two fixed points, called the **foci**, is a constant.

A hyperbola is illustrated in **Figure 14.29a**. In the figure, for every point on the hyperbola, the difference $M - N$ is the same constant. A hyperbola may look like a pair of parabolas. However, the shapes are actually quite different. A hyperbola has two **vertices**. The vertices are the points where the graph crosses the $x$-axis (**Figure 14.29b**) or the $y$-axis (**Figure 14.29c**). The point halfway between the two vertices is the **center** of the hyperbola. The line through the vertices is called the **transverse axis**. In **Figure 14.29b** the transverse axis lies along the $x$-axis, and in **Figure 14.29c**, the transverse axis lies along the $y$-axis.

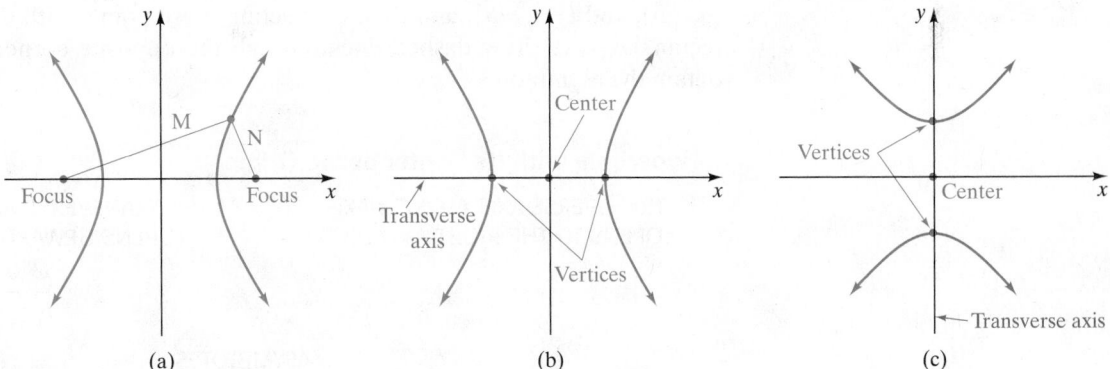

FIGURE 14.29

(a)                    (b)                    (c)

**Figure 14.30** shows the graphs of the standard forms of the equations for both hyperbolas. In **Figure 14.30a**, both vertices are $a$ units from the origin. In **Figure 14.30b**, both vertices are $b$ units from the origin. Note that in the standard form of the equation, the denominator of the $x^2$ term is always $a^2$ and the denominator of the $y^2$ term is always $b^2$.

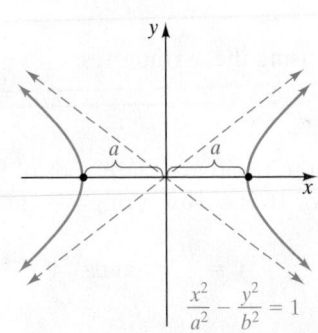

Hyperbola
with transverse axis
along the $x$-axis

$$\frac{x^2}{a^2} - \frac{y^2}{b^2} = 1$$

• Hyperbola centered at origin.
• Vertices $(-a, 0)$ and $(a, 0)$.
• Transverse axis is along the $x$-axis.

(a)

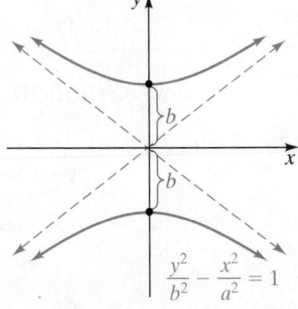

Hyperbola
with transverse axis
along the $y$-axis

$$\frac{y^2}{b^2} - \frac{x^2}{a^2} = 1$$

• Hyperbola centered at origin.
• Vertices $(0, -b)$ and $(0, b)$.
• Transverse axis is along the $y$-axis.

(b)

FIGURE 14.30

When written in standard form, the intercepts will be on the axis indicated by the variable with the positive coefficient. The intercepts will be at the positive and the negative square roots of the denominator of the positive term.

| Examples | Intercepts on | Intercepts |
|---|---|---|
| $\dfrac{x^2}{49} - \dfrac{y^2}{16} = 1$ | $x$-axis | $(-7, 0)$ and $(7, 0)$ |
| $\dfrac{y^2}{16} - \dfrac{x^2}{49} = 1$ | $y$-axis | $(0, -4)$ and $(0, 4)$ |

## Understanding Algebra

Asymptotes are *not* part of the graph, but are used to show values the graph approaches, but does not touch.

The dashed lines in **Figure 14.30** are called **asymptotes**. The asymptotes are not a part of the hyperbola but are used as an aid in graphing hyperbolas. The asymptotes are two straight lines that go through the center of the hyperbola (see **Fig. 14.30**). As the values of $x$ and $y$ get larger, the graph of the hyperbola approaches the asymptotes. The equations of the asymptotes of a hyperbola whose center is the origin are

$$y = \frac{b}{a}x \quad \text{and} \quad y = -\frac{b}{a}x$$

The asymptotes can be drawn quickly by plotting the four points $(a, b)$, $(-a, b)$, $(a, -b)$, and $(-a, -b)$, and then connecting these points with dashed lines to form a rectangle. Next, draw dashed lines through the opposite corners of the rectangle to obtain the asymptotes.

### Hyperbola with Its Center at the Origin

| TRANSVERSE AXIS ALONG $x$-AXIS (OPENS TO THE RIGHT AND LEFT) | TRANSVERSE AXIS ALONG $y$-AXIS (OPENS UPWARD AND DOWNWARD) |
|---|---|
| $\dfrac{x^2}{a^2} - \dfrac{y^2}{b^2} = 1$ | $\dfrac{y^2}{b^2} - \dfrac{x^2}{a^2} = 1$ |

ASYMPTOTES

$$y = \frac{b}{a}x \quad \text{and} \quad y = -\frac{b}{a}x$$

## EXAMPLE 1

**a)** Determine the equations of the asymptotes of the hyperbola with equation

$$\frac{x^2}{9} - \frac{y^2}{16} = 1$$

**b)** Draw the hyperbola using the asymptotes.

*Solution*

**a)** The value of $a^2$ is 9; the positive square root of 9 is 3. The value of $b^2$ is 16; the positive square root of 16 is 4. The asymptotes are

$$y = \frac{b}{a}x \quad \text{and} \quad y = -\frac{b}{a}x$$

or

$$y = \frac{4}{3}x \quad \text{and} \quad y = -\frac{4}{3}x$$

**b)** To graph the hyperbola, we first graph the asymptotes. First plot the points $(3, 4)$, $(-3, 4)$, $(3, -4)$, and $(-3, -4)$ and draw the rectangle as illustrated in **Figure 14.31**. The asymptotes are the dashed lines through the opposite corners of the rectangle.

Since the $x$-term in the original equation is positive, the graph intersects the $x$-axis. Since the denominator of the positive term is 9, the vertices are at $(3, 0)$ and $(-3, 0)$. Now draw the hyperbola by letting the hyperbola approach its asymptotes (**Fig. 14.32**). The asymptotes are drawn using dashed lines since they are not part of the hyperbola. They are used merely to help draw the graph.

FIGURE 14.31

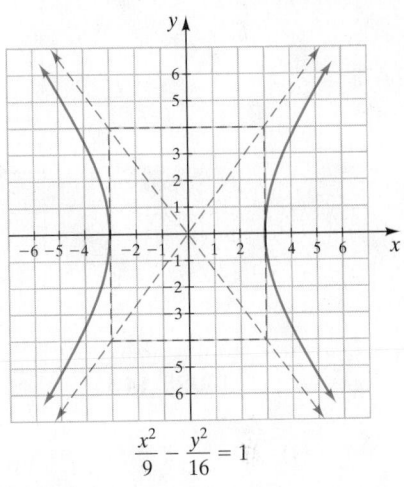

$$\frac{x^2}{9} - \frac{y^2}{16} = 1$$

FIGURE 14.32

Now Try Exercise 21

## EXAMPLE 2

**a)** Show that the equation $-25x^2 + 4y^2 = 100$ is a hyperbola by expressing the equation in standard form.

**b)** Determine the equations of the asymptotes of the graph.

**c)** Draw the graph.

*Solution*

**a)** We divide both sides of the equation by 100 to obtain 1 on the right side of the equation.

$$\frac{-25x^2 + 4y^2}{100} = \frac{100}{100}$$

$$\frac{-25x^2}{100} + \frac{4y^2}{100} = 1$$

$$\frac{-x^2}{4} + \frac{y^2}{25} = 1$$

Rewriting the equation in standard form (positive term first), we get

$$\frac{y^2}{25} - \frac{x^2}{4} = 1$$

**b)** Since $a = 2$ and $b = 5$, the equations of the asymptotes are

$$y = \frac{5}{2}x \quad \text{and} \quad y = -\frac{5}{2}x$$

**c)** The graph intersects the $y$-axis at $(0, 5)$ and $(0, -5)$. **Figure 14.33a** illustrates the asymptotes, and **Figure 14.33b** illustrates the hyperbola.

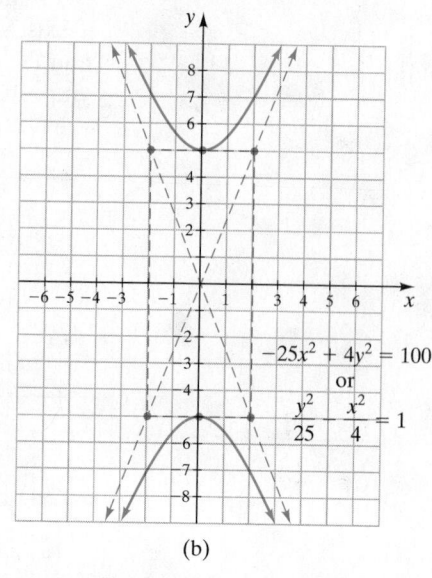

FIGURE 14.33          (a)                                    (b)

Now Try Exercise 29

We have discussed hyperbolas with their centers at the origin. Hyperbolas do not have to be centered at the origin. In this book, we will not discuss such hyperbolas.

## Using Your Graphing Calculator ▪▪▪▪▪

To graph hyperbolas on the graphing calculator, solve the equation for $y$ and graph each part. Consider Example 1,

$$\frac{x^2}{9} - \frac{y^2}{16} = 1.$$

Show that if you solve this equation for $y$ you get $y = \pm \frac{4}{3} \sqrt{x^2 - 9}$. Let $Y_1 = \frac{4}{3} \sqrt{x^2 - 9}$ and $Y_2 = -\frac{4}{3} \sqrt{x^2 - 9}$. **Figures 14.34a, 14.34b, 14.34c,** and **14.34d**, shown below, give the graphs of $Y_1$ and $Y_2$ for different window settings. The window settings used are indicated above each graph.

|  Standard setting | ZOOM: option 5 ZSquare setting | ZOOM: option 4, ZDecimal setting | Set window as shown below figure (called "friendly window settings") |

FIGURE 14.34

$-10, 10, 1, -10, 10, 1$          $\approx -15.2, \approx 15.2, 1, -10, 10, 1$          $-4.7, 4.7, 1, -3.1, 3.1, 1$          $-14.1, 14.1, 1, -9.3, 9.3, 1$

(a)                              (b)                              (c)                              (d)

In part (d), the "friendly window setting," the ratio of the length of the $x$-axis (28.2 units) to the length of the $y$-axis (18.6 units) is about 1.516. This is the same ratio as the length to the width of the display window of the calculator on the TI-84 Plus.

## 2 Review Conic Sections

The following chart summarizes conic sections.

| Parabola | Circle | Ellipse | Hyperbola |
|---|---|---|---|
| $y = a(x - h)^2 + k$ <br> or $y = ax^2 + bx + c$ | $x^2 + y^2 = r^2$ | $\dfrac{x^2}{a^2} + \dfrac{y^2}{b^2} = 1$ | $\dfrac{x^2}{a^2} - \dfrac{y^2}{b^2} = 1$ |

For the parabola: $a > 0$ graph with vertex $(h,k)$; $a < 0$ graph with vertex $(h,k)$; $x = a(y - k)^2 + h$ or $x = ay^2 + by + c$; $a > 0$ graph with vertex $(h,k)$; $a < 0$ graph with vertex $(h,k)$.

For the circle: $(x - h)^2 + (y - k)^2 = r^2$ with center $(h,k)$ and radius $r$.

For the ellipse: $\dfrac{(x - h)^2}{a^2} + \dfrac{(y - k)^2}{b^2} = 1$.

For the hyperbola: $\dfrac{y^2}{b^2} - \dfrac{x^2}{a^2} = 1$.

*Asymptotes*

$y = \dfrac{b}{a}x$ and $y = -\dfrac{b}{a}x$

**EXAMPLE 3** Indicate whether each equation represents a parabola, a circle, an ellipse, or a hyperbola.

**a)** $6x^2 = -6y^2 + 48$      **b)** $x - y^2 = 9y + 3$      **c)** $2x^2 = 8y^2 + 72$

*Solution*

**a)** This equation has an $x$-squared term and a $y$-squared term. Let's place all the squared terms on the left side of the equation.

$$6x^2 = -6y^2 + 48$$

$$6x^2 + 6y^2 = 48 \qquad \text{Add } 6y^2 \text{ to both sides.}$$

Since the coefficients of both squared terms are the same number, we divide both sides of the equation by this number. Divide both sides by 6.

$$\frac{6x^2 + 6y^2}{6} = \frac{48}{6}$$

$$x^2 + y^2 = 8$$

This equation is of the form $x^2 + y^2 = r^2$ where $r^2 = 8$.

The equation $6x^2 = -6y^2 + 48$ represents a circle.

**b)** This equation has a $y$-squared term, but no $x$-squared term. Let's solve the equation for $x$.

$$x - y^2 = 9y + 3$$

$$x = y^2 + 9y + 3 \qquad \text{Add } y^2 \text{ to both sides.}$$

This equation is of the form $x = ay^2 + by + c$ where $a = 1$, $b = 9$, and $c = 3$. The equation $x - y^2 = 9y + 3$ represents a parabola that opens to the right.

**c)** This equation has an $x$-squared term and a $y$-squared term. Let's place all the squared terms on the left side of the equation.

$$2x^2 = 8y^2 + 72$$

$$2x^2 - 8y^2 = 72 \qquad \text{Subtract } 8y^2 \text{ from both sides.}$$

Since the coefficients of both squared terms are different numbers, we want to divide the equation by the constant on the right side. Divide both sides by 72.

$$\frac{2x^2 - 8y^2}{72} = \frac{72}{72}$$

$$\frac{2x^2}{72} - \frac{8y^2}{72} = 1$$

$$\frac{x^2}{36} - \frac{y^2}{9} = 1$$

This equation is of the form $\dfrac{x^2}{a^2} - \dfrac{y^2}{b^2} = 1$ where $a^2 = 36$ (or $a = 6$) and $b^2 = 9$ (or $b = 3$).

The equation $2x^2 = 8y^2 + 72$ represents a hyperbola.

Now Try Exercise 53

---

# EXERCISE SET 14.3    MathXL®    MyMathLab®
MathXL®        MyMathLab

## Warm-Up Exercises

*Fill in the blanks with the appropriate word, phrase, or symbol(s) from the following list.*

| | | |
|---|---|---|
| divide | asymptotes | $x$ |
| horizontal | center | vertical |

| | | |
|---|---|---|
| $y$ | foci | transverse |
| vertices | hyperbola | multiply |

**1.** A set of points, the difference of whose distances from two fixed points is a constant, is a _____.

**2.** The fixed points for a hyperbola are called the _____.

**3.** For the hyperbola $\dfrac{x^2}{36} - \dfrac{y^2}{100} = 1$, the _____ is at $(0, 0)$.

**4.** For the hyperbola $\dfrac{x^2}{9} - \dfrac{y^2}{36} = 1$, the _____ are at the points $(-3, 0)$ and $(3, 0)$.

**5.** The line through the vertices of a hyperbola is the _____ axis.

**6.** For the hyperbola $\dfrac{x^2}{81} - \dfrac{y^2}{4} = 1$, the vertices are on the _____-axis.

**7.** For the hyperbola $\dfrac{y^2}{49} - \dfrac{x^2}{121} = 1$, the vertices are on the _____-axis.

**8.** For the hyperbola $\dfrac{x^2}{4} - \dfrac{y^2}{25} = 1$, the lines $y = \dfrac{5}{2}x$ and $y = -\dfrac{5}{2}x$ are the _____.

**9.** The first step in graphing the hyperbola $25x^2 - 9y^2 = 225$ is to _____ both sides by 225.

**10.** The first step in graphing the hyperbola $-\dfrac{x^2}{2} + \dfrac{y^2}{50} = -\dfrac{1}{2}$ is to _____ both sides by $-2$.

## Practice the Skills

**a)** *Determine the equations of the asymptotes for each equation.* **b)** *Graph the equation.*

**11.** $\dfrac{x^2}{9} - \dfrac{y^2}{4} = 1$

**12.** $\dfrac{y^2}{4} - \dfrac{x^2}{9} = 1$

**13.** $\dfrac{x^2}{4} - \dfrac{y^2}{1} = 1$

**14.** $\dfrac{y^2}{1} - \dfrac{x^2}{4} = 1$

**15.** $\dfrac{x^2}{4} - \dfrac{y^2}{16} = 1$

**16.** $\dfrac{x^2}{4} - \dfrac{y^2}{4} = 1$

**17.** $\dfrac{x^2}{25} - \dfrac{y^2}{16} = 1$

**18.** $\dfrac{y^2}{16} - \dfrac{x^2}{25} = 1$

**19.** $\dfrac{y^2}{25} - \dfrac{x^2}{36} = 1$

**20.** $\dfrac{x^2}{36} - \dfrac{y^2}{25} = 1$

**21.** $\dfrac{y^2}{4} - \dfrac{x^2}{36} = 1$

**22.** $\dfrac{x^2}{16} - \dfrac{y^2}{9} = 1$

**23.** $\dfrac{y^2}{25} - \dfrac{x^2}{4} = 1$

**24.** $\dfrac{x^2}{4} - \dfrac{y^2}{25} = 1$

**25.** $\dfrac{x^2}{81} - \dfrac{y^2}{16} = 1$

**26.** $\dfrac{y^2}{16} - \dfrac{x^2}{81} = 1$

*In Exercises 27–36,* **a)** *write each equation in standard form and determine the equations of the asymptotes.* **b)** *Draw the graph.*

**27.** $25y^2 - x^2 = 25$

**28.** $x^2 - 25y^2 = 25$

**29.** $4y^2 - 16x^2 = 64$

**30.** $16x^2 - 4y^2 = 64$

**31.** $9y^2 - x^2 = 9$

**32.** $x^2 - 9y^2 = 9$

**33.** $25x^2 - 9y^2 = 225$

**34.** $4y^2 - 25x^2 = 100$

**35.** $4y^2 - 36x^2 = 144$

**36.** $25x^2 - 16y^2 = 400$

*In Exercises 37–60, indicate whether the equation represents a parabola, a circle, an ellipse, or a hyperbola. See Example 3.*

**37.** $10x^2 + 10y^2 = 40$

**38.** $30x^2 - 6y^2 = 180$

**39.** $x^2 + 16y^2 = 64$

**40.** $x = 5y^2 + 15y + 1$

**41.** $4x^2 - 4y^2 = 29$

**42.** $1.2x^2 + 1.2y^2 = 24$

**43.** $2y = 12x^2 - 8x + 16$

**44.** $3y^2 - 9x^2 = 54$

**45.** $5x^2 + 12y^2 = 60$

**46.** $9.2x^2 + 9.2y^2 = 46$

**47.** $3x = -2y^2 + 9y - 30$

**48.** $12x^2 - 3y^2 = 48$

**49.** $6x^2 + 6y^2 = 36$

**50.** $11x^2 = -11y^2 + 77$

**51.** $14y^2 = 7x^2 + 35$

**52.** $9x^2 = -18y^2 + 36$

**53.** $x + y = 2y^2 + 6$

**54.** $4x^2 = -4y^2 + 400$

**55.** $13x^2 = 7y^2 + 91$

**56.** $-8x^2 = -9y^2 - 72$

**57.** $y - x + 12 = x^2$

**58.** $17x^2 = -2y^2 + 34$

**59.** $-3x^2 - 3y^2 = -27$

**60.** $x - y^2 = 49$

## Problem Solving

**61.** Determine an equation of the hyperbola whose vertices are $(0, 2)$ and $(0, -2)$ and whose asymptotes are $y = \dfrac{1}{2}x$ and $y = -\dfrac{1}{2}x$.

**62.** Determine an equation of the hyperbola whose vertices are $(0, 6)$ and $(0, -6)$ and whose asymptotes are $y = \dfrac{3}{2}x$ and $y = -\dfrac{3}{2}x$.

**63.** Determine an equation of the hyperbola whose vertices are $(-3, 0)$ and $(3, 0)$ and whose asymptotes are $y = 2x$ and $y = -2x$.

**64.** Determine an equation of the hyperbola whose vertices are $(7, 0)$ and $(-7, 0)$ and whose asymptotes are $y = \dfrac{4}{7}x$ and $y = -\dfrac{4}{7}x$.

**65.** Determine an equation of a hyperbola whose transverse axis is along the $x$-axis and whose equations of the asymptotes are $y = \dfrac{5}{3}x$ and $y = -\dfrac{5}{3}x$. Is this the only possible answer? Explain.

**66.** Determine an equation of a hyperbola whose transverse axis is along the $y$-axis and whose equations of the asymptotes are $y = \dfrac{2}{3}x$ and $y = -\dfrac{2}{3}x$. Is this the only possible answer? Explain.

**67.** Considering the graph of $\dfrac{x^2}{25} - \dfrac{y^2}{4} = 1$, determine the domain and range of the relation.

**68.** Considering the graph of $\dfrac{y^2}{36} - \dfrac{x^2}{9} = 1$, determine the domain and range of the relation.

**69.** Check your answer to Exercise 15 on your graphing calculator.

**70.** Check your answer to Exercise 21 on your graphing calculator.

**71.** Are any hyperbolas of the form $\dfrac{x^2}{a^2} - \dfrac{y^2}{b^2} = 1$ functions? Explain.

**72.** Are any hyperbolas of the form $\dfrac{y^2}{b^2} - \dfrac{x^2}{a^2} = 1$ functions? Explain.

**73.** If the equation $\dfrac{x^2}{a^2} - \dfrac{y^2}{b^2} = 1$, where $a > b$, is graphed, and then the values of $a$ and $b$ are interchanged, and the new

equation is graphed, how will the two graphs compare? Explain your answer.

**74.** If the equation $\dfrac{x^2}{a^2} - \dfrac{y^2}{b^2} = 1$, where $a > b$, is graphed, and then the signs of each term on the left side of the equation are changed, and the new equation is graphed, how will the two graphs compare? Explain your answer.

## Concept/Writing Exercises

**75.** Discuss the graph of $\dfrac{x^2}{a^2} - \dfrac{y^2}{b^2} = 1$ for nonzero real numbers $a$ and $b$. Include the transverse axis, vertices, and asymptotes.

**76.** Discuss the graph of $\dfrac{y^2}{b^2} - \dfrac{x^2}{a^2} = 1$ for nonzero real numbers $a$ and $b$. Include the transverse axis, vertices, and asymptotes.

**77.** Is $\dfrac{x^2}{81} + \dfrac{y^2}{64} = 1$ an equation for a hyperbola? Explain.

**78.** Is $-\dfrac{x^2}{81} - \dfrac{y^2}{64} = 1$ an equation for a hyperbola? Explain.

**79.** Is $4x^2 - 25y^2 = 100$ an equation for a hyperbola? Explain.

**80.** Is $36x^2 - 9y^2 = -324$ an equation for a hyperbola? Explain.

## Cumulative Review Exercises

[4.4] **81.** Write the equation, in slope-intercept form, of the line that passes through the points $(-6, 4)$ and $(-2, 2)$.

[7.4] **82.** Add $\dfrac{3x}{2x - 3} + \dfrac{2x + 4}{2x^2 + x - 6}$.

[8.5] **83.** Let $f(x) = 3x^2 - x + 5$ and $g(x) = 6 - 4x^2$. Find $(f + g)(x)$.

[9.3] **84.** Solve the system of equations.
$$-4x + 9y = 7$$
$$5x + 6y = -3$$

[12.3] **85.** Solve the formula $E = \dfrac{1}{2}mv^2$ for $v$.

[13.6] **86.** Solve the equation $\log(x + 4) = \log 5 - \log x$.

# 14.4 Nonlinear Systems of Equations and Their Applications

**1** Solve nonlinear systems using substitution.

**2** Solve nonlinear systems using addition.

**3** Solve applications.

## 1 Solve Nonlinear Systems Using Substitution

In Chapter 9, we discussed systems of linear equations. Here we discuss nonlinear systems of equations.

### Nonlinear System of Equations

A **nonlinear system of equations** is a system of equations in which at least one equation is not linear (that is, one whose graph is not a straight line).

The solution to a system of equations is the point or points that satisfy all equations in the system. Consider the system of equations

$$x^2 + y^2 = 25$$
$$3x + 4y = 0$$

Both equations are graphed on the same axes in **Figure 14.35**. Note that the graphs appear to intersect at the points $(-4, 3)$ and $(4, -3)$. The check shows that these points satisfy both equations in the system and are therefore solutions to the system.

Check $(-4, 3)$

$$x^2 + y^2 = 25$$
$$(-4)^2 + 3^2 \overset{?}{=} 25$$
$$16 + 9 \overset{?}{=} 25$$
$$25 = 25 \quad \text{True}$$

$$3x + 4y = 0$$
$$3(-4) + 4(3) \overset{?}{=} 0$$
$$-12 + 12 \overset{?}{=} 0$$
$$0 = 0 \quad \text{True}$$

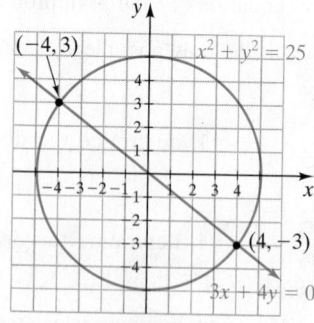

**FIGURE 14.35**

Check $(4, -3)$   $\quad 4^2 + (-3)^2 = 25 \qquad\qquad 3(4) + 4(-3) = 0$

$\qquad\qquad\qquad\qquad 16 + 9 \overset{?}{=} 25 \qquad\qquad\quad 12 - 12 \overset{?}{=} 0$

$\qquad\qquad\qquad\qquad\qquad 25 = 25 \quad$ True $\qquad\qquad\quad 0 = 0 \quad$ True

The graphical procedure for solving a system of equations may be inaccurate since we have to estimate the point or points of intersection. An exact answer may be obtained algebraically.

To solve a system of equations algebraically, we often solve one or more of the equations for one of the variables and then use substitution.

**EXAMPLE 1** Solve the previous system of equations algebraically using the substitution method.

$$x^2 + y^2 = 25$$
$$3x + 4y = 0$$

*Solution*   We first solve the linear equation $3x + 4y = 0$ for either $x$ or $y$. We will solve for $y$.

$$3x + 4y = 0$$
$$4y = -3x$$
$$y = -\frac{3x}{4}$$

Now we substitute $-\dfrac{3x}{4}$ for $y$ in the equation $x^2 + y^2 = 25$ and solve for the remaining variable, $x$.

$$x^2 + y^2 = 25$$

$$x^2 + \left(-\frac{3x}{4}\right)^2 = 25 \qquad \text{Substitute } -\frac{3x}{4} \text{ for } y.$$

$$x^2 + \frac{9x^2}{16} = 25$$

$$16\left(x^2 + \frac{9x^2}{16}\right) = 16(25) \qquad \text{Multiply both sides by 16.}$$

$$16x^2 + 9x^2 = 400$$

$$25x^2 = 400$$

$$x^2 = \frac{400}{25} = 16 \qquad \text{Both sides were divided by 25.}$$

$$x = \pm\sqrt{16} = \pm 4 \quad \text{Square root property}$$

Next, we find the corresponding value of $y$ for each value of $x$ by substituting each value of $x$ (one at a time) into the equation solved for $y$.

$$x = 4 \qquad\qquad\qquad x = -4$$
$$y = -\frac{3x}{4} \qquad\qquad y = -\frac{3x}{4}$$
$$= -\frac{3(4)}{4} \qquad\qquad = -\frac{3(-4)}{4}$$
$$= -3 \qquad\qquad\qquad = 3$$

The solutions are $(4, -3)$ and $(-4, 3)$. This checks with the solution obtained graphically in **Figure 14.35**, on page 862.

Now Try Exercise 9

## Understanding Algebra

Recall that the graph of an equation is the set of ordered pairs that are solutions to the equation. Therefore, the solutions to a system of equations are the ordered pairs that correspond to the points of intersection of the graphs of the equations in the system.

Our objective in using substitution is to obtain a single equation containing only one variable.

> ### Helpful Hint
>
> **Study Tip**
>
> In this section, we will be using the substitution method and addition method to solve nonlinear systems of equations. If you do not remember how to use both methods to solve linear systems of equations, now is a good time to review Chapter 9.
>
> In Examples 1 and 2, we solve systems using the substitution method, while in Examples 3 and 4, we solve systems using the addition method.
>
> You may choose to solve a system by the substitution method if addition of the two equations will not lead to an equation that can be easily solved, as is the case with the systems in Examples 1 and 2.

**EXAMPLE 2**  Solve the system of equations using the substitution method.

$$y = x^2 - 3$$
$$x^2 + y^2 = 9$$

**Solution**  Since both equations contain $x^2$, we will solve one of the equations for $x^2$. We will choose to solve $y = x^2 - 3$ for $x^2$.

$$y = x^2 - 3$$
$$y + 3 = x^2$$

Now substitute $y + 3$ for $x^2$ in the equation $x^2 + y^2 = 9$.

$$x^2 + y^2 = 9$$
$$y + 3 + y^2 = 9 \qquad \text{Substitute } y + 3 \text{ for } x^2.$$
$$y^2 + y + 3 = 9 \qquad \text{Rewrite.}$$
$$y^2 + y - 6 = 0$$
$$(y + 3)(y - 2) = 0 \qquad \text{Factor.}$$
$$y + 3 = 0 \qquad \text{or} \qquad y - 2 = 0$$
$$y = -3 \qquad\qquad y = 2$$

Now find the corresponding values of $x$ by substituting the values found for $y$.

| | |
|---|---|
| $y = -3$ | $y = 2$ |
| $y = x^2 - 3$ | $y = x^2 - 3$ |
| $-3 = x^2 - 3$ | $2 = x^2 - 3$ |
| $0 = x^2$ | $5 = x^2$ |
| $0 = x$ | $\pm\sqrt{5} = x$ |

This system has three solutions: $(0, -3)$, $(\sqrt{5}, 2)$, and $(-\sqrt{5}, 2)$.

Note that the graph of the equation $y = x^2 - 3$ is a parabola and the graph of the equation $x^2 + y^2 = 9$ is a circle. The graphs of both equations are illustrated in **Figure 14.36**.

Now Try Exercise 19

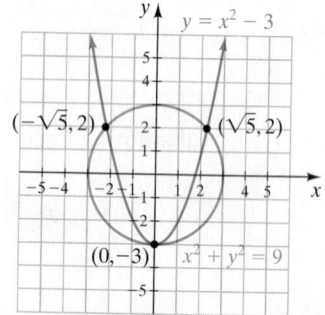

**FIGURE 14.36**

> ### Helpful Hint
>
> Students will sometimes solve for one variable and assume that they have the solution. Remember that the solution, if one exists, to a system of equations in two variables consists of one or more ordered pairs.

## 2 Solve Nonlinear Systems Using Addition

We can often solve systems of equations more easily using the addition method. As with the substitution method, our objective is to obtain a single equation containing only one variable.

**EXAMPLE 3**  Solve the system of equations using the addition method. ⎯

$$x^2 + y^2 = 9$$
$$2x^2 - y^2 = -6$$

*Solution*  If we add the two equations, we will obtain one equation containing only one variable.

$$
\begin{aligned}
x^2 + y^2 &= \phantom{-}9 \\
2x^2 - y^2 &= -6 \\
\hline
3x^2 \phantom{- y^2} &= \phantom{-}3 \\
x^2 &= \phantom{-}1 \\
x &= \pm 1
\end{aligned}
$$

Now solve for the variable $y$ by substituting $x = \pm 1$ into *either* of the original equations.

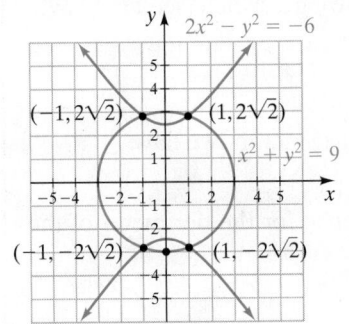

FIGURE 14.37

| $x = 1$ | $x = -1$ |
|---|---|
| $x^2 + y^2 = 9$ | $x^2 + y^2 = 9$ |
| $1^2 + y^2 = 9$ | $(-1)^2 + y^2 = 9$ |
| $1 + y^2 = 9$ | $1 + y^2 = 9$ |
| $y^2 = 8$ | $y^2 = 8$ |
| $y = \pm\sqrt{8}$ | $y = \pm\sqrt{8}$ |
| $= \pm 2\sqrt{2}$ | $= \pm 2\sqrt{2}$ |

There are four solutions to this system of equations:

$$(1, 2\sqrt{2}), (1, -2\sqrt{2}), (-1, 2\sqrt{2}), \text{ and } (-1, -2\sqrt{2})$$

The graphs of the equations in the system are given in **Figure 14.37**. Notice the four points of intersection of the two graphs.

Now Try Exercise 25

It is possible that a system of equations has no real solution (therefore, the graphs do not intersect).

**EXAMPLE 4**  Solve the system of equations using the addition method. ⎯

$$x^2 + 4y^2 = 16 \quad (\text{eq. 1})$$
$$x^2 + y^2 = 1 \quad (\text{eq. 2})$$

*Solution*  Multiply (eq. 2) by $-1$ and add the resulting equation to (eq. 1).

$$
\begin{aligned}
x^2 + 4y^2 &= 16 \\
-x^2 - \phantom{4}y^2 &= -1 \qquad (\text{eq. 2}) \text{ multiplied by } -1 \\
\hline
3y^2 &= 15 \\
y^2 &= 5 \\
y &= \pm\sqrt{5}
\end{aligned}
$$

Now solve for $x$.

$$y = \sqrt{5} \qquad\qquad\qquad y = -\sqrt{5}$$
$$x^2 + y^2 = 1 \qquad\qquad\qquad x^2 + y^2 = 1$$
$$x^2 + (\sqrt{5})^2 = 1 \qquad\qquad x^2 + (-\sqrt{5})^2 = 1$$
$$x^2 + 5 = 1 \qquad\qquad\qquad x^2 + 5 = 1$$
$$x^2 = -4 \qquad\qquad\qquad x^2 = -4$$
$$x = \pm\sqrt{-4} \qquad\qquad\qquad x = \pm\sqrt{-4}$$
$$x = \pm 2i \qquad\qquad\qquad x = \pm 2i$$

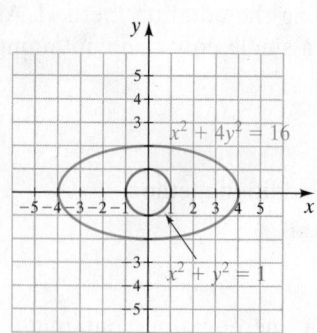

FIGURE 14.38

Since $x$ is an imaginary number for both values of $y$, this system of equations has no real solution. In solving nonlinear systems of equations, we are interested in finding all real number solutions.

The graphs of the equations are shown in **Figure 14.38**. Notice that the two graphs do not intersect; therefore, there is no real solution. This agrees with the answer we obtained algebraically.

*Now Try Exercise 37*

## 3  Solve Applications

EXAMPLE 5  **Flower Garden**  Fred and Judy Vespucci want to build a rectangular flower garden behind their house. Fred went to a local nursery and bought enough topsoil to cover 150 square meters of land. Judy went to the local hardware store and purchased 50 meters of fence for the perimeter of the garden. How should they build the garden to use all the topsoil he bought and all the fence she purchased?

*Solution*    Understand and Translate    We begin by drawing a sketch (see **Fig 14.39**).

Let $x$ = length of garden.

Let $y$ = width of garden.

FIGURE 14.39

Since $A = xy$ and Fred bought topsoil to cover 150 square meters, we have

$$xy = 150$$

Since $P = 2x + 2y$ and Judy purchased 50 meters of fence for the perimeter of the garden, we have

$$2x + 2y = 50$$

The system of equations is

$$xy = 150$$
$$2x + 2y = 50$$

Carry Out    We will solve the system using substitution. The equation $2x + 2y = 50$ is a linear equation. We will solve this equation for $y$.

$$2x + 2y = 50$$
$$2y = 50 - 2x$$
$$y = \frac{50 - 2x}{2} = \frac{50}{2} - \frac{2x}{2} = 25 - x$$

Now substitute $25 - x$ for $y$ in the equation $xy = 150$.

$$xy = 150$$
$$x(25 - x) = 150$$
$$25x - x^2 = 150$$
$$0 = x^2 - 25x + 150$$
$$0 = (x - 10)(x - 15)$$
$$x - 10 = 0 \quad \text{or} \quad x - 15 = 0$$
$$x = 10 \qquad\qquad\qquad x = 15$$

**Answer**  If $x = 10$, then $y = 25 - 10 = 15$. And, if $x = 15$, then $y = 25 - 15 = 10$. Thus, in either case, the dimensions of the flower garden are 10 meters by 15 meters.

*Now Try Exercise 41*

**EXAMPLE 6  Bicycles**  Hike 'n' Bike Company produces and sells bicycles. Its weekly cost equation is $C = 50x + 400, 0 \le x \le 160$, and its weekly revenue equation is $R = 100x - 0.3x^2, 0 \le x \le 160$, where $x$ is the number of bicycles produced and sold each week. Find the number of bicycles that must be produced and sold for Hike 'n' Bike to break even.

*Solution*  Understand and Translate  A company breaks even when its cost equals its revenue. When its cost is greater than its revenue, the company has a loss. When its revenue exceeds its cost, the company makes a profit.

The system of equations is

$$C = 50x + 400$$
$$R = 100x - 0.3x^2$$

For Hike 'n' Bike to break even, its cost must equal its revenue. Thus, we write

$$C = R$$
$$50x + 400 = 100x - 0.3x^2$$

Carry Out  Writing this quadratic equation in standard form, we obtain

$$0.3x^2 - 50x + 400 = 0, \quad 0 \le x \le 160$$

We will solve this equation using the quadratic formula.

$$a = 0.3, \quad b = -50, \quad c = 400$$

$$x = \frac{-b \pm \sqrt{b^2 - 4ac}}{2a}$$

$$= \frac{-(-50) \pm \sqrt{(-50)^2 - 4(0.3)(400)}}{2(0.3)}$$

$$= \frac{50 \pm \sqrt{2020}}{0.6}$$

$$x = \frac{50 + \sqrt{2020}}{0.6} \approx 158.2 \quad \text{or} \quad x = \frac{50 - \sqrt{2020}}{0.6} \approx 8.4$$

Profit region (when revenue exceeds cost)

Cost: $C = 50x + 400$

Revenue: $R = 100x - 0.3x^2$

Number of bicycles

**FIGURE 14.40**

**Answer**  The cost will equal the revenue and the company will break even when approximately 8 bicycles or 158 bicycles are sold. The company will make a profit when between 9 and 158 bicycles are sold. When fewer than 9 or more than 158 bicycles are sold, the company will have a loss (see **Fig. 14.40**).

*Now Try Exercise 53*

# EXERCISE SET 14.4    MathXL®  MyMathLab®
MathXL®   MyMathLab

## Warm-Up Exercises

*Fill in the blanks with the appropriate word, phrase, or symbol(s) from the following list.*

intersection      substitution      1      addition      2      3      nonlinear      no

**1.** A system of equations in which at least one equation is not linear is a _____ system of equations.

**2.** The easiest way to solve the nonlinear system of equations

$$x^2 + y^2 = 100$$
$$6x + 8y = 0$$

algebraically, is to use the _____ method.

**3.** The easiest way to solve the nonlinear system of equations

$$x^2 + y^2 = 25$$
$$4x^2 - y^2 = 100$$

algebraically, is to use the _____ method.

**4.** If the graphs of the equations in the nonlinear system do not intersect, the system has _____ real solutions.

**5.** If the graphs of the equations in the nonlinear system of equations have three points of intersection, the system has _____ real solutions.

**6.** Graphically, the solution to a system of equations occurs at the point or points of _____ of the graphs of the equations.

## Practice the Skills

*Find all real solutions to each system of equations using the substitution method.*

**7.** $x^2 + y^2 = 18$
$x + y = 0$

**8.** $x^2 + y^2 = 18$
$x - y = 0$

**9.** $x^2 + y^2 = 9$
$y = 3x + 9$

**10.** $x^2 + y^2 = 4$
$x - 2y = 4$

**11.** $y = x^2 - 5$
$3x + 2y = 10$

**12.** $x^2 = y^2 + 4$
$x + y = 4$

**13.** $x^2 + y = 6$
$y = x^2 + 4$

**14.** $y - x = 2$
$x^2 - y^2 = 4$

**15.** $x^2 + 2y^2 = 25$
$x^2 - 3y^2 = 25$

**16.** $x + y^2 = 4$
$x^2 + y^2 = 6$

**17.** $x^2 + y^2 = 4$
$y = x^2 - 6$

**18.** $x^2 - 6y^2 = 36$
$x^2 + 2y^2 = 9$

**19.** $x^2 + y^2 = 9$
$y = x^2 - 3$

**20.** $x^2 + y^2 = 16$
$y = x^2 - 4$

**21.** $2x^2 - y^2 = -8$
$x - y = 6$

**22.** $x^2 + y^2 = 1$
$y - x = 3$

*Find all real solutions to each system of equations using the addition method.*

**23.** $x^2 - y^2 = 4$
$2x^2 + y^2 = 8$

**24.** $x^2 + y^2 = 36$
$x^2 - y^2 = 36$

**25.** $x^2 + y^2 = 25$
$x^2 - 2y^2 = -2$

**26.** $x^2 + y^2 = 25$
$x^2 - 2y^2 = 7$

**27.** $3x^2 - y^2 = 4$
$x^2 + 4y^2 = 10$

**28.** $3x^2 + 2y^2 = 30$
$x^2 + y^2 = 13$

**29.** $4x^2 + 9y^2 = 36$
$2x^2 - 9y^2 = 18$

**30.** $x^2 + 4y^2 = 16$
$-9x^2 + y^2 = 4$

**31.** $2x^2 - y^2 = 7$
$x^2 + 2y^2 = 6$

**32.** $5x^2 - 2y^2 = -13$
$3x^2 + 4y^2 = 39$

**33.** $x^2 + y^2 = 25$
$2x^2 - 3y^2 = -30$

**34.** $x^2 - 2y^2 = 7$
$x^2 + y^2 = 34$

**35.** $x^2 + y^2 = 9$
$16x^2 - 4y^2 = 64$

**36.** $3x^2 + 2y^2 = 6$
$4x^2 + 5y^2 = 15$

**37.** $x^2 + y^2 = 4$
$16x^2 + 9y^2 = 144$

**38.** $x^2 + y^2 = 1$
$9x^2 - 4y^2 = 36$

**39.** $x^2 + 4y^2 = 4$
$10y^2 - 9x^2 = 90$

**40.** $-4x^2 + y^2 = -15$
$8x^2 + 3y^2 = -5$

## Problem Solving

**41. Dance Floor** Kris Hundley wants to build a dance floor at her gym. The dance floor is to have a perimeter of 90 meters and an area of 500 square meters. Find the dimensions of the dance floor.

**42. Rectangular Region** Ellen Dupree fences in a rectangular area along a riverbank as illustrated. If 20 feet of fencing encloses an area of 48 square feet, find the dimensions of the enclosed area.

**44. Rectangular Region** A rectangular area is to be fenced along a river as illustrated in Exercise 42. If 20 feet of fencing encloses an area of 50 square feet, find the dimensions of the enclosed area.

**45. Currency** A country's currency includes a bill that has an area of 112 square centimeters with a diagonal of $\sqrt{260}$ centimeters. Find the length and width of the bill.

**43. Vegetable Garden** James Cannon is planning to build a rectangular flower garden in his yard. The garden is to have a perimeter of 54 feet and an area of 170 square feet. Find the dimensions of the vegetable garden.

**46. Ice Rink** A rectangular ice rink has an area of 3000 square feet. If the diagonal across the rink is 85 feet, find the dimensions of the rink.

Rockefeller Plaza, New York City

**47. Piece of Wood** Frank Samuelson, a carpenter, has a rectangular piece of plywood. When he measures the diagonal it measures 34 inches. When he cuts the wood along the diagonal, the perimeter of each triangle formed is 80 inches. Find the dimensions of the original piece of wood.

**48. Sailboat** A sail on a sailboat is shaped like a right triangle with a perimeter of 36 meters and a hypotenuse of 15 meters. Find the length of the legs of the triangle.

**49. Baseball and Football** Paul Martin throws a football upward from the ground. Its height above the ground at any time, $t$, is given by the formula $d = -16t^2 + 64t$. At the same time that the football is thrown, Shannon Ryan throws a baseball upward from the top of an 80-foot-tall building. Its height above the ground at any time, $t$, is given by the formula $d = -16t^2 + 16t + 80$. Find the time at which the two balls will be the same height above the ground.

**50. Tennis Ball and Snowball** Robert Snell throws a tennis ball downward from a helicopter flying at a height of 950 feet. The height of the ball above the ground at any time, $t$, is found by the formula $d = -16t^2 - 10t + 950$. At the instant the ball is thrown from the helicopter, Ramon Sanchez throws a snowball upward from the top of an 750-foot-tall building. The height above the ground of the snowball at any time, $t$, is found by the formula $d = -16t^2 + 80t + 750$. At what time will the ball and snowball pass each other?

**51. Simple Interest** Simple interest is calculated using the simple interest formula, interest = principal · rate · time or $i = prt$. If Seana Hayden invests a certain principal at a specific interest rate for 1 year, the interest she obtains is $7.50. If she increases the principal by $25 and the interest rate is decreased by 1%, the interest remains the same. Find the principal and the interest rate.

**52. Simple Interest** If Claire Brooke invests a certain principal at a specific interest rate for 1 year, the interest she obtains is $72. If she increases the principal by $120 and the interest rate is decreased by 2%, the interest remains the same. Find the principal and the interest rate. Use $i = prt$.

*For the given cost and revenue equations, find the break-even point(s).*

**53.** $C = 10x + 300, R = 30x - 0.1x^2$

**54.** $C = 0.6x^2 + 9, R = 12x - 0.2x^2$

**55.** $C = 12.6x + 150, R = 42.8x - 0.3x^2$

**56.** $C = 80x + 900, R = 120x - 0.2x^2$

*Solve the following systems using your graphing calculator. Round your answers to the nearest hundredth.*

**57.** $3x - 5y = 12$
   $x^2 + y^2 = 10$

**58.** $y = 2x^2 - x + 2$
   $4x^2 + y^2 = 36$

## Concept/Writing Exercises

**59.** Make up your own nonlinear system of equations whose solution is the empty set. Explain how you know the system has no solution.

**60.** If a system of equations consists of an ellipse and a hyperbola, what is the maximum number of points of intersection? Make a sketch to illustrate this.

**61.** Can a nonlinear system of equations have exactly one real solution? If so, give an example. Explain.

**62.** Can a nonlinear system of equations have exactly two real solutions? If so, give an example. Explain.

**63.** Can a nonlinear system of equations have exactly three real solutions? If so, give an example. Explain.

**64.** Can a nonlinear system of equations have no real solutions? If so, give an example. Explain.

## Challenge Problems

**65. Intersecting Roads** The intersection of three roads forms a right triangle, as shown in the figure.

If the hypotenuse is 26 yards and the area is 120 square yards, find the length of the two legs of the triangle.

**66.** In the figure shown, $R$ represents the radius of the larger orange circle and $r$ represents the radius of the smaller orange circles. If $R = 2r$ and if the shaded area is $122.5\pi$, find $r$ and $R$.

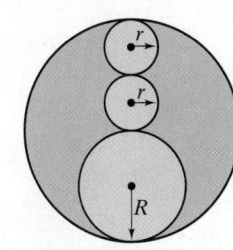

## Cumulative Review Exercises

[1.9]   **67.** List the order of operations we follow when evaluating an expression.

[6.5]   **68.** Factor $(x + 1)^3 + 1$.

[7.8]   **69.** $x$ varies inversely as the square of $P$. If $x = 10$ when $P$ is 6, find $x$ when $P = 20$.

[11.5]   **70.** Simplify $\dfrac{5}{\sqrt{x + 2} - 3}$.

[13.7]   **71.** Solve $A = A_0 e^{kt}$ for $k$.

# Chapter 14 Summary

| IMPORTANT FACTS AND CONCEPTS | EXAMPLES |
|---|---|

### Section 14.1

The four **conic sections** are the parabola, circle, ellipse, and hyperbola, which are obtained by slicing a cone.

Parabola    Circle    Ellipse    Hyperbola

The four different forms for equations of parabolas are summarized below.

**Parabola with Vertex at ($h$, $k$)**

**1.** $y = a(x - h)^2 + k$, $a > 0$ (opens upward)

**2.** $y = a(x - h)^2 + k$, $a < 0$ (opens downward)

**3.** $x = a(y - k)^2 + h$, $a > 0$ (opens to the right)

**4.** $x = a(y - k)^2 + h$, $a < 0$ (opens to the left)

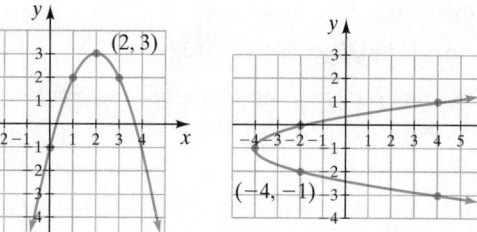

$y = -(x - 2)^2 + 3$          $x = 2(y + 1)^2 - 4$

**Distance Formula**

The distance, $d$, between any two points $(x_1, y_1)$ and $(x_2, y_2)$ can be found by the distance formula:

$$d = \sqrt{(x_2 - x_1)^2 + (y_2 - y_1)^2}$$

The distance between $(-1, 3)$ and $(4, 15)$ is

$$d = \sqrt{[4 - (-1)]^2 + (15 - 3)^2} = \sqrt{5^2 + 12^2} = \sqrt{169} = 13$$

**Midpoint Formula**

Given any two points $(x_1, y_1)$ and $(x_2, y_2)$, the point halfway between the given points can be found by the midpoint formula:

$$\text{midpoint} = \left( \frac{x_1 + x_2}{2}, \frac{y_1 + y_2}{2} \right)$$

The midpoint of the line segment joining $(7, 6)$ and $(-11, 10)$ is

$$\text{midpoint} = \left( \frac{7 + (-11)}{2}, \frac{6 + 10}{2} \right) = \left( \frac{-4}{2}, \frac{16}{2} \right) = (-2, 8)$$

A **circle** is a set of points in a plane that are the same distance, called the **radius**, from a fixed point, called the **center**.

**Circle with Its Center at the Origin and Radius $r$**

$$x^2 + y^2 = r^2$$

Sketch the graph of $x^2 + y^2 = 9$.

The graph is a circle with its center at $(0, 0)$ and radius $r = 3$.

| IMPORTANT FACTS AND CONCEPTS | EXAMPLES |

**Section 14.1 (cont.)**

**Circle with Its center at $(h, k)$ and Radius $r$**

$$(x - h)^2 + (y - k)^2 = r^2$$

Sketch the graph of $(x - 3)^2 + (y + 5)^2 = 25$.
The graph is a circle with its center at $(3, -5)$ and radius $r = 5$.

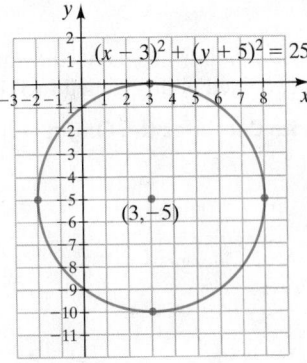

**Section 14.2**

An **ellipse** is a set of points in a plane, the sum of whose distances from two fixed points (called the **foci**) is a constant.

**Ellipse with Its Center at the Origin**

$$\frac{x^2}{a^2} + \frac{y^2}{b^2} = 1$$

where $(a, 0)$ and $(-a, 0)$ are the $x$-intercepts and $(0, b)$ and $(0, -b)$ are the $y$-intercepts.

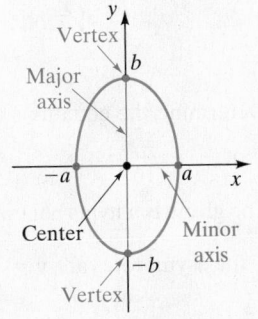

Sketch the graph of $\dfrac{x^2}{25} + \dfrac{y^2}{16} = 1$.

The graph is an ellipse. Since $a = 5$, the $x$-intercepts are $(-5, 0)$ and $(5, 0)$. Since $b = 4$, the $y$-intercepts are $(0, -4)$ and $(0, 4)$.

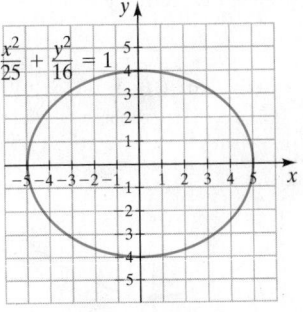

**Ellipse with Its Center at $(h, k)$**

$$\frac{(x - h)^2}{a^2} + \frac{(y - k)^2}{b^2} = 1$$

Sketch the graph of $\dfrac{(x - 2)^2}{9} + \dfrac{(y + 1)^2}{16} = 1$.

The graph is an ellipse with its center at $(2, -1)$, where $a = 3$ and $b = 4$.

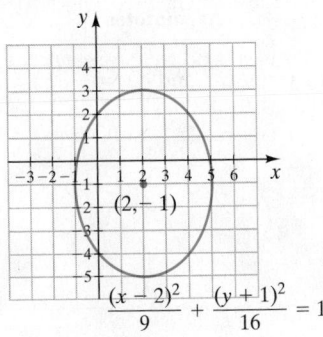

| IMPORTANT FACTS AND CONCEPTS | EXAMPLES |
|---|---|

**Section 14.2 (cont.)**

The area, $A$, of an ellipse is $A = \pi ab$.

The area of the second ellipse from page 871 is

$$A = \pi ab = \pi \cdot 3 \cdot 4 = 12\pi \approx 37.70 \text{ square units.}$$

**Section 14.3**

A **hyperbola** is a set of points in a plane, the difference of whose distances from two fixed points (called **foci**) is a constant.

### Hyperbola with Its Center at the Origin

Hyperbola with transverse axis along the $x$-axis

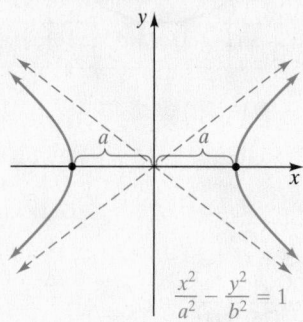

$$\frac{x^2}{a^2} - \frac{y^2}{b^2} = 1$$

**Asymptotes**

$$y = \frac{b}{a}x \quad \text{and} \quad y = -\frac{b}{a}x$$

Hyperbola with transverse axis along the $y$-axis

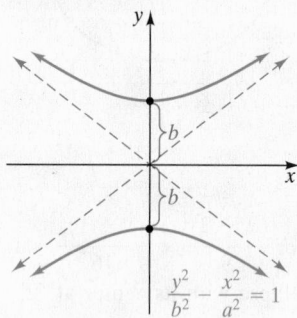

$$\frac{y^2}{b^2} - \frac{x^2}{a^2} = 1$$

**Asymptotes**

$$y = \frac{b}{a}x \quad \text{and} \quad y = -\frac{b}{a}x$$

Determine the equations of the asymptotes and sketch a graph of $\dfrac{x^2}{4} - \dfrac{y^2}{9} = 1$.

The graph is a hyperbola with $a = 2$ and $b = 3$.

The equations for the asymptotes are $y = \dfrac{3}{2}x$ and $y = -\dfrac{3}{2}x$.

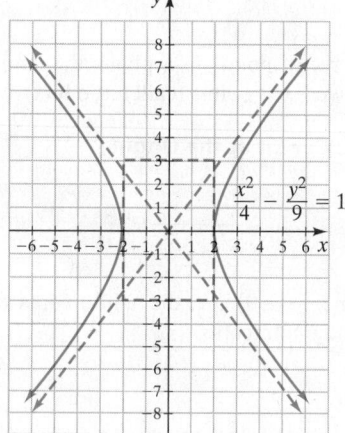

Determine the equations of the asymptotes and sketch a graph of $\dfrac{y^2}{25} - \dfrac{x^2}{16} = 1$.

The graph is a hyperbola with $a = 4$ and $b = 5$. The equations of the asymptotes are $y = \dfrac{5}{4}x$ and $y = -\dfrac{5}{4}x$.

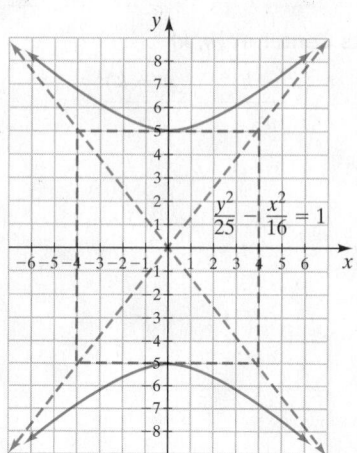

| IMPORTANT FACTS AND CONCEPTS | EXAMPLES |
|---|---|

**Section 14.4**

A **nonlinear system of equations** is a system of equations where at least one equation is not linear. The solution to a nonlinear system of equations is the point or points that satisfy all equations in the system.

Solve the system of equations.

$$x^2 + y^2 = 14$$
$$5x^2 - y^2 = -2$$

We will solve this system using the addition method.

$$x^2 + y^2 = 14$$
$$\underline{5x^2 - y^2 = -2}$$
$$6x^2 \quad\quad = 12$$
$$x^2 = 2$$
$$x = \pm\sqrt{2}$$

To obtain the value(s) for $y$, use the equation $x^2 + y^2 = 14$.

| $x = \sqrt{2}$ | $x = -\sqrt{2}$ |
|---|---|
| $x^2 + y^2 = 14$ | $x^2 + y^2 = 14$ |
| $(\sqrt{2})^2 + y^2 = 14$ | $(-\sqrt{2})^2 + y^2 = 14$ |
| $2 + y^2 = 14$ | $2 + y^2 = 14$ |
| $y^2 = 12$ | $y^2 = 12$ |
| $y = \pm\sqrt{12}$ | $y = \pm\sqrt{12}$ |
| $= \pm 2\sqrt{3}$ | $= \pm 2\sqrt{3}$ |

The system has four solutions:

$$(\sqrt{2}, 2\sqrt{3}), (\sqrt{2}, -2\sqrt{3}), (-\sqrt{2}, 2\sqrt{3}), (-\sqrt{2}, -2\sqrt{3})$$

## Chapter 14 Review Exercises

[14.1]   *Find the length and the midpoint of the line segment between each pair of points.*

**1.** $(0, 0), (5, -12)$      **2.** $(-4, 1), (-1, 5)$      **3.** $(-9, -5), (-1, 10)$      **4.** $(-4, 3), (-2, 5)$

*Graph each equation.*

**5.** $y = (x - 2)^2 + 1$      **6.** $y = (x + 3)^2 - 2$      **7.** $x = (y - 1)^2 + 4$      **8.** $x = -2(y - 4)^2 + 4$

*In Exercises 9–12,* **a)** *write each equation in the form* $y = a(x - h)^2 + k$ *or* $x = a(y - k)^2 + h$. **b)** *Graph the equation.*

**9.** $y = x^2 - 8x + 22$      **10.** $x = -y^2 - 5y - 4$      **11.** $x = y^2 + 5y + 4$      **12.** $y = 2x^2 - 8x - 24$

*In Exercises 13–18,* **a)** *write the equation of each circle in standard form.* **b)** *Draw the graph.*

**13.** Center $(0, 0)$, radius 4      **14.** Center $(-3, 4)$, radius 1      **15.** $x^2 + y^2 - 4y = 0$

**16.** $x^2 + y^2 - 2x + 6y + 1 = 0$      **17.** $x^2 - 8x + y^2 - 10y + 40 = 0$      **18.** $x^2 + y^2 - 4x + 10y + 17 = 0$

*Graph each equation.*

**19.** $y = \sqrt{9 - x^2}$      **20.** $y = -\sqrt{36 - x^2}$

*Determine the equation of each circle.*

**21.**

**22.**

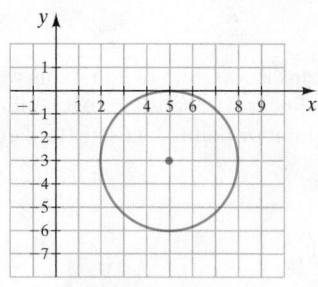

[14.2]   *Graph each equation.*

**23.** $\dfrac{x^2}{4} + \dfrac{y^2}{9} = 1$     **24.** $\dfrac{x^2}{81} + \dfrac{y^2}{49} = 1$     **25.** $9x^2 + 16y^2 = 144$     **26.** $25x^2 + 4y^2 = 100$

**27.** $\dfrac{(x-3)^2}{16} + \dfrac{(y+2)^2}{4} = 1$   **28.** $\dfrac{(x-4)^2}{9} + \dfrac{(y+3)^2}{25} = 1$   **29.** $16(x-2)^2 + 4(y+3)^2 = 16$

**30.** For the ellipse in Exercise 23, find the area.

[14.3]   *In Exercises 31–34,* **a)** *determine the equations of the asymptotes for each equation.* **b)** *Draw the graph.*

**31.** $\dfrac{x^2}{9} - \dfrac{y^2}{25} = 1$     **32.** $\dfrac{y^2}{25} - \dfrac{x^2}{9} = 1$     **33.** $\dfrac{y^2}{9} - \dfrac{x^2}{16} = 1$     **34.** $\dfrac{y^2}{25} - \dfrac{x^2}{16} = 1$

*In Exercises 35–38,* **a)** *write each equation in standard form.* **b)** *Determine the equations of the asymptotes.* **c)** *Draw the graph.*

**35.** $x^2 - 9y^2 = 9$     **36.** $64y^2 - 25x^2 = 1600$

**37.** $9y^2 - 25x^2 = 225$     **38.** $49y^2 - 9x^2 = 441$

[14.1–14.3]   *Identify the graph of each equation as a circle, ellipse, parabola, or hyperbola.*

**39.** $\dfrac{x^2}{49} - \dfrac{y^2}{16} = 1$     **40.** $4x^2 + 8y^2 = 32$     **41.** $6x^2 + 6y^2 = 96$     **42.** $4x^2 - 36y^2 = 36$

**43.** $\dfrac{x^2}{18} + \dfrac{y^2}{9} = 1$     **44.** $y = (x-3)^2 + 4$     **45.** $12x^2 + 9y^2 = 108$     **46.** $x = -y^2 + 8y - 9$

[14.4]   *Find all real solutions to each system of equations using the substitution method.*

**47.** $2x^2 + y^2 = 16$     **48.** $x^2 - y^2 = 4$     **49.** $x^2 + y^2 = 9$     **50.** $x^2 + 2y^2 = 5$
$\quad\ x^2 - y^2 = -4$     $\quad\ x + y = 4$     $\quad\ x + 2y = 3$     $\quad\ x^2 - 4y^2 = 36$

*Find all real solutions to each system of equations using the addition method.*

**51.** $x^2 + y^2 = 36$     **52.** $x^2 + y^2 = 16$     **53.** $x^2 + y^2 = 81$     **54.** $3x^2 + 4y^2 = 35$
$\quad\ x^2 - y^2 = 36$     $\quad\ 2x^2 - 5y^2 = 25$     $\quad\ 25x^2 + 4y^2 = 100$     $\quad\ 2x^2 + 5y^2 = 42$

**55. Pool Table** Jerry and Denise have a pool table in their house. It has an area of 45 square feet and a perimeter of 28 feet. Find the dimensions of the pool table.

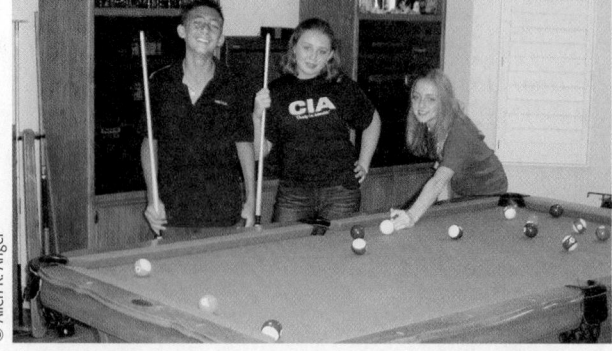

**56. Bottles of Glue** The Dip and Dap Company has a cost equation of $C = 20.3x + 120$ and a revenue equation of $R = 50.2x - 0.2x^2$, where $x$ is the number of bottles of glue sold. Find the number of bottles of glue the company must sell to break even.

**57. Savings Account** If Kien Kempter invests a certain principal at a specific interest rate for 1 year, the interest is $120. If he increases the principal by $2000 and the interest rate is decreased by 1%, the interest remains the same. Find the principal and interest rate. Use $i = prt$.

## Chapter 14 Practice Test

CHAPTER

*Chapter Test Prep Videos provide fully worked-out solutions to any of the exercises you want to review. Chapter Test Prep Videos are available via MyMathLab®, or on* **You Tube** *(search "Angel Elementary and Intermediate Algebra" and click on "Channels")*

1. Why are parabolas, circles, ellipses, and hyperbolas called conic sections?

2. Determine the length of the line segment whose endpoints are $(-1, 8)$ and $(6, 7)$.

3. Determine the midpoint of the line segment whose endpoints are $(-9, 4)$ and $(7, -1)$.

4. Determine the vertex of the graph of $y = -2(x + 3)^2 + 1$, and then graph the equation.

5. Graph $x = y^2 - 2y + 4$.

6. Write the equation $x = -y^2 - 4y - 5$ in the form $x = a(y - k)^2 + h$, and then draw the graph.

7. Write the equation of a circle with center at $(2, 4)$ and radius 3 and then draw the graph of the circle.

8. Find the area of the circle whose equation is $(x + 2)^2 + (y - 8)^2 = 9$.

9. Write the equation of the circle shown.

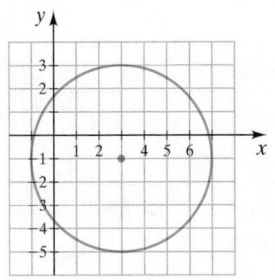

10. Graph $y = -\sqrt{16 - x^2}$.

11. Write the equation $x^2 + y^2 + 2x - 6y + 1 = 0$ in standard form, and then draw the graph.

12. Graph $4x^2 + 25y^2 = 100$.

13. Is the following graph the graph of $\dfrac{(x + 2)^2}{4} + \dfrac{(y + 1)^2}{16} = 1$? Explain your answer.

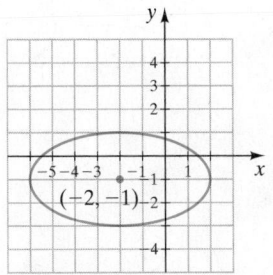

14. Graph $4(x - 4)^2 + 36(y + 2)^2 = 36$.

15. Find the center of the ellipse given by the equation $3(x - 8)^2 + 6(y + 7)^2 = 18$.

16. Explain how to determine whether the transverse axis of a hyperbola lies on the $x$- or $y$-axis.

17. What are the equations of the asymptotes of the graph of $\dfrac{x^2}{16} - \dfrac{y^2}{49} = 1$?

18. Graph $\dfrac{y^2}{25} - \dfrac{x^2}{1} = 1$.

19. Graph $\dfrac{x^2}{4} - \dfrac{y^2}{9} = 1$.

*In Exercises 20 and 21, determine whether the graph of the equation is a parabola, circle, ellipse, or hyperbola.*

20. $4x^2 - 15y^2 = 30$

21. $25x^2 + 4y^2 = 100$

*Solve each system of equations.*

22. $x^2 + y^2 = 7$
    $2x^2 - 3y^2 = -1$

23. $x + y = 8$
    $x^2 + y^2 = 4$

24. **Vegetable Garden** Tom Wilson has a rectangular vegetable garden on his farm that has an area of 1500 square meters. Find the dimensions of the garden if the perimeter is 160 meters.

25. **Truck Bed** Gina Chang owns a truck. The rectangular bed of the truck has an area of 60 square feet, and the diagonal across the bed measures 13 feet. Find the dimensions of the bed of the truck.

## Cumulative Review Test

*Take the following test and check your answers with those given in the back of the book. Review any questions that you answered incorrectly. The section where the material was covered is indicated after the answer.*

1. Solve $4x - 2(3x - 7) = 2x - 5$.

2. Find the solution set: $2(x - 5) + 2x = 4x - 7$.

3. Graph $y = -2x + 2$.

4. Simplify $(9x^2 y^5)(-3xy^4)$.

5. Factor $x^4 - x^2 - 42$.

6. A large triangular sign has a height that is 6 feet less than its base. If the area of the sign is 56 square feet, find the length of the base and the height of the sign.

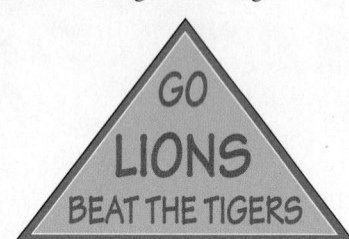

7. Multiply $\dfrac{3x^2 - x - 4}{4x^2 + 7x + 3} \cdot \dfrac{2x^2 - 5x - 12}{6x^2 + x - 12}$.

8. Subtract $\dfrac{x}{x + 3} - \dfrac{x + 5}{2x^2 - 2x - 24}$.

9. Solve $\dfrac{3}{x + 3} + \dfrac{5}{x + 4} = \dfrac{12x + 19}{x^2 + 7x + 12}$.

10. If $f(x) = x^2 + 3x + 9$, find $f(10)$.

11. Solve the system of equations.
$$\frac{1}{2}x - \frac{1}{3}y = 2$$
$$\frac{1}{4}x + \frac{2}{3}y = 6$$

12. Find the solution set: $|3x + 1| > 4$

13. Simplify $\left(\dfrac{18x^{1/2}y^3}{2x^{3/2}}\right)^{1/2}$.

14. Simplify $\dfrac{6\sqrt{x}}{\sqrt{x} - y}$.

15. Solve $3\sqrt[3]{2x + 2} = \sqrt[3]{80x - 24}$.

16. Solve $3x^2 - 4x + 5 = 0$ by the quadratic formula.

17. Solve $\log(3x - 4) + \log 4 = \log(x + 6)$.

18. Solve $35 = 70e^{-0.3t}$.

19. Graph $9x^2 + 4y^2 = 36$.

20. Graph $\dfrac{y^2}{25} - \dfrac{x^2}{16} = 1$.

# 15

# Sequences, Series, and the Binomial Theorem

## Goals of This Chapter

Sequences and series are discussed in this chapter. A sequence is a list of numbers in a specific order and a series is the sum of the numbers in a sequence. In this book, we discuss two types of sequences and series: arithmetic and geometric. Sequences and series can be used to solve many real-life problems as illustrated in this chapter.

In this chapter, we introduce the summation symbol, $\Sigma$, which is often used in statistics and other mathematics courses. We also discuss the binomial theorem for expanding an expression of the form $(a + b)^n$.

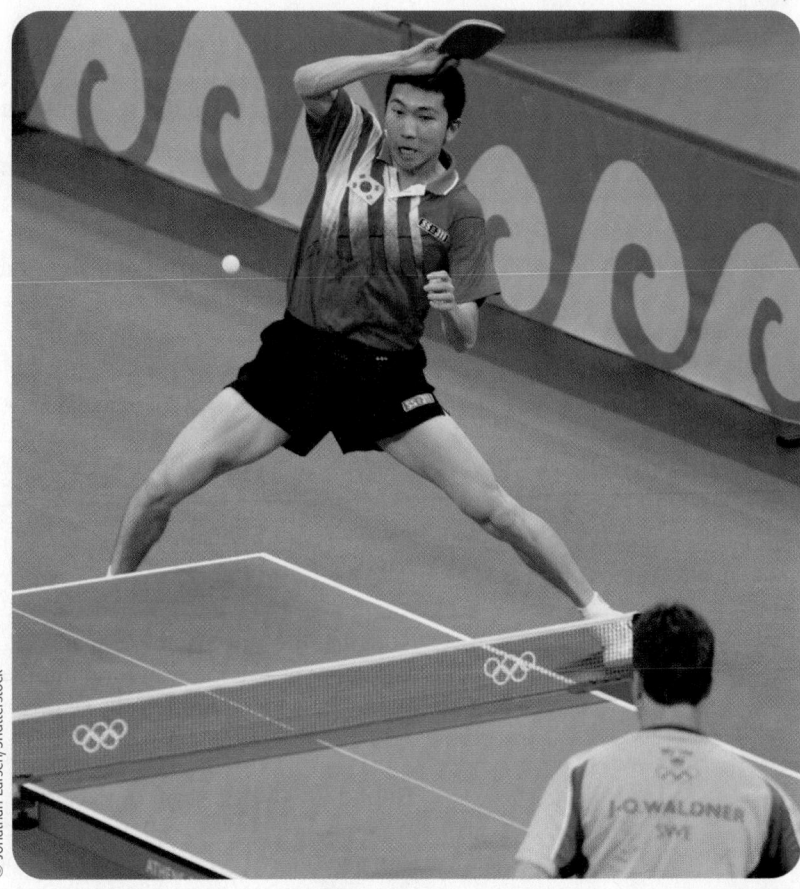

© Jonathan Larsen/Shutterstock

**If a ball rebounds** 9 feet when dropped from 10 feet, it has rebounded 90% of its original height. Theoretically, every rebound will have a rebound and the ball will never stop bouncing. In Exercise 78 on page 891, we determine the height of a ping-pong ball after it falls off a table.

# 15.1 Sequences and Series

1  Find the terms of a sequence.

2  Write a series.

3  Find partial sums.

4  Use summation notation, $\Sigma$.

## 1  Find the Terms of a Sequence

Many times we see patterns in numbers. For example, suppose you are given a job offer with a starting salary of $30,000. You are given two options for your annual salary increases. One option is an annual salary increase of $2000 per year. The salary you would receive under this option is shown below.

| Year | 1 | 2 | 3 | 4 | $\cdots$ |
|------|---|---|---|---|----------|
|      | $\downarrow$ | $\downarrow$ | $\downarrow$ | $\downarrow$ | $\downarrow$ |
| Salary | $30,000 | $32,000 | $34,000 | $36,000 | $\cdots$ |

Each year the salary is $2000 greater than the previous year. The three dots on the right of the lists of numbers indicate that the list continues in the same manner.

The second option is a 5% salary increase each year. The salary you would receive under this option is shown below.

| Year | 1 | 2 | 3 | 4 | $\cdots$ |
|------|---|---|---|---|----------|
|      | $\downarrow$ | $\downarrow$ | $\downarrow$ | $\downarrow$ | $\downarrow$ |
| Salary | $30,000 | $31,500 | $33,075 | $34,728.75 | $\cdots$ |

With this option, the salary in a given year after year 1 is 5% greater than the previous year's salary.

The two lists of numbers that illustrate the salaries are examples of sequences. A **sequence** of numbers is a list of numbers arranged in a specific order. Consider the list of numbers given below, which is a sequence.

$$5, 10, 15, 20, 25, 30, \ldots$$

Each number is called a **term** of the sequence. The first term is 5. We indicate this by writing $a_1 = 5$. Since the second term is 10, $a_2 = 10$, and so on.

$$
\begin{array}{cccccc}
5 & 10 & 15 & 20 & 25 & 30\ldots \\
\downarrow & \downarrow & \downarrow & \downarrow & \downarrow & \downarrow \\
a_1 & a_2 & a_3 & a_4 & a_5 & a_6\ldots
\end{array}
$$

The three dots, called an **ellipsis**, indicate that the sequence continues indefinitely and is an **infinite sequence**.

---

### Infinite Sequence

An **infinite sequence** is a function whose domain is the set of natural numbers.

---

**Understanding Algebra**

An *infinite* sequence continues on forever and never ends.

Consider the infinite sequence $5, 10, 15, 20, 25, 30, 35, \ldots$

$$
\begin{array}{llll}
\text{Domain:} & \{1, \ 2, \ 3, \ 4, \ 5, \ 6, \ 7, \ \ldots, \ n, \ \ldots\} & \text{Natural numbers} \\
& \downarrow \ \downarrow \ \downarrow \ \downarrow \ \downarrow \ \downarrow \ \downarrow \qquad\qquad \downarrow & \\
\text{Range:} & \{5, \ 10, \ 15, \ 20, \ 25, \ 30, \ 35, \ \ldots, \ 5n, \ \ldots\} & \text{Function value}
\end{array}
$$

Note that the terms of the sequence $5, 10, 15, 20, \ldots$ are found by multiplying each natural number by 5. For any natural number, $n$, the corresponding term in the sequence is $5 \cdot n$ or $5n$. The **general term of the sequence**, $a_n$, which defines the sequence, is $a_n = 5n$.

$$a_n = f(n) = 5n$$

To find the twelfth term of the sequence, substitute 12 for $n$ in the general term of the sequence: $a_{12} = 5 \cdot 12 = 60$. Thus, the twelfth term of the sequence is 60. Note that the terms in the sequence are the function values, or the numbers in the range of the function. The general form of a sequence is

$$a_1, a_2, a_3, a_4, \ldots, a_n, \ldots$$

| $n$ | $a_n = 2^n$ |
|---|---|
| 1 | $a_1 = 2^1 = 2$ |
| 2 | $a_2 = 2^2 = 4$ |
| 3 | $a_3 = 2^3 = 8$ |
| 4 | $a_4 = 2^4 = 16$ |
| 5 | $a_5 = 2^5 = 32$ |
| 6 | $a_6 = 2^6 = 64$ |
| $\vdots$ | $\vdots$ $\vdots$ $\vdots$ |

For the infinite sequence $2, 4, 8, 16, 32, \ldots, 2^n, \ldots$ we can write

$$a_n = f(n) = 2^n$$

The table to the left shows the first six terms of this sequence.

### Finite Sequence

A **finite sequence** is a function whose domain includes only the first $n$ natural numbers. A finite sequence has only a finite number of terms.

### Examples of Finite Sequences

| | |
|---|---|
| $5, 10, 15, 20$ | domain is $\{1, 2, 3, 4\}$ |
| $2, 4, 8, 16, 32$ | domain is $\{1, 2, 3, 4, 5\}$ |

**EXAMPLE 1**  Write the finite sequence defined by $a_n = 2n + 3$, for $n = 1, 2, 3, 4$.

*Solution*

$$a_n = 2n + 3$$
$$a_1 = 2(1) + 3 = 5$$
$$a_2 = 2(2) + 3 = 7$$
$$a_3 = 2(3) + 3 = 9$$
$$a_4 = 2(4) + 3 = 11$$

Thus, the sequence is $5, 7, 9, 11$.

Now Try Exercise 13

**Understanding Algebra**

A sequence is *increasing* when each term is greater than the preceding term.

Since each term of the sequence in Example 1 is greater than the preceding term, it is called an **increasing sequence**.

**EXAMPLE 2**  Given $a_n = \dfrac{2n + 3}{n^2}$,

**a)** find the first term in the sequence.     **b)** find the third term in the sequence.
**c)** find the fifth term in the sequence.     **d)** find the tenth term in the sequence.

*Solution*

**a)** When $n = 1$, $a_1 = \dfrac{2(1) + 3}{1^2} = \dfrac{5}{1} = 5$.

**b)** When $n = 3$, $a_3 = \dfrac{2(3) + 3}{3^2} = \dfrac{9}{9} = 1$.

**c)** When $n = 5$, $a_5 = \dfrac{2(5) + 3}{5^2} = \dfrac{13}{25} = 0.52$.

**d)** When $n = 10$, $a_{10} = \dfrac{2(10) + 3}{10^2} = \dfrac{23}{100} = 0.23$.

Now Try Exercise 29

**Understanding Algebra**

A sequence is *decreasing* when each term is less than the preceding term.

Note in Example 2 that since there is no restriction on $n$, $a_n$ is the general term of an infinite sequence.

In Example 2, the first four terms of the sequence are $5, \dfrac{7}{4} = 1.75, 1, \dfrac{11}{16} = 0.6875$.

Since each term of the sequence generated by $a_n = \dfrac{2n + 3}{n^2}$ will be less than the preceding term, the sequence is called a **decreasing sequence**.

**EXAMPLE 3** Find the first four terms of the sequence whose general term is $a_n = (-1)^n(n)$.

*Solution*

$$a_n = (-1)^n(n)$$
$$a_1 = (-1)^1(1) = -1$$
$$a_2 = (-1)^2(2) = 2$$
$$a_3 = (-1)^3(3) = -3$$
$$a_4 = (-1)^4(4) = 4$$

Now Try Exercise 21

If we write the sequence, we get $-1, 2, -3, 4, \ldots, (-1)^n(n), \ldots$. Notice that each term alternates in sign. We call this an **alternating sequence**.

## Understanding Algebra

To *alternate* means to go back and forth with $+$ and $-$ signs. In an *alternating sequence*, the terms alternate in sign.

### 2  Write a Series

**Series**

A **series** is the expressed sum of the terms of a sequence.

A series may be finite or infinite, depending on whether the sequence it is based on is finite or infinite.

## Understanding Algebra

A *series* is obtained by adding the terms from its corresponding *sequence*.

Examples

**Finite Sequence**

$$a_1, a_2, a_3, a_4, a_5$$

**Finite Series**

$$a_1 + a_2 + a_3 + a_4 + a_5$$

**Infinite Sequence**

$$a_1, a_2, a_3, a_4, a_5, \ldots, a_n, \ldots$$

**Infinite Series**

$$a_1 + a_2 + a_3 + a_4 + a_5 + \cdots + a_n + \cdots$$

**EXAMPLE 4** Write the first eight terms of the sequence given by each general term, then write the series that represents the sum of that sequence.

**a)** $a_n = \left(\dfrac{1}{2}\right)^n$       **b)** $a_n = (-2)^n$

*Solution*

**a)** We begin with $n = 1$; thus, the first eight terms of the sequence whose general term is $a_n = \left(\dfrac{1}{2}\right)^n$ are

$$\left(\frac{1}{2}\right)^1, \left(\frac{1}{2}\right)^2, \left(\frac{1}{2}\right)^3, \left(\frac{1}{2}\right)^4, \left(\frac{1}{2}\right)^5, \left(\frac{1}{2}\right)^6, \left(\frac{1}{2}\right)^7, \left(\frac{1}{2}\right)^8$$

or

$$\frac{1}{2}, \frac{1}{4}, \frac{1}{8}, \frac{1}{16}, \frac{1}{32}, \frac{1}{64}, \frac{1}{128}, \frac{1}{256}$$

The series that represents the sum of the sequence is

$$\frac{1}{2} + \frac{1}{4} + \frac{1}{8} + \frac{1}{16} + \frac{1}{32} + \frac{1}{64} + \frac{1}{128} + \frac{1}{256} = \frac{255}{256}$$

**b)** We again begin with $n = 1$; thus, the first eight terms of the sequence whose general term is $a_n = (-2)^n$ are

$$(-2)^1, (-2)^2, (-2)^3, (-2)^4, (-2)^5, (-2)^6, (-2)^7, (-2)^8$$

or

$$-2, 4, -8, 16, -32, 64, -128, 256$$

The series that represents the sum of this sequence is

$$-2 + 4 + (-8) + 16 + (-32) + 64 + (-128) + 256 = 170$$

Now Try Exercise 45

## 3 Find Partial Sums

### Partial Sum

For an infinite sequence with the terms $a_1, a_2, a_3, \ldots, a_n, \ldots$, a **partial sum** is the sum of a finite number of consecutive terms of the sequence, beginning with the first term.

| | |
|---|---|
| $s_1 = a_1$ | First partial sum |
| $s_2 = a_1 + a_2$ | Second partial sum |
| $s_3 = a_1 + a_2 + a_3$ | Third partial sum |
| $\vdots$ | |
| $s_n = a_1 + a_2 + a_3 + \cdots + a_n$ | $n$th partial sum |

The sum of all the terms of the infinite sequence is called an **infinite series** and is given by the following:

$$s = a_1 + a_2 + a_3 + \cdots + a_n + \cdots$$

**EXAMPLE 5** Given the infinite sequence defined by $a_n = \dfrac{3 + n^2}{n}$, find the indicated partial sums.

**a)** $s_1$  **b)** $s_4$

*Solution*

**a)** $s_1 = a_1 = \dfrac{3 + 1^2}{1} = \dfrac{3 + 1}{1} = 4$

**b)** $s_4 = a_1 + a_2 + a_3 + a_4$

$$= \frac{3 + 1^2}{1} + \frac{3 + 2^2}{2} + \frac{3 + 3^2}{3} + \frac{3 + 4^2}{4}$$

$$= 4 + \frac{7}{2} + \frac{12}{3} + \frac{19}{4}$$

$$= \frac{48}{12} + \frac{42}{12} + \frac{48}{12} + \frac{57}{12}$$

$$= \frac{195}{12} \quad \text{or} \quad 16\frac{1}{4}$$

Now Try Exercise 35

## 4 Use Summation Notation, $\Sigma$

When the general term of a sequence is known, the Greek letter **sigma**, $\Sigma$, can be used to write a series. The sum of the first $n$ terms of the sequence whose $n$th term is $a_n$ is represented by

$$\sum_{i=1}^{n} a_i = a_1 + a_2 + a_3 + \cdots + a_n$$

where $i$ is called the **index of summation** or simply the **index**, $n$ is the **upper limit of summation**, and 1 is the **lower limit of summation**. In this illustration, we used $i$ for the index; however, any letter can be used for the index.

Consider the sequence $7, 9, 11, 13, \ldots, 2n + 5, \ldots$. The sum of the first five terms can be represented using **summation notation**.

$$\sum_{i=1}^{5} (2i + 5)$$

This notation is read "the sum as $i$ goes from 1 to 5 of $2i + 5$."

To evaluate the series represented by $\displaystyle\sum_{i=1}^{5} (2i + 5)$,

**1.** First substitute 1 for $i$ in $2i + 5$ to obtain the value 7.

**2.** Then, substitute 2 for $i$ in $2i + 5$ to obtain the value 9.

**3.** Continue to follow this procedure for $i = 3, i = 4$, and $i = 5$.

**4.** Finally, add these values to evaluate the series. The results are summarized as follows:

$$
\begin{array}{ccccc}
i = 1 & i = 2 & i = 3 & i = 4 & i = 5 \\
\downarrow & \downarrow & \downarrow & \downarrow & \downarrow
\end{array}
$$

$$
\sum_{i=1}^{5} (2i + 5) = (2 \cdot 1 + 5) + (2 \cdot 2 + 5) + (2 \cdot 3 + 5) + (2 \cdot 4 + 5) + (2 \cdot 5 + 5)
$$

$$
= 7 + 9 + 11 + 13 + 15
$$

$$
= 55
$$

**EXAMPLE 6** Write out the series $\displaystyle\sum_{i=1}^{6} (i^2 + 1)$ and evaluate it.

*Solution*

$$
\sum_{i=1}^{6} (i^2 + 1) = (1^2 + 1) + (2^2 + 1) + (3^2 + 1) + (4^2 + 1) + (5^2 + 1) + (6^2 + 1)
$$

$$
= 2 + 5 + 10 + 17 + 26 + 37
$$

$$
= 97
$$

Now Try Exercise 57

**EXAMPLE 7** Consider the general term of a sequence $a_n = 2n^2 - 9$. Represent the third partial sum, $s_3$, in summation notation.

*Solution*   The third partial sum will be the sum of the first three terms, $a_1 + a_2 + a_3$.

We can represent the third partial sum as $\displaystyle\sum_{i=1}^{3} (2i^2 - 9)$.

Now Try Exercise 65

**EXAMPLE 8** For the following set of values $x_1 = 3, x_2 = 4, x_3 = 5, x_4 = 6$, and $x_5 = 7$, does $\displaystyle\sum_{i=1}^{5} (x_i)^2 = \left( \sum_{i=1}^{5} x_i \right)^2$?

*Solution*
$$
\sum_{i=1}^{5} (x_i)^2 = (x_1)^2 + (x_2)^2 + (x_3)^2 + (x_4)^2 + (x_5)^2
$$

$$
= 3^2 + 4^2 + 5^2 + 6^2 + 7^2
$$

$$
= 9 + 16 + 25 + 36 + 49 = 135
$$

$$
\left( \sum_{i=1}^{5} x_i \right)^2 = (x_1 + x_2 + x_3 + x_4 + x_5)^2
$$

$$
= (3 + 4 + 5 + 6 + 7)^2 = (25)^2 = 625
$$

Since $135 \neq 625$, $\displaystyle\sum_{i=1}^{5} (x_i)^2 \neq \left( \sum_{i=1}^{5} x_i \right)^2$.

Now Try Exercise 71

When a summation symbol is written without any upper and lower limits, it means that all the given data are to be summed.

EXAMPLE 9  A formula used to find the arithmetic mean, $\bar{x}$ (read $x$ bar), of a set of data is $\bar{x} = \dfrac{\Sigma x}{n}$, where $n$ is the number of pieces of data.

Joan Sally's five test scores are 70, 95, 83, 74, and 92. Find the arithmetic mean of her scores.

*Solution*    $\bar{x} = \dfrac{\Sigma x}{n} = \dfrac{70 + 95 + 83 + 74 + 92}{5} = \dfrac{414}{5} = 82.8$

Now Try Exercise 75

---

# EXERCISE SET 15.1    MathXL®  MyMathLab®

## Warm-Up Exercises

*Fill in the blanks with the appropriate word, phrase, or symbol(s) from the following list.*

| finite | series | greater | upper | sequence | less |
|---|---|---|---|---|---|
| constant | infinite | alternating | index | lower | |

1. A list of numbers arranged in a specific order is a _____.

2. An _____ sequence is a function whose domain is the set of natural numbers.

3. A _____ sequence is a function whose domain includes only the first $n$ natural numbers.

4. In an increasing sequence, each term is _____ than the preceding term.

5. In a decreasing sequence, each term is _____ than the preceding term.

6. A sequence in which each term has the opposite sign of the preceding term is called an _____ sequence.

7. The sum of the terms in a sequence is a _____.

8. Consider the summation $\displaystyle\sum_{k=1}^{50}(k^2 + 3)$. 1 is the _____ limit.

9. Consider the summation $\displaystyle\sum_{k=1}^{50}(k^2 + 3)$. 50 is the _____ limit.

10. Consider the summation $\displaystyle\sum_{k=1}^{50}(k^2 + 3)$. $k$ is the _____.

## Practice the Skills

*Write the first five terms of the sequence whose nth term is shown.*

11. $a_n = 6n$

12. $a_n = -7n$

13. $a_n = 4n - 1$

14. $a_n = 2n + 5$

15. $a_n = \dfrac{9}{n}$

16. $a_n = \dfrac{8}{n^2}$

17. $a_n = \dfrac{n + 2}{n + 1}$

18. $a_n = \dfrac{n - 3}{n + 6}$

19. $a_n = (-1)^n$

20. $a_n = (-1)^{2n}$

21. $a_n = (-2)^{n+1}$

22. $a_n = 3^{n-1}$

*Find the indicated term of the sequence whose nth term is shown.*

23. $a_n = 2n + 7$, twelfth term

24. $a_n = 3n + 2$, sixth term

25. $a_n = \dfrac{n}{4} + 8$, sixteenth term

26. $a_n = \dfrac{n}{2} - 13$, fourteenth term

27. $a_n = (-1)^n$, eighth term

28. $a_n = (-2)^n$, seventh term

29. $a_n = n(n + 2)$, eleventh term

30. $a_n = (n - 1)(n + 4)$, fifth term

31. $a_n = \dfrac{n^2}{2n + 7}$, ninth term

32. $a_n = \dfrac{n(n + 6)}{n^2}$, tenth term

*Find the first and third partial sums, $s_1$ and $s_3$, for each sequence.*

33. $a_n = 3n - 1$

34. $a_n = 2n + 5$

35. $a_n = 2^n + 1$

36. $a_n = 3^n - 8$

37. $a_n = \dfrac{n - 1}{n + 2}$

38. $a_n = \dfrac{n}{n + 4}$

39. $a_n = (-1)^n$

40. $a_n = (-3)^n$

41. $a_n = \dfrac{n^2}{2}$

42. $a_n = \dfrac{n^2}{n + 4}$

*Write the next three terms of each sequence.*

**43.** $2, 4, 8, 16, 32, \ldots$

**44.** $7, 12, 17, 22, 27, \ldots$

**45.** $7, 9, 11, 13, 15, \ldots$

**46.** $\dfrac{1}{2}, \dfrac{1}{3}, \dfrac{1}{4}, \dfrac{1}{5}, \ldots$

**47.** $1, \dfrac{1}{2}, \dfrac{1}{3}, \dfrac{1}{4}, \dfrac{1}{5}, \ldots$

**48.** $\dfrac{2}{3}, \dfrac{3}{4}, \dfrac{4}{5}, \dfrac{5}{6}, \dfrac{6}{7}, \ldots$

**49.** $-1, 1, -1, 1, -1, \ldots$

**50.** $-10, -20, -30, -40, \ldots$

**51.** $1, \dfrac{1}{3}, \dfrac{1}{9}, \dfrac{1}{27}, \ldots$

**52.** $\dfrac{1}{4}, \dfrac{2}{4}, \dfrac{3}{4}, \dfrac{4}{4}, \ldots$

**53.** $1, -\dfrac{1}{2}, \dfrac{1}{4}, -\dfrac{1}{8}, \ldots$

**54.** $\dfrac{1}{3}, -\dfrac{1}{6}, \dfrac{1}{12}, -\dfrac{1}{24}, \ldots$

**55.** $42, 38, 34, 30, \ldots$

**56.** $7, -1, -9, -17, \ldots$

*Write out each series, then evaluate it.*

**57.** $\displaystyle\sum_{i=1}^{5} (3i - 1)$

**58.** $\displaystyle\sum_{i=1}^{4} (4i + 9)$

**59.** $\displaystyle\sum_{i=1}^{6} (i^2 + 1)$

**60.** $\displaystyle\sum_{i=1}^{5} (2i^2 - 7)$

**61.** $\displaystyle\sum_{i=1}^{4} \dfrac{i^2}{3}$

**62.** $\displaystyle\sum_{i=1}^{3} \dfrac{i^2}{4}$

**63.** $\displaystyle\sum_{i=4}^{9} \dfrac{i^2 + i}{i + 1}$

**64.** $\displaystyle\sum_{i=2}^{5} \dfrac{i^3}{i + 1}$

*For the given general term $a_n$, write an expression using $\Sigma$ to represent the indicated partial sum.*

**65.** $a_n = n + 10$, fifth partial sum

**66.** $a_n = n^2 + 5$, fourth partial sum

**67.** $a_n = \dfrac{n^2}{4}$, third partial sum

**68.** $a_n = \dfrac{n^2 + 7}{n + 9}$, third partial sum

*For the set of values $x_1 = 2$, $x_2 = 3$, $x_3 = 5$, $x_4 = -1$, and $x_5 = 4$, find each of the following.*

**69.** $\displaystyle\sum_{i=1}^{5} x_i$

**70.** $\displaystyle\sum_{i=1}^{5} (x_i + 5)$

**71.** $\left(\displaystyle\sum_{i=1}^{5} x_i\right)^2$

**72.** $\displaystyle\sum_{i=1}^{5} 4x_i$

**73.** $\displaystyle\sum_{i=1}^{5} (x_i)^2$

**74.** $\displaystyle\sum_{i=1}^{4} (x_i^2 + 3)$

*Find the arithmetic mean, $\overline{x}$, of the following sets of data.*

**75.** $15, 20, 25, 30, 35$

**76.** $16, 22, 96, 18, 48$

**77.** $72, 83, 4, 60, 18, 20$

**78.** $12, 13, 9, 19, 23, 36, 70$

## Problem Solving

*In Exercises 79 and 80, consider the following rectangles. For the nth rectangle, the length is 2n and the width is n.*

**79. Perimeter**

    **a)** Find the perimeters for the first four rectangles, and then list the perimeters in a sequence.

    **b)** Find the general term for the perimeter of the *n*th rectangle in the sequence. Use $p_n$ for perimeter.

**80. Area**

    **a)** Find the areas for the four rectangles, and then list the areas in a sequence.

    **b)** Find the general term for the area of the *n*th rectangle in the sequence. Use $a_n$ for area.

**81. Write**

    **a)** $\displaystyle\sum_{i=1}^{n} x_i$ as a sum of terms.

    **b)** $\displaystyle\sum_{j=1}^{n} x_j$ as a sum of terms.

    **c)** For a given set of values of $x$, from $x_1$ to $x_n$, will $\displaystyle\sum_{i=1}^{n} x_i = \sum_{j=1}^{n} x_j$? Explain.

**82.** Solve $\overline{x} = \dfrac{\Sigma x}{n}$ for $\Sigma x$.

**83.** Solve $\overline{x} = \dfrac{\Sigma x}{n}$ for $n$.

**84.** Is $\displaystyle\sum_{i=1}^{n} 4x_i = 4\sum_{i=1}^{n} x_i$? Illustrate your answer with an example.

**85.** Is $\displaystyle\sum_{i=1}^{n} \dfrac{x_i}{3} = \dfrac{1}{3}\sum_{i=1}^{n} x_i$? Illustrate your answer with an example.

**86.** Let $x_1 = 3$, $x_2 = 5$, $x_3 = 2$, and $y_1 = 4$, $y_2 = 1$, $y_3 = 6$. Find the following. Note that $\Sigma x = x_1 + x_2 + x_3$, $\Sigma y = y_1 + y_2 + y_3$, and $\Sigma xy = x_1 y_1 + x_2 y_2 + x_3 y_3$.

    **a)** $\Sigma x$

    **b)** $\Sigma y$

    **c)** $\Sigma x \cdot \Sigma y$

    **d)** $\Sigma xy$

    **e)** Is $\Sigma x \cdot \Sigma y = \Sigma xy$?

## Concept/Writing Exercises

**87.** What is the $n$th partial sum of a series?

**88.** Write the following notation in words: $\displaystyle\sum_{i=1}^{5} (i + 4)$.

**89.** Let $a_n = 2n - 1$. Is this an increasing sequence or a decreasing sequence? Explain.

**90.** Let $a_n = -3n + 7$. Is this an increasing sequence or a decreasing sequence? Explain.

**91.** Let $a_n = 1 + (-2)^n$. Is this an alternating sequence? Explain.

**92.** Let $a_n = (-1)^{2n}$. Is this an alternating sequence? Explain.

**93.** Create your own sequence that is an increasing sequence and list the first five terms.

**94.** Create your own sequence that is a decreasing sequence and list the first five terms.

**95.** Create your own sequence that is an alternating sequence and list the first five terms.

## Cumulative Review Exercises

[6.6] **96.** Solve $2x^2 + 15 = 13x$ by factoring.

[12.2] **97.** How many real solutions does the equation $6x^2 - 3x - 4 = 2$ have? Explain how you obtained your answer.

[14.2] **98.** Graph $\dfrac{x^2}{4} + \dfrac{y^2}{1} = 1$.

[14.4] **99.** Solve the system of equations.
$$x^2 + y^2 = 5$$
$$x = 2y$$

# 15.2 Arithmetic Sequences and Series

1 Find the common difference in an arithmetic sequence.

2 Find the $n$th term of an arithmetic sequence.

3 Find the $n$th partial sum of an arithmetic sequence.

## 1 Find the Common Difference in an Arithmetic Sequence

In the previous section, we started our discussion by assuming you got a job with a starting salary of $30,000. One option for salary increases was an increase of $2000 each year. This would result in the sequence

$$\$30,000, \$32,000, \$34,000, \$36,000, \dots$$

This is an example of an arithmetic sequence.

> **Arithmetic Sequence**
>
> An **arithmetic sequence** is a sequence in which each term after the first differs from the preceding term by a constant amount. The constant amount by which each pair of successive terms differs is called the **common difference**, $d$.

### Understanding Algebra

In an *arithmetic sequence*, you can get the next term by adding the *common difference* to the previous term.

The common difference can be found by subtracting any term from the term that directly follows it.

| Arithmetic Sequence | Common Difference |
|---|---|
| $1, 3, 5, 7, 9, \dots$ | $d = 3 - 1 = 2$ |
| $5, 1, -3, -7, -11, -15, \dots$ | $d = 1 - 5 = -4$ |
| $\dfrac{7}{2}, \dfrac{2}{2}, -\dfrac{3}{2}, -\dfrac{8}{2}, -\dfrac{13}{2}, -\dfrac{18}{2}, \dots$ | $d = \dfrac{2}{2} - \dfrac{7}{2} = -\dfrac{5}{2}$ |

Notice that the common difference can be a positive number or a negative number. If the sequence is increasing, then $d$ is a positive number. If the sequence is decreasing, then $d$ is a negative number.

EXAMPLE 1 Write the first five terms of the arithmetic sequence with

**a)** first term 6 and common difference 4.

**b)** first term 3 and common difference $-2$.

**c)** first term 1 and common difference $\dfrac{1}{3}$.

*Solution*

**a)** Start with 6 and keep adding 4. The sequence is 6, 10, 14, 18, 22.

**b)** 3, 1, −1, −3, −5

**c)** $1, \dfrac{4}{3}, \dfrac{5}{3}, 2, \dfrac{7}{3}$

Now Try Exercise 13

### 2 Find the *n*th Term of an Arithmetic Sequence

In general, an arithmetic sequence with first term, $a_1$, and common difference, $d$, has the following terms:

$$a_1 = a_1, \quad a_2 = a_1 + d, \quad a_3 = a_1 + 2d, \quad a_4 = a_1 + 3d, \quad \text{and so on}$$

If we continue this process, we can see that the *n*th term, $a_n$, can be found by the following formula:

> **nth Term of an Arithmetic Sequence**
>
> $$a_n = a_1 + (n - 1)d$$

### EXAMPLE 2

**a)** Write an expression for the general (or *n*th) term, $a_n$, of the arithmetic sequence whose first term is −3 and whose common difference is 2.

**b)** Find the twelfth term of the sequence.

*Solution*

**a)** The *n*th term of the sequence is $a_n = a_1 + (n - 1)d$.

$$
\begin{aligned}
a_n &= a_1 + (n - 1)d && \text{nth term of sequence} \\
&= -3 + (n - 1)2 && \text{Substitute } a_1 = -3 \text{ and } d = 2. \\
&= -3 + 2(n - 1) && \text{Commutative property} \\
&= -3 + 2n - 2 && \text{Distributive property} \\
&= 2n - 5 && \text{Simplify.}
\end{aligned}
$$

Thus, $a_n = 2n - 5$.

**b)** $a_n = 2n - 5$

$$a_{12} = 2(12) - 5 = 24 - 5 = 19$$

The twelfth term in the sequence is 19.

Now Try Exercise 11

### EXAMPLE 3 Find the number of terms in the arithmetic sequence 5, 9, 13, 17, …, 41.

*Solution*    The first term, $a_1$, is 5; the *n*th term is 41, and the common difference, $d$, is 4. Substitute the appropriate values into the formula for the *n*th term and solve for *n*.

$$
\begin{aligned}
a_n &= a_1 + (n - 1)d \\
41 &= 5 + (n - 1)4 \\
41 &= 5 + 4n - 4 \\
41 &= 4n + 1 \\
40 &= 4n \\
10 &= n
\end{aligned}
$$

The sequence has 10 terms.

> **Understanding Algebra**
>
> In an arithmetic sequence, to get the value for the common difference, *d*, subtract a term from the next term in the sequence.

Now Try Exercise 51

## **3** Find the *n*th Partial Sum of an Arithmetic Sequence

An **arithmetic series** is the sum of the terms of an arithmetic sequence. A finite arithmetic series can be written

$$s_n = a_1 + (a_1 + d) + (a_1 + 2d) + (a_1 + 3d) + \cdots + (a_n - 2d) + (a_n - d) + a_n$$

If we consider the last term as $a_n$, the term before the last term will be $a_n - d$, the second before the last term will be $a_n - 2d$, and so on.

A formula for the *n*th partial sum, $s_n$, can be obtained by adding the reverse of $s_n$ to itself.

$$\begin{aligned}
s_n &= \quad a_1 \quad + (a_1 + d) + (a_1 + 2d) + \cdots + (a_n - 2d) + (a_n - d) + \quad a_n \\
s_n &= \quad a_n \quad + (a_n - d) + (a_n - 2d) + \cdots + (a_1 + 2d) + (a_1 + d) + \quad a_1 \\
\hline
2s_n &= (a_1 + a_n) + (a_1 + a_n) + (a_1 + a_n) + \cdots + (a_1 + a_n) + (a_1 + a_n) + (a_1 + a_n)
\end{aligned}$$

Since the right side of the equation contains $n$ terms of $(a_1 + a_n)$, we can write

$$2s_n = n(a_1 + a_n)$$

Now divide both sides of the equation by 2 to obtain the following formula.

---

**nth Partial Sum of an Arithmetic Sequence**

$$s_n = \frac{n(a_1 + a_n)}{2}$$

---

**EXAMPLE 4**  Find the sum of the first 25 natural numbers.

*Solution*    The arithmetic sequence is $1, 2, 3, 4, 5, 6, \ldots, 25$. The first term, $a_1$, is 1; the last term, $a_n$, is 25. There are 25 terms; thus, $n = 25$. Using the formula for the *n*th partial sum, we have

$$\begin{aligned}
s_n &= \frac{n(a_1 + a_n)}{2} \\
&= \frac{25(1 + 25)}{2} \\
&= \frac{25(26)}{2} \\
&= 25(13) = 325
\end{aligned}$$

The sum of the first 25 natural numbers is 325. Thus, $s_{25} = 325$.

Now Try Exercise 57

**EXAMPLE 5**  The first term of an arithmetic sequence is 4, and the last term is 31. If $s_n = 175$, find the number of terms in the sequence and the common difference.

*Solution*    We substitute the appropriate values, $a_1 = 4$, $a_n = 31$, and $s_n = 175$, into the formula for the *n*th partial sum and solve for $n$.

$$\begin{aligned}
s_n &= \frac{n(a_1 + a_n)}{2} \\
175 &= \frac{n(4 + 31)}{2} \\
175 &= \frac{35n}{2} \\
350 &= 35n \\
10 &= n
\end{aligned}$$

There are 10 terms in the sequence. We can now find the common difference by using the formula for the $n$th term of an arithmetic sequence.

$$a_n = a_1 + (n - 1)d$$
$$31 = 4 + (10 - 1)d$$
$$31 = 4 + 9d$$
$$27 = 9d$$
$$3 = d$$

The common difference is 3. The sequence is 4, 7, 10, 13, 16, 19, 22, 25, 28, 31.

Now Try Exercise 31

**EXAMPLE 6** **Salary** Mary Tufts is given a starting salary of $35,000 and is promised a $1200 raise after each of the next 8 years. Find her salary during her eighth year of work.

*Solution*    Understand    Her salaries during the first few years would be

$$\$35{,}000, \$36{,}200, \$37{,}400, \$38{,}600, \ldots$$

Since we are adding a constant amount each year, this is an arithmetic sequence. The general term of an arithmetic sequence is $a_n = a_1 + (n - 1)d$.

Translate    In this example, $a_1 = 35{,}000$ and $d = 1200$. Thus, for $n = 8$, Mary's salary would be

$$a_n = a_1 + (n - 1)\ d$$
$$a_8 = 35{,}000 + (8 - 1)1200$$

Carry Out
$$= 35{,}000 + 7(1200)$$
$$= 35{,}000 + 8400$$
$$= 43{,}400$$

Answer    During her eighth year of work, Mary's salary would be $43,400. If we listed all the salaries for the 8-year period, they would be $35,000, $36,200, $37,400, $38,600, $39,800, $41,000, $42,200, $43,400.

Now Try Exercise 83

**EXAMPLE 7** **Pendulum** Each swing of a pendulum (left to right or right to left) is 3 inches shorter than the preceding swing. The first swing is 8 feet.

**a)** Find the length of the tenth swing.

**b)** Determine the distance traveled by the pendulum during the first 10 swings.

*Solution*

**a)** Understand    Since each swing is decreasing by a constant amount, this problem can be represented as an arithmetic series. Since the first swing is given in feet and the decrease in swing in inches, we will change 3 inches to 0.25 feet ($3 \div 12 = 0.25$). The tenth swing can be considered $a_{10}$. The difference, $d$, is negative since the distance is decreasing with each swing.

Translate
$$a_n = a_1 + (n - 1)d$$
$$a_{10} = 8 + (10 - 1)(-0.25)$$

Carry Out
$$= 8 + 9(-0.25)$$
$$= 8 - 2.25$$
$$= 5.75 \text{ feet}$$

Answer    The tenth swing is 5.75 feet.

**b) Understand and Translate**  The distance traveled during the first 10 swings can be found using the formula for the $n$th partial sum. The first swing, $a_1$, is 8 feet and the tenth swing, $a_{10}$, is 5.75 feet.

$$s_n = \frac{n(a_1 + a_n)}{2}$$

$$s_{10} = \frac{10(a_1 + a_{10})}{2}$$

**Carry Out**  $= \frac{10(8 + 5.75)}{2} = \frac{10(13.75)}{2} = 5(13.75) = 68.75$ feet

**Answer**  The pendulum travels 68.75 feet during its first 10 swings.

NowTry Exercise 75

# EXERCISE SET 15.2  MathXL®  MyMathLab®

## Warm-Up Exercises

*Fill in the blanks with the appropriate word, phrase, or symbol(s) from the following list.*

| | | | | | | | | |
|---|---|---|---|---|---|---|---|---|
| sum | series | constant | 0 | 1 | positive | difference | negative | arithmetic |
| $n$ | adding | subtracting | $d$ | $a_n$ | term | adding | −1 | 2 |

1. A sequence in which each term after the first term differs from the preceding term by a constant amount is an _____ sequence.

2. The constant amount by which each pair of successive terms differs in an arithmetic sequence is called the common _____.

3. The common difference in an arithmetic sequence is identified by the letter _____.

4. The common difference can be found by _____ any term from the next term in the sequence.

5. The formula $a_n = a_1 + (n - 1)d$ is used to find the $n$th _____ of an arithmetic sequence.

6. The sum of the terms of an arithmetic sequence is an arithmetic _____.

7. The formula $s_n = \frac{n(a_1 + a_n)}{2}$ is used to find the $n$th partial _____ of an arithmetic sequence.

8. If an arithmetic sequence is increasing, the value for $d$ is a _____ number.

9. If the terms of an arithmetic sequence are decreasing, the value for $d$ is a _____ number.

10. If all the terms of an arithmetic sequence are the same, the value for $d$ is _____.

## Practice the Skills

*Write the first five terms of the arithmetic sequence with the given first term and common difference. Write the expression for the general (or nth) term, $a_n$, of the arithmetic sequence.*

11. $a_1 = 4, d = 3$

12. $a_1 = -8, d = 4$

13. $a_1 = 7, d = -2$

14. $a_1 = 3, d = -5$

15. $a_1 = \frac{1}{2}, d = \frac{3}{2}$

16. $a_1 = \frac{7}{4}, d = -\frac{3}{4}$

17. $a_1 = 100, d = -5$

18. $a_1 = 50, d = -9$

*Find the indicated quantity of the arithmetic sequence.*

19. $a_1 = 5, d = 3$; find $a_4$

20. $a_1 = -6, d = 5$; find $a_5$

21. $a_1 = -9, d = 4$; find $a_{10}$

22. $a_1 = -1, d = -2$; find $a_{12}$

23. $a_1 = -8, d = \frac{5}{3}$; find $a_{13}$

24. $a_1 = 5, a_8 = 10$; find $d$

25. $a_1 = -\frac{11}{2}, a_9 = \frac{21}{2}$, find $d$

26. $a_1 = \frac{1}{2}, a_7 = \frac{19}{2}$; find $d$

27. $a_1 = 4, a_n = 28, d = 3$; find $n$

28. $a_1 = 1, a_n = -17, d = -3$; find $n$

29. $a_1 = 82, a_n = 42, d = -8$; find $n$

30. $a_1 = -\frac{4}{3}, a_n = -\frac{14}{3}, d = -\frac{2}{3}$; find $n$

*Find the sum, $s_n$, and common difference, d, of each sequence.*

31. $a_1 = 1, a_{10} = 19, n = 10$

32. $a_1 = -8, a_7 = 10, n = 7$

**33.** $a_1 = \dfrac{3}{5}, a_8 = 2, n = 8$

**34.** $a_1 = 12, a_9 = -28, n = 9$

**35.** $a_1 = -5, a_6 = 13.5, n = 6$

**36.** $a_1 = -\dfrac{3}{5}, a_5 = \dfrac{13}{5}, n = 5$

**37.** $a_1 = 7, a_{11} = 67, n = 11$

**38.** $a_1 = 14.25, a_{31} = 18.75, n = 31$

*Write the first four terms of each sequence; then find $a_{10}$ and $s_{10}$.*

**39.** $a_1 = 4, d = 3$

**40.** $a_1 = 1, d = 7$

**41.** $a_1 = -6, d = 2$

**42.** $a_1 = -7, d = -4$

**43.** $a_1 = -8, d = -5$

**44.** $a_1 = 5, d = 4$

**45.** $a_1 = \dfrac{7}{2}, d = \dfrac{5}{2}$

**46.** $a_1 = \dfrac{9}{5}, d = \dfrac{3}{5}$

**47.** $a_1 = 100, d = -7$

**48.** $a_1 = 35, d = 6$

*Find the number of terms in each sequence and find $s_n$.*

**49.** $1, 4, 7, 10, \ldots, 43$

**50.** $-10, -8, -6, -4, \ldots, 40$

**51.** $-9, -5, -1, 3, \ldots, 31$

**52.** $6, 13, 20, 27, \ldots, 55$

**53.** $\dfrac{1}{2}, \dfrac{2}{2}, \dfrac{3}{2}, \dfrac{4}{2}, \dfrac{5}{2}, \ldots, \dfrac{17}{2}$

**54.** $-\dfrac{5}{6}, -\dfrac{7}{6}, -\dfrac{9}{6}, -\dfrac{11}{6}, \ldots, -\dfrac{21}{6}$

**55.** $7, 10, 13, 16, \ldots, 91$

**56.** $-11, -15, -19, \ldots, -51$

## Problem Solving

**57.** Find the sum of the first 50 natural numbers.

**58.** Find the sum of the first 50 even numbers.

**59.** Find the sum of the first 50 odd numbers.

**60.** Find the sum of the first 40 multiples of 5.

**61.** Find the sum of the first 30 multiples of 3.

**62.** Find the sum of the numbers between 50 and 150, inclusive.

**63.** Determine how many numbers between 7 and 1610 are divisible by 6.

**64.** Determine how many numbers between 14 and 1470 are divisible by 8.

*Pyramids occur everywhere. At athletic events, cheerleaders may form a pyramid where the people above stand on the shoulders of the people below. The illustration below shows a pyramid with 1 cheerleader on the top row, 2 cheerleaders in the middle row, and 3 cheerleaders in the bottom row. Notice that $a_1 = 1$, $a_2 = 2$, and $a_3 = 3$. Also, observe that $d = 1$, $n = 3$, and $s_3 = 6$.*

*At a bowling alley, the pins at the end of the bowling lane form a pyramid. The first row has 1 pin, the second row has 2 pins, the third row has 3 pins, and the fourth row has 4 pins. Thus, $a_1 = 1$, $d = 1$, $n = 4$, and $s_4 = 10$.*

*Use the idea of a pyramid to solve Exercises 65–70.*

**65.** **Auditorium**  An auditorium has 20 seats in the first row. Each successive row has two more seats than the previous row. How many seats are in the twelfth row? How many seats are in the first 12 rows?

**66.** **Auditorium**  An auditorium has 22 seats in the first row. Each successive row has four more seats than the previous row. How many seats are in the ninth row? How many seats are in the first nine rows?

**67.** **Logs**  Wolfgang Schmidt stacks logs so that there are 26 logs in the bottom layer, and each layer contains one log less than the layer below it. How many logs are in the pile?

**68.** **Logs**  Suppose Wolfgang, in Exercise 67, stopped stacking the logs after completing the layer containing eight logs. How many logs are in the pile?

**69.** **Glasses in a Stack**  At their fiftieth wedding anniversary, Mr. and Mrs. Carlson are about to pour champagne into the top glass in the photo shown on the next page. The top row has 1 glass, the second row has 3 glasses, the third row has 5 glasses, and so on. Each row has 2 more glasses than the row above it. This pyramid has 14 rows.

**a)** How many glasses are in the fourteenth row (bottom row)?

**b)** How many glasses are there total?

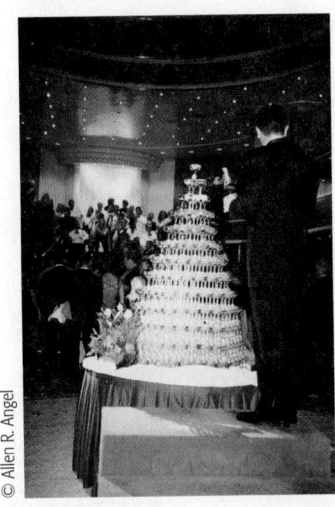

See Exercise 69 on page 890.

**70. Candies in a Stack** Individually wrapped candies are stacked in rows such that the top row has 1 candy, the second row has 3 candies, the third row has 5 candies, and so on. Each row has 2 more candies than the row above it. There are 7 rows of candies.

**a)** How many candies are in the seventh row (bottom row)?

**b)** How many candies are there total?

**71. Sum of Numbers** Karl Friedrich Gauss (1777–1855), a famous mathematician, as a child found the sum of the first 100 natural numbers quickly in his head $(1 + 2 + 3 + \cdots + 100)$. Explain how he might have done this and find the sum of the first 100 natural numbers as you think Gauss might have. (*Hint:* $1 + 100 = 101, 2 + 99 = 101$, etc.)

Karl Friedrich Gauss

**72. Sum of Numbers** Use the same process from Exercise 71 to find the sum of the numbers from 101 to 150.

**73. Sum of Numbers** Find a formula for the sum of the first $n$ consecutive odd numbers starting with 1.

$$1 + 3 + 5 + \cdots + (2n - 1)$$

**74. Sum of Even Numbers** Find a formula for the sum of the first $n$ consecutive even numbers starting with 2.

$$2 + 4 + 6 + 8 + \cdots + 2n$$

**75. Swinging on a Vine** A long vine is attached to the branch of a tree. Sally Wynn swings from the vine, and each swing (left to right or right to left) is $\frac{1}{2}$ foot less than the previous swing. If her first swing is 22 feet, find

**a)** the length of the seventh swing, and

**b)** the distance traveled during her seven swings.

**76. Pendulum** Each swing of a pendulum is 2 inches shorter than the previous swing (left to right or right to left). The first swing is 6 feet. Find

**a)** the length of the eighth swing, and

**b)** the distance traveled by the pendulum during the eight swings.

**77. Bouncing Ball** Frank Holyton drops a ball from a second-story window. Each time the ball bounces, the height reached is 6 inches less than on the previous bounce. If the first bounce reaches a height of 6 feet, find the height attained on the ninth bounce.

**78. Ping-Pong Ball** A ping-pong ball falls from a table and bounces to a height of 3 feet. If each successive bounce is 3 inches less than the previous bounce, find the height attained on the tenth bounce.

**79. Packages** On Monday, March 17, Brian Wallin started a new job at a packing company. On that day, he was able to prepare 105 packages for shipment. His boss expects Brian to be more productive with experience. Each day for six days, Brian is expected to prepare 10 more packages than the previous day's total.

**a)** How many packages is Brian expected to prepare on March 22?

**b)** How many packages is Brian expected to prepare during his first six days of employment?

**80. Salary** Marion Nickelson is making an annual salary of $37,500 at the Thompson Frozen Food Factory. Her boss has promised her an increase of $1500 in her salary each year over the next 10 years.

**a)** What will be Marion's salary 10 years from now?

**b)** What will be her total salary for these 11 years?

**81. Money** If Craig Campanella saves $1 on day 1, $2 on day 2, $3 on day 3, and so on, how much money, in total, will he have saved on day 31?

**82. Money** If Dan Currier saves 50¢ on day 1, $1.00 on day 2, $1.50 on day 3, and so on, how much, in total, will he have saved by the end of 1 year (365 days)?

**83. Money** Carrie Dereshi recently retired and met with her financial planner. She arranged to receive $42,000 the first year. Because of inflation, each year she will get $400 more than she received the previous year.

**a)** What income will she receive in her tenth year of retirement?

**b)** How much money will she have received in total during her first 10 years of retirement?

**84. Salary** Susan Forman is given a starting salary of $25,000 and is told she will receive a $1000 raise at the end of each year.

a) Find her salary during year 12.

b) How much will she receive in total during her first 12 years?

**85. Angles** The sum of the interior angles of a triangle, a quadrilateral, a pentagon, and a hexagon are 180°, 360°, 540°, and 720°, respectively. Use the pattern here to find the formula for the sum of the interior angles of a polygon with $n$ sides.

**86.** Another formula that may be used to find the $n$th partial sum of an arithmetic series is

$$s_n = \frac{n}{2}[2a_1 + (n-1)d]$$

Derive this formula using the two formulas presented in this section.

## Concept/Writing Exercises

**87.** Can an arithmetic sequence consist of only negative numbers? Explain.

**88.** Can an arithmetic sequence consist of only odd numbers? Explain.

**89.** Can an arithmetic sequence consist of only even numbers? Explain.

**90.** Can an alternating sequence be an arithmetic sequence? Explain.

## Group Activity

*In calculus, a topic of importance is limits. Consider* $a_n = \frac{1}{n}$. *The first five terms of this sequence are* $\frac{1}{1}, \frac{1}{2}, \frac{1}{3}, \frac{1}{4}, \frac{1}{5}$. *Since the value of* $\frac{1}{n}$ *gets closer and closer to 0 as n gets larger and larger, we say that the limit of* $\frac{1}{n}$ *as n approaches infinity is 0. We write this as* $\lim\limits_{n \to +\infty} \frac{1}{n} = 0$ *or* $\lim\limits_{n \to +\infty} a_n = 0$. *Notice that* $\frac{1}{n}$ *can never equal 0, but its value approaches 0 as n gets larger and larger.*

a) *Group member 1: Find* $\lim\limits_{n \to +\infty} a_n$ *for Exercises 91 and 92.*

b) *Group member 2: Find* $\lim\limits_{n \to +\infty} a_n$ *for Exercises 93 and 94.*

c) *Group member 3: Find* $\lim\limits_{n \to +\infty} a_n$ *for Exercises 95 and 96.*

d) *Exchange work and check each other's answers.*

**91.** $a_n = \dfrac{1}{n-2}$

**92.** $a_n = \dfrac{n}{n+1}$

**93.** $a_n = \dfrac{1}{n^2 + 2}$

**94.** $a_n = \dfrac{2n+1}{n}$

**95.** $a_n = \dfrac{4n-3}{3n+1}$

**96.** $a_n = \dfrac{n^2}{n+1}$

## Cumulative Review Exercises

[2.6] **97.** Solve $A = P + Prt$ for $r$.

[6.2] **98.** Factor $12n^2 - 6n - 30n + 15$.

[9.2] **99.** Solve the system of equations.

$$y = 2x + 1$$
$$3x - 2y = 1$$

[14.1] **100.** Graph $(x + 4)^2 + y^2 = 25$.

# 15.3 Geometric Sequences and Series

1. Find the common ratio in a geometric sequence.

2. Find the $n$th term of a geometric sequence.

3. Find the $n$th partial sum of a geometric sequence.

4. Identify infinite geometric series.

5. Find the sum of an infinite geometric series.

6. Applications of geometric series.

## 1 Find the Common Ratio in a Geometric Sequence

In Section 15.1, we assumed you got a job with a starting salary of $30,000. We also mentioned that an option for salary increases was a 5% salary increase each year. This would result in the following sequence.

$$\$30,000, \quad \$31,500, \quad \$33,075, \quad \$34,728.75, \ldots$$

This is an example of a geometric sequence.

### Geometric Sequence

A **geometric sequence** is a sequence in which each term after the first term is a common multiple of the preceding term. The common multiple is called the **common ratio**.

**Understanding Algebra**

In a *geometric sequence,* you get the next term by multiplying the previous term by the *common ratio.*

The common ratio, $r$, in any geometric sequence can be found by dividing any term, except the first, by the preceding term. The common ratio of the previous geometric sequence is $\dfrac{31{,}500}{30{,}000} = 1.05$ (or 105%).

Consider the geometric sequence

$$1, 3, 9, 27, 81, \ldots, 3^{n-1}, \ldots$$

The common ratio is 3 since $3 \div 1 = 3, 9 \div 3 = 3, 27 \div 9 = 3$, and so on.

| Geometric Sequence | Common Ratio |
|---|---|
| $4, 8, 16, 32, 64, \ldots, 4(2)^{n-1}, \ldots$ | 2 |
| $3, 12, 48, 192, 768, \ldots, 3(4)^{n-1}, \ldots$ | 4 |
| $7, \dfrac{7}{2}, \dfrac{7}{4}, \dfrac{7}{8}, \dfrac{7}{16}, \ldots, 7\left(\dfrac{1}{2}\right)^{n-1}, \ldots$ | $\dfrac{1}{2}$ |
| $5, -\dfrac{5}{3}, \dfrac{5}{9}, -\dfrac{5}{27}, \dfrac{5}{81}, \ldots, 5\left(-\dfrac{1}{3}\right)^{n-1}, \ldots$ | $-\dfrac{1}{3}$ |

**EXAMPLE 1**  Determine the first five terms of the geometric sequence if $a_1 = 6$ and $r = \dfrac{1}{2}$.

**Solution**  $a_1 = 6,\ a_2 = 6 \cdot \dfrac{1}{2} = 3,\ a_3 = 3 \cdot \dfrac{1}{2} = \dfrac{3}{2},\ a_4 = \dfrac{3}{2} \cdot \dfrac{1}{2} = \dfrac{3}{4},\ a_5 = \dfrac{3}{4} \cdot \dfrac{1}{2} = \dfrac{3}{8}$

Thus, the first five terms of the geometric sequence are

$$6, 3, \dfrac{3}{2}, \dfrac{3}{4}, \dfrac{3}{8}$$

*Now Try Exercise 15*

## 2  Find the *n*th Term of a Geometric Sequence

**Understanding Algebra**

In a geometric sequence, to get the value for the common ratio, $r$, divide a term by the previous term.

In general, a geometric sequence with first term, $a_1$, and common ratio, $r$, has the following terms:

$$a_1, \qquad a_1 r, \qquad a_1 r^2, \qquad a_1 r^3, \qquad a_1 r^4, \ldots, a_1 r^{n-1}, \ldots$$

| $\uparrow$ | $\uparrow$ | $\uparrow$ | $\uparrow$ | $\uparrow$ | $\uparrow$ |
|---|---|---|---|---|---|
| 1st term, $a_1$ | 2nd term, $a_2$ | 3rd term, $a_3$ | 4th term, $a_4$ | 5th term, $a_5$ | $n$th term, $a_n$ |

Thus, we can see that the $n$th term of a geometric sequence is given by the following formula:

**$n$th Term of a Geometric Sequence**

$$a_n = a_1 r^{n-1}$$

**EXAMPLE 2**

**a)** Write an expression for the general (or $n$th) term, $a_n$, of the geometric sequence with $a_1 = 3$ and $r = -2$.

**b)** Find the twelfth term of this sequence.

**Solution**

**a)** The $n$th term of the sequence is $a_n = a_1 r^{n-1}$. Substituting $a_1 = 3$ and $r = -2$, we obtain

$$a_n = a_1 r^{n-1} = 3(-2)^{n-1}$$

Thus, $a_n = 3(-2)^{n-1}$.

**b)**
$$a_n = 3(-2)^{n-1}$$
$$a_{12} = 3(-2)^{12-1} = 3(-2)^{11} = 3(-2048) = -6144$$

The twelfth term of the sequence is $-6144$. The first 12 terms of the sequence are $3, -6, 12, -24, 48, -96, 192, -384, 768, -1536, 3072, -6144$.

Now Try Exercise 35

---

**Helpful Hint**

***Study Tip***

In this chapter, you will be working with exponents and using rules for exponents. The rules for exponents were discussed in Section 5.1 and again in Chapter 11. If you do not remember the rules for exponents, now is a good time to review Section 5.1.

---

**EXAMPLE 3** Find $r$ and $a_1$ for the geometric sequence with $a_2 = 12$ and $a_5 = 324$.

***Solution***    The sequence can be represented with blanks for the missing terms.

$$\underline{\quad}, 12, \underline{\quad}, \underline{\quad}, 324$$

$$\overset{\uparrow}{a_2} \qquad \overset{\uparrow}{a_5}$$

If we assume that $a_2$ is the first term of a sequence with the same common ratio, we obtain

$$12, \underline{\quad}, \underline{\quad}, 324$$

$$\underset{\substack{\text{1st} \\ \text{term}}}{\uparrow} \qquad \underset{\substack{\text{4th} \\ \text{term}}}{\uparrow}$$

Now we use the formula for the $n$th term of a geometric sequence to find $r$. We let the first term, $a_1$, be 12 and the number of terms, $n$, be 4.

$$a_n = a_1 r^{n-1}$$
$$324 = 12 r^{4-1}$$
$$324 = 12 r^3$$
$$\frac{324}{12} = r^3$$
$$27 = r^3$$
$$3 = r$$

Thus, the common ratio is 3.

The first term of the original sequence must be $12 \div 3$ or 4. Thus, $a_1 = 4$. The first term, $a_1$, could also be found using the formula with $a_n = 324$, $r = 3$, and $n = 5$. Find $a_1$ by the formula now.

Now Try Exercise 83

---

## 3  Find the *n*th Partial Sum of a Geometric Sequence

**Geometric Series**

A **geometric series** is the sum of the terms of a geometric sequence.

The sum of the first $n$ terms, $s_n$, of a geometric sequence can be expressed as

$$s_n = a_1 + a_1 r + a_1 r^2 + a_1 r^3 + \cdots + a_1 r^{n-2} + a_1 r^{n-1} \quad \text{(eq. 1)}$$

If we multiply both sides of the equation by $r$, we obtain

$$r s_n = a_1 r + a_1 r^2 + a_1 r^3 + \cdots + a_1 r^{n-1} + a_1 r^n \quad \text{(eq. 2)}$$

Now we subtract the corresponding sides of (*eq.* 2) from (*eq.* 1). The red-colored terms drop out, leaving

$$s_n - rs_n = a_1 - a_1r^n$$

Now we solve the equation for $s_n$.

$$s_n(1 - r) = a_1(1 - r^n) \qquad \text{Factor.}$$

$$s_n = \frac{a_1(1 - r^n)}{1 - r} \qquad \text{Divide both sides by } 1 - r.$$

Thus, we have the following formula for the $n$th partial sum of a geometric sequence.

**$n$th Partial Sum of a Geometric Sequence**

$$s_n = \frac{a_1(1 - r^n)}{1 - r}, \qquad r \neq 1$$

**EXAMPLE 4** Find the seventh partial sum of a geometric sequence whose first term is 16 and whose common ratio is $-\frac{1}{2}$.

**Solution** Substitute the appropriate value for $a_1$, $r$, and $n$.

$$s_n = \frac{a_1(1 - r^n)}{1 - r}$$

$$s_7 = \frac{16\left[1 - \left(-\dfrac{1}{2}\right)^7\right]}{1 - \left(-\dfrac{1}{2}\right)} = \frac{16\left(1 + \dfrac{1}{128}\right)}{\dfrac{3}{2}} = \frac{16\left(\dfrac{129}{128}\right)}{\dfrac{3}{2}} = \frac{\dfrac{129}{8}}{\dfrac{3}{2}} = \frac{129}{8} \cdot \frac{2}{3} = \frac{43}{4}$$

Thus, $s_7 = \dfrac{43}{4}$.

Now Try Exercise 41

**EXAMPLE 5** Given $s_n = 93$, $a_1 = 3$, and $r = 2$, find $n$.

**Solution**

$$s_n = \frac{a_1(1 - r^n)}{1 - r}$$

$$93 = \frac{3(1 - 2^n)}{1 - 2} \qquad \text{Substitute values for } s_n, a_1, \text{ and } r.$$

$$93 = \frac{3(1 - 2^n)}{-1}$$

$$-93 = 3(1 - 2^n) \qquad \text{Both sides were multiplied by } -1.$$

$$-31 = 1 - 2^n \qquad \text{Both sides were divided by 3.}$$

$$-32 = -2^n \qquad \text{1 was subtracted from both sides.}$$

$$32 = 2^n \qquad \text{Both sides were divided by } -1.$$

$$2^5 = 2^n \qquad \text{Write 32 as } 2^5.$$

Therefore, $n = 5$.

Now Try Exercise 65

When working with a geometric series, $r$ can be a positive number as we saw in Example 5 or a negative number as we saw in Example 4.

### 4 Identify Infinite Geometric Series

All the geometric sequences that we have examined thus far have been finite since they have had a last term. The following sequence is an example of an infinite geometric sequence.

$$1, \frac{1}{2}, \frac{1}{4}, \frac{1}{8}, \frac{1}{16}, \ldots, \left(\frac{1}{2}\right)^{n-1}, \ldots$$

Note that the three dots at the end of the sequence indicate that the sequence continues indefinitely.

**Infinite Geometric Series**

An **infinite geometric series** is the sum of the terms in an infinite geometric sequence.

For example,

$$1 + \frac{1}{2} + \frac{1}{4} + \frac{1}{8} + \frac{1}{16} + \cdots + \left(\frac{1}{2}\right)^{n-1} + \cdots$$

is an infinite geometric series. Let's find some partial sums.

| Partial Sum | Series | Sum |
| --- | --- | --- |
| Second | $1 + \frac{1}{2}$ | 1.5 |
| Third | $1 + \frac{1}{2} + \frac{1}{4}$ | 1.75 |
| Fourth | $1 + \frac{1}{2} + \frac{1}{4} + \frac{1}{8}$ | 1.875 |
| Fifth | $1 + \frac{1}{2} + \frac{1}{4} + \frac{1}{8} + \frac{1}{16}$ | 1.9375 |
| Sixth | $1 + \frac{1}{2} + \frac{1}{4} + \frac{1}{8} + \frac{1}{16} + \frac{1}{32}$ | 1.96875 |

With each successive partial sum, the amount being added is less than with the previous partial sum. Also, the sum seems to be getting closer and closer to 2. In Example 6, we will show that the sum of this infinite geometric series is indeed 2.

### 5 Find the Sum of an Infinite Geometric Series

Consider the formula for the sum of the first $n$ terms of an infinite geometric series:

$$s_n = \frac{a_1(1 - r^n)}{1 - r}, \qquad r \neq 1$$

What happens to $r^n$ if $|r| < 1$ and $n$ gets larger and larger? Suppose that $r = \frac{1}{2}$; then

$$\left(\frac{1}{2}\right)^1 = 0.5, \quad \left(\frac{1}{2}\right)^2 = 0.25, \quad \left(\frac{1}{2}\right)^3 = 0.125, \quad \left(\frac{1}{2}\right)^{20} \approx 0.000001$$

We can see that when $|r| < 1$, the value of $r^n$ gets exceedingly close to 0 as $n$ gets larger and larger. Thus, when considering the sum of an infinite geometric series, symbolized $s_\infty$, the expression $r^n$ approaches 0 when $|r| < 1$. Therefore, replacing $r^n$ with 0 in the formula $s_n = \dfrac{a_1(1 - r^n)}{1 - r}$ leads to the following formula.

**Sum of an Infinite Geometric Series**

$$s_\infty = \frac{a_1}{1 - r} \quad \text{where} \quad |r| < 1$$

EXAMPLE 6  Find the sum of the infinite geometric series

$$1 + \frac{1}{2} + \frac{1}{4} + \frac{1}{8} + \cdots + \left(\frac{1}{2}\right)^{n-1} + \cdots$$

**Solution**   $a_1 = 1$ and $r = \frac{1}{2}$. Note that $\left|\frac{1}{2}\right| < 1$.

$$s_\infty = \frac{a_1}{1 - r}$$

$$= \frac{1}{1 - \frac{1}{2}}$$

$$= \frac{1}{\frac{1}{2}}$$

$$= 2$$

Thus, $1 + \frac{1}{2} + \frac{1}{4} + \frac{1}{8} + \frac{1}{16} + \cdots + \left(\frac{1}{2}\right)^{n-1} + \cdots = 2$.

Now Try Exercise 69

EXAMPLE 7  Find the sum of the infinite geometric series

$$3 - \frac{6}{5} + \frac{12}{25} - \frac{24}{125} + \frac{48}{625} - \cdots$$

**Solution**   The terms of the corresponding sequence are $3, -\frac{6}{5}, \frac{12}{25}, -\frac{24}{125}, \ldots$. Note that $a_1 = 3$. To find the common ratio, $r$, we can divide the second term, $-\frac{6}{5}$, by the first term, 3.

$$r = -\frac{6}{5} \div 3$$

$$= -\frac{6}{5} \cdot \frac{1}{3}$$

$$= -\frac{2}{5}.$$

Since $\left|-\frac{2}{5}\right| < 1$,

$$s_\infty = \frac{a_1}{1 - r}$$

$$= \frac{3}{1 - \left(-\frac{2}{5}\right)}$$

$$= \frac{3}{1 + \frac{2}{5}}$$

$$= \frac{3}{\frac{7}{5}}$$

$$= \frac{15}{7}$$

Now Try Exercise 71

**EXAMPLE 8**  Write 0.343434... as a ratio of integers.

*Solution*   We can write this decimal as

$$0.34 + 0.0034 + 0.000034 + \cdots + (0.34)(0.01)^{n-1} + \cdots$$

This is an infinite geometric series with $r = 0.01$. Since $|r| < 1$,

$$s_\infty = \frac{a_1}{1 - r} = \frac{0.34}{1 - 0.01} = \frac{0.34}{0.99} = \frac{34}{99}$$

If you divide 34 by 99, you obtain 0.343434....

Now Try Exercise 81

What is the sum of a geometric series when $|r| > 1$? Consider the geometric sequence in which $a_1 = 1$ and $r = 2$.

$$1, 2, 4, 8, 16, 32, \ldots, 2^{n-1}, \ldots$$

The sum of its terms is

$$1 + 2 + 4 + 8 + 16 + 32 + \cdots + 2^{n-1} + \cdots$$

What is the sum of this series? As $n$ gets larger and larger, the sum gets larger and larger. We therefore say that the sum "does not exist." For $|r| > 1$, the sum of an infinite geometric series does not exist.

## 6 Applications of Geometric Series

**EXAMPLE 9**  **Savings Account**  Gene Simmons invests $1000 at 5% interest compounded annually in a savings account. Determine the amount in his account and the amount of interest earned at the end of 10 years.

*Solution*   **Understand**   Suppose we let $P$ represent any principal invested. At the beginning of the second year, the amount grows to $P + 0.05P$ or $1.05P$. This amount will be the principal invested for year 2. At the beginning of the third year the second year's principal will grow by 5% to $(1.05P)(1.05)$, or $(1.05)^2P$. The amount in Gene's account at the beginning of successive years is

| Year 1 | Year 2 | Year 3 | Year 4 |
|--------|--------|--------|--------|
| $P$ | $1.05P$ | $(1.05)^2P$ | $(1.05)^3P$ |

and so on. This is a geometric series with $r = 1.05$. The amount in his account at the end of 10 years will be the same as the amount in his account at the beginning of year 11. We will therefore use the formula

$$a_n = a_1 r^{n-1}, \quad \text{with} \quad r = 1.05 \quad \text{and} \quad n = 11$$

**Translate**   We have a geometric sequence with $a_1 = 1000$, $r = 1.05$, and $n = 11$. Substituting these values into the formula, we obtain the following.

Savings account
5% interest annually

$$a_n = a_1 r^{n-1}$$
$$a_{11} = 1000(1.05)^{11-1}$$

**Carry Out**
$$= 1000(1.05)^{10}$$
$$\approx 1000(1.62889)$$
$$\approx 1628.89$$

**Answer**   After 10 years, the amount in the account is about $1628.89. The amount of interest is $1628.89 − $1000 = $628.89.

Now Try Exercise 95

**EXAMPLE 10 Money** Suppose someone offered you $1000 a day for each day of a 30-day month. Or, you could elect to take a penny on day 1, 2¢ on day 2, 4¢ on day 3, 8¢ on day 4, and so on. The amount would continue to double each day for 30 days.

**a)** Without doing any calculations, take a guess at which of the two offerings would provide the greater total return for 30 days.

**b)** Calculate the total amount you would receive by selecting $1000 a day for 30 days.

**c)** Calculate the amount you would receive on day 30 by selecting 1¢ on day 1 and doubling the amount each day for 30 days.

**d)** Calculate the total amount you would receive for 30 days by selecting 1¢ on day 1 and doubling the amount each day for 30 days.

*Solution*

**a)** Each of you will have your own answer to part **a)**.

**b)** If you received $1000 a day for 30 days, you would receive 30($1000) = $30,000.

**c)** Understand   Since the amount is doubled each day, this represents a geometric sequence with $r = 2$. The chart that follows shows the amount you would receive in each of the first 7 days. We also show the amounts written with base 2, the common ratio.

| Day | 1 | 2 | 3 | 4 | 5 | 6 | 7 |
|---|---|---|---|---|---|---|---|
| **Amount (cents)** | 1 | 2 | 4 | 8 | 16 | 32 | 64 |
| **Amount (cents)** | $2^0$ | $2^1$ | $2^2$ | $2^3$ | $2^4$ | $2^5$ | $2^6$ |

Notice that for any given day, the exponent on 2 is 1 less than the given day. For example, on day 7, the amount is $2^6$. In general, the amount on day $n$ is $2^{n-1}$.

Translate   To find the amount received on day 30, we evaluate $a_n = a_1 r^{n-1}$ for $n = 30$.

$$a_n = a_1 r^{n-1}$$
$$a_{30} = 1(2)^{30-1}$$

Carry Out
$$a_{30} = 1(2)^{29}$$
$$= 1(536,870,912)$$
$$= 536,870,912$$

Answer   On day 30, the amount that you would receive is 536,870,912 cents or $5,368,709.12

**d)** Understand and Translate   To find the total amount received over the 30 days, we find the thirtieth partial sum.

$$s_n = \frac{a_1(1 - r^n)}{1 - r}$$
$$s_{30} = \frac{1(1 - 2^{30})}{1 - 2}$$

Carry Out
$$= \frac{1(1 - 1,073,741,824)}{-1}$$
$$= 1,073,741,823$$

Answer   Therefore, over 30 days the total amount you would receive by this method would be 1,073,741,823 cents or $10,737,418.23. The amount received by this method greatly surpasses the $30,000 received by selecting $1000 a day for 30 days.

Now Try Exercise 87

**EXAMPLE 11 Pendulum** On each swing (left to right or right to left), a certain pendulum travels 90% as far as on its previous swing. For example, if the swing to the right is 10 feet, the swing back to the left is $0.9 \times 10 = 9$ feet (see **Fig. 15.1**). If the first swing is 10 feet, determine the total distance traveled by the pendulum by the time it comes to rest.

*Solution*   Understand   This problem may be considered an infinite geometric series with $a_1 = 10$ and $r = 0.9$. We can therefore use the formula $s_\infty = \dfrac{a_1}{1 - r}$ to find the total distance traveled by the pendulum.

Translate and Carry Out

$$s_\infty = \frac{a_1}{1 - r} = \frac{10}{1 - 0.9} = \frac{10}{0.1} = 100 \text{ feet}$$

Answer   By the time the pendulum comes to rest, it has traveled 100 feet.

Now Try Exercise 99

**FIGURE 15.1**

# EXERCISE SET 15.3   MathXL®  MyMathLab®
MathXL®    MyMathLab

## Warm-Up Exercises

*Fill in the blanks with the appropriate word, phrase, or symbol(s) from the following list.*

| sum | difference | *d* | *r* | infinite | term | ratio | less |
| 0 | 1 | dividing | multiplying | geometric | series | greater | |

**1.** A sequence in which each term after the first term is a common multiple of the preceding term is a _____ sequence.

**2.** The common multiple of a geometric sequence is called the common _____.

**3.** The common ratio of a geometric sequence is identified by the letter _____.

**4.** The common ratio can be found by _____ any term, except the first term, by the preceding term.

**5.** The formula $a_n = a_1 r^{n-1}$ is used to find the *n*th _____ of a geometric sequence.

**6.** The sum of the terms of a geometric sequence is a geometric _____.

**7.** The formula $s_n = \dfrac{a_1(1 - r^n)}{1 - r}$ is used to find the *n*th partial _____ of a geometric sequence.

**8.** If the terms of a geometric sequence are increasing, the value for *r* is _____ than 1.

**9.** If the terms of a geometric sequence are the same, the value for *r* is _____.

**10.** The formula $s_\infty = \dfrac{a_1}{1 - r}, |r| < 1$, is used to find the sum of an _____ geometric series.

## Practice the Skills

*Determine the first five terms of each geometric sequence.*

**11.** $a_1 = 2, r = 3$

**12.** $a_1 = 5, r = 2$

**13.** $a_1 = 6, r = -\dfrac{1}{2}$

**14.** $a_1 = 6, r = \dfrac{1}{2}$

**15.** $a_1 = 72, r = \dfrac{1}{3}$

**16.** $a_1 = 50, r = \dfrac{1}{5}$

**17.** $a_1 = 90, r = -\dfrac{1}{3}$

**18.** $a_1 = 32, r = -\dfrac{1}{4}$

**19.** $a_1 = -1, r = 3$

**20.** $a_1 = -1, r = -3$

**21.** $a_1 = 7, r = -2$

**22.** $a_1 = -13, r = -1$

**23.** $a_1 = \dfrac{1}{3}, r = \dfrac{1}{2}$

**24.** $a_1 = \dfrac{1}{2}, r = -\dfrac{1}{3}$

**25.** $a_1 = 3, r = \dfrac{3}{2}$

**26.** $a_1 = 60, r = -\dfrac{2}{5}$

*Find the indicated term of each geometric sequence.*

**27.** $a_1 = 4, r = 2$; find $a_6$

**28.** $a_1 = 4, r = -2$; find $a_6$

**29.** $a_1 = -24, r = \dfrac{1}{2}$; find $a_9$

**30.** $a_1 = 27, r = \dfrac{1}{3}$; find $a_7$

**31.** $a_1 = \dfrac{1}{8}, r = 2$; find $a_{10}$

**32.** $a_1 = 3, r = 3$; find $a_5$

**33.** $a_1 = -3, r = -2$; find $a_{12}$

**34.** $a_1 = -10, r = -2$; find $a_4$

**35.** $a_1 = 2, r = \dfrac{1}{2}$; find $a_8$

**36.** $a_1 = 5, r = \dfrac{2}{3}$; find $a_9$

**37.** $a_1 = 50, r = \dfrac{1}{3}$; find $a_7$

**38.** $a_1 = -7, r = -\dfrac{3}{4}$; find $a_7$

*Find the indicated sum.*

**39.** $a_1 = 5, r = 2$; find $s_5$

**40.** $a_1 = 7, r = -3$; find $s_4$

**41.** $a_1 = 2, r = 5$; find $s_6$

**42.** $a_1 = 9, r = \dfrac{1}{2}$; find $s_6$

**43.** $a_1 = 80, r = 2$; find $s_7$

**44.** $a_1 = 2, r = -2$; find $s_{12}$

**45.** $a_1 = -15, r = -\dfrac{1}{2}$; find $s_9$

**46.** $a_1 = \dfrac{3}{4}, r = 3$; find $s_7$

**47.** $a_1 = -9, r = \dfrac{2}{5}$; find $s_5$

**48.** $a_1 = 35, r = \dfrac{1}{5}$; find $s_{12}$

*For each geometric sequence, find the common ratio, r, and then write an expression for the general (or nth) term, $a_n$.*

**49.** $7, \dfrac{7}{2}, \dfrac{7}{4}, \dfrac{7}{8}, \ldots$

**50.** $3, -\dfrac{3}{2}, \dfrac{3}{4}, -\dfrac{3}{8}, \ldots$

**51.** $9, 18, 36, 72, \ldots$

**52.** $2, 6, 18, 54, \ldots$

**53.** $2, -6, 18, -54, \ldots$

**54.** $-1, -4, -16, -64, -256, \ldots$

**55.** $\dfrac{3}{4}, \dfrac{1}{2}, \dfrac{1}{3}, \dfrac{2}{9}, \ldots$

**56.** $\dfrac{2}{3}, \dfrac{10}{3}, \dfrac{50}{3}, \dfrac{250}{3}, \ldots$

*Find the sum of the terms in each geometric sequence.*

**57.** $1, \dfrac{1}{2}, \dfrac{1}{4}, \dfrac{1}{8}, \dfrac{1}{16}, \ldots$

**58.** $1, -\dfrac{1}{2}, \dfrac{1}{4}, -\dfrac{1}{8}, \dfrac{1}{16}, \ldots$

**59.** $1, \dfrac{1}{5}, \dfrac{1}{25}, \dfrac{1}{125}, \dfrac{1}{625}, \ldots$

**60.** $1, -\dfrac{1}{5}, \dfrac{1}{25}, -\dfrac{1}{125}, \dfrac{1}{625}, \ldots$

**61.** $12, 6, 3, \dfrac{3}{2}, \dfrac{3}{4}, \ldots$

**62.** $\dfrac{1}{4}, \dfrac{1}{16}, \dfrac{1}{64}, \dfrac{1}{128}, \ldots$

**63.** $5, 2, \dfrac{4}{5}, \dfrac{8}{25}, \ldots$

**64.** $-\dfrac{4}{3}, -\dfrac{4}{9}, -\dfrac{4}{27}, -\dfrac{4}{81}, \ldots$

*Given $s_n$, $a_1$, and r, find n in each geometric series.*

**65.** $s_n = 93, a_1 = 3$, and $r = 2$

**66.** $s_n = 80, a_1 = 2$, and $r = 3$

**67.** $s_n = \dfrac{189}{32}, a_1 = 3$, and $r = \dfrac{1}{2}$

**68.** $s_n = \dfrac{104}{9}, a_1 = 8$, and $r = \dfrac{1}{3}$

*Find the sum of each infinite geometric series.*

**69.** $2 + 1 + \dfrac{1}{2} + \dfrac{1}{4} + \dfrac{1}{8} + \cdots$

**70.** $16 + 8 + 4 + 2 + \cdots$

**71.** $4 + \dfrac{8}{3} + \dfrac{16}{9} + \dfrac{32}{27} + \cdots$

**72.** $6 - 2 + \dfrac{2}{3} - \dfrac{4}{9} + \cdots$

**73.** $-60 + 20 - \dfrac{20}{3} + \dfrac{20}{9} - \cdots$

**74.** $2 + \dfrac{4}{3} + \dfrac{8}{9} + \dfrac{16}{27} + \cdots$

**75.** $-12 - \dfrac{12}{5} - \dfrac{12}{25} - \dfrac{12}{125} - \cdots$

**76.** $7 - 1 + \dfrac{1}{7} - \dfrac{1}{49} + \cdots$

*Write each repeating decimal as a ratio of integers.*

**77.** $0.242424\ldots$

**78.** $0.454545\ldots$

**79.** $0.7777\ldots$

**80.** $0.375375\ldots$

**81.** $0.515151\ldots$

**82.** $0.742742\ldots$

## Problem Solving

**83.** In a geometric sequence, $a_2 = 15$ and $a_5 = 405$; find $r$ and $a_1$.

**84.** In a geometric sequence, $a_2 = 27$ and $a_5 = 1$; find $r$ and $a_1$.

**85.** In a geometric sequence, $a_3 = 28$ and $a_5 = 112$; find $r$ and $a_1$.

**86.** In a geometric sequence, $a_2 = 12$ and $a_5 = -324$; find $r$ and $a_1$.

**87.** **Loaf of Bread**  A loaf of bread currently costs $1.40. Determine the cost of a loaf of bread after 8 years (the start of the 9th year) if inflation were to grow at a constant rate of 3% per year. *Hint*: After year 1 (the start of year 2), the cost of a loaf of bread is $1.40(1.03)$. After year 2 (the start of year 3), the cost would be $1.40(1.03)^2$, and so on.

**88.** **Bicycle**  A specific bicycle currently costs $300. Determine the cost of the bicycle after 12 years if inflation were to grow at a constant rate of 4% per year.

**89.** **Mass**  A substance loses half its mass each day. If there are initially 600 grams of the substance, find

a) the number of days after which only 37.5 grams of the substance remain.

b) the amount of the substance remaining after 9 days.

**90.** **Bacteria**  The number of a certain type of bacteria doubles every hour. If there are initially 500 bacteria, after how many hours will the number of bacteria reach 32,000?

**91. Population** On July 1, 2008, the population of the United States was about 303.5 million people. If the population grows at a rate of 1.1% per year, find

a) the population after 10 years.

b) the number of years for the population to double.

**92. Farm Equipment** A piece of farm equipment that costs $105,000 decreases in value by 15% per year. Find the value of the equipment after 4 years.

© Jose Ignacio Soto/Shutterstock

**93. Filtered Light** The amount of light filtering through a lake diminishes by one-half for each meter of depth.

a) Write a sequence indicating the amount of light remaining at depths of 1, 2, 3, 4, and 5 meters.

b) What is the general term for this sequence?

c) What is the remaining light at a depth of 7 meters?

**94. Pendulum** On each swing (left to right or right to left), a pendulum travels 80% as far as on its previous swing. If the first swing is 10 feet, determine the total distance traveled by the pendulum by the time it comes to rest.

**95. Investment** You invest $10,000 in a savings account paying 6% interest annually. Find the amount in your account at the end of 8 years.

**96. Injected Dye** A tracer dye is injected into Mark Damion for medical reasons. After each hour, two-thirds of the previous hour's dye remains. How much dye remains in Mark's system after 10 hours?

**97. Bungee Jumping** Shawna Kelly goes bungee jumping off a bridge above water. On the initial jump, the bungee cord stretches to 220 feet. Assume the first bounce reaches a height of 60% of the original jump and that each additional bounce reaches a height of 60% of the previous bounce.

a) What will be the height of the fourth bounce?

b) Theoretically, Shawna would never stop bouncing, but realistically, she will. Use the infinite geometric series to estimate the total distance Shawna travels in a *downward* direction.

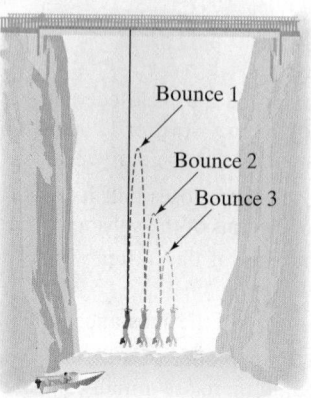

Bounce 1

Bounce 2

Bounce 3

**98. Bungee Jumping** Repeat Exercise 97 **b)**, but this time find the total distance traveled in an *upward* direction.

**99. Ping-Pong Ball** A ping-pong ball falls off a table 30 inches high. Assume that the first bounce reaches a height of 70% of the distance the ball fell and each additional bounce reaches a height of 70% of the previous bounce.

a) How high will the ball bounce on the third bounce?

b) Theoretically, the ball would never stop bouncing, but realistically, it will. Estimate the total distance the ball travels in the *downward* direction.

**100. Ping-Pong Ball** Repeat Exercise 99 **b)**, but this time find the total distance traveled in the *upward* direction.

**101. Stack of Chips** Suppose that you form stacks of blue chips such that there is one blue chip in the first stack and in each successive stack you double the number of chips. Thus, you have stacks of 1, 2, 4, 8, and so on, of blue chips. You also form stacks of red chips, starting with one red chip and then tripling the number in each successive stack. Thus the stacks will contain 1, 3, 9, 27 and so on, red chips. How many more would the sixth stack of red chips have than the sixth stack of blue chips?

**102. Stack of Money** If you start with $1 and double your money each day, how many days will it take to surpass $1,000,000?

**103. Depreciation** One method of depreciating an item on an income tax return is the declining balance method. With this method, a given fraction of the remaining value of the item is depreciated each year. Suppose that an item has a 5-year life and is depreciated using the declining balance method. Then, at the end of its first year, it loses $\frac{1}{5}$ of its value and $\frac{4}{5}$ of its value remains. At the end of the second year it loses $\frac{1}{5}$ of the remaining $\frac{4}{5}$ of its value, and so on. A car has a 5-year life expectancy and costs $15,000.

a) Write a sequence showing the value of the car remaining for each of the first 3 years.

b) What is the general term of this sequence?

c) Find the value of the car at the end of 5 years.

**104. Scrap Value** On page 817, Exercise Set 13.6, Exercise 77, a formula for scrap value was given. The scrap value, $S$, is found by $S = c(1 - r)^n$ where $c$ is the original cost, $r$ is the annual depreciation rate, and $n$ is the number of years the object is depreciated.

a) If you have not already done so, do Exercise 103 above to find the value of the car remaining at the end of 5 years.

b) Use the formula given to find the scrap value of the car at the end of 5 years and compare this answer with the answer found in part **a)**.

**105. Bouncing Ball** A ball is dropped from a height of 10 feet. The ball bounces to a height of 9 feet. On each successive bounce, the ball rises to 90% of its previous height. Find the *total vertical distance* traveled by the ball when it comes to rest.

**106. Wave Action** A particle follows the path indicated by the wave shown. Find the *total vertical distance* traveled by the particle.

**107.** The formula for the $n$th term of a geometric sequence is $a_n = a_1 r^{n-1}$. If $a_1 = 1$, $a_n = r^{n-1}$.

   **a)** How do you think the graphs of $y_1 = 2^{n-1}$ and $y_2 = 3^{n-1}$ will compare?

   **b)** Graph both $y_1$ and $y_2$ and determine whether your answer to part **a)** was correct.

**108.** Use your graphing calculator to approximate the value of $n$ to the nearest hundredth, where $100 = 3 \cdot 2^{n-1}$.

## Concept/Writing Exercises

**109.** In a geometric series, if $|r| < 1$, what does $r^n$ approach as $n$ gets larger and larger?

**110.** Does the sum of an infinite geometric series when $|r| > 1$ exist?

**111.** In a geometric sequence, can the value of $r$ be a negative number?

**112.** In a geometric sequence, can the value of $r$ be a positive number?

**113.** In a geometric series, if $a_1 = 6$ and $r = \dfrac{1}{4}$, does $s_\infty$ exist? If so, what is its value? Explain.

**114.** In a geometric series, if $a_1 = 6$ and $r = -2$, does $s_\infty$ exist? If so, what is its value? Explain.

## Challenge Problem

**115.** Find the sum of the sequence $1, 2, 4, 8, \ldots, 1{,}048{,}576$ and the number of terms in the sequence.

## Cumulative Review Exercises

[5.5] **116.** Multiply $(4x^2 - 3x + 6)(2x - 3)$.

[5.6] **117.** Divide $(16x^2 + 10x - 18) \div (2x + 5)$.

[7.7] **118. Loading a Truck** It takes Mrs. Donovan twice as long to load a truck as it takes Mr. Donovan. If together they can load the truck in 8 hours, how long would it take Mr. Donovan to load the truck by himself?

[11.2] **119.** Evaluate $\left(\dfrac{9}{100}\right)^{-1/2}$.

[11.4] **120.** Simplify $\sqrt[3]{9x^2 y}\left(\sqrt[3]{3x^4 y^6} - \sqrt[3]{8xy^4}\right)$.

**121.** Simplify $x\sqrt{y} - 2\sqrt{x^2 y} + \sqrt{4x^2 y}$.

[11.6] **122.** Solve $\sqrt{a^2 + 9a + 3} = -a$.

## Mid-Chapter Test: 15.1–15.3

*To find out how well you understand the chapter material to this point, take this brief test. The answers, and the section where the material was initially discussed, are given in the back of the book. Review any questions you answered incorrectly.*

**1.** Write the first five terms of the sequence whose $n$th term is $a_n = -3n + 7$.

**2.** If $a_n = n(n + 6)$, find the seventh term.

**3.** Find the first and third partial sums, $s_1$ and $s_3$, for the sequence whose $n$th term is $a_n = 2^n - 1$.

**4.** Write the next three terms of the sequence $5, 1, -3, -7, -11, \ldots$.

**5.** Evaluate the series $\displaystyle\sum_{i=1}^{5} (4i - 3)$.

**6.** If the general term of a sequence is $a_n = \dfrac{1}{4}n + 9$, write an expression using $\Sigma$ to represent the fifth partial sum.

**7.** Write the first four terms of the arithmetic sequence with $a_1 = -6$ and $d = 5$. Find an expression for the general term $a_n$.

**8.** Find $d$ for the arithmetic sequence with $a_1 = \dfrac{11}{2}$ and $a_7 = -\dfrac{1}{2}$.

**9.** Find $n$ for the arithmetic sequence with $a_1 = 22$, $a_n = -3$, and $d = -5$.

**10.** Find the common difference, $d$, and the sum, $s_6$, for the arithmetic sequence with $a_1 = -8$ and $a_6 = 7$.

**11.** Find $s_{10}$ for the arithmetic sequence with $a_1 = \dfrac{5}{2}$ and $d = \dfrac{1}{2}$.

**12.** Find the number of terms in the arithmetic sequence $-7, 0, 7, 14, \ldots, 70$.

**13.** Logs are stacked in a pile with 16 logs on the bottom row, 15 on the next row, 14 on the next row, and so on to the top, with the top row having one log. Each row has one log less than the row below it. How many logs are on the pile?

**14.** Write the first five terms of the geometric series with $a_1 = 100$ and $r = -\dfrac{1}{2}$.

**15.** Find $a_7$ for the geometric sequence with $a_1 = 27$ and $r = \dfrac{1}{3}$.

*(continued next page)*

**16.** Find $s_6$ for the geometric sequence with $a_1 = 5$ and $r = 2$.

**17.** For the geometric sequence $8, -\dfrac{16}{3}, \dfrac{32}{9}, -\dfrac{64}{27}, \ldots$, find $r$.

**18.** Find the sum of the infinite series $12, 4, \dfrac{4}{3}, \dfrac{4}{9}, \ldots$.

**19.** Write the repeating decimal $0.878787\ldots$ as a ratio of two integers.

**20.** **a)** What is a sequence?
  **b)** What is an arithmetic sequence?
  **c)** What is geometric sequence?
  **d)** What is a series?

# 15.4 The Binomial Theorem

1 Evaluate factorials.

2 Use Pascal's triangle.

3 Use the binomial theorem.

## 1 Evaluate Factorials

To understand the binomial theorem, you must have an understanding of what **factorials** are.

> ### *n* Factorial
>
> The notation **n!** is read "**n factorial**."
>
> $$n! = n(n-1)(n-2)(n-3)\cdots(1)$$
>
> for any positive integer $n$.

### Examples

$$6! = 6 \cdot 5 \cdot 4 \cdot 3 \cdot 2 \cdot 1 = 720$$
$$7! = 7 \cdot 6 \cdot 5 \cdot 4 \cdot 3 \cdot 2 \cdot 1 = 5040$$

Note that by definition **0! is 1**.

## Using Your Calculator

### Scientific Calculator

Factorials can be found on calculators that contain an $\boxed{n!}$ or $\boxed{x!}$ key. Often, the factorial key is a second function key.

In the following examples, the answers appear after $\boxed{n!}$.

| | | |
|---|---|---|
| Evaluate 6! | 6 $\boxed{2^{\text{nd}}}$ $\boxed{n!}$ | 720 |
| Evaluate 9! | 9 $\boxed{2^{\text{nd}}}$ $\boxed{n!}$ | 362880 |

### Graphing Calculator

On most graphing calculators, factorials are found under the $\boxed{\text{MATH}}$, Probability function menu (PRB).

On the TI-84 Plus calculator, press $\boxed{\text{MATH}}$, and then scroll to the right using the right arrow key, $\boxed{\blacktriangleright}$, three times, until you get to PRB. The n! (or !) is the fourth item on the menu.

To find 5! or 6!, the keystrokes are as follows.

| KEYSTROKES | ANSWER |
|---|---|
| 5 $\boxed{\text{MATH}}$ $\boxed{\blacktriangleright}$ $\boxed{\blacktriangleright}$ $\boxed{\blacktriangleright}$ 4 $\boxed{\text{ENTER}}$ | 120 |
| 6 $\boxed{\text{MATH}}$ $\boxed{\blacktriangleright}$ $\boxed{\blacktriangleright}$ $\boxed{\blacktriangleright}$ 4 $\boxed{\text{ENTER}}$ | 720 |

## Understanding Algebra

A *two-term* polynomial such as $a + b$ or $2x + 3y$ is called a *binomial*.

## 2 Use Pascal's Triangle

Using polynomial multiplication, we can obtain the following powers of the binomial $a + b$: These multiplications are called expansions of the binomial.

$$(a + b)^0 = 1$$
$$(a + b)^1 = a + b$$
$$(a + b)^2 = a^2 + 2ab + b^2$$
$$(a + b)^3 = a^3 + 3a^2b + 3ab^2 + b^3$$
$$(a + b)^4 = a^4 + 4a^3b + 6a^2b^2 + 4ab^3 + b^4$$
$$(a + b)^5 = a^5 + 5a^4b + 10a^3b^2 + 10a^2b^3 + 5ab^4 + b^5$$
$$(a + b)^6 = a^6 + 6a^5b + 15a^4b^2 + 20a^3b^3 + 15a^2b^4 + 6ab^5 + b^6$$

Note that when expanding a binomial of the form $(a + b)^n$,

1. There are $n + 1$ terms in the expansion.
2. The first term is $a^n$ and the last term is $b^n$.
3. Reading from left to right, the exponents on $a$ decrease by 1 from term to term, while the exponents on $b$ increase by 1 from term to term.
4. The sum of the exponents on the variables in each term is $n$.
5. The coefficients of the terms equidistant from the ends are the same.

If we examine just the variables in $(a + b)^5$, we have $a^5$, $a^4b$, $a^3b^2$, $a^2b^3$, $ab^4$, and $b^5$.

The numerical coefficients of each term in the expansion of $(a + b)^n$ can be found by using **Pascal's triangle**. For example, if $n = 5$, we can determine the numerical coefficients of $(a + b)^5$ as follows.

**Blaise Pascal**
The French mathematician Blaise Pascal (1623–1662) is responsible for developing the array of numbers in Pascal's Triangle.

| Exponent on Binomial | Pascal's Triangle |
|---|---|
| $n = 0$ | 1 |
| $n = 1$ | 1  1 |
| $n = 2$ | 1  2  1 |
| $n = 3$ | 1  3  3  1 |
| $n = 4$ | 1  4  6  4  1 |
| $n = 5$ | 1  5  10  10  5  1 |
| $n = 6$ | 1  6  15  20  15  6  1 |

Examine row 5 ($n = 4$) and row 6 ($n = 5$).

Sum of two numbers

Notice that the first and last numbers in each row are 1, and the inner numbers are obtained by adding the two numbers in the row above (to the right and left). The numerical coefficients of $(a + b)^5$ are 1, 5, 10, 10, 5, and 1. Thus, we can write the expansion of $(a + b)^5$ by using the information in 1–5 above for the variables and their exponents, and by using Pascal's triangle for the coefficients.

$$(a + b)^5 = a^5 + 5a^4b + 10a^3b^2 + 10a^2b^3 + 5ab^4 + b^5$$

This method of expanding a binomial is not practical when $n$ is large.

## 3  Use the Binomial Theorem

We will shortly introduce a more practical method, called the binomial theorem, to expand expressions of the form $(a + b)^n$. However, before we introduce this formula, we need to explain how to find *binomial coefficients* of the form $\binom{n}{r}$.

### Binomial Coefficients

For $n$ and $r$ non-negative integers, $n \geq r$,

$$\binom{n}{r} = \frac{n!}{r! \cdot (n - r)!}$$

The binomial coefficient $\begin{pmatrix} n \\ r \end{pmatrix}$ is read "the number of *combinations* of $n$ items taken $r$ at a time." Combinations are used in many areas of mathematics, including the study of probability.

**EXAMPLE 1** Evaluate $\begin{pmatrix} 6 \\ 2 \end{pmatrix}$.

*Solution*   Using the definition, if we substitute 6 for $n$ and 2 for $r$, we obtain

$$\begin{pmatrix} 6 \\ 2 \end{pmatrix} = \frac{6!}{2! \cdot (6-2)!} = \frac{6!}{2! \cdot 4!} = \frac{6 \cdot 5 \cdot 4 \cdot 3 \cdot 2 \cdot 1}{(2 \cdot 1) \cdot (4 \cdot 3 \cdot 2 \cdot 1)} = 15$$

Thus, $\begin{pmatrix} 6 \\ 2 \end{pmatrix}$ equals 15.

Now Try Exercise 9

**EXAMPLE 2** Evaluate.

**a)** $\begin{pmatrix} 7 \\ 4 \end{pmatrix}$       **b)** $\begin{pmatrix} 8 \\ 8 \end{pmatrix}$       **c)** $\begin{pmatrix} 5 \\ 0 \end{pmatrix}$

*Solution*

**a)** $\begin{pmatrix} 7 \\ 4 \end{pmatrix} = \frac{7!}{4! \cdot (7-4)!} = \frac{7!}{4! \cdot 3!} = \frac{7 \cdot 6 \cdot 5 \cdot 4 \cdot 3 \cdot 2 \cdot 1}{(4 \cdot 3 \cdot 2 \cdot 1)(3 \cdot 2 \cdot 1)} = 35$

**b)** $\begin{pmatrix} 8 \\ 8 \end{pmatrix} = \frac{8!}{8! (8-8)!} = \frac{8!}{8! \cdot 0!} = \frac{1}{1} = 1$       Remember that $0! = 1$

**c)** $\begin{pmatrix} 5 \\ 0 \end{pmatrix} = \frac{5!}{0! \cdot (5-0)!} = \frac{5!}{0! \cdot 5!} = \frac{1}{1} = 1$

Now Try Exercise 17

By studying Examples 2 **b)** and **c)**, you can reason that, for any positive integer $n$,

$$\begin{pmatrix} n \\ n \end{pmatrix} = 1 \quad \text{and} \quad \begin{pmatrix} n \\ 0 \end{pmatrix} = 1$$

## 📱 *Using Your Graphing Calculator* ■ ■ ■ ■ ■

On most graphing calculators the notation $_nC_r$ is used instead of $\begin{pmatrix} n \\ r \end{pmatrix}$. Thus, $\begin{pmatrix} 7 \\ 4 \end{pmatrix}$ would be represented as $_7C_4$.

On the TI-84 Plus calculator, the notation $_nC_r$ can be found under the PRB menu. This time it is item 3, $_nC_r$. To find $_7C_4$ or $_8C_2$, use the following keystrokes:

|  | KEYSTROKES | ANSWER |
|---|---|---|
| $_7C_4$ | [7] [MATH] [▶] [▶] [▶] [3] [4] [ENTER] | 35 |
| $_8C_2$ | [8] [MATH] [▶] [▶] [▶] [3] [2] [ENTER] | 28 |

If you are using a different graphing calculator, consult the manual to learn how to evaluate combinations.

Now we introduce the binomial theorem.

### Binomial Theorem

For any positive integer $n$,

$$(a + b)^n = \begin{pmatrix} n \\ 0 \end{pmatrix}a^n b^0 + \begin{pmatrix} n \\ 1 \end{pmatrix}a^{n-1}b^1 + \begin{pmatrix} n \\ 2 \end{pmatrix}a^{n-2}b^2 + \begin{pmatrix} n \\ 3 \end{pmatrix}a^{n-3}b^3 + \cdots + \begin{pmatrix} n \\ n \end{pmatrix}a^0 b^n$$

Notice in the binomial theorem that the sum of the exponents on the variables in each term is $n$. In the combination, the top number is always $n$ and the bottom number is always the same as the exponent on the second variable in the term.

For example, consider the term $\binom{n}{3} a^{n-3} b^3$.

The top number is always $n$.        The sum of the exponents always equals $n$.

$$\binom{n}{3} \quad a^{n-3} b^3$$

The bottom number always matches the power of the second variable.

Now we will expand $(a + b)^5$ using the binomial theorem and see if we get the same expression as we did when we used polynomial multiplication and Pascal's triangle to obtain the expansion.

$$(a + b)^5 = \binom{5}{0} a^5 b^0 + \binom{5}{1} a^{5-1} b^1 + \binom{5}{2} a^{5-2} b^2 + \binom{5}{3} a^{5-3} b^3 + \binom{5}{4} a^{5-4} b^4 + \binom{5}{5} a^{5-5} b^5$$

$$= \binom{5}{0} a^5 b^0 + \binom{5}{1} a^4 b^1 + \binom{5}{2} a^3 b^2 + \binom{5}{3} a^2 b^3 + \binom{5}{4} a^1 b^4 + \binom{5}{5} a^0 b^5$$

$$= \frac{5!}{0! \cdot 5!} a^5 + \frac{5!}{1! \cdot 4!} a^4 b + \frac{5!}{2! \cdot 3!} a^3 b^2 + \frac{5!}{3! \cdot 2!} a^2 b^3 + \frac{5!}{4! \cdot 1!} ab^4 + \frac{5!}{5! \cdot 0!} b^5$$

$$= a^5 + 5a^4 b + 10a^3 b^2 + 10a^2 b^3 + 5ab^4 + b^5$$

This is the same expression as we obtained earlier.

**EXAMPLE 3**  Use the binomial theorem to expand $(2x + 3)^6$.

*Solution*    If we use $2x$ for $a$ and 3 for $b$, we obtain

$$(2x + 3)^6 = \binom{6}{0}(2x)^6(3)^0 + \binom{6}{1}(2x)^5(3)^1 + \binom{6}{2}(2x)^4(3)^2 + \binom{6}{3}(2x)^3(3)^3 + \binom{6}{4}(2x)^2(3)^4 + \binom{6}{5}(2x)^1(3)^5 + \binom{6}{6}(2x)^0(3)^6$$

$$= 1(2x)^6 + 6(2x)^5(3) + 15(2x)^4(9) + 20(2x)^3(27) + 15(2x)^2(81) + 6(2x)(243) + 1(729)$$

$$= 64x^6 + 576x^5 + 2160x^4 + 4320x^3 + 4860x^2 + 2916x + 729$$

Now Try Exercise 19

**EXAMPLE 4**  Use the binomial theorem to expand $(5x - 2y)^4$.

*Solution*    Write $(5x - 2y)^4$ as $[5x + (-2y)]^4$. Use $5x$ in place of $a$ and $-2y$ in place of $b$ in the binomial theorem.

$$[5x + (-2y)]^4 = \binom{4}{0}(5x)^4(-2y)^0 + \binom{4}{1}(5x)^3(-2y)^1 + \binom{4}{2}(5x)^2(-2y)^2 + \binom{4}{3}(5x)^1(-2y)^3 + \binom{4}{4}(5x)^0(-2y)^4$$

$$= 1(5x)^4 + 4(5x)^3(-2y) + 6(5x)^2(-2y)^2 + 4(5x)(-2y)^3 + 1(-2y)^4$$

$$= 625x^4 - 1000x^3 y + 600x^2 y^2 - 160xy^3 + 16y^4$$

Now Try Exercise 25

# EXERCISE SET 15.4    MathXL®    MyMathLab®
MathXL®    MyMathLab

## Warm-Up Exercises

*Fill in the blanks with the appropriate word, phrase, or symbol(s) from the following list.*

binomial        0              1              2              4              product

Pascal's        10             20             factorial      sum            quotient

**1.** The notation $n!$ is read as "$n$ _____."

**2.** The value of $n!(n \neq 0)$ is the _____ of the whole numbers from 1 to $n$.

**3.** The value of $0!$ is _____.

**4.** The value of $2!$ is _____.

**5.** The number of terms in the expansion of $(a + b)^{19}$ is _____.

**6.** The formula $\binom{n}{r} = \dfrac{n!}{r! \cdot (n - r)!}$ is used to find the _____ coefficients in the expansion of $(a + b)^n$.

**7.** The value of $\binom{5}{3}$ is _____.

**8.** The numerical coefficients of the terms in the expansion of $(a + b)^n$ can be found using _____ triangle.

## Practice the Skills

*Evaluate each combination.*

**9.** $\binom{5}{2}$

**10.** $\binom{6}{3}$

**11.** $\binom{5}{5}$

**12.** $\binom{9}{6}$

**13.** $\binom{7}{0}$

**14.** $\binom{12}{10}$

**15.** $\binom{8}{4}$

**16.** $\binom{11}{8}$

**17.** $\binom{8}{2}$

**18.** $\binom{10}{4}$

*Use the binomial theorem to expand each expression.*

**19.** $(x + 4)^3$

**20.** $(x - 4)^3$

**21.** $(2x - 3)^3$

**22.** $(2x + 3)^3$

**23.** $(a - b)^4$

**24.** $(2r + s^2)^4$

**25.** $(3a - b)^5$

**26.** $(x + 2y)^5$

**27.** $\left(2x + \dfrac{1}{2}\right)^4$

**28.** $\left(\dfrac{2}{3}x + \dfrac{3}{2}\right)^4$

**29.** $\left(\dfrac{x}{2} - 3\right)^4$

**30.** $(3x^2 + y)^5$

*Write the first four terms of each expansion.*

**31.** $(x + 10)^{10}$

**32.** $(2x + 3)^8$

**33.** $(3x - y)^7$

**34.** $(3p - 2q)^{11}$

**35.** $(x^2 - 3y)^8$

**36.** $\left(2x + \dfrac{y}{7}\right)^9$

## Problem Solving

**37.** What are the first, second, next-to-last, and last terms of the expansion $(x + 3)^8$?

**38.** What are the first, second, next-to-last, and last terms of the expansion of $(x + 2)^{10}$?

**39.** What are the first, second, next-to-last, and last terms of the expansion $(2x + 5)^6$?

**40.** What are the first, second, next-to-last, and last terms of the expansion of $(3x - 4)^5$?

## Concept/Writing Exercises

**41.** Is $n!$ equal to $n \cdot (n - 1)!$? Explain and give an example to support your answer.

**42.** Is $(n + 1)!$ equal to $(n + 1) \cdot n!$? Explain and give an example to support your answer.

**43.** Is $(n - 3)!$ equal to $(n - 3)(n - 4)(n - 5)!$ for $n \geq 5$? Explain and give an example to support your answer.

**44.** Is $(n + 2)!$ equal to $(n + 2)(n + 1)(n)(n - 1)!$ for $n \geq 1$? Explain and give an example to support your answer.

**45.** Under what conditions will $\binom{n}{m}$, where $n$ and $m$ are non-negative integers, have a value of 1?

**46.** Can $\binom{n}{m}$ ever have a value of 0? Explain.

**47.** Write the binomial theorem using summation notation.

**48.** Prove that $\binom{n}{r} = \binom{n}{n - r}$ for any whole numbers $n$ and $r$, and $r \leq n$.

## Cumulative Review Exercises

[4.4] **49.** Find the $y$-intercept for the line $2x + y = 10$

[6.6] **50.** Solve $x(x - 11) = -18$.

[9.3] **51.** Solve the system of equations.

$$\frac{1}{5}x + \frac{1}{2}y = 4$$

$$\frac{2}{3}x - y = \frac{8}{3}$$

[11.4] **52.** Simplify $\sqrt{20xy^4}\,\sqrt{6x^5y^7}$.

[13.1] **53.** Find $f^{-1}(x)$ if $f(x) = 3x + 8$.

# Chapter 15 Summary

| IMPORTANT FACTS AND CONCEPTS | EXAMPLES |
|---|---|
| **Section 15.1** | |

| | |
|---|---|
| A **sequence** of numbers is a list of numbers arranged in a specific order. Each number is called a **term** of the sequence. | $2, 6, 10, 14, 18, 22, \ldots$ is a sequence<br>$7, 14, 21, 28, 35, 42, \ldots$ is a sequence |
| An **infinite sequence** is a function whose domain is the set of natural numbers. | Domain: $\{1,\quad 2,\quad 3,\quad 4,\quad \ldots,\quad n,\quad \ldots\}$<br>$\qquad\qquad \downarrow\quad \downarrow\quad \downarrow\quad \downarrow\qquad\qquad \downarrow$<br>Range: $\{7,\quad 14,\quad 21,\quad 28,\quad \ldots\quad 7n,\quad \ldots\}$<br>The infinite sequence is $7, 14, 21, 28, \ldots$. |
| A **finite sequence** is a function whose domain includes only the first $n$ natural numbers. | Domain: $\{1,\quad 2,\quad 3,\quad 4\}$<br>$\qquad\qquad \downarrow\quad \downarrow\quad \downarrow\quad \downarrow$<br>Range: $\{4,\quad 8,\quad 12,\quad 16\}$<br>The finite sequence is $4, 8, 12, 16$. |
| The **general term of a sequence**, $a_n$, defines the sequence. | Let $a_n = n^2 - 3$. Write the first three terms of this sequence<br><br>$$a_1 = 1^2 - 3 = -2$$<br>$$a_2 = 2^2 - 3 = 1$$<br>$$a_3 = 3^2 - 3 = 6$$<br><br>The first three terms of the sequence are $-2, 1, 6$. |
| An **increasing sequence** is a sequence where each term is greater than the preceding term. | $-2, 5, 7, 11$ is an increasing sequence. |
| A **decreasing sequence** is a sequence where each term is less than the preceding term. | $50, 48, 46, 44$ is a decreasing sequence. |
| A **series** is the sum of the terms of a sequence. A series may be finite or infinite. | If the sequence is $1, 3, 5, 7, 9$, then the series is<br>$1 + 3 + 5 + 7 + 9 = 25$.<br><br>If the sequence is $\dfrac{1}{3}, \dfrac{1}{9}, \dfrac{1}{27}, \cdots, \left(\dfrac{1}{3}\right)^n, \ldots$<br><br>then the series is $\dfrac{1}{3} + \dfrac{1}{9} + \dfrac{1}{27} + \cdots + \left(\dfrac{1}{3}\right)^n + \cdots$ |
| A **partial sum**, $s_n$, of an infinite sequence $a_1, a_2, a_3, \ldots, a_n, \ldots$ is the sum of the first $n$ terms. That is,<br><br>$$s_1 = a_1$$<br>$$s_2 = a_1 + a_2$$<br>$$s_3 = a_1 + a_2 + a_3$$<br>$$\vdots$$<br>$$s_n = a_1 + a_2 + a_3 + \cdots + a_n$$ | Let $a_n = \dfrac{5 + n}{n^2}$. Compute $s_1$ and $s_3$.<br><br>$$s_1 = a_1 = \frac{5 + 1}{1^2} = \frac{6}{1} = 6$$<br><br>$$s_3 = a_1 + a_2 + a_3$$<br>$$= \frac{5 + 1}{1^2} + \frac{5 + 2}{2^2} + \frac{5 + 3}{3^2}$$<br>$$= \frac{6}{1} + \frac{7}{4} + \frac{8}{9} = 8\frac{23}{36}$$ |

| IMPORTANT FACTS AND CONCEPTS | EXAMPLES |
|---|---|

### Section 15.1 (cont.)

A series can be written using **summation notation**:

$$\sum_{i=1}^{n} a_1 = a_1 + a_2 + a_3 + \cdots + a_n.$$

$i$ is the **index of summation**, $n$ is the **upper limit of summation**, and 1 is the **lower limit of summation**.

$$\sum_{i=1}^{4}(3i-7) = (3 \cdot 1 - 7) + (3 \cdot 2 - 7) + (3 \cdot 3 - 7) + (3 \cdot 4 - 7)$$
$$= -4 - 1 + 2 + 5 = 2$$

If $a_n = 6n^2 + 11$, the third partial sum, $s_3$, in summation notation, is written as $\sum_{i=1}^{3}(6i^2 + 11)$.

### Section 15.2

An **arithmetic sequence** is a sequence in which each term after the first differs from the preceding term by a **common difference, $d$**.

| Arithmetic Sequence | Common Difference, $d$ |
|---|---|
| $3, 8, 13, 18, 23, \ldots$ | $d = 8 - 3 = 5$ |
| $20, 14, 8, 2, -4, \ldots$ | $d = 14 - 20 = -6$ |

The **nth term, $a_n$,** of an arithmetic sequence is

$$a_n = a_1 + (n-1)d$$

The $n$th term of the arithmetic sequence with $a_1 = 7$ and $d = -5$ is

$$a_n = 7 + (n-1)(-5)$$
$$= 7 - 5n + 5$$
$$= -5n + 12$$

For this sequence, the 20th term is

$$a_{20} = -5(20) + 12 = -100 + 12$$
$$= -88$$

An **arithmetic series** is the sum of the terms of an arithmetic sequence. The sum of the first $n$ terms, $s_n$, of an arithmetic sequence, also known as the **nth partial sum**, is

$$s_n = a_1 + a_2 + a_3 + \cdots + a_n$$

For an arithmetic series, this sum is determined by the formula

$$s_n = \frac{n(a_1 + a_n)}{2}$$

Find the sum of the first 30 natural numbers. That is, find the sum of

$$1 + 2 + 3 + \cdots + 30$$

Since $a_1 = 1$, $a_{30} = 30$, and $n = 30$, the sum is

$$s_{30} = \frac{30(1 + 30)}{2} = \frac{30(31)}{2} = 465$$

### Section 15.3

A **geometric sequence** is a sequence in which each term after the first term is a common multiple of the preceding term. The common multiple is called the **common ratio, $r$**.

| Geometric Sequence | Common Ratio, $r$ |
|---|---|
| $2, 6, 18, 54, 162, \ldots$ | $r = \frac{6}{2} = 3$ |
| $8, -2, \frac{1}{2}, -\frac{1}{8}, \frac{1}{32}, \ldots$ | $r = \frac{-2}{8} = -\frac{1}{4}$ |

The **nth term, $a_n$,** of a geometric sequence is

$$a_n = a_1 r^{n-1}$$

For the geometric sequence with $a_1 = 5$, $r = \frac{1}{2}$, and $n = 6$, $a_6$ is found as follows.

$$a_6 = 5\left(\frac{1}{2}\right)^{6-1} = 5\left(\frac{1}{2}\right)^5 = \frac{5}{32}$$

A **geometric series** is the sum of the terms of a geometric sequence. The sum of the first $n$ terms, $s_n$, of a geometric sequence, also known as the **nth partial sum**, is

$$s_n = a_1 + a_2 + a_3 + \cdots + a_n$$

For a geometric series, this sum is determined by the formula

$$s_n = \frac{a_1(1 - r^n)}{1 - r}, r \neq 1$$

To find the sum of the first six terms of a geometric sequence with $a_1 = 12$ and $r = \frac{1}{3}$, use the formula with $n = 6$ to obtain

$$s_6 = \frac{12\left[1 - \left(\frac{1}{3}\right)^6\right]}{1 - \frac{1}{3}} = \frac{12\left[1 - \frac{1}{729}\right]}{\frac{2}{3}} = \frac{12\left(\frac{728}{729}\right)}{\frac{2}{3}}$$

$$= 12\left(\frac{728}{729}\right)\left(\frac{3}{2}\right) = \frac{1456}{81} \text{ or } 17\frac{79}{81}$$

| IMPORTANT FACTS AND CONCEPTS | EXAMPLES |
|---|---|

**Section 15.3 (cont.)**

The sum of an infinite geometric series is

$$s_\infty = \frac{a_1}{1 - r} \text{ where } |r| < 1$$

To find the sum of the infinite series $4 - 2 + 1 - \frac{1}{2} + \frac{1}{4} + \cdots$,

use the formula with $a_1 = 4$ and $r = -\frac{1}{2}$ to obtain

$$s_\infty = \frac{4}{1 - \left(-\frac{1}{2}\right)} = \frac{4}{\frac{3}{2}} = 4 \cdot \frac{2}{3} = \frac{8}{3} \text{ or } 2\frac{2}{3}$$

**Section 15.4**

**$n$ Factorial**

$$n! = n(n - 1)(n - 2)(n - 3) \cdots (1)$$

for any positive integer $n$.

Note that 0! is defined to be 1.

$5! = 5 \cdot 4 \cdot 3 \cdot 2 \cdot 1 = 120$

$8! = 8 \cdot 7 \cdot 6 \cdot 5 \cdot 4 \cdot 3 \cdot 2 \cdot 1 = 40,320$

**Binomial Coefficients**

For $n$ and $r$ non-negative integers, $n \geq r$,

$$\binom{n}{r} = \frac{n!}{r! \cdot (n - r)!}$$

$$\binom{n}{n} = 1 \text{ and } \binom{n}{0} = 1$$

$$\binom{7}{3} = \frac{7!}{3! \cdot (7 - 3)!} = \frac{7!}{3! \cdot 4!} = \frac{7 \cdot 6 \cdot 5 \cdot 4 \cdot 3 \cdot 2 \cdot 1}{3 \cdot 2 \cdot 1 \cdot 4 \cdot 3 \cdot 2 \cdot 1} = 35$$

$$\binom{10}{10} = 1, \binom{10}{0} = 1$$

**Binomial Theorem**

For any positive integer $n$,

$$(a + b)^n = \binom{n}{0}a^n b^0 + \binom{n}{1}a^{n-1}b^1 + \binom{n}{2}a^{n-2}b^2 +$$
$$\binom{n}{3}a^{n-3}b^3 + \cdots + \binom{n}{n}a^0 b^n$$

$$(x + 2y)^4 = \binom{4}{0}x^4 + \binom{4}{1}x^3(2y) + \binom{4}{2}x^2(2y)^2$$
$$+ \binom{4}{3}x(2y)^3 + \binom{4}{4}(2y)^4$$
$$= 1 \cdot x^4 + 4 \cdot x^3(2y) + 6 \cdot x^2(4y^2) + 4 \cdot x(8y^3) + 1 \cdot 16y^4$$
$$= x^4 + 8x^3 y + 24x^2 y^2 + 32xy^3 + 16y^4$$

## Chapter 15 Review Exercises

[15.1]   *Write the first five terms of each sequence.*

**1.** $a_n = n + 5$
**2.** $a_n = n^2 + n - 3$
**3.** $a_n = \dfrac{12}{n}$
**4.** $a_n = \dfrac{n^2}{n + 4}$

*Find the indicated term of each sequence.*

**5.** $a_n = 3n - 10$, seventh term
**6.** $a_n = (-1)^n + 5$, tenth term

**7.** $a_n = \dfrac{n + 17}{n^2}$, ninth term
**8.** $a_n = (n)(n - 3)$, eleventh term

*For each sequence, find the first and third partial sums, $s_1$ and $s_3$.*

**9.** $a_n = 2n + 5$
**10.** $a_n = n^2 + 11$

**11.** $a_n = \dfrac{n + 3}{n + 2}$
**12.** $a_n = (-1)^n(n + 8)$

*Write the next three terms of each sequence. Then write an expression for the general term, $a_n$.*

**13.** $2, 4, 8, 16, \ldots$
**14.** $-27, 9, -3, 1, \ldots$

**15.** $\dfrac{1}{9}, \dfrac{2}{9}, \dfrac{4}{9}, \dfrac{8}{9}, \ldots$
**16.** $13, 9, 5, 1, \ldots$

*Write out each series. Then find the sum of the series.*

**17.** $\displaystyle\sum_{i=1}^{3} i^2 + 10$
**18.** $\displaystyle\sum_{i=1}^{4} i(i + 5)$

**19.** $\displaystyle\sum_{i=1}^{5} \dfrac{i^2}{6}$
**20.** $\displaystyle\sum_{i=1}^{4} \dfrac{i}{i + 1}$

*For the set of values $x_1 = 3$, $x_2 = 9$, $x_3 = 7$, $x_4 = 10$, evaluate the indicated sum.*

**21.** $\displaystyle\sum_{i=1}^{4} x_i$

**22.** $\displaystyle\sum_{i=1}^{4} (x_i)^2$

**23.** $\displaystyle\sum_{i=2}^{3} (x_i^2 + 1)$

**24.** $\displaystyle\left(\sum_{i=1}^{4} x_i\right)^2$

*In Exercises 25 and 26, consider the following rectangles. For the nth rectangle, the length is n + 3 and the width is n.*

**25. Perimeter**

   **a)** Find the perimeters for the first four rectangles, and then list the perimeters in a sequence.

   **b)** Find the general term for the perimeter of the *n*th rectangle in the sequence. Use $p_n$ for perimeter.

**26. Area**

   **a)** Find the areas for the first four rectangles, and then list the areas in a sequence.

   **b)** Find the general term for the area of the *n*th rectangle in the sequence. Use $a_n$ for area.

**[15.2]**   *Write the first five terms of the arithmetic sequence with the indicated first term and common difference.*

**27.** $a_1 = 5$, $d = 3$

**28.** $a_1 = 5$, $d = -\dfrac{1}{3}$

**29.** $a_1 = \dfrac{1}{2}$, $d = -2$

**30.** $a_1 = -20$, $d = \dfrac{1}{5}$

*For each arithmetic sequence, find the indicated value.*

**31.** $a_1 = 6$, $d = 3$; find $a_9$

**32.** $a_1 = 26$, $a_8 = -9$; find $d$

**33.** $a_1 = -3$, $a_{11} = 2$; find $d$

**34.** $a_1 = 22$, $a_n = -3$, $d = -5$; find $n$

*Find $s_n$ and d for each arithmetic sequence.*

**35.** $a_1 = 7$, $a_8 = 21$, $n = 8$

**36.** $a_1 = -2$, $a_7 = -38$, $n = 7$

**37.** $a_1 = \dfrac{3}{5}$, $a_6 = \dfrac{13}{5}$, $n = 6$

**38.** $a_1 = -\dfrac{10}{3}$, $a_9 = -6$, $n = 9$

*Write the first four terms of each arithmetic sequence. Then find $a_{10}$ and $s_{10}$.*

**39.** $a_1 = -7$, $d = 4$

**40.** $a_1 = 8$, $d = -3$

**41.** $a_1 = \dfrac{5}{6}$, $d = \dfrac{2}{3}$

**42.** $a_1 = -60$, $d = 5$

*Find the number of terms in each arithmetic sequence. Then find $s_n$.*

**43.** $4, 9, 14, \ldots, 64$

**44.** $-7, -4, -1, \ldots, 14$

**45.** $\dfrac{6}{10}, \dfrac{9}{10}, \dfrac{12}{10}, \ldots, \dfrac{36}{10}$

**46.** $-9, -3, 3, 9, \ldots, 45$

**[15.3]**   *Determine the first five terms of each geometric sequence.*

**47.** $a_1 = 6$, $r = 2$

**48.** $a_1 = -56$, $r = \dfrac{1}{2}$

**49.** $a_1 = 20$, $r = -\dfrac{2}{3}$

**50.** $a_1 = -20$, $r = \dfrac{1}{5}$

*Find the indicated term of each geometric sequence.*

**51.** $a_1 = 12$, $r = \dfrac{1}{3}$; find $a_5$

**52.** $a_1 = 15$, $r = 2$; find $a_6$

**53.** $a_1 = -8$, $r = -3$; find $a_4$

**54.** $a_1 = \dfrac{1}{12}$, $r = \dfrac{2}{3}$; find $a_5$

*Find each sum.*

**55.** $a_1 = 7$, $r = 2$; find $s_6$

**56.** $a_1 = -84$, $r = -\dfrac{1}{4}$; find $s_5$

**57.** $a_1 = 9$, $r = \dfrac{3}{2}$; find $s_4$

**58.** $a_1 = 8$, $r = \dfrac{1}{2}$; find $s_7$

*For each geometric sequence, find the common ratio, r, and then write an expression for the general term, $a_n$.*

**59.** $6, 12, 24, \ldots$

**60.** $-6, -30, -150, \ldots$

**61.** $10, \dfrac{10}{3}, \dfrac{10}{9}, \ldots$

**62.** $\dfrac{9}{5}, \dfrac{18}{15}, \dfrac{36}{45}, \ldots$

*Find the sum of the terms in each infinite geometric sequence.*

**63.** $5, \dfrac{5}{2}, \dfrac{5}{4}, \dfrac{5}{8}, \ldots$

**64.** $\dfrac{7}{2}, 1, \dfrac{2}{7}, \dfrac{4}{49}, \ldots$

**65.** $-8, \dfrac{8}{3}, -\dfrac{8}{9}, \dfrac{8}{27}, \ldots$

**66.** $-6, -4, -\dfrac{8}{3}, -\dfrac{16}{9}, \ldots$

*Find the sum of each infinite series.*

**67.** $16 + 8 + 4 + 2 + 1 + \cdots$

**68.** $11 + \dfrac{11}{3} + \dfrac{11}{9} + \dfrac{11}{27} + \cdots$

**69.** $5 - 1 + \dfrac{1}{5} - \dfrac{1}{25} + \cdots$

**70.** $-4, -\dfrac{8}{3}, -\dfrac{16}{9}, -\dfrac{32}{27}, \ldots$

*Write the repeating decimal as a ratio of integers.*

**71.** $0.363636\ldots$

**72.** $0.783783783\ldots$

[15.4]   *Use the binomial theorem to expand the expression.*

**73.** $(3x + y)^4$

**74.** $(2x - 3y^2)^3$

*Write the first four terms of the expansion.*

**75.** $(x - 2y)^9$

**76.** $(2a^2 + 3b)^8$

[15.2]

**77. Sum of Integers** Find the sum of the integers between 101 and 200, inclusive.

**78. Barrels of Oil** Barrels of oil are stacked with 21 barrels in the bottom row, 20 barrels in the second row, 19 barrels in the third row, and so on, to the top row, which has only 1 barrel. How many barrels are there?

**79. Salary** Ahmed Mocanda just started a new job with an annual salary of $36,000. He has been told that his salary will increase by $1000 per year for the next 10 years.

**a)** Write a sequence showing his salary for the first 4 years.

**b)** Write a general term of this sequence.

**c)** What will his salary be 6 years from now?

**d)** How much money will he make in the first 11 years?

[15.3]

**80. Money** You begin with $50, double that to get $100, double that again to get $200, and so on. How much will you have after you perform this process 10 times?

**81. Salary** Gertude Dibble started a new job on January 1, 2010, with a monthly salary of $1600. Her boss has agreed to give her a 4% raise each month for the remainder of the year.

**a)** What is Gertude's salary in July?

**b)** What is Gertude's salary in December?

**c)** How much money does Gertude make in 2010?

**82. Inflation** If the inflation rate was a constant 8% per year (each year the cost of living is 8% greater than the previous year), how much would a product that costs $200 now cost after 12 years?

**83. Pendulum** On each swing (left to right or right to left), a pendulum travels 92% as far as on its previous swing. If the first swing is 12 feet, find the distance traveled by the pendulum by the time it comes to rest.

## Chapter 15 Practice Test

*Chapter Test Prep Videos provide fully worked-out solutions to any of the exercises you want to review. Chapter Test Prep Videos are available via MyMathLab®, or on YouTube (search "Angel Elementary and Intermediate Algebra" and click on "Channels")*

**1.** What is a series?

**2. a)** What is an arithmetic sequence?

   **b)** What is a geometric sequence?

**3.** Write the first five terms of the sequence if $a_n = \dfrac{n-2}{3n}$.

**4.** Find the first and third partial sums if $a_n = \dfrac{2n+1}{n^2}$.

**5.** Write out the following series and find the sum of the series.

$$\sum_{i=1}^{5} (2i^2 + 3)$$

**6.** For $x_1 = 4, x_2 = 2, x_3 = 8,$ and $x_4 = 10$ find $\displaystyle\sum_{i=1}^{4} (x_i)^2$.

7. Write the general term for the following arithmetic sequence.

$$\frac{1}{3}, \frac{2}{3}, \frac{3}{3}, \frac{4}{3}, \ldots$$

8. Write the general term for the following geometric sequence.

$$5, 10, 20, 40, \ldots$$

*In Exercises 9 and 10, write the first four terms of each sequence.*

9. $a_1 = 15, d = -6$

10. $a_1 = \frac{5}{12}, r = \frac{2}{3}$

11. Find $a_{11}$ when $a_1 = 40$ and $d = -8$.

12. Find $s_8$ for the arithmetic sequence with $a_1 = 7$ and $a_8 = -12$.

13. Find the number of terms in the arithmetic sequence $-4, -16, -28, \ldots, -136$.

14. Find $a_6$ when $a_1 = 8$ and $r = \frac{2}{3}$.

15. Find $s_7$ when $a_1 = \frac{3}{5}$ and $r = -5$.

16. Find the common ratio and write an expression for the general term of the sequence $15, 5, \frac{5}{3}, \frac{5}{9}, \ldots$.

17. Find the sum of the following infinite geometric series.

$$4 + \frac{8}{3} + \frac{16}{9} + \frac{32}{27} + \cdots$$

18. Write $0.3939\ldots$ as a ratio of integers.

19. Evaluate $\binom{8}{3}$.

20. Use the binomial theorem to expand $(x + 2y)^4$.

21. **Arithmetic Mean** Paul Misselwitz's five test scores are 76, 93, 83, 87, and 71. Use $\bar{x} = \frac{\Sigma x}{n}$ to find the arithmetic mean of Paul's scores.

22. **A Pile of Logs** Logs are piled with 13 logs in the bottom row, 12 logs in the second row, 11 logs in the third row, and so on, to the top. How many logs are there?

23. **Saving for Retirement** To save for retirement, Jamie Monroe plans to save $1000 the first year, $2000 the second year, $3000 the third year, and to increase the amount saved by $1000 in each successive year. How much will she have saved by the end of her twentieth year of savings?

24. **Earnings** Yolanda Rivera makes $700 per week working at an insurance office. Her boss has guaranteed her an increase of 4% per week for the next 7 weeks. How much is she making in the sixth week?

25. **Culture of Bacteria** The number of bacteria in a culture is tripling every hour. If there are initially 500 bacteria in the culture, how many bacteria will be in the culture by the end of the sixth hour?

## Cumulative Review Test

*Take the following test and check your answers with those given in the back of the book. Review any questions that you answered incorrectly. The section where the material was covered is indicated after the answer.*

1. Solve $A = \frac{1}{2}bh$, for $b$.

2. Find an equation of the line through $(4, -2)$ and $(1, 9)$. Write the equation in slope-intercept form.

3. Multiply $(5x^3 + 4x^2 - 6x + 2)(x + 5)$.

4. Factor $x^3 + 2x - 6x^2 - 12$.

5. Factor $(a + b)^2 + 8(a + b) + 16$.

6. Subtract $5 - \frac{x - 1}{x^2 + 3x - 10}$.

7. $y$ varies directly as the square of $z$. If $y$ is 80 when $z$ is 20, find $y$ when $z$ is 50.

8. Solve the system of equations.

$$x + y + z = 1$$
$$2x + 2y + 2z = 2$$
$$3x + 3y + 3z = 3$$

9. If $f(x) = 2\sqrt[3]{x - 3}$ and $g(x) = \sqrt[3]{5x - 15}$, find all values of $x$ for which $f(x) = g(x)$.

10. Solve $\sqrt{6x - 5} - \sqrt{2x + 6} - 1 = 0$.

11. Solve by completing the square.

$$x^2 + 2x + 15 = 0$$

12. Solve by the quadratic formula.

$$x^2 - \frac{x}{5} - \frac{1}{3} = 0$$

13. **Numbers** Twice the square of a positive number decreased by nine times the number is 5. Find the number.

14. Graph $y = x^2 - 4x$ and label the vertex.

15. Solve $\log_a \frac{1}{64} = 6$ for $a$.

16. Graph $y = 2^x - 1$.

17. Find an equation of a circle with center at $(-6, 2)$ and radius 7.

18. Graph $(x + 3)^2 + (y + 1)^2 = 16$.

19. Graph $9x^2 + 16y^2 = 144$.

20. Find the sum of the infinite geometric series.

$$6 + 4 + \frac{8}{3} + \frac{16}{9} + \frac{32}{27} + \cdots$$

# Appendices

## Appendix A  Review of Decimals and Percent

### Decimals

> **To Add or Subtract Numbers Containing Decimal Points**
>
> 1. Align the numbers by the decimal points.
> 2. Add or subtract the numbers as if they were whole numbers.
> 3. Place the decimal point in the sum or difference directly below the decimal points in the numbers being added or subtracted.

EXAMPLE 1  Add 4.6 + 13.813 + 9.02.

*Solution*

$$
\begin{array}{r}
4.600 \\
13.813 \\
+\ 9.020 \\
\hline
27.433
\end{array}
$$

EXAMPLE 2  Subtract 3.062 from 34.9.

*Solution*

$$
\begin{array}{r}
34.900 \\
-\ 3.062 \\
\hline
31.838
\end{array}
$$

> **To Multiply Numbers Containing Decimal Points**
>
> 1. Multiply as if the factors were whole numbers.
> 2. Determine the total number of digits to the right of the decimal points in the factors.
> 3. Place the decimal point in the product so that the product contains the same number of digits to the right of the decimal as the total found in step 2. For example, if there are a total of three digits to the right of the decimal points in the factors, there must be three digits to the right of the decimal point in the product.

EXAMPLE 3  Multiply 2.34 × 1.9.

*Solution*

$$
\begin{array}{r}
2.34 \\
\times\ \ 1.9 \\
\hline
2106 \\
234\ \ \\
\hline
4.446
\end{array}
$$

2.34 ⟵ two digits to the right of the decimal point

× 1.9 ⟵ one digit to the right of the decimal point

4.446 ⟵ three digits to the right of the decimal point in the product

EXAMPLE 4  Multiply $2.13 \times 0.02$.

*Solution*
$$
\begin{array}{r}
2.13 \\
\times\ 0.02 \\
\hline
0.0426
\end{array}
$$

2.13 ⟵ two digits to the right of the decimal point

× 0.02 ⟵ two digits to the right of the decimal point

0.0426 ⟵ four digits to the right of the decimal point in the product

Note that it was necessary to add a zero preceding the digit 4 in the answer in order to have four digits to the right of the decimal point.

### To Divide Numbers Containing Decimal Points

1. Multiply both the dividend and divisor by a power of 10 that will make the divisor a whole number.
2. Divide as if working with whole numbers.
3. Place the decimal point in the quotient directly above the decimal point in the dividend.

To make the divisor a whole number, multiply *both* the dividend and divisor by 10 if the divisor is given in tenths, by 100 if the divisor is given in hundredths, by 1000 if the divisor is given in thousandths, and so on. Multiplying both the numerator and denominator by the same nonzero number is the same as multiplying the fraction by 1. Therefore, the value of the fraction is unchanged.

EXAMPLE 5  Divide $\dfrac{1.956}{0.12}$.

*Solution*  Since the divisor, 0.12, is twelve-hundredths, we multiply both the divisor and dividend by 100.

$$
\frac{1.956}{0.12} \times \frac{100}{100} = \frac{195.6}{12.}
$$

Now we divide.

$$
\begin{array}{r}
16.3 \\
12\overline{)195.6} \\
\underline{12\phantom{..}} \\
75 \\
\underline{72} \\
3\,6 \\
\underline{3\,6} \\
0
\end{array}
$$

The decimal point in the answer is placed directly above the decimal point in the dividend. Thus, $\dfrac{1.956}{0.12} = 16.3$.

EXAMPLE 6  Divide 0.26 by 10.4.

*Solution*  First, multiply both the dividend and divisor by 10.

$$
\frac{0.26}{10.4} \times \frac{10}{10} = \frac{2.6}{104.}
$$

Now divide.

$$
\begin{array}{r}
0.025 \\
104\overline{)2.600} \\
\underline{2\,08} \\
520 \\
\underline{520} \\
0
\end{array}
$$

Note that a zero had to be placed before the digit 2 in the quotient.

$$\frac{0.26}{10.4} = 0.025$$

## Rounding Decimal Numbers

Now we will explain how to round decimal numbers. The explanation of the procedure will refer to the positional values to the right of the decimal point, as illustrated here:

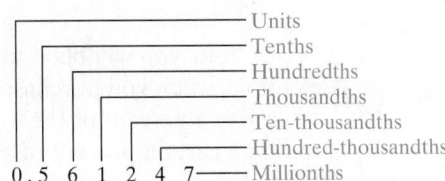

### To Round a Decimal Number

1. Identify the number in the problem to the right of the positional value to which the number is to be rounded. For example, if you wish to round to tenths, you identify the number in the hundredths position. If you wish to round to hundredths, you identify the number in the thousandths position, and so on.

2. **a)** If the number you identified in step 1 is greater than or equal to 5, the number in the preceding position is increased by 1 unit, and all numbers to the right of the number that has been increased by 1 unit are eliminated.

   **b)** If the number you identified in step 1 is less than 5, the number you identified in step 1 and all numbers to its right are eliminated.

**EXAMPLE 7**  Round 4.863 to tenths.

*Solution*   Since we are rounding to tenths, we identify the number in the hundredths position as 6. Because the number 6 is greater than or equal to 5, we increase the number in the preceding position by 1 unit and drop the remaining digits. Therefore, 4.863 when rounded to tenths is 4.9.

**EXAMPLE 8**  Round 5.4738 to hundredths.

*Solution*   Since we are rounding to hundredths, we identify the number in the thousandths position as 3. Because the number 3 is less than 5, we drop the 3 and all numbers to the right of 3. Therefore, 5.4738 when rounded to hundredths is 5.47.

## Percent

The word *percent* means "per hundred." The symbol % means percent. One percent means "one per hundred."

### One Percent

$$1\% = \frac{1}{100} \quad \text{or} \quad 1\% = 0.01$$

**EXAMPLE 9**  Convert 16% to a decimal.

*Solution*   Since 1% = 0.01,

$$16\% = 16(0.01) = 0.16$$

**EXAMPLE 10** Convert 4.7% to a decimal.

*Solution*   4.7% = 4.7(0.01) = 0.047

**EXAMPLE 11** Convert 1.14 to a percent.

*Solution*   To change a decimal number to a percent, we multiply the number by 100%.

$$1.14 = 1.14 \times 100\% = 114\%$$

Often, you will need to find an amount that is a certain percent of a number. For example, when you purchase an item in a state or county that has a sales tax you must often pay a percent of the item's price as the sales tax. Examples 12 and 13 show how to find a certain percent of a number.

**EXAMPLE 12** Find 32% of 300.

*Solution*   To find a percent of a number, use multiplication. Change 32% to a decimal number, then multiply by 300.

$$(0.32)(300) = 96$$

Thus, 32% of 300 is 96.

**EXAMPLE 13** Johnson County charges an 8% sales tax.

**a)** Find the sales tax on a stereo system that cost $580.
**b)** Find the total cost of the system, including tax.

*Solution*

**a)** The sales tax is 8% of 580.

$$(0.08)(580) = 46.40$$

The sales tax is $46.40.

**b)** The total cost is the purchase price plus the sales tax:

$$\text{total cost} = \$580 + \$46.40 = \$626.40$$

# Appendix B Finding the Greatest Common Factor and Least Common Denominator

### Prime Factorization

In Section 1.3, we mentioned that to simplify fractions you can divide both the numerator and denominator by the *greatest common factor* (GCF). One method to find the GCF is to use *prime factorization*. **Prime factorization** is the process of writing a given number as a product of prime numbers. **Prime numbers** are natural numbers, excluding 1, that can be divided by only themselves and 1. The first ten prime numbers are 2, 3, 5, 7, 11, 13, 17, 19, 23, and 29. Can you find the next prime number? If you answered 31, you answered correctly.

To write a number as a product of primes, we can use a *tree diagram*. Begin by selecting any two numbers whose product is the given number. Then continue factoring each of these numbers into prime numbers, as shown in Example 1.

**EXAMPLE 1** Determine the prime factorization of the number 120.

*Solution*    We will use three different tree diagrams to illustrate the prime factorization of 120.

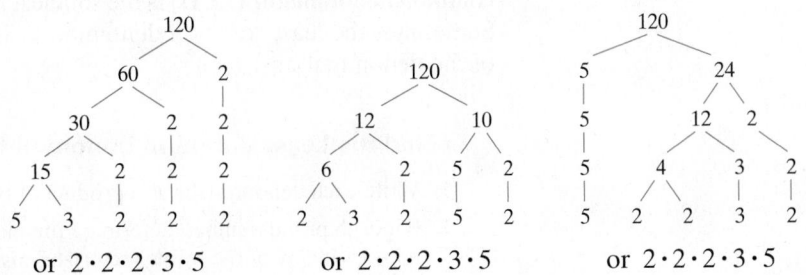

or $2 \cdot 2 \cdot 2 \cdot 3 \cdot 5$          or $2 \cdot 2 \cdot 2 \cdot 3 \cdot 5$          or $2 \cdot 2 \cdot 2 \cdot 3 \cdot 5$

Note that no matter how you start, if you do not make a mistake, you find that the prime factorization of 120 is $2 \cdot 2 \cdot 2 \cdot 3 \cdot 5$. There are other ways 120 can be factored but all will lead to the prime factorization $2 \cdot 2 \cdot 2 \cdot 3 \cdot 5$.

## Greatest Common Factor

The **greatest common factor (GCF)** of two natural numbers is the greatest integer that is a factor of both numbers. We use the GCF when simplifying fractions.

> **To Find the Greatest Common Factor of a Given Numerator and Denominator**
>
> 1. Write both the numerator and the denominator as a product of primes.
> 2. Determine all the prime factors that are common to both prime factorizations.
> 3. Multiply the prime factors found in step 2 to obtain the GCF.

**EXAMPLE 2** Consider the fraction $\dfrac{108}{156}$.

**a)** Find the GCF of 108 and 156.          **b)** Simplify $\dfrac{108}{156}$.

*Solution*

**a)** First determine the prime factorizations of both 108 and 156.

There are two 2s and one 3 common to both prime factorizations; thus

$$\text{GCF} = 2 \cdot 2 \cdot 3 = 12$$

The greatest common factor of 108 and 156 is 12. Twelve is the greatest integer that divides into both 108 and 156.

**b)** To simplify $\dfrac{108}{156}$, we divide both the numerator and denominator by the GCF, 12.

$$\frac{108 \div 12}{156 \div 12} = \frac{9}{13}$$

Thus, $\dfrac{108}{156}$ simplifies to $\dfrac{9}{13}$.

## Least Common Denominator

When adding two or more fractions, you must write each fraction with a common denominator. The best denominator to use is the *least common denominator*. The **least common denominator (LCD)** is the smallest number that each denominator divides into. Sometimes the least common denominator is referred to as the **least common multiple** of the denominators.

> **To Find the Least Common Denominator of Two or More Fractions**
>
> 1. Write each denominator as a product of prime numbers.
> 2. For each prime number, determine the maximum number of times that prime number appears in any of the prime factorizations.
> 3. Multiply all the prime numbers found in step 2. Include each prime number the maximum number of times it appears in any of the prime factorizations. The product of all these prime numbers will be the LCD.

Example 3 illustrates the procedure to determine the LCD.

**EXAMPLE 3** Consider $\dfrac{7}{108} + \dfrac{5}{156}$.

**a)** Determine the least common denominator.

**b)** Add the fractions.

*Solution*

**a)** We found in Example 2 that

$$108 = 2 \cdot 2 \cdot 3 \cdot 3 \cdot 3 \quad \text{and} \quad 156 = 2 \cdot 2 \cdot 3 \cdot 13$$

We can see that the maximum number of 2s that appear in either prime factorization is two (there are two 2s in both factorizations), the maximum number of 3s is three, and the maximum number of 13s is one. Multiply as follows:

$$2 \cdot 2 \cdot 3 \cdot 3 \cdot 3 \cdot 13 = 1404$$

Thus, the least common denominator is 1404. This is the smallest number that both 108 and 156 divide into.

**b)** To add the fractions, we need to write both fractions with a common denominator. The best common denominator to use is the LCD. Since $1404 \div 108 = 13$, we will multiply $\dfrac{7}{108}$ by $\dfrac{13}{13}$. Since $1404 \div 156 = 9$, we will multiply $\dfrac{5}{156}$ by $\dfrac{9}{9}$.

$$\frac{7}{108} \cdot \frac{13}{13} + \frac{5}{156} \cdot \frac{9}{9} = \frac{91}{1404} + \frac{45}{1404} = \frac{136}{1404} = \frac{34}{351}$$

Thus, $\dfrac{7}{108} + \dfrac{5}{156} = \dfrac{34}{351}$.

# Appendix C Geometry

This appendix introduces or reviews important geometric concepts. **Table C.1** gives the names and descriptions of various types of angles.

## Angles

**TABLE C.1**

| Angle | Sketch of Angle |
|---|---|
| An **acute angle** is an angle whose measure is between 0° and 90°. | |
| A **right angle** is an angle whose measure is 90°. | |
| An **obtuse angle** is an angle whose measure is between 90° and 180°. | |
| A **straight angle** is an angle whose measure is 180°. | |
| Two angles are **complementary angles** when the sum of their measures is 90°. Each angle is the complement of the other. Angles $A$ and $B$ are complementary angles. | 60° $A$ 30° $B$ |
| Two angles are **supplementary angles** when the sum of their measures is 180°. Each angle is the supplement of the other. Angles $A$ and $B$ are supplementary angles. | 130° $A$ 50° $B$ |

When two lines intersect, four angles are formed as shown in **Figure C.1**. The pair of opposite angles formed by the intersecting lines are called **vertical angles**.

Angles 1 and 3 are vertical angles. Angles 2 and 4 are also vertical angles. *Vertical angles have equal measures.* Thus, angle 1, symbolized by $\angle 1$, is equal to angle 3, symbolized by $\angle 3$. We can write $\angle 1 = \angle 3$. Similarly, $\angle 2 = \angle 4$.

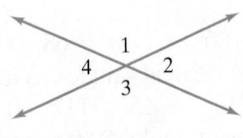

**FIGURE C.1**

## Parallel and Perpendicular Lines

**Parallel lines** are two lines in the same plane that do not intersect (**Fig. C.2**). **Perpendicular lines** are lines that intersect at right angles (**Fig. C.3**).

Parallel lines

Perpendicular lines

**FIGURE C.2**              **FIGURE C.3**

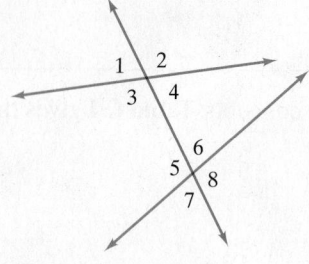

FIGURE C.4

A **transversal** is a line that intersects two or more lines at different points. When a transversal line intersects two other lines, eight angles are formed, as illustrated in **Figure C.4**. Some of these angles are given special names.

**Interior angles:** 3, 4, 5, 6

**Exterior angles:** 1, 2, 7, 8

**Pairs of corresponding angles:** 1 and 5; 2 and 6; 3 and 7; 4 and 8

**Pairs of alternate interior angles:** 3 and 6; 4 and 5

**Pairs of alternate exterior angles:** 1 and 8; 2 and 7

---

**Parallel Lines Cut by a Transversal**

When two parallel lines are cut by a transversal,

1. Corresponding angles are equal ($\angle 1 = \angle 5, \angle 2 = \angle 6, \angle 3 = \angle 7, \angle 4 = \angle 8$).
2. Alternate interior angles are equal ($\angle 3 = \angle 6, \angle 4 = \angle 5$).
3. Alternate exterior angles are equal ($\angle 1 = \angle 8, \angle 2 = \angle 7$).

---

EXAMPLE 1   If line 1 and line 2 are parallel lines and the measure of angle 1 is 112° ($m\angle 1 = 112°$), find the measure of angles 2 through 8.

*Solution*   Angles 1 and 2 are supplementary, so $m\angle 2$ is $180° - 112° = 68°$. The measures of angles 1 and 4 are equal since they are vertical angles. Thus, $m\angle 4 = 112°$. Angles 1 and 5 are corresponding angles. Thus, $m\angle 5 = 112°$. It is equal to its vertical angle, $\angle 8$, so $m\angle 8 = 112°$. The measures of angles 2, 3, 6, and 7 are all equal and measure 68°.

## Polygons

A **polygon** is a closed figure in a plane determined by three or more line segments. Some polygons are illustrated in **Figure C.5**.

A **regular polygon** has sides that are all the same length, and interior angles that all have the same measure. In **Figure C.5**, (b) and (d) are regular polygons.

(a)          (b)          (c)          (d)

FIGURE C.5

## Sum of the Interior Angles of a Polygon

The sum of the interior angles of a polygon can be found by the formula

$$\text{Sum} = (n - 2)180°$$

where $n$ is the number of sides of the polygon.

EXAMPLE 2   Find the sum of the measures of the interior angles of **a)** a triangle; **b)** a quadrilateral (4 sides); **c)** an octagon (8 sides).

*Solution*

**a)** Since $n = 3$, we write

$$\text{Sum} = (n - 2)180°$$
$$= (3 - 2)180° = 1(180°) = 180°$$

The sum of the measures of the interior angles in a triangle is 180°.

**b)**
$$\text{Sum} = (n - 2)180°$$
$$= (4 - 2)180° = 2(180°) = 360°$$

The sum of the measures of the interior angles in a quadrilateral is 360°.

**c)**
$$\text{Sum} = (n - 2)(180°) = (8 - 2)180° = 6(180°) = 1080°$$

The sum of the measures of the interior angles in an octagon is 1080°.

Now we will briefly define several types of triangles in **Table C.2**.

## Triangles

| TABLE C.2 | |
|---|---|
| **Triangle** | **Sketch of Triangle** |
| An **acute triangle** is one that has three acute angles (angles of less than 90°). |  |
| An **obtuse triangle** has one obtuse angle (an angle greater than 90°). |  |
| A **right triangle** has one right angle (an angle equal to 90°). The longest side of a right triangle is opposite the right angle and is called the **hypotenuse**. The other two sides are called the **legs**. |  |
| An **isosceles triangle** has two sides of equal length. The angles opposite the equal sides have the same measure. |  |
| An **equilateral triangle** has three sides of equal length. It also has three equal angles that measure 60° each. |  |

When two sides of a *right triangle* are known, the third side can be found using the **Pythagorean Theorem**, $a^2 + b^2 = c^2$, where $a$ and $b$ are the legs and $c$ is the hypotenuse of the triangle. (See Section 6.7 for examples.)

## Congruent and Similar Figures

If two triangles are **congruent**, it means that the two triangles are identical in size and shape. Two congruent triangles would match up exactly if one were placed on the other.

### Two Triangles Are Congruent If Any One of the Following Statements Is True

1. Two angles of one triangle are equal to two corresponding angles of the other triangle, and the lengths of the sides between each pair of angles are equal. This method of showing that triangles are congruent is called the *angle, side, angle* method.

2. Corresponding sides of both triangles are equal. This is called the *side, side, side* method.

3. Two corresponding pairs of sides are equal, and the angle between them is equal. This is referred to as the *side, angle, side* method.

EXAMPLE 3  Determine whether the two triangles are congruent.

*Solution*   The unknown angle in the figure on the right must measure 100° since the sum of the angles of a triangle is 180°. Both triangles have the same two angles (100° and 50°), with the same length side between them, 8 units. Thus, these two triangles are congruent by the angle, side, angle method.

Two triangles are **similar** if all three pairs of corresponding angles are equal and corresponding sides are in proportion. Similar figures do not have to be the same size but must have the same general shape.

### Two Triangles Are Similar If Any One of the Following Statements Is True

**1.** Two angles of one triangle equal two angles of the other triangle.

$$\angle 1 = \angle 4$$
$$\angle 2 = \angle 5$$
$$(\text{also } \angle 3 = \angle 6)$$

**2.** Corresponding sides of the two triangles are proportional.

$$\frac{a}{x} = \frac{b}{y} = \frac{c}{z}$$

**3.** Two pairs of corresponding sides are proportional, and the angles between them are equal.

$$\frac{a}{x} = \frac{b}{y}$$
$$\text{and } \angle 1 = \angle 2$$

**EXAMPLE 4** Are the triangles $ABC$ and $AB'C'$ similar?

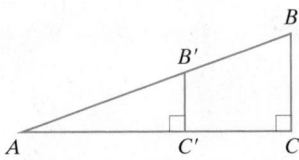

*Solution*   Angle $A$ is common to both triangles. Since angle $C$ and angle $C'$ are equal (both 90°), then $\angle B$ and $\angle B'$ must be equal. Since the three angles of triangle $ABC$ equal the three angles of triangle $AB'C'$, the two triangles are similar.

# Appendix D Review of Exponents, Polynomials, and Factoring

In Chapters 5 and 6 we discussed exponents, polynomials, and factoring. We review these concepts briefly in this Appendix. For a more complete discussion, please see Chapters 5 and 6.

## Exponents

Exponents represent repeated multiplication. For instance, $5 \cdot 5 \cdot 5$ can be written as $5^3$. In the expression $5^3$, the 5 is called the base and the 3 is called the exponent. The number $5^3$ is read "5 cubed" or "5 to the third power" and means $5 \cdot 5 \cdot 5 = 125$. In general, an expression with $n$ factors of $b$ can be written as $b^n$.

In Chapter 5 we discussed the rules of exponents. We summarize them below.

### Summary of the Rules of Exponents

1. **Product Rule**          $x^m \cdot x^n = x^{m+n}$

2. **Quotient Rule**          $\dfrac{x^m}{x^n} = x^{m-n}, \quad x \neq 0$

3. **Zero Exponent Rule**          $x^0 = 1, \quad x \neq 0$

4. **Power Rule**          $(x^m)^n = x^{m \cdot n}$

5. **Power of a Product Rule**          $(xy)^n = x^n y^n$

6. **Power of a Quotient Rule**          $\left(\dfrac{x}{y}\right)^n = \dfrac{x^n}{y^n}, \quad y \neq 0$

7. **Expanded Power Rule**          $\left(\dfrac{ax}{by}\right)^m = \dfrac{a^m x^m}{b^m y^m}, \quad b \neq 0, \quad y \neq 0$

8. **Negative Exponent Rule**          $x^{-m} = \dfrac{1}{x^m}$

In Example 1, we show how these properties can be used to simplify an expression.

**EXAMPLE 1** Simplify $\left(\dfrac{x^4 y^{-5} z^3}{x^2 y z^3}\right)^2 x^4 y^{-8}$.

$$\left(\frac{x^4 y^{-5} z^3}{x^2 y z^3}\right)^2 x^4 y^{-8} = \left(\frac{x^{4\cdot2} y^{-5\cdot2} z^{3\cdot2}}{x^{2\cdot2} y^{1\cdot2} z^{3\cdot2}}\right) x^4 y^{-8} \quad \text{Use expanded power rule.}$$

$$= \left(\frac{x^8 y^{-10} z^6}{x^4 y^2 z^6}\right) x^4 y^{-8} \quad \text{Simplify exponents.}$$

$$= (x^{8-4} y^{-10-2} z^{6-6}) x^4 y^{-8} \quad \text{Use quotient rule.}$$

$$= (x^4 y^{-12} z^0) x^4 y^{-8} \quad \text{Simplify exponents.}$$

$$= \left(\frac{x^4 z^0}{y^{12}}\right) x^4 y^{-8} \quad \text{Use negative exponent rule.}$$

$$= \left(\frac{x^4 \cdot 1}{y^{12}}\right) x^4 y^{-8} \quad \text{Use zero exponent rule.}$$

$$= \left(\frac{x^4}{y^{12}}\right)\left(\frac{x^4}{y^8}\right) \quad \text{Use negative exponent rule.}$$

$$= \left(\frac{x^{4+4}}{y^{12+8}}\right) \quad \text{Use product rule.}$$

$$= \frac{x^8}{y^{20}} \quad \text{Simplify exponents.}$$

Note that this example can be simplified in many different ways but the result will always be the same.

## Polynomials

A *polynomial* in $x$ is an expression containing the sum of a finite number of terms of the form $ax^n$, for any real number $a$ and any whole number $n$. When the exponents on the variable of a polynomial decrease from left to right, the polynomial is written in *descending order*. The parts that are added are called the *terms* of the polynomial. The *degree of a term* of a polynomial in one variable is the value of the exponent on the variable in that term. The *degree of a polynomial* in one variable is the degree of its highest-degree term. The coefficient of the highest-degree term is called the *leading coefficient*.

**EXAMPLE 2** List the terms of the polynomial, $2x^3 + 5 - x + 3x^4$, write it in descending order, determine the degree of the polynomial, and determine the leading coefficient.

*Solution*    The terms of the polynomial are $2x^3$, $5$, $-x$, and $3x^4$. Written in descending order we have $3x^4 + 2x^3 - x + 5$. Because the highest-degree term is $3x^4$, the degree of the polynomial is 4 and the leading coefficient is 3.

When adding or subtracting polynomials, we add or subtract *like terms*. Like terms are terms that have the same variables with the same exponents.

**EXAMPLE 3** Add $(4x^2 - 6x + 3)$ and $(2x^2 + 5x - 1)$.

*Solution*
$$(4x^2 - 6x + 3) + (2x^2 + 5x - 1)$$
$$= 4x^2 - 6x + 3 + 2x^2 + 5x - 1 \qquad \text{Remove the parentheses.}$$
$$= \underbrace{4x^2 + 2x^2} \ \underbrace{-6x + 5x} \ \underbrace{+3 - 1} \qquad \text{Rearrange terms.}$$
$$= \quad 6x^2 \qquad\quad -x \qquad\quad +2 \qquad \text{Combine like terms.}$$

**EXAMPLE 4** Subtract $(-x^2 - 2x + 3)$ from $(x^3 + 4x + 6)$.

*Solution*    $(x^3 + 4x + 6) - (-x^2 - 2x + 3)$
$$= (x^3 + 4x + 6) - 1(-x^2 - 2x + 3) \qquad \text{Insert 1.}$$
$$= x^3 + 4x + 6 + x^2 + 2x - 3 \qquad \text{Distributive property}$$
$$= x^3 + x^2 + 4x + 2x + 6 - 3 \qquad \text{Rearrange terms.}$$
$$= x^3 + x^2 + 6x + 3 \qquad \text{Combine like terms.}$$

When multiplying two polynomials, every term of the first polynomial is multiplied by every term of the second polynomial.

**EXAMPLE 5** Multiply $2xy(3x^2y + 6xy^2 + 4)$.

*Solution*    $2xy(3x^2y + 6xy^2 + 4) = (2xy)(3x^2y) + (2xy)(6xy^2) + (2xy)(4)$
$$= 6x^3y^2 + 12x^2y^3 + 8xy$$

When finding the product of two binomials, we use the FOIL method.

**EXAMPLE 6** Multiply $(3x + 2)(x - 5)$.

*Solution*
$$\qquad\qquad\qquad\quad \text{F} \qquad\quad \text{O} \qquad\quad \text{I} \qquad\quad \text{L}$$
$$(3x + 2)(x - 5) = (3x)(x) + (3x)(-5) + (2)(x) + (2)(-5)$$
$$= 3x^2 - 15x + 2x - 10$$
$$= 3x^2 - 13x - 10$$

When we multiply a binomial by a trinomial, we can multiply vertically as illustrated in Example 7.

**EXAMPLE 7** Multiply $x^2 - 3x + 2$ by $2x - 3$.

*Solution*

$$
\begin{array}{r}
x^2 - 3x + 2 \\
2x - 3 \\
\hline
-3x^2 + 9x - 6 \\
2x^3 - 6x^2 + 4x \\
\hline
2x^3 - 9x^2 + 13x - 6
\end{array}
$$

Multiply top expression by $-3$.

Multiply top expression by $2x$.

Add like terms in columns.

To divide a polynomial by a monomial, we use the fact that

$$\frac{A + B}{C} = \frac{A}{C} + \frac{B}{C}$$

If the polynomial has more than two terms, we expand this procedure.

**To Divide a Polynomial by a Monomial**

Divide each term of the polynomial by the monomial.

**EXAMPLE 8** Divide $\dfrac{4x^2 - 8x - 3}{2x}$.

*Solution*

$$\frac{4x^2 - 8x - 3}{2x} = \frac{4x^2}{2x} - \frac{8x}{2x} - \frac{3}{2x}$$

$$= 2x - 4 - \frac{3}{2x}$$

We divide a polynomial by a binomial in much the same way as we perform long division.

**EXAMPLE 9** Divide $\dfrac{6x^2 - 5x + 5}{2x + 3}$.

*Solution*  In this example we will not show the change of sign in the subtractions.

$$
\begin{array}{r}
3x - 7 \\
2x + 3 \overline{\smash{)}6x^2 - 5x + 5} \\
6x^2 + 9x \\
\hline
-14x + 5 \\
-14x - 21 \\
\hline
26
\end{array}
$$

$\longleftarrow \quad 3x(2x + 3)$

$\longleftarrow \quad -7(2x + 3)$

$\longleftarrow \quad$ remainder

Thus, $\dfrac{6x^2 - 5x + 5}{2x + 3} = 3x - 7 + \dfrac{26}{2x + 3}$.

When dividing a polynomial by a binomial, the answer may be *checked* by multiplying the divisor by the quotient and then adding the remainder. You should obtain the polynomial you began with. To check Example 9, we do the following:

$$(2x + 3)(3x - 7) + 26 = 6x^2 - 5x - 21 + 26$$

$$= 6x^2 - 5x + 5$$

Since we got the polynomial we began with, our division is correct.

*When dividing a polynomial by a binomial, list both the polynomial and binomial in descending order. If a term of any degree is missing, include that term with a numerical coefficient of 0.*

### Factoring

To factor an expression means to write it as the product of other expressions. For a more detailed discussion of factoring, see Chapter 6.

Example 10 illustrates factoring a monomial from a polynomial.

**EXAMPLE 10** Factor $15x^4 - 5x^3 + 20x^2$.

*Solution*   The GCF is $5x^2$. Write each term as the product of the GCF and another product. Then factor out the GCF.

$$15x^4 - 5x^3 + 20x^2 = 5x^2 \cdot 3x^2 - 5x^2 \cdot x + 5x^2 \cdot 4$$
$$= 5x^2(3x^2 - x + 4)$$

Example 11 illustrates factoring by grouping.

**EXAMPLE 11** Factor $x^3 - 5x^2 + 2x - 10$ by grouping.

*Solution*   There are no factors common to all four terms. However, $x^2$ is common to the first two terms and 2 is common to the last two terms.

$$x^3 - 5x^2 + 2x - 10 = x^2(x - 5) + 2(x - 5)$$
$$= (x - 5)(x^2 + 2)$$

Example 12 illustrates factoring a trinomial of the form $x^2 + bx + c$. The key to factoring these kinds of expressions is to find two numbers whose product is $c$ and whose sum is $b$.

**EXAMPLE 12** Factor $x^2 - x - 12$.

*Solution*   We must find two numbers whose product is $-12$ and whose sum is $-1$. We begin by listing the factors of $-12$, trying to find a pair whose sum is $-1$.

| Factors of $-12$ | Sum of Factors |
|---|---|
| $(1)(-12)$ | $1 + (-12) = -11$ |
| $(2)(-6)$ | $2 + (-6) = -4$ |
| $(3)(-4)$ | $3 + (-4) = -1$ |

We can stop here because we found a pair whose product is $-12$ and whose sum is $-1$. Now we factor the trinomial using the 3 and $-4$.

$$x^2 - x - 12 = (x + 3)(x - 4)$$

Example 13 illustrates factoring a trinomial of the form $ax^2 + bx + c, a \neq 1$. This trinomial can be factored trying the various possibilities until you find the one that works.

**EXAMPLE 13** Factor $3x^2 - 13x + 10$.

*Solution*  The three terms have no common factor. The possible factors of the leading coefficient, 3, are 3 and 1. Therefore we write

$$3x^2 - 13x + 10 = (3x \quad)(x \quad)$$

Since the product of the last terms must be positive $(+10)$ and the sum of the products of the outer and inner terms must be negative $(-13)$, the two factors of 10 must both be negative. The negative factors of 10 are $(-1)(-10)$ and $(-2)(-5)$. We look for the factors that give us the correct middle term, $-13x$.

|  Possible Factors  |  Sum of Products of Outer and Inner Terms  |
|---|---|
| $(3x - 1)(x - 10)$ | $-31x$ |
| $(3x - 10)(x - 1)$ | $-13x$ |
| $(3x - 2)(x - 5)$ | $-17x$ |
| $(3x - 5)(x - 2)$ | $-11x$ |

Thus, $3x^2 - 13x + 10 = (3x - 10)(x - 1)$.

Below we list the special factoring formulas.

**Perfect Square Trinomials**

$$a^2 + 2ab + b^2 = (a + b)^2$$
$$a^2 - 2ab + b^2 = (a - b)^2$$

**Difference of Two Squares**

$$a^2 - b^2 = (a + b)(a - b)$$

**Sum of Two Cubes**

$$a^3 + b^3 = (a + b)(a^2 - ab + b^2)$$

**Difference of Two Cubes**

$$a^3 - b^3 = (a - b)(a^2 + ab + b^2)$$

Here is a general factoring procedure.

**To Factor a Polynomial**

1. If all the terms in the polynomial have a greatest common factor (GCF) other than 1, factor it out.
2. If the polynomial has two terms, determine whether it is a difference of two squares or a sum or difference of two cubes. If so, factor using the appropriate formula.
3. If the polynomial has three terms, determine whether it is a perfect square trinomial. If so, factor accordingly. If it is not, factor the trinomial using trial and error or grouping, as explained in Section 6.5.
4. If the polynomial has more than three terms, try factoring by grouping. If that does not work, see if three of the terms are the square of a binomial.
5. As a final step, examine your factored polynomial to see if any factors listed have a common factor and can be factored further. If you find a common factor, factor it out at this point.

The following examples illustrate how to use the procedure.

EXAMPLE  14  Factor $3x^2y^2 - 24xy^2 + 48y^2$. ——————————

*Solution*    Begin by factoring the GCF, $3y^2$, from each term.

$$3x^2y^2 - 24xy^2 + 48y^2 = 3y^2(x^2 - 8x + 16) = 3y^2(x - 4)^2$$

Note that $x^2 - 8x + 16$ is a perfect square trinomial. If you did not recognize this, you would still obtain the correct answer by factoring the trinomial into $(x - 4)(x - 4)$.

EXAMPLE  15  Factor $24x^2 - 6xy + 16xy - 4y^2$. ——————————

*Solution*    As always, begin by determining if all the terms in the polynomial have a common factor. In this example, 2 is common to all terms. Factor out the 2; then factor the remaining four-term polynomial by grouping.

$$
\begin{aligned}
24x^2 - 6xy + 16xy - 4y^2 &= 2(12x^2 - 3xy + 8xy - 2y^2) \\
&= 2[3x(4x - y) + 2y(4x - y)] \\
&= 2(4x - y)(3x + 2y)
\end{aligned}
$$

# Answers

## Chapter 1

**Exercises Set 1.1** **1.** Answers will vary. **3.** Answers will vary. **5.** Answers will vary. **7.** Answers will vary. **9.** Answers will vary. **11.** Do all the homework carefully and completely and preview the new material that is to be covered in class. **13.** At least 2 hours of study and homework time for each hour of class time in generally recommended. **15. a)** You need to do the homework in order to practice what was presented in class. **b)** When you miss class, you miss important information; therefore, it is important that you attend class regularly **17.** Answers will vary.

**Exercise Set 1.2** **1.** Median **3.** Approximately equal to **5.** Checking **7.** Circle graphs **9.** Problem solving **11. a)** 76.4 **b)** 74 **13. a)** $87.32 **b)** $86.57 **15. a)** 11.593 **b)** 11.68 **17.** $470 **19. a)** $1336 **b)** $18,036 **21.** 1,610,000,000 operations **23. a)** 19.375 minutes **b)** 88 minutes **c)** 10 minutes **25.** $\approx$18.49 miles per gallon **27.** $153 **29.** On the 3 on the right **31. a)** 4106.25 gallons **b)** $\approx$ $21.35 **33. a)** $193 **b)** $172 **35. a)** Finland; 550 **b)** Mexico; 405 **c)** 145 **37. a)** 0.275 million or 275,000; 1.05 million or 1,050,000 **b)** 0.775 million or 775,000 **c)** $\approx$3.96 times greater **39. a)** 1.394 million **b)** 0.255 million **c)** 0.051 million **41. a)** 48 **b)** He cannot get a C. **43.** Bachelor's degree **45.** One example is $50, 60, 70, 80, 90$. **47.** Mean

**Exercise Set 1.3** **1.** Added or subtracted **3.** Variables **5.** $\frac{2}{3}$ **7.** Denominator **9.** $\frac{3}{2}$ **11.** $\frac{2}{3}$ **13.** $\frac{1}{4}$ **15.** $\frac{9}{19}$ **17.** $\frac{3}{7}$ **19.** Simplified **21.** Simplified **23.** $\frac{43}{15}$ **25.** $\frac{23}{3}$ **27.** $\frac{59}{18}$ **29.** $\frac{159}{17}$ **31.** $1\frac{3}{4}$ **33.** $3\frac{1}{4}$ **35.** $4\frac{4}{7}$ **37.** $6\frac{1}{7}$ **39.** $\frac{4}{15}$ **41.** $\frac{1}{9}$ **43.** $\frac{3}{2}$ or $1\frac{1}{2}$ **45.** $\frac{1}{2}$ **47.** 6 **49.** $\frac{5}{2}$ or $2\frac{1}{2}$ **51.** $\frac{43}{10}$ or $4\frac{3}{10}$ **53.** $\frac{8}{13}$ **55.** $\frac{5}{8}$ **57.** $\frac{1}{7}$ **59.** $\frac{6}{5}$ or $1\frac{1}{5}$ **61.** $\frac{10}{17}$ **63.** $\frac{7}{12}$ **65.** $\frac{13}{36}$ **67.** $\frac{65}{24}$ or $2\frac{17}{24}$ **69.** $\frac{17}{6}$ or $2\frac{5}{6}$ **71.** $\frac{29}{10}$ or $2\frac{9}{10}$ **73.** $\frac{277}{30}$ or $9\frac{7}{30}$ **75.** $\frac{11}{24}$ mile **77.** $8\frac{15}{16}$ inches **79.** $\frac{9}{55}$ **81.** $\frac{63}{100}$ **83.** 11 feet, $11\frac{1}{4}$ inches or $143\frac{1}{4}$ inches or $\approx$11.94 feet **85.** $2\frac{3}{10}$ minutes **87.** $1\frac{9}{16}$ inches **89.** 5 mg **91.** 40 times **93.** $1\frac{1}{2}$ inches **95.** 6 strips **97. a)** Yes **b)** $82\frac{3}{8}$ in. **99.** Answers will vary **101. a)** $\frac{*\ +\ ?}{a}$ **b)** $\frac{\odot\ -\ \square}{?}$ **c)** $\frac{\triangle\ +\ 4}{\square}$ **d)** $\frac{x\ -\ 2}{3}$ **e)** $\frac{8}{x}$ **103.** 270 pills **105.** Answers will vary. **106.** 16 **107.** 15 **108.** Variables are letters used to represent numbers.

**Exercise Set 1.4** **1.** Irrational numbers **3.** Counting numbers **5.** Real number line **7.** Whole numbers **9.** $\pi$ **11.** $\{\ldots, -3, -2, -1, 0, 1, 2, 3, \ldots\}$ **13.** $\{0, 1, 2, 3, \ldots\}$ **15.** $\{\ldots, -3, -2, -1\}$ **17.** True **19.** True **21.** False **23.** False **25.** True **27.** True **29.** False **31.** True **33.** True **35.** False **37.** True **39.** True **41.** False **43.** True **45.** False **47.** True **49. a)** 13 **b)** $-2, 13$ **c)** $-2, 13$ **d)** 13 **51. a)** 3, 77 **b)** 0, 3, 77 **c)** 0, $-2$, 3, 77 **d)** $-\frac{5}{7}, 0, -2, 3, 6\frac{1}{4}, 1.63, 77$ **e)** $\sqrt{7}, -\sqrt{3}$ **f)** $-\frac{5}{7}, 0, -2,$ $3, 6\frac{1}{4}, \sqrt{7}, -\sqrt{3}, 1.63, 77$ **53.** Answers will vary; three examples are 0, 1, and 2. **55.** Answers will vary; three examples are $-\sqrt{2},$ $-\sqrt{3},$ and $-\sqrt{7}.$ **57.** Answers will vary; three examples are $-\frac{2}{3}, \frac{1}{2},$ and 6.3. **59.** Answers will vary; three examples are $-13, -5,$ and $-1$. **61.** Answers will vary; three examples are $\sqrt{2}, \sqrt{3},$ and $-\sqrt{5}$. **63.** Answers will vary; three examples are $-7, 1,$ and 5. **65.** 87 **67. a)** $\{1, 3, 4, 5, 8\}$ **b)** $\{2, 5, 6, 7, 8\}$ **c)** $\{5, 8\}$ **d)** $\{1, 2, 3, 4, 5, 6, 7, 8\}$ **69. a)** Set $B$ continues beyond 4. **b)** 4 **c)** An infinite number of elements **d)** An infinite set **71.** **a)** An infinite number **b)** An infinite number **73.** $\frac{29}{5}$ **74.** $5\frac{1}{3}$ **75.** $\frac{13}{24}$ **76.** $\frac{4}{45}$

**Exercise Set 1.5** **1.** 0 **3.** $|a|$ **5.** True **7.** $|6 - (-4)|$ **9.** Negative number **11.** 7 **13.** 15 **15.** 0 **17.** $-5$ **19.** $-26$ **21.** > **23.** < **25.** > **27.** < **29.** > **31.** < **33.** < **35.** > **37.** < **39.** > **41.** > **43.** < **45.** > **47.** < **49.** < **51.** < **53.** > **55.** > **57.** > **59.** < **61.** < **63.** < **65.** < **67.** = **69.** = **71.** < **73.** < **75.** $-|-1|, \frac{3}{7}, \frac{4}{9}, 0.46, |-5|$ **77.** $\frac{5}{12}, 0.6, \frac{2}{3}, \frac{19}{25}, |-2.6|$ **79.** 4, $-4$ **81.** Not possible **83.** Answers will vary: one example is $-3, -4$ and $-5$. **85.** Answers will vary: one example is 4, 5, and 6. **87.** Answers will vary: one example is 3, 4, and 5. **89. a)** Does not include the endpoints **b)** Answers will vary: one example is 4.1, 5, and $5\frac{1}{2}$. **c)** No **d)** Yes **e)** True **91. a)** Dietary fiber and thiamin **b)** Vitamin E, niacin, and riboflavin **93.** Yes **95.** No, $|-4| > |-3|$ but $-4 < -3$. **97.** Greater than **99.** No **102.** $\frac{89}{15}$ or $5\frac{14}{15}$ **103.** $\{\ldots, -3, -2, -1, 0, 1, 2, 3, \ldots\}$ **104.** $\{0, 1, 2, 3, \ldots\}$ **105. a)** 5 **b)** 5, 0 **c)** 5, $-2$, 0 **d)** 5, $-2$, 0, $\frac{1}{3}, -\frac{5}{9}, 2.3$ **e)** $\sqrt{3}, \pi$ **f)** 5, $-2$, 0, $\frac{1}{3}, \sqrt{3}, -\frac{5}{9}, 2.3, \pi$

## Mid-Chapter Test: Sections 1.1–1.5[*]

**1.** At least two hours of study and homework for each hour of class time is generally recommended. [1.1]   **2. a)** $80.63   **b)** $83.81 [1.2]   **3.** $824.59 [1.2]   **4. a)** Natwora's   **b)** $24 [1.2]   **5.** $62.35 [1.2]   **6.** $\frac{1}{6}$ [1.3]   **7.** $\frac{39}{80}$ [1.3]   **8.** $\frac{49}{40}$ or $1\frac{9}{40}$ [1.3]   **9.** $\frac{61}{20}$ or $3\frac{1}{20}$ [1.3]   **10.** $54\frac{1}{3}$ feet [1.3]   **11.** False [1.4]   **12.** True [1.4]   **13.** False [1.4]   **14.** True [1.4]   **15.** False [1.4]   **16.** $-\frac{7}{10}$ [1.5]   **17.** $>$ [1.5]   **18.** $>$ [1.5]   **19.** $<$ [1.5]   **20.** $=$ [1.5]

## Exercise Set 1.6

**1.** Negative   **3.** Absolute Value   **5.** Sum   **7.** $-8$   **9.** Common denominator   **11.** Correct   **13.** $-19$   **15.** 28   **17.** 0   **19.** $-\frac{5}{3}$   **21.** $-2\frac{3}{5}$   **23.** $-3.72$   **25.** 21   **27.** 1   **29.** $-6$   **31.** 0   **33.** 0   **35.** $-10$   **37.** $-4$   **39.** $-13$   **41.** 0   **43.** $-8$   **45.** $-27$   **47.** $-64$   **49.** 16   **51.** $-12$   **53.** 3   **55.** $-6$   **57.** $-40$   **59.** $-31$   **61.** $-39$   **63.** 91   **65.** $-9.9$   **67.** $-144.0$   **69.** $-141.91$   **71.** $-53.65$   **73.** $\frac{26}{35}$   **75.** $\frac{107}{84}$ or $1\frac{23}{84}$   **77.** $\frac{4}{55}$   **79.** $-\frac{26}{45}$   **81.** $-\frac{31}{30}$ or $-1\frac{1}{30}$   **83.** $\frac{3}{10}$   **85.** $-\frac{13}{15}$   **87.** $\frac{19}{56}$   **89.** $-\frac{43}{60}$   **91.** $-\frac{23}{21}$ or $-1\frac{2}{21}$   **93. a)** Positive   **b)** 390   **95. a)** Negative   **b)** $-373$   **97. a)** Negative   **b)** $-452$   **99. a)** Negative   **b)** $-1300$   **101. a)** Negative   **b)** $-112$   **103. a)** Negative   **b)** $-3880$   **105. a)** Positive   **b)** 1111   **107. a)** Negative   **b)** $-2050$   **109.** True   **111.** True   **113.** False   **115.** $277   **117.** 21 yards   **119.** 61 feet   **121.** 13,796 feet   **123. a)** $-$$12 thousand   **b)** 2006–2008 surplus, $91 thousand   **125.** $-26$   **127.** 20   **129.** 0   **131.** $\frac{11}{30}$   **133.** 55   **135.** $\frac{19}{14}$ or $1\frac{5}{14}$   **136.** $\frac{43}{16}$ or $2\frac{11}{16}$   **137.** False   **138.** $>$   **139.** $<$

## Exercise Set 1.7

**1.** Minuend   **3.** Difference   **5.** Opposite   **7.** Left to right   **9.** $-a + b$   **11.** Correct   **13.** 6   **15.** 7   **17.** $-1$   **19.** 12   **21.** $-16$   **23.** $-9$   **25.** 0   **27.** $-4$   **29.** $-3$   **31.** 9   **33.** $-20$   **35.** 9.8   **37.** 0.3   **39.** 37   **41.** 4   **43.** 22   **45.** $-11$   **47.** $-131.0$   **49.** $-84$   **51.** 140   **53.** $-7.4$   **55.** $-3.93$   **57.** $-11$   **59.** 0   **61.** 6.1   **63.** 18.2   **65.** $-11$   **67.** $-18.1$   **69.** $\frac{31}{72}$   **71.** $\frac{17}{45}$   **73.** $-\frac{67}{60}$ or $-1\frac{7}{60}$   **75.** $-\frac{5}{12}$   **77.** $-\frac{17}{24}$   **79.** $\frac{1}{2}$   **81.** $\frac{7}{45}$   **83.** $\frac{13}{16}$   **85.** $-\frac{13}{63}$   **87.** $-\frac{7}{60}$   **89. a)** Positive   **b)** 99   **c)** Yes   **91. a)** Negative   **b)** $-619$   **c)** Yes   **93. a)** Positive   **b)** 1588   **c)** Yes   **95. a)** Positive   **b)** 196   **c)** Yes   **97. a)** Negative   **b)** $-448$   **c)** Yes   **99. a)** Positive   **b)** 116.1   **c)** Yes   **101. a)** Negative   **b)** $-69$   **c)** Yes   **103. a)** Negative   **b)** $-1670$   **c)** Yes   **105. a)** Zero   **b)** 0   **c)** Yes   **107.** 4   **109.** 4   **111.** $-15$   **113.** $-2$   **115.** 13   **117.** $-5$   **119.** $-32$   **121.** $-14$   **123.** 9   **125.** $-12$   **127.** 12   **129.** $-18$   **131. a)** 43   **b)** 143   **133.** $1\frac{7}{8}$ inches   **135.** Dropped 100°F   **137. a)** 280   **b)** 14 strokes more   **139.** $x - y = -5, y - x = 5,$ and $x - (-y) = 11$   **141.** $-5$   **143. a)** 8   **b)** $-3 - (-11)$   **145. a)** 9ft   **b)** $-3$ft   **146.** $\{1, 2, 3, \dots\}$   **147.** The set of rational numbers together with the set of irrational numbers form the set of real numbers.   **148.** $>$   **149.** $<$   **150.** $-\frac{1}{24}$

## Exercise Set 1.8

**1.** Negative   **3.** Zero   **5.** $\frac{a}{b}$   **7.** Positive   **9.** 63   **11.** Negative   **13.** Positive   **15.** Negative   **17.** 20   **19.** $-15$   **21.** 160   **23.** $-12.6$   **25.** 42   **27.** $-81$   **29.** 30   **31.** 0   **33.** $-84$   **35.** $-72$   **37.** 1400   **39.** 0   **41.** $-\frac{3}{10}$   **43.** $\frac{7}{27}$   **45.** 4   **47.** $-\frac{1}{10}$   **49.** $-7$   **51.** 4   **53.** 4   **55.** $-18$   **57.** 9.9   **59.** $-10$   **61.** $-33$   **63.** $-4$   **65.** 16.2   **67.** 0   **69.** $-5.5$   **71.** 6   **73.** $-\frac{2}{5}$   **75.** $\frac{5}{36}$   **77.** 1   **79.** $-\frac{144}{5}$ or $-28\frac{4}{5}$   **81.** $-32$   **83.** 20   **85.** $-14$   **87.** $-9.3$   **89.** $-20$   **91.** 1   **93.** 0   **95.** Undefined   **97.** 0   **99.** Undefined   **101. a)** Negative   **b)** $-3496$   **c)** Yes   **103. a)** Negative   **b)** $-16$   **c)** Yes   **105. a)** Negative   **b)** $-9$   **c)** Yes   **107. a)** Positive   **b)** 6174   **c)** Yes   **109. a)** Zero   **b)** 0   **c)** Yes   **111. a)** Undefined   **b)** Undefined   **c)** Yes   **113. a)** Positive   **b)** 3.2   **c)** Yes   **115. a)** Positive   **b)** 226.8   **c)** Yes   **117.** False   **119.** False   **121.** True   **123.** True   **125.** False   **127.** True   **129.** 45 yard loss or $-45$ yards   **131. a)** $104   **b)** $-$$416   **133.** $143.85   **135. a)** 20 point loss or $-20$ points   **b)** 80   **137. a)** 102 to 128 beats per minute   **b)** Answers will vary.   **139.** $-125$   **141.** 1   **143.** Positive   **146.** $>$   **147.** $-\frac{41}{60}$   **148.** $-2$   **149.** $-3$   **150.** 3

## Exercise Set 1.9

**1.** Left to right   **3.** Grouping symbols   **5.** Exponent   **7.** 25   **9.** 1   **11.** $-64$   **13.** 9   **15.** $-1$   **17.** $-100$   **19.** 81   **21.** 27   **23.** 256   **25.** $-16$   **27.** $\frac{9}{16}$   **29.** $-\frac{1}{32}$   **31.** 225   **33.** 576   **35. a)** Positive   **b)** 343   **c)** Yes   **37. a)** Positive   **b)** 1296   **c)** Yes   **39. a)** Negative   **b)** $-243$   **c)** Yes   **41. a)** Positive   **b)** 625   **c)** Yes   **43. a)** Negative   **b)** $-81$   **c)** Yes   **45. a)** Negative   **b)** $-\frac{9}{64}$   **c)** Yes   **47.** 21   **49.** 8   **51.** 57   **53.** 0   **55.** $-16$   **57.** 29   **59.** $-19$   **61.** $-77$   **63.** $\frac{83}{100}$   **65.** 2   **67.** 10   **69.** $-34$   **71.** 103   **73.** 169   **75.** $-23$   **77.** 36.75   **79.** $\frac{5}{8}$   **81.** $\frac{1}{4}$   **83.** $\frac{49}{30}$ or $1\frac{19}{30}$   **85.** $\frac{5}{27}$   **87.** $\frac{32}{53}$   **89.** 9   **91.** $-4$   **93.** 1   **95. a)** 25   **b)** $-25$   **c)** 25   **97. a)** 4   **b)** $-4$   **c)** 4   **99. a)** 36   **b)** $-36$   **c)** 36   **101. a)** $\frac{1}{9}$   **b)** $-\frac{1}{9}$   **c)** $\frac{1}{9}$   **103.** 4   **105.** $-45$

---

*Numbers in blue brackets after the answer indicate the section where the material was discussed.

**107.** 3  **109.** −4  **111.** −1  **113.** $\frac{15}{4}$ or $3\frac{3}{4}$  **115.** 994  **117.** −5  **119.** 193  **121.** −25  **123.** $[(6 \cdot 3) - 4] - 2; 12$

**125.** $\{[(10 \cdot 4) + 9] - 6\} \div 7; \frac{43}{7}$ or $6\frac{1}{7}$  **127.** $\left(\frac{4}{5} + \frac{3}{7}\right) \cdot \frac{2}{3}; \frac{86}{105}$  **129.** All real numbers  **131.** 162.5 miles  **133.** 102 feet

**135.** 50 ft³  **137.** 1.71 inches  **139.** $12 - (4 - 6) + 10 = 24$  **145. a)** 3  **b) Dogs   Number of Houses   c)** 18

**d)** ≈ 1.29 dogs per house  **146.** $6.40  **147.** $-\frac{5}{36}$  **148.** $\frac{10}{3}$ or $3\frac{1}{3}$

| Dogs | Number of Houses |
|------|------------------|
| 0 | 4 |
| 1 | 5 |
| 2 | 3 |
| 3 | 1 |
| 4 | 1 |

## Exercise Set 1.10

**1.** Associative property of addition  **3.** Zero  **5.** Additive inverse  **7.** Distributive property of multiplication over addition  **9.** Multiplicative identity  **11. a)** −6  **b)** $\frac{1}{6}$  **13. a)** 3  **b)** $-\frac{1}{3}$  **15. a)** $-x$  **b)** $\frac{1}{x}$  **17. a)** −1.6  **b)** $\frac{1}{1.6}$ or 0.625  **19. a)** $-\frac{1}{5}$  **b)** 5  **21. a)** $\frac{5}{6}$  **b)** $-\frac{6}{5}$  **23.** Distributive property  **25.** Associative property of addition  **27.** Commutative property of multiplication  **29.** Associative property of multiplication  **31.** Distributive property  **33.** Identity property of multiplication  **35.** Inverse property of multiplication  **37.** $1 + (-4)$  **39.** $(-6 \cdot 4) \cdot 2$  **41.** $-2 \cdot x + -2 \cdot y$ or $-2x - 2y$  **43.** $y \cdot x$  **45.** $3y + 4x$  **47.** $a + (b + 3)$  **49.** $3x + (4 + 6)$  **51.** $(m + n)3$  **53.** $4x + 4y + 12$  **55.** 0  **57.** $\frac{5}{2}n$  **59.** Yes  **61.** Yes  **63.** No  **65.** Yes  **67.** No  **69.** No  **71.** The $(3 + 4)$ is treated as one value.  **73.** Commutative property of addition  **75.** No; Associative property of addition  **77.** $\frac{49}{15}$ or $3\frac{4}{15}$  **78.** $\frac{23}{16}$ or $1\frac{7}{16}$  **79.** −11.2  **80.** $-\frac{7}{8}$

## Chapter 1 Review Exercises

**1.** 28  **2.** $572.45  **3. a)** $74.25  **b)** $974.24  **4.** $100  **5. a)** 78.4  **b)** 79  **6. a)** 11  **b)** 9.5  **7. a)** 30 minutes  **b)** 27 minutes  **8. a)** 225  **b)** 165  **9.** $\frac{1}{2}$  **10.** $\frac{127}{21}$ or $6\frac{1}{21}$  **11.** $\frac{25}{36}$  **12.** $\frac{7}{6}$ or $1\frac{1}{6}$  **13.** $\frac{23}{12}$ or $1\frac{11}{12}$  **14.** $\frac{177}{10}$ or $17\frac{7}{10}$  **15.** $\{1, 2, 3, ...\}$  **16.** $\{0, 1, 2, 3, ...\}$  **17.** $\{..., -3, -2, -1, 0, 1, 2, 3, ...\}$  **18.** The set of all numbers which can be expressed as the quotient of two integers, denominator not zero  **19. a)** $3, 426$  **b)** $3, 0, 426$  **c)** $3, -5, -12, 0, 426$  **d)** $3, -5, -12, 0, \frac{1}{2}, -0.62, 426, -3\frac{1}{4}$  **e)** $\sqrt{7}$  **f)** $3, -5, -12, 0, \frac{1}{2}, -0.62, \sqrt{7}, 426, -3\frac{1}{4}$  **20. a)** 1  **b)** 1  **c)** $-8, -9$  **d)** $-8, -9, 1$  **e)** $-2.3, -8, -9, 1\frac{1}{2}, 1, -\frac{3}{17}$  **f)** $\sqrt{2}, -\sqrt{2}$  **g)** $-2.3, -8, -9, 1\frac{1}{2}, \sqrt{2}, -\sqrt{2}, 1, -\frac{3}{17}$  **21.** <  **22.** >  **23.** <  **24.** >  **25.** <  **26.** >  **27.** =  **28.** >  **29.** −14  **30.** 0  **31.** −3  **32.** −6  **33.** −6  **34.** 2  **35.** 8  **36.** 0  **37.** −5  **38.** 14  **39.** 4  **40.** −12  **41.** $\frac{7}{12}$  **42.** $\frac{11}{10}$ or $1\frac{1}{10}$  **43.** $-\frac{7}{36}$  **44.** $-\frac{19}{56}$  **45.** $-\frac{5}{4}$ or $-1\frac{1}{4}$  **46.** $-\frac{37}{84}$  **47.** $-\frac{7}{90}$  **48.** $\frac{61}{60}$ or $1\frac{1}{60}$  **49.** 14  **50.** −3  **51.** −12  **52.** −7  **53.** 16  **54.** 11  **55.** −63  **56.** 25.42  **57.** −120  **58.** $-\frac{6}{35}$  **59.** $-\frac{6}{11}$  **60.** $\frac{15}{56}$  **61.** 0  **62.** 144  **63.** −15  **64.** −6  **65.** −3.2  **66.** 4.3  **67.** 8  **68.** 9  **69.** $\frac{56}{27}$ or $2\frac{2}{27}$  **70.** $-\frac{35}{9}$ or $-3\frac{8}{9}$  **71.** 0  **72.** 0  **73.** Undefined  **74.** Undefined  **75.** Undefined  **76.** 0  **77.** 25  **78.** −8  **79.** 1  **80.** 3  **81.** −6  **82.** −32  **83.** 6  **84.** −4  **85.** 10  **86.** 1  **87.** 15  **88.** −4  **89.** −36  **90.** 36  **91.** 16  **92.** −27  **93.** −1  **94.** −32  **95.** $\frac{16}{25}$  **96.** $\frac{8}{125}$  **97.** 500  **98.** 4  **99.** 12  **100.** −256  **101.** 7  **102.** 32  **103.** 6.36  **104.** −17  **105.** −39  **106.** −2.3  **107.** 0  **108.** $\frac{9}{7}$ or $1\frac{2}{7}$  **109.** −60  **110.** 10  **111.** 20  **112.** 20  **113.** 14  **114.** 9  **115.** −4  **116.** 50  **117.** 5  **118.** 26  **119.** 45  **120.** 0  **121.** −11  **122.** −3  **123.** −3  **124.** 39  **125.** −215  **126.** 353.6  **127.** −2.88  **128.** 117.8  **129.** 65,536  **130.** −74.088  **131.** Associative property of addition  **132.** Distributive property  **133.** Commutative property of addition  **134.** Commutative property of multiplication  **135.** Distributive property  **136.** Associative property of addition  **137.** Identity property of multiplication  **138.** Inverse property of addition

## Chapter 1 Practice Test

**1. a)** $10.65  **b)** $0.23  **c)** $10.88  **d)** $39.12 [1.2]  **2.** ≈2.5 times greater [1.2]  **3. a)** ≈13 thousand  **b)** During this specific time, half the time KFUN had more than 8.8 thousand listeners and half the time KFUN had less than 8.8 thousand listeners. [1.2]  **4. a)** 42  **b)** 42, 0  **c)** $-6, 42, 0, -7, -1$  **d)** $-6, 42, -3\frac{1}{2}, 0, 6.52, \frac{5}{9}, -7, -1$  **e)** $\sqrt{5}$  **f)** $-6, 42, -3\frac{1}{2}, 0, 6.52, \sqrt{5}, \frac{5}{9}, -7, -1$ [1.4]  **5.** < [1.5]  **6.** > [1.5]  **7.** −15 [1.6]  **8.** −11 [1.7]  **9.** −14 [1.7]  **10.** 8 [1.9]  **11.** −24 [1.8]  **12.** $\frac{16}{63}$ [1.8]  **13.** −3 [1.9]  **14.** $-\frac{53}{56}$ [1.7]  **15.** 12 [1.9]  **16.** $-\frac{32}{243}$ [1.9]  **17.** 100 [1.9]  **18.** $-x^2$ means $-(x^2)$ and $x^2$ will always be positive for any nonzero value of x. Therefore, $-x^2$ will always be negative. [1.9]  **19.** 37 [1.9]  **20.** 11 [1.9]  **21.** 10 [1.9]  **22.** 1 [1.9]  **23.** Commutative property of addition [1.10]  **24.** Distributive property [1.10]  **25.** Associative property of addition [1.10]

## Chapter 2

**Exercise Set 2.1**   **1.** Constant   **3.** Like terms   **5.** Unlike terms   **7.** Terms   **9.** $14x$   **11.** $3x + 6$   **13.** $5y + 3$   **15.** $\frac{9}{44}a$   **17.** $-6x + 7t$   **19.** $-5w + 5$   **21.** $-2x$   **23.** $0$   **25.** $-2t + 21$   **27.** $-12p - 8$   **29.** $10x^2 - 10y^2 - 7$   **31.** $2x - 8$   **33.** $b + \frac{23}{5}$   **35.** $0.8n + 6.42$   **37.** $\frac{1}{2}a + 3b + 1$   **39.** $14.6x + 8.3$   **41.** $x^2 + y$   **43.** $-3x - 5y$   **45.** $-3n^2 - 2n + 13$   **47.** $21.72x - 7.11$   **49.** $-\frac{23}{20}x - 5$   **51.** $5w^3 + 2w^2 + w + 3$   **53.** $-7z^3 - z^2 + 2z$   **55.** $6x^2 - 6xy + 3y^2$   **57.** $4a^2 + 3ab + b^2$   **59.** $5x + 10$   **61.** $5x + 20$   **63.** $3x - 18$   **65.** $-x + 2$   **67.** $x - 4$   **69.** $\frac{4}{5}s - 4$   **71.** $-0.9x^2 - 1.5$   **73.** $-r + 4$   **75.** $1.4x + 0.35$   **77.** $x - y$   **79.** $-2x - 4y + 8$   **81.** $3.41x - 5.72y + 3.08$   **83.** $10x - 45y$   **85.** $r + 3s - 19$   **87.** $3x - 6y - 12$   **89.** $-3x + 1$   **91.** $2x + 1$   **93.** $14x + 18$   **95.** $4x - 2y + 3$   **97.** $5c$   **99.** $7x + 3$   **101.** $\frac{5}{4}x + \frac{1}{3}$   **103.** $\frac{19}{6}x - 2$   **105.** $-4s - 6$   **107.** $2x - 2$   **109.** $0$   **111.** $-y - 6$   **113.** $3x - 5$   **115.** $x + 15$   **117.** $0.2x - 4y - 2.8$   **119.** $-6x + 7y$   **121.** $\frac{3}{2}x + \frac{7}{2}$   **123.** $2\square + 3\ominus$   **125.** $2x + 3y + 2\triangle$   **127.** $1, 2, 3, 6, 9, 18$   **129. a)** The signs of all terms inside the parentheses change when the parentheses are removed. **b)** $-x + 8$   **131.** $22x^2 - 25y^2 - 4x + 3$   **133.** $9x - 39$   **135.** $7$   **136.** $-16$   **137.** $-11$   **138.** Answers will vary.   **139.** $-12$

**Exercise Set 2.2**   **1.** $2$   **3.** Isolate   **5.** Linear   **7.** Equation   **9.** Opposite   **11.** Yes   **13.** No   **15.** Yes   **17.** Yes   **19.** No   **21.** Yes   **23.** $5$   **25.** $-7$   **27.** $-4$   **29.** $43$   **31.** $15$   **33.** $11$   **35.** $-4$   **37.** $-5$   **39.** $-12$   **41.** $-30$   **43.** $0$   **45.** $-57$   **47.** $-4$   **49.** $-65$   **51.** $0$   **53.** $17$   **55.** $-26$   **57.** $28$   **59.** $-46.1$   **61.** $46.5$   **63.** $-8.23$   **65.** $5.57$   **67.** No, the equation is equivalent to $1 = 2$, a false statement.   **69.** Use properties that allow us to get the variable by itself on one side of the equation.   **71.** $x = \square + \triangle$   **73.** $\square = \odot - \triangle$   **76.** $\frac{11}{30}$   **77.** $-\frac{31}{24}$   **78.** $2x - 13$   **79.** $7t - 25$

**Exercise Set 2.3**   **1.** $\frac{x}{3}$   **3.** $3$   **5.** Check   **7.** $-\frac{7}{x}$   **9.** $4$   **11.** $21$   **13.** $-3$   **15.** $-80$   **17.** $20$   **19.** $-3$   **21.** $-\frac{7}{3}$   **23.** $-13$   **25.** $8$   **27.** $30$   **29.** $-\frac{1}{3}$   **31.** $6$   **33.** $\frac{26}{43}$   **35.** $2$   **37.** $\frac{1}{5}$   **39.** $-\frac{3}{40}$   **41.** $-64$   **43.** $240$   **45.** $-45$   **47.** $20$   **49.** $-50$   **51.** $0$   **53.** $0$   **55.** $22.5$   **57.** $6$   **59.** $-20.2$   **61.** $7$   **63.** $9$   **65. a)** In $5 + x = 10$, $5$ is added to the variable, whereas in $5x = 10$, $5$ is multiplied by the variable. **b)** $5$ **c)** $2$   **67.** Multiply by $\frac{3}{2}$; $6$   **69.** Multiply by $\frac{7}{3}$; $\frac{28}{15}$   **71. a)** $\square$ **b)** Divide both sides of the equation by $\triangle$. **c)** $\square = \frac{\odot}{\triangle}$   **73.** $-4$   **74.** $-30$   **75.** $6$   **76.** Associative property of addition   **77.** $-57$

**Exercise Set 2.4**   **1.** Multiplication property of equality   **3.** A power of ten   **5.** Expression   **7.** $5$   **9.** $-5$   **11.** $2$   **13.** $\frac{12}{5}$   **15.** $2$   **17.** $7$   **19.** $\frac{11}{3}$   **21.** $-\frac{19}{16}$   **23.** $-12$   **25.** $2$   **27.** $-\frac{51}{5}$   **29.** $3$   **31.** $6.8$   **33.** $4$   **35.** $12$   **37.** $22$   **39.** $60$   **41.** $-14$   **43.** $-11$   **45.** $-2$   **47.** $\frac{19}{8}$   **49.** $0$   **51.** $-10$   **53.** $-3$   **55.** $6$   **57.** $-21$   **59.** $-\frac{19}{7}$   **61.** $-9$   **63.** $4$   **65.** $5$   **67.** $0.8$   **69.** $-1$   **71.** $\frac{2}{7}$   **73.** $-3.1$   **75.** $-\frac{14}{5}$   **77.** $18$   **79.** $-\frac{1}{15}$   **81.** $\frac{18}{7}$   **83.** $-\frac{16}{21}$   **85.** $10$   **87.** $5$   **89.** $-\frac{39}{4}$   **91.** $2$   **93.** $\frac{25}{3}$   **95.** $\frac{26}{45}$   **97. a)** You will not have to work with fractions. **b)** $x = 3$   **99.** $\frac{35}{6}$   **101.** $-4$   **105.** False   **106.** $64$   **107.** Isolate the variable on one side of the equation.   **108.** Divide both sides of the equation by $-4$ to isolate the variable.

**Mid-Chapter Test: Sections 2.1–2.4**   **1.** $-2x - 5y - 6$ [2.1]   **2.** $-\frac{7}{20}x - \frac{15}{2}$ [2.1]   **3.** $-8a + 12b - 64$ [2.1]   **4.** $3.36x - 5.44y - 8.32$ [2.1]   **5.** $2t - 38$ [2.1]   **6.** Yes [2.2]   **7.** No [2.2]   **8.** $-4$ [2.2]   **9.** $-160$ [2.2]   **10.** $-23$ [2.2]   **11.** Multiply both sides by $4$. [2.3]   **12.** $\frac{1}{3}$ [2.3]   **13.** $24$ [2.3]   **14.** $10$ [2.3]   **15.** $-\frac{3}{7}$ [2.3]   **16.** $\frac{5}{2}$ [2.4]   **17.** $-\frac{3}{2}$ [2.4]   **18.** $\frac{13}{16}$ [2.4]   **19.** $-6$ [2.4]   **20.** $-\frac{20}{9}$ [2.4]

**Exercise Set 2.5**   **1.** Identity   **3.** Least common denominator   **5.** Specific value(s) of the variable   **7.** No real numbers   **9.** False   **11.** $2$   **13.** $1$   **15.** $\frac{3}{5}$   **17.** $3$   **19.** $-2$   **21.** No solution   **23.** $6.5$   **25.** $3.2$   **27.** $3$   **29.** $-2$   **31.** No solution   **33.** $\frac{13}{5}$   **35.** $\frac{7}{9}$   **37.** $5$   **39.** $30$   **41.** $\frac{3}{2}$   **43.** $\frac{5}{2}$   **45.** $25$   **47.** All real numbers   **49.** $23$   **51.** $0$   **53.** All real numbers   **55.** $\frac{21}{20}$   **57.** $14$   **59.** $-\frac{15}{4}$   **61.** $5$   **63.** $4$   **65.** $0$   **67.** $16$   **69.** $-\frac{24}{5}$   **71.** $-\frac{4}{21}$   **73.** $5$   **75.** $30$   **77.** $4$   **79. a)** One example is $x + x + 1 = x + 2$.

**b)** It has a single solution.   **c)** For the example given in part **a)**, $x = 1$.   **81. a)** One example is $x + x + 1 = 2x + 1$.
**b)** Both sides simplify to the same expression.   **c)** All real numbers   **83. a)** One example is $x + x + 1 = 2x + 2$.
**b)** It simplifies to a false statement.   **c)** No solution   **85.** $\ast = -\dfrac{1}{4}$   **87.** All real numbers   **89.** $x = -4$   **91. a)** 4   **b)** 7
**c)** 0   **92.** $x^{a-b}$   **93.** Factors are expressions that are multiplied. Terms are expressions that are added.   **94.** $7x - 10$
**95.** $\dfrac{10}{7}$   **96.** $-3$

## Exercise Set 2.6
**1.** Quadrilateral   **3.** Formula   **5.** Cubic feet   **7.** $\pi$   **9.** Evaluating   **11.** 480   **13.** 96   **15.** 360   **17.** 36
**19.** 314.16   **21.** 82   **23.** 2   **25.** 8   **27.** 6.00   **29.** 127.03   **31.** $\approx$50.27 square feet   **33.** 16.5 square feet   **35.** $\approx$452.39 cubic centimeters
**37.** $C = 10°$   **39.** $F = 77°$   **41.** $P = 40$   **43.** $V = 5$   **45.** $l = A/w$   **47.** $t = d/r$   **49.** $t = i/(pr)$   **51.** $b = 2A/h$
**53.** $w = (P - 2l)/2$   **55.** $r = (-n + 3)/2$   **57.** $b = y - mx$   **59.** $b = d - a - c$   **61.** $y = (-ax - c)/b$   **63.** $h = 3V/(\pi r^2)$
**65.** $m = 2A - d$   **67.** $y = -2x + 8$   **69.** $y = x - 6$   **71.** $y = \dfrac{2}{3}x + \dfrac{4}{3}$   **73.** $y = \dfrac{3}{5}x - 2$   **75.** $y = \dfrac{1}{2}x - \dfrac{5}{2}$   **77.** $y = -\dfrac{1}{2}x + 5$
**79.** $y = -\dfrac{1}{3}x - \dfrac{5}{3}$   **81.** $y = 2x + \dfrac{13}{15}$   **83.** \$1440   **85.** \$5000   **87.** 60 mph   **89.** 7.632 miles   **91.** 48 square inches
**93.** 558 square inches   **95.** $\approx$75.40 feet   **97.** 3 square feet   **99.** 7 square feet   **101.** $\approx$124.1 feet   **103.** $\approx$381.7 cubic inches
**105. a)** $B = \dfrac{703w}{h^2}$   **b)** $\approx$23.91   **107. a)** $V = 18x^3 - 3x^2$   **b)** 6027 cubic centimeters   **c)** $S = 54x^2 - 8x$
**d)** 2590 square centimeters   **109.** Doubles   **111.** 8 times as large   **113.** Circle   **115.** $\dfrac{1}{3}$   **116.** $-6$   **117.** 0   **118.** 8

## Exercise Set 2.7
**1.** Proportion   **3.** Means   **5.** Similar   **7.** Units   **9.** Yes   **11.** No   **13.** $2:3$   **15.** $1:2$   **17.** $8:1$   **19.** $7:4$
**21.** $1:3$   **23.** $6:1$   **25.** $7:30$   **27.** $8:1$   **29. a)** $50:23$   **b)** $\approx 2.17:1$   **31. a)** $7.25:4.75$   **b)** $\approx 1.53:1$   **33. a)** $35.7:2.4$ or $11.9:0.8$
**b)** $5.3:7.1$   **35. a)** $40:32$ or $5:4$   **b)** $15:11$   **37.** 12   **39.** 45   **41.** $-9$   **43.** $-2$   **45.** $-54$   **47.** 6   **49.** 32 inches   **51.** 15.75 inches
**53.** 19.5 inches   **55.** 25 loads   **57.** 403 miles   **59.** 1.5 feet   **61.** 24 teaspoons   **63.** $\approx$0.43 feet   **65.** 3.75 cups   **67.** $\approx$9.49 feet
**69.** 0.55 milliliter   **71.** 570 minutes or 9 hours 30 minutes   **73.** $\approx$360 children   **75.** 6.5 feet   **77.** 2.9 square yards   **79.** 20 inches
**81.** $\approx$22   **83.** \$307   **85.** 2020 pesos   **87.** Less. It is $2.12:1$.   **89.** It must increase.   **91.** $\approx$41,667 miles   **93.** 0.625 cubic centimeters
**96.** Commutative property of addition   **97.** Associative property of multiplication   **98.** Distributive property
**99.** All real numbers.   **100.** $m = (y - b)/x$

## Chapter 2 Review Exercises
**1.** $3x + 24$   **2.** $5x - 10$   **3.** $-2x - 8$   **4.** $-x - 2$   **5.** $-m - 8$   **6.** $-16 + 4x$   **7.** $25 - 5p$
**8.** $24x - 30$   **9.** $-25t + 25$   **10.** $-4x + 12$   **11.** $x + 2$   **12.** $-1 - 2y$   **13.** $-x - 2y + z$   **14.** $-6a + 15b - 21$   **15.** $4q$
**16.** $-3y + 8$   **17.** $5x + 1$   **18.** $-3x + 3y$   **19.** $8m + 8n$   **20.** $9x + 3y + 2$   **21.** $4x + 3y + 6$   **22.** 3   **23.** $-12x^2 + 3$   **24.** 0
**25.** $5x + 7$   **26.** $-3b + 2$   **27.** 0   **28.** $4x - 4$   **29.** $22x - 42$   **30.** $6x^2 - 3x + y$   **31.** $-\dfrac{7}{20}d + 7$   **32.** 3   **33.** $\dfrac{1}{6}x + 2$
**34.** $-\dfrac{7}{12}n$   **35.** 1   **36.** $-13$   **37.** 11   **38.** $-27$   **39.** $\dfrac{11}{5}$   **40.** $\dfrac{11}{2}$   **41.** $-8$   **42.** $-3$   **43.** 12   **44.** 4   **45.** 2   **46.** $-3$   **47.** $\dfrac{3}{2}$   **48.** $-3$
**49.** $-\dfrac{1}{2}$   **50.** $-1$   **51.** All real numbers   **52.** 2   **53.** $-5$   **54.** $-35.5$   **55.** $-1.125$   **56.** 0.6   **57.** $-\dfrac{21}{4}$   **58.** $\dfrac{78}{7}$   **59.** $-\dfrac{3}{2}$   **60.** $-\dfrac{2}{3}$
**61.** 2   **62.** $\dfrac{10}{7}$   **63.** 0   **64.** $-1$   **65.** 10   **66.** No solution   **67.** All real numbers   **68.** $-4$   **69.** No solution   **70.** All real numbers
**71.** $\dfrac{17}{3}$   **72.** $-\dfrac{20}{7}$   **73.** No solution   **74.** 7   **75.** 52   **76.** 32   **77.** 7   **78.** $-18$   **79.** 3   **80.** 48   **81.** 12 square centimeters
**82.** $\approx$33.51 cubic inches   **83.** $l = (P - 2w)/2$   **84.** $m = \dfrac{y - y_1}{x - x_1}$   **85.** $y = \dfrac{1}{3}x + \dfrac{2}{3}$   **86.** 308.5 miles   **87.** 240 square feet
**88.** $\approx$25.13 cubic inches   **89.** $3:4$   **90.** $5:12$   **91.** $6:1$   **92.** 2   **93.** 20   **94.** 9   **95.** $\dfrac{135}{4}$   **96.** $-10$   **97.** $-16$   **98.** $\dfrac{108}{7}$   **99.** 90
**100.** 40 inches   **101.** 1 foot   **102.** 6.3 hours   **103.** 72 dishes   **104.** 440 pages   **105.** $6\dfrac{1}{3}$ inches   **106.** 15.75 feet   **107.** $\approx$\$0.109
**108.** 192 bottles

## Chapter 2 Practice Test
**1.** $6x - 12$ [2.1]   **2.** $-x - 3y + 4$ [2.1]   **3.** $-3x + 4$ [2.1]   **4.** $-x + 10$ [2.1]   **5.** $-5x - y - 6$ [2.1]
**6.** $7a - 8b - 3$ [2.1]   **7.** $2x^2 + 6x - 1$ [2.1]   **8.** 4 [2.4]   **9.** 8 [2.5]   **10.** $-\dfrac{1}{7}$ [2.5]   **11.** No solution [2.5]   **12.** All real numbers [2.5]
**13.** $x = \dfrac{-by - c}{a}$ [2.6]   **14.** $y = \dfrac{6}{5}x - \dfrac{2}{5}$ [2.6]   **15.** 0 [2.5]   **16.** $-45$ [2.7]   **17. a)** Conditional equation   **b)** Contradiction
**c)** Identity [2.5]   **18.** $x = \dfrac{32}{3}$ feet or $10\dfrac{2}{3}$ feet [2.7]   **19.** 4% [2.6]   **20.** $\approx$28.27 inches [2.6]
**21.** 175 minutes or 2 hours 55 minutes [2.6]

## Cumulative Review Test
**1.** $\frac{8}{3}$ [1.3]   **2.** $\frac{15}{16}$ [1.3]   **3.** > [1.5]   **4.** 3 [1.7]   **5.** −1 [1.7]   **6.** 16 [1.9]   **7.** 168 [1.9]   **8.** 12 [1.9]
**9.** Distributive property [1.10]   **10.** $12x + y$ [2.1]   **11.** $\frac{1}{12}x + 25$ [2.1]   **12.** $3x^2 + 8x − 23$ [2.1]   **13.** −1 [2.4]   **14.** −44 [2.3]
**15.** $−\frac{5}{4}$ [2.5]   **16.** $\frac{12}{5}$ [2.5]   **17.** $b = 3A − a − c$ [2.6]   **18.** $\frac{9}{4}$ or 2.25 [2.7]   **19.** ≈380.13 square feet [2.6]   **20.** $42 [2.7]

## Chapter 3

### Exercise Set 3.1
**1.** $0.07c$   **3.** $7 − c$   **5.** $c + 0.07$   **7.** $2c + 7$   **9.** Equals   **11.** $h + 4$   **13.** $a − 5$   **15.** $5h$   **17.** $2d$   **19.** $\frac{1}{2}a$
**21.** $r − 5$   **23.** $12 − m$   **25.** $2w + 8$   **27.** $5a − 4$   **29.** $\frac{1}{3}w − 9$   **31.** $x$ = Sonya's height   **33.** $x$ = length of Jones Beach
**35.** $x$ = number of medals Finland won   **37.** $x$ = cost of the Chevy   **39.** $x$ = Teri's grade   **41.** $x$ = amount Kristen receives or $x$ = amount Yvonne receives   **43.** $x$ = Don's weight or $x$ = Angela's weight   **45.** Let $c$ = cost of chair, then $5c$ = cost of table.
**47.** Let $a$ = area of the kitchen, then $2a + 20$ = area of the living room.   **49.** Let $w$ = width of rectangle, then $6w − 2$ = length of rectangle.   **51.** Let $w$ = number of medals won by Sweden, then $20 − w$ = number of medals won by Brazil, or let $w$ = number of medals won by Brazil, then $20 − w$ = number of medals won by Sweden.   **53.** Let $g$ = George's age, then $\frac{1}{2}g + 2$ = Mike's age.
**55.** Let $m$ = number of miles Jan walked, then $6.4 − m$ = number of miles Edward walked, or let $m$ = number of miles Edward walked, then $6.4 − m$ = number of miles Jan walked.   **57.** $n + 8$   **59.** $\frac{1}{2}x$   **61.** $2a − 1$   **63.** $2t − 30$   **65.** $2p − 2.3$
**67.** $80{,}000 − m$   **69.** $2r − 673$   **71.** $10x$   **73.** $100d$   **75.** $45 + 0.40x$   **77.** $s + 0.20s$   **79.** $e − 0.12e$   **81.** $c + 0.07c$
**83.** $m − 0.307m$   **85.** $f + (f + 15)$   **87.** $(2l − 1) − l$   **89.** $x − (3x − 40)$   **91.** $w + (2w − 3)$   **93.** $r + (479r + 462)$
**95.** $r + (r + 0.394r)$   **97.** $a − (a + 0.112a)$   **99.** $x + 4x = 20$   **101.** $x + (x + 1) = 41$   **103.** $2x − 8 = 12$
**105.** $\frac{1}{5}(x + 10) = 150$   **107.** $x + 2(x + 2) = 22$   **109.** $12.50h = 150$   **111.** $2.99x = 17.94$   **113.** $25q = 175$
**115.** $a + (2a + 1) = 52$   **117.** $(2s + 300) − s = 420$   **119.** $s + (2s − 4) = 890$   **121.** $m + (3m − 2) = 12.6$
**123.** $c + 0.002c = 86{,}460$   **125.** $p − 0.019p = 12{,}087$   **127.** $c + 0.07c = 32{,}600$   **129.** $c + 0.15c = 42.50$
**131. a)** $86{,}400d + 3600h + 60m + s$   **b)** 368,125 seconds   **134.** 4   **135.** 15   **136.** $y = \frac{3x − 6}{2}$ or $y = \frac{3}{2}x − 3$   **137.** 2.52
**138.** 13:32

### Exercise Set 3.2
**1.** $x + (2x + 2) = 20$   **3.** $x + (x + 2) = 20$   **5.** $(2x + 2) − x = 20$   **7.** 5   **9.** 43, 44   **11.** 47, 49
**13.** 8, 19   **15.** 2, 6   **17.** 25, 42   **19.** 65 cards   **21.** 22.2 hours   **23.** 13 weeks   **25.** 9.6 years   **27.** December: 22, June: 226
**29.** $14.74   **31.** Japan: $2.17, U.S.: $0.77   **33.** 12.5 gallons   **35.** 18,100 copies   **37.** 6 movies   **39.** 150 miles   **41.** 6 years
**43.** 4000 pages   **45.** 2000 newsletters   **47.** $276.02   **49.** $23,230.77   **51.** ≈1.21 million   **53.** $25,000   **55.** $39,387.76   **57.** $49,867.37
**59.** $7500   **61.** 500 pages   **63.** $1071.43   **65.** $24.59   **67.** $77,777.78   **69.** 15.3 million gallons   **71.** Answers will vary
**73. a)** $80 = \frac{74 + 88 + 76 + x}{4}$   **b)** 82   **75.** 4   **76.** Commutative property of addition   **77.** $h = \frac{2A}{b}$   **78.** 12

### Mid-Chapter Test: Sections 3.1–3.2
**1.** $6w$ [3.1]   **2.** $3h + 5$ [3.1]   **3.** $c + 0.20c$ [3.1]   **4.** $60 + 0.95m$ [3.1]   **5.** $50n$ [3.1]
**6.** $25 − x$ [3.1]   **7.** $c − 0.25c$ [3.1]   **8.** $x$ = length of Poison Dart Frog [3.1]   **9.** Let $p$ = distance Pedro traveled, then $4p + 6$ = distance Mary traveled. [3.1]   **10.** $v − (v − 0.18v)$ [3.1]   **11.** $p + 0.12p = 38{,}619$ [3.1]   **12.** $x + 3(x + 2) = 26$ [3.1]
**13.** 46, 47 [3.2]   **14.** 4, 11 [3.2]   **15.** 18 days [3.2]   **16.** 15 hours [3.2]   **17.** $700 [3.2]   **18.** Betty: 196 clients, Anita: 404 clients [3.2]
**19.** 52 miles [3.2]   **20.** $5000 [3.2]

### Exercise Set 3.3
**1.** 10   **3.** 360°   **5.** Isosceles   **7.** $l \cdot w$   **9.** 45°   **11.** 46°, 46°, 88°   **13.** 11.5 inches   **15.** $A = 67°$, $B = 23°$
**17.** $A = 47°$, $B = 133°$   **19.** 88°   **21.** 50°, 60°, 70°   **23.** Length is 14 feet, width is 8 feet   **25.** Length is 78 feet, width is 36 feet
**27.** Smaller angles are 50°, larger angles are 130°   **29.** 30°, 30°, 150°, 150°   **31.** 63°, 73°, 140°, 84°   **33.** Width is 4 feet, height is 7 feet   **35. a)** Length: 100 inches, width: 190 inches   **b)** 7.69 inches   **c)** 76 inches   **d)** 53.85 inches   **e)** 19,000 square inches
**37.** Width is 11 feet, length is 16 feet   **39.** $ac + ad + bc + bd$   **41.** <   **42.** >   **43.** −10   **44.** $−2x − 5y + 6$   **45.** $y = −2x + 3$

### Exercise Set 3.4
**1.** Subtracting   **3.** $d = r \cdot t$   **5.** Adding   **7.** 1 hour   **9.** 6 mph   **11.** ≈0.72 hour   **13.** 2.4 hours   **15.** 70 mph
**17.** ≈44.4 seconds   **19.** 35 mph, 40 mph   **21.** ≈0.61 hour   **23.** 1.5 miles/day, 2.25 miles/day   **25.** *Apollo*: 5 mph, *Pythagoras*: 9 mph
**27.** 1.5 hours   **29. a)** 2.4 miles   **b)** 112.0 miles   **c)** 26.2 miles   **d)** 140.6 miles   **e)** 9.23 hours   **31.** Dom: 21 km/hr, Sue: 11 km/hr
**33.** $10,000 at 7%, $2000 at 5%   **35.** $2400 at 6%, $3600 at 4%   **37.** $2000 at 4%, $8000 at 5%   **39.** November   **41.** 8 hours at Home Depot, 10 hours at clinic   **43.** 1200 adult admissions   **45. a)** 41 shares of Nike, 205 shares of Kellogg   **b)** $37   **47.** 11.25 pounds almonds, 18.75 pounds walnuts   **49.** $160: 6 cubic yards, $120: 2 cubic yards   **51.** $2.74 per pound   **53.** ≈11.1%
**55.** $1\frac{2}{3}$ liters   **57.** 18.96%   **59.** ≈11.1%   **61.** 40% pure juice   **63.** ≈3.77 cups Clorox, ≈2.23 cups shock treatment   **65.** 11.25 pints
**67.** ≈5.74 hours   **70. a)** $\frac{22}{13}$ or $1\frac{9}{13}$   **b)** $\frac{35}{8}$ or $4\frac{3}{8}$   **71.** All real numbers   **72.** $\frac{3}{4}$ or 0.75   **73.** 38, 39

## Chapter 3 Review Exercises
**1.** $3n + 7$ **2.** $1.2g$ **3.** $d - 0.25d$ **4.** $16y$ **5.** $200 - x$ **6.** Let $d$ = Dino's age, then $7d + 6$ = Mario's age. **7.** $c - (c - 0.12c)$ **8.** $n - (3n - 24) = 8$ **9.** 33 and 41 **10.** 118 and 119 **11.** 38 and 7 **12.** $21,738.32 **13.** 19 months **14.** $2000 **15.** $618.75 **16.** 10 hours **17.** $\approx$ $171,000 **18.** $684 **19.** 45°, 55°, 80° **20.** 30°, 40°, 150°, 140° **21.** Width is 15.5 feet, length is 19.5 feet **22.** Width is 50 feet, length is 80 feet **23.** 45°, 45°, 135°, 135° **24.** Height is 2 feet, length is 4 feet **25.** 2 hours **26.** 4 hours **27.** $\approx$8.8 feet per second **28.** $4000 at 8%, $8000 at $7\frac{1}{4}$% **29.** $700 at 3%, $3300 at 3.5% **30.** $\approx$0.67 gallon **31.** 9 smaller and 21 larger **32.** 1.2 liters of 10%, 0.8 liters of 5% **33.** 103 and 105 **34.** $450 **35.** $12,000 **36.** 42°, 50°, 88° **37.** 8 years **38.** 70°, 70°, 110°, 110° **39.** 500 copies **40.** 0.6 mph **41.** 60 pounds of $3.50, 20 pounds of $4.10 **42.** Older brother: 55 mph, younger brother: 60 mph **43.** 0.4 liters **44.** Width is 16 feet, length is 24 feet **45.** 1.5 liters

## Chapter 3 Practice Test
**1.** $500 - n$ [3.1] **2.** $2w + 6000$ [3.1] **3.** $60t$ [3.1] **4.** $c + 0.06c$ [3.1] **5.** Let $n$ = number of packages of orange, then $7n - 105$ = number of packages of peppermint. [3.1] **6.** Let $x$ = number of men, then $600 - x$ = number of women, or let $x$ = number of women, then $600 - x$ = number of men. [3.1] **7.** $(2n - 1) - n$ [3.1] **8.** $n + (n + 18)$ [3.1] **9.** $(c + 0.84c) - c$ [3.1] **10.** 56 and 102 [3.2] **11.** 5 and 7 [3.2] **12.** 9 and 33 [3.2] **13.** $2500 [3.2] **14.** $34.78 [3.2] **15.** Peter: $40,000, Julie: $80,000 [3.2] **16.** 6 times [3.2] **17.** 7450 pages [3.2] **18.** 15 inches, 30 inches, 30 inches [3.3] **19.** Width is 6 feet, length is 8 feet [3.3] **20.** 59°, 59°, 121°, 121° [3.3] **21.** Harlene: 0.3 feet per minute, Ellis: 0.5 feet per minute [3.4] **22.** 6 mph [3.4] **23.** Jelly Belly: $\approx$1.91 pounds, Kits: $\approx$1.09 pounds [3.4] **24.** 20 liters [3.4] **25.** 1 liter of 8%, 2 liters of 5% [3.4]

## Cumulative Review Test
**1.** $16,000 [1.2] **2.** **a)** 11.6% **b)** 36 [1.2] **3.** **a)** 7.2 parts per million **b)** 6 parts per million [1.2] **4.** $\frac{5}{9}$ [1.3] **5.** $\frac{13}{24}$ inch [1.3] **6.** **a)** $\{1, 2, 3, 4, \dots\}$ **b)** $\{0, 1, 2, 3, \dots\}$ **c)** A rational number is a quotient of two integers where the denominator is not 0. [1.4] **7.** **a)** 4 **b)** $|-5|$ [1.5] **8.** $-34$ [1.9] **9.** $2x - 28$ [2.1] **10.** 5 [2.5] **11.** $\frac{1}{5}$ [2.5] **12.** No solution [2.5] **13.** $-8$ [2.7] **14.** $\approx$113.10 [2.6] **15.** **a)** $y = -\frac{1}{2}x + 2$ **b)** 4 [2.6] **16.** $w = \dfrac{P - 2l}{2}$ [2.6] **17.** 9 gallons [2.7] **18.** 40 minutes [3.2] **19.** 6, 23 [3.2] **20.** 40°, 45°, 90°, 185° [3.3]

# Chapter 4

## Exercise Set 4.1
**1.** $x$-coordinate, $y$-coordinate **3.** $y$-axis **5.** Line **7.** Roman numerals **9.** II **11.** IV **13.** I **15.** III **17.** III **19.** II **21.** $A(3, 1)$; $B(-3, 0)$; $C(1, -3)$; **23.** **25.** **27.** The points are collinear.

$$D(-2, -3); E(0, 3); F\left(\frac{3}{2}, -1\right)$$

**29.** $(-5, -3)$ is not on the line

**31.** **a)** Point c) does not satisfy the equation. **b)**

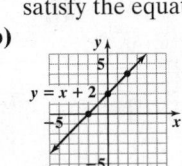

**33.** **a)** Point a) does not satisfy the equation. **b)**

**35.** **a)** Point a) does not satisfy the equation. **b)**

**37.** 2 **39.** $-4$ **41.** 6 **43.** $\frac{1}{2}$ **45.** **a)** Latitude, 16°N; Longitude, 56°W **b)** Latitude, 29°N; Longitude, 90.5°W **c)** Latitude 26°N; Longitude, 80.5°W **d)** Answers will vary. **47.** 0 **49.** To show that the line extends in both directions **55.** $\frac{13}{36}$ **56.** **a)** 9 **b)** $-9$ **c)** 9 **57.** 21 **58.** $y = \dfrac{3x - 4}{2} = \dfrac{3}{2}x - 2$ **59.** 23, 37 **60.** 2 hours

## Exercise Set 4.2
**1.** $x$-axis **3.** $y$-intercept **5.** $(x, 0)$ **7.** Horizontal **9.** 3 **11.** 5 **13.** 3 **15.** 2 **17.** $\frac{8}{3}$ **19.** $-10$

**21.**  **23.**  **25.**  **27.**  **29.**  **31.**

**33.**  **35.**  **37.**  **39.**  **41.**

**43.**  **45.**  **47.**  **49.**  **51.**  **53.**

**55.**  **57.**  **59.**  **61.**  **63.**

**65.** $x = 3$ **67.** $y = 3$ **69.** 5 **71.** 2 **73.** Yes

**75. a)** $C = 0.10n + 15$     **77. a)** $C = m + 40$

**b)**      **b)**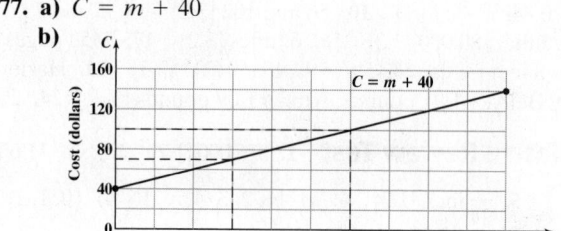

**c)** $19 **d)** 100 minutes     **c)** $100 **d)** 30 miles

**79. a)** 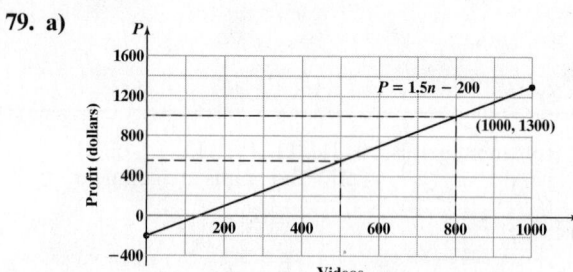     **b)** $550 **c)** 800 videos **81.** 3,2 **83.** 6,4

**85. a)**      **b)** (2,3) **c)** Yes **d)** No

**88.** −57

**89.** $C = 6\pi \approx 18.85$ in.

$A = 9\pi \approx 28.27$ in.$^2$

**90.** 6.67 ounces

**91.** $1400

**Exercise Set 4.3** **1.** Rise **3.** Ratio **5.** Horizontal **7.** Parallel **9.** Equal **11.** 2 **13.** $\frac{1}{2}$ **15.** 0 **17.** 1 **19.** Undefined

**21.** $-\frac{3}{8}$ **23.** $\frac{2}{3}$ **25.** $m = 2$ **27.** $m = -2$ **29.** $m = -\frac{4}{7}$ **31.** $m = \frac{7}{4}$ **33.** $m = 0$ **35.** $m = -\frac{2}{3}$ **37.** Undefined

**39.**  **41.**  **43.**  **45.**  **47.**

**49.** Parallel **51.** Neither

**53.** Perpendicular

**55.** Neither **57.** Neither

**59.** Parallel **61.** Parallel

**63.** Perpendicular **65.** 3

**67.** $\frac{1}{4}$ **69.** First **71. a)** 60 **b)** 75 **73.** −4 **75.** $-\frac{1}{8}$ **77.** 3

**79. a)**  **b)** $AC, m = \frac{3}{5}; CB, m = -2; DB, m = \frac{3}{5}; AD, m = -2$ **c)** Yes, opposite sides are parallel.

**81. a)** $AB, m = 4; BC, m = -2; CD, m = 4$ **b)** $[4 + (-2) + 4]/3 = 2$

**c)** $AD, m = 2$ **d)** Yes **e)** Answers will vary.

**83.** 0 **84. a)** $\frac{5}{2}$ **b)** 0 **85.** $\frac{84}{17}$

**86.** $C = d - a - b$ **87.** $x$-intercept: (6, 0); $y$-intercept: (0, −10)

**Mid-Chapter Test** **1.** IV [4.1] **2.**  [4.1] **3. b)** [4.1]

**4.** −4 [4.1]

**5.** 3 [4.1]

**6.** A graph of an equation in two variables is an illustration of a set of points whose coordinates satisfy the equation. [4.1]

**7.**  [4.2] **8.**  [4.2] **9.**  [4.2] **10.**  [4.2] **11.**  [4.2]

**12.**  [4.2]  **13.** $-\dfrac{2}{7}$ [4.3]   **16.**  [4.3] **17.**  [4.3]  **18.** Neither [4.3]
**19.** Perpendicular [4.3]
**20.** 10 [4.3]

**14.** 0 [4.3]

**15.** Undefined [4.3]

## Exercise Set 4.4
**1.** Point-slope  **3.** Slope-intercept  **5.** $y$-intercept  **7.** Slope  **9.** 3, $(0,1)$  **11.** $\dfrac{4}{3}$, $(0,-7)$

**13.** 1, $(0,-3)$    **15.** 3, $(0,2)$    **17.** 2, $(0,0)$    **19.** 2, $(0,-3)$

**21.** $\dfrac{5}{2}$, $(0,-5)$    **23.** $-\dfrac{1}{2}$, $\left(0,\dfrac{3}{2}\right)$    **25.** 3, $(0,4)$    **27.** $\dfrac{3}{2}$, $(0,2)$

**29.** $y = x - 2$  **31.** $y = -\dfrac{1}{3}x + 2$  **33.** $y = -3x - 5$  **35.** $y = \dfrac{1}{3}x + 5$  **37.** Parallel  **39.** Perpendicular  **41.** Parallel

**43.** Neither  **45.** Perpendicular  **47.** Neither  **49.** $y = 3x + 2$  **51.** $y = -3x - 7$  **53.** $y = \dfrac{1}{2}x - \dfrac{5}{2}$  **55.** $y = \dfrac{2}{3}x + 6$

**57.** $y = 3x + 10$  **59.** $y = -\dfrac{3}{2}x$  **61.** $y = \dfrac{1}{2}x - 2$  **63.** $y = 7.4x - 4.5$  **65. a)** $y = 5x + 60$  **b)** \$210

**67. a)** Slope-intercept form  **b)** Point-slope form  **c)** Point-slope form
**69. a)** No  **b)** $y + 4 = 2(x + 5)$  **c)** $y - 12 = 2(x - 3)$  **d)** $y = 2x + 6$  **e)** $y = 2x + 6$  **f)** Yes
**71. a)** 1.465  **b)** $f = 1.465m$  **c)** $\approx 223.7$ feet per second  **d)** 150 feet per second  **e)** 55 miles per hour

**73.** $y = -2x + 5$  **75.** $y = \dfrac{3}{4}x + 5$    **78.** $<$    **79.** T  **80.** T  **81.** F  **82.** F  **83.** 64  **84.** $r = \dfrac{i}{pt}$

## Chapter 4 Review Exercises
**1.**   **2.** Not collinear

**3. a), b),** and **d)**

**4. a)** $-7$  **b)** $-4$  **c)** 6  **d)** $\dfrac{8}{3}$

**5.** **6.** **7.** **8.** $y = 2x - 1$ **9.** **10.**

**11.** **12.** $5x + 2y + 10 = 0$ **13.** **14.**  **15.** $-\dfrac{9}{5}$  **16.** $-\dfrac{1}{12}$

**17.** $-2$

**18.** 0

**19.** Undefined

**20.** The slope of a line is the ratio of the vertical change to the horizontal change between any two points on the line.  **21.** $-2$

**22.** $\dfrac{1}{4}$  **23.** Neither  **24.** Perpendicular  **25. a)** 18  **b)** 10  **26.** $m = -\dfrac{6}{7}$, $(0,3)$  **27.** Slope is undefined, no $y$-intercept

**28.** $m = 0$, $(0,-3)$  **29.** $y = 3x - 3$  **30.** $y = -\dfrac{1}{2}x + 2$  **31.** Parallel  **32.** Perpendicular  **33.** $y = 3x + 1$  **34.** $y = -\dfrac{2}{3}x + 4$

**35.** $y = 2$  **36.** $x = 4$  **37.** $y = -\dfrac{7}{2}x - 3$  **38.** $x = -5$

## Chapter 4 Practice Test
**1.** A graph is an illustration of the set of points whose coordinates satisfy an equation. [4.1]
**2. a)** IV  **b)** II [4.1]  **3. a)** $ax + by = c$  **b)** $y = mx + b$  **c)** $y - y_1 = m(x - x_1)$ [4.1–4.4]

**4. b) and d)** [4.1]    **5.** $-\dfrac{4}{3}$ [4.3]    **6.** $\dfrac{4}{9}, \left(0, -\dfrac{5}{3}\right)$ [4.4]    **7.** $y = -x - 1$ [4.4]

**8.**  [4.2]    **9.**  [4.2]    **10.** [4.2]    **11. a)** $y = \dfrac{1}{2}x - 2$    **b)**  [4.2]

**12.**  [4.2]    **13.** $y = 4x - 13$ [4.4]    **14.** $y = -\dfrac{3}{7}x + \dfrac{2}{7}$ [4.4]    **15.** The lines are parallel since they have the same slope but different $y$-intercepts. [4.3]

**16.** [4.4]    **17.**  [4.4]

## Cumulative Review Test
**1. a)** $\{1, 2, 3, 4, \dots\}$    **b)** $\{0, 1, 2, 3, \dots\}$ [1.4]    **2.** 18 [1.7]    **3.** 35 [1.9]    **4.** 100 [1.9]
**5. a)** Distributive property    **b)** Commutative property of addition [1.10]    **6.** 20 [2.5]    **7.** All real numbers [2.5]    **8.** 20 [2.7]
**9.** $4.00 [2.7]    **10.** $w = \dfrac{v}{lh}$ [2.6]    **11.** 4 [3.2]    **12.** width: 5 feet, length: 13 feet [3.3]    **13.** 2 hours [3.4]
**14.** Answers will vary. [5.1]    **15.** [5.1]    **16.** [5.2]    **17.** 1 [5.3]    **18.** $-7$ [5.3]    **19.** [5.4]

**20.** $y - 2 = 3(x - 5)$ [5.4]

## Chapter 5

### Exercise Set 5.1
**1.** 0    **3.** Base    **5.** Expanded power rule for exponents    **7.** Product rule for exponents    **9.** $p^3 q^4 r^5$    **11.** $x^{10}$
**13.** $-z^5$    **15.** $y^5$    **17.** $3^5 = 243$    **19.** $z^8$    **21.** 6    **23.** $x^7$    **25.** $3^4 = 81$    **27.** $\dfrac{1}{y^2}$    **29.** 1    **31.** $\dfrac{1}{q^6}$    **33.** 1    **35.** 3    **37.** 4    **39.** $-9$
**41.** $6x^3 y^2$    **43.** $-8r$    **45.** $x^8$    **47.** $x^{25}$    **49.** $x^3$    **51.** $x^{12}$    **53.** $n^{18}$    **55.** $-8w^6$    **57.** $-27t^9$    **59.** $64x^9 y^6$    **61.** $\dfrac{x^2}{9}$    **63.** $\dfrac{y^4}{x^4}$    **65.** $-\dfrac{216}{x^3}$
**67.** $\dfrac{8x^3}{y^3}$    **69.** $\dfrac{16p^2}{25}$    **71.** $\dfrac{27x^{12}}{y^3}$    **73.** $\dfrac{a^7}{b^3}$    **75.** $\dfrac{x^{11}}{2y^7}$    **77.** $\dfrac{6y^4}{z^3}$    **79.** $\dfrac{7}{3x^5 y^3}$    **81.** $-\dfrac{3y^2}{x^3}$    **83.** $-\dfrac{2}{x^3 y^2 z^5}$    **85.** $\dfrac{8}{x^6}$    **87.** $27y^9$    **89.** 1    **91.** $\dfrac{x^4}{y^4}$
**93.** $\dfrac{z^{24}}{16y^{28}}$    **95.** $\dfrac{125}{s^6 t^9}$    **97.** $9x^2 y^8$    **99.** $5ab^4$    **101.** $-6x^2 y^2$    **103.** $15x^3 y^6$    **105.** $-9p^6 q^3$    **107.** $49r^6 s^4$    **109.** $x^2$    **111.** $x^{12}$    **113.** $6.25x^6$
**115.** $\dfrac{x^7}{y^4}$    **117.** $-\dfrac{m^{12}}{n^9}$    **119.** $-216x^9 y^6$    **121.** $-8x^{12} y^6 z^3$    **123.** $729r^{12} s^{15}$    **125.** $108x^5 y^7$    **127.** $1.69x^4 y^8$    **129.** $x^{11} y^{13}$    **131.** $x^8 z^8$
**133.** Cannot be simplified    **135.** Cannot be simplified    **137.** $6z^2$    **139.** Cannot be simplified    **141.** 40    **143.** 1    **145.** The sign will be positive because a negative number with an even exponent will be positive. This is because $(-1)^m = 1$ when $m$ is even.    **147.** $8x^2$
**149.** $ab + a^2 + b^2$    **151.** product rule: $x^m \cdot x^n = x^{m+n}$; power rule: $(x^m)^n = x^{m \cdot n}$    **153.** $576y^8 z^{13}$    **156.** 13    **157.** $x + 10$
**158.** All real numbers    **159. a)** 4 inches, 4 inches, 9 inches, 9 inches    **b)** $w = \dfrac{P - 2l}{2}$

### Exercise Set 5.2
**1.** $\dfrac{1}{x^2}$    **3.** $(-2)^4$    **5.** $x^4$    **7.** $\dfrac{1}{x^4}$    **9.** $4x^4$    **11.** $\dfrac{1}{x^6}$    **13.** $\dfrac{1}{5}$    **15.** $t^3$    **17.** $a$    **19.** 36    **21.** $\dfrac{1}{x^{20}}$    **23.** $\dfrac{1}{y^{20}}$    **25.** $\dfrac{1}{x^8}$
**27.** 9    **29.** $y^2$    **31.** $x^2$    **33.** 9    **35.** $\dfrac{1}{r}$    **37.** $p^3$    **39.** $\dfrac{1}{x^4}$    **41.** 27    **43.** $\dfrac{1}{125}$    **45.** $z^9$    **47.** $p^{24}$    **49.** $y^6$    **51.** $\dfrac{1}{x^4}$    **53.** $\dfrac{1}{x^{15}}$    **55.** $-\dfrac{1}{16}$
**57.** $-\dfrac{1}{16}$    **59.** $-\dfrac{1}{8}$    **61.** $\dfrac{1}{36}$    **63.** $\dfrac{1}{x^{10}}$    **65.** $n^2$    **67.** 1    **69.** 1    **71.** 64    **73.** $x^8$    **75.** 1    **77.** $\dfrac{1}{4}$    **79.** $\dfrac{1}{49}$    **81.** $x^3$    **83.** $\dfrac{1}{16}$    **85.** 125
**87.** $\dfrac{1}{9}$    **89.** 1    **91.** $\dfrac{1}{36x^4}$    **93.** $\dfrac{3y^2}{x^2}$    **95.** 4    **97.** $\dfrac{64}{125}$    **99.** $\dfrac{d^4}{c^8}$    **101.** $-\dfrac{s^4}{r^{16}}$    **103.** $-\dfrac{7}{a^3 b^4}$    **105.** $\dfrac{y^9}{64x^{15}}$    **107.** $\dfrac{18}{z^9}$    **109.** $-8$    **111.** $12x^5$
**113.** $-\dfrac{20z^5}{y}$    **115.** $8d^4$    **117.** $\dfrac{4}{x^2}$    **119.** $\dfrac{x^4}{2y^5}$    **121.** $\dfrac{8x^6}{y^2}$    **123.** $\dfrac{y^{14} z^6}{25x^8}$    **125.** $\dfrac{r^{20} t^{48}}{16s^{36}}$    **127.** $\dfrac{y^{12}}{x^{18} z^6}$    **129.** $\dfrac{1}{16p^4 q^6}$    **131. a)** Yes    **b)** No

**133.** $16\frac{1}{16}$  **135.** $125\frac{1}{125}$  **137.** $\frac{2}{3}$  **139.** $-\frac{7}{8}$  **141.** $-\frac{5}{6}$  **143.** $\frac{22}{9}$  **145.** $-2$  **147.** $-3$  **149.** Answers will vary.  **151.** $-2$  **153.** $2, -3$
**155.** The product rule is $(xy)^m = x^m y^m$, not $(x + y)^m = x^m + y^m$.  **157.** 186 miles  **158.** $94, 96$  **159.** 12 feet by 16 feet
**160.** $3400 at 4%, $5600 at 3%  **161.** $18x^3 y^9$

**Exercise Set 5.3**  **1.** $5.12 \times 10^{-5}$  **3.** Integer  **5.** $5.12 \times 10^{-6}$  **7.** 11  **9.** $2.0 \times 10^4$  **11.** $4.5 \times 10^2$  **13.** $3.5 \times 10^5$
**15.** $7.950 \times 10^3$  **17.** $5.3 \times 10^{-2}$  **19.** $7.26 \times 10^{-4}$  **21.** $5.26 \times 10^9$  **23.** $9.14 \times 10^{-6}$  **25.** $2.203 \times 10^5$  **27.** $5.104 \times 10^{-3}$
**29.** $43,000$  **31.** $0.00000932$  **33.** $0.0000213$  **35.** $625,000$  **37.** $9,000,000$  **39.** $535$  **41.** $0.000000773$  **43.** $10,000$  **45.** $0.000008$ meter
**47.** $125,000,000,000$ watts  **49.** $15,300$ meters  **51.** $0.0482$ meter  **53.** $90,000,000$  **55.** $0.243$  **57.** $0.000064$  **59.** $0.0013$  **61.** $2500$
**63.** $250,000$  **65.** $5.6 \times 10^{12}$  **67.** $1.28 \times 10^{-1}$  **69.** $7.0 \times 10^2$  **71.** $1.75 \times 10^2$  **73.** $3.3 \times 10^{-4}, 5.3, 7.3 \times 10^2, 1.75 \times 10^6$
**75. a)** $\approx 6.415 \times 10^9$  **b)** $\approx 22$  **77.** $8,640,000,000$ cubic feet  **79.** $1.6 \times 10^7$ seconds  **81. a)** $\$1.66 \times 10^8$  **b)** $\approx 1.38$
**83. a)** $5.25 \times 10^{10}$  **b)** $\approx 6.5$  **85. a)** Earth: $5.794 \times 10^{24}$ metric tons; Moon: $7.34 \times 10^{19}$ metric tons; Jupiter: $1.899 \times 10^{27}$ metric tons
**b)** $\approx 7.89 \times 10^4$  **c)** $\approx 3.28 \times 10^2$  **87.** $\$1.36 \times 10^9$  **89. a)** $2.0 \times 10^7$;  **b)** $7.6 \times 10^7$  **c)** $2.2 \times 10^7$  **91.** 17  **93.** A number
greater than or equal to 1 and less than 10 multiplied by some power of 10.  **95. a)** Less  **b)** $1.5 \times 10^{78}$  **97.** Answers will vary.

**99.** $1,000,000$ times greater  **102.** 0  **103. a)** $\frac{3}{2}$  **b)** 0  **104.** 2  **105.** $-\frac{y^{12}}{64x^9}$

**Mid-Chapter Test: Sections 5.1–5.3**  **1.** $y^{31}$ [5.1]  **2.** $x^3$ [5.1]  **3.** $6x^6 y^{13}$ [5.1]  **4.** $\frac{2a^5 b^6}{3}$ [5.1]  **5.** $-64x^6 y^{12}$ [5.1]  **6.** $\frac{t^6}{4s^4}$ [5.1]

**7.** $63x^{11} y^{13}$ [5.1]  **8.** $\frac{1}{p^8}$ [5.2]  **9.** $\frac{1}{x^{10}}$ [5.2]  **10.** 1 [5.2]  **11.** $\frac{49}{9}$ [5.2]  **12.** $\frac{32x}{y}$ [5.2]  **13.** $\frac{3}{m^4 n^6}$ [5.2]  **14.** $\frac{x^{12} y^{10}}{4z^4}$ [5.2]  **15. a)** and
**b)** Answers will vary. [5.3]  **16.** $6.54 \times 10^9$ [5.3]  **17.** $0.0000327$ [5.3]  **18.** $18,900$ meters [5.3]  **19.** $0.0238$ [5.3]  **20.** $3.0 \times 10^{-7}$ [5.3]

**Exercise Set 5.4**  **1.** $-5x^3 + 3x + 4$  **3.** $3x + 4$  **5.** 1  **7.** 5  **9.** True  **11.** 3  **13.** 1  **15.** 4  **17.** 5  **19.** 8  **21.** 3  **23.** 10
**25.** 9  **27.** Trinomial  **29.** Monomial  **31.** Binomial  **33.** Monomial  **35.** Not a polynomial  **37.** Polynomial  **39.** Trinomial
**41.** Not a polynomial  **43.** Already in descending order, 0  **45.** $x^2 - 2x - 4, 2$  **47.** $3x^2 + x - 8, 2$  **49.** Already in descending
order, 1  **51.** Already in descending order, 2  **53.** $4x^3 - 3x^2 + x - 4, 3$  **55.** $-2x^4 + 3x^2 + 5x - 6, 4$  **57.** $13x - 9$

**59.** $-x + 11$  **61.** $-2t - 1$  **63.** $x^2 + 6.6x + 0.8$  **65.** $5m^2 + 4$  **67.** $x^2 + 3x - 3$  **69.** $-3x^2 + x + \frac{17}{2}$  **71.** $8.2n^2 - 4.8n - 0.6$

**73.** $-2x^3 - 3x^2 + 4x - 3$  **75.** $11x^2 - 7x - y + 5$  **77.** $5x^2 y - 3x + 2$  **79.** $11x - 3$  **81.** $7y^2 - 2y + 5$  **83.** $4x^2 + 2x - 4$
**85.** $2x^3 - x^2 + 6x - 2$  **87.** $3n^3 - 11n^2 - n + 2$  **89.** $2x - 6$  **91.** $3x + 4$  **93.** $-3r$  **95.** $-6y^2 + 1.9y - 12.7$  **97.** $8x^2 + x + 4$

**99.** $-6.4n^2 + 6n - 7.6$  **101.** $8x^3 - 7x^2 - 4x - 3$  **103.** $2x^3 - \frac{23}{5}x^2 + 2x - 2$  **105.** $x - 2$  **107.** $2x^2 - 9x + 14$

**109.** $-5c^3 - 4c^2 - 7c + 14$  **111.** $3x + 8$  **113.** $3a^2 - 16a + 28$  **115.** $x^2 - 3x - 3$  **117.** $-4x^2$  **119.** $4x^3 - 7x^2 + x - 2$
**121.** Answers will vary.  **123.** Answers will vary.  **125.** Sometimes  **127.** Sometimes  **129.** Answers will vary; one example is
$x^4 - 2x^3 + x$.  **131.** $a^2 + 2ab + b^2$  **133.** $4x^2 + 3xy$  **135.** No, all three terms would have to be degree 4 or 1. Therefore at least two
of the terms would be like terms.  **137.** $-12x + 18$  **139.** $8x^2 + 28x - 24$  **141.** 4.5  **142.** $n - 5$  **143.**

**144.**    **145.** $\frac{y^3}{27x^{12}}$  **146.** $0.035$

**Exercise Set 5.5**  **1.** Trinomial  **3.** False  **5.** Align  **7.** Difference of two squares  **9.** $15x^5$  **11.** $-30t^{10}$  **13.** $-20x^6$
**15.** $-24x^6$  **17.** $20x^5 y^6$  **19.** $-28x^3 y^{15}$  **21.** $54x^6 y^{14}$  **23.** $3x^6 y$  **25.** $5.94x^8 y^3$  **27.** $9x - 54$  **29.** $-6x^2 + 6x$  **31.** $-16y - 10$
**33.** $-2x^3 + 4x^2 - 10x$  **35.** $-20x^3 + 30x^2 - 20x$  **37.** $0.5x^5 - 3x^4 - 0.5x^2$  **39.** $0.6x^2 y + 1.5x^2 - 1.8xy$  **41.** $x^2 y^4 - 4y^7 - 3y^4$
**43.** $5x^2 + 18x - 8$  **45.** $6t^2 + 3t - 30$  **47.** $4x^2 - 16$  **49.** $-5x^2 - 22x + 48$  **51.** $-12x^2 + 32x - 5$  **53.** $4x^2 - 10x + 4$
**55.** $12k^2 - 30k + 12$  **57.** $x^2 - 4$  **59.** $4x^2 - 12x + 9$  **61.** $-6z^2 + 46z - 28$  **63.** $8x^2 - 50x + 63$  **65.** $xy - 3x + 7y - 21$

**67.** $6x^2 - 5xy - 6y^2$  **69.** $-27x^2 - 3xy + 36x + 4y$  **71.** $x^2 + 0.9x + 0.18$  **73.** $x^2 + \frac{7}{2}x - 2$  **75.** $x^2 - 64$  **77.** $9x^2 - 64$

**79.** $x^2 + 2xy + y^2$  **81.** $x^2 - 0.4x + 0.04$  **83.** $16x^2 + 40x + 25$  **85.** $0.16x^2 + 0.8xy + y^2$  **87.** $16c^2 - 25d^2$  **89.** $4x^2 - 36$
**91.** $49s^2 - 42st + 9t^2$  **93.** $16m^3 - 8m^2 + 9m + 18$  **95.** $12x^3 + 5x^2 + 13x + 10$  **97.** $-14x^3 - 22x^2 + 19x - 3$  **99.** $a^3 + b^3$
**101.** $6t^4 + 5t^3 + 5t^2 + 10t + 4$  **103.** $x^4 - 3x^3 + 5x^2 - 6x$  **105.** $2x^5 + 2x^4 - 23x^3 + x^2 - 12x$  **107.** $b^3 - 3b^2 + 3b - 1$
**109.** $27a^3 - 135a^2 + 225a - 125$  **111.** Yes  **113.** No  **115.** 6, 3, 1  **117. a)** $(x + 2)(x + 1)$ or $2x^2 + 5x + 2$  **b)** 54 square feet
**c)** 1 foot  **119. a)** $3x^2 + 19x + 20$  **b)** $6x^3 + 32x^2 + 2x - 40$  **c)** 864 cubic feet  **d)** 864 cubic feet  **e)** Yes

**121.** $\frac{1}{3}x^2 + \frac{11}{45}x - \frac{4}{15}$  **123.** No solution  **124.** $C = 53°, D = 37°$  **125.** $\frac{x^4}{16y^8}$  **126. a)** $-216$  **b)** $\frac{1}{216}$  **127.** $-5x^2 - 2x + 14$

**Exercise Set 5.6**  **1.** $\frac{6x^2 - 7x - 20}{(2x - 5)} = (3x + 4)$  **3.** Divisor  **5.** Remainder  **7.** $6x^3 + 8x^2 + 0x + 16$  **9.** False

**11.** $\frac{x^2 - x - 42}{x - 7} = x + 6$ or $\frac{x^2 - x - 42}{x + 6} = x - 7$  **13.** $\frac{2x^2 + 5x + 3}{2x + 3} = x + 1$ or $\frac{2x^2 + 5x + 3}{x + 1} = 2x + 3$

**15.** $\frac{4x^2 - 9}{2x + 3} = 2x - 3$ or $\frac{4x^2 - 9}{2x - 3} = 2x + 3$  **17.** $t + 2$  **19.** $n + \frac{5}{2}$  **21.** $\frac{7}{3}x + 2$  **23.** $-3x + 2$  **25.** $3x + 1$  **27.** $\frac{1}{2}x + 4$

**29.** $-1 + \dfrac{5}{2}w$    **31.** $1 + \dfrac{2}{x} - \dfrac{3}{x^2}$    **33.** $-2x^3 + \dfrac{3}{x} + \dfrac{4}{x^2}$    **35.** $x^2 + 3x - \dfrac{3}{x^3}$    **37.** $3x^2 - 2x + 6 - \dfrac{5}{2x}$    **39.** $-2k^2 - \dfrac{3}{2}k + \dfrac{2}{k}$

**41.** $-4x^3 - x^2 + \dfrac{10}{3} + \dfrac{3}{x^2}$    **43.** $x + 3$    **45.** $5y + 1$    **47.** $2x + 4$    **49.** $x + 4$    **51.** $x + 5 - \dfrac{3}{2x - 3}$    **53.** $x + 6$

**55.** $3x^2 + 4x + 5 + \dfrac{4}{3x - 4}$    **57.** $4x - 3 - \dfrac{3}{2x + 3}$    **59.** $7x^2 - 5$    **61.** $2t^2 + \dfrac{12}{t - 2}$    **63.** $w^2 + 3w + 9 + \dfrac{19}{w - 3}$    **65.** $x^2 + 3x + 9$

**67.** $2x^2 + x - 2 - \dfrac{2}{2x - 1}$    **69.** $-m^2 - 7m - 5 - \dfrac{8}{m - 1}$    **71.** $4t^2 - 8t + 15 - \dfrac{26}{t + 2}$    **73.** No; for example $\dfrac{x + 2}{x} = 1 + \dfrac{2}{x}$ which
is not a binomial.    **75.** $2x^2 + 11x + 16$    **77.** 3    **79.** $4x$    **81.** Since the shaded areas minus 2 must equal 3, 1, 0, and $-1$, respectively,
the shaded areas are 5, 3, 2, and 1, respectively.    **83.** $x^2 + \dfrac{2}{3}x + \dfrac{4}{9} - \dfrac{37}{9(3x - 2)}$    **85.** $-3x + 3 + \dfrac{1}{x + 3}$    **88. a)** 2    **b)** 2, 0

**c)** $2, -5, 0, \dfrac{2}{5}, -6.3, -\dfrac{23}{34}$    **d)** $\sqrt{7}, \sqrt{3}$    **e)** $2, -5, 0, \sqrt{7}, \dfrac{2}{5}, -6.3, \sqrt{3}, -\dfrac{23}{34}$    **89. a)** 0    **b)** Undefined    **90.** Parentheses,
exponents, multiplication or division from left to right, addition or subtraction from left to right    **91.** $-\dfrac{2}{3}$    **92.** \$39.50    **93.** $x^{13}$

## Chapter 5 Review Exercises    **1.** $x^8$    **2.** $x^6$    **3.** 243    **4.** 32    **5.** $x^3$    **6.** 1    **7.** 25    **8.** 64    **9.** $\dfrac{1}{x^2}$    **10.** $y^3$    **11.** 1    **12.** 7    **13.** 1

**14.** 1    **15.** $25x^2$    **16.** $27a^3$    **17.** $-27x^3$    **18.** $216s^3$    **19.** $16x^8$    **20.** $t^{24}$    **21.** $p^{32}$    **22.** $\dfrac{4x^6}{y^2}$    **23.** $\dfrac{25y^4}{4b^2}$    **24.** $24x^5$    **25.** $\dfrac{4x}{y}$    **26.** $54x^4y^9$

**27.** $9x^2$    **28.** $24x^7y^7$    **29.** $16x^8y^{11}$    **30.** $6c^6d^3$    **31.** $\dfrac{27a^6}{b^{15}}$    **32.** $27x^{12}y^3$    **33.** $\dfrac{1}{b^9}$    **34.** $\dfrac{1}{27}$    **35.** $\dfrac{1}{25}$    **36.** $z^2$    **37.** $x^7$    **38.** 16    **39.** $\dfrac{1}{y^3}$

**40.** $\dfrac{1}{x^5}$    **41.** $\dfrac{1}{p^2}$    **42.** $\dfrac{1}{a^5}$    **43.** $m^{10}$    **44.** $x^7$    **45.** $\dfrac{1}{x^6}$    **46.** $\dfrac{1}{9x^8}$    **47.** $\dfrac{x^9}{64y^3}$    **48.** $\dfrac{4n^2}{m^6}$    **49.** $12y^2$    **50.** $-\dfrac{125z^3}{y^9}$    **51.** $\dfrac{x^4}{16y^6}$    **52.** $\dfrac{6}{x}$

**53.** $10x^2y^2$    **54.** $\dfrac{12x^2}{y}$    **55.** $\dfrac{24y^2}{x^2}$    **56.** $3y^5$    **57.** $\dfrac{4y}{x^3}$    **58.** $\dfrac{7x^5}{y^4}$    **59.** $\dfrac{x}{2y^5}$    **60.** $\dfrac{4y^{10}}{x}$    **61.** $1.72 \times 10^6$    **62.** $1.53 \times 10^{-1}$    **63.** $7.63 \times 10^{-3}$

**64.** $4.7 \times 10^4$    **65.** $5.76 \times 10^3$    **66.** $3.14 \times 10^{-4}$    **67.** 0.0075    **68.** 0.000652    **69.** 8,900,000    **70.** 51,200    **71.** 0.0000314
**72.** 11,030,000    **73.** 0.092 liter, $9.2 \times 10^{-2}$ liter    **74.** 6,000,000,000 meters, $6.0 \times 10^9$ meters    **75.** 0.0000128 gram, $1.28 \times 10^{-5}$ gram
**76.** 19,200 grams, $1.92 \times 10^4$ grams    **77.** 0.085    **78.** 1260    **79.** 245    **80.** 397,000,000    **81.** 0.00003    **82.** 0.0325    **83.** $3.64 \times 10^9$
**84.** $2.12 \times 10^1$    **85.** $5.0 \times 10^9$    **86.** $5.0 \times 10^{-4}$    **87.** $3.4 \times 10^{-3}$    **88.** $3.4 \times 10^7$    **89.** 50,000 gallons    **90. a)** \$4,300,000,000,000
**b)** $1.075 \times 10^{11}$ dollars    **91.** Not a polynomial    **92.** Monomial, 0    **93.** $x^2 + 3x - 4$, trinomial, 2    **94.** $4x^2 - x - 3$, trinomial, 2
**95.** Not a polynomial    **96.** Binomial, 3    **97.** $-4x^2 + x$, binomial, 2    **98.** Not a polynomial    **99.** $2x^3 + 4x^2 - 3x - 7$, polynomial, 3
**100.** $5x - 3$    **101.** $7d + 4$    **102.** $-3x - 5$    **103.** $-4x^2 + 11x - 7$    **104.** $5m^2 - 10$    **105.** $8.1p + 2.8$    **106.** $-3y - 15$
**107.** $4x^2 - 12x - 15$    **108.** $3a^2 - 5a - 21$    **109.** $4x + 2$    **110.** $-5x^2 + 8x - 19$    **111.** $3x^2 + 3x$    **112.** $-15x^2 - 12x$
**113.** $6x^3 - 12x^2 + 21x$    **114.** $-2c^3 + 3c^2 - 5c$    **115.** $28b^3 + 21b^2 + 35b$    **116.** $x^2 + 9x + 20$    **117.** $-12x^2 - 21x + 6$
**118.** $25x^2 - 30x + 9$    **119.** $-6x^2 + 14x + 12$    **120.** $r^2 - 25$    **121.** $3x^3 + x^2 - 10x + 6$    **122.** $3x^3 + 7x^2 + 14x + 4$

**123.** $-12x^3 + 10x^2 - 30x + 14$    **124.** $x + 2$    **125.** $4y + 6$    **126.** $8x + 4$    **127.** $2x^2 + 3x - \dfrac{4}{3}$    **128.** $2w - \dfrac{5}{3} + \dfrac{1}{w}$

**129.** $4x^5 - 2x^4 - \dfrac{3}{4}x^2 + \dfrac{1}{4x}$    **130.** $-4m + 2$    **131.** $\dfrac{5}{2}x + \dfrac{5}{x} + \dfrac{1}{x^2}$    **132.** $\dfrac{5}{3}x - 2 + \dfrac{5}{x}$    **133.** $x + 4$    **134.** $5x - 2 + \dfrac{2}{x + 6}$
**135.** $n + 3$    **136.** $2x^2 + 3x - 4$    **137.** $2x - 3$

## Chapter 5 Practice Test    **1.** $15x^6$ [5.1]    **2.** $27x^3y^6$ [5.1]    **3.** $8p^5$ [5.1]    **4.** $\dfrac{x^3}{8y^6}$ [5.1]    **5.** $\dfrac{y^4}{4x^6}$ [5.2]    **6.** 4 [5.1]    **7.** $\dfrac{2x^7y}{3}$ [5.2]

**8.** $1.425 \times 10^{10}$ [5.3]    **9.** $2.0 \times 10^{-7}$ [5.3]    **10.** Monomial [5.4]    **11.** Binomial [5.4]    **12.** Not a polynomial [5.4]
**13.** $6x^3 - 2x^2 + 5x - 5$, 3 [5.4]    **14.** $2x^2 + x - 7$ [5.4]    **15.** $-3y^2 - 2y + 5$ [5.4]    **16.** $3x^2 - x + 3$ [5.4]    **17.** $15d^2 - 40d$ [5.5]
**18.** $15x^2 + 4x - 32$ [5.5]    **19.** $-12c^2 + 7c + 45$ [5.5]    **20.** $6x^3 + 2x^2 - 35x + 25$ [5.5]    **21.** $4x^2 + 2x - 1$ [5.6]
**22.** $4x + 2 - \dfrac{5}{3x}$ [5.6]    **23.** $4x + 5$ [5.6]    **24.** $3x - 2 - \dfrac{2}{4x + 5}$ [5.6]    **25. a)** $5.73 \times 10^3$    **b)** $\approx 7.78 \times 10^5$ [5.3]

## Cumulative Review Test    **1.** 17 [1.9]    **2.** $8x + 2$ [2.1]    **3.** $-25$ [1.9]    **4.** $-\dfrac{33}{20}$ [2.4]    **5.** 13 [2.5]    **6.** 0 [2.5]    **7.** $y = 3x + 5$ [2.6]

**8.** $-\dfrac{1}{2}$ [4.3]    **9.** No [4.4]    **10.** $\dfrac{25x^6}{y^{16}}$ [5.1]    **11.** $-7x^2 - 5x + 2$, 2 [5.4]    **12.** $3x^2 + 9x - 2$ [5.4]    **13.** $5a^2 + 6a + 5$ [5.4]

**14.** $10t^2 - 11t + 3$ [5.5]    **15.** $6x^3 - 13x^2 + 9x - 2$ [5.5]    **16.** $\dfrac{5}{2}d + 3 - \dfrac{2}{d}$ [5.6]    **17.** $2x + 5$ [5.6]    **18.** \$3.33 [2.7]
**19.** Bob: 56.5 mph, Nick: 63.5 mph [3.4]    **20.** $l = 10$ feet, $w = 4$ feet [3.3]

## Chapter 6

**Exercise Set 6.1    1.** 6    **3.** $2x^2y^3$    **5.** 2    **7.** 12    **9.** $2^2 \cdot 5$    **11.** $2 \cdot 3^2 \cdot 5$    **13.** $2^3 \cdot 31$    **15.** 8    **17.** 14    **19.** 2    **21.** $x^2$    **23.** $3x$
**25.** $a$    **27.** $qr$    **29.** $x^3y^5$    **31.** 1    **33.** $x^2y^2$    **35.** $x$    **37.** $x - 4$    **39.** $2x - 3$    **41.** $3w + 5$    **43.** $x - 4$    **45.** $x - 1$    **47.** $x - 9$
**49.** $4(x - 20)$    **51.** $5(3x - 1)$    **53.** $7(q + 4)$    **55.** $3x(3x - 4)$    **57.** $x^4(7x - 9)$    **59.** $3x^2(x^3 - 4)$    **61.** $12x^8(3x^4 + 2)$

**63.** $9y^3(3y^{12} - 1)$ **65.** $y(1 + 6x^3)$ **67.** $a^2(7a^2 + 3)$ **69.** $4xy(4yz + x^2)$ **71.** $4x^2yz^3(20x^3y^2z - 9)$ **73.** $25x^2yz(z^2 + x)$
**75.** $x^4y^3z^9(19y^9z^4 - 8x)$ **77.** $4(2c^2 - c - 8)$ **79.** $3(3x^2 + 6x + 1)$ **81.** $4x(x^2 - 2x + 3)$ **83.** $8(5b^2 - 6c + 3)$
**85.** $3(5p^2 - 2p + 3)$ **87.** $3a(3a^3 - 2a^2 + b)$ **89.** $xy(8x + 12y + 5)$ **91.** $(x - 7)(x + 6)$ **93.** $(a - 2)(3b - 4)$
**95.** $(2x + 1)(4x + 1)$ **97.** $(2x + 1)(5x + 1)$ **99.** $(6c + 7)(3c - 2)$ **101.** $6\triangledown(2 - \triangledown)$ **103.** $4\square(3\square^2 - \square + 1)$ **105.** Multiply
to obtain the original expression and be sure all common factors have been factored out. **107.** $2x^2(2x + 7)(3x^3 + 2x - 1)$

**109.** $(x + 2)(x + 3)$ **110.** $-3x + 17$ **111.** $2$ **112.** $y = \dfrac{4x - 20}{5}$ or $y = \dfrac{4}{5}x - 4$ **113.** $\approx 201.06$ cubic inches **114.** $14, 27$

**115.** $\dfrac{9y^2}{4x^6}$

## Exercise Set 6.2
**1.** Factoring by grouping **3.** $(2x - 1)(x + 3)$ **5.** FOIL **7.** $(x + 3)(x + 2)$ **9.** $(t + 5)(t + 4)$
**11.** $(x + 2)(x + 5)$ **13.** $(c - 4)(c + 7)$ **15.** $(2x - 3)(2x + 3)$ **17.** $(x + 3)(3x + 1)$ **19.** $(2x + 1)(3x - 1)$
**21.** $(x + 4)(8x + 1)$ **23.** $(3t - 2)(4t - 1)$ **25.** $(x + 9)(x - 1)$ **27.** $(2p + 5)(3p - 2)$ **29.** $(x + 2y)(x - 3y)$
**31.** $(3x + 2y)(x - 3y)$ **33.** $(5x - 6y)(2x - 5y)$ **35.** $(x - b)(x - a)$ **37.** $(y + 9)(x - 5)$ **39.** $(a + 3)(a + b)$
**41.** $(y - 1)(x + 5)$ **43.** $(3 + 2y)(4 - x)$ **45.** $(z + 5)(z^2 + 1)$ **47.** $(x - 5)(x^2 + 8)$ **49.** $2(x - 6)(x + 4)$
**51.** $4(x + 2)(x + 2) = 4(x + 2)^2$ **53.** $x(2x + 3)(3x - 1)$ **55.** $p(p - 6q)(p + 2q)$ **57.** $(y + 5)(x + 5)$ **59.** $(x + 5)(y + 6)$
**61.** $(a + b)(x + y)$ **63.** $(r + 6)(s - 7)$ **65.** $(c - a)(d + 3)$ **67.** No; $xy + 2x + 5y + 10$ is factorable; $xy + 10 + 2x + 5y$ is
not factorable in this arrangement. **69.** $(\odot + 3)(\odot - 5)$ **71.** $x^2 - 2xy - 3x + 6y$ or $x^2 - 3x - 2xy + 6y$
**73. a)** $2x^2 - 5x - 6x + 15$ **b)** $(2x - 5)(x - 3)$ **75. a)** $2x^2 - 6x - 5x + 15$ **b)** $(x - 3)(2x - 5)$

**77. a)** $4x^2 + 3x - 20x - 15$ **b)** $(4x + 3)(x - 5)$ **79.** $(\odot + 3)(\star + 2)$ **81.** $\dfrac{6}{5}$ **82.** 30 pounds of jelly beans, 20 pounds of

gumdrops **83.** $5x^2 - 2x - 3 + \dfrac{5}{3x}$ **84.** $a - 4$

## Exercise Set 6.3
**1.** Both negative **3.** $2(x - 2)(x - 3)$ **5.** $2(x - 5)(x - 1)$ **7.** Both positive **9.** $y(x - 1)(x - 5)$
**11.** $(x + 3)(x + 8)$ **13.** $(y - 8)(y - 2)$ **15.** $(x - 5)(x - 2)$ **17.** $(x + 2)(x + 4)$ **19.** $(x + 8)(x - 3)$ **21.** Prime
**23.** $(y - 12)(y - 1)$ **25.** $(a - 4)(a + 2)$ **27.** $(r - 5)(r + 3)$ **29.** $(b - 9)(b - 2)$ **31.** Prime **33.** $(q + 9)(q - 5)$
**35.** $(x - 10)(x + 3)$ **37.** $(x + 2)^2$ **39.** $(s - 4)^2$ **41.** $(p - 6)^2$ **43.** $(w - 15)(w - 3)$ **45.** $(x + 13)(x - 3)$
**47.** $(x - 5)(x + 4)$ **49.** $(y + 5)(y + 8)$ **51.** $(x + 16)(x - 4)$ **53.** Prime **55.** $(x - 16)(x - 4)$ **57.** $(a - 9)(a - 11)$
**59.** $(x + 2)(x + 1)$ **61.** $(w + 9)(w - 2)$ **63.** $(x - 3y)(x - 5y)$ **65.** $(m - 3n)^2$ **67.** $(x + 6y)(x + 2y)$
**69.** $(m + 3n)(m - 8n)$ **71.** $6(x - 4)(x - 1)$ **73.** $5(x + 3)(x + 1)$ **75.** $2(x - 4)(x - 5)$ **77.** $b(b - 5)(b - 2)$
**79.** $3z(z - 9)(z + 2)$ **81.** $x(x + 4)^2$ **83.** $7(a - 2b)(a - 3b)$ **85.** $3r(r + 4t)(r - 2t)$ **87.** $x^2(x - 7)(x + 3)$ **89.** Both
negative; one positive and one negative; one positive and one negative; both positive **91.** $x^2 - 12x + 32 = (x - 8)(x - 4)$
**93.** $x^2 - 2x - 35 = (x - 7)(x + 5)$ **95.** The GCF, 2, was not factored out first. **97.** $(x + 0.4)(x + 0.2)$
**99.** $\left(x + \dfrac{1}{5}\right)\left(x + \dfrac{1}{5}\right) = \left(x + \dfrac{1}{5}\right)^2$ **101.** $(x + 8)(x - 32)$ **103.** $9$ **104.** $19.6\%$ **105.**

**106.** $2x^3 + x^2 - 16x + 12$ **107.** $3x + 2 - \dfrac{2}{x - 4}$ **108.** $(5x + 2)(4x - 3)$

## Exercise Set 6.4
**1.** Factoring by grouping **3.** Factor out the GCF **5.** $(2x + 1)(x + 5)$ **7.** $(3x + 2)(x + 4)$
**9.** $(5x + 1)(x - 2)$ **11.** $(3r - 2)(r + 5)$ **13.** $(2z - 3)^2$ **15.** $(2z + 3)(3z - 4)$ **17.** Prime **19.** $(8x + 3)(x + 2)$ **21.** Prime
**23.** $(5y - 1)(y - 3)$ **25.** $(7x + 1)(x + 6)$ **27.** $(2x + 5)(2x - 3)$ **29.** $(7t - 1)^2$ **31.** $(5z + 4)(z - 2)$ **33.** $(4y - 3)(y + 2)$
**35.** $(5x - 1)(2x - 5)$ **37.** $(5d + 4)(2d - 3)$ **39.** $2(4x + 1)(x - 6)$ **41.** $(7t + 3)(t + 1)$ **43.** $2(3x + 5)(x + 1)$
**45.** $x(2x + 1)(3x - 4)$ **47.** $4x(3x + 1)(x + 2)$ **49.** $2x(2x + 3)(x - 2)$ **51.** $8(2c - 1)(3c + 2)$ **53.** $4(2p + 3)(p - 1)$
**55.** $(8c + d)(c + 5d)$ **57.** $(5x + 3y)(3x - 2y)$ **59.** $2(2x - y)(3x + 4y)$ **61.** $(7p + 6q)(p + q)$ **63.** $(3m - 2n)(2m + n)$
**65.** $x(4x + 3y)(2x + y)$ **67.** $x^2(2x + y)(2x + 3y)$ **69.** $3x^2 - 20x - 7$; obtained by multiplying the factors.
**71.** $10x^2 + 35x + 15$; obtained by multiplying the factors. **73.** $3t^4 + 11t^3 - 4t^2$; obtained by multiplying the factors.
**75. a)** Dividing the trinomial by binomial gives the second factor. **b)** $6x + 11$ **77.** Factoring trinomials is the reverse process of
multiplying binomials. **79.** $(6x - 5)(3x + 4)$ **81.** $(5x - 8)(3x - 20)$ **83.** $5(3a - 8)(7a + 4)$ **85.** $2x + 45$, the product of the
three first terms must equal $6x^3$, and the product of the constants must equal 2250. **87.** $49$ **88.** $155.2$ mph
**89.** $12xy^2(3x^3y - 1 + 2x^4y^4)$ **90.** $(b + 12)(b - 8)$

## Mid-Chapter Test: Sections 6.1–6.4
**1.** A factoring problem may be checked by multiplying the factors. [6.1]
**2.** $3xy^2$ [6.1] **3.** $4a^2b(b^2 - 6a)$ [6.1] **4.** $(d - 6)(5c - 3)$ [6.1] **5.** $(2x + 9)(7x + 1)$ [6.1] **6.** $(x + 4)(x + 7)$ [6.2]
**7.** $(x + 5)(x - 3)$ [6.2] **8.** $(2a + 5b)(3a - b)$ [6.2] **9.** $(5x - 2y)(x - 9)$ [6.2] **10.** $4x(2x + 1)(x - 6)$ [6.2]
**11.** $(x - 3)(x - 7)$ [6.3] **12.** $(t + 4)(t + 5)$ [6.3] **13.** Prime [6.3] **14.** $(x + 8)^2$ [6.3] **15.** $(m + 5n)(m - 9n)$ [6.3]
**16.** $(3x + 2)(x + 5)$ [6.4] **17.** $(4z - 3)(z - 2)$ [6.4] **18.** Prime [6.4] **19.** $(3x - 1)^2$ [6.4] **20.** $3(2a - b)(a + b)$ [6.4]

## Exercise Set 6.5
**1.** The difference of two squares **3.** The sum of two squares **5.** Factor out the greatest common factor
**7.** Prime **9.** $9(b^2 + 9)$ **11.** $4(4m^2 + 9n^2)$ **13.** $(y + 5)(y - 5)$ **15.** $(9 + z)(9 - z)$ **17.** $(x + 7)(x - 7)$ **19.** $(x + y)(x - y)$
**21.** $(3y + 5z)(3y - 5z)$ **23.** $4(4a + 3b)(4a - 3b)$ **25.** $(6 + 7x)(6 - 7x)$ **27.** $(z^2 + 9x)(z^2 - 9x)$ **29.** $(5x^2 + 7y^2)(5x^2 - 7y^2)$
**31.** $(6m^2 + 7n)(6m^2 - 7n)$ **33.** $2(x^2 + 5y)(x^2 - 5y)$ **35.** $5(x^2 + 9)(x + 3)(x - 3)$ **37.** $(x + y)(x^2 - xy + y^2)$
**39.** $(x - y)(x^2 + xy + y^2)$ **41.** $(x + 4)(x^2 - 4x + 16)$ **43.** $(x - 3)(x^2 + 3x + 9)$ **45.** $(a + 1)(a^2 - a + 1)$

**47.** $(3x - 1)(9x^2 + 3x + 1)$  **49.** $(3a - 5)(9a^2 + 15a + 25)$  **51.** $(3 - 2y)(9 + 6y + 4y^2)$  **53.** $(4m + 3n)(16m^2 - 12mn + 9n^2)$
**55.** $(2a - 3b)(4a^2 + 6ab + 9b^2)$  **57.** $4(t - 3)^2$  **59.** $2(5x + 2)(5x - 3)$  **61.** $2(d + 4)^2$  **63.** $5(x - 3)(x + 1)$
**65.** $5(x + 2)(x - 2)$  **67.** $2(x + 5)(x - 5)$  **69.** $2y(x + 3)(x - 3)$  **71.** $3y^2(x + 1)(x^2 - x + 1)$  **73.** $2(x - 2)(x^2 + 2x + 4)$
**75.** $2(3x + 5)(3x - 5)$  **77.** $3r(2t^2 - 5t + 7)$  **79.** $2(3x - 2)(x + 4)$  **81.** $2r(s + 3)(s - 8)$  **83.** $(x + 2)(4x - 3)$
**85.** $25(b + 2)(b - 2)$  **87.** $a^3b^2(a + 2b)(a - 2b)$  **89.** $5x^2(x + 1)^2$  **91.** $x(x^2 + 25)$  **93.** $(y^2 + 4)(y + 2)(y - 2)$
**95.** $2(2m + 5)(4m^2 - 10m + 25)$  **97.** $(a + b)(c + 2)$  **99.** $9(1 + y^2)(1 + y)(1 - y)$  **101.** $2\blacklozenge^4(\blacklozenge^2 + 2\maltese^2)$  **103.** Answers will vary.
**105.** $(x^2 - 3y^3)(x^4 + 3x^2y^3 + 9y^6)$  **107.** $(x - 3 + 2y)(x - 3 - 2y)$  **109.** $(x + y + 3)(x - y + 7)$

**110.** $\dfrac{7}{3}$  **111.** 4 inches  **112.** $-9$  **113.** $\dfrac{8x^9}{27y^{12}}$  **114.** $\dfrac{1}{a^{11}}$

## Exercise Set 6.6
**1.** Zero-factor property  **3.** $x = 0$ or $x = 2$  **5.** $x = -2$ or $x = 2$  **7.** $-8, 7$  **9.** $0, 8$  **11.** $-\dfrac{7}{3}, \dfrac{11}{2}$
**13.** $4, -4$  **15.** $0, 12$  **17.** $0, -7$  **19.** $4$  **21.** $-2, -10$  **23.** $-2, -10$  **25.** $-3, 4$  **27.** $1, -24$  **29.** $-5, -6$  **31.** $5, -3$  **33.** $30, -1$
**35.** $-3, -\dfrac{5}{4}$  **37.** $-\dfrac{1}{3}, -2$  **39.** $\dfrac{2}{3}, -5$  **41.** $-4, 3$  **43.** $-\dfrac{3}{4}, \dfrac{1}{2}$  **45.** $8, -8$  **47.** $0, 25$  **49.** $10, -10$  **51.** $-2, 5$  **53.** $-2, \dfrac{1}{3}$  **55.** $\dfrac{3}{2}, 6$

**57.** $x^2 - 2x - 24 = 0$ (other answers are possible)  **59.** $x^2 - 6x = 0$ (other answers are possible)  **61. a)** $(2x - 1)$ and $(3x + 1)$
**b)** $6x^2 - x - 1 = 0$  **63.** They are constant multiples of one another.  **65.** $4, 5$  **67.** $0, 3, -2$  **69.** $\dfrac{17}{45}$  **70. a)** Identity
**b)** Contradiction  **71.** $\approx 738$ people  **72.** $\dfrac{9}{p^8q^2}$  **73.** Monomial  **74.** Binomial  **75.** Not a polynomial  **76.** Trinomial

## Exercise Set 6.7
**1.** Hypotenuse  **3.** $90°$  **5.** $8$  **7.** $13$  **9.** $10$  **11.** $39$  **13.** $13, 17$  **15.** $6, 14$  **17.** $16, 18$  **19.** Width: 3 feet,
length: 12 feet  **21.** Width: 10 feet, length: 15 feet  **23.** 5 meters  **25.** 4 seconds  **27.** Yes  **29.** Yes  **31.** 16 feet  **33.** 41 feet
**35.** 6 feet, 8 feet, 10 feet  **37.** Width: 9 inches, length: 12 inches  **39.** Width: 7 feet, length: 24 feet  **41.** 30 books  **43. a)** 4  **b)** 9
**45.** 100 feet  **47.** 432 square feet  **49.** $0, 8, -4$  **51.** $x^3 - x^2 - 6x = 0$  **53.** 3 and 6  **55.** 20 and 50 units  **57.** $3x - 7$
**58.** $-x^2 + 7x - 4$  **59.** $6x^3 + x^2 - 10x + 4$  **60.** $2x - 3$  **61.** $2x - 3$

## Chapter 6 Review Exercises
**1.** $y^3$  **2.** $3p$  **3.** $6c^2$  **4.** $5x^2y^2$  **5.** $1$  **6.** $s$  **7.** $x - 3$  **8.** $x + 5$  **9.** $7(x - 5)$
**10.** $5(7x - 1)$  **11.** $4y(6y - 1)$  **12.** $5p^2(11p - 4)$  **13.** $12ab(5a - 3b)$  **14.** $9xy(1 - 4x^2y)$  **15.** $4x^3y^2(5 + 2x^6y - 4x^2)$
**16.** Prime  **17.** Prime  **18.** $(5x + 3)(x - 2)$  **19.** $(t - 1)(3t + 4)$  **20.** $(4x - 3)(2x + 1)$  **21.** $(x + 6)(x + 2)$
**22.** $(x - 5)(x + 4)$  **23.** $(y - 6)^2$  **24.** $(y + 1)(3x + 2)$  **25.** $(a - b)(4a - 1)$  **26.** $(x + 6)(2x - 1)$  **27.** $(x + 3)(x - 2)$
**28.** $(5x - y)(x + 4y)$  **29.** $(x + 3y)(4x - 5y)$  **30.** $(3a - 5b)(2a - b)$  **31.** $(p - 3)(q + 4)$  **32.** $(x - 3y)(3x + 2y)$
**33.** $(a + 2b)(7a - b)$  **34.** $(2x - 1)(4x + 3)$  **35.** $(x + 3)(x + 2)$  **36.** Prime  **37.** $(x + 2)(x + 9)$  **38.** $(n + 8)(n - 5)$
**39.** $(b + 5)(b - 4)$  **40.** $(x - 8)(x - 7)$  **41.** Prime  **42.** Prime  **43.** $x(x - 9)(x - 8)$  **44.** $t(t - 9)(t + 4)$
**45.** $(x + 3y)(x - 5y)$  **46.** $4x(x + 5y)(x + 3y)$  **47.** $(2x + 5)(x - 3)$  **48.** $(6x + 1)(x - 5)$  **49.** $(4x - 5)(x - 1)$
**50.** $(5m - 4)(m - 2)$  **51.** $(4y + 3)(4y - 1)$  **52.** $(5x - 2)(x - 6)$  **53.** Prime  **54.** $(5x - 3)(x + 8)$  **55.** $(2s + 1)(3s + 5)$
**56.** $(3x - 2)(2x + 5)$  **57.** $2(3x + 2)(2x - 1)$  **58.** $(5x - 3)^2$  **59.** $x(3x - 2)^2$  **60.** $2x(3x + 4)(3x - 2)$
**61.** $(2a - 3b)(2a - 5b)$  **62.** $(8a + b)(2a - 3b)$  **63.** $(x + 10)(x - 10)$  **64.** $(x + 6)(x - 6)$  **65.** $3(x + 4)(x - 4)$
**66.** $9(3x + y)(3x - y)$  **67.** $(9 + a)(9 - a)$  **68.** $(8 + x)(8 - x)$  **69.** $(4x^2 + 7y)(4x^2 - 7y)$  **70.** $(8x^3 + 7y^3)(8x^3 - 7y^3)$
**71.** $(a + b)(a^2 - ab + b^2)$  **72.** $(x - y)(x^2 + xy + y^2)$  **73.** $(x - 1)(x^2 + x + 1)$  **74.** $(x + 2)(x^2 - 2x + 4)$
**75.** $(a + 3)(a^2 - 3a + 9)$  **76.** $(b - 4)(b^2 + 4b + 16)$  **77.** $(5a + b)(25a^2 - 5ab + b^2)$  **78.** $(3 - 2y)(9 + 6y + 4y^2)$
**79.** $3(x - 4y)(x^2 + 4xy + 16y^2)$  **80.** $3(3x^2 + 5y)(3x^2 - 5y)$  **81.** $(x - 6)(x - 8)$  **82.** $3(x - 3)^2$  **83.** $5(q + 1)(q - 1)$
**84.** $8(x + 3)(x - 1)$  **85.** $4(y + 3)(y - 3)$  **86.** $(x - 9)(x + 3)$  **87.** $(3x - 1)^2$  **88.** $(7x - 3)(x + 4)$  **89.** $6(b - 1)(b^2 + b + 1)$
**90.** $y(x - 3)(x^2 + 3x + 9)$  **91.** $b(a + 3)(a - 5)$  **92.** $3x(2x + 3)(x + 5)$  **93.** $(x - 3y)(x - y)$  **94.** $(3m - 4n)(m + 2n)$
**95.** $(2x + 3y)^2$  **96.** $(5a + 7b)(5a - 7b)$  **97.** $(x + 2)(y - 7)$  **98.** $y^5(4 + 5y)(4 - 5y)$  **99.** $(2x - 3y)(3x + 7y)$

**100.** $2x(2x + 5y)(x + 2y)$  **101.** $x^2(4x + 1)(4x - 3)$  **102.** $(d^2 + 4)(d + 2)(d - 2)$  **103.** $0, -9$  **104.** $2, -6$  **105.** $-5, \dfrac{3}{4}$

**106.** $0, -7$  **107.** $0, -5$  **108.** $0, -3$  **109.** $-3, -6$  **110.** $1, 2$  **111.** $-4, 3$  **112.** $-1, -4$  **113.** $2, 4$  **114.** $3, -5$  **115.** $\dfrac{1}{4}, -\dfrac{3}{2}$

**116.** $-\dfrac{1}{3}, 4$  **117.** $2, -2$  **118.** $\dfrac{10}{7}, -\dfrac{10}{7}$  **119.** $\dfrac{3}{2}, \dfrac{1}{4}$  **120.** $\dfrac{3}{2}, \dfrac{5}{2}$  **121.** $a^2 + b^2 = c^2$  **122.** Hypotenuse  **123.** 10 feet  **124.** 12 meters
**125.** $9, 11$  **126.** $4, 14$  **127.** Width: 12 feet, length: 15 feet  **128.** 8 feet, 15 feet, 17 feet  **129.** 9 inches  **130.** 10 feet
**131.** 1 second  **132.** 80 dozen

## Chapter 6 Practice Test
**1.** $3y^3$ [6.1]  **2.** $8p^2q^2$ [6.1]  **3.** $5x^2y^2(y - 3x^3)$ [6.1]  **4.** $4a^2b(2a - 3b + 7)$ [6.1]
**5.** $(x - 5)(4x + 1)$ [6.2]  **6.** $(a - 4b)(a - 5b)$ [6.2]  **7.** $(r + 8)(r - 3)$ [6.3]  **8.** $(5a - 3b)(5a + 2b)$ [6.4]  **9.** $4(x + 2)(x - 6)$ [6.4]
**10.** $y(2y - 3)(y + 1)$ [6.4]  **11.** $(3x + 2y)(4x - 3y)$ [6.4]  **12.** $(x + 3y)(x - 3y)$ [6.5]  **13.** $(x - 4)(x^2 + 4x + 16)$ [6.5]
**14.** $\dfrac{5}{6}, -3$ [6.6]  **15.** $0, 6$ [6.6]  **16.** $-8, 8$ [6.6]  **17.** $-9$ [6.6]  **18.** $3, 4$ [6.6]  **19.** $-2, -3$ [6.6]  **20.** 24 inches [6.7]  **21.** 34 feet [6.7]
**22.** $4, 9$ [6.7]  **23.** $12, 14$ [6.7]  **24.** Length: 6 meters, width: 4 meters [6.7]  **25.** 10 seconds [6.7]

**Cumulative Review Test**    **1.** $-171$ [1.9]    **2.** 9 [1.9]    **3.** \$90 [3.2]    **4. a)** 7    **b)** $-6, -0.2, \frac{3}{5}, 7, 0, -\frac{5}{9}, 1.34$    **c)** $\sqrt{7}, -\sqrt{2}$

**d)** $-6, -0.2, \frac{3}{5}, \sqrt{7}, -\sqrt{2}, 7, 0, -\frac{5}{9}, 1.34$ [1.4]    **5.** $|-8|$ [1.5]    **6.** 13 [2.5]    **7.** 19.2 [2.7]    **8.** 7 gallons [2.7]

**9.** $y = -\frac{4}{3}x + \frac{7}{3}$ [2.6]    **10.** 6 liters [3.4]    **11.** $\frac{1}{4}$ hour [3.4]    **12.**  [4.4]    **13.** $y = \frac{3}{5}x + \frac{26}{5}$ [4.4]    **14.** $\frac{16y^6}{x^3}$ [5.2]

$y = -\frac{3}{5}x + 1$

**15.** $-3x^3 + 2x^2 + 6x - 12$ [5.4]

**16.** $3x^3 + 13x^2 - 28x + 12$ [5.5]    **17.** $x - 5 + \frac{21}{x+3}$ [5.6]

**18.** $(r+2)(q-8)$ [6.2]    **19.** $(5x+3)(x-2)$ [6.4]    **20.** $7y(y+3)(y-3)$ [6.5]

## Chapter 7

**Exercise Set 7.1**    **1.** Rational expression    **3.** Reduced to lowest terms    **5.** $x - 4$    **7.** $-1$    **9.** $4x - 1$    **11.** All real numbers except 1    **13.** All real numbers except $x = 0$    **15.** All real numbers except $n = 4$    **17.** All real numbers except $x = 2$ and $x = -2$

**19.** All real numbers except $x = \frac{3}{2}$ and $x = 3$    **21.** All real numbers    **23.** All real numbers except $p = \frac{5}{2}$ and $p = -\frac{5}{2}$

**25.** $\frac{x}{3y^4}$    **27.** $\frac{4}{b^5}$    **29.** $\frac{5}{1+y}$    **31.** 5    **33.** $\frac{x^2 + 6x + 7}{2}$    **35.** $r + 1$    **37.** $\frac{x}{x+2}$    **39.** $\frac{z-5}{z+5}$    **41.** $\frac{x+1}{x+2}$    **43.** $-1$

**45.** $-(x+2)$    **47.** $-\frac{x+6}{2x}$    **49.** $-(x+3)$    **51.** $\frac{1}{4m-5}$    **53.** $\frac{x-5}{x+5}$    **55.** $2x - 3$    **57.** $x - 3$    **59.** $\frac{x-4}{x+4}$

**61.** $a^2 + 2a + 4$    **63.** $3s + 4t$    **65.** $\frac{3}{x-y}$    **67.** $\frac{☺}{5}$    **69.** $\frac{\Delta}{2\Delta + 9}$    **71.** $-1$    **73.** $x + 2; (x+2)(x-3) = x^2 - x - 6$

**75.** $x^2 + 9x + 20; (x+5)(x+4) = x^2 + 9x + 20$    **77.** Set the denominator equal to zero and then solve the resulting equation.

**79. a)** $x \neq 0, x \neq -5, x \neq \frac{3}{2}$    **b)** $\frac{1}{x(2x-3)}$    **81.** 1, the numerator and denominator are identical.    **84.** $y = x - 4z$

**85.** $28°, 58°,$ and $94°$    **86.** $\frac{25}{81x^4y^2}$    **87.** $8x^2 - 10x - 19$    **88.** $3(a+4)(a-6)$    **89.** 13 inches

**Exercise Set 7.2**    **1.** Factor    **3.** Divide out    **5.** $x + 4$    **7.** $\frac{3}{19}$    **9.** $-\frac{15}{28}$    **11.** $\frac{5}{16}$    **13.** $\frac{3}{5}$    **15.** $-\frac{13}{48}$    **17.** $\frac{xy}{8}$    **19.** $\frac{70x^4}{y^6}$

**21.** $\frac{36x^9y^2}{25z^7}$    **23.** $\frac{-3x+2}{3x+2}$    **25.** 1    **27.** $\frac{1}{a^2-b^2}$    **29.** 1    **31.** $\frac{x+2}{x+3}$    **33.** $x + 9$    **35.** $4x^2y$    **37.** $\frac{9z}{x}$    **39.** $\frac{11}{6ab^2}$    **41.** $6r^2$    **43.** $x + 9$

**45.** $\frac{x-8}{x+2}$    **47.** $\frac{x+3}{x-1}$    **49.** $-1$    **51.** $\frac{x+1}{x-4}$    **53.** $4x^2y^2$    **55.** $\frac{7c}{5ab^2}$    **57.** $\frac{3y^2}{a^2}$    **59.** $\frac{8mx^7}{3y^2}$    **61.** $\frac{2(x+3)}{x(x-3)}$    **63.** $\frac{7x}{y}$    **65.** $\frac{4}{m^4n^{11}}$

**67.** $\frac{r+2}{r-3}$    **69.** $\frac{x-6}{x-3}$    **71.** $\frac{2w-7}{w+1}$    **73.** $\frac{q-5}{2q+3}$    **75.** $\frac{2n+3}{3n-1}$    **77.** $\frac{1}{6\Delta^3}$    **79.** $\frac{\Delta + ☺}{9(\Delta - ☺)}$    **81.** $x^2 + 5x + 6$    **83.** $x^2 - 4x - 12$

**85.** $x^2 - 3x + 2$    **87.** Answers will vary.    **89.** $\frac{x-1}{x-3}$    **91.** $x^2 - 5x + 6, x^2 - x - 20$    **94.** 1 hour    **95.** $12x^4y^5z^{11}$

**96.** $2x^2 + x - 2 - \frac{2}{2x-1}$    **97.** $6(x-5)(x+2)$    **98.** $5, -2$

**Exercise Set 7.3**    **1.** $\frac{3x+2}{y}$    **3.** True    **5.** $x(x+6)$    **7.** $4x(x+3)$    **9. a)** The negative sign in $-(2x-9)$ was not distributed.

**b)** $\frac{4x-3-2x+9}{5x+4}$    **11. a)** The negative sign in $-(3x^2-4x+5)$ was not distributed.    **b)** $\frac{8x-2-3x^2+4x-5}{x^2-4x+3}$    **13.** $\frac{5}{7}$

**15.** $\frac{5r-1}{4}$    **17.** $\frac{x+6}{x}$    **19.** $\frac{n+8}{n+1}$    **21.** $\frac{5x+9}{x-3}$    **23.** $\frac{t+3}{5t^2}$    **25.** $\frac{1}{x-4}$    **27.** $\frac{1}{m-3}$    **29.** $\frac{p-12}{p-5}$    **31.** $x - 3$    **33.** $\frac{1}{2}$    **35.** 1

**37.** 4    **39.** $\frac{5}{x-2}$    **41.** $\frac{3}{4}$    **43.** $\frac{x-5}{x+2}$    **45.** $\frac{3x+2}{x-4}$    **47.** $\frac{6x+1}{x-8}$    **49.** 5    **51.** $9n$    **53.** $15x$    **55.** $p^3$    **57.** $3m - 4$    **59.** $6t^2$    **61.** $36x^3y$

**63.** $18r^4s^7$    **65.** $m(m+2)$    **67.** $x(x+1)$    **69.** $4n - 1$ or $1 - 4n$    **71.** $4k - 5r$ or $-4k + 5r$    **73.** $18q(q+1)$    **75.** $120x^2y^3$

**77.** $6(x+4)(x+2)$    **79.** $(x+1)(x+8)$    **81.** $(x-8)(x+3)(x+8)$    **83.** $(a-4)^2(a-3)$    **85.** $(x+5)(x+1)(x+3)$

**87.** $(x-3)^2$    **89.** $(x-6)(x-1)$    **91.** $(3t-2)(t+4)(3t-1)$    **93.** $(2x+1)^2(4x+3)$    **95.** $\frac{19}{35}$    **97.** $\frac{35}{36}$    **99.** $\frac{1}{18}$

**101.** $x^2 + x - 9$; the sum of the numerators must be $2x^2 - 5x - 6$    **103.** $x^2 + 9x - 10$; the sum of the numerators must be $5x - 7$

**105.** $5☺$    **107.** $(\Delta + 3)(\Delta - 3)$    **109.** Answers will vary.    **111.** $\frac{-3x^2 + 12x}{x^2 - 25}$    **113.** $30x^{12}y^9$    **115.** $(x-4)(x+3)(x-2)$

**117.** $\frac{92}{45}$ or $2\frac{2}{45}$    **118.** $-\frac{1}{5}$    **119.** 2.25 ounces    **120.** 70 hours    **121.** 1    **122.** $4.0 \times 10^{11}$    **123.** $\frac{3}{2}, -1$

## Exercise Set 7.4

**1.** Opposites  **3.** You are multiplying by 1  **5.** $12z^2$  **7.** $\dfrac{5x + 2y}{xy}$  **9.** $\dfrac{t + 10}{2t^2}$  **11.** $\dfrac{3x + 8}{x}$  **13.** $\dfrac{3x + 10}{5x^2}$

**15.** $\dfrac{45y + 12x}{20x^2y^2}$  **17.** $\dfrac{4y^2 + x}{y}$  **19.** $\dfrac{9a + 7}{6a}$  **21.** $\dfrac{6x^2 + 2y}{xy}$  **23.** $\dfrac{45a^2 - 4b}{5a^2b}$  **25.** $\dfrac{13x - 12}{x(x - 3)}$  **27.** $\dfrac{11p + 6}{p(p + 3)}$  **29.** $\dfrac{-d^2 + 14d + 25}{(d + 1)(3d + 5)}$

**31.** $\dfrac{6}{p - 3}$  **33.** $\dfrac{14}{x + 7}$  **35.** $\dfrac{a + 16}{2(a - 2)}$  **37.** $\dfrac{20x}{(x - 5)(x + 5)}$  **39.** $\dfrac{-7n - 6}{3n(2n + 1)}$  **41.** $\dfrac{15w + 66}{2(w + 5)(w + 2)}$  **43.** $\dfrac{5z - 16}{(z + 4)(z - 4)}$

**45.** $\dfrac{-x + 6}{(x + 2)(x - 2)}$  **47.** $\dfrac{r + 12}{(r - 4)(r - 6)}$  **49.** $\dfrac{6x - 11}{(x + 4)(x - 2)}$  **51.** $\dfrac{x^2 + 3x - 21}{(x + 5)^2}$  **53.** $\dfrac{-a + 16}{(a - 8)(a - 1)(a + 2)}$

**55.** $\dfrac{9x + 17}{(x + 3)^2(x - 2)}$  **57.** $\dfrac{3x^2 - 12x - 5}{(2x + 1)(3x - 2)(x + 3)}$  **59.** $\dfrac{2x^2 - 3x - 4}{(4x + 3)(x + 2)(2x - 1)}$  **61.** $\dfrac{1}{w - 3}$  **63.** $\dfrac{6}{r - 3}$  **65.** $-\dfrac{x + 4}{4x}$

**67.** All real numbers except $x = 0$.  **69.** All real numbers except $x = 4$ and $x = -6$.  **71.** $\dfrac{7}{\Delta - 2}$  **73.** All real numbers except

$a = -b$ and $a = 0$.  **75.** 0  **77.** $\dfrac{2x - 3}{2 - x}$  **79.** $\dfrac{6x + 5}{(x + 2)(x - 3)(x + 1)}$  **82.** $\approx 1.53$ hours  **83.** $-\dfrac{8}{7}$  **84.** $4x - 3 - \dfrac{6}{2x + 3}$  **85.** $-1$

## Mid-Chapter Test

**1.** All real numbers except $x = \dfrac{2}{3}$ [7.1]  **2.** All real numbers except $x = -2, x = 7$ [7.1]  **3.** 9 [7.1]

**4.** $\dfrac{2x + 3}{3x - 1}$ [7.1]  **5.** $5r + 6t$ [7.1]  **6.** $\dfrac{6y^3}{x^3}$ [7.2]  **7.** $-(m + 4)$ or $-m - 4$ [7.2]  **8.** $x - 2$ [7.2]  **9.** $\dfrac{x + 7}{2(x + 1)}$ [7.2]  **10.** $\dfrac{5x + 2}{7x + 3}$ [7.2]

**11.** $x - 6$ [7.3]  **12.** $x - 3$ [7.3]  **13.** $\dfrac{3t + 2}{4t - 1}$ [7.3]  **14.** $3m(2m + 1)$ [7.3]  **15.** $(2x + 3)(x - 4)(x - 5)$ [7.3]  **16.** $\dfrac{13x - 1}{10x}$ [7.4]

**17.** $-\dfrac{a^2 + 13a + 23}{a^2 - a - 12}$ [7.4]  **18.** $\dfrac{4x^2 + 17x - 1}{2x^2 + 13x + 6}$ [7.4]  **19.** $\dfrac{x^2 - 7x - 4}{(x + 1)(x + 2)(x - 3)}$ [7.4]

**20.** Need common denominator of $x(x + 1)$, $\dfrac{15x + 8}{x(x + 1)}$ [7.4]

## Exercise Set 7.5

**1.** $\dfrac{5}{3}$  **3.** 5  **5.** $\dfrac{5}{16}$  **7.** $\dfrac{57}{32}$  **9.** $\dfrac{11}{6}$  **11.** $\dfrac{x^3y^2}{21}$  **13.** $\dfrac{2ab^3}{21c^2}$  **15.** $\dfrac{ab - a}{3 + a}$  **17.** $\dfrac{3}{t}$  **19.** $\dfrac{5x - 1}{4x - 1}$  **21.** $\dfrac{m - n}{m}$

**23.** $-\dfrac{a}{b}$  **25.** $-1$  **27.** 1  **29.** $b - a$  **31.** $\dfrac{a^2 + b}{b(b + 1)}$  **33.** $\dfrac{x^2y}{y - x}$  **35.** $\dfrac{ab^2 + b^2}{a^2(b + 1)}$  **37.** b)–c) $-\dfrac{224}{155}$  **39.** b)–c) $\dfrac{x - y + 6}{2x + 2y - 7}$

**41. a)** $\dfrac{\dfrac{5}{12x}}{\dfrac{8}{x^2} - \dfrac{4}{3x}}$  **b)** $\dfrac{5x}{96 - 16x}$  **43.** A complex fraction is a fraction that contains a fraction in its numerator or its denominator or both.

**45.** $\dfrac{y + x}{3xy}$  **47.** $x + y$  **49. a)** $\dfrac{2}{7}$  **b)** $\dfrac{4}{13}$  **51.** $\dfrac{a^3b + a^2b^3 - ab^2}{a^3 - ab^3 + 3b^2}$  **53.** $\dfrac{17}{2}$  **54.** A polynomial is an expression containing a finite

number of terms of the form $ax^n$ where $a$ is a real number and $n$ is a whole number.  **55.** $(x - 5)(x - 8)$  **56.** $\dfrac{x^2 - 9x + 2}{(3x - 1)(x + 6)(x - 3)}$

## Exercise Set 7.6

**1.** Multiplied  **3.** Rational expression  **5.** Rational equation  **7.** Yes  **9.** 12  **11.** $-6$  **13.** 12  **15.** $-4$

**17.** $-20$  **19.** 30  **21.** 4  **23.** $\dfrac{25}{6}$  **25.** 36  **27.** 8  **29.** $-8$  **31.** 3  **33.** 4  **35.** 2  **37.** 5  **39.** No solution  **41.** 7  **43.** 15

**45.** No solution  **47.** $-3$  **49.** 38  **51.** $-\dfrac{1}{3}, 3$  **53.** $6, -1$  **55.** $4, -4$  **57.** $-4, -5$  **59.** 4  **61.** $-\dfrac{5}{2}$  **63.** No solution  **65.** 24

**67.** $-12$  **69.** $-3$  **71.** 3  **73.** 5  **75.** 0  **77.** $x$ can be any real number since the sum on the left is also $\dfrac{2x - 4}{3}$.  **79.** 15 centimeters

**81.** $-4$  **83.** No, it is impossible for both sides of the equation to be equal.  **85.** More than $6\dfrac{1}{3}$ hours  **86.** 150 minutes  **87.** $40°, 140°$

**88.** $6.8 \times 10^8$  **89.** Linear equation: $ax + b = c, a \neq 0$; quadratic equation: $ax^2 + bx + c = 0, a \neq 0$

## Exercise Set 7.7

**1.** 1 complete task  **3.** $\dfrac{1}{5}$  **5.** length = 11 inches, width = 9 inches  **7.** base = 12 centimeters,

height = 7 centimeters  **9.** Base: 8 feet  **11.** $\dfrac{8}{9}, 8$  **13.** 7  **15.** 9 miles per hour  **17.** 15 miles  **19.** 150 miles per hour, 600 miles per hour

**21.** No wake zone: $\approx 1.83$ miles, zone to Island: $\approx 34.77$ miles  **23.** 3 hours at 600 miles per hour and 2 hours at 550 miles per hour

**25.** 1200 feet  **27.** $3\dfrac{3}{7}$ hours  **29.** 4 hours  **31.** 24 minutes  **33.** 120 minutes or 2 hours.  **35.** 60 minutes or 1 hour  **37.** $8\dfrac{3}{4}$ days

**39.** $\dfrac{6}{5}$ hours or 1 hour, 12 minutes  **41.** 300 hours  **43.** 3 or $\dfrac{1}{2}$  **45.** 8 pints  **47.** $-\dfrac{3}{2}x - \dfrac{27}{2}$  **48.**  **49.** 1

**50.** $\dfrac{3x^2 - 9x - 15}{(2x + 3)(3x - 5)(3x + 1)}$

**Exercise Set 7.8** **1.** Combined **3.** Direct **5.** Constant **7.** Direct **9.** Inverse **11.** Direct **13.** Direct **15.** Direct **17.** Inverse **19.** Direct **21.** Inverse **23.** Inverse **25. a)** $x = ky$ **b)** 72 **27. a)** $y = kR$ **b)** 306 **29. a)** $R = \dfrac{k}{W}$ **b)** $\dfrac{1}{20}$ **31. a)** $A = \dfrac{kB}{C}$ **b)** 9 **33. a)** $x = ky$ **b)** 20 **35. a)** $y = kR^2$ **b)** 20 **37. a)** $S = \dfrac{k}{G}$ **b)** 0.96 **39. a)** $x = \dfrac{k}{P^2}$ **b)** 25 **41. a)** $F = \dfrac{kM_1M_2}{d}$ **b)** 40 **43.** Doubled **45.** Halved **47.** Doubled **49.** Unchanged **51.** Doubled **53.** $y = \dfrac{k}{x}; k = 5$ **55.** \$8814 **57.** 3096 milligrams **59.** 1.05 inches **61.** 6400 cubic centimeters **63.** 3.12 hours **65.** 45 foot-candles **67.** 117.6 feet **69.** 126 cubic meters **71.** 4600 DVDs **73.** $\approx 133.25$ pounds **75.** $\approx 121{,}528$ calls **77.** $\dfrac{1}{49}$ of the light of the flash **79. a)** $P = 14.7 + kx$ **b)** 0.43 **c)** $\approx 337.9$ feet **80.** $2x - 3 + \dfrac{6}{4x + 9}$ **81.** $(z - 2)(y + 8)$ **82.** $-4, 2$ **83.** $x + 8$

**Chapter 7 Review Exercises** **1.** All real numbers except $x = 19$ **2.** All real numbers except $x = 3$ and $x = 5$ **3.** All real numbers except $x = \dfrac{1}{5}$ and $x = -1$ **4.** $\dfrac{1}{x - 8}$ **5.** $x^2 + 5x + 12$ **6.** $3x + y$ **7.** $x + 4$ **8.** $a + 9$ **9.** $-(2x + 1)$ **10.** $\dfrac{b - 2}{b + 2}$ **11.** $\dfrac{x - 3}{x - 2}$ **12.** $\dfrac{x - 8}{2x + 3}$ **13.** $\dfrac{5}{12b^2}$ **14.** $12xz^2$ **15.** $\dfrac{8b^3c^4}{a^2}$ **16.** $-\dfrac{2}{9}$ **17.** $-\dfrac{2}{3}$ **18.** 1 **19.** $\dfrac{36x^2}{y}$ **20.** $\dfrac{20z}{x^3}$ **21.** $\dfrac{6}{a - b}$ **22.** $\dfrac{1}{8(a + 3)}$ **23.** 1 **24.** $6y(x - y)$ **25.** $\dfrac{n - 2}{n + 5}$ **26.** 4 **27.** 5 **28.** $\dfrac{4}{x + 10}$ **29.** $4h - 3$ **30.** $3x + 4$ **31.** 24 **32.** $x + 3$ **33.** $20x^2y^3$ **34.** $x(x - 3)$ **35.** $(n + 5)(n - 4)$ **36.** $x(x + 2)$ **37.** $(r + s)(r - s)$ **38.** $x - 9$ **39.** $(x + 7)(x - 5)(x + 2)$ **40.** $\dfrac{3y^2 + 10}{6y^2}$ **41.** $\dfrac{12x + y}{4xy}$ **42.** $\dfrac{5x^2 - 18y}{3x^2y}$ **43.** $\dfrac{7x + 12}{x + 2}$ **44.** $\dfrac{x^2 - 2xy - y^2}{xy}$ **45.** $\dfrac{9x + 8}{x(x + 4)}$ **46.** $\dfrac{-x - 4}{3x(x - 2)}$ **47.** $\dfrac{z + 14}{(z + 5)^2}$ **48.** $\dfrac{2x - 8}{(x - 3)(x - 5)}$ **49.** $\dfrac{5x + 38}{(x + 6)(x + 2)}$ **50.** $\dfrac{3x - 8}{x - 4}$ **51.** $\dfrac{5ab + 10b}{a - 2}$ **52.** $\dfrac{3x - 1}{(x + 3)(x - 3)}$ **53.** $\dfrac{6p^3}{q}$ **54.** $\dfrac{2x + 2}{(x + 2)(x - 3)(x - 2)}$ **55.** $\dfrac{8x - 29}{(x + 2)(x - 7)(x + 7)}$ **56.** $\dfrac{x}{x + y}$ **57.** $\dfrac{3(x + 3y)}{5(x - 3y)}$ **58.** $a - 3$ **59.** $\dfrac{3a^2 - 8a + 3}{(a + 1)(a - 1)(3a - 5)}$ **60.** $\dfrac{4x - 8}{x}$ **61.** $\dfrac{64}{9}$ **62.** $\dfrac{2}{3}$ **63.** $\dfrac{bc}{3}$ **64.** $\dfrac{8x^3z^2}{y^3}$ **65.** $\dfrac{ab - a}{a + 1}$ **66.** $\dfrac{r^2s + 7}{s^3}$ **67.** $\dfrac{3x + 2}{x(5x - 1)}$ **68.** $\dfrac{4}{x}$ **69.** $x$ **70.** $\dfrac{3a + 1}{4}$ **71.** $\dfrac{-x + 1}{x + 1}$ **72.** $\dfrac{8x^2 - x^2y}{y(y - x)}$ **73.** 13 **74.** 6 **75.** 12 **76.** 20 **77.** $-16$ **78.** $\dfrac{1}{2}$ **79.** $-6$ **80.** 28 **81.** No solution **82.** 2.4 hours **83.** $16\dfrac{4}{5}$ hours **84.** $\dfrac{1}{6}, 1$ **85.** Robert: 2.1 miles per hour, Tran: 5.6 miles per hour **86.** $\dfrac{25}{2}$ **87.** $\approx 426.7$ **88.** 273 milligrams **89.** 1.68 hours

**Chapter 7 Practice Test** **1.** 1 [7.1] **2.** $\dfrac{x^2 + x + 1}{x + 1}$ [7.1] **3.** $\dfrac{8x^2z}{y}$ [7.2] **4.** $a + 3$ [7.2] **5.** $\dfrac{x^2 - 6x + 9}{(x + 3)(x + 2)}$ [7.2] **6.** $-1$ [7.2] **7.** $\dfrac{x - 2y}{5}$ [7.2] **8.** $\dfrac{3}{y + 5}$ [7.2] **9.** $-\dfrac{m + 6}{m - 5}$ [7.2] **10.** $\dfrac{3x - 1}{4y}$ [7.3] **11.** $\dfrac{7x^2 - 6x - 13}{x + 3}$ [7.3] **12.** $\dfrac{2y^2 - 8}{xy^3}$ [7.4] **13.** $-\dfrac{2z + 15}{z - 5}$ [7.4] **14.** $\dfrac{-1}{(x + 4)(x - 4)}$ [7.4] **15.** $\dfrac{25}{28}$ [7.5] **16.** $\dfrac{x^2 + x^2y}{7y}$ [7.5] **17.** $\dfrac{4x + 3}{9 - 5x}$ [7.5] **18.** 2 [7.6] **19.** $-\dfrac{12}{7}$ [7.6] **20.** 12 [7.6] **21.** 6 hours [7.7] **22.** 1 [7.7] **23.** Base: 6 inches, height: 10 inches [7.7] **24.** 2 miles [7.7] **25.** $\approx 1.13$ feet [7.8]

**Cumulative Review Test** **1.** 121 [1.9] **2.** 36 [1.9] **3.** $\dfrac{17}{8}$ [2.5] **4.** $c + (c + 0.018c)$ [3.1] **5.** 0 [4.3] **6.** $y = \dfrac{3}{4}x - 3$ [4.4] **7.** $8x^2 + 5x + 14$ [5.4] **8.** $6n^3 - 23n^2 + 26n - 15$ [5.5] **9.** $\dfrac{1}{2}x - \dfrac{19}{4}$ [5.6] **10.** $(8a - 5)(a - 1)$ [6.2] **11.** $13(x + 3)(x - 1)$ [6.3] **12.** $4, \dfrac{3}{2}$ [6.6] **13.** $\dfrac{x + 4}{2x + 1}$ [7.2] **14.** $\dfrac{r^2 - 8r - 6}{(r + 2)(r - 5)}$ [7.4] **15.** $\dfrac{10x - 18}{(x - 5)(x + 2)(x + 3)}$ [7.4] **16.** $-\dfrac{3}{2}$ [7.6] **17.** No solution [7.6] **18.** \$3000 [3.2] **19.** 20 pounds sunflower seed, 30 pounds premixed assorted seed mix [3.4] **20.** First leg: 3.25 miles, second leg: 9.5 miles [3.4]

# Chapter 8

**Exercise Set 8.1** **1.** $ax + by = c$ **3.** Nonlinear

**5.**
**7.**
**9.**
**11.**
**13.**

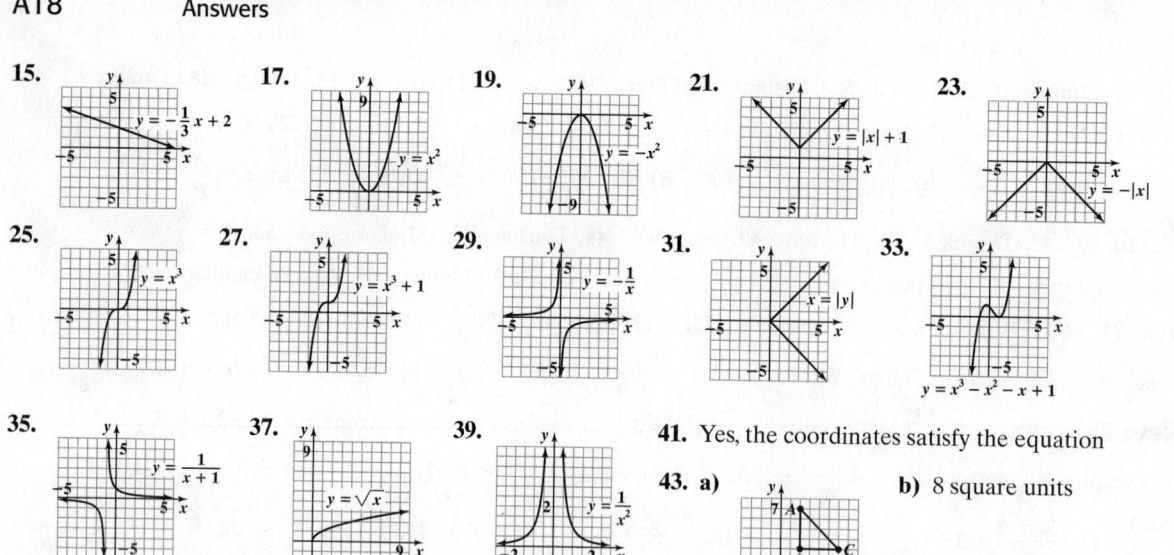

**15.** $y = -\frac{1}{3}x + 2$   **17.** $y = x^2$   **19.** $y = -x^2$   **21.** $y = |x| + 1$   **23.** $y = -|x|$

**25.** $y = x^3$   **27.** $y = x^3 + 1$   **29.** $y = -\frac{1}{x}$   **31.** $x = |y|$   **33.** $y = x^3 - x^2 - x + 1$

**35.** $y = \frac{1}{x + 1}$   **37.** $y = \sqrt{x}$   **39.** $y = \frac{1}{x^2}$

**41.** Yes, the coordinates satisfy the equation

**43. a)** 

**b)** 8 square units

**45. a)** About 800,000 passengers   **b)** About 667,000 passengers   **c)** 1890–1894, 1900–1904, 1905–1909, 1910–1914   **d)** No

**47.** a   **49.** c   **51.** b   **53.** d   **55.** b   **57.** d   **59.** b   **61.** d

**63. a)**    **b)**   $Y_1 = 2X - 3$

**65. a)**    **b)**   $Y_1 = X^2 - 2X - 8$

**67. a)**    **b)**   $Y_1 = X^3 - 2X + 4$

**69.** $y = x + 3$   $y = x + 1$   $y = x - 1$

**a)** Each graph crosses the y-axis at the point corresponding to the constant term in the graph's equation.
**b)** Yes

**71.** The rate of change is 2.   **73.** The rate of change is 3.   **75.** $(4, -3), (5, 1)$, other answers possible.

$y = 2x$      $y = 3x + 2$

**77.** $y = |x - 2|$   **81.** 26   **82.** $\frac{4}{11}$   **83.** $-6$   **84.** $-1$

**Exercise Set 8.2**   **1.** Domain   **3.** Ordered pairs   **5.** Vertical line   **7.** Range   **9.** Function   **11.** Independent
**13. a)** Function   **b)** Domain: {Idaho, Texas, Georgia}, Range: {15, 16, 18}   **15. a)** Function   **b)** Domain: {3, 5, 11},
Range: {6, 10, 22}   **17. a)** Function   **b)** Domain: {Cameron, Tyrone, Vishnu}, Range: {3, 6}   **19. a)** Not a function
**b)** Domain: {1990, 2001, 2002}; Range: {20, 34, 37}   **21. a)** Function   **b)** Domain: {1, 2, 3, 4, 5}, Range: {1, 2, 3, 4, 5}
**23. a)** Function   **b)** Domain: {1, 2, 3, 4, 5, 7},   Range: {−1, 0, 2, 4, 9}   **25. a)** Not a function   **b)** Domain: {1, 2, 3},
Range: {1, 2, 4, 5, 6}   **27. a)** Not a function   **b)** Domain: {0, 1, 2}, Range: {−7, −1, 2, 3}   **29.** $A = \{0\}$   **31.** $C = \{18, 20\}$
**33.** $E = \{0, 1, 2\}$   **35.** $H = \{0, 7, 14, 21, \ldots\}$   **37.** $J = \{1, 2, 3, 4, \ldots\}$ or $J = N$   **39. a)** Set $A$ is the set of all $x$ such that $x$ is a natural
number less than 7   **b)** $A = \{1, 2, 3, 4, 5, 6\}$   **41.**    **43.**   **45.**   **47.**
**49.**    **51.** $\{x | x \geq 1\}$   **53.** $\{x | x < 5 \text{ and } x \in I\}$ or $\{x | x \leq 4 \text{ and } x \in I\}$   **55.** $\{x | -3 < x \leq 5\}$   **57.** $\{x | -2.5 \leq x < 4.2\}$

**59.** $\{x | -3 \leq x \leq 1 \text{ and } x \in I\}$   **61. a)** Function   **b)** Domain: $\mathbb{R}$, Range: $\mathbb{R}$   **c)** 2   **63. a)** Not a function   **b)** Domain:
$\{x | 0 \leq x \leq 2\}$, Range: $\{y | -3 \leq y \leq 3\}$   **c)** ≈1.5   **65. a)** Function   **b)** Domain: $\mathbb{R}$, Range: $\{y | y \geq 0\}$   **c)** −3, −1

**67. a)** Function  **b)** Domain: $\{-1, 0, 1, 2, 3\}$, Range: $\{-1, 0, 1, 2, 3\}$  **c)** 2  **69. a)** Not a function  **b)** Domain: $\{x | x \geq 2\}$, Range: $\mathbb{R}$
**c)** 3  **71. a)** Function  **b)** Domain: $\{x | -2 \leq x \leq 2\}$, Range: $\{y | -1 \leq y \leq 2\}$  **c)** $-2, 2$  **73. a)** 5  **b)** $-11$  **75. a)** $-6$  **b)** $-4$
**77. a)** 2  **b)** 2  **79. a)** 7  **b)** 0  **81. a)** 0  **b)** 3  **83. a)** 1  **b)** Undefined  **85. a)** 16 square feet  **b)** 26 square feet
**87. a)** $A(r) = \pi r^2$  **b)** $\approx 452.4$ square yards  **89. a)** $C(F) = \dfrac{5}{9}(F - 32)$  **b)** $-35°C$  **91. a)** $18.23°C$  **b)** $27.68°C$
**93. a)** $78.32°$  **b)** $73.04°$  **95. a)** 91 oranges  **b)** 204 oranges
**97. a)**  **b)** \$2.65 per bushel

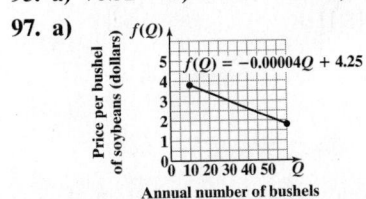

**99.** Answers will vary. One possible interpretation: The person warms up slowly, possibly by walking, for 5 minutes. Then the person begins jogging slowly over a period of 5 minutes. For the next 15 minutes the person jogs. For the next 5 minutes the person walks slowly and his heart rate decreases to his normal heart rate. The rate stays the same for the next 5 minutes.
**101.** Answers will vary. One possible interpretation: The man walks on level ground, about 30 feet above sea level, for 5 minutes. For the next 5 minutes he walks uphill to 45 feet above sea level. For 5 minutes he walks on level ground, then walks quickly downhill for 3 minutes to an elevation of 20 feet above sea level. For 7 minutes he walks on level ground. Then he walks quickly uphill for 5 minutes.
**103.** Answers will vary. One possible interpretation: Driver is in stop-and-go traffic, then gets on highway for about 15 minutes, then drives on a country road for a few minutes, then stops car for a couple of minutes, then stop-and-go traffic.
**105. a)** Yes  **b)** Year  **c)** $\approx\$22,000$  **d)** $\approx\$6,000$  **e)** $\approx 100\%$
**107. a)**  **b)** No. It is not a straight line.  **c)** \$2,300,000

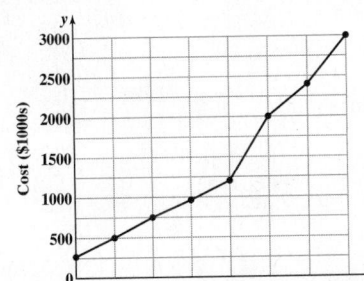

**111.** $\dfrac{1}{2}$  **112.** $p_2 = \dfrac{E - a_1 p_1 - a_3 p_3}{a_2}$

**113.** $y = -\dfrac{1}{2}x + \dfrac{3}{2}, m = -\dfrac{1}{2}, \left(0, \dfrac{3}{2}\right)$

**114.** $\dfrac{3x + 1}{x - 1}$

**Exercise Set 8.3**  **1.** Real numbers  **3.** $x$-intercept  **5.** Linear
**7.**  **9.**  **11.**  **13.**  **15.**

**17.**  **19.**  **21.**  **23.**  **25.**

**27.**  **29.**  **31. a)**  **b)** 1300 bicycles  **c)** 3800 bicycles

    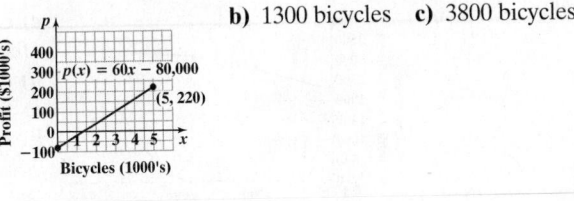

**33. a)** $s(x) = 500 + 0.15x$  **b)**

**c)** \$950  **d)** \$4000
**35. a)** There is only one $y$-value for each $x$-value.
**b)** Independent: length; Dependent: weight  **c)** Yes
**d)** 11.5 kilograms  **e)** 65 centimeters  **f)** 12.0–15.5 kilograms
**g)** Increases; yes, as babies get older their weights vary more.
**37.** When the graph goes through the origin, because at the origin both $x$ and $y$ are zero.

**39.** Answers will vary. One possible answer is $f(x) = 4$.   **41.** Both intercepts will be at 0.   **43.** $(-3.2, 0), (0, 6.4)$   **45.** $(-2, 0), (0, -2.5)$

**49.** $3x^2 - 12x - 88$   **50.** $3(x + 4)(x - 8)$   **51.** $\dfrac{1}{x + 2}$   **52.** $4, 6$

## Mid-Chapter Test   **1.**

[8.1]   **2.**

[8.1]   **3.**

[8.1]

**4.**

[8.1]   **5. a)** A relation is any set of ordered pairs.   **b)** A function is a correspondence between a first set of elements, the domain, and a second set of elements, the range, such that each element of the domain corresponds to exactly one element in the range.   **c)** No   **d)** Yes [8.2]   **6.** Function; Domain: $\{1, 2, 7, -5\}$, Range: $\{5, -3, -1, 6\}$ [8.2]   **7.** Not a function; Domain: $\{x | -2 \le x \le 2\}$, Range: $\{y | -4 \le y \le 4\}$ [8.2]   **8.** Function; Domain: $\{x | -5 \le x \le 3\}$, Range: $\{y | -1 \le y \le 3\}$ [8.2]   **9.** $E = \{9, 18, 27, 36, \ldots\}$ [8.2]   **10.** ←———→ [8.2]   **11.** $-21$ [8.2]   **12.** 105 feet [8.2]

**13.**

[8.3]   **14.**

[8.3]   **15.**
[8.3]

**16. a)**
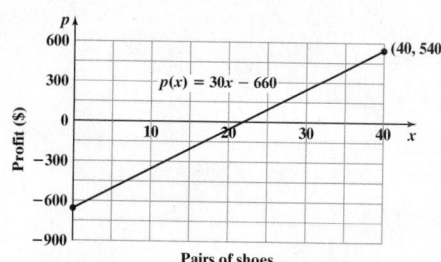
   **b)** 22 pairs of shoes   **c)** 34 pairs of shoes [8.3]

## Exercise Set 8.4   **1.** Slope   **3.** Constant   **5.** Perpendicular   **7.** $m = -3; f(x) = -3x$

**9.** Slope is undefined; $x = -2$   **11.** $m = -\dfrac{1}{3}; f(x) = -\dfrac{1}{3}x + 2$   **13.** $m = -\dfrac{3}{2}; f(x) = -\dfrac{3}{2}x + 15$

**15.**
**17.**
   **19.** Parallel   **21.** Neither   **23.** Perpendicular   **25.** Perpendicular   **27.** Parallel
   **29.** Neither   **31.** Perpendicular   **33.** Parallel   **35.** Neither   **37.** $y = 2x + 1$

**39.** $2x - 5y = 19$   **41.** $y = -\dfrac{5}{3}x + 5$   **43.** $f(x) = -3x + 13$   **45.** $y = -\dfrac{2}{3}x + 6$

**47. a)** $y = 3x$   **b)** $y = 3x - 2$   **49. a)** 1   **b)** $(0, 4)$   **c)** $y = x + 4$   **51.** $y = \dfrac{3}{2}x - 7$   **53.** 0.2   **55. a)** 11.3   **b)** Positive

**c)** 7.075   **57. a–b)**
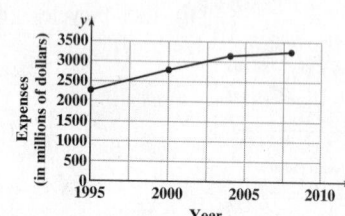
   **c)** 123.8, 64.25, 31.75   **d)** 1995–2000, because its line segment has the greatest slope   **59. a)** $h(x) = -x + 200$
   **b)** 186 beats per minute   **61. a)** $S(t) = 210t + 37,550$   **b)** \$38,180
   **c)** \$39,650   **d)** 12 years   **63. a)** $S(t) = 1853.6t + 37,855$
   **b)** \$43,415.80   **c)** \$56,391   **d)** 8 years

**65. a)** $C(s) = 45.7s + 95.8$   **b)** 324.3 calories   **67. a)** $d(p) = -0.20p + 90$   **b)** 38 iPods   **c)** \$225

**69. a)** $s(p) = 95p - 60$   **b)** 206 kites   **c)** \$3.00   **71. a)** $i(t) = 12.5t$   **b)** \$1500   **c)** 176 tickets   **73. a)** $r(w) = 0.01w + 10$
**b)** \$46.13   **c)** 5000 pounds   **75. a)** $y(a) = -0.865a + 79.25$   **b)** 47.2 years   **c)** 62.7 years old   **77. a)** $w(a) \approx 0.189a + 10.6$
**b)** 14.758 kilograms   **79.** The $y$-intercept is wrong.   **81.** The slope is wrong.   **83.** Height: 14.2 inches, width: 6.4 inches   **87.** 19

**88.** 2.4   **89.** $-\dfrac{92}{5}$   **90. a)** Any set of ordered pairs   **b)** A correspondence where each member of the domain corresponds to a unique member of the range   **c)** Answers will vary.   **91.** Domain: $\{3, 4, 5, 6\}$; Range: $\{-2, -1, 2, 3\}$

**Exercise Set 8.5**  **1.** $x^2 - 2x + 5$  **3.** $x^2 + 2x - 5$  **5.** $2x^3 - 5x^2$  **7.** All real numbers  **9. a)** $4x + 3$  **b)** $4a + 3$  **c)** $11$
**11. a)** $x^3 + x - 4$  **b)** $a^3 + a - 4$  **c)** $6$  **13. a)** $4x^3 - x + 4$  **b)** $4a^3 - a + 4$  **c)** $34$  **15.** $-7$  **17.** $29$  **19.** $-60$
**21.** Undefined  **23.** $13$  **25.** $-\dfrac{3}{4}$  **27.** $2x^2 - 6$  **29.** $-4$  **31.** $18$  **33.** $0$  **35.** $-\dfrac{3}{7}$  **37.** $-\dfrac{1}{45}$  **39.** $-2x^2 + 2x - 6$  **41.** $3$
**43.** $-4$  **45.** $1$  **47.** Undefined  **49.** $0$  **51.** $0$  **53.** $-3$  **55.** $-2$  **57. a)** $2008$  **b)** $\$800$  **c)** $\$7900$  **d)** $\$900$
**59. a)** Summer A, $36$  **b)** Summer B, $10$  **c)** $11$ male students  **d)** $19$ female students  **61. a)** $\approx20$  **b)** $\approx8$  **c)** $\approx12$  **d)** $\approx23$

**63. a)**

**b)**

**a)**

**65. a)**

**b)**

**c)**

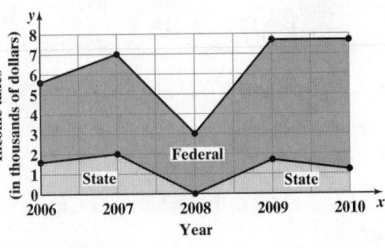

**67.** $g(x) \neq 0$ since division by zero is undefined.  **69.** No, subtraction is not commutative. One example is $5 - 3 = 2$ but $3 - 5 = -2$
**71.** $f(a)$ and $g(a)$ must either be opposites or both be equal to $0$.  **73.** $f(a) = g(a)$  **75.** $f(a)$ and $g(a)$ must have opposite signs.

**78.** $-\dfrac{1}{64}$  **79.** $h = \dfrac{2A}{b}$  **80.** $\$450$  **81.**

**82.** $2.96 \times 10^6$  **83.**

## Chapter 8 Review Exercises

**1.**

**2.**

**3.**

**4.**

**5.**

**6.**

**7.**

**8.**

**9.**

**10.**

**11.** A function is a correspondence where each member of the domain corresponds to exactly one member of the range.
**12.** No, every relation is not a function. $\{(4, 2), (4, -2)\}$ is a relation but not a function. Yes, every function is a relation because it is a set
of ordered pairs.  **13.** Yes, each member of the domain corresponds to exactly one member of the range.  **14.** No, the domain element
$2$ corresponds to more than one member of the range ($-4$ and $-2$).  **15.** $\{3, 4, 5, 6\}$  **16.** $\{6, 7, 8, \ldots\}$  **17.** $(4, \infty)$
**18.** $(-\infty, -2]$  **19.** $(-1.5, 2.7]$  **20. a)** Yes, the relation is a function.  **b)** Domain: $\mathbb{R}$; Range: $\mathbb{R}$  **21. a)** Yes, the relation is a function.
**b)** Domain: $\mathbb{R}$; Range: $\{y \,|\, y \leq 0\}$  **22. a)** No, the relation is not a function.  **b)** Domain: $\{x \,|\, -3 \leq x \leq 3\}$; Range: $\{y \,|\, -3 \leq y \leq 3\}$
**23. a)** No, the relation is not a function.  **b)** Domain: $\{x \,|\, -2 \leq x \leq 2\}$; Range: $\{y \,|\, -1 \leq y \leq 1\}$  **24. a)** $-2$  **b)** $-h^2 + 3h - 4$
**25. a)** $1$  **b)** $2a^3 - 3a^2 + 6$  **26.** Answers will vary. Here is one possible interpretation: The car speeds up to $50$ mph. Stays at this
speed for about $11$ minutes. Speeds up to about $68$ mph. Stays at that speed for about $5$ minutes, then stops quickly. Stopped for about
$5$ minutes. Then in stop and go traffic for about $5$ minutes.  **27. a)** $1020$ baskets  **b)** $1500$ baskets  **28. a)** $180$ feet  **b)** $52$ feet

**29.**

**30.**

**31.**

**32. a)** 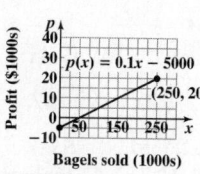   **b)** 50,000 bagels   **c)** 270,000 bagels

**33.**    **34. a)** Yes   **b)** $f(x) = 3$   **35. a)** No   **36. a)** Yes   **b)** $f(x) = -\frac{1}{2}x + 2$

**37. a)** $-2$   **b)** $(0, 1)$   **c)** $y = -2x + 1$   **38.** $(0, 0)$

**39. a)**

**b)** 1970–1980: 16.4; 1980–1990: 4.2; 1990–2000: $-23.5$   **c)** 1970–1980   **40.** $n(t) \approx 0.7t + 35.6$

**41.** Parallel   **42.** Perpendicular   **43.** Neither   **44.** $y = \frac{1}{2}x + 7$   **45.** $y = -x - 2$   **46.** $y = -\frac{2}{3}x + 6$

**47.** $y = \frac{5}{2}x + 3$   **48.** $y = -\frac{5}{3}x - 4$   **49.** $y = -\frac{1}{2}x + 7$   **50.** Neither   **51.** Parallel

**52.** Perpendicular   **53.** Neither   **54. a)** $r(a) = 0.61a - 10.59$   **b)** $13.81

**55. a)** $C(r) = 1.8r + 435$   **b)** 507 calories   **c)** $\approx 91.7$ yards per minute   **56.** $x^2 - x - 1$   **57.** 11

**58.** $-x^2 + 5x - 9$   **59.** $-15$   **60.** $-56$   **61.** 4   **62.** $-\frac{2}{3}$   **63.** $-2$   **64. a)** $\approx 4.6$ billion   **b)** $\approx 2.1$ billion   **c)** $\approx 0.8$ billion   **d)** $\approx 33\%$

**65. a)** $\approx\$47,000$   **b)** $\approx\$28,000$   **c)** $\approx\$3000$

## Chapter 8 Practice Test

**1.** [8.1]    **2.** [8.1]    **3.** [8.1]    **4.** [8.1]

**5.** A function is a correspondence where each member in the domain corresponds with exactly one member in the range. [8.2]
**6.** Yes, because each member in the domain corresponds to exactly one member in the range. [8.2]   **7.** Yes; Domain: $\mathbb{R}$; Range: $\{y | y \leq 4\}$ [8.2]   **8.** No; Domain: $\{x | -3 \leq x \leq 3\}$; Range: $\{y | -2 \leq y \leq 2\}$ [8.2]   **9.** 29 [8.2]   **10.**   [8.3]

**11. a)**   **b)** 4900 books   **c)** 14,700 books [8.3]

**12.** $p(t) = 2.9044t + 274.634$ [8.4]   **13.** Parallel, the slope of both lines is the same, $\frac{2}{3}$. [8.4]

**14.** $y = -2x + 7$ [8.4]   **15. a)** $r(t) = -3t + 266$

**b)** 248 per 100,000   **c)** 206 per 100,000 [8.4]   **16.** 12 [8.5]   **17.** $-\frac{3}{7}$ [8.5]

**18.** $2a^2 - a$ [8.5]   **19. a)** $\approx 44$ million tons   **b)** $\approx 18$ million tons   **c)** $\approx 26$ million tons [8.5]

## Cumulative Review Test

**1.** 126 [1.9]   **2.** 92 [1.9]   **3.** 0 [2.5]   **4.** $R = \dfrac{P - 2E}{3}$ [2.6]   **5.** $y = \frac{5}{2}x - 6$ [4.4]   **6.** $\dfrac{27y^6}{x^9}$ [5.1]

**7.** $8x^2 + 5x - 1$ [5.4]   **8.** $12x^3 - 38x^2 + 39x - 15$ [5.5]   **9.** $(a - 1)(6a - 5)$ [6.2]   **10.** $3, -\frac{5}{2}$ [6.6]   **11.** $\dfrac{x + 3}{2x + 1}$ [7.2]

**12.** $\dfrac{x^2 - 8x - 12}{(x + 4)(x - 5)}$ [7.4]   **13.** $\dfrac{6x + 2}{(x - 5)(x + 2)(x + 3)}$ [7.4]   **14.** $-\frac{3}{2}$ [7.6]   **15.** No solution [7.6]

**16. a)** Not a function   **b)** Domain: $\{x | x \leq 2\}$; Range $\mathbb{R}$ [8.2]   **17.** Neither [8.4]   **18.** $x^2 + 7x - 8$ [8.5]   **19.** 260 [8.5]

**20. a)** $3.052 \times 10^{12}$ cubic feet   **b)** $7.412 \times 10^{12}$ cubic feet   **c)** $2.398 \times 10^{13}$ cubic feet [5.3]

## Chapter 9

**Exercise Set 9.1**   **1.** Ordered pair   **3.** Consistent   **5.** Parallel lines   **7.** c)   **9.** a)   **11.** None   **13.** a), b)   **15.** a)
**17.** Consistent; one solution   **19.** Dependent; infinite number of solutions   **21.** Consistent; one solution   **23.** Inconsistent; no solution
**25.** One solution   **27.** No solution   **29.** One solution   **31.** No solution   **33.** Infinite number of solutions   **35.** No solution

**37.**    **39.**    **41.**    **43.**    **45.**    **47.**

Inconsistent

**49.**    **51.**    **53.**    **55.**    **57.**    **59.**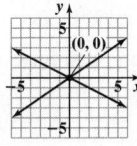

Dependent                      Dependent     Inconsistent

**61.** 6 years   **63.** 7 hours   **65.** Lines are parallel; they have the same slope and different $y$-intercepts.
**67.** The system has an infinite number of solutions. If the two lines have two points in common then they must be the same line.
**69.** The system has no solution. Parallel lines do not intersect.   **71.** One; $(5, 3)$   **79.** $-2x + 18$   **80.** 1   **81.** 0.05
**82. a)** $(6, 0), (0, 4)$   **b)**             **83.** $-(x - 7)$   **84.** $\dfrac{1}{3}, 6$

**Exercise Set 9.2**   **1.** Coefficient   **3.** No solution   **5.** $(2, 1)$   **7.** $(-1, -1)$   **9.** No solution   **11.** $(3, -8)$
**13.** $(-2, -3)$   **15.** $(3, 5)$   **17.** Infinite number of solutions   **19.** $(2, 1)$   **21.** $(5, -3)$   **23.** $(-1, 0)$   **25.** $\left(-\dfrac{12}{7}, \dfrac{6}{35}\right)$   **27.** No solution
**29.** $\left(\dfrac{1}{8}, -\dfrac{1}{2}\right)$   **31.** 42, 21   **33.** Width: 8 feet, length: 17 feet   **35.** 1561 attendees   **37.** \$30,000   **39. a)** 16 months   **b)** Yes
**41. a)** 1.25 hours   **b)** 155 mile marker   **43. a)** $T = 180 - 10t$   **b)** $T = 20 + 6t$   **c)** 10 minutes   **d)** 80°F   **45.** 23 years
**46.** 2   **47.**           **48.** Slope: $\dfrac{3}{5}$, $y$-intercept: $(0, -5)$   **49.** $y = 2x + 4$   **50.** $18x^2 + 9x - 14$

**Exercise Set 9.3**   **1.** Eliminate   **3.** An infinite number of solutions   **5.** $(5, 1)$   **7.** $(-2, -1)$   **9.** $(5, 8)$   **11.** No solution
**13.** $(-8, -26)$   **15.** $\left(-\dfrac{3}{2}, -3\right)$   **17.** $(-2, 2)$   **19.** $(3, -1)$   **21.** Infinite number of solutions   **23.** $(1, 2)$   **25.** No solution
**27.** $(0, 0)$   **29.** $\left(10, \dfrac{15}{2}\right)$   **31.** $(-6, -12)$   **33.** $\left(\dfrac{20}{39}, -\dfrac{16}{39}\right)$   **35.** $\left(\dfrac{14}{5}, -\dfrac{12}{5}\right)$   **37.** No solution
**39.** First number: 49, second number: 11   **41.** First number: 6, second number: 4   **43.** Width: 3 inches, length: 6 inches
**45.** Width: 8 inches, length: 10 inches   **47.** Answers will vary.   **49. a)** $(200, 100)$   **b)** Same solution; dividing both sides of an
equation by a nonzero number does not change the solution.   **51.** $(8, -1)$   **53.** $(1, 2, 3)$   **55.** 125   **56.** 7   **57.** $2x^2y - 9xy + 4y$
**58.** $36a^6b^9c^5$   **59.** $(y + c)(x - a)$   **60.** 14

**Exercise Set 9.4**

**1.** Ordered triple   **3.** Inconsistent   **5.** $(2, 0, 1)$   **7.** $\left(-7, -\dfrac{35}{4}, -3\right)$   **9.** $(0, 3, 6)$   **11.** $(1, 1, 1)$   **13.** $(-3, 15, -7)$

**15.** $(3, 1, -2)$   **17.** $(2, -1, 3)$   **19.** $\left(\dfrac{2}{3}, -\dfrac{1}{3}, 1\right)$   **21.** $(0, -1, 0)$   **23.** $\left(-\dfrac{11}{17}, \dfrac{7}{34}, -\dfrac{49}{17}\right)$   **25.** $(0, 0, 0)$   **27.** $(4, 6, 8)$

**29.** $\left(\dfrac{2}{3}, \dfrac{23}{15}, \dfrac{37}{15}\right)$   **31.** $(1, 1, 2)$   **33.** Inconsistent   **35.** Dependent   **37.** Inconsistent   **39.** $A = 9, B = 6, C = 2; 9x + 6y + 2z = 1$
**41.** Answers will vary. One example is $x + y + z = 10, x + 2y + z = 11, x + y + 2z = 16$
**43. a)** $a = 1, b = 2, c = -4$   **b)** $y = x^2 + 2x - 4$, Substitute 1 for $a$, 2 for $b$, and $-4$ for $c$ into $y = ax^2 + bx + c$
**45.** No point is common to all three planes. Therefore, the system is inconsistent.   **47.** One point is common to all three planes;
therefore, the system is consistent.   **49. a)** Yes, the 3 planes can be parallel   **b)** Yes, the 3 planes can intersect at one point

**c)** No, the 3 planes cannot intersect at exactly two points   **51.** $(1, 2, 3, 4)$   **53. a)** Commutative property of addition   **b)** Associative property of multiplication   **c)** Distributive property   **54.** $\dfrac{1}{2}$   **55. a)** $\dfrac{1}{4}$ hour or 15 minutes   **b)** 1.25 miles   **56.** $w = 3$ ft, $l = 8$ ft

## Mid-Chapter Test

**1.** b) [9.1]   **2.** a) [9.1]   **3.** One solution [9.1]   **4.** Infinite number of solutions [9.1]   **5.** No solution [9.1]

**6.**  [9.1]   **7.**  [9.1]   **8.** $(-2, 4)$ [9.2]   **9.** $\left(1, -\dfrac{1}{3}\right)$ [9.2]   **10.** No solution [9.2]

**11.** Length: 15 feet, width: 7 feet [9.2]   **12.** $(4, -1)$ [9.3]   **13.** $\left(-\dfrac{1}{2}, 2\right)$ [9.3]

**14.** Infinite number of solutions [9.3]

**15.** Solution must have two values, one for $x$ and one for $y$. The solution is $x = 3$ and $y = 5$ or $(3, 5)$. [9.1–9.3]   **16.** $(1, 2, -1)$ [9.4]
**17.** $(2, 0, 3)$ [9.4]

## Exercise Set 9.5

**1.** Ireland: 70,273 square kilometers, Georgia: 69,700 square kilometers   **3.** Hamburger: 21 grams, fries: 67 grams   **5.** Hot dog: $2, soda: $1   **7.** 512 MB: 109 photos, 4 GB: 887 photos   **9.** $25°, 65°$   **11.** $52°, 128°$   **13.** 12.2 miles per hour, 3.4 miles per hour   **15.** $500, 4\%$   **17.** 1.2 ounces of 5%, 1.8 ounces of 30%   **19.** 10 gallons concentrate, 190 gallons water

**21.** $17\dfrac{1}{3}$ pounds birdseed, $22\dfrac{2}{3}$ pounds sunflower seeds   **23.** Adult: $29, child: $18   **25.** $6000 at 5%, $4000 at 6%

**27.** 160 gallons whole, 100 gallons skim milk   **29.** 7 pounds Season's Choice, 13 pounds Garden Mix   **31.** 50 miles per hour, 55 miles per hour   **33.** Cabrina: 8 hours, Dabney: 3.4 hours   **35.** 80 grams $A$, 60 grams $B$   **37.** 200 grams first alloy, 100 grams second alloy
**39.** 2012   **41.** Tom: 60 miles per hour, Melissa: 75 miles per hour   **43.** Personal: 3, bills and statements: 4, advertisements: 17
**45.** Alabama: 54, Tennessee: 47, Texas: 46   **47.** Bandits: 32 wins, Force: 31 wins, Glory: 30 wins   **49.** *Their Greatest Hits 1971–1975*: 29 million albums, *Thriller*: 27 million albums, *The Wall*: 23.5 million albums   **51.** Florida: 15, California: 11, Louisiana: 9
**53.** $30°, 45°, 105°$   **55.** $1500 at 3%, $3000 at 5%, $5500 at 6%   **57.** 4 liters of 10% solution, 2 liters of 12% solution,

2 liters of 20% solution   **59.** 10 children's chairs, 12 standard chairs, 8 executive chairs   **61.** $I_A = \dfrac{27}{38}$; $I_B = -\dfrac{15}{38}$; $I_C = -\dfrac{6}{19}$

**64.** $-\dfrac{35}{8}$   **65.** 4   **66.** $y = x - 10$   **67.** Use the vertical line test.

## Exercise Set 9.6

**1.** Row transformations   **3.** Dimensions   **5.** Inconsistent   **7.** Square matrix   **9.** $\begin{bmatrix} 1 & -2 & | & -5 \\ 3 & -7 & | & -4 \end{bmatrix}$

**11.** $\begin{bmatrix} 1 & 1 & 3 & | & -8 \\ 3 & 2 & 1 & | & -5 \\ 4 & 7 & 2 & | & -1 \end{bmatrix}$   **13.** $\begin{bmatrix} 1 & 3 & | & 12 \\ 0 & 23 & | & 42 \end{bmatrix}$   **15.** $\begin{bmatrix} 1 & 0 & 8 & | & \frac{1}{4} \\ 0 & 2 & -38 & | & -\frac{13}{4} \\ 6 & -3 & 1 & | & 0 \end{bmatrix}$   **17.** $(3, 0)$   **19.** $(-5, 1)$   **21.** $(0, 1)$   **23.** Dependent system

**25.** $\left(-\dfrac{1}{3}, 3\right)$   **27.** Inconsistent system   **29.** $\left(\dfrac{2}{3}, \dfrac{1}{4}\right)$   **31.** $\left(\dfrac{4}{5}, -\dfrac{7}{8}\right)$   **33.** $(1, 2, 0)$   **35.** $(3, 1, 2)$   **37.** $\left(1, -1, \dfrac{1}{2}\right)$

**39.** Dependent system   **41.** $\left(\dfrac{1}{2}, 2, 4\right)$   **43.** Inconsistent system   **45.** $\left(5, \dfrac{1}{3}, -\dfrac{1}{2}\right)$   **47.** $\angle x = 30°, \angle y = 65°, \angle z = 85°$

**49.** Washington: $1100 million (or $1.1 billion), Dallas: $923 million, Houston: $905 million   **51.** Dependent   **53.** No, this is the same as switching the order of the equations.   **55.** $3x - 8 + \dfrac{9}{x + 4}$   **56.** $\dfrac{9}{2}, -4$   **57.** $\dfrac{-2x - 3}{(x + 2)(x - 2)}$   **58.** 30 min

## Exercise Set 9.7

**1.** Determinant   **3.** $D_y$   **5.** Dependent   **7.** 11   **9.** $-8$   **11.** $-12$   **13.** 44   **15.** $(2, 1)$   **17.** $(6, -4)$

**19.** $\left(\dfrac{1}{2}, -1\right)$   **21.** $(-7, -2)$   **23.** Infinite number of solutions   **25.** $(2, -3)$   **27.** No solution   **29.** $(2, 5)$   **31.** $(1, 2, 1)$

**33.** $\left(\dfrac{1}{2}, -\dfrac{1}{2}, 2\right)$   **35.** $\left(\dfrac{1}{2}, -\dfrac{1}{8}, 2\right)$   **37.** $(-1, 0, 2)$   **39.** Infinite number of solutions   **41.** $(1, -1, 2)$   **43.** No solution
**45.** $(3, 4, 1)$   **47.** $(-1, 5, -2)$   **49.** 5   **51.** 6   **53.** adult: $10, student: $5   **55.** It will have the opposite sign. This can be seen by comparing $a_1 b_2 - a_2 b_1$ to $a_2 b_1 - a_1 b_2$   **57.** 0   **59.** 0   **61.** Yes, it will have the opposite sign.   **63.** No, same value as original value

**65.** Yes, value is double original value   **67. a)** $x = \dfrac{c_1 b_2 - c_2 b_1}{a_1 b_2 - a_2 b_1}$   **b)** $y = \dfrac{a_1 c_2 - a_2 c_1}{a_1 b_2 - a_2 b_1}$   **68.** $y = \dfrac{2}{5} x - \dfrac{6}{5}$

**69.**   **70.**   **71.**

## Chapter 9 Review Exercises   **1.** c)   **2.** a)   **3.** Consistent, one solution   **4.** Inconsistent, no solution   **5.** Dependent, infinite number of solutions   **6.** Consistent, one solution   **7.** No solution   **8.** One solution   **9.** Infinite number of solutions   **10.** One solution

**11.**    **12.**    **13.**    **14.**    **15.**    **16.**   **17.**

**18.**    **19.** $(5, 2)$   **20.** $(-3, 2)$   **21.** $(4, 1)$   **22.** $(-15, 5)$   **23.** No solution   **24.** Infinite number of solutions

**25.** $\left(\frac{5}{2}, -1\right)$   **26.** $\left(-\frac{2}{7}, \frac{29}{7}\right)$   **27.** $(-6, -2)$   **28.** $(1, -2)$   **29.** $(-7, 19)$   **30.** $\left(\frac{32}{13}, \frac{8}{13}\right)$   **31.** $\left(-1, \frac{13}{3}\right)$

**32.** No solution   **33.** Infinite number of solutions   **34.** $\left(-\frac{78}{7}, -\frac{48}{7}\right)$   **35.** $(1, 2, -4)$

**36.** $(-1, 3, -2)$   **37.** $(-5, 1, 2)$   **38.** $(3, -2, -2)$   **39.** $\left(\frac{8}{3}, \frac{2}{3}, 3\right)$   **40.** $(0, 2, -3)$   **41.** No solution   **42.** Infinite number of solutions

**43.** Luan: 38, Jennifer: 28   **44.** Airplane: 520 mph, wind: 40 mph   **45.** Combine 2 liters of the 20% acid solution with 4 liters of the 50% acid solution.   **46.** 410 adults' tickets and 240 children's tickets were sold.   **47.** His ages were 41 years and 77 years.
**48.** $20,000 invested at 7%, $15,000 invested at 5%, and $5000 was invested at 3%.   **49.** $(11, -2)$   **50.** $(3, 1)$   **51.** Infinite number of solutions   **52.** $(2, 1, -2)$   **53.** No solution   **54.** $(1, -1, 3)$   **55.** $(2, 3)$   **56.** $(-3, 2)$   **57.** $(-1, 2)$   **58.** $(-2, 3, 4)$   **59.** $(1, 1, 2)$
**60.** No solution

## Chapter 9 Practice Test   **1.** b) [9.1]   **2.** Consistent, one solution [9.1]   **3.** Inconsistent, no solution [9.1]   **4.** Dependent, infinite number of solutions [9.1]   **5.** Consistent; one solution [9.1]   **6.** Dependent; infinite number of solutions [9.1]

**7.** Inconsistent; no solution [9.1]   **8.**    [9.1]   **9.**    [9.1]   **10.** $(1, 1)$ [9.2]   **11.** $(-3, 2)$ [9.2]

**12.** Infinite number of solutions [9.3]   **13.** $\left(\frac{44}{19}, \frac{48}{19}\right)$ [9.3]   **14.** $(1, -1, 2)$ [9.4]   **15.** $\begin{bmatrix} -2 & 3 & 7 & | & 5 \\ 3 & -2 & 1 & | & -2 \\ 1 & -6 & 9 & | & -13 \end{bmatrix}$ [9.6]

**16.** $\begin{bmatrix} 6 & -2 & 4 & | & 4 \\ 0 & 5 & -3 & | & 12 \\ 2 & -1 & 4 & | & -3 \end{bmatrix}$ [9.6]   **17.** $(4, -1)$ [9.6]   **18.** $(3, -1, 2)$ [9.6]   **19.** $-1$ [9.7]   **20.** $165$ [9.7]   **21.** $(-3, 2)$ [9.7]   **22.** $(3, 1, -1)$ [9.7]

**23.** 8 pounds sunflower; 12 pounds bird mix [9.5]   **24.** $6\frac{2}{3}$ liters 6% solution; $3\frac{1}{3}$ liters 15% solution [9.5]   **25.** 4, 9, and 16 [9.5]

## Cumulative Review Test   **1. a)** $9, 1$ **b)** $\frac{1}{2}, -4, 9, 0, -4.63, 1$ **c)** $\frac{1}{2}, -4, 9, 0, \sqrt{3}, -4.63, 1$ [1.4]   **2.** $1$ [1.9]   **3.** $\frac{29}{6}$ [2.5]

**4.** [4.2]   **5.** $y = \frac{2}{5}x + \frac{11}{5}$ [4.4]   **6.** $y = 3x - 2$ [4.4]   **7.** $\frac{1}{7a^5b^6}$ [5.2]   **8.** $(3x + 4)(x^2 + 2)$ [6.2]

**9.** $(x - 14)(x - 2)$ [6.3]   **10.** $0, 5$ [6.6]   **11.** $\frac{3y + 2}{8}$ [7.3]   **12.** $\frac{41}{7}$ [7.6]

**13. a)** Function **b)** Function **c)** Not a function [8.2]   **14. a)** $-\frac{1}{7}$ **b)** $\frac{h + 3}{h^2 - 9}$ **c)** Undefined [8.2]

**15.** $(1, 3)$ [9.2]   **16.** $(7, -1)$ [9.3]   **17.** $(2, 1, 3)$ [9.4]   **18.** $10°, 80°, 90°$ [3.3]   **19.** 1 hour [3.4]   **20.** 600 at $20, 400 at $16 [9.5]

# Chapter 10

**Exercise Set 10.1**   **1.** Compound   **3.** Closed   **5.** Union   **7. a)**   **b)** $(-3, \infty)$   **c)** $\{x | x > -3\}$

**9. a)**   **b)** $(-\infty, \pi]$   **c)** $\{w | w \leq \pi\}$   **11. a)**   **b)** $\left(-3, \frac{4}{5}\right]$   **c)** $\left\{q \middle| -3 < q \leq \frac{4}{5}\right\}$

**13. a)**   **b)** $(-7, -4]$   **c)** $\{x | -7 < x \leq -4\}$   **15.**   **17.**   **19.**

**21.**   **23.**   **25.**   **27.** $\left(-\infty, \frac{3}{2}\right)$   **29.** $[2, \infty)$   **31.** $\left(-\infty, \frac{3}{2}\right]$   **33.** $(-\infty, \infty)$

**35.** $A \cup B = \{5, 6, 7, 8\}; A \cap B = \{6, 7\}$   **37.** $A \cup B = \{-1, -2, -4, -5, -6\}; A \cap B = \{-2, -4\}$

**39.** $A \cup B = \{0, 1, 2, 3\}; A \cap B = \varnothing$   **41.** $A \cup B = \{0, 1, 2, 3, 4, 5, 6, 7, 8\}; A \cap B = \varnothing$

**43.** $A \cup B = \{0.1, 0.2, 0.3, 0.4, \ldots\}; A \cap B = \{0.2, 0.3\}$   **45.** $[-5, 1)$   **47.** $[-4, 5]$   **49.** $\left[4, \frac{11}{2}\right)$   **51.** $\left(-\frac{13}{3}, -4\right]$   **53.** $\{x | 3 \leq x < 7\}$

**55.** $\{x | 0 < x \leq 3\}$   **57.** $\left\{u \middle| 4 \leq u \leq \frac{19}{3}\right\}$   **59.** $\{c | -3 < c \leq 1\}$   **61.** $\varnothing$   **63.** $\{x | -5 < x < 2\}$   **65.** $(-\infty, 2) \cup [7, \infty)$   **67.** $[0, 2]$

**69.** $(-\infty, 0) \cup (6, \infty)$   **71.** $[0, \infty)$   **73. a)** $l + g \leq 130$   **b)** $l + 2w + 2d \leq 130$   **c)** 24.5 inches   **75.** 11 boxes

**77.** 300 text messages   **79.** 1881 books   **81.** 12 ounces   **83.** For sales over \$5,000 per week   **85.** 24   **87.** $76 \leq x \leq 100$

**89. a)** \$12,242.75   **b)** \$78,135.89   **91. a)** $[0, 3)$   **b)** $(3, 10]$   **93. a)** $[0, 5)$   **b)** $(5, 13]$   **95. a)** $[0, 8]$   **b)** None   **97.** $6.97 < x < 8.77$

**99.** Answers will vary.   **101. a)** $[17.5, 23.5]$   **b)** $[23.5, 31]$   **c)** $[27.2, 36.5]$   **103.** $84 \leq x \leq 100$   **105. a)** Answers will vary.

**b)** $(-3, \infty)$   **107. a)** 4   **b)** $0, 4$   **c)** $-3, 4, \frac{5}{2}, 0, -\frac{13}{29}$   **d)** $-3, 4, \frac{5}{2}, \sqrt{7}, 0, -\frac{13}{29}$   **108.** Associative property of addition

**109.** Commutative property of addition   **110.** $V = \dfrac{R - L + Dr}{r}$

**Exercise Set 10.2**   **1.** $|x| \leq 4$   **3.** $|x| = 4$   **5.** $|x| > 4$   **7.** $|x| \geq 5$   **9.** $|x| < 5$   **11.** $|x| > -6$   **13.** $\{-7, 7\}$   **15.** $\left\{-\frac{1}{2}, \frac{1}{2}\right\}$

**17.** $\varnothing$   **19.** $\{-13, 3\}$   **21.** $\{-7\}$   **23.** $\left\{\frac{3}{2}, \frac{11}{6}\right\}$   **25.** $\{-17, 23\}$   **27.** $\{3\}$   **29.** $\varnothing$   **31.** $\{w | -11 < w < 11\}$   **33.** $\{q | -13 \leq q \leq 3\}$

**35.** $\{b | 1 < b < 5\}$   **37.** $\{x | -9 \leq x \leq 6\}$   **39.** $\left\{x \middle| \frac{1}{3} < x < \frac{13}{3}\right\}$   **41.** $\varnothing$   **43.** $\{j | -22 < j < 6\}$   **45.** $\{x | -1 \leq x \leq 7\}$

**47.** $\{y | y < -8 \text{ or } y > 8\}$   **49.** $\{x | x < -9 \text{ or } x > 1\}$   **51.** $\left\{b \middle| b < \frac{2}{3} \text{ or } b > 4\right\}$   **53.** $\{h | h < 1 \text{ or } h > 4\}$   **55.** $\{x | x < 2 \text{ or } x > 6\}$

**57.** $\{x | x \leq -18 \text{ or } x \geq 2\}$   **59.** $\mathbb{R}$   **61.** $\{x | x < 2 \text{ or } x > 2\}$   **63.** $\{-1, 15\}$   **65.** $\{-3, 1\}$   **67.** $\left\{-23, \frac{13}{7}\right\}$   **69.** $\{10\}$   **71.** $\{-9, 9\}$

**73.** $\{q | q < -8 \text{ or } q > -4\}$   **75.** $\{w | -1 \leq w \leq 8\}$   **77.** $\left\{-\frac{8}{5}, 2\right\}$   **79.** $\left\{x \middle| x < -\frac{5}{2} \text{ or } x > -\frac{5}{2}\right\}$   **81.** $\left\{x \middle| -\frac{13}{3} \leq x \leq \frac{5}{3}\right\}$   **83.** $\varnothing$

**85.** $\{w | -16 < w < 8\}$   **87.** $\mathbb{R}$   **89.** $\left\{2, \frac{22}{3}\right\}$   **91.** $\left\{-\frac{3}{2}, \frac{9}{7}\right\}$   **93. a)** $[0.085, 0.093]$   **b)** 0.085 inch   **c)** 0.093 inch

**95. a)** $[132, 188]$   **b)** 132 to 188 feet below sea level, inclusive   **97. a)** Two   **b)** Infinite number   **c)** Infinite number

**99. a)** None   **b)** One   **c)** Two   **101.** $x = -\dfrac{b}{a}; |ax + b|$ is never less than 0, so set $|ax + b| = 0$ and solve for $x$.

**103. a)** Set $ax + b = -c$ or $ax + b = c$ and solve each equation for $x$.   **b)** $x = \dfrac{-c - b}{a}$ or $x = \dfrac{c - b}{a}$

**105. a)** Write $ax + b < -c$ or $ax + b > c$ and solve each inequality for $x$.   **b)** $x < \dfrac{-c - b}{a}$ or $x > \dfrac{c - b}{a}$

**107.** $\mathbb{R}$; Since $4 - x = -(x - 4)$   **109.** $\{x | x \geq 0\}$; by definition of absolute value   **111.** $\{2\}$; set $x + 1 = 2x - 1$ or $x + 1 = -(2x - 1)$

**113.** $\{x | x \leq 4\}$; by definition $|x - 4| = -(x - 4)$ if $x \leq 4$   **115.** $\{4\}$   **117.** $\varnothing$   **119.** $\dfrac{29}{72}$   **120.** 25   **121.** $\approx 1.33$ miles   **122.** $\{x | x \leq 3\}$

**Mid-Chapter Test**   **1.** $[10.1]$   **2.** $[-4, 5)$ $[10.1]$   **3.** $[10.1]$   **4.** $[10.1]$

**5.** $[10.1]$   **6.** $(-1, \infty)$ $[10.1]$   **7.** $(-5, 7]$ $[10.1]$   **8.** $\left[-\frac{1}{6}, 1\right)$ $[10.1]$   **9.** $\{x | -8 \leq x \leq 2\}$ $[10.1]$

**10.** $\{m | -6 < m \leq 7\}$ $[10.1]$   **11.** $\{x | x \leq 0 \text{ or } x > 6\}$ $[10.1]$   **12.** $A \cup B = \{7, 8, 9, 10, 12, 15\}, A \cap B = \{8, 10\}$ $[10.1]$

**13.** $\{-2, 10\}$ [10.2]   **14.** $\{w| -16 \le w \le 26\}$ [10.2]   **15.** $\{z| z \le -8 \text{ or } z \ge 8\}$ [10.2]   **16.** $\mathbb{R}$ [10.2]   **17.** $\left\{-\dfrac{1}{2}, \dfrac{9}{2}\right\}$ [10.2]

**18.** $\left\{m \middle| -1 < m < \dfrac{11}{3}\right\}$ [10.2]   **19.** $\left\{-5, -\dfrac{4}{5}\right\}$ [10.2]   **20.** $\left\{-5, \dfrac{1}{2}\right\}$ [10.2]

## Exercise Set 10.3    **1.** Boundary line   **3.** Solid   **5.** Checkpoint   **7.** Linear programming

**9.**    **11.**    **13.**    **15.**    **17.**    **19.**

**21.**    **23.**    **25.**    **27.**    **29.**    **31.**

**33.**   **35.**    **37.**    **39.**    **41.**   **43.**

**45.**   **47.**    **49.**    **51.**    **53.**    **55.**    **57.**

**59.**    **61.**    **63.**    **65.**    **67.**

**69.** No; the point is not contained in the solution to the $<$ inequality.   **71.** No; the point is not a solution to either inequality.
**73.** There is no solution. Opposite sides of the same line are being shaded and only one inequality includes the line.   **75.** There are an infinite number of solutions. Both inequalities include the line $5x - 2y = 3$.   **77.** There are an infinite number of solutions. The lines are not parallel or identical.

**79.**    **81.**    **83.** $f_2 = \dfrac{f_3 d_3 - f_1 d_1}{d_2}$   **84.** Domain: $\{-1, 0, 4, 5\}$, Range: $\{-5, -2, 2, 3\}$
**85.** Domain: $\mathbb{R}$, Range: $\mathbb{R}$
**86.** Domain: $\mathbb{R}$, Range: $\{y| y \ge -1\}$

## Chapter 10 Review Exercises    **1.**    **2.**    **3.**   **4.**

**5.**   **6.**   **7.**   **8.**    **9.** 6 boxes   **10.** 7 minutes   **11.** $\approx 15.67$ weeks

**12.** $\{x| 81 \le x \le 100\}$   **13.** $A \cup B = \{1, 2, 3, 4, 5\}; A \cap B = \{2, 3, 4, 5\}$   **14.** $A \cup B = \{2, 3, 4, 5, 6, 7, 8, 9\}; A \cap B = \emptyset$
**15.** $A \cup B = \{1, 2, 3, 4, \ldots\}; A \cap B = \{2, 4, 6, \ldots\}$   **16.** $A \cup B = \{3, 4, 5, 6, 9, 10, 11, 12\}; A \cap B = \{9, 10\}$   **17.** $(5, 11)$

**18.** $(-3, 5]$   **19.** $\left(\dfrac{7}{2}, 8\right)$   **20.** $\left(\dfrac{8}{3}, 6\right)$   **21.** $(-3, 1]$   **22.** $(2, 14)$   **23.** $\{h| -3 < h \le 1\}$   **24.** $\mathbb{R}$   **25.** $\{x| x \le -4\}$

**26.** $\{g| g < -6 \text{ or } g \ge 11\}$   **27.** $\{-2, 2\}$   **28.** $\{x| -8 < x < 8\}$   **29.** $\{x| x \le -9 \text{ or } x \ge 9\}$   **30.** $\{-18, 8\}$

**31.** $\{x| x \le -3 \text{ or } x \ge 7\}$   **32.** $\left\{-\dfrac{1}{2}, \dfrac{9}{2}\right\}$   **33.** $\{q| 1 < q < 8\}$   **34.** $\{-1, 4\}$   **35.** $\{x| -14 < x < 22\}$   **36.** $\left\{-5, -\dfrac{4}{5}\right\}$   **37.** $\mathbb{R}$

**38.** $\left[-5, -\dfrac{1}{3}\right]$　**39.** $(4, 8)$　**40.** $\left(-\dfrac{17}{2}, \dfrac{27}{2}\right]$　**41.** $[-2, 6)$　**42.** $(-\infty, \infty)$　**43.** $\left(\dfrac{2}{3}, 10\right]$　**44.** 　**45.**

**46.** 　**47.** 　**48.** 　**49.**

**50.** 　**51.** No solution　**52.** 　**53.** 　**54.** 　**55.**

## Chapter 10 Practice Test

**1.** $A \cup B = \{5, 7, 8, 9, 10, 11, 14\}$; $A \cap B = \{8, 10\}$ [10.1]

**2.** $A \cup B = \{1, 3, 5, 7, \ldots\}$; $A \cap B = \{3, 5, 7, 9, 11\}$ [10.1]　**3.**  [10.1]　**4.** [10.1]　**5.** $\left(\dfrac{9}{2}, 7\right]$ [10.1]

**6.** $[13, 16)$ [10.1]　**7.** $\{-7, 2\}$ [10.2]　**8.** $\left\{-\dfrac{14}{3}, \dfrac{26}{5}\right\}$ [10.2]　**9.** $\{-3\}$ [10.2]　**10.** $\{x \mid x < -1 \text{ or } x > 4\}$ [10.2]　**11.** $\left\{x \mid \dfrac{1}{2} \le x \le \dfrac{5}{2}\right\}$ [10.2]

**12.**  [10.3]　**13.**  [10.3]　**14.**  [10.3]

## Cumulative Review Test

**1.** $-88$ [1.9]　**2.** 9 [1.9]　**3.** $\dfrac{3}{2}x + \dfrac{7}{2}$ [2.1]　**4.** 5 [2.5]　**5.** 208 pounds [2.7]　**6.** \$50,000 [3.2]

**7.** \$1.69 per pound [3.4]　**8.** $\dfrac{1}{4a^{10}b^2}$ [5.2]　**9.** $4.5 \times 10^{-7}$ [5.3]　**10.** $2pq(2p + 3)(6p - 5)$ [6.5]　**11.** $-5$ [7.6]　**12.** No solution [7.6]

**13.**  [8.1]　**14.** a and c are functions [8.2]　**15.** $y = \dfrac{3}{5}x - 2$ [8.4]　**16.** $y = -\dfrac{3}{4}x + \dfrac{1}{4}$ [8.4]

**17. a)** $x^3 - 16x^2 + 85x - 150$　**b)** $x - 6$　**c)** $\{x \mid x \neq 5\}$ [8.5]　**18.** $(2, -1, 3)$ [9.4]　**19.** $\left(-\dfrac{7}{6}, \dfrac{2}{3}\right)$ [10.1]

**20.**  [10.3]

# Chapter 11

**Exercise Set 11.1**　**1.** Radical　**3.** Radicand　**5.** Cube　**7.** Negative　**9.** Even　**11. a)** 3　**b)** $-3$　**c)** Not a real number
**d)** Not a real number　**13.** $-4$　**15.** $-5$　**17.** $-1$　**19.** 1　**21.** Not a real number　**23.** $-7$　**25.** Not a real number

**27.** Not a real number　**29.** $\dfrac{1}{5}$　**31.** $\dfrac{1}{2}$　**33.** $\dfrac{2}{7}$　**35.** $-\dfrac{2}{3}$　**37.** $\approx -2.07$　**39.** 7　**41.** 3　**43.** 119　**45.** 235.23　**47.** 0.06　**49.** $\dfrac{12}{13}$

**51.** $|x - 8|$　**53.** $|x - 3|$　**55.** $|3x^2 - 1|$　**57.** $|6a^3 - 5b^4|$　**59.** $|x^5|$　**61.** $|z^{16}|$　**63.** $|a - 4|$　**65.** $|3a + 2b|$　**67.** $5a$　**69.** $4c^3$

**71.** $x + 2$　**73.** $2x + y$　**75.** 2　**77.** 8　**79.** 9　**81.** $\approx 9.381$　**83.** $\approx 5.290$　**85.** $-3$　**87.** 97　**89.** 11　**91.** 45

**93.** Select a value less than $-\dfrac{1}{2}$.　**95.** d　**97.** a　**99.** 　**101.**

**103.** One answer is $f(x) = \sqrt{x-8}$    **105. a)** $\sqrt{1288} \approx 35.89$ feet per second  **b)** $\sqrt{2576} \approx 50.75$ feet per second
**107.** There is no real number which when squared gives $-81$.    **109.** No; if the radicand is negative, the answer is not a real number.
**111. a)** 1.3 **b)** 1.3   **113.** $x \geq 1$; answers will vary   **115.** $x \geq 3$; answers will vary   **117. a)** All real numbers **b)** $a \geq 0$ **c)** All real
numbers   **119.** If $n$ is even, you are finding an even root of a positive number. If $n$ is odd, the expression is real.   **121.** $x > -5$;
answers will vary   **123. a)** No **b)** Yes, when $x = 0$ **c)** Yes   **127.** $(3a - b)(3x + 4y)$   **128.** $3x(x - 4)(x - 2)$

**129.** $(2x + 1)(2x - 1)(2x^2 + 3)$   **130.** $\left(x - \dfrac{2}{3}y\right)\left(x^2 + \dfrac{2}{3}xy + \dfrac{4}{9}y^2\right)$

**Exercise Set 11.2**   **1.** Numerator   **3.** Index   **5.** Add   **7.** $x^{5/2}$   **9.** $9^{5/2}$   **11.** $z^{5/3}$   **13.** $7^{10/3}$   **15.** $9^{7/4}$   **17.** $y^{14/3}$   **19.** $(a^3 b)^{1/4}$

**21.** $(x^9 z^5)^{1/4}$   **23.** $(3a + 8b)^{1/6}$   **25.** $\left(\dfrac{2x^6}{11y^7}\right)^{1/5}$   **27.** $\sqrt{a}$   **29.** $\sqrt[3]{a^2}$   **31.** $\sqrt[3]{18^5}$   **33.** $\sqrt{24x^3}$   **35.** $\left(\sqrt[5]{11b^2 c}\right)^3$   **37.** $\sqrt[5]{6a + 5b}$

**39.** $\dfrac{1}{\sqrt[3]{b^3 - d}}$   **41.** $a$   **43.** $x^3$   **45.** $\sqrt[3]{y}$   **47.** $\sqrt{y}$   **49.** 19.3   **51.** $x^5 y^{10}$   **53.** $\sqrt{xyz}$   **55.** $\sqrt[4]{x}$   **57.** $\sqrt[8]{y}$   **59.** $\sqrt[9]{x^2 y}$   **61.** $\sqrt[10]{a^9}$

**63.** 3   **65.** 4   **67.** 16   **69.** Not a real number   **71.** $\dfrac{5}{3}$   **73.** $\dfrac{1}{2}$   **75.** $-9$   **77.** $-4$   **79.** $\dfrac{1}{4}$   **81.** $\dfrac{1}{64}$   **83.** $\dfrac{3}{4}$   **85.** Not a real number

**87.** 24   **89.** $\dfrac{11}{28}$   **91.** $x^{9/2}$   **93.** $x^{1/6}$   **95.** $\dfrac{1}{x}$   **97.** 1   **99.** $\dfrac{y^{5/3}}{12}$   **101.** $\dfrac{12}{x^{11/6}}$   **103.** $\dfrac{1}{2x^{1/3}}$   **105.** $\dfrac{121}{x^{1/7}}$   **107.** $\dfrac{64}{a^{66/5}}$   **109.** $\dfrac{x}{y^{20}}$

**111.** $8z^{7/2} - 4$   **113.** $\dfrac{5}{x^5} + \dfrac{20}{x^{3/2}}$   **115.** $12x^{13/6} - 18x^2$   **117.** $\approx 13.42$   **119.** $\approx 3.32$   **121.** $\approx 20.53$   **123.** $\approx 0.03$   **125.** $n$ is odd, or $n$ is

even and $a \geq 0$.   **127.** $(4^{1/2} + 9^{1/2})^2 \neq 4 + 9$; $25 \neq 13$   **129.** $(1^{1/3} + 1^{1/3})^3 \neq 1 + 1$; $8 \neq 2$   **131.** $x^{1/2}(x + 1)$

**133.** $y^{1/3}(1 - y)(1 + y)$   **135.** $\dfrac{1 + y^2}{y^{2/5}}$   **137. a)** $2^{10} = 1024$ bacteria **b)** $2^{10}\sqrt{2} \approx 1448$ bacteria   **139.** 2247 calories   **141.** $\{x \mid x \geq 7\}$

**143.** 2; $z^{\frac{1}{4} \cdot \frac{1}{5} \cdot \frac{1}{a} \cdot \frac{1}{3}} = z^{\frac{1}{60a}}$; $z^{\frac{1}{60a}} = z^{\frac{1}{120}}$, $60a = 120$; $a = 2$   **145. a)** When $n$ is even and $a \geq 0$, or $n$ is odd **b)** $a^{1/n}$   **147. a)** Always real

**b)** $a$ **c)** $a$ **d)** $|a|$   **149. a)** No; $(xy)^{1/2} = x^{1/2} y^{1/2}$ **b)** No; $(xy)^{-1/2} = x^{-1/2} y^{-1/2} = \dfrac{1}{x^{1/2} y^{1/2}}$   **151.** 9   **153.** $\dfrac{b^2 + a^3 b}{a^3 - b}$

**154.** 0, 3   **155.** $\approx 441.67$ miles per hour.   **156. c)** is a function

**Exercise Set 11.3**   **1.** Square   **3.** 3   **5.** 4   **7.** Product   **9.** 7   **11.** $2\sqrt{6}$   **13.** $4\sqrt{2}$   **15.** $5\sqrt{2}$   **17.** $5\sqrt{3}$   **19.** $2\sqrt{10}$   **21.** $2\sqrt[3]{2}$
**23.** $3\sqrt[3]{2}$   **25.** $2\sqrt[3]{4}$   **27.** $2\sqrt[3]{5}$   **29.** $2\sqrt[4]{3}$   **31.** $-2\sqrt[5]{2}$   **33.** $b^3$   **35.** $x^2$   **37.** $x\sqrt{x}$   **39.** $a^2\sqrt{a}$   **41.** $8z^{10}\sqrt[3]{z^2}$   **43.** $b^5\sqrt[4]{b^3}$
**45.** $x\sqrt[6]{x^3}$ or $x\sqrt{x}$   **47.** $3y^4\sqrt[5]{y^3}$   **49.** $10y^4\sqrt{2y}$   **51.** $xy^2\sqrt[3]{y}$   **53.** $ab^4\sqrt[5]{ab^3}$   **55.** $2x^7 y^{10} z^{13}\sqrt{6xz}$   **57.** $3a^2 b^2\sqrt[3]{3b^2}$
**59.** $2x^2 y^2 z^4\sqrt[4]{2yz^3}$   **61.** $3a^2 b^2\sqrt[4]{b}$   **63.** $2a^2 b^2\sqrt[5]{b^2}$   **65.** 3   **67.** $\dfrac{9}{10}$   **69.** 3   **71.** $\dfrac{1}{4}$   **73.** $\dfrac{1}{2}$   **75.** $\dfrac{1}{3}$   **77.** $\dfrac{1}{2}$   **79.** 2   **81.** $\dfrac{x}{3}$   **83.** $\dfrac{4x^2}{5y^5}$

**85.** $\dfrac{c^2}{4}$   **87.** $a^2 b^6\sqrt[3]{a^2 b^2}$   **89.** $2\sqrt{2}$   **91.** $3x^2$   **93.** $2x^2 y\sqrt{2y}$   **95.** $\dfrac{\sqrt[3]{5y}}{2x^4}$   **97.** $\dfrac{y^2\sqrt[3]{5y}}{x^2}$   **99.** $\dfrac{x^3\sqrt[4]{10y}}{3}$

**101. a)** Square the natural numbers. **b)** 1, 4, 9, 16, 25, 36   **103.** If $n$ is even and $a$ or $b$ are negative, the numbers are not real
numbers.   **105.** $(a \cdot b)^{1/2} = a^{1/2} b^{1/2} = \sqrt{a}\sqrt{b}$   **107.** No; One example is $\sqrt{18}/\sqrt{2} = 3$.   **109. a)** No

**b)** When $\sqrt[n]{x}$ is a real number and not equal to 0.   **110.** $C = \dfrac{5}{9}(F - 32)$   **111.** $3x^6 - x^3 + 4$   **112.** $(x - 1)(x^2 - 8x + 19)$

**113.** $4, \dfrac{3}{2}$   **114.** $-28, 32$

**Exercise Set 11.4**   **1.** Like   **3.** Distributive   **5.** FOIL   **7.** 0   **9.** $4\sqrt{5}$   **11.** $-4\sqrt{3} + 5$   **13.** $-7\sqrt[4]{y}$   **15.** $2\sqrt[3]{x} + 9\sqrt{5}$
**17.** $7\sqrt{x} - 6\sqrt{y}$   **19.** $3\sqrt{3}$   **21.** $-30\sqrt{3} + 25\sqrt{5}$   **23.** $-4\sqrt{10}$   **25.** $18y\sqrt{5x}$   **27.** $-16\sqrt{5x}$   **29.** $-27a\sqrt{2}$   **31.** $5\sqrt[3]{4}$   **33.** $-7$
**35.** $6a\sqrt[3]{ab^2}$   **37.** $3r^3 s^2\sqrt{rs}$   **39.** 0   **41.** 4   **43.** $2\sqrt[3]{7}$   **45.** $3m^2 n^5\sqrt{3n}$   **47.** $3x^3 y^4\sqrt[3]{2x^2 y}$   **49.** $x^7 y^7 z^3\sqrt[5]{x^2 y^3 z}$   **51.** $x^2 y^2\sqrt[3]{4y^2}$
**53.** $5 - \sqrt{15}$   **55.** $2\sqrt[3]{y^2} - y^3$   **57.** $4x^5 y^3\sqrt[3]{x} + 4xy^4\sqrt[3]{2x^2 y^2}$   **59.** 59   **61.** $6 - x^2$   **63.** $7 - z$   **65.** $23 + 9\sqrt{3}$   **67.** $16 - 10\sqrt{2}$
**69.** $10 - 3\sqrt{6}$   **71.** $29 - 12\sqrt{5}$   **73.** $18x - \sqrt{3xy} - y$   **75.** $8 - 2\sqrt[3]{18} - \sqrt[3]{12}$   **77.** $x - \sqrt{3x}$   **79.** $x^2 + x\sqrt[3]{x^2}$
**81.** $x\sqrt[4]{27x^2} - x^2\sqrt[4]{3x}$   **83.** $3\sqrt{2}$   **85.** $3\sqrt{5}$   **87.** $-14 + 11\sqrt{2}$   **89.** $5\sqrt{6} - 2\sqrt{3}$   **91.** $15\sqrt{2}$   **93.** $2x^3\sqrt[3]{10x^2}$   **95.** $2b^2 c\sqrt[6]{2ab^5 c^3}$
**97.** $4ab\sqrt[4]{b}$   **99.** $x - 2\sqrt[3]{x^2 y^2} - \sqrt[3]{xy} + 2y$   **101.** $ab\sqrt[3]{12a^2 b^2} - 2a^2 b^2\sqrt[3]{3}$   **103.** $2x - 5$   **105.** $2|r - 4|$   **107.** $P = 14\sqrt{5}, A = 60$
**109.** $P = 12\sqrt{5}, A = 30$   **111. a)** $\approx 45.17$ miles per hour **b)** $\approx 35.33$ miles per hour   **113. a)** 37 inches **b)** $\approx 37.97$ inches

**115. a)**
$y$
$(f + g)(x) = 2 + \sqrt{x}$
$x$

**117. a)**
$y$
$(f - g)(x) = 2, x \geq 0$
$x$

**119.**

$y$
$f(x) = \sqrt{x^2}$
$x$

**b)** Raises the graph 2 units

**b)** $\{x \mid x \geq 0\}$

**121.** No: one example is $\sqrt{9} + \sqrt{16} \neq \sqrt{9+16}$, $3 + 4 \neq 5, 7 \neq 5$.   **123.** No, $-\sqrt{2} + \sqrt{2} = 0$   **125.** A quotient of two integers, denominator not 0.   **126.** A number that can be represented on a real number line.   **127.** A real number that cannot be expressed as a quotient of two integers.   **128.** $|a| = \begin{cases} a, & a \geq 0 \\ -a, & a < 0 \end{cases}$   **129.** $m = \dfrac{2E}{v^2}$   **130. a)** ⟶   **b)** $\left(-\dfrac{1}{2}, 5\right]$   **c)** $\left\{ x \,\middle|\, -\dfrac{1}{2} < x \leq 5 \right\}$

## Mid-Chapter Test

**1.** $11$ [11.1]   **2.** $-\dfrac{3}{4}$ [11.1]   **3.** $16.3$ [11.1]   **4.** $|3a^2 - 4b^3|$ [11.1]   **5.** $3$ [11.1]   **6.** $(7a^4b^3)^{1/5}$ [11.2]   **7.** $20$ [11.2]   **8.** $a^{10}b^{15}c^5$ [11.3]   **9.** $\dfrac{14}{x}$ [11.3]   **10.** $8x + \dfrac{16}{x^{5/2}}$ [11.3]   **11.** $4x^2y^4\sqrt{2y}$ [11.3]   **12.** $2a^2b^3c^2\sqrt[6]{ab^5c^3}$ [11.3]   **13.** $\dfrac{1}{3}$ [11.3]   **14.** $\dfrac{y^2\sqrt{y}}{3x^5}$ [11.3]   **15.** $11\sqrt{x} + 12\sqrt{y}$ [11.4]   **16.** $27x\sqrt{10y}$ [11.4]   **17.** $2x^2 - x\sqrt{5} - 15$ [11.4]   **18.** $18a\sqrt{a} - 20a\sqrt{3}$ [11.4]   **19.** $7ab\sqrt[4]{ab}$ [11.4]   **20.** Part **a)** will have an absolute value. **a)** $|x - 3|$, **b)** $8x$ [11.1]

## Exercise Set 11.5

**1.** Rationalizing   **3.** Power   **5.** Radicand   **7.** $\dfrac{\sqrt{2}}{2}$   **9.** $\dfrac{4\sqrt{5}}{5}$   **11.** $\sqrt{6}$   **13.** $\dfrac{\sqrt{z}}{z}$   **15.** $\dfrac{p\sqrt{2}}{2}$   **17.** $\dfrac{\sqrt{7y}}{7}$
**19.** $3\sqrt{2}$   **21.** $\dfrac{\sqrt{xy}}{y}$   **23.** $\dfrac{\sqrt{10m}}{4}$   **25.** $\dfrac{\sqrt{2n}}{3}$   **27.** $\dfrac{3x^2y\sqrt{yz}}{z^2}$   **29.** $\dfrac{2y^4z\sqrt{15xz}}{3x}$   **31.** $\dfrac{4x^3y^2\sqrt{yz}}{z^2}$   **33.** $\dfrac{\sqrt[3]{4}}{2}$   **35.** $\dfrac{8\sqrt[3]{y^2}}{y}$   **37.** $\dfrac{\sqrt[4]{8}}{2}$
**39.** $\dfrac{a\sqrt[4]{2}}{2}$   **41.** $\dfrac{5\sqrt[4]{z^2}}{z}$   **43.** $\dfrac{10\sqrt[5]{y^2}}{y}$   **45.** $\dfrac{2\sqrt[4]{a^3}}{a}$   **47.** $\dfrac{\sqrt[3]{4x^2}}{2x}$   **49.** $\dfrac{5m\sqrt[4]{8}}{2}$   **51.** $\dfrac{\sqrt[4]{135x}}{3x}$   **53.** $\dfrac{\sqrt[3]{12x^2y}}{2y}$   **55.** $\dfrac{\sqrt[3]{7xy^2z}}{z}$   **57.** $11$
**59.** $62$   **61.** $-6$   **63.** $a - b$   **65.** $4x - 9y$   **67.** $\sqrt{2} - 1$   **69.** $2 - \sqrt{3}$   **71.** $\dfrac{-5\sqrt{2} - 35}{47}$   **73.** $\dfrac{10 + \sqrt{30}}{14}$   **75.** $\dfrac{18 - 3\sqrt{x}}{36 - x}$
**77.** $\dfrac{4x + 4y\sqrt{x}}{x - y^2}$   **79.** $\dfrac{-13 + 3\sqrt{6}}{23}$   **81.** $a + a^3$   **83.** $\dfrac{4\sqrt{x+2} + 12}{x - 7}$   **85.** $\dfrac{a}{5}$   **87.** $\dfrac{\sqrt{2}}{3}$   **89.** $1$   **91.** $\dfrac{2xy^3\sqrt{30xz}}{5z}$   **93.** $\dfrac{\sqrt{14}}{x}$
**95.** $\dfrac{\sqrt{a} - 7}{a - 49}$   **97.** $-\dfrac{\sqrt{2x}}{2}$   **99.** $\dfrac{\sqrt[4]{24x^3y^2}}{2x}$   **101.** $\dfrac{2y^4z^3\sqrt[3]{2x^2z}}{x}$   **103.** $\dfrac{a\sqrt{r} + 2r\sqrt{a}}{a - 4r}$   **105.** $\dfrac{\sqrt[3]{150y^2}}{5y}$   **107.** $\dfrac{y^3z\sqrt[4]{54x^2}}{3x}$   **109.** $\sqrt{2}$
**111.** $\dfrac{3\sqrt{5}}{5}$   **113.** $\dfrac{8\sqrt{6}}{3}$   **115.** $\dfrac{19\sqrt{2}}{2}$   **117.** $\dfrac{21\sqrt{2}}{2}$   **119.** $-\dfrac{301\sqrt{2}}{20}$   **121.** $\dfrac{3\sqrt{6}}{4}$   **123.** $\left(-\dfrac{2}{y} + \dfrac{3}{x}\right)\sqrt{xy}$   **125.** $2\sqrt{a}$   **127.** $\sqrt[3]{(a+b)^5}$
**129.** $\sqrt[15]{(a + 2b)^2}$   **131.** $\sqrt[6]{rs^5}$   **133.** $\sqrt[15]{x^2y^8}$   **135.** $\approx 3.69$ meters   **137.** $\approx 12$ inches   **139. a)** $6.21$ million **b)** $\approx 2.35$ million
**141.** $\dfrac{10}{15 + 3\sqrt{5}}$   **143.** $\dfrac{1}{\sqrt{x+h} + \sqrt{x}}$   **145.** $\dfrac{3}{\sqrt{3}}; \dfrac{2}{\sqrt{2}} = \sqrt{2}, \dfrac{3}{\sqrt{3}} = \sqrt{3}$   **147.** $2 + \sqrt{3}$; rationalize the denominator and compare   **149. a)** $4, 8, 12$ **b)** $9, 18, 27$ **c)** $x^{(3a+2b)/6}$ **d)** $x^{(3a-2b)/6}$   **152.** $b_2 = \dfrac{2A}{h} - b_1$   **153.** $40$ miles per hour, $50$ miles per hour   **154.** $4x^3 + x^2 - 20x + 4$   **155.** $-8, 1$

## Exercise Set 11.6

**1.** Radical   **3.** Original   **5.** Index   **7.** Cube   **9.** $16$   **11.** No real solution   **13.** $-64$   **15.** $11$   **17.** $9$
**19.** $-1$   **21.** $81$   **23.** $71$   **25.** $26$   **27.** No real solution   **29.** $2, 4$   **31.** $8$   **33.** $7$   **35.** $\dfrac{2}{3}$   **37.** $16$   **39.** $2$   **41.** $10$   **43.** $6$   **45.** $8$
**47.** $0$   **49.** $-3$   **51.** $\dfrac{3}{2}$   **53.** No real solution   **55.** $0$   **57.** $5, 8$   **59.** No real solution   **61.** $9$   **63.** $3, 7$   **65.** $-1$   **67.** $2$   **69.** $4$
**71.** $5$   **73.** $v = \dfrac{p^2}{2}$   **75.** $g = \dfrac{v^2}{2h}$   **77.** $F = \dfrac{Mv^2}{R}$   **79.** $m = \dfrac{x^2k}{V_0^2}$   **81.** $A = \pi r^2$   **83.** $\sqrt{87}$   **85.** $2\sqrt{10}$   **87.** $1$   **89.** No solution
**91.** $3$   **93.** $7$   **95.** $1$   **97.** $3$   **99.** $\sqrt{16{,}200} \approx 127.28$ feet   **101.** $\sqrt{2000} \approx 44.7$ feet   **103. a)** $\approx 3.14$ seconds
**b)** $\sqrt{2} \cdot T$; compare $\sqrt{\dfrac{l}{32}}$ with $\sqrt{\dfrac{l}{16}}$ **c)** $\sqrt{24} \approx 4.90$ seconds   **105.** $R = \dfrac{8\mu l}{\pi r^4}$   **107.** $0.2(\sqrt{149.4})^3 \approx 365.2$ days
**109.** $\sqrt{10{,}000} = 100$ pounds   **111.** $\sqrt{320} \approx 17.89$ feet per second   **113.** $\sqrt{1649} \approx 40.61$ meters   **115.** $2, -2$   **117.** $5, -1$
**119.** $30$   **121.** $5, -5$   **123.** All real numbers   **125.** Answers will vary.   **127.** $0$   **129.** $1$; Answers will vary.
**131. a)** $3, 7$; points of intersection **b)** Yes **c)** $3, 7$; yes   **133.** At $x = 4$, $g(x)$ or $y = 0$. Therefore the graph must have an $x$-intercept at 4.
**135.** No solution   **137.** $n = \dfrac{z^2\sigma^2}{(\bar{x} - \mu)^2}$   **139.** $L_1 \approx 0.44, L_2 \approx 0.76$   **142.** $P_2 = \dfrac{P_1P_3}{P_1 - P_3}$   **143.** $x$   **144.** $\dfrac{3a}{2b(2a + 3b)}$
**145.** $t(t - 5)$   **146.** $\dfrac{3}{x + 3}$   **147.** $2$

## Exercise Set 11.7

**1.** Imaginary   **3.** Complex   **5.** Part   **7.** Number   **9.** $-1$   **11.** $4 + 0i$   **13.** $5 + 0i$   **15.** $21 - 6i$
**17.** $0 + 2i\sqrt{6}$   **19.** $8 - 2i\sqrt{3}$   **21.** $3 + 7i\sqrt{2}$   **23.** $12 - 5i$   **25.** $0 + (7 - 3\sqrt{5})i$   **27.** $4 + 4i$   **29.** $2 + 10i$   **31.** $-17 - 12i$
**33.** $(4\sqrt{2} + \sqrt{3}) - 2i\sqrt{2}$   **35.** $11 - 4i\sqrt{2}$   **37.** $-3 - 2i\sqrt{5}$   **39.** $6 - 12i$   **41.** $-9 + 4i$   **43.** $-33 + 18i$   **45.** $28 + 4i\sqrt{3}$
**47.** $9 + 9i$   **49.** $1 + 5i$   **51.** $109$   **53.** $39 - 9i\sqrt{2}$   **55.** $\dfrac{25}{72} + \dfrac{1}{4}i$   **57.** $-\dfrac{2}{3}i$   **59.** $\dfrac{3}{2} - i$   **61.** $\dfrac{12}{5} + \dfrac{6}{5}i$   **63.** $\dfrac{3}{5} + \dfrac{6}{5}i$   **65.** $\dfrac{9}{10} - \dfrac{6}{5}i$
**67.** $\dfrac{3}{5} + \dfrac{1}{5}i$   **69.** $\dfrac{5\sqrt{2}}{37} - \dfrac{2\sqrt{6}}{37}i$   **71.** $\dfrac{3}{8} + \dfrac{7}{8}i$   **73.** $5$   **75.** $\dfrac{\sqrt{2}}{3}$   **77.** $0$   **79.** $4\sqrt{2} + 2i\sqrt{3}$   **81.** $20.8 - 16.64i$   **83.** $37 - 39i$
**85.** $2 - \dfrac{11}{2}i$   **87.** $\dfrac{6\sqrt{3}}{7} + \dfrac{12}{7}i$   **89.** $7 + \dfrac{2}{45}i$   **91.** $\dfrac{1}{4} - \dfrac{31}{50}i$   **93.** $2$   **95.** $-4.33 - 10.91i$   **97.** $-1$   **99.** $1$   **101.** $i$   **103.** $-i$

**105. a)** $-2 - 3i$ **b)** $\dfrac{2}{13} - \dfrac{3}{13}i$   **107.** $-4$   **109.** $16 - 4i$   **111.** $14 + 8i$   **113.** 0   **115.** 1   **117.** Yes   **119.** No   **121.** $\approx 0.83 - 3i$

**123.** $\approx 1.5 - 0.33i$   **125.** $-i$   **127.** $1 + i\sqrt{5}, 1 - i\sqrt{5}$   **129.** $6 + 3i\sqrt{3}$   **131.** $-1 + 7i\sqrt{3}$   **133.** Yes   **135.** No

**137.** True; $(2i)(2i) = -4$   **139.** False; $(1 + i)(1 + 2i) = -1 + 3i$   **141.** Even values; $i^n$ where $n$ is even will either be $-1$ or 1

**143.** $2c - 3 - \dfrac{8}{4c + 9}$   **144.** $\dfrac{a^2}{b(a - b)}$   **145.** 4   **146.** 15 pounds at \$5.50, 25 pounds at \$6.30

## Chapter 11 Review Exercises
**1.** 3   **2.** $-3$   **3.** $-5$   **4.** 4   **5.** 5   **6.** 38.2   **7.** $|x|$   **8.** $|x + 7|$   **9.** $|x - y|$

**10.** $|x^2 - 4x + 12|$   **11.** 7   **12.** 57   **13.** $\approx 2.2$   **14.** 12 meters   **15.** $x^{3/2}$   **16.** $x^{5/3}$   **17.** $y^{13/4}$   **18.** $6^{-2/7}$   **19.** $\sqrt{x}$

**20.** $\sqrt[3]{a^2}$   **21.** $\left(\sqrt[4]{8m^2n}\right)^7$   **22.** $\dfrac{1}{\left(\sqrt[3]{x + y}\right)^5}$   **23.** 16   **24.** $x^6$   **25.** 81   **26.** $\sqrt[4]{a}$   **27.** $-6$   **28.** Not a real number   **29.** $\dfrac{3}{4}$   **30.** $\dfrac{3}{8}$

**31.** $x^{4/15}$   **32.** $\dfrac{4}{y^3}$   **33.** $\dfrac{1}{a^{16/15}}$   **34.** $\dfrac{25x^{10}}{y^7}$   **35.** $5a^2 - 3a^{5/2}$   **36.** $\dfrac{4}{x^{7/6}} + 11$   **37.** $x^{2/5}(1 + x)$   **38.** $\dfrac{1 + a^2}{a^{1/2}}$   **39.** 5   **40.** $\approx 2.668$

**41.**

**42.**

**43.** $3\sqrt{2}$   **44.** $2\sqrt[3]{2}$   **45.** $\dfrac{7}{3}$   **46.** $\dfrac{2}{5}$   **47.** $-\dfrac{9}{7}$   **48.** $-\dfrac{3}{5}$   **49.** 8   **50.** 4   **51.** $3xyz^2\sqrt{2y}$

**52.** $5xy^3\sqrt{3xy}$   **53.** $3a^2b^3\sqrt[3]{2ab}$   **54.** $5x^2y^3z^5\sqrt[3]{x^2z}$   **55.** $x^{14}y^{21}z^{35}$   **56.** $8a^3b^{12}c^{18}$   **57.** $2x^3\sqrt{10}$   **58.** $2x^3y\sqrt[3]{x^2y^2}$   **59.** $2x^2y^3\sqrt[3]{4x^2}$

**60.** $2x^2y^4\sqrt[4]{x}$   **61.** $6x - 2\sqrt{15x}$   **62.** $2x^2y^2\sqrt[3]{y^2} + x\sqrt[3]{18y}$   **63.** $\sqrt[4]{a^3b^2}$   **64.** $\sqrt[6]{x^5y^2}$   **65.** $\dfrac{64r^{9/2}}{p^3}$   **66.** $\dfrac{y^{1/5}}{6xz^{1/3}}$   **67.** $\dfrac{\sqrt{15}}{5}$   **68.** $\dfrac{\sqrt[3]{21}}{3}$

**69.** $\dfrac{\sqrt[4]{20}}{2}$   **70.** $\dfrac{x\sqrt{10}}{10}$   **71.** $\dfrac{8\sqrt{x}}{x}$   **72.** $\dfrac{m\sqrt[3]{5}}{5}$   **73.** $\dfrac{10\sqrt[3]{y}}{y}$   **74.** $\dfrac{9\sqrt[4]{z^3}}{z}$   **75.** $\dfrac{x}{3}$   **76.** $\dfrac{x}{2}$   **77.** $\dfrac{4y^2}{x^3}$   **78.** $2x^2y^3$   **79.** $\dfrac{x^2\sqrt{6y}}{y}$

**80.** $\dfrac{2\sqrt{21ab}}{7b}$   **81.** $\dfrac{x^2y^2\sqrt{6yz}}{z}$   **82.** $\dfrac{5xy^2\sqrt{15yz}}{3z}$   **83.** $3xy\sqrt[3]{2y}$   **84.** $\dfrac{\sqrt[3]{75xy^2}}{5y}$   **85.** $\dfrac{y\sqrt[3]{9x^2}}{x}$   **86.** $\dfrac{y^2\sqrt[3]{25x}}{5x}$   **87.** $\dfrac{b^2\sqrt[4]{2ab^2}}{a}$

**88.** $\dfrac{y\sqrt[4]{6x^3y^2}}{2x}$   **89.** 7   **90.** $-2$   **91.** $x^2 - y$   **92.** $7 + 4\sqrt{3}$   **93.** $x + \sqrt{5xy} - \sqrt{3xy} - y\sqrt{15}$

**94.** $\sqrt[3]{6x^2} - \sqrt[3]{4xy} - \sqrt[3]{9xy} + \sqrt[3]{6y^2}$   **95.** $-12 + 6\sqrt{5}$   **96.** $\dfrac{4x - x\sqrt{x}}{16 - x}$   **97.** $\dfrac{4a + a\sqrt{b}}{16 - b}$   **98.** $\dfrac{x\sqrt{y} + 7x}{y - 49}$   **99.** $\dfrac{x - \sqrt{xy}}{x - y}$

**100.** $\dfrac{x - 2\sqrt{xy} - 3y}{x - y}$   **101.** $\dfrac{2\sqrt{a - 1} + 4}{a - 5}$   **102.** $\dfrac{5\sqrt{y + 2} + 15}{y - 7}$   **103.** $9\sqrt[3]{x}$   **104.** $-4\sqrt{3}$   **105.** $12 - 13\sqrt[3]{2}$   **106.** $\dfrac{45\sqrt{2}}{8}$

**107.** $(9x^2y^3 - 4x^3y^4)\sqrt{x}$   **108.** $(8x^2y^2 - x + 3x^3)\sqrt[3]{xy^2}$   **109.** $3x\sqrt{2} - 3\sqrt{5x}$   **110.** $2x^2 + 2x^2\sqrt[3]{4x}$   **111.** $2x + 7$

**112.** $\sqrt{5}|2a + 5|$   **113.** $\sqrt[6]{x + 5}$   **114.** $\sqrt[12]{b^5}$   **115. a)** $12\sqrt{3}$ **b)** 24   **116. a)** $40\sqrt{2}$ **b)** $\dfrac{2040}{7} \approx 291.4$

**117. a)**
**b)** $x \geq 0$   **118. a)**
**b)** $x \geq 0$   **119.** 16   **120.** No solution   **121.** 64   **122.** $-125$

**123.** 9   **124.** 125   **125.** No solution   **126.** 4   **127.** $-3$   **128.** 3   **129.** 0, 9   **130.** 5   **131.** 4   **132.** 6   **133.** $L = \dfrac{V^2w}{2}$

**134.** $A = \pi r^2$   **135.** $2\sqrt{14}$   **136.** $5\sqrt{3}$   **137.** $\sqrt{29} \approx 5.39$ meters   **138.** $\sqrt{1280} \approx 35.78$ feet per second

**139.** $2\pi\sqrt{2} \approx 2.83\pi \approx 8.89$ seconds   **140.** $\sqrt{\dfrac{90}{0.145}} \approx 24.91$ meters per second   **141.** $m \approx 5m_0$. Thus, it is $\approx 5$ times its original mass.

**142.** $1 + 0i$   **143.** $-8 + 0i$   **144.** $7 - 16i$   **145.** $9 + 4i$   **146.** $13 + i$   **147.** $-3 + 3i$   **148.** $12\sqrt{3} + (\sqrt{5} - \sqrt{7})i$   **149.** $-6 + 6i$

**150.** $17 - 6i$   **151.** $(24 + 3\sqrt{5}) + (4\sqrt{3} - 6\sqrt{15})i$   **152.** $-\dfrac{8i}{3}$   **153.** $-1 - \dfrac{\sqrt{3}}{2}i$   **154.** $\dfrac{12}{13} - \dfrac{8}{13}i$   **155.** $\dfrac{5\sqrt{3}}{31} + \dfrac{3\sqrt{2}}{31}i$   **156.** 0

**157.** 7   **158.** $i$   **159.** $-i$   **160.** 1   **161.** $-1$

## Chapter 11 Practice Test
**1.** $|5x - 3|$ [11.1]   **2.** $\dfrac{1}{x^{12/5}}$ [11.2]   **3.** $\dfrac{1 + x^2}{x^{2/3}}$ [11.2]   **4.**
[11.1]   **5.** $3x^3y^5\sqrt{6x}$ [11.3]

**6.** $5x^3y^3\sqrt[3]{2x^2y}$ [11.4]  **7.** $\dfrac{x^3y\sqrt{14yz}}{4z}$ [11.5]  **8.** $\dfrac{9\sqrt[3]{x^2}}{x}$ [11.5]  **9.** $\dfrac{3-\sqrt{3}}{6}$ [11.5]  **10.** $7\sqrt{6}$ [11.4]  **11.** $(2xy+4x^2y^2)\sqrt[3]{y^2}$ [11.4]

**12.** $6\sqrt{3}-2\sqrt{6}-12+4\sqrt{2}$ [11.4]  **13.** $\sqrt[8]{x^5y^3}$ [11.2]  **14.** $\sqrt[12]{(7x+2)^7}$ [11.5]  **15.** $-5$ [11.6]  **16.** $-3$ [11.6]  **17.** $9$ [11.6]

**18.** $3$ [11.6]  **19.** $g=\dfrac{8w^2}{h}$ [11.6]  **20.** $\sqrt{12{,}880}\approx 113.49$ feet per second [11.6]  **21.** 13 feet [11.6]

**22.** $2\pi\sqrt{\dfrac{1400}{65{,}000}}\approx 0.92$ second [11.6]  **23.** $20+20i$ [11.7]  **24.** $\dfrac{33}{53}-\dfrac{17}{53}i$ [11.7]  **25.** $2$ [11.7]

## Cumulative Review Test    **1.** $\dfrac{57}{9}$ [2.5]  **2.** $-1$ [2.5]  **3.** $\$40$ [3.2]  **4.**

[4.2]  **5.** $25x^2y^2-9$ [5.5]

**6.** $w=2r+1$ [5.6]  **7.** $x(4x-5)(x-1)$ [6.4]  **8.** $(x-2)(x^2+5x+13)$ [6.5]  **9.** $\dfrac{1}{4},-\dfrac{3}{2}$ [6.6]  **10.** $\dfrac{(x+y)y^2}{3x^3}$ [7.2]

**11.** $\dfrac{x+3}{x+5}$ [7.4]  **12.** $18$ [7.6]  **13.** Parallel [8.4]  **14.** $y=-\dfrac{2}{3}x-\dfrac{10}{3}$ [8.4]  **15.** $-x^2+5x-13$ [8.5]  **16.** $\left(2,5,\dfrac{34}{5}\right)$ [9.4]

**17.** $40$ [9.7]  **18.** $\{x\mid -1<x<4\}$ [10.2]  **19.** $3,-3$ [11.6]  **20.** 400 feet [7.8]

## Chapter 12

**Exercise Set 12.1**  **1.** Quadratic  **3.** Complex  **5.** Perfect square trinomial  **7.** Reciprocal  **9.** $25$  **11.** $\pm 9$  **13.** $\pm 5$  **15.** $\pm 7i$
**17.** $\pm 2i\sqrt{6}$  **19.** $\pm i\sqrt{61}$  **21.** $8,0$  **23.** $-3\pm 5i$  **25.** $2\pm 3i\sqrt{5}$  **27.** $-1,\dfrac{1}{3}$  **29.** $\dfrac{2\pm 2i}{3}$  **31.** $0.1,-1.7$  **33.** $\dfrac{5\pm 3\sqrt{2}}{2}$
**35.** $-\dfrac{1}{20},-\dfrac{9}{20}$  **37.** $-1,-5$  **39.** $-3,-5$  **41.** $-2,-4$  **43.** $1,6$  **45.** $-1,\dfrac{1}{2}$  **47.** $-\dfrac{1}{2},4$  **49.** $5,8$  **51.** $-1,7$  **53.** $4,5$  **55.** $7,-4$
**57.** $1,-11$  **59.** $2\pm\sqrt{14}$  **61.** $-4\pm\sqrt{11}$  **63.** $\dfrac{1\pm\sqrt{13}}{2}$  **65.** $\dfrac{-3\pm i\sqrt{15}}{2}$  **67.** $0,1$  **69.** $0,-\dfrac{2}{3}$  **71.** $0,\dfrac{1}{6}$  **73.** $1,-3$  **75.** $8,-4$
**77.** $\dfrac{-9\pm\sqrt{73}}{2}$  **79.** $\dfrac{1}{3},-1$  **81.** $\dfrac{1\pm i\sqrt{39}}{4}$  **83.** $1\pm i$  **85. a)** $21=(x+2)(x-2)$  **b)** $5$  **87. a)** $18=(x+4)(x+2)$
**b)** $-3+\sqrt{19}$  **89.** 12 mph  **91.** $5,7$  **93.** 5 feet by 12 feet  **95.** $\dfrac{12+\sqrt{288}}{2}\approx 14.49$ feet by 14.49 feet  **97.** $\sqrt{200}\approx 14.14$ inches
**99.** $\sqrt{24}\approx 4.90$ feet  **101.** $4\%$  **103.** $\approx 6\%$  **105. a)** $S=32+80\sqrt{\pi}\approx 173.80$ square inches  **b)** $r=\dfrac{4\sqrt{\pi}}{\pi}\approx 2.26$ inches
**c)** $r=-5+\sqrt{\dfrac{80+25\pi}{\pi}}\approx 2.1$ inches  **107.** $2$  **108.** $\$4200$ at $7\%$, $\$5800$ at $6\dfrac{1}{4}\%$  **109.** $0$  **110.** $4x^3+x^2-21x+6$  **111.** $\left\{10,\dfrac{4}{3}\right\}$

**Exercise Set 12.2**  **1.** Completing the square  **3.** Divide  **5.** Two  **7.** No  **9.** Two real solutions  **11.** No real solution
**13.** Two real solutions  **15.** No real solution  **17.** One real solution  **19.** One real solution  **21.** $-2,-5$  **23.** $2,4$  **25.** $1,-7$
**27.** $-2\pm 2\sqrt{6}$  **29.** $\pm 8$  **31.** $\dfrac{2\pm i\sqrt{11}}{3}$  **33.** $0,5$  **35.** $\dfrac{2\pm i\sqrt{2}}{2}$  **37.** $-1$  **39.** $\dfrac{1}{4}$  **41.** $1\pm\sqrt{2}$  **43.** $\dfrac{-3\pm i\sqrt{15}}{2}$  **45.** $-3,\dfrac{1}{2}$
**47.** $\dfrac{1}{2},-\dfrac{5}{3}$  **49.** $4,-6$  **51.** $\dfrac{1}{3},-\dfrac{2}{3}$  **53.** $\dfrac{-6\pm 2\sqrt{6}}{3}$  **55.** $\dfrac{3\pm\sqrt{309}}{30}$  **57.** $\dfrac{3\pm\sqrt{33}}{2}$  **59.** $\dfrac{2\pm i\sqrt{6}}{2}$  **61.** $\dfrac{-1\pm i\sqrt{23}}{4}$
**63.** $\dfrac{-0.6\pm\sqrt{0.84}}{0.2}$ or $-3\pm\sqrt{21}$  **65.** $0,\dfrac{3}{2}$  **67.** $-5,6$  **69.** $\dfrac{7\pm\sqrt{17}}{4}$  **71.** No real number  **73.** $x^2-7x+6=0$
**75.** $x^2+8x-9=0$  **77.** $15x^2-x-6=0$  **79.** $x^2-2=0$  **81.** $x^2+9=0$  **83.** $x^2-6x+7=0$  **85.** $x^2-4x+13=0$
**87. a)** $n(10-0.02n)=450$  **b)** $50$  **89. a)** $n(50-0.4n)=660$  **b)** $15$  **91.** $3$  **93.** $w=3$ feet, $l=8$ feet  **95.** 2 inches
**97. a)** 0.5 second and 2 seconds  **b)** 2.6 seconds  **c)** 28 feet  **99. a)** 1 second and 9 seconds  **b)** 10 seconds  **c)** 66.25 feet
**101.** Yes; if you multiply both sides of one equation by $-1$ you get the other equation.  **103. a)** $b^2-4ac$  **b)** $-84$
**c)** Answers will vary.  **105.** Answers will vary.  **107.** Yes  **109.** $(-0.12+\sqrt{14.3952})/1.2\approx 3.0618$ millimeters
**111. a)** $\approx 1.94$ seconds  **b)** $\approx 2.74$ seconds  **c)** Courtney's  **d)** Yes, at 1.5 seconds  **112.** $5.0\times 10^2$ or $500$  **113.** $7$  **114.** $(2,-1)$
**115.** $\dfrac{x^2+2x\sqrt{y}+y}{x^2-y}$  **116.** No real solution

**Exercise Set 12.3**   **1.** Profit   **3.** One whole task   **5.** $s = \sqrt{A}$   **7.** $i = \sqrt{\dfrac{E}{r}}$   **9.** $t = \dfrac{\sqrt{d}}{4}$   **11.** $c = \sqrt{\dfrac{E}{m}}$   **13.** $r = \sqrt{\dfrac{3V}{\pi h}}$

**15.** $W = \sqrt{d^2 - L^2}$   **17.** $b = \sqrt{c^2 - a^2}$   **19.** $H = \sqrt{d^2 - L^2 - W^2}$   **21.** $t = \sqrt{\dfrac{h - s_0}{-16}}$ or $t = \dfrac{\sqrt{s_0 - h}}{4}$   **23.** $v = \sqrt{\dfrac{2E}{m}}$

**25.** $v_1 = \sqrt{v_2^2 - 2ad}$   **27.** $c = \sqrt{(v')^2 + v^2}$   **29. a)** loss of $2400   **b)** profit of $24,000   **c)** about 1.4 years   **31. a)** loss of $2000

**b)** profit of $21,000   **c)** about 1.5 years   **33. a)** 32°F   **b)** 80.8°F   **c)** ≈2.92 minutes   **35. a)** about 9.64 crashes

**b)** about 12.115 crashes   **c)** about 19 years old and 72 years old   **37. a)** $3,615.25 billion   **b)** $1,114 billion

**c)** 2009 and 2028   **39.** $l = 30$ meters, $w = 20$ meters   **41.** 4 feet per hour   **43.** Going: 6 mph, returning: 8 mph

**45.** Bonita: ≈11.52 hours, Pamela: ≈12.52 hours   **47.** 130 mph   **49.** Chris ≈11.76 hours, John: ≈12.26 hours   **51.** 75 mph

**53.** $l \approx 34.86$ inches, $h \approx 19.61$ inches   **55.** Answers will vary.   **57.** Answers will vary.   **59.** 6 meters by 3 meters or 2 meters by 9 meters

**61.** $-16$   **62.** $R = \dfrac{E - Ir}{I}$   **63.** $\dfrac{8}{r - 4}$   **64.** $\dfrac{x^2}{y^{32}}$   **65.** No solution

**Mid-Chapter Test**   **1.** $\pm 7\sqrt{2}$ [12.1]   **2.** $3 \pm 2i\sqrt{5}$ [12.1]   **3.** $-\dfrac{1}{2}, -\dfrac{13}{2}$ [12.1]   **4.** $-6, 2$ [12.1]   **5.** $2 \pm \sqrt{14}$ [12.1]

**6.** $\dfrac{-1 \pm i\sqrt{143}}{8}$ [12.1]   **7.** $(6 + 6\sqrt{2})$ meters [12.1]   **8. a)** $b^2 - 4ac$   **b)** Two distinct real solutions: $b^2 - 4ac > 0$; single real solution:

$b^2 - 4ac = 0$; no real solution: $b^2 - 4ac < 0$ [12.2]   **9.** Two distinct real solutions [12.2]   **10.** $-\dfrac{5}{3}, \dfrac{3}{2}$ [12.2]   **11.** $-2 \pm 2\sqrt{3}$ [12.2]

**12.** $\dfrac{1 \pm i\sqrt{14}}{3}$ [12.2]   **13.** $x^2 - 5x - 14 = 0$ [12.2]   **14.** $x^2 - 4x - 1 = 0$ [12.2]   **15.** 10 lamps [12.2]   **16.** $r = \sqrt{x^2 - y}$ [12.3]

**17.** $x = \sqrt{\dfrac{3A}{k}}$ [12.3]   **18.** $y = \sqrt{D^2 - x^2}$ [12.3]   **19.** 5 feet by 12 feet [12.3]   **20.** 5 clocks [12.3]

**Exercise Set 12.4**   **1.** Quadratic in form   **3.** $h - 2$   **5.** $\sqrt{v}$   **7.** $(x^2 + 3)(x^2 - 2)$   **9.** $(x^2 + 2)(x^2 + 3)$   **11.** $(2a^2 + 5)(3a^2 - 5)$

**13.** $(2x + 5)(2x + 3)$   **15.** $(3a + 1)(2a + 5)$   **17.** $(ab + 3)(ab + 5)$   **19.** $(3xy - 5)(xy + 1)$   **21.** $(2a - 5)(a - 1)(5 - a)$

**23.** $(2x + 3)(x + 2)(x - 3)$   **25.** $(y^2 + 10)(y^2 + 3)$   **27.** $(x + 2)(x + 1)(x + 3)$   **29.** $a^3 b^2 (5a - 3b)(a - b)$   **31.** $\pm 1, \pm 3$

**33.** $\pm i, \pm 4i$   **35.** $\pm 2, \pm 3$   **37.** $\pm 2, \pm \sqrt{3}$   **39.** $\pm \dfrac{1}{2}, \pm 2$   **41.** $\pm \sqrt{3}, \pm \sqrt{5}$   **43.** $\pm 3, \pm i\sqrt{2}$   **45.** $\pm 1, \pm i\sqrt{5}$   **47.** 4   **49.** 9   **51.** $\dfrac{1}{9}$

**53.** $1, -9$   **55.** $\dfrac{4}{3}, -\dfrac{1}{2}$   **57.** $\pm \sqrt{6}, \pm 1$   **59.** $-6, -\dfrac{5}{2}$   **61.** $\pm \dfrac{5\sqrt{6}}{6}, \pm \dfrac{\sqrt{39}}{3}$   **63.** $-\dfrac{1}{2}$   **65.** $3, 4$   **67.** $2, \dfrac{1}{3}$   **69.** $1, -\dfrac{1}{10}$   **71.** $-\dfrac{1}{2}, \dfrac{1}{6}$

**73.** $1, 27$   **75.** $27, 216$   **77.** $\dfrac{1}{4}$   **79.** $-32, -1$   **81.** $(1, 0), (16, 0)$   **83.** None   **85.** $(-4, 0), \left(\dfrac{1}{5}, 0\right)$   **87.** $(-8, 0), (27, 0)$   **89.** $(-1, 0), (4, 0)$

**91.** $(\pm 2, 0), (\pm 5, 0)$   **93. a) and b)** $\dfrac{1}{5}, -\dfrac{1}{4}$   **95.** $-\dfrac{14}{5}, -\dfrac{8}{3}$   **97.** $2, \dfrac{1}{4}$   **99.** $2, 1$   **101.** $-3, 1, 2, -4$   **103.** $\pm \sqrt{\dfrac{3 \pm \sqrt{15}}{2}}$   **105.** Let $u = x^2$

**107.** Let $u = x^{-1}$   **109.** $x^4 - 5x^2 + 4 = 0$; start with $(x - 2)(x + 2)(x - 1)(x + 1) = 0$

**111.** $x^4 - 7x^2 + 10 = 0$; start with $(x + \sqrt{2})(x - \sqrt{2})(x + \sqrt{5})(x - \sqrt{5}) = 0$   **113.** No; imaginary solutions always occur in pairs.

**115.** $\dfrac{43}{60}$   **116.** 1   **117.** D: $\mathbb{R}$, R: $\{y | y \geq 0\}$   **118.** $2xy^2 \sqrt[3]{2}$   **119.** $9\sqrt{3}$

**Exercise Set 12.5**   **1.** Increase   **3.** Opens downward   **5.** $-\dfrac{b}{2a}$   **7.** $y$-intercept   **9.** Right   **11.** Up   **13.** 3   **15.** $-4$   **17.** $-7$

**19.** $-2.0312$

**21. a)** Upward   **b)** $(0, 15)$
**c)** $(-4, -1)$   **d)** $(-5, 0), (-3, 0)$

**e)** $f(x) = x^2 + 8x + 15$

**23. a)** Upward   **b)** $(0, 3)$
**c)** $(2, -1)$   **d)** $(1, 0), (3, 0)$

**e)**

$f(x) = x^2 - 4x + 3$

**25. a)** Downward   **b)** $(0, 8)$
**c)** $(-1, 9)$   **d)** $(-4, 0), (2, 0)$

**e)**

$f(x) = -x^2 - 2x + 8$

**27. a)** Downward   **b)** $(0, 5)$
**c)** $(2, 9)$   **d)** $(-1, 0), (5, 0)$

**e)**

$g(x) = -x^2 + 4x + 5$

**29. a)** Downward   **b)** $(0, -5)$
**c)** $(2, -1)$   **d)** No $x$-intercepts

**e)**

$t(x) = -x^2 + 4x - 5$

**31. a)** Upward   **b)** $(0, 4)$   **c)** $(2, 0)$
**d)** $(2, 0)$   **e)**

$f(x) = x^2 - 4x + 4$

**33. a)** Upward **b)** $(0, 2)$
**c)** $(0, 2)$ **d)** No $x$-intercepts
**e)**

$y$ $r(x) = x^2 + 2$

**35. a)** Downward **b)** $(0, 5)$
**c)** $(0, 5)$ **d)** $(-\sqrt{5}, 0), (\sqrt{5}, 0)$
**e)**

$l(x) = -x^2 + 5$

**37. a)** Downward **b)** $(0, -8)$
**c)** $(1, -6)$ **d)** No $x$-intercepts
**e)**

$f(x) = -2x^2 + 4x - 8$

**39. a)** Upward **b)** $(0, 3)$
**c)** $\left(-\dfrac{2}{3}, \dfrac{5}{3}\right)$ **d)** No $x$-intercepts
**e)**

$m(x) = 3x^2 + 4x + 3$

**41. a)** Upward **b)** $(0, -6)$
**c)** $\left(-\dfrac{2}{3}, -\dfrac{22}{3}\right)$
**d)** $\left(\dfrac{-2 + \sqrt{22}}{3}, 0\right), \left(\dfrac{-2 - \sqrt{22}}{3}, 0\right)$
**e)**

$y = 3x^2 + 4x - 6$

**43. a)** Upward **b)** $(0, -6)$
**c)** $\left(\dfrac{1}{4}, -\dfrac{49}{8}\right)$ **d)** $\left(-\dfrac{3}{2}, 0\right), (2, 0)$
**e)**

$y = 2x^2 - x - 6$

**45. a)** Downward **b)** $(0, -5)$
**c)** $\left(\dfrac{3}{2}, -\dfrac{11}{4}\right)$ **d)** No $x$-intercepts
**e)**

$f(x) = -x^2 + 3x - 5$

**47.**

$(3, 0)$   $f(x) = (x - 3)^2$

**49.**

$f(x) = (x + 1)^2$   $(-1, 0)$

**51.**

$f(x) = x^2 + 3$   $(0, 3)$

**53.**

$f(x) = x^2 - 1$   $(0, -1)$

**55.**

$f(x) = (x - 2)^2 + 3$   $(2, 3)$

**57.**

$f(x) = (x + 4)^2 + 4$   $(-4, 4)$

**59.**

$(-3, -2)$   $g(x) = -(x + 3)^2 - 2$

**61.**

$(2, 2)$   $y = -2(x - 2)^2 + 2$

**63.**

$(-1, -3)$   $h(x) = -2(x + 1)^2 - 3$

**65. a)** $f(x) = (x - 3)^2 - 1$
**b)**

$f(x) = x^2 - 6x + 8$   $(3, -1)$

**67. a)** $g(x) = \left(x - \dfrac{1}{2}\right)^2 - \dfrac{13}{4}$
**b)**

$y$ $g(x) = x^2 - x - 3$   $\left(\dfrac{1}{2}, -\dfrac{13}{4}\right)$

**69. a)** $f(x) = -(x + 2)^2 - 2$
**b)**

$(-2, -2)$   $f(x) = -x^2 - 4x - 6$

**71. a)** $g(x) = (x - 2)^2 - 5$
**b)**

$(2, -5)$   $g(x) = x^2 - 4x - 1$

**73. a)** $f(x) = 2\left(x + \dfrac{5}{4}\right)^2 - \dfrac{49}{8}$
**b)**

$\left(-\dfrac{5}{4}, -\dfrac{49}{8}\right)$   $f(x) = 2x^2 + 5x - 3$

**75. c)** **77. c)** **79. d)** **81. b)**
**83. a)** $x = 7$ **b)** $A = 121$ **85. a)** $x = 10.5$
**b)** $A = 240.25$ **87. a)** $n = 200$ **b)** $R = \$800$
**89.** 2010 **91.** 4 units **93.** 3 units
**95.** $f(x) = 2(x - 3)^2 - 2$

**97.** $f(x) = -4\left(x + \dfrac{3}{5}\right)^2 - \sqrt{2}$

**99. a)**

$I = -x^2 + 24x - 44, 0 \le x \le 24$

**b)** \$2  **c)** \$22  **d)** \$12  **e)** \$10,000  **101. a)** 100  **b)** \$3800
**103. a)** 40.425 meters  **b)** 2.5 seconds  **c)** $\approx 5.37$ seconds
**105.** 400 square feet  **107.** $-16, 4$ and $-4$  **109.** $900, 30$ and $30$  **111. a)** \$142,400  **b)** 380
**113. a)**   **115.** Minimum

**117. a)** The graphs will have the same $x$-intercepts but $f(x) = x^2 - 8x + 12$ will open upward and $g(x) = -x^2 + 8x - 12$ will open downward.  **b)** Yes, both at $(6, 0)$ and $(2, 0)$  **c)** No; vertex of $f(x)$ at $(4, -4)$, vertex of $g(x)$ at $(4, 4)$

**d)**

$g(x) = -x^2 + 8x - 12$

**119. b)**    **121. a)** $f(t) = -16(t - 1.625)^2 + 45.25$  **b)** 45.25 feet, 1.625 seconds  **c)** Same

**125.** $200\pi$ square feet  **126.** $\dfrac{x}{x + 6}$  **127.** $(-2, 3, 2)$  **128.** $-8$  **129.**

## Exercise Set 12.6  **1.** Solution  **3.** Included  **5.** Open  **7.**    **9.**    **11.**

**13.**    **15.**    **17.**    **19.**    **21.** $(-\infty, -2] \cup [1, 3]$  **23.** $(-\infty, -4) \cup (-2, 3)$

**25.** $\left(-6, -\dfrac{5}{2}\right) \cup (2, \infty)$  **27.** $\left(-\dfrac{5}{3}, -1\right) \cup (3, \infty)$  **29.** $[-2, -2] \cup \left[\dfrac{8}{3}, \infty\right)$  **31.** $(-\infty, 0)$  **33.**

**35.**    **37.**    **39.**    **41.**    **43.** $\{x | -1 \le x < 3\}$  **45.** $\{x | -5 < x < 1\}$

**47.** $\{x | x \le -3 \text{ or } x > 2\}$  **49.** $\{a | -5 < a < 9\}$  **51.** $\{c | c < 4 \text{ or } c > 10\}$  **53.** $\{y | -4 < y \le -2\}$  **55.** $\left\{a \,\middle|\, a \le -2 \text{ or } a > \dfrac{1}{3}\right\}$

**57.** $\left\{x \,\middle|\, -\dfrac{4}{3} < x < \dfrac{1}{2}\right\}$  **59.** $\left\{x \,\middle|\, -\dfrac{8}{3} \le x < 2\right\}$  **61.** $(-\infty, -1) \cup (2, 4)$  **63.** $(-3, 2) \cup (5, \infty)$  **65.** $(-2, 1] \cup [7, \infty)$

**67.** $(-\infty, -8) \cup [0, 3)$  **69.** $(-\infty, -4) \cup (1, 6]$  **71.** $\left[-\dfrac{5}{2}, 3\right] \cup (4, \infty)$  **73.**    **75.**

**77.**    **79.**    **81.**    **83.**    **85. a)** $x < 2$ or $x > 5$  **b)** $2 < x < 5$

**87. a)** $(4, \infty)$; $y > 0$ in this interval  **b)** $(-\infty, 2) \cup (2, 4)$; $y < 0$ in this interval  **89.** $x^2 + 2x - 8 > 0$  **91.** $\dfrac{x + 3}{x - 4} \ge 0$

**93.** All real numbers; for any value of $x$, the expression is $\ge 0$.

**95.** All real numbers except $-2$; for any value of $x$ except $-2$, the expression is $\ge 0$.

**97.** No solution; the graph opens upward and has no $x$-intercepts, so it is always above the $x$-axis.  **99.**

**101.** $x^2 - 3x > 0$; multiply factors containing boundary values.  **103.** $x^2 < 0$; $x^2$ is always $\ge 0$.  **105.** $(-\infty, -3) \cup (-1, 1) \cup (3, \infty)$

**107.** $[-2, -1] \cup [2, \infty)$  **111.** 6 quarts  **112.** $3r + 3s - 9t$  **113.** $\dfrac{x - 3}{x + 1}$  **114.** $-\dfrac{1}{2}$  **115.** $38 - 9i$

## Chapter 12 Review Exercises  **1.** $1, 9$  **2.** $\dfrac{-1 \pm 2\sqrt{15}}{2}$  **3.** $-\dfrac{1}{3}, 1$  **4.** $\dfrac{5}{4}, -\dfrac{3}{4}$  **5.** $2, 6$  **6.** $4, -8$  **7.** $-1 \pm \sqrt{10}$

**8.** $-3 \pm \sqrt{21}$  **9.** $1 \pm 3i$  **10.** $2 \pm 2i\sqrt{7}$  **11. a)** $32 = (x + 1)(x + 5)$  **b)** 3  **12. a)** $63 = (x + 2)(x + 4)$  **b)** 5
**13.** $7, 8$  **14.** $\approx 16.90$ feet by $\approx 16.90$ feet  **15.** Two real solutions  **16.** No real solution  **17.** One real solution  **18.** No real solution
**19.** One real solution  **20.** Two real solutions  **21.** $-\dfrac{5}{2}, \dfrac{10}{3}$  **22.** $2, 9$  **23.** $8, -5$  **24.** $0, \dfrac{9}{7}$  **25.** $\dfrac{3}{2}, -\dfrac{5}{3}$  **26.** $\dfrac{1}{4}, -3$

**27.** $-4 \pm \sqrt{11}$  **28.** $-2 \pm 2\sqrt{3}$  **29.** $\dfrac{-2 \pm \sqrt{10}}{2}$  **30.** $\dfrac{3 \pm \sqrt{33}}{3}$  **31.** $\dfrac{1 \pm i\sqrt{51}}{2}$  **32.** $1 \pm i\sqrt{10}$  **33.** $\dfrac{5}{2}, -\dfrac{5}{3}$  **34.** $\dfrac{1}{4}, -\dfrac{3}{2}$

**35.** $10, -6$  **36.** $\dfrac{2}{3}, -\dfrac{3}{2}$  **37.** $\dfrac{7 \pm \sqrt{89}}{10}$  **38.** $\dfrac{3 \pm 3\sqrt{3}}{2}$  **39.** $x^2 - 2x - 3 = 0$  **40.** $3x^2 + 4x - 4 = 0$  **41.** $x^2 - 11 = 0$

**42.** $x^2 - 6x + 13 = 0$  **43.** 8 feet by 12 feet  **44.** $\sqrt{128} \approx 11.31$  **45.** 4%  **46.** $7, 11$  **47.** 8 inches by 12 inches  **48.** \$540
**49. a)** 10.12 inches  **b)** 11.77 inches or 38.23 inches  **50. a)** 720 feet  **b)** 7 seconds  **51. a)** 40 milliliters  **b)** 150°C
**52.** larger: $\approx 23.51$ hours, smaller: $\approx 24.51$ hours  **53.** 50 miles per hour  **54.** 1.6 miles per hour  **55.** $l = 10$ units, $w = 8$ units

**56.** 20 tables  **57.** $a = \sqrt{c^2 - b^2}$  **58.** $t = \sqrt{\dfrac{c - h}{4.9}}$  **59.** $v_y = \sqrt{v^2 - v_x^2}$  **60.** $v_2 = \sqrt{v_1^2 + 2ad}$  **61.** $(x^2 + 4)(x^2 - 5)$

**62.** $(2x^2 - 1)(2x^2 + 3)$  **63.** $(x + 9)(x + 11)$  **64.** $(4x + 1)(4x + 5)$  **65.** $\pm 2, \pm 3$  **66.** $\pm 4, \pm \sqrt{5}$  **67.** $\pm 2\sqrt{2}, \pm i\sqrt{3}$  **68.** $\dfrac{3}{2}, -\dfrac{1}{6}$

**69.** $\dfrac{1}{9}$  **70.** $\dfrac{27}{8}, 8$  **71.** $4, \dfrac{13}{8}$  **72.** $-\dfrac{1}{5}, -\dfrac{5}{2}$  **73.** $(\pm 1, 0), (\pm 9, 0)$  **74.** $\left(\dfrac{4}{25}, 0\right)$  **75.** None  **76.** $(3 \pm \sqrt{17}, 0), (3 \pm \sqrt{6}, 0)$

**77. a)** Upward  **b)** $(0, 0)$
**c)** $\left(-\dfrac{5}{2}, -\dfrac{25}{4}\right)$  **d)** $(0, 0), (-5, 0)$
**e)**

$\left(-\dfrac{5}{2}, -\dfrac{25}{4}\right)$
$f(x) = x^2 + 5x$

**78. a)** Upward  **b)** $(0, -8)$
**c)** $(1, -9)$  **d)** $(-2, 0), (4, 0)$
**e)**

$y \, f(x) = x^2 - 2x - 8$
$(1, -9)$

**79. a)** Downward  **b)** $(0, -2)$
**c)** $(0, -2)$  **d)** No $x$-intercepts
**e)**

$(0, -2)$
$g(x) = -x^2 - 2$

**80. a)** Downward  **b)** $(0, 15)$
**c)** $\left(-\dfrac{1}{4}, \dfrac{121}{8}\right)$  **d)** $(-3, 0), \left(\dfrac{5}{2}, 0\right)$
**e)**

$\left(-\dfrac{1}{4}, \dfrac{121}{8}\right)$
$g(x) = -2x^2 - x + 15$

**81. a)** $11  **b)** $7600
**82. a)** 2.5 seconds  **b)** 175 feet
**83.**

$y \, f(x) = (x - 3)^2$
$(3, 0)$

**84.**

$(-2, -3)$
$f(x) = -(x + 2)^2 - 3$

**85.**

$(-4, -1)$
$g(x) = -2(x + 4)^2 - 1$

**86.**

$(1, 3)$
$h(x) = \dfrac{1}{2}(x - 1)^2 + 3$

**87.**
**88.**
**89.**   $\dfrac{11 - \sqrt{41}}{2} \quad \dfrac{11 + \sqrt{41}}{2}$
**90.** $-4 \quad \dfrac{4}{3}$
**91.** $-\dfrac{3}{2} \quad \dfrac{3}{2}$
**92.** $-\sqrt{5} \quad \sqrt{5}$

**93.** $\{x \mid x < -1 \text{ or } x > 5\}$  **94.** $\{x \mid -2 < x \le 3\}$  **95.** $\{x \mid x < -3 \text{ or } x \ge 2\}$
**96.** $\left\{x \mid -\dfrac{5}{3} < x < 6\right\}$  **97.** $\{x \mid -4 < x < -1 \text{ or } x > 2\}$  **98.** $\{x \mid x \le 0 \text{ or } 3 \le x \le 6\}$  **99.** $\left[-\dfrac{4}{3}, 1\right] \cup [3, \infty)$
**100.** $(-\infty, -4) \cup (-2, 0)$  **101.** $(-2, 0) \cup (4, \infty)$  **102.** $(-\infty, -3) \cup (2, 8)$  **103.** $(-2, 3] \cup (7, \infty)$  **104.** $(-\infty, -3) \cup [0, 6]$
**105.**  $-9 \quad -4$  **106.**  $-2 \quad 2$  **107.** $\dfrac{5}{3} \quad \dfrac{23}{10}$

## Chapter 12 Practice Test
**1.** $3, -5$ [12.1]  **2.** $3 \pm \sqrt{2}$ [12.1]  **3.** $8, -2$ [12.2]  **4.** $2 \pm i\sqrt{7}$ [12.2]  **5.** $\dfrac{2}{3}, -1$ [12.1–12.2]
**6.** $\dfrac{-7 \pm \sqrt{33}}{2}$ [12.1–12.2]  **7.** $5x^2 - 18x - 8 = 0$ [12.2]  **8.** $v = \sqrt{\dfrac{2K}{m}}$ [12.3]  **9. a)** $121,200  **b)** $\approx 2712.57$ square feet [12.1–12.3]
**10.** 50 mph [12.1–12.3]  **11.** $\pm\dfrac{\sqrt{10}}{2}, \pm i\sqrt{10}$ [12.4]  **12.** $\dfrac{343}{27}, -216$ [12.4]  **13.** $\left(\dfrac{9}{16}, 0\right)$ [12.4]

**14.**

$y \, f(x) = (x - 3)^2 + 2$ [12.5]
$(3, 2)$

**15.**

[12.5]
$(2, -2)$
$h(x) = -\dfrac{1}{2}(x - 2)^2 - 2$

**16.** Two real solutions [12.5]  **17. a)** Upward  **b)** $(0, -8)$
**c)** $(-1, -9)$  **d)** $(-4, 0), (2, 0)$  **e)** [12.5]

$y = x^2 + 2x - 8$

**18.** $2x^2 + 13x - 7 = 0$ [12.5]  **19.** $-6 \quad 7$ [12.6]  **20.** $-5 \quad -1 \quad 4$ [12.6]  **21. a)** $\left[-\dfrac{5}{2}, -2\right)$  **b)** $\left\{x \mid -\dfrac{5}{2} \le x < -2\right\}$ [12.6]
**22.** $w = 5$ feet, $l = 13$ feet [12.5]  **23.** 6 seconds [12.5]  **24. a)** 20  **b)** $490 [12.5]  **25.** 30 [12.5]

## Cumulative Review Test
**1.** 13 [1.9]  **2.** 18 [1.9]  **3.** $3x - 7$ [1.9]  **4.** All real numbers, $\mathbb{R}$ [2.5]
**5. a)** [4.2]  **b)**
  $y = 2$
  $x = -4$
**6.** $y = x - 1$ [4.4]  **7.** $2.54 \times 10^6$ [5.3]  **8.** $\dfrac{2(x - 4)}{(x - 3)(x - 5)}$ [7.4]  **9.** $\dfrac{12}{5}$ [7.6]
**10.** 11.52 watts [7.8]  **11.** 1500 [8.2]
**12. a)** No, the graph does not pass the vertical line test  **b)** Domain: $\{x \mid x \ge -2\}$, Range: $\mathbb{R}$ [8.2]
**13.** $\left(\dfrac{5}{2}, 0\right)$ [9.3]  **14.** 160 [9.7]  **15.** $(-12, 8)$ [10.1]  **16.** $\left\{-\dfrac{1}{2}, \dfrac{9}{2}\right\}$ [10.2]
**17.** $\dfrac{-14 - 23i}{29}$ [11.7]  **18.** $\dfrac{-3 \pm i\sqrt{183}}{8}$ [12.2]  **19.** $r = \sqrt{\dfrac{3V}{\pi h}}$ [12.3]  **20.** $(x + 9)(x + 7)$ [12.4]

# Chapter 13

**Exercise Set 13.1**   **1.** Composition   **3.** Vertical   **5.** Inverse   **7.** Domain   **9. a)** $2x + 1$ **b)** $9$ **c)** $2x + 5$ **d)** $13$

**11. a)** $x^2 + x - 1$ **b)** $19$ **c)** $x^2 + 7x + 8$ **d)** $52$   **13. a)** $\dfrac{1}{2x + 3}$ **b)** $\dfrac{1}{11}$ **c)** $\dfrac{2}{x} + 3$ **d)** $3\dfrac{1}{2}$   **15. a)** $\dfrac{9}{x} + 1$ **b)** $3\dfrac{1}{4}$ **c)** $\dfrac{3}{3x + 1}$ **d)** $\dfrac{3}{13}$

**17. a)** $x^4 + 10x^2 + 26$ **b)** $442$ **c)** $x^4 + 2x^2 + 6$ **d)** $294$   **19. a)** $\sqrt{x + 5} - 4$ **b)** $-1$ **c)** $\sqrt{x + 1}$ **d)** $\sqrt{5}$   **21.** No   **23.** Yes

**25.** Yes   **27.** No   **29.** Yes   **31.** No   **33.** No   **35.** Yes   **37.** Yes   **39.** No   **41.** Yes   **43.** $f(x)$: Domain: $\{-2, -1, 2, 4, 8\}$,

Range: $\{0, 4, 6, 7, 9\}$; $f^{-1}(x)$: Domain: $\{0, 4, 6, 7, 9\}$, Range: $\{-2, -1, 2, 4, 8\}$   **45.** $f(x)$: Domain: $\{-1, 1, 2, 4\}$,

Range: $\{-3, -1, 0, 2\}$; $f^{-1}(x)$: Domain: $\{-3, -1, 0, 2\}$, Range: $\{-1, 1, 2, 4\}$   **47.** $f(x)$: Domain: $\{x|x \geq 2\}$, Range: $\{y|y \geq 0\}$; $f^{-1}(x)$:

Domain: $\{x|x \geq 0\}$, Range: $\{y|y \geq 2\}$   **49. a)** Yes **b)** $f^{-1}(x) = x - 3$   **51. a)** Yes **b)** $h^{-1}(x) = \dfrac{x}{4}$   **53. a)** No   **55. a)** No

**57. a)** Yes **b)** $g^{-1}(x) = \dfrac{1}{x}$   **59. a)** No   **61. a)** Yes **b)** $g^{-1}(x) = \sqrt[3]{x + 6}$   **63. a)** Yes **b)** $g^{-1}(x) = x^2 - 2, x \geq 0$   **65. a)** Yes

**b)** $h^{-1}(x) = \sqrt{x + 4}, x \geq -4$

**67. a)** $f^{-1}(x) = \dfrac{x - 8}{2}$    **69. a)** $f^{-1}(x) = x^2, x \geq 0$    **71. a)** $f^{-1}(x) = x^2 + 1, x \geq 0$    **73. a)** $f^{-1}(x) = x^3$

**b)**     **b)**    **b)**   **b)**

**75. a)** $f^{-1}(x) = \dfrac{1}{x}, x > 0$    **77.** $(f \circ f^{-1})(x) = x, (f^{-1} \circ f)(x) = x$   **79.** $(f \circ f^{-1})(x) = x, (f^{-1} \circ f)(x) = x$

**b)**     **81.** $(f \circ f^{-1})(x) = x, (f^{-1} \circ f)(x) = x$   **83.** $(f \circ f^{-1})(x) = x, (f^{-1} \circ f)(x) = x$

**85.** $f^{-1}(x) = \dfrac{x}{3}$; $x$ is feet and $f^{-1}(x)$ is yards   **87.** $f^{-1}(x) = \dfrac{9}{5}x + 32$.

**89.** $(f \circ g)(x) = 453.6x, x$ is pounds, $(f \circ g)(x)$ is grams   **91.** $(f \circ g)(x) = 0.915x, x$ is

yards, $(f \circ g)(x)$ is meters   **93.** No, composition of functions is not commutative. Let $f(x) = x^2$

and $g(x) = x + 1$. Then $(f \circ g)(x) = x^2 + 2x + 1$, while $(g \circ f)(x) = x^2 + 1$.   **95. a)** $(f \circ g)(x) = x; (g \circ f)(x) = x$

**b)** The domain is $\mathbb{R}$ for all of them.   **97.** The range of $f^{-1}(x)$ is the domain of $f(x)$.   **99. a)** 6 feet **b)** $36\pi \approx 113.10$ square feet

**c)** $A(t) = 4\pi t^2$ **d)** $36\pi \approx 113.10$ square feet **e)** Answers should agree   **102.** $2x + 3y = 10$   **103.** $x^2 + 4x - 2 - \dfrac{4}{x + 2}$

**104.** $\dfrac{18 - 12x}{x^3}$   **105.** $\dfrac{2x\sqrt{2y}}{y}$   **106.** $-1 \pm \sqrt{11}$

**Exercise Set 13.2**   **1.** Exponent   **3.** Rise   **5.** Principal

**7.**    **9.**    **11.**    **13.**    **15.**   **17.**

**19.**    **21.**   **23.**    **25.**    **27. a)** $\approx 36.232$ million **b)** $\approx 187.846$ million

**29.** \$512   **31.** 45   **33.** $\approx$\$6344.93   **35.** \$6166.13   **37.** \$3147.06   **39.** $\approx 10.6$ grams   **41. a)** 5 grams **b)** $\approx 7.28 \times 10^{-11}$ grams

**43. a)** 2400 **b)** $\approx 4977$   **45.** $\approx$\$10,850.92   **47.** $\approx 8.83$ kilometers   **49. a)** \$13,047 thousand **b)** \$282,727 thousand   **51. a)** 14 years

**b)** 10 years **c)** \$25 **d)** Increases it   **53. a)** As $x$ increases, $y$ decreases. **b)** No, $\left(\dfrac{1}{2}\right)^x$ can never be 0. **c)** No, $\left(\dfrac{1}{2}\right)^x$ can never be

negative.   **55. a)** Same; $(0, 1)$. **b)** $y = 3^x$ will be steeper than $y = 2^x$ for $x > 0$.   **57. a)** It is a horizontal line through $y = 1$. **b)** Yes

**c)** No, not a one-to-one function   **59.** $y = a^x - k$ is $y = a^x$ lowered $k$ units.   **61.** The graph of $y = a^{x+2}$ is the graph of $y = a^x$

shifted 2 units to the left.   **63. a)** \$16,384 **b)** \$524,288 **c)** $2^{n-1}$ **d)** $2^{29} = $ \$536,870,912 **e)** $2^0 + 2^1 + 2^2 + \cdots + 2^{29}$

**65. a)** $-6.2x^6y^2 + 9.2x^5y^2 + 2.3x^4y$ **b)** 8 **c)** $-6.2$   **66.** $x^3 + 3x^2 - 6x + 20$   **67.** $|a - 4|$   **68.** $\dfrac{2xy\sqrt[4]{xy^2z^3}}{z}$

## Exercise Set 13.3

**1.** Exponent　**3.** Base　**5.** Domain　**7.** $\log_5 25 = 2$　**9.** $\log_3 9 = 2$　**11.** $\log_{16} 4 = \frac{1}{2}$　**13.** $\log_8 2 = \frac{1}{3}$

**15.** $\log_{1/2} \frac{1}{32} = 5$　**17.** $\log_2 \frac{1}{8} = -3$　**19.** $\log_4 \frac{1}{64} = -3$　**21.** $\log_{64} 4 = \frac{1}{3}$　**23.** $\log_8 \frac{1}{2} = -\frac{1}{3}$　**25.** $\log_{81} \frac{1}{3} = -\frac{1}{4}$

**27.** $\log_{10} 7 = 0.8451$　**29.** $\log_e 7.3891 = 2$　**31.** $\log_a b = n$　**33.** $3^2 = 9$　**35.** $\left(\frac{1}{3}\right)^3 = \frac{1}{27}$　**37.** $5^{-2} = \frac{1}{25}$　**39.** $49^{1/2} = 7$

**41.** $9^{-2} = \frac{1}{81}$　**43.** $10^{-3} = \frac{1}{1000}$　**45.** $6^3 = 216$　**47.** $10^{-0.2076} = 0.62$　**49.** $e^{1.8749} = 6.52$　**51.** $w^{-p} = s$　**53.** 3　**55.** 5　**57.** 27

**59.** $-4$　**61.** $\frac{1}{64}$　**63.** 3　**65.** $\frac{1}{2}$　**67.** 0　**69.** 2　**71.** $-2$　**73.** 4　**75.** 4　**77.** $-4$　**79.** $-2$　**81.** $\frac{1}{2}$　**83.** 1　**85.** $\frac{1}{2}$

**87.** 　**89.** 　**91.** 　**93.** 　**95.** 　**97.**

**99.** 10,000,000　**101.** 10,000　**103.** 　**105. a)** $a > 0$ and $a \neq 1$　**b)** $\{x \mid x > 0\}$　**c)** $\mathbb{R}$

**107.** $\left(\frac{1}{27}, -3\right)\left(\frac{1}{9}, -2\right), \left(\frac{1}{3}, -1\right)$ $(1, 0), (3, 1), (9, 2)$, and $(27, 3)$; the functions $f(x) = a^x$ and $g(x) = \log_a x$ are inverses.

**109.** $f^{-1}(x) = \log_5 x$　**111.** 3 and 4, since 62 lies between $3^3 = 27$ and $3^4 = 81$.　**113.** 2 and 3, since 425 lies between $10^2 = 100$ and $10^3 = 1000$.　**115.** $2^x$; Note that for $x = 10$, $2^x = 1024$ while $\log_{10} x = 1$.　**117.** $2x(x + 3)(x - 6)$　**118.** $(x - 2)(x + 2)(x^2 + 4)$

**119.** $4(2x + 3)(5x - 1)$　**120.** $(3rs - 1)(2rs + 1)$

## Exercise Set 13.4

**1.** Positive　**3.** Sum　**5.** Times　**7.** $\log_2 3 + \log_2 5$　**9.** $\log_8 7 + \log_8 (x + 3)$　**11.** $\log_2 27 - \log_2 11$

**13.** $\frac{1}{2} \log_{10} x - \log_{10} (x - 9)$　**15.** $7 \log_6 x$　**17.** $5 \log_4 (r + 7)$　**19.** $\frac{3}{2} \log_4 a - \frac{1}{2} \log_4 (a + 2)$　**21.** $6 \log_3 d - 4 \log_3 (a - 8)$

**23.** $\log_8 (y + 4) - 2 \log_8 y$　**25.** $\log_{10} 9 + \log_{10} m - \log_{10} 8 - \log_{10} n$　**27.** $\log_2 21$　**29.** $\log_2 \frac{9}{5}$　**31.** $\log_4 64$　**33.** $\log_{10} x(x + 3)$

**35.** $\log_9 \frac{z^2}{z - 2}$　**37.** $\log_5 \left(\frac{p}{3}\right)^4$　**39.** $\log_2 \frac{n(n + 4)}{n - 3}$　**41.** $\log_5 \sqrt{\frac{x - 8}{x}}$　**43.** $\log_9 \frac{16\sqrt[3]{r - 6}}{\sqrt{r}}$　**45.** $\log_6 \frac{81}{(x + 3)^2 x^4}$　**47.** 1　**49.** $-0.3980$

**51.** 1.3980　**53.** 2　**55.** 7　**57.** 3　**59.** 25　**61.** $\log_a (x - 2)$　**63.** $\frac{1}{4} \log_2 x + \frac{1}{4} \log_2 y + \frac{1}{3} \log_2 a - \frac{1}{5} \log_2 (a - b)$　**65.** 0.8640

**67.** 0.1080　**69.** 0.7000　**71.** No, there is no relationship between $\log_{10}(x + y)$ and $\log_{10} xy$ or $\log_{10}\left(\frac{x}{y}\right)$　**73.** Yes, it is an expansion

of property 1.　**75.** Yes, $\log_a (x^2 + 8x + 16) = \log_a (x + 4)^2 = 2 \log_a (x + 4)$　**77.** Yes　**79.** $\log_a \frac{x}{y} = \log_a \left(x \cdot \frac{1}{y}\right) = \log_a x + \log_a \frac{1}{y}$

**82.** $\frac{(2x + 5)^2}{(x - 4)^2}$　**83.** $\frac{(3x + 1)(x + 9)}{(x - 4)(x - 3)(2x + 5)}$　**84.** $2\frac{2}{9}$ days　**85.** $2x^3 y^5 \sqrt[3]{6x^2 y^2}$　**86.** 36

## Mid-Chapter Test

**1. a)** In $f(x)$, replace $x$ by $g(x)$.　**b)** $6x + 18$ [13.1]　**2. a)** $\left(\frac{6}{x}\right)^2 + 5$ or $\frac{36}{x^2} + 5$　**b)** 9　**c)** $\frac{6}{x^2 + 5}$　**d)** $\frac{3}{7}$ [13.1]

**3. a)** Answers will vary.　**b)** No [13.1]　**4. a)** Yes　**b)** $\{(2, -3), (3, 2), (1, 5), (8, 6)\}$ [13.1]　**5. a)** Yes　**b)** $p^{-1}(x) = 3x + 15$ [13.1]

**6. a)** Yes　**b)** $k^{-1}(x) = x^2 + 4$ $x \geq 0$ [13.1]　**7.** $m^{-1}(x) = -\frac{1}{2}x + 2$　[13.1]　**8.** [13.2]

**9.** [13.2]　**10.** [13.3]　

**11. a)** 10　**b)** 320 [13.2]　**12.** $\log_{27} 9 = \frac{2}{3}$ [13.3]　**13.** $2^{-6} = \frac{1}{64}$ [13.3]　**14.** 3 [13.3]　**15.** 2 [13.3]　**16.** 4 [13.3]

 $2 \log_9 x + \log_9 (x - 5)$ [13.4]　**18.** $\log_5 7 + \log_5 m - \frac{1}{2} \log_5 n$ [13.4]　**19.** $\log_2 \frac{x^3(x + 7)}{(x + 1)^4}$ [13.4]　**20.** $\log_7 \sqrt{\frac{x + 2}{x}}$ [13.4]

**Exercise Set 13.5** **1.** Common **3.** $10^y$ **5.** 0 **7.** −1 **9.** −2 **11.** −3 **13.** 1 and 2 **15.** 4 and 5 **17.** −2 and −1 **19.** 2 and 3 **21.** 0 and 1 **23.** −2 and −1 **25.** 1.9345 **27.** 4.2833 **29.** −1.2125 **31.** 2.0004 **33.** 0.5740 **35.** −1.7620 **37.** 3.5514 **39.** −1.1385 **41.** 2.3856 **43.** −2.2277 **45.** −2.0088 **47.** 1.1901 **49.** 1.6357 **51.** 42,491.2982 **53.** 0.0196 **55.** 1.0023 **57.** 578.7620 **59.** 0.0097 **61.** 100 **63.** 2408.2413 **65.** 13,832.4784 **67.** 0.0871 **69.** 0.2389 **71.** 0.7489 **73.** 7 **75.** 7 **77.** 20.8 **79.** 41.5 **81.** 2511.8864 **83.** 501,187.2336 **85. a)** ≈31.62 kilometers **b)** ≈0.50 kilometer **c)** ≈14.68 **87. a)** ≈72% **b)** ≈15% **89.** ≈6310 times more intense **91. a)** ≈$6.31 \times 10^{20}$ **b)** ≈2.19 **93.** ≈6.2 **95.** No; $10^2 = 100$ and since $462 > 100$, log 462 must be greater than 2. **97.** No; $10^0 = 1$ and $10^{-1} = 0.1$ and since $1 > 0.163 > 0.1$, log 0.163 must be between 0 and −1. **99.** No; $\log \frac{y}{4x} = \log y - \log 4 - \log x$ **101.** $I = 10^R$ **103.** $t = 10^{\left(\frac{26-R}{41.9}\right)} - 1$ **106.** $(2, -3)$ **107.** 50 miles per hour, 55 miles per hour **108.** 10 miles per hour **109.**

$y = (x-2)^2 + 1$

**110.**

**Exercise Set 13.6** **1.** Base **3.** Extraneous **5.** Product **7.** 3 **9.** 4 **11.** $\frac{1}{2}$ **13.** 2 **15.** $-\frac{1}{3}$ **17.** 4 **19.** 3 **21.** 3 **23.** 2.01 **25.** 3.56 **27.** 5.59 **29.** 6.34 **31.** 6 **33.** 5 **35.** $\frac{1}{16}$ **37.** 100 **39.** −1 **41.** −6, 4 **43.** 0, −8 **45.** 92 **47.** $\frac{4}{5}$ **49.** $\frac{3}{2}$ **51.** 4 **53.** 2 **55.** 0.87 **57.** 30 **59.** 5 **61.** 4 **63.** 2 **65.** 3 **67.** 9 **69.** ≈3.47 hours **71.** ≈3.19 years **73.** ≈17.07 years **75.** ≈11.81 years **77.** ≈$7112.09 **79.** ≈19.36 **81. a)** 1,000,000,000,000 times greater **b)** 10,000,000 times greater **83.** 8 **85.** $x = 1$ and $x = 2$ **87.** $(3, 1)$ **89.** $(54, 46)$ **91.** 2.8 **93.** No solution **95.** $\log(-2)$ is not a real number. **97.** The box is greater by ≈7.73 cubic feet **98.** −4 **99.**

**100.** $\frac{x\sqrt{y} - y\sqrt{x}}{x - y}$ **101.** $c = \sqrt{\frac{E}{m}}$ **102.** $f(x) = 2(x-3)^2 - 5$

**Exercise Set 13.7** **1.** Base **3.** Exponential **5.** Change **7.** Natural **9.** Growth **11.** 3.2958 **13.** −0.8795 **15.** 3.3201 **17.** 0.5712 **19.** 4.95 **21.** 0.0578 **23.** 4.5850 **25.** −2.8577 **27.** 4.3923 **29.** 1.7297 **31.** 2.7380 **33.** 2.9135 **35.** −0.4784 **37.** 4 **39.** 1 **41.** 4 **43.** 6 **45.** $P = 4757.5673$ **47.** $P_0 = 224.0845$ **49.** $t = 0.7847$ **51.** $k = 0.2310$ **53.** $k = -0.2888$ **55.** $A = 4719.7672$ **57.** $V_0 = \frac{V}{e^{kt}}$ **59.** $t = \frac{\ln P - \ln 150}{7}$ **61.** $k = \frac{\ln A - \ln A_0}{t}$ **63.** $y = xe^{2.3}$ **65.** $y = (x+6)e^5$ **67. a)** ≈$5637.48 **b)** ≈11.55 years **69. a)** ≈$2818.74 **b)** ≈17.33 years **71.** ≈39.98 grams **73. a)** ≈86.47% **b)** ≈34.66 days **75. a)** ≈5.15 feet per second **b)** ≈5.96 feet per second **c)** ≈646,000 **77.** ≈$526,911,558,800,000 **79. a)** ≈7.26 billion people **b)** ≈57.76 years **81. a)** ≈1.39 billion people **b)** ≈115.52 years **83. a)** ≈32.43 inches **b)** ≈35.46 inches **85. a)** ≈999.7 grams **b)** 231,049 years **87. a)** ≈6626.62 years **b)** ≈5752.26 years **89.** ≈$6791.91 **91. a)** Strontium 90, since it has a higher decay rate **b)** ≈31.66% of original amount **93.** $v_0 = \frac{e^{xk} - 1}{kt}$ **95.** $i = Ie^{-t/RC}$ **97.** $-9x^2y^3 + 12x^2y^2 - 3xy^2 + 4xy$ **98.** −20, 20 **99. a)** 0 **b)** $\frac{11}{40}$ or 0.275 **100.** 240 children, 310 adults **101.** $x + x^2$

**Chapter 13 Review Exercises** **1.** $4x^2 - 26x + 44$ **2.** 2 **3.** $2x^2 - 6x + 3$ **4.** 39 **5.** $6\sqrt{x-3} + 7, x \geq 3$ **6.** $\sqrt{6x+4}, x \geq -\frac{2}{3}$ **7.** One-to-one **8.** Not one-to-one **9.** One-to-one **10.** Not one-to-one **11.** One-to-one **12.** Not one-to-one **13.** $f(x)$: Domain: $\{-4, -1, 5, 6\}$, Range: $\{-3, 2, 3, 8\}$; $f^{-1}(x)$: Domain: $\{-3, 2, 3, 8\}$, Range: $\{-4, -1, 5, 6\}$ **14.** $f(x)$: Domain: $\{x|x \geq 0\}$, Range: $\{y|y \geq 4\}$; $f^{-1}(x)$: Domain: $\{x|x \geq 4\}$, Range: $\{y|y \geq 0\}$ **15.** $f^{-1}(x) = \frac{x+2}{4}$;

**16.** $f^{-1}(x) = x^3 + 1$; **17.** $f^{-1}(x) = \frac{x}{36}$, $x$ is inches, $f^{-1}(x)$ is yards.

**18.** $f^{-1}(x) = \frac{x}{4}$, $x$ is quarts, $f^{-1}(x)$ is gallons **19.**

$y = 2^x$

**20.**

$y = \left(\frac{1}{2}\right)^x$

**21.** ≈$1830.29 **22.** $\log_8 64 = 2$

**23.** $\log_{81} 3 = \dfrac{1}{4}$  **24.** $\log_5 \dfrac{1}{125} = -3$  **25.** $2^5 = 32$  **26.** $\left(\dfrac{1}{4}\right)^2 = \dfrac{1}{16}$  **27.** $6^{-2} = \dfrac{1}{36}$  **28.** $4^3 = x; 64$  **29.** $a^4 = 81; 3$

**30.** $\left(\dfrac{1}{5}\right)^{-3} = x; 125$  **31.**   **32.**   **33.** $8\log_5 17$  **34.** $\dfrac{1}{2}\log_3(x-9)$  **35.** $\log 6 + \log(a+1) - \log 19$

**36.** $4\log x - \log 7 - 5\log(2x+3)$  **37.** $\log \dfrac{x^5}{(x+1)^3}$  **38.** $\log \dfrac{(2x)^4}{y}$  **39.** $\ln \dfrac{\sqrt[3]{\dfrac{x}{x+2}}}{2}$  **40.** $\ln \dfrac{x^3\sqrt{x+1}}{(x+4)^6}$  **41.** 10  **42.** 5  **43.** 121

**44.** 3  **45.** 2.9133  **46.** $-3.5720$  **47.** 1442.1154  **48.** 0.0007  **49.** 11,561.1224  **50.** 0.0594  **51.** 5  **52.** 9  **53.** 22.4  **54.** 9.4

**55.** 4  **56.** $-\dfrac{1}{2}$  **57.** 2  **58.** 5  **59.** 2.5818  **60.** 5.9681  **61.** 1.3529  **62.** 2.2402  **63.** 26  **64.** 5  **65.** 1  **66.** 3  **67.** $t \approx 1.1552$

**68.** $A_0 \approx 352.5421$  **69.** $t = \dfrac{\ln A - \ln A_0}{k}$  **70.** $k = \dfrac{\ln 0.25}{t}$  **71.** $y = xe^6$  **72.** $y = 3x + 23$  **73.** 7.6147  **74.** 3.5046

**75.** $\approx\$19{,}126.18$  **76.** $\approx 17.3$ years  **77. a)** $\approx 92.88$ minutes  **b)** $\approx 118.14$ minutes  **78.** $\approx 10.32$ pounds per square inch  **79. a)** 72  **b)** $\approx 61.2$  **c)** $\approx 5$ months

## Chapter 13 Practice Test

**1. a)** Yes  **b)** $\{(2,4),(8,-3),(3,-1),(-7,6)\}$ [13.1]  **2. a)** $x^2 + 4x + 1$  **b)** 61 [13.1]

**3. a)** $\sqrt{x^2+3}$,  **b)** $2\sqrt{13}$ [13.1]  **4. a)** $f^{-1}(x) = -\dfrac{1}{3}(x+5)$ [13.1]  **5. a)** $f^{-1}(x) = x^2 + 1, x \geq 0$

**b)**   **b)**  [13.1]

**6.** $\{x \mid x > 0\}$ [13.3]  **7.** $-4$ [13.3]  **8.**  [13.2]  **9.**  [13.3]  **10.** $\log_2 \dfrac{1}{32} = -5$ [13.3]  **11.** $5^3 = 125$ [13.3]

**12.** $2^4 = x + 3, 13$ [13.3]  **13.** $64^y = 16, \dfrac{2}{3}$ [13.7]  **14.** $3\log_2 x + \log_2(x-4) - \log_2(x+2)$ [13.4]  **15.** $\log_6 \dfrac{(x-4)^7(x+3)^2}{\sqrt{x}}$ [13.4]

**16.** 5 [13.4]  **17. a)** 3.6646  **b)** $-2.6708$ [13.7]  **18.** $\approx 2.68$ [13.6]  **19.** 3 [13.6]  **20.** $\dfrac{17}{5}$ [13.6]  **21.** 16.2810 [13.7]  **22.** 2.0588 [13.7]

**23.** 30.5430 [13.7]  **24.** $\approx\$5211.02$ [13.7]  **25.** $\approx 3364.86$ years old [13.7]

## Cumulative Review Test

**1.** $-1$ [1.9]  **2.** 1 [2.5]  **3.** $y = \dfrac{2x-5}{3}$ [2.6]  **4.** 7.5% [3.2]  **5. a)** 2 hours  **b)** 10 miles [3.4]

**6.** $m = 2, y = 2x + 3$ [4.4]  **7.** $\dfrac{8y^{21}z^6}{x^9}$ [5.2]  **8.** $x^2 + 2x + 3 + \dfrac{1}{x+1}$ [5.6]  **9.** $(4x - 3y)(3x + y)$ [6.4]

**10.** $(x - y + 5)(x - y - 5)$ [6.5]  **11.** $-1$ [7.6]  **12.** 6.25 [7.8]  **13.** $-1$ [8.2]  **14.** $(2,4)$ [9.3]  **15.** $(1,5)$ [9.6]  **16.** 41 [9.7]

**17.** [10.3]  **18.** $g(x) = (x-2)^2 - 9$  **b)** [12.5]  **19.** $[0,1] \cup [5,\infty)$ [12.6]

$g(x) = x^2 - 4x - 5$

**20. a)** $\approx 111.82$ grams  **b)** $\approx 24.8$ years [13.7]

# Chapter 14

**Exercise Set 14.1**   **1.** Midpoint   **3.** Parabola   **5.** $(h, k)$

**7.**    **9.**    **11.**    **13.**    **15.**

**17.**    **19.**    **21.**    **23.**    **25.**    **27. a)** $y = (x + 1)^2 - 1$   **b)**

**29. a)** $y = (x + 3)^2 - 9$   **31. a)** $x = (y + 2)^2 - 4$   **33. a)** $y = \left(x + \dfrac{7}{2}\right)^2 - \dfrac{9}{4}$   **35. a)** $x = -(y - 3)^2$

**b)**    **b)**    **b)**    **b)**

**37. a)** $y = -(x - 2)^2$   **39. a)** $x = -\left(y - \dfrac{3}{2}\right)^2 - \dfrac{7}{4}$   **41.** 5   **43.** 9   **45.** 13   **47.** $\sqrt{90} \approx 9.49$

**b)**    **b)**    **49.** $\sqrt{\dfrac{125}{4}} \approx 5.59$   **51.** $\sqrt{34.33} \approx 5.86$   **53.** $\sqrt{10} \approx 3.16$

**55.** $(3, 5)$   **57.** $(0, 0)$   **59.** $\left(\dfrac{3}{2}, 5\right)$   **61.** $\left(\dfrac{5}{2}, -\dfrac{7}{4}\right)$

**63.** $\left(\dfrac{\sqrt{3} + \sqrt{2}}{2}, \dfrac{7}{2}\right)$   **65.** $x^2 + y^2 = 25$   **67.** $(x - 2)^2 + y^2 = 64$

**69.** $x^2 + (y - 5)^2 = 4$   **71.** $(x - 3)^2 + (y - 4)^2 = 16$   **73.** $(x - 7)^2 + (y + 6)^2 = 144$   **75.** $(x - 1)^2 + (y - 2)^2 = 5$

**77.** $x^2 + y^2 = 16$   **79.** $(x - 3)^2 + (y + 2)^2 = 9$

**81.**    **83.**    **85.**    **87.**    **89.**    **91.**

**93.**    **95. a)** $(x + 4)^2 + y^2 = 1^2$   **97. a)** $(x + 3)^2 + (y - 2)^2 = 3^2$   **99. a)** $(x + 3)^2 + (y - 1)^2 = 2^2$

**b)**    **b)**    **b)**

**101. a)** $(x - 4)^2 + (y + 1)^2 = 2^2$   **103.** $25\pi \approx 78.5$ square units   **105.** $x$-intercept: $(-7, 0)$, $y$-intercepts: $(0, -1)$, $(0, 7)$

**b)**

**107.** $x$-intercept: $(24, 0)$, no $y$-intercepts

**109.** No, different line segments can have the same midpoint   **111.** 10

**113.** $(x + 6)^2 + (y - 2)^2 = 4$   **115. a)** $2\sqrt{2}$   **b)** $(7, 6)$   **c)** $(x - 7)^2 + (y - 6)^2 = 8$

**117.** 4, 0; a parabola opening up or down and a parabola opening right or left can be drawn to have a maximum of 4 intersections, or a minimum of 0 intersections.

**119. a)** 13.6 feet   **b)** 81.8 feet   **c)** $x^2 + (y - 81.8)^2 = 4651.24$   **121. a)** $x^2 + y^2 = 16$   **b)** $(x - 2)^2 + y^2 = 4$
**c)** $(x + 2)^2 + y^2 = 4$   **d)** $8\pi$ square units   **123.** $48\pi$ square units   **125.** Yes; D: $\mathbb{R}$, R: $\{y|y \geq k\}$
**127.** Same vertex; the first graph opens up and the second opens down.   **129.** No   **131.** No   **133.** No   **136.** $\dfrac{y}{3x}$
**137. a)** 1. $a^2$, 2. $ab$, 3. $ab$, 4. $b^2$   **b)** $(a + b)^2$   **138.** 128   **139.** $(0, 7)$   **140.**

# Exercise Set 14.2

**1.** $(0, 0)$   **3.** Ellipse   **5.** $x$-intercepts   **7.** Major   **9.** Divide

**11.**

$\dfrac{x^2}{4} + \dfrac{y^2}{1} = 1$

**13.**

$\dfrac{x^2}{4} + \dfrac{y^2}{9} = 1$

**15.**

$\dfrac{x^2}{25} + \dfrac{y^2}{9} = 1$

**17.**

$\dfrac{x^2}{16} + \dfrac{y^2}{25} = 1$

**19.**

$x^2 + 16y^2 = 16$

**21.**

$49x^2 + y = 49$

**23.**

$4x^2 + 9y^2 = 36$

**25.**

$25x^2 + 100y^2 = 400$

**27.**

$x^2 + 2y^2 = 8$

**29.**

$\dfrac{x^2}{16} + \dfrac{(y - 2)^2}{9} = 1$

**31.**

$\dfrac{(x + 3)^2}{9} + \dfrac{y^2}{25} = 1$

**33.**

$\dfrac{(x + 1)^2}{9} + \dfrac{(y - 2)^2}{4} = 1$

**35.**

$(x + 3)^2 + 9(y + 1)^2 = 81$

**37.**

$(x - 5)^2 + 4(y + 4)^2 = 4$

**39.**

$12(x + 4)^2 + 3(y - 1)^2 = 48$

**41.** $2\pi \approx 6.3$ square units   **43.** $20\pi \approx 62.8$ square units   **45.** $\dfrac{x^2}{25} + \dfrac{y^2}{16} = 1$   **47.** $\dfrac{x^2}{4} + \dfrac{y^2}{9} = 1$

**49.** $\dfrac{(x + 3)^2}{36} + \dfrac{(y + 2)^2}{9} = 1; (-3, -2)$   **51.** 69.5 feet   **53. a)** $\dfrac{x^2}{100} + \dfrac{y^2}{576} = 1$   **b)** $240\pi \approx 753.98$ square feet

**c)** $\approx 376.99$ square feet   **55.** $\sqrt{5} \approx 2.24$ feet, in both directions, from the center of the ellipse, along the major axis.
**57.** Answers will vary.   **59.** Answers will vary.   **61.** If $a = b$, the formula for a circle is obtained.   **63.** No

**65.** It becomes more circular. When $a = b$, the graph is a circle.   **67.** 2   **69.** $\dfrac{(x - 4)^2}{49} + \dfrac{(y - 2)^2}{9} = 1$   **71.** $l = \dfrac{2S - nf}{n}$
**72.** $x + \dfrac{5}{2} + \dfrac{1}{2(2x - 3)}$   **73.** 6   **74.** $\left[-\dfrac{5}{3}, 4\right)$   **75.** $\approx 2.7755$

# Mid-Chapter Test

**1.**

$y = (x - 2)^2 - 1$   [14.1]

**2.**

$y = -(x + 1)^2 + 3$   [14.1]

**3.**

$x = -(y - 4)^2 + 1$   [14.1]

**4.**

$x = 2(y + 3)^2 - 2$   [14.1]

**5.**

$y = x^2 + 6x + 10$   [14.1]

**6.** 13 [14.1]   **7.** $\sqrt{153} \approx 12.37$ [14.1]   **8.** $\left(-1, \dfrac{5}{2}\right)$ [14.1]   **9.** $\left(\dfrac{11}{4}, \dfrac{15}{4}\right)$ [14.1]

**10.** $(x + 3)^2 + (y - 2)^2 = 25$ [14.1]   **11.**

$x^2 + (y - 1)^2 = 16$   [14.1]

**12.**

$y = \sqrt{36 - x^2}$   [14.1]

**13.**

$x^2 + y^2 - 2x + 4y - 4 = 0$

[14.1]   **14.** A circle is a set of points in a plane that are the same distance from a fixed point called its center. [14.1]   **15.**

$\frac{x^2}{4} + \frac{y^2}{9} = 1$

[14.2]   **16.**

$\frac{x^2}{81} + \frac{y^2}{25} = 1$

[14.2]

**17.**

$\frac{(x-1)^2}{49} + \frac{(y+2)^2}{4} = 1$

[14.2]   **18.**

$36(x+3)^2 + (y-4)^2 = 36$

[14.2]   **19.** $6\pi \approx 18.85$ square units [14.2]   **20.** $\frac{x^2}{64} + \frac{y^2}{25} = 1$ [14.2]

## Exercise Set 14.3   **1.** Hyperbola   **3.** Center   **5.** Transverse   **7.** $y$   **9.** Divide

**11. a)** $y = \pm\frac{2}{3}x$   **13. a)** $y = \pm\frac{1}{2}x$   **15. a)** $y = \pm 2x$   **17. a)** $y = \pm\frac{4}{5}x$   **19. a)** $y = \pm\frac{5}{6}x$   **21. a)** $y = \pm\frac{1}{3}x$

**b)**

$\frac{x^2}{9} - \frac{y^2}{4} = 1$

**b)**

$\frac{x^2}{4} - \frac{y^2}{1} = 1$

**b)**

$\frac{x^2}{4} - \frac{y^2}{16} = 1$

**b)**

$\frac{x^2}{25} - \frac{y^2}{16} = 1$

**b)**

$\frac{y^2}{25} - \frac{x^2}{36} = 1$

**b)**

$\frac{y^2}{4} - \frac{x^2}{36} = 1$

**23. a)** $y = \pm\frac{5}{2}x$   **25. a)** $y = \pm\frac{4}{9}x$   **27. a)** $\frac{y^2}{1} - \frac{x^2}{25} = 1, y = \pm\frac{1}{5}x$   **29. a)** $\frac{y^2}{16} - \frac{x^2}{4} = 1, y = \pm 2x$

**b)**

$\frac{y^2}{25} - \frac{x^2}{4} = 1$

**b)**

$\frac{x^2}{81} - \frac{y^2}{16} = 1$

**b)**

$25y^2 - x^2 = 25$

**b)**

$4y^2 - 16x^2 = 64$

**31. a)** $\frac{y^2}{1} - \frac{x^2}{9} = 1, y = \pm\frac{1}{3}x$   **33. a)** $\frac{x^2}{9} - \frac{y^2}{25} = 1, y = \pm\frac{5}{3}x$   **35. a)** $\frac{y^2}{36} - \frac{x^2}{4} = 1, y = \pm 3x$

**b)**

$9y^2 - x^2 = 9$

**b)**

$25x^2 - 9y^2 = 225$

**b)**

$4y^2 - 36x^2 = 144$

**37.** Circle   **39.** Ellipse   **41.** Hyperbola   **43.** Parabola   **45.** Ellipse   **47.** Parabola   **49.** Circle   **51.** Hyperbola   **53.** Parabola

**55.** Hyperbola   **57.** Parabola   **59.** Circle   **61.** $\frac{y^2}{4} - \frac{x^2}{16} = 1$   **63.** $\frac{x^2}{9} - \frac{y^2}{36} = 1$   **65.** $\frac{x^2}{9} - \frac{y^2}{25} = 1$, no, $\frac{x^2}{18} - \frac{y^2}{50} = 1$ and others will

work. The ratio of $\frac{b}{a}$ must be $\frac{5}{3}$.   **67.** D: $(-\infty, -5] \cup [5, \infty)$; R: $\mathbb{R}$   **69.** Answers will vary.   **71.** No, graphs of hyperbolas of this form do not pass the vertical line test.   **73.** The transverse axes of both graphs are along the $x$-axis. The vertices of the second graph will be closer to the origin, and the second graph will open wider.   **75.** It is a hyperbola with vertices at $(a, 0)$ and $(-a, 0)$. The transverse axis is along the $x$-axis. The asymptotes are $y = \pm\frac{b}{a}x$.   **77.** No   **79.** Yes   **81.** $y = -\frac{1}{2}x + 1$   **82.** $\frac{3x + 2}{2x - 3}$   **83.** $-x^2 - x + 11$

**84.** $\left(-1, \frac{1}{3}\right)$   **85.** $v = \sqrt{\frac{2E}{m}}$   **86.** 1

## Exercise Set 14.4   **1.** Nonlinear   **3.** Addition   **5.** 3   **7.** $(3, -3), (-3, 3)$   **9.** $(-3, 0), \left(-\frac{12}{5}, \frac{9}{5}\right)$   **11.** $(-4, 11), \left(\frac{5}{2}, \frac{5}{4}\right)$

**13.** $(-1, 5), (1, 5)$   **15.** $(5, 0), (-5, 0)$   **17.** No real solution   **19.** $(0, -3), (\sqrt{5}, 2)(-\sqrt{5}, 2)$   **21.** $(2, -4), (-14, -20)$

**23.** $(2, 0), (-2, 0)$   **25.** $(4, 3), (4, -3), (-4, 3), (-4, -3)$   **27.** $(\sqrt{2}, \sqrt{2}), (\sqrt{2}, -\sqrt{2}), (-\sqrt{2}, \sqrt{2}), (-\sqrt{2}, -\sqrt{2})$

**29.** $(3, 0), (-3, 0)$   **31.** $(2, 1), (2, -1), (-2, 1), (-2, -1)$   **33.** $(3, 4), (3, -4), (-3, 4), (-3, -4)$

**35.** $(\sqrt{5}, 2), (\sqrt{5}, -2), (-\sqrt{5}, 2), (-\sqrt{5}, -2)$   **37.** No real solution   **39.** No real solution   **41.** 20 meters by 25 meters
**43.** 10 feet by 17 feet   **45.** length: 14 centimeters, width: 8 centimeters   **47.** 16 inches by 30 inches   **49.** $\approx 1.67$ seconds
**51.** $r = 6\%, p = \$125$   **53.** $\approx 16$ and $\approx 184$   **55.** $\approx 5$ and $\approx 95$   **57.** $(-1, -3), (3.12, -0.53)$   **59.** Answers will vary
**61.** Yes, for example    **63.** Yes, for example ⌄↑   **65.** 10 yards, 24 yards
**67.** Parentheses, exponents, multiplication or division, addition or subtraction   **68.** $(x + 2)(x^2 + x + 1)$   **69.** 0.9
**70.** $\dfrac{5\sqrt{x + 2} + 15}{x - 7}$   **71.** $k = \dfrac{\ln A - \ln A_0}{t}$

## Chapter 14 Review Exercises

**1.** $13; \left(\dfrac{5}{2}, -6\right)$   **2.** $5; \left(-\dfrac{5}{2}, 3\right)$   **3.** $17; \left(-5, \dfrac{5}{2}\right)$   **4.** $\sqrt{8} \approx 2.83; (-3, 4)$

**5.**
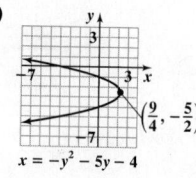

**6.** $y = (x + 3)^2 - 2$

**7.**

**8.**

**9. a)** $y = (x - 4)^2 + 6$
**b)** $y = x^2 - 8x + 22$

**10. a)** $x = -\left(y + \dfrac{5}{2}\right)^2 + \dfrac{9}{4}$
**b)**
$x = -y^2 - 5y - 4$

**11. a)** $x = \left(y + \dfrac{5}{2}\right)^2 - \dfrac{9}{4}$
**b)**
$x = y^2 + 5y + 4$

**12. a)** $y = 2(x - 2)^2 - 32$
**b)**
$y = 2x^2 - 8x - 24$

**13. a)** $x^2 + y^2 = 4^2$
**b)**
$x^2 + y^2 = 4^2$

**14. a)** $(x + 3)^2 + (y - 4)^2 = 1^2$
**b)**
$(x + 3)^2 + (y - 4)^2 = 1^2$

**15. a)** $x^2 + (y - 2)^2 = 2^2$
**b)**
$x^2 + y^2 - 4y = 0$

**16. a)** $(x - 1)^2 + (y + 3)^2 = 3^2$
**b)**
$x^2 + y^2 - 2x + 6y + 1 = 0$

**17. a)** $(x - 4)^2 + (y - 5)^2 = 1^2$
**b)**
$x^2 - 8x + y^2 - 10y + 40 = 0$

**18. a)** $(x - 2)^2 + (y + 5)^2 = (\sqrt{12})^2$
**b)**
$x^2 + y^2 - 4x + 10y + 17 = 0$

**19.**
$y = \sqrt{9 - x^2}$

**20.**
$y = -\sqrt{36 - x^2}$

**21.** $(x + 1)^2 + (y - 1)^2 = 4$   **22.** $(x - 5)^2 + (y + 3)^2 = 9$   **23.**
$\dfrac{x^2}{4} + \dfrac{y^2}{9} = 1$

**24.**
$\dfrac{x^2}{81} + \dfrac{y^2}{49} = 1$

**25.**
$9x^2 + 16y^2 = 144$

**26.**
$25x^2 + 4y^2 = 100$

**27.**
$\dfrac{(x - 3)^2}{16} + \dfrac{(y + 2)^2}{4} = 1$

**28.**
$\dfrac{(x - 4)^2}{9} + \dfrac{(y + 3)^2}{25} = 1$

**29.**

$16(x - 2)^2 + 4(y + 3)^2 = 16$

**30.** $6\pi \approx 18.85$ square units

**31. a)** $y = \pm\dfrac{5}{3}x$

**b)**

$\dfrac{x^2}{9} - \dfrac{y^2}{25} = 1$

**32. a)** $y = \pm\dfrac{5}{3}x$

**b)**

$\dfrac{y^2}{25} - \dfrac{x^2}{9} = 1$

**33. a)** $y = \pm\dfrac{3}{4}x$

**b)**

$\dfrac{y^2}{9} - \dfrac{x^2}{16} = 1$

**34. a)** $y = \pm\dfrac{5}{4}x$

**b)**

$\dfrac{y^2}{25} - \dfrac{x^2}{16} = 1$

**35. a)** $\dfrac{x^2}{9} - \dfrac{y^2}{1} = 1$   **b)** $y = \pm\dfrac{1}{3}x$

**c)**

$x^2 - 9y^2 = 9$

**36. a)** $\dfrac{y^2}{25} - \dfrac{x^2}{64} = 1$   **b)** $y = \pm\dfrac{5}{8}x$

**c)**

$64y^2 - 25x^2 = 1600$

**37. a)** $\dfrac{y^2}{25} - \dfrac{x^2}{9} = 1$   **b)** $y = \pm\dfrac{5}{3}x$

**c)**

$9y^2 - 25x^2 = 225$

**38. a)** $\dfrac{y^2}{9} - \dfrac{x^2}{49} = 1$   **b)** $y = \pm\dfrac{3}{7}x$

**c)**

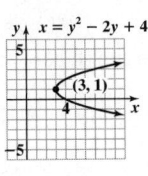

$49y^2 - 9x^2 = 441$

**39.** Hyperbola
**40.** Ellipse
**41.** Circle
**42.** Hyperbola
**43.** Ellipse

**44.** Parabola   **45.** Ellipse   **46.** Parabola
**47.** $(2, 2\sqrt{2}), (2, -2\sqrt{2}), (-2, 2\sqrt{2}),$
$(-2, -2\sqrt{2})$
**48.** $\left(\dfrac{5}{2}, \dfrac{3}{2}\right)$   **49.** $(3, 0), \left(-\dfrac{9}{5}, \dfrac{12}{5}\right)$
**50.** No real solution   **51.** $(6, 0), (-6, 0)$

**52.** $(\sqrt{15}, 1), (-\sqrt{15}, 1), (\sqrt{15}, -1), (-\sqrt{15}, -1)$   **53.** No real solution   **54.** $(1, 2\sqrt{2}), (-1, 2\sqrt{2}), (1, -2\sqrt{2}), (-1, -2\sqrt{2})$
**55.** 5 feet by 9 feet   **56.** $\approx 4$ and $\approx 145$   **57.** $r = 3\%$, $p = \$4000$

## Chapter 14 Practice Test   **1.** They are formed by cutting a cone or a pair of cones. [14.1]   **2.** $\sqrt{50} \approx 7.07$ [14.1]

**3.** $\left(-1, \dfrac{3}{2}\right)$ [14.1]   **4.** $(-3, 1)$ [14.1]   **5.**  [14.1]   **6.** $x = -(y + 2)^2 - 1$ [14.1]   **7.** $(x - 2)^2 + (y - 4)^2 = 9$ [14.1]

$y = -2(x + 3)^2 + 1$

**8.** $9\pi \approx 28.27$ square units [14.1]   **9.** $(x - 3)^2 + (y + 1)^2 = 16$ [14.1]   **10.** 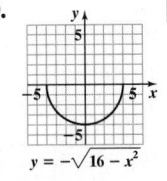 [14.1]   **11.** $(x + 1)^2 + (y - 3)^2 = 9$ [14.1]

$y = -\sqrt{16 - x^2}$

$x^2 + y^2 + 2x - 6y + 1 = 0$

**12.**  [14.2]

$4x^2 + 25y^2 = 100$

**13.** No, the major axis should be along the $y$-axis. [14.2]
**14.**  [14.2]

$4(x - 4)^2 + 36(y + 2)^2 = 36$

**15.** $(8, -7)$ [14.2]
**16.** The transverse axis lies along the axis corresponding to the positive term of the equation in standard form. [14.3]
**17.** $y = \pm\dfrac{7}{4}x$ [14.3]

**18.**  [14.3]

$\dfrac{y^2}{25} - \dfrac{x^2}{1} = 1$

**19.**  [14.3]

$\dfrac{x^2}{4} - \dfrac{y^2}{9} = 1$

**20.** Hyperbola, divide both sides of the equation by 30. [14.3]
**21.** Ellipse, divide both sides of the equation by 100. [14.3]
**22.** $(2, \sqrt{3}), (2, -\sqrt{3}), (-2, \sqrt{3}), (-2, -\sqrt{3})$ [14.4]
**23.** No real solution [14.4]
**24.** 30 meters by 50 meters [14.4]
**25.** 5 feet by 12 feet [14.4]

**Cumulative Review Test**   **1.** $\frac{19}{4}$ [2.5]   **2.** $\emptyset$ [2.5]   **3.**  [4.2]   **4.** $-27x^3y^9$ [5.1]   **5.** $(x^2 + 6)(x^2 - 7)$ [12.4]

**6.** base: 14 feet, height: 8 feet [6.7]   **7.** $\frac{x - 4}{4x + 3}$ [7.2]   **8.** $\frac{2x^2 - 9x - 5}{2(x + 3)(x - 4)}$ [7.4]   **9.** 2 [7.6]   **10.** 139 [8.2]   **11.** $(8, 6)$ [9.2]

**12.** $\left\{x \mid x < -\frac{5}{3} \text{ or } x > 1\right\}$ [10.2]   **13.** $\frac{3y^{3/2}}{x^{1/2}}$ [11.2]   **14.** $\frac{6x + 6y\sqrt{x}}{x - y^2}$ [11.5]   **15.** 3 [11.6]   **16.** $\frac{2 \pm i\sqrt{11}}{3}$ [12.2]   **17.** 2 [13.6]

**18.** $\approx 2.31$ [13.2]   **19.**  [14.2]   **20.**  [14.3]

## Chapter 15

**Exercise Set 15.1**   **1.** Sequence   **3.** Finite   **5.** Less   **7.** Series   **9.** Upper   **11.** $6, 12, 18, 24, 30$   **13.** $3, 7, 11, 15, 19$
**15.** $9, \frac{9}{2}, \frac{9}{3}, \frac{9}{4}, \frac{9}{5}$   **17.** $\frac{3}{2}, \frac{4}{3}, \frac{5}{4}, \frac{6}{5}, \frac{7}{6}$   **19.** $-1, 1, -1, 1, -1$   **21.** $4, -8, 16, -32, 64$   **23.** 31   **25.** 12   **27.** 1   **29.** 143   **31.** $\frac{81}{25}$
**33.** $2, 15$   **35.** $3, 17$   **37.** $0, \frac{13}{20}$   **39.** $-1, -1$   **41.** $\frac{1}{2}, 7$   **43.** $64, 128, 256$   **45.** $17, 19, 21$   **47.** $\frac{1}{6}, \frac{1}{7}, \frac{1}{8}$   **49.** $1, -1, 1$
**51.** $\frac{1}{81}, \frac{1}{243}, \frac{1}{729}$   **53.** $\frac{1}{16}, -\frac{1}{32}, \frac{1}{64}$   **55.** $26, 22, 18$   **57.** $2 + 5 + 8 + 11 + 14 = 40$   **59.** $2 + 5 + 10 + 17 + 26 + 37 = 97$
**61.** $\frac{1}{3} + \frac{4}{3} + \frac{9}{3} + \frac{16}{3} = \frac{30}{3} = 10$   **63.** $4 + 5 + 6 + 7 + 8 + 9 = 39$   **65.** $\sum_{i=1}^{5}(i + 10)$   **67.** $\sum_{i=1}^{3}\frac{i^2}{4}$   **69.** 13   **71.** 169   **73.** 55
**75.** 25   **77.** $\approx 42.83$   **79. a)** $6, 12, 18, 24$   **b)** $p_n = 6n$   **81. a)** $x_1 + x_2 + x_3 + \cdots + x_n$   **b)** $x_1 + x_2 + x_3 + \cdots + x_n$   **c)** Yes
**83.** $n = \frac{\sum x}{x}$   **85.** Yes   **87.** The $n$th partial sum of a series is the sum of the first $n$ consecutive terms of a series.
**89.** Increasing sequence   **91.** Yes   **93.** Answers will vary.   **95.** Answers will vary.   **96.** $5, \frac{3}{2}$   **97.** two; $b^2 - 4ac > 0$
**98.**    **99.** $(2, 1), (-2, -1)$

**Exercise Set 15.2**   **1.** Arithmetic   **3.** $d$   **5.** Term   **7.** Sum   **9.** Negative   **11.** $4, 7, 10, 13, 16; a_n = 3n + 1$
**13.** $7, 5, 3, 1, -1; a_n = -2n + 9$   **15.** $\frac{1}{2}, 2, \frac{7}{2}, 5, \frac{13}{2}; a_n = \frac{3}{2}n - 1$   **17.** $100, 95, 90, 85, 80; a_n = -5n + 105$   **19.** 14   **21.** 27   **23.** 12
**25.** 2   **27.** 9   **29.** 6   **31.** $s_{10} = 100, d = 2$   **33.** $s_8 = \frac{52}{5}, d = \frac{1}{5}$   **35.** $s_6 = 25.5, d = 3.7$   **37.** $s_{11} = 407, d = 6$
**39.** $4, 7, 10, 13; a_{10} = 31, s_{10} = 175$   **41.** $-6, -4, -2; 0; a_{10} = 12, s_{10} = 30$   **43.** $-8, -13, -18, -23; a_{10} = -53, s_{10} = -305$
**45.** $\frac{7}{2}, 6, \frac{17}{2}, 11; a_{10} = 26, s_{10} = 147.5$   **47.** $100, 93, 86, 79; a_{10} = 37, s_{10} = 685$   **49.** $n = 15, s_{15} = 330$   **51.** $n = 11, s_{11} = 121$
**53.** $n = 17, s_{17} = \frac{153}{2}$   **55.** $n = 29, s_{29} = 1421$   **57.** 1275   **59.** 2500   **61.** 1395   **63.** 267   **65.** $42; 372$   **67.** 351   **69. a)** 27   **b)** 196
**71.** $101 \cdot 50 = 5050$   **73.** $s_n = n^2$   **75. a)** 19 feet   **b)** 143.5 feet   **77.** 2 feet   **79. a)** 155   **b)** 780   **81.** $496   **83. a)** $45,600
**b)** $438,000   **85.** $a_n = 180°(n - 2)$   **87.** Yes   **89.** Yes   **97.** $r = \frac{A - P}{Pt}$   **98.** $3(2n - 5)(2n - 1)$   **99.** $(-3, -5)$
**100.**

**Exercise Set 15.3**  **1.** Geometric  **3.** $r$  **5.** Term  **7.** Sum  **9.** 1  **11.** $2, 6, 18, 54, 162$  **13.** $6, -3, \dfrac{3}{2}, -\dfrac{3}{4}, \dfrac{3}{8}$  **15.** $72, 24, 8, \dfrac{8}{3}, \dfrac{8}{9}$

**17.** $90, -30, 10, -\dfrac{10}{3}, \dfrac{10}{9}$  **19.** $-1, -3, -9, -27, -81$  **21.** $7, -14, 28, -56, 112$  **23.** $\dfrac{1}{3}, \dfrac{1}{6}, \dfrac{1}{12}, \dfrac{1}{24}, \dfrac{1}{48}$  **25.** $3, \dfrac{9}{2}, \dfrac{27}{4}, \dfrac{81}{8}, \dfrac{243}{16}$

**27.** 128  **29.** $-\dfrac{3}{32}$  **31.** 64  **33.** 6144  **35.** $\dfrac{1}{64}$  **37.** $\dfrac{50}{729}$  **39.** 155  **41.** 7812  **43.** 10,160  **45.** $-\dfrac{2565}{256}$  **47.** $-\dfrac{9279}{625}$

**49.** $r = \dfrac{1}{2}, a_n = 7\left(\dfrac{1}{2}\right)^{n-1}$  **51.** $r = 2, a_n = 9(2)^{n-1}$  **53.** $r = -3, a_n = 2(-3)^{n-1}$  **55.** $r = \dfrac{2}{3}, a_n = \dfrac{3}{4}\left(\dfrac{2}{3}\right)^{n-1}$  **57.** 2  **59.** $\dfrac{5}{4}$  **61.** 24

**63.** $\dfrac{25}{3}$  **65.** 5  **67.** 6  **69.** 4  **71.** 12  **73.** $-45$  **75.** $-15$  **77.** $\dfrac{8}{33}$  **79.** $\dfrac{7}{9}$  **81.** $\dfrac{17}{33}$  **83.** $r = 3, a_1 = 5$  **85.** $r = 2$ or $r = -2, a_1 = 7$

**87.** $\approx \$1.77$  **89. a)** 4 days  **b)** $\approx 1.172$ grams  **91. a)** $\approx 338.59$ million  **b)** $\approx 63.4$ years  **93. a)** $\dfrac{1}{2}, \dfrac{1}{4}, \dfrac{1}{8}, \dfrac{1}{16}, \dfrac{1}{32}$

**b)** $a_n = \dfrac{1}{2}\left(\dfrac{1}{2}\right)^{n-1} = \left(\dfrac{1}{2}\right)^{n}$  **c)** $\dfrac{1}{128} \approx 0.78\%$  **95.** $\approx \$15,938.48$  **97. a)** 28.512 feet  **b)** 550 feet  **99. a)** 10.29 inches  **b)** 100 inches

**101.** 211  **103. a)** $\$12,000, \$9600, \$7680$  **b)** $a_n = 12,000\left(\dfrac{4}{5}\right)^{n-1}$  **c)** $\approx \$4915.20$  **105.** 190 feet  **107. a)** $y_2$ goes up more steeply.

**b)**

**109.** 0  **111.** Yes  **113.** Yes, $s_\infty = 8$  **115.** $n = 21, s_n = 2,097,151$  **116.** $8x^3 - 18x^2 + 21x - 18$

**117.** $8x - 15 + \dfrac{57}{2x + 5}$

**118.** 12 hours  **119.** $\dfrac{10}{3}$  **120.** $3x^2y^2\sqrt[3]{y} - 2xy\sqrt[3]{9y^2}$  **121.** $x\sqrt{y}$  **122.** $-\dfrac{1}{3}$

**Mid-Chapter Test**  **1.** $4, 1, -2, -5, -8$ [15.1]  **2.** 91 [15.1]  **3.** $1, 11$ [15.1]  **4.** $-15, -19, -23$ [15.1]  **5.** 45 [15.1]

**6.** $\displaystyle\sum_{i=1}^{5}\left(\dfrac{1}{4}i + 9\right)$ [15.1]  **7.** $-6, -1, 4, 9; a_n = -11 + 5n$ [15.2]  **8.** $-1$ [15.2]  **9.** 6 [15.2]  **10.** $3, -3$ [15.2]  **11.** $47\dfrac{1}{2}$ [15.2]

**12.** 12 [15.2]  **13.** 136 [15.2]  **14.** $100, -50, 25, -\dfrac{25}{2}, \dfrac{25}{4}$ [15.3]  **15.** $\dfrac{1}{27}$ [15.3]  **16.** 315 [15.3]  **17.** $-\dfrac{2}{3}$ [15.3]  **18.** 18 [15.3]

**19.** $\dfrac{29}{33}$ [15.3]  **20. a)** A sequence is a list of numbers arranged in a specific order.  **b)** An arithmetic sequence is a sequence where each term differs by a constant amount.  **c)** A geometric sequence is a sequence where the terms differ by a common multiple. **d)** A series is the sum of the terms of a sequence. [15.1–15.3]

**Exercise Set 15.4**  **1.** Factorial  **3.** 1  **5.** 20  **7.** 10  **9.** 10  **11.** 1  **13.** 1  **15.** 70  **17.** 28  **19.** $x^3 + 12x^2 + 48x + 64$
**21.** $8x^3 - 36x^2 + 54x - 27$  **23.** $a^4 - 4a^3b + 6a^2b^2 - 4ab^3 + b^4$  **25.** $243a^5 - 405a^4b + 270a^3b^2 - 90a^2b^3 + 15ab^4 - b^5$
**27.** $16x^4 + 16x^3 + 6x^2 + x + \dfrac{1}{16}$  **29.** $\dfrac{1}{16}x^4 - \dfrac{3}{2}x^3 + \dfrac{27}{2}x^2 - 54x + 81$  **31.** $x^{10} + 100x^9 + 4500x^8 + 120,000x^7$
**33.** $2187x^7 - 5103x^6y + 5103x^5y^2 - 2835x^4y^3$  **35.** $x^{16} - 24x^{14}y + 252x^{12}y^2 - 1512x^{10}y^3$  **37.** $x^8, 24x^7, 17,496x, 6561$
**39.** $64x^6, 960x^5, 37,500x, 15,625$  **41.** Yes, $4! = 4 \cdot 3!$  **43.** Yes, $(7 - 3)! = (7 - 3)(7 - 4)(7 - 5)! = 4 \cdot 3 \cdot 2!$  **45.** $m = n$ or $m = 0$
**47.** $(a + b)^n = \displaystyle\sum_{i=0}^{n}\binom{n}{i}a^{n-i}b^i$  **49.** $(0, 10)$  **50.** $2, 9$  **51.** $(10, 4)$  **52.** $2x^3y^5\sqrt{30y}$  **53.** $f^{-1}(x) = \dfrac{x - 8}{3}$

**Chapter 15 Review Exercises**  **1.** $6, 7, 8, 9, 10$  **2.** $-1, 3, 9, 17, 27$  **3.** $12, 6, 4, 3, \dfrac{12}{5}$  **4.** $\dfrac{1}{5}, \dfrac{2}{3}, \dfrac{9}{7}, 2, \dfrac{25}{9}$  **5.** 11  **6.** 6  **7.** $\dfrac{26}{81}$

**8.** 88  **9.** $s_1 = 7, s_3 = 27$  **10.** $s_1 = 12, s_3 = 47$  **11.** $s_1 = \dfrac{4}{3}, s_3 = \dfrac{227}{60}$  **12.** $s_1 = -9, s_3 = -10$  **13.** $32, 64, 128; a_n = 2^n$

**14.** $-\dfrac{1}{3}, \dfrac{1}{9}, -\dfrac{1}{27}; a_n = (-1)^n(3^{4-n})$  **15.** $\dfrac{16}{9}, \dfrac{32}{9}, \dfrac{64}{9}; a_n = \dfrac{2^n - 1}{9}$  **16.** $-3, -7, -11; a_n = 17 - 4n$  **17.** $11 + 14 + 19 = 44$

**18.** $6 + 14 + 24 + 36 = 80$  **19.** $\dfrac{1}{6} + \dfrac{4}{6} + \dfrac{9}{6} + \dfrac{16}{6} + \dfrac{25}{6} = \dfrac{55}{6}$  **20.** $\dfrac{1}{2} + \dfrac{2}{3} + \dfrac{3}{4} + \dfrac{4}{5} = \dfrac{163}{60}$  **21.** 29  **22.** 239  **23.** 132  **24.** 841

**25. a)** $10, 14, 18, 22$  **b)** $p_n = 4n + 6$  **26. a)** $4, 10, 18, 28$  **b)** $a_n = n(n + 3) = n^2 + 3n$  **27.** $5, 8, 11, 14, 17$  **28.** $5, \dfrac{14}{3}, \dfrac{13}{3}, 4, \dfrac{11}{3}$

**29.** $\dfrac{1}{2}, -\dfrac{3}{2}, -\dfrac{7}{2}, -\dfrac{11}{2}, -\dfrac{15}{2}$  **30.** $-20, -\dfrac{99}{5}, -\dfrac{98}{5}, -\dfrac{97}{5}, -\dfrac{96}{5}$  **31.** 30  **32.** $-5$  **33.** $\dfrac{1}{2}$  **34.** 6  **35.** $s_8 = 112, d = 2$

**36.** $s_7 = -140, d = -6$  **37.** $s_7 = \dfrac{48}{5}, d = \dfrac{2}{5}$  **38.** $s_9 = -42, d = -\dfrac{1}{3}$  **39.** $-7, -3, 1, 5; a_{10} = 29, s_{10} = 110$

**40.** $8, 5, 2, -1; a_{10} = -19, s_{10} = -55$  **41.** $\dfrac{5}{6}, \dfrac{3}{2}, \dfrac{13}{6}, \dfrac{17}{6}; a_{10} = \dfrac{41}{6}, s_{10} = \dfrac{115}{3}$  **42.** $-60, -55, -50, -45; a_{10} = -15, s_{10} = -375$

**43.** $n = 13, s_{13} = 442$  **44.** $n = 8, s_8 = 28$  **45.** $n = 11, s_{11} = \dfrac{231}{10}$  **46.** $n = 10, s_{10} = 180$  **47.** $6, 12, 24, 48, 96$

48. $-28, -14, -7, -\frac{7}{2}, -\frac{7}{14}$  49. $20, -\frac{40}{3}, \frac{80}{9}, -\frac{160}{27}, \frac{320}{81}$  50. $-20, -4, -\frac{4}{5}, -\frac{4}{25}, -\frac{4}{125}$  51. $\frac{4}{27}$  52. 480  53. 216

54. $\frac{4}{243}$  55. 441  56. $-\frac{4305}{64}$  57. $\frac{585}{8}$  58. $\frac{127}{8}$  59. $r=2; a_n=6(2)^{n-1}$  60. $r=5, a_n=-6(5)^{n-1}$  61. $r=\frac{1}{3}, a_n=10\left(\frac{1}{3}\right)^{n-1}$

62. $r=\frac{2}{3}, a_n=\frac{9}{5}\left(\frac{2}{3}\right)^{n-1}$  63. 10  64. $\frac{49}{10}$  65. $-6$  66. $-18$  67. 32  68. $\frac{33}{2}$  69. $\frac{25}{6}$  70. $-12$  71. $\frac{4}{11}$  72. $\frac{29}{37}$

73. $81x^4 + 108x^3y + 54x^2y^2 + 12xy^3 + y^4$  74. $8x^3 - 36x^2y^2 + 54xy^4 - 27y^6$  75. $x^9 - 18x^8y + 144x^7y^2 - 672x^6y^3$

76. $256a^{16} + 3072a^{14}b + 16,128a^{12}b^2 + 48,384a^{10}b^3$  77. 15,050  78. 231  79. a) \$36,000, \$37,000, \$38,000, \$39,000

b) $a_n = \$35,000 + 1000n$  c) \$41,000  d) \$451,000  80. \$51,200  81. a) $\approx\$2024.51$  b) $\approx\$2463.13$  c) $\approx\$24,378.78$

82. $\approx\$503.63$  83. 150 feet

## Chapter 15 Practice Test

1. A series is the sum of the terms of a sequence. [15.1]  2. a) An arithmetic sequence is one whose terms differ by a constant amount.  b) A geometric sequence is one whose terms differ by a common multiple. [15.2–15.3]

3. $-\frac{1}{3}, 0, \frac{1}{9}, \frac{1}{6}, \frac{1}{5}$ [15.1]  4. $s_1 = 3, s_3 = \frac{181}{36}$ [15.1]  5. $5 + 11 + 21 + 35 + 53 = 125$ [15.1]  6. 184 [15.1]

7. $a_n = \frac{1}{3} + \frac{1}{3}(n - 1) = \frac{1}{3}n$ [15.2]  8. $a_n = 5(2)^{n-1}$ [15.3]  9. $15, 9, 3, -3$ [15.2]  10. $\frac{5}{12}, \frac{5}{18}, \frac{5}{27}, \frac{10}{81}$ [15.3]  11. $-40$ [15.2]

12. $-20$ [15.2]  13. 12 [15.2]  14. $\frac{256}{243}$ [15.3]  15. $\frac{39,063}{5}$ [15.3]  16. $r = \frac{1}{3}, a_n = 15\left(\frac{1}{3}\right)^{n-1}$ [15.3]  17. 12 [15.3]  18. $\frac{13}{33}$ [15.3]

19. 56 [15.4]  20. $x^4 + 8x^3y + 24x^2y^2 + 32xy^3 + 16y^4$ [15.4]  21. 82 [15.1]  22. 91 [15.2]  23. \$210,000 [15.2]  24. $\approx\$851.66$ [15.3]

25. 364,500 [15.3]

## Cumulative Review Test

1. $b = \frac{2A}{h}$ [2.6]  2. $y = -\frac{11}{3}x + \frac{38}{3}$ [4.4]  3. $5x^4 + 29x^3 + 14x^2 - 28x + 10$ [5.5]

4. $(x^2 + 2)(x - 6)$ [6.2]  5. $(a + b + 4)^2$ [12.4]  6. $\frac{5x^2 + 14x - 49}{(x + 5)(x - 2)}$ [7.4]  7. 500 [7.8]  8. Infinite number of solutions [9.4]

9. 3 [11.6]  10. 5 [11.6]  11. $-1 \pm i\sqrt{14}$ [12.1]  12. $\frac{3 \pm \sqrt{309}}{30}$ [12.2]  13. 5 [6.7]  14.  [14.1]  15. $\frac{1}{2}$ [13.3]

16.  [13.2]  17. $(x + 6)^2 + (y - 2)^2 = 49$ [14.1]

18.  [14.1]  19. 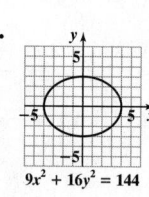 [14.2]  20. 18 [15.3]

# Applications Index

# Subject Index

## A

Absolute value equations, 452, 603, 606

Absolute values
 addition using, 45–47, 86
 in equations, 452, 603, 606
 to evaluate radicals, 628–629
 explanation of, 37–38, 85, 600
 inequalities containing, 600–606, 615–616
 negative of, 38

Acute angles, 921

Acute triangles, 923

Addend, 42

Addition
 associative property of, 77, 87
 commutative property of, 76, 78, 87
 of complex numbers, 676, 685
 of decimals, 915
 distributive property of multiplication over, 77–78, 87
 of fractions, 23–24, 43–45, 84–85, 442
 identity property of, 79, 87
 inverse property of, 79, 87
 of polynomials, 292–293, 316, 927
 of radicals, 647–649, 659, 684
 of rational expressions, 394–396, 401–406, 418
 of real numbers, 42–47
 using absolute values, 45–47, 86
 words and phrases for, 166

Addition method
 explanation of, 533
 to solve nonlinear systems of equations, 865–866
 to solve systems of equations, 533–538, 541–542, 578

Addition property of equality
 explanation of, 103, 157
 mentally using, 105–106
 to solve equations, 103–105, 112, 123

Additive inverse
 explanation of, 45, 79, 86
 expression divided by, 384

Algebra
 applications for, 166
 arithmetic vs., 20
 explanation of, 103, 108
 of functions, 499–503

Algebraic expressions. *See also* Expressions
 explanation of, 8, 83, 92

in quadratic form, 723–724
 words and phrases translated into, 166–167

Alternating sequence, 880

Angles
 acute, 921
 alternate exterior angles, 922
 alternate interior angles, 922
 complementary, 194, 556
 corresponding, 921
 exterior, 922
 formed by transversal intersecting other lines, 922
 interior, 922
 obtuse, 921
 right, 921
 straight, 921
 supplementary, 194, 556, 921
 vertical, 194, 921

Antilogarithm (antilog), 809

Application problems. *See also* Applications index; Problem solving
 exponential equations to solve, 815
 exponential functions to solve, 784, 786–788
 functions to solve, 466–468, 478–479
 geometric series to solve, 898–900
 graphs to solve, 229–230
 involving geometric shapes, 191–193
 involving mixtures, 200–204, 552
 involving money, 182–183, 198–200, 210
 involving motion, 197–198, 210, 423–425, 443, 714–715
 involving numbers, 180–182, 365
 involving percent, 184–185
 involving variation, 432–436
 involving work, 425–428, 443, 715–716
 logarithmic functions to solve, 797
 natural base and natural logarithms to solve, 822–824
 nonlinear systems of equations to solve, 866–867
 quadratic equations to solve, 365–366, 707–709, 713–718, 762

radical equations to solve, 668–669
 systems of equations to solve, 548–555, 579

Application problem solving. *See also* Problem solving
 expressing relationship between two related quantities, 167–169
 procedure for, 180, 209
 suggestions for, 185
 translating applications into equations for, 172–174
 translating phrases into mathematical expressions, 166–167
 writing expressions involving multiplication for, 169–172

Approximately equal to, 12

Approximation
 of common logarithms, 807–809, 821
 decimal, 655
 of natural logarithms, 820, 821

Area
 of circles, 191, 472
 explanation of, 134, 159
 of parallelograms, 134
 of quadrilaterals, 134, 191
 of rectangles, 134, 191, 472, 738
 of squares, 134
 of three-dimensional figures, 134
 of trapezoids, 134
 of triangles, 134, 191

Argument, of logarithm, 799

Arithmetic
 algebra vs., 20
 operations of, 42

Arithmetic sequence
 common difference in, 885–886
 explanation of, 885
 $n$th partial sum of, 887–889, 909, 910
 $n$th term of, 886, 910

Arithmetic series, 887, 910

Associative property
 of addition, 77, 87
 function of, 77, 78
 of multiplication, 77, 78, 87

Asymptotes, 856–858

Augmented matrices
 explanation of, 561, 579
 to solve systems of linear equations, 562–564, 566

Averages, 14, 84

Axis of symmetry, of parabola, 733, 734, 741–742

## B

Bar graphs, 11–12, 83

Base, 67, 315

Binomial coefficients, 905–906, 911

Binomials. *See also* Polynomials
 dividing polynomials by, 309–311, 929
 expansions of, 904–905
 explanation of, 292, 316, 904
 square of, 303–304, 316, 694
 use of distributive property to multiply, 299–300
 use of FOIL method to multiply, 300–303, 316
 use of formulas for special products to multiply, 302–304

Binomial theorem, 905–907, 911

Boundary lines, of linear inequalities, 610

Boundary values, 751, 753–754, 756

Braces , 69

Brackets, 69

## C

Calculators. *See* Graphing calculators; Scientific calculators

Cartesian coordinate system. *See also* Graphs/graphing
 explanation of, 216, 255
 plotting points in, 217–221, 224–226, 256

Celsius temperature, 472

Center
 of circle, 842–845, 870
 of ellipse, 849, 871
 of hyperbola, 855, 856, 872

Change of base formula, 821, 832

Circle graphs
 explanation of, 13, 84
 problem solving involving, 13–14

Circles
 area of, 191, 472
 with centers at origin, 842–843, 870
 circumference of, 136, 159, 191, 433
 concentric, 848
 explanation of, 842, 870
 formulas for, 191
 on graphing calculator, 844
 graphs of, 842–845

Circumference, 136, 159, 191, 433